PRINCIPLES OF MATHEMATICAL PHYSICS

For BSc [Hons]/MSc, BE/ME students of all leading universities in India and abroad.

PRINCIPLES OF MATHEMATICAL PHYSICS

Prof. (Dr) S P Kuila MSc, PhD [IIT, Kharagpur]

Ex-Professor and Head, Department of Applied Physical Sciences
Chairman, Research and Development Cell
Bengal Institute of Technology and Management, Santiniketan, WB
and
Former Professor of Physics, Abhedananda Mahavidyalya,
Sainthia, Birbhum, WB

Author of

Perspective of Quantum Mechanics
Vector Analysis and its Application, Principles of Engineering Physics [Vols. I, II]
Fundamentals of Quantum Mechanics, Statistical Mechanics and Solid State Physics,
Essentials of Solid State Physics
Advanced Engineering Physics, Concept of Nuclear Physics, etc.

New Central Book Agency (P) Ltd

LONDON
HYDERABAD ERNAKULAM BHUBANESWAR
NEW DELHI KOLKATA PUNE GUWAHATI

NCBA

REGD OFFICE
8/1 Chintamoni Das Lane, Kolkata 700 009, India
email: ncbapvtltd@eth.net

OVERSEAS
NCBA (UK) Ltd, 149 Park Avenue North, Northampton, NN3 2HY, UK
email: ncbauk@yahoo.com

EXPORT
NCBA Exports Pvt. Ltd, 212 Shahpur Jat, New Delhi 110 049
email: ncbaexports@ncbapvtltd.com

BRANCHES

2/27 Ansari Road, 1st Floor
Daryaganj, **New Delhi** 110 002
email: ncbadel@ncbapvtltd.com

House No. 3-1-315, 1st Floor
Nimboliadda, Kachiguda, **Hyderabad** 500 027
email: ncbahydb@ncbapvtltd.com

Shop Nos. 3 & 4, Vinayak Towers
681/B Budhwar Peth, Appa Balwant Chowk
Pune 411 002
email: ncbapune@ncbapvtltd.com

GSS Shopping Complex, 1st Floor
Opposite College Post Office, Convent Road
Ernakulam 682 035
0484-4021054 / 53

Shop No. 15, Bharati Towers, Ground Floor
Forest Park
Bhubaneswar 751 001
email: ncbabub@ncbapvtltd.com

Radhamayee Complex [Second Floor]
Opp. Axis Bank, Panbazar, Jaswanta Road
Guwahati 781 001
email: ncbaguwahati@ncbapvtltd.com

PRINCIPLES OF MATHEMATICAL PHYSICS ● S P Kuila

First Published: July 2015
Reprinted: April 2016, January 2017
Second Edition: January 2020

PUBLISHER
New Central Book Agency (P) Ltd
8/1 Chintamoni Das Lane, Kolkata 700 009

TYPESETTER
Anin, Kolkata

PRINTER
New Central Book Agency (P) Ltd
Web-Offset Division, Dhulagarh, Sankrail, Howrah

TECHNICAL EDITOR
Dipan Roy

PROJECT TEAM
Prodip Baidya, Pradip Biswas

ISBN: 978 81 7381 977 3

Price ₹ 835.00

Dedicated to

the sweet memory of my

loving mother

Contents

Preface to the Second Edition

The second edition of the book '**Principles of Mathematical Physics**' has been thoroughly revised and upgraded in view of new UGC syllabus prescribed by UGC for the students pursuing B.Sc. (Honours/Major) of Indian Universities. The book entirely covers the new UGC syllabus in the subject and large number of questions and problems have been incorporated to judge the depth of understanding of the students.

The author will deem his labour amply rewarded if the book receives over whelming popularity as in its first edition.

S P Kuila

Preface to the First Edition

Mathematics is an essential tool to the workers in the fields of physical sciences and technology. The author all along has been conscious regarding the requirements of the students while teaching them in various branches of physics and technology in different institutions and as such feels the need of a suitable textbook with minimum guidance. The book '**Principles of Mathematical Physics**' attempts to provide a thorough and comprehensive knowledge to the students pursuing B.Sc. (Hons.)/M.Sc., B.E. and M.E. courses under all leading universities in India and abroad.

Various questions and problems set in different Indian universities and competitive examinations like GATE, AMIETE, etc. have been selected so as to make the reader familiar with the type of questions usually set therein. Apart from that objective type questions, short answer type questions and broad answer type questions have been added at the end of each chapter so as to judge the depth of understanding of the subject by the students.

The author of this treatise is indebted to many eminent authors whose works have been freely consulted, the faculty members and other friends for their advice and encouragement during the preparation of the manuscript.

The author would like to acknowledge Mr. Amitabh Sen, the Director, New Central Book Agency (P) Ltd. for bringing out the book in time.

Thanks are due to the editorial staff of the above organisation for their valuable help and active cooperation.

Any suggestion for the improvement of the text as well as intimations of errors and misprints will be most gratefully received and necessary steps will be taken in future. Finally, the author will deem his labour amply rewarded if the book proves useful to those for whom it is intended.

18 July 2015 **S P Kuila**
Kolkata

CHAPTER 1

Determinants

1.1 Introduction

A **determinant** (or **matrix**) is a two dimensional collection of numbers which may be a simple sequence like $a_1, a_2, \ldots, a_n$ or two dimensional characters arranged in rectangular quadratic or even triangular form which obeys certain laws such as :

$$\begin{vmatrix} a_1 & a_2 & \cdots & a_n \\ b_1 & b_2 & \cdots & b_n \\ \cdots & \cdots & \cdots & \cdots \\ g_1 & g_2 & \cdots & g_n \end{vmatrix} \qquad \begin{vmatrix} a_1 & a_2 & a_3 & a_4 \\ b_1 & b_2 & b_3 & b_4 \\ c_1 & c_2 & c_3 & c_4 \\ d_1 & d_2 & d_3 & d_4 \end{vmatrix}$$

rectangular form quadratic form

The collection is usually enclosed by bars in case of determinant and by brackets in case of Matrix. The individual numbers are called **elements**. The horizontal array of elements is called **row** and vertical array is called **column** of the matrix. The individual element is referred to by a pair of indices (e.g., a_{ij}), the first index (i) represents its position in the row and the second index represents its position in the column. Thus, the element a_{ij} belongs to ith row and jth column of the determinant.

If a determinant contains m rows and n columns, then the total number of elements in it is $(m \times n)$ and the determinant is said to be of order $(m \times n)$. If $m = n$, then the determinant is said to be quadratic (or square) and if $m \neq n$, then the determinant is rectangular.

In physics and engineering, the knowledge of determinant is necessary in solving problems on mechanical and electrical systems. A large number of physical phenomena expressed in terms of set of linear differential equations can be solved through the application of the principle of determinant.

1.2 Definition of Determinant

A determinant may be defined as an array of real or complex quantities in two dimensions. A determinant of order $(m \times n)$ is expressed as

$$\text{Det } a = |a| = \begin{vmatrix} a_{11} & a_{12} & a_{13} & \cdots & a_{1m} \\ a_{21} & a_{22} & a_{23} & \cdots & a_{2m} \\ a_{31} & a_{32} & a_{33} & \cdots & a_{3m} \\ \cdots & \cdots & \cdots & \cdots & \cdots \\ a_{m1} & a_{m2} & a_{m3} & \cdots & a_{mn} \end{vmatrix} \tag{1.1}$$

There are $m \times m$ elements in the determinant. The diagonal through the left-hand top corner containing the elements $a_{11}, a_{22}, a_{33}, \ldots, a_{mm}$ is called the leading or **principal diagonal** and $a_{11}, a_{22}, a_{33}, \ldots, a_{mm}$ is the **leading term**.

The origin of determinant may be attributed to the linear equations. For example, consider the three linear equations,

$$a_1x + b_1y + c_1z = 0 \tag{1.2}$$

$$a_2x + b_2y + c_2z = 0 \tag{1.3}$$

$$a_3x + b_3y + c_3z = 0. \tag{1.4}$$

Solving the above equations,

$$\frac{x}{b_2c_3 - c_2b_3} = \frac{y}{c_2a_3 - a_2c_3} = \frac{z}{a_2b_3 - b_2a_3} \tag{1.5}$$

Substituting the proportionate values of x, y, z from Eqn. (1.5) in Eqn. (1.2), we get

$$a_1(b_2c_3 - c_2b_3) + b_1(c_2a_3 - a_2c_3) + c_1(a_2b_3 - b_2a_3) = 0. \tag{1.6}$$

Eqn. (1.6) can be expressed in the determinant form as

$$\begin{vmatrix} a_1 & b_1 & c_1 \\ a_2 & b_2 & c_2 \\ a_3 & b_3 & c_3 \end{vmatrix}$$

and is called the **expansion of the determinant.**

1.3 Minor and Co-factors

(i) The Minor : The minor m_{ij} of an element a_{ij} of a determinant Δ is obtained by deleting the ith row and the jth column in which the element a_{ij} appears in the determinant Δ.

For example, we consider the determinant

$$\Delta = \begin{vmatrix} a_{11} & a_{12} & a_{13} \\ a_{21} & a_{22} & a_{23} \\ a_{31} & a_{32} & a_{33} \end{vmatrix}. \tag{1.7}$$

The minor M_{12} of the element a_{12} is

$$M_{12} = \begin{vmatrix} a_{21} & a_{23} \\ a_{31} & a_{33} \end{vmatrix} = a_{21}a_{33} - a_{31}a_{23} \tag{1.8}$$

and that of a_{33} is

$$M_{33} = \begin{vmatrix} a_{11} & a_{12} \\ a_{21} & a_{22} \end{vmatrix} = a_{11}a_{22} - a_{21}a_{12}. \tag{1.9}$$

(ii) The Co-factor : The Co-factor A_{ij} of an element a_{ij} in the determinant Δ is the $(-1)^{i+j}$ times the minor M_{ij}, i.e.,

$$A_{ij} = (-1)^{i+j} M_{ij}$$

For example, the co-factor of the element a_{12} in the determinant Δ of Eqn. (1.6) is

$$A_{12} = (-1)^{1+2} M_{12} = -M_{12} = -\begin{vmatrix} a_{21} & a_{23} \\ a_{31} & a_{33} \end{vmatrix}. \tag{1.10}$$

Similarly, the co-factor of a_{33} is

$$A_{33} = (-1)^{3+3} \begin{vmatrix} a_{11} & a_{12} \\ a_{21} & a_{22} \end{vmatrix} = \begin{vmatrix} a_{11} & a_{12} \\ a_{22} & a_{23} \end{vmatrix}. \tag{1.11}$$

Note that the determinants representing M_{12}, M_{33}, A_{12} and A_{33} are all homogeneous second order polynomials.

1.4 Laplace's Expansion of Determinant

For evaluation of a determinant, Laplace utilised the principle of expanding the determinant in terms of co-factors, i.e.,

$$\Delta = \sum a_{ij} A_{ij}, \tag{1.12}$$

where A_{ij} is the co-factor of the element a_{ij} in Δ.

The working rule is to multiply each element of the row (or, column) in terms of which the determinant is to be expanded by its co-factor and then to be added all these terms with the sign of the product alternately positive and negative.

To illustrate Laplace's principle we consider the determinant,

$$\Delta = \begin{vmatrix} a_{11} & a_{12} & a_{13} \\ a_{21} & a_{22} & a_{23} \\ a_{31} & a_{32} & a_{33} \end{vmatrix}.$$

The expansion of Δ in terms of 1st row is given by

$$\begin{aligned} \Delta &= a_{11}A_{11} + a_{12}A_{12} + a_{13}A_{13} \\ &= (-1)^{1+1} a_{11} \begin{vmatrix} a_{22} & a_{23} \\ a_{32} & a_{33} \end{vmatrix} + (-1)^{1+2} a_{12} \begin{vmatrix} a_{21} & a_{23} \\ a_{31} & a_{33} \end{vmatrix} + (-1)^{1+3} a_{13} \begin{vmatrix} a_{21} & a_{22} \\ a_{31} & a_{32} \end{vmatrix} \\ &= a_{11}(a_{22}a_{33} - a_{32}a_{23}) - a_{12}(a_{21}a_{33} - a_{31}a_{23}) + a_{13}(a_{21}a_{32} - a_{31}a_{22}). \end{aligned} \tag{1.13}$$

Eqn. (1.13) is a polynomial of order three in a_{ij}.

Similarly, Δ can be expanded in terms of 2nd and 3rd rows.

If we take the sum of the products of the element of any row (or column) by the co-factors of another row (or, column), then the result of expansion of Δ becomes zero, i.e.,

$$\begin{aligned} \Delta &= a_{31}A_{21} + a_{32}A_{22} + a_{33}A_{23} \\ &= (-1)^{2+1} a_{31} \begin{vmatrix} a_{12} & a_{13} \\ a_{32} & a_{33} \end{vmatrix} + (-1)^{2+2} a_{32} \begin{vmatrix} a_{11} & a_{13} \\ a_{31} & a_{33} \end{vmatrix} + (-1)^{2+3} a_{33} \begin{vmatrix} a_{11} & a_{12} \\ a_{31} & a_{32} \end{vmatrix} \\ &= a_{11}(a_{22}a_{33} - a_{32}a_{23}) - a_{12}(a_{21}a_{33} - a_{31}a_{23}) + a_{13}(a_{21}a_{32} - a_{31}a_{22}) \\ &= 0 \end{aligned} \tag{1.14}$$

1.5 Properties of Determinant

The properties of determinant help us to simplify and evaluate a given determinant without expanding it. The properties hold good for determinant of any order.

1. **The value of the determinant remains unaltered if the rows and columns are interchanged.**

   Let $\Delta = \begin{vmatrix} a_1 & b_1 & c_1 \\ a_2 & b_2 & c_2 \\ a_3 & b_3 & c_3 \end{vmatrix}$

   By changing the rows into columns and columns into rows, we get

$$\Delta' = \begin{vmatrix} a_1 & a_2 & a_3 \\ b_1 & b_2 & b_3 \\ c_1 & b_2 & c_3 \end{vmatrix}.$$

$$\begin{aligned}\text{Now, } \Delta' &= a_1(b_2c_3 - c_2b_3) - a_2(b_1c_3 - c_1b_3) + a_3(b_1c_2 - c_1b_2)\\ &= a_1(b_2c_3 - b_3c_2) - b_1(a_2c_3 - c_2a_3) + c_1(a_2b_3 - b_2a_3)\\ &= \Delta \end{aligned} \tag{1.15}$$

N.B. Any theorem concerning the rows of a determinant is equally applicable to its column and vice versa.

2. **If any two rows (or, columns) are interchanged, then the value of the determinant remains unchanged, but its sign is changed.**

$$\text{Let } \Delta = \begin{vmatrix} a_1 & b_1 & c_1 \\ a_2 & b_2 & c_2 \\ a_3 & b_3 & c_3 \end{vmatrix}.$$

Interchanging the 2nd and 3rd columns, we have

$$\Delta' = \begin{vmatrix} a_1 & c_1 & b_1 \\ a_2 & c_2 & b_2 \\ a_3 & c_3 & b_3 \end{vmatrix}.$$

Now expanding Δ', we get

$$\begin{aligned}\Delta' &= a_1(b_3c_2 - c_3b_2) + c_1(a_2b_3 - b_2a_3) + b_1(a_2c_3 - c_2a_3)\\ &= -a_1(b_2c_3 - b_3c_2) + b_1(a_2c_3 - a_3c_2) - c_1(a_2b_3 - a_3b_2)\\ &= \begin{vmatrix} a_1 & b_1 & c_1 \\ a_2 & b_2 & c_2 \\ a_3 & b_3 & c_3 \end{vmatrix}\\ &= -\Delta \end{aligned} \tag{1.16}$$

N.B. If a column (or, row) of a determinant passes over two columns (or, rows), then

$$\begin{vmatrix} b_1 & c_1 & a_1 \\ b_2 & c_2 & a_2 \\ b_3 & c_3 & a_3 \end{vmatrix}.$$

In that case, $\Delta' = (-1)^2\Delta$.

In general, if m columns (or, rows) are passed over by any column (or, row) of Δ, then

$$\Delta' = (-1)^m\Delta. \tag{1.17}$$

3. **If any two columns (or, rows) of a determinant are identical, then the determinant vanishes.**

Let the 1st and the 2nd columns of a determinant be identical, i.e.,

$$\Delta = \begin{vmatrix} a_1 & a_1 & c_1 \\ a_2 & a_2 & c_2 \\ a_3 & a_3 & c_3 \end{vmatrix}.$$

interchanging the **1st** and **2nd** columns, we get

$$\Delta' = \begin{vmatrix} a_1 & a_1 & c_1 \\ a_2 & a_2 & c_2 \\ a_3 & a_3 & c_3 \end{vmatrix}.$$

Now by property 2,

$$\Delta' = -\Delta, \quad \text{i.e.,} \quad \Delta = -\Delta$$

or, $\quad 2\Delta = 0, \quad$ i.e., $\quad \Delta = 0. \qquad (1.18)$

4. **If each element of a column (or, row) of a determinant is multiplied by a common factor (say, q), then the resulting determinant is multiplied by that factor (i.e., q)**

Let $\Delta = \begin{vmatrix} a_1 & b_1 & c_1 \\ a_2 & b_2 & c_2 \\ a_3 & b_3 & c_3 \end{vmatrix}.$

Multiplying the third column of Δ by q, we get

$$\Delta' = \begin{vmatrix} a_1 & b_1 & qc_1 \\ a_2 & b_2 & qc_2 \\ a_3 & b_3 & qc_3 \end{vmatrix}.$$

Now, expanding Δ', we have

$$\begin{aligned} \Delta' &= qa_1(b_2c_3 - b_3c_2) - qb_1(a_2c_3 - a_3c_2) + qc_1(a_2b_3 - a_3b_2) \\ &= q[a_1(b_2c_3 - b_3c_2) - b_1(a_2c_3 - a_3c_2) + c_1(a_2b_3 - a_3b_2)] \\ &= q\begin{vmatrix} a_1 & b_1 & c_1 \\ a_2 & b_2 & c_2 \\ a_3 & b_3 & c_3 \end{vmatrix} \end{aligned}$$

or, $\Delta' = q\Delta. \qquad (1.19)$

Similarly, $\begin{vmatrix} a_1 & b_1 & c_1 \\ a_2 & b_2 & c_2 \\ qa_3 & qb_3 & qc_3 \end{vmatrix} = q\begin{vmatrix} a_1 & b_1 & c_1 \\ a_2 & b_2 & c_2 \\ a_3 & b_3 & c_3 \end{vmatrix}.$

N.B. $\begin{vmatrix} a_1 & b_1 & pb_1 \\ a_2 & b_2 & pb_2 \\ a_3 & b_3 & pb_3 \end{vmatrix} = p\begin{vmatrix} a_1 & b_1 & b_1 \\ a_2 & b_2 & b_2 \\ a_3 & b_3 & b_3 \end{vmatrix} = p \times 0 = 0.$

5. **If each element of a column (or, row) consists of m terms, then the determinant can be expressed as the sum of m determinants.** This is known as the **sum and difference rule** of determinants.

Let $\Delta = \begin{vmatrix} a_1 & b_1 & c_1 + d_1 - e_1 \\ a_2 & b_2 & c_2 + d_2 - e_2 \\ a_3 & b_3 & c_3 + d_3 - e_3 \end{vmatrix}.$

Expanding Δ by third column, we have

$$\begin{aligned} \Delta &= (c_1 + d_1 - e_1)(a_2b_3 - a_3b_2) - (c_2 + d_2 - e_2)(a_1b_3 - a_3b_1) \\ &\qquad + (c_3 + d_3 - e_3)(a_1b_2 - a_2b_1) \\ &= [c_1(a_2b_3 - a_3b_2) - c_2(a_1b_3 - a_3b_1) + c_3(a_1b_2 - a_2b_1)] \\ &\qquad + [d_1(a_2b_3 - a_3b_2) - d_2(a_1b_3 - a_3b_1) + d_3(a_1b_2 - a_2b_1)] \\ &\qquad - [e_1(a_2b_3 - a_3b_2) - e_2(a_1b_3 - a_3b_1) + e_3(a_1b_2 - a_2b_1)] \end{aligned}$$

$$= \begin{vmatrix} a_1 & b_1 & c_1 \\ a_2 & b_2 & c_2 \\ a_3 & b_3 & c_3 \end{vmatrix} + \begin{vmatrix} a_1 & b_1 & d_1 \\ a_2 & b_2 & d_2 \\ a_3 & b_3 & d_3 \end{vmatrix} - \begin{vmatrix} a_1 & b_1 & e_1 \\ a_2 & b_2 & e_2 \\ a_3 & b_3 & e_3 \end{vmatrix}.$$

It can be shown that,

$$\begin{vmatrix} a_1 & b_1+p_1 & c_1+q_1 \\ a_2 & b_2+p_2 & c_2+q_2 \\ a_3 & b_3+p_3 & c_3+q_3 \end{vmatrix} = \begin{vmatrix} a_1 & b_1 & c_1 \\ a_2 & b_2 & c_2 \\ a_3 & b_3 & c_3 \end{vmatrix} + \begin{vmatrix} a_1 & p_1 & c_1 \\ a_2 & p_2 & c_2 \\ a_3 & p_3 & c_3 \end{vmatrix} + \begin{vmatrix} a_1 & b_1 & q_1 \\ a_2 & b_2 & q_2 \\ a_3 & b_3 & q_3 \end{vmatrix} + \begin{vmatrix} a_1 & p_1 & q_1 \\ a_2 & q_2 & q_2 \\ a_3 & p_3 & q_3 \end{vmatrix}. \tag{1.20}$$

6. **If a multiple of one column is added (row by row) to another column or, if a multiple of one row is added (column by column) to another row, then the determinant remains unaltered.**

Let $\Delta = \begin{vmatrix} a_1 & b_1 & c_1 \\ a_2 & b_2 & c_2 \\ a_3 & b_3 & c_3 \end{vmatrix}$.

Then, $\Delta' = \begin{vmatrix} a_1+pb_1-qc_1 & b_1 & c_1 \\ a_2+pb_2-qc_2 & b_2 & c_2 \\ a_3+pb_3-qc_3 & b_3 & c_3 \end{vmatrix}$

$$= \begin{vmatrix} a_1 & b_1 & c_1 \\ a_2 & b_2 & c_2 \\ a_3 & b_3 & c_3 \end{vmatrix} + \begin{vmatrix} pb_1 & b_1 & c_1 \\ pb_2 & b_2 & c_2 \\ pb_3 & b_3 & c_3 \end{vmatrix} + \begin{vmatrix} -qc_1 & b_1 & c_1 \\ -qc_2 & b_2 & c_2 \\ -qc_3 & b_3 & c_3 \end{vmatrix}$$

$= \Delta + 0 + 0$ [by property 4 (cf. N.B.)]

$= \Delta.$ (1.21)

7. **If each element of a column (or, row) is zero, then the value of the determinant is zero.**

Let $\Delta = \begin{vmatrix} a_1 & b_1 & 0 \\ a_2 & b_2 & 0 \\ a_3 & b_3 & 0 \end{vmatrix} = - \begin{vmatrix} a_1 & 0 & b_1 \\ a_2 & 0 & b_2 \\ a_3 & 0 & b_3 \end{vmatrix}$

$= \begin{vmatrix} 0 & a_1 & b_1 \\ 0 & a_2 & b_2 \\ 0 & a_3 & b_3 \end{vmatrix}$ [by property 2]

$= 0(a_2b_3 - a_3b_2) - a_1(0 \times b_3 - 0 \times b_2) + b_1(0 \times a_3 - 0 \times a_2)$

$= 0.$ (1.22)

8. **The Factor Theorem : If the elements of a determinant Δ be rational integral function of x and two columns (or, rows) becomes identical for $x = a$, then $(x - a)$ is a factor of Δ.**

Let $\Delta = f(x)$.

Since $\Delta = 0$ for $x = a$, $f(a) = 0$, i.e., $(x - a)$ is a factor of $f(x)$.

$\therefore$ $(x - a)$ is a factor of Δ.

1.6 Symmetric and Skew-symmetric Determinants

1. Symmetric Determinant : If the elements of a given determinant Δ and the elements of Δ', obtained by a complete interchange of rows and columns of Δ are identical, then Δ is called a **symmetric determinant**. For a symmetric determinant, $\Delta = |a|$, $a_{ij} = a_{ji}$.

Skew Determinant and Skew-symmetric Determinant

If the elements of a given determinant $\Delta = |a|$ are equal to the negative of the elements of the determinant Δ', obtained by complete interchange of rows and columns of Δ, then Δ is a **skew determinant**. For a skew determinant, $a_{ij} = -a_{ij}$.

In addition to the above, if $a_{11} = a_{22} = a_{33} = 0$ and $a_{12} = -a_{21}$, $a_{13} = -a_{31}$, $a_{23} = -a_{32}$, then the corresponding determinant is called a **skew symmetric** determinant.

Example : Symmetric determinant, $\begin{vmatrix} a & h & g \\ h & b & f \\ g & f & c \end{vmatrix}$.

Skew determinant, $\begin{vmatrix} a & h & -g \\ h & b & -f \\ g & f & c \end{vmatrix}$.

Skew-symmetric determinant, $\begin{vmatrix} 0 & a & b \\ -a & 0 & c \\ -b & -c & 0 \end{vmatrix}$.

1.7 Multiplication of determinants

The result of multiplication of two determinants of the same order is itself a determinant of that order.

Let Δ_1 and Δ_2 be two 3×3 determinants as under :

$$\Delta_1 = \begin{vmatrix} a_1 & b_1 & c_1 \\ a_2 & b_2 & c_2 \\ a_3 & b_3 & c_3 \end{vmatrix} \quad \text{and} \quad \Delta_2 = \begin{vmatrix} l_1 & m_1 & n_1 \\ l_2 & m_2 & n_2 \\ l_3 & m_3 & n_3 \end{vmatrix};$$

$$\text{then} \quad \Delta_1 \times \Delta_2 = \begin{vmatrix} a_1 & b_1 & c_1 \\ a_2 & b_2 & c_2 \\ a_3 & b_3 & c_3 \end{vmatrix} \times \begin{vmatrix} l_1 & m_1 & n_1 \\ l_2 & m_2 & n_2 \\ l_3 & m_3 & n_3 \end{vmatrix}$$

$$= \begin{vmatrix} a_1l_1 + b_1m_1 + c_1n_1 & a_1l_2 + b_1m_2 + c_1n_2 & a_1l_3 + b_1m_3 + c_1n_3 \\ a_2l_1 + b_2m_1 + c_2n_1 & a_2l_2 + b_2m_2 + c_2n_2 & a_2l_3 + b_2m_3 + c_2n_3 \\ a_3l_1 + b_3m_1 + c_3n_1 & a_3l_2 + b_3m_2 + c_3n_2 & a_3l_3 + b_3m_3 + c_3n_3 \end{vmatrix}. \quad (1.23)$$

1.8 Properties of Symmetric and Skew-symmetric Determinants

1. The square of any determinant is a symmetric determinant.

Let $\Delta = \begin{vmatrix} a_1 & b_1 & c_1 \\ a_2 & b_2 & c_2 \\ a_3 & b_3 & c_3 \end{vmatrix}$.

$$\therefore \quad \Delta^2 = \begin{vmatrix} a_1 & b_1 & c_1 \\ a_2 & b_2 & c_2 \\ a_3 & b_3 & c_3 \end{vmatrix}^2 = \begin{vmatrix} a_1 & b_1 & c_1 \\ a_2 & b_2 & c_2 \\ a_3 & b_3 & c_3 \end{vmatrix} \times \begin{vmatrix} a_1 & b_1 & c_1 \\ a_2 & b_2 & c_2 \\ a_3 & b_3 & c_3 \end{vmatrix}$$

$$\text{or, } D = \Delta^2 = \begin{vmatrix} a_1^2+b_1^2+c_1^2 & a_1a_2+b_1b_2+c_1c_2 & a_1a_3+b_1b_3+c_1c_3 \\ a_2a_1+b_2b_1+c_2c_1 & a_2^2+b_2^2+c_2^2 & a_2a_3+b_2b_3+c_2c_3 \\ a_3a_1+b_3b_1+c_3c_1 & a_3a_2+b_3b_2+c_3c_2 & a_3^2+b_3^2+c_3^2 \end{vmatrix}$$

$$\therefore \quad D' = \begin{vmatrix} a_1^2+b_1^2+c_1^2 & a_2a_1+b_2b_1+c_2c_1 & a_3a_1+b_3b_1+c_3c_1 \\ a_1a_2+b_1b_2+c_1c_2 & a_2^2+b_2^2+c_2^2 & a_3a_2+b_3b_2+c_3c_2 \\ a_1a_3+b_1b_3+c_1c_3 & a_2a_3+b_2b_3+c_2c_3 & a_3^2+b_3^2+c_3^2 \end{vmatrix}$$

From the expressions of D and D' it is clear that $a_{ij} = a_{ji}$. Hence, D is a symmetric determinant.

2. A skew-symmetric determinant of even order is a perfect square.

$D = \begin{vmatrix} 0 & a \\ -a & 0 \end{vmatrix}$ is a skew-symmetric determinant of order 2 (even order). Then,

$$D = 0 + a^2 = a^2, \tag{1.24}$$

which is a perfect square.

3. The value of a skew-symmetric determinant of odd order is zero.

$\Delta = \begin{vmatrix} 0 & a & b \\ -a & 0 & c \\ -b & -c & 0 \end{vmatrix}$ is a skew-symmetric determinant of odd order.

$$\text{Then, } \Delta' = \begin{vmatrix} 0 & -a & -b \\ a & 0 & -c \\ b & c & 0 \end{vmatrix} = (-1)^3 \begin{vmatrix} 0 & a & b \\ -a & b & c \\ -b & -c & 0 \end{vmatrix} = -\Delta$$

or, $\quad \Delta = -\Delta \quad (\because \Delta' = \Delta)$

or, $\quad 2\Delta = 0, \quad \text{i.e., } \quad \Delta = 0 \quad (1.25)$

1.9 Reduction in Order of a Determinant

In evaluating the numerical determinants, the reduction in order of the determinant sometimes appears to be very useful. **F. Chio** reduced the order of the determinant as explained below :

Let Δ be a fourth order determinant as,

$$\Delta = \begin{vmatrix} a_1 & b_1 & c_1 & d_1 \\ a_2 & b_2 & c_2 & d_2 \\ a_3 & b_3 & c_3 & d_3 \\ a_4 & b_4 & c_4 & d_4 \end{vmatrix}$$

First the elements of a row (or column) are made unity. The elements of Δ in the second row may be made unity as follows :

$$\Delta = a_2b_2c_2d_2 \begin{vmatrix} \frac{a_1}{a_2} & \frac{b_1}{b_2} & \frac{c_1}{c_2} & \frac{d_1}{d_2} \\ 1 & 1 & 1 & 1 \\ \frac{a_3}{a_2} & \frac{b_3}{b_2} & \frac{c_3}{c_2} & \frac{d_3}{d_2} \\ \frac{a_4}{a_2} & \frac{b_4}{b_2} & \frac{c_4}{c_2} & \frac{d_4}{d_2} \end{vmatrix} = a_2b_2c_2d_2 \begin{vmatrix} A_1 & B_1 & C_1 & D_1 \\ 1 & 1 & 1 & 1 \\ A_3 & B_3 & C_3 & D_3 \\ A_4 & B_4 & C_4 & D_4 \end{vmatrix},$$

where $\frac{a_i}{a_2} = A_i$, $\frac{b_i}{b_2} = B_i$, $\frac{c_i}{c_2} = C_i$ and $\frac{d_i}{d_2} = D_i \quad i = 1, 2, 3, 4;\ A_2 = B_2 = C_2 = D_2 = 1$

$$\text{or,}\qquad \Delta = a_2b_2c_2d_2 \begin{vmatrix} (A_1 - B_1) & (B_1 - C_1) & (C_1 - D_1) & D_1 \\ 0 & 0 & 0 & 1 \\ (A_3 - B_3) & (B_3 - C_3) & (C_3 - D_3) & D_3 \\ (A_4 - B_4) & (B_4 - C_4) & (C_4 - D_4) & D_4 \end{vmatrix}$$

$$= a_2b_2c_2d_2(-1)^{2+4} \begin{vmatrix} (A_1 - B_1) & (B_1 - C_1) & (C_1 - D_1) \\ (A_3 - B_3) & (B_3 - C_3) & (C_3 - D_3) \\ (A_4 - B_4) & (B_4 - C_4) & (C_4 - D_4) \end{vmatrix}. \tag{1.26}$$

This is a determinant of order three.

Repeating the above process, the order of a higher order determinant can be reduced to any desired lower order.

1.10 Adjoint and Reciprocal Determinant

The adjoint, or adjugate of a determinant Δ is the determinant Δ' whose elements are the co-factors of the corresponding elements of Δ, i.e.,

$$\text{If } \Delta = \begin{vmatrix} a_{11} & a_{12} & a_{13} & a_{14} \\ a_{21} & a_{22} & a_{23} & a_{24} \\ a_{31} & a_{32} & a_{33} & a_{34} \\ a_{41} & a_{42} & a_{43} & a_{44} \end{vmatrix}, \text{ then the adjoint or adjugate of } \Delta \text{ is}$$

$$\Delta' = \begin{vmatrix} A_{11} & A_{12} & A_{13} & A_{14} \\ A_{21} & A_{22} & A_{23} & A_{24} \\ A_{31} & A_{32} & A_{33} & A_{34} \\ A_{41} & A_{42} & A_{43} & A_{44} \end{vmatrix},$$

where A_{ij} $(i, j = 1, 2, 3, 4$ are the co-factors of a_{ij} $(i, j = 1, 2, 3, 4)$ in Δ.

The **reciprocal** or **inverse** of Δ $(\neq 0)$ is

$$\begin{vmatrix} \frac{A_{11}}{\Delta} & \frac{A_{12}}{\Delta} & \frac{A_{13}}{\Delta} & \frac{A_{14}}{\Delta} \\ \frac{A_{21}}{\Delta} & \frac{A_{22}}{\Delta} & \frac{A_{23}}{\Delta} & \frac{A_{24}}{\Delta} \\ \frac{A_{31}}{\Delta} & \frac{A_{32}}{\Delta} & \frac{A_{33}}{\Delta} & \frac{A_{34}}{\Delta} \\ \frac{A_{41}}{\Delta} & \frac{A_{42}}{\Delta} & \frac{A_{43}}{\Delta} & \frac{A_{44}}{\Delta} \end{vmatrix}. \tag{1.27}$$

1.11 Jacobi's Theorem

(a) If Δ $(\neq 0)$ be a determinant of order n and Δ' be its adjoint, then $\Delta' = \Delta^{n-1}$.

We consider a determinant of order n, i.e.,

$$\Delta = \begin{vmatrix} a_{11} & a_{12} & \cdots & a_{1n} \\ a_{21} & a_{22} & \cdots & a_{2n} \\ \cdots & \cdots & \cdots & \cdots \\ \cdots & \cdots & \cdots & \cdots \\ a_{n1} & a_{n2} & \cdots & a_{nn} \end{vmatrix},$$

then the adjoint of Δ is

$$\Delta' = \begin{vmatrix} A_{11} & A_{12} & \cdots & A_{1n} \\ A_{21} & A_{22} & \cdots & A_{2n} \\ \cdots & \cdots & \cdots & \cdots \\ \cdots & \cdots & \cdots & \cdots \\ A_{n1} & A_{n2} & \cdots & A_{nn} \end{vmatrix},$$

where A_{ij} is the co-factor of a_{ij} in Δ.

Now consider the product,

$$\Delta\Delta' = \begin{vmatrix} a_{11} & a_{12} & \cdots & a_{1n} \\ a_{21} & a_{22} & \cdots & a_{2n} \\ \cdots & \cdots & \cdots & \cdots \\ \cdots & \cdots & \cdots & \cdots \\ a_{n1} & a_{n2} & \cdots & a_{nn} \end{vmatrix} \times \begin{vmatrix} A_{11} & A_{12} & \cdots & A_{1n} \\ A_{21} & A_{22} & \cdots & A_{2n} \\ \cdots & \cdots & \cdots & \cdots \\ \cdots & \cdots & \cdots & \cdots \\ A_{n1} & A_{n2} & \cdots & A_{nn} \end{vmatrix}$$

$$= \begin{vmatrix} \sum a_{1k}A_{1k} & \sum a_{1k}A_{2k} & \cdots & \sum a_{1k}A_{nk} \\ \sum a_{2k}A_{1k} & \sum a_{2k}A_{2k} & \cdots & \sum a_{2k}A_{nk} \\ \cdots & \cdots & \cdots & \cdots \\ \cdots & \cdots & \cdots & \cdots \\ \sum a_{nk}A_{1k} & \sum a_{nk}A_{2k} & \cdots & \sum a_{nk}A_{nk} \end{vmatrix}$$

$$= \begin{vmatrix} \Delta & 0 & \cdots & 0 \\ 0 & \Delta & \cdots & 0 \\ \cdots & \cdots & \cdots & \cdots \\ \cdots & \cdots & \cdots & \cdots \\ 0 & 0 & \cdots & \Delta \end{vmatrix} \qquad \left[\because \sum a_{ik}A_{jk} = 0, \text{ when } i \neq j \atop = \Delta, \text{ when } i = j\right]$$

$$= \Delta^n.$$

$$\therefore \quad \Delta' = \frac{\Delta^n}{\Delta} = \Delta^{n-1} \quad [\text{if } \Delta \neq 0] \tag{1.28}$$

N.B. (1) If $\Delta = 0$, then it can be shown that $\Delta' = 0$, hence the relation $\Delta' = \Delta^{n-1}$ follows.

(2) The value of the reciprocal determinant of Δ ($\neq 0$) is $\dfrac{\Delta'}{\Delta^n} = \dfrac{\Delta^{n-1}}{\Delta^n} = \dfrac{1}{\Delta}$.

(3) The reciprocal of Δ' is Δ, i.e., $\dfrac{1}{\Delta'} = \dfrac{1}{\frac{1}{\Delta}} = \Delta$.

(b) If Δ' be the adjoint of a determinant Δ ($\neq 0$) of order n, then any minor M'_r of Δ' of order r is equal to the product of Δ^{r-1} and the algebraic component $\overline{M}_{n-r}$ of the corresponding minor M_r of Δ, i.e., $M'_r = \Delta^{r-1}\overline{M}_{n-r}$.

Consider the determinant Δ for $n = 4$, i.e., $\Delta = \begin{vmatrix} a_1 & b_1 & c_1 & d_1 \\ a_2 & b_2 & c_2 & d_2 \\ a_3 & b_3 & c_3 & d_3 \\ a_4 & b_4 & c_4 & d_4 \end{vmatrix} \neq 0$, then

$$\begin{vmatrix} A_1 & B_1 \\ A_2 & B_2 \end{vmatrix} = \Delta \begin{vmatrix} c_3 & d_3 \\ c_4 & d_4 \end{vmatrix}, \quad \begin{vmatrix} A_1 & B_1 & C_1 \\ A_2 & B_2 & C_2 \\ A_3 & B_3 & C_3 \end{vmatrix} = \Delta^2 d_4, \text{ etc.} \tag{1.29}$$

1.12 Cramer's Rule

A system of linear simultaneous equations can be easily solved by using the method of **Gabriel Cramer** using determinants.

Consider the set of non-homogeneous linear equations as below :

$$\left.\begin{aligned} a_1x + b_1y + c_1z &= d_1 \\ a_2x + b_2y + c_2z &= d_2 \\ a_3x + b_3y + c_3z &= d_3 \end{aligned}\right\}. \tag{1.30}$$

If the determinant of the coefficient of Eqns. (1.30) be $\Delta = \begin{vmatrix} a_1 & b_1 & c_1 \\ a_2 & b_2 & c_2 \\ a_3 & b_3 & c_3 \end{vmatrix}$, then

$$x\Delta = \begin{vmatrix} xa_1 & b_1 & c_1 \\ xa_2 & b_2 & c_2 \\ xa_3 & b_3 & c_3 \end{vmatrix} = \begin{vmatrix} a_1x + b_1y + c_1z & b_1 & c_1 \\ a_2x + b_2y + c_2z & b_2 & c_2 \\ a_3x + b_3y + c_3z & b_3 & c_3 \end{vmatrix} \quad \text{[by Eqn. (1.30)]}$$

$$= \Delta_1 \quad \text{(say)}.$$

$$\therefore \qquad x = \frac{\Delta_1}{\Delta}. \tag{1.31}$$

$$\text{Similarly, } y\Delta = \begin{vmatrix} a_1 & b_1y & c_1 \\ a_2 & b_2y & c_2 \\ a_3 & b_3y & c_3 \end{vmatrix} = \begin{vmatrix} a_1 & a_1x + b_1y + c_1z & c_1 \\ a_2 & a_2x + b_2y + c_2z & c_2 \\ a_3 & a_3x + b_3y + c_3z & c_3 \end{vmatrix} = \begin{vmatrix} a_1 & d_1 & c_1 \\ a_2 & d_2 & c_2 \\ a_3 & d_3 & c_3 \end{vmatrix}$$

$$= \Delta_2 \quad \text{(say)}.$$

$$\therefore \qquad y = \frac{\Delta_2}{\Delta} \tag{1.32}$$

$$\text{and} \qquad z\Delta = \begin{vmatrix} a_1 & b_1 & c_1z \\ a_2 & b_2 & c_2z \\ a_3 & b_3 & c_3z \end{vmatrix} = \begin{vmatrix} a_1 & b_1 & a_1x + b_1y + c_1z \\ a_2 & b_2 & a_2x + b_2y + c_2z \\ a_3 & b_3 & a_3x + b_3y + c_3z \end{vmatrix} = \begin{vmatrix} a_1 & b_1 & d_1 \\ a_2 & b_2 & d_2 \\ a_3 & b_3 & d_3 \end{vmatrix}$$

$$= \Delta_3 \quad \text{(say)}.$$

$$\therefore \qquad z = \frac{\Delta_3}{\Delta}. \tag{1.33}$$

1.13 Consistency of Equations

If a system of linear equations has one (or more) solutions, the equations are said to be consistent or compatible, i.e., the equations exist, otherwise they are said to be inconsistent or incompatible.

Let a set of equations be

$$\left.\begin{aligned} 2x + 3y &= 3 \\ 4x + 6y &= 7 \end{aligned}\right\}. \tag{1.34}$$

There is no set of values of x and y which satisfies both these equations an hence are **inconsistent**.

Again, the set of simultaneous equations

$$\left.\begin{aligned} 2x + 3y &= 3 \\ 4x + 6y &= 6 \end{aligned}\right\} \tag{1.35}$$

are **consistent**, for there exist values of x and y which satisfy both the equations.

Here, Δ for both equations (1.34) and (1.35) is zero, i.e., $\Delta = 0$.

By **Cramer's rule**, the equations,

$$a_i x + b_i y + c_i z = d_i \quad (i = 1, 2, 3) \tag{1.36}$$

have a unique solution :

$$x = \frac{\Delta_1}{\Delta}, \quad y = \frac{\Delta_2}{\Delta}, \quad z = \frac{\Delta_3}{\Delta}, \quad \text{if } \Delta \neq 0,$$

where Δ_i is the determinant obtained by replacing the ith column in the coefficient by d_1, d_2, d_3.

The equations (1.36) are **consistent**, if $\Delta = 0$. But for $\Delta = 0$ and $\Delta_1 \neq 0$, $\Delta_2 \neq 0$, $\Delta_3 \neq 0$, these equations are **inconsistent**, since they do not possess any solutions in this condition.

N.B. The concept of consistency is of great practical importance specially in dealing with problems in electrical networks and frequency of oscillatory system.

In a more general way consider the system of equations,

$$a_{i1}x_1 + a_{i2}x_2 + \cdots a_{in}x_n = d_i \quad (i = 1, 2, \ldots, n). \tag{1.37}$$

The unique solution, by Cramer's rule,

$$\boldsymbol{x_1 = \frac{\Delta_1}{\Delta}, \; x_2 = \frac{\Delta_2}{\Delta}, \; \ldots, \; x_n = \frac{\Delta_n}{\Delta}, \quad \text{if } \Delta \neq 0} \tag{1.38}$$

and Δ_i is the determinant obtained by replacing the ith column in the coefficient determinant by the column consisting the elements, d_1, d_2, $\ldots$, d_n.

Equations (1.37) are **consistent**, if $\Delta \neq 0$.

If $\Delta = 0$ and at least one of the determinants, Δ_1, Δ_2, $\ldots$, Δ_n be non-zero, then these equations do not have any solutions and hence, **inconsistent**.

The system of Eqns. (1.36), however, admits an infinite number of solutions, if $\Delta = 0$ alongwith $\Delta_1 = \Delta_2 = \cdots = \Delta_n = 0$.

1.14 Homogeneous Linear Equations

We consider a system of homogeneous linear equations,

$$\left.\begin{aligned} a_1x + b_1y + c_1z &= d_1 \\ a_2x + b_2y + c_2z &= d_2 \\ a_3x + b_3y + c_3z &= d_3 \end{aligned}\right\}. \tag{1.39}$$

By **Cramer's rule**, the solution is

$$x = \frac{0}{\Delta}, \quad y = \frac{0}{\Delta}, \quad z = \frac{0}{\Delta}. \tag{1.40}$$

If $\Delta \neq 0$, then $\boldsymbol{x = y = z = 0}$. Such a solution for which all unknowns, i.e., x, y, z are zero is called **a trivial solution.**

If not all of x, y, z are zero simultaneously, then the solution is called **non-trivial**. The system of Eqns. (1.36) are then said to be consistent or compatible.

Condition that Eqns. (1.36) have non-trivial solutions : Let $z \neq 0$. Then

$$a_1\left(\frac{x}{z}\right) + b_1\left(\frac{y}{z}\right) + c_1 = 0$$

$$a_2\left(\frac{x}{z}\right) + b_2\left(\frac{y}{z}\right) + c_2 = 0$$

$$a_3\left(\frac{x}{z}\right) + b_3\left(\frac{y}{z}\right) + c_3 = 0.$$

Eliminating the unknowns x/z and y/z form the above equations, we get

$$\Delta = \begin{vmatrix} a_1 & b_1 & c_1 \\ a_2 & b_2 & c_2 \\ a_3 & b_3 & c_3 \end{vmatrix} = 0. \tag{1.41}$$

This is the required condition.

1.15 Solved Problems

Example 1.1. *Evaluate the given determinant Δ as below and find its value, if $x+y+z=0$.*

$$\Delta = \begin{vmatrix} 1 & 1 & 1 \\ x & y & z \\ x^3 & y^3 & z^3 \end{vmatrix} = 0$$

Solution : The given determinant is

$$\Delta = \begin{vmatrix} 1 & 1 & 1 \\ x & y & z \\ x^3 & y^3 & z^3 \end{vmatrix} = \begin{vmatrix} 1 & 0 & 0 \\ x & y-x & z-x \\ x^3 & y^3-x^3 & z^3-x^3 \end{vmatrix}.$$

The 2nd column $(C_2') =$ [2nd column (C_2) − 1st column (C_1)] of the given determinant Similarly, $C_3' = C_3 - C_1$.

$$\text{or,}\quad \Delta = \begin{vmatrix} (y-x) & (z-x) \\ (y-x)(y^2+xy+x^2) & (z-x)(z^3+xz+x^2) \end{vmatrix}$$

$$= (y-x)(z-x)\begin{vmatrix} 1 & (z-x) \\ (x^2+xy+y^2) & (x^2+xz+z^2) \end{vmatrix}$$

$$= (y-x)(z-x)(x^2+xz+z^2-x^2-xy-y^2)$$

$$= (y-x)(z-x)\{x(z-y)+(z^2-y^2)\}$$

$$\text{or,}\quad \boldsymbol{\Delta = (y-x)(z-x)(z-y)(x+y+z)}.$$

If now we put $x+y+z=0$, then the value of $\Delta = (y-x)(z-x)(z-y) \times 0 = 0$.

Example 1.2. *If $1+\omega+\omega^2=0$, then show that* $\Delta = \begin{vmatrix} 1 & \omega & \omega^2 \\ \omega & \omega^2 & 1 \\ \omega^2 & 1 & \omega \end{vmatrix} = 0.$

Solution : We have $\Delta = \begin{vmatrix} 1 & \omega & \omega^2 \\ \omega & \omega^2 & 1 \\ \omega^2 & 1 & \omega \end{vmatrix} = \begin{vmatrix} 1+\omega+\omega^2 & \omega & \omega^2 \\ 1+\omega+\omega^2 & \omega^2 & 1 \\ 1+\omega+\omega^2 & 1 & \omega \end{vmatrix}.$

By adding the 2nd column (C_2) and third column (C_3) to the first (C_1), i.e.,

$$C_1' = C_2 + C_3 \quad \text{or,}\quad \Delta = \begin{vmatrix} 0 & \omega & \omega^2 \\ 0 & \omega^2 & 1 \\ 0 & 1 & \omega \end{vmatrix} = 0, \quad \textbf{(Proved)} \quad [\because\ 1+\omega+\omega^2=0]$$

Example 1.3. *If $A + B + C = \pi$, then show that* $\Delta = \begin{vmatrix} -1 & \cos C & \cos B \\ \cos C & -1 & \cos A \\ \cos B & \cos A & -1 \end{vmatrix} = 0.$

Solution : We have

$$\Delta = \begin{vmatrix} -1 & \cos C & \cos B \\ \cos C & -1 & \cos A \\ \cos B & \cos A & -1 \end{vmatrix} = \begin{vmatrix} -1 & \cos C & \cos B \\ 0 & \cos^2 C & \cos B \cos C + \cos A \\ 0 & \cos B \cos C + \cos A & \cos^2 B - 1 \end{vmatrix}$$

$$R_2' = R_2 + R_1 \cos C$$
$$R_3' = R_3 + R_1 \cos B$$

R_1, R_2, R_3 stand for the 1st, 2nd and 3rd rows respectively of Δ and R_2' and R_3' the 2nd and 3rd rows of the latter determinant,

or, $\Delta = -\begin{vmatrix} -\sin^2 C & \cos A + \cos B \cos C \\ \cos A + \cos B \cos C & -\sin^2 B \end{vmatrix}$

$$= -[\sin^2 B \sin^2 C - (\cos A + \cos B \cos C)^2]$$
$$= -[\sin^2 B \sin^2 C - \{\cos(\pi - \overline{B + C}) + \cos B \cos C\}^2] \quad [\because A + B + C = \pi]$$
$$= -[\sin^2 B \sin^2 C - \{-\cos(B + C) + \cos B \cos C\}^2]$$
$$= \mathbf{-[(\sin B \sin C)^2 - (\sin B \sin C)^2] = 0} \quad \textbf{(Proved).}$$

Example 1.4. *Prove that,* $\begin{vmatrix} -2a & a+b & a+c \\ b+a & -2b & b+c \\ c+a & c+b & -2c \end{vmatrix} = 4(a+b)(b+c)(c+a).$

Solution : Let $\Delta = \begin{vmatrix} -2a & a+b & a+c \\ b+a & -2b & b+c \\ c+a & c+b & -2c \end{vmatrix}$.

Putting $a = -b$ in Δ, we get

$$\Delta = \begin{vmatrix} 2b & 0 & c-b \\ 0 & -2b & b+c \\ c-b & b+c & -2c \end{vmatrix} = 2b \begin{vmatrix} -2b & b+c \\ b+c & -2c \end{vmatrix} = (c-b) \begin{vmatrix} 0 & -2b \\ c-b & b+c \end{vmatrix}$$

$$= 2b\{4bc - (b+c)^2\} + (c-b) \cdot 2b \cdot (c-a) = -2b(b-c)^2 + 2b(b-c)^2$$
$$= 0.$$

Hence, $(a + b)$ is a factor of Δ.

In a similar way we can show that $(b + c)$ and $(c + a)$ are also factors of Δ.

$$\therefore \quad \boldsymbol{\Delta = k(a+b)(b+c)(c+a)} \tag{1}$$

As the degree of Δ is three in a, b, c and the right-hand side of the above equation without k is of degree 3 in a, b, c, we can say that k is independent of a, b, c.

Now putting, $a=0$, $b=1$, $c=1$, in Δ, we have

$$\Delta = \begin{vmatrix} 0 & 1 & 1 \\ 1 & -2 & 2 \\ 1 & 2 & -2 \end{vmatrix} = \begin{vmatrix} 0 & 1 & 1 \\ 1 & -4 & 2 \\ 1 & 4 & -2 \end{vmatrix}, \quad [C_2' = C_2 - C_3]$$

$$= 4 + 4 = 8.$$

From Eqn (1),

$$8 = k(0+1)(1+1)(1+0) = 2k$$

$\therefore \quad k = 4.$

$\therefore \quad \boldsymbol{\Delta = 4(a+b)(b+c)(c+a)}$ **(Proved).**

Example 1.5. *Prove that,* $\begin{vmatrix} 1 & 1 & 1 \\ a & b & c \\ a^3 & b^3 & c^3 \end{vmatrix} = (a-b)(b-c)(c-a)(a+b+c).$

Solution :

$$\text{Let } \boldsymbol{\Delta} = \begin{vmatrix} 1 & 1 & 1 \\ a & b & c \\ a^3 & b^3 & c^3 \end{vmatrix} = \begin{vmatrix} 0 & 0 & 1 \\ (a-b) & (b-c) & c \\ (a^3-b^3) & (b^3-c^3) & c^3 \end{vmatrix} = \begin{vmatrix} (a-b) & (b-c) \\ (a^3-b^3) & (b^3-c^3) \end{vmatrix}$$

$$= (a-b)(b^3-c^3) - (b-c)(a^3-b^3) = (a-b)(b-c)(b^2+bc+c^2-a^2-ab-b^3)$$

$$= (a-b)(b-c)\{b(c-a)+(c-a)(c+a)\}$$

$$= \boldsymbol{(a-b)(b-c)(c-a)(a+b+c)} \quad \textbf{(Proved)}.$$

Example 1.6. *Eliminate l, m and n form the equations :*

$$al + cm + bn = 0$$
$$cl + bm + an = 0$$
$$bl + am + cn = 0$$

and express the result in the simplest form. **[BU(H) 1964]**

Solution : The determinant of the coefficient of the system of equation is

$$0 = \begin{vmatrix} a & c & b \\ c & b & a \\ b & a & c \end{vmatrix}$$

$$= \begin{vmatrix} a+b+c & c & b \\ a+b+c & b & a \\ a+b+c & a & c \end{vmatrix}, \quad [c_1' = c_1 + c_2 + c_3]$$

$$= (a+b+c)\begin{vmatrix} 1 & c & b \\ 1 & b & a \\ 1 & a & c \end{vmatrix}$$

$$= (a+b+c)\begin{vmatrix} 1 & c & b \\ 0 & (b-c) & (a-b) \\ 0 & (a-c) & (c-b) \end{vmatrix}, \quad [R_2' = R_2 - R_1,\ R_3' = R_3 - R_1]$$

$$= (a+b+c)[-(b-c)^2-(a-b)(a-c)] = (a+b+c)(-b^2-c^2+2bc-a^2+ab+ca-bc)$$
$$= -(a+b+c)(a^2+b^2+c^2-bc-ca-ab) = -(a^3+b^3+c^3-3abc)$$

$\therefore \quad \boldsymbol{a^3+b^3+c^3=3abc}.$

Example 1.7. *Show that,* $\begin{vmatrix} x-\alpha & l-\beta & m-\gamma & 1 \\ 0 & x-\beta & n-\gamma & 1 \\ 0 & 0 & x-\gamma & 1 \\ 0 & 0 & 0 & 1 \end{vmatrix} = (x-\alpha)(x-\beta)(x-\gamma).$

[BU(H) 1981]

Solution : Let $\Delta = \begin{vmatrix} x-\alpha & l-\beta & m-\gamma & 1 \\ 0 & x-\beta & n-\gamma & 1 \\ 0 & 0 & x-\gamma & 1 \\ 0 & 0 & 0 & 1 \end{vmatrix}$

$$= \begin{vmatrix} x-\alpha & l-\beta & m-\gamma & 1 \\ 0 & x-\beta & n-\gamma & 1 \\ 0 & 0 & x-\gamma & 1 \\ 0 & 0 & 0 & 1 \end{vmatrix}, \quad \begin{bmatrix} C_1' = C_1 - \alpha C_4 \\ C_2' = C_2 - \beta C_4 \\ C_3' = C_3 - \gamma C_4 \end{bmatrix}$$

$$= (x-\alpha)\begin{vmatrix} x-\beta & n-\gamma & 1 \\ 0 & x-\gamma & 1 \\ 0 & 0 & 1 \end{vmatrix} = (x-\alpha)(x-\beta)\begin{vmatrix} x-\gamma & 1 \\ 0 & 1 \end{vmatrix}$$

$$= \boldsymbol{(x-\alpha)(x-\beta)(x-\gamma)} \quad \textbf{(Proved)}.$$

Example 1.8. *Show that* $\begin{vmatrix} a^2 & 2ab & b^2 \\ b^2 & a^2 & 2ab \\ 2ab & b^2 & a^2 \end{vmatrix}$ *is a perfect square.*

Solution : We consider the product

$$\begin{vmatrix} a & b & 0 \\ 0 & a & b \\ b & 0 & a \end{vmatrix} \times \begin{vmatrix} a & b & 0 \\ 0 & a & b \\ b & 0 & a \end{vmatrix} = \begin{vmatrix} a & b & 0 \\ 0 & a & b \\ b & 0 & a \end{vmatrix} \times \begin{vmatrix} a & 0 & b \\ b & a & 0 \\ 0 & b & a \end{vmatrix}$$

$$= \begin{vmatrix} a^2 & ab+ab & b^2 \\ b^2 & a^2 & ab+ab \\ ab+ab & b^2 & a^2 \end{vmatrix} = \begin{vmatrix} a^2 & 2ab & b^2 \\ b^2 & a^2 & 2ab \\ 2ab & b^2 & a^2 \end{vmatrix}.$$

$$\begin{vmatrix} a^2 & 2ab & b^2 \\ b^2 & a^2 & 2ab \\ 2ab & b^2 & a^2 \end{vmatrix} = \begin{vmatrix} a & b & 0 \\ 0 & a & b \\ b & 0 & a \end{vmatrix}^2 \textbf{ is a perfect square.}$$

Example 1.9. *Prove that,* $\begin{vmatrix} 1+a & 1 & 1 & 1 \\ 1 & 1+b & 1 & 1 \\ 1 & 1 & 1+c & 1 \\ 1 & 1 & 1 & 1+d \end{vmatrix} = abcd\left(1+\frac{1}{a}+\frac{1}{b}+\frac{1}{c}+\frac{1}{d}\right).$

[AU 1986; RU 1982]

Solution : We have $\boldsymbol{\Delta} = \begin{vmatrix} 1+a & 1 & 1 & 1 \\ 1 & 1+b & 1 & 1 \\ 1 & 1 & 1+c & 1 \\ 1 & 1 & 1 & 1+d \end{vmatrix}$

$$= abcd \begin{vmatrix} \left(\frac{1}{a}+1\right) & \frac{1}{a} & \frac{1}{a} & \frac{1}{a} \\ \frac{1}{b} & \left(\frac{1}{b}+1\right) & \frac{1}{b} & \frac{1}{b} \\ \frac{1}{c} & \frac{1}{c} & \left(\frac{1}{c}+1\right) & \frac{1}{c} \\ \frac{1}{d} & \frac{1}{d} & \frac{1}{d} & \left(\frac{1}{d}+1\right) \end{vmatrix}$$

$$= abcd \begin{vmatrix} \left(1+\frac{1}{a}+\frac{1}{b}+\frac{1}{c}+\frac{1}{d}\right) & \left(1+\frac{1}{a}+\frac{1}{b}+\frac{1}{c}+\frac{1}{d}\right) & \left(1+\frac{1}{a}+\frac{1}{b}+\frac{1}{c}+\frac{1}{d}\right) & \left(1+\frac{1}{a}+\frac{1}{b}+\frac{1}{c}+\frac{1}{d}\right) \\ \frac{1}{b} & \left(\frac{1}{b}+1\right) & \frac{1}{b} & \frac{1}{b} \\ \frac{1}{c} & \frac{1}{c} & \left(\frac{1}{c}+1\right) & \frac{1}{c} \\ \frac{1}{d} & \frac{1}{d} & \frac{1}{d} & \left(\frac{1}{d}+1\right) \end{vmatrix}$$

$[R_1' = R_1 + (R_2 + R_3 + R_4)]$

$$= abcd\left(1+\frac{1}{a}+\frac{1}{b}+\frac{1}{c}+\frac{1}{d}\right) \begin{vmatrix} 1 & 1 & 1 & 1 \\ \frac{1}{b} & \left(\frac{1}{b}+1\right) & \frac{1}{b} & \frac{1}{b} \\ \frac{1}{c} & \frac{1}{c} & \left(\frac{1}{c}+1\right) & \frac{1}{c} \\ \frac{1}{d} & \frac{1}{d} & \frac{1}{d} & \left(\frac{1}{d}+1\right) \end{vmatrix}$$

$$= abcd\left(1+\frac{1}{a}+\frac{1}{b}+\frac{1}{c}+\frac{1}{d}\right) \begin{vmatrix} 1 & 0 & 0 & 0 \\ \frac{1}{b} & 1 & 0 & 0 \\ \frac{1}{c} & 0 & 1 & 0 \\ \frac{1}{d} & 0 & 0 & 1 \end{vmatrix}, \begin{bmatrix} C_2' = C_2 - C_1 \\ C_3' = C_3 - C_1 \\ C_4' = C_4 - C_1 \end{bmatrix}$$

$$= \boldsymbol{abcd\left(1+\frac{1}{a}+\frac{1}{b}+\frac{1}{c}+\frac{1}{d}\right)} \quad \textbf{(Proved)}.$$

Example 1.10. *Show that,* $\begin{vmatrix} A_1 & B_1 & C_1 \\ A_2 & B_2 & C_2 \\ A_3 & B_3 & C_3 \end{vmatrix} = \begin{vmatrix} a_1 & b_1 & b_1 \\ a_2 & b_2 & b_2 \\ a_3 & b_3 & b_3 \end{vmatrix}^2$, *where A, B, C are the co-factors of a, b, c respectively in the determinant $(a_1 b_2 c_3)$.* **[RU 1889; NBU 1988; CU 2000]**

Solution : Let $\Delta = \begin{vmatrix} a_1 & b_1 & c_1 \\ a_2 & b_2 & c_2 \\ a_3 & b_3 & c_3 \end{vmatrix}$ and $\Delta' = \begin{vmatrix} A_1 & B_1 & C_1 \\ A_2 & B_2 & C_2 \\ A_3 & B_3 & C_3 \end{vmatrix}$.

Then, $\Delta\Delta' = \begin{vmatrix} a_1 & b_1 & c_1 \\ a_2 & b_2 & c_2 \\ a_3 & b_3 & c_3 \end{vmatrix} \times \begin{vmatrix} A_1 & B_1 & C_1 \\ A_2 & B_2 & C_2 \\ A_3 & B_3 & C_3 \end{vmatrix}$

$$= \begin{vmatrix} a_1A_1+b_1B_1+c_1C_1 & a_1A_2+b_1B_2+c_1C_2 & a_1A_3+b_1B_3+c_1C_3 \\ a_2A_1+b_2B_1+c_2C_1 & a_2A_2+b_2B_2+c_2C_2 & a_2A_3+b_2B_3+c_2C_3 \\ a_3A_1+b_3B_1+c_3C_1 & a_3A_2+b_3B_2+c_3C_2 & a_3A_3+b_3B_3+c_3C_3 \end{vmatrix}$$

$$= \begin{vmatrix} \Delta & 0 & 0 \\ 0 & \Delta & 0 \\ 0 & 0 & \Delta \end{vmatrix} = \Delta^3.$$

$$\therefore \quad \boldsymbol{\Delta' = \frac{\Delta^3}{\Delta} = \Delta^2} \quad \textbf{(Proved).}$$

N.B. Δ' is called the reciprocal (or, adjugate) of Δ.

Example 1.11. *Express* $\begin{vmatrix} 2bc-a^2 & c^2 & b^2 \\ c^2 & 2ca-b^2 & a^2 \\ b^2 & a^2 & 2ab-c^2 \end{vmatrix}$ *as the square of a determinant and find its value.* **[GU 1984; BU 1999]**

Solution : We have the given determinant

$$\begin{vmatrix} 2bc-a^2 & c^2 & b^2 \\ c^2 & 2ca-b^2 & a^2 \\ b^2 & a^2 & 2ab-c^2 \end{vmatrix} = \begin{vmatrix} a(-a)+b\cdot c+c\cdot b & a(-b)+b\cdot a+c\cdot c & a(-c)+b\cdot b+c\cdot a \\ b(-a)+c\cdot c+a\cdot b & b(-b)+c\cdot a+a\cdot c & b(-c)+c\cdot b+a\cdot a \\ c(-a)+a\cdot c+b\cdot b & c(-b)+a\cdot a+b\cdot c & c(-c)+a\cdot b+b\cdot a \end{vmatrix}$$

$$= \begin{vmatrix} a & b & c \\ b & c & a \\ c & a & b \end{vmatrix} \times \begin{vmatrix} -a & c & b \\ -b & a & c \\ -c & b & a \end{vmatrix} = \begin{vmatrix} a & b & c \\ b & c & a \\ c & a & b \end{vmatrix} \times (-1)^2 \begin{vmatrix} a & b & c \\ b & c & a \\ c & a & b \end{vmatrix}$$

$$= \Delta^2,$$

where $\Delta = \begin{vmatrix} a & b & c \\ b & c & a \\ c & a & b \end{vmatrix} = -(a^3+b^3+c^3-3abc).$

$\therefore$ the given determinant, $\boldsymbol{\Delta^2 = (a^3+b^3+c^3-3abc)^2}$.

Example 1.12. *Solve :* $\begin{vmatrix} x^3-a^3 & x^2 & x \\ b^3-a^3 & b^2 & b \\ c^3-a^3 & c^2 & c \end{vmatrix} = 0$ **[BU(H) 1966]**

Solution : We have $\begin{vmatrix} x^3-a^3 & x^2 & x \\ b^3-a^3 & b^2 & b \\ c^3-a^3 & c^2 & c \end{vmatrix} = 0$

or, $\begin{vmatrix} x^3 & x^2 & x \\ b^3 & b^2 & b \\ c^3 & c^2 & c \end{vmatrix} - \begin{vmatrix} a^3 & x^2 & x \\ a^3 & b^2 & b \\ a^3 & c^2 & c \end{vmatrix} = 0$ or, $xbc \begin{vmatrix} x^2 & x & 1 \\ b^2 & b & 1 \\ c^2 & c & 1 \end{vmatrix} - a^3 \begin{vmatrix} 1 & x^2 & x \\ 1 & b^2 & b \\ 1 & c^2 & c \end{vmatrix} = 0$

or, $xbc \begin{vmatrix} x^2 & x & 1 \\ b^2 & b & 1 \\ c^2 & c & 1 \end{vmatrix} - a^3 \begin{vmatrix} x^2 & x & 1 \\ b^2 & b & 1 \\ c^2 & c & 1 \end{vmatrix} = 0$ or, $(xbc - a^3) \begin{vmatrix} x^2 & x & 1 \\ b^2 & b & 1 \\ c^2 & c & 1 \end{vmatrix} = 0$

or, $(xbc - a^3)\begin{vmatrix} x^2 - b^2 & x - b & 0 \\ b^2 - c^2 & b - c & 0 \\ c^2 & c & 1 \end{vmatrix} = 0$ [operating $R_1' = R_1 - R_2$, $R_2' = R_2 - R_3$]

or, $(xbc - a^2)(x - b)(b - c)\begin{vmatrix} x + b & 1 \\ b + c & 1 \end{vmatrix} = 0$ or, $(xbc - a^2)(x - b)(b - c)(x + b - b - c) = 0$

or, $\boldsymbol{(x - b)(x - c)(xbc - a^2)(b - c) = 0}$.

Hence, $x = b,\ c,\ \dfrac{a^2}{bc}$, if $b \neq c$.

Example 1.13. *Show that,* $\begin{vmatrix} 2ab & a^2 & b^2 \\ a^2 & b^2 & 2ab \\ b^2 & 2ab & a^2 \end{vmatrix} = -(a^3 + b^3)^2$. **[CU 1999]**

Solution : We have the given determinant,

$$\begin{vmatrix} 2ab & a^2 & b^2 \\ a^2 & b^2 & 2ab \\ b^2 & 2ab & a^2 \end{vmatrix} = -\begin{vmatrix} a^2 & 2ab & b^2 \\ b^2 & a^2 & 2ab \\ 2ab & b^2 & a^2 \end{vmatrix} \quad \text{[by interchanging } c_1 \text{ and } c_2]$$

$$= -\begin{vmatrix} a^2 + 2ab + b^2 & a^2 + 2ab + b^2 & a^2 + 2ab + b^2 \\ b^2 & a^2 & 2ab \\ 2ab & b^2 & a^2 \end{vmatrix} \quad [R_1' = R_1 + R_2 + R_3]$$

$$= -(a + b)^2\begin{vmatrix} 1 & 1 & 1 \\ b^2 & a^2 & 2ab \\ 2ab & b^2 & a^2 \end{vmatrix}$$

$$= -(a + b)^2\begin{vmatrix} 1 & 0 & 0 \\ b^2 & a^2 - b^2 & 2ab - b^2 \\ 2ab & b^2 - 2ab & a^2 - 2ab \end{vmatrix} \quad \begin{bmatrix} C_2' = C_2 - C_1 \\ C_3' = C_2 - C_1 \end{bmatrix}$$

$$= -(a + b)^2\begin{vmatrix} a^2 - b^2 & 2ab - b^2 \\ b^2 - 2ab & a^2 - 2ab \end{vmatrix}$$

$$= -(a + b)^2[(a^2 - b^2)(a^2 - 2ab) - (2ab - b^2)(b^2 - 2ab)]$$

$$= -(a + b)^2[a^4 - a^2b^2 - 2a^3b + 2ab^3 + 4a^2b^2 + b^4 - 4ab^3]$$

$$= -(a + b)^2[a^4 + a^2b^2 + b^4 - 2a^3b - 2ab^2 + 2a^2b^2]$$

$$= -(a + b)^2(a^2 - ab + b^2)^2 = -[(a + b)(a^2 - ab + b^2)]^2$$

$$= \boldsymbol{-(a^3 + b^3)^2} \quad \textbf{(Proved)}.$$

Example 1.14. *Prove that,* $\begin{vmatrix} 1 + a & 1 & 1 \\ 1 & 1 + b & 1 \\ 1 & 1 & 1 + c \end{vmatrix} = abc\left(1 + \dfrac{1}{a} + \dfrac{1}{b} + \dfrac{1}{c}\right)$. **[NBU 1989]**

Solution : We have the determinant,

$$\begin{vmatrix} 1+a & 1 & 1 \\ 1 & 1+b & 1 \\ 1 & 1 & 1+c \end{vmatrix} = \begin{vmatrix} a & 0 & 1 \\ -b & b & 1 \\ 0 & -c & 1+c \end{vmatrix} \quad \begin{bmatrix} C_1' = C_1 - C_2 \\ C_2' = C_2 - C_3 \end{bmatrix}$$

$$= a\begin{vmatrix} b & 1 \\ -c & 1+c \end{vmatrix} + \begin{vmatrix} -b & b \\ 0 & -c \end{vmatrix} = a(b + bc + c) + bc = abc + bc + ca + ab$$

$$= \boldsymbol{abc\left(1 + \frac{1}{a} + \frac{1}{b} + \frac{1}{c}\right)} \quad \textbf{(Proved).}$$

Example 1.15. *Prove without expanding that,* $\begin{vmatrix} 1 & a & a^2 - bc \\ 1 & b & b^2 - ca \\ 1 & c & c^2 - ab \end{vmatrix} = 0.$

[CU 2000; VU 2000; KU 1995]

Solution : We have the determinant,

$$\begin{vmatrix} 1 & a & a^2 - bc \\ 1 & b & b^2 - ca \\ 1 & c & c^2 - ab \end{vmatrix} = \begin{vmatrix} 1 & a & a^2 \\ 1 & b & b^2 \\ 1 & c & c^2 \end{vmatrix} - \begin{vmatrix} 1 & a & bc \\ 1 & b & ca \\ 1 & c & ab \end{vmatrix} = \begin{vmatrix} 1 & a & a^2 \\ 1 & b & b^2 \\ 1 & c & c^2 \end{vmatrix} - \frac{1}{a}\frac{1}{b}\frac{1}{c}\begin{vmatrix} a & a^2 & abc \\ b & b^2 & abc \\ c & c^2 & abc \end{vmatrix}$$

$$= \begin{vmatrix} 1 & a & a^2 \\ 1 & b & b^2 \\ 1 & c & c^2 \end{vmatrix} - \frac{abc}{abc}\begin{vmatrix} a & a^2 & 1 \\ b & b^2 & 1 \\ c & c^2 & 1 \end{vmatrix} = \begin{vmatrix} 1 & a & a^2 \\ 1 & b & b^2 \\ 1 & c & c^2 \end{vmatrix} - \begin{vmatrix} 1 & a & a^2 \\ 1 & b & b^2 \\ 1 & c & c^2 \end{vmatrix}$$

$$= \mathbf{0} \quad \textbf{(Proved).}$$

Example 1.16. *Prove that,* $\begin{vmatrix} (b+c)^2 & a^2 & a^2 \\ b^2 & (c+a)^2 & b^2 \\ c^2 & c^2 & (a+b)^2 \end{vmatrix} = 2abc(a+b+c)^2.$

[BU(H) 1966, 1983]

Solution : We have $\begin{vmatrix} (b+c)^2 & a^2 & a^2 \\ b^2 & (c+a)^2 & b^2 \\ c^2 & c^2 & (a+b)^2 \end{vmatrix}$

$$= \begin{vmatrix} (b+c)^2 & a^2 - (b+c)^2 & a^2 - (b+c)^2 \\ b^2 & (c+a)^2 - b^2 & 0 \\ c^2 & 0 & (a+b)^2 - c^2 \end{vmatrix} \quad [C_2' = C_2 - C_1,\ C_3' = C_3 - C_1]$$

$$= (a+b+c)^2\begin{vmatrix} (b+c)^2 & a-b-c & a-b-c \\ b^2 & c+a-b & 0 \\ c^2 & 0 & a+b-c \end{vmatrix}$$

$$= (a+b+c)^2\begin{vmatrix} 2bc & -2c & -2b \\ b^2 & c+a-b & 0 \\ c^2 & 0 & a+b-c \end{vmatrix} \quad [R_1' = R_1 - (R_2 + R_3)]$$

$$\cdot b+c)^2\frac{1}{b}\cdot\frac{1}{c}\begin{vmatrix} bc & -bc. & -bc \\ b^2 & bc+ab-b^2 & 0 \\ c^2 & 0 & ac+bc-c^2 \end{vmatrix}$$

[The 2nd and 3rd columns are multiplied by b and c respectively].

$$= \frac{2}{bc}(a+b+c)^2\begin{vmatrix} bc & 0 & 0 \\ b^2 & bc+ab & b^2 \\ c^2 & c^2 & ac+bc \end{vmatrix} \quad [C_2' = C_2 + C_1;\ C_3' = C_3 + C_1]$$

$$= \frac{2}{bc}(a+b+c)^2 bc\begin{vmatrix} b(a+c) & b^2 \\ c^2 & c(a+b) \end{vmatrix} \quad [C_2' = C_2 + C_1;\ C_3' = C_3 + C_1]$$

$$= 2(a+b+c)^2 bc\begin{vmatrix} a+c & b \\ c & a+b \end{vmatrix} = 2bc(a+b+c)^2(a^2+ab+bc+ca-bc)$$

$$= \mathbf{2abc(a+b+c)^3} \quad \textbf{(Proved)}.$$

Example 1.17. *Prove that,* $\begin{vmatrix} b_1+c_1 & c_1+a_1 & a_1+b_1 \\ b_2+c_2 & c_2+a_2 & a_2+b_2 \\ b_3+c_3 & c_3+a_3 & a_3+b_3 \end{vmatrix} = 2\begin{vmatrix} a_1 & b_1 & c_1 \\ a_2 & b_2 & c_2 \\ a_3 & b_3 & c_3 \end{vmatrix}$. [BU 2000]

Solution : The given determinant is

$$\begin{vmatrix} b_1+c_1 & c_1+a_1 & a_1+b_1 \\ b_2+c_2 & c_2+a_2 & a_2+b_2 \\ b_3+c_3 & c_3+a_3 & a_3+b_3 \end{vmatrix} = \begin{vmatrix} b_1 & c_1 & a_1 \\ b_2 & c_2 & a_2 \\ b_3 & c_3 & a_3 \end{vmatrix} + \begin{vmatrix} c_1 & a_1 & b_1 \\ c_2 & a_2 & b_2 \\ c_3 & a_3 & b_3 \end{vmatrix}$$

$$= \begin{vmatrix} a_1 & b_1 & c_1 \\ a_2 & b_2 & c_2 \\ a_3 & b_3 & c_3 \end{vmatrix} + (-1)\begin{vmatrix} a_1 & c_1 & b_1 \\ a_2 & c_2 & b_2 \\ a_3 & c_3 & b_3 \end{vmatrix}$$

$$= \begin{vmatrix} a_1 & b_1 & c_1 \\ a_2 & b_2 & c_2 \\ a_3 & b_3 & c_3 \end{vmatrix} + (-1)^2\begin{vmatrix} a_1 & b_1 & c_1 \\ a_2 & b_2 & c_2 \\ a_3 & b_3 & c_3 \end{vmatrix}$$

$$= \mathbf{2}\begin{vmatrix} \mathbf{a_1} & \mathbf{b_1} & \mathbf{c_1} \\ \mathbf{a_2} & \mathbf{b_2} & \mathbf{c_2} \\ \mathbf{a_3} & \mathbf{b_3} & \mathbf{c_3} \end{vmatrix} \quad \textbf{(Proved)}.$$

Example 1.18. *Find the value of* $\begin{vmatrix} (a-x)^2 & (a-y)^2 & (a-z)^2 \\ (b-x)^2 & (b-y)^2 & (b-z)^2 \\ (c-x)^2 & (c-y)^2 & (c-z)^2 \end{vmatrix}$ *by expressing it as the product of two determinant.*

Solution : The given determinant,

$$\Delta = \begin{vmatrix} (a-x)^2 & (a-y)^2 & (a-z)^2 \\ (b-x)^2 & (b-y)^2 & (b-z)^2 \\ (c-x)^2 & (c-y)^2 & (c-z)^2 \end{vmatrix} = \begin{vmatrix} a^2-2ax+x^2 & a^2-2ay+y^2 & a^2-2az+z^2 \\ b^2-2bx+x^2 & b^2-2by+y^2 & b^2-2bz+z^2 \\ c^2-2cx+x^2 & c^2-2cy+y^2 & c^2-2cz+z^2 \end{vmatrix}.$$

Now since $a^2 - 2ax + x^2 = a^2 \cdot 1 + (-2a) \cdot x + 1 \cdot x^2$,

$$\therefore \quad \Delta = \begin{vmatrix} a^2 & -2a & 1 \\ b^2 & -2b & 1 \\ c^2 & -2c & 1 \end{vmatrix} \times \begin{vmatrix} 1 & x & x^2 \\ 1 & y & y^2 \\ 1 & z & z^2 \end{vmatrix} = -2 \begin{vmatrix} a^2 & a & 1 \\ b^2 & b & 1 \\ c^2 & c & 1 \end{vmatrix} \times \begin{vmatrix} 1 & x & x^2 \\ 1 & y & y^2 \\ 1 & z & z^2 \end{vmatrix}$$

$$= -2 \begin{vmatrix} a^2 - b^2 & a - b & 0 \\ b^2 - c^2 & b - c & 0 \\ c^2 & c & 1 \end{vmatrix} \times \begin{vmatrix} 0 & x - y & x^2 - y^2 \\ 0 & y - z & y^2 - z^2 \\ 1 & z & z^2 \end{vmatrix}$$

$[R_1' = R_1 - R_2,\ R_2' = R_2 - R_3$ for both the determinants$]$

$$= -2(a-b)(b-c)(x-y)(y-z) \begin{vmatrix} a+b & 1 \\ b+c & 1 \end{vmatrix} \times \begin{vmatrix} 1 & x+y \\ 1 & y+z \end{vmatrix}$$

$$= -2(a-b)(b-c)(x-y)(y-z)(a+b-b-c)(y+z-x-y)$$

$$= \mathbf{2(a-b)(b-c)(c-a)(x-y)(y-z)(z-x)}.$$

Example 1.19. *Show that,* $\begin{vmatrix} a & b & c \\ b+c & c+a & a+b \\ a^2 & b^2 & c^2 \end{vmatrix} = -(a-b)(b-c)(c-a)(a+b+c).$

[RU 1990; Andhra U 1981]

Solution : The given determinant is

$$\begin{vmatrix} a & b & c \\ b+c & c+a & a+b \\ a^2 & b^2 & c^2 \end{vmatrix} = \begin{vmatrix} a+b+c & a+b+c & a+b+c \\ b+c & c+a & a+b \\ a^2 & b^2 & c^2 \end{vmatrix} \quad [R_1' = R_1 + R_2]$$

$$= (a+b+c) \begin{vmatrix} 1 & 1 & 1 \\ b+c & c+a & a+b \\ a^2 & b^2 & c^2 \end{vmatrix}$$

$$= (a+b+c) \begin{vmatrix} 0 & 0 & 1 \\ b-a & c-b & a+b \\ a^2-b^2 & b^2-c^2 & c^2 \end{vmatrix}, \quad \begin{bmatrix} C_1' = C_1 - C_2 \\ C_2' = C_2 - C_3 \end{bmatrix}$$

$$= (a+b+c) \begin{vmatrix} b-a & c-b \\ a^2-b^2 & b^2-c^2 \end{vmatrix}$$

$$= (a+b+c)[(b-a)(b^2-c^2) - (a^2-b^2)(c-b)]$$

$$= (a+b+c)[(b-a)(b+c)(b-c) - (a+b)(a-b)(c-b)]$$

$$= (a+b+c)[-(a-b)(b+c)(b-c) + (a+b)(a-b)(b-c)]$$

$$= (a+b+c)[(a-b)(b-c)[-(b+c)+a+b]$$

$$= \mathbf{-(a-b)(b-c)(c-a)(a+b+c)} \quad \textbf{(Proved)}.$$

Example 1.20. *Evaluate the determinant product,*

$$\begin{vmatrix} a^2+\lambda^2 & ab+c\lambda & ca-b\lambda \\ ab-c\lambda & b^2+\lambda^2 & bc+a\lambda \\ ca+b\lambda & bc-a\lambda & c^2+\lambda^2 \end{vmatrix} \times \begin{vmatrix} \lambda & c & -b \\ -c & \lambda & a \\ b & -a & \lambda \end{vmatrix}.$$

Solution : We have $\Delta = \begin{vmatrix} a^2+\lambda^2 & ab+c\lambda & ca-b\lambda \\ ab-c\lambda & b^2+\lambda^2 & bc+a\lambda \\ ca+b\lambda & bc-a\lambda & c^2+\lambda^2 \end{vmatrix} \times \begin{vmatrix} \lambda & c & -b \\ -c & \lambda & a \\ b & -a & \lambda \end{vmatrix}$

or, $\quad \Delta = \begin{vmatrix} x_{11} & x_{12} & x_{13} \\ x_{21} & x_{22} & x_{23} \\ x_{31} & x_{32} & x_{33} \end{vmatrix}$, [by the rule of multiplication of determinants]

where $x_{11} = (a^2+\lambda^2)\lambda + (ab+c\lambda)c + (ca-b\lambda)(-b) = \lambda(a^2+b^2+c^2+\lambda^2)$

$x_{12} = (a^2+\lambda^2)(-c) + (ab+c\lambda)\lambda + (ca-b\lambda)a = 0$

$x_{13} = 0$

$x_{21} = 0$

$x_{22} = \lambda(a^2+b^2+c^2+\lambda^2)$

$x_{23} = 0$

$x_{31} = 0$

$x_{32} = 0$

$x_{33} = \lambda(a^2+b^2+c^2+\lambda^2)$.

$$\therefore \quad \boldsymbol{\Delta} = \begin{vmatrix} \lambda(a^2+b^2+c^2+\lambda^2) & 0 & 0 \\ 0 & \lambda(a^2+b^2+c^2+\lambda^2) & 0 \\ 0 & 0 & \lambda(a^2+b^2+c^2+\lambda^2) \end{vmatrix}$$

$$= \boldsymbol{\lambda^3(a^2+b^2+c^2+\lambda^2)^3}.$$

Example 1.21. *Show that,* $\begin{vmatrix} a & b & c & d \\ -b & a & d & -c \\ -c & -d & a & b \\ -d & c & -b & a \end{vmatrix} = (a^2+b^2+c^2+d^2)^2.$

[NBU(H) 2004; BU(H) 1990; VU(H) 1989; CU(H) 1981]

Solution : We have $\Delta = \begin{vmatrix} a & b & c & d \\ -b & a & d & -c \\ -c & -d & a & b \\ -d & c & -b & a \end{vmatrix}$

$$= \begin{vmatrix} \frac{1}{a}(a^2+b^2+c^2+d^2) & b & c & d \\ 0 & a & d & -c \\ 0 & -d & a & b \\ 0 & c & -b & a \end{vmatrix}, \quad C_1' = \frac{1}{a}(aC_1+bC_2+cC_3+dC_4)$$

$$= \frac{1}{a}(a^2+b^2+c^2+d^2)\begin{vmatrix} a & d & -c \\ -d & a & b \\ c & -b & a \end{vmatrix}$$

$$= (a^2+b^2+c^2+d^2)\begin{vmatrix} 1 & \frac{d}{a} & -\frac{c}{a} \\ -d & a & b \\ c & -b & a \end{vmatrix}$$

$$= (a^2+b^2+c^2+d^2)\begin{vmatrix} 1 & \frac{d}{a} & -\frac{c}{a} \\ 0 & a+\frac{d^2}{a} & b-\frac{cd}{a} \\ 0 & -b-\frac{cd}{a} & a+\frac{c^2}{a} \end{vmatrix},$$

where $R_2' = R_2 + dR_1$ and $R_3' = R_3 - cR_1$

$$= (a^2+b^2+c^2+d^2)\left[\left(a+\frac{d^2}{a}\right)\left(a+\frac{c^2}{a}\right)+\left(b-\frac{cd}{a}\right)\left(b+\frac{cd}{a}\right)\right]$$

$$= (a^2+b^2+c^2+d^2)\left(a^2+c^2+d^2+\frac{c^2d^2}{a^2}+b^2-\frac{c^2d^2}{a^2}\right)$$

$$= (a^2+b^2+c^2+d^2)(a^2+b^2+c^2+d^2)$$

$$= \mathbf{(a^2+b^2+c^2+d^2)^2} \quad \textbf{(Proved)}.$$

Example 1.22. *If a, b, c be all different and* $\begin{vmatrix} a & a^2 & a^3-1 \\ b & b^2 & b^3-1 \\ c & c^2 & c^3-1 \end{vmatrix} = 0$, *then prove that,* $abc = 1$.

Solution : Let $\Delta = \begin{vmatrix} a & a^2 & a^3-1 \\ b & b^2 & b^3-1 \\ c & c^2 & c^3-1 \end{vmatrix} = \begin{vmatrix} a & a^2 & a^3 \\ b & b^2 & b^3 \\ c & c^2 & c^3 \end{vmatrix} + \begin{vmatrix} a & a^2 & -1 \\ b & b^2 & -1 \\ c & c^2 & -1 \end{vmatrix}$

$$= abc\begin{vmatrix} 1 & a & a^2 \\ 1 & b & b^2 \\ 1 & c & c^2 \end{vmatrix} + \begin{vmatrix} a-b & a^2-b^2 & 0 \\ b-c & b^2-c^2 & 0 \\ c & c^2 & -1 \end{vmatrix}$$

$$= abc\begin{vmatrix} 0 & a-b & a^2-b^2 \\ 0 & b-c & b^2-c^2 \\ 1 & c & c^2 \end{vmatrix} - \begin{vmatrix} a-b & a^2-b^2 \\ b-c & b^2-c^2 \end{vmatrix} \quad \left[\because \begin{matrix} R_1' = R_1 - R_2 \\ R_2' = R_2 - R_3 \end{matrix}\right]$$

$$= abc\begin{vmatrix} a-b & a^2-b^2 \\ b-c & b^2-c^2 \end{vmatrix} - \begin{vmatrix} a-b & a^2-b^2 \\ b-c & b^2-c^2 \end{vmatrix} = \begin{vmatrix} a-b & a^2-b^2 \\ b-c & b^2-c^2 \end{vmatrix}(abc-1).$$

By the given condition, $\Delta = 0$.

$\therefore \quad \begin{vmatrix} a-b & a^2-b^2 \\ b-c & b^2-c^2 \end{vmatrix}(abc-1) = 0$ or, $(abc-1) = 0$ $(\because a \neq b \neq c)$

or, $\mathbf{abc = 1}$ **(Proved).**

Example 1.23. *Find the adjoint and the reciprocal determinant of* $\begin{vmatrix} 0 & 1 & 2 \\ 2 & 0 & 1 \\ 1 & 2 & 0 \end{vmatrix}$.

Solution : Let $\Delta = \begin{vmatrix} 0 & 1 & 2 \\ 2 & 0 & 1 \\ 1 & 2 & 0 \end{vmatrix}$.

The co-factors of the elements : 0, 1, 2, 2, 0, 1, 1, 2, 0 are −2, 1, 4, 4, −2, 1, 1, 4, −2 respectively.

$\therefore$ the adjoint of Δ is, $\mathbf{\Delta'} = \begin{vmatrix} \mathbf{-2} & \mathbf{1} & \mathbf{4} \\ \mathbf{4} & \mathbf{-2} & \mathbf{1} \\ \mathbf{1} & \mathbf{4} & \mathbf{-2} \end{vmatrix}$.

Again, $\Delta = 0(0-2) + 1(1-0) + 2(4-0) = 9, \quad \therefore \quad \Delta \neq 0.$

Hence, the reciprocal of Δ is $\dfrac{\Delta'}{\Delta} = \dfrac{1}{9}\begin{vmatrix} -2 & 1 & 4 \\ 4 & -2 & 1 \\ 1 & 4 & -2 \end{vmatrix} = \begin{vmatrix} -\frac{2}{9} & \frac{1}{9} & \frac{4}{9} \\ \frac{4}{9} & -\frac{2}{9} & \frac{1}{9} \\ \frac{1}{9} & \frac{4}{9} & -\frac{2}{9} \end{vmatrix}$.

Example 1.24. *If a determinant is given by* $\Delta = \begin{vmatrix} x+1 & 2 & 3 \\ 3 & x+2 & 4 \\ 4 & 4 & x+3 \end{vmatrix}$, *then show that* $(x-1)$ *is a factor of* Δ.

Solution : We put $x = 1$ in Δ, then $\Delta = \begin{vmatrix} 2 & 2 & 3 \\ 3 & 3 & 4 \\ 4 & 4 & 4 \end{vmatrix}$.

Since, the 1st and 2nd rows of Δ are equal, therefore, from the properties of determinant, $\Delta = 0$.

Hence, **$(x-1)$ is a factor of the determinant Δ.**

Example 1.25. *Solve by Cramer's rule, the following equations :*

$$x + y + z = 1$$
$$ax + by + cz = d$$
$$a^2x + b^2y + c^2z = d^2, \quad a \neq b \neq c.$$

[VU, 2000]

Solution : The determinant of the coefficients of x, y, z is

$$\Delta = \begin{vmatrix} 1 & 1 & 1 \\ a & b & c \\ a^2 & b^2 & c^2 \end{vmatrix} = \begin{vmatrix} 1 & 0 & 0 \\ a & b-a & c-a \\ a^2 & b^2-a^2 & c^2-a^2 \end{vmatrix} \quad [C_2' = C_2 - C_1 \text{ and } C_3' = C_3 - C_1]$$

$$= (b-a)(c-a)\begin{vmatrix} 1 & 1 \\ b+a & c+a \end{vmatrix} = (a-b)(b-c)(c-a).$$

Now, $\dfrac{x}{\Delta_1} = \dfrac{y}{\Delta_2} = \dfrac{z}{\Delta_3} = \dfrac{-1}{\Delta}$,

where $\Delta_1 = \begin{vmatrix} 1 & 1 & -1 \\ b & c & -k \\ b^2 & c^2 & -k^2 \end{vmatrix}$, $\Delta_2 = \begin{vmatrix} 1 & -1 & 1 \\ a & -k & c \\ a^2 & -k^2 & c^2 \end{vmatrix}$, $\Delta_3 = \begin{vmatrix} 1 & 1 & -1 \\ a & b & -k \\ a^2 & b^2 & -k^2 \end{vmatrix}$.

On expanding we can show that,

$$\Delta_1 = -(b-c)(c-k)(k-b) \qquad \Delta_2 = (a-c)(c-k)(k-a) \qquad \Delta_3 = (b-a)(b-k)(k-a).$$

$$\therefore \quad \boldsymbol{x} = -\frac{\Delta_1}{\Delta} = \frac{(b-c)(c-k)(k-b)}{(a-b)(b-c)(c-a)} = -\boldsymbol{\frac{(k-c)(k-b)}{(a-b)(c-a)}}$$

$$\boldsymbol{y} = -\frac{\Delta_2}{\Delta} = -\frac{(a-c)(c-k)(k-a)}{(a-b)(b-c)(c-a)} = -\boldsymbol{\frac{(k-c)(k-a)}{(a-b)(b-c)}}$$

$$\boldsymbol{z} = -\frac{\Delta_3}{\Delta} = -\frac{(b-a)(k-b)(k-a)}{(a-b)(b-c)(c-a)} = -\boldsymbol{\frac{(k-b)(k-a)}{(b-c)(c-a)}}.$$

Example 1.26. *Apply Cramer's rule, to solve the equations,*

$$3x + y + 2z = 3$$
$$2x - 3y - z = -3$$
$$x + 2y + z = 4.$$

[RU 1985; AU 1983]

Solution : The determinant of the coefficients of x, y, z is

$$\Delta = \begin{vmatrix} 3 & 1 & 2 \\ 2 & -3 & -1 \\ 1 & 2 & 1 \end{vmatrix} = 3(-3+2) - 2(1-4) + (-1+6) = 8.$$

$$\therefore \quad x = \frac{\Delta_1}{\Delta}, \quad y = \frac{\Delta_2}{\Delta} \quad \text{and} \quad z = \frac{\Delta_3}{\Delta}.$$

But, $\Delta_1 = \begin{vmatrix} 3 & 1 & 2 \\ -3 & -3 & -1 \\ 4 & 2 & 1 \end{vmatrix} = 3(-3+2) - 1(-3+4) + 2(-6+12) = -3 - 1 + 12 = 8$

$$\Delta_2 = \begin{vmatrix} 3 & 3 & 2 \\ 2 & -3 & -1 \\ 1 & 4 & 1 \end{vmatrix} = 3(-3+4) + 3(2+1) + 2(8+3) = 3 - 9 + 22 = 16$$

$$\Delta_3 = \begin{vmatrix} 3 & 1 & 3 \\ 2 & -3 & -3 \\ 1 & 2 & 4 \end{vmatrix} = 3(-12+6) - 1(8+3) + 3(4+3) = -18 - 11 + 21 = -8.$$

$$\therefore \quad x = \frac{8}{8} = 1, \quad y = \frac{16}{8} = 2 \quad \text{and} \quad z = \frac{-8}{8} = -1.$$

$$\therefore \quad \boldsymbol{x = 1, \; y = 2, \; z = -1.}$$

Example 1.27. *Apply Cramer's rule, to solve the equations,*

$$x + 2y + 3z = 6$$
$$2x + 4y + z = 7$$
$$3x + 2y + 9z = 14.$$

Solution : The determinant of the coefficients of x, y, z is

$$\Delta = \begin{vmatrix} 1 & 2 & 3 \\ 2 & 4 & 1 \\ 3 & 2 & 9 \end{vmatrix} = 1(36-2) + 2(3-18) + 3(4-12) = 34 - 30 - 24 = -20.$$

$\therefore \Delta \neq 0.$

By Cramer's rule, $x = \dfrac{\Delta_1}{\Delta}, \quad y = \dfrac{\Delta_2}{\Delta} \quad z = \dfrac{\Delta_3}{\Delta},$

$$\text{where } \Delta_1 = \begin{vmatrix} 6 & 2 & 3 \\ 7 & 4 & 1 \\ 14 & 2 & 9 \end{vmatrix} = 6(36-2) + 2(14-63) + 3(14-56) = 6 \times 34 + 2(-49) + 3(-42)$$
$$= 204 - 98 - 126 = 204 - 224 = -20$$

$$\Delta_2 = \begin{vmatrix} 1 & 6 & 3 \\ 2 & 7 & 1 \\ 3 & 14 & 9 \end{vmatrix} = 1(63-14) + 6(3-18) + 3(28-21) = 49 + 6(-15) + 3 \times 7$$
$$= 49 - 90 + 21 = 70 - 90 = -20$$

$$\Delta_3 = \begin{vmatrix} 1 & 2 & 6 \\ 2 & 4 & 7 \\ 3 & 2 & 14 \end{vmatrix} = 1(56-14) + 2(21-28) + 6(4-12) = 42 + 2 \times (-7) + 6(-8)$$
$$= 42 - 14 - 48 = 42 - 62 = -20.$$

$$\therefore \quad x = \frac{-20}{-20} = 1, \quad y = \frac{-20}{-20} = 1 \quad \text{and} \quad z = \frac{-20}{-20} = 1.$$

$\therefore \quad \boldsymbol{x = 1, y = 1, z = 1.}$

Exercises

1(A) : Objective Type Questions

Select the correct answers

1.1 If any two rows (or columns) of a determinant are identical then the value of the determinant is

(a) 1 **(b)** 0 **(c)** ∞ **(d)** unchanged.

1.2 If each element of a row (or column) of a determinant (Δ) be multiplied by a common factor (k), then the resulting determinant Δ' is given by

(a) $\Delta' = \dfrac{\Delta}{k}$ **(b)** $\Delta' = k\Delta$ **(c)** $\Delta' = \dfrac{k}{\Delta}$ **(d)** none of these.

1.3 If any two columns (or rows) of a determinant (Δ) have proportional elements, then

(a) $\Delta = 1$ (b) $\Delta = 10$ (c) $\Delta = 0$ (d) $\Delta = \infty$.

1.4 If each element of a row (or column) of a determinant (Δ) is zero, then

(a) $\Delta = \infty$ (b) $\Delta = 0$ (c) $\Delta = -1$ (d) $\Delta = 10$.

1.5 For a symmetric determinant,

(a) $a_{ij} = \dfrac{1}{a_{ji}}$ (b) $a_{ij} \neq a_{ji}$ (c) $a_{ij} = a_{ji}$ (d) none of these.

1.6 For a skew-symmetric determinant,

(a) $a_{ij} = a_{ji}$ (b) $a_{ij} = -a_{ji}$ (c) $a_{ij} = \dfrac{1}{a_{ji}}$ (d) none of these.

1.7 The square of a determinant is a determinant which

(a) is symmetric (b) is skew-symmetric (c) may be both (a) and (b) (d) is none of these.

1.8 Any skew symmetric determinant of odd order is

(a) ∞ (b) 0 (c) 1 (d) -1.

1.9 A square skew-symmetric determinant of even order is

(a) a perfect cube (b) a perfect square (c) of any order (d) none of these.

1.10 If $\Delta = \begin{vmatrix} a_1 & b_1 & c_1 \\ a_2 & b_2 & c_2 \\ a_3 & b_3 & c_3 \end{vmatrix}$ be a determinant, then the co-factor of b_1 is

(a) $B_1 = (-1)^{1+2} M_1$ (b) $B_1 = M_1$ (c) $B_1 = (-1)^{-(1+2)} M_1$ (d) $B_1 = \dfrac{1}{(-1)^{1+2} M_1}$.

M_1 is the minor of b_1

1.11 If a, b, c be the elements and A, B, C the corresponding co-factors of a determinant (Δ), then $a_i A_j + b_i B_j + c_i C_j$ is equal to

(a) Δ, when $i = j$ (b) 0, when $i \neq j$ (c) Δ, when $i \neq 0$ (d) 0, when $i = j$.

1.12 The value of $\Delta = \begin{vmatrix} 5 & 6 \\ 3 & 8 \end{vmatrix}$ is

(a) 20 (b) 22 (c) 10 (d) 27.

Answers

1.1 (b)	**1.2** (b)	**1.3** (c)	**1.4** (b)	**1.5** (c)	**1.6** (b)
1.7 (a)	**1.8** (b)	**1.9** (b)	**1.10** (a)	**1.11** (a), (b)	**1.12** (b)

1(B) : Questions and Problems

1.1 Define 'minor' and 'co-factor' of a determinant. Find the relation between them.

If $\Delta = \begin{vmatrix} a_{11} & a_{12} & a_{13} \\ a_{21} & a_{22} & a_{23} \\ a_{31} & a_{32} & a_{33} \end{vmatrix}$, then find the minor and co-factor of the element a_{12}.

1.2 Discuss the important properties of a determinant.

If $\Delta = \begin{vmatrix} a_1 & b_1 & c_1 \\ a_2 & b_2 & c_2 \\ a_3 & b_3 & c_3 \end{vmatrix}$, then show that $a_1A_1+b_1B_1+c_1C_1 = \Delta$ and $a_2A_1+b_2B_1+c_2C_1 = 0$.

1.3 Expand $\Delta = \begin{vmatrix} a & h & g \\ h & b & f \\ g & f & c \end{vmatrix}$ by Laplace's method. [**Ans:** $abc + 2fgh - af^2 - bg^2 - ch^2$]

1.4 Define symmetric and skew-symmetric determinant with example. Show that the square of any determinant is s symmetric determinant.

1.5 Show that a skew-symmetric determinant of odd order is zero and of even order is a perfect square.

1.6 What are adjoint (or adjugate) and reciprocal determinants?

If $\Delta(\neq 0)$ be a determinant of order n and Δ' be its adjoint, then show that $\Delta' = \Delta^{n-1}$.

1.7 If the adjugate of $\Delta = \begin{vmatrix} a & h & g \\ h & b & f \\ g & f & c \end{vmatrix}$ be $\Delta' = \begin{vmatrix} A & H & G \\ H & B & F \\ G & F & C \end{vmatrix}$, then prove that,

(i) $\dfrac{BC - F^2}{a} = \dfrac{CA - G^2}{b} = \dfrac{AB - H^2}{c} = \Delta.$ **[CU(H) 2002]**

(ii) $\dfrac{GH - F^2}{a} = \dfrac{HF - BG}{g} = \dfrac{FG - CH}{c} = \Delta.$ **[CU(H) 1970]**

1.8 How can you test by the method of determinant if there is any non-trivial solution of a given set of homogeneous equations :

$$a_1x + b_1y + c_1z = 0$$
$$a_2x + b_2y + c_2z = 0$$
$$a_3x + b_3y + c_3z = 0.$$

1.9 Discuss the method of finding solution of non-homogeneous equations by Cramer's rule.

1.10 What do you mean by consistency and inconsistency of equations. State whether the following sets of equations are consistent or inconsistent,

(i) $\left.\begin{matrix} 2x + 3y = 3 \\ 4x + 6y = 7 \end{matrix}\right\}$ (ii) $\left.\begin{matrix} 2x + 3y = 3 \\ 4x + 6y = 6 \end{matrix}\right\}$ [**Ans:** (i) inconsistent (ii) consistent]

1.11 Discuss with example how the order of a higher order determinant may be reduced.

1.12 Prove without expanding that, $\begin{vmatrix} 1 & a & a^2 - bc \\ 1 & b & b^2 - ca \\ 1 & c & c^2 - ab \end{vmatrix} = 0.$ **[CU 2000]**

1.13 Prove without expanding that, $\begin{vmatrix} 5 & 2 & 3 \\ 7 & 3 & 4 \\ 9 & 4 & 5 \end{vmatrix} = 0.$ **[VU 2000]**

1.14 Prove that, $\begin{vmatrix} 2 + 3i & 4 + 5i \\ -4 + 5i & 2 - 3i \end{vmatrix} = 2^2 + 3^2 + 4^2 + 5^2$ and express $1^2 + 3^2 + 5^2 + 7^2$ in the form of a determinant **[TU 1992]**

1.15 Without expanding the determinant prove that, $\begin{vmatrix} a & b & c \\ x & y & z \\ p & q & r \end{vmatrix} = \begin{vmatrix} y & b & q \\ x & a & p \\ z & c & r \end{vmatrix}$. **[NBU 1989]**

1.16 Prove without expanding that,

(i) $\begin{vmatrix} 0 & c & b \\ -c & 0 & a \\ -b & -a & 0 \end{vmatrix} = 0$ **[NBU 1995]** **(ii)** $\begin{vmatrix} z & 0 & y \\ 0 & -z & x \\ -x & -y & 0 \end{vmatrix} = 0$ **[NBU 1995]**

(iii) $\begin{vmatrix} 0 & (x-y)^3 & (x-z)^3 \\ (y-x)^3 & 0 & (y-z)^3 \\ (z-x)^3 & (z-y)^3 & 0 \end{vmatrix} = 0$

(iv) $\begin{vmatrix} a+b+2c & a & b \\ c & b+c+2a & b \\ c & a & c+a+2b \end{vmatrix} = 2(a+b+c)^3.$

1.17 Prove without expanding that, $\begin{vmatrix} -a^2 & ab & ac \\ ba & -b^2 & bc \\ ac & bc & -c^2 \end{vmatrix} = (2abc)^2.$

1.18 Show without expanding the determinant $\begin{vmatrix} \dfrac{a^2+b^2}{c} & c & c \\ a & \dfrac{b^2+c^2}{a} & a \\ b & b & \dfrac{c^2+a^2}{b} \end{vmatrix} = 4abc.$

1.19 If $\omega =$ the cube root of unity, show that $(a + b\omega + c\omega^2)$ is a factor of the determinant

$$\begin{vmatrix} a & b & c \\ b & c & a \\ c & a & b \end{vmatrix}.$$

Hence, show that the above determinant is equal to $-(a^3 + b^3 + c^3 - 3abc)$.

1.20 Show that $(x + a + b + c)$ is one of the factor of the determinant : $\begin{vmatrix} x+a & b & c \\ b & x+c & a \\ c & a & x+b \end{vmatrix}$.

1.21 **(a)** Show that, $\begin{vmatrix} 2 & a+b+c+d & ab+cd \\ a+b+c+d & 2(a+b)(c+d) & ab(c+d)+cd(a+b) \\ ab+cd & ab(c+d)+cd(a+b) & 2abcd \end{vmatrix} = 0.$

(b) Show that, $\begin{vmatrix} \cos(A-P) & \cos(A-Q) & \cos(A-R) \\ \cos(B-P) & \cos(B-Q) & \cos(B-R) \\ \cos(C-P) & \cos(C-Q) & \cos(C-R) \end{vmatrix} = 0$, for all values of A, B, C and P, Q, R. **[NBU(H), 1999]**

1.22 If $ax + by + cz = 1$, $cx + ay + bz = 0$, $bx + cy + az = 0$, then prove that,

$$\begin{vmatrix} x & y & z \\ z & x & y \\ y & z & x \end{vmatrix} \text{ is the reciprocal of } \begin{vmatrix} a & c & b \\ b & a & c \\ c & b & a \end{vmatrix}.$$

1.23 If $\Delta = \begin{vmatrix} ap & aq & bp & bq \\ ar & as & br & bs \\ cp & cq & dp & dq \\ cr & cs & dr & ds \end{vmatrix} = \begin{vmatrix} a & b & 0 & 0 \\ 0 & 0 & a & b \\ c & d & 0 & 0 \\ 0 & 0 & c & d \end{vmatrix} \times A$, then find the value of A.

Hence, evaluate the determinant Δ. **[CU(H) 2001]**

1.24 Prove that, $\begin{vmatrix} 1+a^2-b^2 & 2ab & -2b \\ 2ab & 1-a^2+b^2 & 2a \\ 2b & -2a^2 & 1-a^2-b^2 \end{vmatrix}$ is a perfect cube. **[AU 1982]**

1.25 Find the value of the determinant (M) if $M = 3A^2 + AB + B^2$, where $A = \begin{vmatrix} 2 & 1 & 1 \\ 1 & 2 & 1 \\ 0 & -1 & 0 \end{vmatrix}$,

$B = \begin{vmatrix} \frac{1}{2} & 0 & 1 \\ -1 & 0 & 1 \\ 1 & 2 & 3 \end{vmatrix}$ without evaluating A and B independently. **[RU 1986]**

1.26 Show that, $\begin{vmatrix} 0 & z & y \\ z & 0 & x \\ y & x & 0 \end{vmatrix}^2 = \begin{vmatrix} y^2+z^2 & xy & zx \\ xy & z^2+x^2 & yz \\ zx & yz & x^2+y^2 \end{vmatrix}$.

1.27 Show that, $\begin{vmatrix} a & b & c \\ c & a & b \\ b & c & a \end{vmatrix} = a^3 + b^3 + c^3 - 3abc$

and deduce that, $\begin{vmatrix} a^2-bc & b^2-ca & c^2-ab \\ c^2-ab & a^2-bc & b^2-ca \\ b^2-ca & c^2-ab & a^2-bc \end{vmatrix} = (a^3+b^3+c^3-3abc)^2$. **[VU 2000]**

1.28 Show that $(x-1)$ is s factor of the determinant $\Delta = \begin{vmatrix} x+1 & 2 & 3 \\ 3 & x+2 & 4 \\ 4 & 4 & x+3 \end{vmatrix}$.

1.29 Factorize : $\begin{vmatrix} (a-x)^2 & (a-y)^2 & (a-z)^2 \\ (b-x)^2 & (b-y)^2 & (b-z)^2 \\ (c-x)^2 & (c-y)^2 & (c-z)^2 \end{vmatrix}$.

[Ans: $2(a-b)(b-c)(c-a)(x-y)(y-z)(z-x)$**]**

1.30 Show that $(b-c)(c-a)(a-b)$ is a factor of $\begin{vmatrix} 1 & 1 & 1 \\ a & b & c \\ a^{n+2} & b^{n+2} & c^{n+2} \end{vmatrix}$ and the remaining factor is a homogeneous product of a, b, c of degree n, where n is an integer.

1.31 Factorize : $\begin{vmatrix} a & a^2 & a^3+bc \\ b & b^2 & b^3+ca \\ c & c^2 & c^3+ab \end{vmatrix}$. **[Ans:** $(a-b)(b-c)(c-a)(ab+bc+ca+abc)$**]**

1.32 Solve the equations :

(i) $\begin{vmatrix} x+p & q & r \\ q & x+r & p \\ r & p & x+q \end{vmatrix} = 0$ **[CU(H) 1962]**

(ii) $\begin{vmatrix} a-x & c & b \\ c & b-x & a \\ b & a & c-x \end{vmatrix} = 0, \text{ if } a+b+c=0$ **[BU(H) 1970]**

[Ans: (i) $-(p+q+r) \pm \sqrt{p^2+q^2+r^2-qr-rp-pq}$ **(ii)** $0, \pm\sqrt{\frac{3}{2}(a^2+b^2+c^2)}$**]**

1.33 Solve by Cramer's rule the equations :

$$x+y+z=1$$
$$ax+by+cz=d$$
$$a^2x+b^2y+c^2z=d^2, \quad a \neq b \neq c$$

[VU 2000]

1.34 Solve by Cramer's rule :

$$2x-y=3$$
$$3y-2z=5$$
$$x-2z=4.$$

[BU(H) 1967] [Ans: $x=1,\ y=2,\ z=3$**]**

1.35 Solve by Cramer's rule :

$$2x+3y+4z=20$$
$$x-5y+6z=9$$
$$3x+4y-5z=-4.$$

[BU 2000] [Ans: $x=1,\ y=2,\ z=3$**]**

1.36 Show that $x=1$ is a solution of the following equation $\begin{vmatrix} x+1 & 3 & 5 \\ 2 & x+2 & 5 \\ 2 & 3 & x+4 \end{vmatrix} = 0.$

1.37 Using determinants prove that there is no solution of the equations :

$$x+4y-2z=3$$
$$3x+y+5z=7$$
$$2x+3y+z=5.$$

1.38 Examine whether the following system of equations is consistent or not :

$$3x-2y-5=0$$
$$4x-3y-6=0$$
$$x+y-5=0.$$

[Ans: Yes]

1.39 Solve by Cramer's rule :

$$x+y+z=1$$
$$ax+by+cz=1$$
$$bcx+cay+abz=1, \quad (a \neq b \neq c).$$

[Ans: $x = \frac{a(b+c-1)-1}{(c-a)(a-b)},\ y = \frac{b(a+c-1)-1}{(a-b)(b-c)},\ z = \frac{c(a+b-1)-1}{(b-c)(c-a)}$**]**

1.40 Test for consistency of the following system and find the solution if it exists :

$$3x + 3y + 2z = 1$$
$$x + 2y = 4$$
$$10y + 3z = -2$$
$$2x - 3y - z = 5.$$

[AMIE 1981]

1.41 For what values of t will the system :

$$3x + ty + 5z = 0$$
$$4tx - 2y - tz = 0$$
$$18x - 4y + 7z = 0$$

have non-trivial solutions? Also find these solutions.

[**Ans:** $t = 1,\ 3$; $230x = 90y = 207z$ and $y = -69x,\ z = -42x$]

1.42 Solve by Cramer's rule :

$$x - y + z = 0$$
$$2x + 3y - 5z = 7$$
$$3x - 4y - 2z = -1.$$

[**Ans:** $x = \frac{11}{8},\ y = \frac{21}{16},\ z = -\frac{1}{16}$]

1.43 Solve by Cramer's rule :

$$2x + y + 2z = 2$$
$$3x + 2y + z = 2$$
$$-x + y + 3z = 6.$$

[**Ans:** $x = -1,\ y = 2,\ z = 1$]

1.44 Solve the equations by Cramer's rule :

$$x - y = 3$$
$$y - z = 1$$
$$x + y + z = 7.$$

[VU 1999]

1.45 Solve by Cramer's rule :

$$x - 3z = 1$$
$$2x - y - 4z = 2$$
$$y + z = 4.$$

[CU 1999]

1.46 Solve by Cramer's rule :

$$2x + y + 2z = 2$$
$$3x + 2y + z = -2$$
$$-x + y + 3z = 6.$$

[**Ans:** $x = -1,\ y = 2,\ z = 1$]

CHAPTER 2

Matrices

2.1 Introduction

Matrices were discovered in 1860 by a French mathematician, **Caley** and was much developed in twentieth century. It has great importance in many branches of physics and engineering such as **algebraic and differential equations, mechanics, theory of electric circuits, nuclear physics, astronomy**, etc.

N.B. : (i) A matrix is a single entity with a number of components such as the coordinates of a point $P(x, y, z)$ in three dimension.

(ii) Just like determinant, a matrix cannot be reduced to a single number and hence the value of a matrix does not arise.

(iii) The interchange of rows and columns in a determinant does not alter the determinant but in case of matrix, the change of rows and columns gives an entirely new matrix.

2.2 Definition

The arrangement of a system of numbers mn in rectangular form along m rows and n columns and bounded by a [ ] bracket is called a matrix. It is written as $m \times n$ matrix and as such called an m by n matrix.

Thus, a matrix or order mn having m rows and n columns is represented as

$$A = \begin{bmatrix} a_{11} & a_{12} & \cdots & a_{1j} & \cdots & a_{1n} \\ a_{21} & a_{22} & \cdots & a_{ij} & \cdots & a_{2n} \\ \cdots & \cdots & \cdots & \cdots & \cdots & \cdots \\ a_{i1} & a_{i2} & \cdots & a_{ij} & \cdots & a_{in} \\ a_{m1} & a_{m2} & \cdots & a_{mj} & \cdots & a_{mn} \end{bmatrix}. \tag{2.1}$$

Each of the numbers of mn is called an element and in particular the element a_{ij} belongs to i^{th} row and j^{th} column. If $m = n$, then A is called a **square matrix** and if $m \neq n$, then A is a **rectangular matrix**.

2.3 Types of Matrices

(1) Row matrix : A matrix having a single row is called a row matrix, e.g.,

$$A = [a_1,\ a_2, \ldots, a_n].$$

(2) Column matrix : A matrix having only a single column is called as column matrix, e.g.,

$$B = \begin{bmatrix} a_1 \\ a_2 \\ \vdots \\ a_m \end{bmatrix}$$

B is also written as $B = \{a_1, a_2, \ldots, a_m\}$.

(3) Square matrix : A matrix in which the number of rows(m) is equal to the number of columns(n), is called a **square matrix**. For example,

$$A = \begin{bmatrix} a_{11} & a_{12} & a_{13} \\ a_{21} & a_{22} & a_{23} \\ a_{31} & a_{32} & a_{33} \end{bmatrix} \text{ is a square matrix of order 3.}$$

$$|A| = \begin{vmatrix} a_{11} & a_{12} & a_{13} \\ a_{21} & a_{22} & a_{23} \\ a_{31} & a_{32} & a_{33} \end{vmatrix} \text{ is called the determinant of the matrix } A.$$

(4) Diagonal matrix : A square matrix with all the non-diagonal elements zero is called a diagonal matrix, e.g.,

$$A = \begin{bmatrix} d_1 & 0 & 0 \\ 0 & d_2 & 0 \\ 0 & 0 & d_3 \end{bmatrix} \text{ is a \textbf{diagonal matrix}.}$$

So for a diagonal matrix $a_{ij} = 0$ for $i \neq j$ and $a_{ij} \neq 0$ for $i = j$.

(5) Singular and Non-singular matrices : A square matrix (A) is said to be **singular** if its determinant $|A| = 0$. For example, the matrix

$$A = \begin{bmatrix} 1 & 2 & 3 \\ 2 & 3 & 4 \\ 3 & 4 & 5 \end{bmatrix} \text{ is a singular matrix as } |A| = 0.$$

If, however, $|A| \neq 0$, then the matrix is **non-singular**, e.g.,

$$B = \begin{bmatrix} 1 & 4 & 3 \\ 4 & 9 & 6 \\ 5 & 6 & 2 \end{bmatrix}, \ |B| \neq 0 \text{ is a non-singular matrix.}$$

(6) Unit or identity matrix : If all the diagonal elements of a square matrix are unity and non-diagonal elements are zero then it is a **unit**, or, **identity matrix**. For example,

$$I = \begin{bmatrix} 1 & 0 & 0 \\ 0 & 1 & 0 \\ 0 & 0 & 1 \end{bmatrix} \text{ is a unit or identity matrix.}$$

(7) Null matrix or zero matrix : A matrix in which all the elements are zero, is called a **zero matrix** or **null matrix**, e.g.,

$$\begin{bmatrix} 0 & 0 & 0 \\ 0 & 0 & 0 \\ 0 & 0 & 0 \end{bmatrix} \text{ is a \textbf{null} or \textbf{zero matrix}.}$$

(8) Symmetric matrix : A square matrix is called a symmetric matrix if for all values of i and j, $a_{ij} = a_{ji}$ or, $A' = A$. For example,

$$A = \begin{bmatrix} a & h & g \\ h & p & f \\ g & h & c \end{bmatrix} \text{ is a \textbf{symmetric matrix}.}$$

(9) Skew-symmetric matrix : A skew-symmetric matrix is a square matrix in which $a_{ij} = -a_{ji}$ for all values of i and j so that all the **leading diagonal elements** are zero or, $A' = -A$. For example,

$$\begin{bmatrix} 0 & -h & -g \\ h & 0 & -f \\ g & f & 0 \end{bmatrix} \text{ is a skew-symmetric matrix.}$$

(10) Triangular matrix : If all the elements below the leading diagonal of a square matrix are zero, then it is called an **upper triangular matrix**, e.g.,

$$\begin{bmatrix} 1 & 3 & 2 \\ 0 & 4 & 1 \\ 0 & 0 & 6 \end{bmatrix} \text{ is an upper triangular matrix.}$$

If all thc elements above the leading diagonal of a square matrix are zero, then it is callcd a lower triangular matrix, e.g.,

$$\begin{bmatrix} 2 & 0 & 0 \\ 4 & 1 & 0 \\ 5 & 6 & 7 \end{bmatrix} \text{ is a lower triangular matrix.}$$

(11) Transpose of a matrix : If the rows and the corresponding columns of a matrix A are interchanged, then the resulting matrix is called the transpose of the matrix A and is denoted by A' (or, A^T). For example, if

$$A = \begin{bmatrix} 2 & 3 & 4 \\ 1 & 0 & 5 \\ 6 & 7 & 8 \end{bmatrix}, \quad \text{then } A^T = \begin{bmatrix} 2 & 1 & 6 \\ 3 & 0 & 7 \\ 4 & 5 & 8 \end{bmatrix}.$$

It is to be noted that for a symmetric matrix, $A^T = A$ and for a skew-symmetric matrix $A^T = -A$. For any matrix A, $(A^T)^T = A$.

(12) Orthogonal matrix : A square matrix A is said to be **orthogonal**, if $A^T A = I$, where A^T is he transpose of A and I is a unit matrix. For example, let

$$A = \begin{bmatrix} \cos\alpha & -\sin\alpha \\ \sin\alpha & \cos\alpha \end{bmatrix} \text{ be a square matrix.}$$

Then $$A^T = \begin{bmatrix} \cos\alpha & \sin\alpha \\ -\sin\alpha & \cos\alpha \end{bmatrix}$$

and $$A^T A = \begin{bmatrix} \cos\alpha & \sin\alpha \\ -\sin\alpha & \cos\alpha \end{bmatrix} \begin{bmatrix} \cos\alpha & -\sin\alpha \\ \sin\alpha & \cos\alpha \end{bmatrix} = \begin{bmatrix} 1 & 0 \\ 0 & 1 \end{bmatrix} = I.$$

$\therefore$ A is an orthogonal matrix. It is to note that $\begin{bmatrix} 1 & 0 \\ 0 & 1 \end{bmatrix}$ and $\begin{bmatrix} 1 & 0 \\ 0 & -1 \end{bmatrix}$ are orthogonal matrices.

(13) Conjugate matrix : If the elements of a matrix A are complex, then the conjugate matrix of A is the matrix (A^* or $\overline{A}$), whose elements are the complex conjugate of the elements of A.

For example, if $A = \begin{bmatrix} 1+i & 2-3i & 4 \\ 7+2i & -i & 3-2i \end{bmatrix}$,

then $A^* = \begin{bmatrix} 1-i & 2+3i & 4 \\ 7-2i & i & 3+2i \end{bmatrix}$.

Note that, **if the elements of the matrix A are real numbers, then $A^* = A$.**

Also $[A^*]^* = [A]$.

(14) Hermitian matrix : A square matrix $A = [a_{ij}]$ is called Hermitian matrix if every ijth element of A is equal to complex conjugate of jith element of A. That is, $a_{ij} = a_{ij}^*$.

The necessary and sufficient condition for a matrix A to be Hermitian is that,

$$[A] = [A^*]^T = [A^\dagger],$$

where $[A^*]^T$ is the conjugate transpose of A and $[A^\dagger]$ is the Hermitian matrix.

Example If $A = \begin{bmatrix} 0 & -i \\ i & 0 \end{bmatrix}$, then $[A^*]^T = A^\dagger = \begin{bmatrix} 0 & -i \\ i & 0 \end{bmatrix} = A$.

$\therefore$ A is **Hermitian.**

(15) Hermitian conjugate : The Hermitian conjugate (written as $A^\dagger$ or A^θ) of a square matrix A is the transpose of the complex conjugate of A. So, if $A^\dagger = [A^*]^T$, then $A^\dagger$ is the Hermitian conjugate of A. If $A = \begin{bmatrix} 1+i & 2-3i \\ 2 & 3i \end{bmatrix}$ be a square matrix, then its Hermitian conjugate is,

$$A^\dagger = [A^*]^T = \text{transpose of} \begin{bmatrix} 1-i & 2+3i \\ 2 & -3i \end{bmatrix} = \begin{bmatrix} 1-i & 2 \\ 2+3i & -3i \end{bmatrix}.$$

(16) Skew-Hermitian matrix : A square matrix $A = [a_{ij}]$ will be a skew Hermitian matrix if every ijth element of A is equal to the negative conjugate complex of jith element of A, i.e.,

$$a_{ij} = -a_{ji}^* \quad \text{or,} \quad a_{ij} + a_{ji}^* = 0.$$

If $a_{ii} = a + ib$, then $a_{ii}^* = a - ib$, then from the above relation,

$$(a+ib) + (a-ib) = 0 \quad \text{or,} \quad 2a = 0, \quad \text{i.e., } a = 0.$$

So, **a_{ii} is either pure imaginary or $a_{ii} = 0$. Hence, all the diagonal elements of a skew-Hermitian matrix are either zero or pure imaginary.**

Example Let $A = \begin{bmatrix} i & 2-3i & 4+5i \\ -(2+3i) & 0 & 2i \\ -(4-5i) & 2i & -3i \end{bmatrix}$.

The necessary and sufficient condition for A to be skew-symmetric is

$$[\boldsymbol{A^*}]^{\boldsymbol{T}} = \boldsymbol{A^\dagger} = -\boldsymbol{A}.$$

(17) Equal matrices : Two matrices A and B are said to be equal, if

(i) A and B are of the same order,

(ii) the elements of the corresponding positions are equal.

Example If $A = \begin{bmatrix} 2 & 3 \\ 1 & 4 \end{bmatrix}$ and $B = \begin{bmatrix} 2 & 3 \\ 1 & 4 \end{bmatrix}$, then $A = B$.

(18) Scalar matrix : A scalar matrix is a diagonal matrix whose diagonal elements are equal, e.g.,

$$A = \begin{bmatrix} u & 0 & 0 \\ 0 & u & 0 \\ 0 & 0 & u \end{bmatrix} \text{ is a scalar matrix}$$

(19) Sub-matrix : If some rows or some columns or some of both rows and columns of a matrix are deleted, then the resulting matrix is a sub-matrix. Thus,

$$\begin{bmatrix} a_1 & b_1 \\ a_2 & b_2 \end{bmatrix} \text{ is a sub-matrix of } \begin{bmatrix} a_1 & b_1 & c_1 \\ a_2 & b_2 & c_c \\ a_3 & b_3 & c_3 \end{bmatrix}$$

(20) Periodic matrix : A matrix A is a periodic matrix, if $A^{k+1} = A$, where k is a positive integer. If k be the least positive integer, for which $A^{k+1} = A$, then k is called the period of A.

If $k = 1$, then $A^2 = A$ and A is then called the *idempotent matrix.*

(21) Involuntary matrix : A matrix A is involuntary matrix, if $A^2 = I$, where I is a unit matrix. Since $I^2 = I$, a unit matrix is involuntary.

2.4 Addition of Matrices

If A and B be two matrices of the same order, then their sum $A + B$ is defined as the matrix whose elements are the sum of the corresponding elements of A and B.

If $$A = \begin{bmatrix} 1 & 2 & 3 \\ 2 & 3 & 4 \\ 3 & 4 & 5 \end{bmatrix} \quad \text{and} \quad B = \begin{bmatrix} 2 & 3 & 4 \\ 3 & 4 & 5 \\ 4 & 5 & 6 \end{bmatrix},$$

then $$A + B = \begin{bmatrix} 1+2 & 2+3 & 3+4 \\ 2+3 & 3+4 & 4+5 \\ 3+4 & 4+5 & 5+6 \end{bmatrix} = \begin{bmatrix} 3 & 5 & 7 \\ 5 & 7 & 9 \\ 7 & 9 & 11 \end{bmatrix}.$$

If $A = [a_{ij}]$ and $B = [b_{ij}]$, then $A + B = [a_{ij} + b_{ij}]$.

2.4.1 Properties of Matrix Addition

(a) Matrix addition follows commutative law : If A and B be two matrices of the same order then, $A + B = B + A$.

Proof : Let $A = [a_{ij}]$, $B = [b_{ij}]$ and $C = [C_{ij}]$ be the matrices of the same order.

Then, $A + B = [a_{ij} + b_{ij}]$, $\quad B + A = [b_{ij} + a_{ij}]$.

But, $[a_{ij} + b_{ij}] = [b_{ij} + a_{ij}]$.

Hence, $\mathbf{A + B = B + A}$.

This is known as **commutative law** of matrix addition.

(b) Matrix addition is associative : If A, B and C be three matrices of the same order, then $A + (B + C) = (A + B) + C$.

Proof : Let $A = [a_{ij}]$, $B = [b_{ij}]$ and $C = [c_{ij}]$ be the matrices of the same order, then

$$A + (B + C) = [a_{ij} + (b_{ij} + c_{ij})]$$

$$(A + B) + C = [(b_{ij} + a_{ij}) + c_{ij}].$$

But, $[a_{ij} + (b_{ij} + c_{ij})] = [(a_{ij} + b_{ij}) + c_{ij}]$.

$\therefore \quad \boldsymbol{A + (B + C) = (A + B) + C}$.

This is the **associative law** of matrix addition.

2.5 Subtraction of Matrices

If A and B be two matrices of the same order, then their difference $A - B$ is a matrix whose elements are obtained by subtracting the elements of B from the corresponding elements of A.

If $\quad A = [a_{ij}]$ and $B = [b_{ij}]$, then $A - B = [a_{ij} - b_{ij}]$.

If $\quad A = \begin{bmatrix} a_1 & b_1 \\ a_2 & b_2 \end{bmatrix}$ and $B = \begin{bmatrix} c_1 & d_1 \\ c_2 & d_2 \end{bmatrix}$,

then $\quad A - B = \begin{bmatrix} a_1 & b_1 \\ a_2 & b_2 \end{bmatrix} - \begin{bmatrix} c_1 & d_1 \\ c_2 & d_2 \end{bmatrix} = \begin{bmatrix} a_1 - c_1 & b_1 - d_1 \\ a_2 - c_2 & b_2 - d_2 \end{bmatrix}$.

N.B. : (i) **Subtraction of B from A is the same as the addition of A and $-B$.**
(ii) Subtraction of matrices is associative, i.e., $\boldsymbol{(A+B)-C = A+(B-C) = B+(A-C)}$.

2.5.1 Multiplication of Matrix by a Scalar

The product of a matrix A by a scalar k is a matrix whose each element is k times the corresponding elements of A.

If $\quad A = [a_{ij}]$, then $kA = [ka_{ij}]$.

Thus, if $A = \begin{bmatrix} a_1 & b_1 & c_1 \\ a_2 & b_2 & c_2 \\ a_3 & b_3 & c_3 \end{bmatrix}$, then $kA = \begin{bmatrix} ka_1 & kb_1 & kc_1 \\ ka_2 & kb_2 & kc_2 \\ ka_3 & kb_3 & kc_3 \end{bmatrix}$.

N.B. : (i) **Distribution law holds for such multiplication**, i.e., $k(A + B) = kA + kB$.
(ii) All laws of ordinary algebra hold for the multiplication of a matrix by a scalar.

2.6 Matrix Multiplication

Two matrices can be multiplied, only if the number of columns in the first is equal to the number of rows in the second. Such matrices are called **conformable**.

If $\quad A = \begin{bmatrix} a_1 & b_1 & c_1 \\ a_2 & b_2 & c_2 \\ a_3 & b_3 & c_3 \\ a_4 & b_4 & c_4 \end{bmatrix}$ and $B = \begin{bmatrix} l_1 & l_2 \\ m_1 & m_2 \\ n_1 & n_2 \end{bmatrix}$,

then the product $AB = \begin{bmatrix} a_1 & b_1 & c_1 \\ a_2 & b_2 & c_2 \\ a_3 & b_3 & c_3 \\ a_4 & b_4 & c_3 \end{bmatrix} \times \begin{bmatrix} l_1 & l_2 \\ m_1 & m_2 \\ n_1 & n_2 \end{bmatrix}$

$$= \begin{bmatrix} a_1l_1 + b_1m_1 + c_1n_1 & a_1l_2 + b_1m_2 + c_1n_2 \\ a_2l_1 + b_2m_1 + c_2n_1 & a_2l_2 + b_2m_2 + c_2n_2 \\ a_3l_1 + b_3m_1 + c_3n_1 & a_3l_2 + b_3m_2 + c_3n_2 \\ a_4l_1 + b_4m_1 + c_4n_1 & a_4l_2 + b_4m_2 + c_4n_2 \end{bmatrix}.$$

In general, if

$$A = \begin{bmatrix} a_{11} & a_{12} & \cdots & a_{1n} \\ a_{21} & a_{22} & \cdots & a_{2n} \\ \cdots & \cdots & \cdots & \cdots \\ a_{m1} & a_{m2} & \cdots & a_{mn} \end{bmatrix} \quad \text{and} \quad B = \begin{bmatrix} b_{11} & b_{12} & \cdots & b_{1p} \\ b_{21} & b_{22} & \cdots & b_{2p} \\ \cdots & \cdots & \cdots & \cdots \\ b_{n1} & b_{n2} & \cdots & b_{np} \end{bmatrix}$$

be two $m \times n$ and $n \times p$ conformable matrices, then their product is defined as the $m \times p$ matrix.

$$AB = \begin{bmatrix} C_{11} & C_{12} & \cdots & C_{1p} \\ C_{21} & C_{22} & \cdots & C_{2p} \\ \cdots & \cdots & \cdots & \cdots \\ C_{m1} & C_{m2} & \cdots & C_{mp} \end{bmatrix}$$

where $C_{ij} = a_{i1}b_{1j} + a_{i2}b_{2j} + a_{i3}b_{3j} + \cdots + a_{in}b_{nj}$.

The expression for C_{ij} is known as the **inner product** of the ith row with the jth column.

2.7 Idempotent Matrix

A matrix A is called an idempotent matrix if $A^2 = A$.

Let $A = \begin{bmatrix} 2 & -2 & 4 \\ -1 & 3 & 4 \\ 1 & -2 & 3 \end{bmatrix}$, then it can be easily verified that $A^2 = A \cdot A = A$. So, A is an **idempotent matrix.**

2.8 Properties of Matrix Multiplication

1. Matrix multiplication is not commutative, i.e., $AB \neq BA$.
2. If conformability is assured, then matrix multiplication is associative, i.e.,
$$A(BC) = (AB)C.$$
3. With respect to addition, matrix multiplication $A(B + C) = AB + AC$.
4. Multiplication of matrix A by unit matrix does not change the matrix, i.e.,
$$AI = IA = A.$$
5. Multiplicative inverse of a matrix exists if $|A| \neq 0$.
$$A \cdot A^{-1} = A^{-1} \cdot A = 1.$$
6. **The transpose of the product of two matrices is the product of their transposes taken in the reverse order.**

   If A and B be two matrices conformable for product AB, then $[AB]^T = B^T A^T$, where T represents the transpose of a matrix.

Proof : Let $A = [a_{ij}]$ be an $m \times n$ matrix and $B = [b_{ij}]$ be an $n \times p$ matrix. Then AB is an $m \times p$ matrix and $[AB]^T$ is a $p \times m$ matrix.

Again, B^T is a $p \times n$ matrix and A' is an $n \times p$ matrix. So, $B^T A^T$ is a $p \times m$ matrix. Thus, $[AB]^T$ and $B^T A^T$ are matrices of the **same order.**

Now, the jith element of $[AB]^T = ij$th element of $AB = \sum_{k=1}^{n} a_{ik}b_{kj}$.

Also, the jth row of $B^T = b_{1j}, b_{2j}, \ldots$ and the ith column of $A^T = a_{i1}, a_{i2}, a_{i3}, \ldots$. So, the jith element of $B^T A^T = \sum_{k=1}^{n} b_{kj} a_{ik}$.

$\therefore$ the jith element of $[AB]^T$ = jith element of $B^T A^T$. Hence, $[AB]^T$ and $B^T A^T$ are of the same order and their corresponding elements are equal.

$\therefore$ $\boldsymbol{[AB]^T = B^T A^T}$.

2.9 Trace of a Matrix

The sum of the diagonal elements of a matrix is called the **trace of a matrix**.

If $A = \begin{bmatrix} a_1 & a_2 & a_3 \\ b_1 & b_2 & b_3 \\ c_1 & c_2 & c_3 \end{bmatrix}$, then the trace of the matrix A is, $T_r A = a_1 + b_2 + c_3$.

Note : The trace of the product of two matrices A and B is independent of multiplication, i.e.,

$$T_r[AB] = T_r[BA].$$

2.10 Rank of a Matrix

A matrix is said to be or rank r, when (i) it has at least one non-zero minor of order r, and (ii) every minor of order higher than r vanishes.

Hence, **the rank of a matrix is the largest order of any non-vanishing minor of the matrix.**

If a matrix has a non-zero minor of order r, its rank is $\geq r$. If all minors of a matrix of order $r+1$ are zero, its rank is $\leq r$.

Example If $A = \begin{bmatrix} 1 & 2 & 3 \\ 2 & 3 & 4 \\ 3 & 4 & 5 \end{bmatrix}$, then $|A| = 0$. But there is at least one 2×2 sub matrix, viz., $\begin{bmatrix} 3 & 4 \\ 4 & 5 \end{bmatrix}$, whose determinant is not zero. So, the rank of A is 2.

Note : (i) The rank of a non-singular square matrix of order n is n.

(ii) The rank of a singular square matrix of order n is less than n.

(iii) The rank of a null matrix is zero.

(iv) The rank of the transpose of a matrix A is the same as the rank of A.

In section 2.3 we have discussed about symmetric and skew-symmetric matrices. Here the properties of both the types of matrices will be discussed.

2.11 Properties of Symmetric and Skew-symmetric Matrices

(a) The product of a matrix and its transpose is a symmetric matrix.

Let A be a matrix of order $m \times n$. Then A^T is a matrix of order $n \times m$.

$\therefore$ $AA^T = a$ square matrix of order $m \times m$.

Now, $[AA^T]^T = [A^T]^T A^T = AA^T$.

Hence, $\boldsymbol{AA^T}$ **is a symmetric matrix. Similarly, $\boldsymbol{A^T A}$ is also a symmetric matrix.**

(b) The product of two symmetric matrices of the same order is symmetric, if the product be commutative.

Let A and B be two symmetric matrices of the same order. Also $AB = BA$.

$\therefore$ $[AB]^T = B^T A^T = BA = AB$. **Hence, AB is symmetric.**

(c) If A be a square matrix, then $(A + A^T)$ is symmetric and $(A - A^T)$ is skew-symmetric.

We have, $[A + A^T]^T = A^T + [A^T]^T = A^T + A$, $\therefore$ **$(A + A^T)$ is symmetric.**

Again, $[A - A^T]^T = A^T - A = -(A - A^T)$, $\therefore$ **$[A - A^T]$ is skew-symmetric.**

(d) Every square matrix can be expressed as a sum of a symmetric matrix and a skew-symmetric matrix uniquely.

Let A be a square matrix.

Then, $A = \frac{1}{2}(A + A^T) + \frac{1}{2}(A - A^T)$.

Now, $$\left[\frac{1}{2}(A + A^T)\right]^T = \frac{1}{2}[A^T + (A^T)^T] = \frac{1}{2}(A^T + A) = \frac{1}{2}(A + A^T).$$

Again, $$\left[\frac{1}{2}(A - A^T)\right]^T = \frac{1}{2}[A^T - (A^T)^T] = \frac{1}{2}[A^T - A] = -\frac{1}{2}(A - A^T).$$

$\therefore$ **$\frac{1}{2}(A + A^T)$ is symmetric** and **$\frac{1}{2}(A - A^T)$ is skew-symmetric**

The above decomposition may be proved as below :

We assume, $A = B + C$, where B is symmetric and C is skew-symmetric.

Then, $A^T = B^T + C^T = B - C$.

$\therefore$ $(A + A^T) = B + C + B - C = 2B$ and $(A - A^T) = 2C$.

$\therefore$ $B = \frac{1}{2}(A + A^T)$ and $C = \frac{1}{2}(A - A^T)$.

Thus, the decomposition is unique.

We have already discussed about orthogonal matrices in section 2.3. Here we shall describe the properties of such matrices.

2.12 Properties of Orthogonal Matrices

(a) Orthogonal matrices are non-singular.

Let A be an orthogonal matrix, then $A^T \cdot A = I$.

$\therefore$ $|A^T \cdot A| = |A^T|\,|A| = |A|^2 = |I|^2 = 1$.

This shows that $|A| \neq 0$, i.e., A is **non-singular.**

(b) Unit matrices are orthogonal.

We have, $I^T I = I \cdot I = 1$. This proves the statement.

(c) The value of the determinant of an orthogonal matrix is either 1 or (−1).

If A be an orthogonal matrix and A^T its transpose, then $A^T \cdot A = 1$.

$\therefore$ $|A^T \cdot A| = |A^T|\,|A| = |A|^2 = |I| = 1$.

$\therefore$ $\mathbf{|A| = \pm 1}$.

(d) The product of two orthogonal matrices is orthogonal.

If A and B be two orthogonal matrices, then

$$A^T A = I \quad \text{and} \quad B^T B = I.$$

$\therefore \quad [AB]^T \cdot [AB] = [B^T \cdot A^T] \cdot [A \cdot B] = B^T[A^T A]B = B^T IB = B^T B = I.$

[$\because$ the matrix multiplication is associative]

$\therefore$ **the matrix AB is orthogonal.**

(e) The transpose of an orthogonal matrix is orthogonal.

Let A be an orthogonal matrix. Then $A^T A = I$.

$\therefore \quad A^T \cdot AA^{-1} = IA^{-1} = A^{-1} \quad \text{or,} \quad A^T I = A^T = A^{-1}.$

Hence, $AA^T = AA^{-1} = I \quad \{\because\ [A^T]^T = A,\ [A^T] \cdot A^T = AA^T = I$, i.e., A^T is orthogonal$\}$

Note that if A be an orthogonal matrix, then $\boldsymbol{A^T = A^{-1}}$.

(f) The inverse of an orthogonal matrix is orthogonal.

If A be an orthogonal matrix, then, $A^T A = I$.

Also, $AA^{-1} = A^{-1}A = I$.

$$\therefore \quad [AA^{-1}]^T[AA^{-1}] = [A^{-1}]^T A^T AA^{-1} = [A^{-1}]^T 1 A^{-1}$$
$$= [A^{-1}]^T A^{-1}, \quad \text{where, } [A^{-1}]^T A^{-1} = I^T I = I$$

$\therefore$ **A^{-1} is orthogonal.**

2.13 Inverse of a Matrix

If A and B be two square matrices of the same order such that,

$$AB = BA = I, \quad [I = \text{ unit matrix }]$$

then B is called the inverse of A, i.e., $B = A^{-1}$; also A is the inverse of B, i.e., $A = B^{-1}$.

2.13.1 Properties of the Inverse of a Matrix

(a) The inverse of a matrix, if exists, is unique.

Let B and C be two inverse matrices of A.

Then, $A \cdot B = B \cdot A = I$ and $A \cdot C = C \cdot A = I$ [where I = unit matrix]

As matrix multiplication is associative,

$\therefore \quad C(A \cdot B) = (C \cdot A)B \quad \text{or,} \quad C \cdot I = I \cdot B, \quad \therefore \quad C = B.$

This proves that the inverse of a matrix is unique.

(b) The necessary and sufficient condition that the inverse of a square matrix A exists, is that A is non-singular.

Let the inverse of A exists and let that inverse be equal to B.

$\therefore \quad A \cdot B = B \cdot A = I \quad \text{or,} \quad |A \cdot B| = |A|\,|B| = |I| = 1.$

So, $|A| \neq 0$, this means A is **non-singular**.

Hence, the condition is necessary.

Again, let A is non-singular, i.e., $|A| \neq 0$, then

$$A \cdot \left[\frac{\text{adj}.A}{|A|}\right] = I = \left[\frac{\text{adj}.A}{|A|}\right] \cdot A.$$

Hence, inverse of A exists and is equal to $\dfrac{\text{adj}.A}{|A|}$, which is equal to $\dfrac{1}{A}$.

Thus, for the existence of A the condition $|A| \neq 0$ is necessary as well as sufficient.

N.B. : If A^{-1} exists then A is called **invertible.**

(c) The inverse of the product of two matrices is the product of their inverses in the reverse order.

If A and B be two **non-singular square matrices of the same order, then**

$$(AB)^{-1} = B^{-1}A^{-1}.$$

By definition, $[B^{-1}A^{-1}] \cdot [AB] = B^{-1}[A^{-1}A]B = B^{-1}IB = B^{-1}B = I$

and $[AB] \cdot [B^{-1}A^{-1}] = A[BB^{-1}]A^{-1} = AIA^{-1} = AA^{-1} = I.$

So, $[B^{-1}A^{-1}] \cdot [AB] = [AB] \cdot [B^{-1}A^{-1}] = I.$

Hence, $B^{-1}A^{-1}$ is inverse of AB, i.e., $\mathbf{[AB]^{-1} = B^{-1}A^{-1}}$.

This is known as the **law of reversal for inverses.**

N.B. : (i) Since $[AB]^{-1}$ exists, $[AB]$ is non-singular.

(ii) In general, $[ABC \cdots K]^{-1} = K^{-1} \cdots C^{-1}B^{-1}A^{-1}$.

(d) If A be an invertible matrix, then A^{-1} is invertible and $[A^{-1}]^{-1} = A$.

As A is invertible, A^{-1} exists and $AA^{-1} = A^{-1}A = I$ also $|A| \neq 0$.

Now, $|A|\,|A^{-1}| = |AA^{-1}| = |I| = 1$, $\therefore$ $|A^{-1}| \neq 0$, hence, A^{-1} **is invertible.**

Again, $A^{-1}A = AA^{-1} = I$, $\therefore$ A is the inverse of A^{-1}, i.e., $\mathbf{[A^{-1}]^{-1} = A}$.

2.14 In section 2.3 we have defined the transpose of a matrix in which the rows of matrix (say, A) are interchanged to the corresponding columns resulting the transpose of the matrix A which is A^T (or A').

2.14.1 Properties of the Transpose of a Matrix

(a) If the transpose of a matrix A be represented by A^T, then $(A^T)^T = A$.

Suppose, $A = [a_{ij}]$, then $A^T = [a_{ji}]$, $\therefore$ $\mathbf{[A^T]^T = A}$.

(b) If k be a scalar, then $(kA)^T = kA^T$.

Let $A = [a_{ij}]$ or, $kA = k[a_{ij}]$.

$\therefore$ $[kA]^T = [ka_{ij}]^T = [ka_{ji}] = k[a_{ji}] = \mathbf{kA^T}$.

(c) If A and B be two matrices of the same order then $(A+B)^T = A^T + B^T$.

Let $A = [a_{ij}]$, then $A^T = [a_{ji}]$ and $B = [b_{ij}]$, then $B^T = [b_{ji}]$.

$\therefore$ $[A + B] = [a_{ij}] + [b_{ij}] = [a_{ij} + b_{ij}]$.

The transpose of matrix $[A + B]$ is

$$\mathbf{[A+B]^T} = [a_{ij} + b_{ij}]^T = [a_{ji}] + [b_{ji}] = \mathbf{A^T + B^T}.$$

(d) A and B be two conformable matrices product AB, then $[AB]^T = B^T + A^T$.

Let $A = \begin{bmatrix} a_{11} & a_{12} \\ a_{21} & a_{22} \end{bmatrix}$ and $B = \begin{bmatrix} b_{11} & b_{12} \\ b_{21} & b_{22} \end{bmatrix}$;

then $A^T = \begin{bmatrix} a_{11} & a_{21} \\ a_{12} & a_{22} \end{bmatrix}$ and $B^T = \begin{bmatrix} b_{11} & b_{21} \\ b_{12} & b_{22} \end{bmatrix}$.

Now, $AB = \begin{bmatrix} a_{11}b_{11} + a_{12}b_{21} & a_{11}b_{12} + a_{12}b_{22} \\ a_{21}b_{11} + a_{22}b_{21} & a_{21}b_{12} + a_{22}b_{22} \end{bmatrix}$

and $[AB]^T = \begin{bmatrix} a_{11}b_{11} + a_{12}b_{21} & a_{21}b_{11} + a_{22}b_{21} \\ a_{11}b_{12} + a_{12}b_{22} & a_{21}b_{12} + a_{22}b_{22} \end{bmatrix}$.

Again, $B^T A^T = \begin{bmatrix} b_{11} & b_{21} \\ b_{12} & b_{22} \end{bmatrix} \begin{bmatrix} a_{11} & a_{21} \\ a_{12} & a_{22} \end{bmatrix} = \begin{bmatrix} b_{11}a_{11} + b_{21}a_{12} & b_{11}a_{21} + b_{21}a_{22} \\ b_{12}a_{11} + b_{22}a_{12} & b_{12}a_{21} + b_{22}a_{22} \end{bmatrix}$.

Hence, $[AB]^T = B^T A^T$.

(e) If A be a square matrix, then the determinant of A is equal to the determinant of A^T i.e., $|A| = |A^T|$.

Let $A = \begin{bmatrix} a_1 & b_1 \\ a_2 & b_2 \end{bmatrix}$ be a square matrix. Then, $A^T = \begin{bmatrix} a_1 & a_2 \\ b_1 & b_2 \end{bmatrix}$

Now, $|A| = \begin{vmatrix} a_1 & b_1 \\ a_2 & b_2 \end{vmatrix}$ and $|A^T| = \left[\begin{matrix} a_1 & a_2 \\ b_1 & b_2 \end{matrix} \right|$

or, $|A^T| = \begin{vmatrix} a_1 & b_1 \\ a_2 & b_2 \end{vmatrix}$, [interchanging rows and columns]

$\therefore$ $|A| = |A^T|$.

2.15 Calculus of Matrices

Derivative of a Matrix : If the elements of a matrix are functions of a variable x, then the derivative of the matrix with respect to x is equal to the matrix whose elements are derivatives of each element separately.

Thus, if $A = [a_{ij}]$ be a matrix and a_{ij} be a function of the variable x, then

$$\frac{dA}{dx} = \left[\frac{d}{dx}(a_{ij})\right].$$

Example Let $A = \begin{bmatrix} \log x & e^x & x \\ 3x & x^3 & 0 \end{bmatrix}$; then $\frac{dA}{dx} = \begin{bmatrix} \frac{1}{x} & e^x & 1 \\ 3 & 3x^2 & 0 \end{bmatrix}$.

Integration of a Matrix : The integration of a matrix with respect to a variable x is the matrix whose elements are integral of each element with respect to x separately.

If $A = [a_{ij}]$ be a matrix, then the integral of A is given by

$$\int A dx = \left[\int a_{ij}\, dx\right].$$

Example If $A = \begin{bmatrix} x & e^x & x^3 \\ e^{-x} & 0 & x^2 \end{bmatrix}$, then

$$\int A\,dx = \begin{bmatrix} \frac{x^2}{2} & e^x & \frac{x^4}{4} \\ -e^{-x} & c & \frac{x^3}{3} \end{bmatrix} + \begin{bmatrix} a_1 & a_2 & a_3 \\ a_4 & c & a_5 \end{bmatrix},$$

where a_1, a_2, a_3 and c are constants of integration.

2.16 Elementary Transformations

The term 'elementary transformation' refers to any of the following operations on a matrix.

(i) Interchange of any two rows or columns.

Example

$$\begin{bmatrix} a_1 & b_1 & c_1 \\ a_2 & b_2 & c_2 \\ a_3 & b_3 & c_3 \end{bmatrix} \xrightarrow{R_1 \leftrightarrow R_2} \begin{bmatrix} a_2 & b_2 & c_2 \\ a_1 & b_1 & c_1 \\ a_3 & b_3 & c_3 \end{bmatrix}$$

$$\begin{bmatrix} a_1 & b_1 & c_1 \\ a_2 & b_2 & c_2 \\ a_3 & b_3 & c_3 \end{bmatrix} \xrightarrow{C_2 \leftrightarrow C_3} \begin{bmatrix} a_1 & c_1 & b_1 \\ a_2 & c_2 & b_2 \\ a_3 & c_3 & b_3 \end{bmatrix}.$$

Here, R and C refer to row and column respectively.

(ii) Multiplication of the ith row (or, ith column) by a non-zero number.

Example

$$\begin{bmatrix} a & b \\ c & d \end{bmatrix} \xrightarrow{R_1' \to kR_1} \begin{bmatrix} ka & kb \\ c & d \end{bmatrix}$$

$$\begin{bmatrix} a & b \\ c & d \end{bmatrix} \xrightarrow{C_1' \to lC_1} \begin{bmatrix} la & b \\ lc & d \end{bmatrix}.$$

(iii) Addition of k times the jth row (or, column) to the ith row (or, column).

$$\begin{bmatrix} a & b \\ c & d \end{bmatrix} \xrightarrow{R_2' \to R_2 + kR_1 R_1} \begin{bmatrix} a & b \\ c + ka & d + kb \end{bmatrix}$$

$$\begin{bmatrix} a & b \\ c & d \end{bmatrix} \xrightarrow{C_2' \to C_2 + lC_1} \begin{bmatrix} a & b + la \\ c & d + lc \end{bmatrix}.$$

These operations are called *elementary transformations* as they transform the matrices from one form to another.

Elementary matrices : A square matrix obtained from a unit matrix by any one of the elementary row or column operations is called an elementary matrix.

Example Let $I = \begin{bmatrix} 1 & 0 & 0 \\ 0 & 1 & 0 \\ 0 & 0 & 1 \end{bmatrix}$. The matrix obtained from it by $R_2' = R_2 + 2R_1$ is

$$\begin{bmatrix} 1 & 0 & 0 \\ 2 & 1 & 0 \\ 0 & 0 & 1 \end{bmatrix}$$

which is an **elementary matrix.**

Theorem : Every elementary row transformation of a matrix can be affected by pre-multiplication with the corresponding elementary matrix.

Let a matrix $A = \begin{bmatrix} 2 & 1 & 3 \\ 4 & 2 & 5 \\ 2 & 5 & 6 \end{bmatrix}$, then by the transformation $R_3' = R_3 + 2R_1$, we get

$$B = \begin{bmatrix} 2 & 1 & 3 \\ 4 & 2 & 5 \\ 6 & 7 & 12 \end{bmatrix}.$$

We shall now show that pre-multiplication of A by corresponding elementary matrix $R_3' = R_3 + 2R_1$ will give B.

Now, if $I = \begin{bmatrix} 1 & 0 & 0 \\ 0 & 1 & 0 \\ 0 & 0 & 1 \end{bmatrix} \xrightarrow[\text{Elementary matrix}]{R_3' = R_3 + 2R_1} \begin{bmatrix} 1 & 0 & 0 \\ 0 & 1 & 0 \\ 2 & 0 & 1 \end{bmatrix}$.

$\therefore$ **elem. matrix** $\times$ $\boldsymbol{A} = \begin{bmatrix} 1 & 0 & 0 \\ 0 & 1 & 0 \\ 2 & 0 & 1 \end{bmatrix} \times \begin{bmatrix} 2 & 1 & 3 \\ 4 & 2 & 5 \\ 2 & 5 & 6 \end{bmatrix} = \begin{bmatrix} 2 & 1 & 3 \\ 4 & 2 & 5 \\ 6 & 7 & 12 \end{bmatrix} = \boldsymbol{B}$.

Similarly, every elementary column transformation of a matrix can be affected by post multiplication with the corresponding elementary matrix.

Equivalent matrix : Two matrices A and B are said to be equivalent if one can be obtained from the other by a sequence of elementary transformations. Two equivalent matrices have the same order and same rank. The equivalence of A and B matrices is represented as $A \sim B$.

2.17 Row-Echelon Normal Form (RENF)

The elementary row operations may be applied to any matrix A so that the ultimate transformed matrix has the following properties :

(a) All the zero rows are its last rows.

(b) The first non-zero element in a non-zero row is 1. Each column containing this 1 will have all other elements zero.

(c) The leading 1 in ith row will be the left of the leading 1 in jth row, if $i > j$.

A matrix A having the above properties is called the row-echelon normal form (RENF).

Example The matrix, $A = \begin{bmatrix} 1 & 3 & 4 \\ 0 & 1 & 0 \\ 0 & 0 & 0 \end{bmatrix}$ is in RENF.

2.18 Linear Equations

An equation containing n unknowns $x_1, x_2, \ldots, x_n$ in the form,

$$a_1x_1 + a_2x_2 + \cdots + a_nx_n = b \tag{2.1}$$

is called a linear equation.

If $b = 0$, in the above equation, i.e.,

$$a_1x_1 + a_2x_2 + \cdots + a_nx_n = 0, \tag{2.2}$$

then it is called a **homogeneous linear equation**.

But if $b \neq 0$, then the equation is a **non-homogeneous equation**.

2.18.1 Consistency of a System of Linear Equations

Now, consider a set of m non-homogeneous linear equations in n unknowns :

$$\left.\begin{array}{l} a_{11}x_1 + a_{12}x_2 + \cdots + a_{1n}x_n = b_1 \\ a_{21}x_1 + a_{22}x_2 + \cdots + a_{2n}x_n = b_2 \\ \cdots\cdots\cdots\cdots\cdots\cdots\cdots\cdots\cdots\cdots \\ a_{m1}x_1 + a_{m2}x_2 + \cdots + a_{mn}x_n = b_m \end{array}\right\} \quad (2.3)$$

The above, if there be at least one set of values of $x_1, x_2, \ldots, x_n$ satisfying all the equations, then the equations are called **consistent**. If no such set of $x_1, x_2, \ldots, x_n$ be possible, then the equations are called **inconsistent**.

Example

(1) Consistent equations : A system of equations is said to be consistent if they have one or more solutions, i.e.,

$$\left.\begin{array}{r} 2x + 5y = 9 \\ x - y = 1 \end{array}\right\}. \quad (2.4)$$

$$\left.\begin{array}{r} x + 2y = 7 \\ 4x + 8y = 28 \end{array}\right\}. \quad (2.5)$$

Eqns. (2.4) have a unique solution ($x = 2, y = 1$) and Eqns. (2.5) have an infinite number of solutions ($x = 7 - 2c, y = c$), where c is arbitrary. Hence, the above sets of equations are **consistent**.

(2) Inconsistent equations : A system of equations having no solutions is called **inconsistent**, i.e.,

$$\left.\begin{array}{r} x + 2y = 4 \\ 3x + 6y = 5 \end{array}\right\}. \quad (2.6)$$

2.18.2 Consistency of a System of Linear Equations

The system of linear equations as mentioned in equation (2.3) is

$$\left.\begin{array}{l} a_{11}x_1 + a_{12}x_2 + \cdots + a_{1n}x_n = b_1 \\ a_{21}x_1 + a_{22}x_2 + \cdots + a_{2n}x_n = b_2 \\ \cdots\cdots\cdots\cdots\cdots\cdots\cdots\cdots\cdots\cdots \\ a_{m1}x_1 + a_{m2}x_2 + \cdots + a_{mn}x_n = b_m \end{array}\right\} \quad (2.7)$$

which can be represented as

$$\begin{bmatrix} a_{11} & a_{12} & \cdots & a_{1n} \\ a_{21} & a_{22} & \cdots & a_{2n} \\ \cdots & \cdots & \cdots & \cdots \\ a_{m1} & a_{m2} & & a_{mn} \end{bmatrix} \begin{bmatrix} x_1 \\ x_2 \\ \vdots \\ x_n \end{bmatrix} = \begin{bmatrix} b_1 \\ b_2 \\ \vdots \\ b_m \end{bmatrix} \quad (2.8)$$

or, $\quad AX = B \quad (2.9)$

and $$A_b = [A, B] = \begin{bmatrix} a_{11} & a_{12} & \cdots & a_{1n} & \cdots & b_1 \\ a_{21} & a_{22} & \cdots & a_{2n} & \cdots & b_2 \\ \cdots & \cdots & \cdots & \cdots & \cdots & \cdots \\ a_{m1} & a_{m2} & \cdots & a_{mn} & \cdots & b_m \end{bmatrix} \quad (2.10)$$

is called the **augmented matrix** which is of order $n \times (m + 1)$.

Now, (a) The equations are **consistent**, if Rank A = Rank A_b.

(i) The solutions are unique, if Rank A = Rank $A_b = n$, where n = number of unknowns.

(ii) Infinite solutions, if Rank A = Rank $A_b = r$, $r < n$.

(b) The equations are **inconsistent**, if Rank $A \neq$ Rank A_b.

2.19 Solutions of Simultaneous Linear Equations

Matrices play an important role in solving the systems of linear algebraic equations.

2.19.1 Solutions of Homogeneous Equations

For homogeneous system, Eqn. (2.9) can be written as

$$AX = 0. \tag{2.11}$$

If $|A| \neq 0$, then A^{-1} exists.

Premultiplying both sides of Eqn. (2.11) by A^{-1}, we get

$$A^{-1}AX = A^{-1}0 \quad \text{or,} \quad IX = 0,$$

which shows that $X = 0$, i.e., $x_1 = x_2 = \cdots = x_n = 0$. This is called the **trivial solution**.

If however, the matrix A is singular, i.e., $|A| = 0$, then the equations under consideration will have infinite solutions where some solutions may be non-zero.

Example Let the system of homogeneous equations be represented by

$$\begin{aligned} x + 2y + 3z &= 0 \\ 2x + 3y + 4z &= 0 \\ 3x + 4y + 5z &= 0. \end{aligned}$$

Here, $A = \begin{bmatrix} 1 & 2 & 3 \\ 2 & 3 & 4 \\ 3 & 4 & 5 \end{bmatrix}$.

$\therefore \quad |A| = 1 \times (-1) + 2 \times (2) + 3(-1) = 0.$

$\therefore$ ***A* is a singular matrix.**

Now making the elementary row operations $R_2' \to R_2 - 2R_1$ and $R_3' \to R_3 - 3R_1$, we have

$$A = \begin{bmatrix} 1 & 2 & 3 \\ 0 & -1 & -2 \\ 0 & -2 & -4 \end{bmatrix}.$$

$\therefore$ the given set of equations can be written as

$$\begin{bmatrix} 1 & 2 & 3 \\ 0 & -1 & -2 \\ 0 & -2 & -4 \end{bmatrix} \begin{bmatrix} x \\ y \\ z \end{bmatrix} = \begin{bmatrix} 0 \\ 0 \\ 0 \end{bmatrix}.$$

This gives
$$\begin{aligned} x + 2y + 3z &= 0 \\ y + 2z &= 0 \\ 4y + 4z &= 0. \end{aligned}$$

The last two equations are the same giving $y = -2z$. Putting this value of y in the first equation, we get $x - z = 0$, i.e., $x = z$ giving infinite number of solutions.

Hence, if (i) A is of order $m \times n$ and rank r, then the number of linearly independent solutions of equation $AX = 0$ is $n = r$.

(ii) A is of order $m \times n$, then $AX = 0$ has non-zero solutions, if and only if rank $A < n$.

2.19.2 Solutions of Non-homogeneous System of Equations

A system of non-homogeneous equations is represented as

$$AX = B. \tag{2.12}$$

Now, A^{-1} exists if we assume that matrix A is non-singular.

Pre-multiplying both sides of equation (2.12) by A^{-1}, we have

$$A^{-1}AX = A^{-1}B \quad \text{or,} \quad IX = A^{-1}B$$

$$\therefore \quad \mathbf{X = A^{-1}B}. \tag{2.13}$$

Thus, finding the value of A^{-1}, we can find out the solution using eqn. (2.13).

N.B. : (i) This method of finding out the solution, however, fails if A is singular, i.e., A^{-1} does not exist.

(ii) The solutions given by eqn. (2.13) are unique.

Proof : If the solutions are not unique, let X_1 and X_2 be two solutions.

Then $\quad AX_1 = B \quad$ and $\quad AX_2 = B$

so that, $\quad AX_1 = AX_2$.

Pre-multiplying by A^{-1} on both sides of the above equation, we get

$$A^{-1}AX_1 = A^{-1}AX_2 \quad \text{or,} \quad IX_1 = IX_2 \quad \text{or,} \quad X_1 = X_2.$$

Hence, the solution is unique.

Example Solve the given set of equations :

$$\begin{aligned} x + z &= 4 \\ 2x + y + z &= 7 \\ x + y + 2z &= 9. \end{aligned}$$

Solution : In this case, $A = \begin{bmatrix} 1 & 0 & 1 \\ 2 & 1 & 1 \\ 1 & 1 & 2 \end{bmatrix}$.

$$\therefore \quad |A| = 1 \times (2-1) + 1(2-1) = 2.$$

Co-factors of the elements of A are

$$A_{11} = (-1)^{1+1}(2-1) = 1, \; A_{12} = 1, \; A_{13} = -1$$

$$A_{21} = (-1)^{1+2}(4-1) = -3, \; A_{22} = 1, \; A_{23} = 1$$

$$A_{31} = (-1)^{1+3}(2-1) = 1, \; A_{32} = -1, \; A_{33} = 1.$$

$$\therefore \quad A^{-1} = \frac{\text{Adj}A}{|A|} = \frac{1}{2}\begin{bmatrix} 1 & 1 & -1 \\ -3 & 1 & 1 \\ 1 & -1 & 1 \end{bmatrix}.$$

Hence, $\begin{bmatrix} x \\ y \\ z \end{bmatrix} = \frac{1}{2}\begin{bmatrix} 1 & 1 & -1 \\ -3 & 1 & 1 \\ 1 & -1 & 1 \end{bmatrix}\begin{bmatrix} 4 \\ 7 \\ 9 \end{bmatrix} = \frac{1}{2}\begin{bmatrix} 2 \\ 4 \\ 6 \end{bmatrix}$ or, $\begin{bmatrix} x \\ y \\ z \end{bmatrix} = \begin{bmatrix} 1 \\ 2 \\ 3 \end{bmatrix}$.

$\therefore$ the solution is $x = 1$, $y = 2$, $z = 3$.

2.20 Eigenvalues or Characteristic Roots

(a) Characteristic matrix : If A be a square matrix and λ, a scalar, then the matrix $A - \lambda I$ is called the **characteristic matrix** of A, where I is the unit matrix.

Example We consider the matrix $A = \begin{bmatrix} 2 & 2 & 1 \\ 1 & 3 & 1 \\ 1 & 2 & 2 \end{bmatrix}$.

The characteristic matrix of A is

$$A - \lambda I = \begin{bmatrix} 2 & 2 & 1 \\ 1 & 3 & 1 \\ 1 & 2 & 2 \end{bmatrix} - \lambda\begin{bmatrix} 1 & 0 & 0 \\ 0 & 1 & 0 \\ 0 & 0 & 1 \end{bmatrix} = \begin{bmatrix} 2-\lambda & 2 & 1 \\ 1 & 3-\lambda & 1 \\ 1 & 2 & 2-\lambda \end{bmatrix}.$$

(b) Characteristic polynomial : The determinant when expanded gives a polynomial called the **characteristic polynomial of matrix *A*.**

Example $\begin{vmatrix} 2-\lambda & 2 & 1 \\ 1 & 3-\lambda & 1 \\ 1 & 2 & 2-\lambda \end{vmatrix}$

$$= (2-\lambda)[(3-\lambda)(2-\lambda) - 2] - 2[(2-\lambda) - 1] + 1[2 - (3-\lambda)]$$
$$= (2-\lambda)[(4 - 5\lambda + \lambda^2) + 2\lambda - 2 + \lambda - 1] = 8 - 14\lambda + 7\lambda^2 - \lambda^3 + 3\lambda - 3$$
$$= -\lambda^3 + 7\lambda^2 - 11\lambda + 5.$$

This gives the **characteristic polynomial** of the matrix A.

(c) Characteristic equation : The equation $|A - \lambda I| = 0$ is called characteristic equation of the matrix A.

Example $\begin{vmatrix} 2-\lambda & 2 & 1 \\ 1 & 3-\lambda & 1 \\ 1 & 2 & 2-\lambda \end{vmatrix} = 0$

or, $-\lambda^3 + 7\lambda^2 - 11\lambda + 5 = 0$ or, $\lambda^3 - 7\lambda^2 + 11\lambda + 5 = 0$,

is the characteristic equation of the matrix A.

(d) Eigenvalues or characteristic values : The roots of the characteristic equation $|A - \lambda I| = 0$ are called the eigenvalues or characteristic roots of the matrix A.

Now, $\lambda^3 - 7\lambda^2 + 11\lambda + 5 = 0$ or, $(\lambda - 5)(\lambda - 1)(\lambda - 1) = 0$

i.e., $\lambda_1 = 5$, $\lambda_2 = 1$, $\lambda_2 = 1$ are the characteristic roots of A.

2.20.1 Properties of Eigenvalues

(a) Any square matrix *A* and its transpose *A′* have the same eigenvalues.

(b) The sum of the elements on the principal diagonal of a matrix is called the trace of the matrix.

(c) The sum of the eigenvalues of a matrix is equal to the trace of the matrix.

(d) The product of the eigenvalues of a matrix A is equal to the determinant of A.

(e) If $\lambda_1, \lambda_2, \ldots, \lambda_n$ be the eigenvalues of the matrix A, then the eigenvalues of

(i) kA **are** $k\lambda_1, k\lambda_2, \ldots, k\lambda_n$

(ii) A^m **are** $\lambda_1^m, \lambda_2^m, \ldots, \lambda_n^m$

(iii) A^{-1} **are** $\dfrac{1}{\lambda_1}, \dfrac{1}{\lambda_2}, \ldots, \dfrac{1}{\lambda_n}$.

2.21 Caley-Hamilton Theorem

Statement : Every square matrix satisfies its own characteristic equation.

If A be a square matrix of order n, then the characteristic polynomial is

$$|A - \lambda I| = a_0 + a_1\lambda + a_2\lambda^2 + \cdots + a_n\lambda^n.$$

The matrix equation is

$$a_0 I + a_1 X + a_2 X^2 + \cdots + a_n X^n = 0$$

is satisfied by $X = A$.

Proof : We have the characteristic polynomial,

$$|A - \lambda I| = a_0 + a_1\lambda + a_2\lambda^2 + \cdots + a_n\lambda^n. \tag{2.14}$$

The Characteristic equation of A is

$$|A - \lambda I| = a_0 + a_1\lambda + a_2\lambda^2 + \cdots + a_n\lambda^n = 0. \tag{2.15}$$

The matrix equation is

$$a_0 I + a_1 X + a_2 X^2 + \cdots + a_n X^n = 0. \tag{2.16}$$

If this equation is satisfied by A, then we have to show that,

$$a_0 I + a_1 A + a_2 A^2 + \cdots + a_n A^n = 0. \tag{2.17}$$

As each element of the characteristic matrix $[A - \lambda I]$ is an ordinary polynomial of degree n (at most), the co-factor of every element of $|A - \lambda I|$ is at most an ordinary polynomial of degree $(n-1)$.

So, each element of

$$B = \text{adj } [A - \lambda I] \tag{2.18}$$

is at most an ordinary polynomial of degree $(n-1)$.

$$\therefore \quad B = \text{adj}\,[A - \lambda I] = B_0 + B_1\lambda + B_2\lambda^2 + \cdots + B_n\lambda^{n-1}, \tag{2.19}$$

where $B_1, B_2, \ldots, B_{n-1}$ are all square matrices of order n, their elements being polynomials in λ.

We know that,

$$[A - \lambda I]\,\text{adj}\,[A - \lambda I] = |A - \lambda I| I$$

$$\text{or,} \quad [A - \lambda I] = [B_0 + B_1\lambda + B_2\lambda^2 + \cdots B_n\lambda^{n-1}]$$

$$= (a_0 + a_1\lambda + a_2\lambda^2 + \cdots + a_n\lambda^n)I. \tag{2.20}$$

Equating the coefficients of like powers of λ on both sides, we get

$$\begin{aligned} AB_0 &= a_0 I \\ AB_1 - B_0 &= a_1 I \\ AB_2 - B_1 &= a_2 I \\ &\cdots\cdots \\ AB_{n-1} - B_{n-2} &= a_{n-1} I \\ -B_{n-1} &= a_{n-1} I. \end{aligned}$$

Multiplying these equations by A^n, A^{n-1}, ..., I respectively and then adding, we get

$$0 = a_0 I + a_1 A + a_2 A^2 + \cdots + a_n A^n.$$

This is equation (2.16) with $X = A$. This proves the theorem.

N.B. : If A be a square matrix with characteristic equation,

$$\lambda^3 - 2\lambda^2 + 3\lambda - 4 = 0, \tag{2.21}$$

then according to **Cayley-Hamilton theorem**, equn. (2.21) is satisfied by A.

$$\therefore \quad A^3 - 2A^2 + 3A - 4 = 0. \tag{2.22}$$

A^{-1} can be found out from equation (2.22) by multiplying it with A^{-1}, i.e.,

$$A^{-1}A^3 - 2A^{-1}A^2 + 3A^{-1}A - 4A^{-1} = 0 \quad \text{or,} \quad A^2 - 2A + 3I - 4A^{-1} = 0$$

$$\text{or,} \quad \mathbf{A^{-1} = \frac{1}{4}[A^2 - 2A + 3I]}. \tag{2.23}$$

2.22 Characteristic Vectors or, Eigenvectors

For each value of the eigenvalues λ, there is a a non-zero vector X satisfying the equation,

$$|A - \lambda I|X = 0.$$

The non-zero value X is called the **Characteristic Vector or Eigenvector**. So, the eigenvector X is not unique and is of the form μX.

Orthogonal vectors : The eigenvectors X_1 and X_2 are said to be orthogonal, if $X_1 \cdot X_2 = 0$.

The condition of orthogonality of two vectors $X_1 = \begin{bmatrix} x_1 \\ x_2 \\ x_3 \end{bmatrix}$ and $X_2 = \begin{bmatrix} x_4 \\ x_5 \\ x_6 \end{bmatrix}$ is

$$x_1 x_4 + x_2 x_5 + x_3 x_6 = 0.$$

Normalised form of vectors : Let $A = \begin{bmatrix} a \\ b \\ c \end{bmatrix}$ be a matrix.

To find the normalised form of A we divide each element of A by $\sqrt{a^2 + b^2 + c^2}$.

Example Suppose, we have to find the normalised form of the matrix

$$A = \begin{bmatrix} 2 \\ \sqrt{3} \\ 3 \end{bmatrix}.$$

Here, $\sqrt{a^2+b^2+c^2} = \sqrt{2^2+\sqrt{3}^2+3^2} = 4.$

$\therefore$ the normalised form of A is $\begin{bmatrix} \frac{1}{2} \\ \frac{\sqrt{3}}{4} \\ \frac{3}{4} \end{bmatrix}$.

2.23 Diagonalisation of a Matrix

Theorem : If a square matrix A of order n has n linearly independent eigenvectors, then a matrix P can be found out such that $P^{-1}AP$ is a diagonal matrix.

Proof : Let A be a square matrix of order $n = 3$ whose eigenvalues are λ_1, λ_2, λ_3 and eigenvectors are

$$X_1 = \begin{bmatrix} x_1 \\ y_1 \\ z_1 \end{bmatrix}, \; X_2 = \begin{bmatrix} x_2 \\ y_2 \\ z_2 \end{bmatrix}, \; X_3 = \begin{bmatrix} x_3 \\ y_3 \\ z_3 \end{bmatrix}.$$

If $A = \begin{bmatrix} a_1 & b_1 & c_1 \\ a_2 & b_2 & c_2 \\ a_3 & b_3 & c_3 \end{bmatrix}$, then for the eigenvalues λ_1, the eigenvector is given by

$$\begin{aligned} (a_1-\lambda_1)x_1 + b_1y_1 + c_1z_1 &= 0 \\ a_2x_1 + (b_2-\lambda_1)y_1 + c_2z_1 &= 0 \\ a_3x_1 + b_3y_1 + (c_3-\lambda_1)z_1 &= 0. \end{aligned}$$

$$\therefore \quad \left.\begin{aligned} a_1x_1 + b_1y_1 + c_1z_1 &= \lambda_1x_1 \\ a_2x_1 + b_2y_1 + c_2z_1 &= \lambda_1y_1 \\ a_3x_1 + b_3y_1 + c_3z_1 &= \lambda_1z_1 \end{aligned}\right\}. \tag{2.24}$$

Likewise for λ_2 and λ_3,

$$\left.\begin{aligned} a_1x_2 + b_1y_2 + c_1z_2 &= \lambda_2x_2 \\ a_2x_2 + b_2y_2 + c_2z_2 &= \lambda_2y_2 \\ a_3x_2 + b_3y_2 + c_3z_2 &= \lambda_2z_2 \end{aligned}\right\}. \tag{2.25}$$

and
$$\left.\begin{aligned} a_1x_3 + b_1y_3 + c_1z_3 &= \lambda_3x_3 \\ a_2x_3 + b_2y_3 + c_2z_3 &= \lambda_3y_3 \\ a_3x_3 + b_3y_3 + c_3z_3 &= \lambda_3z_3 \end{aligned}\right\}. \tag{2.26}$$

We now consider a matrix P such that,

$$P = \begin{bmatrix} x_1 & x_2 & x_3 \\ y_1 & y_2 & y_3 \\ z_1 & z_2 & z_3 \end{bmatrix},$$

whose columns are the eigenvectors of A, then

$$AP = \begin{bmatrix} a_1 & b_1 & c_1 \\ a_2 & b_2 & c_2 \\ a_3 & b_3 & c_3 \end{bmatrix} \begin{bmatrix} x_1 & x_2 & x_3 \\ y_1 & y_2 & y_3 \\ z_1 & z_2 & z_3 \end{bmatrix}$$

$$= \begin{bmatrix} a_1x_1 + b_1y_1 + c_1z_1 & a_1x_2 + b_1y_2 + c_1z_2 & a_1x_3 + b_1y_3 + c_1z_3 \\ a_2x_1 + b_2y_1 + c_2z_1 & a_2x_2 + b_2y_2 + c_2z_2 & a_2x_3 + b_2y_3 + c_2z_3 \\ a_3x_1 + b_3y_1 + c_3z_1 & a_3x_2 + b_3y_2 + c_3z_2 & a_3x_3 + b_3y_3 + c_3z_3 \end{bmatrix}$$

$$= \begin{bmatrix} \lambda_1x_1 & \lambda_2x_2 & \lambda_3x_3 \\ \lambda_1y_1 & \lambda_2y_2 & \lambda_3y_3 \\ \lambda_1z_1 & \lambda_2z_2 & \lambda_3z_3 \end{bmatrix} \quad \text{[by using equations (2.24), (2.25) and (2.26)]}$$

$$= \begin{bmatrix} x_1 & x_2 & x_3 \\ y_1 & y_2 & y_3 \\ z_1 & z_2 & z_3 \end{bmatrix} \begin{bmatrix} \lambda_1 & 0 & 0 \\ 0 & \lambda_2 & 0 \\ 0 & 0 & \lambda_3 \end{bmatrix}$$

$$= PD, \text{ where } D \text{ is the diagonal matrix given by } D = \begin{bmatrix} \lambda_1 & 0 & 0 \\ 0 & \lambda_2 & 0 \\ 0 & 0 & \lambda_3 \end{bmatrix}.$$

$\therefore \quad AP = PD$

or, $\quad P^{-1}AP = P^{-1}AD = D. \qquad (2.27)$

N.B. : 1. The matrix P which diagonises A is called the model matrix of A and the resulting diagonal matrix D is known as the spectral matrix of A.

The matrix P is constructed by grouping the eigenvectors of A into square matrix and the diagonal matrix has the eigenvalues of A as its elements.

2. The transformation of the matrix A to $P^{-1}AP$ is known as a similarity transformation.

3. The reduction of A into a diagonal matrix is obviously, a particular case of similarity transformation.

2.24 Linear Dependence and Independence of Vectors

A set of vectors (matrices), $X_1, X_2, X_3, \ldots, X_n$ are **linearly dependent** if the following conditions are satisfied :

(1) All the vectors are of the same order.

(2) If there exist n scalars $\lambda_1, \lambda_2, \ldots, \lambda_n$ (not all zero) such that,

$$\sum_{i=1}^{n} \lambda_i X_i = \theta,$$

where θ is a null vector.

But if, $\sum_{i=1}^{n} \lambda_i X_i = \theta$ implies $\lambda_i = 0$ for all i, then the vectors $X_1, X_2, \ldots, X_n$ are said to be **linearly independent**.

Example The vectors [1 0 0], [0 1 0] and [0 0 1] **linearly independent** as

$$\lambda_1[1\ 0\ 0] + \lambda_2[1\ 0\ 0] + \lambda_3[0\ 0\ 1] = [0\ 0\ 0]$$

implies $\lambda_1 = \lambda_2 = \lambda_3 = 0$.

On the other hand, the vectors $[1\ -1\ 1]$, $[-1\ 1\ 1]$ and $[1\ -1\ -1]$ are **linearly dependent**.

2.25 Transformation Matrix

We consider two sets of variables $x_1, x_2, \ldots, x_n$ and $x'_1, x'_2, \ldots, x'_n$ which are related by he equations :

$$x'_1 = a_{11}x_1 + a_{12}x_2 + \cdots + a_{1n}x_n$$
$$x'_2 = a_{21}x_1 + a_{22}x_2 + \cdots + a_{2n}x_n$$
$$\cdots\cdots\cdots\cdots\cdots\cdots\cdots\cdots\cdots\cdots\cdots$$
$$x'_n = a_{n1}x_1 + a_{n2}x_2 + \cdots + a_{nn}x_n$$

The above set of equations can be expressed in the matrix form as below :

$$\begin{bmatrix} x'_1 \\ x'_2 \\ \vdots \\ x'_n \end{bmatrix} = \begin{bmatrix} a_{11} & a_{12} & \cdots & a_{1n} \\ a_{21} & a_{22} & \cdots & a_{2n} \\ \cdots & \cdots & \cdots & \cdots \\ a_{n1} & a_{n2} & \cdots & a_{nn} \end{bmatrix} \begin{bmatrix} x_1 \\ x_2 \\ \vdots \\ x_n \end{bmatrix}$$

or, $\quad X' = AX,$

$$\text{where } X' = \begin{bmatrix} x'_1 \\ x'_2 \\ \vdots \\ x'_n \end{bmatrix}, A = \begin{bmatrix} a_{11} & a_{12} & \cdots & a_{1n} \\ a_{21} & a_{22} & \cdots & a_{2n} \\ \vdots & & & \\ a_{n1} & a_{n2} & \cdots & a_{nn} \end{bmatrix}, X = \begin{bmatrix} x_1 \\ x_2 \\ \vdots \\ x_n \end{bmatrix}.$$

The matrix A is known as the transformation matrix.

$\therefore$ the identical relations $x'_1 = x_1$, $x'_2 = x_2$, $x'_3 = x_3$ can be expressed through the unit matrix I, where

$$I = \begin{bmatrix} 1 & 0 & \cdots & 0 \\ 0 & 1 & \cdots & 0 \\ \cdots & \cdots & \cdots & \cdots \\ 0 & 0 & \cdots & 1 \end{bmatrix}.$$

2.26 Solved Problems

Example 2.1. Given $A = \begin{bmatrix} 1 & -1 & -1 \\ -3 & 2 & -1 \\ -2 & 1 & 0 \end{bmatrix}$ and $B = \begin{bmatrix} 1 & 2 & 3 \\ 2 & 4 & 6 \\ 1 & 2 & 3 \end{bmatrix}$; compute AB and BA and show that $AB \neq BA$.

Solution : Since both A and B are 3×3 matrices, the product AB and BA are 3×3 matrices.

Now, $\quad AB = \begin{bmatrix} 1 & -1 & 1 \\ -3 & 2 & -1 \\ -2 & 1 & 0 \end{bmatrix} \begin{bmatrix} 1 & 2 & 3 \\ 2 & 4 & 6 \\ 1 & 2 & 3 \end{bmatrix}$

$$= \begin{bmatrix} 1\times1+(-1)\times2+1\times1 & 1\times2+(-1)\times4+1\times2 & 1\times3+(-1)\times6+1\times3 \\ (-3)\times1+2\times2+(-1)\times1 & (-3)\times2+2\times4+(-1)\times2 & (-3)\times3+2\times6+(-1)\times3 \\ (-2)\times1+1\times2+0\times1 & (-2)\times2+1\times4+0\times2 & -2\times3+1\times6+0\times3 \end{bmatrix}$$

$$= \begin{bmatrix} 0 & 0 & 0 \\ 0 & 0 & 0 \\ 0 & 0 & 0 \end{bmatrix}$$

and $BA = \begin{bmatrix} 1 & 2 & 3 \\ 2 & 4 & 6 \\ 1 & 2 & 3 \end{bmatrix} \begin{bmatrix} 1 & -1 & 1 \\ -3 & 2 & -1 \\ -2 & 1 & 0 \end{bmatrix}$

$$= \begin{bmatrix} 1\times 1+2\times(-3)+3\times(-2) & 1\times(-1)+2\times 2+3\times 1 & 1\times 1+2\times(-1)+3\times 0 \\ 2\times 1+4\times(-3)+6\times(-2) & 2\times(-1)+4\times 2+6\times 1 & 2\times 1+4\times(-1)+6\times 0 \\ 1\times 1+2\times(-3)+3\times(-2) & 1\times(-1)+2\times 2+3\times 1 & 1\times 1+2\times(-1)+3\times 0 \end{bmatrix}$$

$$= \begin{bmatrix} -11 & 6 & -1 \\ -22 & 12 & -2 \\ -11 & 6 & -1 \end{bmatrix}.$$

Hence, $\boldsymbol{AB \neq BA}$.

Example 2.2. *Determine the matrices A and B, where*

$$A + 2B = \begin{bmatrix} 1 & 2 & 0 \\ 6 & -3 & 3 \\ -5 & 3 & 1 \end{bmatrix} \text{ and } 2A - B = \begin{bmatrix} 2 & -1 & 5 \\ 2 & -1 & 6 \\ 0 & 1 & 2 \end{bmatrix}.$$

Solution : We have $= 4A - 2B = \begin{bmatrix} 4 & -2 & 10 \\ 4 & -2 & 12 \\ 0 & 2 & 4 \end{bmatrix}$

$\therefore \quad (A+2B)+(4A-2B) = 5A = \begin{bmatrix} 1 & 2 & 0 \\ 6 & -3 & 3 \\ -5 & 3 & 1 \end{bmatrix} + \begin{bmatrix} 4 & -2 & 10 \\ 4 & -2 & 12 \\ 0 & 2 & 4 \end{bmatrix}$

$$= \begin{bmatrix} 5 & 0 & 10 \\ 10 & -5 & 15 \\ -5 & 5 & 5 \end{bmatrix} = 5 \begin{bmatrix} 1 & 0 & 2 \\ 2 & -1 & 3 \\ -1 & 1 & 1 \end{bmatrix}.$$

$\therefore \quad \boldsymbol{A} = \begin{bmatrix} \mathbf{1} & \mathbf{0} & \mathbf{2} \\ \mathbf{2} & \mathbf{-1} & \mathbf{3} \\ \mathbf{-1} & \mathbf{1} & \mathbf{1} \end{bmatrix}.$

Again, $2B = (A+2B) - A = \begin{bmatrix} 1 & 2 & 0 \\ 6 & -3 & 3 \\ -5 & 3 & 1 \end{bmatrix} - \begin{bmatrix} 1 & 0 & 2 \\ 2 & -1 & 3 \\ -1 & 1 & 1 \end{bmatrix}$

$$= \begin{bmatrix} 0 & 2 & -2 \\ 4 & -2 & 0 \\ -4 & 2 & 0 \end{bmatrix} = 2 \begin{bmatrix} 0 & 1 & -1 \\ 2 & -1 & 0 \\ -2 & 1 & 0 \end{bmatrix}$$

or, $\quad \boldsymbol{B} = \begin{bmatrix} \mathbf{0} & \mathbf{1} & \mathbf{-1} \\ \mathbf{2} & \mathbf{-1} & \mathbf{0} \\ \mathbf{-2} & \mathbf{1} & \mathbf{0} \end{bmatrix}.$

Example 2.3. *Show by means of an example that in matrices $A+B = 0$ does not necessarily mean that either $A = 0$ or $B = 0$, where 0 stands for the null matrix.*

Solution : Consider matrices A and B such that, $A = \begin{bmatrix} 1 & -1 & 1 \\ -3 & 2 & -1 \\ -2 & 1 & 0 \end{bmatrix}$ and $B = \begin{bmatrix} 1 & 2 & 3 \\ 2 & 4 & 6 \\ 1 & 2 & 3 \end{bmatrix}$.

$$\therefore \quad AB = \begin{bmatrix} 1-2+1 & 2-4+2 & 3-6+3 \\ -3+4-1 & -6+8+-2 & -9+12-3 \\ -2+2+0 & -4+4+0 & -6+6+0 \end{bmatrix} = \begin{bmatrix} 0 & 0 & 0 \\ 0 & 0 & 0 \\ 0 & 0 & 0 \end{bmatrix} \quad \text{or,} \quad AB = 0.$$

But neither $A = 0$ nor $B = 0$. This proves the statement.

Example 2.4. *If* $M_x = \frac{1}{\sqrt{2}}\begin{bmatrix} 0 & 1 & 0 \\ 1 & 0 & 1 \\ 0 & 1 & 0 \end{bmatrix}$, $M_y = \frac{1}{\sqrt{2}}\begin{bmatrix} 0 & -i & 0 \\ i & 0 & -i \\ 0 & i & 0 \end{bmatrix}$ *and* $M_z = \begin{bmatrix} 1 & 0 & 0 \\ 0 & 0 & 0 \\ 0 & 0 & -1 \end{bmatrix}$, *then show that,* $\frac{1}{2}M^2 = I$, *where* $M = M_x^2 + M_y^2 + M_z^2$ *and* $i = \sqrt{-1}$.

Solution : We have $2M_x^2 = \begin{bmatrix} 0 & 1 & 0 \\ 1 & 0 & 1 \\ 0 & 1 & 0 \end{bmatrix} \times \begin{bmatrix} 0 & 1 & 1 \\ 1 & 0 & 1 \\ 0 & 1 & 0 \end{bmatrix} = \begin{bmatrix} 1 & 0 & 1 \\ 0 & 2 & 0 \\ 1 & 0 & 1 \end{bmatrix}$

$$2M_y^2 = \begin{bmatrix} 0 & -i & 0 \\ i & 0 & -i \\ 0 & i & 0 \end{bmatrix} \times \begin{bmatrix} 0 & -i & 0 \\ i & 0 & -i \\ 0 & i & 0 \end{bmatrix} = \begin{bmatrix} 1 & 0 & -1 \\ 0 & 2 & 0 \\ -1 & 0 & 1 \end{bmatrix}$$

and $2M_z^2 = 2\begin{bmatrix} 1 & 0 & 0 \\ 0 & 0 & 0 \\ 0 & 0 & -1 \end{bmatrix} \times \begin{bmatrix} 1 & 0 & 0 \\ 0 & 0 & 0 \\ 0 & 0 & -1 \end{bmatrix} = \begin{bmatrix} 2 & 0 & 0 \\ 0 & 0 & 0 \\ 0 & 0 & 2 \end{bmatrix}$.

$$\therefore \quad 2(M_x^2 + M_y^2 + M_z^2) = \begin{bmatrix} 1 & 0 & 1 \\ 0 & 2 & -1 \\ 1 & 0 & 1 \end{bmatrix} + \begin{bmatrix} 1 & 0 & -1 \\ 0 & 2 & 0 \\ -1 & 0 & 1 \end{bmatrix} + \begin{bmatrix} 2 & 0 & 0 \\ 0 & 0 & 0 \\ 0 & 0 & 2 \end{bmatrix}$$

$$= \begin{bmatrix} 4 & 0 & 0 \\ 0 & 4 & 0 \\ 0 & 0 & 4 \end{bmatrix} = 4\begin{bmatrix} 1 & 0 & 0 \\ 0 & 1 & 0 \\ 0 & 0 & 1 \end{bmatrix} = 4I.$$

$$\therefore \quad \mathbf{\frac{1}{2}M^2 = I.}$$

Example 2.5. *If* $A = \begin{bmatrix} 1 & 2 & 2 \\ 2 & 1 & 2 \\ 2 & 2 & 1 \end{bmatrix}$, *show that,* $A^2 - 4A - 5I = O$, *where* I *and* O *are the unit matrix and the null matrix of order 3 respectively. Also use this result to find* A^{-1}.

[AMIE, Summer 2004]

Solution : We have $A = \begin{bmatrix} 1 & 2 & 2 \\ 2 & 1 & 2 \\ 2 & 2 & 1 \end{bmatrix}$.

$$\therefore \quad A^2 = \begin{bmatrix} 1 & 2 & 2 \\ 2 & 1 & 2 \\ 2 & 2 & 1 \end{bmatrix} \times \begin{bmatrix} 1 & 2 & 2 \\ 2 & 1 & 2 \\ 2 & 2 & 1 \end{bmatrix} = \begin{bmatrix} 9 & 8 & 8 \\ 8 & 9 & 8 \\ 8 & 8 & 9 \end{bmatrix}.$$

$$\therefore \quad A^2 - 4A - 5I = \begin{bmatrix} 9 & 8 & 8 \\ 8 & 9 & 8 \\ 8 & 8 & 9 \end{bmatrix} - 4\begin{bmatrix} 1 & 2 & 2 \\ 2 & 1 & 2 \\ 2 & 2 & 1 \end{bmatrix} - 5\begin{bmatrix} 1 & 0 & 0 \\ 0 & 1 & 0 \\ 0 & 0 & 1 \end{bmatrix}$$

$$= \begin{bmatrix} 9-4-5 & 8-8-0 & 8-8-0 \\ 8-8-0 & 9-4-5 & 8-8-0 \\ 8-8-0 & 8-8-0 & 9-4-5 \end{bmatrix} = \begin{bmatrix} 0 & 0 & 0 \\ 0 & 0 & 0 \\ 0 & 0 & 0 \end{bmatrix}.$$

$\therefore \quad \mathbf{A^2 - 4A - 5I = 0}$ or, $5I = A^2 - 4A$.

Now multiplying by A^{-1}, we get

$$5A^{-1} = A - 4I = \begin{bmatrix} 1 & 2 & 2 \\ 2 & 1 & 2 \\ 2 & 2 & 1 \end{bmatrix} - 4\begin{bmatrix} 1 & 0 & 0 \\ 0 & 1 & 0 \\ 0 & 0 & 1 \end{bmatrix} = \begin{bmatrix} -3 & 2 & 2 \\ 2 & -3 & 2 \\ 2 & 2 & -3 \end{bmatrix}$$

$$\therefore \quad A^{-1} = \frac{1}{5}\begin{bmatrix} -3 & 2 & 2 \\ 2 & -3 & 2 \\ 2 & 2 & -3 \end{bmatrix}.$$

Example 2.6. *Express* $\begin{bmatrix} 1 & 2 & 3 \\ 3 & 4 & 5 \\ 5 & 6 & 7 \end{bmatrix}$ *as the sum of a symmetric and a skew-symmetric matrix.*

Solution : We consider the matrix $A = \begin{bmatrix} 1 & 2 & 3 \\ 3 & 4 & 5 \\ 5 & 6 & 7 \end{bmatrix}$;

$\therefore$ the transpose of A is $A' = \begin{bmatrix} 1 & 3 & 5 \\ 2 & 4 & 6 \\ 3 & 5 & 7 \end{bmatrix}$.

Now, $A + A' = \begin{bmatrix} 1 & 2 & 3 \\ 3 & 4 & 5 \\ 5 & 6 & 7 \end{bmatrix} + \begin{bmatrix} 1 & 3 & 5 \\ 2 & 4 & 6 \\ 3 & 5 & 7 \end{bmatrix} = \begin{bmatrix} 2 & 5 & 8 \\ 5 & 8 & 11 \\ 8 & 11 & 14 \end{bmatrix}$ and $A - A' = \begin{bmatrix} 0 & -1 & -2 \\ 1 & 0 & -1 \\ 2 & 1 & 0 \end{bmatrix}$

$$X = \frac{1}{2}(A + A') = \begin{bmatrix} 1 & \frac{5}{2} & 4 \\ \frac{5}{2} & 4 & \frac{11}{2} \\ 4 & \frac{11}{2} & 7 \end{bmatrix} \quad \text{and} \quad Y = \frac{1}{2}(A - A') = \begin{bmatrix} 0 & -\frac{1}{2} & -1 \\ \frac{1}{2} & 0 & -\frac{1}{2} \\ 1 & \frac{1}{2} & 0 \end{bmatrix},$$

where X is symmetric and Y is skew symmetric and $\mathbf{A = X + Y}$ **(Proved)**.

Example 2.7. *Find the inverse of the matrix,* $A = \begin{bmatrix} 5 & -2 & 4 \\ -2 & 1 & 1 \\ 4 & 1 & 0 \end{bmatrix}$.

Solution : We have $|A| = \begin{vmatrix} 5 & -2 & 4 \\ -2 & 1 & 1 \\ 4 & 1 & 0 \end{vmatrix} = 5 \times (-1) + 2(-4) + 4(-2-4) = -37$

and transpose of A, $A^T = \begin{bmatrix} 5 & -2 & 4 \\ -2 & 1 & 1 \\ 4 & 1 & 0 \end{bmatrix}$.

Now the co-factors are :

$$A_{11} = \begin{vmatrix} 1 & 1 \\ 1 & 0 \end{vmatrix} = -1 \qquad A_{21} = -\begin{vmatrix} -2 & 4 \\ 1 & 0 \end{vmatrix} = 4 \qquad A_{31} = +\begin{vmatrix} -2 & 4 \\ 1 & 1 \end{vmatrix} = -6$$

$$A_{12} = -\begin{vmatrix} -2 & 1 \\ 4 & 0 \end{vmatrix} = 4 \qquad A_{22} = +\begin{vmatrix} 5 & 4 \\ 4 & 0 \end{vmatrix} = -16 \qquad A_{32} = -\begin{vmatrix} 5 & 4 \\ -2 & 1 \end{vmatrix} = -13$$

$$A_{13} = +\begin{vmatrix} -2 & 1 \\ 4 & 1 \end{vmatrix} = -6 \qquad A_{23} = -\begin{vmatrix} 5 & -2 \\ 4 & 1 \end{vmatrix} = -13 \qquad A_{33} = +\begin{vmatrix} 5 & -2 \\ -2 & 1 \end{vmatrix} = 1$$

$$\therefore \text{ Adj } A = \begin{bmatrix} -1 & 4 & -6 \\ 4 & -16 & -13 \\ -6 & -13 & 1 \end{bmatrix}. \qquad \therefore A^{-1} = -\frac{1}{37}\begin{bmatrix} -1 & 4 & -6 \\ 4 & -16 & -13 \\ -6 & -13 & 1 \end{bmatrix}.$$

Example 2.8. *If* $A = \begin{bmatrix} 1 & 2 & -1 \\ 2 & 0 & 3 \\ 0 & 1 & 2 \end{bmatrix}$ *and* $B = \begin{bmatrix} 3 & -1 & 1 \\ 0 & 0 & 2 \\ 4 & -3 & 2 \end{bmatrix}$, *then prove that,*

$$(A+B)^2 = A^2 + AB + BA + B^2.$$

Solution : We have $A + B = \begin{bmatrix} 1 & 2 & -1 \\ 2 & 0 & 3 \\ 0 & 1 & 2 \end{bmatrix} + \begin{bmatrix} 3 & -1 & 1 \\ 0 & 0 & 2 \\ 4 & -3 & 2 \end{bmatrix} = \begin{bmatrix} 1+3 & 2+(-1) & -1+1 \\ 2+0 & 0+0 & 3+2 \\ 0+4 & 1+(-3) & 2+2 \end{bmatrix}$

$$= \begin{bmatrix} 4 & 1 & 0 \\ 2 & 0 & 5 \\ 4 & -2 & 4 \end{bmatrix}.$$

Then, $(A+B)^2 = (A+B)(A+B) = \begin{bmatrix} 4 & 1 & 0 \\ 2 & 0 & 5 \\ 4 & -2 & 4 \end{bmatrix}\begin{bmatrix} 4 & 1 & 0 \\ 2 & 0 & 5 \\ 4 & -2 & 4 \end{bmatrix}$

$$= \begin{bmatrix} 4\times4+1\times2+0\times4 & 4\times1+1\times0+0\times(-2) & 4\times0+1\times5+0\times4 \\ 2\times4+0\times2+5\times4 & 2\times1+0\times0+5\times(-2) & 2\times0+0\times5+5\times4 \\ 4\times4+(-2)\times2+4\times4 & 4\times1+(-2)\times0+4\times(-2) & 4\times0+(-2)\times5+4\times4 \end{bmatrix}$$

$$= \begin{bmatrix} 18 & 4 & 5 \\ 28 & -8 & 20 \\ 28 & -4 & 6 \end{bmatrix}.$$

Now, $A^2 = \begin{bmatrix} 1 & 2 & -1 \\ 2 & 0 & 3 \\ 0 & 1 & 2 \end{bmatrix}\begin{bmatrix} 1 & 2 & -1 \\ 2 & 0 & 3 \\ 0 & 1 & 2 \end{bmatrix} = \begin{bmatrix} 1+4+0 & 2+0-1 & -1+6-2 \\ 2+0+0 & 4+0+3 & -2+0+6 \\ 0+2+0 & 0+0+2 & 0+3+4 \end{bmatrix} = \begin{bmatrix} 5 & 1 & 3 \\ 2 & 7 & 4 \\ 2 & 2 & 7 \end{bmatrix}.$

$$AB = \begin{bmatrix} 1 & 2 & -1 \\ 2 & 0 & 3 \\ 0 & 1 & 2 \end{bmatrix}\begin{bmatrix} 3 & -1 & 1 \\ 0 & 0 & 2 \\ 4 & -3 & 2 \end{bmatrix} = \begin{bmatrix} 3+0-4 & -1+0+3 & 1+4-2 \\ 6+0+12 & -2+0-9 & 2+0+6 \\ 0+0+8 & 0+0-6 & 0+2+4 \end{bmatrix}$$

$$= \begin{bmatrix} -1 & 2 & 3 \\ 18 & -11 & 8 \\ 8 & -6 & 6 \end{bmatrix}$$

$$BA = \begin{bmatrix} 3 & -1 & 1 \\ 0 & 0 & 2 \\ 4 & -3 & 2 \end{bmatrix} \begin{bmatrix} 1 & 2 & -1 \\ 2 & 0 & 3 \\ 0 & 1 & 2 \end{bmatrix} = \begin{bmatrix} 3-2+0 & 6+0+1 & -3-3+2 \\ 0+0+0 & 0+0+2 & 0+0+4 \\ 4-6+0 & 8+0+2 & -4-9+4 \end{bmatrix}$$

$$= \begin{bmatrix} 1 & 7 & -4 \\ 0 & 2 & 4 \\ -2 & 10 & -9 \end{bmatrix}.$$

$$B^2 = \begin{bmatrix} 3 & -1 & 1 \\ 0 & 0 & 2 \\ 4 & -3 & 2 \end{bmatrix} \begin{bmatrix} 3 & -1 & 1 \\ 0 & 0 & 2 \\ 4 & -3 & 2 \end{bmatrix} = \begin{bmatrix} 9+0+4 & -3+0-3 & 3-2+2 \\ 0+0+8 & 0+0-6 & 0+0+4 \\ 12-0+8 & -4+0-6 & 4-6+4 \end{bmatrix}$$

$$= \begin{bmatrix} 13 & -6 & 3 \\ 8 & -6 & 4 \\ 20 & -10 & 2 \end{bmatrix}.$$

$$\therefore \quad A^2 + AB + BA + B^2 = \begin{bmatrix} 5 & 1 & 3 \\ 2 & 7 & 4 \\ 2 & 2 & 7 \end{bmatrix} + \begin{bmatrix} -1 & 2 & 3 \\ 18 & -11 & 8 \\ 8 & -6 & 6 \end{bmatrix} + \begin{bmatrix} 1 & 7 & -4 \\ 0 & 2 & 4 \\ -2 & 10 & -9 \end{bmatrix} + \begin{bmatrix} 13 & -6 & 3 \\ 8 & -6 & 4 \\ 20 & -10 & 2 \end{bmatrix}$$

$$= \begin{bmatrix} 5-1+1+13 & 1+2+7-6 & 3+3-4+3 \\ 2+18+0+8 & 7-11+2-6 & 4+8+4+4 \\ 2+8-2+20 & 2-6+10-10 & 7+6-9+2 \end{bmatrix}$$

$$= \begin{bmatrix} 18 & 4 & 5 \\ 28 & -8 & 20 \\ 28 & -4 & 6 \end{bmatrix}.$$

Hence, $\boldsymbol{A^2 + AB + BA + B^2} = \begin{bmatrix} 18 & 4 & 5 \\ 28 & -8 & 20 \\ 28 & -4 & 6 \end{bmatrix} = \boldsymbol{(A+B)^2}$ **(Proved).**

Example 2.9. *If* $A = \begin{bmatrix} 1 & 1 & 1 \\ 2 & 4 & 1 \\ 2 & 3 & 1 \end{bmatrix}$ *and* $B = \begin{bmatrix} 2 & 3 \\ 3 & 4 \end{bmatrix}$, *then for what value of* C, $BCA = \begin{bmatrix} 1 & 0 & 1 \\ 0 & 1 & 0 \end{bmatrix}$?

Solution : Here, $BCA = \begin{bmatrix} 1 & 0 & 1 \\ 0 & 1 & 0 \end{bmatrix}$ or, $\begin{bmatrix} 2 & 3 \\ 3 & 4 \end{bmatrix} C \begin{bmatrix} 1 & 1 & 1 \\ 2 & 4 & 1 \\ 2 & 3 & 1 \end{bmatrix} = \begin{bmatrix} 1 & 0 & 1 \\ 0 & 1 & 0 \end{bmatrix}$

or, $C \begin{bmatrix} 1 & 1 & 1 \\ 2 & 4 & 1 \\ 2 & 3 & 1 \end{bmatrix} = \begin{bmatrix} 2 & 3 \\ 3 & 4 \end{bmatrix}^{-1} \begin{bmatrix} 1 & 0 & 1 \\ 0 & 1 & 0 \end{bmatrix}$

$$\therefore \quad C = -\begin{bmatrix} 4 & -3 & 4 \\ -3 & 2 & -3 \end{bmatrix} \begin{bmatrix} 1 & 1 & 1 \\ 2 & 4 & 1 \\ 2 & 3 & 1 \end{bmatrix}^{-1} = \begin{bmatrix} 4 & -3 & 4 \\ -3 & 2 & -3 \end{bmatrix} \begin{bmatrix} 1 & 2 & -3 \\ 0 & -1 & 1 \\ -2 & -1 & 2 \end{bmatrix}$$

$$= \begin{bmatrix} -4 & 7 & -7 \\ 3 & -5 & 5 \end{bmatrix}.$$

Example 2.10. *Verify that,* $[AB]^T = B^T A^T$, *where*

$$A = \begin{bmatrix} 1 & 2 & 3 \\ 6 & 7 & 8 \\ 6 & -3 & 4 \end{bmatrix} \text{ and } B = \begin{bmatrix} 1 & 2 & 3 \\ 3 & 4 & 2 \\ 5 & 6 & 1 \end{bmatrix}.$$

Solution : We have $AB = \begin{bmatrix} 1 & 2 & 3 \\ 6 & 7 & 8 \\ 6 & -3 & 4 \end{bmatrix} \times \begin{bmatrix} 1 & 2 & 3 \\ 3 & 4 & 2 \\ 5 & 6 & 1 \end{bmatrix} = \begin{bmatrix} 22 & 28 & 10 \\ 67 & 88 & 40 \\ 17 & 24 & 16 \end{bmatrix}.$

$$\therefore \quad [AB]^T = \begin{bmatrix} 22 & 67 & 17 \\ 28 & 88 & 24 \\ 10 & 40 & 16 \end{bmatrix}.$$

Again, $A^T = \begin{bmatrix} 1 & 6 & 6 \\ 2 & 7 & -3 \\ 3 & 8 & 4 \end{bmatrix}$ and $B^T = \begin{bmatrix} 1 & 3 & 5 \\ 2 & 4 & 6 \\ 3 & 2 & 1 \end{bmatrix}.$

$$\therefore \quad B^T A^T = \begin{bmatrix} 1 & 3 & 5 \\ 2 & 4 & 6 \\ 3 & 2 & 1 \end{bmatrix} \times \begin{bmatrix} 1 & 6 & 6 \\ 2 & 7 & -3 \\ 3 & 8 & 4 \end{bmatrix} = \begin{bmatrix} 22 & 67 & 17 \\ 28 & 88 & 24 \\ 10 & 40 & 16 \end{bmatrix}.$$

Hence, $\boldsymbol{[AB]^T = B^T A^T}$ **(Proved).**

Example 2.11. *Show that,* $[AB]^n = A^n B^n$ *provided* $AB = BA$. **[BU(H) 2008]**

Solution : We have $[AB] = AB = [A][B]$

$$[AB]^2 = [AB][AB] = [ABA]B = \{A[AB]\}B = [A^2B]B = A^2[BB] = A^2B^2.$$

Again, since $[AB]^n = A^n B^n$,

$$\begin{aligned} \therefore \quad \boldsymbol{[AB]^{n+1}} &= [AB]^n AB = [A^n B^n]AB = A^n(B^n A)B = A^n(B^{n-1}BA]B \\ &= A^n[B^{n-1}AB]B \quad [\because AB = BA] \\ &= A^n[B^{n-2}BAB]B = A^n[B^{n-2}ABB]B = A^n[B^{n-2}AB^2]B = A^n[AB^n]B \\ &= \boldsymbol{A^n[AB^{n+1}] = A^{n+1}B^{n+1}} \quad \textbf{(Proved).} \end{aligned}$$

Example 2.12. *If A and B are idempotent matrices, then* $A + B$ *will be idempotent, if and only if* $AB = BA = 0$. **[Banaras university 1970]**

Solution : Since matrices A and B are idempotent, we must have $A^2 = A$ and $B^2 = B$.

If $(A + B)$ is idempotent, then $(A + B)^2 = A + B$.

We have $(A + B)^2 = (A + B)(A + B) = A^2 + AB + BA + B^2$

$$= A + AB + BA + B \quad [\because A^2 = A, B^2 = B]$$

$$= A + B + (AB + BA).$$

$\therefore$ $\boldsymbol{(A + B)^2}$ **will be idempotent, if**

$$\boldsymbol{AB + BA = 0} \quad \textbf{or,} \quad \boldsymbol{AB = BA = 0}.$$

Example 2.13. *If A, B and C are matrices such that the products AC and BC obey the commutative law, prove that, $[AB + BA]C = C[AB + BA]$.*

Solution : We have

$$\begin{aligned}
[AB + BA]C &= [AB]C + [BA]C, && \text{Distributive law}\\
&= A[BC] + B[AC], && \text{Associative law}\\
&= A[CB] + B[CA], && BC \text{ and } AC \text{ obey commutative law}\\
&= [AC]B + [BC]A, && \text{Associative law}\\
&= [CA]B + [CB]A, && \text{Commutative law}\\
&= C[AB] + C[BA], && \text{Associative law}\\
&= C[AB + BA], && \text{Distributive law.}
\end{aligned}$$

Example 2.14. *Given,* $A = \begin{bmatrix} 0 & 1+2i \\ -1+2i & 0 \end{bmatrix}$*; show that,* $U = [I - A][1 + A]^{-1}$ *is unitary.* **[BU(old) 2002]**

Solution : We have $I - A = \begin{bmatrix} 1 & 0 \\ 0 & 1 \end{bmatrix} - \begin{bmatrix} 0 & 1+2i \\ -1+2i & 0 \end{bmatrix} = \begin{bmatrix} 1 & -1-2i \\ 1-2i & 1 \end{bmatrix}$

and $I + A = \begin{bmatrix} 1 & 1+2i \\ -1+2i & 1 \end{bmatrix}$.

$\therefore \quad |I + A| = 1 - (1+2i)(-1+2i) = 6$ and $\text{Adj}\,(I + A) = \begin{bmatrix} 1 & -1-2i \\ 1-2i & 1 \end{bmatrix}$.

$\therefore \quad [I + A]^{-1} = \dfrac{1}{6}\begin{bmatrix} 1 & -1-2i \\ 1-2i & 1 \end{bmatrix}$.

Now, $[I - A][I + A] = \dfrac{1}{6}\begin{bmatrix} 1 & -1-2i \\ 1-2i & 1 \end{bmatrix} \times \begin{bmatrix} 1 & -1-2i \\ 1-2i & 1 \end{bmatrix}$

or, $U = \dfrac{1}{3}\begin{bmatrix} -2 & -1-2i \\ 1-2i & -2 \end{bmatrix}$ and $U^* = \dfrac{1}{3}\begin{bmatrix} -2 & -1+2i \\ -1+2i & -2 \end{bmatrix}$.

Now, $UU^* = \dfrac{1}{9}\begin{bmatrix} -2 & -1-2i \\ 1-2i & -2 \end{bmatrix}\begin{bmatrix} -2 & 1+2i \\ -1+2i & -2 \end{bmatrix} = \begin{bmatrix} 1 & 0 \\ 0 & 1 \end{bmatrix} = 1.$

Hence, U is unitary.

Example 2.15. *Show that the inverse of a unitary matrix is unitary and the product of two unitary matrices is unitary.* **[BU(H) 2002]**

Solution : Let U be a unitary matrix. Then $UU^* = I$, where U^* is the conjugate transpose of U.

$\therefore \quad U^* = U^{-1}$ and $(U^{-1})^* = (U^*)^* = U$,

as the conjugate transpose of he transpose conjugate of a matrix is the matrix itself.

$\therefore \quad (U^{-1})^*(U^{-1}) = UU^{-1} = I$ and $(U^{-1})(U^{-1})^* = U^{-1}U = I$.

$\therefore \quad$ **U^{-1} is unitary.**

Again, if U and V be two unitary matrices of the same order, then $U^*U = I$ and $V^*V = I$.

Now, $(UV)^*(UV) = V^*U^*UV = V^*IV = V^*V = I$

and $(UV)(UV)^* = UVV^*U^* = UIU = UU^* = I,$

which show that **UV is also an unitary matrix.**

Example 2.16. *Given,* $A = \begin{bmatrix} i & 2 & 3+i \\ -2 & 2i & 0 \\ -3+i & 0 & -i \end{bmatrix}$; *find its* (i) *transpose* (ii) *Hermitian conjugate.* **[BU(H) 1993]**

Solution : The transpose of A is $A^T = \begin{bmatrix} i & -2 & -3+i \\ 2 & 2i & 0 \\ 3+i & 0 & -i \end{bmatrix}$.

Then Hermitian conjugate is $A^* = \begin{bmatrix} -i & -2 & -3-i \\ 2 & -2i & 0 \\ 3-i & 0 & i \end{bmatrix}$.

Example 2.17. *Express the given matrix as the sum of a symmetric and an antisymmetric matrices :*

$$\begin{bmatrix} 1 & 1 & 3 \\ 2 & 3 & 2 \\ 1 & 2 & 4 \end{bmatrix}.$$ **[NBU(H) 2006]**

Solution : Let $A = \begin{bmatrix} 1 & 1 & 3 \\ 2 & 3 & 2 \\ 1 & 2 & 4 \end{bmatrix}$.

The transpose of A is $A^T = \begin{bmatrix} 1 & 2 & 1 \\ 1 & 3 & 2 \\ 3 & 2 & 4 \end{bmatrix}$.

$$\therefore \quad X = \frac{1}{2}[A + A^T] = \frac{1}{2}\begin{bmatrix} 2 & 3 & 4 \\ 3 & 6 & 4 \\ 4 & 4 & 8 \end{bmatrix} = \begin{bmatrix} 1 & \frac{3}{2} & 2 \\ \frac{3}{2} & 3 & 2 \\ 2 & 2 & 4 \end{bmatrix}.$$

$$Y = \frac{1}{2}[A - A^T] = \frac{1}{2}\begin{bmatrix} 0 & -1 & 2 \\ 1 & 0 & 0 \\ -2 & 0 & 0 \end{bmatrix} = \begin{bmatrix} 0 & -\frac{1}{2} & 1 \\ \frac{1}{2} & 0 & 0 \\ -1 & 0 & 0 \end{bmatrix}.$$

Here X is a symmetric and Y is an antisymmetric matrices, and $A = X + Y$.

(Proved).

Example 2.18. *Determine the value of x, y, z and t for which the matrices.*

$$\begin{bmatrix} x+y & y-z \\ 5-t & 7+x \end{bmatrix} \text{ and } \begin{bmatrix} t-x & z-t \\ z-y & y+z+t \end{bmatrix} \textit{ may be equal.}$$

Solution : Two matrices A and B are equal if their corresponding elements are equal, i.e.,

$$x + y = t - x, \qquad 5 - t = z - y$$
$$y - z = z - t, \qquad 7 + x = y + z + t.$$

Solving for x, y, z and t, we get

$$\boldsymbol{x = 1, \quad y = 2, \quad z = 3 \quad \text{and} \quad t = 4.}$$

Example 2.19. *Show that the following matrices are orthogonal,*

(i) $\begin{bmatrix} \cos\theta & -\sin\theta \\ \sin\theta & \cos\theta \end{bmatrix}$ *and* (ii) $\begin{bmatrix} 0 & 1 & 0 \\ 1 & 0 & 0 \\ 0 & 0 & 1 \end{bmatrix}$.

Solution : (i) Let $A = \begin{bmatrix} \cos\theta & -\sin\theta \\ \sin\theta & \cos\theta \end{bmatrix}$, then the transpose of A is $A^T = \begin{bmatrix} \cos\theta & \sin\theta \\ -\sin\theta & \cos\theta \end{bmatrix}$.

$$\therefore \quad A^T A = \begin{bmatrix} \cos\theta & \sin\theta \\ -\sin\theta & \cos\theta \end{bmatrix} \times \begin{bmatrix} \cos\theta & -\sin\theta \\ \sin\theta & \cos\theta \end{bmatrix} = \begin{bmatrix} 1 & 0 \\ 0 & 1 \end{bmatrix} = I.$$

Hence, matrix A is orthogonal.

(ii) Let $B = \begin{bmatrix} 0 & 1 & 0 \\ 1 & 0 & 0 \\ 0 & 0 & 1 \end{bmatrix}$, then $B^T = \begin{bmatrix} 0 & 1 & 0 \\ 1 & 0 & 0 \\ 0 & 0 & 1 \end{bmatrix}$.

$$\therefore \quad B^T B = \begin{bmatrix} 0 & 1 & 0 \\ 1 & 0 & 0 \\ 0 & 0 & 1 \end{bmatrix} \begin{bmatrix} 0 & 1 & 0 \\ 1 & 0 & 0 \\ 0 & 0 & 1 \end{bmatrix} = \begin{bmatrix} 1 & 0 & 0 \\ 0 & 1 & 0 \\ 0 & 0 & 1 \end{bmatrix} = I.$$

Hence, B is orthogonal.

Example 2.20. *Show that the matrix* $\begin{bmatrix} \frac{1}{\sqrt{2}} & \frac{i}{\sqrt{2}} \\ -\frac{i}{\sqrt{2}} & -\frac{1}{\sqrt{2}} \end{bmatrix}$ *is unitary.*

Solution : Let $A = \begin{bmatrix} \frac{1}{\sqrt{2}} & \frac{i}{\sqrt{2}} \\ -\frac{i}{\sqrt{2}} & -\frac{1}{\sqrt{2}} \end{bmatrix}$

Now the conjugate transpose of A is, $A^* = \begin{bmatrix} \frac{1}{\sqrt{2}} & \frac{i}{\sqrt{2}} \\ -\frac{i}{\sqrt{2}} & -\frac{1}{\sqrt{2}} \end{bmatrix}$

Now $A^* A = \begin{bmatrix} \frac{1}{\sqrt{2}} & \frac{i}{\sqrt{2}} \\ -\frac{i}{\sqrt{2}} & -\frac{1}{\sqrt{2}} \end{bmatrix} \begin{bmatrix} \frac{1}{\sqrt{2}} & \frac{i}{\sqrt{2}} \\ -\frac{i}{\sqrt{2}} & -\frac{1}{\sqrt{2}} \end{bmatrix} = \begin{bmatrix} 1 & 0 \\ 0 & 1 \end{bmatrix} = I.$

Hence, the given matrix is unitary.

Example 2.21. *Construct a* 3×3 *matrix* A *such that,*

$$A \begin{bmatrix} 1 \\ 0 \\ -1 \end{bmatrix} = \begin{bmatrix} 2 \\ 0 \\ 0 \end{bmatrix}, \; A \begin{bmatrix} 0 \\ 1 \\ 0 \end{bmatrix} = \begin{bmatrix} 1 \\ 0 \\ 1 \end{bmatrix} \text{ and } A \begin{bmatrix} 1 \\ 0 \\ 1 \end{bmatrix} = \begin{bmatrix} 1 \\ 0 \\ -1 \end{bmatrix}.$$ **[BU(H) 1996]**

Solution : Let $A = \begin{bmatrix} a_1 & a_2 & a_3 \\ b_1 & b_2 & b_3 \\ c_1 & c_2 & c_3 \end{bmatrix}$.

Then, $\begin{bmatrix} a_1 & a_2 & a_3 \\ b_1 & b_2 & b_3 \\ c_1 & c_2 & c_3 \end{bmatrix} \begin{bmatrix} 1 \\ 0 \\ -1 \end{bmatrix} = \begin{bmatrix} 2 \\ 0 \\ 0 \end{bmatrix}$

implies : $a_1 - a_3 = 2,\ b_1 - b_3 = 0$ or, $b_1 = b_3$ and $c_1 - c_3 = 0$ or, $c_1 = c_3$.

Again, $\begin{bmatrix} a_1 & a_2 & a_3 \\ b_1 & b_2 & b_3 \\ c_1 & c_2 & c_3 \end{bmatrix} \begin{bmatrix} 0 \\ 1 \\ 0 \end{bmatrix} = \begin{bmatrix} 1 \\ 0 \\ 1 \end{bmatrix}$ implies : $a_1 = 1,\ b_2 = 0,\ c_2 = 1$.

and $\begin{bmatrix} a_1 & a_2 & a_3 \\ b_1 & b_2 & b_3 \\ c_1 & c_2 & c_3 \end{bmatrix} \begin{bmatrix} 1 \\ 0 \\ 1 \end{bmatrix} = \begin{bmatrix} 1 \\ 0 \\ -1 \end{bmatrix}$

implies : $a_1 + a_3 = 1,\ b_1 + b_3 = 0$ or, $b_1 = -b_3$ and $c_1 + c_3 = 1$.

Solving the above equations, we get

$$\left.\begin{aligned} a_1 &= \frac{3}{2} \\ a_2 &= 1 \\ a_3 &= -1 \end{aligned}\right\} \qquad \left.\begin{aligned} b_1 &= 0 \\ b_2 &= 0 \\ b_3 &= 0 \end{aligned}\right\} \text{ and } \left.\begin{aligned} c_1 &= -\frac{1}{2} \\ c_2 &= 1 \\ c_3 &= -\frac{1}{2} \end{aligned}\right\}$$

$$\therefore \qquad A = \begin{bmatrix} \frac{3}{2} & 1 & -\frac{1}{2} \\ 0 & 0 & 0 \\ -\frac{1}{2} & 1 & -\frac{1}{2} \end{bmatrix}.$$

[Alternatively, Let $x_1 = \begin{bmatrix} 1 \\ 0 \\ -1 \end{bmatrix}$, $x_2 = \begin{bmatrix} 0 \\ 1 \\ 0 \end{bmatrix}$, $x_3 = \begin{bmatrix} 1 \\ 0 \\ 1 \end{bmatrix}$ and $y_1 = \begin{bmatrix} 2 \\ 0 \\ 0 \end{bmatrix}$, $y_2 = \begin{bmatrix} 1 \\ 0 \\ 1 \end{bmatrix}$, $y_3 = \begin{bmatrix} 1 \\ 0 \\ -1 \end{bmatrix}$.

then we may write,

$$AX = Y, \quad \text{where, } X = [(x_1)\ (x_2)\ (x_3)], \quad Y = [(y_1)\ (y_2)\ (y_3)],$$

then $\quad AXX^{-1} = YX^{-1}$ or, $AI = YX^{-1}$ or, $A = YX^{-1}$.

But here, $X = \begin{bmatrix} 1 & 0 & 1 \\ 0 & 1 & 0 \\ -1 & 0 & 1 \end{bmatrix}$, $\therefore \quad X = 1(1) + 1(1) = 2.$

$$\text{Adj } X = \begin{bmatrix} 1 & 0 & -1 \\ 0 & 2 & 0 \\ 1 & 0 & 1 \end{bmatrix}, \quad \therefore \quad X^{-1} = \frac{1}{2}\begin{bmatrix} 1 & 0 & -1 \\ 0 & 2 & 0 \\ 1 & 0 & 1 \end{bmatrix}.$$

$$\therefore \qquad A = YX^{-1} = \frac{1}{2}\begin{bmatrix} 2 & 1 & 1 \\ 0 & 0 & 0 \\ 0 & 1 & -1 \end{bmatrix}\begin{bmatrix} 1 & 0 & -1 \\ 0 & 2 & 0 \\ 1 & 0 & 1 \end{bmatrix} = \frac{1}{2}\begin{bmatrix} 3 & 2 & -1 \\ 0 & 0 & 0 \\ -1 & 2 & -1 \end{bmatrix}$$

$$\therefore \qquad A = \begin{bmatrix} \frac{3}{2} & 1 & -\frac{1}{2} \\ 0 & 0 & 0 \\ -\frac{1}{2} & 1 & -\frac{1}{2} \end{bmatrix}.$$

Example 2.22. *Determine the values of* α, β, γ, *when* $\begin{bmatrix} 0 & 2\beta & \gamma \\ \alpha & \beta & -\gamma \\ \alpha & -\beta & \gamma \end{bmatrix}$ *is orthogonal.*

[AMIE(Summer) 1991]

Solution : Let $A = \begin{bmatrix} 0 & 2\beta & \gamma \\ \alpha & \beta & -\gamma \\ \alpha & -\beta & \gamma \end{bmatrix}$

On transposing A, $A^T = \begin{bmatrix} 0 & \alpha & \alpha \\ 2\beta & \beta & -\beta \\ \gamma & -\gamma & \gamma \end{bmatrix}$.

Now if A is orthogonal, then $AA^T = I$.

$$\text{Now,} \quad \begin{bmatrix} 0 & 2\beta & \gamma \\ \alpha & \beta & -\gamma \\ \alpha & -\beta & \gamma \end{bmatrix} \begin{bmatrix} 0 & \alpha & \alpha \\ 2\beta & \beta & -\beta \\ \gamma & -\gamma & \gamma \end{bmatrix} = \begin{bmatrix} 1 & 0 & 0 \\ 0 & 1 & 0 \\ 0 & 0 & 1 \end{bmatrix}$$

$$\text{or,} \quad \begin{bmatrix} 4\beta^2+\gamma^2 & 2\beta^2-\gamma^2 & -2\beta^2+\gamma^2 \\ 2\beta^2-\gamma^2 & \alpha^2+\beta^2+\gamma^2 & \alpha^2-\beta^2-\gamma^2 \\ -2\beta^2+\gamma^2 & \alpha^2-\beta^2-\gamma^2 & \alpha^2+\beta^2+\gamma^2 \end{bmatrix} = \begin{bmatrix} 1 & 0 & 0 \\ 0 & 1 & 0 \\ 0 & 0 & 1 \end{bmatrix}$$

$$\text{Now,} \quad \left.\begin{matrix} 4\beta^2+\gamma^2 = 1 \\ 2\beta^2-\gamma^2 = 0 \end{matrix}\right\} \quad \text{or,} \quad \beta = \pm\frac{1}{\sqrt{6}}, \quad \gamma = \pm\frac{1}{\sqrt{3}}.$$

But, $\alpha^2+\beta^2+\gamma^2 = 1$

$$\therefore \quad \boldsymbol{\beta = \pm\frac{1}{\sqrt{6}}, \ \gamma = \pm\frac{1}{\sqrt{3}}, \ \alpha = \pm\frac{1}{\sqrt{2}}} \quad \textbf{[Ans.]}$$

Example 2.23. *Reduce the matrix to normal form and find its rank :* $A = \begin{bmatrix} 2 & 6 & 5 \\ 2 & 7 & 6 \\ 6 & 16 & 13 \end{bmatrix}$.

Solution : Given matrix is $A = \begin{bmatrix} 2 & 6 & 5 \\ 2 & 7 & 6 \\ 6 & 16 & 13 \end{bmatrix} = \begin{bmatrix} 2 & 6 & 5 \\ 0 & 1 & 1 \\ 2 & 4 & 3 \end{bmatrix}$.

[by $R_3' \to (R_3 - 2R_1)$ and $R_2' \to (R_2 - R_1)$]

$$= \begin{bmatrix} 2 & 4 & 3 \\ 0 & 1 & 1 \\ 2 & 6 & 5 \end{bmatrix}, \text{ by } R_1 \longleftrightarrow R_3$$

$$= \begin{bmatrix} 1 & 4 & 3 \\ 0 & 1 & 1 \\ -2 & -2 & -1 \end{bmatrix}, \text{ by } R_3' \to (R_3 - 2R_1)$$

$$= \begin{bmatrix} 1 & 4 & 3 \\ 0 & 1 & 1 \\ -2 & 0 & 1 \end{bmatrix}, \text{ by } R_3' \to (R_3 + 2R_2)$$

$$= \begin{bmatrix} 1 & 4 & 0 \\ 0 & 1 & 0 \\ 0 & 0 & 1 \end{bmatrix}, \text{ by } R_3' \to (R_1 - 3R_3) \text{ and } R_2' \to (R_2 - R_3)$$

$$= \begin{bmatrix} 1 & 0 & 0 \\ 0 & 1 & 0 \\ 0 & 0 & 1 \end{bmatrix}, \text{ by } R_1' \to (R_1 - 4R_2)$$

which is in normal form and its rank is 3.

Example 2.24. *Find the rank of the matrix* $A = \begin{bmatrix} 1 & 2 & 3 \\ 3 & 4 & 5 \\ 4 & 6 & 8 \end{bmatrix}$. **[BU(H) 1965]**

Solution : We have $A = \begin{bmatrix} 1 & 2 & 3 \\ 3 & 4 & 5 \\ 4 & 6 & 8 \end{bmatrix} = \begin{bmatrix} 1 & 2 & 3 \\ 0 & -2 & -4 \\ 0 & 0 & 0 \end{bmatrix}$,

by $R_2' \to (R_2 - 3R_1)$ and $R_3' \to [R_3 - (R_1 + R_2)]$

$$= \begin{bmatrix} 1 & 2 & 3 \\ 0 & -2 & -4 \end{bmatrix} = \begin{bmatrix} 1 & 0 & -1 \\ 0 & -2 & -4 \end{bmatrix}, \text{ by } R_1' \to (R_1 + R_2).$$

Thus, $A = \begin{bmatrix} 1 & 0 & -1 \\ 0 & 1 & 2 \end{bmatrix} = \begin{bmatrix} 1 & 0 & 0 \\ 0 & 1 & 0 \end{bmatrix}$, [by $C_3 \to \{(C_3 + C_1) - 2C_2\}$]

Hence, **the rank of the matrix A is 2.**

Example 2.25. *Find the inverse of the matrix,* $A = \begin{bmatrix} 1 & 2 & -2 \\ -1 & 3 & 0 \\ 0 & -2 & 1 \end{bmatrix}$

[AMIE, Summer 1990]

Solution : We have $|A| = 1(3-0) - 2(-1-0) - 2(2-0) = 3 + 2 - 4 = 1$.

Adjoint of A, which is the matrix of the co-factors of A is Adj $A = \begin{bmatrix} 3 & 1 & 2 \\ 2 & 1 & 2 \\ 6 & 2 & 5 \end{bmatrix}$.

$\therefore$ **inverse of the matrix,** $A^{-1} = \dfrac{\text{adj } A}{|A|} = \begin{bmatrix} 3 & 2 & 6 \\ 1 & 1 & 2 \\ 2 & 2 & 5 \end{bmatrix}$ **[Ans.]**

Example 2.26. *If A and B are non-singular matrices of the same order, then*

$$[AB]^{-1} = B^{-1}A^{-1}.$$ **[AMIE(Summer) 1996]**

Proof : We know that,

$$[AB][B^{-1}A^{-1}] = \{[AB]B^{-1}\}A^{-1} = \{A[BB^{-1}]\}A^{-1} = [AI]A^{-1} = AA^{-1} = I.$$

Also, $B^{-1}A^{-1}[AB] = B^{-1}\{A^{-1}[AB]\} = B^{-1}\{[A^{-1}A]B\} = B^{-1}[IB] = B^{-1}B = I$.

Now by definition of inverse of a matrix,

$B^{-1}A^{-1}$ is the inverse of AB or $B^{-1}A^{-1} = (AB)^{-1}$.

Example 2.27. *Solve the following system of equations by the matrix method :*

$$2x + 3y + z = 9$$
$$x + 2y + 3z = 6$$
$$3x + y + 2z = 8$$

[BU(H) 1994]

Solution : From the coefficients of the given sets of equations, we get

$$A = \begin{bmatrix} 2 & 3 & 1 \\ 1 & 2 & 3 \\ 3 & 1 & 2 \end{bmatrix} \text{ and } B = \begin{bmatrix} 9 \\ 6 \\ 8 \end{bmatrix}. X = \begin{bmatrix} x_1 \\ x_2 \\ x_3 \end{bmatrix}.$$

So, the system of equations can be expressed as, $AX = B$.

Multiplying the above equation by A^{-1} from left, we get

$$A^{-1}AX = A^{-1}B \quad \text{or,} \quad X = A^{-1}B.$$

Here, $|A| = 2(4-3) + 3(9-2) + 1(1-6) = 18$.

Now, the matrix of the co-factors of A is $C = \begin{bmatrix} 1 & 7 & -5 \\ -5 & 1 & 7 \\ 7 & -5 & 1 \end{bmatrix}$.

Then $\text{adj } A = C^T = \begin{bmatrix} 1 & -5 & 7 \\ 7 & 1 & -5 \\ -5 & 7 & 1 \end{bmatrix}$.

$\therefore$ inverse of matrix A is, $A^{-1} = \dfrac{\text{adj } A}{|A|} = \dfrac{1}{18}\begin{bmatrix} 1 & -5 & 7 \\ 7 & 1 & -5 \\ -5 & 7 & 1 \end{bmatrix}$

Now, $X = A^{-1}B = \dfrac{1}{18}\begin{bmatrix} 1 & -5 & 7 \\ 7 & 1 & -5 \\ -5 & 7 & 1 \end{bmatrix}\begin{bmatrix} 9 \\ 6 \\ 8 \end{bmatrix} = \dfrac{1}{18}\begin{bmatrix} 35 \\ 29 \\ 5 \end{bmatrix}$ or, $\begin{bmatrix} x_1 \\ x_2 \\ x_3 \end{bmatrix} = \begin{bmatrix} \frac{35}{18} \\ \frac{29}{18} \\ \frac{5}{18} \end{bmatrix}$.

$\therefore \quad x_1 = \dfrac{35}{18}, x_2 = \dfrac{29}{18}, x_3 = \dfrac{5}{18}$ i.e., $\boldsymbol{x = \dfrac{35}{18}, y = \dfrac{29}{18}}$ **and** $\boldsymbol{z = \dfrac{5}{18}}$.

Example 2.28. *Solve with the help of matrices, the simultaneous equations :*

$$x + y + z = 3$$
$$x + 2y + 3z = 4$$
$$x + 4y + 9z = 6.$$

[AMIE(Summer) 1995]

Solution : The matrix form of the given set of equations is

$$\begin{bmatrix} 1 & 1 & 1 \\ 1 & 2 & 3 \\ 1 & 4 & 9 \end{bmatrix}\begin{bmatrix} x \\ y \\ z \end{bmatrix} = \begin{bmatrix} 3 \\ 4 \\ 6 \end{bmatrix} \quad \text{or,} \quad AX = B,$$

where $A = \begin{bmatrix} 1 & 1 & 1 \\ 1 & 2 & 3 \\ 1 & 4 & 9 \end{bmatrix}, X = \begin{bmatrix} x \\ y \\ z \end{bmatrix}, B = \begin{bmatrix} 3 \\ 4 \\ 6 \end{bmatrix}$.

Multiplying the equation by A^{-1} from left, we get

$$A^{-1}AX = A^{-1}B \quad \text{or,} \quad X = A^{-1}B,$$

Now, $|A| = 1 \times 6 + 1 \times (-6) + 1 \times 2 = 6 - 6 + 2 = 2.$

The matrix of the co-factors of A is $C = \begin{bmatrix} 6 & -6 & 2 \\ -5 & 8 & -3 \\ 1 & -2 & 1 \end{bmatrix}.$

Now, $\text{adj } A = \begin{bmatrix} 6 & -5 & 1 \\ -6 & 8 & -2 \\ 2 & -3 & 1 \end{bmatrix}.$

$$\therefore \quad A^{-1} = \frac{\text{adj } A}{|A|} = \frac{1}{2}\begin{bmatrix} 6 & -5 & 1 \\ -6 & 8 & -2 \\ 2 & -3 & 1 \end{bmatrix}.$$

Now, $X = A^{-1}B = \frac{1}{2}\begin{bmatrix} 6 & -5 & 1 \\ -6 & 8 & -2 \\ 2 & -3 & 1 \end{bmatrix}\begin{bmatrix} 3 \\ 4 \\ 6 \end{bmatrix}$

or, $$\begin{bmatrix} x \\ y \\ z \end{bmatrix} = \frac{1}{2}\begin{bmatrix} 18 - 20 + 6 \\ -18 + 32 - 12 \\ 6 - 12 + 6 \end{bmatrix} = \frac{1}{2}\begin{bmatrix} 4 \\ 2 \\ 0 \end{bmatrix} = \begin{bmatrix} 2 \\ 1 \\ 0 \end{bmatrix}.$$

$\therefore \quad$ $\boldsymbol{x = 2, \; y = 1, \; z = 0.}$

Example 2.29. *Solve the system of equations by reducing the matrix to normal form :*

$$x + z = 0$$
$$2x - 3y + 4z = -4$$
$$-y + 4z = 2.$$

Solution : The system of equations can be expressed in the matrix form as

$$\begin{bmatrix} 1 & 0 & 1 \\ 2 & -3 & 4 \\ 0 & -1 & 4 \end{bmatrix}\begin{bmatrix} x \\ y \\ z \end{bmatrix} = \begin{bmatrix} 0 \\ -4 \\ 2 \end{bmatrix}$$

or, $$\begin{bmatrix} 1 & 0 & 1 \\ 0 & -3 & 2 \\ 0 & -1 & 4 \end{bmatrix}\begin{bmatrix} x \\ y \\ z \end{bmatrix} = \begin{bmatrix} 0 \\ -4 \\ 2 \end{bmatrix}, \text{ by } R_2' \to (R_2 - 2R_1)$$

or, $$\begin{bmatrix} 1 & 0 & 1 \\ 0 & 1 & 6 \\ 0 & -1 & 4 \end{bmatrix}\begin{bmatrix} x \\ y \\ z \end{bmatrix} = \begin{bmatrix} 0 \\ 8 \\ 2 \end{bmatrix}, \text{ by } R_2' \to -R_2 \text{ and } R_2' \to (R_2 + 2R_3)$$

or, $$\begin{bmatrix} 1 & 0 & 1 \\ 0 & 1 & 6 \\ 0 & 0 & 1 \end{bmatrix}\begin{bmatrix} x \\ y \\ z \end{bmatrix} = \begin{bmatrix} 0 \\ 8 \\ 1 \end{bmatrix}, \text{ by } R_3' \to (R_3 + R_2) \text{ then } R_3' \to \tfrac{1}{10}R_3$$

or, $$\begin{bmatrix} 1 & 0 & 0 \\ 0 & 1 & 0 \\ 0 & 0 & 1 \end{bmatrix}\begin{bmatrix} x \\ y \\ z \end{bmatrix} = \begin{bmatrix} -1 \\ 2 \\ 1 \end{bmatrix}, \text{ by } R_1' \to (R_1 - R_3) \text{ and } R_2' \to (R_2 - 6R_3)$$

Hence, **the solution is $\boldsymbol{x = -1, \; y = 2, \; z = 1}$.**

Example 2.30. *Determine the characteristic roots and the characteristic vector of the matrix,* $A = \begin{bmatrix} 2 & 3 \\ 1 & 0 \end{bmatrix}$.

Solution : The characteristic equation of the matrix A is

$$\begin{bmatrix} 2-\lambda & 3 \\ 1 & -\lambda \end{bmatrix} = 0 \quad \text{or,} \quad -\lambda(2-\lambda) - 3 = 0.$$

$$\therefore \qquad (\lambda+1)(\lambda-3) = 0.$$

Hence, **the characteristic roots are 3, −1.**

Now, for $\lambda = 3$,

$$\begin{bmatrix} 2-3 & 3 \\ 1 & 0-3 \end{bmatrix} \begin{bmatrix} x_1 \\ x_2 \end{bmatrix} = 0, \quad \text{where } X_1 = \begin{bmatrix} x_1 \\ x_2 \end{bmatrix}.$$

$$\therefore \qquad -x_1 + 3x_2 = 0 \quad \text{or,} \quad x_1 - 3x_2 = 0.$$

If we assign any arbitrary value of x_1 (say, 3), then $\mathbf{X_1} = \begin{bmatrix} \mathbf{3} \\ \mathbf{1} \end{bmatrix}$,

which is the characteristic vector of A associated with $\lambda = 3$, satisfying $Ax_i = \lambda_i x_i$.

Again, **for $\lambda = -1$, we have $X_2 = \begin{bmatrix} 1 \\ -1 \end{bmatrix}$.**

Example 2.31. *Determine the eigenvalues and the normalised eigenvectors of the following matrix :*

$$\begin{bmatrix} 3 & 5 & 1 \\ -2 & -2 & 0 \\ 2 & 7 & 3 \end{bmatrix}.$$ **[BU(H) 1993]**

Solution : The characteristic equation is $\begin{vmatrix} 3-\lambda & 5 & 1 \\ -2 & -2-\lambda & 0 \\ 2 & 7 & 3-\lambda \end{vmatrix} = 0.$

Expanding, $\lambda^3 - 4\lambda^2 + 5\lambda - 2 = 0$ or, $(\lambda-1)^2(\lambda-2) = 0$

$\therefore$ **the eigenvalues are $\lambda_1 = 1$, $\lambda_2 = 1$, $\lambda_3 = 2$**

For λ, we have $AX = X$ where $X = [x_1 \ x_2 \ x_3]$

$$\text{or,} \qquad \begin{bmatrix} 3 & 5 & 1 \\ -2 & -2 & 0 \\ 2 & 7 & 3 \end{bmatrix} \begin{bmatrix} x_1 \\ x_2 \\ x_3 \end{bmatrix} = \begin{bmatrix} x_1 \\ x_2 \\ x_3 \end{bmatrix}$$

from which $3x_1 + 5x_2 + x_3 = x_1$ (i)

$-2x_1 - 2x_2 = x_2$ (ii)

$2x_1 + 7x_2 + 3x_3 = x_3.$ (iii)

From eqn. (ii),

$$2x_1 = -3x_2 \quad \text{or,} \quad x_1 = -\frac{3}{2}x_2.$$

Choosing $x_2 = -2$, we get $x_1 = 3$ and $x_3 = 4$.

So, the eigenvector corresponding to λ_1 is $\begin{bmatrix} 3 \\ -2 \\ 4 \end{bmatrix}$, then

$$\phi^* = [3 \quad -2 \quad 4] = \phi \text{ (say)}.$$

$$\therefore \qquad \phi^*\phi = [3 \quad -2 \quad 4]\begin{bmatrix} 3 \\ -2 \\ 4 \end{bmatrix} = 29.$$

The normalised eigenvector is $\dfrac{1}{\sqrt{29}}\begin{bmatrix} 3 \\ -2 \\ 4 \end{bmatrix}$.

For $\lambda_3 = 2$, the eigenvector X satisfied $AX = 2X$,

$$\text{or,} \qquad \begin{bmatrix} 3 & 5 & 1 \\ -2 & -2 & 0 \\ 2 & 7 & 3 \end{bmatrix}\begin{bmatrix} x_1 \\ x_2 \\ x_3 \end{bmatrix} = 2\begin{bmatrix} x_1 \\ x_2 \\ x_3 \end{bmatrix}$$

$$\text{or,} \qquad \begin{aligned} x_1 + 5x_2 + x_3 &= 0 \\ x_1 + 2x_2 &= 0 \\ 2x_1 + 7x_2 + x_3 &= 0. \end{aligned}$$

Taking $x_2 = -1$, we get $x_1 = 2$, $x_3 = 3$.

$\therefore$ the eigenvector corresponding to $\lambda_2 = 2$ is $\begin{bmatrix} 2 \\ -1 \\ 3 \end{bmatrix}$.

The normalised eigenvector is $\dfrac{1}{\sqrt{14}}\begin{bmatrix} \mathbf{2} \\ \mathbf{-1} \\ \mathbf{3} \end{bmatrix}$.

Example 2.32. *Find the eigenvalues and the eigenvectors of the matrix,*

$$\begin{bmatrix} 1 & 0 & 0 \\ 0 & 3 & -1 \\ 0 & -1 & 3 \end{bmatrix}. \qquad \textbf{[BU(H) 2002]}$$

Solution : The characteristic polynomial is

$$\phi(\lambda) = \begin{vmatrix} 1-\lambda & 0 & 0 \\ 0 & 3-\lambda & -1 \\ 0 & -1 & 3-\lambda \end{vmatrix}$$

and the characteristic equation is given by

$$(1-\lambda)[(3-\lambda)^2 - 1] = 0 \quad \text{or,} \quad (\lambda - 1)(\lambda - 2)(\lambda - 4) = 0.$$

So, **the eigenvalues of the given matrix are : $\lambda = 1$, $\lambda = 2$ and $\lambda = 4$.**

Now, for $\lambda = 1$, we have $AX = X$, where $X = [x \ \ y \ \ z]^T$

$$\text{or,} \qquad \begin{bmatrix} 1 & 0 & 0 \\ 0 & 3 & -1 \\ 0 & -1 & 3 \end{bmatrix}\begin{bmatrix} x \\ y \\ z \end{bmatrix} = \begin{bmatrix} x \\ y \\ z \end{bmatrix}$$

$\therefore \quad x = x,$ i.e., $x = C$ (constant). As C is arbitrary, we choose $C = 1$.

$\therefore \quad x = 1$

Again, $3y - z = y$ or, $z = 2y$ and $-y + 3z = z$ or, $y = 2z$.

The above two equations are satisfied for $y = z = 0$.

$\therefore$ **the eigenvector corresponding to $\lambda = 1$ is** $\begin{bmatrix} 1 \\ 0 \\ 0 \end{bmatrix}$.

This is also a normalised eigenvector. For the eigenvalue $\lambda = 2$, we have $AX = 2X$.

$$\text{or,} \quad \begin{bmatrix} 1 & 0 & 0 \\ 0 & 3 & -1 \\ 0 & -1 & 3 \end{bmatrix} \begin{bmatrix} x \\ y \\ z \end{bmatrix} = 2 \begin{bmatrix} x \\ y \\ z \end{bmatrix}.$$

From the above equation, we get

$$x = 2x \quad \text{or,} \quad x = 0$$

$$3y - z = 2y \quad \text{or,} \quad y = z$$

$$\text{and} \quad -y + 3z = 2z \quad \text{or,} \quad y = z.$$

Since the last two equations (i.e., $y = z$) are identical, we write $y = z = c$ will satisfy both the equations. Choosing $C = 1$, an arbitrary constant, we get $y = z = 1$.

$\therefore$ **the eigenvector corresponding to $\lambda = 2$ is** $\begin{bmatrix} 0 \\ 1 \\ 1 \end{bmatrix}$.

Similarly, for $\lambda = 4$, we have $AX = 4X$.

$$\text{or,} \quad \begin{bmatrix} 1 & 0 & 0 \\ 0 & 3 & -1 \\ 0 & -1 & 3 \end{bmatrix} \begin{bmatrix} x \\ y \\ z \end{bmatrix} = 4 \begin{bmatrix} x \\ y \\ z \end{bmatrix}.$$

From the above equation, we get $x = 0$, $y = 1$ and $z = 1$.

So, **the eigenvector corresponding to $\lambda = 4$ is** $\begin{bmatrix} 0 \\ 1 \\ -1 \end{bmatrix}$.

Example 2.33. *Find the eigenvalues and eigenvectors of the Hermitian matrix :*

$$A = \begin{bmatrix} 0 & 1 & 1 \\ 1 & 0 & 1 \\ 1 & 1 & 0 \end{bmatrix}.$$

[BU(H)(Old) 2001]

Solution : Since, $A = \begin{bmatrix} 0 & 1 & 1 \\ 1 & 0 & 1 \\ 1 & 1 & 0 \end{bmatrix}$, the characteristic equation for the matrix is $|A - \lambda I| = 0$.

$$\text{or,} \quad \begin{vmatrix} -\lambda & 1 & 1 \\ 1 & -\lambda & 1 \\ 1 & 1 & -\lambda \end{vmatrix} = 0 \quad \text{or,} \quad (\lambda + 1)(\lambda + 1)(\lambda - 2) = 0.$$

$\therefore$ **the eigenvalues of A are $\lambda_1 = -1$, $\lambda_2 = -1$, $\lambda_3 = 2$.**

For $\lambda_1 = -1$, we have $AX = X$, where $X = [x \ y \ z]$.

or, $$\begin{bmatrix} 0 & 1 & 1 \\ 1 & 0 & 0 \\ 1 & 1 & 0 \end{bmatrix} \begin{bmatrix} x \\ y \\ z \end{bmatrix} = - \begin{bmatrix} x \\ y \\ z \end{bmatrix}.$$

$\therefore$ $\quad y + z + x = 0;\ x + y + z = 0;\ x + y + z = 0.$

The above three equations are equivalent to $x + y + z = 0$. Through arbitrary choice, we may get infinite number of independent solutions.

Choosing any two values of x and y, wet get the convenient form of eigenvectors for $\lambda_1 = -1$, $\lambda_2 = -1$ as $\begin{bmatrix} 1 \\ 0 \\ -1 \end{bmatrix}$ **and** $\begin{bmatrix} 1 \\ -1 \\ 0 \end{bmatrix}$**.**

For $\lambda_3 = 2$, $AX = X$ or, $$\begin{bmatrix} 0 & 1 & 1 \\ 1 & 0 & 1 \\ 1 & 1 & 0 \end{bmatrix} \begin{bmatrix} x \\ y \\ z \end{bmatrix} = 2 \begin{bmatrix} x \\ y \\ z \end{bmatrix}.$$

Then, $y + z = 2x$, $z + x = 2y$, $x + y = 2z$ from which $x = y = z$.

Choosing $x = 1$, we get the eigenvector corresponding to $\lambda_3 = 2$ as $\begin{bmatrix} 1 \\ 1 \\ 1 \end{bmatrix}$.

Hence, **the normalised eigenvectors for the matrix A are :**

$$\frac{1}{\sqrt{2}} \begin{bmatrix} 1 \\ 0 \\ -1 \end{bmatrix}, \frac{1}{\sqrt{2}} \begin{bmatrix} 1 \\ -1 \\ 0 \end{bmatrix}, \frac{1}{\sqrt{2}} \begin{bmatrix} 1 \\ 1 \\ 1 \end{bmatrix}.$$

Example 2.34. *Verify Caley-Hamilton theorem for the matrix* $A = \begin{bmatrix} 1 & 2 \\ 2 & -1 \end{bmatrix}$. *Find* A^{-1} *and determine* A^8.

Solution : The characteristic equation of the given matrix A is

$$|A - \lambda I| = \begin{vmatrix} 1-\lambda & 2 \\ 2 & -1-\lambda \end{vmatrix} = 0 \quad \text{or,} \quad (\lambda - 1)(\lambda + 1) - 4 = 0 \quad \text{or,} \quad \lambda^2 - 5 = 0.$$

Now, $$A^2 = A \cdot A = \begin{bmatrix} 1 & 2 \\ 2 & -1 \end{bmatrix} \begin{bmatrix} 1 & 2 \\ 2 & -1 \end{bmatrix} = \begin{bmatrix} 5 & 0 \\ 0 & 5 \end{bmatrix} = 5 \begin{bmatrix} 1 & 0 \\ 0 & 1 \end{bmatrix} = 5I \quad \text{or,} \quad A^2 - 5I = 0.$$

Thus, A satisfies the characteristic equation.
Now, to find A^{-1} we multiply both sides of $A^2 - 5I = 0$ by A^{-1} from the left.

$\therefore$ $\quad A^{-1}A^2 - 5A^{-1}I = 0$ or, $A - 5A^{-1} = 0$.

So, $$A^{-1} = \frac{1}{5}A = \frac{1}{5} \begin{bmatrix} 1 & 2 \\ 2 & -1 \end{bmatrix}.$$

To find, A^8 we multiply $A^2 - 5I = 0$ by A^6.

$\therefore$ $\quad A^6 \cdot A^2 - 5IA^6 = 0$ or, $\mathbf{A^8 = 5A^6 = 5A^2 \cdot A^2 \cdot A^2} = \mathbf{5(5I)(5I)(5I)} = \mathbf{625I}$.

Example 2.35. *Verify Caley-Hamilton theorem for the matrix* $A = \begin{bmatrix} 1 & 2 & 3 \\ 2 & 4 & 5 \\ 3 & 5 & 6 \end{bmatrix}$ *and hence find the inverse of A. Express $B = A^8 - 11A^7 - 4A^6 + A^5 + A^4 - 11A^3 - 3A^2 + 2A + I$ as a quadratic polynomial in A. Find B.*

Solution : Here, the characteristic equation of A is

$$|A-\lambda I| = \begin{vmatrix} 1-\lambda & 2 & 3 \\ 2 & 4-\lambda & 5 \\ 3 & 5 & 6-\lambda \end{vmatrix}$$

$$= (1-\lambda)[(4-\lambda)(6-\lambda)-25] - 2[2(6-\lambda)-15] + 3[10-3(4-\lambda)] = 0$$

or, $\lambda^3 - 11\lambda^2 - 4\lambda + I = 0.$

Caley-Hamilton theorem is satisfied if A satisfies the above equation, i.e.,

$$A^3 - 11A^2 - 4A + I = 0.$$

Now, $A^2 = A \cdot A = \begin{bmatrix} 1 & 2 & 3 \\ 2 & 4 & 5 \\ 3 & 5 & 6 \end{bmatrix} \begin{bmatrix} 1 & 2 & 3 \\ 2 & 4 & 5 \\ 3 & 5 & 6 \end{bmatrix} = \begin{bmatrix} 14 & 25 & 31 \\ 25 & 45 & 56 \\ 31 & 56 & 70 \end{bmatrix}.$

$\therefore \quad A^3 = A \cdot A^2 = \begin{bmatrix} 1 & 2 & 3 \\ 2 & 4 & 5 \\ 3 & 5 & 6 \end{bmatrix} \begin{bmatrix} 14 & 25 & 31 \\ 25 & 45 & 56 \\ 31 & 56 & 70 \end{bmatrix} = \begin{bmatrix} 157 & 283 & 353 \\ 283 & 510 & 636 \\ 353 & 636 & 793 \end{bmatrix}.$

Now, $A^3 - 11A^2 - 4A + I$

$$\begin{bmatrix} 157 & 283 & 353 \\ 283 & 510 & 636 \\ 353 & 636 & 793 \end{bmatrix} - 11 \begin{bmatrix} 14 & 25 & 31 \\ 25 & 45 & 56 \\ 31 & 56 & 70 \end{bmatrix} - 4 \begin{bmatrix} 1 & 2 & 3 \\ 2 & 4 & 5 \\ 3 & 5 & 6 \end{bmatrix} + \begin{bmatrix} 1 & 0 & 0 \\ 0 & 1 & 0 \\ 0 & 0 & 1 \end{bmatrix} = \begin{bmatrix} 0 & 0 & 0 \\ 0 & 0 & 0 \\ 0 & 0 & 0 \end{bmatrix}.$$

$\therefore$ A satisfies the characteristic equation.

To find A^{-1} : We have the characteristic equation, $A^3 - 11A^2 - 4A + I = 0.$

Multiplying from left by A^{-1}, we get

$$A^{-1}A^3 - 11A^{-1}A^2 - 4A^{-1}A + A^{-1}I = 0 \quad \text{or,} \quad A^2 - 11A - 4I + A^{-1} = 0$$

or, $\quad A^{-1} = -A^2 + 11A + 4I$

or, $\quad A^{-1} = -\begin{bmatrix} 14 & 25 & 31 \\ 25 & 45 & 56 \\ 31 & 56 & 70 \end{bmatrix} + 11 \begin{bmatrix} 1 & 2 & 3 \\ 2 & 4 & 5 \\ 3 & 5 & 6 \end{bmatrix} + 4 \begin{bmatrix} 1 & 0 & 0 \\ 0 & 1 & 0 \\ 0 & 0 & 1 \end{bmatrix} = \begin{bmatrix} \mathbf{1} & \mathbf{-3} & \mathbf{2} \\ \mathbf{-3} & \mathbf{3} & \mathbf{-1} \\ \mathbf{2} & \mathbf{-1} & \mathbf{0} \end{bmatrix}.$

To find A^4 : From Caley-Hamilton's theorem,

$$A^3 - 11A^2 - 4A + I = 0 \quad \text{or,} \quad A^3 = 11A^2 + 4A - I.$$

Multiplying both sides by A, we get

$$A^4 = A \cdot A^3 = A[11A^2 + 4A - I] = 11A^3 + 4A^2 - A$$

$$= 11 \begin{bmatrix} 157 & 283 & 353 \\ 283 & 510 & 636 \\ 353 & 636 & 793 \end{bmatrix} + 4 \begin{bmatrix} 14 & 25 & 31 \\ 25 & 45 & 56 \\ 31 & 56 & 70 \end{bmatrix} - \begin{bmatrix} 1 & 2 & 3 \\ 2 & 4 & 5 \\ 3 & 5 & 6 \end{bmatrix}$$

$$= \begin{bmatrix} \mathbf{1782} & \mathbf{3211} & \mathbf{4004} \\ \mathbf{3211} & \mathbf{5786} & \mathbf{7215} \\ \mathbf{4004} & \mathbf{7215} & \mathbf{8997} \end{bmatrix}.$$

To find B : We have $B = A^8 - 11A^7 - 4A^6 + A^5 + A^4 - 11A^3 - 3A^2 + 2A + I$

$$= A^5[A^3 - 11A^2 - 4A + I] + A[A^3 - 11A^2 - 4A + I] + A^2 + A + I$$
$$= A^5 \cdot 0 + A \cdot 0 + A^2 + A + I$$

$\because$ $A^3 - 11A^2 - 4A + I = 0$, as A satisfies the characteristic equation.

$$\therefore \quad B = A^2 + A + I = \begin{bmatrix} 14 & 25 & 31 \\ 25 & 45 & 56 \\ 31 & 56 & 70 \end{bmatrix} + \begin{bmatrix} 1 & 2 & 3 \\ 2 & 4 & 5 \\ 3 & 5 & 6 \end{bmatrix} + \begin{bmatrix} 1 & 0 & 0 \\ 0 & 1 & 0 \\ 0 & 0 & 1 \end{bmatrix} \text{ or, } B = \begin{bmatrix} \mathbf{16} & \mathbf{27} & \mathbf{34} \\ \mathbf{27} & \mathbf{50} & \mathbf{61} \\ \mathbf{34} & \mathbf{61} & \mathbf{77} \end{bmatrix}.$$

Example 2.36. *Test whether the set of equations is consistent :*

$$x - y + z = 2$$
$$3x - y + 2z = -6$$
$$3x + y + z = -18$$

and solve, if consistent.

Solution : The given set of equations expressed in matrix form is

$$AX = B \quad \text{or,} \quad \begin{bmatrix} 1 & -1 & 1 \\ 3 & -1 & 2 \\ 3 & 1 & 1 \end{bmatrix} \begin{bmatrix} x \\ y \\ z \end{bmatrix} = \begin{bmatrix} 2 \\ -6 \\ -18 \end{bmatrix}.$$

The augmented matrix

$$[A, B] = \begin{bmatrix} 1 & -1 & 1 & 2 \\ 3 & -1 & 2 & -6 \\ 3 & 1 & 1 & -18 \end{bmatrix}$$

$$= \begin{bmatrix} 1 & -1 & 1 & 2 \\ 0 & 2 & -1 & -12 \\ 0 & 4 & -2 & -24 \end{bmatrix}, \quad \text{by } R_2' \to (R_2 - 3R_1);\ R_3' \to (R_3 - 3R_1)$$

$$= \begin{bmatrix} 1 & -1 & 1 & 2 \\ 0 & 1 & -\frac{1}{2} & -6 \\ 0 & 1 & -\frac{1}{2} & -6 \end{bmatrix}, \quad \text{by } R_2' \to \frac{1}{2}R_2;\ R_3' \to \frac{1}{4}R_3$$

$$= \begin{bmatrix} 1 & -1 & 1 & 2 \\ 0 & 1 & -\frac{1}{2} & -6 \\ 0 & 0 & 0 & 0 \end{bmatrix}, \quad \text{by } R_3' \to (R_3 - R_2)$$

Hence, rank of $[A, B]$ = rank $A = 2$, which is less than the number of unknowns.

$\therefore$ **the given system of equations are consistent and as such it has infinite number of solutions.**

Let us now find the solutions. The reduced form of the equations is

$$\begin{bmatrix} 1 & -1 & 1 \\ 0 & 1 & -\frac{1}{2} \\ 0 & 0 & 0 \end{bmatrix} \begin{bmatrix} x \\ y \\ z \end{bmatrix} = \begin{bmatrix} 2 \\ -6 \\ 0 \end{bmatrix}.$$

$$\therefore \quad x - y + z = 2, \quad y - \frac{1}{2}z = -6.$$

Let us put $z = 2k$, then $y = k - 6$ and $x = -(k + 4)$.

Hence, $[-(k+4),\ (k-6),\ 2k]$, where k is any arbitrary real number, are solutions of the given system of equations.

Example 2.37. *Find the value of λ so that the set of equations :*

$$\begin{aligned} x+y+3z &= 0\\ 4x+3y+\lambda z &= 0\\ 2x+y+2z &= 0 \end{aligned}$$

have a non-trivial solution.

Solution : In matrix form the given set of equations can be expressed as

$$AX = B \quad \text{or,} \quad \begin{bmatrix} 1 & 1 & 3\\ 4 & 3 & \lambda\\ 2 & 1 & 2 \end{bmatrix} \begin{bmatrix} x\\ y\\ z \end{bmatrix} = \begin{bmatrix} 0\\ 0\\ 0 \end{bmatrix}.$$

The augmented matrix,

$$[A, B] = \begin{bmatrix} 1 & 1 & 3 & 0\\ 4 & 3 & \lambda & 0\\ 2 & 1 & 2 & 0 \end{bmatrix}$$

$$\sim \begin{bmatrix} 1 & 1 & 3 & 0\\ 0 & -1 & \lambda-12 & 0\\ 0 & -1 & -4 & 0 \end{bmatrix}, \quad R_2' \to (R_2 - 4R_1);\ R_3' \to (R_3 - 2R_1)$$

$$\sim \begin{bmatrix} 1 & 1 & 3 & 0\\ 0 & -1 & \lambda-12 & 0\\ 0 & 0 & 8-\lambda & 0 \end{bmatrix} \quad R_3' \to (R_3 - R_2)$$

$$\sim \begin{bmatrix} 1 & 1 & 3 & 0\\ 0 & 1 & 12-\lambda & 0\\ 0 & 0 & 8-\lambda & 0 \end{bmatrix}, \quad R_2' \to (-R_2).$$

If $8 - \lambda = 0$ or, $\lambda = 8$, then the system of equations have non-trivial solutions.

Example 2.38. *If λ is an eigenvalue of an orthogonal matrix, then show that $\dfrac{1}{\lambda}$ is also an eigenvalue.* **[BU(H) 2002]**

Solution : The characteristic equation for the matrix A is given by,

$$|A - \lambda I| = 0 \quad \text{or,} \quad |A^T - \lambda I| = 0,$$

where λ is the eigenvalue of the matrix A. As matrix A is orthogonal,

$$\therefore \quad A^T = A^{-1}.$$

$$\text{Then,} \quad |A^{-1} - \lambda I| = 0 \quad \text{or,} \quad \left|\frac{1}{\lambda}A^{-1} - I\right| = 0$$

$$\text{or,} \quad \left|\frac{1}{\lambda}A^{-1} - A^{-1}A\right| = 0, \qquad [\because I = A^{-1}A]$$

$$\text{or,} \quad \left|A^{-1}\left[\frac{1}{\lambda} - A\right]\right| = 0 \quad \text{or,} \quad |A^{-1}|\,\left|A - \frac{1}{\lambda}I\right| = 0$$

or, $\left[A - \frac{1}{\lambda}I\right] = 0, \quad [\because A^{-1} \neq 0].$

This shows that $\frac{1}{\lambda}$ is also an eigenvalue of the matrix A.

Example 2.39. *Consider the matrix $H = \begin{bmatrix} 3 & i \\ -i & 3 \end{bmatrix}$. Find the eigenvalues and the eigenvectors and hence, construct the unitary matrix U which diagonalizes H. Compute $H^{-1} = U^{-1}HU$.*

Solution : The characteristic equation of the matrix H is

$$\phi(X) = |H - \lambda I| = \begin{vmatrix} 3-\lambda & i \\ -i & 3-\lambda \end{vmatrix} = 0$$

or, $\quad \lambda^2 - 6\lambda + 8 - 0 \quad$ or, $\quad (\lambda-2)(\lambda-4) = 0.$

Hence, the eigenvalues of the matrix H is $\lambda_1 = 2,\ \lambda_2 = 4$.

Now, for $\lambda_1 = 2$, we have

$$HX = 2X, \quad X = \begin{bmatrix} x \\ y \end{bmatrix} \quad \text{or,} \quad \begin{bmatrix} 3 & i \\ -i & 3 \end{bmatrix}\begin{bmatrix} x \\ y \end{bmatrix} = 2\begin{bmatrix} x \\ y \end{bmatrix}$$

or, $\quad x + iy = 0 \quad$ and $\quad -ix + y = 0.$

The above two equations are equivalent.

Taking $x = 1$, we get the convenient eigenvector $X_1 = \begin{bmatrix} 1 \\ i \end{bmatrix}$.

Similarly, for $\lambda_2 = 4$, we can show that the eigenvector corresponding to λ_2 is $X_2 = \begin{bmatrix} 1 \\ -i \end{bmatrix}$.

Thus, the normalised eigenvectors for H are $\frac{1}{\sqrt{2}}\begin{bmatrix} 1 \\ i \end{bmatrix}$ and $\frac{\cdot 1}{\sqrt{2}}\begin{bmatrix} 1 \\ -i \end{bmatrix}$.

Now, the diagonalisable matrix is

$$U = [x_1\ x_2] \quad \text{or,} \quad U = \frac{1}{\sqrt{2}}\begin{bmatrix} 1 & 1 \\ i & -i \end{bmatrix}.$$

$$\therefore \quad U^* = \frac{1}{\sqrt{2}}\begin{bmatrix} 1 & -i \\ 1 & i \end{bmatrix}.$$

$$\text{So,} \quad U^*U = \left(\frac{1}{\sqrt{2}}\right)^{2\cdot}\begin{bmatrix} 1 & -i \\ 1 & i \end{bmatrix}\begin{bmatrix} 1 & 1 \\ i & -i \end{bmatrix} = \frac{1}{2}\begin{bmatrix} 2 & 0 \\ 0 & 2 \end{bmatrix} = \begin{bmatrix} 1 & 0 \\ 0 & 1 \end{bmatrix} = I.$$

This shows that U is a unitary matrix.

Now, $\quad |U| = -\sqrt{2}i.$

$$\therefore \quad U^{-1} = -\frac{1}{\sqrt{2}i}\begin{bmatrix} -i & -1 \\ -i & 1 \end{bmatrix} = \frac{1}{\sqrt{2}}\begin{bmatrix} 1 & -i \\ 1 & i \end{bmatrix}.$$

$$\text{Hence,} \quad H' = U^{-1}HU = \frac{1}{2}\begin{bmatrix} 1 & -i \\ -i & i \end{bmatrix}\begin{bmatrix} 3 & i \\ -i & 3 \end{bmatrix}\begin{bmatrix} 1 & 1 \\ i & -i \end{bmatrix} = \begin{bmatrix} 1 & -i \\ 2 & 2i \end{bmatrix}\begin{bmatrix} 1 & 1 \\ i & -i \end{bmatrix} = \begin{bmatrix} 2 & 0 \\ 0 & 4 \end{bmatrix}.$$

Hence, H' is a diagonal matrix having diagonal elements as the eigenvalues of H.

Example 2.40. *Consider the matrix* $[A] = \begin{bmatrix} a & h \\ h & b \end{bmatrix}$. *Suppose it is transformed to the diagonal form* $[B] = T_\theta A T_\theta^{-1}$, *where* $T_\theta = \begin{bmatrix} \cos\theta & \sin\theta \\ -\sin\theta & \cos\theta \end{bmatrix}$. *Show that,* $\theta = \frac{1}{2}\tan^{-1}\left(\frac{2h}{a-b}\right)$.

[BU(H) 2003]

Solution : $\because\ T_\theta = \begin{bmatrix} \cos\theta & \sin\theta \\ -\sin\theta & \cos\theta \end{bmatrix}$ $\quad [\because\ T_\theta = \cos^2\theta + \sin^2\theta = 1]$

$$T_\theta^{-1} = \begin{bmatrix} \cos\theta & -\sin\theta \\ \sin\theta & \cos\theta \end{bmatrix}$$

Now, $[B] = T_\theta[A]T_\theta^{-1} = \begin{bmatrix} \cos\theta & \sin\theta \\ -\sin\theta & \cos\theta \end{bmatrix}\begin{bmatrix} a & h \\ h & b \end{bmatrix}\begin{bmatrix} \cos\theta & -\sin\theta \\ \sin\theta & \cos\theta \end{bmatrix}$

$$= \begin{bmatrix} a\cos\theta + h\sin\theta & h\cos\theta + b\sin\theta \\ -a\sin\theta + h\cos\theta & -h\sin\theta + b\cos\theta \end{bmatrix}\begin{bmatrix} \cos\theta & -\sin\theta \\ \sin\theta & \cos\theta \end{bmatrix}$$

$\because$ $[B]$ is a diagonal matrix, $B_{21} = B_{12} = 0$.

$\therefore$ $\quad \cos\theta(-a\sin\theta + h\cos\theta) + \sin\theta(-h\sin\theta + b\cos\theta) = 0$

or, $\quad h(\cos^2\theta - \sin^2\theta) = (a-b)\sin\theta\cos\theta$ or, $h\cos 2\theta = \frac{1}{2}(a-b)\sin 2\theta$ or, $\tan 2\theta = \frac{2h}{a-b}$

or, $\quad \boldsymbol{\theta = \frac{1}{2}\tan^{-1}\left(\frac{2h}{a-b}\right)}$ **(Proved).**

Example 2.41. *Consider the following transformation in three dimensions,*

$$x' = x\cos\theta + y\sin\theta$$
$$y' = -x\sin\theta + y\cos\theta$$
$$z' = z.$$

(i) Write down the transformation matrix $A(\theta)$.

(ii) Show that, $A(\theta_1)A(\theta_2) = A(\theta_1 + \theta_2)$.

(iii) Show that $A(\theta)$ is unitary. **[BU(H)(Old) 2001]**

Solution : (i) The transformation in this case is $\begin{bmatrix} x' \\ y' \\ z' \end{bmatrix} = \begin{bmatrix} \cos\theta & \sin\theta & 0 \\ -\sin\theta & \cos\theta & 0 \\ 0 & 0 & 1 \end{bmatrix}\begin{bmatrix} x \\ y \\ z \end{bmatrix}$

$\therefore$ **the transformation matrix is** $\boldsymbol{A(\theta) = \begin{bmatrix} \cos\theta & \sin\theta & 0 \\ -\sin\theta & \cos\theta & 0 \\ 0 & 0 & 1 \end{bmatrix}}$.

(ii) The elements of $A(\theta)$ are the coefficients of x, y and z.

$$\boldsymbol{A(\theta_1)A(\theta_2)} = \begin{bmatrix} \cos\theta_1 & \sin\theta_1 & 0 \\ -\sin\theta_1 & \cos\theta_1 & 0 \\ 0 & 0 & 1 \end{bmatrix}\begin{bmatrix} \cos\theta_2 & \sin\theta_2 & 0 \\ -\sin\theta_2 & \cos\theta_2 & 0 \\ 0 & 0 & 1 \end{bmatrix}$$

$$= \begin{bmatrix} \cos(\theta_1+\theta_2) & \sin(\theta_1+\theta_2) & 0 \\ -\sin(\theta_1+\theta_2) & \cos(\theta_1+\theta_2) & 0 \\ 0 & 0 & 1 \end{bmatrix} = \boldsymbol{A(\theta_1+\theta_2)}$$

(iii) $A^T(\theta) = \begin{bmatrix} \cos\theta & -\sin\theta & 0 \\ \sin\theta & \cos\theta & 0 \\ 0 & 0 & 1 \end{bmatrix}$

$$\therefore \quad A(\theta)A^T(\theta) = \begin{bmatrix} \cos\theta & \sin\theta & 0 \\ -\sin\theta & \cos\theta & 0 \\ 0 & 0 & 1 \end{bmatrix} \begin{bmatrix} \cos\theta & -\sin\theta & 0 \\ \sin\theta & \cos\theta & 0 \\ 0 & 0 & 1 \end{bmatrix} = \begin{bmatrix} 1 & 0 & 0 \\ 0 & 1 & 0 \\ 0 & 0 & 1 \end{bmatrix} = I.$$

$\therefore$ **$A(\theta)$ is unitary.**

Example 2.42. *Find the matrix $A(\theta)$ representing a counter-clockwise rotation through an angle θ of the Cartesian x_1x_2 coordinate system. The axes of rotation is at right angles to the x_1x_2-plane and passes through the origin.*

Show that, $[A(\theta)]^2 = A(2\theta)$. **[BU(H) 1998]**

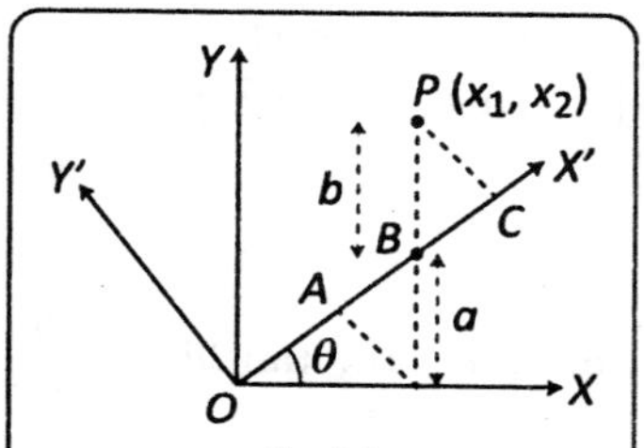

Fig. 2.1

Solution : Let $P(x_1, x_2')$ be a point in the coordinate system OX and OY. Let the coordinate system rotate in anticlockwise direction by an angle θ so that OX', OY' are he new positions of the axes as shown in Fig. 2.1.

If x_1', x_2' be the new coordinates of P then from Fig. 2.1,

$$x_1' = OA + AB + BC = x\cos\theta + a\sin\theta + b\sin\theta$$
$$= x\cos\theta + y\sin\theta \quad [\because\ a + b = y]$$
$$x_2' = a\cos\theta - x\sin\theta = -x\sin\theta + a\cos\theta.$$

In matrix form,

$$\begin{bmatrix} x_1' \\ x_2' \end{bmatrix} = \begin{bmatrix} \cos\theta & \sin\theta \\ -\sin\theta & \cos\theta \end{bmatrix} \begin{bmatrix} x_1 \\ x_2 \end{bmatrix} \quad \text{or,} \quad \begin{bmatrix} x_1' \\ x_2' \end{bmatrix} = A(\theta) \begin{bmatrix} x_1 \\ x_2 \end{bmatrix},$$

where $A(\theta) = \begin{bmatrix} \cos\theta & \sin\theta \\ -\sin\theta & \cos\theta \end{bmatrix}$ is the rotation matrix or transformation matrix.

Again, $[\boldsymbol{A(\theta)}]^2 = [A(\theta)][A(\theta)] = \begin{bmatrix} \cos\theta & \sin\theta \\ -\sin\theta & \cos\theta \end{bmatrix} \begin{bmatrix} \cos\theta & \sin\theta \\ -\sin\theta & \cos\theta \end{bmatrix}$

$$= \begin{bmatrix} \cos^2\theta - \sin^2\theta & \cos\theta\sin\theta + \sin\theta\cos\theta \\ -\sin\theta\cos\theta - \cos\theta\sin\theta & -\sin^2\theta + \cos^2\theta \end{bmatrix} = \begin{bmatrix} \cos 2\theta & \sin 2\theta \\ -\sin 2\theta & \cos 2\theta \end{bmatrix}$$

$= \boldsymbol{A(2\theta)}$ **(Proved).**

Example 2.43. *Show that trace is invariant under a similarity transformation.* **[BU(H) 2002]**

Solution : In case of similarity transformation from matrix A to matrix B,

$$C^{-1}AC = B \quad [|C| \neq 0]$$

Now, trace B, i.e., tr $[B] = \sum_i [B]_{ii} = \sum (C^{-1}AC)_{ii} = \sum_i \sum_{j,k} C_{ij}^{-1} A_{jk} C_{ki}$

$$= \sum_j \sum_k \left(\sum_i C_{ki} C_{ij}^{-1} \right) A_{jk} = \sum_j \sum_k (CC^{-1})_{kj} A_{jk}$$

$$= \sum_j (ACC^{-1})_{jj} = \sum_j (A_{jj}) = \text{tr}\,[A].$$

Thus, the traces of two similar matrices are equal. The above treatment shows that traces of two similar matrices are equal.

Example 2.44. *If A is non-singular, show that the eigenvalues of A^{-1} are reciprocals of those of A and every eigenvector of A is also an eigenvector of A^{-1}.* **[BU(H) 1998]**

Solution : If λ be the eigenvalue of the given non-singular matrix, then its characteristic equation is

$$|A - \lambda I| = 0 \quad \text{or,} \quad |AI - \lambda AA^{-1}| = 0$$

or, $$|A(I - \lambda A^{-1})| = 0 \quad \text{or,} \quad |A|\,|\lambda A^{-1} - I| = 0.$$

or, $$|\lambda|\left|A^{-1} - \frac{1}{\lambda}I\right| = 0 \qquad [\because |A| \neq 0]$$

or, $$\left|A^{-1} - \frac{1}{\lambda}I\right| = 0.$$

i.e., $\dfrac{1}{\lambda}$ is the eigenvalue of A^{-1}.

Again, $AX = \lambda_1 X$, where X is the eigenvector corresponding to eigenvalue λ_1 of A.

Now multiplying by A^{-1} from the left we get from the above equation :

$$A^{-1}AX = A^{-1}\lambda_1 X = \lambda_1 A^{-1}X, \quad [\because AX = \lambda_1 X]$$

or, $$IX = \lambda_1 A^{-1}X \quad \text{or,} \quad \lambda_1 A^{-1}X = X$$

or, $$A^{-1}X = \frac{1}{\lambda_1}X.$$

This shows that X is the eigenvector of A^{-1} corresponding to the eigenvalue $\frac{1}{\lambda_1}$ of A^{-1}.

Example 2.45. *Solve the differential equation by matrix method :*

$$\frac{d^2x}{dt^2} - \frac{5dx}{dt} + 6x = 0, \; x(0) = 1, \; x'(0) = 2.$$

Solution : Let $x = x_1$ and $\dfrac{dx}{dt} = x_2$.

Then from the equation,

$$\frac{d^2x}{dt^2} - \frac{5dx}{dt} + 6x = 0, \tag{i}$$

we get

$$\frac{dx_2}{dt} = 5x_2 - 6x_1 \quad \text{or,} \quad \frac{dx_2}{dt} = -6x_1 + 5x_2. \tag{ii}$$

Equations (i) and (ii) can be expressed by a single matrix equation,

$$\frac{d}{dx}\begin{bmatrix} x_1 \\ x_2 \end{bmatrix} = \begin{bmatrix} 0 & 1 \\ -6 & 5 \end{bmatrix}\begin{bmatrix} x_1 \\ x_2 \end{bmatrix}.$$

The characteristic equation is

$$\begin{vmatrix} 0 - \lambda & 1 \\ -6 & 5 - \lambda \end{vmatrix} = 0, \quad \text{where } \lambda \text{ is the eigenvalue.}$$

or, $-\lambda(5-\lambda)+6=0$ or, $\lambda^2-5\lambda+6=0$ or, $\lambda=2,3$.

Eigenvectors for $\lambda = 2$ and $\lambda = 3$ are $\begin{bmatrix}1\\2\end{bmatrix}$ and $\begin{bmatrix}1\\3\end{bmatrix}$. Now if B be the matrix of eigenvectors, then

$$B=\begin{bmatrix}1 & 1\\2 & 3\end{bmatrix} \quad \text{and} \quad B^{-1}=\begin{bmatrix}3 & -1\\-2 & 1\end{bmatrix}$$

$$\therefore \quad Be^{\lambda t}B^{-1}=\begin{bmatrix}1 & 1\\2 & 3\end{bmatrix}\begin{bmatrix}e^{2t} & 0\\0 & e^{3t}\end{bmatrix}\begin{bmatrix}3 & -1\\-2 & 1\end{bmatrix}=\begin{bmatrix}e^{2t} & e^{3t}\\2e^{2t} & 3e^{3t}\end{bmatrix}\begin{bmatrix}3 & -1\\-2 & 1\end{bmatrix}$$
$$=\begin{bmatrix}3e^{2t}-2e^{3t} & -e^{2t}+e^{3t}\\6e^{2t}-6e^{3t} & -2e^{2t}+3e^{3t}\end{bmatrix}.$$

By the given conditions $x(0) = 1$ and $x'(0) = 2$,

$$\begin{bmatrix}x_1\\x_2\end{bmatrix}=\begin{bmatrix}3e^{2t}-2e^{3t} & -e^{-2t}+e^{3t}\\6e^{2t}-6e^{36t} & -2e^{2t}+3e^{3t}\end{bmatrix}=\begin{bmatrix}1\\2\end{bmatrix}=\begin{bmatrix}3e^{2t}-2e^{3t} & -e^{2t}+2e^{3t}\\6e^{2t}-6e^{3t} & -4e^{2t}+6e^{3t}\end{bmatrix}$$
$$=\begin{bmatrix}e^{2t}\\2e^{2t}\end{bmatrix}.$$

$$\therefore \quad \boldsymbol{x_1=x=e^{2t} \quad \text{and} \quad x_2=\frac{dx}{dt}=2e^{2t}}.$$

Exercises

2(A) : Objective Type Questions

Select the correct answers

2.1 The elements of a matrix may be

(a) real numbers (b) imaginary numbers (c) complex numbers (d) none of these.

2.2 In a null matrix,

(a) the diagonal elements are zero (b) the off-diagonal elements are zero
(c) all the elements are zero (d) elements of the first row are zero.

2.3 By changing the rows into columns and columns into rows of a matrix A, we get

(a) transpose of A (b) inverse of A (c) adjoint of A (d) neither of these

2.4 If m and n be respectively the number of rows and number of columns, then for a square matrix,

(a) $m > n$ (b) $m = n$ (c) $m < n$ (d) $m \times n = 1$.

2.5 If the elements of a diagonal matrix are all equal, then it is a

(a) unit matrix (b) reciprocal matrix (c) scalar mtrix (d) square matrix.

2.6 If all the elements of a diagonal matrix are equal to unity, then the matrix is

(a) a scalar matrix (b) a unit matrix (c) a square matrix (d) none of these.

2.7 A square matrix A is called a singular matrix, if

(a) $|A| \neq 0$ (b) $|A| =$ a constant (c) $|A| = 0$ (d) $|A| = \infty$

2.8 If A be a symmetric matrix, then

(a) $A^T = A^{-1}$ (b) $A^T = A$ (c) $A^{-1} = A$ (d) $A = A^T A^{-1}$.

2.9 If A be a skew symmetric matrix, then

(a) $A = -A$ (b) $A^{-1} = A^T$ (c) $A^T = -A$ (d) $A^T = -A^{-1}$.

2.10 If A and B be two matrices conformable to multiplication, then

(a) $AB = BA$ (b) $AB \neq BA$ (c) $AB = A^{-1}B^{-1}$ (d) $AB = B^{-1}A^{-1}$.

2.11 A square matrix A is orthogonal, if

(a) $A^T A = A^{-1}$ (b) $A^{-1}A^T = A$ (c) $A^T A = I$ (d) none of these.

2.12 The reciprocal of matrix A is

(a) $A^{-1} = \dfrac{\text{adj } A}{|A|}$ (b) $A^{-1} = \dfrac{|A|}{\text{adj } A}$ (c) $A^{-1} = \text{adj } A \cdot |A|$ (d) none of these.

2.13 If A and B be two matrices of the same order, then

(a) $[AB]^{-1} = A^{-1}B^{-1}$ (b) $[AB]^{-1} = B^{-1}A^{-1}$ (c) $[AB]^{-1} = BA^{-1}$ (d) $[AB]^{-1} = B^{-1}A$.

2.14 If $AB = 0$ and $BA \neq 0$, then necessarily

(a) $A = 0$ (b) $B = 0$ (c) $A = 0,\ B = 0$ (d) $A \neq 0,\ B \neq 0$.

2.15 If A and B be two matrices such that $A = 0$, $AB = 0$, the equation always implies that

(a) $B = 0$ (b) $B \neq 0$ (c) $B = -A$ (d) $B = A'$.

2.16 If $A = \begin{bmatrix} 2 & 1 \\ 2 & 3 \end{bmatrix}$, $B = \begin{bmatrix} -3 & 1 \\ 2 & 0 \end{bmatrix}$, then AB is

(a) $\begin{bmatrix} 2 & 4 \\ 0 & 2 \end{bmatrix}$ (b) $\begin{bmatrix} -4 & 2 \\ 1 & 2 \end{bmatrix}$ (c) $\begin{bmatrix} -4 & 2 \\ 0 & 2 \end{bmatrix}$ (d) $\begin{bmatrix} 4 & 2 \\ 1 & 2 \end{bmatrix}$.

2.17 Inverse of $A = \begin{bmatrix} 4 & 3 \\ -7 & 1 \end{bmatrix}$ is

(a) $\begin{bmatrix} \frac{1}{4} & \frac{1}{3} \\ -\frac{1}{7} & 1 \end{bmatrix}$ (b) $\dfrac{1}{25}\begin{bmatrix} 4 & 3 \\ -7 & 1 \end{bmatrix}$ (c) $\dfrac{1}{25}\begin{bmatrix} 1 & 0 \\ 0 & 1 \end{bmatrix}$ (d) $\dfrac{1}{25}\begin{bmatrix} 1 & -3 \\ 7 & 4 \end{bmatrix}$.

2.18 If $f(x) = x^2 + 3x + 4$, then $f(x) =$

(a) $A^2 + 3A + 4I$ (b) $A^2 + 3A + 4$ (c) $A^2 + 3A$ (d) $(A + 4)(A + I)$.

2.19 The rank of the matrix $\begin{bmatrix} 4 & 0 & 0 \\ 0 & 3 & 0 \\ 0 & 0 & 5 \end{bmatrix}$ is equal to

(a) 4 (b) 3 (c) 5 (d) 1.

2.20 An $n \times n$ homogeneous system of equation $AX = 0$ is given. The rank of A is $r < n$. Then the system has

(a) $(n - r)$ independent solutions (b) r independent solutions

(c) no solution (d) $(n - 2r)$ independent solutions.

[AMIETE, Summer 1993]

2.21 If λ_1, λ_2, λ_3 be the eigenvalues of the matrix, $\begin{bmatrix} -2 & -9 & 5 \\ -5 & -10 & 7 \\ -9 & -21 & 14 \end{bmatrix}$, then $\lambda_1 + \lambda_2 + \lambda_3$ is equal to

(a) -16 (b) 2 (c) -6 (d) -14. **[AMIETE, Winter 1997]**

2.22 If $A = \begin{bmatrix} 1 & 0 \\ 2 & 4 \end{bmatrix}$, then the eigenvalues of $4A^{-1} + 3A + 2I$ are

(a) 6, 15 (b) 9, 12 (c) 9, 15 (d) 7, 15. **[AMIETE, Winter 1996]**

2.23 An eigenvalue of a square matrix A is $\lambda = 0$. Then

(a) $|A| \neq 0$ (b) A is symetric (c) A is singular
(d) A is skew-symmetric (e) A is an even order matrix (f) A is an odd order matrix.

[AMIETE, Summer 1996, 1994; Winter 1993]

2.24 The eigenvalues of matrix A are 1, -2, 3. The eigenvalues of $3I - 2A + A^2$ are

(a) 2, 11, 6 (b) 3, 11, 18 (c) 2, 3, 6 (d) 6, 3, 11 **[AMIETE, Summer 1995]**

2.25 If the matrix is $A = \begin{bmatrix} -1 & 2 & 3 \\ 0 & 3 & 5 \\ 0 & 0 & -2 \end{bmatrix}$, then the eigenvalues of $A^3 + 5A + 8I$ are

(a) $-1, 27, -8$ (b) $-1, 3, -2$ (c) $2, 50, -10$ (d) $2, 50, 10$ **[AMIETE, Winter 1994]**

2.26 If a square matrix A has an eigenvalue λ, then an eigenvalue of the matrix $(kA)^T$, where $k \neq 0$ is a scalar, is

(a) $\dfrac{\lambda}{k}$ (b) $\dfrac{k}{\lambda}$ (c) $k\lambda$ (d) none of these.

2.27 A matrix A has eigenvalues $\lambda_i \neq 0$. Then $A^{-1} - 2I + A$ has eigenvalues

(a) $1 + 2\lambda_i + \lambda_i^2$ (b) $\dfrac{1}{\lambda_i} = 2 + \lambda_i$ (c) $1 - 2\lambda_i + \lambda_i^2$ (d) $1 - \dfrac{2}{\lambda_i} + \dfrac{1}{\lambda_i^2}$.

[AMIETE, Winter 1990]

2.28 If I be the unit matrix of order n and adj $(2I) = 2^k I$, then k is equal to

(a) 1 (b) 2 (c) $n - 1$ (d) n. **[AMIETE, Summer 1992]**

2.29 If A is a non-zero column vector $(n \times 1)$, then the rank of the matrix AA^T is

(a) 0 (b) 1 (c) $n - 1$ (d) n.

2.30 The value of k such that the system of equations,

$$2x + 3y - 2z = 0$$
$$3x - y + 3z = 0$$
$$7x + ky - z = 0$$

has non-trivial solutions is

(a) 2 (b) 5 (c) 3 (d) 0. **[AMIETE, Winter 1996]**

2.31 If A and B be two matrices, then $(A + B)^2$ is

(a) $A^2 + 2AB + B^2$ (b) $A^2 + B^2$ (c) $\neq A^2 + 2AB + B^2$ (d) $A^2 + 2BA + B^2$.

2.32 The normalised form of the matrix $\begin{bmatrix} 1 \\ 2 \\ 0 \end{bmatrix}$ is

(a) $\begin{bmatrix} \frac{1}{\sqrt{5}} \\ \frac{2}{\sqrt{5}} \\ 0 \end{bmatrix}$ (b) $\begin{bmatrix} \frac{1}{\sqrt{3}} \\ \frac{2}{\sqrt{3}} \\ 0 \end{bmatrix}$ (c) $\begin{bmatrix} \frac{1}{\sqrt{7}} \\ \frac{2}{\sqrt{7}} \\ 0 \end{bmatrix}$ (d) $\begin{bmatrix} 1 \\ 2 \\ 0 \end{bmatrix}$.

2.33 If a square matrix A of order n has n linearly independent eigenvectors, then a matrix P can be found such that $P^{-1}AP$ is a

(a) triangular matrix (b) symmetric matrix (c) diagonal matrix (d) unit matrix.

2.34 The vectors $X_1 = [1\ 2\ 3]$ and $X_2 = [2\ \ -2\ \ 6]$ are linearly

(a) dependent (b) independent (c) both (a) and (b) (d) neither of these.

2.35 The eigenvalues of the matrix $A = \begin{bmatrix} 1 & 0 & 0 \\ 0 & 1 & 0 \\ 0 & 0 & 1 \end{bmatrix}$ are

(a) $\lambda = 1, 1, -1$ (b) $0, 1, -1$ (c) $1, 0, -1$ (d) $1, 2, -1$.

2.36 The matrix $A = \begin{bmatrix} \frac{1}{\sqrt{2}} & \frac{i}{\sqrt{2}} \\ -\frac{i}{\sqrt{2}} & -\frac{1}{\sqrt{2}} \end{bmatrix}$ is

(a) symmetric (b) unitary (c) singular (d) triangular.

2.37 The matrix $A = \begin{bmatrix} 0 & -1 & 0 \\ 1 & 0 & 0 \\ 0 & 0 & 1 \end{bmatrix}$ is

(a) orthogonal (b) Hermitian (c) symmetric (d) antisymmetric [GATE 1994]

2.38 The eigenvalues of the matrix $\begin{bmatrix} 0 & 1 & 0 \\ 0 & 0 & 1 \\ 1 & 0 & 0 \end{bmatrix}$ are given by λ_1, λ_2, λ_3 and λ_4. Which of the following system is not true?

(a) $\lambda_1 + \lambda_2 + \lambda_3 + \lambda_4 = 0$ (b) $\lambda_1\lambda_2 + \lambda_2\lambda_3 + \lambda_3\lambda_1 = 0$

(c) all eigenvalues are real (d) product of the eigenvalues is 1. [GATE 1998]

2.39 The eigenvalues of the matrix $\begin{bmatrix} 1 & 1 \\ 1 & 1 \end{bmatrix}$ are

(a) 1, 0 (b) 1, 1 (c) 1, 2 (d) 0, 2. [GATE 2000]

2.40 If two matrices A and B can be diagonalised simultaneously, which of the following is true?

(a) $A^2B = B^2A$ (b) $A^2B^2 = B^2A$ (c) $AB = BA$ (d) $AB^2AB = BABA^2$. [GATE 2002]

2.41 Which of the following matrices is the inverse of the matrix $\begin{bmatrix} 1 & -1 \\ 0 & 1 \end{bmatrix}$?

(a) $\begin{bmatrix} 1 & 1 \\ -1 & 1 \end{bmatrix}$ (b) $\begin{bmatrix} 1 & -1 \\ 0 & 1 \end{bmatrix}$ (c) $\begin{bmatrix} 1 & 1 \\ 0 & 1 \end{bmatrix}$ (d) $\begin{bmatrix} -1 & 1 \\ 0 & -1 \end{bmatrix}$ [GATE 2002]

2.42 A 3×3 matrix has eigenvalues 0, $2 + i$ and $2 - i$. Which of the following statements is correct?

(a) The matrix is Hermitian. (b) The matrix is unitary.

(c) The inverse of the matrix exists. (d) The determinant of the matrix is zero.

[GATE 2003]

2.43 A real traceless 4×4 unitary matrix has two eigenvalues -1 and $+1$. The other eigenvalues are,

(a) 0 and + 2 (b) -1 and $+1$ (c) 0 and + 1 (d) + 1 and -1. **[GATE 2004]**

2.44 If every minor of order r of a matrix A is zero, then the rank of A is

(a) $> r$ (b) $= r$ (c) $\le r$ (d) $< r$.

2.45 The rank of the diagonal matrix $\begin{bmatrix} 1 & & & & & \\ & 0 & & & & \\ & & 1 & & & \\ & & & 0 & & \\ & & & & 0 & \\ & & & & & 4 \end{bmatrix}$ is

(a) 1 (b) 2 (c) 3 (d) 4.

Answers

2.1 (a), (c)	**2.2** (c)	**2.3** (a)	**2.4** (b)	**2.5** (c)	**2.6** (b)	**2.7** (c)	**2.8** (b)	**2.9** (c)
2.10 (b)	**2.11** (c)	**2.12** (a)	**2.13** (b)	**2.14** (d)	**2.15** (b)	**2.16** (c)	**2.17** (d)	**2.18** (a)
2.19 (b)	**2.20** (a)	**2.21** (b)	**2.22** (c)	**2.23** (c)	**2.24** (a)	**2.25** (c)	**2.26** (c)	**2.27** (b)
2.28 (c)	**2.29** (a)	**2.30** (b)	**2.31** (c)	**2.32** (a)	**2.33** (c)	**2.34** (b)	**2.35** (a)	**2.36** (b).

2(B) : Review Questions

2.1 (a) Define the term 'matrix' with example.

(b) Define : Unit matrix, column matrix, diagonal matrix, square matrix, singular and non-singular matrices, symmetric matrix, orthogonal matrix and Hermitian matrix with examples.

2.2 What do you mean by conformable matrices? How matrix multiplication is defined? Discuss the properties of matrix multiplication.

2.3 What do you mean by transpose of a matrix? Prove that the product of two matrices is the product of their transposes taken in reverse order.

2.4 (a) What is symmetric matrix? Prove that the product of two symmetric matrices of same order is symmetric.

(b) Show that $[A][B]$ is Hermitian, if and only if $[AB] = [BA]$.

2.5 Show that if a matrix A possesses its inverse, then it is a square matrix and its determinant is not equal to zero.

2.6 Prove that

(a) $A + (B + C) = (A + B) + C$, where A, B and C are the matrices of the same order.

(b) $k(A + B) = kA + kB$, where k is scalar.

2.7 (a) Prove that, the product of two symmetric matrices of the same order is symmetric, if the product be commutative.

(b) Prove that every square matrix can be expressed as a sum of a symmetric matrix and a skew-symmetric matrix uniquely.

2.8 What do you mean by the inverse of a matrix? State and prove the necessary and sufficient condition for the existence of the inverse of a matrix. Show that the inverse of a matrix is unique.

2.9 (a) Show that the inverse of a unitary matrix is unitary and the product of two unitary matrices is also unitary.

(b) Define Hermitian matrix and anti-Hermitian matrix.

Show that an arbitrary matrix can be uniquely decomposed into the sum of a Hermitian and an anti-Hermitian matrix. **[BU(H) 2003]**

2.10 (a) What do you mean by trace of a matrix? Show that $T_r(AB) = T_r(BA)$.

(b) Extend this result to the case of three matrices and generalise it for any number of matrices.

(c) Show that trace is invariant under a similarity transformation. **[BU(H) 2002]**

2.11 (a) If $\lambda_1, \lambda_2, \ldots, \lambda_n$ be the eigenvalues of a $n \times n$ matrix A, show that,

(i) $\lambda_1 + \lambda_2 + \cdots + \lambda_n = T_r A$ (ii) $\lambda_1 \lambda_2 \cdots \lambda_n = \det A$.

What restriction is imposed on λ_i's, if the matrix is invertible? **[BU(H) 2000]**

2.12 (a) If λ_1, λ_2, ..., λ_n be the eigenvalues of an $(n \times n)$ unitary matrix, what is the value of $\sum_{i=1}^{n} |\lambda_i|$? **[BU(H) 2006]**

(b) Find the matrix representing a counterclockwise rotation through 60° in the $x-y$ plane. **[BU(H) 2000]**

2.13 Show that, (i) the eigenvalues of a Hermitian matrix are real and (ii) eigen matrices corresponding to distinct eigenvalues are mutually orthogonal. **[NBU(H) 2000]**

2.14 (a) Establish that any two eigenvectors corresponding to two distinct eigenvalues of a Hermitian matrix are orthogonal. **[NBU(H) 2006]**

(b) $[A, B] = iC$.

If A and B are Hermitian matrices, then show that C is also a Hermitian matrix. **[NBU(H) 2008]**

2.15 (a) If A and B are two Hermitian matrices, prove that AB is Hermitian, only if A and B commute. **[NBU(H) 2009]**

(b) Let $H' = UHU^*$, where U is a unitary matrix. Show that if H is Hermitian, then H' is also Hermitian. **[BU(H) 2003]**

2.16 (a) Show that the three vectors $[1\ 1\ 1]$, $[1\ 0\ -1]$ and $[1\ -2\ 1]$ constitute a set of linearly independent vectors. Verify that they are also orthogonal vectors. Express the vector $[1\ 0\ 0]$ as a linear combination of the three given vectors.

(b) Prove that the eigenvalues of a unitary matrix have unit magnitude **[BU(H) 1999]**

2.17 (a) Show that the diagonal elements of a skew-Hermitian matrix are either zero or imaginary.

(b) Show by an example that if $[A][B] = 0$, it does not necessarily mean that $[A] = 0$ or, $[B] = O$ where O stands for the null matrix.

2.18 (a) Define adjoint of a square matrix and find the inverse matrix with the help of adjoint matrix.

(b) What do you mean by **elementary transformation** and **elementary matrices**.

2.19 (a) What is Caley-Hamilton theorem? Find the inverse of a matrix with its help.

(b) Prove the theorem of diagonalisation of matrix. What is similarity transformation?

2.20 Explain the principle of solving simultaneous linear equations using the matrix method.

Discuss linear dependence and independence of vectors.

2.21 (a) Show that the eigenvalues of an orthogonal matrix are unimodulus (i.e., of unit modulus).

(b) The eigenvalues of a diagonal matrix are precisely the elements in the diagonal.

2.22 (a) If A and B are orthogonal matrices, show that AB is also orthogonal.

(b) If $\lambda_1, \lambda_2, \lambda_3$ are the eigenvalues of a matrix M and X_1, X_2, X_3 are the corresponding eigenvectors, find the eigenvalues and eigenvectors of the product MM.

2.23 (a) If P and Q be two non-zero square matrices such that PQ is zero, show that P and Q are singular.

(b) Prove that any two eigenvectors of a real asymmetric matrix are orthogonal, provided the corresponding eigenvalues are different.

2.24 (a) What is unitary transformation? How does it affect the eigenvalues and eigenvectors of a matrix?

(b) If T be the trace of a matrix, find its trace after a similarity transformation has been performed on it. Show that the eigenvalues and the eigenfunctions will be affected by the similarity transformation.

2.25 (a) If A and B are Hermitian matrices, then show that $A + B$ is also Hermitian.

(b) Prove that any matrix can be expressed as the sum of symmetric and antisymmetric matrices.

2(C) : Problems

2.1 Given $A = \begin{bmatrix} 1 & -1 & 1 \\ -3 & 2 & -1 \\ -2 & 1 & 0 \end{bmatrix}$ and $B = \begin{bmatrix} 1 & 2 & 3 \\ 2 & 4 & 6 \\ 1 & 2 & 3 \end{bmatrix}$. Compute AB and BA and hence, show that $AB \neq BA$. [**Ans:** $AB = \begin{bmatrix} 0 & 0 & 0 \\ 0 & 0 & 0 \\ 0 & 0 & 0 \end{bmatrix}$, $BA = \begin{bmatrix} -11 & 6 & -1 \\ -22 & 12 & -2 \\ -11 & 6 & -1 \end{bmatrix}$, $\therefore AB \neq BA$]

2.2 Show that, $\begin{vmatrix} 1 & \cos(\alpha-\beta) & \cos(\gamma-\alpha) \\ \cos(\alpha-\beta) & 1 & \cos(\beta-\gamma) \\ \cos(\gamma-\alpha) & \cos(\beta-\gamma) & 1 \end{vmatrix} = 0.$

2.3 For what value of x, the matrix, $\begin{bmatrix} 3-x & 2 & 2 \\ 2 & 4-x & 1 \\ -2 & -4 & -1-x \end{bmatrix}$ is singular? [**Ans:** $x = 0, 3$]

2.4 If $A = \begin{bmatrix} 2 & 3 & 4 \\ 4 & 5 & 6 \\ 6 & 7 & 9 \end{bmatrix}$, then find $3A$. [**Ans:** $\begin{bmatrix} 6 & 9 & 12 \\ 12 & 15 & 18 \\ 18 & 21 & 27 \end{bmatrix}$]

2.5 If matrices A and B are such that, $3A - 2B = \begin{bmatrix} 2 & 1 \\ -2 & -1 \end{bmatrix}$ and $-4A + B = \begin{bmatrix} -1 & 2 \\ -4 & 3 \end{bmatrix}$, find A and B. **[AMIE, Summer 1992]** [**Ans:** $A = \begin{bmatrix} 0 & -1 \\ 2 & -1 \end{bmatrix}$, $B = \begin{bmatrix} -1 & -2 \\ 4 & -1 \end{bmatrix}$]

2.6 If $A = \begin{bmatrix} 1 & 2 & -2 \\ 4 & 1 & 5 \\ -3 & 2 & 4 \end{bmatrix}$, find B, when $A^T + 2B = 3I$. [**Ans:** $\frac{1}{2}\begin{bmatrix} 2 & -4 & 3 \\ -2 & 2 & -2 \\ -2 & -5 & -1 \end{bmatrix}$]

2.7 If $A = \begin{bmatrix} 0 & 2 & 0 \\ 1 & 0 & 3 \\ 1 & 1 & 2 \end{bmatrix}$ and $B = \begin{bmatrix} 1 & 2 & 1 \\ 2 & 1 & 0 \\ 0 & 0 & 3 \end{bmatrix}$, then find $2A + 3B$ and $3A - 4B$.

[**Ans:** $2A + 3B = \begin{bmatrix} 3 & 10 & 3 \\ 8 & 3 & 6 \\ 2 & 2 & 13 \end{bmatrix}$ and $3A - 4B = \begin{bmatrix} -4 & -2 & -4 \\ -5 & -4 & 9 \\ 3 & 3 & -6 \end{bmatrix}$]

2.8 If $A = \begin{bmatrix} 0 & 1 & 0 \\ 0 & 0 & 1 \\ 0 & 0 & 0 \end{bmatrix}$, $B = \begin{bmatrix} 0 & 0 & 0 \\ 1 & 0 & 0 \\ 0 & 1 & 0 \end{bmatrix}$,

(i) calculate AB and BA. Hence, evaluate $A^2B + B^2A$.

(ii) show that for any number k, $(A + kB^2)^3 = kI$, where I is the unit matrix.

[AMIE, Winter 1990]

2.9 If $A = \begin{bmatrix} 1 & 2 \\ -2 & 3 \end{bmatrix}$, $B = \begin{bmatrix} 2 & 1 \\ 2 & 3 \end{bmatrix}$ and $C = \begin{bmatrix} -3 & 1 \\ 2 & 0 \end{bmatrix}$, then verify that $(AB)C = A(BC)$ and $A(B + C) = AB + AC$.

2.10 If $A = \begin{bmatrix} 3 & 1 \\ -1 & 2 \end{bmatrix}$, show that $A^2 - 5A + 7I = 0$, where I is a unit matrix of the second order. **[AMIE, 1990]**

2.11 If $A = \begin{bmatrix} \cos\alpha & \sin\alpha \\ -\sin\alpha & \cos\alpha \end{bmatrix}$, show that, $A^n = \begin{bmatrix} \cos n\alpha & \sin n\alpha \\ -\sin n\alpha & \cos n\alpha \end{bmatrix}$, where n is a positive integer.

2.12 Show that, $\begin{bmatrix} \cos\theta & -\sin\theta \\ \sin\theta & \cos\theta \end{bmatrix} = \begin{bmatrix} 1 & -\tan\frac{\theta}{2} \\ \tan\frac{\theta}{2} & 1 \end{bmatrix} \begin{bmatrix} 1 & -\tan\frac{\theta}{2} \\ -\tan\frac{\theta}{2} & 1 \end{bmatrix}^{-1}$

[AMIE, Summer 1991]

2.13 Find the adjoint and inverse of the following matrices :

(a) $\begin{bmatrix} 1 & 0 & -1 \\ 3 & 4 & 5 \\ 0 & -6 & -7 \end{bmatrix}$ **(b)** $\begin{bmatrix} 1 & 1 & 2 \\ 1 & 9 & 2 \\ 1 & 4 & 2 \end{bmatrix}$.

[**Ans:** (a) $\frac{1}{20}\begin{bmatrix} 2 & 6 & 4 \\ 21 & -7 & -8 \\ -18 & 6 & 4 \end{bmatrix}$, (b) $-\frac{1}{3}\begin{bmatrix} 6 & 6 & -15 \\ 1 & 0 & -1 \\ -5 & -3 & 8 \end{bmatrix}$]

2.14 Given, $A = \begin{bmatrix} 3 & 2 & 2 \\ 1 & 3 & 1 \\ 5 & 3 & 4 \end{bmatrix}$, compute $|A|$, A^{-1} and the matrix B such that

$$AB = \begin{bmatrix} 3 & 4 & 2 \\ 1 & 6 & 1 \\ 5 & 6 & 4 \end{bmatrix}.$$

[Ans: $5, \frac{1}{5}\begin{bmatrix} 9 & -2 & -4 \\ 1 & 2 & -1 \\ -12 & 1 & 7 \end{bmatrix}, B = \begin{bmatrix} 1 & 0 & 0 \\ 0 & 2 & 0 \\ 0 & 0 & 1 \end{bmatrix}, AB \neq BA$]

2.15 If $A = \begin{bmatrix} 0 & 1 & 0 \\ 0 & 0 & 2 \\ p & q & r \end{bmatrix}$, then show that, $A^3 = pI + qA + rA^2$.

2.16 If $A = \begin{bmatrix} 1 & 0 & -1 \\ 1 & 0 & 0 \\ 0 & 1 & 0 \end{bmatrix}$ and $I = \begin{bmatrix} 1 & 0 & 0 \\ 0 & 1 & 0 \\ 0 & 0 & 1 \end{bmatrix}$, then show that, $(A^{-1})^2 + A = I$.

2.17 Show that the matrix $\frac{1}{3}\begin{bmatrix} -1 & 2 & -2 \\ -2 & 1 & 2 \\ 2 & 2 & 1 \end{bmatrix}$ is orthogonal. Hence, find A^{-1}.

[NBU(Math. Hons) 1990; BU(Math. Hons.) 1994]

2.18 Show that, for the skew-symmetric matrix $A = \begin{bmatrix} 0 & 1 & 2 \\ -1 & 0 & 3 \\ -2 & -3 & 0 \end{bmatrix}$, $(I - A)(I + A)^{-1}$ is an orthogonal matrix. **[KU(Math. Hons.) 1988]**

2.19 Prove that, $(A^{-1})^T = (A^T)^{-1}$. **[AMIE, Summer 1993]**

2.20 Solve : $x + 2y + 3z = 1$

$3x - 2y + z = 2$

$4x + 2y + z = 3$. **[AMIE, Summer 1991]** **[Ans:** $x = \frac{7}{10}$, $y = \frac{3}{4}$, $z = \frac{1}{40}$]

2.21 Prove that the matrix $\begin{bmatrix} \frac{1}{3} & \frac{2}{3} & \frac{2}{3} \\ -\frac{2}{3} & -\frac{1}{3} & \frac{2}{3} \\ \frac{2}{3} & -\frac{2}{3} & \frac{1}{3} \end{bmatrix}$ is orthogonal, and use it to solve the equations,

$$x + 2y + 2z = 2$$
$$-2x - y + 2z = 1$$
$$2x - 2y + z = 7$$

[BU (Math. Hons.) 2003]

2.22 Express $A = \begin{bmatrix} 2 & 3 & 1 \\ 3 & 5 & 6 \\ 4 & 6 & 7 \end{bmatrix}$ as the sum of symmetric and skew symmetric matrices.

[BU (Math. Hons.) 2003]

2.23 Find the rank of the matrices : **(a)** $\begin{bmatrix} 1 & 2 & 1 \\ -1 & 0 & 2 \\ 2 & 1 & -3 \end{bmatrix}$ **(b)** $\begin{bmatrix} 1 & 2 & 3 \\ 3 & 4 & 5 \\ 4 & 6 & 8 \end{bmatrix}$. **[Ans:** (a) 3 (b) 2]

2.24 Find the value of λ for which the following system of equations has non-trivial solution :

$$3x_1 + x_2 - \lambda x_3 = 0$$
$$2x_1 + 4x_2 + \lambda x_3 = 0$$
$$8x_1 - 4x_2 - 6x_3 = 0.$$

[AMIETE, Winter 1992] **[Ans:** $\lambda = 1$]

2.25 Show that the following system of equations are consistent :

(i) $x + y + z = 9$
$3x - 2y + 47 = 3$

(ii) $2x + y + 11 = 0$
$6x + 20y - 6z = 0$
$6y - 18z + 1 = 0.$

2.26 Determine the values of α, β, γ, when the matrix $\begin{bmatrix} 0 & 2\beta & \gamma \\ \alpha & \beta & -\gamma \\ \alpha & -\beta & \gamma \end{bmatrix}$ is orthogonal.

[Ans: $[\alpha = \pm\frac{1}{\sqrt{2}},\ \beta = \pm\frac{1}{\sqrt{6}},\ \gamma = \pm\frac{1}{\sqrt{3}}]$

2.27 Find the characteristic and corresponding eigenvectors of the following matrices :

(a) $\begin{bmatrix} 1 & 1 & 1 \\ 1 & 2 & 1 \\ 3 & 2 & 3 \end{bmatrix}$ **[AMIETE, Winter 1996; Summer 1993]**

(b) $\begin{bmatrix} 3 & 10 & 5 \\ -2 & -3 & -4 \\ 3 & 5 & 7 \end{bmatrix}$ **[AMIETE 1996; Summer 1991]**

[Ans: (a) Eigenvalues 1, 0, 5; Eigenvectors $[-1\ 0\ 1]$, $[0\ -1\ 1]$, $[4\ 5\ 11]$
(b) Eigenvalues 2, 2, 3; Eigenvectors $[-1\ -1\ 2]$, $[-5\ -2\ 5]$]

2.28 One of the eigenvalues of the matrix $\begin{bmatrix} 2 & 3 & 0 \\ 3 & 2 & 0 \\ 0 & 0 & 1 \end{bmatrix}$ is 5. What are the other two eigenvalues?

Show that the normalised eigenvector corresponding to the eigenvalue 5 is $\frac{1}{\sqrt{5}}\begin{bmatrix} 1 \\ 0 \end{bmatrix}$.

[GATE 2006]

[Ans: $\lambda_1 = 5,\ \lambda = 1,\ \lambda_3 = -1$

Now, $\begin{bmatrix} 2-5 & 3 & 0 \\ 3 & 2-5 & 0 \\ 0 & 0 & 1-5 \end{bmatrix}\begin{bmatrix} x_1 \\ x_2 \\ x_3 \end{bmatrix} = 0$

or,
$$-3x_1 + 3x_2 = 0$$
$$3x_1 - 3x_2 = 0$$
$$-4x_3 = 0$$

$\therefore$ $x_1 = x_2;\ x_3 = 0.$

$\therefore$ the normalised eigenvector corresponding to the eigenvalue λ is $X_1 = \frac{1}{\sqrt{2}}\begin{bmatrix} 1 \\ 1 \\ 0 \end{bmatrix}$.]

2.29 Diagonalise the matrix, **(a)** $\begin{bmatrix} 1 & 0 & -1 \\ 1 & 2 & 1 \\ 2 & 2 & 3 \end{bmatrix}$. **[Ans:** $\begin{bmatrix} 1 & 0 & 0 \\ 0 & 2 & 0 \\ 0 & 0 & 3 \end{bmatrix}$]

2.30 Find the eigenvalues and eigenvectors of the matrix $A = \begin{bmatrix} 2 & 0 & -2 \\ 0 & 4 & 0 \\ -2 & 0 & 5 \end{bmatrix}$. Is it a Hermitian matrix? **[VU(H) 2009]**

2.31 Solve by matrix method the following equations :

$$-x_1 + 3x_3 + 1 = 0$$
$$-2x_1 + x_2 - 4x_3 - 2 = 0$$
$$x_2 - 2x_3 - 4 = 0.$$

[VU(H) 2008]

2.32 Find the eigenvalues and the eigenvectors of the matrix $\begin{bmatrix} 0 & -i \\ i & 0 \end{bmatrix}$, where $i = \sqrt{-1}$. Check the orthogonality of the eigenvectors corresponding to different eigenvalues. **[VU(H) 2007]**

2.33 **(a)** Express the given matrix as the sum of a symmetric and antisymmetric matrix :

$$\begin{bmatrix} 1 & 1 & 3 \\ 2 & 3 & 2 \\ 1 & 2 & 4 \end{bmatrix}.$$

(b) Find the inverse of the matrix $\begin{bmatrix} 1 & -1 & 3 \\ -1 & 1 & 2 \\ 3 & 2 & 1 \end{bmatrix}$.

(c) Find the rank of the matrix $\begin{bmatrix} 1 & 2 & 3 \\ 1 & 2 & 5 \\ 2 & 4 & 8 \end{bmatrix}$. **[NBU(H) 2006]**

2.34 **(a)** Show that the eigenvalues of Hermitian matrix are real and

(b) eigen matrices corresponding to distinct eigenvalues are mutually orthogonal.

(c) Find the eigenvalues and eigenvectors of the matrix $\begin{bmatrix} 0 & 1 & 0 \\ 1 & 0 & 1 \\ 0 & 1 & 0 \end{bmatrix}$. **[NBU(H) 2007]**

2.35 Diagonalise the following matrix, $M = \dfrac{1}{\sqrt{2}} \begin{bmatrix} 0 & -i & 0 \\ i & 0 & i \\ 0 & i & 0 \end{bmatrix}$. **[NBU(H) 2008]**

2.36 **(a)** If $A = \begin{bmatrix} i & 2 & 3+i \\ -2 & 2i & 0 \\ -3+i & 0 & -i \end{bmatrix}$, find $A^\dagger$.

(b) Solve the following equations by matrix method :

$$x + 5y + 3z = 1$$
$$3x + y + 2z = 1$$
$$x + 2y + z = 0.$$

[NBU(H) 2009]

2.37 Find the eigenvalues and eigenvectors of the matrix, $\begin{bmatrix} 1 & -2 \\ -2 & -2 \end{bmatrix}$. **[CU(H) 2009]**

2.38 **(a)** Find the eigenvalues and eigenvectors of the matrix $\begin{bmatrix} 1 & 3 \\ 3 & 1 \end{bmatrix}$.

(b) If a matrix is both Hermitian and unitary, show that all its eigenvalues are ± 1.

[CU(H) 2008]

2.39 (a) If A and B are two commutating matrices, show that they must be square matrices of the same order.

(b) If S and A are unitary matrices, show that $S^{-1}AS$ is also a unitary matrix.

(c) Find the eigenvalues and the normalised eigenvectors of the matrix $\begin{bmatrix} 1 & 1 \\ 1 & 1 \end{bmatrix}$.

[CU(H) 2007]

2.40 (a) A Cartesian coordinate axes is rotated about the z-axis in anti-clockwise direction through an angle θ, such that a point $P(x, y, z)$ changes to (x', y', z') and

$$\begin{bmatrix} x' \\ y' \\ z' \end{bmatrix} = R_z(\theta) \begin{bmatrix} x \\ y \\ z \end{bmatrix},$$

find $R_z(\theta)$

(b) Given $A = \begin{bmatrix} 1 & -2 \\ -2 & 2 \end{bmatrix}$, determine the eigenvalues and normalised eigenvectors of A.

[CU(H) 2006]

2.41 (a) Find the eigenvalues and eigenvectors of the matrix $\begin{bmatrix} 1 & 0 & 0 \\ 0 & 0 & 2 \\ 0 & 2 & 0 \end{bmatrix}$.

Is this matrix unitary? [CU(H) 2005]

(b) Find the eigenvalues and eigenvector of the following matrices :

(i) $A = \begin{bmatrix} 1 & 1 \\ 4 & 1 \end{bmatrix}$ [CU(H), 2004]

(ii) $A = \begin{bmatrix} 10 & 0 & 0 \\ 0 & 6 & -4 \\ 0 & -4 & 6 \end{bmatrix}$ [CU(H) 2003]

(iii) $A = \begin{bmatrix} 5 & 2 \\ 2 & 2 \end{bmatrix}$. [CU(H) 2002]

2.42 (a) Find the trace and determinant of the matrix $A = \begin{bmatrix} 0 & -i \\ i & 0 \end{bmatrix}$; find the eigenvalues using these values.

(b) Show that A does not commute with $B = \begin{bmatrix} 0 & 1 \\ 1 & 0 \end{bmatrix}$.

(c) If a matrix A is Hermitian and $A^2 = I$, show that A is also unitary. [CU(H) 2001]

2.43 Find the eigenvalues and eigenvectors of the following matrices :

(i) $\begin{bmatrix} 4 & -1 & -4 \\ -1 & 4 & -1 \\ -1 & -1 & 4 \end{bmatrix}$ [CU(H) 2000]

(ii) $\begin{bmatrix} 2 & -1 & -1 \\ -1 & 2 & -1 \\ -1 & -1 & 2 \end{bmatrix}$ [CU(H) 1999]

(iii) $\begin{bmatrix} 1 & 0 & 0 \\ 0 & 1 & 0 \\ 0 & 0 & 1 \end{bmatrix}$. Is this matrix unitary? **[CU(H) 1998]**

(iv) $A = \begin{bmatrix} 1 & 1 \\ 0 & 1 \end{bmatrix}$. **[CU(H) 1997]**

2.44 Consider an orthogonal transformation of coordinate system $OXYZ$ to $OXYZ'$ with unit vectors $\hat{i}, \hat{j}, \hat{k}; \hat{i}', \hat{j}', \hat{k}'$. A position vector $\hat{r}$ in these systems can be expressed as

$$\hat{r} = \hat{i}x + \hat{j}y + \hat{k}z = \hat{i}'x + \hat{j}'y + \hat{k}'z.$$

Show that, $x' = (\hat{i}' \cdot \hat{i})x + (\hat{i}' \cdot \hat{j})y + (\hat{i}' \cdot \hat{k})z$

Hence, establish : $\begin{bmatrix} x' \\ y' \\ z' \end{bmatrix} \begin{bmatrix} a_{11} & a_{12} & a_{13} \\ a_{21} & a_{22} & a_{23} \\ a_{31} & a_{32} & a_{33} \end{bmatrix} \begin{bmatrix} x \\ y \\ z \end{bmatrix}$, where $a_{11} = \hat{i}' \cdot \hat{i}$, $a_{12} = \hat{i}' \cdot \hat{j}$, etc.

[CU(H) 1997]

2.45 (a) Solve the following set of equations by the matrix method :

$$2x + 3y + z = 9$$
$$x + 2y + 3z = 6$$
$$3x + y + 2z = 8.$$

(b) Prove that $[AB]' = A'B'$, if $AB = BA$.

(c) Find eigenvalues and eigenvectors of the matrix, $A = \begin{bmatrix} 1 & 0 & -1 \\ 1 & 2 & 1 \\ 2 & 2 & 3 \end{bmatrix}$. **[BU(H) 2008]**

2.46 (a) Find the eigenvalues and the normalised eigenvectors of the matrix,

$$A = \begin{bmatrix} 1 & 1 & 3 \\ 1 & 1 & -3 \\ 3 & -3 & -3 \end{bmatrix}.$$

(b) If H be a Hermitian matrix, prove that, $U = (H - il)(H + il)^{-1}$ is a unitary matrix.

[BU(H) 2006]

2.47 Consider the matrix, $H = \begin{bmatrix} 3 & i \\ -i & 3 \end{bmatrix}$. Find the eigenvalues and eigenvectors and hence, construct the unitary matrix U which diagonalises H. Compute $H' = U^{-1}HU$.

[BU(H) 2003]

2.48 (a) Given $A = \begin{bmatrix} 0 & 1+2i \\ -1+2i & 0 \end{bmatrix}$. Show that, $U = (1 - A)(1 + A)^{-1}$ is unitary.

(b) Find the eigenvalues and normalised eigenvectors of the matrix $\sigma_y = \begin{bmatrix} 0 & i \\ -i & 0 \end{bmatrix}$.

Find the value of σ_y^n, where n is an integer. **[BU(H) 2002]**

2.49 (a) When do you call a set of vectors linearly dependent? Are the following vectors linearly dependent? If so, find a relation between them.

$x_1 = [3\ 2\ 7]$; $x_2 = [4\ 2\ 1]$, $x_3 = [1\ -2\ 6]$

(b) (i) Show that $T_r(AB) = T_r(BA)$.

(ii) Extend this result to the case of three matrices and generalise it for any number of matrices.

(iii) Show that trace is invariant under a similarity transformation. **[BU(H) 2002]**

2.50 (a) Find the eigenvalues and eigenvectors of the matrix, $\begin{bmatrix} 1 & 0 & 0 \\ 0 & 3 & -1 \\ 0 & -1 & 3 \end{bmatrix}$.

(b) Solve the following set of equations by the matrix method :

$$x + 3y + 3z = 1$$
$$x + 4y + 3z = 0$$
$$x + 3y + 4z = 2.$$

[BU(H) 2002]

2.51 Solve the following set of equations by the matrix method :

$$x + 5y + 3z = 1$$
$$3x + y + 2z = 1$$
$$x + 2y + z = 0.$$

[BU(H) 2001]

2.52 (a) Find the inverse of the following matrix if it exists, $A = \begin{bmatrix} -3 & 1 & -1 \\ 15 & -6 & 5 \\ -5 & 2 & -2 \end{bmatrix}$.

(b) Find the value of k for which the following equations possess a non-trivial solution :

$$x - 2y + 2z = 0$$
$$3x + 2y + z = 0$$
$$x + 2y + kz = 0.$$

[BU(H) 1999]

2.53 (a) If A is non-singular, show that the eigenvalues of A^{-1} are reciprocals of those of A and every eigenvector of A is also an eigenvector of A^{-1}.

(b) Find the eigenvalues and the normalised eigenvectors of the matrix :

$$A = \begin{bmatrix} -5 & 2 \\ 2 & -2 \end{bmatrix}.$$

[BU(H) 1998]

2.54 (a) Given matrices : $A_1 = \begin{bmatrix} 0 & 1 \\ 1 & 0 \end{bmatrix}$, $A_2 = \begin{bmatrix} 0 & -i \\ i & 0 \end{bmatrix}$, $A_3 = \begin{bmatrix} 1 & 0 \\ 0 & -1 \end{bmatrix}$.

Show that (i) $A_1^2 = A_2^2 = A_3^2 = I$, the 2×2 unit matrix

(ii) $A_1A_2 - A_2A_1 = 2iA_2$ (iii) A_1, A_2, A_3 are Hermitian.

(b) Find the eigenvalues and eigenvectors of A_1, A_2, A_3 given above. **[BU(H) 1997]**

2.55 (a) Find the inverse of the matrix, $A = \begin{bmatrix} \cos\theta & \sin\theta \\ -\sin\theta & \cos\theta \end{bmatrix}$.

(b) Use the matrix method to solve the following set of equations :

$$x + 5y + 3z = 1$$
$$3x + y + 2z = 1$$
$$x + 2y + z = 1.$$

(c) Construct a 3×3 matrix A such that,

$$A\begin{bmatrix} 1 \\ 0 \\ -1 \end{bmatrix} = \begin{bmatrix} 2 \\ 0 \\ 0 \end{bmatrix}, \quad A\begin{bmatrix} 0 \\ 1 \\ 0 \end{bmatrix} = \begin{bmatrix} 1 \\ 0 \\ 1 \end{bmatrix} \quad \text{and} \quad A\begin{bmatrix} 1 \\ 0 \\ 1 \end{bmatrix} = \begin{bmatrix} 1 \\ 0 \\ 1 \end{bmatrix}.$$

[BU(H) 1996]

2.56 **(a)** Given two matrices, $A = \begin{bmatrix} i & 2 & 3+i \\ -2 & 2i & 0 \\ -3+i & 0 & -i \end{bmatrix}$ and $B = \begin{bmatrix} 1 & 2+i & 3-i \\ 2-i & 5 & 5+9i \\ 3+i & 5-9i & 2 \end{bmatrix}$.

Find the Hermitian conjugates examine whether any of them is Hermitian.

(b) Given $\sigma_1 = \begin{bmatrix} 0 & 1 \\ 1 & 0 \end{bmatrix}$ and $\sigma_2 = \begin{bmatrix} 0 & -i \\ i & 0 \end{bmatrix}$. Find σ_1^2 and σ_2^2 and hence, infer the eigenvalues of σ_1 and σ_2. **[BU(H) (old) 1996]**

2.57 **(a)** Find A such that $A^{-1} = \begin{bmatrix} 3 & 2 \\ 1 & 6 \end{bmatrix}$.

(b) Let A be a matrix $\begin{bmatrix} \frac{7}{3} & -\frac{2}{3} \\ \frac{2}{3} & \frac{2}{3} \end{bmatrix}$.

(i) Find the eigenvalues and eigenvectors of A. **(ii)** Compute A^4.

(c) Ascertain whether $A = \begin{bmatrix} 1 & 0 \\ 0 & 0 \end{bmatrix}$ is a singular matrix or not.

(d) Given that $B(\theta) = \begin{bmatrix} \cos\theta & \sin\theta \\ -\sin\theta & \cos\theta \end{bmatrix}$; prove that, $B(\theta_1)B(\theta_2) = B(\theta_1 + \theta_2)$.

[BU(Hons.) 1994]

2.58 **(a)** Given, $A = \begin{bmatrix} i & 2 & 3+i \\ -2 & 2i & 0 \\ -3+i & 0 & -i \end{bmatrix}$; find its **(i)** transpose and **(ii)** Hermitian conjugate.

(b) Solve the following equations :

$$\begin{aligned} x_1 + x_2 - 2x_3 &= 1 \\ 2x_1 + \lambda x_2 + 2x_3 &= 2 \\ 3x_1 + 2x_2 - 4x_3 &= 3. \end{aligned}$$

Is the method valid for all values of λ?

(c) Determine the eigenvalues and the normalised eigenvectors of the following matrix :

$$\begin{bmatrix} 3 & 5 & 1 \\ -2 & -2 & 0 \\ 2 & 7 & 3 \end{bmatrix}.$$

[BU(Hons.) 1993]

$$a_7 = \frac{a_5}{7.6} = \frac{a_1+1}{7\cdot 6\cdot 5!} = \frac{a_1+1}{7!} \text{ and so on.}$$

$$\therefore \quad y = a_0 + a_1 x + \frac{1}{2}a_0 x^2 + \frac{a_1+1}{3!}x^3 + \frac{a_0}{4!}x^4 + \frac{(a_1+1)}{5!}x^5 + \cdots$$

$$= \pm a_0\left(1 + \frac{1}{2!}x^2 + \frac{1}{4!}x^4 + \cdots\right) + a_1 x + (a+1)\left(\frac{x^3}{3!} + \frac{x^5}{5!} + \frac{x^7}{7!} + \cdots\right)$$

$$= a_0 \cos hx + (a_1+1)\left(x + \frac{x^3}{3!} + \frac{x^5}{5!} + \cdots\right) - x$$

$$= a_0 \cos hx + (a_1+1)\sin hx - x$$

with a_0 and a_1 are arbitrary constants.

2.59 (a) Find the eigenvalues and normalised eigenvectors of the matrix $\begin{bmatrix} 5 & 2 \\ 2 & 5 \end{bmatrix}$.

[CU(H) 2014]

(b) Consider the matrices $\sigma_x = \begin{bmatrix} 0 & 1 \\ 1 & 0 \end{bmatrix}$ and $\sigma_y = \begin{bmatrix} 0 & -i \\ i & 0 \end{bmatrix}$. Which of the matrices is/are Hermitian? Which of the matrices are unitary? Find $(\sigma_x\sigma_y - \sigma_y\sigma_x)$. [CU(H) 2013]

2.60 (a) Prove that $e^{i\theta\sigma} = \cos\theta 1 + i\sin\theta\sigma$, where σ is a matrix $\sigma^2 = 1$ and θ is real.

(b) Show that a Hermitian matrix remains Hermitian under a unitary transformation.

[CU(H) 2010]

(c) Find the eigenvalues and normalised eigenvectors of the following matrix.

$$A = \begin{bmatrix} 1 & 0 & 0 \\ 0 & 0 & 1 \\ 0 & 1 & 0 \end{bmatrix}$$

[BU(H) 2009]

2.61 (a) Given $M = \begin{bmatrix} I_{xx} & I_{xy} & I_{xz} \\ I_{yx} & I_{yy} & I_{yz} \\ I_{zx} & I_{zy} & I_{zz} \end{bmatrix}$ and $U = \begin{bmatrix} l \\ m \\ n \end{bmatrix}$. Find I, if $I = U^{-1}MU$.

[VU(H) 2014]

(b) A is a square matrix and A^T is its transpose. Show that $A + A^T$ is a symmetric matrix. [VU(H) 2013]

2.62 (a) Establish that any two eigenvectors corresponding to two distinct eigenvalues of a Hermitian matrix are orthogonal.

(b) Express the given matrix as the sum of a symmetric and anti-symmetric matrix.

$$\begin{bmatrix} 1 & 1 & 3 \\ 2 & 3 & 2 \\ 1 & 2 & 4 \end{bmatrix}$$

(c) Find the inverse of the matrix $\begin{bmatrix} 1 & -1 & 3 \\ -1 & 1 & 2 \\ 3 & 2 & 1 \end{bmatrix}$.

(d) Find the rank of the matrix $\begin{bmatrix} 1 & 2 & 3 \\ 1 & 2 & 5 \\ 2 & 4 & 8 \end{bmatrix}$. [NBU(H) 2006]

2.63 (a) Prove that orthogonal matrices are non-singular.

(b) Find the value of λ for which the matrix $A = \begin{bmatrix} \cos\psi & -\sin\psi & 0 \\ \sin\psi & \cos\psi & 0 \\ 0 & 0 & \lambda \end{bmatrix}$ will be orthogonal.

2.64 (a) Pauli matrices are given by $\sigma_x = \begin{bmatrix} 0 & 1 \\ 1 & 0 \end{bmatrix}$, $\sigma_y = \begin{bmatrix} 0 & -i \\ i & 0 \end{bmatrix}$ and $\sigma_z = \begin{bmatrix} 1 & 0 \\ 0 & -1 \end{bmatrix}$. Which of the Pauli matrices is/are Hermitian?

(b) Find the values of $\sigma_x\sigma_y - \sigma_y\sigma_x$, $\sigma_{yz} - \sigma_{zy}$ and $\sigma_z\sigma_x - \sigma_x\sigma_z$. [VU(H) 2011]

2.65 Show that Trace $[ABC]$ = Trace $[CBA]$, if any three of the two matrices commute. **[MGU (Model questions) 2012]**

2.66 Prove that the set of all matrices $\begin{bmatrix} \cos\theta & \sin\theta \\ -\sin\theta & \cos\theta \end{bmatrix}$ for $0 \leq \theta \leq 2\pi$ form a lie group.

2.67 Find the characteristic equation of the matrix

$\begin{bmatrix} 1 & 2 & 3 \\ 2 & -1 & 4 \\ 3 & 1 & 1 \end{bmatrix}$ and verify Cayley-Hamilton theorem for it. **[MGU (MSc) 2012]**

2.68 (a) If a matrix A is given by $A = \frac{1}{2}\begin{bmatrix} 3 & 1 \\ 1 & 3 \end{bmatrix}$, find e^A and 4^A.

(b) Prove that if A is an orthogonal matrix, then A^{-1} is also orthogonal.

2.69 (a) Determine the eigenvalues and corresponding eigenvectors of the matrix

$A = \begin{bmatrix} a & b & b \\ b & a & b \\ b & b & a \end{bmatrix}$, where a and b are scalars and $b \neq 0$.

(b) Show that $A = \begin{bmatrix} 1 & 1 \\ 0 & 1 \end{bmatrix}$ is not diagonalisable. **[DU 2015]**

2.70 Let $A = \begin{bmatrix} 4 & 1 & -1 \\ 2 & 5 & -2 \\ 1 & 1 & 2 \end{bmatrix}$. Find all eigenvalues of A. **[Pune(MSc)U 2013]**

2.71 (a) Diagonalise the following matrix : $A = \begin{bmatrix} 2 & -2 \\ -2 & 5 \end{bmatrix}$.

(b) Let $A = \begin{bmatrix} 1 & -3 & 3 \\ 3 & -5 & 3 \\ 6 & -6 & 4 \end{bmatrix}$. Find eigenvalues and eigenvectors of A.

[Pune (MSc) U 2014]

2.72 Diagonalise the matrix $\begin{bmatrix} 1 & 0 & 1 \\ 0 & 1 & 0 \\ 1 & 0 & 1 \end{bmatrix}$, find eigenvalues and eigenvectors.

[Pune U (MSc) 2014]

2.73 (a) Diagonalise the following matrix $A = \begin{bmatrix} 1 & 4 \\ 2 & 3 \end{bmatrix}$.

(b) Let $A = \begin{bmatrix} 1 & 2 \\ 3 & 4 \end{bmatrix}$ and let T be the linear operator on R^2 defined by $T(V) = AV$, where V is written as column matrix of T in each of the following bases: (i) $\{e_1 = (1,0),\ e_2 = (0,1)$, i.e., usual basis; (ii) $\{f_1 = (1,3),\ f_2 = (2,5)\}$.

[Pune U(MSc), 2014]

2.74 (a) Show that any square matrix can be written as the sum of a symmetric and antisymmetric matrix.

(b) Find the eigenvalues and the normalised eigenvector of the matrix $M = \begin{bmatrix} 2 & 2 \\ 2 & -1 \end{bmatrix}$.

(c) A normal matrix N is defined by the relation $NN^+ = N^+N$. If N is written as $(A + iB)$, where A and B are Hermitian, show that A and B commute.

[CU(H) 2017]

2.75 **(a)** If a matrix A has eigenvalue λ with respect to an eigenvector Y, then show that e^A has eigenvalue e^λ with respect to the same eigenvector Y.

(b) Find out eigenvalue λ and normalised orthogonal eigenvectors of matrix $\begin{bmatrix} 1 & 1 \\ 0 & 1 \end{bmatrix}$.

(c) Consider the matrices $\sigma_1 = \begin{bmatrix} 0 & 1 \\ 1 & 0 \end{bmatrix}$, $\sigma_2 = \begin{bmatrix} 0 & -i \\ i & 0 \end{bmatrix}$, $\sigma_3 = \begin{bmatrix} 1 & 0 \\ 0 & -1 \end{bmatrix}$.

Which of the matrices is/are Hermitian? Which is/are unitary? **[CU(H) 2016]**

2.76 **(a)** Find the eigenvalues and the normalised eigenvectors of the matrix

$$\begin{bmatrix} \cos\theta & \sin\theta \\ -\sin\theta & \cos\theta \end{bmatrix}.$$

(b) Prove that all eigenvalues of a Hermitian matrix are real. **[CU(H) 2015]**

2.77 **(a)** Find the eigenvalues and normalised eigenvectors of the matrix $\begin{bmatrix} 1 & 1 \\ 4 & 1 \end{bmatrix}$.

(b) If a matrix is both Hermitian and unitary, show that all its eigenvalues are ± 1.

(c) Show that the product of two symmetric matrices is symmetric, only if they commute. **[CU(H) 2014]**

CHAPTER 3

Linear Vector Space

3.1 Definition of Vector Space

A linear vector space V_n is a set over field F such that elements of V_n obey certain mathematical operations like addition and multiplication by a scalar as follows :

If α, β, γ be the elements of V_n and a, b are scalars in F, then

(i) $\alpha + \beta$ will also be in V_n

(ii) $\alpha + \beta = \beta + \alpha$

(iii) $\alpha + (\beta + \gamma) = (\alpha + \beta) + \gamma$

(iv) $a(\alpha + \beta) = a\alpha + a\beta$

(v) $\alpha + (-\alpha) = 0$, where 0 is the null vector

(vi) $(ab)\alpha = a(b\alpha)$

(vii) $1 \cdot \alpha = \alpha$, where 1 is unity in field F.

The elements of V, i.e., α, β, γ are called **vectors** and those of F, i.e., a, b, c are called **scalars**.

3.2 Vector Space of Ordered n-Tuples

We assume that α, β and γ are three elements of non-empty space V such that,

$$\alpha = (a_1,\ a_2, \ldots, a_n), \quad \beta = (b_1,\ b_2, \ldots, b_n) \quad \text{and} \quad \gamma = (c_1,\ c_2, \ldots, c_n),$$

where a_i, b_i, c_i are all in non-empty space F.

Let a be an element of F. Then we define addition of α and β as

$$\alpha + \beta = (a_1,\ a_2, \ldots, a_n) + (b_1,\ b_2, \ldots, b_n) = (a_1 + b_1,\ a_2 + b_2, \ldots, a_n + b_n). \tag{3.1}$$

Eqn. (3.1) represents **vector addition.**

The multiplication of α by a :

$$a\alpha = a(a_1, a_2, \ldots, a_n) = (aa_1, aa_2, \ldots, aa_n). \tag{3.2}$$

This represents multiplication of a vector (α) by a scalar (a).

3.2.1 Verification of the Postulates of Vector Space V

(i) For $\alpha, \beta \in V$, we have

$$\alpha + \beta = (a_1, a_2, \ldots, a_n) + (b_1, b_2, \ldots, b_n) = (a_1 + b_1,\ a_2 + b_2, \ldots, a_n + b_n)$$

(since a_i and b_i are commutative in F)

$$= (b_1 + a_1,\ b_2 + a_2, \ldots, b_n + a_n) = (b_1, b_2, \ldots, b_n) + (a_1, a_2, \ldots, a_n) = \beta + \alpha.$$

$$\therefore \quad \alpha + \beta = \beta + \alpha. \tag{3.3}$$

(ii) Let α, β, $\gamma \in V$, then

$$\alpha + (\beta + \gamma) = (a_1, a_2, \ldots, a_n) + \{(b_1, b_2, \ldots, b_n) + (c_1, c_2, \ldots, c_n)\}$$

$$= (a_1, a_2, \ldots, a_n) + (b_1 + c_1,\ b_2 + c_2, \ldots, b_n + c_n)$$
$$= [a_1 + (b_1 + c_1),\ a_2 + (b_2 + c_2), \ldots, a_n + (b_n + c_n)]$$
$$= [(a_1 + b_1) + c_1,\ (a_2 + b_2) + c_2, \ldots, (a_n + b_n) + c_n]$$

(as $a_i,\ b_i,\ c_i$ are associative in F)

$$= [(a_1 + b_1,\ a_2 + b_2, \ldots, a_n + b_n) + (c_1, c_2, \ldots, c_n)]$$
$$= [(a_1, a_2, \ldots, a_n) + (b_1, b_2, \ldots, b_n)] + (c_1, c_2, \ldots, c_n) = (\alpha + \beta) + \gamma.$$

$$\therefore \quad \alpha + (\beta + \gamma) = (\alpha + \beta) + \gamma. \tag{3.4}$$

(iii) $a(\alpha + \beta) = a[(a_1, a_2, \ldots, a_n) + (b_1, b_2, \ldots, b_n)] = a[a_1 + b_1,\ a_2 + b_2, \ldots, a_n + b_n]$

$$= [a(a_1 + b_1),\ a(a_2 + b_2), \ldots a(a_n + b_n)]$$
$$= [aa_1 + ab_1,\ aa_2 + ab_2, \ldots, aa_n + ab_n]$$
$$= a(a_1, a_2, \ldots, a_n) + a(b_1, b_2, \ldots, b_n) = a\alpha + a\beta$$

$$\therefore \quad a(\alpha + \beta) = a\alpha + a\beta. \tag{3.5}$$

(iv) $\alpha + (-\alpha) = (a_1,\ a_2, \ldots, a_n) + (-a_1, -a_2, \ldots, a_n)$

$$= (a_1 - a_1,\ a_2 - a_2, \ldots, a_n - a_n) = (0, 0, \ldots, 0)$$

$$\therefore \quad \alpha + (-\alpha) = 0. \tag{3.6}$$

(v) If a, b are in F, then

$(ab)\alpha = (ab)(a_1, a_2, \ldots, a_n) = [a(ba_1), a(ba_2), \ldots, a(ba_n)] = a[ba_1, ba_2, \ldots, ba_n]$

$$= a[b(a_1, a_2, \ldots, a_n)] = a(b\alpha).$$

$$\therefore \quad (ab)\alpha = a(b\alpha). \tag{3.7}$$

(iv) If 1 is in F, then

$1 \cdot \alpha = 1(a_1, a_2, \ldots, a_n) = (1 \cdot a_1,\ 1 \cdot a_2, \ldots, 1 \cdot a_n) = (a_1, a_2, \ldots, a_n) = \alpha.$

$$\therefore \quad 1 \cdot \alpha = \alpha. \tag{3.8}$$

Hence, **$V = (a_1, a_2, \ldots, a_n)$ with all a_i in F represents a vector space.**

3.2.2 Some Properties of Vector Spaces

(1) For each element, say α in V, there exists a null vector 0 in V such that, $\alpha + 0 = \alpha$ and $0 + \alpha = \alpha$.

We have $\alpha + 0 = (a_1, a_2, \ldots, a_n) + (0, 0, \ldots, 0) = (a_1 + 0,\ a_2 + 0, \ldots, a_n + 0)$

$$= (a_1,\ a_2, \ldots, a_n) = \alpha.$$

$$\therefore \quad \alpha + 0 = \alpha. \tag{3.9}$$

Similarly, we can show that, $0 + \alpha = \alpha$.

$$\text{Hence,} \quad 0 + \alpha = \alpha + 0 = \alpha. \tag{3.10}$$

(2) If $a \neq 0$, then $a\alpha = 0$ implies $\alpha = 0$.

Since, $a \neq 0$, $\therefore$ a^{-1} exists.

So, $(a^{-1}a)\alpha = 0$ or, $1 \cdot \alpha = 0$

$$\text{or,} \quad \alpha = 0. \tag{3.11}$$

(3) If α be a non-zero vector element in V and a, b are scalars in F, then $a\alpha = b\alpha$ implies $a = b$.

We have $a\alpha = b\alpha$ or, $a\alpha - b\alpha = 0$ or, $(a-b)\alpha = 0$,

$\therefore$ $a - b = 0$ $[\because \alpha \neq 0]$

or, $a = b$.

3.3 Linear Independence and Dependence of Vectors

A set of vectors $\{\alpha_i\}$, $i = 1, 2, \ldots, n$ in a vector space V is said to be **linearly dependent**, if thee exists a corresponding set of scalars a_i, $i = 1, 2, \ldots, n$, not all zero, such that

$$\sum_{i=1}^{n} a_i\alpha_i = 0. \tag{3.12}$$

On he other hand, a set of vectors $\{\alpha_i\}$, $i = 1, 2, \ldots, n$ in a vector space V is said to be **linearly independent**, if

$$\sum_{i=1}^{n} a_i\alpha_i = 0, \quad \text{where } a_i = 0 \text{ for all } i\text{'s}.$$

Example We consider the vectors, $\alpha_1 = (1, 0, 1)$, $\alpha_2 = (0, 1, 1)$, $\alpha_3 = (1, 1, 1)$.

If there be three scalars a_1, a_2 and a_3, then if

$$a_1\alpha_1 + a_2\alpha_2 + a_3\alpha_3 = (0, 0, 0) \quad \text{or,} \quad a_1(1, 0, 1) + a_2(0, 1, 1) + a_3(1, 1, 1) = (0, 0, 0).$$

That is, if $a_1 + a_3 = 0$, $a_2 + a_3 = 0$, $a_1 + a_2 + a_3 = 0$, then

$$a_1 = a_2 = a_3 = 0.$$

Therefore, the set of vectors $\{\alpha_i\}$, $i = 1, 2, 3$, is **linearly independent**.

But if $\alpha_1 = (1, -1, 1)$, $\alpha_2 = (-1, 1, 1)$ and $\alpha_3 = (1, -1, -1)$, then we can show that the set of vector $\{a_i\}$, $i = 1, 2$, is **linearly dependent**.

Dimension of a vector space : A vector space containing n linearly independent vectors is called **n dimensional.**

If the vector space contains a large number of linearly independent vectors, then it is called **infinite dimensional.**

3.4 A Few Deductions

N.B. (1) If α and β be two linearly independent vectors, then one of them is a scalar multiple of the other.

Proof : We consider two linearly independent vectors α and β in a vector space V. Then $a\alpha + b\beta = 0$, where a and b are two scalars such that $a \neq 0$, $b \neq 0$.

or, $a\alpha = -b\beta$ or, $\alpha = \left(-\frac{b}{a}\right)\beta$ $[\because a \neq 0]$.

Thus, α is a scalar multiple of β.

Also, $b\beta = -a\alpha$ or, $\beta = \left(-\frac{a}{b}\right)\alpha$ $[\because b \neq 0]$.

This shows that β is a scalar multiple of α.

(2) The set of vectors $e_1 = (1, 0, 0, \ldots, 0)$, $e_2 = (0, 1, 0, \ldots, 0) \cdots e_n(0, 0, 0, \ldots, 1)$ defined over the field F is linearly independent. Here 1 denotes unity in the field F.

Proof : Taking $a_i(i = 1, 2, \ldots, n)$ as scalars in the field F,

$$\sum_{i=1}^{n} a_i e_i = 0, \text{ implying that } (a_1, a_2, \ldots, a_n) = (0, 0, \ldots, 0).$$

$\therefore \quad a_1 = a_2 = \cdots = a_n = 0.$

Hence, the set of vectors $e_1, e_2, \ldots, e_n$ is linearly independent.

(3) Any sub-set of a linearly independent set of vectors in a vector space is also linearly independent.

Proof : Let $\{\alpha_i\}$, $i = 1, 2, \ldots, n$ be a set of linearly independent vectors. There exists a set of scalars $a_i(i = 1, 2, \ldots, n)$ such that,

$$\sum_{i=1}^{n} a_i \alpha_i = 0.$$

This implies that $a_i = 0$, $i = 1, 2, \ldots, n$.

Let $\{\alpha_i\}$, $i = 1, 2, \ldots, r$, $(r < n)$ be another set of vectors which must be a subset of $\{\alpha_i\}$, $i = 1, 2, \ldots, n$.

Then $\sum_{i=1}^{r} a_i \alpha_i = 0$, which implies that $a_i(i = 1, 2, \ldots, r)$ is zero.

Hence, **the subset of a linearly independent set of vectors is also linearly independent**.

3.5 Theorems on Linear Dependence

Theorem 3.1 : A set of non-zero vectors in a vector field V is linearly dependent, if and only if some elements of that set is a linear combination of others.

Proof : Let $\{\alpha_i\}$, $i = 1, 2, \ldots, n$ be a set of vectors in the vector space V over the field F, which are linearly dependent. Then there exists a set of scalars $a_i, i = 1, 2, \ldots, n$ in F such that

$$\sum_{i=1}^{n} a_i \alpha_i = 0. \tag{3.13}$$

As the set of vectors is linearly dependent, at least one of the scalars a_i, say $a_k, \neq 0$ $(k < n)$.

Then from equation (3.13),

$$a_k \alpha_k = -\sum_{i=1}^{n} a_i \alpha_i \ (i \neq k). \tag{3.14}$$

But a_k^{-1} exists s $a_k \neq 0$.

$$\therefore \quad a_k = -\sum_{i=1}^{n} (a_i a_k^{-1}) \alpha_i \quad (i \neq k).$$

This shows that **the vector α_k can be expressed as a linear combination of others.**

Conversely, let α_m of the set of vectors $\{\alpha_i\}$, $i = 1, 2, \ldots, n$ which can be expressed as a linear combination of other elements of $\{\alpha_i\}$, i.e.,

$$\alpha_m = \sum_{i=1}^{n} b_i \alpha_i \quad (i \neq m), \tag{3.15}$$

where the scalars b_i's are in F.

Now, from the postulate of vector space,

$$\alpha_m + (-1)\alpha_m = 0. \quad \text{or,} \quad \sum b_i\alpha_i + (-1)\alpha_m = 0.$$

Hence, the set of vectors $\{\alpha_i\}$, $i = 1, 2, \ldots, n$ is linearly dependent.

Theorem 3.2 : A set of non-zero vectors in a vector space is linearly dependent, if and only if one of them be a linear combination of the preceding vectors.

Proof : Let V be a vector space over a field F.

Let the vectors $\{\alpha_i\}$, $i = 1, 2, \ldots, n$ be non-zero and that belonging to V be linearly dependent. Then there exist scalars a_i $(i = 1, 2, \ldots, n)$ in F, not all a_i's are zero such that,

$$\sum_{i=1}^{n} a_i\alpha_i = 0 \tag{3.16}$$

holds good.

Now we assume that the scalars, $a_{k+1}, a_{k+2}, a_{k+3}, \ldots, a_n$ are all zero. Then from eqn. (3.16) we write

$$\sum_{i=1}^{k} a_i\alpha_i + \sum_{i=k+1}^{n} a_i\alpha_i = 0 \tag{3.17}$$

$$\text{or,} \quad a_k\alpha_k + \sum_{i=1}^{k-1} a_i\alpha_i = 0 \quad \left[\because \sum_{i=k+1}^{n} a_i\alpha_i = 0\right]$$

$$\text{or,} \quad \alpha_k = -\sum_{i=1}^{k-1} a_k^{-1} a_i\alpha_i. \tag{3.18}$$

ak^{-1} exists as $a_k \neq 0$.

Eqn. (3.18) shows that α_k is a linear combination of the vectors preceding to it.

Conversely, let a_m be a vector of the set $\{\alpha_i\}$, $i = 1, 2, \ldots, n$, which can be expressed as a linear combination of its preceding vectors.

$$\text{Then,} \quad a_m = \sum_{i=1}^{m-1} b_i\alpha_i, \quad \text{for all } b_i\text{'s in } F.$$

From the postulate of vector space, $\alpha_m + (-1)\alpha_m = 0$

$$\text{or,} \quad \sum_{i=1}^{m-1} b_i\alpha_i + (-1)\alpha_m = 0. \tag{3.19}$$

$$\text{Thus,} \quad \sum_{i=1}^{m-1} b_i\alpha_I + (-1)\alpha_m + \sum_{i=k+1}^{n} a_i\alpha_i = 0.$$

$\therefore$ **the set of vectors $\{\alpha_i\}$, $i = 1, 2, \ldots, n$ is linearly dependent.**

Example 3.1. *Show that the vectors $(-1,2,1)$, $(3,0,-1)$ and $(-5,4,3)$ are linearly dependent in V_3.*

Solution : Let a_1, a_2, a_3 be scalars such that,

$$a_1(-1,2,1) + a_2(3,0,-1) + a_3(-5,4,3) = (0,0,0)$$

or, $\quad (-a_1 + 3a_2 - 5a_3,\ 2a_1 + 4a_3,\ a_1 - a_2 + 3a_3) = (0,0,0)$

i.e., $\quad -a_1 + 3a_2 - 5a_3 = 0, \quad 2a_1 + 4a_3 = 0, \quad a_1 - a_2 + 3a_3 = 0.$

These equations will have non-zero solutions in V_3, if the rank of he coefficient matrix be less than 3, which is the number of unknowns, viz., a_1, a_2, a_3. For rank 3, the only solution will be $a_1 = a_2 = a_3 = 0$.

The coefficient matrix of the above equations is $\begin{bmatrix} -1 & 3 & -5 \\ 2 & 0 & 4 \\ 1 & -1 & 3 \end{bmatrix} = A$ (say).

Then, $|A| = -1(0+4) - 3(6-4) - 5(-2-0) = 0$.

So, the rank is less than 3 and as such the equations will have non-zero solution which is $a = -2$, $b = 1$, $c = 1$. **Hence, the given vectors are linearly dependent.**

Example 3.2. *Show that the three-dimensional vectors $x^1 = (2,0,-2)$, $x^2 = (0,2,0)$, $x^3 = (2,0,2)$ are linearly independent.*

Express $(0,0,3)$ as a linear combination of vectors x^i given above **[BU(H) 1993]**

Solution : Let a_1, a_2, a_3 be three constants such that,

$$a_1x_1 + a_2x_2 + a_3x_3 = 0 \quad \text{or,} \quad a_1(2,0,-2) + a_2(0,2,0) + a_3(2,0,2) = (0,0,0).$$

Then, $\quad 2a_1 + 2a_3 = 0, \quad 2a_2 = 0 \quad \text{and} \quad -2a + 2a_3 = 0.$

or, $\quad a_1 + a_3 = 0, \quad a_2 = 0, \quad a_1 - a_3 = 0.$

Solving we get $a_1 = 0$, $a_2 = 0$, $a_3 = 0$. Hence, the given vectors are linearly dependent.

Again, let for a_1, a_2, a_3,

$$a_1(2,0,-2) + a_2(0,2,0) + a_3(2,0,2) = (0,0,3).$$

$\therefore \quad a_1 + a_3 = 0, \quad a_2 = 0, \quad 2a_3 - 2a_1 = 3.$

Solving we get

$$\boldsymbol{a_1 = -\frac{3}{4}, \ a_2 = 0, \ a_3 = \frac{3}{4}.}$$

$$\therefore \quad \boldsymbol{(0,0,3) = -\frac{3}{4}(2,0,-2) + 0(0,2,0) + \frac{3}{4}(2,0,2).}$$

Example 3.3. *Show that the vectors $(2,-5,3)$ cannot be expressed as a linear combination of the vectors $\alpha_i = (1,-3,2)$, $\alpha_2 = (2,-4,-1)$, $a_3 = (1,-5,7)$.*

Solution : First we assume that the vector $(2,-5,3)$ can be expressed as a linear combination of α_1, α_2, α_3. Then there should exist scalars a_1, a_2, a_3 not all zero, so that

$$a_1\alpha_1 + a_2\alpha_2 + a_3\alpha_3 = (2,-5,3)$$

or, $\quad a_1(1,-3,2) + a_2(2,-4,-1) + a_3(1,-5,7) = (2,-5,3).$

Thus, $a_1 + 2a_2 + a_3 = 2$

$-3a_1 - 4a_2 - 5a_3 = -5$

$2a_1 - a_2 + 7a_3 = 3.$

The matrix of the coefficients is $A = \begin{bmatrix} 1 & 2 & 1 \\ -3 & -4 & -5 \\ 2 & -1 & 7 \end{bmatrix} = 0$

$\therefore \quad |A| = 1[(-4 \times 7) - (-1) \times (-5)] + 2[2(-5) - (-3) \times 7] + 1[(-3)(-1) - 2(-4)]$

$= (-28 - 5) + 2(-10 + 21) + (3 + 8) = -33 + 22 + 11 = -33 + 33 = 0.$

$\therefore |A| = 0$, the solution of the system of equations does not exist. Hence, the given vector cannot be expressed as a linear combination of α_1, α_2 and α_3.

3.6 Basis of a Vector Space

If an arbitrary vector ψ in a vector space V over the field F can be expressed as a linear combination of the vectors $\{\alpha_i\}$ in V so that,

$$\psi = \sum_{i=1}^{n} a_i \alpha_i, \tag{3.20}$$

where a_i are scalars, then α_i is said to span the vector space V.

Thus, a set of linearly independent vectors in V that spans the vector space is called a basis for V. Hence, if there be a subset S of V over F so that, S contains linearly independent vectors, and each vector in V is a linear combination of finite number of elements in S, then S is the basis set or simply basis of V.

Example 3.4. *Let $S = \{(1,0,0),\ (0,1,0),\ (0,0,1)\}$ be a subset in V_3 over the real numbers. The set S is linearly independent. Also ψ spans V_3 because any vector (ψ_1, ψ_2, ψ_3) of V_3 can be expressed as a linear combination of the vectors of ψ, i.e.,*

$$(\psi_1,\ \psi_2,\ \psi_3) = \psi_1(1,0,0) + \psi_3(0,1,0) + \psi_3(0,0,1).$$

Hence, ψ is a basis of V_3 over the real number.

Example 3.5. *Show that the vectors $(1,2,1)$, $(2,1,0)$, $(1,-1,2)$ from a basis of the vector space V_3 over the field of real numbers.* **[V.U.Math(H) 2001]**

Solution : If $V(F)$ be a finite dimensional vector space of dimension n, then any set of n linearly independent vectors in V form is a basis of V.

The set $(1,0,0)$, $(0,1,0)$, $(0,0,1)$ forms a basis of a vector space V_3 over the field of real numbers. Hence, its dimension is 3.

Now, if the set, $S = \{(1,2,1),\ (2,1,0),\ (1,-1,2)\}$ is linearly independent, then S will form a basis of V_3.

Now, the set will be linearly independent, if $a_1(1,2,1) + a_2(2,1,0) + a_3(1,-1,2) = (0,0,0)$, where a_1, a_2, a_3 are scalars, would produce $a_1 = a_2 = a_3 = 0$. The above equation can be written as

$$(a_1 + 2a_2 + a_3,\ 2a_1 + a_2 - a_3,\ a_1 + 2a_3) = (0,0,0).$$

$\therefore \quad a_1 + 2a_2 + a_3 = 0$

$$2a_1 + a_2 - a_3 = 0$$
$$a_1 + 2a_3 = 0.$$

The coefficient matrix is $\begin{bmatrix} 1 & 2 & 1 \\ 2 & 1 & -1 \\ 1 & 0 & 2 \end{bmatrix} = A$ (say).

$|A| = -9 \neq 0$. Thus, the rank of A is 3.

Solving the equations, we get the only solution $a_1 = a_2 = a_3 = 0$.

$\therefore$ the set S is linearly independent and forms a basis of V_3 in the field of real numbers.

Theorem 3.3 : Every vector in a finite dimensional vector space V over the field F can be uniquely expressed as a linear combination of the vectors of its basis.

Proof : Let $S = \{a_i\} = \{\alpha_1, \alpha_2, \dots, \alpha_n\}$ be a basis of the vector space V over F. Then any vector α in V can be expressed as a linear combination of the vectors S. That is,

$$\alpha = \sum_{i=1}^{n} a_i\alpha_i = a_1\alpha_1 + a_2\alpha_2 + \dots + a_n\alpha_n, \tag{3.21}$$

where a_i, $i = 1, 2, \dots, n$, are scalars in F. We have to prove that the above expression is unique. If possible, let b_i, $i = 1, 2, \dots n$, be another set of scalars in F. Then,

$$\alpha = \sum_{i=1}^{n} b_i\alpha_i = b_1\alpha_1 + b_2\alpha_2 + \dots + b_n\alpha_n. \tag{3.22}$$

Subtracting eqn. (3.22) from eqn. (3.21), we get

$$\sum a_i\alpha_i - \sum b_i\alpha_i = \alpha - \alpha = 0 \quad \text{or,} \quad \sum (a_i - b_i)\alpha_i = 0.$$

Since α_i is a set of linearly independent vectors,

$$\therefore \quad a_i - b_i = 0 \quad \text{or,} \quad a_i = b_i.$$

Hence, the expression for α as combination of $\{\alpha_i\}$ is unique.

Theorem 3.4 : If $\{\alpha_i\} = \alpha_1, \alpha_{2j}$ of n independent vectors and β be a non-null vector such that, $\beta = a_1\alpha_1 + a_2\alpha_2 + \dots + a_n\alpha_n$, $a_j \neq 0$, where $a_j \neq 0$ and $a_i (i = 1, 2, \dots n)$ scalars, then the vectors $\alpha_1, \alpha_2, \dots, \alpha_{j-1}, \beta, \alpha_{j+1}, \dots, \alpha_n$ are linearly independent.

Proof : Let $b_1, b_2, \dots, b_n$ be a set of vectors such that

$$b_1\alpha_1 + b_2\alpha_2 + \dots + b_{j-1}\alpha_{j-1} + b_j\beta + b_{j+1}\alpha_{j+1} + \dots + b_n\alpha_n = 0.$$

Hence, $b_1\alpha_1 + \dots + b_{j-1}\alpha_{j-1} + b_j(a_1\alpha_1 + \dots + a_n\alpha_n) + b_{j+1}\alpha_{j+1} + \dots + b_n\alpha_n = 0$

or,
$$(b_1 + b_ja_1)\alpha_1 + (b_2 + b_ja_2)\alpha_2 + \dots + (b_{j-1} + b_ja_{j-1})\alpha_{j-1} + b_ja_j\alpha_j + (b_{j+1} + b_ja_{j+1})\alpha_{j+1} + \dots + (b_n + b_ja_n)\alpha_n = 0.$$

Since, $\alpha_1, \alpha_2, \dots, \alpha_n$ are linearly independent,

$$b_1 + a_1b_j = 0$$
$$b_2 + a_2b_j = 0$$
$$\dots\dots\dots\dots\dots$$
$$b_{j-1} + a_{j-1}b_j = 0$$
$$a_jb_j = 0$$

$$b_{j+1} + a_{j+1}b_j = 0$$
$$\ldots\ldots\ldots\ldots\ldots$$
$$b_n + a_n b_j = 0.$$

From these we have $b_j a_j = 0$. Since $a_j \neq 0$, we get $b_j = 0$. Putting $b_j = 0$ into other equations of the above, wet get $b_1 = b_2 = \cdots = b_n = 0$.

This proves that, $\alpha_1, \alpha_2, \ldots, \alpha_{j-1}, \beta, \alpha_{j+1}, \ldots, \alpha_n$ are linearly independent.

Theorem 3.5 : (The Replacement Theorem) : If $\{\alpha_i\}$, $i = 1, 2, \ldots n$ be a basis of vector space V over the field F and a non-null vector β can be expressed as a linear combination of these vectors in the form :

$$\beta = \sum_{i=1}^{n} a_i \alpha_i = a_1\alpha_1 + a_2\alpha_2 + \cdots + a_n\alpha_n,$$

where a_1, a_2, ..., a_n are scalars in F, then $\{\alpha_1, \alpha_2, \ldots \alpha_{j-1}, \beta, a_{j+1}, \ldots, a_n\}$ is also a basis of V.

Proof : We have $\beta = a_1\alpha_1 + \cdots + a_{j-1}\alpha_{j-1} + a_j\alpha_j + a_{j+1}\alpha_j + \cdots + a_n\alpha_n$.

$$\therefore \quad a_j\alpha_j = \beta - a_1\alpha_1 - \cdots - a_{j-1}\alpha_{j-1} - a_{j+1}\alpha_{j+1} - \cdots - a_n\alpha_n. \tag{3.23}$$

As $a_j \neq 0$, a_j exists, where a_j^{-1} is in F.

Now from eqn. (3.23),

$$\alpha_j = a_j^{-1}\beta - a_j^{-1}a_1\alpha_1 - \cdots - a_j^{-1}a_{j-1}\alpha_{j-1} - a_j^{-1} \cdot a_{j+1}\alpha_{j+1} - \cdots a_j^{-1}a_n\alpha_n.$$

Putting $b_j = a_j^{-1}$ and $b_i = -a_j^{-1}a_i \quad (i \neq j)$, we get

$$\alpha_j = b_1\alpha_1 + b_2\alpha_2 + \cdots + b_{j-1}\alpha_{j-1} + b_j\beta + b_{j+1}\alpha_{j+1} + \cdots + b_n\alpha_n. \tag{3.24}$$

Since, $\{\alpha_1, \alpha_2, \ldots, \alpha_n\}$ is a basis of V and if γ be any arbitrary vector in V over F, then

$$\gamma \sum_{i=1}^{n} c_i\alpha_i = c_1\alpha_1 + c_2\alpha_2 + \cdots + c_n\alpha_n. \tag{3.25}$$

Replacing α_j by eqn. (3.24),

$$\gamma = c_1\alpha_1 + \cdots + c_{j-1}\alpha_{j-1} + c_j(b_1\alpha_1 + \cdots + b_{j-1}\alpha_{j-1} + b_j\beta + b_{j+1}\alpha_{j+1} + \cdots b_n\alpha_n)$$
$$+ c_j + \alpha_{j+1} + \cdots + c_n\alpha_n$$
$$= d_1\alpha_1 + \cdots + d_{j-1}\alpha_{j-1} + d_2\beta + d_{j+1}\alpha_{j+1} + \cdots + d_n\alpha_n,$$

where $\quad d_i = c_i + c_j b_i,\; d_j = c_j b_j \quad (i \neq j)$.

So, $\quad \gamma \in L\{\alpha_1, \alpha_2, \ldots, \alpha_{j-1}, \beta, \alpha_{j+1}, \ldots, \alpha_n\}$.

$$\text{Hence,} \quad V \subseteq L\{\alpha_1,\ \alpha_2, \ldots, \alpha_{j-1}, \ldots, \alpha_n\}. \tag{3.26}$$

Also, $L\{\alpha_1,\ \alpha_2, \ldots, \alpha_{j-1},\ \beta,\ \alpha_{j+1}, \ldots, \alpha_n\}$ is the smallest sub-space containing the set,

$$\{\alpha_1,\ \alpha_2, \ldots, \alpha_{j-1}, \beta, \alpha_{j+1}, \ldots, \alpha_n\}.$$

$$\therefore \quad L\{\alpha_1,\ \alpha_2, \ldots, \alpha_{j-1}, \beta,\ \alpha_{j+1}, \ldots, \alpha_n\} \subseteq V. \tag{3.27}$$

From eqns. (3.26) and (2.27), we get

$$L\{\alpha_1,\ \alpha_2, \ldots, \alpha_{j-1}, \beta, \alpha_{j+1}, \ldots \alpha_n\} = V.$$

Again, from theorem (3.4),

$\alpha_1, \alpha_2, \ldots, \alpha_{j-1}, \ \beta, \alpha_{j+1}, \ldots \alpha_n$ are independent.

Hence, $\{\alpha_1, \alpha_2, \ldots, \alpha_{j-1}, \beta, \alpha_{j+1}, \ldots, \alpha_n = \{\alpha_i\}$, $i = 1, 2, \ldots, n$ **is a basis of** V.

3.7 Coordinates

Let V be a finite-dimensional vector space defined over the field F and let $B = \{\alpha_1, \alpha_2, \ldots, \alpha_n\}$ be an ordered basis of V, whose vectors are enumerated in some well-defined way, i.e., ordered.

Let α be a vector in V. Then there exists a unique n-tuples $(a_1, a_2, \ldots, a_n)$ of scalars such that,

$$\alpha = a_1\alpha_1 + a_2\alpha_2 + \cdots + a_n\alpha_n.$$

This n-tuple $(a_1, a_2, \ldots, a_n)$ is called the **n-tuple of coordinates** of α relative to the ordered basis B.

If the order vector B is changed, then the coordinates of α will also change.

Example Find the coordinates of the vector $\alpha = (2, 3, 1)$ relative to the basis,

$$B = \{(1, 1, 1), \ (1, 1, 0), \ (0, 1, 0)\}.$$

Solution : Let the scalars a_1, a_2, a_3 be such that $a_1(1, 1, 1) + a_2(1, 1, 0) + a_3(0, 1, 0) = (2, 3, 1)$.

This gives $a_1 + a_2 = 2$

$$a_1 + a_2 + a_3 = 0$$

$$a_1 + 0 + 0 = 1.$$

Solving we get, $a_1 = a_2 = a_3 = 1$. Then the coordinates of the given vector are $(1, 1, 1)$.

3.8 Norms and Inner Products

Let V be a vector space defined over the field F of real or complex numbers. The inner product of the pair of vector α and β in V is denoted by.

$$(\alpha, \beta) = a_1b_1 + a_2b_2 + \cdots + a_nb_n,$$

where $\alpha = (a_1, a_2, \ldots, a_n)$ and $\beta = (b_1, b_2, \ldots, b_n)$.

This is known as scalar product.

The inner product has the important properties :

(i) $(\alpha, \beta) = (\beta, \alpha)^*$, α, β in V

(ii) $(\alpha, \beta + \gamma) = (\alpha, \beta) + (\alpha, \gamma)$

(iii) $(\alpha + \beta, \gamma) = (\alpha, \gamma) + (\beta, \gamma)$

and (iv) $(\alpha, \alpha) > 0$, unless $\alpha = 0$.

In (i), $(\beta, \alpha)^*$ denotes the complex conjugate of he number (β, α).

The inner product space V in F is called the **inner product space** or **Euclidean space**, if F be a field R of real numbers. It is called **unitary space**, if F be a field C of complex numbers.

Norm : The **norm or length** of a vector α is defined as the positive square root of the inner product (α, α) and is denoted as $\|\alpha\|$, i.e.,

$$\|\alpha\| = (\alpha, \alpha)^{\frac{1}{2}}. \qquad (3.27)$$

In n-dimensional space, $\alpha = (a_1, a_2, \ldots, a_n)$ and the norm is

$$\|\alpha\| = \left(\sum_{i=1}^{n} a_i^2\right)^{\frac{1}{2}}. \tag{3.28}$$

Examples of inner product :

(i) **Proof of $(\alpha, \beta) = (\beta, \alpha)$**

If R^n be the n-dimensional vector space over the real field R, then for the vectors α and β such that, $\alpha = (a_1, a_2, \ldots, a_n)$, $\beta = (b_1, b_2, \ldots, b_n)$ in R^n, we define

$$(\alpha, \beta) = \{(a_1, a_2, \ldots, a_n),\ (b_1, b_2, \ldots, b_n)\} = a_1b_1 + a_2b_2 + \cdots + a_nb_n.$$

This is the **standard inner product** or the det product of the vectors α and β.

Now, $(\alpha, \beta) = a_1b_1 + a_2b_2 + \cdots + a_nb_n = b_1a_1 + b_2a_2 + \cdots + b_na_n$

$$= (\beta, \alpha) \text{ (symmetry)} \tag{3.29}$$

(ii) **Proof of $(a\alpha + b\beta, \gamma) = a(\alpha, \gamma) + b(\beta, \gamma)$**

Let $\gamma = (c_1, c_2, \ldots, c_n)$ in R^n and let a, b in R.

Then, $(a\alpha + b\beta) = a(a_1, a_2, \ldots, a_n) + b(b_1, b_2, \ldots, b_n) = (aa_1 + bb_1, \ldots, aa_n + bb_n)$

Hence, $(a\alpha + b\beta, \gamma) = (aa_1 + bb_1)c_1 + \cdots + (aa_n + bb_n)c_n$ (by symmetry)

$$= (aa_1c_1 + \cdots + aa_nc_n) + \cdots + (bb_1c_1 + \cdots + bb_nc_n)$$

$$= a(\alpha, \gamma) + b(\beta, \gamma), \text{ (linearly)}. \tag{3.30}$$

(iii) **Proof of $(\alpha, \alpha) \geq 0$**

We have $(\alpha, \alpha) = a_1a_1 = a_2a_2 + \cdots + a_na_n = |a_1|^2 + |a_2|^2 + \cdots + |a_n|^2$

$= $ sum of n non-negative real numbers.

$$\therefore \quad (\alpha, \alpha) \geq 0. \tag{3.31}$$

Also, $(\alpha, \alpha) = 0$ or, $|a_1|^2 + |a_2|^2 + \cdots + |a_n|^2 = 0$

or, each $|a_1|^2 = 0$ and so $a_1 = 0$ or, $\alpha = 0$ (non-negatively).

(iv) **Proof of $(\alpha, \beta + \gamma) = (\alpha, \beta) + (\alpha, c)$**

Let $\alpha = (a_1, a_2, \ldots, a_n)$, $\beta = (b_1, b_2, \ldots b_n)$ and $\gamma = (c_1, c_2, \ldots, c_n)$; then,

$$\beta + \gamma = (b_1 + c_1,\ b_2 + c_2, \ldots, b_n + c_n)$$

and $(\alpha, \beta + \gamma) = a_1(b_1 + c_1) + a_2(b_2 + c_2) + \cdots + a_n(b_n + c_n)$

$$= (a_1b_1 + a_1c_1) + (a_2b_2 + a_2c_2) + \cdots + (a_nb_n + a_nc_n)$$

$$= (\alpha, \beta) + (\alpha, c). \tag{3.32}$$

Similarly, we can prove that, $(\alpha + \beta, \gamma) = (\alpha, \gamma) + (\beta, \gamma)$.

3.9 Important Theorems

Theorem 3.6 : If V be an inner product space over a field F, then

(i) $(a\alpha - b\beta, \gamma) = a(\alpha, \gamma) - b(\beta, \gamma)$ (ii) $(\alpha, a\beta + b\gamma) = a^*(\alpha, \beta) + b^*(\alpha, \gamma)$

where α, β, γ are in V, a, b are in F, a^*, b^* are complex conjugate of a, b.

Proof : (i) We have $(a\alpha - b\beta, \gamma) = \{a\alpha + (-b\beta), \gamma\} = a(\alpha,\gamma) + (-b)(\beta,\gamma)$

$$= a(\alpha,\gamma) - b(\beta,\gamma). \quad (3.33)$$

(ii) We have $(\alpha, a\beta + b\gamma) = (a\beta + b\gamma, \alpha)^*$ [by property (i)]

$$= [a(\beta,\alpha) + b(\gamma,\beta)]^* = a^*(\beta,\alpha)^* + B^*(\gamma,\alpha)^*$$

$$= a^*(\alpha,\beta) + b^*(\alpha,\gamma). \quad (3.34)$$

Theorem 3.7 : If α and β be any two vectors in an inner product space V over field F, then

$$|(\alpha,\beta)| \leq \|\alpha\|\ \|\beta\|, \quad (3.35)$$

where $|(\alpha,\beta)|$ denotes the modulus of the numbers (α,β), real or complex. This is the **Schwarz's inequality**.

Proof : Let $\alpha = 0$, then $\|\alpha\| = 0$.

In this case, $(\alpha,\beta) = (0,\beta) = (00,\beta) = 0(0,\beta) = 0$.

$\therefore \quad |(\alpha,\beta)| = 0$.

Thus, if $\alpha = 0$, both sides being zero, the inequality holds good.

Now let $\alpha = 0$.

Then $\dfrac{1}{\|\alpha\|^2}$ is a positive real number, since $\|a\| > 0$, for $\alpha \neq 0$.

We now consider the vector, $\gamma = \beta - \dfrac{(\beta,\alpha)}{\|\alpha\|^2}\alpha$.

$$\text{Then, } (\gamma,\gamma) = \left[\beta - \frac{(\beta,\alpha)}{\|\alpha\|^2}\alpha,\ \beta - \frac{(\beta-\gamma)}{\|\alpha\|^2}\alpha\right]$$

$$= \left[\beta, \beta - \frac{(\beta,\alpha)}{\|\alpha\|^2}\alpha\right] - \frac{(\beta,\alpha)}{\|\alpha\|^2}\left[\alpha, \beta - \frac{(\beta,\alpha)}{\|\alpha\|^2}\right] \text{ (linearly)}$$

$$= (\beta,\beta) - \frac{(\beta,\alpha)^*}{\|\alpha\|^2}(\beta,\alpha) - \frac{(\beta,\alpha)}{\|\alpha\|^2}(\alpha,\beta) + \frac{(\beta,\alpha)(\beta,\alpha)^*}{\|\alpha\|^2\|\alpha\|^2}(\alpha,\alpha)$$

$$= \|\beta\|^2 - \frac{(\beta,\alpha)^*(\beta,\alpha)}{\|\alpha\|^2} - \frac{(\alpha,\beta)^*(\alpha,\beta)}{\|\alpha\|^2} + \frac{(\beta,\alpha)^*(\beta,\alpha)}{\|\alpha\|^2\|\alpha\|^2}\|\alpha\|^2$$

$$= \|\beta\|^2 - \frac{(\alpha,\beta)(\alpha,\beta)^*}{\|\alpha\|^2}$$

$$= \|\beta\|^2 - \frac{|(\alpha,\beta)|^2}{\|\alpha\|^2}, \quad \because (\alpha,\beta),\ (\alpha,\beta)^* \text{ are conjugate complex.}$$

Again, $(\gamma,\gamma) = \|\gamma\|^2 \geq 0$.

$$\therefore \quad \|\beta\|^2 - \frac{|(\alpha,\beta)|^2}{\|\alpha\|^2} \geq 0 \quad \text{or,} \quad |(\alpha,\beta)|^2 \leq \|\beta\|^2\ \|\alpha\|^2$$

or, $\quad |\alpha,\beta| \leq \|\beta\|\ \|\alpha\|$.

This is **Schwarz's inequality**.

Theorem 3.8 : If α and β be any two vectors in an inner product space V, then,

$$\|\alpha + \beta\| \leq \|\alpha\| + \|\beta\|$$

[BU(H) 1996]

Proof : From the definition of norm, we have

$$\|\alpha+\beta\|^2 = (\alpha+\beta,\ \alpha+\beta) = (\alpha,\ \alpha+\beta+\beta,\ \alpha+\beta),\ \text{(linearity)}$$
$$= (\alpha,\alpha)+(\alpha,\beta)+(\beta,\alpha)+(\beta,\beta)$$
$$= \|\alpha\|^2+(\alpha,\beta)+(\alpha,\beta)^*+\|\beta\|^2 \quad [\because\ (\beta,\alpha)=(\alpha,\beta)^*]$$
$$= \|\alpha\|^2+2\text{Re}\,(\alpha,\beta)+\|\beta\|^2$$
$$\leq \|\alpha\|^2+2|\alpha,\beta|+\|\beta\|^2 \quad [\because\ \text{Re}\,(\alpha,\beta)\leq|\alpha,\beta|]$$
$$\leq \|\alpha\|^2+2\|\alpha\|\cdot\|\beta\|+\|\beta\|^2 \quad \text{[by Schwarz's inequality]}$$
$$\leq (\|\alpha\|+\|\beta\|)^2.$$

$$\therefore\quad \|\alpha+\beta\| \leq \|\alpha\|+\|\beta\|.$$

This is known as **triangle inequality**.

Example 3.6. *If α and β be vectors in an inner product space, then show that*

$$\|\alpha+\beta\|^2+\|\alpha-\beta\|^2 = 2\|\alpha\|^2+2\|\beta\|^2.$$

Solution : We have $\|\alpha+\beta\|^2 = (\alpha+\beta,\ \alpha+\beta)$ by definition

$$= (\alpha,\alpha+\beta)+(\beta,\alpha+\beta)\quad \text{(linearity)}$$
$$= (\alpha,\alpha)+(\alpha,\beta)+(\beta,\alpha)+(\beta,\beta)$$
$$= \|\alpha\|^2+(\alpha,\beta)+(\beta,\alpha)+|\beta\|^2.$$

Also, $\|\alpha-\beta\|^2 = (\alpha-\beta,\ \alpha-\beta) = (\alpha,\ \alpha-\beta)-(\beta,\alpha-\beta)$

$$= (\alpha,\alpha)-(\alpha,\beta)-(\beta,\alpha)+(\beta,\beta) = |\alpha\|^2-(\alpha,\beta)-(\beta,\alpha)+|\beta\|^2.$$

$$\therefore\quad \|\alpha+\beta\|^2+\|\alpha-\beta\|^2 = 2\|\alpha\|^2+2\|\beta\|^2.$$

Example 3.7. *If in an inner product space $\|\alpha+\beta\| = \|\alpha\|+\|\beta\|$, then show that α and β are linearly dependent.*

Solution : We have $\|\alpha+\beta\| = \|\alpha\|+\|\beta\|$.

$$\therefore\quad \|\alpha+\beta\|^2 = \|\alpha\|^2+\|\beta\|^2+2\|\alpha\|\ \|\beta\|.$$

Thus, $\|\alpha\|^2+2\text{Re}\,(\alpha,\beta)+\|\beta\|^2 = \|\alpha\|^2+\|\beta\|^2+2\|\alpha\|\ \|\beta\|$.

This shows that

$$\text{Re}\ (\alpha,\beta) = \|\alpha\|\ \|\beta\|.$$

$$\therefore\quad \|\alpha\|\ \|\beta\| \leq |\alpha,\beta|.$$

But by Schwarz's inequality, we have

$$\|\alpha\|\ \|\beta\| \geq |\alpha,\beta|.$$

From the last two inequalities, we write

$$\|\alpha\|\ \|\beta\| = |(\alpha,\beta)|. \tag{3.36}$$

This equality shows that α and β are linearly dependent.

3.10 Orthogonal Set of Vectors

Two vectors α, β in an inner-product space are said to be orthogonal if

$$(\alpha, \beta) = 0. \tag{3.37}$$

Let S be a set of vectors in an inner-product space. If any two distinct vectors in S be orthogonal, then the set is orthogonal.

Suppose, α and β are two orthogonal vectors in a Euclidean space, then

$$\begin{aligned}\|\alpha + \beta\|^2 &= (\alpha + \beta,\ \alpha + \beta) = (\alpha, \alpha) + 2(\alpha, \beta) + (\beta, \beta)\\ &= \|\alpha\|^2 = \|\beta\|^2 \qquad [\because\ (\alpha, \beta) = 0].\end{aligned}$$

This is called **Pythagorus theorem.**

N.B. Orthogonal set of non-null vectors is linearly independent.

3.11 Orthonormal Set of Vectors

An **orthonormal** set of vectors ns an **orthogonal** set of vectors in which each element has unit norm. Thus, a set $\{\alpha_i\}$ of vectors in an inner-product space is orthonormal if

(i) $\|\alpha_i\|^2 = 1$ and

(ii) $(\alpha_i, \alpha_j) = 0$ for $i \neq j$, i.e., $(\alpha_i, \alpha_j) = \delta_{ij}$,

$$\text{where} \quad \delta_{ij} = \begin{cases} 1, & \text{for } i = j \\ 0, & \text{for } i \neq j. \end{cases} \tag{3.38}$$

δ_{ij} is the **Kronecker delta.**

So, if $\{\beta_i\}$, $i = 1, 2, \ldots, m$, be a set of non-null orthogonal vectors, then the set of vectors $\{\gamma_i\}$, $i = 1, 2, \ldots, m$, will be orthonormal if

$$\gamma_i = \frac{\beta_i}{\|\beta_i\|}. \tag{3.39}$$

N.B. Any orthonormal set of vectors in an inner-product space is linearly independent.

3.12 Gram-Schmidt Process of Orthonormalisation

An orthonormal set $\{\alpha_1, \alpha_2, \ldots, \alpha_n\}$ can be constructed from a linearly independent set $\{\psi_1, \psi_2, \ldots, \psi_n\}$. The set $\{\psi_1, \psi_2, \ldots, \psi_n\}$ being linearly independent, none of the vectors is null vector.

Let us define, $\gamma_1 = \psi_1$.

$$\text{Then,} \quad \alpha_1 = \frac{\gamma_1}{\|\gamma_1\|} = \frac{\psi_1}{\|\psi_1\|} \text{ is of unit norm.} \tag{3.40}$$

$$\text{Now we define,} \quad \gamma_2 = \psi_2 - (\alpha_1, \psi_2)\alpha_1. \tag{3.41}$$

Then $\alpha_2 = \dfrac{\gamma_2}{\|\gamma_2\|}$ is of unit norm.

$$\text{Again, we define,} \quad \gamma_3 = \psi_3 - (\alpha_1, \psi_3)\alpha_1 - (\alpha_2, \psi_3)\alpha_2. \tag{3.42}$$

Then $\alpha_3 = \dfrac{\gamma_3}{\|\gamma_3\|}$ is of unit norm.

Thus, in general, $\gamma_i = \psi_i - \sum_{k=1}^{i-1} (a_k, \psi_i) a_k, \; i > 1$ (3.43)

Then $\alpha_i = \dfrac{\gamma_i}{\|\gamma_i\|}$ is of unit norm.

So, as long as $\psi_i \neq 0$, the continued process will give vectors of unit norm.

Again, $(\alpha_2, \alpha_1) = \dfrac{1}{\|\gamma_2\|}(\gamma_2, \alpha_1) = \dfrac{1}{\|\gamma_2\|}|(\psi_2, \alpha_1) - (\alpha_1, \psi_2) \cdot (\alpha_1, \alpha_1)| = 0.$

Thus, $\boldsymbol{\alpha_2}$ **is orthonormal to** $\boldsymbol{\alpha_1}$.

Likewise we can prove that $\boldsymbol{(\alpha_3, \alpha_2) = 0}$ **and** $\boldsymbol{(\alpha_3, \alpha_1) = 0}$.

Example 3.8. *Show that the set of vectors*

(i) $S = \{(3,0,4),\ (-4,0,3),\ (0,9,0)\}$ is orthogonal and

(ii) $S_1 = \left\{\frac{1}{5}(3,0,4),\ \frac{1}{5}(-4,0,3),\ (0,1,0)\right\}$ is orthonormal.

Solution : (i) Let $S = \{\alpha_1, \alpha_2, \alpha_3\}$, then,

$$(\alpha_1, \alpha_2) = -12 + 12 = 0, \quad (\alpha_2, \alpha_3) = 0 \quad \text{and} \quad (\alpha_3, \alpha_1) = 0.$$

As $(\alpha_1, \alpha_1) \neq 1$, the set S is **orthogonal**.

(ii) Again let $S_1 = \{\beta_1, \beta_2, \beta_3\}$, then

$$(\beta_i, \beta_j) = 0 \quad \text{for } i \neq 0 \quad \text{and} \quad i, j = 1, 2, 3.$$

But $(\beta_1, \beta_1) = 1$, $(\beta_2, \beta_2) = 1$ and $(\beta_3, \beta_3) = 1$.

Hence, the set $\boldsymbol{S_1}$ **is orthonormal.**

Example 3.9. *Given,* $\psi_1 = \begin{bmatrix} 1 \\ 1 \\ 1 \end{bmatrix}$, $\psi_2 = \begin{bmatrix} 1 \\ 0 \\ 1 \end{bmatrix}$, $\psi_3 = \begin{bmatrix} -1 \\ -2 \\ 1 \end{bmatrix}$ *; construct an orthonormal set.*

Solution : We have $\psi_1 = \begin{bmatrix} 1 \\ 1 \\ 1 \end{bmatrix}$;

$$\therefore \quad \|\psi_1\| = \sqrt{3}, \quad \text{so, } \alpha_1 = \frac{1}{\sqrt{3}} \begin{bmatrix} 1 \\ 1 \\ 1 \end{bmatrix}.$$

$$\gamma_2 = \psi_2 - (\alpha_1, \psi_2)\alpha_1 = \begin{bmatrix} 1 \\ 0 \\ 1 \end{bmatrix} - \begin{bmatrix} \frac{1}{\sqrt{3}} \\ \frac{1}{\sqrt{3}} \\ \frac{1}{\sqrt{3}} \end{bmatrix} \begin{bmatrix} 1 \\ 0 \\ 1 \end{bmatrix} \alpha_1 = \begin{bmatrix} 1 \\ 0 \\ 1 \end{bmatrix} - \frac{2}{\sqrt{3}} \cdot \frac{1}{\sqrt{3}} \begin{bmatrix} 1 \\ 1 \\ 1 \end{bmatrix}$$

$$= \begin{bmatrix} 1 \\ 0 \\ 1 \end{bmatrix} - \frac{2}{3} \begin{bmatrix} 1 \\ 1 \\ 1 \end{bmatrix} = \frac{1}{3} \begin{bmatrix} 1 \\ -1 \\ 1 \end{bmatrix}$$

$$\therefore \quad \|\gamma_2\| = \frac{1}{\sqrt{3}}; \quad \text{so,} \quad \alpha_2 = \frac{1}{\sqrt{3}} \begin{bmatrix} 1 \\ -1 \\ 1 \end{bmatrix}.$$

Again, $\gamma_3 = \psi_3 - (\alpha_1, \psi_3)\alpha_1 - (\alpha_2, \psi_3)\alpha_2$.

$$(\alpha_1, \psi_3) = \begin{bmatrix} \frac{1}{\sqrt{3}} \\ \frac{1}{\sqrt{3}} \\ \frac{1}{\sqrt{3}} \end{bmatrix} \cdot \begin{bmatrix} -1 \\ -2 \\ 1 \end{bmatrix} = -\frac{2}{\sqrt{3}}.$$

$$\therefore \quad (\alpha_1, \psi_3)\alpha_1 = -\frac{2}{\sqrt{3}} \begin{bmatrix} \frac{1}{\sqrt{3}} \\ \frac{1}{\sqrt{3}} \\ \frac{1}{\sqrt{3}} \end{bmatrix} = -\frac{2}{3} \begin{bmatrix} 1 \\ 1 \\ 1 \end{bmatrix}.$$

Again, $(\alpha_2, \psi_3) = \dfrac{1}{\sqrt{3}} \begin{bmatrix} -1 \\ -1 \\ 1 \end{bmatrix} \cdot \begin{bmatrix} -1 \\ -2 \\ 1 \end{bmatrix} = \dfrac{2}{\sqrt{3}}.$

$$\therefore \quad (\alpha_2, \psi_3)\alpha_2 = \frac{1}{\sqrt{3}} \cdot \frac{2}{\sqrt{3}} \begin{bmatrix} 1 \\ -1 \\ 1 \end{bmatrix} = \frac{2}{3} \begin{bmatrix} 1 \\ -1 \\ 1 \end{bmatrix}.$$

So, $\gamma_3 = \begin{bmatrix} -1 \\ -2 \\ 1 \end{bmatrix} + \dfrac{2}{3} \begin{bmatrix} 1 \\ 1 \\ 1 \end{bmatrix} - \dfrac{2}{3} \begin{bmatrix} 1 \\ -1 \\ 1 \end{bmatrix} \begin{bmatrix} -1 \\ -\frac{2}{3} \\ 1 \end{bmatrix}.$

$$\therefore \quad \|\gamma_3\| = \frac{\sqrt{13}}{3}. \quad \text{So,} \quad \alpha_3 = \frac{\gamma_3}{\|\gamma_3\|} = \frac{3}{\sqrt{13}} \begin{bmatrix} -1 \\ -\frac{2}{3} \\ 1 \end{bmatrix}.$$

So, the required orthonormal set is

$$(\alpha_1, \alpha_2, \alpha_3) = \left\{ \frac{1}{\sqrt{3}} \begin{bmatrix} 1 \\ 1 \\ 1 \end{bmatrix} \cdot \frac{1}{\sqrt{3}} \begin{bmatrix} 1 \\ -1 \\ 1 \end{bmatrix}, \frac{2}{\sqrt{13}} \begin{bmatrix} -1 \\ -\frac{2}{3} \\ 1 \end{bmatrix} \right\}.$$

3.13 Operator

An operator is a device for obtaining a vector V_i in vector space V_n for each $\vec{U}_i$ in V_n.

If $\hat{A}$ be an operator, then $\hat{A}\vec{U}_i = \vec{V}_i$. (3.44)

Linear operator : An operator is said to be linear if

$$\hat{A}(\alpha\vec{U} + \beta\vec{V}) = \alpha\hat{A}\vec{U} + \beta\hat{A}\vec{V}, \tag{3.45}$$

where α and β are scalars.

We consider the operator $A(x, y) = (x, y)$. Then, for any scalars α and β,

$$A[\alpha(x_1, y_1) + \beta(x_2, y_2)] = A[\alpha x_1 + \beta x_2,\ \alpha y_1 + \beta y_2] = (\alpha x_1 + \beta x_2,\ \alpha y_1 + \beta y_2)$$
$$= \alpha(x_1, y_1) + \beta(x_2, y_2) = \alpha A(x_1, y_1) + \beta A(x_2, y_2).$$

Hence, the operator $\hat{A}$ is **linear**.

Unit operator ($\hat{1}$) : The operator which maps any vector $\vec{U}$ to itself is called the unit operator, i.e., $\hat{1}\,\vec{U} = \vec{U}$.

Now, $\hat{1}(\alpha\vec{U}+\beta\vec{V})=\alpha\hat{1}\,\vec{U}+\beta\hat{1}\,\vec{V}=\alpha\vec{U}+\beta\vec{V}$. (3.46)

This shows that the unit operator $\hat{1}$ is **linear**.

Null operator ($\hat{0}$) : An operator which maps any vector $\vec{U}$ into a null vector ($\vec{0}$) is called a null operator, i.e., $\hat{0}\,\vec{U}=\vec{0}$.

Also, $\hat{0}(\alpha\vec{U}+\beta\vec{V})=\vec{0}=\alpha\hat{0}\,\vec{U}+\beta\hat{0}\,\vec{V}$. (3.47)

Hence, a null operator is also a **linear operator**.

3.14 Linear Differential Operators

The operators in terms of derivatives of a function is called a differential operator.

Example The expression, $L(u)\equiv\dfrac{d^2u}{dx^2}+p(x)\dfrac{du}{dx}+q(x)u$ (3.48)

always defines a function $R(x)$ for each permissible input function $u(x)$. Hence, L is a differential operator.

Let $u_1(x)$ and $u_2(x)$ be two functions of x and a_1, a_2 two scalars, then we can show that, $L[a_1u_1(x)+a_2u_2(x)]$ is a linear operator.

Proof : We have $L[a_1u_1(x)+a_2u_2(x)]=\dfrac{d^2}{dx^2}[a_1u_1(x)+a_2u_2(x)+p(x)\dfrac{d}{dx}[a_1u_1(x)$

$$+a_2u_2(x)]+q(x)[a_1u_1(x)+a_2u_2(x)]$$

$$=a_1\left[\frac{d^2u_1}{dx^2}+p(x)\frac{du_1}{dx}+q(x)u_1\right]+a_2\left[\frac{d^2u_2}{dx^2}+p(x)\frac{du_2}{dx_2}+q(x)u_2\right]$$

$$=a_1L(u_1)+a_2L(u_2). \tag{3.49}$$

Eqn. (3.49) shows that L is a **linear operator**.

N.B. From eqn. (3.49), we get

(i) $L[u+v]=L[u]+L[v]$ (ii) $L[\gamma u]=\gamma L[u]$, where γ is a constant.

3.15 Algebra of Linear Operators

We consider two linear operators $\widehat{A}$ and $\widehat{B}$ in the vector space V.

(i) Sum of operators : The sum or difference of the operators $\widehat{A}$ and $\widehat{B}$ is represented by $\widehat{A}\pm\widehat{B}$ and is defined as

$$(A\pm B)\psi=A\psi\mp B\psi, \tag{3.50}$$

where ψ is any vector in vector space V.

(ii) Product of operators : The product of the operators $\widehat{A}$ and $\widehat{B}$ is denoted by $\widehat{A}\,\widehat{B}$ and is defined as

$$\widehat{A}\,\widehat{B}\psi=\widehat{A}(\widehat{B}\psi). \tag{3.51}$$

Generally, $\widehat{A}\widehat{B}\neq\widehat{B}\widehat{A}$ does not commute.

$(\widehat{A}\,\widehat{B}-\widehat{B}\,\widehat{A})$ is called the commutator of $\widehat{A}$ and $\widehat{B}$ and symbolically,

$$[\widehat{A},\widehat{B}]=\widehat{A}\,\widehat{B}-\widehat{B}\,\widehat{A}. \tag{3.52}$$

If $\widehat{A}\,\widehat{B}-\widehat{B}\,\widehat{A}=0$, then the operators $\widehat{A}$ and $\widehat{B}$ commute each other.

N.B. If $\widehat{A}$ and $\widehat{B}$ are linear, then the product $\widehat{A}\,\widehat{B}$ is also a linear operator.

(iii) Inverse operator : Let ϕ and ψ be two vectors in vector space V, then

$$\widehat{A}\phi = \psi. \tag{3.53}$$

The inverse operator of $\widehat{A}$ is $\widehat{A}^{-1}$ and

$$\widehat{A}^{-1}\varphi = \phi. \tag{3.54}$$

It can be shown that A^{-1} is linear if $\widehat{A}$ is linear.

Proof : If $\widehat{A}$ is a linear operator, then

$$\widehat{A}(\alpha_1\phi_1 + \alpha_2\phi_2) = \alpha_1\widehat{A}\phi_1 + \alpha_2\widehat{A}\phi_2.$$

By definition,

$$\alpha_1\phi_1 + \alpha_2\phi_2 = \widehat{A}^{-1}(\alpha_1\widehat{A}\phi_1 + \alpha_2\widehat{A}\phi_2). \tag{3.55}$$

Now, corresponding to vectors ϕ_1 and ϕ_2, let there be two vectors ψ_1 and ψ_2 such that,

$$\widehat{A}\phi_1 = \psi_1 \quad \text{and} \quad \widehat{A}\phi_2 = \varphi_2.$$

$$\therefore \quad \phi_1 = \widehat{A}^{-1}\psi_1 \quad \text{and} \quad \phi_2 = \widehat{A}^{-1}\varphi_2.$$

So LHS of Eqn. 3.55 is

$$\alpha_1\phi_1 + \alpha_2\phi_2 = \alpha_1\widehat{A}^{-1}\psi_1 + \alpha_2\widehat{A}^{-1}\psi_2$$

and RHS of Eqn. 3.55 is

$$\widehat{A}^{-1}(\alpha_1\psi_1 + \alpha_2\psi_2).$$

$$\therefore \quad \widehat{A}^{-1}(\alpha_1\psi_1 + \alpha_2\psi_2) = \alpha_1\widehat{A}^{-1}\psi_1 + \alpha_2\widehat{A}^{-1}\psi_2.$$

Hence, $\widehat{A}^{-1}$ is also a linear operator.

N.B. (i) $\widehat{A}(A^{-1}\phi) = \phi$

Let $\widehat{A}\psi = \phi, \quad \therefore \quad \widehat{A}^{-1}\phi = \psi, \quad \text{then } \widehat{A}(\widehat{A}^{-1}\phi) = \widehat{A}\psi = \phi.$

(ii) $(AB)^{-1} = B^{-1}\,A^{-1}$

Let $(\widehat{A}\widehat{B})^{-1}\phi = \psi, \quad \therefore \quad \phi = (\widehat{A}\widehat{B})\psi.$

Now, $B^{-1}A^{-1}\phi = B^{-1}A^{-1}\widehat{A}\,\widehat{B}\phi = f$ (say).

$\therefore \quad A^{-1}\widehat{A}\widehat{B}\psi = \widehat{B}f \quad \text{or,} \quad \widehat{A}\widehat{B}\psi = \widehat{A}(\widehat{B}f) = \widehat{A}\widehat{B}f.$

$\therefore \quad \psi = f.$

So, $B^{-1}A^{-1}\phi = \psi, \quad \therefore \quad (\widehat{A}\widehat{B})^{-1}\phi = \widehat{B}^{-1}\widehat{A}^{-1}\phi,$

i.e.,

$$(\widehat{\boldsymbol{A}}\widehat{\boldsymbol{B}})^{-1} = \widehat{\boldsymbol{B}}^{-1}\widehat{\boldsymbol{A}}^{1}. \tag{3.56}$$

Example 3.10. *If $\widehat{A}$ and $\widehat{B}$ be the linear operators in the vector space V, then show that their product $\widehat{A}\widehat{B}$ is also a linear operator.*

Solution : If a and b any two scalars, then

$$\widehat{A}\widehat{B}(a\alpha + b\beta) = \widehat{A}[\widehat{B}(a\alpha + b\beta] = \widehat{A}[a\widehat{B}\alpha + b\widehat{B}\beta], \quad [\because \widehat{B} \text{ is linear}]$$

$$= a\widehat{A}(\widehat{B}\alpha) + b\widehat{A}(\widehat{B}\beta) = a\widehat{A}\widehat{B}(\alpha) + b\widehat{A}\widehat{B}(\beta). \tag{3.57}$$

This shows that $\widehat{A}\widehat{B}$ is a linear operator.

3.16 Matrix Representation of Linear Operators

We consider two vector spaces V and W over the field F. Let the vectors X and Y be in V and W respectively. Let $\widehat{A}$ be a linear operator such that $\widehat{A}: V \to W$.

If $X: \{x_1, x_2, \ldots, x_n\}$ and $Y: \{y_1, y_2, \ldots, y_n\}$ be two sets of independent vectors, then for every vector $x_i (i = 1, 2, \ldots, n)$ in V, $\widehat{A}x_i$ will be in W. As $(y_1, y_2, \ldots, y_n)$ are the basis vectors, we have

$$\widehat{A}x_i = a_{1i}y_1 + a_{2i}y_2 + \cdots + a_{mi}y_m. \tag{3.58}$$

For any vector $x = b_1x_1 + b_2x_2 + \cdots + b_nx_n$ in V, there corresponds y in W, such that,

$$\begin{aligned} y = \widehat{A}x = b_1\widehat{A}x_1 + b_2\widehat{A}x_2 + \cdots + b_n\widehat{A}x_n &= b_1(a_{11}y_1 + \cdots + a_{m1}y_m) \\ &\quad + b_2(a_{12}y_1 + \cdots + a_{m2}y_m) + \cdots + b_n(a_{1n}y_1 + \cdots + a_{mn}y_m) \\ &= b_1a_{11} + b_2a_{12} + \cdots + b_na_{1n})y_1 + \cdots + (b_1a_{m1} + b_2a_{m2} + \cdots + b_na_{mn})y_m \\ &= c_1y_1 + c_2y_2 + \cdots + c_my_m, \end{aligned}$$

where $c_i = b_1a_{i1} + b_2a_{i2} + \cdots + b_na_{in} \quad (i = 1, 2, \ldots, m)$. (3.59)

Eqn. (3.59) can be expressed in matrix form as

$$\begin{bmatrix} C_1 \\ C_2 \\ \vdots \\ C_m \end{bmatrix} = \begin{bmatrix} a_{11} & a_{12} & \cdots & a_{1n} \\ a_{21} & a_{22} & \cdots & a_{2n} \\ \cdots & \cdots & \cdots & \cdots \\ a_{m1} & a_{m2} & \cdots & a_{mn} \end{bmatrix} \begin{bmatrix} b_1 \\ b_2 \\ \vdots \\ b_n \end{bmatrix}. \tag{3.60}$$

In conclusion, any linear operator $\widehat{A}: V \to W$, V and W having bases X and Y respectively, any $y = \widehat{A}x$ in W can be determined through the matrix $A = (a_{ij})$. The elements a_{ij} of A can be uniquely determined through equation (3.58) with the help of the bases X and Y and the coefficient vector $u = (b_1b_2b_3 \cdots b_n)^T$ of x. After knowing A and u the vector $v = (c_1c_2 \cdots c_m)^T$ of y can be determined by equation (3.60). **Hence, the computation of linear transformation $y = \widehat{A}x$ is nearly similar to the matrix relation $v = Au$.**

Example 3.11. *The vectors $x_1 = (1, 1, 0)$, $x_2 = (1, 0, 1)$, $x_3 = (0, 1, 1)$ form a basis of R^3 and $\widehat{A}$ is an operator such that, $\widehat{A}: R^3 \to R^2$ and $\widehat{A}x_1 = (-1, 0)$, $\widehat{A}x_2(0, 1)$, $\widehat{A}x_3 = (1, -1)$, find $\widehat{A}y$. [R^3 represents a vector space of real numbers in three dimensions]*

Solution : Let $y = (y_1, y_2, y_3)$ be a vector in R^3 and α, β, γ any three scalars, then

$$y = \alpha x_1 + \beta x_2 + \gamma x_3 = \alpha(1, 1, 0) + \beta(1, 0, 1) + \gamma(0, 1, 1) = [\alpha\ \beta\ \gamma]\begin{bmatrix} 1 & 1 & 0 \\ 1 & 0 & 1 \\ 0 & 1 & 1 \end{bmatrix}.$$

If $$A = \begin{bmatrix} 1 & 1 & 0 \\ 1 & 0 & 1 \\ 0 & 1 & 1 \end{bmatrix}, \quad \text{then } A^{-1} = \frac{1}{2}\begin{bmatrix} 1 & 1 & -1 \\ 1 & -1 & 1 \\ -1 & 1 & 1 \end{bmatrix}.$$

Then $$[\alpha\ \beta\ \gamma] = \frac{1}{2}[y_1\ y_2\ y_3]\begin{bmatrix} 1 & 1 & -1 \\ 1 & -1 & 1 \\ -1 & 1 & 1 \end{bmatrix}.$$

$$\therefore \quad \alpha = \frac{1}{2}(y_1 + y_2 - y_3), \quad \beta = \frac{1}{2}(y_1 - y_2 + y_3), \quad \gamma = \frac{1}{2}(-y_1 + y_2 + y_3).$$

$$\therefore \quad \widehat{A}y = \widehat{A}(\alpha x_1 + \beta x_2 + x_3) = \alpha\widehat{A}x_1 + \beta\widehat{A}x_2 + \gamma\widehat{A}x_3 = \alpha(-1,0) + \beta(0,1) + \gamma(1,-1)$$
$$= (\gamma - \alpha,\ \beta - \gamma) = (y_3 - y_1,\ y_1 - y_2).$$

3.17 Solved Problems

Example 3.1. *Show that the vectors $(1,2,3)$ and $(4,-2,7)$ are linearly independent in V_3 over the field F of real numbers.*

Solution : Let a, b be in F such that,

$$a(1,2,3) + b(4,-2,7) = 0 = (0,0,0).$$
$$\therefore \quad a + 4b = 0, \qquad 2a - 2b = 0 \quad \text{and} \quad 3a + 7b = 0.$$

From the first two equations, we have $a = b = 0$, which also satisfies the last equation.

$\therefore \quad a(1,2,3) + b(4,-2,7) = 0$, if and only if $a = b = 0$.

Hence, the given vectors are linearly independent.

Example 3.2. *Find the value of μ such that the vectors $x^{(1)} = (\mu,2,1)$, $x^{(2)} = (-1,\mu,1)$, $x^{(3)} = (0,2,1)$ are linearly independent.* **[BU(H) 1994]**

Solution : Let $a_1(\mu,2,1) + a_2(-1,\mu,1) + a_3(0,2,1) = 0$, where a_1, a_2, a_3 are arbitrary constants.

$$\therefore \quad a_1\mu - a_2 = 0 \qquad \text{(i)}$$
$$2a_1 + \mu a_2 + 2a_3 = 0 \qquad \text{(ii)}$$
$$a_1 + a_2 + a_3 = 0. \qquad \text{(iii)}$$

From eqn. (i), $a_2 = a_1\mu$.

Putting $a_2 = a_1\mu$ in (iii), we get $a_3 = -(\mu+1)a_1$.

Then from (ii), $\mu(\mu-2)a_1 = 0$.

$\therefore \quad a_1 = 0, \quad$ if $\mu \neq 0,\ \mu \neq 2$.

Again, if $\mu \neq 0$ and $\mu \neq 2$, we get $\boldsymbol{a_1 = a_2 = a_3 = 0}$.

So, **in order to be independent vectors, $\boldsymbol{\mu}$ may possess any value other than 0 and 2.**

Example 3.3. *Are the following vectors linearly dependent? If so, find a relation between them : $x_1 = (3,2,7)$, $x_2 = (2,4,1)$, $x_3 = (1,-2,6)$* **[BU(H) 2002]**

Solution : We assume that, the given vectors are linearly dependent, then

$$a_1(3,2,7) + a_2(2,4,1) = (1,-2,6).$$

Then, $\quad 3a_1 + 2a_2 = 1, \quad 2a_1 + 4a_2 = -2, \quad 7a_1 + a_2 = 6.$

From the first two equations on solving, we get $a_1 = -a_2 = 1$. These values of a_1 and a_2 satisfy the third equation. **Hence, we conclude that the given vectors are linearly dependent and $\boldsymbol{x_3 = x_1 - x_2}$.**

Example 3.4. *Examine whether the following set of vectors are linearly dependent or independent : $(2,-3,1)$, $(3,-1,5)$, $(1,-4,3)$* **[CU(H) 1987]**

Solution : Let a_1, a_2, a_3 be three scalars, such that

$$a_1(2,-3,1)+a_2(3,-1,5)+a_3(1,-4,3)=(0,0,0)$$

or, $[(2a_1+3a_2+a_3),\ (-3a_1-a_2-4a_3),\ (a_1+5a_2+3a_3)]=(0,0,0)$.

This gives $2a_1+3a_2+a_3=0$

$$-3a_1-a_2-4a_3=0$$
$$a_1+5a_2+3a_3=0.$$

The matrix of the coefficients is $A=\begin{bmatrix} 2 & 3 & 1 \\ -3 & -1 & -4 \\ 1 & 5 & 3 \end{bmatrix}$

or, $|A|=2(-3+20)+3(9-5)+(-12+1)=35$, i.e., $|A|\neq 0$.

$\therefore$ the rank of $A=3$, the number of unknowns.

Hence, $a_1=a_2=a_3=0$ is the only solution.

So, the given set of vectors is linearly independent.

Example 3.5. *Find a basis of R^3 containing vectors $(1,1,2)$ and $(3,5,2)$.*

Solution : For any two scalars a_1 and a_2 the given vectors are linearly independent as

$$a_1(1,1,2)+a_2(3,5,2)=(0,0,0).$$

or, $\quad a_1+3a_2=0,\quad a_1+5a_2=0,\quad 2a_1+2a_2=0,$

which gives $a_1=0,\ a_2=0$.

Now, we consider the fundamental basis of R^3

$$\{(1,0,0),\ (0,1,0),\ (0,0,1)$$

For any three scalars x_1, x_2, x_3 we consider the linear relation as

$$x_1(1,1,2)+x_2(3,5,2)+x_3(1,0,0)=(0,0,0).$$

This gives $\quad x_1+3x_2+x_3=0$

$$x_1+5x_2=0$$
$$2x_1+2x_2=0.$$

On solving, we get $x_1=x_2=x_3=0$.

Thus, the vectors $(1,1,2)$, $(2,5,2)$, $(1,0,0)$ are linearly independent and form a basis of R^3.

Example 3.6. *Find a basis of R^3 containing the vectors $(1,1,0)$.* **[CU(H) 1985]**

Solution : The vectors $(1,1,0)$, $(1,1,1)$ are linearly independent. For any linear combination,

$$x_1(1,1,0)+x_2(1,1,1)=(0,0,0).$$

This gives, $x_1=x_2=0$.

Let $\quad S=\{(1,1,0),\ (1,1,1)\}$.

Considering the fundamental basis in R^3, we have $A=\{(1,0,0),\ (0,1,0),\ (0,0,1)$.

For any three scalars x_1, x_2, x_3, we consider the linear relation as

$$x_1(1,1,0)+x_2(1,1,1)+x_3(1,0,0)=(0,0,0)$$

This gives, $x_1 + x_2 + x_3 = 0$

$x_1 + x_2 = 0$

and $x_2 = 0.$

$\therefore$ $x_1 = x_2 = x_3 = 0.$

Thus, the vectors $(1, 1, 0)$, $(1, 1, 1)$, $(1, 0, 0)$ are linearly independent and form a basis of R^3.

Example 3.7. *Show that,* $\begin{pmatrix} 1 & 0 \\ 0 & 1 \end{pmatrix}$ *and* $\begin{pmatrix} 0 & 1 \\ 1 & 1 \end{pmatrix}$ *cannot form a basis.*

Solution : If a and b be two scalars, then let $a\begin{pmatrix} 1 & 0 \\ 0 & 1 \end{pmatrix} + b\begin{pmatrix} 0 & 1 \\ 1 & 1 \end{pmatrix} = \begin{pmatrix} 0 & 0 \\ 0 & 0 \end{pmatrix}$.

This gives $a = b = 0$.

So, the matrices are linearly independent.

Again, let $a\begin{pmatrix} 1 & 0 \\ 0 & 1 \end{pmatrix} + b\begin{pmatrix} 0 & 1 \\ 1 & 1 \end{pmatrix} = \begin{pmatrix} x & y \\ z & w \end{pmatrix}$.

This gives $\boldsymbol{a = x}$, $\boldsymbol{b = y}$, $\boldsymbol{b = z}$, $\boldsymbol{a + b = w}$.

Since x, y, z, w are arbitrary, the matrices do not span the space. Hence, the given matrices do not form a basis.

Example 3.8. *Determine whether or not the set of vectors* $\{(2, -1, 1),\ (2, 0, 3),\ (1, 1, -2)\}$ *is a basis of the vector space* R^3. **[BU(H) 2000]**

Solution : The given set of vectors can form a basis of the vector space R^3, if the set is linearly independent.

The set will be linearly independent if for scalars a_1, a_2, a_3,

$$a_1(2, -1, 1) + a_2(2, 0, 3) + a_3(1, 1, -2) = (0, 0, 0)$$

or, $2a_1 + 2a_2 + a_3 = 0$

$-a_1 + a_3 = 0$

$a_1 + 3a_2 - 2a_3 = 0.$

Solving these equations, we get $a_1 = a_2 = a_3 = 0$.

Thus, the given set is linearly independent and hence, form a basis of R^3.

Example 3.9. *The vectors $\vec{r}$ in real three-dimensional space are transformed to $\vec{r}'$ by an operator A as follows :*

$$\vec{r}' = A\vec{r} = \vec{a} \times \vec{r},$$

where $\vec{a}$ is a constant vector. Show that $A^3 + |\vec{a}|^2 A = 0$. **[BU(H) 1995, 1997]**

Solution : Given, $A\vec{r} = \vec{a} \times \vec{r}$.

$$\therefore \quad A^2\vec{r} = A(A\vec{r}) = A(\vec{a} \times \vec{r}) = \vec{a} \times (\vec{a} \times \vec{r}) = \vec{a}(\vec{a} \cdot \vec{r}) - \vec{r}(\vec{a} \cdot \vec{a})$$

$$= \vec{a}(\vec{a} \cdot \vec{r}) - |\vec{a}|^2\vec{r}.$$

Again, $A^3\vec{r} = A(A^2\vec{r}) = \vec{a} \times \vec{a}(\vec{a} \cdot \vec{r}) - |\vec{a}|^2\vec{a} \times \vec{r}$

$$= -|\vec{a}|^2 A\vec{r} \quad [\because\ \vec{a} \times \vec{a} = 0].$$

$\therefore \qquad A^3\vec{r} + |\vec{a}|^2 A\vec{r} = 0 \quad \text{or,} \quad (A^3 + |\vec{a}|^2 A)\vec{r} = 0.$

This shows that $A^3 + |\vec{a}|A^2 = 0$ is a null operator.

Example 3.10. *If $X = (x_1, x_2)$ and $Y = (y_1, y_2)$ be two vectors, then the operator $\widehat{A}(x, y)$ $= (x - y,\ x + y)$ is linear.*

Solution : Let a_1 and a_2 be two scalars, then

$$\begin{aligned}\widehat{A}(a_1X + a_2Y) &= \widehat{A}(a_1x_1 + a_2y_1,\ a_1x_2 + a_2y_2)\\ &= [a_1(x_1 - x_2) + a_2(y_1 - y_2),\ a_1(x_1 + x_2) + a_2(y_1 + y_2)]\\ &= a_1(x_1 - x_2,\ x_1 + x_2) + a_2(y_1 - y_2,\ y_1 + y_2)\\ &= a_1\widehat{A}(x_1, x_2) + a_2\widehat{A}(y_1, y_2) = a_1\widehat{A}X + a_2\widehat{A}Y.\end{aligned}$$

Hence, $\widehat{A}$ is a **linear operator.**

Example 3.11. *If $\widehat{A}(x_1, x_2, x_3) = (x_1, x_2)$, then show that $\widehat{A}$ is a linear operator.*

Solution : Let X and Y be two vectors such that, $X = (x_1, x_2, x_3)$ and $Y = (y_1, y_2, y_3)$.

For any two scalars a_1 and a_2,

$$\begin{aligned}\widehat{A}(a_1X + A_2Y) &= \widehat{A}(a_1x_1 + a_2y_1,\ a_1x_2 + a_2y_2,\ a_1x_3 + a_2y_3) = (a_1x_1 + a_2y_1,\ a_1x_2 + a_2y_2)\\ &= a_1(x_1, x_2) + a_2(y_1, y_2) = a_1\widehat{A}(x_1, x_2, x_3) + a_2\widehat{A}(y_1, y_2, y_3)\\ &= a_1\widehat{A}X + a_2\widehat{A}Y.\end{aligned}$$

This shows that the operator $\widehat{A}$ is linear.

Example 3.12. *Examine if the following operators are linear :*

(i) $Af(x) = x^2 f(x)$ (ii) $Bf(x) = f^*(x)$ (iii) $Cf(x) = f^2(x)$ (iv) $Df(x) = \dfrac{\alpha}{dx} f(x)$.

[BU(H) 1995]

Solution : (i) We have $Af(x) = x^2 f(x)$.

$$\begin{aligned}\therefore \quad A[\alpha f(x) + \beta\phi(x)] &= x^2[\alpha f(x) + \beta\phi(x)] \qquad (\alpha \text{ and } \beta \text{ are two arbitrary constants})\\ &= \alpha x^2 f(x) + \beta x^2\phi(x) = \alpha Af(x) + \beta A\phi(x).\end{aligned}$$

Hence, A is a linear operator.

(ii) $Bf(x) = f^*(x)$, [* represents complex conjugation.]

$$\begin{aligned}\therefore \quad B[\alpha f(x) + \beta\phi(x) &= [\alpha f(x) + \beta\phi(x)]^* \qquad (\alpha \text{ and } \beta \text{ are two arbitrary constants})\\ &= \alpha^* f^*(x) + \beta^*\phi^*(x) = \alpha^* Bf(x) + \beta^* B\phi(x).\end{aligned}$$

Hence, B is a linear operator.

(iii) $\because \quad cf(x) = f^2(x)$,

$$\begin{aligned}\therefore \quad c[\alpha f(x) + \beta\phi(x)] &= [\alpha f(x) + \beta\phi(x)]^2 \qquad (\text{where } \alpha \text{ and } \beta \text{ are two arbitrary constants})\\ &= \alpha^2 f^2(x) + \beta^2\phi^2(x) + 2\alpha\beta f(x)\phi(x)\\ &\neq \alpha f^2(x) + \beta\phi^2(x).\end{aligned}$$

$\therefore$ c is not a linear operator.

(iv) $\because \quad Df(x) = \dfrac{d}{dx} f(x)$

$$\therefore \quad D[\alpha f(x) + \beta\phi(x) = \frac{d}{dx}[\alpha f(x) + \beta\phi(x)] \qquad (\alpha \text{ and } \beta \text{ are two arbitrary constants})$$

$$= \alpha\frac{d}{dx} f(x) + \beta\frac{d}{dx}\phi(x) = \alpha Df(x) + \beta d\phi(x).$$

Hence, ***D* is a linear operator**.

Example 3.13. *The vectors $\vec{r}$ in three-dimensional space are subject to the transformation $\vec{r}\,' = \hat{A}\vec{r} = \vec{a} \times \vec{r}$. Show that $\hat{A}$ is a linear operator.*

Solution : Consider the vector, $\vec{a} = a_1\hat{i} + a_2\hat{j} + a_3\hat{k}$, $\vec{r} = x\hat{i} + y\hat{j} + z\hat{k}$, $\vec{r} = x'\hat{i} + y'\hat{j} + z'\hat{k}$.

$$\because \quad \vec{r}\,' = \hat{A}\vec{r} = \vec{a} \times \vec{r} = \hat{i}(a_2 z - a_3 y) + \hat{j}(a_3 x - a_1 z) + \hat{k}(a_1 y - a_2 x) = \hat{i}x' + \hat{j}y' + \hat{k}z'.$$

$$\therefore \quad x' = a_2 z - a_3 y, \quad y' = a_3 x - a_1 z, \quad z' = a_1 y - a_2 x.$$

Consider the vector, $\vec{S} = \alpha\vec{r}_1 + \beta\vec{r}_2$, where α and β are constants.

Then, $S_x = \alpha x_1 + \beta x_2, \quad S_y = \alpha y_1 + \beta y_2$ and $S_z = \alpha z_1 + \beta z_2$.

$$\therefore \quad \vec{S}\,' = \hat{A}\vec{S} = \vec{a} \times \vec{S}.$$

$$S'_x = a_2 S_z - a_3 S_y = a_2(\alpha z_1 + \beta z_2) - a_3(\alpha y_1 + \beta y_2) = \alpha(a_2 z_1 - a_3 y_1) + \beta(a_2 z_2 - a_3 y_2)$$

$$= \alpha x'_1 + \beta x'_2$$

Likewise, $S'_y = \alpha y'_1 + \beta y'_2$ and $S'_z = \alpha z'_1 + \beta z'_2$

$$\therefore \quad \hat{A}\vec{S} = \vec{S}\,' = \hat{i}S_x + \hat{j}S_y u + \hat{k}S_z = \hat{i}(\alpha x'_1 + \beta x'_2) + \hat{j}(\alpha y'_1 + \beta y'_2) + \hat{k}(\alpha z'_1 + \beta z'_2)$$

$$= \alpha\vec{r}\,'_1 + \beta\vec{r}\,'_2$$

or, $\quad \hat{A}(\alpha\vec{r}_1 + \beta\vec{r}\,'_2) = \alpha\hat{A}\vec{r}_1 + \beta\hat{A}\vec{r}_2$

Hence, **$\hat{A}$ is a linear operator**.

Example 3.14. *Examine the linearity of the operators :*

(i) $\hat{A}f(x) = xf'(x) + f(x)$ (ii) $= \hat{C}f(x,y,z) = \left(\dfrac{\partial f}{\partial x}\right)^2 + \left(\dfrac{\partial f}{\partial y}\right)^2 + \left(\dfrac{\partial f}{\partial z}\right)^2$

(iii) $\hat{A}(a + ib) = (a, b)$ (iv) $\hat{B}(x, y) = (x + a, \; y + b)$.

Solution : (i) Given, $\hat{A}f(x) = xf'(x) + f(x)$.

If α and β be two scalars, then

$$\hat{A}[\alpha f(x) + \beta\phi(x)] = x\frac{d}{dx}[\alpha f(x) + \beta\phi(x)] + \alpha f(x) + \beta\phi(x)$$

$$= x[\alpha f'(x) + \beta\phi'(x)] + \alpha f(x) + \beta\phi(x)$$

$$= \alpha[xf'(x) + f(x)] + \beta[x\phi'(x) + \phi(x)]$$

$$= \alpha\hat{A}f(x) + \beta\hat{A}\phi(x).$$

$\therefore$ $\hat{A}$ is a linear operator.

(ii) Given, $\hat{C}f(x,y,z) = \left(\frac{\partial f}{\partial x}\right)^2 + \left(\frac{\partial f}{\partial y}\right)^2 + \left(\frac{\partial f}{\partial z}\right)^2.$

If α and β be two scalars, then

$$\hat{C}(\alpha f + \beta\phi) = \left(\alpha\frac{\partial f}{\partial x} + \beta\frac{\partial \phi}{\partial x}\right)^2 + \left(\alpha\frac{\partial f}{\partial y} + \beta\frac{\partial \phi}{\partial y}\right)^2 + \left(\alpha\frac{\partial f}{\partial z} + \beta\frac{\partial \phi}{\partial z}\right)^2$$

$$= \alpha\left[\left(\frac{\partial f}{\partial x}\right)^2 + \left(\frac{\partial f}{\partial y}\right)^2 + \left(\frac{\partial f}{\partial z}\right)^2\right] + \beta\left[\left(\frac{\partial \phi}{\partial x}\right)^2 + \left(\frac{\partial \phi}{\partial y}\right)^2 + \left(\frac{\partial \phi}{\partial z}\right)^2\right]$$

$$+ 2\alpha\beta\left(\frac{\partial f}{\partial x}\right)\left(\frac{\partial \phi}{\partial x}\right) + 2\alpha\beta\left(\frac{\partial f}{\partial y}\right)\left(\frac{\partial \phi}{\partial z}\right) + 2\alpha\beta\left(\frac{\partial f}{\partial z}\right)\left(\frac{\partial \phi}{\partial z}\right).$$

$$\therefore\quad \hat{C}(\alpha f + \beta\phi) \neq \alpha\left[\left(\frac{\partial f}{\partial x}\right)^2 + \left(\frac{\partial f}{\partial y}\right)^2 + \left(\frac{\partial f}{\partial z}\right)^2\right] + \beta\left[\left(\frac{\partial \phi}{\partial x}\right)^2 + \left(\frac{\partial \phi}{\partial y}\right)^2 + \left(\frac{\partial \phi}{\partial z}\right)^2\right].$$

This shows that $\hat{C}$ is not a linear operator.

(iii) We have $\widehat{A}(a+ib) = (a,b)$.

Now, let $x_1 = a_1 + ib_1$, $x_2 = a_2 + ib_2$.

Then $\widehat{A}(ax_1 + bx_2) = \widehat{A}[a(a_1+ib_1) + b(a_2+ib_2)] = \widehat{A}[(aa_1+ba_2) + i(ab_1+bb_2)$

$$= (aa_1 + ba_2,\ ab_1 + bb_2) = a(a_1,b_1) + b(a_2,b_2)$$

$$= a\widehat{A}(x_1) + b\widehat{A}(x_2).$$

This shows that the operator $\widehat{A}$ is linear.

(iv) Given, $\widehat{B}(x,y) = (x+a,\ y+b)$.

Then $\widehat{B}[\alpha(x_1,y_1) + \beta(x_2,y_2)]$, α and β are scalars

$$= \hat{\beta}(\alpha x_1 + \beta x_2 + a,\ \alpha y_1 + \beta y_2 + b) = \alpha(x_1+a,\ y_1+b) + \beta(x_2+a,\ y_2+b) - (a,b)$$

$$= \alpha\widehat{B}(x_1,y_1) + \beta\widehat{B}(x_2,y_2) - (a,b).$$

Hence, $\widehat{B}$ **is not a linear operator.**

Exercises

3(A) : Objective Type Questions

Select the correct answers

3.1 A linear vector space V is a set over field F, such that the elements of V_n obey certain mathematical operations like

(a) addition (b) multiplication (c) both (a) and (b) (d) none of these.

3.2 If α and β be two elements of non-empty space V, then

(a) $\alpha+\beta = \alpha\beta$ (b) $\alpha+\beta = \beta+\alpha$ (c) $\alpha+\beta = 1$ (d) $\alpha+\beta = (\alpha+\beta)^{-1}$.

3.3 If a, b are in F, then

(a) $(ab)\alpha = a(b\alpha)$ (b) $(ab)\alpha = b(a\alpha)$ (c) $(ab)\alpha = (ab)\alpha^{-1}$ (d) $(ab)\alpha = \frac{a\alpha}{b}$.

3.4 If α be a non-zero vector element in V and a, b are scalars in F, then $a\alpha = b\alpha$ implies

(a) $a \neq b$ (b) $a = b$ (c) $a = \frac{1}{b}$ (d) none of these.

3.5 A set of vectors $\{\alpha_i\}$, $i = 1, 2, \ldots, n$ in a vector space V is said to be linearly independent, if $\sum_{i=1}^{n} a_i\alpha_i = 0$

(a) all a_i's $= 0$ (b) at least one $a_i \neq 0$ (c) no $a_i = 0$ (d) none of these.

3.6 For a set of vectors $\{\alpha_i\}$, $i = 1, 2, \ldots, n$ in a vector field V, there exists a set of scalars a_i, $i = 1, 2, \ldots, n$ such that some of a_i's are zero, then the set of vectors $\{\alpha_i\}$ is linearly

(a) independent (b) dependent (c) may be both (a) and (b) (d) none of these.

3.7 If the set of vectors $\{\alpha_i\}$, $i = 1, 2, \ldots, n$ be the basis of a vector space V, then $\{\alpha_i\}$ are linearly

(a) dependent (b) independent (c) may be both (a) and (b) (d) none of these.

3.8 The inner-product of vectors α and β in a vector space has the property,

(a) $(\alpha, \beta) = (\beta, \alpha)$ (b) $(\alpha, \beta) = (\beta, \alpha)^*$ (c) $(\alpha, \beta) = \alpha^*\beta^*$ (d) $(\alpha, \beta) = \alpha\beta^*$.

3.9 If α and β be any two vectors in an inner-product space V over field F, then $|\alpha, \beta|$

(a) $= \|\alpha\| \, \|\beta\|$ (b) $< \|\alpha\| \, \|\beta\|$ (c) $\leq \|\alpha\| \, \|b\|$ (d) $\|\alpha\| + \|\beta\|$.

3.10 If α and β be any two vectors in an inner-product space V, then $\|\alpha + \beta\|$ is

(a) $< \|\alpha\| + \|\beta\|$ (b) $= \|\alpha\| + \|\beta\|$ (c) $\leq \|\alpha\| + \|\beta\|$ (d) $= \|\alpha\| \, \|\beta\|$.

3.11 If two vectors α and β in $V(F)$ be orthogonal, then

(a) $(\alpha, \beta) = 1$ (b) $(\alpha, \beta) = 0$ (c) $(\alpha, \beta) < 1$ (d) $(\alpha, \beta) = \infty$.

3.12 A set of vectors $\{\alpha_i\}$ in an inner-product space is orthonormal, if

(a) $\|\alpha_i\| = 1$ (b) $\|\alpha_i\|^2 = 1$ (c) $|\alpha_i\|^2 = 0$ (d) $\|\alpha_i\|^2 = \infty$.

3.13 The operator $\hat{A}$ is said to be linear, if

(a) $\hat{A}(\alpha\vec{U} + \beta\vec{V}) = \alpha\hat{A}\vec{u} + \beta\hat{A}\vec{V}$ (b) $\hat{A}(\alpha\vec{U} + \beta\vec{V}) = \alpha\hat{A}\vec{U} - \beta\hat{A}\vec{V}$

(c) $\hat{A}(\alpha\vec{U} + \beta\vec{V}) = \vec{U}\alpha\hat{A} - \vec{V}\beta\hat{A}$ (d) $\hat{A}(\alpha\vec{U} + \beta\vec{V}) = \alpha\hat{A}\vec{U} + \vec{V}\beta\hat{A}$

where α and β are scalars and $\vec{U}$ and $\vec{V}$ are two vectors.

3.14 If $\hat{A}$ and $\hat{B}$ be two operators which commute, then

(a) $\hat{A}\hat{B} - \hat{B}\hat{A} = 1$ (b) $\hat{A}\hat{B} - \hat{B}\,\hat{A} = 0$ (c) $\hat{A}\hat{B} + \hat{A}\hat{B} = 0$ (d) $(\hat{A}\hat{B})(\hat{A}\hat{B}) = 1$.

3.15 If $\hat{A}$ and $\hat{B}$ be two operators, then

(a) $(\hat{A}\hat{B})^{-1} = \hat{B}^{-1}\hat{A}^{-1}$ (b) $(\hat{A}\hat{B})^{-1} = \hat{A}^{-1}\hat{B}^{-1}$

(c) $(\hat{A}\hat{B})^{-1} = \hat{A}\hat{B}$ (d) $(\hat{A}\hat{B})^{-1} = \hat{A}^{-1} + \hat{B}^{-1}$.

Answers

3.1 (c) **3.2** (b) **3.3** (a) **3.4** (b) **3.5** (a) **3.6** (b) **3.7** (b) **3.8** (b) **3.9** (c) **3.10** (c) **3.11** (b) **3.12** (b) **3.13** (a) **3.14** (b) **3.15** (a).

3(B) : Review Questions

3.1 **(a)** Define a linear vector space.

(b) Write down the postulates of the linear vector space.

(c) Verify that, $\alpha+(\beta+\gamma)=(\alpha+\beta)+\gamma$, where α,β,γ are the three elements of vector space V.

3.2 **(a)** Prove that, $(ab)\alpha=a(b\alpha)$, where α is a vector and a,b are scalars.

(b) Discuss the properties of vector space.

3.3 **(a)** What do you mean by linearly dependence and independence of vectors? Discuss with examples.

(b) Define dimension of a vector space.

3.4 **(a)** If α and β be two linearly independent vectors, then prove that one of them is a scalar multiple of the other.

(b) Prove that any subset of a linearly independent set of vectors in a vector space is also linearly independent.

3.5 **(a)** Prove that a set of non-zero vectors in a vector field V is linearly dependent, if and only if some elements of that set is a linear combination of others.

(b) Prove that a set of non-zero vectors in a vector space is linearly dependent, if and only if one of them be a linear combination of the preceding vectors.

3.6 **(a)** Define 'basis' of a vector space.

(b) Prove that every vector in a finite-dimensional vector space V over the field F can be uniquely expressed as a linear combination of the vectors of its basis.

3.7 **(a)** Define inner-product of a pair of vectors.

(b) What is 'norm' of a vector?

(c) If V be an inner-product space over a field F, then show that,

$$(\alpha, a\beta+b\gamma)=a^*(\alpha,\beta)+b^*(\alpha,\gamma),$$

whose α,β,γ are vectors and a,b are scalars, $*$ represents complex conjugation.

3.8 **(a)** State and obtain Schwarz's inequality.

(b) If α and β be two vectors, then prove that, $\|\alpha\|\;\|\beta\|=|(\alpha,\beta)|$.

3.9 **(a)** What are (i) orthogonal set of vectors and (ii) orthonormal set of vectors?

(b) Discuss Gram-Schmidt process of orthonormalisation.

3.10 **(a)** What is an operator?

(b) What is a linear operator?

(c) Describe matrix representation of operator.

3.11 **(a)** The orthogonal set of vectors $\hat{e_1},\hat{e_2},\hat{e_3}$ form a basis in a linear vector space. An operator A is such that it transforms $\hat{e_1}$ to $\hat{e_2}$, $\hat{e_2}$ to $\hat{e_3}$ and $\hat{e_3}$ to $\hat{e_1}$. Find the matrix representing the operator in the given basis. Show that the operator A is unitary. **[BU(H) 1999, 2006]**

(b) The orthogonal set of vectors ϕ_1,ϕ_2 and ϕ_3 form a basis in a linear vector space. An operator is such that it transforms ϕ_1 to ϕ_2, ϕ_2 to ϕ_3 and ϕ_3 to ϕ_1.

(i) Find the matrix representing the operator in the given basis.

(ii) Is the operator Hermitian? Unitary?

(iii) What are the eigenvalues of the operator? [BU(H) 2001]

3.12 (a) Let u, v, w be three linearly independent vectors in a real vector space. Show that the vectors $u+v$, $v+w$ and $w+u$ are also linearly independent. [BU(H) 2006, 2008]

(b) Show that any system of vectors containing two equal vectors must be linearly dependent [BU(H) 1998, 2008]

3(C) : Problems

3.1 If $V(F)$ be a vector space, hen show that the following relations hold :

(i) $(-1)\alpha = -\alpha$

(ii) $\beta + (\alpha - \beta) = \alpha$

(iii) if $a\alpha = 0$, then either $a = 0$ or, $\alpha = 0$, where α, β are in V and a, b are in F.

3.2 Examine whether the set of vectors $\alpha_1 = (2, -3, 1)$, $\alpha_2 = (3, -1, 5)$, $\alpha_3 = (1, -4, 3)$ are linearly dependent or independent. [**Ans:** linearly independent]

3.3 Show that the vector $(2, -5, 3)$ cannot be expressed as a linear combination of vectors,

$$\alpha_1 = (1, -3, 2),\ \alpha_2 = (2, -4, -1),\ \alpha_3 = (1, -5, 7).$$

3.4 Show that the coordinates of the vector $(-1, 3, 1)$ relative to the ordered basis,

$$B = \{(2, 1, 0),\ (2, 1, 1),\ (2, 2, 1)\} \text{ are } \left(-\frac{3}{2}, -\frac{5}{2}, \frac{7}{2}\right).$$

3.5 Construct an orthonormal base for R^3 out of the vectors, $\alpha_1 = (1, 2, 2)$, $\alpha_2 = (1, -1, 2)$, $\alpha_3 = (1, 0, 1)$. [**Ans:** $\{\left(\frac{1}{3}, \frac{2}{3}, \frac{2}{3}\right), \frac{1}{\sqrt{5}}\left(\frac{2}{3}, -\frac{5}{3}, \frac{4}{3}\right), \frac{1}{\sqrt{5}}(2, 0, -1)\}$]

3.6 Examine if the following operator A is linear :

$$Af(x) = (1 - x^2)\frac{d^2 f}{dx^2} - 5x\frac{df}{dx} + 7f(x). \qquad \text{[BU(H) 1993]}$$

[**Solution :** If α and β be two scalars, then

$$\begin{aligned} A[\alpha f(x) + \beta\phi(x)] &= (1 - x^2)\frac{d^2}{dx^2}[\alpha f(x) + \beta\phi(x)] \\ &\quad - 5x\frac{d}{dx}[\alpha f(x) + \beta\phi(x)] + 7[\alpha f(x) + \beta\phi(x)] \\ &= \alpha\left[(1 - x^2)\frac{d^2}{dx^2}f(x) - 5x\frac{d}{dx}f(x) + 7f(x)\right] \\ &\quad + \beta\left[(1 - x^2)\frac{d^2}{dx^2}\phi(x) - 5x\frac{d}{dx}\phi(x) + 7\phi(x)\right] \\ &= \alpha Af(x) + \beta A\phi(x). \end{aligned}$$

Hence, the operator A is linear.]

3.7 Write the vector $\alpha = \begin{bmatrix} 3 & -1 \\ 1 & -2 \end{bmatrix}$ in the vector space 2×2 matrices as a linear combination of $\alpha_1 = \begin{bmatrix} 1 & 1 \\ 0 & -1 \end{bmatrix}$, $\alpha_2 = \begin{bmatrix} 1 & 1 \\ -1 & 0 \end{bmatrix}$, $\alpha_3 = \begin{bmatrix} 1 & -1 \\ 0 & 0 \end{bmatrix}$. [**Ans:** $\alpha = 2\alpha_1 - \alpha_2 + 2\alpha_3$]

3.8 Determine k so that the vectors $(1,2,1)$, $(k,1,1)$ and $(1,1,2)$ are linearly independent $(k \in R)$.

3.9 Prove that the vectors (a,b) and (c,d) in $R^2(R)$ are linearly dependent, if $ad - bc = 0$.

3.10 If α, β, γ be a basis of a vector space R^3, then show that $\{\alpha+\beta,\ \beta+\gamma,\ \gamma+\alpha\}$ is also a basis set of R^3. **[CU(H) 1994]**

3.11 For all α, β in a Euclidean space V, prove that, $\|\alpha+\beta\|^2 = \|\alpha\|^2 + \|\beta\|^2$, if $\alpha, \beta = 0$.

3.12 If α, β be vectors in a real inner product space and $\|\alpha\| = \|\beta\|$, then show that $(\alpha+\beta,\ \alpha-\beta) = 0$. **[CU(H) 1989]**

3.13 Show that the standard inner-product between $\alpha = (1,2,x,y)$ and $\beta = (x,y,2,1)$ is $3x+3y$.

3.14 Use the Gram-Schmidt orthonormalisation process to find the orthonormal basis for R^3 generated by the set of vectors, $\{(1,-1,1),\ (2,0,1),\ (0,1,1)\}$. **[VU(H) 1990]**

3.15 Construct an orthonormal base for R^3 out of the vectors $(1,2,2)$, $(1,-1,2)$, $(1,0,1)$. **[CU(H) 1996]**

3.16 **(a)** Define vector space and subspace.

(b) For what value of k is $(1,k,5)$ a linear combination of $u = (1,-3,2)$ and $v = (2,-1,1)$ **[Pune U(MSc) 2014]**

3.17 **(a)** Let $V = R^3$. Show that W is a subspace of V, where $W = \{(a,b,0)/a,b \in R\}$.

(b) Write the vector $V = (1,-2,5)$ as a linear combination of the vectors

$e_1 = (1,1,1)$, $e_2 = (1,2,3)$ and $e_3 = (2,-1,1)$.

(c) Consider the following basis of Euiclidean space R^3 :

$\{V_1 = (1,1,1),\ V_2 = (0,1,1),\ V_3 = (0,0,1)\}$. Using the Gram-Schmidt orthogonalisation process, transform $\{v_i\}$ into an orthonormal basis $\{u_i\}$. **[Pune U(MSc) 2014]**

3.18 **(a)** Define subspace W of vector space V. Let V be the vector space of all 2×2 matrices over R. Show that W is not a subspace of V, where W consists of all matrices with zero determinant.

(b) Find whether the following vectors in R^3 are linearly dependent or independent : $\{(1,-2,1),\ (2,1,-1),\ (7,-4,1)\}$.

(c) Let $W = \{(a,b,c) : a = b = c\}$ and $W = \{0,b,c\}$. Show that $R^3 = \text{U} \oplus W$. **[Pune U(MSc) 2014]**

3.19 **(a)** Find the basis and dimension of subspace W of $V = M_{2\times 2}$ spanned by

$$A = \begin{pmatrix} 1 & -5 \\ -4 & 2 \end{pmatrix},\ B = \begin{pmatrix} 1 & 1 \\ -1 & 5 \end{pmatrix},\ C = \begin{pmatrix} 2 & -4 \\ -5 & 7 \end{pmatrix},\ D = \begin{pmatrix} 1 & -7 \\ -5 & 1 \end{pmatrix}.$$

(b) Let $V = R^3$. Determine whether W is a subspace of V, where

(i) $W = \{(a,b,0) : a,b \in R\}$ **(ii)** $W = \{(a,b,c) : a \geq 0\}$.

(c) Determine whether or not the following form a basis for the vector space R^3 : $\{(1,1,1),\ (2,0,0),\ (0,2,0), (0,0,2)\}$.

3.20 **(a)** Define basis and dimension of a vector space defined by $T(x,y,z) = (2y+z, x-4y, 3x)$.

(b) Find the matrix T in the basis $\{f_1 = (1,1,1),\ f_2 = (1,1,0),\ f_3(1,0,0)\}$.

(c) Verify that $[T]_f[V]_f = [T(V)]_f$. **[Pune U(MSc) 2014]**

3.21 (a) Apply Gram-Schmidt orthogonalisation process to find an orthogonal basis for the subspace U of R^4 spanned by $V_1 = (1, 1, 1, 1)$, $V_2 = (1, 2, 4, 5)$, $V_3 = (1, -3, -4, 2)$.

(b) Find if set $S = \{(1, 1, 1),\ (1, 1, 0),\ (1, 0, 0)\}$ can form the basis of R^3. If yes, find the coordinates of (i) $v = (4, -3, 2)$, (ii) (a, b, c) relative to S.

3.22 (a) Define inner product space. Verify that $\langle u, v\rangle = x_1y_1 - x_1y_2 - x_2y_1 + 3x_2y_2$, where $u = (x_1, y_1)$ and $v = (x_2, y_2)$ is an inner product in R^3. **[Pune U(MSc) 2008]**

(b) What is subspace of vector space? Show that W is subspace of $V = R^3$, where W consists those vectors each whose sum of component is zero.

CHAPTER 4

Infinite Series

4.1 Introduction

In many problems of physics and engineering infinite series occur frequently and as such studying their convergence or divergence becomes essential. A series employed in any problem, if not convergent then it may lead to absurd result. It is, therefore, essential that the students should have a sound knowledge in applying infinite series to physical problems.

4.2 Sequence

A sequence is an ordered set of numbers $a_1, a_2, \ldots, a_n$, formed according to some definite rule, such that corresponding to every positive integer n there exists a number a_n of the set. The individual numbers are called the **elements of the sequence.** The sequence with nth element a_n is denoted by $\{a_n\}$.

Example of sequence :

1. $\{n\}$ **represents 1, 2, 3, 4, ...**
2. $\{\frac{1}{n}\}$ **represents 1, $\frac{1}{2}, \frac{1}{3}, \frac{1}{4}, \ldots$**
3. $(2n-1)$ **represents 1, 3, 5, 7, ...**
4. $[(-1)^n n]$ **represents −1, +2, −3, +4, ...**
5. $\left\{\frac{n}{n+1}\right\}$ **represents $\frac{1}{2}, \frac{2}{3}, \frac{3}{4}, \frac{4}{5}, \ldots$**

A **finite sequence** has finite number of terms whereas an **infinite sequence** has unlimited number of terms in which each term is followed by another term.

Example *The sequence $\{3,5,7\}$ is finite but the sequence $\{2,3,5,7,11,\ldots\}$ is infinite sequence and is represented by $\{a_n\}$, where a_n is the nth prime number.*

4.3 Bounded Sequence

If in a sequence $\{a_n\}$ there exists a number k such that k is greater than or equal to a_n of the sequence, i.e., $a_n \leq k$, for every n, then the sequence is called a **bound sequence.** Here, k is called the **rough upper bound** and the least of all the upper bound is called the **exact upper bound** of the sequence.

Example *The sequence $\{1+\frac{1}{n}\}$ is bounded above and the upper bound is 2.*

If in a sequence $\{a_n\}$, there exists a number which is less than or equal to any number of the sequence (i.e., $a_n \geq k$), then the sequence is said to be **bounded below.** k is called the **rough lower bound.** If k is the least of all rough lower bounds then it is called the **exact-lower bound.**

Example *The sequence $\{\frac{1}{n}\}$ is bounded below, the lower bound being 0.*

A sequence $\{a_n\}$, bounded both above and below is called a **bounded sequence.**

Example *The sequence* $\{2+(-1)^n \frac{1}{n}\}$ *the upper and lower bounds being* $2\frac{1}{2}$ *and* 1.

N.B. The sequence $\{1+(-1)^n\} \to 0, 3, -2, 5, -4, 7, \ldots$ is unbounded as it has neither upper or lower bound.

4.4 Monotonic Sequence

A sequence is said to be monotonic sequence, if it be monotonic increasing or monotonic decreasing.

Monotonic Increasing Sequence

A sequence $\{a_n\}$ is said to be monotonic increasing, if $a_n \leq a_{n+1}$, for all n.

Example *The sequence* 1, 2, 3, 4, ... *is a monotonic increasing sequence.*

Monotonic Decreasing Sequence

A sequence $\{a_n\}$ is said to be monotonic increasing, if $a_n \geq a_{n+1}$, for all n.

Example *The sequence* $-1, -2, -3, -4, \ldots$ *is a monotonic decreasing sequence.*

N.B.

(i) If a sequence $\{a_n\}$ be monotonic increasing, then the sequence $\{-a_n\}$ is monotonic decreasing.

(ii) A sequence can neither be monotonic increasing nor decreasing, e.g., $\{1, 0, 1, 0, 1, 0, \ldots\}$.

Limit : An infinite sequence $\{a_n\}$ is said to be convergent and has the limit l, if corresponding to any arbitrary small positive number ϵ we can find out positive integer m (depending on ϵ) such that

$$|a_n - 1| < \epsilon, \quad \text{when } n \geq m$$

or, $1 - \epsilon < a_n < 1 + t$, when $n \geq m$.

If this holds, then a_n tends to l as n tends to infinity i.e., $\underset{n\to\infty}{\text{Lt}}\ a_n = l$.

Example *The limit of the sequence* $\{\frac{1}{n}\}$ *as* $n \to \infty$ *is* 0.

To prove it, we have

$$\left|\frac{1}{n} - 0\right| < \epsilon, \text{ if } \frac{1}{n} < \epsilon.$$

This happens, if $n > \dfrac{1}{\epsilon}$.

If $m = \frac{1}{\epsilon}$ or the integral part of $\frac{1}{\epsilon}$, we have

$$\left|\frac{1}{n} - 0\right| < \epsilon, \text{ if } n \geq m.$$

$$\therefore \quad \underset{n\to\infty}{\text{Lt}} \frac{1}{n} = 0.$$

Hence, $\left\{\dfrac{1}{n}\right\}$ converges to 0.

Note that, limit of a sequence may or may not be a term of the sequence. For example, consider the convergent series $\{\frac{1}{n}\} \equiv 1, \frac{1}{2}, \frac{1}{3}, \ldots, \frac{1}{n}$. It has the limit 0 but no member of the sequence is 0.

Convergent, Divergent, Oscillatory Sequences

A sequence $\{a_n\}$ is said to be **convergent**, if it has a finite limit.

If the sequence $\{a_n\}$ is not convergent, then it is a **divergent sequence**, diverging either to $+\infty$ or $-\infty$.

If, however, a sequence is neither convergent nor divergent, then it is called an **oscillatory sequence**.

Thus, $\{n^2\} \equiv 1^2, 2^2, 3^3, \ldots$ diverges to $+\infty$

$\{-n\} \equiv -1, -2, -3, \ldots$ diverges to $-\infty$

$\{(-1)^n\} \equiv -1, 1, -1, \ldots$ oscillates.

If for an oscillatory sequence $\{a_n\}$, a constant c exists such that $|a_n| < c$, for all n, then the sequence is said to **oscillate finitely**, otherwise it is said to **oscillate infinitely**.

Example $\{(-1)^n \frac{n+2}{n}\} \equiv -3, -\frac{4}{2}, -\frac{5}{2}, \frac{6}{4}, -\frac{7}{2}, \ldots$ *oscillates finitely*

and $\{(-1)^n \sqrt{n}\} \equiv -1, \sqrt{2}, -\sqrt{3}, 2, -\sqrt{5}, \ldots$ *oscillates infinitely.*

N.B.

(i) A monotonic sequence cannot oscillate as it always tends to either a finite or infinite limit.

(ii) A monotonic and bounded sequence is also convergent.

4.5 Cauchy's General Principle for Convergence of a Sequence

A necessary and sufficient condition for the convergence of the sequence $\{a_n\}$ is that corresponding to any pre-assigned positive number ϵ, however small, there exists a positive integer m such that, for $n \geq m$,

$$|a_{n+p} - a_n| < \epsilon, \text{ for all positive integral values of } p.$$

This is known as **Cauchy's criterion** and the sequence satisfying this criterion is called Cauchy's sequence.

Proof : Suppose, the sequence $\{a_n\}$ is convergent having a finite limit, say, l. Then for given ϵ, however small, there exists a positive integer m such that,

$$|a_n - l| < \frac{1}{2}\epsilon, \text{ for all } n \geq m.$$

Then $|a_{n+1} - l| < \frac{1}{2}\epsilon$, for $n \geq m$ and $p \geq 0$.

$$\therefore \quad |a_{n+p} - a_n| = |a_{n+p} - l + l - a_n| \leq |a_{n+p} - l| + |l - a_n| < \frac{1}{2}\epsilon + \frac{1}{2}\epsilon.$$

So $|a_{n+p} - a_n| < \epsilon$, for $n \geq m$ and $p \geq 0$.

This, **condition is necessary.**

Now, let $|a_{n+p} - a_n| < \epsilon$, when $n \geq m$ and $p \geq 0$.

Then, $a_n - \epsilon < a_{n+p} < a_n + \epsilon$, for all positive integral values of p.

$$\therefore \quad \underset{n\to\infty}{\text{Lt}} \frac{1}{n} = 0.$$

Hence, $\left\{\frac{1}{n}\right\}$ **converges to 0.**

Thus, $\{a_{n+p}\}$ is bounded as $p \to \infty$.

If L and M be the lower and upper bounds respectively of $\{a_n\}$, then

$$L \le a_n - \epsilon \quad \text{and} \quad M \le a_n + \epsilon.$$

$$\therefore \quad M - L \le (a_n - \epsilon) - (a_n + \epsilon) = 2\epsilon.$$

This implies that in the limit $M - L = 0$ as ϵ is arbitrary.

So, $\boldsymbol{M - \epsilon \le a_{n+p} \le M + \epsilon}$.

It follows that, $a_{n+p} \to M$ for all positive integral values of p.

Hence, **the sequence $\{a_n\}$ is convergent and the condition is also sufficient.**

4.6 Series

Definition : If a set of real numbers $u_1, u_2, \ldots, u_n, \ldots$ occur according to some definite law, then

$$u_1 + u_2 + \cdots + u_n + \cdots \infty$$

is called a series.

If the number of terms is limited, then the series is called a **finite series**. If the number of terms is infinitely large, it is called an infinite series.

A series is, thus, the sum of a sequence.

An infinite series is denoted by

$$u_1 + u_2 + \cdots + u_n + \cdots \infty \quad \text{or} \quad \sum_{n=1}^{\infty} u_n \quad \text{or} \quad \sum u_n.$$

The sum of its first n terms is expressed by S_n, where $S_n = u_1 + u_2 + \cdots + u_n$.

S_n is called the partial sum of the series.

4.7 Convergence, Divergence and Oscillation of a Series

We consider the infinite series $\sum u_n = u_1 + u_2 + \cdots + u_n + \cdots \infty$.

Let the sum of the first n terms be $S_n = u_1 + u_2 + \cdots + u_n$.

$\therefore$ S_n is a function of n. As n increases infinitely, three possibilities may arise :

(i) If S_n tends to a finite limit as $n \to \infty$, the series $\sum u_n$ is said to be **convergent**.

(ii) If S_n tends to $\pm\infty$, as $n \to \infty$, the series $\sum u_n$ is said to be **divergent**.

(iii) If S_n does not tend to a unique limit as $n \to \infty$, then the series $\sum u_n$ is oscillatory.

A divergent or oscillating series is also called non-convergent series.

Example 4.1. *Show that the following series is convergent :*

$$\frac{1}{1\cdot 2} + \frac{1}{2\cdot 3} + \frac{1}{3\cdot 4} + \cdots + \frac{1}{n(n+1)} + \cdots \infty.$$

Solution : $S_n = \left(1 - \frac{1}{2}\right) + \left(\frac{1}{2} - \frac{1}{3}\right) + \cdots + \left(\frac{1}{n-1} - \frac{1}{n}\right) + \left(\frac{1}{n} - \frac{1}{n+1}\right)$

$$= 1 - \frac{1}{n+1} \to 1 \text{ as } n \to \infty$$

$\therefore$ $\boldsymbol{S_n = 1}$ **as** $\boldsymbol{n \to \infty}$

Hence, the series is **convergent** with **limit** 1.

Example 4.2. *Show that the series $1+2+3+\cdots+n+\cdots\infty$ is divergent.*

Solution : Here, $S_n = 1+2+3+\cdots+n = \dfrac{n(n+1)}{2}$.

$\therefore \quad \underset{n\to\infty}{\text{Lt}} S_n = \underset{n\to\infty}{\text{Lt}} \frac{1}{2}n(n+1) \to \infty.$

Hence, the given series is **divergent**.

Example 4.3. *Show that the series $5-4-1+5-4-1+5-4-1+\cdots\infty$ is oscillatory.*

Solution : Here, $S_n = 5-4-1+5-4-1+5-4-1+\cdots n$ terms

$= 0$, if the number of terms is $3m$

$= 5$, if the number of terms is $3m+1$

$= 1$, if the number of terms is $3m+2$.

As S_n tends to no unique limit, the series is oscillatory.

Example 4.4. **Geometric series:** *Show that the series $1+r+r^2+r^3+\cdots+\infty$ (i) converges, if $|r|<1$, (ii) diverges, if $r \geq 1$ and (iii) oscillates, if $r \leq -1$.*

Solution : Let $S_n = 1+r+r^2+\cdots+r^{n-1}$.

(i) When $|r|<1$, $\underset{n\to\infty}{\text{Lt}} = 0$.

Also, $S_n = \dfrac{1-r^n}{1-r} = \dfrac{1}{r-1} - \dfrac{r^n}{1-r}$.

$\therefore \quad \underset{n\to\infty}{\text{Lt}} S_n = \dfrac{1}{1-r}.$

Thus, the series is **convergent**.

(ii) (a) When $r>1$, $\underset{n\to\infty}{\text{Lt}} r^n \to \infty$.

Also, $S_n = \dfrac{r^n-1}{r-1} = \dfrac{r^n}{r-1} - \dfrac{1}{r-1}$.

$\therefore \quad \underset{n\to\infty}{\text{Lt}} S_n \to \infty.$

$\therefore$ the series is **divergent**.

(b) When $r=1$, $S_n = 1+1+1+\cdots+1 = n$ and $\underset{n\to\infty}{\text{Lt}} S_n \to \infty$.

$\therefore$ the series is divergent.

(iii) (a) when $r=-1$, the series becomes $1-1+1-1+1-1+\cdots$

The series is oscillatory.

(b) when $r<-1$, let $r=-p$ so that $p>1$.

Then, $r^n = (-1)^n p^n$ and $S_n = \dfrac{1-r^n}{1-r} = \dfrac{1-(-1)^n p^n}{1+p}$.

As $\underset{n\to\infty}{\text{Lt}} p^n \to \infty$,

$\therefore \quad$ **$\underset{n\to\infty}{\text{Lt}} S_n \to -\infty$ or $+\infty$ according as n is even or odd.**

Hence, the series oscillates.

4.8 Addition and Substraction of Series

Let us assume that $\sum u_n$ converges to s and $\sum v_n$ converges to t, then

(i) $\sum(u_n + v_n)$ converges to $(s+t)$ and (ii) $\sum(u_n - v_n)$ converges to $(s-t)$.

Proof : (i) $\text{Lt}\{(u_1+v_1)+(u_2+v_2)+\cdots+(u_n+v_n)\}$

$$= \text{Lt}\{(u_1+u_2+\cdots+u_n)+(v_1+v_2+\cdots+v_n)\}$$
$$= \text{Lt}(u_1+u_2+\cdots+u_n)+\text{Lt}(v_1+v_2+\cdots+v_n)$$
$$= \text{Lt}\sum u_n + \text{Lt}\sum v_n = s+t.$$

(ii) **The proof is similar to (i).**

4.9 General Properties of Series

1. **A series $\sum u_n$ of positive terms is either convergent or, divergent but cannot oscillate.**

   We suppose, $S_n = u_1 + u_2 + \cdots + u_n$.

   Since all the elements $u_1, u_2, \ldots$ are positive, the sequence $\{S_n\}$ is monotonic increasing. So, $\underset{n\to\infty}{\text{Lt}} S_n$ exists either as a finite quantity or as an infinite quantity with a definite sign. In the former case the series is convergent and in the latter it is divergent. So, it cannot oscillate.

2. **An infinite series in which all the terms after some particular term are positive, is s positive term series.**

   Example *The series, $-7-5-2+2+7+13+20+\cdots$ is a positive term series as all its terms after the third are positive.*

3. **The convergence or divergence of an infinite series remains unaffected by multiplying each term by a finite number.**

   Let $\underset{n\to\infty}{\text{Lt}} S_n = s$, where $S_n = u_1+u_2+\cdots+u_n$.

   $\therefore \quad \underset{n\to\infty}{\text{Lt}}(cu_1+cu_2+\cdots+cu_n)$ [here c is a constant]

   $$= \underset{n\to\infty}{\text{Lt}} c(u_1+u_2+\cdots+u_n) = c\underset{n\to\infty}{\text{Lt}}(u_1+u_2+\cdots+u_n) = cs.$$

   If $c \neq 0$ and $S_n \to \pm\infty$, then cS_n also $\to \pm\infty$.

   If S_n tends to no fixed limit finite or infinite, so does cS_n. This establishes the property (3).

   N.B.

   (a) **If $\sum u_n$ and $\sum v_n$ converge and their sums be s and t respectively, then the series $\sum(au_n + bv_n)$, a, b being constant, converges and its sum is $(as+bt)$.**

   (b) **The convergence or divergence is not affected, if the signs of all the terms be changed.**

4. **The convergence or divergence of a series is not affected by the inclusion or removal of a finite number of terms.**

   The sum of a finite number of terms in an infinite series is finite and definite. So the inclusion or removal of a finite number of terms only changes the value of the limiting sum ($\text{Lt}\, S_n$) by a finite and definite quantity.

5. **If the terms of a convergent series of positive terms be arranged, then the series remains convergent and its sum remains the same.**

Let $S_n = u_1 + u_2 + \cdots + u_n$.

Suppose, $\sum v_n$ be another series whose terms are the same as those of $\sum u_n$ in a different order so that, $S'_n = v_1 + v_2 + \cdots + v_n$.

For every u there is a v and vice versa.

Let the first m terms of $\sum v_n$ contains the first n terms of $\sum u_n$ and first $(n+p)$ terms of $\sum u_n$ contains first m terms of $\sum v_n$.

Then, $n \le m \le n+p$, where $m \to \infty$ as $n \to \infty$.

Also, $S_n \le S'_m \le S_{n+p}$.

Now, let $S_n = \text{Lt}\, S_{n+p} = s$.

$\therefore \text{Lt}\, S'_m = s$.

Hence, $\sum v_n$ is convergent and has the same sum as $\sum u_n$.

4.10 An Important Series

The infinite series, $\sum_{p=1}^{n} \frac{1}{p^2} = \frac{1}{1^p} + \frac{1}{2^p} + \frac{1}{3^p} + \cdots + \frac{1}{n^p} + \cdots \infty$

is convergent if $p > 1$ and divergent if $p \le 1$.

As the series is of positive terms, its nature is not affected by grouping of terms in any way. The above series is called the harmonic series of order of p.

(i) (i) Let $p > 1$.

The first term $= \frac{1}{1^p}$.

As $p > 1$, $3^p > 2^p$, $\therefore\ \frac{1}{3^p} < \frac{1}{2^p}$.

Thus, $\frac{1}{2^p} < \frac{1}{3^p} < \frac{2}{2^p}$.

Similarly, $\frac{1}{4^p} + \frac{1}{5^p} + \frac{1}{6^p} + \frac{1}{7^p} < \frac{4}{4^p}$; and so on.

Hence, the series is less than the comparing series, $1 + \frac{2}{2^p} + \frac{4}{4^p} + + \frac{8}{8^p} + \cdots$

This is a geometric series whose common ratio is $\frac{2}{2^p}$ which is less than 1 as $p > 1$.

Hence, the comparing series is convergent, the given series is also convergent, if $p > 1$.

(ii) Let $p = 1$.

Then the series becomes $1 + \frac{1}{2} + \frac{1}{3} + \frac{1}{4} + \cdots + \infty$.

The 3rd and 4th terms together is greater than $\frac{4}{8}$ or $\frac{1}{2}$, the following eight terms are together greater than $\frac{8}{16} = \frac{1}{2}$; and so on.

Hence, the series is greater than the series $1 + \frac{1}{2} + \frac{1}{2} + \frac{1}{2} + \cdots$, and is divergent.

6. Necessary condition for convergence.

If a positive term series $\sum u_n$ is convergent, then $\underset{n\to\infty}{\text{Lt}}\ u_n = 0$.

Proof : Let $S_n = u_1 + u_2 + \cdots + u_n$.

$\because \sum u_n$ is convergent, $\underset{n\to\infty}{\text{Lt}}\ S_n = k$ (k is a finite quantity).

Also, $\underset{n\to\infty}{\text{Lt}}\ S_{n-1} = k$.

But, $u_n = S_n - S_{n-1}$

$\therefore \quad \underset{n\to\infty}{\textbf{Lt}}\ \boldsymbol{u_n} = \underset{n\to\infty}{\textbf{Lt}}\ \boldsymbol{(S_n - S_{n-1}) = 0}$.

N.B.

(i) The convergence of this result is not true.

For instance, consider the series, $1 + \frac{1}{\sqrt{2}} + \frac{1}{\sqrt{3}} + \cdots + \frac{1}{\sqrt{n}} + \cdots \infty$.

Since the terms go on descending,

$$S_n = 1 + \frac{1}{\sqrt{2}} + \frac{1}{\sqrt{3}} + \cdots + \frac{1}{\sqrt{n}} > \frac{1}{\sqrt{n}} + \frac{1}{\sqrt{n}} + \frac{1}{\sqrt{n}} + \cdots + \frac{1}{\sqrt{n}} = \frac{n}{\sqrt{n}} = \sqrt{n}.$$

$\therefore \quad \underset{n\to\infty}{\textbf{Lt}}\ \boldsymbol{S_n} = \underset{n\to\infty}{\textbf{Lt}}\ \boldsymbol{\sqrt{n} \to \infty}$.

Hence, the series is divergent although, $\underset{n\to\infty}{\text{Lt}}\ u_n = \underset{n\to\infty}{\text{Lt}}\ \frac{1}{\sqrt{n}} = 0$.

So, $\underset{n\to\infty}{\text{Lt}}\ u_n = 0$ is a necessary but not sufficient condition for convergence $\sum n_n$.

(ii) The above result leads to a simple test of divergence.

If $\underset{n\to\infty}{\text{Lt}}\ u_n \neq 0$, the series $\sum u_n$ must be divergent.

Hence, for $p = 1$, the given series is divergent.

(iii) Let $p < 1$.

Now, each term after the first is greater than the corresponding term in (ii) above.

Hence, the series is divergent for $p < 1$.

4.11 Necessary Condition for Convergence

If a positive term series $\sum u_n$ is convergent, then $\underset{n\to\infty}{\text{Lt}}\ u_n = 0$.

Proof : Let the sum of the series $\sum u_n$ be represented by S_n, then

$$S_n = \sum u_n = u_1 + u_2 + \cdots + u_n.$$

Since $\sum u_n$ is convergent,

$\therefore \quad \underset{n\to\infty}{\text{Lt}}\ S_n =$ a finite quantity k (say).

Also, $\text{Lt}_{n\to\infty} S_{n-1} = k$.

Now, $u_n = S_n - S_{n-1}$.

$\therefore \quad \underset{n\to\infty}{\textbf{Lt}}\ \boldsymbol{u_n} = \underset{n\to\infty}{\textbf{Lt}}\ \boldsymbol{(S_n - S_{n-1}) = 0}$ **(Proved).**

N.B.

(i) The convergence is not true.

We consider the series $1 + \frac{1}{\sqrt{2}} + \frac{1}{\sqrt{3}} + \cdots + \frac{1}{\sqrt{n}} + \cdots \infty$.

As the terms go on descending,

$$S_n = 1 + \frac{1}{\sqrt{2}} + \frac{1}{\sqrt{3}} + \cdots + \frac{1}{\sqrt{n}} > \frac{1}{\sqrt{n}} + \frac{1}{\sqrt{n}} + \cdots + \frac{1}{\sqrt{n}} = \frac{n}{\sqrt{n}} = \sqrt{n}.$$

$$\therefore \quad \underset{n\to\infty}{\text{Lt}}\, S_n = \underset{n\to\infty}{\text{Lt}}\, \sqrt{n} = \infty.$$

This shows that, **though** $\underset{n\to\infty}{\text{Lt}}\, u_n = \underset{n\to\infty}{\text{Lt}}\, \frac{1}{\sqrt{n}} = 0$, **the series** $\sum u_n$ **is divergent.**

Thus, we may say that, $\underset{n\to\infty}{\text{Lt}}\, u_n = 0$ **is a necessary but not sufficient condition for the convergence of** $\sum u_n$.

(ii) If $\underset{n\to\infty}{\text{Lt}}\, u_n \neq 0$, **then** $\sum_n u_n$ **must be divergent.**

This result provides a simple test of divergence of a series.

4.12 Comparison Tests

(a) If $\sum u_n$ **and** $\sum v_n$ **be two positive term series such that, (i)** $\sum v_n$ **converges (ii)** $u_n \leq v_n$, **for all values of** n, **then** $\sum u_n$ **also converges.**

Proof : As $\sum v_n$ is a positive term convergent series, we have

$$\underset{n\to\infty}{\text{Lt}} \sum v_n = \underset{n\to\infty}{\text{Lt}}\, (v_1 + v_2 + \cdots + v_n) = k, \text{ a finite quantity.}$$

Also, $u_1 \leq v_1$

$u_2 \leq v_1$

....

$u_n \leq v_n$.

We get by adding $(u_1 + u_2 + \cdots + u_n) \leq (v_1 + v_2 + \cdots + v_n)$.

$$\therefore \quad \underset{n\to\infty}{\text{Lt}} \sum u_n \leq \underset{n\to\infty}{\text{Lt}} \sum v_n = k.$$

Hence, the series $\sum u_n$ **is convergent, since** $\sum v_n$ **is convergent.**

N.B. If the relation $u_n \leq v_n$ **holds for values of** n **greater than** m, **a fixed number, then the first** m **terms of both the series can be ignored without affecting their convergence or divergence.**

(b) If the positive term series $\sum u_n$ **and** $\sum v_n$ **be such that, (i)** $\sum v_n$ **diverges (ii)** $u_n \geq v_n$, **for all values of** n, **then** $\sum u_n$ **also diverges.**

This can be proved easily in the same way as the test (a).

(c) Limit Form : If $\sum u_n$ **and** $\sum v_n$ **be two positive term series such that** $\underset{n\to\infty}{\text{Lt}} \frac{u_n}{v_n} = l$, **a finite quantity** $(l \neq 0)$, **then** $\sum u_n$ **and** $\sum v_n$ **converge or diverge together.**

Proof : We have $\underset{n\to\infty}{\text{Lt}} \dfrac{u_n}{v_n} = l$, l being a finite number $(l \neq 0)$.

According to the definition of limit, there exists a positive number ϵ, however small, such that

$$\left|\frac{u_n}{v_n} - l\right| < \epsilon, \text{ for } n \geq m$$

$$\text{or,} \quad -\epsilon < \left(\frac{u_n}{v_n} - l\right) < \epsilon, \text{ for } n \geq m$$

$$\text{or,} \quad (l - \epsilon) < \left(\frac{u_n}{v_n}\right) < (l + \epsilon), \text{ for } n \geq m.$$

Omitting the first m terms of both the series, we have

$$(l - \epsilon) < \frac{u_n}{v_n} < (l + \epsilon), \text{ for all } n. \tag{4.1}$$

Case 1 : If $\sum v_n$ is convergent, then

$$\underset{n\to\infty}{\text{Lt}} \sum v_n = \underset{n\to\infty}{\text{Lt}} (v_1 + v_2 + \cdots + v_n) = k, \text{ where } k \text{ is a finite number.} \tag{4.2}$$

Again, from Eqn. (4.1),

$$\frac{u_n}{v_n} < l + \epsilon \quad \text{or,} \quad u_n < (l + \epsilon)v_n, \text{ for all } n.$$

$$\therefore \quad \underset{n\to\infty}{\text{Lt}} (u_1 + u_2 + \cdots + u_n) < (l + \epsilon) \underset{n\to\infty}{\text{Lt}} (v_1 + v_2 + \cdots + v_n)$$
$$= (l + \epsilon)k \quad \text{[by Eqn. 4.2]}.$$

$\therefore$ **u_n is also convergent.**

Case 2 : When v_n is divergent.

$$\underset{n\to\infty}{\text{Lt}} (v_1 + v_2 + \cdots + v_n) \to \infty \tag{4.3}$$

Also from Eqn. (4.1),

$$(l - \epsilon) < \frac{u_n}{v_n} \quad \text{or,} \quad u_n > (l - \epsilon)v_n, \text{ for all } n.$$

$$\therefore \quad \text{Lt}_{n\to\infty}(u_1 + u_2 + \cdots + u_n) \geq (l - \epsilon) \underset{n\to\infty}{\text{Lt}} (v_1 + v_2 + \cdots + v_n)$$
$$\to \infty \text{ [by Eqn. (4.3)]}$$

$$\text{or,} \quad \sum_{n\to\infty} \sum u_n > (l - \epsilon) \underset{n\to\infty}{\text{Lt}} \sum v_n \to \infty. \tag{4.4}$$

$\therefore$ **$\sum u_n$ is also divergent.**

4.13 Comparison of Ratios

(a) If $\sum u_n$ and $\sum v_n$ be two positive term series, then $\sum u_n$ converges, if

(i) $\sum v_n$ converges, and (ii) from and after some particular term $\frac{u_{n+1}}{u_n} \leq \frac{v_{n+1}}{v_n}$.

Proof : If $\frac{u_2}{u_1} < \frac{v_2}{v_1}, \frac{u_3}{u_2} < \frac{v_3}{v_2}, \ldots,$ then

$$u_1 + u_2 + u_3 + \cdots = u_1\left(1 + \frac{u_2}{u_1} + \frac{u_3}{u_1} + \cdots\right) = u_1\left(1 + \frac{u_2}{u_1} + \frac{u_2}{u_1}\cdot\frac{u_3}{u_2} + \cdots\right)$$

$$> u_1\left(1 + \frac{v_2}{v_1} + \frac{v_2}{v_1}\cdot\frac{v_3}{v_2} + \cdots\right) > \frac{u_1}{v_1}(v_2 + v_2 + v_3 + \cdots). \qquad (4.5)$$

$\therefore \sum u_n$ **converges, if** $\sum v_n$ **converges.**

A convenient form of the above test is to apply it as follows :

(i) $\sum u_n$ **converges if** $\sum v_n$ **converges and (ii) from and after a particular term** $\frac{u_n}{u_{n+1}} > \frac{v_n}{v_{n+1}}$.

(b) If $\sum u_n$ **and** $\sum v_n$ **be two positive term series, then** $\sum u_n$ **is divergent, if** $\frac{u_{n+1}}{u_n} \geq \frac{v_{n+1}}{v_n}$ **for all values of** n.

Proof : If $\frac{u_2}{u_1} \geq \frac{v_2}{v_1}, \frac{u_3}{u_2} \geq \frac{v_3}{v_2}, \ldots,$ then

$$u_1 + u_2 + u_3 + \cdots = u_1\left(1 + \frac{u_2}{u_1} + \frac{u_3}{u_1} + \cdots\right) = u_1\left(1 + \frac{u_2}{u_1} + \frac{u_2}{u_1}\cdot\frac{u_3}{u_2} + \cdots\right)$$

$$\geq u_1\left(1 + \frac{v_2}{v_1} + \frac{v_2}{v_1}\cdot\frac{v_3}{v_2} + \cdots\right) \geq \frac{u_1}{v_1}(v_1 + v_2 + v_3 + \cdots). \qquad (4.6)$$

$\therefore \sum u_n$ **is divergent, if** $\sum \frac{u_{n+1}}{u_n} \geq \frac{v_{n+1}}{v_n}$.

4.14 D'Alembert's Ratio Test

If $\sum u_n$ be a positive term series for which $\underset{n\to\infty}{\text{Lt}} \frac{u_{n+1}}{u_n} = \lambda$, then the series converges for $\lambda < 1$ and diverges for $\lambda > 1$.

Case 1 : If $\underset{n\to\infty}{\text{Lt}} \frac{u_{n+1}}{n} = \lambda < 1$.

From the definition of limit, there may be a positive r $(r < 1)$ such that,

$$\frac{u_{n+1}}{u_n} < r, \text{ for all } n > m.$$

Omitting the first m terms, let the series be $u_1 + u_2 + u_3 + \cdots$ so that, $\frac{u_2}{u_1} < r$, $\frac{u_3}{u_2} < r$, $\frac{u_1}{u_3} < 1$, and so on.

Then, $u_1 + u_2 + u_3 + \cdots + \infty = u_1\left(1 + \frac{u_2}{u_1} + \frac{u_3}{u_2}\cdot\frac{u_2}{u_1} + \frac{u_4}{u_3}\cdot\frac{u_3}{u_2}\cdot\frac{u_2}{u_1} + \cdots\infty\right)$

$$< u_1(1 + r + r^2 + r^3 + \cdots\infty)$$

$$= \frac{u_1}{(1-r)}, \quad \left[\because r < 1, \frac{u_1}{1-r} \text{ is finite}\right]. \qquad (4.7)$$

$\therefore \sum u_n$ **is convergent.**

Case 2 : If $\underset{n\to\infty}{\text{Lt}} \frac{u_{n+1}}{n} = \lambda > 1$.

From the definition of limit, we can find m such that,

$$\frac{u_{n+1}}{u_n} \geq 1, \text{ for all } n \geq m.$$

Omitting the first m terms, let the series be $u_1 + u_2 + u_3 + \cdots$ so that, $\frac{u_2}{u_1} \geq 1$, $\frac{u_3}{u_2} \geq 1$, $\frac{u_4}{u_3} \geq 1$, and so on.

$$\therefore \quad u_1 + u_2 + u_3 + u_4 + \cdots + u_n = u_1 \left(1 + \frac{u_2}{u_1} + \frac{u_3}{u_2} \cdot \frac{u_2}{u_1} + \cdots\right)$$

$$\geq u_1(1 + 1 + 1 + \cdots \text{ up to } n \text{ terms}) = nu_1.$$

$$\therefore \quad \operatorname*{Lt}_{n\to\infty} (u_1 + u_2 + \cdots + u_n) \geq \operatorname*{Lt}_{n\to\infty} nu_1 \to \infty. \tag{4.8}$$

Hence, $\sum u_n$ is divergent.

N.B.

(i) If $\operatorname*{Lt}_{n\to\infty} \frac{u_{n+1}}{n} = \lambda = 1$, then the limit does not exist and as such the ratio test fails for $\lambda = 1$.

For example, we consider the series $\sum u_n = \frac{1}{r^n}$.

$$\text{Then, } \lambda = \operatorname*{Lt}_{n\to\infty} \frac{u_{n+1}}{u_n} = \operatorname*{Lt}_{n\to\infty} \left[\frac{1}{(n+r)^p} \cdot \frac{n^p}{1}\right] = \operatorname*{Lt}_{n\to\infty} \frac{1}{\left(1+\frac{1}{n}\right)^p} = 1. \tag{4.9}$$

For all values of p, $\lambda = 1$. But $\sum \frac{1}{r^n}$ converges for $p > 1$ and diverges for $p < 1$.

This leads to the conclusion that $\sum u_n$ both converges and diverges for $\lambda = 1$, which is absurd.

(ii) **The test does not make any reference to the magnitude of $u_n + \frac{1}{u_n}$ but concerns only with the limit of the ratio.**

For example, we consider a series, $1 + \frac{1}{2} + \frac{1}{3} + \frac{1}{4} + \cdots + \frac{1}{n} + \cdots$

$$\text{Here, the ratio, } \frac{u_{n+1}}{u_n} = \frac{n}{n+1} < 1. \tag{4.10}$$

for all values of n but tends to unity as $n \to \infty$. **So, as $\operatorname*{Lt}_{n\to\infty} \frac{u_{n+1}}{u_n} = 1$ or the limit does not exist, no conclusion can be drawn.**

4.15 Raabe's Test

We have seen that **D'Alembert's ratio test** fails for $\lambda = 1$. When the ratio test fails, we apply '**Raabe's test**'.

In a positive term series $\sum u_n$, if $\operatorname*{Lt}_{n\to\infty} n\left(\frac{u_n}{u_{n+1}} - 1\right) = k$, then if $k > 1$, the series converges and for $k < 1$, the series diverges. The test, however fails for $k = 1$.

When $k > 1$, we choose a number p such that $k > p > 1$ and compare $\sum u_n$ with a series $v_n = \frac{1}{n^p}$, which is convergent for $p > 1$.

$\therefore \sum u_n$ will converge from and after some term,

$$\text{if } \frac{u_n}{u_{n+1}} > \frac{v_n}{v_{n+1}} \quad \text{or, if } \frac{u_n}{u_{n+1}} > \frac{(n+1)^p}{n^p} = \left(1 + \frac{1}{n}\right)^p$$

$$\text{or, if } \frac{u_n}{u_{n+1}} > 1 + \frac{p}{n} + \frac{p(p-1)}{2!}\frac{1}{n^2} + \cdots \quad \text{or, if } n\left(\frac{u_n}{u_{n+1}} - 1\right) > p + \frac{p(p-1)}{2!}\frac{1}{n} + \cdots$$

$$\text{or, if } \underset{n\to\infty}{\text{Lt}} \left\{ n \left(\frac{u_n}{u_{n+1}} - 1 \right) \right\} > \underset{n\to\infty}{\text{Lt}} \left\{ p + \frac{p(p-1)}{2n} + \cdots \right\} \quad \text{i.e., if } k > p. \tag{4.11}$$

Hence, $\sum u_n$ is convergent.

When $k < 1$, (i.e., $p < 1$), it can be proved that the series diverges.

N.B. If however, $\underset{n\to\infty}{\text{Lt}} \left\{ n \left(\frac{u_n}{u_{n+1}} - 1 \right) \right\} = 1$, then Raabe's test also fails.

4.16 Logarithmic Test

The series u_n of positive terms is convergent or divergent according as

$$\text{Lt} \left(n \log \frac{u_n}{u_{n+1}} \right) > 1 \quad \text{or,} \quad < 1.$$

Proof : Let any number p be such that $k > p > 1$. We now compare the given series $\sum u_n$ with the series $\sum v_n$, where $v_n = \frac{1}{n^p}$ which converges as $p > 1$.

Now, $\sum u_n$ will be convergent, if after some particular terms,

$$\frac{u_n}{u_{n+1}} > \left(\frac{n+1}{n} \right)^p \quad \text{or, if} \quad \log \frac{u_n}{u_{n+1}} > p \log \left(1 + \frac{1}{n} \right)$$

$$\text{or, if} \quad \log \frac{u_n}{u_{n+1}} > p \left(\frac{1}{n} - \frac{1}{2n^2} + \frac{1}{3n^2} - \cdots \right)$$

$$\text{or, if} \quad n \log \frac{u_n}{u_{n+1}} > p + \frac{p}{n} \left(-\frac{1}{2} + \frac{1}{3n} - \cdots \right). \tag{4.12}$$

For large value of n, the LHS approaches k while the RHS approaches p. But $k > p$. Hence, the inequality is satisfied for large values of n.

So, the given series $\sum u_n$ is convergent if Lt $\left(n \log \frac{u_n}{u_{n+1}} \right) > 1$.

Similarly, we can also prove that, $\sum u_n$ is divergent, if Lt $\left(n \log \frac{u_n}{u_{n+1}} \right) < 1$.

N.B. The test (logarithmic test) is a substitute of Raabe's test. It should be applied when either n occurs as an exponent of $\frac{u_n}{u_{n+1}}$ or evaluation of $\underset{n\to\infty}{\text{Lt}}$ becomes easier taking logarithm of $\left(\frac{u_n}{u_{n+1}} \right)$.

4.17 Gauss' Test

This test may be used instead of beginning with **D'Alembert's** and **Raabe's** test and as such is very much useful.

If $\sum u_n$ is a series of positive terms such that,

$$\frac{u_n}{u_{n+1}} = \alpha + \frac{\beta}{m} + \frac{\gamma_n}{n^p} \tag{4.13}$$

in which $\alpha > 0$, $\beta > 1$ and $\{\gamma_n\}$ is a bounded sequence, then

(i) **for $\alpha \neq 1$ and for any value of β, $\sum u_n$ converges, if $\alpha > 1$ and $\sum n$ diverges, if $\alpha < 1$.**

(ii) **for $\alpha = 1$, $\sum u_n$ converges, if $\beta > 1$ and diverges, if $\beta \leq 1$.**

Proof : (i) If $\alpha \neq 1$, then $\underset{n\to\infty}{\text{Lt}} \dfrac{u_n}{u_{n+1}} = \alpha$.

Thus, by **D'Alembert's ratio test**, $\sum u_n$ converges, if $\alpha > 1$ and diverges, if $\alpha < 1$.

(ii) If $\alpha = 1$, then as $p > 1$,

$$\underset{n\to\infty}{\text{Lt}}\ n\left(\frac{u_n}{u_{n+1}} - 1\right) = \beta. \tag{4.14}$$

Thus, by **Raabe's test**, $\sum u_n$ converges, if $\beta > 1$ and diverges, if $\beta < 1$.

When $\beta = 1$, we have

$$\frac{u_n}{u_{n+1}} = 1 + \frac{1}{n} + \frac{\gamma_n}{n^p};\quad p > 1.$$

Let us now compare the given series $\sum u_n$ with the series $\sum v_n$ which is divergent, where $v_n = \frac{1}{n\log n}$ so that,

$$v_{n+1} = \frac{1}{(n+1)\log(n+1)}.$$

Now, $$\frac{u_n}{u_{n+1}} - \frac{v_n}{v_{n+1}} = 1 + \frac{1}{n} + \frac{\gamma_n}{n^p} - \frac{(n+1)\log(n+1)}{n\log n} = \frac{\gamma_n}{n^p} - \frac{(n+1)}{n}\left\{\frac{\log(n+1)}{\log n} - 1\right\}$$

$$= \frac{\gamma_n}{n^p} - \frac{(n+1)}{n}\left\{\frac{\log(n+1)}{\log n} - 1\right\} = \frac{\gamma_n}{n^p}\left\{\gamma_n - (n+1)\log\left(1+\frac{1}{n}\right)\frac{n^{p-1}}{\log n}\right\}.$$

But, $$\underset{n\to\infty}{\text{Lt}}(n+1)\log\left(1+\frac{1}{n}\right) = \underset{n\to\infty}{\text{Lt}}\left\{\log\left(1+\frac{1}{n}\right)^n + \log\left(1+\frac{1}{n}\right)\right\} = 1$$

and $$\frac{n^{p-1}}{\log n} \to \infty \text{ as } n \to \infty \quad (\because\ p > 1).$$

For sufficiently large n, $\left\{\gamma_n(n+1)\log\left(1+\frac{1}{n}\right)\frac{n^{p-1}}{\log n}\right\}$ remains negative as γ_n is bounded.

Thus for all $n \geq m$, m being a positive integer,

$$\gamma_n - (n+1)\log\left(1+\frac{1}{n}\right)\frac{n^{p-1}}{\log n} < 0.$$

Hence, for $n \geq m$, $\dfrac{u_n}{u_{n+1}} < \dfrac{v_n}{v_{n+1}}$

As v_n is divergent by **comparison test**, $\sum u_n$ is also divergent.

4.18 Alternating Series

If the signs of consecutive terms of a series are alternatively positive and negative, then the series is called an alternating series.

Example *$u_1 - u_2 + u_3 - u_4 + u_5 + \cdots$, where $u_1 > u_2 > u_3 > u_4 > \cdots$ is an alternating series.*

Leibnitz's Rule for Alternating Series

An infinite series in which the terms are alternatively positive and negative is convergent if (i) each term be numerically less than the proceeding term and (ii) $\underset{n\to\infty}{\text{Lt}}\ u_n = 0$.

Let the series be

$$\sum_{n=1}^{\infty} u_n = u_1 - u_2 + u_3 - u_4 + u_5 + \cdots \tag{4.15}$$

We suppose, $u_1 > u_2 > u_3 > u_4 > \cdots$ (4.16)

and $\underset{n\to\infty}{\text{Lt}}\ u_n = 0.$ (4.17)

Now, the sum of $2n$ terms is

$$S_{2n} = (u_1 - u_2) + (u_3 - u_4) + \cdots + (u_{2n-1} - u_{2n}) \tag{4.18}$$

or, $$S_{2n} = u_1 - (u_2 - u_3) - (u_4 - u_5) - \cdots - \text{ up to } 2n \text{ terms.} \tag{4.19}$$

By the condition (4.16), the expressions within brackets of Eqns (4.18) and (4.19) are all positive.

So, according to Eqn. (4.18), S_{2n} is positive and increases with n. Eqn. (4.19) shows that S_{2n} is always less than u_1.

∴ S_{2n} must tend towards a finite limit.

Furthermore, $\underset{n\to\infty}{\text{Lt}}\ S_{2n+1} = \underset{n\to\infty}{\text{Lt}}\ (S_{2n} + u_{2n+1})$

$$= \underset{n\to\infty}{\text{Lt}}\ S_{2n} + 0 \quad \text{[by Eqn. (4.17)].} \tag{4.20}$$

Thus, $\underset{n\to\infty}{\text{Lt}}\ S_n$ tends to the same finite value whether n is even or odd. Hence, the given series [Eqn. 4.15] is convergent.

If $\underset{n\to\infty}{\text{Lt}}\ u_n \neq 0$, $\underset{n\to\infty}{\text{Lt}}\ S_{2n} \neq \underset{n\to\infty}{\text{Lt}}\ S_{2n} + 1$, the given series is oscillatory.

4.19 Power Series

(i) General Form of Power Series : The general form of a power series is expressed as

$$\sum_{n=0}^{\infty} a_n (x - x_0)^n = a_0 + a_1(x - x_0) + a_2(x - x_0)^2 + \cdots, \tag{4.21}$$

where x_0 is a fixed number and the coefficients a_0, a_1, a_2, ... are constants (depending on n but not on x) in x about x_0, x being a real variable.

In particular for $x_0 = 0$, the power series is

$$\sum_{n=0}^{\infty} a_n x^n = a_0 + a_1 x + a_2 x^2 + \cdots + a_n x^n + \cdots. \tag{4.22}$$

(ii) Interval of Convergence : From Eqn. (4.22), $u_n = a_n x^n$.

$$\therefore \quad \underset{n\to\infty}{\text{Lt}}\ \frac{u_{n+1}}{u_n} = \underset{n\to\infty}{\text{Lt}}\ \frac{a_{n+1}x^{n+1}}{a_n x^n} = \underset{n\to\infty}{\text{Lt}}\ \left(\frac{a_{n+1}}{a_n}\right) x. \tag{4.23}$$

If $\underset{n\to\infty}{\text{Lt}}\ \left(\frac{a_{n+1}}{a_n}\right) = l$, then by ratio test the series of Eqn. (4.22) **converges**, when lx is numerically less than 1, i.e., $|x| < \frac{1}{l}$ and **diverges** for other values.

So, the series of Eqn. (4.22) has an interval given by

$$-\frac{1}{l} < x < \frac{1}{l},$$

within which the series converges or diverges for values of x outside the above interval. The interval is called the **interval of convergence of the power series.**

4.20 Convergence of Binomial Series

The binomial series,

$$1 + nx + \frac{n(n-1)x^2}{2!} + \cdots + \frac{n(n-1)\cdots(n-r+1)}{r!}x^r + \cdots\infty \tag{4.24}$$

converges for $|x| < 1$.

Proof : Here the rth term of the series is,

$$u_r = \frac{n(n-1)\cdots(n-r)}{(r-1)!}x^{r-1}$$

and $(r+1)$th term of the series is

$$u_{r+1} = \frac{n(n-1)\cdots(n-r+1)}{r!}x^r$$

$$\therefore \qquad \underset{r\to\infty}{\text{Lt}} \frac{u_{r+1}}{u_r} = \underset{r\to\infty}{\text{Lt}} \frac{n-r+1}{r}x = \underset{r\to\infty}{\text{Lt}} \left(\frac{n+1}{r} + 1\right)x$$

$$= -x, \text{ for } r > (n+r) \tag{4.25}$$

Hence, the binomial series converges for $|x| < 1$.

4.21 Convergence of Exponential Series

An exponential series in x is expressed as

$$1 + x + \frac{x^2}{2} + \cdots + \frac{x^n}{n!} + \cdots\infty \tag{4.26}$$

is convergent for all values of x.

Proof : We have $\underset{n\to\infty}{\text{Lt}} \dfrac{u_{n+1}}{u_n} = \underset{n\to\infty}{\text{Lt}} \left[\dfrac{x^n}{n!} + \dfrac{x^{n-1}}{(n-1)!}\right] = \underset{n\to\infty}{\text{Lt}} \dfrac{x}{n} = 0.$ (4.27)

Hence, for any value of x, the exponential series of Eqn. (4.26) converges.

4.22 Convergence of Logarithmic Series

A logarithmic series in x is expressed as

$$x - \frac{x^2}{2} + \frac{x^3}{3} + \cdots + (-1)^n\frac{x^n}{n!} + \cdots\infty. \tag{4.28}$$

The series converges, whatever be the value of x.

Proof : Here, $\dfrac{u_{n+1}}{u_n} = \underset{n\to\infty}{\text{Lt}} \dfrac{(-1)^n x^{n+1}}{n+1} \cdot \dfrac{n}{(-1)^n x} = -x \underset{n\to\infty}{\text{Lt}} \dfrac{n}{n+1} = -x \underset{n\to\infty}{\text{Lt}} \left(\dfrac{1}{1+\frac{1}{n}}\right)$

$$= -x. \tag{4.29}$$

$\therefore$ the series converges for $|x| < 1$ and diverges for $|x| > 1$.

(i) When $x = 1$, the series becomes $1 - \frac{1}{2} + \frac{1}{3} - \frac{1}{4} + \cdots$, which is convergent.

(ii) When $x = -1$, the series becomes $-\left(1 + \frac{1}{2} + \frac{1}{3} + \frac{1}{4} + \cdots\right)$, which is divergent.

So, the series converges for $-1 < x \leq 1$.

4.23 Absolute and Conditional Convergence

4.23.1 How to Test the Convergency of a Series

Procedure : First check whether the given series is

(a) a series with alternatively positive or negative terms,

(b) a series of positive terms excluding power series,

(c) a series is a power series.

For alternating series (a), apply Leibnitz's rule. For series (b), first find u_n and, if possible, evaluate $\text{Lt}\, u_n$. If $\text{Lt}\, u_n \neq 0$, the series is divergent.

If $\text{Lt}\, u_n = 0$, compare $\sum u_n$ with $\sum \frac{1}{n^2}$ and then apply comparison test.

If the comparison tests are not applicable, apply the ratio test.

If $\underset{n\to\infty}{\text{Lt}} \frac{u_n}{u_{n+1}} = 1$, i.e., the ratio test fails, apply Raabe's test. If Raabe's test fails for a similar reason, then apply logarithmic test. If logarithmic test fails apply Cauchy's root test.

For power series (c), apply ratio test. If the ratio test fails, then examine the given series following the same procedure as in (b).

Different tests for convergence or divergence of various series.

Table 1

Type of Series	Name of Test	Condition for Convergence or Divergence	Remark
	1. Cauchy's fundamental test	If $\underset{n\to\infty}{\text{Lt}}\, u_n \neq 0$, then the series $\sum u_n$ is divergent.	
	2. Comparison test	If $\underset{n\to\infty}{\text{Lt}} \frac{u_n}{v_n} = k$, then $\sum u_n$ and $\sum v_n$ are of the same nature.	
Positive term series	3. D'Alembert's ratio test	$\underset{n\to\infty}{\text{Lt}} \frac{u_{n+1}}{u_n} = k$. $\sum u_n$ is convergent, if $k < 1$ $\sum v_n$ is divergent, if $k > 1$.	This test is to be applied, if comparison test is not applicable.
	4. Raabe's test	$\underset{n\to\infty}{\text{Lt}}\, n\left(\frac{u_n}{u_{n+1}} - 1\right) = k$. $\sum u_n$ is convergent, if $k > 1$ and divergent, if $k < 1$.	Apply Raabe's test in case D'Alembert's ratio test fails.
	5. Cauchy's root test	$\underset{n\to\infty}{\text{Lt}} (u_n)^{\frac{1}{n}} = k$ If $k < 1$, $\sum u_n$ is convergent and if $k > 1$, $\sum u_n$ is divergent.	Apply logarithmic test, if root test fails.

(Contd.)

Type of Series	Name of Test	Condition for Convergence or Divergence	Remark
Positive term series	6. Gauss' test	If $\sum u_n$ is a series of positive terms so that $\frac{u_n}{u_{n+1}} = \alpha + \frac{\beta}{n} + \frac{\gamma_n}{n^p}$ in which $\alpha > 0$, $\beta > 1$ and $\{\gamma_n\}$ is a bounded sequence. Then (i) for $\alpha \neq 1$ and for any value of β. $\sum u_n$ converges, if $\alpha > 1$ and diverges, if $\alpha < 1$. (ii) for $\alpha = 1$, $\sum u_n$ converges, if $\beta > 1$ and diverges, if $\beta \leq 1$.	
Alternating series	7. Leibnitz test	If (i) $\lvert u_{n+1}\rvert > \lvert u_n\rvert$ and (ii) $\underset{n\to\infty}{\text{Lt}}\, u_n = 0$, then the series is convergent.	
	8. Cauchy's condensation test	If the function $f(x)$ be positive for integral values of n and continually decreases as n increases, then the two infinite series $f(1)+f(2)+\cdots+f(n)+\cdots$ and $af(a)+a^2f(a^2)+\cdots+a^nf(a^n)+\cdots$ are both convergent or divergent, where $a > 1$.	
	9. D'Morgan and Bertrand's test	The series $\sum u_n$ of positive terms is convergent or divergent according as $\text{Lt}\left[\log n\left\{n\left(\frac{u_n}{u_{n+1}} - 1\right) - 1\right\}\right] >$ 1 or < 1.	

4.24 Solved Problems

Example 4.1. *Test the divergence of the series* $\sum_{n=1}^{\infty} \frac{2^n+1}{3^n+n}$.

Solution : We have $u_n = \frac{2^n+1}{3^n+n}$, $u_{n+1} = \sum_{n=1}^{\infty} \frac{2^{n+1}+1}{3^{n+1}+(n+1)}$.

$$\therefore \underset{n\to\infty}{\text{Lt}} \left|\frac{u_{n+1}}{u_n}\right| = \underset{n\to\infty}{\text{Lt}} \left|\left(\frac{2^{n+1}+1}{3^{n+1}+(n+1)}\right)\left(\frac{3^n+n}{2^n+1}\right)\right|$$

$$= \underset{n\to\infty}{\text{Lt}} \frac{2^n\left(2+\frac{1}{2^n}\right)}{3^n\left(3+\frac{n}{3^n}+\frac{1}{3^n}\right)} \times \underset{n\to\infty}{\text{Lt}} \frac{3^n\left(1+\frac{n}{3^n}\right)}{2^n\left(1+\frac{n}{2^n}\right)}$$

$$= \frac{2+0}{3+0+0} \times \frac{1+0}{1+0} = \frac{2}{3} < 1.$$

$\therefore$ **the series is convergent.**

Example 4.2. *Show that the series* $\sum_{n=1}^{\infty} n^4 e^{-n^2}$ *is convergent.*

Solution : Here, $u_n = \sum_{n=1}^{\infty} n^4 e^{-n^2}$ and $u_{n+1} = \sum_{n=1}^{\infty} (n+1)^4 e^{-(n+1)^2}$.

$$\therefore \quad \underset{n\to\infty}{\mathrm{Lt}} \left|\frac{u_{n+1}}{u_n}\right| = \underset{n\to\infty}{\mathrm{Lt}} \left\{\frac{(n+1)^4 + e^{-(n+1)^2}}{n^4 e^{-n^2}}\right\} = \underset{n\to\infty}{\mathrm{Lt}} \left\{\left(1+\frac{1}{n}\right)^4 \frac{e^{n^2}}{e^{n^2} e^{2n} e}\right\}$$

$$= \underset{n\to\infty}{\mathrm{Lt}} \frac{\left(1+\frac{1}{n}\right)^4}{e^{2n}\cdot e} = \frac{1}{\infty} = 0 < 1.$$

$\therefore$ **the series is convergent.**

Example 4.3. *Show that the given series* $\{a_n\}$ *is a bounded monotonic increasing sequence, where* $a_n = \frac{1}{1\cdot 2} + \frac{1}{2\cdot 3} + \frac{1}{3\cdot 4} + \frac{1}{4\cdot 5} + \cdots + \frac{1}{n(n+1)}$.

Solution : We have $a_n = \frac{1}{1\cdot 2} + \frac{1}{2\cdot 3} + \frac{1}{3\cdot 4} + \frac{1}{4\cdot 5} + \cdots + \frac{1}{n(n+1)}$

$$= \left(1-\frac{1}{2}\right) + \left(\frac{1}{2}-\frac{1}{3}\right) + \left(\frac{1}{3}-\frac{1}{4}\right) + \cdots + \left(\frac{1}{n}-\frac{1}{n+1}\right)$$

$$= 1 - \frac{1}{n+1} = \frac{n}{n+1}.$$

$$\therefore \quad a_{n+1} - a_n = \frac{n+1}{n+2} - \frac{n}{n+1} = \frac{1}{(n+1)(n+2)} > 0,$$

if n is a positive integer and hence a_n is a **monotonic increasing sequence.**

Again, for all positive integers, $n < n+1$.

$$\therefore \quad \frac{n}{n+1} < 1$$

For positive n, $\frac{1}{n+1} > 0$.

$\therefore \quad 0 < a_n < 1$, for all positive n.

This shows that the given sequence is bounded.

Example 4.4. *Test whether the series* $\{a_n\}$ *is convergent, where* $a_n = 1+2+3+\cdots+\infty$.

Solution : We have $a_n = 1+2+3+\cdots+\infty$.

$\therefore$ sum of n terms of the series is $S_n = 1+2+3+\cdots+n = \frac{n(n+1)}{2}$.

$$\therefore \quad \underset{n\to\infty}{\mathrm{Lt}} \frac{n(n+1)}{2} = \frac{1}{2} \underset{n\to\infty}{\mathrm{Lt}}\, n(n+1) \to \infty.$$

Hence, a_n is divergent.

Example 4.5. *Show that the series* $a_n = 1^2 + 2^2 + 3^2 + \cdots + n^2 + \cdots \infty$ *is divergent.*

Solution : We have $S_n = 1^2 + 2^2 + 3^2 + \cdots + n^2 = \left\{\frac{1}{2}n(n+1)\right\}^2$.

$$\therefore \quad \underset{n\to\infty}{\text{Lt}} \left\{\frac{1}{2}n(n+1)\right\}^2 = \frac{1}{4}\underset{n\to 0}{\text{Lt}}\{n(n+1)\}^2 \to \infty.$$

Hence, the series a_n is divergent.

Example 4.6. *Show that the sequence, $a_n = \frac{1}{n}\sin\left(\frac{n\pi}{2}\right)$ is convergent. Find the limit.*

Solution : It is known that $\sin\left(\frac{n\pi}{2}\right)$ always lies between (-1) and $(+1)$.

In this case,

$$|a_n - a_0| = \left|\frac{1}{n}\sin\left(\frac{n\pi}{2}\right) - 0\right| = \left|\frac{1}{n}\sin\left(\frac{n\pi}{2}\right)\right| < \frac{1}{n} < \epsilon, \quad \text{if } n > \frac{1}{\epsilon}.$$

We choose $m =$ a positive integral part of $\frac{1}{\epsilon}$, if it is not an integer.

Example 4.7. *Show that, $\underset{n\to\infty}{\text{Lt}}\ n^{\frac{1}{n}} = 1$ and hence, show that,*

$$\underset{n\to\infty}{\text{Lt}} \frac{1}{n}\left(1 + 2^{\frac{1}{2}} + 3^{\frac{1}{3}} + \cdots + n^{\frac{1}{n}}\right) = 1.$$

Solution : Let $a_n = n$, then $\frac{a_{n+1}}{a_n} = \frac{n+1}{n} = 1 + \frac{1}{n}$.

$$\therefore \quad \underset{n\to\infty}{\text{Lt}} \frac{a_{n+1}}{a_n} = 1.$$

By Cauchy's theorem,

$$\underset{n\to\infty}{\text{Lt}}\ a_n^{\frac{1}{n}} = \underset{n\to\infty}{\text{Lt}}\ n^{\frac{1}{n}} = \underset{n\to\infty}{\text{Lt}} \frac{a_{n+1}}{a_n} = 1.$$

Then, $\displaystyle \underset{n\to\infty}{\textbf{Lt}} \frac{1}{n}\left(1 + 2^{\frac{1}{2}} + 3^{\frac{1}{3}} + \cdots + n^{\frac{1}{n}}\right) = \underset{n\to\infty}{\textbf{Lt}} \frac{n}{n} = 1$ **(Proved).**

Example 4.8. *Test for convergence of the series $\frac{1}{1\cdot 2\cdot 3} + \frac{3}{2\cdot 3\cdot 4} + \frac{5}{3\cdot 4\cdot 5} + \cdots\infty$*

Solution : Let $u_n = \frac{2n-1}{n(n+1)(n+2)} = \frac{1}{n^2}\frac{2-\frac{1}{n}}{\left(1+\frac{1}{n}\right)\left(1+\frac{2}{n}\right)}$.

Now, putting $v_n = \frac{1}{n^2}$, we have

$$\underset{n\to\infty}{\text{Lt}} \frac{u_n}{v_n} = \underset{n\to\infty}{\text{Lt}} \frac{2-\frac{1}{n}}{\left(1+\frac{1}{n}\right)\left(1+\frac{2}{n}\right)} = \frac{2-0}{(1+0)(1+0)} = 2,$$

which is finite and non-zero.

So, $\sum u_n$ and $\sum v_n$ converge or diverge together.

But, $\sum v_n = \sum \frac{1}{n}$ is converging.

$\therefore \sum u_n$ **is also convergent.**

Example 4.9. *Discuss the convergence of the series, (i) $\sum_{r=1}^{\infty} \frac{n^2}{3^n}$ (ii) $1 + \frac{2!}{2^2} + \frac{3!}{3^3} + \frac{4!}{4^4} + \cdots$.*

Solution : (i) Let $u_n = \frac{n^2}{3^n}$ and $u_{n+1} = \frac{(n+1)^2}{3^{n+1}}$.

$$\therefore \quad \operatorname*{Lt}_{n\to\infty} \frac{u_{n+1}}{u_n} = \operatorname*{Lt}_{n\to\infty} \frac{(n+1)^2}{3^{n+1}} \times \frac{3^n}{n^2} = \operatorname*{Lt}_{n\to\infty} \frac{1}{3}\left(1+\frac{1}{n}\right)^2 = \frac{1}{3} < 1.$$

$\therefore$ the series $\sum_{n=1}^{\infty} \frac{n^2}{3^n}$ **is convergent.**

$$\text{(ii)} \quad \frac{u_{n+1}}{u_n} = \frac{(n+1)!}{(n+1)^{n+1}} \cdot \frac{n^n}{n!} = \frac{n^n}{(n+1)^n} = \frac{1}{\left(1+\frac{1}{n}\right)^n}.$$

$$\therefore \quad \operatorname*{Lt}_{n\to\infty} \frac{u_{n+1}}{u_n} = \frac{1}{\operatorname*{Lt}_{n\to\infty}\left(1+\frac{1}{n}\right)} = \frac{1}{e} < 1.$$

Hence, the series (ii) is convergent.

Example 4.10. *If the series, $\sum_{n=1}^{\infty} u_n^2 = u_1^2 + u_2^2 + u_3^2 + \cdots$ and $\sum_{n=1}^{\infty} v_n^2 = v_1^2 + v_2^2 + v_3^2 + \cdots$ be convergent, then prove that $u_1v_1 + u_2v_2 + u_3v_3 + \cdots$ is also convergent.*

Solution : We know that, the geometric mean is less than the arithmetic mean.

$$\therefore \quad u_1v_1 < \frac{1}{2}(u_1^2 + v_1^2) \quad \text{and} \quad u_2v_2 < \frac{1}{2}(u_2^2 + v_2^2) \text{ and so on.}$$

Hence, every term of the series, $u_1v_1 + u_2v_2 + \cdots$ is less than the corresponding term of the series $\frac{1}{2}\sum(u_n^2 + v_n^2)$.

But, $\sum u_n^2$ and $\sum v_n^2$ are both convergent.

$\therefore$ $\frac{1}{2}\sum(u_n^2 + v_n^2)$ **is also convergent.**

This proves that, the series $\sum u_n v_n$ is also convergent.

Example 4.11. *Use Cauchy's criterion to show that $\{a_n\}$ diverges to $(+\infty)$, where $a_n = 1 + \frac{1}{2} + \frac{1}{3} + \cdots + \frac{1}{n}$.*

Solution : Let $\{a_n\}$ converges to a finite limit. Then by Cauchy's criterion,

$$|a_{n+p} - a_n| < \epsilon, \text{ for all positive integral values of } p.$$

Assuming $\epsilon = \frac{1}{2}$, $n \geq m$, a positive integer.

$$\left|\frac{1}{n+1} + \frac{1}{n+2} + \cdots + \frac{1}{n+p}\right| < \frac{1}{2}, \text{ for } (n+p) > n \geq m.$$

When $p = n$, $\left|\frac{1}{n+1} + \frac{1}{n+2} + \cdots + \frac{1}{n+p}\right| < \frac{1}{2}$.

Again, $\frac{1}{n+1} + \frac{1}{n+2} + \cdots + \frac{1}{2n} > n \cdot \frac{1}{2n}$ or, $\frac{1}{n+1} + \frac{1}{n+2} + \cdots + \frac{1}{2n} > \frac{1}{2}$.

This shows that $\{a_n\}$ is diverging, which is a contradiction to the assumption. So, the series $\{a_n\}$ cannot converge.

Further, $a_{n+1} > a_n$.

Hence, the sequence is monotonic increasing. The sequence diverges to $(+\infty)$.

Example 4.12. *Test the convergency of the series :* $\sum_{n=1}^{\infty} \frac{1}{\sqrt{n}+\sqrt{n+1}}$.

Let $u_n = \frac{1}{\sqrt{n}+\sqrt{n+1}} = \frac{\sqrt{n+1}-\sqrt{n}}{(\sqrt{n+1}+\sqrt{n})(\sqrt{n+1}-\sqrt{n})} = \sqrt{n+1}-\sqrt{n}$

$$= \sqrt{n}\left[\left(1+\frac{1}{n}\right)^{\frac{1}{2}} - 1\right] \quad \text{[by binomial theorem]}$$

$$= \sqrt{n}\left\{\left(1+\frac{1}{2n}-\frac{1}{8n^2}+\cdots\right)-1\right\} = \sqrt{n}\left(\frac{1}{2n}-\frac{1}{8n^2}+\cdots\right)$$

$$= \frac{1}{\sqrt{n}}\left(\frac{1}{2}-\frac{1}{8n}+\cdots\right) = v_n\left(\frac{1}{2}-\frac{1}{8n}+\cdots\right), \quad \text{where } v_n = \frac{1}{\sqrt{n}}.$$

$\therefore \frac{u_n}{v_n} = \left(\frac{1}{2}-\frac{1}{8n}+\cdots\right)$ or, $\underset{n\to\infty}{\text{Lt}} \frac{u_n}{v_n} = \frac{1}{2}$, which is finite and non-zero.

$\therefore \sum u_n$ and $\sum v_n$ converge or diverge together.

But, $\sum v_n = \sum \frac{1}{\sqrt{n}}$ is known to be divergent.

$\therefore$ **$\sum u_n$ is also divergent.**

Example 4.13. *Test the convergence of the series :* $\frac{1}{2\sqrt{1}}+\frac{x^2}{3\sqrt{2}}+\frac{x^4}{4\sqrt{3}}+\frac{x^6}{5\sqrt{4}}+\cdots\infty$

Solution : Let $u_n = \frac{x^{2n-1}}{(n+1)\sqrt{n}}$, then $u_{n+1} = \frac{x^{2n}}{(n+2)\sqrt{n+1}}$.

$$\therefore \underset{n\to\infty}{\text{Lt}} \frac{u_{n+1}}{u_n} = \underset{n\to\infty}{\text{Lt}} \frac{x^{2n}}{(n+2)\sqrt{n+1}} \times \frac{(n+1)\sqrt{n}}{x^{2n-2}} = \underset{n\to\infty}{\text{Lt}} \left[\frac{n+1}{n+2}\left(\frac{n}{n+1}\right)^{\frac{1}{2}}\right] x^2$$

$$= \underset{n\to\infty}{\text{Lt}} \left[\frac{1+\frac{1}{n}}{1+\frac{2}{n}} \cdot \frac{1}{\sqrt{1+\frac{1}{n}}}\right] x^2 = x^2.$$

So, $\sum u_n$ converges for $x^2 < 1$ and diverges for $x^2 > 1$.

Now, if $x^2 = 1$, then

$$u_n = \frac{2}{(n+1)\sqrt{n}} = \frac{1}{n^{\frac{3}{2}}} \cdot \frac{1}{1+\frac{1}{n}} = v_n \frac{1}{1+\frac{1}{n}}, \quad \text{where } v_n = \frac{1}{n^{\frac{3}{2}}}.$$

$$\therefore \underset{n\to\infty}{\text{Lt}} \frac{u_n}{v_n} = \underset{n\to\infty}{\text{Lt}} = \frac{1}{1+n} = 1, \text{ a finite quantity.}$$

$\therefore \sum u_n$ and $\sum v_n$ converge or diverge together.

But, $\sum v_n = \frac{1}{n^{\frac{3}{2}}}$ is a converging series.

Hence, $\sum u_n$ is also a converging series.

Example 4.14. *Test the convergence of the series :*

$$1+\frac{2}{5}x+\frac{6}{9}x^2+\frac{14}{17}x^3+\cdots+\frac{2^n-2}{2^n+1}x^{n-1}+\cdots,\ x>0.$$

Solution : Let $u_n = \dfrac{2^n-2}{2^n+1}x^{n-1}$.

$$\therefore\quad u_n = \frac{2^{n+1}-2}{2^{n+1}}x^n.$$

$$\therefore\quad \frac{u_{n+1}}{u_n} = \frac{2^{n+1}-2}{2^{n+1}+1}x^n \times \frac{2^n+1}{2^n-2}\frac{1}{x^{n-1}} = \frac{2-\frac{2}{2^n}}{2+\frac{1}{2^n}} \times \frac{1+\frac{1}{2^n}}{1-\frac{2}{2^n}}x.$$

$$\therefore\quad \operatorname*{Lt}_{n\to\infty}\frac{u_{n+1}}{u_n} = \frac{2-0}{2+0}\times\frac{1+0}{1-0}x = x.$$

By Ratio test $\sum u_n$ converges for $x<1$ and diverges for $x>1$. But the Ratio test fails for $x=1$.

When $x=1$,

$$\operatorname*{Lt}_{n\to\infty} u_n = \operatorname*{Lt}_{n\to\infty}\frac{2^n-2}{2^n+1} = \operatorname*{Lt}_{n\to\infty}\frac{1-\frac{2}{2^n}}{1+\frac{1}{2^n}} = 1 \neq 0.$$

$\therefore$ **$\sum u_n$ diverges for $x=1$.**

So, **the given series converges for $x<1$ and diverges for $x \geq 1$.**

Example 4.15. *Test for the convergence of the series* $\sum \dfrac{4\cdot 7\cdot(3n+1)}{1\cdot 2\cdots n}x^n$.

Solution : Let $u_n = \dfrac{4\cdot 7\cdots(3n+1)}{1\cdot 2\cdots n}x^n$.

$$\therefore\quad u_{n+1} = \frac{4\cdot 7\cdots(3n+4)}{1\cdot 2\cdots(n+1)}x^{n+1}$$

$$\text{or,}\quad \frac{u_n}{u_{n+1}} = \frac{4\cdot 7\cdots(3n+1)}{1\cdot 2\cdots n}x^n \Big/ \frac{4\cdot 7\cdots(3n+4)}{1\cdot 2\cdots(n+1)}x^{n+1} = \frac{n+1}{3n+4}\cdot\frac{1}{x} = \left[\frac{1+\frac{1}{n}}{3+\frac{4}{n}}\right]\frac{1}{x}.$$

$$\therefore\quad \operatorname*{Lt}_{n\to\infty}\frac{u_n}{u_{n+1}} = \frac{1}{3x}.$$

By Ratio test, $\sum u_n$ converges for $\frac{1}{3x}>1$, i.e., $x<\frac{1}{3}$ and diverges for $x>\frac{1}{3}$. The Ratio test fails for $x=\frac{1}{3}$.

We now try the Raabe's test as follows :

$$\frac{u_n}{u_{n+1}} = \left(1+\frac{1}{n}\right)\left(1+\frac{4}{3n}\right)^{-1}$$

$$= \left(1+\frac{1}{n}\right)\left(1-\frac{4}{3n}+\frac{16}{9n^2}-\cdots\right) \quad \text{[Expanding by Binomial Theorem]}$$

$$= 1-\frac{1}{3n}+\frac{4}{9n^2}+\cdots.$$

$$\therefore\quad \operatorname*{Lt}_{n\to\infty} n\left(\frac{u_n}{u_{n+1}}-1\right) = -\frac{1}{3},\ \text{i.e., } <1.$$

Thus Raabe's test shows that the series diverges.
Here, the given series $\sum u_n$ converges for $x < 1$ and diverges for $x \geq \frac{1}{3}$.

Example 4.16. *Apply D'Alembert's ratio test to show that the series (i) $\frac{n^2}{2^n}$ converges and (ii) $\frac{2^n}{n^3}$ diverges.*

Solution : (i) Let $u_n = \frac{n^2}{2^n}$.

$$\therefore \quad u_{n+1} = \frac{(n+1)^2}{2^{n+1}}.$$

$$\therefore \quad \frac{u_{n+1}}{u_n} = \frac{(n+1)^2}{2^{n+1}} \cdot \frac{2^n}{n^2} \text{ or, } \underset{n\to\infty}{\text{Lt}} \frac{u_{n+1}}{u_n} = \underset{n\to\infty}{\text{Lt}} \frac{(n+1)^2}{2^{n+1}} \cdot \frac{2^n}{n^2} = \underset{n\to\infty}{\text{Lt}} \frac{1}{2}\left(1+\frac{1}{n}\right)^2 = \frac{1}{2} < 1.$$

Thus, by D'Alembert's Ratio test, the series $\sum u_n$ is convergent.

(ii) We have, $u_n = \frac{2^n}{n^3}$, $\therefore \; u_{n+1} = \frac{2^{n+1}}{(n+1)^3}$.

$$\text{Now, } \frac{u_{n+1}}{u_n} = \frac{2^{n+1}}{(n+1)^3} \cdot \frac{n^3}{2^n} = 2\left(1 - \frac{1}{n+1}\right)^3$$

$$\therefore \quad \underset{n\to\infty}{\text{Lt}} \frac{u_{n+1}}{u_n} = \underset{n\to\infty}{\text{Lt}} 2\left(1 - \frac{1}{(n+1)}\right)^3 = 2 > 1.$$

Hence, by D'Alembert's Ratio test, the series is divergent.

Example 4.17. *Discuss the convergence of the series :* $x + \frac{2^2x^2}{2!} + \frac{3^3x^3}{3!} + \frac{4^4x^4}{4!} + \frac{5^5x^5}{5!} + \cdots \infty.$

Solution : Let $\sum u_n = x + \frac{2^2x^2}{2!} + \frac{3^3x^3}{3!} + \frac{4^4x^4}{4!} + \frac{5^5x^5}{5!} + \cdots \infty = \sum \frac{n^nx^n}{n!}$;

then $u_n = \frac{n^nx^n}{n!}$ and $u_{n+1} = \frac{(n+1)^{n+1}x^{n+1}}{(n+1)!}$.

$$\therefore \quad \frac{u_n}{u_{n+1}} = \frac{n^nx^n}{n!} \Big/ \frac{(n+1)^{n+1}x^{n+1}}{(n+1)!} = \frac{n^n}{(n+1)^nx} = \frac{1}{\left(1+\frac{1}{n}\right)^n}\frac{1}{x} \text{ or, } \underset{n\to\infty}{\text{Lt}} \frac{u_n}{u_{n+1}} = \frac{1}{e^x}.$$

Thus, by Ratio test the given series converges for $x < \frac{1}{e}$ and diverges for $x > \frac{1}{e}$, but it fails for $x = \frac{1}{e}$.

Example 4.18. *Test for the convergence of the series :* $\sum \left(1 + \frac{1}{\sqrt{n}}\right)^{-n^{3/2}}$.

Solution : Let $u_n = \left(1 + \frac{1}{\sqrt{n}}\right)^{-n^{3/2}}$.

$$\therefore \quad u_n^{\frac{1}{n}} = \left[\frac{1}{\left(1+\frac{1}{\sqrt{n}}\right)^{3/2}}\right]^{1/n} = \frac{1}{\left(1+\frac{1}{\sqrt{n}}\right)^n}$$

$$\therefore \quad \underset{n\to\infty}{\text{Lt}} (u_n)^{1/n} = \underset{n\to\infty}{\text{Lt}} \frac{1}{\left(1+\frac{1}{\sqrt{n}}\right)^{\sqrt{n}}} = \frac{1}{e} < 1.$$

Hence, the given series is convergent.

Example 4.19. *Show that the series :* $\sum_{n=1}^{\infty} \frac{(-1)^{n-1}n}{2n-1}$ *is oscillatory.*

Solution : Let $u_n = \sum_{n=1}^{\infty} \frac{(-1)^{n-1}n}{2n-1} = 1 - \frac{2}{3} + \frac{3}{5} - \frac{4}{7} + \frac{5}{9} + \cdots + \frac{(-1)^{n-1}n}{2n-1} + \cdots\infty.$

The terms of the series are alternately positive and negative and each term is numerically smaller than its preceding term.

Again, $\underset{n\to\infty}{\text{Lt}}\ u_n = \underset{n\to\infty}{\text{Lt}} \frac{n}{2n-1} = \underset{n\to\infty}{\text{Lt}} \frac{1}{2-\frac{1}{n}} = \frac{1}{2} \neq 0.$

Hence, the given series is oscillatory.

Example 4.20. *Discuss the convergence of the following series :*

$$1 - \frac{1}{\sqrt{2}} + \frac{1}{\sqrt{3}} - \frac{1}{\sqrt{4}} + \frac{1}{\sqrt{5}} - \frac{1}{\sqrt{6}} + \cdots.$$

Solution : It is evident that each term of the given series are alternately positive and negative.

Again, each term is numerically less than its preceding term because

$$u_n - u_{n-1} = \frac{1}{\sqrt{n}} - \frac{1}{\sqrt{n-1}} < 0.$$

But, $\underset{n\to\infty}{\text{Lt}}\ u_n = \underset{n\to\infty}{\text{Lt}} \left(\frac{1}{\sqrt{n}}\right) = 0.$

Hence, by Leibnitz rule the givèn series is **convergent**.

Example 4.21. *Show that the series* (i) $\frac{1}{a\cdot 1^2+b} + \frac{2}{a\cdot 2^2+b} + \frac{3}{a\cdot 3^2+b} + \cdots + \frac{n}{a\cdot n^2+b} +$ $\cdots$ *is divergent and* (ii) *the series whose* n*th term is* $\frac{1}{n}\sin\frac{1}{n}$ *is convergent.*

Solution : We suppose, $\sum v_n$ be such that $v_n = \frac{1}{n}$.

$$\therefore \quad \underset{n\to\infty}{\text{Lt}} \frac{u_n}{v_n} = \underset{n\to\infty}{\text{Lt}} \frac{n}{a\cdot n^2+b}\cdot n = \frac{n^2}{an^2} = \frac{1}{a}, \text{ is a finite quantity.}$$

But, $\sum v_n$ is divergent, ($\because$ $p = 1$) hence, the given series = $\sum u_n$ is divergent.

(ii) The nth term of the series is

$$u_n = \frac{1}{n}\sin\frac{1}{n} = \frac{1}{n}\left(\frac{1}{n} - \frac{1}{3!}\frac{1}{n^3} + \frac{1}{5!}\frac{1}{n^3} - \cdots\right) \simeq \frac{1}{n^2}, \text{ when } n \text{ is large.}$$

Let $v_n = \frac{1}{n^2}$.

$\therefore \quad \underset{n\to\infty}{\text{Lt}} \dfrac{u_n}{v_n} = 1$, a finite quantity.

The series $\sum v_n$ is convergent, so the series $\sum u_n$ is also convergent.

Example 4.22. *Test the convergency or divergency of the following series :*

(i) $1 + \dfrac{x}{2} + \dfrac{x^2}{5} + \dfrac{x^3}{10} + \cdots + + \dfrac{x^n}{n^2+1} + \cdots$ (ii) $\dfrac{1}{2} + \dfrac{2}{3} + \dfrac{3}{4} + \dfrac{4}{5} + \cdots$.

Solution : (i) From after the first term, $u_n = \dfrac{x^n}{n^2+1}$.

$\therefore \quad u_{n+1} = \dfrac{x^{n+1}}{(n+1)^2+1}$.

$\therefore \quad \underset{n\to\infty}{\text{Lt}} \dfrac{u_n}{u_{n+1}} = \underset{n\to\infty}{\text{Lt}} \dfrac{n^2+2n+2}{n^2+1} \cdot \dfrac{1}{x} = \underset{n\to\infty}{\text{Lt}} \dfrac{1+\frac{2}{n}+\frac{2}{n^2}}{1+\frac{1}{n^2}} \cdot \dfrac{1}{x} = \dfrac{1}{x}$.

For $\frac{1}{x} > 1$ or, $x < 1$, the series is convergent and for $\frac{1}{x} < 1$ or, $x > 1$, the series is divergent by D'Alembert's Ratio test.

If $x = 1$, then $u_n = \dfrac{1}{n^2+1}$.

Let $v_n = \dfrac{1}{n^2}$, a convergent series.

$\therefore \quad \underset{n\to\infty}{\text{Lt}} \dfrac{u_n}{v_n} = \underset{n\to\infty}{\text{Lt}} \dfrac{n^2}{n^2+1} = 1$, a finite quantity.

So, the given series is convergent for $x \le 1$ and divergent for $x > 1$.

(ii) Here $u_n = \dfrac{n}{n+1}$, $\quad \therefore \quad u_{n+1} = \dfrac{n+1}{n+2}$.

So, $\underset{n\to\infty}{\text{Lt}} \dfrac{u_{n+1}}{u_n} = \dfrac{(n+1)^2}{n(n+2)} = \dfrac{\left(1+\frac{1}{n}\right)^2}{\left(1+\frac{2}{n}\right)} \to 1$.

Thus, the D'Alembert's ratio test fails. Now, we try Raabe's test.

$$\underset{n\to\infty}{\text{Lt}} \left\{ n\left(\frac{u_n}{u_{n+1}} - 1\right)\right\} = \underset{n\to\infty}{\text{Lt}} \left[n \left\{ \frac{n(n+2)}{(n+1)^2} - 1 \right\}\right] = 0 < 1.$$

Hence, the given series is divergent.

Example 4.23. *Test the convergence of the series, if $(y - x) \neq 1$,*

$$1 + \frac{1+x}{1+y} + \frac{(1+x)(2+x)}{(1+y)(2+y)} + \frac{(1+x)(2+x)(3+x)}{(1+y)(2+y)(3+y)} + \cdots$$

Solution : Here, $u_n = \dfrac{(1+x)(2+x)\cdots(n+x)}{(1+y)(2+y)\cdots(n+y)}$

Again, $u_{n+1} = \dfrac{(1+x)(2+x)\cdots(n+x)(n+x+1)}{(1+y)(2+y)\cdots(n+y)(n+y+1)}$.

$\therefore \quad \dfrac{u_n}{u_{n+1}} = \dfrac{1+n+y}{1+n+x}$.

Now, $\underset{n\to\infty}{\text{Lt}} \dfrac{u_n}{u_{n+1}} = \underset{n\to\infty}{\text{Lt}} \dfrac{1+n+y}{1+n+x} = \underset{n\to\infty}{\text{Lt}} \dfrac{\frac{1}{n}+1+\frac{y}{n}}{\frac{1}{n}+1+\frac{x}{n}} = 1.$

Again, $\dfrac{u_n}{u_{n+1}} - 1 = \dfrac{y-x}{1+n+x}.$

$$\therefore \quad \underset{n\to\infty}{\text{Lt}}\ n\left(\frac{u_n}{u_{n+1}} - 1\right) = \underset{n\to\infty}{\text{Lt}} \frac{n(y-x)}{n+1+x} = \underset{n\to\infty}{\text{Lt}} \frac{y-x}{1+\frac{(1+x)}{n}} = y - x.$$

Hence, for $(y-x) > 1$, the series is convergent.
If $(y-x) < 1$, the series is divergent by Raabe's test.

Example 4.24. *Test the convergence of the series :* $\dfrac{1}{(\log 2)^r} + \dfrac{1}{(\log 3)^r} + \cdots + \dfrac{1}{(\log n)^r} + \cdots.$

Solution : We have $u_n = \dfrac{1}{[\log(n+1)]^r}$ and $u_{n+1} = \dfrac{1}{[\log(n+2)]^r}.$

$$\frac{u_n}{u_{n+1}} = \left[\frac{\log(n+2)}{\log(n+1)}\right]^r = \left[\frac{\log(n+1) + \log\left(1+\frac{1}{n+1}\right)}{\log(n+1)}\right]^r$$

$$= \left[1 + \frac{1}{\log(n+1)}\left\{\frac{1}{n+1} - \frac{1}{2(n+1)^2} + \cdots\right\}\right]^r$$

$$= \left[1 + \frac{1}{(n+1)\log(n+1)} - \frac{1}{2(n+1)^2\log(n+1)} + \cdots\right]^r.$$

$$\therefore \quad \underset{n\to\infty}{\text{Lt}} \frac{u_n}{u_{n+1}} = 1, \text{ the test fails.}$$

Now, $\log \dfrac{u_n}{u_{n+1}} = r \log\left[1 + \dfrac{1}{(n+1)\log(n+1)} - \dfrac{1}{2(n+1)^2\log(n+1)} + \cdots\right]$

$$= r\left[\frac{1}{(n+1)\log(n+1)} + \cdots\right].$$

$$\therefore \quad \underset{n\to\infty}{\text{Lt}}\left(n\log\frac{u_n}{u_{n+1}}\right) = \underset{n\to\infty}{\text{Lt}}\ r\left[\frac{n}{(n+1)\log(n+1)} + \cdots\right] = 0, \text{ for all values of } r.$$

Hence, by logarithm test, the series is divergent for all values of r.

Example 4.25. *Test the convergence of the series :* $\dfrac{x^2}{2\log 2} + \dfrac{x^3}{3\log 3} + \dfrac{x^4}{4\log 4} + \cdots \infty.$

Solution : We have $u_n = \dfrac{x^{n+1}}{(n+1)\log(n+1)}.$

$$u_n = \frac{x^{n+2}}{(n+2)\log(n+2)}$$

$$\therefore \quad \underset{n\to\infty}{\text{Lt}}\left|\frac{u_n}{u_{n+1}}\right| = \underset{n\to\infty}{\text{Lt}} \frac{x^{n+2}}{(n+2)\log(n+2)} \cdot \frac{(n+1)\log(n+1)}{x^{n+1}}$$

$$= \underset{n\to\infty}{\text{Lt}} \frac{n\left(1+\frac{1}{n}\right)\log\left(1+\frac{1}{n}\right)}{n\left(1+\frac{2}{n}\right)\log\left(1+\frac{2}{n}\right)} x = x.$$

The series is convergent, if $x < 1$ and divergent, if $x > 1$. But for $x = 1$, the Ratio test fails. Now, for $x = 1$,

$$\frac{u_n}{u_{n+1}} = \frac{(n+1)\log(n+1)}{(n+2)\log(n+2)} = \frac{(n+2-1)}{(n+2)}\frac{\log\left\{n\left(1+\frac{1}{n}\right)\right\}}{\log\left\{n\left(1+\frac{2}{n}\right)\right\}}$$

$$= \left(1-\frac{1}{n+2}\right)\left\{\frac{\log n + \frac{1}{n} - \frac{1}{2}\frac{1}{n^2} + \cdots}{\log n + \frac{2}{n} - \frac{1}{2}\frac{4}{n^2} + \cdots}\right\}$$

$$= \left(1-\frac{1}{n+2}\right)\frac{\log n\left(1+\frac{1}{n\log n} - \frac{1}{2n^2\log n} + \cdots\right)}{\log n\left(1+\frac{2}{n\log n} - \frac{2}{n^2\log n} + \cdots\right)}$$

$$= \left(1-\frac{1}{n+2}\right)\frac{\left(1+\frac{1}{n\log n} + \cdots\right)}{\left(1+\frac{2}{n\log n} + \cdots\right)}$$

$$= \left(1-\frac{1}{n+2}\right)\left(1+\frac{1}{n\log n} - \cdots\right)\left(1+\frac{2}{n\log n} - \cdots\right)^{-1}$$

$$= \left(1-\frac{1}{n+2}\right)\left(1+\frac{1}{n\log n} - \frac{2}{n\log n} + \cdots\right)$$

$$= \left(1-\frac{1}{n+2}\right)\left(1-\frac{1}{n\log n} + \cdots\right)$$

$$= 1 - \frac{1}{n\log n} + \frac{1}{(n+2)n\log n} + \cdots = 1 - \frac{1}{n\log n}.$$

By Raabe's test

$$\underset{n\to\infty}{\text{Lt}}\, n\left(1-\frac{1}{n\log n}\right) = \underset{n\to\infty}{\text{Lt}}\, n\left(1-1+\frac{1}{n\log n}\right) = \underset{n\to\infty}{\text{Lt}}\frac{n}{n\log n} = \underset{n\to\infty}{\text{Lt}}\frac{1}{\log n} = 0 < 1.$$

$\therefore$ **the series is divergent.**

Hence, if $x < 1$, the series convergent, if $x \geq 1$, the series is divergent.

Example 4.26. *Test whether the series :* $\frac{2^2 4^2}{3^2 5^2} + \frac{2^2 4^2 6^2}{3^2 5^2 7^2} + \frac{2^2 4^2 6^2 8^2}{3^2 5^2 7^2 9^2} + \cdots$ *is convergent or divergent.*

Solution : Here, $u_n = \left[\frac{2\cdot 4\cdot 6\cdots(2n)^2}{3\cdot 5\cdot 7\cdots(2n+1)^2}\right]^2$ and $u_n = \left[\frac{2\cdot 4\cdot 6\cdots(2n)(2n+2)}{3\cdot 5\cdot 7\cdots(2n+1)(2n+3)}\right]^2$.

$$\therefore \quad \frac{u_{n+1}}{u_n} = \left(\frac{2n+2}{2n+3}\right)^2 \quad \text{or,} \quad \underset{n\to\infty}{\text{Lt}}\left|\frac{u_{n+1}}{u_n}\right| = \underset{n\to\infty}{\text{Lt}}\frac{n^2\left(2+\frac{2}{n}\right)^2}{n^2\left(2+\frac{3}{n}\right)^2} = 1.$$

Hence, the Ratio test fails.

We now try with Raabe's test,

$$\underset{n\to\infty}{\text{Lt}}\, n\left(1-\left|\frac{u_{n+1}}{u_n}\right|\right) = \underset{n\to\infty}{\text{Lt}}\, n\left[1-\frac{(2n+2)^2}{(2n+3)^2}\right] = \underset{n\to\infty}{\text{Lt}}\frac{n^2\left(4+\frac{5}{n}\right)}{n^2\left(2+\frac{3}{n}\right)} = 1.$$

The Raabe's test also fails.

Now we apply Gauss' test. According to this test if for series, u_n,

$$\left|\frac{u_{n+1}}{u_n}\right| = 1 - \frac{L}{n} + \frac{C_n}{n^2},$$

where L is independent of n and C_n is finite, then the series is divergent. In the present case,

$$\left|\frac{u_{n+1}}{u_n}\right| = \frac{4n^2+8n+4}{4n^2+12n+9} = 1 - \frac{1}{n} + \frac{7+\frac{9}{n}}{4n^2+12n+9} = 1 - \frac{L}{n} + \frac{c_n}{n^2},$$

where $L = 1$ and $C_n = 7 + \dfrac{9}{n} = 7$ (finite) as $n \to \infty$.

Hence, by Gauss' test, the given series is divergent.

Example 4.27. *Show by Cauchy's root test the following series is convergent :*

$$\left(\frac{2^2}{1^2} - \frac{2}{1}\right)^{-1} + \left(\frac{3^3}{2^3} - \frac{3}{2}\right)^{-2} + \left(\frac{4^4}{3^4} - \frac{4}{3}\right)^{-3} + \cdots.$$

Solution : We have $u_n = \left[\dfrac{(n+1)^{n+1}}{n^{n+1}} - \dfrac{n+1}{n}\right]^{-n}$.

$$\therefore \operatorname*{Lt}_{n\to\infty} (u_n)^{\frac{1}{n}} = \operatorname*{Lt}_{n\to\infty}\left[\left\{\frac{(n+1)^{n+1}}{n^{n+1}} - \frac{n+1}{n}\right\}\right]^{-1} = \operatorname*{Lt}_{n\to\infty}\left[\left\{\left(1+\frac{1}{n}\right)^{n+1} - 1\right\} - \left(1+\frac{1}{n}\right)\right]^{-1}$$

$$= (1+0)(e-1)^{-1} = \frac{1}{e-1} < 1 \quad [\because\ e > 2].$$

$$\left[\because \operatorname*{Lt}_{n\to\infty}\left(1+\frac{1}{n}\right)^n = e\right]$$

Hence, by Gauss' root test the given series is convergent.

Example 4.28. *Examine whether the given series is convergent or divergent :*

$$x + \frac{2^2x^2}{2!} + \frac{3^3x^3}{3!} + + \frac{4^4x^4}{4!} + \cdots \infty.$$

Solution : Here, $u_n = \dfrac{n^n x^n}{n!}$ and $u_{n+1} = \dfrac{(n+1)^{n+1}x^{n+1}}{(n+1)!}$.

$$\therefore \frac{u_n}{u_{n+1}} = \frac{n^n x^n}{n!} \cdot \frac{(n+1)!}{(n+1)^{n+1}x^{n+1}} = \frac{n^n(n+1)n!}{(n+1)^n(n+1)n!\frac{1}{x}} = \frac{1}{\left(1+\frac{1}{n}\right)^n}\frac{1}{x}.$$

$$\therefore \operatorname*{Lt}_{n\to\infty}\frac{u_n}{u_{n+1}} = \operatorname*{Lt}_{n\to\infty}\frac{1}{\left(1+\frac{1}{n}\right)^n}\cdot\frac{1}{x} = \frac{1}{ex}.$$

Thus, the series is convergent, if $\frac{1}{ex} > 1$ or $x < \frac{1}{e}$; and divergent, if $\frac{1}{ex} < 1$ or, $x > \frac{1}{e}$.

Now, if $ex = 1$ or, $x = \frac{1}{e}$, the test fails.

Let us now try with logarithmic test as follows :

$$\log\frac{u_n}{u_{n+1}} = \log\left\{\frac{1\cdot e}{\left(1+\frac{1}{n}\right)^n}\right\} = \log e - n\log\left(1+\frac{1}{n}\right) = 1 - n\left\{\frac{1}{n} - \frac{1}{2}\frac{1}{n^2} + \frac{1}{3}\frac{1}{n^2} - \cdots\right\}$$

$$= 1 - 1 + \frac{1}{2n} - \frac{1}{3n^2} + \cdots = \frac{1}{2n} - \frac{1}{3n^2}.$$

$$\therefore \quad \underset{n\to\infty}{\text{Lt}}\; n\log\left(\frac{u_n}{u_{n+1}}\right) = \underset{n\to\infty}{\text{Lt}}\; n\left(\frac{1}{2n}-\frac{1}{3n^2}\right) = \underset{n\to\infty}{\text{Lt}}\left(\frac{1}{2}-\frac{1}{3n}\right) = \frac{1}{2} < 1.$$

Thus, the series is divergent.

Finally, the series is convergent, if $x < \frac{1}{e}$ and divergent, if $x \geq \frac{1}{e}$.

Exercises

4(A) : Objective Type Questions

Select the correct answers

4.1 A sequence is a set of real numbers formed

(a) arbitrarily (b) according to some definite law
(c) with or without any definite law (d) none of these.

4.2 The sequence $[1+(-1)^n n] \to 0,\ 3,\ -2,\ 5,\ -4,\ 7\cdots$ has

(a) upper bound (b) lower bound (c) neither upper nor lower bound (d) none of these.

4.3 A sequence $\{a_n\}$ is said to be monotonic increasing, if

(a) $a_n > a_{n+1}$ (b) $a_n < a_{n+1}$ (c) $a_n \leq a_{n+1}$ (d) $a_n = a_{n+1}$.

4.4 A sequence $\{a_n\}$ is monotonic decreasing, if

(a) $a_n = a_{n+1}$ (b) $a_n > a_{n+1}$ (c) $a_n \geq a_{n+1}$ (d) $a_n < a_{n+1}$.

4.5 The limit of the sequence $\left\{\frac{1}{n}\right\}$ as $n \to \infty$ is

(a) ∞ (b) 100 (c) 0 (d) 50.

4.6 The sequence $\{n^2\} \to 1^2,\ 2^2,\ 3^2,\ \ldots$ is

(a) convergent (b) divergent (c) oscillatory (d) none of these.

4.7 The sequence $\{(-1)^n\} = -1,\ 1,\ -1,\ \ldots$ is

(a) convergent (b) divergent (c) oscillatory (d) none of these.

4.8 A series is a

(a) sum of sequences (b) product of sequences
(c) difference of sequences (d) division of sequences.

4.9 If the sum of the first n terms S_n of a series $\sum u_n$ tends to $\pm\infty$, then the series is

(a) convergent (b) divergent (c) oscillatory (d) none of these.

4.10 If the sum of the first n terms S_n of a series $\sum u_n$ does not tend to a unique limit as $n \to \infty$, then the series is

(a) divergent (b) convergent (c) oscillatory (d) none of these.

4.11 If each term of an infinite series is multiplied by a finite number, then the convergence or divergence is

(a) affected (b) unaffected (c) both (a) and (b) (d) none of these.

4.12 If a finite number of terms be included or removed from an infinite series, the convergence or divergence is

(a) affected (b) unaffected (c) both (a) and (b) (d) none of these.

4.13 The necessary condition for convergence of a series is

(a) $\underset{n\to 0}{\text{Lt}}\ u_n = \infty$ (b) $\underset{n\to 0}{\text{Lt}}\ u_n = \infty$ (c) $\underset{n\to 0}{\text{Lt}}\ u_n =$ any finite value (d) $\underset{n\to 0}{\text{Lt}}\ u_n = 1$.

4.14 If $\sum u_n$ be a positive term series for which $\underset{n\to\infty}{\text{Lt}}\ \dfrac{u_{n+1}}{u_n} = \lambda$, then series converges for

(a) $\lambda = 0$ (b) $\lambda > 1$ (c) $\lambda < 1$ (d) $\lambda = \infty$.

4.15 According to Raabe's test if in a positive term series $\sum u_n$, $\underset{n\to\infty}{\text{Lt}}\ n\left(\dfrac{u_n}{u_{n+1}} - 1\right) = k$, then the series $\sum u_n$ diverges for

(a) $k > 1$ (b) $k = 1$ (c) $k < 1$ (d) $k = 0$.

4.16 If $\sum u_n$ be a positive term series such that, $\dfrac{u_n}{u_{n+1}} = \alpha + \dfrac{\beta}{m} + \dfrac{\gamma_n}{n^p}$, then for $\alpha = 1$, $\sum u_n$ converges, if

(a) $\beta = 0$ (b) $\beta \leq 1$ (c) $\beta > 1$ (d) $\beta = \infty$.

4.17 A binomial series, $1 + \dfrac{nx}{1!} + \dfrac{n(n-1)x^2}{2!} + \cdots + n(n-1)\cdots(n-r+1)x^n + \cdots\infty$ converges for

(a) $|x| = 0$ (b) $|x| < 1$ (c) $|x| > 1$ (d) $|x| = \infty$.

4.18 An exponential series in x expressed as $1 + x + \dfrac{x^2}{2!} + \cdots + \dfrac{x^n}{n!} + \cdots\infty$ converges for

(a) $x = 0$ (b) $x = -1$ (c) $x = x$ (d) all values of x.

4.19 A logarithmic series in x as $x - \dfrac{x^2}{2} + \dfrac{x^3}{3} + \cdots + (-1)^n\dfrac{x^n}{n!} + \cdots\infty$ converges for

(a) $|x| < 1$ (b) $|x| > 1$ (c) $-1 < x \leq 1$ (d) $|x| = 1$.

4.20 The series $\sum_{n=0}^{\infty}(2x)^n$ converges, if

(a) $-1 \leq x \leq 1$ (b) $-\dfrac{1}{2} \leq x \leq \dfrac{1}{2}$ (c) $-2 \leq x \leq 2$ (d) none of these.

4.21 If $u_n > 0$ and the series $\sum u_n$ be convergent, then the series $\sum \dfrac{1}{n}u_n$ is

(a) divergent (b) convergent (c) oscillating (d) none of these.

4.22 The series $1 + x + x^2 + x^3 + \cdots$ is convergent, if

(a) $0 < x < 1$ (b) $-1 < x < 1$ (c) $1 < x < 2$ (d) $-2 < x < 2$.

4.23 The series $\sum\left(1 + \dfrac{1}{\sqrt{n}}\right)^{-n^{3/2}}$ is convergent, if $\underset{n\to\infty}{\text{Lt}}\ (u_n)^{\frac{1}{n}}$ is equal to

(a) 1 (b) 2 (c) -1 (d) $\dfrac{1}{e}$.

4.24 The harmonic series of order p : $\sum_{n=1}^{\infty} \frac{1}{n^p}$ converges, if

(a) $p > 1$ (b) $p = 1$ (c) $p < 1$ (d) $p = 0$.

4.25 The geometric series, $1 + r + r^2 + r^3 + \cdots \infty$ oscillates, if

(a) $r = -1$ (b) $r = 1$ (c) $r = 0$ (d) $r \leq -1$.

Answers

4.1 (b) **4.2** (c) **4.3** (c) **4.4** (c) **4.5** (c) **4.6** (b) **4.7** (c) **4.8** (a) **4.9** (b)
4.10 (b) **4.11** (b) **4.12** (b) **4.13** (b) **4.14** (c) **4.15** (c) **4.16** (c) **4.17** (b) **4.18** (d)
4.19 (c) **4.20** (a) **4.21** (b) **4.22** (b) **4.23** (d) **4.24** (a) **4.25** (d).

4(B) : Short Answer Type Questions

4.1 Define a sequence and give an example. What do you mean by finite sequence and infinite sequence? Give one example of each.

4.2 When a sequence is said to be bounded? What do you mean by rough upper bound and exact upper bound of a sequence.

Is the sequence $\left\{2 + (-1)^n \frac{1}{n}\right\}$ bounded? What are its upper and lower bounds?

4.3 What is a monotonic sequence? What are monotonic increasing and monotonic decreasing sequences? Give one example of each.

4.4 Explain what is meant by the limit of an infinite sequence. State the nature of the sequence $\{1, 0, 1, 0, 1, 0, \ldots\}$.

4.5 What do you mean by convergent, divergent and oscillating sequences? Give one example of each.

4.6 Define a series. When the series is finite and infinite? Give examples.

4.7 Discuss the convergence, divergence and oscillation of a series.

4.8 If $\sum u_n$ converges to s and $\sum v_n$ converges to t, then show that $\sum(u_n + v_n)$ converges to $(s + t)$.

4.9 Discuss with example, the absolute and conditional convergences.

4.10 State and prove D'Alembert's Ratio test for the convergence of a series.

4.11 State and prove Raabe's test and Gauss' test for the convergence of a series.

4.12 What is an alternating series? State and prove Leibnitz's rule for alternating series.

4.13 Discuss the convergence of **(a)** Binomial series **(b)** Exponential series and **(c)** Power series.

4.14 Discuss the convergence or divergence of a logarithmic series.

4.15 Discuss the procedure to check whether a series is convergent or divergent.

4(C) : Broad Answer Type Questions

4.1 Show that the sequence $\{a_n\}$ is monotonic increasing and bounded, where

(a) $a_n = \dfrac{3n+1}{n+2}$ (b) $a_n = \dfrac{1}{n+1} + \dfrac{1}{n+2} + \cdots + \dfrac{1}{2n}$

(c) $a_n = \dfrac{1}{1\cdot 3} + \dfrac{1}{3\cdot 5} + \dfrac{1}{5\cdot 7} + \cdots + \dfrac{1}{(2n-1)(2n+1)}$.

4.2 (a) Show that the sequence $\sqrt{2}, \sqrt{2+\sqrt{2}}, \sqrt{2+\sqrt{2+\sqrt{2}}}, \ldots$ tends to a definite finite limit and find the limit. **[Ans: $-1 < a_n < 2$]**

(b) Show that the supper and lower limits of $\left\{\dfrac{n-1}{2n}\right\}$ are $\dfrac{1}{2}$ and 0.

4.3 (a) A sequence $\{x_n\}$ is defined as $x_n = \sqrt{2x_{n-1}}$, $x_1 = \sqrt{2}$. Show that the sequence is monotonic increasing and bounded above.

(b) Show that the sequence $a_n = 2^n$ is strictly monotonic increasing.

4.4 Evaluate the limit, as $n \to \infty$ of (a) $\dfrac{5n}{n+5\sqrt{n}}$ (b) $\dfrac{\log n}{n}$ (c) $\dfrac{(2n+1)^2}{2n^2}$.

4.5 Show that the sequence

(a) $\sqrt{5}, \sqrt{5+\sqrt{5}}, \sqrt{5+\sqrt{5+\sqrt{5}}}, \ldots$ converges to $\dfrac{1}{2}(\sqrt{21}+1)$.

(b) $\sqrt{2}, \sqrt{2\sqrt{2}}, \sqrt{2\sqrt{2\sqrt{2}}}, \ldots$ converges to 2.

(c) If $S_n > 1$ and $S_{n+1} = \dfrac{2}{S_{n+1}}$, then prove that the sequence $\{S_n\}$ converge to 1.

4.6 Show that the following series are convergent :

(a) $2 - \dfrac{3^2}{2} + \dfrac{3^3}{5} - \dfrac{3^4}{4} + \cdots$ (b) $1 + \dfrac{2}{3} + \dfrac{4}{9} + \dfrac{8}{27} + \dfrac{16}{81} + \cdots$ (c) $1 + \dfrac{2^2}{2!} + \dfrac{3^3}{3!} + \dfrac{4^4}{4!} + \cdots\infty$.

4.7 Show that the following series are divergent :

(a) $\dfrac{1}{1\cdot 3} + \dfrac{2}{3\cdot 5} + \dfrac{3}{5\cdot 7} + \cdots\infty$ (b) $1^2 + 2^2 + 3^2 + \cdots + n^2 + \cdots\infty$ (c) $u_n = \dfrac{1}{n^{1+\frac{1}{n}}}$.

4.8 Test the convergence or divergence of the following series :

(a) $1 + \dfrac{1}{3} + \dfrac{1}{3^3} + \cdots$ **[Ans: convergent]**

(b) $\dfrac{1}{1\cdot 2\cdot 3} + \dfrac{3}{2\cdot 3\cdot 4} + \dfrac{5}{3\cdot 4\cdot 5}$ **[Ans: convergent]**

(c) $\dfrac{\sqrt{2}-1}{3^3-1} + \dfrac{\sqrt{3}-1}{4^3-1} + \dfrac{\sqrt{4}-1}{5^3-1} + \cdots$ **[Ans: convergent]**

(d) $\displaystyle\sum_{n=1}^{\infty} \frac{3^n}{n^3}$ **[Ans: divergent]**

(e) $\displaystyle\sum_{1}^{\infty} \frac{\sqrt[3]{3n^2+1}}{\sqrt[4]{4n^3+2n+7}}$. **[Ans: divergent]**

4.9 Show that the series

(a) $1+\dfrac{1}{2\cdot1^2+1}+\dfrac{1}{2\cdot2^2+1}+\dfrac{1}{2\cdot3^2+1}+\cdots$ is convergent

(b) $1+\dfrac{1+2}{2^3}+\dfrac{1+2+3}{3^3}+\dfrac{1+2+3+4}{4^3}+\cdots$ is divergent.

4.10 Test the convergence or divergence of the following series :

(a) $x+\dfrac{3}{5}x^2+\dfrac{8}{10}x^3+\dfrac{15}{17}x^4+\cdots+\dfrac{n^2-1}{n^2+1}x^n+\cdots\infty$

[**Ans:** convergent, if $x<1$, divergent, if $x\geq 1$]

(b) $1+\dfrac{\alpha\cdot\beta}{1\cdot\gamma}x+\dfrac{\alpha(\alpha+1)\beta(\beta+1)}{1\cdot2\cdot\gamma(\gamma+1)}x^2+\dfrac{\alpha(\alpha+1)(\alpha+2)\beta(\beta+1)(\beta+2)}{1\cdot2\cdot3\cdot\gamma(\gamma+1)(\gamma+2)(\gamma+3)}x^3+\cdots$

[**Ans:** convergent, if $x<1$, divergent, if $x>1$. If $x=1$, then the series converges, if $\gamma>\alpha+\beta$ and diverges, if $\gamma\leq\alpha+\beta$]

(c) $1+2x+3^2\dfrac{x^2}{2!}+4^3\dfrac{x^3}{3!}+5^4\dfrac{x^4}{4!}+\cdots.$

[**Ans:** converges, if $0<x<\frac{1}{e}$, divergent if $\frac{1}{e}<x<\infty$]

4.11 Apply ratio test and logarithmic test to discuss the convergence of the following series :

$$x+\frac{2^2x^2}{2!}+\frac{3^3x^3}{3!}+\frac{4^4x^4}{4!}+\cdots$$

[**Ans:** converges, if $x<\frac{1}{e}$, diverges, if $x\geq\frac{1}{e}$]

4.12 Test the convergence of the following series :

(a) $\dfrac{a}{b}+\dfrac{a(a+d)}{b(b+d)}x+\dfrac{a(a+d)(a+2d)}{b(b+d)(b+2d)}x^2+\cdots \quad (a>0,\ b>0,\ x>0)$

[**Ans:** convergence, if $x<1$ or, $x=1$ and $b>a+d$; divergence, if $x>1$ or, $x=1$ and $b<a+d$]

(b) $\displaystyle\sum_{n=1}^{\infty}\frac{n!}{x(x+1)(x+2)\cdots(x+n-1)}$ [**Ans:** converges, if $x>2$, diverges, if $x\leq 2$]

4.13 For what values of x are the following series convergent?

(a) $x-\dfrac{x^2}{\sqrt{2}}+\dfrac{x^3}{\sqrt{3}}-\dfrac{x^4}{\sqrt{4}}+\cdots\infty$ [**Ans:** $-1\leq x\leq 1$]

(b) $\dfrac{1}{1-x}+\dfrac{1}{2(1-x)^2}+\dfrac{1}{3(1-x)^3}+\cdots\infty.$ [**Ans:** $x<0$ and $x\geq 2$]

4.14 Test the convergence of the following series :

(a) $1-0.1+0.001-0.001+\cdots$ [**Ans:** convergent]

(b) $\dfrac{1}{xy}-\dfrac{1}{(x+1)(y+1)}+\dfrac{1}{(x+2)(y+2)}-\dfrac{1}{(x+3)(y+3)}+\cdots$ [**Ans:** convergent]

4.15 Test the convergence of the series $\sum u_n$, where

(a) $u_n=\dfrac{n^{n^2}}{(1+n)^{n^2}}$ [**Ans:** convergent]

(b) $u_n = \frac{n+1}{n^3}x^n$ [**Ans:** convergent for $x \le 1$, divergent for $x > 1$]

(c) $u_n = \frac{1}{n}\left(\frac{n+\sqrt{n}}{2n}\right)^n$. [**Ans:** convergent]

4.16 Test the convergence of the following series :

(a) $\sum_{n=1}^{\infty} u_n = \sum_{n=1}^{\infty} \frac{\sqrt{n}x^n}{\sqrt{n^2+1}}$, $x > 0$ [**Ans:** convergent for $x < 1$ and divergent for $x \ge 1$]

(b) $\sum_{n=1}^{\infty} u_n = \sum_{n=1}^{\infty} \left(1+\frac{1}{\sqrt{n}}\right)^{-n^{\frac{3}{2}}}$. [**Ans:** convergent]

4.17 Test for uniform convergence of the following series :

(a) $\sum \frac{\cos n\pi}{n^p}$, $(p > 1)$ [**Ans:** uniformly convergent for all x]

(b) $\frac{ax}{2} + \frac{a^2x^2}{5} + \frac{a^3x^3}{10} + \cdots + \frac{a^nx^n}{n^2+1} + \cdots \infty$. [**Ans:** uniformly convergent for all x]

4.18 Test the following series for conditional convergence :

(a) $\sum_{n=1}^{\infty} \frac{(-1)^{n-1}n}{n^2+1}$ [**Ans:** conditionally convergent]

(b) $\sum_{n=1}^{\infty} \frac{(-1)^{n-1}}{\sqrt{n}}$. [**Ans:** conditionally convergent]

4.19 Test the convergence of the following series :

(a) $\sum_{n=1}^{\infty} \frac{(2n)!}{(n!)}x^n$ [**Ans:** convergent for $x < \frac{1}{4}$; divergent for $x \ge \frac{1}{4}$]

(b) $\sum_{n=1}^{\infty} \frac{(n+1)x^n}{(n+2)(n+3)}$, $(x > 0)$. [**Ans:** convergent for $x \le 1$, divergent for $x > 1$]

4.20 (a) Show that the expansion of $\log(1+x)$ in ascending powers of x is convergent when $|x| < 1$.

(b) Test the series $1^2 + 2^2x + 3^2x^2 + 4^2x^3 + \cdots$ for convergence for $x < 1$, $x > 1$ and $x = 1$. **[CU(H) 2007]**

4.21 (a) Write down Taylor series expansion of a function $f(x, y)$ about the values x_0, y_0.

(b) Prove that the series $x + \frac{x^2}{2} + \frac{x^3}{3} + \cdots$ is convergent for $0 < x < 1$.

(c) Expand $\frac{1}{x-2}$ in a Taylor series about the point $x = 1$. What is the radius of convergence of this series? **[CU(H) 2015]**

4.22 (a) Show that the infinite series $\sum \frac{1}{n^p}$ is convergent, if $p > 1$.

(b) Find the first four terms in the Maclaurian series expansion of $(1+x)\ln(1+x)$. **[CU(H) 2014]**

4.23 (a) Expand $\dfrac{1}{x+3}$ in a Taylor series about the point $x = 1$. What is the radius of convergence of this series?

(b) Find the first three non-zero coefficients in the Taylor series expansion for $\dfrac{1}{\cos x}$ about $x = 0$.

(c) Show that by appropriate choice of n, one series at a time may be converted into a polynomial. **[CU(H) 2013]**

4.24 (a) Prove that series $x + \dfrac{x^2}{2} + \dfrac{x^3}{3} + \cdots$ is convergent for $0 < x < 1$.

(b) Use Maclaurian series expansion to find an upper limit of θ (in degrees) for which we can use the approximation $\sin\theta = \theta$ up to 3 decimal places.

(c) Find the condition on the parameter α such that an infinite series solution becomes a polynomial. What is the order of this polynomial? **[CU(H) 2012]**

4.25 (a) Test the series $1^2 + 2^2x + 3^2x^2 + 4^2x^3 + \cdots$ for convergence for $|x| < 1$ and $|x| > 1$. **[CU(H) 2011]**

(b) Expand $\dfrac{1}{x-2}$ in Taylor series about the point $x = 1$. What is the radius of convergence of this series? **[CU(H) 2010]**

CHAPTER 5

Complex Variables

5.1 Definition of a Complex Number

A complex number z is defined as an ordered pair (x, y) where both x and y are real numbers and is written as

$$z = x + iy, \quad \text{where} \quad i = \sqrt{-1}. \tag{5.1}$$

The numbers x and y are called **real** and **imaginary parts** of the complex number z and are denoted as

$$\left.\begin{aligned} x &= \operatorname{Re}(z) \\ y &= \operatorname{Im}(z) \end{aligned}\right\} \tag{5.2}$$

The era of complex number began with the introduction of the symbol $i = \sqrt{-1}$ by **Euler**.

A pair of complex numbers $z_1 = x + iy$ and $z_2 = x - iy$ are called *complex conjugate* of each other. The complex conjugate of a complex number z is denoted by $\overline{z}$.

Thus, $\overline{z}_1 = z_2$ and $\overline{z}_2 = z_1$.

N.B. If $x = 0$, then $z = iy$, the complex number is purely imaginary. But if $y = 0$, then $z = x$, the complex number is purely real.

5.2 Algebra of Complex Numbers

1. **Equality of complex numbers : Two complex numbers $z_1 = x_1 + iy_1$ and $z_2 = x_2 + iy_2$ are said to be equal, if and only if their real and imaginary parts of each are equal.**

   If $x_1 + iy_1 = x_2 + iy_2$, then $(x_1 - x_2) + i(y_1 - y_2) = 0$.

   This is possible, if $x_1 - x_2 = 0$ and $y_1 - y_2 = 0$.

   i.e., $\boldsymbol{x_1 = x_2}$ and $\boldsymbol{y_1 = y_2}$. (5.3)

2. **Addition (or subtraction) : The addition or subtraction of two complex numbers is the sum or difference of real and imaginary parts of the numbers separately, i.e.,** if $z_1 = x_1 + iy_1$ and $z_2 = x_2 + iy_2$, then

   $$\boldsymbol{z_1 + z_2 = (x_1 \pm x_2) + i(y_1 \pm y_2)}. \tag{5.4}$$

   The addition of complex numbers obeys commutative and associative laws.

   This, $z_1 + z_2 = (x_1 + x_2) + i(y_1 + y_2) = (x_2 + x_1) + i(y_2 + y_1)$

   $$= (x_2 + iy_2) + (x_1 + iy_1) = z_2 + z_1. \tag{5.5}$$

   If $z_3 = x_3 + iy_3$ be another complex number, then

   $$\begin{aligned} z_1 + (z_2 + z_3) &= (x_1 + iy_1) + (x_2 + iy_2 + x_3 + iy_3) \\ &= (x_1 + iy_1 + x_2 + iy_2) + x_3 + iy_3. \end{aligned}$$

   $$\therefore \quad \boldsymbol{z_1 + (z_2 + z_3) = (z_1 + z_2) + z_3}. \tag{5.6}$$

   N.B. If z be a complex number, then $-z$ is called its additive inverse and $z + (-z) = 0$.

3. **Multiplication : Multiplication of a complex number by a scalar λ is equal to λ times z, i.e.,**

$$\boldsymbol{\lambda z = \lambda(x + iy) = (\lambda x + i\lambda y).} \tag{5.7}$$

Multiplication of two complex number is always a complex number.

$$\boldsymbol{z_1 z_2 = (x_1 + iy_1)(x_2 + iy_2) = (x_1 x_2 + y_1 y_2) + i(x_1 y_2 + x_2 y_1).} \tag{5.8}$$

The multiplication of complex numbers follows the **distributive law**.

Thus, if z_1, z_2, z_3 be three complex numbers, then $(z_1 + z_2)z_3 = z_1 z_3 + z_2 z_3$.

$$\begin{aligned}
\text{We have}\quad (z_1 + z_2)z_3 &= (x_1 + iy_1 + x_2 + iy_2)(x_3 + iy_3) \\
&= \{(x_1 + x_2) + i(y_1 + y_2)\}(x_3 + iy_3) \\
&= \{(x_1 + x_2)x_3 - (y_1 + y_2)y_3\} + i\{(x_1 + x_2)y_3 - (y_1 + y_3)x_3\} \\
&= \{(x_1 x_3 - y_1 y_3) + i(x_1 y_3 + x_3 y_1)\} \\
&\qquad + \{(x_2 x_3 - y_2 y_3) + i(x_2 y_3 + x_3 y_2)\} \\
&= (x_1 + iy_1)(x_3 + iy_3) + (x_2 + iy_2)(x_3 + iy_3)
\end{aligned}$$

or,
$$\boldsymbol{(z_1 + z_2)z_3 = z_1 z_3 + z_2 z_3.} \tag{5.9}$$

N.B. The complex number $1 + i0 = (1, 0)$ is called unity of the system of the complex numbers. The complex number, $0 + i1 = (0, 1)$ is denoted by i, so that

$$i^2 = (0 + i1)(0 + i1) = -1.$$

$$\therefore \quad i = \sqrt{-1}.$$

4. **Inverse of a complex number : If z_1 and z_2 be two complex numbers such that,**

$$\boldsymbol{z_1 \cdot z_2 = (1 + i \cdot 0) = 1,} \tag{5.10}$$

then z_2 is called the inverse of the complex number z_1.

Thus, if $z_1 = x_1 + iy_1$ and $z_2 = x_2 + iy_2$, then from Eqn. (5.10),

$$\{x_1 x_2 - y_1 y_2\} + i\{x_1 y_2 + x_2 y_1\} = 1 + i \cdot 0.$$

$$\left.\begin{aligned}
\therefore \quad & x_1 x_2 - y_1 y_2 = 1 \\
\text{and} \quad & x_1 y_2 + x_2 y_1 = 0
\end{aligned}\right\} \tag{5.11}$$

Solving Eqns. (5.11), we get

$$\left.\begin{aligned}
x_2 &= \frac{x_1}{x_1^2 + y_1^2} \\
y_2 &= \frac{-y_1}{x_1^2 + y_1^2}.
\end{aligned}\right\} \tag{5.12}$$

Eqns. (5.12) exist, if and only if $x_1^2 + y_1^2 = 0$.

i.e., $|z_1| \neq 0$, i.e., $z \neq 0$.

This implies that every non-zero complex number has a unique inverse.

5. **Division : Division of a complex number is the inverse of the operation of multiplication.**

If $z_1 = x_1 + iy_1$ and $z_2 = x_2 + iy_2$, then if $z_2 \neq 0$,

$$\frac{z_1}{z_2} = z_1 z_2^{-1} = (x_1 + iy_1)\left(\frac{x_2}{x_2^2+y_2^2} - i\frac{y_2}{x_2^2+y_2^2}\right) \text{ (by Eqn. 5.12)}$$

$$= \left[\left\{x_1\left(\frac{x_2}{x_2^2+y_2^2}\right) + y_1\left(\frac{y_2}{x_2^2+y_2^2}\right)\right\} + i\left\{x_1\left(\frac{-y_2}{x_2^2+y_2^2}\right) + y_1\left(\frac{x_2}{x_2^2+y_2^2}\right)\right\}\right]. \quad (5.13)$$

Thus, the division of two complex numbers is also a complex number.

6. **The sum and product of a complex number and its conjugate are both real.**
   For, $z + \overline{z} = (x + iy)(x - iy) - 2x =$ real.
   Also, $z\overline{z} = (x + iy)(x - iy) = x^2 - (iy)^2 = \boldsymbol{x^2 + y^2}$ = **real.**

5.3 Complex Conjugate

The complex number resulting by changing the sign of the imaginary part of $z = x + iy$, i.e., $x - iy$ is called the **complex conjugate** or conjugate of z.

So, if $z = x + iy$ then $\overline{z} = x - iy$.

N.B. The conjugate of $\overline{z}$ is z.

Properties : 1. $z_1 + z_2 = \overline{z}_1 + \overline{z}_2$

2. If $z_1 = x_1 + iy_1$ and $z_2 = x_2 + iy_2$, then,

$$z_1 + z_2 = (x_1 + iy_1) + i(y_1 + y_2)$$

$$\therefore \quad \overline{z_1 + z_2} = (x_1 + iy_1) - i(y_1 + y_2) = (x_1 - iy_1) + (x_2 - iy_2) = \overline{z_1} + \overline{z_2}.$$

Hence, $\boldsymbol{\overline{z_1 + z_2} = \overline{z_1} + \overline{z_2}}$. (5.14)

Again, $\overline{z_1 z_2} = (x_1x_2 - y_1y_2) - i(x_1y_2 + x_2y_1) \quad [\because z_1z_2 = (x_1x_2 - y_1y_2) + i(x_1y_2 + x_2y_1)]$

$$= (x_1 - iy_1)(x_2 - iy_2) = z_1 z_2.$$

$$\therefore \quad \boldsymbol{\overline{z_1 z_2} = z_1 z_2} \quad (5.15)$$

and $$\left(\frac{z_1}{z_2}\right) = \frac{\overline{z_1}}{\overline{z_2}}. \quad (5.16)$$

4. $z_1 \cdot \overline{z_1} = (x + iy)(x - iy) = x^2 + y^2 = |z|^2 = |\overline{z}|^2$. (5.17)
5. $z + \overline{z} = (x + iy) + (x - iy) = 2x = 2\,\mathrm{Re}\,(z)$. (5.18)
6. $z - \overline{z} = (x + iy) - (x - iy) - 2iy = 2\,\mathrm{Im}\,(z)$. (5.19)

5.4 Geometrical Representation of a Complex Number : Argand Diagram

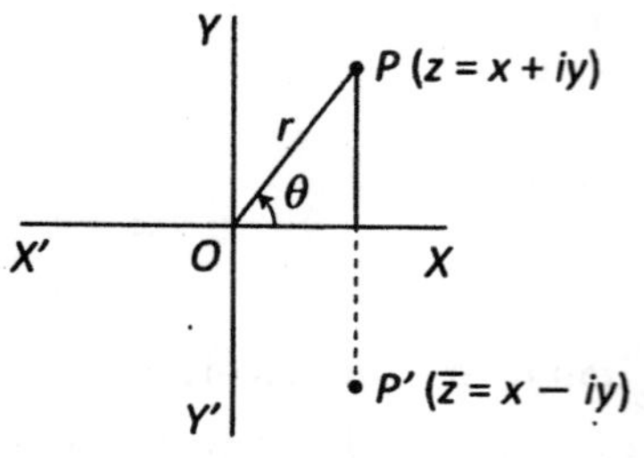

Fig. 5.1: Argand Diagram

We take two mutually perpendicular straight lines XOX' and YOY' as x-axis and y-axis respectively, O being the origin. Let $P(x, y)$ be a point in the X-Y plane as shown in Fig. 5.1 which represents a complex number $z = x + iy$.

If $y = 0$, then z is real and lies on the x-axis. So, XOX' is called the real axis. If $x = 0$, then z is purely imaginary and the point lies on the y-axis. So, YOY' is called the imaginary axis. The plane X-Y is called the **Argand plane** and the diagram showing points representing complex number is called the **Argand diagram**.

The point $P'(x, -y)$ represents the complex number $\overline{z} = x - iy$, which is the complex conjugate of z.

5.5 Modulus and Amplitude of Complex Number

In Fig. 5.1, let (r, θ) be the polar coordinates of P whose Cartesian coordinates are (x, y). Then,

$$x = r\cos\theta \quad \text{and} \quad y = r\sin\theta,$$

$$r = \sqrt{x^2 + y^2} \quad \text{and} \quad \theta = \tan^{-1}\frac{y}{x}.$$

$$\therefore \quad \boldsymbol{z = x + iy = r(\cos\theta + i\sin\theta).} \tag{5.20}$$

This is known as the **polar form** of complex number, where $r = OP$, called the **modulus** or **magnitude** of z and is denoted by $|z|$ and is essentially positive.

Thus, $\boldsymbol{|z| = |x + iy| = r = \pm\sqrt{x^2 + y^2}}$.

The angle θ is called the **amplitude** or **argument** of z and is denoted by **amp** z. $\theta = n\times 2\pi$, where $n = 0, 1, 2, \ldots$. θ has infinite number of values differing by 2π. For the **principal value** of θ, $-\pi < \theta < \pi$. Unless otherwise stated, we shall always mean the principal value.

5.6 Euler's Formula

For a real variable x, we have

$$e^x = 1 + \frac{x}{1!} + \frac{x^2}{2!} + \cdots. \tag{5.21}$$

If the power series expansion of Eqn. (5.21) is also valid for a complex number z, then

$$e^z = 1 + \frac{z}{1!} + \frac{z^2}{2!} + \cdots$$

Putting, $z = i\theta$, we have

$$e^{i\theta} = 1 + i\theta - \frac{\theta^2}{2!} - \frac{i\theta^3}{3!} + \cdots = \left(1 - \frac{\theta^2}{2!} + \frac{\theta^4}{4!} - \cdots\right) + i\left(\theta - \frac{\theta^3}{3!} + \frac{\theta^5}{5!} - \cdots\right)$$

or, $$\boldsymbol{e^{i\theta} = \cos\theta + i\sin\theta}. \tag{5.22}$$

This is **Euler's formula**.

If $x = r\cos\theta$ and $y = r\sin\theta$, then

$$z = x + iy = r(cos\theta + i\sin\theta) = re^{i\theta}. \tag{5.23}$$

5.7 De Moivre's Theorem

Statement : If n be (i) an integer positive or negative, then

$$\boldsymbol{(\cos\theta + i\sin\theta)^n = \cos n\theta + i\sin n\theta}. \tag{5.24(a)}$$

(ii) a fraction positive or negative, then one of the values of

$$\boldsymbol{(\cos\theta + i\sin\theta)^n = \cos n\theta + i\sin n\theta}. \tag{5.24(b)}$$

Proof : Case I : When n is a positive integer.

We have $(\cos\theta + i\sin\theta)^1 = \cos\theta + i\sin\theta$

$$(\cos\theta + i\sin\theta)^2 = \cos^2\theta - \sin^2\theta + 2i\sin\theta\cos\theta = \cos 2\theta + i\sin 2\theta.$$

Thus, the theorem is valid for $n = 1$ and 2.

Now, we assume that the theorem is true for $n = m$, m being a positive integer.

Then, $(\cos\theta + i\sin\theta)^m = \cos m\theta + i\sin m\theta$.

$$\therefore \quad (\cos\theta + i\sin\theta)^{m+1} = (\cos\theta + i\sin\theta)(\cos m\theta + i\sin m\theta)$$
$$= \cos m\theta\cos\theta - \sin m\theta\sin\theta + i(\sin m\theta\cos\theta + \cos m\theta\sin\theta)$$
$$= \cos(m+1)\theta + i\sin(m+1)\theta. \qquad (5.25)$$

Thus, if the theorem is valid for $n = m$, it is also valid for $m = m + 1$.

But it was found to be true for $n = 2$, so, it is true for $n = 3 + 1 = 4$ and so on.

Hence, the theorem is true for **all positive integral values of n.**

Case II : When n is a negative integer.

Let $n = -m$, m being a positive integer.

$$\therefore \quad (\cos\theta + i\sin\theta)^n = (\cos\theta + i\sin\theta)^{-m} = \frac{1}{(\cos\theta + i\sin\theta)^m}$$
$$= \frac{\cos m\theta - i\sin m\theta}{(\cos m\theta + i\sin m\theta)(\cos m\theta - i\sin m\theta)}$$
$$= \frac{\cos m\theta - i\sin m\theta}{\cos^2 m\theta + \sin^2 m\theta} = \cos m\theta - i\sin m\theta$$
$$= \cos(-n)\theta - i\sin(-n)\theta, \quad [\because\ m = -n]$$
$$= \cos n\theta + i\sin n\theta. \qquad (5.26)$$

Hence, the theorem is true for **all negative integral values of n.**

Case III : When n is a fraction, positive or negative.

Let $n = \frac{p}{q}$, where q is a positive number and p is any integer—positive or negative.

$$\text{Then by case I,} \quad \left(\cos\frac{\theta}{q} + i\sin\frac{\theta}{q}\right)^q = \cos q\cdot\frac{\theta}{q} + i\sin q\cdot\frac{\theta}{q} = \cos\theta + i\sin\theta. \qquad (5.27)$$

Extracting the qth root of both sides, we see that one of the values of $(\cos\theta + i\sin\theta)^{\frac{1}{q}}$ is

$$\left(\cos\frac{\theta}{q} + i\sin\frac{\theta}{q}\right).$$

Now, raising both sides to p-th power, we say that one of the values of $[(\cos\theta + i\sin\theta)^{\frac{1}{q}}]^p$ i.e., of $(\cos\theta + i\sin\theta)^{\frac{1}{q}}$, i.e., of $(\cos\theta + \sin\theta)^n$ is

$$\left(\cos\frac{\theta}{q} + i\sin\frac{\theta}{q}\right)^p = \cos\frac{p}{q}\theta + i\sin\frac{p}{q}\theta = \cos n\theta + i\sin n\theta. \qquad (5.28)$$

(by case I and case II)

Hence, one of the values of $(\cos\theta + i\sin\theta)^n$ is $(\cos\theta + i\sin\theta)$.

N.B. De Moivre's theorem is valid for all rational values of n.

5.8 Complex Functions

Some definitions

(i) **Set of points :** A collection of points (finite or infinite in number) in the complex plane is called a **set of points**.

A set of points is said to be **bounded**, if there exists a constant M, such that $|z| \leq M$, for every point z of the set.

(ii) Neighbourhood of a point : The neighbourhood of a point z_0 in the complex plane is the set of all points z such that $|z - z_0| < \epsilon$, where ϵ is an arbitrary small positive number and is called the **radius of the neighbourhood**.

If, from the neighbourhood of a point z_0, the point z_0 is excluded, then it is a **deleted neighbourhood**.

The region outside the circle $|z| = R$, R being a positive real large number, is called the **neighbourhood of a point at infinity**.

(iii) Limiting point : A point z_0 of the set of points S is called **limiting point**, if every neighbourhood of z_0 contains a point of S distinct from z_0.

For example, every point on the circumference of the circle $|z| = R$ is a limiting point of the set of points $|z| < R$.

The limiting points may be either interior and boundary points. All the points inside the circle $|z| = R$ are interior points of the set defined by $|z| < R$, but all the points on the circumference of the circle are boundary points of the set $|z| < R$. The set of points defined by $|z| < R$ is **open** but the set of points $|z| \leq R$ is **closed**.

(iv) Domain: If every pair of points of a set S, in a complex plain, can be joined by a polynomial arc which consists of only one of the points of the set, then the set of points is said to be **connected**.

An open connected set of points is called a **domain**. If the limiting points of a sets are added to a domain, then the resulting set is called a **closed region**. Two domains are said to be **separated**, if they do not have common point.

5.9 Functions of a Complex Variable

(a) Complex variable : If (x, y) be an ordered pair of real variables such that, $z = x + iy$, where $i = \sqrt{-1}$ represents a point in the **Argand diagram**, then z is called a **complex variable**.

(b) Function of a complex variable : If for each value of the complex variable $z(= x + iy)$, in a domain D, we have one more value $w(= u + iv)$, then w is said to be a **complex function of z**, i.e., $w = f(z) = u + iv$, where u and v are the real and imaginary parts of $f(z)$.

If for each value of z, there is only one value of w, then w is said to be a **single valued function**. Otherwise, w is said to be a **multivalued function** of z. For example, $w = \frac{1}{z}$ is a single valued and $w = \sqrt{z}$ is a multivalued function of z.

5.9.1 Exponential Function of a Complex Variable

For a complex variable $z = x + iy$, the exponential function is defined as

$$e^z = 1 + \frac{z}{1!} + \frac{z^2}{2!} + \cdots + \frac{z^n}{n!} + \cdots \tag{5.29}$$

Substituting $x = 0$, i.e., $z = iy$ in the above equation, we get

$$\begin{aligned} e^{iy} &= 1 + \frac{iy}{1!} + \frac{(iy)^2}{2!} + \frac{(iy)^3}{3!} + \frac{(iy)^4}{4!} + \cdots = \left(1 - \frac{y^2}{2!} + \frac{y^4}{4!} \cdots\right) + i\left(y - \frac{y^3}{3!} + \frac{y^5}{5!} - \cdots\right) \\ &= \cos y + i \sin y. \end{aligned} \tag{5.30}$$

$\therefore \qquad e^z = e^x \cdot e^{iy} = e^x(\cos y + i \sin y).$

Also, $x + iy = r(\cos\theta + i\sin\theta) = re^{i\theta}$. (5.31)

Hence, the exponential form of the complex variable z may be written as

$$z = re^{i\theta}. \tag{5.32}$$

5.9.2 Circular Function of a Complex Variable

We have $e^{iy} = \cos y + i \sin y$ and $e^{-iy} = \cos y - i \sin y$.

So, for the **circular function** of real angles, $\cos y = \dfrac{e^{iy} + e^{-iy}}{2}$; $\sin y = \dfrac{e^{iy} - e^{-iy}}{2}, \ldots$

Thus, the different circular functions of the **complex variable z** are as follows :

$$\sin z = \frac{e^{iz} - e^{-iz}}{2i}, \quad \cos z = \frac{e^{iz} + e^{-iz}}{2i}, \quad \tan z = \frac{e^{iz} - e^{-iz}}{e^{iz} + e^{-iz}}.$$

N.B. (i) **Euler's theorem for complex number :** By definition,

$$\cos z + i \sin z = \frac{e^{iz} + e^{-iz}}{2} + i\frac{e^{iz} - e^{-iz}}{2} = e^{iz}, \; (z = x + iy).$$

Also, $e^{iy} = \cos y + i \sin y$, **where y is real**

and $e^{i\theta} = \cos\theta + i \sin\theta$, **where θ is real or complex.** This is **Euler's theorem.**

(ii) **De Moivre's theorem for complex numbers :** For θ is real or complex, we have

$$(\cos\theta + i\sin\theta)^n = (e^{i\theta})^n = e^{in\theta}.$$

Hence, **De Moivre's theorem** is true for all values of θ—real or complex.

5.9.3 Hyperbolic Functions

For all values of x (real or complex), the hyperbolic functions are defined as below :

(i) $\sinh x = \dfrac{e^x - e^{-x}}{2}$,

(ii) $\cosh x = \dfrac{e^x + e^{-x}}{2}$,

(iii) $\tanh x = \dfrac{\sinh x}{\cosh x} = \dfrac{e^x - e^{-x}}{e^x + e^{-x}}$

(iv) $\coth x = \dfrac{e^x + e^{-x}}{e^x - e^{-x}}$

(v) $\text{sech}\, x = \dfrac{2}{e^x + e^{-x}}$

(vi) $\text{cosech}\, x = \dfrac{2}{e^x - e^{-x}}$

(vii) $\cosh x + \sinh x = \dfrac{e^x + e^{-x}}{2} + \dfrac{e^x - e^{-x}}{2} = e^x$

(viii) $\cosh x - \sinh x = \dfrac{e^x - e^{-x}}{2} - \dfrac{e^x - e^{-x}}{2} = e^{-x}$

(ix) $(\cosh x + i \sinh x)^n = \cos nx + \sin nx.$

5.9.4 Relation between Circular and Hyperbolic Functions

Circular functions	Hyperbolic functions	Hyperbolic functions	Circular functions
$\sin x$	$= -i\sinh(ix)$	$\sinh x$	$= -i\sin(ix)$
$\cos x$	$= \cosh(ix)$	$\cosh x$	$= \cos(ix)$
$\tan x$	$= -i\tanh(ix)$	$\tanh x$	$= -i\tan(ix)$
$\cot x$	$= -i\coth(ix)$	$\coth x$	$= i\cot(ix)$
$\sec x$	$= \operatorname{sech}(ix)$	$\operatorname{sech} x$	$= \sec(ix)$
$\operatorname{cosec} x$	$= i\operatorname{cosech}(ix)$	$\operatorname{cosech} x$	$= i\operatorname{cosec}(ix)$

5.9.5 Inverse Hyperbolic Functions

If $\sinh u = z$, then $u = \sinh^{-1} z$; u is called the **hyperbolic sine inverse of z**. Similarly, $u = \sinh^{-1} z$.

It can be shown that, $\sinh^{-1} z = \log\left[z + (1+z^2)^{\frac{1}{2}}\right]$ and $\cosh^{-1} z = \log\left[z + (z^2-1)^{\frac{1}{2}}\right]$.

5.9.6 Logarithmic Function of a Complex Variable

If $z = (x+iy)$ and $w = u + iv$ are so related that $e^w = z$, then w is defined as the logarithm of z to the base e, i.e.,

$$w = \log_e z.$$

Also, $e^{w+2\pi ni} = e^w \cdot e^{2\pi ni} = e^w \cdot 1 = e^w = z$.

[$\because$ $e^{2\pi ni} = \cos 2\pi n + i\sin 2\pi n = 1$, (for $n = 0$, or any integer)].

$$\therefore \quad \boldsymbol{\log z = w + 2\pi ni}. \tag{5.33}$$

Thus, $\log z$ has infinite number of values and is called a **multivalued functions**.

For $n = 0$, $\log z$ is called the principal value. For $n \neq 0$, $\log z$ is written as $\log z$.

So, $$\boldsymbol{\log z = \log(x+iy) = w + i2\pi n = \log(x+iy) + i2\pi n}. \tag{5.34}$$

5.10 Calculus of Complex Function

(a) **Limit of a function of complex variable :** If $f(z)$ be a single valued function defined at all points in some neighbourhood of z_0, then the limit of $f(z)$ as z approaches z_0 is w_0, i.e.,

$$\underset{z\to z_0}{\text{Lt}}\ f(z) = w_0.$$

If for each value of z, there is only one value of w, then w is **single valued function of z**, otherwise w is **multivalued**.

(b) **Continuity of functions :** A complex function $f(z)$, $z = x + iy$, is said to be continuous at a point $z = z_0$, if $\underset{z\to z_0}{\text{Lt}}\ f(z) = f(z_0)$. $f(z_0)$ is the value of the function $f(z)$ at $z = z_0$.

N.B. If $w = f(z) = u(x,y) + iv(x,y)$ is continuous at $z = z_0$, then $u(x,y)$ and $v(x,y)$ are also continuous at $z = z_0$, i.e., at $x = 0$ and $y = 0$.

(c) **Differentiability of functions :** If $f(z)$ be a single valued function of the variable z, then

$$f'(z) = \underset{\delta z\to 0}{\text{Lt}}\ \frac{f(z+\delta z) - f(z)}{\delta z},$$

provided that the limit exists and is independent of the path along which $\delta z \to 0$. This unique limit is called the **differential coefficient** or **derivative of $f(z)$ at $z = z_0$**.

5.11 The Cauchy-Riemann Equations

We consider two functions $u(x, y)$ and $v(x, y)$ of real variables x, y such that,

$$f(z) = u(x, y) + iv(x, y) \tag{5.35}$$

is a function of complex variable $z = x + iy$.

Let us assume that $f(z)$ is analytic at any point z in its domain. Then,

$$f'(z) = \underset{h \to 0}{\text{Lt}} \frac{f(z+h) - f(z)}{h}. \tag{5.36}$$

If h is real then from Eqn. (5.36),

$$\underset{h \to 0}{\text{Lt}} \frac{u(x+h, y) - u(x, y)}{h} + i\frac{v(x+h, y) - v(x, y)}{h} \tag{5.37}$$

must exist and must be equal to $f'(z)$.

From Eqn. (5.37),

$$f'(z) = \frac{\partial u}{\partial x} + i\frac{\partial v}{\partial x}. \tag{5.38}$$

If h be purely imaginary, then

$$f'(z) = \frac{\partial v}{\partial y} - i\frac{\partial u}{\partial y}. \tag{5.39}$$

By definition the right-hand side of Eqn. (5.38) and (5.39) must be equal, i.e.,

$$\frac{\partial u}{\partial x} + i\frac{\partial v}{\partial x} = \frac{\partial v}{\partial y} - i\frac{\partial u}{\partial y} \tag{5.40}$$

equating the real and imaginary parts, we get

$$\left.\begin{aligned} \frac{\partial u}{\partial x} &= \frac{\partial v}{\partial y} \\ \frac{\partial v}{\partial x} &= -\frac{\partial u}{\partial y} \end{aligned}\right\}. \tag{5.41}$$

The equations (5.41) are known as **Cauchy-Reimann equations.**

5.12 Analytic Functions

A single valued function $f(z)$, differentiable at $z = z_0$, is said to be analytic at $z = z_0$ in the domain D.

The point at which the function $f(z)$ is not differentiable is called a **singular point** of the function.

5.12.1 The Necessary Conditions for a Function $f(z)$ to be Analytic

Theorem : The necessary conditions for a function $f(z) = u + iv$ to be analytic at all points in domain D are : (i) $\frac{\partial u}{\partial x} = \frac{\partial v}{\partial y}$ (ii) $\frac{\partial u}{\partial y} = -\frac{\partial v}{\partial x}$, provided $\frac{\partial u}{\partial x}, \frac{\partial u}{\partial y}, \frac{\partial v}{\partial x}, \frac{\partial v}{\partial y}$ exist.

Proof : Let $f(z)$ be an analytic function in domain D, $f(z) = u + iv$, where u and v are the functions of x and y.

If δu and δv be the increments of u and v respectively corresponding to the increments δx and δy of x and y, then

$$f(z + \delta z) = (u + \delta u) + i(v + \delta v).$$

Now, $\dfrac{f(z+\delta z)-f(z)}{\delta z} = \dfrac{(u+\delta u)+i(v+\delta v)-(u+v)}{\delta z} = \dfrac{\delta u + i\delta v}{\delta z} = \dfrac{\delta u}{\delta z} + i\dfrac{\delta v}{\delta z}.$

$$\therefore \quad \underset{\delta z\to 0}{\text{Lt}} \frac{f(z+\delta z)-f(z)}{\delta z} = \underset{\delta z\to 0}{\text{Lt}} \left(\frac{\partial u}{\partial z} + i\frac{\partial v}{\partial z}\right) \tag{5.42}$$

as δz can approach to zero along any path.

(i) **Along real axis :** When δz is wholly real, then $\delta y = 0$ and $\delta z = \delta x$, then from Eqn. (5.42)

$$f'(z) = \underset{\delta x\to 0}{\text{Lt}} \left(\frac{\delta u}{\delta x} + i\frac{\delta v}{\delta x}\right). \tag{5.43}$$

(ii) **Along imaginary axis :** When δz is wholly imaginary, $\delta x = 0$, $\delta z = i\delta y$, then

$$f'(z) = \underset{\delta y\to 0}{\text{Lt}} \left(\frac{\delta u}{i\delta y} + i\frac{\delta v}{i\delta y}\right) = \frac{1}{i}\frac{\partial u}{\partial y} + \frac{\partial v}{\partial y}. \tag{5.44}$$

Now $f'(z)$ exists, if $f'(z)$ of Eqns. (5.43) and (5.44) must be equal, i.e.,

$$\frac{\partial u}{\partial x} + i\frac{\partial v}{\partial x} = -i\frac{\partial u}{\partial y} + \frac{\partial v}{\partial y}. \tag{5.45}$$

Equating the real and imaginary parts of above equation, we get

$$\left.\begin{aligned} \frac{\partial u}{\partial x} &= \frac{\partial v}{\partial y} \\ \frac{\partial v}{\partial x} &= -\frac{\partial u}{\partial y} \end{aligned}\right\} \tag{5.46}$$

Necessary condition for $f(z)$ to be analytic.

These are **Cauchy-Rieman equations.**

From (5.46), $\dfrac{\partial^2 u}{\partial x \partial y} = \dfrac{\partial^2 v}{\partial y^2}$ and $\dfrac{\partial^2 u}{\partial y \partial x} = -\dfrac{\partial^2 v}{\partial x^2}$

As, $\frac{\partial u}{\partial x}, \frac{\partial u}{\partial y}$ etc. are continuous,

$$\frac{\partial^2 u}{\partial x \partial y} = \frac{\partial^2 u}{\partial y \partial x} \qquad \therefore \quad \frac{\partial v}{\partial x^2} + \frac{\partial^2 v}{\partial y^2} = 0$$

Similarly, $\dfrac{\partial^2 u}{\partial x^2} + \dfrac{\partial^2 u}{\partial y^2} = 0$

These are **Laplace's equations.**

5.12.2 Harmonic Functions

The function which satisfy Laplace's equation is known as harmonic function. So, if $f(z)$ is analytic, then both u and v are harmonic functions.

Conjugate Functions

If $f(z) = u(x,y) + iv(x,y)$ is analytic then $u(x,y)$ and $v(x,y)$ are said to be **conjugate harmonic functions** or **conjugate functions.**

5.12.3 Sufficient Conditions for $f(z)$ to be Analytic

Theorem : The sufficient condition for a function $f(z)$ to be analytic at all points in a domain D are : (i) $\dfrac{\partial u}{\partial x} = \dfrac{\partial v}{\partial y}$; $\dfrac{\partial u}{\partial y} = -\dfrac{\partial v}{\partial y}$ (ii) $\dfrac{\partial u}{\partial x}, \dfrac{\partial u}{\partial y}, \dfrac{\partial v}{\partial x}, \dfrac{\partial v}{\partial y}$ are continuous functions of x and y in domain D.

Proof : Let $f(z)$ be a single valued function having $\frac{\partial u}{\partial x}, \frac{\partial u}{\partial y}, \frac{\partial v}{\partial x}, \frac{\partial v}{\partial y}$ at each point in the domain D. Then **Cauchy-Rieman equations** are satisfied.

By **Taylor's theorem** for a function of two variables, we have

$$\begin{aligned} f(z+\delta z) &= u(x+\delta x, y+\delta y) + iv(x+\delta x, y+\delta y) \\ &= u(x,y) + \left(\frac{\partial u}{\partial x}\delta x + \frac{\partial u}{\partial y}\delta y\right) + \cdots + i\left[v(x,y) + \left(\frac{\partial v}{\partial x}\delta x + \frac{\partial v}{\partial y}\delta y\right) + \cdots\right] \\ &= [u(x,y) + iv(x,y)] + \left[\frac{\partial u}{\partial x}\delta x + i\frac{\partial v}{\partial x}\delta x\right] + \left[\frac{\partial u}{\partial y}\delta y + i\frac{\partial v}{\partial y}\delta y\right] \\ &= f(z) + \left(\frac{\partial u}{\partial x} + i\frac{\partial v}{\partial y}\right)\delta x + \left(\frac{\partial u}{\partial y} + i\frac{\partial v}{\partial y}\right)\delta y + \cdots \end{aligned}$$

(ignoring the terms of second and higher powers)

or, $$f(z+\delta z) - f(z) = \left(\frac{\partial u}{\partial x} + i\frac{\partial v}{\partial x}\right)\delta x + \left(\frac{\partial u}{\partial y} + i\frac{\partial v}{\partial y}\right)\delta y. \quad (5.47)$$

By C-R equations, i.e., $\frac{\partial u}{\partial x} = \frac{\partial v}{\partial y}$ and $\frac{\partial u}{\partial y} = -\frac{\partial v}{\partial x}$.

$\therefore$ from Eqn. (5.47), we get

$$f(z+\delta z) - f(z) = \left(\frac{\partial u}{\partial x} + i\frac{\partial v}{\partial x}\right)\delta x + \left(-\frac{\partial v}{\partial x} + i\frac{\partial u}{\partial x}\right)\delta y$$

$$= \left(\frac{\partial u}{\partial x} + i\frac{\partial v}{\partial x}\right)\delta x + \left(i\frac{\partial v}{\partial x} + \frac{\partial u}{\partial x}\right)i\delta y = \left(\frac{\partial u}{\partial x} + i\frac{\partial v}{\partial x}\right)(\delta x + \delta y) = \left(\frac{\partial u}{\partial x} + i\frac{\partial v}{\partial y}\right)\delta z$$

or, $$\frac{f(z+\delta z) - f(z)}{\delta z} = \left(\frac{\partial u}{\partial x} + i\frac{\partial v}{\partial y}\right) \quad \text{or,} \quad \underset{\delta z \to 0}{\text{Lt}} \frac{f(z+\delta z) - f(z)}{\delta z} = \frac{\partial u}{\partial x} + i\frac{\partial v}{\partial y}$$

or, $$\boldsymbol{f'(z) = \frac{\partial u}{\partial x} + i\frac{\partial v}{\partial x}} \quad \text{or,} \quad \boldsymbol{f'(z) = \frac{\partial v}{\partial y} - i\frac{\partial u}{\partial y}} \quad \textbf{(Proved).}$$

5.13 Cauchy-Rieman's Equation in Polar Coordinates

The Cauchy-Rieman's equations are $\frac{\partial u}{\partial x} = \frac{\partial v}{\partial y}$ and $\frac{\partial u}{\partial y} = -\frac{\partial v}{\partial x}$.

In polar coordinates,

$$\left.\begin{aligned} & x = r\cos\theta \text{ and } y = r\sin\theta \\ \therefore\ & r = (x^2+y^2)^{\frac{1}{2}},\ \theta = \tan^{-1}(y/x) \end{aligned}\right\} \quad (5.48)$$

$$\left.\begin{aligned} \frac{\partial r}{\partial x} &= \frac{1}{2\sqrt{x^2+y^2}}2x = \frac{x}{r} = \cos\theta \\ \frac{\partial r}{\partial y} &= \frac{y}{r} = \sin\theta \\ \frac{\partial \theta}{\partial x} &= \frac{1}{1+\frac{y^2}{x^2}}\left(-\frac{y}{x^2}\right) = \frac{-y}{x^2+y^2} = -\frac{\sin\theta}{r} \\ \frac{\partial \theta}{\partial y} &= \frac{1}{1+\frac{y^2}{x^2}}\left(\frac{1}{x}\right) = \frac{x}{x^2+y^2} = \frac{\cos\theta}{r} \end{aligned}\right\} \quad (5.49)$$

Now, $\dfrac{\partial u}{\partial x} = \dfrac{\partial u}{\partial r}\dfrac{\partial r}{\partial x} + \dfrac{\partial u}{\partial \theta}\dfrac{\partial \theta}{\partial x} = \cos\theta\dfrac{\partial u}{\partial x} - \dfrac{\sin\theta}{r}\dfrac{\partial u}{\partial \theta}$ (5.50)

[Using equation (5.49)]

and $\dfrac{\partial v}{\partial y} = \dfrac{\partial v}{\partial r}\dfrac{\partial r}{\partial y} + \dfrac{\partial v}{\partial \theta}\dfrac{\partial \theta}{\partial y} = \sin\theta\dfrac{\partial v}{\partial r} + \dfrac{\cos\theta}{r}\dfrac{\partial v}{\partial \theta}.$ (5.51)

But, $\dfrac{\partial u}{\partial x} = \dfrac{\partial v}{\partial y}.$

$\therefore$ from equations (5.50) and (5.51), we get

$$\cos\theta\frac{\partial u}{\partial r} - \frac{\sin\theta}{r}\frac{\partial u}{\partial \theta} = \sin\theta\frac{\partial v}{\partial r} + \frac{\cos\theta}{r}\frac{\partial v}{\partial \theta}. \tag{5.52}$$

Again, $\dfrac{\partial u}{\partial y} = \dfrac{\partial u}{\partial r}\cdot\dfrac{\partial r}{\partial y} + \dfrac{\partial u}{\partial \theta}\cdot\dfrac{\partial \theta}{\partial y} = \sin\theta\dfrac{\partial u}{\partial r} + \dfrac{\cos\theta}{r}\dfrac{\partial u}{\partial \theta}$ (5.53)

and $\dfrac{\partial v}{\partial x} = \dfrac{\partial v}{\partial r}\cdot\dfrac{\partial r}{\partial x} + \dfrac{\partial v}{\partial \theta}\cdot\dfrac{\partial \theta}{\partial x} = \cos\theta\dfrac{\partial v}{\partial r} - \dfrac{\sin\theta}{r}\dfrac{\partial v}{\partial \theta}.$ (5.54)

By Cauchy-Rieman's equation,

$$\frac{\partial u}{\partial r} = -\frac{\partial v}{\partial x}.$$

$\therefore$ from equations (5.52) and (5.53), we have

$$\sin\theta\frac{\partial u}{\partial r} + \frac{\cos\theta}{r}\frac{\partial u}{\partial \theta} = -\cos\theta\frac{\partial v}{\partial r} + \frac{\sin\theta}{r}\frac{\partial v}{\partial \theta}. \tag{5.55}$$

Multiplying Eqn. (5.52) by $\cos\theta$ and (5.55) by $\sin\theta$ and then adding, we get

$$\frac{\partial u}{\partial r} = \frac{1}{r}\frac{\partial v}{\partial \theta}. \tag{5.56}$$

Again, multiplying Eqns. (5.55) by $\sin\theta$ and (5.56) by $\cos\theta$ and then adding, we get

$$\frac{\partial u}{\partial r} = -\frac{1}{r}\frac{\partial u}{\partial \theta}. \tag{5.57}$$

Eqns. (5.56) and (5.57) are the Cauchy-Riemann's Eqns. in polar coordinates.

5.14 Complex Line Integral

A function $f(z)$ of complex variable z may or may not be analytic but it has a definite finite value on an arc L given by $z = x(t) + iy(t)$ for $t_0 \le t \le T$.

The arc L is sub-divided into n number of smaller arcs by points $z_0, z_1, \ldots, z_n$ corresponding to the values of the parameter t as $t_0, t_1, \ldots, t_n = T$ satisfying the equation

$$t_0 < t_1 < t_2 \cdots < t_n = T.$$

Let ξ_r be a point between z_r and z_{r-1} as shown in Fig. 5.2.

Now, if the sum $S = \displaystyle\sum_{r=1}^{n} f(\xi_r)(z_r - z_{r-1})$ (5.58)

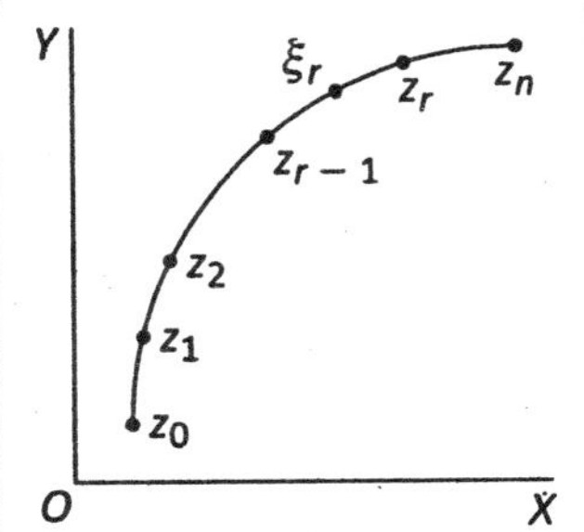

Fig. 5.2

tends to a unique limit J as $n \to \infty$ and the greatest of the numbers $(t_r - t_{r-1})$ tends to zero, then $f(z)$ is integrable from z_0 to z_n of the arc L. Then,

$$J = \int_z f(z)dz \tag{5.59}$$

The complex integral of equation (5.59) depends on the initial and final points on L as well as on the actual form of L. The sufficient condition for the existence of J is that $f(z)$ is continuous on the arc L.

The integration in Eqn. (5.59) has the following properties :

(i) If A and B be the starting and end points of the arc L and P, a point between A and B, then

$$\int_{AB} f(z)dz = \int_{AP} f(z)dz + \int_{PB} f(z)dz. \tag{5.60}$$

(ii) If the arc L is traversed in opposite direction defined by arc L_1, then

$$\int_{L_1} f(z)dz = -\int_L f(z)dz. \tag{5.61}$$

(iii) If λ be a constant, then

$$\int_L \lambda f(z)dz = \lambda \int_L f(z)dz. \tag{5.62}$$

Example *We consider the integral* $\int_L f(z)dz = \int_L z\,dz$. *The arc* L *is defined between* $z_0 = a$ *and* $z_n = b$. *The integral being continuous, the integral exists.*

Now, $\int_L z\,dz = \underset{n\to\infty}{\text{Lt}} \sum_{r=1}^{n} z_r(z_r - z_{r-1})$

Also, $\int_L z\,dz = \underset{n\to\infty}{\text{Lt}} \sum_{r=1}^{n} z_{r-1}(z_r - z_{r-1})$.

Adding we get

$$\int_L z\,dz = \frac{1}{2}\underset{n\to\infty}{\text{Lt}} \sum_{r=1}^{n}(z_r - z_{r-1})(z_r + z_{r-1}) = \frac{1}{2}\underset{n\to\infty}{\text{Lt}} \sum_{r=1}^{n}(z_r^2 - z_{r-1}^2)$$

$$= \frac{1}{2}\underset{n\to\infty}{\text{Lt}} \{(z_1^2 - z_0^2) + (z_2^2 - z_1^2) + \cdots + (z_n^2 - z_{n-1}^2)\}$$

$$= \frac{1}{2}(z_1^2 - z_0^2) = \frac{1}{2}(b^2 - a^2). \tag{5.63}$$

If the curve is closed one then z_0 and z_n coincide and $a = b$.

$$\therefore \quad \oint f(z)dz = \oint z\,dz = 0. \tag{5.64}$$

Again, we consider the integral $\int_L |z|dz$.

By definition,

$$\int_L f(z)dz = \underset{n\to\infty}{\text{Lt}} \sum_{r=1}^{n} |z_r - z_{r-1}| = \underset{n\to\infty}{\text{Lt}} \; \|z_1 - z_1\| + |z_2 - z_1| + \cdots$$

Each term represents the length of the chord between corresponding points and hence for $n \to a$ the sum of the chords equal to l, the length of the curve L,

$$\therefore \quad \int_L |z| dz = l \tag{5.65}$$

5.15 Absolute Value of a Complex Integral

If a complex function $f(z)$ is continuous on a contour C of length l, such that $|f(z)| \leq M$, then

$$\left| \int_C f(z) dz \right| \leq Ml. \tag{5.66}$$

The above statement can be proved in the following way :

By definition we have

$$\int_C f(z) dz = \underset{n \to \infty}{\text{Lt}} \sum_{r=1}^{n} f(\xi_r)(z_r - z_{r-1})$$

$$\therefore \quad \left| \int_0 f(z) dz \right| = \underset{n \to \infty}{\text{Lt}} \left| \sum_{r=1}^{n} f(\xi_r)(z_r - z_{r-1}) \right|.$$

But the modulus of sum of n complex numbers is less than or equal to the sum of their moduli.

$$\therefore \quad \left| \int_0 f(z) dz \right| \leq \underset{n \to \infty}{\text{Lt}} \sum_{r=1}^{n} |f(\xi_r)(z_r - z_{r-1})|$$

$$\leq \underset{n \to \infty}{\text{Lt}} \sum_{r=1}^{n} |f(\xi_r)|\, |z_r - z_{r-1}|, \quad [\because |z_1 z_2| = |z_1||z_2|]$$

$$\leq \underset{n \to \infty}{\text{Lt}}\ \underset{n \to \infty}{\text{Lt}} \sum_{r=1}^{n} |z_r - z_{r-1}|.$$

Hence, $\left| \int_C f(z) dz \right| \leq \int_C |z| dz \leq Ml, \quad \left[\because \int_C |z| dz = l\right]$ **(Proved)**.

5.16 Green's Theorem

If $M(x, y)$ and $N(x, y)$ be two functions of x and y having continuous first order derivatives $\frac{\partial M}{\partial y}$ and $\frac{\partial N}{\partial x}$ at every point within and on the closed contour C, then

$$\int_C (M dx + N dy) = \int_S \left(\frac{\partial N}{\partial x} - \frac{\partial M}{\partial y} \right) dx dy,$$

where S is the area enclosed by the curve C.

5.17 Cauchy's Integral Theorem

The theorem states that if a function $f(z)$ is analytic and $f'(z)$ is continuous at any point inside and on a simple closed curve C, then

$$\int_C f(z) dz = 0.$$

Proof : Let R be the region enclosed by the simple close curve C, and $f(z) = u(x, y) + iv(x, y)$, $z = x + iy$ and $dz = dx + idy$.

$$\therefore \quad \int_C f(z)dz = \int_C (u+iv)(dx+idy) = \int_C (udx - vdy) + i\int_C (vdx + udy). \tag{5.67}$$

As $f'(x)$ is continuous, $\dfrac{\partial u}{\partial x}, \dfrac{\partial u}{\partial y}, \dfrac{\partial v}{\partial x}, \dfrac{\partial v}{\partial y}$ are also continuous in R.

Hence, by Green's theorem

$$\int_C f(z)dz = \iint\limits_R \left(-\frac{\partial v}{\partial x} - \frac{\partial u}{\partial y}\right) dxdy + i\iint\limits_R \left(\frac{\partial u}{\partial x} - \frac{\partial v}{\partial y}\right) dxdy. \tag{5.68}$$

Since $f(z)$ is analytic, u and v satisfy **Cauchy-Riemann** equation. So the integrands of the two integrals of Eqn. (5.68) reduce to zero independently.

$$\therefore \quad \int_C f(z)dz = 0.$$

N.B. A simple closed curve is that which does not cross itself and encloses one separate region R, [Fig. 5.3].

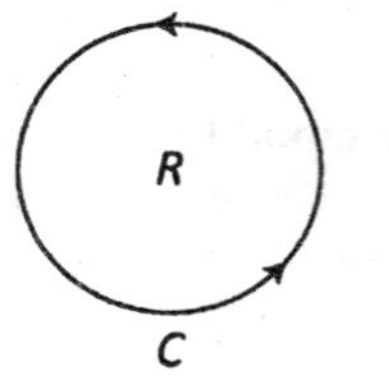

Fig. 5.3: Simple Curve C

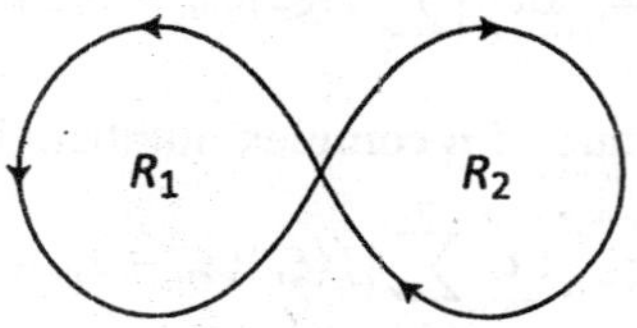

Fig. 5.4: Multiple Curve enclosing two regions R_1 and R_2

If a curve crosses itself and encloses more than one separate region (R_1, R_2) is called a multiple curve [Fig. 5.4].

5.18 Cauchy's Integral Formula

The theorem states that, if $f(z)$ be an analytic function within and on any closed curve C and a is any point within C, then

$$f(a) = \frac{1}{2\pi i}\int \frac{f(z)}{z-a}dz. \tag{5.69}$$

Proof : Let the function $f(z)/(z-a)$ is analytic at all points within C, except $z = a$.

If we encircle the point a with radius r, then by previous theorem,

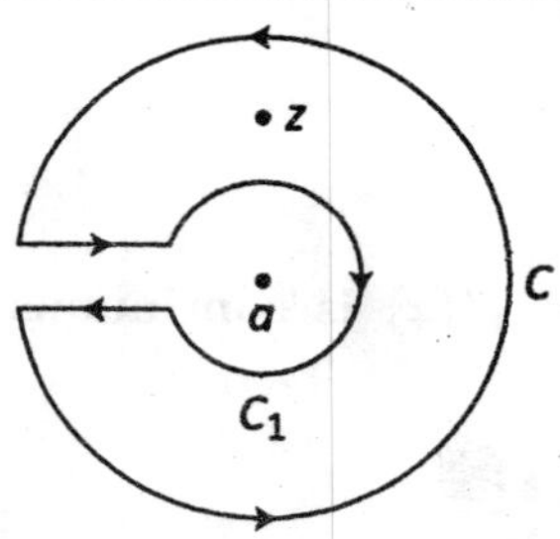

Fig. 5.5

$$\int_C \frac{f(z)}{(z-a)}dz = \int_{C_1} \frac{f(z)}{z-a}dz = \int_{C_1} \frac{f(a+re^{i\theta})}{re^{i\theta}} ire^{i\theta}d\theta$$

[$\because$ for any point on C_1, $z - a = re^{i\theta}$ and $dz = ir\,e^{i\theta}d\theta$]

$$= i\int_{C_1} f(r + re^{i\theta})d\theta. \tag{5.70}$$

In the limit form, as the circle C_1 shrinks to the point a, i.e., as $r \to 0$, the integral of equation (5.70) approaches to

$$i\int_{C_1} f(a)d\theta = if(a)\int_0^{2\pi} d\theta = 2\pi i f(a).$$

$$\therefore \quad f(a) = \frac{1}{2\pi i}\int_C \frac{f(z)}{z-a}dz. \tag{5.71}$$

This is the **Cauchy's integral formula.**

N.B. Differentiating Eqn. (5.71) with respect to a

$$f'(a) = \frac{1}{2\pi i}\int_C \frac{\partial}{\partial a}\left[\frac{f(z)}{z-a}\right]dz = \frac{1}{2\pi i}\int_C \frac{f(z)}{(z-a)^2}dz.$$

Likewise, $$f''(a) = \frac{2!}{2\pi i}\int_C \frac{f(z)}{(z-a)^3}dz.$$

In general, $$f^n(a) = \frac{n!}{2\pi i}\int_C \frac{f(z)}{(z-a)^{n+1}}dz.$$

5.19 Cauchy's Residue Theorem

If a function $f(z)$ is analytic within and on a closed contour C, except at a finite number of poles within C, then

$$\int_C f(z)dz = 2\pi i \sum R, \tag{A}$$

where $\sum R$ is the sum of residues at poles within C, the integration being taken in positive sense.

Proof : Let C be a contour within which there are finite number of poles $z_1, z_2, \ldots, z_n$. The circles $C_1, C_2, \ldots, C_n$ are drawn with $z_1, z_2, \ldots, z_n$ as centres such that all of them lie within C without overlapping as shown in Fig. 5.6. The function $f(z)$ would be analytic within the region bounded by $C_1, C_2, \ldots, C_n$.

Then by Cauchy's theorem on **multiple-connected region,**

$$\int_C f(z)dz = \int_{C_1} f(z)dz + \int_{C_2} f(z)dz + \cdots + \int_{C_n} f(z)dz, \tag{B}$$

all the integrations being taken in **anti-clockwise direction.**.

But from definition,

$$\frac{1}{2\pi i}\int_{C_1} f(z)dz = \text{residue at } z = z_1, \text{ which is (say) } R_1 \text{ and so on.}$$

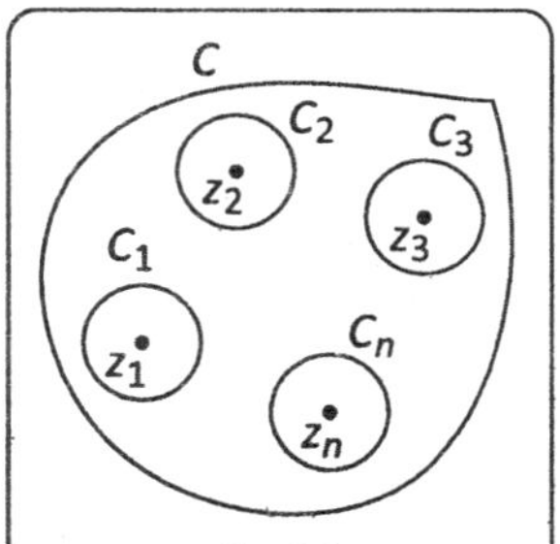

Fig. 5.6

$$\therefore \quad \int_C f(z) = 2\pi i(R_1 + R_2 + \cdots + R_n) = 2\pi i \sum R, \text{ [from Eqn. (A)]}$$

where $R_1, R_2, \ldots, R_n$ represent residues of $f(z)$ at $z_1, z_2, \ldots, z_n$ respectively. This proves the theorem.

Again, let the analytic function $f(z)$ has finite number of singularities within a closed curve C. Then by this theorem,

$$\int_C f(z)dz = 2\pi i \sum R \quad \text{or,} \quad \sum R = \frac{1}{2\pi i}\int_C f(z)dz. \tag{C}$$

Let $f(z)$ have a singularity at $z = \infty$ also.

Then from definition, residue at $z = \infty$ is

$$-\frac{1}{2\pi i}\int_C f(z)dz. \tag{D}$$

Hence, the sum of the residues at all the singularities of the function $f(z)$ is

$$\frac{1}{2\pi i}\int_C f(z)dz + \left(-\frac{1}{2\pi i}\int_C f(z)dz\right) = 0.$$

A circle C_1 is drawn with circle at O enclosing the point z. If w be a point on C_1, then

$$\frac{1}{w-z} = \frac{1}{(w-a)-(z-a)} = \frac{1}{(w-a)}\left(1-\frac{z-a}{w-a}\right)^{-1}$$

$$= \frac{1}{(w-a)}\left[1+\frac{z-a}{w-a}+\left(\frac{z-a}{w-a}\right)^2+\cdots+\left(\frac{z-a}{w-a}\right)^n+\cdots\right]. \quad (5.74)$$

[by Binomial theorem]

Since $|z-a| < |w-a|$, i.e., $|(z-a)/(w-a)| < 1$, the series converges uniformly and is integrable over C_1.

Now multiplying both sides of Eqn. (5.74) by $f(w)$ and then integrating with respect to w, we get

$$\int_{C_1}\frac{f(w)}{(w-z)}dw = \int_{C_1}\frac{f(w)}{(w-a)}dw + (z-a)\int_{C_1}\frac{f(w)}{(w-a)^2}dw$$

$$+\cdots+(z-a)^n\int_{C_1}\frac{f(w)}{(w-a)^{n+1}}dw. \quad (5.75)$$

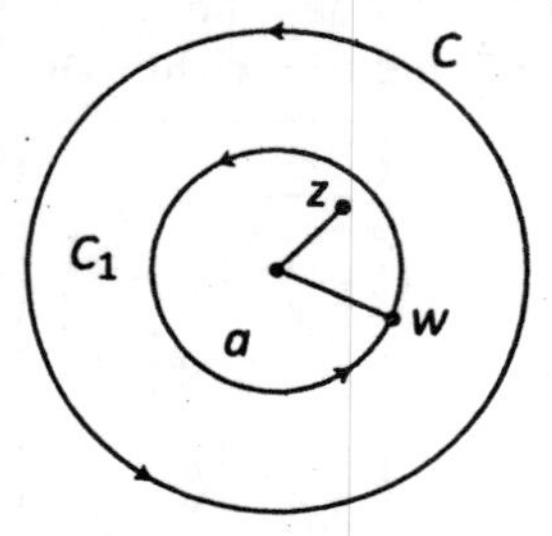

Fig. 5.7

But Cauchy's integral formula is

$$\int_{C_1}\frac{f(w)}{w-z}dw = 2\pi i f(z) \quad \text{or,} \quad \int_{C_1}\frac{f(w)}{(w-z)}dw = 2\pi i f(a)$$

$$f'(a) = \frac{1}{2\pi i}\int_{C_1}\frac{f(w)}{(w-a)^2}dw$$

$$f''(a) = \frac{2!}{2\pi i}\int_{C_1}\frac{f(w)}{(w-a)^3}dw, \cdots$$

Using these values in Eqn. (5.75), we get

$$\boldsymbol{f(z) = f(a) + (z-a)f'(a) + (z-a)^2\frac{f''(a)}{2!} + \cdots + (z-a)^n\frac{f^n(a)}{n!} + \cdots}$$

This is the **Taylor's series.**

5.20 Solved Problems

Example 5.1. *Find the conjugate of the complex number* $\frac{2-i}{(1+2i)^2}$.

Solution : We have $z = \frac{2-i}{(1+2i)^2} = \frac{2-i}{1+4i+4i^2} = \frac{2-i}{-3+4i}$

or, $\quad z = -\frac{(2-i)(3+4i)}{(3-4i)(3+4i)} = \frac{-(6+8i-3i-4i^2)}{9-16i^2} = \frac{-5(2+i)}{9+16} = -\frac{1}{5}(2+i).$

$\therefore$ the conjugate of $z = \boldsymbol{-\frac{1}{5}(2-i)}.$

Example 5.2. *If ω be the cube root of unity, show that, $(1-\omega)(1-\omega^2)(1-\omega^4)(1-\omega^8) = 9$.*

Solution : Since ω is the cube root of unity.

$\therefore \quad \omega^3 = 1$ and $1+\omega+\omega^2 = 0$.

Now, $1-\omega^4 = 1-\omega\cdot\omega^3 = 1-\omega$

and $1-\omega^8 = (1-\omega^4)(1+\omega^4) = (1-\omega\cdot\omega^3)(1+\omega\cdot\omega^3) = (1-\omega)(1+\omega) = 1-\omega^2$.

Hence $(1-\omega)(1-\omega^2)(1-\omega^4)(1-\omega^4) = (1-\omega)(1-\omega^2)(1-\omega)(1-\omega^2)$

$$= (1-\omega^2-\omega+\omega^3)^2 = (1+1+1)^2 = \mathbf{9}.$$

Example 5.3. *Find the value of* (i) $(\sqrt{3}-i)^{\frac{1}{7}}$ (ii) *Solve* $x^5 = 1$.

Solution : (i) We have $(\sqrt{3}-i) = 2\left(\frac{\sqrt{3}}{2}-\frac{1}{2}i\right) = 2\left(\cos\frac{\pi}{6} - i\sin\frac{\pi}{6}\right)$

$$= 2\left\{\cos\left(2m\pi+\frac{\pi}{6}\right) - i\sin\left(2m\pi+\frac{\pi}{6}\right)\right\},$$

where $m = 0$ or any integer.

$$\therefore \quad (\sqrt{3}-i)^{\frac{1}{7}} = 2^{\frac{2}{7}}\left\{\cos\left(2m\pi+\frac{\pi}{6}\right) - i\sin\left(2m\pi+\frac{\pi}{6}\right)\right\}^{\frac{1}{7}}$$

$$= 2^{\frac{1}{7}}\left\{\cos\left(2m+\frac{1}{6}\right)\frac{\pi}{7} - i\sin\left(2m+\frac{1}{6}\right)\frac{\pi}{7}\right\}, \text{ where } m = 0,\ 1,\ 3,\ 4,\ 5,\ 6.$$

(ii) Given, $x^5 = 1$.

$\therefore \quad x^5 = \cos 0 + i\sin 0 = \cos 2k\pi + i\sin 2k\pi$ where $k = 0$ or any integer

$\therefore \quad x = (\cos 2k\pi + i\sin 2k\pi)^{\frac{1}{5}}$ or, $\boldsymbol{x = \cos\frac{2k\pi}{5} + i\sin\frac{2k\pi}{5}}$ **where $\boldsymbol{k = 0, 1, 2, 3, 4}$.**

Example 5.4. *Find the cube root of* (-1)

Solution : Let $x^3 = -1 = \cos\pi + i\sin\pi = \cos(2m\pi+\pi) + i\sin(2m\pi+\pi)$

$$= \cos(2m+1)\pi + i\sin(2m+1)\pi, \quad \text{where } m = 0 \text{ or any integer.}$$

$$\therefore \quad x = [\cos(2m+1)\pi + i\sin(2m+1)\pi]^{\frac{1}{3}}$$

$$= \cos\frac{1}{3}(2m+1)\pi + i\sin\frac{1}{3}(2m+1)\pi, \quad [m = 0, 1, 2, \ldots]$$

or, $x = \cos\left(\frac{\pi}{3}\right) + i\sin\left(\frac{\pi}{3}\right)$; $\cos\pi + i\sin\pi$; $\cos\frac{4\pi}{3} + i\sin\frac{4\pi}{3}$.

$$= \mathbf{\frac{1}{2}(1+\sqrt{3});\ -1;\ \frac{1}{2}(1-\sqrt{3})}.$$

Example 5.5. *If z be a variable complex number such that,* (i) amp$\frac{(z-i)}{(z+i)} = \frac{\pi}{4}$ *and* (ii) $|z^2+z| < 1$, *find the loci of the points in both cases.*

Solution : (i) Let the complex number be $z = x+iy$, then

$$\text{amp}\,\frac{z-i}{z+i} = \frac{\pi}{4} \quad \text{or,} \quad \text{amp}\,\{x+iy-i\} - \text{amp}\,(x+iy+i) = \frac{\pi}{4}$$

or, $\quad \operatorname{amp}\{x+i(y-i)\} - \operatorname{amp}\{x+i(y-1)\} = \frac{\pi}{4}$

or, $\quad \tan\frac{y-1}{x} - \tan^{-1}\frac{y+1}{x} = \frac{\pi}{4}$ or, $\tan^{-1}\left[\frac{\frac{y-1}{x}-\frac{y+1}{x}}{1+\frac{y^2-1}{x^2}}\right] = \frac{\pi}{4}$.

$\therefore \quad \frac{\frac{-2}{x}}{1+\frac{y^2-1}{x^2}} = \tan\frac{\pi}{4} = 1$ or, $x^2+y^2-1=-2x$

or, $\boldsymbol{x^2+y^2+2x-1=0}$.

This is the equation of a circle.

(ii) $|z^2+z| = |(x+iy)^2+(x+iy)| = |(x^2-y^2+x)+i(2xy+y)|$.

Hence, $|z^2+z|<1$ or, $(x^2-y^2+x)^2+(2xy+y)^2<1$

or, $\quad x^4+y^4+x^2-2x^2y^2+2x^3-2y^2x+4x^2y^2+y^2+4y^2x<1$

or, $\quad x^4+y^4+2x^2y^2+2x^3+x^2+y^2+2y^2x<1$

or, $\quad (x^2+y^2)(x^2+y^2+2x+1)<1$.

Hence, the locus of the point is interior of the curve,

$$\boldsymbol{(x^2+y^2)(x^2+y^2+2x+1)=1.}$$

Example 5.6. *If z_1 and z_2 be two complex numbers, then show that, $|z_1+z_2| \le |z_1|+|z_2|$.*

Solution : We know that, $z\overline{z} = |z|^2$.

$\therefore \quad (z_1+\overline{z}_2)(\overline{z}_1+\overline{z}_2) = |z_1+z_2|^2$.

$\therefore \quad |z_1+z_2|^2 = z_1\overline{z}_1 + z_1\overline{z}_2 + z_2\overline{z}_1 + z_2\overline{z}_2 = |z_1|^2+|z_2|^2+z_1\overline{z}_2+z_2\overline{z}_1$.

Now, $\quad z_1 = x_1+iy_1$ and $z_2 = x_2+iy_2$.

$\therefore \quad z_1\overline{z}_2 + z_2\overline{z_1} = (x_1+iy_1)(x_2-iy_2)+(x_2+iy_2)(x_1-iy_1)$

$$= 2(x_1x_2+y_1y_2) = 2\operatorname{Re}(z_1\overline{z_2}).$$

Again, $\operatorname{Re}(z) \le |x|$ for all x, $\quad \therefore \operatorname{Re}(z_1\overline{z}_2) \le |z_1\overline{z}_2|$.

Also, $\quad |z_1\overline{z}_2| = |(x_1+iy_1)(x_2-iy_2)| = [(x_1x_2+y_1y_2)^2+(x_2y_1-x_1y_2)^2]^{\frac{1}{2}}$.

$$= \sqrt{(x_1^2+y_1^2)(x_2^2+y_2^2)} = |z_1z_2|.$$

$\therefore \quad \operatorname{Re}(z_1\overline{z}_2) \le |z_1z_2|$ or, $|z_1+z_2|^2 \le |z_1|^2+|z_2|^2+2|z_1z_2|$

or, $\quad |z_1+z_2|^2 \le (|z_1|+|z_2|)^2$ or, $\boldsymbol{|z_1+z_2| \le |z_1|+|z_2|}$ **(Proved).**

[**Alternative Proof :** Let $z_1 = r_1(\cos\theta_1+i\sin\theta_1)$, $\quad z_2 = r_2(\cos\theta_2+i\sin\theta_2)$.

$\therefore \quad z_1+z_2 = (r_1\cos\theta_1+r_2\cos\theta_2)+i(r_1\sin\theta_1+r_2\sin\theta_2)$

$\therefore \quad |z_1+z_2|^2 = (r_1\cos\theta_1+r_2\cos\theta_2)^2+(r_1\sin\theta_1+r_2\sin\theta_2)^2$

$$= r_1^2+r_2^2+2r_1r_2\cos(\theta_1-\theta_2)$$

$$\le r_1^2+r_2^2+2r_1r_2, \quad [\because \cos(\theta_1-\theta_2)\le 1]$$

$$\le (r_1+r_2)^2 = (|z_1|+|z_2|)^2.$$

$\therefore \quad \boldsymbol{(z_1+z_2) \le |z_1|+|z_2|}$]

Example 5.7. *If z_1 and z_2 be two complex numbers, then show that, $|z_1 - z_2| \geq |z_1| \sim |z_2|$.*

Solution : We have $|z_1 - z_2|^2 = (z_1 - z_2)(\overline{z}_1 - \overline{z}_2) = z_1\overline{z}_1 + z_2\overline{z}_2 - z_1\overline{z}_2 - z_2\overline{z}_1$.

But since $z_1\overline{z}_2 + z_2\overline{z}_1 \leq 2|z_1 z_2|$,

$$\therefore \quad |z_1 - \overline{z}_2|^2 \geq |z_1|^2 + |z_2|^2 - 2|z_1|\,|z_2| \quad \text{or,} \quad |z_1 - z_2|^2 \geq (|z_1| - |z_2|)^2,$$

$$\therefore \quad \boldsymbol{|z_1 - z_2| \geq |z_1| - |z_2|} \quad \textbf{(Proved)}.$$

[**Alternative proof :** Let $z_1 = r_1(\cos\theta_1 - i\sin\theta_1)$, $\quad z_2 = r_2(\cos\theta_2 - i\sin\theta_2)$.

$$\therefore \quad |z_1 - z_2|^2 = (r_1\cos\theta_1 - r_2\cos\theta_2)^2 + (r_1\sin\theta_1 - r_2\sin\theta_2)^2$$

$$= r_1^2 + r_2^2 - 2r_1r_2\cos(\theta_1 - \theta_2) \geq r_1^2 + r_2^2 - 2r_1r_2, \quad [\because \cos(\theta_1 - \theta_2) \leq 1]$$

$$\geq (|z_1|^2 \sim |z_2|^2), \qquad \therefore \ \boldsymbol{|z_1 + z_2| \geq |z_1| - |z_2|}].$$

Example 5.8. *If z_1 and z_2 be two complex numbers such that, $|z_2 + z_1| = |z_2 - z_1|$, then show that, z_2/z_1 is a purely imaginary number.*

Solution : Let $z_1 = x_1 + iy_1$, $z_2 = x_2 + iy_2$, then

$$|z_2 + z_1| = \sqrt{(x_2 + x_1)^2 + (y_2 + y_1)^2} \quad \text{and} \quad |z_2 - z_1| = \sqrt{(x_2 - x_1)^2 + (y_2 - y_1)^2},$$

$\because |z_2 + z_1| = |z_2 - z_1|$, $\quad \therefore \quad x_1x_2 + y_1y_2 = 0.$

$$\text{Now,} \quad \frac{z_2}{z_1} = \frac{x_2 + iy_2}{x_1 + iy_1} = \frac{(x_2 + iy_2)(x_1 - iy_1)}{x_1^2 + y_1^2} = \frac{(x_1x_2 + y_1y_2) + i(x_1y_2 - y_1x_2)}{x_1^2 + y_1^2}$$

$$\text{or,} \quad \boldsymbol{\frac{z_2}{z_1} = \frac{i(x_1y_2 - y_1x_2)}{x_1^2 + y_1^2}}, \textbf{ which is a purely imaginary number.}$$

Example 5.9. *Show that* $\sin\left[i\dfrac{a - ib}{a + ib}\right] = \dfrac{2ab}{a^2 + b^2}$.

Solution : Let $a = r\cos\theta$, $b = r\sin\theta$, so that $\tan\theta = \dfrac{b}{a}$.

$$\therefore \quad \frac{a - ib}{a + ib} = \frac{r(\cos\theta - i\sin\theta)}{r(\cos\theta + i\sin\theta)} = \frac{e^{-i\theta}}{e^{i\theta}} = e^{-2i\theta}.$$

$$\therefore \quad \log\frac{a - ib}{a + ib} = -2i\theta$$

$$\text{or,} \quad \boldsymbol{\sin\left[i\log\frac{a - ib}{a + ib}\right]} = \sin[i(-2i\theta)] = \sin 2\theta = \frac{2\tan\theta}{1 + \tan^2\theta} = \frac{\frac{2b}{a}}{1 + \frac{b^2}{a^2}} = \boldsymbol{\frac{2ab}{a^2 + b^2}} \quad \textbf{(Proved)}.$$

Example 5.10. *If $z = x + iy$, find the real and imaginary parts of e^{z^2}.*

Solution : We have $z = x + iy$, $\quad \therefore \quad z^2 = (x + iy)^2 = x^2 - y^2 + 2ixy$.

$$\therefore \quad e^{z^2} = e^{x^2 - y^2 + 2ixy} = e^{x^2 - y^2} \cdot e^{i(2xy)} = e^{x^2 - y^2}[\cos 2xy + i\sin 2xy]$$

$$= e^{x^2 - y^2}(\cos 2xy) + ie^{x^2 - y^2}(\sin 2xy) = \operatorname{Re} e^{z^2} + \operatorname{Im} e^{z^2},$$

where, the real part of e^{z^2} is $\boldsymbol{\operatorname{Re} e^{z^2} = e^{x^2 - y^2}(\cos 2xy)}$, and the imaginary part of e^{z^2} is $\boldsymbol{\operatorname{Im} e^{z^2} = e^{x^2 - y^2}(\sin 2xy)}$.

Example 5.11. *For any complex number z, show that,* $|z| \geq \frac{1}{\sqrt{2}}\{|\text{Re}\,(z)| + |\text{Im}\,(z)|\}$.

Solution : Let $z = x + iy$, then $\text{Re}\,(z) = x$, $\text{Im}\,(z) = y$ and $|z| = \sqrt{x^2 + y^2}$.

$\therefore \quad 2|z|^2 = 2(x^2 + y^2), \qquad |\text{Re}\,(z)| + |\text{Im}\,(z)| = |x| + |y|.$

Now, $2|z|^2 - \{|\text{Re}\,(z)| + |\text{Im}\,(z)|\}^2 = 2(x^2 + y^2) - (x + y)^2 = (x - y)^2 \geq 0$,

for all values of x and y.

So, $|z|^2 \geq \frac{1}{2}\{|\text{Re}\,(z)| + |\text{Im}\,(z)|\}^2$ or, $\boldsymbol{|z| \geq \frac{1}{\sqrt{2}}\{|\text{Re}\,(z)| + |\text{Im}\,(z)|\}}$. **(Proved).**

Example 5.12. *If* $\sin(\alpha + \beta) = x + iy$, *prove that,*

$$\frac{x^2}{\sin^2\alpha} - \frac{y^2}{\cos^2\alpha} = 1 \quad \text{and} \quad \frac{x^2}{\cosh^2\beta} + \frac{y^2}{\sinh^2\beta} = 1.$$

Solution : We have $x + iy = \sin(\alpha + i\beta) = \sin\alpha\cos(i\beta) + \cos\alpha\sin(i\beta)$

$$= \sin\alpha\cosh\beta + i\cos\alpha\sinh\beta.$$

Similarly, $x - iy = \sin\alpha\cosh\beta - i\cos\alpha\sinh\beta$.

$\therefore \quad (x + iy) + (x - iy) = 2\sin\alpha\cosh\beta$ or, $x = \sin\alpha\cosh\beta$

and $\quad (x + iy) - (x - iy) = 2i\cos\alpha\sinh\beta$ or, $y = \cos\alpha\sinh\beta$,

$$\therefore \quad \boldsymbol{\frac{x^2}{\sin^2\alpha} - \frac{y^2}{\cos^2\alpha} = \cosh^2\beta - \sinh^2\beta = 1} \quad \textbf{(Proved)}.$$

$$\text{and} \quad \boldsymbol{\frac{x^2}{\cosh^2\beta} + \frac{y^2}{\sinh^2\beta} = \sin^2\alpha + \cos^2\alpha = 1} \quad \textbf{(Proved)}.$$

Example 5.13. *If* $u = \log\tan\left(\frac{\pi}{4} + \frac{\theta}{2}\right)$, *prove that,* $\tan h\frac{u}{2} = \tan\frac{\theta}{2}$.

[Marathawada U 1990; Rewa U 1986]

Solution : Given $u = \log\tan\left(\frac{\pi}{4} + \frac{\theta}{2}\right)$,

$$\therefore \quad e^u = \tan\left(f\pi 4 + \frac{\theta}{2}\right) \quad \text{or,} \quad \frac{e^{u/2}}{e^{-u/2}} = \frac{1 + \tan\frac{\theta}{2}}{2 - \tan\frac{\theta}{2}}.$$

By componendo and dividend, we get

$$\frac{e^{\frac{u}{2}} - e^{-\frac{u}{2}}}{e^{\frac{u}{2}} + e^{-\frac{u}{2}}} = \tan\frac{\theta}{2} \quad \text{or,} \quad \boldsymbol{\tanh\frac{u}{2} = \tan\frac{\theta}{2}} \quad \textbf{(Proved)}.$$

Example 5.14. *If* $\tan(\theta + i\phi) = e^{i\alpha}$, *show that,* $\theta = \left(n + \frac{1}{2}\right)\frac{\pi}{2}$ *and* $\phi = \frac{1}{2}\log\tan\left(\frac{\pi}{4} + \frac{\alpha}{2}\right)$.

[Pune U 1990; Andhra U 1986]

Solution : We have $\tan(\theta + i\phi) = \cos\alpha + i\sin\alpha$.

$\therefore \quad \tan(\theta - i\phi) = \cos\alpha - i\sin\alpha$.

$$\therefore \quad \tan 2\theta = \tan[(\theta + i\phi) + (\theta - i\phi)] = \frac{\tan(\theta + i\phi) + \tan(\theta - i\phi)}{1 - \tan(\theta + i\phi)\tan(\theta - i\phi)}$$

$$= \frac{2\cos\alpha}{1 - (\cos^2\alpha + \sin^2\alpha)} = \frac{2\cos\alpha}{0} \to \infty$$

i.e., $2\theta = n\pi + \frac{\pi}{2}$ or, $\boldsymbol{\theta = \left(n + \frac{1}{2}\right)\frac{\pi}{2}}$ **(Proved).**

Also, $\tan 2i\phi = \tan[(\theta + i\phi) - (\theta - i\phi)] = \dfrac{\tan(\theta + i\phi) - \tan(\theta - i\phi)}{1 + \tan(\theta + i\phi)\tan(\theta - i\phi)}$

or, $\quad i\tan 2i\phi = \dfrac{2i\sin\alpha}{1 + (\cos^2\alpha + \sin^2\alpha)} = i\sin\alpha$ or, $\dfrac{e^{2\phi} - e^{-2\phi}}{e^{2\phi} + e^{-2\phi}} = \dfrac{\sin\alpha}{1}$.

By componendo and dividendo,

$$\frac{e^{2\phi}}{e^{-2\phi}} = \frac{1 + \sin\alpha}{1 - \sin\alpha} = \frac{\cos^2\frac{\alpha}{2} + \sin^2\frac{\alpha}{2} + 2\sin\frac{\alpha}{2}\cos\frac{\alpha}{2}}{\cos^2\frac{\alpha}{2} + \sin^2\frac{\alpha}{2} + 2\sin\frac{\alpha}{2}\cos\frac{\alpha}{2}}$$

or, $\quad e^{4\phi} = \dfrac{(\cos\frac{\alpha}{2} + \sin\frac{\alpha}{2})^2}{(\cos\frac{\alpha}{2} - \sin\frac{\alpha}{2})^2} = \dfrac{(1 + \tan\frac{\alpha}{2})^2}{(1 - \tan\frac{\alpha}{2})^2}$ or, $e^{2\phi} = \dfrac{1 + \tan\frac{\alpha}{2}}{1 - \tan\frac{\alpha}{2}} = \tan\left(\dfrac{\pi}{4} + \dfrac{\alpha}{2}\right)$.

$\therefore \quad \boldsymbol{\phi = \frac{1}{2}\log\tan\left(\frac{\pi}{4} + \frac{\alpha}{2}\right)}$ **(Proved).**

Example 5.15. *Prove that,* $\log\left(\dfrac{a + ib}{a - ib}\right) = 2i\tan^{-1}\left(\dfrac{b}{a}\right)$.

Hence, evaluate $\cos\left[i\log\left(\dfrac{a + ib}{a - ib}\right)\right]$. **[Rewa U 1990]**

Solution : Let $a = r\cos\theta$, $b = r\sin\theta$, $\therefore \quad \tan\theta = \tan^{-1}\left(\dfrac{b}{a}\right)$.

Now, $\boldsymbol{\log\left(\frac{a + ib}{a - ib}\right)} = \log\dfrac{r(\cos\theta + i\sin\theta)}{r(\cos\theta - i\sin\theta)} = \log(e^{i\theta} + e^{-i\theta}) = \log e^{2i\theta} = 2i\theta$

$$\boldsymbol{= 2i\tan^{-1}\left(\frac{b}{a}\right)} \quad \textbf{(Proved)}.$$

$$\therefore \quad \boldsymbol{\cos\left[i\log\left(\frac{a + ib}{a - ib}\right)\right]} = \cos[i(2i\theta)] = \cos 2\theta = \frac{1 - \tan^2\theta}{1 + \tan^2\theta} = \frac{1 - \left(\frac{b}{a}\right)^2}{1 + \left(\frac{b}{a}\right)^2} = \boldsymbol{\frac{a^2 - b^2}{a^2 + b^2}}.$$

Example 5.16. *Show that,* $\sin\left[i\log\dfrac{a - ib}{a + ib}\right] = \dfrac{2ab}{a^2 + b^2}$.

Solution : Put $a = r\cos\theta$, $b = r\sin\theta$, so that $\tan\theta = \dfrac{b}{a}$.

$$\therefore \quad \frac{a - ib}{a + ib} = \frac{r(\cos\theta - i\sin\theta)}{r(\cos\theta + i\sin\theta)} = \frac{e^{-i\theta}}{e^{i\theta}} = e^{-2i\theta}.$$

$$\therefore \quad \log\frac{a - ib}{a + ib} = -2i\theta.$$

So, $\boldsymbol{\sin\left[i\log\frac{a - ib}{a + ib}\right]} = \sin[i(-2\theta)] = \sin 2\theta = \dfrac{2\tan\theta}{1 + \tan^2\theta} = \dfrac{\frac{2b}{a}}{1 + \frac{b^2}{a^2}} = \boldsymbol{\dfrac{2ab}{a^2 + b^2}}$ **(Proved).**

Example 5.17. *Show that $f(z) = z^2$ is analytic.*

Solution : We have $f(z) = z^2 = (x+iy)^2 = x^2 - y^2 + 2ixy$.

$$\therefore \quad u(x,y) = x^2 - y^2 \quad \text{and} \quad v(x,y) = 2xy.$$

Now, $\dfrac{\partial u}{\partial x} = 2x,\ \dfrac{\partial u}{\partial y} = -2y,\ \dfrac{\partial v}{\partial x} = 2y,\ \dfrac{\partial v}{\partial y} = 2x.$

$$\therefore \quad \frac{\partial u}{\partial x} = \frac{\partial v}{\partial y} \quad \text{and} \quad \frac{\partial u}{\partial x} = -\frac{\partial v}{\partial y}.$$

Hence, ***f*(*x*) is analytic.**

Example 5.18. *Examine whether* $\sin z$ *is an analytic function of* z. **[BU(H) 1996]**

Solution : Let $z = x + iy$, then $\sin(z) = \sin(x+y) = \sin x \cos(iy) + \cos x \sin(iy)$.

But, $\cos(iy) = \dfrac{1}{2}(e^y + e^{-y}) = \cosh y$ and $\sin(iy) = -\dfrac{1}{2i}(e^y - e^{-y}) = i \sinh y$.

$$\therefore \quad \sin(x+iy) = \sin x \cosh x + i \cos x \sinh y \quad \text{or,} \quad u(x,y) = \sin \cosh x,\ v(x,y) = \cos \sinh y$$

Now, $\dfrac{\partial u}{\partial x} = \cos x \cos hy = \dfrac{\partial v}{\partial y}$ and $\dfrac{\partial u}{\partial y} = \sin x \sin hy = -(-\sin x \sin hy) = -\dfrac{\partial v}{\partial x}$.

Hence, **$\sin z$ is analytic.**

Example 5.19. *Examine whether the function,*

$$f(z) = \frac{x^2y^3(x+iy)}{x^4+y^6}, \quad \text{for } z \neq 0$$
$$= 0, \quad \text{for } z = 0$$

is analytic or not.

Solution : We have here, $y(x,y) = \frac{x^3y^3}{x^4+y^6}$ and $v(x,y) = \frac{x^2y^4}{x^4+y^6}$ with the condition $u(0,0) = 0$ and $v(0,0) = 0$.

It is obvious that $\frac{\partial u}{\partial x}, \frac{\partial u}{\partial y}, \frac{\partial v}{\partial x}, \frac{\partial v}{\partial y}$ are all zero at the origin. Hence, Cauchy-Rieman equations are satisfied at the origin.

$$\text{Now} \quad f'(0) = \underset{z\to 0}{\text{Lt}} \frac{f(z)-f(0)}{z} = \underset{\substack{x\to 0\\ y\to 0}}{\text{Lt}} \frac{x^2y^3(x+iy)}{(x^4+y^6)(x+iy)} = \underset{\substack{x\to 0\\ y\to 0}}{\text{Lt}} \frac{x^2y^3}{x^4+y^6}.$$

Let us assume that $z \to 0$ along the curve $y^3 = mx^2$

$$\text{Then,}\ f'(0) = \underset{x\to 0}{\text{Lt}} \frac{mx^4}{x^4+m^2x^4} \quad \text{or,} \quad f'(0) = \underset{x\to 0}{\text{Lt}} \frac{mx^4}{(1+m^2)x^4} = \frac{m}{1+m^2},$$

which is different for different values of m.

$\therefore$ $f'(0)$ does not exist. Thus, **the function is not analytic at the origin.**

Example 5.20. *Examine whether* $\sin z$ *is an analytic function of* z*. Determine the constants* a *and* b *so that the function* $f(z) = (x^2 + ay^2 - 2xy) + i(bx^2 - y^2 + 2xy)$ *becomes analytic.*

[BU(H) 1998]

Solution : $\sin z = \sin(x+iy) = \sin x \cos iy + \cos x \sin iy$

We know, $\cos iy = \frac{1}{2}(e^y + e^{-y}) = \cosh y$ and $\sin iy = -\frac{1}{2i}(e^y - e^{-y}) = i \sinh y$

$\therefore \quad \sin(x+iy) = \sin x \cosh y + i \cos x \sinh y$

$\therefore \quad u(x,y) = \sin x \cosh y, \quad v(x,y) = \cos x \sinh y$

Now, $\frac{\partial u}{\partial x} = \cos x \cosh y = \frac{\partial v}{\partial y}$ and $\frac{\partial u}{\partial y} = \sin x \sinh y = -(-\sin x \sinh y) = -\frac{\partial v}{\partial x}$

Hence, $\sin z$ is analytic.

The condition for $f(z)$ to be analytic is that $u(x,y)$ and $v(x,y)$ should be harmonic.

Here, $u(x,y) = x^2 + ay^2 - 2xy$ and $v(x,y) = bx^2 - y^2 + 2xy$.

$$\frac{\partial u}{\partial x} - 2x - 2y; \quad \frac{\partial u}{\partial y} = 2ay - 2x; \quad \frac{\partial^2 u}{\partial x^2} = 2; \quad \frac{\partial^2 u}{\partial y^2} = 2a.$$

$$\therefore \quad \frac{\partial^2 u}{\partial x^2} + \frac{\partial^2 u}{\partial y^2} = 2 + 2a = 0, \quad \therefore \quad a = -1.$$

Similarly, $b = 1$, $\quad \therefore$ **$f(z)$ is analytic for $a = -1$ and $b = 1$.**

Example 5.21. *Obtain the analytic function whose imaginary part is*

$$v(x,y) = -\cos x \cosh y.$$

Solution : $\frac{\partial v}{\partial x} = \sin x \cosh y; \quad \frac{\partial v}{\partial y} = -\cos x \sinh y.$

Now, $f'(z) = \frac{\partial u}{\partial x} + i\frac{\partial v}{\partial x} = \frac{\partial v}{\partial y} + i\frac{\partial v}{\partial x} = -\cos x \sinh y + i \sin x \cosh y$

$$= i(\sin x \cosh y + i \cos x \sinh y) = i \sin(x+iy) = i \sin z.$$

$\therefore \quad$ **$f'(z) = -i \cos z + c$, is the required analytic function.**

Example 5.22. *Show that $e^x(\cos y + i \sin y)$ is an analytic function and find its derivative. If $w = \ln z$, find $\frac{dw}{dz}$ and determine whether w is non-analytic.*

Solution : Let $e^x(\cos y + i \sin y) = u + iv$, then $e^x \cos y = u$ and $e^x \sin y = v$.

$$\therefore \quad \frac{\partial u}{\partial x} = e^x \cos y; \quad \frac{\partial v}{\partial x} = e^x \sin y$$

$$\frac{\partial u}{\partial y} = -e^x \sin y; \quad \frac{\partial v}{\partial y} = e^x \cos y.$$

$$\therefore \quad \frac{\partial u}{\partial x} = \frac{\partial v}{\partial y} \text{ and } \frac{\partial u}{\partial y} = -\frac{\partial v}{\partial x}.$$

Thus, **the Cauchy-Riemann equation is satisfied and hence, $e^x(\cos y + i \sin y)$ is analytic.**

$\therefore \quad f(z) = u + iv = e^x(\cos y + i \sin y)$

and $f'(z) = \frac{\partial u}{\partial x} + i\frac{\partial v}{\partial x} = e^x \cos y + ie^x \sin y = e^x(\cos y + i \sin y) = e^x \cdot e^{iy} = e^{x+iy} e^z.$

Also, $w = u + iv = \ln(x+iy) = \frac{1}{2}\ln(x^2+y^2) + i\tan^{-1}\frac{y}{x}$,

$\therefore \quad u = \frac{1}{2}\ln(x^2+y^2); \quad v = \tan^{-1}\frac{y}{x}$.

So, $\frac{\partial u}{\partial x} = \frac{x}{x2+y^2} = \frac{\partial v}{\partial y}$ and $\frac{\partial u}{\partial y} = \frac{y}{x^2+y^2} = \frac{y}{x^2+y^2} = -\frac{\partial v}{\partial x}$.

Thus, the Cauchy-Riemann equations are satisfied. Also, the partial derivatives are continuous except at $(0,0)$.

Hence, $\boldsymbol{w}$ **is analytic except at** $\boldsymbol{z = 0}$.

Now, $\frac{dw}{dz} = \frac{\partial u}{\partial x} + i\frac{\partial v}{\partial x} = \frac{x}{x^2+y^2} + i\frac{-y}{x^2+y^2} = \frac{x-iy}{x^2+y^2} = \frac{1}{x+iy} = \frac{1}{z}$.

Example 5.23. *If $f(z)$ is analytic, then show that,*

$$\left(\frac{\partial}{\partial x}|f(z)|\right)^2 + \left(\frac{\partial}{\partial y}|f(z)|\right)^2 = |f'(z)|^2 \qquad \textbf{[BU(H) 1998]}$$

Solution : Let $f(z) = u(x,y) + iv(x,y)$.

$\therefore \quad |f(z)| = \sqrt{u^2+v^2}, \quad \therefore \quad \frac{\partial}{\partial x}|f(z)| = \frac{1}{\sqrt{u^2+v^2}}\left(u\frac{\partial u}{\partial x} + v\frac{\partial v}{\partial x}\right)$.

Now Cauchy-Riemann equations for analytic function are :

$$\frac{\partial v}{\partial x} = -\frac{\partial u}{\partial y}; \quad \frac{\partial v}{\partial y} = \frac{\partial u}{\partial x}.$$

$$\therefore \quad \frac{\partial}{\partial x}|f(z)| = \frac{1}{\sqrt{u^2+v^2}}\left(u\frac{\partial u}{\partial x} - v\frac{\partial u}{\partial y}\right)$$

$$\text{and} \quad \frac{\partial}{\partial y}|f(z)| = \frac{1}{\sqrt{u^2+v^2}}\left(u\frac{\partial u}{\partial y} + v\frac{\partial v}{\partial y}\right) = \frac{1}{\sqrt{u^2+v^2}}\left(u\frac{\partial u}{\partial y} + v\frac{\partial u}{\partial x}\right).$$

$$\therefore \quad \left(\frac{\partial}{\partial x}|f(z)|\right)^2 + \left(\frac{\partial}{\partial y}|f(z)|\right)^2 = \frac{1}{(u^2+v^2)}\left[\left(u\frac{\partial u}{\partial x} - v\frac{\partial u}{\partial y}\right)^2 + \left(u\frac{\partial u}{\partial y} + v\frac{\partial u}{\partial x}\right)^2\right]$$

$$= \left[\left(\frac{\partial u}{\partial x}\right)^2 + \left(\frac{\partial u}{\partial y}\right)^2\right] = |f'(z)|^2 \quad \textbf{(Proved).} \quad \left[\because f'(z) = \frac{\partial u}{\partial x} + i\frac{\partial v}{\partial x} = \frac{\partial u}{\partial x} - i\frac{\partial u}{\partial y}\right]$$

Example 5.24. *If $f(z)$ is a regular function of z, prove that,*

$$\left(\frac{\partial^2}{\partial x^2} + \frac{\partial^2}{\partial z^2}\right)|f(z)|^2 = 4|f'(z)|^2.$$

Solution : We have $z = x + iy$ and $\overline{z} = x - iy$.

$$\therefore \quad x = \frac{1}{2}(z + \overline{z}) \quad \text{and} \quad y = -\frac{i}{2}(z - \overline{z}).$$

$$\therefore \quad \frac{\partial}{\partial z} = \frac{\partial}{\partial x}\cdot\frac{\partial x}{\partial z} + \frac{\partial}{\partial y}\cdot\frac{\partial y}{\partial z} = \frac{1}{2}\left(\frac{\partial}{\partial x} - i\frac{\partial}{\partial y}\right)$$

and $\dfrac{\partial}{\partial z} = \dfrac{\partial}{\partial x}\dfrac{\partial x}{\partial \overline{z}} + \dfrac{\partial}{\partial y}\dfrac{\partial y}{\partial \overline{z}} = \dfrac{1}{2}\left(\dfrac{\partial}{\partial x} + i\dfrac{\partial}{\partial y}\right).$

$$\therefore \quad \frac{\partial}{\partial z}\cdot\frac{\partial}{\partial z} = \frac{1}{4}\left(\frac{\partial}{\partial x} - i\frac{\partial}{\partial y}\right)\left(\frac{\partial}{\partial x} + i\frac{\partial}{\partial y}\right) = \frac{1}{4}\left(\frac{\partial^2}{\partial x^2} + \frac{\partial^2}{\partial y^2}\right).$$

$$\therefore \quad \left(\frac{\partial^2}{\partial x^2} + \frac{\partial^2}{\partial y^2}\right)|f(z)|^2 = 4\frac{\partial^2}{\partial z\partial\overline{z}}[f(z)f(\overline{z})] = 4\frac{\partial}{\partial z}[f(z)f'(\overline{z})]$$

$$= \mathbf{4f'(z)f'(\overline{z}) = 4|f'(z)|^2} \quad \textbf{(Proved).}$$

Example 5.25. *Given, $u(x,y) = x^3 - 3x^2$. Under what condition would you expect a conjugate function $v(x,y)$ to exist such that the function $f(z) = u(x,y) + iv(x,y)$ is analytic? If $u(x,y)$ satisfies the condition, then find $v(x,y)$.* **[BU(H) 1995]**

Solution : We have $u(x,y) = x^3 - 3xy^2$.

$$\therefore \quad \frac{\partial u}{\partial x} = 3x^2 - 3y^2, \quad \frac{\partial u}{\partial y} = -6xy.$$

$$\therefore \quad \frac{\partial^2 u}{\partial x^2} + \frac{\partial^2 u}{\partial y^2} = 6x - 6x = 0.$$

Hence, $u(x,y)$ will have its conjugate function $v(x,y)$.

Now, $f'(z) = \dfrac{\partial u}{\partial x} + i\dfrac{\partial v}{\partial y} = \dfrac{\partial u}{\partial x} - i\dfrac{\partial u}{\partial y} = 3(x^2 - y^2) + i6xy = 3(x^2 - y^2 + 2ixy)$

$$= 3(x+iy)^2 = 3z^2.$$

$$\therefore \quad f(z) = z^3 + c = (x+iy)^2 + c,$$

$$\therefore \quad \boldsymbol{v(x,y) = (3x^2y - y^3).}$$

Example 5.26. *Evaluate $\displaystyle\int_C \frac{dz}{z^n}$, n positive integer > 1, where C is unit circle with centre at origin.*

Solution : The unit circle may be represented by $z = e^{i\theta}$, $\therefore\ dz = ie^{i\theta}d\theta$.

$$\therefore \quad \int_C \frac{dz}{z^n} = \int_0^{2\pi} \frac{ie^{i\theta}}{e^{in\theta}}d\theta = \int_0^{2\pi} ie^{(1-n)i\theta}d\theta = \frac{ie^{(1-n)i\theta}}{i(1-n)}\Big]_0^{2\pi} = \frac{1}{1-n}(1-1) = \mathbf{0}.$$

Example 5.27. *Evaluate $\displaystyle\int_0^{1+i}(x^2+iy)dz$ along the path $y = x^2$.* **[BU(H) 2001]**

Solution : We have $\displaystyle\int_C (x^2+iy)dz = \int_C (x^2+iy)(dx+idz)$

$$= \int_C x^2dx - y\,dy + i\int_C y\,dy + x^2dy.$$

Now, the parametric equation of the curve is $x = t$, $y = t^2$ and the curve moves from $t = 0$ to $t = 1$.

$\therefore$ the line integral $= \displaystyle\int_{t=0}^1 t^2dt - t^2(2t)dt + i\int_{t=0}^1 t^2dt + t^2(2t)dt$

$$= \int_0^1 (t^2 - 2t^3)dt + i\int_0^1 (t^2 + 2t^3)dt = \left(\frac{t^3}{3} - \frac{t^4}{2}\right)_0^t + i\left(\frac{t^3}{3} + \frac{t^4}{2}\right)_0^1$$

$$= \frac{1}{6}(-1 + 5i).$$

Example 5.28. *Find the value of* $\int_0^{2+i} z^2 dz$ *along the following contours* (i) *the straight line from* 0 *to* $2+i$; (ii) polygonal line from 0 to 2 and from 2 to $2+i$.

Solution : We have $\int_0^{2+i} z^2 dz = \int_0^{2+i} (x^2 - y^2 + i2xy)(dx + idy)$

$$= \int_0^{2+i} [(x^2 - y^2)dx - 2xy\,dy] + i\int_0^{2+i} [2xy\,dx + (x^2 - y^2)dy]$$

(i) The straight line from 0 to $2+i$ has parametric equation $x = 2t,\ y = t,\ 0 \le t \le 1$.

$\therefore \quad dx = 2dt,\ dy = dt.$

$\therefore$ the integral becomes

$$\int_0^{2+i} z^2 dz = \int_0^1 [(4t^2 - t^2)(2dt) - 2(2t)dt] + i\int_0^1 [4t^2(2\alpha t) + (4t^2 - t^2)dt]$$

$$= \int_0^1 2t^2 dt + i\int_0^1 11t^2 dt = \frac{2}{3} + i\frac{11}{3}.$$

(ii) When we have two paths one along real axis, $x = x$ and $y = 0$ and the other along the imaginary axis, when $x = 2 + iy$ and $dx = 0$. Thus,

$$\int_0^{2+i} z^2 dz = \int_0^2 x^2 dx + i\int_0^1 (2 + iy)^2 dy = \frac{8}{3} + i\int_0^1 (4 - y^2 + 4iy)dy$$

$$= \frac{8}{3} + i\left(4 - \frac{1}{3} + 2i\right) = \frac{2}{3} + i\frac{11}{3}.$$

Example 5.29. *Evaluate* $\int_{1-i}^{2+i} (2x+iy+1)dz$ *along the straight line joining the points* $(1-i)$ *and* $(2+i)$. **[BU(H) 2001]**

Solution : We have $\int_C (2x + iy + 1)dz = \int_C (2x + 1 + iy)(dx + idy)$

$$= \int_C (2x + 1)dx - y\,dy + i\int_C y\,dx + (2x + 1)dy.$$

The numbers $1-i$ and $2+i$ represent the point having coordinates $(1, -1)$ and $(2, 1)$ respectively. The equation to the line joining the points $(1, -1)$ and $(2, 1)$ is $y = 2x - 3$. Changing y to x and integrating from $x = 1$ to $x = 2$, we get the integral as

$$\int_{x=1}^2 (2x + 1)dx - (2x - 3)2dx + i\int_{x=1}^2 (2x - 3)dx + (2x + 1)2dx$$

$$= \int_1^2 (7 - 2x)dx + i\int_1^2 (6x - 1)dx = (7x - x^2)_1^2 + i(3x^2 - x)_1^2 = \mathbf{4 + 8i}.$$

Example 5.30. *Evaluate the integral* $\int_C \frac{dz}{z-a}$, *where* C *is a cardioid given by the equation* $r = a(1+\cos\theta)$ *and* $z = a$ *is a point inside* C.

Solution : Let us imagine a circle with centre at the point a and radius equal to ρ.

Then, $f(z) = \frac{1}{2-a}$

is continuous in the closed annulus and regular in the open annulus bound by C and C_1.

$$\therefore \quad \int_C \frac{dz}{z-a} = \int_{C_1} \frac{dz}{C_1}$$

But on C_1, $z = a + \rho(\cos\theta + i\sin\theta) = a + \rho e^{i\theta}$, $0 \le \theta \le 2\pi$,

so that $dz = \rho i e^{i\theta}$

$$\text{Hence,} \quad \int_{C_1} \frac{dz}{z-a} = \int_0^{2\pi} \frac{\rho i e^{i\theta}}{\rho e^{i\theta}} d\theta = \int_0^{2\pi} i d\theta = 2\pi i.$$

If the point $z = a$ be outside the boundary of C, then the function $f(z) = \frac{1}{z-a}$ becomes analytic at any point within and on C. In that case

$$\int_C \frac{dz}{z-a} = 0.$$

Example 5.31. *If* $f(z)$ *is analytic in a simple connected region* R, *prove that* $\int_a^b f(z)dz$ *is independent of the path joining any two points in* R. **[BU(H) 2001]**

Solution : Let the two points $A(a)$ and $B(b)$ be connected by curves $ACB(C_1)$ and $ADB(C_2)$. Then by Cauchy's theorem,

$$\int_{ADBCA} f(z)dz = 0$$

$$\text{or,} \quad \int_{ADB} f(z)dz + \int_{BCA} f(z)dz = 0$$

$$\text{or,} \quad \int_{ADB} f(z)dz = -\int_{BCA} f(z)dz = \int_{ACB} f(z)dz$$

$$\text{or,} \quad \int_{C_2} f(z)dz = \int_{C_1} f(z)dz = \int_a^b f(z)dz \quad \textbf{(Proved)}.$$

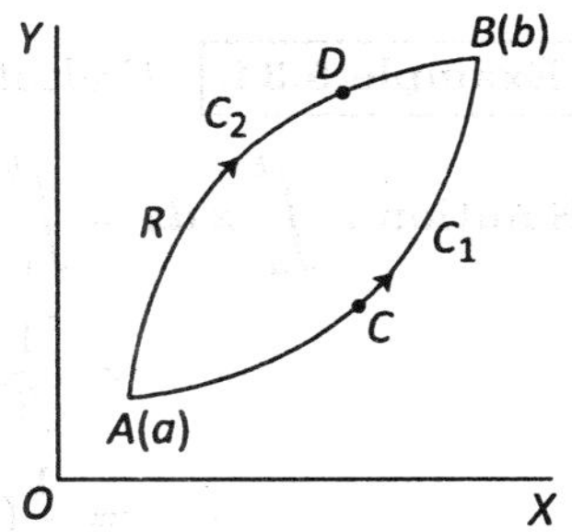

Fig. 5.8

$\therefore$ $\int_a^b$ ***f(z)dz* is independent of the path joining any two points in *R*.**

Example 5.32. *Evaluate the integral* $\oint \frac{dz}{z-a}$ *in the conventional positive sense, where* C *is any simply closed curve and* $z = a$ *is a point inside* C. *What is the value of the integral, if* $z = a$ *is outside* C. **[BU(H) 1996]**

Solution : Let C be a closed boundary and a a point inside C [Fig. 5.9]. We encircle the point a with radius z_0 and the boundary is C_1.

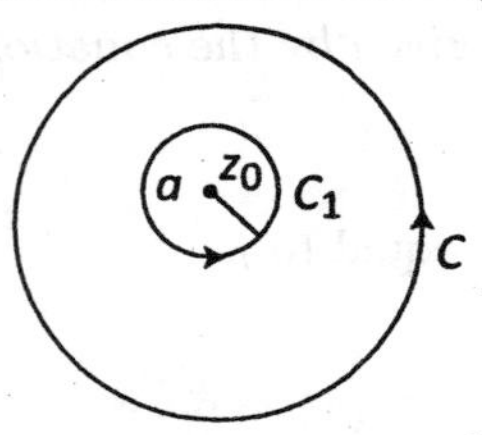
Fig. 5.9

Then $\oint_C \frac{dz}{z-a} = \oint_{C_1} \frac{dz}{z-a}$.

Again, on C_1,

$$z = a + Z_0(\cos\theta + i\sin\theta) = a + z_0 e^{i\theta},\ 0 \le \theta \le 2\pi.$$

$$\therefore \quad dz = z_0 i e^{i\theta}\, d\theta.$$

$$\therefore \quad \oint_{C_1} \frac{dz}{z-a} = \int_0^{2\pi} \frac{1}{z_0 e^{i\theta}} z_0 i e^{i\theta}\, d\theta = i\int_0^{2\pi} d\theta = 2\pi i.$$

When the point a lies outside the boundary C, then clearly $f(z) = \frac{1}{z-a}$ is analytic at any point on C.

$\therefore$ by Cauchy's theorem, $\oint_C \frac{dz}{z-a} = 0.$

Example 5.33. *Evaluate the integral* $\int_C \frac{e^z(z^2+1)}{(z-1)^2}dz$, *where C is the circle $|z| = 2$.*

Solution : As C is given by $|z| = 2$, the point $z = 1$ lies within C, then according to the theorem,

$f'(a) = \frac{1}{2\pi i}\int_C \frac{f(z)dz}{(z-a)^2}$, we have in the present case,

$$\int_C \frac{e^z(z^2+1)}{(z-1)^2}dz = \int_C \frac{f(z)}{(z-1)^2}dz = 2\pi i f'(1).$$

Now, $f(z) = e^z(z^2+1)$, $\quad\therefore\quad f'(z) = e^z(z^2+2z+1) = 4e.$

$$\therefore \quad \int_C \frac{f(z)}{(z-1)^2}dz = 2\pi i(4e) = 8\pi e i.$$

Example 5.34. *Evaluate from the first principle* $\int_a^b z\,dz$. **[BU(H) 1994]**

Solution : $\int_a^b z\,dz = \int_a^b (x+iy)(dx+idy) = \int_a^b x\,dx - y dy + i\int_a^b x\,dy + y\,dx$

$$= \left[\frac{1}{2}x^2 - \frac{1}{2}y^2 + ixy\right]_a^b = \frac{1}{2}[x^2 + (iy)^2 + 2xiy]_a^b = \frac{1}{2}[(x+iy)]_a^b = \frac{1}{2}[z^2]_a^b$$

$$= \frac{1}{2}(b^2 - a^2).$$

Example 5.35. *Evaluate* $\oint_C \frac{e^{2z}}{(z+1)^4}dz$, *where C is the circle within $|z| = 3$.* **[BU(H) 1995]**

Solution : Since C is $|z| = 3$, $z = -1$ must be a point within C.

$$\therefore \quad \oint_C \frac{f(z)}{(z+1)^4} = \frac{2\pi i f^3(-1)}{3!}.$$

whose $f^3(z)$ is the 3rd derivative of $f(z)$.

Again, $f(z) = e^{2z}$, $\quad\therefore\quad f^3(z) = 8e^{2x}$.

$\therefore \quad f^3(-1) = 8e^{-2}.$

Hence, $\displaystyle\oint \frac{f(z)}{(z+1)^4} = 2\pi i \frac{8e^{-2}}{6} = \frac{\mathbf{8\pi i}}{\mathbf{3e^2}}.$

Example 5.36. *Using Cauchy's integral formula evaluate the integral.*

$$\int_C \frac{2z+1}{z^2+z}dz, \text{ where } C \text{ is } |x| = \frac{1}{2}. \qquad \textbf{[BU(H) 2002]}$$

Solution : The given integral is $\displaystyle\int_C \frac{2z+1}{z^2+z}dz = \int_C \frac{2z+1}{z(z+1)}dz = \int_C \frac{1}{z}dz + \int_C \frac{1}{z+1}dz.$

In the first integral of the right side $a = 0$ is a point within C; $|z| = \frac{1}{2}$ and in the second integral $a = -1$ is a point outside C.

So, in $\displaystyle\int_C \frac{dz}{z}$, $f(z) = 1$.

$\therefore \quad f(0) = 1$, as $a = 0$ and $\displaystyle\int_C \frac{dz}{z} = 2\pi i f(0) = 2\pi i.$

In $\displaystyle\int_C \frac{dz}{z+1}$, the point $a = -1$ is outside C.

$\therefore \quad f(z) = \dfrac{1}{z+1}$ is analytic everywhere within C, $\because \quad |z| = \dfrac{1}{2}.$

So, by Cauchy's theorem, $\displaystyle\int_C \frac{dz}{z+1} = 0.$

$$\therefore \quad \int_0 \frac{\mathbf{2z+1}}{\mathbf{z^2+z}}\mathbf{dz} = \mathbf{2\pi i}.$$

Example 5.37. *Evaluate the integral,* $\displaystyle\int_C \frac{z-3}{z^2+2z+3}dz$, *where C is the circle* $|z+1+i| = 2$.

[BU(H) 2001]

Solution : In this case the circle C is $|z+1+i| = 2$.

or, $\quad |(x+1) + i(y+1)| = 2$ or, $(x+1)^2 + (y+1)^2 = 4$.

This equation represents a circle with centre $(-1, -1)$ and radius 2.

$$\text{Again, } \int_C \frac{(z-3)dz}{x^2+2z+3} = \int_C \frac{(z-3)}{(z+1+i\sqrt{2})(z+1-i\sqrt{2})}dz$$

$$= \int_C \frac{(z-3)}{\{z-(-1-i\sqrt{2})\}\{z-(-1+i\sqrt{2})\}}dz.$$

The point $(-1+i\sqrt{2})$, lies outside the circle C.

For the outside point $(-1-i\sqrt{2})$, we consider $f(z)$ as $\dfrac{(z-3)}{(z+1-i\sqrt{2})}$.

$$\text{Then } \oint_C \frac{\mathbf{(z-3)}}{\mathbf{z^2+2z+3}}\mathbf{dz} = 2\pi i f(-1-i\sqrt{2}) = 2\pi i \frac{(-1-i\sqrt{2}-3)}{(-1-i\sqrt{2}+1-i\sqrt{2})} = \pi\frac{(4+i\sqrt{2})}{\sqrt{2}}$$

$$= \boldsymbol{\pi(2\sqrt{2}+i)}.$$

Example 5.38. *Discuss the singularities of the function* $f(z) = \frac{z^2-2}{z^3}\cos\left(\frac{1}{1+z}\right)$.

Solution : The zero's of the function $f(z)$ are given by

$$(z^2-2)\cos\left(\frac{1}{1+z}\right) = 0$$

or, $(z^2-2) = 0$ and $\cos\left(\frac{1}{1+z}\right) = 0$

or, $z = \pm\sqrt{2}$ and $\frac{1}{1+z} = 2n\pi + \frac{\pi}{2}$ $(n = 0, \pm 1, \ldots)$

or, $z = \pm\sqrt{2}$ and $z = 1 - \frac{1}{2n\pi + \frac{\pi}{2}}$ $(n = 0, \pm 1, \pm 2, \ldots)$.

This shows that the function has two zeros at $z = \sqrt{2}$ and $z = -\sqrt{2}$ and the limit point of zeros are given by

$$z = 1 - \frac{1}{2n\pi + \frac{\pi}{2}} \text{ is } 1.$$

Hence, $z = 1$ is an isolated essential singularity.

The poles of the function are given by $z^3 = 0$. Thus, $z = 0$ is a pole of $f(z)$ of order three.

Example 5.39. *Find the residue of* $f(z) = \frac{e^2}{z^2 + a^2}$ *at its singularities.*

Solution : Here, $f(z) = \frac{e^z}{z^2 + a^2} = \frac{e^2}{(z+ia)(z-ia)}$.

The singularities of the function $f(z)$ are given by

$$(z+ia)(z-ia) = 0, \quad \text{i.e., } z = \pm ia.$$

Thus, the given function $f(z)$ has two simple poles one at $z = ia$ and another at $z = -ia$.

The residue of $f(z)$ at $z = ia$ is given by

$$\operatorname*{Res}_{z \to ia} f(z) = \operatorname*{Lt}_{z \to ia} (z - i\alpha) f(z) = \operatorname*{Lt}_{z \to ia} (z - ia)\frac{e^z}{(z+ia)(z-ia)} = \operatorname*{Lt}_{z \to ia} \frac{e^z}{(z+ia)} = \frac{e^{ia}}{zia}.$$

Similarly, the residue of $f(z)$ at $z = ia$ is given by

$$\operatorname*{\mathbf{Res}}_{z \to -ia} f(z) = \operatorname*{Lt}_{z \to -ia} (z+ia) f(z) = \operatorname*{Lt}_{z \to -ia} (z+ia)\frac{e^z}{(z+ia)(z-ia)} = \operatorname*{Lt}_{z = -ia} \frac{e^z}{z - ia}$$

$$= \frac{e^{-ia}}{-2ai} = \frac{e^{-ia}}{2ia}.$$

Example 5.40. *Find the residue of* $\frac{z^3 - z^2 + 1}{z^3}$ *at infinity.*

Solution : The function $f(z) = \frac{z^3 - z^2 + 1}{z^3}$ behaves like $\frac{1}{2}$ for large values of z and is, therefore, analytic at $z = \infty$. Now, the residue of $f(z)$ at $z = \infty$ is given by

$$\operatorname*{Res}_{z = \infty} f(z) = \operatorname*{Lt}_{z \to \infty} [-z f(z)] = \operatorname*{Lt}_{z \to \infty} \left[\frac{z^2}{(z-a)(z-b)}\right] = \operatorname*{Lt}_{z \to \infty} \left[-\frac{1}{\left(1 - \frac{a}{x}\right)\left(1 - \frac{b}{x}\right)}\right].$$

Thus, the function $f(z)$ is analytic at ∞ and still has residue at ∞.

Example 5.41. *Find the residue of* (i) $\dfrac{z}{(z-1)(z+1)^2}$ *at all its singularities* (ii) $e^{-\frac{1}{z}}$ *at its singularity.* **[BU(H) 1997]**

Solution : (i) The function $f(z) = \frac{z}{(z-1)(z+1)^2}$ has the pole of order one at $z = 1$ and order two at $z = -1$ respectively.

Now, for residue at $z = 1$, $\phi(z) = z(z+1)^{-2}$. Since the pole is of order one, $m = 1$.

So, the residue at $z = 1$, is $\dfrac{\phi(1)}{0!} = 1 \cdot (2)^{-2} = \dfrac{1}{4}$.

For residue at $z = -1$, $\phi(z) = z(z-1)^{-1}$.

The pole is of order 2, $\therefore$ $m = 2$.

$\therefore$ **the residue at $z = 2$ is** $\dfrac{\phi'(-1)}{1!} = \left[\dfrac{1}{z-1} - \dfrac{z}{(z-1)^2}\right]_{z=1} = \dfrac{1}{-2} + \dfrac{1}{4} = -\dfrac{1}{4}$.

(ii) Residue can also be found by taking the coefficient of $\frac{1}{z}$ in the expression of $f(z)$.

Here, $f(z) = e^{-\frac{1}{z}} = 1 - \dfrac{1}{z} + \dfrac{1}{2!z^2} - \dfrac{1}{3!z^3} + \cdots$.

Since $z = 0$ is the singularity of $f(z)$, the residue at $z = 0$ is the coefficient of $\frac{1}{z}$ in the expression of $f(z) = -1$.

Example 5.42. *Find the residue of* $f(z) = \dfrac{z\cos z}{(z-\pi)^3}$ *at* $z = \pi$.

Solution : Here $z = \pi$ is a pole of order 3.

$$\therefore \quad \phi(z) = z\cos z$$
$$\phi'(z) = \cos z - z\sin z$$
$$\phi''(z) = -\sin z - (\sin z + z\cos z) = -2\sin z - z\cos z.$$

The residue at $z = \pi$ is

$$\left.\frac{\phi''(z)}{2!}\right|_{z=\pi} = \frac{1}{2!}[-2\sin z - z\cos z]_{z=\pi} = \frac{\pi}{2}.$$

Example 5.43. *Apply calculus of residue to show that,*

(a) $\displaystyle\int_0^{2\pi} \frac{d\theta}{a + b\cos\theta} = \frac{2\pi}{\sqrt{a^2 - b^2}}, \; a > b > 0$ (b) $\displaystyle\int_0^{2\pi} \frac{d\theta}{1 + \in\cos\theta} = \frac{2\pi}{\sqrt{1 - \in^2}}, \; 0 < \in < 1$

(c) $\displaystyle\int_0^{2\pi} \frac{d\theta}{25 - 24\cos\theta} = \frac{2\pi}{7}$ (d) $\displaystyle\int_0^{2\pi} \frac{d\theta}{2 + \cos\theta} = \frac{2\pi}{\sqrt{3}}$.

Solution : (a) Let $I = \displaystyle\int_0^{2\pi} \frac{d\theta}{a + b\cos\theta}$.

Putting $z = e^{i\theta}$, $d\theta = \dfrac{dz}{iz}$ and $\cos\theta = \dfrac{1}{2}\left(z + \dfrac{1}{z}\right)$;

we get $I = \displaystyle\int_C \frac{\frac{dz}{iz}}{a + \frac{b}{2}\left(z + \frac{1}{z}\right)}$, where C denotes the unit circle $|z| = 1$

or, $$I = \frac{2}{ib}\int_C \frac{dz}{z^2 + \frac{2a}{b}z + 1} = \int_C f(z)dz, \tag{i}$$

where $$f(z) = \frac{2}{ib\left(z^2 + \frac{2a}{b}z + 1\right)}. \tag{ii}$$

The poles of $f(z)$ are given by

$$z^2 + \frac{2a}{b}z + 1 = 0 \tag{iii}$$

or, $$(z - \alpha)(z - \beta) = 0,$$

where α and β are the roots of Eqn. (iii) and are given by

$$\alpha = \frac{-a\sqrt{a^2 + b^2}}{b}; \quad \beta = \frac{-a - \sqrt{a^2 - b^2}}{b}.$$

We have, $\alpha + \beta = -\frac{2a}{b}$ and $\alpha\beta = 1$.

As $a > b > 0$, $\therefore$ $|\beta| > 1$.

Also $|\alpha|\,|\beta| = 1$. This implies that $|\alpha| < 1$. That is, α lies inside and β lies outside the contour C. Thus, the only pole lying inside the contour is the simple pole at $z = \alpha$.

The residue of $f(z) = \dfrac{2}{ib\left(z^2 + \frac{2a}{b}z + 1\right)} = \dfrac{2}{ib(z - \alpha)(z - \beta)}$.

Its simple pole at $z = \alpha$ is given by

$$\underset{z=\alpha}{\text{Res}}\ f(z) = \underset{z\to\alpha}{\text{Lt}}\ (z - \alpha)f(z) = \underset{z\to\alpha}{\text{Lt}}\ (z - \alpha)\frac{2}{ib(z - \alpha)(z - \beta)}$$

$$= \underset{z\to\alpha}{\text{Lt}}\ \frac{2}{ib(z - \beta)} = \frac{2}{ib(\alpha - \beta)} = \frac{2b}{ib2\sqrt{a^2 - b^2}}$$

$$= \frac{1}{ib\sqrt{a^2 - b^2}} \quad \left[\because\ (\alpha - \beta) = \frac{2\sqrt{a^2 - b^2}}{b}\right].$$

Hence, by residue theorem,

$$I = \int_C f(z)dz = 2\pi i \times (\text{sum of residues at poles within } C)$$

$$= 2\pi i \times \frac{1}{i(\sqrt{a^2 - b^2})} = \frac{2\pi}{\sqrt{a^2 - b^2}}.$$

Hence, $$\int_0^{2\pi} \frac{d\theta}{a + b\cos\theta} = \frac{2\pi}{\sqrt{a^2 - b^2}}, \quad a > b > 0 \quad \textbf{(Proved)}.$$

(b) $\displaystyle\int_0^{2\pi} \frac{d\theta}{1 + \epsilon\cos\theta}$ is a particular case of (i).

Here $a = 1$, $b = \epsilon$.

$\therefore$ $$\int_0^{2\pi} \frac{d\theta}{1 + \epsilon\cos\theta} = \frac{2\pi}{\sqrt{1 - \epsilon^2}} \quad \textbf{(Proved)}.$$

(c) Putting $a = 25,\ b = 24$ in (i), we get

$$\int_0^{2\pi} \frac{d\theta}{25 - 24\cos\theta} = \frac{2\pi}{\sqrt{(25)^2 - (24)^2}} = \frac{2\pi}{7} \quad \textbf{(Proved)}.$$

(d) Putting $a = 2,\ b = 1$ in (i), we get

$$\int_0^{2\pi} \frac{d\theta}{2 + \cos\theta} = \frac{2\pi}{\sqrt{2^2 - 1^2}} = \frac{2\pi}{\sqrt{3}} \quad \textbf{(Proved)}.$$

Example 5.44. *Evaluate* $\displaystyle\int_0^{2\pi} \frac{\cos 3\theta\, d\theta}{5 - 4\cos\theta}$. **[BU(H) 1996]**

Solution : Let $z = e^{i\theta}$, $\therefore\ d\theta = \dfrac{dz}{iz}$, $\cos\theta = \dfrac{1}{2}\left(z + \dfrac{1}{z}\right)$.

Again, $z^3 = e^{3i\theta} = \cos 3\theta + i\sin 3\theta$, $\quad\therefore\quad z^{-3} = \cos 3\theta - i\sin 3\theta$.

$$\therefore \qquad \cos 3\theta = \frac{1}{2}\left(z^3 + \frac{1}{z^3}\right).$$

Now,
$$\int_0^{2\pi} \frac{\cos 3\theta}{5 - 4\cos\theta} d\theta = \frac{1}{2}\oint \frac{(z^3 + \frac{1}{z^3})dz}{iz\left[5 - 2\left(z + \frac{1}{z}\right)\right]} = -\frac{1}{2i}\oint_C \frac{(z^6 + 1)dz}{z^3(2z^3 - 5z + 2)}$$
$$= -\frac{1}{4i}\oint_C \frac{(z^6 + 1)dz}{z^3\left(z - \frac{1}{2}\right)(z - 2)}.$$

C is a circle $|z| = 1$.

Here, the pole is of order 3 at $z = 0$, two simple poles at $z = \frac{1}{2}$ and $z = 2$ of which the pole $z = 2$ is outside the circle.

Now residue at $z = \dfrac{1}{2}$ is $\dfrac{\phi\left(\frac{1}{2}\right)}{0!} = \phi\left(\dfrac{1}{2}\right)$, where, $\phi(z) = \dfrac{z^6 + 1}{z^3(z - 2)}$,

$$\therefore \qquad \phi\left(\frac{1}{2}\right) = \frac{\left(1 + \frac{1}{64}\right)}{\frac{1}{8}\left(-\frac{3}{2}\right)} = -\frac{65}{12}.$$

Pole of order 3 at $z = 0$,

$$\phi(z) = \frac{z^6 + 1}{\left(z - \frac{1}{2}\right)(z - 2)} = \frac{2(z^6 + 1)}{2z^2 - 5z + 2}$$

$$\phi'(z) = \phi(z)\left[\frac{6z^5}{z^6 - 1} - \frac{2}{2z - 1} - \frac{1}{z - 2}\right].$$

Now, $\quad \phi(0) = \dfrac{2}{2} = 1, \quad \therefore \quad \phi'(0) = 1\left[2 + \dfrac{1}{2}\right] = \dfrac{5}{2}.$

Then,
$$\phi''(z) = \phi'(z)\left[\frac{6z^5}{z^6 + 1} - \frac{2}{2z - 1} - \frac{1}{z - 2}\right]$$
$$+ \phi(z)\left[\frac{(z^6 + 1)30z^4 - (6z^5)^2}{(z^6 + 1)} + \frac{4}{(2z - 1)^2} + \frac{1}{(z - 2)^2}\right].$$

$$\therefore \qquad \phi''(z) = \frac{5}{2}\left(2 + \frac{1}{2}\right) + 1\left(4 + \frac{1}{4}\right) = \frac{21}{2}.$$

Residue at $z = 0$ is $\dfrac{\phi''(0)}{2!} = \dfrac{21}{4}$.

Then $$\int_0^{2\pi} \frac{\cos 3\theta}{5 - 4\cos\theta} d\theta = \frac{1}{4!} \oint_C \frac{(2^6 + 1)dz}{z^3 \left(z - \frac{1}{2}\right)(z - 2)} = 2\pi i \left(-\frac{1}{4i}\right)\left(\frac{21}{4} - \frac{65}{12}\right) = \frac{\pi}{12}.$$

Example 5.45. *Evaluate* $\displaystyle\int_0^\infty \frac{\cos m\pi}{x^2 + 1} dx.$ **[BU(H) 1993]**

Solution : We consider the integral $\displaystyle\oint \frac{e^{imx}}{z^2 + 1} dz$, where the contour C represents the real axis $-R$ to $+R$ and semicircular are C_1. Then the integral has the pole at e

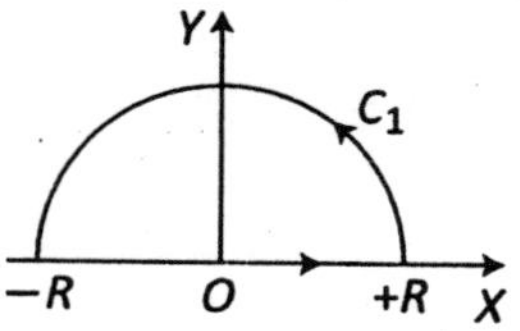

Fig. 5.10

$$z = i \text{ (inside)} \quad \text{and} \quad z = -i \text{ (outside)}.$$

Now residue at $z = i$ is

$$\underset{z \to i}{\text{Lt}} \left[(z - i) \frac{e^{imz}}{(z - i)(z + i)}\right] = \frac{e^{i^2 m}}{2i} = \frac{e^{-m}}{2i}.$$

$$\therefore \quad \oint \frac{e^{imz}}{z^2 + 1} dz = 2\pi i \frac{e^{-m}}{2i} = \pi e^{-m} \quad \text{or,} \int_{-R}^{+R} \frac{e^{imx}}{x^2 + 1} dx + \int_{C_1} \frac{e^{imx}}{x^2 + 1} dz = \pi e^{-m}$$

$$\text{or,} \quad \int_{-R}^{+R} \frac{\cos mx}{x^2 + 1} dx + i \int_{-R}^{+R} \frac{\sin mx}{x^2 + 1} dx + \int_{C_1} \frac{e^{imz}}{z^2 + 1} dz = \pi e^{-m}$$

$$\text{or,} \quad 2 \int_0^R \frac{\cos mx}{x^2 + 1} dx + \int_{C_1} \frac{e^{imx}}{z^2 + 1} dz = \pi e^{-m}.$$

$$\left[\because \frac{\sin mx}{x^2 + 1} \text{ is an odd function on } \int_{-R}^{R} \frac{e^{imx}}{x^2 + 1} dx = 0\right]$$

$$\text{Hence,} \int_0^\infty \frac{\cos m\pi}{x^2 + 1} dx = \frac{\pi}{2} e^{-m}, \quad \left[\because \underset{R \to \infty}{\text{Lt}} \int_{C_1} \frac{e^{imz}}{z^2 + 1} dz = 0\right]$$

Example 5.46. *Prove by Contour Integration Method :* $\displaystyle\int_0^\pi \frac{a d\theta}{a^2 + \sin^2\theta} = \frac{\pi}{\sqrt{1 + a^2}},\ a > 0.$

[Rohilkhand U 1987; Agra U 1983; Meerut U 1980]

Solution : Let $\displaystyle I = \int_0^\pi \frac{a d\theta}{a^2 + \sin^2\theta} = \int_0^\pi \frac{2a\, d\theta}{2a^2 + (1 - \cos\theta)}.$

Put $2\theta = \phi$, $2d\theta = d\phi$, $\displaystyle\therefore \quad I = \int_0^{2\pi} \frac{a d\theta}{2a^2 + (1 - \cos\phi)}.$

Substituting, $z = e^{i\phi}$, $\displaystyle\therefore \quad \cos\theta = \frac{1}{2}\left(z + \frac{1}{z}\right)$ and $\displaystyle d\phi = \frac{dz}{iz},$

we get $\displaystyle I = \int_C \frac{a(dz/iz)}{2a^2 + 1 - \frac{1}{2}\left(z + \frac{1}{z}\right)}$ [where C denotes the unit circle $|z| = 1$]

$$= 2ai \int_C \frac{dz}{z^2 - 2(2a^2 + 1)z + 1} = \int_c f(z) dz,$$

where $f(z) = \dfrac{2ai}{z^2 - 2(2a^2+1)z + 1} = \dfrac{2ai}{(z-\alpha)(z-\beta)}$,

where $\alpha = 2a^2 + 1 + 2a\sqrt{a^2+1}$ and $\beta = 2a^2 + 1 - 2a\sqrt{a^2+1}$.

The singularities of $f(z)$ are given by $(z-\alpha)(z-\beta) = 0$, i.e., the $f(z)$ is at simple poles $z = \alpha$, and $z = \beta$, where

$$\alpha + \beta = 2(2a^2+1) \quad \text{and} \quad \alpha - \beta = 4a\sqrt{a^2+1},\ \alpha\beta = 1.$$

We note that $\alpha > 1$ and $\beta < 1$ for $a > 0$.

$\therefore$ the only pole lying within the contour of unit radius is $z = \beta$.

Now, the residue of $f(z)$ at $z = \beta$ is

$$\operatorname*{Res}_{z=\beta} f(z) = \operatorname*{Lt}_{z\to\beta} (z-\beta)\frac{2ai}{(z-\alpha)(z-\beta)} = \frac{2ai}{(\beta-\alpha)} = -\frac{i}{2\sqrt{1+a^2}}.$$

Hence, by Cauchy's residue theorem,

$$I = \int_C f(z)dz = 2\pi i \times (\text{sum of the residues at poles within } C)$$

$$\therefore \quad I = \int_0^{\pi} \frac{a\,d\theta}{a^2 + \sin^2\theta} = 2\pi i \times \left(-\frac{i}{2\sqrt{1+a^2}}\right) = \frac{\pi}{\sqrt{1+a^2}} \quad \textbf{(Proved)}.$$

Example 5.47. *Show that* $\displaystyle\int_{-\infty}^{+\infty} \frac{x^2dx}{(x^2+1)(x^2+4)} = \frac{\pi}{3}$, *where the contour C represents the real axis* $-R$ *to* $+R$ *and semicircular arc* C_1 *as shown in Fig. (5.10).*

Solution : Here, the poles are at $\pm i$ and $\pm 2i$ as the contour is on the upper half of the plane. Hence, $z = -i$ and $z = -2i$ are outside the contour.

$$\text{Residue at } z = +i \text{ is } \operatorname*{Lt}_{z\to i} (z-i)f(z) = \frac{i^2}{(2i)(i^2+4)} = \frac{i}{6}$$

$$\text{and} \quad \text{residue at } z = +2i \text{ is } \frac{4i^2}{(4i^2+1)4i} = -\frac{i}{3}.$$

$\therefore$ sum of the residues at poles $= \dfrac{i}{6} - \dfrac{i}{3} = -\dfrac{i}{6}$, then

$$\oint_C \frac{z^2}{(z^2+1)(z^2+4)}dz = 2\pi i\left(-\frac{i}{6}\right) = \frac{\pi}{6}$$

$$\text{or,} \quad \int_{-R}^{R} \frac{x^2}{(x^2+1)(x^2+4)}dx + \int_{C_1} \frac{z^2}{(z^2+1)(z^2+4)}dz = \frac{\pi}{3}.$$

$$\text{For } r \to \infty, \text{ we get } \int_{-\infty}^{+\infty} \frac{x^2}{(x^2+1)(x^2+4)}dx = \frac{\pi}{3} \quad \textbf{(Proved)}.$$

Example 5.48. *Find the value of the following integral by the method of residue :*

$$\int_0^{\infty} \frac{dx}{x^4 + a^4},\ a > 0.$$ **[BU(H) 2001]**

Solution : The integral being an even function, we write :

$$I = \int_0^{\pi} \frac{dx}{x^4 + a^4} = \frac{1}{2} \int_{-\infty}^{+\infty} \frac{dx}{x^4 + a^4}. \tag{i}$$

The integral in Eqn.(i) represents integration of

$$f(z) = \frac{1}{x^4 + a^4} \tag{ii}$$

along the entire real axis. The poles of $f(z)$ are given by

$$z^4 + a^4 = 0 \quad \text{or,} \quad z = a(-1)^4, \tag{iii}$$

where $(-1)^{\frac{1}{9}} = [\cos(2n\pi + \pi) + i \sin(2\pi n + \pi)]^{\frac{1}{4}} = \cos\left(\frac{2n\pi + \pi}{4}\right) + i \sin\left(\frac{2n\pi + \pi}{4}\right),$

where $n = 0, 1, 2, 3$.

In this case the poles are, $ae^{i\frac{\pi}{4}}$, $ae^{i\frac{3\pi}{4}}$, $ae^{i\frac{5\pi}{4}}$ and $ae^{i\frac{7\pi}{4}}$.

Out of the four, only first two poles remain within the semicircle, the other two having amplitude $> \pi$, therefore, remain outside.

$\therefore$ $f(z)$ is analytic in the upper half of z-plane except these poles, $f(z)$ has no pole on real axis.

$$\underset{z \to \infty}{\text{Lt}}\, z f(z) = \underset{z \to \infty}{\text{Lt}} \frac{z}{z^4 + z^4} = 0.$$

So,
$$\int_{-\infty}^{+\infty} \frac{dx}{x^4 + a^4} = 2\pi i \sum R^*, \tag{iv}$$

where R^* is the sum of the residues of $f(z)$ at its poles which are

$$\alpha = ae^{i\frac{\pi}{4}} \quad \text{and} \quad \beta = ae^{i\frac{3\pi}{4}}. \tag{v}$$

Now, residue at $z = \alpha$ is

$$\underset{z \to \alpha}{\text{Lt}} (z - \alpha) f(z) = \underset{z \to \alpha}{\text{Lt}} (z - \alpha) \frac{1}{z^4 + a^4} = \underset{z \to \alpha}{\text{Lt}} (z - \alpha) \frac{1}{z^4 - \alpha^4} \text{ [by eqn. (iii)]}$$

$$= \underset{z \to \alpha}{\text{Lt}} \frac{1}{4z^3} \text{ [by L' Hospital's rule]}$$

$$= \frac{1}{4\alpha^3} = -\frac{\alpha}{4a^4} = -\frac{1}{4a^4} ae^{i\frac{\pi}{4}}.$$

Similarly, the residue at $z = \beta$ is

$$-\frac{1}{4a^4} ae^{i\frac{2\pi}{4}}.$$

$\therefore$ sum of the residues is

$$\sum R^* = -\frac{1}{4a^3}\left(e^{i\frac{\pi}{4}} + e^{i\frac{3\pi}{4}}\right) = -\frac{1}{4a^3}\left(e^{i\frac{\pi}{4}} - e^{-i\frac{\pi}{4}}\right) = -\frac{1}{4a^3}\left(2i \sin\frac{\pi}{4}\right) = -\frac{\sqrt{2}i}{4a^3}. \tag{vi}$$

Hence, from (i), (ii) and (vi), we get

$$I = \int_0^{\infty} \frac{dx}{x^4 + a^4} = \frac{1}{2} \int_{-\infty}^{+\infty} \frac{dx}{x^4 + a^4} = \frac{1}{2}(2\pi i)\left(\frac{-\sqrt{2}i}{4a^3}\right) = \frac{\sqrt{2}\pi}{4a^3}.$$

Example 5.49. *Evaluate* $\displaystyle\int_{-\infty}^{+\infty} \frac{x \sin \pi x \, dx}{x^2 + 2x + 5}$. **[BU(H) 1997]**

Solution : We consider the integral $\displaystyle\oint_C \frac{ze^{i\pi z}}{z^2+2z+5}dz$, where C is the contour from $-R$ to R along the real axis [Fig. (5.10)] and semicircular arc C_1. The integrand has the simple poles at $z = -1 \pm 2i$ of which $z = -i - 2i$ is outside the contour.

$\therefore$ the residue at $z = -1 + 2i$ is

$$\operatorname*{Lt}_{z\to -1+2i} (z+1-2i)\frac{ze^{i\pi z}}{(z+1-2i)(z+1+2i)} = \frac{(-1+2i)}{4i}e^{-(i+2)\pi}.$$

$$\therefore \quad \oint_C \frac{ze^{i\pi z}}{z^2+2z+5}dz = 2\pi i\frac{(-i+2i)}{4i}e^{-(i+2)\pi} = -\frac{\pi}{2}(-1+2i)e^{-2\pi}$$

$$[\because e^{-i\pi} = \cos\pi - i\sin\pi = -1]$$

$$\text{or,} \quad \int_{-R}^{+R} \frac{xe^{i\pi x}}{x^2+2x+5}dx + \int_{C_1} \frac{ze^{i\pi z}}{z^2+2z+5}dz = -\frac{\pi}{2}(-1+2i)e^{-2\pi}.$$

Now, taking $R \to \infty$, we have $\displaystyle\oint_{C_1} f(z)dz = 0$.

$$\therefore \quad \int_{-\infty}^{+\infty} \frac{xe^{i\pi x}}{x^2+2x+5}dx = -\frac{\pi}{2}(-1+2i)e^{-2\pi}$$

$$\text{or,} \quad \int_{-\infty}^{+\infty} \frac{x\cos\pi x\,dx}{x^2+2x+5} + i\int_{-\infty}^{+\infty} \frac{x\sin\pi x\,dx}{x^2+2x+5} = \frac{\pi}{2}e^{-2\pi} - i\pi e^{-2\pi}.$$

Equating the real and imaginary parts, $\displaystyle\int_{-\infty}^{+\infty} \frac{x\sin\pi x}{x^2+2x+5}dx = -\pi e^{-2\pi}$.

Example 5.50. *Evaluate the following integrals by contour integration :*

(i) $\displaystyle\int_0^{\infty} \frac{\sin mx}{x}dx;\ m > 0$, **[BU(H) 1996]** (ii) $\displaystyle\int_{-\infty}^{\infty} \frac{\sin x}{x}dx$ (iii) $\displaystyle\int_0^{\infty} \frac{\sin x}{x}dx$.

Solution : We consider the function $f(z) = \dfrac{e^{imz}}{z}$.

The function $f(z)$ has simple pole at $z = 0$ and none in the upper half plane. Let us now choose the indented semicircular contour C, shown in Fig. 5.11.

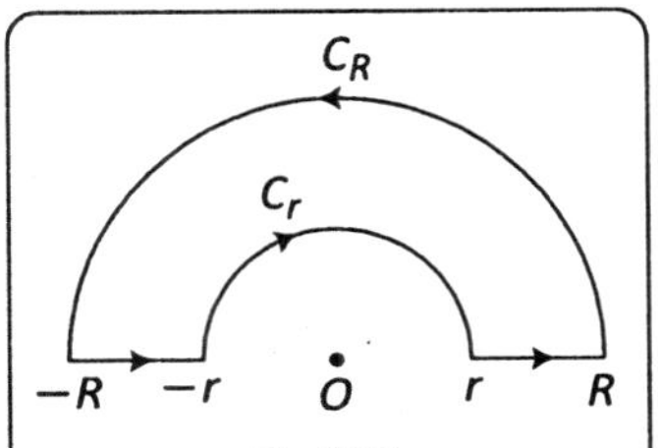

Fig. 5.11

Fig. 5.11 consists of (a) the semicircle C_R with centre at the origin and a large radius R in the upper half plane given by $|z| = R$.

(b) The line segment of the real axis from $-R$ to $-r$.

(c) the semi-circle C_r with cenre at the pole $z = 0$ and of infinitesimal radius r given by $|z| = r$.

(d) the line segment of the real axis from r to R. By Cauchy's theorem,

$$\int_C f(z)dz = \int_{CR} f(z)dz + \int_{-R}^{-r} f(x)dx + \int_{Cr} f(z)dz + \int_r^R f(x)dx = 0 \qquad \text{(i)}$$

as there is no singularity in the interior of C.

As $|f(z)| = \left|\frac{e^{imz}}{z}\right| = \frac{1}{|z|} = \frac{1}{R} \to 0$ as $R \to \infty$

$$\underset{R\to\infty}{\text{Lt}} \int_{C_R} f(z)dz = \underset{R\to 0}{\text{Lt}} \int_{C_R} \frac{e^{imz}}{z} dz = 0. \quad \text{(ii)}$$

On semi-circle C_r, we have $z = re^{i\theta}$; so that $\frac{dz}{z} = id\theta$ and so

$$\underset{r\to 0}{\text{Lt}} \int_{C_r} f(z)dz = \underset{r\to 0}{\text{Lt}} \int_{\pi}^{0} e^{imr}(\cos\theta + i\sin\theta)d\theta$$

$$= \underset{r\to 0}{\text{Lt}} \int_{\pi}^{0} e^{-mr\sin\theta} \times e^{imr\cos\theta} id\theta = -i\pi. \quad \text{(iii)}$$

If $R \to \infty$ and $r \to 0$ then, from Eqn. (i),

$$0 + \int_{-\infty}^{0} f(x)dx - i\pi + \int_{0}^{\infty} f(x)dx = 0 \text{ or, } \int_{+\infty}^{-\infty} f(x)dx = i\pi \text{ or, } \int_{-\infty}^{+\infty} \frac{e^{imx}}{x} = i\pi$$

$$\text{or, } \int_{-\infty}^{+\infty} \frac{1}{x}(\cos mx + i\sin mx) = i\pi. \quad \text{(iv)}$$

(i) Equating imaginary parts on both sides,

$$\int_{-\infty}^{+\infty} \frac{\sin mx}{x} dx = \pi. \quad \text{(v)}$$

As $\frac{\sin mx}{x}$ is an even function of x, we have

$$\int_{0}^{\infty} \frac{\sin mx}{x} dx = \frac{\pi}{2}. \quad \text{(vi)}$$

(ii) Substituting $m = 1$ in Eqn. (v), we get

$$\int_{-\infty}^{+\infty} \frac{\sin x}{x} dx = \pi. \quad \text{(vii)}$$

(iii) Substituting $m = 1$ in Eqn. (vi), we get

$$\int_{0}^{\infty} \frac{\sin x}{x} dx = \frac{\pi}{2}.$$

Example 5.51. *Evaluate* $\int_0^\infty \frac{dx}{x^4+1}$. **[BU(H) 1996]**

Solution : We consider the integral $\int_C \frac{dz}{z^4+1}$, where C is the contour as in Fig. 5.10.

$$\therefore \quad \int_C f(z)dz = \int_{-R}^{R} f(x)dx + \int_{C_1} f(z)dz.$$

$$\because \quad \underset{z\to\infty}{\text{Lt}}\ z\,f(z) = 0,$$

$$\therefore \quad \underset{R\to\infty}{\text{Lt}} \int_{C_1} f(z)dz = 0 \text{ or, } \int_C f(z)dz = \underset{R\to\infty}{\text{Lt}} \int_{-R}^{+R} f(x)dx = \int_{-\infty}^{\infty} f(x)dx.$$

Now, $f(z)$ has simple poles at $z = e^{(2n+1)\frac{\pi i}{4}}$, $n = 0, 1, 2, 3$.

So, $z = e^{\pi i/4}, e^{3\pi i/4}, e^{5\pi i/4}, e^{7\pi i/4}$,

of which only the first two are lying within the upper half plane.

Putting $z = \alpha$, we get $\alpha^4 + 1 = 0$, then residue at $z = \alpha$ is

$$\underset{z\to\infty}{\text{Lt}} \frac{z-\alpha}{z^4-\alpha^4} = \frac{1}{4\alpha^3}.$$

$\therefore$ sum of the residues for $\alpha = e^{\pi i/4}, e^{3\pi i/4}$ is

$$\frac{1}{4}\left(e^{-3\pi i/4} + e^{-9\pi i/4}\right) = \frac{1}{4}\left(\frac{3\pi}{4} - i\sin\frac{3\pi}{4} + \cos\frac{9\pi}{4} - i\sin\frac{9\pi}{4}\right) = -\frac{\sqrt{2}i}{4}.$$

$$\therefore \quad \oint_C f(z)dz = \int_{-\infty}^{+\infty} \frac{dx}{x^4+1} = 2\pi i\left(-\frac{\sqrt{2}i}{4}\right) = \frac{\sqrt{2}\pi}{2}.$$

So, $$\int_0^{\infty} \frac{dx}{x^4+1} = \frac{1}{2}\int_{-\infty}^{+\infty} \frac{dx}{x^4+1} = \frac{\pi}{2\sqrt{2}}.$$

Example 5.52. *Evaluate* $\int_0^{2\pi} \frac{d\theta}{(1|a^2) - 2a\cos\theta}$, $0 < a < 1$

Solution : We put $z = e^{i\theta}$,

$$\therefore \quad \cos\theta = \frac{1}{2}\left(z + \frac{1}{z}\right), \quad dz = ie^{i\theta}d\theta.$$

$$\therefore \quad I = \int_{\theta}^{2\pi} \frac{d\theta}{(1+a)^2 - 2a\cos\theta} = \int_C \frac{1}{(1+a^2) - a\left(z+\frac{1}{z}\right)}\frac{dz}{z} = -\frac{1}{ia}\int_C f(z)dz, \quad \text{(i)}$$

where $$f(z) = z^2 - z\left(a + \frac{1}{a}\right) + 1 = (z-a)\left(z - \frac{1}{a}\right).$$

The poles of $f(z)$ are $z = a$ and $z = \frac{1}{a}$. But only the point $z = a$ is within the circle $|z| < 1$ as $a < 1$.

The residue of $f(z)$ at $z = a$ is

$$R_1 = \underset{z\to a}{\text{Lt}}(z-a)f(z) = \underset{z\to a}{\text{Lt}}(z-a)\frac{1}{(z-a)\left(z-\frac{1}{a}\right)} = \frac{1}{a-\frac{1}{a}} = \frac{a}{a^2-1}.$$

Now, by Cauchy's residue theorem,

$$\int_C f(z)dz = 2\pi i\left(\frac{a}{a^2-1}\right). \quad \text{(ii)}$$

$\therefore$ from Eqns. (i) and (ii) we get,

$$I = -\frac{1}{ai}\int_C f(z)dz = -\frac{1}{ai}2\pi i\left(\frac{a}{a^2-1}\right) \quad \text{or,} \quad I = \frac{2\pi}{1-a^2}.$$

Exercises

5(A) : Objective Type Questions

Select the correct answer

5.1 If $z = x + iy$ be a complex number, then its complex conjugate is

(a) $\overline{z} = ix + y$ (b) $\overline{z} = x - iy$ (c) $\overline{z} = -ix + y$ (d) $\overline{z} = -ix - iy$.

5.2 If $x + iy = \sqrt{3} + i\sqrt{2}$, then $x^2 + y^2$ is equal to

(a) 5 (b) 13 (c) 7 (d) none of these.

5.3 If z_1 and z_2 be two complex numbers, then $|z_1 \pm z_2|$ is

(a) $\leq |z_1| + |z_2|$ (b) $< |z_1| - |z_2|$ (c) $> |z_1| + |z_2|$ (d) $\geq |z_1| + |z_2|$.

5.4 If ω be the imaginary cube root of unity, then ω^2 is equal to

(a) $\frac{1}{2}(-i + \sqrt{3})$ (b) $\frac{1}{2}(-1 + i\sqrt{3})$ (c) $\frac{1}{2}(-1 - i\sqrt{3})$ (d) $\frac{1}{2}(-\sqrt{3} + i)$.

5.5 The argument of $(-1 - i)$ is

(a) $-\frac{\pi}{4}$ (b) $-\frac{4}{3\pi}$ (c) $-\frac{3\pi}{4}$ (d) $\frac{3\pi}{4}$.

5.6 If $f(z) = u(x, y) + iv(x, y)$ is differentiable at a point $z = x + iy$ in a domain, then

(a) $\frac{\partial u}{\partial x} = \frac{\partial v}{\partial y}, \frac{\partial u}{\partial y} = -\frac{\partial v}{\partial x}$ (b) $\frac{\partial u}{\partial x} = -\frac{\partial v}{\partial y}, \frac{\partial u}{\partial y} = -\frac{\partial v}{\partial x}$

(c) $\frac{\partial u}{\partial y} = \frac{\partial v}{\partial y}; \frac{\partial u}{\partial x} = \frac{\partial v}{\partial x}$ (d) none of these.

5.7 If $f(z)$ is analytic and $f(z) = u(x, y) + iv(x, y)$, then $f'(z)$ equals

(a) $\frac{\partial u}{\partial x} - i\frac{\partial u}{\partial y}$ (b) $\frac{\partial u}{\partial x} - i\frac{\partial v}{\partial x}$ (c) $\frac{\partial v}{\partial y} - i\frac{\partial v}{\partial x}$ (d) none of these.

5.8 If C be a simple closed curve enclosing the origin, then

(a) $\oint_c \frac{dz}{z} = 0$ (b) $\oint_c \frac{dz}{z} = 2\pi i$ (c) $\oint_c \frac{dz}{z} = \text{const.}$ (d) none of these.

5.9 The imaginary part of $e^{i\theta}$ is

(a) $\cos\theta$ (b) $\sin\theta$ (c) $\sin\theta + \cos\theta$ (d) $\sin\theta - \cos\theta$.

5.10 At the point of singularity of an analytic function $f(z)$, it is

(a) analytic (b) not analytic (c) may or may not be analytic (d) none of these.

5.11 A simply connected domain contains

(a) no point inside (b) all points inside (c) some points outside (d) none of these.

5.12 If $f(z)$ is analytic within and on any closed contour C and if a be a point within C, then

(a) $f(a) = \frac{1}{2\pi i}\oint_C \frac{f(z)}{z - a}dz$ (b) $f(a) = \frac{i}{2\pi}\oint_C \frac{f(z)}{z - a}dz$

(c) $f(a) = \frac{1}{2\pi i}\oint \frac{f(z)}{z + a}dz$ (d) $f(a) = -\frac{1}{2\pi i}\oint_C \frac{f(z)}{z - a}dz$.

Answers

5.1 (b) **5.2** (a) **5.3** (a) **5.4** (b) **5.5** (c) **5.6** (a) **5.7** (a) **5.8** (b) **5.9** (b) **5.10** (b) **5.11** (b) **5.12** (a).

5(B) : Short Answer type questions

5.1 Find the smallest positive integer for which $\left(\frac{1+i}{1-i}\right)^n = 1$. **[BU(H) 2008]**

5.2 Write the complex number $z = -1 - i\sqrt{3}$ in polar form. **[BU(H) 1994]**

5.3 Find the complex number z, which satisfies the condition $|z - 2 + 2i| = 1$ and has the least absolute value. **[BU(H) 1998]**

5.4 Evaluate $f(2)$ and $f(3)$, where $f(z_0) = \oint_C \frac{2az^2 - z - 2}{z - z_0} dz$ and C is the circle $|z| = 25$. **[BU(H) 2008]**

5.5 For what value(s) of k will be the function $f(z) = e^x(\cos ky + i \sin ky)$ be analytic? **[BU(H) 2006]**

5.6 What is the nature of singularity of the functionsin $\left(\frac{1}{z}\right)$ at $z = 0$? **[BU(H) 2006]**

5.7 **(a)** Find the residue of $\frac{\sin z}{z^4}$ at $z = 0$.

(b) What is the value of the integral? **[BU(H) 2006]**

5.8 Show that, $(1+i)^n + (1-i)^n = 2^{\frac{n}{2}+1} \cos \frac{n\pi}{4}$. **[BU(H) 2002]**

5.9 Verify that $f(z) = ze^{-z}$ is an analytic function.

5.10 Locate and name the singularities of the following function in the finite z-plane :

(i) $\sec\left(\frac{1}{z^2}\right)$ **(ii)** $\frac{e^z - 1}{z\sqrt{z+1}}$. **[BU(H) 2000]**

5.11 Does the function $f(z) = \frac{e^z - 1}{z}$ have a pole at the origin? **[BU(H) 2001]**

5.12 Express the following expression in the modulus amplitude form : $1 + \sin x + i \cos \alpha$. **[BU(H) 2001]**

5.13 Evaluate the integral $\int_0^{1+i} (x^2 + iy)dz$ along the path $y = x^2$. **[BU(H) 2001]**

5.14 If $|z_1 + z_2| = |z_1 - z_2|$, show that the difference of the amplitudes of z_1 and z_2 is $\frac{\pi}{2}$. **[BU(H) 2001]**

5.15 If $f(z)$ is analytic in a simply connected region R, prove that, $\int_a^b f(z)dz$ is independent of the path joining any two points in R. **[BU(H) 1999, 2001]**

5.16 Show that the derivative of the function $f(z) = \overline{z}$ (bar denotes complex conjugate) does not exist anywhere. **[BU(H) 1999]**

5.17 Write down the necessary and sufficient conditions for a function $f(z)$ of the complex variable z to be analytic where $z = x + iy$. **[BU(H) 1997]**

5.18 Show that if $f(z) = u(x, y) + iv(x, y)$ is analytic, then u and v are harmonic. **[BU(H) 1997]**

5.19 Examine whether the following functions are analytic :

(i) $\bar{z} = x - iy$ (ii) $\dfrac{1+z}{1-z}$ (iii) e^z (iv) $\sin z$. **[BU(H) 1997]**

5.20 Show that if $f(z)$ be analytic in a region R and on its boundary C, then $\oint f(z)dz = 0$. **[BU(H) 1997]**

5.21 Define an analytic function $f(z)$ of a complex variable z mentioning when it is called analytic (i) in a region R, (ii) at a point $z = z_0$. **[BU(H) 1996]**

5.22 State the necessary and sufficient conditions for $f(z)$ to be analytic. **[BU(H) 1996]**

5.23 Show that the function $w = e^z$ is analytic in the entire complex plane. **[BU(H) 1994]**

5.24 Show that, the sum, difference, product, ratio of any two complex numbers are always complex numbers.

5.25 Prove the following for a complex variable :

(i) $\tanh x = \dfrac{e^x - e^{-x}}{e^x + e^{-x}}$ (ii) $\sin ix = i \sinh x$.

5.26 The only singularities of a single-valued function $f(z)$ are poles of order 1 and 2 at $z = -1$ and $z = 2$ with residues at these poles 1 and 2 respectively. If $f(0) = \dfrac{7}{4}$, $f(1) = \dfrac{5}{2}$, determine the function.

5.27 Examine whether the function $f(z) = \sin z$ is analytic or not. **[BU(H) 2014]**

5.28 Show that $\left(\dfrac{1+\sqrt{3}i}{1-\sqrt{3}i}\right)^{10} = -\dfrac{1}{2} + \dfrac{\sqrt{3}}{2}i$ **[BU(H) 2014]**

5(C) : Broad Answer Type Questions :

5.1 (a) State Cauchy's residue theorem.

(b) (i) Find the residue of $f(z) = \dfrac{z^2 + 2z}{(z+1)^2(z^2+4)}$ at all its pole in the finite plane.

(ii) Evaluate the integral $\displaystyle\int_0^{2\pi} \frac{d\theta}{2 + \sin\theta}$ with the help of residue theorem. **[BU(H) 2008]**

5.2 (a) Show that in polar form the Cauchy's-Riemann equations can be written as

$$\frac{\partial u}{\partial r} = \frac{1}{r}\frac{\partial u}{r\theta}; \quad \frac{\partial v}{\partial r} = -\frac{1}{r}\frac{\partial u}{\partial \theta}.$$

(b) Given, $u = 3x^2y + 2x^2 - y^3 - 2y^2$, find v such that $f(z) = u + v$ is an analytic function of z.

(c) Prove that, $\displaystyle\int_0^{2\pi} \cos^{2n}\theta\, d\theta = \frac{1 \cdot 3 \cdot 5, \ldots (2n-1)}{2 \cdot 4 \cdot 6, \ldots 2n} 2\pi$, where n is a natural number. **[BU(H) 2006]**

5.3 **(a)** **(i)** State Cauchy's theorem and Cauchy's integral formula.

(ii) Using Cauchy's integral formula evaluate the integral

$$\int_C \frac{(2z+1)}{z^2+z}dz, \text{ where } C \text{ is } |z| = \frac{1}{2}.$$

(b) **(i)** State residue theorem,

(ii) Using the residue theorem, evaluate the integral, $\int_0^{\infty} \frac{\cos x\, dx}{x^2+a^2}$, $a > 0$.

[BU(H) 2002]

5.4 **(a)** What is an analytic function? State the necessary and sufficient conditions for a function to be analytic.

(b) Evaluate $\int_{1-i}^{2+i} (2x + iy + 1)dz$ along the straight line joining the points $(1 - i)$ and $(2 + i)$.

(c) Evaluate $\int_C \frac{z-3}{z^2+2z+3}dz$, where C is the circle $|z + 1 + i| = 2$. **[BU(H) 2001]**

5.5 **(a)** Find the analytic function $f(z)$ whose real part is $\cosh y \cos x$.

(b) Find the value of the following integrals by the method of residue :

(i) $\int_0^{\infty} \frac{dx}{x^4+a^4}$, $a > 0$ **[BU(H) 2001]** **(ii)** $I = \int_0^{2\pi} \frac{d\theta}{(5-3\sin\theta)^2}$. **[BU(H) 2008]**

5.6 **(a)** Evaluate $\oint_C \frac{\sin \pi z^2 + \cos \pi z^2}{(z-2)(z-3)}dz$, where C is the circle $|z| = \pi$.

(b) Find the residue of $\frac{\sec \pi z}{z^3}$ at $z = 0$.

(c) Using the residue theorem evaluate $\int_0^{2\pi} \frac{d\theta}{(3 - 2\cos\theta + \sin\theta)}$. **[BU(H) 2000]**

5.7 Prove that,

(a) $\int_0^{2\pi} \frac{a\theta}{a+b\sin\theta} = \frac{2\pi}{\sqrt{a^2-b^2}}$, $a > |b|$ **(b)** $\int_0^{2\pi} \frac{\sin 3\theta\, d\theta}{5-3\cos\theta} = 0$

(c) $\int_{-\infty}^{+\infty} \frac{x^2 dx}{(x^2+1)^2(x^2+2x+2)} = \frac{7\pi}{50}$ **[BU(H) 1997]**

(d) $\int_0^{\infty} \frac{dx}{(1+x^2)^2} = \frac{\pi}{2}$ **(e)** $\int_0^{\infty} \frac{x^2 dx}{x^6+1} = \frac{\pi}{6}$ **(f)** $\int_0^{\pi} \frac{a d\theta}{a^2+\sin^2\theta} = \frac{\pi}{\sqrt{1+a^2}}$, $a > 0$.

5.8 **(a)** Identify singularities of the function $f(z) = \frac{\operatorname{cosec} z}{z}$ and find the residues of the function at its poles.

(b) Evaluate $\int_C \tan z\, dz$, where C is the circle $|z| = 2$.

(c) Using the method of residues, show that $\int_0^{2\pi} \frac{d\theta}{1-2a\cos\theta+a^2} = \frac{2\pi}{1-a^2}$, if $|a| < 1$.

[BU(H) 1998]

5.9 **(a)** Define an analytic function $f(z)$ of a complex variable z and state the necessary and sufficient conditions for the analyticity.

(b) Examine whether $\sin z$ is an analytic function of z.

(c) State Cauchy's integral theorem and apply it to evaluate the integral, $\int_C e^{-2z} dz$, over any path C joining the points $(1 - \pi i)$ and $(2 + 3\pi i)$.

(d) Show that, $\int_C \bar{z}\, dz - 10 - \frac{8i}{3}$, where C is the curve given by $z = t^2 + it$ and joining the points $z = 0$ and $z = 4 + 2i$, $\bar{z}$ is the complex conjugate of z. **[BU(H) 1996]**

5.10 **(a)** Locate and state the nature of the singularities of the function,

$$f(z) = \frac{1}{(z-1)^2(z+4)}.$$

(b) Evaluate the following integrals :

(i) $\int_0^{2\pi} \frac{d\theta}{a + b\sin\theta}$, $a > |b|$ (ii) $\int_0^{\infty} \frac{dx}{x^4 + 1}$

(iii) $\int_0^{\infty} \frac{\sin mx}{x} dx$, $m > 0$ (iv) $\int_{-\infty}^{+\infty} \frac{x \sin \pi x\, dx}{(x^2 + 2x + 5)}$. **[BU(H) 1996]**

5.11 **(a)** Prove Cauchy's integral formula, $f(z_0) = \frac{1}{2\pi i}\int_C \frac{f(z)}{z - z_0} dz$,

where $f(z)$ is analytic on a positively oriented simple closed path C and in the interior of $C \cdot z_0$ is any point in the interior of C.

(b) Prove that,

(i) $\int_{-\infty}^{+\infty} \frac{dx}{(x^2+4)^2} = \frac{\pi}{16}$ (ii) $\int_0^{\infty} \frac{x \sin x}{x^2 + 4} dx = \frac{\pi e^{-2}}{2}$ (iii) $\int_0^{2\pi} \frac{d\theta}{\frac{5}{4} + \sin\theta} = \frac{8\pi}{3}$.

5.12 Use contour integration to show that,

(a) $\int_0^{\infty} \frac{x \sin mx}{x^4 + a^4} dx = \frac{\pi}{2a^2} e^{-\frac{ma}{\sqrt{2}}}$

(b) $\int_{-\infty}^{+\infty} \frac{\cos x\, dx}{(x^2 + a^2)(x^2 + b^2)} = \frac{\pi}{a^2 - b^2}\left(\frac{e^{-b}}{b} - \frac{e^{-a}}{a}\right)^{\sin \frac{ma}{\sqrt{2}}}$, for $a > b > 0$.

(c) $\int_{-\infty}^{+\infty} \frac{dx}{(x^2 + b^2)(x^2 + c^2)^2} = \frac{\pi(b + 2c)}{2bc^3(b+c)^2}$, for $b > 0$, $c > 0$.

5.13 State and prove Cauchy's Residue Theorem. Explain how it is extended for the case of an isolated first order pole lying on the contour of integration. Using this theorem show that,

$$\int_{-\infty}^{+\infty} \frac{e^{ax}}{1 + e^x} dx = \frac{\pi}{\sin \pi a}, \quad \text{where } 0 < a < 1.$$

5.14 By contour integration prove that,

(a) $\int_0^{\infty} \frac{dx}{(a + bx^2)^n} = \frac{\pi}{2^n\sqrt{b}} \cdot \frac{1 \cdot 3 \cdot 5 \cdots (2n-3)}{1 \cdot 2 \cdot 3 \cdots (n-1)} \frac{1}{a^{\frac{2n-1}{2}}}$

(b) $\displaystyle\int_0^{\infty} \frac{\cos mx}{x^2+a^2}dx = \frac{\pi}{2a}e^{-ma} \ (m \geq 0)$

(c) $\displaystyle\int_{-\infty}^{+\infty} \frac{a\cos x + x\sin x}{x^2+a^2}dx = 2\pi e^{-a}$ [**Hints :** Take $\phi(z) = \dfrac{e^{iz}}{z-ai}$, $a > 0$]

(d) $\displaystyle\int_0^{\infty} \frac{\cos mx}{x^4+x^2+1}dx = \frac{\pi}{\sqrt{3}}e^{\frac{-m\sqrt{3}}{2}} \sin\left(\frac{m}{2}+\frac{\pi}{6}\right).$

5.15 Prove that, **(a)** $\displaystyle\int_0^{\infty} \frac{\sin x}{x^2+4x+5}dx = -\frac{\pi}{e}\sin 2$ **(b)** $\displaystyle\int_0^{\infty} \frac{\cos x}{x^2+4x+5}dx = \frac{\pi}{e}\cos 2.$

5.16 Show that,

(a) $\displaystyle\int_0^{2\pi} \frac{\cos^2 3\theta \, d\theta}{1-2a\cos 2\theta + a^2} = \frac{\pi(1-a+a^2)}{1-a}, \ 0 < a < 1$

(b) $\displaystyle\int_0^{2\pi} e^{\cos\theta}\cdot\cos(\sin\theta - n\theta)d\theta = \frac{2\pi}{n!}$, where n is a positive number.

5.17 **(a)** Using Cauchy's residue theorem, evaluate the integral $\displaystyle\int_C^{\infty} \frac{\log x}{(1+x^2)^2}dx$. [**Ans:** $\frac{-\pi}{2}$]

(b) Applying the method of contour integration, show that,

$$\int_{-\infty}^{+\infty} \frac{e^{\alpha x}}{1+e^x}dx = \frac{\pi}{\sin\pi\alpha}, \quad \text{where } 0 < \alpha < 1.$$

5.18 Integrating around rectangular contour, show that, for $-\pi < a < \pi$,

(a) $\displaystyle\int_0^{\infty} \frac{\sinh ax}{\sinh \pi x} = \frac{1}{2}\tan\frac{a}{2}$ **(b)** $\displaystyle\int_0^{\infty} \frac{\cosh ax}{\cos \pi x} = \frac{1}{2}\sec\frac{a}{2}.$

5.19 **(a)** Evaluate $\displaystyle\int_C \frac{z-3}{z^2+2z+3}dz$, where C is the circle $|z+1+i| = 2$. **[BU(H) 2001]**

(b) Evaluate $\displaystyle\int_0^{2\pi} \frac{d\theta}{a+b\cos\theta+c\sin\theta}$, $a^2 > h^2 + c^2$.

5.20 Using the method of contour integration, show that,

(a) $\displaystyle\int_0^{\infty} \frac{x-\sin x}{x^2(a^2+x^2)}dx = \frac{\pi}{2a^4}\left(\frac{a^2}{2} - a + 1 - e^{-a}\right)$, (for $a > 0$)

(b) $\displaystyle\int_0^{\infty} \frac{\sin^2 mx \, dx}{x^2(a^2+x^2)} = \frac{\pi}{4a^2}(e^{-2ma} - 1 + 2ma)$, $[m > 0, a > 0]$.

5.21 Show that, **(a)** $\displaystyle\int_0^{\infty} \frac{dx}{\sqrt{x}(1+x)} = \pi$ **(b)** $\displaystyle\int_{\infty}^{+\infty} \frac{dx}{(1+x^2)^2} = \frac{\pi}{2}$ **(c)** $\displaystyle\int_0^{\infty} \frac{dx}{x^4+1} = \frac{\pi}{8}.$

5.22 **(a)** State Cauchy's integral formula.

Evaluate $\displaystyle\oint_C \frac{e^z(z^2+1)}{(z-1)^2}dz$, where C is the circle $|z| = 2$.

(b) What do you mean by removable and essential singularities of the function $f(z)$?

(c) Locate and classify the singularities of the function of complex variable z :

(i) $\dfrac{\ln(z-2)}{(z^4+2z+z)^4}$ **(ii)** z^4

(d) Determine the poles and residue at each pole of the function $f(z)=\cot z$.

[BU(H) 2014]

5.23 (a) State Cauchy's integral theorem.

Evaluate $\oint_C \dfrac{dz}{z-a)}dz$, where C is any simple closed curve and $z=a$ is

(i) outside C (ii) inside C.

(b) Locate the name of singularity of the function $f(z)=\dfrac{\sin\sqrt{z}}{\sqrt{2}}$.

(c) Determine the poles of the following function and residue at each pole :

$$f(z)=\frac{4-3z}{z(z-1)(z-2)}$$

and hence evaluate $\oint_C = \dfrac{4-3z}{z(z-1)(z-2)}dz$ where C is the circle $|z|=\dfrac{3}{2}$.

(d) Prove that the eigenvectors corresponding to different eigenvalues of a Hermitian matrix are orthogonal. **[BU(H) 2011]**

5.24 (a) Simplify $\left(\cos\dfrac{2\pi}{3}+i\sin\dfrac{2\pi}{3}\right)$.

(b) Determine the poles of the following function and residue at each pole :

$$f(z)=\frac{z^2}{(z^2-1)^2(z+2)}.$$

Hence evaluate, $\oint_C \dfrac{z^2dz}{(z-1)^2(z+2)}$, where C is the circle $|z|=3$.

(c) For any complex number z show that $|z|\geq\dfrac{1}{\sqrt{2}}\{|\text{Re}(z)|+|\text{Im}(z)|\}$. **[BU(H) 2009]**

5.25 State and prove Cauchy-Riemann equations for a function to be analytic.

[Pune U(MSc) 2008]

5.26 (a) Check whether $f(z)=i^z$ is analytic.

(b) If $f(z)=\sin z$, shows that $f^*(z)=f(z^*)$ for complex z.

5.27 (a) Find the polar representation of $z=-2\sqrt{3}-zi$ in terms of r and θ.

(b) Find the residue of the function $f(z)=\dfrac{z^3}{z(z^2+4)}$ at $z=0$ and $z=2i$.

(c) Obtain all values of $\ln(\sqrt{3}-i)$ and find its principal values.

5.28 (a) Prove that $u=e^{-x}(x\sin y-y\cos y)$ is harmonic such that the function $f(z)=u+iv$ is analytic.

(b) Show that if $f(z)$ is analytic in a region R and on its boundary then $\oint f(z)dz=0$.

5.29 If $f(z)$ is analytic in entire z-plane and $z=a$ is a point in the z-plane, then prove that,

$$f'(a)=\frac{1}{2\pi i}\oint\frac{f(z)}{(z-a)^2}dz.$$

5.30 (a) What is the argument of $-3i$?
(b) Express the number $-\sqrt{6}-\sqrt{2}i$ in polar form.
(c) Find the number $e^{\left(1+\frac{\pi}{4}i\right)}$.
(d) Find the real part of $\dfrac{1+z}{1-z}$. **[Assam U(Major) 2014]**

5.31 (a) Find and plot all roots of $(1-i)^{\frac{1}{4}}$.
(b) Prove that, $\arg\left(\dfrac{z_1}{z_2}\right)=\arg z_1-\arg z_2$ **[Assam U(Major) 2014]**

5.32 (a) Check the analyticity and hence find derivative of the function $f(z)=\sin z$.
(b) Use Cauchy's integral formula to evaluate $\oint \dfrac{z^2}{(z-1)^3}dz$, where C is the circle given by $|z|=2$. **[Assam U(Major) 2014]**

5.33 (a) Derive Cauchy's integral formulas.
(b) Calculate the residues of $f(z)=\dfrac{z^2}{(1+z^2)^2}$ and evaluate, $\displaystyle\int_0^{\infty}\frac{x^2dx}{(1+x^2)^2}$. **[Assam U(Major) 2014]**

5.34 (a) Evaluate : $\displaystyle\int_0^{2\pi}\frac{d\theta}{(2+\cos\theta)}$, using residues theorem.
(b) Find the residues of $\dfrac{z^2}{z^2+a^2}$ at $z=ia$ and $\dfrac{z}{(z-a)(z-b)}$ at infinity. **[Gujrat U(MSc) 2017]**

5.35 (a) Evaluate : $\displaystyle\int_c\frac{2z-1}{z(z+1)(z+3)}dz$, where c is $|z|=2$, using residue theorem.
(b) State and prove Cauchy's integral theorem. Deduce Cauchy's integral formula from it.
Evaluate : $\displaystyle\oint_c\frac{dz}{\sinh 2z}$.

5.36 (a) Find all roots of the equation $\log(-1)$.
(b) State whether $f(z)=\overline{z}$ is analytic or not.
(c) Evaluate : $\displaystyle\int_c z^2\,dz$, where c is any curve joining 0 to $1+i$.
(d) Find the residue of $f(z)=\dfrac{z^2-1}{z^2+z}$ at $z=0$.

5.37 (a) Prove that the real and imaginary parts of an analytic function are harmonic.
(b) Show that an analytic function is constant if its modulus is constant.
(c) Find the residue of $f(z)=\dfrac{1}{(z-1)^3}$ at its poles.
(d) Evaluate $\displaystyle\int_c(y-x-3x^2i)dz$, where $z=x+iy^C$ and C is the straight line joining 0 to $1+i$.

5.38 (a) Show that the function $f(z)=\begin{cases}\frac{(\overline{z})^2}{z}; & z\neq 0\\ 0, & z=0\end{cases}$ is not differentiable even though Cauchy-Riemann equations are satisfied there.
(b) If $u-v=(x-y)(x^2+4xy+y^2)$, find an analytic function $f(z)=u+iv$ in terms of z.

CHAPTER 6

Beta and Gamma Functions

6.1 Introduction

The Beta, Gamma and Error functions find important applications in various branches of mathematical analysis like Physics, Mechanics, Statistics, etc.

(a) The Beta Function : The beta function is defined as

$$\beta(m, n) = \int_0^1 x^{m-1}(1-x)^{n-1}dx \quad \begin{cases} m > 0 \\ n > 0 \end{cases} \tag{6.1}$$

This is known as the Eulerian integral of the first kind.

(b) The Gamma Function : The second Eulerian function is known as Gamma function and is defined by

$$\Gamma(n) = \int_0^\infty e^{-x}x^{n-1}dx, \quad n > 0. \tag{6.2}$$

(c) The Error Function : The error function, $erf(x)$ or the probability integral is defined as

$$erf(x) = \frac{2}{\sqrt{\pi}}\int_0^x e^{-y^2}dy. \tag{6.3}$$

This is a particular form of incomplete Gamma function which has extensive use in the theory of probability.

The complementary error function is defined as

$$erfc(x) = 1 - erf(x). \tag{6.4}$$

6.2 Convergence of Beta Function

We have $\beta(m, n) = \int_0^1 x^{m-1}(1-x)^{n-1}dx \quad \begin{cases} m > 0 \\ n > 0 \end{cases}$

To test the convergence of $\beta(m, n)$, we split the interval $(0, 1)$ into two intervals which are $(0, \frac{1}{2})$ and $(\frac{1}{2}, 1)$.

$$\text{So, } \int_0^1 x^{m-1}(1-x)^{n-1}dx = \int_0^{\frac{1}{2}} x^{m-1}(1-x)^{n-1}dx + \int_{\frac{1}{2}}^0 x^{m-1}(1-x)^{n-1}dx. \tag{6.5}$$

Convergence at 0 of $\int_0^{\frac{1}{2}} x^{m-1}(1-x)^{n-1}dx$

Let $f(x) = x^{m-1}(1-x)^{n-1} = \dfrac{(1-x)^{n-1}}{x^{1-m}}$ and $g(x) = \dfrac{1}{x^{1-m}}$.

Then, $f(x)g(x) = (1-x)^{n-1} \to 1$ as $x \to 0$:

Again, $\int_0^{\frac{1}{2}} g(x)dx = \int_0^{\frac{1}{2}} \frac{1}{x^{1-m}}dx$ is convergent at 0, if and only if $(1-m) < 1$, i.e., $m > 0$.

Hence, $\int_0^{\frac{1}{2}} x^{m-1}(1-x)^{n-1}dx$ is also convergent at 0, for $m > 0$.

Convergence of $\int_0^{\frac{1}{2}} x^{m-1}(1-x)^{n-1}dx$ at 1

Now, let $f(x) = \frac{x^{m-1}}{(1-x)^{1-n}}$ and $g(x) = \frac{1}{(1-x)^{1-n}}$.

$\therefore \quad \frac{f(x)}{g(x)} = x^{m-1} \to 1$ as $m \to 1$.

So, $\int_{\frac{1}{2}}^1 g(x)dx = \int_{\frac{1}{2}}^1 \frac{1}{(1-x)^{1-n}}dx$ converges at 1 for, $1-n < 1$, i.e., $n > 0$.

So, $\int_{\frac{1}{2}}^1 x^{m-1}(1-x)^{n-1}dx$ converges at 1, if and only if $n > 0$.

$\therefore \quad$ **$\int_0^1 x^{m-1}(1-x)^{n-1}dx$ exists for all $m, n > 0$.**

6.3 Convergence of Gamma Function

We have $\Gamma(n) = \int_0^\infty e^{-x}x^{n-1}dx,\ n > 0$.

To test the convergence of $\Gamma(n)$ we split the interval $(0, \infty)$ into $(0, 1)$ and $(1, \infty)$.

$$\therefore \quad \int_0^\infty e^{-x}x^{n-1}dx = \int_0^1 e^{-x}x^{n-1}dx + \int_1^\infty e^{-x}x^{n-1}dx. \tag{6.6}$$

Convergence of $\int_0^1 e^{-x}x^{n-1}dx$ at 0

Let $f(x) = e^{-x}x^{n-1} = \frac{e^{-x}}{x^{1-n}}$ and $g(x) = \frac{1}{x^{1-n}}$.

$\therefore \quad \frac{f(x)}{g(x)} = e^{-x} \to 1$ as $x \to 0$.

Now, as $\int_0^1 \frac{1}{x^{n-1}}dx$ converges at 0, if and only if $1-n < 1$, i.e., $n > 0$.

$\therefore \quad$ **$\int_0^1 e^{-x}x^{n-1}dx$ converges at 0 for $n > 0$.**

Convergence of $\int_1^\infty e^{-x}x^{n-1}dx$ at $n > 0$

$e^x > x^{n+1}$ for any value of n, if x is very large.

$\therefore \quad x^{n+1} < e^x \quad$ or, $\quad e^{-x}x^{n-1} < \frac{1}{x^2}$.

But, $\int_1^\infty \frac{1}{x^2}dx$ converges.

$\therefore \quad \int_1^\infty e^{-x}x^{n-1}dx$ also converges, if and only if $n > 0$.

Hence, $\boldsymbol{\int_0^\infty e^{-x}x^{n-1}dx}$ **converges for** $\boldsymbol{n > 0}$.

6.4 Properties of Beta Function

1. The Beta function is symmetrical in m and n, i.e., $\beta(m,n) = \beta(n,m)$.

Proof : $\beta(m,n) = \beta(n,m) = \int_0^1 x^{m-1}(1-x)^{n-1}dx.$

Let $1 - x^0 = p$, $\therefore$ for $x = 0$, $p = 1$ and $t = 1$, $p = 0$.

$$\therefore \quad \beta(m,n) = \int_0^1 (1-p)^{m-1}p^{n-1}dp = \int_0^1 p^{n-1}(1-p)^{m-1}dp = \beta(n,m)$$

i.e., $\boldsymbol{\beta(m,n) = \beta(n,m)}$. (6.7)

2. $\beta(m,n) = \frac{m-1}{m+n-1}\beta(m-1,n) = \frac{n-1}{m+n-1}\beta(m,n-1).$

Proof : We have $\beta(m,n) = \int_0^1 x^{m-1}(1-x)^{n-1}dx.$

Integrating by parts,

$$\beta(m,n) = (1-x)^{n-1}\frac{x^m}{m}\bigg|_0^1 + \int_0^1 \frac{x^m}{n}(n-1)(1-x)^{n-2}dx = \frac{n-1}{m}\int_0^1 x^m(1-x)^{n-2}dx.$$

Now, $x^m = x^{m-1} - x^{m-1}(1-x)$.

$$\therefore \quad \beta(m,n) = \frac{n-1}{m}\int_0^1 x^{m-1}(1-x)^{n-2}dx - \frac{n-1}{m}\int_0^1 x^{m-1}(1-x)^{n-1}dx$$

$$= \frac{n-1}{m}\beta(m,n-1) - \frac{n-1}{m}\beta(m,n).$$

$$\therefore \quad \beta(m,n) = \frac{n-1}{m+n-1}\beta(m,n-1). \tag{6.8}$$

Interchanging m and n,

$$\beta(n,m) = \frac{m-1}{m+n-1}\beta(n,m-1). \tag{6.9}$$

$$\therefore \quad \boldsymbol{\beta(m,n) = \frac{m-1}{m+n-1}\beta(m-n,n) = \frac{n-1}{m+n-1}\beta(m,n-1).}$$

3. $\beta(m,n) = \beta(n,m) = \frac{(n-1)!(m-1)!}{(m+n-1)!}$, where m, n are integers.

Proof : We have $\beta(m,n) = \frac{(n-1)}{(m+n-1)}\beta(m,n-1)$, n being integers

$$= \frac{n-1}{m+n-1}\cdot\frac{n-2}{m+n-2}\cdots\frac{1}{m+1}\beta(m,1). \tag{6.10}$$

But, $\beta(m,1)=\int_0^1 x^{m-1}dx=\frac{1}{m}$.

$$\therefore \quad \beta(m,n)=\frac{1\cdot 2\cdots(n-1)}{m(m+1)(m+2)\cdots(m+n-1)}. \tag{6.11}$$

If m and n both are integers, then from Eqn. (6.10), we write by the successive use of Eqn. (6.9).

$$\beta(m,n)=\beta(n,m)=\frac{1\cdot 2\cdots(n-1)}{(m+1)(m+2)\cdots(m+n-1)}\beta(m,1)$$

$$=\frac{1\cdot 2\cdots(n-1)}{(m+1)\cdots(m+n-1)}\cdot\frac{(m-1)(m-2)\cdots 2\cdot 1}{m(m-1)\cdots 2}\beta(1,1).$$

But $\beta(1,1)=1$.

$$\therefore \quad \boldsymbol{\beta(m,n)=\beta(n,m)=\frac{(n-1)!(m-1)!}{(m+n-1)!}}. \tag{6.12}$$

4. $\beta(m,n)=\frac{1}{2^{2m-1}}\beta\left(\frac{1}{2},m\right)$

Proof : We have $\beta(m,n)=\int_0^1 x^{m-1}(1-x)^{n-1}dx$.

Putting $n=m$ in the above integral,

$$\beta(m,n)=\int_0^1 x^{m-1}(1-x)^{n-1}dx$$

$$\text{or,}\quad \beta(m,n)=2\cdot\int_0^{\frac{1}{2}}\left[\frac{1}{4}-\left(\frac{1}{2}-x\right)^2\right]^{m-1}dx. \tag{6.13}$$

[as the integrand is symmetrical about the line $x=\frac{1}{2}$]

Now putting, $\frac{1}{2}-x=\frac{\sqrt{y}}{2}$ in Eqn. (6.13), we get

$$\beta(m,n)=\frac{1}{2^{2m-1}}\int_0^1 y^{-\frac{1}{2}}(1-y)^{m-1}dy$$

$$\text{or,}\quad \boldsymbol{\beta(m,n)=\frac{1}{2^{2m-1}}\beta\left(\frac{1}{2},m\right)}. \tag{6.14}$$

6.5 Others Form of Beta Function

By definition, $\beta(m,n)=\int_0^1 x^{m-1}(1-x)^{n-1}dx$.

(a) Putting $x=\frac{y}{1+y}$, $dx=\frac{dy}{(1+y)^2}$ and $(1-x)=\frac{1}{1+y}$.

$$\therefore \quad \beta(m,n)=\int_0^\infty \frac{y^{m-1}}{(1+y)^{m-1}}\cdot\frac{1}{(1+y)^{n-1}}\cdot\frac{dy}{(1+y)^2}=\int_0^\infty \boldsymbol{\frac{y^{m-1}}{(1+y)^{m+n}}dy}. \tag{6.15}$$

(b) Since, $\beta(m,n)=\beta(n,m)$,

$$\therefore \quad \beta(m,n)=\int_0^\infty \frac{y^{n-1}}{(1+y)^{m+n}}dy. \tag{6.16}$$

Eqn. (6.16) can be obtained directly by putting $x = \frac{1}{1+y}$. Eqns. (6.15) and (6.16) are the other (or, transformed) forms of Beta function.

6.6 Incomplete Beta Functions

A Beta function is said to be incomplete if the upper limit is a variable quantity.

The incomplete Beta function is defined as

$$\beta_\xi(m,n) = \int_0^\xi x^{m-1}(1-x)^{n-1}dx, \text{ for } 0 \le \xi \le 1,\ m>0,\ n>0 \tag{6.17}$$

Its series expansion is given by

$$\beta_\xi(m,n) = \xi^m\left[\frac{1}{m} + \frac{1-n}{m+1}\xi + \frac{(1-n)(2-n)}{2!(m+2)}\xi^2 + \cdots + \frac{(1-n)(2-n)\cdots(p-n)}{p!(m+p)}\right]. \tag{6.18}$$

$\therefore$ for $\xi = 1$, $\beta_\xi(m,n) = \beta(m,n)$.

$$\text{So, } \boldsymbol{\beta(m,n) = \frac{1}{m} + \frac{1-n}{m+1} + \frac{(1-n)(2-n)}{2!(m+2)} + \cdots + \frac{(1-n)(2-n)\cdots(p-n)}{p!(m+p)}}. \tag{6.19}$$

6.7 Convergence of Gamma Function

By definition, $\Gamma(n) = \int_0^\infty e^{-x}x^{n-1}dx,\ n>0.$

We split the interval $(0,\infty)$ to $(0,1)$ and $(1,\infty)$.

$$\text{Then, } \Gamma(n) = \int_0^1 e^{-x}x^{n-1}dx + \int_1^\infty e^{-x}x^{n-1}dx. \tag{6.20}$$

Convergence of $\int_0^1 e^{-x}x^{n-1}dx$ at 0

Let $f(x) = e^{-x}x^{n-1} = \dfrac{e^{-x}}{x^{1-n}}$ and $g(x) = \dfrac{1}{x^{1-n}}$.

So, $\dfrac{f(x)}{g(x)} = e^{-x} \to 1$ as $x \to 0$.

$\therefore \displaystyle\int_0^1 \frac{1}{x^{n-1}}dx$ converges at 0, if and only if $1-n<1$ i.e., $n>0$.

Hence, $\displaystyle\int_0^1 e^{-x}x^{n-1}dx$ **converges at 0 for** $\boldsymbol{n>0}$.

Convergence of $\int_0^1 e^{-x}x^{n-1}dx$ at ∞

We have $e^x > x^{n+1}$, for any value of n when x is very large.

$\therefore\ x^{n+1} < e^x$ or, $e^{-x}x^{n-1} < \dfrac{1}{x^2}$.

But $\displaystyle\int_0^\infty \frac{1}{x^2}dx$ converges, so $\displaystyle\int_0^\infty e^{-x}x^{n-1}dx$ also converges, if and only if $n>0$.

$\therefore \displaystyle\int_0^\infty e^{-x}x^{n-1}dx$ **converges for** $\boldsymbol{n>0}$.

6.8 Properties of Gamma Functions

1. **Recursion formula for Gamma Function**

We have $\Gamma(n) = \int_0^\infty e^{-x}x^{n-1}dx$

$$\therefore \quad \boldsymbol{\Gamma(n+1) = \int_0^\infty e^{-x}x^n dx.} \tag{6.19}$$

Integrating by parts the right side of Eqn. (6.19) becomes

$$\Gamma(n+1) = \left[-x^n e^{-x}\right]_0^\infty + n\int_0^\infty e^{-x}x^{n-1}dx.$$

$$\therefore \quad \boldsymbol{\Gamma(n+1) = n\Gamma(n).} \quad \textbf{[BU(H) 2002]} \tag{6.20}$$

Eqn. (6.20) is known as the **recursion formula for Gamma function**.

2. If n be an integer, then $\Gamma(n) = n!$.

$\because \ \Gamma(n+1) = n\Gamma(n)$

$$\begin{aligned}\therefore \quad \Gamma(n) &= n\Gamma(n-1)\\ &= n(n-1)\Gamma(n-2)\\ &\cdots\cdots\cdots\cdots\\ &= n(n-1)\cdots 3\cdot 2\cdot 1\Gamma(1).\end{aligned}$$

But, $\Gamma(1) = \int_0^\infty e^{-1}x^{1-0}dx = \left[e^{-1}\right]_0^\infty = 1.$

$$\therefore \quad \Gamma(n) = n(n-1)(n-2)\cdots 3\cdot 2\cdot 1 \tag{6.21}$$

or, $\boldsymbol{\Gamma(n) = n!}$

$$\Gamma(n) = \frac{\Gamma(n+m)}{n(n+1)\cdots(n+m-1)}.$$

3. We have, $\Gamma(n+1) = n\Gamma(n)$

$$\therefore \quad \Gamma(n) = \frac{\Gamma(n+1)}{n} = \frac{\Gamma(n+2)}{n(n+1)} = \frac{\Gamma(n+3)}{n(n+1)(n+2)} = \cdots$$

$$= \frac{\Gamma(n+m)}{n(n+1)\cdots(n+m-1)} \tag{6.22}$$

4. $\boldsymbol{\Gamma\left(\frac{1}{2}\right) = \sqrt{\pi} = 1.772.}$

[Bhopal U 1991; Coimbatore U 1990; Andhra U 1987; Calicut U 1987]

Proof : We have $\Gamma\left(\frac{1}{2}\right) = \int_0^t e^{-x}x^{\frac{1}{2}-1}dx = \int_0^\infty e^{-x}x^{-\frac{1}{2}}dx.$

Putting $x = y^2$, $dx = 2ydy$.

$$\therefore \quad \Gamma\left(\frac{1}{2}\right) = \int_0^\infty e^{-y^2}y^{-1}2ydy = 2\int_0^\infty e^{-y^2}dy.$$

$$\therefore \quad \left[\Gamma\left(\frac{1}{2}\right)\right]^2 = 4\int_0^\infty e^{-y^2}dy\int_0^\infty e^{-z^2}dz = 4\int_0^\infty e^{-(y^2+z^2)}dydz.$$

Putting $y = r\cos\theta$, $z = r\sin\theta$,

$$\left[\Gamma\left(\frac{1}{2}\right)\right]^2 = 4\int_{\theta=0}^{\pi/2}\left[\int_{r=0}^{\infty} e^{-r^2} r dr\right] d\theta = 4\int_0^{\pi/2}\left[-\frac{1}{2}e^{-r^2}\right]_0^{\infty} d\theta = 2\int_0^{\pi/2} d\theta = \pi.$$

$$\therefore \quad \Gamma\left(\frac{1}{2}\right) = \sqrt{\pi} = 1.772. \tag{6.23}$$

6.9 Relation between Beta and Gamma Functions

The relation between Beta and Gamma Functions is expressed as

$$\beta(m,n) = \frac{\Gamma(m)\Gamma(n)}{\Gamma(m+n)}.$$ **[BU(H) 2004]**

Proof : We have $\Gamma(m) = \int_0^{\infty} e^{-x}x^{m-1}dx$.

Putting $x = ty$, $t > 0$, we get

$$\Gamma(m) = t^m \int_0^{m} e^{-ty}y^{m-1}dy.$$

Replacing $m \to m+n$ and $t \to 1+t$, we get

$$\frac{\Gamma(m+1)}{(1+t)^{m+n}} = \int_0^{\infty} y^{m+n-1}e^{-(1+t)y}dy. \tag{6.24}$$

Multiplying both sides of the above equation by t^{m-1} and integrating the result on t in the interval $(0,\infty)$, we get

$$\Gamma(m+n)\int_0^{\infty}\frac{t^{m-1}}{(1+t)^{m+n}}dt = \int_0^{\infty}\left[\int_0^{\infty} y^{m+n-1}e^{-(1+t)y}dy\right] t^{m-1}dt. \tag{6.25}$$

Integrating the RHS of the above equation, we get

$$\int_0^{\infty}\left[\int_0^{\infty}(ty)^{m-1}e^{-ty}d(ty)\right] y^{n-1}e^{-y}dy = \int_0^{\infty}\left[\int_0^{\infty} x^{m-1}e^{-x}dx\right] y^{n-1}e^{-y}dy \ [\because\ x = ty]$$

$$= \Gamma(m)\int_0^{\infty} e^{-y}y^{n-1}dy = \Gamma(m)\Gamma(n).$$

The LHS of equation (6.25) can be written as

$$\Gamma(m+n)\beta(m,n), \quad \left[\because \int_0^{\infty}\frac{t^{m-1}}{(1+t)^{m+n}}dt = \beta(m,n) \text{ (by Eqn. (6.16))}\right]$$

Hence, $\Gamma(m+n)\beta(m,n) = \Gamma(m)\Gamma(n)$

or, $$\boldsymbol{\beta(m,n) = \frac{\Gamma(m)\Gamma(n)}{\Gamma(m+n)}}. \tag{6.26}$$

6.10 Dirichlet's Integral

The volume integral $$I = \iiint_R x^{a-1}y^{b-1}z^{c-1}dxdydz = \frac{\Gamma(a)\Gamma(b)\Gamma(c)}{\Gamma(a+b+c+1)}, \tag{6.27}$$

where the region R is given by $x \geq 0$, $y \geq 0$; $z \geq 0$, $x+y+z \leq 1$ is known as the **Dirichlet's integral**.

The value of the integral can be expressed in terms of Gamma function.

The integral in Eqn. (6.27) represents the volume enclosed within coordinate planes and the plane

$$\left.\begin{array}{l} x+y+z=1 \\ \text{or,} \quad z=1-x-y \end{array}\right\}. \tag{6.28}$$

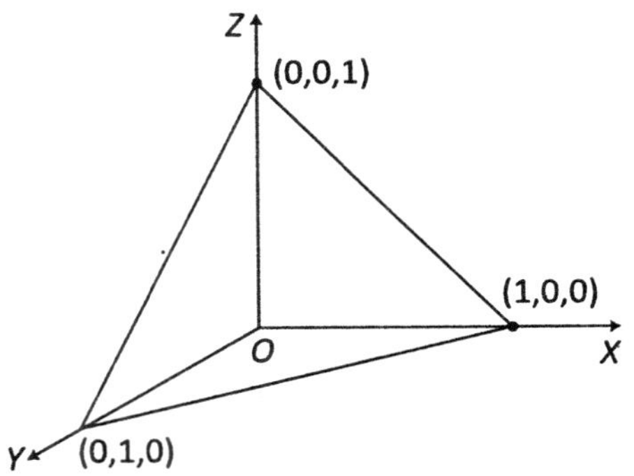

Fig. 6.1

$$\therefore \quad I=\int_{x=0}^{1}\int_{y=0}^{1}\int_{z=0}^{1-x-y} x^{a-1}y^{b-1}z^{c-1}dxdydz$$

$$=\int_{x=0}^{1}\int_{y=0}^{1-x} x^{a-1}y^{b-1}\left[\frac{z^c}{c}\right]_0^{1-x-y}dxdy$$

$$=\int_{x=0}^{1}\frac{1}{c}\int_{y=0}^{1-x} x^{a-1}y^{b-1}(1-x-y)^c dxdy$$

Now, let $y=(1-x)\xi$, i.e., $dy=(1-x)d\xi$; for $y=0$, $\xi=0$; for $y=1-x$, $\xi=1$.

$$\therefore \quad I=\int_{z=0}^{1}\frac{1}{c}\int_{\xi=0}^{1} x^{a-1}(1-x)^{b-1}\xi^{b-1}(1-x)^c(1-\xi)^c(1-x)dxd\xi$$

$$=\frac{1}{c}\int_{z=0}^{1} x^{a-1}(1-x)^{b+c}dx\int_{\xi=0}^{1}\xi^{b-1}(1-\xi)^c d\xi$$

$$=\frac{1}{c}\beta(a,b+c+1)\beta(b,c+1)=\frac{1}{c}\frac{\Gamma(a)\Gamma(b+c+1)}{\Gamma(a+b+c+1)}\cdot\frac{\Gamma(b)\Gamma(c+a)}{\Gamma(b+c+a)}$$

$$=\frac{\Gamma(a)\Gamma(b)\Gamma(c)}{\Gamma(a+b+c)}, \quad [\because\ \Gamma(c+1)=c\Gamma(c)].$$

Hence, $$\iiint_R \boldsymbol{x^{a-1}y^{b-1}z^{c-1}dxdydz=\frac{\Gamma(a)\Gamma(b)\Gamma(c)}{\Gamma(a+b+c+1)}}.$$

6.11 Error Function

The error function is defined in Art 6.1(c) which is $erf(x)=\frac{2}{\sqrt{\pi}}\int_0^x e^{-y^2}dy$.

Properties of Error Function

1. $\boldsymbol{erf(-x)=-erf(x)}$

   Proof : $erf(-x)=\frac{2}{\sqrt{\pi}}\int_0^{-x}e^{-y^2}dy$

   Putting $y=-z$, $dy=-dz$ for $y=0$, $z=0$; $y=-x$, $z=x$.

   $$\therefore \quad \boldsymbol{erf(-x)}=-\frac{2}{\sqrt{\pi}}\int_0^x e^{-z^2}dz=\boldsymbol{-erf(x)}. \tag{6.29}$$

2. $\boldsymbol{erf(0)=0}$ (6.30)

   Proof : $\boldsymbol{erf(0)}=\int_0^0 e^{-y^2}dy=\mathbf{0}.$

   Also from Eqn. (6.29),

   $$erf(0)=-erf(0) \quad \text{or,} \quad 2erf(0)=0 \quad \text{or,} \quad \boldsymbol{erf(0)=0}.$$

3. $erf(\infty) = 1$

Proof : $$erf(\infty) = \frac{2}{\sqrt{\pi}}\int_0^\infty e^{-y^2}dy = \frac{2}{\sqrt{\pi}}\frac{\sqrt{\pi}}{2} = 1. \tag{6.31}$$

This is why the normalisation factor in error function is chosen as $\frac{2}{\sqrt{\pi}}$.

4. For $|x|$ is large, $erfc(x) = 1 - erf(c)$

Proof : When $|x|$ is large, the error function $erf(x)$ may be computed by Maclaurin series.

$$e^{-y^2} = 1 - \frac{y^2}{1!} + \frac{y^4}{2!} - \frac{y^6}{3!} + \frac{y^8}{4!} - \cdots.$$

$$\therefore \quad erf(x) = \frac{2}{\sqrt{\pi}}\int_0^x e^{-y^2}dy = \frac{2}{\sqrt{\pi}}\int_0^x \left(1 - \frac{y^2}{1!} + \frac{y^4}{2!} - \frac{y^6}{3!} + \frac{y^8}{4!} - \cdots\right)dy$$

$$\text{or,} \quad erf(x) = \frac{2}{\sqrt{\pi}}\left[x - \frac{x^3}{1!3} + \frac{x^5}{2!5} - \frac{x^7}{3!7} + \cdots\right]. \tag{6.32}$$

The complementary error function is

$$erf(c) = \frac{2}{\sqrt{\pi}}\int_{x=0}^\infty e^{-y^2}dy \tag{6.33}$$

$$= \frac{2}{\sqrt{\pi}}\int_0^x e^{-y^2}dy + \frac{2}{\sqrt{\pi}}\int_x^\infty e^{-y^2}dy.$$

From Eqns. (6.31), (6.32) and (6.33), we get

$$1 = erf(x) + erfc(x)$$

$$\text{or,} \quad \boldsymbol{erfc(x) = 1 - erf(x)}. \tag{6.34}$$

6.12 Incomplete Gamma Functions

The Gamma functions with variable limit are defined as **Incomplete Gamma Functions.** There are two types of incomplete Gamma Functions, which are

$$\gamma(n, x) = \int_0^x e^{-z}z^{n-1}dz \tag{6.35}$$

$$\text{and} \quad \Gamma(n, x) = \int_x^\infty e^{-z}z^{n-1}dz. \tag{6.36}$$

$$\text{So,} \quad \gamma(n, x) + \Gamma(n, x) = \int_0^x e^{-z}z^{n-1}dz + \int_x^\infty e^{-z}z^{n-1}dz = \int_x^\infty e^{-z}z^{n-1}dz = \Gamma(n). \tag{6.37}$$

The series expansion of $\gamma(n, x)$ and $\Gamma(n, x)$ are :

(i) **When n is a positive integer**

$$\gamma(n, x) = (n-1)!\left(1 - e^{-x}\sum_{r=0}^{n-1}\frac{x^r}{r!}\right) \tag{6.38}$$

$$\text{and} \quad \gamma(n, x) = (n-1)!e^{-x}\sum_{r=0}^{n-1}\frac{x^r}{r!}. \tag{6.39}$$

(ii) **When n is non-integer**

$$\gamma(n,x) = x^n \sum_{r=0}^{\infty} (-1)^r \frac{x^n}{r!(n+r)} \tag{6.40}$$

and
$$\gamma(n,x) = x^{n-1} e^{-x} \sum_{r=0}^{\infty} (-1)^r \frac{(r-n)!}{\Gamma(1-n)x^r}. \tag{6.41}$$

Eqn. (6.41) is true for large values of x.

6.13 Series Expansion of Error Function

We have the error function $erf(x) = \frac{2}{\sqrt{\pi}} \int_0^x e^{-y^2} dy$.

Putting $y^2 = z$; $y = \sqrt{z}$ and $dy = \frac{y^{-\frac{1}{2}}}{2} dy$.

$$\therefore \quad erf(x) = \frac{2}{\sqrt{\pi}} \int_0^{x^2} \frac{1}{2} e^{-y} y^{-\frac{1}{2}} dy = \frac{1}{\sqrt{\pi}} \int_0^{x^2} e^{-y} y^{-\frac{1}{2}} dy$$

or,
$$erf(x) = \frac{1}{\sqrt{\pi}} \gamma\left(\frac{1}{2}, x^2\right). \tag{6.42}$$

Again,
$$erfc(x) = 1 - erf(x) = \frac{2}{\sqrt{\pi}} \int_0^{\infty} e^{-y^2} dy - \frac{2}{\sqrt{\pi}} \int_0^x e^{-y^2} dy = \frac{2}{\sqrt{\pi}} \int_x^{\infty} e^{-y^2} dy. \tag{6.43}$$

Again, putting $y^2 = z$ in Eqn. (6.43), we get

$$erfc(x) = \frac{2}{\sqrt{\pi}} \frac{1}{2} \int_{x^2}^{\infty} e^{-z} z^{-\frac{1}{2}} dz = \frac{1}{\sqrt{\pi}} \Gamma\left(\frac{1}{2}, x^2\right). \tag{6.44}$$

$\therefore$ from Eqn. (6.40),

$$\gamma\left(\frac{1}{2}, x^2\right) = (x^2)^{\frac{1}{2}} \left[2 - \frac{x^2}{\frac{3}{2}} + \frac{x^4}{2!\left(\frac{5}{2}\right)} - \frac{x^6}{3!\left(\frac{7}{2}\right)} + \cdots\right] = 2x\left[1 - \frac{x^2}{3} + \frac{x^4}{5 \cdot 2!} - \frac{x^6}{7 \cdot 3!} + \cdots\right]$$

$\therefore$ **the series expansion of $erf(x)$ is**

$$\boldsymbol{erf(x) = \frac{2}{\sqrt{\pi}} \left[x - \frac{x^3}{3} + \frac{x^5}{5 \cdot 2!} - \frac{x^7}{7 \cdot 3!} + \cdots\right]}. \tag{6.45}$$

6.14 Stirling's Formula

Stirling's formula gives an approximate value of Gamma function $\Gamma(n+1)$ or $n!$ when n is very large.

From definition,
$$\Gamma(n+1) = \int_0^{\infty} e^{-x} x^n dx = \int_0^{\infty} e^{-x} e^{\log x^n} dx = \int_0^{\infty} e^{\log x^n - x} dx \tag{6.46}$$

Putting $x = n + \sqrt{n}y$, (6.47)

$$dx = \sqrt{n}dy \quad \text{and} \quad \text{for } x = 0, \quad y = -\sqrt{n}. \tag{6.48}$$

Again, $\log x^n - x = n \log x - x$

and
$$\log x = \log(n + \sqrt{n}y) = \log n + \log\left(1 + \frac{y}{\sqrt{n}}\right) = \log n + \frac{y}{\sqrt{n}} - \frac{y^2}{2n} + \cdots$$

or, $$\log x \simeq \log x + \frac{y}{\sqrt{n}} - \frac{y^2}{2n} \text{ [for } n \text{ large].} \tag{6.49}$$

Putting relation (6.47)–(6.49) in Eqn. (6.46), we get

$$\Gamma(n+1) = \int_{-\sqrt{n}}^{\infty} e^{\left\{n\log n + y\sqrt{n} - \frac{y^2}{2} - n - \sqrt{n}y\right\}} \sqrt{n}dy = e^{(n\log n - n)}\sqrt{n}\int_{-\sqrt{n}}^{\infty} e^{-\frac{y^2}{2}} dy$$

$$= \sqrt{n}n^n e^{-n}\left[\int_{-\infty}^{\infty} e^{-\frac{y^2}{2}} dy - \int_{-\infty}^{+\infty} e^{-\frac{y^2}{2}} dy\right]$$

$$= \sqrt{n}n^n e^{-n}\int_{-\infty}^{+\infty} e^{-\frac{y^2}{2}} dy. \tag{6.50}$$

From Eqn. (6.50), we get

$$\Gamma(n+1) = 2\sqrt{n}n^n e^{-n}\int_{0}^{\infty} e^{-\frac{y^2}{2}} dy = 2\sqrt{n}n^n e^{-n}\frac{1}{2}\sqrt{2n}$$

or, $$\boldsymbol{n! = \sqrt{2n\pi}n^n e^{-n}}, \quad [\because \Gamma(n+1) = n!]. \tag{6.51}$$

Eqn. (6.51) is the **Stirling's formula**.

6.15 Solved Problems

Example 6.1. *Show that* $\beta(x,y) = 2\int_0^{\frac{\pi}{2}} \sin^{2x-1}\theta\cos^{2y-1}\theta d\theta.$

Solution : We have $\beta(x,y) = 2\int_0^1 x^{x-1}(1-x)^{y-1}dx.$

We put $x = \sin^2\theta$, $\therefore$ $dx = 2\sin\theta\cos\theta d\theta.$

For $x = 0$, $\theta = 0$ and $x = 1$, $\theta = \frac{\pi}{2}$.

$$\therefore \beta(x,y) = \int_0^{\frac{\pi}{2}} (\sin\theta)^{2y-y}(\cos\theta)^{2y-2} 2\sin\theta\cos\theta d\theta = \mathbf{2}\int_0^{\frac{\pi}{2}} \sin^{2x-1}\theta\cos^{2y-1}\theta d\theta$$

(Proved).

Example 6.2. *Write the integral* $\int_0^1 \frac{x^3}{\sqrt{1-x^2}}dx$ *in the form of a Beta function and hence, evaluate it.* **[BU(H) 1993]**

Solution : Let $I = \int_0^1 \frac{x^3}{\sqrt{1-x^2}}dx.$

Putting $x = \sin^2\theta$, $dx = 2\sin\theta\cos\theta d\theta.$

For $x = 0$, $\theta = 0$ and $x = 1$, $\theta = \frac{\pi}{2}$.

$$\therefore I = 2\int_0^{\frac{\pi}{2}} \sin^6\theta(\cos\theta)^{-1}\sin\theta\cos\theta d\theta = 2\int_0^{\frac{\pi}{2}} \sin^7\theta d\theta = 2\int_0^{\frac{\pi}{2}} \sin^7\theta(\cos\theta)^0 d\theta = \beta\left(4, \frac{1}{2}\right)$$

$$= \frac{\Gamma(4)\Gamma\left(\frac{1}{2}\right)}{\Gamma\left(4\frac{1}{2}\right)} = \frac{3!\Gamma\left(\frac{1}{2}\right)}{\frac{7}{2}\cdot\frac{5}{2}\cdot\frac{3}{2}\cdot\frac{1}{2}\Gamma\left(\frac{1}{2}\right)} = \frac{6\times 16}{105} = \frac{\mathbf{32}}{\mathbf{35}}.$$

$$\therefore \int_0^1 \frac{\boldsymbol{x^3}}{\sqrt{\mathbf{1}-\boldsymbol{x}}}\boldsymbol{dx} = \frac{\mathbf{32}}{\mathbf{35}}.$$

Example 6.3. *Show that,* $\int_0^{\frac{\pi}{2}} \sin^p\theta\cos^q\theta d\theta = \frac{\Gamma\left(\frac{p+1}{2}\right)\Gamma\left(\frac{q+1}{2}\right)}{2\Gamma\left(\frac{p+q+2}{2}\right)}$.

Hence, evaluate (a) $\int_0^{\frac{\pi}{2}} \sin^p\theta d\theta$ *and* (b) $\int_0^{\frac{\pi}{2}} \cos^q\theta d\theta$.

Solution : From the definition of β-function, $\beta(m,n) = \int_0^1 x^{m-1}(1-x)^{n-1}dx$.

Putting $x = \sin^2\theta$; $dx = 2\sin\theta\cos\theta d\theta$ and $1 - x = 1 - \sin^2\theta = \cos^2\theta$, we get

$$\beta(m,n) = \int_0^{\frac{\pi}{2}} \sin^{2m-2}\theta\cos^{2n-2}\theta\cdot 2\sin\theta\cos\theta d\theta = 2\int_0^{\frac{\pi}{2}} \sin^{2m-1}\theta\cos^{2n-1}\theta d\theta \qquad \text{(i)}$$

Again, $\beta(m,n) = \frac{\Gamma(m)\Gamma(n)}{\Gamma(m+n)}$. (ii)

Comparing Eqns. (i) and (ii), we get

$$\int_0^{\frac{\pi}{2}} \sin^{2m-1}\theta\cos^{2n-1}\theta d\theta = \frac{\Gamma(m)\Gamma(n)}{\Gamma(m+n)}. \qquad \text{(iii)}$$

Replacing $(2m-1)$ by p and $(2n-1)$ by q, the above equation becomes

$$\int_0^{\frac{\pi}{2}} \sin^p\theta\cos^q\theta d\theta = \frac{\Gamma\left(\frac{p+1}{2}\right)\Gamma\left(\frac{q+1}{2}\right)}{2\Gamma\left(\frac{p+q+2}{2}\right)} \quad \textbf{(Proved)}. \qquad \text{(iv)}$$

(a) Putting $q = 0$ in equation (iv), we get

$$\int_0^{\frac{\pi}{2}} \sin^p\theta d\theta = \frac{\Gamma\left(\frac{p+1}{2}\right)\Gamma\left(\frac{1}{2}\right)}{2\Gamma\left(\frac{p+2}{2}\right)} = \frac{\Gamma\left(\frac{q+1}{2}\right)\sqrt{\pi}}{2\Gamma\left(\frac{q}{2}+1\right)}.$$

Now, if $p = 1$, then $\int_0^{\frac{\pi}{2}} \sin\theta d\theta = \frac{\Gamma(1)\sqrt{\pi}}{2\sqrt{\frac{3}{2}}} = \frac{\sqrt{\pi}}{2\cdot\frac{1}{2}\sqrt{\pi}} = 1$.

(b) Also putting $p = 0$ in equation (iv),

$$\int_0^{\frac{\pi}{2}} \cos^q\theta d\theta = \frac{\Gamma\left(\frac{1}{2}\right)\Gamma\left(\frac{q+1}{2}\right)}{2\Gamma\left(\frac{q+2}{2}\right)} = \frac{\Gamma\left(\frac{q+1}{2}\right)\sqrt{\pi}}{2\Gamma\left(\frac{q}{2}+1\right)}.$$

Now, if $q = 1$, then $\int_0^{\frac{\pi}{2}} \cos\theta d\theta = \frac{\Gamma(1)\sqrt{\pi}}{2\sqrt{\frac{3}{2}}} = 1$.

Example 6.4. *If* $\Gamma(x) = \int_0^{\infty} e^{-t}t^{z-1}dt$, *evaluate* $\int_0^{\frac{\pi}{2}} \cos^{m-1}x\sin^{n-1}x\,dx$ *in terms of Gamma function.*

Solution : We have $\int_0^{\frac{\pi}{2}} \sin^p\theta\cos^q\theta d\theta = \frac{\Gamma\left(\frac{p+1}{2}\right)\Gamma\left(\frac{q+1}{2}\right)}{2\Gamma\left(\frac{p+q+2}{2}\right)}$.

Replacing p by $n-1$, q by $m-1$ and θ by x, we get

$$\int_0^{\frac{\pi}{2}} \cos^{m-1}x\sin^{n-1}xdx = \frac{\Gamma\left(\frac{n}{2}\right)\Gamma\left(\frac{m}{2}\right)}{2\Gamma\left(\frac{m+n}{2}\right)}. \qquad \text{(i)}$$

But $\Gamma\left(\frac{m}{2}\right)\Gamma\left(\frac{n}{2}\right) = \int_0^\infty e^{-t}t^{\frac{m}{2}-1}dt\int_0^\infty e^{-u}u^{(\frac{n}{2})-1}du$

$$= \int_0^\infty\int_0^\infty e^{-(t+u)}t^{\frac{m}{2}-1}u^{\frac{n}{2}-1}dtdu. \quad \left[\because \Gamma(z) = \int_0^\infty e^{-t}t^{z-1}dt\right] \qquad \text{(ii)}$$

Now, let $t = r\cos^2\theta$, $u = r\sin^2\theta$.

$$\therefore \quad \frac{\partial(t,u)}{\partial(r,\theta)} = \begin{vmatrix} \frac{\partial t}{\partial r} & \frac{\partial u}{\partial \theta} \\ \frac{\partial u}{\partial r} & \frac{\partial u}{\partial \theta} \end{vmatrix} = \begin{vmatrix} \cos^2\theta & -2r\cos\theta\sin\theta \\ \sin^2\theta & 2r\sin\theta\cos\theta \end{vmatrix} = 2r\sin\theta\cos\theta.$$

Then $\Gamma\left(\frac{m}{2}\right)\Gamma\left(\frac{n}{2}\right) = 2\int_0^\infty\int_0^{\frac{\pi}{2}} e^{-r}(\cos^2\theta)^{\frac{m}{2}-1}(r\sin^2\theta)^{\frac{n}{2}-1} \times r\sin\theta\cos\theta drd\theta$

$$= 2\int_0^\infty e^{-r}r^{\frac{m+n}{2}-1}dr\int_0^{\frac{\pi}{2}}\cos^{m-1}\theta\sin^{n-1}\theta d\theta$$

$$= 2\Gamma\left(\frac{m+n}{2}\right)\int_0^{\frac{\pi}{2}}\cos^{m-1}\theta\sin^{n-1}\theta\, d\theta. \left[\because \int_0^\infty e^{-r}r^{\frac{m+n}{2}-1}dr = \Gamma\left(\frac{m+n}{2}\right)\right]$$

$$\therefore \quad \int_0^{\frac{\pi}{2}}\cos^{m-1}\theta\sin^{n-1}\theta d\theta = \int_0^{\frac{\pi}{2}}\cos^{m-1}x\sin^{n-1}x dx = \frac{\Gamma\left(\frac{m}{2}\right)\Gamma\left(\frac{n}{2}\right)}{2\Gamma\left(\frac{m+n}{2}\right)}.$$

Example 6.5. *Evaluate* : (a) $\int_0^1 \frac{dx}{\sqrt{1-x^n}}$ (b) $\int_0^1 \frac{dx}{(1-x^n)^{\frac{1}{n}}}$.

Solution : Let $x^n = \sin^2\theta$.

$\therefore \quad x = \sin^{\frac{2}{n}}\theta$ and $dx = \frac{2}{n}\sin^{(\frac{2}{n})-1}\cos\theta d\theta$, then

$$\text{(i)} \int_0^1 \frac{dx}{\sqrt{1-x^n}} = \int_0^{\frac{\pi}{2}} \frac{\frac{2}{n}\sin^{(\frac{2}{n}-1)}\theta\cos\theta d\theta}{\cos\theta} = \frac{2}{n}\int_0^{\frac{\pi}{2}}\sin^{(\frac{2}{n}-1)}\theta d\theta$$

$$= \frac{2}{n}\frac{\Gamma\left(\frac{1}{n}\right)\sqrt{\pi}}{2\Gamma\left(\frac{1}{n}+\frac{1}{2}\right)} \quad \left[\because \int_0^{\frac{\pi}{2}}\sin^p\theta d\theta = \frac{\Gamma\left(\frac{p+1}{2}\right)\sqrt{\pi}}{2\Gamma\left(\frac{p}{2}+1\right)}\right]$$

$$= \frac{\frac{\sqrt{\pi}}{n}\Gamma\left(\frac{1}{n}\right)}{\Gamma\left(\frac{1}{n}+\frac{1}{2}\right)}.$$

$$\text{(ii)} \int_0^1 \frac{dx}{(1-x^n)^{\frac{1}{n}}} = \int_0^{\frac{\pi}{2}} \frac{\frac{2}{n}\sin^{(\frac{2}{n}-1)}\theta\cos\theta d\theta}{\cos^{\frac{2}{n}}\theta} = \frac{2}{n}\int_0^{\frac{\pi}{2}}\sin^{(\frac{2}{n}-1)}\theta\cos^{(1-\frac{2}{n})}\theta d\theta$$

$$= \frac{2}{n}\frac{\Gamma\left(\frac{1}{n}\right)\Gamma\left(1-\frac{1}{2}\right)}{2\Gamma(1)} \quad \left[\because \int_0^{\frac{\pi}{2}}\sin^p\theta\cos^q\theta d\theta = \frac{\Gamma\left(\frac{p+1}{2}\right)\Gamma\left(\frac{q+1}{2}\right)}{2\Gamma\left(\frac{p+q+2}{2}\right)}\right]$$

$$= \frac{\pi}{\pi\sin\left(\frac{\pi}{n}\right)}. \quad \left[\because \Gamma\left(\frac{1}{n}\right)\Gamma\left(1-\frac{1}{n}\right) = \frac{\pi}{\sin\frac{\pi}{n}}\right]$$

Example 6.6. *Show that,* $\beta(p,q+1) = \dfrac{\beta(p+1,q)}{p} = \dfrac{\beta(p,q)}{p+q}$. **[BU(H) 1994]**

Solution : We have $\dfrac{\beta(p,q+1)}{q} = \dfrac{\Gamma(p)\Gamma(q+1)}{q\Gamma(p+q+1)} = \dfrac{\Gamma(p)q\Gamma(q)}{q(p+q)\Gamma(p+q)} = \dfrac{1}{p+q}\dfrac{\Gamma(p)\Gamma(q)}{\Gamma(p+q)}$

$$= \frac{\beta(p,q)}{p+q}.$$

Similarly, $\dfrac{\beta(p+1,q)}{p} = \dfrac{\beta(p,q)}{p+q}$ **(Proved)**.

Example 6.7. *Prove that,*

(a) $\displaystyle\int_0^{\frac{\pi}{2}} \frac{d\theta}{\sqrt{\sin\theta}} \cdot \int_0^{\frac{\pi}{2}} \sqrt{\sin\theta}\,d\theta = \pi$ (b) $\displaystyle\int_0^{\frac{\pi}{2}} \sqrt{\tan\theta}\,d\theta = \frac{\Gamma\left(\frac{1}{4}\right)\Gamma\left(\frac{3}{4}\right)}{2}$. **[BU(H) 2002]**

Solution : (a) **LHS** $\displaystyle = \int_0^{\frac{\pi}{2}} (\sin\theta)^{-\frac{1}{2}}d\theta \cdot \int_0^{\frac{\pi}{2}} \sqrt{\sin\theta}\,d\theta$

$$= \frac{\Gamma\left(\frac{1}{2}\right)\Gamma\left(\frac{1}{4}\right)}{2\Gamma\left(\frac{3}{4}\right)}\,\frac{\Gamma\left(\frac{1}{2}\right)\Gamma\left(\frac{1}{4}\right)}{2\Gamma\left(\frac{5}{2}\right)}, \quad \left[\because \int_0^{\pi} \sin^p d\theta = \frac{\Gamma\left(\frac{p+1}{2}\right)\Gamma\left(\frac{1}{2}\right)}{2\Gamma\left(\frac{p}{2}+1\right)}\right]$$

$$= \frac{\sqrt{\pi}\Gamma\left(\frac{1}{4}\right)}{2\Gamma\left(\frac{3}{4}\right)} \cdot \frac{\sqrt{\pi}\Gamma\left(\frac{3}{4}\right)}{2\times\frac{1}{4}\Gamma\left(\frac{1}{4}\right)} \quad \left[\because \Gamma\left(\frac{1}{2}\right) = \sqrt{\pi},\ \Gamma\left(\frac{5}{2}\right) = \frac{1}{4}\Gamma\left(\frac{1}{4}\right)\right]$$

$= \pi$ **(Proved)**.

(b) **LHS** $\displaystyle = \int_0^{\frac{\pi}{2}} (\tan\theta)^{\frac{1}{2}}d\theta = \int_0^{\frac{\pi}{2}} (\sin\theta)^{\frac{1}{2}}(\cos\theta)^{-\frac{1}{2}}d\theta = \frac{1}{2}\frac{\Gamma\left(\frac{-\frac{1}{2}+1}{2}\right)\Gamma\left(\frac{\frac{1}{2}+1}{2}\right)}{\Gamma\left(\frac{\frac{1}{2}-\frac{1}{2}}{2}\right)+1}$

$$= \frac{\Gamma\left(\frac{3}{4}\right)\Gamma\left(\frac{1}{4}\right)}{2\Gamma(1)} = \frac{\Gamma\left(\frac{1}{4}\right)\Gamma\left(\frac{3}{4}\right)}{2} \quad \textbf{(Proved)}.$$

Example 6.8. *Prove that,* $\displaystyle\int_0^1 x^m(\log x)^n dx = \frac{(-1)^n n!}{(m+n)^{m+1}}$, *where n is a positive integer and $m > -1$.* **[BU(H) 1996]**

Solution : Let $\displaystyle I = \int_0^1 x^m(\log x)^n dx$ and $\log x = -y$.

$\therefore \quad x = e^{-y}, \quad \dfrac{1}{x}dx = -dy.$

So, $\displaystyle I = -\int_\infty^0 e^{-my}(-y)^n x dx = (-1)^n \int_0^\infty y^n e^{-my}e^{-y}dy = (-1)^n \int_0^\infty y^n e^{-(m+1)y}dy.$

Putting $(m+1)y = z$,

$$I = (-1)^n \int_0^\infty \frac{z^n}{(m+1)^n}e^{-z}\frac{dz}{(m+1)} = \frac{(-1)^n}{(m+1)^{n+1}}\int_0^\infty z^n e^{-z}dz = \frac{(-1)^n}{(m+1)^{n+1}}\Gamma(n+1)$$

$$= \frac{(-1)^n n!}{(m+1)^{n+n}} \quad \textbf{(Proved)}.$$

Example 6.9. *Express* $\int_0^{\frac{\pi}{2}} \sqrt{\cot\theta}\,d\theta$ *in terms of Gamma function.*

Solution : $$\int_0^{\frac{\pi}{2}} \sqrt{\cot\theta}\,d\theta = \int_0^{\frac{\pi}{2}} (\cos\theta)^{\frac{1}{2}}(\sin\theta)^{-\frac{1}{2}}d\theta = \frac{1}{2}\frac{\Gamma\left(\frac{\frac{1}{2}+1}{2}\right)\Gamma\left(\frac{-\frac{1}{2}+1}{2}\right)}{\Gamma\left(\frac{\frac{1}{2}-\frac{1}{2}}{2}+1\right)}$$

$$= \frac{1}{2}\Gamma\left(\frac{1}{4}\right)\Gamma\left(\frac{3}{4}\right). \quad [\because\ \Gamma(1)=1]$$

Example 6.10. *Given* $\Gamma\left(\frac{1}{2}\right) = \pi$, *prove that,* $\Gamma\left(-\frac{1}{2}\right) = -2\sqrt{\pi}$ *and* $\Gamma\left(-\frac{3}{2}\right) = \frac{4\sqrt{\pi}}{3}$.

[BU(H) 2001]

Solution : We know that, $\Gamma(n+1) = n\Gamma(n)$.

Putting $n = -\frac{1}{2}$,

$$\Gamma\left(\frac{1}{2}\right) = -\frac{1}{2}\Gamma\left(-\frac{1}{2}\right) \quad \text{or,} \quad \pi = -\frac{1}{2}\Gamma\left(-\frac{1}{2}\right).$$

$$\therefore\ \Gamma\left(-\frac{1}{2}\right) = -2\sqrt{\pi} \quad \textbf{(Proved).}$$

Again, putting $n = -\frac{3}{2}$, we get

$$\Gamma\left(-\frac{3}{2}+1\right) = -\frac{3}{2}\Gamma\left(-\frac{1}{2}\right) = -\frac{3}{2} \times -\frac{1}{2}\sqrt{\pi} = \frac{3}{4}\sqrt{\pi} \quad \textbf{(Proved).}$$

Example 6.11. *Prove that,*

(i) $\Gamma\left(\frac{1}{2}\right) = \sqrt{\pi}$ **[BU(H) 2004, 2005]** (ii) $2\cdot5\cdot8\cdots(3n-1) = \frac{3^n\Gamma\left(n+\frac{2}{3}\right)}{\Gamma\left(\frac{2}{3}\right)}$. **[BU(H) 2002]**

Solution : (i) See Art 6.8.

(ii) We know that, $\Gamma(n+1) = n\Gamma(n)$.

$$\therefore\ \Gamma\left(n+\frac{2}{3}\right) = \left(n-\frac{1}{3}\right)\Gamma\left(n-\frac{1}{3}\right)$$

$$= \left(n-\frac{1}{3}\right)\left(n-\frac{4}{3}\right)\Gamma\left(n-\frac{4}{3}\right)$$

$$\cdots\cdots\cdots\cdots\cdots\cdots\cdots\cdots$$

$$= \left(n-\frac{1}{3}\right)\left(n-\frac{4}{3}\right)\left(n-\frac{7}{3}\right)\cdots\frac{8}{3}\cdot\frac{5}{3}\cdot\frac{2}{3}\Gamma\left(\frac{2}{3}\right)$$

$$= \frac{(3n-1)(3n-4)(3n-7)\cdots8\cdot5\cdot2\Gamma\left(\frac{2}{3}\right)}{3^n}$$

$$\therefore\ 2\cdot5\cdot8\cdots(3n-1) = \frac{3^n\Gamma\left(n+\frac{2}{3}\right)}{\Gamma\left(\frac{2}{3}\right)} \quad \textbf{(Proved).}$$

Example 6.12. *Evaluate* $\int_0^{\frac{\pi}{2}} \sqrt{\tan\theta}d\theta$.

Solution : Let $I = \int_0^{\frac{\pi}{2}} \sqrt{\tan\theta}d\theta = \int_0^{\frac{\pi}{2}} \sin^{\frac{1}{2}}\theta\cos^{-\frac{1}{2}}\theta d\theta = \frac{1}{2}\Gamma\left(\frac{3}{4}\right)\Gamma\left(\frac{1}{4}\right)$.

Putting $n = \frac{1}{4}$ in $\sqrt{\pi}\Gamma(2m) = 2^{2m-1}\Gamma(m)\Gamma\left(m+\frac{1}{2}\right)$, we get

$$\Gamma\left(\frac{3}{4}\right)\Gamma\left(\frac{1}{4}\right) = \sqrt{2}\sqrt{\pi}\Gamma\left(\frac{1}{2}\right) = \sqrt{2}\pi.$$

$\therefore \quad I = \frac{1}{2}(\sqrt{2}\pi) = \frac{\pi}{\sqrt{2}}$.

Example 6.13. *Show that,* $\Gamma\left(\frac{1}{3}\right)\Gamma\left(\frac{2}{3}\right) = \frac{2\pi}{\sqrt{3}}$.

Solution : Let $P = \Gamma\left(\frac{1}{3}\right)\Gamma\left(\frac{2}{3}\right)$ and $p = \Gamma\left(\frac{2}{3}\right)\Gamma\left(\frac{1}{3}\right)$.

Multiplying, we get $P^2 = \Gamma\left(\frac{1}{3}\right)\Gamma\left(\frac{2}{3}\right)\Gamma\left(\frac{2}{3}\right)\Gamma\left(\frac{1}{3}\right) = \Gamma\left(\frac{1}{3}\right)\Gamma\left(1-\frac{1}{3}\right)\Gamma\left(\frac{2}{3}\right)\Gamma\left(1-\frac{2}{3}\right)$

$$= \frac{\pi}{\sin\frac{\pi}{3}}\cdot\frac{\pi}{\sin\frac{2\pi}{3}} \quad \left[\because\ \Gamma(m)\Gamma(1-m) = \frac{\pi}{\sin m\pi}\right]$$

$$= \frac{4\pi^2}{3}$$

$\therefore \quad P = \frac{2\pi}{\sqrt{3}}$ **(Proved).**

Example 6.14. *Show that,* $1\cdot 3\cdot 5\cdots(2n-1) = \frac{2^n}{\sqrt{\pi}}\Gamma\left(n+\frac{1}{2}\right)$ *and hence, show that,* $\Gamma\left(n+\frac{1}{2}\right) = \frac{(2n)!}{2^{2n}n!}\sqrt{\pi}$.

Solution : We have $\beta(n,m) = \beta(m,n) = \frac{(n-1)!}{m(m+1)\cdots(m+n-1)}$ when n is a positive integer.

Putting $n = \frac{1}{2}$, we get

$$\beta\left(n,\frac{1}{2}\right) = \frac{2^n(n-1)!}{1\cdot 3\cdots(2n-1)} \quad \text{or,} \quad \frac{\Gamma(n)\Gamma\left(\frac{1}{2}\right)}{\Gamma\left(n+\frac{1}{2}\right)} = \frac{2^n(n-1)!}{1\cdot 3\cdots(2n-1)}$$

or, $\frac{(n-1)!\sqrt{\pi}}{\Gamma\left(n+\frac{1}{2}\right)} = \frac{2^n(n-1)!}{1\cdot 3\cdots(2n-1)}$.

$\therefore \quad \mathbf{1\cdot 3\cdots(2n-1) = \frac{2^n}{\sqrt{\pi}}\Gamma\left(n+\frac{1}{2}\right)}$ **(Proved).**

Again, $1\cdot 3\cdots(2n-1) = 1\cdot 3\cdot(2n-1) = \frac{2\cdot 4\cdot 6\cdots 2n}{2\cdot 4\cdot 6\cdots 2n} = \frac{(2n)!}{2^n n!}$.

Hence, $\frac{2^n}{\sqrt{\pi}} = \Gamma\left(n+\frac{1}{2}\right) = \frac{(2n)!}{2^n n!}$.

$\therefore \quad \mathbf{\Gamma\left(n+\frac{1}{2}\right) = \frac{(2n)!}{2^n n!}\sqrt{\pi}}$ **(Proved).**

Example 6.15. *Evaluate* $\int_0^\infty \frac{x^a}{a^x}dx.$

Solution : Here $a^x = e^{x\log a}$.

$$\therefore \quad I = \int_0^\infty \frac{x^a}{a^x}dx = \int_0^\infty x^a e^{-x\log a}dx.$$

Putting $x\log a = z$, so that $dx = \frac{dz}{\log a}$.

$$\therefore \quad I = \int_0^\infty \left(\frac{z}{\log a}\right)^a e^{-z}\frac{dz}{\log a} = \frac{1}{(\log a)^{a+1}}\int_0^\infty z^a e^{-z}dz = \frac{\Gamma(a+1)}{(\log a)^{a+1}}.$$

Example 6.16. *Evaluate* $\Gamma\left(\frac{1}{n}\right)\Gamma\left(\frac{2}{n}\right)\Gamma\left(\frac{3}{n}\right)\cdots$, *where* n *is a positive integer.*

Solution : Let $R = \Gamma\left(\frac{1}{n}\right)\Gamma\left(\frac{2}{n}\right)\Gamma\left(\frac{3}{n}\right)\cdots\Gamma\left(\frac{n-1}{n}\right)$

$$= \Gamma\left(\frac{1}{n}\right)\Gamma\left(\frac{2}{n}\right)\Gamma\left(\frac{3}{n}\right)\cdots\Gamma\left(1-\frac{1}{n}\right). \quad \text{(i)}$$

Reversing the order of the factors, we get

$$R = \Gamma\left(1-\frac{1}{n}\right)\Gamma\left(1-\frac{2}{n}\right)\Gamma\left(1-\frac{3}{n}\right)\cdots\Gamma\left(\frac{2}{n}\right)\Gamma\left(\frac{1}{n}\right). \quad \text{(ii)}$$

Multiplying (i) and (ii), we get

$$R^2 = \left[\Gamma\left(\frac{1}{n}\right)\Gamma\left(1-\frac{1}{n}\right)\right]\left[\Gamma\left(\frac{2}{n}\right)\Gamma\left(1-\frac{2}{n}\right)\right]\cdots\left[\Gamma\left(1-\frac{1}{n}\right)\Gamma\left(\frac{1}{n}\right)\right]$$

$$= \left[\Gamma\left(\frac{1}{n}\right)\Gamma\left(1-\frac{1}{n}\right)\right]\left[\Gamma\left(\frac{2}{n}\right)\Gamma\left(1-\frac{2}{n}\right)\right]\cdots\left[\Gamma\left(\frac{n-1}{n}\right)\Gamma\left(1-\frac{n-1}{n}\right)\right]$$

$$= \frac{\pi}{\sin\frac{\pi}{n}}\cdot\frac{\pi}{\sin\frac{2\pi}{n}}\cdots\frac{\pi}{\sin\frac{(n-1)\pi}{n}} \quad \left[\because \Gamma(m)\Gamma(m-1) = \frac{\pi}{\sin m\pi}\right]$$

$$= \frac{\pi^{n-1}}{\sin\frac{\pi}{n}\sin\frac{2\pi}{n}\cdots\sin\frac{(n-1)\pi}{n}}.$$

Again, we know that, $\frac{\sin n\theta}{\theta} = 2^{n-1}\sin\left(\theta+\frac{\pi}{n}\right)\sin\left(\theta+\frac{2\pi}{n}\right)\cdots\sin\left(\theta+\frac{(n-1)\pi}{n}\right).$

Putting $\theta = 0$,

$$\underset{\theta\to 0}{\text{Lt}}\frac{\sin n\theta}{\sin\theta} = \underset{\theta\to 0}{\text{Lt}}\left[\frac{n\sin n\theta}{n\theta}\cdot\frac{\theta}{\sin\theta}\right] = n.$$

$$\therefore \quad n = 2^{n-1}\sin\frac{\pi}{n}\sin\frac{2\pi}{n}\cdots\sin\frac{(n-1)\pi}{n}.$$

$$\therefore \quad \sin\frac{\pi}{n}\sin\frac{2\pi}{n}\cdots\sin\frac{(n-1)\pi}{n} = \frac{n}{2^{n-1}}.$$

$$\therefore \quad R^2 = \frac{\pi^{n-1}}{\left(\frac{n}{2^{n-1}}\right)} = \frac{(2\pi)^{n-1}}{n} \quad \text{or,} \quad R = \Gamma\left(\frac{1}{n}\right)\Gamma\left(\frac{2}{n}\right)\Gamma\left(\frac{3}{n}\right)\cdots\Gamma\left(\frac{n-1}{n}\right) = \frac{(2\pi)^{\frac{n-1}{2}}}{\sqrt{n}}.$$

Example 6.17. *Evaluate* $\Gamma\left(-\frac{5}{2}\right)$ *using* $\Gamma\left(\frac{1}{2}\right)=\sqrt{\pi}$. [BU(H) 1996]

Solution : We have $\Gamma(n+1)=n\Gamma(n)$.

$$\therefore \quad \Gamma(n)=\frac{\Gamma(n+1)}{n}.$$

Putting $n=-\frac{5}{2}$,

$$\Gamma\left(-\frac{5}{2}\right)=\frac{\Gamma\left(-\frac{5}{2}+1\right)}{-\frac{5}{2}}=\frac{\Gamma\left(-\frac{3}{2}\right)}{-\frac{5}{2}}=\frac{\Gamma\left(-\frac{1}{2}\right)}{\left(-\frac{5}{2}\right)\left(-\frac{3}{2}\right)}=\frac{\Gamma\left(\frac{1}{2}\right)}{\left(-\frac{5}{2}\right)\left(-\frac{3}{2}\right)\left(-\frac{1}{2}\right)}$$

$$=-\frac{8}{15}\sqrt{15}\quad\left[\because\ \Gamma\left(\frac{1}{2}\right)=\sqrt{\pi}\right].$$

Example 6.18. *Find the value of*

(i) $\Gamma\left(\frac{1}{9}\right)\Gamma\left(\frac{2}{9}\right)\Gamma\left(\frac{3}{9}\right)\cdots\Gamma\left(\frac{8}{9}\right)$ (ii) $\Gamma(0.1)\Gamma(0.2)\cdots\Gamma(0.9)$.

Solution : We know that, $\Gamma\left(\frac{1}{n}\right)\Gamma\left(\frac{2}{n}\right)\Gamma\left(\frac{3}{n}\right)\cdots\Gamma\left(\frac{n-1}{n}\right)=\frac{(2\pi)^{\frac{n-1}{2}}}{\sqrt{n}}$.

(i) Putting $n=9$,

$$\therefore \quad \Gamma\left(\frac{1}{9}\right)\Gamma\left(\frac{2}{9}\right)\Gamma\left(\frac{3}{9}\right)\cdots\Gamma\left(\frac{8}{9}\right)=\frac{(2\pi)^{\frac{9-1}{2}}}{\sqrt{9}}=\frac{16}{3}\pi^4.$$

(ii) Putting $n=10$,

$$\Gamma\left(\frac{1}{10}\right)\Gamma\left(\frac{2}{10}\right)\Gamma\left(\frac{3}{10}\right)\cdots\Gamma\left(\frac{9}{10}\right)=\frac{(2\pi)^{\frac{9}{2}}}{\sqrt{10}}$$

or, $$\Gamma(0.1)\Gamma(0.2)\Gamma(0.3)\cdots\Gamma(0.9)=\frac{2^5\sqrt{5}\pi^{\frac{9}{2}}}{10}.$$

Example 6.19. *For any positive number n, show that*

$$\Gamma\left(-n+\frac{1}{2}\right)=\frac{(-1)^n2^n\sqrt{\pi}}{(2n-1)(2n-3)\cdots3\cdot1}.$$

Solution : We have $\Gamma(x)\Gamma(1-x)=\dfrac{\pi}{\sin\pi x}$ $\left[\because\ \Gamma(m)\Gamma(1-m)=\dfrac{\pi}{\sin m\pi}\right]$

and $\Gamma(1-x)=-x\Gamma(-x)$ $[\because\ \Gamma(m+1)=m\Gamma(m)$ and putting $m=-x]$

From the above two relations, we write :

$$\Gamma(x)[-x\Gamma(-x)]=\frac{\pi}{\sin\pi x}$$

or, $$\Gamma(-x)=-\frac{\pi}{x\sin\pi x}\cdot\frac{1}{\Gamma(x)}. \qquad \text{(i)}$$

Putting $x=\left(n-\frac{1}{2}\right)$ in the above equation,

$$\Gamma\left(-n+\frac{1}{2}\right)=-\frac{\pi}{\left(n-\frac{1}{2}\right)\sin\left(n-\frac{1}{2}\right)\pi}\cdot\frac{1}{\Gamma\left(n-\frac{1}{2}\right)}$$

or, $$\Gamma\left(-n+\frac{1}{2}\right) = -\frac{\pi}{\sin\left(n-\frac{1}{2}\right)\pi}\cdot\frac{1}{\Gamma\left(n+\frac{1}{2}\right)}. \quad \text{(ii)}$$

Again, $$\Gamma\left(n+\frac{1}{2}\right) = \left(n-\frac{1}{2}\right)\left(n-\frac{3}{2}\right)\cdots\frac{3}{2}\cdot\frac{1}{2}\cdot\Gamma\left(\frac{1}{2}\right) = \frac{(2n-1)(2n-3)\cdots3\cdot1}{2^n}\sqrt{\pi}. \quad \text{(iii)}$$

From Eqns. (ii) and (iii),

$$\Gamma\left(-n+\frac{1}{2}\right) = \frac{(-1)^n 2^n\sqrt{\pi}}{(2n-1)(2n-3)\cdots3\cdot1}. \quad \text{(iv)}$$

Special Cases : For, $n = 1$, $\Gamma\left(-\frac{1}{2}\right) = -2\sqrt{\pi}$; $n = 2$, $\Gamma\left(-\frac{3}{2}\right) = \frac{4}{3}\sqrt{\pi}$; $n = 3$, $\Gamma\left(-\frac{5}{2}\right) = -\frac{8}{15}\sqrt{\pi}$, etc.

Exercises

6(A) : Objective Type Questions

Select the correct answer

6.1 The Beta function is defined as $\beta(m,n) = \int_0^1 x^{m-1}(1-x)^n dx$, where

(a) $m > 0, n < 0$ (b) $m = 0, n > 0$ (c) $m > 0, n > 0$ (d) $m < 0, n < 0$.

6.2 The Gamma function is defined as $\Gamma(n) = \int_0^\infty e^{-x}x^{n-1}dx$, where

(a) $n < 0$ (b) $n > 0$ (c) $n > -1$ (d) $n = 1$.

6.3 The complementary error function is defined as

(a) $erfc(x) = 1 - erf(x)$ (b) $erfc(x) = 1 + erf(x)$

(c) $erf(x) = -erf(x)$ (d) $erfc(x) = erf(-x)$.

6.4 If $m > 0$, $n > 0$ then

(a) $\beta(m,n) = -\beta(n,m)$ (b) $\beta(m,n) = \beta(n,m)$

(c) $\beta(m,n) = -\beta(n+m)$ (d) $\beta(m,n) = \beta(-m,n)$.

6.5 If $m > 0$, $n > 0$, then $\beta(m,n)$ is given by

(a) $\frac{(n-1)!(m-1)!}{(m+n+1)!}$ (b) $\frac{(n-1)!(m+1)!}{(m+n-1)!}$ (c) $\frac{(n+1)!(m+1)!}{(m+n-1)!}$ (d) $\frac{(n-1)!(m-1)!}{(m+n-1)!}$.

6.6 $\frac{\beta(m,n)}{\beta\left(\frac{1}{2},m\right)}$ is equal to

(a) 2^{2m-1} (b) $\frac{1}{2^{2m-1}}$ (c) $\frac{1}{2^{-m-1}}$ (d) 2^{-2m+1}.

6.7 $\Gamma(n) = \int_0^1 e^{-x}x^{n-1}dx$ converges at 0 for

(a) $n = 0$ (b) $n = 1$ (c) $n > 0$ (d) $n < 0$.

6.8 $\Gamma(n) = \int_0^1 e^{-x}x^{n-1}dx$ converges at ∞ for

(a) $n = \infty$ **(b)** $n = -1$ **(c)** $n > 0$ **(d)** $n < 0$.

6.9 $\Gamma(n+1)$ is equal to

(a) $\Gamma(n-1)$ **(b)** $n\Gamma(n-1)$ **(c)** $n\Gamma(n+1)$ **(d)** $n\Gamma(n)$.

6.10 $\Gamma\left(\frac{1}{2}\right)$ is equal to

(a) π **(b)** $\sqrt{\pi}$ **(c)** $\frac{1}{\sqrt{\pi}}$ **(d)** $\frac{1}{\pi}$.

6.11 $\beta(m,n)$ is equal to

(a) $\frac{\Gamma(m)\Gamma(n)}{\Gamma(m+n)}$ **(b)** $\frac{\Gamma(-m)\Gamma(n)}{\Gamma(m+n)}$ **(c)** $\frac{\Gamma(-m)\Gamma(-n)}{\Gamma(m+n)}$ **(d)** $\frac{\Gamma(m)\Gamma(n)}{\Gamma(m+n)}$.

6.12 $erf(0)$ is equal to

(a) ∞ **(b)** 0 **(c)** 1 **(d)** $-\infty$.

6.13 $erf(\infty)$ is equal to

(a) 0 **(b)** 1 **(c)** -1 **(d)** ∞.

6.14 When n is very large, $n!$ is given by

(a) $\sqrt{2n\pi}\, n^{-n}e^{n}$ **(b)** $\sqrt{2n\pi}\, n^{n}e^{n}$ **(c)** $\sqrt{2n\pi}\, n^{n}e^{-n}$ **(d)** $2n\pi n^{n}e^{-n}$.

6.15 $\int^{\frac{\pi}{2}} \cos\theta d\theta$ is equal to

(a) -1 **(b)** 1 **(c)** 0 **(d)** ∞.

6.16 $\Gamma\left(-\frac{3}{2}\right)$ is equal to,

(a) $\frac{3\pi}{4}$ **(b)** $\frac{-3\pi}{4}$ **(c)** $\frac{3\pi}{4}\sqrt{\pi}$ **(d)** $\frac{3\pi}{2}\sqrt{\pi}$.

6.17 $\Gamma\left(-\frac{5}{2}\right)$ is equal to

(a) $\frac{8}{15}\sqrt{\pi}$ **(b)** $-\frac{8}{15}\sqrt{\pi}$ **(c)** $\frac{8\pi}{15}$ **(d)** $\frac{3\sqrt{\pi}}{15}$.

6.18 The value of $\int_1^{\infty} \frac{1}{1+x^2}dx$ is

(a) 1 **(b)** $\frac{\pi}{2}$ **(c)** $\frac{\pi}{4}$ **(d)** 0.

6.19 The value of $\int_{-\infty}^{0} \frac{dx}{(1-x)^2}$ is

(a) 1 **(b)** 2 **(c)** $\frac{1}{2}$ **(d)** $\frac{1}{3}$.

6.20 The value of $\int_{-\infty}^{\infty} \frac{dx}{a^2+x^2}$ $(a>0)$ is

(a) π **(b)** $\frac{\pi}{2}$ **(c)** $\frac{\pi}{a}$ **(d)** $-\frac{\pi}{a}$.

6.21 The value of $\frac{\Gamma\left(\frac{7}{2}\right)}{\Gamma\left(\frac{1}{2}\right)}$ is equal to

(a) $\frac{3}{4}$ **(b)** $\frac{3}{8}$ **(c)** $\frac{15}{a}$ **(d)** none of these.

6.22 The value of $\int_0^{\infty} \frac{\sin t}{t} dt$ is equal to

(a) $\frac{\pi}{3}$ **(b)** $\frac{\pi}{6}$ **(c)** $\frac{\pi}{4}$ **(d)** $\frac{\pi}{2}$. **[NBUT 2009]**

6.23 $\beta\left(\frac{1}{2},\frac{1}{2}\right)$ is equal to

(a) $\sqrt{\pi}$ **(b)** $\frac{\sqrt{\pi}}{2}$ **(c)** π **(d)** $\frac{\pi}{2}$.

6.24 $\int_0^{\infty} e^{-x^2} dx$ is equal to

(a) π **(b)** $\frac{\pi}{2}$ **(c)** $\sqrt{\pi}$ **(d)** $\frac{\sqrt{\pi}}{2}$.

6.25 $\beta\left(\frac{3}{2},2\right)$ is equal to

(a) $\frac{4}{15}\pi$ **(b)** $\frac{4}{15}$ **(c)** $\frac{4}{15}\sqrt{\pi}$ **(d)** $\frac{15}{4}\pi$.

6.26 $\Gamma\left(\frac{1}{2}\right)\Gamma\left(\frac{3}{2}\right)\Gamma\left(\frac{5}{2}\right)\Gamma\left(\frac{7}{2}\right)$ is equal to

(a) $\frac{45}{32}\pi^2$ **(b)** $\frac{45}{32}\pi$ **(c)** $\frac{45}{64}\pi$ **(d)** $\frac{45}{64}\pi^2$.

6.27 $\Gamma(3,5)$ is equal to

(a) $\sqrt{\pi}$ **(b)** $\frac{15}{8}\sqrt{\pi}$ **(c)** $\frac{15}{16}\sqrt{\pi}$ **(d)** none of these.

6.28 The integral $\int_0^2 \frac{dx}{2-x}$

(a) does not exist **(b)** exists **(c)** none of these.

6.29 The integral $\int_0^{\infty} \frac{dx}{x^n}$ $(a>0)$ exists, if

(a) $n>1$ **(b)** $n=1$ **(c)** $n<1$ **(d)** none of these.

6.30 If $\underset{x\to\infty}{\text{Lt}}\ x^n f(x)$ exists, then $\int_0^{\infty} f(x)dx$ converges absolutely, if

(a) $n<1$ **(b)** $n>1$ **(c)** $n=1$ **(d)** none of these.

6.31 The integral $\int_0^{\frac{\pi}{2}} \frac{\sqrt{x}}{\sin x} dx$

(a) diverges to $+\infty$ (b) converges (c) oscillatory (d) none of these.

Answers

6.1 (a)	**6.2** (b)	**6.3** (a)	**6.4** (b)	**6.5** (d)	**6.6** (b)	**6.7** (c)	**6.8** (c)	**6.9** (d)
6.10 (b)	**6.11** (d)	**6.12** (b)	**6.13** (b)	**6.14** (c).	**6.15** (b)	**6.16** (c)	**6.17** (b)	**6.18** (c)
6.19 (a)	**6.20** (c)	**6.21** (c)	**6.22** (d)	**6.23** (c)	**6.24** (d)	**6.25** (b)	**6.26** (d)	**6.27** (b)
6.28 (a)	**6.29** (a)	**6.30** (c)	**6.31** (b).					

6(B) : Short Answer Type Questions

6.1 Define **(a)** Beta function, **(b)** Gamma function and **(c)** Error function.

6.2 Using $\Gamma\left(\frac{1}{2}\right) = \sqrt{\pi}$, evaluate $\Gamma\left(-\frac{5}{2}\right)$. **[BU(H) 1996]**

6.3 Define Beta function $\beta(m,n)$. Show that $\beta(m,n) = \beta(n,m)$. **[BU(H) 1996]**

6.4 Define the gamma function $\Gamma(z)$, where z is not necessarily an integer and hence, find the recurrence formula $\Gamma(z+1) = z\Gamma(z)$. **[BU(H) 1998]**

6.5 Prove that, $\beta(m,n) = \frac{\Gamma(m)\Gamma(n)}{\Gamma(m+n)}$. **[BU(H) 1993]**

6.6 The gamma function may be defined as $\Gamma(n) = \int_0^{\infty} e^{-x}x^{n-1}dx$.

Is there any restriction on the value of n? Why? **[BU(H) 2002]**

6.7 Show that, $\Gamma(n+1) = n\Gamma(n)$. **[BU(H) 2002]**

6.8 Can you define Gamma function as $\Gamma(n) = \int_0^{\infty} e^{-x}x^{n-1}dx$ for $x < 0$? **[BU(H) 2003]**

6.9 Express $\int_0^{\frac{\pi}{2}} \sqrt{\cot\theta}\, d\theta$ in terms of Gamma function.

6.10 Show that, $\frac{\beta(p,q+1)}{q} = \frac{\beta(p+1,q)}{p} = \frac{\beta(p,q)}{p+q}$.

6.11 Prove that, $erf(\infty) = 1$.

6(C) : Broad Answer Type Questions

6.1 **(a)** Define Beta function and Gamma function.

(b) Show that, $\beta(m,n) = \frac{\Gamma(m)\Gamma(n)}{\Gamma(m+n)}$ for $m > 0$, $n > 0$. **[BU(H) 2001]**

(c) $\Gamma(m)\Gamma\left(m+\frac{1}{2}\right) = \frac{\Gamma(\pi)}{2^{2m-1}}\Gamma(2m)$.

6.2 **(a)** Prove that the Gamma function $\Gamma(z)$ satisfies the relation, $\Gamma(z)\Gamma(1-z) = \frac{\pi}{\sin \pi z}$.

(b) Show, using Beta function, that, $\int_t^z = \frac{dx}{(z-x)^{1-\alpha}(x-t)^{\alpha}} = \frac{\pi}{\sin \pi x}$; $0 < \alpha < 1$.

6.3 Compute (a) $\beta(1,2)$, (b) $\beta(2,4)$, (c) $\beta(1.9,0.6)$ **[Ans:** (a) 0.5, (b) 0.05, (c) 1.077**]**

6.4 (a) Show that, $\displaystyle\int_0^1 \frac{dx}{\sqrt{1-x^n}} = \frac{\sqrt{\pi}}{n}\frac{\Gamma\left(\frac{1}{n}\right)}{\Gamma\left(\frac{1}{n}+\frac{1}{2}\right)}$.

(b) Evaluate $\displaystyle\int_{-1}^{+1} (1+x)^{2m-1}(1-x)^{2n-1}(1+x^2)^{-m-n}$; $m>0$, $n>0$.

6.5 (a) Show that, $\displaystyle\int_0^{\frac{\pi}{2}} \cos^{2n}\phi d\phi = \frac{\pi}{2}\frac{1\cdot3\cdot5\cdots(2n-1)}{2\cdot4\cdot6\cdots(2n)}$.

$\left[\text{Given, } \dfrac{\Gamma(m)\Gamma(n)}{\Gamma(m+n)} = \beta(m,n) = 2\displaystyle\int_0^{\frac{\pi}{2}} \sin^{2m-1}\phi\cos^{2m-1}\phi d\phi\right]$ **[BU(H) 2005]**

(b) Given, $\Gamma\left(\dfrac{1}{2}\right) = \sqrt{\pi}$, prove that, $\Gamma\left(-\dfrac{3}{2}\right) = \dfrac{4\sqrt{\pi}}{3}$. **[BU(H) 2001]**

6.6 (a) Prove that, $2\cdot5\cdot8\cdots(3n-1) = \dfrac{3^n\Gamma\left(n+\frac{2}{3}\right)}{3^n\Gamma\left(\frac{2}{3}\right)}$. **[BU(H) 2002]**

(b) Show that, $\displaystyle\int_0^{\frac{\pi}{2}} (\cos\theta)^r d\theta = \frac{\Gamma\left(\frac{r+1}{2}\right)}{\Gamma\left(\frac{r}{2}+1\right)}\frac{\sqrt{\pi}}{2}$; $r>-1$. **[BU(H) 1997]**

6.7 (a) Prove that, $\beta(m,n) = \displaystyle\int_0^1 \frac{x^{m-1}+x^{n-1}}{(1+x)^{m+n}}dx$.

(b) Show that, $\displaystyle\int_0^{\frac{\pi}{2}} \sqrt{\sec\theta}d\theta = \frac{\sqrt{\pi}}{\Gamma\left(\frac{3}{4}\right)}$.

6.8 (a) Express $\displaystyle\int_0^{\frac{\pi}{2}} \sqrt{\tan\theta}d\theta$ in terms of Gamma function. **[BU(H) 2000]**

(b) Prove that, $\displaystyle\int_0^{\frac{\pi}{2}} \sqrt{\sin\theta}d\theta = \int_0^{\frac{\pi}{2}} \frac{d\theta}{\sqrt{\sin\theta}} = \pi$. **[BU(H) 1995]**

6.9 Show that (a) $\displaystyle\int_0^{\frac{\pi}{2}} \sin^2\phi+\cos^2\phi d\phi = \frac{\pi}{16}$ (b) $\displaystyle\int_0^{\frac{\pi}{2}} \sin^5\phi\cos^5\phi d\phi = \frac{1}{120}$.

6.10 Prove that (a) $\displaystyle\int_0^{\frac{\pi}{2}} (\cot\theta)^{\frac{1}{2}}d\theta = \frac{\pi}{\sqrt{2}}$ (b) $\displaystyle\int_0^{\frac{\pi}{2}} \cos^{\frac{4}{3}}\theta\sin^{\frac{5}{3}}\theta d\theta = \frac{2\left(\frac{1}{6}!\frac{1}{3}\right)}{3\sqrt{\pi}!}$.

6.11 (a) Write the integral $\displaystyle\int_0^1 \frac{x^3}{\sqrt{1-x}}dx$ in the form of Beta function and hence, evaluate it. **[BU(H) 1993]**

(b) Define the Gamma function $\Gamma(z)$, where z is not necessarily an integer. Find the value of $\Gamma(1)$ from your definition and show that when $z=n$, a positive integer, $\Gamma(n)=(n-1)!$.

Does the definition of $\Gamma(z)$ hold for $z=0,-1,-2,\ldots$, etc.? **[BU(H) 1998]**

6.12 (a) Define the Gamma function $\Gamma(n)$. Evaluate $\Gamma\left(-\dfrac{5}{2}\right)$, using $\Gamma\left(\dfrac{1}{2}\right) = \sqrt{\pi}$.

(b) Define the Beta function $\beta(m,n)$. Show that, $\beta(m,n)=\beta(n,m)$.

(c) Show that, $\beta(m,n)$ can be expressed as $\int_0^\infty \frac{y^{n-1}dy}{(1+y)^{m+n}}$.

(d) Prove that, $\int_0^1 x^m(\log_e x)^n dx = \frac{(-1)^n n!}{(m+1)^{m+1}}$, where n is a positive integer and $m > -1$. **[BU(H) 1996]**

6.13 (a) Prove that, $\int_0^1 \frac{35x^3}{38(1-x)}dx = 1$. (b) Evaluate $\int_0^\infty \frac{x^a}{x^x}dx$.

6.14 (a) Show that, $\Gamma\left(\frac{1}{m}\right)\Gamma\left(\frac{2}{m}\right)\cdots\Gamma\left(\frac{m-1}{m}\right) = \left\{\frac{(2\pi)^{m-1}}{m}\right\}^{\frac{1}{2}}$.

(b) Prove that, $\Gamma(a+1,x) = a\Gamma(a,x) + x^a e^{-x}$.

6.15 (a) Prove that $\iiint\limits_R x^{a-1}y^{b-1}z^{c-1}dxdydz = \frac{\Gamma(a)\Gamma(b)\Gamma(c)}{\Gamma(a+b+c+1)}$.

(b) Show that, $\int_0^1 (1-x^n)^{\frac{1}{n}}dx = \frac{1}{n}\frac{\{\Gamma(\frac{1}{n})\}^2}{2\Gamma(\frac{2}{n})}$.

6.16 (a) Evaluate $\int_0^1 x^2\left(\ln\frac{l}{x}\right)^3 dx$ in terms of gamma function. **[BU(H) 2013]**

(b) Evaluate $\int_0^\infty x^{\frac{5}{4}}e^{-\sqrt{x}}dx$ by using gamma function. **[BU(H) 2012]**

6.17 The beta function is defined as $\beta(m,n) = \int_0^1 x^{m-1}(1-x)^{n-1}dx$ where $m > 0$, $n > 0$.

From this prove that,

(i) $\beta(m,n) = \frac{m-1}{m+n-1}\beta(m-1,n)$ (ii) $\beta(m,n) = \int_0^1 \frac{y^{m-1}+y^{n-1}}{(1+y)^{m+n}}dy$ **[BU(H) 2010]**

6.18 Establish Dirichlet's integral,

$$\iiint\limits_V x^{l-1}y^{m-1}z^{n-1}dx\,dy\,dz = \frac{\Gamma(l)\Gamma(m)\Gamma(n)}{\Gamma(l+m+n+1)},$$

where V is the region $x \geq 0$, $y \geq 0$, $z \geq 0$ and $x+y+z \leq 1$. **[BU(H) 2009]**

6.19 (a) Evaluate $\int_0^\infty e^{-ax}x^{m-1}\sin bx\,dx$ in terms of gamma function.

Show that $\Gamma(n) = \int_0^1 \left[\ln\left(\frac{1}{y}\right)\right]^{n-1} dy$, $(n > 0)$.

(b) Prove that, $\Gamma(n+1) = n\Gamma(n)$. **[BU(H) 2008]**

6.20 (a) Prove the identity $\frac{\sqrt{m}\sqrt{n}}{\sqrt{m+n}} = \beta(m,n)$, where all the symbols have their usual meaning.

(b) Show that $\beta(m,n) = 2\int_0^{\frac{\pi}{2}} (\sin\theta)^{2m-1}(\cos\theta)^{2n-1}d\theta$.

6.21 **(a)** Evaluate the value of $\dfrac{\Gamma(3)\Gamma\left(\frac{3}{2}\right)}{\Gamma\left(\frac{9}{2}\right)}$.

(b) Show that $\Gamma\left(\dfrac{1}{2}\right) = \sqrt{\dfrac{\pi}{2}}$.

6.22 **(a)** Establish that $\beta(m,n) = \dfrac{\Gamma(m)\Gamma(n)}{\Gamma(m+n)}$; $m, n > 0$.

(b) Evaluate the value of $\Gamma\left(-\dfrac{5}{2}\right)$, given that $\Gamma\left(\dfrac{1}{2}\right) = \sqrt{\pi}$. **[Assam U(Major) 2014]**

(c) Define Gamma function and establish $\Gamma(n+1) = n\Gamma(n)$. **[Assam U(Major) 2014]**

Chapter 7 Fourier Series

7.1 Introduction

Fourier analysis is an important mathematical technique which is widely used in the diverse fields of physics, viz., **acoustics, optics, electricity** and **quantum mechanics** as well as in **engineering**, viz., **electronics engineering, communication science, electrical engineering**, etc. **Fourier synthesis** deals with the addition of a number of sinusoidal oscillations to get the resultant whereas **Fourier analysis** determines the sine and cosine oscillations which are compounded to obtain a given resultant.

The French mathematician **Jaques Fourier** (1768–1830) proposed a systematic study of the topic in his monograph "**Theorie Analytic de la Chaleur**". In any branch of science and technology involving sine and cosine oscillations, the use of Fourier analysis is a must.

7.2 Periodic Functions

A function $f(x)$ is said to be periodic, if for a constant number $T > 0$,

$$f(x+T) = f(x) \tag{7.1}$$

for any value of x. Here, T is called the period of the function. It is obvious that for a periodic function $f(x)$ of period T,

$$\left.\begin{aligned} f(x) &= f(x+T) = f(x+2T) = \cdots \\ \text{and} \quad f(x) &= f(x-T) = f(x-2T) = \cdots \end{aligned}\right\} \tag{7.2}$$

For example, $\sin x = \sin(x+2\pi) = \sin(x+4x) = \cdots$ so $\sin x$ is a periodic function having period 2π. $f(x) = \sin x$ is called a sinusoidal periodic function and is depicted in Fig. 7.1.

The integral of a periodic function $f(x)$ does not change its value when the range of integration is equal to its period, i.e.,

$$\int_a^b f(x)dx = \int_{a+T}^{b+T} f(x)dx \tag{7.3}$$

for arbitrary values of a and b.

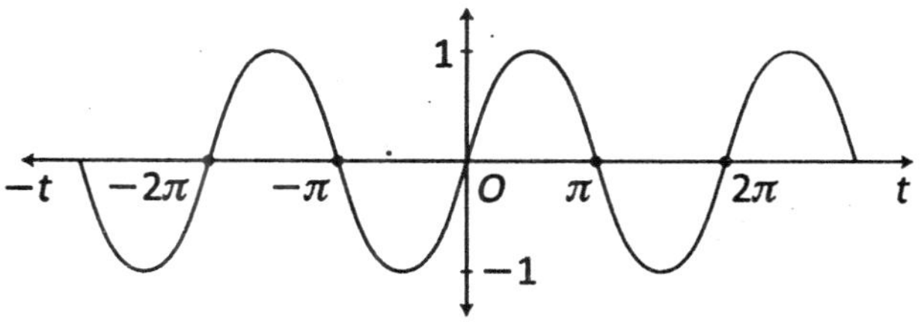

Fig. 7.1: Graphical representation of $f(x) = \sin x$

7.3 Fourier Series

Statement of Fourier's Theorem

Any single valued periodic function $f(x)$ defined in the closed interval $(-\pi, \pi)$ may be expressed as a series of sines and cosines in the form :

$$f(x) = \frac{a_0}{2} + \sum_{n=1}^{\infty}(a_n \cos nx + b_n \sin nx); \tag{7.4}$$

where the expansion coefficients a and b are determined by Euler's formulae as :

$$a_n = \frac{1}{\pi}\int_{-\pi}^{+\pi} f(x)\cos nx\, dx, \quad (n = 0, 1, 2) \tag{7.5}$$

and $$b_n = \frac{1}{\pi}\int_{-\pi}^{+\pi} f(x)\sin nx\, dx, \quad (n = 1, 2, 3). \tag{7.6}$$

The integrals of Eqns. (7.5) and (7.6) are known as **Fourier integrals.**

7.4 Dirichlet's Theorem and Dirichlet's Conditions

Dirichlet's Theorem establishes the convergence of Fourier series for a very large class of functions.

Statement : *If a function $f(x)$ is well defined and bounded in the interval $-\pi < x < \pi$ and has only a finite number of maxima and minima and a finite number of discontinuities and also for other values of x, it satisfies the periodicity condition,*

$$f(x + 2\pi) = f(x),$$

then the function $f(x)$ may be expanded in a Fourier series which converges to

$$\frac{1}{2}[f(x^+) + f(x^-)]$$

at every value of x (and hence, it converges to $f(x)$ at points where $f(x)$ is continuous).

Dirichlet's Conditions

The conditions imposed on $f(x)$ in Dirichlet Theorem are called **Dirichlet's conditions**. The conditions are as below :

(i) *The function must be well defined and bounded in the interval $-\pi < x < \pi$,*

(ii) *The function must be continuous or piecewise continuous, i.e., the function may have only a finite number of discontinuities and only a finite number of maxima and minima,*

(iii) *The function $f(x)$ must satisfy the periodicity condition, $f(x + 2\pi) = f(x)$.*

7.5 Functions Having Points of Discontinuities

The function $f(x)$ may have a finite number of points of finite discontinuity, i.e., its graph may consist of a finite number of different curves with different equations. Inspite of this, a function can be expressed as a Fourier series.

Example *If in the interval $(\alpha, \alpha + 2\pi)$, $f(x)$ is defined by,*

$$\begin{aligned} f(x) &= \phi(x), \quad \alpha < x < c \\ &= \psi(x), \quad c < x < \alpha + 2\pi \end{aligned}$$

i.e., c is the point of discontinuity, then

$$a_0 = \frac{1}{\pi}\left[\int_{\alpha}^{c} \phi(x)dx + \int_{c}^{\alpha+2\pi} \psi(x)dx\right] \tag{7.7}$$

$$a_n = \frac{1}{\pi}\left[\int_{\alpha}^{c} \phi(x)\cos nx dx + \int_{c}^{\alpha+2\pi} \psi(x)\cos nx dx\right] \tag{7.8}$$

and $$b_n = \frac{1}{\pi}\left[\int_{\alpha}^{c} \phi(x)\sin nx dx + \int_{c}^{\alpha+2\pi} \psi(x)\sin nx dx\right]. \tag{7.9}$$

At the point c of finite discontinuity there is a finite jump in the graph (Fig. 7.2). Here, both the limit on the left, i.e., $f(c-0)$ and the limit on the right, i.e., $f(c+0)$ are different and exist. At such point c, the value of $f(x)$ according to Fourier series is given by

$$f(x) = \frac{1}{2}[f(c-0) + f(c+0)]. \tag{7.10}$$

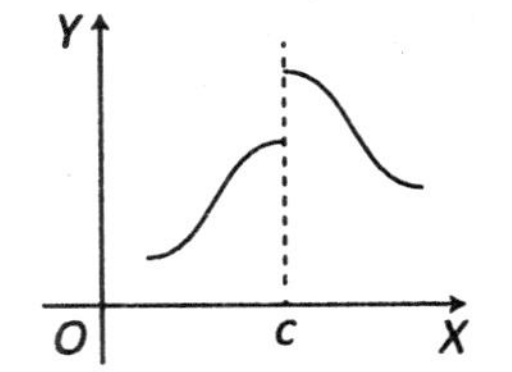

Fig. 7.2: Discontinuity at c

7.6 Evaluation of the Fourier's Coefficients

The Fourier coefficients a_0, a_n and b_n in equation (7.4) may be evaluated. In this connections the following results will be used :

1. $$\int_{-\pi}^{\pi} \cos nx \cos mx\, dx = \begin{cases} 0, & \text{for } n \neq m \\ 2\pi, & \text{for } n = m = 0 \\ \pi, & \text{for } n = m > 0 \end{cases} \tag{7.11}$$

2. $$\int_{-\pi}^{\pi} \sin nx \sin mx\, dx = \begin{cases} 0, & \text{for } n \neq m \\ 0, & \text{for } n = m = 0 \\ \pi, & \text{for } n = m > 0 \end{cases} \tag{7.12}$$

3. $$\int_{-\pi}^{\pi} \sin nx \cos mx = 0, \text{ for all } n \text{ and } m \tag{7.13}$$

4. $$\int_{-\pi}^{\pi} \sin nx\, dx = 0, \text{ for all } n \tag{7.14}$$

5. $$\int_{-\pi}^{\pi} \cos nx\, dx = \begin{cases} 0, & \text{for } n > 0 \\ 2\pi, & \text{for } n = 0 \end{cases} \tag{7.15}$$

Evaluation of a_0

We have the Fourier series in the interval $-\pi \leq x \leq \pi$ as

$$f(x) = \frac{a_0}{2} + \sum_{n=1}^{\infty}(a_n \cos nx + b_n \sin nx). \tag{7.16}$$

Integrating both sides of the above equation with respect to x in the interval $(-\pi, \pi)$, we get

$$\int_{-\pi}^{\pi} f(x)dx = \frac{1}{2}a_0 \int_{-\infty}^{\pi} dx + \sum_{n=1}^{\infty} a_n \int_{-\pi}^{\pi} \cos nx\, dx + \sum_{n=1}^{\infty} b_n \int_{-\pi}^{\pi} \sin nx\, dx = \frac{1}{2}a_0 \cdot 2\pi = a_0\pi.$$

$$\therefore \quad a_0 = \frac{1}{\pi}\int_{-\pi}^{+\pi} f(x)dx. \tag{7.17}$$

Evaluation of a_n

Multiplying both sides of Eqn. (7.16), by $\cos mx$ and then integrating within $(-\pi, \pi)$, we get

$$\int_{-\pi}^{\pi} f(x)\cos mx\, dx = \frac{1}{2}a_0 \int_{-\pi}^{\pi} \cos mx\, dx + \sum_{n=1}^{\infty} a_n \int_{-\pi}^{\pi} \cos nx \cos mx\, dx$$
$$+ \sum_{n=1}^{\infty} b_n \int_{-\pi}^{\pi} \sin nx \cos mx\, dx$$
$$= 0 + a_n \int_{-\pi}^{\pi} \cos^2 nx\, dx = a_n\pi.$$

[$\because$ the first integral is zero. Within summation, the first integral is zero except $m = n$ and the second integral in the summation also vanishes even for $m = n$.]

$$\therefore \quad a_n = \frac{1}{\pi}\int_{-\pi}^{\pi} f(x)\cos nx\, dx. \tag{7.18}$$

Evaluation of b_n

Multiplying both sides of Eqn. (7.16), by $\sin mx$ and then integrating within $(-\pi, \pi)$, we get

$$\int_{-\pi}^{\pi} f(x)\sin mx\, dx = \frac{a_0}{2}\int_{-\pi}^{\pi} \sin mx\, dx + \sum_{n=1}^{\infty} a_n \int_{-\pi}^{\pi} \cos nx \sin mx\, dx + \sum_{n=1}^{\infty} b_n \int_{-\pi}^{\pi} \sin nx \sin mx\, dx$$

$$= 0 + 0 + \pi b_n.$$

$$\therefore \quad b_n = \frac{1}{\pi}\int_{-\pi}^{\pi} f(x)\sin nx\, dx. \tag{7.19}$$

N.B. As the RHS of Eqn. (7.4) is periodic the interval $(-\pi, \pi)$ may be replaced by any of the interval of length 2π, viz., $(0, 2\pi)$ or $(\alpha, \alpha + 2\pi)$.

7.7 Even and Odd Functions

Even function symmetry : A function $f(x)$ is said to be an **even function**, if

$$f(x) = f(-x) \tag{7.20}$$

in the range $-\pi < x \leq \pi$.

Example $|x|$, $\cos x$, x^2, etc.

Graphically these functions are symmetrical about the Y-axis as shown in Fig. (7.3). Moreover,

$$\int_{-\pi}^{\pi} f(x)dx = 2\int_{0}^{\pi} f(x)dx, \tag{7.21}$$

when $f(x)$ is defined in $-\pi < x < \pi$.

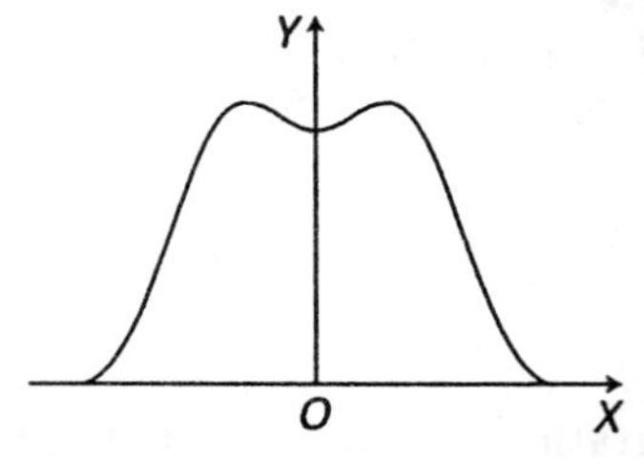

Fig. 7.3: Even function (Symmetric)

The Fourier coefficients in this case are :

$$a_0 = \frac{1}{\pi}\int_{-\pi}^{\pi} f(x)dx = \frac{2}{\pi}\int_{0}^{\pi} f(x)dx \tag{7.22}$$

$$a_n = \frac{1}{\pi}\int_{-\pi}^{\pi} f(x)\cos nx\, dx = \frac{2}{\pi}\int_{0}^{\pi} f(x)\cos nx\, dx \tag{7.23}$$

$$b_n = \frac{1}{\pi}\int_{-\pi}^{\pi} f(x)\sin nx\, dx = \frac{2}{\pi}\int_{0}^{\pi} f(x)\sin nx\, dx. \tag{7.24}$$

If $f(x)$ be an even function, its Fourier series reduces to cosine series with the constant $\frac{a_0}{2}$, i.e.,

$$f(x) = \frac{a_0}{2} + \sum_{n=1}^{\infty} a_n \cos nx, \tag{7.25}$$

all sine terms vanish.

Odd function symmetry : A function $f(x)$ is said to be an **odd function**, if

$$f(x) = -f(-x) \tag{7.26}$$

in the interval $-\pi < x < \pi$.

Example x, x^3, $\sin x$, *etc.*

The graph of this type of function is symmetrical about the origin as shown in Fig. (7.4). Moreover,

$$\int_{-\pi}^{\pi} f(x)dx = 0. \tag{7.27}$$

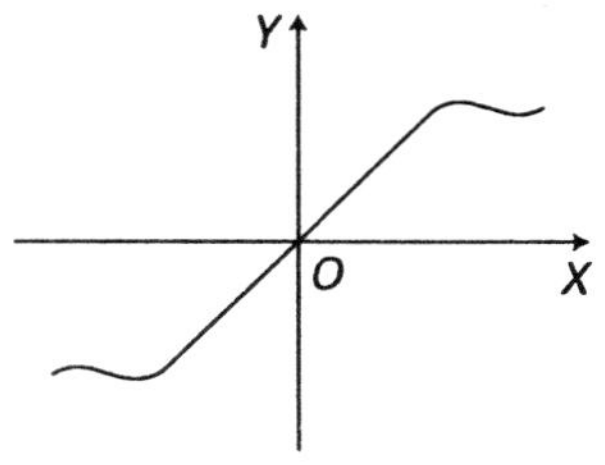

Fig. 7.4: Odd function (Antisymmetric)

The Fourier coefficients are :

$$a_0 = \frac{1}{\pi}\int_{-\pi}^{\pi} f(x)dx = 0 \tag{7.28}$$

$$a_n = \frac{1}{\pi}\int_{-\pi}^{\pi} f(x)\cos nx\, dx = 0 \textit{ for all } n \tag{7.29}$$

$$b_n = \frac{1}{\pi}\int_{-\pi}^{\pi} f(x)\sin nx\, dx = \frac{2}{\pi}\int_{0}^{\pi} f(x)\sin nx\, dx \quad (n = 1, 2, 3, \ldots). \tag{7.30}$$

Thus, if $f(x)$ is an odd function, the Fourier series reduces to sine series, i.e.,

$$f(x) = \sum_{n=1}^{\infty} b_n \sin nx, \tag{7.31}$$

all cosine terms and constant vanishes.

7.8 Complex Form of Fourier Series

The Fourier expansion of a single valued function defined in the interval $(-\pi, \pi)$ can be expressed as

$$f(x) = \frac{a_0}{2} + \sum_{n=1}^{\infty}(a_n \cos nx + b_n \sin nx), \tag{7.32}$$

where $$a_n = \frac{1}{\pi}\int_{-\pi}^{+\pi} f(x)\cos nx\, dx \quad (n = 0, 1, 2, \ldots) \tag{7.33}$$

and $$b_n = \frac{1}{\pi}\int_{-\pi}^{+\pi} f(x)\sin nx\, dx \quad (n = 1, 2, 3, \ldots). \tag{7.34}$$

But
$$\left.\begin{aligned} \cos nx &= \frac{1}{2}(e^{inx} + e^{-inx}) \\ \sin nx &= \frac{i}{2}(e^{-inx} - e^{inx}) \end{aligned}\right\} \tag{7.35}$$

Now substituting equations (7.35) into equation (7.32), we get

$$f(x) = \frac{a_0}{2} + \sum_{n=1}^{\infty}\left[\frac{a_n - ib_n}{2}e^{inx} + \frac{a_n + ib_n}{2}e^{-inx}\right]$$

or, $$f(x)=\sum_{n=-\infty}^{\infty} C_n e^{inx}, \tag{7.36}$$

where $$\left.\begin{aligned} C_0 &= \frac{1}{2}a_0 \\ C_n &= \frac{1}{2}(a_n - ib_n) \\ C_{-n} &= \frac{1}{2}(a_n + ib_n) \end{aligned}\right\} \tag{7.37}$$

Eqn. (7.36) is the **complex form of Fourier series**. C_0, C_n and C_{-n} are called **complex coefficients of Fourier series** for the function $f(x)$.

Multiplying both sides of Eqn. (7.36) by e^{-inx} and integrating the result with respect to x from $-\pi$ to π, we get

$$\int_{-\infty}^{+\infty} f(x)e^{-inx}dx = \sum_{n=-\infty}^{\infty} C_n \int_{-\pi}^{+\pi} e^{ix(n-m)}dx = \sum_{n=-\infty}^{\infty} C_n \times 2\pi\delta_{mn} = 2\pi C_n,$$

where $$\delta_{mn} = \begin{cases} 1, & \text{when } n=m \\ 0, & \text{when } n \neq m. \end{cases}$$

$$\therefore \quad C_n = \frac{1}{2\pi}\int_{-\pi}^{\pi} f(x)e^{-inx}dx. \tag{7.38}$$

Eqn. (7.38) represents the general formula for obtaining the coefficients C_n.

7.9 Change of Interval

Let the function $f(x)$ be defined in the interval $(-l, l)$. We want to change the function to the period of 2π so that we can use the formulae of a_n, b_n.

$\because$ $2l$ is the interval for the variable x,

$\therefore$ 1 is the interval for the variable $\frac{x}{2l}$.

$\therefore$ 2π is the interval for the variable $\frac{x}{2l}2\pi = \frac{\pi x}{l}$.

Now we put $z = \frac{\pi x}{l}$ or, $x = \frac{zl}{\pi}$.

Thus, the function $f(x)$ of period $2l$ is transformed to the function $f\left(\frac{lz}{\pi}\right)$ or $F(z)$ of period 2π.

Now, we can expand $F(z)$ in the Fourier series,

$$F(z) = f\left(\frac{lz}{\pi}\right) = \frac{a_0}{2} + \sum_{n=1}^{\infty} a_n \cos nz + \sum_{n=1}^{\infty} b_n \sin nz, \tag{7.39}$$

where $$a_0 = \frac{1}{\pi}\int_0^{2\pi} F(z)dz = \frac{1}{\pi}\int_0^{2\pi} f\left(\frac{lz}{\pi}\right)dz = \frac{1}{\pi}\int_0^{2l} f(x)d\left(\frac{\pi x}{l}\right)$$

$$= \frac{1}{l}\int_0^{2l} f(x)dx \quad \left[\text{putting } z = \frac{\pi x}{l}\right]. \tag{7.40}$$

$$a_n = \frac{1}{\pi}\int_0^{2\pi} \cos nz\,dz$$

$$= \frac{1}{\pi}\int_0^{2\pi} f\left(\frac{lz}{\pi}\right) \cos nx\,dz \quad \left[\text{putting } z = \frac{\pi x}{l}\right]$$

$$= \frac{1}{\pi}\int_0^{2l} f(x) \cos \frac{n\pi x}{l} d\left(\frac{\pi x}{l}\right)$$

$$= \frac{1}{l}\int_0^{2l} f(x) \cos \frac{n\pi x}{l} dx. \tag{7.41}$$

Similarly, $$b_n = \frac{1}{l}\int_0^{2l} f(x) \sin \frac{n\pi x}{l} dx. \tag{7.42}$$

Half Range Series

Cosine series : $$f(x) = \frac{a_0}{2} + \sum_{n=1}^{\infty} a_n \cos \frac{n\pi x}{l}, \tag{7.43}$$

where $$a_0 = \frac{2}{l}\int_0^{l} f(x)dx \tag{7.44}$$

and $$a_n = \frac{2}{l}\int_0^{l} f(x) \cos \frac{n\pi x}{l} dx. \tag{7.45}$$

Sine series : $$f(x) = \sum_{n=1}^{\infty} b_n \sin \frac{n\pi x}{l}, \tag{7.46}$$

where $$b_n = \frac{2}{l}\int_0^{l} f(x) \sin \frac{n\pi x}{l} dx. \tag{7.47}$$

7.10 Parseval's Relations

Parseval's relation is the relation of Fourier coefficients of a function $f(x)$ and the average of the square of the function in its interval of its definition.

We shall find out three such relations as follows :

(i) Let a function $f(x)$ be defined in $(-\pi, \pi)$ so that its Fourier expansion is

$$\boldsymbol{f(x) = \frac{a_0}{2} + \sum_{n=1}^{\infty}(a_n \cos nx + b_n \sin nx)}$$

The average of the square of $f(x)$ in $(-\pi, \pi)$ is

$$\frac{1}{2\pi}\int_{-\pi}^{+\pi} |f(x)|^2 dx = \frac{1}{2\pi}\int_{-\pi}^{+\pi} \left[\frac{a_0}{2} + \sum_{n=1}^{\infty}(a_n \cos nx + b_n \sin nx)\right]^2 dx$$

$$= \frac{1}{2\pi}\int_{-\pi}^{+\pi} \left[\left(\frac{a_0}{2}\right)^2 + a_n^2 \cos^2 nx + b_n^2 \sin^2 nx\right] dx. \tag{7.48}$$

[due to orthogonality property all other terms vanish]

Now, $$\int_{-\pi}^{+\pi} \cos^2 nx dx = \frac{1}{2}\int_{-\pi}^{\pi}(1 + \cos 2nx)dx = \frac{1}{2}\left[x + \frac{\sin 2nx}{2n}\right]_{-\pi}^{+\pi} = \pi$$

and $\int_{-\pi}^{+\pi} \sin^2 nx dx = \frac{1}{2}\int_{-\pi}^{\pi}(1-\cos 2nx)dx = \pi.$

Now, from Eqn. (7.48),

$$\frac{1}{\pi}\int_{-\pi}^{\pi}[f(x)]^2 dx = \frac{1}{2}a_0^2 + \sum_{n=1}^{\infty}(a_n^2+b_n^2). \tag{7.49}$$

Eqn. (7.47) is one of the **Parseval relation.**

(ii) If $f(x)$ is defined in arbitrary interval $(-l, l)$, then

$$f(x) = \frac{a_0}{2} + \sum_{n=1}^{\infty}\left(a_n \cos\frac{n\pi x}{l} + b_n \sin\frac{n\pi x}{l}\right).$$

The average of the square of $f(x)$ in $(-l, l)$ is

$$\frac{1}{l}\int_{-l}^{l}[f(x)]^2 dx = \frac{1}{2}a_0^2 + \sum_{n=1}^{\infty}(a_n^2+b_n^2). \quad \text{(as before)} \tag{7.50}$$

Eqn. (7.50) represents another form of Parseval relation when the interval is $(-l, l)$.

(iii) If $f(x)$ is defined in $(0, l)$ and expanded in a half-range sine series like

$$f(x) = \sum_{n=1}^{\infty} b_n \sin\left(\frac{n\pi x}{l}\right).$$

Then Eqn. (7.50) reduces to $\frac{2}{l}\int_0^l [f(x)]^2 dx = \frac{1}{2}a_0^2 + \sum_{n=1}^{\infty} a_n^2.$ (7.51)

Eqn. (7.51) is still another form of **Parseval relation.**

7.11 Some Practical Applications of Fourier Series

7.11.1 Half-wave Rectifier

Expansion of the function $f(x) = \sin x$, when $0 \le x \le \pi$ and $f(x) = 0$, when $\pi < x < 2\pi$.

The process of converting alternating current into direct current is called rectification and the corresponding device, the rectifier. Fig. 7.5 shows the graphical representation of such a function.

Here, $a_0 = \frac{1}{\pi}\int_0^{\pi} f(x)dx + \frac{1}{\pi}\int_{\pi}^{2\pi} f(x)dx = \frac{1}{\pi}\int_0^{\pi}\sin x\, dx = \frac{2}{\pi}.$

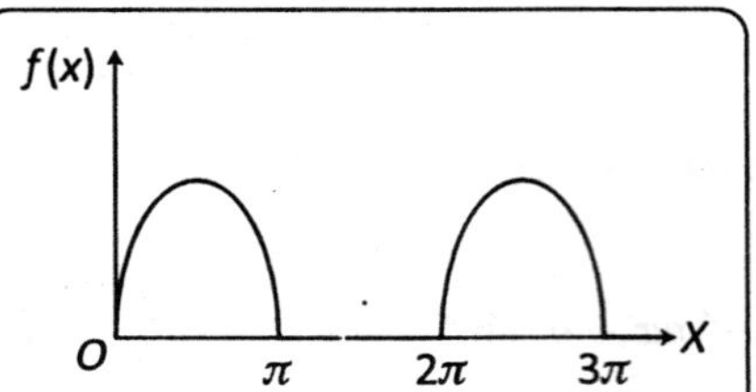

Fig. 7.5: Plot of f(x) against x

Also for $n > 1$,

$$\begin{aligned} a_n &= \frac{1}{\pi}\int_0^{\pi} f(x)\cos nx\, dx + \frac{1}{\pi}\int_0^{2\pi} f(x)\cos nx\, dx \\ &= \frac{1}{\pi}\int_0^{\pi}\sin x \cos nx\, dx \\ &= \begin{cases} 0 & \text{, when } n \text{ is odd} \\ -\frac{2}{\pi}\cdot\frac{1}{n^2-1}, & \text{for } n \text{ even.} \end{cases} \end{aligned}$$

For $n > 1$, .

$$b_n = \frac{1}{\pi}\int_0^{\pi} f(x)\sin nx\,dx + \frac{1}{\pi}\int_0^{2\pi} f(x)\sin nx\,dx = \frac{1}{\pi}\int_0^{\pi}\sin x\sin nx\,dx$$

$$= \begin{cases} 0, & \text{for all } n \text{ except } 1 \\ \frac{1}{2}, & \text{for } n = 1. \end{cases}$$

$$\therefore\quad f(x) = \frac{1}{2}a_0 + \sum_{n=1}^{\infty}(a_n\cos nx + b_n\sin nx) = \frac{1}{\pi} + \frac{1}{2}\sin x - \frac{2}{\pi}\sum_{n(\text{even})}\frac{\cos nx}{n^2-1}. \quad (7.52)$$

Here, $f(x)$ refers to **half wave rectifier**.

7.11.2 Full-wave Rectifier

Expansion of $f(x) = \sin x$, when $0 \le x \le \pi$ and $f(x) = -\sin x$, when $-\pi \le x \le 0$.

Fig. 7.6 represents the graphical representation of $f(x)$ in the interval $(-\pi, \pi)$ and the periodic extension outside $(-\pi, \pi)$.

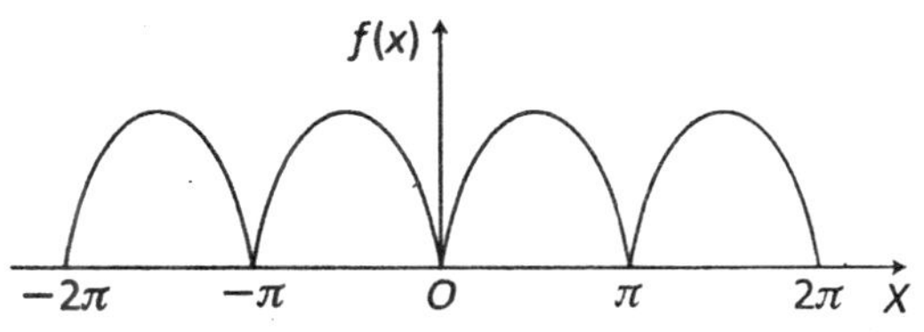

Fig. 7.6: Plot of $f(x)$ in the interval $(-\pi,\pi)$

$$\text{Here, } a_0 = \frac{1}{\pi}\int_{-\pi}^{0} f(x)dx + \frac{1}{\pi}\int_0^{\pi} f(x)dx$$

$$= -\frac{1}{\pi}\int_{-\pi}^{0}\sin x\,dx + \frac{1}{\pi}\int_0^{\pi}\sin x\,dx$$

$$= \frac{2}{\pi}\int_0^{\pi}\sin x\,dx = \frac{4}{\pi}.$$

Also for $n > 0$,

$$a_n = \frac{1}{\pi}\int_{-\pi}^{0} -\sin nx\cos nx\,dx + \frac{1}{\pi}\int_0^{\pi}\sin x\cos nx\,dx = \frac{2}{\pi}\int_0^{\pi}\sin x\cos nx\,dx$$

$$= \begin{cases} 0 & , \text{ for } n \text{ odd} \\ -\frac{4}{\pi}\cdot\frac{1}{n^2-1}, & \text{for } n \text{ even} \end{cases}$$

and for $n > 0$,

$$b_n = \frac{2}{\pi}\int_0^{\pi}\sin x\sin nx\,dx = 0.$$

The Fourier expansion of $f(x)$ is, thus,

$$f(x) = \frac{1}{2}a_0 + \sum_{n=1}^{\infty}(a_n\cos nx + b_n\sin nx) = \frac{2}{\pi} - \frac{4}{\pi}\sum_{n=0}^{\infty}\frac{\cos nx}{n^2-1}\ (n\text{ even}). \quad (7.53)$$

Eqn. (4.53) corresponds to **full-wave rectifier**.

7.11.3 Sawtooth Wave

Expansion of $f(x) = x$, $-\pi \le x \le \pi$

Fig. 7.7 represents graphically the function $f(x) = x$ in the interval $(-\pi, \pi)$ and its periodic extension outside $(-\pi, \pi)$.

Here, $a_0 = \frac{1}{\pi}\int_{-\pi}^{\pi} f(x)dx + .\frac{1}{\pi}\int_{-\pi}^{\pi} f(x)dx = 0.$

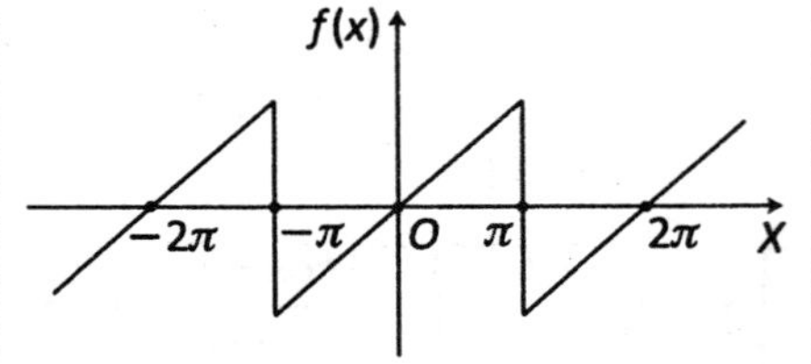

Fig. 7.7: Plot of $f(x) = x$ in the interval $(-\pi,\pi)$ and its periodic extension outside $(-\pi,\pi)$

For $n > 0$,

$$a_n = \frac{1}{\pi}\int_{-\pi}^{\pi} f(x)\cos nx\,dx = \frac{1}{\pi}\int_{-\pi}^{\pi} x\cos nx\,dx = 0$$

and for $n > 0$,

$$b_n = \frac{1}{\pi}\int_{-\pi}^{\pi} f(x)\sin nx\,dx = \frac{1}{\pi}\int_{-\pi}^{\pi} x\sin nx\,dx$$

$$= -\frac{2}{n}\cos nx = \frac{2}{n}(-1)^{n+1}.$$

Thus, the Fourier expansion of $f(x) = x,\ -\pi \leq x \leq \pi$ is given by

$$f(x) = x = \frac{1}{2}a_0 + \sum_{n=1}^{\infty}(a_n\cos nx + b_n\sin nx) = 2\sum_{n=1}^{\infty}\frac{(-1)^{n+1}\sin nx}{n}. \tag{7.54}$$

The above function $f(x)$ corresponds to **sawtooth wave.**

7.11.4 Square Wave

Expansion of $f(x) = \begin{cases} 0, \text{ for } -\pi < x < 0 \\ h, \text{ for } 0 \leq x < \pi \end{cases}$

Fig. 7.8 represents graphically $f(x)$ in $(-\pi,\pi)$ and its periodic extension outside $(-\pi,\pi)$.

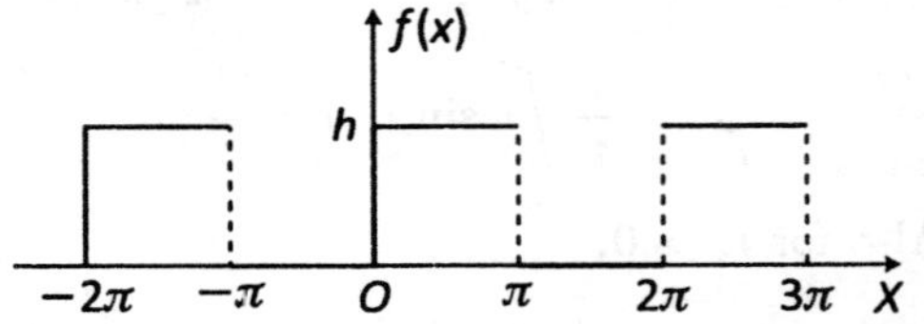

Fig. 7.8: Plot of $f(x)$ in $(-\pi,\pi)$ and its periodic extension outside $(-\pi,\pi)$

$$\text{Here, } a_0 = \frac{1}{\pi}\int_{-\pi}^{0} f(x)dx + \frac{1}{\pi}\int_{0}^{\pi} f(x)dx$$

$$= \frac{1}{\pi}\int_{0}^{\pi} h\,dx = h.$$

For $n > 1$,

$$a_n = \frac{1}{\pi}\int_{-x}^{0} f(x)\cos nx\,dx + \frac{1}{\pi}\int_{0}^{\pi} f(x)\cos nx\,dx = \frac{1}{\pi}\int_{0}^{\pi} h\cos nx\,dx$$

$$= \frac{h}{\pi}\int_{0}^{\pi} f(x)\cos nx\,dx = 0.$$

Also, for $n > 1$,

$$b_n = \frac{1}{\pi}\int_{-\pi}^{0} f(x)\sin nx\,dx + \frac{1}{\pi}\int_{0}^{\pi} f(x)\sin nx\,dx = \frac{h}{\pi}\int_{0}^{\pi}\sin nx\,dx = \frac{h}{\pi}\sin nx\,dx$$

$$= \frac{h}{\pi}\left[-\frac{\cos nx}{n}\right]_0^{\pi} = \begin{cases} 0 \text{ , for } n \text{ even} \\ \frac{2h}{n\pi}, \text{ for } n \text{ odd.} \end{cases}$$

The Fourier expansion is, thus,

$$f(x) = \frac{1}{2}a_0 + \sum_{n=1}^{\infty}(a_n\cos nx + b_n\sin nx) = \frac{h}{2} + \frac{2h}{\pi}\sum_{n=1}^{\infty}\frac{\sin nx}{n} \text{ (for } n \text{ odd).} \tag{7.55}$$

Eqn. (7.55) corresponds to **square wave.**

7.11.5 Rectangular Wave

Expansion of $f(x) = \begin{cases} h, \text{ for } 0 \le x \le \pi \\ -h, \text{ for } \pi \le x \le 2\pi \end{cases}$

The function $f(x)$ here represents a rectangular wave as shown in Fig. 7.9.

The Fourier expansion of $f(x)$ is

$$f(x) = \frac{a_0}{2} + \sum_{n=1}^{\infty} (a_n \cos nx + b_n \sin nx) dx,$$

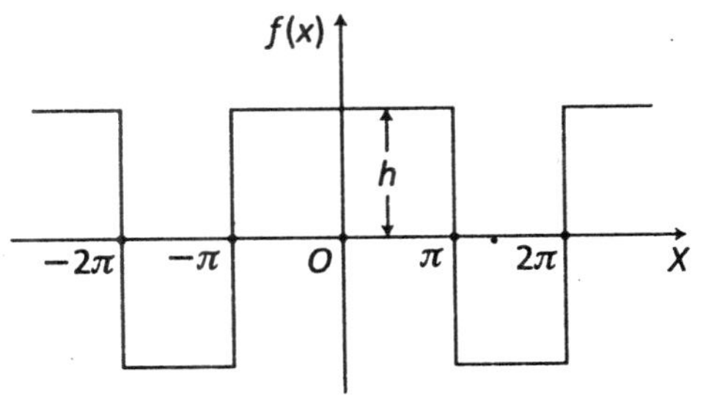

Fig. 7.9: The function $f(x)$ representing a rectangular wave

where $a_0 = \frac{1}{\pi}\int_0^{2\pi} f(x)dx$

$$= \frac{1}{\pi}\int_0^{\pi} h\,dx + \frac{1}{\pi}\int_{\pi}^{2\pi} -h\,dx$$

$$= \frac{h}{\pi}[\pi - \pi] = 0.$$

For $n > 0$,

$$a_n = \frac{1}{\pi}\int_0^{2\pi} f(x)\cos nx\,dx = \frac{1}{\pi}\left[\int_0^{\pi} h\cos nx\,dx - \int_{\pi}^{2\pi} h\cos nx\,dx\right]$$

$$= \frac{h}{\pi}\left[\int_0^{\pi} \cos nx\,dx - \int_{\pi}^{2\pi} \cos nx\,dx\right] = 0$$

and $$b_n = \frac{1}{\pi}\int_0^{2\pi} f(x)\sin nx\,dx = \frac{h}{\pi}\left[\int_0^{\pi} h\sin nx\,dx - \int_{\pi}^{2\pi} \sin nx\,dx\right]$$

$$= \frac{h}{n\pi}[-\cos nx]_0^{\pi} + [\cos nx]_{\pi}^{2\pi} = \frac{2h}{n\pi}(1 - \cos n\pi).$$

$$\therefore \quad b_n = \begin{cases} 0\ , & \text{for } n \text{ even} \\ \frac{4h}{n\pi}, & \text{for } n \text{ odd.} \end{cases}$$

Thus, the Fourier expansion of $f(x) = x$ is given by

$$f(x) = \frac{4h}{\pi}\sum_{n=1}^{\infty}\frac{\sin nx}{n}, \ (n \text{ odd})$$

$$= \frac{4h}{\pi}\left[\sin x + \frac{\sin 3x}{3} + \frac{\sin 5x}{5} + \cdots\right]. \tag{7.56}$$

Eqn. (7.56) shows that,

(i) the frequency of the component vibrations are odd multiples of the fundamental frequency,

(ii) all components are of same phase, and

(iii) the amplitude decreases in the ratio $1:3:5:\cdots$ as the frequency increases in the ratio $1:3:5:\cdots$.

7.12 Two Important Formulae

The following two important formulae are very useful in discussing the convergence of Fourier series.

(1) $$\frac{1}{2}+\sum_{n=1}^{n}\cos nx=\frac{\sin\left(n+\frac{1}{2}\right)x}{2\sin\frac{x}{2}} \tag{7.57}$$

(2) $$\frac{1}{\pi}\int_0^{\infty}\frac{\sin\left(n+\frac{1}{2}\right)x}{2\sin\frac{x}{2}}=\frac{1}{2}. \tag{7.58}$$

Proof of (1) : Let $S=\frac{1}{2}+\sum_{n=1}^{n}\cos nx.$

Then, $$2S\sin\frac{x}{2}=\sin\frac{x}{2}+2\sin\frac{x}{2}\sum_{n=1}^{n}\cos nx=\sin\frac{x}{2}+2\sin\frac{x}{2}\cos x+2S\sin\frac{x}{2}\cos 2x+\cdots$$

$$=\sin\frac{x}{2}+\left(\sin\frac{3x}{2}-\sin\frac{x}{2}\right)+\left(\sin\frac{5x}{2}-\sin\frac{3x}{2}\right)+\cdots+\left[\sin\left(n+\frac{1}{2}\right)x-\sin\left(n-\frac{1}{2}\right)x\right]$$

$$=\sin\left(n+\frac{1}{2}\right)x.$$

$$\therefore\quad S=\frac{1}{2}+\sum_{n=1}^{n}\cos x=\frac{\sin\left(n+\frac{1}{2}\right)x}{2\sin\left(\frac{x}{2}\right)}. \tag{7.59}$$

Proof of (2) : Integrating Eqn. (7.59) within the limit $-\pi$ to π, we get

$$\int_{-\pi}^{\pi}\frac{\sin\left(n+\frac{1}{2}\right)x}{\sin\left(\frac{x}{2}\right)}dx=\pi+0=\pi$$

or, $$\frac{1}{\pi}\int_{-\pi}^{\pi}\frac{\sin\left(n+\frac{1}{2}\right)x}{\sin\left(\frac{x}{2}\right)}=1 \tag{7.60}$$

or, $$\frac{1}{\pi}\int_{0}^{\pi}\frac{\sin\left(n+\frac{1}{2}\right)x}{2\sin\left(\frac{x}{2}\right)}=\frac{1}{2} \tag{7.61}$$

[∵ the integrand of Eqn. [7.60] is an even function].

7.13 Solved Problems

Example 7.1. *A periodic function $f(x)$ with period 2π is defined as*

$$f(x)=x^2,\quad(-\pi\le x\le\pi).$$

Expand $f(x)$ in a Fourier series and hence show that, $\sum_{n=1}^{\infty}\frac{1}{n^2}=\frac{\pi^2}{6}.$

Show the nature of the function graphically.

[BU(H) 1993, CU(H) 2000, VU(H) 2009, CU(H) 2007]

Solution : According to Euler's formula, $a_0=\frac{1}{\pi}\int_{-\pi}^{\pi}x^2dx=\frac{1}{3\pi}\left[x^3\right]_{-\pi}^{+\pi}=\frac{2}{3}\pi^2,$

$$a_n = \frac{1}{\pi}\int_{-\pi}^{\pi} x^2 \cos nx\, dx = \frac{1}{\pi}\left[x^2 \frac{\sin nx}{n}\right]_{-\pi}^{+\pi} - \frac{2}{n\pi}\int_{-\pi}^{\pi} x \sin nx\, dx$$

$$= \frac{2}{n\pi}\left[x\frac{\cos nx}{n}\right]_{-\pi}^{+\pi} - \frac{2}{n^2\pi}\int_{-\pi}^{\pi} x\cos nx\, dx = \frac{2}{n^2\pi}\cdot 2\pi\cos n\pi - \frac{2}{n^3\pi}\left[\sin n\pi\right]_{-\pi}^{+\pi}$$

$$= \frac{4}{n^2}(-1)^n.$$

$$b_n = \frac{1}{\pi}\int_{-\pi}^{\pi} x^2 \sin nx\, dx = \frac{1}{n\pi}\left[-x^2\cos nx\right]_{-\pi}^{\pi} - \frac{2}{n\pi}\int_{-\pi}^{\pi} x\cos nx\, dx$$

$$= -\frac{2}{n\pi}\left[x\frac{\sin nx}{n}\right]_{-\pi}^{\pi} + \frac{2}{n^2\pi}\int_{-\pi}^{\pi}\sin nx\, dx = 0.$$

Thus, $f(x) = \frac{1}{2}a_0 + \sum_{n=1}^{\infty}(a_n\cos nx + b_n \sin nx)$

$$= \frac{\pi^2}{3} + 4\sum_{n=1}^{\infty}(-1)^n\frac{\cos nx}{n^2} \quad (\text{for } -\pi \le x \le \pi).$$

Putting $x = \pi$ on both sides of the above equation,

$$\pi^2 = \frac{\pi^2}{3} + 4\sum_{n=1}^{\infty}\frac{(-1)^{2n}}{n^2}, \quad [\because\ f(\pi) = \pi^2].$$

$$\therefore \quad \sum_{n=0}^{\infty}\frac{1}{n^2} = \frac{\pi^2}{6}.$$

Here $f(x) = x^2$ is an even function. The nature of the function is shown in Fig. 7.10 graphically.

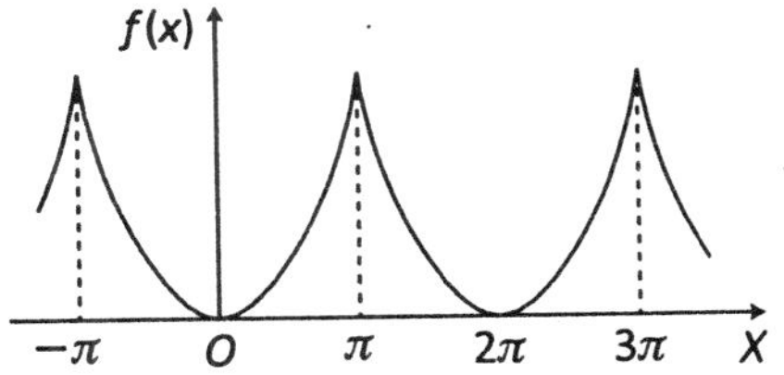

Fig. 7.10: Plot of $f(x) = x^2$ against x

Example 7.2. *Expand in Fourier series the function $f(x) = x$, for $0 < x < 2\pi$.*

Solution : The function is neither odd nor even. The coefficients of the Fourier expansion of $f(x)$ are

$$a_0 = \frac{1}{\pi}\int_0^{2\pi} x dx = \frac{1}{\pi}\left(\frac{x^2}{2}\right)_0^{2\pi} = 2\pi,$$

$$a_n = \frac{1}{\pi}\int_0^{2\pi} x\cos nx\, dx = -\frac{1}{n\pi}\int_0^{2\pi} x\sin nx\, dx + \left[\frac{x}{n\pi}\sin nx\right]_0^{2\pi},$$

$$b_n = \frac{1}{\pi}\int_0^{2\pi} x\sin(nx)dx = -\frac{1}{nx}\left[x\cos nx\right]_0^{2\pi} + \frac{1}{n\pi}\int_0^{2\pi}\cos nx\, dx = -\frac{2}{n}$$

Hence, for $0 < x < 2\pi$, $\boldsymbol{x = \pi - 2\left(\sin x + \frac{\sin 2x}{2} + \frac{\sin 3x}{3} + \cdots\right)}$.

Example 7.3. *Find the Fourier series for the function defined by*

$$f(x) = 1, \quad \text{for } -\pi \le x < 0$$
$$= 0, \quad \text{for } x = 0$$
$$= +1, \quad \text{for } 0 < x \le \pi.$$

[BU(H) 2002]

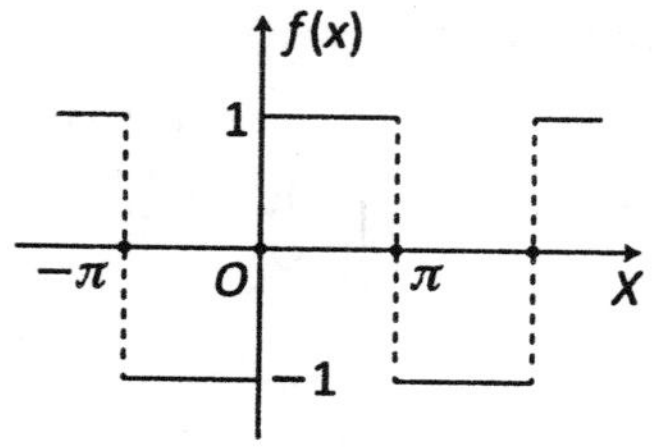

Fig. 7.11: Plot of f(x) against x along with its periodic extension

Solution : Here $f(x)$ is an odd function. Its graph is shown in Fig. 7.11 along with its periodic extension.

The coefficients of expansion are :

$$a_n = 0, \quad b_n = \frac{2}{\pi}\int_0^{\pi} \sin nx\, dx = \begin{cases} 0\ , & \text{for } n \text{ even} \\ \dfrac{4}{n\pi}, & \text{for } n \text{ odd.} \end{cases}$$

Thus, the Fourier series is :

$$\boldsymbol{f(x) = \frac{4}{\pi}\left[\sin x + \frac{\sin 3x}{3} + \frac{\sin 5x}{5} + \cdots\right].}$$

Example 7.4. *Find the Fourier series of the function*

$$f(x) = \begin{cases} -k, & \text{if } -\pi < x < 0 \\ k, & \text{if } 0 < x < \pi. \end{cases}$$

$$\text{and} \quad f(x + 2\pi) = f(x)$$

Hence, show that, $\sum_{n=0}^{\infty} = \frac{1}{n^2} = \frac{\pi^2}{6}$. **[BU(H) 1995]**

Solution : Here the Fourier coefficients of $f(x)$ are :

$$a_0 = \frac{1}{\pi}\int_{-\pi}^{\pi} f(x)dx = \frac{1}{\pi}\left[\int_{-\pi}^{0}(-k)dx + \int_0^{\pi} kdx\right] = \frac{1}{\pi}[-k\pi + k\pi] = 0,$$

$$a_n = \frac{1}{\pi}\left[\int_{-\pi}^{0} -k\cos nx\, dx + \int_0^{\pi} k\cos nx\, dx\right] = \frac{k}{\pi}\left[\left(-\frac{\sin nx}{n}\right)_{-\pi}^{0} + \left(\frac{\sin nx}{n}\right)_0^{\pi}\right] = 0$$

$$\text{and} \quad b_n = \frac{1}{\pi}\left[\int_{-\pi}^{0} -k\sin nx\, dx + \int_0^{\pi} k\sin nx\, dx\right] = \frac{k}{\pi}\left[\left(-\frac{\cos nx}{n}\right)_{-\pi}^{0} - \left(\frac{\cos nx}{n}\right)_0^{\pi}\right]$$

$$= \frac{2k}{n\pi}[1 - (-1)^n].$$

For even n, $b_n = 0$

odd n, $b_n = \frac{4k}{n\pi}$.

$$\therefore \quad \boldsymbol{f(x)} = \frac{4k}{\pi}\sum_{n=1}^{\infty}\frac{\sin n\pi}{n}, \text{ (for odd } n)$$

$$\boldsymbol{= \frac{4k}{\pi}\sum_{n=0}^{\infty}\frac{\sin(2n+1)x}{(2n+1)} = \frac{4k}{\pi}\left[\frac{\sin x}{1} + \frac{\sin 3x}{3} + \frac{\sin 5x}{5} + \cdots\right].}$$

For $x = \frac{\pi}{2}$, $f\left(\frac{\pi}{2}\right) = k$.

$$\therefore \quad k = \frac{4k}{\pi}\left[\sin\frac{\pi}{2} + \frac{1}{3}\sin\frac{3\pi}{2} + \frac{1}{5}\sin\frac{5\pi}{2} + \cdots\right] = \frac{4k}{\pi}\sum_{n=0}^{\infty}\frac{(-1)^n}{2n+1}.$$

$$\therefore \quad \boldsymbol{\sum_{n=0}^{\infty}\frac{(-1)^n}{2n+1} = \frac{\pi}{2}} \quad \textbf{(Proved).}$$

Example 7.5. *A periodic sawtooth wave has the displacement falling off linearly with time t from $y = c$ to $y = 0$ when $t = T$ (time period as shown). Find $f(t)$ into Fourier series.*

[VU(H) 1991]

Solution : The displacement at any time may be represented by $f(t) = y(t)$. Then from the similar triangles PRT and OQT,

$$\frac{y}{a} = \frac{T-t}{T} \quad \text{or,} \quad y = a\left[\frac{T-t}{T}\right] = a\left[1 - \frac{t}{T}\right].$$

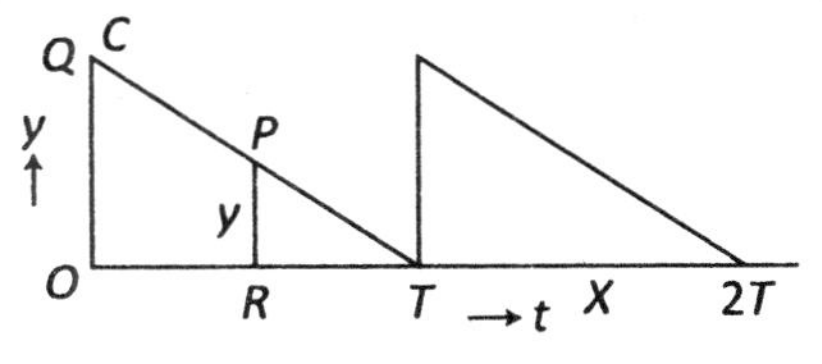

Fig. 7.12: Plot of $f(x) = y(t)$ against t

When the time axis passes through the lowest point of the periodic curve, we have

$$a_0 = \frac{1}{T}\int_0^T y\,dt = \frac{a}{T}\int_0^T \left[1 - \frac{t}{T}\right] dt = \frac{a}{T}\left[t - \frac{t^2}{2T}\right]_a^T = \frac{a}{T}\left[T - \frac{T^2}{2T}\right] = \frac{a}{2},$$

$$a_n = \int_0^T y\cos n\omega t\,dt = \frac{2a}{T}\left[\int_0^T \left\{1 - \frac{1}{T}\right\}\cos n\omega t\,dt\right] = 0.$$

$\therefore$ all the cosine terms of the Fourier series disappear.

Also, $b_n = \int_0^T y\sin n\omega t\,dt = \frac{2a}{T}\int_0^T \left\{1 - \frac{1}{T}\right\}\sin n\omega t\,dt = \frac{2a}{T}\left[\int_0^T \sin n\omega t\,dt - \frac{1}{T}\int_0^T t\sin n\omega t\,dt\right]$

or, $b_n = \frac{2a}{T}\left[\left\{-\frac{\cos n\omega t}{n\omega}\right\}_0^T - \frac{1}{T}\left\{\left(-t\frac{\cos n\omega t}{n\omega}\right)_0^T + \int_0^T \frac{\cos n\omega t}{n\omega}\right\}\right]$ [integrating by parts]

$$= \frac{2a}{T}\left[\left\{-\frac{\cos n\omega t}{n\omega} + \frac{\cos n\omega t}{n\omega}\right\} - \frac{1}{T}\left\{-\frac{T\cos n\omega t}{n\omega} + 0 + \left(\frac{\sin n\omega t}{n^2\omega^2}\right)_0^T\right\}\right]$$

$$= \frac{2a}{T}\cdot\frac{1}{n\omega} = \frac{2a}{2\pi n} = \frac{a}{\pi n}.$$

$\therefore$ the Fourier's series in this case is given by

$$\boldsymbol{f(t) = y(t) = \frac{a}{2} + \frac{a}{\pi}\left(\sin\omega t + \frac{1}{2}\sin 2\omega t + \frac{1}{3}\sin 3\omega t + \cdots + \frac{1}{n}\sin n\omega t\right)}$$

$$\boldsymbol{= \frac{a}{2} + \frac{a}{\pi}\sum_{n=1}^{\infty}\frac{\sin n\omega t}{n}.}$$

Example 7.6. *Expand the function $f(x) = |x|$ in Fourier series in the interval $-\pi \le x \le \pi$.*

Solution : As the given function is an even function it can be expanded only in cosine series.

Thus, $a_0 = \frac{1}{\pi}\int_{-\pi}^{\pi} x\,dx = \frac{2}{\pi}\int_0^{\pi} x\,dx = \frac{2}{\pi}\left[\frac{x^2}{2}\right]_0^{\pi} = \pi.$

$$a_n = \frac{1}{\pi}\int_{-\pi}^{\pi} x\cos nx\,dx = \frac{2}{\pi}\int_0^{\pi} x\cos nx\,dx = -\frac{2}{n\pi}\int_0^{\pi} x\sin nx\,dx = \frac{2}{\pi n^2}[\cos nx]_0^{\pi}$$

$$= \frac{2}{\pi n^2}(\cos n\pi - 1) = \frac{2}{\pi n^2}[(-1)^n - 1] = \begin{cases} 0, & \text{for } n \text{ even} \\ -\frac{4}{\pi n^2}, & \text{for } n \text{ odd.} \end{cases}$$

$$\therefore \quad |x| = \frac{\pi}{2} - \frac{4}{\pi}\left(\cos x + \frac{\cos 3x}{3^2} + \frac{\cos 5x}{5^2} + \cdots\right) \text{ for } -\pi \le x \le \pi.$$

Example 7.7. *Find the Fourier series for the periodic function*

$$f(x) = \begin{cases} 0, & -\pi < x < 0 \\ x, & 0 < x < \pi \end{cases}$$

$$f(x+2\pi) = f(x).$$

From this expansion show that, $1 + \frac{1}{3^2} + \frac{1}{5^2} + \cdots = \frac{\pi^2}{8}$.

[VU(H) 2013, CU(H) 2005, AMIETE (Winter) 1992]

Solution : Let $f(x) = \frac{a_0}{2} + a_1 \cos x + a_2 \cos 2x + \cdots + b_1 \sin x + b_2 \sin 2x + \cdots$. (i)

Then $a_0 = \frac{1}{\pi}\int_{-\pi}^{0} 0\,dx + \frac{1}{\pi}\int_0^{\pi} x\,dx = \frac{1}{\pi}\left[\frac{x^2}{2}\right]_0^{\pi} = \frac{1}{\pi}\left(\frac{\pi^2}{2}\right) = \frac{\pi}{2}$.

$$a_n = \frac{1}{\pi}\int_0^{\pi} x\cos nx\,dx = \frac{1}{\pi}\left[x\frac{\sin nx}{x} - (1)\left(-\frac{\cos nx}{n}\right)\right]_0^{\pi} = \frac{1}{\pi}\left[\frac{(-1)^n}{n^2} - \frac{1}{n^2}\right]$$

$$= -\frac{2}{\pi n^2}, \text{ (for } n \text{ odd)}$$

$$= 0, \text{ (for } n \text{ even)}.$$

$$b_n = \frac{1}{\pi}\int_0^{\pi} x\sin nx\,dx = \frac{1}{\pi}\left[x\left(-\frac{\cos nx}{x}\right) - (1)\left(-\frac{\sin nx}{n^2}\right)\right]_0^{\pi} = \frac{1}{\pi}\left[-\pi\frac{(-1)^n}{n}\right] = -\frac{(-1)^n}{n}.$$

Substituting the values of $a_0, a_1, a_2, \ldots, b_1, b_2$ in Eqn. (i), we get

$$f(x) = \frac{\pi}{4} - \frac{2}{\pi}\left[\frac{\cos x}{1^2} + \frac{\cos 3x}{3^2} + \frac{\cos 5x}{5^2} + \cdots\right] + \left[\frac{\sin x}{1} + \frac{\sin 2x}{2} + \frac{\sin 3x}{3} + \cdots\right].$$

Putting $x = 0$ in the above equation, we get

$$f(0) = 0 = \frac{\pi}{4} - \frac{2}{\pi}\left[\frac{1}{1^2} + \frac{1}{3^2} + \frac{1}{5^2} + \cdots\right] + 0 \quad \text{or,} \quad 1 + \frac{1}{3^2} + \frac{1}{5^2} + \cdots = \frac{\pi^2}{8}.$$

Example 7.8. *Find the Fourier series expansion of the periodic function of period* 2π *defined by*

$$f(x) = x, \qquad \text{if, } -\frac{\pi}{2} < x < \frac{\pi}{2}$$

$$= (\pi - x), \text{ if, } \frac{\pi}{2} < x < \frac{3\pi}{2}.$$

[AMIETE (Summer) 1995, Winter 1994]

Solution : Let $f(x) = \frac{a_0}{2} + a_1 \cos x + a_2 \cos 2x + \cdots + b_1 \sin x + b_2 \sin 2x + \cdots$,

where $a_0 = \frac{1}{\pi}\int_{-\frac{\pi}{2}}^{\frac{\pi}{2}} x\,dx + \frac{1}{\pi}\int_{\frac{\pi}{2}}^{\frac{3\pi}{2}} (\pi - x)\,dx = \frac{1}{\pi}\left(\frac{x^2}{2}\right)_{-\frac{\pi}{2}}^{\frac{\pi}{2}} + \frac{1}{\pi}\left[\pi x - \frac{x^2}{2}\right]_{\frac{\pi}{2}}^{\frac{3\pi}{2}}$

$$= \frac{1}{8}\left[\frac{\pi^2}{8} - \frac{\pi^2}{8}\right] + \frac{1}{\pi}\left[\frac{3\pi^2}{2} - \frac{9\pi^2}{8} - \frac{\pi^2}{2} + \frac{\pi^2}{8}\right] = \pi\left[\frac{3}{2} - \frac{9}{8} - \frac{1}{2} + \frac{1}{8}\right] = 0,$$

$$a_n = \frac{1}{\pi}\int_{-\frac{\pi}{2}}^{\frac{\pi}{2}} x\cos nx\,dx + \frac{1}{\pi}\int_{\frac{\pi}{2}}^{\frac{3\pi}{2}} (\pi - x)\cos nx\,dx$$

$$= \frac{1}{\pi}\left[x\frac{\sin nx}{n} - (1)\left(-\frac{\cos nx}{n^2}\right)\right]_{-\frac{\pi}{2}}^{\frac{\pi}{2}} + \frac{1}{\pi}\left[(\pi - x)\frac{\sin nx}{n} - (-1)\left(-\frac{\cos nx}{n^2}\right)\right]_{\frac{\pi}{2}}^{\frac{3\pi}{2}}$$

$$= \frac{1}{\pi}\left[\frac{\pi}{2}\frac{\sin\frac{n\pi}{2}}{n} + \frac{\cos\frac{n\pi}{2}}{n^2} - \frac{\pi}{2}\frac{\sin\frac{n\pi}{2}}{n} - \frac{\cos\frac{n\pi}{2}}{n^2}\right]$$

$$+ \frac{1}{\pi}\left[-\frac{\pi}{2}\frac{\sin\frac{3n\pi}{2}}{n} - \frac{\cos\frac{3n\pi}{2}}{n^2} - \frac{\pi}{2}\frac{\sin\frac{n\pi}{2}}{n} + \frac{\cos\frac{n\pi}{2}}{n^2}\right]$$

$$= \frac{1}{\pi}\left[-\frac{\pi}{2n}\left(\sin\frac{3n\pi}{2} + \sin\frac{n\pi}{2}\right) - \frac{1}{n^2}\left(\cos\frac{3n\pi}{2} - \cos\frac{n\pi}{2}\right)\right]$$

$$= \frac{1}{\pi}\left[-\frac{\pi}{n}\sin n\pi\cos\frac{n\pi}{2} + \frac{2}{n^2}\sin\frac{n\pi}{2}\sin n\pi\right] = 0.$$

$\therefore \qquad a_n = 0.$

$$b_n = \frac{1}{\pi}\int_{-\frac{\pi}{2}}^{\frac{\pi}{2}} x\sin nx\,dx + \frac{1}{\pi}\int_{\frac{\pi}{2}}^{\frac{3\pi}{2}} (\pi - x)\sin nx\,dx$$

$$= \frac{2}{\pi}\int_{0}^{\frac{\pi}{2}} x\sin nx\,dx + \frac{1}{\pi}\int_{\frac{\pi}{2}}^{\frac{3\pi}{2}} (\pi - x)\sin nx\,dx$$

$$= \frac{2}{\pi}\left[x\left(-\frac{\cos nx}{n}\right) - (1)\left(-\frac{\sin nx}{n^2}\right)\right]_0^{\frac{\pi}{2}} + \frac{1}{\pi}\left[(\pi - x)\left(-\frac{\cos nx}{n}\right) - (-1)\left(-\frac{\sin nx}{n^2}\right)\right]_{\frac{\pi}{2}}^{\frac{3\pi}{2}}$$

$$= \frac{2}{\pi}\left[-\frac{\pi}{2}\frac{\cos\frac{n\pi}{2}}{n} + \frac{\sin\frac{n\pi}{2}}{n^2}\right] + \frac{1}{\pi}\left[\frac{\pi}{2}\frac{\cos\frac{3n\pi}{2}}{n} - \frac{\sin\frac{3n\pi}{2}}{n^2} + \frac{\pi}{2}\frac{\cos\frac{n\pi}{2}}{n} + \frac{\sin\frac{n\pi}{2}}{n^2}\right]$$

$$= \frac{1}{\pi}\left[-\frac{\pi}{2}\frac{\cos\frac{n\pi}{2}}{n} + \frac{3\sin\frac{n\pi}{2}}{n^2} + \frac{\pi}{2}\frac{\cos\frac{3n\pi}{2}}{n} - \frac{\sin\frac{3n\pi}{2}}{n^2}\right]$$

$$= \frac{1}{\pi}\left[\frac{\pi}{2n}\left(\cos\frac{3n\pi}{2} - \cos\frac{n\pi}{2}\right) + \frac{3}{n^2}\sin\frac{n\pi}{2} - \frac{1}{n^2}\sin\frac{3n\pi}{2}\right]$$

$$= \frac{1}{\pi}\left[-\frac{\pi}{n}\sin\frac{n\pi}{2}\sin n\pi + \frac{3}{n^2}\sin\frac{n\pi}{2} - \frac{1}{n^2}\sin\frac{3n\pi}{2}\right] = \frac{1}{n^2\pi}\left[3\sin\frac{n\pi}{2} - \sin\frac{3n\pi}{2}\right].$$

Using the values of $a_0, a_1, a_2, \ldots, b_1, b_2, \ldots$, we get

$$\boldsymbol{f(x) = \frac{4}{\pi}\left[\frac{\sin x}{1^2} - \frac{\sin 3x}{3^2} + \frac{\sin 5x}{5^2}\cdots\right]}.$$

Example 7.9. *A periodic function $f(x)$ is defined as*

$$f(x) = x, \quad -\frac{\pi}{2} < x < \frac{\pi}{2}$$

$$= 0, \quad \frac{\pi}{2} < x < \frac{3\pi}{2}.$$

Find the Fourier series expansion of $f(x)$. **[AMIETE (Winter) 1993]**

Solution : Here, $a_0 = \frac{1}{\pi}\int_{-\frac{\pi}{2}}^{\frac{\pi}{2}} x\,dx = \frac{1}{\pi}\left[\frac{x^2}{2}\right]_{-\frac{\pi}{2}}^{\frac{\pi}{2}} = \frac{1}{\pi}\left[\frac{\pi^2}{8} - \frac{\pi^2}{8}\right] = 0.$

$$a_n = \frac{1}{\pi}\int_{-\frac{\pi}{2}}^{\frac{\pi}{2}} x\cos nx\,dx = \frac{1}{\pi}\left[x\frac{\sin nx}{n} - (1)\left(-\frac{\cos nx}{n^2}\right)\right]_{-\frac{\pi}{2}}^{\frac{\pi}{2}}$$

$$= \frac{1}{\pi}\left[\frac{\pi}{2}\frac{\sin\frac{n\pi}{2}}{n} + \frac{\cos\frac{n\pi}{2}}{n^2} - \frac{\pi}{2}\frac{\sin\frac{n\pi}{2}}{n} - \frac{\cos\frac{n\pi}{2}}{n^2}\right] = 0.$$

$$b_n = \frac{1}{\pi}\int_{-\frac{\pi}{2}}^{\frac{\pi}{2}} x\sin nx\,dx = \frac{1}{\pi}\left[x\left(-\frac{\cos nx}{n}\right) - (1)\left(\frac{\sin nx}{n^2}\right)\right]_{-\frac{\pi}{2}}^{\frac{\pi}{2}}$$

$$= \frac{1}{\pi}\left[-\frac{\pi}{2}\frac{\cos\frac{n\pi}{2}}{n} + \frac{\sin\frac{n\pi}{2}}{n^2} - \frac{\pi}{2}\frac{\cos\frac{n\pi}{2}}{n} + \frac{\sin\frac{n\pi}{2}}{n^2}\right] = \frac{2}{\pi}\left[-\frac{\pi}{2}\frac{\cos\frac{n\pi}{2}}{n} + \frac{\sin\frac{n\pi}{2}}{n^2}\right].$$

$$\therefore\quad b_1 = \frac{2}{\pi}\left[0 + \frac{1}{n^2}\right] = \frac{2}{\pi}\frac{1}{1^2}, \qquad b_2 = \frac{2}{\pi}\left[-\frac{\pi}{2}\left(-\frac{1}{2}\right) + 0\right] = \frac{1}{2},$$

$$b_3 = \frac{2}{\pi}\left[0 - \frac{1}{3^2}\right] = \frac{2}{\pi}\left(-\frac{1}{3^2}\right),\quad b_4 = \frac{2}{\pi}\left[-\frac{\pi}{2}\left(\frac{1}{4}\right) + 0\right] = -\frac{1}{4}.$$

$\therefore$ the Fourier series of the periodic function $f(x)$ is

$$f(x) = \frac{a_0}{2} + a_1\cos x + a_2\cos 2x + \cdots + a_n\cos nx + \cdots$$
$$+ b_1\sin x + b_2\sin 2x + \cdots + b_n\sin nx + \cdots$$

or, $$\boldsymbol{f(x) = \frac{2}{\pi}\sin x + \frac{1}{2}\sin 2x - \frac{2}{9\pi}\sin 3x - \frac{1}{4}\sin 4x + \cdots.}$$

Example 7.10. *Find the Fourier series of the function $f(x) = e^x$ in the interval $-\pi < x < \pi$.*

Solution : Let the Fourier series expansion of $f(x) = e^x$ be

$$f(x) = \frac{a_0}{2} + \sum_{n=1}^{\infty} a_n\cos nx + \sum_{n=1}^{\infty} b_n\sin nx, \tag{i}$$

where $$a_0 = \frac{1}{\pi}\int_{-\pi}^{\pi} f(x)dx = \frac{1}{\pi}\int_{-\pi}^{\pi} e^x\,dx = \frac{1}{\pi}[e^x]_{-\pi}^{\pi} = \frac{1}{\pi}[e^\pi - e^{-\pi}] = \frac{2\sin h\pi}{\pi},$$

$$a_n = \frac{1}{\pi}\int_{-\pi}^{\pi} f(x)dx = \frac{1}{\pi}\int_{-\pi}^{\pi} e^x\cos nx\,dx = \frac{4}{1+n^2}\frac{\sin h\pi}{\pi}\cos n\pi$$

and $$b_n = \frac{1}{\pi}\int_{-\pi}^{\pi} e^x\sin nx\,dx = -\frac{4}{1+n^2}\frac{\sin h\pi}{\pi}\cos n\pi$$

Hence, $$\boldsymbol{e^x = \frac{\sin h\pi}{\pi} + \frac{4}{\pi}\sum_{n=1}^{\infty}\sin h\pi\frac{\cos n\pi}{1+n^2}\cos nx + \frac{4}{\pi}\sum_{n=1}^{\infty}\frac{n}{1+n^2}\sin h\pi\cos n\pi\sin n\pi.}$$

Example 7.11. *Obtain Fourier series for $f(x) = e^{-x}$ in the interval $0 < x < 2\pi$.*

[Poona U 1987, Sambalpur U 1987]

Solution : Let $$e^{-x} = \frac{a_0}{2} + \sum_{n=1}^{\infty} a_n\cos nx + \sum_{n=1}^{\infty} b_n\sin nx, \tag{i}$$

where $a_0 = \frac{1}{\pi}\int_0^{2\pi} f(x)dx = \frac{1}{\pi}\int_0^{2\pi} e^{-x}\,dx = \frac{1}{\pi}\left[-e^{-x}\right]_0^{2\pi} = \frac{1-e^{-2\pi}}{\pi}$,

$$a_n = \frac{1}{\pi}\int_0^{2\pi} f(x)\cos nx\,dx = \frac{1}{\pi}\int_0^{2\pi} e^{-x}\cos nx\,dx = \frac{1}{\pi(n^2+1)}\left[e^{-x}(-\cos nx + n\sin nx)\right]_0^{2\pi}$$

$$= \left(\frac{1-e^{-2\pi}}{\pi}\right)\cdot\frac{1}{n^2+1}.$$

Putting $n = 1$, $a_1 = \left(\frac{1-e^{-2\pi}}{\pi}\right)\cdot\frac{1}{2}$,

$n = 2$, $a_2 = \left(\frac{1-e^{-2\pi}}{\pi}\right)\cdot\frac{1}{5}$, etc.

and $\quad b_n = \frac{1}{\pi}\int_0^{2\pi} f(x)\sin nx\,dx = \frac{1}{\pi}\int_0^{2\pi} e^{-x}\sin nx\,dx = \frac{1}{\pi(n^2+1)}\left[e^{-x}(-\sin nx - n\cos nx)\right]_0^{2\pi}.$

$$= \left(\frac{1-e^{-2\pi}}{\pi}\right)\cdot\frac{n}{n^2+1}.$$

$\therefore \quad b_1 = \frac{1-e^{-2\pi}}{\pi}\cdot\frac{1}{2}$, $\quad b_2 = \frac{1-e^{-2\pi}}{\pi}\cdot\frac{2}{5}$, etc.

Substituting the values of a_0, a_n, b_n in Eqn. (1), we get

$$\boldsymbol{e^{-x} = \frac{1-e^{-2x}}{\pi}\left[\frac{1}{2} + \left(\frac{1}{2}\cos x + \frac{1}{5}\cos 2x + \frac{1}{10}\cos 3x + \cdots\right) + \left(\frac{1}{2}\sin x + \frac{2}{5}\sin 2x + \frac{3}{10}\sin 3x + \cdots\right)\right].}$$

Example 7.12. *A function $f(x)$ is defined as* $f(x) = \begin{cases} \sin x, & \text{for } 0 \le x \le \pi \\ -\sin x, & \text{for } -\pi \le x \le 0. \end{cases}$

Expand $f(x)$ in Fourier series.

Solution : The $f(x)$ in Fourier expansion is

$$f(x) = \frac{a_0}{2} + \sum_{n=1}^{\infty} a_n \cos nx\,dx + \sum_{n=1}^{\infty} b_n \sin nx\,dx, \tag{1}$$

where $a_0 = \frac{1}{\pi}\int_{-\pi}^{0}(-\sin x)dx + \int_0^{\pi}\sin x\,dx = \frac{2}{\pi}\int_0^{\pi}\sin x\,dx = \frac{4}{\pi}$,

$$a_n = \frac{1}{\pi}\int_{-\pi}^{0}(-\sin x)\cos nx\,dx + \int_0^{\pi}\sin x\cos nx\,dx = \frac{2}{\pi}\int_0^{\pi}\sin x\cos nx\,dx$$

$$= \begin{cases} -\frac{4}{\pi}\cdot\frac{1}{n^2-1}, & \text{when } n \text{ even} \\ 0 & , \text{when } n \text{ odd} \end{cases}$$

$b_n = 0$ as the function is an even function.

$\therefore$ from Eqn. (1), $\boldsymbol{f(x) = \frac{2}{\pi} - \frac{4}{\pi}\sum_{n=1}^{\infty}\frac{\cos nx}{n^2-1}}$ **(when n even).**

Example 7.13. *Find the Fourier series to represent $x - x^2$ from $x = -\pi$ to $x = \pi$.*

Solution : Let $f(x) = x - x^2 = \frac{a_0}{2} + \sum_{n=1}^{\infty} a_n \cos nx + \sum_{n=1}^{\infty} b_n \sin nx$,

where $a_0 = \frac{1}{\pi}\int_{-\pi}^{\pi}(x - x^2)dx = \frac{1}{\pi}\left[\frac{x^2}{2} - \frac{x^3}{3}\right]_{-\pi}^{\pi} = -\frac{2\pi^2}{3}$,

$$a_n = \frac{1}{\pi}\int_{-\pi}^{\pi}(x - x^2)\cos nx\, dx$$

$$= \frac{1}{\pi}\left[(x - x^2)\frac{\sin nx}{n} - (1 - 2x)\left(-\frac{\cos nx}{n^2}\right) + (-2)\left(-\frac{\sin nx}{n^3}\right)\right]_{-\pi}^{\pi}$$

[integrating by parts]

$$= -\frac{4(-1)^n}{n^2} \quad [\because \cos N\pi = (-1)^n].$$

$\therefore \quad a_1 = \frac{4}{1^2},\ a_2 = -\frac{4}{2^2},\ a_3 = \frac{4}{3^2},\ a_4 = -\frac{4}{4^2}, \ldots$, etc.

Also, $b_n = \frac{1}{\pi}\int_{-\pi}^{\pi}(x - x^2)\sin nx\, dx$

$$= \frac{1}{\pi}\left[(x - x^2)\left(-\frac{\cos nx}{n}\right) - (1 - 2x)\left(-\frac{\sin nx}{n^2}\right) + (-2)\left(\frac{\cos nx}{n^3}\right)\right]_{-\pi}^{\pi}$$

$$= \frac{-2(-1)^n}{n} \quad [\because \sin n\pi = 0].$$

$\therefore \quad b_1 = \frac{2}{1},\ b_2 = -\frac{2}{2},\ b_3 = \frac{2}{3},\ b_4 = -\frac{2}{4}$, etc.

Hence, $\boldsymbol{f(x) = x - x^2}$

$$= -\frac{\pi^2}{3} + 4\left[\frac{\cos x}{2^2} + \frac{\cos 3x}{3^2} - \cdots\right] + 2\left[\frac{\sin x}{1} - \frac{\sin 2x}{2} + \frac{\sin 3x}{3} - \cdots\right].$$

Example 7.14. *A function $f(x)$ is defined as*

$$f(x) = \begin{cases} -\frac{2}{\pi}(x + \pi), & \text{when } -\pi \le x \le -\frac{\pi}{2} \\ \frac{2}{\pi}x, & \text{when } -\frac{\pi}{2} \le x \le \frac{\pi}{2} \\ -\frac{2}{\pi}(\pi - x), & \text{when } -\infty < x < \infty. \end{cases}$$

Find the Fourier expansion of $f(x)$.

Solution : Here $f(x)$ is a continuous periodic function having a period $-\pi$ to π.

$$\therefore \quad f(x) = \frac{a_0}{2} + \sum_{n=1}^{\infty} a_n \cos nx + \sum_{n=1}^{\infty} b_n \sin nx.$$

Here, $a_0 = \frac{1}{\pi}\int_{-\pi}^{\pi} f(x)dx = \frac{1}{\pi}\int_{-\pi}^{\pi} -\frac{2}{\pi}(x + \pi)dx + \frac{1}{\pi}\int_{-\frac{\pi}{2}}^{\frac{\pi}{2}} \frac{2}{\pi}x\, dx + \frac{1}{\pi}\int_{\frac{\pi}{2}}^{\pi} -\frac{2}{\pi}(x - \pi)dx$

$$= \frac{2}{\pi^2}\left[\int_{-\pi}^{-\frac{\pi}{2}} -(x+\pi)dx + \int_{-\frac{\pi}{2}}^{\frac{\pi}{2}} x\,dx + \int_{\frac{\pi}{2}}^{\pi} -(x-\pi)dx\right]$$

$$= \frac{2}{\pi^2}\left[-\frac{(x+\pi)^2}{2}\right]_{-\pi}^{-\frac{\pi}{2}} + \frac{2}{\pi}\left[\frac{x^2}{2}\right]_{-\frac{\pi}{2}}^{\frac{\pi}{2}} + \frac{2}{\pi^2}\left[-\frac{(x-\pi)^2}{2}\right]_{\frac{\pi}{2}}^{\pi}$$

$$= \frac{1}{\pi^2}\left[-\frac{\pi^2}{4} + \frac{\pi^2}{4} - \frac{\pi^2}{4} + \frac{\pi^2}{4}\right] = 0.$$

$\therefore \quad a_0 = 0.$

$$a_n = \frac{1}{\pi}\int_{-\pi}^{\pi} f(x)\cos nx\,dx$$

$$= \frac{1}{\pi}\int_{-\pi}^{-\frac{\pi}{2}} -\frac{2}{\pi}(x+\pi)\cos nx\,dx + \frac{1}{\pi}\int_{-\frac{\pi}{2}}^{\frac{\pi}{2}} \frac{2}{\pi}x\cos nx\,dx + \frac{1}{\pi}\int_{\frac{\pi}{2}}^{\pi} -\frac{2}{\pi}(x-\pi)\cos nx\,dx$$

$= 0.$

Again, $b_n = \frac{1}{\pi}\int_{-\pi}^{\pi} f(x)\sin nx\,dx$

$$= \frac{1}{\pi}\int_{\pi}^{-\frac{\pi}{2}} -\frac{2}{\pi}(x+\pi)\sin nx\,dx + \frac{1}{\pi}\int_{-\frac{\pi}{2}}^{\frac{\pi}{2}} \frac{2}{\pi}x\sin nx\,dx + \frac{1}{\pi}\int_{\frac{\pi}{2}}^{\pi} -\frac{2}{\pi}(x-\pi)\sin nx\,dx$$

$$= \begin{cases} 0 & , \text{ for } n \text{ even} \\ \dfrac{8}{n^2\pi^2}\sin\dfrac{n\pi}{2}, & \text{for } n \text{ odd.} \end{cases}$$

Hence, the Fourier expansion of the function

$$\boldsymbol{f(x) = \frac{8}{\pi^2}\left[\sin\pi - \frac{1}{3^2}\sin 3\pi + \frac{1}{5^2}\sin 5\pi - \cdots\right].}$$

Example 7.15. *Expand in Fourier series the periodic function $f(x)$ with period $2l$ which in the interval $(-l, l)$ is given by $f(x) = |x|$.* **[BU(H) 1995]**

Solution : Here $f(x)$ is defined as below :

$$f(x) = -x, \quad -l < x < l$$
$$= x, \quad 0 \le x < l.$$

Now, the Fourier constant,

$$a_0 = \frac{1}{l}\int_{-l}^{+l} f(x)dx = \frac{1}{l}\left[\int_{-l}^{0} -x\,dx + \int_{0}^{l} x\,dx\right] = \frac{1}{2l}\left[\left(-x^2\right)_{-l}^{0} + \left(x^2\right)_{0}^{l}\right] = \frac{2l^2}{2l} = l,$$

$$a_n = \frac{1}{l}\int_{-l}^{0} -x\cos\frac{n\pi x}{l}dx = \frac{1}{l}\int_{-l}^{0} -x\cos\frac{n\pi x}{l}dx + \frac{1}{l}\int_{0}^{l} x\cos\frac{n\pi x}{l}dx$$

$$= -\frac{1}{n^2\pi^2}\left[\cos\frac{n\pi x}{l}\right]_{-l}^{0} + \frac{1}{n^2\pi^2}\left[\cos\frac{n\pi x}{l}\right]_{0}^{l} = \frac{2l}{n^2\pi^2}[(-1)^n - 1].$$

For n even, $a_n = 0$.

For n odd, $a_n = -\dfrac{4l}{n^2\pi^2}$

and $b_n = \frac{1}{l}\int_{-l}^{+l} f(x)\sin\frac{n\pi x}{l}dx = \frac{1}{l}\int_{-l}^{0} -x\sin\frac{n\pi x}{l}dx + \frac{1}{l}\int_{0}^{l} x\sin\frac{n\pi x}{l}dx = 0.$

$$\therefore \quad f(x) = \frac{l}{2} - \frac{4l}{\pi^2}\left[\cos\frac{\pi x}{l} + \frac{1}{9}\cos\frac{3\pi x}{l} + \frac{1}{25}\cos\frac{5\pi x}{l} + \cdots\right].$$

Example 7.16. *Expand in Fourier series the function $f(x) = x$ in the interval $-1 < x < 1$.*

Solution : It is a case of change of interval.

Let $f(x) = x = \frac{a_0}{2}\sum_{n=1}^{\infty}(a_n\cos n\pi x + b_n\sin n\pi x)$,

where $a_0 = \int_{-1}^{1} f(x)dx = \int_{-1}^{1} x\,dx = \left[\frac{x^2}{2}\right]_{-1}^{1} = 0,$

$$a_n = \int_{-1}^{1} f(x)\cos n\pi x\,dx = \int_{-1}^{1} x\cos n\pi x\,dx = 0, \text{ for all } n$$

and $b_n = \int_{-1}^{1} f(x)\sin n\pi x\,dx = -\frac{2}{\pi n}(-1)^n.$

$$\therefore \quad f(x) = \frac{2}{\pi}\left[\sin\pi x - \frac{1}{2}\sin 2\pi x + \frac{1}{3}\sin 3\pi x + \cdots\right] = -\frac{2}{\pi}\sum_{n=1}^{\infty}(-1)^n\frac{\sin n\pi x}{n}.$$

Example 7.17. *Expand e^{-x} as a Fourier series in the interval $(-l, l)$.* **[Calicut U 1989]**

Solution : The required Fourier series is of the form :

$$e^{-x} = \frac{a_0}{2} + \sum_{n=1}^{\infty} a_n\cos\frac{n\pi x}{l} + \sum_{n=1}^{\infty} b_n\sin\frac{n\pi x}{l}, \qquad \text{(i)}$$

where $a_0 = \frac{1}{l}\int_{-l}^{l} e^{-x}dx = \frac{1}{l}\left[e^{-x}\right]_{-l}^{l} = \frac{1}{l}(e^l - e^{-l}) = \frac{2\sin hl}{l},$

$$a_n = \frac{1}{l}\int_{-l}^{l} e^{x}\cos\frac{n\pi x}{l}dx = \frac{1}{l}\left[\frac{e^{-x}}{1+(\frac{n\pi}{l})^2}\left(-\cos\frac{n\pi x}{l} + \frac{n\pi}{l}\sin\frac{n\pi x}{l}\right)\right]_{-l}^{l}$$

$$\left[\because \int e^{ax}\cos bx\,dx = \frac{e^{ax}}{a^2+b^2}(a\cos bx + b\sin bx)\right]$$

$$= \frac{2l(-1)^n\sin hl}{l^2+(n\pi)^2} \quad [\because \cos nx = (-1)^n].$$

$$\therefore \quad a_1 = -\frac{2l\sin nhl}{l^2+\pi^2},\ a_2 = \frac{2l\sin nhl}{l^2+2^2\pi^2},\ a_3 = \frac{2l\sin nhl}{l^2+3^2\pi^2}, \text{ etc.}$$

and $b_n = \frac{1}{l}\int_{-l}^{+l} e^{-x}\sin\frac{n\pi x}{l}dx = \frac{1}{l}\left[\frac{e^{-x}}{1+(\frac{n\pi}{l})^2}\left(-\sin\frac{n\pi x}{l} - \frac{n\pi}{l}\cos\frac{n\pi x}{l}\right)\right]_{-l}^{l}$

$$= \frac{2n\pi(-1)^n\sin n\pi hl}{l^2+(n\pi)^2}.$$

$$\therefore \quad b_1 = -\frac{2\pi\sin hl}{l^2+\pi^2},\ b_2 = \frac{4\pi\sin hl}{l^2+2^2\pi^2},\ b_3 = \frac{-6\pi\sin hl}{l^2+3^2\pi^2}, \text{ etc.}$$

Substituting the values of a_1, a_2, a_3, ... and b_1, b_2, b_3, ... in equation (i), we get

$$e^{-x} = \sinh hl\left[\frac{1}{l} - 2l\left(\frac{1}{l^2+\pi^2}\cos\frac{\pi x}{l} - \frac{1}{l^2+2^2\pi^2}\cos\frac{2\pi x}{l} + \frac{1}{l^2+3^2\pi^2}\cos\frac{3\pi x}{l} - \cdots\right)\right.$$
$$\left. -2\pi\left(\frac{1}{l^2+\pi^2}\sin\frac{n\pi}{l} - \frac{2}{l^2+2^2\pi^2}\sin\frac{2\pi x}{l} + \frac{3}{l^2+3^2\pi^2}\sin\frac{3\pi x}{l} - \cdots\right)\right].$$

Example 7.18. *Expand $f(x) = \sin x$, $0 < x < \pi$ in Fourier cosine series.* [BU(H) 1997]

Solution : The plot of $f(x)$ against x is shown in the Fig. 7.13 where the portion ABC is in $(0, \pi)$. The extension of the curve of $f(x)$ in $(-\pi, 0)$ is represented by ADE. Hence, the curve is symmetrical about the y-axis. Hence, $f(x)$ is an even function within $(-\pi, \pi)$.

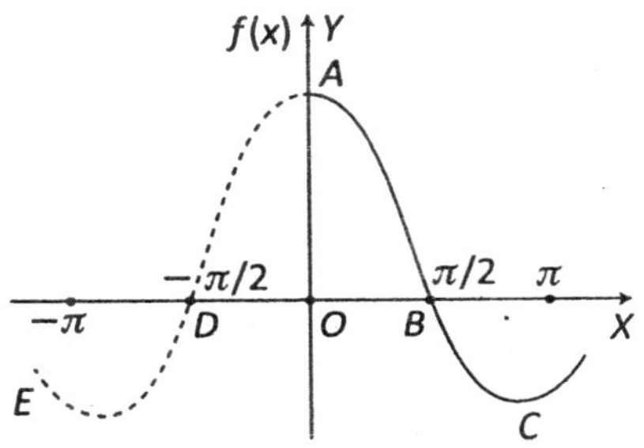

Fig. 7.13: Plot of $f(x)$ against x

Hence, the Fourier expansion of $\sin x$ in $(-\pi, \pi)$ will contain only cosine terms.

$$\therefore \quad \sin x = \frac{1}{2}a_0 + \sum_{n=1}^{\infty} a_n \cos nx, \tag{i}$$

where $a_0 = \dfrac{2}{\pi}\displaystyle\int_0^{\pi} \sin x\, dx + \frac{2}{\pi}[-\cos x]_0^{\pi} = \frac{2}{\pi}[1+1] = \frac{4}{\pi}$,

$$a_n = \frac{2}{\pi}\int_0^{\pi} \sin x \cos nx\, dx = \frac{1}{\pi}\int_0^{\pi}[\sin(n+1)x - \sin(n-1)x]dx$$
$$= \frac{1}{\pi}\left[\frac{\cos(n-1)x}{n-1} - \frac{\cos(n+1)x}{n+1}\right]_0^{\pi} = \frac{1}{\pi}\left[\frac{1}{n+1} - \frac{1}{n-1} - \frac{(-1)^{n+1}}{n+1} + \frac{(-1)^{n-1}}{n-1}\right].$$

For odd n, $a_n = 0$.

For even n, $a_n = -\dfrac{4}{\pi}\dfrac{1}{(n^2-1)}$

$$\therefore \quad f(x) = \frac{2}{\pi} - \frac{4}{\pi}\sum_{n=1}^{\infty}\frac{\cos 2n\pi}{4n^2-1}.$$

Example 7.19. *Expand the function $y = \cos 2x$ in a series of sines in the interval $(0, \pi)$.* [BU(H) 1992]

Solution : For the Fourier expansion of $y = \cos 2x$, we have to find out b_n for the odd extension of $\cos 2x$.

$$\therefore \quad b_n = \frac{2}{\pi}\int_0^{\pi} \cos 2x \sin nx\, dx = \frac{1}{\pi}\int_0^{\pi}[\sin(n+2)x + \sin(n-2)x]dx$$
$$= \frac{1}{\pi}\left[-\frac{\cos(n+2)x}{n+2} - \frac{\cos(n-2)x}{n-2}\right]_0^{\pi} = \frac{1}{\pi}\left[\frac{1}{n+2} + \frac{1}{n-2} + \frac{(-1)^{n+3}}{n+2} + \frac{(-1)^{n-1}}{n-2}\right].$$

For odd n, $b_n = \dfrac{4}{\pi}\dfrac{n}{(n^2-4)}$.

For even n, $b_n = 0$

$$\therefore \quad f(x) = \cos 2x = \frac{4}{\pi}\sum_{n=1}^{\infty}\frac{(2n-1)\sin(2n-1)x}{(2n-1)^2-4}$$

$$= \frac{4}{\pi}\left[-\frac{\sin x}{3} + \frac{3}{5}\sin 3x + \frac{5}{21}\sin 5x + \cdots\right].$$

Example 7.20. *Find Fourier half-range even expansion of the function,*

$$f(x) = -\left(\frac{x}{l}\right) + l,\ 0 \le x \le l. \qquad \textbf{[AMIETE (Summer) 1991]}$$

Solution : Here, $a_0 = \frac{2}{l}\int_0^l f(x)dx = \frac{2}{l}\int_0^l \left(-\frac{x}{l}+1\right)dx = \frac{2}{l}\left[-\frac{x^2}{2l}+x\right]_0^l$

$$= \frac{2}{l}\left[-\frac{l^2}{2l}+l\right] = \frac{2l}{l}\left[-\frac{1}{2}+l\right] = 1.$$

$$a_n = \frac{2}{l}\int_0^l f(x)\cos\frac{n\pi x}{l}dx = \frac{2}{l}\int_0^l\left(-\frac{x}{l}+1\right)\cos\frac{n\pi x}{l}dx$$

$$= \frac{2}{l}\left[\left(-\frac{x}{l}+1\right)\left(\frac{l}{n\pi}\sin\frac{n\pi x}{l}\right) - \left(-\frac{1}{l}\right)\left(-\frac{l^2}{n^2\pi^2}\cos\frac{n\pi x}{l}\right)\right]_0^l$$

$$= \frac{2}{l}\left[0 - \frac{l}{n^2\pi^2}\cos n\pi + \frac{l}{n^2\pi^2}\right] = \frac{2}{l}\frac{l}{n^2\pi^2}[-(-1)^n+1] = \frac{2}{n^2\pi^2}[1-(-1)^n]$$

$$= \frac{4}{n^2\pi^2}, \text{ for } n \text{ odd}$$

$$= 0, \text{ for } n \text{ even.}$$

$$\therefore \quad f(x) = \frac{1}{2} + \frac{4}{\pi^2}\left[\frac{1}{1^2}\cos\frac{\pi x}{l} + \frac{1}{3^2}\cos\frac{3\pi x}{l} + \frac{1}{5^2}\cos\frac{5\pi x}{l} + \cdots\right].$$

Example 7.21. *Express the function $f(x) = x$ as a half range sine series in the interval* $0 < x < 2$.

Solution : We have $f(x) = x = \sum_{n=1}^{\infty}\sin\frac{n\pi x}{2}$,

where $b_n = \frac{2}{2}\int_0^2 f(x)\sin\frac{n\pi x}{2}dx = \int_0^2 x\sin\frac{n\pi x}{2}dx = \left[x\cdot\frac{-\cos\left(\frac{n\pi x}{2}\right)}{\frac{n\pi}{2}} - \frac{-\sin\left(\frac{n\pi x}{2}\right)}{\left(\frac{n\pi}{2}\right)^2}\right]_0^2$

$$= -\frac{4}{n\pi}\cos n\pi = \begin{cases} \frac{4}{n\pi}, & \text{for } n \text{ odd} \\ -\frac{4}{n\pi}, & \text{for } n \text{ even.}\end{cases}$$

$$\therefore \quad f(x) = x = -\frac{4}{\pi}\sum_{n=1}^{\infty}\frac{(-1)^n}{n}\sin\frac{n\pi x}{2} = \frac{4}{\pi}\left[\sin\frac{\pi x}{2} - \frac{1}{2}\sin\pi x + \frac{1}{3}\sin\frac{3\pi x}{2} - \cdots\right].$$

Example 7.22. *Express $f(x) = x$ as a half-range sine series in $0 < x < 2$.*

Solution : The plot of $f(x)$ against x is shown in Fig. 7.14. If we extend $f(x)$ in the interval $-2 < x < 0$, then it is found that the new function is symmetrical about the origin and hence represents an odd function in $(-2, 2)$.

The Fourier series for $f(x)$ in $(-2, 2)$ contain will sine terms, i.e.,

$$f(x) = \sum_{n=1}^{\infty} b_n \sin \frac{n\pi x}{2},$$

where $b_n = \frac{2}{2}\int_0^2 f(x) \sin \frac{n\pi x}{2} dx = \int_0^2 x \sin \frac{n\pi x}{2} dx$

$$= \left[\frac{2\pi}{n\pi} \cos \frac{n\pi x}{2} + \frac{4}{n^2\pi^2} \sin \frac{n\pi x}{2}\right]_0^2 = -\frac{4(-1)^n}{n\pi}.$$

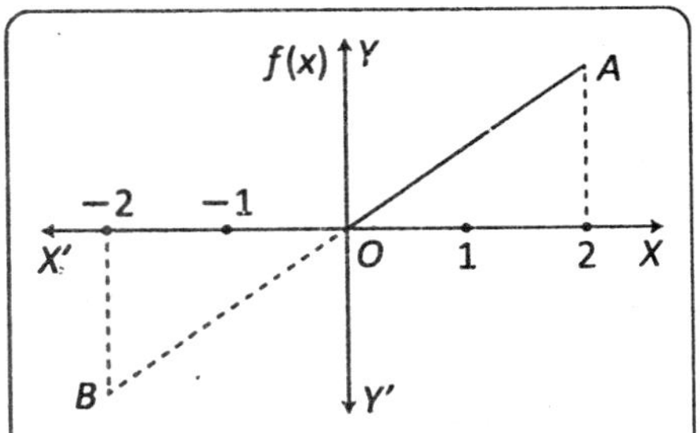

Fig. 7.14: Plot of $f(x)$ against x

$\therefore \quad b_1 = \frac{4}{\pi}, b_2 = \frac{4}{2\pi}, b_3 = \frac{4}{3\pi}, b_4 = -\frac{4}{4\pi}$, etc.

Hence, $\boldsymbol{f(x) = x = \frac{4}{\pi}\left[\sin \frac{\pi x}{2} - \frac{1}{2}\sin \frac{2\pi x}{2} + \frac{1}{3}\sin \frac{3\pi x}{2} - \frac{1}{4}\sin \frac{4\pi x}{2} + \cdots\right]}$.

Example 7.23. *Express $f(x) = x$ as a half-range cosine series in $0 < 2 < x$.*

Solution : The plot of $f(x)$ against x is shown in Fig. 7.14. Extension of $f(x)$ in the interval $(-2, 0)$ shows that the new function is symmetrical about the Y-axis. Hence, $f(x)$ represents an even function in $(-2, 2)$.

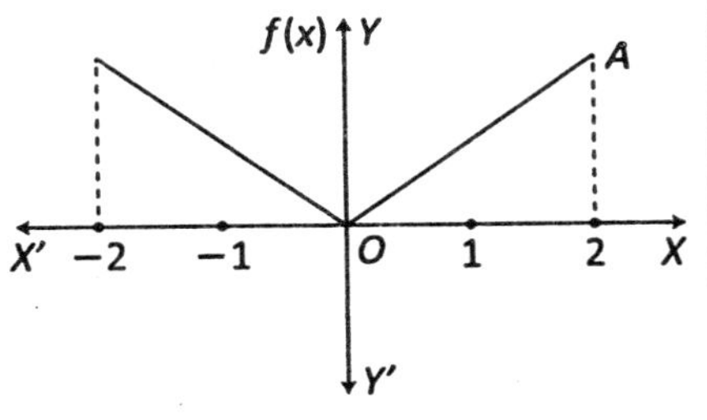

Fig. 7.15: Plot of $f(x)$ against x

$\therefore$ the Fourier series for $f(x)$ in $(-2, 2)$ will contain only cosine terms, i.e.,

$$f(x) = \frac{a_0}{2} + \sum_{n=1}^{\infty} a_n \cos \frac{n\pi x}{2},$$

where $a_0 = \frac{2}{2}\int_0^2 f(x) dx = \int_0^2 x\, dx = 2,$

$$a_n = \frac{2}{2}\int_0^2 f(x) \cos \frac{n\pi x}{2} dx = \int_0^2 x \sin \frac{n\pi x}{2} dx = \left[\frac{2\pi}{nx} \sin \frac{n\pi x}{2} + \frac{4}{n^2\pi^2} \cos \frac{n\pi x}{2}\right]_0^2$$

$$= \frac{4}{n^2\pi^2}[(-1)^n - 1].$$

$\therefore \quad a_1 = -\frac{8}{\pi^2}, a_2 = 0, a_3 = -\frac{8}{3^2\pi^2}, a_4 = 0, a_5 = -\frac{8}{5^2\pi^2}$, etc.

$$\therefore \quad \boldsymbol{f(x) = 1 - \frac{8}{\pi^2}\left[\frac{\cos\left(\frac{\pi x}{2}\right)}{1^2} + \frac{\cos\left(\frac{3\pi x}{2}\right)}{3^2} + \frac{\cos\left(\frac{5\pi x}{2}\right)}{5^2} + \cdots\right]}.$$

Example 7.24. *Find the Fourier series expansion of the periodic function of period 2π, $f(x) = x^2, -\pi \le x \le \pi$.*

Hence, find the sum of $\frac{1}{1^2} - \frac{1}{2^2} + \frac{1}{3^2} - \frac{1}{4^2} + \cdots$. **[AMIETE (Winter) 1996, (Summer) 1993]**

Solution : The function $f(x) = x^2, -\pi \le x \le \pi$ is an even function,

$\therefore \quad b_n = 0.$

$$a_0 = \frac{2}{\pi}\int_0^{\pi} f(x)dx = \frac{2}{\pi}\int_0^{\pi} x^2\,dx = \frac{2}{\pi}\left[\frac{x^3}{3}\right]_0^{\pi} = \frac{2\pi^2}{3}.$$

$$a_n = \frac{2}{\pi}\int_0^{\pi} f(x)dx = \frac{2}{\pi}\int_0^{\pi} x^2\cos nx\,dx$$

$$= \frac{2}{\pi}\left[x^2\left(\frac{\sin nx}{n}\right) - (2x)\left(-\frac{\cos nx}{n^2}\right) + 2\left(-\frac{\sin nx}{n^3}\right)\right]_0^{\pi}$$

$$= \frac{2}{\pi}\left[\frac{x^2\sin nx}{n} + \frac{2\pi\cos nx}{n^2} - \frac{2\sin nx}{n^3}\right] = \frac{4(-1)^n}{n^2}.$$

$\therefore$ Fourier series is $f(x) = \frac{a_0}{2} + a_1\cos x + a_2\cos 2x + a_3\cos 3x + \cdots + a_n\cos nx$
$+ b_1\sin x + b_2\sin 2x + b_3\sin 3x + \cdots + b_n\sin nx + \cdots$

or, $\boldsymbol{x^2 = \frac{\pi^2}{3} - 4\left[\frac{\cos x}{1^2} - \frac{\cos 2x}{2^2} + \frac{\cos 3x}{3^2} - \frac{\cos 4x}{4^2} + \cdots\right]}$.

For $x = 0$,

$$0 = \frac{\pi^2}{3} - 4\left[\frac{1}{1^2} - \frac{1}{2^2} + \frac{1}{3^2} - \frac{1}{4^2} + \cdots\right] \quad \text{or,} \quad \boldsymbol{\frac{1}{1^2} - \frac{1}{2^2} + \frac{1}{3^2} - \frac{1}{4^2} = \frac{1}{12}}.$$

Example 7.25. *If* $f(x) = \frac{a_0}{2} + \sum_{n=1}^{\infty}\left(a_n\cos\frac{n\pi x}{l} + b_n\sin\frac{n\pi x}{l}\right)$, *prove that,*

$$\int_{-l}^{+l}[f(x)]^2dx = l\left[\frac{a_0^2}{2}\sum_{n=1}^{\infty}(a_n^2 + b_n^2)\right].$$ **[BU(H) 2004]**

Solution : $\because$ the average value of the square of $f(x)$ in $(-l, l)$ is

$$\frac{1}{2l}\int_{-l}^{l}[f(x)]^2dx = \frac{1}{2l}\int_{-l}^{l}\left[\frac{a_0}{2} + \sum_{n=1}^{\infty}\left(a_n\cos\frac{n\pi x}{l} + b_n\sin\frac{n\pi x}{l}\right)\right]dx$$

or, $$\int_{-l}^{l}[f(x)]^2dx = \frac{a_0^2}{4}\int_{-l}^{l}dx + \left[\sum_{n=1}^{\infty}\int_{-l}^{l}a_n\cos\frac{n\pi x}{l}dx\right]^2 + \left[\sum_{n=1}^{\infty}\int_{-l}^{l}b_n\sin\frac{n\pi x}{l}dx\right]^2.$$

($\because$ all other terms vanish due to orthogonality property)

Now, $$\int_{-l}^{l}\cos^2\frac{n\pi x}{l}dx = \frac{1}{2}\int_{-l}^{l}\left[1 + \cos\frac{2n\pi x}{l}\right]dx = \frac{1}{2}\left[x + \frac{\sin\frac{2n\pi x}{l}}{\frac{2n\pi}{l}}\right]_{-l}^{l} = \frac{1}{2}2l = l.$$

Similarly, $$\int_{-l}^{l}\sin^2\frac{n\pi x}{l}dx = \frac{1}{2}\int_{-l}^{l}\left(1 - \cos\frac{2n\pi x}{l}\right)dx = l.$$

$$\therefore \quad \boldsymbol{\int_{-l}^{l}[f(x)]^2dx = l\left[\frac{a_0^2}{2} + \sum_{n=1}^{\infty}(a_n^2 + b_n^2)\right]} \quad \textbf{(Proved).}$$

Example 7.26. *A function* $f(x)$ *has the Fourier expansion in the interval* $0 < x < L$, $f(x) = \frac{a_0}{2} + \sum_{k=1}^{\infty}a_n\cos\frac{k\pi x}{l}$. *Prove that,* $\int_0^L [f(x)]^2dx = \frac{L}{4}a_0^2 + \frac{L}{2}\sum_{k=1}^{\infty}a_k^2$. **[BU(H) 1998]**

Solution : Here $f(x)$ is defined in $0 < x < L$ and expanded in half-range cosine series like

$$f(x) = \frac{a_0}{2} + \sum_{k=1}^{\infty} a_n \cos\frac{k\pi x}{l} dx.$$

The average value of $f(x)$ in $0 < x < L$ is

$$\frac{1}{L}\int_0^L [f(x)]^2 dx = \frac{1}{L}\int_0^L \left[\frac{a_0}{2} + \sum_{k=1}^{\infty} a_k \cos\frac{k\pi x}{L}\right]^2 dx$$

$$\text{or,} \quad \int_0^L [f(x)]^2 dx = \frac{La_0^2}{4} + \left[\int_0^L \sum_{k=1}^{\infty} a_k \cos\frac{k\pi x}{L}\right]^2 + 2\frac{a_0}{2}\int_0^L \cos\frac{k\pi x}{L} dx$$

$$\therefore \quad \int_0^L [f(x)]^2 dx = \frac{La_0^2}{4} + \frac{1}{2}\sum_{k=1}^{\infty} a_k^2 \quad \left[\because \int_0^L \cos\frac{k\pi x}{L} dx = 0\right] \quad \textbf{(Proved)}.$$

Example 7.27. *A function $f(x)$ has the Fourier expansion in the interval $0 < x < l$, $f(x) = \sum_{k=1}^{\infty} b_k \sin\frac{k\pi x}{l} dx$, then prove that, $\int_0^l [f(x)]^2 dx = \frac{l}{2}\sum_{k=1}^{\infty} b_k^2$.*

Solution : We have the half-range sine series in $0 < x < l$ is

$$f(x) = \sum_{k=1}^{\infty} b_k \sin\frac{k\pi x}{l} dx.$$

The average value of $f(x)$ in $0 < x < l$ is

$$\frac{1}{l}\int_0^l [f(x)]^2 dx = \frac{1}{l}\left[\int_0^l \sum_{k=1}^{\infty} b_k \sin\frac{k\pi x}{l}\right]^2$$

$$\text{or,} \quad \int_0^l [f(x)]^2 dx = \int_0^l \sum_{k=1}^{\infty} b_k^2 \sin^2\frac{k\pi x}{l} dx = \frac{1}{2}\sum_{k=1}^{\infty} b_k^2 \int_0^l \left(1 - 2\cos\frac{2k\pi x}{L}\right) dx$$

$$= \frac{1}{2}\sum_{k=1}^{\infty} b_k^2 \int_0^l dx - \frac{1}{2}\sum_{k=1}^{\infty} b_k^2 \left[-2\frac{\sin\frac{2k\pi x}{l}}{\frac{2k\pi}{l}}\right]_0^l$$

$$= \frac{l}{2}\sum_{k=1}^{\infty} b_k^2 + \frac{1}{2}2\frac{l}{2\pi k}\sum_{k=1}^{\infty}\left[\sin\frac{2k\pi x}{l}\right]_0^l = \frac{l}{2}\sum_{k=1}^{\infty} b_k^2 + 0.$$

$$\therefore \quad \int_0^l [f(x)]^2 dx = \frac{l}{2}\sum_{k=1}^{\infty} b_k^2 \quad \textbf{(Proved)}.$$

Example 7.28. *Show that the complex Fourier coefficients are pure imaginary and those for an even function are real.* **[BU(H) 2000]**

Solution : Consider a function $f(x)$ defined in $(-l, l)$ with period $2l$ in the form,

$$f(x) = \frac{1}{2}a_0 + \sum_{n=1}^{\infty}\left(a_n \cos\frac{n\pi x}{l} + b_n \sin\frac{n\pi x}{l}\right).$$

We know, $\cos\theta = \frac{1}{2}(e^{i\theta} + e^{-i\theta})$ and $\sin\theta = \frac{1}{2i}(e^{i\theta} - e^{-i\theta})$.

$$\therefore \quad f(x) = \frac{1}{2}a_0 + \frac{1}{2}\sum_{n=1}^{\infty} a_n \left[e^{i\frac{n\pi x}{l}} + e^{-i\frac{n\pi x}{l}}\right] + \frac{1}{2i}\sum_{n=1}^{\infty} b_n \left[e^{i\frac{n\pi x}{l}} - e^{-i\frac{n\pi x}{l}}\right]$$

$$= \frac{1}{2}a_0 + \frac{1}{2}\sum_{n=1}^{\infty}(a_n - ib_n)e^{i\frac{n\pi x}{l}} + \frac{1}{2}\sum_{n=1}^{\infty}(a_n + ib_n)e^{-i\frac{n\pi x}{l}}$$

$$= C_0 + \sum_{n=1}^{\infty} C_n e^{i\frac{n\pi x}{l}} + \sum_{n=1}^{\infty} C_{-n} e^{-i\frac{n\pi x}{l}},$$

where $C_0 = \frac{1}{2}a_0,$ (i)

$$C_n = \frac{1}{2}(a_n - ib_n) \tag{ii}$$

$$= \frac{1}{2l}\int_{-l}^{l} f(x)\cos\frac{n\pi x}{l}dx - \frac{1}{2l}i\int_{-l}^{l} f(x)\sin\frac{n\pi x}{l}dx$$

$$= \frac{1}{2l}\int_{-l}^{l} f(x)\left[\cos\frac{n\pi x}{l} - i\sin\frac{n\pi x}{l}\right]dx = \frac{1}{2l}\int_{-l}^{l} f(x)e^{-i\frac{n\pi x}{l}}dx$$

and $$C_{-n} = \frac{1}{2}(a_n + ib_n) = \frac{1}{2l}\int_{-l}^{l} f(x)e^{i\frac{n\pi x}{l}}dx. \tag{iii}$$

Now putting $x = -y$, $dx = -dy$ for $x = l$, $y = -l$ and $x = -l$, $y = l$.

$\therefore$ from Eqn. (ii),

$$C_n = \frac{1}{2l}\int_{l}^{-l} f(-y)e^{i\frac{n\pi y}{l}}(-dy) = \frac{1}{2l}\int_{-l}^{l} f(-y)e^{i\frac{n\pi y}{l}}dy. \tag{iv}$$

For even function, $f(-y) = f(y)$.

$\therefore$ from Eqn. (iv),

$$C_n = \frac{1}{2l}\int_{-l}^{l} f(y)e^{i\frac{n\pi y}{l}}dy = C_{-n} \quad \text{or,} \quad \frac{1}{2}(a_n - ib_n) = \frac{1}{2}(a_n + ib_n) \quad \text{or,} \quad -ib_n = ib_n,$$

i.e., $b_n = 0.$

$$\therefore \quad C_n = \frac{1}{2}a_n,$$

which shows that the complex Fourier coefficients of an even function are real.

Again, for odd function, $f(-y) = -f(y)$.

$\therefore$ From Eqn. (iii),

$$C_n = \frac{1}{2l}\int_{-l}^{l} -f(y)e^{i\frac{n\pi y}{l}}dy = -C_{-n} \quad \text{or,} \quad \frac{1}{2}(a_n - ib_n) = \frac{1}{2}(-a_n - b_n)$$

or, $a_n = 0.$

$$\therefore \quad C_n = -\frac{1}{2}(ib_n),$$

which shows that the complex Fourier coefficients of an odd function are purely imaginary.

Example 7.29. *Find the Fourier series of the function $f(x) = e^{i\lambda x}$, where λ is an integer. Hence, prove that,*

$$\frac{\pi \sin \lambda x}{2 \sin \lambda x} = \frac{\sin x}{1^2 - \lambda^2} - \frac{2 \sin 2x}{2^2 - \lambda^2} + \cdots \quad \text{and} \quad \frac{\pi \cos \lambda x}{2 \sin \lambda x} = \frac{1}{2\lambda} + \frac{\lambda \cos x}{1^2 - \lambda^2} - \frac{\lambda \cos 2x}{2^2 - \lambda^2} + \cdots$$

Solution : Let $e^{i\lambda x} = \frac{a_0}{2} + \sum_{n=1}^{\infty}(a_n \cos nx + b_n \sin nx)$.

The coefficients of the Fourier expansion of $e^{i\lambda x}$ are :

$$a_0 = \frac{1}{\pi}\int_{-\pi}^{\pi} e^{i\lambda x}\,dx = \frac{2}{\lambda\pi}\sin \lambda x, \tag{i}$$

$$a_n = \frac{1}{\pi}\int_{-\pi}^{\pi} e^{i\lambda x}\cos nx\,dx$$

or, $$\pi a_n = \left[e^{i\lambda x}\frac{\sin nx}{n}\right]_{-\pi}^{\pi} - \int \frac{i\lambda}{n} e^{i\lambda x}\sin nx\,dx = \left[i\lambda e^{i\lambda x}\frac{\cos nx}{n^2}\right]_{-\pi}^{\pi} + \lambda^2 \int e^{i\lambda x}\frac{\cos kx}{n^2}dx$$

$$= -\frac{2\lambda}{n^2}\sin \lambda x \cos n\pi + \frac{\lambda^2}{n^2}(\pi a_n) \text{ [from Eqn. (i)]}.$$

$\therefore$ $$\pi a_n\left(1 - \frac{\lambda^2}{n^2}\right) = -\frac{2\lambda}{n^2}(-1)^n \sin \lambda\pi.$$

So, $$a_n = -(-1)^n \frac{2\lambda \sin \lambda x}{\pi n^2 - \lambda^2}. \tag{ii}$$

Also, $$b_n = \frac{1}{\pi}\int_{-\pi}^{\pi} e^{i\lambda x}\sin nx\,dx, \tag{iii}$$

i.e., $$\pi b_n = -\left[e^{i\lambda x}\frac{\cos nx}{n}\right]_{-\pi}^{\pi} + \frac{i\lambda}{n}\int_{-\pi}^{\pi} e^{i\lambda x}\cos nx\,dx$$

$$= -\frac{2i}{n}\cos n\pi \sin \lambda x + \frac{\lambda^2}{n^2}\int_{-\pi}^{\pi} e^{i\lambda x}\sin nx\,dx$$

$$= -\frac{2i}{n}(-1)^n \sin \lambda\pi + \frac{\lambda^2}{n^2}(\pi b_n) \text{ [from Eqn. (iii)]}$$

or, $$\pi b_n\left(1 - \frac{\lambda^2}{n^2}\right) = -\frac{2i(-1)^n}{n}\sin \lambda x \quad \text{or,} \quad b_n = -\frac{2i}{\pi}(-1)^n \frac{n \sin \lambda x}{n^2 - \lambda^2}. \tag{iv}$$

Hence, $$\boldsymbol{e^{i\lambda x} = \frac{2}{\pi}\sin \lambda x\left[\frac{1}{2\lambda} + \sum_{n=1}^{\infty}(-1)^{n+1}\left(\frac{n \cos nx}{n^2 - \lambda^2} + \frac{in \sin nx}{n^2 - \lambda^2}\right)\right]}. \tag{v}$$

Equating the imaginary and real parts of Equation (v), we get

$$\frac{\pi \sin \lambda x}{2 \sin \lambda x} = \sum_{n=1}^{\infty}(-1)^{n+1}\frac{n \sin nx}{n^2 - \lambda^2} \quad \text{and} \quad \frac{\pi \cos \lambda x}{2 \sin \lambda x} = \frac{1}{2\lambda} + \sum_{n=1}^{\infty}(-1)^{n+1}\frac{\lambda \sin nx}{n^2 - \lambda^2}$$

or, $$\boldsymbol{\frac{\pi \sin \lambda x}{2 \sin \lambda x} = \frac{\sin x}{1^2 - \lambda^2} - \frac{2 \sin 2x}{2^2 - \lambda^2} + \cdots}$$

and $$\boldsymbol{\frac{\pi \cos \lambda x}{2 \sin \lambda x} = \frac{1}{2\lambda} + \frac{\lambda \cos x}{1^2 - \lambda^2} - \frac{\lambda \cos 2x}{2^2 - \lambda^2} + \cdots}.$$

Example 7.30. *Find the Fourier series expansion for the function,* $f(x) = x^4 - 2\pi x^2$, $(-\pi \le x \le \pi)$; *hence show that,* $\sum_{n=1}^{\infty} \frac{1}{n^4} = \frac{\pi^4}{90}$.

Solution : The given function $f(x) = x^4 - 2\pi x^2$ is even. The Fourier expansion is

$$x^4 - 2\pi x^2 = \frac{a_0}{2} + \sum_{n=1}^{\infty} a_n \cos nx. \tag{i}$$

Then, the coefficients are :

$$a_0 = \frac{1}{\pi}\int_{-\pi}^{\pi}(x^4 - 2\pi x^2)dx = 2\left(\frac{\pi^4}{5} - \frac{2\pi^3}{3}\right), \tag{ii}$$

$$a_n = \frac{1}{\pi}\int_{-\pi}^{\pi}(x^4 - 2\pi x^2)\cos nx\, dx.$$

Now, $$I = \int_{-\pi}^{\pi} x^2 \cos nx\, dx = \left[x^2\frac{\sin nx}{n}\right]_{-\pi}^{\pi} - \int_{-\pi}^{\pi} 2x\frac{\sin nx}{n}dx$$

$$= -\frac{2}{n}\left[\left(-x\frac{\cos nx}{n}\right)_{-\pi}^{\pi} + \int_{-\pi}^{\pi}\frac{\cos nx}{n}dx\right] = (-1)^n\frac{4\pi}{n^2} \tag{iii}$$

and $$\int_{-\pi}^{\pi} x^4 \cos nx\, dx = \left[x^4\frac{\sin nx}{n}\right]_{-\pi}^{\pi} - \int_{-\pi}^{\pi} 4x^3\frac{\sin nx}{n}dx = \frac{4}{n^2}\left(x^3\cos nx\right)_{-\pi}^{\pi} - \frac{12}{n^2}I$$

$$= \frac{8\pi^3}{n^2}(-1)^n - \frac{12}{n^2}\left(\frac{4\pi}{n^2}\right)(-1)^n \text{ [by Eqn. (iii)]}$$

$$= (-1)^n\left(\frac{8\pi^3}{n^2} - \frac{48\pi}{n^4}\right). \tag{iv}$$

Now, from Eqns. (i), (ii) and (iii), we have

$$a_n = \frac{(-1)^n}{\pi}\left[\frac{8\pi^3}{n^2} - \frac{48\pi}{n^4} - 2\pi\left(\frac{4\pi}{n^2}\right)\right] = (-1)^n\left[\frac{8\pi^2 - 8\pi}{n^2} - \frac{48}{n^4}\right] \tag{v}$$

$\therefore$ from Eqn. (i), we get

$$\boldsymbol{x^4 - 2\pi x^2 = \left(\frac{\pi^4}{5} - \frac{2\pi^3}{3}\right) + (8\pi^2 - 8\pi)\sum_{n=1}^{\infty}(-1)^n\frac{\sin nx}{n^2}}$$

$$\boldsymbol{- 48\sum_{n=1}^{\infty}(-1)\frac{\cos n\pi}{n^4}}. \tag{vi}$$

Putting $x = 0$ on both sides of Eqn. (6), we get, the Fourier expansion of $f(x)$,

$$0 = \frac{\pi^4}{5} - \frac{2\pi^3}{3} - (8\pi^2 - 8\pi)\left[\frac{1}{1^2} - \frac{1}{2^2} + \frac{1}{3^2} - \cdots\right] + 48\left[\frac{1}{1^4} - \frac{1}{2^4} + \frac{1}{3^4} - \cdots\right]$$

$$= \frac{\pi^4}{5} - \frac{2\pi^3}{3} - (8\pi^2 - 8\pi)\left[\left(\frac{1}{1^2} + \frac{1}{2^2} + \frac{1}{3^2} + \cdots\right) - \frac{2}{2^2}\left(\frac{1}{1^2} + \frac{1}{2^2} + \frac{1}{3^2} + \cdots\right)\right]$$

$$+ 48\left[\left(\frac{1}{1^4} + \frac{1}{2^4} + \cdots\right) - \frac{2}{2^4}\left(\frac{1}{1^4} + \frac{1}{2^4} + \cdots\right)\right]$$

or, $$0 = \frac{\pi^4}{5} - \frac{2\pi^3}{3} - (8\pi^2 - 8\pi)\frac{1}{2}\sum_{n=1}^{\infty}\frac{1}{n^2} + 48\left(\frac{7}{8}\right)\sum_{n=1}^{\infty}\frac{1}{n^4}$$

$$= \frac{\pi^4}{5} - \frac{2\pi^3}{3} - (8\pi^2 - 8\pi)\frac{1}{2}\left(\frac{\pi^2}{6}\right) + 42\sum_{n=1}^{\infty}\frac{1}{n^4} \quad \left[\because \sum_{n=1}^{\infty}\frac{1}{n^2} = \frac{\pi^2}{6}\right].$$

Hence, $42\sum_{n=1}^{\infty}\frac{1}{n^4} = \frac{2}{3}\pi^4 - \frac{1}{5}\pi^4 = \frac{7}{15}\pi^4$

or, $$\sum_{n=1}^{\infty}\frac{1}{n^4} = \frac{\pi^4}{90} \quad \textbf{(Proved)}.$$

Example 7.31. *Expand $f(x) = e^x$ in a cosine series over (0, 1).*

Solution : We have $f(x) = e^x$ and $l = 1$.

$$\therefore \quad a_0 = \frac{2}{l}\int_0^l f(x)dx = \frac{2}{1}\int_0^1 e^x dx = 2(e-1),$$

$$a_n = \frac{2}{l}\int_0^l f(x)\cos\frac{n\pi x}{l}dx = \frac{2}{1}\int_0^1 e^x\cos\frac{n\pi x}{1}dx = 2\left[\frac{e^x}{n^2\pi^2+1}(n\pi\sin n\pi x + \cos n\pi x)\right]_0^1$$

$$= 2\left[\frac{e^x}{n^2\pi^2+1}(n\pi\sin n\pi + \cos n\pi) - \frac{1}{n^2\pi^2+1}\right] = \frac{2}{n^2\pi^2+1}[(-1)^n e - 1].$$

$$\therefore \quad f(x) = e^x = \frac{a_0}{2} + a_1\cos\pi x + a_2\cos 2\pi x + a_3\cos 3\pi x + \cdots$$

or, $$e^x = e - 1 + 2\left[\frac{-e-1}{\pi^2+1}\cos\pi x + \frac{e-1}{4\pi^2+1}\cos 2\pi x + \frac{-e-1}{9\pi^2+1}\cos 3\pi x + \cdots\right].$$

Exercises

7(A) : Objective Type Questions

Select the correct answer

7.1 A function $f(x)$ can be expressed in terms of Fourier series, if it is

(a) single valued and periodic **(b)** single valued and bounded

(c) single valued, periodic and bounded **(d)** periodic and bounded.

7.2 The system of functions $\sin x \ \sin 3x \ \sin 5x \cdots \sin(2n+1)x$ is orthogonal in the interval

(a) $(0, \pi)$ **(b)** $\left(0, \frac{\pi}{2}\right)$ **(c)** $(0, 2\pi)$ **(d)** $(-\pi, \pi)$.

7.3 A function $f(x)$ is even, if

(a) $f(x) = -f(x)$ **(b)** $f(-x) = f(x)$ **(c)** $f(-x) = -f(x)$ **(d)** $f(x) = \pm f(x)$.

7.4 An even function $f(x)$ is symmetrical about the

(a) y-axis **(b)** x-axis **(c)** origin **(d)** none of these.

7.5 An odd function $f(x)$ is symmetrical about the

(a) y-axis **(b)** origin **(c)** x-axis **(d)** none of these.

7.6 Product of two odd or two even functions is an

(a) odd function **(b)** even function **(c)** neither odd nor even function **(d)** none of these.

7.7 The product of an odd and an even function is an

(a) even function **(b)** odd function **(c)** neither odd nor even function **(d)** none of these.

7.8 The sum of the series $1 - \frac{1}{2^2} + \frac{1}{3^2} - \frac{1}{4^2} + \cdots$ is

(a) $\frac{\pi^2}{6}$ **(b)** $\frac{\pi^2}{8}$ **(c)** $\frac{\pi}{4}$ **(d)** $\frac{\pi^2}{12}$.

7.9 The sum of the series $1 + \frac{1}{3^2} + \frac{1}{5^2} + \cdots$ is

(a) $\frac{\pi^2}{8}$ **(b)** $\frac{\pi^2}{12}$ **(c)** $\frac{\pi^2}{6}$ **(d)** $\frac{\pi^2}{10}$.

7.10 $\int_{-\pi}^{\pi} \sin nx \sin mx dx = \pi$, if

(a) $n \neq m$ **(b)** $n = m = 0$ **(c)** $n = m > 0$ **(d)** $n = -m$.

7.11 For an odd function, the Fourier coefficients are

(a) $a_0 = 0$, $a_n = 0$ for all n, $b_n \neq 0$ **(b)** $a_0 \neq 0$, $a_n = 0$, $b_n = 0$

(c) $a_0 = 0$, $a_n \neq 0$, $b_n \neq 0$ **(d)** $a_0 = 0$, $a_n \neq 0$, $b_n = 0$.

7.12 For an even function, the Fourier coefficients are

(a) $a_0 = 0$, $a_n \neq 0$, $b_n \neq 0$ **(b)** $a_0 = 0$, $a_n \neq 0$, $b_n = 0$

(c) $a_0 \neq 0$, $a_n \neq 0$, $b_n = 0$ **(d)** $a_0 \neq 0$, $a_n = 0$, $b_n \neq 0$.

7.13 To satisfy the Dirichlet conditions, a function $f(x)$ should be

(a) single valued and bounded **(b)** piecewise continuous

(c) a finite number of extrema **(d)** all of these.

7.14 Identify the odd function,

(a) x^2 **(b)** $\sec x$ **(c)** $\tan x$ **(d)** $\cos x$.

7.15 The value of the sum of the series, $\sum_{n=1}^{\infty} \frac{1}{n^4}$ is

(a) $\frac{\pi^2}{90}$ **(b)** $\frac{\pi^4}{90}$ **(c)** $\frac{90}{\pi^4}$ **(d)** $\frac{\pi^4}{180}$.

Answers

7.1 (c) **7.2** (b) **7.3** (b) **7.4** (a) **7.5** (b) **7.6** (b) **7.7** (b) **7.8** (d) **7.9** (a) **7.10** (c) **7.11** (a) **7.12** (c) **7.13** (d) **7.14** (c) **7.15** (b).

7(B) : Short Answer Type Questions

7.1 What is a Fourier series? How can Fourier series be applied to periodic functions?

7.2 State the conditions under which function can be expanded into a convergent series.

7.3 What are odd and even functions? Explain.

7.4 Can you expand $f(x) = \tan x$ in Fourier series? Explain.

7.5 Is the system of functions $\sin x$, $\sin 2x$, $\ldots$, $\sin nx$, $\ldots$ orthogonal in the interval $(0, \pi)$? **[BU(H) 2004]**

7.6 State Fourier's theorem and Dirichlet condition. **[BU(H) 1992, '98, 2000]**

7.7 If $f(x)$ is an even function then the Fourier series reduces to a cosine series—prove it.

7.8 Show that for an odd function the Fourier series is a sine series.

7.9 What is a half-range series? Is $f(x) = \frac{a_0}{2} + \sum_{n=1}^{\infty} a_n \cos \frac{n\pi x}{l}$, write down the expressions of a_0 and a_n.

7.10 Which of the following statements are true or false?

(a) The graph of an odd function $f(x)$ in $(-\pi, \pi)$ is symmetric about the x-axis.

(b) If $f(x)$ is an even function, then the graph is symmetrical about the origin.

(c) If $f(x)$ is defined in $(0, l)$, then the period of $f(x)$ to expand it is a half-range sine series of period $2l$.

(d) x^2, $\cos x$, $\sec x$ are all even functions.

(e) x^3, $\sin x$, $\tan x$ are all odd functions.

(f) $\int_{-\pi}^{\pi} \cos nx \cos mx = 2\pi$, for $n \neq m$.

(g) If $f(x) = |x|$, for $-l < x < l$, then $\frac{1}{1^2} + \frac{1}{3^2} + \frac{1}{5^2} + \cdots = \frac{\pi^2}{8}$.

7.11 What is a complex form of Fourier series? Write down the expressions of the coefficient.

7(C) : Review Questions

7.1 A periodic function $f(x)$ with period 2π is defined as $f(x) = x^2$, $-\pi \leq x \leq \pi$.

Expand $f(x)$ in Fourier series and hence, show that, $\sum_{n=1}^{\infty} \frac{1}{n^2} = \frac{\pi^2}{6}$.

[VU(H) 2009, CU(H) 2007]

7.2 Expand the following periodic function in a Fourier series :

$$\begin{aligned} f(t) &= 0, \qquad \text{for } -\frac{T}{2} < t \leq 0 \\ &= E_0 \sin \omega t, \quad \text{for } 0 < t \leq \frac{T}{2}, \end{aligned}$$

where $T = \frac{2\pi}{\omega}$. **[VU(H) 1999]**

7.3 Obtain the Fourier series of the periodic function $f(x)$ defined as

$$\begin{aligned} f(x) &= 0, \quad \text{for } -\pi < x \leq 0 \\ &= x, \quad \text{for } 0 < x \leq \pi. \end{aligned}$$

Hence, show that, $\frac{\pi^2}{8} = \sum_{n=1}^{\infty} \frac{1}{(2n-1)^2}$. **[BU(H) 2000, NBU(H) 2008]**

7.4 Expand in Fourier series of the function $f(x) = x$ in the interval $-1 < x < 1$.

[NBU(H) 2009]

7.5 Find the Fourier series expansion for the function $f(x)$ defined by

$$\begin{aligned} f(x) &= 0, \quad \text{for } 0 < x < \pi \\ &= -x, \quad \text{for } -\pi < x < 0 \\ f(x+2\pi) &= f(x). \end{aligned}$$

[CU(H) 2008]

7.6 The function $f(x)$ is expanded in Fourier series :

$$f(x) = \frac{a_0}{2} + \sum_{n=1}^{\infty} \left(a_n \cos \frac{n\pi x}{l} + b_n \sin \frac{n\pi x}{l} \right).$$

(i) State the Dirichlet conditions imposed upon function $f(x)$.

(ii) Evaluate b_n's when $f(x)$ is an even function. **[CU(H) 2002]**

7.7 (a) If the Fourier series expansion of a function $f(x)$ contains only sine terms, what kind of function is $f(x)$?

(b) Let $f(x)$ have a Fourier series expansion $f(x) = \sum_{n=1}^{\infty} a_n \cos nx + \sum_{n=1}^{\infty} b_n \sin nx$, then prove that,

$$\langle f^2(x) \rangle = \frac{1}{2\pi} \int_{-\pi}^{\pi} f^2(x) dx = \sum_{n=1}^{\infty} \frac{a_n^2 + b_n^2}{2}.$$

[CU(H) 2001]

7.8 Find a series of sine and cosine of multiple of x which will represent $f(x)$ in the interval $(-\pi, \pi)$, where

$$\begin{aligned} f(x) &= 0 \quad ; \quad -\pi < x < 0 \\ &= \frac{n\pi}{4}; \quad 0 < x < \pi \end{aligned}$$

and hence, deduce that, $\frac{\pi^2}{8} = \left[1 + \frac{1}{3^2} + \frac{1}{5^2} + \cdots \right]$.

7.9 Find the Fourier series of the function $f(x) = \begin{cases} -2; & -4 < x < -2 \\ x; & -2 < x < 2 \\ 2; & 2 < x < 4. \end{cases}$

[Ans: $\frac{4}{\pi} + \frac{8}{\pi^2} \sin \frac{\pi x}{4} \dot{-} \frac{2}{\pi} \sin \frac{2\pi x}{4} + \left(\frac{4}{3\pi} - \frac{8}{3^2\pi} \right) \sin \frac{3\pi x}{4} - \frac{2}{2\pi} \sin \frac{4\pi x}{4} + \cdots$**]**

7.10 A periodic square wave has a period 4. The function generating the square is

$$\begin{aligned} f(t) &= 0, \text{ for } -2 < t < -1 \\ &= k, \text{ for } -1 < t < 1 \\ &= 0, \text{ for } 1 < t < 2. \end{aligned}$$

Find the Fourier series of the function.

[AMIETE (Summer) 1995] **[Ans:** $f(t) = \frac{k}{2} + \frac{2k}{\pi} \left[\cos \frac{\pi t}{2} - \frac{1}{3} \cos \frac{3\pi t}{2} + \cdots \right]$**]**

7.11 Find the Fourier series to represent, $f(x) = x^2 - 2$, from $-2 < x < 2$.

[Ans: $-\frac{2}{3} - \frac{16}{\pi^2} \left[\cos \frac{\pi x}{2^2} - \frac{1}{4} \cos \pi x + \frac{1}{9} \cos \frac{3\pi x}{2} - \cdots \right]$**]**

7.12 Find the Fourier series of the function $f(x) = \begin{cases} -\frac{\pi}{4}; \text{ for } -\pi < x < 0 \\ \frac{\pi}{4}; \text{ for } 0 < x < \pi \end{cases}$

and $f(-\pi) = f(0) = f(\pi) = 0$, $f(x) = f(x + 2\pi)$ for all x.

Deduce that, $\frac{\pi}{4} = 1 - \frac{1}{3} + \frac{1}{5} - \frac{1}{7} + \cdots$.

7.13 Find the displacement $y(x, t)$ of the string, if it is set into vibration by striking at the centre so as to give it a velocity which varies linearly from zero at $x = 0$ to v_0 at $x = \frac{L}{2}$, keeping the remaining portion of the string initially at rest. **[BU(H) (Old) 1995]**

7.14 Find the Fourier series of the function

$$f(x) = 0, \quad -5 < x < 0$$
$$\text{and} \quad f(x) = 3, \quad 0 < x < 5.$$

Period 10. **[BU(H) (Old) 1996]**

7.15 Expand $f(x) = x^2$ in the interval $0 < x < 2\pi$ in a Fourier series when the period is 2π to show that,

$$f(x) = \frac{4\pi^2}{3} + \sum_{n=1}^{\infty} \left(\frac{4}{n^2} \cos nx - \frac{4\pi}{n} \sin nx \right).$$

7.16 Use Fourier series expansion to show that,

(a) $\sum_{n=1}^{\infty} \frac{1}{n^2} = \frac{\pi^2}{6}$, **(b)** $\sum_{n=1}^{\infty} \frac{1}{n^4} = \frac{\pi^2}{90}$, **(c)** $\sum_{n=1}^{\infty} \frac{1}{(2n-1)^2} = \frac{\pi^2}{8}$,

(d) Develop the Fourier expansion for the function, $f(t) = \begin{cases} 0 \ , \text{ for } -\pi \le \omega t \le 0 \\ \sin \omega t, \text{ for } 0 \le \omega t \le -\pi. \end{cases}$

7.17 An alternating current passing through a rectifier has the form

$$I(\theta) = \begin{cases} I_0 \sin \theta, \text{ for } 0 < \theta < \pi \\ 0 \ , \text{ for } \pi < \theta < 2\pi. \end{cases}$$

Find the Fourier series of the function $I(\theta)$.

[Ans: $\frac{I_0}{\pi} - \frac{2I_0}{\pi} \left(\frac{\cos 2\theta}{3} + \frac{\cos 4\theta}{15} + \cdots \right) + \frac{I}{2} \sin \theta$**]**

7.18 Obtain a Fourier expansion for $f(x) = x^3$, $-\pi < x < \pi$.

[Ans: $x^3 = 2\left[-\left(-\frac{\pi^2}{1} + \frac{6}{1^3} \right) \sin x + \left(-\frac{\pi^2}{2} + \frac{6}{2^3} \right) \sin 2x - \left(-\frac{\pi^2}{3} + \frac{6}{3^3} \right) \sin 3x \right]$**]**

7.19 Expand the function $f(x) = |x|$, for $-\pi \le x \le \pi$ in Fourier series.

[Ans: $|x| = \frac{\pi}{2} - \frac{4}{\pi} \left(\cos x + \frac{\cos 3x}{3^2} + \frac{\cos 5x}{5^2} + \cdots \right)$**]**

7.20 Find the Fourier series of the function $f(x) = \begin{cases} -1, \text{ for } -\pi < x < 0 \\ 0, \text{ for } x = 0 \\ 1, \text{ for } 0 < x < \pi. \end{cases}$

[Ans: $f(x) = \frac{4}{\pi} \left(\sin x + \frac{\sin 3x}{3} + \frac{\sin 5x}{5} + \cdots \right)$**]**

7.21 If $f(x) = x + 1$, for $0 < x < \pi$, find its Fourier (i) Sine series (ii) Cosine series.

Hence, deduce that (i) $1 - \frac{1}{3} + \frac{1}{5} - \frac{1}{7} + \cdots = \frac{\pi}{4}$ (ii) $1 + \frac{1}{3^2} + \frac{1}{5^2} + \frac{1}{7^2} + \cdots = \frac{\pi^2}{8}$.

[**Ans:** (i) $\frac{2}{\pi}\left[(\pi - 2)\sin x - \frac{\pi}{2}\sin 2x + \frac{1}{3}(\pi + 2)\sin 3x - \frac{\pi}{4}\sin 4x + \cdots\right]$

(ii) $\frac{\pi}{2} + 1 - 4\left(\cos x + \frac{\cos 3x}{3^2} + \frac{\cos 5x}{5^2} + \cdots\right)$]

7.22 Obtain the Fourier cosine series expansion of the periodic function defined by

$$f(t) = \sin\left(\frac{\pi t}{l}\right),\ 0 < t < l$$ **[AMIETE (Winter) 1994]**

[**Ans:** $f(t) = \frac{2}{\pi} - \frac{4}{\pi}\left[\frac{1}{3}\cos\frac{2\pi t}{l} + \frac{1}{15}\cos\frac{4\pi t}{l} + \frac{1}{35}\cos\frac{6\pi t}{l} + \cdots\right]$]

7.23 Find the Fourier series to represent $f(x) = x^2 - 2$, from $-2 < x < 2$.

[**Ans:** $-\frac{2}{3} - \frac{16}{\pi^2}\left[\cos\frac{\pi x}{2^2} - \frac{1}{4}\cos\pi x + \frac{1}{9}\cos\frac{3\pi x}{2} - \cdots\right]$]

7.24 A sawtooth wave is given by $f(x) = x,\ -\pi < x < \pi$.

Show that, $f(x) = 2\sum_{n=1}^{\infty}\frac{(-1)^{n+1}}{1}\sin n\pi$.

7.25 Show that, for x in $[0, \pi]$, $\sum_{k=1}^{\infty}\frac{\cos(2k-1)}{(2k-1)^4} = \frac{1}{96}\pi(\pi - 2x)(\pi^2 + 2\pi x - 2x^2)$.

7.26 Show that a constant c can be expanded in an infinite series

$$\frac{4c}{\pi}\left[\sin x + \frac{\sin 3x}{3} + \frac{\sin 5x}{5} + \cdots\right] \text{ in the range } 0 < x < \pi.$$

7.27 Obtain a half-range cosine series for

$$f(x) = kx, \qquad \text{for } 0 \le x \le \frac{l}{2}$$

$$f(x) = k(l - x), \quad \text{for } \frac{l}{2} \le x \le l.$$

[**Ans:** $f(x) = \frac{kl}{3} - \frac{8kl}{\pi^2}\left[\frac{1}{2^2}\cos\frac{2\pi x}{l} + \frac{1}{6^2}\cos\frac{6\pi x}{l} + \frac{1}{10^2}\cos\frac{10\pi x}{l} + \cdots\right]$]

7.28 Expand a Fourier series of the following function : $f(x) = \begin{cases} 0, & -\pi < x < 0 \\ 1, & 0 < x < \pi \end{cases}$

[BU(H) 2012]

7.29 (i) State the Dirichlet conditions in the context of Fourier series expansion.

(ii) Find the Fourier coefficients corresponding to the function,

$$f(x) = \begin{cases} 0, & \text{for } -5 < x < 0 \\ 3, & \text{for } 0 < x < 5 \end{cases}$$

having period 10. Hence, write the corresponding Fourier series. How should $f(x)$ be defined at $x = -5$, $x = 0$ and $x = 5$ in order that the Fourier series will converge to $f(x)$ for $-5 \le x \le 5$? **[BU(H) 2011]**

7.30 Find the Fourier series of the function $f(x) = x + x^2$ for $-\pi < x < \pi$ and show that,

$$\frac{\pi^2}{6} = \frac{1}{1^2} + \frac{1}{2^2} + \frac{1}{3^2} + \frac{1}{4^2} + \cdots$$ **[BU(H) 2010]**

7.31 A periodic function $f(x)$ in the period 2π is defined as $f(x) = -x^2;\ -x \leq x \leq \pi$. Express $f(x)$ in a Fourier series and show that $\sum_{n=1}^{\infty} \frac{1}{n^2} = \frac{\pi^2}{6}$. **[NBU(H) 2008]**

7.32 Find the Fourier series for the function shown graphically

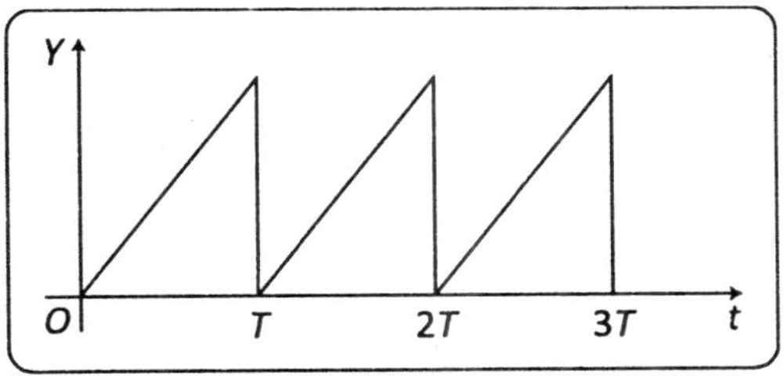

[NBU(H) 2007]

7.33 Find the Fourier series expansion for the function $f(x)$ defined by

$$\begin{aligned} f(x) &= x, && \text{for } 0 < x < \pi \\ f(x) &= -x, && \text{for } -\pi < x < 0 \\ f(x+2\pi) &= f(x). \end{aligned}$$

[CU(H) 2014]

7.34 Expand the function

$$\begin{aligned} f(x) &= x^2, && \text{for } -\pi < x < \pi \\ f(x+2\pi) &= f(x). \end{aligned}$$

in Fourier series. Hence deduce, $\frac{\pi^2}{6} = \sum_{n=1}^{\infty} \frac{1}{n^2}$. **[CU(H) 2013]**

7.35 Expand the function

$$\begin{aligned} f(x) &= x, && \text{for } -\pi < x < \pi \\ f(x+2\pi) &= f(x). \end{aligned}$$

in Fourier series. Hence show that, $1 - \frac{1}{3} + \frac{1}{5} - \frac{1}{7} + \cdots = \frac{\pi}{4}$. **[CU(H) 2010]**

7.36 **(a)** State the Dirichlet condition for Fourier series expansion, $f(x)$.

(b) Let $f(x)$ have a convergent Fourier series expansion,

$$f(x) = \frac{1}{2}a_0 + \sum_{n=1}^{\infty}(a_n \cos nx + b_n \sin nx)$$

Show that, $\frac{1}{\pi}\int_{-\pi}^{\pi}[f(x)]^2 dx = \frac{1}{2}a_0 + \sum_{n=1}^{\infty}(a_n^n + b_n^2)$.

(c) Expand the function

$$\begin{aligned} f(x) &= x, && \text{for } -\pi < x < \pi \\ f(x+2\pi) &= f(x). \end{aligned}$$

in Fourier series. Is this series differentiable everywhere? **[CU(H) 2007]**

7.37 Express $f(x) = \frac{x}{2}$ as a Fourier series in the interval $-\pi < x < \pi$. **[VU(H) 2012]**

7.38 Expand $f(x) = x^2$ for $-\pi \leq x \leq \pi$ in a Fourier series.

Now show that, $\sum_{n=1}^{\infty} \frac{(-1)^{n-1}}{n^2} = \frac{\pi^2}{12}$ **[VU(H) 2011]**

7.39 (a) State and prove Parseval's identity for Fourier series.

(b) What is half-way fourier sine and half-way Fourier cosine series?

(c) Expand $f(x) = x^2$, $0 < x < 2\pi$ in Fourier series, if the period is 2π.

[Pune U(MSc) 2014]

7.40 (a) State and explain Dirichlet conditions.

(b) What are odd and even functions? Determine the Fourier coefficients for odd functions.

(c) Obtain the Fourier series corresponding to the functions

$f(x) = 0, \quad -5 < x < 0$

$f(x) = 3, \quad 0 < x < 5$

period 10. **[Pune U(MSc) 2014]**

7.41 (a) State the Dirichlet conditions that the function $f(x)$ should satisfy in order to expand in terms of Fourier series.

(b) Expand $F(x)$ as a Fourier series, where

$$\begin{aligned} F(x) &= x, \quad 0 < x < \pi \\ &= -x, \quad -\pi < x < 0. \end{aligned}$$

[Pune U(MSc) 2014]

7.42 (a) State and prove the Parseval's identity for Fourier series.

(b) Find the Fourier coefficients a_n and b_n in the interval $(-L, +L)$ for odd and even functions.

(c) Obtain the Fourier series corresponding to the function :

$f(x) = 0, \quad -5 < x < 0$

$f(x) = 3, \quad 0 < x < 5$

period 10. **[Pune U(MSc), 2008]**

7.43 (a) Define Fourier series. Determine the Fourier coefficients for even functions.

(b) Find the Fourier coefficients corresponding to the function

$$f(x) = \begin{cases} 0, & -5 < x < 0 \\ 3, & 0 < x < 5 \end{cases}$$

and write corresponding Fourier series. **[Pune U(MSc) (2013 Pattern)]**

7.44 (a) State Dirichlet condition for expansion of a periodic function in Fourier series.

(b) Classify the following function as even, odd or neither even nor odd and expand in a Fourier series :

$$f(x) = \begin{cases} -x, & -2 < x < 0 \\ x, & 0 < x < 2 \end{cases}$$

period 4.

(c) Making use of Parseval's identity corresponding to the Fourier series of the above function, show that,

$$\frac{1}{1^4} + \frac{1}{3^4} + \frac{1}{5^4} + \cdots = \frac{\pi^4}{96}.$$

(d) Find fourier series expansion of the output of a full-wave rectifier.

7.45 (a) What is the standard form of Fourier series for an even function?

(b) Find the Fourier of (i) $f(x) = x;\ 0 < x < 2\pi$ (ii) $f(x) = |x|;\ -2 < x < 2$.

(c) Obtain the Fourier series of $f(x) = x - x^2$ in $(-\pi, \pi)$.

Hence, deduce, $\frac{1}{1^2} - \frac{1}{2^2} + \frac{1}{3^2} - \frac{1}{4^2} + \cdots = \frac{\pi^2}{12}$ **[Kerala U(Model question)]**

7.46 **(a)** Find the half-range cosine series of $f(x) = \begin{cases} \frac{2K}{L}x; & 0 < x < \frac{L}{2} \\ \frac{2K}{L}(L-x); & \frac{L}{2} < x < L. \end{cases}$

(b) Find the Fourier of $f(x) = \frac{x^2}{2}$; $-\pi < x < \pi$.

Hence, deduce $\sum_{n=1}^{\infty} \frac{1}{n^2}$. [Kerala U(Model question)]

7.47 **(a)** Find the Fourier series expansion of the output of a half wave rectifier. Draw the rectifier wave function between 0 to T, where T is the time period.

(b) Obtain a cosine series expansion of the function $f(x) = 1 + x$ valid in the interval $0 < x < 2$ and hence sum the series $\frac{1}{1^2} + \frac{1}{3^2} + \frac{1}{5^2} + \cdots$.

7.48 Find the Fourier series expansion of the function

$$f(x) = \begin{array}{ll} -k, & \text{if } -\pi < x < 0 \\ k, & \text{if } 0 < x < \pi \end{array}$$

and $f(x + 2\pi) = f(x)$. Hence, find the value of $\sum_{n=0}^{\infty} \frac{(-1)^n}{2n+1}$.

7.49 **(a)** Expand the function $\begin{aligned} f(x) &= x^2, \quad \text{for } -\pi < x < \pi \\ f(x+2\pi) &= f(x) \end{aligned}$

in a Fourier series. Hence prove, $\frac{\pi^2}{6} = \sum_{n=1}^{\infty} \frac{1}{n^2}$. [CU(H) 2017]

(b) Consider the function, $\begin{aligned} f(x) &= \sin x \quad \text{for } 0 < x < \pi \\ &= -\sin x \quad \text{for } -\pi < x < 0, \end{aligned}$

$f(x + 2\pi) = f(x)$. Find the Fourier series expansion of $f(x)$. [CU(H) 2016]

7.50 **(a)** Plot the function $f(x)$ defined by

$f(x) = 0$, for $-\pi < x < 0$,
$f(x) = 1$, for $0 < x < \pi$,
$f(x + 2\pi) = f(x)$ and find the Fourier series expansion. [CU(H) 2015]

(b) Find the Fourier expansion for the function defined by

$f(x) = x$ for $0 < x < \pi$,
$f(x) = -x$ for $-\pi < x < 0$
$f(x + 2\pi) = f(x)$.

(i) Show that $\delta(kx) = \frac{1}{k}\delta(x)$, where k is a constant > 0. [CU(H) 2014]

7.51 **(a)** State the Dirichlet condition for Fourier series expansion of a function $f(x)$.

(b) Expand the function $f(x + 2\pi) = f(x)$ in a Fourier series. [CU(H) 2012]

(c) Expand the function $\begin{aligned} f(x) &= x \quad \text{for } -\pi < x < \pi \\ f(x+2\pi) &= f(x) \end{aligned}$

in a Fourier series. Hence, show that $1 - \frac{1}{3} + \frac{1}{5} - \frac{1}{7} + \cdots = \frac{\pi}{4}$. [CU(H) 2010]

CHAPTER 8

Fourier Transformation

8.1 Introduction

Fourier transformation is a type of integral transformation which affords mathematical devices through which solutions of many boundary value problems in various fields of science and technology are obtained. The fundamental idea which make the transformation a very powerful technique is that **it replaces operation of calculus by the operation of algebra**. The number of independent variables is reduced by one when an integral transformation is applied to a partial differential equation.

8.2 Definition

The integral transformation of a function $f(x)$ is defined by

$$\overline{f}(s) = \int_{x_1}^{x_2} f(x)K(s,x)dx, \tag{8.1}$$

where $K(s,x)$ is called the **Kernel** of the transformation and is a function of s and x.

Here $f(x)$ is the inverse transform of $\overline{f}(s)$.

Depending on the choice of the Kernel and the range of integration a large number of integral transformations exist as for example :

1. **If $K(s,x) = e^{isx}$, then**

$$F(s) = \int_{-\infty}^{\infty} f(x)e^{isx}dx, \quad \text{[Fourier Transform of } f(x)]$$

2. **If $K(s,x) = e^{-sx}$, then**

$$\overline{f}(s) = \int_{0}^{\infty} f(x)e^{-sx}dx, \quad \text{[Laplace's Transform of } f(x)]$$

3. **If $K(s,x) = x^{s-1}$, then**

$$M(s) = \int_{0}^{\infty} f(x)e^{s-1}dx, \quad \text{[Mellin Transform of } f(x)]$$

4. **If $K(s,x) = xJ_n(sx)$, then**

$$H(s) = \int_{0}^{\infty} f(x)xJ_n(sx), \quad \text{[Henkel Transform of } f(x)]$$

5. **If $K(s,x) = \sin sx$, then**

$$F_s(x) = \int_{0}^{\infty} f(x)\sin sx\, dx, \quad \text{[Fourier Sine Transform of } f(x)]$$

The integral transformation are used in solving the partial differential equation with boundary conditions.

8.3 Fourier Integral Theorem

The theorem states that, under certain condition a function $f(x)$ may be expressed as

$$f(x) = \frac{1}{2\pi} \int_0^{\infty} \int_{-\infty}^{\infty} f(t) \cos u(t-x) dt\, du. \tag{8.2}$$

Proof : Fourier series of the function $f(x)$ in $(-l, l)$ is given by

$$f(x) = \frac{a_0}{2} + \sum_{n=1}^{\infty} a_n \cos\frac{n\pi x}{l} + \sum_{n=1}^{\infty} b_n \sin\frac{n\pi x}{l}, \tag{8.3}$$

where a_0, a_n and b_n are given by

$$a_0 = \frac{1}{l}\int_{-l}^{l} f(t)dt, \qquad a_n = \frac{1}{l}\int_{-l}^{l} f(t)\cos\frac{n\pi t}{l}dt, \qquad b_n = \frac{1}{l}\int_{-l}^{l} f(t)\sin\frac{n\pi t}{l}dt.$$

Substituting the values of a_0, a_n and b_n in Eqn. (8.3), we get

$$f(x) = \frac{1}{2l}\int_{-l}^{l} f(t)dt + \sum_{n=1}^{\infty}\frac{1}{l}\int_{-l}^{l} f(t)\cos\frac{n\pi t}{l}\cos\frac{n\pi x}{l}dt + \sum_{n=1}^{\infty}\frac{1}{l}\int_{-l}^{l} f(t)\sin\frac{n\pi t}{l}\sin\frac{n\pi x}{l}dt$$

$$= \frac{1}{2l}\int_{-l}^{l} f(t)dt + \sum_{n=1}^{\infty}\frac{1}{l}\int_{-l}^{+l} f(t)\left[\cos\frac{n\pi t}{l}\cos\frac{n\pi x}{l} + \sin\frac{n\pi t}{l}\sin\frac{n\pi x}{l}\right]dt$$

$$= \frac{1}{2l}\int_{-l}^{l} f(t)dt + \sum_{n=1}^{\infty}\frac{1}{l}\int_{-l}^{l} f(t)\cos\frac{n\pi}{l}(t-x)dt$$

or, $$\boldsymbol{f(x) = \frac{1}{2l}\int_{-l}^{l} f(t)\left\{1 + \sum_{n=1}^{\infty}\cos\frac{n\pi}{l}(t-x)\right\}dt.} \tag{8.4}$$

[$\because$ $\cos(-\theta) = \cos\theta$, as cosine functions are even functions,

$$\therefore\ 1 + 2\sum_{n=1}^{\infty}\cos\frac{n\pi}{l}(t-x) = \sum_{-\infty}^{\infty}\cos\frac{n\pi}{l}(y-x)]$$

$\therefore$ Eqn. (8.4) becomes,

$$f(x) = \frac{1}{2l}\int_{-l}^{l} f(t)\left\{\sum_{-\infty}^{\infty}\cos\frac{n\pi}{l}(t-x)\right\}dt = \frac{1}{2\pi}\int_{-l}^{+l} f(t)\left\{\frac{\pi}{l}\sum_{-\infty}^{\infty}\cos\frac{n\pi}{l}(t-x)\right\}dt. \tag{8.5}$$

We now assume that l increases indefinitely, so that, we may write $\frac{n\pi}{l} = u$ and $\frac{\pi}{l} = .du$.

Then $$\boldsymbol{\underset{l\to\infty}{\text{Lt}}\left\{\frac{\pi}{l}\sum_{-\infty}^{\infty}\cos\frac{n\pi}{l}(t-x)\right\} = \int_{-\infty}^{+\infty}\cos u(t-x)du.} \tag{8.6}$$

From Eqn. (8.5),

$$f(x) = \frac{1}{2\pi}\int_{-\infty}^{\infty} f(t)\left\{2\int_0^{\infty}\cos u(t-x)du\right\}dt. \tag{8.7}$$

So, $$\boldsymbol{f(x) = \frac{1}{\pi}\int_0^{\infty}\int_{-\infty}^{\infty} f(t)\cos u(t-x)du\,dt.} \tag{8.8}$$

The above treatment assumes that,

(a) $f(x)$ is defined as a single-valued function except at finite points in $(-l, l)$.

(b) $f(x)$ is periodic outside $(-l, l)$ with period $2l$.

(c) $f(x)$ and $f'(x)$ are sectionally continuous in $(-l, l)$.

(d) $\int_{-\infty}^{+\infty} |f(x)|dx$ converges, this means $f(x)$ is absolutely integrable in $(-\infty, \infty)$.

8.4 Fourier Sine and Cosine Integrals

Fourier Sine Integral : $f(x) = \frac{2}{\pi}\int_0^\infty \sin ux\, du \int_0^\infty f(x) \sin ut\, dt.$

Fourier Cosine Integral : $f(x) = \frac{2}{\pi}\int_0^\infty \cos ux\, du \int_0^\infty f(t) \cos ut\, dt.$

Proof : We have $\cos u(t-x) = \cos(ut - ux) = \cos ut \cos ux + \sin ut \sin ux.$

Now, Eqn. (8.8) may be written as

$$f(x) = \frac{1}{\pi}\int_0^\infty \int_{-\infty}^\infty f(t)[\cos ut \cos ux + \sin ut \sin ux] du\, dt$$

or, $$\boldsymbol{f(x) = \frac{1}{\pi}\int_0^\infty \int_{-\infty}^\infty f(t) \cos ut \cos ux\, du\, dt + \frac{1}{\pi}\int_0^\infty \int_{-\infty}^\infty f(t) \sin ut \sin ux\, du\, dt.} \quad (8.9)$$

Case I : When $f(t)$ is odd. $f(x) \cos ut$ is odd.

$$\therefore \quad \int_0^\infty \int_{-\infty}^\infty f(t) \cos ut\, \cos ux\, du\, dt = 0.$$

$$\left[\int_{-a}^a f(x)dx = 0, \text{ if } f(x) \text{ is odd}; \int_{-a}^a f(x)dx = 2\int_0^a f(x)dx, \text{ if } f(x) \text{ is even}\right]$$

$\therefore$ from Eqn. (8.9),

$$\boldsymbol{f(x) = \frac{2}{\pi}\int_0^\infty \sin ux\, du \int_0^\infty f(t) \sin ut\, dt.} \quad (8.10)$$

Eqn. (8.10) is known as **Fourier sine integral.**

Case II : When $f(x)$ is even.

$\therefore$ $f(t) \sin ut$ is odd.

So, $$\int_0^\infty \int_{-\infty}^\infty f(t) \sin ut \sin ux\, du\, dx = 0.$$

$\therefore$ $f(t) \cos ut$ is even.

Hence, from Eqn. (8.9),

$$\boldsymbol{f(x) = \frac{2}{\pi}\int_0^\infty \cos ux\, dx \int_0^\infty f(t) \cos ut\, dt.} \quad (8.11)$$

Eqn. (8.11) is known as **Fourier cosine integral.**

8.5 Fourier Complex Integral

We have the Fourier integral theorem (Eqn. 8.9),

$$f(x) = \frac{1}{\pi}\int_0^\infty\int_{-\infty}^\infty f(t)\cos u(t-x)du\,dt = \frac{1}{\pi}\int_0^\infty f(t)\frac{e^{iu(t-x)}+e^{-iu(t-x)}}{2}du\,dt, \quad i=\sqrt{-1}$$

$$= \frac{1}{2\pi}\int_0^\infty\int_{-\infty}^\infty f(t)[e^{iux}\cdot e^{-iut}+e^{-iux}\cdot e^{iut}]du\,dt$$

$$= \frac{1}{2\pi}\int_{-\infty}^\infty f(t)\left[\int_0^\infty e^{iux}\cdot e^{-iut}du\right]dt + \frac{1}{2\pi}\int_{-\infty}^\infty f(t)\left[\int_0^\infty e^{-iux}\cdot e^{iut}du\right]dt.$$

Change u by $-u$ in the second integral,

$$f(x) = \frac{1}{2\pi}\int_{-\infty}^\infty f(t)\left[\int_0^\infty e^{-iux}\cdot e^{iut}du\right]dt + \frac{1}{2\pi}\int_{-\infty}^\infty f(t)\left[\int_0^{-\infty} e^{-iux}\cdot e^{iut}(-du)\right]dt$$

$$= \frac{1}{2\pi}\int_{-\infty}^\infty f(t)\left[\int_0^\infty e^{-iux}e^{iut}du + \int_{-\infty}^0 e^{-iux}e^{iut}du\right]dt$$

or, $$\boldsymbol{f(x) = \frac{1}{2\pi}\int_{-\infty}^\infty\int_{-\infty}^\infty f(t)e^{-iux}\cdot e^{iut}du\,dt.} \tag{8.12}$$

Eqn. (8.12) is called **Fourier complex integral**.

8.6 Fourier Transform

We have $f(x) = \frac{1}{2\pi}\int_{-\infty}^\infty\int_{-\infty}^\infty f(t)e^{-iux}\cdot e^{iut}du\,dt = \frac{1}{\sqrt{2\pi}}\int_{-\infty}^\infty e^{-iux}\left[\frac{1}{\sqrt{2\pi}}\int_{-\infty}^\infty f(t)e^{iut}dt\right]du$

$$= \frac{1}{\sqrt{2\pi}}\int_{-\infty}^\infty e^{-iux}F(u)du, \tag{8.13}$$

where $$\boldsymbol{F(u) = \frac{1}{\sqrt{2\pi}}\int_{-\infty}^\infty f(t)e^{iut}dt.} \tag{8.14}$$

The function $F(u)$ is called the **Fourier transform of $f(t)$** and $f(t)$ is called the **Fourier inverse transform of $F(u)$** .

8.7 Fourier Sine and Cosine Transforms

The fourier transform of the function $f(x)$ is given by

$$F(u) = \frac{1}{\sqrt{2\pi}}\int_{-\infty}^\infty f(t)e^{iut}dt = \frac{1}{\sqrt{2\pi}}\left[\int_{-\infty}^0 f(t)e^{iut}dt + \int_0^\infty f(t)e^{iut}dt\right]$$

$$= \frac{1}{\sqrt{2\pi}}\int_0^\infty f(-t)e^{-iut}(-dt) + \frac{1}{\sqrt{2\pi}}\int_0^\infty f(t)e^{iut}dt.$$

In the first integral, t has been replaced by $-t$.

Again, $f(t) = \begin{cases} f(-t), & \text{for even function} \\ -f(-t), & \text{for odd function.} \end{cases}$

$$\therefore \quad F(u) = \frac{1}{\sqrt{2\pi}} \int_0^\infty f(t)[e^{iut} + e^{-iut}]dt, \text{ for even function}$$

$$\text{or,} \quad F(u) = \sqrt{\frac{2}{\pi}} \int_0^\infty f(t) \cos ut\, dt, \text{ for even function} \tag{8.15}$$

$$\text{or,} \quad F(u) = \frac{1}{\sqrt{2\pi}} \int_0^\infty f(t)[e^{-iut} - e^{iut}], \text{ for } f(t) \text{ odd}$$

$$F(u) = \sqrt{\frac{2}{\pi}} \int_0^\infty f(t) \sin ut\, dt, \text{ for } f(t) \text{ odd.} \tag{8.16}$$

Eqns. (8.15) and (8.16) are known as **Fourier's cosine and sine transforms** respectively and are denoted by

$$F_c(u) = \sqrt{\frac{2}{\pi}} \int_0^\infty f(t) \cos ut\, dt \quad \text{and} \quad F_s(u) = \sqrt{\frac{2}{\pi}} \int_0^\infty f(t) \sin ut\, dt.$$

8.8 Properties of Fourier Transforms

1. **Linear property : The Fourier transform of a linear combination of a given number of functions is the linear combination of their Fourier transform.**

   If $F(s)$ and $G(s)$ be the Fourier transforms of $f(x)$ and $g(x)$ respectively, then

$$F[af(x) + bg(x)] = aF(s) + bG(s),$$

   where a and b are constants.

   Proof : We have $F(s) = \int_{-\infty}^\infty e^{isx} f(x)dx$ and $G(s) = \int_{-\infty}^\infty e^{isx} g(x)dx$.

$$\therefore \quad F[af(x) + bg(x)] = \int_{-\infty}^{+\infty} e^{isx}[af(x) + bg(x)] = a\int_{-\infty}^\infty e^{isx} f(x)dx + b\int_{-\infty}^{+\infty} e^{isx} g(x)dx$$

$$\text{or,} \quad \boldsymbol{F[af(x) + bg(x)] = aF(s) + bG(s)} \quad \textbf{(Proved).} \tag{8.17}$$

2. **Change of scale property :** If $F(s)$ is the complex Fourier transformation of $f(x)$, then

$$F\{f(ax)\} = \frac{1}{a} F\left(\frac{s}{a}\right).$$

   Proof : We know that, $F(s) = \int_{-\infty}^\infty e^{isx} f(x)dx.$ (8.18)

$$\therefore \quad F\{f(ax)\} = \int_{-\infty}^\infty e^{isx} f(ax)dx.$$

   Put $ax = t$, then $dx = \dfrac{dt}{a}$.

$$\therefore \quad F\{f(ax)\} = \int_{-\infty}^\infty e^{isx} f(t)\frac{dt}{a} = \int_{-\infty}^\infty e^{is\frac{t}{a}} f(t)\frac{dt}{a} = \frac{1}{a}\int_{-\infty}^\infty e^{i\frac{s}{a}t} f(t)dt$$

$$\text{or,} \quad \boldsymbol{F\{f(ax)\} = \frac{1}{a} F\left(\frac{s}{a}\right)} \quad \text{[using Eqn. (8.17)]} \tag{8.18}$$

Cor. : If $F_s(s)$ and $F_c(s)$ be the Fourier sine and cosine transforms of $f(x)$ respectively, then

$$\boldsymbol{F_s\{f(ax)\} = \frac{1}{a}F_s\left(\frac{s}{a}\right)} \tag{8.19}$$

and $$\boldsymbol{F_c\{f(ax)\} = \frac{1}{a}F_c\left(\frac{s}{a}\right)}. \tag{8.20}$$

3. **Shifting property** : If $F(s)$ be the complex fourier transform of $f(x)$, then

$$F\{f(x-a)\} = e^{isx}F(s).$$

Proof : We know that, $F(s) = \int_{-\infty}^{\infty} e^{isx} f(x)dx.$

$$\therefore \quad F\{f(x-a)\} = \int_{-\infty}^{\infty} e^{isx} f(x-a)dx = \int_{-\infty}^{\infty} e^{is(t+a)} f(t)dt$$

[by putting $x - a = t$, $\therefore$ $dx = dt$]

$$= e^{isa}\int_{-\infty}^{\infty} e^{ist} f(t)dt$$

or, $$\boldsymbol{F\{f(x-a)\} = e^{isa}F(s)} \quad \textbf{(Proved)}. \tag{8.21}$$

4. **Modulation theorem** : If $F(s)$ be the complex Fourier transform of $f(x)$, then

$$F\{f(x)\cos ax\} = \frac{1}{2}\{F(s+a) + F(s-a)\}.$$

Proof : We have $F(s) = \int_{-\infty}^{\infty} e^{isx} f(x)dx.$

$$\therefore \quad F\{f(x)\cos ax\} = \int_{-\infty}^{\infty} e^{isx} f(x)\cos ax\, dx = \int_{-\infty}^{\infty} e^{isx} f(x)\frac{e^{isx}+e^{-isx}}{2}dx$$

$$= \frac{1}{2}\left[\int_{-\infty}^{\infty} e^{i(s+a)x} f(x)dx + \int_{-\infty}^{\infty} e^{i(s-a)x} f(x)dx\right]$$

or, $$\boldsymbol{F\{f(x)\cos ax\} = \frac{1}{2}\{F(s+a) + F(s-a)\}} \quad \textbf{(Proved)}. \tag{8.22}$$

Cor. If $F_s(x)$ and $F_c(s)$ be the Fourier sine and cosine transforms of $f(x)$ respectively, then

(a) $F_s\{f(x)\cos ax\} = \frac{1}{2}\{F_s(s+a) + F_s(s-a)\}$

(b) $F_c\{f(x)\sin ax\} = \frac{1}{2}\{F_s(s+a) - F_s(s-a)\}$

(c) $F_s\{f(x)\sin ax\} = \frac{1}{2}\{F_c(s-a) - F_c(s+a)\}.$

Proof of (a) : We have, $F_s\{f(x)\cos ax\} = \int_0^{\infty} f(x)\cos ax \sin sx\, dx$

$$= \int_0^{\infty} f(x)\cdot\frac{1}{2}[\sin(s+a)x + \sin(s-a)x]dx$$

$$= \frac{1}{2}\left[\int_0^\infty f(x)\sin(s+a)x\,dx + \int_0^\infty f(x)\sin(s-a)x\,dx\right]$$

or, $\boldsymbol{F_s\{f(x)\cos ax\} = \frac{1}{2}\{F_s(s+a) + F_s(s-a)\}}$ **(Proved).**

5. **Conjugate property : If $F(u)$ be the Fourier's transform of $f(t)$, then the Fourier's transform of the complex conjugate of $f(t)$ is given by $F^*(-u)$, where $*$ represents the complex conjugate of the corresponding complex function.**

   Proof : We know that, $F(s) = \frac{1}{\sqrt{2\pi}}\int_{-\infty}^{+\infty} f(t)e^{-ist}dt.$

   $$\therefore \quad F^*(s) = \frac{1}{\sqrt{2\pi}}\int_{-\infty}^{+\infty} f^*(t)e^{ist}dt$$

   or, $F^*(-s) = \frac{1}{\sqrt{2\pi}}\int_{-\infty}^{\infty} f^*(t)e^{-ist}dt.$ [replacing s by $-s$]

   $$\therefore \quad \mathbf{F^*(-s) = F.T.\ [f^*(t)]} \quad \textbf{(Proved).} \tag{8.23}$$

6. **Convolution property : The transform of a product of two functions is given by a convolution integral.**

   If $F(s)$ and $G(s)$ be the Fourier transforms of $f(t)$ and $g(t)$ respectively, then the Fourier transform of the convolution of $f(t)$ and $g(t)$ is the product of their Fourier transforms.

   So, $\mathbf{F.T.\ [f(t) * g(t)] = F.T.\ [f(t)] \cdot F.T.\ [g(t)] = F(s)G(s)}$.

   Proof : The convolution of two functions $f(t)$ and $g(t)$ over the interval $(-\infty, \infty)$ is defined as

   $$f * g = \frac{1}{\sqrt{2\pi}}\int_{-\infty}^{\infty} f(t)g(x-t)dt. \tag{8.24}$$

   Now, the Fourier transform of $f * g$ is

   $$\frac{1}{\sqrt{2\pi}}\int_{-\infty}^{\infty}\left[\frac{1}{\sqrt{2m}}\int_{-\infty}^{\infty} f(t)g(x-t)dt\right]e^{ist}dt = \frac{1}{2\pi}\int_{-\infty}^{\infty} f(t)\left[\int_{-\infty}^{\infty} g(x-t)e^{ist}dx\right]dt$$

   $$= \frac{1}{\sqrt{2\pi}}\int_{-\infty}^{\infty} f(t)[e^{ist}G(s)]dt = G(s)\cdot\frac{1}{\sqrt{2\pi}}\int_{-\infty}^{+\infty} f(t)e^{ist}dt$$

   $$\therefore \quad \boldsymbol{f * g = G(s)\cdot F(s)} \quad \textbf{(Proved).} \tag{8.25}$$

8.9 Perseval Theorem

The Fourier transform of a convolution integral is given by the product of transformations of the convolving functions.

If $F(s)$ and $G^*(s)$ be he Fourier transforms of the functions $f(x)$ and $g^*(x)$ respectively, then

(1) $\int_{-\infty}^{\infty} f(x)g^*(x)dx = \int_{-\infty}^{\infty} F(s)G^*(s)ds.$

(2) $\int_{-\infty}^{\infty} |f(x)|^2 dx = \int_{-\infty}^{\infty} |F(s)|^2 ds.$

Here $*$ indicates the complex conjugate.

Proof of (1) : We have $\int_{-\infty}^{\infty} f(x)g^*(x)dx \int_{-\infty}^{+\infty} \left[\frac{1}{\sqrt{2\pi}} \int_{-\infty}^{\infty} f(s)e^{-isx}ds\right]$

$$\times \left[\frac{1}{\sqrt{2\pi}} \int_{-\infty}^{\infty} G^*(k)e^{ikx}dk\right] dx$$

$$= \int_{-\infty}^{\infty} F(s)ds \int_{-\infty}^{\infty} G^*(k)dk \left[\frac{1}{2\pi} \int_{-\infty}^{\infty} e^{ix(k-s)}dx\right]$$

$$= \int_{-\infty}^{\infty} F(s)ds \int_{-\infty}^{\infty} G^*(k)dx\delta(k-\omega) = \int_{-\infty}^{\infty} F(s)G^*(k)dk \tag{8.26}$$

$$\left[\because \frac{1}{2\pi} \int_{-\infty}^{\infty} e^{ix(k-s)}dx = \delta(k-s)\right]$$

$$\therefore \quad \int_{-\infty}^{\infty} \boldsymbol{f(x)g^*(x)dx} = \int_{-\infty}^{\infty} \boldsymbol{F(s)G^*(k)dk}$$

Proof of (2) : If $f(x) = g(x)$, then from Eqn. (8.26), we get

$$\int_{-\infty}^{\infty} F(s)F^*(s)ds = \int_{-\infty}^{\infty} f(x)f^*(x)dx$$

or,
$$\int_{-\infty}^{\infty} |\boldsymbol{F(s)}|^2\boldsymbol{ds} = \int_{-\infty}^{\infty} |\boldsymbol{f(s)}|^2\boldsymbol{dx}. \tag{8.27}$$

8.10 Fourier Transforms of Derivatives

If $F(k)$ be the Fourier transform of a function $f(x)$, then the Fourier transform of $\frac{df}{dx}$ is $-ikF(k)$.

Proof : We know that, $F(k) = \frac{1}{\sqrt{2\pi}} \int_{-\infty}^{\infty} f(x)e^{ikx}dx$, then the Fourier transform of $f'(x) = \frac{df}{dx}$ is given by

$$F_1(k) = \frac{1}{\sqrt{2\pi}} \int_{-\infty}^{\infty} f'(x)e^{ikx}dx. \tag{8.28}$$

Now, $f'(x)e^{ikx} = \underset{h\to 0}{\text{Lt}} \frac{f(x+h)-f(x)}{h}e^{ikx} = \underset{h\to 0}{\text{Lt}} \left[\frac{f(x+h)}{h}e^{ik(x+h)}e^{-ikh} - \frac{f(x)}{h}e^{ikx}\right]$. (8.29)

Using Eqn. (8.29) in Eqn. (8.28), we get

$$F_1(k) = \underset{h\to 0}{\text{Lt}} \left[\frac{F(k)e^{-ikh}}{h} - \frac{F(k)}{h}\right] = F(k) \underset{h\to 0}{\text{Lt}} \left[\frac{e^{-ikh}-1}{h}\right]$$

$$= F(k) \underset{h\to 0}{\text{Lt}} (-ike^{-ikh}) \quad \text{[using L' Hospital's rule]}$$

or,
$$\boldsymbol{F_1(k) = -ikF(k)}. \tag{8.30}$$

For nth derivative, $\boldsymbol{F_n(k) = (-ik)^n F(k)}$. [8.30(a)]

8.11 Application of Fourier Transform to Solve Differential Equations

We consider the motion of a damped harmonic oscillator whose differential equation of motion is

$$\frac{d^2x}{dt^2} + 2\alpha\frac{dx}{dt} + \xi^2 x = f(t), \tag{8.31}$$

where $f(t) = \frac{1}{\sqrt{2\pi}} \int_{-\infty}^{\infty} F(\omega) e^{-i\omega t} d\omega$ (8.32)

and $F(\omega) = \frac{1}{\sqrt{2\pi}} \int_{-\infty}^{\infty} f(t) e^{i\omega t} dt.$ (8.33)

Now taking Fourier transform of both sides of Eqn. (8.31)

$$\frac{1}{\sqrt{2\pi}} \int_{-\infty}^{\infty} \frac{d^2x}{dt^2} e^{i\omega t} dt + \frac{2\alpha}{\sqrt{2\pi}} \int_{-\infty}^{\infty} \frac{dx}{dt} e^{i\omega t} dt + \frac{\xi^2}{\sqrt{2\pi}} \int_{-\infty}^{\infty} x(t) e^{i\omega t} dt = \frac{1}{\sqrt{2\pi}} \int_{-\infty}^{\infty} f(t) e^{i\omega t} dt.$$

Using Eqns. [8.30(a)] and (8.33) in the above equation, we get

$$-\omega^2 X(\omega) - 2\alpha i\omega X(\omega) + \xi^2 X(\omega) = F(\omega)$$

or, $X(\omega) = \frac{F(\omega)}{\xi^2 - \omega^2 - 2\alpha i\omega},$ (8.34)

where $X(\omega) = \frac{1}{\sqrt{2\pi}} \int_{-\infty}^{\infty} x(t) e^{i\omega t} dt.$ (8.35)

The Fourier inverse transform is

$$x(t) = \frac{1}{\sqrt{2\pi}} \int_{-\infty}^{\infty} X(\omega) e^{-i\omega t} d\omega = \frac{1}{\sqrt{2\pi}} \int_{-\infty}^{\infty} \frac{F(\omega) e^{-i\omega t}}{\xi^2 - \omega^2 - 2\alpha i\omega} d\omega. \quad (8.36)$$

[using Eqn. (8.34)]

Eqn. (8.36) is the desired solution of Eqn. (8.31).

8.12 Evaluation of Integrals

Fourier transforms may be used to evaluate certain integrals as follows :

(i) $\int_0^{\infty} \frac{\cos nx}{a^2 + n^2} dx$ (ii) $\int_0^{\infty} \frac{n \sin nx}{a^2 + n^2} dx$ (iii) $\int_0^{\infty} \frac{\sin^2 na}{n^2} dx = \pi \frac{a}{2}.$

Evaluation of (i) and (ii) :

$$I_1 = \int_0^{\infty} e^{-ax} \cos nx \, dx \quad (1)$$

and $I_2 = \int_0^{\infty} e^{-ax} \sin nx \, dx.$ (2)

Integrating by parts,

$$I_1 = \left[-\frac{1}{a} e^{-ax} \cos nx \right]_0^{\infty} - \frac{1}{a} \int_0^{\infty} e^{-ax} \sin nx \, dx = \frac{1}{a} - \frac{n}{a} I_2. \quad (3)$$

Similarly, $I_2 = \left[-\frac{1}{a} e^{-ax} \sin nx \right]_0^{\infty} + \frac{n}{a} \int_0^{\infty} e^{-ax} \cos nx \, dx = \frac{n}{a} I_1.$ (4)

Solving Eqns. (3) and (4), we get

$$I_1 = \frac{a}{a^2 + n^2} \quad \text{and} \quad I_2 = \frac{n}{a^2 + n^2}. \quad (5)$$

Taking $f(x) = e^{-ax}$, the cosine and sine transforms of $f(x)$ are :

$$F_c(n) = \sqrt{\frac{2}{\pi}} \int_0^\infty e^{-ax} \cos nx \, dx = \sqrt{\frac{2}{\pi}} \cdot \frac{a}{a^2 + n^2} \tag{6}$$

and $$F_s(n) = \sqrt{\frac{2}{\pi}} \int_0^\infty e^{-ax} \sin nx \, dx = \sqrt{\frac{2}{\pi}} \cdot \frac{n}{a^2 + n^2}. \tag{7}$$

$\therefore$ Fourier inverse transform is

$$f(x) = e^{-ax} = \sqrt{\frac{2}{\pi}} \int_0^\infty F_c(n) \cos nx \, dx = \frac{2}{\pi} \int_0^\infty \frac{a}{a^2 + n^2} \cos nx \, dx \tag{8}$$

and $$f(x) = e^{-ax} = \sqrt{\frac{2}{\pi}} \int_0^\infty F_s(n) \sin nx \, dx = \frac{2}{\pi} \int_0^\infty \frac{a}{a^2 + n^2} \sin nx \, dx. \tag{9}$$

$\therefore$ from Eqn. (8), **(i)** : $$\int_0^\infty \frac{\cos nx}{a^2 + n^2} dn = \frac{\pi}{2a} e^{-ax} \tag{10}$$

and from Eqn. (9), **(ii)** : $$\int_0^\infty \frac{\sin nx}{a^2 + n^2} dn = \frac{\pi}{2} e^{-ax}. \tag{11}$$

Proof of (iii) : If $f(x) = \begin{matrix} 1, & |x| < a \\ 0, & |x| > a \end{matrix}$, then Fourier transform of $f(x)$ is

$$F(n) = \frac{1}{\sqrt{2\pi}} \int_{-\infty}^\infty f(x) e^{inx} dx = \frac{1}{\sqrt{2\pi}} \int_{-a}^a 1 \cdot e^{inx} dx = \frac{1}{\sqrt{2a}} \cdot \left[\frac{e^{inx}}{in} \right]_{-a}^a$$

$$= \frac{1}{\sqrt{2\pi}} \cdot \frac{2}{n} \frac{e^{ina} - e^{-ina}}{2i} = \frac{1}{\sqrt{2\pi}} \frac{2 \sin na}{n} = \frac{2}{\sqrt{\pi}} \frac{\sin na}{n}.$$

Now, from Parseval relation, we have

$$\int_{-\infty}^\infty |f(x)|^2 dx = \int_{-\infty}^\infty |F(n)|^2 dn \quad \text{or,} \quad \int_{-a}^a 1^2 dx = \frac{2}{\pi} \int_{-\infty}^\infty \frac{\sin^2 na}{n^2} dn$$

or, $$2a = \frac{4}{\pi} \int_0^\infty \frac{\sin^2 na}{n^2} dn.$$

$\therefore$ $$\int_0^\infty \frac{\sin^2 na}{n^2} dn = \frac{\pi}{2} a \quad \textbf{(Proved)}.$$

8.13 Solution of Boundary Value Problems

(a) One dimensional heat flow : Solve $\frac{\partial u(x,t)}{\partial t} = \frac{\partial^2 u(x,t)}{\partial x^2}$, $x > 0$, $t > 0$, subject to the conditions :

(i) $u(0, t) = 0$

(ii) $u(x, 0) = \begin{cases} 1, & 0 < x < 1 \\ 0, & x \geq 1 \end{cases}$

(iii) $u(x, t)$ is bounded.

Solution : The given partial differential equation is

$$\frac{\partial u}{\partial t} = \frac{\partial^2 u}{\partial x^2}. \tag{1}$$

Taking Fourier sine transform on both sides, we get

$$\sqrt{\frac{2}{\pi}}\int_0^\infty \frac{\partial u}{\partial t}\sin sx\,dx = \sqrt{\frac{2}{\pi}}\int_0^\infty \frac{\partial^2 u}{\partial x^2}\sin sx\,dx$$

or, $$\frac{2}{\sqrt{\pi}}\frac{\partial}{\partial t}\int_0^\infty u\sin sx\,dx = \sqrt{\frac{2}{\pi}}\left[\left\{\frac{\partial u}{\partial x}\sin sx\right\}_0^\infty - s\int_0^\infty \frac{\partial u}{\partial x}\cos sx\,dx\right]$$

or, $$\frac{\partial \overline{u}}{\partial t} = -\sqrt{\frac{2}{\pi}}s\int_0^\infty \frac{\partial u}{\partial x}\cos sx\,dx,$$

where $\overline{u} = \sqrt{\frac{2}{\pi}}\int_0^\infty u\sin sx\,dx$ (2)

and $\frac{\partial u}{\partial x} \to 0$ as $x \to \infty$

or, $$\frac{\partial \overline{u}}{\partial t} = -\sqrt{\frac{2}{\pi}}s[(u\cos sx)_0^\infty + s\int_0^\infty u\sin sx\,dx]$$

$$= \sqrt{\frac{2}{\pi}}\left[u(0,t) - s^2\sqrt{\frac{2}{\pi}}\int_0^\infty u\sin sx\,dx\right], \quad [\text{for } u \to 0 \text{ as } x \to \infty]$$

$$= -\omega^2\overline{u}, \quad [\text{using boundary condition (i)}]$$

or, $$\frac{\partial \overline{u}}{\partial t} + s^2\overline{u} = 0 \quad (3)$$

or, $$\frac{\delta \overline{u}}{\overline{u}} = -s^2\delta t.$$

Integrating, $\log\overline{u} = -s^2t\log A$, ($A$ being a constant)

or, $\overline{u} = Ae^{-s^2t}$. (4)

At $t = 0$, $\overline{u}(s,0) = A$. (5)

$$A = \overline{u}(s,0) = \sqrt{2\pi}\int_0^\infty u(x,0)\sin sx\,dx, \quad [\text{using Eqn. (2)}]$$

$$= \sqrt{\frac{2}{\pi}}\left[\int_0^1 1\cdot\sin sx\,dx + \int_0^\infty 0\cdot\sin sx\,dx\right] = -\sqrt{\frac{2}{\pi}}\left[\frac{\cos sx}{s}\right]_0^1 + 0$$

or, $$A = \sqrt{\frac{2}{\pi}}\frac{1-\cos sx}{s}. \quad (6)$$

$\therefore$ Eqn. (4) becomes $\overline{u} = \sqrt{\frac{2}{\pi}}\left(\frac{1-\cos s}{s}\right)e^{-s^2t}$. (7)

Applying **inverse Fourier sine transform**, we get

$$u(x,t) = \sqrt{\frac{2}{\pi}}\int_0^\infty \overline{u}\sin sx\,ds = \sqrt{\frac{2}{\pi}}\int_0^\infty \sqrt{\frac{2}{\pi}}\left(\frac{1-\cos s}{s}\right)\sin sx\,ds$$

or, $$\boldsymbol{u(x,t) = \frac{2}{\pi}\int_0^\infty \left(\frac{1-\cos s}{s}\right)\sin sx\,ds} \quad (8)$$

Eqn. (8) represents the required solution.

8.14 Transverse Vibration of an Elastic String Fixed at Both Ends

Let us consider the transverse vibration of a string fixed at both ends. The differential equation of the wave equation is

$$\frac{\partial^2 y}{\partial t^2} = c^2 \frac{\partial^2 y}{\partial x^2}, \tag{1}$$

where $y(x,t)$ is the displacement at any point of the string at any instant of time, such that,

$$y(x,t) \to 0 \quad \text{and} \quad \frac{\partial y}{\partial x} \to 0 \text{ as } x \to \pm\infty. \tag{2}$$

The initial conditions of motion are

$$y(x,0) = f(x) \quad \text{and} \quad \left.\frac{\partial y}{\partial t}\right|_{t=0} = 0. \tag{3}$$

Let us define the Fourier transform,

$$F(k,t) = \frac{1}{\sqrt{2\pi}} \int_{-\infty}^{\infty} y(x,t) e^{kx} dx. \tag{4}$$

Now, taking Fourier transform of the wave equation (1), we get

$$\frac{1}{\sqrt{2\pi}} \frac{\partial^2}{\partial t^2} \int_{-\infty}^{\infty} y(x,t) e^{ikx} dx = c^2 \frac{1}{\sqrt{2\pi}} \int_{-\infty}^{\infty} \frac{\partial^2 y}{\partial x^2} e^{ikx} dx$$

or, $$\frac{d^2}{dt^2} F(k,t) = -c^2 k^2 F(k,t). \tag{5}$$

The general solution of Eqn. (5) is

$$F(k,t) = A_1 e^{ikct} + A_2 e^{-ikct}, \tag{6}$$

where A_1 and A_2 are the constants of integration.

Using the condition $y(x,0) = f(x)$, Eqn. (4) becomes

$$F(k,0) = \frac{1}{\sqrt{2\pi}} \int_{-\infty}^{\infty} y(x,0) e^{ikx} dx$$

or, $$A_1 + A_2 = \frac{1}{\sqrt{2\pi}} \int_{-\infty}^{\infty} f(x) e^{ikx} dx. \tag{7}$$

From the second boundary condition,

$$\left.\frac{dF}{dt}\right|_{t=0} = \frac{1}{\sqrt{2\pi}} \int_{-\infty}^{\infty} \left(\frac{\partial y}{\partial t}\right)_{t=0} e^{ikx} dx = 0,$$

i.e., $$ikc(A_1 - A_2) = 0. \tag{8}$$

Solving Eqns. (7) and (8), we get

$$A_1 = \frac{1}{2} F(k,0), \ A_2 = \frac{1}{2} F(k,0). \tag{9}$$

Now, the solution Equation (6) can be written as

$$F(k,t) = \frac{1}{2} F(k,0) \left[e^{ikct} + e^{-ikct}\right]. \tag{10}$$

Taking the Fourier transform of Eqn. (4), we get the solution of the wave equation as

$$y(x,t) = \frac{1}{\sqrt{2\pi}} \int_{-\infty}^{\infty} F(k,t) e^{-ikt} dk = \frac{1}{\sqrt{2\pi}} \int_{-\infty}^{\infty} \frac{1}{2} F(k,0) [e^{ikct} + e^{-ikct}] e^{-ikx} dk$$

or, $y(x,t) = \frac{1}{\sqrt{2\pi}} \int_{-\infty}^{\infty} \frac{1}{2} F(k,0) e^{ik(ct-x)} dk + \frac{1}{\sqrt{2\pi}} \int_{-\infty}^{\infty} \frac{1}{2} F(k,0) e^{-ik(ct+x)} dk$

or, $\boldsymbol{y(x,t) = \frac{1}{2}[f(x-ct) + f(x+ct)]}$, [using Eqn. (4) for $t = 0$] (11)

$f(x - ct)$ represents a wave travelling in the positive x-direction with speed c. At time $t + \Delta t$, x becomes $x + \Delta x$

$\therefore$ $\boldsymbol{y(x + c\Delta t, t + \Delta t) = f[x + c\Delta t - c(t + \Delta t)] = f(x - ct)}$. (12)

Thus, the wave repeats itself at $t + \Delta t$ and $x + \Delta x$. Likewise, $f(x + ct)$ represents a wave travelling in the $-x$ direction with speed c and repeats at $t + \Delta t$ and $x - \Delta x$.

Eqn. (11) represents the solution of the wave Eqn. (1).

8.15 The Temperature u in a Semi-infinite Bar ($0 < x < \infty$) is Determined by $\frac{\partial u}{\partial t} = c^2 \frac{\partial^2 u}{\partial x^2}$

$u(x,0) = 0$, $\frac{\partial u(0,t)}{\partial x} = k$, $\frac{\partial u(x,t)}{\partial x} = 0$ as $x \to \infty$.

Show that, $u = \frac{2k}{\pi} \int_0^{\infty} (e^{-s^2c^2t} - 1) \frac{\cos sx}{s^2} ds$.

Solution : Here, we shall apply fourier cosine transformation, since $\frac{\partial u(0,t)}{\partial x} = u$.

$$\frac{\partial u}{\partial t} = c^2 \frac{\partial u}{\partial x^2} \tag{1}$$

$$u(x,0) = 0 \tag{2}$$

$$\frac{\partial u(0,t)}{\partial x} = k \tag{3}$$

$$\frac{\partial u(x,t)}{\partial x} = 0 \tag{4}$$

as $x \to \infty$.

Multiplying Eqn. (1) by $\cos sx$ and then integrating with respect to x between 0 and ∞, we get

$$\int_0^{\infty} \frac{\partial u}{\partial t} \cos sx \, dx = c^2 \int_0^{\infty} \frac{\partial^2 u}{\partial x^2} \cos sx \, dx$$

$$\frac{du_c}{dt} = c^2 \left[\cos sx \frac{\partial u}{\partial x} + s \sin sx \cdot u\right]_0^{\infty} - c^2 \int_0^{\infty} s^2 \cos sx \cdot u \, dx$$

$$= c^2 \left[\cos sx \frac{\partial u}{\partial x}\right]_0^{\infty} + c^2 [s \sin sx \, u]_0^{\infty} - c^2 s^2 \overline{u}_c = c^2(0 - k) + c^2 s(0 - 0) - c^2 s^2 \overline{u}_c$$

or, $$\frac{d\overline{u}_c}{dt} + s^2c^2\overline{u}_c = -c^2k^2. \tag{5}$$

Solution of Eqn. (5) is

$$\overline{u}_c e^{s^2c^2t} = A - \frac{k}{s^2} e^{s^2c^2t}. \tag{6}$$

Now $\because$ $\overline{u}_c = 0$ at $t = 0$, $\therefore$ $A = \frac{k}{s^2}$.

Using this value of A in Eqn. (6), we get

$$\overline{u}_c e^{s^2c^2t} = \frac{k}{s^2} - \frac{k}{s^2}e^{s^2c^2t} = \frac{k}{s^2}\,[1 - e^{s^2c^2t}] \quad \text{or,} \quad \overline{u}_c = \frac{k}{s^2}\,[e^{-s^2c^2t} - 1].$$

Taking the inverse transformation, we get

$$\boldsymbol{u = \frac{2k}{\pi}\int_0^\infty (e^{-s^2c^2t} - 1)\frac{\cos sx}{s^2}ds.}$$

8.16 Solved Problems

Example 8.1. *Find the Fourier transform of* $f(x) = \begin{matrix} 1, & \textit{for } |x| < a \\ 0, & \textit{for } |x| > a. \end{matrix}$

[AMIETE (Summer) 1987]

Solution : The Fourier transform of $f(x)$ is given by

$$F(s) = \int_{-\infty}^{\infty} f(x)e^{isx}dx.$$

Using the values of $f(x)$, we get

$$\boldsymbol{F(s)} = \int_{-a}^{a} 1\cdot e^{isx}dx = \left[\frac{e^{isx}}{is}\right]_{-a}^{a} = \frac{1}{(is)}[e^{ias} - e^{-ias}] = \frac{2}{s}\frac{e^{ias} - e^{-ias}}{2i} = \boldsymbol{\frac{2\sin sa}{s}}.$$

Example 8.2. *Find the Fourier transform of the function* $f(x) = e^{-\frac{x^2}{2}}$.

Solution : The Fourier transform of $f(x)$ is

$$F(s) = \frac{1}{\sqrt{2\pi}}\int_{-\infty}^{\infty} e^{-\frac{x^2}{2}}e^{isx}dx \quad \left[\because\ f(x) = e^{-\frac{x^2}{2}}\right]$$

$$= \frac{1}{\sqrt{2\pi}}\int_{-\infty}^{\infty} e^{-\frac{1}{2}(x-is)^2-\frac{s^2}{2}}dx$$

$$= \frac{e^{-\frac{s^2}{2}}}{\sqrt{2\pi}}\int_{-\infty}^{\infty} e^{-z^2}\cdot\sqrt{2}dz \quad \left[\text{where } z = \frac{1}{\sqrt{2}}(x - is)\right]$$

or, $\boldsymbol{F(s) = e^{\frac{s^2}{2}}\frac{1}{\sqrt{\pi}}\sqrt{\pi} = e^{-\frac{s^2}{2}}}$.

Example 8.3. *Solve the integral equation,* $\displaystyle\int_0^\infty f(\theta)\cos\alpha\theta\,d\theta = \begin{cases} 1-\alpha, & 0 \le \alpha \le 1 \\ 0, & \alpha > 1 \end{cases}$

Hence, evaluate $\displaystyle\int_0^\infty \frac{\sin^2 t}{t^2}dt = \frac{\pi}{2}$.

Solution : We have $\displaystyle\int_0^\infty f(\theta)\cos\alpha\theta\,d\theta = F_c(\alpha)$

$$F_c(\alpha) = \begin{cases} 1-\alpha, & 0 \le \alpha \le 1 \\ 0, & \alpha > 1. \end{cases} \tag{1}$$

By the inversion formula,

$$f(\theta) = \frac{2}{\pi}\int_0^\infty F_c(\alpha)\cos\alpha\theta\,d\alpha = \frac{2}{\pi}\int_0^1 (1-\alpha)\cos\alpha\theta\,d\alpha$$

$$= \frac{2}{\pi}\left[\left|(1-\alpha)\frac{\sin\alpha\theta}{\theta}\right|_0^1 - \int_0^1 (-1)\frac{\sin\alpha\theta}{\theta}d\alpha\right] \quad \text{[integrating by parts]}$$

$$= \frac{2}{\pi\theta}\left|-\frac{\cos\alpha\theta}{\theta}\right|_0^l = \frac{2(1-\cos\theta)}{\pi\theta^2}.$$

Now, $F_c(\alpha) = \int_0^\infty f(\theta)\cos\alpha\theta\,d\theta = \int_0^\infty \frac{2(1-\cos\theta)}{\pi\theta^2}\cos\alpha\theta\,d\theta.$ (2)

From Eqns. (1) and (2), we get

$$\frac{2}{\pi}\int_0^\infty \frac{1-\cos\theta}{\theta^2}\cos\alpha\theta\,d\theta = \begin{cases} 1-\alpha, & 0 \le \alpha \le 1 \\ 0, & \alpha > 1. \end{cases}$$

If now $\alpha \to 0$, then

$$\frac{2}{\pi}\int_0^\infty \frac{1-\cos\theta}{\theta^2}d\theta = 1 \quad \text{or,} \quad \int_0^\infty \frac{2\sin^2\frac{\theta}{2}}{\theta^2}\cdot d\theta = \frac{\pi}{2}$$

Putting $\frac{\theta}{2} = t$, so that, $d\theta = 2dt$,

or, $$\int_0^\infty \frac{\sin^2 t}{t^2}dt = \frac{\pi}{2}.$$

Example 8.4. *Find Fourier sine and cosine transforms of $f(x) = e^{-ax}$, $(a > 0, x \ge 0)$.*

Solution : The Fourier sine transform of $f(x)$ is given by

$$F_s(s) = \sqrt{\frac{2}{\pi}}\int_0^\infty f(x)\sin sx\,dx = \sqrt{\frac{2}{\pi}}\int_0^\infty e^{-ax}\sin sx\,dx$$

$$= \sqrt{\frac{2}{\pi}}\left[\frac{e^{-ax}}{a^2+s^2}\{-a\sin sx - s\cos sx\}\right]_0^\infty = \sqrt{\frac{2}{\pi}}\left[-\frac{e^{-ax}}{a^2+s^2}\{a\sin sx + s\sin\cos sx\}\right]_0^\infty$$

$$= \sqrt{\frac{2}{\pi}}\left[-0 + \frac{1}{a^2+s^2}s\right] = \sqrt{\frac{2}{\pi}}\frac{s}{a^2+s^2}.$$

Again, Fourier cosine transform is given by

$$F_c(s) = \sqrt{\frac{2}{\pi}}\int_0^\infty e^{-ax}\cos sx\,dx = \sqrt{\frac{2}{\pi}}\left[\frac{e^{-ax}}{a^2+s^2}\{-a\cos sx + s\sin sx\}\right]_0^\infty$$

$$= \sqrt{\frac{2}{\pi}}\left[0 + \frac{a}{a^2+s^2}\right] = \sqrt{\frac{2}{\pi}}\frac{a}{a^2+s^2}.$$

Example 8.5. *Find the Fourier transform of the function $f(x) = \frac{e^{-ax}}{x}$.*

Solution : Fourier transform of $f(x)$ is given by

$$F(s) = \sqrt{\frac{2}{\pi}}\int_0^\infty f(x)\sin sx\,dx \quad \text{or,} \quad F(s) = \sqrt{\frac{2}{\pi}}\int_0^\infty \frac{e^{-ax}}{x}\sin sx\,dx.$$

Differentiating both sides w.r.t. s,

$$\frac{d}{dx}[F(s)] = \sqrt{\frac{2}{\pi}}\int_0^\infty \frac{e^{-ax}}{x}(x\cos sx)dx = \sqrt{\frac{2}{\pi}}\int_0^\infty e^{-ax}\cos sx\,dx = \sqrt{\frac{2}{\pi}}\frac{a}{s^2+a^2}.$$

Now, integrating w.r.t. s, we get $F(s) = \sqrt{\frac{2}{\pi}} \int \frac{a}{s^2+a^2} ds = \sqrt{\frac{2}{\pi}} \tan^{-1}\left(\frac{s}{a}\right) + C.$

For $s = 0,\ F(s) = 0, \quad \therefore \quad C = 0.$

Hence, $\boldsymbol{F(s) = \sqrt{\frac{2}{\pi}} \tan^{-1} \frac{s}{a}}.$

Example 8.6. *Find the Fourier transform for the function* $f(x) = \begin{cases} x^2, & 0 \le x < \pi \\ 0, & \textit{elsewhere} \end{cases}$

Solution : The Fourier transform for $f(x)$ is given by

$$F(s) = \sqrt{\frac{2}{\pi}} \int_0^\pi x^2 \sin sx\, dx = \sqrt{\frac{2}{\pi}} \left[\left(-\frac{x^2 \cos sx}{s} \right)_0^\pi + \int_0^\pi \frac{2x}{s} \cos sx \right]$$

$$= \sqrt{\frac{2}{x}} \left[-\frac{\pi^2}{s} \cos s\pi - \frac{2}{s^2} \int_0^\pi \sin s\pi dx \right] = \sqrt{\frac{2}{\pi}} \left[-\frac{\pi^2}{s} \cos s\pi + \frac{2}{s^3} \cos sx \Big|_0^\pi \right]$$

or, $$\boldsymbol{F(s) = \sqrt{\frac{2}{\pi}} \left[-\frac{\pi^2}{s}(-1)^s - \frac{2}{s^3}\{1-(-1)^s\} \right]}.$$

Example 8.7. *Find the fourier cosine transform of the function* $f(x) = e^{-x^2}$.

Solution : The fourier cosine transform of the given function is given by

$$F_c(s) = \sqrt{\frac{2}{\pi}} \int_0^\infty f(x) \cos sx\, dx = \sqrt{\frac{2}{\pi}} \int_0^\infty e^{-x^2} \cos sx\, dx. \tag{1}$$

Now, let $I = \int_0^\infty e^{-x^2} \cos sx\, dx.$ (2)

Differentiating both sides w.r.t. s,

$$\frac{dI}{ds} = \int_0^\infty -xe^{-x^2} \sin sx\, dx = \frac{1}{2} \int_0^\infty (-2xe^{-x^2}) \sin sx\, dx = \frac{1}{2} \int_0^\infty d(e^{-x^2}) \sin sx\, dx.$$

Integrating by parts, we get

$$\frac{dI}{ds} = \frac{1}{2} \left[\left\{ \sin sx(e^{-x^2}) \right\}_0^\infty - \int_0^\infty se^{-x^2} \cos sx\, dx \right] = -\frac{1}{2} s \int_0^\infty e^{-x^2} \cos x\, dx = -\frac{s}{2} I$$

or, $$\frac{dI}{I} = -\frac{s}{2} ds.$$

Integrating, $I = Ae^{-\frac{s^2}{4}}.$ (3)

Putting $s = 0$ in Eqn. (2), we get

$$I = \int_0^\infty e^{-x^2} dx = \frac{\sqrt{\pi}}{2}. \tag{4}$$

From Eqns. (3) and (4), $A = \frac{\sqrt{\pi}}{2}.$

$\therefore \quad I = \frac{\sqrt{\pi}}{2} e^{-\frac{s^2}{4}}.$

Hence, $F_c(s) = \sqrt{\frac{2}{\pi}} I = \sqrt{\frac{2}{\pi}} \frac{\sqrt{\pi}}{2} e^{-\frac{s^2}{4}}$ or, $\boldsymbol{F_c(s) = \frac{1}{\sqrt{2}} e^{-\frac{s^2}{4}}}.$

Example 8.8. *Express the function,* $f(x) = \begin{cases} 1, & \text{for } |x| \le 1 \\ 0, & \text{for } |x| > 1 \end{cases}$ *as a Fourier integral.*

Hence, evaluate, $\displaystyle\int_0^\infty \frac{\sin\lambda\cos\lambda x}{\lambda} d\lambda$ **[AMIETE (Summer) 1991]**

Solution : We know that the Fourier integral for $f(x)$ is

$$f(x) = \frac{1}{\pi}\int_0^\infty \int_{-\infty}^\infty f(t)\cos\lambda(t-x)dt\,d\lambda$$

$$= \frac{1}{\pi}\int_0^\infty \int_{-1}^1 \cos\lambda(t-x)dt\,d\lambda \quad [\because f(t)=1]$$

$$= \frac{1}{\pi}\int_0^\infty \left[\frac{\sin\lambda(t-x)}{\lambda}\right]_{-1}^1 d\lambda = \frac{1}{\pi}\int_0^\infty \frac{\sin\lambda(1-x)+\sin\lambda(1+x)}{\lambda}d\lambda$$

$$= \frac{2}{\pi}\int_0^\infty \frac{\sin\lambda\cos\lambda x}{\lambda}d\lambda.$$

Now, $\displaystyle\int_0^\infty \frac{\sin\lambda\cos\lambda x}{\lambda}d\lambda = \frac{\pi}{2}f(x).$

$$\therefore \qquad \int_0^\infty \frac{\sin\lambda\cos\lambda x}{\lambda} = \begin{cases} \frac{\pi}{2}, & \text{for } |x| < 1 \\ 0, & \text{for } |x| > 1. \end{cases}$$

For $|x| = 1$, which is a point of discontinuity of $f(x)$.

The value of the integral is $\dfrac{\frac{\pi}{2}+0}{2} = \dfrac{\pi}{4}$ **[Ans.]**

Example 8.9. *Find Fourier cosine transform of* $f(x)$, *where* $f(x) = \begin{cases} x, & \text{for } 0 < x < \frac{1}{2} \\ 1-x, & \text{for } \frac{1}{2} < x < 1 \\ 0, & \text{for } x > 1. \end{cases}$

Write the inverse transform. **[AMIETE 1987]**

Solution : Fourier cosine transform of $f(x)$ is given by

$$F(s) = \int_0^\infty f(x)\cos x\,dx = \int_0^{\frac{1}{2}} x\cos x\,dx + \int_{\frac{1}{2}}^1 (1-x)\cos sx\,dx$$

$$= \left[x\frac{\sin sx}{s}\left(-\frac{\cos sx}{s^2}\right)\right]_0^{\frac{1}{2}} + \left[(1-x)\frac{\sin sx}{s} - (-1)\frac{(-\cos sx)}{s^2}\right]_{\frac{1}{2}}^1$$

$$= \left[\frac{1}{2}\frac{\sin\frac{s}{2}}{s} + \frac{\cos\frac{s}{2}}{s^2} - \frac{1}{s^2}\right] + \left[-\frac{\cos x}{s^2} - \frac{1}{2}\frac{\sin\frac{s}{2}}{s} + \frac{\cos\frac{s}{2}}{s^2}\right]$$

or, $\displaystyle \boldsymbol{F(s) = -\frac{\cos s}{s^2} + \frac{2\cos\frac{s}{2}}{s^2} - \frac{1}{s^2}}.$

Example 8.10. *Using Parseval's identity evaluate the integral,* $\displaystyle\int_0^\infty \frac{dx}{(x^2+1)^2}.$

Solution : If $f(x) = e^{-x}$, then $F_c(s) = \dfrac{1}{1+s^2}.$

Applying parseval's identity cosine transformation, we get

$$\frac{2}{\pi}\int_0^{\infty}[F_c(s)]^2 ds = \int_0^{\infty}|f(x)|^2 dx$$

or, $$\frac{2}{\pi}\int_0^{\infty}\left|\frac{1}{(1+s^2)^2}\right|^2 ds = \int_0^{\infty}|e^{-x}|^2 dx = \int_0^{\infty}|e^{-2x}|dx = \left.\frac{e^{-2x}}{-2}\right|_0^{\infty} = \frac{1}{2}.$$

Hence, $$\int_0^{\infty}\left|\frac{1}{(1+s^2)^2}\right|^2 ds = \frac{1}{2}\frac{\pi}{2} = \frac{\pi}{4}.$$

Example 8.11. *Using Fourier integral representation of an appropriate function, show that,*

$$\int_0^{\infty}\frac{\cos\omega x}{k^2+\omega^2}d\omega = \frac{\pi e^{-kx}}{2k},\quad x>0,\ k>0.$$ **[AMIETE (Winter) 1992]**

Solution : We have $f(x) = \frac{2}{\pi}\int_0^{\infty}\cos ux\,du\int_0^{\infty}f(t)\cos ut\,dt.$

Putting value of $f(t)$ and replacing u by ω,

$$\begin{aligned} e^{-kx} &= \frac{2}{\pi}\int_0^{\infty}\cos\omega x d\omega\int_0^{\infty}e^{-kt}\cos\omega t\,dt \\ &= \frac{2}{\pi}\int_0^{\infty}\cos\omega x\,d\omega\left[\frac{e^{-kt}}{k^2+\omega^2}\{-k\cos\omega t+\omega\sin\omega t\}_0^{\infty}\right] \\ &= \frac{2}{\pi}\int_0^{\infty}\cos\omega x\,d\omega\left[0+\frac{k}{k^2+\omega^2}\right] = \frac{2k}{\pi}\int_0^{\infty}\frac{\cos\omega x\,d\omega.}{k^2+\omega^2} \end{aligned}$$

or, $$\int_0^{\infty}\frac{\cos\omega x}{k^2+\omega^2}d\omega = \frac{\pi e^{-kx}}{2k}.$$

Example 8.12. *If a function is defined as $f(x) = e^{-\beta x}$, $(\beta>0)$. Find the Fourier sine integral for $f(x)$. Also show that, $\frac{\pi}{2}e^{-\beta x} = \int_0^{\infty}\frac{\lambda\sin\lambda x}{\beta^2+\lambda^2}d\lambda$.*

Solution : Fourier sine transform of $f(x)$ is given by

$$f(x) = \frac{2}{\pi}\int_0^{\infty}\sin\lambda x\,d\lambda\int_0^{\infty}f(t)\sin\lambda t\,dt.$$

Putting the value of $f(x)$ in the above equation,

$$\begin{aligned} e^{-\beta x} &= \frac{2}{\pi}\int_0^{\lambda}\sin\lambda x\,d\lambda\int_0^{\lambda}e^{-\beta t}\sin\lambda t\,dt = \frac{2}{\pi}\int_0^{\infty}\sin\lambda x\,dx\left[\frac{e^{-\beta t}}{(\beta^2+\lambda^2)}\{-\beta\sin\lambda t-\lambda\cos\lambda t\}_0^{\infty}\right] \\ &= \frac{2}{\pi}\int_0^{\infty}\sin\lambda x\,d\lambda\left[0+\frac{\lambda}{\beta^2+\lambda^2}\right] = \frac{2}{\pi}\int_0^{\infty}\frac{\lambda\sin\lambda x d\lambda}{\beta^2+\lambda^2}d\lambda \end{aligned}$$

or, $$\frac{\pi}{2}e^{-\beta x} = \int_0^{\infty}\frac{\lambda\sin\lambda x}{\beta^2+\lambda^2}d\lambda$$ **(Proved).**

Exercises

8(A) Short Answer Type Questions

8.1 Define Fourier transform and Fourier inverse transform

8.2 Write down Fourier sine transform and cosine transform.

8.3 State and prove the properties of the following Fourier transforms :

(a) Linearity property (b) change of scale property (c) Shifting property

(d) Convolution property and (e) conjugate property.

8.4 If $F(s)$ be the Fourier transform of a function $f(x)$, then show that the Fourier transform of

$$\frac{d}{dx}[f(x)] = -isF(s).$$

8.5 State and prove Parseval identity of Fourier transform.

8.6 Find the Fourier transform of $f(x) = \dfrac{1}{1+x^2}$.

8.7 Find the Fourier transform of $f(x) = \dfrac{e^{\lambda x}}{x}$.

8(B) problems

8.1 Calculate the Fourier transform of $f(t) = \begin{cases} e^{-i\omega t}, & \text{for } -\frac{a}{2} \le t \le \frac{a}{2} \\ 0, & \text{for } |t| > \frac{a}{2} \end{cases}$

[**Ans:** $\frac{a \sin \xi}{\sqrt{2\pi}\xi}$, $\xi = (k-\omega)\frac{a}{2}$]

8.2 Find the fourier transform of $e^{-|t|}$. [**Ans:** $\frac{2}{\pi}\left(\frac{1}{1+s^2}\right)$]

8.3 Find the Fourier transform of $e^{-\frac{r^2}{a^2}}$, where a is a constant and $r = \sqrt{x^2+y^2+z^2}$.

[**Ans:** $F(u,v,\omega) = \frac{a^3}{2^{\frac{3}{2}}} \cdot e^{-(u^2+v^2+\omega^2)\frac{a^2}{4}}$]

8.4 Find finite cosine transform of $f(x) = \dfrac{\pi}{3} - x + \dfrac{x^2}{2\pi}$. [**Ans:** $F_c(n) = \frac{1}{n^2}$, if $n = 1, 2, 3.$]

8.5 (a) Show that the finite sine transform of $\dfrac{x}{\pi}$ is $(-1)^{n+1} \cdot \dfrac{1}{n}$.

(b) Show that the finite cosine transform of $\left(1 - \dfrac{x}{\pi}\right)^2$ is $\begin{cases} \frac{\pi}{3}, & \text{if } n = 0 \\ \frac{2}{\pi n^2}, & \text{if } n = 1, 2, 3. \end{cases}$

8.6 If the Fourier transform of $f(x)$ be $\dfrac{1-\cos nx}{n^2\pi^2}$ $(0 < x < \pi)$, find $f(x)$.

[**Ans:** $\dfrac{2}{\pi^3} \displaystyle\sum_{\pi^3}^{\infty} \dfrac{\cos n\pi}{n^2} \sin n\pi$]

8.7 Find the Fourier transform of $e^{-a|x|}$. Hence, prove that $\displaystyle\int_0^\infty \frac{x \sin mx}{1+x^2}dx = \frac{\pi e^{-m}}{2}$, $m > 0$.

[**Ans:** $\frac{2a}{(a^2+s^2)}$]

8.8 Solve for $f(x)$ from the integral equation $\displaystyle\int_0^\infty f(x)\cos x\, dx = e^{-s}$. [**Ans:** $\frac{2}{\pi}\frac{1}{1+x^2}$]

8.9 Show that, $\int_0^{\infty} \frac{dt}{(a^2+t^2)(b^2+t^2)} = \frac{\pi}{2ab(a+b)}$ (use Parseval identity.)

8.10 Using Fourier integral representation, show that, $\int_0^{\infty} \frac{\sin\omega\cos\lambda\omega}{\omega} d\omega = \frac{\pi}{2}$, if $(0 \leq x \leq 1)$.

8.11 Prove that, $\int_0^{\infty} \frac{\sin\pi\alpha\sin a\theta}{1-\alpha^2} d\alpha = \begin{cases} \frac{1}{2}\pi\sin\alpha, & 0 \leq \theta \leq \pi \\ 0, & \theta > \pi. \end{cases}$

8.12 Solve the integral equation, $\int_0^{\infty} f(x)\sin tx\, dx = \begin{cases} 1, & 0 \leq t < 1 \\ 2, & 1 \leq t < 2 \\ 3, & t \geq 2. \end{cases}$

8.13 If $F_s(\alpha)$ and $G_s(\alpha)$ be the Fourier sine transform of $f(x)$ and $g(x)$ respectively, show that,

$$\int_0^{\infty} F_s(\alpha)G_s(\alpha)d\alpha = \int_0^{\infty} f(x)g(x)dx.$$

8.14 Use the complex form of the Fourier transform to show that,

$$V = \frac{1}{2\sqrt{xt}} \int_{-\infty}^{\infty} \overline{f}(u)e^{\left[-\frac{(x-u)}{ut}\right]}du$$

is the solution of the boundary value problem,

$$\frac{\partial V}{\partial t} = \frac{\partial^2 V}{\partial x^2}, \ -\infty < x < \infty, \ t > 0;$$

$V = f(x)$, when $t = 0$. **[Delhi 1990]**

8.15 Use Fourier transform to solve, $2\frac{\partial\theta}{\partial t} = \frac{\partial^2\theta}{\partial x^2}$ $(x < 0)$, if $\theta_x(0,t) = 0$, $\theta(x,0) = e^{-x}$ and $\theta(x,t)$ is finite as $x \to \infty$.

8.16 Find the steady temperature $V(x,y)$ in a long square bar of side π when one face is kept at constant temperature v_0 and the face at zero temperature. Also $V(x,y)$ is bounded.

$$V(x,y) = \frac{4V_0}{\pi} \sum \frac{\sin h(2n+1)y \sin(2n+1)x}{(2n+1)\sin h(2n+1)\pi}.$$

8.17 Using Fourier transform, solve the equation, $\frac{\partial T}{\partial t} = h^2\frac{\partial^2 T}{\partial x^2}$ assuming that, $T(x,y) \to 0$ and $\frac{\partial T}{\partial x} \to 0$ as $x \to \pm\infty$ and initial condition, $T(x,0) = e^{-\alpha^2 x^2}$, $\alpha =$ a constant > 0.

8.18 Consider $f(x) = \frac{1}{2a}$, for $|x| < a$

$= 0$, for $|x| > a$

Calculate the Fourier transform of $f(x)$ and state its limiting value as $a \to 0$.

[CU(H) 2014, 2009]

8.19 Consider $f(x) \to 0$ for $x \to \pm\infty$. Let Fourier transform of $f(x)$ be

$$g(K) = \frac{1}{\sqrt{2\pi}} \int_{-\infty}^{+\infty} f(x)e^{iks}dx.$$

If $g_1(K)$ is Fourier Transform of $\frac{df}{dx}$, then show that, $g_1(K) = -iKg(K)$. **[CU(H) 2010]**

8.20 Define Fourier transform of a function. Obtain the fourier transform of a function

$$f(t) = e^{-\alpha t^2} \ (\alpha = \text{constant}).$$

Given : $\int_{-\infty}^{+\infty} e^{-(\alpha x^2 - \beta x)} dx = \sqrt{\frac{\pi}{\alpha}} e^{\frac{\beta^2}{4\alpha}}$ **[CU(H) 2007]**

8.21 Write down the expression for the Fourier transform of a function $f(x)$ defined in the domain $-\infty < x < \infty$. Calculate the Fourier transform when $f(x)$ is the Dirac delta function $\delta(x)$. **[CU(H) 2005]**

8.22 Find the Fourier transform of the function $f(x) = \begin{cases} 1, & \text{for } |x| < 1 \\ 0, & \text{for } |x| > 1. \end{cases}$

Show that, $\int_0^{\infty} \frac{\sin \alpha}{\alpha} dx = \frac{\pi}{2}$. **[BU(H) 2014]**

8.23 Find the Fourier transform of the function $f(x) = 1, \quad \text{for } |x| < 1$
$= 0, \quad \text{for } |x| > 1$

and hence calculate the integral $\int_0^{\infty} \frac{\sin x}{x}$. **[BU(H) 2008]**

8.24 **(a)** Write down the expression for the Fourier transform of a function $f(x)$. Obtain the Fourier transform of $\delta(x)$. **[CU(H) 2013]**

(b) Show that $\delta(kx) = \frac{1}{k}\delta(x)$, where k is a constant. **[CU(H) 2014]**

8.25 **(a)** Shows that a Fourier transform of a Gaussian function is itself a Gaussian function.

(b) Find the Fourier cosine transform of $f(t) = e^{-pt}$, $p > 0$.

Hence, evaluate the integral, $\int_0^{\alpha} \frac{\cos(\omega t)}{p^2 + \omega^2} d\omega$.

8.26 **(a)** What are the sufficient conditions for the existence of Fourier transform.

(b) Derive the Fourier transform of $f(x)$, the derivative of $f(x)$.

(c) Find the Fourier transform of $e^{-\frac{x^2}{2}}$. What is your inference?

[Kerala U (Model questions)]

8.27 **(a)** Show that the Fourier transform is a linear operator.

(b) Find the Fourier transform of $f(x)$, where

$$f(x) = \begin{cases} e^{x}, & x < 0 \\ e^{-x}, & x > 0. \end{cases}$$ **[Kerala U (Model questions)]**

8.28 **(a)** Define Fourier transform of a function $f(x)$. What is the Fourier transform of its first derivative.

(b) Show that the Fourier transform of a Gaussian function is another Gaussian.

8.29 **(a)** Deduce the necessary expression of Fourier, Fourier sine and cosine transforms by deriving necessary relations.

(b) Find the Fourier transform of

$$f(t) = 0; \quad t < 0$$
$$= e^{-\frac{t}{\tau}} \sin \omega_0 t; \quad t > 0.$$ **[Gujrat U(MSc) 2017]**

8.30 **(a)** State and prove the convolution theorem for Fourier transform.

(b) Find the Fourier cosine transform of $f(x) = e^{-2x} + 4e^{-3x}$. **[Pune U(MSc) 2013]**

8.31 (a) Consider the function $f(x) = 1, \quad \text{for} \quad -1 \leq x \leq 1$
$\qquad\qquad = 0, \quad \text{for} \quad |x| > 0.$

Find the Fourier transform of $f(x)$.

(b) Consider the function, $f(x) = \sin x \quad \text{for} \quad 0 < x < \pi$
$\qquad\qquad = -\sin x \quad \text{for} \quad -\pi < x < 0$

$f(x + 2\pi) = f(x)$.

Find the Fourier series expansion for $f(x)$. **[CU(H) 2016]**

(c) Show that the Fourier transform of $f(x)e^{-|x|}$ is $F(K) = \sqrt{\frac{2}{\pi}} \frac{1}{1+K^2}$. **[CU(H) 2017]**

8.32 (a) Find the Fourier transform of the function $f(x) = Ne^{-\alpha x}$, where N and α are constants. Calculate the standard deviation of both $f(x)$ and $F(K)$. Plot $F(K)$ against K and $f(x)$ against x. **[CU(H) 2015]**

8.33 (a) Consider $f(x) = \frac{1}{2a} \quad \text{for} \quad |x| < a$
$\qquad\qquad = 0 \quad \text{for} \quad |x| > a$

where $a > 0$. Calculate the Fourier transform of $f(x)$ and state its limiting value as $a \to 0$. **[CU(H) 2014]**

(b) Write down the expression for Fourier transform of a function $f(x)$. Obtain Fourier transform of $\delta(x)$. **[CU(H) 2013]**

8.34 Calculate the Fourier transform $F(K)$ of the box function $f(x)$, where

$f(x) = 1 \quad \text{for} \quad -a \leq x \leq a$
$\qquad = 0 \quad \text{for} \quad |x| > 0.$

Sketch $F(K)$ against K. Write down its value at $K = 0$. **[CU(H) 2011]**

CHAPTER 9

Laplace Transforms

9.1 Introduction

Laplace transformation method provides an easy and effective way for the solution of many problems in science and technology. The method helps in solving differential equations with boundary values where the general solution as well as the values of arbitrary constants need not be found.

9.2 Definition of Laplace Transform

If $f(t)$ be a function defined for all positive values of t, then

$$F(s) = \int_0^\infty e^{-st} f(t)dt, \tag{9.1}$$

provided the integral exists, is called the **Laplace transform** of $f(t)$. It is denoted as

$$F(s) = L[f(t)]. \tag{9.2}$$

Here 's' is a parameter—a real or complex number.

The Laplace transform of $f(t)$ exists if the function satisfies the following conditions :

(a) The function $f(t)$ should be an arbitrary piecewise continuous function in every finite interval and that $f(t) = 0$, for all negative values of t.

(b) The function $f(t)$ should be of exponential order. That is there exist positive constants M and a, such that,

$$|f(t)| \leq Me^{at}, \quad \text{for all } t > 0. \tag{9.3}$$

9.3 Properties of Laplace's Transform

(1) Linearity property : A transformation T is called linear if for every pair of functions $f_1(t)$ and $f_2(t)$ and for every pair of constants, a and b, we have

$$T\{af_1(t) + bf_2(t)\} = aT\{f_1(t)\} + bT\{f_2(t)\}. \tag{9.4}$$

Theorem 1 : Laplace transform is a linear transformation, i.e.,

$$L\{af_1(t) + bf_2(t)\} = aL\{f_1(t)\} + bL\{f_2(t)\}, \tag{9.5}$$

where a and b are constants.

Proof : We have by definition, $L\{f(t)\} = \int_0^\infty e^{-st} f(t)dt$.

$$\therefore \; L\{af_1(t) + bf_2(t)\} = \int_0^\infty e^{-st}\{af_1(t) + bf_2(t)\}dt = a\int_0^\infty e^{-st} f_1(t)dt + b\int_0^\infty e^{-st} f_2(t)dt$$
$$= aL\{f_1(t)\} + bL\{f_2(t)\}.$$

Existence of Laplace theorem

Theorem 2 : Let a function $f(t)$ satisfy the following conditions :

(i) For every $b > 0$, the interval $(0, b)$ can be broken up into a finite number of subintervals in each of which $f(t)$ is continuous and $f(t)$ approaches definite limit as t approaches either end points of these subintervals from the interior.

(ii) $|e^{-at}f(t)| \leq \mu$ or, $|f(t)| \leq \mu e^{at}$, for all $t \geq 0$ and for some constants $\mu > 0$.

Then the Laplace transform of $f(t)$ exists i.e., converges in the domain $\text{Re}(s) > a$.

Proof : We have $L\{f(t)\} = \int_0^\infty e^{-st}f(t)dt.$

$$\text{Now, } \left|\int_0^\infty \boldsymbol{e^{-st}f(t)dt}\right| \leq \int_0^\infty \left|e^{-st}f(t)\right| dt = \int_0^\infty |e^{-st}|\ |f(t)|dt$$

$$\leq \int_0^\infty \left|e^{-(x+iy)}\right| \mu e^{at}dt \quad [\text{putting } s = x + iy]$$

$$= \mu \int_0^\infty \boldsymbol{e^{-t(x-a)}dt}. \tag{9.6}$$

If $x > 0$, i.e., $\text{Re}\,(s) > a$, then the above is convergent.

Theorem 3 : (First shifting property)

If $L\{f(t)\} = f(s)$, then $L\{e^{at}f(t)\} = f(s-a)$.

Proof : By definition, $L\{f(t)\} = \int_0^\infty e^{-st}e^{at}f(t)dt = \overline{f}(s).$

$$\therefore\quad \boldsymbol{L\{e^{at}f(t)\}} = \int_0^\infty e^{-st}e^{at}f(t)dt = \int_0^\infty e^{-(s-a)t}f(t)dt$$

$$= \int_0^\infty e^{-pt}f(t)dt,\ (p = s - a)$$

$$= \overline{f}(t) = \boldsymbol{\overline{f}(s-a)} \quad [\because p = s - a]. \tag{9.7}$$

N.B. If the Laplace transform $\overline{f}(s)$ of $f(t)$ be known then the Laplace transform of $e^{\alpha t}f(t)$ can be written simply by replacing s by $(s-a)$ to get $f(s-a)$.

Theorem 4 : (Second Shifting property)

If $L\{f(t)\} = f(s)$ and $g(t) = \begin{cases} f(t-a), & t > a \\ 0, & t < (a) \end{cases}$, then $L\{g(t)\} = e^{-as}\overline{f}(s)$.

$$\therefore\quad \boldsymbol{L\{g(t)\}} = \int_0^\infty e^{-st}g(t)dt = \int_0^a e^{-st}\cdot 0dt + \int_a^\infty e^{-st}f(t-a)dt$$

$$= \int_0^\infty e^{-st}f(t-a)dt = \underset{B\to 0}{\text{Lt}} \int_0^B e^{-st}f(t-a)dt$$

$$= \underset{B\to\infty}{\text{Lt}} \int_0^{B-a} e^{-s(x+a)}f(x)dx\,[\text{by putting } t - a = x, \text{ then } t = x + a, \text{ and } dt = dx]$$

$$= e^{-as}\int_0^\infty e^{-sx}f(x)dx = e^{-as}\int_0^\infty e^{-sa}f(t)dt = \boldsymbol{e^{-as}\overline{f}(s)}. \tag{9.8}$$

Theorem 5 : (Change of scale)

If $L\{f(t)\} = \overline{f}(s)$, then $L\{f(at\} = \frac{1}{a}\overline{f}\left(\frac{s}{a}\right)$, $(a > 0)$

Proof : By definition, $L\{f(t)\} = \int_0^\infty e^{-st} f(t)dt = f(s)$.

$$\therefore \quad L\{f(at)\} = \int_0^\infty e^{-st} f(at)dt = \underset{B\to\infty}{\text{Lt}} \int_0^B e^{-st} s(at)dt$$

$$= \underset{B\to 0}{\text{Lt}} \int_0^{aB} e^{-s\frac{x}{a}} f(x)\frac{dx}{a} \quad \left[\text{putting } at = x, \quad \therefore \quad dt = \frac{dx}{a}\right]$$

$$= \frac{1}{a}\int_0^\infty e^{-\frac{s}{a}x} f(x)dx \quad [\text{Here, } a > 0 \ \therefore \ aB \to \infty \text{ as } B \to \infty]$$

$$= \frac{1}{a}\overline{f}\left(\frac{s}{a}\right). \tag{9.9}$$

9.4 Laplace Transform of Elementary Functions

The following transforms of elementary functions follow directly from the definition of Laplace transform :

(i) $L(1) = \frac{1}{s}$, $s > 0$

(ii) $L(t^n) = \frac{\Gamma(n+1)}{s^{n+1}}$, $n > -1$, $s > 0$

(iii) $L(e^{at}) = \frac{1}{s-a}$, $s > 0$

(iv) $L(\sin at) = \frac{a}{s^2+a^2}$, $s > 0$

(v) $L(\cos at) = \frac{s}{s^2+a^2}$, $s > 0$

(vi) $L \sinh at) = \frac{a}{s^2-a^2}$, $x > |a|$

(vii) $L(\cosh at) = \frac{s}{s^2-a^2}$, $s > |a|$.

Proof : (i) $L(1) = \int_0^\infty e^{-at}\cdot 1\, dt = \underset{B\to\infty}{\text{Lt}} \int_0^B e^{-st}dt = \underset{B\to\infty}{\text{Lt}} \left[\frac{e^{-st}}{-s}\right]_0^B = \underset{B\to\infty}{\text{Lt}} \frac{1}{s}(1 - e^{-sB})$

$$= \frac{1}{s} \quad [\because s > 0].$$

(ii) $L(t^n) = \int_0^\infty e^{-st} t^n dt = \int_0^\infty e^{-st} t^{(n+1)-1} dt = \frac{\Gamma(n+1)}{n+1}$, $n > -1$, $s > 0$.

$$\left[\because \int_0^\infty e^{-ax} x^{n-1} dx = \frac{\Gamma(n)}{a^n}, \ a > 0, \ n > 0\right]$$

(iii) $L(e^{at}) = \int_0^\infty e^{-st} e^{at} dt = \int_0^\infty e^{-(s-a)} dt = \underset{B\to\infty}{\text{Lt}} \int_0^B e^{-(s-a)t} dt$

$$= \underset{B\to\infty}{\text{Lt}} \left[\frac{e^{-(s-a)t}}{-(s-a)}\right]_0^B = \underset{B\to\infty}{\text{Lt}} \frac{1}{(s-a)} [1 - e^{-(s-a)B}]$$

$$= \frac{\mathbf{1}}{(\boldsymbol{s-a})}, \quad \textbf{for } \boldsymbol{s > a}.$$

(iv) $\boldsymbol{L(\sin at)} = \int_0^\infty e^{-st} \sin at\, dt = \underset{B\to\infty}{\text{Lt}} \int_0^B e^{-st} \sin at\, dt$

$$= \underset{B\to\infty}{\text{Lt}} \left[\frac{e^{-st}}{s^2+a^2}(-s\sin at - a\cos at)\right]_0^B$$

$$= \underset{B\to 0}{\text{Lt}} \left[\frac{a}{s^2+a^2} - \frac{e^{-sB}}{s^2+a^2}(s\sin aB + a\cos aB)\right]$$

$$= \frac{\boldsymbol{a}}{\boldsymbol{s^2+a^2}} \quad [\because e^{-sB} \to 0 \text{ as } B\to\infty, \ \because s > 0].$$

(v) $\boldsymbol{L(\cos at)} = \int_0^\infty e^{-st} \cos at\, dt = \underset{B\to\infty}{\text{Lt}} \int_0^B e^{-st} \cos at\, dt$

$$= \underset{B\to\infty}{\text{Lt}} \left[\frac{e^{-st}}{s^2+a^2}(-s\cos at + a\sin at)\right]_0^B$$

$$= \frac{\boldsymbol{s}}{\boldsymbol{s^2+a^2}} \quad [\because e^{-sB} \to 0 \text{ as } B\to 0, \ \because s > 0]$$

(vi) $\boldsymbol{L(\sin at)} = \int_0^\infty e^{-st} \sin h\, at\, dt = \int_0^\infty e^{-st} \frac{1}{2}(e^{at} - e^{-at})dt$

$$= \underset{B\to\infty}{\text{Lt}} \frac{1}{2} \int_0^B \left\{e^{-(s-a)t} - e^{-(s+a)t}\right\} dt = \underset{B\to\infty}{\text{Lt}} \frac{1}{2} \left[\frac{e^{-(s-a)t}}{-(s-a)} - \frac{e^{-(s+a)t}}{-(s+a)}\right]_0^B$$

$$= \underset{B\to\infty}{\text{Lt}} \frac{1}{2} \left[\frac{1}{s-a}\left\{1 - e^{-(s-a)B}\right\} - \frac{1}{s+a}\left\{1 - e^{-(s+a)B}\right\}\right]$$

$$= \frac{1}{2}\left(\frac{1}{s-a} - \frac{1}{s+a}\right) \quad [\because e^{-(s-a)B} \to 0, e^{-(s+a)B} \to 0 \text{ as } B\to\infty \text{ as } s > |a|]$$

$$= \frac{\boldsymbol{a}}{\boldsymbol{s^2-a^2}}, \ \textbf{for } \boldsymbol{s > |a|}.$$

(vii) $\boldsymbol{L(\cos at)} = \int_0^\infty e^{-st} \cos h\, at\, dt = \int_0^\infty e^{-st} \frac{1}{2}(e^{at} + e^{-at})dt$

$$= \underset{B\to\infty}{\text{Lt}} \frac{1}{2} \int_0^B \left\{e^{-(s-a)t} + e^{-(s+a)t}\right\} dt = \underset{B\to\infty}{\text{Lt}} \frac{1}{2} \left[\frac{e^{-(s-a)t}}{-(s-a)} + \frac{e^{-(s+a)t}}{-(s+a)}\right]_0^B$$

$$= \underset{B\to\infty}{\text{Lt}} \frac{1}{2} \left[\frac{1}{s-a}\left\{1 - e^{-(s-a)B}\right\} + \frac{1}{s+a}\{1 - e^{-(s+a)B}\}\right]$$

$$= \frac{1}{2}\left(\frac{1}{s-a} + \frac{1}{s+a}\right)$$

$$[\because e^{-(s-a)B} \to 0, \ e^{-(s+a)B} \to 0 \text{ as } B\to\infty \text{ because } s > |a|]$$

$$= \frac{\boldsymbol{s}}{\boldsymbol{s^2-a^2}}, \ \textbf{for } \boldsymbol{s > |a|}.$$

The following useful results may be obtained using properties (ii), (iv), (v), (vi) and (vii) :

(i) $L(e^{at}) = \frac{1}{s-a}, \; s > a$

(ii) $L(e^{at}\sin bt) = \frac{b}{(s-a)^2+b^2}; \; s-a > 0$

(iii) $L(e^{at}\cos bt) = \frac{s-a}{(s-a)^2+b^2}, \; s-a > 0$

(iv) $L(e^{at}\sin h\, bt) = \frac{b}{(s-a)^2-b^2}, \; s-a > |b|$

(v) $L(e^{at}\cos h\, bt) = \frac{s-a}{(s-a)^2-b^2}, \; s-a > |b|.$

9.5 Laplace Transform of Derivative of $f(t)$

$L[f'(t)] = sL[f(t) - f(0)]$, where $L[f(t)] = F(s)$.

Proof : We know that, $L[f'(t)] = \int_0^\infty e^{-st} f'(t)dt.$

Integrating by parts, $L[f'(t)] = \left[e^{-st}f(t)\right]_0^\infty - \int_0^\infty (-se^{-st})f(t)dt$

$$= -f(0) + s\int_0^\infty e^{-st} f(t)dt \quad [\because\ e^{-st}f(t) = 0, \text{ when } t = \infty]$$

$$= -f(0) + sLf(t).$$

$$\therefore\ L[f'(t)] = sLf(t) - f(0). \tag{9.10}$$

9.6 Laplace Transform of Derivative of Order n

$L[f^n(t)] = s^n L[f(t) - s^{n-1}f(0) - s^{n-2}f'(0) - s^{n-3}f''(0) - \cdots - f^{n-1}(0).$

Proof : We have $l[f'(t)] = sL[f(t)] - f(0).$ (9.11)

Replacing $f(t)$ by $f'(t)$ and $f'(t)$ by $f''(t)$ in the above equation, we get

$$L[f''(t)] = sL[f'(t) - f'(0)]. \tag{9.12}$$

Putting the value of $L[f'(t)]$ from (9.10) in (9.11), we get

$$L[f''(t)] = s[sL[f(t)] - f(0)] - f'(0) \quad \text{or,} \quad L[f''(t)] = s^2L[f(t)] - sf(0) - f'(0).$$

Similarly, $L[f'''(t)] = s^3L[f(t)] - s^2f(0) - sf(0) - f''(0)$

$$L[f^{\text{iv}}(t)] = s^4L[f(t)] - s^3f(0) - s^2f'(0) - sf''(0) - f'''(0)$$

$$\cdots \quad \cdots \quad \cdots \quad \cdots \quad \cdots \quad \cdots$$

$$L[f^n(t)] = s^nL[f(t)] - s^{n-1}f(0) - s^{n-2}f'(0) - s^{n-3}f''(0) \cdots - f^{n-1}(0). \tag{9.13}$$

9.7 Laplace Transform of Integration of $f(t)$

$L\left[\int_0^t f(t)dt\right] = \frac{1}{s}F(s)$, where $L[f(t)] = F(s)$. **[Gujarat U. 1991]**

Proof : Let $\phi(t) = \int_0^t f(t)dt$ and $\phi(0) = 0$, then $\phi'(t) = f(t)$.

The Laplace transforms of $\phi'(t)$ are

$$L[\phi'(t)] = sL[\phi(t)] - \phi(0) \quad \text{or,} \quad L[\phi'(t)] = sL[\phi(t)] \quad [\phi(0) = 0]$$

or, $L[\phi(t)] = \dfrac{1}{s}L[f'(t)]$.

Using the values of $\phi(t)$ and $\phi'(t)$, we get

$$L[\int_0^t f(t)dt] = \frac{1}{s}L[f(t)]$$

or, $$\boldsymbol{L\left[\int_0^t f(t)dt\right] = \frac{1}{s}F(s)}. \tag{9.14}$$

N.B. (i) This shows that Laplace transform of integration of $f(t)$ corresponds to the direction of the Laplace transform of $f(t)$ by s.

(ii) $$\boldsymbol{\int_0^t f(t) = L^{-1}\left[\frac{1}{s}F(s)\right]}.$$

9.8 Laplace Transform of $t^n f(t)$, where n is a Positive Number

Theorem 6 : If $L[f(t)] = \overline{f}(s)$ then $L[t^n f(t)] = (-1)^n \dfrac{d^n}{ds^n}\overline{f}(s)$, where n is a positive integer.

[WBUT 2005]

Proof : By definition, $\overline{f}(s) = L[f(t)] = \displaystyle\int_0^\infty e^{-st} f(t)dt$.

$$\therefore \quad \frac{d}{ds}\overline{f}(s) = \frac{d}{ds}\int_0^\infty e^{-st} f(t)dt = \int_0^\infty \frac{\partial}{\partial s}[e^{-st} f(t)]dt \quad \text{[By Leibnitz's rule]}$$

$$= \int_0^\infty -te^{-st} f(t)dt.$$

$$\therefore \quad \int_0^\infty e^{-st}[tf(t)]dt = -\frac{d}{ds}\overline{f}(s). \tag{9.15}$$

The theorem is thus true for $n = 1$. Now we assume that the theorem is true for a positive integer m.

$$\therefore \quad \int_0^\infty e^{-st}[t^m f(t)]dt = (-1)^m \frac{d^m}{ds^m}\overline{f}(s).$$

Then, $$\frac{d}{ds}\int_0^\infty e^{-st}[t^m f(t)]dt = (-1)^m \frac{d^{m+1}}{ds^{m+1}}\overline{f}(s)$$

or, $$\int_0^\infty \frac{\partial}{\partial s}[e^{-st}t^m f(t)]dt = (-1)^m \frac{d^{m+1}}{ds^{m+1}}\overline{f}(s) \quad \text{[by Leibnitz's rule]}$$

or, $$\int_0^\infty (-te^{-st})t^m f(t)dt = (-1)^m \frac{d^{m+1}}{ds^{m+1}}\overline{f}(s).$$

$$\therefore \quad \int_0^\infty e^{-st}[t^{m+1} f(t)]dt = (-1)^{m+1} \frac{d^{m+1}}{ds^{m+1}} f(s). \tag{9.16}$$

Hence, the theorem is true for a positive integer m, then it is also true for $m + 1$. But the theorem is true for $n = 1$, hence it is true for $1 + 1 = 2$, $2 + 1 = 3$, $3 + 1 = 4$ and so on.

Thus, the theorem is true for all positive integral values of n.

9.9 Laplace Transform of $\frac{1}{t}f(t)$

$L\left[\frac{1}{t}f(t)\right] = \int_s^\infty \overline{f}(s)ds$, where $L[f'(t)] = \overline{f}(s)$.

Proof : $L[f(t)] = \overline{f}(s) = \int_0^\infty e^{-st}f(t)dt$ (9.17)

Integrating Eqn. (9.17) w.r.t. s, we get

$$\int_s^\infty \overline{\boldsymbol{f}}(\boldsymbol{s})\boldsymbol{ds} = \int_s^\infty \left[\int_0^\infty e^{-st}f(t)dt\right]ds = \int_0^\infty \left[\int_s^\infty e^{-st}f(t)ds\right]dt = \int_0^\infty \left[\frac{e^{-st}f(t)}{-t}\right]_s^\infty dt$$

$$= \int_0^\infty \frac{f(t)}{t}\left[e^{-st}\right]_0^\infty ds = \int_0^\infty \frac{-f(t)}{t}[0-e^{-st}]dt \int_0^\infty e^{-st}\left[\frac{1}{t}f(t)\right]dt$$

$$= \boldsymbol{L}\left[\frac{\boldsymbol{1}}{\boldsymbol{t}}\boldsymbol{f}(\boldsymbol{t})\right]. \tag{9.18}$$

9.10 Laplace Transforms of Some Special Functions

1. **Periodic function :** If $f(t)$ be a periodic function of period $T(>0)$, i.e., $f(t+T) = f(t)$, then

$$Lf(t) = \frac{1}{1-e^{-st}}\int_0^T e^{-st}f(t)dt.$$

Proof : By definition, $Lf(t) = \int_0^\infty e^{-st}f(t)dt$

$$= \int_0^T e^{-st}f(t)dt + \int_T^{2T} e^{-st}f(t)dt + \int_{2T}^{3T} e^{-st}f(t)dt + \cdots \text{ to } \infty.$$

Putting $t = u + T$ in the second integral and $t = 1 + 2T$ in the integral and so on, then

$$L[f(t)] = \int_0^T e^{-st}f(t)dt + \int_0^T e^{-s(u+T)}f(u+T)du + \int_0^T e^{-s(u+2T)}f(y+2T)du + \cdots \text{ to } \infty$$

$$= \int_0^T e^{-st}f(t)dt + e^{-sT}\int_0^T e^{-su}f(u)du + e^{-2sT}\int_0^T e^{-su}f(u)du + \cdots \text{ to } \infty$$

$[\because f(u) = f(u+T) = f(u+2T)$, etc.]

$$= \int_0^T e^{-st}f(t)dt + e^{-st}\int_0^T e^{-st}f(t)dt + e^{-2sT}\int_0^T e^{-st}f(t)dt + \cdots \text{ to } \infty$$

$$= (1 + e^{-st} + e^{-2st} + \cdots \text{ to } \infty)\int_0^T e^{-st}f(t)dt$$

$$\therefore\ \boldsymbol{L[f(t)]} = \frac{\boldsymbol{1}}{\boldsymbol{1-e^{-st}}}\int_0^T \boldsymbol{e^{-st}f(t)dt}. \tag{9.19}$$

2. **Gamma function :** By definition, $\Gamma(n) = \int_0^\infty e^{-t}t^{n-1}dt$, $(n > 0)$

We have $L(t^n) = \int_0^\infty e^{-st}t^n dt = \int_0^\infty e^{-x}\left(\frac{x}{s}\right)^n \frac{dx}{s}$, putting $st = x$

$$= \frac{1}{s^{n+1}}\int_0^\infty x^n e^{-x}dx = \frac{\Gamma(n+1)}{s^{n+1}}.$$

$$\therefore \quad L^{-1}\left[\frac{\Gamma(n+1)}{s^{n+1}}\right] = t^n$$

$$\text{or,} \quad \boldsymbol{L^{-1}\left[\frac{1}{s^{n+1}}\right] = \frac{t^n}{\Gamma(n+1)}}. \tag{9.20}$$

N.B. For $n = -\frac{1}{2}$, $L[t^{-\frac{1}{2}}] = \frac{\Gamma\left(\frac{1}{2}\right)}{\sqrt{s}} = \sqrt{\frac{\pi}{s}} \quad \left(\because \Gamma\left(\frac{1}{2}\right) = \sqrt{\pi}\right)$

$$\therefore \quad \boldsymbol{L\left[\frac{1}{\sqrt{t}}\right] = \sqrt{\frac{\pi}{s}}}. \tag{9.21}$$

3. Bessel functions $J_0(x)$ and $J_1(x)$: We have

$$J_0(t) = \left[1 - \frac{t^2}{2^2} + \frac{t^4}{2^2\cdot 4^2} - \frac{t^6}{2^2\cdot 4^2\cdot 6^2} + \cdots\right].$$

$$\begin{aligned}
\therefore \quad L[J_0(t)] &= \frac{1}{s} - \frac{1}{2^2}\cdot\frac{2!}{s^3} + \frac{1}{2^2\cdot 4^2}\frac{4!}{s^5} - \frac{1}{2^2\cdot 4^2\cdot 6^2}\frac{6!}{s^7} + \cdots \\
&= \frac{1}{s}\left[1 - \frac{1}{2}\left(\frac{1}{s^2}\right) + \frac{1\cdot 3}{2\cdot 4}\left(\frac{1}{s^4}\right) - \frac{1\cdot 3\cdot 5}{2\cdot 4\cdot 6}\left(\frac{1}{s^6}\right) + \cdots\right] \\
&= \frac{1}{s}\left[1 + \left(-\frac{1}{2}\right)\left(\frac{1}{s^2}\right) + \frac{(-\frac{1}{2})(-\frac{3}{2})}{2!}\left(\frac{1}{s^2}\right)^2 + \frac{(-\frac{1}{2})(-\frac{3}{2})(-\frac{5}{2})}{3!}\left(\frac{1}{s^2}\right)^3 + \cdots\right] \\
&= \frac{1}{s}\left[1 + \frac{1}{s^2}\right]^{-\frac{1}{2}}, \quad \text{[using Binomial theorem]}
\end{aligned}$$

$$\therefore \quad \boldsymbol{L[J_0(t)] = \frac{1}{\sqrt{s^2+1}}}. \tag{9.22}$$

Again, $\because \; Lf(at) = \frac{1}{a}F\left(\frac{s}{a}\right)$,

$$\therefore \quad \boldsymbol{LJ_0(at) = \frac{1}{a}\frac{1}{\sqrt{\frac{s^2}{a^2}+1}} = \frac{1}{\sqrt{s^2+a^2}}}$$

$$\begin{aligned}
\text{and} \quad \boldsymbol{LJ_1(x)} &\boldsymbol{= -LJ_0'(x) = -[sLJ_0(x) - J_0(x)] = -\left[s\cdot\frac{1}{\sqrt{s^2+1}} - 1\right]} \\
&\boldsymbol{= 1 - \frac{s}{\sqrt{s^2+1}}}.
\end{aligned} \tag{9.23}$$

9.11 Convolution Theorem

If the Laplace transforms $L[f_1(t)] = F_1(s)$ and $L[f_2(t)] = F_2(s)$, then

$$L\left\{\int_0^t f_1(x)f_2(t-x)dx\right\} = F_1(s)\cdot F_2(s) \quad \text{or,} \quad F_1(s)\cdot F_2(s) = L^{-1}\int_0^t f_1(x)f_2(t-x)dx.$$

Proof : We have $L\left\{\int_0^\infty f_1(x)f_2(t-x)dx\right\} = \int_0^\infty e^{-st}\int_0^t f_1(x)f_2(t-x)dx\,dt$

$$= \int_0^\infty \int_0^t e^{-st}f_1(x)f_2(t-x)dx\,dt.$$

The double integral is over the infinite region in the first quadrant lying between the lines $x = 0$ and $x = t$.

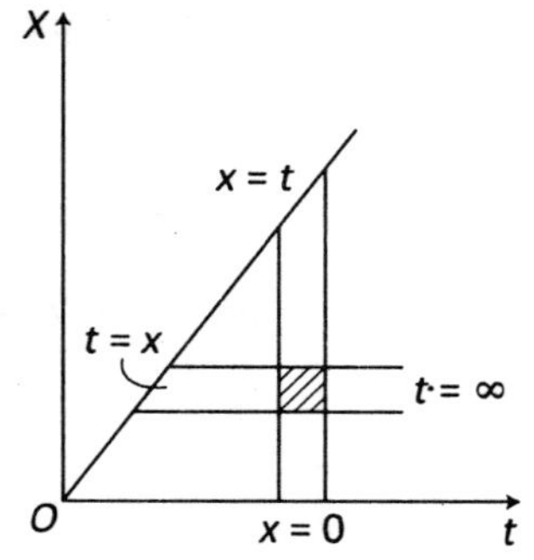

Fig. 9.1

Changing the order of integration, we get

$$\int_0^\infty \int_x^t e^{-st} f_1(x) f_2(t-x) dt\, dx$$

$$= \int_0^\infty e^{-sx} f_1(x) dx \int_x^\infty e^{-s(t-x)} f_2(t-x) dt$$

$$= \int_0^\infty e^{-sx} f_1(x) dx \int_0^\infty e^{-sz} f_2(z) dz \quad [\text{putting } t - x = z]$$

$$= \int_0^\infty e^{-sx} f_1(x) F_2(s) dx = \left[\int_0^\infty e^{-sx} f_1(x) dx\right] F_2(s)$$

$$= \boldsymbol{F_1(s) F_2(s)}. \tag{9.24}$$

9.12 Inverse Laplace Transform

Definition : If $L[f(t)] = \overline{f}(s)$, then $f(t)$ is called the **inverse Laplace transform** of the function $f(s)$ and is expressed as

$$\boldsymbol{f(t) = L^{-1}[\overline{f}(s)]}. \tag{9.25}$$

N.B. L^{-1} is known as the **inverse Laplace transformation operator**.

List of some important and useful results :

1. $\boldsymbol{L(1) = \frac{1}{s}}$, $\quad \therefore \boldsymbol{L^{-1}\left(\frac{1}{s}\right) = 1}$.
2. $\boldsymbol{L(e^{at}) = \frac{1}{s-a}}$, $\quad \therefore \boldsymbol{L^{-1}\left(\frac{1}{s-a}\right) = e^{at},\ s > a}$.
3. $\boldsymbol{L(t^n) = \frac{\Gamma(n+1)}{s^{n+1}}}$, $\quad \therefore \boldsymbol{L^{-1}\left(\frac{1}{s^{n+1}}\right) = \frac{t^n}{\Gamma(n+1)},\ n > -1}$.
4. $\boldsymbol{L(\sin at) = \frac{a}{s^2+a^2}}$, $\quad \therefore \boldsymbol{L^{-1}\left(\frac{1}{s^2+a^2}\right) = \frac{1}{a}\sin at}$.
5. $\boldsymbol{L(\cos at) = \frac{s}{s^2+a^2}}$, $\quad \therefore \boldsymbol{L^{-1}\left(\frac{s}{s^2+a^2}\right) = \cos at}$.
6. $\boldsymbol{L(\sinh at) = \frac{a^2}{s^2-a^2}}$, $\quad \therefore \boldsymbol{L^{-1}\left(\frac{1}{s^2-a^2}\right) = \frac{1}{a}\sinh at,\ s > |a|}$.
7. $\boldsymbol{L(\cosh at) = \frac{s}{s^2-a^2}}$, $\quad \therefore \boldsymbol{L^{-1}\left(\frac{s}{s^2-a^2}\right) = \cosh at,\ s > |a|}$.

9.13 Properties of Inverse Laplace Transform

1. **Linear property :** If $L[f_1(t)] = \overline{f}_1(s)$ and $L[f_2(t)] = \overline{f_2}(s)$, then

$$L^{-1}[af_1(s) + bf_2(s)] = aL^{-1}[\overline{f}_1(s)] + bL^{-1}[\overline{f}_2(s)],$$

where a and b are two constants.

Proof : Since, $L[f_1(t)] = \overline{f}_1(s)$, $L[f_2(t)] = \overline{f}_2(s)$,

$\therefore \quad f_1(t) = L^{-1}[f_1(s)], \quad f_2(t) = L^{-1}[\overline{f}_2(s)]$.

By the linear property Laplace transform, we have

$$L[af_1(t) + bf_2(t)] = aL[f_1(t)] + bL[f_2(t)] = a\overline{f}(s) + b\overline{f}_2(s).$$

$$\therefore \quad \boldsymbol{L^{-1}[a\overline{f_1}(s) + b\overline{f_2}(s) = af_1(t) + bf_2(t) = aL^{-1}[\overline{f_1}(s)] + bL^{-1}[\overline{f_2}(s)]}. \quad (9.26)$$

The above result can also be generalized.

2. **Shifting property of inverse Laplace transform : First shifting property :**

$$L^{-1}[\overline{f}(s-a)] = e^{at}L^{-1}[\overline{f}(s)].$$

Proof : Let $L^{-1}[\overline{f}(s)] = f(t)$.

$\therefore$ by the first shifting property of Laplace transform, we have $L[e^{at}f(t)] = \overline{f}(s-a)$.

$$\therefore \quad \boldsymbol{L^{-1}[f(s-a)] = e^{at}f(t) = e^{at}L^{-1}[\overline{f}(s)]}. \quad (9.27)$$

3. **Second shifting property :** If $L^{-1}[\overline{f}(s)] = f(t)$, then

$$\boldsymbol{L^{-1}[e^{-as}\overline{f}(s)] = \begin{cases} f(t-a), & t > a \\ 0, & t < a \end{cases} \doteq f(t-a)u(t-a)},$$

where $u(t-a)$ is a unit step function.

Proof : $\because \; L^{-1}[\overline{f}(s)] = f(t)$, so $L[f(t)] = \overline{f}(s)$.

Let $g(t) = f(t-a)u(t-a) = \begin{cases} f(t-a), & t > a \\ 0, & t < a. \end{cases}$

$\therefore$ by the second shifting property of Laplace transform, we have

$$L[g(t)] = e^{-as}\overline{f}(s).$$

$$\therefore \quad \boldsymbol{L^{-1}[e^{-as}\overline{f}(s)] = g(t) = f(t-a)u(t-a) = \begin{cases} f(t-a), & t > a \\ 0, & t < a. \end{cases}} \quad (9.28)$$

4. **Change of scale property :** If $L[f(t)] = \overline{f}(s)$ denotes the Laplace transform of the function $f(t)$, then

$$L^{-1}[f(as)] = \frac{1}{a}f\left(\frac{t}{a}\right), \; a > 0. \qquad \textbf{[WBUT 2002]}$$

Proof :
$$L\left[\frac{1}{a}f\left(\frac{t}{a}\right)\right] = \frac{1}{a}L\left[f\left(\frac{t}{a}\right)\right]$$
$$= \frac{1}{a}\frac{1}{\frac{1}{a}}\overline{f}\left(\frac{s}{\frac{1}{a}}\right), \quad \text{[by the scale property of Laplace transform]}$$
$$= \overline{f}(as).$$

$$\therefore \quad \boldsymbol{L^{-1}[f(as)] = \frac{1}{a}f\left(\frac{t}{a}\right), \; a > 0}. \quad (9.29)$$

9.14 Unit Step Function and its Transformation

There are many functions whose **inverse Laplace transforms** cannot be found with the methods discussed earlier. In those cases **unit step function** is of great help. Now what is a unit step-function?

The unit step function may be defined as

$$u(t-a)=\begin{cases}0, & \text{for } t<0\\ 1, & \text{for } t\geq a\end{cases}$$

where $a \geq 0$ (Fig. 9.2).

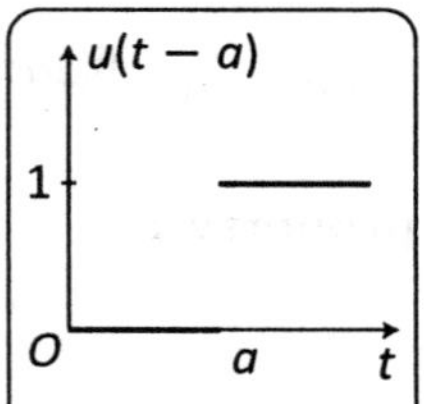

Fig. 9.2

The Laplace transform of unit function is $\boldsymbol{L[u(t-a)]=\frac{e^{-as}}{s}}$.

Proof : We have $L[(u-t)]=\int_0^{\infty} e^{-st}u(t-a)dt$

$$=\int_0^a e^{-at}\cdot 0\,dt+\int_a^{\infty} e^{-st}\cdot 1\cdot dt=0+\left[\frac{e^{-st}}{-s}\right]_a^{\infty}.$$

$$\therefore \quad \boldsymbol{L[u(t-a)=\frac{e^{-as}}{s}}. \tag{9.30}$$

9.15 Second Shifting Theorem

If $L[f(t)]=F(s)$, then $L[f(t-a)]\cdot u(t-a)]=e^{-as}F(s)$.

Proof : $\boldsymbol{L[f(t-a),u(t-a)]}=\int_0^{\infty} e^{-st}f(t-a)\cdot 1\cdot dt$

$$=\int_0^a e^{-st}f(t-a)\cdot 0dt+\int_0^{\infty} e^{-st}f(t-a)\cdot 1\cdot dt=\int_0^{\infty} e^{-st}f(t-a)dt$$

$$=\int_0^{\infty} e^{-s(u+a)}f(u)du,\ [u=t-a]$$

$$=e^{-st}\int_0^{\infty} e^{-su}f(u)du=\boldsymbol{e^{-sa}F(s)}. \tag{9.31}$$

An important theorem : $\boldsymbol{L[f(t)\cdot u(t-a)]=e^{-sa}L[f(t+a)]}$.

Proof : $\boldsymbol{L[f(t-a)\cdot u(t-a)]}=\int_0^{\infty} e^{-sx}[f(t)\cdot u(t-a)]dt$

$$=\int_0^a e^{-sx}[f(t)\cdot u(t-a)]dt+\int_a^{\infty} e^{sx}[f(t)\cdot u(t-a)]dx=0+\int_a^{\infty} e^{-sa}[f(t)\cdot 1\cdot dt]$$

$$=\int_0^{\infty} e^{-s}(y+a)f(y+a)dy \quad [\text{putting } t-a=y]$$

$$=\int_0^{\infty} e^{-sa}\int_0^{\infty} e^{-sy}f(y+a)dy=e^{-sa}\int_0^{\infty} e^{-st}f(t+a)dt=\boldsymbol{e^{-sa}L[f(t+a)]}. \tag{9.32}$$

9.16 (a) Impulse Function

In applied mechanics, impulse is the product of a large force acting for a short time and the time of action. The unit impulse function is the limiting function,

$$\boldsymbol{g(t-1)=\begin{cases}\frac{1}{\epsilon}, & a<t<a+\epsilon\\ 0, & \text{otherwise.}\end{cases}}$$

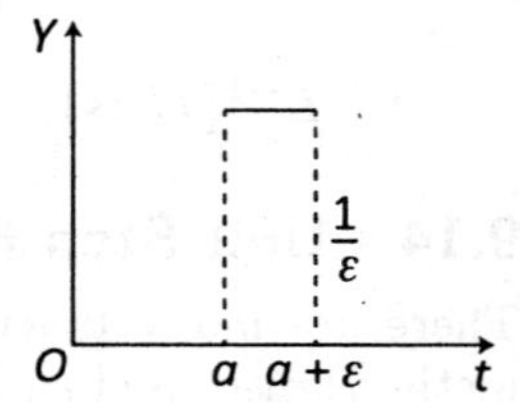

Fig. 9.3

The value of the function becomes infinite as $\epsilon \to 0$ and the area of the rectangle is unity [Fig. (9.3)].

(b) The unit impulse function is defined as follows :

$$\boldsymbol{\delta(t-a) = \begin{cases} \infty, & \text{for } t = a \\ 0, & \text{for } t \neq a \end{cases}}$$

and $\displaystyle\int_0^\infty \boldsymbol{\delta(t-a)\cdot dt = 1}$ **(area of the strip).**

(c) Laplace transform of unit impulse function :

$$\begin{aligned} \int_0^\infty f(t)\delta(t-a)dt &= \int_a^{a+\epsilon} f(t)\cdot\frac{1}{\epsilon}dt \\ &= (a+\epsilon-a)f(\eta),\frac{1}{\epsilon}, \quad \text{where } a<\eta<a+\epsilon \\ &= f(\eta) \end{aligned}$$

$$\int_0^\infty \boldsymbol{f(t)\delta(t-a)dt = f(a)}, \quad [\textbf{as } \epsilon \to \mathbf{0}].$$

N.B. If $f(t) = e^{-st}$, then $L[\delta(t-a)] = e^{-as}$.

9.17 Use of Partial Fractions

If $g(s)$ and $h(s)$ are polynomials in s and $\overline{f}(s) = \frac{g(s)}{h(s)}$, then break $\overline{f}(s)$ in partial fractions is as given below :

(i) The degree of the polynomial $h(s)$ must be less than the degree of the polynomial in the denominator, i.e., $g(s)$. If this is not so, then the process of division is to be used to obtain the necessary requirement.

(ii) Then the denominator is to be factorized into its ultimate real factors and these factors must be one or other of the following four types :

1. Linear but not repeated, of the type $(ax+b)$.
2. Linear and repeated, of the type $(ax+b)^n$.
3. Quadratic but not repeated, such as (ax^2+bx+c).
4. Repeated quadratic, such as $(ax^2+bx+c)^n$.

(iii) Then the given fraction is to be written as he sum of a set of simple fractions according to the rules as :

(a) For each factor of the type $(ax+b)$ there should be a single fraction of the form

$$\frac{A}{ax+b},$$

A being a constant.

(b) For each factor of the type $(ax+b)^n$, there should be fractions of the form :

$$\frac{A_1}{ax+b} + \frac{A_2}{(ax+b)^2} + \cdots + \frac{A_n}{(ax+b)^n},$$

$A_1, A_2, \ldots, A_n$ are constants.

(c) Likewise for ax^2+bx+c, take $\dfrac{Ax+B}{ax^2+bx+c}$

and for $(ax^2+bx+c)^n$, take $\dfrac{A_1x+B_1}{ax^2+bx+c} + \dfrac{A_2x+B_2}{(ax^2+bx+c)^2} + \cdots + \dfrac{A_nx+B_n}{(ax^2+bx+c)^n}$.

(iv) Finally the numerator of the sum of component fractions are to be made identically equal to the numerator of the given fraction to determine A, B; A_1, B_1; ...

9.18 Solution of Differential Equations Using Laplace Transform

Laplace transforms offer a very convenient method for solving ordinary differential equations with constant or variable coefficients, partial differential equations, integral equations and simultaneous differential equations.

(a) **Ordinary differential equations :** We consider a linear differential equation with constant coefficients as

$$a_0\frac{d^2x}{dt^2}+a_1\frac{dx}{dt}+a_2x=F(t), \tag{9.33}$$

where a_0, a_1 and a_2 are constants.

Taking Laplace transform of both sides

$$a_0L\left[\frac{d^2x}{dt^2}\right]+a_1L\left[\frac{dx}{dt}\right]+a_2L[x]=L[F(t)]. \tag{9.34}$$

Using the property of Laplace transform of derivatives,

$$L\left[\frac{dx}{dt}\right]=sL[x]-[x]_{t=0} \quad \text{or,} \quad L\left[\frac{d^2x}{dt^2}\right]=s^2L[x]_{t=0}-s[x]_{t=0}$$

and $L[F(t)]=f(s)$.

Then from Eqn. (9.34),

$$a_0\left\{s^2L(x)-s[x]_{t=0}-\left(\frac{dx}{dt}\right)_{t=0}\right\}+a_1[sL[x]-[x]_{t=0}+a_2L[x]-f(s)$$

or, $$(a_0s^2+a_1s+a_2)L[x]=f(s)+(a_0s+a_1)[x]_{t=0}+a_0\left(\frac{dx}{dt}\right)_{t=0}$$

or, $$L[x]=\frac{f(s)+(a_0s+a_1)[x]+a_1\left(\frac{dx}{dt}\right)_{t=0}}{a_0s^2+a_1s+a_2}. \tag{9.35}$$

The R.H.S. of Eqn. (9.35) is of the form $\left[\frac{g(s)}{h(s)}\right]$.

∴ the solution of the differential Eqn. (9.33) becomes

$$\boldsymbol{x(t)=L^{-1}\left[\frac{g(s)}{h(s)}\right]}. \tag{9.36}$$

Procedure of solving differential equations :

(i) Replace D^nx by $s^nL[x]$, $D^{n-1}x$ by $s^{n-1}L[x]$, ..., D^2x by $L[x]$, Dx by $sL[x]$ and $xsL[x]$.

(ii) Then for D^nx in the L.H.S. add $s^{n-1}x_0+s^{n-2}x_1+\cdots+x_{n-1}$ in the R.H.S., where

$$x_0=(x)_{t=0} \quad \text{and} \quad x_r=(D^rx)_{t=0}=\left(\frac{d^rx}{dx^r}\right)_{t=0}.$$

(iii) Replace the function (if any) in the R.H.S. by its Laplace transform.

(iv) Finally, express $L[x]$ in the form $\left[\frac{g(s)}{h(s)}\right]$ and then take the inverse Laplace transform to get the solution.

(b) Ordinary differential equation with variable coefficients : Suppose we have to solve the differential equation,

$$t\frac{d^2x}{dt^2} + \frac{dx}{dt} + 4tx = 0, \tag{9.34}$$

when $x(0) = 3$ and $x_1(0) = 0$.

Solution : Taking Laplace transform of Eqn. (9.37),

$$L[tx''] + L[x'] + 4L[tx] = L(0) = 0$$

or, $$-\frac{d}{ds}[s^2L[x] - s(x)_{t=0} - (x')_{t=0}] + [sL[x] - (x)_{t=0}] - 4\frac{d}{ds}\{L[x]\} = 0$$

or, $$\frac{d}{dx}\{s^2L[x] - 3s\} - sL\{[x] - 3\} + 4\frac{d}{ds}\{L[x]\} = 0$$

$$[\because\ (x)_{t=0} = x(0) = 3 \quad \text{and} \quad (x')_{t=0} = x_1(0) = 0]$$

or, $(s^2 + 4)\frac{d}{ds}[L[x]\} + sL[x] = 0$ or, $\frac{d\{L[x]\}}{L[x]} + \frac{sds}{s^2+4} = 0.$

Integrating, we get

$$\log_e\{L[x]\} + \frac{1}{2}\log_e(s^2+4) = \log_e k, \quad [k = \text{a constant}]$$

or, $$\boldsymbol{L[x] = \frac{k}{\sqrt{s^2+4}}}.$$

Taking inverse Laplace transform, we get $x(t) = L^{-1}\left\{\frac{k}{\sqrt{s^2+4}}\right\} = kJ_0(2t)$.

Now, at $t = 0$, $x = 3$, $\therefore\ 3 = kJ_0(0) = k$.

$$\therefore \quad \boldsymbol{x(t) = 3J_0(2t)}. \tag{9.38}$$

(c) Solution of integral equations : An integral equation contains a dependent variable under an integral sign.

Suppose, we have to solve the following integral equation,

$$F(t) = 1 + 2\int_0^t F(t-x)e^{-2x}dx. \tag{9.39}$$

Taking Laplace transform of both sides, we get

$$L[F(t)] = L[1] + 2L\left\{\int_0^t F(t-x)e^{-2x}dx\right\}. \tag{9.40}$$

Putting, $L[F(t)] = f(s)$; $L(1) = \frac{1}{s}$ and $L\left\{\int_0^t F(t-x)e^{-2x}dx\right\} = \frac{f(s)}{s+2}$.

From Eqn. (9.40),

$$f(s) = \frac{1}{s} + \frac{2f(s)}{s+2}. \quad \text{or,} \quad f(s) = \frac{1}{s} + \frac{2}{s^2}.$$

$$\therefore \quad \boldsymbol{F(t)} = L^{-1}[f(s)] = L^{-1}\left[\frac{1}{s} + \frac{2}{s^2}\right] = \boldsymbol{1 + 2t}. \tag{9.42}$$

9.19 Table for Laplace Transform Theorems

Sl. No.	Operation	$f(t)$	$L[f(t)] = \bar{f}(s)$
1.	Linearity property	$a_1 f_1(t) + a_2 f_2(t)$	$a_1 L[f_1(t)] + a_2 L[f_2(t)]$
2.	First shifting property	$e^{at} f(t)$	$\bar{f}(s-a)$
3.	Second shifting property	(i) $g(t) = \begin{cases} f(t-a), & t > a \\ 0, & t < a \end{cases}$ (ii) $f(t-a)u(t-a)$ (iii) $f(t)u(t-a)$	(i) $e^{-at}\bar{f}(s)$ (ii) $e^{-as}\bar{f}(s)$ (iii) $e^{-as}L[f(t+a)]$
4.	Change of scale property	$f(at)$	$\frac{1}{a}\bar{f}\left(\frac{s}{a}\right)$, $a > 0$
5.	Multiplication theorem	(i) $tf(t)$ (ii) $t^n f(t)$	(i) $-\frac{d}{ds}\bar{f}(s)$ (ii) $(-1)^n \frac{d^n}{ds^n}\bar{f}(s)$
6.	Division theorem	$\frac{1}{t} f(t)$	$\int_0^\infty \bar{f}(s)ds$
7.	Differentiation theorem	(i) $f'(t) = \frac{d}{dt} f(t)$ (ii) $f^n(t) = \frac{d^n}{dt^n} f(t)$	(i) $-f(0) + s\bar{f}(s)$ (ii) $-f^{n-1}(0) - sf^{n-2}(0) \cdots - s^{n-1}f(0) + s^n \bar{f}(s)$
8.	Integral theorem	$\int_0^t f(u)du$	$\frac{1}{s}\bar{f}(s)$
9.	Periodic function	$f(t)$ is periodic with period T	$\frac{1}{1-e^{-sT}}\int_0^T e^{-st} f(t)dt$
10.	Unit step function	(i) If $f(t) = \begin{cases} f_1(t), & t < a \\ f_2(t), & t > a \end{cases}$ then $f(t) = f_1(t) + [f_2(t) - f_1(t)]$ $u(t-a)$ (ii) If $f(t) = \begin{cases} f_1(t), & t < a_1 \\ f_2(t), & a_1 < t < a_2 \\ f_3(t), & t > a_2 \end{cases}$ then $f(t) = f_1(t) + [f_2(t) - f_1(t)]$ $u(t-a_1) + [f_3(t) - f_2(t)]u(t-a_2)$	(i) $L[f(t)] = L[f_1(t)] + e^{-as}L[f_2(t+a)] - f_1(t+a)$ (ii) $L[f(t)] = L[f_1(t)] + e^{-a_1 s}L[f_2(t+a_1) - f_1(t+a_1)] + e^{-a_2 s}L[f_3(t+a_2) - f_2(t+a_2)]$

9.20 Table of Inverse Laplace Theorem

Sl. No.	Operation	$f(s)$	$L^{-1}[f(s)] = f(t)$
1.	Linear property	$a_1 f_1(s) + a_2 f_2(s)$	$a_1 L^{-1}[f_1(s)] + a_2 L^{-1}[f_2(s)]$
2.	First shifting property	$f(s - a)$	$e^{at} L^{-1}[f(s)]$
3.	Second shifting property	$e^{-as} f(s)$	$f(t-a)u(t-a) = \begin{cases} f(t-a), & t > a \\ 0, & t < a \end{cases}$
4.	Change of scale	$f(as),\ a > 0$	$\frac{1}{a} f\left(\frac{t}{a}\right)$
5.	Differentiation theorem	$f^{-n}(s)$	$(-1)^n t^n f(t)$
6.	Multiplication theorem	(i) $sf(s)$ (ii) $s^n f(s)$	(i) $f'(t)$, if $f(0) = 0$ (ii) $f^n(t)$, if $f(0) = f'(0) = \cdots = f^{n-1}(0) = 0$
7.	Division theorem	(i) $\frac{f(s)}{s}$ (ii) $\frac{f(s)}{s^n}$	(i) $\int_0^t f(u)du$ (ii) $\int_0^t \int_0^t \cdots \int_0^t f(u)du\,du \cdots du$ (n times integration)
8.	Integration theorem	$\int_0^\infty \overline{f}(u)du$	$\frac{f(t)}{t}$
9.	Convolution theorem	$f(s)g(s)$	$\int_0^t fu)g(t-u)du.$

9.21 Solved problems

Example 9.1. *Find the Laplace transforms of the function* (i) $\sin at$ (ii) $\cosh(at)$.

Solution : (i) $$L[\sin(at)] = \int_0^\infty e^{-st}\sin(at)dt = \left[\frac{e^{-st}}{a^2+s^2}\{-s\sin(at) - a\cos(at)\}\right]_0^\infty$$
$$= \frac{a}{a^2+s^2},\ s > 0$$

(ii) $$L[\cosh(at)] = L\left[\frac{1}{2}(e^{at}+e^{-at})\right] = \frac{1}{2}\left[\frac{1}{s-a}+\frac{1}{s+a}\right] = \frac{s}{s^2-a^2},\ s > |a|.$$

Example 9.2. *Find the Laplace transform of* $(1 + \cos 2t)$. **[AMIETE(Winter) 1994]**

Solution : $$L(1+\cos 2t) = \int_0^\infty e^{-st}(1+\cos 2t)dt = \int_0^\infty e^{-st}\left(1+\frac{e^{2it}+e^{-2it}}{2}\right)dt$$
$$= \frac{1}{2}\int_0^\infty \left[2e^{-st} + e^{(-s+2i)t} + e^{(-s-2i)t}\right]dt$$

$$= \frac{1}{2}\left[\frac{2e^{-st}}{-s} + \frac{e^{(-s+2i)t}}{-s+2i} + \frac{e^{(-s-2i)t}}{-s-2i}\right]_0^\infty$$

$$= \frac{1}{2}\left[\left(0+\frac{2}{s}\right) + \frac{1}{-s+2i}(0-1) + \frac{1}{-s-2i}(0-1)\right]$$

$$= \frac{1}{2}\left[\frac{2}{s} + \frac{1}{s-2i} + \frac{1}{s+2i}\right] = \frac{1}{2}\left[\frac{2}{s} + \frac{2s}{s^2+4}\right]$$

$$= \frac{1}{s} + \frac{s}{s^2+4} = \boldsymbol{\frac{2s^2+4}{s(s^2+4)}}.$$

Example 9.3. *Find the Laplace Transform of $f(t)$ given by* $f(t) = \begin{cases} 1, & \text{if } 0 < t < 2 \\ 2, & \text{if } t > 2 \end{cases}$

[WBUT 2002]

Solution : By definition,

$$\boldsymbol{L[f(t)]} = \int_0^\infty e^{-st} f(t)dt = \int_0^2 e^{-st}\cdot 1\cdot dt + \int_2^\infty e^{-st}2dt = \left[\frac{e^{-st}}{-s}\right]_0^2 + \underset{B\to\infty}{\text{Lt}}\ 2\int_0^B e^{-st}dt$$

$$= \frac{e^{-2s}}{-s} + \frac{1}{s} + 2\underset{B\to\infty}{\text{Lt}}\left[\frac{e^{-st}}{-s}\right]_2^B = -\frac{e^{-2s}}{s} + \frac{1}{s} + 2\underset{B\to\infty}{\text{Lt}}\left[\frac{e^{-sB}}{-s} + \frac{e^{-2s}}{s}\right]$$

$$= -\frac{e^{-2s}}{s} + \frac{1}{s} + 2\left[0 + \frac{e^{-2s}}{s}\right] \quad [\because\ e^{-sB} \to 0 \text{ as } B \to \infty \text{ for } s > 0]$$

$$= \boldsymbol{\frac{1}{s} + \frac{e^{-2s}}{s}}.$$

Example 9.4. *Find the Laplace transform of the function $f(t)$, where*

$$f(t) = \begin{cases} \sin\left(t - \frac{\pi}{3}\right), & t > \frac{\pi}{3} \\ 0, & t \le \frac{\pi}{3}. \end{cases}$$

[WBUT 2004]

Solution : We put $\phi(t) = \sin t$

$$\therefore \quad f(t) = \begin{cases} \phi\left(t - \frac{\pi}{3}\right), & t > \frac{\pi}{3} \\ 0, & t \le \frac{\pi}{3}. \end{cases}$$

We have $L[\phi(t)] = L[\sin t] = \dfrac{1}{s^2+1} = \phi(s)$.

$\therefore$ by the use of second shifting property,

$$\boldsymbol{L[f(t)] = e^{-\left(\frac{\pi}{3}\right)s}\bar{\phi}(s) = \frac{e^{-\frac{\pi s}{3}}}{s^2+1},\ s > 0.}$$

Example 9.5. *Find the Laplace transform of $f(t)$, where $f(t) = \dfrac{\sin(at)}{t}$,*

given that, $L\left[\dfrac{\sin at}{t}\right] = \tan^{-1}\left(\dfrac{1}{s}\right)$.

[WBUT 2007]

Solution : According to the change of scale property, if $L[f(t)] = \overline{f}(s)$, then

$$\boldsymbol{L[f(at)] = \frac{1}{a}\overline{f}\left(\frac{s}{a}\right).}$$

$$\because \quad L\left[\frac{\sin t}{t}\right] = \tan^{-1}\frac{1}{s},$$

$$\therefore \quad L\left[\frac{\sin at}{at}\right] = \frac{1}{a}\tan^{-1}\frac{1}{\frac{s}{a}}... $$

$$\therefore \quad L\left[\frac{\sin at}{at}\right] = \frac{1}{a}\tan^{-1}\frac{1}{s} = \frac{1}{a}\tan^{-1}\frac{a}{s} \quad \text{or,} \quad \frac{1}{a}L\left[\frac{\sin at}{t}\right] = \frac{1}{a}\tan^{-1}\frac{a}{s},$$

$$\therefore \quad \boldsymbol{L\left[\frac{\sin at}{t}\right] = \tan^{-1}\frac{a}{s}}$$

Example 9.6. *Find the Laplace transform of $e^{-3t}(2\cos 5t - 3\sin 5t)$.*

Solution : We have $L[2\cos 5t - 3\sin 5t] = 2L(\cos 5t) - 3L(\sin 5t)$

$$= 2\frac{s}{s^2+5} - 3\frac{5}{s^2+5} = \frac{2s-15}{s^2+25}.$$

$\therefore$ by the first shifting property we have

$$\boldsymbol{L\left[e^{-3t}(2\cos 5t - 3\sin 5t)\right]} = \frac{2(s+3)-15}{(s+3)^2+25} = \boldsymbol{\frac{2s-9}{s^2+6s+34}}$$

Example 9.7. *If $L[f(t)] = \dfrac{p^2-p+1}{(2p+1)^2(p-1)}$, apply the change of scale property to show that,*

$$L[f(2t)] = \frac{p^2-2p+4}{4(p+1)^2(p-2)}.$$ **[WBUT 2002]**

Solution : Change of scale property states that, if $L[f(t)] = f(s)$, then

$$L[f(at)] = \frac{1}{a}\overline{f}\left(\frac{s}{a}\right).$$

$$\because \quad L[f(t)] = \frac{p^2-p+1}{(2p+1)^2(p-1)}.$$

$$\therefore \quad \boldsymbol{L[f(2t)]} = \frac{1}{2}\frac{\left(\frac{p}{2}\right)^2 - \left(\frac{p}{2}\right) + 1}{\left(2\cdot\frac{p}{2}+1\right)^2\left(\frac{p}{2}-1\right)} = \boldsymbol{\frac{p^2-2p+4}{4(p+1)^2(p-2)}} \quad \textbf{(Proved)}$$

Example 9.8. *Find he Laplace transform of $f(t)$, where $f(t) = \begin{cases} \sin t, & 0 < t < \pi \\ 0, & t > \pi. \end{cases}$*

[WBUT 2008]

Solution : By definition,

$$L[f(t)] = \int_0^\infty e^{-st}f(t)dt = \int_0^\pi e^{-st}\sin t\,dt + \int_\pi^\infty e^{-st}\cdot 0\cdot dt = \int_0^\pi e^{-st}\sin t\,dt = I \text{ (say)}.$$

$$\therefore \quad I = [e^{-st}\cos t]_0^\pi - \int_0^\pi e^{-st}(-s)(-\cos t)dt \quad \text{[integrating by parts]}$$

$$= e^{-s\pi} + 1 - s\int_0^\infty e^{-st}\cos t\,dt = e^{-s\pi} + 1 - s\left(e^{-st}\sin t\right)_0^\pi - \int_0^\pi e^{-st}(-s)\sin t\,dt.$$

$\therefore\quad I = e^{-s\pi} + 1 - s^2 I \quad$ or, $\quad (1+s^2)I = 1 + e^{-s\pi}$.

$$\therefore\quad I = L[f(t)] = \frac{1+e^{-s\pi}}{1+s^2}.$$

Example 9.9. *Find the Laplace transform of* $t^{-\frac{1}{2}}$.

Solution : We have $L(t^n) = \dfrac{\Gamma(n+1)}{s^{n+1}}$.

Putting $n = -\dfrac{1}{2}$, $L(t^{-\frac{1}{2}}) = \dfrac{\Gamma\left(-\frac{1}{2}+1\right)}{s^{-\frac{1}{2}+1}} = \dfrac{\Gamma\left(\frac{1}{2}\right)}{\sqrt{s}} = \dfrac{\sqrt{\pi}}{\sqrt{s}} \quad \left[\because \Gamma\left(\dfrac{1}{2}\right) = \sqrt{\pi}\right]$

Example 9.10. *Find the Laplace transform of* $t \sin at$.

Solution : We have $L(t\sin at) = L\left(t\dfrac{e^{iat}-e^{-iat}}{2i}\right) = \dfrac{1}{2i}\left[L(te^{iat}) - L(te^{-iat})\right]$

$$= \frac{1}{2i}\left[\frac{1}{(s-ia)^2} - \frac{1}{(s+ia)^2}\right] = \frac{1}{2i}\left[\frac{(s+ia)^2-(s-ia)^2}{(s-ia)(s+ia)}\right]$$

$$= \frac{1}{2i}\frac{(s^2+2ias-a^2)-(s^2-2ias-a^2)}{(s^2+a^2)^2} = \frac{1}{2i}\frac{4ias}{(s^2+a^2)^2} = \frac{2as}{(s^2+a^2)^2}.$$

Example 9.11. *Evaluate* (i) $L^{-1}\left[\dfrac{s^2}{(s+1)^2}\right]$ $(a^2 \neq b^2)$ **[WBUT 2006]**

(ii) $L^{-1}\left[\dfrac{1}{(s^2+a^2)(s^2+b^2)}\right]$ **[WBUT 2006]** *and* (iii) $L^{-1}\left[\dfrac{4p+5}{(p-4)^2(p+3)}\right]$ **[WBUT 2003]**

Solution : (i) We know, $L^{-1}\left(\dfrac{1}{s^5}\right) = \dfrac{t^4}{4!}$, $\because L^{-1}\left(\dfrac{1}{s^{n+1}}\right) = \dfrac{t^n}{n!}$, [$n$ is a positive integer.]

Using first shifting property, we have

$$L^{-1}\left[\frac{1}{(s+1)^5}\right] = e^{-t}\frac{t^4}{4!} = f(t), \text{ say.}$$

$$\therefore\quad f'(t) = \frac{df}{dt} = \frac{1}{4!}(e^{-t}t^4 + 4t^3e^{-t}) = \frac{e^{-t}}{24}t^3(4-t), \quad \therefore\ f(0) = f'(0) = 0.$$

Hence, $L^{-1}\left[\dfrac{s^2}{(s+1)^2}\right] = \dfrac{d^2f}{dt^2} = \dfrac{d}{dt}\left[\dfrac{e^{-t}}{24}t^3(4-t)\right] = \dfrac{1}{24}\left[e^{-t}(4t^3-t^4) + e^{-t}(12t^2-4t^3)\right]$

$$= \frac{t^2e^{-t}}{24}(t^2-8t+12).$$

(ii) We have $L^{-1}\left[\dfrac{1}{(s^2+a^2)(s^2+b^2)}\right] = \dfrac{1}{a^2-b^2}L^{-1}\left[\dfrac{(s^2+a^2)-(s^2+b^2)}{(s^2+a^2)(s^2+b^2)}\right]$

$$= \frac{1}{a^2-b^2}L^{-1}\left[\frac{1}{s^2+b^2} - \frac{1}{s^2+a^2}\right]$$

$$= \frac{1}{a^2-b^2}\left[\frac{\sin bt}{b} - \frac{\sin at}{a}\right], \quad \left[\because L^{-1}\left(\frac{1}{s^2+a^2}\right) = \frac{1}{a}\sin at\right] \text{ where } a^2 \neq b^2.$$

Example 9.12. *Find the Laplace transform of* $\dfrac{4p+5}{(p-4)^2(p+3)}$.

Solution : Let $\dfrac{4p+5}{(p-4)^2(p+3)} = \dfrac{A}{p-4} + \dfrac{B}{(p-4)^2} + \dfrac{C}{p+3}$.

$$\therefore \quad \frac{4p+5}{(p-4)^2(p+3)} = \frac{A(p-4)(p+3) + B(p+3) + C(p-4)^2}{(p-4)^2(p+3)}$$

or, $4p+5 = A(p-4)(p+3) + B(p+3) + C(p-4)^2$.

This being an identity, is true for all values of p.

Putting $p = -3$, $-7 = 49C$, $\therefore$ $C = -\dfrac{1}{7}$.

$p = 4$, $21 = 7B$, $\therefore$ $B = 3$.

$p = 0$, $5 = -12A + 3B + 16C = -12A + 9 - \dfrac{16}{C}$.

$$\therefore \quad \frac{4p+5}{(p-4)^2(p+3)} = \frac{1}{7}\cdot\frac{1}{p-4} + \frac{3}{(p-4)^2} - \frac{1}{7}\cdot\frac{1}{p+3}$$

$$\therefore \quad \boldsymbol{L^{-1}\left[\frac{4p+5}{(p-4)^2(p+3)}\right]} = \frac{1}{7}L^{-1}\left(\frac{1}{p-4}\right) + 3L^{-1}\left[\frac{1}{(p-4)^2}\right] - \frac{1}{7}L^{-1}\left(\frac{1}{p+3}\right)$$

$$= \frac{1}{7}e^{4t} + 3e^{4t}L^{-1}\left(\frac{1}{p^2}\right) - \frac{1}{7}e^{-3t} \quad \left[\because\ L^{-1}\left(\frac{1}{p-a}\right) = e^{at} \text{ and by first siifting property}\right]$$

$$= \boldsymbol{\frac{1}{7}e^{4t} + 3te^{4t} - \frac{1}{7}e^{-3t}}, \quad \because\ L^{-1}\left(\frac{1}{p^{n+1}}\right) = \frac{t^n}{n!} \quad [\text{where } n \geq 0].$$

Example 9.13. *Find the Laplace transform of* $\dfrac{s^2+2s-4}{(s^2+2s+5)(s^2+2s+2)}$.

Solution : Let $\dfrac{s^2+2S-4}{(s^2+2s+5)(s^2+2s+2)} = \dfrac{As+B}{s^2+2s+5} + \dfrac{Cs+D}{s^2+2s+2}$.

$$\therefore \quad s^2+2s+4 = (As+B)(s^2+2s+2) + (cs+D)(s^2+2s+5). \qquad \text{(i)}$$

Equating the coefficients of s^3, s^2, s and the constant terms from both sides of Eqn. (i), we get

$$A + C = 0, \quad B + 2A + D + 2C = 1, \quad 2A + 2B + 5C + 2D = 2 \quad \text{and} \quad 2B + 5D = -4.$$

Solving the above equations, we get $A = 0$, $B = 3$, $C = 0$, $D = -2$.

$$\therefore \quad \frac{s^2+2s-4}{(s^2+2s+5)(s^2+2s+2)} = \frac{3}{s^2+2s+5} - \frac{2}{s^2+2s+2} = \frac{3}{(s+1)^2+2^2} - \frac{2}{(s+1)^2+1^2}.$$

Hence, $\boldsymbol{L^{-1}\left[\frac{s^2+2s-4}{(s^2+2s+5)(s^2+2s+2)}\right]} = 3L^{-1}\left[\dfrac{1}{(s+1)^2+2^2}\right] - 2L^{-1}\left[\dfrac{1}{(s+1)^2+1^2}\right]$

$$= \boldsymbol{3e^{-t}L^{-1}\left[\frac{1}{s^2+2}\right] - 2e^{-t}L^{-1}\left(\frac{1}{s^2+1}\right)} \quad [\text{by shifting property}].$$

Example 9.14. *Evaluate :* $\displaystyle\int_0^\infty \frac{\sin t}{t}dt$ **[WBUT 2005, '06]**

Hence, show that $\displaystyle\int_0^\infty \frac{\sin t}{t} = \frac{\pi}{2}$.

Solution : Let $F(t) = \sin t$.

$\therefore \quad L[F(t)] = L[\sin t]$ or, $f(s) = \dfrac{1}{s^2+1}$.

$$\text{Now, } \int_0^\infty \frac{\sin t}{t} dt = \int_0^\infty \frac{F(t)}{t} dt = \int_0^\infty f(x)dx$$

$$= \int_0^\infty \frac{dx}{x^2+1} \quad \left[\because \int_0^\infty \frac{f(t)}{t} dt = \int_0^\infty f(x)dx\right]$$

$$= \left[\tan^{-1} x\right]_0^\infty = \frac{\pi}{2}.$$

Example 9.15. *Evaluate :* $\int_0^\infty e^{-3t} t \sin t \, dt$

Solution : We have $\int_0^\infty e^{-st} t \sin t \, dt = L(t \sin t) = -\dfrac{d}{ds}[L(\sin t)] = -\dfrac{d}{ds}\left(\dfrac{1}{s^2+1}\right) = \dfrac{2s}{(s^2+1)^2}$.

Putting $s = 3$, we get $\int_0^\infty e^{-3t} t \sin t \, dt = \dfrac{6}{(3+1)^2} = \dfrac{3}{50}$.

Example 9.16. *Evaluate :* $\int_0^\infty e^{-4t} \dfrac{\sin^2 t}{t} dt$

Solution : $\int_0^\infty e^{-st} \dfrac{1}{t} \sin^2 t \, dt = L\left[\dfrac{1}{t}\sin^2 t\right]$

$$\sin^2 t = \frac{1}{2}(1 - \cos 2t).$$

$$\therefore \quad L(\sin^2 t) = \frac{1}{2} L(1 - \cos 2t) = \frac{1}{2}[L(1) - L(\cos 2t)] = \frac{1}{2}\left[\frac{1}{s} - \frac{s}{s^2+1}\right].$$

$$\therefore \quad L\left[\frac{1}{t}\sin^2 t\right] = \int_0^\infty \frac{1}{2}\left[\frac{1}{s} - \frac{s}{s^2+4}\right] ds = \underset{B\to\infty}{\text{Lt}} \int_0^B \frac{1}{2}\left[\frac{1}{s} - \frac{s}{s^2+4}\right] ds$$

$$= \underset{B\to\infty}{\text{Lt}} \frac{1}{2}\left[\log s - \frac{1}{2}\log(s^2+4)\right]_s^B = \underset{B\to\infty}{\text{Lt}} \frac{1}{4}\left[\log\left(\frac{s^2}{s^2+4}\right)\right]_s^B$$

$$= \underset{B\to\infty}{\text{Lt}} \frac{1}{4}\left[\log\left(\frac{B^2}{B^2+4}\right) - \log\left(\frac{s^2}{s^2+4}\right)\right]$$

$$= -\frac{1}{4}\log\left(\frac{s^2}{s^2+4}\right) \quad \left[\because \log\left(\frac{B^2}{B^2+4}\right) = \log\left(\frac{1}{1+\frac{4}{B^2}}\right) \to \log 1 = 0 \text{ as } B \to \infty\right]$$

$$= \frac{1}{4}\log\left(\frac{s^2+4}{s^2}\right).$$

$$\therefore \quad \int_0^\infty e^{-st} \frac{1}{z} \sin^2 t \, dt = \frac{1}{4}\log\left(\frac{s^2+4}{s^2}\right).$$

Putting $s = 4$, $\int_0^\infty e^{-4t} \dfrac{1}{t} \sin^2 t \, dt = \dfrac{1}{4}\log\left(\dfrac{4^2+4}{4^2}\right) = \dfrac{1}{4}\log\dfrac{5}{4}$.

Example 9.17. *Find the Laplace transform of* $\dfrac{s+1}{s^2(s+4)}$.

Solution : We have $L^{-1}\left[\dfrac{s}{s^2(s+4)}\right] = L^{-1}\left[\dfrac{1}{s(s+4)} + \dfrac{1}{s^2(s+4)}\right]$

$$= L^{-1}\left[\frac{1}{s(s+4)}\right] + L^{-1}\left[\frac{1}{s^2(s+4)}\right].$$

But, $L^{-1}\left[\dfrac{1}{s+4}\right] = e^{-4t}$.

$$\therefore \quad L^{-1}\left[\frac{1}{s}\cdot\frac{1}{s+4}\right] = \int_0^t e^{-4u}du = \left[\frac{e^{-4u}}{-4}\right]_0^t = \frac{1}{4}(1-e^{-4t}).$$

$$\therefore \quad L^{-1}\left[\frac{1}{s^2(s+4)}\right] = L^{-1}\left[\frac{1}{s}\cdot\frac{1}{s(s+4)}\right] = \int_0^t \frac{1}{4}(1-e^{-4u})du$$

$$= \frac{1}{4}\left[u - \frac{e^{-4u}}{-4}\right]_0^t = \frac{1}{4}\left[t + \frac{1}{4}e^{-4t} - \frac{1}{4}\right].$$

$$\therefore \quad \boldsymbol{L^{-1}\left[\frac{s+1}{s^2(s+4)}\right]} = \frac{1}{4}(1-e^{-4t}) + \frac{1}{4}\left(t + \frac{1}{4}e^{-4t} - \frac{1}{4}\right) = \boldsymbol{\frac{1}{4}\left(\frac{3}{4} + t - \frac{3}{4}e^{-4t}\right)}.$$

Example 9.18. *Find inverse laplace transform of* (i) $\log\left(1+\dfrac{a^2}{s^2}\right)$ (ii) $\dfrac{s^2}{(s+a)^3}$.

Solution : (i) Let $\overline{f}(s) = \boldsymbol{\log\left(1+\frac{a^2}{s^2}\right)} = \log\left(\dfrac{s^2+a^2}{s^2}\right) = \log(s^2+a^2) - 2\log s$

$$= \frac{d}{ds}\overline{f}(s) = \frac{2s}{s^2+a^2} - 2\frac{1}{s} = \mathbf{2\cos at - 2}. \qquad \text{(i)}$$

Again, $\boldsymbol{L^{-1}[\overline{f}(s)]} = -tL^{-1}[\overline{f}(s)] = \boldsymbol{\frac{2}{t}(1-\cos at)}$ [using Eqn. (i)].

(ii) We have $L^{-1}\left[\dfrac{1}{(s+a)^3}\right] = e^{-at}L^{-1}\left(\dfrac{1}{s^3}\right)$ [by first shifting property]

$$= e^{-at}\frac{t^2}{2}, \quad \because \quad L^{-1}\left(\frac{1}{s^{n+1}}\right) = \frac{t^n}{n!} \quad \text{[where } n \text{ is a positive integer]}.$$

Here, $g'(t) = te^{-at} - \dfrac{at^2}{2}e^{-at}$, $\therefore$ $g(0) = g'(0) = 0$.

$$\therefore \quad \boldsymbol{L^{-1}\left[\frac{s^2}{(s+a)^3}\right]} = \frac{d^2}{dt^2}L^{-1}\left[\frac{1}{(s+a)^2}\right] = \frac{d^2}{dt^2}g(t) = e^{-at} - ate^{-at} - ate^{-at} + \frac{a^2t^2}{2}e^{-at}$$

$$= \boldsymbol{\frac{1}{2}(a^2t^2 - 4at + 2)e^{-at}}.$$

Example 9.19. *Find the inverse Laplace transform of*

(i) $\dfrac{1}{(s^2+2s+5)}$ **[WBUT 2008]** (ii) $\dfrac{p}{(p^2+1)^2}$, **[WBUT 2005, '06]**

(iii) $\dfrac{1}{(s-1)^2(s-2)^3}$, **[WBUT 2009]** (iv) $\dfrac{s^2}{(s^2+a^2)(s^2+b^2)}$, $a^2 \neq b^2$

Solution : (i) We have $L^{-1}\left[\dfrac{1}{s^2+2s+5}\right] = L^{-1}\left[\dfrac{1}{(s+1)^2+2^2}\right]$

$$= e^{-t}L^{-1}\left[\frac{1}{s^2+2^2}\right], \quad \text{[by first shifting property]}$$

$$= e^{-t}\frac{1}{2}\sin 2t = \frac{1}{2}e^{-t}\sin 2t = f(t) \text{ (say)}.$$

(ii) $\because$ $L^{-1}\left[\dfrac{p}{p^2+1}\right] = \cos t$ and $L^{-1}\left[\dfrac{1}{p^2+1}\right] = \sin t.$

$\therefore$ by convolution theorem,

$$L^{-1}\left[\frac{p}{(p^2+1)^2}\right] = L^{-1}\left[\frac{p}{p^2+1}\cdot\frac{1}{p^2+1}\right] = \int_0^t \cos u \sin(t-u)du$$

$$= \frac{1}{2}\int_0^t [\sin t - \sin(2u-t)]du = \frac{1}{2}\left[u\sin t + \frac{1}{2}\cos(2u-t)\right]_{u=0}^{t}$$

$$= \frac{1}{2}\left[t\sin t + \frac{1}{2}\cos t - \frac{1}{2}\cos t\right] = \frac{1}{2}t\sin t.$$

(iii) We have $L^{-1}\left[\dfrac{1}{(s-1)^2}\right] = e^t L^{-1}\left(\dfrac{1}{s^2}\right)$ [by first shifting property]

$$L^{-1}\left(\frac{1}{s^{n+1}}\right) = \frac{t^n}{n!} \quad \text{[where } n \text{ is a positive integer]}.$$

$$= g(t) \text{ (say)}.$$

Similarly, $L^{-1}\left[\dfrac{1}{(s-2)^3}\right] = e^{2t}L^{-1}\left(\dfrac{1}{s^3}\right) = e^{2t}\cdot\dfrac{t^2}{2!} = \dfrac{1}{2}t^2e^{2t} = f(t)$ (say).

$\therefore$ by convolution theorem,

$$L^{-1}\left[\frac{1}{(s-1)^2(s-2)^3}\right] = \int_0^t f(u)g(t-u)du = \int_0^t \frac{1}{2}u^2e^{2u}e^{t-u}(t-u)du$$

$$= \frac{1}{2}e^t\int_0^t e^u(-tu^2-u^3)du.$$

Integrating by parts, we have

$$\int e^u(tu^2-u^3)du = (tu^2-u^3)\int e^u du - \int\left[\frac{d}{du}(tu^2-u^3)\right]\left[\int e^u du\right]du$$

$$= (tu^2-u^3)e^u - \int(2tu-3u^2)e^u du$$

$$= (tu^2-u^3)e^u - \left[(2tu-3u^2)e^u - \int(2t-6u)e^u du\right]$$

$$= (tu^2-u^3)e^u - (2tu-3u^2)e^u + [(2t-6u)e^u + 6e^u]$$

$$= (tu^2-u^3-2tu+3u^2+2t-6u+6)e^u$$

$$\therefore \quad \int_0^t e^u(tu^2-u^3)du = (t^2-4t+6)e^t - 2t - 6.$$

Hence, $$L^{-1}\left[\frac{1}{(s-1)^2(s-2)^3}\right] = (t^2-4t+6)e^t - 2t - 6.$$

(iv) We have $L^{-1}\left(\frac{s}{s^2+a^2}\right) = \cos at$ and $L^{-1}\left(\frac{s}{s^2+a^2}\right) = \cos bt.$

$\therefore$ by convolution theorem,

$$L^{-1}\left[\frac{s^2}{(s^2+a^2)(s^2+b^2)}\right] = L^{-1}\left[\frac{s}{s^2+a^2}\cdot\frac{s}{s^2+b^2}\right] = \int_0^t \cos au \cos b(t-u)du$$

$$= \frac{1}{2}\int_0^t [\cos\{(a-b)u+bt\} + \cos\{(a+b)u-bt\}]du$$

$$= \frac{1}{2}\left[\frac{\sin\{(a-b)u+bt\}}{a-b} + \frac{\sin\{(a+b)u+bt\}}{a+b}\right]_{u=0}^{t}$$

$$= \frac{1}{2}\left[\frac{\sin at - \sin bt}{a-b} + \frac{\sin at + \sin bt}{a+b}\right]$$

$$= \frac{a\sin at - b\sin bt}{a^2-b^2}\ (a^2 \neq b^2).$$

Example 9.20. *Find the inverse transform of the following :* $\frac{p+2}{p^2-4p+13}$.

Solution : We have $$L^{-1}\left[\frac{p+2}{p^2-4p+13}\right] = L^{-1}\left[\frac{p+2}{(p-2)^2+9}\right] = L^{-1}\left[\frac{p-2+4}{(p-2)^2+9}\right]$$

$$= L^{-1}\left[\frac{p-2}{(p-2)^2+3^2}\right] + 4L^{-1}\left[\frac{1}{(p-2)^2+3^2}\right]$$

$$= e^{2t}\cos 3t + \frac{4}{3}e^{2t}\sin 3t.$$

Example 9.21. *Find the Laplace transform of the half-wave rectifier function,*

$$f(t) = \begin{cases} \sin \omega t, & 0 < t < \frac{\pi}{\omega} \\ 0, & \frac{\pi}{\omega} < t < \frac{2\pi}{\omega}. \end{cases}$$

Solution : The time period of the function $f(t)$ is $T = \frac{2\pi}{\omega}$.

So, $$L[f(t)] = \frac{1}{1-e^{-sT}}\int_0^T e^{-st}f(t)dt = \frac{1}{1-e^{-2\pi s/\omega}}\int_0^{\frac{2\pi}{\omega}} e^{-st}f(t)dt$$

$$= \frac{1}{1-e^{-\frac{2\pi s}{\omega}}}\int_0^{\frac{\pi}{\omega}} \sin\omega t\, dt \quad \left[\because f(t) = 0, \text{ in the range } \frac{\pi}{\omega} \text{ to } \frac{2\pi}{\omega}\right]$$

$$= \frac{1}{1-e^{-\frac{2\pi s}{\omega}}}\left[e^{-st}\frac{(-s\sin\omega t - \omega\cos\omega t)}{s^2+\omega^2}\right]_0^{\frac{\pi}{\omega}}$$

$$= \frac{1}{(s^2+\omega^2)(1-e^{-\frac{2\pi s}{\omega}})}\left[e^{-s\pi\omega}(-p\sin\pi - \omega\cos\pi)+\omega\right]$$

$$\left[\because \int e^{ax}\sin bx\,dx = e^{ax}\cdot\frac{a\sin bx - b\sin bx}{a^2+b^2}\right]$$

$$= \frac{\omega(1+e^{-\frac{s\pi}{\omega}})}{(s^2+\omega^2)(1-e^{\frac{2\pi s}{\omega}})} = \frac{\omega}{(s^2+\omega^2)(1-e^{-\frac{s\pi}{\omega}})}$$

Example 9.22. *Find the Laplace transform of*

(i) $f(t) = k\frac{t}{T},\ 0<t<T$ and (ii) $f(t) = \begin{cases} 1, & 0 \le t < a \\ -1, & 0 < t < 2a \end{cases}$

$f(t)$ is periodic with period $2a$.

Solution : (i) Since $f(t)$ is periodic with period T, we have

$$L[f(t)] = \frac{1}{1-e^{-st}}\int_0^T e^{-st}f(t)dt = \frac{1}{1-e^{-sT}}\int_0^T e^{-st}\frac{kt}{T}dt = \frac{1}{1-e^{-sT}}\cdot\frac{k}{T}\int_0^T te^{-st}dt$$

$$= \frac{1}{1-e^{-st}}\,\frac{k}{T}\int_0^T\left[\frac{te^{-st}}{-s}-\frac{e^{-st}}{s^2}\right]_0^T = \frac{1}{1-e^{-sT}}\cdot\frac{k}{T}\left[-\frac{Te^{-sT}}{s}-\frac{e^{-sT}}{s^2}+\frac{1}{s^2}\right]$$

$$= \frac{1}{1-e^{-sT}}\cdot\frac{k}{T}\left[\frac{1}{s^2}(1-e^{-sT})-\frac{Te^{-sT}}{s}\right] = \boldsymbol{k\left[\frac{1}{Ts^2}-\frac{e^{-sT}}{s(1-e^{-sT})}\right]}.$$

(ii) $\because f(t)$ is periodic with period $2a$,

$$\therefore\ L[f(t)] = \frac{1}{1-e^{-2as}}\int_0^{2a}e^{-st}f(t)dt = \frac{1}{1-e^{-2as}}\left[\int_0^a e^{-st}\cdot 1\cdot dt + \int_a^{2a}e^{-st}(-1)dt\right]$$

$$= \frac{1}{1-e^{-2as}}\left[\left(\frac{e^{-st}}{-s}\right)_0^a + \left(\frac{e^{-st}}{s}\right)_a^{2a}\right] = \frac{1}{s(1-e^{-2as})}\left(e^{-as}+1+e^{-2as}-e^{-as}\right)$$

$$= \frac{(1-e^{-as})}{s(1-e^{-as})(1+e^{as})} = \frac{1}{s}\left(\frac{1-e^{-as}}{1+e^{-as}}\right) = \frac{1}{s}\left(\frac{e^{\frac{as}{2}}-e^{-\frac{as}{2}}}{e^{\frac{as}{2}}+e^{-\frac{as}{2}}}\right) = \boldsymbol{\frac{1}{2}\tanh\left(\frac{as}{2}\right)}.$$

Example 9.23. *Solve the differential equation, $(D^2+a^2)y = \sin bx$, where a and b are constants and $a \ne b,\ a \ne 0$.*

Solution : We have the Laplace transform of the given equation,

$$s^2\cdot\overline{y}(s) - sy(0) - y'(0) + a^2\overline{y}(s) = \frac{b}{s^2+b^2}$$

or, $$\overline{y}(s) = \frac{b}{(s^2+a^2)(s^2+b^2)} + \frac{sy(0)}{s^2+a^2} + \frac{y'(0)}{s^2+a^2},$$

where $y(0)$ and $y'(0)$ represent the values of y and $\frac{dy}{dx}$ at $x = 0$.

Using partial fraction,

$$\overline{y}(s) = \frac{b}{b^2-a^2}\left[\frac{1}{s^2+a^2}-\frac{1}{(s^2+a^2)}\right] + y(0)\left(\frac{s}{s^2+a^2}\right) + \frac{y'(0)}{a}\left(\frac{a}{s^2+a^2}\right).$$

$$\therefore \quad \boldsymbol{y(x)} = L^{-1}[\overline{y}(p)] = \frac{b\sin ax}{a(b^2-a^2)} - \frac{\sin bx}{b^2-a^2} + y(0)\cos ax + \frac{y'(0)\sin ax}{a}$$

$$= \boldsymbol{y(0)\cos ax + A\sin ax + \frac{\sin bx}{a^2-b^2}}, \quad \text{where } A = \frac{1}{a}\left[y'(0) + \frac{b}{b^2-a^2}\right].$$

Example 9.24. *Solve* $y''(t) + y(t) = \sin 2t$, *when* (i) $y(0) = 0$ *and* $y'(0) = 1$ (ii) $y(0) = y'(0) = 1$. **[WBUT 2003]**

Solution : Let $L(y) = y$.

Taking Laplace transform on both sides of the given equation, we get

$$L(y'') + L(y) = L(\sin 2t) \quad \text{or,} \quad s^2\overline{y} - sy(0) - y''(0) + y\frac{2}{s^2+2^2}$$

$$\text{or,} \quad (s^2+1)\overline{y} = \frac{2}{s^2+4} + sy(0) + y'(0).$$

(i) When $y(0) = 0$ and $y'(0) = 1$, we get

$$\overline{y} = \frac{s^2+6}{(s^2+4)(s^2+1)} = \frac{1}{3}\frac{3s^2+18}{(s^2+4)(s^2+1)} = \frac{1}{3}\frac{5(s^2+4)-2(s^2+1)}{(s^2+4)(s^2+1)}$$

$$= \frac{5}{3}\cdot\frac{1}{s^2+1} - \frac{1}{3}\frac{2}{s^2+4}.$$

$$\therefore \quad \boldsymbol{y(t)} = \frac{5}{3}L^{-1}\left(\frac{1}{s^2+1}\right) - \frac{1}{3}L^{-1}\left(\frac{2}{s^2+2^2}\right) = \frac{5}{3}\sin t - \frac{1}{3}\sin 2t = \boldsymbol{\frac{1}{3}(5\sin t - \sin 2t)}.$$

(ii) When $y(0) = y'(0) = 1$, we have

$$\overline{y} = \frac{2}{(s^2+4)(s^2+1)} + \frac{s+1}{s^2+1} = \frac{2}{3}\left(\frac{1}{s^2+1} - \frac{1}{s^2+4}\right) + \frac{s}{s^2+1} + \frac{1}{s^2+1}$$

$$= \frac{2}{3}\left(\frac{1}{s^2+1} - \frac{1}{s^2+4}\right) = \frac{s}{s^2+1} + \frac{1}{s^2+1} = \frac{5}{3}\cdot\frac{1}{s^2+1} - \frac{2}{3}\frac{1}{s^2+4} + \frac{s}{s^2+1}.$$

$$\therefore \quad \boldsymbol{y(t)} = \frac{5}{3}L^{-1}\left(\frac{1}{s^2+1}\right) - \frac{1}{3}L^{-1}\left(\frac{2}{s^2+2^2}\right) + L^{-1}\left(\frac{s}{s^2+1}\right) = \frac{5}{3}\sin t - \frac{1}{3}\sin 2t + \cos t$$

$$= \boldsymbol{\frac{1}{3}(5\sin t - \sin 2t + 3\cos t)}.$$

Example 9.25. *Using Laplace transformation find the solution of the initial value problem,*

$y'' + 9y = 6\cos 3t$, $y(0) = 2$, $y'(0) = 0$. **[AMIETE(Winter) 1997, (Summer) 1993]**

Solution : $y'' + 9y = 6\cos 3t$

$y(0) = 2, \ y'(0) = 0$

Taking Laplace transform on both sides of the given equation,

$$[s^2\overline{y} - sy(0) - y''(0)] + 9\overline{y} = 6\frac{s}{s^2+9}.$$

$$\therefore \quad s^2\overline{y} - 2s + 9\overline{y} = \frac{6s}{s^2+9} \quad \text{or,} \quad (s^2+9)\overline{y} = 2s + \frac{6s}{s^2+9} \quad \text{or,} \quad \overline{y} = \frac{2s}{s^2+9} + \frac{6s}{(s^2+9)^2}$$

or, $$y = L^{-1}\left(\frac{2s}{s^2+9}\right) + L^{-1}\left[\frac{6s}{(s^2+9)^2}\right] = 2\cos 3t + 3L^{-1}\frac{d}{ds}\left[\frac{-3}{(s^2+9)}\right]$$

$$= 2\cos 3t - L^{-1}\frac{d}{ds}\frac{3}{s^2+9} = \mathbf{2\cos 3t + t\sin 3t}.$$

Example 9.26. *Solve the differential equation,*

$$\frac{d^2y}{dx^2} - \lambda^2 y = 0,\ \lambda \neq 0,\ \textit{when } y(0) = 0,\ y'(0) = 1.$$

Solution : We have $\frac{d^2y}{dx^2} - \lambda^2 y = 0$

Taking Laplace transform on both sides of the above equation, we get

$$L\left(\frac{d^2y}{dx^2}\right) - \lambda^2 L(y) = 0. \quad \text{(i)}$$

Let $L(y) = f(s)$. (ii)

Then from equations (i) and (ii), we have

$$[s^2 f(s) - sy(0) - y'(0)] - \lambda^2 f(s) = 0. \quad \text{(iii)}$$

Applying the given boundary conditions,

$$f(s)[s^2 - \lambda^2] = 1.$$

Hence, $$f(s) = \frac{1}{s^2-\lambda^2} = \frac{1}{(s-\lambda)(s+\lambda)}$$

or, $$f(s) = \frac{1}{2\lambda}\left[\frac{1}{s-\lambda} - \frac{1}{s+\lambda}\right]. \quad \text{(iv)}$$

Now, taking inverse Laplace transform on both sides,

$$\boldsymbol{y(x)} = \frac{1}{2\lambda}\left[L^{-1}\left(\frac{1}{s-\lambda}\right) - L^{-1}\left(\frac{1}{s+\lambda}\right)\right] = \boldsymbol{\frac{1}{2\lambda}[e^{\lambda x} - e^{-\lambda x}]}.$$

Example 9.27. *Solve the differential equation by Laplace transform :*

$$\frac{d^2y}{dt^2} + 2\frac{dy}{dt} + 5y = e^{-t}\sin t,\ \textit{when } y(0) = 0,\ y'(0) = L.$$

Solution : Let $L(y) = \overline{y}$.

Taking Laplace transform on both sides of the given equation, we get

$$L\left(\frac{d^2y}{dt^2}\right) + 2L\left(\frac{dy}{dt}\right) + 5L(y) = L(e^{-t}\sin t)$$

or; $$s^2\overline{y} - sy(0) - y'(0) + 2\{sy - y(0)\} + 5y = \frac{1}{(s+1)^2+1}, \quad \because\ L(e^{at}\sin bt) = \frac{b}{(s-a)^2+b^2}$$

or, $$s^2\overline{y} - 1 + 2s\overline{y} + 5\overline{y} = \frac{1}{s^2+2s+2} \quad \text{or,} \quad (s^2+2s+5)\overline{y} = \frac{1}{s^2+2s+2} + 1$$

or, $$\overline{y} = \frac{s^2+2s+3}{(s^2+2s+2)(s^2+2s+5)} = \frac{(s+1)^2+2}{[(s+1)^2+1][(s+1)^2+4]}.$$

$$\therefore \quad y = L^{-1}\left[\frac{(s+1)^2+2}{\{(s+1)^2+1\}\{(s+1)^2+4\}}\right] = e^{-t}L^{-1}\left[\frac{s^2+2}{(s^2+1)^2(s^2+4)}\right]$$

[by the first shifting property]

$$= \frac{e^{-t}}{3}\left[L^{-1}\left(\frac{1}{s^2+1}\right)+L^{-1}\left(\frac{2}{s^2+2}\right)\right] = \frac{\mathbf{e^{-t}}}{\mathbf{3}}[\mathbf{\sin t + \sin 2t}].$$

Example 9.28. *Solve the differential equation by Laplace transform :*

$$\frac{d^2y}{dt^2} - 2\frac{dy}{dt} - 3y = t\cos t \text{ when, } y(0) = y'(0) = 0. \quad \textbf{[WBUT 2010]}$$

Solution : Let $L(y) = \overline{y}$.

Taking Laplace transform on both sides of the given differential equation, we get

$$L\left(\frac{d^2y}{dt^2}\right) - 2L\left(\frac{dy}{dt}\right) - 3L(y) = L(t\cos t)$$

or, $\quad s^2y - sy(0) - y'(0) - 2[s\overline{y} - y(0)] - 3\overline{y} = -\frac{d}{ds}[L(\cos t)].$

$$\therefore \quad (s^2-2s-3)\overline{y} = -\frac{d}{ds}\left(\frac{s}{s^2+1}\right) = -\frac{1}{s^2+1} + \frac{2s^2}{(s^2+1)^2} \quad [\because y(0) = y'(0) = 0]$$

$$\therefore \quad \overline{y} = \frac{s^2-1}{(s+1)(s+3)(s^2+1)^2} = \frac{s-1}{(s-3)(s^2+1)^2} = \frac{A}{s-3} + \frac{Bs-c}{s^2+1} + \frac{Ds+F}{(s^2+1)^2} \text{ (say)}.$$

$$\therefore \quad s-1 = A(s^2+1)^2 + (Bs+c)(s-3)(s^2+1) + (Ds+E)(s-3). \quad \text{(i)}$$

The above equation is true for all values of s.

Putting $s = 3$, we get $A = \frac{1}{50}$.

Equating the coefficients of s^4, s^3, s^2 and the constant terms from both sides of Eqn. (i), we get

$$A+B=0 \quad -3B+C=0 \quad 2A-3C+B+D=0 \quad A-3C-3E=-1.$$

Solving, $B = -\frac{1}{50} \quad C = -\frac{3}{50} \quad D = -\frac{1}{5} \quad E = \frac{2}{5}$.

$$\therefore \quad \overline{y} = \frac{1}{50}\cdot\frac{1}{s-3} - \frac{1}{50}\cdot\frac{s}{s^2+1} - \frac{3}{50}\cdot\frac{1}{s^2+1} - \frac{1}{5}\cdot\frac{s}{(s^2+1)^2} + \frac{2}{5}\cdot\frac{1}{(s^2+1)}.$$

$$\therefore \quad y(t) = \frac{1}{50}L^{-1}\left(\frac{1}{s-3}\right) - \frac{1}{50}L^{-1}\left(\frac{s}{s^2+1}\right)$$

$$- \frac{3}{50}L^{-1}\left(\frac{1}{s^2+1}\right) - \frac{1}{5}L^{-1}\left(\frac{s}{(s^2+1)^2}\right) + \frac{2}{5}L^{-1}\left(\frac{1}{(s^2+1)^2}\right)$$

$$= \frac{\mathbf{e^{3t}}}{\mathbf{50}} - \frac{\mathbf{1}}{\mathbf{50}}\mathbf{\cos t} - \frac{\mathbf{3}}{\mathbf{50}}\mathbf{\sin t} - \frac{\mathbf{1}}{\mathbf{5}}\mathbf{L^{-1}}\left[\frac{\mathbf{s}}{\mathbf{(s^2+1)^2}}\right] + \frac{\mathbf{2}}{\mathbf{5}}\mathbf{L^{-1}}\left[\frac{\mathbf{1}}{\mathbf{(s^2+1)^2}}\right]. \quad \text{(ii)}$$

But we know that, $L^{-1}[\overline{f}'(s)] = -tL^{-1}[\overline{f}(s)]$.

Put $\overline{f}(s) = \frac{1}{s^2+1}$, $\quad \therefore \quad L^{-1}\left[\frac{2s}{(s^2+1)^2}\right] = -tL^{-1}\left(\frac{1}{s^2+1}\right) = -t\sin t.$

$$\therefore \quad L^{-1}\left[\frac{s}{(s^2+1)^2}\right] = \frac{1}{2}t\sin t. \qquad \text{(iii)}$$

Now, $L^{-1}\left(\frac{1}{s^2+1}\right) = \sin t = f(t)$ (say).

Now, by convolution theorem,

$$L^{-1}\left[\frac{1}{(s^2+1)^2}\right] = L^{-1}\left[\frac{1}{s^2+1}\cdot\frac{1}{s^2+1}\right] = \int_0^t f(u)f(t-u)du$$

$$= \int_0^t \sin u \sin(t-u)du = \frac{1}{2}\int_0^t [\cos(2u-t) - \cos t]du$$

$$= \frac{1}{2}\left[\frac{1}{2}\sin(2u-t) - u\cos t\right]_{u=0}^{t} = \frac{1}{2}\left[\frac{1}{2}\sin t - t\cos t + \frac{1}{2}\sin t\right]$$

$$= \frac{1}{2}(\sin t - t\cos t). \qquad \text{(iv)}$$

Subtracting Eqn. (iv) from Eqn. (ii), we get

$$y(t) = \frac{e^{3t}}{50} - \frac{1}{50}\cos t - \frac{3}{50}\sin t - \frac{1}{10}t\sin t + \frac{1}{50}(\sin t - t\cos t)$$

$$= \boldsymbol{\frac{e^{3t}}{50} - \frac{1}{50}(1+10t)\cos t + \frac{1}{50}(7-5t)\sin t.}$$

Example 9.29. *A particle moves on a vertical z-axis under the force of gravity, the resistance of the medium being proportional to its velocity. Its equation of motion is* $m\frac{d^2z}{dt^2} = mg - k\frac{dz}{dt}$. *Solve the equation under the condition* $z(0) = 0$ *and* $\left.\frac{dz}{dt}\right|_{t=0} = V_0$.

Solution : We have the equation of motion of the particle,

$$m\frac{d^2z}{dt^2} = mg - k\frac{dz}{dt}$$

$$\text{or,} \quad \frac{d^2z}{dt^2} + b\frac{dz}{dt} = g, \quad \text{where } b = \frac{k}{m}. \qquad \text{(i)}$$

Taking Laplace transform of both sides,

$$[s^2f(s) - sz(0) - z'(0)] + b[sf(s) - z(0)] = \frac{g}{s}, \quad \text{where } f(s) = L[z(t)].$$

$$\therefore \quad f(s) = \frac{g+V_0s}{s^2(s+b)} \quad \text{[by the initial conditions]}$$

$$= \frac{A}{s} + \frac{B}{s^2} + \frac{C}{s+b} \text{ (say).}$$

$$\text{Then } C = \left.\frac{g+V_0s}{s^2}\right|_{s=-b} = -\frac{bV_0 - g}{b^2} \qquad \text{(ii)}$$

$$B = \left.\frac{g+V_0s}{s+b}\right|_{s=0} = \frac{g}{b}$$

$$A = \frac{d}{ds}\left[\frac{g+V_0 s}{s+b}\right]_{s=0} = \left.\frac{1(1+V_0)+bV_0-g}{(s+b)^2}\right|_{s=0} = \frac{bV_0-g}{b^2}. \tag{iii}$$

$$\therefore \quad f(s) = \frac{bV_0-g}{b^2}\left[\frac{1}{s}-\frac{1}{s+b}\right]+\frac{g}{b}\cdot\frac{1}{s^2}.$$

Taking the inverse transform, we get

$$\boldsymbol{L^{-1}[f(s)] = z(t) = \frac{bV_0-g}{b^2}(1-e^{-bt})+\frac{gt}{b}}, \text{ where } b=\frac{k}{m}.$$

Example 9.30. *A resistance R in series with an inductance L is connected with e.m.f. E(t). The current I is given by* $L\dfrac{dI}{dt}+RI=E(t)$.

If the switch is connected at t = 0 and disconnected at t = a, find the current in terms of t.

Solution : The conditions of current flow are :

$$I = 0 \text{ at } t = 0 \quad\text{and}\quad E(t) = \begin{cases} E, & 0 < t < a \\ 0, & t > a. \end{cases}$$

The given differential equation is

$$L\frac{dI}{dt}+RI = E(t). \tag{i}$$

Taking Laplace transform of Eqn., we get

$$L[s\overline{I}(t)-E(0)]+R\overline{I}(t) = \int_0^\infty e^{-st}E(t)dt$$

or, $\quad Ls\overline{I}(t)+R\overline{I}(t) = \int_0^\infty e^{-st}E(t)dt \quad [\because\ I(0)=0]$

or, $\quad (Ls+R)\overline{I} = \int_0^a e^{-st}E\,dt+\int_a^\infty e^{-st}E\,dt = E\left[\frac{e^{-st}}{-s}\right]_0^a+0 = \frac{E}{s}(1-e^{-sa}) = \frac{E}{s}-\frac{E}{s}e^{-sa}.$

$$\therefore \quad \overline{I} = \frac{E}{s(Ls+R)}-\frac{Ee^{-sa}}{s(Ls+R)}.$$

By inverse Laplace transform,

$$I = L^{-1}\left[\frac{E}{s(Ls+R)}\right]-L^{-1}\left[\frac{Ee^{-sa}}{s(Ls+R)}\right].$$

But, $L^{-1}\left[\dfrac{E}{s(Ls+R)}\right] = \dfrac{E}{L}L^{-1}\left[\dfrac{1}{s\left(s+\frac{R}{L}\right)}\right] = \dfrac{E}{L}\cdot\dfrac{L}{R}L^{-1}\left[\dfrac{1}{s}-\dfrac{1}{s+\frac{R}{L}}\right]$ [by partial fraction]

$$= \frac{E}{R}L^{-1}\left[\frac{1}{s}-\frac{1}{s+\frac{R}{L}}\right] = \frac{E}{R}\left(1-e^{-\frac{R}{L}t}\right)$$

and $L^{-1}\left[\dfrac{Ee^{-sa}}{s(Ls+R)}\right] = \dfrac{E}{R}\left[1-e^{-\frac{R}{L}(t-a)}\right]u(t-a)$, by the second shifting property.

$\therefore \quad I = \dfrac{E}{R}\left[1-e^{-\frac{R}{L}t}\right]$, [for $0 < t < a$, $\because\ u(t-a)=0$].

Also, $I = \frac{E}{R}\left[1-e^{-\frac{R}{L}t}\right] = \frac{E}{R}\left[1-e^{-\frac{R(t-a)}{L}}\right]$, for $t > a$ since, $u(t-a) = 1$

$$= \frac{E}{R}\left[-e^{-\frac{Rt}{L}} + e^{-\frac{R(t-a)}{L}}\right]$$

or, $\boldsymbol{I = \frac{E}{R}e^{-\frac{R}{L}t}\left[e^{\frac{Ra}{L}} - 1\right]}$, **for $t > 0$.**

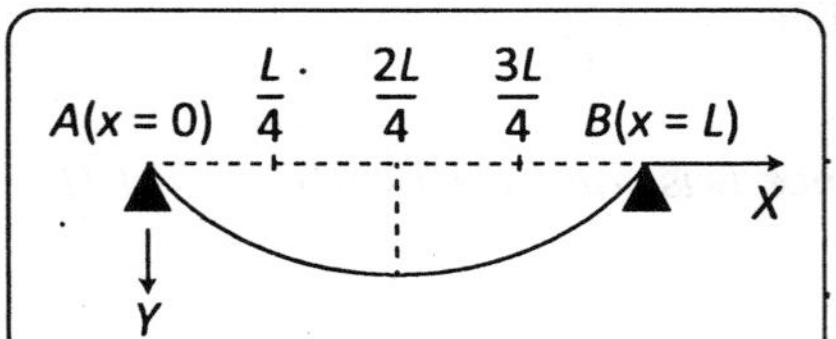

Fig. 9.4

Example 9.31. *A beam is simply supported at its end $x = 0$ and is clamped at other end $x = L$. It carries a load W at $x = \frac{L}{4}$. Find the resulting deflection at any point if the differential equation for deflection is*

$$\frac{d^4y}{dx^4} = \frac{W}{EI}\delta\left(x - \frac{L}{4}\right).$$ **[Coimbatore U 1985]**

Solution : Here, $\frac{d^4y}{dx^4} = \frac{W}{EI}\delta\left(x - \frac{L}{4}\right).$ (i)

Taking Laplace transform of Eqn. (i), we get

$$s^4\overline{y} - s^3y(0) - s^2y'(0) - sy''(0) - y'''(0) = \frac{W}{EI}e^{-\frac{1}{4}x}. \quad \text{(ii)}$$

$$\left[\because\ L\delta\left(x - \frac{L}{4}\right) = e^{-\frac{L}{4}x}\right]$$

The initial conditions are : $y(0) = 0$, $y'(0) = C_1$, $y''(0) = 0$, $y'''(0) = C_2$.

Putting the values of $y(0)$, $y'(0)$, $y''(0)$, $y'''(0)$, we obtain

$$s^4y - C_1s^2 - C_2 = \frac{W}{EI}e^{-\frac{L}{4}x} \quad \text{or,} \quad s^4\overline{y} = C_1s^2 + C_2 + \frac{W}{EI}e^{-\frac{L}{4}x}$$

or, $\overline{y} = \frac{C_1}{s^2} + \frac{C_2}{s^4} + \frac{W}{EI}\frac{e^{-\frac{L}{4}x}}{s^4}$

Taking inverse Laplace transform, we get

$$y = C_1x + C_2\frac{x^3}{3!} + \frac{W}{EI}\frac{\left(x-\frac{L}{4}\right)^3}{3!}u\left(x - \frac{L}{4}\right). \quad \text{(unit step function)}$$

$$\therefore \quad y = C_1x + C_2\frac{x^3}{6}, \ 0 < x < \frac{1}{4}$$

$$y = C_1x + C_2\frac{x^3}{6} + \frac{W}{6EI}\left(x - \frac{L}{4}\right)^3, \ \frac{L}{4} < x < L. \quad \text{(iii)}$$

$$\because \quad y = 0 \text{ at } x = L, \quad \therefore \quad 0 = C_1L + C_2\frac{L^3}{6} + \frac{W}{6EI}\left(\frac{27L^3}{64}\right)$$

or, $0 = C_1 + \frac{C_2L^2}{6} + \frac{9WL^3}{128EI}.$ (iii)

Differentiating Eqn. (ii), we get

$$y' = C_1 + 3C_2\frac{x^2}{6} + \frac{3W}{6EI}\left(x - \frac{L}{4}\right)^2. \quad \text{(iv)}$$

Again since $y' = 0$ at $L = 0$, the above equation becomes

$$0 = C_1 + C_2\frac{L^2}{2} + \frac{9WL^2}{32EI}, \qquad \text{(v)}$$

Solving Eqns. (iii) and (v), we get

$$C_1 = \frac{45WL^2}{128EI}, \quad C_2 = -\frac{81W}{128EI}.$$

Putting these values of C_1 and C_2 in Eqn. (ii), we get

$$y = \frac{45WL^2}{128EI}x - \frac{27W}{256EL}x^3, \; 0 < x < \frac{1}{4}$$

and $\boldsymbol{y = \dfrac{45WL^2}{128EI}x - \dfrac{27Wx^3}{256EI} + \dfrac{W}{6EI}\left(x - \dfrac{L}{4}\right)^3, \; \dfrac{L}{4} < x < L.}$

Example 9.32. *Calculate the maximum deflection of an encastre beam of length L and carrying a uniformly distributed load W on its central half length.*

Solution : Let A be the origin [Fig. 9.4], then

$$EI\frac{d^4y}{dx^4} = W(x), \quad \text{where } W(x) = W\left[u\left(x - \frac{L}{4}\right) - u\left(x - \frac{3L}{4}\right)\right].$$

Taking Laplace transform of both sides, we get

$$EI[s^4\overline{y} - s^3y(0) - s^2y'(0) - sy''(0) - y'''(0)] = W\left[\frac{e^{-Ls/4}}{s} - \frac{e^{-3Ls/4}}{s}\right].$$

The conditions are $y(0) = y'(0) = 0$ and putting $y''(0) = C_1$ and $y'''(0) = C_2$, we have

$$EL\overline{y} = W\left[\frac{e^{-Ls/4}}{s^5} - \frac{e^{-3Ls/4}}{s^5}\right] + \frac{C_1}{s^3} + \frac{C_2}{s^4}.$$

Taking inversion, we get

$$ELy = \frac{W}{24}\left[\left(x - \frac{L}{4}\right)^4 u\left(x - \frac{1}{4}\right) - \left(x - \frac{3L}{4}\right)^4 u\left(x - \frac{3L}{4}\right)\right] + \frac{1}{2}C_1x^2 + \frac{1}{6}C_2x^3. \qquad \text{(i)}$$

Now, when $x > \frac{3L}{4}$,

$$ELy = \frac{W}{24}\left[\left(x - \frac{L}{4}\right)^4 - \left(x - \frac{3L}{4}\right)^4\right] + \frac{1}{2}C_1x^2 + \frac{1}{6}C_2x^3$$

and $$ELy' = \frac{W}{6}\left[\left(x - \frac{L}{4}\right)^3 - \left(x - \frac{3L}{4}\right)^3\right] + C_1x + \frac{1}{2}C_2x^3.$$

Using the conditions, $y(L) = 0$ and $y'(L) = 0$, we get

$$0 = \frac{W}{24}\left[\left(\frac{3L}{4}\right)^4 - \left(\frac{L}{4}\right)^4\right] + \frac{1}{2}C_1L^2 + \frac{1}{6}C_2L^3$$

and $$0 = \frac{W}{6}\left[\left(\frac{3L}{4}\right)^3 - \left(\frac{L}{4}\right)^3\right] + C_1L + \frac{1}{2}C_2L^2, \quad \text{where } C_1 = \frac{11WL}{192} \text{ and } C_2 = -\frac{WL}{4}.$$

Hence, for $\frac{L}{4} < x < \frac{3L}{4}$, we get from Eqn. (i),

$$ELy = \frac{W}{24}\left(x + \frac{L}{4}\right)^4 + \frac{11WL^2}{384}x^2 - \frac{WL}{24}x^3.$$

Thus, the maximum deflection at y_m at $\frac{L}{2}$ is

$$y_{\max}\left(\frac{L}{2}\right) = \frac{13WL^4}{6144EI}.$$

Example 9.33. *Solve the wave equation,* $\frac{\partial^2 y}{\partial t^2} = c^2\frac{\partial^2 y}{\partial x^2}$ *under the conditions that,*

$$y(x,0) = 0, \quad y'(x,0) = 0, \quad x > 0 \quad \textit{and} \quad y(0,t) = t, \quad \operatorname*{Lt}_{x\to\infty} y(x,t) = 0, \quad t \geq 0.$$

Solution : We have $\frac{\partial^2 y}{\partial t^2} = c^2\frac{\partial^2 y}{\partial x^2}$. (i)

Taking Laplace transform on both sides,

$$s^2 f(x,s) - sy(x,0) - y'(x,0) = c^2\frac{d^2 f}{dx^2}$$

or, $\frac{d^2 f}{dx^2} = \frac{s^2}{c^2}f$ [by using the initial conditions]. (ii)

Its general solution is

$$f(x,s) = Ae^{-ux} + Be^{-ux}, \quad \text{where } u = \frac{s}{c}. \tag{iii}$$

$\because \quad \operatorname*{Lt}_{x\to\infty} y(x,t) = 0, \; f(x,s) = 0$ at $x = \infty$.

$\therefore \quad A = 0. \qquad \therefore \quad f(x,s) = Be^{-ux}.$ (iv)

To find the constant B, we use the boundary condition,

$$y(0,t) = t, \quad \text{i.e.,} \; L[y(0,t)] = \frac{1}{s^2} \quad \text{or,} \; f(0,s) = \frac{1}{s^2}. \tag{v}$$

Putting $x = 0$ in Eqn. (iv) and comparing with Eqn. (v), we get

$$B = \frac{1}{s^2}. \tag{vi}$$

$$\therefore \quad f(x,s) = \frac{1}{s^2}e^{-ux}. \tag{vii}$$

Taking inverse Laplace transform,

$$y(x,t) = L^{-1}\left[\frac{1}{s^2}e^{-\frac{s}{c}x}\right].$$

Using the second shifting property, we get

$$y(x,t) = \begin{cases} t - \dfrac{x}{c}, & \text{for } t > \dfrac{x}{c} \\ 0, & \text{for } t < \dfrac{x}{c}. \end{cases}$$

Example 9.34. *A string is stretched between two fixed points $(0,0)$ and $(l,0)$ and is released from position $y = y_0 \sin\frac{\pi x}{l}$, then show that the expression for its subsequent displacement $y(x,t)$ is*

$$y(x,t) = y_0 \sin\left(\frac{\pi x}{l}\right)\cos\left(\frac{h\pi t}{T}\right), \quad \text{where } h^2 \text{ is diffusivity.}$$

Solution : The equation of vibration of the string is given by

$$\frac{\partial^2 y}{\partial t^2} = h^2 \frac{\partial^2 u}{\partial x^2}. \tag{i}$$

where the boundary conditions are

$$y(0,t) = 0, \ y(l,t) = 0 \tag{ii}$$

and initial conditions are

$$y(x,0) = y_0 \sin\left(\frac{\pi x}{l}\right) \quad \text{and} \quad \left(\frac{\partial y}{\partial t}\right)_{t=0} = y_t(x,0) = 0. \tag{iii}$$

The Laplace transform of Eqn. (i) gives

$$s^2 f(x,s) - sy(x,0) - y_t(x,0) = h^2 \frac{\partial^2 f}{\partial x^2}. \tag{iv}$$

where $f = f(x,s) = L[y(x,t)]$

Using Eqn. (3) in Eqn. (4), we get

$$\frac{\partial^2 f}{\partial x^2} - \frac{s^2}{h^2} f = \frac{y_0}{h^2} \sin\frac{\pi x}{l}. \tag{v}$$

Laplace transform of Eqn. (2) is

$$f(0,s) = 0; \ f(l,s) = 0. \tag{vi}$$

$\therefore$ the solution of Eqn. (4) is

$$f = Ae^{+\frac{x}{h}} + Be^{-\frac{sx}{h}} + \frac{y_0 s \sin\frac{\pi x}{l}}{s^2 + \frac{\pi^2 h^2}{l}}. \tag{vii}$$

Now, $\quad f(0,s) = 0$, then $A + B = 0$

and $\quad f(l,s) = 0$ implies $Ae^{\frac{sx}{h}} + Be^{-\frac{sx}{h}} = 0$.

Solving, we get $A = B = 0$.

$\therefore$ Eqn. (vii) becomes

$$f(x,s) = \frac{y_0 s \sin\frac{\pi x}{l}}{s^2 + \frac{\pi^2 h^2}{l}}. \tag{viii}$$

Taking inverse Laplace transform of Eqn. (viii), we get

$$\boldsymbol{y(x,t)} = L^{-1}[f(x,s)] = y_0 \sin\frac{\pi x}{l} L^{-1}\left[\frac{s}{s^2 + \frac{\pi^2 h^2}{l^2}}\right] = \boldsymbol{y_0 \sin\frac{\pi x}{l} \cos\frac{\pi h t}{l}}. \tag{ix}$$

Example 9.35. *Find a relation between Fourier and Laplace transforms.*

Solution : If $f(t) = \begin{cases} e^{-st} g(t), & t > 0 \\ 0, & t < 0, \end{cases}$ (i) then $F[f(t)] = \int_{-\infty}^{0} e^{ist} \cdot 0 \cdot dt + \int_{0}^{\infty} e^{ist} \cdot e^{-xt} \cdot g(t)dt = \int_{0}^{\infty} e^{(is-x)t} g(t)dt$

$$= \int_{0}^{\infty} e^{-pt} g(t)dt, \quad \text{[where } p = (x - is),$$

$$= L[g(t)].$$

Hence, the Fourier transform of $f(t)$ is the Laplace transform of $g(t)$.

Exercises

9(A) : Objective Type Questions

Select the correct answer

9.1 The Laplace transform of a function $f(t)$ $(t \geq 0)$ is given by $F(s)$ equal to

(a) $\int_{0}^{\infty} e^{st} f(t)dt$ (b) $\int_{-\infty}^{+\infty} e^{-st} f(t)dt$ (c) $\int_{0}^{\infty} e^{-st} f(t)dt$ (d) $\int_{-\infty}^{+\infty} e^{st} f(t)dt$

provided the integral exists.

9.2 The Laplace transform of the function $f(t)$ is $L[f(t)]$ given by

(a) s (b) $\frac{1}{s^2}$ (c) $\frac{1}{s}$ (d) s^3

where $s > 0$.

9.3 Laplace transform of the function $f(t) = e^{at}$ is

(a) $\frac{1}{s}$ (b) $\frac{1}{s-a}$ (c) $(s-a)$ (d) $\frac{1}{(s-a)^2}$

where $s > 0$.

9.4 Laplace transform of $t^n (n > -1)$ is

(a) $\frac{\Gamma(n+1)}{s^{n+1}}$ (b) $\frac{\Gamma(n)}{s^n}$ (c) $\frac{\Gamma(n+1)}{s^n}$ (d) $\frac{\Gamma(n-1)}{s^{n-1}}$

where $s > 0$.

9.5 Laplace transform of $\sin bt$ is

(a) $\frac{b^2}{s^2+b^2}$ (b) $\frac{b}{s^2+b^2}$ (c) $\frac{b}{s^2-b^2}$ (d) $\frac{b^2}{s^2-b^2}$

where $s > 0$.

9.6 Laplace transform of $\cos bt$ is

(a) $\frac{s}{s^2-b^2}$ (b) $\frac{s^2}{s^2-b^2}$ (c) $\frac{s^2}{s^2+a^2}$ (d) $\frac{s}{s^2+b^2}$

where $s > 0$.

9.7 If $L[f(t)] = f(s)$ then $L[e^{at} f(t)]$ is equal to

(a) $f(s+a)$ (b) $\overline{f}(s+a)$ (c) $f(s-a)$ (d) $\overline{f}(s-a)$.

9.8 If $L[f(t)] = f(s)$ then $L[f(at)]$ is

(a) $\frac{1}{a}f\left(\frac{s}{a}\right)$ (b) $\frac{1}{a}\overline{f}\left(\frac{s}{a}\right)$ (c) $\frac{1}{a}\overline{f}\left(\frac{a}{s}\right)$ (d) $\frac{1}{a}f\left(\frac{a}{s}\right)$

where $a > 0$.

9.9 $L[\sin 3t \cos 2t]$ is

(a) $\frac{3s^2+15}{(s^2+25)(s^2+1)}$ (b) $\frac{3s^2+15}{(s^2+25)(s^2-1)}$ (c) $\frac{3s^2-15}{(s^2+25)(s^2+1)}$ (d) $\frac{3s^2+15}{(s^2-25)(s^2+1)}$.

9.10 $L[e^{2t}\cos^2 t]$ is

(a) $\frac{1}{2}\left[\frac{1}{s-2} + \frac{s-2}{(s-2)^2+4}\right]$ (b) $\frac{1}{2}\left[\frac{1}{s+2} + \frac{s-2}{(s-2)^2+4}\right]$

(c) $\left[\frac{1}{s-2} + \frac{s-2}{(s+2)^2+4}\right]$ (d) $\frac{1}{2}\left[\frac{1}{s-2} + \frac{s+2}{(s-2)^2+4}\right]$.

9.11 If $L[f(t)] = f(s)$, then $L[t^n f(t)]$ (n is a positive number) is equal to

(a) $\frac{d^n}{ds^n}\overline{f}(s)$ (b) $\frac{d^n}{ds^n}f(s)$ (c) $(-1)^n\frac{d^n}{ds^n}\overline{f}(s)$ (d) $(-1)^n\frac{d^n}{ds^n}f(s)$.

9.12 If $L[f(t)] = \overline{f}(s)$, then $L\left[\frac{1}{t}f(t)\right]$ is equal to

(a) $\int_0^\infty \overline{f}(s)ds$ (b) $\int_s^\infty \overline{f}(s)ds$ (c) $\int_s^\infty f(s)ds$ (d) $\int_0^\infty f(s)ds$.

9.13 $L\left[\frac{1}{t}(1-e^t)\right]$ is given by

(a) $\log\frac{s-1}{s}$ (b) $\log\left|\frac{s-1}{s}\right|$ (c) $\log\left|\frac{s+1}{s}\right|$ (d) $\log\left|\frac{s}{s-1}\right|$.

9.14 $L[f(t)]$, where $f(t) = \begin{cases} 1, & \text{for } 0 \le t < a \\ -1, & \text{for } a < t < 2a \end{cases}$

(a) $\frac{1}{2}\coth\left(\frac{as}{2}\right)$ (b) $\frac{1}{2}\tanh\left(\frac{as}{2}\right)$ (c) $\coth\left(\frac{as}{2}\right)$ (d) $\tanh\left(\frac{as}{2}\right)$.

9.15 $L^{-1}\left(\frac{1}{s}\right)$ is

(a) 1 (b) 0 (c) ∞ (d) none of these.

9.16 $L^{-1}\left(\frac{s}{s^2+a^2}\right)$ is

(a) $\sin at$ (b) $a\sin t$ (c) $a\cos at$ (d) $\cos at$.

9.17 If $L[f(t)] = \overline{f}(s)$ denotes the Laplace transform of the function $f(t)$, then $L^{-1}[\overline{f}(as)]$ is

(a) $f\left(\frac{t}{a}\right)$ (b) $\frac{1}{a}f\left(\frac{t}{a}\right)$ (c) $\frac{1}{a}f\left(\frac{a}{t}\right)$ (d) $\frac{1}{a}f(at)$

where $a > 0$.

9.18 $L^{-1}\left[\dfrac{s}{(s^2+a^2)^2}\right]$ is

(a) $\dfrac{1}{2a}\sin at$ (b) $\dfrac{1}{2a}t\sin at$ (c) $\dfrac{1}{2a}t\cos at$ (d) $2at\sin at$.

9.19 $L^{-1}\left[\dfrac{s^2}{(s^2+a^2)(s^2+b^2)}\right]$ is

(a) $\dfrac{a\sin at+b\sin bt}{a^2-b^2}$ (b) $\dfrac{a\sin at-b\sin bt}{a^2-b^2}$ (c) $\dfrac{a\sin at-b\sin bt}{a^2+b^2}$ (d) $\dfrac{\sin at+\sin bt}{ab}$.

9.20 $L^{-1}\left[\dfrac{1}{s(s^2+1)^2}\right]$ is

(a) $\dfrac{1}{2}[-2\cos t+t\sin t+2]$ (b) $\dfrac{1}{2}[-2\sin t+t\cos t+2]$

(c) $\dfrac{1}{2}[-2\cos t-t\sin t+2]$ (d) $\dfrac{1}{2}[-2\cos t-t\sin t-2]$.

Fill up the blanks in the followings :

9.21 $L[t^n]=\ldots\ldots$ [**Ans:** $\frac{\Gamma(n+1)}{s^{n+1}}$, $n>-1$, $s>0$]

9.22 $L[\sinh at]=\ldots\ldots$ [**Ans:** $\frac{a}{s^2-a^2}$, $s>|a|$]

9.23 $L[\cosh at]=\ldots\ldots$ [**Ans:** $\frac{s}{s^2-a^2}$, $s>|a|$]

9.24 $L^{-1}\left[\dfrac{1}{s^2-a^2}\right]=\ldots\ldots$ [**Ans:** $\frac{1}{a}\sinh at$, $s>|a|$]

9.25 $L^{-1}\left[\dfrac{1}{(s+3)(s-1)}\right]=\ldots\ldots$ [**Ans:** $\frac{e^t}{4}(1-e^{-4t})$]

9.26 $L^{-1}\left[\dfrac{1}{s^3(s^2+1)}\right]=\ldots\ldots$ [**Ans:** $\frac{t^2}{2}+\cos t-1$]

9.27 $L^{-1}\left[\dfrac{1}{s(s^2+a^2)}\right]=\ldots\ldots$ [**Ans:** $\frac{1}{a^2}(1-\cos at)$]

9.28 If $L\left[\dfrac{\sin t}{t}\right]=\tan^{-1}\left(\dfrac{1}{s}\right)$, then $L\left[\dfrac{\sin at}{t}\right]$ is

(a) $\tan^{-1}\left(\dfrac{1}{s^2}\right)$ (b) $\tan^{-1}\left(\dfrac{a}{s}\right)$ (c) $\tan^{-1}\left(\dfrac{1}{as}\right)$ (d) $\tan^{-1}\left(\dfrac{1}{s^2+a^2}\right)$. [WBUT 2006]

9.29 $L^{-1}\left[\dfrac{s}{s^2+a^2}\right]$ is equal to

(a) $\dfrac{1}{2a}t\sin at$ (b) $\dfrac{1}{a}t\sin at$ (c) $\dfrac{1}{a^2}\sin at$ (d) $\cos at$. [WBUT 2006]

9.30 Laplace transform of the function $\cos(at)$ is

(a) $\dfrac{s}{s^2-a^2}$ (b) $\dfrac{a}{s^2+a^2}$ (c) $\dfrac{s}{s^2+a^2}$ (d) $\dfrac{1}{s^2-a^2}$. [WBUT 2007]

9.31 $L[te^{2t}]$ is equal to

(a) $\frac{1}{s-2}$ (b) $2(s-2)^2$ (c) $\frac{1}{(s-2)^2}$ (d) $\frac{2}{s^2}$. [WBUT 2006]

9.32 If $L[f(t)] = \tan^{-1}\left(\frac{1}{p}\right)$, then $L[tf(t)]$ is

(a) $\tan^{-1}\left(\frac{1}{p^2}\right)$ (b) $\frac{1}{1+p^2}$ (c) $\frac{1}{1+p}$ (d) $\tan^{-1}\left(\frac{2}{\pi p}\right)$. [WBUT 2007]

9.33 The value of $\int_0^\infty \frac{\sin t}{t} dt$ is equal to

(a) $\frac{\pi}{3}$ (b) $\frac{\pi}{6}$ (c) $\frac{\pi}{4}$ (d) $\frac{\pi}{2}$. [WBUT 2008]

9.34 $L[e^{-2t}\cos t]$ is equal to

(a) $\frac{p}{p^2+2p+5}$ (b) $\frac{p-1}{p^2+4p+5}$ (c) $\frac{p+1}{p^2+4p+5}$ (d) $\frac{p+2}{p^2+4p+5}$ [WBUT 2010]

9.35 $L^{-1}\left[\int_s^\infty \frac{du}{u+a}\right] =$

(a) $\frac{e^{-at}}{t}$ (b) e^{-at} (c) $\frac{e^{at}}{t}$ (d) e^{at}.

Answers

9.1 (c)	**9.2** (c)	**9.3** (b)	**9.4** (a)	**9.5** (b)	**9.6** (d)	**9.7** (d)
9.8 (b)	**9.9** (a)	**9.10** (a)	**9.11** (c)	**9.12** (b)	**9.13** (b)	**9.14** (b)
9.15 (a)	**9.16** (b)	**9.17** (b)	**9.18** (b)	**9.19** (b)	**9.20** (c)	**9.28** (b)
9.29 (d)	**9.30** (c)	**9.31** (c)	**9.32** (b)	**9.33** (d)	**9.34** (d)	**9.35** (a).

9(B) Short Answer Type Questions

9.1 Define Laplace transform of a function $f(t)$. State the conditions under which Laplace transform exists.

9.2 State the properties of Laplace transform. Show that Laplace transform is a linear transformation.

9.3 State and prove first and second shifting properties of Laplace transform.

9.4 State and prove change of scale property of Laplace transform.

9.5 Find the Laplace transform of the derivative of a function $y = f(x)$. What form does it take when the order of the derivative is n.

9.6 Find the Laplace transform of $\frac{1}{t}f(t)$.

9.7 Find the Laplace transform of integration of $f(t)$.

9.8 Find the Laplace transform of $t^n f(t)$, where n is a positive integer.

9.9 Find the Laplace transform of gamma function, $\Gamma(n) = \int_0^\infty e^{-t}t^{n-1}dt$, $(n > 0)$.

9.10 Find the Laplace transform of Bessel functions $J_0(x)$ and $J_1(x)$.

9.11 State and prove convolution theorem.

9.12 What is inverse Laplace transform? Find the inverse Laplace transform of $\frac{s}{s^2+a^2}$.

[**Ans:** $\cos at$]

9.13 State and prove first shifting property of inverse Laplace transform.

9.14 State and prove second shifting property of inverse Laplace transform.

9.15 Find the inverse Laplace transform of $\frac{1}{(s-1)^2(s-2)^3}$ [**Ans:** $(t^2-4t+6)e^t-2t-6$]

9(C) Broad Answer Type Questions

9.1 Find the Laplace transform of the following :

(a) $\cos at \sinh at$ [**Ans:** $\frac{1}{2}\left\{\frac{s-a}{(s-a)^2+a^2}-\frac{s+a}{(s+a)^2+a^2}\right\}$]

(b) $e^{-t}\cos^2 t$ [**Ans:** $\frac{1}{2s+1}+\frac{s+1}{2s^2+4s+10}$]

(c) $t^2e^{-3t}\sin 2t$ [**Ans:** $\frac{2(s^3+6s^2+9s+2)}{(s^2+4s+5)^3}$]

9.2 Find the Laplace transform of the following :

(a) $\frac{1}{t}(e^{-at}-e^{-bt})$ [**Ans:** $\ln\frac{s+b}{s+a}$]

(b) $\frac{1}{t}(1-\cos at)$ [**Ans:** $-\frac{1}{2}\ln\frac{s^2}{s^2+a^2}$]

(c) $\frac{1}{t}(\cos at-\cos bt)$ [**Ans:** $-\frac{1}{2}\ln\frac{p^2+a^2}{p^2+b^2}$]

9.3 Find the Laplace transform of

(a) $e^{-2t}u_\pi(t)$, where $u_\pi(t)=\begin{cases}0, & t<\pi\\ 1; & t>\pi.\end{cases}$ [AMIETE(Summer) 1994] [**Ans:** $\frac{e^{-\pi(s+2)}}{s+2}$]

(b) Find the Laplace transform of $t^2u(t-3)$. [**Ans:** $e^{-3s}\left[\frac{2}{s^3}+\frac{6}{s^2}+\frac{9}{s}\right]$]

(c) Find the Laplace transform of $f(t)=\begin{cases}\frac{t}{a}, & 0<t\le a\\ \frac{1}{a}(2a-t), & a<t<2a.\end{cases}$ [**Ans:** $\frac{1}{as^2}\tanh\left(\frac{as}{2}\right)$]

9.4 Find the inverse Laplace transform of

(a) $\log\left[\frac{s^2-4}{(s-3)^2}\right]$ [**Ans:** $\frac{2}{t}(e^{3t}-\cosh 2t)$]

(b) $\frac{(s+2)^2}{(s^2+4s+8)^2}$. [**Ans:** $\frac{1}{4}e^{-2t}(2t\cos 2t+\sin 2t)$]

(c) $\frac{4s+12}{s^2+8s+16}$. [**Ans:** $4e^{-4t}(1-t)$]

9.5 Use convolution theorem to evaluate the inverse Laplace transform of

(a) $\frac{1}{s(s+4)}$ [**Ans:** $\frac{1}{4}(1-\cos 2t)$]

(b) $\dfrac{s^2}{(s^2+a^2)(a^2+b^2)},\ a \neq b$ [**Ans:** $\frac{a\sin at - b\sin bt}{a^2-b^2}$]

(c) $\dfrac{s^2}{(s^2-a^2)^2}$ [**Ans:** $\frac{1}{2a}(\sinh at \cosh at)$]

9.6 Show that,

(a) $L[erf(\sqrt{t})] = \dfrac{1}{s\sqrt{s+1}}$

(b) $L\left[erfc\left(\dfrac{a}{2\sqrt{t}}\right)\right] = \dfrac{1}{s}e^{-a\sqrt{s}}$

(c) $L[J_1(t)] = 1 - \dfrac{s}{\sqrt{s^2+1}}$ [$J_1(t)$ = Bessel function of order one].

9.7 Using the convolution theorem, show that

(a) $\beta(m,n) = \dfrac{\Gamma(m)\Gamma(n)}{\Gamma(m+n)}$ (b) $L^{-1}\left[\dfrac{f(s)}{(s+a)^2}\right] = \displaystyle\int_0^t xe^{-ax}F(t-x)dx.$

9.8 (a) If $L^{-1}\left[\dfrac{s^2-1}{(s+1)^2}\right] = t\cot t$, then evaluate $L^{-1}\left[\dfrac{9s^2-1}{9(s^2+1)^2}\right]$

[**Hint :** Use change of scale property]

(b) Find $L^{-1}\left[\log\left(1+\dfrac{\omega^2}{s^2}\right)\right]$ [**Ans:** $\frac{2}{t}(1-\cos\omega t)$]

9.9 (a) Show that inverse Laplace transform of $\dfrac{1}{s^2(s^2+1)}$ is $\dfrac{t^2}{2}+\cos t - 1$.

(b) Show that, $L^{-1}\left[\dfrac{1}{1+\sqrt{1+s^2}}\right] = -1 + \text{erf}\,(\sqrt{t}) + \dfrac{e^{-t}}{\sqrt{\pi t}}$.

9.10 Find the Laplace transform of the following periodic functions :

(a) $F(t) = \begin{cases} \dfrac{t}{T}, & 0 < t < T \\ 1, & t > T \end{cases}$ [**Agra U 1992**] [**Ans:** $-\frac{1}{s^2T}(1-e^{-sT})$]

(b) $F(t) = \begin{cases} t+1, & 0 \le t \le 2 \\ 3, & t > 2 \end{cases}$ [**Ans:** $\frac{1}{s^2}(s+1-e^{-2s})$]

(c) $F(t) = \begin{cases} \dfrac{h}{a}t, & 0 < t < a \\ \dfrac{h}{a}(2a-t), & a < t < 2a \end{cases}$ [**Ans:** $\frac{h}{as^2}\tanh\left(\frac{as}{2}\right)$]

(d) $F(t) = \begin{cases} 1, & 0 < t < \dfrac{T}{2} \\ 0, & \dfrac{T}{2} < t < T. \end{cases}$ [**Ans:** $\frac{1-e^{-2s}}{s(1-e^{-4s})}$]

9.11 Evaluate the following proper integrals by using Laplace transform :

(a) $\displaystyle\int_0^\infty \frac{\cos 6t - \cos 4t}{t}dt$ [**Ans:** $\log\left(\frac{2}{3}\right)$]

(b) $\displaystyle\int_0^\infty \frac{\cos at - \cos bt}{t}\,dt$ **[Ans:** $\frac{1}{2}\log\left(\frac{b^2}{a^2}\right)$**]**

(c) $\displaystyle\int_0^\infty \frac{e^{-t} - e^{-4t}}{t}\,dt.$ **[Ans:** $\log 4$**]**

9.12 Solve the following differential equations using Laplace transform :

(a) $\dfrac{d^2y}{dx^2} + 2\dfrac{dy}{dx} - 3y = \sin x$, where $y(0) = y'(0) = 0$.
[WBUT 2004] [Ans: $y = -\frac{1}{10}\cos x - \frac{1}{5}\sin x - \frac{1}{40}e^{-3x} + \frac{1}{8}e^{x}$**]**

(b) $\dfrac{d^2x}{dt^2} + 4x = \sin 3t$, $x(0) = x'(0) = 0$ **[WBUT 2007] [Ans:** $y = \frac{1}{10}(3\sin 2t - 2\sin 3t)$**]**

(c) $\dfrac{d^2y}{dx^2} + y = 0$, where $y = 1$, $\dfrac{dy}{dx} = 1$ at $x = 0$. **[Ans:** $y = \sin x + \cos x$**]**

(d) $(D^2 - D - 2)x = 20\sin 2t$, $x_0 = -1$, $x_1 = 2$.
[AMIETE(Summer) 1989] [Ans: $x = 2e^{2t} - 4e^{-t} + \cos 2t - 3\sin 2t$**]**

(e) $(D^3 + D^2)x = 6t^2 + 4$, $x(0) = 0$, $x'(0) = 2$, $x''(0) = 0$. **[AMIETE(Summer) 1988]**

(f) $(D^2 + n^2)x = a\sin(nt + \alpha)$, where $x = Dx$ at $t = 0$.
[Ans: $x = an\cos\alpha(\sin nt - nt\sin nt) + \frac{a\sin 2\alpha}{2n}(t\sin nt)$**]**

(g) $y'' + 2y' + y = te^{-t}$ if $y(0) = 1$, $y'(0) = -2$.
[Gujarat U 1992] [Ans: $y = \left(1 - t + \frac{t^3}{6}\right)e^{-t}$**]**

(h) $y'' + 4y' + 3y = t$, $t > 0$, given that $y(0) = 0$, $y'(0) = 1$.
[AMIETE(Winter) 1990] [Ans: $y = -\frac{4}{9} + \frac{t}{6} + e^{-t} - \frac{5}{9}e^{-3t}$**]**

(i) $\dfrac{d^2y}{dx^2} + y = x\cos 2x$, where $y = \dfrac{dy}{dx} = 0$ at $x = 0$.
[Ans: $y = \frac{4}{9}\sin 2x - \frac{5}{9}\sin x - \frac{x}{3}\cos 2x$**]**

(j) $y'' + 2y = r(t)$, $y(0) = 0$, $y'(0) = 0$, where $r(t) = \begin{cases} 0, & t > 1 \\ 1, & 0 \le t < 1. \end{cases}$
[Ans: $y = \frac{1}{2} - \frac{1}{2}\cos\sqrt{2}t$**]**

(k) $\dfrac{d^2x}{dt^2} + \dfrac{dx}{dt} - 2x = 2(1 + t - t^2)$, $x(0) = 0$, $x''(0) = 3$. **[Ans:** $x = t^2 + e^t - e^{-2t}$**]**

(l) $x\dfrac{d^2y}{dx^2} + (2x + 3)\dfrac{dy}{dx} + (x + 3)y = ae^{-x}$ **[Ans:** $y(x) = \left(A + \frac{a}{3}t\right)e^{-t}$, $A =$ Const.**]**

(m) $\dfrac{d^2\phi}{dt^2} + k^2\phi = A\sin\omega t$, $\phi(0) = \phi'(0) = 0$, $\omega \neq k$.
[Ans: $\phi(t) = \frac{A}{\omega^2 - k^2}\left(\frac{\omega}{k}\sin kt - \sin\omega t\right)$**]**

(n) $\dfrac{d^2y}{dt^2} + 7\dfrac{dy}{dt} + 12y = 0$, $y(0) = 2$, $y'(0) = 0$. **[Ans:** $y(t) = 2e^{-3t}$**]**

9.13 Solve the partial differential equations by Laplace transform method :

(a) $2\dfrac{\partial\theta}{\partial t}=\dfrac{\partial^2\theta}{\partial t^2}$ $(x<0)$, if $\theta_x(0,t)=0$, $\theta(x,0)=e^{-x}$ and $\theta(x,t)$ is finite as $x\to\infty$.

[**Ans:** $\theta(x,t)=e^{t-x}-\dfrac{2e^t}{\sqrt{\pi}}\displaystyle\int_0^{\sqrt{t}}e^{-[u^2+(x/2u^2)^2]}du$]

(b) $\dfrac{\partial^2\phi}{\partial x^2}-\dfrac{\partial^2\phi}{\partial t^2}=xt$, under the conditions, $\phi=\dfrac{\partial\phi}{\partial t}=0$ for $t=0$. [**Ans:** $\phi(x,t)=-\frac{xt^3}{6}$]

9.14 Solve the integral equations :

(a) $F(t)=t+2\displaystyle\int_0^t\cos(t-u)F(u)du$ [**Ans:** $F(t)=2e^t(t-1)+2+t$]

(b) $F(t)=at+\displaystyle\int_0^t\Gamma(x)\sin(t-x)dx.$ [**Ans:** $a+t\left(1+\frac{t^2}{6}\right)$]

9.15 If $L\{f(t)\}=\frac{p^2-p+1}{(2p^2+1)^2(p-1)}$, then apply the change of scale property to show that,

$$L\{f(2t)\}=\frac{p^2-2p+4}{4(p+1)^2(p-2)}.$$ [**WBUT 2002**]

9.16 If $L\{f(t)\}=\overline{f}(s)$, then $L\{t^nf(t)\}=(-1)^n\dfrac{d^n}{ds^n}\overline{f}(s)$, where n is a positive integer.

[**WBUT 2005**]

9.17 Find the Laplace transform of $\dfrac{\sin at}{t}$. Hence, show that, $\displaystyle\int_0^\infty\frac{\sin t}{t}dt=\frac{\pi}{2}$.

[**WBUT 2005, '07**] [**Ans:** $\cos^{-1}\left(\frac{s}{a}\right)$]

9.18 Find the inverse Laplace's transform of $\dfrac{1}{s(s^2+1)}$.

[**AMIETE(Winter) 1994**] [**Ans:** $-\cos t+1$]

9.19 Find the Laplace transform of $t\sinh at$. [**Ans:** $\frac{2as}{(s^2-a^2)^2}$]

9.20 Find the inverse of the following by convolution theorem.

(i) $\dfrac{1}{(s^2+1)^3}$, **(ii)** $xJ_1(x)$. [**Ans:** (i) $\frac{1}{8}[(3-t^2)\sin t-3t\cos t$, (ii) $\frac{1}{(s^2+1)^{3/2}}$]

9.21 Prove that, $L^{-1}\left[\dfrac{e^{-a\sqrt{s}}}{\sqrt{s}}\right]=\dfrac{1}{\sqrt{\pi}}t\,e^{-\frac{a^2}{4t}}$.

9.22 Find the Laplace transform of the periodic function (saw tooth wave)

$$f(t)=\frac{k}{T}\text{ for }0<t<T,\ f(t+T)=f(t).$$

[**AMIETE(Winter) 1996**] [**Ans:** $-\frac{k\,e^{-sT}}{s(1-e^{-sT})}+\frac{k}{Ts^2}$]

9.23 **(a)** Prove that inverse Laplace transform operator L^{-1} is linear.

(b) Find Laplace transform of $\cos(\omega t)$. [**Pune U(MSc) 2014**]

9.24 **(a)** Prove that the Laplace transform operator is linear.

(b) Find $L^{-1}\left[\dfrac{5S^2-15S+Z}{(S+1)(S-2)^3}\right]$.

(c) Find Laplace transform of $\sin(\omega t)$. [**Pune U(MSc) 2014**]

9.25 (a) Find the Laplace transform of
(i) $L\{6\sin 2t - 5\cos 2t\}$ (ii) $L\{e^{-4t}\cosh 2t\}$.

(b) Solve using Laplace transform :

$$x'' + 4x' + 4x = 6e^{-2t}$$
$$x(0) = 2,\; x'(0) = 8$$

[Pune U(MSc) 2013]

9.26 (a) Find Laplace transform of $\sinh(at)$.

(b) Find $L^{-1}\left\{\dfrac{5s^2 - 15s + 7}{(s+1)(s-2)^3}\right\}$. **[Pune U(MSc) 2008]**

9.27 (a) Find $L\{F(t)\}$, if

$$F(t) = \begin{cases} 5, & 0 < t < 3 \\ 0, & t > 3. \end{cases}$$

[Pune U(MSc) 2013]

(b) State and prove convolution theorem of Laplace transform.

9.28 (a) Solve the differential equation of coupled pendulums using Laplace transform with initial conditions :

$x_1 = x_2 = 0,\; x_1' = v$ and $x_2' = 0$ at $t = 0$.

(b) Prove that $L[f''(x)] = s^2 L[f(x) - sf(0) - f'(0)$.

(c) Find $L[\sin \lambda x]$. **[Gujrat U(MSc) 2017]**

9.29 (a) State and prove convolution theorem in Laplace transforms.

(b) Obtain the Laplace transform of $f(x) = x^n$.

9.30 (a) Solve $y'(t) + y(t) = t,\; y(0) = -2$, using Laplace transforms.

(b) Given that $f(t)$ is a periodic function with period T, find its Laplace transform.

CHAPTER 10

Vector, Scalar, the Dot and Cross Product

10.1 Introduction

All measurable physical quantities can be divided into two classes : (a) Scalars and (b) Vectors.

Scalars

Scalars are those physical quantities which have magnitude and unit without any reference to the idea of direction.

Example *Length, mass, time, temperature, volume, work, electric charge, pressure, entropy, etc. All these quantities have magnitude and unit, e.g.,* 5 m, 150 kg, 50 s, 30 °C, 500 m^3, *etc. and hence, are scalar quantities.*

Vectors

The physical quantities which have magnitude, unit and direction and obey the parallelogram law of addition of vectors are called vector quantities.

Example *Displacement, acceleration, momentum, electric intensity, electric current, density, weight, force, etc. All these quantities have magnitude, unit, direction and obey the parallelogram law of addition of vectors, e.g.,* 10 m *towards north,* 50 kg-m/s *towards east,* 2 C/m^2/s, *etc. and hence are vector quantities.*

Notes : Physical quantities having magnitude, unit and direction only are not vectors. If the physical quantities have magnitude, unit and direction but do not obey the parallelogram law of vector addition are not vectors.

For example, electric current has magnitude, unit and direction but does not obey the parallelogram law of vector addition and hence, is not a vector quantity. Similarly, the angular displacement of a rigid body is not a vector quantity though it has magnitude, unit and direction. This is due to the fact that two successive angular displacements are not given by the parallelogram law of vectors.

Representation of a Vector

Analytically a vector is represented by placing an arrow on the corresponding letter (e.g., $\vec{A}$) and its magnitude is represented by $|\vec{A}|$.

Graphically, a vector A is represented by $\vec{A}$ and drawn parallel to the direction of $\vec{A}$ on a definite scale and putting an arrow at the terminal point of the line (Fig. 10.1). The arrow-head represents the direction of the vector. Here O is called the *initial point* and P, the *terminal point.*

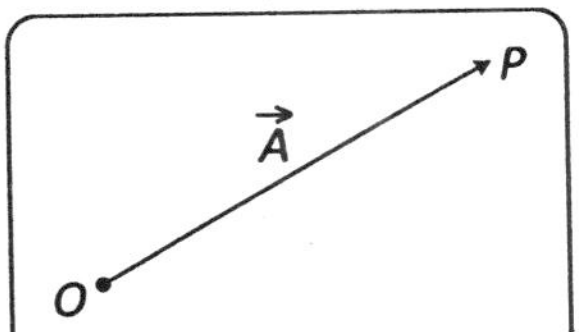

Fig. 10.1

10.2 Kinds of Vectors

(a) **Equal Vectors :** Two vectors $\vec{A}$ and $\vec{B}$ are said to be equal if they have the same magnitude and direction regardless of their initial points (Fig. 10.2).

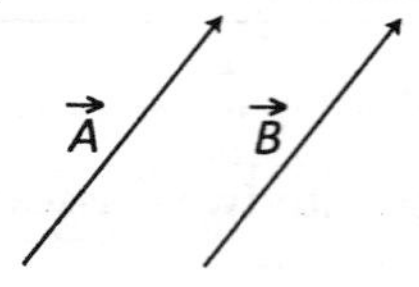

Fig. 10.2

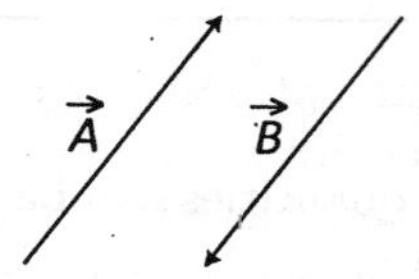

Fig. 10.3

(b) **Unlike Vectors :** If two vectors $\vec{A}$ and $\vec{B}$ are collinear but have opposite sense, then $\vec{A}$ and $\vec{B}$ are called *unlike vectors* (Fig. 10.3).

(c) **Null Vectors :** If the initial point O and the final point P of a line segment OP coincide, then the line segment OP becomes a point losing direction. For the sake of generality of the rules of vector algebra such a pair of coincident points is regarded as a vector, called the *null (or zero) vector.* A null vector is represented by the symbol $\vec{O}$.

All vectors, except the null vector, are called the *proper vectors.* If $\vec{A}$ is a proper vector, then $|\vec{A}| \neq 0$.

(d) **Unit Vectors :** A vector whose magnitude (or, length) is unity is called a *unit vector.* A unit vector indicates direction and is represented by putting a cap (∧) on the corresponding letter representing the vector.

Thus, the unit vector of $\vec{A}$ is $\hat{A} = \dfrac{\vec{A}}{|\vec{A}|} = \dfrac{\vec{A}}{A}$, where $A = |\vec{A}|$

$\hat{A}$ is pronounced as 'A-cap' or 'A-hat'.

(e) **Collinear Vectors :** Two or more vectors parallel or antiparallel to the same line are called collinear vectors. Thus, two equal vectors or two like vectors are *collinear vectors.*

(f) **Like Vectors :** Vectors are directed in the same sense irrespective of their magnitudes are like vectors or codirectional vectors.

(g) **Coplanar Vectors :** If a system of vectors lie in the same plane, then they are called *coplanar vectors.* The plane on which the coplanar vectors lie is the plane of the vectors.

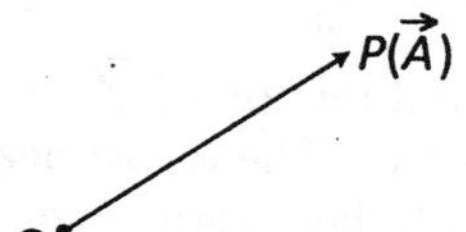

Fig. 10.4

(h) **Position Vectors :** The vector used to specify the position of a point P with reference to a fixed point O is represented by $\overrightarrow{OP}$ and is called the position vector of P with respect to O. The origin O is arbitrarily chosen and is called the *origin of the reference.*

The point P is sometimes referred to as the point $P(\vec{r})$ Fig. (10.4).

(i) **Reciprocal Vectors :** A vector $\vec{a}'$ having direction the same as that of a given vector $\vec{a}$ but magnitude as the reciprocal of the magnitude of $\vec{a}$ is termed as the *reciprocal vector* of $\vec{a}$. Thus,

$$\vec{a}' = \frac{1}{a}\hat{a} = \frac{\vec{a}}{a^2}.$$

(j) Axial Vectors : The line vectors involving some rotational action about an axis and drawn parallel to the axis of rotation such that the magnitude of the quantity is represented by the length of the vector and the direction by the rule of right-handed screw is called *axial vector.*

The angular velocity of a particle P, rotating in the horizontal plane is represented by a line AB which is perpendicular to the plane of rotation and has magnitude equal to the magnitude of angular velocity (Fig. 10.5). The direction of $\vec{\omega}$ and the rotation of P are related by the rule of right-handed screw. Here, $\vec{\omega}$ is an axial vector.

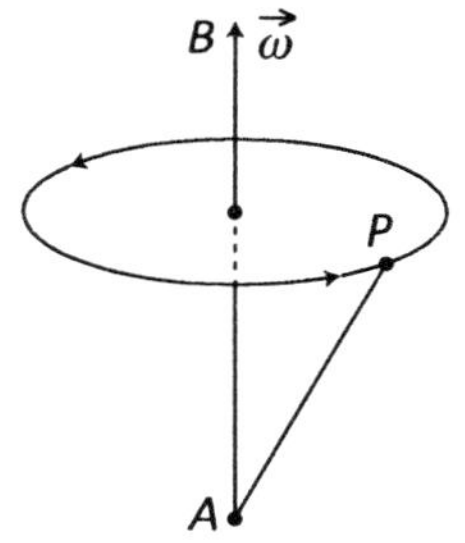

Fig. 10.5

10.3 Laws of Vector Algebra

If $\vec{A}$, $\vec{B}$, $\vec{C}$ are vectors and p and q are scalars, then the following laws hold good :

(i) Commutative law of addition : $\vec{A}+\vec{B}=\vec{B}+\vec{A}$

(ii) Associative law of addition : $\vec{A}+(\vec{B}+\vec{C})=(\vec{A}+\vec{B})+\vec{C}$

(iii) Commutative law for multiplication : $p\vec{A}=\vec{A}p$

(iv) Associative law for multiplication : $p(q\vec{A})=q(p\vec{A})$

(v) Distributive law for multiplication : $p(\vec{A}+\vec{B})=p\vec{A}+p\vec{B}$

10.4 Resolution of Vectors

(a) Co-planar Vectors : *If $\vec{a}$ and $\vec{b}$ be two non-collinear vectors and any vector $\vec{R}$, which is co-planar with $\vec{a}$ and $\vec{b}$, then $\vec{R}$ can be uniquely represented as a linear combination of $\vec{a}$ and $\vec{b}$, i.e., $\vec{R}=m\vec{a}+n\vec{b}$.*

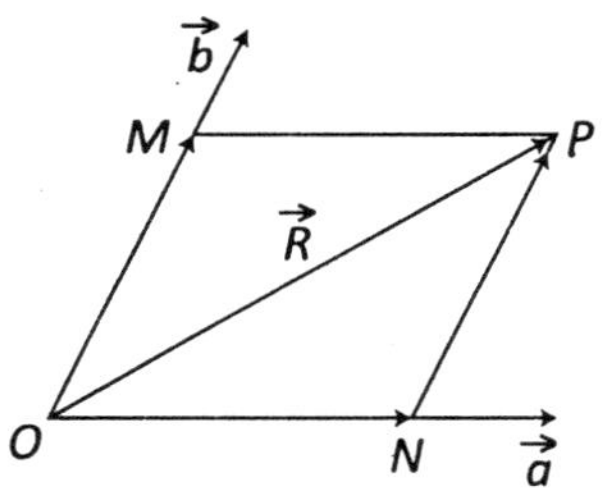

Fig. 10.6

Proof : Let vector $\vec{R}$ is coplanar with $\vec{a}$ and $\vec{b}$, where $\vec{a}$ and $\vec{b}$ are non-coplanar as shown in Fig. 10.6.

PN and PM are drawn parallel to $\vec{b}$ and $\vec{a}$ respectively. Since $\overrightarrow{ON}$ and $\vec{a}$ are collinear, $\overrightarrow{ON}=m\vec{a}$.

Similarly, $\overrightarrow{OM}=n\vec{b}$, where m and n are appropriate scalars. But,

$$\vec{R}=\overrightarrow{ON}+\overrightarrow{NP}=\overrightarrow{ON}+\overrightarrow{OM}$$

or, $$\vec{R}=m\vec{a}+n\vec{b}. \tag{10.1}$$

The statement regarding $\vec{R}$ is unique in that, only one parallelogram can be drawn with OP as diagonal and ON and OM as adjacent sides. Here $\vec{a}$ and $\vec{b}$ are called **base vectors**.

(b) Space Vector : *If $\vec{a}$, $\vec{b}$ and $\vec{c}$ are three given non-coplanar vectors and m, n, l any appropriate scalars then, $\vec{R}$ can be expressed uniquely by the relation,*

$$\vec{R}=m\vec{a}+n\vec{b}+l\vec{c}. \tag{10.2}$$

Proof : In Fig. 10.7, the position vector of the point P with respect to O is $\vec{R}$. We construct a parallelopiped with edges OA, OC and OD along $\vec{a}$, $\vec{b}$ and $\vec{c}$ respectively (Fig. 10.7).

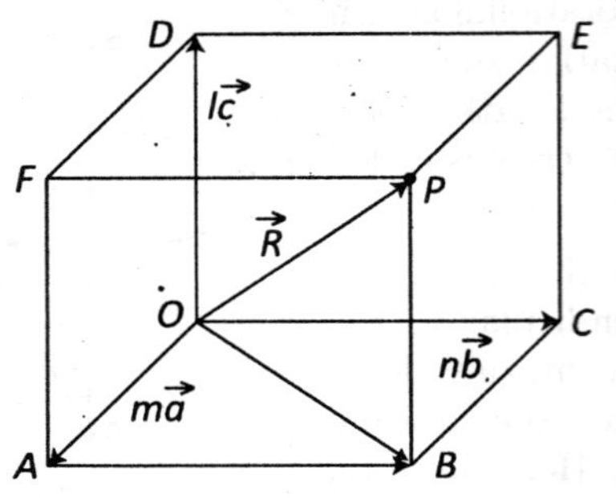

Fig. 10.7

Let $\overrightarrow{OA} = m\vec{a}$, $\overrightarrow{OC} = n\vec{b}$ and $\overrightarrow{OD} = l\vec{c}$.

Then, $\vec{R} = \overrightarrow{OB} + \overrightarrow{BP} = \overrightarrow{OA} + \overrightarrow{AB} + \overrightarrow{BP}$

$$= \overrightarrow{OA} + \overrightarrow{OC} + \overrightarrow{OD} = m\vec{a} + n\vec{b} + l\vec{c}.$$

Here, $\vec{a}$, $\vec{b}$ and $\vec{c}$ are called the **base vectors** and l, m, n are the **measure number**.

If the base vectors $\vec{a}$, $\vec{b}$ and $\vec{c}$ are known and m, n, l are given, then $\vec{R}$ is uniquely determined as only one parallelopiped can be constructed with a given diagonal and known edges.

If, $\vec{R}$ as expressed by Eqn. (10.2), is not unique, then there must exist m', n' and l' such that,

$$\vec{R} = m'\vec{a} + n'\vec{b} + l'\vec{c}. \tag{10.3}$$

From Eqns. (10.2) and (10.3), we have

$$m\vec{a} + n\vec{b} + l\vec{c} = m'\vec{a} + n'\vec{b} + l'\vec{c}$$

or, $$(m - m')\vec{a} + (n - n')\vec{b} + (l - l')\vec{c} = 0. \tag{10.4}$$

But, $\vec{a}$, $\vec{b}$ and $\vec{c}$ are non-coplanar vectors, so for relation $(1, 4)$ to be true,

$$m - m' = 0,\ n - n' = 0 \text{ and } l - l' = 0. \quad \text{i.e., } m = m',\ n = n' \text{ and } l = l'$$

∴ the representation of the vector $\vec{R}$ is unique.

10.5 Rectangular Unit Vectors

We consider a rectangular Cartesian coordinate system X, Y and Z. The rectangular unit vectors $\hat{i}$, $\hat{j}$ and $\hat{k}$ are mutually perpendicular and have directions along the positive X-, Y- and Z-axes respectively (Fig. 10.8).

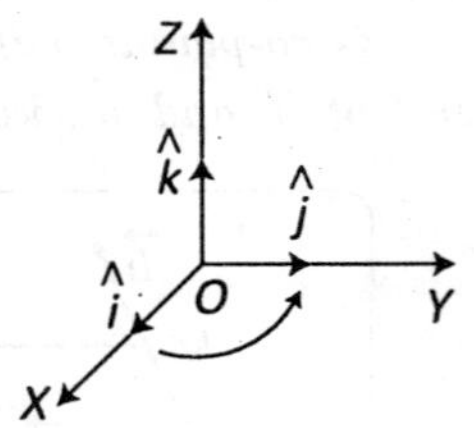

Fig. 10.8

Such a system is right-handed because, if a right-handed screw be rotated in the anticlockwise directions from OX to OY, the screw head will advance in the positive Z-direction (Fig. 10.8).

10.6 Components of a Vector

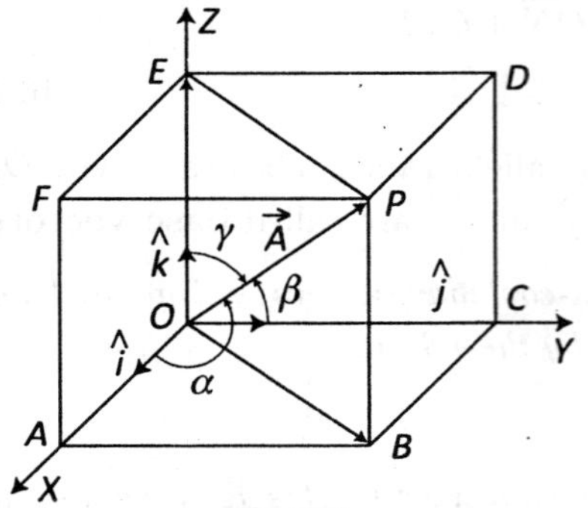

Fig. 10.9

Let $\overrightarrow{OP} = \vec{A}$ be any vector with initial point O, the origin of a Cartesian coordinate system OX, OY and OZ.

To find out the Cartesian components of $\vec{A}$ along X-, Y- and Z-axes, a perpendicular from P is dropped on the XY-plane. Let B be the foot of the perpendicular. Then lines BC and AB are drawn parallel to X- and Y-axes respectively.

If $\hat{i}$, $\hat{j}$ and $\hat{k}$ be the unit vectors along X-, Y- and Z-directions respectively, then

$$\overrightarrow{OA} = \hat{i}A_1, \quad \overrightarrow{OC} = \hat{j}A_2 \quad \text{and} \quad \overrightarrow{OE} = \hat{k}A_3,$$

where A_1, A_2 and A_3 are the magnitudes of the components of $\vec{A}$ along X-, Y- and Z-axes respectively.

Now, $\overrightarrow{OP} = \overrightarrow{OB} + \overrightarrow{BP} = (\overrightarrow{OA} + \overrightarrow{OC}) + \overrightarrow{BP}$ · or, $\vec{A} = \hat{i}A_1 + \hat{j}A_2 + \hat{k}A_3$.

Equation (10.5) below represents the vector $\vec{A}$ in terms of rectangular components. Geometrically, the magnitude of $\vec{A}$ may be represented as

$$A = |\vec{A}| = \sqrt{(OA)^2 + (OC)^2 + (OE)^2} \quad \text{or,} \quad A = \sqrt{A_1^2 + A_2^2 + A_3^2}. \tag{10.5}$$

The unit vector along $\vec{A}$ is $\hat{n}$, where

$$\hat{n} = \frac{\vec{A}}{A} = \hat{i}\left(\frac{A_1}{A}\right) + \hat{j}\left(\frac{A_2}{A}\right) + \hat{k}\left(\frac{A_3}{A}\right). \tag{10.6}$$

Now, $\frac{A_1}{A} = \cos\alpha$, $\frac{A_2}{A} = \cos\beta$ and $\frac{A_3}{A} = \cos\gamma$, are called the *direct cosines* of $\vec{A}$ with respect to X-, Y- and Z-axes and α, β and γ are the angles as shown in Fig. (1.9).

If $\cos\alpha = l$, $\cos\beta = m$ and $\cos\gamma = n$, then

$$\hat{n} = \hat{i}l + \hat{j}m + \hat{k}n. \tag{10.7}$$

Taking modulus on both sides of Eqn. (10.7), we get

$$|\hat{n}|^2 = l^2 + m^2 + n^2$$

or, $$l^2 + m^2 + n^2 = 1 \tag{10.8}$$

or, $$\mathbf{\cos^2\alpha + \cos^2\beta + \cos^2\gamma = 1}. \tag{10.9}$$

Hence, the sum of the squares of the direction cosines of any vector is unity.

10.7 Linearly Dependent and Independent Vectors

A set of n vectors $\vec{a}_1, \vec{a}_2, \ldots, \vec{a}_n$ is said to be linearly dependent, if there exists a set of n scalars $k_1, k_2, \ldots, k_n$ of which at least one is non-zero, such that,

$$k_1\vec{a}_1 + k_2\vec{a}_2 + \cdots + k_n\vec{a}_n = 0.$$

If $k_1 = k_2 = \cdots = k_n = 0$, then the set of n vectors $\vec{a}_1, \vec{a}_2, \ldots, \vec{a}_n$ is said to be *linearly independent*, if

$$\boldsymbol{k_1\vec{a}_1 + k_2\vec{a}_2 + \cdots + k_n\vec{a}_n = 0}.$$

10.8 Solved Problems (A)

Example 10.1. *Find the position vector of the point P which divides the line joining two given points A and B in a given ratio, say, $m : n$.*

Solution : Let the position vectors of the points A and B with respect to the origin be $\vec{a}$ and $\vec{b}$. Then, $\overrightarrow{OA} = \vec{a}$ and $\overrightarrow{OB} = \vec{b}$.

Let the point P divide AB such that,

$$\frac{AP}{PB} = \frac{m}{n} \quad \text{or,} \quad nAP = mPB \quad \text{or,} \quad n\overrightarrow{AP} = m\overrightarrow{PB}$$

or, $$n(\overrightarrow{OP} - \overrightarrow{OA}) = m(\overrightarrow{OB} - \overrightarrow{OP}) \quad \text{or,} \quad (n+m)\overrightarrow{OP} = m\overrightarrow{OB} + n\overrightarrow{OA}$$

or, $$\overrightarrow{OP} = \frac{m\overrightarrow{OB} + n\overrightarrow{OA}}{m+n} \quad \text{or,} \quad \boldsymbol{\overrightarrow{OP} = \frac{m\vec{b} + n\vec{a}}{m+n}}.$$

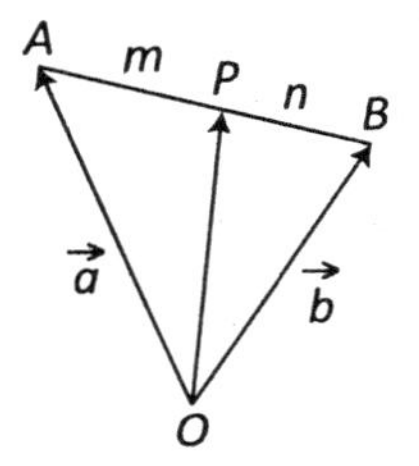

Fig. 10.10

Middle Point : If P be the middle point of AB, then the position vector of P is $\frac{\vec{a} + \vec{b}}{2}$.

Example 10.2. *If $\vec{a}$ and $\vec{b}$ are two non-collinear vectors and $x\vec{a} + y\vec{b} = 0$, then show that $x = 0$, $y = 0$.*

Solution : Let $x \neq 0$, then $\vec{a} = -\frac{y}{x}\vec{b}$.

The above equation is of the type, $\vec{a} = m\vec{b}$ and hence, $\vec{a}$ and $\vec{b}$ are collinear where m is a scalar. But this is contrary to hypothesis.

$\therefore$ **$x = 0$. Similarly, it can be shown that $y = 0$.**

Example 10.3. *If $\vec{a}$, $\vec{b}$ and $\vec{c}$ are non-coplanar vectors and $x\vec{a} + y\vec{b} + z\vec{c} = 0$, then show that $x = 0$, $y = 0$, $z = 0$.*

Solution : Let $x \neq 0$.

Now, $\vec{a} = -\frac{y}{x}\vec{b} - \frac{z}{x}\vec{c}$.

From the above equation we conclude that, $\vec{a}$ is coplanar with $\vec{b}$ and $\vec{c}$, which is contrary to the hypothesis.

$\therefore$ **$x = 0$. Similarly, we can show that, $y = 0$ and $z = 0$.**

Example 10.4. *Show that the line joining one vertex of a parallelogram to the mid-point of an opposite side trisects the diagonal and is itself trisected thereat.*

Solution : We consider parallelogram $PQRS$, whose P is the origin. Let the position vectors of Q, R and S be respectively $\vec{A}$, $\vec{B}$ and $\vec{C}$.

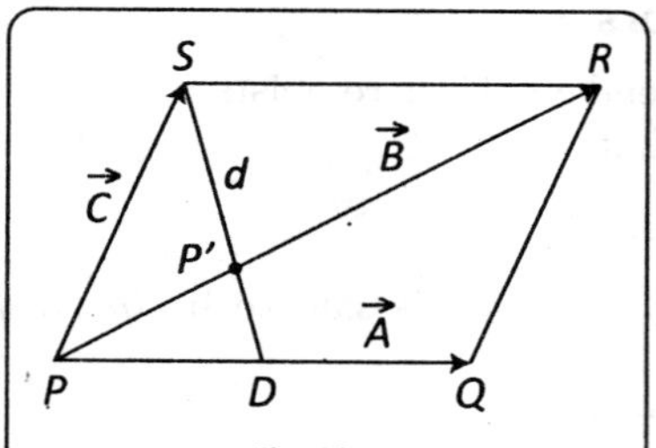

Fig. 10.11

The position vector of the mid-point D is $\frac{\vec{A}}{2}$.

$\because$ PQ is equal and parallel to SR,

$\therefore \quad \overrightarrow{PQ} = \overrightarrow{SR}$ or, $\vec{A} = \vec{B} - \vec{C}$ or, $\vec{A} + \vec{C} = \vec{B}$

or, $\dfrac{2\left(\frac{\vec{A}}{2}\right) + 1\vec{C}}{2+1} = \dfrac{\vec{B}}{3} = \overrightarrow{PP'}$.

$\therefore$ **P' trisects DS and PR.**

Example 10.5. *Prove that, the line joining the mid-points of two sides of a triangle is parallel and half to the third.*

Solution : Let ABC be triangle and $\vec{a}$, $\vec{b}$, $\vec{c}$, the position vectors of A, B, C respectively.

If the mid-points of AB and AC be D and E respectively, then the position vector of D is $\frac{\vec{a}+\vec{b}}{2}$.

The position vector of E is $\frac{\vec{a}+\vec{c}}{2}$.

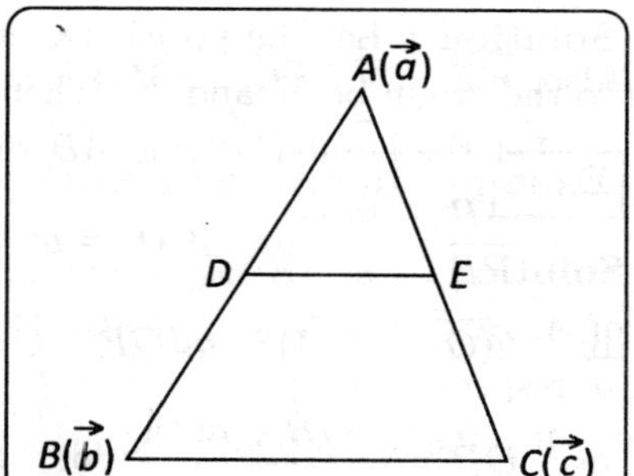

Fig. 10.12

$\therefore \quad \overrightarrow{DE}$ = (position vector of $\vec{E}$) – (position vector of $\vec{D}$)

$$= \frac{\vec{a}+\vec{c}}{2} - \frac{\vec{a}+\vec{b}}{2} = \frac{\vec{c}-\vec{b}}{2} = \frac{1}{2}\overrightarrow{BC}.$$

Hence, **DE is parallel to BC and half of BC.**

Example 10.6. *Find the position vector of the centroid of a triangle ABC, when the position vector of A, B and C are $\vec{a}$, $\vec{b}$ and $\vec{c}$.* **[AMIETE (Summer) 1991]**

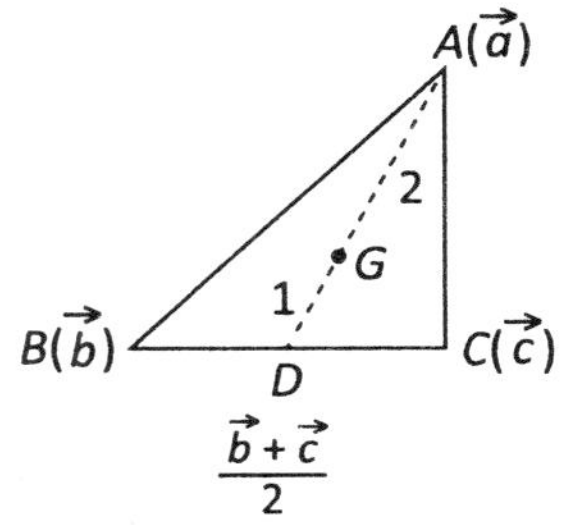

Fig. 10.13

Solution : The centroid G of $\triangle ABC$ divides the medians of the triangle in the ratio 2 : 1.

$$\text{Position vector of } D = \frac{\vec{b}+\vec{c}}{2}.$$

$$\text{Position vector of } \boldsymbol{G} = \frac{1\cdot\vec{a}+2\cdot\frac{\vec{b}+\vec{c}}{2}}{1+2} = \frac{\vec{a}+\vec{b}+\vec{c}}{3}.$$

Example 10.7. *In a trapezium prove that, the straight line joining the mid-points of the diagonals is parallel to the parallel sides and half their difference.*

Solution : Let $ABCD$ be a trapezium and AB and DC sides are parallel to each other. Let the position vectors of A, B, C and D be $\vec{a}$, $\vec{b}$, $\vec{c}$ and $\vec{d}$ respectively.

Now, DC is parallel to AB.

$\therefore \quad \overrightarrow{DC} = m\overrightarrow{AB}$ or, $(\vec{c}-\vec{d}) = m(\vec{b}-\vec{a})$.

If E be the mid-point of AC, then its position vector $= \dfrac{\vec{a}+\vec{c}}{2}$.

Similarly, the position vector of the mid-point F of $BD = \dfrac{\vec{b}+\vec{d}}{2}$.

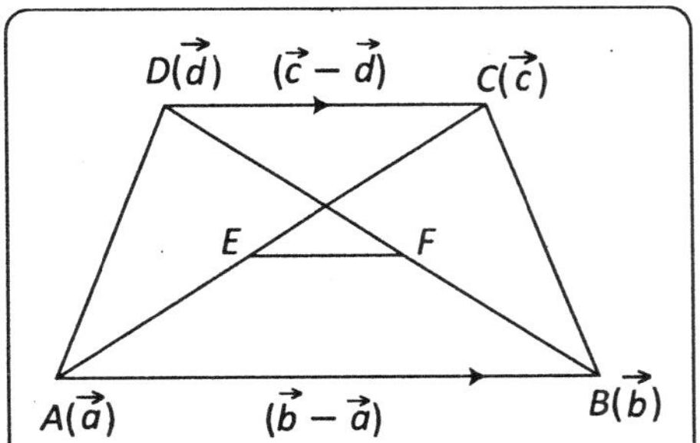

Fig. 10.14

$$\text{Now, } \overrightarrow{EF} = \frac{\vec{b}+\vec{d}}{2} - \frac{\vec{a}+\vec{c}}{2}$$

$$= \frac{1}{2}[(\vec{b}-\vec{a}) - (\vec{c}-\vec{d}) \qquad (10.10)$$

$$= \frac{1}{2}[(\vec{b}-\vec{a}) - m(\vec{b}-\vec{a})$$

$$= \frac{1}{2}(1-m)(\vec{b}-\vec{a}) \qquad (10.11)$$

$$= \frac{1}{2}(1-m)\overrightarrow{AB}. \qquad (10.12)$$

$\therefore$ **$\overrightarrow{EF}$ is parallel to $\overrightarrow{AB}$.**

Also, $\overrightarrow{EF} = \frac{1}{2}(\overrightarrow{AB} - \overrightarrow{DC})$ [from (10.10)]

Example 10.8. *Prove that the diagonals of a parallelogram bisect each other.*

Solution : Let $ABCD$ be a parallelogram and the position vectors of A, B, C and D with respect to a fixed origin be $\vec{a}$, $\vec{b}$, $\vec{c}$ and $\vec{d}$ respectively.

If possible, suppose the mid-points of AC and BD be M and N respectively.

Now, the position vectors of $M = \frac{\vec{a}+\vec{c}}{2}$ and that of $N = \frac{\vec{b}+\vec{d}}{2}$.

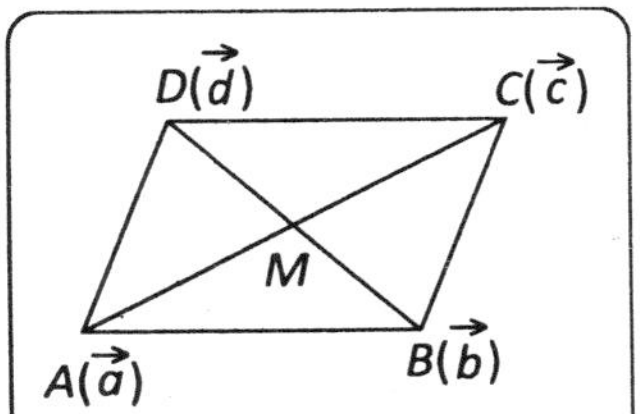

Fig. 10.15

But, $\overrightarrow{AB} = \overrightarrow{DC}$

$\therefore \quad \vec{b} - \vec{a} = \vec{c} - \vec{d}$ or, $\vec{b} + \vec{d} = \vec{a} + \vec{c}$ or, $\frac{1}{2}(\vec{a} + \vec{c}) = \frac{1}{2}(\vec{b} + \vec{d})$.

This shows that the points M and N are coincident and hence, the result follows.

Example 10.9. *If $\vec{A}$ and $\vec{B}$ are the vectors determined by two adjacent sides of a regular hexagon, what are the vectors represented by the other sides taken in order?*

Solution : In $ABCDEF$ regular hexagon, let $\overrightarrow{AB} = \vec{A}$ and $\overrightarrow{BC} = \vec{B}$.

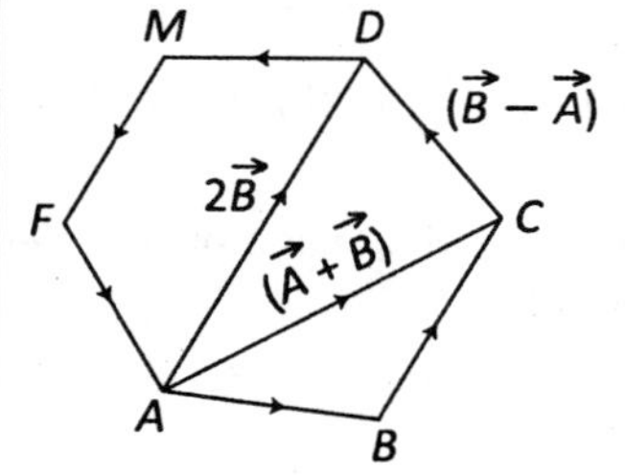

Fig. 10.16

$$\therefore \quad \overrightarrow{AC} = \overrightarrow{AB} + \overrightarrow{BC} = \vec{A} + \vec{B} \quad \text{and} \quad \overrightarrow{AD} = 2\overrightarrow{BC} = 2\vec{B},$$

$$\overrightarrow{CD} = \overrightarrow{AD} - \overrightarrow{AC} = 2\vec{B} - (\vec{A} + \vec{B}) = \vec{B} - \vec{A}.$$

Now, $\overrightarrow{DE} = -\overrightarrow{AB} = -\vec{A}$ [$\because$ AB is equal and parallel to ED]

$\overrightarrow{EF} = -\overrightarrow{BC} = -\vec{B}$ [$\because$ BC is equal and parallel to FE]

and $\overrightarrow{FA} = -\overrightarrow{CD} = -(\vec{B} - \vec{A})$

$= \vec{A} - \vec{B}$. [$\because$ CD is equal and parallel to AF]

Example 10.10. *If the mid-points of consecutive sides of any quadrilateral are connected by straight lines, prove that the resulting quadrilateral is a parallelogram.*

Solution : Let $ABCD$ be a quadrilateral and K, L, M, N be the mid-points of AB, BC, CD and DA. Let the position vectors of the points, A, B, C, D be $\vec{a}$, $\vec{b}$, $\vec{c}$ and $\vec{d}$ respectively.

Now, the position vectors of K, L, M, N are :

$\vec{K} = \frac{\vec{a}+\vec{b}}{2}$, $\vec{L} = \frac{\vec{b}+\vec{c}}{2}$, $\vec{M} = \frac{\vec{c}+\vec{d}}{2}$ and $\vec{N} = \frac{\vec{a}+\vec{d}}{2}$ respectively.

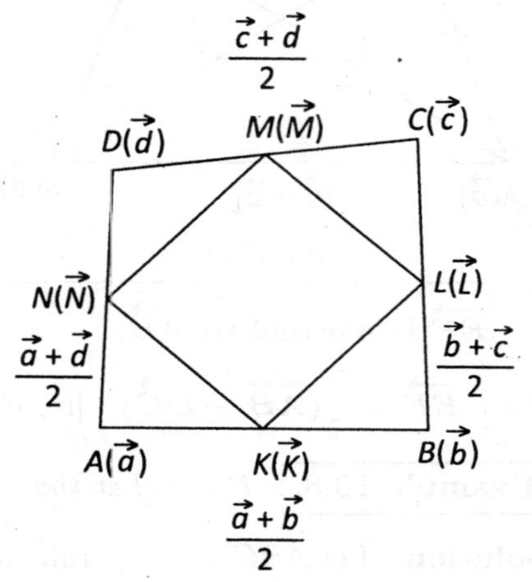

Fig. 10.17

$$\text{Now, } \overrightarrow{KL} = \vec{L} - \vec{K} = \frac{\vec{b} + \vec{c}}{2} - \frac{\vec{a} + \vec{b}}{2}$$

$$= \frac{\vec{c} - \vec{a}}{2} \qquad (10.13)$$

$$\overrightarrow{NM} = \vec{M} - \vec{N} = \frac{\vec{c} + \vec{d}}{2} - \frac{\vec{a} + \vec{d}}{2}$$

$$= \frac{\vec{c} - \vec{a}}{2}. \qquad (10.14)$$

From (10.13) and (10.14), $\overrightarrow{KL} = \overrightarrow{NM}$.

$\therefore$ $KL = NM$ and KL is parallel to NM.

$\therefore$ **$KLMN$ is a parallelogram.**

10.9 Product of Two Vectors

Two vectors when multiplied may give rise to either a scalar or a vector. As such two types of vector products result, viz., *Scalar product* and *Vector product*.

10.10 Scalar or Dot Product

The scalar or dot product of two vectors $\vec{A}$ and $\vec{B}$ is defined as the scalar $AB\cos\theta$, where θ is the angle between $\vec{A}$ and $\vec{B}$ (Fig. 1.18) i.e.,

$$\vec{A}\cdot\vec{B} = |\vec{A}||\vec{B}|\cos\theta = AB\cos\theta. \qquad (10.15)$$

where $|\vec{A}| = A$ and $|\vec{B}| = B$ represent the magnitudes of the vectors $\vec{A}$ and $\vec{B}$ respectively.

Geometrical Interpretation

The product $\vec{A}\cdot\vec{B}$ represents the product of the length of one vector and the length of the projection of the other in the direction of the former.

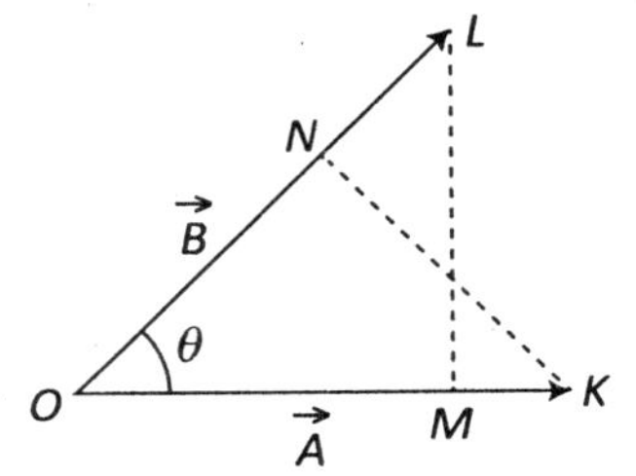

Fig. 10.18

If $\overrightarrow{OK} = \vec{A}$ and $\overrightarrow{OL} = \vec{B}$, then

$$\vec{A}\cdot\vec{B} = AB\cos\theta = A(OL\cos\theta) = A(OM)$$

$$= |\vec{A}|, \textbf{projection of } |\vec{B}| \textbf{in the direction of } \vec{A}.$$

Similarly, $\vec{A}\cdot\vec{B} = |\vec{B}|(OK\cos\theta) = |\vec{B}|(ON)$

$$= |\vec{B}|, \textbf{projection of } |\vec{A}| \textbf{along } \vec{B}.$$

It is clear that, $\vec{A}\cdot\vec{B} \le |A||B| \quad (\because \cos\theta \le 1)$.

10.11 Properties and Some Important Results

1. The scalar product of two products obeys commutative law of multiplication, i.e.,
$$\vec{A}\cdot\vec{B} = \vec{B}\cdot\vec{A}.$$
Because, $\vec{A}\cdot\vec{B} = AB\cos\theta = BA\cos(-\theta) = \vec{B}\cdot\vec{A}$.
2. If $\vec{A}$ and $\vec{B}$ are not null vectors, then $\vec{A}\cdot\vec{B} = 0$ implies $\vec{A}$ is perpendicular to $\vec{B}$.
3. $n(\vec{A}\cdot\vec{B}) = (n\vec{A})\cdot\vec{B} = \vec{A}\cdot(n\vec{B}) = (\vec{A}\cdot\vec{B})n$, n is a scalar.
4. $\vec{A}\cdot\vec{A} = AA\cos 0° = A^2$. Thus, the dot product of two equal vectors is a scalar and is equal to the square of the modulus of any of them.

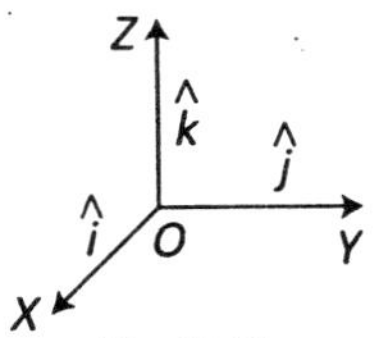

Fig. 10.19

5. If $\hat{i}$, $\hat{j}$ and $\hat{k}$ be mutually perpendicular unit vectors, then
$$\hat{i}\cdot\hat{i} = \hat{j}\cdot\hat{j} = \hat{k}\cdot\hat{k} = 1$$
and $\quad \hat{i}\cdot\hat{j} = \hat{j}\cdot\hat{k} = \hat{k}\cdot\hat{i} = 0.$

The system of three unit vectors, any two of which are mutually orthogonal, is called an *orthogonal trial of vectors*.

6. If $\vec{A}$ and $\vec{B}$ be two vectors whose orthogonal projections along X-, Y- and Z-axes are (A_1, A_2, A_3) and (B_1, B_2, B_3) respectively, then
$$\vec{A} = \hat{i}A_1 + \hat{j}A_2 + \hat{k}A_3 \qquad \vec{B} = \hat{i}B_1 + \hat{j}B_2 + \hat{k}B_3.$$
Then, $\vec{A}\cdot\vec{B} = (\hat{i}A_1 + \hat{j}A_2 + \hat{k}A_3)\cdot(\hat{i}B_1 + \hat{j}B_2 + \hat{k}B_3) = A_1B_1 + A_2B_2 + A_3B_3$.

If $\vec{A} = \vec{B}$, then
$$\vec{A}\cdot\vec{A} = A^2 = A_1^2 + A_2^2 + A_3^2 \quad \text{and} \quad \vec{B}\cdot\vec{B} = B^2 = B_1^2 + B_2^2 + B_3^2.$$

7. If θ be the angle between $\vec{A}$ and $\vec{B}$, then

$$\vec{A}\cdot\vec{B} = |\vec{A}||\vec{B}|\cos\theta \quad \text{or,} \quad \cos\theta = \frac{\vec{A}\cdot\vec{B}}{|\vec{A}||\vec{B}|} = \frac{\vec{A}\cdot\vec{B}}{\sqrt{A_1^2+A_2^2+A_3^2}\sqrt{B_1^2+B_2^2+B_3^2}}.$$

$$\therefore \quad \boldsymbol{\theta = \cos^{-1}\left[\frac{A_1B_1+A_2B_2+A_3B_3}{\sqrt{A_1^2+A_2^2+A_3^2}\sqrt{B_1^2+B_2^2+B_3^2}}\right]}.$$

8. If we have $\vec{A} = \hat{i}A_1 + \hat{j}A_2 + \hat{k}A_3$, then $\vec{A}\cdot\hat{i} = A_1$, $\vec{A}\cdot\hat{j} = A_2$, $\vec{A}\cdot\hat{k} = A_3$.

$$\therefore \quad \boldsymbol{\vec{A} = \hat{i}(\vec{A}\cdot\hat{i}) + \hat{j}(\vec{A}\cdot\hat{j}) + \hat{k}(\vec{A}\cdot\hat{k})}.$$

Here, $\vec{A}\cdot\hat{i}$, $\vec{A}\cdot\hat{j}$ and $\vec{A}\cdot\hat{k}$ are the rectangular components of $\vec{A}$ along X-, Y- and Z-axes respectively.

9. Scalar product of vectors is distributive, i.e., $\vec{A}\cdot(\vec{B}+\vec{C}) = \vec{A}\cdot\vec{B} + \vec{A}\cdot\vec{C}$.

Also, $\vec{A}\cdot(\vec{B}-\vec{C}) = \vec{A}\cdot\vec{B} - \vec{A}\cdot\vec{C}$.

10. (i) $\boldsymbol{(\vec{A}+\vec{B})^2} = (\vec{A}+\vec{B})\cdot(\vec{A}+\vec{B}) = \vec{A}\cdot\vec{A} + \vec{A}\cdot\vec{B} + \vec{B}\cdot\vec{A} + \vec{B}\cdot\vec{B}$

$= \boldsymbol{A^2 + 2\vec{A}\cdot\vec{B} + B^2}$.

(ii) $\boldsymbol{(\vec{A}-\vec{B})^2} = (\vec{A}-\vec{B})\cdot(\vec{A}-\vec{B}) = \boldsymbol{A^2 - 2\vec{A}\cdot\vec{B} + B^2}$

(iii) $\boldsymbol{(\vec{A}+\vec{B})\cdot(\vec{A}-\vec{B})} = \vec{A}\cdot\vec{A} - \vec{A}\cdot\vec{B} + \vec{B}\cdot\vec{A} - \vec{B}\cdot\vec{B} = \boldsymbol{A^2 - B^2}$.

11. For any three vectors $\vec{A}$, $\vec{B}$ and $\vec{C}$, if $\vec{A}\cdot\vec{B} = \vec{A}\cdot\vec{C}$, then it does not imply that $\vec{B} = \vec{C}$.
$\because \vec{A}(\vec{B}-\vec{C}) = 0$ as $\vec{A}\cdot\vec{B} = \vec{A}\cdot\vec{C}$, it implies either $(\vec{B}-\vec{C})$ is perpendicular to $\vec{A}$, or, $\boldsymbol{(\vec{B}-\vec{C}) = \vec{O}}$, **a null vector**.

Illustrative Examples

(a) In mechanics the work done by a force $(\vec{F})$ causing a displacement $d\vec{r}$ (Fig. 10.20) is given by

$$dW = (\text{Component of } \vec{F} \text{ along } OA)|\overrightarrow{OP}| = |\vec{F}|\cos\theta|\overrightarrow{OP}| = \vec{F}\cdot d\vec{r}. \tag{10.17}$$

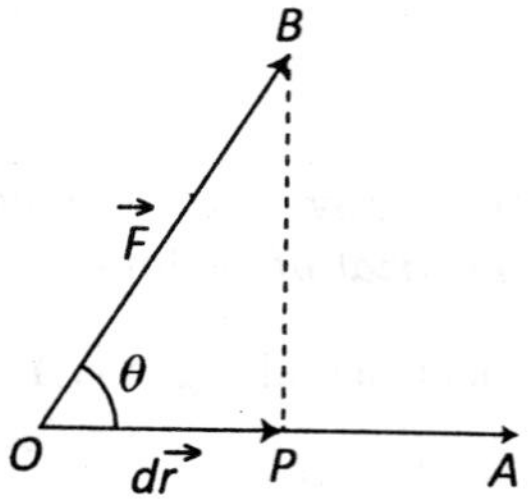

Fig. 10.20

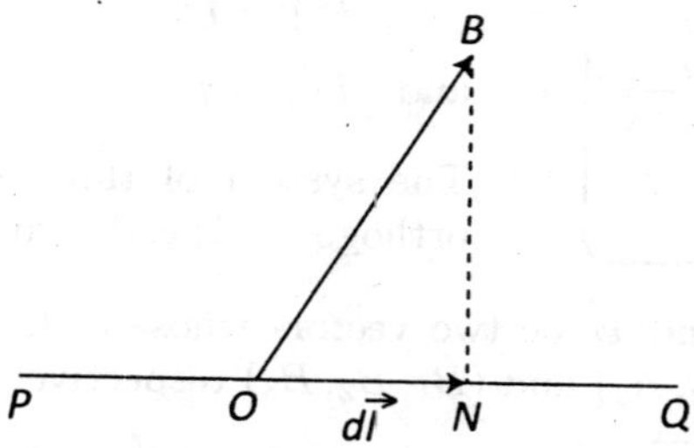

Fig. 10.21

(b) In a magnetic field $\vec{B}$, the work done in displacing $\vec{dl}$ a unit north pole in direction θ with the field is written as

$$\boldsymbol{dW = |\vec{B}|\cos\theta|\vec{dl}| = \vec{B}\cdot\vec{dl}}. \tag{10.18}$$

The work done for a finite displacement is obtained as the *line integral* of the magnetic field as

$$W = \int_P^Q \vec{B} \cdot \vec{dl}. \tag{10.19}$$

10.12 Vector Product or Cross Product of Two Vectors

The vector product of two vectors $\vec{A}$ and $\vec{B}$ is defined as a vector whose modulus is equal to the product of the magnitudes of the vectors and the sine of the angle between their directions and whose direction is perpendicular to the plane of $\vec{A}$ and $\vec{B}$, i.e.,

$$\vec{A} \times \vec{B} = |\vec{A}||\vec{B}| \sin\theta \hat{n}, \tag{10.20}$$

$\hat{n}$ is a unit vector perpendicular to the plane of $\vec{A}$ and $\vec{B}$. If the angle θ is described in the anticlockwise direction, then $\hat{n}$ is positive. Here, $\vec{A}$, $\vec{B}$ and $\hat{n}$ form a right-handed system.

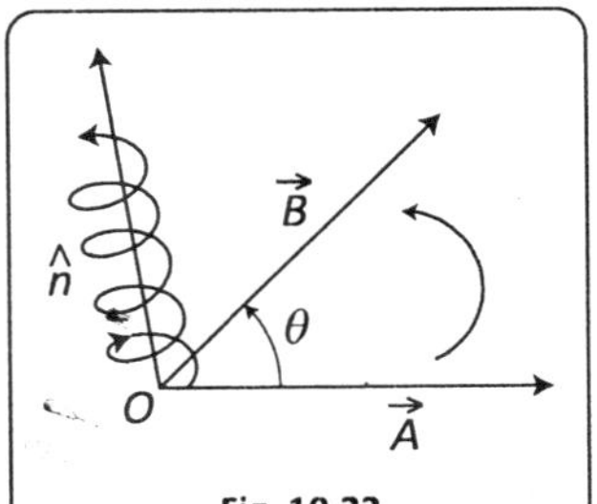

Fig. 10.22

Geometrical Interpretation

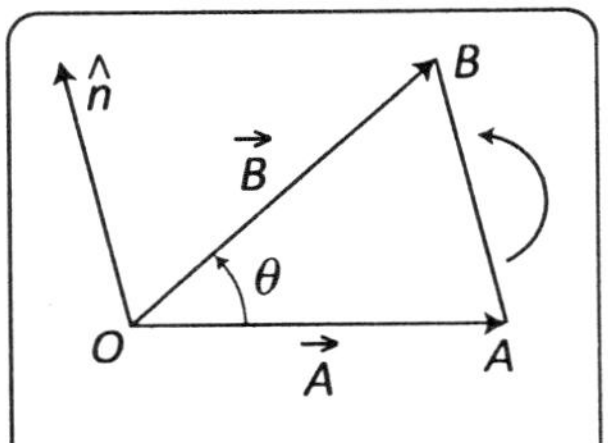

Fig. 10.23

If $\vec{A}$ and $\vec{B}$ are represented by the two adjacent sides of a triangle OAB, then $\vec{A} \times \vec{B}$ represents twice the vector area of the triangle.

Proof : If $\hat{n}$ be the unit vector normal to the plane of the triangle OAB, then

$$\vec{A} \times \vec{B} = |\vec{A}||\vec{B}| \sin\theta \cdot \hat{n} = 2\left[\frac{1}{2}|\vec{A}||\vec{B}|\sin\theta\right]\hat{n} = 2\triangle OAB \cdot \hat{n}$$

or, $$\vec{A} \times \vec{B} = 2\triangle\overrightarrow{OAB}. \tag{10.21}$$

10.13 Properties and Other Results

1. **Commutative Law fails in case of vector product.**

$$\vec{A} \times \vec{B} \neq \vec{B} \times \vec{A}; \quad \text{in fact,} \quad \vec{A} \times \vec{B} = -(\vec{B} \times \vec{A}).$$

$$\because \vec{A} \times \vec{B} = |\vec{A}||\vec{B}| \sin\theta \cdot \hat{n} \text{ and } \vec{B} \times \vec{A} = |\vec{A}||\vec{B}| \sin(-\theta) \cdot \hat{n} = -|\vec{A}||\vec{B}| \sin\theta \cdot \hat{n}.$$

2. **Distribute Law for cross product holds good.**

$$\vec{A} \times (\vec{B} + \vec{C}) = \vec{A} \times \vec{B} + \vec{A} \times \vec{C}.$$

3. $m(\vec{A} \times \vec{B}) = (m\vec{A}) \times \vec{B} = (\vec{A} \times \vec{B})m$, m is a scalar.

4. **Condition for two collinear vectors.**

Two vectors $\vec{A}$ and $\vec{B}$ are collinear, if the angle between them $\theta = 0°$ or, π, i.e., $\sin\theta = 0$.

Thus, for two collinear vectors,

$$\vec{A} \times \vec{B} = |\vec{A}||\vec{B}| \sin\theta \cdot \hat{n} = 0, \quad [\text{since } \sin\theta = 0]$$

Thus, two non-zero vectors will be parallel or antiparallel, if and only if $\vec{A} \times \vec{B} = 0$.

Conversely, if $\vec{A} \times \vec{B} = 0$ $|\vec{A}||\vec{B}| \sin\theta = 0$ or, $\sin\theta = 0$.

If $|\vec{A}| \neq 0$, $|\vec{B}| = 0$, then $\theta = 0$ or π.

In particular, if $\vec{A} = \vec{B}$, then $\vec{A} \times \vec{A} = 0$.

5. For orthogonal vector triad $\hat{\imath}$, $\hat{\jmath}$ and $\hat{k}$,

$$\hat{\imath}\times\hat{\imath} = \hat{\jmath}\times\hat{\jmath} = \hat{k}\times\hat{k} = 0$$

$$\hat{\imath}\times\hat{\jmath} = \hat{k}, \quad \hat{\jmath}\times\hat{k} = \hat{\imath}, \quad \hat{k}\times\hat{\imath} = \hat{\jmath}$$

and $\hat{\jmath}\times\hat{\imath} = -\hat{k}, \quad \hat{k}\times\hat{\jmath} = -\hat{\imath}, \quad \hat{\imath}\times\hat{k} = -\hat{\jmath}.$

6. **Vector product in terms of components.**

If $\vec{A} = \hat{\imath}A_1 + \hat{\jmath}A_2 + \hat{k}A_3$ and $\vec{B} = \hat{\imath}B_1 + \hat{\jmath}B_2 + \hat{k}B_3$, then

$$\vec{A}\times\vec{B} = \begin{vmatrix} i & j & k \\ A_1 & A_2 & A_3 \\ B_1 & B_2 & B_3 \end{vmatrix}.$$

7. The unit vector ($\hat{n}$) perpendicular to $\vec{A}$ and $\vec{B}$ is

$$\hat{\boldsymbol{n}} = \frac{\vec{\boldsymbol{A}}\times\vec{\boldsymbol{B}}}{|\vec{\boldsymbol{A}}||\vec{\boldsymbol{B}}|\sin\theta} = \frac{\vec{\boldsymbol{A}}\times\vec{\boldsymbol{B}}}{|\vec{\boldsymbol{A}}\times\vec{\boldsymbol{B}}|}.$$

8. **Relation between Scalar and Vector Product.**

$$(\vec{A}\cdot\vec{B})^2 = |\vec{A}|^2|\vec{B}|^2\cos^2\theta = |\vec{A}|^2|\vec{B}|^2 - |\vec{A}|^2|\vec{B}|^2\sin^2\theta = A^2B^2 - (\vec{A}\times\vec{B})^2.$$

$$\therefore\quad \boldsymbol{(\vec{A}\times\vec{B})^2 = A^2B^2 - (\vec{A}\cdot\vec{B})^2}.$$

Illustrative Examples

10.13.1 Moment of a Force about a Point

The moment of a force (or, torque) $\vec{F}$ about the point O is a vector $\vec{\wedge} = \vec{r}\times\vec{F}$, where $\vec{r}$ is the position vector of any point A relative to O on the line of action of $\vec{F}$ (Fig. 10.24).

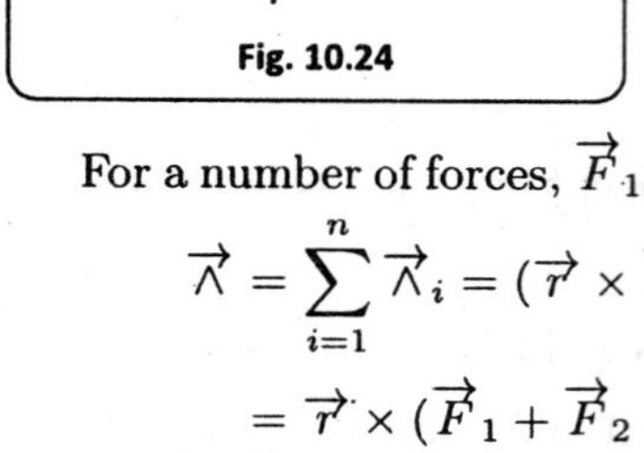

Fig. 10.24

We drop a perpendicular OP from O on the line of $\vec{F}$. Then

$$|\vec{\wedge}| = (|\overrightarrow{OP}|)(|\vec{F}|).$$

If $\overrightarrow{OA} = \vec{r}$, then

$$|\overrightarrow{OP}| = |\vec{r}|\sin\theta, \quad \theta \text{ being the angle between } \vec{r} \text{ and } \vec{F}.$$

$$|\vec{\wedge}| = |\vec{r}||\vec{F}|\sin\theta = |\vec{r}\times\vec{F}|$$

or,
$$|\vec{\wedge}| = |\vec{r}\times\vec{F}|. \tag{10.22}$$

For a number of forces, $\vec{F}_1, \vec{F}_2, \ldots, \vec{F}_n$ passing through A we have the sum of the moments,

$$\vec{\wedge} = \sum_{i=1}^{n}\vec{\wedge}_i = (\vec{r}\times\vec{F}_1) + (\vec{r}\times\vec{F}_2) + \cdots + (\vec{r}\times\vec{F}_n) \tag{10.23}$$

$$= \vec{r}\times(\vec{F}_1 + \vec{F}_2 + \cdots + \vec{F}_n) \tag{10.24}$$

$$= \vec{r}\times\vec{F}, \tag{10.25}$$

where
$$\vec{\boldsymbol{F}} = \sum_{i=1}^{n}(\vec{\boldsymbol{F}}_1 + \vec{\boldsymbol{F}}_2 + \cdots + \vec{\boldsymbol{F}}_n). \tag{10.26}$$

$|\vec{\lambda}| = |\vec{r} \times \vec{F}|$ shows that, $\vec{\lambda}$ does not change, if $\vec{F}$ is displaced along the line of action of the force.

10.13.2 Angular Velocity of a Rigid Body

Let a rigid body rotates in the anticlockwise direction about the axis OA with an angular velocity $\vec{\omega}$. $\vec{\omega}$ is a vector whose magnitude is the angular speed in radian per second and direction along the axis of rotation $\overrightarrow{OA}$.

If $\vec{r}$ be the position vector of a point P on the rigid body, then its linear velocity

$$v = \omega a \quad \text{or,} \quad v = \omega(r \sin\theta).$$

Since $\vec{v}$ is normal to the plane of $\vec{r}$ and $\vec{\omega}$, the vectors $\vec{\omega}$, $\vec{r}$ and $\vec{v}$ form a right-handed system.

$\therefore \quad \vec{v} = \omega \times \vec{r}$.

If $\vec{v} = \hat{i}v_1 + \hat{j}v_2 + \hat{k}v_3$, $\vec{\omega} = \hat{i}\omega_1 + \hat{j}\omega_2 + \hat{k}\omega_3$ and $\vec{r} = \hat{i}x + \hat{j}y + \hat{k}z$, then

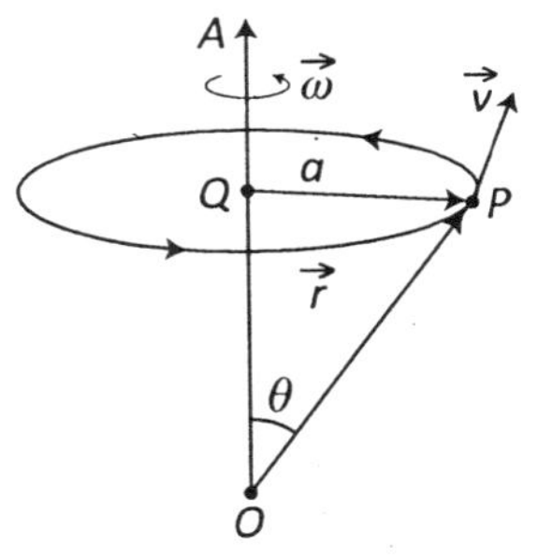

Fig. 10.25

$$\vec{v} = \begin{vmatrix} \hat{i} & \hat{j} & \hat{k} \\ \omega_1 & \omega_2 & \omega_3 \\ x & y & z \end{vmatrix}$$

or, $\quad \hat{i}v_1 + \hat{j}v_2 + \hat{k}v_3 = \hat{i}(z\omega_2 - y\omega_3) + \hat{j}(x\omega_3 - z\omega_1) + \hat{k}(y\omega_1 - x\omega_2)$.

$$\left.\begin{aligned} \text{So,} \quad v_1 &= z\omega_2 - y\omega_3 \\ v_2 &= x\omega_3 - z\omega_1 \\ \text{and} \quad v_3 &= y\omega_1 - x\omega_2 \end{aligned}\right\} \tag{10.27}$$

10.13.3 Angular Momentum

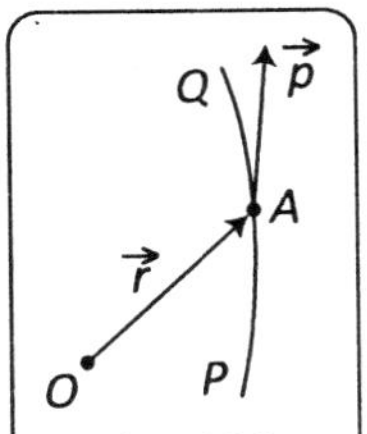

Fig. 10.26

Let a particle move along the curve PAQ and A be its position at any instant on it. If $\vec{r}$ be the position vector of A with respect to O, then the angular momentum of the particle about O is

$$\vec{\Omega} = \vec{r} \times \vec{p}, \tag{10.28}$$

where $\vec{p} = m\vec{v}$ is the linear momentum of the particle of mass m and moving with linear velocity $\vec{v}$.

10.13.4 Force on a Charged Particle in a Uniform Magnetic Field

If a charge q moves with velocity $\vec{v}$ in a uniform magnetic field $\vec{B}$, then the force experienced by the charge is

$$\vec{F} = q(\vec{v} \times \vec{B}). \tag{10.29}$$

$\vec{F}$ is generally called the *Lorentz force*.

10.14 Solved Problems (B)

Example 10.11. *Using vector method prove that the angle in a semicircle is a right angle.*

Solution : Let ACB be a semi-circle with centre at D and radius R.

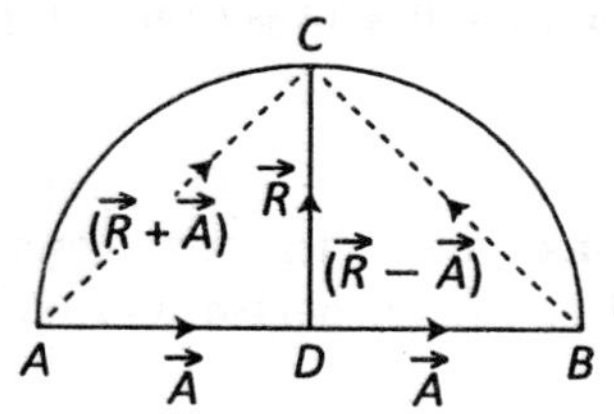

Fig. 10.27

From Fig. 10.27,

$$\overrightarrow{AC} = \overrightarrow{AD} + \overrightarrow{DC} = \vec{A} + \vec{R} = \vec{R} + \vec{A}$$

and $\overrightarrow{BC} = \overrightarrow{BD} + \overrightarrow{DC} = -\vec{A} + \vec{R} = \vec{R} - \vec{A}$.

$$\therefore \quad \overrightarrow{AC} \cdot \overrightarrow{BC} = (\vec{R} + \vec{A}) \cdot (\vec{R} - \vec{A})$$

$$= R^2 - A^2 = 0 \quad [\because \ |\vec{R}| = |\vec{A}|].$$

So, $\overrightarrow{AC}$ is perpendicular to BC and $\angle ACB = 90°$.

Example 10.12. *Prove that the diagonals of a rhombus are at right angles.*

Solution : Let $PQRS$ represent a rhombus whose diagonals PR and QS intersect at O. We have to prove that PR and SQ are at right angle (Fig. 10.28).

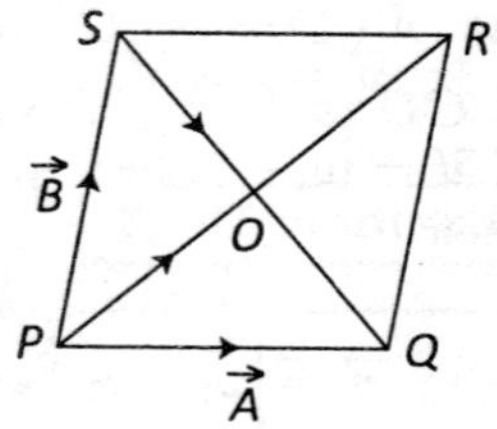

Fig. 10.28

Now, $\overrightarrow{PR} = \vec{A} + \vec{B} \quad [\because \ \overrightarrow{PS} = \overrightarrow{QR} = \vec{B}]$

$$\overrightarrow{SQ} = \vec{A} - \vec{B}.$$

$$\therefore \quad \overrightarrow{PR} \cdot \overrightarrow{SQ} = (\vec{A} + \vec{B}) \cdot (\vec{A} - \vec{B}) = A^2 - B^2 = 0 \quad [\because \ A = B].$$

$\therefore$ $\overrightarrow{PR}$ and $\overrightarrow{SQ}$ are at right angles to each other.

Example 10.13. *Prove that the perpendicular bisectors of the sides of a triangle are concurrent.*

Solution : ABC is a triangle and DO, EO, FO are the perpendicular bisectors of the sides AB, AC and BC, which meet at O (Fig. 10.29).

If the position vectors of the vertices be $\vec{A}$, $\vec{B}$, $\vec{C}$ with respect to an origin, then the position vectors of D, F and E are respectively: $\frac{\vec{A}+\vec{B}}{2}$, $\frac{\vec{B}+\vec{C}}{2}$ and $\frac{\vec{A}+\vec{C}}{2}$.

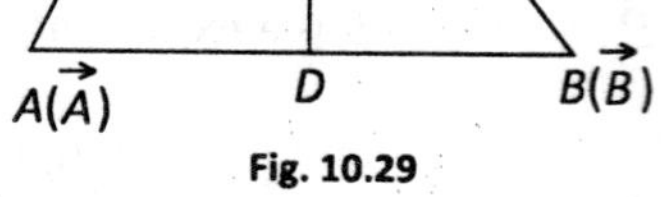

Fig. 10.29

But $\overrightarrow{OD}$ is $\perp$ $\overrightarrow{AB}$,

$$\therefore \quad \frac{(\vec{A} + \vec{B})}{2} \cdot (\vec{B} - \vec{A}) = 0$$

$$\text{or,} \quad B^2 = A^2. \tag{10.30}$$

Again, $\overrightarrow{OE} \perp \overrightarrow{AC}$,

$$\therefore \quad \frac{1}{2}(\vec{C} + \vec{A}) \cdot (\vec{C} - \vec{A}) = 0 \quad \text{or,} \quad |\vec{C}|^2 = |\vec{A}|^2$$

$$\text{or,} \quad C^2 = A^2. \tag{10.31}$$

Now, the position vector of F is $\dfrac{\vec{B} + \vec{C}}{2}$.

$$\overrightarrow{OF} \cdot \overrightarrow{BC} = \frac{1}{2}(\vec{B} + \vec{C}) \cdot (\vec{C} - \vec{B}) = 0 \quad \text{[from (10.30) and (10.31)]}$$

$$\text{or,} \quad C^2 = B^2. \tag{10.32}$$

Thus, the right bisector of BC passes through O, and hence the result.

Example 10.14. *Show that the perpendiculars drawn from the vertices to the opposite sides of a triangle are concurrent.*

Solution : In a triangle ABC, let the perpendiculars AD and BE meet at O. CO is joined and produced to meet AB at F.

Let $\overrightarrow{OA} = \vec{A}$, $\overrightarrow{OB} = \vec{B}$ and $\overrightarrow{OC} = \vec{C}$.

Now, $\overrightarrow{BC} = \vec{C} - \vec{B}$, $\overrightarrow{CA} = \vec{A} - \vec{C}$ and $\overrightarrow{AB} = \vec{B} - \vec{A}$.

$\because$ AO is $\perp$ to BC, $\vec{A}\cdot(\vec{C} - \vec{B}) = 0$ or, $\vec{A}\cdot\vec{C} = \vec{A}\cdot\vec{B}$.

BO is $\perp$ to CA, $\vec{B}\cdot(\vec{A} - \vec{C}) = 0$ or, $\vec{B}\cdot\vec{A} = \vec{B}\cdot\vec{C}$.

$\therefore$ $\vec{A}\cdot\vec{C} = \vec{B}\cdot\vec{C}$ or, $\vec{C}\cdot(\vec{B} - \vec{A}) = 0$.

$\therefore$ ***CO* is $\perp$ to *AB*. Hence the result.**

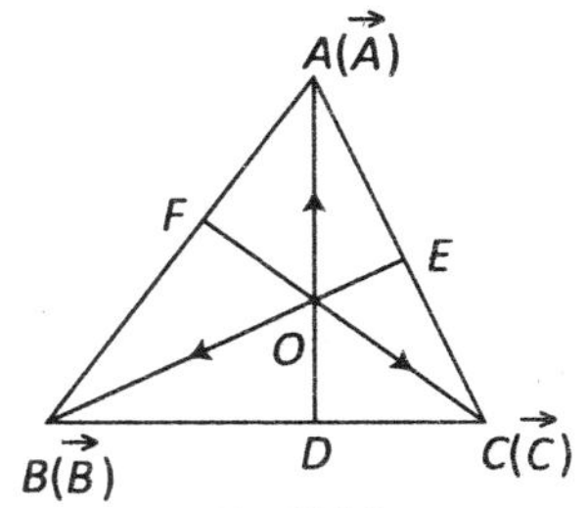

Fig. 10.30

Example 10.15. *In a triangle ABC, prove that, $a = b\cos C + c\cos B$.*

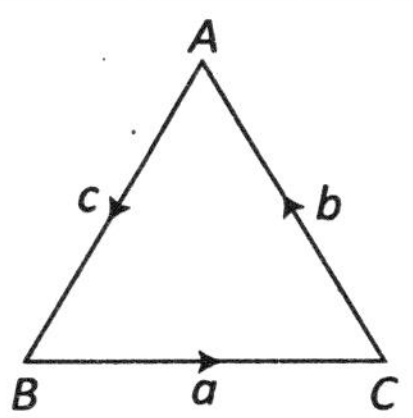

Fig. 10.31

Solution : Let ABC be a triangle.

Then, $\overrightarrow{AB} + \overrightarrow{BC} + \overrightarrow{CA} = 0$.

$\therefore$ $\overrightarrow{BC}\cdot\overrightarrow{AB} + \overrightarrow{BC}\cdot\overrightarrow{DC} + \overrightarrow{BC}\cdot\overrightarrow{CA} = 0$

or, $ac\cos(\pi - B) + a^2 + ab\cos(\pi - C) = 0$

or, $a - c\cos B - b\cos C = 0$

or, $\boldsymbol{a = b\cos C + c\cos B}$.

Example 10.16. *Prove that, in triangle ABC, $\cos A = \dfrac{b^2 + c^2 - a^2}{2bc}$.*

Solution : Let ABC be any triangle, Fig. 10.32.

Then, $\overrightarrow{BC} + \overrightarrow{CA} = \overrightarrow{BA}$ or, $\overrightarrow{BC} = \overrightarrow{BA} - \overrightarrow{CA}$.

Now, $\overrightarrow{BC}\cdot\overrightarrow{BC} = (\overrightarrow{BA} - \overrightarrow{CA})\cdot(\overrightarrow{BA} - \overrightarrow{CA})$

$= \overrightarrow{BA}\cdot\overrightarrow{BA} - 2\overrightarrow{BA}\cdot\overrightarrow{CA} + \overrightarrow{CA}\cdot\overrightarrow{CA}$

or, $a^2 = c^2 + b^2 - 2cb\cos A$ [$\because$ $BC = a$, $CA = b$ and $AB = c$]

or, $$\boldsymbol{\cos A = \frac{b^2 + c^2 - a^2}{2bc}} \quad \textbf{(Proved)}. \qquad (10.39)$$

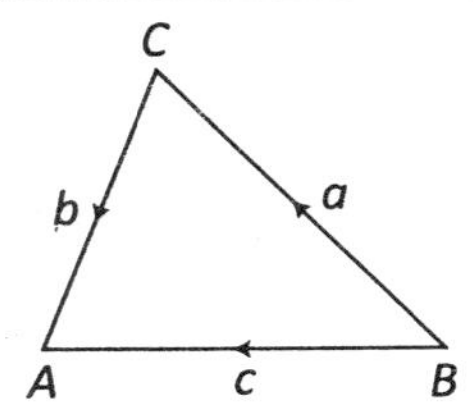

Fig. 10.32

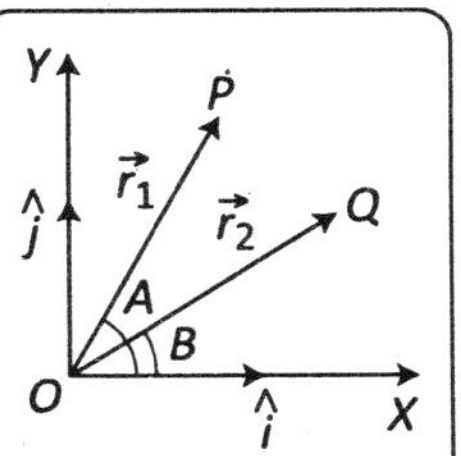

Fig. 10.33

Example 10.17. *Using dot product of two vectors, prove that,*

$$\cos(A - B) = \cos A\cos B + \sin A\sin B.$$

Solution : Let the position vectors of P and Q with respect to the origin O of a Cartesian coordinate system be $\vec{r_1}$ and $\vec{r_2}$ respectively. $\vec{r_1}$ and $\vec{r_2}$ make angles A and B with OX (Fig. 10.33).

Then, $\vec{r_1} = (r\cos A)\hat{\imath} + (r\sin A)\hat{\jmath}$ and $\vec{r_2} = (r\cos B)\hat{\imath} + (r\sin B)\hat{\jmath}$, where $\hat{\imath}$ and $\hat{\jmath}$ are unit vectors along OX and OY respectively (Fig. 10.33).

$$\therefore \quad \vec{r_1} \cdot \vec{r_2} = [(r\cos A)\hat{i} + (r\sin A)\hat{j}] \cdot [(r\cos B)\hat{i} + (r\sin B)\hat{j}]$$

$$= r^2(\cos A\cos B + \sin A\sin B). \qquad (10.34)$$

Again, $\vec{r_1} \cdot \vec{r_2} = |\vec{r_1}||\vec{r_2}|\cos(A-B) = r^2\cos(A-B). \qquad (10.35)$

From eqns. (10.34) and (10.35)

$$\mathbf{\cos(A - B) = \cos A\cos B + \sin A\sin B.}$$

Example 10.18. *Given n fixed point masses $m_1, m_2, \ldots, m_n$, whose position vectors with respect to an origin O are $\vec{r_1}, \vec{r_2}, \ldots, \vec{r}_n$ respectively. The position vector of the centroid of this system of masses with respect to O is* $\vec{r} = \dfrac{\sum_{k=1}^{n} m_k \vec{r}_k}{\sum_{k=1}^{n} m_k}$.

Show that the position of the centroid is independent of the choice of the origin.

[BU(H) 1999]

Solution : The position vectors of the point masses $m_1, m_2, \ldots, m_k, \ldots m_n$ are $\vec{r_1}, \vec{r_2}, \ldots, \vec{r}_k, \ldots, \vec{r}_n$ respectively, with respect to the origin O (Fig. 10.34).

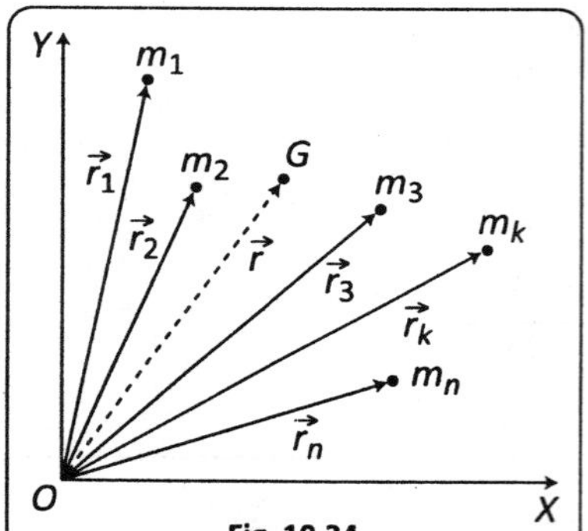

Fig. 10.34

Then the position vectors of the centroid (G) with respect to O,

$$\vec{r} = \overrightarrow{OG} = \frac{m_1\vec{r_1} + m_2\vec{r_2} + \cdots + m_n\vec{r}_n}{m_1 + m_2 + \cdots + m_n} = \frac{\sum_{k=1}^{n} m_k \vec{r}_k}{\sum_{k=1}^{n} m_k}.$$

If now O' be any other point in space such that $\overrightarrow{OO'} = \vec{n}$ and G' be the new centroid (if possible). Then the position vectors of $m_1, m_2, \ldots, m_n$ with respect to O' are :

$$(\vec{r_1} - \vec{n}), (\vec{r_2} - \vec{n}), (\vec{r}_3 - \vec{n}) \cdots (\vec{r}_n - \vec{n}).$$

Then the position vector of G' is

$$\overrightarrow{O'G'} = \frac{m_1(\vec{r_1} - \vec{n}) + m_2(\vec{r_2} - \vec{n}) + \cdots + m_n(\vec{r}_n - \vec{n})}{m_1 + m_2 + \cdots + m_n}$$

$$\text{or,} \quad \overrightarrow{O'G'} = \frac{m_1\vec{r_1} + m_2\vec{r_2} + \cdots + m_n\vec{r}_n}{m_1 + m_2 + \cdots + m_n} - \vec{n} = \overrightarrow{OG} - \overrightarrow{OO'} = \overrightarrow{O'G}.$$

So, G and G' are in coincident. Hence, G is unique and is independent of the coordinate system.

Example 10.19. *If $|\vec{A} + \vec{B}| = |\vec{A} - \vec{B}|$, then show that $\vec{A}$ and $\vec{B}$ are perpendicular.*

[BU(H) 2003]

Solution : We have $|\vec{A} + \vec{B}| = |\vec{A} - \vec{B}|$.

Squaring both sides,

$$|\vec{A} + \vec{B}|^2 = |\vec{A} - \vec{B}|^2 \quad \text{or,} \quad A^2 + B^2 + 2\vec{A} \cdot \vec{B} = A^2 + B^2 - 2\vec{A} \cdot \vec{B}$$

or, $4\vec{A}\cdot\vec{B}=0$ or, $\vec{A}\cdot\vec{B}=0$.

Hence, $\vec{A}$ and $\vec{B}$ are perpendicular to each other.

Example 10.20. *A particle under the action of two forces $\vec{F}_1=4\hat{i}+3\hat{j}$ and $\vec{F}_2=3\hat{i}+2\hat{j}$ is displaced from the point $(\hat{i}+3\hat{j})$ to the point $(4\hat{i}+8\hat{j})$. Calculate the work done by the force.* **[BU(H) 1992]**

Solution : The resultant of the forces $\vec{F}_1$ and $\vec{F}_2$ is

$$\vec{F}=\vec{F}_1+\vec{F}_2=(4\hat{i}+3\hat{j})+(3\hat{i}+2\hat{j})=7\hat{i}+5\hat{j}.$$

This displacement of the particle is

$$\vec{r}=(4\hat{i}+8\hat{j})-(\hat{i}+3\hat{j})=3\hat{i}+5\hat{j}.$$

$\therefore$ work done by the forces is

$$W=\vec{F}\cdot\vec{r}=(7\hat{i}+5\hat{j})\cdot(3\hat{i}+5\hat{j})=21+25=\mathbf{46\ units}.$$

Example 10.21. *Prove that, $\begin{vmatrix}\vec{A}\cdot\vec{A} & \vec{A}\cdot\vec{B}\\ \vec{A}\cdot\vec{B} & \vec{B}\cdot\vec{B}\end{vmatrix}=|\vec{A}\times|\vec{B}|^2$, where $\vec{A}$ and $\vec{B}$ are any two vectors.* **[BU(H) 1992]**

Solution : We have $\begin{vmatrix}\vec{A}\cdot\vec{A} & \vec{A}\cdot\vec{A}\\ \vec{A}\cdot\vec{A} & \vec{A}\cdot\vec{A}\end{vmatrix}=(\vec{A}\cdot\vec{A})(\vec{B}\cdot\vec{B})-(\vec{A}\cdot\vec{B})(\vec{A}\cdot\vec{B})$

$$=A^2B^2-(\vec{A}\cdot\vec{B})^2=A^2B^2-A^2B^2\cos^2\theta$$

$$=A^2B^2-A^2B^2+A^2B^2\sin^2\theta=(AB\sin\theta)^2$$

$$=|\vec{\boldsymbol{A}}\times\vec{\boldsymbol{B}}|^2.$$

Example 10.22. *Given two vectors, $\vec{A}=a_1\hat{i}+a_2\hat{j}+a_3\hat{k}$*

$$\vec{B}=b_1\hat{i}+b_2\hat{j}+b_3\hat{k}.$$

Find the condition that they are parallel. **[BU(H) 1998]**

Solution : If $\vec{A}$ and $\vec{B}$ are parallel, then

$\vec{A}=\lambda\vec{B}$, where λ is a constant.

or, $a_1\hat{i}+a_2\hat{j}+a_3\hat{k}=\lambda(b_1\hat{i}+b_2\hat{j}+b_3\hat{k})$ or, $(a_1-\lambda b_1)\hat{i}+(a_2-\lambda b_2)\hat{j}+(a_3-\lambda b_3)\hat{k}=0.$

$\because$ $\hat{i}$, $\hat{j}$ and $\hat{k}$ are independent of each other, the above condition will be satisfied, if

$$a_1-\lambda b_1=0,\quad a_2-\lambda b_2=0,\quad a_3-\lambda b_3=0\quad\text{or,}\quad \lambda=\frac{a_1}{b_1},\quad \lambda=\frac{a_2}{b_2},\quad \lambda=\frac{a_3}{b_3}.$$

So, the desired condition is $\boldsymbol{\dfrac{a_1}{b_1}=\dfrac{a_2}{b_2}=\dfrac{a_3}{b_3}}$.

Example 10.23. *A particle is acted upon by constant forces $\vec{P}=6\hat{i}-\hat{j}+3\hat{k}$ and $\vec{Q}=3\hat{i}-2\hat{j}+\hat{k}$. Find the work done in displacing the particle from the point $(1,-2,3)$ to the point $(5,4,-2)$.* **[BU(H) 1995]**

Solution : The resultant force, $\vec{R}=\vec{P}+\vec{Q}$

or, $\vec{R}=(6\hat{i}-\hat{j}+3\hat{k})+(3\hat{i}-2\hat{j}+\hat{k})=9\hat{i}-3\hat{j}+4\hat{k}.$

The displacement of the particle is

$$\vec{r} = (5-1)\hat{i} + (4+2)\hat{j} + (-2-3)\hat{k}) = 4\hat{i} + 6\hat{j} - 5\hat{k}.$$

Now the required work done is

$$W = \vec{R}\cdot\vec{r} = (9\hat{i} - 3\hat{j} + 4\hat{k}]\cdot(4\hat{i} + 6\hat{j} - 5\hat{k}) = 36 - 18 - 20 = \mathbf{-2\ units}.$$

∴ **the work done is 2 units.**

Example 10.24. *If $\vec{A}\cdot\vec{B} = \vec{A}\cdot\vec{C}$, does it necessarily follow that $\vec{B}$ and $\vec{C}$ are equal?* **[BU(H) 2001]**

Solution : If $\vec{A}\cdot\vec{B} = \vec{A}\cdot\vec{C}$, then

$$\vec{A}\cdot\vec{B} - \vec{A}\cdot\vec{C} = 0 \quad \text{or,} \quad \vec{A}\cdot(\vec{B}-\vec{C}) = 0.$$

i.e., either $(\vec{B}-\vec{C})$ is perpendicular to $\vec{A}$ or, $(\vec{C}-\vec{C}) = \vec{O}$, where $\vec{O}$ is a null vector. Hence, it does not necessarily follow that $\vec{B}$ and $\vec{C}$ are equal.

Example 10.25. *Two non-zero vectors $\vec{P}$ and $\vec{Q}$ are such that $|\vec{P}+\vec{Q}| = |\vec{P}-\vec{Q}|$. Find the angle between them.* **[BU(H) 2003]**

Solution : Since $|\vec{P}+\vec{Q}| = |\vec{P}-\vec{Q}|$,

∴ $|\vec{P}+\vec{Q}|^2 = |\vec{P}-\vec{Q}|^2$ or, $P^2 + Q^2 + 2\vec{P}\cdot\vec{Q} = P^2 + Q^2 - 2\vec{P}\cdot\vec{Q}$

or, $4\vec{P}\cdot\vec{Q} = 0$ or, $\vec{P}\cdot\vec{Q} = 0$.

∴ $\vec{P}$ **and** $\vec{Q}$ **are perpendicular to each other and the angle between them is 90°.**

Example 10.26. *When are two vectors $\vec{A}$ and $\vec{B}$ called linearly independent?* **[BU(H) 2005, AU 1995]**

Solution : The two vectors $\vec{A}$ and $\vec{B}$ are said to be linearly independent, if there are two scalars λ and μ such that,

$$\lambda\vec{A} + \mu\vec{B} \neq 0.$$

Example 10.27. *If $\vec{a}$ and $\vec{b}$ are two vectors, then show that,*

$$|\vec{a}\times\vec{b}| = \sqrt{(\vec{a}\cdot\vec{a})(\vec{b}\cdot\vec{b}) - (\vec{a}\cdot\vec{b})^2},$$

where the notations have usual meanings. **[BU(H) 2005]**

Solution : We have $|\vec{a}\times\vec{b}| = \sqrt{(\vec{a}\cdot\vec{a})(\vec{b}\cdot\vec{b}) - (\vec{a}\cdot\vec{b})^2}$.

$$\textbf{R.H.S.} = \sqrt{(\vec{a}\cdot\vec{a})(\vec{b}\cdot\vec{b}) - (\vec{a}\cdot\vec{b})^2}$$

$$= \sqrt{a^2b^2 - a^2b^2\cos^2\theta} \quad [\text{where } \theta \text{ is the angle between } \vec{a} \text{ and } \vec{b}]$$

$$= \sqrt{a^2b^2 - a^2b^2 + a^2b^2\sin\theta} = \sqrt{a^2b^2\sin^2\theta} = |\vec{a}\times\vec{b}| = \textbf{L.H.S.}$$

Example 10.28. *If $|\vec{A}| = |\vec{B}|$, prove that $(\vec{A}+\vec{B})$ is perpendicular to $(\vec{A}-\vec{B})$.* **[AU 1994]**

Solution : We have $|\vec{A}| = |\vec{B}|$.

∴ $|\vec{A}|^2 = |\vec{B}|^2$ or, $\vec{A}\cdot\vec{A} = \vec{B}\cdot\vec{B}$.

Now, $(\vec{A}+\vec{B})\cdot(\vec{A}-\vec{B}) = \vec{A}\cdot\vec{A} - \vec{A}\cdot\vec{B} + \vec{B}\cdot\vec{A} - \vec{B}\cdot\vec{B} = 0 \quad (\because\ \vec{A}\cdot\vec{A} = \vec{B}\cdot\vec{B})$

$\therefore$ $(\vec{A}+\vec{B})$ is perpendicular to $(\vec{A}-\vec{B})$.

Example 10.29. *Show that the vectors, $\vec{A} = 3\hat{i} - 2\hat{j} + \vec{k}$, $\vec{B} = \hat{i} - 3\hat{j} + 5\vec{k}$ and $\vec{C} = 2\hat{i} + \hat{j} - 4\vec{k}$ form a right-angled triangle.*

Solution : The vectors $\vec{A}$, $\vec{B}$ and $\vec{C}$ form a triangle if any one of the vector, is equal to the vector sum of the remaining two vectors.

Now, $\vec{B}+\vec{C} = \hat{i} - 3\hat{j} + 5\hat{k} + 2\hat{i} + \hat{j} - 4\hat{k} = 3\hat{i} - 2\hat{j} + \hat{k} = \vec{A}$.

Hence, the three vectors for a triangle.

Again, $\vec{A}\cdot\vec{B} = (3\hat{i} - 2\hat{j} + \hat{k})\cdot(\hat{i} - 3\hat{j} + 5\hat{k}) = 14$

$\vec{A}\cdot\vec{C} = (3\hat{i} - 2\hat{j} + \hat{k})\cdot(2\hat{i} + \hat{j} - 4\hat{k}) = 0$

$\vec{B}\cdot\vec{C} = (\hat{i} - 3\hat{j} + 5\hat{k})\cdot(2\hat{i} + \hat{j} + 4\hat{k}) = -21.$

As $\vec{A}\cdot\vec{C} = 0$, the vectors $\vec{A}$ and $\vec{C}$ are orthogonal and the triangle formed is a right-angled triangle.

Example 10.30. *A particle moves from a point $(3,-4,-2)m$ to a point $(-2,3,5)m$ under the influence of a force $\vec{F} = (-2\hat{i} + 3\hat{j} + 4\hat{k})N$.*

Calculate the work done by the force. **[AU 1993]**

Solution : Here force $\vec{F} = (-2\hat{i} + 3\hat{j} + 4\hat{k})N$

and displacement $\vec{r} = (-2-3)\hat{i} + (3+4)\hat{j} + (5+2)\hat{k} = (-5\hat{i} + 7\hat{j} + 7\hat{k})m.$

$\therefore$ work done, $\boldsymbol{W} = \vec{F}\cdot\vec{r} = (-2\hat{i} + 3\hat{j} + 4\hat{k})\cdot(-5\hat{i} + 7\hat{j} + 7\hat{k}) = (10 + 21 + 28) =$ **59 J**

Example 10.31. *Define direction cosine of a vector. If l_1, m_1, n_1 and l_2, m_2, n_2 be the direction cosines of two vectors, show that the angle θ between them is*

$$\cos\theta = l_1 l_2 + m_1 m_2 + n_1 n_2.$$ **[AU 1993, KU 1994]**

Solution : The cosines of the angles made by a vector $\vec{A}$ with X-, Y- and Z-axes are known as X-direction cosine, Y-direction cosine and Z-direction cosine respectively. These are denoted by l, m and n respectively. If α, β and γ be the angles made by $\vec{A}$ with X-, Y- and Z-axes respectively, then

$$l = \cos\alpha = \frac{A_x}{A}, \quad m = \cos\beta = \frac{A_y}{A}, \quad n = \cos\gamma = \frac{A_z}{A}.$$

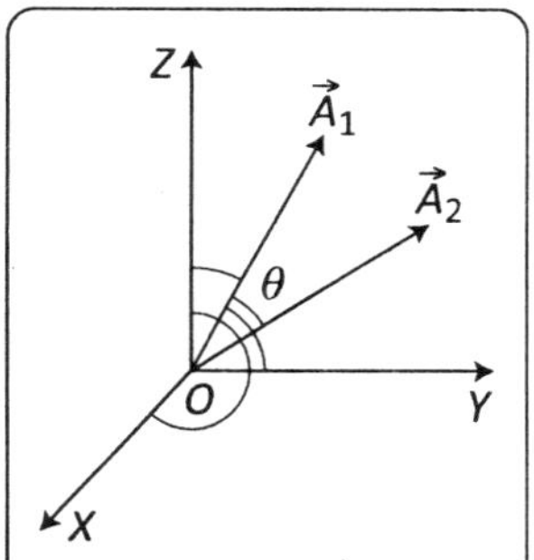

Fig. 10.35

Here, A_x, A_y and A_z are the X, Y and Z components of $\vec{A}$.

$$\therefore \quad \cos^2\alpha + \cos^2\beta + \cos^2\gamma = \left(\frac{A_x}{A}\right)^2 + \left(\frac{A_y}{A}\right)^2 + \left(\frac{A_z}{A}\right)^2 = \frac{A^2}{A^2} = 1$$

or, $l^2 + m^2 + n^2 = 1.$

If there be two vectors $\vec{A}_1$ and $\vec{A}_2$ such that

$$\vec{A}_1 = A_{1x}\hat{i} + A_{1y}\hat{j} + A_{1z}\hat{k}$$

or, $\frac{\vec{A}_1}{A_1}\left(\frac{A_{1x}}{A_1}\right)\hat{i}+\left(\frac{A_{1y}}{A_1}\right)\hat{j}+\left(\frac{A_{1z}}{A_1}\right)\hat{k}=l_1\hat{i}+m_1\hat{j}+n_1\hat{k}.$

$\therefore\quad \vec{A}_1=A_1(l_1\hat{i}+m_1\hat{j}+n_1\hat{k},$ where $\frac{A_{1x}}{A_1}=l_1,\ \frac{A_{1y}}{A_1}=m_1$ and $\frac{A_{1z}}{A_1}=n_1.$

Similarly, $\vec{A}_2=A_2(l_2\hat{i}+m_2\hat{j}+n_2\hat{k}).$

If θ be the angle between $\vec{A}_1$ and $\vec{A}_2$ [Fig. (10.35)], then

$$\cos\theta=\frac{\vec{A}_1\cdot\vec{A}_2}{\vec{A}_1\vec{A}_2}=\frac{A_1(l_1\hat{i}+m_1\hat{j}+n_1\hat{k})\cdot A_2(l_2\hat{i}+m_2\hat{j}+n_2\hat{k})}{A_1A_2}$$

$$=\boldsymbol{l_1l_2+m_1m_2+n_1n_2}\quad \textbf{(Proved)}.$$

Example 10.32. *Find the projection of* $\vec{A}=\hat{i}-2\hat{j}+\hat{k}$ *on* $\vec{B}=4\hat{i}-4\hat{j}+7\hat{k}.$

Solution : Unit vector in the direction of $\vec{B}$ is

$$\hat{n}=\frac{\vec{B}}{B}=\frac{4\hat{i}-4\hat{j}+7\hat{k}}{\sqrt{4^2+(-4)^2+7^2}}=\frac{4}{9}\hat{i}-\frac{4}{9}\hat{j}+\frac{7}{9}\hat{k}.$$

Now, the projection of $\vec{A}$ on $\vec{B}$ is

$$\vec{\boldsymbol{A}}\cdot\hat{\boldsymbol{n}}=(\hat{i}-2\hat{j}+\hat{k})\left(\frac{4}{9}\hat{i}-\frac{4}{9}\hat{j}+\frac{7}{9}\hat{k}\right)=(1)\frac{4}{9}+(-2)\left(-\frac{4}{9}\right)+(1)\frac{7}{9}=\boldsymbol{\frac{19}{9}}.$$

Example 10.33. *Determine the unit vector perpendicular to the plane of* $\vec{A}=2\hat{i}-6\hat{j}-3\hat{k}$ *and* $\vec{B}=4\hat{i}+3\hat{j}-\hat{k}.$

Solution : Let the vector $\vec{N}=n_1\hat{i}+n_2\hat{j}+n_3\hat{k}$ be perpendicular to the plane of $\vec{A}$ and $\vec{B}$. Then $\hat{n}$ is perpendicular to $\vec{A}$ and $\vec{B}$.

$\therefore\quad \vec{N}\cdot\vec{A}=(n_1\hat{i}+n_2\hat{j}+n_3\hat{k})\cdot(2\hat{i}-6\hat{j}-3\hat{k})=2n_1-6n_2-3n_3=0$

or, $\quad 2n_1-6n_2=3n_3.$ (i)

Similarly, $\vec{N}\cdot\vec{B}=4n_1+3n_3=0$

or, $\quad 4n_1+3n_2=n_3.$ (ii)

Solving (i) and (ii), we get $n_1=\frac{n_3}{2},\ n_2=-\frac{n_3}{2}.$

$\therefore\quad \vec{N}=\left(\frac{1}{2}\hat{i}-\frac{1}{3}\hat{j}+\hat{k}\right)n_3.$

$\therefore$ the unit vector along $\vec{\boldsymbol{N}}=\frac{\vec{N}}{N}=\frac{n_3\left(\frac{1}{2}\hat{i}-\frac{1}{3}\hat{j}+\hat{k}\right)}{\sqrt{n_3^2\left[\left(\frac{1}{2}\right)^2+\left(-\frac{1}{3}\right)^2+1^2\right]}}=\pm\left(\boldsymbol{\frac{3}{7}\hat{i}-\frac{2}{7}\hat{j}+\frac{6}{7}\hat{k}}\right).$

Example 10.34. (a) *Find an equation for the plane perpendicular to the vector* $\vec{A}=2\hat{i}+3\hat{j}+6\hat{k}$ *and passing through the terminal point of the vector* $\vec{B}=\hat{i}+5\hat{j}+3\hat{k}.$

(b) *Find the distance of the plane from the origin.*

Solution : (a) In Fig. (10.36), $\vec{r}$ is the position vector of the point P. The terminal point of $\vec{B}$ is Q.

Since, $\overrightarrow{PQ} = \overrightarrow{B} - \overrightarrow{r}$ is perpendicular to $\overrightarrow{A}$,

$$(\overrightarrow{B} - \overrightarrow{r}) \cdot \overrightarrow{A} = 0 \quad \text{or,} \quad \overrightarrow{r} \cdot \overrightarrow{A} = \overrightarrow{B} \cdot \overrightarrow{A}$$

is the equation of the plane in vector form

or, $(\hat{i}x + \hat{j}y + \hat{k}z) \cdot (2\hat{i} + 3\hat{j} + 6\hat{k}) = (\hat{i} + 5\hat{j} + 3\hat{k}) \cdot (2\hat{i}x + 3\hat{j} + 6\hat{k})$

or, $\mathbf{2x + 3y + 6z = 2 + 15 + 18 = 35}$

is the required equation of the plane.

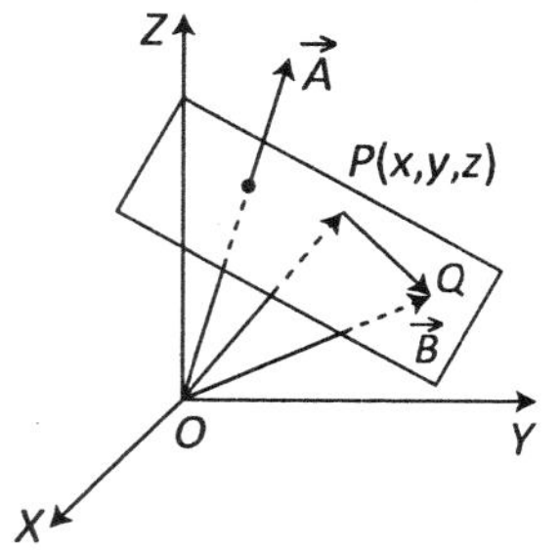

Fig. 10.36

(b) The distance of the plane from the origin is the projection of $\overrightarrow{B}$ on $\overrightarrow{A}$. The unit vector in the direction of $\overrightarrow{A}$ is, $\hat{n} = \frac{2\hat{i}+3\hat{j}+6\hat{k}}{\sqrt{2^2+3^2+6^2}}$

$$\hat{n} = \frac{2}{7}\hat{i} + \frac{3}{7}\hat{j} + \frac{6}{7}\hat{k}.$$

The projection of $\overrightarrow{B}$ on $\overrightarrow{A}$ is,

$$\overrightarrow{\boldsymbol{B}} \cdot \hat{\boldsymbol{n}} = (\hat{i} + 5\hat{j} + 3\hat{k}) \cdot \left(\frac{2}{7}\hat{i} + \frac{3}{7}\hat{j} + \frac{6}{7}\hat{k}\right) = \frac{2}{7} + \frac{15}{7} + \frac{18}{7} = \mathbf{5\ units}.$$

Example 10.35. *Prove that in a triangle ABC,* $\dfrac{\sin A}{a} = \dfrac{\sin B}{b} = \dfrac{\sin C}{c}$.

Solution : Let in the $\triangle ABC$, $\overrightarrow{BC} = \overrightarrow{a}$, $\overrightarrow{CA} = \overrightarrow{b}$ and $\overrightarrow{AB} = \overrightarrow{c}$.

Now, for the triangle, $\overrightarrow{a} + \overrightarrow{b} + \overrightarrow{c} = 0$.

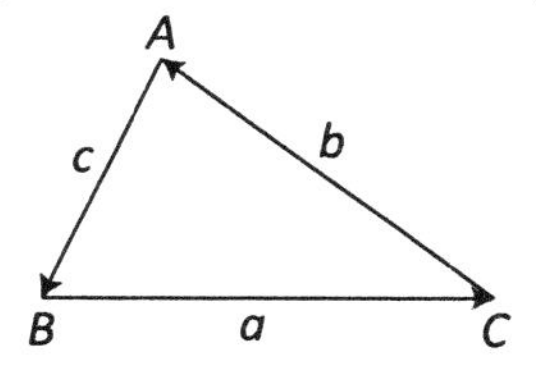

Fig. 10.37

$\therefore \quad \overrightarrow{a} \times \overrightarrow{a} + \overrightarrow{a} \times \overrightarrow{b} + \overrightarrow{a} \times \overrightarrow{c} = 0$

or, $\quad \overrightarrow{a} \times \overrightarrow{b} + \overrightarrow{a} \times \overrightarrow{c} = 0 \quad (\because\ \overrightarrow{a} \times \overrightarrow{a} = 0)$

or, $\quad \overrightarrow{a} \times \overrightarrow{b} = \overrightarrow{c} \times \overrightarrow{a}$ or, $ab\sin(\pi - C)\hat{n} = ca\sin(\pi - B)\hat{n}$,

where $\hat{n}$ is a unit vector normal to the plane of the triangle

or, $\quad b\sin C = c\sin B$ or, $\dfrac{\sin B}{b} = \dfrac{\sin C}{c}$. (i)

Again, $\overrightarrow{b} \times \overrightarrow{a} + \overrightarrow{b} \times \overrightarrow{b} + \overrightarrow{b} \times \overrightarrow{c} = 0$

or, $\quad \overrightarrow{b} \times \overrightarrow{a} = \overrightarrow{c} \times \overrightarrow{b} \quad (\because\ \overrightarrow{b} \times \overrightarrow{b} = 0)$

or, $\quad ba\sin(\pi - C)\hat{n} = cb\sin(\pi - A)\hat{n}$.

or, $\quad a\sin C = c\sin A$ or, $\dfrac{\sin A}{a} = \dfrac{\sin C}{c}$. (ii)

From (i) and (ii),

$$\frac{\sin \boldsymbol{A}}{\boldsymbol{a}} = \frac{\sin \boldsymbol{B}}{\boldsymbol{b}} = \frac{\sin \boldsymbol{C}}{\boldsymbol{c}} \quad \textbf{(Proved)}.$$

Example 10.36. *Prove that,* $\sin(A - B) = \sin A\cos B - \cos A\sin B$.

Solution : Let $\overrightarrow{OP} = \overrightarrow{r_1}$ and $\overrightarrow{OQ} = \overrightarrow{r_2}$ be two vectors making angles A and B respectively with the X-axis. PM and PN are dropped perpendiculars from P on OX and OY respectively. Then from Fig. 10.38,

$$\overrightarrow{OM} = \overrightarrow{r_1}\cos A \quad \text{and} \quad \overrightarrow{ON} = \overrightarrow{r_1}\sin A.$$

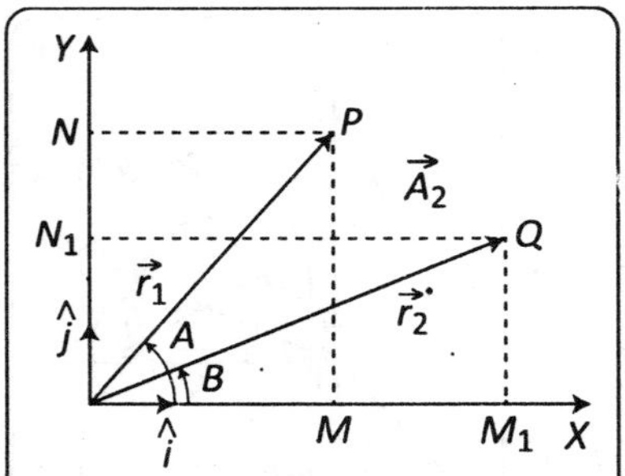

Fig. 10.38

$$\therefore \quad \vec{r_1} = \overrightarrow{OM} + \overrightarrow{ON} = \vec{r_1}\cos A + \vec{r_1}\sin A$$
$$= (r_1 \cos A)\hat{\imath} + (r_1 \sin A)\hat{\jmath},$$

where $\hat{\imath}$ and $\hat{\jmath}$ are unit vectors along OX and OY respectively.

Similarly, if M_1 and N_1 be the feet of the perpendicular dropped form Q on OX and OY respectively, then

$$\vec{r_2} = \vec{r_2}\cos B + \vec{r_2}\sin B = (r_2 \cos B)\hat{\imath} + (r_2 \sin B)\hat{\jmath}.$$

$$\therefore \quad \vec{r_1} \times \vec{r_2} = [(r_1 \cos A)\hat{\imath} + (r_1 \sin A)\hat{\jmath}]$$
$$\times [(r_1 \cos B)\hat{\imath} + (r_1 \sin B)\hat{\jmath}]$$

or, $\quad \vec{r_1} \times \vec{r_2} = r_1 r_2 \cos A \sin B(\hat{k}) + r_1 r_2 \sin A \cos B(\hat{k}) = r_1 r_2[\cos A \sin B - \sin A \cos B]\hat{k} \quad$ (i)

Again, $\vec{r_1} \times \vec{r_2} = r_1 r_2(-\hat{k})\sin(A - B) = r_1 r_2 \sin(A - B)\hat{k} \quad$ (ii)

From eqns. (i) and (ii), we get $\mathbf{\sin(A - B) = \sin A \cos B - \cos A \sin B}$.

Example 10.37. *If three co-planar forces acting at at point be in equilibrium, then prove that each is proportional to the sine of the angle between the other two.*

Solution : Let three forces $\vec{P}$, $\vec{Q}$ and $\vec{R}$ acting at O are in equilibrium.

Let $\hat{n}_1$, $\hat{n}_2$ and $\hat{n}_3$ be the unit vectors along $\vec{P}$, $\vec{Q}$ and $\vec{R}$ respectively.

Since the forces are in equilibrium,

$$P\hat{n}_1 + Q\hat{n}_2 + R\hat{n}_3 = 0, \quad \text{(i)}$$

where $|\vec{P}| = P$, $|\vec{Q}| = Q$ and $|\vec{R}| = R$.

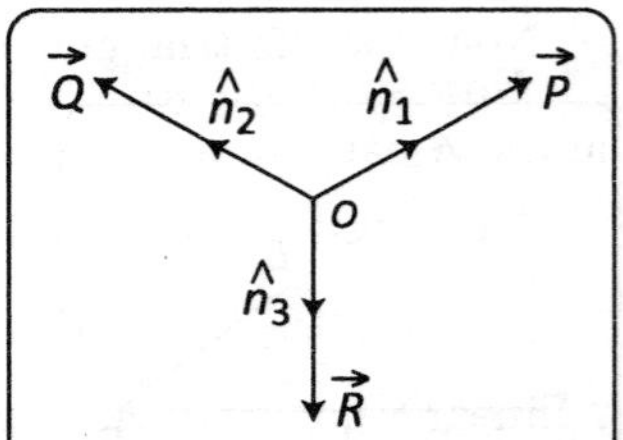

Fig. 10.39

Taking cross-product by $\hat{n}_1$ on both sides of (i),

$$P(\hat{n}_1 \times \hat{n}_1) + Q(\hat{n}_2 \times \hat{n}_1) + R(\hat{n}_3 \times \hat{n}_1) = 0$$

or, $\quad Q(\hat{n}_2 \times \hat{n}_1) + R(\hat{n}_3 \times \hat{n}_1) = 0 \quad$ or, $\quad Q(\hat{n}_2 \times \hat{n}_1) = R(\hat{n}_1 \times \hat{n}_3)$

$$\text{or,} \quad \frac{Q}{|\hat{n}_1 \times \hat{n}_3|} = \frac{R}{|\hat{n}_2 \times \hat{n}_1|}. \quad \text{(ii)}$$

$$\text{Similarly, we get } \frac{P}{|\hat{n}_2 \times \hat{n}_3|} = \frac{Q}{|\hat{n}_3 \times \hat{n}_1|}. \quad \text{(iii)}$$

From (ii) and (iii), we get

$$\frac{P}{|\hat{n}_2 \times \hat{n}_3|} = \frac{Q}{|\hat{n}_3 \times \hat{n}_1|} = \frac{R}{|\hat{n}_1 \times \hat{n}_2|}.$$

$$\because \quad |\hat{n}_1| = |\hat{n}_2| = |\hat{n}_3| = 1.$$

$$\therefore \quad \frac{\mathbf{P}}{\sin(\hat{n}_2, \hat{n}_3)} = \frac{\mathbf{Q}}{\sin(\hat{n}_3, \hat{n}_1)} = \frac{\mathbf{R}}{\sin(\hat{n}_1, \hat{n}_2)}$$

This result is known as **Lami's theorem** in statics.

Example 10.38. *Find the unit vector perpendicular to the plane of vectors,* $\vec{A} = 3\hat{\imath} - 2\hat{\jmath} + 4\hat{k}$ *and* $\vec{B} = \hat{\imath} + \hat{\jmath} - 2\hat{k}$.

Solution : $\vec{A} \times \vec{B}$ is a vector perpendicular to the plane of $\vec{A}$ and $\vec{B}$.

Now, $\vec{A} \times \vec{B} = \begin{vmatrix} \hat{\imath} & \hat{\jmath} & \hat{k} \\ 3 & -2 & 4 \\ 1 & 1 & -2 \end{vmatrix} = \hat{\imath}(4-4) + \hat{\jmath}(-6-4) + \hat{k}(3+2) = -10\hat{\jmath} + 5\hat{k}.$

The unit vector parallel to $\vec{A} \times \vec{B}$ is $\hat{\boldsymbol{n}} = \dfrac{\vec{A} \times \vec{B}}{|\vec{A} \times \vec{B}|} = \pm\dfrac{-10\hat{\jmath} + 5\hat{k}}{\sqrt{100+25}} = \pm\left(\dfrac{\mathbf{2}}{\sqrt{\mathbf{5}}}\hat{\boldsymbol{\imath}} + \dfrac{\mathbf{1}}{\sqrt{\mathbf{5}}}\hat{\boldsymbol{k}}\right).$

The $\pm$ sign indicates that $\hat{n}$ has two values, viz., $\hat{n_1} = +\left(\frac{2}{\sqrt{5}}\hat{\imath} + \frac{1}{\sqrt{5}}\hat{k}\right)$ along the upward direction and $\hat{n_2} = -\left(\frac{2}{\sqrt{5}}\hat{\imath} + \frac{1}{\sqrt{5}}\hat{k}\right)$ along the downward direction to the plane of $\vec{A}$ and $\vec{B}$.

Example 10.39. *A rigid body is rotating with an angular velocity 5 radians per second about an axis parallel to $\hat{\imath} - 2\hat{\jmath} + 2\hat{k}$ passing through the point $2\hat{\imath} - 3\hat{\jmath} + \hat{k}$. Find the velocity of the particle at the point $4\hat{\imath} - 2\hat{\jmath} - \hat{k}$.*

Solution : The axis of rotation is parallel to $\hat{\imath} - 2\hat{\jmath} + 2\hat{k}$.

The unit vector along the axis of rotation is $\hat{n} = \dfrac{\hat{\imath} - 2\hat{\jmath} + 2\hat{k}}{\sqrt{1+4+4}} = \dfrac{1}{3}(\hat{\imath} - 2\hat{\jmath} + 2\hat{k}).$

So, the angular velocity is $\vec{\omega} = 5\hat{n} = \dfrac{5}{3}(\hat{\imath} - 2\hat{\jmath} + 2\hat{k}).$

Now since the axis of rotation passes through $2\hat{\imath} - 3\hat{\jmath} + \hat{k}$ and the particle is at $4\hat{\imath} - 2\hat{\jmath} - \hat{k}$.

The position vector of P with respect to Q (Fig. 10.40) is

$\vec{r} = (4\hat{\imath} - 2\hat{\jmath} - \hat{k}) - (2\hat{\imath} - 3\hat{\jmath} + \hat{k}) = 2\hat{\imath} + \hat{\jmath} - 2\hat{k}.$

$\therefore$ linear velocity of the particle at P is

$$\vec{v} = \vec{\omega} \times \vec{r} = \frac{5}{3}\begin{vmatrix} \hat{\imath} & \hat{\jmath} & \hat{k} \\ 1 & -2 & 2 \\ 2 & 1 & -2 \end{vmatrix} = \frac{5}{3}(2\hat{\imath} + 6\hat{\jmath} + 5\hat{k}).$$

$$\therefore \quad |\vec{v}| = \frac{5}{3}\sqrt{2^2 + 6^2 + 5^2} = \frac{\mathbf{5}}{\mathbf{3}}\sqrt{\mathbf{65}} \textbf{ units.}$$

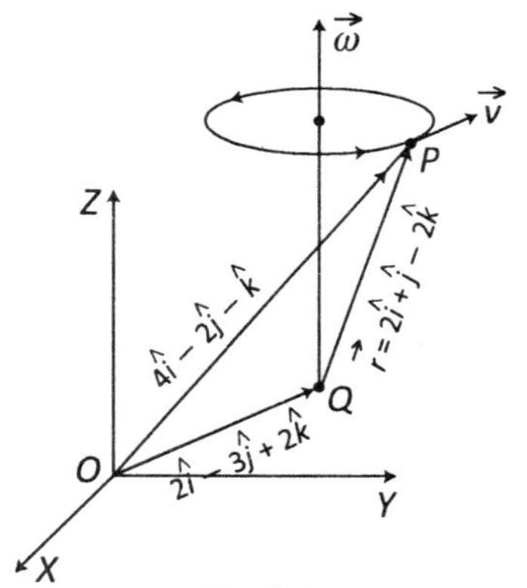

Fig. 10.40

Example 10.40. *A rigid body is spinning with an angular velocity 5 units about an axis parallel to $4\hat{\imath} - 3\hat{\jmath}$ and passing through the point $\hat{\imath} + 2\hat{\jmath} - 3\hat{k}$. Find the velocity of the particle at the point $3\hat{\imath} - 2\hat{\jmath} + \hat{k}$.* **[BU(H) 1998]**

Solution : The unit vector along the axis of rotation is : $\hat{n} = \dfrac{4\hat{\imath} - 3\hat{\jmath}}{\sqrt{4^2 + 3^2}} = \pm\dfrac{1}{5}(4\hat{\imath} - 3\hat{\jmath}).$

$\therefore$ the angular velocity $\vec{\omega} = 5\hat{n} = \dfrac{5}{5}(4\hat{\imath} - 3\hat{\jmath})$ or, $\vec{\omega} = 4\hat{\imath} - 3\hat{\jmath}.$

Now the position vector of the particle at $3\hat{\imath} - 2\hat{\jmath} + \hat{k}$ (say P) with respect to the point $\hat{\imath} + 2\hat{\jmath} - 3\hat{k}$ (say Q) is

$\vec{r} = (3\hat{\imath} - 2\hat{\jmath} + \hat{k}) - (\hat{\imath} + 2\hat{\jmath} - 3\hat{k}) = 2\hat{\imath} - 4\hat{\jmath} + 4\hat{k}.$

$\therefore$ linear velocity of the particle at P is

$$\vec{v} = \vec{\omega} \times \vec{r} = (4\hat{i} - 3\hat{j}) \times (2\hat{i} - 4\hat{j} + 4\hat{k}) = 2[(4\hat{i} - 3\hat{j} + 0\hat{k}) \times (\hat{i} - 2\hat{j} + 2\hat{k})]$$

$$= 2\begin{vmatrix} \hat{i} & \hat{j} & \hat{k} \\ 4 & -3 & 0 \\ 1 & -2 & 2 \end{vmatrix} = 2[\hat{i}(-6+0) + \hat{j}(0-8) + \hat{k}(-8+3)] = 2(-6\hat{i} - 8\hat{j} - 5\hat{k}).$$

$$\therefore \quad |\vec{r}| = 2\sqrt{(-6)^2 + (-8)^2 + (-5)^2} = 2 \times \sqrt{125} = \mathbf{10\sqrt{5}\ units}.$$

Example 10.41. *Express $\vec{r}$ in terms of $\vec{b}$ and $\vec{c}$ from the following equations :*

$$\vec{r} \times \vec{b} = \vec{c} \times \vec{b} \text{ and } \vec{r} \cdot \vec{a} = 0, \ (\vec{a} \cdot \vec{b}) \neq 0.$$ **[BU(H) 1996]**

Solution : We have $\vec{r} \times \vec{b} = \vec{c} \times \vec{b}$.

$$\therefore \quad \vec{a} \times (\vec{r} \times \vec{b}) = \vec{a} \times (\vec{c} \times \vec{b})$$

or, $(\vec{r} \cdot \vec{a}) \cdot \vec{b} - (\vec{b} \cdot \vec{a}) \cdot \vec{r} = -[\vec{a} \times (\vec{b} \times \vec{c})]$ or, $-(\vec{b} \cdot \vec{a})\vec{r} = -[\vec{a} \times (\vec{b} \times \vec{c})]$

or, $$\vec{r} = \frac{\vec{a} \times (\vec{b} \times \vec{c})}{\vec{a} \cdot \vec{b}} \quad [\because \vec{r} \cdot \vec{a} = 0 \text{ and } \vec{a} \cdot \vec{b} \neq 0].$$

Example 10.42. *The magnetic force on a particle of charge e moving in a static magnetic field $\vec{B}$ with a velocity $\vec{V}$ is given by $\vec{F} = e\vec{V} \times \vec{B}$. Show that no work is done by this force.* **[BU(H) 1999]**

Solution : The force on the charged particle is, $\vec{F} = e\vec{V} \times \vec{B}$.

The rate of work done by the force on the particle is

$$\vec{F} \cdot \vec{V} = e(\vec{V} \times \vec{B}) \cdot \vec{V} = 0.$$

Hence the result.

Example 10.43. *Find a unit vector parallel to the xy-plane and perpendicular to the vector* $4\hat{i} - 3\hat{j} + \hat{k}$.

Solution : Let $\vec{r} = \hat{i}x + \hat{j}y$ be a vector in the xy plane. Then unit vector along $\vec{r}$ is

$$\hat{n} = \pm \frac{\hat{i}x + \hat{j}y}{\sqrt{x^2 + y^2}}.$$

$\because \hat{n}$ is perpendicular to $4\hat{i} - 3\hat{j} + \hat{k}$, we have

$$\hat{n} \cdot (4\hat{i} - 3\hat{j} + \hat{k}) = 0 \quad \text{or,} \quad \frac{\hat{i}x + \hat{j}y}{\sqrt{x^2 + y^2}} \cdot (4\hat{i} - 3\hat{j} + \hat{k}) = 0$$

or, $4x - 3y = 0$ or, $\dfrac{x}{y} = \dfrac{3}{4}$.

Taking $x = 3$ and $y = 4$, we have

$$\hat{n} = \pm \frac{3\hat{i} + 4\hat{j}}{\sqrt{3^2 + 4^2}} = \pm \left(\frac{3\hat{i} + 4\hat{j}}{5} \right).$$

Example 10.44. (a) *$\vec{A}$, $\vec{B}$, $\vec{C}$ are three vectors. If $\vec{A}+\vec{B}+\vec{C}=0$ show that,*

$$\vec{A}\times\vec{B}=\vec{B}\times\vec{C}=\vec{C}\times\vec{A}.$$

(b) *Is the converse true, that is, if $\vec{A}\times\vec{B}=\vec{B}\times\vec{C}=\vec{C}\times\vec{A}$, is $\vec{A}+\vec{B}+\vec{C}=0$? Justify your answer.* **[BU(H) 1999]**

Solution : (a) If $\vec{A}+\vec{B}+\vec{C}=0$, then $\vec{A}+\vec{B}=-\vec{C}$

$$\vec{B}\times(\vec{A}+\vec{B})=-\vec{B}\times\vec{C} \quad \text{or,} \quad \vec{B}\times\vec{A}+\vec{B}\times\vec{B}=-\vec{B}\times\vec{C}$$

or, $-\vec{A}\times\vec{B}=-\vec{B}\times\vec{C}$ or, $\vec{A}\times\vec{B}=\vec{B}\times\vec{C}$.

Similarly, $\vec{B}\times\vec{C}=\vec{C}\times\vec{B}$.

$\therefore \quad \vec{A}\times\vec{B}=\vec{B}\times\vec{C}=\vec{C}\times\vec{B}$.

(b) If $\vec{A}\times\vec{B}=\vec{B}\times\vec{C}=\vec{C}\times\vec{A}$, then we are to prove that $\vec{A}+\vec{B}+\vec{C}=0$.

$\because \quad \vec{A}\times\vec{B}=\vec{B}\times\vec{C}$

$\therefore \quad -\vec{A}\times\vec{B}=-\vec{B}\times\vec{C}$ or, $\vec{B}\times\vec{A}=\vec{C}\times\vec{B}$ or, $\vec{B}\times\vec{A}+\vec{B}\times\vec{B}=-\vec{B}\times\vec{C}$

or, $\vec{B}\times\vec{A}+\vec{B}\times\vec{B}+\vec{B}\times\vec{C}=0$ or, $\vec{B}\times(\vec{A}+\vec{B}+\vec{C})=0$.

$\therefore \quad \boldsymbol{\vec{A}+\vec{B}+\vec{C}=0}$ $(\because\ \vec{B}\neq 0)$.

Example 10.45. *Find the unit vector perpendicular to both the vectors $\vec{A}=2\hat{\imath}+\hat{\jmath}+3\hat{k}$ and $\vec{B}=\hat{\imath}-2\hat{\jmath}-\hat{k}$. Also find the angle between $\vec{A}$ and $\vec{B}$.* **[BU(H) 1994]**

Solution : $\vec{A}\times\vec{B}=\begin{vmatrix}\hat{\imath} & \hat{\jmath} & \hat{k}\\ 2 & 1 & 3\\ 1 & -2 & -1\end{vmatrix}=\hat{\imath}(-1+6)+\hat{\jmath}(3+2)+\hat{k}(-4-1)=5\hat{\imath}+5\hat{\jmath}-5\hat{k}$

and $|\vec{A}\times\vec{B}|=\sqrt{5^2+5^2+(-5)^2}=\sqrt{75}=5\sqrt{3}$.

$\therefore$ unit vector perpendicular to $\vec{A}$ and $\vec{B}$ is

$$\hat{n}=\frac{\vec{A}\times\vec{B}}{|\vec{A}\times\vec{B}|}=\frac{5(\hat{\imath}+\hat{\jmath}-\hat{k})}{5\sqrt{5}}=\frac{1}{\sqrt{5}}(\hat{\imath}+\hat{\jmath}-\hat{k}).$$

Also, $|\vec{A}|=\sqrt{2^2+1^2+3^2}=\sqrt{14}$ and $|\vec{B}|=\sqrt{1^2+(-2)^2+(-1)^2}=\sqrt{6}$.

If θ be the angle between $\vec{A}$ and $\vec{B}$, then $|\vec{A}\times\vec{B}|=|\vec{A}||\vec{B}|\sin\theta$.

$$\therefore \quad \sin\theta=\frac{|\vec{A}\times\vec{B}|}{|\vec{A}||\vec{B}|}=\frac{5\sqrt{3}}{\sqrt{14}\sqrt{6}}=\frac{5\sqrt{3}}{\sqrt{84}}. \qquad \therefore \quad \boldsymbol{\theta=\sin^{-1}\left(\frac{5\sqrt{3}}{\sqrt{84}}\right)}.$$

Example 10.46. *Find the torque about the point $O(3,-1,3)$ of a force $\vec{F}(4,2,1)$ passing through the point $A(5,2,4)$.*

Solution : If $\hat{\imath}$, $\hat{\jmath}$, $\hat{k}$ be the unit vectors along the three mutually perpendicular directions x, y, z then the position vector of $\vec{A}$ relative to O is

$$\vec{r}=(5\hat{\imath}+2\hat{\jmath}+4\hat{k})-(3\hat{\imath}-\hat{\jmath}+3\hat{k})=2\hat{\imath}+3\hat{\jmath}+\hat{k}$$

Again, force $\vec{F} = 4\hat{i} + 2\hat{j} + \hat{k}$.

Hence, the required moment is

$$\vec{\wedge} = \vec{r} \times \vec{F} = \begin{vmatrix} \hat{i} & \hat{j} & \hat{k} \\ 2 & 3 & 1 \\ 4 & 2 & 1 \end{vmatrix} = \hat{i} + 2\hat{j} - 8\hat{k}.$$

Example 10.47. *Show that the components of a vector $\vec{G}$ along and perpendicular to a vector $\vec{H}$ in the plane of $\vec{G}$ and $\vec{H}$ are respectively given by $\dfrac{(\vec{H}\cdot\vec{G})\vec{H}}{|\vec{H}|^2}$ and $\dfrac{(\vec{H}\times\vec{G})\times\vec{H}}{|\vec{H}|^2}$.*

[Rohilkkand U 1984, Agra U 1984, Gulbarga 1984, cf. BU(H) 1999]

Solution : Let $\overrightarrow{OA} = \vec{H}$, $\overrightarrow{OB} = \vec{G}$ and OM, the projection of $\vec{G}$ on $\vec{H}$.

The component of $\vec{G}$ along $\vec{H} = \vec{G}\cdot\hat{H} = \dfrac{\vec{G}\cdot\vec{H}}{H}$, where $\hat{H}$ is the unit vector along $\overrightarrow{OA}$.

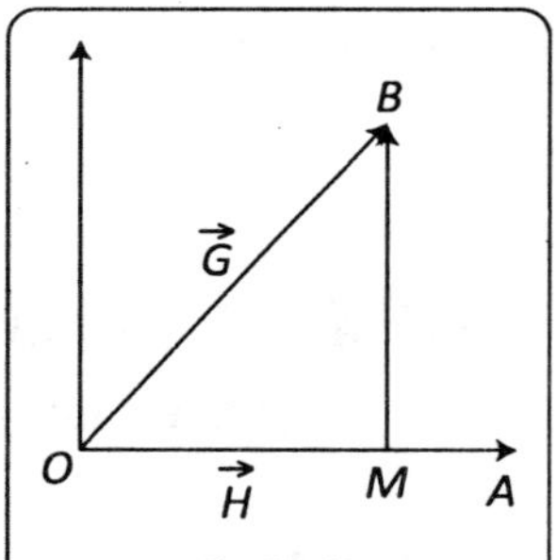

Fig. 10.41

$\therefore$ vector component of $\vec{G}$ along $\vec{H}$ is

$$\frac{\vec{G}\cdot\hat{H}}{|\vec{H}|}\hat{H} = \frac{(\vec{H}\cdot\vec{G})}{|\vec{H}|^2}\vec{H}.$$

Vector component of $\vec{G}$ perpendicular to $\vec{H}$ is

$$\overrightarrow{OB} - \overrightarrow{OM} = \vec{G} - \frac{(\vec{H}\cdot\vec{G})\vec{H}}{|\vec{H}|^2} = \frac{(\vec{H}\cdot\vec{H})\vec{G} - (\vec{H}\cdot\vec{G})\vec{H}}{|\vec{H}|^2}$$

$$= \frac{(\vec{H}\times\vec{G})\times\vec{H}}{|\vec{H}|^2}$$

10.15 Triple Product

If $\vec{B}$ and $\vec{C}$ be any two vectors, then the vector product $\vec{B}\times\vec{C}$ is a vector. This may again be multiplied by a third vector $\vec{A}$ scalarly or, vectorially. Such expression known as triple product appears in two forms, viz., *scalar triple product* $\vec{A}\cdot(\vec{B}\times\vec{C})$ and *vector triple product* $\vec{A}\times(\vec{B}\times\vec{C})$.

Scalar Triple Product

The scalar triple product of three vectors $\vec{A}$, $\vec{B}$ and $\vec{C}$ is a scalar and is represented as

$$\vec{A}\cdot(\vec{B}\times\vec{C}).$$

According to **H.G. Grassmann**, this is also represented as $[\vec{A}\,\vec{B}\,\vec{C}]$ or, $[\vec{A}, \vec{B}, \vec{C}]$.

10.15.1 Geometrical Interpretation of Scalar Triple Product

The scalar triple product of three vectors $\vec{A}$, $\vec{B}$ and $\vec{C}$ represents numerically the volume of the parallelopiped having $\vec{A}$, $\vec{B}$ and $\vec{C}$ as coterminous edges.

Let $\overrightarrow{OK} = \vec{A}$, $\overrightarrow{OR} = \vec{B}$ and $\overrightarrow{OM} = \vec{C}$ as the coterminous edges of the parallelopiped $OKLMNPQRN$.

Let $|\vec{B} \times \vec{C}| = S$ be the area of each of the two faces parallel to the vectors $\vec{B}$ and $\vec{C}$ and ρ be the perpendicular distance between these faces, (Fig. 10.42), i.e.,

$$|\vec{A}| \cos\theta = \rho \text{ (or, } -\rho)$$

Now, $\vec{A} \cdot (\vec{B} \times \vec{C}) = \vec{A} \cdot$ area of the parallelogram $ORQM\hat{n}$

$= S \times$ perpendicular distance $(\pm\rho)$ between the parallel faces $ORQM$ and $KNPL$

$= \pm S \times \rho =$ volume of the parallelopiped $(\pm V)$,

where, $\hat{n} = \dfrac{\vec{B} \times \vec{C}}{|\vec{B} \times \vec{C}|}$, the unit vector perpendicular to the plane of $\vec{B}$ and $\vec{C}$.

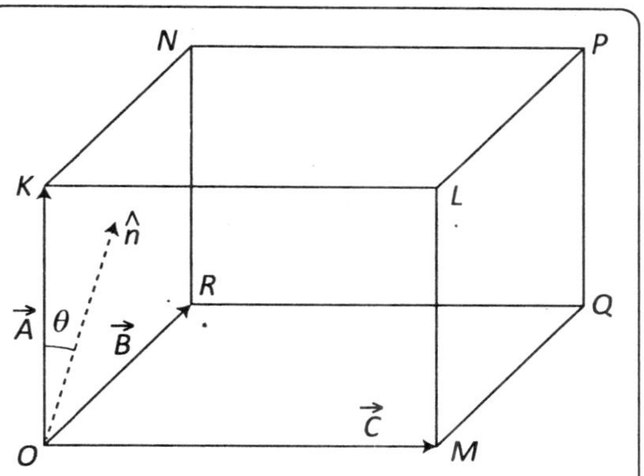

Fig. 10.42

Thus, $[\vec{A}\,\vec{B}\,\vec{C}] = +V$ or $-V$ according as $\vec{A}$, $\vec{B}$, $\vec{C}$ form a right-handed or left-handed triad of vectors.

10.15.2 Characteristics of Scalar Triple Product

1. If $\hat{i}$, $\hat{j}$ and $\hat{k}$ be a set of orthonormal right-handed vector triad, then

$$[\hat{i}\hat{j}\hat{k}] = \hat{i} \cdot (\hat{j} \times \hat{k}) = \hat{i} \cdot \hat{i} = 1.$$

Similarly, $[\hat{j}\hat{k}\hat{i}] = [\hat{k}\hat{i}\hat{j}] = 1$ and $[\hat{i}\hat{k}\hat{j}] = [\hat{k}\hat{j}\hat{i}] = [\hat{j}\hat{i}\hat{k}] = -1$.

2. $\vec{A} \cdot (\vec{B} \times \vec{C}) = \vec{B} \cdot (\vec{C} \times \vec{A}) = \vec{C} \cdot (\vec{A} \times \vec{B})$

We have $\vec{A} \cdot (\vec{B} \times \vec{C}) = (\hat{i}A_1 + \hat{j}A_2 + \hat{k}A_3) \cdot \begin{vmatrix} \hat{i} & \hat{j} & \hat{k} \\ B_1 & B_2 & B_3 \\ C_1 & C_2 & C_3 \end{vmatrix}$,

where $\quad \vec{A} = \hat{i}A_1 + \hat{j}A_2 + \hat{k}A_3, \quad \vec{B} = \hat{i}B_1 + \hat{j}B_2 + \hat{k}B_3, \quad \vec{C} = \hat{i}C_1 + \hat{j}C_2 + \hat{k}C_3$

or, $\quad \vec{A} \cdot (\vec{B} \times \vec{C}) = A_1(B_2C_3 - C_2B_3) + A_2(B_3C_1 - C_3B_1) + A_3(B_1C_2 - C_1B_2)$

$$= \begin{vmatrix} A_1 & A_2 & A_3 \\ B_1 & B_2 & B_3 \\ C_1 & C_2 & C_3 \end{vmatrix}.$$

Similarly, $\vec{A}\cdot(\vec{B}\times\vec{C}) = \begin{vmatrix} B_1 & B_2 & B_3 \\ C_1 & C_2 & C_3 \\ A_1 & A_2 & A_3 \end{vmatrix}$ and $\vec{C}\cdot(\vec{A}\times\vec{B}) = \begin{vmatrix} C_1 & C_2 & C_3 \\ A_1 & A_2 & A_3 \\ B_1 & B_2 & B_3 \end{vmatrix}$.

$$\because \begin{vmatrix} A_1 & A_2 & A_3 \\ B_1 & B_2 & B_3 \\ C_1 & C_2 & C_3 \end{vmatrix} = \begin{vmatrix} B_1 & B_2 & B_3 \\ C_1 & C_2 & C_3 \\ A_1 & A_2 & A_3 \end{vmatrix} = \begin{vmatrix} C_1 & C_2 & C_3 \\ A_1 & A_2 & A_3 \\ B_1 & B_2 & B_3 \end{vmatrix}.$$

$$\therefore \quad \vec{\boldsymbol{A}}\cdot(\vec{\boldsymbol{B}}\times\vec{\boldsymbol{C}}) = \vec{\boldsymbol{B}}\cdot(\vec{\boldsymbol{C}}\times\vec{\boldsymbol{A}}) = \vec{\boldsymbol{C}}\cdot(\vec{\boldsymbol{A}}\times\vec{\boldsymbol{B}}).$$

Also, $\vec{\boldsymbol{A}}\cdot(\vec{\boldsymbol{B}}\times\vec{\boldsymbol{C}}) = \vec{\boldsymbol{B}}\cdot(\vec{\boldsymbol{C}}\times\vec{\boldsymbol{A}}) = \vec{\boldsymbol{C}}\cdot(\vec{\boldsymbol{A}}\times\vec{\boldsymbol{B}}) = \pm\boldsymbol{V}.$

3. If any two of the non-vanishing vectors $\vec{A}$, $\vec{B}$, $\vec{C}$ be parallel or equal, then $[\vec{A}\,\vec{B}\,\vec{C}] = 0$.

4. If the three vectors $\vec{A}$, $\vec{B}$, $\vec{C}$ be coplanar, then $\vec{B}\times\vec{C}$ is perpendicular to $\vec{A}$.

   $\therefore \quad \vec{A}\cdot(\vec{B}\times\vec{C}) = 0$

   Conversely, if $\vec{A}\cdot(\vec{B}\times\vec{C}) = 0$, then $\vec{A}$ is perpendicular to $\vec{B}\times\vec{C}$ and $\vec{A}$ lies in the plane of $\vec{B}$ and $\vec{C}$. So, $\vec{A}$, $\vec{B}$, $\vec{C}$ are coplanar and the volume of the parallelopiped formed by these vectors as coterminous edges is zero.

5. If λ, μ, ν be three scalars, then any vector $\vec{R}$ can be expressed as a linear combination of three non-coplanar vectors as

$$\vec{R} = \lambda\vec{A} + \mu\vec{B} + \nu\vec{C}.$$

$$\therefore \quad \vec{R}\cdot(\vec{B}\times\vec{C}) = \lambda\vec{A}\cdot(\vec{B}\times\vec{C}) + \mu\vec{B}\cdot(\vec{B}\times\vec{C}) + \nu\vec{C}\cdot(\vec{B}\times\vec{C})$$

$$= \lambda\vec{A}\cdot(\vec{B}\times\vec{C}) \quad [\because\ \vec{B}\cdot(\vec{B}\times\vec{C}) = \vec{C}\cdot(\vec{B}\times\vec{C}) = 0].$$

$$\therefore \quad \lambda = \frac{\vec{R}\cdot(\vec{B}\times\vec{C})}{\vec{A}\cdot(\vec{B}\times\vec{C})} = \frac{[\vec{R}\cdot(\vec{B}\times\vec{C})]}{[\vec{A}\,\vec{B}\,\vec{C}]} = \frac{[\vec{R}\,\vec{B}\,\vec{C}]}{[\vec{A}\,\vec{B}\,\vec{C}]}.$$

Similarly, $\mu = \dfrac{[\vec{R}\,\vec{C}\,\vec{A}]}{[\vec{A}\,\vec{B}\,\vec{C}]}$ and $\nu = \dfrac{[\vec{R}\,\vec{A}\,\vec{C}]}{[\vec{A}\,\vec{B}\,\vec{C}]}$.

$$\therefore \quad [\vec{\boldsymbol{A}}\,\vec{\boldsymbol{B}}\,\vec{\boldsymbol{C}}]\vec{\boldsymbol{R}} = [\vec{\boldsymbol{R}}\,\vec{\boldsymbol{B}}\,\vec{\boldsymbol{C}}]\vec{\boldsymbol{A}} + [\vec{\boldsymbol{R}}\,\vec{\boldsymbol{C}}\,\vec{\boldsymbol{A}}]\vec{\boldsymbol{B}} + [\vec{\boldsymbol{R}}\,\vec{\boldsymbol{A}}\,\vec{\boldsymbol{B}}]\vec{\boldsymbol{C}}.$$

6. If $\vec{A}$, $\vec{B}$, $\vec{C}$ be any three non-coplanar vectors such that $\vec{A} = \vec{B}\times\vec{C}$, $\vec{B} = \vec{C}\times\vec{A}$ and $\vec{C} = \vec{A}\times\vec{B}$, then $\vec{A}$, $\vec{B}$ and $\vec{C}$ are perpendicular to each other.

   We have $\vec{A} = \vec{B}\times\vec{C}$.

   $\therefore \quad \vec{A}\cdot\vec{B} = \vec{A}\cdot(\vec{B}\times\vec{C}) = 0.$

   Also, $\vec{A}\cdot\vec{C} = \vec{A}\cdot(\vec{A}\times\vec{B}) = 0.$

   $\therefore$ **$\vec{A}$ is perpendicular to $\vec{B}$ and $\vec{C}$.**

   Similarly, it can be shown that **$\vec{B}$ is perpendicular to $\vec{C}$ and $\vec{A}$, also $\vec{C}$ is perpendicular to $\vec{A}$ and $\vec{B}$.**

Vector Triple Product

The vector triple product of three vectors is a vector product of any vector with the vector product of remaining two vectors.

If $\vec{A}$, $\vec{B}$ and $\vec{C}$ be three vectors, then their vector triple product is given by

$$\vec{A}\times(\vec{B}\times\vec{C})$$

This is in fact a vector product of a polar vector and an axial vector.

It can be proved that,

$$\vec{A}\times(\vec{B}\times\vec{C})=(\vec{A}\cdot\vec{C})\vec{B}-(\vec{A}\cdot\vec{B})\vec{C}. \qquad \text{(i)}$$

Proof : Let $\vec{P}=\vec{A}\times(\vec{B}\times\vec{C})$ and $\vec{D}=(\vec{B}\times\vec{C})$. The vector $\vec{D}$ is normal to the plane of $\vec{B}$ and $\vec{C}$. Also $\vec{P}=\vec{A}\times\vec{D}$ is normal to the plane of $\vec{A}$ and $\vec{D}$. So, $\vec{P}$ is in the plane of $\vec{B}$ and $\vec{C}$. Hence, $\vec{P}$ can be expressed as

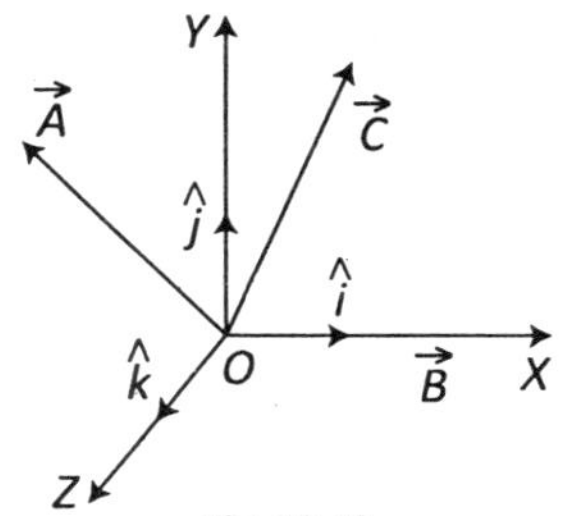

Fig. 10.43

$$\vec{P}=\alpha\vec{B}+\beta\vec{C}.$$

Again, $\vec{P}\cdot\vec{A}=0$ as $\vec{P}$ is normal to $\vec{A}$

or, $\quad \alpha(\vec{A}\cdot\vec{B})+\beta(\vec{A}\cdot\vec{C})=0.$

$\therefore \quad \beta=-\dfrac{(\vec{A}\cdot\vec{B})}{(\vec{A}\cdot\vec{C})}\alpha$

or, $$\vec{P}=\vec{A}\times(\vec{B}\times\vec{C})=\frac{\alpha}{(\vec{A}\cdot\vec{C})}[(\vec{A}\cdot\vec{C})\vec{B}-(\vec{A}\cdot\vec{B})\vec{C}]. \qquad \text{(ii)}$$

To find the value of α, let $\vec{B}$ be along X-axis, Y-axis lying in the plane of $\vec{B}-\vec{C}$ and Z-axis along $\vec{B}\times\vec{C}$.

Then, $\quad \vec{A}=\hat{\imath}A_1+\hat{\jmath}A_2+\hat{k}A_3, \quad \vec{B}=\hat{\imath}B_1, \quad \vec{C}=\hat{\imath}C_1+\hat{\jmath}C_2.$

$\therefore \quad \vec{B}\times\vec{C}=\hat{\imath}B_1\times(\hat{\imath}C_1+\hat{\jmath}C_2)=B_1C_2(\hat{\imath}\times\hat{\jmath})=B_1C_2\hat{k}.$

Hence, $$\vec{A}\times(\vec{B}\times\vec{C})=(\hat{\imath}A_1+\hat{\jmath}A_2+\hat{k}A_3)\times\hat{k}B_1C_2. \qquad \text{(iii)}$$

Again, $$\vec{A}\cdot\vec{C}=A_1C_1+A_2C_2 \qquad \text{(iv)}$$

and $$\vec{A}\cdot\vec{B}=A_1B_1 \qquad \text{(v)}$$

$$\therefore \quad \vec{A}\times(\vec{B}\times\vec{C})=\hat{\imath}A_2B_1C_2-\hat{\jmath}A_1B_1C_2=\hat{\imath}B_1(A_2C_2+A_1C_1)-A_1B_1C_1\hat{\imath}-A_1B_2C_2\hat{\jmath}$$

$$=(A_1C_1+A_2C_2)\vec{B}-A_1B_1(\hat{\imath}C_1+\hat{\jmath}C_2)$$

$$=(\vec{A}\cdot\vec{C})\vec{B}-(\vec{A}\cdot\vec{B})\vec{C}. \qquad \text{(vi)}$$

Comparing equations (ii) and (vi), we get $\dfrac{\alpha}{\vec{A}\cdot\vec{C}}=1.$

$$\therefore \quad \boldsymbol{\vec{A}\times(\vec{B}\times\vec{C})=(\vec{A}\cdot\vec{C})\vec{B}-(\vec{A}\cdot\vec{B})\vec{C}}. \qquad \text{(vii)}$$

But, $$(\vec{A}\times\vec{B})\times\vec{C}=\vec{B}(\vec{A}\cdot\vec{C})-\vec{A}(\vec{B}\cdot\vec{C}) \qquad \text{(viii)}$$

for, $(\vec{A}\times\vec{B})\times\vec{C}=-\vec{C}\times(\vec{A}\times\vec{B})=-[\vec{A}(\vec{C}\cdot\vec{B})-\vec{B}(\vec{C}\cdot\vec{A})]=\vec{B}(\vec{A}\cdot\vec{C})-\vec{A}(\vec{B}\cdot\vec{C}).$

So, $\vec{A} \times (\vec{B} \times \vec{C}) \neq (\vec{A} \times \vec{B}) \times \vec{C}$.

i.e., the associative law in the context of vector cross products for $\vec{A}$, $\vec{B}$ and $\vec{C}$ is not valid.

10.15.3 Characteristics of Vector Triple Product

1. We have shown that $\vec{A} \times (\vec{B} \times \vec{C}) \neq (\vec{A} \times \vec{B}) \times \vec{C}$.

   If follows that, $\vec{A} \times (\vec{B} \times \vec{C})$ represents a vector coplanar with $\vec{B}$ and $\vec{C}$, but $(\vec{A} \times (\vec{B}) \times \vec{C}$ represents a vector lying in the plane of $\vec{A}$ and $\vec{B}$. Hence, the two products do not represent the same vector.

2. If $\vec{A}$, $\vec{B}$, $\vec{C}$ be three non-zero vectors and satisfy the condition

$$\vec{A} \times (\vec{B} \times \vec{C}) = (\vec{A} \times \vec{B}) \times \vec{C}$$

   then $\vec{A}$ and $\vec{C}$ must be collinear or $\vec{B}$ must be perpendicular to $\vec{A}$ and $\vec{C}$.

   Proof : Since $\vec{A} \times (\vec{B} \times \vec{C}) = (\vec{A} \times \vec{B}) \times \vec{C}$,

$$\therefore \quad \vec{B}(\vec{A} \cdot \vec{C}) - \vec{C}(\vec{A} \cdot \vec{B}) = \vec{B}(\vec{A} \cdot \vec{C}) - \vec{A}(\vec{B} \cdot \vec{C})$$

$$\text{or,} \quad \vec{C}(\vec{A} \cdot \vec{B}) = \vec{A}(\vec{B} \cdot \vec{C}). \tag{ix}$$

   Since $\vec{A}$, $\vec{B}$, $\vec{C}$ are non-zero vectors it follows from (i) that,

$$\text{(i) } \vec{B} \cdot \vec{C} = 0, \quad \vec{A} \cdot \vec{B} = 0 \quad \text{or,} \quad \text{(ii) } \vec{B} \cdot \vec{C} \neq 0, \quad \vec{A} \cdot \vec{B} \neq 0.$$

   From (i), **we conclude that $\vec{B}$ is perpendicular to both $\vec{A}$ and $\vec{C}$.**

   From equation (ix), $\vec{C} = \dfrac{\vec{B} \cdot \vec{C}}{\vec{A} \cdot \vec{B}} \vec{A} = m\vec{A}, \quad m \neq 0.$

   $\therefore$ **$\vec{A}$ and $\vec{C}$ are parallel. Hence the result.**

3. For three vectors $\vec{A}$, $\vec{B}$, $\vec{C}$,

$$\vec{A} \times (\vec{B} \times \vec{C}) + \vec{B} \times (\vec{C} \times \vec{A}) + \vec{C} \times (\vec{A} \times \vec{B}) = 0.$$

   For, $\vec{A} \times (\vec{B} \times \vec{C}) + \vec{B} \times (\vec{C} \times \vec{A}) + \vec{C} \times (\vec{A} \times \vec{B})$

$$= \vec{B}(\vec{A} \cdot \vec{C}) - \vec{C}(\vec{A} \cdot \vec{B}) + \vec{C}(\vec{B} \cdot \vec{A}) - \vec{A}(\vec{B} \cdot \vec{C}) + \vec{A}(\vec{C} \cdot \vec{B}) - \vec{B}(\vec{C} \cdot \vec{A}) = 0.$$

10.16 Product of Four Vectors

10.16.1 Scalar Product of Four Vectors

If $\vec{A}$, $\vec{B}$, $\vec{C}$, $\vec{D}$ be four vectors, then the product $(\vec{A} \times \vec{B}) \cdot (\vec{C} \times \vec{D})$ is called the scalar product of four vectors. **[BU(H) 1996]**

$$\text{Now, } (\vec{A} \times \vec{B}) \cdot (\vec{C} \times \vec{D}) = \vec{A} \cdot \vec{B} \times (\vec{C} \times \vec{D}) \tag{i}$$

[since the position of dot and cross in a scalar triple product can be interchanged]

$$= \vec{A} \cdot [\vec{C}(\vec{B} \cdot \vec{D}) - \vec{D}(\vec{B} \cdot \vec{C})] = (\vec{A} \cdot \vec{C})(\vec{B} \cdot \vec{D}) - (\vec{A} \cdot \vec{D})(\vec{B} \cdot \vec{C})$$

$$= \begin{vmatrix} \vec{A} \cdot \vec{C} & \vec{B} \cdot \vec{C} \\ \vec{A} \cdot \vec{D} & \vec{B} \cdot \vec{D} \end{vmatrix}.$$

10.16.2 Vector Product of Four Vectors

If $\vec{A}$, $\vec{B}$, $\vec{C}$, $\vec{D}$ be four vectors, then the product $(\vec{A} \times \vec{B}) \times (\vec{C} \times \vec{D})$ is the vector product of four vectors.

It can be shown that,

$$(\vec{A} \times \vec{B}) \times (\vec{C} \times \vec{D}) = \vec{B}(\vec{A} \cdot \vec{C} \times \vec{D}) - \vec{A}(\vec{B} \cdot \vec{C} \times \vec{D})$$
$$= \vec{C}(\vec{A} \cdot \vec{B} \times \vec{D}) - \vec{D}(\vec{A} \cdot \vec{B} \times \vec{C}.$$

Let $\vec{P} = \vec{A} \times \vec{B}$, then

$$(\vec{A} \times \vec{B}) \times (\vec{C} \times \vec{D}) = \vec{P} \times (\vec{C} \times \vec{D}) = \vec{C}(\vec{P} \cdot \vec{D}) - \vec{D}(\vec{P} \cdot \vec{C})$$
$$= \vec{C}(\vec{A} \times \vec{B} \cdot \vec{D}) - \vec{D}(\vec{A} \times \vec{B} \cdot \vec{C})$$
$$= \vec{C}(\vec{A} \cdot \vec{B} \times \vec{D}) - \vec{D}(\vec{A} \cdot \vec{B} \times \vec{C}).$$

Again, let $\vec{Q} = \vec{C} \times \vec{D}$, then

$$(\vec{A} \times \vec{B}) \times (\vec{C} \times \vec{D}) = (\vec{A} \times \vec{B}) \times \vec{Q} = \vec{B}(\vec{A} \cdot \vec{Q}) - \vec{A}(\vec{B} \cdot \vec{Q})$$
$$= \vec{B}(\vec{A} \cdot \vec{B} \times \vec{D}) - \vec{A}(\vec{B} \cdot \vec{C} \times \vec{D}).$$

10.17 Reciprocal System of Vectors

If $\vec{a}$, $\vec{b}$, $\vec{c}$ be a set of non-coplanar vectors, then the set $\vec{a}'$, $\vec{b}'$, $\vec{c}'$ will be reciprocal to the above set of vectors if the following conditions be satisfied.

$$\vec{a} \cdot \vec{a}' = \vec{b} \cdot \vec{b}' = \vec{c} \cdot \vec{c}' = 1$$

and $\vec{a} \cdot \vec{b}' = \vec{a} \cdot \vec{c}' = \vec{b} \cdot \vec{a}' = \vec{b} \cdot \vec{c}' = \vec{c} \cdot \vec{a}' = \vec{c} \cdot \vec{b}' = 0.$

It can be shown that,

$$\vec{a}' = \frac{\vec{b} \times \vec{c}}{\vec{a} \cdot \vec{b} \times \vec{c}}, \quad \vec{b}' = \frac{\vec{c} \times \vec{a}}{\vec{a} \cdot \vec{b} \times \vec{c}} \quad \text{and} \quad \vec{c}' = \frac{\vec{a} \times \vec{b}}{\vec{a} \cdot \vec{b} \times \vec{c}}.$$

Since, $\vec{a} \cdot (\vec{b} \times \vec{c}) \neq 0.$

The above relations can be obtained in the following way.

$\because$ $\vec{a}' \cdot \vec{b} = \vec{a}' \cdot \vec{c} = 0$, so $\vec{a}'$ is normal to both $\vec{b}$ and $\vec{c}$. So $\vec{a}'$ and $\vec{b} \times \vec{c}$ are parallel.

$\therefore$ $\vec{a}' = p(\vec{b} \times \vec{c})$, where p is a scalar quantity.

Again, $\vec{a} \cdot \vec{a}' = 1$ or, $\vec{a} \cdot p(\vec{b} \times \vec{c}) = 1$ or, $p = \dfrac{1}{\vec{a} \cdot \vec{b} \times \vec{c}}$.

$$\therefore \quad \vec{a}' = \frac{\vec{b} \times \vec{c}}{\vec{a} \cdot \vec{b} \times \vec{c}}.$$

Similarly, we can show that,

$$\vec{b}' = \frac{\vec{c} \times \vec{a}}{\vec{a} \cdot \vec{b} \times \vec{c}} \quad \text{and} \quad \vec{c}' = \frac{\vec{a} \times \vec{b}}{\vec{a} \cdot \vec{b} \times \vec{c}}.$$

10.18 Solved Problems (C)

Example 10.48. *If* $\vec{A} = x_1\vec{a} + y_1\vec{b} + z_1\vec{c}$, $\vec{B} = x_2\vec{a} + y_2\vec{b} + z_2\vec{c}$ *and* $\vec{C} = x_3\vec{a} + y_3\vec{b} + z_3\vec{c}$, *then prove that,* $\vec{A}\cdot\vec{B}\times\vec{C} = \begin{vmatrix} x_1 & y_1 & z_1 \\ x_2 & y_2 & z_2 \\ x_3 & y_3 & z_3 \end{vmatrix}(\vec{a}\cdot\vec{b}\times\vec{c})$.

Solution : We have
$$\begin{aligned}\vec{B}\times\vec{C} &= (x_2\vec{a} + y_2\vec{b} + z_2\vec{c})\times(x_3\vec{a} + y_3\vec{b} + z_3\vec{c})\\ &= x_2y_3(\vec{a}\times\vec{b}) + x_2z_3(\vec{a}\times\vec{c}) + y_2x_3(\vec{b}\times\vec{a})\\ &\quad + y_2z_3(\vec{b}\times\vec{c}) + z_2x_3(\vec{c}\times\vec{a}) + z_2y_3(\vec{c}\times\vec{b})\\ &= (x_2y_3 - y_2x_3)(\vec{a}\times\vec{b}) + (y_2z_3 - z_2y_3)(\vec{b}\times\vec{c})\\ &\quad + (z_2x_3 - x_2z_3)(\vec{c}\times\vec{a}).\end{aligned}$$

$$\begin{aligned}\therefore\ \vec{A}\cdot(\vec{B}\times\vec{C}) &= (x_1\vec{a} + y_1\vec{b} + z_1\vec{c})\cdot[(x_2y_3 - y_2x_3)(\vec{a}\times\vec{b})]\\ &\quad + (y_2z_3 - z_2y_3)(\vec{b}\times\vec{c}) + (z_2x_3 - x_2z_3)(\vec{c}\times\vec{a})\\ &= x_1(x_2y_3 - y_2x_3)(\vec{a}\cdot\vec{a}\times\vec{b}) + x_1(y_2z_3 - z_2y_3)(\vec{a}\cdot\vec{b}\times\vec{c})\\ &\quad + x_1(z_2x_3 - x_2z_3)(\vec{a}\cdot\vec{c}\times\vec{a}) + y_1(x_2y_3 - y_2x_3)(\vec{b}\cdot\vec{a}\times\vec{b})\\ &\quad + y_1(y_2z_3 - z_2y_3)(\vec{b}\cdot\vec{b}\times\vec{c}) + y_1(z_2x_3 - x_2z_3)(\vec{b}\cdot\vec{c}\times\vec{a})\\ &\quad + z_1(x_2y_3 - y_2x_3)(\vec{c}\cdot\vec{a}\times\vec{b}) + z_1(y_2z_3 - z_2y_3)(\vec{c}\cdot\vec{b}\times\vec{c})\\ &\quad + z_1(z_2x_3 - x_2z_3)(\vec{c}\cdot\vec{c}\times\vec{a})\\ &= x_1(y_2z_3 - z_2y_3)(\vec{a}\cdot\vec{b}\times\vec{c}) + y_1(z_2x_3 - x_2z_3)(\vec{b}\cdot\vec{c}\times\vec{a})\\ &\quad + z_1(x_2y_3 - y_2x_3)(\vec{c}\cdot\vec{a}\times\vec{b})\\ &= \begin{vmatrix} x_1 & y_1 & z_1 \\ x_2 & y_2 & z_2 \\ x_3 & y_3 & z_3 \end{vmatrix}(\vec{a}\cdot\vec{b}\times\vec{c})\quad \textbf{(Proved)}\end{aligned}$$

$$[\because\ \vec{a}\cdot\vec{b}\times\vec{c} = \vec{b}\cdot\vec{c}\times\vec{a} = \vec{c}\cdot\vec{a}\times\vec{b}].$$

Example 10.49. *If* $\vec{A}$, $\vec{B}$, $\vec{C}$, $\vec{D}$ *are such that* $\vec{A}\times\vec{B} = \vec{C}\times\vec{D}$ *and* $\vec{A}\times\vec{C} = \vec{B}\times\vec{D}$, *then show that the vectors* $(\vec{A} - \vec{D})$ *and* $(\vec{B} - \vec{C})$ *are collinear.*

Solution : We have
$$\begin{aligned}(\vec{A} - \vec{D})\times(\vec{B} - \vec{C}) &= (\vec{A} - \vec{D})\times\vec{B} - (\vec{A} - \vec{D})\times\vec{C}\\ &= (\vec{A}\times\vec{B}) - (\vec{D}\times\vec{B}) - (\vec{A}\times\vec{C}) + (\vec{D}\times\vec{C}) = 0.\end{aligned}$$

$\because\ \vec{A}\times\vec{B} = \vec{C}\times\vec{D}$ and $\vec{A}\times\vec{C} = \vec{B}\times\vec{D}$.

Hence, $\vec{A} - \vec{D}$ **and** $\vec{B} - \vec{C}$ **are collinear.**

Example 10.50. *Prove that,*

(a) $(\vec{A}\times\vec{B})\cdot(\vec{C}\times\vec{D}) = (\vec{A}\cdot\vec{C})(\vec{B}\cdot\vec{D}) - (\vec{A}\cdot\vec{D})(\vec{B}\cdot\vec{C})$ **[BU(H) 1996]**

(b) $(\vec{A}\times\vec{B})\cdot(\vec{C}\times\vec{D}) + (\vec{B}\times\vec{C})\cdot(\vec{A}\times\vec{D}) + (\vec{C}\times\vec{A})\cdot(\vec{B}\times\vec{D}) = 0.$ **[BU(H) 1993]**

Solution : (a) Let $\vec{P} = \vec{A} \times \vec{B}$, then

$$(\vec{A} \times \vec{B}) \cdot (\vec{C} \times \vec{D}) = \vec{P} \cdot (\vec{C} \times \vec{D}) = (\vec{C} \times \vec{D}) \cdot \vec{P} = \vec{C} \cdot (\vec{D} \times \vec{P}) = \vec{C} \cdot (\vec{D} \times \vec{A} \times \vec{B})$$
$$= \vec{C} \cdot [\vec{A}(\vec{D} \cdot \vec{B}) - \vec{B}(\vec{D} \cdot \vec{A})] = (\vec{C} \cdot \vec{A})(\vec{D} \cdot \vec{B}) - (\vec{C} \cdot \vec{B})(\vec{D} \cdot \vec{A})$$
$$= (\vec{A} \cdot \vec{C})(\vec{B} \cdot \vec{D}) - (\vec{A} \cdot \vec{D})(\vec{B} \cdot \vec{C}) \quad \textbf{(Proved)}.$$

(b) Now, $(\vec{A} \times \vec{B}) \cdot (\vec{C} \times \vec{D}) = (\vec{A} \cdot \vec{C})(\vec{B} \cdot \vec{D}) - (\vec{A} \cdot \vec{D})(\vec{B} \cdot \vec{C})$.

Similarly, $(\vec{B} \times \vec{C}) \cdot (\vec{A} \times \vec{D}) = (\vec{B} \cdot \vec{A})(\vec{C} \cdot \vec{D}) - (\vec{B} \cdot \vec{D})(\vec{A} \cdot \vec{C})$

$$(\vec{C} \times \vec{A}) \cdot (\vec{B} \times \vec{D}) = (\vec{C} \cdot \vec{B})(\vec{A} \cdot \vec{D}) - (\vec{C} \cdot \vec{D})(\vec{A} \cdot \vec{B}).$$

$$\therefore \quad \boldsymbol{(\vec{A} \times \vec{B}) \cdot (\vec{C} \times \vec{D}) + (\vec{B} \times \vec{C}) \cdot (\vec{A} \times \vec{D}) + (\vec{C} \times \vec{A}) \cdot (\vec{B} \times \vec{D}) = 0} \quad \textbf{(Proved)}.$$

Example 10.51. *If $\vec{A}$, $\vec{B}$, $\vec{C}$ and $\vec{D}$ are four coplanar but non-parallel vectors, then show that, $(\vec{A} \times \vec{B}) \times (\vec{C} \times \vec{D}) = 0$.* **[BU(H) 1998]**

Solution : Let $\vec{P} = \vec{A} \times \vec{B}$, then

$$(\vec{A} \times \vec{B}) \cdot (\vec{C} \times \vec{D}) = \vec{P} \times (\vec{C} \times \vec{D}) = \vec{C}(\vec{P} \cdot \vec{D}) - \vec{D}(\vec{P} \cdot \vec{C})$$
$$= \vec{C}(\vec{A} \times \vec{B} \cdot \vec{D}) - \vec{D}(\vec{A} \times \vec{B} \cdot \vec{C})$$
$$= \vec{C}(\vec{A} \cdot \vec{B} \times \vec{D}) - \vec{D}(\vec{A} \cdot \vec{B} \times \vec{C}).$$

Now, since $\vec{A}$, $\vec{B}$, $\vec{C}$, $\vec{D}$ are coplanar vectors,

$\therefore \quad \vec{A} \cdot \vec{B} \times \vec{D} = \vec{A} \cdot \vec{B} \times \vec{C} = 0;$

$\therefore \quad \boldsymbol{(\vec{A} \times \vec{B}) \times (\vec{C} \times \vec{D}) = 0}$ **(Proved).**

Example 10.52. *Prove that, $[(\vec{A} + \vec{B}) \cdot \{(\vec{B} + \vec{C}) \times (\vec{C} + \vec{A})\}] = 2(\vec{A} \cdot \vec{B} \times \vec{C})$.* **[RU 1989, BU 1990]**

Solution : We have,

$$\text{LHS} = [(\vec{A} + \vec{B}) \cdot \{(\vec{B} + \vec{C}) \times (\vec{C} + \vec{A})\}]$$
$$= (\vec{A} + \vec{B}) \cdot \{(\vec{B} \times \vec{C}) + (\vec{B} \times \vec{A}) + (\vec{C} \times \vec{C}) + (\vec{C} \times \vec{A})\}$$
$$= \vec{A} \cdot (\vec{B} \times \vec{C}) + \vec{A} \cdot (\vec{B} \times \vec{A}) + \vec{A} \cdot (\vec{C} \times \vec{C}) + \vec{A} \cdot (\vec{C} \times \vec{A}) + \vec{B} \cdot (\vec{B} \times \vec{C})$$
$$+ \vec{B} \cdot (\vec{B} \times \vec{A}) + \vec{B} \cdot (\vec{C} \times \vec{C}) + \vec{B} \cdot (\vec{C} \times \vec{A})$$
$$= \vec{A} \cdot (\vec{B} \times \vec{C}) + \vec{B} \cdot (\vec{C} \times \vec{A}) = \vec{A} \cdot (\vec{B} \times \vec{C}) + \vec{A} \cdot (\vec{B} \times \vec{C})$$
$$= \boldsymbol{2(\vec{A} \cdot \vec{B} \times \vec{C}) = \textbf{RHS}} \quad \textbf{(Proved)}.$$

$$\because \vec{A} \cdot (\vec{B} \times \vec{A}) = \vec{A} \cdot (\vec{C} \times \vec{C}) = \vec{A} \cdot (\vec{C} \times \vec{A})$$
$$= \vec{B} \cdot (\vec{B} \times \vec{C}) + \vec{B} \cdot (\vec{B} \times \vec{A}) + \vec{B} \cdot (\vec{C} \times \vec{C}) = 0.$$

Example 10.53. *Prove that,* $[(\vec{B} \times \vec{C}) \cdot \{(\vec{C} \times \vec{A}) \times (\vec{A} \times \vec{B})\}] = [\vec{A} \cdot (\vec{B} \times \vec{C})]^2$.

[Karnataka U 1988, Bhopal U 1991, WBUT (Math) 2006]

Solution : LHS $= (\vec{B} \times \vec{C}) \cdot \{(\vec{C} \times \vec{A}) \times (\vec{A} \times \vec{B})\}$

$= (\vec{B} \times \vec{C}) \cdot \{[(\vec{C} \times \vec{A}) \cdot \vec{B}] \cdot \vec{A} - [(\vec{C} \times \vec{A}) \cdot \vec{A}]\vec{B}\}$

$= (\vec{B} \times \vec{C}) \cdot \{[\vec{B} \cdot (\vec{C} \times \vec{A})]\vec{A}\} \quad [\because (\vec{C} \times \vec{A}) \cdot \vec{A} = 0]$

$= [(\vec{B} \times \vec{C}) \cdot \vec{A}][(\vec{B} \times \vec{C}) \cdot \vec{A}] = [(\vec{B} \times \vec{C}) \cdot \vec{A}]^2 = [\vec{A} \cdot (\vec{B} \times \vec{C})]^2$

= **RHS (Proved).**

Example 10.54. *Show that the volume of the tetrahedron having* $(\vec{a} + \vec{b})$, $(\vec{b} + \vec{c})$, $(\vec{c} + \vec{a})$ *as concurrent edges is twice the volume of the tetrahedron having* $\vec{a}$, $\vec{b}$, $\vec{c}$ *as concurrent edges.*

Solution : We have the volume of the tetrahedron

$$= \frac{1}{6}[(\vec{a} + \vec{b}) \cdot \{(\vec{b} + \vec{c}) \times (\vec{c} + \vec{a})\}]$$

$$= \frac{1}{6}[(\vec{a} + \vec{b}) \cdot \{\vec{b} \times \vec{c} + \vec{b} \times \vec{a} + \vec{c} \times \vec{c} + \vec{c} \times \vec{a}\}]$$

$$= \frac{1}{6}[(\vec{a} + \vec{b}) \cdot \{\vec{b} \times \vec{c} + \vec{b} \times \vec{a} + \vec{c} \times \vec{a}\}]$$

$$= \frac{1}{6}[\vec{a} \cdot (\vec{b} \times \vec{c}) + \vec{a} \cdot (\vec{b} \times \vec{a}) + \vec{a} \cdot (\vec{c} \times \vec{a}) + \vec{b} \cdot (\vec{b} \times \vec{c}) + \vec{b} \cdot (\vec{b} \times \vec{a}) + \vec{b} \cdot (\vec{c} \times \vec{a})$$

$$= \frac{1}{6}[\vec{a} \cdot (\vec{b} \times \vec{c}) + \vec{b} \cdot (\vec{c} \times \vec{a})] = \frac{1}{3}[\vec{a} \cdot (\vec{b} \times \vec{c})] = \mathbf{2} \times \frac{\mathbf{1}}{\mathbf{6}}[\vec{a} \cdot (\vec{b} \times \vec{c})]$$

= **2 × volume of tetrahedron with $\vec{a}$, $\vec{b}$, $\vec{c}$ as concurrent edges**

Example 10.55. *If* $\vec{A}$, $\vec{B}$, $\vec{C}$, $\vec{D}$ *be the position vectors of four points lying in the same plane, then show that,* $\vec{A} \cdot \vec{B} \times \vec{C} = \vec{A} \cdot \vec{B} \times \vec{D} + \vec{A} \cdot \vec{D} \times \vec{C} + \vec{D} \cdot \vec{B} \times \vec{C}$.

Solution : Let K, L, M, N be four points whose position vectors are $\vec{A}$, $\vec{B}$, $\vec{C}$, $\vec{D}$ respectively.

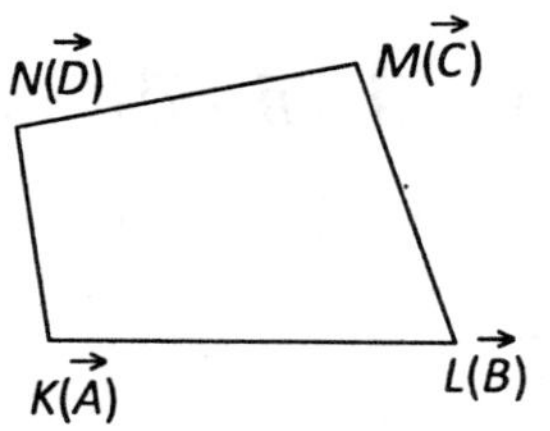

Fig. 10.44

Then, $\overrightarrow{KN} = \vec{D} - \vec{A}$, $\overrightarrow{LN} = \vec{D} - \vec{B}$ and $\overrightarrow{MN} = \vec{D} - \vec{C}$.

If $\overrightarrow{KN}$, $\overrightarrow{LN}$ and $\overrightarrow{MN}$ are coplanar, then

$$\overrightarrow{KN} \cdot (\overrightarrow{LN} \times \overrightarrow{MN}) = 0$$

or, $(\vec{D} - \vec{A}) \cdot [(\vec{D} - \vec{B}) \times (\vec{D} - \vec{C})] = 0$

or, $(\vec{D} - \vec{A}) \cdot [\vec{D} \times \vec{D} - \vec{D} \times \vec{C} - \vec{B} \times \vec{D} + \vec{B} \times \vec{C}] = 0$

or, $(\vec{D} - \vec{A}) \cdot [-\vec{D} \times \vec{C} - \vec{B} \times \vec{D} + \vec{B} \times \vec{C}] = 0$

or, $-\vec{D} \cdot (\vec{D} \times \vec{C}) - \vec{D} \cdot (\vec{B} \times \vec{D}) + \vec{D} \cdot (\vec{B} \times \vec{C}) + \vec{A} \cdot (\vec{D} \times \vec{C}) + \vec{A} \cdot (\vec{B} \times \vec{D}) - \vec{A} \cdot (\vec{B} \times \vec{C}) = 0$

or, $\quad 0+0+(\vec{D}\cdot\vec{B}\times\vec{C})+(\vec{A}\cdot\vec{D}\times\vec{C})+(\vec{A}\cdot\vec{B}\times\vec{D})-(\vec{A}\cdot\vec{B}\times\vec{C})=0$

or, $\quad \vec{A}\cdot(\vec{B}\times\vec{C})=(\vec{A}\cdot\vec{B}\times\vec{D})+(\vec{A}\cdot\vec{D}\times\vec{C})+(\vec{D}\cdot\vec{B}\times\vec{C})$ **(Proved).**

Example 10.56. *If $\vec{a}$, $\vec{b}$, $\vec{c}$ and $\vec{a}'$, $\vec{b}'$, $\vec{c}'$ are reciprocal system of vectors, then prove that, $(\vec{a}\times\vec{a}')+(\vec{b}\times\vec{b}')+(\vec{c}\times\vec{c}')=0$.*

Solution : Here, $\vec{a}'=\dfrac{\vec{b}\times\vec{c}}{\vec{a}\cdot(\vec{b}\times\vec{c})}$, $\vec{b}'=\dfrac{\vec{c}\times\vec{a}}{\vec{a}\cdot(\vec{b}\times\vec{c})}$, $\vec{c}'=\dfrac{\vec{a}\times\vec{b}}{\vec{a}\cdot(\vec{b}\times\vec{c})}$.

$$\therefore \quad \vec{a}\times\vec{a}'=\frac{\vec{a}\times(\vec{b}\times\vec{c})}{\vec{a}\cdot(\vec{b}\times\vec{c})}=\frac{\vec{b}(\vec{a}\cdot\vec{c})-\vec{c}(\vec{a}\cdot\vec{b})}{\vec{a}\cdot(\vec{b}\times\vec{c})},$$

$$\vec{b}\times\vec{b}'=\frac{\vec{b}\times(\vec{c}\times\vec{a})}{\vec{a}\cdot(\vec{b}\times\vec{c})}=\frac{\vec{c}(\vec{b}\cdot\vec{a})-\vec{a}(\vec{b}\cdot\vec{c})}{\vec{a}\cdot(\vec{b}\times\vec{c})}$$

$$\text{and} \quad \vec{c}\times\vec{c}'=\frac{\vec{c}\times(\vec{a}\times\vec{b})}{\vec{a}\cdot(\vec{b}\times\vec{c})}=\frac{\vec{a}(\vec{c}\cdot\vec{b})-\vec{b}(\vec{c}\cdot\vec{c})}{\vec{a}\cdot(\vec{b}\times\vec{c})}.$$

$\therefore \quad (\vec{a}\times\vec{a}')+(\vec{b}\times\vec{b}')+(\vec{c}\times\vec{c}')=0$ **(Proved).**

Example 10.57. *Show that the orthonormal vector triad $(\hat{i},\hat{j},\hat{k})$ is self-reciprocal.*

[Bang. U 1990]

Solution : If $\hat{i}'$, $\hat{j}'$, $\hat{k}'$ be the set of vectors reciprocal to $(\hat{i},\hat{j},\hat{k})$, then

$$\hat{i}'=\frac{\hat{j}\times\hat{k}}{\hat{i}\cdot\hat{j}\times\hat{k}}=\frac{\hat{i}}{\hat{i}\cdot\hat{i}}=\frac{\hat{i}}{1}=\hat{i}.$$

Similarly $\hat{j}'=\hat{j}$ and $\hat{k}'=\hat{k}$.

$\therefore$ **the orthonormal set of vector triad $\hat{i}$, $\hat{j}$, $\hat{k}$ is self-reciprocal.**

Example 10.58. *If $\vec{a}$, $\vec{b}$, $\vec{c}$ and $\vec{a}'$, $\vec{b}'$, $\vec{c}'$ be set of reciprocal vectors, then show that,*

$$\vec{a}'\cdot(\vec{b}'\times\vec{c}')=\frac{1}{\vec{a}\cdot(\vec{b}\times\vec{c})}.$$

[Bang. U 1985, Calicut U 1983]

Solution : We have $\vec{a}'\cdot(\vec{b}'\times\vec{c}')=\vec{a}'\dfrac{(\vec{c}\times\vec{a})\times(\vec{a}\times\vec{b})}{[\vec{a}\cdot(\vec{b}\times\vec{c})]^2}$

$$=\frac{\vec{a}'\cdot[\vec{a}\cdot(\vec{b}\times\vec{c})]\vec{a}}{[\vec{a}\cdot(\vec{b}\times\vec{c})]^2}=\frac{\vec{a}'\times\vec{a}}{[\vec{a}\cdot(\vec{b}\times\vec{c})]^2}$$

$$=\frac{1}{[\vec{a}\cdot(\vec{b}\times\vec{c})]} \quad [\because \ \vec{a}'\times\vec{a}=1].$$

Example 10.59. *If $\vec{a}'=\dfrac{\vec{b}\times\vec{c}}{\vec{a}\cdot(\vec{b}\times\vec{c})}$, $\vec{b}'=\dfrac{\vec{c}\times\vec{a}}{\vec{a}\cdot(\vec{b}\times\vec{c})}$, $\vec{c}'=\dfrac{\vec{a}\times\vec{b}}{\vec{a}\cdot(\vec{b}\times\vec{c})}$,*

prove that $\vec{a}=\dfrac{\vec{b}'\times\vec{c}'}{\vec{a}'\cdot(\vec{b}'\times\vec{c}')}$, $\vec{b}=\dfrac{\vec{c}'\times\vec{a}'}{\vec{a}'\cdot(\vec{b}'\times\vec{c}')}$, $\vec{c}=\dfrac{\vec{a}'\times\vec{b}'}{\vec{a}'\cdot(\vec{b}'\times\vec{c}')}$.

Solution : We have $$\frac{\vec{b}' \times \vec{c}'}{\vec{a}' \cdot (\vec{b}' \times \vec{c}')} = \frac{\frac{\vec{c}\times\vec{a}}{\vec{a}\cdot(\vec{b}\times\vec{c})} \times \frac{\vec{a}\times\vec{b}}{\vec{a}\cdot(\vec{b}\times\vec{c})}}{\frac{\vec{b}\times\vec{c}}{\vec{a}\cdot(\vec{b}\times\vec{c})} \cdot \frac{\vec{c}\times\vec{a}}{\vec{a}\cdot(\vec{b}\times\vec{c})} \times \frac{\vec{a}\times\vec{b}}{\vec{a}\cdot(\vec{b}\times\vec{c})}}$$

$$= \frac{(\vec{c}\times\vec{a})\times(\vec{a}\times\vec{b})}{\frac{(\vec{b}\times\vec{c})\cdot(\vec{c}\times\vec{a})\times(\vec{a}\times\vec{b})}{\vec{a}\cdot(\vec{b}\times\vec{c})}} = \frac{(\vec{a}\times\vec{b})\times(\vec{a}\times\vec{c})}{\frac{(\vec{a}\times\vec{b})\cdot(\vec{b}\times\vec{c})\times(\vec{c}\times\vec{a})}{\vec{a}\cdot(\vec{b}\times\vec{c})}}$$

$$= \frac{(\vec{a}\cdot\vec{b}\times\vec{c})\vec{a}}{\frac{(\vec{a}\cdot\vec{b}\times\vec{c})^2}{\vec{a}\cdot(\vec{b}\times\vec{c})}} = \vec{a}$$

$$\therefore \quad \vec{a} = \frac{\vec{b}' \times \vec{c}'}{\vec{a}' \cdot (\vec{b}' \times \vec{c}')}, \quad \because \quad (\vec{A}\times\vec{B})\times(\vec{A}\times\vec{C}) = (\vec{A}\cdot\vec{B}\times\vec{C})\vec{A}$$

and $(\vec{A}\times\vec{B})\cdot(\vec{B}\times\vec{C})\times(\vec{C}\times\vec{A}) = (\vec{A}\cdot\vec{B}\times\vec{C})^2$.

Similarly, we can prove that, $\vec{b} = \dfrac{\vec{c}' \times \vec{a}'}{\vec{a}' \cdot (\vec{b}' \times \vec{c}')}$ and $\vec{c} = \dfrac{\vec{a}' \times \vec{b}'}{\vec{a}' \cdot (\vec{b}' \times \vec{c}')}$.

Example 10.60. *If $\vec{a}'$, $\vec{b}'$ and $\vec{c}'$ be a set of reciprocal vectors of $\vec{a}$, $\vec{b}$, $\vec{c}$ and $\vec{r}$ is any vector, then show that, $\vec{r} = (\vec{r}\cdot\vec{a}')\vec{a} + (\vec{r}\cdot\vec{b}')\vec{b} + (\vec{r}\cdot\vec{c}')\vec{c}$.*

Solution : We know that, if $\vec{A}$, $\vec{B}$, $\vec{C}$, $\vec{D}$ be any four vectors, then

$$\vec{B}(\vec{A}\cdot\vec{C}\times\vec{D}) - \vec{A}(\vec{B}\cdot\vec{C}\times\vec{D}) = \vec{C}(\vec{A}\cdot\vec{B}\times\vec{D}) - \vec{D}(\vec{A}\cdot\vec{B}\times\vec{C}).$$

$$\therefore \quad \vec{D} = \frac{\vec{A}(\vec{B}\cdot\vec{C}\times\vec{D})}{\vec{A}\cdot\vec{B}\times\vec{C}} - \frac{\vec{B}(\vec{A}\cdot\vec{C}\times\vec{D})}{\vec{A}\cdot\vec{B}\times\vec{C}} + \frac{\vec{C}(\vec{A}\cdot\vec{B}\times\vec{D})}{\vec{A}\cdot\vec{B}\times\vec{C}}.$$

Now putting $\vec{A} = \vec{a}$, $\vec{B} = \vec{b}$, $\vec{C} = \vec{c}$ and $\vec{D} = \vec{r}$, we get from the above relation,

$$\vec{r} = \frac{\vec{r}\cdot\vec{b}\times\vec{c}}{\vec{a}\cdot\vec{b}\times\vec{c}}\vec{a} + \frac{\vec{r}\cdot\vec{c}\times\vec{a}}{\vec{a}\cdot\vec{b}\times\vec{c}}\vec{b} + \frac{\vec{r}\cdot\vec{a}\times\vec{b}}{\vec{a}\cdot\vec{b}\times\vec{c}}\vec{c}$$

$$= \vec{r}\cdot\frac{\vec{b}\times\vec{c}}{\vec{a}\cdot\vec{b}\times\vec{c}}\vec{a} + \vec{r}\cdot\frac{\vec{c}\times\vec{a}}{\vec{a}\cdot\vec{b}\times\vec{c}}\vec{b} + \vec{r}\cdot\frac{\vec{a}\times\vec{b}}{\vec{a}\cdot\vec{b}\times\vec{c}}\vec{c}$$

$$= (\vec{r}\cdot\vec{a}')\vec{a} + (\vec{r}\cdot\vec{b}')\vec{b} + (\vec{r}\cdot\vec{c}')\vec{c} \quad \textbf{(Proved)}.$$

Example 10.61. *$\vec{a}$, $\vec{b}$, $\vec{c}$ are three non-coplanar vectors and three reciprocal vectors are defined by $\vec{a}^* = \dfrac{\vec{b}\times\vec{c}}{\vec{a}\cdot\vec{b}\times\vec{c}}$, $\vec{b}^* = \dfrac{\vec{a}\times\vec{b}}{\vec{a}\cdot\vec{b}\times\vec{c}}$, $\vec{c}^* = \dfrac{\vec{a}\times\vec{c}}{\vec{a}\cdot\vec{b}\times\vec{c}}$.*

Show that

(i) *$\vec{a}^*$, $\vec{b}^*$, $\vec{c}^*$ describe a parallelopiped whose volume is the reciprocal of the volume described by $\vec{a}$, $\vec{b}$, $\vec{c}$.*

(ii) *Any vector $\vec{H} = h\vec{a}^* + k\vec{b}^* + l\vec{c}^*$ is normal to the plane that intersects $\vec{a}$, $\vec{b}$, $\vec{c}$ at the points $\frac{\vec{a}}{h}$, $\frac{\vec{b}}{k}$, $\frac{\vec{c}}{l}$ respectively, where h, k, l are constants and it is assumed that the vectors $\vec{a}$, $\vec{b}$, $\vec{c}$ are drawn from a common origin.* **[BU(H) 1998]**

Solution : (i) Let the volume of the parallelopiped described by the vectors $\vec{a}$, $\vec{b}$, $\vec{c}$ be V, then

$$V = \vec{a} \cdot \vec{b} \times \vec{c}.$$

But, $\vec{b}^* = \dfrac{\vec{c} \times \vec{a}}{\vec{a} \cdot \vec{b} \times \vec{c}}$, $\vec{c}^* = \dfrac{\vec{a} \times \vec{b}}{\vec{a} \cdot \vec{b} \times \vec{c}}$.

$$\therefore \quad \vec{b}^* \times \vec{c}^* = \frac{(\vec{c} \times \vec{a}) \times (\vec{a} \times \vec{b})}{(\vec{a} \cdot \vec{b} \times \vec{c})^2} = \frac{(\vec{c} \cdot \vec{a} \times \vec{b})\vec{a} - (\vec{a} \cdot \vec{a} \times \vec{b})\vec{c}}{(\vec{a} \cdot \vec{b} \times \vec{c})^2}$$

$$= \frac{\vec{a}}{\vec{a} \cdot (\vec{b} \times \vec{c})} \quad [\because \vec{a} \cdot \vec{a} \times \vec{b} = 0].$$

$$\therefore \quad \vec{a}^* \cdot \vec{b}^* \times \vec{c}^* = \frac{\vec{a}^* \cdot \vec{a}}{\vec{a} \cdot (\vec{b} \times \vec{c})} = \frac{1}{\vec{a} \cdot (\vec{b} \times \vec{c})} = \frac{1}{V}.$$

(ii) Let AB be a plane which intersects $\vec{a}$, $\vec{b}$, $\vec{c}$ at P, Q, R respectively whose position vectors w.r.t. O (Fig. 10.45) are $\frac{\vec{a}}{h}$, $\frac{\vec{b}}{k}$, $\frac{\vec{c}}{l}$.

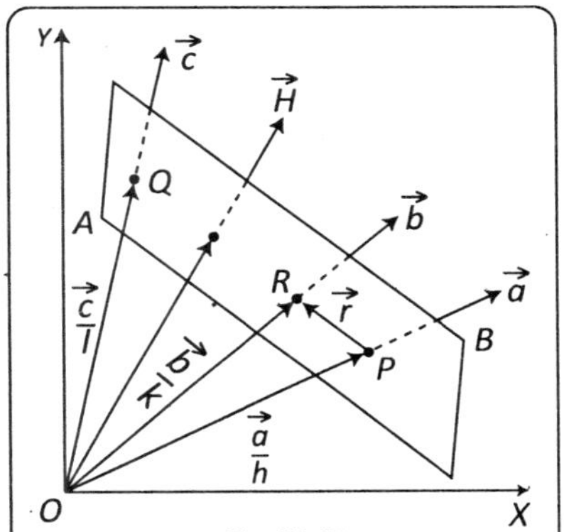

Fig. 10.45

Now, from $\triangle OPR$,

$$\overrightarrow{PR} = \overrightarrow{OR} - \overrightarrow{OP} \quad \text{or,} \quad \vec{r} = \frac{\vec{b}}{k} - \frac{\vec{a}}{h}$$

and $\vec{H} = h\vec{a}^* + k\vec{b}^* + l\vec{c}^*$

$$= h\frac{\vec{b} \times \vec{c}}{\vec{a} \cdot \vec{b} \times \vec{c}} + k\frac{\vec{c} \times \vec{a}}{\vec{a} \cdot \vec{b} \times \vec{c}} + l\frac{\vec{a} \times \vec{b}}{\vec{a} \cdot \vec{b} \times \vec{c}}$$

$$= \frac{1}{(\vec{a} \cdot \vec{b} \times \vec{c})}[h(\vec{b} \times \vec{c}) + k(\vec{c} \times \vec{a}) + l(\vec{a} \times \vec{b})].$$

$$\therefore \quad \vec{H} \cdot \vec{r} = \frac{1}{(\vec{a} \cdot \vec{b} \times \vec{c})}[h(\vec{b} \times \vec{c}) + k(\vec{c} \times \vec{a}) + l(\vec{a} \times \vec{b})] \cdot \left[\left(\frac{\vec{b}}{k} - \frac{\vec{a}}{h}\right)\right]$$

$$= \frac{1}{(\vec{a} \cdot \vec{b} \times \vec{c})}[\{0 + (\vec{c} \times \vec{a}) \cdot \vec{b} + 0\} - \{\vec{a} \cdot (\vec{b} \times \vec{c}) - 0 - 0\}]$$

$$= \frac{1}{(\vec{a} \cdot \vec{b} \times \vec{c})}[\vec{a} \cdot (\vec{b} \times \vec{c}) - \vec{a} \cdot (\vec{b} \times \vec{c})] = 0.$$

∴ $\vec{H}$ is perpendicular to the plane AB as $\vec{r}$ lies in the plane (AB).

Example 10.62. *Let the points P, Q and R have position vectors $\vec{r_1} = 3\hat{i} - 2\hat{j} - \hat{k}$, $\vec{r_2} = \hat{i} + 3\hat{j} + 4\hat{k}$ and $\vec{r_3} = 2\hat{i} + \hat{j} - 2\hat{k}$ relative to an origin O. Find the distance of P to the plane OQR.*

Solution : Since the position vectors $\vec{r}$, $\vec{r_2}$ and $\vec{r_3}$ are in the same plane Fig. 10.45,

$$\therefore \quad \vec{r} \cdot (\vec{r_2} \times \vec{r_3}) = 0 \quad \text{or,} \quad \begin{vmatrix} x & y & z \\ 1 & 3 & 4 \\ 2 & 1 & -2 \end{vmatrix} = 0$$

$\because \vec{r} = \hat{i}x + \hat{j}y + \hat{k}z$ is the position vector of the points and lying in the plane OQR (Fig. 10.46) which is the foot of the perpendicular drawn from P on the plane OQR.

$\therefore \quad x(-6-4) - y(-2-8) + z(1-6) = 0 \quad$ or, $\quad -10x + 10y - 5z = 0$

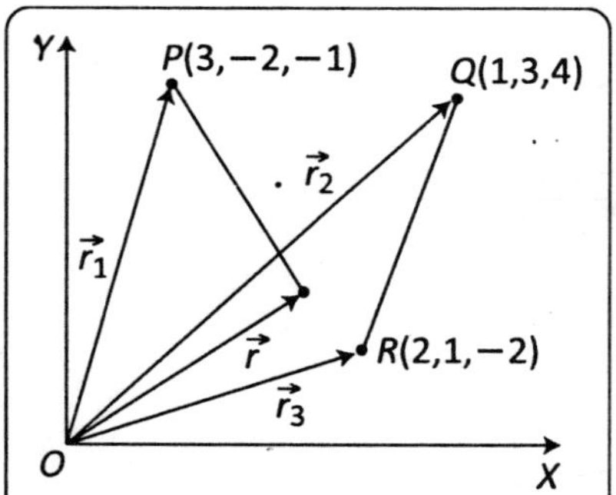

Fig. 10.46

or, $\quad -2x + 2y - z = 0$ is the equation of the plane OQR.

Now, $\vec{r_1} = 3\hat{i} - 2\hat{j} - \hat{k}$ is the position vector of P having coordinates $(3, -2, -1)$.

$\therefore$ the perpendicular distance from P to the plane OQR is

$$PS = \frac{-2 \times 3 + 2(-2) - 1 \times (-1)}{\sqrt{2^2 + 2^2 + 1^2}} = \frac{-6 - 4 + 1}{\pm 3}.$$

$$= \frac{-9}{\pm 3} = \pm 3.$$

$\therefore \quad |\boldsymbol{PS}| = \mathbf{3}$

Exercises

10(A) : Objective Type Questions

Select the correct answer

10.1 If $\hat{n}$ is a unit vector in the direction of $\vec{A}$, then

(a) $\hat{n} = \dfrac{\vec{A}}{|\vec{A}|}$ **(b)** $\hat{n} = \vec{A}|\vec{A}|$ **(c)** $\hat{n} = \dfrac{|\vec{A}|}{\vec{A}}$ **[WBUT 2005]**

10.2 Three vectors $\vec{a}$, $\vec{b}$ and $\vec{c}$ will form a triangle, if

(a) $\vec{a} + \vec{b} > \vec{c}$ **(b)** $\vec{a} - \vec{b} < \vec{c}$ **(c)** $\vec{a} + \vec{b} = \vec{c}$ **(d)** $\vec{a} + \vec{b} = -\vec{c}$.

10.3 Two vectors $\vec{A}$ and $\vec{B}$ are parallel, when

(a) $\vec{A} \times \vec{B} = 0$ **(b)** $\vec{A} \cdot \vec{B} = 0$ **(c)** $\vec{A} \cdot \vec{B} = 1$ **[WBUT 2005]**

10.4 The vectors $\vec{a} = -2\hat{i} - 2\hat{j}$ and $\hat{b} = \hat{i} - 2\hat{k}$ are

(a) parallel **(b)** orthogonal **(c)** linearly dependent **(d)** none of these.

10.5 The vectors $\vec{a} = 2\hat{i} + 3\hat{j} - 4\hat{k}$ and $\vec{b} = a\hat{i} + b\hat{j} + c\hat{k}$ are perpendicular when

(a) $a = 2, b = 3, c = -4$ **(b)** $a = 4, b = 4, c = 5$
(c) $a = 4, b = 4, c = -5$ **(d)** none of these.

10.6 A vector $\hat{n}$, perpendicular to $\vec{b}$ and $\vec{a}$ such that $\vec{b}, \vec{n}, \vec{a}$ form a right-handed system, is given by

(a) $\vec{a} \times \vec{b}$ **(b)** $\vec{a} \cdot \vec{b}$ **(c)** $\vec{b} \times \vec{a}$ **(d)** $|\vec{a} \times \vec{b}|$. **[AMIETE(Winter) 1989]**

10.7 If $|\vec{A}| = |\vec{B}|$, then $(\vec{A} + \vec{B})$ and $(\vec{A} - \vec{B})$ are

(a) parallel **(b)** orthogonal **(c)** equal **(d)** none of these.

10.8 If $\vec{A} = 3\hat{i} - 2\hat{j} + \hat{k}$, $\vec{B} = \hat{i} - 3\hat{j} + 5\hat{k}$ and $\vec{C} = 2\hat{i} + \hat{j} - 4\hat{k}$, then $\vec{A}$, $\vec{B}$, $\vec{C}$ form a

(a) right-angled triangle (b) equilateral triangle
(c) acute-angled triangle (d) isosceles triangle.

10.9 If $\vec{A} \cdot \vec{B} = \vec{A} \cdot \vec{C}$ and $\vec{A} \times \vec{B} = \vec{A} \times \vec{C}$, then

(a) $\vec{A}$ is $\perp$ to $(\vec{B} - \vec{C})$ (b) $\vec{A}$ is $\parallel$ to $(\vec{B} - \vec{C})$
(c) either $\vec{A} = 0$ or $\vec{B} = \vec{C}$ (d) none of these.

10.10 The volume of the parallelopiped with $\vec{A}$, $\vec{B}$, $\vec{C}$ as the coterminous edges is given by

(a) $\vec{A} \times (\vec{B} \times \vec{C})$ (b) $\vec{A} + \vec{B} \times \vec{C}$ (c) $\vec{A} \cdot (\vec{B} \times \vec{C})$ (d) $\vec{A} + \vec{B} + \vec{C}$

10.11 Three vectors $\vec{a}$, $\vec{b}$ and $\vec{c}$ are linearly independent, if

(a) $\lambda\vec{a} + \mu\vec{b} + \nu\vec{c} = 0$ (b) $\lambda\vec{a} + \mu\vec{b} + \nu\vec{c} \neq 0$
(c) $\lambda\vec{a} + \mu\vec{b} + \nu\vec{c} = 1$ (d) none of these.

10.12 If $\vec{A}$, $\vec{B}$, $\vec{C}$ be non-zero vectors, then they will be coplanar, if

(a) $\vec{A} \times (\vec{B} \times \vec{C}) = 1$ (b) $\vec{A} \cdot (\vec{B} \times \vec{C}) = 1$ (c) $\vec{A} \times (\vec{B} \times \vec{C}) = 0$ (d) $\vec{A} \cdot (\vec{B} \times \vec{C}) \neq 0$.

10.13 If $\hat{i}$, $\hat{j}$, $\hat{k}$ be triad of unit vectors, then

(a) $\hat{i} \cdot (\vec{j} \times \vec{k}) = 0$ (b) $\hat{i} \cdot (\vec{j} \times \vec{k}) = 1$ (c) $(\hat{i} \times \vec{j}) \cdot \vec{k} = 0$ (d) $\hat{i} + \vec{j} + \vec{k} = 3$.

10.14 The projection of $\vec{B}$ along $\vec{A}$ is

(a) $\vec{B} \cdot \dfrac{\vec{A}}{|\vec{A}|}$ (b) $\vec{B} \cdot \dfrac{\vec{A}}{|\vec{B}|}$ (c) $\dfrac{\vec{B}}{|\vec{B}|} \cdot \vec{A}$ (d) $\dfrac{\vec{A} \cdot \vec{B}}{|\vec{A} \cdot \vec{B}|}$.

10.15 If $0.5\hat{i} + 0.8\hat{j} + a\hat{k}$ is a unit vector, then a is

(a) $\sqrt{0.89}$ (b) 0.2 (c) 0.3 (d) $\sqrt{0.11}$.

10.16 Which of the following is not correct?

(a) $\vec{A} \times \vec{B} = -\vec{B} \times \vec{A}$ (b) $\vec{A} \times \vec{B} \neq \vec{B} \times \vec{A}$
(c) $\vec{A} \times (\vec{B} \times \vec{C}) \neq (\vec{A} \times \vec{B})$ (d) $\vec{A} \times (\vec{B} + \vec{C}) = \vec{A} \times \vec{B} + \vec{C}$.

10.17 If a body is in equilibrium under the action of three forces two of which are $\vec{F_1} = 5\hat{i}$ and $\vec{F_2} = 7\hat{j}$, then force is

(a) $5\hat{i} + 7\vec{j}$ (b) $5\hat{i} - 7\vec{j}$ (c) $-5\hat{i} - 7\vec{j}$ (d) $-5\hat{i} + 7\vec{j}$.

10.18 If $|\vec{a} + \vec{b}| = |\vec{a} - \vec{b}|$, then

(a) $\vec{a} \cdot \vec{b} \neq 0$ (b) $\vec{a} \cdot \vec{b} \neq \vec{b} \cdot \vec{a}$ (c) $\vec{a} \cdot \vec{b} = 0$ (d) $\vec{a} \times \vec{b} = 0$.

10.19 If $\vec{\alpha} = 3\hat{i} - 2\hat{j} + \hat{k}$, $\vec{\beta} = 2\hat{i} - \hat{k}$, then $(\vec{\alpha} \times \vec{\beta}) \cdot \vec{\alpha}$ is equal to

(a) $\vec{i} + \vec{j} + \vec{k}$ (b) $\vec{i} + \vec{k}$ (c) 0 (d) > 2. [WBUT(Math) 2006]

10.20 If $\phi = 3x^2y - y^3z^2$, then grad ϕ at $(1, -2, -1)$ is

(a) $-16\hat{i} + 12\vec{j} - 9\vec{k}$ (b) $-9\hat{i} - 12\vec{j} + 16\vec{k}$

(c) $-12\hat{i} - 9\vec{j} - 16\vec{k}$ (d) $12\hat{i} + 16\vec{j} - 9\vec{k}$. **[WBUT(Math) 2006]**

Answers

10.1 (a)	**10.2** (b)	**10.3** (a)	**10.4** (d)	**10.5** (b)	**10.6** (c)	**10.7** (b)	**10.8** (a)
10.9 (d)	**10.10** (c)	**10.11** (b)	**10.12** (c)	**10.13** (b)	**10.14** (a)	**10.15** (c)	**10.16** (d)
10.17 (c)	**10.18** (c)	**10.19** (c)	**10.20** (c).				

10(B) : Short Answer Type Questions

10.1 Prove that, $\begin{vmatrix} \vec{A}\cdot\vec{A} & \vec{A}\cdot\vec{B} \\ \vec{A}\cdot\vec{B} & \vec{B}\cdot\vec{B} \end{vmatrix} = |\vec{A}\times\vec{B}|$.

10.2 Prove that, $\hat{A} = \hat{i}\cos(\vec{A}, x) + \hat{j}\cos(\vec{A}, y) + \hat{k}\cos(\vec{A}, z)$, where $\hat{A}$ is a unit vector and $\cos(\vec{A}, x)$, $\cos(\vec{A}, y)$ and $\cos(\vec{A}, z)$ are direction cosines of $\vec{A}$. **[AU 1994]**

10.3 What are **(i)** polar vector **(ii)** pseudovector and **(iii)** pseudoscalar?

Solution : (i) *Polar vector* : If there be two vectors $\vec{r}(x, y, z)$ and $\vec{r}(x', y', z')$ such that $\vec{r}(x', y', z')$ can be obtained just by transformation of x' to $-x$, y' to $-y$ and z' to $-z$ in the reverse direction, then $\vec{r}(x, y, z)$ is called the *polar vector.*

For example, displacement, velocity, force, etc. are polar vectors.

(ii) *Pseudovector (or axial vectors)* : If there be two polar vectors $\vec{A}$ and $\vec{B}$ such that $\hat{A} = \hat{i}\vec{A_1} + \hat{j}\vec{A_2} + \hat{k}\vec{A_3}$ and $\hat{B} = \hat{i}\vec{B_1} + \hat{j}\vec{B_2} + \hat{k}\vec{B_3}$, then

$$\vec{C} = \vec{A}\times\vec{B} = \hat{i}(A_2B_3 - A_3B_2) + \hat{j}(A_3B_1 - A_1B_3) + \hat{k}(A_1B_3 - B_1A_3)$$

is called a *pseudovector* or an *axial vector.* So, a vector obtained through the cross product of two vectors is a *pseudo* or an *axial vector.* For example, angular velocity (ω), angular momentum ($\vec{\Omega}$), torque ($\vec{\Lambda}$), etc. are pseudovectors.

(iii) *Pseudoscalar* : If $\vec{A}$, $\vec{B}$ and $\vec{C}$ be three polar vectors, then $\vec{A}\cdot\vec{B}\times\vec{C} = V$ is a scalar. If $V' = -V$ can be obtained by reversed transformation through reversal axes, then V is called a *pseudoscalar.* In general, $V' = |p|$, where the inversion determinant

$$|p| = \begin{vmatrix} -1 & 0 & 0 \\ 0 & -1 & 0 \\ 0 & 0 & -0 \end{vmatrix} = -1$$

10.4 Angular momentum is defined by $\vec{L} = \vec{r}\times\vec{p}$, where the symbols have their usual meaning. Is $\vec{L}$ a polar vector or, pseudovector? **[BU(H) 2003] [Ans:** Pseudovector**]**

10.5 Given, $\vec{A}$, $\vec{B}$, $\vec{C}$ which are all unequal and not mutually perpendicular. Is it possible to have $\vec{A}\cdot\vec{B} = \vec{A}\cdot\vec{C}$? **[BU(H) 2003]**

[**Hint :** Let $\vec{A} = \hat{i} + \hat{j} + \hat{k}$, $\vec{B} = \hat{i}a + \hat{j}b + \hat{k}c$ and $\vec{C} = \hat{i}b + \hat{j}c + \hat{k}a$, then

$$\vec{B} - \vec{C} = \hat{i}(a-b) + \hat{j}(b-c) + \hat{k}(c-a).$$

$\therefore \quad \vec{A}\cdot(\vec{B}-\vec{C}) = a-b+b-c+c-a$ and $\vec{A}\cdot\vec{B} = \vec{A}\cdot\vec{C} = a+b+c.$

Since $\vec{A}\cdot\vec{B} \neq 0, \vec{B}\cdot\vec{C} \neq 0, \vec{C}\cdot\vec{A} \neq 0,$

$\therefore \vec{A}, \vec{B}, \vec{C}$ are not mutually perpendicular.

Also, $\because \vec{A}\cdot(\vec{B}-\vec{C}) = 0,$

$\therefore \quad (\vec{B}-\vec{C})$ is $\perp$ to $\vec{A}$ or, $\vec{A}\cdot\vec{B} = \vec{A}\cdot\vec{C}.$

So, it is possible to have $\vec{A}\cdot\vec{B} = \vec{A}\cdot\vec{C}.$

10.6 Does the commutative law apply to scalar product and vector product? **[BU(H) 2003]**

10.7 Is electric current a vector quantity? **[BU(H) 2002]**

10.8 If $\vec{A} = \lambda\vec{B}$, prove that $\dfrac{A_x}{B_x} = \dfrac{A_y}{B_y} = \dfrac{A_z}{B_z}.$ **[AU 1995]**

10.9 If $\vec{A} = 2\hat{\imath}+3\hat{\jmath}+4\hat{k}$ and $\vec{B} = 3\hat{\imath}+4\hat{\jmath}+5\hat{k}$, find the angle between $\vec{A}$ and $\vec{B}$.

[Ans: $\cos^{-1}\left(\frac{38}{\sqrt{1450}}\right)$**]**

10.10 If $\vec{A}\times\vec{B} = 0$, then can we conclude that $\vec{A} = \vec{B}$.

10.11 If a vector $\vec{A}$ points towards vertically upward and $\vec{B}$ towards west, then what is the direction of $\vec{B}\times\vec{A}$?

10.12 If $\vec{a}^*, \vec{b}^*, \vec{c}^*$ be a reciprocal system of vectors, then prove that $\vec{a}^*\cdot\vec{a} = 1$, but $\vec{a}^*\cdot\vec{b} = 0.$

10.13 If the magnitude of the vector field $\vec{A}(t)$ is constant with respect to t, show that $\dfrac{d\vec{A}}{dt}$, is perpendicular to $\vec{A}$. **[CU(H) 2013, 2010]**

10.14 For two sets of vectors $(\vec{a}, \vec{b}, \vec{c})$ and $(\vec{a}', \vec{b}', \vec{c}')$ if $\vec{a}\cdot\vec{a}' = 1$, $\vec{b}\cdot\vec{a}' = \vec{c}\cdot\vec{a} = 0$, then show that $\vec{a}' = \dfrac{\vec{b}\times\vec{c}}{\vec{a}\cdot(\vec{b}\times\vec{c})}.$ **[CU(H) 2011]**

10.15 If $\vec{A}\times\vec{B} = \vec{A}\times\vec{C}$, then explain whether it always implies $\vec{B}\times\vec{C}$. **[CU(H) 2005]**

10.16 The position vectors $\vec{r_1}$ and $\vec{r_2}$ have direction cosines (l_1, m_1, n_1) and (l_2, m_2, n_2) respectively. Find the angle between $\vec{r_1}$ and $\vec{r_2}$. **[VU(H) 2010]**

10.17 If $\vec{A}+\vec{B} = 5\hat{\imath}+3\hat{\jmath}+2\hat{k}$ and $\vec{A}\times\vec{B} = 7\hat{\imath}+11\hat{\jmath}+\hat{k}$, then find $\vec{A}$ and $\vec{B}$. **[VU(H) 2007]**

10.18 If $\vec{A}+\vec{B}+\vec{C} = 0$, prove that $\vec{A}\times\vec{B} = \vec{B}\times\vec{C} = \vec{C}\times\vec{A}$, also interpret it geometrically. **[VU(H) 2006]**

10.19 If $\hat{a}, \hat{b}, \hat{c}$ be three unit vectors such that $\hat{a}\times(\hat{b}\times\hat{c}) = \frac{1}{2}\hat{b}$, find the angles which $\hat{a}$ makes with $\hat{b}$ and $\hat{c}$. **[BU(H) 2010]**

10.20 Determine λ and μ such that the points $(-1,3,2)$, $(-4,2,-2)$ and $(5,\lambda,\mu)$ are collinear. **[BU(H) 2010]**

10.21 Find the volume of the parallelopiped with sides $\vec{a} = \hat{\imath}+2\hat{\jmath}+3\hat{k}$, $\vec{b} = 4\hat{\imath}+5\hat{\jmath}+6\hat{k}$ and $\vec{c} = 7\hat{\imath}+8\hat{\jmath}+10\hat{k}$. **[BU(H) 2008]**

10.22 A point P divides the line segment AB in the ratio $\lambda : \mu$ (see fig). If the position vectors of the points A and B are $\vec{a}$ and $\vec{b}$ respectively, find the position of vector $\vec{p}$ of the point P.

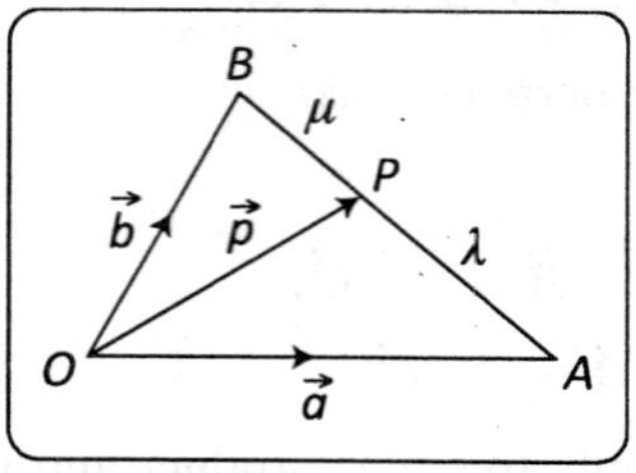

[BU(H) 2008]

10.23 If $\vec{A}(t)$ is a vector function of the scalar variable t and is of constant length, show that $\dfrac{d\vec{A}}{dt}$ is a vector perpendicular to $\vec{A}$. [NBU(H) 2008]

10(C) : Broad Answer Type Questions

10.1 Prove that, $\vec{A}\times(\vec{B}\times\vec{C})+\vec{B}\times(\vec{C}\times\vec{A})+\vec{C}\times(\vec{A}\times\vec{B})=0$, the symbols have their usual meaning. [WBUT 2004]

10.2 Find the work done to move an object along a vector $\vec{r}=3\hat{i}+2\hat{j}-5\hat{k}$, if the applied force is $\vec{F}=2\hat{i}-\hat{j}-\hat{k}$. [WBUT 2005]

10.3 A force of 15 units acts along $\hat{i}-2\hat{j}+2\hat{k}$ and passes through the point $2\hat{i}-2\hat{j}+2\hat{k}$. Find the moment of the force about the point $(1,1,1)$.

10.4 Find the volume of the tetrahedron whose vertices are the points, $A(2,1,8)$, $B(3,2,9)$, $C(2,1,4)$ and $D(3,3,10)$. [Ans: $\frac{2}{3}$]

10.5 Show that the acute angle between two diagonals of a cube has cosine $\dfrac{1}{3}$.

10.6 Prove vectorially that,

(a) $\sin(A+B)=\sin A\cos B+\cos A\sin B$ (b) $\cos(A+B)=\cos A\cos B-\sin A\sin B$.

10.7 Find the value of λ such that the scalar product of the vector $\hat{i}+\hat{j}+\hat{k}$ with the unit vector parallel to the sum of the vectors $\lambda\hat{i}+2\hat{j}+3\hat{k}$ and $2\hat{i}+4\hat{j}-5\hat{k}$ is equal to one. [Ans: $\lambda=-\frac{38}{7}$]

10.8 In a tetrahedron, if two pairs of opposite edges are perpendicular, the third pair is also perpendicular to each other and the sum of the squares on two opposite edges is the same for each pair. [AMIE(Summer) 1987]

10.9 A rigid body is rotating at the rate of 2.5 radians per second about an axis AB, where A and B are the points $(1,-2,1)$ and $(3,-4,2)$. Find the velocity of the point P at $(5,-1,-1)$ of the body. [AMIE(Summer) 1992]

10.10 If $\vec{A}$, $\vec{B}$, $\vec{C}$ be three non-collinear vectors, then show that $\vec{A}=5\vec{a}+6\vec{b}+7\vec{c}$, $\vec{B}=7\vec{a}-8\vec{b}+9\vec{c}$ and $\vec{C}=3\vec{a}+20\vec{b}+5\vec{c}$ are coplanar. [AMIE(Summer) 1992]

10.11 If $\vec{A}$, $\vec{B}$, $\vec{C}$ are coplanar vectors and $\vec{A}$ perpendicular to $\vec{B}$, then prove that,

$$\vec{C}=\frac{\begin{vmatrix}\vec{C}\cdot\vec{A} & \vec{B}\cdot\vec{A}\\ \vec{C}\cdot\vec{B} & \vec{B}\cdot\vec{B}\end{vmatrix}\vec{A}+\begin{vmatrix}\vec{A}\cdot\vec{A} & \vec{C}\cdot\vec{A}\\ \vec{A}\cdot\vec{B} & \vec{C}\cdot\vec{B}\end{vmatrix}\vec{B}}{A^2B^2-(\vec{A}\cdot\vec{B})^2}.$$

[PU 1986]

10.12 If $\vec{A}$ and $\vec{B}$ are two given vectors then prove that,

(a) $|\vec{A}+\vec{B}| \le |\vec{A}|+|\vec{B}|$ (b) $|\vec{A}-\vec{B}| \ge |\vec{A}|-|\vec{B}|$

(c) $|\vec{A}+\vec{B}+\vec{C}| \le |\vec{A}|+|\vec{B}|+|\vec{C}|$. **[BU(H) 1998]**

10.13 Prove that, $(\vec{A}\cdot\vec{B}\times\vec{C})(\vec{a}\cdot\vec{b}\times\vec{c}) = \begin{vmatrix} \vec{A}\cdot\vec{a} & \vec{A}\cdot\vec{b} & \vec{A}\cdot\vec{c} \\ \vec{B}\cdot\vec{a} & \vec{B}\cdot\vec{b} & \vec{B}\cdot\vec{c} \\ \vec{C}\cdot\vec{a} & \vec{C}\cdot\vec{b} & \vec{C}\cdot\vec{c} \end{vmatrix}$.

10.14 Find a unit vector normal to the plane through the point $(-1,2,3)$, $(1,1,1)$ and $(2,-1,3)$. **[Calicut U 1984]** **[Ans:** $\frac{1}{3}(2\hat{i}+2\hat{j}+\hat{k})$**]**

10.15 Show that, $\hat{i}\times(\vec{R}\times\hat{i})+\hat{j}\times(\vec{R}\times\hat{j})+\hat{k}\times(\vec{R}\times\hat{k}) = 2\vec{R}$. **[AU 1990, RU 1989]**

10.16 Find the volume of the tetrahedron, three of whose coterminous edges are $(3\hat{i}-\hat{j}+2\hat{k})$, $(2\hat{i}+\hat{j}-\hat{k})$ and $(\hat{i}-2\hat{j}+2\hat{k})$. **[KU 1984]**

10.17 Prove that, $(\vec{B}\times\vec{C})\times(\vec{A}\times\vec{D})+(\vec{C}\times\vec{A})\times(\vec{B}\times\vec{D})+(\vec{A}\times\vec{B})\times(\vec{C}\times\vec{D}) = -2(\vec{A}\cdot\vec{B})\times\vec{C})\vec{D}$. **[RU 1987]**

10.18 If $\vec{A}=\hat{i}+\hat{j}-\hat{k}$, $\vec{B}=\hat{i}-\hat{j}+\hat{k}$, $\vec{C}=\hat{i}-\hat{j}+\hat{k}$, then find (a) $\vec{A}\cdot\vec{B}\times\vec{C}$ (b) $\vec{A}\times(\vec{B}\times\vec{C})$.

10.19 If $\vec{a}$, $\vec{b}$, $\vec{c}$ and $\vec{a}^*$, $\vec{b}^*$, $\vec{c}^*$ be the sets of reciprocal vectors, then prove that,

(a) $\vec{a}^*\times\vec{b}^*+\vec{b}^*\times\vec{c}^*+\vec{c}^*\times\vec{a}^* = \dfrac{\vec{a}^*+\vec{b}^*+\vec{c}^*}{(\vec{a}+\vec{b}+\vec{c})}$

(b) $\vec{a}\times\vec{a}^*+\vec{b}\times\vec{b}^*+\vec{c}\times\vec{c}^* = 3$.

10.20 Prove that, $(\vec{B}\times\vec{c})\cdot|\vec{A}\times(\vec{B}\times\vec{C})| = 0$. **[AU 1993]**

10.21 If $\vec{R_1} = \hat{i}x_1+\hat{j}y_1+\hat{k}z_1$, $\vec{R_2} = \hat{i}x_2+\hat{j}y_2+\hat{k}z_2$, $\vec{R_3} = \hat{i}x_3+\hat{j}y_3+\hat{k}z_3$, be the position vector of points $P_1(x_1,y_1,z_1)$, $P_2(x_2,y_2,z_2)$ and $P_3(x_3,y_3,z_3)$, then show that, the equation for the plane passing through P_1, P_2 and P_3 is

$$(\vec{r}-\vec{r_1})\cdot(\vec{r_2}-\vec{r_1})\times(\vec{r_3}-\vec{r_1}) = 0 \quad \text{or,} \quad \begin{vmatrix} x-x_1 & y-y_1 & z-z_1 \\ x_2-x_1 & y_2-y_1 & z_2-z_1 \\ x_3-x_1 & y_3-y_1 & z_3-z_1 \end{vmatrix} = 0,$$

where $\vec{r} = \hat{i}x+\hat{j}y+\hat{k}z$ is the position vector of any point $P(x,y,z)$ in the same plane.

10.22 (a) Prove the identity $\nabla\cdot(\vec{A}\times\vec{B}) = \vec{B}\cdot(\nabla\times\vec{A}) - \vec{A}(\nabla\times\vec{B})$.

(b) Evaluate : $\nabla^2\left[\nabla\cdot\frac{r}{r^2}\right]$, where $\vec{r} = x\hat{i}+y\hat{i}+z\hat{k}$.

(c) Prove that $\nabla\left[\vec{M}\cdot\nabla\left(\frac{1}{r}\right)\right] = \dfrac{3[\vec{M}\cdot\vec{r}]\vec{r}}{r^5} - \dfrac{\vec{M}}{r^3}$.

10.23 (a) Show that, $\vec{a}\times(\vec{b}\times\vec{c}) = (\vec{a}\cdot\vec{c}) - (\vec{b}\cdot\vec{a})\vec{c}$.

(b) Prove that, $\nabla^2\vec{r}^n = r^{n-2}\vec{r}$.

10.24 (a) Give the vector diagram representation of $\vec{A}\times\vec{B}=\vec{C}$ and $\vec{B}\times\vec{A}=\vec{D}$. Name a vector physical quantity which is the product of two vectors. **[Assam U(Major) 2014]**

10.25 **(a)** Deduce $\nabla \times (\phi \vec{A}) = (\nabla\phi) \times \vec{A} + \phi(\nabla \times \vec{A})$ for the scalar field ϕ and vector field $\vec{A}$. **[CU(H) 2017]**

(b) Deduce $\nabla \cdot (\phi \vec{A}) = (\nabla\phi) \cdot \vec{A} + \phi(\nabla \cdot \vec{A})$. **[CU(H) 2012, '13]**

(c) If $\nabla \times \vec{A} = 0$, show that $\nabla(\vec{A} \times \vec{r}) = 0$. **[CU(H) 2012]**

10.26 For two sets of vectors $[\vec{a}, \vec{b}, \vec{c}]$ and $[\vec{a}', \vec{b}', \vec{c}']$, show that

$$\vec{a}' = \frac{\vec{b} \times \vec{c}}{\vec{a} \cdot (\vec{b} \times \vec{c})}.$$

[CU(H) 2011]

CHAPTER 11

Vector Differentiation

11.1 Vector Differentiation

The function $\vec{R}(t)$ is said to be a continuous function of t, if $\vec{R}(t)$ varies continuously as a scalar variable t changes.

Let a particle moves from $P(\vec{R})$ to $Q(\vec{R}+\Delta\vec{R})$ along the curve PQ in time Δt. Then from Fig. 11.1,

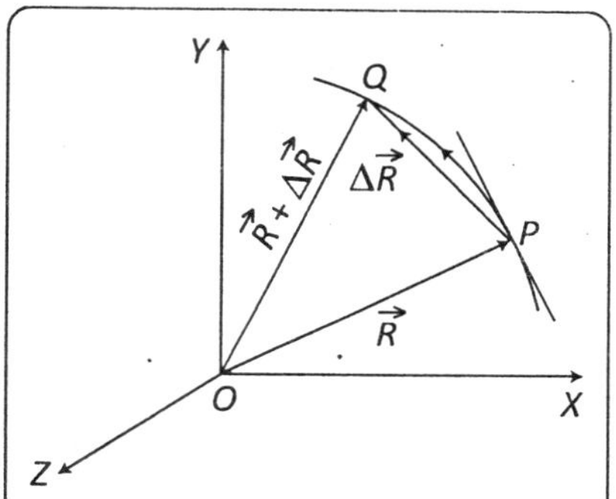

Fig. 11.1

$$\Delta\vec{R} = \overrightarrow{OQ} - \overrightarrow{OP} = \vec{R}(t+\Delta t) - \vec{R}(t)$$

$$\therefore \quad \frac{\Delta\vec{R}}{\Delta t} = \frac{\vec{R}(t+\Delta t) - \vec{R}(t)}{\Delta t}$$

Now, in the limit when $\Delta t \to 0$,

$$\mathop{\text{Lt}}_{\Delta t\to 0} \frac{\Delta\vec{R}}{\Delta t} = \mathop{\text{Lt}}_{\Delta t\to 0} \frac{\vec{R}(t+\Delta t) - \vec{R}(t)}{\Delta t} = \frac{d\vec{R}}{dt} = \vec{R}'(t)$$

or, $\boldsymbol{d\vec{R} = \vec{R}'(t)dt}$.

$\frac{d\vec{R}}{dt}$ is called the *derivative* (or *differential coefficient*) of $\vec{R}(t)$ w.r.t. t and $d\vec{R}$ is called the *differential.*

Now, as $\Delta t \to 0$, $Q \to P$ and the chord becomes the tangent at P to the curve $\frac{d\vec{R}}{dt}$ is a vector in the direction of the tangent at P. $\frac{d^2\vec{R}}{dt^2}$ is *the second order derivative* of $\vec{R}(t)$ w.r.t. t.

Here, $\frac{d\vec{R}}{dt}$ is the velocity of the particle and $\frac{d^2\vec{R}}{dt^2}$ is the acceleration.

If $\vec{R} = R_1(t)\hat{i} + R_2(t)\hat{j} + R_3(t)\hat{k}$, then

$$\boldsymbol{\frac{d\vec{R}}{dt} = \frac{dR_1(t)}{dt}\hat{i} + \frac{dR_2(t)}{dt}\hat{j} + \frac{dR_3(t)}{dt}\hat{k}}.$$

11.2 Some Important Theorems

(i) Any vector function $\vec{R}(t)$ is derivable provided $R_1(t), R_2(t), R_3(t)$ are derivable.

Now,
$$\frac{d\vec{R}}{dt} = \mathop{\text{Lt}}_{\Delta t\to 0} \frac{\vec{R}(t+\Delta t) - \vec{R}(t)}{\Delta t}$$
$$= \mathop{\text{Lt}}_{\Delta t\to 0} \frac{R_1(t+\Delta t) - R_1(t)}{\Delta t}\hat{i} + \mathop{\text{Lt}}_{\Delta t\to 0} \frac{R_2(t+\Delta t) - R_2(t)}{\Delta t}\hat{j} + \mathop{\text{Lt}}_{\Delta t\to 0} \frac{R_3(t+\Delta t) - R_3(t)}{\Delta t}\hat{k}.$$

Thus, for $\frac{d\vec{R}}{dt}$ to exist,

$$\mathop{\text{Lt}}_{\Delta t\to 0} \frac{R_1(t+\Delta t) - R_1(t)}{\Delta t}, \quad \mathop{\text{Lt}}_{\Delta t\to 0} \frac{R_2(t+\Delta t) - R_2(t)}{\Delta t} \text{ and } \frac{R_3(t+\Delta t) - R_3(t)}{\Delta t}$$

must exist.

(ii) If $\vec{A}$ and $\vec{B}$ be two derivable vectors and each is function of the scalar variable t, then

$$\frac{d}{dt}(\vec{A} \pm \vec{B}) = \frac{d\vec{A}}{dt} \pm \frac{d\vec{B}}{dt}.$$

Let $\vec{G} = \vec{A} + \vec{B}$.

Then $\vec{G}(t) = \vec{A}(t) \pm \vec{B}(t)$.

$$\therefore \quad \frac{d\vec{G}}{dt} = \underset{h \to 0}{\text{Lt}} \frac{\vec{G}(t+h) - \vec{G}(t)}{h} = \underset{h \to 0}{\text{Lt}} \frac{[\vec{A}(t+h) \pm \vec{B}(t+h)] - [\vec{A}(t) \pm \vec{B}(t)]}{h}$$

$$= \underset{h \to 0}{\text{Lt}} \frac{\vec{A}(t+h) - \vec{A}(t)}{h} \pm \underset{h \to 0}{\text{Lt}} \frac{\vec{B}(t+h) - \vec{B}(t)}{h}$$

$$\text{or,} \quad \boldsymbol{\frac{d}{dt}(\vec{A} \pm \vec{B}) = \frac{d\vec{A}}{dt} \pm \frac{d\vec{B}}{dt}}.$$

(iii) If ϕ be a scalar function and $\vec{A}$, a vector function of the scalar variable t, then

$$\frac{d}{dt}(\phi\vec{A}) = \phi\frac{d\vec{A}}{dt} + \vec{A}\frac{d\phi}{dt}.$$

We have $\frac{d}{dt}(\phi A) = \underset{\Delta t \to 0}{\text{Lt}} \frac{[(\phi + \Delta\phi)(\vec{A} + \Delta\vec{A}) - \phi\vec{A}]}{\Delta t}$

$$= \underset{\Delta t \to 0}{\text{Lt}} \frac{[\phi\Delta\vec{A} + (\Delta\phi)\vec{A} + \Delta\phi\Delta\vec{A}]}{\Delta t}$$

$$= \underset{\Delta t \to 0}{\text{Lt}} \left[\phi\left(\frac{\Delta\vec{A}}{\Delta t}\right) + \left(\frac{\Delta\phi}{\Delta t}\right)\vec{A} + \left(\frac{\Delta\phi}{\Delta t}\right)\Delta\vec{A}\right]$$

$$\therefore \quad \boldsymbol{\frac{d}{dt}(\phi\vec{A}) = \phi\frac{d\vec{A}}{dt} + \frac{d\phi}{dt}\vec{A}}.$$

(iv) If $\vec{A}$ and $\vec{B}$ are two vector functions of the scalar variable t, then

$$\frac{d}{dt}(\vec{A} \cdot \vec{B}) = \vec{A} \cdot \frac{d\vec{B}}{dt} + \frac{d\vec{A}}{dt} \cdot \vec{B}.$$

Let $\vec{A} = \hat{i}A_1 + \hat{j}A_2 + \hat{k}A_3$ and $\vec{B} = \hat{i}B_1 + \hat{j}B_2 + \hat{k}B_3$.

$$\vec{A} \cdot \frac{d\vec{B}}{dt} + \frac{d\vec{A}}{dt} \cdot \vec{B} = (\hat{i}A_1 + \hat{j}A_2 + \hat{k}A_3) \cdot \left(\hat{i}\frac{dB_1}{dt} + \hat{j}\frac{dB_2}{dt} + \hat{k}\frac{dB_3}{dt}\right)$$

$$+ \left(\hat{i}\frac{dA_1}{dt} + \hat{j}\frac{dA_2}{dt} + \hat{k}\frac{dA_3}{dt}\right) \cdot (\hat{i}B_1 + \hat{j}B_2 + \hat{k}B_3)$$

$$= \left(A_1\frac{dB_1}{dt} + A_2\frac{dB_2}{dt} + A_3\frac{dB_3}{dt}\right) + \left(\frac{dA_1}{dt}B_1 + \frac{dA_2}{dt}B_2 + \frac{dA_3}{dt}B_3\right)$$

$$= \left(A_1\frac{dB_1}{dt} + \frac{dA_1}{dt}B_1\right) + \left(A_2\frac{dB_2}{dt} + \frac{dA_2}{dt}B_2\right) + \left(A_3\frac{dB_3}{dt} + \frac{dA_3}{dt}B_3\right)$$

$$= \frac{d}{dt}(A_1B_2) + \frac{d}{dt}(A_2B_2) + \frac{d}{dt}(A_3B_3)$$

$$= \frac{d}{dt}(A_1B_1 + A_2B_2 + A_3B_3) = \frac{d}{dt}(\vec{A} \cdot \vec{B}).$$

Hence, $\boldsymbol{\frac{d}{dt}(\vec{A} \cdot \vec{B}) = \vec{A} \cdot \frac{d\vec{B}}{dt} + \frac{d\vec{A}}{dt} \cdot \vec{B}}$

(v) If $\vec{A}$ and $\vec{B}$ be two vector functions of the scalar variable t, then

$$\frac{d}{dt}(\vec{A} \times \vec{B}) = \frac{d\vec{A}}{dt} \times \vec{B} + \vec{A} \times \frac{d\vec{B}}{dt}.$$

Let $\vec{A} = \hat{i}A_1 + \hat{j}A_2 + \hat{k}A_3$ and $\vec{B} = \hat{i}B_1 + \hat{j}B_2 + \hat{k}B_3$.

Then, $\frac{d}{dt}(\vec{A} \times \vec{B}) = \frac{d}{dt}\begin{vmatrix} \hat{i} & \hat{j} & \hat{k} \\ A_1 & A_2 & A_3 \\ B_1 & B_2 & B_3 \end{vmatrix} = \begin{vmatrix} \hat{i} & \hat{j} & \hat{k} \\ \frac{dA_1}{dt} & \frac{dA_2}{dt} & \frac{dA_3}{dt} \\ B_1 & B_2 & B_3 \end{vmatrix} + \begin{vmatrix} \hat{i} & \hat{j} & \hat{k} \\ A_1 & A_2 & A_3 \\ \frac{dB_1}{dt} & \frac{dB_2}{dt} & \frac{dB_3}{dt} \end{vmatrix}$

$$= \frac{d\vec{A}}{dt} \times \vec{B} + \vec{A} \times \frac{d\vec{B}}{dt}.$$

$$\therefore \quad \boldsymbol{\frac{d}{dt}(\vec{A} \times \vec{B}) = \frac{d\vec{A}}{dt} \times \vec{B} + \vec{A} \times \frac{d\vec{B}}{dt}}.$$

Since the vector product does not commute, the order of differentiation cannot be changed.

If $\vec{B} = \frac{d\vec{A}}{dt}$, then

$$\boldsymbol{\frac{d}{dt}(\vec{A} \times \vec{B})} = \frac{d}{dt}\left(\vec{A} \times \frac{d\vec{A}}{dt}\right) = \frac{d\vec{A}}{dt} \times \frac{d\vec{A}}{dt} + \vec{A} \times \frac{d^2\vec{A}}{dt^2}$$

$$= \boldsymbol{\vec{A} \times \frac{d^2 A}{dt^2}} \quad \left[\because \frac{d\vec{A}}{dt} \times \frac{d\vec{A}}{dt} = 0\right].$$

(vi) The necessary and sufficient condition for a vector function $\vec{R}(t)$ to be constant is

$$\frac{d}{dt}\vec{R}(t) = 0.$$

Now, if the scalar variable t changes by a small amount, then

$$\frac{d\vec{R}}{dt} = \underset{h \to 0}{\text{Lt}} \frac{\vec{R}(t+h) - \vec{R}(t)}{h} = 0.$$

Thus, the condition $\boldsymbol{\frac{d}{dt}\vec{R}(t) = 0}$ **for** $\boldsymbol{\vec{R}(t)}$ **to be constant is necessary.**

Again, to prove that the condition is sufficient we have

$$\vec{R}(t) = R_1(t)\hat{i} + R_2(t)\hat{j} + R_3(t)\hat{k} = 0,$$

where $R_1(t), R_2(t)$ and $R_3(t)$ are three scalar functions of t.

$$\therefore \quad \frac{d\vec{R}}{dt} = \hat{i}\frac{dR_1}{dt} + \hat{j}\frac{dR_2}{dt} + \hat{k}\frac{dR_3}{dt} = 0.$$

$$\therefore \quad \frac{dR_1}{dt} = \frac{dR_2}{dt} = \frac{dR_3}{dt} = 0.$$

So, $R_1(t)$, $R_2(t)$, $R_3(t)$ are constants, $\therefore \quad \vec{R}(t) = \text{constant}.$

Hence, the condition $\boldsymbol{\frac{d}{dt}\vec{R}(t) = 0}$ **is sufficient for $\vec{R}(t)$ to be constant.**

(vii) The necessary and sufficient condition that a vector $\vec{R}(t)$ have constant magnitude is

$$\vec{R} \cdot \frac{d\vec{R}}{dt} = 0.$$

We have $\vec{R}(t) \cdot \vec{R}(t) = |\vec{R}(t)|^2$, $\therefore \quad \vec{R}(t) \cdot \frac{d\vec{R}(t)}{dt} = |\vec{R}(t)|\frac{d}{dt}|\vec{R}(t)|.$

$$\therefore \quad \vec{R} \cdot \frac{d\vec{R}}{dt} = 0, \text{ if and only if } \frac{d}{dt}|\vec{R}(t)| = 0 \text{ and } |\vec{R}(t)| \text{ not necessarily zero}$$

$$\therefore \quad \frac{d}{dt}|\vec{R}(t)| = 0 \quad \text{or,} \quad |\vec{R}(t)| = \textbf{constant}.$$

This condition is necessary.

Again, $\frac{d\vec{R}(t)}{dt} = 0$, if $\vec{R}(t)$ has a constant magnitude. So, $\vec{R}(t) \cdot \frac{d}{dt}\vec{R}(t) = 0.$

Thus, the condition $\boldsymbol{\vec{R}(t) \cdot \frac{d}{dt}\vec{R}(t) = 0}$ **is sufficient for $\vec{R}(t)$ to have a constant magnitude.**

(viii) If $\vec{R} \times \frac{d\vec{R}}{dt} = 0$, then the vector $\vec{R}(t)$ has a constant direction.

Let $\vec{R}(t) = R(t)\hat{n}(t)$, where $\hat{n}(t)$ is a unit vector along $\vec{R}(t)$.

Then, $\frac{d\vec{R}}{dt} = R\frac{d\hat{n}}{dt} + \frac{dR}{dt}\hat{n}$

and
$$\vec{R} \times \frac{d\vec{R}}{dt} = R\hat{n} \times \left[R\frac{d\hat{n}}{dt} + \frac{dR}{dt}\hat{n}\right] = R^2\hat{n} \times \frac{d\hat{n}}{dt} + R\frac{dR}{dt}(\hat{n} \times \hat{n})$$
$$= R^2\hat{n} \times \frac{d\hat{n}}{dt} \quad [\because \ \hat{n} \times \hat{n} = 0]$$
$$= \mathbf{0}, \quad \left[\frac{d\hat{n}}{dt} = 0, \ \text{if } \hat{n} = \text{ constant}\right].$$

$\therefore$ **$\vec{R}(t)$ has a constant direction.**

(ix) If $\vec{A}$, $\vec{B}$, $\vec{C}$ are three vector functions of the scalar variable t, then

$$\frac{d}{dt}[\vec{A}, \vec{B}, \vec{C}] = \frac{d\vec{A}}{dt} \cdot (\vec{B} \times \vec{C}) + \vec{A} \cdot \left(\frac{d\vec{B}}{dt} \times \vec{C}\right) + \vec{A} \cdot \left(\vec{B} \times \frac{d\vec{C}}{dt}\right).$$

The cyclic order of the factors in each term is maintained.

(x) If $\vec{A}, \vec{B}, \vec{C}$ be any three vector functions of the scalar variable t, then

$$\frac{d}{dt}[\vec{A} \times (\vec{B} \times \vec{C})] = \frac{d\vec{A}}{dt} \times (\vec{B} \times \vec{C}) + \vec{A} \times \left(\frac{d\vec{B}}{dt} \times \vec{C}\right) + \vec{A} \times \left(\vec{B} \times \frac{d\vec{C}}{dt}\right)$$

the order of the factors in each term is maintained.

11.2.1 Partial Derivatives of Vectors

If a vector $\vec{A}$ be a function of more than one scalar variable x, y, z, then $\vec{A} = \vec{A}(x, y, z)$. The derivative of $\vec{A}$ with respect to x keeping y and z constant is given by

$$\frac{\partial \vec{A}}{\partial x} = \underset{\Delta x \to 0}{\text{Lt}} \frac{\vec{A}(x + \Delta x, y, z) - \vec{A}(x, y, z)}{\Delta x}$$

provided the limit exists.

In a similar way the partial derivatives of $\vec{A}$ with respect to y and z are

$$\frac{\partial \vec{A}}{\partial y} = \underset{\Delta y \to 0}{\text{Lt}} \frac{\vec{A}(x, y + \Delta y, z) - \vec{A}(x, y, z)}{\Delta y} \quad \text{and} \quad \frac{\partial \vec{A}}{\partial z} = \underset{\Delta z \to 0}{\text{Lt}} \frac{\vec{A}(x, y, z + \Delta z) - \vec{A}(x, y, z)}{\Delta z}$$

respectively provided these limits exist.

The second and the third partial derivatives are given by

$$\frac{\partial^2 \vec{A}}{\partial x^2} = \frac{\partial}{\partial x}\left(\frac{\partial \vec{A}}{\partial x}\right), \frac{\partial^2 \vec{A}}{\partial y^2} = \frac{\partial}{\partial y}\left(\frac{\partial \vec{A}}{\partial y}\right), \frac{\partial^2 \vec{A}}{\partial z^2} = \frac{\partial}{\partial z}\left(\frac{\partial \vec{A}}{\partial z}\right).$$

$$\frac{\partial^2 \vec{A}}{\partial x \partial y} = \frac{\partial}{\partial x}\left(\frac{\partial \vec{A}}{\partial y}\right), \frac{\partial^2 \vec{A}}{\partial y \partial x} = \frac{\partial}{\partial y}\left(\frac{\partial \vec{A}}{\partial x}\right), \frac{\partial^2 \vec{A}}{\partial y \partial z} = \frac{\partial}{\partial y}\left(\frac{\partial \vec{A}}{\partial z}\right)$$

$$\frac{\partial^3 \vec{A}}{\partial x \partial z^2} = \frac{\partial}{\partial_x}\left(\frac{\partial^2 \vec{A}}{\partial z^2}\right).$$

When $\vec{A}$ has continuous partial derivatives, $\frac{\partial^2 \vec{A}}{\partial x \partial y} = \frac{\partial^2 \vec{A}}{\partial y \partial x}$, then order of differential is unimportant.

Rules for Partial Differentiation of Vectors

If $\vec{A} = \vec{A}(x, y, z)$ and $\vec{B} = \vec{B}(x, y, z)$, then

(i) $\frac{\partial}{\partial x}(\vec{A} \pm \vec{B}) = \frac{\partial \vec{A}}{\partial x} + \frac{\partial \vec{B}}{\partial x}$

(ii) $\frac{\partial}{\partial x}(\vec{A} \cdot \vec{B}) = \vec{A} \cdot \frac{\partial \vec{B}}{\partial x} + \frac{\partial \vec{A}}{\partial x} \cdot \vec{B}$

(iii) $\frac{\partial}{\partial x}(\vec{A} \times \vec{B}) = \vec{A} \times \frac{\partial \vec{B}}{\partial x} + \frac{\partial \vec{A}}{\partial x} \times \vec{B}$

(iv) $\frac{\partial^2}{\partial y \partial x}(\vec{A} \cdot \vec{B}) = \frac{\partial}{\partial y}\left[\frac{\partial}{\partial x}(\vec{A} \cdot \vec{B})\right]$

$$= \frac{\partial}{\partial y}\left[\vec{A} \cdot \frac{\partial \vec{B}}{\partial x} + \frac{\partial \vec{A}}{\partial x} \cdot \vec{B}\right] \vec{A} \cdot \frac{\partial^2 \vec{B}}{\partial y \partial x} + \frac{\partial \vec{A}}{\partial y} \cdot \frac{\partial \vec{B}}{\partial x} + \frac{\partial^2 \vec{A}}{\partial y \partial x} \cdot \vec{B}, \text{ etc.}$$

Differentials of Vectors

(i) If $\vec{A} = \hat{\imath}A_1 + \hat{\jmath}A_2 + \hat{k}A_3$, then the differential of $\vec{A}$ is $d\vec{A} = \hat{\imath}dA_1 + \vec{\jmath}\,dA_2 + \hat{k}dA_3$

(ii) $d(\vec{A}\cdot\vec{B}) = \vec{A}\cdot d\vec{B} + d\vec{A}\cdot\vec{B}$

(iii) $d(\vec{A}\times\vec{B}) = \vec{A}\times d\vec{B} + d\vec{A}\times\vec{B}$

(iv) $d\vec{A} = \frac{\partial\vec{A}}{\partial x}dx + \frac{\partial\vec{A}}{\partial y}dy + \frac{\partial\vec{A}}{\partial z}dz$, if $\vec{A} = \vec{A}(x,y,z)$.

11.3 Space Curve

(1) Tangent

Let a particle moves along the curve P_0PQ in space and P and Q be the positions of it at time t and $t+\Delta t$ respectively. Let $\vec{R}(t)$ and $\vec{R}(t+\Delta t)$ be the position vectors of P and Q with respect to the origin O of the Cartesian co-ordinate system X, Y, Z. Then,

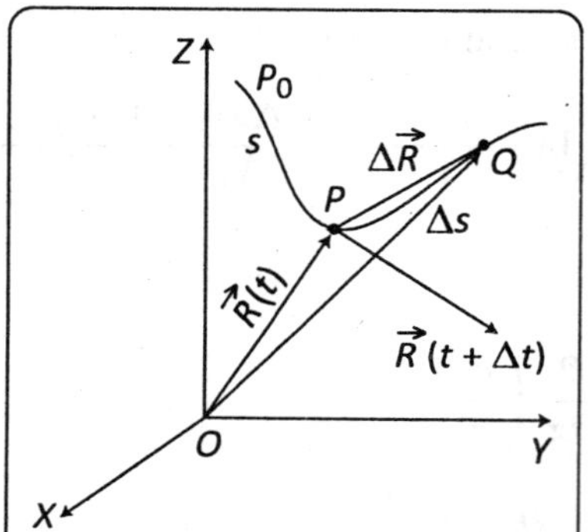

Fig. 11.2

$$\Delta\vec{R} = \vec{R}(t+\Delta t) - \vec{R}(t).$$

$\frac{\Delta\vec{R}}{\Delta t}$ is directed along the chord PQ. In the limit when $\Delta t\to 0$, $\frac{\Delta\vec{R}}{\Delta t}$ becomes the tangent to the curve at P, if it exists. Thus,

$$\underset{\Delta t\to 0}{\text{Lt}}\frac{\Delta\vec{R}}{\Delta t} = \frac{d\vec{R}}{dt}$$

is a tangent vector to the space curve $\vec{R} = \vec{f}(t)$.

If P_0 be the initial point corresponding to $t = t_0$ and s be the length of the arc P_0P, then

$$\frac{\Delta s}{\Delta t} = \frac{\Delta s}{|\Delta\vec{R}|}\cdot\frac{|\Delta\vec{R}|}{\Delta t} = \frac{\text{arc }PQ}{\text{chord }PQ}\left|\frac{\Delta\vec{R}}{\Delta t}\right|,$$

where $\Delta s = \text{arc }PQ$.

But, when $\Delta t\to 0, Q$ approaches the point P along the curve QP, then $\frac{\text{arc }PQ}{\text{chord }PQ} = 1$, so,

$$\frac{ds}{dt} = \underset{\Delta t\to 0}{\text{Lt}}\frac{\Delta s}{\Delta t} = \underset{\Delta t\to 0}{\text{Lt}}\left|\frac{\Delta\vec{R}}{\Delta t}\right| = \left|\frac{d\vec{R}}{dt}\right|.$$

Now, if $\frac{d\vec{R}}{dt}$ is is continuous, then P_0P is given by

$$s = \int_{t=t_0}^{t}\left|\frac{d\vec{R}}{dt}\right|dt = \int_{t=t_0}^{t}\sqrt{\left(\frac{dx}{dt}\right)^2 + \left(\frac{dy}{dt}\right)^2 + \left(\frac{dz}{dt}\right)^2}\,dt.$$

Taking the arc length s as the parameter, i.e., $s = t$, then $\frac{ds}{dt} = 1$ and $\frac{d\vec{R}}{dt} = \frac{d\vec{R}}{ds}$. Then the magnitude of the tangent vector $(\vec{T})$ is $\left|\frac{d\vec{R}}{ds}\right| = 1$. Thus, the unit tangent vector is

$$\hat{T} = \frac{d\vec{R}}{ds}. \qquad \text{(i)}$$

(2) Principal Normal

As $\widehat{T}$ is a unit vector,

$$\widehat{T} \cdot \frac{d\widehat{T}}{ds} = 0,$$

this means, $\frac{d\widehat{T}}{ds}$ perpendicular to $\widehat{T}$ or, $\frac{d\widehat{T}}{ds} = 0$, i.e., $\widehat{T}$ is a constant vector with respect to s, which means $\widehat{T}$ has constant direction and hence the curve is a straight line.

If we denote $\frac{d\widehat{T}}{ds}$ by $\widehat{N}$, then

$$\frac{d\widehat{T}}{ds} = \kappa \widehat{N}, \; |\widehat{N}| = 1,$$

where the scalar κ is called the *curvature*. The unit vector $\widehat{N} = \frac{d\widehat{T}}{ds}$ is called the *principal normal* to the space curve at P. The plane containing $\widehat{T}$ and $\widehat{N}$ is called the *Osculating plane* of the curve at P (cf. Fig. 11.3).

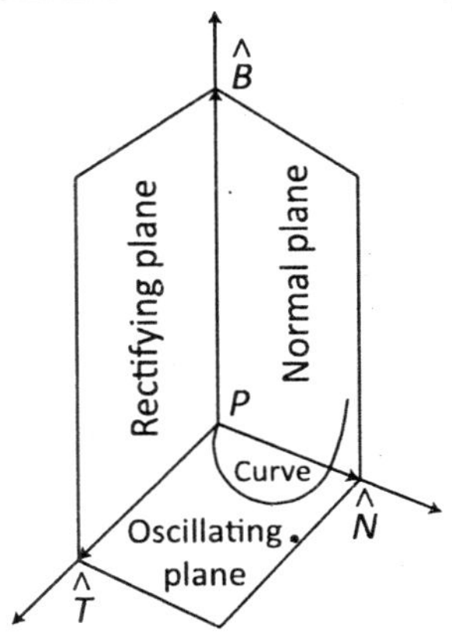

Fig. 11.3

(3) Binormal

The third unit vector given by $\widehat{B} = \widehat{T} \times \widehat{N}$, is called the **binormal** at P. $\widehat{B}$ is perpendicular to $\widehat{T}$ and $\widehat{N}$ and therefore, normal to the osculating plane at P of the space curve.

Therefore, at each point P of the space curve there are three mutually perpendicular unit vectors $\widehat{T}$, $\widehat{N}$ and $\widehat{B}$. $\widehat{T}$, $\widehat{N}$ and $\widehat{B}$ form a right-handed set of unit vectors. Hence,

$$\left.\begin{aligned} \widehat{T} &= \widehat{N} \times \widehat{B} \\ \text{and} \quad \widehat{N} &= \widehat{B} \times \widehat{T} \\ \widehat{B} &= \widehat{T} \times \widehat{N} \end{aligned}\right\} \quad \text{(ii)}$$

This $\widehat{T}$, $\widehat{N}$, $\widehat{B}$ constitute a moving tetrahedral which determines the following three fundamental planes at each point of the space curve $\vec{R}(t)$ [cf. Fig. (11.3)]

(a) The osculating plane containing $\widehat{T}$ and $\widehat{N}$.

(b) The normal plane containing $\widehat{N}$ and $\widehat{B}$.

(c) The rectifying plane containing $\widehat{B}$ and $\widehat{T}$.

(4) Curvature

The curvature κ is the arc rate of turning of the tangent, given by $\frac{d\widehat{T}}{ds}$, which is in the direction of the principal normal $\widehat{N}$.

$$\therefore \quad \frac{d\widehat{T}}{ds} = \kappa \widehat{N}. \quad \text{(iii)}$$

$\rho = \dfrac{1}{\kappa}$ is called the *radius of curvature* of the space curve.

(5) Torsion

We have $\frac{d\widehat{B}}{ds} \cdot \widehat{B} = 0$, since $\widehat{B}$ is a unit vector.

Again, $\widehat{B} \cdot \widehat{T} = 0$.

$$\therefore \quad \frac{d\widehat{B}}{ds}\cdot\widehat{T}+\widehat{B}\cdot\frac{d\widehat{T}}{ds}=0 \quad \text{or,} \quad \frac{d\widehat{B}}{ds}\cdot\widehat{T}+\widehat{B}\cdot(\kappa\widehat{N})=0$$

$$\text{or,} \quad \frac{d\boldsymbol{\widehat{B}}}{\boldsymbol{ds}}\cdot\boldsymbol{\widehat{T}}=\mathbf{0}. \quad [\because\ \widehat{B}\cdot\widehat{N}=0]$$

Hence, $\frac{d\widehat{B}}{ds}$ is perpendicular to both $\widehat{B}$ and $\widehat{T}$ and therefore, parallel to $\widehat{N}$.

Torsion τ is the arc rate of turning of the binormal and is given by $\frac{d\widehat{B}}{ds}$ and is directed along the negative direction of $\widehat{N}$.

$$\therefore \quad \frac{d\widehat{B}}{ds}=-\tau\widehat{N}. \tag{iv}$$

The quantity, $\sigma=\frac{1}{\tau}$, is called the *radius of the torsion.*

Now, since $\widehat{N}=\widehat{B}\times\widehat{T}$,

$$\therefore \quad \frac{d\widehat{N}}{ds}=\frac{d\widehat{B}}{ds}\times\widehat{T}+\widehat{B}\times\frac{d\hat{T}}{ds}=-\tau\widehat{N}\times\widehat{T}+\widehat{B}\times\kappa\widehat{N}=-\tau(-\widehat{B})+\kappa(-\widehat{T})$$

$$\text{or,} \quad \frac{d\boldsymbol{\widehat{N}}}{\boldsymbol{ds}}=\boldsymbol{\tau\widehat{B}-\kappa\widehat{T}}. \tag{v}$$

$$[\because\ \widehat{N}\times\widehat{T}=-\widehat{B} \quad \text{and} \quad \widehat{B}\times\widehat{N}=-\widehat{T}] \tag{vi}$$

Equations (iii), (iv) and (v) are the *Frenet-Serret formulae* of differential geometry.

11.4 Velocity and Acceleration

If the position of a moving particle at any time be represented by

$$\vec{R}(t)=\hat{i}x(t)+\hat{j}y(t)+\hat{k}z(t),$$

then the position of the particle at a later time $(t+\Delta t)$ is represented by

$$\vec{R}(t+\Delta t)=\hat{i}x(t+\Delta t)+\hat{j}y(t+\Delta t)+\hat{k}z(t+\Delta t).$$

$\therefore$ the displacement of the particle in time Δt is

$$\vec{R}(t+\Delta t)-\vec{R}(t)=\Delta\vec{R} \quad (say).$$

The velocity of a moving particle is given by the rate of change of displacement.

$$\therefore \quad \vec{v}=\underset{\Delta t\to 0}{\mathbf{Lt}}\ \frac{\boldsymbol{\Delta\vec{R}}}{\boldsymbol{\Delta t}}=\frac{\boldsymbol{d\vec{R}}}{\boldsymbol{dt}}. \tag{i}$$

$\vec{v}$ is along the tangent to the path of the particle at time t, the magnitude of which is he speed of the partial, i.e., $|\vec{v}|=v$.

The time rate of change of velocity $\vec{v}$ of a moving particle is defined as he acceleration $\vec{a}$ of the particle, i.e.,

$$\vec{a}=\underset{\Delta t\to 0}{\text{Lt}}\ \frac{\Delta\vec{v}}{\Delta t}=\frac{d\vec{v}}{dt}=\frac{d}{dt}\left(\frac{d\vec{R}}{dt}\right)$$

$$\text{or,} \quad \boldsymbol{\vec{a}}=\frac{\boldsymbol{d^2\vec{R}}}{\boldsymbol{dt^2}}. \tag{ii}$$

11.4.1 Relative Velocity and Relative Acceleration

Let $\overrightarrow{R_1}$ and $\overrightarrow{R_2}$ be the position vectors of the particles P_1 and P_2 with respect to 0 moving along C_1 and C_2 at any time t [Fig. 11.4]. Then,

$$\vec{R} = \overrightarrow{P_1P_2} = \overrightarrow{R_2} - \overrightarrow{R_1}$$

$$\therefore \quad \frac{d\vec{R}}{dt} = \frac{d\overrightarrow{R_2}}{dt} - \frac{d\overrightarrow{R_1}}{dt} \quad \text{or,} \quad \vec{v} = \overrightarrow{v_2} - \overrightarrow{v_1} = \overrightarrow{v_{21}}. \qquad \text{(i)}$$

$\overrightarrow{v_{21}}$ is the relative velocity vector of P_2 w.r.t. P_1.

$$\text{Again,} \quad \frac{d\vec{v}}{dt} = \frac{d\overrightarrow{v_2}}{dt} - \frac{d\overrightarrow{v_1}}{dt} \quad \text{or,} \quad \vec{a} = \overrightarrow{a_2} - \overrightarrow{a_1} = \overrightarrow{a_{21}}. \qquad \text{(ii)}$$

$\overrightarrow{a_{21}}$ gives the relative acceleration of P_2 w.r.t. P_1 and is equal to the acceleration of P_2 − acceleration of P_1.

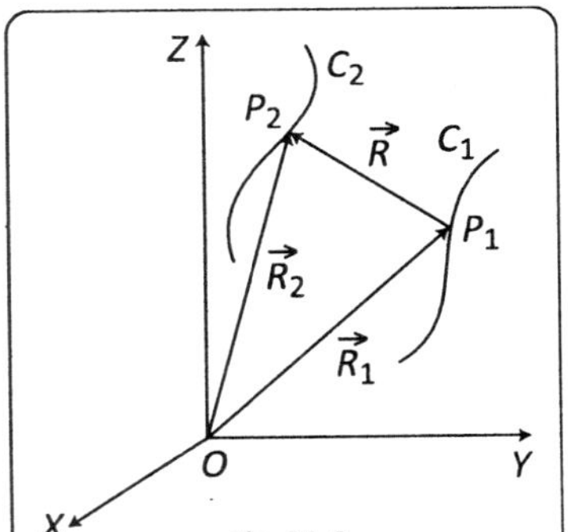

Fig. 11.4

11.5 Scalar and Vector Point Function

11.5.1 Field

A field is a region of space at every point of which a function is defined. The function must be single valued, continuous and must have continuous first derivatives w.r.t. space coordinates in the region.

Example *Gravitational field, electric field, magnetic field, electromagnetic field, etc.*

11.5.2 Scalar Point Function

If in a region R, a scalar function $f(\vec{r})$ is associated with each point $P(\vec{r})$, then $f(\vec{r})$ is called a *scalar point function*. The region R, so defined, is called the *scalar field*.

Example *The temperature at any instant, density of a body, potential due to gravitational matter, etc.*

11.5.3 Vector Point Function

If in a region R, a vector function $\vec{A}(\vec{r})$ is associated with each point $P(\vec{r})$, then $\vec{A}(\vec{r})$ is called a *vector point function*. The region R so defined is called a *vector field*.

Example *Velocity of a moving fluid at any instant, the gravitational intensity of force, the electrostatic and magnetic intensities, etc.*

11.5.4 Level Surface

If in a region of space R, a scalar point function $\phi(x, y, z)$ has the same value at each point $P(\vec{r})$, i.e., $\phi(x, y, z) = C$ (constant), then R is called a **level surface**.

Example *Free surface of a liquid in a vessel, the surface of an insulated spherical charged conductor, etc.*

11.6 Vector Differential Operator (Del, i.e., ∇)

The vector differential operator Del is defined as

$$\nabla = \hat{i}\frac{\partial}{\partial x} + \hat{j}\frac{\partial}{\partial y} + \hat{k}\frac{\partial}{\partial z},$$

where $\hat{i}$, $\hat{j}$, $\hat{k}$ are the triad of unit vectors.

11.6.1 Gradient of a Scalar Point Function

Definition : The gradient of a scalar point function $\phi(x, y, z)$ is defined as $\nabla\phi$ and is written as

$$\text{Grad } \phi = \nabla\phi = \left(\hat{i}\frac{\partial}{\partial x} + \hat{j}\frac{\partial}{\partial y} + \hat{k}\frac{\partial}{\partial z}\right)\phi = \hat{i}\frac{\partial\phi}{\partial x} + \hat{j}\frac{\partial\phi}{\partial y} + \hat{k}\frac{\partial\phi}{\partial z}. \tag{11.3}$$

Grad ϕ is a *vector quantity.*

Geometrical Interpretation of $\nabla\phi$

We consider two level surfaces S_1 and S_2 in the scalar field which are very close to each other and characterised by the scalar functions ϕ and $\phi + \Delta\phi$ respectively. Let $\vec{r}$ and $\vec{r} + \Delta\vec{r}$ be the position vectors of P in S_1 and Q in S_2 relative to any arbitrary origin O. Then,

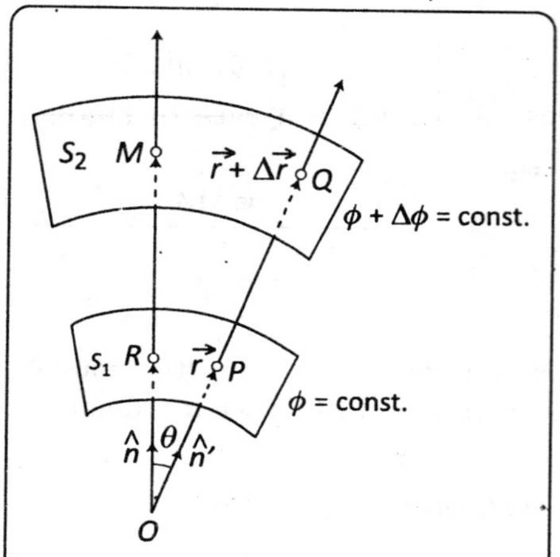

Fig. 11.5

$$\overrightarrow{PQ} = (\vec{r} + \Delta\vec{r}) - \vec{r} = \Delta\vec{r}.$$

If the Cartesian coordinates of P and Q w.r.t. 0 be (x, y, z) and $(x + \Delta x, y + \Delta y, z + \Delta z)$ respectively, then

$$\Delta\vec{r} = \hat{i}\Delta x + \hat{j}\Delta y + \hat{k}\Delta z.$$

$$\therefore \quad \Delta\phi = \frac{\partial\phi}{\partial x}\Delta x + \frac{\partial\phi}{\partial y}\Delta y + \frac{\partial\phi}{\partial z}\Delta z$$

$$= \left(\hat{i}\frac{\partial\phi}{\partial x} + \hat{j}\frac{\partial\phi}{\partial y} + \hat{k}\frac{\partial\phi}{\partial z}\right) \cdot (\hat{i}\Delta x + \hat{j}\Delta y + \hat{k}\Delta z)$$

$$= \nabla\phi \cdot \Delta\vec{r}.$$

If we consider that the point Q lies on the level surface S_1, then $\Delta\phi = 0$.

So $\nabla\phi \cdot \Delta\vec{r} = 0$. (11.4)

This shows that $\nabla\phi$ is normal to the level surface S_1.

If $\hat{n}$ be a unit vector along the normal to the level surface S_1, then

$$\nabla\phi = |\nabla\phi|\hat{n}.$$

If the perpendicular distance RM between the surfaces through P and Q be ΔN, then the rate of change of ϕ normal to the surface through P is

$$\frac{d\phi}{dN} = \underset{\Delta N \to 0}{\text{Lt}} \frac{\Delta\phi}{\Delta N} = \underset{\Delta N \to 0}{\text{Lt}} \nabla\phi \cdot \frac{\Delta\vec{r}}{\Delta N} = |\nabla\phi| \underset{\Delta N \to 0}{\text{Lt}} \frac{\hat{n} \cdot \Delta\vec{r}}{\Delta N}$$

$$= |\nabla\phi| \underset{\Delta N \to 0}{\text{Lt}} \frac{\Delta N}{\Delta N} \quad [\because \hat{n} \cdot \Delta\vec{r} = |\Delta\vec{r}|\cos\theta = \Delta N]$$

$$= |\nabla\phi|.$$

Hence, the magnitude of $\nabla\phi$ is $\frac{d\phi}{dN}$. Since, $\nabla\phi = \frac{d\phi}{dN}\hat{n}$ is the greatest rate of charge of ϕ with respect to the space variables, grad ϕ may be defined as follows :

The gradient of a scalar function ϕ is a vector whose magnitude is equal to the maximum rate of change of ϕ with respect to the space variables and whose direction is along that change.

Directional Derivative

If $|\Delta\vec{r}|$ be the length of PQ [Fig. 11.5] and $\hat{n}'$ be a unit vector along PQ, then the limiting value of $\frac{\Delta\phi}{\Delta r}$ as $\Delta r \to 0$, i.e., $\frac{d\phi}{dr}$ is called the *directional derivative* of ϕ at P along the direction PQ.

$$\because \quad \Delta r = \frac{\Delta N}{\cos\theta} = \frac{\Delta N}{\hat{n}\cdot\hat{n}'},$$

$$\therefore \quad \frac{d\phi}{dr} = \underset{\Delta r\to 0}{\text{Lt}} \left[\hat{n}\cdot\hat{n}'\frac{\Delta\phi}{\Delta N}\right] \qquad \left[\because \; \frac{\Delta\phi}{\Delta r} = \frac{\Delta\phi}{\frac{\Delta N}{\hat{n}\cdot\hat{n}'}} = \hat{n}\cdot\hat{n}'\frac{\Delta\phi}{\Delta N}\right]$$

$$= \hat{n}'\cdot\left(\underset{\Delta r\to 0}{\text{Lt}} \frac{\Delta\phi}{\Delta N}\right)\hat{n}.$$

$$\therefore \quad \frac{d\phi}{dr} = \hat{n}'\cdot\nabla\phi.$$

Thus, the directional derivative of ϕ in the direction $\hat{n}'$ is the resolved part of $\nabla\phi$ along $\hat{n}'$.

Since, $\nabla\phi\cdot\hat{n}' = |\nabla\phi|\cos\theta \leq |\nabla\phi|$, **hence, $\nabla\phi$ gives the maximum rate of change of ϕ.**

11.6.2 Lamellar Vector (or non-curl) Vector Field

If a vector field $\vec{A}$ can be expressed as the gradient of a scalar field, then $\vec{A}$ is called a *lamellar (or, non-curl) vector field.*

A pure electric field $\vec{E}$ can be expressed as (negative) gradient of a scalar potential ϕ, i.e.,

$$\vec{E} = -\nabla\phi.$$

Hence, $\vec{E}$ is a lamellar vector field.

11.7 Divergence of a Vector Function

The divergence of a vector field at any point is defined as the amount of flux diverging through the surface enclosing unit volume.

Let there be a vector field $\vec{A}$ in a certain region and V be he infinitesimal volume element enclosed by an infinitely small closed surface S surrounding a point $P(x, y, z)$, then

$$\text{div}\,\vec{A} = \underset{V\to 0}{\text{Lt}} \iint_S \frac{\vec{A}\cdot d\vec{S}}{V} \qquad \text{(iv)}$$

The divergence of a vector function is a *scalar quantity.*

Physical Significance of div $\vec{A}$

If $\vec{A}$ represents the velocity of a moving fluid at any point P, then div $\vec{A}$ gives the rate at which the fluid is diverging per unit volume from the point P.

If div $\vec{A} > 0$ at any point P, then either the fluid is expanding or its density at P is decreasing with time or the point P is the *source* of the fluid. If div $\vec{A} < 0$, then either the fluid is contracting or its density is increasing at P or the point is a *sink.*

If div $\vec{A} = 0$, then flux of $\vec{A}$ entering any element of space is exactly balanced by the flux leaving it. Then there is no source or sink in the field nor its density is changing (i.e., the fluid is **incompressible**).

A vector $\vec{A}$, which satisfies the condition div $\vec{A} = 0$, is called a *solenoidal* vector. Thus, a solenoidal vector field is without source and sink. The equation div $\vec{A} = 0$, when $\vec{A}$ represents the velocity of a liquid, is also called the **equation of continuity** for an in compressible fluid.

11.7.1 Divergence in Cartesian Coordinates

We consider an infinitesimal parallelopiped with sides Δx, Δy and Δz parallel to the coordinate axes X, Y and Z in a vector field [Fig. 11.6]. Let $\vec{A}$ be the vector at the centre P of the parallelopiped, then

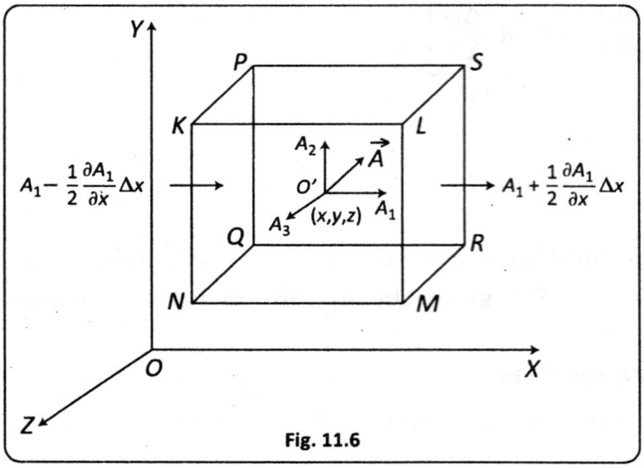

Fig. 11.6

$$\vec{A} = \hat{i}A_1 + \hat{j}A_2 + \hat{k}A_3.$$

Now, the x-component of $\vec{A}$ at the centre of the face $PQNK$ is

$$A_1 - \frac{1}{2}\frac{\partial A_1}{\partial x}\Delta x.$$

Similarly, the value of the x-component of $\vec{A}$ at the centre O' of the face SRML

$$= A_1 + \frac{1}{2}\frac{\partial A_1}{\partial x}\Delta x.$$

$\therefore$ the flux entering the face $PQNK$ is $= \left(A_1 - \frac{1}{2}\frac{\partial A_1}{\partial x}\Delta x\right)\Delta y \Delta z$

and that leaving the face $SRML$ is $= \left(A_1 + \frac{1}{2}\frac{\partial A_1}{\partial x}\Delta x\right)\Delta y \Delta z.$

Thus, excess of the flux leaving the parallelopiped over that entering along the x-direction is

$$\left[\left(A_1 + \frac{1}{2}\frac{\partial A_1}{\partial x}\Delta x\right) - \left(A_1 - \frac{1}{2}\frac{\partial A_1}{\partial x}\Delta x\right)\right]\Delta y \Delta z = \frac{\partial A_1}{\partial x}\Delta x \Delta y \Delta z.$$

Similarly, the net flux leaving the parallelopiped along y- and z-directions are $\frac{\partial A_2}{\partial y}\Delta x \Delta y \Delta z$ and $\frac{\partial A_3}{\partial z}\Delta x \Delta y \Delta z$ respectively.

Hence, the total flux diverging from the parallelopiped is

$$\iint_s \vec{A} \cdot \vec{dS} = \left(\frac{\partial A_1}{\partial x} + \frac{\partial A_2}{\partial y} + \frac{\partial A_3}{\partial z}\right)\Delta x \Delta y \Delta z = \left(\frac{\partial A_1}{\partial x} + \frac{\partial A_2}{\partial y} + \frac{\partial A_3}{\partial z}\right)\Delta V,$$

where $\Delta V = \Delta x \Delta y \Delta z$ is the volume of the parallelopiped. Thus,

$$\text{div}\,\vec{A} = \underset{\Delta V \to 0}{\text{Lt}} \frac{\iint_s \vec{A} \cdot \vec{dS}}{\Delta V}$$

or, $$\textbf{div}\,\vec{A} = \frac{\partial A_1}{\partial x} + \frac{\partial A_2}{\partial y} + \frac{\partial A_3}{\partial z}.$$

The above equation represents the divergence of $\vec{A}$ in Cartesian coordinates.

Again, $\text{div}\,\vec{A} = \frac{\partial A_1}{\partial x} + \frac{\partial A_2}{\partial y} + \frac{\partial A_3}{\partial z}$

$$\left(\hat{i}\frac{\partial}{\partial x} + \hat{j}\frac{\partial}{\partial y} + \hat{k}\frac{\partial}{\partial z}\right) \cdot (\hat{i}A_1 + \hat{j}A_2 + \hat{k}A_3)$$

or, $$\textbf{div}\,\vec{A} = \nabla \cdot \vec{A}$$

11.8 Line Integral of a Lamellar Vector Field

We consider a lamellar vector field defined by $\vec{A} = \nabla\phi$, where ϕ is a continuously differentiable function of x, y, z. In such a field, let $KLMN$ be a closed path and $\vec{A}$ makes an angle θ with an element of the path $\vec{dl}$ as shown in Fig. 11.7.

The line integral of $\vec{A}$ over $\vec{dl}$ is $\vec{A}\cdot\vec{dl} = \nabla\phi\cdot\vec{dl} = d\phi$ which is the variation of ϕ over the path length dl.

$\therefore$ line integral of $\vec{A}$ from K to M is

$$\int_K^M \vec{A}\cdot\vec{dl} = \int_K^M d\phi = \phi(M) - \phi(K).$$

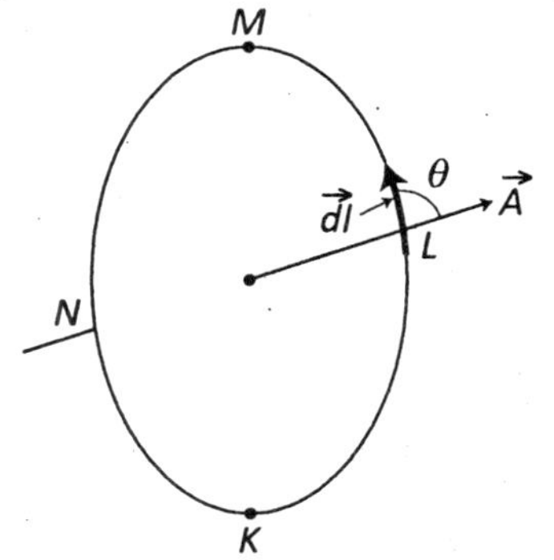

Fig. 11.7

Hence, the line integral of $\vec{A}$ over the closed path $KLMN$ is

$$\begin{aligned}\oint \vec{A}\cdot\vec{dl} &= \int_K^M \vec{A}\cdot\vec{dl} + \int_M^K \vec{A}\cdot\vec{dl}\\ &= \phi(M) - \phi(K) + \phi(K) - \phi(M)\\ &= 0.\end{aligned}$$

Thus, we conclude that, the line integral of a vector field derived from the gradient of a scalar field between two points, is independent of path, but depends on the values of the scalar at the initial and final points of the path and the line integral around a closed path vanishes. Such a vector field ($\vec{A}$) is called the *lamellar vector field.*

Example *Electric field $\vec{E}$ is derived from the negative gradient of potential V, i.e., $\vec{E} = -\nabla V$. The line integral of $\vec{E}$ around a closed path is zero, i.e., $\oint_C \vec{E}\cdot\vec{dl} = 0$, where C represents a closed path.* **Hence, $\vec{E}$ is a lamellar vector field.**

The curl of $\vec{A}$: *The curl of a vector field at a point is defined as the maximum value of the line integral of the vector expressed per unit area surrounding the point and is directed along the normal to the plan of the area.*

If $\vec{A}$ be the vector at any point P and ΔS an infinitesimal area surrounding P, then according to the definition,

$$\text{curl}\,\vec{A} = \underset{\Delta S\to 0}{\text{Lt}} \frac{\left[\oint_c \vec{A}\cdot\vec{dl}\right]_{\text{maximum}}}{\Delta S}\hat{n}$$

and $$|\text{curl}\,\vec{A}| = \underset{\Delta S\to 0}{\text{Lt}} \frac{\left[\oint_c \vec{A}\cdot\vec{dl}\right]_{\text{maximum}}}{\Delta S},$$

where $\hat{n}$ is a unit vector normal to the plane of the area.

11.8.1 Curl of a Vector Function

The maximum value of line integral. We know that for a **lamellar vector field** the line integral around a closed path is zero. In other words, for a **non-lamellar vector field** the line integral around a closed path is not zero and has a definite value.

In Fig. 11.8, a non-lamellar vector field is represented by flux lines. We consider, a rectangular plane lamina $PQRS$ be placed in the vector field. If the area $PQRS$ is perpendicular to the flux lines, then each side of it is at right angles to the field. Hence, the line integral of $\vec{A}$ around the boundary is zero. But if the area $P'Q'R'S'$ of the same rectangle be parallel to the field, then the line integrals along $P'S'$ and $Q'R'$ will vanish but the line integrals $P'Q'$ and $S'R'$ have some finite value.

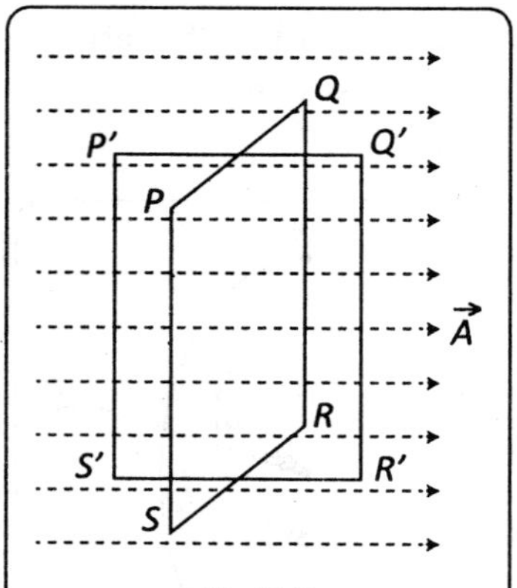

Fig. 11.8

As the value of $\vec{A}$ along $P'Q'$ and $S'R'$ are different, the line integral of $\vec{A}$ around the boundary of the rectangle must have some finite value. So, the value of the line integral around the closed boundary depends on the orientation of the area $PQRS$ in the vector field and as such there is a definite orientation of the area with respect to the vector field, for which the line integral of $\vec{A}$ is the maximum.

11.8.2 Curl in Cartesian Coordinates

To obtain the x-component of curl $\vec{A}$ at a point P, we consider a rectangular area $KLMN$ of sides Δy and Δz in the y-z plane as shown in Fig. 11.9. The area of the rectangle is $\Delta y \Delta z$.

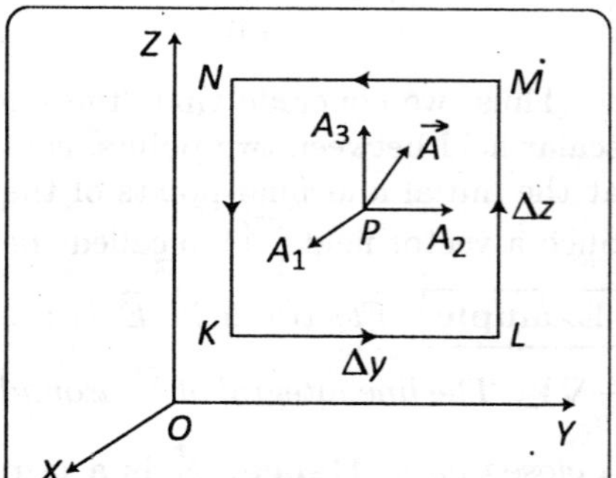

Fig. 11.9

Now at P, $\vec{A} = \hat{\imath}A_1 + \hat{\jmath}A_2 + \hat{k}A_3$ and the components $\vec{A}$ along y- and z-directions are as shown below :

Along KL, the y-component of $\vec{A} = A_2 - \dfrac{\partial A_2}{\partial z}\dfrac{\Delta z}{2}$.

The y-component along $MN = -\left(A_2 + \dfrac{\partial A_2}{\partial z}\dfrac{\Delta z}{2}\right)$.

Similarly, the z-component of $\vec{A}$ along $LM = A_3 + \dfrac{\partial A_3}{\partial y}\dfrac{\Delta y}{2}$

and the z-component of $\vec{A}$ along $NK = -\left(A_3 - \dfrac{\partial A_3}{\partial y}\dfrac{\Delta y}{2}\right)$.

Hence,
$$\int_{KL} \vec{A}\cdot\vec{dl} = \left(A_2 - \frac{\partial A_2}{\partial z}\frac{\Delta z}{2}\right)\Delta y \qquad \int_{LM} \vec{A}\cdot\vec{dl} = \left(A_3 + \frac{\partial A_3}{\partial y}\frac{\Delta y}{2}\right)\Delta z$$

$$\int_{MN} \vec{A}\cdot\vec{dl} = -\left(A_2 + \frac{\partial A_2}{\partial z}\frac{\Delta z}{2}\right)\Delta y \quad \text{and} \quad \int_{NK} \vec{A}\cdot\vec{dl} = -\left(A_3 - \frac{\partial A_3}{\partial y}\frac{\Delta y}{2}\right)\Delta z.$$

$\therefore$
$$\int_{KLMN} \vec{A}\cdot\vec{dl} = \left(\frac{\partial A_3}{\partial y} - \frac{\partial A_2}{\partial z}\right)\Delta y \Delta z.$$

So,
$$(\text{curl } \vec{A})_x = \underset{S_x \to 0}{\text{Lt}} \frac{\left(\int_{KLMN} \vec{A}\cdot\vec{dl}\right)}{S_x} = \left(\frac{\partial A_3}{\partial y} - \frac{\partial A_2}{\partial z}\right),$$

where S_x is the area perpendicular to x-axis.

Similarly, $(\text{curl}\,\vec{A})_y = \left(\dfrac{\partial A_1}{\partial z} - \dfrac{\partial A_3}{\partial x}\right)$ and $(\text{curl}\,\vec{A})_z = \left(\dfrac{\partial A_2}{\partial x} - \dfrac{\partial A_1}{\partial y}\right)$.

Hence, $\text{curl}\,\vec{A} = \hat{i}(\text{curl}\,\vec{A})_x + \hat{j}(\text{curl}\,\vec{A})_y + \hat{k}(\text{curl}\,\vec{A})_z$

$$= \hat{i}\left(\frac{\partial A_3}{\partial y} - \frac{\partial A_2}{\partial z}\right) + \hat{j}\left(\frac{\partial A_1}{\partial z} - \frac{\partial A_3}{\partial x}\right) + \hat{k}\left(\frac{\partial A_2}{\partial x} - \frac{\partial A_1}{\partial y}\right) \quad (1)$$

or,
$$\textbf{curl}\,\vec{A} = \begin{vmatrix} \hat{i} & \hat{j} & \hat{k} \\ \frac{\partial}{\partial x} & \frac{\partial}{\partial y} & \frac{\partial}{\partial z} \\ A_1 & A_2 & A_3 \end{vmatrix}. \quad (2)$$

Also from equation (1),

$$\text{curl}\,\vec{A} = \left(\hat{i}\frac{\partial}{\partial x} + \hat{j}\frac{\partial}{\partial y} + \hat{k}\frac{\partial}{\partial z}\right) \times (\hat{i}A_1 + \hat{j}A_2 + \hat{k}A_3)$$

or,
$$\textbf{curl}\,\vec{A} = \nabla \times \vec{A}. \quad (3)$$

11.8.3 Physical Significance of Curl

Curl in hydrodynamics, represents rotation of a fluid, that is why sometimes curl $\vec{A}$ is written as rot $\vec{A}$. The curl of a vector field is also known as *circulation* or *rotation*.

If $\vec{v}$ represents the fluid velocity, then curl $\vec{v}$ represents that the velocity field, over and above the general motion in a certain direction, has motion like 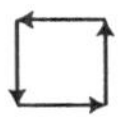 or, 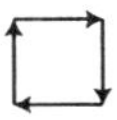 superposed on it in the direction of $\vec{v}$. This is because of the existence of curl of $\vec{v}$ or **vorticity** at that point of the fluid flow.

Thus, if a vector field $\vec{A}$ is derived from the scalar field ϕ, i.e., $\vec{A} = \nabla\phi$, then we have seen the $\oint_c \vec{A}\cdot\vec{dl} = 0$ and hence, $\nabla \times \vec{A} = 0 \cdot \vec{A}$ is called the *lamellar vector field.*

Example *The electric field $\vec{E}$ is given by, $\vec{E} = -\nabla V$, whose V is the potential. Then*

$$\text{curl}\,\vec{E} = \nabla \times \vec{E} = \nabla \times (\nabla V) = \left(\hat{i}\frac{\partial}{\partial x} + \hat{j}\frac{\partial}{\partial y} + \hat{k}\frac{\partial}{\partial z}\right) \times \left(\hat{i}\frac{\partial V}{\partial x} + \hat{i}\frac{\partial V}{\partial y} + \hat{k}\frac{\partial V}{\partial z}\right)$$

$$= \begin{vmatrix} \hat{i} & \hat{j} & \hat{k} \\ \frac{\partial}{\partial x} & \frac{\partial}{\partial y} & \frac{\partial}{\partial z} \\ \frac{\partial V}{\partial x} & \frac{\partial V}{\partial y} & \frac{\partial V}{\partial z} \end{vmatrix}$$

or,
$$\text{curl}\,\vec{E} = \hat{i}\left(\frac{\partial^2 V}{\partial y\partial z} - \frac{\partial^2 V}{\partial z\partial y}\right) + \hat{j}\left(\frac{\partial^2 V}{\partial z\partial x} - \frac{\partial^2 V}{\partial x\partial z}\right) + \hat{k}\left(\frac{\partial^2 V}{\partial x\partial y} - \frac{\partial^2 V}{\partial y\partial x}\right) = 0.$$

Hence, $\nabla \times \vec{E} = 0$, i.e., $\vec{E}$ is a lamellar vector field.

N.B. When electric field $\vec{E}$ generated by changing magnetic field generated by electric currents are considered, $\nabla \times \vec{E} \neq 0$.

If $\nabla \times A = 0$, then the vector field $\vec{E}$ is called *irrotational.* Otherwise, if $\nabla \times \vec{A} \neq 0$, then $\vec{A}$ is rotational.

It is to be noted that curl of a vector is a *vector* quantity.

11.8.4 Physical Meaning of Curl

It can be shown that curl of a vector field is related to the rotational motion.

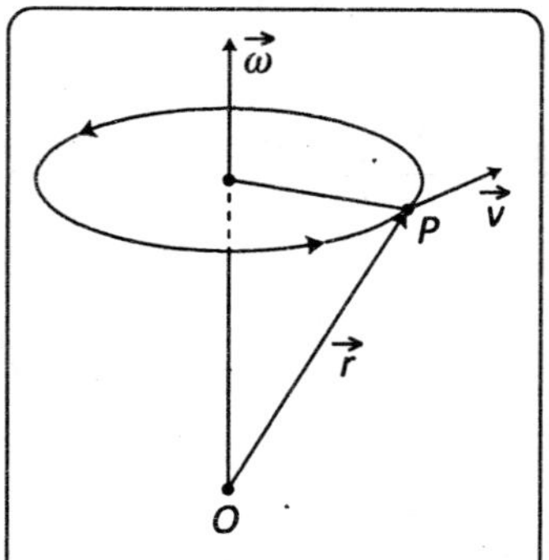

Fig. 11.10

Let $\vec{V}$ be the linear velocity and $\vec{\omega}$ the angular velocity of a particle in a rigid body rotating about an axis passing through origin 0 [Fig. 11.10].

Then, $\vec{V} = \vec{\omega} \times \vec{r}$.

Let $\vec{\omega} = \hat{i}\omega_1 + \hat{j}\omega_2 + \hat{k}\omega_3$ and $\vec{r} = \hat{i}x + \hat{j}y + \hat{k}z$.

$$\text{So, } \vec{V} = \vec{\omega} \times \vec{r} = \begin{vmatrix} \hat{i} & \hat{j} & \hat{k} \\ \omega_1 & \omega_2 & \omega_3 \\ x & y & z \end{vmatrix}$$

$$= \hat{i}(\omega_2 z - \omega_8 y) + \hat{j}(\omega_3 x - \omega_1 z) + \hat{k}(\omega_1 y - \omega_2 x).$$

$$\therefore \quad \text{curl } \vec{v} = \nabla \times \vec{V} = \begin{vmatrix} \hat{i} & \hat{j} & \hat{k} \\ \dfrac{\partial}{\partial x} & \dfrac{\partial}{\partial y} & \dfrac{\partial}{\partial z} \\ (\omega_2 z - \omega_3 y) & (\omega_3 x - \omega_1 z) & (\omega_1 y - \omega_2 x) \end{vmatrix}$$

$$= \hat{i}(\omega_1 + \omega_1) + \hat{j}(\omega_2 + \omega_2) + \hat{k}(\omega_3 + \omega_3) = 2(\hat{i}\omega_1 + \hat{j}\omega_2 + \hat{k}\omega_3)$$

$$\therefore \quad \textbf{curl } \vec{V} = 2\vec{\omega}.$$

This shows that, the curl of a vector field is associated with rotational properties of the vector field and justifies the name *rotation* used for curl.

11.9 Invariance

Let us consider two rectangular coordinate systems, X, Y, Z and X', Y', Z' with the origin O. The axes of the two coordinate systems are rotated with respect to each others as shown in Fig. 11.11. .

If the coordinates of the point P in the two coordinate systems be (x, y, z) and (x', y', z') respectively, the coordinate transformations are represented by the following ;

$$\left.\begin{aligned} x' &= l_{11}x + l_{12}y + l_{13}z \\ y' &= l_{21}x + l_{22}y + l_{23}z \\ z' &= l_{31}x + l_{32}y + l_{33}z \end{aligned}\right\} \qquad (1)$$

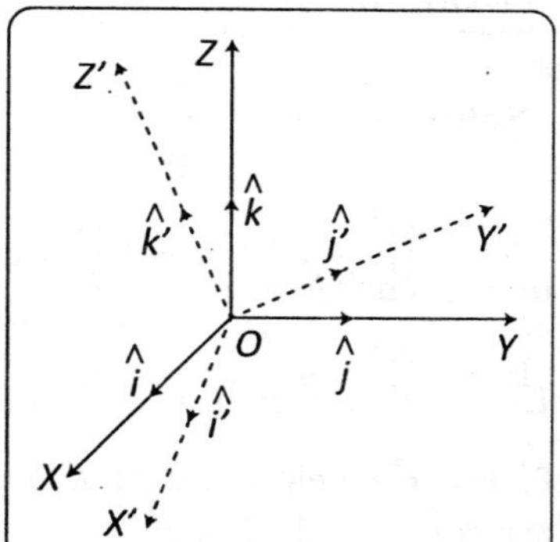

Fig. 11.11

where l_{jk} with j, $k = 1, 2, 3$ are direction cosine of X', Y' and Z' axes with respect to X, Y and Z axes.

If, however, the origin of X, Y, Z be $O(0, 0, 0)$ and that of X', Y' and Z' be $O'(a_1', a_2', a_3')$, then the transformation equations are as follows :

$$\left.\begin{aligned} x' &= l_{11}x + l_{12}y + l_{13}z + a_1' \\ y' &= l_{21}x + l_{22}y + l_{23}z + a_2' \\ z' &= l_{31}x + l_{32}y + l_{33}z + a_3' \end{aligned}\right\} \qquad (2)$$

Equations (1) and (2) represent pure rotation and rotation plus translation respectively. The transformation represented by Equation (1) is called *orthogonal transformation*.

It is to be noted that,

(i) $x^2+y^2+z^2$ is invariant, since $x^2+y^2+z^2=x'^2+y'^2+z'^2$

(ii) A scalar point function $\phi(x,y,z)$ is invariant, if $\phi(x,y,z)=\phi'(x',y',z')$

(iii) A vector point function $A(x,y,z)$ is invariant, if $\vec{A}(x,y,z)=\vec{A}'(x',y',z')$, where x, y, z and x', y', z' are related by equations (1) and (2).

11.10 Solved Problems

Example 11.1. *If $\vec{A}(t)$ has a constant magnitude, then show that $\frac{d\vec{A}}{dt}$ is perpendicular to $\vec{A}$.* **[CU(H) 2013; '10; NBU(H) 2008; GU 1984, MU 1983]**

Solution : We have $\vec{A}(t)\cdot\vec{A}(t)=|\vec{A}(t)|^2=$ constant.

$$\therefore\quad 2\vec{A}\cdot\frac{d\vec{A}}{dt}=0 \quad\text{or,}\quad \vec{A}\cdot\frac{d\vec{A}}{dt}=0 \quad\therefore\quad \frac{d\vec{A}}{dt} \textbf{ is perpendicular to } \vec{A}.$$

Example 11.2. *If $\vec{A}(t)$ has fixed direction then show that,*

$$\vec{A}(t)\times\frac{d\vec{A}(t)}{dt}=0,\quad i.e.,\quad \vec{A}\times\frac{d\vec{A}}{dt}=0.$$ **[NBU 1988]**

Solution : Let $\hat{n}(t)$ be a unit vector along $\vec{A}(t)$.

$$\therefore\quad \vec{A}(t)=A(t)\hat{n}(t) \quad\text{or,}\quad \vec{A}=A\hat{n},\quad \text{where } A(t)=|\vec{A}(t)|.$$

$$\therefore\quad \frac{d\vec{A}}{dt}=A\frac{d\hat{n}}{dt}+\frac{dA}{dt}\hat{n}$$

and
$$\vec{A}\times\frac{d\vec{A}}{dt}=A\hat{n}\times\left[A\frac{d\hat{n}}{dt}+\frac{dA}{dt}\hat{n}\right]=A^2\hat{n}\times\frac{d\hat{n}}{dt}+A\frac{dA}{dt}(\hat{n}\times\hat{n})$$

$$=A^2\hat{n}\times\frac{d\hat{n}}{dt}=\mathbf{0},\quad \left[\because\ \hat{n}\text{ is a constant, } \frac{d\hat{n}}{dt}=0 \text{ and } \hat{n}\times\hat{n}=0\right]$$

Example 11.3. *If $\hat{r}$ is a unit vector in the direction of $\vec{R}$, then show that, $\hat{r}\times d\hat{r}=\frac{\vec{R}\times d\vec{R}}{|\vec{R}|^2}$.*

Solution : We have $\hat{r}=\frac{\vec{R}}{|\vec{R}|}$ or, $\vec{R}=\hat{r}|\vec{R}|$.

$$\therefore\quad \frac{d\vec{R}}{dt}=|\vec{R}|\frac{d\hat{r}}{dt}+\hat{r}\frac{d}{dt}|\vec{R}|.$$

But,
$$d\vec{R}=\frac{d\vec{R}}{dt}dt=\left[|\vec{R}|\frac{d\hat{r}}{dt}+\hat{r}\frac{d}{dt}|\vec{R}|\right]dt=|\vec{R}|\frac{d\hat{r}}{dt}dt+\hat{r}\frac{d}{dt}|\vec{R}|dt=|\vec{R}|d\hat{r}+\hat{r}d|\vec{R}|.$$

$$\therefore\quad \vec{R}\times d\vec{R}=\vec{R}\times[|\vec{R}|d\hat{r}+\hat{r}d|\vec{R}|]=|\vec{R}|\vec{R}\times d\hat{r}+\vec{R}\times\hat{r}d|\vec{R}|$$

$$=|\vec{R}|(|\vec{R}|\hat{r})\times d\hat{r}+(|\vec{R}|\hat{r})\times\hat{r}d|\vec{R}|=|\vec{R}|^2\hat{r}\times d\hat{r}\quad[\because\ \hat{r}\times\hat{r}=0]$$

$$\therefore\quad \hat{\boldsymbol{r}}\times d\hat{\boldsymbol{r}}=\frac{\vec{\boldsymbol{R}}\times d\vec{\boldsymbol{R}}}{|\vec{\boldsymbol{R}}|^2}.$$

Example 11.4. *If* $\vec{A} = 5t^2\hat{i} + t\hat{j} - t^3\hat{k}$ *and* $\vec{B} = \sin t\hat{i} - \cos t\hat{j}$, *then show that,*

(i) $\frac{d}{dt}(\vec{A} \cdot \vec{B}) = 5t^2 \cos t + 11t \sin t - \cos t$ *and*

(ii) $\frac{d}{dt}(\vec{A} \times \vec{B}) = (t^3 \sin t - 3t^2 \cos t)\hat{i} - t^2(t \cos t + 3 \sin t)\hat{j} + [(5t^2 - 1) \sin t - 11t \cos t]\hat{k}$.

Solution : (i) We have $\frac{\boldsymbol{d}}{\boldsymbol{dt}}(\vec{\boldsymbol{A}} \cdot \vec{\boldsymbol{B}}) = \vec{A} \cdot \frac{d\vec{B}}{dt} + \frac{d\vec{A}}{dt} \cdot \vec{B}$

$$= (5t^2\hat{i} + t\hat{j} - t^3\hat{k}) \cdot (\cos t\hat{i} + \sin t\hat{j}) + (10t\hat{i} + \hat{j} - 3t^2\hat{k}) \cdot (\sin t\hat{i} - \cos t\hat{j})$$

$$= (5t^2 \cot t + t \sin t) + (10t \sin t - \cos t)$$

$$= \boldsymbol{5t^2 \cos t + 11t \sin t - \cos t}.$$

(ii) $\frac{\boldsymbol{d}}{\boldsymbol{dt}}(\vec{\boldsymbol{A}} \times \vec{\boldsymbol{B}}) = \vec{A} \times \frac{d\vec{B}}{dt} + \frac{d\vec{A}}{dt} \times \vec{B}$

$$= (5t^2\hat{i} + t\hat{j} - t^2\hat{k}) \times (\cos t\hat{i} + \sin t\hat{j}) + (10t\hat{i} + \hat{j} - 3t^3\hat{k}) \times (\sin t\hat{i} - \cos t\hat{j})$$

$$= 5t^2 \sin t\hat{k} + t \cos t(-k) - t^3 \cos t\hat{j} - t^3 \sin t(-\hat{i}) - 10t \cos t\hat{k} + \sin t(-\hat{k}) - 3t^2 \sin t\hat{j} + 3t2 \cos t(-\hat{i})$$

$$= \boldsymbol{(t^3 \sin t - 3t^2 \cos t)\hat{i} - t^2(t \cos t + 3 \sin t)\hat{j} + [(5t^2 - 1) \sin t - 11t \cos t]\hat{k}}.$$

Example 11.5. *If* $\vec{A}$, $\vec{B}$ *and* $\vec{C}$ *are vector functions of t such that,* $\frac{d\vec{A}}{dt} = \vec{C} \times \vec{A}$ *and* $\frac{d\vec{B}}{dt} = \vec{C} \times \vec{B}$, *then show that,* $\frac{d}{dt}(\vec{A} \times \vec{B}) = \vec{C} \times (\vec{A} \times \vec{B})$.

Solution : $\frac{\boldsymbol{d}}{\boldsymbol{dt}}(\vec{\boldsymbol{A}} \times \vec{\boldsymbol{B}}) = \frac{d\vec{A}}{dt} \times \vec{B} + \vec{A} \times \frac{d\vec{B}}{dt} = (\vec{C} \times \vec{A}) \times \vec{B} + \vec{A} \times (\vec{C} \times \vec{B})$

$$= (\vec{C} \cdot \vec{B})\vec{A} - (\vec{A} \cdot \vec{B})\vec{C} + (\vec{A} \cdot \vec{B})\vec{C} - (\vec{A} \cdot \vec{C})\vec{B}$$

$$= (\vec{C} \cdot \vec{B})\vec{A} - (\vec{A} \cdot \vec{C})\vec{B} = \vec{\boldsymbol{C}} \times (\vec{\boldsymbol{A}} \times \vec{\boldsymbol{B}}).$$

Example 11.6. *If a particle moves in a circle with constant angular velocity, then show that,*

(a) *its linear velocity* $\vec{v}$ *is at right angles to its position vector* $\vec{r}$.

(b) *the acceleration* $\vec{f}$ *is directed towards the centre of the circle and is proportional to* $\vec{r}$.

(c) *the vector* $(\vec{r} \times \vec{v})$ *is of constant magnitude.*

Solution : If r and ω be the radius of the circular path and, the angular velocity of a particle moving in the circular path, the motion of the particle is represented by $\vec{r} = r(\hat{i}\cos\omega t + \hat{j}\sin\omega t)$.

(a) $\therefore \quad \frac{d\vec{r}}{dt} = \frac{dr}{dt}(\hat{i}\cos\omega t + \hat{j}\sin\omega t) + r\omega(-\hat{i}\sin\omega t + \hat{j}\cos\omega t)$

$$\therefore \quad \vec{r} \cdot \frac{d\vec{r}}{dt} = \vec{r} \cdot \vec{v} = r\frac{d\vec{r}}{dt}[(\hat{i}\cos\omega t + \hat{j}\sin\omega t) \cdot (\hat{i}\cos\omega t + \hat{j}\sin\omega t)] + r\omega(\hat{i}\cos\omega t + \hat{j}\sin\omega t) \cdot (-\hat{i}\sin\omega t + \hat{j}\cos\omega t)$$

$$= r\frac{dr}{dt}[\cos^2\omega t + \sin^2\omega t + r\omega(-\sin\omega t\cos\omega t + \sin\omega t\cos\omega t)]$$

$$= r\frac{dr}{dt} = 0 \quad \left[\because\ r \text{ is constant, } \frac{dr}{dt} = 0\right]$$

Hence, **r and $\frac{dr}{dt}$ are at right angles to each other.**

(b) $\because\ \vec{r} = r(\hat{i}\cos\omega t + \hat{j}\sin\omega t)$,

$$\therefore \quad \frac{d\vec{r}}{dt} = \omega r(-\hat{i}\sin\omega t + \hat{j}\cos\omega t) \quad [\because\ r \text{ is constant}]$$

$$\therefore \quad \frac{d^2\vec{r}}{dt^2} = \omega^2 r(-\hat{i}\cos\omega t - \hat{j}\sin\omega t) = -\omega^2 r(\vec{i}\cos\omega t + \hat{j}\sin\omega t) = -\omega^2 r\frac{\vec{r}}{r} = -\omega^2\vec{r} = \vec{f}.$$

$$\therefore \quad \vec{f} \propto \vec{r} \quad [\because\ \omega = \text{constant}]$$

The negative sign shows that the acceleration $\vec{f}$ is directed towards the centre of the circular path.

(c) We have $\vec{r} \times \vec{v} = \vec{r} \times \frac{d\vec{r}}{dt} = r(\hat{i}\cos\omega t + \hat{j}\sin\omega t) \times r\omega(-i\sin\omega t + \hat{j}\cos\omega t)$

$$= \boldsymbol{r^2\omega(\hat{k}\cos^2\omega t + \hat{k}\sin^2\omega t) = r^2\omega\hat{k} = \textbf{a constant.}}$$

$\therefore$ the vector $(\vec{r} \times \vec{v})$ has a constant magnitude.

Example 11.7. *Show that, $\vec{r} = \dot{e}^{-t}(\vec{C}_1\cos 2t + \vec{C}_2\sin 2t)$, where $\vec{C}_1$ and $\vec{C}_2$ are constant vectors, is a solution of the differential equation, $\frac{d^2\vec{r}}{dt^2} + 2\frac{d\vec{r}}{dt} + 5\vec{r} = 0$.*

Solution : We have $\vec{r} = e^{-t}(\vec{C}_1\cos 2t + \vec{C}_2\sin 2t)$.

$$\therefore \quad \frac{d\vec{r}}{dt} = (-1)e^{-t}(\vec{C}_1\cos 2t + \vec{C}_2\sin 2t) + e^{-t}(-2\vec{C}_1\sin 2t + 2\vec{C}_2\cos 2t)$$

$$= e^{-t}[\vec{C}_2(2\cos 2t - \sin 2t) - \vec{C}_1(2\sin 2t + \cos 2t)]$$

and
$$\frac{d^2\vec{r}}{dt^2} = \frac{d}{dt}\left(\frac{d\vec{r}}{dt}\right) = e^{-t}[\vec{C}_2(2\cos 2t - \sin 2t) - \vec{C}_1(2\sin 2t + \cos 2t)]$$
$$+ e^{-t}[\vec{C}_2(-4\sin 2t - 2\cos 2t) - \vec{C}_1(4\cos 2t - 2\sin 2t)]$$
$$= e^{-t}[\vec{C}_2(-2\cos 2t + \sin 2t - 4\sin 2t - 2\cos 2t)$$
$$+ \vec{C}_1(2\sin 2t + \cos 2t - 4\cos 2t + 2\sin 2t)]$$
$$= e^{-t}[\vec{C}_1(4\sin 2t - 3\cos 2t) - \vec{C}_2(4\cos 2t + 3\sin 2t)].$$

Now,
$$\frac{d^2\vec{r}}{dt^2} + 2\frac{d\vec{r}}{dt} + 5\vec{r} = e^{-t}[\vec{C}_1(4\sin 2t - 3\cos 2t) - \vec{C}_1(2\sin 2t + \cos 2t)]$$
$$+ 2e^{-t}[-\vec{C}_1(2\sin 2t + \cos 2t) + \vec{C}_2(2\cos 2t - \sin 2t)]$$
$$+ 5e^{-t}(\vec{C}_1\cos 2t + \vec{C}_2\sin 2t)$$
$$= e^{-t}[\vec{C}_1(4\sin 2t - 3\cos 2t - 4\sin 2t - 2\cos 2t + 5\cos 2t)]$$

$$= e^{-t}[\vec{C}_2(-4\cos 2t - 3\sin 2t + 4\cos 2t - 2\sin 2t + 5\sin 2t)$$
$$= e^{-t}\cdot 0 = 0.$$

$\therefore$ $\vec{r}$ **satisfies the equation,** $\dfrac{d^2\vec{r}}{dt^2} + 2\dfrac{d\vec{r}}{dt} + 5\vec{r} = 0.$

Example 11.8. *Find a unit tangent vector to any point on the curve, $x = a\cos\omega t$, $y = a\sin\omega t$, $z = bt$, where a, b, ω are constant quantities.*

Solution : We have $\vec{r} = \hat{i}r + \hat{j}y + \hat{k}z = \hat{i}(a\cos\omega t) + \hat{j}(a\sin\omega t) + \hat{k}bt.$

$$\therefore \quad \frac{d\vec{r}}{dt} = \hat{i}(-a\omega\sin\omega t) + \hat{j}(a\omega\cos\omega t) + \hat{k}b$$

$$\frac{ds}{dt} = \left|\frac{d\vec{r}}{dt}\right| = \sqrt{\frac{d\vec{r}}{dt}\cdot\frac{d\vec{r}}{dt}} = \sqrt{(-a\omega\sin\omega t)^2 + (a\omega\cos\omega t)^2 + b^2} = \sqrt{a^2\omega^2 + b^2}.$$

$\therefore$ unit vector to any point on the curve is

$$\hat{T} = \frac{d\vec{r}}{ds} = \frac{\frac{d\vec{r}}{dt}}{\frac{ds}{dt}} = \frac{\hat{i}(-a\omega\sin\omega t) + \hat{j}a\omega\cos\omega t + b\vec{k}}{\sqrt{a^2\omega^2 + b^2}}.$$

Example 11.9. *Solve* $\dfrac{d\vec{Y}}{dt} = \vec{X},\ \dfrac{d\vec{X}}{dt} = -\vec{Y}.$

Solution : We have $\dfrac{d\vec{X}}{dt} = -\vec{Y},$

$$\therefore \quad \frac{d^2\vec{X}}{dt^2} = -\frac{d\vec{Y}}{dt} = -\vec{X}$$

$$\text{or,} \quad \frac{d^2\vec{X}}{dt^2} + \vec{X} = 0. \tag{1}$$

The solution of equation (1) is

$$\vec{X} = \vec{C}_1\cos t + \vec{C}_2\sin t \quad [\vec{C}_1 \text{ and } \vec{C}_2 \text{ are arbitrary constant vectors}]$$

Again, since $\dfrac{d\vec{Y}}{dt} = \vec{X},$

$$\therefore \quad \frac{d^2\vec{Y}}{dt^2} = \frac{d\vec{X}}{dt} = -\vec{Y} \tag{2}$$

$$\text{or,} \quad \frac{d^2\vec{Y}}{dt^2} + \vec{Y} = 0. \tag{3}$$

The solution of equation (2) is $\boldsymbol{\vec{Y} = \vec{C}'_1\sin t - \vec{C}'_2\cos t}$.

Example 11.10. *If $\vec{C}_1$ and $\vec{C}_2$ are constant vectors and λ is a constant scalar, then show that, $H = e^{-\lambda x}(\vec{C}_1\sin\lambda y + C_2\cos\lambda y)$ satisfies the partial differential equation,*

$$\frac{\partial^2\vec{H}}{\partial x^2} + \frac{\partial^2\vec{H}}{\partial y^2} = 0.$$

Solution : We have $\vec{H} = e^{-\lambda x}(\vec{C}_1 \sin \lambda y + \vec{C}_2 \cos \lambda y)$

$$\therefore \quad \frac{\partial \vec{H}}{\partial x} = e^{-\lambda x}(-\lambda)(\vec{C}_2 \sin \lambda y + \vec{C}_2 \cos \lambda y)$$

and $$\frac{\partial^2 \vec{H}}{\partial x^2} = \frac{\partial}{\partial x}\left(\frac{\partial \vec{H}}{\partial x}\right) = \lambda^2 e^{-\lambda x}(\vec{C}_1 \sin \lambda y + \vec{C}_2 \cos \lambda y) = \lambda^2 \vec{H}.$$

Again, $$\frac{\partial \vec{H}}{dy} = e^{-\lambda x}[\lambda \vec{C}_1 \cos \lambda y - \lambda \vec{C}_2 \sin \lambda y] = \lambda e^{-\lambda x}(\vec{C}_1 \cos \lambda y - \vec{C}_2 \sin \lambda y)$$

and $$\frac{\partial^2 \vec{H}}{dy^2} = \lambda^2 e^{-\lambda x}(-\vec{C}_1 \sin \lambda y - \vec{C}_2 \cos \lambda y) = \lambda^2 \vec{H}.$$

$$\therefore \quad \boldsymbol{\frac{\partial^2 \vec{H}}{\partial x^2} + \frac{\partial^2 \vec{H}}{\partial y^2} = \lambda^2 \vec{H} = 0.}$$ **Hence the result.**

Example 11.11. *Show the general solution of the differential equation,*

$$\frac{d^2 \vec{r}}{dt^2} + 2\alpha \frac{d\vec{r}}{dt} + \omega^2 \vec{r} = 0,$$

where α and ω are constants is

(i) $\vec{r} = e^{-\alpha t}[\vec{C}_1 e^{\sqrt{\alpha^2-\omega^2}t} + \vec{C}_2 e^{\sqrt{-\alpha^2-\omega^2}t}]$, *if* $\alpha^2 - \omega^2 > 0$

(ii) $\vec{r} = e^{-\alpha t}[\vec{C}_1 \sin \sqrt{\omega^2 - \alpha^2}t + \vec{C}_2 \cos \sqrt{\alpha^2 - \omega^2}t]$, *if* $\alpha^2 - \omega^2 < 0$

(iii) $\vec{r} = e^{-\alpha t}(\vec{C}_1 + \vec{C}_2 t)$, *if* $\alpha^2 - \omega^2 = 0$, *where $\vec{C}_1$ and $\vec{C}_2$ are arbitrary constant vectors.*

Solution : (i) Let $\vec{r} = \vec{C} e^{mt}$ be a trial solution of the given equation. Then,

$$\frac{d\vec{r}}{dt} = \vec{C} m e^{mt} \quad \text{and} \quad \frac{d^2 \vec{r}}{dt^2} = m^2 \vec{C} e^{mt}.$$

Substituting these values in the given equation, we get

$$m^2 \vec{C} e^{mt} + 2\alpha \vec{C} m e^{mt} + \omega^2 \vec{C} e^{mt} = 0 \quad \text{or,} \quad \vec{C} e^{mt}[m^2 + 2m\alpha + \omega^2] = 0.$$

$$\because \quad \vec{C} e^{mt} \neq 0, \quad \therefore \quad m^2 + 2\alpha m + \omega^2 = 0.$$

$$\therefore \quad m = \frac{-2\alpha \pm \sqrt{4\alpha^2 - 4\omega^2}}{2} = -\alpha \pm \sqrt{\alpha^2 - \omega^2}.$$

$$\therefore \quad \vec{r} = \vec{C}_1 e^{(-\alpha + \sqrt{\alpha^2 - \omega^2})t} + \vec{C} 2 e^{(-\alpha - \sqrt{\alpha^2 - \omega^2})}t$$

$$= \boldsymbol{e^{-\alpha t}[\vec{C}_1 e^{\sqrt{\alpha^2-\omega^2}t} + \vec{C}_2 e^{-\sqrt{\alpha^2-\omega^2}t}]}, \text{ when } \alpha^2 - \omega^2 > 0.$$

(ii) When, $\alpha^2 - \omega^2 < 0$, $\sqrt{\alpha^2 - \omega^2} = i\sqrt{\omega^2 - \alpha^2}$.

$$\therefore \quad \vec{r} = e^{-\alpha t}[\vec{C}'_1 e^{i\sqrt{\omega^2-\alpha^2}t} + \vec{C}'_2 e^{-i\sqrt{\omega^2-\alpha^2}t}]$$

$$= e^{-\alpha t}[\vec{C}'_1 \cos \sqrt{\omega^2 - \alpha^2}t + i\vec{C}'_1 \sin \sqrt{\omega^2 - \alpha^2}t + \vec{C}'_2 \cos \sqrt{\omega^2 - \alpha^2}t - i\vec{C}'_2 \sin \sqrt{\omega^2 - \alpha^2}t]$$

$$= e^{-\alpha t}\left[(\vec{C}'_1 + \vec{C}'_2) \cos \sqrt{\omega^2 - \alpha^2}t + i(\vec{C}'_1 - \vec{C}'_2) \sin \sqrt{\omega^2 - \alpha^2}t\right]$$

$$= \boldsymbol{e^{-\omega t}[\vec{C}_1 \sin \sqrt{\omega^2 - \alpha^2}t + \vec{C}_2 \cos \sqrt{\omega^2 - \alpha^2}t]}, \text{ where } \alpha^2 - \omega^2 < 0$$

(iii) If $(\alpha^2 - \omega^2) = 0$, then

$$\vec{r} = \vec{C}_1 e^{-\alpha t} + \vec{C}_2 e^{-\alpha t} = (\vec{C}_1 + \vec{C}_2) e^{-\alpha t} \quad [\because \alpha^2 - \omega^2 = 0]$$
$$= \vec{C} e^{-\alpha t}.$$

This cannot be a solution of the equation which is a second order linear differential equation. According to the theory of differential equation a second order linear differential equation must contain two constants in its solution. Hence, the accepted solution of the given equation, when $\alpha^2 - \omega^2 = 0$ is

$$\vec{r} = e^{-\alpha t}(\vec{C}_1 + \vec{C}_2 t).$$

Here $\vec{C}_1$ and $\vec{C}_2$ are arbitrary constant vectors.

Example 11.12. *Prove that,* $\vec{A} = \vec{P}_0 \frac{e^{i\omega(t-r/c)}}{r}$, *where $\vec{P}_0$ is a constant vector $\vec{\omega}$ and C are constant scalars and $i = \sqrt{-1}$, satisfies the equation,* $\dfrac{\partial^2 \vec{A}}{\partial r^2} + \dfrac{2}{r}\dfrac{\partial \vec{A}}{\partial r} - \dfrac{1}{C^2}\dfrac{\partial^2 \vec{A}}{\partial t^2} = 0.$

Solution : We have $\vec{A} = \vec{P}_0 \frac{e^{i\omega(t-r/c)}}{r}$.

$$\therefore \quad \frac{\partial \vec{A}}{\partial r} = \vec{P}_0 \left[\left(-\frac{1}{r^2}\right) e^{i\omega(t-r/c)} + \frac{1}{r}\left(\frac{-i\omega}{c}\right) e^{i\omega(t-r/c)}\right] = -\vec{P}_0 \left[\frac{1}{r^2} + \frac{i\omega}{cr}\right] e^{i\omega(t-r/c)}$$

and
$$\frac{\partial^2 \vec{A}}{\partial r^2} = -\vec{P}_0 \left[\left(-\frac{2}{r^3} - \frac{i\omega}{cr^2}\right) e^{i\omega(t-r/c)} + \left(\frac{1}{r^2} + \frac{i\omega}{cr}\right) e^{i\omega(t-r/c)} \left(-\frac{i\omega}{c}\right)\right]$$
$$= -\vec{P}_0 \left[-\frac{2}{r^3} - \frac{2i\omega}{cr^2} + \frac{\omega^2}{c^2 r}\right] e^{i\omega(t-r/c)}.$$

Also,
$$\frac{\partial \vec{A}}{\partial t} = \vec{P}_0 (i\omega) \frac{e^{i\omega(t-r/c)}}{r} \quad \text{and} \quad \frac{\partial^2 \vec{A}}{\partial t^2} = -\omega^2 \vec{P}_0 \frac{e^{i\omega(t-r/c)}}{r}.$$

Now, substituting the values of $\frac{\partial \vec{A}}{\partial r} \cdot \frac{\partial^2 \vec{A}}{\partial r^2}$ and $\frac{\partial^2 \vec{A}}{\partial t}$ in the expression, $\frac{\partial^2 \vec{A}}{\partial r^2} + \frac{2}{r}\frac{\partial \vec{A}}{\partial r} - \frac{1}{c^2}\frac{\partial^2 \vec{A}}{\partial t^2}$, we get

$$\frac{\partial^2 \vec{A}}{\partial r^2} + \frac{2}{r}\frac{\partial \vec{A}}{\partial r} - \frac{1}{c^2}\frac{\partial^2 \vec{A}}{\partial t^2} = -\vec{P}_0 \left(-\frac{2}{r^3} - \frac{2i\omega}{cr^2} + \frac{\omega^2}{c^2 r}\right) e^{i\omega(t-r/c)}$$
$$- \vec{P}_0 \left(\frac{2}{r^3} + \frac{2i\omega}{cr^2}\right) e^{i\omega(t-r/c)} + \frac{\omega^2}{c^2 r} \vec{P}_0 e^{i\omega(t-r/c)}$$
$$= -\vec{P}_0 \left(\frac{\omega^2}{c^2 r}\right) e^{i\omega(t-r/c)} + \vec{P}_0 \left(\frac{\omega^2}{c^2 r}\right) e^{i\omega(t-r/c)} = \mathbf{0},$$

the result follows.

Example 11.13. *If $\vec{A} = \vec{A}(x, y, z)$ and $x = x(t) \cdot y = y(t)$, $z = z(t)$, then prove that,*

$$\frac{d\vec{A}}{dt} = \frac{\partial \vec{A}}{\partial t} + \frac{\partial \vec{A}}{\partial x}\frac{dx}{dt} + \frac{\partial \vec{A}}{\partial y}\frac{dy}{dt} + \frac{\partial \vec{A}}{\partial z}\frac{dz}{dt}.$$

Solution : Let $\vec{A} = \hat{i}A_1(x, y, z, t) + \hat{j}A_2(x, y, z, t) + \hat{k}A_3(x, y, z, t)$.

Then, $d\vec{A} = \hat{i}dA_1 + \hat{j}dA_2 + \hat{k}dA_3$

$$= \hat{i}\left[\frac{\partial A_1}{\partial t}dt + \frac{\partial A_1}{\partial x}dx + \frac{\partial A_1}{\partial y}dy + \frac{\partial A_1}{\partial z}dz\right] + \hat{j}\left[\frac{\partial A_2}{\partial t}dt + \frac{\partial A_2}{\partial x}dx + \frac{\partial A_2}{\partial y}dy + \frac{\partial A_2}{\partial z}dz\right]$$
$$+ \hat{k}\left[\frac{\partial A_3}{\partial t}dt + \frac{\partial A_3}{\partial x}dx + \frac{\partial A_3}{\partial y}dy + \frac{\partial A_3}{\partial z}dz\right]$$

or, $$d\vec{A} = \left[\hat{i}\frac{\partial A_1}{\partial t} + \hat{j}\frac{\partial A_2}{\partial t} + \hat{k}\frac{\partial A_3}{\partial t}\right]dt + \left[\hat{i}\frac{\partial A_1}{\partial x} + \hat{j}\frac{\partial A_2}{\partial x} + \hat{k}\frac{\partial A_3}{\partial x}\right]dx$$
$$+ \left[\hat{i}\frac{\partial A_1}{\partial y} + \hat{j}\frac{\partial A_2}{\partial y} + \hat{k}\frac{\partial A_3}{\partial y}\right]dy + \left[\hat{i}\frac{\partial A_1}{\partial z} + \hat{j}\frac{\partial A_2}{\partial z} + \hat{k}\frac{\partial A_3}{\partial z}\right]dz$$

or, $$d\vec{A} = \frac{\partial \vec{A}}{\partial t}dt + \frac{\partial \vec{A}}{\partial x}dx + \frac{\partial \vec{A}}{\partial y}dy + \frac{\partial \vec{A}}{\partial z}dz$$

or, $$\boldsymbol{\frac{d\vec{A}}{dt} = \frac{\partial \vec{A}}{\partial t} + \frac{\partial \vec{A}}{\partial x}\frac{dx}{dt} + \frac{\partial \vec{A}}{\partial y}\frac{dy}{dt} + \frac{\partial \vec{A}}{\partial z}\frac{dz}{dt}}$$ **(Proved).**

Example 11.14. *If $\vec{A} = \vec{A}(t)$, then show that,*

$$\frac{d}{dt}\left[\vec{A}\cdot\left(\frac{d\vec{A}}{dt}\times\frac{d^2\vec{A}}{dt^2}\right)\right] = \vec{A}\left[\frac{d\vec{A}}{dt}\times\frac{d^3\vec{A}}{dt^3}\right].$$

Solution : We have $$\frac{d}{dt}\left[\vec{A}\cdot\left(\frac{d\vec{A}}{dt}\times\frac{d^2\vec{A}}{dt^2}\right)\right] = \frac{d\vec{A}}{dt}\cdot\left(\frac{d\vec{A}}{dt}\times\frac{d^2\vec{A}}{dt^2}\right) + \vec{A}\cdot\frac{d}{dt}\left(\frac{d\vec{A}}{dt}\times\frac{d^2\vec{A}}{dt^2}\right)$$
$$= \vec{A}\cdot\frac{d}{dt}\left(\frac{d\vec{A}}{dt}\times\frac{d^2\vec{A}}{dt^2}\right) \quad \left[\because \frac{d\vec{A}}{dt}\cdot\left(\frac{d\vec{A}}{dt}\times\frac{d^2\vec{A}}{dt^2}\right) = 0\right]$$
$$= \vec{A}\cdot\left[\frac{d^2\vec{A}}{dt^2}\times\frac{d^2\vec{A}}{dt^2} + \frac{d\vec{A}}{dt}\times\frac{d^3\vec{A}}{dt^3}\right]$$
$$= \boldsymbol{\vec{A}\cdot\left(\frac{d\vec{A}}{dt}\times\frac{d^3\vec{A}}{dt^3}\right)} \quad \left[\because \frac{d^2\vec{A}}{dt^2}\times\frac{d^2\vec{A}}{dt^2} = 0\right]$$ **(Proved).**

Example 11.15. *Find the radial and transverse velocity and acceleration of a particle moving in a plane. Find the cartesian components of the acceleration o the particle.*

[NBU 1989, AU 1990S]

Solution : Let $\vec{r}$ be the position vector of a moving particle at any instant of time and $\hat{r}$, the unit vector along $\vec{r}$. If $P(r,\theta)$ be the position of the particle at t, then

$$\vec{r} = r\hat{r} = r(\hat{i}\cos\theta + \hat{j}\sin\theta),$$

where $\hat{i}$ and $\hat{j}$ are the unit vectors along X- and Y-directions respectively (Fig. 11.12).

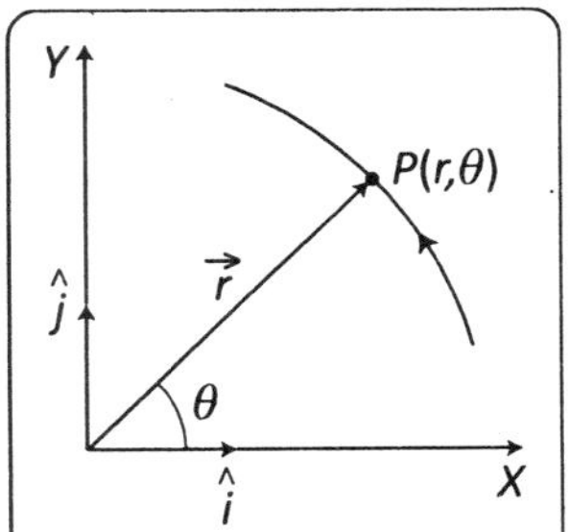

Fig. 11.12

The velocity of the particle at t is

$$\vec{v} = \frac{d\vec{r}}{dt} = \frac{d\vec{r}}{dt}\hat{r} + r\frac{d\vec{r}}{dt} \quad [\because \vec{r} = r\hat{r}].$$

Now, $\hat{r} = \hat{\imath}\cos\theta + \hat{\jmath}\sin\theta$.

$$\because \quad \frac{d\hat{r}}{dt} = (-\imath\sin\theta + \hat{\jmath}\cos\theta)\frac{d\theta}{dt},$$

$$\therefore \quad \vec{v} = \frac{d\vec{r}}{dt}\hat{r} + r\frac{d\theta}{dt}(-\hat{\imath}\sin\theta + \hat{\jmath}\cos\theta). \tag{1}$$

Again, unit vector transverse to OP is $\hat{\theta} = -\hat{\imath}\sin\theta + \hat{\jmath}\cos\theta$.

$$\therefore \quad \frac{d\hat{\theta}}{dt} = (-\hat{\imath}\cos\theta - \hat{\jmath}\sin\theta)\frac{d\theta}{dt} = -\hat{r}\frac{d\theta}{dt}.$$

$$\therefore \quad \vec{v} = \frac{dr}{dt}\hat{r} + r\frac{d\theta}{dt}\hat{\theta} = \dot{r}\hat{r} + r\dot{\theta}\hat{\theta}, \quad \left[\text{where } \frac{dr}{dt} = \dot{r} \text{ and } \frac{d\theta}{dt} = \dot{\theta}\right]$$

$$= \boldsymbol{v_r\hat{r} + v_\theta\hat{v}}, \tag{2}$$

where $v_r = \dot{r}$ and $v_\theta = r\dot{\theta}$ are respectively the radial and transverse component of velocity of the particle moving in the X-Y plane.

The acceleration of the particle is given by

$$\vec{a} = \frac{d\vec{v}}{dt} = \frac{d}{dt}(\dot{r}\hat{r} + r\dot{\theta}\hat{\theta})$$

$$= \ddot{r}\hat{r} + \dot{r}\frac{d\hat{r}}{dt} + \dot{r}\dot{\theta}\hat{\theta} + r\ddot{\theta}\hat{\theta} + r\dot{\theta}\frac{d\hat{\theta}}{dt}, \quad \left[\text{where } \ddot{r} = \frac{d^2t}{dt^2} \text{ and } \ddot{\theta} = \frac{d^2\theta}{dt^2}\right]$$

$$= \ddot{r}\hat{r} + \dot{r}(-\hat{\imath}\sin\theta + \hat{\jmath}\cos\theta)\frac{d\theta}{dt} + \dot{r}\dot{\theta}\hat{\theta} + r\ddot{\theta}\hat{\theta} + r\dot{\theta}\left(-\hat{r}\frac{d\theta}{dt}\right)$$

$$\left[\text{using the values of } \frac{d\hat{r}}{dt} \text{ and } \frac{d\hat{\theta}}{dt}\right]$$

$$= \boldsymbol{(\ddot{r} - r\dot{\theta}^2)\hat{r} + (2\dot{r}\dot{\theta} + r\ddot{\theta})\hat{\theta} = a_r\hat{r} + a_\theta\hat{\theta}}, \tag{3}$$

where $a_r = (\ddot{r} - r\dot{\theta}^2)$ and $a_\theta = (2\dot{r}\dot{\theta} + r\ddot{\theta})$ are respectively the radial and transverse components of accelerations of the particle moving in the XY-plane.

Cartesian Components of Acceleration ($\vec{a}$)

We have position vector $\vec{r}$ of the particle at any time t is

$$\vec{r} = \hat{\imath}x + \hat{\jmath}y = \hat{\imath}(r\cos\theta) + \hat{\jmath}(r\sin\theta).$$

$\therefore$ the velocity of the particle at t is

$$\vec{v} = \frac{d\vec{r}}{dt} = \hat{\imath}\left(\frac{dx}{dt}\right) + \hat{\jmath}\left(\frac{dy}{dt}\right) \quad \text{or,} \quad \dot{\vec{r}} = \hat{\imath}\dot{x} + \hat{\jmath}\dot{y} = \hat{\imath}v_x + \hat{\jmath}v_y$$

Now, the x-component of velocity

$$v_x = \dot{x} = \frac{d}{dt}(r\cos\theta) = \frac{dr}{dt}\cos\theta - r\sin\theta\frac{d\theta}{dt} = \boldsymbol{(\dot{r}\cos\theta - r\sin\theta\dot{\theta})} \tag{4}$$

and the y-component of velocity of the particle is

$$v_y = \dot{y} = \frac{d}{dt}(r\sin\theta) = \frac{dr}{dt}\sin\theta + r\cos\theta\frac{d\theta}{dt} = \boldsymbol{\dot{r}\sin\theta + (r\dot{\theta}\cos\theta)} \tag{5}$$

$$\therefore \quad \dot{\vec{r}} = \hat{r}(\dot{r}\cos\theta - r\dot{\theta}\sin\theta) + \hat{\jmath}(\dot{r}\sin\theta + r\dot{\theta}\cos\theta) = \vec{v}. \tag{6}$$

Now the acceleration of the particle at t is

$$\vec{a} = \frac{d\vec{r}}{dt} = \frac{d}{dt}(\vec{v}) = \frac{d}{dt}[\hat{\imath}(\dot{r}\cos\theta - r\dot{\theta}\sin\theta)] + \frac{d}{dt}[\hat{\jmath}(\dot{r}\sin\theta + r\dot{\theta}\cos\theta)]$$

$$= [\ddot{r}\cos\theta + \dot{r}\dot{\theta}(-\sin\theta) - \dot{r}\dot{\theta}\sin\theta - r\ddot{\theta}\sin\theta - r\dot{\theta}^2\cos\theta]\hat{\imath}$$
$$+ [\ddot{r}\sin\theta + \dot{r}\dot{\theta}\cos\theta + \dot{r}\dot{\theta}\cos\theta + r\ddot{\theta}\cos\theta - r\theta^2\sin\theta]\hat{\jmath}$$

$$= [(\ddot{r} - r\dot{\theta}^2)\cos\theta - 2(\dot{r}\dot{\theta} + r\theta)\sin\theta]\hat{\imath} + [(\ddot{r} - r\dot{\theta}^2)\sin\theta + 2(\dot{r}\dot{\theta} + r\ddot{\theta})\cos\theta]\hat{\jmath}$$

$$= \boldsymbol{a_x\hat{\imath} + \hat{\jmath}a_y}, \tag{7}$$

where $\boldsymbol{a_x = (\ddot{r} - r\dot{\theta}^2)\cos\theta - (2\dot{r}\dot{\theta} + r\ddot{\theta})\sin\theta}$

and $\boldsymbol{a_y = (\ddot{r} - r\dot{\theta}^2)\sin\theta + (2\dot{r}\dot{\theta} + r\ddot{\theta})\cos\theta}$

are respectively the x- and y-components of acceleration of the particle moving in the XY plane.

Example 11.16. *Find the curvature and torsion of the curve,*

$$x = a\cos t, \quad y = a\sin t, \quad z = bt.$$ **[Poona U 1987]**

Solution : The position vector of any point on the curve with respect to the origin is

$$\vec{r} = (a\cos t)\hat{\imath} + (a\sin t)\hat{\jmath} + (bt)\hat{k},$$

$$\therefore \quad \frac{d\vec{r}}{dt} = (-a\sin t)\hat{\imath} + (a\cos t)\hat{\jmath} + b\hat{k}.$$

The arc length of the curve from P_0 at $t = 0$ to any point P at $t = t$ is given by

$$s = \int_{t=0}^{t} \left|\frac{d\vec{r}}{dt}\right| dt = \int_{t=0}^{t} \sqrt{(-a\sin t)^2 + (a\cos t)^2 + b^2}dt = \sqrt{a^2+b^2}t.$$

$$\therefore \quad \frac{ds}{dt} = \sqrt{a^2+b^2}.$$

The unit tangent to the curve at $P(t)$ is

$$\hat{T} = \frac{d\vec{r}}{ds} = \frac{d\vec{r}}{dt}\Big/\frac{ds}{dt} = \frac{(-a\sin t)\hat{\imath} + (a\cos t)\hat{\jmath} + b\hat{k}}{\sqrt{a^2+b^2}}$$

$$\therefore \quad \frac{d\hat{T}}{ds} = \frac{d\hat{T}}{dt}\Big/\frac{ds}{dt} = \frac{-a(\cos t\hat{\imath} + \sin t\hat{\jmath})}{a^2+b^2}.$$

But, $\dfrac{d\hat{T}}{ds} = \kappa\hat{N}$ or, $\boldsymbol{\kappa = \left|\dfrac{d\hat{T}}{ds}\right| = \dfrac{a}{a^2+b^2}}$. (1)

$\therefore$ $\hat{N} = -(\cos t\hat{\imath} + \sin t\hat{\jmath})$ is the unit principal normal to the curve at $P(t)$.

Again, $\because$ the bi-normal $\hat{B} = \hat{T} \times \hat{N}$ or, $\hat{B} = (b\sin t\hat{\imath} - b\cos t\hat{\jmath} + a\hat{k})$.

$$\therefore \quad \frac{d\hat{B}}{ds} = \frac{d\vec{B}}{dt}\Big/\frac{ds}{dt} = \frac{b\cos t\hat{\imath} + b\sin t\hat{\jmath}}{a^2+b^2}.$$

But, $\dfrac{d\hat{B}}{ds} = -\tau\hat{N} = \tau(\cos t\hat{\imath} + \sin t\hat{\jmath})$.

$\therefore$ **Torsion** $\boldsymbol{\tau = \dfrac{b}{a^2+b^2}}$. (2)

Example 11.17. *The equation of an ellipse is given by $\vec{r} = a\cos t\hat{i} + b\sin t\hat{j}$. Find the curvature.*

Solution : $\because \quad \vec{r} = a\cos t\hat{i} + b\sin t\hat{j}$,

$$\therefore \quad \frac{d\vec{r}}{dt} = -a\sin t\hat{i} + b\cos t\hat{j}; \quad \therefore \quad \frac{ds}{dt} = \left|\frac{d\vec{r}}{dt}\right| = \sqrt{a^2\sin^2 t + b^2\cos^2 t}.$$

Now the unit tangent vector to the curve is

$$\widehat{T} = \frac{d\vec{r}}{ds} = \frac{d\vec{r}}{dt}\Big/\frac{ds}{dt} = \frac{1}{\sqrt{a^2\sin^2 t + b^2\cos t}}(-a\sin t\hat{i} + b\cos\hat{j}).$$

$$\therefore \quad \frac{d\widehat{T}}{dt} = \frac{d}{dt}\left[\frac{-a\sin t}{\sqrt{a^2\sin^2 t + b^2\cos^2 t}}\right]\hat{i} + \frac{d}{dt}\left[\frac{b\cos t}{\sqrt{a^2\sin^2 t + b^2\cos^2 t}}\right]\hat{j}$$

$$= \frac{-ab^2\cos t}{(a^2\sin^2 t + b^2\cos^2 t)^{3/2}}\hat{i} - \frac{a^2 b\sin t}{(a^2\sin^2 t + b^2\cos^2 t)^{3/2}}\hat{j}.$$

$$\therefore \quad \frac{d\widehat{T}}{ds} = \frac{d\widehat{T}}{dt}\Big/\frac{ds}{dt} = \frac{-ab(b\cos t\hat{i} + a\sin t\hat{i})}{(a^2\sin^2 tb^2\cos^2 t)^2}.$$

But, $\dfrac{d\widehat{T}}{ds} = \kappa\widehat{N}$.

$$\textbf{Curvature, } \kappa = \left|\frac{d\widehat{T}}{ds}\right| = \frac{+ab\sqrt{a^2\sin^2 t + b^2\cos^2 t}}{(a^2\sin^2 t + b^2\cos^2 t)^{3/2}} = \frac{ab}{(a^2\sin^2 t + b^2\cos^2 t)^{3/2}}$$

$$= \frac{1}{\rho}. \quad [\rho = \text{the radius of curvature}]$$

Example 11.18. *Find the equation for the* (a) *tangent* (b) *principal normal and* (c) *binormal to the curve $x = 3\cos t$, $y = 3\sin t$, $z = 4t$ at the point where $t = \pi$.*

Solution : The equation to the curve is $\vec{r} = \hat{i}x + \hat{j}y + \hat{k}z = \hat{i}(3\cos t) + \hat{j}(3\sin t) + \hat{k}4t$.

$$\therefore \quad \frac{d\vec{r}}{dt} = -3\sin t\hat{i} + 3\cos t\hat{j} + 4\hat{k}.$$

The unit tangent vector is

$$\widehat{T} = \frac{d\vec{r}}{ds} = \frac{d\vec{r}}{dt}\Big/\frac{ds}{dt} = -\frac{3\sin t\hat{i} + 3\cos t\hat{j} + 4\hat{k}}{\sqrt{9\sin^2 t + 9\cos^2 t + 4^2}}$$

or, $$\boldsymbol{\widehat{T} = -\frac{3}{5}\sin t\hat{i} + \frac{3}{5}\cos t\hat{j} + \frac{4}{5}\hat{k}.}$$

Again, $$\frac{d\widehat{T}}{dt} = -\frac{3}{5}\cos t\hat{i} - \frac{3}{5}\sin t\hat{j}, \quad \therefore \quad \frac{d\widehat{T}}{ds} = \frac{d\widehat{T}}{dt}\Big/\frac{ds}{dt} = -\frac{3}{25}\cos t\hat{i} - \frac{3}{25}\sin t\hat{j}.$$

$$\because \quad \frac{d\widehat{T}}{ds} = \kappa\widehat{N},$$

$$\therefore \quad \textbf{curvature, } k = \left|\frac{\widehat{T}}{ds}\right| = \sqrt{\left(\frac{3}{25}\cos t\right)^2 + \left(\frac{3}{25}\sin t\right)^2} = \frac{\mathbf{3}}{\mathbf{25}}.$$

$\therefore$ unit principal normal to the curve at t is $\boldsymbol{\widehat{N} = -\cos t\hat{i} - \sin t\hat{j}}$.

$$\text{Again, } \widehat{B} = \widehat{T} \times \widehat{N} = \begin{vmatrix} \hat{i} & \hat{j} & \hat{k} \\ -\frac{3}{5}\sin t & \frac{3}{t}\cos t & \frac{4}{5} \\ -\cos t & -\sin t & 0 \end{vmatrix} = \mathbf{\frac{4}{5}\sin t\hat{i} - \frac{4}{5}\cos t\hat{j} + \frac{3}{5}\hat{k}}$$

at $t = \pi$, let the vectors $\widehat{T}$, $\widehat{N}$ and $\widehat{B}$ be $\widehat{T}_0$, $\widehat{N}_0$ and $\widehat{B}_0$ respectively.

Then, $\widehat{T}_0 = -\frac{3}{5}\hat{j} + \frac{4}{5}\hat{k}$, $\quad \widehat{N}_0 = \hat{i}$, $\quad \widehat{B}_0 = \frac{4}{5}\hat{j} + \frac{3}{5}\hat{k}$.

If the vector equation to the tangent line be $\vec{A}$ and $\vec{r}$ be arbitrary point on the curve, then $(\vec{r} - \vec{r}_0)$ is parallel to $\vec{A}$.

So, $(\vec{r} - \vec{r}_0)$ is parallel to $\widehat{T}_0$

or, $\quad \vec{r} - \vec{r}_0 = t\widehat{T}$ [where $\vec{r}_0 = -3\hat{i} + 4\pi\hat{k}$ is the initial point on the tangent at $t = \pi$]

or, $\quad \vec{r} = \vec{r}_0 + t\widehat{T}$.

Hence, the equation to the tangent line is

$$\vec{r} = -3\hat{i} + 4\pi\hat{k} + t\left(-\frac{3}{5}\hat{j} + \frac{4}{5}\hat{k}\right) \quad \text{or,} \quad \vec{r} = \mathbf{-3\hat{i} - \frac{3}{5}t\hat{j} + \left(\frac{4}{5}t + 4\pi\right)\hat{k}}.$$

Similarly, principal normal is

$$\vec{r} = \vec{r}_0 + t\widehat{N}_0 = -3\hat{i} + 4\pi\hat{k} + t\hat{i} = \mathbf{(t-3)\hat{i} + 4\pi\hat{k}}$$

and binormal is

$$\vec{r} = \vec{r}_0 + t\widehat{B}_0 = -3\hat{i} + 4\pi\hat{k} + \frac{4}{5}t\hat{j} + \frac{3}{5}t\hat{k} = \mathbf{-3\hat{i} + \frac{4}{5}t\hat{j} + \left(4\pi + \frac{3}{5}t\right)\hat{k}}.$$

Example 11.19. *The parametric equation of a curve C is $x = x(s)$, $y = y(s)$, $z = z(s)$, where s is the arc length of C measured from a fixed point on C. If $\vec{r}$ be the position vector of any point on C, show that $\frac{d\vec{r}}{ds}$ is a unit tangent vector to C. Find the radius of curvature of the curve.*

Solution : We have the equation to the curve C is $\vec{r} = \hat{i}x + \hat{j}y + \hat{k}z$,

$$\therefore \quad \frac{d\vec{r}}{ds} = \hat{i}\frac{dx}{ds} + \hat{j}\frac{dy}{ds} + \hat{k}\frac{dz}{ds}. \qquad \frac{d\vec{r}}{ds} \text{ is a tangent to } C.$$

$$\text{Now, } \left|\frac{d\vec{r}}{ds}\right| = \sqrt{\left(\frac{dx}{ds}\right)^2 + \left(\frac{dy}{ds}\right)^2 + \left(\frac{dz}{ds}\right)^2} = \sqrt{\frac{(dx)^2 + (dy)^2 + (dz)^2}{(ds)^2}} = \frac{(ds)^2}{(ds)^2} = 1.$$

$\because (ds)^2 = (dx)^2 + (dy)^2 + (dz)^2$, so, $\frac{d\vec{r}}{ds}$ is a unit tangent vector to the curve C.

To find the radius of curvature (ρ) of the curve, we have

$$\widehat{T} = \frac{d\vec{r}}{ds} = \hat{i}\frac{dx}{ds} + \hat{j}\frac{dy}{ds} + \hat{k}\frac{dz}{ds}.$$

$$\therefore \quad \frac{d\widehat{T}}{ds} = \hat{i}\frac{d^2x}{ds^2} + \hat{j}\frac{d^2y}{ds^2} + \hat{k}\frac{d^2z}{ds^2}.$$

But $\dfrac{d\widehat{T}}{ds} = \kappa\widehat{N}$, where κ is the curvature and $\widehat{N}$ is the unit principal normal to C.

$$\therefore \quad \kappa = \left|\frac{d\widehat{T}}{ds}\right| = \sqrt{\left(\frac{d^2x}{ds^2}\right)^2 + \left(\frac{d^2y}{ds^2}\right)^2 + \left(\frac{d^2z}{ds}\right)^2}.$$

Hence, the radius of curvature of C is $\rho = \dfrac{1}{\kappa} = \dfrac{1}{\sqrt{\left(\frac{d^2x}{ds^2}\right)^2 + \left(\frac{d^2y}{ds^2}\right)^2 + \left(\frac{d^2z}{ds^2}\right)^2}}$.

Example 11.20. *The equation to the trajectory of a particle is given by $\vec{r} = \hat{i}e^t\cos t + \hat{j}e^t\sin t$. Find out an expression for the acceleration of the particle and show that it is at right angles to $\vec{r}$.*

Solution : We have the equation to the trajectory of the moving particle,

$$\vec{r} = \hat{i}e^t\cos t + \hat{j}e^t\sin t.$$

$\therefore$ the velocity of the particle is

$$\vec{v} = \frac{d\vec{r}}{dt} = \hat{i}(e^t\cos t) - \hat{i}(e^t\sin t) + \hat{j}(e^t\sin t) + \hat{j}(e^t\cos t)$$

$$= (\hat{i}e^t\cos t + \hat{j}e^t\sin t) - \hat{i}e^t\sin t + \hat{j}e^t\cos t = \vec{r} - \hat{i}e^t\sin t + \hat{j}e^t\cos t.$$

Acceleration of the particle is

$$\vec{a} = \frac{\vec{v}}{dt} = \frac{d\vec{r}}{dt} - \hat{i}(e^t\sin t + e^t\cos t) + \hat{j}(e^t\cos t - e^t\sin t)$$

$$= \vec{r} - \hat{i}e^t\sin t + \hat{j}e^t\cos t - \hat{i}e^t\sin t - \hat{i}e^t\cos t - \hat{j}e^t\sin t + \hat{j}e^t\cos t$$

$$= \vec{r} - (\hat{i}e^t\cos t + \hat{j}e^t\sin t) - 2\hat{i}e^t\sin t + 2\hat{j}e^t\cos t$$

or, $\quad \vec{a} = \vec{r} - \vec{r} - \hat{i}2e^t\sin t + \hat{j}2e^t\cos t = -\hat{i}2e^t\sin t + \hat{j}2e^t\cos t.$

Now, $\quad \vec{a}\cdot\vec{r} = (-\hat{i}2e^t\sin t + \hat{j}2e^t\cos t)\cdot(\hat{i}e^t\cos t + \hat{j}e^t\sin t)$

$$= -2e^t\sin t\cos t + 2e^t\cos t\sin t = \mathbf{0}.$$

Hence, the acceleration $\vec{a}$ is at right angles to $\vec{r}$.

Example 11.21. *The parametric equation to the curve of a moving particle at t is $x = (t^3 - 4t)$, $y = (t^2 + 4t)$, $z = (8t^3 - 3t^3)$. Find the magnitude of the*

(a) *tangential components of its acceleration and*

(b) *normal components of its acceleration at $t = 2$.*

Solution : The equation to the curve of the moving particle at t is

$$\vec{r} = \hat{i}x + \hat{j}y + \hat{k}z = \hat{i}(t^3 - 4t) + \hat{j}(t^2 + 4t) + \hat{k}(8t^2 - 3t^3).$$

$$\therefore \quad \frac{d\vec{r}}{dt} = \hat{i}(3t^2 - 4) + \hat{j}(2t + 4) + \hat{k}(16t - 9t^2).$$

$$\therefore \quad \frac{ds}{dt} = \left|\frac{d\vec{r}}{dt}\right| = \sqrt{(3t^2 - 4)^2 + 4(t+2)^2 + (16t - 9t^2)^2}.$$

The tangential components of the acceleration is

$$\frac{d^2s}{dt^2} = \frac{2(3t^2-4)6t + 8(t+2) + 2(16t-9t^2)(16-18t)}{2\sqrt{(3t^2-4)^2 + 4(t+2)^2 + (16t-9t^2)^2}}.$$

For $t = 2$, $\dfrac{d^2s}{dt^2} = 16$ units.

Again, $\widehat{T} = \dfrac{d\vec{r}}{ds} = \dfrac{d\vec{r}}{dt}\Big/\dfrac{ds}{dt} = \dfrac{\hat{\imath}(3t^2-4) + \hat{\jmath}(2t+4) + \hat{k}(16t-9t^2)}{\sqrt{(3t^2-4)^2 + 4(t+2)^2 + (16t-9t^2)^2}}.$

$$\therefore \quad \frac{d\widehat{T}}{ds} = \frac{d\widehat{T}}{dt}\Big/\frac{ds}{dt} = \frac{2f(t)[\hat{\imath}a'(t) + \hat{\jmath}b'(t) + \hat{k}c'(t)] - f'(t)[\hat{\imath}a(t) + \hat{\jmath}b(t) + \hat{k}c(t)]}{[f(t)]^2},$$

where $a(t) = 3t^2 - 4$, $b(t) = (2t+4)$, $c(t) = 16t - 9t^2$

and $f(t) = (3t^2-4)^2 + 4(t+2)^2 + (16t-9t^2)^2$.

$$\therefore \quad \frac{d\widehat{T}}{ds} \text{ (at } t = 2) = \frac{2\hat{\imath} - 13\hat{\jmath} - 22\hat{k}}{216}.$$

Hence, **the curvature,** $\boldsymbol{\kappa} = \left|\left(\dfrac{d\widehat{T}}{ds}\right)_{t=2}\right| = \dfrac{\sqrt{73}}{72}$.

$\therefore$ **the normal component of acceleration** $= \kappa\left|\left(\dfrac{ds}{dt}\right)_{t=2}\right|^2 = \dfrac{\sqrt{73}}{72} \times 144 = \mathbf{2\sqrt{73}}$ **units**.

Example 11.22. *The equation to the curve of a moving particle is represented by*

$$\vec{r} = \hat{\imath}t + \hat{\jmath}t^2 + \hat{k}\frac{2}{3}t^3.$$

Find,

(a) *the curvature (κ) and the torsion (τ) of the curve.*

(b) *the equations in vector and rectangular form of the tangent, principal normal and binormal to the curve at $t = 1$.*

Solution : (a) We have the equation to the curve $\vec{r} = \hat{\imath}t + \hat{\jmath}t^2 + \hat{k}\dfrac{2}{3}t^3$,

$$\therefore \quad \frac{d\vec{r}}{dt} = \hat{\imath} + \hat{\jmath}2t + \hat{k}2t^2.$$

$$\frac{ds}{dt} = \left|\frac{d\vec{r}}{dt}\right| = \sqrt{1^2 + (2t)^2 + (2t^2)^2} = 1 + 2t^2$$

and $$\widehat{T} = \frac{d\vec{r}}{ds} = \frac{d\vec{r}}{dt}\Big/\frac{ds}{dt} = \frac{\hat{\imath} + \hat{\jmath}2t + \hat{k}2t^2}{1+2t^2}. \qquad (1)$$

$$\therefore \quad \frac{d\widehat{T}}{dt} = \frac{(1+2t^2)(2\hat{\jmath} + \hat{k}4t) - (\hat{\imath} + \hat{\jmath}2t + \hat{k}2t^2)(4t)}{(1+2t^2)^2} = \frac{-\hat{\imath}4t + \hat{\jmath}(2-4t^2) + \hat{k}4t}{(1+2t^2)^2}$$

$$= \frac{-\hat{\imath}4t + \hat{\jmath}(2-4t^2) + \hat{k}4t}{(1+2t^2)^2}.$$

Now, $$\frac{d\widehat{T}}{ds} = \frac{d\widehat{T}}{dt}\Big/\frac{ds}{dt} = \frac{-\hat{\imath}4t + \hat{\jmath}(2-4t^2) + \hat{k}4t}{(1+2t^2)^3}.$$

Again, $\frac{d\widehat{T}}{ds} = \kappa \widehat{N}$.

$$\therefore \quad \kappa = \text{curvature} = \left|\frac{d\widehat{T}}{ds}\right| = \frac{\sqrt{(-4t)^2 + (2-4t^2)^2 + (4t)^2}}{(1+2t^2)^3} = \frac{\mathbf{2}}{\mathbf{(1+2t^2)^3}}.$$

To find out torsion, we have

$$\widehat{\mathbf{N}} = \frac{1}{\kappa}\frac{d\widehat{T}}{ds} = \frac{-\hat{\imath}2t + \hat{\jmath}(1-2t^2) + \hat{k}2t}{1+2t^2}. \tag{2}$$

$$\therefore \quad \widehat{B} = \widehat{T} \times \widehat{N} = \begin{vmatrix} \hat{\imath} & \hat{\jmath} & \hat{k} \\ \frac{1}{1+2t^2} & \frac{2t}{1+2t^2} & \frac{2t^2}{1+2t^2} \\ \frac{-2t}{1+2t^2} & \frac{1-2t^2}{1+2t^2} & \frac{2t}{1+2t^2} \end{vmatrix} = \frac{\hat{\imath}2t^2 - \hat{\jmath}2t + \hat{k}}{1+2t^2}. \tag{3}$$

$$\frac{d\widehat{B}}{dt} = \frac{\hat{\imath}4t + \hat{\jmath}(4t^2-2) - \hat{k}4t}{(1+2t^2)^2} \tag{4}$$

$$\text{and} \quad \frac{d\widehat{B}}{ds} = \frac{d\widehat{B}}{dt}\bigg/\frac{ds}{dt} = \frac{\hat{\imath}4t + \hat{\jmath}(4t^2-2) - \hat{k}4t}{(1+2t^2)^3}.$$

Again, $\because \frac{d\widehat{B}}{ds} = -\tau \widehat{N}$,

$$\therefore \quad \textbf{Torsion } \tau = \left|\frac{d\widehat{B}}{ds}\right| = \frac{\sqrt{(4t^2) + (4t^2-2)^2 + (-4t)^2}}{(1+2t^2)^3} = \frac{2\sqrt{1+4t+4t^4}}{(1+2t^2)^3}$$

$$= \frac{2\sqrt{(1+2t^2)^2}}{(1+2t^2)^3} = \frac{\mathbf{2}}{\mathbf{(1+2t^2)^2}}.$$

(b) Let $\widehat{T} = \widehat{T}_0$, $\widehat{N} = \widehat{N}_0$ and $\widehat{B} = \widehat{B}_0$ at $t = 1$.

$\therefore$ from equations (1), (2) and (3), we get

$$\widehat{T}_0 = \frac{\hat{\imath} + \hat{\jmath}2 + \hat{k}2}{3}, \quad \widehat{N}_0 = \frac{-\hat{\imath}2 - \hat{\jmath} + \hat{k}2}{3} \quad \text{and} \quad \widehat{B}_0 = \frac{\hat{\imath}2 - \hat{\jmath}2 + \hat{k}}{3}.$$

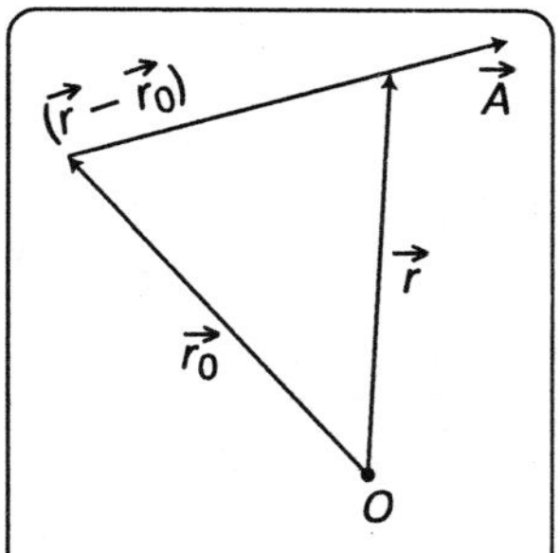

Fig. 11.13

Let $\vec{A}$ be any vector and $\vec{r}_0$ and $\vec{r}$ represent respectively the initial and an arbitrary point on $\vec{A}$, then $(\vec{r} - \vec{r}_0)$ is parallel to $\vec{A}$.

Thus, the equation of $\vec{A}$ is $(\vec{r} - \vec{r}_0) \times \vec{A} = 0$,

and the equation to the tangent, principal normal and binormal are respectively,

$$(\vec{r} - \vec{r}_0) \times \widehat{T}_0 = 0 \tag{5}$$

$$(\vec{r} - \vec{r}_0) \times \widehat{N}_0 = 0 \tag{6}$$

$$\text{and} \quad (\vec{r} - \vec{r}_0) \times \widehat{B}_0 = 0. \tag{7}$$

Equations (5), (6) and (7) represent respectively the vector equations of tangent, principal normal and binormal.

In rectangular system, let

$$\vec{r} = \hat{i}x + \hat{j}y + \hat{k}z \quad \text{and} \quad \vec{r}_0 = \hat{i}x_0 + \hat{j}y_0 + \hat{k}z_0 = \hat{i} + \hat{j} + \hat{k}\frac{2}{3}$$

then, $$\left[(\hat{i}x + \hat{j}y + \hat{k}z) - \left(\hat{i} + \hat{j} + \hat{k}\frac{2}{3}\right)\right] \times \frac{(\hat{i} + 2\hat{j} + 2\hat{k})}{3} = 0$$

or, $$\left[\hat{i}(x-1) + \hat{j}(y-1) + \hat{k}\left(z - \frac{2}{3}\right)\right] \times [\hat{i} + 2\hat{j} + 2\hat{k}] = 0$$

or, $$\begin{vmatrix} \hat{i} & \hat{j} & \hat{k} \\ (x-1) & (y-1) & \left(z - \frac{2}{3}\right) \\ 1 & 2 & 2 \end{vmatrix} = 0$$

or, $$\hat{i}\left[2(y-1) - 2\left(z - \frac{2}{3}\right)\right] + \hat{j}\left[\left(z - \frac{2}{3}\right) 2(x-1)\right] + \hat{k}[2(x-1) - (y-1)] = 0.$$

$\because$ $\hat{i}$, $\hat{j}$, $\hat{k}$ are triad of unit vectors.

$$\therefore \quad 2(y-1) - 2\left(z - \frac{2}{3}\right) = 0,$$

$$\therefore \quad (y-1) = \left(z - \frac{2}{3}\right) \quad \text{or,} \quad \left(z - \frac{2}{3}\right) - 2(x-1) = 0 \quad \text{or,} \quad (x-1) = \frac{z - \frac{2}{3}}{2}$$

and $$[2(x-1) - (y-1)] = 0 \quad \text{or,} \quad (x-1) = \frac{y-1}{2}.$$

$$\therefore \quad \frac{(x-1)}{1} = \frac{(y-1}{2} = \frac{(z - \frac{2}{3})}{2}. \tag{8}$$

Similarly, $$\frac{x-1}{-2} = \frac{y-1}{-1} = \frac{z - \frac{2}{3}}{2} \tag{9}$$

and $$\frac{x-1}{2} = \frac{y-1}{-2} = \frac{z - \frac{2}{3}}{1}. \tag{10}$$

Equations (8), (9) and (10) represent respectively the equation of tangent, principal normal and binormal to the curve.

Example 11.23. *The equation to a plane curve in the xy-plane is given by $y = f(x)$, $z = 0$. Find the radius of curvature of the curve.*

Solution : Now, the equation to the plane curve in the xy-plane is, $y = f(x), z = 0$, so

$$\vec{r} = \hat{i}x + \hat{j}y \quad [\text{for } z = 0].$$

$$\therefore \quad \frac{d\vec{r}}{dx} = \hat{i} + \hat{j}y' \quad \left[\text{where } y' = \frac{dy}{dx}\right].$$

$$\therefore \quad \frac{ds}{dx} = \left|\frac{d\vec{r}}{dx}\right| = \sqrt{1 + y'^2}.$$

The unit tangent vector to the curve is $\hat{T} = \dfrac{d\vec{r}}{ds} = \dfrac{d\vec{r}}{dx}\dfrac{dx}{ds} = \dfrac{\hat{i} + \hat{j}y'}{\sqrt{1 + y'^2}}$.

But, $\dfrac{d\widehat{T}}{dx} = \dfrac{d}{dx}\left[\dfrac{\hat{i}+\hat{j}y'}{\sqrt{1+y'^2}}\right] = \dfrac{d}{dx}\left[\dfrac{\hat{i}}{\sqrt{1+y'^2}} + \hat{j}\dfrac{y'}{\sqrt{1+y'^2}}\right]$

$$= \left[\hat{i}\left(-\frac{1}{2}\right)\frac{2y'y''}{(1+y'^2)^{3/2}} + \hat{j}\frac{y''\sqrt{1+y'^2} - y'\frac{1}{2}(1+y'^2)^{-\frac{1}{2}}2y'y''}{(1+y'^2)}\right]$$

$$= \left[-\hat{i}\frac{y'y''}{(1+y'^2)^{3/2}} + \hat{j}y''\frac{(1+y'^2) - y'^2 y''}{(1+y'^2)^{3/2}}\right] = \frac{y''}{(1+y'^2)^{3/2}}[-\hat{i}y' + \hat{j}].$$

$\therefore \quad \dfrac{d\widehat{T}}{ds} = \dfrac{d\widehat{T}}{dx}\dfrac{dx}{ds} = \dfrac{y''}{(1+y'^2)^{3/2}}(-\hat{i}y'+\hat{j})\dfrac{1}{(1+y'^2)^{\frac{1}{2}}} = \dfrac{y''}{(1+y'^2)^2}(-\hat{i}y'+\hat{j}).$

Now, $\dfrac{d\widehat{T}}{ds} = \kappa\widehat{N}.$

$\therefore \quad \kappa$ (the curvature) $= \left|\dfrac{d\widehat{T}}{ds}\right| \quad [\because |\widehat{N}| = 1]$

or, $\quad \kappa = \dfrac{|y''|}{(1+y'^2)^2}\sqrt{y'^2+1} = \dfrac{|y''|}{(1+y'^2)^{3/2}}.$

$\therefore$ the radius of curvature of the curve is $\boldsymbol{\rho = \dfrac{1}{\kappa} = \dfrac{(1+y'^2)^{3/2}}{|y''|}}$.

Example 11.24. *Show that the differential arc length on the surface, $\vec{r} = \vec{r}(u,v)$ is given by $ds^2 = Edu^2 + 2F\,dudv + Gdv^2$, where*

$$E = \frac{\partial\vec{r}}{\partial u}\cdot\frac{\partial\vec{r}}{\partial u} = \left(\frac{\partial\vec{r}}{\partial u}\right)^2; \quad F = \frac{\partial\vec{r}}{\partial u}\cdot\frac{\partial\vec{r}}{\partial v}; \quad G = \frac{\partial\vec{r}}{\partial v}\cdot\frac{\partial\vec{r}}{\partial v} = \left(\frac{\partial\vec{r}}{\partial v}\right)^2.$$

Solution : The equation to the surface is, $\vec{r} = \vec{r}(u,v)$,

$\therefore \quad d\vec{r} = \dfrac{\partial\vec{r}}{\partial u}du + \dfrac{\partial\vec{r}}{\partial v}dv.$

The differential arc length is given by

$$ds^2 = d\vec{r}\cdot d\vec{r} = \left(\frac{\partial\vec{r}}{\partial u}du + \frac{\partial\vec{r}}{\partial v}dv\right)\cdot\left(\frac{\partial\vec{r}}{\partial u}du + \frac{\partial\vec{r}}{\partial v}dv\right)$$

$$= \frac{\partial\vec{r}}{\partial u}\cdot\frac{\partial\vec{r}}{\partial u}du^2 + 2\frac{\partial\vec{r}}{\partial u}\frac{\partial\vec{r}}{\partial v}du\,dv + \frac{\partial\vec{r}}{\partial v}\frac{\partial\vec{r}}{\partial v}dv^2$$

$$= \boldsymbol{Edu^2 + 2Fdudv + Gdv^2}, \quad \textbf{(Proved)}$$

where $E = \dfrac{\partial\vec{r}}{\partial u}\cdot\dfrac{\partial\vec{r}}{\partial u}, \quad F = \dfrac{\partial\vec{r}}{\partial u}\cdot\dfrac{\partial\vec{r}}{\partial v}, \quad G = \dfrac{\partial\vec{r}}{\partial v}\cdot\dfrac{\partial\vec{r}}{\partial v}.$

Example 11.25. *Find the acceleration of a particle moving along a space curve with velocity $\vec{v}$.* **[BU(H) 1997]**

Solution : If $\widehat{T}$ be the unit tangent vector at any point on the space curve, then the velocity of the particle at that point is

$$\vec{v} = v\widehat{T}, \quad \text{where} \quad v = |\vec{v}|.$$

The acceleration of the particle at that point is

$$\vec{a} = \frac{d\vec{v}}{dt} = \frac{d}{dt}(v\widehat{T}) = \frac{dv}{dt}\widehat{T} + v\frac{d\widehat{T}}{dt}.$$

But $\displaystyle \frac{d\widehat{T}}{dt} = \frac{d\widehat{T}}{ds}\frac{ds}{dt} = \kappa\widehat{N}\frac{ds}{dt} = \kappa v\widehat{N} = \frac{v\widehat{N}}{\rho}, \quad \left[\because \frac{d\widehat{T}}{ds} = \kappa\widehat{N} \text{ and } \frac{1}{\kappa} = \rho\right]$

Then from equation (1),

$$\vec{a} = \frac{dv}{dt}\widehat{T} + v\left(\frac{v\widehat{N}}{\rho}\right) \tag{2}$$

or, $$\boldsymbol{\vec{a} = \left(\frac{dv}{dt}\right)\widehat{T} + \left(\frac{v^2}{\rho}\right)\widehat{N}}. \tag{3}$$

Here, $\left(\frac{dv}{dt}\right)$ and $\left(\frac{v^2}{\rho}\right)$ are the components of the acceleration $\vec{a}$ along $\widehat{T}$ and $\widehat{N}$ respectively.

Example 11.26. *Find the unit normal to the surface,*

$$\vec{r} = \hat{i}a\cos u\sin v + \hat{j}a\sin u\sin v + \hat{k}a\cos v, \quad \textit{where } a > 0.$$

Solution : The vector normal to the surface

$$\vec{r} = \hat{i}a\cos u\sin v + \hat{j}a\sin u\sin v + \hat{k}a\cos v = \vec{r}(u,v) \text{ is } \left(\frac{\partial\vec{r}}{\partial u} \times \frac{\partial\vec{r}}{\partial v}\right).$$

But $\displaystyle \frac{\partial\vec{r}}{\partial u} = -\hat{i}a\sin u\sin v + \hat{j}a\cos u\sin v$

and $\displaystyle \frac{\partial\vec{r}}{\partial v} = \hat{i}a\cos u\cos v + \hat{j}a\sin u\cos v - \hat{k}a\sin v.$

$$\therefore \quad \frac{\partial\vec{r}}{\partial u} \times \frac{\partial\vec{r}}{\partial v} = \begin{vmatrix} \hat{i} & \hat{j} & \hat{k} \\ -a\sin u\sin v & a\cos u\sin v & 0 \\ +a\cos u\cos v & a\sin u\cos v & -a\sin v \end{vmatrix}$$

$$= -\hat{i}a^2\cos u\sin^2 v - \hat{j}a^2\sin u\sin^2 v - \hat{k}a^2\sin v\cos v.$$

$$\therefore \quad \left|\frac{\partial\vec{r}}{\partial u} \times \frac{\partial\vec{r}}{\partial v}\right| = \pm\sqrt{a^4\cos^2 u\sin^4 v + a^4\sin^2 u\sin^4 v + a^4\sin^2 v\cos^2 v}$$

$$= \sqrt{a^4\sin^2 v(\sin^2 v + \cos^2 v)} = \begin{cases} a^2\sin v, & \text{if } \sin v > 0 \\ -a^2\sin u, & \text{if } \sin v < 0 \end{cases}.$$

The unit normal vector to the surface, $\vec{r} = \vec{r}(u,v)$ is

$$\hat{\boldsymbol{n}} = \pm\frac{\frac{\partial\vec{r}}{\partial u} \times \frac{\partial\vec{r}}{\partial v}}{\left|\frac{\partial\vec{r}}{\partial u} \times \frac{\partial\vec{r}}{\partial v}\right|} = \pm\frac{-\hat{i}a^2\cos u\sin^2 v - \hat{j}a^2\sin u\sin^2 v - \hat{k}a^2\sin v\cos v}{a^2\sin v}$$

$$= \pm[\hat{\boldsymbol{i}}\cos\boldsymbol{u}\sin\boldsymbol{v} + \hat{\boldsymbol{j}}\sin\boldsymbol{u}\sin\boldsymbol{v} + \hat{\boldsymbol{k}}\cos\boldsymbol{v}].$$

The '+' and '−' signs in the expression of $\hat{n}$ represent respectively the outward drawn and inward drawn unit normal to the surface.

Example 11.27. *If $\phi(x,y,z) = 2xz^4 - x^2y$, find $\nabla\phi$ and $|\nabla\phi|$ at the point $(2,-2,-1)$.*

Solution : $\nabla\phi = \left(\hat{i}\frac{\partial}{\partial x} + \hat{j}\frac{\partial}{\partial y} + \hat{k}\frac{\partial}{\partial z}\right)(2xz^4 - x^2y)$

$$= \hat{i}\frac{\partial}{\partial x}(2xz^4 - x^2y) + \hat{j}\frac{\partial}{\partial y}(2xz^4 - x^2y) + \hat{k}\frac{\partial}{\partial z}(2xz - x^2y)$$

$$= \hat{i}(2z^4 - 2xy) + \hat{j}(-x^2) + \hat{k}(8xz^3).$$

$\therefore \quad \boldsymbol{\nabla\phi}|_{2,-2,-1} = \hat{i}[2(-1)^4 - 2(2)(-2)] + \hat{j}(-2)^2 + \hat{k}[8(2)(-1)] = \mathbf{10\hat{i} - 4\hat{j} - 16\hat{k}}$

and $\quad |\boldsymbol{\nabla\phi}|_{2,-2,-1} = \sqrt{10^2 + (-4)^2 + (-16)^2} = \sqrt{372} = \mathbf{2\sqrt{93}}.$

Example 11.28. *Find a unit vector normal to the surface, $x^2 + 3y^2 + 2z^2 = 6$ at $P(2,0,1)$.*

[AMIETE(Summer) 1995]

Solution : Let $\phi(x,y,z) = x^2 + 3y^2 + 2z^2 - 6 = 0$

The vector normal to the surface $\phi(x,y,z)$ is

$$\nabla\phi = \left(\hat{i}\frac{\partial}{\partial x} + \hat{j}\frac{\partial}{\partial y} + \hat{k}\frac{\partial}{\partial z}\right)(x^2 + 3y^2 + 2z^2 - 6) = \hat{i}2x + \hat{j}6y + 4z\hat{k}.$$

The unit normal vector at (2, 0, 1) $= \left[\frac{\nabla\phi}{|\nabla\phi|}\right]_{2,0,1} = \frac{\hat{i}4 + 4\hat{k}}{\sqrt{16+16}} = \frac{\mathbf{1}}{\sqrt{\mathbf{2}}}(\hat{i} + \hat{j}).$

Example 11.29. *The temperature at any point in space is given by $T = xy + yz + zx$. Determine the derivative of T in the direction of the vector $3\hat{i} - 4\hat{k}$ at the point $(1,1,1)$.*

Solution : We have $T = xy + yz + zx$.

$\because \quad \nabla T = \left(\hat{i}\frac{\partial}{\partial x} + \hat{j}\frac{\partial}{\partial y} + \hat{k}\frac{\partial}{\partial z}\right)(xy + yz + zx) = \hat{i}(y+z) + \hat{j}(x+z) + \hat{k}(y+x).$

∇T at $(1,1,1)$ is $2\hat{i} + 2\hat{j} + 2\hat{k}$.

The directional derivative at $(1,1,1)$ in the direction $(\hat{i} - 4\hat{k})$ is

$$\Delta T \cdot \hat{n}, \quad \text{where } \hat{n} = \frac{3\hat{i} - 4\hat{k}}{\sqrt{9+16}} = \frac{3\hat{i} - 4\hat{k}}{5}.$$

$\therefore \quad \boldsymbol{\nabla T \cdot \hat{n}} = (2\hat{i} + 2\hat{j} + 2\hat{k}) \cdot \frac{(3\hat{i} - 4\hat{k})}{5} = \frac{1}{5}(6-8) = \frac{\mathbf{-2}}{\mathbf{5}}.$

Example 11.30. *Find the directional derivative of $f(x,y,z) = xy^2 + yz^3$ at the point $(2,-1,1)$ in the direction of vector $\hat{i} + 2\hat{j} + 2\hat{k}$.* **[MU 1994, RU 1982]**

Solution : $\because \; f(x,y,z) = xy^2 + yz^2,$

$\therefore \quad \nabla f = \left(\hat{i}\frac{\partial}{\partial x} + \hat{j}\frac{\partial}{\partial y} + \hat{k}\frac{\partial}{\partial z}\right)(xy^2 + yz^2) = \hat{i}y^2 + \hat{j}(2xy + z^2) + \hat{k}(3yz^2).$

$\therefore \quad \nabla f$ at $(2,-1,1) = \hat{i}(-1)^2 + \hat{j}[2(2)(-1) + (1)^3] + \hat{k}[3(-1)(1)^2] = \mathbf{\hat{i} - 3\hat{j} - 3\hat{k}}.$

Now unit vector in the direction of $\hat{i}+2\hat{j}+2\hat{k}$ is

$$\hat{n}=\frac{\hat{i}+2\hat{j}+2\hat{k}}{\sqrt{1^2+2^2+2^3}}=\frac{1}{\sqrt{9}}(\hat{i}+2\hat{j}+2\hat{k})=\frac{\hat{i}+2\hat{j}+2\hat{k}}{3}.$$

Directional derivative of $f(x,y,z)$ along the vector $\hat{i}+2\hat{j}+2\hat{k}$ is

$$\boldsymbol{\nabla f}\cdot\boldsymbol{\hat{n}}=(\hat{i}-3\hat{j}-3\hat{k})\cdot\frac{(\hat{i}+2\hat{j}+2\hat{k})}{3}=\frac{1}{3}(1-6-6)=\frac{11}{3}=\boldsymbol{-3\frac{2}{3}}.$$

Example 11.31. *Find a unit vector perpendicular to the surface, $x^2+y^2-z^2=11$ at the point $(4,2,3)$.* **[BU(H) 1997]**

Solution : Let $\phi(x,y,z)=x^2+y^2-z^2-11$.

$$\therefore\quad \nabla\phi=\hat{i}\frac{\partial\phi}{\partial x}+\hat{j}\frac{p\phi}{\partial y}+\hat{k}\frac{\partial\phi}{\partial z}=\hat{i}2x+\hat{j}2y+\hat{k}2z.\qquad \therefore\quad \nabla\phi|_{(4,2,3)}=8\hat{i}+4\hat{j}-6\hat{k}.$$

Now the unit vector normal to the surface $\phi(x,y,z)$ at $(4,2,3)$ is

$$\boldsymbol{\hat{n}}=\frac{8\hat{i}+4\hat{j}-6\hat{k}}{\sqrt{64+16+36}}=\boldsymbol{\frac{1}{\sqrt{29}}(4\hat{i}+2\hat{j}-3\hat{k})}.$$

Example 11.32. *For the function $\phi(x,y)=\dfrac{x}{x^2+y^2}$, find the magnitude of the directional derivative along a line making an angle 30° with the positive x-axis at $(0,2)$.*

[AMIETE(Winter) 1993]

Solution : We have $\phi(x,y)=\dfrac{x}{x^2+y^2}$.

$$\therefore\quad \nabla\phi=\left(\hat{i}\frac{\partial}{\partial x}+\hat{j}\frac{\partial}{\partial y}+\hat{z}\frac{\partial}{\partial z}\right)\frac{x}{x^2+y^2}=\hat{i}\frac{\partial}{\partial x}\left(\frac{x}{x^2+y^2}\right)+\hat{j}\left(\frac{x}{x^2+y^2}\right)+\hat{k}\left(\frac{x}{x^2+y^2}\right)$$

$$=\hat{i}\left[\frac{1}{x^2+y^2}-\frac{x(2x)}{(x^2+y^2)^2}\right]-\hat{j}\frac{x(2y)}{(x^2+y^2)^2}.$$

$\therefore$ $\nabla\phi$ at the point $(0,2)$ is $\nabla\phi\Big|_{0,2}=\hat{i}\dfrac{(4-0)}{(0+4^2)^2}-\hat{j}\dfrac{2(0)(2)}{(0+4)^2}=\dfrac{\hat{i}}{4}.$

Now as the line makes an angle 30° with the x-axis,

$$\vec{r}=\hat{i}(r\cos 30°)+\hat{j}r\sin 30°=\left[\hat{i}\frac{\sqrt{3}}{2}+\hat{j}\frac{1}{2}\right]\quad\text{or,}\quad \frac{\vec{r}}{r}=\hat{n}=\hat{i}\frac{\sqrt{3}}{2}+\hat{j}\frac{1}{2}.$$

$\therefore$ **the directional derivative of $\boldsymbol{f(x,y,z)}$ along $\vec{r}$ is** $\dfrac{df}{dr}=\dfrac{\hat{i}}{4}\cdot\left(\hat{i}\dfrac{\sqrt{3}}{2}+\hat{j}\dfrac{1}{2}\right)=\boldsymbol{\dfrac{\sqrt{3}}{8}}.$

Example 11.33. *Find a unit normal vector to the surface, $z^2=x^2+y^2$ at the point $(1,0,-1)$.* **[AMIETE(Summer) 1990]**

Solution : Let $\phi(x,y,z)=x^2+y^2-z^2$.

$$\therefore\quad \nabla\phi=\left(\hat{i}\frac{\partial}{\partial x}+\hat{j}\frac{\partial}{\partial y}+\hat{k}\frac{\partial}{\partial z}\right)(x^2+y^2-z^2)$$

$$= \hat{i}\frac{\partial}{\partial x}(x^2 + y^2 - z^2) + \hat{j}\frac{\partial}{\partial x}(x^2 + y^2 - z^2) + \hat{k}\frac{\partial}{\partial x}(x^2 + y^2 - z^2)$$

$$= \hat{i}2x + \hat{j}2y + \hat{k}(-2z) = 2(\hat{i}x + \hat{j}y - \hat{k}z).$$

$$\therefore \quad \nabla\phi\Big|_{(1,0,-1)} = 2(\hat{i} + \hat{k}).$$

Now, the unit normal to $\phi(x, y, z)$ at $(1, 0, -1)$ is

$$\hat{\boldsymbol{n}} = \frac{\nabla\phi|_{1,0,-1}}{|(\nabla\phi)_{1,0,-1}|} = \frac{2(\hat{i} + \hat{k})}{\sqrt{2^2 + 2^2}} = \frac{2(\hat{i} + \hat{j})}{2\sqrt{2}} = \frac{(\hat{\boldsymbol{i}} + \hat{\boldsymbol{j}})}{\sqrt{\mathbf{2}}}.$$

Example 11.34. *Find the directional derivative of $\phi = x^2yz + 4xz^2$ at the point $(1, -2, -1)$ in the direction of the vector $2\hat{i} - \hat{j} - 2\hat{k}$.* **[BU(H) 2006]**

Solution : We have $\phi(x, y, z) = x^2yz + 4xz^2$.

$$\therefore \quad \nabla\phi = \left(\hat{i}\frac{\partial}{\partial x} + \hat{j}\frac{\partial}{\partial y} + \hat{k}\frac{\partial}{\partial z}\right)(x^2yz + 4xz^2)$$

$$= \hat{i}\frac{\partial}{\partial x}(x^2yz + 4xz^2) + \hat{j}\frac{\partial}{\partial y}(x^2yz + 4xz^2) + \hat{k}\frac{\partial}{\partial z}(x^2yz + 4xz^2)$$

$$= \hat{i}(2xyz + 4z^2) + \hat{j}(x^2z) + \hat{k}(x^2y + 8xz).$$

$$\therefore \quad \nabla\phi|_{1,-2,-1} = \hat{i}[\{2(1)(-2)(-1) + 4(-1)^2\} + \hat{j}(1)^2(-1) + \hat{k}\{(1)^2(-2) + 8(1)(-1)\}]$$

$$= \hat{i}(4 + 4) - \hat{j} - \hat{k}10 = 8\hat{i} - \hat{j} - 10\hat{k}.$$

Let $\vec{r} = 2\hat{i} - \hat{j} - 2\hat{k}$, then the unit vector along $\vec{r}$ is

$$\hat{n} = \frac{\vec{r}}{|\vec{r}|} = \frac{2\hat{i} - \hat{j} - 2\hat{k}}{\sqrt{2^2 + (-1)^2 + (-2)^2}} = \frac{1}{3}(2\hat{i} - \hat{j} - 2\hat{k})$$

$\therefore$ directional derivative of $\phi(x, y, z)$ along $\vec{r}$ is

$$\left(\frac{\boldsymbol{d\phi}}{\boldsymbol{dr}}\right) = \nabla\phi|_{1,-2,-1}\hat{n} = (8\hat{i} - \hat{j} - 10\hat{k}) \cdot \frac{1}{3}(2\hat{j} - \hat{j} - 2\hat{k}) = \frac{1}{3}(16 + 1 + 20) = \frac{37}{3} = \mathbf{12\frac{1}{3}}.$$

Example 11.35. *Find a unit vector perpendicular to the surface, $x^2 + y^2 - z^2 = 11$ at the point $(4, 2, 3)$.* **[BU(H) 1997]**

Solution : Let $\phi(x, y, z) = x^2 + y^2 - z^2 = 11$.

$$\therefore \quad \nabla\phi = \left(\hat{i}\frac{\partial}{\partial x} + \hat{j}\frac{\partial}{\partial y} + \hat{k}\frac{\partial}{\partial z}\right)(x^2 + y^2 - z^2 - 11) = 2x\hat{i} + 2y\hat{j} - 2z\hat{k}.$$

$$\therefore \quad \nabla\phi|_{4,2,3} = 8\hat{i} + 4\hat{j} - 6\hat{k}.$$

$\therefore$ the unit vector perpendicular to the surface $\phi(x, y, z) =$ const. is

$$= \frac{8\hat{i} + 4\hat{j} - 6\hat{k}}{\sqrt{64 + 16 + 36}} = \pm\frac{\mathbf{1}}{\sqrt{\mathbf{29}}}(\mathbf{4\hat{i} + 2\hat{j} - 3\hat{k}}).$$

Example 11.36. *Find the unit vector normal to the surface, $x^2 + y - z = 1$ at the point* $(1, 0, 0)$. **[BU(H) 1996]**

Solution : Let $\phi(x, y, z) = x^2 + y - z - 1 = 0$.

Then $\nabla\phi = \left(\hat{\imath}\frac{\partial}{\partial x} + \hat{\jmath}\frac{p}{\partial y} + \hat{k}\frac{p}{\partial z}\right)(x^2 + y^2 - z - 1) = \hat{\imath}2x + \hat{\jmath} - \hat{k}$.

$\therefore \quad \nabla\phi\big|_{1,0,1} = 2\hat{\imath} + \hat{\jmath} - \hat{k}$.

$\therefore$ unit vector normal to $\phi(x, y, z) = 0$ is

$$\hat{\boldsymbol{n}} = \frac{\nabla\phi\big|_{1,0,0}}{|\nabla\phi|_{1,0,0}|} = \pm\frac{2\hat{\imath} + \hat{\jmath} - \hat{k}}{\sqrt{2^2 + 1^2 + (-1)^2}} = \pm\frac{\mathbf{2\hat{\imath} + \hat{\jmath} - \hat{k}}}{\sqrt{\mathbf{6}}}.$$

Example 11.37. *Find the constants a and b so that the surface $ax^2 - byz - (a + 2)x = 0$ will be orthogonal to the surface $4x^2y + z^3 - 4 = 0$ at the point $(1, -1, 2)$.* **[BU(H) 1997]**

Solution : Let $\phi_1(x, y, z) = ax^2 - byz - (a + z)x = 0$ and $\phi_2(x, y, z) = 4x^2y + z^3 - 4 = 0$.

$$\therefore \quad \nabla\phi_1 = \hat{\imath}\frac{\partial\phi_1}{\partial x} + \hat{\jmath}\frac{\partial\phi_1}{\partial y} + \hat{k}\frac{\partial\phi_1}{\partial z}$$

$$= \hat{\imath}\frac{\partial}{\partial x}[ax^2 - byz - (a+z)x] + \hat{\jmath}\frac{\partial}{\partial y}[ax^2 - byz - (a+z)x] + \hat{k}\frac{\partial}{\partial z}[ax^2 - byz - (a+z)x]$$

$$= \hat{\imath}[2ax - (a + z) + \hat{\jmath}(-bz) + \hat{k}(-by).$$

$$\therefore \quad \nabla\phi\big|_{(1,-1,2)} = \hat{\imath}(a - 2) - \hat{\jmath}2b + \hat{k}b$$

and $\nabla\phi_2 = \hat{\imath}\frac{\partial\phi_2}{\partial x} + \hat{\jmath}\frac{\partial\phi_2}{\partial y} + \hat{k}\frac{\partial\phi_2}{\partial z} = \hat{\imath}8xy + \hat{\jmath}4x^2 + \hat{k}3x^2$.

$$\therefore \quad \nabla\phi_2\big|_{(1,-1,2)} = -8\hat{\imath} + 4\hat{\jmath} + 12\hat{k}.$$

Now, the angle between the surfaces is equal to the angle between the normals to the surfaces at their point of intersection. The surface being orthogonal to each other we have

$$\nabla\phi_1 \cdot \nabla\phi_2 = 0 \quad \text{or,} \quad [(a - 2)\hat{\imath} - 2b\hat{\jmath} + b\hat{k}] \cdot [-8\hat{\imath} + 4\hat{\jmath} + 12\hat{k}] = 0$$

or, $\quad -8(a - 2) - 8b + 12b = 0 \quad$ or, $\quad 2a - b = 4$.

Again, as $(1, -1, 2)$ is a point on the surface $\phi_1 = 0$, so,

$$a + 2b - a - 2 = 0 \quad \text{or,} \quad \boldsymbol{b = 1}, \quad \therefore \ \boldsymbol{a = \frac{5}{2}}.$$

Example 11.38. *$\phi(\overrightarrow{X})$ is a scalar field. Find the rate of change of $\phi(\overrightarrow{X})$ with position along the vector $2\hat{\imath} + 3\hat{\jmath} + \sqrt{3}\hat{k}$, if $\phi = x^2 + y^2 + z^2$.* **[BU(H) 1999]**

Solution : We have to find the directional derivative $\frac{d\phi}{dX}$ along the vector $\overrightarrow{r} = 2\hat{\imath} + 3\hat{\jmath} + \sqrt{3}\hat{k}$.

Now, $\phi(x, y, z) = x^2 + y^2 + z^2$.

$$\therefore \quad \nabla\phi = \left(\hat{\imath}\frac{\partial}{\partial x} + \hat{\jmath}\frac{\partial}{\partial y} + \hat{k}\frac{\partial}{\partial z}\right)(x^2 + y^2 + z^2) = 2x\hat{\imath} + 2y\hat{\jmath} + 2z\hat{k}.$$

The unit vector along $\vec{r}$ is

$$\hat{n} = \frac{r}{|\vec{r}|} = \frac{2\hat{i} + 3\hat{j} + \sqrt{3}\hat{k}}{\sqrt{4+9+3}} = \frac{1}{4}(2\hat{i} + 3\hat{j} + \sqrt{3}\hat{k}).$$

$$\therefore \quad \frac{d\phi}{dr} = \nabla\phi \cdot dr = (2x\hat{i} + 2y\hat{j} + 2z\hat{k}) \cdot \frac{1}{4}(2\hat{i} + 3\hat{j} + \sqrt{3}\hat{k})$$

or, $$\frac{d\phi(\vec{X})}{dr} = \frac{1}{2}(2x + 3y + \sqrt{3}z).$$

Example 11.39. *Find the angle between the surfaces $x^2 + y^2 + z^2 = 4$ and $z = x^2 + y^2 - 5$ at the point $(1, -1, 2)$.* **[BU(H) 1995]**

Solution : Let $\phi_1(x, y, z) = x^2 + y^2 + z^2 - 4 = 0$ and $\phi_2(x, y, z) = x^2 + y^2 - z - 5 = 0$.

Now, $$\nabla\phi_1 = \left(\hat{i}\frac{\partial}{\partial x} + \hat{j}\frac{\partial}{\partial y} + \hat{k}\frac{\partial}{\partial z}\right)(x^2 + y^2 + z - 4) = 2x\hat{i} + 2y\hat{j} + 2z\hat{k} = 2(x\hat{i} + y\hat{j} + z\hat{k}).$$

$$\therefore \quad \nabla\phi\big|_{(1,-1,2)} = 2[\hat{i} - \hat{j} + 2\hat{k}]$$

and $$\nabla\phi_2 = \left(\hat{i}\frac{\partial}{\partial x} + \hat{j}\frac{\partial}{\partial y} + \hat{k}\frac{\partial}{\partial z}\right)(x^2 + y^2 - z - 5) = 2x\hat{i} + 2y\hat{j} - \hat{k}.$$

$$\therefore \quad \nabla\phi_2\big|_{(1,-1,2)} = 2\hat{i} - 2\hat{j} - \hat{k}.$$

Now, the angle between two surfaces is equal to the angle between their normals.

$$\therefore \quad \nabla\phi_1\big|_{(1,-1,2)} \cdot \nabla\phi_2\big|_{(1,-1,2)} = \left|\nabla\phi_1\big|_{(1,-1,2)}\right| \left|\nabla\phi_2\big|_{(1,-1,2)}\right| \cos\theta.$$

where θ is the angle between the two surfaces at $(1, -1, 2)$ given by

$$\cos\theta = \frac{\nabla\phi_1\big|_{(1,-1,2)} \cdot \nabla\phi_2\big|_{(1,-2,2)}}{\left|\nabla\phi_1\big|_{(1,-1,2)}\right| \left|\nabla\phi_2\big|_{(1,-1,2)}\right|} = \frac{2(\hat{i} - \hat{j} + 2\hat{k}) \cdot (2\hat{i} - 2\hat{j} - \hat{k})}{\sqrt{2^2 + (-2)^2 + 2^2}\sqrt{2^2 + (-2)^2 + (-1)^2}}$$

$$= \frac{2(2+2-2)}{\sqrt{4+4+4}\sqrt{9}} = \frac{4}{\sqrt{12} \times 3} = \frac{2}{3\sqrt{3}} = \frac{2\sqrt{3}}{3 \times 3} = \frac{2\sqrt{3}}{9}$$

$$\therefore \quad \theta = \cos^{-1}\left(\frac{2\sqrt{3}}{9}\right).$$

Example 11.40. *Find the directional derivative of the scalar function, $\phi = 2xz^4 - x^2y$ at the point $(2, 1, -1)$ in the direction $(2\hat{i} - \hat{j} + 2\hat{k})$. In which direction is the directional derivative a maximum? What is the magnitude of this maximum?* **[BU(H) 1992]**

Solution : We have $\phi(x, y, z) = 2xz^4 - x^2y$.

$$\therefore \quad \nabla\phi = \left(\hat{i}\frac{\partial}{\partial x} + \hat{j}\frac{\partial}{\partial y} + \hat{k}\frac{\partial}{\partial z}\right)(2xz^4 - x^2y) = (2z^4 - 2xy)\hat{i} + (-x^2)\hat{j} + 8xz^3\hat{k}.$$

$$\therefore \quad \nabla\phi\big|_{(2,1,-1)} = [2(-1)^4 - 2(2)(1)]\hat{i} - (2)^2\hat{j} + 8(2)(-1)^3\hat{k} = -2\hat{i} - 4\hat{j} - 16\hat{k}.$$

$\because$ the directional derivative at a point is the maximum along the direction of $\nabla\phi$ at that point, so the directional derivative is the maximum along $\boldsymbol{-2\hat{i} - 4\hat{j} - 16\hat{k}}$.

The magnitude of the maximum directional derivative is

$$\left|\nabla\phi|_{(2,1,-1)}\right| = \sqrt{4+16+256} = \sqrt{276} = \mathbf{2\sqrt{69}}.$$

Example 11.41. *Find the directional derivative of $\phi = x^2yz^3$ at $(2,1,-1)$. In which direction the directional derivative is the maximum and what is the value of this maximum directional derivative?*

Solution : We have $\phi(x,y,z) = x^2yz^3$.

$$\therefore \quad \nabla\phi = \left(\hat{i}\frac{\partial}{\partial x} + \hat{j}\frac{\partial}{\partial y} + \hat{k}\frac{\partial}{\partial z}\right) x^2yz^3 = 2xyz^3\hat{i} + x^2z^3\hat{j} + 3x^2yz^2\hat{k}.$$

$$\therefore \quad \nabla\phi|_{(2,1,-1)} = 2(2)(1)(-1)^3\hat{i} + (2)^2(-1)^3\hat{j} + 3(2)^2(1)(-1)^2\hat{k} = -4\hat{i} - 4\hat{j} + 12\hat{k}.$$

The directional derivative is the maximum along the direction of $\nabla\phi|_{(2,1,-1)} = \mathbf{-4\hat{i} - 4\hat{j} + 12\hat{k}}$.

The magnitude of this maximum is

$$\left|\nabla\phi|_{(2,1,-1)}\right| = \sqrt{(-4)^2 + (-4)^2 + (12)^2} = \sqrt{176} = \mathbf{4\sqrt{11}}.$$

Example 11.42. *The surfaces $x^2 + y^2 + z^2 = 9$ and $z = x^2 + y^2 = 3$ intersect at $(2,-1,2)$. Find the angle between the surfaces at the point of intersection.*

Solution : We know that the angle between the surfaces is equal to the angle between their normals.

Again we know that, a normal to $x^2 + y^2 + z^2 = 9$ at $(2,-1,2)$ is

$$\nabla\phi_1 = \nabla(x^2 + y^2 + z^2 - 9) = \hat{i}2x + \hat{j}2y + 2\hat{k}z.$$

$$\therefore \quad \nabla\phi_1 \text{ at } (2,-1,2) = 4\hat{i} - 2\hat{j} + 4\hat{k}.$$

Similarly, a normal to the surface $\phi_2 = (x^2 + y^2 - z - 3)$ is

$$\nabla\phi_2 = \nabla(x^2 + y^2 - z - 3) = \hat{i}2x + \hat{j}2y - \hat{k}.$$

$$\therefore \quad \nabla\phi_2 \text{ at } (2,-1,2) = 4\hat{i} - 2\hat{j} - \hat{k}.$$

If θ be the angle between the surfaces $\phi_1 =$ const. and $\phi_2 =$ const. then,

$$\nabla\phi_1 \cdot \nabla\phi_2 = |\nabla\phi_1||\nabla\phi_2|\cos\theta \text{ at } (2,-1,2)$$

$$\text{or,} \quad (4\hat{i} - 2\hat{j} + 4\hat{k}) \cdot (4\hat{i} - 2\hat{j} - \hat{k}) = |4\hat{i} - 2\hat{j} + 4\hat{k}|\,|4\hat{i} - 2\hat{j} - \hat{k}|\cos\theta$$

$$\text{or,} \quad 16 + 4 - 4 = \sqrt{4^2 + (-2)^2 + 4^2}\sqrt{4^2 + (-2)^2 + 1^2}\cos\theta \quad \text{or,} \quad 16 = 6\sqrt{21}\cos\theta$$

$$\text{or,} \quad \cos\theta = \frac{16}{6\sqrt{21}} = \frac{8\sqrt{21}}{63} = 0.5819 = \cos 54°25'.$$

$$\therefore \quad \boldsymbol{\theta = 54°25'}.$$

Example 11.43. *Find the acute angle between the surfaces $xy^2z = 3x + z^2$ and $3x^2 - y^2 + 2z = 1$ at the point $(1,-2,1)$.*

Solution : Let $\phi_1 = xy^2z - 3x - z^2$ and $\phi_2 = 3x^2 - y^2 + 2z - 1$.

$$\therefore \quad \nabla\phi_1 = \left(\hat{i}\frac{\partial}{\partial x} + \hat{j}\frac{\partial}{\partial y} + \hat{k}\frac{\partial}{\partial z}\right)(xy^2z - 3x - z^2) = \hat{i}(y^2z - 3) + \hat{j}(2xyz) + \hat{k}(xy^2 - 2z)$$

and $\quad \nabla\phi_1\big|_{(1,-2,1)} = \hat{i}[(-2)^2(1)-3]+\hat{j}[2(1)(-2)(1)]+\hat{k}[(1)(-2)^2-2(1)] = \hat{i}-4\hat{j}+2\hat{k}.$

Again, $\phi_2 = 3x^2-y^2+2z-1.$

$$\therefore \quad \nabla\phi_2 = \left(\hat{i}\frac{\partial}{\partial x}+\hat{j}\frac{\partial}{\partial y}+\hat{k}\frac{\partial}{\partial z}\right)(3x^2-y^2+2z-1) = \hat{i}6x-\hat{j}2y+\hat{k}2$$

and $\quad \nabla\phi_2\big|_{(1,-2,1)} = \hat{i}6(1)-\hat{j}2(-2)+2\hat{k} = 6\hat{i}+4\hat{j}+2\hat{k}.$

If θ be the angle between the surfaces, the it is also the angle between their normals at $(1,-2,1)$.

So, $\quad \nabla\phi_2\big|_{(1,-2,1)}\cdot\nabla\phi_2\big|_{(1,-2,1)} = \left|\nabla\phi_1\big|_{(1,-2,1)}\right|\left|\nabla\phi_2\big|_{(1,-2,1)}\right|\cos\theta$

or, $\quad (\hat{i}-4\hat{j}+2\hat{k})\cdot(6\hat{i}+4\hat{j}+2\hat{k}) = \sqrt{1+16+4}\sqrt{36+16+4}\cos\theta$

or, $\quad 6-16+4 = \sqrt{21}\;2\sqrt{14}\cos\theta \quad$ or, $\quad \cos\theta = \dfrac{-6}{2\sqrt{21}\sqrt{14}} = \dfrac{-3}{\sqrt{14}\sqrt{21}}.$

Hence the acute angle between the given surfaces is $\boldsymbol{\theta = \cos^{-1}\left(\dfrac{3}{\sqrt{14}\sqrt{21}}\right) = 79°55'}$.

Example 11.44. *If $\vec{A}$ is a constant vector, then prove that, $\nabla(\vec{r}\cdot\vec{A}) = \vec{A}$.*

Solution : Let $\vec{r} = \hat{i}x+\hat{j}y+\hat{k}z$ and $\vec{A} = \hat{i}A_1+\hat{j}A_2+\hat{k}A_3$.

$$\therefore \quad \vec{r}\cdot\vec{A} = (\hat{i}x+\hat{j}y+\hat{k}z)\cdot(\hat{i}A_1+\hat{j}A_2+\hat{k}A_3) = xA_1+yA_2+zA_3.$$

Now, $\boldsymbol{\nabla(\vec{r}\cdot\vec{A})} = \left(\hat{i}\frac{\partial}{\partial x}+\hat{j}\frac{\partial}{\partial y}+\hat{k}\frac{\partial}{\partial z}\right)(xA_1+yA_2+zA_3) = \hat{i}A_1+\hat{j}A_2+\hat{k}A_3$

$$= \boldsymbol{\vec{A}}, \quad [\because \vec{A} \text{ is a constant vector}]$$

Example 11.45. *Find $\phi(r)$ such that $\nabla\phi = \dfrac{\vec{r}}{r^5}$ and $\phi(1) = 0$.*

Solution : We have $\nabla\phi = \dfrac{\vec{r}}{r^5}$ or, $\dfrac{d}{dr}\hat{r}\phi = \dfrac{r}{r^5}\hat{r}$ or, $d\phi = r^{-4}\,dr$.

Integrating, $\phi(r) = -\dfrac{1}{3r^3}+A$, where A is an integration constant.

$$\therefore \quad \phi(1) = -\frac{1}{3}+A = 0, \quad \therefore \quad A = \frac{1}{3}.$$

Hence, $\boldsymbol{\phi(r)} = -\dfrac{1}{3r^3}+\dfrac{1}{3} = \boldsymbol{\dfrac{1}{3}\left(1-\dfrac{1}{r^3}\right)}$.

Example 11.46. *Find the directional derivative of $P = 4e^{2x-y+z}$ at the point $(1,1,-1)$ in a direction towards the point $(-3,5,6)$.*

Solution : We have $P = 4e^{2x-y+z}$.

$$\therefore \quad \nabla P = \left(\hat{i}\frac{\partial}{\partial x}+\hat{j}\frac{\partial}{\partial y}+\hat{k}\frac{\partial}{\partial z}\right)4e^{2x-y+z}$$

$$= 4\left[\hat{i}\frac{\partial}{\partial x}(e^{2x-y+z})+\hat{j}\frac{\partial}{\partial y}(e^{2x-y+z})+\hat{k}\frac{\partial}{\partial x}(e^{2x-y+z})\right]$$

$$= 4\left[\hat{i}e^{2x-y+z}(2) + \hat{j}e^{2x-y+z}(-1) + \hat{k}e^{2x-y+z}(1)\right] = 4e^{2x-y+z}[2\hat{i} - \hat{j} + \hat{k}].$$

$$\therefore \quad \nabla P\big|_{(1,1,-1)} = 4e^{2-1-1}[2\hat{i} - \hat{j} + \hat{k}] = 4(2\hat{i} - \hat{j} + \hat{k}).$$

The direction $(1, 1, -1)$ to $(-3, 5, 6)$ is represented by the vector,

$$\vec{r} = (-3-1)\hat{i} + (5-1)\hat{j} + (6+1)\hat{k} = -4\hat{i} + 4\hat{j} + 7\hat{k}.$$

The unit vector in the direction $\vec{r}$ is

$$\hat{n} = \frac{\vec{r}}{|\vec{r}|} = \frac{-4\hat{i} + 4\hat{j} + 7\hat{k}}{\sqrt{(-4)^2 + 4^2 + 7^2}} = \frac{-4\hat{i} + 4\hat{j} + 7\hat{k}}{\sqrt{81}} = \frac{1}{9}(-4\hat{i} + 4\hat{j} + 7\hat{k}).$$

The directional derivative in a direction towards the point $(-3, 5, 6)$ is

$$\frac{dP}{dr} = \nabla P\big|_{(1,1,-1)} \cdot \hat{n} = \frac{1}{9}4(2\hat{i} - \hat{j} + \hat{k}) \cdot (-4\hat{i} + 4\hat{j} + 7\hat{k}) = \frac{4}{9}(-8 - 4 + 7) = -\frac{20}{9}.$$

Example 11.47. *If $\vec{R}$ be the distance from a fixed point $A(a, b, c)$ to any point $P(x, y, z)$, then show that, ∇R is a unit vector in the direction $\overrightarrow{AP} = \vec{R}$.*

Solution : Let $\vec{r}_A$ and $\vec{r}_P$ be the position vectors of the points $A(a, b, c)$ and $P(x, y, z)$ respectively. Then from Fig. 11.14.

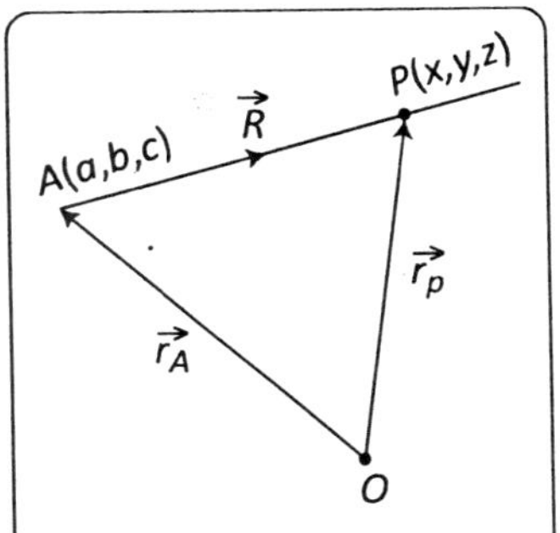

Fig. 11.14

$$\vec{R} = \vec{r}_P - \vec{r}_A = (x-a)\hat{i} + (y-b)\hat{j} + (z-c)\hat{k},$$

or, $\quad R = \sqrt{(x-a)^2 + (y-b)^2 + (z-c)^2}.$

Then, $\quad \nabla R = \dfrac{(x-a)\hat{i} + (y-b)\hat{j} + (z-c)\hat{k}}{\sqrt{(x-a)^2 + (y-b)^2 + (z-c)^2}} = \dfrac{\vec{R}}{R} = \hat{n},$

where **$\nabla R = \hat{n}$ is the unit vector in the direction $\overrightarrow{AP} = \vec{R}$.**

Example 11.48. *Show that, $\nabla\phi$ is a vector perpendicular to the surface $\phi(x, y, z) = C$, where C is a constant.*

Solution : Let $P(x, y, z)$ be a point on the surface, then its position vector with respect to the origin is

$$\vec{r} = x\hat{i} + y\hat{j} + z\hat{k}, \qquad \therefore \quad d\vec{r} = dx\hat{i} + dy\hat{j} + dz\hat{k}.$$

Now, $\quad \nabla\phi = \left(\hat{i}\dfrac{\partial}{\partial x} + \hat{j}\dfrac{\partial}{\partial y} + \hat{k}\dfrac{\partial}{\partial z}\right)\phi(x, y, z) = \left(\hat{i}\dfrac{\partial\phi}{\partial x} + \hat{j}\dfrac{\partial\phi}{\partial y} + \hat{k}\dfrac{\partial\phi}{\partial z}\right).$

$$\therefore \quad \nabla\phi \cdot d\vec{r} = \left(\hat{i}\frac{\partial\phi}{\partial x} + \hat{j}\frac{\partial\phi}{\partial y} + \hat{k}\frac{\partial\phi}{\partial z}\right) \cdot (\hat{i}dx + \hat{j}dy + \hat{k}dz) = \frac{\partial\phi}{\partial x}dx + \frac{\partial\phi}{\partial y}dy + \frac{\partial\phi}{\partial z}dz = d\phi$$

$$= 0, \quad \text{as } \phi(x, y, z) = C.$$

$\therefore$ $\nabla\phi$ is perpendicular to $d\vec{r}$ which lies on the surface $\phi = C$, i.e., $\nabla\phi$ is perpendicular to the surface.

Example 11.49. *Find $\nabla\phi$, if (i) $\phi = \ln|\vec{r}|$ and (ii) $\phi = \dfrac{1}{r}$.*

Solution : (i) We have $\vec{r} = \hat{i}x + \hat{j}y + \hat{k}z$.

$\therefore \quad |\vec{r}| = \sqrt{x^2+y^2+z^2}.$

$\therefore \quad \phi = \ln|\vec{r}| = \ln[\sqrt{x^2+y^2+z^2}] = \frac{1}{2}\ln(x^2+y^2+z^2).$

$$\therefore \quad \nabla\phi = \left(\hat{i}\frac{\partial}{\partial x} + \hat{j}\frac{\partial}{\partial y} + \hat{k}\frac{\partial}{\partial z}\right)\frac{1}{2}\ln(x^2+y^2+z^2)$$

$$= \frac{1}{2}\frac{\partial}{\partial x}[\ln(x^2+y^2+z^2)\hat{i} + \frac{1}{2}\frac{\partial}{\partial y}[\ln(x^2+y^2+z^2)]\hat{j} + \frac{1}{2}\frac{\partial}{\partial z}[\ln(x^2+y^2+z^2)]\hat{k}$$

$$= \frac{1}{2}\left[\frac{2x}{x^2+y^2+z^2}\hat{i} + \frac{2y}{x^2+y^2+z^2}\hat{j} + \frac{2z}{x^2+y^2+z^2}\hat{k}\right] = \frac{\hat{i}x+\hat{j}y+\hat{k}z}{x^2+y^2+z^2} = \frac{\vec{r}}{r^2}.$$

$$\therefore \quad \boldsymbol{\nabla\phi = \frac{\vec{r}}{r^2}}$$

(b) $\because$ we have $\phi = \frac{1}{r}$,

$$\therefore \quad \nabla\phi = \left(\hat{i}\frac{\partial}{\partial x} + \hat{j}\frac{\partial}{\partial y} + \hat{k}\frac{\partial}{\partial z}\right)\frac{1}{r} = \left(\hat{i}\frac{\partial}{\partial x} + \hat{j}\frac{\partial}{\partial y} + \hat{k}\frac{\partial}{\partial z}\right)(x^2+y^2+z^2)^{-\frac{1}{2}}$$

$$\text{or,} \quad \nabla\phi = \hat{i}\frac{\partial}{\partial x}(x^2+y^2+z^2)^{-\frac{1}{2}} + \hat{j}\frac{\partial}{\partial y}(x^2+y^2+z^2)^{-\frac{1}{2}} + \hat{k}\frac{\partial}{\partial x}(x^2+y^2+z^2)^{-\frac{1}{2}}$$

$$= \hat{i}\left[-\frac{1}{2}(x^2+y^2+z^2)^{-\frac{3}{2}}2x\right] + \hat{j}\left[-\frac{1}{2}(x^2+y^2+z^2)^{-\frac{3}{2}}2y\right] + \hat{k}\left[-\frac{1}{2}(x^2+y^2+z^2)^{-\frac{3}{2}}2z\right]$$

$$= \frac{-(\hat{i}x+\hat{j}y+\hat{k}z)}{(x^2+y^2+z^2)^{3/2}} = -\frac{\vec{r}}{r^3}.$$

$$\therefore \quad \boldsymbol{\nabla\phi = -\frac{\vec{r}}{r^3}}.$$

Example 11.50. *If $\vec{r} = \hat{i}x + \hat{j}y + \hat{k}z$, then prove that,*

$$\nabla r^n = nr^{n-2}\vec{r}, \quad \text{where } r = |\vec{r}| = \sqrt{x^2+y^2+z^2}.$$

Solution : We have $\nabla r^n = \left(\hat{i}\frac{\partial}{\partial x} + \hat{j}\frac{\partial}{\partial y} + \hat{k}\frac{\partial}{\partial z}\right)(\sqrt{x^2+y^2+z^2})^n$

$$\text{or,} \quad \nabla r^n = \hat{i}\frac{\partial}{\partial x}(x^2+y^2+z^2)^{n/2} + \hat{j}\frac{\partial}{\partial y}(x^2+y^2+z^2)^{n/2} + \hat{k}\frac{\partial}{\partial z}(x^2+y^2+z^2)^{n/2}$$

$$= \hat{i}\left[\frac{n}{2}(x^2+y^2+z^2)^{\frac{n}{2}-1}2x\right] + \hat{j}\left[\frac{n}{2}(x^2+y^2+z^2)^{\frac{n}{2}-1}2y\right] + \hat{k}\left[\frac{n}{2}(x^2+y^2+z^2)^{\frac{n}{2}-1}2z\right]$$

$$= n(x^2+y^2+z^2)^{\frac{n}{2}-1}(\hat{i}x+\hat{j}y+\hat{k}z) = n(r^2)^{\frac{n}{2}-1}\vec{r} = nr^{n-2}\vec{r}.$$

$$\therefore \quad \boldsymbol{\nabla r^n = nr^{n-2}\vec{r}}.$$

Example 11.51. *If $\phi(x,y,z)$ is a single-valued and continuously differentiable function of (x,y,z) and $\vec{r} = \hat{i}x+\hat{j}y+\hat{k}z$, then show that,*

(i) $\frac{d\phi}{ds} = \nabla\phi\cdot\frac{d\vec{r}}{ds}$, *where ds is an elementary length of a space curve and*

(ii) *the greatest rate of change of ϕ i.e., the maximum directional derivative, takes place in the direction of, and has a magnitude of, the vector $\nabla\phi$.*

Solution : (i) We have, $\dfrac{d\phi}{ds} = \dfrac{\partial\phi}{\partial x}\dfrac{dx}{ds} + \dfrac{\partial\phi}{\partial y}\dfrac{dy}{ds} + \dfrac{\partial\phi}{\partial z}\dfrac{dz}{ds}$

$$= \left(\hat{i}\frac{\partial}{\partial x} + \hat{j}\frac{\partial}{\partial y} + \hat{k}\frac{\partial}{\partial z}\right)\phi \cdot \left(\hat{i}\frac{dx}{ds} + \hat{j}\frac{dy}{ds} + \hat{k}\frac{dz}{ds}\right) = \nabla\phi \cdot \frac{d\vec{r}}{ds}.$$

Here, $\frac{d\vec{r}}{ds}$ is a unit vector and hence, $\nabla\phi \cdot \frac{d\vec{r}}{ds}$ is the component of $\nabla\phi$ in the direction of $\frac{d\vec{r}}{ds}$.

(ii) We have $\nabla\phi \cdot \frac{d\vec{r}}{ds}$ is a projection of $\nabla\phi$ in the direction of the unit vector $\frac{d\vec{r}}{ds}$. If $\nabla\phi$ and $\frac{d\vec{r}}{ds}$ are in the same direction, then $\nabla\phi \cdot \frac{d\vec{r}}{ds}$ will be a maximum. Under this condition, the maximum value of $\frac{d\phi}{ds}$ takes place in the direction of $\nabla\phi$. **The maximum value of $\frac{d\vec{r}}{ds}$ is $|\nabla\phi|$.**

Example 11.52. *If $\phi(x, y, z) = 2x^3y^2z^4$, then find* (i) $\nabla \cdot \nabla\phi$, (ii) *show that*, $\nabla \cdot \nabla\phi = \nabla^2\phi$, *where* $\nabla^2 = \dfrac{\partial^2}{\partial x^2} + \dfrac{\partial^2}{\partial y^2} + \dfrac{\partial^2}{\partial z^2}$ *is the Laplacian operator.*

Solution : (i) We have $\nabla\phi = \left(\hat{i}\dfrac{\partial}{\partial x} + \hat{j}\dfrac{\partial}{\partial y} + \hat{k}\dfrac{\partial}{\partial z}\right) 2x^3y^2z^4$

or, $\quad \nabla\phi = \hat{i}\dfrac{\partial}{\partial x}(2x^3y^2z^4) + \hat{j}\dfrac{\partial}{\partial x}(2x^3y^2z^4) + \hat{k}\dfrac{\partial}{\partial x}(2x^3y^2z^4)$

$$= 6x^2y^2z^4\hat{i} + 4x^3yz^4\hat{j} + 8x^3y^2z^3\hat{k}.$$

Now, $\quad \nabla \cdot \nabla\phi = \left(\hat{i}\dfrac{\partial}{\partial x} + \hat{j}\dfrac{\partial}{\partial y} + \hat{k}\dfrac{\partial}{\partial z}\right) \cdot (6x^2y^2z^4\hat{i} + 4x^3yz^4\hat{j} + 8x^3y^2z^3\hat{k})$

$$= \frac{\partial}{\partial x}(6x^2y^2z^4) + \frac{\partial}{\partial y}(4x^3yz^4) + \frac{\partial}{\partial z}(8x^3y^2z^3)$$

$$= 12xy^2z^4 + 4x^3z^4 + 24x^3y^2z^2.$$

(ii) $\boldsymbol{\nabla \cdot \nabla\phi} = \left[\hat{i}\dfrac{\partial}{\partial x} + \hat{j}\dfrac{\partial}{\partial y} + \hat{k}\dfrac{\partial}{\partial z}\right] \cdot \left(\hat{i}\dfrac{\partial\phi}{\partial x} + \hat{j}\dfrac{\partial\phi}{\partial y} + \hat{k}\dfrac{\partial\phi}{\partial z}\right)$

$$= \frac{\partial}{\partial x}\left(\frac{\partial\phi}{\partial x}\right) + \frac{\partial}{\partial y}\left(\frac{\partial\phi}{\partial y}\right) + \frac{\partial}{\partial z}\left(\frac{\partial\phi}{\partial z}\right) = \frac{\partial^2\phi}{\partial x^2} + \frac{\partial^2\phi}{\partial y^2} + \frac{\partial^2\phi}{\partial z^2}$$

$$= \left(\frac{\partial^2}{\partial x^2} + \frac{\partial^2}{\partial y^2} + \frac{\partial^2}{\partial z^2}\right)\phi = \boldsymbol{\nabla^2\phi} \quad \textbf{(Proved)}.$$

Example 11.53. *Prove that*, $\nabla \cdot (\phi\vec{A}) = (\nabla\phi) \cdot \vec{A} + \phi(\nabla \cdot \vec{A})$. *Use this result to prove that,*

(i) $\nabla \cdot (u\nabla V - V\nabla U) = U\nabla^2 V - V\nabla^2 u$ and (ii) $\nabla \cdot \left(\dfrac{\vec{r}}{r^3}\right) = 0$.

Solution : We have $\nabla(\phi\vec{A}) = \nabla \cdot (\phi A_1\hat{i} + \phi A_2\hat{j} + \phi A_3\hat{k}) = \dfrac{\partial}{\partial x}(\phi A_1) + \dfrac{\partial}{\partial y}(\phi A_2) + \dfrac{\partial}{\partial z}(\phi A_3)$

$$= \left(\frac{\partial\phi}{\partial x}A_1 + \phi\frac{\partial A_1}{\partial x}\right) + \left(\frac{\partial\phi}{\partial y}A_2 + \phi\frac{\partial A_2}{\partial y}\right) + \left(\frac{\partial\phi}{\partial z}A_3 + \phi\frac{\partial A_3}{\partial z}\right)$$

$$= \left(\frac{\partial \phi}{\partial x}A_1 + \frac{\partial \phi}{\partial y}A_2 + \frac{\partial \phi}{\partial z}A_3\right) + \phi\left(\frac{\partial A_1}{\partial x} + \frac{\partial A_2}{\partial y} + \frac{\partial A_3}{\partial z}\right)$$

$$= \left(\hat{i}\frac{\partial \phi}{\partial x} + \hat{j}\frac{\partial \phi}{\partial y} + \hat{k}\frac{\partial \phi}{\partial z}\right)\cdot(\hat{i}A_1 + \hat{j}A_2 + \hat{k}A_3)$$

$$+ \phi\left(\hat{i}\frac{\partial}{\partial x} + \hat{i}\frac{\partial}{\partial y} + \hat{k}\hat{i}\frac{\partial}{\partial z}\right)\cdot(\hat{i}A_1 + \hat{j}A_2 + \hat{k}A_3)$$

or, $\boldsymbol{\nabla(\phi\vec{A}) = \nabla\phi\cdot\vec{A} + \phi(\nabla\cdot\vec{A})}$ (4)

(Hence proved)

(i) Let $\phi = U$ and $\vec{A} = \nabla V$ in the above equation (4), then

$\nabla\cdot(U\nabla V) = (\nabla U)\cdot(\nabla V) + U(\nabla\cdot\nabla V) = (\nabla U)\cdot(\nabla V) + U\nabla^2 V.$

Interchanging U and V in the above equation, we get

$\boldsymbol{\nabla\cdot(V\nabla U)} = (\nabla V)\cdot(\nabla U) + V\nabla^2 U$

$$\therefore\quad \nabla\cdot(U\nabla V) - \nabla\cdot(V\nabla U) = \nabla\cdot(U\nabla V - V\nabla U)$$

$$= [(\nabla U)\cdot(\nabla V) + U\nabla^2 V] - [(\nabla V)\cdot(\nabla U) + V\nabla^2 U]$$

$$= (\nabla U)\cdot(\nabla V) + U\nabla^2 V - (\nabla V)\cdot(\nabla U) - V\nabla^2 U$$

$$= \boldsymbol{U\nabla^2 V - V\nabla^2 U} \quad \textbf{(Proved)}.$$

(ii) We put $\phi = r^{-3}$ and $\vec{A} = \vec{r}$ in eqn. (4), then

$$\nabla\cdot(r^{-3}\vec{r}) = (\nabla r^{-3})\cdot\vec{r} + (r^{-3})\nabla\cdot\vec{r}$$

$$= (-3r^{-5}\vec{r})\cdot\vec{r} + 3r^{-3} \qquad [\because\ \nabla r^n = nr^{n-2}\vec{r} \text{ and } \nabla\cdot\vec{r} = 3]$$

$$\therefore\quad \boldsymbol{\nabla\cdot(r^{-3}\vec{r})} = -3r^{-3} + 3r^{-3} = \mathbf{0}.$$

Example 11.54. *Prove that,* $\nabla\cdot\vec{r} = 3$. **[MU 1992]**

Solution : We have $\boldsymbol{\nabla\cdot\vec{r}} = \left(\hat{i}\frac{\partial}{\partial x} + \hat{j}\frac{\partial}{\partial y} + \hat{k}\frac{\partial}{\partial z}\right)\cdot(\hat{i}x + \hat{j}y + \hat{k}z) = \frac{\partial}{\partial x}(x) + \frac{\partial}{\partial y}(y) + \frac{\partial}{\partial z}(z)$

$$= 1 + 1 + 1 = \mathbf{3} \quad \textbf{(Proved)}.$$

Example 11.55. *Prove that* $\nabla^2(\ln r) = \frac{1}{r^2}$.

Solution : We have $\vec{r} = \hat{i}x + \hat{j}y + \hat{k}z$.

$$\therefore\quad r^2 = \vec{r}\cdot\vec{r} = x^2 + y^2 + z^2 \text{ and } r = (x^2 + y^2 + z^2)^{\frac{1}{2}}.$$

$$\therefore\quad \ln r = \ln(x^2 + y^2 + z^2)^{\frac{1}{2}} = \frac{1}{2}\ln(x^2 + y^2 + z^2).$$

Now, $\nabla \ln r = \frac{1}{2}\nabla\ln(x^2 + y^2 + z^2) = \frac{1}{2}\left(\hat{i}\frac{\partial}{\partial x} + \hat{j}\frac{\partial}{\partial y} + \hat{k}\frac{\partial}{\partial z}\right)\ln(x^2 + y^2 + z^2)$

$$= \frac{1}{2}\left[\hat{i}\frac{2x}{x^2 + y^2 + z^2} + \hat{j}\frac{2y}{x^2 + y^2 + z^2} + \hat{k}\frac{2z}{x^2 + y^2 + z^2}\right]$$

$$= \frac{1}{2}\cdot\frac{1}{(x^2 + y^2 + z^2)}2(\hat{i}x + \hat{j}y + \hat{k}z) = \frac{\vec{r}}{r^2}$$

and $\quad \nabla^2(\ln r) = \nabla \cdot \nabla \ln r = \nabla \cdot \left(\frac{\vec{r}}{r^2}\right) = \nabla \cdot (r^{-2}\vec{r}) = (\nabla r^{-2})\vec{r} + r^{-2}(\nabla \cdot \vec{r})$

$$= -2r^{-2-2}(\vec{r} \cdot \vec{r}) + \frac{3}{r^2}$$

$$= -2r^{-4}r^2 + \frac{3}{r^2} \quad [\because \nabla r^n = nr^{n-2}\vec{r} \text{ and } \nabla \cdot \vec{r} = 3]$$

or, $\quad \boldsymbol{\nabla^2 \ln r = -\frac{2}{r^2} + \frac{3}{r^2} = \frac{1}{r^2}}$ **(Proved).**

Example 11.56. *Prove that,* $\nabla^2 \left(\frac{1}{r}\right) = 0.$

Solution : $\nabla^2 \left(\frac{1}{r}\right) = \left(\frac{\partial^2}{\partial x^2} + \frac{\partial^2}{\partial y^2} + \frac{\partial^2}{\partial z^2}\right) \frac{1}{\sqrt{x^2+y^2+z^2}}$

Now, $\quad \frac{\partial}{\partial x}\left(\frac{1}{\sqrt{x^2+y^2+z^2}}\right) = \frac{\partial}{\partial x}(x^2+y^2+z^2)^{-\frac{1}{2}} = \left(-\frac{1}{2}\right)(x^2+y^2+z^2)^{-\frac{3}{2}}2x$

$$= -x(x^2+y^2+z^2)^{-\frac{3}{2}}$$

and $\quad \frac{\partial^2}{\partial x^2}\frac{1}{(\sqrt{x^2+y^2+z^2})} = \frac{\partial}{\partial x}\left(\frac{\partial}{\partial x}\frac{1}{\sqrt{x^2+y^2+z^2}}\right) = \frac{\partial}{\partial x}[-x(x^2+y^2+z^2)^{-\frac{3}{2}}]$

$$= 3x^2(x^2+y^2+z^2)^{-5/2} - (x^2+y^2+z^2)^{-3/2} = \frac{2x^2-y^2-z^2}{(x^2+y^2+z^2)^{5/2}}.$$

Similarly, $\quad \frac{\partial^2}{\partial y^2}\frac{1}{\sqrt{x^2+y^2+z^2}} = \frac{2y^2-z^2-x^2}{(x^2+y^2+z^2)^{5/2}}$

and $\quad \frac{\partial^2}{\partial z^2}\frac{1}{\sqrt{x^2+y^2+z^2}} = \frac{2z^2-x^2-y^2}{(x^2+y^2+z^2)^{5/2}}.$

$\therefore \quad \left(\frac{\partial^2}{\partial x^2} + \frac{\partial^2}{\partial y^2} + \frac{\partial^2}{\partial z^2}\right)\frac{1}{\sqrt{x^2+y^2+z^2}}$ or, $\boldsymbol{\nabla^2\left(\frac{1}{r}\right) = 0}$ **(Proved).**

Example 11.57. *If* $\vec{A} = \frac{\vec{r}}{r}$, *find grad div.* $\vec{A}$ **[BU(H) 1997]**

Solution : We have **grad div** $\vec{A} = \nabla\left(\nabla \cdot \frac{\vec{r}}{r}\right)$

$$= \nabla[\nabla \cdot (r^{-1}\vec{r})] = \nabla[\nabla r^{-1} \cdot \vec{r} + r^{-1}(\nabla \cdot \vec{r})]$$

$$= \nabla\left[\left(-\frac{\vec{r}}{r^3}\right) \cdot \vec{r} + \frac{3}{r}\right] \quad \left[\because \nabla r^{-1} = \nabla\left(\frac{1}{r}\right) = -\frac{\vec{r}}{r^3} \text{ and } \nabla \cdot \vec{r} = 3\right]$$

$$= \nabla\left[-\frac{r^2}{r^3} + \frac{3}{r}\right] = \nabla\left[-\frac{1}{r} + \frac{3}{r}\right]$$

$$= \nabla\left(\frac{2}{r}\right) = 2\nabla\left(\frac{1}{r}\right) = 2\left(\frac{-\vec{r}}{r^3}\right) = -\frac{\boldsymbol{2\vec{r}}}{\boldsymbol{r^3}}.$$

Example 11.58. *If $\phi = 3x^2y$, $\psi = xz^2 - 2y$, then evaluate $\nabla(\nabla\phi \cdot \nabla\psi)$.*

Solution : $\because \ \phi = 3x^2y, \quad \therefore \ \nabla\phi = \left(\hat{i}\frac{\partial}{\partial x} + \hat{j}\frac{\partial}{\partial y} + \hat{k}\frac{\partial}{\partial z}\right) 3x^2y$

or, $\quad \nabla\phi = \hat{i}(6xy) + \hat{j}(3x^2) + \hat{k}(0) = 6xy\hat{i} + 3x^2\hat{j}$

and $\quad \because \ \psi = xz^2 - 2y, \quad \therefore \ \nabla\psi = \left(\hat{i}\frac{\partial}{\partial x} + \hat{j}\frac{\partial}{\partial y} + \hat{k}\frac{\partial}{\partial z}\right)(xz^2 - 2y)$

or, $\quad \nabla\psi = \hat{i}(z^2) + \hat{j}(-2) + \hat{k}(2xz) = z^2\hat{i} - 2\hat{j} + 2xz\hat{k}$

$$\nabla\phi \cdot \nabla\psi = (6xy\hat{i} + 3x^2\hat{j}) \cdot (z^2\hat{i} - 2\hat{j} + 2xz\hat{k}) = (6xyz^2 - 6x^2).$$

$$\therefore \quad \boldsymbol{\nabla(\nabla\phi \cdot \nabla\psi)} = \left(\hat{i}\frac{\partial}{\partial x} + \hat{j}\frac{\partial}{\partial y} + \hat{k}\frac{\partial}{\partial z}\right)(6xyz^2 - 6x^2)$$

$$= \boldsymbol{(6yz^2 - 12x)\hat{i} + 6xz^2\hat{j} + 12xyzk^2}.$$

Example 11.59. *Prove that $\nabla^2(\phi\psi) = \phi\nabla^2\psi + 2\nabla\phi \cdot \nabla\psi + \psi\nabla^2\phi$.*

Solution : We have $\nabla(\phi\psi) = \phi\nabla\psi + \psi\nabla\phi$.

$$\therefore \quad \boldsymbol{\nabla^2(\phi\psi)} = \nabla \cdot \nabla(\phi\psi) = \nabla \cdot (\phi\nabla\psi + \psi\nabla\phi) = \nabla \cdot (\phi\nabla\psi) + \nabla \cdot (\psi\nabla\phi)$$

$$= \nabla\phi \cdot \nabla\psi + \phi\nabla \cdot (\nabla\phi) + \nabla\psi \cdot \nabla\phi + \psi\nabla \cdot (\nabla\phi)$$

$$= \boldsymbol{\phi\nabla^2\psi + 2\nabla\phi \cdot \nabla\psi + \psi\nabla^2\phi}. \quad \textbf{(Proved)}$$

Example 11.60. *Show that, the vector field $\vec{V} = -\dfrac{x\hat{i} - y\hat{j}}{\sqrt{x^2 + y^2}}$ is a "sink" field.*

Solution : We know that if $\nabla \cdot \vec{A}$ is negative at a point in the vector field, then the vector field $\vec{A}$ is called a sink field.

Now, $\vec{V} = \dfrac{-x\hat{i} - y\hat{j}}{\sqrt{x^2 + y^2}}$.

$$\therefore \quad \nabla \cdot \vec{V} = -\nabla \cdot \left[\frac{x\hat{i} + y\hat{j}}{\sqrt{x^2 + y^2}}\right] = -\left[\hat{i}\frac{\partial}{\partial x} + \hat{j}\frac{\partial}{\partial y} + \hat{k}\frac{\partial}{\partial z}\right] \cdot \left[\frac{x\hat{i} + y\hat{j}}{\sqrt{x^2 + y^2}}\right]$$

$$= -\left[\frac{\partial}{\partial x}\left(\frac{x}{\sqrt{x^2 + y^2}}\right) + \frac{\partial}{\partial y}\left(\frac{y}{\sqrt{x^2 + y^2}}\right)\right]$$

$$= -\left[\frac{\partial}{\partial x}(x^2 + y^2)^{-\frac{1}{2}}x + \frac{\partial}{\partial y}(x^2 + y^2)^{-\frac{1}{2}}y\right]$$

$$= -\left[-\frac{1}{2}(x^2 + y^2)^{-\frac{3}{2}}2x \cdot x + (x^2 + y^2)^{-\frac{1}{2}} \cdot 1 \right.$$

$$\left. + \left(-\frac{1}{2}\right)(x^2 + y^2)^{-\frac{3}{2}}2y \cdot y + (x^2 + y^2)^{-\frac{1}{2}} \cdot 1\right]$$

$$= \frac{x^2}{(x^2 + y^2)^{3/2}} - \frac{1}{(x^2 + y^2)^{\frac{1}{2}}} + \frac{y^2}{(x^2 + y^2)^{3/2}} - \frac{1}{(x^2 + y^2)^{\frac{1}{2}}}$$

$$= \frac{x^2 - (x^2+y^2) + y^2 - (x^2+y^2)}{(x^2+y^2)^{3/2}} = -\frac{(x^2+y^2)}{(x^2+y^2)^{3/2}} = -\frac{1}{\sqrt{x^2+y^2}}.$$

$\therefore$ $\nabla \cdot \vec{V}$ is negative, $\therefore$ $\vec{V}$ **is a sink field.**

Example 11.61. *Show that,* $\nabla^2(r^n) = n(n+1)r^{n-2}$. **[BU 1985S, MU 1985, GU 1984]**

Solution : We have $\nabla^2 r^n = \nabla \cdot (\nabla r^n) = \nabla \cdot \left(n r^{n-1} \frac{\vec{r}}{r}\right) = n\nabla \cdot (r^{n-2}\vec{r})$

or, $\boldsymbol{\nabla^2 r^n} = n[(\nabla r^{n-2}) \cdot \vec{r} + r^{n-2}(\nabla \cdot \vec{r})] = n[(n-2)r^{n-3}\frac{\vec{r}}{r} \cdot \vec{r} + r^{n-2}(3)]$

$$= n[(n-2)r^{n-4}(r^2) + (3r^{n-2})] = \boldsymbol{n(n+1)r^{n-2}}.$$

Alternatively, $\nabla^2(r^n) = \frac{\partial^2(r^n)}{\partial x^2} + \frac{\partial^2(r^n)}{\partial y^2} + \frac{\partial^2(r^n)}{\partial z^2}.$

Now, $\frac{\partial^2(r^n)}{\partial x^2} = n\left[r^{n-2} + (n-2)r^{n-3}\frac{\partial r}{\partial x}x\right] = n\left[r^{n-2} + (n-2)r^{n-3}\frac{x}{r}x\right]$

$$= n[r^{n-2} + (n-2)r^{n-4}x^2].$$

Similarly, we can show that,

$$\frac{\partial^2(r^n)}{\partial y^2} = n[r^{n-2} + (n-2)r^{n-4}y^2] \quad \text{and} \quad \frac{\partial^2(r^n)}{\partial z^2} = n[r^{n-2} + (n-2)r^{n-4}z^2].$$

$$\therefore \quad \boldsymbol{\frac{\partial^2(r^n)}{\partial x^2} + \frac{\partial^2(r^n)}{\partial y^2} + \frac{\partial^2(r^n)}{\partial z^2}} = n[3r^{n-2} + (n-2)r^{n-4}(x^2+y^2+z^2)]$$

$$= n[3r^{n-2} + (n-2)r^{n-4}(r^2)] = n[3r^{n-2} + (n-2)r^{n-2}]$$

$$= \boldsymbol{n(n+1)r^{n-2}} \quad \textbf{(Proved)}.$$

Example 11.62. *Prove that,* (a) $\nabla^2 f(r) = \frac{d^2 f}{dr^2} + \frac{2}{r}\frac{df}{dr}$ (b) *Find* $f(r)$ *such that* $\nabla^2 f(r) = 0$.

Solution : (a) We have $\boldsymbol{\nabla^2 f(r)} = \nabla \cdot \nabla f(r) = \nabla \cdot \left[\frac{df(r)}{dr}\hat{r}\right]$

$$= \nabla \cdot \left[\frac{1}{r}\frac{df(r)}{dr}\vec{r}\right], \quad \left[\because \nabla f(r) = \frac{df(r)}{dt}\hat{r}\right]$$

$$= \frac{1}{r}\frac{df(r)}{dr}(\nabla \cdot \vec{r}) + \vec{r} \cdot \nabla\left(\frac{1}{r}\frac{df(r)}{dr}\right)$$

$$= \frac{3}{r}\frac{df(r)}{dt} + \frac{d^2 f(r)}{dr^2} - \frac{1}{r}\frac{df(r)}{dr}$$

$$\therefore \quad \boldsymbol{\nabla^2 f(r) = \frac{d^2 f(r)}{dr^2} + \frac{2}{r}\frac{df(r)}{dr}} \quad \textbf{(Proved)}.$$

(b) $\nabla^2 f(r) = 0$, we have $\frac{d^2 f(r)}{dr^2} + \frac{2}{r}\frac{df(r)}{dr} = 0$ or, $\frac{1}{r^2}\frac{d}{dr}\left(r^2\frac{df(r)}{dr}\right) = 0.$

Integrating, $r^2\frac{df(r)}{dr} = B'$ [B' is an integration constant] or, $\dfrac{df(r)}{dr} = \dfrac{B'}{r^2}$

or, $f(r) = -\dfrac{B'}{r} + A = A + \dfrac{B}{r}$ [where $B = -B'$] [A and B are arbitrary constant]

Example 11.63. *Prove that, the vector, $\vec{A} = 3y^4z^2\hat{\imath} + 4x^3z^2\hat{\jmath} - 3x^2y^2\hat{k}$ is solenoidal.*

Solution : If $\nabla\cdot\vec{A} = 0$, then $\vec{A}$ is said to be solenoidal.

Now, $\nabla\cdot\vec{A} = \left(\hat{\imath}\dfrac{\partial}{\partial x} + \hat{\jmath}\dfrac{\partial}{\partial y} + \hat{k}\dfrac{\partial}{\partial z}\right)[3y^4z^2\hat{\imath} + 4x^3z^2\hat{\jmath} - 3x^2y^2\hat{k}]$

$$= \frac{\partial}{\partial x}(3y^4z^2) + \frac{\partial}{\partial y}(4x^3z^2) + \frac{\partial}{\partial z}(-3x^2y^2) = 0 + 0 + 0 = 0.$$

$\therefore$ $\vec{A}$ **is solenoidal.**

Example 11.64. *For what value of a the vector $\vec{A} = (x+3y)\hat{\imath} + (y-2z)\hat{\jmath} + (x+az)\hat{k}$ is solenoidal?*

Solution : $\vec{A}$ is solenoidal if $\nabla\cdot\vec{A} = 0$.

Now, $\nabla\cdot\vec{A} = \left(\hat{\imath}\dfrac{\partial}{\partial x} + \hat{\jmath}\dfrac{\partial}{\partial y} + \hat{k}\dfrac{\partial}{\partial z}\right)\cdot[(x+3y)\hat{\imath} + (y-2z)\hat{\jmath} + (x+az)\hat{k}]$

$$= \frac{\partial}{\partial x}(x+3y) + \frac{\partial}{\partial y}(y-2x) + \frac{\partial}{\partial z}(x+az) = 1 + 1 + a = 2 + a.$$

If $\vec{A}$ is solenoidal, then $\nabla\cdot\vec{A} = 0$ or, $2 + a = 0$, or, $\boldsymbol{a = -2}$.

Example 11.65. *If $\vec{A} = 2xz^2\hat{\imath} - yz\hat{\jmath} + 3xz^2\hat{k}$, then find* (a) $\nabla\times\vec{A}$ *and* (b) $\nabla\times(\nabla\times\vec{A})$ *at the point* $(1,1,1)$.

Solution : We have $\vec{A} = 2xz^2\hat{\imath} - yz\hat{\jmath} + 3xz^3\hat{k}$.

$$\therefore\quad \nabla\times\vec{A} = \left(\hat{\imath}\frac{\partial}{\partial x} + \hat{\jmath}\frac{\partial}{\partial y} + \hat{k}\frac{\partial}{\partial z}\right)\times(2xz^2\hat{\imath} - yz\hat{\jmath} + 3xz^3\hat{k}) = \begin{vmatrix} \hat{\imath} & \hat{\jmath} & \hat{k} \\ \dfrac{\partial}{\partial x} & \dfrac{\partial}{\partial y} & \dfrac{\partial}{\partial z} \\ 2xz^2 & -yz & 3xz^3 \end{vmatrix}$$

$$= \left[\frac{\partial}{\partial y}(3xz^2) - \frac{\partial}{\partial z}(-yz)\right]\hat{\imath} + \left[\frac{\partial}{\partial z}(2xz^2) - \frac{\partial}{\partial x}(3xz^3)\right]\hat{\jmath} + \left[\frac{\partial}{\partial x}(-yz) - \frac{\partial}{\partial y}(2xz^3)\right]\hat{k}$$

$\therefore$ $\boldsymbol{\nabla\times\vec{A} = y\hat{\imath} + (4xz - 3z^2)\hat{\jmath}}$

$\therefore$ $\boldsymbol{(\nabla\times\vec{A})_{1,1,1} = \hat{\imath} + \hat{\jmath}}$.

(b) $\because$ $\nabla\times\vec{A} = y\hat{\imath} + (4xz - 3z^3)\hat{\jmath}$

$$\therefore\quad \nabla\times(\nabla\times\vec{A}) = \begin{vmatrix} \hat{\imath} & \hat{\jmath} & \hat{k} \\ \dfrac{\partial}{\partial x} & \dfrac{\partial}{\partial y} & \dfrac{\partial}{\partial z} \\ y & (4xz - 3z^3) & 0 \end{vmatrix}$$

$$= \left[\frac{\partial}{\partial x}(0) - \frac{\partial}{\partial z}(4xz - 3z^3)\right]\hat{\imath} + \left[\frac{\partial}{\partial z}(y) - \frac{\partial}{\partial x}(0)\right]\hat{\jmath} + \left[\frac{\partial}{\partial x}(4xz - 3z^3) - \frac{\partial}{\partial y}(y)\right]\hat{k}$$

$$= -(4x - 9y^2)\hat{\imath} + 0\hat{\jmath} + (4z - 1)\hat{k}$$

$$\therefore \quad [\boldsymbol{\nabla} \times (\boldsymbol{\nabla} \times \vec{A})]_{1,1,1} = -(4 - 9)\hat{\imath} + 3\hat{k} = \mathbf{5\hat{\imath} + 3\hat{k}}.$$

Example 11.66. *If $\vec{A} = \nabla\phi$ and $\phi = xy + yz + zx$, then find* (i) $\nabla \cdot \vec{A}$ *and* (ii) $\nabla \times \vec{A}$.

Solution : (i) We have $\phi = xy + yz + zx$.

$$\therefore \quad \nabla\phi = \hat{\imath}\frac{\partial\phi}{\partial x} + \hat{\jmath}\frac{\partial\phi}{\partial y} + \hat{k}\frac{\partial\phi}{\partial z} \quad \text{or,} \quad \vec{A} = \hat{\imath}(y+z) + \hat{\jmath}(z+x) + \hat{k}(x+y).$$

Now, $\boldsymbol{\nabla} \cdot \vec{A} = \left(\hat{\imath}\frac{\partial}{\partial x} + \hat{\jmath}\frac{\partial}{\partial y} + \hat{k}\frac{\partial}{\partial z}\right) \cdot [\hat{\imath}(y+z) + \hat{\jmath}(z+x) + \hat{k}(x+y)]$

$$= \frac{\partial}{\partial x}(y+z) + \frac{\partial}{\partial y}(z+x) + \frac{\partial}{\partial z}(x+y) = 0 + 0 + 0 = \mathbf{0}.$$

$$\therefore \quad \boldsymbol{\nabla} \cdot \vec{A} = \mathbf{0}$$

(ii) Again, $\nabla \times \vec{A} = \left(\hat{\imath}\frac{\partial}{\partial x} + \hat{\jmath}\frac{\partial}{\partial y} + \hat{k}\frac{\partial}{\partial z}\right) \cdot [\hat{\imath}(y+z) + \hat{\jmath}(z+x) + \hat{k}(x+y)]$

$$= \begin{vmatrix} \hat{\imath} & \hat{\jmath} & \hat{k} \\ \frac{\partial}{\partial x} & \frac{\partial}{\partial y} & \frac{\partial}{\partial z} \\ y+z & z+x & x+y \end{vmatrix} = (1-1)\hat{\imath} + (1-1)\hat{\jmath} + (1-1)\hat{k} = 0.$$

$$\therefore \quad \boldsymbol{\nabla} \times \vec{A} = \mathbf{0}$$

Example 11.67. *If a rigid body rotates about an axis passing through the origin with angular velocity $\vec{\omega}$ and with linear velocity $\vec{v} = \vec{\omega} \times \vec{r}$, then prove that,*

$$\vec{\omega} = \frac{1}{2}(\nabla \times \vec{v}).$$ **[BU(H) 1996]**

Solution : Let $\vec{\omega} = \hat{\imath}\omega_1 + \hat{\jmath}\omega_2 + \hat{k}\omega_3$ and $\vec{r} = \hat{\imath}x + \hat{\jmath}y + \hat{k}z$

Then, $\vec{v} = \vec{\omega} \times \vec{r} = (\hat{\imath}\omega_1 + \hat{\jmath}\omega_2 + \hat{k}\omega_3) \times (\hat{\imath}x + \hat{\jmath}y + \hat{k}z)$

$$\text{and} \quad \nabla \times \vec{v} = \begin{vmatrix} \hat{\imath} & \hat{\jmath} & \hat{k} \\ \frac{\partial}{\partial x} & \frac{\partial}{\partial y} & \frac{\partial}{\partial z} \\ (\omega_2 z - \omega_3 y) & (\omega_3 x - \omega_1 z) & (\omega_1 y - \omega_2 x) \end{vmatrix}$$

$$= (\omega_1 + \omega_1)\hat{\imath} + (\omega_2 + \omega_2)\hat{\jmath} + (\omega_3 + \omega_3)\hat{k} = 2\omega_1\hat{\imath} + 2\omega_2\hat{\jmath} + 2\omega_3\hat{k}$$

$$= 2(\omega_1\hat{\imath} + \omega_2\hat{\jmath} + \omega_3\hat{k}) = 2\vec{\omega}.$$

$$\therefore \quad \vec{\omega} = \frac{1}{2}(\boldsymbol{\nabla} \times \vec{v}) \quad \textbf{(Proved)}.$$

N.B. The above result may be thought as the significance of curl.

Example 11.68. *A vector field is defined by $\vec{F} = \dfrac{\vec{r}}{r^2}$, where $\vec{r} = \hat{i}x+\hat{j}y+\hat{k}z$, then evaluate $\nabla \cdot \vec{F}$ and $\nabla \times \vec{F}$ and state whether the field is solenoidal or irrotational.* **[BU(H) 1992]**

Solution : We have $\vec{F} = \dfrac{\vec{r}}{r} = \dfrac{\hat{i}x+\hat{j}y+\hat{k}z}{\sqrt{x^2+y^2+z^2}}$.

Now, $\nabla \cdot \vec{F} = \nabla \cdot \left(\dfrac{\vec{r}}{r}\right) = \left(\hat{i}\dfrac{\partial}{\partial x}+\hat{j}\dfrac{\partial}{\partial y}+\hat{k}\dfrac{\partial}{\partial z}\right)\cdot\left(\dfrac{\hat{i}x+\hat{j}y+\hat{k}z}{\sqrt{x^2+y^2+z^2}}\right)$

$$= \frac{\partial}{\partial x}\left(\frac{x}{x^2+y^2+z^2}\right)+\frac{\partial}{\partial y}\left(\frac{y}{x^2+y^2+z^2}\right)+\frac{\partial}{\partial z}\left(\frac{z}{x^2+y^2+z^2}\right)$$

or, $$\nabla \cdot \vec{F} = \frac{1}{(x^2+y^2+z^2)} - \frac{2x^2}{(x^2+y^2+z^2)^2} + \frac{1}{x^2+y^2+z^2} - \frac{2y^2}{(x^2+y^2+z^2)^2} + \frac{1}{x^2+y^2+z^2} - \frac{2z^2}{(x^2+y^2+z^2)^2}$$

$$= \frac{y^2+z^2-x^2}{(x^2+y^2+z^2)^2} + \frac{x^2+z^2-y^2}{(x^2+y^2+z^2)^2} + \frac{x^2+y^2-z^2}{(x^2+y^2+z^2)^2}$$

$$= \frac{y^2+z^2-x^2+x^2+z^2-y^2+x^2+y^2-z^2}{(x^2+y^2+z^2)^2} = \frac{x^2+y^2+z^2}{(x^2+y^2+z^2)^2}$$

$$= \frac{1}{x^2+y^2+z^2} = \frac{1}{r^2}.$$

$\because$ $\nabla \cdot \vec{F} \neq 0$, $\vec{F}$ **is not solenoidal.**

Now, $$\nabla \times \vec{F} = \left(\hat{i}\frac{\partial}{\partial x}+\hat{j}\frac{\partial}{\partial y}+\hat{k}\frac{\partial}{\partial z}\right)\cdot\left(\frac{\hat{i}x+\hat{j}y+\hat{k}z}{r^2}\right) = \begin{vmatrix} \hat{i} & \hat{j} & \hat{k} \\ \dfrac{\partial}{\partial x} & \dfrac{\partial}{\partial y} & \dfrac{\partial}{\partial z} \\ \dfrac{x}{r^2} & \dfrac{y}{r^2} & \dfrac{z}{r^2} \end{vmatrix}$$

$$= \hat{i}\left[\frac{\partial}{\partial y}\left(\frac{z}{r^2}\right)-\frac{\partial}{\partial z}\left(\frac{y}{r^2}\right)\right]+\hat{j}\left[\frac{\partial}{\partial z}\left(\frac{x}{r^2}\right)-\frac{\partial}{\partial x}\left(\frac{z}{r^2}\right)\right]+\hat{k}\left[\frac{\partial}{\partial x}\left(\frac{y}{r^2}\right)-\frac{\partial}{\partial y}\left(\frac{x}{r^2}\right)\right].$$

The coefficient of $\hat{i} = \dfrac{\partial}{\partial y}\left(\dfrac{z}{r^2}\right)-\dfrac{\partial}{\partial z}\left(\dfrac{y}{r^2}\right) = -z\dfrac{2y}{(x^2+y^2+z^2)^2}+y\dfrac{2z}{(x^2+y^2+z^2)^2} = 0.$

Similarly, we can show that the coefficient of $\hat{j}$ and $\hat{k}$ are separately zero.

$\therefore$ $\quad \boldsymbol{\nabla \times \vec{F} = 0.}$

This suggests that the field $\vec{F}$ **is irrotational but not solenoidal.**

Example 11.69. *If $\vec{A} = x^2y\hat{i} - 2xz\hat{j} + 2yz\hat{k}$, find $\nabla \times (\nabla \times \vec{A})$.*

Solution : We have $$\nabla \times (\nabla \times \vec{A}) = \nabla \times \begin{vmatrix} \hat{i} & \hat{j} & \hat{k} \\ \dfrac{\partial}{\partial x} & \dfrac{\partial}{\partial y} & \dfrac{\partial}{\partial z} \\ x^2y & -2xz & 2yz \end{vmatrix}$$

$$= \nabla \times [(2x+2z)\hat{\imath} - (x^2+2z)\hat{k}] = \begin{vmatrix} \hat{\imath} & \hat{\jmath} & \hat{k} \\ \dfrac{\partial}{\partial x} & \dfrac{\partial}{\partial y} & \dfrac{\partial}{\partial z} \\ (2x+2z) & 0 & -x^2-2z \end{vmatrix}$$

$\therefore \quad \boldsymbol{\nabla \times (\nabla \times \vec{A}) = (2x+2)\hat{\jmath}}.$

Example 11.70. *If $\vec{A} = xz^2\hat{\imath} + 2y\hat{\jmath} - 3xz\hat{k}$ and $\vec{B} = 3xz\hat{\imath} + 2yz\hat{\jmath} - z^2\hat{k}$, then evaluate $(\vec{A} \times \nabla) \times \vec{B}$ at the point $(1,-1,2)$.* **[BU(H) 2001]**

Solution : We have $\vec{A} \times \nabla = (xz^2\hat{\imath} + 2y\hat{\jmath} - 3xz\hat{k}) \times \left(\hat{\imath}\dfrac{\partial}{\partial x} + \hat{\jmath}\dfrac{\partial}{\partial y} + \hat{k}\dfrac{\partial}{\partial z}\right)$

$$= \begin{vmatrix} \hat{\imath} & \hat{\jmath} & \hat{k} \\ xz^2 & 2y & -3xz \\ \dfrac{\partial}{\partial x} & \dfrac{\partial}{\partial y} & \dfrac{\partial}{\partial z} \end{vmatrix} = \hat{\imath}\left[2y\frac{\partial}{\partial z} + 3xz\frac{\partial}{\partial y}\right] - \hat{\jmath}\left[xz^2\frac{\partial}{\partial z} + 3xz\frac{\partial}{\partial x}\right] + \hat{k}\left[xz^2\frac{\partial}{\partial y} - 2y\frac{\partial}{\partial x}\right]$$

$$\therefore \quad (\vec{A} \times \nabla) \times \vec{B} = \begin{vmatrix} \hat{\imath} & \hat{\jmath} & \hat{k} \\ \left(2y\dfrac{\partial}{\partial z} + 3xz\dfrac{\partial}{\partial y}\right) & -\left(xz^2\dfrac{\partial}{\partial z} + 3xz\dfrac{\partial}{\partial x}\right) & \left(xz^2\dfrac{\partial}{\partial y} - 2y\dfrac{\partial}{\partial x}\right) \\ 3xz & 2yz & -z^2 \end{vmatrix}$$

$$= (2xz^3 - 2xz^3)\hat{\imath} + (-6yz + 4yz)\hat{\jmath} + (4y^2 + 6xz^2 + 3x^2z^2 + 9xz^2)$$

$$= -2yz\hat{\jmath} + (3x^2z^2 + 15xz^2 + 4y^2)\hat{k}.$$

At the point $(1,-1,2)$,

$$\boldsymbol{(\vec{A} \times \nabla) \times \vec{B}} = -2(-1)(2)\hat{\jmath} + [3(1)^2(2)^2 + 15(1)(2)^2 + 4(-1)^2]\hat{k} = \boldsymbol{4j + 76\hat{k}}.$$

Example 11.71. *If $\vec{r}$ be the position vector with magnitude r, then find $(\nabla \times \nabla r^n)$.*

Solution : We have $\nabla \times \nabla r^n = \nabla \times nr^{n-2}\vec{r} \quad [\because r^n = nr^{n-2}\vec{r}]$.

$$\left(\hat{\imath}\frac{\partial}{\partial x} + \hat{\jmath}\frac{\partial}{\partial y} + \hat{k}\frac{\partial}{\partial z}\right) \times nr^{n-2}(\hat{\imath}x + \hat{\jmath}y + \hat{k}z) = \begin{vmatrix} \hat{\imath} & \hat{\jmath} & \hat{k} \\ \dfrac{\partial}{\partial x} & \dfrac{\partial}{\partial y} & \dfrac{\partial}{\partial z} \\ nr^{n-2}x & nr^{n-2}y & nr^{n-2}z \end{vmatrix}.$$

The coefficient of $\hat{\imath} = \dfrac{\partial}{\partial y}(nr^{n-2}z) - \dfrac{\partial}{\partial y}(nr^{n-2}y) = n(n-2)r^{n-3}\dfrac{\partial r}{\partial y}z - n(n-2)r^{n-3}\dfrac{\partial r}{\partial z}y$

$$= n(n-2)r^{n-3} \cdot \left(\frac{\partial r}{\partial y}z - \frac{\partial r}{\partial z}y\right) = n(n-2)r^{n-3}\left(\frac{y}{r}z - \frac{z}{r}y\right) = 0.$$

$$\because r = \sqrt{x^2+y^2+z^2}, \quad \therefore \frac{\partial r}{\partial y} = \frac{y}{\sqrt{x^2+y^2+z^2}} = \frac{y}{r} \text{ and } \frac{\partial r}{\partial z} = \frac{z}{r}.$$

Similarly, the coefficients of $\hat{j}$ and $\hat{k}$ are separately zero.

$\therefore \quad \nabla \times \nabla r^n = 0.$

Example 11.72. *Prove that,* $\nabla \times \nabla\phi = 0$.

Solution : $\nabla \times \nabla\phi = \left(\hat{i}\frac{\partial}{\partial x} + \hat{j}\frac{\partial}{\partial y} + \hat{k}\frac{\partial}{\partial z}\right) \times \left(\hat{i}\frac{\partial}{\partial x} + \hat{j}\frac{\partial}{\partial y} + \hat{k}\frac{\partial}{\partial z}\right)\phi$

$$= \begin{vmatrix} \hat{i} & \hat{j} & \hat{k} \\ \frac{\partial}{\partial x} & \frac{\partial}{\partial y} & \frac{\partial}{\partial z} \\ \frac{\partial\phi}{\partial x} & \frac{\partial\phi}{\partial y} & \frac{\partial\phi}{\partial z} \end{vmatrix} = \hat{i}\left(\frac{\partial^2\phi}{\partial y\partial z} - \frac{\partial^2\phi}{\partial z\partial y}\right) + \hat{j}\left(\frac{\partial^2\phi}{\partial z\partial x} - \frac{\partial^2\phi}{\partial x\partial z}\right) + \hat{k}\left(\frac{\partial^2\phi}{\partial x\partial y} - \frac{\partial^2\phi}{\partial y\partial x}\right).$$

Now, if $\phi(x, y, z)$ is a perfect differential, then $\frac{\partial^2\phi}{\partial y\partial z} = \frac{\partial^2\phi}{\partial z\partial y}$, etc.

$\therefore \quad \nabla \times \nabla\phi = 0$ **(Proved).**

Example 11.73. *Prove that, curlcurl* $\vec{A}$ = grad div $\vec{A} - \nabla^2\vec{A}$ *or,* $\nabla \times (\nabla \times \vec{A}) = \nabla(\nabla \cdot \vec{A}) - \nabla^2\vec{A}$.

[BU(H) 1992, WBUT 2002]

Solution : We have curlcurl $\vec{A} = \nabla \times (\nabla \times \vec{A}) = \nabla \times \begin{vmatrix} \hat{i} & \hat{j} & \hat{k} \\ \frac{\partial}{\partial x} & \frac{\partial}{\partial y} & \frac{\partial}{\partial z} \\ A_1 & A_2 & A_3 \end{vmatrix}$

$$= \nabla \times \left[\hat{i}\left(\frac{\partial A_3}{\partial y} - \frac{\partial A_2}{\partial z}\right) + \hat{j}\left(\frac{\partial A_1}{\partial z} - \frac{\partial A_3}{\partial x}\right) + \hat{k}\left(\frac{\partial A_2}{\partial x} - \frac{\partial A_1}{\partial y}\right)\right]$$

$$= \begin{vmatrix} \hat{i} & \hat{j} & \hat{k} \\ \frac{\partial}{\partial x} & \frac{\partial}{\partial y} & \frac{\partial}{\partial z} \\ \left(\frac{\partial A_3}{\partial y} - \frac{\partial A_2}{\partial z}\right) & \left(\frac{\partial A_1}{\partial z} - \frac{\partial A_3}{\partial x}\right) & \left(\frac{\partial A_2}{\partial x} - \frac{\partial A_1}{\partial y}\right) \end{vmatrix}.$$

Now considering the x-component of $\nabla \times (\nabla \times \vec{A})$, we get

$$(\text{curl})_x\text{curl } \vec{A} = \left[\frac{\partial}{\partial y}\left(\frac{\partial A_2}{\partial x} - \frac{\partial A_1}{\partial y}\right) - \frac{\partial}{\partial z}\left(\frac{\partial A_1}{\partial z} - \frac{\partial A_3}{\partial x}\right)\right]\hat{i}$$

$$= \left[\frac{\partial^2 A_2}{\partial y\partial x} - \frac{\partial^2 A_1}{\partial y^2} - \frac{\partial^2 A_1}{\partial z^2} + \frac{\partial^2 A_3}{\partial z\partial x}\right]\hat{i}$$

$$= \left(\frac{\partial^2 A_1}{\partial x^2} + \frac{\partial^2 A_2}{\partial x\partial y} + \frac{\partial^2 A_3}{\partial z\partial x}\right)\hat{i} - \left(\frac{\partial^2 A_1}{\partial x^2} + \frac{\partial^2 A_1}{\partial y^2} + \frac{\partial^2 A_1}{\partial z^2}\right)\hat{i}$$

[Adding and subtracting the term $\frac{\partial^2 A_1}{\partial x^2}$]

$$= \frac{\partial}{\partial x}\left(\frac{\partial A_1}{\partial x} + \frac{\partial A_2}{\partial y} + \frac{\partial A_3}{\partial z}\right)\hat{i} - \left(\frac{\partial^2 A_1}{\partial x^2} + \frac{\partial^2 A_1}{\partial y^2} + \frac{\partial^2 A_1}{\partial z^2}\right)\hat{i}$$

$$= \frac{\partial}{\partial x}\hat{i}(\nabla \cdot \vec{A}) - \nabla^2(A_1\hat{i}).$$

Similarly, $(\text{curl})_y \text{curl } \vec{A} = \frac{\partial}{\partial y}\hat{j}(\nabla \cdot \vec{A}) - \nabla^2(A_2\hat{j})$

and $(\text{curl})_z \text{curl } \vec{A} = \frac{\partial}{\partial z}\hat{k}(\nabla \cdot \vec{A}) - \nabla^2(A_3\hat{k}).$

$\therefore$ $\text{curl curl } \vec{A} = \left(\hat{i}\frac{\partial}{\partial x} + \hat{j}\frac{\partial}{\partial y} + \hat{k}\frac{\partial}{\partial z}\right)(\nabla \cdot \vec{A}) - \nabla^2(\hat{i}A_1 + \hat{j}A_2 + \hat{k}A_3)$

$$= \nabla(\nabla \cdot \vec{A}) - \nabla^2\vec{A}$$

or, **curl curl $\vec{A}$ = grad div $\vec{A} - \nabla^2\vec{A}$ (Proved).**

Example 11.74. *Prove that, div curl $\vec{A} = \nabla \cdot (\nabla \times \vec{A}) = 0$.* **[CU(H) 2008]**

Solution : We have div curl $\vec{A} = \nabla \cdot (\nabla \times \vec{A}) = \nabla \cdot \begin{vmatrix} \hat{i} & \hat{j} & \hat{k} \\ \frac{\partial}{\partial x} & \frac{\partial}{\partial y} & \frac{\partial}{\partial z} \\ A_1 & A_2 & A_3 \end{vmatrix}$

$$= \left(\hat{i}\frac{\partial}{\partial x} + \hat{j}\frac{\partial}{\partial y} + \hat{k}\frac{\partial}{\partial z}\right) \cdot \left[\hat{i}\left(\frac{\partial A_3}{\partial y} - \frac{\partial A_2}{\partial z}\right) + \hat{j}\left(\frac{\partial A_1}{\partial z} - \frac{\partial A_3}{\partial x}\right) + \hat{k}\left(\frac{\partial A_2}{\partial x} - \frac{\partial A_1}{\partial y}\right)\right]$$

$$= \frac{\partial}{\partial x}\left(\frac{\partial A_3}{\partial y} - \frac{\partial A_2}{\partial z}\right) + \frac{\partial}{\partial y}\left(\frac{\partial A_1}{\partial z} - \frac{\partial A_3}{\partial x}\right) + \frac{\partial}{\partial z}\left(\frac{\partial A_2}{\partial x} - \frac{\partial A_1}{\partial y}\right)$$

$$= \frac{\partial^2 A_3}{\partial x \partial y} - \frac{\partial^2 A_2}{\partial x \partial z} + \frac{\partial^2 A_1}{\partial y \partial z} - \frac{\partial^2 A_3}{\partial y \partial x} + \frac{\partial^2 A_2}{\partial z \partial x} - \frac{\partial^2 A_1}{\partial z \partial y} = 0,$$

if $\vec{A}(x, y, z)$ is a perfect differential. $\therefore$ $\boldsymbol{\nabla \cdot (\nabla \times \vec{A}) = 0}$

Example 11.75. *Find the constants a, b, c, so that $\nabla \times \vec{V} = 0$, where $\vec{V} = (x + 2y + az)\hat{i} + (bx - 3y - z)\hat{j} + (4x + cy + 2z)\hat{k}$. Show that, $\vec{V}$ can be expressed as the gradient of a scalar function.* **[BU(H) 1997; AMIETE(Winter) 1986]**

Solution : We have $\nabla \times \vec{V} = \left(\hat{i}\frac{\partial}{\partial x} + \hat{j}\frac{\partial}{\partial y} + \hat{k}\frac{\partial}{\partial z}\right)[(x + 2y + z)\hat{i} + (bx - 3y - z)\hat{j} + (4x + cy + 2z)\hat{k}]$

$$= \begin{vmatrix} \hat{i} & \hat{j} & \hat{k} \\ \frac{\partial}{\partial x} & \frac{\partial}{\partial y} & \frac{\partial}{\partial z} \\ (x + 2y + az) & (bx - 3y - z) & (4x + cy + 2z) \end{vmatrix}$$

$$= \hat{i}(c + 1) + \hat{j}(a - 4) + \hat{k}(b - 2) = 0.$$

$\therefore \quad c+1=0,\ a-4=0,\ b-2=0 \quad$ or, $\quad c=-1,\ a=4,\ b=2.$

So, $\vec{V} = (x+2y+4z)\hat{i} + (2x-3y-z)\hat{j} + (4x-y+2z)\hat{k}.$ (1)

If $\phi(x,y,z)$ be the scalar function in terms of whose gradient $\vec{V}$ can be expressed, i.e.,

$$\vec{V} = \nabla\phi = \hat{i}\frac{\partial\phi}{\partial x} + \hat{j}\frac{\partial\phi}{\partial y} + \hat{k}\frac{\partial\phi}{\partial z}. \tag{2}$$

Comparing eqns. (1) and (2), we get

$$\frac{\partial\phi}{\partial x} = x+2y+4z \tag{3}$$

$$\frac{\partial\phi}{\partial y} = 2x-3y+z \tag{4}$$

$$\frac{\partial\phi}{\partial z} = 4x-y+2z. \tag{5}$$

Integrating eqns. (3), (4) and (5), we get

$$\phi = \frac{x^2}{2} + 2xy + 4zx + f(y,z)$$

$$\phi = 2xy - \frac{3}{2}y^2 - yz + g(z,x)$$

$$\phi = 4xz - yz + z^2 + h(x,y).$$

If we choose $f(y,z) = -yz, g(z,x) = 4zx$ and $h(x,y) = 2xy$, then

$$\phi(x,y,z) = \frac{x^2}{2} - \frac{3}{2}y^2 + z^2 + 2xy - yz + 4zx + C.$$

This shows that the vector $\vec{V}$ **can be expressed as the gradient of a scalar function.**

Example 11.76. *If $\vec{r}$ and $\vec{R}$ have their usual meanings and $\vec{A}$ is a constant vector, then prove that,* $\nabla \times \left(\frac{\vec{A}\times\vec{R}}{r^n}\right) = \frac{2-n}{r^n}\vec{A} + \frac{n(\vec{A}\cdot\vec{R})}{r^{n+2}}\vec{R}.$

Solution : We have $\nabla \times [r^{-n}(\vec{A}\times\vec{R})] = r^{-n}[\nabla\times(\vec{A}\times\vec{R})] + \nabla r^{-n}\times(\vec{A}\times\vec{R})]$

$$= r^{-n}[(\nabla\cdot\vec{R})\vec{A} - (\vec{A}\cdot\nabla)\vec{R}] + \left(-nr^{(-n+1)}\frac{\vec{R}}{r}\right)\times(\vec{A}\times\vec{R})$$

$$= r^{-n}(3\vec{A}-\vec{A}) - nr^{(-n+2)}\vec{R}\times(\vec{A}\times\vec{R}) \quad [\because\ \nabla\cdot\vec{R}=3,\ (\vec{A}\cdot\nabla)\vec{R}=\vec{A}]$$

$$= 2\vec{A}r^{-n} - nr^{-(n+2)}[(\vec{R}\cdot\vec{R})\vec{A} - (\vec{A}\cdot\vec{R})\vec{R}] = \frac{2\vec{A}}{r^n} - \frac{n}{r^{n+2}}[r^2\vec{A} - (\vec{A}\cdot\vec{R})\vec{R}]$$

or, $\quad \boldsymbol{\nabla\times[r^{-n}(\vec{A}\times\vec{r})] = \frac{2-n}{r^n}\vec{A} + \frac{n(\vec{A}\cdot\vec{R})}{r^{n+2}}\vec{R}}$ **(Proved).**

Example 11.77. *If $\vec{A}$ and $\vec{B}$ are irrotational, show that $\vec{A}\times\vec{B}$ is solenoidal.* **[BU(H) 2002]**

Solution : Since $\vec{A}$ and $\vec{B}$ are irrotational, $\nabla\times\vec{A}=0$ and $\nabla\times\vec{B}=0$.

Now, $\nabla\cdot(\vec{A}\times\vec{B})=\sum\hat{i}\cdot\frac{\partial}{\partial x}(\vec{A}\times\vec{B})\quad\left[\because\ \nabla\cdot\vec{A}=\sum\hat{i}\frac{\partial\vec{A}}{\partial x}\right]$

$$=\sum\hat{i}\cdot\left[\left(\frac{\partial\vec{A}}{\partial x}\times\vec{B}\right)+\left(\vec{A}\times\frac{\partial\vec{B}}{\partial x}\right)\right]=\sum\hat{i}\cdot\frac{\partial\vec{A}}{\partial x}\times\vec{B}-\sum\hat{i}\cdot\left(\frac{\partial\vec{B}}{\partial x}\times\vec{A}\right)$$

$$=\left(\sum\hat{i}\times\frac{\partial\vec{A}}{\partial x}\right)\cdot\vec{B}-\left(\sum\hat{i}\times\frac{\partial\vec{B}}{\partial x}\right)\cdot\vec{A}=\vec{B}\cdot(\nabla\times\vec{A})-\vec{A}\cdot(\nabla\times\vec{B})=0.$$

$\therefore\ \vec{A}\times\vec{B}$ is solenoidal as $\nabla\cdot(\vec{A}\times\vec{B})=0$.

Example 11.78. *Prove that,* $\nabla\times(\vec{A}\times\vec{B})=(\vec{B}\cdot\nabla)\vec{A}-(\vec{A}\cdot\nabla)\vec{B}+\vec{A}(\nabla\cdot\vec{B})-\vec{B}(\nabla\cdot\vec{A})$.

Solution : We have $\nabla\times(\vec{A}\times\vec{B})=\sum\hat{i}\times\frac{\partial}{\partial x}(\vec{A}\times\vec{B})$

$$=\sum\hat{i}\times\left[\vec{A}\times\frac{\partial\vec{B}}{\partial x}+\frac{\partial\vec{A}}{\partial x}\times\vec{B}\right]=\sum\hat{i}\times\left(\vec{A}\times\frac{\partial\vec{B}}{\partial x}\right)+\sum\hat{i}\left(\frac{\partial\vec{A}}{\partial x}\times\vec{B}\right)$$

$$=\sum\left[\left(\hat{i}\cdot\frac{\partial\vec{B}}{\partial x}\right)\vec{A}-(\hat{i}\cdot\vec{A})\frac{\partial\vec{B}}{\partial x}\right]+\sum\left[(\hat{i}\cdot\vec{B})\frac{\partial\vec{A}}{\partial x}-\left(\hat{i}\cdot\frac{\partial\vec{A}}{\partial x}\right)\vec{B}\right]$$

$$=\vec{A}\left(\sum\hat{i}\cdot\frac{\partial\vec{B}}{\partial x}\right)-\left(\vec{A}\cdot\sum\hat{i}\frac{\partial}{\partial x}\right)\vec{B}+\left(\vec{B}\cdot\sum\hat{i}\frac{\partial}{\partial x}\right)\vec{A}-\vec{B}\left(\sum\hat{i}\cdot\frac{\partial\vec{A}}{\partial x}\right)$$

$$=\vec{A}(\nabla\cdot\vec{B})-(\vec{A}\cdot\nabla)\vec{B}+(\vec{B}\cdot\nabla)\vec{A}-\vec{B}(\nabla\cdot\vec{A})$$

or, $\boldsymbol{\nabla\times(\vec{A}\times\vec{B})=(\vec{B}\cdot\nabla)\vec{A}-(\vec{A}\cdot\nabla)\vec{B}+\vec{A}(\vec{B}\cdot\nabla)-\vec{B}(\nabla\cdot\vec{A})}$ **(Proved).**

Example 11.79. *Prove that,* $\nabla(\vec{A}\cdot\vec{B})=(\vec{B}\cdot\nabla)\vec{A}+(\vec{A}\cdot\nabla)\vec{B}+\vec{B}\times(\nabla\times\vec{A})+\vec{A}\times(\nabla\times\vec{B})$.

Solution : We have $\nabla(\vec{A}\cdot\vec{B})=\frac{\partial}{\partial x}(\vec{A}\cdot\vec{B})\hat{i}+\frac{\partial}{\partial y}(\vec{A}\cdot\vec{B})\hat{j}+\frac{\partial}{\partial z}(\vec{A}\cdot\vec{B})\hat{k}$

$$=\left(\vec{A}\cdot\frac{\partial\vec{B}}{\partial x}+\frac{\partial\vec{A}}{\partial x}\cdot\vec{B}\right)\vec{i}+\left(\vec{A}\cdot\frac{\partial\vec{B}}{\partial y}+\vec{B}\cdot\frac{\partial\vec{A}}{\partial y}\right)\hat{j}+\left(\vec{A}\cdot\frac{\partial\vec{B}}{\partial z}+\vec{B}\cdot\frac{\partial\vec{A}}{\partial z}\right)\hat{k}$$

$$=\left(\vec{A}\cdot\frac{\partial\vec{B}}{\partial x}\right)\hat{i}+\left(\vec{A}\cdot\frac{\partial\vec{B}}{\partial y}\right)\hat{j}+\left(\vec{A}\cdot\frac{\partial\vec{B}}{\partial z}\right)\hat{k}+\left(\vec{B}\cdot\frac{\partial\vec{A}}{\partial x}\right)\hat{i}$$

$$+\left(\vec{B}\cdot\frac{\partial\vec{A}}{\partial y}\right)\hat{j}+\left(\vec{B}\cdot\frac{\partial\vec{A}}{\partial z}\right)\hat{k}$$

$$=\sum\hat{i}\left(\vec{A}\cdot\frac{\partial\vec{B}}{\partial x}\right)+\sum\hat{i}\left(\vec{B}\cdot\frac{\partial\vec{A}}{\partial x}\right)\qquad(1)$$

Now, $\vec{A}\times\left(\hat{i}\cdot\frac{\partial\vec{B}}{\partial x}\right)=\hat{i}\left(\vec{A}\cdot\frac{\partial\vec{B}}{\partial x}\right)-\frac{\partial\vec{B}}{\partial x}(\vec{A}\cdot\hat{i})$

$$\therefore \quad \hat{i}\left(\vec{A}\cdot\frac{\partial\vec{B}}{\partial x}\right)=\vec{A}\times\left(\hat{i}\times\frac{\partial\vec{B}}{\partial x}\right)+\frac{\partial\vec{B}}{\partial x}(\vec{A}\cdot\hat{i})$$

$$\therefore \quad \sum\hat{i}\left(\vec{A}\cdot\frac{\partial\vec{B}}{\partial x}\right)=\vec{A}\times\sum\left(\hat{i}\times\frac{\partial\vec{B}}{\partial x}\right)+\sum\frac{\partial\vec{B}}{\partial x}(\vec{A}\cdot\hat{i})=\vec{A}\times(\nabla\times\vec{B})+(\vec{A}\cdot\nabla)\vec{B} \quad (2)$$

Similarly we can show that,

$$\sum\hat{i}\left(\vec{B}\cdot\frac{\partial\vec{A}}{\partial x}\right)=\vec{B}\times\left(\sum\hat{i}\times\frac{\partial\vec{A}}{\partial x}\right)+\sum(\vec{B}\cdot\hat{i})\frac{\partial\vec{A}}{\partial x}=\vec{B}\times(\nabla\times\vec{A})+(\vec{B}\cdot\nabla)\vec{A} \quad (3)$$

Using eqns. (2) and (3) in eqn. (1), we get

$$\boldsymbol{\nabla(\vec{A}\cdot\vec{B})=(\vec{B}\cdot\nabla)\vec{A}+(\vec{A}\cdot\nabla)\vec{B}+\vec{B}\times(\nabla\times\vec{A})+\vec{A}\times(\nabla\times\vec{B})} \quad \textbf{(Proved)}.$$

Example 11.80. *Prove that,* $\nabla\cdot(\phi\nabla\psi)=\phi\nabla^2\psi+\nabla\phi\cdot\nabla\psi$.

Solution : We have $\nabla\cdot(\phi\nabla\psi)=\nabla\cdot\left(\phi\frac{\partial\psi}{\partial x}\hat{i}+\phi\frac{\partial\psi}{\partial y}\hat{j}+\phi\frac{\partial\psi}{\partial z}\hat{k}\right)$

$$=\left(\hat{i}\frac{\partial}{\partial x}+\hat{j}\frac{\partial}{\partial y}+\hat{k}\frac{\partial}{\partial z}\right)\cdot\left(\phi\frac{\partial\psi}{\partial x}\hat{i}+\psi\frac{\partial\psi}{\partial y}\hat{j}+\phi\frac{\partial\psi}{\partial z}\hat{k}\right)$$

$$=\frac{\partial}{\partial x}\left(\phi\frac{\partial\psi}{\partial x}\right)+\frac{\partial}{\partial y}\left(\phi\frac{\partial\psi}{\partial y}\right)+\frac{\partial}{\partial z}\left(\phi\frac{\partial\psi}{\partial z}\right)$$

$$=\phi\left(\frac{\partial^2\psi}{\partial x^2}+\frac{\partial^2\psi}{\partial y^2}+\frac{\partial^2\psi}{\partial z^2}\right)+\frac{\partial\phi}{\partial x}\frac{\partial\psi}{\partial x}+\frac{\partial\phi}{\partial y}\frac{\partial\psi}{\partial y}+\frac{\partial\phi}{\partial z}\frac{\partial\psi}{\partial z}$$

$$=\phi\nabla^2\psi+\left(\frac{\partial\phi}{\partial x}\hat{i}+\frac{\partial\phi}{\partial y}\hat{j}+\frac{\partial\phi}{\partial z}\hat{k}\right)\cdot\left(\frac{\partial\psi}{\partial x}\hat{i}+\frac{\partial\psi}{\partial y}\hat{j}+\frac{\partial\psi}{\partial z}\hat{k}\right)$$

or, $\boldsymbol{\nabla\cdot(\phi\nabla\psi)=\phi\nabla^2\psi+\nabla\phi\cdot\nabla\psi}$ **(Proved)**.

Example 11.81. *Prove that,* $\nabla\cdot(\phi\vec{A})=\phi(\nabla\cdot\vec{A})+\vec{A}\cdot\nabla\phi$.

Solution : We have $\nabla\cdot(\phi\vec{A})=\left(\hat{i}\frac{\partial}{\partial x}+\hat{j}\frac{\partial}{\partial y}+\hat{k}\frac{\partial}{\partial z}\right)\cdot(\phi A_1\hat{i}+\phi A_2\hat{j}+\phi A_3\hat{k})$

$$=\frac{\partial}{\partial x}(\phi A_1)+\frac{\partial}{\partial y}(\phi A_2)+\frac{\partial}{\partial z}(\phi A_3)$$

$$=\left(\phi\frac{\partial A_1}{\partial x}+A_1\frac{\partial\phi}{\partial x}\right)+\left(\phi\frac{\partial A_2}{\partial y}+A_2\frac{\partial\phi}{\partial y}\right)+\left(\phi\frac{\partial A_3}{\partial z}+A_1\frac{\partial\phi}{\partial z}\right)$$

$$=\phi\left(\frac{\partial A_1}{\partial x}+\frac{\partial A_2}{\partial y}+\frac{\partial A_3}{\partial z}\right)+\left(A_1\frac{\partial\phi}{\partial x}+A_2\frac{\partial\phi}{\partial y}+A_3\frac{\partial\phi}{\partial z}\right)$$

$$=\phi(\nabla\cdot\vec{A})+\left[(\hat{i}A_1+\hat{j}A_2+\hat{k}A_3)\cdot\left(\frac{\partial\phi}{\partial x}\hat{i}+\frac{\partial\phi}{\partial y}\hat{j}+\frac{\partial\phi}{\partial z}\hat{k}\right)\right]$$

or, $\boldsymbol{\nabla\cdot(\phi\vec{A})=\phi(\nabla\cdot\vec{A})+\vec{A}\cdot(\nabla\phi)}$ **(Proved)**.

Example 11.82. *Prove that,* $(\vec{A}\cdot\nabla)\vec{A} = \frac{1}{2}\nabla A^2 - \vec{A}\times(\nabla\times\vec{A})$.

Solution : We have $\vec{A}\times(\nabla\times\vec{A}) = \nabla(\vec{A}\cdot\vec{A}) - \vec{A}(\vec{A}\cdot\nabla)$

or, $\boldsymbol{(\vec{A}\cdot\nabla)\vec{A} = \frac{1}{2}\nabla A^2 - \vec{A}\times(\nabla\times\vec{A})}$ **(Proved).**

Example 11.83. *If* $\nabla\cdot\vec{E} = 0$, $\nabla\cdot\vec{H} = 0$, $\nabla\times\vec{E} = \frac{\partial\vec{H}}{\partial t}\cdot\nabla\times\vec{H} = \frac{\partial\vec{E}}{\partial t}$, *show that* $\vec{E}$ *and* $\vec{H}$ *satisfy the wave equation,* $\nabla^2 u = \frac{\partial^2\vec{u}}{\partial t^2}$. **[BU(H) 1994]**

Solution : We have $\nabla\times(\nabla\times\vec{E}) = \nabla(\nabla\cdot\vec{E}) - \nabla^2\vec{E} = -\nabla^2\vec{E}$ (as $\nabla\cdot\vec{E} = 0$).

Again, $\nabla\times(\nabla\times\vec{E}) = \nabla\times\left(-\frac{\partial\vec{H}}{\partial t}\right) = -\frac{\partial}{\partial t}(\nabla\times\vec{H}) = -\frac{\partial}{\partial t}\left(\frac{\partial\vec{E}}{\partial t}\right) = -\frac{\partial^2\vec{E}}{\partial t^2}$

or, $\boldsymbol{\nabla^2\vec{E} = \frac{\partial^2\vec{E}}{\partial t^2}}$.

Replacing $\vec{E}$ by $\vec{H}$ in the above we can show that,

$$\boldsymbol{\nabla^2\vec{H} = \frac{\partial^2\vec{H}}{\partial t^2}}.$$

Hence, the result follows.

Example 11.84. *If* $\vec{r} = \hat{i}x + \hat{j}y + \hat{k}z$, *then find* $\nabla\cdot\nabla r^m$ *and verify that* $\nabla\times\nabla r^m = 0$.

Solution : We have $\nabla r^m = \left(\hat{i}\frac{\partial}{\partial x} + \hat{j}\frac{\partial}{\partial y} + \hat{k}\frac{\partial}{\partial z}\right) r^m = \left(\hat{i}\frac{\partial}{\partial x} + \hat{j}\frac{\partial}{\partial y} + \hat{k}\frac{\partial}{\partial z}\right)(x^2+y^2+z^2)^{\frac{m}{2}}$

$$= m[\hat{i}x + \hat{j}y + \hat{k}z][x^2+y^2+z^2]^{\frac{m}{2}-1} = mr^{m-2}(\hat{i}x + \hat{j}y + \hat{k}z).$$

$$\therefore \quad \nabla\cdot\nabla r^m = \left(\hat{i}\frac{\partial}{\partial x} + \hat{j}\frac{\partial}{\partial y} + \hat{k}\frac{\partial}{\partial z}\right)\cdot(\hat{i}x + \hat{j}y + \hat{k}z)mr^{m-2}$$

$$= \frac{\partial}{\partial x}(mxr^{m-2}) + \frac{\partial}{\partial y}(myr^{m-2} + \frac{\partial}{\partial z}(mzr^{m-2}).$$

Now, $\frac{\partial}{\partial x}(mxr^{m-2}) = mr^{m-2} + m(m-2)r^{m-3}\frac{\partial r}{\partial x}x$

$$= mr^{m-2} + m(m-2)r^{m-4}x^2 \quad \left[\because \frac{\partial r}{\partial x} = \frac{x}{r}\right].$$

Similarly, $\frac{\partial}{\partial y}(myr^{m-2}) = mr^{m-2} + m(m-2)r^{m-4}y^2$

and $\frac{\partial}{\partial y}(mzr^{m-2}) = mr^{m-2} + m(m-2)r^{m-4}z^2$.

$$\therefore \quad \nabla\cdot\nabla r^m = 3mr^{m-2} + m(m-2)r^{m-4}(x^2+y^2+z^2)$$

$$= 3mr^{m-2} + m(m-2)r^{m-2} = m(m+1)r^{m-2}.$$

Again, $\nabla\times\nabla r^m = \left(\hat{i}\frac{\partial}{\partial x} + \hat{j}\frac{\partial}{\partial y} + \hat{k}\frac{\partial}{\partial z}\right)\times[mr^{m-2}x\hat{i} + mr^{m-2}y\hat{j} + mr^{m-2}z\hat{k}]$

$$= \begin{vmatrix} \hat{i} & \hat{j} & \hat{k} \\ \dfrac{\partial}{\partial x} & \dfrac{\partial}{\partial y} & \dfrac{\partial}{\partial z} \\ mr^{m-2}x & mr^{m-2}y & mr^{m-2}z \end{vmatrix}$$

$$= \hat{i}\left[\frac{\partial}{\partial y}(mr^{m-2}z) - \frac{\partial}{\partial z}(mr^{m-2}y)\right] + \hat{j}\left[\frac{\partial}{\partial z}(mr^{m-2}x) - \frac{\partial}{\partial x}(mr^{m-r}z)\right]$$
$$+ \hat{k}\left[\frac{\partial}{\partial x}(mr^{m-2}y) - \frac{\partial}{\partial y}(mr^{m-2}x)\right]$$

$$= \hat{i}\left[m(m-2)r^{m-3}\frac{\partial r}{\partial y}z - m(m-2)r^{m-3}\frac{\partial r}{\partial z}y\right]$$
$$+ \hat{j}\left[m(m-2)r^{m-3}\frac{\partial r}{\partial z}x - m(m-2)r^{m-3}\frac{\partial r}{\partial x}z\right]$$
$$+ \hat{k}\left[m(m-2)r^{m-3}\frac{\partial r}{\partial x}y - m(m-2)r^{m-3}\frac{\partial r}{\partial y}x\right]$$

$$= m(m-2)r^{m-3}\left[\left(\frac{yz}{r} - \frac{zy}{r}\right)\hat{i} + \left(\frac{zx}{r} - \frac{xz}{r}\right)\hat{j} + \left(\frac{xy}{r} - \frac{yx}{r}\right)\hat{k}\right]$$

$$= m(m-2)r^{m-3}[0+0+0] = 0$$

Hence, it is verified that $\boldsymbol{\nabla \times \nabla r^m = 0}$.

Example 11.85. *What do you mean by an exact differential? show that a necessary and sufficient condition that, $F_1dx + F_2dy + F_3dz$ be an exact differential is that, $\nabla \times \vec{F} = 0$, where $\vec{F} = \hat{i}F_1 + \hat{j}F_2 + \hat{k}F_3$.* **[BU(H) 1998]**

Solution : If $P = P(x, y, z)$, $Q = Q(x, y, z)$, $R = R(x, y, z)$, then $Pdx + Qdy + Rdz$ is an exact differential if there exists a function $\phi(x, y, z)$, such that

$$d\phi = \frac{\partial \phi}{\partial x}dx + \frac{\partial \phi}{\partial y}dy + \frac{\partial \phi}{\partial z}dz \text{ is equal to } Pdx + Qdy + Rdz.$$

For exact differential the following condition must hold good,

$$\left(\frac{\partial Q}{\partial z} - \frac{\partial R}{\partial y}\right) = \left(\frac{\partial R}{\partial x} - \frac{\partial P}{\partial z}\right) = \left(\frac{\partial P}{\partial y} - \frac{\partial Q}{\partial x}\right) = 0.$$

The necessary condition that $F_1dx + F_2dy + F_3dz$ is an exact differential :

Let $F_1dx + F_2dy + F_3dz$ be an exact differential, then

$$F_1dx + F_2dy + F_3dz = d\phi = \frac{\partial \phi}{\partial x}dx + \frac{\partial \phi}{\partial y}dy + \frac{\partial \phi}{\partial z}dz.$$

$$\therefore \quad F_1 = \frac{\partial \phi}{\partial x} \cdot F_2 = \frac{\partial \phi}{\partial y}, \; F_3 = \frac{\partial \phi}{\partial z}.$$

$$\therefore \quad \vec{F} = \hat{i}F_1 + \hat{j}F_2 + \hat{k}F_3 = \hat{i}\frac{\partial \phi}{\partial x} + \hat{j}\frac{\partial \phi}{\partial y} + \hat{j}\frac{\partial \phi}{\partial z} = \nabla \cdot \phi.$$

$$\therefore \quad \boldsymbol{\nabla \times \vec{F} = \nabla \times \nabla \phi = 0}.$$

The sufficient condition that $F_1dx + F_2dy + F_3dz$ is an exact differential :

Let $\nabla \times \vec{F} = 0$ or, $\begin{vmatrix} \hat{i} & \hat{j} & \hat{k} \\ \frac{\partial}{\partial x} & \frac{\partial}{\partial y} & \frac{\partial}{\partial z} \\ F_1 & F_2 & F_3 \end{vmatrix} = 0$

or, $\left(\frac{\partial F_3}{\partial y} - \frac{\partial F_2}{\partial z}\right) = \left(\frac{\partial F_1}{\partial z} - \frac{\partial F_3}{\partial x}\right) = \left(\frac{\partial F_2}{\partial x} - \frac{\partial F_1}{\partial y}\right) = 0.$

Then, $\vec{F} = \nabla\psi$, so, $\vec{F} \cdot d\vec{r} = \nabla\psi \cdot d\vec{r}$.

Here $\vec{F} \cdot d\vec{r} = F_1dx + F_2dy + F_3dz$

$$\nabla\psi \cdot d\vec{r} = (\hat{i}F_1 + \hat{j}F_2 + \hat{k}F_3) \cdot (\hat{i}dx + \hat{j}dy + \hat{k}dz) = \frac{\partial\psi}{\partial x}dx + \frac{\partial\psi}{\partial y}dy + \frac{\partial\psi}{\partial z}dz$$

$= \boldsymbol{d\psi}$ **is an exact differential.**

Example 11.86. *Prove that, curl* $\left(\frac{\vec{A} \times \vec{r}}{r^3}\right) = -\frac{\vec{A}}{r^3} + \frac{3\vec{r}}{r^5}(\vec{A} \cdot \vec{r})$, *where* $\vec{A}$ *is a constant vector and* $\vec{r} = \hat{i}x + \hat{j}y + \hat{k}z$. **[BU(H) 2001]**

Solution : We have **curl** $\left(\frac{\vec{A} \times \vec{r}}{r^3}\right) = \nabla \times \left(\frac{\vec{A} \times \vec{r}}{r^3}\right) = \nabla \times [r^{-3}(\vec{A} \times \vec{r})]$

$$= r^{-3}[\nabla \times (\vec{A} \times \vec{r})] + \nabla r^{-3} \times (\vec{A} \times \vec{r})$$

$$= r^{-3}[(\nabla \cdot \vec{r})\vec{A} - (\vec{A} \cdot \nabla)\vec{r}] + \left(-3r^{-(3+1)}\frac{\vec{r}}{r}\right) \times (\vec{A} \times \vec{r})$$

$$= r^{-3}(3\vec{A} - \vec{A}) - 3r^{-(3+2)}\vec{r} \times (\vec{A} \times \vec{r}) \quad [\because \nabla \cdot \vec{r} = 3, (\vec{A} \cdot \nabla)\vec{r} = \vec{A}]$$

$$= 2\vec{A}r^{-3} - 3r^{-(3+2)}[(\vec{r} \cdot \vec{r})\vec{A} - (\vec{A} \cdot \vec{r})\vec{r}] = \frac{2}{r^3}\vec{A} - \frac{3}{r^5}[r^2\vec{A} - (\vec{A} \cdot \vec{r})\vec{r}]$$

$$= \frac{(2-3)}{r^3}\vec{A} + \frac{3(\vec{A} \cdot \vec{r})\vec{r}}{r^5} = -\frac{\vec{A}}{r^3} + \frac{3\vec{r}}{r^5}(\vec{A} \cdot \vec{r}) \quad \textbf{(Proved)}.$$

Example 11.87. *Show that,* $r^n\vec{r}$ *is solenoidal only when* $n = -3$, $(r \neq 0)$. *We know that* $\vec{A}$ *is solenoidal if* $\nabla \cdot \vec{A} = 0$. **[BU(H) 2000, 1994]**

Solution : Now $\nabla \cdot (r^n\vec{r}) = \left(\hat{i}\frac{\partial}{\partial x} + \hat{j}\frac{\partial}{\partial y} + \hat{k}\frac{\partial}{\partial z}\right) \cdot [r^n(\hat{i}x + \hat{j}y + \hat{k}z)]$

$$= \frac{\partial}{\partial x}[r^n x] + \frac{\partial}{\partial y}[r^n y] + \frac{\partial}{\partial z}[r^n y] + \frac{\partial}{\partial z}[r^n z]$$

$$= \left[x\frac{\partial r^n}{\partial x} + 1 \cdot r^n\right] + \left[y\frac{\partial r^n}{\partial y} + 1 \cdot r^3\right] + \left[z\frac{\partial r^n}{\partial z} + 1 \cdot r^n\right]$$

$$= 3r^n + \left[x\frac{\partial r^n}{\partial x} + y\frac{\partial r^n}{\partial y} + z\frac{\partial r^n}{\partial z}\right]$$

$$= 3r^n + \left[x \cdot \frac{n}{2}(x^2+y^2+z^2)^{\frac{n}{2}-1} \cdot 2x + y \cdot \frac{n}{2}(x^2+y^2+z^2)^{\frac{n}{2}-1} 2y + z \cdot \frac{n}{2}(x^2+y^2+z^2)^{\frac{n}{2}-1} \cdot 2z\right]$$

$$= 3r^n + \left[n(x^2+y^2+z^2)(x^2+y^2+z^2)^{\frac{n}{2}-1}\right]$$

$$= 3r^n + \left[n(x^2+y^2+z^2)^{\frac{n}{2}}\right] = 3r^n + nr^n = (3+n)r^n.$$

So, $\nabla \cdot (r^n \vec{r}) = 0$, if $n = -3[|\vec{r}| = r \neq 0]$

$\therefore$ **$r^n \vec{r}$ is solenoidal, if $n = -3$ (Proved).**

Example 11.88. *Show that, $\vec{F} = (2xy + z^3)\hat{i} + x^2\hat{j} + 3xz^2\hat{k}$ is a conservative force field.*

[WBUT 2003, 2004, 2006]

Solution : If $\vec{F}$ is a conservative force field then it must be irrotational i.e., $\nabla \times \vec{F} = 0$.

We have, $$\nabla \times \vec{F} = \begin{vmatrix} \hat{i} & \hat{j} & \hat{k} \\ \frac{\partial}{\partial x} & \frac{\partial}{\partial y} & \frac{\partial}{\partial z} \\ (2xy+z^3) & x^2 & 3xz^2 \end{vmatrix}$$

$$= \hat{i}\left[\frac{\partial}{\partial y}(3xz^2) - \frac{\partial}{\partial z}(x^2)\right] + \hat{j}\left[(2xy+z^3) - \frac{\partial}{\partial x}(3xz^3)\right] + \hat{k}\left[\frac{\partial}{\partial y}(x^2) - \frac{\partial}{\partial y}(2xy+z^3)\right]$$

$$= \hat{i}(0-0) + \hat{j}(3z^2 - 3z^2) + \hat{k}(2x - 2x) = 0$$

$\because$ $\nabla \times \vec{F} = 0$ $\therefore$ **$\vec{F}$ is a conservative force field.**

Example 11.89. (a) *If $\nabla \times \vec{A} = 0$ then show that, $\nabla \cdot (\vec{A} \times \vec{r}) = 0$*

[BU(H) 1996; KU 1986; MU 1984]

(b) *Prove that $\nabla \times (\vec{A} \times \vec{r}) = 2\vec{A}$,* **[AMIETE(Summer), 1997]**

(c) *Prove that $(\vec{A} \times \nabla) \times \vec{r} = -2\vec{A}$, where $\vec{r} = \hat{i}x + \hat{j}y + \hat{k}z$.* **[BU(H) 1995]**

Solution : (a) We have $\nabla \cdot (\vec{A} \times \vec{r}) = \vec{A} \cdot (\nabla \times \vec{r}) - \vec{r} \cdot (\nabla \times \vec{A})$.

Since, $$\nabla \times \vec{r} = \begin{vmatrix} \hat{i} & \hat{j} & \hat{k} \\ \frac{\partial}{\partial x} & \frac{\partial}{\partial y} & \frac{\partial}{\partial z} \\ x & y & z \end{vmatrix} = \hat{i}\left(\frac{\partial}{\partial y}z - \frac{\partial}{\partial z}y\right) + \hat{j}\left(\frac{\partial}{\partial z}x - \frac{\partial}{\partial x}z\right) + \hat{k}\left(\frac{\partial}{\partial x}y - \frac{\partial}{\partial y}x\right)$$

$$= \hat{i}0 + \hat{j}0 + \hat{k}0 = 0$$

and $\nabla \times \vec{A} = 0$.

$\therefore$ $\boldsymbol{\nabla \times (\vec{A} \times \vec{r}) = \vec{A} \cdot 0 = 0}$ **(Proved).**

(b) We have $\vec{A} \times \vec{r} = (\hat{i}A_1 + \hat{j}A_2 + \hat{k}A_3) \times (\hat{i}x + \hat{j}y + \hat{k}z) = \begin{vmatrix} \hat{i} & \hat{j} & \hat{k} \\ A_1 & A_2 & A_3 \\ x & y & z \end{vmatrix}$

$$= \hat{i}(zA_2 - yA_3) + \hat{j}(xA_3 - zA_1) + \hat{k}(A_1y - xA_2).$$

$$\therefore \quad \nabla \times (\vec{A} \times \vec{r}) = \begin{vmatrix} \hat{i} & \hat{j} & \hat{k} \\ \dfrac{\partial}{\partial x} & \dfrac{\partial}{\partial y} & \dfrac{\partial}{\partial z} \\ (zA_2 - yA_3) & (xA_3 - zA_1) & (A_1y - xA_2) \end{vmatrix}$$

$$= \hat{i}\left[\frac{\partial}{\partial y}(A_1y - xA_2) - \frac{\partial}{pz}(xA_3 - zA_1)\right] + \hat{j}\left[\frac{\partial}{\partial z}(zA_2 - yA_3) - \frac{\partial}{\partial x}(A_1y - xA_2)\right] + \hat{k}\left[\frac{\partial}{\partial x}(xA_3 - zA_1) - \frac{\partial}{\partial y}(zA_2 - yA_3)\right]$$

$$= \hat{i}(2A_1) + \hat{j}(2A_2) + \hat{k}(2A_3) = 2(\hat{i}A_1 + \hat{j}A_2 + \hat{k}A_3) = \mathbf{2}\vec{\boldsymbol{A}}$$

or, $\quad \boldsymbol{\nabla \times (\vec{A} \times \vec{r}) = 2\vec{A}}$ **(Proved).**

(c) We have $\vec{A} \times \vec{\nabla} = \begin{vmatrix} \hat{i} & \hat{j} & \hat{k} \\ A_1 & A_2 & A_3 \\ \dfrac{\partial}{\partial x} & \dfrac{\partial}{\partial y} & \dfrac{\partial}{\partial z} \end{vmatrix}$

$$= \hat{i}\left(A_2\frac{\partial}{\partial z} - A_3\frac{\partial}{\partial y}\right) + \hat{j}\left(A_3\frac{\partial}{\partial x} - A_1\frac{\partial}{\partial z}\right) + \hat{k}\left(A_1\frac{\partial}{\partial y} - A_2\frac{\partial}{\partial z}\right)$$

$$\therefore \quad (\vec{A} \times \vec{\nabla}) \times \vec{r} = \begin{vmatrix} \hat{i} & \hat{j} & \hat{k} \\ \left(A_2\dfrac{\partial}{\partial z} - A_3\dfrac{\partial}{\partial y}\right) & \left(A_3\dfrac{\partial}{\partial x} - A_1\dfrac{\partial}{\partial z}\right) & \left(A_1\dfrac{\partial}{\partial y} - A_2\dfrac{\partial}{\partial z}\right) \\ x & y & z \end{vmatrix}$$

$$= \hat{i}\left[\left(A_3\frac{\partial}{\partial x} - A_1\frac{\partial}{\partial z}\right)z - \left(A_1\frac{\partial}{\partial y} - A_2\frac{\partial}{\partial x}\right)y\right] + \hat{j}\left[\left(A_1\frac{\partial}{\partial z} - A_2\frac{\partial}{\partial x}\right)x - \left(A_2\frac{\partial}{\partial z} - A_3\frac{\partial}{\partial y}\right)z\right] + \hat{k}\left[\left(A_2\frac{\partial}{\partial z} - A_3\frac{\partial}{\partial y}\right)y - \left(A_3\frac{\partial}{\partial x} - A_1\frac{\partial}{\partial z}\right)x\right]$$

$$= \hat{i}(-2A_1) + \hat{j}(-2A_2) + \hat{k}(-2A_3) = -2(\hat{i}A_1 + \hat{j}A_2 + \hat{k}A_3) = -\mathbf{2}\vec{\boldsymbol{A}}.$$

or, $\quad \boldsymbol{(\vec{A} \times \vec{\nabla}) \times \vec{r} = -2\vec{A}}$ **(Proved).**

Example 11.90. *Prove that,* $\sum_{p=1}^{3} l_{pm}l_{pm} \begin{matrix} = 1, & \text{if } m = n \\ = 0, & \text{if } m \neq n \end{matrix}$, *where m and n can assume any of the values* 1, 2, 3.

Solution : We have $\hat{i} = l_{11}\hat{i}' + l_{21}\hat{j}' + l_{31}\hat{k}'$

$$\therefore \quad \hat{i}\cdot\hat{i} = (l_{11}\hat{i}' + l_{21}\hat{j}' + l_{31}\hat{k}')\cdot(l_{11}\hat{i}' + l_{21}\hat{j}" + l_{31}\hat{k}') = l_{11}^2 + l_{21}^2 + l_{31}^2 = l_{11}l_{11} + l_{21}l_{21} + l_{31}l_{31}$$

$$= \sum_{p=1}^{3} l_{p1}\cdot l_{p1}.$$

$$\hat{i}\cdot\hat{j} = 0 = (l_{11}\hat{i}' + l_{21}\hat{j}' + l_{31}\hat{k}')\cdot(l_{12}\hat{i}' + l_{22}\hat{j}' + l_{32}\hat{k}') = l_{11}l_{12} + l_{21}l_{22} + l_{31}l_{32}$$

$$= \sum_{p=1}^{3} l_{p1}l_{p2}.$$

$$\hat{i}\cdot\hat{k} = 0 = (l_{11}\hat{i}' + l_{21}\hat{j}' + l_{31}\hat{k}')\cdot(l_{13}\hat{i}' + l_{23}\hat{j}' + l_{33}\hat{k}') = l_{11}l_{13} + l_{21}l_{23} + l_{31}l_{33}$$

$$= \sum_{p=1}^{3} l_{p1}l_{p3}.$$

$$\therefore \quad \sum_{p=1}^{3} l_{pm}l_{pn} \left.\begin{matrix} = 1, & \text{if } m = n \\ = 0, & \text{if } m \neq n \end{matrix}\right\}$$

$\therefore \quad \sum_{p=1}^{3} l_{pm}l_{pn} = \delta_{mn};$ **δ_{mn} is called the Kronecker's delta. (Proved)**

Example 11.91. *If $\phi(x, y, z)$ is a scalar invariant with respect to rotation axes, prove that grand ϕ is a vector invariant under this transformation.*

Solution : According to the condition of the problem, $\phi(x, y, z) = \phi'(x', y', z')$.

We have to prove that, $\nabla\phi = \nabla'\phi'$ or, $\hat{i}\frac{\partial\phi}{\partial x} + \hat{j}\frac{\partial\phi}{\partial y} + \hat{k}\frac{\partial\phi}{\partial z} = \hat{i}'\frac{\partial\phi'}{\partial x'} + \hat{j}'\frac{\partial\phi'}{\partial y'} + \hat{k}'\frac{\partial\phi'}{\partial z'}$.

$$\text{Now,} \quad \frac{\partial\phi}{\partial x} = \frac{\partial\phi}{\partial x'}\frac{\partial x'}{\partial x} + \frac{\partial\phi}{\partial y'}\frac{\partial y'}{\partial x} + \frac{\partial\phi}{\partial z'}\frac{\partial z'}{\partial x}$$

$$\because \quad x' = l_{11}x + l_{12}y + l_{13}z, \quad y' = l_{21}x + l_{22}y + l_{23}z, \quad z' = l_{31}x + l_{32}y + l_{33}z.$$

$$\therefore \quad \frac{\partial x'}{\partial x} = l_{11}, \quad \text{Similarly,} \ \frac{\partial y'}{\partial x} = l_{21}, \ \frac{\partial z'}{\partial x} = l_{31}$$

$$\frac{\partial\phi}{\partial x} = \frac{\partial\phi'}{\partial x'}l_{11} + \frac{\partial\phi'}{\partial y'}l_{21} + \frac{\partial\phi'}{\partial z'}l_{31}$$

$$\frac{\partial\phi}{\partial y} = \frac{\partial\phi'}{\partial x'}\frac{\partial x'}{\partial y} + \frac{\partial\phi'}{\partial y'}\frac{\partial y'}{\partial y} + \frac{\partial\phi'}{\partial z'}\frac{\partial z'}{\partial y} = \frac{\partial\phi'}{\partial x'}l_{12} + \frac{\partial\phi'}{\partial h'}l_{22} + \frac{\partial\phi'}{\partial z'}l_{32}$$

$$\text{and} \quad \frac{\partial\phi}{\partial z} = \frac{\partial\phi'}{\partial x'}\frac{\partial x'}{\partial z} + \frac{\partial\phi'}{\partial y'}\frac{\partial y'}{\partial z} + \frac{\partial\phi'}{\partial z'}\frac{\partial z'}{\partial z} = \frac{\partial\phi'}{\partial x'}l_{13} + \frac{\partial\phi'}{\partial y'}l_{23} + \frac{\partial\phi'}{\partial z'}l_{33}.$$

$$\therefore \quad \hat{i}\frac{\partial \phi}{\partial x}+\hat{j}\frac{\partial \phi}{\partial y}+\hat{k}\frac{\partial \phi}{\partial z}=\hat{i}\left(\frac{\partial \phi'}{\partial x'}l_{11}+\frac{\partial \phi'}{\partial y'}l_{21}+\frac{\partial \phi'}{\partial z'}l_{31}\right)+\hat{j}\left(\frac{\partial \phi'}{\partial x'}l_{12}+\frac{\partial \phi'}{\partial y'}l_{22}+\frac{\partial \phi'}{\partial z'}l_{32}\right)$$
$$+\hat{k}\left(\frac{\partial \phi'}{\partial x'}l_{13}+\frac{\partial \phi'}{\partial y'}l_{23}+\frac{\partial \phi'}{\partial z'}l_{33}\right)$$
$$=\frac{\partial \phi'}{\partial x'}(\hat{i}l_{11}+\hat{j}l_{12}+\hat{k}l_{13})+\frac{\partial \phi'}{\partial y'}(\hat{i}l_{21}+\hat{j}l_{22}+\hat{k}l_{23})$$
$$+\frac{\partial \phi'}{\partial z'}(\hat{i}l_{31}+\hat{j}l_{32}+\hat{k}l_{33})$$
$$=\frac{\partial \phi'}{\partial x'}\hat{i}'+\frac{\partial \phi'}{\partial y'}\hat{j}'+\frac{\partial \phi'}{\partial z'}\hat{k}'=\nabla'\phi'.$$

$\therefore$ $\boldsymbol{\nabla\phi = \nabla'\phi'}$. **$\therefore$ $\nabla\phi$ is a vector invariant under transformation. (Proved)**

Exercises

11(A) : Objective Type Questions

11.1 If $\vec{A}$ is a constant vector then $\frac{d\vec{A}}{dt}$ is

(a) parallel to $\vec{A}$ **(b)** perpendicular to $\vec{A}$ **(c)** $m\vec{A}$ **(d)** $\frac{\vec{A}}{m}$.

11.2 If $\vec{A}=\hat{i}e^{-t}+\hat{j}\ln(t^2+1)-\hat{k}\tan t$, then $\left|\frac{d^2\vec{A}}{dt^2}\right|$ at $t=0$ is

(a) $\sqrt{4\cdot 5}$ **(b)** $\sqrt{8}$ **(c)** $\sqrt{5}$ **(d)** $(2\cdot 6)^{\frac{1}{3}}$.

11.3 The unit normal vector to the surface $z=x^2+y^2$ at the point $(1,1,2)$ is

(a) $2\hat{i}+2\hat{j}-\hat{k}$ **(b)** $2\hat{i}+2\hat{j}+\hat{k}$ **(c)** $(\hat{i}+\hat{j}+2\hat{k})|\sqrt{6}$ **(d)** $(2\hat{i}+2\hat{j}-\hat{k})|3$.

[AMIE(Winter) 1990]

11.4 The angle between the tangents to the curve $\vec{r}=\hat{i}t^2+\hat{j}2t-t^3\hat{k}$ at the point $t=\pm 1$ is

(a) $\cos^{-1}\left(\frac{9}{17}\right)$ **(b)** $\cos^{-1}\left(\frac{1}{3}\right)$ **(c)** $\cos^{-1}\left(\frac{2}{5}\right)$ **(d)** $\cos^{-1}\left(\frac{8}{9}\right)$.

11.5 A particle moves along a curve $x=2t^2, y=t^2-4t$ and $z=3t-5$, where t is the time. The velocity of the particle in the direction $\hat{i}-3\hat{j}+2\hat{k}$ at $t=1$ is

(a) $\sqrt{149/9}$ **(b)** $\frac{8\sqrt{14}}{7}$ **(c)** $\sqrt{7/3}$ **(d)** $\sqrt{\frac{15}{13}}$.

11.6 If $\vec{P}=\hat{i}\cos xy+\hat{j}(3xy-2x^2)-\hat{k}(3x+2y)$, then $\frac{\partial^2\vec{P}}{\partial y\partial x}$ is

(a) $(xy\sin xy+\cos xy)\hat{i}-2\hat{j}$ **(b)** $(\cos xy+\sin xy)\hat{i}+2\hat{j}$

(c) $-(xy\cos xy+\sin xy)\hat{i}+3\hat{j}$ **(d)** $\hat{i}e^{xy}+\hat{j}e^{-xy}$.

11.7 If $\vec{F}=\hat{i}x^2yz-\hat{j}2xz^3+xz^2\hat{k}$ and $\vec{G}=\hat{i}2z+\hat{j}y-\hat{k}x^2$, then $\frac{\partial^2}{\partial x\partial y}(\vec{F}\times\vec{G})$ at $(1,0,-2)$ is

(a) $4\hat{i}+8\hat{j}$ **(b)** $3\hat{i}+5\hat{j}$ **(c)** $6\hat{i}+13\hat{j}$ **(d)** $-(4\hat{i}+8\hat{j})$.

11.8 If $\vec{r} = \hat{i}x + \hat{j}y + \hat{k}z$, then, $\nabla \cdot \vec{r}$ is equal to

(a) 2 (b) 3 (c) 5 (d) 1. **[cf. WBUT 2004]**

11.9 If $r^2 = x^2 + y^2 + z^2$, then $\nabla \cdot \left(\frac{\vec{r}}{r}\right)$ is equal to

(a) $\frac{2}{r}$ (b) r^2 (c) 0 (d) $3r$.

11.10 If $\phi = \ln(x^2 + y^2 + z^2)$, then $\nabla\phi$ is equal to

(a) $\frac{2(\hat{i}x + \hat{j}y + \hat{k}z)}{(x^2 + y^2 + z^2)}$ (b) $(\hat{i}x + \hat{j}y + \hat{k}z)$ (c) $\frac{1}{(x^2 + y^2 + z^2)}$ (d) $\frac{\hat{i}x + \hat{j}y + \hat{k}z}{(x^2 + y^2 + z^2)}$.

11.11 If $\phi(x, y, z) = 2xy + z^2$, then the directional derivative of ϕ at $(1, -1, 3)$ in the direction $\hat{i} + 2\hat{j} + 2\hat{k}$ is

(a) $\frac{11}{3}$ (b) $\frac{14}{3}$ (c) $\frac{3}{14}$ (d) $\frac{9}{14}$. **[PU 1984, AU 1986, KU 1990]**

11.12 If $\vec{A}$ is a constant vector, then $\nabla(\vec{r} \cdot \vec{A})$ is equal to

(a) $3\vec{A}$ (b) $2\vec{A}$ (c) $(\vec{r} \cdot \vec{A})$ (d) $\vec{A}$.

11.13 If $\vec{A} = \frac{\vec{r}}{r}$, then $\nabla(\nabla \cdot \vec{A})$ is equal to

(a) 5 (b) 2 (c) 9 (d) 6.

11.14 If $\nabla \cdot \vec{A} = 0$, then $\vec{A}$ is

(a) irrotational (b) solenoidal (c) axial (d) rotational.

11.15 If ϕ and ψ are differentiable scalar functions, then $\nabla\phi \times \nabla\psi$ is

(a) solenoidal (b) rotational (c) not solenoidal (d) axial.

11.16 If ϕ is a continuously differentiable function of x, y, z, then $\nabla \times (\phi\nabla\phi)$ is equal to

(a) 3 (b) ∞ (c) 0 (d) 6.

11.17 If $\nabla \times \vec{A} = 0$, then $\vec{A}$ is

(a) rotational (b) irrotational (c) a constant vector (d) solenoidal.

11.18 $f(r)$ is differentiable, then $f(r)\vec{r}$ is

(a) solenoidal (b) rotational (c) irrotational (d) neither of these.

11.19 If $\vec{A} = \hat{i}e^x \sin y + \hat{j}e^x \cos y$, then $\vec{A}$ is

(a) rotational (b) solenoidal (c) rotational in part of space (d) irrotational.

[AMIETE(Winter) 1994]

11.20 The value of m for which $\vec{A} = \hat{i}m(x + y) + \hat{j}4y + \hat{k}3$ is solenoidal is equal to

(a) 4 (b) -4 (c) 0 (a) -3. **[AMIETE(Winter) 1991]**

11.21 If $\vec{A}$ is a constant vector, then $\nabla \times (\vec{A} \cdot \vec{r})\vec{A}$ is equal to

(a) $\vec{A}$ (b) $(\vec{A} \cdot \vec{r})$ (c) $\nabla \cdot \vec{A}$ (d) 0.

11.22 If $\vec{A}$ is solenoidal, then $\nabla \times \nabla \times \nabla \times \nabla \times \vec{A}$ is equal to

(a) $\nabla^4 \vec{A}$ (b) $\nabla^3 \vec{A}$ (c) $\nabla(\nabla \times \vec{A})$ (d) $\nabla \vec{A}$.

11.23 If $\nabla \times \vec{A} = 0$, then $\nabla \cdot (\vec{A} \times \vec{r})$ is equal to

(a) 0 (b) 3 (c) 2 (d) $\vec{A} \cdot \vec{r}$.

11.24 If $\vec{A}$ and $\vec{b}$ are irrotational, then $\vec{A} \times \vec{B}$ is

(a) rotational (b) irrotational (c) solenoidal (d) neither of these.

11.25 The vector $r^n \vec{r}$ is solenoidal, if

(a) $n = 3$ (b) $n = 1$ (c) $n = -3$ (d) $n = 0$.

Answers

11.1 (b) **11.2** (c) **11.3** (d) **11.4** (a) **11.5** (c) **11.6** (c) **11.7** (d) **11.8** (b) **11.9** (a) **11.10** (a) **11.11** (b) **11.12** (a) **11.13** (d) **11.14** (b) **11.15** (a) **11.16** (c) **11.17** (b) **11.18** (c) **11.19** (d) **11.20** (b) **11.21** (d) **11.22** (a) **11.23** (a) **11.24** (c) **11.25** (c).

11(B) Short Answer Type Questions

11.1 Find the velocity and acceleration of a particle moving in a curve whose equation is $\vec{r} = \hat{i}e^{-t} + \hat{j}2\cos 3t + \hat{k}2\sin 3t$ at time $t = 1$.

[**Ans:** $\vec{v} = -\frac{\hat{i}}{e} - \hat{j}6\sin 3 + \hat{k}6\cos 3$, $\vec{a} = \frac{\hat{i}}{e} - \hat{j}18\cos 3 - \hat{k}18\sin 3$]

11.2 If $\vec{A} = \hat{i}t^2 - \hat{j}t + \hat{k}(2t+1)$ and $\vec{B} = \hat{i}(2t-3) + \hat{j} - \hat{k}t$, then find $\frac{d}{dt}(\vec{A} \times \vec{B})$ at $t = 1$.

[**Ans:** $(7\hat{j} + 3\hat{k})$]

11.3 Show that, $(\vec{A} \times \vec{\nabla}) \times \vec{r} = -2\vec{A}$, where $\vec{r} = \hat{i}x + \hat{j}y + \hat{k}z$. [**BU(H) 1995**]

11.4 If $\nabla \times \vec{A} = \dfrac{\partial \vec{B}}{\partial t}$, the show that $\nabla \cdot \vec{B}$ is independent of t. [**BU(H) 1999**]

11.5 Find a unit vector perpendicular to the surface $x^2 + y^2 - z = 11$ at the point $(1, -1, 2)$. [**BU(H) 1997**]

11.6 If $\vec{A}$ and $\vec{B}$ are irrotational, show that $\vec{A} \times \vec{B}$ is solenoidal. [**WBUT 2006; BU(H) 2002**]

11.7 If $u\vec{A} = \vec{\nabla}_v$, where u and v are scalar fields and $\vec{A}$ is a vector field, prove that,

$$\vec{A} \cdot \vec{\nabla} \times \vec{A} = 0.$$ [**BU(H) 2002**]

11.8 If $\phi(x, y, z) = 3(x^2 y - y^2 z)$, calculate grad ϕ at the point $(1, -2, 1)$. [**WBUT 2005**]

11.9 If $\vec{r} = \sqrt{x^2 + y^2 + z^2}$, then find $\nabla^2 \left(\dfrac{1}{r}\right)$. Is the result you obtain for $r = 0$ valid? [**BU(H) 1998**]

11.10 Find $\nabla \cdot \vec{A}$ and $\nabla \times \vec{A}$, if $\vec{A} = \dfrac{\hat{i}x + \hat{j}y + \hat{k}z}{\sqrt{x^2 + y^2 + z^2}}$. [**Ans:** 0]

11.11 If $\vec{A}$ is a constant vector, show that, $\vec{A} \times (\nabla \times \vec{r}) = \nabla(\vec{A} \cdot \vec{r}) - (\vec{A} \cdot \nabla)\vec{r}$.

11.12 If $\vec{A}$ is a constant vector, then prove that $\dfrac{\nabla \times (\vec{A} \times \vec{r})}{r^3} = -\dfrac{\vec{A}}{r^3} + \dfrac{3\vec{r}}{r^5}(\vec{A} \cdot \vec{r})$.

[CU Math(H) 1999]

11.13 The motion of a particle can be represented in terms of the following parametric equation, $x = 5t^2 - 7$, $y = 7\cos t$ and $z = 3\sin t$. Find the magnitude of instantaneous velocity and acceleration of the particle.

[HPU 1986] [**Ans:** $|v| = 9 + 40\sin^2 t + 100t^2$, $|a| = \sqrt{109 + 40\cos^2 t}$]

11.14 Find the maximum value of the directional derivative of $\phi = 2xy - y^2$ at $(1, 2, 3)$. Find the direction in which this occurs. [**Ans:** $2\sqrt{14}$, $6\hat{i} - 4\hat{j} + 2\hat{k}$]

11.15 What is the greatest rate of increase of $\phi = x^2 + yz^2$ at the point $(1, -1, 3)$?

[MU 1983] [**Ans:** $2\hat{i} + 9\hat{j} - 6\hat{k}$]

11.16 In what direction of $(3, 1, -2)$ is the directional derivative of $\phi = x^2y^2z^4$ maximum? Find the magnitude of this maximum.

[BU 1991, KU 1987] [**Ans:** $96(\hat{i} + 3\hat{j} - 3\hat{k})$; $96\sqrt{19}$]

11.17 If $\vec{A} = \nabla(x^3 + y^3 + z^3 - 3xyz)$, find $\nabla \cdot \vec{A}$ and $\nabla \times \vec{A}$. **[MU 1990, KU 1987]**

11.18 Show that for a plane curve, the torsion $\tau = 0$.

11.19 Prove that for a vector field $\vec{A}$ and scalar field ϕ, $\nabla \cdot (\phi\vec{A}) = \nabla\phi \cdot \vec{A} + \phi\nabla \cdot \vec{A}$.

[CU(H) 2012, VU(H) 2010]

11.20 Show that, $\nabla \times (\phi\vec{A}) = \nabla\phi \times \vec{A} + \phi\nabla \times \vec{A}$ for a scalar field ϕ and a vector field $\vec{A}$.

[CU(H) 2014]

11.21 The position vector of a particle at time t is given by $\vec{r}(t) = (6 - 2t)\hat{i} + (3 + 4t - 6t^2)\hat{j} - (4 + 3t - 2t^2)\hat{k}$ with distances given in metres and time in seconds. Find velocity and speed of the particle at time $t = 3$ sec. **[CU(H) 2014]**

11.22 If the magnitude of a vector field $\vec{A}(t)$ is constant with respect to time, show that $\dfrac{d\vec{A}}{dt}$ is perpendicular to $\vec{A}$. **[CU(H) 2010, 2013]**

11.23 For two sets of vectors $(\vec{a}, \vec{b}, \vec{c})$ and $(\vec{a'}, \vec{b'}, \vec{c'})$ if $\vec{a} \cdot \vec{a} = 1$, and $\vec{b} \cdot \vec{a'} = \vec{c} \cdot \vec{a'} = 0$ then show that $\vec{a'} = \dfrac{\vec{b} \times \vec{c}}{\vec{a} \cdot (\vec{b} \times \vec{c})}$. **[CU(H) 2011]**

11.24 A particle moves along a curve $x = 2t^2$, $y = t^2 - 4t$ and $z = 3t - 5$ where t is time. Find its component of velocity at time $t = 1$ in the direction $\hat{i} - 2\hat{j} + 2\hat{k}$.

[CU(H) 2008, VU(H) 2008]

11.25 Find a unit vector perpendicular to the surface $x^2y + 2xz = 4$ at the point $(2, -2, 3)$.

[VU(H) 2009]

11.26 A surface in three dimension is described by $\phi(x, y, z) = c$ where c is a constant. Show that $\nabla\phi = 0$ on this surface. **[CU(H) 2008]**

11.27 Verify whether the force $\vec{F} = (3x - 4y - 3z^2)\hat{i} + (4x + 2y + 2z)\hat{j} + (xz - 4y^2 + 2x^3)\hat{k}$ is conservative. **[VU(H) 2014]**

11.28 The coordinates of a moving point mass are given by $\gamma = 2t + 3$ and $\theta = 3t - t^2$ at any instant t. Calculate the instantaneous radial and transverse components of velocity of the point mass. **[NBU(H) 2007]**

11.29 A vector field is given by $\vec{F} = 2xz^2\hat{i} + 2yz^2\hat{j} + (2x^2z + 2y^2z - 1)\hat{k}$. Show that $\vec{F}$ is a conservative field. Find the corresponding scalar function. **[BU(H) 2013]**

11.30 If $\vec{r} = \hat{i}x + \hat{j}y + \hat{k}z$ evaluate the following quantities **(i)** $\nabla(\ln r)$, **(ii)** $\nabla^2 \ln r$. **[BU(H) 2013]**

11(C) Broad Answer Type Questions

11.1 A particle moves along the curve, $\vec{r} = \hat{i}(t^3 - 4t) + \hat{j}(t^2 + 4t) + \hat{k}(8t^2 - 3t^3)$, where t is time. Find the magnitudes of the tangential and normal component of its acceleration for $t = 2$. **[Ans:** Tangential, 16; normal, $2\sqrt{73}$**]**

11.2 A particle of constant mass describes the parabola $r = a\sec^2(\theta/2)$ such that the cross-radial component of the velocity is constant. Show that $\frac{d^2r}{dt^2}$ is constant. **[CU 1984]**

11.3 A particle moves with a velocity $\vec{v}$ and acceleration $\vec{a}$ along a space curve. Prove that the curvature of the path is given by $\kappa = \frac{|\vec{v} \times \vec{a}|}{v^3}$.

11.4 **(a)** Define vector field. Give an example.

(b) What do you mean by a scalar field?

If $\phi(x, y, z) = 3x^2y - y^3z^2$, find $\nabla\phi$ at the point $(1, -2, 1)$. **[WBUT 2004]**

11.5 A particle moves in a field of force given by

$$\vec{F} = \hat{i}2xz(1 - 6xyz) + \hat{j}2xz(1 - 6xyz) + \hat{k}2xy(1 - 6xyz).$$

Can you define a potential function $V(x, y, z)$ for this force field? **[BU(H) 2004]**

11.6 **(a)** What do you mean by gradient of a scalar field?

(b) What is its physical significance?

(c) Find the gradient of $\phi = 3x^2 + 2y^2 + z^2 = 3$ at $(2, 1, 3)$.

(d) What is directional derivative? Find the directional derivative of ϕ at $(2, 1, 3)$ in the direction $2\hat{i} + \hat{j} - 3\hat{k}$.

11.7 Prove that, $\nabla \times (\nabla \times \vec{A}) = \nabla(\nabla \cdot \vec{A}) - \nabla^2\vec{A}$. **[BU(H) 1999, 1992]**

11.8 Find the values of the constants a, b, c so that the directional derivative of $\phi = axy^2 + byz + cz^2x^3$ at $(1, 2, -1)$ has a maximum magnitude 64 in a direction parallel to the z-axis. **[Ans:** $a = 6$, $b = 24$, $c = -8$**]**

11.9 Find the value of $\nabla \cdot \left[r\nabla\left(\frac{1}{r^3}\right)\right]$. **[Ans:** $3r^{-4}$**]**

11.10 Prove that, $\nabla(\vec{A} \cdot \vec{B}) = (\vec{B} \cdot \nabla)\vec{A} + (\vec{A} \cdot \nabla)\vec{B} + \vec{B} \times (\nabla \times \vec{A}) + \vec{A} \times (\nabla \times \vec{B})$.

11.11 If $\phi(x, y, z)$ is any solution of Laplace's equation, then show that $\nabla\phi$ is a vector which is both solenoidal and irrotational.

11.12 For a particle moving in a circular path $\vec{r} = \hat{i}r\cos\omega t + \hat{j}r\sin\omega t$,

(a) Evaluate $\vec{r} \times \dot{\vec{r}}$. **[Ans.** $\hat{k}\omega r^2$**]**

(b) Show that $\ddot{\vec{r}} + \omega^2\vec{r} = 0$, where the radius r and the angular velocity ω are constant.

11.13 If $\vec{A} = \hat{i}3xyz^2 + \hat{j}2xy^3 - \hat{k}x^2yz$ and $\phi = 3x^2 - yz$, find :

(a) $\nabla \cdot \vec{A}$ (b) $\vec{A} \cdot \nabla\phi$ (c) $\nabla \cdot (\phi\vec{A})$ (d) $\nabla \cdot \nabla\phi$ at $(1, -1, 1)$.

[**Ans:** (a) 4, (b) -15, (c) 1, (d) 6]

11.14 Prove that

(a) $\nabla \times (\vec{A} \times \vec{B}) = (\vec{B} \cdot \nabla)\vec{A} - \vec{B}(\nabla \cdot \vec{A}) - (\vec{A} \cdot \nabla)\vec{B} + \vec{A}(\nabla \cdot \vec{B})$

(b) $\nabla \cdot (\vec{A} \times \vec{B}) = \vec{B} \cdot (\nabla \times \vec{A}) - \vec{A} \cdot (\nabla \times \vec{B})$.

11.15 Show that the solutions of Maxwell's equations,

$$\nabla \times \vec{H} = \frac{1}{c}\frac{\partial \vec{E}}{\partial t}$$

$$\nabla \times \vec{E} = -\frac{1}{c}\frac{\partial \vec{H}}{\partial t}$$

$$\nabla \cdot \vec{H} = 0$$

$$\nabla \cdot \vec{E} = 4\pi\rho$$

where ρ is a function of (x, y, z) and c is the velocity of light assumed constant, are given by

$$\vec{E} = -\nabla\phi - \frac{1}{c}\frac{\partial \vec{A}}{\partial t}, \quad \vec{H} = \nabla \times \vec{A},$$

where $\vec{A}$ and ϕ called the *vector and scalar potentials* respectively, satisfy the equations :

$$\nabla \cdot \vec{A} + \frac{1}{c}\frac{\partial \phi}{\partial t} = 0$$

$$\nabla^2\phi - \frac{1}{c^2}\frac{\partial^2 \phi}{\partial t^2} = 4\pi\rho$$

$$\nabla^2\vec{A} = \frac{1}{c^2}\frac{\partial^2 \vec{A}}{\partial t^2}.$$

11.16 Show that the force field $\vec{F}$ defined by $\vec{F} = \hat{i}(y^2z^3 - 6xz^2) + \hat{j}2xyz^3 + \hat{k}(3xy^2z^2 - 6x^2z)$ is a conservative force field.

Solution : The force field $\vec{F}$ is conservative if and only if there exists a scalar function $\phi(x, y, z)$ such that,

$$\vec{F} = -\nabla\phi.$$

Then, $\vec{F} = \hat{i}(y^2z^3 - 6xz^2) + \hat{j}(2xyz^3) + \hat{k}(3xy^2z^2 - 6x^2z) = -\hat{i}\dfrac{\partial \phi}{\partial x} - \hat{j}\dfrac{\partial \phi}{\partial y} - \hat{k}\dfrac{\partial \phi}{\partial z}$

$\therefore \quad \dfrac{\partial \phi}{\partial x} = 6xz^2 - y^2z^3, \quad \dfrac{\partial \phi}{\partial y} = -2xyz^3, \quad \dfrac{\partial \phi}{\partial z} = 6x^2z - 3x^2yz^2.$

Integrating $\frac{\partial \phi}{\partial x} = 6xz^2 - y^2z^3$ keeping y and z constant, we get

$$\phi = 3x^2z^2 - xy^2z^3 + g(y, z).$$

Similarly, $\phi = -xy^2z^3 + h(x, z)$ and $\phi = 3x^2z^2 - xy^2z^3 + k(x, y)$.

If we choose $g(y,z) = c$, $h(x,z) = 3x^2z^2 + c$ and $k(x,y) = C$, then there will be a common value of

$$\phi = 3x^2z^2 - xy^2z^3 + C, \quad \text{where } C \text{ is an arbitrary constant.}$$

Since this ϕ satisfies $\vec{F} = -\nabla\phi$, $\vec{F}$ is conservative.

Alternatively : $\vec{F}$ is conservative if $\nabla \times \vec{F} = 0$. In the present case, $\nabla \times \vec{F} = 0$,

$\therefore$ $\vec{F}$ is conservative.

11.17 Show that the force of attraction between two particles of masses m_1 and m_2 separated by a distance $\vec{r}$ is conservative.

Solution : We have the force of attraction between m_1 and m_2 separated by $|\vec{r}|$ is

$$\vec{F} = -G\frac{m_1m_2}{r_2}, \quad \text{where } G \text{ is the gravitational constant.}$$

In cartesian coordinates, $r^2 = x^2 + y^2 + z^2$.

$$\therefore \quad \frac{\partial r}{\partial x} = \frac{x}{r}, \quad \frac{\partial r}{\partial y} = \frac{y}{r}, \quad \frac{\partial r}{\partial z} = \frac{z}{r}.$$

Now, $\quad F^2 = F_1^2 + F_2^2 + F_z^2, \quad \therefore \quad \dfrac{\partial F}{\partial F_1} = \dfrac{F_1}{F} = \dfrac{x}{r}.$

Since $\vec{F}$ is along $\vec{r}$,

$$\therefore \quad F_1 = \frac{xF}{r} = -Gm_1m_2\left(\frac{x}{r^3}\right) = -k\left(\frac{x}{r^3}\right), \quad \text{where } k = Gm_1m_2.$$

Similarly, $F_2 = -k\left(\dfrac{y}{r^3}\right)$ and $F_3 = -k\left(\dfrac{z}{r^3}\right)$.

$$\text{Now,} \quad \nabla \times \vec{F} = \begin{vmatrix} \hat{i} & \hat{j} & \hat{k} \\ \dfrac{\partial}{\partial x} & \dfrac{\partial}{\partial y} & \dfrac{\partial}{\partial z} \\ -k\left(\dfrac{x}{r^3}\right) & -k\left(\dfrac{y}{r^3}\right) & -k\left(\dfrac{z}{r^3}\right) \end{vmatrix}$$

$$= \hat{i}\left[\frac{\partial}{\partial x}\left(-\frac{kz}{r^3}\right) - \frac{\partial}{\partial z}\left(-\frac{ky}{r^3}\right)\right] + \hat{j}\left[\frac{\partial}{\partial z}\left(-\frac{kx}{r^3}\right) - \frac{\partial}{\partial x}\left(-\frac{kz}{r^3}\right)\right]$$
$$+ \hat{k}\left[\frac{\partial}{px}\left(-\frac{ky}{r^3}\right) - \frac{\partial}{\partial y}\left(-\frac{kx}{r^3}\right)\right].$$

$$\text{Now} \quad \frac{\partial}{\partial x}\left(-\frac{ky}{r^3}\right) = -ky\frac{\partial}{\partial x}(x^2+y^2+z^2)^{-\frac{3}{2}} = ky\frac{3x}{r^5}.$$

$$\text{Likewise,} \quad \frac{\partial}{\partial y}\left(-\frac{ky}{r^3}\right) = kz\frac{3y}{r^5}, \text{ etc.}$$

$\therefore \quad \nabla \times \vec{F} = 0$ and hence $\vec{F}$ is conservative.

11.18 If $\hat{n}$ be a unit vector, then show that,

(a) $\nabla \cdot (\hat{n} \cdot \vec{r})\hat{n} = 1$ **(b)** $\nabla \times (\hat{n} \cdot \vec{r})\hat{n} = 0$

(c) $\nabla \cdot [(\hat{n} \times \vec{r}) \times \hat{n}] = 2$ **(d)** $\nabla \times [(\hat{n} \times \vec{r}) \times \hat{n}] = 0.$

11.19 In electromagnetic theory, we have

$$\nabla \cdot \vec{D} = \rho, \quad \nabla \cdot \vec{H} = 0, \quad \nabla \times \vec{D} = -\frac{1}{c}\frac{\partial \vec{H}}{\partial t}, \quad \nabla \times \vec{H} = \frac{1}{c}\left(\rho\vec{v} + \frac{\partial \vec{D}}{\partial t}\right).$$

Prove that, $\nabla^2\vec{D} - \frac{1}{c^2}\frac{\partial^2 \vec{D}}{\partial t^2} = \nabla\rho + \frac{1}{c^2}\frac{\partial}{\partial t}(\rho\vec{v})$

and $\nabla^2\vec{H} - \frac{1}{c^2}\frac{\partial^2\vec{H}}{\partial t^2} = -\frac{1}{c}\nabla \times (\rho\vec{v})$ **[RU 1990]**

[**Hint** : $\nabla \times (\nabla \times \vec{V}) = \nabla(\nabla \cdot \vec{V}) - \nabla^2\vec{V}$]

11.20 If $\vec{A}$ is a constant vector, then prove that,

(a) $\nabla A^2 = 2(\vec{A} \cdot \nabla)\vec{A} + 2\vec{A} \times (\nabla \times \vec{A})$

(b) $\nabla \times (\vec{R} \times \vec{V}) = \vec{R}(\nabla \cdot \vec{V}) - 2\vec{V} - (\vec{R} \cdot \nabla)\vec{V}$.

11.21 Find the value of a if the vector $[\hat{i}(ax^2y + yz) + \hat{j}(xy^2 - xz^2) + \hat{k}(2xyz - 2x^2y^2)]$ has zero divergence. Find the curl of the above vector which has zero divergence.

$[a = -2, \hat{i}4x(z - xy) + \hat{j}(y - 2yz + 4xy^2) + \hat{k}(2x^2 + y^2 - z^2 - 2)]$ **[AU 1990S]**

(a) Find a unit vector perpendicular to the surface $(x-2)^2 + 5y^2 + 2z^2 = 8$ at the point $(1, 1, 1)$.

(b) If $\nabla \times \vec{A} = 0$, show that $\nabla \cdot (\vec{A} \times \vec{r}) = 0$ **[CU(H) 2012]**

11.22 **(a)** Define conservative force. Show that for a conservative force $\vec{F}$,

(i) $\vec{F} = -\nabla\phi$, $\phi = \phi(x, y, z)$ and **(ii)** $\nabla \times \vec{F} = 0$.

(b) Show that for a system of particles, angular momentum about a point is equal to the angular momentum of a single particle of total mass $\mu(= \sum m_i)$ situated at the centre of mass together with the angular momentum of the system of particles about the centre of mass.

(c) A particle of mass m moves in the X-Y plane so that its position vector is, $\vec{r} = a\cos\omega t\hat{i} + b\sin\omega t\hat{j}$ where a, b and ω are constants. Show that the force acting on the particle is always directed towards the origin.

[CU(H) 2010, BU(H) 2000]

11.23 If the points P, Q and R not all lying in the same straight line, have position vectors $\vec{a}$, $\vec{b}$ and $\vec{c}$ relative to a given origin. Show that, $\vec{a} \times \vec{b} + \vec{b} \times \vec{c} + \vec{c} \times \vec{a}$ is a vector perpendicular to the plane containing P, Q and R. **[VU(H) 2014]**

11.24 **(a)** A fluid motion is given by $\vec{v} = (y + z)\hat{i} + (z + x)\hat{j} + (x + y)\hat{k}$.

Is this motion irrotational? If so find the velocity potential.

(b) Find the unit vector perpendicular to the surface $x^2 + y^2 + z^2 = 3$ at point $(1, 1, 1)$.

[VU(H) 2012]

11.25 **(a)** The position vectors $\vec{r}_1$ and $\vec{r}_2$ have direction cosines (l_1, m_1, n_1) and (l_2, m_2, n_2) respectively. Find the angle between $\vec{r}_1$ and $\vec{r}_2$.

(b) If $\vec{a'} = \dfrac{\vec{b} \times \vec{c}}{\vec{a} \cdot \vec{b} \times \vec{c}}$, $\vec{b'} = \dfrac{\vec{c} \times \vec{a}}{\vec{a} \cdot \vec{b} \times \vec{c}}$, $\vec{c'} = \dfrac{\vec{a} \times \vec{b}}{\vec{a} \cdot \vec{b} \times \vec{c}}$ and then $\vec{a} \cdot \vec{b} \times \vec{c} = V$, show that

$$\vec{a'} \cdot \vec{b'} \times \vec{c'} = \frac{1}{V}.$$

[VU(H) 2010]

11.26 (a) Explain the physical significance of gradient, divergence and curl of a vector $\vec{A}$.

(b) Time evolution of a vector $\vec{A}$ is given by $\dfrac{d\vec{A}}{dt} = \vec{A} \times \vec{B}$, $\vec{B}$ is a constant vector. Show that,

(i) the magnitude of $\vec{A}$ and

(ii) the angle between $\vec{A}$ and $\vec{B}$ do not change with time.

(c) $\vec{E} = \vec{E}_0 e^{i\vec{k}\cdot\vec{r}}$ and $\vec{B} = \nabla \times \vec{E}$ are two vector fields with $\vec{E}_0$ and $\vec{k}$ constant vectors. Show that $\vec{B}$ is perpendicular to both $\vec{E}$ and $\vec{k}$. **[NBU(H) 2007]**

11.27 (a) Find the derivative of the function $\phi = x^2 - 2xy + z^2$ at the point $(2, -1, 1)$ in the direction of the vector $\vec{A} = 2\hat{i} - 4\hat{j} + 4\hat{k}$. **[BU(H) 2013]**

(b) Find the directional derivative of $f(x, y, z) = xy^2 + yz^2$ at the point $(2, -1, 1)$ in the direction $\hat{i} + 2\hat{j} + \hat{k}$. **[BU(H) 2008]**

(c) (i) If $\vec{a'}, \vec{b'}, \vec{c'}$ are the reciprocal vector triads for the vectors $\vec{a}, \vec{b}, \vec{c}$ then

$$[\vec{a} \cdot (\vec{b} \times \vec{c})][\vec{a'} \cdot (\vec{b'} \times \vec{c'})] = 1.$$

(ii) Show that $\nabla \times (\nabla \times \vec{f}) = \nabla(\nabla \cdot \vec{f}) - \nabla^2 \vec{f}$, $\vec{f}$ being an arbitrary vector. **[BU(H) 2006]**

11.28 (a) Is the direction of a vector an absolute concept? Explain using the idea of derivative of a vector.

(b) The Euclidean space cannot be considered as a vector space. Why? Show how can you get a scalar field from a vector field. **[Assam U Phys(Major) 2004]**

11.29 (a) Show that the gradient of any scalar field $\phi(r)$ is irrotational and the curl of any vector field $\vec{V}(r)$ is solenoidal.

(b) Prove that, $\nabla \cdot (\vec{A} \times \vec{B}) = \vec{B} \cdot (\nabla \times \vec{A}) - \vec{A} \cdot (\nabla \times \vec{B})$.

If $\vec{A}$ and $\vec{B}$ are irrotational, show that $\vec{A} \times \vec{B}$ is solenoidal. **[Assam U Phys.(Major) 2014]**

11.30 (a) Let $\vec{r}(t)$ be a vector of fixed magnitude. Show that $\dfrac{d\vec{r}(t)}{dt}$ is perpendicular to $\vec{r}(t)$.

(b) If $\vec{v}(t)$ is a time dependent vector, show that, $\vec{v} \cdot \dfrac{d\vec{v}}{dt} = v\dfrac{dv}{dt}$, where v is the magnitude. How is this relation relevant in connection with the motion of a particle in a circular orbit? Explain. **[Assam U Phys.(Major) 2014]**

11.31 (a) Explain the physical meaning of the direction and magnitude of the gradient of a scalar function.

(b) Find $\text{div}(r^n \vec{r})$.

(c) Find the angle between the surfaces, $x^2+y^2+z^2=9$ and $x^2+y^2-z^2=3$ at the point $(2,-1,2)$. **[Mahatma Gandhi U (PG Model questions) 2012]**

11.32 (a) Show that $\vec{A}=(6xy+z^3)\hat{\imath}+(3x^2-z)\hat{\jmath}+(3xz^2-y)\hat{k}$ is irrotational. Find ϕ such that $\vec{A}=\nabla\phi$. **[CU(H) 2016]**

(b) Consider the spherical polar coordinate system as given by
$\vec{r}=r\sin\theta\cos\phi\hat{\imath}+r\sin\theta\sin\phi\hat{\jmath}+r\cos\theta\hat{k}$.
Find the unit vectors $\hat{e}_r, \hat{e}_\theta$ and $\hat{e}_\phi$ in this system. **[CU(H) 2015]**

(c) A particle moves along the curve $x=2t^2$, $y=4t^2-1$, $z=5t-3$, where t denotes time. Find the component of acceleration at time $t=1$ in the direction $(\hat{\imath}-2\hat{\jmath}+2\hat{k})$. **[CU(H) 2014]**

11.33 (a) If the magnitude of a vector field $\vec{A}(t)$ is constant with respect to time t, show that $\dfrac{d\vec{A}}{dt}$ is perpendicular to $\vec{A}$. **[CU(H) 2013]**

(b) Consider a vector $\vec{A}(t)$ in cylindrical coordinates (ρ,ϕ,Z). Show that,

$$\frac{d\vec{A}}{dt}=\left(\frac{dA_\rho}{dt}-A_\phi\dot{\phi}\right)\hat{\rho}+\left(\frac{dA_\phi}{dt}+A_\rho\dot{\phi}\right)\hat{\phi}+\frac{dA_z}{dt}\hat{z}.$$

If $\vec{A}$ in the above expression is the position vector of a particle, determine the corresponding expression of $\dfrac{dA}{dt}$, i.e., the velocity of the particle. Hence, by the proper choice of $\vec{A}$, determine the acceleration of the particle. **[CU(H) 2012]**

(c) A particle of mass m moves in the XY-plane so that its position vector is $\vec{r}=a\cos\omega t\hat{\imath}+b\sin\omega t\hat{\jmath}$, where a,b,ω are constant. Show that the force acting on the particle is always directed towards the origin. **[CU(H) 2010]**

CHAPTER 12

Vector Integration

12.1 Ordinary Integration of Vectors

If $\vec{P}(t)$ be a vector function of single variable t, then it can be integrated in the same way as a scalar function.

If $\vec{P}(t) = \hat{i}P_1(t) + \hat{j}P_2(t) + \hat{k}P_3(t)$, then

$$\int \vec{P}(t)dt = \hat{i}\int P_1(t)dt + \hat{j}\int P_2(t)dt + \hat{k}\int P_3(t)dt,$$

where $P_1(t)$, $P_2(t)$, $P_3(t)$ are continuous in a certain interval. This is *indefinite integral* of $\vec{P}(t)$.

If $\vec{P}(t) = \frac{d}{dt}\vec{Q}(t)$, where $\vec{Q}(t)$ is another vector function of single variable t, then

$$\int \vec{P}(t)dt = \int \frac{d}{dt}\vec{Q}(t)dt = \int d\vec{Q}(t)$$

or, $$\int \vec{P}(t)dt = \vec{Q}(t) + \vec{C}, \tag{12.1}$$

where $\vec{C}$ is an arbitrary constant vector.

The *definite integral* of $\vec{P}(t)$ between limits $t = a$ and $t = b$ is expressed as

$$\int_a^b \vec{P}(t)dt = \int_a^b \frac{d}{dt}\vec{Q}(t)dt = \left[\vec{Q}(t) + \vec{C}\right]_a^b = \vec{Q}(b) - \vec{Q}(a). \tag{12.2}$$

12.2 Line Integral

Any integral which is to be evaluated along a curve is called a *line integral.*

Let $\vec{A}(\vec{r})$ be a continuous vector function acting on a particle at P, which moves along a space curve C. The curve is represented by a continuously differentiable function $\vec{r} = f(t)$ of t (Fig. 12.1).

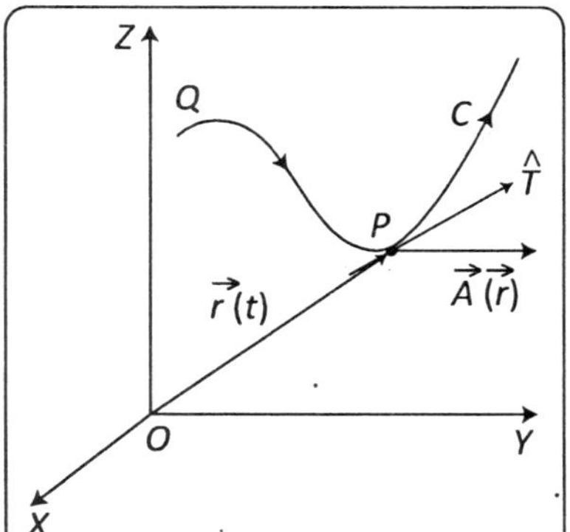

Fig. 12.1

The unit vector along the tangent to the curve at P is

$$\vec{T} = \frac{d\vec{r}}{ds},$$

where s is the length of the curve from initial point Q to P.

The component $\vec{A}(\vec{r})$ along $\vec{T}$ is

$$\vec{A}\cdot\vec{T} = \vec{A}\cdot\frac{d\vec{r}}{ds}$$

Then, $$\int_C \vec{A}\cdot\frac{d\vec{r}}{ds}ds = \int \vec{A}\cdot d\vec{r} \tag{12.3}$$

is the *line integral* of the vector function $\vec{A}(\vec{r})$ along the curve C.

Note: (1) *Work* : If $A(\vec{r})$ be the variable force acting on a particle moving along an arc QP, then the work done for small displacement $d\vec{r}$ is $dW = \vec{A}\cdot d\vec{r}$. The total work done by $\vec{A}(\vec{r})$ during the displacement from Q to P is

$$W = \int_Q^P \vec{A}\cdot d\vec{r}.$$

(2) *Circulation* : If $A(\vec{r})$ represents the velocity of a liquid, then

$$\oint \vec{A}\cdot d\vec{r}$$

is called the *circulation* of $\vec{A}(\vec{r})$ around the closed curve C.

(3) If $\vec{A} = \nabla\phi$, where ϕ is a scalar function of position, then

(i) $\int_Q^P \vec{A}\cdot d\vec{r}$ is independent of path Q to P (ii) $\oint \vec{A}\cdot d\vec{r} = \oint_C \nabla\phi\cdot d\vec{r} = \oint_C d\phi = 0.$

$\vec{A}(\vec{r})$ is then called the *conservative vector field.* Gravitational field, electrostatic field are the examples of conservative field.

If $\vec{A}$ is conservative, then $\nabla\times\vec{A} = 0$ and $\vec{A} = \nabla\phi$, then

$$\vec{A}\cdot d\vec{r} = A_1dx + A_2dy + A_3dz = d\phi \text{ is an } \textit{exact differential.}$$

12.3 Surface Integral

Let S be the outward surface of a body placed in a vector field $\vec{A}$. If $\hat{n}$ be an outward drawn unit normal to an element of surface dS such that $\vec{dS} = \hat{n}dS$, where $\vec{dS}$ is the vector area of dS, i.e., $|dS| = dS$ (Fig. 12.2).

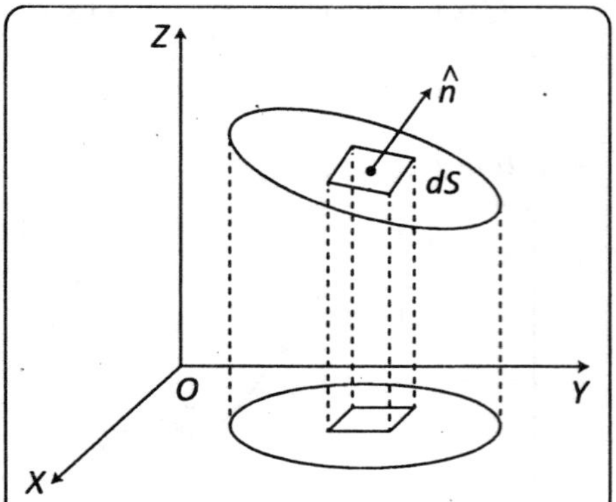

Fig. 12.2

The outward flux of the continuous vector function $\vec{A}$ through dS is

$$dN = \vec{A}\cdot\vec{dS}.$$

The outward flux through the whole surface S is

$$N = \iint_S \vec{A}\cdot\vec{dS} = \iint_S \vec{A}\cdot\hat{n}dS. \qquad (12.4)$$

The other type of surface integrals are

$$\iint_S \phi dS,\quad \iint_S \phi\hat{n}dS,\quad \iint_S \vec{A}\times\vec{dS}.$$

where ϕ is a scalar function of position.

12.4 $\iint_S \vec{A}\cdot\hat{n}\,dS$ as the Limit of a Sum

We consider the outward surface S of a body in a vector field $\vec{A}$. Let ΔS_p be an element of surface S surrounding the point P_p, where the vector field is $\vec{A}_p$ and the outward unit normal is $\hat{n}_p$ (Fig. 12.3).

The normal flux of $\vec{A}_p$ at P_p is $\sum_{p=1}^{M} \vec{A}_p\cdot\hat{n}_p\Delta S_p$, where $p = 1, 2, \ldots, M$.

When $M \to \infty$, the largest $\Delta S_p \to 0$, if the limit exists then, $\iint_S \vec{A} \cdot \hat{n}\, dS$ is called the surface integral of the normal component of $\vec{A}$ over S.

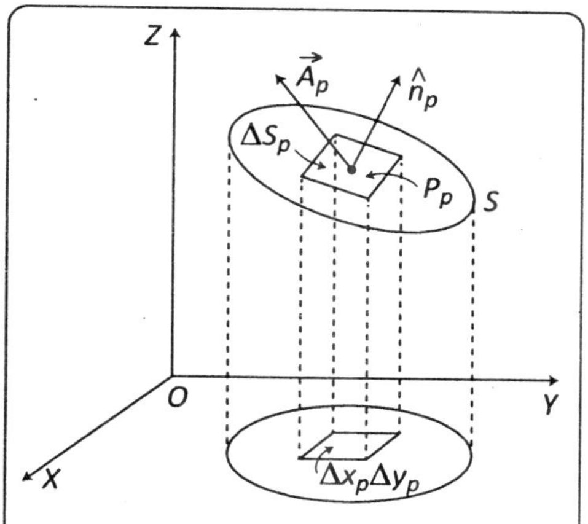

Fig. 12.3

12.5 To Show that $\iint_S \vec{A} \cdot \hat{n}\, dS = \iint_R \vec{A} \cdot \hat{n} \frac{dxdy}{|\hat{n}_p \cdot \hat{k}|}$

If the surface S has projection R on the xy-plane, then

$$\iint_S \vec{A} \cdot \hat{n} dS = \iint_R \vec{A} \cdot \hat{n} \frac{dxdy}{|\hat{n}_p \cdot \hat{k}|}.$$

Proof : From Fig. 12.3, the projection of ΔS_p on the xy-plane is $|(\hat{n}_p \Delta S) \cdot \hat{k}| = |\hat{n}_p \cdot \hat{k}| \Delta S_p$.

But, $[(\hat{n}_p \cdot \hat{k}|\Delta S_p = \Delta x_p \Delta y_p, \quad \therefore \quad \Delta S_p = \dfrac{\Delta x_p \Delta y_p}{|\hat{n}_p \cdot \hat{k}|}$.

Using this value of ΔS_p in the result of section 12.4, we get

$$\sum_{p=1}^{M} \vec{A}_p \cdot \hat{n}_p \frac{\Delta x_p \Delta y_p}{|\hat{n}_p \cdot \hat{k}|}.$$

In the limit, when $M \to \infty$ largest Δx_p and Δy_p tend to zero at the same time, we have

$$\sum_{p=1}^{M} \vec{A}_p \cdot \hat{n}_p \frac{\Delta x_p \Delta y_p}{|\hat{n}_p \cdot \hat{k}|} = \iint_R \vec{A} \cdot \hat{n} \frac{dxdy}{|\hat{n}_p \cdot \hat{k}|}$$

or, $$\iint_S \vec{A} \cdot \hat{n} dS = \iint_R \vec{A} \cdot \hat{n} \frac{dxdy}{|\hat{n}_p \cdot \hat{k}|}. \qquad (12.5)$$

12.6 Volume Integral

If $\vec{A}$ be a single valued continuous vector function in volume V enclosed by a surface S, then the volume integral of $\vec{A}$ is given by

$$\iiint_V \vec{A} dV = \hat{i} \iiint_V A_1 dV + \hat{j} \iiint_V A_2 dV + \hat{k} \iiint_V A_3 dV, \text{ where } \vec{A} = \hat{i}A_1 + \hat{j}A_2 + \hat{k}A_3. \quad (12.6)$$

12.7 Solved Problems (A)

Example 12.1. *A force field $\vec{F} = 3x^2\hat{i} + (2xz - y)\hat{j} + z\hat{k}$ acts on a particle displacing it from $(0,0,0)$ to $(2,1,3)$ in 1 second. Find the work done.*

Solution : We have the work done $= \int_C = \vec{F}\cdot d\vec{r}$

$$= \int_C [3x^2\hat{i} + (2xz - y)\hat{j} + z\hat{k}]\cdot(\hat{i}dx + \hat{j}dy + \hat{k}dz) = \int_C 3x^2dx + (2xz - y)dy + zdz.$$

Now, the equation of the straight line from $(0,0,0)$ to $(2,1,3)$ is $\frac{x}{2} = \frac{y}{1} = \frac{z}{3} = t$ (say).

$\therefore x = 2t,\ y = t,\ z = 3t$. The point $(0,0,0)$ and $(2,1,3)$ corresponds to $t = 0$ to $t = 1$.

$$\therefore \textbf{work done} = \int_C \vec{F}\cdot d\vec{r} = \int_0^t [3(2t)^2 2dt + (2\cdot 2t\cdot 3t - t)dt + 3t\cdot 3dt] = \int_0^1 (36t^2 + 8t)dt$$

$$= \left[36\frac{t^3}{3} + 8\frac{t^2}{2}\right]_0^1 = \textbf{16 units.}$$

Example 12.2. *Calculate the work done when a force $\vec{F} = 3xy\hat{i} - y^2\hat{j}$ moves a particle in the xy-plane from $(0,0)$ to $(1,2)$ along the parabola $y = 2x^2$.* **[BU(H) 1994]**

Solution : We have $\vec{F} = 3xy\hat{i} - y^2\hat{j} = 3x(2x^2)\hat{i} - (2x^2)^2\hat{j}$ $\quad [\because y = 2x^2]$

$$= 6x^3\hat{i} - 4x^4\hat{j}.$$

Also, $d\vec{r} = \hat{i}dx + \hat{j}dy + \hat{k}dz = \hat{i}dx + \hat{j}dy$ $\quad (\because dz = 0)$

$$\therefore \quad \int_C \vec{F}\cdot d\vec{r} = \int_0^1 (6x^3\hat{i} - 4x^4\hat{j})\cdot(\hat{i}dx + \hat{j}dy) = \int_0^1 (6x^3\hat{i} - 4x^4\hat{j})\cdot(\hat{i}dx + 4x\hat{j})$$

$$= \int_0^1 \left[\frac{6}{4}x^3 - \frac{16}{6}x^6\right]dx = \left[\frac{3}{2}x^4 - \frac{8}{3}x^6\right]_0^1 = -\frac{7}{6} \text{ units.}$$

$\therefore$ **Work done = −7/6 units.**

Example 12.3. *A particle of constant mass m is moving in a conservative force field, $\vec{F} = -\nabla\phi$. If A and B be two points in space, prove that, $\phi(A) + \frac{1}{2}mv_A^2 = \phi(B) + \frac{1}{2}mv_B^2$, where v_A and v_B are the magnitudes of velocities of the particle at A and B respectively.* **[BU(H) 2000]**

Solution : If $\vec{r}$ be the position vector of the particle at any time t, then its acceleration is $\frac{d^2\vec{r}}{dt^2}$.

The force acting on the particle is $\vec{F} = m\frac{d^2\vec{r}}{dt^2}$.

The work done by $\vec{F}$ is $dW = \vec{F}\cdot d\vec{r} = m\frac{d^2\vec{r}}{dt^2}\cdot d\vec{r} = m\frac{d}{dt}\left(\frac{d\vec{r}}{dt}\right)\cdot\left(\frac{d\vec{r}}{dt}\right)dt$

$$= \frac{1}{2}m\frac{d}{dt}\left(\frac{d\vec{r}}{dt}\right)^2 dt = \frac{1}{2}md(v^2).$$

$$\therefore \quad \text{total work done, } W = \int_A^B \frac{1}{2}md(v^2) = \frac{1}{2}m(v_B^2 - v_A^2). \qquad \text{(i)}$$

But, $\vec{F} = -\nabla\phi$,

$$\therefore \quad W = \int_A^B \vec{F}\cdot d\vec{r} = -\int_A^B \nabla\phi\cdot d\vec{r} = \int_B^A d\phi = \phi_A - \phi_B.$$

$$\text{So, } \frac{1}{2}mv_B^2 - \frac{1}{2}mv_A^2 = \phi_A - \phi_B. \qquad \text{(ii)}$$

$$\therefore \quad \boldsymbol{\phi_B + \frac{1}{2}mv_B^2 = \phi_A + \frac{1}{2}mv_A^2},$$

where ϕ_A and $\frac{1}{2}mv_A^2$ are respectively potential energy and kinetic energy of the particle at A. Thus, the total energy at A and B are equal (conservation of energy). This is known as the *work-energy principle* in Mechanics.

Example 12.4. *Find the work done in going around a unit circle in the xy-plane.*

(i) counter clockwise from 0 to π (ii) clockwise from 0 to $-\pi$

against a force field given by $\vec{F} = -\frac{y}{x^2+y^2}\hat{i} + \frac{x}{x^2+y^2}\hat{j}$.

Is the field conservative? **[BU(H) 1993]**

Solution : We have $\vec{F} = \frac{-y\hat{i}+x\hat{j}}{x^2+y^2} = -y\hat{i}+x\hat{j} \quad [\because\ x^2+y^2=1]$

and $\quad d\vec{r} = \hat{i}dx + \hat{j}dy$.

$$\therefore \text{ work done is } \int \vec{F}\cdot d\vec{r} = \int(-y\hat{i}+x\hat{j})\cdot(\hat{i}dx+\hat{j}dy) = \int(-ydx+xdy).$$

(i) Let the counterclockwise path be C_1 (from 0 to π).

$\therefore$ work done along the counter clockwise path is

$$\int_{C_1}(-\boldsymbol{ydx} + \boldsymbol{xdy}) = \int_0^{-\pi}(-\sin\theta \times -\sin\theta d\theta + \cos\theta d\theta + \cos\theta\cos\theta)d\theta,$$

$$\left[\begin{array}{lll}\because & x=\cos\theta, & \therefore\ dx = -\sin\theta d\theta \\ & y = \sin\theta, & \therefore\ dy = -\cos\theta d\theta\end{array}\right]$$

$$= \int_0^{\pi}(\sin^2\theta + \cos^2\theta)d\theta = \int_0^{\pi} d\theta = \boldsymbol{\pi}.$$

(ii) The work done along the clockwise path (C_2) is

$$\int_{C_2}(-\boldsymbol{ydx} + \boldsymbol{xdy}) = \int_0^{-\pi}(\sin^2\theta+\cos^2\theta)d\theta = \int_0^{-\pi} d\theta = \boldsymbol{-\pi}.$$

$$\therefore \quad \oint \vec{F}\cdot d\vec{r} = \int_0^{\pi}\vec{F}\cdot d\vec{r} + \int_{\pi}^{0}\vec{F}\cdot d\vec{r} = \pi - (-\pi) = \pi + \pi = 2\pi.$$

$\therefore$ $\vec{F}$ **is not conservative.**

Example 12.5. *If $\vec{A} = (y-2x)\hat{i} + (3x+2y)\hat{j}$, compute the circulation of $\vec{A}$ about a circle C in the xy-plane with centre at the origin and radius 2, if C is traversed in the positive direction.*

Solution : We have, $\vec{A} \cdot d\vec{r} = [(y-2x)\hat{i} + (3x+2y)\hat{j}] \cdot (\hat{i}dx + \hat{j}dy)$
$= (y-2x)dx + (3x+2y)dy.$

Now, $x = 2\cos\theta$ and $y = 2\sin\theta$, where $\theta = 0$ to $\theta = 2\pi$.

$$\therefore \text{ circulation of } \vec{A} = \int_C \vec{A} \cdot A\vec{r}$$
$$= \int_0^{2\pi} [(2\sin\theta - 4\cos\theta) \times (-2\sin\theta)d\theta + (6\cos\theta + 4\sin\theta) \times (2\cos\theta d\theta)]$$
$$= \mathbf{8\pi}.$$

Example 12.6. *If $\vec{F} = (2x+y^2)\hat{i} + (3y-4x)\hat{j}$, evaluate $\oint \vec{F} \cdot d\vec{r}$ around the right-angled triangle OAB having vertices $O(0,0)$, $A(2,0)$, $B(2,1)$ and right angle at A.*

Solution : We have $\vec{F} \cdot d\vec{r} = [(2x+y^2)\hat{i} + (3y-4x)\hat{j}] \cdot (\hat{i}dx + \hat{j}dy)$
$= (2x+y^2)dx + (3y-4x)dy.$

$$\therefore \quad \oint_C \vec{F} \cdot d\vec{r} = \oint_{OA} \vec{F} \cdot d\vec{r} + \oint_{AB} \vec{F} \cdot d\vec{r} + \oint_{BO} \vec{F} \cdot d\vec{r}.$$

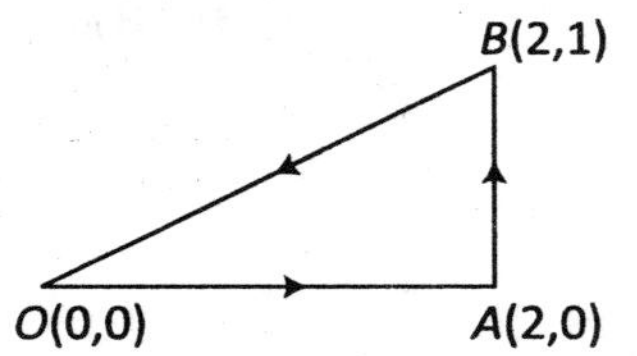

Fig. 12.4

From Fig. 12.4, along OA, $y = 0$, $dy = 0$ and x varies from 0 to 2.

$$\therefore \quad \oint_{OA} \vec{F} \cdot d\vec{r} = \int_0^2 2x dx = \left[x^2\right]_0^2 = 4.$$

Along AB, $x = 2$, $\therefore$ $dx = 0$ and y varies from 0 to 1.

$$\therefore \quad \oint_{AB} \vec{F} \cdot d\vec{r} = \int_0^1 (3y-8)dy = \left[\frac{3}{2}y^2 - 8y\right]_0^1 = \frac{3}{2} - 8 = -\frac{13}{2}.$$

Equation of BO is $\dfrac{y-1}{1-0} = \dfrac{x-2}{2-0}$ or, $2y - 2 = x - 2$ or, $x = 2y$.

$$\therefore \quad \oint_{BO} \vec{F} \cdot d\vec{r} = \int_2^0 \left[\left(2x + \frac{x^2}{4}\right)dx + \left(\frac{3x}{2} - 4x\right) \times \frac{1}{2}dx\right] = \int_2^0 \left(\frac{x^2}{4} + \frac{3x}{4}\right)dx$$
$$= \left[\frac{x^3}{12} + \frac{3}{8}x^2\right]_2^0 = \frac{1}{12}(-8) + \frac{3}{8}(-4) = -\frac{13}{6}.$$

$$\therefore \quad \oint_C \vec{\boldsymbol{F}} \cdot d\vec{\boldsymbol{r}} = 4 - \frac{13}{2} - \frac{13}{6} = -\frac{\mathbf{14}}{\mathbf{3}}.$$

Example 12.7. *If C be the boundary of the rectangle (in the xy-plane) defined by $y = 0$, $x = a$; $y = b$, $x = 0$, evaluate the line integral $\int_C \vec{F} \cdot d\vec{r}$ around C, where $\vec{F} = (x^2+y^2)\hat{i} - 2xy\hat{j}$.*

[BU(H) 1996(Old)]

Solution : We have from Fig. 12.5,

$$\therefore \quad \oint_C \vec{F} \cdot d\vec{r} = \oint_{OA} \vec{F} \cdot d\vec{r} + \oint_{AB} \vec{F} \cdot d\vec{r} + \oint_{BC} \vec{F} \cdot d\vec{r} + + \oint_{CO} \vec{F} \cdot d\vec{r}.$$

Now, along OA, $y = 0$, $dy = 0$ and x varies from 0 to a, $d\vec{r} = \hat{i}dx$.

$\therefore \quad \vec{F} \cdot d\vec{r} = [(x^2 + y^2)\hat{i} - 2xy\hat{j}] \cdot (\hat{i}dx).$

$\therefore \quad \oint_{OA} \vec{F} \cdot d\vec{r} = \int_0^a x^2 dx = \dfrac{a^3}{3}.$

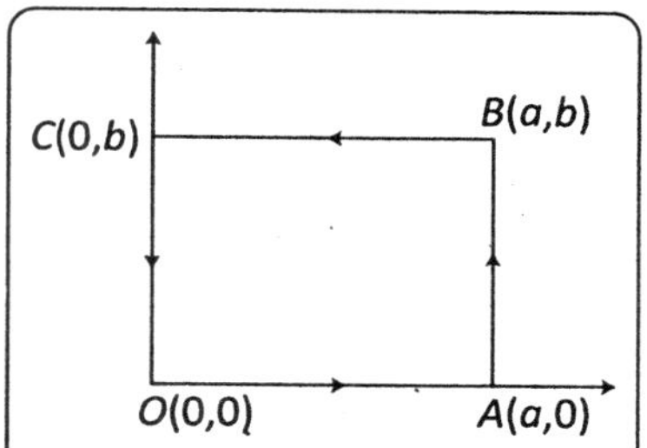

Fig. 12.5

Along AB, $x = a$, $dx = 0$, y varies from 0 to b, $d\vec{r} = \hat{j}dy$.

$\therefore \quad \vec{F} \cdot d\vec{r} = [(x^2 + y^2)\hat{i} - 2xy\hat{j}] \cdot (\hat{j}dy) = -2aydy.$

$\therefore \quad \int_{AB} \vec{F} \cdot d\vec{r} = -2a\int_0^b y\, dy = -2a\left[\dfrac{y^2}{2}\right]_0^b = -\dfrac{2ab^2}{2} = -ab^2.$

Along BC, $y = b$, $dy = 0$, x varies from a to 0, $d\vec{r} = -\hat{i}dx$.

$\therefore \quad \vec{F} \cdot d\vec{r} = [(x^2 + y^2)\hat{i} - 2xy\hat{j}] \cdot (-\hat{i}dx) = -(x^2 + y^2)dx.$

$\therefore \quad \int_{BC} \vec{F} \cdot d\vec{r} = -\int_a^0 y(x^2 + b^2)dx = -\int_a^0 x^2 dx - b^2\int_a^0 dx = -\left[\dfrac{x^3}{3}\right]_a^0 - b^2\,[x]_a^0 = \dfrac{a^3}{3} + ab^2.$

Along CO, $x = 0$, $\therefore$ $dx = 0$, y varies from b to 0, $d\vec{r} = -\hat{j}dy$.

$\therefore \quad \vec{F} \cdot d\vec{r} = 2(0)(-b)dy = 0, \qquad \therefore \quad \int_{CO} \vec{F} \cdot d\vec{r} = 0.$

$\therefore \quad \oint_C \boldsymbol{\vec{F} \cdot d\vec{r}} = \dfrac{a^3}{3} - ab^2 + \dfrac{a^3}{3} + ab^2 = \dfrac{\mathbf{2a^3}}{\mathbf{3}}.$

Example 12.8. *Evaluate* $\oint_C \vec{F} \cdot d\vec{r}$, *where* $\vec{F} = (x - 3y)\hat{i} + (y - 2x)\hat{j}$ *and* C *is the closed curve in the xy-plane,* $x = 2\cos t$, $y = 3\sin t$ *from* $t = 0$ *to* 2π.

Solution : We have $\vec{F} \cdot d\vec{r} = [(x - 3y)\hat{i} + (y - 2x)\hat{j}] \cdot (\hat{i}dx + \hat{j}dy) = (x - 3y)dx + (y - 2x)dy.$

$$\therefore \quad \oint_C \vec{F} \cdot d\vec{r} = \oint_C [(x - 3y)dx + (y - 2x)dy]$$

$$= \int_0^{2\pi} [(2\cos t - 9\sin t)(-2\sin t dt) + (3\sin t - 4\cos t)(3\cos t dt)]$$

$$= \int_0^{2\pi} [-4\cos t\sin t + 18\sin^2 t + 9\sin t\cos t - 12\cos^2 t]dt$$

$$= 5\int_0^{2\pi} [\sin t\cos t dt + 18\int_0^{2\pi} \sin^2 t dt - 12\int_0^{2\pi} \cos^2 t dt$$

$$= 5 \times 0 + 18 \times \frac{2\pi}{2} - 12 \times \frac{2\pi}{2} = 6\pi.$$

$$\therefore \quad \oint_C \boldsymbol{\vec{F} \cdot d\vec{r} = 6\pi}$$

Example 12.9. *If* $\vec{F} = (3x^2 + 6y)\hat{i} - 14yz\hat{j} + 20xa^2\hat{k}$, *evaluate* $\int \vec{F} \cdot d\vec{r}$ *from* $(0, 0, 0)$ *to* $(1, 1, 1)$ *along the following paths* C *:*

(i) $x = t$, $y = t^2$, $z = t^3$.

(ii) *the straight lines from* $(0,0,0)$ *to* $(1,0,0)$, *then to* $(1,1,0)$ *and then to* $(1,1,1)$.

(iii) *the straight lines from* $(0,0,0)$ *to* $(1,1,1)$.

Solution : (i) $\int_C \vec{F}\cdot d\vec{r} = \int_C [(3x^2+6y)dx - 14yzdy + 20xz^2dz]$

$$= \int_C [(3x^2+6x^2)dx - 14x^5 2xdx + 20x^7\cdot 3x^2dx]$$

$$[\because\ x=t,\ y=t^2=x^2,\ z=t^3=x^3]$$

$$= \int_0^1 (9x^2 - 28x^6 + 60x^9)dx = \mathbf{5}$$

(ii) Along the line from $(0,0,0)$ to $(1,0,0)$ we get $y=0$, $z=0$, $dy=0$, $dz=0$, x varies from 0 to 1.

$$\therefore\quad \int_0^1 (3x^2)dx = \left[x^3\right]_0^1 = 1.$$

Along, the line from $(1,0,0)$ to $(1,1,0)$ we have $z=0$, $dz=0$, $x=1$, $dx=0$ and y varies from 0 to 1.

$$\therefore\quad \int_C \vec{F}\cdot d\vec{r} = \int_0^1 -14yzdy = -\int_0^1 0dy = 0$$

Along the line from $(1,0,0)$ to $(1,1,0)$ we have, $z=0, dz=0, x=1, dx=0$ and y varies from 0 to 1.

$$\therefore\quad \int \vec{F}\cdot d\vec{r} = \int_0^1 20z^2dz = \frac{20}{3},\qquad \therefore\quad \int_C \vec{F}\cdot d\vec{r} = 1+0+\frac{20}{3} = \frac{23}{3}.$$

(iii) Along the straight line joining $(0,0,0)$ and $(1,1,1)$, $x=t$, $y=t$, $z=t$.

Then, $\int_C \vec{F}\cdot d\vec{r} = \int_{t=0}^1 (3t^2+6t)dt - 14(t)(t)dt + 20(t)(t)^2dt = \int_{t=0}^1 (3t^2+6t-14t^2+20t^3)dt$

$$= \int_{t=0}^1 (6t - 11t^2 + 20t^3)dt = \frac{13}{3}.$$

Example 12.10. *If* $\vec{F} = \nabla\phi$ *everywhere in a region* R *and* ϕ *is a single-valued function and has continuous derivatives in* R, *then show that,* $\int_A^B \vec{F}\cdot d\vec{r}$ *is independent of path* C *in* R *joining the points* $A(x_1,y_1,z_1)$ *and* $B(x_2,y_2,z_2)$. **[BU(H) 1992]**

Solution : We have $\int_A^B \vec{F}\cdot d\vec{r} = \int_A^B \nabla\phi\cdot d\vec{r} = \int_A^B d\phi = \phi(B)-\phi(A)$

$$= \phi(x_2,y_2,z_2) - \phi(x_1,y_1,z_1),$$

assuming ϕ to be single-valued and continuous scalar function. Hence the result.

Example 12.11. *If* $\int_P^Q \vec{A}\cdot d\vec{r}$ *is independent of path joining any two points* P *and* Q *in a region* R, *then prove that,* $\oint \vec{A}\cdot d\vec{r} = 0$ *for all closed paths in the same region. Prove that, the converse is also true.*

Solution : Let $PXQYP$ be a closed curve as shown in Fig. 12.6. Then,

$$\oint_{PXQYP} \vec{A}\cdot d\vec{r} = \oint_{PXQ} \vec{A}\cdot d\vec{r} + \oint_{QYP} \vec{A}\cdot d\vec{r}$$

$$= \oint_{PXQ} \vec{A}\cdot d\vec{r} - \oint_{PYQ} \vec{A}\cdot d\vec{r} = 0,$$

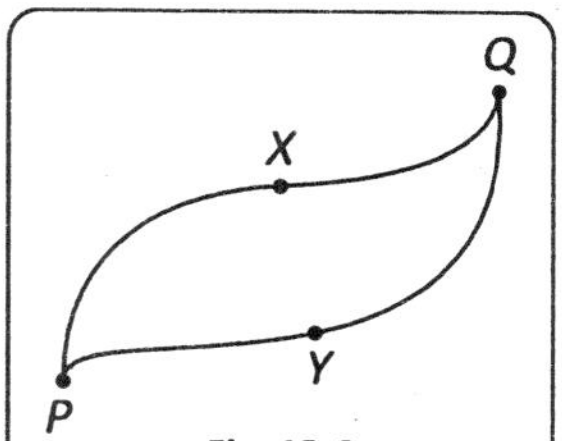

Fig. 12.6

since, $\oint_{PXQ} \vec{A}\cdot d\vec{r} = \oint_{PYQ} \vec{A}\cdot d\vec{r}$ by the condition of the problem.

Conversely, if $\oint \vec{A}\cdot d\vec{r} = 0$,

then $\int_{PXQYP} \vec{A}\cdot d\vec{r} = \int_{PXQ} \vec{A}\cdot d\vec{r} + \oint_{QYP} \vec{A}\cdot d\vec{r} = \int_{PXQ} \vec{A}\cdot d\vec{r} - \oint_{PYQ} \vec{A}\cdot d\vec{r} = 0.$

$\therefore \quad \int_{PXQ} \vec{A}\cdot d\vec{r} = \oint_{PYQ} \vec{A}\cdot d\vec{r}$, i.e., $\int_P^Q \vec{A}\cdot d\vec{r}$ **is independent of path.**

Example 12.12. *If $\int_C \vec{A}\cdot d\vec{r}$ is independent of path C joining any two points P and Q in a region R, show that, there exists a scalar function ϕ, such that $\vec{A} = \nabla\phi$.*

Solution : If $\vec{A}$ be the force field and $d\vec{r}$ the elementary displacement of a particle by the application of $\vec{A}$, then the work done for displacing the particle from P to Q is

$$\phi(x,y,z) = \int_P^Q \vec{A}\cdot d\vec{r} = \int_P^Q \vec{A}\cdot \frac{d\vec{r}}{ds}ds \quad \text{is independent of path.}$$

$$\therefore \quad \frac{d\phi}{ds} = \vec{A}\cdot\frac{d\vec{r}}{ds}.$$

Again, since $d\phi = \vec{A}\cdot d\vec{r}$,

$$\therefore \quad \frac{d\phi}{ds} = \frac{\nabla\phi\cdot d\vec{r}}{ds} = \nabla\phi\cdot\frac{d\vec{r}}{ds}.$$

$$\therefore \quad \frac{d\phi}{ds} = \vec{A}\cdot\frac{d\vec{r}}{ds} = \nabla\phi\cdot\frac{d\vec{r}}{ds} \quad \text{or,} \quad \vec{A}\cdot\frac{d\vec{r}}{ds} = \nabla\phi\cdot\frac{d\vec{r}}{ds} \quad \text{or,} \quad (\vec{A}-\nabla\phi) = \frac{d\vec{r}}{ds} = 0.$$

$$\therefore \quad \vec{A} - \nabla\phi = 0 \quad \left[\because \frac{d\vec{r}}{ds} \neq 0\right]$$

or, $\boldsymbol{\vec{A} = \nabla\phi}$.

Example 12.13. (a) *Prove that if $\vec{A}$ is a conservative field, then curl $\vec{A} = \nabla\times\vec{A} = 0$.*

(b) *Conversely, if $\nabla\times\vec{A} = 0$, then prove that, $\vec{A}$ is conservative.*

Solution : (a) If the vector field $\vec{A}$ be conservative, then it can be shown that it is associated with a scalar potential ϕ such that $\vec{A} = \nabla\phi$.

$$\therefore \quad \nabla\times\vec{A} = \nabla\times\nabla\phi = 0.$$

So, **curl $\vec{A} = 0$ i.e., $\vec{A}$ is irrotational.**

(b) If $\nabla \times \vec{A} = 0$, then

$$\begin{vmatrix} \hat{i} & \hat{j} & \hat{k} \\ \dfrac{\partial}{\partial x} & \dfrac{\partial}{\partial y} & \dfrac{\partial}{\partial z} \\ A_1 & A_2 & A_3 \end{vmatrix} = 0, \quad \text{where } \vec{A} = \hat{i}A_1 + \hat{j}A_2 + \hat{k}A_3$$

or, $$\hat{i}\left[\frac{\partial A_3}{\partial y} - \frac{\partial A_2}{\partial z}\right] + \hat{j}\left[\frac{\partial A_1}{\partial z} - \frac{\partial A_3}{\partial x}\right] + \hat{k}\left[\frac{\partial A_2}{\partial x} - \frac{\partial A_1}{\partial y}\right] = 0$$

or, $$\frac{\partial A_3}{\partial y} = \frac{\partial A_2}{\partial z}, \quad \frac{\partial A_1}{\partial z} = \frac{\partial A_3}{\partial x} \quad \text{and} \quad \frac{\partial A_2}{\partial x} = \frac{\partial A_1}{\partial y}.$$

Now, we have to prove that, $\int_C \vec{A} \cdot d\vec{r}$ is independent of path, where $\vec{A}$ will be conservative.

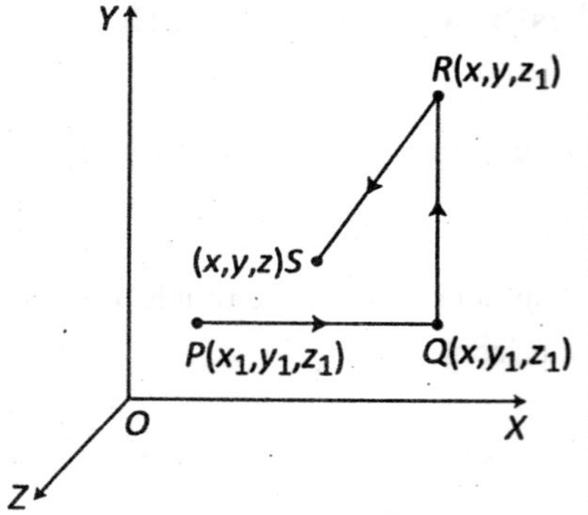

Fig. 12.7

Let us now consider a path $PQRS$, where PQ, QR and RS are parallel to OX, OY and OZ respectively (Fig. 12.7).

The work done $\phi(x, y, z)$ in moving a particle from P to S in the force field $\vec{A}$ is

$$\phi(x, y, z) = \int_P^S \vec{A} \cdot d\vec{r}$$

$$= \int_P^Q \vec{A} \cdot d\vec{r} + \int_Q^R \vec{A} \cdot d\vec{r} + \int_R^S \vec{A} \cdot d\vec{r}.$$

Along PQ, y and z are constants, so $dy = dz = 0$.

$$\therefore \quad \vec{A} \cdot d\vec{r} = A_1(x_1, y_1, z_1)dx.$$

Similarly, along QR and RS,

$$\vec{A} \cdot d\vec{r} = F_2(x, y, z_1)dy \quad \text{and} \quad \vec{A} \cdot d\vec{r} = F_3(x, y, z_1)dz.$$

$$\therefore \quad \phi(x, y, z) = \int_{x_1, y_1, z_1}^{x, y_1, z_1} A_1(x_1, y_1, z_1)dx + \int_{x, y_1, z_1}^{x, y, z_1} A_2(x, y, z_1)dy + \int_{x, y, z_1}^{x, y, z} A_3(x, y, z)dz.$$

$$\therefore \quad \frac{\partial \phi}{\partial z} = A_3(x, y, z)dz,$$

$$\frac{\partial \phi}{\partial y} = A_2(x, y, z_1) + \int_{z_1}^{z} \frac{\partial A_3}{\partial y}(x, y, z)dz$$

$$= A_2(x, y, z_1) + \int_{z_1}^{z} \frac{\partial A_2}{\partial y}(x, y, z)dz \quad \left[\frac{\partial A_3}{\partial y} = \frac{\partial A_2}{\partial z}, \text{ from problem 12.13}\right]$$

$$= A_2(x, y, z_1) + [A_2(x, y, z)]_{z_1}^{z} = A_2(x, y, z_1) + A_2(x, y, z) - A_2(x, y, z_1) = A_2(x, y, z)$$

and $$\frac{\partial \phi}{\partial x} = A_1(x, y_1, z_1) + \int_{y_1}^{y} \frac{\partial A_1}{\partial y}(x, y, z)dy + \int_{z_1}^{z} \frac{\partial A_1}{\partial z}(x, y, z)dz$$

$$\left[\because \frac{\partial A_2}{\partial x} = \frac{\partial A_1}{\partial y} \text{ and } \frac{\partial A_3}{\partial x} = \frac{\partial A_1}{\partial z}, \text{ from problem 12.13}\right]$$

$$\therefore \quad \frac{\partial \phi}{\partial x} = A_1(x, y_1, z_1) + [A_1(x, y, z_1)]_{y_1}^{y} + [A_1(x, y, z)]_{z_1}^{z}$$

$$= A_1(x, y_1, z_1) + A_1(x, y, z_1) - A_1(x, y_1, z_1) + A_1(x, y, z) - A_1(x, y, z_1) = A_1(x, y, z).$$

$$\therefore \quad \vec{A} = \hat{i}A_1 + \hat{j}A_2 + \hat{k}A_3 = \hat{i}\frac{\partial \phi}{\partial x} + \hat{j}\frac{\partial \phi}{\partial y} + \hat{k}\frac{\partial \phi}{\partial z} \quad \text{or,} \quad \vec{A} = \nabla \phi.$$

Hence, **if $\vec{A}$ be conservative, then $\vec{A} = \nabla \phi$. (Proved)**

Example 12.14. *If $\phi = 2xyz^2$, $\vec{F} = xy\hat{i} - \hat{j}z + x^2 k$ and C is the curve, $x = t^2$, $y = 2t$, $z = t^3$ from $t = 0$ to $t = 1$. Evaluate the line integral,* (a) $\int_C \phi \, d\vec{r}$, (b) $\int_C \vec{F} \times d\vec{r}$.

Solution : (a) Along the curve C, $\phi = 2xyz^2 = 2(t)^2(2t)(t^3)^2 = 4t^9$.

$$\therefore \quad \vec{r} = \hat{i}x + \hat{j}y + \hat{k}z = \hat{i}t^2 + \hat{j}2t + \hat{k}t^3 \quad \text{and} \quad d\vec{r} = (\hat{i}2t + \hat{j}2 + \hat{k}3t^2)dt.$$

$$\text{Then,} \int_C \phi d\vec{r} = \int_{t=0}^{1} 4t^9(2t\hat{i} + 2\hat{j} + 3t^2\hat{k})dt = \hat{i}\int_0^1 8t^{10}dt + \hat{j}\int_0^1 8t^9 dt + \hat{k}\int_0^1 12t^{11}dt$$

$$= \frac{8}{11}\hat{i} + \frac{4}{5}\hat{j} + \hat{k}.$$

(b) Along C, $\vec{F} = xy\hat{i} - z\hat{j} + x^2\hat{k} = 2t^3\hat{i} - t^3\hat{j} + t^4\hat{k}$.

$$\text{Then, } \vec{F} \times d\vec{r} = \begin{vmatrix} \hat{i} & \hat{j} & \hat{k} \\ 2t^3 & -t^3 & t^4 \\ 2t & 2 & 3t^2 \end{vmatrix} = [(-3t^5 - 2t^4)\hat{i} + (2t^5 - 6t^5)\hat{j} + (4t^3 + 2t^4)\hat{k}].$$

$$\therefore \quad \int_C \vec{F} \times d\vec{r} = \hat{i}\int_0^1 (-3t^5 - 2t^4)dt + \hat{j}\int_0^1 (2t^5 - 6t^5)dt + \hat{k}\int_0^1 (4t^3 + 2t^4)dt$$

$$= \hat{i}\left[\left(-3\frac{t^6}{6} - 2\frac{t^5}{5}\right)\right]_0^1 + \hat{j}\left[\left(-4\frac{t^6}{6}\right)\right]_0^1 + \hat{k}\left[\left(4\frac{t^4}{4} + 2\frac{t^5}{5}\right)\right]_0^1$$

$$= \hat{i}\left(-\frac{3}{6} - \frac{2}{5}\right) + \hat{j}\left(-\frac{4}{6}\right) + \hat{k}\left(1 + \frac{2}{5}\right)$$

$$= -\frac{9}{10}\hat{i} - \frac{2}{3}\hat{j} + \frac{7}{5}\hat{k}.$$

Example 12.15. *Show that, $\iint_S \vec{r} \cdot \hat{n} dS = 3$, over the entire surface S of the unit cube bounded by coordinate planes and the planes $x = 1$, $y = 1$, $z = 1$.*

Solution : Let $OABCDEFG$ be a unit cube bounded by the surfaces $x = 0$, $x = 1$; $y = 0$, $y = 1$; $z = 0$, $z = 1$ as shown in Fig. 12.8.

(i) On the surface, $OABC$, unit normal $\hat{n} = -\hat{i}$, $x = 0$.

$$\therefore \quad \vec{r} \cdot \hat{n} = (\hat{i}x + \hat{j}y + \hat{k}z) \cdot (-\hat{i}) = -x = 0, \quad dS_1 = dydz.$$

$$\therefore \quad \iint_{S_1} \vec{r} \cdot \hat{n} \, dS_1 = \int_{z=0}^{1} \int_{y=0}^{1} 0 \, dydz = 0.$$

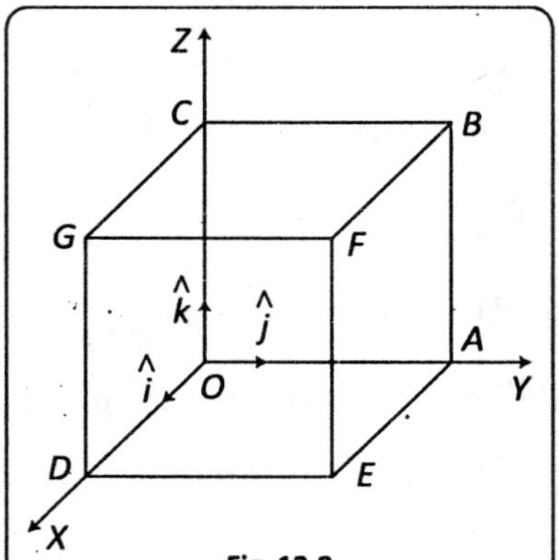

Fig. 12.8

(ii) On the surface, $DEFG$, $\hat{n} = \hat{i}$, $x = 1$.

$\therefore \quad \vec{r} \cdot \hat{n} = x = 1; \quad dS_2 = dydz$

$$\therefore \quad \iint_{S_2} \vec{r} \cdot \hat{n}\, dS_2 = \int_{z=0}^{1} \int_{y=0}^{1} (1)\, dydz = \int_{z=0}^{1} dz = 1.$$

(iii) On the surface, $ODEA$, $\hat{n} = -\hat{k}$, $z = 0$, $dS_3 = dxdy$.

$\therefore \quad \vec{r} \cdot \hat{n} = -z = 0.$

$$\therefore \quad \iint_{S_3} \vec{r} \cdot \hat{n}\, dS_3 = \int_{z=0}^{1} \int_{x=0}^{1} 0\, dxdy = 0.$$

(iv) On the surface, $GFBC$, $\hat{n} = \hat{k}$, $z = 1$, $\hat{r} \cdot \hat{n} = z = 1$, $dS_4 = dxdy$.

$$\therefore \quad \iint_{S_4} \vec{r} \cdot \hat{n}\, dS_4 = \int_{y=0}^{1} \int_{x=0}^{1} (1)\, dxdy = 1.$$

(v) On the surface, $EABF$, $\hat{n} = \hat{j}$, $y = 1$, $\hat{r} \cdot \hat{n} = y = 1$, $dS_5 = dxdz$.

$$\therefore \quad \iint_{S_5} \vec{r} \cdot \hat{n}\, dS_5 = \int_{z=0}^{1} \int_{x=0}^{1} (1)\, dxdz = 1.$$

(vi) On the surface, $ODGC$, $\hat{n} = -\hat{j}$, $y = 0$, $\hat{r} \cdot \hat{n} = -y = 0$, $dS_6 = dxdz$.

$$\therefore \quad \iint_{S_6} \vec{r} \cdot \hat{n}\, dS_5 = \int_{z=0}^{1} \int_{x=0}^{1} 0\, dxdz = 0.$$

Hence, $\displaystyle\iint_{S} \vec{r} \cdot \hat{n}\, dS = \mathbf{0 + 1 + 0 + 1 + 1 + 0 = 3.}$

Example 12.16. *Evaluate,* $\displaystyle\iint_S \vec{A} \cdot \hat{n}\, dS$, *where S is the surface of unit cube bounded by* $x = 0$, $x = 1$, $y = 0$, $y = 1$, $z = 0$, $z = 1$ *and* $\vec{A} = 4xz\hat{i} - y^2\hat{j} + yz\hat{k}$.

Solution : $OABCDEFG$ is a unit cube bounded by the surfaces $x = 0$, $x = 1$; $y = 0$, $y = 1$, $z = 0$, $z = 1$ as shown in Fig. 12.9.

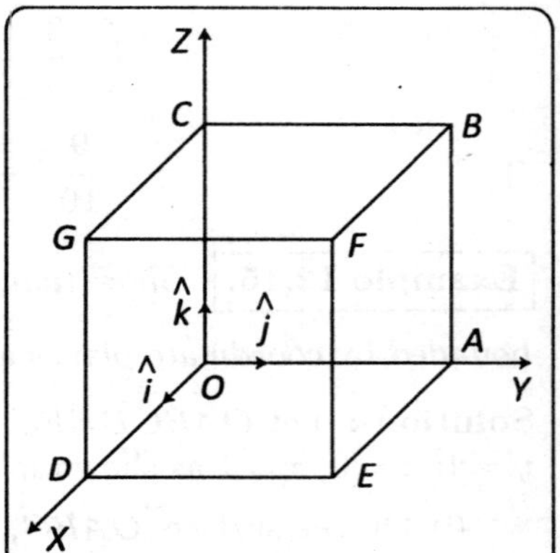

Fig. 12.9

(i) For the surface, $OABC$, unit normal $\hat{n} = -\hat{i}$, $x = 0$ and

$$\vec{A} \cdot \hat{n} = (4xz\hat{i} - y^2\hat{j} + yz\hat{k}) \cdot (-\hat{i}) = -4xz = 0.$$

$$\therefore \iint_S \vec{A} \cdot \hat{n}\, dS = \int_{z=0}^{1} \int_{y=0}^{1} 0\, dydz = 0.$$

(ii) For the surface, $DEFG$, $\hat{n} = \hat{i}$, $x = 1$, $dS = dydz$

$$\vec{A} \cdot \hat{n} = (4xz\hat{i} - y^2\hat{j} + yz\hat{k}) \cdot (\hat{i}) = 4xz = 4z.$$

$$\therefore \iint_S \vec{r} \cdot \hat{n}\, dS = \int_{z=0}^{1} \int_{y=0}^{1} 4(1)z\, dydz = 4\int_{z=0}^{1} \int_{y=0}^{1} z\, dzdy = 4\int_{z=0}^{1} z\, dz = 4\left[\frac{z^2}{2}\right]_0^1 = 2.$$

(ii) For the surface, $ODEA$, $\hat{n} = -\hat{k}$, $z = 0$, $dS = dxdy$.

$$\vec{A} \cdot \hat{n} = (4xz\hat{i} - y^2\hat{j} + yz\hat{k}) \cdot (-\hat{k}) = -yz = 0.$$

$$\therefore \iint_S \vec{A} \cdot \hat{n}\, dS = \int_{y=0}^{1} \int_{x=0}^{1} 0\, dxdy = 0.$$

(iii) For the surface, $CGFB$, $\hat{n} = \hat{k}$, $z = 1$, $dS = dxdy$.

$$\vec{A} \cdot \hat{n} = (4xz\hat{i} - y^2\hat{j} + yz\hat{k}) \cdot (\hat{k}) = yz.$$

$$\therefore \iint_S \vec{A} \cdot \hat{n}\, dS = \int_{y=0}^{1} \int_{x=0}^{1} yz\, dxdy = \int_{y=0}^{1} \int_{x=0}^{1} y\, dxdy = \frac{1}{2}.$$

(iv) For the surface, $DOCG$, $\hat{n} = -\hat{j}$, $y = 0$, $dS = dxdz$.

$$\vec{A} \cdot \hat{n} = (4xz\hat{i} - y^2\hat{j} + yz\hat{k}) \cdot (-\hat{j}) = y^2,\ dS = dxdz.$$

$$\therefore \iint_S \vec{A} \cdot \hat{n}\, dS = \int_{z=0}^{1} \int_{x=0}^{1} y^2\, dxdz = \int_{z=0}^{1} \int_{x=0}^{1} 0\, dxdz = 0.$$

(v) For the surface, $ABFE$, $\hat{n} = \hat{j}$, $y = 1$, $dS = dxdz$.

$$\vec{A} \cdot \hat{n} = (4xz\hat{i} - y^2\hat{j} + yz\hat{k}) \cdot (\hat{j}) = -y^2.$$

$$\therefore \iint_S \vec{A} \cdot \hat{n}\, dS = \int_{z=0}^{1} \int_{x=0}^{1} -y^2\, dxdz = -\int_{z=0}^{1} \int_{x=0}^{1} dxdz = -1.$$

$$\therefore \iint_S \vec{\boldsymbol{A}} \cdot \hat{\boldsymbol{n}}\, d\boldsymbol{S} = 0 + 2 + 0 + \frac{1}{2} + 0 - 1 = \frac{\mathbf{3}}{\mathbf{2}}.$$

Example 12.17. *Evaluate,* $\iint_S \vec{A} \cdot \hat{n}\, dS$, *over the entire surface* S *of the region bounded by the cylinder* $x^2 + z^2 = 9$, $x = 0$, $y = 0$, $z = 0$ *and* $y = 8$, *if* $\vec{A} = 6z\hat{i} + (2x + y)\hat{j} - x\hat{k}$.

Solution : We have, from Fig. 12.10,

$$\iint_S \vec{A} \cdot \hat{n}\, dS = \iint_{S_1} \vec{A} \cdot \hat{n}_1\, dS + \iint_{S_2} \vec{A} \cdot \hat{n}_2\, dS + \iint_{S_3} \vec{A} \cdot \hat{n}_3\, dS. \quad (1)$$

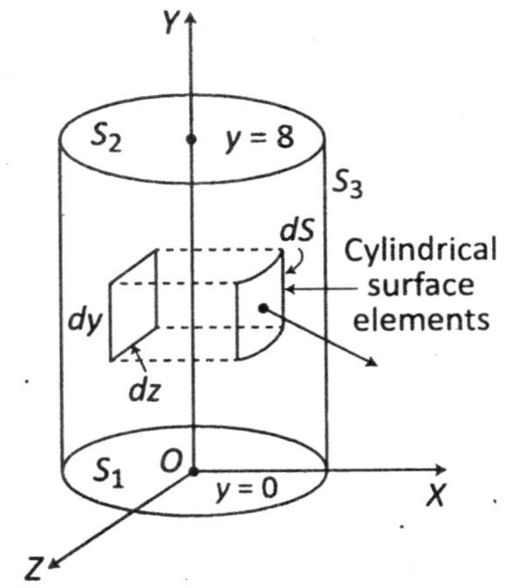

Fig. 12.10

Now over the surface S_1,

$$\iint_{S_1} \vec{A} \cdot \hat{n_1}\, dS$$

$$= \int_{x=0}^{3} \int_{z=0}^{\sqrt{9-x^2}} [6z\hat{i} + (2x + 0)\hat{j} - x\hat{k}] \cdot (-\hat{j}) dxdz$$

$$= \int_{x=0}^{3} \int_{z=0}^{\sqrt{9-x^2}} -2x\, dxdz = -2\int_{x=0}^{3} (\sqrt{9 - x^2}) x\, dx$$

$$= -2\left[x\int\sqrt{9-x^2}dx - \int 1\cdot\int\sqrt{9-x^2}dx\right]_0^3$$

$$= -2\left[x\left\{\frac{\sqrt{9-x^2}}{2}+\frac{9}{2}\sin^{-1}\frac{x}{3}\right\}_0^3 + 2\left\{\frac{x\sqrt{9-x^2}}{2}+\frac{9}{2}\sin^{-1}\frac{x}{3}\right\}_0^3\right]$$

$$= -2\times 3\times\frac{9}{2}\times\frac{\pi}{2}+2\times\frac{9}{2}\times\frac{\pi}{2} = -\frac{27\pi}{2}+\frac{9\pi}{2} = -9\pi.$$

$$\therefore \iint_{S_1}\vec{A}\cdot\hat{n}_1\,dS = -9\pi. \tag{2}$$

Again, $\iint_{S_2}\vec{A}\cdot\hat{n}_2\,dS = \iint(2x+8)dxdz \quad [\because\ dS = dxdz \text{ and } \hat{n}_2 = \hat{j}]$

$$= 2\int_{x=0}^{3}\int_{z=0}^{\sqrt{9-x^2}} x\,dxdz + 8\int_{x=0}^{3}\int_{z=0}^{\sqrt{9-x^2}} dxdz$$

$$= 2\int_{x=0}^{3}\sqrt{9-x^2}x\,dx + 8\int_{x=0}^{3}\sqrt{9-x^2}\,dx$$

$$= 9\pi + 8\left[\frac{x\sqrt{9-x^2}}{2}+\frac{9}{2}\sin^{-1}\frac{x}{3}\right] = 9\pi+8\left[\frac{9}{2}\times\frac{\pi}{2}\right]$$

$$= 9\pi + 18\pi = 27\pi.$$

$$\therefore \qquad \iint_{S_2}\vec{A}\cdot\hat{n}_1\,dS = 27\pi. \tag{3}$$

Also, $\iint_{S_3}\vec{A}\cdot\hat{n}_1\,dS = \iint_R(\vec{A}\cdot\hat{n})\frac{dydz}{|\hat{n}\cdot\hat{i}|} \qquad [\because\ \hat{n}_3 = \hat{n}]$

Now, $\phi(x,z) = (x^2+z^2), \quad \therefore\ \nabla\phi = 2(x\hat{i}+z\hat{k}).$

$$\therefore \qquad \hat{n} = \frac{\nabla\phi}{|\nabla\phi|} = \frac{2(x\hat{i}+z\hat{k})}{2\sqrt{9}} = \frac{x\hat{i}+z\hat{k}}{3} \quad \text{and} \quad |\hat{n}\cdot\hat{i}| = \frac{x}{3}.$$

$$\therefore \qquad \vec{A}\cdot\hat{n} = [6z\hat{i}+(2x+y)\hat{j}-x\hat{k}]\cdot\frac{1}{3}(x\hat{i}+z\hat{k}) = \frac{1}{3}[6zx-xz] = \frac{5}{3}xz.$$

Again, $\iint_{S_3}\vec{A}\cdot\hat{n}_3\,dS = \int_{z=-3}^{3}\int_{y=0}^{8}\frac{5}{3}xz\frac{dydz}{\frac{x}{3}} = 5\int_{z=-3}^{3}\int_{y=0}^{8}z\,dydz = 40\int_{-3}^{3}z\,dz$

$$= 40\left[\frac{z^2}{2}\right]_{-3}^{3} = 20\times 0 = 0.$$

$$\therefore \qquad \iint_{S_3}\vec{A}\cdot\hat{n}\,dS = 0. \tag{4}$$

Using eqns. (2), (3) and (4) in eqn. (1), we get

$$\iint_S \boldsymbol{\vec{A}\cdot\hat{n}\,dS} = -9\pi+27\pi+0 = \mathbf{18\pi}.$$

Example 12.18. *Let R be the projection of the surface S on the xy-plane, prove that the surface area of S is given by* $\iint_S \sqrt{1+\left(\frac{\partial z}{\partial x}\right)^2+\left(\frac{\partial z}{\partial y}\right)^2}\,dxdy$, *if the equation for S is* $z = f(x,y)$.

Solution : The equation for S is $z = f(x,y)$.

$\therefore \quad \phi(x,y,z) = z - f(x,y)$.

$$\therefore \quad \nabla\phi = \hat{i}\frac{\partial\phi}{\partial x} + \hat{j}\frac{\partial\phi}{\partial y} + \hat{k}\frac{\partial\phi}{\partial z} = \hat{i}\left(-\frac{\partial f}{\partial x}\right) + \hat{j}\left(-\frac{\partial f}{\partial y}\right) + \hat{k}.$$

Unit normal to S is

$$\hat{n} = \frac{\nabla\phi}{|\nabla\phi|} \quad \text{or,} \quad \hat{n} = \frac{-\hat{i}\frac{\partial f}{\partial x} - \hat{j}\frac{\partial f}{\partial y} + \hat{k}}{\sqrt{1+\left(\frac{\partial f}{\partial x}\right)^2+\left(\frac{\partial f}{\partial y}\right)^2}}.$$

$\therefore$ the surface area S is given by

$$\iint_R \nabla\phi\cdot\hat{n}\,dxdy = \iint_R \left(-\hat{i}\frac{\partial f}{\partial x} - \hat{j}\frac{\partial f}{\partial y} + \hat{k}\right)\cdot\frac{-\hat{i}\frac{\partial f}{\partial x} - \hat{j}\frac{\partial f}{\partial y} + \hat{k}}{\sqrt{1+\left(\frac{\partial f}{\partial x}\right)^2+\left(\frac{\partial f}{\partial y}\right)^2}}dxdy$$

$$= \iint_R \frac{\left[1+\left(\frac{\partial f}{\partial x}\right)^2+\left(\frac{\partial f}{\partial y}\right)^2\right]}{\sqrt{1+\left(\frac{\partial f}{\partial x}\right)^2+\left(\frac{\partial f}{\partial y}\right)^2}}dxdy = \iint_R \sqrt{1+\left(\frac{\partial f}{\partial x}\right)^2+\left(\frac{\partial f}{\partial y}\right)^2}\,dxdy$$

$$= \iint_R \sqrt{1+\left(\frac{\partial x}{\partial x}\right)^2+\left(\frac{\partial z}{\partial y}\right)^2}\,dxdy, \quad [\because z = f(x,y)].$$

Example 12.19. *Given,* $\vec{A} = (2x-y)\hat{i} - yz^2\hat{j} - y^2z\hat{k}$, *evaluate* $\iint_S (\nabla\times\vec{A})\cdot\hat{n}\,dS$, *where S is the upper surface of the sphere,* $x^2+y^2+z^2 = a^2$ *above the xy-plane.*

Solution : We have $\nabla\times\vec{A} = \left(\hat{i}\frac{\partial}{\partial x} + \hat{j}\frac{\partial}{\partial y} + \hat{k}\frac{\partial}{\partial z}\right)\times[(2x-y)\hat{i} - yz^2\hat{j} - y^2z\hat{k}]$

$$= \begin{vmatrix} \hat{i} & \hat{j} & \hat{k} \\ \frac{\partial}{\partial x} & \frac{\partial}{\partial y} & \frac{\partial}{\partial z} \\ (2x-y) & -yz^2 & -y^2z \end{vmatrix}$$

$$= \hat{i}\left[\frac{\partial}{\partial x}(-yz^2) - \frac{\partial}{\partial y}(-y^2z)\right] + \hat{j}\left[\frac{\partial}{\partial z}(2x-y)\frac{\partial}{\partial x}(-y^2z)\right] + \hat{k}\left[\frac{\partial}{\partial x}(-yz^2) - \frac{\partial}{\partial y}(2x-y)\right]$$

$$= \hat{i}[-2yz+2yz] + \hat{j}[0-0] + \hat{k}[0+1] = \hat{k}.$$

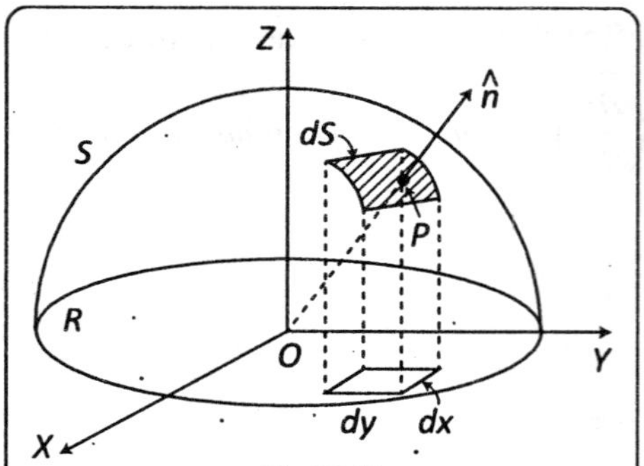

Fig. 12.11

Now, normal to $x^2+y^2+z^2=a^2$ or, $\phi(x,y,z)=a^2$ is

$$\nabla\phi=\hat{i}2x+\hat{j}2y+\hat{k}2z=2(\hat{i}x+\hat{j}y+\hat{k}z).$$

$\therefore$ outward unit normal to the surfaced is

$$\hat{n}=\frac{\nabla\phi}{|\nabla\phi|}=\frac{2(x\hat{i}+y\hat{j}+z\hat{k})}{\sqrt{4x^2+4y^2+4z^2}}$$

$$=\frac{x\hat{i}+y\hat{j}+z\hat{k}}{a}\quad [\because\ x^2+y^2+z^2=a^2].$$

The projection of S on xy-plane is the region R, whose boundary is the circle $x^2+y^2=a^2$, $z=0$.

Then, $$\iint_S(\nabla\times\vec{A})\cdot\hat{n}\frac{dxdy}{|\hat{n}\cdot\hat{k}|}=\iint_R\hat{k}\cdot\frac{(x\hat{i}+y\hat{j}+z\hat{k})}{a}\frac{dxdy}{\frac{z}{a}}$$

$$\left[\because\ \hat{n}=\frac{x\hat{i}+y\hat{j}+z\hat{k}}{a}\text{ and }|\hat{n}\cdot\hat{k}=\frac{z}{a}\right]$$

or, $$\iint_S(\nabla\times\vec{A})\cdot\hat{n}\,dS=\iint_R dxdy.$$

Now $x=\rho\cos\phi$, $y=\rho\sin\phi$, $dxdy=\rho d\rho d\phi$.

$$\therefore\quad \iint_R dxdy=\int_{\rho=0}^{1}\int_{\phi=0}^{2\pi}\rho d\rho d\phi=\int_{\rho=0}^{1}\left(\int_{\phi=0}^{2\pi}d\phi\right)\rho d\phi=2\pi\int_{\rho=0}^{1}\rho d\rho=2\pi\left[\frac{\rho^2}{2}\right]_0^1=\pi.$$

$$\therefore\quad \iint_S(\nabla\times\vec{A})\cdot\hat{n}\,dS=\pi.$$

Example 12.20. *Find the value of $\iint_S\vec{A}\cdot\hat{n}\,dS$ over the surface, $2x+y+2z=6$, in the first octant, when $\vec{A}=(x+y^2)\hat{i}-2x\hat{j}+2yz\hat{k}$.*

Solution : The given surface S as well as its projection on the xy-plane is shown in Fig. 12.12.

Now, $$\iint_S\vec{A}\cdot\hat{n}\,dS=\iint_R\vec{A})\cdot\hat{n}\frac{dxdy}{|\hat{n}\cdot\hat{k}|}.$$

Here, $\phi(x,y,z)=2x+y+2z-6$.

$$\therefore\quad \nabla\phi=\left(\hat{i}\frac{\partial}{\partial x}+\hat{j}\frac{\partial}{\partial y}+\hat{k}\frac{\partial}{\partial z}\right)(2x+y+2z-6)$$

$$=2\hat{i}+\hat{j}+2\hat{k}.$$

$\therefore$ the outward unit normal to the surface S is

$$\hat{n}=\frac{\nabla\phi}{|\nabla\phi|}=\frac{2\hat{i}+\hat{j}+2\hat{k}}{\sqrt{4+1+4}}=\frac{1}{3}(2\hat{i}+\hat{j}+2\hat{k})$$

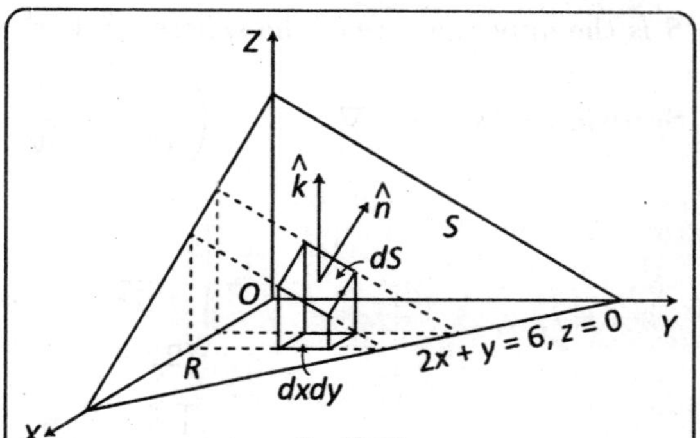

Fig. 12.12

$$\therefore\ \vec{A}\cdot\hat{n}=[(x+y^2)\hat{i}-2x\hat{j}+2yz\hat{k}]\cdot\frac{1}{3}(2\hat{i}+\hat{j}+2\hat{k})=\frac{1}{3}[2(x+y^2)-2x+4yz]$$

$$=\frac{1}{3}(2y^2+4yz).$$

$$\therefore \iint_S \vec{A}\cdot\hat{n}\,dS = \iint_R \vec{A}\cdot\hat{n}\frac{dxdy}{|\hat{n}\cdot\hat{k}|} = \iint_R \frac{1}{3}(2y^2+4yz)\frac{dxdy}{|\hat{n}\cdot\hat{k}|}$$

$$= \frac{2}{3}\iint_R (y^2+2yz)\frac{dxdy}{\frac{2}{3}} \qquad \left[\because\ \hat{n}\cdot\hat{k} = \frac{2}{3}\right]$$

$$= \iint_R (y^2+2yz)dxdy = \iint_R y(y+2z)dxdy$$

$$= \iint_R y(6-2x)dxdy \qquad [\because\ y+2z = 6-2x]$$

$$= \int_{x=0}^{3}\int_{y=0}^{(6-2x)} (6y-2yx)dxdy = \int_{x=0}^{3}\left(6\frac{y^2}{2} - 2x\frac{y^2}{2}\right)_0^{6-2x} dx$$

$$= \int_{x=0}^{3} [3y^2 - xy^2]_0^{6-2x} dx = \int_{x=0}^{3} [3(6-2x)^2 - x(6-2x)^2]dx$$

$$= 4\int_{x=0}^{3}(3-x)(3-x)^3 dx = 4\int_{x=0}^{3}(3-x)^3 dx = 4\int_{x=0}^{3}[27-27x+9x-x^3]dx$$

$$= 4\int_0^3 [27-18x-x^3]dx = 4\left[27x - 18\frac{x^2}{2} - \frac{1}{4}x^4\right]_0^3 = 4\left[27x - 9x^2 - \frac{x^4}{4}\right]_0^3$$

$$= 4\left[27\times 3 - 9\times 9 - \frac{1}{4}3^4\right]_0^3 = 4\times -\frac{1}{4}81 = -81.$$

$$\therefore \iint_S \vec{A}\cdot\hat{n}\,dS = -81$$

Example 12.21. *Evaluate the integrals,* (a) $\iiint\limits_V \nabla\cdot\vec{A}dV$ (b) $\iiint\limits_V \nabla\times\vec{F}dV$, *where V is the region bounded by the planes $x = 0$, $y = 0$, $z = 0$ and $2x+2y+z = 4$ and $\vec{A} = (2x^2-3z)\hat{i} - 2xy\hat{j} - 4x\hat{k}$.*

Solution : (a) We have $\nabla\times\vec{A} = \left(\hat{i}\frac{\partial}{\partial x} + \hat{j}\frac{\partial}{\partial y} + \hat{k}\frac{\partial}{\partial z}\right)\cdot[(2x^2-3z)\hat{i} - 2xy\hat{j} - 4x\hat{k}]$

$$= \frac{\partial}{\partial x}(2x^3-3z) + \frac{\partial}{\partial x}(-2xy) + \frac{\partial}{\partial z}(-4x) = 4x - 2x - 0 = 2x.$$

Limits are $z = 0$ to $4-2x-2y$; $y = 0$ to $\frac{4-2x}{2} = (2-x)$; $x = 0$ to $\frac{4}{2} = 2$.

$$\therefore \iiint\limits_V \nabla\cdot\vec{A}dV = \iiint 2x\,dxdydz = \iint 2x[z]_0^{4-2x-2y}dxdy = \iint 2x(4-2x-2y)dydx$$

$$= \int [8xy - 4x^2y - 2xy^2]_0^{2-x} dx = \int_0^2 [8x(2-x) + 4x^2(2-x) - 2x(2-x^2)]dx$$

$$= \int_0^2 [2x^3 - 8x^2 + 8x]dx = \left[2\frac{2^4}{4} - 8\frac{2^3}{3} + 8\frac{2^2}{2}\right] = \frac{8}{3}$$

$$\therefore \iiint\limits_V \nabla\cdot\vec{A}dV = \frac{8}{3}$$

(b) Now, $\nabla \times \vec{A} = \begin{vmatrix} \hat{i} & \hat{j} & \hat{k} \\ \dfrac{\partial}{\partial x} & \dfrac{\partial}{\partial y} & \dfrac{\partial}{\partial z} \\ 2x^2 - 3z & -2xy & -4x \end{vmatrix} = \hat{j} - 2\hat{k}y$, also $dV = dx\,dy\,dz$.

$$\therefore \iiint_V (\nabla \cdot \vec{A})dV = \int_{x=0}^{2}\int_{y=0}^{(2-x)}\int_{z=0}^{(4-2x-2y)} (\hat{j} - 2\hat{k}y)dx\,dy\,dz$$

$$= \int_{x=0}^{2}\int_{y=0}^{2-x} [\hat{j}(4-2x-2y) - 2\hat{k}(4-2x-2y)]dxdy$$

$$= \int_{x=0}^{2} \left[\hat{j}(4y - 2xy - y^2) - 2\hat{k}(4y - 2xy - y^2)\right]_0^{2-x} dx$$

$$= \int_{x=0}^{2} \left[\hat{j}(2-x)^2 - 2\hat{k}\frac{(2-x)^3}{3}\right] dx = \left[\hat{j}\frac{(2-x)^2}{3} + \frac{2\hat{k}}{3}\frac{(x-2)^4}{4}\right]_0^2$$

$$= \left[-\hat{j}\left(-\frac{8}{3}\right) - \frac{2\hat{k}}{3}\cdot\frac{2^4}{4}\right] = \frac{8}{3}\hat{j} - \frac{8}{3}\hat{k} = \frac{\mathbf{8}}{\mathbf{3}}(\hat{\boldsymbol{j}} - \hat{\boldsymbol{k}}).$$

Example 12.22. *Evaluate* $\int_V x^2 y\,dV$, *where* V *is the closed region bounded by the planes* $4x + 2y + z = 8$, $x = 0$, $y = 0$ *and* $z = 0$. **[BU(H) 1995]**

Solution : In the problem, z varies from 0 to $8 - 4x - 2y$, y varies from 0 to $4 - 2x$ and x varies from 0 to 2.

$\therefore$ the given integral is

$$\iiint_V x^2 y dV = \int_{x=0}^{2}\int_{y=0}^{4-2x}\int_{z=0}^{8-4x-2y} x^2 y\,dxdydz$$

$$= \int_{x=0}^{2}\int_{y=0}^{4-2x} x^2 y(8-4x-2y)dxdy = \int_{x=0}^{2}\left[\int_{y=0}^{4-2x} x^2 y(8-4x-2y)dy\right]dx$$

$$= \int_{x=0}^{2}\left[8x^2\frac{y^2}{2} - 4x^3\frac{y^2}{2} - 2x^2\frac{y^3}{3}\right]_0^{4-2x} dx = \int_{x=0}^{2}\left[4x^2y^2 - 2x^3y^2 - \frac{2}{3}x^2y^3\right]_0^{4-2x} dx$$

$$= \int_{x=0}^{2}\left[4x^2(4-2x)^2 - 2x^3(4-2x)^3 - \frac{2}{3}x^2(4-2x)^3\right] dx$$

$$= \int_{x=0}^{2}\left[\frac{64}{3}x^2 - 32x^3 + 16x^4 - \frac{8}{3}x^5\right] dx = \left[\frac{64}{3}\cdot\frac{x^3}{3} - 32\frac{x^4}{4} + 16\frac{x^5}{5} - \frac{8}{3}\frac{x^6}{6}\right]_0^2$$

$$= \frac{2560 - 1152 + 4608 - 1280}{45} = \frac{128}{45} = \mathbf{2\frac{38}{45}}.$$

Example 12.23. *Evaluate* $\iiint_V (2x + y)dV$, *where* V *is the closed region bounded by the cylinder* $z = 4 - x^2$ *and the planes,* $x = 0$, $y = 0$, $y = 2$ *and* $z = 0$.

Solution : We have $\iiint\limits_V (\boldsymbol{2x+y})\boldsymbol{dV} = \int_{y=0}^{2}\int_{x=0}^{2}\left(\int_{z=0}^{(4-x^2)}(2x+y)dz\right)dxdy$

$$=\int_{y=0}^{2}\int_{x=0}^{2}[2xz+yz]_0^{4-x^2}dxdy=\int_{y=0}^{2}\int_{x=0}^{2}[2x(4-x^2)+y(4-x^2)]dxdy$$

$$=\int_{y=0}^{2}\int_{x=0}^{2}[8x-2x^3+4y-x^2y]dxdy=\int_{y=0}^{2}\left[4x^2-\frac{x^4}{2}+4xy-\frac{1}{3}yx^3\right]_0^2 dy$$

$$=\int_{y=0}^{2}\left[16-\frac{1}{2}\times 2^4+4\cdot 2y-\frac{y}{3}8\right]dy=\int_{y=0}^{2}\left[8+\frac{16}{3}y\right]_0^2 dy$$

$$=\left[8y+\frac{16}{3}\frac{y^2}{2}\right]_0^2 dy=16+\frac{32}{3}=\frac{\mathbf{80}}{\mathbf{3}}.$$

Integral Theorems

12.8 Stoke's Theorem

Statement

The line integral of the tangential component of a vector taken around a simple closed curve (curve that does not intersect) is equal to the surface integral of the normal component of the curl of the vector taken over any surface having the curve as the boundary.

Mathematically, the theorem states that, if S is an open two sided surface bounded by a simple closed curve C, and if $\vec{A}$ be any continuously differentiable vector point function, then

$$\oint_C \vec{A}\cdot d\vec{r}=\iint_S(\nabla\times\vec{A})\cdot\hat{n}\,dS=\iint_S(\nabla\times\vec{A})\cdot\vec{dS},$$

where the boundary is traversed in the positive (counterclockwise) direction. $\hat{n}$ is the outward drawn unit normal to the element of surface dS.

Stoke's theorem thus expresses the relation between line integral and surface integral of a vector.

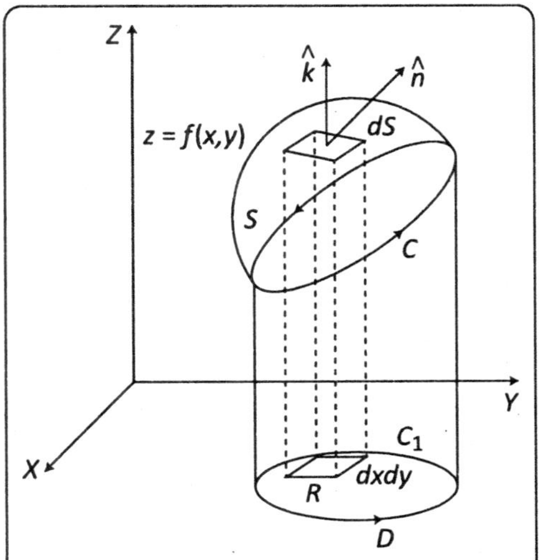

Fig. 12.13

Proof : Let C be the boundary of a surface S whose projection on the xy-plane is R having C_1 as the boundary (Fig. 12.13). Let the surface S be represented by the function $Z=f(x,y)$, where f is the single valued continuous and differential functions of x and y.

If $\vec{r}=\hat{i}x+\hat{j}y+\hat{k}z$ be any point on the surface S, then

$$\frac{\partial\vec{r}}{\partial y}=\hat{j}\frac{\partial z}{\partial y}\hat{k}\quad[\because\ z=f(x,y)].$$

$$\therefore\quad \hat{n}\cdot\frac{\partial\vec{r}}{\partial y}=\hat{n}\cdot\hat{j}+\frac{\partial z}{\partial y}\hat{n}\cdot\hat{k},$$

where $\hat{n}$ is a unit normal to the surface S and $\frac{\partial\vec{r}}{\partial y}$ is tangential to S.

$$\therefore \quad \hat{n}\cdot\frac{\partial\vec{r}}{\partial y}=0 \quad \text{or,} \quad \hat{n}\cdot\hat{j}+\frac{\partial z}{\partial y}\hat{n}\cdot\hat{k}=0$$

$$\text{or,} \quad \hat{n}\cdot\hat{j}=-\frac{\partial z}{\partial y}\hat{n}\cdot\hat{k}. \tag{i}$$

If $\vec{A}=\hat{i}A_1+\hat{j}A_2+\hat{k}A_3$, then

$$\nabla\times A_1\hat{i}=\begin{vmatrix}\hat{i} & \hat{j} & \hat{k}\\ \dfrac{\partial}{\partial x} & \dfrac{\partial}{\partial y} & \dfrac{\partial}{\partial z}\\ A_1 & 0 & 0\end{vmatrix}=\frac{\partial A_1}{\partial z}\hat{j}-\frac{\partial A_1}{\partial y}\hat{k}.$$

$$\nabla\times A_1\hat{i}\cdot\hat{n}\,dS=\frac{\partial A_1}{\partial z}\hat{n}\cdot\hat{j}\,dS-\frac{\partial A_1}{\partial y}\hat{n}\cdot\hat{k}\,dS=\left(\frac{\partial A_1}{\partial z}\hat{n}\cdot\hat{j}-\frac{\partial A_1}{\partial y}\hat{n}\cdot\hat{k}\right)dS. \tag{ii}$$

Using eqns. (i) and (ii), we get

$$\nabla\times A_1\hat{i}\cdot\hat{n}\,dS=-\left(\frac{\partial A_1}{\partial z}\frac{\partial z}{\partial y}+\frac{\partial A_1}{\partial y}\right)\hat{n}\cdot\hat{k}dS. \tag{iii}$$

Again, z being a function of x, y, i.e., $z=f(x,y)$ and $A_1=A_1(x,y,z)$ we can express A_1 as a function of two variables viz. x and y.

Let $A_1(x,y,z)=\phi(x,y)$. Then,

$$\frac{\partial\phi}{\partial y}=\frac{\partial A_1}{\partial y}+\frac{\partial A_1}{\partial z}\frac{\partial z}{\partial y}. \tag{iv}$$

$$\therefore \quad \nabla\times A_1\hat{i}\cdot\hat{n}\,dS=-\frac{\partial\phi}{\partial y}\hat{n}\cdot\hat{k}\frac{dxdy}{\hat{n}\cdot\hat{k}}, \quad \left[\because\ dS=\frac{dxdy}{\hat{n}\cdot\hat{k}}\right]$$

$$=-\frac{\partial\phi}{\partial y}dxdy. \tag{v}$$

According to **Green's theorem in the plane,**

$$\oint_C M(x,y)dx+N(x,y)dy=\iint_S\left(\frac{\partial N}{\partial x}-\frac{\partial M}{\partial y}\right)dxdy.$$

Putting $M=\phi$ and $N=0$, we get

$$\oint_D \phi dx=\iint_R-\frac{\partial\phi}{\partial y}dxdy. \tag{vi}$$

Since, at each point (x,y) on the boundary D, $\phi(x,y)=A_1(x,y,z)$ on the contour C,

$$\therefore \quad \int_C\phi(x,y)dx=\int_C A_1(x,y,z)dx. \tag{vi}$$

$$\therefore \quad \iint_S\nabla\times A_1\hat{i}\cdot\hat{n}\,dS=\int_C A_1dx. \tag{vii}$$

Considering the projection of S on yz-and zx-plane it can be proved that,

$$\iint_S\nabla\times A_2\hat{j}\cdot\hat{n}\,dS=\int_C A_2dy \tag{viii}$$

and $\iint_S \nabla \times A_3\hat{k} \cdot \hat{n}\, dS = \int_C A_3 dz.$ (ix)

From eqns. (vii), (viii) and (ix) we get by addition

$$\iint_S \nabla \times A_1\hat{i} \cdot \hat{n}\, dS + \iint_S \nabla \times A_2\hat{j} \cdot \hat{n}\, dS + \iint_S \nabla \times A_3\hat{k} \cdot \hat{n}\, dS$$
$$= \oint_C A_1\, dx + \oint_C A_2\, dy + \oint_C A_3\, dz$$

or, $\iint_S \nabla \times (\hat{i}A_1 + \hat{j}A_2 + \hat{k}A_3) \cdot \hat{n}\, dS = \oint_C A_1 dx + A_2 dy + A_3 dz$

or, $\iint_S (\nabla \times \mathbf{A}) \cdot \hat{\mathbf{n}}\, d\mathbf{S} = \oint_C \vec{A} \cdot d\vec{r}.$ (x)

This proves Stoke's theorem.

12.9 Gauss' Divergence Theorem

Statement

The theorem states that *the surface integral of the normal component of a vector taken around a closed surface is equal to the integral of the divergence of the vector taken over the volume enclosed by the surface.*

If V be the volume bounded by a closed surface S and $\vec{A}$ is a vector function of position with continuous derivatives, then according to the divergence theorem of Gauss,

$$\oiint_S \vec{A} \cdot \hat{n}\, dS = \iiint_V (\nabla \cdot \hat{A}) dV,$$

where $\hat{n}$ is the outward unit normal to S, Gauss' divergence theorem thus expresses the relation between surface and volume integrals.

Proof : We consider a closed surface S of any arbitrary shape drawn in a vector field $\vec{A}$ enclosing a volume V (Fig. 12.14). A line drawn parallel to the z-axis cuts S in two points, say, $P_1(x, y, z_1)$ and $P_2(x, y, z_2)$ $(z_1 \le z_2)$ and meets at $Q(x, y, 0)$ on the projection R of V on the xy-plane.

Now, $\iiint_V (\nabla \cdot \vec{A}) dV = \iiint_V \left(\frac{\partial A_1}{\partial x} + \frac{\partial A_2}{\partial y} + \frac{\partial A_3}{\partial z}\right) dxdydz$, where $\vec{A} = \hat{i}A_1 + \hat{j}A_2 + \hat{k}A_3$

or, $$\iiint_V (\nabla \cdot \vec{A}) dV = \iiint_V \frac{\partial A_1}{\partial x} dxdydz + \iiint_V \frac{\partial A_2}{\partial y} dxdydz + \iiint_V \frac{\partial A_3}{\partial z} dxdydz. \quad \text{(i)}$$

Now, $$\iiint_V \frac{\partial A_3}{\partial z} dxdydz = \iint_R \left[\int_{z=f_1(x,y)}^{z=f_2(x,y)} \frac{\partial A_3}{\partial z} dz\right] dxdy$$

$$= \iint_R [A_3(x, y, f_2) - A_3(x, y, f_1)] dxdy. \quad \text{(ii)}$$

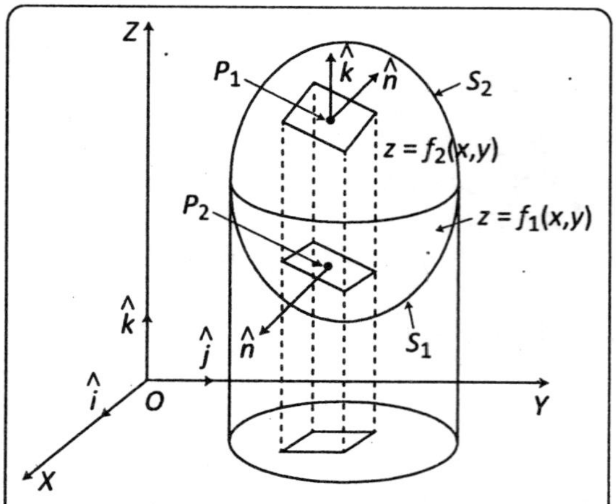

Fig. 12.14

Let S_1 and S_2 be respectively the lower and upper parts of S corresponding to P_1 and P_2 and $\hat{n}$ an outward drawn unit normal to S.

$$\therefore \quad dS = \frac{dxdy}{\hat{n}\cdot\hat{k}}, \quad \text{for } S_2$$

$$\text{and } dS = -\frac{dxdy}{\hat{n}\cdot\hat{k}}, \quad \text{for } S_1$$

for $\hat{n}$ makes an acute angle with $\hat{k}$ and an obtuse angle at S_1 with $\hat{k}$.

$$\therefore \quad \iint_R A_3(x, y, f_2)dxdy = \int_{S_2} A_3(\hat{n}\cdot\hat{k})dS$$

$$\text{and } \iint_R A_3(x, y, f_1)dxdy = -\int_{S_1} A_3(\hat{n}\cdot\hat{k})dS.$$

$$\therefore \quad \iiint_V \frac{\partial A_3}{\partial z}dxdydz = \int_{S_2} A_3(\hat{n}\cdot\hat{k})dS + \int_{S_1} A_3(\hat{n}\cdot\hat{k})dS = \int_{S_2} A_3(\hat{n}\cdot\hat{k})dS. \tag{iii}$$

Likewise we can show,

$$\iiint_V \frac{\partial A_2}{\partial y}dxdydz = \int_S A_2(\hat{n}\cdot\hat{j})dS \tag{iv}$$

$$\text{and} \quad \iiint_V \frac{\partial A_1}{\partial x}dxdydz = \int_S A_1(\hat{n}\cdot\hat{i})dS. \tag{v}$$

Now, adding eqns. (iii), (iv) and (v), we get

$$\iiint_V \left(\frac{\partial A_1}{\partial x} + \frac{\partial A_2}{\partial y} + \frac{\partial A_3}{\partial z}\right) dxdydz = \int_S (\hat{i}A_1 + \hat{j}A_2 + \hat{k}A_3)\cdot\hat{n}\, dS$$

$$\text{or,} \quad \iiint_V (\nabla\cdot\vec{A})dV = \oiint_S \vec{A}\cdot\hat{n}\, dS. \tag{vi}$$

This proves Gauss' divergence theorem.

12.10 Green's Theorem in the Plane

Statement

If $M(x, y)$ and $N(x, y)$ be two continuous functions of x, y having continuous derivatives in region R of the xy-plane bounded by a simple closed curve C, then

$$\oint_C M(x, y)dx + N(x, y)dy = \iint_R \left(\frac{\partial N}{\partial x} - \frac{\partial N}{\partial y}\right) dxdy.$$

where C is traversed in the counter-clockwise direction.

Proof : We consider a region R in the xy-plane bounded by a simple closed curve C such that a line parallel to either x-axis or y-axis cuts the curve C in atmost two points (Fig. 12.15).

The curves PKQ and PLQ are represented by equations $y = f_1(x)$ and $y = f_2(x)$ respectively. Then,

$$\iint_R \frac{\partial M}{\partial y} dxdy = \int_{x=a}^{b} \left[\int_{y=f_1(x)}^{y=f_2(x)} \frac{\partial M}{\partial y} dy \right] dx$$

$$= \int_{x=a}^{b} [M(x, y)]_{f_1(x)}^{f_2(x)} dx$$

$$= \int_{x=a}^{b} [M(x, f_2(x)) - M(x, f_1(x))] dx$$

$$= -\int_{x=b}^{a} [M(x, f_2(x))] dx - \int_{x=a}^{b} M[x, f_1(x)] dx$$

$$= -\oint M\, dx. \qquad \text{(i)}$$

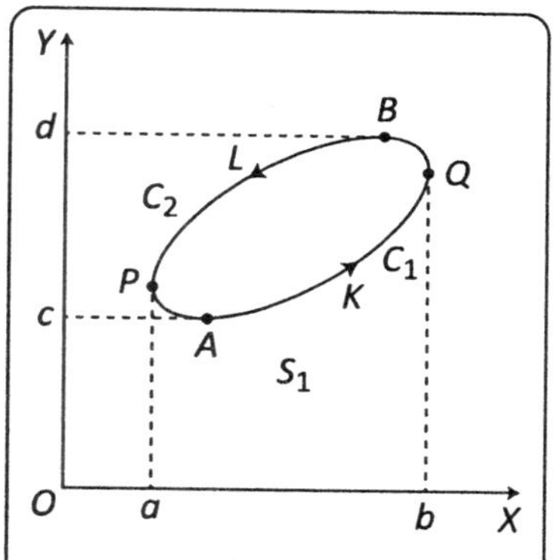

Fig. 12.15

Now, let the equations to the curves APB and AQB be $x = f_1(y)$ and $x = f_2(y)$ respectively. Then,

$$\iint_R \frac{\partial N}{\partial x} dxdy = \int_{y=c}^{d} \left[\int_{x=f_1(y)}^{f_2(y)} \frac{\partial N}{\partial x} dx \right] dy = \int_{y=c}^{d} [N(x, y)]_{f_1(y)}^{f_2(y)} dy$$

$$= \int_{y=c}^{d} [N(f_2(y), y) - N(f_1(y), y)] dy = \int_c^d [N(f_2(y), y) dy + \int_d^c N(f_1(y), y) dy$$

$$= \oint N(x, y)\, dy. \qquad \text{(ii)}$$

Adding eqns. (i) and (ii), we get

$$\oint_C \boldsymbol{M(x, y)dx + N(x, y)dy = \iint_R \left(\frac{\partial N}{\partial x} - \frac{\partial N}{\partial y} \right) dxdy.}$$

This proves Green's theorem in the plane. Green's theorem in the plane is a special case of Stokes' theorem. Also Gauss' divergence theorem is the Green's theorem in space.

12.11 Green's Theorem in the Plane in Vector Form

Let $\vec{A} = M\hat{i} + N\hat{j}$ and $\vec{r} = \hat{i}x + \hat{j}y$.

$\therefore \quad d\vec{r} = \hat{i}dx + \hat{j}dy$ and $\vec{A} \cdot d\vec{r} = (M\hat{i} + N\hat{j}) \cdot (\hat{i}dx + \hat{j}dy) = M\,dx + N\,dy.$

$$\nabla \times \vec{A} = \begin{vmatrix} \hat{i} & \hat{j} & \hat{k} \\ \frac{\partial}{\partial x} & \frac{\partial}{\partial y} & \frac{\partial}{\partial z} \\ M & N & O \end{vmatrix} = -\frac{\partial N}{\partial x}\hat{i} + \frac{\partial M}{\partial z}\hat{j} + \left(\frac{\partial N}{\partial x} - \frac{\partial M}{\partial y} \right) \hat{k}$$

$$\therefore \quad (\nabla \times \vec{A}) \cdot \hat{k} = \frac{\partial N}{\partial x} - \frac{\partial M}{\partial y}.$$

Now, Green's theorem in plane is

$$\oint_C M\, dx + N\, dy = \iint_R \left(\frac{\partial N}{\partial x} - \frac{\partial M}{\partial y} \right) dxdy.$$

In vector form, $\displaystyle \oint_C \boldsymbol{\vec{A} \cdot d\vec{r} = \iint_R \left(\frac{\partial N}{\partial x} - \frac{\partial M}{\partial y} \right) dxdy.}$

12.12 Solved Problems (B)

Example 12.24. *Verify Stokes' theorem for* $\vec{A} = (2x-y)\hat{i} - yz^2\hat{j} - y^2z\hat{k}$ *considering the upper half surface of sphere* $x^2+y^2+z^2=1$. **[BU(H) 1997]**

Solution : We have $\phi(x,y,z) = x^2+y^2+z^2-1=0$.

$\therefore$ outward drawn unit normal to the surface of the sphere is

$$\hat{n} = \frac{\nabla\phi}{|\nabla\phi|} = \frac{2(\hat{i}x+\hat{j}y+\hat{k}z)}{2\sqrt{x^2+y^2+z^2}} = \hat{i}x+\hat{j}y+\hat{k}z, \quad [\because\ x^2+y^2+z^2=1].$$

Now, $\displaystyle\iint_S (\nabla\times\vec{A})\cdot d\vec{S} = \iint_S \vec{B}\cdot dS = \iint_R \vec{B}\cdot\hat{n}\frac{dxdy}{|\hat{n}\cdot\hat{k}|}$, [where $\vec{B} = \nabla\times\vec{A}$]

and $$\vec{B} = \begin{vmatrix} \hat{i} & \hat{j} & \hat{k} \\ \frac{\partial}{\partial x} & \frac{\partial}{\partial y} & \frac{\partial}{\partial z} \\ (2x-y) & -yz^2 & -y^2z \end{vmatrix} = \hat{i}(-2yz+2yz) - \hat{j}0 + \hat{k}(0+1) = \hat{k}.$$

$$\therefore \quad \iint_S (\nabla\times\vec{A})\cdot d\vec{S} = \iint_R \frac{z\,dxdy}{z} = \iint_R dxdy, \quad [\because\ (\nabla\times\vec{A})\cdot\hat{n} = \vec{B}\cdot\hat{n} \text{ and } \hat{k}\cdot\hat{n} = z].$$

In polar coordinates (Fig. 12.16), $dxdy = r\,d\theta dr$.

$$\therefore \quad \iint_S (\nabla\times\vec{A})\cdot d\vec{S} = \int_0^1 r\,dr \int_0^{2\pi} d\theta = \pi.$$

Also, $\vec{A}|_{z=0} = (2x-y)\hat{i}$ and $d\vec{r} = \hat{i}\,dx + \hat{j}\,dy$.

$$\therefore \quad \oint_C \vec{A}\cdot d\vec{r} = \int (2x-y)dx.$$

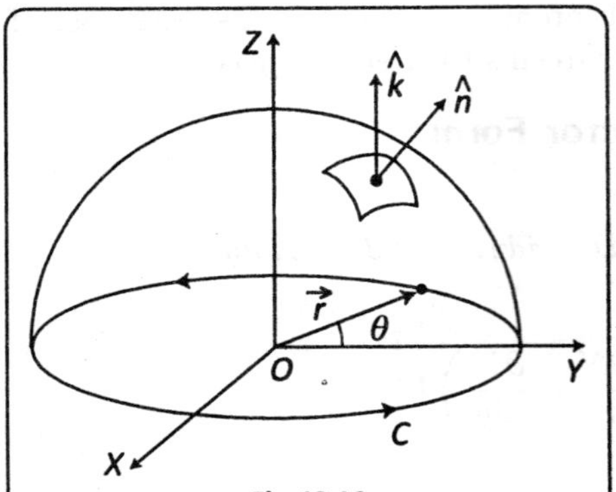

Fig. 12.16

But, $x=\cos\theta$ and $y=\sin\theta$, $dx = -\sin\theta\,d\theta$

$$\therefore \quad \oint_C \vec{A}\cdot d\vec{r} = \int_0^{2\pi} (2\cos\theta - \sin\theta)(-\sin\theta)d\theta$$

$$= 2\int_0^{2\pi} \sin\theta\, d(\sin\theta) + \int_0^{2\pi} \sin^2\theta\, d\theta$$

$$= 2\times\left[\frac{1}{2}\sin^2\theta\right]_0^{2\pi} + \int_0^{2\pi} \frac{1-\cos 2\theta}{2}d\theta$$

$$= 0 + \frac{1}{2}\left[2\pi - \left(\frac{1}{2}\sin 2\theta\right)_0^{2\pi}\right] = \pi.$$

$$\therefore \quad \oint_C \vec{A}\cdot d\vec{r} = \iint_S (\nabla\times\vec{A})\cdot d\vec{S}.$$

Hence, Stokes' theorem is verified.

Example 12.25. *Show that,* $\vec{F} = \dfrac{-\hat{i}x+\hat{j}y}{x^2+y^2}$ *is irrotational and that* $\oint_C \vec{F}\cdot d\vec{r} = 2\pi$, *where C is a circle containing the origin.*

Solution : We have $\vec{F} = \dfrac{-\hat{i}x + \hat{j}y}{\sqrt{x^2+y^2}} = -\hat{i}\dfrac{x}{\sqrt{x^2+y^2}} + \hat{j}\dfrac{y}{\sqrt{x^2+y^2}}$

$= \nabla\phi$, where $\phi = \tan^{-1}\left(\dfrac{y}{x}\right)$.

$\therefore \quad \nabla \times \vec{F} = \nabla \times \nabla\phi = 0$, i.e., $\vec{F}$ **is irrotational.**

$$\therefore \quad \oint_C \vec{F} \cdot d\vec{r} = \oint_C \nabla\phi \cdot d\vec{r} = \oint_C d\phi == \oint_C d\left(\tan^{-1}\frac{y}{x}\right) = \oint_C d\theta = 2\pi.$$

Hence, the result follows.

Example 12.26. *If $\vec{A} = -\hat{i}y + \hat{j}x$ and C, a closed curve in the xy-plane, then applying Stokes' theorem prove that, $\oint_C \vec{A} \cdot d\vec{r} = 2S$, where S is the area enclosed in the xy-plane by C. Hence, prove that area of an ellipse is πab.*

Solution : We have, $\oint_C \vec{A} \cdot d\vec{r} = \oint_C (-\hat{i}y + \hat{j}x) \cdot (\hat{i}dx + \hat{j}dy) = \oint_C (x\,dy - y\,dx)$.

By Stokes' theorem, $\oint_C \vec{A} \cdot d\vec{r} = \iint_S (\nabla \times \vec{A}) \cdot \vec{dS}$.

$$\text{Now, } \nabla \times \vec{A} = \begin{vmatrix} \hat{i} & \hat{j} & \hat{k} \\ \dfrac{\partial}{\partial x} & \dfrac{\partial}{\partial y} & \dfrac{\partial}{\partial z} \\ -y & x & 0 \end{vmatrix} = \hat{i}(0-0) + \hat{j}(0-0) + \hat{k}(1+1) = 2\hat{k}.$$

$$\therefore \quad (\nabla \times \vec{A}) \cdot d\vec{S} = 2\hat{k} \cdot \hat{k}\,dxdy = 2dxdy$$

$$\therefore \quad \oint_C \vec{A} \cdot d\vec{r} = \oint_C (x\,dy - y\,dx) = 2S,$$

where $S = \dfrac{1}{2}\oint_C (x\,dy - y\,dx)$ is the area of the region in xy-plane enclosed by C.

Now, for an ellipse, $x = a\cos\theta, \quad \therefore \quad dx = -a\sin\theta\,d\theta$

$y = a\sin\theta, \quad \therefore \quad dy = a\cos\theta\,d\theta.$

$$\therefore \quad S = \frac{1}{2}\oint_C (ab\cos^2\theta + ab\sin^2\theta)d\theta = \frac{1}{2}ab\int_0^{2\pi} d\theta = \frac{1}{2}ab2\pi = \pi ab. \quad \textbf{(Proved)}$$

Example 12.27. *Verify Stokes' theorem for the function $\vec{F} = x^2\hat{i} - xy\hat{j}$ integrated round the square in the plane $z = 0$ and bounded by the lines $x = 0$, $y = 0$, $x = a$, $y = a$.*

[AMIETE(Summer) 1989]

Solution : Since $\vec{F} = x^2\hat{i} - xy\hat{j}$,

$$\therefore \quad \nabla \times \vec{F} = \begin{vmatrix} \hat{i} & \hat{j} & \hat{k} \\ \dfrac{\partial}{\partial x} & \dfrac{\partial}{\partial y} & \dfrac{\partial}{\partial z} \\ x^2 & -xy & 0 \end{vmatrix} = \hat{i}(0-0) + \hat{j}(0-0) + \hat{k}(-y) = -y\hat{k}.$$

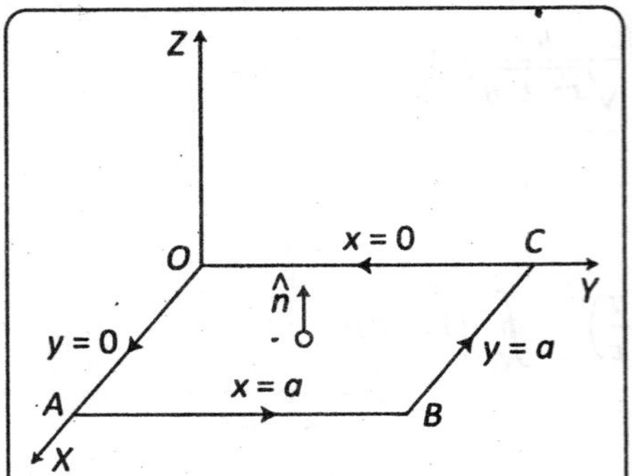

Fig. 12.17

$$\therefore \iint_S (\nabla \times \vec{F}) \cdot \hat{n} dS = \iint_S (-y\hat{k}) \cdot \hat{n}\, dx dy$$

$$= \iint_S (-y\hat{k}) \cdot \hat{k}\, dx dy \quad [\because \hat{n} = \hat{k}]$$

$$= \int_{x=0}^{a} dx \int_0^a -y\, dy$$

$$= \int_0^1 dx \left[-\frac{y^2}{2}\right]_0^a$$

$$= -\frac{a^3}{2}. \quad \text{(i)}$$

Again, $\oint_C \vec{F} \cdot d\vec{r} = \oint_C (x^2\hat{\imath} - xy\hat{\jmath}) \cdot (\hat{\imath}dx + \hat{\jmath}dy) = \int_C (x^2\, dx - xy\, dy),$

where C is the boundary $OABCO$ (Fig. 12.17).

$$\therefore \oint_C \vec{F} \cdot d\vec{r} = \int_{OA} \vec{F} \cdot d\vec{r} + \int_{AB} \vec{F} \cdot d\vec{r} + \int_{BC} \vec{F} \cdot d\vec{r} + \int_{CO} \vec{F} \cdot d\vec{r}.$$

Now, considering the path along OA, $y = 0$, $\therefore dy = 0$.

$$\int_{OA} \vec{F} \cdot d\vec{r} = \int_{OA} (x^2 dx - xy\, dy) = \int_0^a x^2 dx = \left[\frac{x^3}{3}\right]_0^a = \frac{a^3}{3}.$$

Along AB, $x = a$, $dx = 0$.

$$\int_{AB} \vec{F} \cdot d\vec{r} = \int_{AB} (x^2 dx - xy\, dy) = \int_0^a -ay\, dy = -a\left[\frac{y^2}{2}\right]_0^a = -\frac{a^3}{2}.$$

Along BC, $y = a$, $dy = 0$.

$$\int_{BC} \vec{F} \cdot d\vec{r} = \int_{BC} (x^2 dx - xy\, dy) = \int_a^0 x^2 dx = \left[\frac{x^3}{3}\right]_a^0 = -\frac{a^3}{2}.$$

Along CO, $x = 0$, $dx = 0$.

$$\therefore \int_{CO} \vec{F} \cdot d\vec{r} = \int_{CO} (x^2 dx - xy\, dy) = 0.$$

$$\therefore \oint_C \vec{F} \cdot d\vec{r} = \frac{a^3}{3} - \frac{a^3}{2} - \frac{a^3}{3} + 0 = -\frac{a^3}{2}. \quad \text{(ii)}$$

$\therefore$ from eqns. (i) and (ii),

$$\iint_S (\nabla \times \vec{F}) \cdot \vec{dS} = \int_C \vec{F} \cdot d\vec{r}.$$

$\therefore$ Stokes' theorem is verified.

Example 12.28. *Verify Stokes' theorem for $\vec{A} = (y - z + 2)\hat{\imath} + (yz + 4)\hat{\jmath} - zx\hat{k}$ over the surface of the cube, $x = 0$, $y = 0$, $z = 0$, $x = 2$, $y = 2$, $z = 2$ above the xy-plane.* **[BU(H) 2000]**

Solution : We consider the surface S of the cube as shown in Fig. 12.18, the boundary C being $OKLMO$.

Now, $\oint_C \vec{F} \cdot d\vec{r} = \int [(y - z + 2)\hat{i} + (yz + 4)\hat{j} - xz\hat{k}] \cdot [\hat{i}dx + \hat{j}dy + \hat{k}dz]$

$$= \int [(y - z + 2)dx + (yz + 4)dy - xz\,dz.$$

Now, $\oint_C \vec{F} \cdot d\vec{r} = \int_{OK} \vec{F} \cdot d\vec{r} + \int_{KL} \vec{F} \cdot d\vec{r} + \int_{LM} \vec{F} \cdot d\vec{r} + \int_{MO} \vec{F} \cdot d\vec{r}.$ (i)

Along the path OK, $y = 0$, $dy = 0$, $z = 0$, $dz = 0$.

$\therefore \quad \int_{OK} \vec{F} \cdot d\vec{r} = \int_0^2 2x\,dx = 2[x]_0^2 = 4.$

Along the path KL, $x = 2$, $dx = 0$, $z = 0$, $dz = 0$.

$\therefore \quad \int_{KL} \vec{F} \cdot d\vec{r} = \int_0^2 4\,dy = 4[y]_0^2 = 8.$

Along the path LM, $y = 2$, $dy = 0$, $z = 0$, $dz = 0$.

$\therefore \quad \int_{LM} \vec{F} \cdot d\vec{r} = \int_2^0 (2 - 0 + 2)dx = -8.$

Along the path MO, $x = 0$, $dx = 0$, $z = 0$, $dz = 0$.

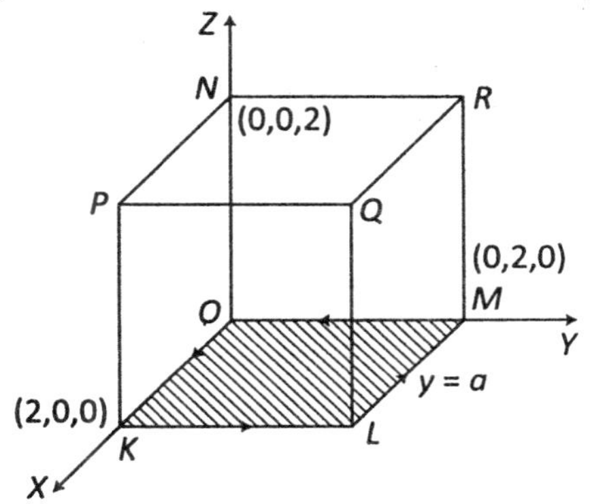

Fig. 12.18

$\therefore \quad \int_{MO} \vec{F} \cdot d\vec{r} = \int (y - 0 + 2)(0) + (0 + 4)dy - 0 = 4\int_2^0 dy = -8.$

$\therefore \quad \oint_C \vec{F} \cdot d\vec{r} = 4 + 8 - 8 - 8 = -4.$ (ii)

Again, $\nabla \times \vec{F} = \begin{vmatrix} \hat{i} & \hat{j} & \hat{k} \\ \dfrac{\partial}{\partial x} & \dfrac{\partial}{\partial y} & \dfrac{\partial}{\partial z} \\ y - z + 2 & yz + 4 & -xz \end{vmatrix}$ (iii)

$$= (0 - y)\hat{i} - (-z + 1)\hat{j} + (0 - 1)\hat{k} \quad \text{(iv)}$$

$$= -y\hat{i} + (z - 1)\hat{j} - \hat{k}. \quad \text{(v)}$$

We have to integrate over five surfaces $KLQP$, $OMRN$, $LMRQ$, $OKPM$, $QPNR$.

Considering the surface $KLQP$, $x = 2$, $\hat{n} = \hat{i}$, $dS = \hat{i}\,dydz$.

$\therefore \quad \iint_S (\nabla \times \vec{F}) \cdot \hat{n}\,dS = \iint [-y\hat{i} + (z - 1)\hat{j} - \hat{k}] \cdot \hat{i}\,dydz = \iint -y\,dydz$

$$= -\int_0^2 y\,dy \int_0^2 dz = -4.$$

Over the surface $OMRN$, $x = 0$, $\hat{n} = -\hat{i}$, $dS = \hat{i}\,dydz$.

$\therefore \quad \iint_S (\nabla \times \vec{F}) \cdot \hat{n}\,dS = \iint [-y\hat{i} + (z - 1)\hat{j} - \hat{k}] \cdot (-\hat{i})\,dydz = \iint y\,dydz$

$$= \int_0^2 y\,dy \int_0^2 dz = 4.$$

Over the surface $LMRQ$, $y = 2$, $\hat{n} = \hat{j}$, $dS = dxdz$.

$$\therefore \quad \iint (\nabla \times \vec{F}) \cdot dS = \iint [-y\hat{i} + (z-1)\hat{j} - \hat{k}] \cdot \hat{j}\, dxdz = \iint (z-1)\, dxdz$$

$$= \int_0^2 dx \int_0^2 (z-1)dz = [x]_0^2 \left[\frac{z^2}{2} - z\right]_0^2 = 0.$$

Over the surface $OKPN$, $y = 0$, $\hat{n} = -\hat{j}$, $dS = dxdz$.

$$\therefore \quad \iint (\nabla \times \vec{F}) \cdot \hat{n}\, dS = \iint [-y\hat{i} + (z-1)\hat{j} - \hat{k}] \cdot (-\hat{j})\, dxdz = -\iint (z-1)\, dxdz$$

$$= -\int_0^2 dx \int_0^2 (z-1)dz = [x]_0^2 \left[\frac{z^2}{2} - z\right]_0^2 = 0.$$

Over the surface $QPNR$, $z = 2$, $\hat{n} = \hat{k}$, $dS = dxdy$.

$$\therefore \quad \iint_S (\nabla \times \vec{F}) \cdot \hat{n}\, dS = \iint [-y\hat{i} + (z-1)\hat{j} - \hat{k}] \cdot \hat{k}\, dx\, dy = -\iint dx\, dy = -\int_0^2 dx \int_0^2 dy = -4.$$

$$\therefore \quad \iint_S (\nabla \times \vec{F}) \cdot \vec{dS} = -4 + 4 + 0 + 0 = -4 \qquad \text{(vi)}$$

From eqns. (ii) and (vi),

$$\iint_S (\nabla \times \vec{F}) \cdot \vec{dS} = \int_C \vec{F} \cdot d\vec{r} = -4.$$

$\therefore$ Stokes' theorem is verified.

Example 12.29. *Verify Stokes' theorem for $\vec{F} = (x^2+y^2)\hat{i} - 2xy\hat{j}$ taken around the rectangle bounded by lines $x = \pm a$, $y = 0$, $y = b$.*

Solution : $ABCD$ is a rectangle with $x = \pm a$, $y = 0$, $y = b$ (Fig. 12.19) in the vector field $\vec{F} = (x^2 + y^2)\hat{i} - 2xy\hat{j}$.

The line integral of $\vec{F}$ around the rectangular contour is

$$\int_{ABCDA} \vec{F} \cdot d\vec{r} = \int_{AB} \vec{F} \cdot d\vec{r} + \int_{BC} \vec{F} \cdot d\vec{r} + \int_{CD} \vec{F} \cdot d\vec{r} + \int_{DA} \vec{F} \cdot d\vec{r}.$$

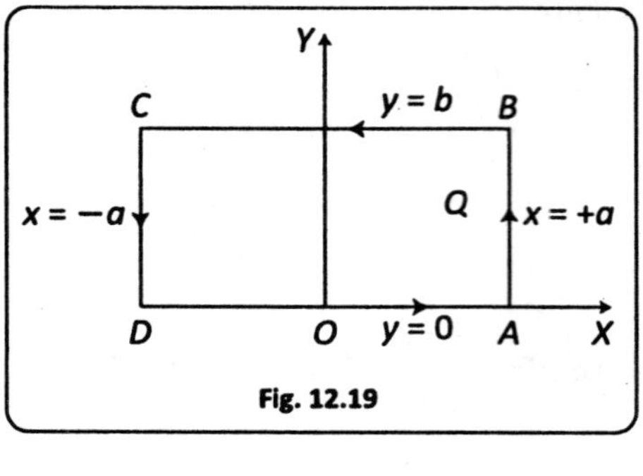

Fig. 12.19

Now, $\vec{F} \cdot d\vec{r} = [(x^2 + y^2)\hat{i} - 2xy\hat{j}] \cdot (\hat{i}\, dx + \hat{j}\, dy)$

$= (x^2 + y^2)dx - 2xy\, dy.$

Along AB, $x = a$, $dx = 0$, y varies from 0 to b.

$$\therefore \quad \int_{AB} \vec{F} \cdot d\vec{r} = -2a \int_0^b y\, dy = -2a\frac{b^2}{2} = -ab^2.$$

Along BC, $y = b$, $dy = 0$ and x varies from $+a$ to $-a$.

$$\therefore \quad \int_{BC} \vec{F} \cdot d\vec{r} = \int_a^{-a} (x^2 + y^2)\, dx = -\frac{2a^3}{3} - 2ab^2.$$

Along CD, $x = -a$, $dx = 0$, y varies from b to 0.

$$\therefore \quad \int_{CD} \vec{F} \cdot d\vec{r} = 2a \int_b^0 y\, dy = -ab^2.$$

Along DA, $y = 0$, $dy = 0$, x varies from $-a$ to $+a$.

$$\therefore \quad \int_{DA} \vec{F} \cdot d\vec{r} = \int_{-a}^{+a} x^2\, dx = \frac{2a^3}{3}.$$

Thus, $$\int_{ABCDA} \vec{F} \cdot d\vec{r} = -ab^2 - \frac{2a^3}{3} - 2ab^2 - ab^2 + \frac{2a^3}{3} = -4ab^2. \quad \text{(i)}$$

Again, $$\nabla \times \vec{F} = \begin{vmatrix} \hat{i} & \hat{j} & \hat{k} \\ \dfrac{\partial}{\partial x} & \dfrac{\partial}{\partial y} & \dfrac{\partial}{\partial z} \\ (x^2 + y^2) & -2xy & 0 \end{vmatrix} = \hat{i}(0-0) + \hat{j}(0-0) + \hat{k}[(-2y) - (2y)] = -4y\hat{k}.$$

$$\therefore \quad \iint_S (\nabla \times \vec{F}) \cdot \vec{dS} = \int_0^b \int_{-a}^{+a} -4y\hat{k} \cdot \hat{k}\, dx dy \qquad [\because\ \vec{dS} = \hat{k}\, dx dy]$$

$$= -4 \int_0^b \int_{-a}^{+a} y\, dx dy = -4 \int_0^b [x]_{-a}^{+a}\, y\, dy = -8a \left[\frac{y^2}{2}\right]_0^b = -4ab^2. \quad \text{(ii)}$$

$\therefore$ from eqns. (i) and (ii),

$$\int_{ABCDA} \vec{F} \cdot d\vec{r} = \iint_S (\nabla \times \vec{F}) \cdot \vec{dS} = -4ab^2.$$

$\therefore$ Stokes' theorem is verified.

Example 12.30. *Applying Stokes' theorem to find the value of, $\oint_C (y\,dx + z\,dy + x\,dz)$, where C is the curve of intersection of $x^2 + y^2 + z^2 = a^2$ and $x + z = a$.* **[Coim.U 1984]**

We have $$\oint_C (y\,dx + z\,dy + x\,dz) = \oint_C (y\hat{i} + z\hat{j} + x\hat{k}) \cdot (\hat{i}dx + \hat{j}dy + \hat{k}dz)$$

$$= \oint_C (y\hat{i} + z\hat{j} + x\hat{k}) \cdot d\vec{r}$$

$$= \iint_S \nabla \times (y\hat{i} + z\hat{j} + x\hat{k}) \cdot \hat{n}\, dS \quad \text{[using Stokes' theorem]}$$

$$= \iint_S \left(\hat{i}\frac{\partial}{\partial x} + \hat{j}\frac{\partial}{\partial y} + \hat{k}\frac{\partial}{\partial z}\right) \times (y\hat{i} + z\hat{j} + x\hat{k}) \cdot \hat{n}\, dS$$

$$= \iint_S -(\hat{i} + \hat{j} + \hat{k}) \cdot \hat{n}\, dS.$$

C is the circle formed by the intersection of $x^2 + y^2 + z^2 = a^2$ and $x + z = a$. S is the area bounded by C.

$$\therefore \quad \hat{n} = \text{outward drawn unit normal to } S = \frac{\nabla\phi}{|\nabla\phi|} = \left(\hat{i}\frac{\partial}{\partial x} + \hat{j}\frac{\partial}{\partial y} + \hat{k}\frac{\partial}{\partial z}\right)(x + z - a)$$

$$= \frac{\hat{i} + \hat{k}}{\sqrt{1+1}} = \frac{\hat{i} + \hat{k}}{\sqrt{2}}.$$

$$\therefore \iint_S (-\hat{i}+\hat{j}+\hat{k})\cdot\frac{\hat{i}+\hat{k}}{\sqrt{2}}\,dS=\iint_S -\left(\frac{1}{\sqrt{2}}+\frac{1}{\sqrt{2}}\right)dS=-\frac{2}{\sqrt{2}}\iint_S dS=-\frac{2}{\sqrt{2}}\pi\left(\frac{a}{\sqrt{2}}\right)^2=-\frac{\pi a^2}{\sqrt{2}}.$$

$$\therefore \oint (\boldsymbol{y\,dx+z\,dy+x\,dz})=-\frac{\boldsymbol{\pi a^2}}{\boldsymbol{\sqrt{2}}}.$$

Example 12.31. *Using Stokes' theorem evaluate* $\int_C [(x+y)\,dx+(2x-z)\,dy+(y+z)\,dz]$ *where, C is the boundary of the triangle with vertices* $(2,0,0)$, $(0,3,0)$ *and* $(0,0,6)$.

[AMIETE(Winter) 1996, RU 1990]

Solution : We have $\oint_C \vec{F}\cdot d\vec{r}=\oint_C [(x+y)dx+(2x-z)dy+(y+z)dz]$

$$=\oint_C [(x+y)\hat{i}+(2x-z)\hat{j}+(y+z)\hat{k}]\cdot(\hat{i}\,dx+\hat{j}\,dy+\hat{k}\,dz),$$

where $\vec{F}=(x+y)\hat{i}+(2x-z)\hat{j}+(y+z)\hat{k}$.

$$\therefore \quad \nabla\times\vec{F}=\begin{vmatrix}\hat{i} & \hat{j} & \hat{k}\\ \frac{\partial}{\partial x} & \frac{\partial}{\partial y} & \frac{\partial}{\partial z}\\ (x+y) & (2x-z) & (y+z)\end{vmatrix}=\hat{i}[1+1]+\hat{j}[0-0]+\hat{k}[2-1]=2\hat{i}+\hat{k}.$$

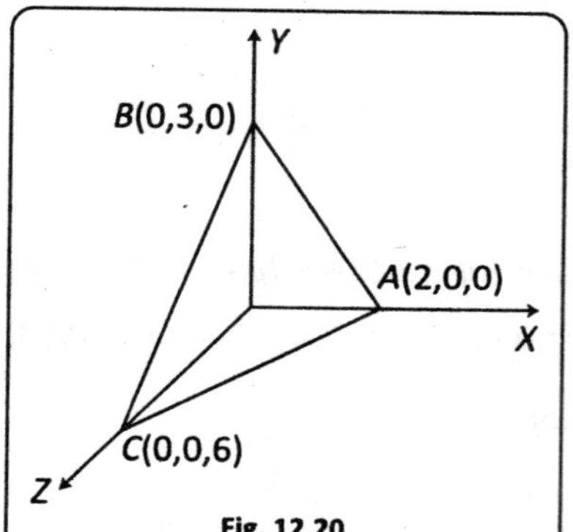

Fig. 12.20

If S be the surface of the plane ABC, then outward drawn unit normal to S is

$$\hat{n}=\frac{\nabla\phi}{|\nabla\phi|}=\frac{\left(\hat{i}\frac{\partial}{\partial x}+\hat{j}\frac{\partial}{\partial y}+\hat{k}\frac{\partial}{\partial z}\right)\left(\frac{x}{2}+\frac{y}{3}+\frac{z}{6}-1\right)}{\nabla\phi}$$

$$\left[\because\ \phi=\frac{x}{2}+\frac{y}{3}+\frac{z}{6}-1\right]$$

$$=\frac{\frac{\hat{i}}{2}+\frac{\hat{j}}{3}+\frac{\hat{k}}{6}}{|\nabla\phi|}=\frac{\frac{1}{6}(3\hat{i}+2\hat{j}+\hat{k})}{\frac{1}{6}\sqrt{3^2+2^2+1^2}}$$

$$=\frac{1}{\sqrt{14}}(3\hat{i}+2\hat{j}+\hat{k}).\quad\left[\because\ \nabla\phi=\frac{\hat{i}}{2}+\frac{\hat{j}}{3}+\frac{\hat{k}}{6}\right]$$

$$\therefore\quad (\nabla\times\vec{F})\cdot\hat{n}=(2\hat{i}+\hat{k})\cdot\frac{1}{\sqrt{14}}(3\hat{i}+2\hat{j}+\hat{k})=\frac{1}{\sqrt{14}}(6+1)=\frac{7}{\sqrt{14}}$$

$$\therefore\quad \oint_C \vec{F}\cdot d\vec{r}=\iint_S(\nabla\times\vec{F})\cdot\hat{n}\,dS=\frac{7}{\sqrt{14}}\iint\frac{dxdy}{\hat{k}\cdot\frac{1}{\sqrt{14}}(3\hat{i}+2\hat{j}+\hat{k})}=7\iint dxdy$$

$$=7\times\text{area of }\triangle ABC.$$

Example 12.32. *Prove using Stokes' theorem that,*

(a) $\oint \phi\,d\vec{r}=\iint_S(\hat{n}\times\nabla\phi)\,dS$ (b) $\int_C d\vec{r}\times\vec{F}=\iint_S(\hat{n}\times\nabla)\times\vec{F}\,dS.$

Solution : (a) Let $\vec{A}=\vec{a}\phi$, where $\vec{a}$ is an arbitrary constant vector and ϕ, a continuously differentiable scalar point function. Then,

$$\int_C \vec{A} \cdot d\vec{r} = \iint_S (\nabla \times \vec{A}) \cdot \hat{n}\, dS$$

or, $$\int_C \vec{a}\phi \cdot d\vec{r} = \iint_S (\nabla \times \vec{a}\phi) \cdot \hat{n}\, dS. \qquad \text{(i)}$$

Now, $\nabla \times (\vec{a}\phi) = \phi(\nabla \times \vec{a}) + \nabla\phi \times \vec{a}$

$$= \nabla\phi \times \vec{a} \quad [\because \vec{a} = \text{a constant vector, } \nabla \times \vec{a} = 0] \qquad \text{(ii)}$$

From eqn. (i), $\vec{a} \cdot \int_C \phi\, d\vec{r} = \iint_S \nabla\phi \times \vec{a} \cdot \hat{n}\, dS = -\vec{a} \cdot \iint_S \nabla\phi \times \hat{n}\, dS$

or, $$\vec{a} \cdot \left[\int_C \phi\, d\vec{r} + \iint_S (\nabla\phi \times \vec{n})\, dS\right] = 0$$

or, $$\boldsymbol{\int_C \phi\, d\vec{r} = \iint_S (\hat{n} \times \nabla\phi)\, dS = 0}. \quad [\because \vec{a} \text{ is a constant vector}]$$

(b) Let $\vec{B} = \vec{a} \times \vec{F}$, where $\vec{a}$ is any arbitrary constant vector.

$$\therefore \quad \int_C B \cdot d\vec{r} = \iint_S (\nabla \times \vec{B}) \cdot \hat{n}\, dS. \qquad \text{(iii)}$$

But, $\nabla \times \vec{B} = \nabla \times (\vec{a} \times \vec{F}) = \vec{a}(\nabla \cdot \vec{F}) - \vec{F}(\nabla \cdot \vec{a}) + (\vec{F} \cdot \nabla)\vec{a} - (\vec{a} \cdot \nabla)\vec{F}$

$$= \vec{a}(\nabla \cdot \vec{F}) - (\vec{a} \cdot \nabla)\vec{F}$$

and $(\vec{a} \cdot \nabla)\vec{F} \cdot \hat{n} = \hat{a} \cdot \nabla(\vec{F} \cdot \hat{n})$.

$\therefore$ from eqn. (iii), $\int_C \vec{a} \cdot \vec{F} \times d\vec{r} = \iint_S \vec{a}(\nabla \cdot \hat{F}) \cdot \hat{n} - \vec{a} \cdot \nabla(\vec{F} \cdot \hat{n})\, dS$

or, $$\vec{a} \cdot \int_C \vec{F} \times d\vec{r} = \vec{a} \cdot \iint_S [(\nabla \cdot \hat{F})\hat{n} - \nabla(\vec{F} \cdot \hat{n})]\, dS = -\vec{a} \cdot \iint_S (\hat{n} \times \nabla) \times \hat{F}\, dS$$

or, $$\vec{a} \cdot \left[\int_C \vec{F} \times d\vec{r} + \iint_S \{(\hat{n} \times \nabla) \times \hat{F}\, dS\}\right] = 0.$$

$\because \vec{a}$ is a constant vector, $\boldsymbol{\int_C d\vec{r} \times \vec{F} = \iint_S (\hat{n} \times \nabla) \times \hat{F}\, dS}$.

Example 12.33. *Evaluate $\iint_S \vec{r} \times \hat{n}\, dS$, where S is a closed surface.* **[BU(H) 1997]**

Solution : We have $\vec{r} = \hat{i}x + \hat{j}y + \hat{k}z$.

$$\therefore \quad \nabla \cdot \vec{r} = \left(\hat{i}\frac{\partial}{\partial x} + \hat{j}\frac{\partial}{\partial y} + \hat{k}\frac{\partial}{\partial z}\right) \cdot (\hat{i}x + \hat{j}y + \hat{k}z) = 1 + 1 + 1 = 3.$$

Now, $\boldsymbol{\iint_S \vec{r} \times \hat{n}\, dS} = \iiint_V \nabla \cdot \vec{r}\, dV$ [by Gauss' divergence theorem]

$$= 3\iiint_V dV = \mathbf{3V}. \qquad \left[\because \iiint_V dV = V\right]$$

Example 12.34. *Prove the following :*

(a) $$\iiint_V (\phi\nabla^2\psi - \psi\nabla^2\phi)\, dV = \iint_S (\phi\nabla\psi - \psi\nabla\phi)\cdot d\vec{S}$$

where V is the volume bounded by the surface S and ϕ, ψ are scalar fields.

(b) $$\iiint_V (\nabla \times \vec{B})\, dV = \iint_S (\hat{n} \times \vec{B})\, dS.$$

Solution : (a) By Gauss' divergence theorem, $\iiint_V (\nabla\cdot\vec{A})\, dV = \iint_S \vec{A}\cdot\hat{n}\, dS.$

Let $\vec{A} = \phi\nabla\psi$, then

$$\iiint_V \nabla\cdot(\phi\nabla\psi)\, dV = \iint_S (\phi\nabla\psi)\cdot\hat{n}\, dS. \qquad \text{(i)}$$

But, $\nabla\cdot(\phi\nabla\psi) = \phi(\nabla\cdot\nabla\psi) + \nabla\phi\cdot\nabla\psi = \phi\nabla^2\psi + \nabla\phi\cdot\nabla\psi.$

Using this value of $\nabla\cdot(\phi\nabla\psi)$ in eqn. (i), we get

$$\iiint_V [\phi\nabla^2\psi + \nabla\phi\cdot\nabla\psi]\, dV = \iint_S (\phi\nabla\psi)\cdot d\vec{S}. \qquad \text{(ii)}$$

Now, interchanging ϕ and ψ, we get

$$\iiint_V [\psi\nabla^2\phi + \nabla\psi\cdot\nabla\phi]\, dV = \iint_S (\psi\nabla\phi)\cdot d\vec{S}. \qquad \text{(iii)}$$

Subtracting eqn. (iii) from eqn. (ii), we get

$$\iiint_V [\phi\nabla^2\psi - \psi\nabla^2\phi]\, dV = \iint_S (\phi\nabla\psi - \psi\nabla\phi)\cdot d\vec{S}. \qquad \text{(iv)}$$

(b) We have from Gauss' divergence theorem, $\iiint_V (\nabla\cdot\vec{A})\, dV = \iint_S \vec{A}\cdot\hat{n}\, dS$

Let $\vec{A} = \vec{C}\times\vec{B}$, where $\vec{C}$ is a constant vector.

$$\therefore \quad \iiint_V \nabla\cdot(\vec{C}\times\vec{B})\, dV = \iint_S (\vec{C}\times\vec{B})\cdot\hat{n}\, dS. \qquad \text{(i)}$$

But, $\vec{C}\times\vec{B}\cdot\hat{n} = \vec{C}\cdot\vec{B}\times\hat{n}$

and $\nabla\cdot(\vec{C}\times\vec{B}) = \vec{B}\cdot(\nabla\times\vec{C}) - \vec{C}\cdot(\nabla\times\vec{B})$

$= -\vec{C}\cdot(\nabla\times\vec{B})$ [$\because$ $\nabla\times\vec{C} = 0$, as $\vec{C}$ is a constant vector]

Hence from eqn. (i),

$$\therefore \quad -\iiint_V \vec{C}\cdot(\nabla\times\vec{B})\, dV = \iint_S \vec{C}\cdot\vec{B}\times\hat{n}\, dS$$

or, $$-\vec{C}\cdot\iiint_V (\nabla\times\vec{B})\, dV = \vec{C}\cdot\iint_S (\vec{B}\times\hat{n})\, dS$$

or, $\vec{C} \cdot \iint_S (\vec{B} \times \hat{n})\, dS + \vec{C} \cdot \iiint_V (\nabla \times \vec{B})\, dV = 0$

or, $$\iiint_V (\nabla \times \vec{B})\, dV = \iint_S (\hat{n} \times \vec{B})\, dS. \qquad \text{(ii)}$$

Example 12.35. *A vector $\vec{B}$ is always normal to a given closed surface S. Show that,*

$$\iiint_V (\nabla \times \vec{B})\, dV = 0, \text{ where } V \text{ is the volume bounded by } S.$$

Solution : According to problem (12.34)

$$\iiint_V (\nabla \times \vec{B})\, dV = \iint_S (\hat{n} \times \vec{B})\, dS$$

$$\therefore \quad \iiint_V (\nabla \times \vec{B})\, dV = -\iint_S \vec{B} \times \hat{n}\, dS = -\iint_S \vec{B} \times d\vec{S}.$$

$\because$ $\vec{B}$ is normal to $|d\vec{S}|$, $\therefore$ $\vec{B}$ and $d\vec{S}$ are parallel and hence, $\vec{B} \times d\vec{S} = 0$.

$$\therefore \quad \iiint_V (\nabla \times \vec{B})\, dV = 0 \quad \textbf{(Proved)}.$$

Example 12.36. *Prove that,* $\oint_C \phi d\vec{r} = \iint_S d\vec{S} \times \nabla\phi.$

Solution : Let $\vec{A} = \vec{C}\phi$, where $\vec{C}$ is an arbitrary constant vector and ϕ, a continuously differential scalar point function. Then,

$$\oint_C \vec{C}\phi \cdot d\vec{r} = \iint_S \nabla \times (\vec{C}\phi) \cdot \hat{n}\, dS. \qquad \text{(i)}$$

But, $\nabla \times (\vec{C}\phi) = \phi(\nabla \times \vec{C}) + \nabla\phi \times \vec{C}$

$= \nabla\phi \times \vec{C}$ $\quad [\because \nabla \times \vec{C} = 0, \vec{C}$ being a constant vector$]$

$$\therefore \quad \vec{C} \cdot \left(\oint_C \phi d\vec{r}\right) = \iint_S (\nabla\phi \times \vec{C}) \cdot \hat{n}\, dS = -\iint_S \vec{C} \cdot \nabla\phi \times \hat{n}\, dS = -\vec{C} \cdot \iint_S \nabla\phi \times \hat{n}\, dS$$

$[\because (\nabla\phi \times \vec{C}) \cdot \hat{n} = -(\vec{C} \times \nabla\phi) \cdot \hat{n} = -\vec{C} \cdot \nabla\phi \times \hat{n}]$

$$\therefore \quad \vec{C} \cdot \left[\oint_C \phi d\vec{r} + \iint_S \nabla\phi \times \hat{n}\, dS\right] = 0.$$

$$\therefore \quad \left[\oint_C \phi d\vec{r} + \iint_S \nabla\phi \times \hat{n}\, dS\right] = 0, \quad [\because \vec{C} \text{ is a constant vector}]$$

or, $$\oint_C \phi\, d\vec{r} = \iint_S \hat{n} \times \nabla\phi\, dS = \iint_S d\vec{S} \times \nabla\phi \quad \textbf{(Proved)}.$$

Example 12.37. *Prove that,* $\iint_S (\nabla\phi \times \nabla\psi)\cdot \vec{dS} = 0.$ **[BU(H) 1994]**

Solution : If we put $\vec{A} = \nabla\phi \times \nabla\psi$, then

$$\iint_S \vec{A}\cdot\vec{dS} = \iiint_V (\nabla\cdot\vec{A})\,dV \quad \text{(by Gauss' divergence theorem)}$$

$$= \iiint_V \nabla\cdot(\nabla\phi\times\nabla\psi)\,dV = \iiint_V [\nabla\psi\cdot(\nabla\times\nabla\phi) - \nabla\phi\cdot(\nabla\times\nabla\psi)]\,dV$$

$$\therefore \quad \iint_S (\nabla\phi\times\nabla\psi)\cdot\vec{dS} = 0, \quad [\because\ \nabla\times\nabla\phi = 0,\ \ \nabla\times\nabla\psi = 0] \quad \textbf{(Proved)}.$$

Example 12.38. *If* $\oint_C \vec{E}\cdot d\vec{r} = -\frac{1}{c}\frac{\partial}{\partial t}\iint_S \vec{H}\cdot\vec{dS}$, *where S is any surface bounded by the curve C, show that,* $\nabla\times\vec{E} = -\frac{1}{c}\frac{\partial\vec{H}}{\partial t}$.

Solution : We have $\oint \vec{E}\cdot d\vec{r} = \iint_S (\nabla\times\vec{E})\cdot\vec{dS}$ [by Stokes' theorem] (i)

But, $\oint_C \vec{E}\cdot d\vec{r} = \iint_S \left(-\frac{1}{c}\frac{\partial\vec{H}}{\partial t}\right)\cdot\vec{dS}$ (ii)

as given in the problem.

$$\therefore \quad \iint_S (\nabla\times\vec{E})\cdot\vec{dS} = \iint_S \left(-\frac{1}{c}\frac{\partial\vec{H}}{\partial t}\right)\cdot\vec{dS}.$$

$\because$ surface is arbitrary, $\boldsymbol{\nabla\times\vec{E} = -\frac{1}{c}\frac{\partial\vec{H}}{\partial t}}$ **(Proved).**

Example 12.39. *If S is any closed surface enclosing a volume V and* $\vec{A} = ax\hat{i} + by\hat{j} + cz\hat{k}$, *prove that,* $\iint_S \vec{A}\cdot\hat{n}\,\vec{dS} = (a+b+c)V$.

Solution : We have $\iint_S \vec{A}\cdot\hat{n}\,\vec{dS} = \iiint_V (\nabla\cdot\vec{A})\,dV$ [by Gauss' divergence theorem]

But, $\nabla\cdot\vec{A} = \left(\hat{i}\frac{\partial}{\partial x} + \hat{j}\frac{\partial}{\partial y} + \hat{k}\frac{\partial}{\partial z}\right)\cdot(ax\hat{i} + by\hat{j} + cz\hat{k}) = (a+b+c)$.

$$\therefore \quad \iint_S \vec{A}\cdot\hat{n}\,\vec{dS} = \iiint_V (a+b+c)\,dV = (a+b+c)\iiint_V dV = \mathbf{(a+b+c)V}.$$

Example 12.40. *By using the divergence theorem determine the value of the integral*

$$\int_S [x\,dydz + y\,dzdx + z\,dxdy].$$

where S is the surface of the region bounded by the cylinder $x^2 + y^2 = 9$ *and the planes* $z = 0$ *and* $z = 3$. **[BU(H) 1996]**

Solution : We have $\int_S [x\,dydz + y\,dzdx + z\,dxdy] = \int_{-S} \vec{r} \cdot \vec{dS}$.

where $\vec{r} = \hat{i}x + \hat{j}y + \hat{k}z$ and $\vec{dS} = \hat{i}\,dydz + \hat{j}\,dzdx + \hat{k}\,dxdy$.

By Gauss' divergence theorem,

$$\iint_S \vec{r} \cdot \vec{dS} = \iiint_V (\nabla \cdot \vec{r})\,dV = \iiint_V 3dV \quad [\because\ \nabla \cdot \vec{r} = 3]$$

$$= 3\int_{x=-3}^{3}\int_{y=-\sqrt{9-x^2}}^{\sqrt{9-x^2}}\int_{z=0}^{3} dxdydz = (3 \times 3)\int_{x=-3}^{3}\left[\int_0^{\sqrt{9-x^2}} dy\right] dx$$

$$= 18\int_{x=-3}^{3} \sqrt{9-x^2}dx = 36\int_0^3 \sqrt{9-x^2} = 36\left[\frac{x}{2}\sqrt{9-x^2} + \frac{9}{2}\sin^{-1}\frac{x}{3}\right]_0^3$$

$$= 36 \cdot \frac{9}{2} \cdot \frac{\pi}{2} = \mathbf{81\pi}.$$

Example 12.41. *If $\vec{B} = \nabla \times \vec{A}$, prove that $\iint_S \vec{B} \cdot \hat{n}\,dS = 0$ for any closed surface S.*

[BU(H) 1992, 1997]

Solution : We have $\iint_S \vec{B} \cdot \hat{n}\,dS = \iiint_V (\nabla \cdot \vec{B})\,dV = \iiint_V [\nabla \cdot (\nabla \times \vec{A})]dV$

$$= \iiint_V [(\nabla \times \nabla) \cdot \vec{A}]dV = 0 \quad \textbf{Proved).}$$

Example 12.42. *By using the divergence theorem evaluate $\iint_S \vec{F} \cdot \hat{n}\,dS$, where $\vec{F} = 4xz\hat{i} - y^2\hat{j} + yz\hat{k}$ and S is the surface of the cube bounded by $x = 0$, $x = 1$; $y = 0$, $y = 1$; $z = 0$, $z = 1$.*

Solution : Since $\vec{F} = 4xz\hat{i} - y^2\hat{j} + yz\hat{k}$,

$$\therefore\ \nabla \cdot \vec{F} = \left(\hat{i}\frac{\partial}{\partial x} + \hat{j}\frac{\partial}{\partial y} + \hat{k}\frac{\partial}{\partial z}\right) \cdot (4xz\hat{i} - y^2\hat{j} + yz\hat{k}) = 4z - 2y + y = 4z - y.$$

$$\therefore\ \iiint_V (\nabla \cdot \vec{F})\,dV = \iiint (4z - y)dxdydz = \int_0^1 dx \int_0^1 dy \int_0^1 (4z - y)dz$$

$$= \int_0^1 dx \int_0^1 dy\,(2z^2 - yz)_0^1 = \int_0^1 dx \int_0^1 dy(2 - y)$$

$$= \int_0^1 dx \left[2y - \frac{y^2}{2}\right]_0^1 = \frac{3}{2}[x]_0^1 = \frac{3}{2}. \qquad \text{(i)}$$

Again, $\iint_S \vec{F} \cdot \hat{n}\,dS = \int_{OKLP} \vec{F} \cdot \hat{n}\,dS + \int_{LPQM} \vec{F} \cdot \hat{n}\,dS + \int_{QMNR} \vec{F} \cdot \hat{n}\,dS$

$$+ \int_{OPQR} \vec{F} \cdot \hat{n}\,dS + \int_{KORN} \vec{F} \cdot \hat{n}\,dS + \int_{KLMN} \vec{F} \cdot \hat{n}\,dS.$$

Now over the face $OKLP$, $z = 0$, $dz = 0$, $\hat{n} = -\hat{k}\,dS$, $dS = dxdy$.

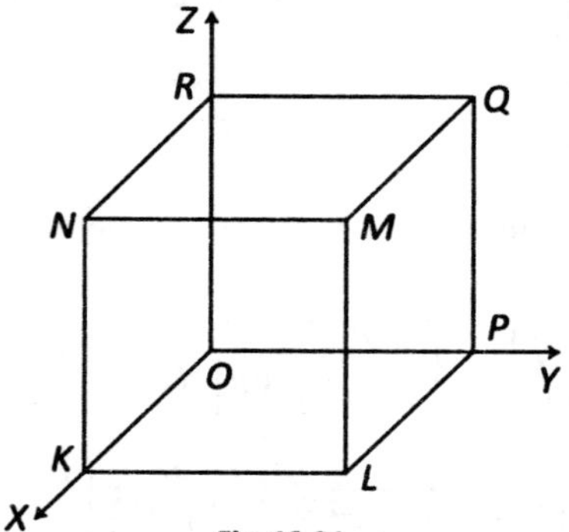

Fig. 12.21

$$\therefore \quad \iint \vec{F}\cdot\hat{n}\,dS = \int_0^1\int_0^1 (-y^2\hat{j})\cdot(-\hat{k})\,dxdy = 0.$$

Over the face $LPQM$, $y = 1$, $dy = 0$, $\hat{n} = \hat{j}$, $dS = dxdz$.

$$\therefore \quad \iint \vec{F}\cdot\hat{n}\,dS = \int_0^1\int_0^1 (4xz\hat{i} - \hat{j} + z\hat{k})\cdot\hat{j}\,dxdz$$

$$= -\int_0^1\int_0^1 -dxdz = -\int_0^1 dx\int_0^z dz$$

$$= -(1)(1) = -1.$$

Over the face $QMNR$, $z = 1$, $dz = 0$, $\hat{n} = -\hat{k}$, $dS = dxdy$.

$$\therefore \quad \iint \vec{F}\cdot\hat{n}\,dS = \int_0^1\int_0^1 (4x(1)\hat{i} - y^2\hat{k} + y(1)\cdot\hat{k})\,dxdy = \int_0^1\int_0^1 y\,dxdy = \int_0^1 dx\int_0^1 y\,dy$$

$$= [x]_0^1\left[\frac{y^2}{2}\right]_0^1 = \frac{1}{2}.$$

Over the face $OPQR$, $x = 0$, $dx = 0$, $\hat{n} = -\hat{i}$, $dS = dydz$.

$$\therefore \quad \iint \vec{F}\cdot\hat{n}\,dS = \int_0^1\int_0^1 (0\hat{i} - y^2\hat{j} + yz\hat{k})\cdot(-\hat{i})\,dydz = 0.$$

Over the face $KORN$, $y = 0$, $dy = 0$, $\hat{n} = -\hat{j}$, $dS = dxdz$.

$$\therefore \quad \iint \vec{F}\cdot\hat{n}\,dS = \int_0^1\int_0^1 (4xz\hat{i})\cdot(-\hat{j})\,dxdz = 0.$$

Over the face $KLMN$, $x = 1$, $dx = 0$, $\hat{n} = \hat{i}$, $dS = dydz$.

$$\therefore \quad \iint \vec{F}\cdot\hat{n}\,dS = \int_0^1\int_0^1 [4z\hat{i} - y^2\hat{j} + yz\hat{k}]\cdot\hat{i}\,dydz = \int_0^1\int_0^1 4z\,dydz = \int_0^1 dy\int_0^1 4z\,dz$$

$$= \int_0^1 dy\,[2z^2]_0^1 = 2\int_0^1 dy = 2.$$

$$\therefore \quad \iint \vec{F}\cdot\hat{n}\,dS = 0 - 1 + \tfrac{1}{2} + 0 + 0 + 2 = \frac{3}{2}. \qquad \text{(ii)}$$

Also, from eqns. (i) and (ii),

$$\iiint_V (\nabla\cdot\vec{F})\,dV = \iint_S \vec{F}\cdot\hat{n}\,dS.$$

So, divergence theorem is verified.

Example 12.43. *Verify divergence theorem for $\vec{A} = 4x\hat{i} - 2y^2\hat{j} + z^2\hat{k}$ taken over the region bounded by $x^2 + y^2 = 4$, $z = 0$, $z = 3$.*

Solution : We have $\vec{A} = 4x\hat{i} - 2y^2\hat{j} + z^2\hat{k}$.

$\therefore\ \nabla \cdot \vec{A} = \left(\hat{i}\frac{\partial}{\partial x} + \hat{j}\frac{\partial}{\partial y} + \hat{k}\frac{\partial}{\partial z}\right) \cdot (4x\hat{i} - 2y^2\hat{j} + z^2\hat{k}) = (4 - 4y + 2z).$

i.e., volume integral is

$$\iiint_V (\nabla \cdot \vec{A})\, dV = \int_{x=-2}^{2}\int_{z=0}^{3}\int_{y=-\sqrt{4-x^2}}^{\sqrt{4-x^2}} (4 - 4y + 2z)\, dx\, dy\, dz$$

$$= \int_{x=-2}^{2}\int_{z=0}^{3} \left[4y - 2y^2 + 2yz\right]_{y=-\sqrt{4-x^2}}^{\sqrt{4-x^2}} dx\, dy\, dz$$

$$= 2\int_{-2}^{2}\int_{0}^{3} [(4+2z)y]_0^{\sqrt{4-x^2}}\, dx\, dz = 2\int_{-2}^{2}\int_{0}^{3}(4+2z)\sqrt{4-x^2}\, dx\, dz$$

$$= 4\left[\{4z + z^2\}_0^3 \left\{\frac{x}{2}\sqrt{4-x^2} + 2\sin^{-1}\frac{x}{2}\right\}_0^2\right] = 4[(12+9)\pi] = \mathbf{84\pi}.$$

To evaluate the surface integral, we note that the surface of the cylinder consists of base S_1 $(z = 0)$, the top S_2 $(z = 3)$ and curved surface S_3 $(x^2+y^2 = 4)$.

$$\therefore \quad \iint_S \vec{A} \cdot \hat{n}\, dS = \iint_{S_1} \vec{A} \cdot \hat{n_1}\, dS_1 + \iint_{S_2} \vec{A} \cdot \hat{n_2}\, dS_2 + \iint_{S_3} \vec{A} \cdot \hat{n_3}\, dS_3.$$

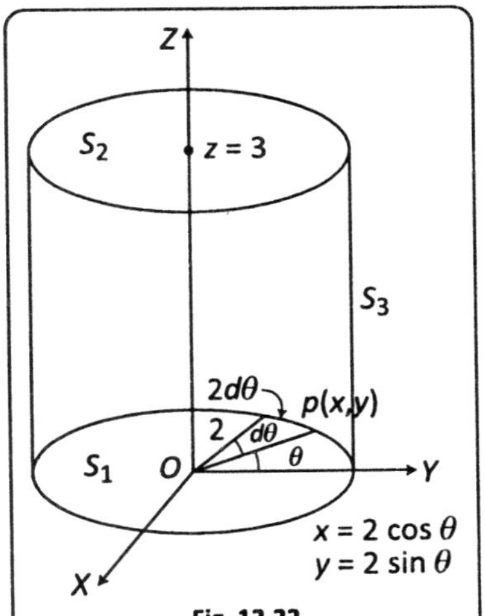

Fig. 12.22

For S_1, $\hat{n_1} = -\hat{k}$ and $z = 0$,

$$\therefore \quad \iint_{S_1} \vec{A} \cdot \hat{n}_1\, dS_1 = \iint (4x\hat{i} - 2y^2\hat{j}) \cdot (-\hat{k})\, dS_1 = \mathbf{0} \quad \text{(i)}$$

$$\iint_{S_2} \vec{A} \cdot \hat{n_2}\, dS_2 = \iint_{S_2} (4x\hat{i} - 2y^2\hat{j} + 3^2\hat{k}) \cdot \hat{k}\, dS_2$$

$$= 9\iint_{S_2} dS_2 = 9S_2 = 9(4\pi) = 36\pi. \quad \text{(ii)}$$

$$\because \quad \mathbf{S_2 = \pi(2^2) = 4\pi} \quad \text{(iii)}$$

and $$\iint_{S_3} \vec{A} \cdot \hat{n}_3\, dS_3 = \iint_{S_3} \left[\{4x\hat{i} - 2y^2\hat{j} + z^2\hat{k}\} \cdot \left\{\frac{2x\hat{i} + 2y\hat{j}}{\sqrt{2^2+2^2}}\right\}\right] dS_3$$

$$\left[\because\ \hat{n_3} = \frac{\nabla\phi}{|\nabla\phi|} = \frac{\nabla(x^2+y^2-4)}{|\nabla\phi|} = \frac{2x\hat{i} + 2y\hat{j}}{\sqrt{2^2+2^2}}\right]$$

$$= \iint_{S_3} \left\{\frac{8x^2 - 4y}{4}\right\} dS_3 = \iint_{S_3} (2x^2 - y^3)2d\theta\, dz$$

$$= \int_{\theta=0}^{2\pi}\int_{z=0}^{3} \{2(2\cos\theta)^2 - 2^3\sin\theta\}(2d\theta\, dz) = 48\int_{\theta=0}^{2\pi}(\cos^2\theta - \sin^3\theta)d\theta$$

$$= 48\int_0^{2\pi}\cos^2\theta\, d\theta = 48(4)\int_0^{\frac{\pi}{2}}\cos^2\theta\, d\theta = 48(4)\left(\frac{\pi}{4}\right) = \mathbf{48\pi}.$$

$$\therefore \quad \iint_S \vec{A} \cdot \hat{n}\, ds = 36\pi + 48\pi = 84\pi.$$

$$\therefore \quad \iiint_V (\nabla \cdot \vec{A})\, dV = \iint_S (\vec{A} \cdot \hat{n})\, dS.$$

$\therefore$ the divergence theorem is verified.

Example 12.44. *Applying Gauss' Divergence theorem, evaluate* $\iint_S \frac{\vec{r} \cdot \hat{n}}{r^3}\, dS$, *where* S *represents any closed surface enclosing volume* V.

Solution : **Case (i) :** The origin is outside S.

The vector function, $\vec{A} = \frac{\vec{r}}{r^3}$ is continuously differentiable throughout the volume enclosed by S. By Gauss' divergence theorem,

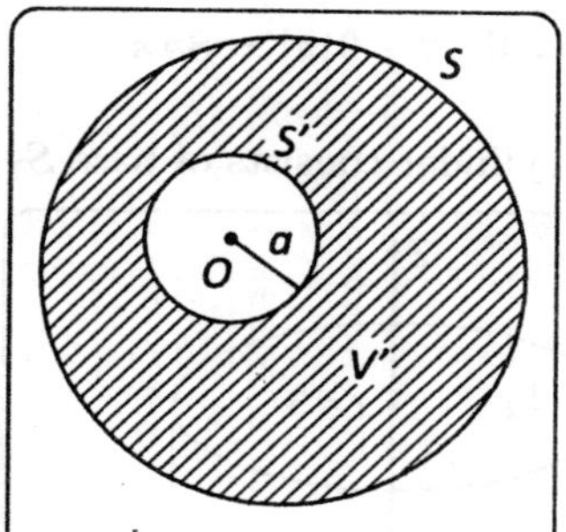

Fig. 12.23

$$\iint_S \vec{A} \cdot \hat{n}\, dS = \iiint_V \nabla \cdot \vec{A}\, dV$$

or, $$\iint_S \frac{\vec{r} \cdot \hat{n}}{\vec{r}^3}\, dS = \iiint_V \left(\nabla \cdot \frac{\vec{r}}{\vec{r}^3}\right) dV.$$

Now, $$\nabla \cdot \left(\frac{\vec{r}}{\vec{r}^3}\right) = \nabla \cdot (r^{-3}\vec{r}) = (\nabla r^{-3}) \cdot \vec{r} + r^{-3}(\nabla \cdot \vec{r})$$
$$= -3r^{-5}\vec{r} \cdot \vec{r} + 3r^{-3} \quad [\because \nabla \cdot \vec{r} = 3]$$
$$= -3r^{-3} + 3r^{-3} = 0.$$

$$\therefore \quad \iint_S \frac{\vec{r} \cdot \hat{n}}{r^3} = \iiint_V \nabla \cdot \left(\frac{\vec{r}}{\vec{r}^3}\right) dV = \mathbf{0}.$$

Case (ii) : The origin is inside S.

Here, the vector point function $\vec{A} = \frac{\vec{r}}{\vec{r}^3}$ has a point of discontinuity and as such Gauss' divergence theorem cannot be applied.

Let the origin be enclosed by small sphere S' of radius a. The outward normal at each point on S' will be towards O. Then $\vec{A}$ becomes continuously differentiable at all points of the region V' enclosed by S' and S.

Now, by Gauss' divergence theorem,

$$\iint_S \frac{\vec{r} \cdot \hat{n}}{r^3}\, dS + \iint_{S'} \frac{\vec{r} \cdot \hat{n}}{r^3}\, dS = \iint_{V'} \nabla \cdot \frac{\vec{r} \cdot \hat{n}}{r^3}\, dV. \quad \text{(i)}$$

Now, $$\iint_{S'} \frac{\vec{r} \cdot \hat{n}}{r^3}\, dS = -\iint_{S'} \frac{\vec{r} \cdot \vec{r}}{r^3 a}\, dS \quad \left[\because \text{ for } S',\ \hat{n} = -\frac{\vec{r}}{|\vec{r}|} = -\frac{\vec{r}}{a}\right]$$

$$= \iint_{S'} \frac{r^2}{r^3 a}\, dS = -\frac{1}{a^2}\iint_{S'} dS = -\frac{1}{a^2}4\pi a^2 = -4\pi. \quad \text{(ii)}$$

$\therefore$ from eqns. (i) and (ii),

$$\iint_{S'} \frac{\vec{r} \cdot \hat{n}}{r^3}\, dS = -\iint_{S'} \frac{\vec{r} \cdot \hat{n}}{r^3}\, dS = -(-4\pi) = \mathbf{4\pi}.$$

Example 12.45. *If $\vec{V}(x, y, z, t)$ be the velocity of a moving fluid of density $\rho(x, y, z, t)$, then in absence of sources and sinks prove that, $\nabla \cdot \vec{J} + \frac{\partial \rho}{\partial t} = 0$, where $\vec{J} = \rho\vec{V}$.* **[BU(H) 1999]**

Solution : We consider a volume V of the moving fluid enclosed by a surface S at any time t. Then, the mass of the fluid at any time t in V is

$$M = \iiint_V \rho\, dV.$$

The rate of increase of mass of the fluid is

$$\frac{\partial M}{\partial t} = \frac{\partial}{\partial t}\iiint_V \rho\, dV = \iiint_V \frac{\partial \rho}{\partial t} dV.$$

As the fluid under consideration is flowing one, and there being no accumulation of it is in V, $\frac{\partial M}{\partial t}$ must be equal to the fluid entering V through S in unit time, i.e.,

$$-\iint_S \rho\vec{V} \cdot \hat{n} dS.$$

Now, $-\iint_S \rho\vec{V} \cdot \hat{n}\, dS = -\iiint_V \nabla \cdot (\rho\vec{V})\, dV$ [by the Gauss' divergence theorem]

$\therefore \quad \iiint_V \frac{\partial \rho}{\partial t} dV = -\iiint_V \nabla \cdot (\rho\vec{V})\, dV$ or, $\iiint_V \left[\nabla \cdot (\rho\vec{V}) + \frac{\partial \rho}{\partial t}\right] dV = 0.$

Since V is arbitrary,

$$\nabla \cdot (\rho\vec{V}) + \frac{\partial \rho}{\partial t} = 0$$

or, $\quad \boldsymbol{\nabla \cdot \vec{J} + \frac{\partial \rho}{\partial t} = 0} \quad [\because \vec{J} = \rho\vec{V}]$

This is known as the *equation of continuity.*

Example 12.46. *If $\hat{n}$ be the unit outward drawn normal to any closed surface S, then show that, $\iiint_V \nabla \cdot \hat{n}\, dV = S$.*

Solution : By Divergence theorem of Gauss,

$$\iiint_V (\nabla \cdot \hat{n})\, dV = \iint_S \hat{n} \cdot \hat{n}\, dS = \iint_S dS = S.$$

Example 12.47. *If there be a distribution of charges $q_1, q_2, q_3, \ldots, q_n$ having position vectors $\vec{r_1}, \vec{r_2}, \vec{r_3}, \ldots, \vec{r_n}$ and $\phi(P) = \sum_{m=1}^{n} \frac{q_m}{r_m}$, the potential at any point $P(x, y, z)$, then prove that, $\iint_S \vec{E} \cdot \vec{dS} = 4\pi Q$, where $\vec{E} = -\nabla\phi$ is the electric field intensity, S is the surface enclosing all the charges and $Q = \sum_{m=1}^{n} q_m$ is the total charge within S.*

Solution : We have $\vec{E}\cdot\vec{dS} = -\nabla\phi\cdot\vec{dS}$

$$\text{or, } \vec{E}\cdot\vec{dS} = -\nabla\left(\sum_{m=1}^{n}\frac{q_m}{r_m}\right)\cdot\vec{dS} = -\nabla\left(\sum_{m=1}^{n} q_m \sum_{m=1}^{n}\frac{1}{r_m}\right)\cdot\vec{dS}$$

$$= -Q\nabla\sum_{m=1}^{n}\frac{1}{r_m}\cdot\vec{dS} \qquad \left[\because Q = \sum_{m=1}^{n} q_m\right]$$

$$= -Q\sum_{m=1}^{n}\nabla\left(\frac{1}{r_m}\right)\cdot\vec{dS}$$

$$= -Q\sum_{m=1}^{n} -\frac{\vec{r_m}\cdot\vec{dS}}{r_m^3} \qquad \left[\because \nabla\left(\frac{1}{r}\right) = -\frac{\vec{r}}{r^3}\right]$$

$$\text{or, } \vec{E}\cdot\vec{dS} = Q\sum_{m=1}^{n}\frac{\vec{r_m}\cdot\vec{dS}}{r_m^3} = Q\sum_{m=1}^{n}\frac{dS\cos\theta}{r_m^2}$$

$$= Q\sum_{m=1}^{n}\frac{dS\cos\theta}{r_m^2} = Q\,d\omega, \quad \text{where } \sum_{m=1}^{n}\frac{dS\cos\theta}{r_m^2} = d\omega \text{ is the elementary solid angle.}$$

$$\therefore \quad \iint_S \vec{E}\cdot\vec{dS} = Q\int_{\omega=0}^{4\pi} d\omega = 4\pi Q.$$

This is known as the *Gauss' law* in electrostatics.

Example 12.48. *Apply Green's theorem in plane to evaluate the integral,*

$$\oint_C [(xy - x^2)dx + x^2y\,dy]$$

over the triangle bounded by the lines, $y = 0$, $x = 1$ and $y = x$. **[BU(H) 1994, 2001, 2014]**

Solution : The triangle bounded by the lines, $y = 0$, $x = 1$ and $y = x$ is shown in Fig. 12.24.

Fig. 12.24

Along the line OB, $y = 0$, $\therefore\ dy = 0$.

$$\int_{OB}[(xy - x^2)dx + x^2y\,dy] = \int_0^1 -x^2dx = -\left[\frac{x^3}{3}\right]_0^1 = -\frac{1}{3}.$$

Along BA, $x = 1$, $\therefore\ dx = 0$.

$$\int_{BA}[(xy - x^2)dx + x^2y\,dy] = \int_0^1 y\,dy = \frac{1}{2}.$$

Along AO, $\dfrac{y}{x} = \dfrac{1}{1}$ or, $y = x$, $\therefore\ dy = dx$.

$$\int_{AO}[(xy - x^2)dx + x^2y\,dy] = \int_1^0 x^3dx = \left[\frac{x^4}{4}\right]_1^0 = -\frac{1}{4}.$$

$$\therefore \quad \oint_C [(xy - x^2)dx + x^2y\,dy] = -\frac{1}{3} + \frac{1}{2} - \frac{1}{4} = -\frac{1}{12}. \qquad \text{(i)}$$

Again, $\iint_R \left(\frac{\partial N}{\partial x} - \frac{\partial M}{\partial y}\right) dxdy = \iint \left[\frac{\partial}{\partial x}(x^2 y) - \frac{\partial}{\partial y}(xy - x^2)\right] dxdy$ (ii)

$$= \iint_R (2xy - x)dxdy = \int_{x=0}^{1} \left[\int_{y=0}^{x} (2xy - x)dy\right] dx$$

$$= \int_{x=0}^{1} \left[xy^2 - xy\right]_0^x dx = \int_{x=0}^{1} (x^3 - x^2)dx$$

$$= \frac{1}{4} - \frac{1}{3} = -\frac{1}{12}. \quad \text{(iii)}$$

From eqns. (i) and (ii),

$$\oint_C [(xy - x^2)dx + x^2 y\, dy] = \iint_R \left(\frac{\partial M}{\partial x} - \frac{\partial N}{\partial y}\right) dxdy.$$

Hence, Green's theorem is verified.

Example 12.49. *Using Green's theorem in plane evaluate the integral,*

$$\oint_C [(x^2 + xy)dx + (x^2 + y^2)\, dy],$$

where C is the square formed by the line, $y = \pm 1$, $x = \pm 1$.

Solution : By Green's theorem,

$$\oint_0 [(x^2 + xy)dx + (x^2 + y^2)\, dy] = \int_{-1}^{+1}\int_{-1}^{+1} \left[\frac{\partial}{\partial x}(x^2 + y^2) - \frac{\partial}{\partial y}(x^2 + xy)\right]$$

$$= \int_{-1}^{+1}\int_{-1}^{+1} (2x - x)dxdy = \int_{-1}^{+1}\int_{-1}^{+1} x\, dxdy$$

$$= \int_{-1}^{+1} x\, dx \int_{-1}^{+1} dy = \int_{-1}^{+1} x\, dx[y]_{-1}^{+1} = \int_{-1}^{+1} 2x\, dx$$

$$= \left[x^2\right]_{-1}^{+1} = 1 - 1 = \mathbf{0}.$$

Example 12.50. *Evaluate $\oint_C [(y - \sin x)dx + \cos x\, dy]$, where C is the triangle of the Fig. 12.25 by using Green's theorem in the plane.*

Solution : We have by Green's theorem in the plane,

$$\oint_C M\, dx + N\, dy = \iint_R \left(\frac{\partial N}{\partial x} - \frac{\partial M}{\partial y}\right) dydx.$$

Here, $M = y - \sin x$, $N = \cos x$.

$$\therefore \quad \frac{\partial N}{\partial x} = -\sin x, \quad \frac{\partial M}{\partial y} = 1.$$

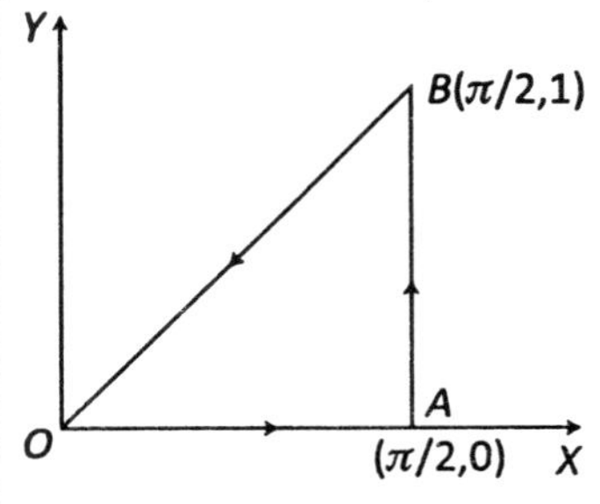

Fig. 12.25

Now, $\oint_C \boldsymbol{M\, dx + N\, dy} = \iint_R \left(\frac{\partial N}{\partial x} - \frac{\partial M}{\partial y}\right) dydx$

$$= \iint_R (-\sin x - 1)dydx = \int_{x=0}^{\frac{\pi}{2}} \left[\int_{y=0}^{\frac{2x}{\pi}} (-\sin x - 1)dydx\right]$$

$$= \int_0^{\frac{\pi}{2}} [-y\sin x - y]_0^{\frac{2x}{\pi}} dx = \int_0^{\frac{\pi}{2}} \left(-\frac{2x}{\pi}\sin x - \frac{2x}{\pi}\right) dx$$

$$= -\frac{2}{\pi}\left[(-x\cos x + \sin x) - \frac{x^2}{\pi}\right]_0^{\frac{\pi}{2}} = -\frac{2}{\pi} - \frac{\pi}{4}.$$

Example 12.51. *Evaluate the integral,* $\int_C (xy + y^2)dx + x^2dy$, *where* C *is the closed curve of the region bounded by* $y = x$ *and* $y = x^2$ *and verify Green's theorem in the plane.*

Solution : The curves $y = x$ and $y = x^2$ are shown in Fig. 12.26, which intersect at $O(0,0)$ and $A(1,1)$. Integrating along the curve $y = x^2$, we have

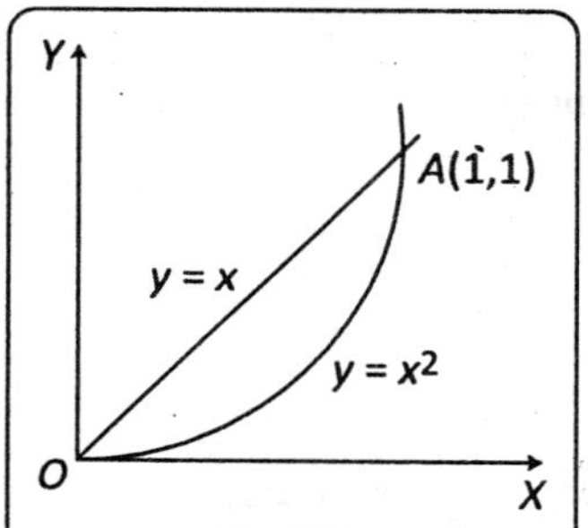

Fig. 12.26

$$\int_0 [(x)(x^2) + x^4]\,dx + (x^2)(2x)\,dx \quad [\text{putting } y = x^2,\ dy = (2x)dx]$$

$$= \int_0^1 (3x^3 + x^4)dx = \frac{19}{20}.$$

Integrating along $y = x$ from A to O, we get

$$\int_1^0 [(x)(x) + x^2]\,dx + x^2\,dx \quad [\text{putting } y = x,\ \therefore\ dy = dx]$$

$$= \int_1^0 3x^2\,dx = -1.$$

$$\therefore \quad \oint_C (xy + y^2)dx + x^2dy = \frac{19}{20} - 1 = -\frac{1}{20}. \qquad \text{(i)}$$

To verify Green's theorem we have

$$\iint_R \left(\frac{\partial N}{\partial x} - \frac{\partial M}{\partial y}\right) dydx = \iint_R \left[\frac{\partial}{\partial x}(x^2) - \frac{\partial}{\partial y}(xy + y^2)\right] dxdy = \iint_R (x - 2y)dxdy$$

$$= \int_{x=0}^1 \int_{y=x^2}^x (x - 2y)dxdy = \int_0^1 \left[\int_{x^2}^x (x - 2y)dy\right] dx$$

$$= \int_0^1 (xy - y^2)_{x^2}^x\,dx = -\frac{1}{20}. \qquad \text{(ii)}$$

From eqns. (i) and (ii),

$$\int_0^1 (xy + y^2)dx + x^2dy = \iint_R \left[\frac{\partial}{\partial x}(x^2) - \frac{\partial}{\partial y}(xy + y^2)\right].$$

$\therefore$ Green's theorem is verified.

Example 12.52. *Evaluate the integral,* $\int_C (3x + 4y)dx + (2x - 3y)dy$, *where* C *is the circle of radius two with centre at the origin of the xy-plane, is traversed in the positive sense.*

Solution : We have, $\oint_C (3x+4y)dx + (2x-3y)dy$

$$= \iint \left[\frac{\partial}{\partial x}(2x-3y) - \frac{\partial}{\partial y}(3x+4y)\right] dxdy$$

[by Green's theorem in the plane]

$$= \iint (2-4)dxdy = -2\iint dxdy.$$

Now, $dxdy = r\,drd\theta$ in polar coordinates,

$$\therefore \oint_C [(3x+4y)dx + (2x-3y)dy] = -2\int_{r=0}^{2}\int_{\theta=0}^{2\pi} r\,drd\theta$$

$$= -2\int_{r=0}^{2} r\,dr \int_{\theta=0}^{2\pi} d\theta = -2\left[\frac{r^2}{2}\right]_0^2 [\theta]_0^{2\pi} = -8\pi.$$

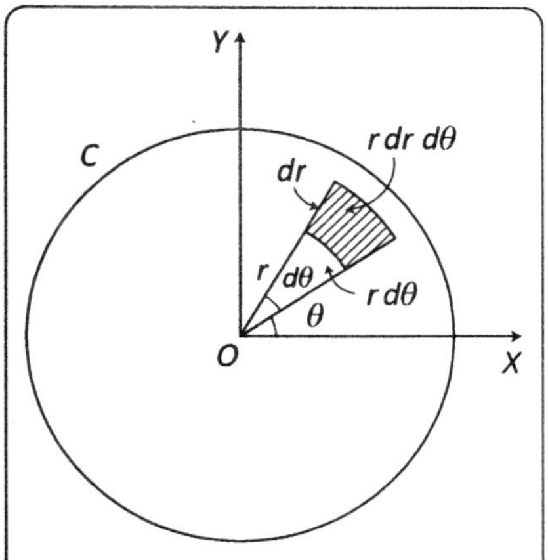

Fig. 12.27

Example 12.53. *Verify Green's theorem in the plane for* $\int_C [(3x^2-8y^2)dx + (4y-6xy)dy]$, *where C is the boundary of the region defined by* $y=\sqrt{x}$, $y=x^2$.

Solution : We have, along the path OKA, (Fig. 12.28)

$$\int_C [(3x^2-8y^2)dx + (4y-6xy)dy] = \int_0^1 (3x^2-8x^4)dx + (4x^2-6x^3)2x\,dx$$

$[\because\ y=x^2,\ \therefore\ dy = 2x\,dx]$

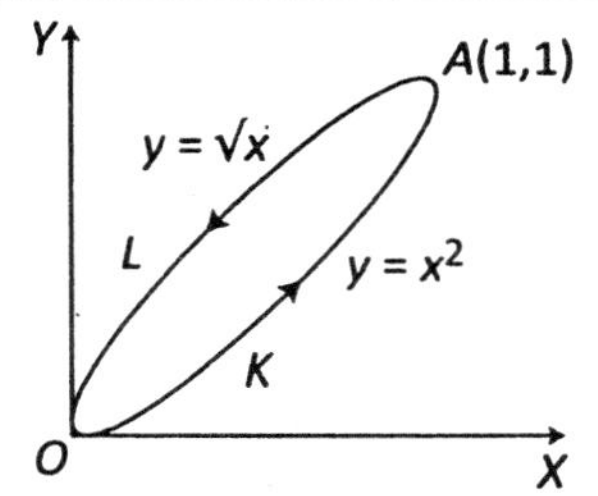

Fig. 12.28

$$= \int_0^1 (3x^2+8x^3-20x^4)dx$$

$$= 3\left[\int_0^1 x^2dx\right] + 8\left[\int_0^1 x^3dx\right] - 20\left[\int_0^1 x^4dx\right]$$

$$= 3\times\frac{1}{3} + 8\times\frac{1}{4} - 20\times\frac{1}{5} = 1+2-4 = -1.$$

Along the path ALO, $y = x^{\frac{1}{2}}$, $\therefore\ dy = \frac{1}{2}x^{-\frac{1}{2}}dx.$

$$\int_{ALO} [(3x^2-8y^2)dx + (4y-6xy)dy] = \int_1^0 \left[(3x^2-8x)dx + \left(4x^{\frac{1}{2}} - 6x^{\frac{3}{2}}\right)\frac{1}{2}x^{-\frac{1}{2}}dx\right]$$

$$= \int_1^0 [3x^2-8x+2-3x]dx = \int_1^0 [2-11x+3x^2]dx$$

$$= 2[x]_1^0 - \frac{11}{2}\left[x^2\right]_1^0 + \frac{3}{3}\left[x^2\right]_1^0 = -2 + \frac{11}{2} - 1$$

$$= \frac{11}{2} - 3 = \frac{5}{2}.$$

$$\therefore \int_C [(3x^2-8y^2)dx + (4y-6xy)dy] = \frac{5}{2} - 1 = \frac{3}{2}. \quad \text{(i)}$$

By Green's theorem in the plane,

$$\oint_C [(3x^2 - 8y^2)dx + (4y - 6xy)dy] = \iint_R \left[\frac{\partial}{\partial x}(4y - 6xy) - \frac{\partial}{\partial y}(3x^2 - 8y^2)\right] dxdy$$

$$= \int_{x=0}^{1}\int_{y=x^2}^{\sqrt{x}} (-6y + 16y)dxdy = \int_{x=0}^{1}\int_{y=x^2}^{\sqrt{x}} 10y\, dxdy$$

$$= 10\int_0^1 \left[\frac{y^2}{2}\right]_{x^2}^{\sqrt{x}} dx = 5\int_0^1 (x - x^4)dx = 5\left[\frac{x^2}{2}\right]_0^1 - 5\left[\frac{x^5}{5}\right]_0^1$$

$$= 5 \times \frac{1}{2} - 5 \times \frac{1}{5} = \frac{5}{2} - 1 = \frac{3}{2}. \quad \text{(ii)}$$

From eqns. (i) and (ii),

$$\oint_C [(3x^2 - 8y^2)dx + (4y - 6xy)dy] = \int_R \left[\frac{\partial}{\partial x}(4y - 6xy) - \frac{\partial}{\partial y}(3x^2 - 8y^2)\right] dxdy.$$

Hence, Green's theorem is verified.

Exercises

12(A) : Objective Type Questions

Select the correct answer

12.1 If $\hat{A} = ax\hat{i} + by\hat{j} + cz\hat{k}$, where a, b, c are constants, then $\iint_S \vec{A} \cdot \vec{dS}$, where S is the surface of a unit sphere, is

(a) $\frac{4}{3}\pi(a+b+c)^2$ (b) $\frac{4}{3}\pi(a+b+c)$ (c) 0 (d) $\frac{4}{3}\pi(a^2+b^2+c^2)$.

12.2 A necessary and sufficient condition that the line integral $\int_C \vec{A} \cdot d\vec{r} = 0$ for every closed curve C is that,

(a) $\nabla \times \vec{A} = 0$ (b) $\nabla \times \vec{A} \neq 0$ (c) $\nabla \cdot \vec{A} \neq 0$ (d) $\nabla \cdot \vec{A} = 0$.

12.3 If $\oint_C \vec{A} \cdot d\vec{r} = 0$ for a closed curve, then

(a) $\vec{A} = 0$ (b) $\vec{A} = \nabla \times \vec{r}$ (c) $\vec{A} = -\nabla\phi$ (d) $d\vec{r} = 0$.

12.4 The area bounded by a simple closed curve is

(a) $\oint_C (x\,dx - y\,dy)$ (b) $\frac{1}{2}\oint_C (x\,dx - y\,dy)$ (c) $\oint_C (x\,dy - y\,dx)$ (d) $\oint_C (xy\,dydx)$.

12.5 The value of $\oint_C (y^2 dx + x^2 dy)$, where C is any closed counter representing the boundary of the square $-1 \leq x < 1$, $-1 \leq y < 1$ is

(a) $\frac{4}{3}$ (b) 0 (c) 4 (d) 2.

12.6 The value of $\iint_R \vec{r} \cdot \hat{n}\, dS$ where S is a closed surface is

(a) V (b) $6V$ (c) $3V$ (d) 0.

12.7 The equation of continuity for an incompressible fluid in absence of source or sink is

(a) $\nabla \cdot \vec{J} = -\frac{\partial \rho}{\partial t}$ (b) $\frac{\partial \rho}{\partial t} = 0$ (c) $\nabla \cdot \vec{v} = 0$ (d) $\nabla \vec{J} = 0$.

12.8 If $\vec{A} = \nabla \times \vec{F}$, then $\iint_S \vec{A} \cdot \hat{n}\, dS$ for any closed surface S is

(a) 0 (b) $4S$ (c) $3V$ (d) none of these.

12.9 If the surface S encloses the change q, then $\iint_S \vec{E} \cdot \vec{dS}$ is equal to

(a) 0 (b) $4\pi q$ (c) q (d) $-q$.

12.10 If $\vec{r} = \hat{i}x + \hat{j}y + \hat{k}z$ and $\hat{n}$ be an outward drawn unit normal to the surface enclosing unit volume, then $\iint_S \vec{r} \cdot \hat{n}\, dS$ is

(a) 1 (b) 4 (c) 3 (d) 6.

12.11 If $\oint \vec{E} \cdot \vec{dr} = -\frac{1}{c}\frac{\partial}{\partial t}\int \vec{H} \cdot \vec{dS}$, where S is the surface bounded by the curve C, then $\nabla \times \vec{E}$ is

(a) $\nabla \cdot \vec{H}$ (b) $\frac{\partial \vec{H}}{\partial t}$ (c) $-\frac{1}{c}\frac{\partial \vec{H}}{\partial t}$ (d) $\frac{\partial}{\partial t}(\nabla \times \vec{H})$.

12.12 If $\vec{A} = ax\hat{i} + by\hat{j} + cz\hat{k}$ and S is any closed surface enclosing volume V, then $\iint_S \vec{A} \cdot \hat{n}\, dS$ is equal to

(a) $3V$ (b) $(a+b+c)V$ (c) V (d) $(abc)V$.

12.13 If $\vec{r} = \hat{i}x + \hat{j}y + \hat{k}z$ be the position vector of a point $P(x, y, z)$ measured from O, then $\iint_S \frac{\hat{n} \cdot \vec{r}}{r^3}\, dS$, where S is a closed surface enclosing O, is equal to

(a) 4π (b) 2π (c) 0 (d) 4.

12.14 If $\oint (M\, dx + N\, dy) = 0$, around any closed curve C in a simple connected region, then

(a) $\frac{\partial M}{\partial y} \neq \frac{\partial N}{\partial y}$ (b) $\frac{\partial N}{\partial x} = \frac{\partial M}{\partial y}$ (c) $\frac{\partial N}{\partial x} \neq \frac{\partial M}{\partial y}$ (d) $\frac{\partial M}{\partial y} = \frac{\partial N}{\partial y}$.

12.15 If $\nabla \cdot \vec{A} = 0$, then $\iint_S \vec{A} \cdot \hat{n}\, dS$ where S' is a closed surface enclosing volume V is

(a) S (b) $3V$ (c) 0 (d) none of these.

12.16 If $\vec{A}$ is solenoidal, then which is not correct?

(a) $\nabla \cdot \vec{A} = 0$ (b) $\iint_S \vec{A} \cdot \hat{n}\, dS = 0$ (c) $\nabla \times \vec{A} = 0$ (d) $\vec{A} = \nabla \times \vec{B}$.

Answers

12.1 (b)	**12.2** (a)	**12.3** (c)	**12.4** (c)	**12.5** (b)	**12.6** (c)	**12.7** (c)	**12.8** (a)
12.9 (b)	**12.10** (c)	**12.11** (c)	**12.12** (b)	**12.13** (a)	**12.14** (d)	**12.15** (c)	**12.16** (c).

12(B) : Short Answer Type Questions

12.1 What do you mean by ordinary integration of vectors?

12.2 What is line integral of vector? Illustrate.

12.3 If $\vec{A} = \nabla\phi$, then show that $\oint \vec{A} \cdot \vec{dr} = 0$.

12.4 Define surface integral of vector.

12.5 Prove that, $\phi(A) + \frac{1}{2}mv_A^2 = \phi(B) + \frac{1}{2}mv_B^2$, the symbols have their usual meaning.

12.6 State and explain Stokes' theorem.

12.7 State and explain Gauss' divergence theorem.

12.8 State Green's theorem in the plane.

12.9 Prove that $\nabla \cdot \vec{r} = 3$, where $\vec{r} = \hat{i}x + \hat{j}y + \hat{k}z$.

12.10 Prove that the area of the ellipse $\frac{x^2}{a^2} + \frac{y^2}{b^2} = 1$ is πab.

12.11 If S is a closed surface enclosing a volume V and $\vec{A} = ax\hat{i} + by\hat{j} + cz\hat{k}$, prove that,

$$\iint_S \vec{A} \cdot \hat{n}\, dS = (a + b + c)V.$$

12.12 If a force $\vec{F} = 2x^2y\hat{i} + 3xy\hat{j}$ displaces a particle in the xy-plane from $(0, 0)$ to $(1, 4)$ along a curve $y = 4x^2$, find the work done. **[Ans:** $\frac{104}{5}$**]**

12.13 If $\vec{A} = \nabla\phi$, then show that the work done in moving a particle in the force field $\vec{A}$ from $A(x_1, y_1, z_1)$ to $B(x_2, y_2, z_2)$ is independent of path.

12.14 If $\vec{B} = \nabla \times \vec{A}$, then prove that, $\iint_S \vec{B} \cdot \hat{n}\, dS = 0$ for any closed surface S.

12.15 Evaluate $\int_C \vec{F} \cdot d\vec{r}$ where, $\vec{F} = xy\hat{i} + yz\hat{j} + zx\hat{k}$ where C is the curve consisting of points having position vectors $\vec{r} = t\hat{i} + t^2\hat{j} + t^3\hat{k}$, with t varying from -1 to 1.

[BU(H) 2014]

12.16 Using Stoke's theorem show that $\oint_C \vec{r} \times d\vec{r} = 2\int_S\int d\vec{S}$, where C is the closed perimeter curve bounding the open surface S. **[BU(H) 2014]**

12.17 Find the unit vector at the point $(a/\sqrt{3}, b\sqrt{3}, c/\sqrt{3})$ normal to the surface of the ellipsoid $\frac{x^2}{a^2} + \frac{y^2}{b^2} + \frac{z^2}{c^2} = 1$ **[BU(H) 2014]**

12.18 A vector field is given by $\vec{F} = xy^2\hat{i} + 2\hat{j} + x\hat{k}$ and L is a path parameterised by $x = ct, y = \frac{c}{t}, z = d$ for the range $k < t \leq 2$.

Evaluate the following integrals :

(i) $\int_L \vec{F} \cdot dt$ (ii) $\int \vec{F} dy$ and (iii) $\int_L \vec{F} \cdot d\vec{r}$ where $\vec{r} = \hat{i}x + \hat{j}y + \hat{k}z$ **[BU(H) 2013]**

12.19 An object is moved from $(3, 2, -4)$ to $(5, 0, 6)$. If the force acting on the object is $\vec{F} = 3\hat{i} - 10\hat{j}$ Newtons, determine the work done, where the distance is measured in metre. **[BU(H) 2012]**

12.20 State Gauss divergence theorem for a vector field $\vec{A} = \vec{A}(x, y, z)$ having continuous first derivatives. **[CU(H) 2009]**

12.21 State Stokes theorem. Give an example where this theorem is not valid. **[CU(H) 2007]**

12.22 Show that the area bounded by a simple closed curve C in a plane is given by

$$A = \frac{1}{2} \oint_C (x\,dy - y dx)$$

[VU(H) 2014]

12.23 State Gauss divergence theorem if $\vec{F} = 2y\hat{i} - z\hat{j} + x^2\hat{k}$ and S is the surface of the parabolic cylinder $y^2 = 8x$ in the first octant bounded by the planes $y = -4$ and $z = 6$. Evaluate $\iint \vec{F} \cdot \hat{n} dS$. **[VU(H) 2013]**

12.24 A vector field is given by $\vec{A} = \sin y\hat{i} + (1 + \cos y)\hat{j}$. Evaluate $\oint_C \vec{A} \cdot d\vec{r}$ over a circular path C given by $x^2 + y^2 = a^2$, $z = 0$ **[VU(H) 2012]**

12.25 State stokes theorem. Using the theorem prove that $\oint_C \phi d\vec{r} = \iint_S d\vec{S} \times \nabla\phi$, where ϕ is a scalar function of r, S is an open surface bounded by a closed curve. **[VU(H) 2012]**

12(C) : Broad Answer Type Questions

12.1 Apply Stokes' theorem to evaluate $\oint_C [(x + y)dx + (2x - z)dy + (y + z)dz]$, where C is the boundary of the triangle with vertices $(2, 0, 0)$, $(0, 3, 0)$ and $(0, 0, 6)$.

[RU 1990; AMIETE(Winter) 1996] **[Ans: 21]**

12.2 If $\vec{A} = y\hat{i} + xz^3\hat{j} - zy^3\hat{k}$, evaluate $\oint_C \vec{A} \cdot d\vec{r}$, where C is the circle $x^2 + y^2 = 4$, $z = 1.5$.

[DU 1992] **[Ans: $(\frac{19\pi}{2})$]**

12.3 Evaluate $\oint_C (x^2 + y^2)dx + 3xy^2 dy$, where C is the circle of radius two with centre at the origin of the xy-plane. **[Ans: 12π]**

12.4 Apply Stokes' theorem to evaluate $\int_C (y\,dx + z\,dy + x\,dz)$, where C is the curve of intersection of the sphere $x^2 + y^2 + z^2 = a^2$ and $x + y = a$. **[Ans: $\pi a^2/\sqrt{2}$]**

12.5 Prove that if $\int \vec{F} \cdot d\vec{r}$ is independent of the path connecting two points P and Q in a given region, then $\oint \vec{F} \cdot d\vec{r} = 0$ for all paths in the region.

12.6 Show that, $\oint_S \vec{B}\cdot d\vec{S} = 0$, when $\vec{B}$ is the magnetic field and S is a closed surface. State the theorem that you have used. **[WBUT 2006]**

[**Hint:** $\oint_S \vec{B}\cdot d\vec{S} = \iint_R (\nabla \times \vec{B})\, dV$.

But, $\nabla\cdot\vec{B} = 0$, $\therefore$ $\oint_S \vec{B}\cdot d\vec{S} = \iint_R 0\, dV = 0$.

Here, Gauss' divergence theorem has been used.]

12.7 Using Divergence theorem evaluate $\iint_S \vec{u}\cdot\hat{n}\, dS$, where $\vec{u} = x\hat{i} + y\hat{j} + z\hat{k}$ and S is the sphere $x^2 + y^2 + z^2 = 9$ and $\hat{n}$ is an outward normal to S. **[WBUT(Math) 2006]**

12.8 Verify Stokes' theorem for $\vec{F} = (2x - y)\hat{i} - yz^2\hat{j} - y^2z\hat{k}$, where S is the upper half surface of the sphere $x^2 + y^2 + z^2 = 1$ and C is its boundary. **[WBUT(Math) 2006]**

12.9 Apply Divergence theorem to evaluate $\iint_S (lx^2 + my^2 + nz^2)dS$ taken over the sphere $(x - a)^2 + (y - b)^2 + (z - c)^2 - \rho^2$; l, m, n being the direction cosines of the external normal to the sphere. **[Ans:** $\frac{8\pi}{3}(a + b + c)\rho^3$**]**

12.10 Calculate, $\oint_C \vec{A}\cdot d\vec{r}$, where $\vec{F} = -\dfrac{y}{x^2 + y^2}\hat{i} + \dfrac{x}{x^2 + y^2}\hat{j}$ and C is a closed curve and explain the results. **[Ans:** 2π or 0 as C surrounds the origin or not**]**

12.11 Prove that, $\iint_S \dfrac{\vec{r}\cdot\hat{n}}{r^2}\, dS = \iiint_V \dfrac{dV}{r^2}$.

12.12 Verify Stokes' theorem for the vector field $\vec{A} = -y^3\hat{i} + x^3\hat{j}$ and the closed curve C is the ellipse $\dfrac{x^2}{a^2} + \dfrac{y^2}{b^2} = 1$.

12.13 Verify Stokes' theorem for $\vec{A} = (y - z + 2)\hat{i} + (yz + 4)\hat{j} - xz\hat{k}$, where S is the surface of the cube $x = 0$, $y = 0$, $z = 0$; $x = 2$, $y = 2$, $z = 2$ above the xz-plane.

Find the line integral $\oint_C \vec{A}\cdot d\vec{r}$ along the boundary of the square on the xy-plane and $\iint_S (\nabla\times\vec{A})\cdot d\vec{S}$ over the five surfaces except that in the xy-plane. The common value of the integrals is -4.

12.14 The potential at a point P, in a region of volume V bounded by a surface S and having continuous charge distribution of density ρ, is $\phi(P) = \iiint_V \dfrac{\rho\, dV}{r}$.

Deduce

(a) Poisson's equation, $\nabla^2\phi = -4\pi\rho$ [when $\rho \neq 0$]

(b) Laplace's equation, $\nabla^2\phi = 0$ [when $\rho = 0$].

12.15 If $\vec{A} = (y^2 + z^2 - x^2)\hat{i} + (z^2 + x^2 - y^2)\hat{j} + (x^2 + y^2 - z^2)\hat{k}$, then verify Stokes' theorem over the hemisphere $x^2 + y^2 + z^2 - 2ax - 2ay - 2az = 0$ on the xy-plane.

12.16 Evaluate $\iint_S \vec{A} \cdot \vec{dS}$, where $\vec{A} = \hat{i}x + \hat{j}y + \hat{k}z$ and S is the closed surface bounded by $x^2 + y^2 = z^2$ and the plane $z = 4$. **[Ans:** $\frac{128\pi}{3}$**]**

12.17 Evaluate $\phi(P) = \iiint_V \vec{A}\, dV$, where $\vec{A} = (2xz\hat{i} - x\hat{j} + \hat{k}y^2)$ and where V is the region bounded by the surfaces $x = 0$, $y = 0$, $y = 6$, $z = x^2$, $z = 4$. **[Ans:** $128\hat{i} - 24\hat{j} + 384\hat{k}$**]**

12.18 For a given vector $\vec{A}$, prove by the application of divergence theorem and Stokes' theorem that, $\nabla \cdot \nabla \times \vec{A} = 0$.

12.19 Verify Divergence theorem for $\vec{A} = (x^2 - yz)\hat{i} + (y^2 - zx)\hat{j} + (z^2 - xy)\hat{k}$ over the rectangular parallelopiped $0 \le x \le a$, $0 \le y \le b$, $0 \le z \le c$.

[Ans: Common value $= abc(a + b + c)$**]**

12.20 Verify Green's theorem in the plane for, $\oint_C (2x - y^3)dx - xy\, dy$, where C is the boundary of the region enclosed by the circles $x^2 + y^2 = 1$ and $x^2 + y^2 = 9$.

[Ans: Common value $= 60\pi$**]**

12.21 Find the area of the ellipse $x = a\cos\theta$, $y = a\sin\theta$. **[Ans:** πab**]**

12.22 **(a)** State Stokes' theorem of vector calculus.

(b) Given the vector $\vec{A} = (x^2 - y^2)\hat{i} + 2xy\hat{j}$

(i) Evaluate $\int_S (\nabla \times \vec{A}) \cdot d\vec{S}$ over the rectangle in the xy-plane bounded by the lines $x = 0, x = a, y = 0, y = b$.

(ii) Evaluate $\int \vec{A} \cdot d\vec{r}$ around the boundary of the rectangle and thus verify Stokes' theorem. **[BU(H) 2012]**

12.23 **(a)** State divergence theorem of Gauss.

(b) Verify it for the vector $\vec{A} = 4xz\hat{i} - y^2\hat{j} + yz\hat{k}$ taken over the cube bounded by $0 \le x \le 1$, $0 \le y \le 1$ and $0 \le z \le 1$. **[BU(H) 2010]**

(c) If $\vec{F} = 3xy\hat{i} - y^2\hat{j}$ evaluate $\int_C \vec{F} \cdot d\vec{r}$ where C is the segment of the curve connecting the points $(0, 0)$ and $(1, 2)$. **[BU(H) 2008]**

12.24 **(a)** Show that $\frac{1}{3}\int_S \vec{r} \cdot d\vec{S} = V$, where V is the volume enclosed by the closed surface S.

(b) By use of Stokes' theorem prove that $\nabla \times \nabla\phi = 0$.

(c) Show that the line integral of $\vec{F} = -\hat{i}y + \hat{j}x$ around a continuous closed curve in the xy-plane is twice the area enclosed by the curve. **[CU(H) 2014]**

12.25 **(a)** $ABCD$ is a square in the xy-plane. The x, y coordinates of A, B, C and D are $(0, 0)$, $(0, a)$, (a, a), $(a, 0)$, respectively. Show that the line integral of $\vec{F} = (2xy + z^2)\hat{i} + x^2\hat{j} + 3xz^2\hat{k}$ is zero over $ABCDA$ **[CU(H) 2010]**

(b) Evaluate $\oint \vec{A}\cdot d\vec{r}$ around the right-angled triangle PQR having vertices at $P(0,0)$, $Q(2,0)$, and $R(2,1)$, where $\vec{A} = \hat{\imath}(2x+y^2)+\hat{\jmath}(3y-4x)$.

(c) Evaluate $\int_S (yz\hat{\imath}+zx\hat{\jmath}+xy\hat{k})\cdot d\vec{S}$, where S is the surface of the sphere $x^2+y^2+z^2 = a^2$ in the first octant. **[VU(H) 2006]**

12.26 (a) Evaluate $\oint \vec{F}\cdot d\vec{r}$ along a closed curve surrounding the origin and lying in the xy-plane in the following cases,

(i) $\vec{F} = \dfrac{\hat{\imath}x+\hat{\jmath}y}{x^2+y^2}$ (ii) $\dfrac{-\hat{\imath}y+\hat{\jmath}x}{x^2+y^2}$.

(b) If $\vec{r}$ be the position vector of a point on a closed curve C, prove that the line integral $\oint_C \vec{r}\cdot d\vec{r} = 0$. **[VU(H) 2011]**

12.27 (a) Verify Green's theorem in the plane for $\oint_C [(xy+y^2)dx + x^2dy]$, where C is the closed curve of the region bounded by $y=x$ and $y=x^2$. **[VU(H) 2009]**

(b) If $\vec{F} = (5xy-6x^3)\hat{\imath}+(2y-4x)\hat{\jmath}$, then evaluate $\int_C \vec{F}\cdot d\vec{r}$ along the curve C in the xy-plane given by $y=x^3$ from the point $(1,1)$ to $(2,8)$.

12.28 (a) Verify Green's theorem in the plane for $\left[\int_C (3x^2-8y^2)dx + (4y-6xy)\right] dy$, where C is is the boundary of the region defined by $x=0, y=0, x+y=1$.

(b) State and prove Stokes' theorem.

(c) Evaluate $\int\int\int \sqrt{x^2+y^2}\,dx\,dy\,dz$, where V is the region bounded by $Z = x^2+y^2$ and $Z = 8-x^2-y^2$.

12.29 (a) State and prove Green's theorem in the plane. **[Mahatma Gandhi U(MSc) 2012]**

(b) If $\vec{V} = \vec{\omega}\times\vec{r}$, prove that $\vec{\omega} = \frac{1}{2}\text{Curl}\vec{V}$, where ω is a constant vector.

12.30 (a) If $\vec{R}(u) = \dfrac{d}{du}\vec{S}(u)$, find $\int_a^b \vec{R}(u)du$.

(b) The surface integral of $\vec{A}$ over a closed surface S vanishes, evaluate $\nabla\cdot\vec{A}$.

12.31 (a) If $\vec{E} = -\nabla\phi$, evaluate $\oint_C \vec{E}\cdot d\vec{r}$, where ϕ is a scalar function of $\vec{r}$.

(b) Show that, $\int_V (\phi\nabla^2\psi - \psi\nabla^2\phi)dV = \int_S (\phi\nabla\psi - \psi\nabla\phi)d\vec{S}$.

12.32 (a) Using Green's theorem in plane, find $\oint_C [(xy+y^2)dx + x^2dy]$ where C is the closed curve of the region bounded by $y=x$ and $y=x^2$.

(b) Prove that $\oint_C \vec{A}\cdot d\vec{\lambda} = \int_S (\nabla\times\vec{A})\cdot d\vec{S}$, where C is the curve bounding the surface S. Hence, find $\oint_C \vec{r}\cdot d\vec{r}$. **[Assam U Phys(Major) 2014]**

(c) Prove that, $\int_V (\nabla\times\vec{B})dV = \int_S (\hat{n}\times\vec{B})dS$, where V is the volume enclosed by the surface S and $\hat{n}$ is the unit normal vector to the plane of dS. **[Assam U Phys.(Major) 2014]**

12.33 (a) State Stokes' theorem for a vector $\vec{V}$. Let $\vec{V} = \phi\vec{c}$, where $\vec{c}$ is a constant vector. Deduce that $\int \nabla\phi\times d\vec{s} = \oint \phi d\vec{r}$. **[Use $\nabla\times(\phi\vec{A}) = (\nabla\phi)\times\vec{A} + \phi(\nabla\times\vec{A})$]**

(b) Given $\vec{A} = (x^2-y)\hat{i} + 2x\hat{j} + 2\hat{k}$, evaluate $\oint \vec{A}\cdot d\vec{r}$ around the boundary of the circle $x^2+y^2=1$.

12.34 (a) Show that $\vec{A} = (6xy+z^3)\hat{i} + (3x^2-z)\hat{j} + (3xz^2-y)\hat{k}$ is irrotational. Find ϕ such that $\vec{A} = \nabla\phi$.

(b) If $\vec{F} = 3xy\hat{i} - y^2\hat{j}$, evaluate $\int_C \vec{F}\cdot d\vec{r}$, where C is the curve $y = 2x^2$ in the xy-plane, from $(0,0)$ to $1,2)$.

(c) Use divergence theorem to prove that $\oint \frac{\vec{r}}{r^2}\cdot d\vec{S} = \int \frac{dV}{r^2}$. **[CU(H) 2016]**

12.35 (a) For a force defined by $\vec{F} = \dfrac{-x\hat{i}+y\hat{j}}{x^2+y^2}$, evaluate the integral along the semi-circular path connecting the points $(-1,0)$ and $(1,0)$ with the origin as centre.

(b) Use the divergence theorem to evaluate $\oint_S \vec{A}\cdot d\vec{S}$, where $\vec{A} = (2x-z)\hat{i}+x^2y\hat{j}+xz^2\hat{k}$ and S is the surface of the cube bounded by $x=0$, $x=1$, $y=0$, $y=1$, $z=0$, $z=1$.

(c) Show that $|\oint d\vec{r}\times\vec{r}|$ taken around a curve in the XY-plane is twice the area enclosed by the curve. **[CU(H) 2015]**

12.36 (a) Show that $\frac{1}{3}\oint_S \vec{r}\cdot d\vec{s} = V$, where V is the volume enclosed by the closed surface S.

(b) By the use of Stokes' theorem prove that $\nabla\times\nabla\phi = 0$.

(c) A particle moves along the curve $x = 2t^2$, $y = 4t^2-1$, $z = 5t-3$, where t denotes the time. Find the component of acceleration at time $t=1$ in the direction$(\hat{i}-2\hat{j}+2\hat{k})$.

(d) Show that the line integral of $\vec{F} = -\hat{i}y + \hat{j}x$ around a continuous closed curve in the XY-plane is twice the area enclosed by the curve. **[CU(H) 2014]**

12.37 (a) Let $\vec{V} = \phi\vec{C}$ where $\vec{C}$ is a constant vector in the divergence theorem, deduce that $\int (\nabla\phi)d\vec{r} = \int \phi d\vec{\sigma}$.

Use : $\nabla\cdot(\phi\vec{A}) = (\nabla\phi)\cdot\vec{A} + \phi(\nabla\cdot\vec{A})$.

12.38 (a) If $\vec{V} = 2xy\hat{i} - y^2\hat{j} + (x+xy)\hat{k}$, evaluate $\int_\sigma \vec{V}\cdot d\vec{\sigma}$, where σ is the surface bounded by $x^2+y^2=9$, $z=0$ and $z=5$. **[CU(H) 2012]**

(b) Using Green's theorem for a plane prove that the area bounded by the curve in XY-plane is given by $\frac{1}{2}\oint(xdy - ydx)$. Hence, calculate the area of the ellipse $\frac{x^2}{a^2} + \frac{y^2}{b^2} = 1$.

(c) Use Stokes' theorem to compute the line integral of the vector $\vec{A} = uy\hat{x} - x^2\hat{y} - 2\hat{z}$ around the contour $x^2 + y^2 = 4,\ z = 3$. **[CU(H) 2011]**

(d) If $\vec{F} = (2x+y)\hat{\imath} + (3y-x)\hat{\jmath}$, evaluate $\int_C \vec{F}\cdot d\vec{r}$, where C is the curve in XY-plane consisting of the straight line segment from $(0,0)$ to $(2,0)$ and then to $(3,2)$. **[CU(H) 2011]**

CHAPTER 13

Curvilinear Coordinates

13.1 Curvilinear Coordinates

The curvilinear coordinate system is the most general coordinate system which can easily be transformed to any special coordinate system like *Cartesian, spherical, cylindrical, paraboloid, ellipsoidal, bi-polar, etc., coordinate system* which have important applications in physical problems.

In rectangular Cartesian coordinate system a point P is represented by (x, y, z). If there be a set of variables u, v, w such that,

$$x = f_1(u, v, w), \quad y = f_2(u, v, w), \quad z = f_3(u, v, w) \tag{13.1}$$

we can solve eqn. (13.1) for u, v, w in terms of x, y, z, i.e.,

$$u = f_4(x, y, z), \quad v = f_5(x, y, z), \quad w = f_6(x, y, z). \tag{13.2}$$

If the functions in eqns. (13.1) and (13.2) are *single-valued* and have *continuous derivatives,* then there is a unique correspondence between (x, y, z) and (u, v, w) except in certain special cases. Thus, we can associate a unique set of coordinates (u, v, w) to (x, y, z). The set (u, v, w) is called *Curvilinear Coordinates* of the point P.

13.2 Orthogonal Curvilinear Coordinates

The level surfaces $u = u_0$, $v = v_0$, $w = w_0$, where u_0, v_0, w_0 are constants, are called *coordinate surfaces* through the point $P(u_0, v_0, w_0)$. The surfaces $u = u_0$ and $v = v_0$ intersect in curves called the w-curve as only w changes along this curve. Similarly, we get v-curve and w-curve due to the intersection of $w = w_0$, $u = u_0$ and $v = v_0$, $w = w_0$ respectively, along which only v and u-changes (Fig. 13.1).

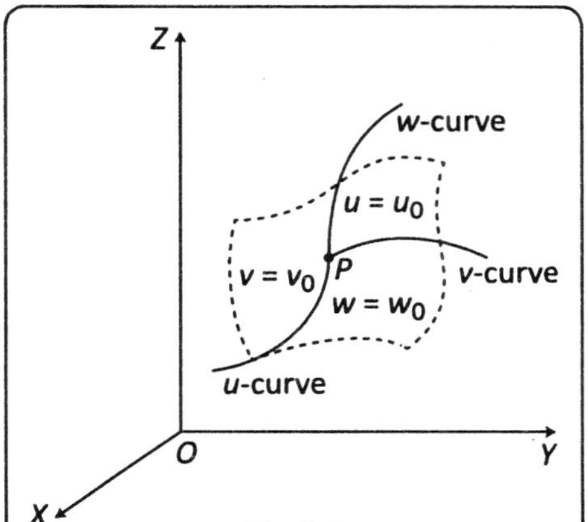

Fig. 13.1

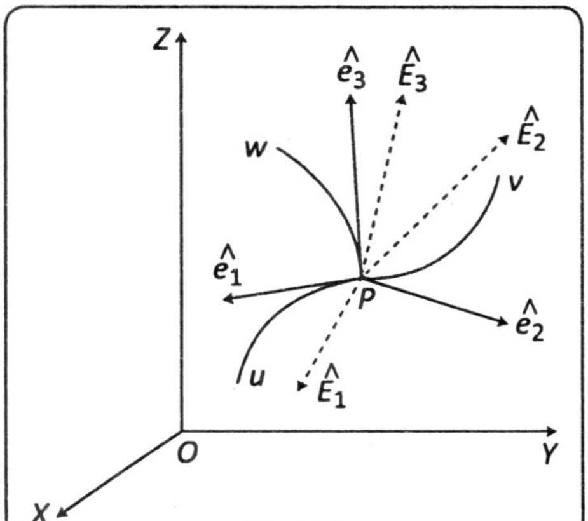

Fig. 13.2

If the coordinate surfaces intersect at right angles, then the curvilinear coordinate system is orthogonal and u, v, w are called the *orthogonal curvilinear coordinates.* In that case u, v, w

are analogous to the rectangular coordinates x, y, z. The coordinate axes are determined by the tangents to the coordinate lines at the point of intersection of the $u = u_0$, $v = v_0$ and $w = w_0$ surfaces. Unlike Cartesian system, these are not in a fixed direction in space.

13.3 Two Sets of Units Vectors

If $\vec{r}$ be the position vector of a point $P(x, y, z)$, then $\vec{r} = \hat{i}x + \hat{j}y + \hat{k}z$

or, $\quad \vec{r} = \hat{i}f_1(u, v, w) + \hat{j}f_2(u, v, w) + \hat{k}f_3(u, v, w)$

$$\therefore \quad d\vec{r} = \frac{\partial \vec{r}}{\partial u}du + \frac{\partial \vec{r}}{\partial v}dv + \frac{\partial \vec{r}}{\partial w}dw.$$

Now, $\frac{\partial \vec{r}}{\partial u}$ is a tangent vector to $u = u_0$ curve. If $\hat{e}_1$ be a unit vector along the tangent $\frac{\partial \vec{r}}{\partial u}$, then

$$\hat{e}_1 = \frac{\partial \vec{r}}{\partial u} \Big/ \left|\frac{\partial \vec{r}}{\partial u}\right|, \quad \therefore \quad \frac{\partial \vec{r}}{\partial u} = h_1\hat{e}_1,\ h_1 = \left|\frac{\partial \vec{r}}{\partial u}\right|.$$

Similarly, $$\hat{e}_2 = \frac{\partial \vec{r}}{\partial v} \Big/ \left|\frac{\partial \vec{r}}{\partial v}\right|, \quad \therefore \quad \frac{\partial \vec{r}}{\partial v} = h_2\hat{e}_2,\ h_2 = \left|\frac{\partial \vec{r}}{\partial v}\right|$$

and $$\hat{e}_3 = \frac{\partial \vec{r}}{\partial w} \Big/ \left|\frac{\partial \vec{r}}{\partial w}\right|, \quad \therefore \quad \frac{\partial \vec{r}}{\partial w} = h_3\hat{e}_3,\ h_3 = \left|\frac{\partial \vec{r}}{\partial w}\right|.$$

Here h_1, h_2, h_3 are three scalars called the *scale factors*. The unit vectors $\hat{e}_1$, $\hat{e}_2$, $\hat{e}_3$ are directed along the directions of increasing u, v, w respectively.

Again, ∇u, ∇v and ∇w are the vectors along the direction of normal to the $u = u_0$, $v = v_0$ and $w = w_0$ surfaces respectively. The unit vectors along these normals are

$$\widehat{E}_1 = \frac{\nabla u}{|\nabla u|},\ \widehat{E}_2 = \frac{\nabla v}{|\nabla v|},\ \widehat{E}_3 = \frac{\nabla w}{|\nabla w|}.$$

Thus, two sets of vectors $\hat{e}_1$, $\hat{e}_2$, $\hat{e}_3$ and $\widehat{E}_1$, $\widehat{E}_2$, $\widehat{E}_3$ exist at each point P in curvilinear coordinate system. In case of orthogonal curvilinear coordinate system both sets become identical. These are called *unit base vectors*.

13.4 Contravariant and Co-variant Components of a Vector $\vec{A}$

If A_1, A_2, A_3 and a_1, a_2, a_3 represent respectively the components of $\vec{A}$ in *unit base vectors* $\hat{e}_1$, $\hat{e}_2$, $\hat{e}_3$ and $\widehat{E}_1$, $\widehat{E}_2$, $\widehat{E}_3$, then

$$\vec{A} = A_1\hat{e}_1 + A_2\hat{e}_2 + A_3\hat{e}_3 \quad \text{and} \quad \vec{A} = A_1\widehat{E}_1 + A_2\widehat{E}_2 + A_3\widehat{E}_3.$$

$\vec{A}$ can also be represented by sets of *unitary* base vectors $\frac{\partial \vec{r}}{\partial u}$, $\frac{\partial \vec{r}}{\partial v}$, $\frac{\partial \vec{r}}{\partial w}$ and ∇u, ∇v, ∇w as follows :

$$\vec{A} = C_1\frac{\partial \vec{r}}{\partial u} + C_2\frac{\partial \vec{r}}{\partial v} + C_3\frac{\partial \vec{r}}{\partial w} = C_1\vec{\alpha_1} + C_2\vec{\alpha_2} + C_3\vec{\alpha_3}$$

and $$\vec{A} = c_1\nabla u + c_2\nabla v + c_3\nabla w = c_1\vec{\beta_1} + c_2\vec{\beta_2} + c_3\vec{\beta_3},$$

where $\vec{\alpha_1} = \frac{\partial \vec{r}}{\partial u}$, $\vec{\alpha_2} = \frac{\partial \vec{r}}{\partial v}$, $\vec{\alpha_3} = \frac{\partial \vec{r}}{\partial w}$ and $\vec{\beta_1} = \nabla u$, $\vec{\beta_2} = \nabla v$, $\vec{\beta_3} = \nabla w$

Here, C_1, C_2, C_3 are called the *co-variant components of* $\vec{A}$ and c_1, c_2, c_3 are the *contra-variant components of* $\vec{A}$.

13.5 Arc Length and Volume Elements

We consider an infinitesimal parallelopiped $PABCDEFG$ in orthogonal curvilinear coordinate space whose faces coincide with the planes $u = u_0$, $v = v_0$ and $w = w_0$ (Fig. 13.3).

Since $\vec{r} = \vec{r}(u, v, w)$

$$\therefore \quad d\vec{r} = \frac{\partial \vec{r}}{\partial u}du + \frac{\partial \vec{r}}{\partial v}du + \frac{\partial \vec{r}}{\partial w}dw$$

$$= h_1 du\hat{e}_1 + h_2 dv\hat{e}_2 + h_3 dw\hat{e}_3.$$

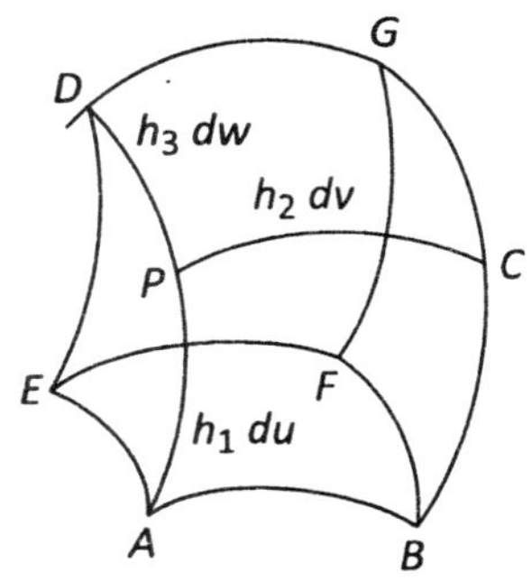

Fig. 13.3

$\therefore$ the square of the differential arc length is

$$ds^2 = d\vec{r} \cdot d\vec{r}$$

$$= (h_1 du\hat{e}_1 + h_2 dv\hat{e}_2 + h_3 dw\hat{e}_3) \cdot (h_1 du\hat{e}_1 + h_2 dv\hat{e}_2 + h_3 dw\hat{e}_3)$$

$$\boldsymbol{ds^2 = h_1^2 du^2 + h_2^2 dv^2 + h_3^2 dw^2}.$$

$\because \ \hat{e}_1 \cdot \hat{e}_1 = \hat{e}_2 \cdot \hat{e}_2 = \hat{e}_3 \cdot \hat{e}_3 = 1$ and $\hat{e}_1 \cdot \hat{e}_2 = \hat{e}_2 \cdot \hat{e}_3 = \hat{e}_3 \cdot \hat{e}_1 = 0$

or, $\quad ds^2 = ds_1^2 + ds_2^2 + ds_3^2$.

where $ds_1 = h_1 du$, $ds_2 = h_2 dv$, $ds_3 = h_3 dw$ are the differential arc lengths along u, v, w respectively.

Volume Element

The volume element in orthogonal curvilinear coordinates is given by

$$dv = |(h_1 du\hat{e}_1) \cdot (h_2 dv\hat{e}_2) \times (h_3 dw\hat{e}_3)| = h_1 h_2 h_3\, du\, dv\, dw |\hat{e}_1 \cdot \hat{e}_2 \times \hat{e}_3|$$

or, $\quad \boldsymbol{dv = h_1 h_2 h_3\, du\, dv\, dw}, \quad \because \ |\hat{e}_1 \cdot \hat{e}_2 \times \hat{e}_3| = 1|$

13.6 Differential Operators in Terms of Orthogonal Curvilinear Coordinates

Let $\phi(u, v, w)$ be a scalar function of the variables u, v and w and

$$\nabla\phi = f_1\hat{e}_1 + f_2\hat{e}_2 + f_3\hat{e}_3, \tag{i}$$

where f_1, f_2, f_3 are scalars, whose values are to be determined.

Again, $\vec{r} = \vec{r}(u, v, w)$,

$$\therefore \quad d\vec{r} = \frac{\partial \vec{r}}{\partial u}du + \frac{\partial \vec{r}}{\partial v}dv + \frac{\partial \vec{r}}{\partial w}dw = h_1\hat{e}_1 du + h_2\hat{e}_2 dv + h_3\hat{e}_3 dw,$$

Now, $\quad d\phi = \nabla\phi \cdot d\vec{r} = (f_1\hat{e}_1 + f_2\hat{e}_2 + f_3\hat{e}_3) \cdot (h_1\hat{e}_1 du + h_2\hat{e}_2 dv + h_3\hat{e}_3 dw)$

$$= h_1 f_1 du + h_2 f_2 dv + h_3 f_3 d_2. \tag{ii}$$

$$\text{Again, } d\phi = \frac{\partial \phi}{\partial u}du + \frac{\partial \phi}{\partial v}dv + \frac{\partial \phi}{\partial w}dw. \tag{iii}$$

From eqns. (ii) and (iii) we can write,

$$h_1 f_1 du + h_2 f_2 dv + h_3 f_3 dw = \frac{\partial \phi}{\partial u}du + \frac{\partial \phi}{\partial v}dv + \frac{\partial \phi}{\partial w}dw.$$

$$\therefore \quad h_1 f_1 du = \frac{\partial \phi}{\partial u}du \quad \text{or,} \quad f_1 = \frac{1}{h_1}\frac{\partial \phi}{\partial u}.$$

$$h_2 f_2 dv = \frac{\partial \phi}{\partial v} dv \quad \text{or,} \quad f_2 = \frac{1}{h_2}\frac{\partial \phi}{\partial v},$$

and $$h_3 f_3 dw = \frac{\partial \phi}{\partial w} dw \quad \text{or,} \quad f_3 = \frac{1}{h_3}\frac{\partial \phi}{\partial w}. \tag{iv}$$

Now, substituting the values of f_1, f_2 and f_3 in eqn. (i) from eqn. (iv), we get

$$\boldsymbol{\nabla}\phi = \frac{1}{h_1}\frac{\partial \phi}{\partial u}\hat{\mathbf{e}}_1 + \frac{1}{h_2}\frac{\partial \phi}{\partial v}\hat{\mathbf{e}}_2 + \frac{1}{h_3}\frac{\partial \phi}{\partial w}\hat{\mathbf{e}}_3. \tag{v}$$

Also, $$\boldsymbol{\nabla} = \frac{1}{h_1}\frac{\partial}{\partial u}\hat{\mathbf{e}}_1 + \frac{1}{h_2}\frac{\partial}{\partial v}\hat{\mathbf{e}}_2 + \frac{1}{h_3}\frac{\partial}{\partial w}\hat{\mathbf{e}}_3. \tag{vi}$$

[Mang.U 1985, Myso.U 1987(S)]

13.6.1 Divergence

Let $\vec{A} = A_1\hat{e}_1 + A_2\hat{e}_2 + A_3\hat{e}_3$.

Then, $$\nabla \cdot \vec{A} = \nabla \cdot (A_1\hat{e}_1 + A_2\hat{e}_2 + A_3\hat{e}_3) = \nabla \cdot (A_1\hat{e}_1) + \nabla \cdot (A_2\hat{e}_2) + \nabla(A_3\hat{e}_3). \tag{i}$$

Now, $$\nabla \cdot (A_1\hat{e}_1) = \nabla \cdot [A_1(h_2h_3\nabla v \times \nabla w)]$$

$$\left[\because\ \nabla v \times \nabla w = \frac{\hat{e}_2}{h_2} \times \frac{\hat{e}_3}{h_3} = \frac{\hat{e}_1}{h_2h_3}, \quad \therefore\ \hat{e}_1 = h_2h_3\nabla u \times \nabla w\right]$$

$$= \nabla(A_1h_2h_3) \cdot \nabla v \times \nabla w + A_2h_2h_3\nabla \cdot (\nabla v \times \nabla w)$$

$$= \nabla(A_1h_2h_3) \cdot \frac{\hat{e}_2}{h_2} \times \frac{\hat{e}_3}{h_3} + 0 = \nabla(A_1h_2h_3) \cdot \frac{\hat{e}_1}{h_2h_3}$$

$$= \left[\frac{1}{h_1}\frac{\partial}{\partial u}(A_1h_2h_3)\hat{e}_1 + \frac{1}{h_2}\frac{\partial}{\partial v}(A_1h_2h_3)\hat{e}_2 + \frac{1}{h_3}\frac{\partial}{\partial w}(A_1h_2h_3)\hat{e}_3\right] \cdot \frac{\hat{e}_1}{h_2h_3}.$$

$$\therefore \quad \nabla \cdot (A_1\hat{e}_1) = \frac{1}{h_1h_2h_3}\frac{\partial}{\partial u}(A_1h_2h_3). \tag{ii}$$

Similarly, $$\nabla \cdot (A_2\hat{e}_2) = \frac{1}{h_1h_2h_3}\frac{\partial}{\partial v}(A_2h_3h_1) \tag{iii}$$

and $$\nabla \cdot (A_3\hat{e}_3) = \frac{1}{h_1h_2h_3}\frac{\partial}{\partial w}(A_3h_1h_2). \tag{iv}$$

Substituting eqns. (ii), (iii), (iv) in eqn. (i), we get

$$\nabla \cdot \vec{A} = \frac{1}{h_1h_2h_3}\frac{\partial}{\partial u}(A_1h_2h_3) + \frac{1}{h_1h_2h_3}\frac{\partial}{\partial r}(A_2h_3h_1) + \frac{1}{h_1h_2h_3}\frac{\partial}{\partial w}(A_3h_1h_2)$$

or, $$\boldsymbol{\nabla} \cdot \vec{\boldsymbol{A}} = \frac{1}{h_1h_2h_3}\left[\frac{\partial}{\partial u}(A_1h_2h_3) + \frac{\partial}{\partial v}(A_2h_3h_1) + \frac{\partial}{\partial w}(A_3h_1h_2)\right]$$

[AU 1995(S), Bang.U 1990]

Curl

Let $\vec{A} = A_1\hat{e}_1 + A_2\hat{e}_2 + A_3\hat{e}_3$.

$$\therefore \quad \nabla \times \vec{A} = \nabla \times (A_1\hat{e}_1 + A_2\hat{e}_2 + A_3\hat{e}_3). \tag{i}$$

Now, $\nabla \times A_1\hat{e}_1 = \nabla \times (A_1 h_1 \nabla u) \quad \left[\because \ \nabla u = \dfrac{\hat{e}_1}{h_1}\right]$

$$= \nabla(A_1h_1) \times \nabla u + A_1 h \nabla \times \nabla u = \nabla(A_1 h_1) \times \frac{\hat{e}_1}{h_1} \quad [\because \ \nabla \times \nabla u = 0]$$

$$= \left[\frac{1}{h_1}\frac{\partial}{\partial u}(A_1h_1)\hat{e}_1 + \frac{1}{h_2}\frac{\partial}{\partial v}(A_1h_1)\hat{e}_2 + \frac{1}{h_3}\frac{\partial}{\partial w}(A_1h_1)\hat{e}_3\right] \times \frac{\hat{e}_1}{h_1}$$

or, $$\nabla \times A_1\hat{e}_1 = \frac{1}{h_3h_1}\frac{\partial}{\partial w}(A_1h_1)\hat{e}_2 - \frac{1}{h_1h_2}\frac{\partial}{\partial v}(A_1h_1)\hat{e}_3. \quad \text{(ii)}$$

Similarly, $$\nabla \times A_2\hat{e}_2 = \frac{1}{h_1h_2}\frac{\partial}{\partial u}(A_2h_2)\hat{e}_3 - \frac{1}{h_2h_3}\frac{\partial}{\partial w}(A_2h_2)\hat{e}_1 \quad \text{(iii)}$$

and $$\nabla \times A_3\hat{e}_3 = \frac{1}{h_2h_3}\frac{\partial}{\partial v}(A_3h_3)\hat{e}_1 - \frac{1}{h_3h_1}\frac{\partial}{\partial u}(A_3h_3)\hat{e}_2. \quad \text{(iv)}$$

Substituting eqns. (ii), (iii) and (iv) in eqn. (ii), we get

$$\nabla \times \vec{A} = \nabla \times A_1\hat{e}_1 + \nabla \times A_2\hat{e}_2 + \nabla \times A_3\hat{e}_3$$

$$= \left[\frac{1}{h_3h_1}\frac{\partial}{\partial w}(A_1h_1)\hat{e}_2 - \frac{1}{h_1h_2}\frac{\partial}{\partial v}(A_1h_1)\hat{e}_3\right] + \left[\frac{1}{h_1h_2}\frac{\partial}{\partial u}(A_2h_2)\hat{e}_3 - \frac{1}{h_2h_3}\frac{\partial}{\partial w}(A_2h_2)\hat{e}_1\right] + \left[\frac{1}{h_2h_3}\frac{\partial}{\partial v}(A_3h_3)\hat{e}_1 - \frac{1}{h_3h_1}\frac{\partial}{\partial u}(A_3h_3)\hat{e}_2\right]$$

or, $$\nabla \times \vec{A} = \frac{1}{h_2h_3}\left[\frac{\partial}{\partial v}(A_3h_3) - \frac{\partial}{\partial w}(A_2h_2)\right]\hat{e}_1 + \frac{1}{h_3h_1}\left[\frac{\partial}{\partial w}(A_1h_1) - \frac{\partial}{\partial u}(A_3h_3)\right]\hat{e}_2 + \frac{1}{h_1h_2}\left[\frac{\partial}{\partial u}(A_2h_2) - \frac{\partial}{\partial v}(A_1h_1)\right]\hat{e}_3.$$

$$\therefore \quad \boldsymbol{\nabla \times \vec{A} = \frac{1}{h_1h_2h_3}\begin{vmatrix} h_1\hat{e}_1 & h_2\hat{e}_2 & h_3\hat{e}_3 \\ \dfrac{\partial}{\partial u} & \dfrac{\partial}{\partial v} & \dfrac{\partial}{\partial w} \\ A_1h_1 & A_2h_2 & A_3h_3 \end{vmatrix}}. \quad \textbf{[Bang.U 1990(S)]}$$

Laplacian (∇^2)

If $\phi(u, v, w)$ be a scalar function of u, v, w, then we have

$$\nabla\phi = \frac{1}{h_1}\frac{\partial\phi}{\partial u}\hat{e}_1 + \frac{1}{h_2}\frac{\partial\phi}{\partial v}\hat{e}_2 + \frac{1}{h_3}\frac{\partial\phi}{\partial w}\hat{e}_3. \quad \text{(i)}$$

If $\vec{A} = \nabla\phi$, then

$$A_1\hat{e}_1 + A_2\hat{e}_2 + A_2\hat{e}_3 = \frac{1}{h_1}\frac{\partial\phi}{\partial u}\hat{e}_1 + \frac{1}{h_2}\frac{\partial\phi}{\partial v}\hat{e}_2 + \frac{1}{h_3}\frac{\partial\phi}{\partial w}\hat{e}_3.$$

Comparing the coefficients of $\hat{e}_1, \hat{e}_2$ and $\hat{e}_3$ on both sides, we get.

$$A_1 = \frac{1}{h_1}\frac{\partial\phi}{\partial u}, \ A_2 = \frac{1}{h_2}\frac{\partial\phi}{\partial v}, \ A_3 = \frac{1}{h_3}\frac{\partial\phi}{\partial w}.$$

Also we have

$$\nabla\cdot\vec{A}=\frac{1}{h_1h_2h_3}\left[\frac{\partial}{\partial u}(A_1h_2h_3)+\frac{\partial}{\partial v}(A_2h_3h_1)+\frac{\partial}{\partial w}(A_3h_1h_2)\right]$$

$$=\frac{1}{h_1h_2h_3}\left[\frac{\partial}{\partial u}\left(\frac{h_2h_3}{h_1}\frac{\partial\phi}{\partial u}\right)+\frac{\partial}{\partial v}\left(\frac{h_3h_1}{h_2}\frac{\partial\phi}{\partial v}\right)+\frac{\partial}{\partial w}\left(\frac{h_1h_2}{h_3}\frac{\partial\phi}{\partial w}\right)\right]$$

or, $$\nabla\cdot\nabla\phi=\nabla^2\phi=\frac{1}{h_1h_2h_3}\left[\frac{\partial}{\partial u}\left(\frac{h_2h_3}{h_1}\frac{\partial\phi}{\partial u}\right)+\frac{\partial}{\partial v}\left(\frac{h_3h_1}{h_2}\frac{\partial\phi}{\partial v}\right)+\frac{\partial}{\partial w}\left(\frac{h_1h_2}{h_3}\frac{\partial\phi}{\partial w}\right)\right].$$

$$\therefore\quad \nabla^2=\frac{1}{h_1h_2h_3}\left[\frac{\partial}{\partial u}\left(\frac{h_2h_3}{h_1}\frac{\partial}{\partial u}\right)+\frac{\partial}{\partial v}\left(\frac{h_3h_1}{h_2}\frac{\partial}{\partial v}\right)+\frac{\partial}{\partial w}\left(\frac{h_1h_2}{h_3}\frac{\partial}{\partial w}\right)\right].$$

13.7 Spherical Polar Coordinates

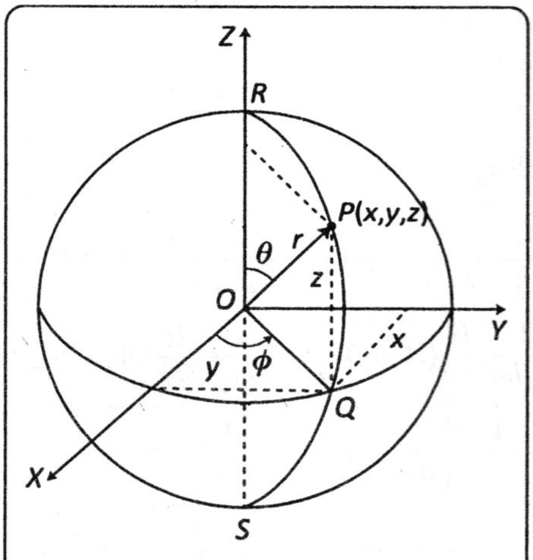

Fig. 13.4

Let P be a point in space whose Cartesian coordinates are (x, y, z). To locate the point P in spherical polar coordinates we draw a sphere of radius r so that the point P lies on the surface of the sphere. Next we draw a vertical great circle through P which cuts the horizontal great circle at Q such that $\angle XOQ=\phi$, the **azimuthal angle**. As the point P may be anywhere on the vertical great circle for given values of r and ϕ, we need to know the value of the angle $\angle POZ=\theta$. Thus r, θ, ϕ, called the *spherical polar coordinates*, are necessary to specify the point P (Fig. 13.4). Hence, the polar coordinates of the point P are r, θ, ϕ.

From Fig. 13.4 the transformations from (x, y, z) to (r, θ, ϕ) are as follows :

$$x=r\sin\theta\cos\phi,\quad y=r\sin\theta\sin\phi \quad\text{and}\quad z=r\cos\theta$$

$$\therefore\quad x=x(r,\theta,\phi),\quad y=x(r,\theta,\phi)\quad\text{and}\quad z=z(r,\theta).$$

$$\therefore\quad dx=\frac{\partial x}{\partial r}dr+\frac{\partial x}{\partial\theta}d\theta+\frac{\partial r}{\partial\phi}d\phi=\sin\theta\cos\phi dr+r\cos\theta\cos\phi d\theta+r\sin\theta\sin\phi d\phi.$$

Similarly, $dy=\sin\theta\sin\phi dr+r\cos\theta\sin\phi d\theta+r\sin\theta\cos\phi d\phi$ (ii)

and $dz=\cos\theta dr-r\sin\theta d\theta.$ (iii)

Now, the line element in Cartesian coordinates is

$$ds^2=dx^2+dy^2+dz^2. \quad\text{(iv)}$$

Substituting the values of dx, dy and dz from eqns. (i), (ii) and (iii) in eqn. (iv), we get after simplification :

$$ds^2=(dr)^2+r^2(d\theta)^2+r^2\sin^2\theta(d\phi)^2. \quad\text{(v)}$$

Comparing this equation with $ds^2=h_1^2du^2+h_2^2dv^2+h_3^2dw^2$, we get

$$h_1=h_r=1,\quad h_2=h_\theta=r,\quad h_3=h_\phi=r\sin\theta.$$

Differential Operators in Spherical Polar Coordinates

(a) Gradient : We have in curvilinear coordinates,

$$\nabla\phi = \frac{1}{h_1}\frac{\partial\phi}{\partial u}\hat{e}_1 + \frac{1}{h_2}\frac{\partial\phi}{\partial v}\hat{e}_2 + \frac{1}{h_3}\frac{\partial\phi}{\partial w}\hat{e}_3.$$

If $\hat{r}, \hat{\theta}$ and $\hat{\phi}$ be unit vectors along r, θ, ϕ, then

$$\nabla\phi = \frac{\partial\phi}{\partial r}\hat{r} + \frac{1}{r}\frac{\partial\phi}{\partial\theta}\hat{\theta} + \frac{1}{r\sin\theta}\frac{\partial\phi}{\partial\phi}\hat{\phi},$$

$[\because\ u = r,\ v = \theta,\ w = \phi$ and $h_1 = 1,\ h_2 = r,\ h_3 = r\sin\theta.]$

$$\therefore \quad \boldsymbol{\nabla} = \frac{\partial}{\partial r}\hat{r} + \frac{1}{r}\frac{\partial}{\partial\theta}\hat{\theta} + \frac{1}{r\sin\theta}\frac{\partial}{\partial\phi}\hat{\phi}. \tag{i}$$

(b) Divergence : The expression for $\nabla\cdot\vec{A}$ in orthogonal curvilinear coordinates is

$$\nabla\cdot\vec{A} = \frac{1}{h_1h_2h_3}\left[\frac{\partial}{\partial u}(A_1h_2h_3) + \frac{\partial}{\partial v}(A_2h_3h_1) + \frac{\partial}{\partial w}(A_3h_1h_2)\right].$$

Putting $h_1 = 1,\ h_2 = r,\ h_3 = r\sin\theta$ and $u = r,\ v = \theta,\ w = \phi,\ A_1 = A_r,\ A_2 = A_\theta,\ A_3 = A_\phi$, we get

$$\nabla\cdot\vec{A} = \frac{1}{r^2\sin\theta}\left[\frac{\partial}{\partial r}(r^2\sin\theta Ar) + \frac{\partial}{\partial\theta}(r\sin\theta A\theta) + \frac{\partial}{\partial\phi}(rA_\phi)\right]$$

or, $$\boldsymbol{\nabla}\cdot\vec{A} = \frac{1}{r^2}\frac{\partial}{\partial r}(r^2A_r) + \frac{1}{r\sin\theta}\frac{\partial}{\partial\theta}(\sin\theta A_\theta) + \frac{1}{r\sin\theta}\left(\frac{\partial A_\phi}{\partial\phi}\right).$$

(c) Curl : In orthogonal curvilinear coordinates,

$$\nabla\times\vec{A} = \frac{1}{h_1h_2h_3}\begin{vmatrix} h_1\hat{e}_1 & h_2\hat{e}_2 & h_3\hat{e}_3 \\ \dfrac{\partial}{\partial u} & \dfrac{\partial}{\partial v} & \dfrac{\partial}{\partial w} \\ A_1h_1 & A_2h_2 & A_3h_3 \end{vmatrix}.$$

Using $u = r,\ v = \theta,\ w = \phi,\ h_1 = 1,\ h_2 = r,\ h_3 = r\sin\theta,\ \hat{e}_1 = \hat{r},\ \hat{e}_2 = \hat{\theta},\ \hat{e}_1 = \hat{\phi}$, in the above expression, we get

$$\boldsymbol{\nabla}\times\vec{A} = \frac{1}{r^2\sin\theta}\begin{vmatrix} \hat{r} & r\hat{\theta} & r\sin\theta\hat{\phi} \\ \dfrac{\partial}{\partial r} & \dfrac{\partial}{\partial\theta} & \dfrac{\partial}{\partial\phi} \\ A_r & rA_\theta & r\sin\theta A_\phi \end{vmatrix}$$

$$= \frac{1}{r\sin\theta}\left[\frac{\partial}{\partial\theta}(\sin\theta A_\phi) - \frac{\partial}{\partial\phi}(A_\theta)\right]\hat{r} + \frac{1}{r}\left[\frac{1}{\sin\theta}\frac{\partial A_r}{\partial\phi} - \frac{\partial(rA_\phi)}{\partial r}\right]\hat{\theta}$$

$$+ \frac{1}{r}\left[\frac{\partial}{\partial r}(rA_\theta) - \frac{\partial A_r}{\partial\theta}\right]\hat{\phi}.$$ **[BU(H) 2009]**

(d) Laplacian : $\nabla^2\phi$ in orthogonal curvilinear coordinates is given by

$$\nabla^2\phi = \frac{1}{h_1h_2h_3}\left[\frac{\partial}{\partial u}\left(\frac{h_2h_3}{h_1}\frac{\partial\phi}{\partial u}\right) + \frac{\partial}{\partial v}\left(\frac{h_3h_1}{h_2}\frac{\partial\phi}{\partial v}\right) + \frac{\partial}{\partial w}\left(\frac{h_1h_2}{h_3}\frac{\partial\phi}{\partial w}\right)\right].$$

Using $u = r$, $v = \theta$, $w = \phi$ and $h_1 = 1$, $h_2 = r$, $h_3 = r\sin\theta$, we have in spherical polar coordinates,

$$\nabla^2\phi = \frac{1}{r^2\sin\theta}\left[\frac{\partial}{\partial r}\left(r^2\sin\theta\frac{\partial\phi}{\partial r}\right) + \frac{\partial}{\partial\theta}\left(\sin\theta\frac{\partial\phi}{\partial\theta}\right) + \frac{\partial}{\partial\phi}\left(\frac{1}{\sin\theta}\frac{\partial\phi}{\partial\phi}\right)\right].$$

$\therefore$ Laplacian in spherical polar coordinates is

$$\boldsymbol{\nabla}^2 = \frac{1}{r^2\sin\theta}\left[\frac{\partial}{\partial r}\left(r^2\sin\theta\frac{\partial}{\partial r}\right) + \frac{\partial}{\partial\theta}\left(\sin\theta\frac{\partial}{\partial\theta}\right) + \frac{\partial}{\partial\phi}\left(\frac{1}{\sin\theta}\frac{\partial}{\partial\phi}\right)\right]$$

$$= \frac{1}{r^2}\frac{\partial}{\partial r}\left(r^2\frac{\partial}{\partial r}\right) + \frac{1}{r^2\sin\theta}\frac{\partial}{\partial\theta}\left(\sin\theta\frac{\partial}{\partial\theta}\right) + \frac{1}{r^2\sin^2\theta}\frac{\partial^2}{\partial\phi^2}.$$

13.8 Cylindrical Coordinates

To find the position coordinates of a point $P(x, y, z)$ in cylindrical coordinates we draw a radius vector r from O' to P as shown in Fig. 13.5.

Then PQ is drawn parallel to Z-axis and OQ is parallel to $O'P$. The angle $\angle QOX = \theta$. Now the position of the point P is specified by (r, θ, z) in cylindrical coordinates.

Here r represents the distance of P from O' in the XY-plane, P being on the surface of the cylinder of radius r, z is the distance of P from XY-plane.

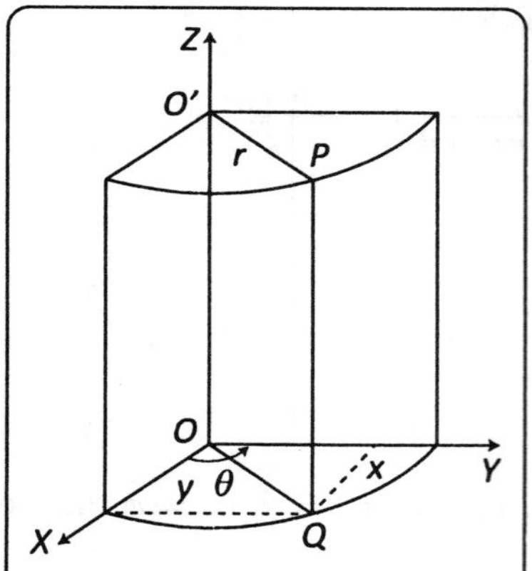

Fig. 13.5

The transformations between rectangular Cartesian coordinates (x, y, z) and cylindrical coordinates from Fig. 13.5 are as follows :

$$\left.\begin{aligned} x &= r\cos\theta \\ y &= r\sin\theta \\ z &= z. \end{aligned}\right\} \tag{i}$$

From $x = x(r, \theta)$, $y = y(r, \theta)$ and $z = z(z)$, we have

$$dx = \frac{\partial x}{\partial r}dr + \frac{\partial x}{\partial\theta}d\theta = \cos\theta dr - r\sin\theta d\theta \tag{ii}$$

$$dy = \frac{\partial y}{\partial r}dr + \frac{\partial y}{\partial\theta}d\theta = \sin\theta dr + r\cos\theta d\theta \tag{iii}$$

and $\quad dz = dz. \qquad$ (iv)

The line element in Cartesian coordinates is given by

$$ds^2 = dx^2 + dy^2 + dz^2$$

or, $\quad ds^2 = (\cos\theta dr - r\sin\theta d\theta)^2 + (\sin\theta dr + r\cos\theta d\theta)^2 + dz^2 = dr^2 + r^2 d\theta^2 + dz^2. \qquad$ (v)

In orthogonal carvilinear coordinates the arc length is given by

$$\boldsymbol{ds^2 = h_1^2 du^2 + h_2^2 dv^2 + h_3^2 dw^2}. \tag{vi}$$

Comparing eqns. (v) and (vi), we get

$$\left.\begin{aligned} u &= r, \quad v = \theta, \quad w = z \\ h_1 &= 1, \quad h_2 = r, \quad h_3 = 1 \end{aligned}\right\} \tag{vii}$$

(a) Gradient : In $\phi(u,v,w)$ be a scalar function in orthogonal curvilinear coordinates, then

$$\nabla\phi = \frac{1}{h_1}\frac{\partial\phi}{\partial u}\hat{e}_1 + \frac{1}{h_2}\frac{\partial\phi}{\partial v}\hat{e}_2 + \frac{1}{h_3}\frac{\partial\phi}{\partial w}\hat{e}_3.$$

In cylindrical coordinates,

$$\boldsymbol{\nabla\phi = \frac{\partial\phi}{\partial r}\hat{r} + \frac{1}{r}\frac{\partial\phi}{\partial\theta}\hat{\theta} + \frac{\partial\phi}{\partial z}\hat{z}}, \quad \text{[by eqn. (vii)]} \tag{viii}$$

where $\hat{r}$, $\hat{\theta}$, $\hat{z}$ are the unit vectors along r, θ and z axes respectively.

$$\therefore \quad \boldsymbol{\nabla = \frac{\partial}{\partial r}\hat{r} + \frac{1}{r}\frac{\partial}{\partial\theta}\hat{\theta} + \frac{\partial}{\partial z}\hat{z}}. \tag{ix}$$

(b) Divergence : If $\vec{A}$ is a vector in orthogonal curvilinear coordinates such that, $\vec{A} = A_1\hat{e}_1 + A_2\hat{e}_2 + A_3\hat{e}_3$, then

$$\nabla\cdot\vec{A} = \frac{1}{h_1h_2h_3}\left[\frac{\partial}{\partial u}(A_1h_2h_3) + \frac{\partial}{\partial v}(A_2h_3h_1) + \frac{\partial}{\partial w}(A_3h_1h_2)\right]$$

In cylindrical coordinates,

$$\boldsymbol{\nabla\cdot\vec{A} = \frac{1}{r}\frac{\partial}{\partial r}(rA_r) + \frac{1}{r}\frac{\partial A_\theta}{\partial\theta} + \frac{\partial A_z}{\partial z}}. \quad \text{[by eqn. (vii)]} \tag{x}$$

(c) Curl : If $\vec{A} = A_1\hat{e}_1 + A_2\hat{e}_2 + A_3\hat{e}_3$ be a vector in orthogonal curvilinear coordinates, then

$$\nabla\times\vec{A} = \frac{1}{h_1h_2h_3}\begin{vmatrix} h_1\hat{e}_1 & h_2\hat{e}_2 & h_3\hat{e}_3 \\ \dfrac{\partial}{\partial u} & \dfrac{\partial}{\partial v} & \dfrac{\partial}{\partial w} \\ A_1h_1 & A_2h_2 & A_3h_3 \end{vmatrix}.$$

In cylindrical coordinates, $\nabla\times\vec{A} = \begin{vmatrix} \hat{r} & \hat{\theta} & \hat{z} \\ \dfrac{\partial}{\partial r} & \dfrac{\partial}{\partial\theta} & \dfrac{\partial}{\partial z} \\ A_r & rA_\theta & A_2 \end{vmatrix}$.

Since in cylindrical coordinates, $\vec{A} = A_r\hat{r} + A_\theta\hat{\theta} + A_z\hat{z}$.

$$\therefore \quad \boldsymbol{\nabla\times\vec{A} = \left(\frac{1}{r}\frac{\partial A_z}{\partial\theta} - \frac{\partial A_\theta}{\partial z}\right)\hat{r} + \left(\frac{\partial A_r}{\partial z} - \frac{\partial A_z}{\partial r}\right)\hat{\theta} + \frac{1}{r}\left(\frac{\partial}{\partial r}(rA_\theta) - \frac{\partial A_r}{\partial\theta}\right)\hat{z}}. \tag{xi}$$

(d) Laplacian : If $\phi = \phi(u,v,w)$ be a scalar function in orthogonal curvilinear coordinates, then

$$\nabla^2\phi = \frac{1}{h_1h_2h_3}\left[\frac{\partial}{\partial u}\left(\frac{h_2h_3}{h_1}\frac{\partial\phi}{\partial u}\right) + \frac{\partial}{\partial v}\left(\frac{h^3h_1}{h_2}\frac{\partial\phi}{\partial v}\right) + \frac{\partial}{\partial w}\left(\frac{h_1h_2}{h_3}\frac{\partial\phi}{\partial w}\right)\right]$$

In cylindrical coordinates,

$$\nabla^2\phi = \frac{1}{r}\left[\left\{\frac{\partial}{\partial r}\left(r\frac{\partial\phi}{\partial r}\right)\right\} + \left\{\frac{\partial}{\partial\theta}\left(\frac{1}{r}\frac{\partial\phi}{\partial\theta}\right)\right\} + \left\{\frac{\partial}{\partial z}\left(r\frac{\partial\phi}{\partial z}\right)\right\}\right] \quad \text{[by the use of eqn. (vii)]}$$

$$= \frac{\partial^2\phi}{\partial r^2} + \frac{1}{r}\frac{\partial\phi}{\partial r} + \frac{1}{r^2}\frac{\partial^2\phi}{\partial\theta^2} + \frac{\partial^2\phi}{\partial z^2}. \tag{xii}$$

$$\therefore \quad \boldsymbol{\nabla^2 = \left(\frac{\partial^2}{\partial r^2} + \frac{1}{r}\frac{\partial}{\partial r}\right) + \frac{1}{r^2}\frac{\partial^2}{\partial\theta^2} + \frac{\partial^2}{\partial z^2}}. \tag{xiii}$$

13.9 Volume Element in (a) Cylindrical and (b) Spherical Coordinates

13.9.1 (a) Cylindrical Coordinates

In Fig. 13.6, $P(x, y, z)$ is a point in rectangular coordinates. Here $OQ = r$, $\angle XOQ = \theta$, $QP = z$.

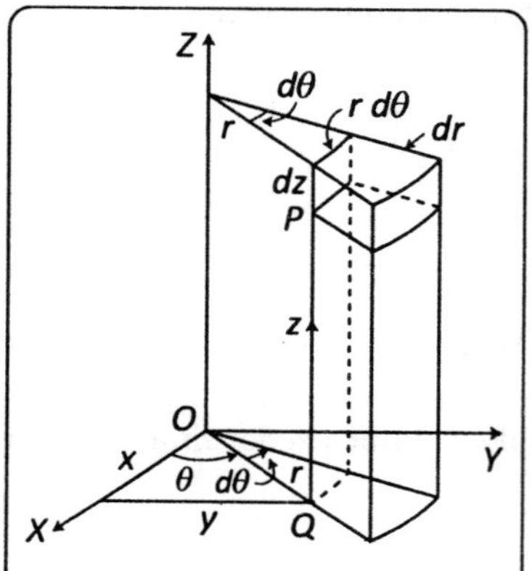

Fig. 13.6

$\therefore$ the cylindrical coordinates of P are (r, θ, z).

From Fig. 13.6, $x = r\cos\theta$, $y = r\sin\theta$, $z = z$.

$\therefore$ the line element in cylindrical coordinates is given by

$$(dS)^2 \ (dr)^2 + (dy)^2 + (dz)^2$$

$$= (dr)^2 + r^2(d\theta)^2 + (dz)^2. \tag{i}$$

Comparing eqn. (i) with the equation :

$$\boldsymbol{(dS)^2 = h_1^2 du^2 + h_2^2 dv^2 + h_3^2 dw^2}$$

we get the scale factors, $h_1 = 1$, $h_2 = r$, $h_3 = 1$.

The volume element in cylindrical coordinates is

$$\boldsymbol{dV} = dr \times rd\theta \times dz = \boldsymbol{r\, dr\, d\theta\, dz}. \tag{ii}$$

(b) Spherical Polar Coordinates

In Fig. 13.7, $P(x, y, z)$ is a point in Cartesian coordinate space.

Here $OP = r$, $\angle ZOP = \theta$, $\angle XOP = \phi$.

$\therefore$ the spherical polar coordinates of P are (r, θ, ϕ) and

$$x = OQ\cos\phi = OP\cos(90° - \theta)\cos\phi = r\sin\theta\cos\phi$$

$$y = OQ\sin\phi = r\sin\theta\sin\phi$$

and $z = r\cos\theta$.

$\therefore$ the element of length in Cartesian coordinates is given by

$$(dS)^2 = (dx)^2 + (dy)^2 + (dz)^2.$$

$\therefore$ $\boldsymbol{dS^2}$ in spherical polar coordinates

$$\boldsymbol{= (dr)^2 + (rd\theta)^2 + (r\sin\theta)^2(d\phi)^2}. \tag{i}$$

$\therefore$ the scale factors are $h_1 = 1$, $h_2 = r$, $h_3 = r\sin\theta$.

Then the volume element in spherical polar coordinate is

$$\boldsymbol{dV = dr \times rd\theta \times r\sin\theta d\phi}$$

$$\boldsymbol{= r^2\sin\theta dr\, d\theta d\phi}. \tag{ii}$$

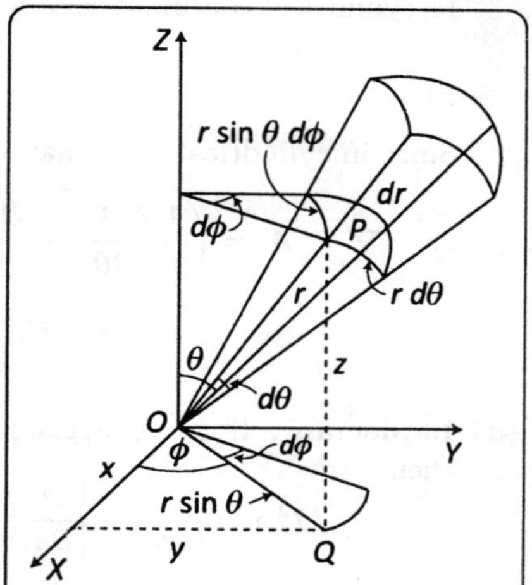

Fig. 13.7

13.10 Solved Problems

Example 13.1. *If $\vec{r} = \vec{r}(u, v)$ represents a surface, then show that*

(a) *the square of the element of arc length on the surface is $ds^2 = Edu^2 + 2F du\, dv + G dv^2$ and*

(b) *the element of surface area is $dS = \sqrt{EG - F^2} du\, dv$, where $E = \frac{\partial \vec{r}}{\partial u} \cdot \frac{\partial \vec{r}}{\partial u}$, $F = \frac{\partial \vec{r}}{\partial u} \cdot \frac{\partial \vec{r}}{\partial v}$ and $G = \frac{\partial \vec{r}}{\partial v} \cdot \frac{\partial \vec{r}}{\partial v}$.*

Solution : (a) We have the square of the element of arc length, $ds^2 = d\vec{r} \cdot d\vec{r}$.

But, $\vec{r} = \vec{r}(u, v)$

$$d\vec{r} = \frac{\partial \vec{r}}{\partial u} du + \frac{\partial \vec{r}}{\partial v} dv.$$

$$\therefore \quad \boldsymbol{ds^2} = \left(\frac{\partial \vec{r}}{\partial u} du + \frac{\partial \vec{r}}{\partial u} dv\right) \cdot \left(\frac{\partial \vec{r}}{\partial u} du + \frac{\partial \vec{r}}{\partial v} dv\right)$$

$$= \left(\frac{\partial \vec{r}}{\partial u} \cdot \frac{\partial \vec{r}}{\partial u}\right) du^2 + 2\frac{\partial \vec{r}}{\partial u} \cdot \frac{\partial \vec{r}}{\partial v} dudv + \left(\frac{\partial \vec{r}}{\partial v} \cdot \frac{\partial \vec{r}}{\partial u}\right) dv^2$$

$$= \boldsymbol{Edu^2 + 2Fdudv + Gdv^2}$$

(b) The element of surface area is

$$\boldsymbol{dS} = \left|\frac{\partial \vec{r}}{\partial u} du \times \frac{\partial \vec{r}}{\partial v} dv\right| = \left|\frac{\partial \vec{r}}{\partial u} \times \frac{\partial \vec{r}}{\partial v}\right| du\, dv = \sqrt{\left(\frac{\partial \vec{r}}{\partial u} \times \frac{\partial \vec{r}}{\partial v}\right) \cdot \left(\frac{\partial \vec{r}}{\partial u} \times \frac{\partial \vec{r}}{\partial v}\right)}\, du\, dv$$

$$= \sqrt{\left(\frac{\partial \vec{r}}{\partial u} \cdot \frac{\partial \vec{r}}{\partial u}\right)\left(\frac{\partial \vec{r}}{\partial v} \cdot \frac{\partial \vec{r}}{\partial v}\right) - \left(\frac{\partial \vec{r}}{\partial u} \cdot \frac{\partial \vec{r}}{\partial v}\right)\left(\frac{\partial \vec{r}}{\partial v} \cdot \frac{\partial \vec{r}}{\partial u}\right)}\, du\, dv$$

$$= \boldsymbol{\sqrt{EG - F^2}\, du\, dv}, \quad [\because \ (\vec{A} \times \vec{B}) \cdot (\vec{C} \times \vec{D}) = (\vec{A} \cdot \vec{C})(\vec{B} \cdot \vec{D}) - (\vec{A} \cdot \vec{D})(\vec{B} \cdot \vec{C})]$$

Example 13.2. *Represent of vector $\vec{A} = 2y\hat{i} - z\hat{j} + 3x\hat{k}$ in cylindrical coordinates (ρ, ϕ, z) and find A_ρ, A_ϕ, A_z.* **[BU(H) 1993, 2006]**

Solution : The transformation from rectangular to cylindrical coordinates are

$$x = \rho\cos\phi, \quad y = \rho\sin\phi, \quad z = z.$$

The position vector of any point $P(x, y, z)$ is $\vec{r} = x\hat{i} + y\hat{j} + z\hat{k}$.

In cylindrical coordinates, $\vec{r} = \rho\cos\phi\hat{i} + \rho\sin\phi\hat{j} + z\hat{k}$.

$$\therefore \quad \hat{\rho} = \frac{\frac{\partial \vec{r}}{\partial \rho}}{\left|\frac{\partial \vec{r}}{\partial \rho}\right|} = \frac{\cos\phi\hat{i} + \sin\phi\hat{j}}{\sqrt{\cos^2\phi + \sin^2\phi}}$$

$$\text{or,} \quad \hat{\rho} = \cos\phi\hat{i} + \sin\phi\hat{j} \tag{i}$$

$$\hat{\phi} = \frac{\frac{\partial \vec{r}}{\partial \phi}}{\left|\frac{\partial \vec{r}}{\partial \phi}\right|} = \frac{-\rho\sin\phi\hat{i} + \rho\cos\phi\hat{j}}{\sqrt{\rho^2\sin^2\phi + \rho^2\cos^2\phi}} = -\sin\phi\hat{i} + \cos\phi\hat{j} \tag{ii}$$

$$\hat{z} = \frac{\frac{\partial \vec{r}}{\partial z}}{\left|\frac{\partial \vec{r}}{\partial z}\right|} = \hat{k}. \tag{iii}$$

Solving (i) and (ii), we get

$$\hat{i} = \cos\phi\hat{\rho} - \sin\phi\hat{\phi} \qquad \hat{j} = \sin\phi\hat{\rho} + \cos\phi\hat{\phi}.$$

$$\therefore \quad \vec{A} = 2y\hat{i} - z\hat{j} + 3x\hat{k} = 2\rho\sin\phi(\cos\phi\hat{\rho} - \sin\phi\hat{\phi}) - z(\sin\phi\hat{\rho} + \cos\phi\hat{\phi}) + 3\rho\cos\phi\hat{k}$$

$$= (2\rho\sin\phi\cos\phi - z\sin\phi)\hat{\rho} - (2\rho\sin^2\phi + z\cos\phi)\hat{\phi} + 3\rho\cos\phi\hat{k} = A_\rho\hat{\rho} + A_\phi\hat{\phi} + A_z\hat{z},$$

where $\boldsymbol{A_\rho = 2\rho\sin\phi\cos\phi - z\sin\phi}$, $\boldsymbol{A_\rho = -(2\rho\sin^2\phi\cos\phi + z\cos\phi)}$, $\boldsymbol{A_z = 3\rho\cos\phi}$.

Example 13.3. *Prove that a cylindrical coordinate system is orthogonal.* **[Bang.U 1990]**

Solution : The transformations from rectangular to cylindrical coordinates system are

$$x = r\cos\theta, \quad y = r\sin\theta \quad \text{and} \quad z = z.$$

The position vector of any point in cylindrical coordinate system is

$$\vec{r} = \hat{i}x + \hat{j}y + \hat{k}z = r\cos\theta\hat{i} + r\sin\theta\hat{j} + z\hat{k}.$$

Now,

$$\frac{\partial \vec{r}}{\partial r} = \cos\theta\hat{i} + \sin\theta\hat{j}$$

$$\frac{\partial \vec{r}}{\partial \theta} = -r\sin\theta\hat{i} + r\cos\theta\hat{j}$$

$$\frac{\partial \vec{r}}{\partial z} = \hat{k}.$$

$$\therefore \quad \hat{e}_1 = \hat{r} = \frac{\frac{\partial \vec{r}}{\partial r}}{\left|\frac{\partial \vec{r}}{\partial r}\right|} = \frac{\cos\theta\hat{i} + \sin\theta\hat{j}}{\sqrt{\cos^2\theta + \sin^2\theta}} = \cos\theta\hat{i} + \sin\theta\hat{j}$$

$$\hat{e}_2 = \hat{\theta} = \frac{\frac{\partial \vec{r}}{\partial \theta}}{\left|\frac{\partial \vec{r}}{\partial \theta}\right|} = \frac{-r\sin\theta\hat{i} + r\cos\theta\hat{j}}{\sqrt{r^2\sin^2\theta + r^2\cos^2\theta}} = -\sin\theta\hat{i} + \cos\theta\hat{j}$$

$$\hat{e}_3 = \hat{z} = \frac{\frac{\partial \vec{r}}{\partial z}}{\left|\frac{\partial \vec{r}}{\partial z}\right|} = \hat{k}.$$

Now,

$$\hat{r}\cdot\hat{\theta} = (\cos\theta\hat{i} + \sin\theta\hat{j})\cdot(-\sin\theta\hat{i} + \cos\theta\hat{j}) = 0$$

$$\hat{\theta}\cdot\hat{k} = (-\sin\theta\hat{i} + \cos\theta\hat{j})\cdot\hat{k} = 0$$

$$\hat{k}\cdot\hat{r} = \hat{k}\cdot(\cos\theta\hat{i} + \sin\theta\hat{j}) = 0.$$

Thus, $\hat{\boldsymbol{r}}$, $\hat{\boldsymbol{\theta}}$, $\hat{\boldsymbol{k}}$ **are being mutually perpendicular, the cylindrical coordinate system is orthogonal.**

Example 13.4. *If u_1, u_2, u_3 are orthogonal curvilinear coordinates and x, y, z are rectangular Cartesian coordinates, then show that the Jacobian, $J\left(\frac{x, y, z}{u_1, u_2, u_3}\right) = h_1h_2h_3$, where h_1, h_2, h_3 are the scale factors.* **[BU(H) 1994; Bang.U 1990S]**

Solution : We have $J\left(\dfrac{x, y, z}{u_1, u_2, u_3}\right) = \begin{vmatrix} \dfrac{\partial x}{\partial u_1} & \dfrac{\partial y}{\partial u_1} & \dfrac{\partial z}{\partial u_1} \\ \dfrac{\partial x}{\partial u_2} & \dfrac{\partial y}{\partial u_2} & \dfrac{\partial z}{\partial u_2} \\ \dfrac{\partial x}{\partial u_3} & \dfrac{\partial y}{\partial u_3} & \dfrac{\partial z}{\partial u_3} \end{vmatrix}$

$$= \left(\frac{\partial x}{\partial u_1}\hat{i} + \frac{\partial y}{\partial u_1}\hat{j} + \frac{\partial z}{\partial u_1}\hat{k}\right) \cdot \left(\frac{\partial x}{\partial u_2}\hat{i} + \frac{\partial y}{\partial u_2}\hat{j} + \frac{\partial z}{\partial u_2}\hat{k}\right) \times \left(\frac{\partial x}{\partial u_3}\hat{i} + \frac{\partial y}{\partial u_3}\hat{j} + \frac{\partial z}{\partial u_3}\hat{k}\right)$$

$$= \frac{\partial \vec{r}}{\partial u_1} \cdot \frac{\partial \vec{r}}{\partial u_2} \times \frac{\partial \vec{r}}{\partial u_3} \quad \left[\because \; \frac{\partial x}{\partial u_1}\hat{i} + \frac{\partial y}{\partial u_1}\hat{j} + \frac{\partial z}{\partial u_1}\hat{k} = \frac{\partial}{\partial u_1}(\hat{i}x + \hat{j}y + \hat{k}z) = \frac{\partial \vec{r}}{\partial u_1}\right]$$

$$= h_1\hat{e}_1 \cdot h_2\hat{e}_2 \times h_3\hat{e}_3 \quad \left[\because \; \hat{e}_1 = \frac{\partial \vec{r}}{\partial u_1}\Big/ h_1\right]$$

$$= h_1 h_2 h_3 (\hat{e}_1 \cdot \hat{e}_2 \times \hat{e}_3)$$

$$= \boldsymbol{h_1 h_2 h_3} \quad [\because \; \hat{e}_1 \cdot \hat{e}_2 \times \hat{e}_3 = 1 \text{ for orthogonal curvilinear coordinates}]$$

Example 13.5. *Represent the vector $\vec{A} = 2y\hat{i} - z\hat{j} + 3x\hat{k}$ in spherical polar coordinates.*

[Manga.U 1985]

Solution : We have the transformations from rectangular to spherical polar coordinates

$$x = r\sin\theta\cos\phi, \quad y = r\sin\theta\sin\phi, \quad z = r\cos\theta.$$

$\therefore$ the position vector of any point P in spherical polar coordinates is

$$\vec{r} = \hat{i}x + \hat{j}y + \hat{k}z = r\sin\theta\cos\phi\hat{i} + r\sin\theta\sin\phi\hat{j} + r\cos\theta\hat{k}.$$

If $\hat{r}$, $\hat{\theta}$, $\hat{\phi}$ be the unit vectors along the tangents to r, θ, ϕ curves respectively, then

$$\hat{r} = \frac{\frac{\partial \vec{r}}{\partial r}}{\left|\frac{\partial \vec{r}}{\partial r}\right|} = \frac{\sin\theta\cos\phi\hat{i} + \sin\theta\sin\phi\hat{j} + \cos\theta\hat{k}}{\sqrt{(\sin\theta\cos\phi)^2 + (\sin\theta\sin\phi)^2 + \cos^2\theta}}$$

$$= \sin\theta\cos\phi\hat{i} + \sin\theta\sin\phi\hat{j} + \cos\theta\hat{k}.$$

$$\hat{\theta} = \frac{\frac{\partial \vec{r}}{\partial \theta}}{\left|\frac{\partial \vec{r}}{\partial \theta}\right|} = \frac{r\cos\theta\cos\phi\hat{i} + r\cos\theta\sin\phi\hat{j} - r\sin\theta\hat{k}}{\sqrt{(r\cos\theta\cos\phi)^2 + (r\cos\theta\sin\phi)^2 + (-r\sin^2\theta)^2}}$$

$$= \cos\theta\cos\phi\hat{i} + \cos\theta\sin\phi\hat{j} - \sin\theta\hat{k}.$$

$$\hat{\phi} = \frac{\frac{\partial \vec{r}}{\partial \phi}}{\left|\frac{\partial \vec{r}}{\partial \phi}\right|} = \frac{-r\sin\theta\sin\phi\hat{i} + r\sin\theta\cos\phi\hat{j}}{\sqrt{(-r\sin\theta\sin\phi)^2 + (r\sin\theta\cos\phi)^2}} = -\sin\phi\hat{i} + \cos\phi\hat{j}$$

Now as $\vec{A} = 2y\hat{i} - z\hat{j} + 3x\hat{k} = A_r\hat{r} + A_\theta\hat{\theta} + A_\phi\hat{\phi}$, (i)

where $A_r = \hat{A} \cdot \hat{r}$

$$= (2r\sin\theta\cos\phi\hat{i} - r\cos\theta\hat{j} + 3r\sin\theta\cos\phi\hat{k}) \cdot (\sin\theta\cos\phi\hat{i} + \sin\theta\sin\phi\hat{j} + \cos\theta\hat{k})$$

$$= 2r\sin^2\theta\sin\phi\cos\phi - r\sin\theta\cos\theta\sin\phi + 3r\sin\theta\cos\theta\cos\phi. \qquad \text{(ii)}$$

$A_\theta = \vec{A} \cdot \hat{\theta}$

$$= (2r\sin\theta\sin\phi\hat{i} - r\cos\theta\hat{j} + 3r\sin\theta\cos\phi\hat{k})\cdot(\cos\theta\cos\phi\hat{i} + \cos\theta\sin\phi\hat{j} - \sin\theta\hat{k})$$

$$= 2r\sin\theta\cos\theta\sin\phi\cos\phi - r\cos^2\theta\sin\phi - 3r\sin^2\theta\cos\phi \quad \text{(iii)}$$

and $A_\phi = \vec{A}\cdot\hat{\phi} = (2r\sin\theta\sin\phi\hat{k} - r\cos\theta\hat{j} + 3r\sin\theta\cos\phi\hat{k})(-\sin\phi\hat{i} + \cos\phi\hat{j}).$

$$= -2r\sin\theta\sin^2\phi - r\cos\theta\cos\phi. \quad \text{(iv)}$$

Substituting the values of A_r, A_θ and A_ϕ from eqns. (ii), (iii) and (iv) in eqn. (i) we get the given vector $\vec{A}$ in spherical polar coordinates.

Example 13.6. *Prove that, $\frac{\partial\vec{r}}{\partial u_1}, \frac{\partial\vec{r}}{\partial u_2}, \frac{\partial\vec{r}}{\partial u_3}$ and ∇u_1, ∇u_2, ∇u_3 are reciprocal systems of vectors, $\vec{r}$ being the position vector.* **[BU(H) 1994]**

Solution : For $\frac{\partial\vec{r}}{\partial u_1}, \frac{\partial\vec{r}}{\partial u_2}, \frac{\partial\vec{r}}{\partial u_3}$ and ∇u_1, ∇u_2, ∇u_3 to be two reciprocal systems of vectors the following relations must hold good.

$$\frac{\partial\vec{r}}{\partial u_i}\cdot\nabla u_j = \begin{cases}1, & \text{if } i = j\\ 0, & \text{if } i \neq j.\end{cases} \quad \text{(v)}$$

Now, $\vec{r} = \vec{r}(u_1, u_2, u_3).$

$$\therefore \quad d\vec{r} = \frac{\partial\vec{r}}{\partial u_1}du_1 + \frac{\partial\vec{r}}{\partial u_2}du_2 + \frac{\partial\vec{r}}{\partial u_3}du_3.$$

Again, $\nabla u_1\cdot d\vec{r} = du_1 = \left(\nabla u_1\cdot\frac{\partial\vec{r}}{\partial u_1}\right)du_1 + \left(\nabla u_1\cdot\frac{\partial\vec{r}}{\partial u_2}\right)du_2 + \left(\nabla u_1\cdot\frac{\partial\vec{r}}{\partial u_3}\right)du_3$

or, $\nabla u_1\cdot\frac{\partial\vec{r}}{\partial u_1} = 1, \quad \nabla u_1\cdot\frac{\partial\vec{r}}{\partial u_2} = 0, \quad \nabla u_1\cdot\frac{\partial\vec{r}}{\partial u_3} = 0,$

and $\nabla u_2\cdot d\vec{r} = du_2 = \left(\nabla u_2\cdot\frac{\partial\vec{r}}{\partial u_1}\right)du_1 + \left(\nabla u_2\cdot\frac{\partial\vec{r}}{\partial u_2}\right)du_2 + \left(\nabla u_2\cdot\frac{\partial\vec{r}}{\partial u_3}\right)du_3$

or, $\nabla u_2\cdot\frac{\partial\vec{r}}{\partial u_1} = 0, \quad \nabla u_2\cdot\frac{\partial\vec{r}}{\partial u_2} = 1, \quad \nabla u_2\cdot\frac{\partial\vec{r}}{\partial u_3} = 0,$

also, $\nabla u_3\cdot d\vec{r} = du_3 = \left(\nabla u_3\cdot\frac{\partial\vec{r}}{\partial u_1}\right)du_1 + \left(\nabla u_3\cdot\frac{\partial\vec{r}}{\partial u_2}\right)du_2 + \left(\nabla u_3\cdot\frac{\partial\vec{r}}{\partial u_3}\right)du_3$

or, $\nabla u_3\cdot\frac{\partial\vec{r}}{\partial u_1} = 0, \quad \nabla u_3\cdot\frac{\partial\vec{r}}{\partial u_2} = 0, \quad \nabla u_3\cdot\frac{\partial\vec{r}}{\partial u_3} = 1,$

i.e., $$\boldsymbol{\frac{\partial\vec{r}}{\partial u_i}\cdot\nabla u_j = \begin{cases}1, & \text{if } i = j\\ 0, & \text{if } i \neq j.\end{cases}}$$

Hence, the result.

Example 13.7. *Prove that spherical polar coordinate system is orthogonal.* **[Bang.U 1990]**

Solution : The transformations from rectangular to spherical polar coordinates are

$$x = r\sin\theta\cos\phi, \quad y = r\sin\theta\sin\phi, \quad z = r\cos\theta.$$

$$\therefore \quad \vec{r} = \hat{i}x + \hat{j}y + \hat{k}z = r\sin\theta\cos\phi\hat{i} + r\sin\theta\sin\phi\hat{j} + r\cos\theta\hat{k}.$$

If $\hat{r}$, $\hat{\theta}$ and $\hat{\phi}$ be the unit vectors along the tangents to r, θ and ϕ curves respectively, then

$$\hat{r} = \frac{\frac{\partial \vec{r}}{\partial r}}{\left|\frac{\partial \vec{r}}{\partial r}\right|} = \frac{\sin\theta\cos\phi\hat{i} + \sin\theta\sin\phi\hat{j} + \cos\theta\hat{k}}{\sqrt{(\sin\theta\cos\phi)^2 + (\sin\theta\sin\phi)^2 + (\cos\theta)^2}}$$

$$= \sin\theta\cos\phi\hat{i} + \sin\theta\sin\phi\hat{j} + \cos\theta\hat{k}$$

$$\hat{\theta} = \frac{\frac{\partial \vec{r}}{\partial \theta}}{\left|\frac{\partial \vec{r}}{\partial \theta}\right|} = \frac{r\cos\theta\cos\phi\hat{i} + r\cos\theta\sin\phi\hat{j} - r\sin\theta\hat{k}}{\sqrt{(r\cos\theta\cos\phi)^2 + (r\cos\theta\sin\phi)^2 + (-r\cos\theta)^2}}$$

$$= \cos\theta\cos\phi\hat{i} + \cos\theta\sin\phi\hat{j} - \sin\theta\hat{k}$$

and $$= \frac{\frac{\partial \vec{r}}{\partial \theta}}{\left|\frac{\partial \vec{r}}{\partial \phi}\right|} = \frac{-r\sin\theta\sin\phi\hat{i} + r\sin\theta\cos\phi\hat{j}}{\sqrt{(-r\sin\theta\sin\phi)^2 + (r\sin\theta\cos\phi)^2}} = -\sin\phi\hat{i} + \cos\phi\hat{j}.$$

$\therefore$ $$\hat{r}\cdot\hat{\theta} = (\sin\theta\cos\phi\hat{i} + \sin\theta\sin\phi\hat{j} + \cos\theta\hat{k})\cdot(\cos\theta\cos\phi\hat{i} + \cos\theta\sin\phi\hat{j} - \sin\theta\hat{k})$$

$$= \sin\theta\cos\theta\cos^2\phi + \sin\theta\cos\theta\sin^2\phi - \sin\theta\cos\theta = 0$$

$$\hat{\theta}\cdot\hat{\phi} = (\cos\theta\cos\phi\hat{i} + \cos\theta\cos\phi\hat{j} - \sin\theta\hat{k})\cdot(-\sin\phi\hat{i} + \cos\phi\hat{j})$$

$$= -\cos\theta\sin\phi\cos\phi + \cos\theta\sin\phi\cos\phi = 0$$

$$\hat{\phi}\cdot\hat{r} = (-\sin\phi\hat{i} + \cos\phi\hat{j})\cdot(\sin\theta\cos\phi\hat{i} + \sin\theta\sin\phi\hat{j} + \cos\theta\hat{k})$$

$$= -\sin\theta\sin\phi\cos\phi + \sin\theta\sin\phi\cos\phi = 0.$$

$\therefore$ the spherical polar coordinate system is orthogonal.

Example 13.8. (a) *Calculate the components of velocity of a particle in cylindrical coordinates.* **[DU(H) 1997]**

(b) *Calculate the components of acceleration of a particle in cylindrical coordinates.*

Solution : (a) The transformations from Cartesian to cylindrical coordinates are

$$x = r\cos\theta, \qquad y = r\sin\theta, \qquad z = z.$$

Now the position vector of a particle in rectangular Cartesian coordinates is

$$\vec{r} = \hat{i}x + \hat{j}y + \hat{k}z$$

or, $$\vec{r} = (r\cos\theta)(\cos\theta\hat{r} - \sin\theta\hat{\theta}) + r\sin\theta(\sin\theta\hat{r} + \cos\theta\hat{\theta}) + z\hat{z} = r\hat{r} + z\hat{z},$$

since $\hat{i} = \cos\theta\hat{r} - \sin\theta\hat{\theta}$, $\quad \hat{j} = \sin\theta\hat{r} + \cos\theta\hat{\theta}$, $\quad \hat{k} = \hat{z}$.

Now velocity of the particle in cylindrical coordinates (r, θ, z) is

$$\vec{v} = \frac{d\vec{r}}{dt} = \frac{dr}{dt}\hat{r} + r\frac{d\hat{r}}{dt} + \frac{dz}{dt}\hat{z} = \dot{r}\hat{r} + r\dot{\theta}\hat{\theta} + \dot{z}\hat{z} = v_r\hat{r} + v_\theta\hat{\theta} + v_z\hat{z},$$

where $v_r = \dot{r}$, $v_\theta = r\dot{\theta}$ and $v_z = \dot{z}$ are the components of velocity of a particle in cylindrical coordinates.

(b) The acceleration of the particle in cylindrical coordinates (r, θ, z) is

$$\vec{a} = \frac{d}{dt}\left(\frac{d\vec{r}}{dt}\right) = \frac{d}{dt}(\dot{r}\hat{r} + r\dot{\theta}\hat{\theta} + \dot{z}\hat{z})$$

or, $\vec{a} = \dot{r}\frac{d\hat{r}}{dt} + \ddot{r}\hat{r} + r\dot{\theta}\frac{d\theta}{dt} + r\ddot{\theta}\hat{\theta} + r\dot{\phi}\hat{\theta} + \hat{\theta}\ddot{z}\hat{z} = \dot{r}\dot{\phi}\hat{\theta} + \ddot{r}\hat{r} + r\dot{\theta}(-\dot{\theta}\hat{r}) + r\ddot{\theta}\hat{\theta} + \dot{r}\dot{\theta}\hat{\theta} + \ddot{z}\hat{z}$

$$= (\ddot{r} - r\dot{\theta}^2)\hat{r} + (r\ddot{\theta} + 2\dot{r}\dot{\theta})\hat{\theta} + \ddot{z}\hat{z} = a_r\hat{r} + a_\theta\hat{\theta} + a_z\hat{z}.$$

where $a_r = (\ddot{r} - r\dot{\theta}^2)$, $a_\theta = (r\ddot{\theta} + 2\dot{r}\dot{\theta})$ **and** $a_z = \ddot{z}$ **are the components of accelerations in cylindrical coordinates along** $\hat{r}, \hat{\theta}$ **and** $\hat{z}$ **directions respectively.**

Exercises

13(A) : Objective Type Questions

Select the correct answer

13.1 Curvilinear coordinate system is

(a) the most general form of coordinate system

(b) a special type of coordinate system

(c) a mixture of both (a) and (b)

(d) neither of these.

13.2 Number of set of unit vectors in curvilinear coordinates is

(a) 1 **(b)** 3 **(c)** 2 **(d)** 4.

13.3 The two sets of unit vector $\hat{e}$, $\hat{e}_2$, $\hat{e}_3$ and $\widehat{E}$, $\widehat{E}_2$, $\widehat{E}$ are identical in

(a) Cartesian coordinates **(b)** cylindrical coordinates

(c) spherical coordinates **(d)** orthogonal curvilinear coordinates.

13.4 If a vector $\vec{A} = C_1\frac{\partial \vec{r}}{\partial u} + C_2\frac{\partial \vec{r}}{\partial v} + C_3\frac{\partial \vec{r}}{\partial w}$, then C_1, C_2, C_3 are called the

(a) contra-variant components of $\vec{A}$ **(b)** co-variant components of $\vec{A}$

(c) curvilinear components of $\vec{A}$ **(d)** none of these.

13.5 If $\vec{A} = C_1\nabla u + C_2\nabla v + C_3\nabla w$, then C_1, C_2, C_3 are called the

(a) contra-variant components of $\vec{A}$ **(b)** co-variant components of $\vec{A}$

(c) rectangular components of $\vec{A}$ **(d)** none of these.

13.6 The scale factors are represented by

(a) $h_i = \frac{\partial \vec{r}}{\partial u_i}$ **(b)** $h_i = \left|\frac{\partial \vec{r}}{\partial u_i}\right|$ **(c)** $h_i = \frac{\partial \vec{r}}{\partial v_i}$ **(d)** $h_i = \left|\frac{\partial \vec{r}}{\partial v_i}\right|$.

13.7 Volume element in curvilinear coordinates is

(a) $h_1h_2h_3$ **(b)** $du\,dv\,dw$ **(c)** $h_1h_2h_3\,du\,dv\,dw$ **(d)** $h_1h_2h_3 + du\,dv\,dw$.

13.8 Laplacian in curvilinear coordinates is

(a) $\nabla = \frac{\partial \phi}{\partial u}\hat{e}_1 + \frac{\partial \phi}{\partial v}\hat{e}_2 + \frac{\partial \phi}{\partial w}\hat{e}_3$ **(b)** $\nabla = \frac{1}{h_1}\frac{\partial \phi}{\partial u} + \frac{1}{h_2}\frac{\partial \phi}{\partial v} + \frac{1}{h_3}\frac{\partial \phi}{\partial w}$

(c) $\nabla = \frac{1}{h_1}\frac{\partial \phi}{\partial u}\hat{e}_1 + \frac{1}{h_2}\frac{\partial \phi}{\partial v}\hat{e}_2 + \frac{1}{h_3}\frac{\partial \phi}{\partial w}\hat{e}_3$ **(d)** none of these.

13.9 Laplacian in spherical coordinates is

(a) $\nabla = \frac{\partial}{\partial r} + \frac{1}{r}\frac{\partial}{\partial \theta} + \frac{1}{\sin\theta}\frac{\partial}{\partial \phi}$ (b) $\nabla = \frac{\partial}{\partial r}\hat{r} + \frac{1}{r}\frac{\partial}{\partial \theta}\hat{\theta} + \frac{1}{r\sin\theta}\frac{\partial}{\partial \phi}\hat{\phi}$

(c) $\nabla = \frac{\partial}{\partial r}\hat{r} + \frac{\partial}{\partial \theta}\hat{\theta} + \frac{\partial}{\partial \phi}\hat{\theta}$ (d) $\nabla = \frac{\partial}{\partial r}\hat{r} + \frac{1}{r\sin\theta}\frac{\partial}{\partial \theta}\hat{\theta} + \frac{1}{r^2\sin^2\theta}\frac{\partial}{\partial \phi}$.

13.10 Scale factors in spherical polar coordinates are

(a) $h_1 = 1,\ h_2 = r,\ h_3 = 1$ (b) $h_1 = r,\ h_2 = 1,\ h_3 = 1$

(c) $h_1 = 1,\ h_2 = 1,\ h_3 = r$ (d) none of these.

13.11 Volume element in cylindrical coordinates

(a) $dV = r\,dr\,d\theta\,dz$ (b) $dV = dr\,d\theta\,dz$ (c) $dV = \frac{dr\,d\theta\,dz}{r}$ (d) none of these.

13.12 Cylindrical coordinate system is

(a) non-orthogonal (b) orthogonal (c) both (a) and (b) (d) none of these.

Answers

13.1 (a) **13.2** (c) **13.3** (d) **13.4** (b) **13.5** (a) **13.6** (b), (d) **13.7** (c) **13.8** (c)
13.9 () **13.10** (a) **13.11** (a) **13.12** (b).

13(B) : Short Answers Type Questions

13.1 What do you mean by orthogonal curvilinear coordinates? [BU(H) 2001]

13.2 What are curvilinear coordinates? When are they called orthogonal? [BU(H) 2002]

13.3 Obtain the expressions for the orthogonal unit vectors in the cylindrical coordinate system in terms of those in the Cartesian coordinate system. [BU(H) 2005]

13.4 Express elements of area in orthogonal curvilinear coordinates.

13.5 Show that $|\nabla u_i| = \frac{1}{h_i}$, $i = 1, 2, 3$.

13.6 Prove that $\hat{e}_1 = h_2 h_3(\nabla u_2 \times \nabla u_3)$.

13.7 Express the vector $2y\hat{i} - z\hat{j} + 3x\hat{k}$ in cylindrical coordinates. [BU(H) 2006]

13(C) : Broad Answers Type Questions

13.1 Find the expression for the divergence of a vector in orthogonal curvilinear coordinates and write its value in spherical polar coordinates. [DU(H) 1997]

13.2 Obtain expression for $\nabla \cdot \vec{A}$ and $\nabla \times \vec{A}$ in orthogonal curvilinear coordinates. [BU(H) 1992]

13.3 Express $\nabla^2\phi$ in cylindrical coordinates. [BU(H) 1996]

13.4 Derive expressions for divergence and curl of a vector in orthogonal curvilinear coordinates. [BU(H) 2001]

13.5 Prove that spherical polar coordinate system is orthogonal. [Bang.U 1990]

13.6 Express the vector $2y\hat{i} - z\hat{j} + 3x\hat{k}$ in cylindrical coordinates. [BU(H) 2006]

13.7 Express $z\hat{\imath} - 2x\hat{\jmath} + y\hat{k}$ in cylindrical coordinates. **[Bang.U 1990S]**

13.8 Prove that a vector length p, which is everywhere normal to the surface $\vec{r}(u, v)$ is given by

$$\vec{A} = \pm p \frac{\left(\frac{\partial \vec{r}}{\partial u} \times \frac{\partial \vec{r}}{\partial v}\right)}{\sqrt{EG - F^2}}.$$

13.9 Prove that in orthogonal curvilinear coordinates, (i) $\nabla \cdot (\nabla \times \vec{A}) = 0$ (ii) $\nabla \times \nabla\phi = 0$.

13.10 Prove that, $\left[\frac{\partial \vec{r}}{\partial u} \cdot \left(\frac{\partial \vec{r}}{\partial v} \times \frac{\partial \vec{r}}{\partial w}\right)\right] [\nabla u \cdot (\nabla v \times \nabla w)] = 1.$

13.11 (a) Find unit vectors $\hat{r}, \hat{\theta}, \hat{\phi}$ in spherical coordinates in terms of $\hat{\imath}, \hat{\jmath}, \hat{k}$.

(b) Find $\hat{\imath}, \hat{\jmath}, \hat{k}$ in terms of $\hat{r}, \hat{\theta}, \hat{\phi}$.

13.12 Show that $\nabla(\ln r)$ and $\nabla\phi$ are solenoidal vectors, where r, θ, z are cylindrical coordinates.

13.13 Using the expression

$$\nabla^2\psi = \frac{1}{h_1h_2h_3}\left[\frac{\partial}{\partial u_1}\left(\frac{h_2h_3}{h_1}\frac{\partial\psi}{\partial u_1}\right) + \frac{\partial}{\partial u_2}\left(\frac{h_3h_1}{h_2}\frac{\partial\psi}{\partial u_2}\right) + \frac{\partial}{\partial u_3}\left(\frac{h_1h_2}{h_3}\frac{\partial\psi}{\partial u_3}\right)\right]$$

(the symbols have their usual meaning) find the expression of $\nabla^2\psi$ in spherical polar coordinates. **[VU(H) 2014]**

13.14 The position vector of a particle in cylindrical coordinates (ρ, ϕ, z) is given by

$$\vec{r} = \rho\hat{\rho} + z\hat{z}.$$

(i) Prove that $\dot{\hat{\rho}} = \dot{\phi}\hat{\phi}$ and $\dot{\hat{\phi}} = -\dot{\phi}\hat{\rho}$.

(ii) Show that acceleration of the particle is $\vec{a} = (\ddot{\rho} - \rho\dot{\phi}^2)\hat{\rho} + (\rho\ddot{\phi} + 2\dot{\rho}\dot{\phi})\hat{\phi} + \ddot{z}\hat{z}$. **[VU(H) 2012]**

13.15 Show that

$$\begin{aligned} \dot{\hat{e}}_r &= \dot{\theta}\hat{e}_\theta + \sin\theta\dot{\phi}\hat{e}_\phi \\ \dot{\hat{e}}_\theta &= -\dot{\theta}\hat{e}_r + \cos\theta\dot{\phi}\hat{e} \\ \dot{\hat{e}}_r &= -\sin\theta\dot{\phi}\hat{e}_r - \cos\theta\dot{\phi}\hat{e}_\theta \end{aligned}$$

where r, θ, ϕ and $\hat{e}_r$, $\hat{e}_\theta$, $\hat{e}_\phi$ are respectively the coordinates and unit vectors in spherical polar coordinates. **[VU(H) 2010]**

13.16 (a) Find an expression for the divergence of a vector in cylindrical polar coordinates. Express the vector $\vec{A} = yz\hat{\imath} - y\hat{\jmath} + xz^2\hat{k}$ in cylindrical polar coordinates and hence calculate the divergence of $\vec{A}$.

(b) Express the vector $\vec{R} = 2x\hat{\imath} - z\hat{\jmath} + y\hat{k}$ in spherical polar coordinates. **[BU(H) 2013]**

13.17 (a) How does curvilinear coordinate system differ from rectangular Cartesian coordinate system?

(b) Find expression for volume element in spherical polar coordinates.

(c) Express divergence of $\vec{A}$ in spherical polar coordinates. **[BU(H) 2012]**

13.18 (a) Derive an expression for ∇^2 in orthogonal curvilinear co-ordinates.

(b) Prove that cylindrical coordinate system is orthogonal. **[Mahatma Gandhi U 2012]**

13.19 (a) Derive the expression for the divergence of a vector in curvilinear co-ordinates.

(b) Find out arc length and volume element in curvilinear co-ordinates.

13.20 (a) For an orthogonal co-ordinate system show that $\hat{e_2} = h_3 h_1 \vec{\nabla} u_3 \times \vec{\nabla} u_1$, where the symbols stand for usual meanings. **[Assam U Phys(Major) 2014]**

(b) Obtain general solution of Laplace's equation in circular cylindrical co-ordinates.

13.21 (a) Write down the expression for a volume element in spherical polar co-ordinates. Hence, find the volume of a hemispherical shell bounded by hemispherical surfaces of radii r_1 and $r_2 (r_1 > r_2)$ using this expression. **[CU(H) 2007]**

(b) Express the divergence operator in spherical polar co-ordinates starting from the expression of the same in Cartesian co-ordinates. **[C.U.(H), 2014]**

(c) Consider the spherical polar co-ordinate system as given by

$$\vec{r} = r \sin\theta \cos\phi \hat{i} + r \sin\theta \sin\phi \hat{j} + r \cos\theta \hat{k}.$$

Find out the unit vectors $\hat{e}_r, \hat{e}_v, \hat{e}_\phi$ in this system. **[CU(H) 2015]**

13.22 In cylindrical polar co-ordinate system (r, θ, z) express an elementary volume bounded by r to $r + dr$, θ to $\theta + d\theta$, z to $z + dz$. Draw a neat diagram to indicate the volume. **[CU(H) 2006]**

CHAPTER 14

Application of Vector

14.1 Hydrodynamics

The branch of science which deals with the motion of fluid is known as **Hydrodynamics**. Fluid is that state of matter which has the capacity of changing shape when flowing, e.g., gases and liquids.

Following five physical properties are possessed by an *actual fluid*:

(i) density (ii) velocity (iii) temperature (iv) pressure and (v) viscosity.

The property by virtue of which a slower moving layer offers a retarding tangential stress to the faster moving layer is called *viscosity* and is a general property of fluids. If a fluid is devoid of viscosity (i.e., shearing stress), then it is called a *perfect fluid*. A perfect fluid is, however, only an ideal abstraction and is supposed to be an aggregate of particles which yield at once to the slightest effect to separate them.

Let $F(x, y, z, t)$ represent any general property of a particle of the moving fluid at (x, y, z) and time t. The fluid particle may be supposed to be a geometrical mass point.

$\because\ F = F(x, y, z, t)$, we have the total differentiation of F as

$$dF = \frac{\partial F}{\partial x}dx + \frac{\partial F}{\partial y}dy + \frac{\partial F}{\partial z}dz + \frac{\partial F}{\partial t}dt.$$

Now the rate of change of the property of the same particle with time is

$$\frac{dF}{dt} = \frac{\partial F}{\partial x}\frac{dx}{dt} + \frac{\partial F}{\partial y}\frac{dy}{dt} + \frac{\partial F}{\partial z}\frac{dz}{dt} + \frac{\partial F}{\partial t}; \qquad \text{(ii)}$$

also, $\nabla F = \hat{i}\frac{\partial F}{\partial x} + \hat{j}\frac{\partial F}{\partial y} + \hat{k}\frac{\partial F}{\partial z}.$

If $\vec{v}$ be the velocity of the particle at t, then

$$\vec{v} = \hat{i}v_1 + \hat{j}v_2 + \hat{k}v_3,$$

where v_1, v_2, v_3 are the components of $\vec{v}$ along X, Y and Z axes respectively,

$$\therefore \quad \vec{v}\cdot\nabla\vec{F} = (\hat{i}v_1 + \hat{j}v_2 + \hat{k}v_3)\cdot\left(\hat{i}\frac{\partial F}{\partial x} + \hat{j}\frac{\partial F}{\partial y} + \hat{k}\frac{\partial F}{\partial z}\right) = \frac{\partial F}{\partial x}v_1 + \frac{\partial F}{\partial y}v_2 + \frac{\partial F}{\partial z}v_3$$

$$= \frac{\partial F}{\partial x}\frac{dx}{dt} + \frac{\partial F}{\partial y}\frac{dy}{dt} + \frac{\partial F}{\partial z}\frac{dz}{dt}. \qquad \text{(iii)}$$

Substituting eqn. (iii) in eqn. (ii), we get

$$\frac{dF}{dt} = \frac{\partial F}{\partial t} + \vec{v}\cdot\nabla F. \qquad \text{(iv)}$$

If F represents the velocity $\vec{v}$ of the particle, then we get

$$\frac{d\vec{v}}{dt} = \frac{d\vec{v}}{\partial t} + (\vec{v}\cdot\nabla)\vec{v}. \qquad \text{(v)}$$

14.1.1 Equation of Continuity

According to the equation of continuity, *the rate of generation of the fluid mass within a given volume is equal to the net outflow of fluid mass from that volume (i.e., the mass of fluid is neither created nor destroyed).*

We consider a fluid of density $\rho(x, y, z, t)$ enclosed by a surface S of volume V. If $\rho(x, y, z, t)$ be the density of the fluid at a point $P(x, y, z)$ at any time t, then the mass of the fluid in volume V is

$$M(t) = \iiint_V \rho dV.$$

$\therefore$ the rate of net outflow of the fluid from V is

$$\frac{dM}{dt} = \frac{\partial}{\partial t}\iiint_V \rho dV = \iiint_V \frac{\partial \rho}{\partial t} dV. \qquad \text{(i)}$$

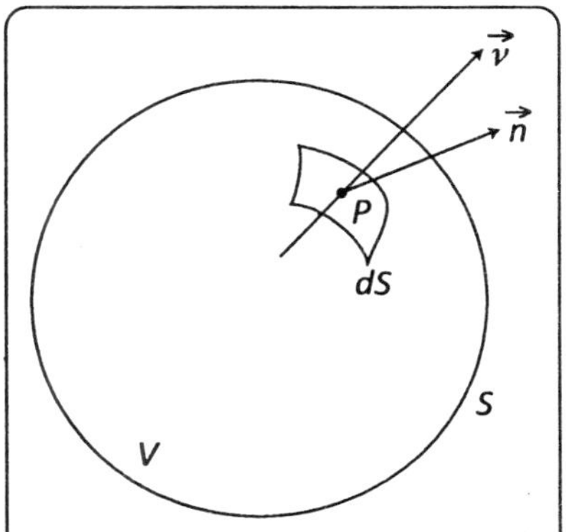

Fig. 14.1

Now let $\vec{v}$ be the velocity of the fluid at any point P surrounded by an element of area dS and $\hat{n}$ be the outward drawn normal to dS, then the component of $\vec{v}$ along $\hat{n}$ is given by $(\vec{v} \cdot \hat{n})$.

$\therefore$ the inward normal velocity of the fluid at P is $-(\vec{v} \cdot \hat{n})$.

So, the rate of fluid mass entering through dS is

$$= \rho dS(-\vec{v} \cdot \hat{n}) = -\rho(\vec{v} \cdot \hat{n} dS) = -\rho \vec{v} \cdot d\vec{S}.$$

$\therefore$ the total mass of fluid entering through S per unit time is

$$M' = -\iint_S \rho \vec{v} \cdot d\vec{S} = -\iiint_V \nabla \cdot (\rho \vec{v}) dV, \quad \text{[by Gauss' divergence theorem]} \qquad \text{(ii)}$$

If there is no source or sink of fluid, then

$$\frac{d}{dt} M = M' \quad \text{or,} \quad \iiint_V \frac{\partial \rho}{\partial t} dV = -\iiint_V \nabla \cdot (\rho \vec{v}) dV \quad \text{or,} \quad \iiint_V \left[\nabla \cdot (\rho \vec{v}) + \frac{\partial \rho}{\partial t}\right] dV = 0.$$

Since, the integrand is continuous and the volume V is arbitrary,

$$\boldsymbol{\nabla \cdot (\rho \vec{v}) + \frac{\partial \rho}{\partial t} = 0.} \qquad \text{(iii)}$$

This is the *equation of continuity.*

For *incompressible fluid,* $\rho =$ constant and $\frac{\partial \rho}{\partial t} = 0$

$\therefore \quad \nabla \cdot (\rho \vec{v}) = 0.$

or, $\quad \boldsymbol{\nabla \cdot \vec{v} = 0}. \qquad$ (iv)

Equation (iv) represents the equation of continuity for an *incompressible fluid.*

Equation of continuity for incompressible irrotational flow :

If the fluid is **incompressible**, then we have seen that $\nabla \cdot \vec{v} = 0$.

If the fluid is **irrotational**, then $\nabla \times \vec{v} = 0$.

So, there exists a scalar function ϕ (called the **velocity potential**) such that, $\vec{v} = -\nabla \phi$.

$\therefore \quad \nabla \cdot \vec{v} = -\nabla \cdot (\nabla\phi) = -\nabla^2\phi = 0.$

i.e., $\boldsymbol{\nabla^2\phi = 0},$ (v)

which is the *Laplace's equation.*

Hence, for **incompressible irrotational fluid** there exists a *velocity potential* ϕ which satisfies the Laplace's equation.

14.2 Euler's Equation of Motion

We consider a volume V of perfect fluid enclosed by the surface S. This volume of fluid is assumed to be moving and contains the same fluid particles at every time.

The forces acting on the fluid particles in V are

(a) the body force (i.e., weight of the body)

(b) the surface force due to pressure

(c) the viscous force.

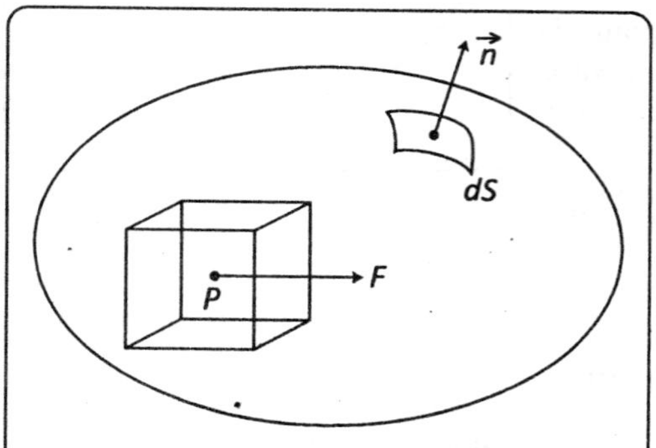

Fig. 14.2

Since, the fluid under consideration is *perfect* the viscous force is zero.

Now, let the density of the fluid at P within dV is ρ, so that the mass within dV is ρdV, which is constant as the same fluid particles are contained in S.

The momentum of the mass ρdV is $= (\rho dV)\vec{v}$, where $\vec{v}$ is the velocity of the fluid.

The total momentum of the fluid in volume V is

$$\vec{p} = \iiint_V (\rho dV)\vec{v}.$$

$\therefore$ rate of change of momentum,

$$\frac{d\vec{p}}{dt} = \frac{d}{dt}\iiint_V \vec{v}(\rho dV) = \iiint_V \frac{d\vec{v}}{dt}(\rho dV) + \iiint_V \vec{v}\frac{d}{dt}(\rho dV)$$

$$= \iiint_V \frac{d\vec{v}}{dt}(\rho dV) \quad \left[\because \frac{d}{dt}(\rho dV) = 0,\ \rho dV \text{ being constant}\right].$$

Here, $\frac{d\vec{p}}{dt}$ is equal to the total force acting on the mass of the fluid.

Body force : If the external force acting per unit mass of the fluid at P be $\vec{F}$, then the total external force acting on the fluid in V is

$$\boldsymbol{\vec{F_b} = \iiint_V \vec{F}(\rho dV).}$$

Surface force : If P be the pressure along the outward drawn normal at dS, then the surface force on dS is $-P\hat{n}dS$, the negative sign is due to the fact that the surface force acts in the inward direction and $\hat{n}$ is the outward unit normal to dS.

$\therefore$ total surface force is $\vec{F_S} = -\iint_S P\hat{n}dS = -\iiint_S \nabla P dV.s$

Thus, the total force acting on the fluid in V is

$$\vec{F_b} + \vec{F_S} = \iiint_V (\rho\vec{F} - \nabla P)dV.$$

Now, $\displaystyle\iiint_V \frac{d\vec{v}}{dt}\rho dV = \iiint_V (\rho\vec{F} - \nabla P)dV$ or, $\displaystyle\iiint_V \left(\rho\frac{d\vec{v}}{dt} - \rho\vec{F} + \nabla P\right) dV = 0.$

Since V is arbitrary, $\rho\dfrac{d\vec{v}}{dt} - \rho\vec{F} + \nabla P = 0$

or,
$$\boldsymbol{\frac{d\vec{v}}{dt} = \vec{F} - \frac{1}{\rho}\nabla P.} \tag{ii}$$

This is known as the *Euler's equation of motion.*

Again, $\because \dfrac{d}{dt} = \dfrac{\partial}{\partial t} + (\vec{v}\cdot\nabla)$,

$$\therefore \quad \boldsymbol{\frac{\partial\vec{v}}{\partial t} + (\vec{v}\cdot\nabla)\vec{v} = \vec{F} - \frac{1}{\rho}\nabla P.} \tag{iii}$$

14.3 Cartesian Form of Euler's Equation

If F_1, F_2, F_3 and u, v, w be the Cartesian components of $\vec{F}$ and $\vec{v}$ respectively, then

$$\vec{v} = \hat{i}u + \hat{j}v + \hat{k}\omega$$

$$\vec{F} = \hat{i}F_1 + \hat{j}F_2 + \hat{k}F_3 \quad \text{and} \quad \nabla P = \hat{i}\frac{\partial P}{\partial x} + \hat{j}\frac{\partial P}{\partial y} + \hat{k}\frac{\partial P}{\partial z}.$$

Putting these values in eqn. (iii), we get

$$\hat{i}\frac{\partial u}{\partial t} + \hat{j}\frac{\partial v}{\partial t} + \hat{k}\frac{\partial w}{\partial t} + \left[(\hat{i}u + \hat{j}v + \hat{k}w)\cdot\left(\hat{i}\frac{\partial}{\partial x} + \hat{j}\frac{\partial}{\partial y} + \hat{k}\frac{\partial}{\partial z}\right)\right](\hat{i}u + \hat{j}v + \hat{k}w)$$

$$= \hat{i}F_1 + \hat{j}F_2 + \hat{k}F_3 - \frac{1}{\rho}\left(\hat{i}\frac{\partial P}{\partial x} + \hat{j}\frac{\partial P}{\partial y} + \hat{k}\frac{\partial P}{\partial z}\right) \tag{iv}$$

or,
$$\hat{i}\left(\frac{\partial u}{\partial t} + u\frac{\partial u}{\partial x} + v\frac{\partial u}{\partial y} + w\frac{\partial u}{\partial z}\right) + \hat{j}\left(\frac{\partial v}{\partial t} + u\frac{\partial v}{\partial x} + v\frac{\partial v}{\partial y} + w\frac{\partial u}{\partial z}\right)$$
$$+ \hat{k}\left(\frac{\partial w}{\partial t} + u\frac{\partial w}{\partial x} + v\frac{\partial w}{\partial y} + w\frac{\partial w}{\partial z}\right)$$

$$= \hat{i}\left(F_1 - \frac{1}{\rho}\frac{\partial P}{\partial x}\right) + \hat{j}\left(F_2 - \frac{1}{\rho}\frac{\partial P}{\partial y}\right) + \hat{k}\left(F_3 - \frac{1}{\rho}\frac{\partial P}{\partial z}\right).$$

Equating the coefficients of like unit vectors, we get

$$\boldsymbol{\frac{\partial u}{\partial t} + u\frac{\partial u}{\partial x} + v\frac{\partial u}{\partial y} + w\frac{\partial u}{\partial z} = F_1 - \frac{1}{\rho}\frac{\partial P}{\partial x}}$$

$$\boldsymbol{\frac{\partial v}{\partial t} + u\frac{\partial v}{\partial x} + v\frac{\partial v}{\partial y} + w\frac{\partial v}{\partial z} = F_2 - \frac{1}{\rho}\frac{\partial P}{\partial y}}$$

$$\boldsymbol{\frac{\partial w}{\partial t} + u\frac{\partial w}{\partial x} + v\frac{\partial w}{\partial y} + w\frac{\partial w}{\partial z} = F_3 - \frac{1}{\rho}\frac{\partial P}{\partial z}.}$$

These are *Euler's equations in Cartesian component form.*

14.4 Bernoulli's Theorem

Statement : *In case of a steady flow of an incompressible fluid the sum of potential energy, Kinetic energy and the pressure energy per unit mass is constant along any streamline.*

Proof : We have the vector identity,

$$\nabla(\vec{A}\cdot\vec{B}) = (\vec{A}\cdot\nabla)\vec{B} + (\vec{B}\cdot\nabla)\vec{A} + \vec{A}\times(\nabla\times\vec{B}) + \vec{B}\times(\nabla\times\vec{A})$$

If $\vec{A} = \vec{B} = \vec{v}$, then

$$\nabla(\vec{v}\cdot\vec{v}) = (\vec{v}\cdot\nabla)\vec{v} + (\vec{v}\cdot\nabla)\vec{v} + \vec{v}\times(\nabla\times\vec{v}) + \vec{v}\times(\nabla\times\vec{v})$$

or, $$\nabla(\vec{v}\cdot\vec{v}) = 2[(\vec{v}\cdot\nabla)\vec{v} + \vec{v}\times(\nabla\times\vec{v})]$$

or, $$(\vec{v}\cdot\nabla)\vec{v} = \frac{1}{2}\nabla\vec{v} + (\nabla\times\vec{v})\times\vec{v}. \tag{v}$$

Substituting eqn. (v) in eqn. (iii), we get

$$\frac{\partial\vec{v}}{\partial t} + \frac{1}{2}\nabla v^2 + (\nabla\times\vec{v})\times\vec{v} = \vec{F} - \frac{1}{\rho}\nabla P. \tag{vi}$$

If the external force is conservative, then $\vec{F} = -\nabla U$, U is the related potential.

Also, if the liquid is irrotational, then $\nabla\times\vec{v} = 0$ and $\vec{v} = \nabla\phi$, where ∇ is the velocity potential.

Using the above two conditions, eqn. (v) reduces to

$$\frac{\partial}{\partial t}(\nabla\phi) + \frac{1}{2}\nabla v^2 = -\nabla U - \frac{1}{\rho}\nabla P \quad \text{or,} \quad \nabla\left[\frac{\partial\phi}{\partial t} + \frac{1}{2}v^2 + U + \frac{P}{\rho}\right] = 0$$

or, $$\nabla\psi = 0, \quad \text{where } \nabla\psi = \frac{\partial\phi}{\partial t} + \frac{1}{2}v^2 + U + \frac{P}{\rho}$$

If, $d\vec{r}$ be an arbitrary path in the fluid at any instant of time, then

$$\nabla\psi\cdot d\vec{r} = d\psi = \left(\hat{i}\frac{\partial\psi}{\partial x} + \hat{j}\frac{\partial\psi}{\partial y} + \hat{k}\frac{\partial\psi}{\partial z}\right)\cdot(\hat{i}dx + \hat{j}dy + \hat{k}dz) = \frac{\partial\psi}{\partial x}dx + \frac{\partial\psi}{\partial y}dy + \frac{\partial\psi}{\partial z}dz.$$

Now, $d\psi = 0$ as $\nabla\psi = 0$ or, $d\left[\frac{\partial\phi}{\partial t} + \frac{1}{2}v^2 + U + \frac{P}{\rho}\right] = 0$

or, $$\boldsymbol{\frac{\partial\phi}{\partial t} + \frac{1}{2}v^2 + U + \frac{P}{\rho} = k(t).} \tag{vii}$$

where $k(t)$ is an arbitrary function of time. Equation (vii) is known as the *Bernoulli's equation.*

In case of steady motion of the fluid, $\frac{\partial\vec{v}}{\partial t} = \frac{\partial\phi}{\partial t} = 0$. Equation (vii) then takes the form,

$$\boldsymbol{\frac{1}{2}v^2 + U + \frac{P}{\rho} = k,} \tag{viii}$$

where k is constant for a particular *stream line.* Dividing both sides of eqn. (viii) by acceleration due to gravity (g), we get

$$\boldsymbol{\frac{v^2}{2g} + h + \frac{P}{\rho g} = k'} \text{ (constant)}, \tag{xi}$$

where $U = gh$. Here $\frac{v^2}{2g}$, h and $\frac{P}{\rho g}$ are called respectively *velocity head, gravity head* and *pressure head*. Hence, in engineering language, Bernoulli's theorem is expressed as **the sum of the three heads is constant**.

14.5 Fourier's Heat Flow Equation

We consider a solid of volume V enclosed by a surface S, the temperature of which is not same at all points of the body. Due to the temperature difference heat flows from the places of higher temperature to the places of lower temperature by the conduction process.

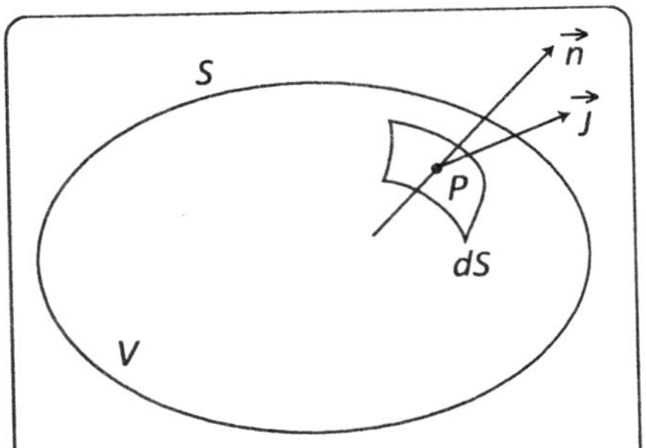

Fig. 14.3

Experiment reveals that the thermal current $\vec{J}$ is proportional to the thermal conductivity K as well as $-\nabla\theta$, where $\theta(x, y, z, t)$ is temperature, i.e.,

$$\vec{J} = -k\nabla\theta.$$

Now, if $\hat{n}$ be the outward drawn normal to a surface element dS, then the rate of flow of heat through dS is

$$\vec{J} \cdot \hat{n}dS = \vec{J} \cdot \vec{dS}$$

Hence, the rate of flow of heat through the surface S is given by

$$-\frac{\partial Q}{\partial t} = \iint_S \vec{J} \cdot \vec{dS} = \iiint_V (\nabla \cdot \vec{J})dV \quad \text{[by Gauss' divergence therorem]}$$

$$= \iiint_V \nabla \cdot (-K\nabla\theta)dV = -K \iiint_V \nabla^2\theta dV. \tag{i}$$

Again, the amount of heat in the body is

$$Q = \iiint_V (\rho dV)c\theta = \iiint_V pc\theta dV,$$

where ρ is the density and c, the specific heat of the body. The rate of decrease of heat of the body is

$$-\frac{\partial Q}{\partial t} = -\frac{\partial}{\partial t} \iiint_V \rho c \cdot \theta dV. \tag{ii}$$

In absence of any source or sink of heat within the body, eqns. (i) and (ii) are equal.

$$\therefore \quad \frac{\partial}{\partial t} \iiint_V \rho c\theta dV = K \iiint_V \nabla^2\theta dV \quad \text{or,} \quad \iiint_V \rho c\frac{\partial\theta}{\partial t}dV = K \iiint_V \nabla^2\theta dV$$

$$\text{or,} \quad \iiint_V \left(\rho c\frac{\partial\theta}{\partial t} - K\nabla^2\theta\right) dV = 0.$$

Since the volume is arbitrary,

$$\rho c\frac{\partial\theta}{\partial t} - k\nabla^2\theta = 0 \quad \text{or,} \quad \frac{\partial\theta}{\partial t} = \frac{K}{\rho c}\nabla^2\theta$$

$$\text{or,} \quad \boldsymbol{\frac{\partial\theta}{\partial t} = h\nabla^2\theta} \tag{iii}$$

Equation (iii) is called the *three-dimensional Fourier's heat flow equation*. Here, $h = \frac{K}{\rho c}$ is the *thermal diffusivity* or *thermometric conductivity* of the solid.

14.6 Gauss' Law of Gravitation

Statement : *The surface integral of gravitational field strength $\vec{E}$ over any closed surface S drawn in the field is equal to -4π times the total mass enclosed by the surface, i.e.,*

$$\iint_S \vec{E} \cdot \vec{dS} = -4\pi \times (\text{total mass enclosed by } S)$$

Proof : Let mass m at O is within a volume V enclosed by the surface S. $\hat{n}$ is an outward drawn normal to the surface element dS surrounding the point P on the surface. The gravitational intensity at P due to mass m at 0 is

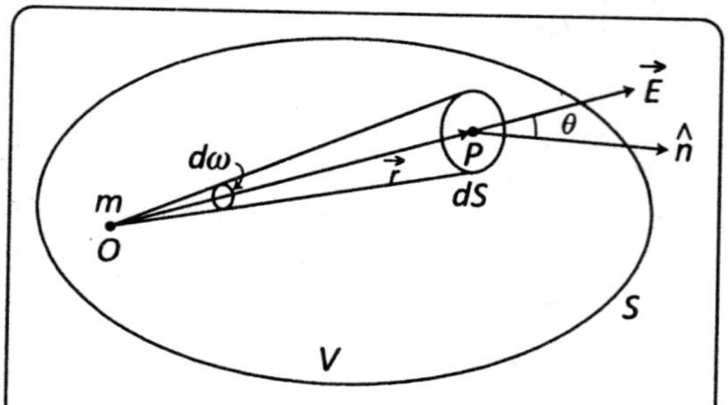

Fig. 14.4

$$\vec{E} = \frac{Gm}{r^2}\hat{r},$$

where $\overrightarrow{OP} = \vec{r}$, and $\hat{r}$ is a unit vector along $\vec{r}$.

$$\therefore \quad \vec{E} \cdot \hat{n} = -\frac{Gm}{r^2}\hat{r} \cdot \hat{n} = -\frac{Gm}{r^2}\cos\theta$$

where, θ is the angle as shown in Fig. 14.4.

$$\therefore \quad \iint_S \vec{E} \cdot \hat{n} dS = -\iint_S Gm\frac{dS\cos\theta}{r^2} \quad \text{or,} \quad \iint_S \vec{E} \cdot \vec{dS} = -Gm\int_0^{4\pi} d\omega = -Gm4\pi$$

where $d\omega = \frac{dS\cos\theta}{r^2}$ is the element of *solid angle* subtended by dS at O

$$\text{or,} \quad \iint_S \vec{E} \cdot \vec{dS} = -4\pi Gm. \qquad \text{(i)}$$

If there are number of masses $m_1, m_2, \ldots, m_i, \ldots, m_n$ within S, then

$$\iint_S \vec{E} \cdot \vec{dS} = -\sum_{t=1}^{n} 4\pi Gm_i = -\iiint_V 4\pi G\rho dV, \qquad \text{(ii)}$$

where ρ = density of the mass distribution.

14.7 Poisson's Equation in Gravitational Field

We have $\iint_S \vec{E} \cdot \vec{dS} = \iiint_V (\nabla \cdot \vec{E})dV$

$$\therefore \quad \iiint_V (\nabla \cdot \vec{E})dV = -\iiint_V 4\pi G\rho dV \quad \text{[by eqn. (ii)]}$$

$$\text{or,} \quad \iiint_V (\nabla \cdot \vec{E} + 4\pi G\rho)dV = 0.$$

This relation is true for any volume,

$$\therefore \quad \nabla \cdot \vec{E} + 4\pi G\rho = 0 \quad \text{or,} \quad \nabla \cdot \vec{E} = -4\pi G\rho.$$

But, $\vec{E} = -\nabla V$ where V is the potential at P.

$\therefore \quad \nabla(-\nabla V) = -4\pi G\rho$

or, $\quad \boldsymbol{\nabla^2 V = 4\pi G\rho}.$ (iii)

This is known as the *Poisson's equation in Gravitational field.*

If, however, the volume does not contain any matter, then $\rho = 0$. Equation (iii) then reduces to

$$\boldsymbol{\nabla^2 V = 0}. \qquad \text{(iv)}$$

Equation (iv) is known as the *Laplace's equation in Gravitational field.*

14.8 Gauss' Law in Electrostatics

Statement : *The total electric flux linked with a closed surface is $\frac{1}{\epsilon_0}$ times the charge or zero according as the closed surface encloses the charge or not.*

If $\vec{E}$ be the electric intensity due to a point charge q placed in a medium of permittivity ϵ_0, being enclosed by a surface S, then

$$\oint_S \vec{E}\cdot\vec{dS} = \frac{1}{\epsilon_0}q \quad \text{(if } S \text{ encloses } q)$$
$$= 0 \quad \text{(otherwise)}.$$

Proof : Consider a point charge q placed at O enclosed by the surface S, and P is a point on the surface at a distance r from O. If $\hat{n}$ be the outward drawn unit normal to an element of surface dS surrounding P, then the normal flux through dS is (Fig. 14.5)

$$\vec{E}\cdot\hat{n}dS = \vec{E}\cdot\vec{dS} = \frac{q}{4\pi\epsilon_0 r^2}\hat{r}\cdot\hat{n}dS$$
$$\left[\because \vec{E} = \frac{q}{4\pi\epsilon_0 r^2}\hat{r} \text{ and } \hat{r} = \frac{\vec{r}}{r}\right]$$

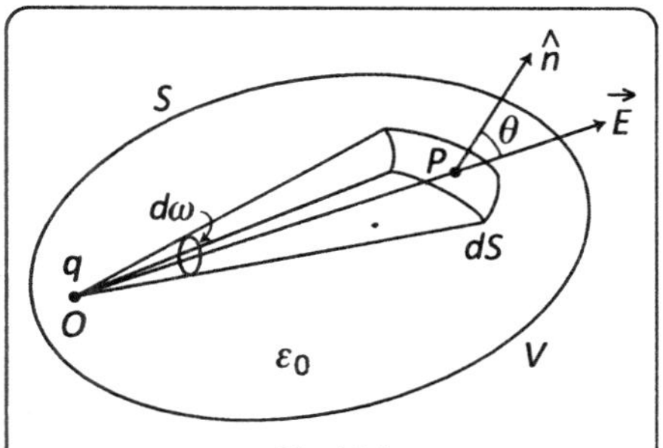

Fig. 14.5

If θ be the angle between $\hat{r}$ and $\hat{n}$, then $\hat{r}\cdot\hat{n} = \cos\theta$.

$$\therefore \quad \vec{E}\cdot\vec{dS} = \frac{q}{4\pi\epsilon_0}\frac{dS\cos\theta}{r^2} = \frac{q}{4\pi\epsilon_0}d\omega,$$
$$\left[\because d\omega = \frac{dS\cos\theta}{r^2}\right]$$

where $d\omega$ is the solid angle subtended by dS at 0.

$\therefore$ the total normal flux over the entire surface S is

$$\oint_S \vec{E}\cdot\vec{dS} = \frac{q}{4\pi\epsilon_0}\int_0^{4\pi} d\omega. \qquad \text{(i)}$$

Now, $$\int_0^{4\pi} d\omega = \begin{cases} 4\pi, & \text{(if } q \text{ is enclosed by } S) \\ 0, & \text{(otherwise)} \end{cases}$$

or, $$\oint_S \boldsymbol{\vec{E}\cdot\vec{dS}} = \begin{cases} \dfrac{q}{\epsilon_0}, & \textbf{if } q \textbf{ lies inside } S \\ 0, & \textbf{if } q \textbf{ lies outside } S. \end{cases} \qquad \text{(ii)}$$

This proves Gauss' law is electrostatic for a single point charge q. For a **discrete set of charges** $q_1, q_2, \ldots, q_n$ the law takes the form,

$$\oint_S \vec{E} \cdot \vec{dS} = \frac{1}{\epsilon_0} \sum_{i=1}^{n} q_i. \tag{iii}$$

The summation is only over those charges which are enclosed by S.

For a **continuous charge distribution** of density ρ, Gauss' law is expressed as

$$\oint_S \vec{E} \cdot \vec{dS} = \frac{1}{\epsilon_0} \iiint_V \rho dV. \tag{iv}$$

This represents the *integral form of Gauss' law in electrostatics.*

14.9 Differential Form of Gauss' Law

We have by Gauss' divergence theorem,

$$\oint_S \vec{E} \cdot \vec{dS} = \iiint_V \nabla \cdot \vec{E} dV \tag{v}$$

$$\therefore \quad \iiint_V (\nabla \cdot \vec{E}) dV = \frac{1}{\epsilon_0} \iiint_V \rho dV \quad \text{[from eqns. (iv) and (v)]}$$

$$\text{or,} \quad \iiint_V \left(\nabla \cdot \vec{E} - \frac{\rho}{\epsilon_0}\right) dV = 0.$$

Since volume is arbitrary, the integrand must vanish.

$$\therefore \quad \left(\nabla \cdot \vec{E} - \frac{\rho}{\epsilon_0}\right) = 0 \quad \text{or,} \quad \nabla \cdot \vec{E} - \frac{\rho}{\epsilon_0}. \tag{vi}$$

This is known as the *differential form of Gauss' law.*

14.10 Biot-Savart Law From Ampere's Law of Force

On the basis of a series of experiments Ampere formulated a law of force between two current carrying conductors separated by a distance. If C_1 and C_2 be two circuits carrying currents I_1 and I_2 respectively placed in a medium of permeability μ_0 (Fig. 14.6); then, according to ampere, the force exerted by the circuit C_1 on C_2 is given by

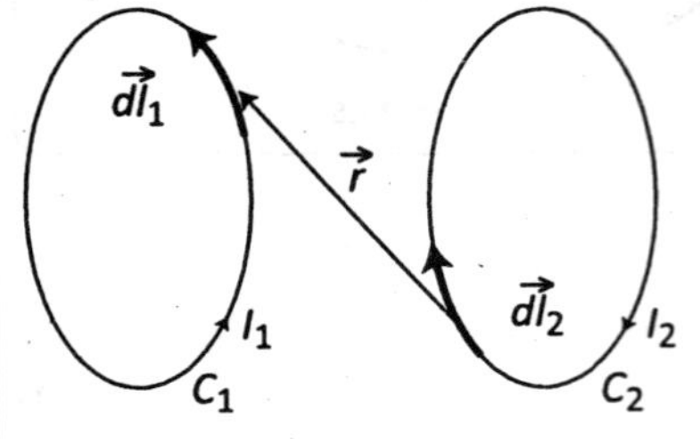

Fig. 14.6

$$\vec{F_{12}} = \frac{\mu_0}{4\pi} I_1 I_2 \oint_{C_1} \oint_{C_2} \frac{\vec{dl_1} \times (\vec{dl_2} \times \vec{r})}{r^3}, \tag{i}$$

where $I_1\vec{dl_1}$ and $I_2\vec{dl_2}$ are the current elements in C_1 and C_2 respectively separated by a distance $\vec{r}$.

Equation (i) can be written as

$$\vec{F_{12}} = I_1 \oint_{C_1} \vec{dl_1} \times \left[\frac{\mu_0}{4\pi} I_2 \oint_{C_2} \frac{\vec{dl_2} \times \vec{r}}{r^3}\right] = \oint_{C_1} I_1 \vec{dl_1} \times \vec{B_2}, \tag{ii}$$

where $\vec{B_2} = \frac{\mu_0}{4\pi} \oint_{C_2} \frac{I_2 \vec{dl_2} \times \vec{r}}{r^3}$. (iii)

$\vec{B_2}$ is the *magnetic induction* at the position of $I_1\vec{dl_1}$ due to current I_2 in C_2.

Hence, from eqn. (iii) it is evident that the magnetic induction $d\vec{B}$ due to a current element $I\vec{dl}$ at a distance $\vec{r}$ is (Fig. 14.7)

$$d\vec{B} = \frac{\mu_0}{4\pi}\left(\frac{I\vec{dl} \times \vec{r}}{r^3}\right).$$

For whole of the closed circuit C,

$$\vec{B} = \frac{\mu_0}{4\pi}\oint_C \frac{I\vec{dl} \times \vec{r}}{r^3}. \quad \text{(iv)}$$

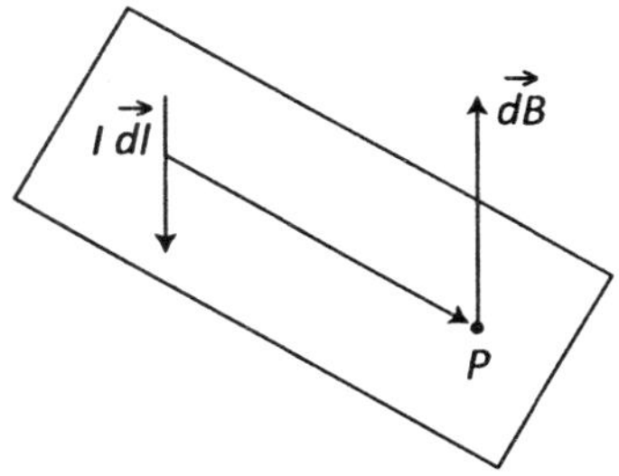

Fig. 14.7

Equation (iv) is called *Biot-Savart law* or *Laplace's formula.*

Case I : If I is distributed in space with a current density $\vec{J}$, then since $I\vec{dl} = \vec{J}dV$,

$$\vec{B} = \frac{\mu_0}{4\pi}\int \frac{\vec{J} \times \vec{r}}{r^3} dV. \quad \text{(v)}$$

Case II : If single charge q moves with velocity $\vec{v}$ in the circuit, then

$$I\vec{dl} = \frac{dq}{dt}\vec{dl} = dq\frac{\vec{dl}}{dt} = \vec{v}\,dq,$$

then, $\vec{B} = \frac{\mu_0}{4\pi}\int dq\frac{\vec{v} \times \vec{r}}{r^3}$ [from equn. (iv)]

$$= \frac{\mu_0}{4\pi} q \frac{\vec{v} \times \vec{r}}{r^3}$$

or, $\vec{B} = \mu_0\epsilon_0 \vec{v} \times \frac{q\vec{r}}{4\pi\epsilon_0 r^3}$

or, $\vec{B} = \frac{\vec{v} \times \vec{E}}{c^2}$, (vi)

where $\mu_0\epsilon_0 = \frac{1}{c^2}$ and $\vec{E} = \frac{q\vec{r}}{4\pi\epsilon_0 r^3}$.

Equation (vi) gives a compact relation between $\vec{E}$ and $\vec{B}$ for a uniformly moving charge q.

14.11 Divergence of $\vec{B}$

According to **Biot-Savart's law**, magnetic induction due to a current element at a distance $\vec{r}$ is

$$d\vec{B} = \frac{\mu_0 I}{4\pi}\left[\frac{\vec{dl} \times \vec{r}}{r^3}\right].$$

For the whole circuit,

$$d\vec{B} = \frac{\mu_0 I}{4\pi}\left[\oint_C \frac{\vec{dl} \times \vec{r}}{r^3}\right].$$

So, $\nabla \cdot \vec{B} = \nabla \cdot \left[\frac{\mu_0 I}{4\pi} \oint_C \frac{\vec{dl} \times \vec{r}}{r^3}\right] = \frac{\mu_0 I}{4\pi}\left[\oint_C \nabla \cdot \left(\frac{\vec{dl} \times \vec{r}}{r^3}\right)\right].$ (i)

But, $\nabla \cdot (\vec{A} \times \vec{C}) = \vec{C} \cdot (\nabla \times \vec{A}) - \vec{A} \cdot (\nabla \times \vec{C})$

i.e., $\nabla \cdot \left(\frac{\vec{dl} \times \vec{r}}{r^3}\right) = \frac{\vec{r}}{r^3}(\nabla \times \vec{dl}) - \vec{dl} \cdot \left(\nabla \times \frac{\vec{r}}{r^3}\right)$ $\left[\text{where } \vec{dl} = \vec{A} \text{ and } \frac{\vec{r}}{r^3} = \vec{B}\right]$

Now eqn. (i) reduces to

$$\nabla \cdot \vec{B} = \frac{\mu_0 I}{4\pi} \oint_C \left[\frac{\vec{r}}{r^3} \cdot (\nabla \times \vec{dl}) - \vec{dl} \cdot \left(\nabla \times \frac{\vec{r}}{r^3}\right)\right].$$

Now, $\nabla \times \vec{dl} = 0$ since $\vec{dl}$ is a constant vector, and

$$\nabla \times \left(\frac{\vec{r}}{r^3}\right) = -\nabla \times \nabla\left(\frac{1}{r}\right) = 0, \quad \therefore \ \boldsymbol{\nabla \cdot \vec{B} = 0}.$$

This indicates that *magnetic field is solenoidal* and as such *magnetic lines of force are closed loops*. This implies the *non-existence of magnetic monopole*.

14.12 Ampere's Circuited Law

Statement : *The line integral of magnetic induction vector $\vec{B}$ round a closed path is equal to μ_0 times the total current crossing any surface bounded by the line integral path, i.e.,*

$$\oint_C \vec{B} \cdot \vec{dl} = \mu_0 \iint_S \vec{J} \cdot \vec{dS} = \mu_0 I. \quad \left[\because \iint_S \vec{J} \cdot \vec{dS} = I\right]$$

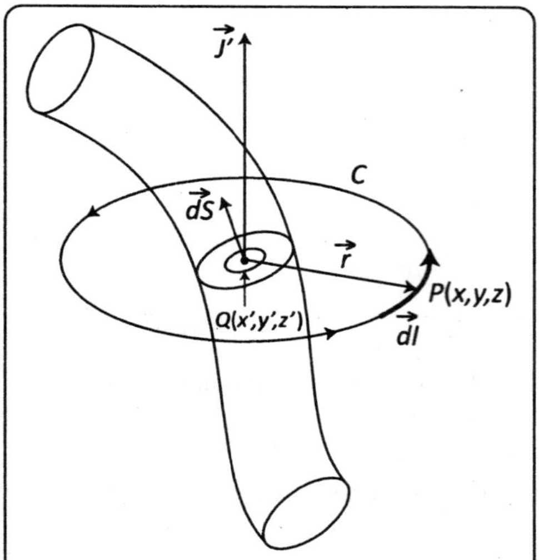

Fig. 14.8

Proof : We consider a current element $\vec{J} \cdot \vec{dS}$ in a conductor and a closed path linking the currents element as shown in Fig. 14.8. The magnetic induction at P due to current at Q by **Biot-Savart law** is

$$d\vec{B} = \frac{\mu_0}{4\pi} \frac{\vec{J}' \times \vec{r}}{r^3},$$

where $\vec{J}'$ is the current density at $Q(x', y', z')$ of the conductor.

Also, $r^2 = (x - x')^2 + (y - y')^2 + (z - z')^2$.

$\therefore$ the total induction at P is

$$\vec{B} = \frac{\mu_0}{4\pi} \iiint_{\tau'} \frac{\vec{J}' \times \vec{r}}{r^3} d\tau', \quad \text{(i)}$$

where $d\tau'$ is the volume element at Q,

or, $\vec{B} = \frac{\mu_0}{4\pi} \iiint_{\tau'} \vec{J}' \times \nabla\left(-\frac{1}{r}\right) d\tau'.$ $\left[\because -\frac{\vec{r}}{r^3} = \nabla\left(\frac{1}{r}\right)\right]$ (ii)

Again, $\nabla \times \phi\vec{A} = \phi(\nabla \times \vec{A}) - \vec{A} \times \nabla\phi.$

$$\therefore \quad \nabla \times \frac{\vec{J}'}{r} = \frac{1}{r}(\nabla \times \vec{J}') - \vec{J}' \times \nabla\left(\frac{1}{r}\right) = -\vec{J}' \times \nabla\left(\frac{1}{r}\right). \quad [\because \nabla \times \vec{J}' = 0]$$

$$\therefore \quad \vec{B} = \frac{\mu_0}{4\pi} \iiint_{\tau'} \nabla \times \left(\frac{\vec{J}'}{r}\right) d\tau'. \tag{iii}$$

So, $$\nabla \times \vec{B} = \frac{\mu_0}{4\pi} \iiint_{\tau'} \nabla \times \left[\nabla \times \left(\frac{\vec{J}'}{r}\right)\right] d\tau'. \tag{iv}$$

But, $\nabla \times (\nabla \times \vec{A}) = \nabla(\nabla \cdot \vec{A}) - \nabla^2 \vec{A}$.

$$\therefore \quad \nabla \times \vec{B} = \frac{\mu_0}{4\pi} \iiint_{\tau'} \left[\nabla\left\{\nabla \cdot \left(\frac{\vec{J}'}{r}\right)\right\} - \nabla^2\left(\frac{\vec{J}'}{r}\right)\right] d\tau'$$

$$= \frac{\mu_0}{4\pi} \iiint_{\tau'} -\nabla^2\left(\frac{\vec{J}'}{r}\right) d\tau' + \frac{\mu_0}{4\pi}\nabla \iiint_{\tau'} \nabla \cdot \left(\frac{\vec{J}'}{r}\right) d\tau'. \tag{v}$$

Now, $$-\frac{\mu_0}{4\pi} \iiint_{\tau'} \nabla^2\left(\frac{\vec{J}'}{r}\right) d\tau' = -\frac{\mu_0}{4\pi} \iiint_{\tau'} \vec{J}'\nabla^2\left(\frac{1}{r}\right) d\tau' \quad \left[\because \vec{J}' \text{ is independent of } x, y, z\right]$$

$$= -\frac{\mu_0}{4\pi} \iiint_{\tau'} \vec{J}'[-4\pi\delta(\vec{r} - \vec{r}')d\tau' \quad \left[\because \nabla^2\left(\frac{1}{r}\right) = -4\pi\delta(\vec{r} - \vec{r}')\right]$$

$$= \mu_0 \vec{J}' \quad \left[\text{as } \iiint_{\tau'} \delta(\vec{r} - \vec{r}')d\tau' = 1\right] \tag{vi}$$

and $$\frac{\mu_0}{4\pi} \iiint_{\tau'} \nabla \cdot \left(\frac{\vec{J}'}{r}\right) d\tau' = \frac{\mu_0}{4\pi} \iiint_{\tau'} \left[\frac{1}{r}(\nabla \cdot \vec{J}') + \vec{J}' \cdot \nabla\left(\frac{1}{r}\right)\right] d\tau'$$

$$= \frac{\mu_0}{4\pi} \iiint_{\tau'} \vec{J}' \cdot \nabla\left(\frac{1}{r}\right) d\tau' \quad [\because \nabla \cdot \vec{J}' = 0]$$

$$= -\frac{\mu_0}{4\pi} \iiint_{\tau'} \vec{J}' \cdot \nabla\left(\frac{1}{r}\right) d\tau' \quad \left[\because \nabla\left(\frac{1}{r}\right) = -\nabla'\left(\frac{1}{r}\right)\right]$$

$$= -\frac{\mu_0}{4\pi} \iiint_{\tau'} \nabla' \cdot \left(\frac{\vec{J}'}{r}\right) d\tau'$$

$$= -\frac{\mu_0}{4\pi} \oint_S \left(\frac{\vec{J}'}{r}\right) \cdot d\vec{S}. \quad [\text{by Gauss' divergence theorem}] \tag{vii}$$

But, $$\oint_S \left(\frac{\vec{J}'}{r}\right) \cdot d\vec{S} = 0$$

as in case of steady current either $\vec{J}' = 0$ or, tangential at the boundary.

So, $$\frac{\mu_0}{4\pi}\nabla \iiint_{\tau'} \nabla \cdot \left(\frac{\vec{J}'}{r}\right) d\tau' = 0. \tag{viii}$$

Using the results of eqn. (vi) and (viii) in eqn. (v), we get

$$\nabla \times \vec{B} = \mu_0 \vec{J}. \quad \text{(omitting the prime)} \tag{ix}$$

Equation (xiv) is known as the *differential form of Ampere's law.*

Again, $$\iint_S (\nabla \times \vec{B}) \cdot \vec{dS} = \mu_0 \iint_S \vec{J} \cdot \vec{dS} \tag{x}$$

[Integrating eqn. (ix) over the whose cross-section (S) of the conductor].

According to Stokes' law,

$$\iint_S (\nabla \times \vec{B}) \cdot \vec{dS} = \iint_S \vec{B} \cdot \vec{dl} \tag{xi}$$

where C is the closed curve enclosing S.

From eqns. (x) and (xi) we get,

$$\oint_C \vec{B} \cdot \vec{dl} = \mu_0 \iint_S \vec{J} \cdot \vec{dS} = \mu_0 I \tag{xii}$$

where $\iint_S \vec{J} \cdot \vec{dS} = I$, the total current.

Equation (xii) is known as the *integral form of Ampere's circuital law.*

14.13 Faraday's Law of Electromagnetic Induction

Statement : *Wherever there is any change in the magnetic flux linked with a circuit, an induced electromotive force is set up in the circuit, the magnitude of the e.m.f. is proportional to the rate of change of magnetic flux and its direction is such as to oppose the very cause for which that is due.*

14.13.1 Integral Form of Faraday's Law

Let us consider a closed circuit C enclosing area S be placed in a magnetic field of flux density $\vec{B}$ (Fig. 14.9).

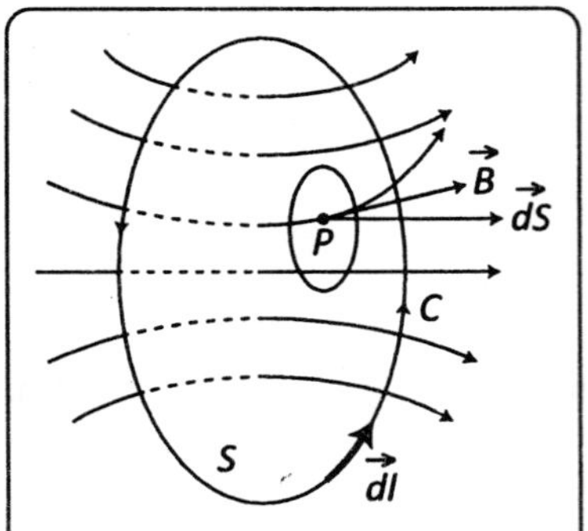

Fig. 14.9

The magnetic flux ϕ linked with the circuit is

$$\phi = \iint_S \vec{B} \cdot \vec{dS}.$$

In the magnetic flux ϕ changes through the circuit then,

$$\frac{d\phi}{dt} = \iint_S \frac{\partial \vec{B}}{\partial t} \cdot \vec{dS}$$

or according to Faradays law of electromagnetic induction, induced e.m.f. in the circuit is

$$\varepsilon = -\frac{d\phi}{dt} = -\iint_S \frac{\partial \vec{B}}{\partial t} \cdot \vec{dS}. \tag{i}$$

Equation (i) is known as the *integral form* of Faraday's law of electromagnetic induction.

14.13.2 Differential Form of Faraday's Law

We have the induced e.m.f.

$$\varepsilon = \oint \vec{E} \cdot \vec{dl}, \tag{ii}$$

where $\vec{E}$ is the electric field due to the induced e.m.f. and $\vec{dl}$ is an elementary length of the circuit. The integral $\oint \vec{E} \cdot \vec{dl}$ is the work done in taking a unit positive charge round the circuit.

From eqns. (i) and (ii), we get

$$\oint \vec{E} \cdot \vec{dl} = -\int_S \int \frac{\partial \vec{B}}{\partial t} \cdot \vec{dS}.$$

But, $\oint \vec{E} \cdot \vec{dl} = -\int_S \int (\nabla \times \vec{E}) \cdot \vec{dS}$ [by Stokes' theorem]

$$\therefore \quad \iint_S (\nabla \times \vec{E}) \cdot \vec{dS} = -\iint_S \frac{\partial \vec{B}}{\partial t} \cdot \vec{dS} \quad \text{or,} \quad \iint_S \left[(\nabla \times \vec{E}) + \frac{\partial \vec{B}}{\partial t}\right] \cdot \vec{dS} = 0.$$

Since S is arbitrary, $\nabla \times \vec{E} + \dfrac{\partial \vec{B}}{\partial t} = 0$

$$\text{or,} \quad \boldsymbol{\nabla \times \vec{E} = -\frac{\partial \vec{B}}{\partial t}}. \tag{iii}$$

Equation (iii) is known as the *differential form* of Faraday's law of electromagnetic induction.

Equation (iii) shows that $\nabla \times \vec{E} \neq 0$, i.e., $\vec{E} \neq -\nabla \phi$ i.e., $\vec{E}$ is not conservative. In this case, $\vec{E}$ is produced due to changing magnetic field.

But for an electric field due to stationary charge, $\nabla \times \vec{E} = 0$ and $\vec{E} = -\nabla \phi$. In this case, $\vec{E}$ is conservative.

Thus, **there is a fundamental difference between the electric fields produced due to changing magnetic field and due to stationary charge.**

14.14 Coefficient of Self-induction (or, Self-inductance of a Coil)

The coefficient of self-induction of a coil is numerically equal to the induced e.m.f. in the coil due to unit rate of change of current in it.

To find out an expression for the self-inductance of a coil C, we consider two elementary lengths $\vec{dl_1}$ and $\vec{dl_2}$ separated by a vector distance $\vec{r}$, placed in a medium of magnetic permeability μ_0 (Fig. 14.10).

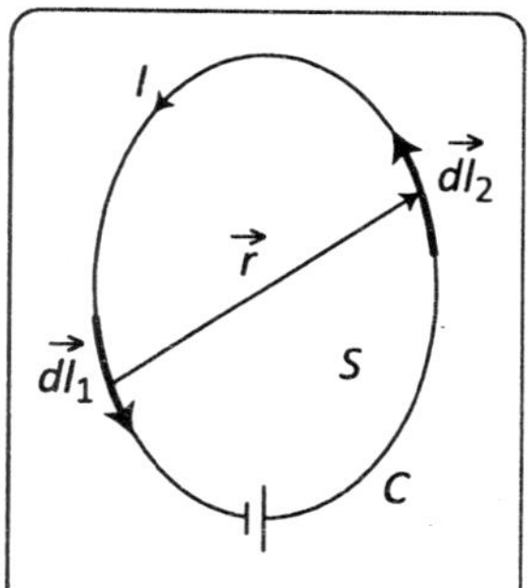

Fig. 14.10

According to **Biot-Savart's law**, the magnetic field produced by a current element $I\vec{dl_1}$ at a point on $I\vec{dl_2}$ is

$$d\vec{B} = \frac{\mu_0}{4\pi} \frac{I\vec{dl_1} \times \hat{r}}{r^2}.$$

So, the magnetic field $\vec{B}$ at $I\, dl_2$ due to the entire coil C is

$$\vec{B} = \frac{\mu_0}{4\pi} \oint_C \frac{I\vec{dl_1} \times \hat{r}}{r^2} = \frac{\mu_0}{4\pi} \oint_C \frac{I\vec{dl_1} \times \vec{r}}{r^3} \quad \left[\text{where } \hat{r} = \frac{\vec{r}}{r}\right]$$

But, $\frac{\vec{r}}{r^3} = -\nabla\left(\frac{1}{r}\right)$.

$$\therefore \quad \vec{B} = -\frac{\mu_0}{4\pi}\oint_C I\vec{dl_1} \times \nabla\left(\frac{1}{r}\right) = \frac{\mu_0}{4\pi}\oint_C \nabla\left(\frac{1}{r}\right) \times I\vec{dl_1}.$$

Again, $\nabla \times \frac{I\vec{dl_1}}{r} = \nabla\left(\frac{1}{r}\right) \times I\vec{dl_1} + \frac{1}{r}\nabla \times I\vec{dl_1}$

$$= \nabla\left(\frac{1}{r}\right) \times I\vec{dl_1}. \quad [\because\ \nabla \times I\vec{dl_1} = 0 \text{ as } I\vec{dl_1} \text{ is not a function of } (x, y, z)]$$

$$\therefore \quad \vec{B} = \frac{\mu_0}{4\pi}\oint_C \nabla \times \frac{I\vec{dl_1}}{r}. \tag{iv}$$

The magnetic flux linked with the coil C is

$$\phi = \iint_S \vec{B} \cdot \vec{dS} = \iint_S \vec{dS} \cdot \vec{B} = \iint_S \vec{dS} \cdot \frac{\mu_0}{4\pi}\oint_C \nabla \times \left(\frac{I\vec{dl_1}}{r}\right)$$

$$= \frac{\mu_0}{4\pi}\oint_C \iint_S \nabla \times \left(\frac{I\vec{dl_1}}{r}\right) \cdot \vec{dS}.$$

But according to **Stokes' theorem,**

$$\iint_S \nabla \times \left(\frac{I\vec{dl_1}}{r}\right) \cdot \vec{dS} = \oint_C \frac{I\vec{dl_1}}{r} \cdot \vec{dl_2}.$$

$$\therefore \quad \phi = \frac{\mu_0}{4\pi}\oint_C \oint_C \frac{I\vec{dl_1}}{r} \cdot \vec{dl_2}.$$

According to **Faraday's Law of electromagnetic induction**, the induced e.m.f. is

$$\varepsilon = -\frac{d\phi}{dt} = -\frac{\mu_0}{4\pi}\oint_C \oint_C \frac{dI}{dt}\frac{\vec{dl_1} \cdot \vec{dl_2}}{r}. \tag{v}$$

If L be the self-inductance of the coil, then

$$\varepsilon = -L\frac{dI}{dt}. \tag{vi}$$

Comparing eqns. (v) and (vi), we get

$$L = \frac{\mu_0}{4\pi}\oint_C \oint_C \frac{\vec{dl_1} \cdot \vec{dl_2}}{r}. \tag{vii}$$

Eqn. (vii) is known as *Neumann's formula.*

14.15 Reciprocity Theorem of Mutual Inductance

If there be two magnetically coupled circuits C_1 and C_2 carrying currents I_1 and I_2 respectively, then the mutual inductance (M_{12}) of circuit C_1 w.r.t. C_2 is equal to the mutual inductance (M_{21}) of circuit C_2 w.r.t. C_1, i.e., $M_{12} = M_{21}$.

If $\vec{A}$ be the magnetic vector potential related to magnetic induction $\vec{B}$, then $\vec{B} = \nabla \times \vec{A}$.

But according to Stokes' law,

$$\oint \vec{A}\cdot\vec{dl} = \iint_S (\nabla\times\vec{A})\cdot\vec{dS}.$$

$$\text{Flux, } \phi = \iint_S \vec{B}\cdot\vec{dS} = \int_C \vec{A}\cdot\vec{dl}. \qquad \text{(viii)}$$

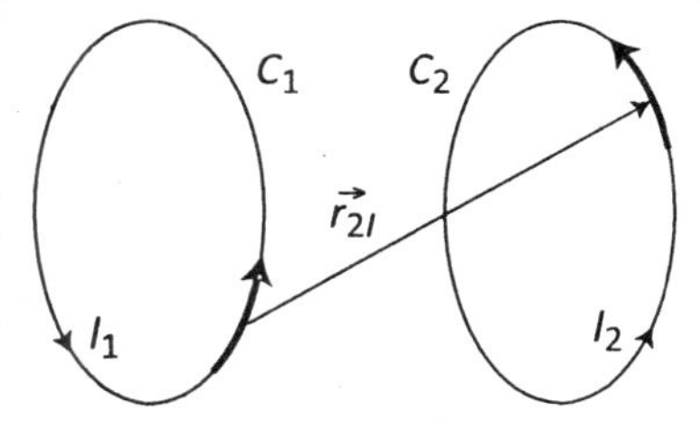

Fig. 14.11

Thus the line integral of the vector potential $\vec{A}$ around a closed loop C is equal to the flux ϕ through the surface S enclosed by C.

Now, the magnetic vector potential,

$$\vec{A} = \frac{\mu_0}{4\pi}\iiint_\tau \frac{\vec{J}\,d\tau}{r},$$

where $\vec{J}$ is the current density and $d\tau$ is the volume element. $\vec{r}$ is the vector distance of the point of observation from the current element.

If α be the cross-sectional area of the conductors, $\vec{dl}$ is the elemental length and I, the current flowing through it, then $I = \vec{J}\cdot\vec{\alpha}$ and $d\tau = \alpha\cdot\vec{dl}$.

$$\therefore \quad I\vec{\alpha}\cdot\vec{dl} = \text{[illegible]}\cdot\vec{dl} = \vec{J}\cdot\vec{\alpha}\,d\tau \quad \text{or,} \quad I\vec{dl}\cdot\vec{\alpha} = \vec{J}\,d\tau\cdot\vec{\alpha} \quad \text{or,} \quad I\vec{dl} = \vec{J}\,d\tau.$$

$$\therefore \quad \vec{A} = \frac{\mu_0}{4\pi}\oint_C \frac{I\vec{dl}}{r} = \frac{\mu_0 I}{4\pi}\oint_C \frac{\vec{dl}}{r}. \qquad \text{(ix)}$$

Now, the vector potential $\overrightarrow{A_{21}}$ at a point P due to current I_1 in C_1 is

$$\overrightarrow{A_{21}} = \frac{\mu_0}{4\pi}I_1\oint_{C_1}\frac{\overrightarrow{dl_1}}{r_{21}}, \qquad \text{(x)}$$

where r_{21} is the magnitude of the distance from $\overrightarrow{dl_1}$ to P.
Since, P is on the boundary of C_2, the flux through C_2 due to I_1 in C_1 is

$$\phi_{21} = \oint_{C_2}\overrightarrow{A_{21}}\cdot\overrightarrow{dl_2}\oint_{C_2}\overrightarrow{dl_2}\cdot\overrightarrow{A_{21}} = \oint_{C_2}\overrightarrow{dl_2}\cdot\frac{\mu_0 I_1}{4\pi}\oint_{C_1}\frac{\overrightarrow{dl_1}}{r_{21}}.$$

If I_1 changes at the rate $\frac{dI_1}{dt}$, then ϕ_{21} will also change,

$$\therefore \quad \frac{d\phi_{21}}{dt} = \frac{\mu_0}{4\pi}\frac{dI_1}{dt}\oint_{C_2}\overrightarrow{dl_2}\cdot\oint_{C_1}\frac{\overrightarrow{dl_1}}{r_{21}}. \qquad \text{(xi)}$$

So, the induced e.m.f. ε_{21} in C_2 is

$$\varepsilon_{21} = -\frac{d\phi_{21}}{dt} = -M_{21}\frac{dI_1}{dt}, \qquad \text{(xii)}$$

where M_{21} is the coefficient of mutual induction of coil C_2 w.r.t. C_1.

$$\text{Also,} \quad \varepsilon_{21} = -\frac{d\phi_{21}}{dt} = -\frac{\mu_0}{4\pi}\frac{dI_1}{dt}\oint_{C_2}\overrightarrow{dl_2}\cdot\oint_{C_1}\frac{\overrightarrow{dl_1}}{r_{21}} = -M_{21}\frac{dI_1}{dt},$$

$$\text{where } M_{21} = \frac{\mu_0}{4\pi}\oint_{C_2}\overrightarrow{dl_2}\cdot\oint_{C_1}\frac{\overrightarrow{dl_1}}{r_{21}} = \frac{\mu_0}{4\pi}\oint_{C_1}\oint_{C_2}\frac{\overrightarrow{dl_2}\cdot\overrightarrow{dl_1}}{r_{21}}. \qquad \text{(xiii)}$$

Similarly, current I_2 flowing through C_2 produces magnetic flux ϕ_{12} through the coil C_1, i.e.,

$$\phi_{12} = \oint_{C_1} \overrightarrow{dl_1} \cdot \frac{\mu_0}{4\pi} I_2 \oint_{C_1} \frac{\overrightarrow{dl_2}}{r_{12}}$$

and proceeding as above it can be show that,

$$M_{12} = \frac{\mu_0}{4\pi} \oint_{C_1} \oint_{C_2} \frac{\overrightarrow{dl_1} \cdot \overrightarrow{dl_2}}{r_{12}}, \qquad \text{(xiv)}$$

where M_{12} is the coefficient of mutual induction of coil C_1 w.r.t. coil C_2.

Since $r_{12} = r_{21}$, $\therefore$ $\dfrac{\mu_0}{4\pi} \oint_{C_1} \oint_{C_2} \dfrac{\overrightarrow{dl_2} \cdot \overrightarrow{dl_1}}{r_{21}} = \dfrac{\mu_0}{4\pi} \oint_{C_1} \oint_{C_2} \dfrac{\overrightarrow{dl_1} \cdot \overrightarrow{dl_2}}{r_{12}}$.

$$\therefore \quad M_{12} = M_{21} \quad [\because \overrightarrow{dl_1} \cdot \overrightarrow{dl_2} = \overrightarrow{dl_2} \cdot \overrightarrow{dl_1}] \qquad \text{(xv)}$$

Thus the *reciprocity theorem* of mutual inductance is proved.

Maxwell's Equations and Electromagnetic Waves

14.16 Displacement Current

Ampere's circuital law in most general form is given by

$$\oint_C \vec{B} \cdot \vec{dl} = \mu_0 I = \mu_0 \iint_S \vec{J} \cdot \vec{dS}, \qquad \text{(i)}$$

where $\vec{B}$ is the magnetic induction vector due to a current I in a conductor, C is a closed path linking current I (Fig. 14.12). $\vec{J}$ is the current density and S is the area enclosed by C.

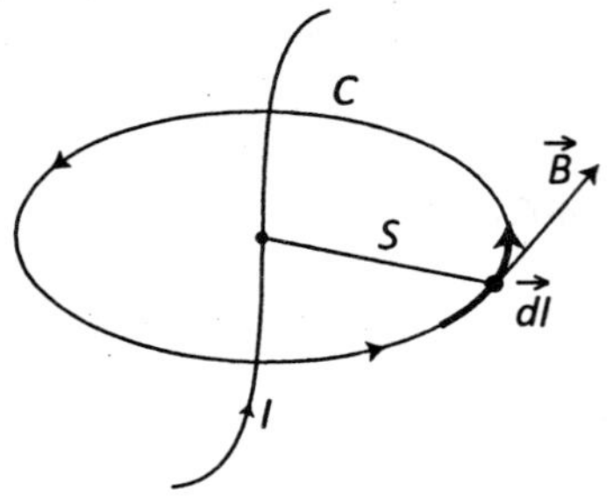

Fig. 14.12

By Stokes' law,

$$\oint_C \vec{B} \cdot \vec{dl} = \iint_S (\nabla \times B) \cdot \vec{dS} \qquad \text{(ii)}$$

$$\therefore \quad \iint_S (\nabla \times B) \cdot \vec{dS} = \mu_0 \iint_S \vec{J} \cdot \vec{dS}$$

$$\text{or,} \quad \nabla \times \vec{B} = \mu_0 \vec{J}. \qquad \text{(iii)}$$

Equation (iii) is true in case of steady fields.

To examine whether eqn. (iii) is true in case of time varying fields, we take divergence on both sides of the eqn. (iii),

$\therefore \quad \nabla \cdot (\nabla \times \vec{B}) = \mu_0 (\nabla \cdot \vec{J})$. Since $\nabla \cdot (\nabla \times \vec{B}) = 0$ and $\mu_0 \neq 0$,

$$\therefore \quad \nabla \cdot \vec{J} = 0. \qquad \text{(iv)}$$

The equation of continuity is

$$\nabla \cdot \vec{J} + \frac{\partial \rho}{\partial t} = 0, \quad \text{where } \rho \text{ is the charge density}$$

$$\text{or,} \quad \frac{\partial \rho}{\partial t} = 0, \quad \text{i.e., } \rho = \text{constant of time.}$$

This means **Ampere's circuital law** is valid in case where charge density is static, i.e., steady state condition of charge flow, so that ρ = constant. Thus, Ampere's circuital law as stated above does not hold good in case where $\frac{\partial \rho}{\partial t} \neq 0$, i.e., in case of time dependent fields.

This led Maxwell to assume that eqn. (i) is not complete and something is to be added to it.

Let the quantity to be added to the right-hand side of eqn. (iv) be $\vec{J_d}$, then

$$\nabla \times \vec{B} = \mu_0(\vec{J} + \vec{J_d}). \quad \text{(v)}$$

To investigate the physical significance of $\vec{J_d}$, we calculate the divergence of eqn. (v),

$$\nabla \cdot (\nabla \times \vec{B}) = \nabla \cdot \mu_0(\vec{J} + \vec{J_d}).$$

But, $\nabla \cdot (\nabla \times \vec{B}) = 0,$

$$\therefore \quad \mu_0(\nabla \cdot \vec{J} + \nabla \cdot \vec{J_d}) = 0$$

or, $\nabla \cdot \vec{J_d} = -\nabla \cdot \vec{J} \quad [\because \mu_0 \neq 0]$

$$= \frac{\partial \rho}{\partial t} \quad \left[\because \nabla \cdot \vec{J} + \frac{\partial \rho}{\partial t} = 0\right]$$

$= \frac{\partial}{\partial t}(\nabla \cdot \vec{D})$, as $\nabla \cdot \vec{D} = \rho$, $\vec{D}$ being the *electric displacement vector.*

or, $\nabla \cdot \vec{J_d} - \frac{\partial}{\partial t}(\nabla \cdot \vec{D}) = 0$ or, $\nabla \cdot \left[\vec{J_d} - \frac{\partial \vec{D}}{\partial t}\right] = 0.$

The above holds for any arbitrary volume,

$$\therefore \quad \vec{J_d} = \frac{\partial \vec{D}}{\partial t}. \quad \text{(vi)}$$

Hence, the *modified* form of Ampere's circuital law takes the form,

$$\nabla \times \vec{B} = \mu_0 \left(J + \frac{\partial \vec{D}}{\partial t}\right). \quad \text{(vii)}$$

This term $\frac{\partial \vec{D}}{\partial t}$ is known as the displacement current density $(\vec{J_d})$, which is different from *charge transported current density* $(\vec{J})$.

14.16.1 Displacement Current

Let a parallel plate capacitor be connected to a d.c. voltage through a resistance R as shown in Fig. 14.13. Then plates A and K of the capacitor will be charged with positive and negative charges. If the voltage source is removed, then the capacitor will discharge through R and a varying current will flow through the circuit, when key K_1 is opened and K_2 is closed. During charging the capacitor an electric field $\vec{E}$ is established between the plates whose magnitude is

$$E = \frac{\sigma}{\epsilon_0} \quad \text{or,} \quad E = \frac{1}{\epsilon_0}\frac{q}{\alpha},$$

where σ is the surface density of charge on the plates of area α. During discharge, a current (I) flows through R given by

$$I = \frac{dq}{dt} = \epsilon_0 \alpha \frac{dE}{dt} \quad [\because q = \epsilon_0 \propto E]$$

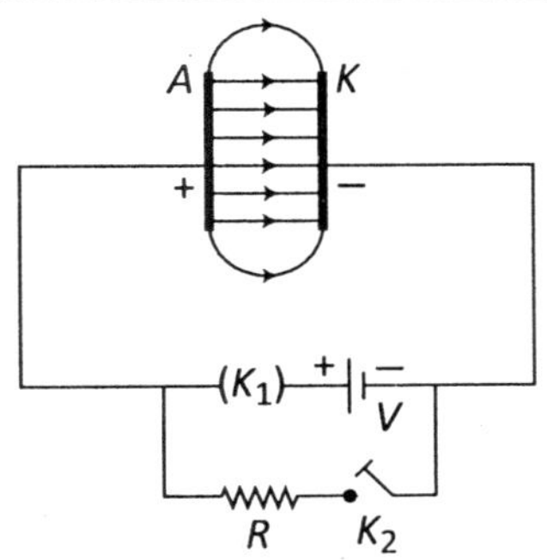

Fig. 14.13

$$\therefore \quad J_d = \frac{I}{\alpha} = \epsilon_0 \frac{dE}{dt} = \frac{dD}{dt} \quad [\because D = \epsilon_0 E]. \tag{viii}$$

Equation (viii) shows that, when the capacitor discharges through R, a varying current flows from positive to negative plate giving rise to the displacement current J_d. Hence, *a changing electric field gives rise to a displacement current which produces a magnetic field* $\vec{B}$ *in the same way as a charged transported current does.*

As no charge flows through the plates of the capacitor, the conduction current is itself not continuous through the gap of the capacitor plates. The continuity of current through the circuit, is however, maintained by the **displacement current**.

14.17 Maxwell's Electromagnetic Equations

The following four fundamental equations concerning electromagnetic theory are known as Maxwell's differential form of electromagnetic equations :

1. $\nabla \cdot \vec{D} = \rho$ [Differential form of Gauss' law in electrostatics]
2. $\nabla \cdot \vec{B} = 0$ [Differential form of Gauss' law in magnetostatics]
3. $\nabla \times \vec{E} = -\frac{\partial \vec{B}}{\partial t}$ [Differential form of Faradays' law of electromagnetic induction]
4. $\nabla \times \vec{B} = \mu(\vec{J} + \vec{J_d})$ [Maxwell's modification of Ampere's circuital law]

where $\vec{D}$ = electric displacement vector in coulomb/m^3, ρ = charge density in coulomb/m^3, $\vec{B}$ = magnetic induction in Weber/m^2, $\vec{E}$ = electric field intensity in volt/coulomb, μ, ϵ = the magnetic permeability and electric permittivity of the medium respectively.

In *charge-free space*, $\mu = \mu_0$, $\epsilon = \epsilon_0$, $\rho = 0$ and $\vec{J} = 0$. Therefore, the Maxwell's equations in *free space* are as follows :

1. $\nabla \cdot D = 0$
2. $\nabla \cdot B = 0$
3. $\nabla \times \vec{E} = -\frac{\partial \vec{B}}{\partial t}$
4. $\nabla \times \vec{B} = \mu_0 \frac{\partial \vec{D}}{\partial t}$.

14.17.1 Integral Form of Maxwell's Equations

(1) Integrating Maxwell's first equation $\nabla \cdot \vec{D} = \rho$ over an arbitrary volume V, we get

$$\iiint_V (\nabla \cdot \vec{D}) dV = \iiint_V \rho dV$$

or,
$$\iint_S \vec{D} \cdot d\vec{S} = q, \tag{i}$$

where S is the surface which bounds volume V and $\iiint_V dV = q$, the total charge in V.

The above equation is the *integral form of Gauss' law in electrostatics.*

(2) Integrating Maxwell's second equation $\nabla \cdot \vec{B} = 0$ over volume V, we get $\iiint_V \nabla \cdot \vec{B} = 0$.

By Gauss' divergence theorem,

$$\iint_S \vec{B} \cdot \vec{dS} = 0, \tag{ii}$$

where S is the surface enclosing volume V. Equation (ii) is the *integral form of Gauss' divergence theorem in magnetostatics.*

Thus, the net magnetic flux through a closed surface is zero, i.e., the magnetic field lines cannot begin or end at any point. Hence, *isolated magnetic monopoles do not exist.*

N.B. Recently a group of physicists of American Institute of Physics and the University of California jointly claimed that they have observed magnetic monopoles. If the reported observation be confirmed, then it must have a major impact on existing physics as well as technology.

(3) Integrating the third equation of Maxwell over a surface S bounded by a closed curve C,

$$\iint_S (\nabla \times \vec{E}) \cdot \vec{dS} = \iint_S \frac{\partial \vec{B}}{\partial t} \cdot \vec{dS}$$

$$\text{or,} \quad \oint_C \vec{E} \cdot \vec{dl} = \frac{\partial}{\partial t} \iint_S \vec{B} \cdot \vec{dS} \quad \text{[by Stokes' theorem]}$$

$$\text{or,} \quad \oint_C \vec{E} \cdot \vec{dl} = \frac{\partial \phi}{\partial t} \quad \left[\because \phi = \iint_S \vec{B} \cdot \vec{dS}\right]$$

This is the *integral form of Faraday's law of electromagnetic induction.*

(4) Maxwell's fourth equations is

$$\nabla \times \vec{B} = \mu(\vec{J} + \vec{J_D}) = \mu \left(\vec{J} + \frac{\partial \vec{D}}{\partial t}\right).$$

Integrating over the surface S bounded by the closed curve C, we get,

$$\iint_S (\nabla \times \vec{B}) \cdot \vec{dS} = \mu \iint_S \left(\vec{J} + \frac{\partial \vec{D}}{\partial t}\right) \cdot \vec{dS}$$

$$\text{or,} \quad \oint_C \vec{B} \cdot \vec{dl} = \mu \iint_S \left(\vec{J} + \frac{\partial \vec{D}}{\partial t}\right) \cdot \vec{dS}. \tag{iii}$$

Equation (iii) is the *Maxwell's modification of Ampere's circuital law.* Equation (iii) can also be written as

$$\oint_C \vec{H} \cdot \vec{dl} = \iint_S \left(\vec{J} + \frac{\partial \vec{D}}{\partial t}\right) \cdot \vec{dS}, \quad \text{where } \vec{B} = \mu \vec{H}. \tag{iv}$$

Equation (iv) shows that the **magnetomotive force** $\left(\oint_C \vec{H} \cdot \vec{dl}\right)$ around a closed path is equal to the conduction current plus displacement current through any surface bounded by the path.

14.18 Derivations of Maxwell's Equations

(a) *Derivation of* $\nabla \cdot \vec{D} = \rho$

We consider a dielectric of volume V enclosed by a surface S containing no net charge. If now the dielectric be placed in an electric field, then it will be polarised and contain polarised charge of density ρ'. If we place some real charge of density ρ, then by Gauss' law in electrostatics,

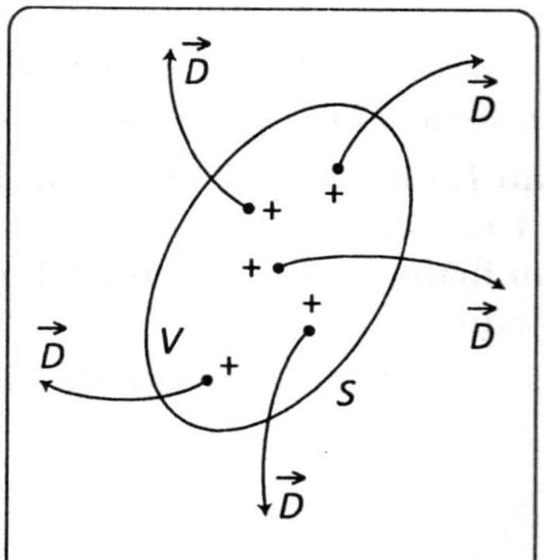

Fig. 14.14

$$\oint_S \vec{E} \cdot d\vec{S} = \frac{1}{\epsilon_0} \int\int_V \int (\rho + \rho') dV$$

or, $$\epsilon_0 \iiint_V (\nabla \cdot \vec{E}) dV = \iiint_V \rho dV - \iiint_V \nabla \cdot \vec{P} dV$$

$[\because \rho' = -\nabla \cdot \vec{P}, \vec{P}$ being the polarisation vector$]$

or, $$\epsilon_0 \iiint_V (\nabla \cdot \vec{E}) dV + \iiint_V (\nabla \cdot \vec{P}) dV = \iiint_V \rho dV$$

or, $$\iiint_V \nabla \cdot (\epsilon_0 \vec{E} + \vec{P}) dV = \iiint_V \rho dV$$

or, $$\iiint_V \nabla \cdot \vec{D} = \iiint_V \rho dV \quad [\because \vec{D} = \epsilon_0 \vec{E} + \vec{P}]$$

or, $$\iiint_V (\nabla \cdot \vec{D} - \rho) dV = 0$$

Since this is true for all volumes, $\nabla \cdot \vec{D} = \rho$. (i)

(b) *Derivation of* $\nabla \cdot \vec{B} = 0$

The experimental finding that magnetic monopole does not exist, leads to the fact that magnetic lines of force are closed group or, extend up to infinity. So the number of magnetic lines of force entering or leaving a closed surface are equal, i.e.,

$$\oint_S \vec{B} \cdot d\vec{S} = 0$$

or, $$\iiint_V \nabla \cdot \vec{B} = 0 \quad \text{[by Gauss' divergence theorem]}$$

Since the volume is arbitrary, $\nabla \cdot \vec{B} = 0$ (ii)

The is *Maxwell's second electromagnetic equation.*

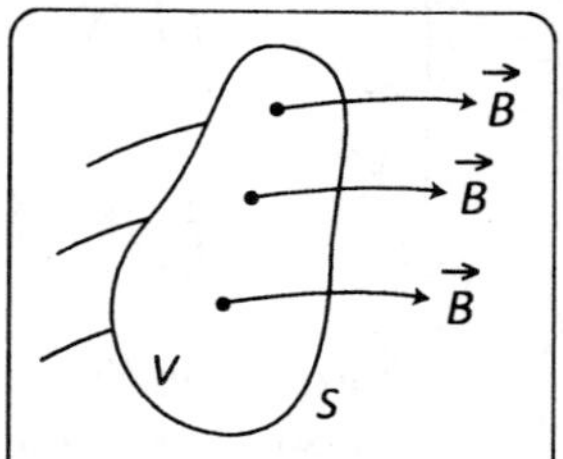

Fig. 14.15

(c) *Derivation of* $\nabla \times \vec{E} = -\dfrac{\partial \vec{B}}{\partial t}$

According to Faraday's law of electromagnetic induction, the induced e.m.f. ε is given by

$$\varepsilon = -\frac{\partial \phi}{\partial t}. \quad \text{[where } \phi \text{ is the magnetic flux linking the coil]}.$$

If $\vec{E}$ be the electric field intensity at a point, then the work done in moving a unit charge through a small distance $\vec{dl}$ is $\vec{E}\cdot\vec{dl}$. The total work done in moving the unit charge around a closed path C is

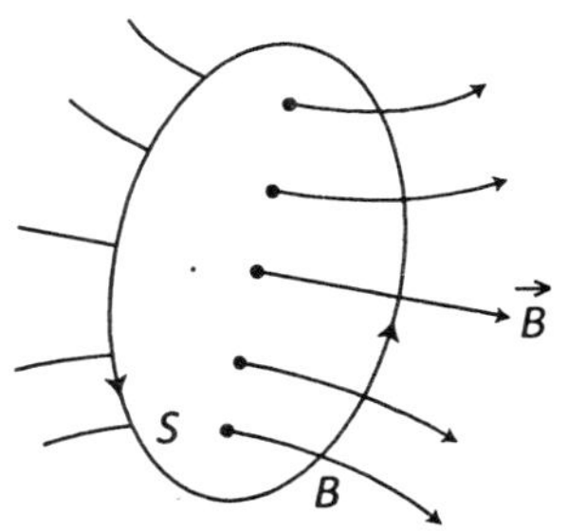

Fig. 14.16

$$\oint_C \vec{E}\cdot\vec{dl}.$$

Since, this is equal to the induced e.m.f.,

$$\therefore \quad \varepsilon = \oint_C \vec{E}\cdot\vec{dl}.$$

So, $\displaystyle\oint_C \vec{E}\cdot\vec{dl} = -\frac{\partial}{\partial t}\iint_S \vec{B}\cdot\vec{dS}$

or, $\displaystyle\iint_S\left(\nabla\times\vec{E}+\frac{\partial\vec{B}}{\partial t}\right)\cdot d\vec{S}=0 \quad \left[\because \oint\vec{E}\cdot\vec{dl}=\iint_S(\nabla\times\vec{E})\cdot\vec{dS}\right]$

As the surface S is arbitrary,

$$\nabla\times\vec{E}+\frac{\partial\vec{B}}{\partial t}=0$$

or, $$\nabla\times\vec{E}=-\frac{\partial\vec{B}}{\partial t}=0 \qquad \text{(iii)}$$

Equation (iii) represents *Maxwell's third electromagnetic equation connecting* $\vec{E}$ *and* $\vec{B}$.

(d) *Derivation of* $\nabla\times\vec{B}=\mu\left(\vec{J}+\dfrac{\partial\vec{D}}{\partial t}\right)$

If $\vec{B}$ be the magnetic field linked with the current I in a conductor, then according to Ampere's circuital law,

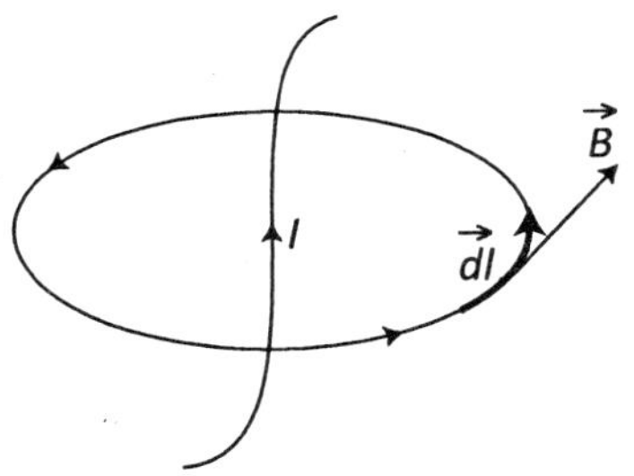

Fig. 14.17

$$\oint_C \vec{B}\cdot\vec{dl}=I=\iint_S \vec{J}\cdot\vec{dS},$$

where $\vec{J}$ is the current density in the conductor of cross-sectional area S. By Stokes' law,

$$\iint_S(\nabla\times\vec{B})\cdot\vec{dS}=\iint_S\vec{J}\cdot\vec{dS}.$$

Since S is arbitrary,

$$\nabla\times\vec{B}=\vec{J}. \qquad \text{(iv)}$$

For time varying electric field, Maxwell showed that equation (iv) is incomplete. To make it complete he assumed that a quantity $\vec{J_D}=\frac{\partial\vec{D}}{\partial t}$ is to be added to the right-hand side of equation (iv) so that,

$$\nabla\times\vec{B}=\vec{J}+\vec{J_D}. \qquad \text{(v)}$$

This is *Maxwell's fourth electromagnetic equation.*

14.19 Plane Electromagnetic Wave in Free Space

Maxwell's electromagnetic equations in free space (i.e., $\rho = 0,\ \sigma = 0,\ \mu = \mu_0,\ \epsilon = \epsilon_0$) are as follows :

1. $\nabla \cdot \vec{E} = 0$ (i)
2. $\nabla \cdot \vec{H} = 0$ (ii)
3. $\nabla \times \vec{E} = -\mu_0 \dfrac{\partial \vec{H}}{\partial t}$ (iii)
4. $\nabla \times \vec{H} = -\epsilon_0 \dfrac{\partial \vec{E}}{\partial t}$. (iv)

Now taking curl on both sides of equation (iii), we get

$$\nabla \times (\nabla \times \vec{E}) = -\mu_0 \frac{\partial}{\partial t}(\nabla \times \vec{H}) = -\mu_0\epsilon_0 \frac{\partial^2 \vec{E}}{\partial t} \quad \text{[using eqn. (iv)]}$$

or, $$\nabla(\nabla \cdot \vec{E}) - \nabla^2 \vec{E} = -\mu_0\epsilon_0 \frac{\partial^2 \vec{E}}{\partial t^2}$$

or, $$\nabla^2 \vec{E} - \mu_0\epsilon_0 \frac{\partial^2 \vec{E}}{\partial t^2} \quad [\because\ \nabla \cdot \vec{E} = 0]$$

or, $$\nabla^2 \vec{E} - \frac{1}{c^2}\frac{\partial^2 \vec{E}}{\partial t^2} = 0 \quad \left[\text{where } \mu_0\epsilon_0 = \frac{1}{c^2}\right] \tag{v}$$

Again, taking curl on both sides of eqn. (iv), we get

$$\nabla \times \nabla \times \vec{H} = \epsilon_0 \frac{\partial}{\partial t}(\nabla \times \vec{E}) = -\mu_0\epsilon_0 \frac{\partial^2 \vec{H}}{\partial t^2} \quad \left[\because\ \nabla \times \vec{E} = -\mu_0 \frac{\partial \vec{H}}{\partial t}\right]$$

or, $$\nabla(\nabla \cdot \vec{H}) - \nabla^2 \vec{H} = -\mu_0\epsilon_0 \frac{\partial^2 \vec{H}}{\partial t^2}$$

or, $$\nabla^2 \vec{H} - \frac{1}{c^2}\frac{\partial^2 \vec{H}}{\partial t^2} = 0 \quad [\because\ \nabla(\nabla \cdot \vec{H}) = 0]. \tag{vi}$$

The vector eqns. (v) and (vi) are of identical form. So each of the six components of $\vec{E}$ and $\vec{H}$ separately satisfies the same scalar equation of the form,

$$\nabla^2 \psi - \frac{1}{v^2}\frac{\partial^2 \psi}{\partial t^2} = 0. \tag{vii}$$

This is the three-dimensional wave equation without damping and travelling with velocity v. Hence, $\vec{E}$ and $\vec{H}$ vectors propagate in free space as waves with velocity c.

Now, $$c = \frac{1}{\sqrt{\mu_0\epsilon_0}} = \sqrt{\frac{4\pi}{\mu_0 4\pi\epsilon_0}} = \sqrt{\frac{4\pi}{4\pi \times 10^{-7}} \times 9 \times 10^9},$$

where $\mu_0 = 4\pi \times 10^{-7}$, $\epsilon_0 = 8.8542 \times 10^{-12}$ farad/m.

But, $\dfrac{\mu_0}{4\pi} = 10^{-7}$ weber/A-m and $\dfrac{1}{4\pi\epsilon_0} = 9 \times 10^9$ m/farad.

or, $c = 3 \times 10^8$ m/s which is the velocity of light in free space.

14.20 Plane Wave Solution

A plane wave is that whose amplitude of vibration is the same at any point in a plane perpendicular to the specified direction.

Now, since the plane progressive solution of eqn. (vii) is

$$\psi = \psi_0 e^{-i(\omega t - k\vec{r})} \quad (i = \sqrt{-1}), \tag{viii}$$

the solutions of eqns. (v) and (vii) will be of the form

$$\vec{E} = \vec{E_0} e^{-i(\omega t - \vec{k}\cdot\vec{r})}$$

$$\vec{H} = \vec{H_0} e^{-i(\omega t - \vec{k}\cdot\vec{r})}, \tag{ix}$$

where $\vec{E_0}$ and $\vec{H}$ are the complex amplitudes which are constant in space and time and $\vec{k}$ is the propagation vector given by

$$\vec{k} = k\hat{n} = \frac{2\pi}{\lambda}\hat{n} = \frac{2\pi\nu}{c}\hat{n} = \frac{\omega}{c}\hat{n}, \tag{x}$$

$\hat{n}$ being a unit vector in the direction of propagation of the wave.

Now, $$\nabla\cdot\vec{E} = \left(\hat{i}\frac{\partial}{\partial x} + \hat{j}\frac{\partial}{\partial y} + \hat{k}\frac{\partial}{\partial z}\right)\vec{E_0}e^{i\vec{k}\cdot\vec{r} - i\omega t}$$

$$= \left(\hat{i}\frac{\partial}{\partial x} + \hat{j}\frac{\partial}{\partial y} + \hat{k}\frac{\partial}{\partial z}\right)\cdot[\hat{i}E_0x + \hat{j}E_0y + \hat{k}E_0z) \times e^{i(k_xx + k_yy + k_zz) - i\omega t}].$$

$\because$ $$\vec{k}\cdot\vec{r} = (\hat{i}k_x + \hat{j}k_y + \hat{k}k_z)\cdot[(\hat{i}x + \hat{j}y + \hat{k}z)].$$

$\therefore$ $$\nabla\cdot\vec{E} = (E_{0x}ik_x + E_{0y}ik_y + E_{0z}ik_z)e^{i\vec{k}\cdot\vec{r} - i\omega t}$$

$$= i(\hat{i}k_x + \hat{j}k_y + \hat{k}k_z)\cdot(\hat{i}E_{0x} + \hat{j}E_{0y} + \hat{k}E_{0z})e^{i\vec{k}\cdot\vec{r} - i\omega t} = i\vec{k}\cdot\vec{E_0}e^{i(\vec{k}\cdot\vec{r} - \omega t)}$$

$$\nabla\cdot\vec{E} = i\vec{k}\cdot\vec{E}. \tag{x}$$

Similarly, we can show that,

$$\nabla\cdot\vec{H} = i\vec{k}\cdot\vec{H}. \tag{xi}$$

From equations (x) and (xi) we see that ∇ is equivalent to $i\vec{k}$.

Also, $$\frac{\partial\vec{E}}{\partial t} = \frac{\partial}{\partial t}[\vec{E_0}e^{-i(\omega t - \vec{k}\cdot\vec{r})}] = -i\omega\vec{E_0}e^{i(\vec{k}\cdot\vec{r} - \omega t)}$$

or, $$\frac{\partial\vec{E}}{\partial t} = -i\omega\vec{E}. \tag{xii}$$

Equation (xii) shows that $\frac{\partial}{\partial t}$ is equivalent to $(-i\omega)$. Thus, Maxwell's equations in free space in terms of the operators (ik) and $(-i\omega)$ can be written as

$$\left.\begin{aligned} \vec{k}\cdot\vec{E} &= 0 && \text{(a)}\\ \vec{k}\cdot\vec{H} &= 0 && \text{(b)}\\ \vec{k}\times\vec{E} &= \omega\mu_0\vec{H} && \text{(c)}\\ -\vec{k}\times\vec{H} &= \omega\epsilon_0\vec{E} && \text{(d)} \end{aligned}\right\} \tag{xiii}$$

From eqns. xiii(a) and xiii(b), we note that $\vec{E}$ is perpendicular to $\vec{k}$ and $\vec{H}$ is also perpendicular to $\vec{k}$. Hence, both of the wave vectors $\vec{E}$ and $\vec{H}$ are perpendicular to $\vec{k}$. In other words **electromagnetic waves are transverse in nature.**

Again, eqns. xiii(c) and xiii(d) show that $\vec{H}$ is perpendicular to both $\vec{E}$, $\vec{k}$ and $\vec{E}$ is perpendicular to both $\vec{H}$, $\vec{k}$. Thus, **in a plane electromagnetic wave vectors $\vec{E}$, $\vec{H}$ and $\vec{k}$ are all orthogonal as shown in Fig. (14.18).**

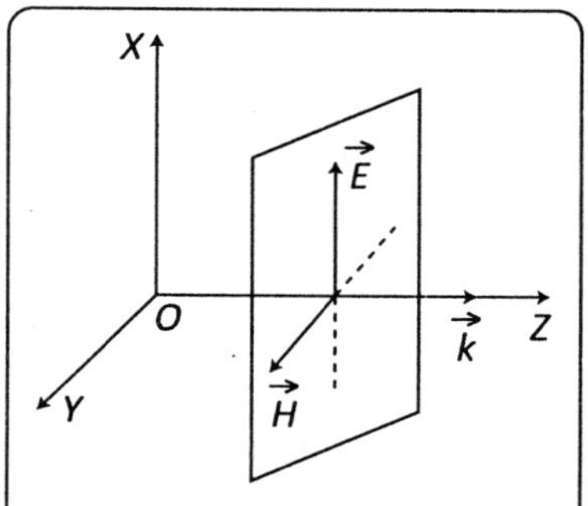

Fig. 14.18

Again since, $\vec{k} \times \vec{E} = \omega\mu_0\vec{H}$

$$\therefore \quad \vec{H} = \frac{k}{\omega\mu_0}(\hat{n} \times \vec{E}) \quad [\therefore \ \vec{k} = \hat{n}k]$$

$$= \frac{\hat{n} \times \vec{E}}{c\mu_0} = c\epsilon_0(\hat{n} \times \vec{E}) \quad \left[\because \ k = \frac{\omega}{c} \text{ and } \epsilon_0\mu_0 = \frac{1}{c^2}\right]$$

So, $\vec{B} = \dfrac{\hat{n} \times \vec{E}}{c}$

$$\therefore \quad \left|\frac{\vec{E}}{\vec{H}}\right| = \frac{E_0}{H_0} = e\mu_0 = \frac{1}{c\epsilon_0} = \sqrt{\frac{\mu_0}{\epsilon_0}} = z_0.$$

Thus, the ration $\left|\frac{\vec{E}}{\vec{H}}\right|$ is real and positive, $\vec{E}$ and $\vec{H}$ are in the same phase of vibration, i.e., when the amplitude of $\vec{E}$ is maximum, the amplitude of $\vec{H}$ is also maximum (Fig. 14.19).

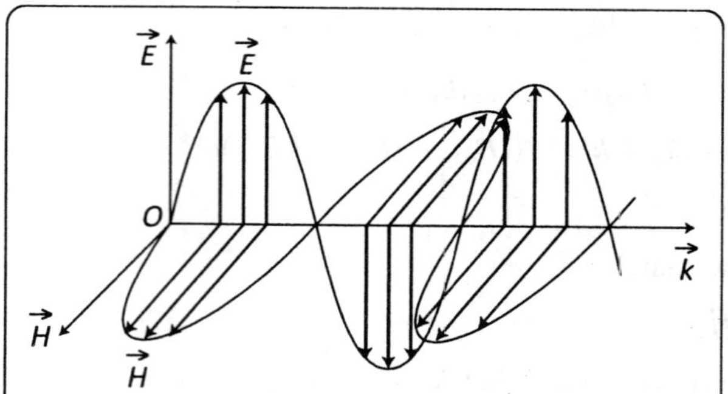

Fig. 14.19

14.21 Poynting's Theorem and Poynting Vector

Poynting Theorem

The theorem state that, *the rate at which electromagnetic energy in a finite volume decreases with time is equal to the rate of dissipation of energy in the form of joule heat plus the rate at which energy flows out of the volume.*

We have from Maxwell's third and fourth electromagnetic equations,

$$\nabla \times \vec{E} = -\frac{\partial \vec{B}}{\partial t} \tag{i}$$

and $$\nabla \times \vec{H} = \vec{J} + \frac{\partial \vec{D}}{\partial t}. \tag{ii}$$

Now, $$-\vec{H}\cdot(\nabla\times\vec{E}) = \vec{H}\cdot\frac{\partial\vec{B}}{\partial t} \tag{iii}$$

and $$-\vec{E}\cdot(\nabla\times\vec{H}) = \vec{E}\cdot\vec{J} + \vec{E}\cdot\frac{\partial\vec{D}}{\partial t}. \tag{iv}$$

Adding eqns. (iii) and (iv), we get

$$-\vec{H}\cdot(\nabla\times\vec{E}) + \vec{E}\cdot(\nabla\times\vec{H}) = \vec{E}\cdot\vec{J} + \left[\vec{H}\cdot\frac{\partial\vec{B}}{\partial t} + \vec{E}\cdot\frac{\partial\vec{D}}{\partial t}\right]$$

$$-[\vec{H}\cdot(\nabla\times\vec{E}) - \vec{E}\cdot(\nabla\times\vec{H}) = \vec{E}\cdot\vec{J} + \left[\vec{H}\cdot\frac{\partial\vec{B}}{\partial t} + \vec{E}\cdot\frac{\partial\vec{D}}{\partial t}\right]$$

or, $$-\nabla\cdot(\vec{E}\times\vec{H}) = \vec{E}\cdot\vec{J} + \left[\vec{H}\cdot\frac{\partial\vec{B}}{\partial t} + \vec{E}\cdot\frac{\partial\vec{D}}{\partial t}\right]. \tag{v}$$

$[\because\ \vec{B}\cdot(\nabla\times\vec{A}) - \vec{A}\cdot(\nabla\times\vec{B}) = \nabla\cdot(\vec{A}\times\vec{B})]$

Now, $$\vec{H}\cdot\frac{\partial\vec{B}}{\partial t} = \mu\vec{H}\cdot\frac{\partial\vec{H}}{\partial t} = \frac{1}{2}\mu\frac{\partial}{\partial t}(\vec{H}\cdot\vec{H}) = \frac{1}{2}\frac{\partial}{\partial t}(\vec{H}\cdot\vec{B})$$

and $$\vec{E}\cdot\frac{\partial\vec{D}}{\partial t} = \epsilon\vec{E}\cdot\frac{\partial\vec{E}}{\partial t} = \frac{1}{2}\epsilon\frac{\partial}{\partial t}(\vec{E}\cdot\vec{E}) = \frac{1}{2}\frac{\partial}{\partial t}(\vec{E}\cdot\vec{D}).$$

Equation (v) reduces to the form,

$$\nabla\cdot(\vec{E}\times\vec{H}) = -\vec{E}\cdot\vec{J} - \frac{\partial}{\partial t}\frac{1}{2}(\vec{E}\cdot\vec{D} + \vec{H}\cdot\vec{B}).$$

Integrating over a volume V, we get

$$\iiint_V \nabla\cdot(\vec{E}\times\vec{H})dV = -\iiint_V(\vec{E}\cdot\vec{J})dV - \frac{\partial}{\partial t}\iiint_V\frac{1}{2}(\vec{E}\cdot\vec{D} + \vec{H}\cdot\vec{B})dV.$$

If S be the surface enclosing volume V, then

$$-\frac{\partial}{\partial t}\iiint_V\frac{1}{2}(\vec{E}\cdot\vec{D} + \vec{H}\cdot\vec{B})dV = \iiint_V(\vec{J}\cdot\vec{E})dV + \oint_S(\vec{E}\times\vec{H})\cdot d\vec{S}. \tag{vi}$$

Now, the term

$$-\frac{\partial}{\partial t}\iiint_V\frac{1}{2}[(\vec{E}\cdot\vec{D}) + (\vec{H}\cdot\vec{B})dV = -\frac{\partial}{\partial t}\left[\iiint_V\frac{1}{2}(\vec{E}\cdot\vec{D})dV + \iiint_V\frac{1}{2}(\vec{H}\cdot\vec{B})dV\right]$$

$$= -\frac{\partial}{\partial t}[U_e + U_m], \quad \text{where } U_e = \iiint_V\frac{1}{2}(\vec{E}\cdot\vec{D})dV$$

and $$U_m = \iiint_V\frac{1}{2}(\vec{H}\cdot\vec{B})dV$$

are respectively electric energy and magnetic energy stored in volume V.

$\therefore$ the left-hand side of eqn. (vi) represents the time rate of decrease of electromagnetic energy in volume V.

The term $\iiint_V (\vec{J} \cdot \vec{E}) dV$ represents the time rate of dissipation of energy in the form of joule heat in V.

The term $\oint_S (\vec{E} \times \vec{H}) \cdot \vec{dS}$ represents the time rate of flow of energy from V through the surface S. Now, $\vec{E} \times \vec{H} = \vec{P}$ (say) has the dimension of

$$\left(\frac{\text{volt}}{\text{metre}}\right)\left(\frac{\text{ampere}}{\text{metre}}\right) = (\text{volt} \times \text{ampere})/\text{metre} = \text{joule/m}^2/\text{s}.$$

So, $\vec{P} = \vec{E} \times \vec{H}$ is the energy flowing through unit area and unit time and is known as the *Poynting vector.*

Hence, *the Poynting vector* ($\vec{P}$) *may be defined as the amount of electromagnetic field energy flowing through unit area of the surface in a direction perpendicular to the plane containing* $\vec{E}$ *and* $\vec{H}$ *per unit time.*

14.22 Solved Problems

Example 14.1. *Find out an expression for the velocity of a liquid which rushes out of a hole in the wall of the container.*

Solution : Let h = depth of the hole below the level of liquid in the container, h_1 and h_2 = the heights of the liquid level and the hole from the bottom of the container respectively.

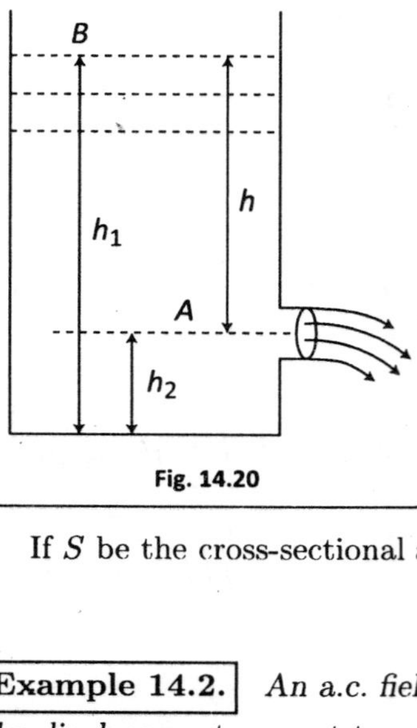

Fig. 14.20

Now, according to Bernoulli's theorem at A and B, we get

$$\frac{P_A}{\rho} + \frac{1}{2}v_A^2 + h_2 g = \frac{P_B}{\rho} + \frac{1}{2}v_B^2 + h_1 g,$$

where $P_A = P_B = P$, the atmospheric pressure and v_A and v_B are the velocity of the liquid at A and B respectively.

$\because \; v_B = 0,$

$$\therefore \quad \frac{P_A}{\rho} + \frac{1}{2}v_A^2 + h_2 g = \frac{P_B}{\rho} + h_1 g \quad \text{or,} \quad (h_1 - h_2)g = \frac{1}{2}v_A^2$$

or, $\; v_A^2 = 2(h_1 - h_2)g = 2hg \quad$ [where $h_1 - h_2 = h$]

or, $\; v_A = \sqrt{2hg}.$

This result is known as the *Torricelli's theorem.*

If S be the cross-sectional area of the hole, then the efflux of the liquid is

$$\boldsymbol{S v_A = S\sqrt{2hg}}.$$

Example 14.2. *An a.c. field* $\vec{E} = \vec{E_0} \cos\omega t$ *is applied to a conductor. Find out the ratio of the displacement current to conduction current.*

Solution : The conduction current density in the conductor of conductivity σ is given by

$$\vec{J} = \sigma\vec{E} = \sigma\vec{E_0} \cos\omega t. \tag{i}$$

The displacement current density through the conductor is

$$\vec{J_d} = \frac{\partial \vec{D}}{\partial t} = \frac{\partial}{\partial t}(\epsilon \vec{E}) \quad (\because \vec{D} = \epsilon \vec{E}).$$

$$= \epsilon \frac{\partial}{\partial t}(\vec{E_0} \cos \omega t) = -\epsilon \vec{E_0} \omega \sin \omega t = \omega \epsilon \vec{E_0} \cos\left(\omega t + \frac{\pi}{2}\right). \quad \text{(ii)}$$

$$\therefore \quad \left|\frac{\vec{J_d}}{\vec{J}}\right| = \left|\frac{\omega \epsilon E_0 \cos\left(\omega t + \frac{\pi}{2}\right)}{\sigma E_0 \cos \omega t}\right| = \frac{\omega \epsilon}{\sigma}.$$

For a good conductor, $\sigma \sim 10^7$ mho/metre and $\epsilon = \epsilon_r \epsilon_0 \simeq \epsilon_0$ $[\because \epsilon_r \simeq 1]$

$$\therefore \quad \left|\frac{\vec{J_d}}{\vec{J}}\right| \simeq \frac{2\pi f \times \epsilon_0}{10^7} = \frac{2\pi f \times 9 \times 10^{-12}}{10^7} \simeq f \times 10^{-17}.$$

But, the optical frequency is $\sim 10^{15}$ hertz.

$$\therefore \quad \left|\frac{\vec{J_d}}{\vec{J}}\right| = 10^{15} \times 10^{-17} = \frac{1}{100}.$$

$$\therefore \quad |\vec{J}| \simeq 100 \times |\vec{J_d}|. \quad \text{(iii)}$$

Example 14.3. (a) *If the earth receives* 2 cal/min/cm^2 *solar energy, what are the amplitudes of electric and magnetic fields of radiation?*

(b) *Assuming that all the energy from a* 100 W *lamp is radiated uniformly, calculate the average values of the intensities of electric and magnetic fields and of radiation at a distance of* 2 m *from the lamp.*

Solution : The energy flux, i.e., the flow of energy per unit area per unit time according to Poynting's theorem is given by

$$|\vec{P}| = |\vec{E} \times \vec{H}| = EH \sin 90° = EH. \quad \text{(iv)}$$

Again, energy flux per unit area per second is $= 2 \text{ cal/min/cm}^2 = \dfrac{2 \times 4.2 \times 10^4}{60}$ J/m^2/s. (v)

Comparing eqns. (iv) and (v), we get

$$EH = \frac{2 \times 4.2 \times 10^4}{60} = 1400. \quad \text{(vi)}$$

Again, $$\frac{E}{H} = \sqrt{\frac{\mu_0}{H_0}} = \sqrt{\frac{4\pi \times 10^{-7}}{8.85 \times 10^{-12}}} = 376.6. \quad \text{(vii)}$$

$\therefore \quad E^2 = 1400 \times 376.6$ or, $E = \sqrt{1400 \times 376.6} = 726.1$ volt/m.

Now from eqn. (vi),

$$H = \frac{1400}{E} = \frac{1400}{726.1} = 1.928 \text{ A-turn/m}.$$

Now, the amplitudes of electric and magnetic field of the radiation are given by,

$$\boldsymbol{E_0 = E\sqrt{2} = 726.1 \times \sqrt{2} = 1027 \text{ volt/m}}$$

$$\boldsymbol{H_0 = H\sqrt{2} = 1.928\sqrt{2} = 2.730 \text{ A-turn/m}}.$$

(b) The flux of energy of a sphere with the lamp as the centre is 1000 W = 1000 J/s

$\therefore$ the energy flux per unit area per sec is $= \dfrac{1000}{16\pi^2}$ W/m^2

Hence, from Poynting's theorem,

$$|\vec{P}| = |\vec{E} \times \vec{H}| = EH = \frac{1000}{16\pi^2} \tag{viii}$$

and $$E/H = \sqrt{\frac{\mu_0}{\epsilon_0}} = 376.6. \tag{ix}$$

$$\therefore \quad E^2 = \frac{1000}{16\pi^2} \times 376.6 \quad \text{or,} \quad \boldsymbol{E} = \left[\frac{1000}{16\pi^2} \times 376.6\right]^{\frac{1}{2}} = \mathbf{48.87\ volt/m.}$$

Now from eqn. (viii),

$$\boldsymbol{H} = \frac{1000}{16\pi^2 E} = \frac{1000}{16\pi^2 \times 48.87} = \mathbf{0.1297\ A\text{-}turn/m.}$$

Example 14.4. *If a point charge is fixed at the origin, then show that the magnetic field is constant of time.*

Solution : According to Maxwell's first electromagnetic equation,

$$\nabla \cdot \vec{D} = \rho. \tag{i}$$

But, the electric field $\vec{E}$ at a point distance $|\vec{r}|$ apart from the charge q is

$$\vec{E}(\vec{r}) = \frac{q}{4\pi\epsilon_0} \frac{\vec{r}}{r^3}. \tag{ii}$$

From eqn. (ii),

$$\nabla \times \vec{E} = 0. \tag{iii}$$

But according to Maxwell's 3rd electromagnetic equation,

$$\nabla \times \vec{E} = -\frac{\partial \vec{B}}{\partial t}. \tag{iv}$$

From (iii) and (iv), we get $\dfrac{\partial \vec{B}}{\partial t} = 0.$

So, $\vec{B} = \mu\vec{H}$ is constant of time or, $\vec{H}$ = constant of time as $\mu \neq 0$.

Again, since $\vec{J} = 0$

$$\therefore \quad \nabla \times \vec{H} = -\frac{\partial \vec{B}}{\partial t} = \vec{0} \quad [\because \ \vec{D} = \epsilon_0 \vec{E} \text{ is constant of time}]$$

Also, $\nabla \cdot \vec{B} = 0$ gives $\nabla \cdot \vec{H} = 0$.

$$\therefore \quad \vec{H} = \textbf{constant w.r.t., } \boldsymbol{x, y, z}. \tag{v}$$

Equations (ii) and (v) represent the solutions of Maxwell's equations.

Example 14.5. *A charged particle moves in a uniform magnetic field, find out an expression for the force experienced by it if it moves along the magnetic field. What will be the work done by the magnetic field if it moves normal to the magnetic field?*

Solution : The force experienced by a charged particle of charge q moving with velocity $\vec{v}$ in a magnetic field $\vec{B}$ is

$$\vec{F} = q(\vec{v} \times \vec{B}).$$

If q moves along $\vec{B}$, then

$$\vec{F} = q(\vec{v} \times \vec{B}) = qvB \sin 0 \hat{n} = 0.$$

So, **the force acting on the particle will be zero if it moves along $\vec{B}$.**

The work done per second on the particle by the magnetic field if it moves in a direction perpendicular to $\vec{B}$ is

$$W = \vec{F} \cdot \vec{v} = q(\vec{v} \times \vec{B}) \cdot \vec{v} = 0.$$

According to Newton's 2nd law of motion,

$$\vec{F} = m\frac{d\vec{v}}{dt}.$$

$$\therefore \quad m\frac{d\vec{v}}{dt} \cdot \vec{v} = \frac{d}{dt}\left(\frac{1}{2}mv^2\right) = 0.$$

Hence, $\frac{1}{2}mv^2 =$ K.E. remains constant during the motion. **Thus, $\vec{B}$ produces no change in the K.E. of the particle and performs no work on q so long it moves in a direction perpendicular to $\vec{B}$.**

Example 14.6. *A straight conductor of length l is moving with a constant velocity $\vec{v}$ at right angles to a uniform magnetic field $\vec{B}$. What is the induced e.m.f. between the ends of the conductor?*

Solution : Let the conductor PQ moves with velocity $\vec{v}$ parallel to Y-axis in a uniform magnetic field $\vec{B}$ along Z-axis.

Each charged particle (electron) inside the conductor will experience a Lorentz force,

$$\vec{F} = -e(\vec{v} \times \vec{B}). \qquad \text{(vi)}$$

The direction of $\vec{F}$ is along negative X-axis if the charge e is negative.

In the steady state an equal and opposite force $\vec{F}'$ must develop at every point of the conductor to balance $\vec{F}$. This internal force $\vec{F}'$ can develop only from an electric field $\vec{E}$ developed inside PQ due to the charge separation as shown in Fig. 14.21. So in steady state,

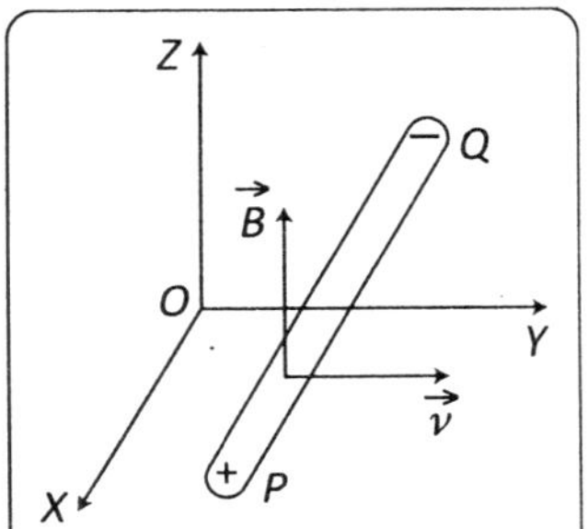

Fig. 14.21

$$-e\vec{E} - e(\vec{v} \times \vec{B}) = 0$$

$$\text{or,} \quad \vec{E} = -(\vec{v} \times \vec{B}). \qquad \text{(vii)}$$

The e.m.f. induced between two ends of the conductor is

$$\varepsilon = \int_0^l \vec{E} \cdot \vec{dl} = \left| -\int_0^l (\vec{v} \times \vec{B}) \cdot \vec{dl} \right| = \boldsymbol{vBl}. \qquad \text{(viii)}$$

ε is sometimes called the *motional e.m.f.*

Example 14.7. *Find the element of force dF_{12} caused by the current element $I_2 dl_2$ on $I_1 dl_1$ as also the force dF_{21} on the current element $I_2\, dl_2$ produced by $I_1\, dl_1$. Show that they are unequal. How do you recoincide your results with Newton's third law?*

[MU 1983, BU(H) 2014]

Solution : According to Ampere's force law the force on current element $I_1 dl_1$ of circuit 1 due to circuit 2 will be

$$\overrightarrow{dF_{12}} = I_1 \overrightarrow{dl_1} \times \overrightarrow{B_2} \tag{i}$$

But according to Biot-Savart law,

$$\vec{B} = \oint_2 \frac{\mu_0}{4\pi} I_2 \frac{\overrightarrow{dl_2} \times \vec{r}}{r^3} \tag{ii}$$

So, from eqns. (i) and (ii)

$$\overrightarrow{dF_{12}} = \frac{\mu_0}{4\pi} I_1 I_2 \overrightarrow{dl_1} \times \oint_2 \frac{\overrightarrow{dl_2} \times \vec{r}}{r^3}$$

i.e., $$\overrightarrow{dF_{12}} = \frac{\mu_0}{4\pi} I_1 I_2 \oint_2 (\overrightarrow{dl_1} \cdot \vec{r})\overrightarrow{dl_2} - (\overrightarrow{dl_1} \cdot \overrightarrow{dl_2})\vec{r} \tag{iii}$$

[as $\vec{A} \times (\vec{B} \times \vec{C}) = (\vec{A} \cdot \vec{C})\vec{B} - (\vec{A} \cdot \vec{B})\vec{C}$]

Similarly, the force on current element $I_2\overrightarrow{dl_2}$ of circuit 2 due to circuit 1 will be,

$$\overrightarrow{dF_{12}} = I_2 \overrightarrow{dl_2} \times \overrightarrow{B_1} = I_2 \overrightarrow{dl_2} \times \oint_1 \frac{\mu_0}{4\pi} \frac{I_1 \overrightarrow{dl_1} \times (-\vec{r})}{r^3}$$

i.e., $$\overrightarrow{dF_{12}} = \frac{\mu_0}{4\pi} I_1 I_2 \oint_1 \frac{\overrightarrow{dl_2} \times (\vec{r} \times \overrightarrow{dl_1})}{r^3} \quad [\text{as } \vec{A} \times \vec{B} = -(\vec{B} \times \vec{A})]$$

i.e., $$\overrightarrow{dF_{12}} = \frac{\mu_0}{4\pi} I_1 I_2 \oint_1 \frac{(\overrightarrow{dl_2} \cdot \overrightarrow{dl_1})\vec{r} - (\overrightarrow{dl_2} \cdot \vec{r})\overrightarrow{dl_1}}{r^3} \tag{iv}$$

Comparing eqn. (iii) and (iv) we find that $\overrightarrow{dF_{12}} \neq -\overrightarrow{dF_{21}}$.

This is quite distressing since from Newton's third law we expect, $\overrightarrow{dF_{12}} = -\overrightarrow{dF_{21}}$.

However, if the force due to one complete circuit is evaluated on the other hand we find

$$\overrightarrow{F_{12}} = \oint_1 \overrightarrow{dF_{12}} \quad \text{i.e.,} \quad \oint_1 \overrightarrow{dF_{12}} = \frac{\mu_0}{4\pi} I_1 I_2 \oint_1 \oint_2 \frac{(\overrightarrow{dl_1} \cdot \vec{r})\overrightarrow{dl_2} - (\overrightarrow{dl_1} \cdot \overrightarrow{dl_2})\vec{r}}{r^3}$$

Now, $$\oint_1 \oint_2 \frac{(\overrightarrow{dl_1} \cdot \vec{r})\overrightarrow{dl_2}}{r^3} = \oint_2 \overrightarrow{dl_2} \oint_1 \frac{(\overrightarrow{dl_1} \cdot \vec{r})}{r^3}$$

$$= \oint_2 \overrightarrow{dl_2} \oint_1 \left[-\nabla\left(\frac{1}{r}\right)\right] \cdot \overrightarrow{dl_1} \quad \left[\text{as } \nabla\left(\frac{1}{r}\right) = -\frac{\vec{r}}{r^3}\right]$$

$$= -\oint_2 dl_2 \int_S \text{curl grad}\left(\frac{1}{r}\right) \cdot dS_1 \quad \left[\text{as } \oint \vec{A} \cdot \vec{dl} = \int_S (\text{curl } \vec{A}) \cdot \vec{dS}\right]$$

$$= 0 \quad [\text{as curl grad of a scalar} = 0]$$

$$\therefore \quad \overrightarrow{F_{12}} = -\frac{\mu_0}{4\pi} I_1 I_2 \oint_1 \oint_2 \frac{(\overrightarrow{dl_1} \cdot \overrightarrow{dl_2})\vec{r}}{r^3} \qquad \text{(v)}$$

Similarly, $\overrightarrow{F_{21}} = -\frac{\mu_0}{4\pi} I_1 I_2 \oint_1 \oint_2 \frac{(\overrightarrow{dl_1} \cdot \overrightarrow{dl_2})\vec{r} - \overrightarrow{dl_2} \cdot \vec{r_1})\overrightarrow{dl_1}}{r^3}$

$$\text{i.e.,} \quad \overrightarrow{F_{21}} = -\frac{\mu_0}{4\pi} I_1 I_2 \oint_1 \oint_2 \frac{(\overrightarrow{dl_2} \cdot \overrightarrow{dl_1})\vec{r}}{r^3} \qquad \text{(vi)}$$

as $\oint_1 \oint_2 \frac{(\overrightarrow{dl_2} \cdot \vec{r})\overrightarrow{dl_1}}{r^3} = -\oint_1 \overrightarrow{dl_1} \oint_2 \overrightarrow{dl_2} \cdot \text{grad}\left(\frac{1}{r}\right) = -\oint_1 \overrightarrow{dl_1} \int_{S_2} \text{curl grad}\left(\frac{1}{r}\right) \cdot dS_2 = 0$

So, from eqns. (v) and (vi) it is apparent that $\overrightarrow{\boldsymbol{F_{12}}} = -\overrightarrow{\boldsymbol{F_{21}}}$.

Exercises

14(A) : Objective Type Questions

Select the correct answers

14.1 The equation of continuity for an incompressible fluid is

(a) $\nabla \cdot (\rho\vec{r}) + \frac{\partial \rho}{\partial t} = 0$ (b) $\nabla \cdot \vec{v} = 0$ (c) $\frac{\partial \rho}{\partial t}$ (d) $\nabla \cdot (\rho\vec{v}) = \frac{\partial \rho}{\partial t}$.

14.2 Gauss' law in electrostatics is valid if the electric field $\vec{E}$ varies as

(a) $\frac{1}{r^3}$ (b) $\frac{1}{r}$ (c) $\frac{1}{r^2}$ (d) r^2.

14.3 The magnetic induction $\vec{B}$ and electric field $\vec{E}$ are related as

(a) $\vec{B} = \frac{\vec{v} \times \vec{E}}{c^2}$ (b) $\vec{B} = \frac{\vec{E} \times \vec{B}}{c^2}$ (c) $\vec{B} = \frac{\vec{E} \times \vec{B}}{c}$ (d) $\vec{B} = \frac{\vec{B} \times \vec{E}}{c^2}$.

14.4 Displacement current arises, if

(a) $\frac{\partial \rho}{\partial t} = 0$ (b) $\frac{\partial \rho}{\partial t} \neq 0$ (c) $\rho =$ a constant (d) $\frac{\partial \rho}{\partial t}$ is $-$ve.

14.5 The displacement current in a conductor is given by,

(a) $\epsilon_0 \frac{\partial \vec{E}}{\partial t}$ (b) $\mu_0 \epsilon_0 \frac{\partial \vec{E}}{\partial t}$ (c) $\epsilon_0 \frac{\partial \phi}{\partial t}$ (d) $\mu_0 \epsilon_0 \frac{\partial \phi}{\partial t}$. **[HPU 1993]**

14.6 An electromagnetic field satisfies

(a) Ampere's law (b) Gauss' law (c) Faraday's law (d) All of them.

14.7 The direction of induced e.m.f. in a circuit is given by

(a) Fleming's left hand rule (b) Faraday's law (c) Lenz' law (d) none of these. **[HPU 1996]**

14.8 For a conductor $|\vec{J}/\vec{J_d}|$ is equal to

(a) 10 (b) 50 (c) 80 (d) 100

Answers

14.1 (b) 14.2 (c) 14.3 (a) 14.4 (b) 14.5 (a) 14.6 (d) 14.7 (c) 14.8 (d)..

14(B) : Short Answers Type Questions

14.1 Express Euler's equation in Cartesian form.

14.2 What are velocity head, gravity head and pressure head.

14.3 State Gauss' law in gravitation.

14.4 State Gauss' law in differential form. **[CU(H) 2005]**

14.5 State and explain Gauss' law in electrostatics, obtain its differential form. **[WBUT 2002, BU(H) 2000]**

14.6 Extend Gauss' law in electrostatics to Poisson's equation. When does it reduce to Laplace's law? **[CU(H) 1999]**

14.7 State and prove Poisson's equation in electrostatics. **[CU(H) 2003, BU(H) 2003, 2001]**

14.8 State Ampere's circuital law. **[WBUT 2002, BU(H) 2002]**

14.9 Write down Faraday's law of electromagnetic induction. **[BU(H) 2001]**

14.10 Bring out the significance of $\nabla \cdot \vec{B} = 0$, where $\vec{B}$ is the magnetic induction vector.

14.11 What is the limitation of Ampere's circuital law?

14.12 State the reciprocity theorem of mutual inductance.

14.13 State Biot-Savart's law **[WBUT 2002, BU(H) 2001]**

14.14 Give an example of an electrical circuit carrying non-steady current where Ampere's circuital law is not applicable. **[WBUT 2006]**

14.15 State the law expressed by $\nabla \times \vec{H} = \vec{J} + \dot{\vec{D}}$. Explain the reason for the $\dot{\vec{D}}$ term in the equation. **[BU(H) 2003]**

14.16 Explain with an example the meaning of motional e.m.f. **[BU(H) 2001]**

14.17 State Poynting theorem and define Poynting vector. **[cf. BU(H) 2000]**

14.18 Maxwell's equations are based on existing laws in electricity and magnetism. What is then Maxwell's contribution? **[BU(H) 2001]**

14.19 Explain the physical significance of the Poynting vector. **[BU(H) 2001]**

14(C) : Numerical Problems

14.1 Calculate the value of Poynting vector at the surface of the sum if the power radiated by the sum is 3.8×10^{26} watts, while its radius is 7×10^{8} m.

[MU 1969] **[Ans:** 6.175×10^{7} watt/m^2**]**

14.2 A 240 monochromatic point source radiates energy uniformly in all directions in free space and the radiation is detected at a distance of 2 m from the source. What is the amplitude of the electric field at the detector? ($\mu_0 = 4\pi \times 10^{-7}$ H/m) **[Ans:** 60 V/m**]**

14.3 Find from Poynting flow the mean value of intensity of the magnetic field in air at a distance of 100 cm from a radiating source of power 10 kW.

[CU 1987] **[Ans:** 1.44 A/m**]**

14.4 If a 2 kW laser beam is concentrated by a lens into cross-sectional area about 10^{-6} square cm, what is the value of Poynting vector and the amplitude of the electric field?

[Ans: 2×10^{13} Watt/m^2, $\sim 1.2 \times 10^8$ volt/m]

14.5 An electron circulates around a nucleus in a path of radius 5.1×10^{-11} m at a frequency of 6.8×10^5 revolutions/sec. Find $\vec{B}$ at the centre. **[CU(H) 2003]**

14.6 A 10 kW oscillator is radiating isotropically, calculate the amplitudes of electric and magnetic fields at a distance 2 km form it. **[Ans:** $\frac{1}{4\pi\epsilon_0} = 9 \times 10^9$ volt-m C^{-1}]

14.7 Prove that $\nabla \cdot \vec{D} = \rho_f$, where D = electric displacement vector and ρ_f = free charge density. **[NBU(H) 2009]**

14(D) Broad Answer Type Questions ......

14.1 **(a)** Write down Maxwell's field equations, explaining the terms used. Show that in vacuum both electric and magnetic vectors obey wave equation. Assuming plane wave solution, show that the magnetic field is always orthogonal to the electric field.

(b) Find the displacement current within a parallel plate capacitor in series with a resistor which carries current I. Area of the capacitor plates is A and the dielectric is vacuum. **[WBUT 2006]**

14.2 **(a)** Write down Maxwell's equation for an electromagnetic field mentioning the laws represented by each of these equations.

(b) Obtain plane wave solution of the wave equation representing an electromagnetic wave travelling in the x-direction. **[BU(H) 2002, 2006]**

14.3 Write down Maxwell's equations for an electromagnetic field and explain physical significance of each. **[BU(H) 2000, 2001]**

14.4 **(a)** Write down Maxwell's equations for an electromagnetic field. Show by integration of these equations that there is a well-known law behind each.

(b) Starting from Maxwell's equations in free space that the magnetic field $\vec{H}$ and the electric field $\vec{E}$ in an electromagnetic wave travel with the same speed.

[BU(H) 1998]

14.5 **(a)** Write down Maxwell's field equations.

(b) The electric vector $\vec{E}$ of an electromagnetic wave in free space is given by,

$$E_x = E_z = 0$$

$$E_y = Ae^{i\omega(t-\frac{x}{y})}.$$

Using Maxwell's equations for free space conditions obtain expressions for the components of the magnetic vector $\vec{H}$.

(c) Starting from Maxwell's field equations prove that for a plane electromagnetic wave propagating in an isotropic dielectric medium, the electric field $\vec{E}$ and the magnetic field $\vec{B}$ are normal to each other and also normal to the direction of wave propagation. **[BU(H) 1996]**

14.6 **(a)** Write down Maxwell's field equations for an electromagnetic field giving the physical interpretation of each. State the law expressed by curl $\vec{H} = \vec{D} + \vec{J}$. Explain the reason for the $\vec{D}$ term in the equation.

(b) Define Poynting vector. **[BU(H) 2003]**

14.7 (a) State and obtain Ampere's circuital law. What was its limitation and how was it overcome?

(b) Explain the physical significance of displacement current. How is it different from conduction current?

14.8 (a) In case of a magnetic interaction of two current elements $I_1\overrightarrow{dl_1}$ and $I_2\overrightarrow{dl_2}$, show that $\overrightarrow{dF_{12}} \neq -\overrightarrow{dF_{21}}$ but for entire loop i.e., if the force of one entire circuit on the other is considered then $\overrightarrow{F_{12}} = -\overrightarrow{F_{21}}$ in accordance with Newton's third law.

[BU(H) 2014]

(b) State Biot-Savart's law and discuss its application in magnetostatics.

[BU(H) 2010]

CHAPTER 15

Ordinary Differential Equations—I

15.1 Introduction

Equations involving differential coefficients or differentials of an unknown function are called **differential equations**.

Differential equations play an important role in science and technology. For example, differential equations have indispensable role in the study of oscillations of mechanical and electrical problems, conduction of heat, etc.

Examples

(i) $$\frac{d^2x}{dt^2} + \omega^2 x = 0 \tag{15.1}$$

(ii) $$\frac{dy}{dx} + y = \sin x \tag{15.2}$$

(iii) $$\left.\begin{aligned} \frac{dy}{dt} - \omega y = a\cos pt \\ \frac{dy}{dt} + \omega y = a\sin pt \end{aligned}\right\} \tag{15.3}$$

(iv) $$y = x\left(\frac{dy}{dx}\right) + \frac{x}{\left(\frac{dy}{dx}\right)} \tag{15.4}$$

(v) $$\frac{\left[1 + \left(\frac{dy}{dx}\right)^2\right]^{\frac{3}{2}}}{\frac{d^2y}{dx^2}} = C \tag{15.5}$$

(vi) $$x^2 dx + e^y dy = 0 \tag{15.6}$$

(vii) $$x\frac{\partial u}{\partial x} + y\frac{\partial u}{\partial y} = 2u \tag{15.7}$$

(viii) $$\frac{\partial^2 y}{\partial t^2} = c^2\frac{\partial^2 y}{\partial x^2}. \tag{15.8}$$

These are all differential equations.

Differential equations are classified into two main types :

(i) Ordinary differential equation (ODE) : An ordinary differential equation contains only one independent variable and hence, total derivative. For example, differential equations (i) to (vi) in section 15.1 are all ordinary differential equations.

(ii) Partial differential equations (PDE) : A differential equation involving differentials with respect to two or more independent variables is called a partial differential equation. For example, equations (vii) and (viii) in setion 15.1 are partial differential equations.

15.2 Some Terminologies

(i) Order of differential equation : The order of the highest order derivative appearing in the equation is called the order of the differential equation. For example, the order of each of the differential equations (15.2), (15.3), (15.6) is one whereas the order of each of the equations (15.1), (15.4), (15.5) is two.

(ii) Degree of differential equation : The degree of differential equation is the highest power of the highest ordered derivative after it has been expressed rational and integral as far as derivatives are concerned. For example, the degree of each of the differential equations (15.2), (15.3) and (15.6) is **one** whereas that of equations (15.4), (15.5) is **two**.

(iii) Solution of differential equations : A solution (or integral) of a differential equation is a relation between the variables which satisfy the given differential equation.

Example *The solution of the differential equation,*

$$\frac{d^2y}{dx^2} + y = 0 \tag{15.9}$$

is $y = A\cos x + B\sin x.$ (15.10)

The order of the equation (15.9) is two and the solution (15.10) contains two arbitrary constants. Thus, ***the number of arbitrary constants in the solution is equal to the order of the differential equation.***

Hence, the solution of a differential equation is an equation containing dependent variable (y), independent variable (x) and free from derivative and satisfying the differential equation.

(iv) Linear differential equations : A differential equation of first degree in the dependent variable and its derivatives (there cannot be any term involving the product of the dependent variable and its derivatives) is called a **linear differential equation.**

The general form of a linear ordinary differential equation is

$$a_0\frac{d^ny}{dx^n} + a_1\frac{d^{n-1}y}{dx^{n-1}} + \cdots + a_{n-1}\frac{dy}{dx} + a_ny = f(x), \tag{15.11}$$

where $a_0, a_1, \ldots, a_n$ are either given functions of x or, constants. The ordinary differential equations which cannot be expressed in the form of equation (15.11) are known as *non-linear ordinary differential equations.*

For example,

The equation : $\dfrac{d^3y}{dx^3} + 3\dfrac{dy}{dx} + 2y = x$ is a linear ODE.

But, $\left(\dfrac{dy}{dx}\right)^2 + y = 0$ is a non-linear ODE.

15.3 Derivation of a Differential Equation

Derivation of a differential equation is the inverse process of finding the solution.

To derive a differential equation for the equation of a curve, say, $f(x, y, z, \ldots, c_1, c_2, \ldots) = 0$ we have to eliminate the constants.

If there be a single constant in the equation to the curve, then the differential equation will be of first order. The number of order of the differential equation will be equal to the number of constants in the equation to the curve.

Example 1. *Form the differential equation for the following :*

(i) $xy = ae^x + be^{-x}$ (ii) $y = bx + b - b^3$ (iii) (a) $y = Ax + A^2$

(b) $y = A\cos x + B\sin x$ (c) $y^2 = Ax^2 + Bx + C$ **[AMIETE (Winter) 1990, (Summer) 1994]**

(iv) $y = ax^2 + bx$ (v) $y = e^{ax}$.

Solution : (i) We have $xy = ae^x + be^{-x}$. (i)

$$\therefore \quad x\frac{dy}{dx} + y = ae^x - be^{-x}. \tag{ii}$$

Adding (i) and (ii),

$$xy + x\frac{dy}{dx} + y = 2ae^x. \tag{iii}$$

Differentiating again w.r.t. x,

$$x\frac{dy}{dx} + y + x\frac{d^2y}{dx^2} + \frac{dy}{dx} + \frac{dy}{dx} = 2ae^x. \tag{iv}$$

Equating equations (iii) and (iv), we get

$$\boldsymbol{x\frac{d^2y}{dx^2} + 2\frac{dy}{dx} - xy = 0}. \tag{v}$$

The differential equation is of order two obtained by eliminating two constants.

(ii) Here, $y = bx + b - b^3$.

$$\therefore \quad \frac{dy}{dx} = b.$$

$$\therefore \quad \boldsymbol{y = x\left(\frac{dy}{dx}\right) + \left(\frac{dy}{dx}\right) - \left(\frac{dy}{dx}\right)^3}.$$

The equation is of order one obtained by eliminating one constant.

(iii) (a) We have $y = Ax + A^2$. (i)

$$\therefore \quad \frac{dy}{dx} = A.$$

Putting the value of A in equation (i),

$$\therefore \quad y = x\left(\frac{dy}{dx}\right) + \left(\frac{dy}{dx}\right)^2.$$

Thus, eliminating one constant we get the differential equation of order 1.

(b) $y = A\cos x + B\sin x$

$$\therefore \quad \frac{dy}{dx} = -A\sin x + B\cos x \quad \text{and} \quad \frac{d^2y}{dx^2} = -(A\cos x + B\sin x) = -y \quad \text{or,} \quad \boldsymbol{\frac{d^2y}{dx^2} + y = 0}.$$

This is a differential equation of order two, obtained by eliminating two constants A and B.

(c) We have $y^2 = Ax^2 + Bx + C$.

Differentiating w.r.t. x,

$$2y\frac{dy}{dx} = 2Ax + B.$$

Again differentiating w.r.t. x,

$$2y\frac{d^2y}{dx^2} + 2\left(\frac{dy}{dx}\right)^2 = 2A$$

Differentiating again w.r.t. x,

$$y\frac{d^3y}{dx^3} + \frac{dy}{dx}\cdot\frac{d^2y}{dx^2} + 2\frac{dy}{dx}\cdot\frac{d^2y}{dx^2} = 0 \quad \text{or,} \quad \boldsymbol{y\frac{d^3y}{dx^3} + 3\frac{dy}{dx}\cdot\frac{d^2y}{dx^2} = 0.}$$

This is a differential equation of order three obtained by eliminating three constants A, B and C.

(iv) $y = ax^2 + bx$

Differentiating w.r.t. x, $\frac{dy}{dx} = 2ax + b.$

$$\therefore \quad b = \frac{dy}{dx} - 2ax. \tag{ii}$$

Differentiating again w.r.t. x, $0 = \frac{d^2y}{dx^2} - 2a.$

$$\therefore \quad a = \frac{1}{2}\frac{d^2y}{dx^2}. \tag{iii}$$

Putting (ii) and (iii) in Eqn. (i),

$$y = \frac{1}{2}\frac{d^2y}{dx^2}x^2 + \left(\frac{dy}{dx} - \frac{d^2y}{dx^2}x\right)x = -\frac{x^2}{2}\cdot\frac{d^2y}{dx^2} + x\frac{dy}{dx}$$

$$\text{or,} \quad \boldsymbol{x^2\frac{d^2y}{dx^2} - 2x\frac{dy}{dx} + 2y = 0.} \tag{iv}$$

Equation (iv) is a differential equation of order two obtained by eliminating two constants a and b.

15.4 Ordinary Differential Equations of First Order and First Degree

An ordinary differential equation of first order and first degree is written as

$$\frac{dy}{dx} = F(x, y), \tag{15.12}$$

where $F(x, y)$ is a given function of (x, y).

Analytical solution of Eqn. (15.12) can be obtained if $F(x, y)$ is expressible in some simple forms. Four such forms are as below :

(i) Equations solvable by separation of variables.

(ii) Homogeneous equations.

(iii) Exact equations.

(iv) Linear equations of first order.

15.4.1 Variables Separable

Let $F(x, y) = f(x)g(y)$, where $f(x)$ and $g(y)$ are functions of x and y only. Then from equation (15.12),

$$\frac{dy}{dx} = f(x)g(y) \quad \text{or,} \quad \int \frac{dy}{g(y)} = \int f(x)dx. \tag{15.13}$$

Special Cases :

(i) If y is absent in F, then Eqn. (15.12) becomes

$$\frac{dy}{dx} = F(x).$$

The solution of the above equation is

$$y = \int F(x)dx. \tag{15.14}$$

(ii) If x is absent in F, then

$$\frac{dy}{dx} = F(y) \quad \text{or,} \quad \int dx = \int \frac{dy}{F(y)}. \tag{15.15}$$

(iii) If both x and y are absent in F, then

$$F(x, y) = \frac{dy}{dx} = \text{a constant } (k).$$

On integration,

$$\boldsymbol{y = kx + A} \text{ (constant)}. \tag{15.16}$$

Example 15.1. *Solve the differential equation* $y\dfrac{dy}{dx} + xy^2 = x$.

Solution : The given equation can be written as

$$y\,dy + x(y^2 - 1)\,dx = 0 \quad \text{or,} \quad \frac{y\,dy}{y^2 - 1} + x\,dx = 0.$$

Integrating, $\dfrac{1}{2}\ln(y^2 - 1) + \dfrac{1}{2}x^2 = C$ or, $\boldsymbol{(y^2 - 1)e^{x^2} = \text{constant}}$.

15.4.2 Homogeneous Equations

If $f(x, y)$ be a homogeneous function of degree n, then for any λ, the identity is true.

A differential equation of the form

$$\frac{dy}{dx} = \frac{f(x, y)}{g(x, y)}$$

where $f(x, y)$ and $g(x, y)$ are homogeneous functions of x and y, then the above equation becomes

$$\frac{dy}{dx} = \frac{f(x, y)}{g(x, y)} = \frac{x^n \phi\left(\frac{y}{x}\right)}{x^n \psi\left(\frac{y}{x}\right)} = \theta\left(\frac{y}{x}\right), \tag{15.17}$$

where ϕ, ψ and θ are all functions of $\left(\frac{y}{x}\right)$. Equations of the type of (15.17) are called **first order homogeneous equations.**

Such equation can be converted to separable variables form by making the substitution

$$y = vx \tag{15.18}$$

where v is a function of x.

Differentiating Eqn. (15.18) w.r.t. x, we get

$$\frac{dy}{dx} = v + x\frac{dy}{dx}. \tag{15.19}$$

From Eqns. (15.17) and (15.19), we have

$$v + x\frac{dy}{dx} = \theta(v).$$

Integrating,

$$\int \frac{dv}{\theta(v) - v} = \int \frac{dx}{x}. \tag{15.20}$$

Eqn. (15.20) shows that v and y can be expressed in terms of x.

Example 15.2. *Solve* : $\dfrac{dy}{dx} = \dfrac{x^2 + y^2}{2xy}$.

Solution : The given equation is homogeneous in x and y. Let $y = vx$, where v is a new variable.

Then, $\dfrac{dy}{dx} = v + x\dfrac{dv}{dx}$.

$\therefore$ given equation becomes

$$v + x\frac{dv}{dx} = \frac{x^2 + (vx)^2}{2x(vx)} = \frac{1 + v^2}{2v} \quad \text{or,} \quad x\frac{dv}{dx} = \frac{1 + v^2}{2v} - v = \frac{1 - v^2}{2v} \quad \text{or,} \quad \frac{2v\,dv}{1 - v^2} = \frac{dx}{x}.$$

Integrating,

$$\int \frac{2v\,dv}{1 - v^2} = \int \frac{dx}{x} \quad \text{or,} \quad -\ln(1 - v^2) = \ln x - \ln C \quad \text{or,} \quad \ln x(1 - v^2) = \ln C$$

or, $$x\left(1 - \frac{y^2}{x^2}\right) = C.$$

15.4.3 Non-homogeneous Equations Reducible to Homogeneous Equations

A non-homogeneous equation of the form,

$$\frac{dy}{dx} = \frac{ax + by + c}{Ax + By + c} \tag{15.21}$$

can be reduced to homogeneous form by putting $x = X + h$, $y = Y + k$ (h, k are constants).

$$\therefore \quad \frac{dy}{dx} = \frac{dY}{dX}.$$

Then Eqn. (15.21) reduces to

$$\frac{dY}{dX} = \frac{a(X + h) + b(Y + k) + c}{A(X + h) + B(Y + k) + C} = \frac{aX + bY + ah + bk + c}{AX + BY + Ah + Bk + C}.$$

If h, k be so chosen that,

$$\left.\begin{aligned} ah + bk + c = 0 \\ ah + bk + c = 0 \end{aligned}\right\} \tag{15.22}$$

then Eqn. (15.21) becomes homogeneous

$$\frac{dY}{dX} = \frac{aX + bY}{AX + BY} \tag{15.23}$$

$$\text{or,}\quad \frac{dY}{dX}=\frac{a+b\left(\frac{Y}{X}\right)}{A+B\left(\frac{Y}{X}\right)}. \tag{15.24}$$

Eqn. (15.24) is a homogeneous equation and can be readily solved.

N.B. h and k can be determined from Eqn. (15.22) only if $aB - bA \neq 0$. Before proceeding this should be checked properly.

Example 15.3. *Find the solution of the differential equation*

$$(x+2y)(dy-2dx)+dx+2dy=0. \text{ [AMIETE(Winter) 1994]}$$

Solution : We have $(x+2y)(dy-2\,dx)+dx+2\,dy=0$

$$\text{or,}\quad (x+2y)\left(\frac{dy}{dx}-2\right)+1+\frac{2dy}{dx}=0 \quad\text{or,}\quad (x+2y+2)\frac{dy}{dx}=2x+4y-1$$

$$\text{or,}\quad \frac{dy}{dx}=\frac{2x+4y-1}{x+2y+2}.$$

Hence, $\dfrac{a}{A}=\dfrac{b}{B}$, i.e., $\dfrac{2}{1}=\dfrac{4}{2}$ (case of failure)

Let $x+2y=v$.

$$\therefore\quad 1+\frac{2dy}{dx}=\frac{dv}{dx} \quad\text{or,}\quad \frac{1}{2}\left(\frac{dv}{dx}-1\right)=\frac{2v-1}{v+2} \quad\text{or,}\quad \frac{dv}{dx}-1=\frac{4v-2}{v+2}$$

$$\text{or,}\quad \frac{dv}{dx}=\frac{4v-2}{v+2}+1=\frac{5v}{v+2} \quad\text{or,}\quad \frac{v+2}{5v}dv=dx \quad\text{or,}\quad \frac{1}{5}\int\left(1+\frac{2}{v}\right)dv=dx$$

$$\text{or,}\quad \frac{1}{5}(v+2\log v)=x+C \quad\text{or,}\quad v+2\log v=5x+5C$$

$$\text{or,}\quad x+2y+2\log(x+2y)=5x+5C \quad\text{or,}\quad -4x+2y+2\log(x+2y)=5C$$

$$\text{or,}\quad \mathbf{-2x+y+\log(x+2y)=\frac{5}{2}C.}$$

15.4.4 Exact Equations

If the function $\phi(x,y)=$ constant be continuous along with its first order partial derivatives, then the total differential of $\phi(x,y)$ is

$$d\phi=\frac{\partial\phi}{\partial x}dx+\frac{\partial\phi}{\partial y}dy=0$$

$$\text{or,}\quad d\phi=M(x,y)dx+N(x,y)dy, \tag{15.25}$$

where $M(x,y)=\dfrac{\partial\phi}{\partial x}$ and $N(x,y)=\dfrac{\partial\phi}{\partial y}$.

Since $\phi(x,y)$ has continuous first order derivative, we can write,

$$\frac{\partial^2\phi}{\partial y\partial x}=\frac{\partial^2\phi}{\partial x\partial y} \quad\text{or,}\quad \frac{\partial M}{\partial y}=\frac{\partial N}{\partial x} \tag{15.26}$$

Eqn. (15.26) represents the necessary and sufficient condition that relation (15.25) be an exact differential equation. The solution of the differential equation will be of the form $\phi(x,y)=$ constant.

Example 15.4. *Solve* : $e^x \sin y\, dx + (e^x + 1)\cos y\, dy = 0.$ **[WBUT 2006]**

Solution : The given equation is of the form

$$M\,dx + N\,dy = 0, \quad \text{where } M = e^x \sin y \text{ and } N = (e^x + 1)\cos y.$$

$$\therefore \quad \frac{\partial M}{\partial y} = e^x \cos y = \frac{\partial N}{\partial x}.$$

Hence, the given equation is exact.

So, $\displaystyle\int M\,dx = \int e^x \sin y\,dx = e^x \sin y$ [assuming y = constant]

and $\displaystyle\int N\,dx = \int (e^x + 1)\cos y\,dy = (e^x + 1)\sin y = e^x \sin y + \sin y.$ [assuming y = constant]

Hence, the required solution is $\boldsymbol{e^x \sin y + \sin y = C}$, where C is an arbitrary constant.

15.4.5 Integrating Factor

Let us consider an equation

$$M(x,y)dx + N(x,y)dy = 0. \tag{15.27}$$

which is not exact. If there exists a function $f(x,y)$ such that, $f(x,y)(M\,dx + N\,dy) = d\phi$ for some function $\phi(x,y)$, then $f(x,y)$ is called an integrating factor (I.F.) of the Eqn. (15.27).

For example the equation $y\,dx - x\,dy = 0$ is not exact.

Multiplying it by $\frac{1}{y^2}$, the equation becomes

$$\frac{y\,dx - x\,dy}{y^2} = 0 \quad \text{or,} \quad d\left(\frac{x}{y}\right) = 0,$$

which is exact and has the general solution

$$\frac{x}{y} = C, \quad C \text{ being an arbitrary constant.}$$

Thus I.F. is a multiplying factor by which the equation $M(x,y)dx + N(x,y)dy = 0$, if not exact, becomes exact.

Rules to determine integrating factors : Let the equation $M\,dx + N.dy = 0$ be not exact, i.e., $\frac{\partial M}{\partial y} \neq \frac{\partial N}{\partial x}$.

(1) **Integrating factor by inspection :** We know $y\,dx - x\,dy = 0$ is not exact. But multiplying it by $\frac{1}{y^2}$, the equation

$$\frac{y\,dx - x\,dy}{y^2} = 0 \text{ becomes exact} \quad \text{or,} \quad d\left(\frac{x}{y}\right) = 0 \text{ is exact}$$

$\therefore$ in this particular case, $\frac{1}{y^2}$ is the integrating factor.

(2) If $M\,dx + N\,dy = 0$ and $M(x,y)$, $N(x,y)$ are both homogeneous function in x, y of the same degree, then $\frac{1}{Mx+Ny}$ is an integrating factor of the equation $M\,dx + N\,dy = 0$.

(3) If $M\,dx+N\,dy \neq 0$, then $\frac{1}{Mx-Ny}$ is an integrating factor of the equation $M\,dx+N\,dy = 0$, where $M = yf(x,y)$ and $N = xg(x,y)$.

(4) If $\frac{1}{N}\left(\frac{\partial M}{\partial y} - \frac{\partial N}{\partial x}\right) = f(x)$, i.e., a function of x only, then $e^{\int f(x)dx}$ is an integrating factor of $M\,dx + N\,dy = 0$.

(5) If $\frac{1}{M}\left(\frac{\partial M}{\partial y}-\frac{\partial N}{\partial x}\right)=g(y)$, then $e^{-\int g(y)dy}$ is an integrating factor of $M\,dx+N\,dy=0$.

(6) If $M\,dx+N\,dy=0$ is of the form $x^a y^b(my\,dx+nx\,dy)+x^{a_1}y^{b_1}(m_1y\,dx+n_1x\,dy)=0$, where a_1, b_1, m, n, m_1, n_1 are constants and $mn_1-m_1n\neq 0$, then $x^h y^k$ is an integrating factor of $M\,dx+N\,dy=0$, where $\frac{a+h+1}{m}=\frac{b+k+1}{n}$ and $\frac{a_1+h+1}{m_1}=\frac{b_1+k+1}{n_1}$.

To determine h, k following equations are to be solved,

$$nh-mk=(m-n)+(mb-na)$$

and $\quad n_1h-m_1k=(m_1-n_1)+(m_1b_1-n_1a_1)$.

h, k can be determined if $mn_1-m_1n\neq 0$.

Example 15.5. *Solve* : $(x^2y-2xy^2)dx+(3x^2y-x^3)dy=0$. **[WBUT 2007]**

Solution : The given differential equation is of the form :

$$M\,dx+N\,dy=0,\quad \text{where } M=x^2y-2xy^2 \text{ and } N=3x^2y-x^3.$$

$$\therefore\quad \frac{\partial M}{\partial y}=x^2-4xy,\quad \frac{\partial N}{\partial x}=6xy-3x^2.$$

So, $\dfrac{\partial M}{\partial y}\neq\dfrac{\partial N}{\partial x}$.

$\therefore$ the given equation is not exact. Here both M and N are homogeneous function in x, y of same degree.

$$Mx+Ny=(x^2y-2xy^2)x+(3x^2y-x^3)y=x^2y^2\neq 0.$$

$\therefore\quad \dfrac{1}{x^2y^2}$ is an I.F.

Multiplying both sides of the given equation by $\frac{1}{x^2y^2}$, we get

$$\left(\frac{1}{y}-\frac{2}{x}\right)dx+\left(\frac{3}{y}-\frac{x}{y^2}\right)dy=0$$

or, $$\frac{y\,dx-x\,dy}{y^2}-2\frac{dx}{x}+3\frac{dy}{y}=0$$

or, $$d\left(\frac{x}{y}\right)-2d(\log x)+3d(\log y)=0.$$

Integrating, we get

$$\boldsymbol{\frac{x}{y}-2\log x+3\log y=C},\quad \text{where } C \text{ is an arbitrary constant.}$$

15.4.6 Linear Differential Equations

A differential equation of the form :

$$\frac{dy}{dx}+Py=Q, \tag{15.28}$$

where P and Q are functions of x only or constants, is called a **first order linear differential equation.**

The Eqn. (15.28) can be expressed as

$$(Py - Q)dx + dy = 0$$

or, $M\,dx + N\,dy = 0$, where $M = (Py - Q)$ and $N = 1$.

$$\therefore \quad \frac{\partial M}{\partial y} = P \quad \text{and} \quad \frac{\partial N}{\partial x} = 0.$$

$$\because \quad \frac{\partial M}{\partial y} \neq \frac{\partial N}{\partial x}, \text{ Eqn. (15.28) is exact.}$$

Here, $\frac{1}{N}\left(\frac{\partial M}{\partial y} - \frac{\partial N}{\partial x}\right) = P$ is a function of x only or a constant.

$$\therefore \quad \text{I.F.} = e^{\int P\,dx}.$$

Now multiplying both sides by $e^{\int P\,dx}$, we get

$$e^{\int P\,dx}\frac{dy}{dx} + Pye^{\int P\,dx} = Qe^{\int P\,dx} \quad \text{or,} \quad \frac{d}{dx}\left(ye^{\int P\,dx}\right) = Qe^{\int P\,dx}$$

or, $d\left(ye^{\int P\,dx}\right) = Qe^{\int P\,dx}dx.$

Integrating both sides w.r.t. x, we get

$$ye^{\int P\,dx} = \int Qe^{\int P\,dx}dx + C \tag{15.29}$$

or,

$$y = e^{-\int P\,dx}\left(\int Qe^{\int P\,dx}dx + C\right), \tag{15.30}$$

where C is an arbitrary constant.

Eqn. (15.29) shows that, the required solution is

$$y \times [\text{I.F.}] = \int Q\,(\text{I.F.})dx + C. \tag{15.31}$$

Example 15.6. *Find the solution of the first order ordinary differential equation*

$$(x^3 + 3)\frac{dy}{dx} + 4xy = x.$$

[AMIETE 1995]

Solution : We have $(x^3 + 3)\frac{dy}{dx} + 4xy = x$

or, $\frac{dy}{dx} + \frac{4x}{x^2 + 3}y = \frac{x}{x^2 + 3}.$

Here, integrating factor

$$\text{I.F.} = e^{\int \frac{4x}{x^2+3}dx} = e^{2\log(x^2+3)} = e^{\log(x^2+3)^2} = (x^2 + 3)^2.$$

$\therefore$ the solution is

$$y(x^2 + 3)^2 = \int \left(\frac{x}{x^2 + 3}\right)(x^2 + 3)^2dx + C = \int x(x^2 + 3)dx + C = \int (x^3 + 3x)dx + C$$

$$= \frac{x^4}{4} + \frac{3x^2}{2} + C.$$

15.5 Bernoulli's Equation

We consider the differential equation of the type

$$\frac{dy}{dx} + Py = Qy^n, \tag{15.32}$$

where P and Q are functions of x only or constants. Eqn. (15.32) is called **Bernoulli's equation**.

Multiplying both sides of Eqn. (15.32) by y^{-n}, we get

$$y^{-n}\frac{dy}{dx} + Py^{1-n} = Q. \tag{15.33}$$

We put, $v = y^{1-n}$, $\quad\therefore\quad \dfrac{dv}{dx} = (1-n)y^{-n}\dfrac{dy}{dx}$.

Then Eqn. (15.33) becomes

$$\frac{dv}{dx} + (1-n)Pv = (1-n)Q. \tag{15.34}$$

Eqn. (15.34) is a linear differential equation and can be solved by usual method of solving linear differential equation.

15.5.1 Ordinary Differential Equations of First Order and Higher Degree

An ODE of first order and higher degree involves $\frac{dy}{dx}$ in higher degree. An ODE of first order and nth degree is of the form :

$$p^n + P_1p^{n-1} + P_2p^{n-2} + +P_{n-1}p + P_n = 0, \tag{15.35}$$

where $p = \frac{dy}{dx}$ and $P_1, P_2, \ldots, P_n$ are functions of x and y or constants.

The following four special types of such equations will be discussed :

(a) Equations solvable for p

(b) Equations solvable for y

(c) Clairaut's form of equations and

(d) Equations solvable for x.

(a) Equations solvable for p : Expressing Eqn. (15.35) into n linear factors, we have

$$[p - f_1(x,y)][p - f_2(x,y)]\cdots[p - f_n(x,y)] = 0. \tag{15.36}$$

$\therefore$ each factor equated to zero gives a solution of Eqn. (15.36).

$\therefore\quad p = f_1(x,y);\ p = f_2(x,y);\ \cdots;\ p = f_n(x,y)$.

Solving each of these equations, we get

$$F_1(x,y,C_1);\quad F_2(x,y,C_2);\ \cdots;\ F_n(x,y,C_n) = 0.$$

Since Eqn. (15.36) is of order one, its general solution will contain only one arbitrary constant.

Hence, the general solution is

$$\boldsymbol{F_1(x,y,C);\quad F_2(x,y,C);\ \cdots;F_n(x,y,C) = 0.} \tag{15.37}$$

Example 15.7. *Solve* : $\dfrac{dy}{dx} - \dfrac{dx}{dy} = \dfrac{x}{y} - \dfrac{y}{x}$. **[Rewa U 1990, WBUT 2006]**

Solution : The given equation can be written as $p - \dfrac{1}{p} = \dfrac{x}{y} - \dfrac{y}{x}$, where $p = \dfrac{dy}{dx}$.

$$\therefore \quad \frac{p^2-1}{p} = \frac{x^2-y^2}{xy} \quad \text{or,} \quad p^2xy - (x^2-y^2)p - xy = 0$$

$$\text{or,} \quad p^2xy - x^2p + y^2p - xy = 0 \quad \text{or,} \quad (px+y)(py-x) = 0$$

If $px + y = 0$, then

$$x\frac{dy}{dx} + y = 0 \quad \text{or,} \quad \frac{dy}{y} + \frac{dx}{x} = 0.$$

Integrating, $\int \frac{dy}{y} + \int \frac{dx}{x} = 0$ or, $\log y + \log x = \log C$

or, $\boldsymbol{xy = C}$, where C is a constant.

If $px - x = 0$, then

$$y\frac{dy}{dx} - x = 0 \quad \text{or,} \quad y\,dy - x\,dx = 0.$$

Integrating, $\int y\,dy - \int x\,dx = 0$ or, $\frac{y^2}{2} - \frac{x^2}{2} = C$ or, $\boldsymbol{y^2 - x^2 = 2C}$.

$\therefore$ the general solution of the given equation is

$$\boldsymbol{(xy - C)(y^2 - x^2 - 2C) = 0}, \quad \text{where } C \text{ is an arbitrary constant.}$$

(b) Equations solvable for y : If the equation, $p^n + P_1p^{n-1} + P_2p^{n-2} + \cdots + + P_{n-1}p + P_n = 0$, is solvable for y, then it can be written as

$$y = f(x, p). \tag{15.38}$$

Differentiating the above equation w.r.t. x, we get

$$\frac{dy}{dx} = p = \phi\left(x, p, \frac{dp}{dx}\right) \tag{15.39}$$

This equation involves two variables viz., x, p having a solution of the form

$$F(x, p, C) = 0 \tag{15.40}$$

Solving Eqns. (15.38) and (15.39) for x, y, we get

$$\left.\begin{aligned} \boldsymbol{x = f_1(p, C)} \\ \boldsymbol{y = f_2(p, C)} \end{aligned}\right\} \tag{15.41}$$

where p is a parameter and c is an arbitrary constant. The above two equations (15.41) represent the solution in parametric form.

Example 15.8. *Solve : $y = 2px + p^n$.*

Solution : Here, $y = 2px + p^n$. (i)

Differentiating both sides of Eqn. (i) w.r.t. x and putting $p = \frac{dy}{dx}$, we get

$$p = 2p + 2x\frac{dp}{dx} + np^{n-1}\frac{dp}{dx} \quad \text{or,} \quad 0 = p\frac{dx}{dp} + 2x + np^{n-1}.$$

$$\text{or,} \quad \frac{dx}{dp} + \frac{2x}{p} = -np^{n-2} \tag{ii}$$

Eqn. (ii) is a linear equation in x having integrating factor, I.F. $= e^{\int \frac{2dp}{p}} = e^{\log p^2} = p^2$.

Now, multiplying both sides of Eqn. (ii) by p^2, we get

$$p^2\frac{dx}{dp} + 2xp = -np^n \quad \text{or,} \quad \frac{d}{dp}(xp^2) = -np^n \quad \text{or,} \quad d(xp^2) = -np^n dp.$$

Integrating, $\int d(xp^2) = -n\int p^n dp$

or, $\quad xp^2 = -\dfrac{np^{n+1}}{n+1} + C,$ where C is an arbitrary constant

or, $$x = -\frac{np^{n-1}}{n+1} + Cp^{-2}. \tag{iii}$$

Using this value of x in Eqn. (i), we get

$$y = \frac{1-n}{1+n}p^n + \frac{2C}{p}. \tag{iv}$$

Eqn. (iii) and (iv) together give the general solution of Eqn. (i).

(c) Clairaut's Equation : An *ODE* of the form

$$y = px + f(p) \tag{15.38}$$

is known as the *Clairaut's equation* after the name of *A. C. Clairaut*, who first solved this equation.

Differentiating Eqn. (15.38) w.r.t. x, we get

$$\frac{dy}{dx} = p + x\frac{dp}{dx} + f'(p)\frac{dp}{dx}$$

or, $$p = p + x\frac{dp}{dx} + f'(p)\frac{dp}{dx} \quad \text{or,} \quad 0 = x\frac{dp}{dx} + f'(p)\frac{dp}{dx}$$

or, $$[x + f'(p)]\frac{dp}{dx} = 0 \quad \left[\text{where } p = \frac{dy}{dx}\right]$$

Now either $\dfrac{dp}{dx} = 0$

or, $$p = C \text{ (a constant)} \tag{15.39}$$

or, $$x + f'(p) = 0. \tag{15.40}$$

Eliminating p form (15.38) and (15.40), we get

$$y = Cx + f(C). \tag{15.41}$$

Eqn. (15.41) represents the general solution of the given equation.

Eliminating p from (15.38) and (15.40) we get another solution free from any arbitrary constant which cannot be deduced from the general solution (15.41). This solution is the **singular solution** of Eqn. (15.38).

N.B.

(i) A singular solution of a differential equation cannot be deduced from the general solution.

(ii) The general solution of Clairaut's equation can be directly written by replacing p in the equation by an arbitrary constant C.

Example 15.9. *Solve* : $y = x\dfrac{dy}{dx} + \left(\dfrac{dy}{dx}\right)^2$. **[WBUT 2006]**

Solution : The given equation can be written as

$$y = px + p^2 \quad \left[\text{where } p = \frac{dy}{dx}\right] \tag{i}$$

This is a Clairaut's equation.

Differentiating both sides of Eqn.(i) w.r.t. x, we get

$$p = p + x\frac{dp}{dx} + 2p\frac{dp}{dx}$$

or, $$(2p + x)\frac{dp}{dx} = 0. \tag{ii}$$

$\therefore$ either $\dfrac{dp}{dx} = 0$ or, $p = C$, an arbitrary constant

or, $2p + x = 0$

or, $$p = -\frac{x}{2}. \tag{iii}$$

Putting $p = C$, the general solution is

$$y = Cx + C^2 \tag{iv}$$

Putting $p = -\frac{x}{2}$ in Eqn. (i), we get

$$y = -\frac{x^2}{2} + \frac{x^2}{4} = -\frac{x^2}{4}$$

or, $$4y = -x^2. \tag{v}$$

This is the singular solution of Eqn. (i).

(d) Equations solvable for x : If the given equation is solvable for x, then it can be expressed as

$$x = f(y, p). \tag{i}$$

Differentiating w.r.t. y, we get

$$\frac{dx}{dy} = \frac{1}{p} = \phi\left(y, p, \frac{dp}{dy}\right). \tag{ii}$$

Eqn. (i) may be possible to solve in y and p. Let this solution be

$$F(y, p, C) = 0. \tag{iii}$$

Eliminating p from (i) and (iii) we get the required solution. If elimination is not feasible, (i) and (iii) may be expressed in terms of p and p may be regarded as a parameter.

Example 15.10. *Find the general solution and singular solution of* $y = 2px + y^2p^3$.

Solution : The given equation can be written as

$$2x = \frac{y}{p} - y^2p^2. \tag{i}$$

Differentiating w.r.t. y and putting $\frac{1}{p} = \frac{dx}{dy}$, we get

$$\frac{2}{p} = \frac{1}{p} - \frac{y}{p^2}\frac{dy}{dy} - 2yp^2 - 2y^2p\frac{dp}{dy} \quad \text{or,} \quad \frac{1}{p} + 2yp^2 + \left(\frac{1}{p} + 2yp^2\right)\frac{y}{p}\frac{dp}{dy} = 0$$

or, $$\left(\frac{1}{p} + 2yp^2\right)\left(1 + \frac{y}{p}\frac{dp}{dy}\right) = 0.$$

$\therefore$ either $1 + \frac{y}{p}\frac{dp}{dy} = 0$ or, $\frac{1}{p} + 2yp^2 = 0$ or, $yp^3 = -\frac{1}{2}$.

(a) When $1 + \frac{y}{p}\frac{dp}{dy} = 0$, $\frac{dy}{y} + \frac{dp}{p} = 0$.

Integrating, $\int \frac{dy}{y} + \int \frac{dp}{p} = 0$. or, $\log y + \log p = \log C$ or, $\log(yp) = \log C$

or, $py = C$ or, $p = \frac{C}{y}$.

Putting this value of p in the given eqn. we get

$$y = 2\frac{C}{y}x + y\left(\frac{C^2}{y^2}\right)$$

or, $\boldsymbol{y^2 = 2Cx + C^3}$, C is an arbitrary constant. (ii)

Eqn. (ii) is the general solution of the given equation.

(b) When $yp^3 = -\frac{1}{2}$, we get from the given equation

$$y = 2px - \frac{1}{2}y \quad \text{or,} \quad 2px = \frac{3}{2}y \quad \text{or,} \quad 4px = 3y \quad \text{or,} \quad 64p^3x^3 = 27y^3$$

or, $$64\left(-\frac{1}{2y}\right)x^3 = 27y^3 \quad \left[\because\ yp^3 = -\frac{1}{2}\right]$$

or, $\boldsymbol{32x^3 + 27y^4 = 0}$. (iii)

Eqn. (iii) is the required singular solution, it cannot be deduced from the general solution.

15.6 Lagrange's Equation

A differential equation of the form

$$y = x\phi(p) + \psi(p), \tag{15.42}$$

where $p = \frac{dy}{dx}$ and $\phi(p)$, $\psi(p)$ are functions of p, is known as **Lagrange's equation**. For $\phi(p) = p$, the above equation reduces to **Clairaut's equation**. Let $\phi(p) \neq p$.

Differentiating equation (15.42) w.r.t. x, we get

$$p = \phi(p) + \{x\phi'(p) + \psi'(p)\}\frac{dp}{dx} \quad \text{or,} \quad \{\phi(p) - p\}\frac{dx}{dp} + x\phi'(p) + \psi'(p) = 0$$

or, $$\frac{dx}{dp} + \frac{\phi'(p)}{\phi(p) - p}x = -\frac{\psi'(p)}{\phi(p) - p}. \tag{15.43}$$

Eqn. (15.43) can be integrated following the previous methods and we get

$$\boldsymbol{x = Cf(p) + g(p)}. \tag{15.44}$$

15.7 Solved Problems

Example 15.1. *Solve the following* : $(x+y)^2\left(x\dfrac{dy}{dx}+y\right)=xy\left(1+\dfrac{dy}{dx}\right)$ **[Pune U. Dec. 1991]**

Solution : We have $(x+y)^2\left(x\dfrac{dy}{dx}+y\right)=xy\left(1+\dfrac{dy}{dx}\right)$ or, $\dfrac{x\frac{dy}{dx}+y}{xy}=\dfrac{1+\frac{dy}{dx}}{(x+y)^2}$.

Integrating, $\boldsymbol{\log xy = -\dfrac{1}{x+y} + C}$.

Example 15.2. *Solve* : $(2x^2+3y^2-7)x\,dx-(3x^2+2y^2-8)y\,dy=0$.

Solution : The given equation may be written as

$$\frac{y\,dy}{x\,dx}=\frac{2x^2+3y^2-7}{3x^2+2y^2-8}$$

or, $$\frac{y\,dy+x\,dx}{y\,dy-x\,dx}=\frac{5(x^2+y^2-3)}{-x^2+y^2+1}\quad\text{(by componendo and dividendo)}$$

or, $$\frac{x\,dx+y\,dy}{x^2+y^2-3}=5\frac{x\,dx-y\,dy}{x^2-y^2-1}\quad\text{or,}\quad\frac{2x\,dx+2y\,dy}{x^2+y^2-3}=\frac{5(2x\,dx-2y\,dy)}{x^2-y^2-1}$$

or, $$\frac{d(x^2+y^2-3)}{x^2+y^2-3}=5\frac{d(x^2-y^2-1)}{x^2-y^2-1}.$$

Integrating, $$\int\frac{d(x^2+y^2-3)}{x^2+y^2-3}=5\int\frac{d(x^2-y^2-1)}{x^2-y^2-1}$$

or, $\log(x^2+y^2-3)=5\log(x^2-y^2-1)+\log C$

or, $\boldsymbol{x^2+y^2-3=C(x^2-y^2-1)^5}$, **where C is an arbitrary constant.**

Example 15.3. *Solve* : $(1+y^2)x+(1+x^2)\dfrac{dy}{dx}=0$.

Solution : The given equation can be written as $\dfrac{x\,dx}{1+x^2}+\dfrac{dy}{1+y^2}=0$.

Integrating, $\dfrac{1}{2}\log(1+x^2)+\tan^{-1}y=C$ or, $\log(1+x^2)^{\frac{1}{2}}+\tan^{-1}y=C$

or, $$\tan^{-1}y=C-\log(1+x^2)^{\frac{1}{2}}=\log\frac{C}{(1+x^2)^{\frac{1}{2}}}$$

or, $$\boldsymbol{y=\tan^{-1}\left[\log\frac{C}{(1+x^2)^{\frac{1}{2}}}\right]}.$$

Example 15.4. *Solve* : $y(2xy+e^x)dx=e^x dy$.

Solution : The given equation can be written as $(ye^x dx-e^x dy)+2xy^2dx=0$

or, $$\frac{ye^x dx-e^x dy}{y^2}+2x\,dx=0\quad\text{or,}\quad d\left(\frac{e^x}{y}\right)+d(x^2)=0\quad\text{or,}\quad d\left(\frac{e^x}{y}+x^2\right)=0.$$

Integrating, $\boldsymbol{\dfrac{e^x}{y}+x^2=C}$, where C is an arbitrary constant.

Example 15.5. *Solve* : $(x^2+y^2+2xy)dx+xy\,dy=0$.

Solution : The given equation can be written as

$$(x^2+2x)dx+y^2dx+xy\,dy=0 \quad \text{or,} \quad (x^2+2x)dx+y(y\,dx+x\,dy)=0$$

or, $(x^2+2x)dx+y\,d(xy)=0$ or, $(x^3+2x^2)dx+xy\,d(xy)=0$.

Integrating, $\int(x^3+2x^2)dx+\int xy\,d(xy)=0$

or, $\dfrac{x^4}{4}+\dfrac{2x^3}{3}+\dfrac{(xy)^2}{2}=C$, where C is an arbitrary constant.

Example 15.6. *Solve* : $y(y^2-2x^2)dx+x(2y^2-x^2)dy=0$.

Solution : The given equation is of the form

$M\,dx+N\,dy=0$, where $M=y(y^2-2x^2)$ and $N=x(2y^2-x^2)$.

$$\therefore \quad \frac{\partial M}{\partial y}=3y^2-2x^2, \quad \frac{\partial N}{\partial x}=2y^2-3x^2.$$

As $\dfrac{\partial M}{\partial y}\neq\dfrac{\partial N}{\partial x}$, the given equation is not exact.

Here both M and N are homogeneous functions of x and y and are of the same degree. Also,

$$Mx+Ny=xy(y^2-2x^2)+xy(2y^2-x^2)=3xy(y^2-x^2)\neq 0, \text{ unless } x=y.$$

$\therefore$ the integrating factor I.F. $=\dfrac{1}{3xy(y^2-x^2)}$.

Now multiplying both sides of the given equation by $\dfrac{1}{3xy(y^2-x^2)}$, we get

$$\frac{y(y^2-2x^2)}{3xy(y^2-x^2)}dx+\frac{x(2y^2-x^2)}{3xy(y^2-x^2)}=0 \quad \text{or,} \quad \frac{(y^2-x^2)-x^2}{x(y^2-x^2)}dx+\frac{y^2+(y^2-x^2)}{y(y^2-x^2)}dy=0$$

or, $\dfrac{dx}{x}-\dfrac{x\,dx}{y^2-x^2}+\dfrac{y\,dy}{y^2-x^2}+\dfrac{dy}{y}=0$ or, $d(\log x)-\dfrac{1}{2}\dfrac{(2y\,dy-2x\,dx)}{y^2-x^2}+d\log(y)=0$

or, $d(\log xy)+\dfrac{1}{2}\dfrac{d(y^2-x^2)}{y^2-x^2}=0$.

Integrating, $\int d(\log xy)+\dfrac{1}{2}\int\dfrac{d(y^2-x^2)}{y^2-x^2}=0$

or, $\log(xy)+\dfrac{1}{2}\log(y^2-x^2)=\log C$ or, $\log[x^2y^2(y^2-x^2)]=2\log C$.

$\therefore \quad \boldsymbol{x^2y^2(y^2-x^2)=C^2}$ [C is an arbitrary constant]

Example 15.7. *Solve* : $y\cos x\,dx-\sin x\,dy=0$ *after converting the equation to an exact one.*

Solution : We have $y\cos x\,dx-\sin x\,dy=0$.

It can be written in the form $M\,dx+N\,dy=0$, where $M=y\cos x$ and $N=-\sin x$.

$$\therefore \quad \frac{\partial M}{\partial y}=\cos x \quad \text{and} \quad \frac{\partial N}{\partial x}=-\cos x$$

$$\therefore \quad \frac{\partial M}{\partial y} - \frac{\partial N}{\partial x} = 2\cos x \quad \text{or,} \quad \frac{1}{M}\left(\frac{\partial M}{\partial y} - \frac{\partial N}{\partial x}\right) = \frac{2}{y}.$$

$$\therefore \quad \text{I.F.} = e^{-\int \frac{2}{y}dy} = \frac{1}{y^2}.$$

Multiplying both sides of the given equation by I.F. $\left(\frac{1}{y^2}\right)$, we get

$$\frac{\cos x}{y}dx - \frac{\sin x}{y^2}dy = 0, \text{ which is an exact equation.}$$

$\therefore$ the general solution is $\boldsymbol{\sin x + Cy = 0}$.

Example 15.8. *Solve* : $(1 + xy)y\,dx + (1 - xy)x\,dx = 0$.

Solution : The given equation is of the form $M\,dx + N\,dy = 0$, where $M = (1 + xy)y$ and $N = (1 - xy)x$.

$$\therefore \quad \frac{\partial M}{\partial y} = 1 + 2xy, \quad \frac{\partial N}{\partial x} = 1 - 2xy.$$

$$\therefore \quad \frac{\partial M}{\partial y} \neq \frac{\partial N}{\partial x} \text{ and hence, the given equation is not exact.}$$

Here M and N are both of the forms $yf(x, y)$, $xg(x, y)$ respectively.

Also, $Mx - Ny = (1 + xy)xy - (1 - xy)xy = 2x^2y^2 \neq 0$.

$$\therefore \quad \frac{1}{2x^2y^2} \text{ is an I.F.}$$

Multiplying both sides of the given equation by $\frac{1}{2x^2y^2}$, we get

$$\frac{1}{2}\left(\frac{1+xy}{x^2y}\right)dx + \frac{1}{2}\left(\frac{1-xy}{xy^2}\right)dy = 0 \quad \text{or,} \quad \left(\frac{1}{x^2y}dx + \frac{1}{xy^2}dy\right) + \left(\frac{1}{x}dx - \frac{1}{y}dy\right) = 0$$

$$\text{or,} \quad \frac{y\,dx + x\,dy}{x^2y^2} + \frac{y\,dx - x\,dy}{xy} = 0 \quad \text{or,} \quad \frac{d(xy)}{(xy)^2} + \frac{y\,dx - x\,dy}{y^2} \cdot \frac{y^2}{xy} = 0$$

$$\text{or,} \quad \frac{d(xy)}{(xy)^2} + \frac{d\left(\frac{x}{y}\right)}{\frac{x}{y}} = 0.$$

$$\text{Integrating,} \int \frac{d(xy)}{(xy)^2} + \int \frac{d\left(\frac{x}{y}\right)}{\frac{x}{y}} = 0$$

$$\text{or,} \quad -\frac{1}{xy} + \log\left|\frac{x}{y}\right| = C, \quad C \text{ is an arbitrary constant.}$$

Example 15.9. *Solve* : $(2x\log x - xy)dy + 2y\,dx = 0$.

Solution : The equation is of the form $M\,dx + N\,dy = 0$, where $M = 2y$ and $N = 2x\log x - xy$.

$$\therefore \quad \frac{\partial M}{\partial y} = 2, \quad \frac{\partial N}{\partial x} = 2(1 + \log x) - y.$$

$$\because \frac{\partial M}{\partial y} \neq \frac{\partial N}{\partial x}, \text{ the given equation is not exact.}$$

Here, $\frac{1}{N}\left(\frac{\partial M}{\partial y}-\frac{\partial N}{\partial x}\right)=\frac{-2\log x+y}{x(2\log x-y)}=-\frac{1}{x}$, a function of x only.

$\therefore$ I.F. $=e^{\int-\frac{1}{x}dx}=e^{-\log x}=e^{\log\frac{1}{x}}=\frac{1}{x}$.

Multiplying both sides of the given equation by $\frac{1}{x}$, we get

$$2\frac{y}{x}dx+(2\log x-y)dy=0 \qquad \text{(i)}$$

or, $M_1dx+N_1dy=0$, where $M_1=\frac{2y}{x}$ and $N_1=2\log x-y$.

$\therefore$ $\frac{\partial M_1}{\partial y}=\frac{2}{x}=\frac{\partial N_1}{\partial x}$. $\therefore$ Eqn. (i) is exact.

Here, $\int M_1dx=\int\frac{2y}{x}dx=2y\log x$, if y is constant

and $\int N_1dy=\int(2\log x-y)dy=2y\log x-\frac{y^2}{2}$, if x is constant.

$\therefore$ the required solution is $\mathbf{2y\log x-\frac{y^2}{2}+C}$, C being an arbitrary constant.

Example 15.10. *Solve* : $(3x+2y^2)y\,dx+2x(2x+3y^2)dy=0$

Solution : The given equation is of the form $M\,dx+N\,dy=0$, where $M=3xy+2y^3$ and $N=4x^2+6xy^2$.

$\therefore$ $\frac{\partial M}{\partial y}=3x+6y^2$, $\frac{\partial N}{\partial x}=8x+6y^2$

As $\frac{\partial M}{\partial y}\neq\frac{\partial N}{\partial x}$, the given equation is not exact.

Again, the given equation can be written as

$$x(3y\,dx+4x\,dy)+y^2(2y\,dx+6x\,dy)=0 \qquad \text{(i)}$$

$$x^ay^b(my\,dx+nx\,dy)+x^{a_1}y^{b_1}(m_1y\,dx+n_1x\,dy)=0. \qquad \text{(ii)}$$

Comparing equations (i) and (ii), we get

$$a=1,\ b=0,\ m=3,\ n=4,\ a_1=0,\ b_1=2,\ m_1=2,\ n_1=6.$$

$\therefore$ I.F. $=x^hy^k$,

where $\frac{a+h+1}{m}=\frac{b+k+1}{n}$, $\frac{a_1+h+1}{m_1}=\frac{b_1+k+1}{n_1}$

i.e., $\frac{1+h+1}{3}=\frac{0+k+1}{4}$, $\frac{0+h+1}{2}=\frac{2+k+1}{6}$ or, $4h-3k=-5$, $3h=k$.

Solving we get $h=1$, $k=3$.

$\therefore$ I.F. $=xy^3$.

Now multiplying both sides of the given equation by xy^3, we get

$$xy^4(3x+2y^2)dx+2x^2y^3(2x+3y^2)dy=0. \qquad \text{(iii)}$$

Eqn. (iii) is of the form

$$M_1dx + N_1dy = 0, \quad \text{where } M_1 = 3x^2y^4 + 2xy^6 \text{ and } N_1 = 4x^3y^3 + 6x^2y^5.$$

$$\therefore \quad \frac{\partial M_1}{\partial y} = 12x^2y^3 + 12xy^5 = \frac{\partial N_1}{\partial x}. \quad \therefore \text{ Eqn. (iii) is exact.}$$

Now, $\int M_1dx = (3x^2y^4 + 2xy^6)dx = x^3y^4 + x^2y^6$, assuming $y =$ constant

and $\int N_1dy = (4x^3y^3 + 6x^2y^5)dy = x^3y^4 + x^2y^6$, if x is constant

Hence, the general solution of the given equation is

$$\boldsymbol{x^3y^4 + x^2y^6 = C}, \quad C \text{ being an arbitrary constant.}$$

Example 15.11. *Solve : $xy^2dx + (2 + x^2y)dy = 0$ with the condition that $y(1) = 2$.*

Solution : In this case, $M(x, y) = xy^2$, $\quad \therefore \quad \dfrac{\partial M}{\partial y} = 2xy.$

$$N(x, y) = 2 + x^2y, \quad \therefore \quad \frac{\partial N}{\partial x} = 2xy.$$

$$\therefore \quad \frac{\partial M}{\partial y} = \frac{\partial N}{\partial x}, \text{ so the given equation is exact.}$$

Then the solution of the equation is $\phi(x, y) =$ constant.

$$\therefore \quad \frac{\partial \phi}{\partial x} = M(x, y) = xy^2, \quad \text{i.e.,} \quad \phi(x, y) = \frac{x^2y^2}{2} + f(y) \qquad \text{(i)}$$

$$\text{and} \quad \frac{\partial \phi}{\partial y} = N(x, y) = 2 + x^2y, \quad \text{i.e.,} \quad \phi(x, y) = 2y + \frac{x^2y^2}{2} + g(x). \qquad \text{(ii)}$$

For the consistency of (i) and (ii), $f(y) = 2y$ and $g(x) = 0$.

$$\therefore \quad \phi(x, y) = \frac{x^2y^2}{2} + 2y.$$

Hence, the solution of the differential equation is

$$\frac{x^2y^2}{2} + 2y = C \quad [C = \text{an arbitrary constant}]$$

By the given condition, for $x = 1$, $y = 2$, $\therefore C = 6$.

Thus, the solution of the equation is $\boldsymbol{\dfrac{x^2y^2}{2} + 2y = 6}$.

Example 15.12. *Solve :* $\dfrac{dy}{dx} + \dfrac{y \log y}{x} = \dfrac{y(\log y)^2}{x^2}$ **[WBUT 2007]**

Solution : Dividing both sides of the given equation by $y(\log y)^2$, we get

$$\frac{1}{y(\log y)^2}\frac{dy}{dx} + \frac{1}{\log y}\frac{1}{x} = \frac{1}{x^2}. \qquad \text{(i)}$$

Put $v = \dfrac{1}{\log y}$, $\quad \therefore \quad \dfrac{dv}{dx} = -\dfrac{1}{y(\log y)^2}\dfrac{dy}{dx}.$

Then Eqn. (i) reduces to

$$-\frac{dv}{dx}+\frac{v}{x}=\frac{1}{x^2} \quad \text{or,} \quad \frac{dv}{dx}-\frac{v}{x}=-\frac{1}{x^2}. \qquad \text{(ii)}$$

Eqn. (ii) is a linear equation in v.

$$\text{I.F.} = e^{\int -\frac{1}{x}dx} = e^{-\log x} = e^{\log \frac{1}{x}} = \frac{1}{x}.$$

Multiplying both sides of Eqn. (ii) by $\frac{1}{x}$,

$$\frac{1}{x}\frac{dv}{dx}-\frac{v}{x^2}=-\frac{1}{x^3} \quad \text{or,} \quad \frac{d}{dx}\left(\frac{v}{x}\right)=-\frac{1}{x^3} \quad \text{or,} \quad d\left(\frac{v}{x}\right)=-\frac{1}{x^3}dx.$$

Integrating, $\int d\left(\frac{v}{x}\right) = -\int \frac{1}{x^3}dx$ or, $\frac{v}{x}=\frac{1}{2x^2}+C$

$$\text{or,} \quad \frac{1}{x\log y}=\frac{1}{2x^2}+C, \quad \left[\because v=\frac{1}{\log y}\right] \qquad \text{(iii)}$$

where C = an arbitrary constant.

Eqn. (iii) is the general solution of the given equation.

Example 15.13. *Solve* : $2xy\frac{dy}{dx}=x^2+y^2$.

Solution : The given equation may be written as

$$\frac{dy}{dx}=\frac{x^2+y^2}{2xy}. \qquad \text{(i)}$$

Eqn. (i) is homogeneous.

Let its solution be $y = vx$, where v is a function of x.

$$\therefore \quad \frac{dy}{dx}=v+x\frac{dv}{dx}. \qquad \text{(ii)}$$

Using equation (ii) in (i), we get

$$v+x\frac{dv}{dx}=\frac{x^2+y^2}{2xy}=\frac{x^2+v^2x^2}{2x\cdot vx}=\frac{1+v^2}{2v}$$

$$\text{or,} \quad x\frac{dv}{dx}=\frac{1+v^2}{2v}-v=\frac{1-v^2}{2v} \quad \text{or,} \quad \frac{dx}{x}=\frac{2v}{1-v^2}dv.$$

On integration, we get

$$\int\frac{dx}{x}=\int\frac{2v}{1-v^2}dv+C'$$

$$\text{or,} \quad \ln x = -\ln(1-v^2)+\ln C \qquad \text{(iii)}$$

$$\text{or,} \quad x(1-v^2)=C.$$

Substituting $v=\frac{y}{x}$ in Eqn. (iii), we get

$$\boldsymbol{x^2-y^2=Cx.}$$

This is the required general solution.

Example 15.14. *Solve* : $\dfrac{dy}{dx} = \dfrac{x+y+3}{x-y-5}$.

Solution : The given equation is inhomogeneous. Changing the variables it can be put in homogeneous form.

Let $x = X + h$, $y = Y + k$, where h and k are constants.

So, the equation becomes $\dfrac{dY}{dX} = \dfrac{X+Y+h+k+3}{X-Y+h-k+5}$.

If now, $h + k + 3 = 0$ and $h - k - 5 = 0$, then on solving, $h = 1$, $k = -4$.

$$\therefore \quad \frac{dY}{dX} = \frac{X+Y}{X-Y}.$$

It is a homogeneous equation.

Now putting $Y = vX$, $\dfrac{dY}{dX} = v + X\dfrac{dv}{dx}$

$$\therefore \quad v + X\frac{dv}{dX} = \frac{X+vX}{X-vX} = \frac{1+v}{1-v}. \qquad \therefore \ X\frac{dv}{dX} = \frac{1+v}{1-v} - v = \frac{1+v^2}{1-v}.$$

$$\therefore \quad \frac{dX}{X} = \frac{dv}{1+v^2} - \frac{v\,dv}{1+v^2}.$$

Integrating, $\log X = \tan^{-1} v - \dfrac{1}{2}\log(1+v^2) + \log C.$ (i)

$$\therefore \quad X = \frac{Ce^{\tan^{-1} v}}{\sqrt{1+v^2}}.$$

Using this value of X in Eqn. (i) and simplifying,

$$X^2 + Y^2 = Ce^{2\tan^{-1}\left(\frac{Y}{X}\right)}.$$

Again substituting for X and Y, we get

$$(x-1)^2 + (y+4)^2 = Ce^{2\tan^{-1}\left[\frac{(y+4)}{(x-1)}\right]}. \qquad \text{(ii)}$$

This is the required general solution of the given equation.

Example 15.15. *Solve* : $\dfrac{dy}{dx} = \dfrac{6x+2y-7}{3x+y-6}$.

Solution : If we put $x = X + h$ and $y = Y + k$ then the set of equations giving h and k cannot be solved as the determinant of the coefficients of the variables, $\begin{vmatrix} 6 & 2 \\ 3 & 1 \end{vmatrix} = 0$.

However, it can be reduced to 'variable separable' form as below.

Let $3x + y = v$.

$$\therefore \quad 3 + \frac{dy}{dx} = \frac{dv}{dx}.$$

Then the given equation reduces to

$$\frac{dy}{dx} = \frac{2v-7}{v-6} \quad \text{or,} \quad \frac{dv}{dx} - 3 = \frac{2v-7}{v-6} \quad \text{or,} \quad \frac{dv}{dx} = 3 + \frac{2v-7}{v-6} = \frac{5v-25}{v-6} = \frac{5(v-5)}{v-6}.$$

$$\therefore \quad dx = \frac{v-6}{5(v-6)}dv = \frac{1}{5}\left(1 - \frac{1}{v-5}\right)dv.$$

Integrating we get $5x = v - \log(v-5) + C$.

Substituting the value of v and simplifying,

$$\mathbf{2x - y + \log(3x + y - 5) = C}, \quad \text{where } C \text{ is an arbitrary constant.} \qquad \text{(i)}$$

Eqn. (i) represents the general solution of the given equation.

Example 15.16. *Solve* : $3x(1-x^2)y^2\dfrac{dy}{dx} + (2x^2-1)y^3 = kx^3$.

Solution : We put $v = y^3$, then, $\dfrac{dv}{dx} = 3y^2\dfrac{dy}{dx}$.

The given equation then reduces to

$$x(1-x^2)\frac{dv}{dx} + (2x^2-1)v = kx^3$$

$$\text{or,} \quad \frac{dv}{dx} + \frac{(2x^2-1)}{x(1-x^2)}v = \frac{kx^2}{1-x^2}. \qquad \text{(i)}$$

which is linear in v.

$$\text{I.F.} = e^{\int \frac{2x^2-1}{x(1-x^2)}dx}.$$

$$\text{Now,} \quad \frac{2x^2-1}{x(1-x^2)} = \frac{2x^2-1}{x(1-x)(1+x)} = \frac{A}{x} + \frac{B}{1-x} + \frac{C}{1+x}.$$

$$\therefore \quad 2x^2 - 1 = A(1-x)(1+x) + B(1+x)x + Cx(1-x).$$

This being an identity, is true for all x.

Put $x = 1$, then $B = \dfrac{1}{2}$; $x = -1$, then $C = -\dfrac{1}{2}$; $x = 0$, then $A = -1$.

$$\therefore \quad \frac{2x^2-1}{x(x^2-1)} = -\frac{1}{x} + \frac{1}{2}\cdot\frac{1}{1-x} - \frac{1}{2}\cdot\frac{1}{1+x}.$$

$$\therefore \quad \int \frac{(2x^2-1)}{x(1-x^2)}dx = -\log x - \frac{1}{2}\log(1-x) - \frac{1}{2}\log(1+x) = -\log x(x\sqrt{1-x^2})$$

$$= \log(x\sqrt{1-x^2})^{-1}.$$

$$\therefore \quad \text{I.F.} = (x\sqrt{1-x^2})^{-1}.$$

Multiplying both sides of Eqn. (i) by I.F., we get

$$\frac{d}{dx}\left[\frac{v}{x\sqrt{1-x^2}}\right] = kx(1-x^2)^{-\frac{3}{2}} \quad \text{or,} \quad d\left[\frac{v}{x\sqrt{1-x^2}}\right] = kx(1-x^2)^{-\frac{3}{2}}dx.$$

$$\text{Integrating,} \quad \frac{v}{x\sqrt{1-x^2}} = k(1-x^2)^{-\frac{1}{2}} + C.$$

$$\therefore \quad \mathbf{y^3 = kx + Cx\sqrt{1-x^2}} \quad [\because v = y^3] \qquad \text{(ii)}$$

where C = an arbitrary constant, Eqn. (ii) is the required general solution of the given equation.

Example 15.17. *Solve* : $\dfrac{dy}{dx} = \dfrac{y}{x+\sqrt{xy}}$.

Solution : The given differential equation can be written as

$$y\frac{dx}{dy} = x + \sqrt{xy} \quad \text{or,} \quad \frac{dx}{dy} - \frac{x}{y} = x^{\frac{1}{2}}y^{-\frac{1}{2}}.$$

This is a Bernoulli's equation in x.

$$\therefore \quad x^{-\frac{1}{2}}\frac{dx}{dy} - \frac{x^{\frac{1}{2}}}{y} = y^{-\frac{1}{2}}. \tag{i}$$

Put $z = x^{\frac{1}{2}}$. $\therefore \quad \dfrac{dz}{dy} = \dfrac{1}{2}x^{-\frac{1}{2}}\dfrac{dx}{dy}$.

Eqn. (i) now becomes $2\dfrac{dz}{dy} - \dfrac{z}{y} = y^{-\frac{1}{2}}$

$$\text{or,} \quad \frac{dz}{dy} - \frac{z}{2y} = \frac{1}{2}y^{-\frac{1}{2}}. \tag{ii}$$

Eqn. (ii) is a **linear equation** in z.

$$\therefore \quad \text{I.F.} = e^{-\frac{1}{2}\int \frac{dy}{y}} = e^{-\frac{1}{2}\log y} == e^{\log y^{-\frac{1}{2}}} = y^{-\frac{1}{2}}.$$

Now multiplying both sides of Eqn. (ii) by $y^{-\frac{1}{2}}$,

$$y^{-\frac{1}{2}}\frac{dz}{dy} - \frac{1}{2}y^{-\frac{3}{2}}z = \frac{1}{2y} \quad \text{or,} \quad \frac{d}{dy}(zy^{-\frac{1}{2}}) = \frac{1}{2y} \quad \text{or,} \quad d(zy^{-\frac{1}{2}}) = \frac{1}{2y}dy.$$

Integrating, $\displaystyle\int d(zy^{-\frac{1}{2}}) = \frac{1}{2}\int \frac{dy}{y}$ or, $zy^{-\frac{1}{2}} = \dfrac{1}{2}\log y + C$

$$\text{or,} \quad \boldsymbol{\sqrt{x} = \sqrt{y}(\log\sqrt{y} + C)} \quad [\because\ z = \sqrt{x}] \tag{iii}$$

where C is an arbitrary constant.

Eqn. (iii) is the general solution of the given equation.

Example 15.18. *Solve* : $\cos^2 x\dfrac{dy}{dx} + y = \tan x$. **[BU(SP) 1995]**

Solution : We have $\cos^2 x\dfrac{dy}{dx} + y = \tan x$

$$\text{or,} \quad \frac{dy}{dx} + y\sec^2 x = \tan x \sec^2 x. \tag{i}$$

Here, $\displaystyle\int P\,dx = \int \sec^2 x\,dx = \tan x$

$$\therefore \quad \text{I.F.} = e^{\tan x}$$

Multiplying both sides of equation (i) by I.F. and then integrating, we get

$$\begin{aligned} ye^{\tan x} &= \int e^{\tan x}\tan x \sec^2 x\,dx \\ &= \int ze^{z}dz \quad [\text{where } z = \tan x \text{ and } dz = \sec^2 x\,dx] \\ &= ze^{z} - e^{z} + C \end{aligned}$$

or, $\boldsymbol{y = \tan x e^{\tan x} - e^{\tan x} + C}$, where C is an arbitrary constant. (ii)

Eqn. (ii) is the general solution of the given equation.

Example 15.19. *Solve* : $\frac{dy}{dx} + \frac{3}{x}y = x^2$.

Solution : The given differential equation is a linear differential equation with

$$\text{I.F.} = e^{\int \frac{3}{x}dx} = e^{3\log x} = e^{\log x^3} = x^3.$$

Multiplying both sides of the given equation by x^3, we get

$$x^3\frac{dy}{dx} + 3x^2y = x^5 \quad \text{or,} \quad \frac{d}{dx}(yx^3) = x^5.$$

Integrating, $yx^3 = \int x^5 dx + C = \frac{x^6}{6} + C$

or, $\boldsymbol{y = \frac{x^3}{6} + \frac{C}{x^3}}$, where C is an arbitrary constant. (i)

Eqn. (i) is the general solution of the given equation.

Example 15.20. *Solve* : $y = xp^2 + p$.

Solution : Differentiating the given equation w.r.t. x and putting $p = \frac{dy}{dx}$, we get

$$p = p^2 + 2px\frac{dp}{dx} + \frac{dp}{dx} = p^2 + (2px + 1)\frac{dp}{dx} \quad \text{or,} \quad (p^2 - p)\frac{dx}{dp} + 2px = -1$$

or, $\frac{dx}{dp} + \frac{2p}{p(p-1)} = -\frac{1}{p(p-1)}$. (i)

Eqn. (i) is a linear equation in x.

$\therefore$ $\text{I.F.} = e^{2\int \frac{dp}{p-1}} = e^{2\log(p-1)} = e^{\log(p-1)^2} = (p-1)^2$.

Multiplying both sides of equation (i) by $(p-1)^2$, we get

$$(p-1)^2\frac{dx}{dp} + 2x(p-1) = \frac{1-p}{p} \quad \text{or,} \quad \frac{d}{dp}[x(p-1)^2] = \frac{1}{p} - 1$$

or, $d[x(p-1)^2] = \left(\frac{1}{p} - 1\right)dp$.

Integrating, $\int d[x(p-1)^2] = \int \left(\frac{1}{p} - 1\right)dp$.

$\therefore$ $x(p-1)^2 = \log p - p + C$ or, $\boldsymbol{x = (\log p - p + C)(p-1)^{-2}}$. (ii)

where C is an arbitrary constant.

Also, $y = xp^2 + p$

or, $\boldsymbol{y = p[p(\log p - p + C)(p-1)^{-2} + 1]}$ [using Eqn. (ii)] (iii)

Eqns. (ii) and (iii) taken together or the p-eliminants of (ii) and (iii) give the general solution.

Example 15.21. *Find the general solution and the singular solution of the given differential equation* $y = px + \sqrt{1+p^2}$.

Solution : The given equation is a **Clairaut's** form of equation.

Differentiating the given equation w.r.t. x and putting $p = \frac{dy}{dx}$, we get

$$p = p + x\frac{dp}{dx} + \frac{p}{\sqrt{1-p^2}}\frac{dp}{dx} \quad \text{or,} \quad \left[x + \frac{p}{\sqrt{1-p^2}}\right]\frac{dp}{dx} = 0.$$

$$\therefore \quad \text{either } \frac{dp}{dx} = 0 \quad \text{or,} \quad p = C, \text{ an arbitrary constant} \tag{i}$$

$$\text{or,} \quad x + \frac{p}{\sqrt{1-p^2}} = 0 \quad \text{or,} \quad x = -\frac{p}{\sqrt{1-p^2}} \tag{ii}$$

Using (i) in (ii), we get the general solution as $y = Cx + \sqrt{1+C^2}$.

Using Eqn. (ii) in the given equation, we get

$$y = -\frac{p}{\sqrt{1+p^2}} + \sqrt{1+p^2} = \frac{1}{\sqrt{1+p^2}} \quad \text{or,} \quad x^2 + y^2 = \frac{p^2}{1+p^2} + \frac{1}{1+p^2} = \frac{p^2+1}{1+p^2} = 1.$$

Thus, the singular solution is $\boldsymbol{x^2 + y^2 = 1}$.

Example 15.22. *Solve* : (a) $y = x\frac{dy}{dx} + \left(\frac{dy}{dx}\right)^2$ **[WBUT 2006]**

(b) $(y - px)(p - 1) = p$. *Obtain the singular solution, where* $p = \frac{dy}{dx}$. **[WBUT 2009]**

(c) $y = px + \sqrt{a^2p^2 + p^2}$, $p = \frac{dy}{dx}$. **[WBUT 2005]**

Solution : (a) The given equation is

$$y = px + p^2, \quad \text{where } p = \frac{dy}{dx}, \tag{i}$$

which is **Clairaut's equation**.

Differentiating Eqn. (i) w.r.t. x and then putting $p = \frac{dy}{dx}$, we get

$$p = p + x\frac{dp}{dx} + 2p\frac{dp}{dx} \quad \text{or,} \quad (2p + x)\frac{dp}{dx} = 0,$$

$$\text{Now, either } \frac{dp}{dx} = 0 \quad \text{or,} \quad p = C, \text{ an arbitrary constant} \tag{ii}$$

$$\text{or,} \quad (2p + x) = 0 \quad \text{or,} \quad p = -\frac{x}{2}. \tag{iii}$$

Using Eqn. (ii) in Eqn. (i) we get the general solution as

$$y = Cx + C^2. \tag{iv}$$

Again, putting $p = -\frac{x}{2}$ in Eqn. (i), we get

$$y = -\frac{x^2}{2} + \frac{x^2}{4} = -\frac{x^2}{4} \quad \text{or,} \quad \boldsymbol{4y = -x^2}. \tag{v}$$

This is the singular solution of the given differential equation.

(b) The given equation may be written as

$$y - px = \frac{p}{p-1} \quad \text{or,} \quad y = px + \frac{p}{1-p}. \tag{i}$$

This is **Clairaut's equation**.

Differentiating equation (i), w.r.t. x and then putting $p = \frac{dy}{dx}$, we get

$$p = p + x\frac{dp}{dx} - \frac{1}{(p-1)^2}\frac{dp}{dx} \qquad \left[\because\ \frac{p}{p-1} = \frac{p-1+1}{p-1} = 1 + \frac{1}{p-1}\right]$$

or, $$\left[x - \frac{1}{(p-1)^2}\right]\frac{dp}{dx} = 0.$$

Now, either $\dfrac{dp}{dx} = 0$, i.e., $p = C$ (constant)

or, $$x - \frac{1}{(p-1)^2} = 0 \quad \text{or,} \quad p = \frac{1+\sqrt{x}}{\sqrt{x}}.$$

Putting $p = C$ in Eqn. (i) we get the general solution as

$$\boldsymbol{y = Cx + \frac{C}{C-1}} \qquad [C \text{ an arbitrary constant}]$$

Again putting $p = \frac{1+\sqrt{x}}{\sqrt{x}}$ in Eqn. (i) we get the singular solution as

$$y = (1+\sqrt{x})\sqrt{x} + 1 + \sqrt{x} = x + 1 + 2\sqrt{x} \quad \text{or,} \quad y - x - 1 = 2\sqrt{x}$$

or, $\boldsymbol{(y - x - 1)^2 = 4x}$, which cannot be deduced from the general solution.

(c) The given equation is $y = px + \sqrt{a^2p^2 + b^2}$. (i)

This is *Clairaut's equation.*

Differentiating equation (i) w.r.t. x and putting $p = \frac{dy}{dx}$, we get

$$p = p + x\frac{dp}{dx} + \frac{a^2p}{\sqrt{a^2p^2+b^2}}\frac{dp}{dx} \quad \text{or,} \quad \left[x + \frac{a^2p}{\sqrt{a^2p^2+b^2}}\right]\frac{dp}{dx} = 0.$$

$\therefore$ either $\dfrac{dp}{dx} = 0$ or, $p = C$ (a constant) (ii)

or, $$x + \frac{a^2p}{\sqrt{a^2p^2+b^2}} = 0 \quad \text{or,} \quad x = -\frac{a^2p}{\sqrt{a^2p^2+b^2}}. \qquad \text{(iii)}$$

Using Eqn. (ii) in Eqn. (i) we get the general solution as

$$\boldsymbol{y = Cx + \sqrt{a^2C^2 + b^2}}, \quad [C \text{ is an arbitrary constant}]. \qquad \text{(iv)}$$

Using Eqn. (iii) in Eqn. (i), we get

$$y = -\frac{a^2p^2}{\sqrt{a^2p^2+b^2}} + \sqrt{a^2p^2+b^2} = \frac{b^2}{\sqrt{a^2p^2+b^2}}. \qquad \text{(v)}$$

From Eqns. (iii) and (v), we get

$$\left(\frac{x}{a}\right)^2 + \left(\frac{y}{b}\right)^2 = \left(-\frac{ap}{\sqrt{a^2p^2+b^2}}\right)^2 + \left(\frac{b}{\sqrt{a^2p^2+b^2}}\right)^2 = \frac{a^2p^2+b^2}{a^2p^2+b^2} = 1$$

or, $$\boldsymbol{\left(\frac{x}{a}\right)^2 + \left(\frac{y}{b}\right)^2 = 1}. \qquad \text{(vi)}$$

Eqn. (vi) cannot be deduced from the general solution and hence, it is a singular solution.

Example 15.23. *Solve* : (a) $\frac{dy}{dx} + \frac{x}{\sqrt{1-x}}y = x\sqrt{y}$ (b) $y^3\frac{dy}{dx} + \frac{y^4}{x} = x.$

Solution : (a) The given equation is in **Bernoulli's** form.

Dividing both sides of the given equation by $\sqrt{y}$, we get

$$y^{-\frac{1}{2}}\frac{dy}{dx} + \frac{x}{\sqrt{1-x^2}}y^{\frac{1}{2}} = x \tag{i}$$

Put $y^{\frac{1}{2}} = v$, then $\frac{1}{2}y^{-\frac{1}{2}}\frac{dy}{dx} = \frac{dv}{dx}$

or, $$\frac{dv}{dx} + \frac{x}{2(1-x^2)}v = \frac{x}{2}. \tag{ii}$$

Eqn. (ii) is a linear differential equation.

$\therefore$ $$\int p\,dx = \int \frac{x\,dx}{2(1-x^2)} = -\frac{1}{4}\log(1-x^2) = \log(1-x^2)^{-\frac{1}{4}}.$$

$\therefore$ $$\text{I.F.} = e^{\log(1-x^2)^{-\frac{1}{4}}} = (1-x^2)^{-\frac{1}{4}}.$$

Multiplying both sides of Eqn. (ii) by $(1-x^2)^{-\frac{1}{4}}$, we get

$$\frac{v}{(1-x^2)^{\frac{1}{4}}} = \int \frac{x}{2(1-x^2)^{\frac{1}{4}}}dx = -\frac{1}{3}(1-x^2)^{\frac{3}{4}} + C \quad [C \text{ is a constant}]. \tag{iii}$$

The required solution is obtained by using Eqn. (iii) in Eqn. (ii) as

$$\sqrt{y} = -\frac{1}{3}(1-x^2) + C(1-x^2)^{\frac{1}{4}}. \tag{iv}$$

(b) The given equation after arranging, we get

$$\frac{dy}{dx} + \frac{1}{x}y = xy^{-3}. \tag{i}$$

Eqn. (i) is in *Bernoulli's* form. It can be put in linear form by putting

$$y^4 = v \quad \text{or,} \quad 4y^3\frac{dy}{dx} = \frac{dv}{dx}. \tag{ii}$$

Using Eqn. (ii) in the given equation and simplifying, we get

$$\frac{dv}{dx} + \frac{4}{x}v = 4x. \tag{iii}$$

This is a linear equation in v whose I.F. is $e^{\int \frac{4}{x}dx} = e^{4\log x} = e^{\log x^4} = x^4$.

Multiplying both sides of Eqn. (iii) by x^4, $x^4\frac{dv}{dx} + 4x^3v = 4x^5$ or, $\frac{d}{dx}(vx^4) = 4x^5$.

Integrating, $vx^4 = \frac{4x^6}{6} + C = \frac{2}{3}x^6 + C$ [C is a constant].

Now restoring the value of v in the above equation we get the general solution as

$$y^4x^4 = \frac{2}{3}x^6 + C$$

or, $$y = \left(\frac{2}{3}x^2 + \frac{C}{x^4}\right)^{\frac{1}{4}}.$$

Example 15.24. *Solve* : $\dfrac{dy}{dx}-\dfrac{\tan y}{1+x}=(1+x)e^x\sec y.$ [BU(SP) 1993]

Solution : We have $\dfrac{dy}{dx}-\dfrac{\tan y}{1+x}=(1+x)e^x\sec y$

or, $\cos y\dfrac{dy}{dx}-\dfrac{\sin y}{1+x}=(1+x)e^x.$ (i)

Putting $v=\sin y$, $\dfrac{dv}{dx}=\cos y\dfrac{dy}{dx}$,

Eqn. (i) becomes $\dfrac{dv}{dx}-\dfrac{v}{1+x}=(1+x)e^x.$

Now, $\displaystyle\int P\,dx=\int -\frac{1}{1+x}dx=\log\left(\frac{1}{1+x}\right).$

$\therefore$ I.F. $=e^{\int P\,dx}=\dfrac{1}{1+x}.$

$\therefore$ the general solution is $\displaystyle v\cdot\frac{1}{1+x}=\int\frac{1+x}{1+x}e^x\,dx=e^x+C.$

Restoring the value of v, $\boldsymbol{\sin y=(1+x)(e^x+C)}$, where C is an arbitrary constant.

Example 15.25. *Solve* : $y=x(p+a)-\frac{1}{4}(p+a)^2$, *where* $p=\dfrac{dy}{dx}.$

Solution : Differentiating the given equation w.r.t. x, we get

$$p=p+a+x\frac{dp}{dx}-\frac{1}{2}(p+a)\frac{dp}{dx}\quad\text{or}\quad a+\left[x-\frac{1}{2}(p+a)\right]\frac{dp}{dx}=0 \qquad \text{(i)}$$

or, $\dfrac{dx}{dp}+\dfrac{x}{a}=\dfrac{p+a}{2a}.$ (ii)

The integrating factor I.F. $=e^{\int\frac{1}{a}dp}=e^{\frac{p}{a}}.$

$\therefore$ the solution of the given differential equation is

$$x=e^{-\frac{p}{a}}\left[\int\frac{p+a}{2a}e^{\frac{p}{a}}dp\right]+Ce^{-\frac{p}{a}}\quad\text{or,}\quad x=\frac{e^{-\frac{p}{a}}}{2a}\cdot a[(p+a)e^{\frac{p}{a}}-ae^{\frac{p}{a}}]+Ce^{-\frac{p}{a}}$$

or, $\boldsymbol{x=Ce^{-\frac{p}{a}}+\frac{p}{a}}$ [where C = a constant] (iii)

Using Eqn. (iii) into the given differential equation, we get

$$y=C(p+a)e^{-\frac{p}{a}}+\frac{1}{2}p(p+a)-\frac{1}{4}p(p+a)^2.$$

If $C=0$, then

$$y=\frac{1}{2}p(p+a)-\frac{1}{4}(p+a)^2=\frac{1}{4}(p^2-a^2)\quad\text{or,}\quad p^2=4y+a^2.$$

$$\therefore\quad \frac{dy}{dx}=\pm\sqrt{4y+a^2}\quad\text{or,}\quad\int\frac{dy}{\pm\sqrt{4y+a^2}}=\int dx\quad\text{or,}\quad\pm2\sqrt{4y+a^2}=x$$

or, $\boldsymbol{\dfrac{x^2}{4}-4y-a^2=0.}$ (iv)

15.8 Applications of First Order Differential Equations

The first order differential equations have important applications in numerous physical problems. Here we propose to develop the technique of constructing appropriate differential equation out of physical problems and to solve the same.

15.8.1 Falling Body under Gravity

We consider a body of mass m falling under the influence of gravity. The resistance to the motion of the body is proportional to its velocity at any instant of motion. The equation of motion of the body is

$$\frac{dv}{dt} = g - kv, \tag{i}$$

the mass of the falling body being unity and k is a constant of proportionality

or, $$\frac{dv}{g - kv} = k\,dt.$$

Integrating, $\log(g - kv) = -kt + \log C$.

$$\therefore \quad g - kv = Ce^{-kt} \quad \text{or,} \quad v = \frac{1}{k}(g - Ce^{kt}). \tag{ii}$$

At $t = 0$, $v = 0$ (as the particle starts from rest).

So, $$0 = \frac{1}{k}(g - C) \quad \text{or,} \quad C = g.$$

$$\therefore \text{ from Eqn. (ii), } \quad v = \frac{g}{k}(1 - e^{-kt}). \tag{iii}$$

Eqn. (iii) shows that for $t = \infty$, $v = g/k$. Particle can never actually attain this velocity and v is called the **terminal velocity**.

Putting $v = \frac{dx}{dt}$ in Eqn. (iii), we get

$$\frac{dx}{dt} = \frac{g}{k}(1 - e^{-kt}).$$

Integrating, $$x = \frac{g}{k}\left(t + \frac{1}{k}e^{-kt}\right) + C. \tag{iv}$$

At $t = 0$, $x = 0$;

$$\therefore \quad 0 = \frac{g}{k}\left(0 + \frac{1}{k}e^{-k0}\right) + C \quad \text{or,} \quad 0 = \frac{g}{k^2} + C \quad \text{or,} \quad C = -\frac{g}{k^2}.$$

Putting this value of C in Eqn. (iv), we get

$$x = \frac{g}{k}\left(t + \frac{1}{k}e^{-kt} - \frac{1}{k}\right) \quad \text{or,} \quad x = \frac{g}{k}\left[t + \frac{1}{k}\left(e^{-kt} - 1\right)\right]. \tag{v}$$

Eqn. (v) gives the required equation which gives the distance traversed by the body after a time t.

N.B. If the resistance to the motion of the body be proportional to v^2 (instead of v), then the equation of motion is

$$\frac{dv}{dt} = g - kv^2 = k\left(\frac{g}{k} - v^2\right) \tag{vi}$$

$$= k(v_0^2 - v^2), \quad \text{where } v_0 = \sqrt{\frac{g}{k}} = \text{constant}.$$

By breaking into partial fraction, we get

$$\frac{1}{v_0^2 - v^2} = \frac{1}{2v_0}\left(\frac{1}{v_0 - v} + \frac{1}{v_0 + v}\right). \qquad \text{(vii)}$$

Now from Eqns. (vi) and (vii), we get

$$\left(\frac{1}{v_0 + v} + \frac{1}{v_0 - v}\right) dv = 2v_0 k\, dt.$$

Integrating, $\log\left(\frac{v_0 + v}{v_0 - v}\right) = 2v_0 kt + \log C$ or, $\frac{v_0 + v}{v_0 - v} = Ce^{2v_0 kt}$. (viii)

15.8.2 Maximum Height Attained by a Rocket

Let a rocket is projected vertically upward under gravity with an initial velocity V_0, from which matter in the form of exhaust, is continuously ejected downward with a constant velocity u.

Let v be the velocity and x the height at any instant t. Also the mass of the rocket at t is

$$\frac{M}{g}(1 + e^{-kt}) = \frac{M}{g}\left[1 + \left(1 - kt + \frac{k^2t^2}{2!} - \cdots\right)\right]$$
$$= \frac{M}{g}(2 - kt) \quad [\text{neglecting } k^2,\ k^3, \text{ etc.}]$$

Now, the momentum imparted per second by the rocket to the ejected matter is

$$-\frac{d}{dt}(\text{mass}) \times u = -\frac{d}{dt}\left[\frac{M}{g}(2 - kt)\right] \cdot u = \frac{kMu}{g}.$$

The reactional force acts in the upward direction to the rocket whose magnitude is $\frac{kMu}{g}$.

$\therefore$ the force acting on the rocket in the direction of motion is $\frac{kMu}{g} - \frac{M}{g}(2 - kt)g$.

Now, the equation to the motion of the rocket is

$$\frac{kMu}{g} - \frac{M}{g}(2 - kt)g = \frac{M}{g}(2 - kt)\frac{dv}{dt}$$

or, $$\frac{dv}{dt} = \frac{ku}{2 - kt} - g = \frac{ku}{2}\left(1 - \frac{kt}{2}\right)^{-1} - g$$
$$= \frac{ku}{2}\left(1 + \frac{kt}{2}\right) - g \quad [\text{neglecting } k^2, \text{ etc.}]$$
$$= \frac{ku}{2} - g = -\left(g - \frac{ku}{2}\right).$$

Thus, the retardation acting on the rocket is $\left(g - \frac{ku}{2}\right)$.

Now, the maximum height attained by a rocket is

$$H = \frac{V_0^2}{2 \times \text{retardation}} = \frac{V_0^2}{2\left(g - \frac{ku}{2}\right)} = \frac{V_0^2}{2g\left(1 - \frac{ku}{2}\right)}.$$

$$\boldsymbol{H = \frac{V_0^2}{2g}\left(1 + \frac{ku}{2g}\right)}.$$

15.8.3 *The opposing force to the motion of a body per unit mass is Cx and the resistance per unit mass is bv^2, where x and v are the displacement and velocity of the particle at any instant t. If the body starts from rest, what is the velocity of the body in terms of x?*

Solution : According to the Newton's second law of motion, the equation of motion of the body is

$$v\frac{dv}{dt} = -Cx - bv^2 \quad \text{or,} \quad v\frac{dv}{dt} + bv^2 = -Cx.$$

Putting $v^2 = z$, $2v\,dv = \dfrac{dz}{dx}$.

$$\therefore \quad \frac{1}{2}\frac{dz}{dt} + bz = -Cx \quad \text{or,} \quad \frac{dz}{dt} + 2bz = -2Cx.$$

$$\therefore \quad \text{I.F.} = \int e^{2b\,dx} = e^{2bx}.$$

$$\therefore \quad ze^{2bx} = \int -2Cxe^{2bx}\,dx + C' = -2C\left[\frac{xe^{2bx}}{2b} - \int \frac{e^{2bx}}{2b}dx\right] + C' = -\frac{C}{b}xe^{2bx} + \frac{C}{b}\frac{e^{2bx}}{2b} + C'$$

$$= -\frac{C}{b}xe^{2bx} + \frac{C}{2b^2}e^{2bx} + C'$$

$$\text{or,} \quad z = -\frac{C}{b}x + \frac{C}{2b^2} + C'e^{-2bx} \quad \text{or,} \quad v^2 = -\frac{Cx}{b} + \frac{C}{2b^2} + C'e^{-2bx}.$$

At $t = 0$, $x = 0$ and $v = 0$,

$$\therefore \quad 0 = \frac{C}{2b^2} + C', \quad \therefore \; C' = -\frac{C}{2b^2}.$$

$$\therefore \quad v^2 = -\frac{Cx}{b} + \frac{C}{2b^2} - \frac{C}{2b^2}e^{-2bx}$$

$$\text{or,} \quad v = \left(-\frac{Cx}{b} + \frac{C}{2b^2} - \frac{C}{2b^2}e^{-2bx}\right)^{\frac{1}{2}} = \left(\frac{C}{b}\right)^{\frac{1}{2}}\left[-x + \frac{1}{2b}(1 - e^{-2bx})\right]^{\frac{1}{2}}.$$

15.8.4 *A spherical rain drop of radius r falls from rest through a vertical height h. As it falls, it accumulates condensed vapour at the rate k/cm^2/s. If there is no other vertical force except gravity, show that when the rain drop reaches the ground, its radius 'a' is given by*

$$a = k\sqrt{\frac{2h}{g}}\left(1 + \sqrt{1 + \frac{gr^2}{2k^2h}}\right)$$

[BU(H) Model Q]

Solution : Suppose that after falling through a vertical distance z at time t, the radius and mass of the rain drop become x and m respectively.

If g be the acceleration due to gravity, the force acting on the rain drop at time t is

$$mg = \frac{d}{dt}(mv) = \frac{d}{dt}\left(m\frac{dz}{dt}\right) \qquad \left[\because \; v = \frac{dz}{dt}\right]$$

$$\text{or,} \quad \frac{4}{3}\pi x^3 \rho g = \frac{d}{dt}\left(\frac{4}{3}\pi x^3 \rho \frac{dz}{dt}\right), \quad \rho \text{ being the density of the rain drop.}$$

$$\therefore \quad x^3 g = \frac{d}{dt}\left(x^3\frac{dz}{dt}\right) \quad \text{or,} \quad g\frac{d}{dx}\left(\frac{x^4}{4}\right) = \frac{d}{dt}\left(x^3\frac{dz}{dt}\right)$$

or, $$g\frac{d}{dt}\left(\frac{x^4}{4}\right)\frac{dt}{dx} = \frac{d}{dt}\left(x^3\frac{dz}{dt}\right). \tag{i}$$

But the rate of increase of mass of the rain drop is

$$\frac{dm}{dt} = 4\pi x^2 k\rho \quad [\because \text{ the rain drop accumulates vapour at the rate k/cm}^2\text{/s}]$$

or, $$\frac{d}{dt}\left(\frac{4}{3}\pi x^3\rho\right) = 4\pi x^2 k\rho \quad \text{or,} \quad 4\pi x^2\rho\frac{dx}{dt} = 4\pi x^2 k\rho \quad \text{or,} \quad k = \frac{dx}{dt}.$$

Using Eqn. (i), $$\frac{d}{dt}\left(x^3\frac{dz}{dt}\right) = \frac{g}{k}\frac{d}{dt}\left(\frac{x^4}{4}\right). \tag{ii}$$

Integrating w.r.t., t, we get $x^3\frac{dz}{dt} = \frac{g}{4k} + C.$

But at $t = 0$, $x = r$ and $\frac{dz}{dt} = v = 0$.

$$\therefore \quad 0 = \frac{gr^4}{4k} + C \quad \text{or,} \quad C = -\frac{gr^4}{4k}.$$

$$\therefore \quad x^3\frac{dz}{dt} = \frac{g}{4k}(x^4 - r^4) \tag{iii}$$

or, $$\frac{dz}{dt} = \frac{g}{4k}\left(x - \frac{r^4}{x^3}\right) \quad \text{or,} \quad \frac{dz}{dx}\cdot\frac{dx}{dt} = \frac{g}{4k}\left(x - \frac{r^4}{x^3}\right)$$

or, $$\frac{dz}{dx} = \frac{g}{4k^2}\left(x - \frac{r^4}{x^3}\right) \quad \left[\because \frac{dx}{dt} = k\right] \tag{iv}$$

or, $$dz = \frac{g}{4k^2}\left(x - \frac{r^4}{x^3}\right)dx.$$

Now, integrating, $$z = \frac{g}{4k^2}\left(\frac{x^2}{2} + \frac{r^4}{2x^2}\right) + A. \tag{v}$$

At $t = 0$, $x = r$ and $z = 0$.

$$\therefore \quad 0 = \frac{g}{4k^2}\left(\frac{x^2}{2} + \frac{r^4}{2r^2}\right) + A, \quad \therefore \ A = -\frac{gr^2}{4k^2}.$$

$$\therefore \quad z = \frac{g}{4k^2}\left(\frac{x^2}{2} + \frac{r^4}{2x^2} - r^2\right) = \frac{g}{8k^2}\left(\frac{x^2 - r^2}{x}\right)^2.$$

When the rain drop reaches just before the ground, $z = h$ and $x = a$.

$$\therefore \quad h = \frac{g}{8k^2}\left(\frac{a^4 - r^2}{a}\right)^2 \quad \text{or,} \quad \frac{a^4 - r^2}{a} = \sqrt{\frac{8k^2h}{g}} = 2k\sqrt{\frac{2h}{g}} \quad \text{or,} \quad a^2 - 2k\sqrt{\frac{2h}{g}}\cdot a - r^2 = 0.$$

This quadratic equation having solution $a = \pm k\sqrt{\frac{2h}{g}}\left(1 + \sqrt{1 + \frac{gr^2}{2k^2h}}\right)$.

Since, the radius of the rain drop cannot be negative,

$$\boldsymbol{a = k\sqrt{\frac{2h}{g}}\left(1 + \sqrt{1 + \frac{gr^2}{2k^2h}}\right)} \quad \textbf{(Proved)}. \tag{vi}$$

15.8.5 *If the population of a country doubles in 50 years, in how many years will it treble assuming that the rate of increase is proportional to the number of inhabitants?*

Solution : If y be the population of the country in t years, then

$$\frac{dy}{dt} \propto y \quad \text{or,} \quad \frac{dy}{dt} = ky \quad [k \text{ is a constant}]$$

or, $k\,dt = \dfrac{dy}{y}$.

Integrating, $kt = \log y + C$. (i)

If P be the population at $t = 0$, i.e., at $t = 0$, $y = P$. By the condition of the problem, when $t = 50$, $y = 2P$.

$\therefore \quad 50k = \log 2P + C$.

$\therefore$ from (i), for $t = 0$, $0 = \log P + C$ and for $t = 50$, $50k = \log 2P + C$.

Solving, $C = -\log P$.

$\therefore \quad 50k = \log 2P - \log P = \log 2 \quad \text{or,} \quad k = \dfrac{1}{50}\log 2.$

Let $t = T$ be the time in years when population is trebled.

$\therefore \quad kT = \log 3P + C = \log 3P - \log P = \log 3.$

$$\therefore \quad T = \frac{1}{k}\log 3 = \frac{50}{\log 2} \times \log 3 = \mathbf{50 \times \frac{\log 3}{\log 2}} \textbf{ years.}$$

15.8.6 Centrifuge

We consider a cylinder of height h filled with a fluid and rotates about its axis. We have to find out the pressure at a point distance r apart from the axis.

Considering a cylindrical shell of fluid of thickness dr, distance r from the axis of rotation, the net force pushing inward in the shell is $(2\pi rh)dP$. The effective centrifugal force due to angular velocity ω is $m\omega^2 r$, where m is the mass of the fluid.

$\therefore \quad (2\pi rh\,dP) = m\omega^2 r = (2\pi rhdr\rho)\omega^2 r$

or, $dP = \rho\omega^2 r\,dr$, where ρ is the density of the liquid. (i)

(a) For incompressible liquids : Since ρ is the density of incompressible fluid we have from equation (i)

$$\int dP = \rho\omega^2 \int r\,dr + C \quad \text{or,} \quad P = \frac{1}{2}\rho\omega^2 r^2 + C.$$

If $P = P_0$ at $r = r_0$, then $C = P_0$.

$$\therefore \quad \mathbf{P = P_0 + \frac{1}{2}\rho\omega^2 r^2}.$$

(b) For a compressible gas : Here $P \propto \rho$ or, $P = C\rho$.

$$\therefore \quad dP = \frac{P}{C}\omega^2 r\,dr \quad \text{or,} \quad \frac{dP}{P} = \frac{\omega^2}{C} r\,dr.$$

Integrating, $\ln P = \dfrac{1}{2}\dfrac{\omega^2}{C} r^2 + A.$

$\because P = P_0$ at $r = r_0$, $\quad\therefore\ A = \log P_0$.

$$\therefore \quad \log\left(\frac{P}{P_0}\right) = \frac{\omega^2 r^2}{2C} \quad \text{or,} \quad \boldsymbol{P = P_0 e^{\frac{\omega^2 r^2}{2C}}}. \tag{iii}$$

15.8.7 Pressure Variation for a Static Compressible Fluid

For a fluid the hydrostatic pressure varies only in the vertical direction. If dP be the change of pressure between two points separated by dz then,

$$dP = -\rho g\, dz \quad \text{or,} \quad \frac{dP}{dz} = -\rho g, \tag{i}$$

where ρ is the density and g the acceleration due to gravity. For a compressible fluid (viz., air),

$$P = \rho RT, \tag{ii}$$

where T is the temperature and R is the universal gas constant.

From Eqns. (i) and (ii),

$$\frac{dP}{P} = -\frac{g}{RT}dz. \tag{iii}$$

Now, the temperature variation in atmosphere is given by

$$T = T_0 - \lambda z \tag{iv}$$

where T_0 = temperature at sea level and λ is a constant called **lapse rate**.

$$\therefore \quad dT = -\lambda\, dz \quad \text{or,} \quad dz = -\frac{dT}{\lambda}. \tag{v}$$

From Eqns. (iii) and (v),

$$\frac{dP}{P} = \frac{g}{\lambda R}\frac{dT}{T}.$$

Integrating, $\log P = \dfrac{g}{\lambda R}\log T + C,\quad (C = \text{a constant})$ (vi)

If at sea level, $P = P_0$ and $T = T_0$, then

$$\log P_0 = \frac{g}{\lambda R}\log T_0 + C. \tag{vii}$$

$$\therefore \quad \log\frac{P}{P_0} = \frac{g}{\lambda R}\log\left(1 - \frac{\lambda z}{T_0}\right) \quad \text{or,} \quad \frac{P}{P_0} = \left(1 - \frac{\lambda z}{T_0}\right)^{\frac{g}{\lambda R}}.$$

Now from Eqns. (ii) and (iv), we get

$$\frac{P}{P_0} = \frac{\rho T}{\rho_0 T_0} \quad [\rho_0 = \text{density at sea level}]$$

so that, $\dfrac{\rho}{\rho_0} = \dfrac{P}{P_0}\dfrac{T_0}{T} = \dfrac{P}{P_0}\left(1 - \dfrac{\lambda z}{T_0}\right)^{-1}$.

$$\therefore \quad \boldsymbol{\frac{\rho}{\rho_0} = \left(1 - \frac{\lambda z}{T_0}\right)^{\frac{g}{\lambda R} - 1}}.$$

15.8.8 Fluid Flow through An Aperture

We consider a vertical cylinder of cross section A be filled with water up to a height h_0. Water is flowing out through an orifice of area α. If v be the velocity of efflux, then the volume of water flowing through the orifice in time dt is $\alpha v\,dt$.

If dh be the decrease in height of water in the tank due to this volume, then

$$\alpha v\,dt = -A\,dh.$$

According to Torricelli's formula,

$$v = k\sqrt{2gh}, \quad \text{where } k \text{ is a constant.}$$

$$\therefore \quad \alpha k\sqrt{2gh}dt = -A\,dh \quad \text{or,} \quad \frac{dh}{\sqrt{h}} = -\frac{k\alpha}{A}\sqrt{2g}\,dt.$$

Integrating, $2\sqrt{h} = -\dfrac{k\alpha}{A}\sqrt{2g}\,t + C.$ (i)

But at $t = 0$, $h = h_0$, $\quad \therefore\ C = 2\sqrt{h_0}$.

Hence, from Eqn. (i),

$$\sqrt{h} = \sqrt{h_0} - \frac{k\alpha}{2A}\sqrt{2g}\,t. \tag{ii}$$

15.8.9 *A man and a parachute weighing W fall from rest under gravity. If the resistance of air is proportional to the speed v, find the speed as a function of t. What is the limiting speed?*

Solution : Since the air resistance is proportional to v the net force acting on the system is

$$W - kv = \frac{W}{g}\frac{dv}{dt}, \quad [\text{where } k = \text{a constant}]$$

$$\text{or,} \quad \frac{dv}{W - kv} = \frac{g}{W}dt.$$

Since $v = 0$ at $t = 0$ and $v = v$ at $t = t$, we get

$$\int_0^v \frac{dv}{W - kv} = \frac{g}{W}\int_0^t dt \quad \text{or,} \quad -\frac{1}{k}\log\left(\frac{W - kv}{W}\right) = \frac{g}{W}t \quad \text{or,} \quad -\frac{1}{k}\log\left(\frac{W - kv}{W}\right) = \frac{g}{W}t$$

$$\text{or,} \quad \left(1 - \frac{kv}{W}\right)^{-\frac{1}{k}} = e^{\frac{g}{W}t} \quad \text{or,} \quad 1 - \frac{kv}{W} = e^{-\left(\frac{gk}{W}\right)t}$$

$$\therefore \quad v = \frac{W}{k}\left[1 - e^{-\left(\frac{gk}{W}\right)t}\right] \tag{i}$$

Eqn. (i) gives v as a function of t. Putting $t = \infty$ in Eqn. (i), we get

$$v_{\max} = \frac{W}{k} = \text{limiting speed.} \tag{ii}$$

15.8.10 Time of Cooling

We consider a body of mass m and specific heat s. The surrounding temperature is $\theta_0\,^\circ$C and the body cools due to radiation from $\theta_1\,^\circ$C to $\theta_2\,^\circ$C in time t.

If the temperature of the body decreases by a small amount $d\theta$ in time dt, then by Newton's law of cooling,

$$\frac{dQ}{dt} = k(\theta - \theta_0). \tag{i}$$

Here Q is the total heat of the body and k is a constant depending on the surface area of the body.

$$\frac{dQ}{dt} = -ms\frac{d\theta}{dt}, \quad [\because\ Q = ms\theta] \tag{ii}$$

the negative sign denotes decrease of θ with t.

$$\therefore \quad -ms\frac{d\theta}{dt} = k(\theta - \theta_0) \quad \text{or,} \quad \frac{d\theta}{\theta - \theta_0} = -\frac{k}{ms}dt.$$

Integrating, $-\int_{\theta_1}^{\theta_2} \frac{d\theta}{\theta - \theta_0} = \frac{k}{ms}\int_0^t dt$

$$\text{or,} \quad \boldsymbol{t = \frac{ms}{k}\log\left(\frac{\theta_1 - \theta_0}{\theta_2 - \theta_0}\right)}. \tag{iii}$$

15.8.11 *The rate at which ice melts is proportional to the amount of ice at the instant. Find the amount of ice left after* 2 *hours if half of the quantity melts in* 30 *minutes.*

Solution : If m be the amount of ice at time t, then

$$\frac{dm}{dt} = km \quad \text{or,} \quad \frac{dm}{m} = k\,dt \quad \text{or,} \quad \int\frac{dm}{m} = k\int dt + C$$

$$\text{or,} \quad \log m = kt + C \quad [C = \text{a constant}]. \tag{i}$$

Let $m = M$ at $t = 0$, then

$$\log M = C \quad \text{or,} \quad C = \log M.$$

$\therefore$ from Eqn. (i),

$$\log m = kt + \log M. \tag{ii}$$

But $m = \frac{M}{2}$ at $t = \frac{1}{2}$ hour.

$$\therefore \quad \log\frac{M}{2} = \frac{k}{2} + \log M \quad \text{or,} \quad \log\frac{M}{2M} = \frac{k}{2} \quad \text{or,} \quad \log\frac{1}{2} = \frac{k}{2}.$$

$$\therefore \quad k = 2\log\frac{1}{2}. \tag{iii}$$

$\therefore$ from Eqn. (ii),

$$\log m = \left(2\log\frac{1}{2}\right)t + \log M. \tag{iv}$$

Putting $t = 2$ hours in Eqn. (iv), we get

$$\log m = 4\log\frac{1}{2} + \log M \quad \text{or,} \quad \log\frac{m}{M} = \log\left(\frac{1}{2}\right)^4 \quad \text{or,} \quad \frac{m}{M} = \frac{1}{16} \quad \text{or,} \quad m = \frac{M}{16}.$$

$\therefore$ **the amount of ice left after 2 hours is** $\boldsymbol{\frac{M}{16}}$.

15.8.12 *A cylinder of volume V contains a gas at a constant temperature. The gas slowly leaks out of a small orifice of area α. Show that the pressure at any instant t is given by* $p = p_0e^{-kt}$, *where* p_0 = *initial pressure,* $k = \frac{c\alpha}{4V}$, c = *the average velocity of the gas molecules. [Assume that no gas molecule backs into the cylinder]*

Solution : For isothermal process, pV = constant.

$$\therefore \quad p\,dV + V\,dp = 0 \quad \text{or,} \quad \frac{dp}{p} = -\frac{dV}{V}.$$

From kinetic theory of gas, we know that the number of gas molecules striking unit area of the vessel wall is $\frac{1}{4}nc$, n = number density of the gas.

$$\therefore \quad \text{the velocity of flow of the gas} = \frac{c}{4}.$$

$$\therefore \quad \frac{dp}{p} = -\frac{c\alpha}{4V}dt \quad \left[\because\ dV = \frac{c\alpha dt}{4}\right]$$

$$\text{or,} \quad \log p = -\frac{c\alpha}{4V}t + A.$$

But at $t = 0$, $p = p_0$, $\therefore\ A = \log p_0$

$$\text{Now,} \quad \log p = -\frac{c\alpha}{4V}t + \log p_0 \quad \text{or,} \quad \log\left(\frac{p}{p_0}\right) = -\frac{c\alpha}{4V}t \quad \text{or,} \quad p = p_0e^{-\frac{c\alpha t}{4V}}$$

$$\text{or,} \quad \boldsymbol{p = p_0e^{-kt}}, \quad \left[\text{where } k = \frac{c\alpha}{4V}\right]$$

15.8.13 Growth of Current in an L-R Circuit

A circuit containing an inductance L and a resistance R in series with a voltage source E is shown in Fig. 15.1.

If i be the current at any time t, then by voltage law,

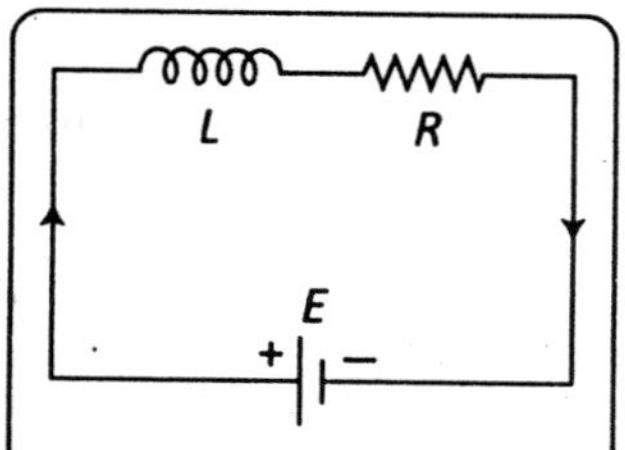

Fig. 15.1

$$Ri + L\frac{di}{dt} = E \quad \text{or,} \quad \frac{di}{dt} + \frac{R}{L}i = \frac{E}{L}. \tag{i}$$

This is a linear differential equation.

$$\therefore \quad \text{I.F.} = e^{\int \frac{R}{L}dt} = e^{\frac{R}{L}t}.$$

Multiplying both sides of Eqn. (i) by $e^{\frac{R}{L}t}$, we get

$$\frac{di}{dt}e^{\frac{R}{L}t} + \frac{R}{L}ie^{\frac{R}{L}t} = \frac{E}{L}e^{\frac{R}{L}t} \quad \text{or,} \quad \frac{d}{dt}\left(ie^{\frac{R}{L}t}\right) = \frac{E}{L}e^{\frac{R}{L}t}.$$

$$\text{Integrating,} \quad ie^{\frac{R}{L}t} = \frac{E}{L}e^{\frac{R}{L}t} \quad \text{or,} \quad i = e^{-\frac{R}{L}t}\left[\frac{E}{L}\frac{L}{R}e^{\frac{R}{L}t}\right] + Ae^{\frac{R}{L}t}.$$

$$i = \frac{E}{R} + Ae^{\frac{R}{L}t}. \tag{ii}$$

At $t = 0$, $i = 0$, $\therefore$ from Eqn. (ii), $A = -\frac{E}{R}$.

$$\therefore \quad i = \frac{E}{R}\left(1 - e^{\frac{R}{L}t}\right)$$

$$\text{or,} \quad \boldsymbol{i = i_0\left(1 - e^{\frac{R}{L}t}\right)}, \quad \text{where } i_0 = \frac{E}{R}, \text{ the maximum value of current.} \tag{iii}$$

Eqn. (iii) gives the growth of current in an 'L-R' circuit.

15.8.14 a.c. Voltage Applied to an 'L-R' Circuit

Let an a.c. voltage $E_0 \sin \omega t$ be applied to a circuit containing an inductance L and a resistance R in series. Then the voltage equation is

$$L\frac{di}{dt} + Ri = E_0 \sin \omega t, \quad \text{where } i \text{ is the current at any instant } t.$$

$$\therefore \quad \frac{di}{dt} + \frac{R}{L}i = \frac{E_0}{L}\sin \omega t.$$

$$\text{I.F.} = e^{\int \frac{R}{L}dt} = e^{\frac{R}{L}t}.$$

Solution is : $i \cdot e^{\frac{R}{L}t} = \dfrac{E_0}{L}\displaystyle\int e^{\frac{R}{L}t} \sin \omega t\, dt + A = E_0 \frac{e^{\frac{R}{L}t}}{\sqrt{\frac{R^2}{L^2} + \omega^2}} \sin\left(\omega t - \tan^{-1}\frac{\omega L}{R}\right) + A$

$$= \frac{E_0 L}{\sqrt{R^2 + \omega^2 L^2}} \sin\left(\omega t - \tan^{-1}\frac{\omega L}{R}\right) + Ae^{\frac{-R}{L}t}.$$

As $t \to \infty$, $Ae^{\frac{-R}{L}t} \to 0$.

$$\therefore \quad \boldsymbol{i = \frac{E_0 L}{\sqrt{R^2 + \omega^2 L^2}} \sin\left(\omega t - \tan^{-1}\frac{\omega L}{R}\right)}.$$

Exercises

15(A) : Objective Type Questions

Select the correct answer

15.1 The order and degree of the differential equation $\left(\frac{d^2y}{dx^2}\right)^{\frac{2}{3}} + \frac{d^2y}{dx^2} - \frac{dy}{dx} = 0$ are respectively

(a) 2, 2 (b) 1, 3 (c) 2, 3 (d) 3, 2.

15.2 The order, degree and linearity of the differential equation $\left(1 + \frac{d^2y}{dx^2}\right)^{\frac{2}{3}} = y\frac{d^2y}{dx^2}$ are respectively

(a) 2, 2, non-linear (b) 2, 3, non-linear (c) 2, 3, linear (d) 3, 2, linear

15.3 The order and degree of the differential equation $\frac{d^2y}{dx^2} + \left(\frac{dy}{dx}\right)^2 = y$ are respectively

(a) 2, 2 (b) 2, 1 (c) 1, 2 (d) 1, 1. **[WBUT 2007]**

15.4 The degree and order of the differential equation $\left(\frac{d^2y}{dx^2} + 2\right)^{\frac{3}{2}} = x\frac{dy}{dx}$ are respectively

(a) $\frac{3}{2}$, 2 (b) 2, 3 (c) 3, 2 (d) 2, 1. **[WBUT 2006, 2010]**

15.5 The condition that a differential equation $M(x, y)dy + N(x, y)dx = 0$ to be exact is that

(a) $\frac{\partial M}{\partial x} = \frac{\partial N}{\partial y}$ (b) $\frac{\partial^2 M}{\partial x^2} = \frac{\partial^2 N}{\partial y^2}$ (c) $\frac{\partial M}{\partial y} = \frac{\partial N}{\partial x}$ (d) $M(x, y) = N(x, y)$.

15.6 $\dfrac{dy}{dx}+P(x)y=Q(x)y^n \quad (n\neq 1)$ is

(a) Clairaut's equation (b) Bernoulli's equation

(c) Lagrange's equation (d) homogeneous equation.

15.7 $\dfrac{x\,dy-y\,dx}{x^2+y^2}$ is equal to

(a) $d\left\{\tan^{-1}\left(\frac{y}{x}\right)\right\}$ (b) $d\left\{\log\left(\frac{y}{x}\right)\right\}$ (c) $d\left\{\tan^{-1}\left(\frac{x}{y}\right)\right\}$ (d) $d\left(\frac{y}{x}\right)$.

15.8 The integrating factor of $\dfrac{dy}{dx}+y=1$ is

(a) e^x (b) x^2 (c) x (d) $\dfrac{1}{x}$. **[WBUT 2007]**

15.9 The equation $x(x^2+2y^2)dx+y(2x^2+y^2)dy=0$ is

(a) exact (b) inexact (c) may be both (a) and (b) (d) none of these.

15.10 The equation $y=xp+\phi(p)$, where $p=\dfrac{dy}{dx}$ is

(a) Bernoulli's equation (b) Lagrange's equation

(c) Clairaut's equation (d) none of these.

15.11 The solution of the differential equation $x\dfrac{dy}{dx}-y=x\sqrt{x^2+y^2}$ is

(a) $y=x\sin(x+C)$ (b) $y=x\sin h(x+C)$

(c) $y=x\cos(x+C)$ (d) $y=x\cos h(x+C)$.

15.12 The solution of $y(2xy+e^x)dx=e^x dy$ is

(a) $x^2+\dfrac{y}{e^x}=C$ (b) $x^2e^x+y=C$ (c) $\dfrac{e^x}{x^2}+y=C$ (d) $\dfrac{e^x}{y}+x^2=C$.

15.13 The equation $y^2dx+(x^2-xy-y^2)dy=0$ is

(a) exact (b) inexact (c) both (a) and (b) (d) none of these.

15.14 The integrating factor of the equation $x\dfrac{dy}{dx}+y=y^2\log x$ is

(a) x (b) x^2 (c) $\dfrac{1}{x}$ (d) e^x.

15.15 The solution of equation $y=px+\sqrt{1+p^2}$ is

(a) $x+y=1$ (b) $x^2+y^2=1$ (c) $y^2=x^2$ (d) $x-y=1$.

15.16 The solution of the equation $y=px+\sqrt{a^2p^2+b^2}$, $p=\dfrac{dy}{dx}$ is

(a) $x^2y^2=1$ (b) $x+y=1$ (c) $x^2+y^2=a^2+b^2$ (d) $\dfrac{x^2}{a^2}+\dfrac{y^2}{b^2}=1$.

15.17 The general solution of $p=\log(px-y)$, $p=\dfrac{dy}{dx}$ is

(a) $y=Cx-C$ (b) $y=Cx-e^C$ (c) $y=C^2x-e^{-C}$ (d) none of these. **[WBUT 2009]**

15.18 Solution of the equation $\frac{dy}{dx} + y = 0$ is

(a) $y = e^{-x} + C$ (b) $y = \frac{C}{e^{-x}}$ (c) $y = Ce^{-x}$ (d) none of these.

15.19 Solution of $\frac{2xy}{x^2 + y^2} = \frac{dy}{dx}$ is

(a) $x^2 + y^2 = Cy$ (b) $x^2 - y^2 = Cy$ (c) $\frac{x}{y} = Cy$ (d) $\frac{y}{x} = y + C$.

15.20 General solution of $p = \cos(y - px)$ is

(a) $y = Cx + \cos^{-1} C$ (b) $y = x + \cos^{-1} C$ (c) $y = \cos^{-1} \frac{x}{C}$ (d) none of these.

15.21 Solution of $y\,dx + (x + \cos y)dy = 0$ is

(a) $xy + \sin y = C$ (b) $x + y + \sin y = C$ (c) $\frac{x}{y} + \sin y = C$ (d) $\frac{y}{x} + \sin x = C$.

Answers

15.1 (a) **15.2** (b) **15.3** (b) **15.4** (c) **15.5** (c) **15.6** (b) **15.7** (a) **15.8** (a)
15.9 (a) **15.10** (c) **15.11** (b) **15.12** (d) **15.13** (b) **15.14** (c) **15.15** (b) **15.16** (d)
15.17 (b) **15.18** (c) **15.19** (b) **15.20** (a) **15.21** (a).

15(B) : Short Answer Type Questions

15.1 What are differential equations? Explain the importance of differential equations in physical problems. What do you mean by the term 'solution' of differential equations?

15.2 What are general solutions and singular solutions of linear differential equations of first order? Explain with suitable illustration.

15.3 What are 'ordinary' and 'partial' differential equations? Distinguish between 'homogeneous' and 'non-homogeneous' differential equations with example.

15.4 What do you mean by 'order' and 'degree' of differential equations? Explain with examples.

15.5 Discuss the methods of construction of differential equation with example. Form the differential equations of the following : **(a)** $y = Ax + A^2$ **(b)** $y = e^{ax}$.

15.6 Mention the different forms of ordinary differential equation of first order and first degree. Give example in each case.

15.7 How an inhogeneous equation can be reduced to homogeneous equation? Discuss with example.

15.8 What are 'exact' and 'inexact' differential equations? Establish the condition of an 'exact' equation.

15.9 Explain the rules to determine integrating factors. What is the integrating factor of the equation $\frac{y\,dx - x\,dy}{y^2} = 0$?

15.10 Show that the solution of a linear differential equation of the form : $\frac{dy}{dx} + Py = Q$ is

$$y \times [\text{I.F.}] = \int Q\ (\text{I.F.})dx + C,$$

where P and Q are functions of x only or constant.

15.11 Write the form of the following differential equations :

(a) Bernoulli's equation **(b)** Clairaut's equation **(c)** Lagrange's equation.

15.12 Discuss the growth and decay of current in a series L-R circuit with a steady e.m.f.

15.13 Obtain radioactive decay law : $N = N_0 e^{-\lambda t}$, the different letters having their usual significance.

15.14 Solve the linear differential equation : $\frac{dy}{dx} + \frac{3}{x}y = x^2$. **[Ans:** $\frac{x^3}{6} + \frac{C}{x^3}$**]**

15.15 Solve the differential equation : $\frac{dy}{dx} - \frac{y}{x} = -x$. **[Ans:** $y = -x^2 + Cx$**]**

15.16 State the order and degree of the differential equation : $\frac{d^2y}{dx^2} + \left(\frac{dy}{dx}\right)^2 + xy = 0$ **[CU(H) 2014]**

15.17 Show that, $(y^2x^3 \cos x - 4x^3 z)dx + 2z^3 y \sin dy + 3y^2z^2 \sin x - x^4 dz$ is an exact differential of a function ϕ. Find the form of ϕ. **[BU(H) 2011]**

15.18 Is the differential equation of $\frac{dy}{dx} = 1 + x$ linear or non-linear and homogeneous or non-homogeneous? **[CU(H) 2013]**

15.19 Show that, $(x^2z^3 \cos x - 4x^3 z)dx + 2z^2 y \sin x\, dy + (3y^2z^2 \sin x - x^4)dz$ is an exact differential of a function ϕ. Find the form of ϕ. **[BU(H) 2011]**

15(C) : Broad Answer Type Questions

15.1 Solve the following differential equations :

(i) $y\,dx + (x + e^x)dy = 0$ **[Ans:** $xy + e^y = C$**]**

(ii) $y\,dx + (x + \cos y)dy = 0$ **[Ans:** $xy + \sin y = C$**]**

(iii) $(4x^3y^3 - 2xy)dx + (3x^4y^2 - x^2)dy$ **[Ans:** $x^4y^3 - x^2y = C$**]**

(iv) $x(1+y)\frac{dy}{dx} + (1-x)y = 0$ **[Ans:** $y - x + \log xy - C = 0$**]**

(v) $(1+x)\frac{dy}{dx} + (1-y) = 0$ **[Ans:** $y = -\frac{x}{C} + (1 - \frac{1}{C})$**]**

(vi) $\frac{dx}{x} = \tan y\, dy$ **[Ans:** $x \cos y = C$**]**

(vii) $y \sec^2 x + (x+7) \tan x \frac{dy}{dx}$ **[Ans:** $y^7 \tan x = Ce^{-y}$**]**

(viii) $\frac{dy}{dx} = e^{x-y} + x^2 e^{-y}$ **[Ans:** $e^y = e^x + \frac{x^3}{3} + C$**]**

(ix) $\frac{dy}{dx} = (4x + y + 1)^2$ **[AMIETE(Summer) 1995]** **[Ans:** $\tan^{-1} \frac{4x+y+1}{2} = 2x + C$**]**

(x) $\frac{dy}{dx} = 1 + x + y + xy$ **[Ans:** $y = Ce^{x+\frac{x^2}{2}} - 1$**]**

(xi) $\left(1 + e^{\frac{x}{y}}\right) dx + e^{\frac{x}{y}} \left(1 - \frac{x}{y}\right) dy = 0$ **[Ans:** $e^x \sin y + xe^{-y} = C$**]**

(xii) $(e^x \sin y + e^{-y})dx + (e^x \cos y - xe^{-y})dy = 0$. **[Ans:** $(e^x + 2) \cos x + C = 0$**]**

15.2 Solve the following differential equations :

(i) $x\,dy - y\,dx = \cos\frac{1}{x}dx$ [**Ans:** $\frac{y}{x} + \sin\frac{1}{x} = C$]

(ii) $y^2 + (3xy + x^2)\frac{dy}{dx} = 0$ [**Ans:** $y^4 = C\left(\frac{4y}{x+1}\right)$]

(iii) $\frac{dy}{dx} = \frac{7y - 3x - 7}{-3y + 7x + 3}$ [**Ans:** $(y - x - 1)^2 = C(y + x - 1)^5$]

(iv) $\frac{dy}{dx} = \frac{x + y + 1}{x - y - 7}$ [**Ans:** $C\{(x-3)^2 + (y+4)^2\}^{\frac{1}{2}} = \tan^{-1}\frac{y+4}{x-3}$]

(v) $\frac{dy}{dx} = \frac{y + x - 2}{y - x - 4}$ [**Ans:** $x^2 + 2xy - y^2 - 4x + 8y - 14 = C$]

(vi) $\frac{dy}{dx} - \frac{2y}{x+1} = (x+1)^3$ [**Ans:** $y = \frac{1}{2}(x+1)^4 + C(x+1)^2$]

(vii) $\left(\frac{e^{-2\sqrt{x}}}{\sqrt{x}} - \frac{y}{\sqrt{x}}\right) = \frac{dy}{dx} = 1$ [**Ans:** $ye^y = 2\sqrt{x} + C$]

(viii) $e^{-y}\sec^2 y\,dy = dx + x\,dy.$ [**Ans:** $xe^y = C + \tan y$]

15.3 Solve the following differential equations :

(i) $2(x+y)dx + 2(x-y)dy = 0$ [**Ans:** $x^2 + 2xy - y^2 = C$]

(ii) $x\,dx + y\,dy = \frac{a^2(x\,dy - y\,dx)}{x^2 + y^2}$ [**Ans:** $x^2 + y^2 - 2a^2\tan^{-1}\left(\frac{y}{x}\right) = 0$]

(iii) $(x + \log y)dy + y\,dx = 0$ [**Ans:** $y(\log y - 1 + x) = C$]

(iv) $(x - y^2)dx + 2xy\,dy = 0$ [**Ans:** $\frac{y^2}{x} + \log x = C$]

(v) $(xy^2 - e^{\frac{1}{x^3}})dx - x^2y\,dy = 0$ [**Ans:** $2e^{x-3} - 3\left(\frac{y}{x}\right)^2 = C$]

(vi) $(y^4 + 2y)dx + (xy^3 + 2y^4 - 4x)dy = 0$ [**Ans:** $x\left(y + \frac{2}{y^3}\right) + y^2 = C$]

(vii) $(x^2 + y^2 + x)dx + y\,dy$ **[AMIETE(Summer) 1995]** [**Ans:** $\frac{x^4}{4} + \frac{x^2y^2}{2} + \frac{x^3}{3} = C$]

(viii) $y(1 + xy)dx + x(1 + xy + x^2y^2)dy = 0$ [**Ans:** $\frac{1}{2x^2y^2} + \frac{1}{xy} - \log y = C$]

(ix) $(y^2 + 2x^2y)dx + (2x^3 - xy)dy = 0.$ [**Ans:** $4(xy)^{\frac{1}{3}} - \frac{2}{3}\left(\frac{x}{y}\right)^{\frac{3}{2}} = C$]

15.4 Solve the following differential equations :

(i) $\frac{dy}{dx} + \log x^y = 0$ [**Ans:** $y = Cx^{-x}e^x$]

(ii). $\frac{dy}{dx} - ny = e^{ax}$ [**Ans:** $y = \frac{1}{a-n}e^{ax} + Ce^{nx}$]

(iii) $\frac{dy}{dx} + \frac{3y}{x} = x^2$ [**Ans:** $y = \frac{x^3}{6} + \frac{C}{x^3}$]

(iv) $(x - y^2)dx + 2xy\,dy = 0$ [**Ans:** $\frac{y^2}{x} + \log x = C$]

(v) $e^y\left(\frac{dy}{dx} + 1\right) = e^x$ [**Ans:** $e^{x+y} = \frac{e^{2x}}{2} + C$]

(vi) $xy\frac{dy}{dx} + (x^2 + x + y^2)$ [AMIETE(Summer) 1999] [Ans: $x^2y^2 = -\frac{x^4}{2} - \frac{2}{3}x^3 + C$]

(vii) $\frac{dy}{dx} + y\log x = e^{-x\log x}$ [Ans: $y = \left(\frac{Ce^x}{x^x}\right) - \frac{1}{x^x}$]

(viii) $\frac{dy}{dx} + \frac{1}{x} = \frac{e^y}{x^2}$ [Ans: $2x = e^y(1 + Cx^2)$]

(ix) $\frac{dy}{dx} + \frac{y}{x} = \frac{y^2}{x^2}$ [Ans: $2x - y = Cx^2y$]

(x) $\frac{dy}{dx} + y = 3e^xy^3$ [Ans: $\frac{1}{y^2} = 6e^x + Ce^{2x}$]

(xi) $x^2y - x^3\frac{dy}{dx} = y^4\cos x$ [Ans: $x^3 = y^3(3\sin x - C)$]

(xii) $3\frac{dy}{dx} + \frac{2}{x+1}y = \frac{x^2}{y^2}$ [Ans: $y^3(x+1)^2 = \frac{x^5}{5} + \frac{x^4}{2} + \frac{x^3}{3} + C$]

(xiii) $(2x - 10y^3)\frac{dy}{dx} + y = 0$ [Ans: $x = 2y^3 + Cy^{-2}$]

(xiv) $\frac{dy}{dx} + x^2y = 0.$ [Ans: $y = Ce^{-\frac{x^3}{3}}$]

(xv) $\frac{dy}{x} + x\sin 2y = x^2\cos^3 y$ [VU(H) 2006]

15.5 Solve the following differential equations :

(i) $x(1 + p^2) = 1$ [Ans: $y - C = \sqrt{(x - x^2)} - \tan^{-1}\frac{\sqrt{1-x}}{x}$]

(ii) $y = px + p^2$ [Ans: $y = ax + a^3$]

(iii) $x^2p^2 + xyp - 6y^2 = 0$ [Ans: $y = \frac{C}{x^3}$, $y = C_1x^2$]

(iv) $y = px + \frac{a}{p}$ [Ans: $y = Cx + \frac{a}{C}$, $z = 4ax$]

(v) $y = 2px + p^2$ [Ans: $(3xy + 2x^3 + C)^2 = 4(x^2 + y)^3$]

(vi) $xp = y + \frac{p^2}{4}$ [Ans: $y = C_1x + C_2$, $y = x^2 + C_3$]

(vii) $y = px + p^3$ [Ans: $y^2 = Cx^2 + C^2$]

(viii) $xp^2 + x = 2yp$ [Ans: $2Cy = C^2x^2 + 1$]

(ix) $x^2(1 - px) = yp^2$ [Ans: $y^2 = Cx^2 + C^2$]

(x) $y = 2px + yp^2$; $y(0) = 1$ [Ans: $y = 2x + 1$]

(xi) $p^3 - p(x^2 + xy + y^2) + x^2y + xy^2 = 0$

[Ans: $(2y - x^2 - C)(y - Ce^x)(y + x - 1 - Ce^{-x}) = 0$]

(xii) $e^y - p^3 - p = 0$ [Ans: $y = \log(p^3 + p)$, $x = 2\tan^{-1}p - p^{-1} + C$]

(xiii) $p = \cos(y - px)$ [WBUT 2007]

[Ans: General solution : $y = Cx + \cos^{-1}C$

Singular solution : $y = \sqrt{x^2 - 1} + \cos^{-1}\left(\frac{1}{x}\sqrt{x^2 - 1}\right)$]

(xiv) $y = xp^2 + p$ [Ans: $y = C - \frac{1}{1+p^2}$, $x = \frac{p}{1+p^2} + \tan^{-1}p$]

(xv) Taking $u = x^2$ and $v = y^2$, solve the differential equation,

$$xyp^2 - (x^2 + y^2 - 1)p + xy = 0$$ [**Ans:** $y^2 = Cx^2 + \frac{C}{C-1}$]

(xvi) $x^2(y - px) = yp^2$ [**Ans:** $y = Cx^2 + C^2$]

(xvii) $y + 2\left(\frac{dy}{dx}\right)^2 = (x+1)\frac{dy}{dx}$ [**Ans:** $y = Cx + (C - 2C^2)$]

(xviii) $p = \log(px - y)$ [**Ans:** $C = \log(Cx - y)$.]

(xix) $\frac{dy}{dx} = (4y + y + 1)^2$ if $y(0) = 1$, [**BU(H) 2008**]

15(D) : Physical Problems

15.1 Find the initial velocity of a particle which is fired in radial direction from the earth's centre and is supposed to escape from the earth. Suppose that it is acted upon by the gravitational attraction of the earth only. [**Ans:** $v_0 = \sqrt{2gR}$]

15.2 A projectile is fired vertically with an initial velocity v_0. Show that its speed v at any height is given by

$$v^2 = v_0^2 - 2gR\left(1 - \frac{R}{r}\right),$$

where R = the radius of the earth and r = the distance traversed.

15.3 A body falling vertically under gravity encounters the resistance of the atmosphere. The resistance varies as the velocity. Show that the velocity-time and space-time relations are given respectively by

$$\text{(a) } u = \frac{g}{k} - Ce^{-kt} \quad \text{(b) } x = \frac{gt}{k} - \frac{g}{k^2}(1 - e^{-kt}).$$

where k and C are constant and g is the acceleration due to gravity.

15.4 A cylinder of radius a and height h is full of water. Show that the maximum angular velocity attainable before water is lost through overflow is

$$\sqrt{\left(\frac{2gh}{a}\right)}.$$

15.5 Solve the following equations by the use of a suitable integrating factor.

$$\text{(a) } x\,dy - y\,dx + x^2 dx = 0 \quad \text{(b) } L\frac{di}{dt} + R_i = E.$$

where L, R, E are the inductance, resistance and a constant EMF respectively constituting an electric circuit all connected in series. Hence, find the growth of current in the circuit. [**Ans:** $y = Cx - x^2$; $i = \frac{E}{R} + Ce^{-\frac{Rt}{L}}$; $u = \frac{E}{R}\left(1 - e^{-\frac{R}{L}t}\right)$]

15.6 A particle is projected with velocity v along a smooth horizontal plane in the medium whose resistance per unit mass is μ times the cube of the velocity. Show that the distance it has described in time t is

$$\frac{1}{\mu v}(\sqrt{(1 + 2\mu v^2 t)} - 1).$$

15.7 A man and a parachute weighing w fall from rest under the force of gravity. If the resistance of air is proportional to the speed v, find the sped as a function of time. Find also the limiting velocity v_0. [**Ans:** $v = \frac{gw}{k}\left(1 - e^{-\frac{k}{w}t}\right)$; $v_0 = \frac{gw}{k}$]

15.8 Find the atmospheric pressure p at a height z above the sea level, both when the temperature is constant or variable.

[**Ans:** $p = p_0 e^{-\frac{gz}{k}}$, when temp. is constant; $p = gk^{-\frac{1}{n}}z$, when temp. varies]

15.9 A particle of mass m is attached to the lower end of a light spring (whose upper end is fixed) and is released. Express the velocity v as a function of the stretch x.

[**Ans:** $v^2 = 2gx - \frac{\lambda}{m}x^2$]

15.10 The rate at which a body cools is proportional to the difference between the temperature of the body and that of the surrounding air. If a body in air at 25 °C cools from 100 °C to 75 °C in one minute, find the temperature at the end of three minutes.

[**Ans:** 47.22 °C]

15.11 A particle is subjected to a frictional force so that its velocity decreases with time according to $v = c(t - t_1)^2$, where t_1 is the time when it stops and C is a constant. Find the force $F(v)$ and the distance traversed before it stops.

[**Hint :** $F = \dfrac{d}{dt}(mv) = m\dfrac{d}{dt}[C(t - t_1)^2]$ and $m\dfrac{dv}{dt} = -kv$.

Boundary condition : $v = 0$ at $t = t_1$]

15.12 Two friends A and B order coffee and receive cups of equal temperature at the same time. A adds a small amount of cool cream immediately but does not drink his coffee until 10 minutes later, B waits for 10 minutes and adds the same amount of cool cream and begins to drink. Assuming the Newton's law of cooling, decide who drinks the hotter coffee. [**Ans:** B drinks hotter coffee]

15.13 Uranium disintegrates at a rate proportional to the amount then present at any instant. If M_1 and M_2 grams of uranium are preset at times T_1 and T_2 respectively, show that the half-life of uranium is

$$\frac{(T_1 - T_2)\log 2}{\log\left(\frac{M_1}{M_2}\right)}.$$

15.14 The current in a coil, containing a resistance R and inductance L and a constant emf E at time t is given by

$$i = \frac{E}{R}\left(1 - e^{-\frac{R}{L}t}\right).$$

Obtain suitable approximation to i when t is very small.

[**Ans:** When t is very small, $i = \frac{E}{R}$]

15.15 When a resistance R ohms and a capacitance C farads are connected in series with an emf E volts, the current i amperes is given by

$$R\frac{di}{dt} + \frac{i}{C} = \frac{dE}{dt}.$$

If $R = 1000$ ohms, $C = 50 \times 10^{-1}$ farads, $i = 10$ amperes and $t = 0$, find the current for $t = 1$ second and $E = 100 \sin 120\pi t$ volts. [**Ans:** 8.187 amperes]

15.16 Set up the equation of motion of a spherical body of radius r moving vertically downward through a viscous fluid. Calculate the terminal velocity v_t. Draw v_t vs. r curve.

[VU(H) 2013]

15.17 Set up the equation of vertical motion of a rocket of initial mass m_0 near earth surface assuming the velocity (u) and rate (α) of ejection of the fuel to be constant. Solve it to show that the condition of soft landing (velocity at the time of touching the ground is to be zero) in time T of a rocket, which has initial downward velocity V given by :

$$v - u \ln \frac{m_0}{m_0 - \alpha t} - gT = 0.$$

[VU(H) 2011]

15.18 A spherical rain drop falling in a constant gravitational field grows by absorption of moisture from its surroundings at a rate which is proportional to the instantaneous surface area. Assuming that, it starts with radius zero, determine its acceleration.

[VU(H) 2010]

15.19 **(a)** Find whether $d\phi$ is an exact differential, where $d\phi = (x^3 - 4xy)dx + (y^2 - 2x)dy$.

(b) A particle moves in a plane such that $\dfrac{dx}{dt} = -50y$ and $\dfrac{dx}{dt} = 18x$.

Given that it passes through the point $(2, 0)$, find the equation of the path.

[CU(H) 2017]

(c) Check that $df = x^2ydx + 3x^3dy$ is a perfect differential. **[CU(H) 2009]**

CHAPTER 16

Ordinary Differential Equations—II

16.1 Introduction

The general linear differential equation of the nth order :

The differential equations, in which the dependent variable and its derivatives occur only in the first degree and are not multiplied together, are called **linear differential equations.**

A general linear differential equation of the nth degree has the form :

$$\frac{d^n y}{dx^n} + P_1 \frac{d^{n-1} y}{dx^{n-1}} + P_2 \frac{d^{n-2} y}{dx^{n-2}} + \cdots + P_n = X, \tag{16.1}$$

where P_1, P_2, $\ldots$, P_n and X are functions of x only is known as **Euler's homogeneous linear equation** or, **Cauchy-Euler equation**. Eqn. (16.1) is said to be **non-homogeneous** if $X \neq 0$. On the other hand, if $X = 0$, then Eqn. (16.1) is **homogeneous**.

Linear differential equation with constant coefficients are of the form :

$$\frac{d^n y}{dx^n} + k_1 \frac{d^{n-1} y}{dx^{n-1}} + k_2 \frac{d^{n-2} y}{dx^{n-2}} + \cdots + k_n y = 0, \tag{16.2}$$

where k_1, k_2, $\ldots$, k_n are constants.

16.2 Linear Differential Equations of First Order with Constant Coefficients

A linear differential equation of first order with constant coefficients is of the form :

$$\frac{dy}{dx} + Py = Q. \tag{16.3}$$

The solution of the equation is

$$ye^{\int P dx} = \int (Qe^{\int P dx}) dx + C \quad \text{or,} \quad y = Ce^{-\int P dx} + e^{-\int P dx} \int (Qe^{\int P dx}) dx$$

$$= Cu + v \quad \text{(say)}, \tag{16.4}$$

where $u = e^{-\int P dx}$ and $v = e^{-\int P dx} \int Qe^{\int P dx} dx$.

(i) Differentiating $u = e^{-\int P dx}$ w.r.t. x, we get

$$\frac{du}{dx} = -Pe^{-\int P dx} = -Pu \quad \text{or,} \quad \frac{du}{dx} + Pu = 0 \quad \text{or,} \quad \frac{d}{dx}(Cu) + P(Cu) = 0 \cdot$$

$$\therefore \quad \boldsymbol{y = Cu} \text{ is a solution of equation, } \frac{dy}{dx} + Py = 0. \tag{16.5}$$

(ii) Differentiating $v = e^{-\int P dx} \int (Qe^{\int P dx}) dx$ w.r.t. x, we get

$$\frac{dv}{dx} = -Pe^{-\int P dx} \int (Qe^{\int P dx}) dx + e^{-\int P dx} Qe^{\int P dx} = -Pv + Q$$

$$\text{or,} \quad \frac{dv}{dx} + Pv = Q. \tag{16.6}$$

Eqn. (16.4) is the solution of the differential equation (16.3) consisting of two parts viz., Cu and v. Cu is the solution of the differential equation whose RHS = 0. The solution Cu is known as the **complementary function** (C.F.).

The solution v is free from any arbitrary constant and is known as **particular integral** (P.I.).

∴ **complete solution (C.S.) = complementary function (C.F.)**
+ particular integral (P.I.)

16.3 Operator D

The operator D stands for $\frac{d}{dx}$, i.e., $D \equiv \frac{d}{dx}$, $D^2 \equiv \frac{d^2}{dx^2}$, $D^3 \equiv \frac{d^3}{dx^3}$, ..., $D^n \equiv \frac{d^n}{dx^n}$, etc.

$$\therefore \quad \frac{dy}{dx} = Dy$$

$$\frac{d^2y}{dx^2} = D^2y$$

....

$$\frac{d^ny}{dx^n} = D^ny.$$

Then Eqn. (16.1) may be written as

$$(D^n + k_1D^{n-1} + \cdots + k_n)y = X \quad \text{or,} \quad f(D) = X,$$

where $f(D) = D^n + k_1D^{n-1} + \cdots + k_n$ **is a polynomial in** D. (16.7)

The symbolic operator D can be treated as an algebraic quantity, i.e., $f(D)$ can be factorised by ordinary rules of algebra. For example,

$$\frac{d^2y}{dx^2} + 2\frac{dy}{dx} - 3y = (D^2 - 2D - 3)y = (D + 3)(D - 1)y = (D - 1)(D + 3)y.$$

16.3.1 Properties of D-operator

· Taking u and v as two differentiable functions of x,

(i) $D(u + v) = Du + Dv$ **[Distributive law]**

(ii) $(D + m)y = (m + D)y$
Also, $(D^2 + lD + m)y = (D^2 + m + lD)y = (lD + D^2 + m)y$ **i.e., order of operation with D is immaterial.**

(iii) $D^nD^mu = D^{n+m}u = D^mD^ny$

(iv) $D(ku) = kDu$, **where k is a constant**

(v) **But** $D(v, v) \neq uDv$ **i.e., D does not commute with u.**

(vi) **Inverse of D (i.e., D^{-1})** : If $u = Dv$, $v = D^{-1}u$.
∴ D^{-1} (i.e., inverse of D) is identified with integration.
∴ $DD^{-1}u = u = Dv$.
It can be shown that $(D - k)^{-1}$, where k is a constant, is given by

$$(D-k)^{-1} = \frac{1}{D-k} = \frac{1}{-k\left(1-\frac{D}{k}\right)} = -\frac{1}{k}\left(1-\frac{D}{k}\right)^{-1} = -\frac{1}{k}\left(1+\frac{D}{k}+\frac{D^2}{k^2}+\cdots\right).$$

Proof : We have $(D-k)[(D-k)^{-1}u] = u$.

Let $(D-k)v = u, \quad \therefore \quad v = (D-k)^{-1}u$.

But $(D-k)^{-1} = \dfrac{1}{D-k} = \dfrac{1}{-k\left(1-\frac{D}{k}\right)} = -\dfrac{1}{k}\left(1-\dfrac{D}{k}\right)^{-1}$

or, $$(D-k)^{-1} = -\frac{1}{k}\left(1+\frac{D}{k}+\frac{D^2}{k^2}+\cdots\right). \tag{16.8}$$

N.B. It can be proved that, $I = \dfrac{1}{D} = D^{-1}$ [where $Iy = \int y\,dx$]

$$D = \frac{1}{I} = I^{-1};\ I^2 = D^{-2};\ I^n = D^{-n}, \text{ etc.}$$

16.4 Second Order Linear Differential Equations

The general form of a second order linear differential equation with constant coefficient is

$$\frac{d^2y}{dx^2} + P_1\frac{dy}{dx} + P_2 y = X, \tag{16.9}$$

where P_1 and P_2 are constants and X is a function of x or constant.

or, $$(D^2 + P_1 D + P_2)y = X \tag{16.10}$$

or, $f(D)y = X$, where $f(D) = D^2 + P_1 D + P_2$.

When $X = 0$, $f(D)y = 0$

or, $$\frac{d^2y}{dx^2} + P_1\frac{dy}{dx} + P_2 y = 0 \tag{16.11}$$

is called the **homogeneous equation.**

Linearly independent solutions : If $y_1(x)$ and $y_2(x)$ are the two solutions of Eqn. (16.11) such that, $C_1y_1 + C_2y_2 = 0$ for arbitrary constants C_1, C_2, i.e., $C_1 = C_2 = 0$, then these two solutions are called **linearly independent solutions.**

(i) **The two solutions $y_1(x)$, $y_2(x)$ of (16.11) are linearly independent, if and only if their Wronskian $W(y_1, y_2) = \begin{vmatrix} y_1 & y_2 \\ y_1' & y_2' \end{vmatrix} \neq 0$.**

Proof : Let y_1 and y_2 are linearly dependent solutions of Eqn. (16.11). By definition, there exist constants C_1 and C_2, not all zero, so that,

$$C_1y_1 + C_2y_2 = 0,$$

$$\therefore \quad C_1y_1' + C_2y_2' = 0.$$

Now, as this system of two equations in C_1, C_2 has **non-trivial solution** (i.e., C_1 and C_2 are not all zero), we must have

$$\begin{vmatrix} y_1 & y_2 \\ y_1' & y_2' \end{vmatrix} = 0 \quad \text{or,} \quad W(y_1, y_2) = 0.$$

Conversely, let $W(y_1, y_2) = 0$, then there exists a non-trivial solution (C_1, C_2 are not all zero) to the following homogeneous equations :

$$C_1y_1 + C_2y_2 = 0,$$

$$\therefore \quad C_1y_1' + C_2y_2' = 0.$$

$\therefore$ C_1, C_2, not all zero, such that $C_1y_1 + C_2y_2 = 0$.

$\therefore$ y_1 and y_2 are **linearly dependent**.

Thus, the two solutions of Eqn. (16.11) $y_1(x)$ and $y_2(x)$ are **linearly independent**, if

$$\boldsymbol{W(y_1, y_2) = \begin{vmatrix} y_1 & y_2 \\ y_1' & y_2' \end{vmatrix} \neq 0}, \text{ where } C_1 = C_2 = 0$$

and linearly dependent, if

$$\boldsymbol{W(y_1, y_2) = \begin{vmatrix} y_1 & y_2 \\ y_1' & y_2' \end{vmatrix} = 0}, \text{ where } C_1, C_2 \text{ are not all zero.}$$

(ii) **If $y_1(x)$ and $y_2(x)$ be two linearly independent solutions of homogeneous differential equation (16.11), then its general solution is given by**

$$\boldsymbol{y(x) = C_1y_1(x) + C_2y_2(x)},$$

where C_1 and C_2 are two arbitrary constants.

Proof : If $y_1(x)$ and $y_2(x)$ be two solutions of Eqn. (16.11), then

$$\frac{d^2y_1}{dx^2} + P_1\frac{dy_1}{dx} + P_2y_1 = 0 \tag{16.12}$$

and $$\frac{dy_2}{dx^2} + P_1\frac{dy_2}{dx} + P_2y_2 = 0. \tag{16.13}$$

Adding Eqns. (16.12) and (16.13), we get

$$\frac{d^2}{dx^2}(C_1y_1 + C_2y_2) + P_1\frac{d}{dx}(C_1y_1 + C_2y_2) + P_2(C_1y_1 + C_2y_2)$$

$$= C_1\left(\frac{d^2y_1}{dx^2} + P_1\frac{dy_1}{dx} + P_2y_1\right) + C_2\left(\frac{d^2y_2}{dx^2} + P_1\frac{dy_2}{dx} + P_2y_2\right)$$

$$= C_1 \cdot 0 + C_2 \cdot 0 = 0, \quad \text{where } C_1 \text{ and } C_2 \text{ are two arbitrary constants.}$$

$\therefore$ $\boldsymbol{y(x) = C_1y_1(x) + C_2y_2(x)}$ is the general solution of Eqn. (16.11). This is the **principle of superposition.**

N.B. The principle of superposition is not applicable to non-homogeneous and non-linear differential equations.

(iii) **If $u(x)$ and $v(x)$ (free from arbitrary constants) be respectively the general solution and particular solution of the homogeneous differential equation (16.9) i.e.,**

$$\boldsymbol{\frac{d^2y}{dx^2} + P_1\frac{dy}{dx} + P_2y = X},$$

then the general solution of the differential equation is $y = u(x) + v(x)$.

Proof : Since $u(x)$ is the general solution of Eqn. (16.11),

$$\therefore \quad \frac{d^2u}{dx^2} + P_1\frac{du}{dx} + P_2u = 0. \tag{16.14}$$

Again, $v(x)$ is a solution of Eqn. (16.9),

$$\therefore \quad \frac{d^2v}{dx^2} + P_1\frac{dv}{dx} + P_2v = X. \tag{16.15}$$

Adding Eqns. (16.14) and (16.15), we get

$$\frac{d^2(u+v)}{dx^2} + P_1\frac{d}{dx}(u+v) + P_2(u+v) = X \quad \text{or,} \quad \frac{d^2y}{dx^2} + P_1\frac{dy}{dx} + P_2y = X.$$

Hence, $\boldsymbol{y = u(x)+v(x)}$ is the general solution of Eqn. (16.9). Here, $u(x)$ is the complementary function (C.F.) and $v(x)$ is the particular integral (P.I.) of Eqn. (16.9).

Procedure to solve the differential equation,

$$\frac{d^2y}{dx^2} + P_1\frac{dy}{dx} + P_2y = X \tag{16.9}$$

involves the following steps :

(i) Put $X = 0$ and then find the complete solution $\boldsymbol{u = c_1u_1 + c_2u_2}$ of the reduced equation, i.e., (C.F.)

(ii) Find the particular solution of Eqn. (16.9), i.e., (P.I.). Then the **complete solution of Eqn. (16.9) is**

$$\textbf{C.S. = C.F.+P.I.}$$

16.5 Method for Finding the Complementary Function

(1) To find the complementary function the RHS of the given equation is replaced by zero.

(2) Let $y = C_1e^{mx}$ be the C.F. of

$$\frac{d^2y}{dx^2} + P_1\frac{dy}{dx} + P_2y = 0$$

or, $$(D^2 + P_1D + P_2) = 0. \tag{16.16}$$

where $D = \dfrac{d}{dy}$ and $D^2 = \dfrac{d^2}{dx^2}$ and P_1 and P_2 are constants.

Substituting $y = C_1e^{mx}$ in equation (16.16), we get,

$$(m^2 + P_1m + P_2)C_1e^{mt} = 0$$

or, $$m^2 + P_1m + P_2 = 0 \quad [\because C_1e^{mt} = 0]. \tag{16.17}$$

Eqn. (16.17) is known as the **characteristic or auxiliary equation of Eqn. (16.16).**

(3) **Solve the auxiliary equation :**

Case I : Roots are real and different.

If m_1 and m_2 be two real and distinct roots of Eqn. (16.17), then the C.F.

$$y = C_1e^{m_1x} + C_2e^{m_2x}, \quad \text{where } C_1 \text{ and } C_2 \text{ are two arbitrary constants.}$$

Example 16.1. *Solve :* $\dfrac{d^2y}{dx^2} - \dfrac{dy}{dx} - 6y = 0.$

Solution : The given equation can be written as

$$(D^2 - D - 6)y = 0. \quad \left[\text{where } D \equiv \frac{d}{dy}\right]$$

Let $y = e^{mx}$ be a trial solution. Then the auxiliary equation is

$$(m^2 - m - 6) = 0 \quad \text{or,} \quad (m-3)(m-2) = 0.$$

$\therefore$ the roots of the auxiliary equation are $m_1 = 3$ and $m_2 = 2$, which are real and different.

$\therefore$ the general solution is $\boldsymbol{y = C_1e^{3x} + C_2e^{2x}}$.

Case II : Roots are real and equal.

If the roots of the auxiliary equation (16.17) be equal, i.e., $m_1 = m_2 = m$, then Eqn. (16.16) can be written as

$$(D-m)^2y = 0 \quad \text{or,} \quad (D^2 - 2mD + m^2)y = 0$$

$$\text{or,} \quad \frac{d^2y}{dx^2} - 2m\frac{dy}{dx} + m^2y = 0. \tag{16.18}$$

Let the solution of the equation be $y = ve^{mx}$, where v is a function of x. Substituting $y = ve^{mx}$ in Eqn. (16.18), we get

$$m^2ve^{mx} + 2me^{mx}\frac{dv}{dx} + e^{mx}\frac{d^2v}{dx^2} - 2m\left(mv\,e^{mx} + e^{mx}\frac{dv}{dx}\right) + m^2ve^{mx} = 0$$

$$\text{or,} \quad \frac{d^2v}{dx^2} = 0 \quad [\because\ e^{mx} \neq 0].$$

The solution of this equation is $v = C_1x + C_2$

$$\therefore \quad \textbf{C.F.} = (C_1x + C_2)e^{mx}. \tag{16.19}$$

where C_1 and C_2 are arbitrary constants.

Example 16.2. *Solve* : $\dfrac{d^2y}{dx^2} - 4\dfrac{dy}{dx} + 4y = 0.$

Solution : The given equation can be written as

$$(D^2 - 4D + 4)y = 0. \quad \left[\text{where } D \equiv \frac{d}{dx}\right]$$

Let $y = e^{mx}$ be a trial solution. Then the auxiliary equation is

$$m^2 - 4m + 4 = 0 \quad \text{or,} \quad (m-2)^2 = 0.$$

$\therefore m = 2, 2$. The roots are real and equal.

Hence, the general solution is $\boldsymbol{y = (C_1x + C_2)e^{2x}}$, where C_1 and C_2 are arbitrary constants.

Case III : Roots are complex.

Let the solution of Eqn. (16.17) has complex roots $\alpha \pm i\beta$ (α, β are real), then

$$\begin{aligned}\textbf{C.F.} &= C_1e^{(\alpha+i\beta)x} + C_2e^{(\alpha-i\beta)x}\\ &= e^{\alpha x}[C_1(\cos\beta x + i\sin\beta x) + C_2(\cos\beta x - i\sin\beta x)]\\ &= e^{\alpha x}[(C_1 + C_2)\cos\beta x + i(C_1 - C_2)\sin\beta x]\\ &= \boldsymbol{e^{\alpha x}(A\cos\beta x + B\sin\beta x)}, \text{ where } C_1 + C_2 = A \text{ and } i(C_1 - C_2) = B.\end{aligned}$$

Example 16.3. *Solve* : $\dfrac{d^2y}{dx^2} + 4\dfrac{dy}{dx} + 5y = 0.$

Solution : The given equation is $(D^2 + 4D + 5)y = 0$, $\because\ D \equiv \dfrac{d}{dx}$.

Let $y = e^{-mx}$ be a trial solution.

Then the auxiliary equation is $m^2 + 4m + 5 = 0$, $\quad \therefore \quad m = -2 \pm i$.

Hence, the C.F. $\boldsymbol{y = e^{-2x}(C_1\cos x + C_2\sin x)}$, where C_1 and C_2 are arbitrary constants.

16.6 Rules to Find Particular Integral

The particular integral of the differential equation $\frac{d^n y}{dx^n}+k_1\frac{d^{n-1}y}{dx^{n-1}}+k_2\frac{d^{n-2}y}{dx^{n-2}}+\cdots+k_n y = X$ is

$$\text{P.I.} = \frac{1}{f(D)}X. \tag{16.20}$$

Case I : When $X = e^{ax}$.

We have $De^{ax} = ae^{ax}$

$$D^2e^{ax} = a^2e^{ax}$$

....

$$D^ne^{ax} = a^ne^{ax}$$

$\therefore \quad (D^n + k_1D^{n-1} + \cdots + k_n)e^{ax} = a^n + k_1e^{n-1} + \cdots + k_n)e^{ax}$

or, $\quad f(D)e^{ax} = f(a)e^{ax}. \quad (16.21)$

Operating both sides of Eqn. (16.21) by $\frac{1}{f(D)}$, we get

$$\frac{1}{f(D)}f(D)e^{ax} = \frac{1}{f(D)}f(a)e^{ax} \quad \text{or,} \quad e^{ax} = f(a)\frac{e^{ax}}{f(D)}.$$

Dividing by $f(a)$, $\quad \boldsymbol{\frac{1}{f(D)}e^{ax} = \frac{1}{f(a)}e^{ax}}$, $\quad$ if $f(a) \neq 0. \quad (16.22)$

If $f(a) = 0$, the above rule fails.

Then, since a is a root of the auxiliary equation,

$\therefore \quad f(D) = D^n + k_1D^{n-1} + \cdots + k_n = 0.$

$\therefore \quad (D - a)$ is a factor of $f(D)$.

Let $f(D) = (D - a)\phi(D)$, $\phi(a) \neq 0$.

Then $\quad \frac{1}{f(D)}e^{ax} = \frac{1}{D-a}\cdot\frac{1}{\phi(D)}e^{ax} = \frac{1}{D-a}\cdot\frac{1}{\phi(a)}e^{ax}$, by Eqn. 16.26

$$= \frac{1}{\phi(a)}\frac{1}{D-a}e^{ax} = \frac{1}{\phi(a)}e^{ax}\int e^{ax}\cdot e^{-ax}dx \quad \left[\because \frac{1}{D-a}X = e^{ax}\int Xe^{-ax}dx\right]$$

$$= \frac{1}{\phi(a)}e^{ax}\int dx = x\frac{1}{\phi(a)}e^{ax}.$$

$\therefore \quad \boldsymbol{\frac{1}{f(D)}e^{ax} = x\frac{1}{f'(a)}e^{ax}}$, $\quad$ if $f'(a) \neq 0 \quad (16.23)$

$$[\because f'(D) = (D-a)\phi'(D) + 1\cdot\phi(D), \quad \therefore f'(a) = 0\times\phi'(a) + \phi(a)]$$

If $f'(a) = 0$, then

$$\boldsymbol{\frac{1}{f(D)}e^{ax} = x^2\frac{1}{f''(a)}e^{ax}}, \textbf{ if } \boldsymbol{f''(a) \neq 0}. \tag{16.24}$$

Example 16.4. *Solve* : $\frac{d^2y}{dx^2} - \frac{dy}{dx} - 2y = \sin x$. **[BU(H) 1999]**

Solution : Here, the auxiliary equation is

$$m^2 - m - 2 = 0 \quad \text{or,} \quad (m-2)(m+1) = 0, \quad \therefore \quad m = 2, -1.$$

$\therefore$ C.F. $= Ae^{2x} + Be^{-x}$.

$$\text{P.I.} = \frac{1}{D^2 - D - 2}\sin x = \frac{1}{(-1) - D - 2}\sin x = -\frac{1}{D+3}\sin x = -\frac{D-3}{D^2-9}\sin x$$

$$= -\frac{D-3}{(-1)-9}\sin x = \frac{1}{10}(D-3)\sin x = \frac{1}{10}(\cos x - 3\sin x).$$

$\therefore$ general solution is $\boldsymbol{y = Ae^{2x} + Be^{-x} + \frac{1}{10}(\cos x - 3\sin x)}$.

A and B are two arbitrary constants.

Case II : When $\boldsymbol{X = \sin(ax+b)}$ **or,** $\boldsymbol{\cos(ax+b)}$

We have $D\sin(ax+b) = a\cos(ax+b)$

$$D^2\sin(ax+b) = -a^2\sin(ax+b)$$

$$D^3\sin(ax+b) = a^3\cos(ax+b)$$

$$D^4\sin(ax+b) = -a^4\sin(ax+b).$$

So, $$D^2\sin(ax+b) = (-a^2)\sin(ax+b)$$

$$(D^2)^2\sin(ax+b) = (-a^2)^2\sin(ax+b).$$

In general, $(D^2)^r\sin(ax+b) = (-a^2)^r\sin(ax+b)$.

$\therefore$ $$f(D^2)\sin(ax+b) = f(-a^2)\sin(ax+b).$$

Operating by $\frac{1}{f(D^2)}$, we get

$$\frac{1}{f(D^2)}f(D^2)\sin(ax+b) = \frac{1}{f(D^2)}f(-a^2)\sin(ax+b)$$

or, $$\sin(ax+b) = f(-a^2)\frac{1}{f(D^2)}\sin(ax+b).$$

Now, dividing by $f(-a^2)$,

$$\boldsymbol{\frac{1}{f(D^2)}\sin(ax+b) = \frac{1}{f(-a^2)}\sin(ax+b)} \text{ [provided } f(-a^2) \neq 0]. \quad (16.25)$$

For $f(-a^2) = 0$, the above rule fails. Then,

$\because$ $\cos(ax+b) + i\sin(ax+b) = e^{i(ax+b)}$,

$\therefore$ $$\frac{1}{f(D^2)}\sin(ax+b) = \text{I.P. of } \frac{1}{f(D^2)}e^{i(ax+b)} \quad [\because f(-a)^2 = 0]$$

$$= \text{I.P. of } x\frac{1}{f'(D^2)}e^{i(ax+b)}, \quad \text{where } D^2 = -a^2.$$

$\therefore$ $$\boldsymbol{\frac{1}{f(D^2)}\sin(ax+b) = x\frac{1}{f'(-a^2)}\sin(ax+b)} \text{ [provided } f'(-a^2) \neq 0]. \quad (16.26)$$

If $f'(-a)^2 = 0$,

$\therefore$ $$\boldsymbol{\frac{1}{f(D^2)}\sin(ax+b) = x^2\frac{1}{f''(-a^2)}\sin(ax+b)} \text{ [provided } f''(-a^2) \neq 0] \text{ and so on.}$$

In a similar way we can show that,

$$\frac{1}{f(D^2)}\cos(ax+b)=\frac{1}{f(-a^2)}\cos(ax+b),\ [\text{if } f(-a)^2\neq 0] \tag{16.27}$$

For $f(-a^2)=0$,

$$\frac{1}{f(D^2)}\cos(ax+b)=x\frac{1}{f'(-a^2)}\cos(ax+b),\ [\text{if } f'(-a^2)\neq 0]$$

If $f'(-a^2)=0$, then

$$\frac{1}{f(D^2)}\cos(ax+b)=x^2\frac{1}{f''(-a^2)}\cos(ax+b),\ [\text{if } f''(-a^2)\neq 0] \tag{16.28}$$

and so on.

Example 16.5. *Solve :* $\dfrac{d^2y}{dx^2}-2\dfrac{dy}{dx}+2y=x+e^x\cos x.$

Solution : Putting $D\equiv\frac{d}{dx}$, the given differential equation can be written as

$$(D^2-2D+2)y=x+e^x\cos x.$$

Let $y=e^{mx}$ be a trial solution, then the A.E. is

$$m^2-2m+2=0,\qquad\therefore\quad m=1\pm i.$$

$\therefore$ the C.F. $=e^x(A\cos x+B\sin x)$, where A and B are arbitrary constants.

The particular integral is

$$\begin{aligned}\text{P.I.}&=\frac{1}{D^2-2D+2}(x+e^x\cos x)=\frac{1}{D^2-2D+2}x+\frac{1}{D^2-2D+2}(e^x\cos x)\\&=\frac{1}{2}\left\{1-\left(D-\frac{D^2}{2}\right)\right\}^{-1}x+e^x\frac{1}{(D+1)^2-2(D+1)+2}\cos x\\&=\frac{1}{2}\left(1+D-\frac{D^2}{2}+\cdots\right)x+e^x\frac{1}{D^2+1}\cos x.\end{aligned} \tag{i}$$

In $\frac{1}{D^2+1}\cos x$, if we replace D^2 by -1^2, the method fails.

Now, $$\begin{aligned}\frac{1}{D^2+1}(\cos x+i\sin x)&=\frac{1}{D^2+1}e^{ix}=e^{ix}\frac{1}{(D+i)^2+1}1=e^{ix}\frac{1}{D^2+2iD}1\\&=e^{ix}\frac{1}{2iD}\left(1+\frac{D}{2i}\right)^{-1}1=e^{ix}\frac{1}{2iD}\left(1-\frac{D}{2i}+\cdots\right)1\\&=e^{ix}\cdot\frac{1}{2i}\cdot\frac{1}{D}1=\frac{e^{ix}}{2i}x=\frac{(\cos x+i\sin x)x}{2i}=\frac{1}{2}x\sin x-\frac{i}{2}x\cos x.\end{aligned}$$

$$\therefore\quad \frac{1}{D^2+1}(\cos x+i\sin x)=\frac{1}{2}x\sin x-\frac{i}{2}x\cos x.$$

Now equating the real parts from both sides,

$$\frac{1}{D^2+1}\cos x=\frac{1}{2}x\sin x.$$

$\therefore$ from equation (i), we get

$$\text{P.I.} = \frac{1}{2}(x+1) + \frac{1}{2}xe^x \sin x.$$

Now, the general solution is

$$\boldsymbol{y = e^x(A\cos x + B\sin x) + \frac{1}{2}(x+1) + \frac{1}{2}xe^x \sin x}$$

Case III : When $X = x^m$; m is a positive integer.

$$\text{Here, P.I.} = \frac{1}{f(D)}x^m = [f(D)]^{-1}x^m. \tag{16.29}$$

Now, expand $[f(D)]^{-1}$ in ascending powers of D and operate on x^m. The term of this expansion beyond mth power of D need not be considered as the result of their operation on x^m vanishes.

N.B. This result is true for nth ($n > 2$) order differential equation also.

Example 16.6. *Solve* : $\dfrac{d^2y}{dx^2} + 4\dfrac{dy}{dx} + 3y = x^2 + 3.$ **[B.U.(H) 1992]**

Solution : Let $y = e^{mx}$ be the trial solution; then the auxiliary equation is

$$m^2 + 4m + 3 = 0 \quad \text{or,} \quad (m+1)(m+3) = 0.$$

$$\therefore \quad m = -1,\ 3.$$

So, the C.F. $= Ae^{-x} + Be^{-3x}$.

$$\text{Now,}\quad \text{P.I.} = \frac{1}{D^2+4D+3}(x^2+3) = \frac{1}{(D+1)(D+3)}(x^2+3)$$

$$= \frac{1}{3}(1+D)^{-1}\left(1+\frac{D}{3}\right)^{-1}(x^2+3) = \frac{1}{3}(1+D)^{-1}\left[\left(1-\frac{D}{3}+\frac{D^2}{9}-\cdots\right)(x^2+3)\right]$$

$$= \frac{1}{3}\left(1-D+\frac{D^2}{2}-\cdots\right)\left(x^2+3-\frac{2x}{3}+\frac{2}{9}\right) = \frac{1}{3}\left[x^2-\frac{2}{3}x+\frac{29}{9}-2x+\frac{2}{3}+1\right]$$

$$= \frac{1}{3}\left(x^2-\frac{8}{3}x+\frac{44}{9}\right) = \boldsymbol{\frac{1}{27}(9x^2-24x+44)}.$$

$\therefore$ The general solution is, $y =$ C.F. + P.I.

$$\text{i.e.,}\quad \boldsymbol{y = Ae^{-x} + Be^{-3x} + \frac{1}{27}(9x^2-24x+44)}.$$

Case IV : When $X = e^{ax}V$, V being a function of x.

$$\text{P.I.} = \frac{1}{f(D)}e^{ax}V = e^{ax}\frac{1}{f(D+a)}V$$

Proof : Let u is a function of x, then

$$D(e^{ax}u) = e^{ax}Du + ae^{ax}u = e^{ax}(D+a)u.$$

$$D^2(e^{ax}u) = e^{ax}D^2u + 2ae^{ax}Du + A^2e^{ax}u = e^{ax}(D+a)^2u.$$

$$\therefore \quad (D^2 + P_1D + P_2)e^{ax}u = e^{ax}[(D+a)^2 + P_1(D+a) + P_2]u.$$

$$\therefore \quad f(D)e^{ax}u = e^{ax}f(D+a)u.$$

Operating on both sides by $\frac{1}{f(D)}$, we have

$$\frac{1}{f(D)}[f(D)e^{ax}u] = \frac{1}{f(D)}[e^{ax}f(D+a)u] \quad \text{or,} \quad e^{ax}u = \frac{1}{f(D)}[e^{ax}f(D+a)u].$$

Now, we put $f(D+a)u = V$, i.e., $u = \dfrac{1}{f(D+a)}V$.

$$\therefore \qquad e^{ax} = \frac{1}{f(D+a)}V = \frac{1}{f(D)}(e^{ax}V)$$

$$\text{or,} \qquad \boldsymbol{\frac{1}{f(D)}(e^{ax}V) = e^{ax}\frac{1}{f(D+a)}V.} \tag{16.30}$$

N.B. The above result is true also for nth ($n > 2$) order linear differential equation.

Example 16.7. *Solve* : $\dfrac{d^2y}{dx^2} + 2\dfrac{dy}{dx} + 2y = xe^{-x}$. **[BU(H) 1999]**

Solution : The given equation may be written as $(D^2 + 2D + 2)y = xe^{-x}$.

The auxiliary equation is $D^2 + 2D + 2 = 0$.

$\therefore$ C.F. $= e^{-x}(A\cos x + B\sin x)$

$$\text{P.I.} = \frac{1}{D^2+2D+2}(xe^{-x}) = e^{-x}\frac{1}{(D-1)^2+2(D-1)+2}(x) \ [\text{replacing } D \text{ by } (D-1)]$$

$$= e^{-x}\frac{1}{D^2+1}(x) = e^{-x}(1+D^2)^{-1}(x) = e^{-x}(1-D^2+D^4-\cdots)x = xe^{-x}.$$

$\therefore$ the general solution is $\boldsymbol{y = e^{-x}(A\cos x + B\sin x) + xe^{-x}}$.

Case V : When $X = xV$, where V is any function of x.

$$\text{Then P.E.} = \frac{1}{f(D)}\, xV = \left[x - \frac{1}{f(D)}f(D)\right]\frac{1}{f(D)}V.$$

Proof : Let $V_1 = \dfrac{1}{f(D)}V$.

Now, $D(xV_1) = xDV_1 + V_1$

$$D^2(xV_1) = DV_1 + xD^2V_1 + DV_1 = xD^2V_1 + 2DV_1.$$

$$\therefore \quad (D^2 + P_1D + P_2)xV_1 = xD^2V_1 + 2DV_1 + P_1xDV_1 + P_1V_1 + P_2xV_1$$

$$= x(D^2 + P_1D + P_2)V_1 + (2D + P_1)V_1.$$

So, $f(D)xV_1 = xf(D)V_1 + f'(D)V_1$

$$\text{or,} \quad f(D)x\left[\frac{1}{f(D)}V\right] = xf(D)\left[\frac{1}{f(D)}V\right] + f'(D)\left[\frac{1}{f(D)}V\right] \quad \left[\because V_1 = \frac{1}{f(D)}V\right]$$

$$\text{or,} \quad f(D)x\left[\frac{1}{f(D)}V\right] = xV + f'(D)\left[\frac{1}{f(D)}V\right]$$

$$\text{or,} \quad x\left[\frac{1}{f(D)}V\right] = \frac{1}{f(D)}xV + \frac{1}{f(D)}f'(D)\left[\frac{1}{f(D)}V\right].$$

$$\therefore \quad \frac{1}{f(D)}xV = \left[x - \frac{1}{f(D)}f'(D)\right]\frac{1}{f(D)}V. \tag{16.31}$$

N.B. The above result is true for nth ($n > 2$) order linear differential equations with constant coefficient.

Example 16.8. *Solve* : $\dfrac{d^2y}{dx^2} - 5\dfrac{dy}{dx} + 6y = xe^x$.

Solution : The given differential equation may be written as

$$(D^2 - 5D + 6)y = xe^x, \quad \text{where } D = \frac{d}{dx}.$$

The auxiliary equation is : $m^2 - 5m + 6 = 0, \quad \therefore \; m = 3,\ 2.$

The complementary function $y = Ae^{2x} + Be^{3x}$.

The particular integral,

$$\text{I.P.} = \frac{1}{D^2 - 5D + 6}xe^x = e^x \frac{1}{(D+1)^2 - 5(D+1) + 6}x \quad [\text{replacing } D \text{ by } (D+1)]$$

$$= e^x \frac{1}{D^2 - 3D + 2}x = e^x \frac{1}{(D-1)(D-2)}e^x = \frac{1}{2}e^x(1-D)^{-1}\left(1 - \frac{D}{2}\right)^{-1}x$$

$$= \frac{1}{2}e^x(1 + D + D^2 + \cdots)\left(1 + \frac{D}{2} + \frac{1}{4}D^2 + \cdots\right)x$$

$$= \frac{1}{2}e^x\left(1 + D + \frac{D}{2} + D^2 + \frac{D^2}{2} + \frac{D}{4} + \cdots\right)x$$

$$= \frac{1}{2}e^x\left(x + \frac{3}{2}\right) \quad [\text{terms higher than } D^2 \text{ are neglected}]$$

$\therefore$ the complete solution is $\boldsymbol{y = Ae^{2x} + Be^{3x} + \frac{1}{2}e^x\left(x + \frac{3}{2}\right)}$.

Case VI : When X is any other function of x

The particular integral is P.I. $= \dfrac{1}{f(D)}x$.

If $f(D) = (D - m_1)(D - m_2)\cdots(D - m_n)$, then

$$\frac{1}{f(D)} = \frac{A_1}{(D - m_1)} + \frac{A_2}{D - m_2} + \cdots + \frac{A_n}{D - m_n}.$$

$$\therefore \quad \text{P.I.} = \left[\frac{A_1}{D - m_1} + \frac{A_2}{D - m_2} + \cdots + \frac{A_n}{D - m_n}\right]X$$

$$= A_1\frac{1}{D - m_1}X + A_2\frac{1}{D - m_2}X + \cdots + A_n\frac{1}{D - m_n}X$$

$$\text{or,} \quad \text{P.I.} = \boldsymbol{A_1 e^{m_1 x}\int Xe^{-m_1 x}dx + A_2 e^{m_2 x}\int Xe^{-m_2 x}dx + \cdots}$$

$$\boldsymbol{+ A_n e^{m_n x}\int Xe^{-m_n x}dx}. \tag{16.32}$$

16.7 Method of Variation of Parameters

This is a very potential method to find the particular solution of a **Linear non-homogeneous** differential equation of second order with variable coefficients.

It can be stated as, if

$$\frac{d^2y}{dx^2} + P_1\frac{dy}{dx} + P_2y = X, \tag{16.33}$$

where P_1, P_2 and X are functions of x or constants, then the particular integral of the differential equation is

$$\text{P.I.} = -y_1\int \frac{y_2X}{W}dx + y_2\int \frac{y_1X}{W}dx,$$

where y_1 and y_2 are two linearly independent solutions of

$$\frac{d^2y}{dx^2} + P_1\frac{dy}{dx} + P_2y = 0 \tag{16.34}$$

and $W = \begin{vmatrix} y_1 & y_2 \\ y_1' & y_2' \end{vmatrix}$ is called the **Wronskian** of y_1 and y_2.

To prove the above statement we consider

$$y_p = u_1y_1 + u_2y_2, \tag{16.35}$$

where u_1 and u_2 are unknown functions of x, as the general solution of Eqn. (16.34).

Differentiating Eqn. (16.35) w.r.t. x we get

$$y_p' = u_1y_1' + u_2y_2' + u_1'y_1 + u_2'y_2. \tag{16.36}$$

Now, we choose u_1 and u_2 such that,

$$u_1'y_1 + u_2'y_2 = 0. \tag{16.37}$$

Then Eqn. (16.36) reduces to

$$y_p' = u_1y_1' + u_2y_2'. \tag{16.38}$$

$$\therefore \quad y_p'' = u_1y_1'' + u_2y_2'' + u_1'y_1' + u_2'y_2'. \tag{16.39}$$

Using the values of y_p, y_p', y_p'' in Eqn. (16.33), we get

$$u_1(y_1'' + P_1y_1' + P_2y_1) + u_2(y_2'' + P_1y_2' + P_2y_2) + u_1'y_1' + u_2'y_2' = X$$

or, $\quad u_1'y_1' + u_2'y_2' = X, \quad [\because y_1, y_2$ are the solutions of Eqn. (16.34)] $\quad$ (16.40)

Solving Eqns. (16.37) and (16.40), we get

$$u_1' = \frac{\begin{vmatrix} 0 & y_2 \\ X & y_2' \end{vmatrix}}{\begin{vmatrix} y_1 & y_2 \\ y_1' & y_2' \end{vmatrix}} = \frac{-y_2X}{W} \tag{16.41}$$

and

$$u_2' = \frac{\begin{vmatrix} y_1 & 0 \\ y_1' & X \end{vmatrix}}{\begin{vmatrix} y_1 & y_2 \\ y_1' & y_2' \end{vmatrix}} = \frac{y_1X}{W}. \tag{16.42}$$

Here, $W(y_1, y_2) = \begin{vmatrix} y_1 & y_2 \\ y_1' & y_2' \end{vmatrix} \neq 0$ as y_1 and y_2 are linearly independent functions.

Integrating, $u_1 = -\int \frac{y_2 X}{W} dx$ (16.43)

and $u_2 = \int \frac{y_1 X}{W} dx.$ (16.44)

Using the values of u_1 and u_2 in Eqn. (16.35), we get

$$\textbf{P.I.} = -y_1 \int \frac{y_2 X}{W} dx + y_2 \int \frac{y_1 X}{W} dx. \tag{16.45}$$

Example 16.9. *Solve by the method of variation of parameters :* $\frac{d^2y}{dx^2} + a^2 y = \sec ax.$ **[WBUT 2010]**

Solution : The given equation is $(D^2 + a^2)y = \sec ax, \quad D \equiv \frac{d}{dx}.$

Let $y = e^{mx}$ be a trial solution, then the A.E. becomes

$m^2 + a^2 = 0 \quad \therefore \; m = \pm ia.$

So, C.F. $= A\cos ax + B\sin ax.$

To find out P.I., let $y_1 = \cos ax$, $y_2 = \sin ax$ and $X = \sec ax$.

$$\therefore \quad W = \begin{vmatrix} y_1 & y_2 \\ y_1' & y_2' \end{vmatrix} = \begin{vmatrix} \cos ax & \sin ax \\ -a\sin ax & a\cos ax \end{vmatrix} = a \neq 0.$$

$\therefore$ y_1 and y_2 are two linearly independent solutions of the corresponding homogeneous equation.

$$\therefore \quad \text{P.I.} = -y_1 \int \frac{y_2 X}{W} dx + y_2 \int \frac{y_1 X}{W} dx$$

$$= -\cos ax \int \frac{\sin ax \sec ax}{a} dx + \sin ax \int \frac{\cos ax \sec ax}{a} dx$$

$$= -\frac{1}{a^2} \cos ax \log(\sec ax) + \frac{1}{a} x \sin ax.$$

Hence, $y = \text{C.F.} + \text{P.I.}$

or, $\quad \boldsymbol{y = A\cos ax + B\sin ax - \frac{1}{a^2}\cos ax \log(\sec ax) + \frac{1}{a} x \sin ax},$

if $a \neq 0$, A and B are two arbitrary constants.

16.8 Solved Problems

Example 16.1. *Solve :* $\frac{d^2y}{dx^2} + 9y = 0$, given $y = 3$, $\frac{dy}{dx} = 0$, when $x = 0$. **[BU(H) 2000]**

Solution : Let $y = e^m x$ be a trial solution. Then the auxiliary equation is

$m^2 + 9 = 0 \quad$ or, $\quad m = \pm 3i.$

The general solution is $y = A\cos 3x + B\sin 3x.$ (i)

Putting $y = 3$ when $x = 0$, we get $A = 3$.

Again, $\frac{dy}{dx} = -3A\sin 3x + 3B\cos 3x.$

But, $\frac{dy}{dx} = 0$ for $x = 0$, $\therefore \quad B = 0$.

So, the particular solution is $\boldsymbol{y = 3\cos 3x}$. (ii)

Example 16.2. *Solve :* $\frac{d^2y}{dx^2} - 4\frac{dy}{dx} + 4y = 0$ and find the particular solution if $y = 1$, $\frac{dy}{dx} = -2$, when $x = 0$.

Solution : The given equation can be written as $(D^2 - 4D + 4)y = 0$, where $D = \frac{dy}{dx}$.

Let $y = e^{mx}$ be a trial solution. Then the auxiliary equation is

$$m^2 - 4m + 4 = 0 \quad \text{or,} \quad (m-2)^2 = 0, \quad \therefore \; m = 2, 2.$$

The general solution is $y = (Ax + B)e^{2x}$, where A and B are two arbitrary constants.

$$\therefore \quad \frac{dy}{dx} = Ae^{2x} + 2(Ax + B)e^{2x} = (2Ax + A + 2B)e^{2x}.$$

When $x = 0$, $y = 1$,

$$\therefore \quad B = 1 \text{ and } A + 2B = -2 \text{ or, } A = -4.$$

$\therefore$ the particular solution of the given equation is $\boldsymbol{y = (1 - 4x)e^{2x}}$.

Example 16.3. *Solve the initial value problem,* $y'' - 4y' + 4y = 0$, $y(0) = 0$, $y'(0) = -3$.

[AMIETE(Winter) 1991]

Solution : We have the given equation, $D^2y - 4Dy + 4y = 0$, where $D \equiv \frac{d}{dx}$

or, $\quad (D^2 - 4D + 4)y = 0$.

The auxiliary equation is

$$D^2 - 4D + 4 = 0 \quad \text{or,} \quad (D-2)^2 = 0, \quad D = 2, 2.$$

The complementary function (C.F.) is

$$y = (C_1 + C_2x)e^{2x}. \qquad \text{(i)}$$

Putting $x = 0$ and $y = 0$, we get $C_1 = 0$.

$$\therefore \quad y = C_2xe^{2x} \quad \text{or,} \quad \frac{dy}{dx} = C_2e^{2x} + 2C_2xe^{2x}.$$

Putting $\frac{dy}{dx} = -3$, $x = 0$ in the above equation, we get $C_2 = -3$.

$\therefore$ substituting the values of C_1 and C_2 in Eqn. (i), we get

$$\boldsymbol{y = -3xe^{3x}}.$$

Example 16.4. *Solve the following differential equation,*

$$\frac{d^2y}{dx^2} + 6\frac{dy}{dx} + 13y = 0; \;\; y(0) = 2, \; \frac{dy}{dx} = 1.$$

Solution : Taking $y = e^{mx}$, we have the auxiliary equation, $m^2 + 6m + 13 = 0$.

$$\therefore \quad m = \frac{-6 \pm \sqrt{36 - 4 \times 13}}{2} = \frac{1}{2}(-6 \pm \sqrt{-16}) = \frac{1}{2}(-6 \pm 4i).$$

$\therefore \quad m_1 = -3+2i, \quad m_2 = -3-2i, \quad i = \sqrt{-1}.$

$\therefore$ the general solution is $y = e^{-3x}(A\cos 2x + B\sin 2x)$. (i)

$$\therefore \quad \frac{dy}{dx} = e^{-3x}(-2A\sin 2x + 2B\cos 2x) - 3e^{-3x}(A\cos 2x + B\sin 2x). \quad \text{(ii)}$$

Now, putting $y(0) = 2$ in Eqn. (i), we get $A = 2$ and $\left.\dfrac{dy}{dx}\right|_{x=0} = 1$, we get $B = \dfrac{7}{2}$.

$\therefore$ the required solution is $\boldsymbol{y = e^{-3x}\left(2\cos 2x + \frac{7}{2}\sin 2x\right)}$.

Example 16.5. *Obtain the general solution of the equation,* $\dfrac{d^2y}{dx^2} - 3\dfrac{dy}{dx} + 2y = x^3$.

Solution : The given equation can be written as,

$$(D^2-3D+2)y = x^3, \quad D \equiv \frac{d}{dx}.$$

Taking $y = e^{mx}$ as the trial solution, the auxiliary equation is $m^2 - 3m + 2 = 0$.

$\therefore \quad m_1 = 1,\ m_2 = 2.$

The complementary function is $y = Ae^x + Be^{2x}$.

The particular integral is $y = \dfrac{1}{D^2-3D+2}x^3 = \dfrac{1}{(D-2)(D-1)}x^3$.

$$\frac{1}{D-1}x^3 = -(1-D)x^3 = -(1+D+D^2+D^3+\cdots)x^3 = -(x^3+3x^2+6x+6).$$

$$\therefore \quad y = \frac{1}{2-D}(x^3+3x^2+6x+6) = \frac{1}{2}\left(1-\frac{D}{2}\right)^{-1}(x^3+3x^2+6x+6)$$

$$= \frac{1}{2}\left(1+\frac{D}{2}+\frac{D^2}{4}+\frac{D^3}{8}+\cdots\right)(x^3+3x^2+6x+6) = \frac{1}{2}\left(x^3+\frac{9}{2}x^2+\frac{21}{2}x+\frac{45}{4}\right).$$

$\therefore$ the complete solution of the given equation is

$$\boldsymbol{y = Ae^x + Be^{2x} + \frac{1}{2}\left(x^3+\frac{9}{2}x^2+\frac{21}{2}x+\frac{45}{4}\right)}.$$

Example 16.6. $(D^2+4)y = x\sin^2 x$, *where* $D \equiv \dfrac{d}{dx}$. **[WBUT 2008, 2010]**

Solution : To find the complementary function (C.F.) we have, $(D^2+4)y = 0$.

Let $y = e^{mx}$ be a trial solution. Then the auxiliary equation is $m^2+4=0$. $\therefore\ m = \pm 2i$.

So the C.F. $= A\cos 2x + B\sin 2x$.

The particular integral (P.I.) is given by

$$\text{P.I.} = \frac{1}{D^2+4}x\sin^2 x = \frac{1}{2}\frac{1}{D^2+4}x(1-\cos 2x) = \frac{1}{2}\frac{1}{D^2+4}x - \frac{1}{2}\frac{1}{D^2+4}x\cos 2x. \quad \text{(i)}$$

$$\text{Now,} \quad \frac{1}{D^2+4}x = \frac{1}{4}\left(1+\frac{D^2}{4}\right)^{-1}x = \frac{1}{4}\left(1-\frac{D^2}{4}+\cdots\right)x = \frac{x}{4} \quad \text{(ii)}$$

and $$\frac{1}{D^2+4}x\cos 2x = \left(x - \frac{2D}{D^2+4}\right) = \frac{1}{D^2+4}\cos 2x. \tag{iii}$$

In $\frac{1}{D^2+4}\cos 2x$, if D^2 is replaced by -2^2, then the method fails.

$$\frac{1}{D^2+4}(\cos 2x + i\sin 2x) = \frac{1}{D^2+4}e^{2ix} = e^{2ix}\frac{1}{(D+2i)^2+4}1 = e^{2ix}\frac{1}{D^2+4iD}1$$

$$= \frac{1}{4iD}\left(1+\frac{D}{4i}\right)^{-1}\cdot 1 = e^{2ix}\frac{1}{4i}\frac{1}{D}\left(1-\frac{D}{4i}+\cdots\right)1 = e^{2ix}\frac{1}{4i}\frac{1}{D}1$$

$$= \frac{e^{2ix}}{4i}x = \frac{\cos 2x + i\sin 2x}{4i} = \frac{1}{4}x\sin 2x - \frac{i}{4}x\cos 2x.$$

$$\therefore \quad \frac{1}{D^2+4}(\cos 2x + i\sin 2x) = \frac{1}{4}x\sin 2x - \frac{i}{4}x\cos 2x.$$

Equating real and imaginary parts from both sides, we get

$$\left.\begin{aligned} \frac{1}{D^2+4}\cos 2x &= \frac{1}{4}x\sin 2x \\ \frac{1}{D^2+4}\sin 2x &= -\frac{1}{4}x\cos 2x \end{aligned}\right\} \tag{iv}$$

From Eqn. (iii),

$$\frac{1}{D^2+4}x\cos 2x = \left(x - \frac{2D}{D^2+4}\right)\frac{1}{4}x\sin 2x = \frac{1}{4}x^2\sin 2x - \frac{1}{2}\frac{1}{D^2+4}D(x\sin 2x)$$

$$= \frac{1}{4}x^2\sin 2x - \frac{1}{2}\frac{1}{D^2+4}(\sin 2x + 2x\cos 2x)$$

$$= \frac{1}{4}x^2\sin 2x - \frac{1}{2}\frac{1}{D^2+4}\sin 2x - \frac{1}{D^2+4}x\cos 2x.$$

$$\therefore \quad 2\frac{1}{D^2+4}x\cos 2x = \frac{1}{4}x^2\sin 2x - \frac{1}{2}\frac{1}{D^2+4}\sin 2x = \frac{1}{4}x^2\sin 2x + \frac{1}{8}x\cos 2x.$$

$$\therefore \quad \frac{1}{D^2+4}x\cos 2x = \frac{1}{8}x^2\sin 2x + \frac{1}{16}x\cos 2x. \tag{v}$$

From Eqns. (i), (ii), (iii) and (v), we get

$$\textbf{P.I.} = \frac{x}{8} - \frac{x^2\sin 2x}{16} - \frac{x\cos 2x}{32}.$$

$\therefore$ the general solution is

$$y = \text{C.F.} + \text{P.I.} \quad \text{or,} \quad y = A\cos 2x + B\sin 2x + \frac{x}{8} - \frac{x^2\sin 2x}{16} - \frac{x\cos 2x}{32}.$$

Example 16.7. *Solve* : $(D^2 - 2D)y = e^x \sin x,\ D \equiv \frac{d}{dx}$. **[WBUT 2007]**

Solution : The C.F. may be found out by writing $(D^2 - 2D)y = 0$.

Taking $y = e^{mx}$ as a trial solution, the AE equation is

$$m^2 - 2m = 0 \quad \text{or,} \quad m(m-2) = 0, \quad \text{i.e.,} \quad m = 0, 2.$$

$\therefore$ C.F. $= Ae^{0x} + Be^{2x} = A + Be^{2x}$.

$$\text{P.I.} = \frac{1}{D^2 - 2D} e^x \sin x = e^x \frac{1}{(D+1)^2 - 2(D+1)} \sin x = e^x \frac{1}{D^2 - 1} \sin x$$

$$= e^x \frac{\sin x}{-1^2 - 1} = -\frac{1}{2} e^x \sin x.$$

$\therefore$ the general solution of the given equation is

$y =$ C.F. + P.I. or, $\boldsymbol{y = A + Be^{2x} - \frac{1}{2} e^x \sin x}$.

A and B are two arbitrary constants.

Example 16.8. *Find the solution of the differential equation* $y'' + 4y' + 5y = 65 \sin 2x$.

[AMIETE(Summer) 1994]

Solution : We have, $y'' + 4y' + 5y = 65 \sin 2x$ or, $\dfrac{d^2y}{dx^2} + 4\dfrac{dy}{dx} + 5y = 65 \sin 2x$

or, $(D^2 + 4D + 5)y = 65 \sin 2x$, where $D \equiv \dfrac{d}{dx}$.

The auxiliary equation is $D^2 + 4D + 5 = 0$.

Let $y = e^{mx}$ be a trial solution. Then,

$$m^2 + 4m + 5 = 0 \quad \text{or,} \quad m = \frac{-4 \pm \sqrt{16 - 20}}{2} = \frac{-4 \pm 2i}{2} = -2 \pm i.$$

$\therefore$ C.F. $= e^{-2x}(A \cos x + B \sin x)$.

$$\text{P.I.} = \frac{1}{D^2 + 4D + 5} 65 \sin 2x = 65 \frac{1}{(-2^2) + 4D + 5} \sin 2x = 65 \frac{4D - 1}{16D^2 - 1} \sin 2x$$

$$= 65 \frac{4D - 1}{16(-2^2) - 1} \sin 2x = 65 \frac{4D - 1}{-65} \sin 2x = -(4D - 1) \sin 2x = -(8 \cos 2x - \sin 2x)$$

$\therefore$ the general solution is

$y =$ C.F. + P.I. or, $\boldsymbol{y = e^{-2x}(A \cos x + B \sin x) - 8 \cos 2x + \sin 2x}$.

A and B are two arbitrary constants.

Example 16.9. *Solve* : $\dfrac{d^2y}{dx^2} - 4\dfrac{dy}{dx} + 4y = e^{2x}$, when $y = 0$, $\dfrac{dy}{dx} = 1$ for $x = 0$.

Solution : The given equation is $(D^2 - 4D + 4)y = e^{2x}$, where $D \equiv \dfrac{d}{dx}$.

The AE is $D^2 - 4D + 4 = 0$ or, $(D - 2)^2 = 0$ or, $D = 2, 2$.

$\therefore$ C.F. $= (A + Bx)e^{2x}$.

$$\text{P.I.} = \frac{1}{D^2 - 4D + 4} e^{2x} = \frac{1}{(D-2)^2} e^{2x}$$

$$= x \frac{1}{2D - 4} e^{2x} \qquad [\because f(D) = (D-2)^2, \quad \therefore f(2) = 0]$$

$$= \frac{x}{2} \frac{1}{D - 2} e^{2x} = \frac{x}{2} x e^{2x} = \frac{x^2}{2} e^{2x}.$$

[P.I. may be found in an alternative way as follows :

$$\text{P.I.} = \frac{1}{(D-2)^2}e^{2x} = e^{2x}\frac{1}{D^2}(1) = \frac{x^2}{2}e^{2x} \quad [\because f(2)=0,\ D \text{ is replaced by } (D+2)]$$

So, the general solution is

$$\boldsymbol{y = (A + Bx)e^{2x} + \frac{x^2}{2}e^{2x}}. \quad \text{(i)}$$

Now, $\dfrac{dy}{dx} = Be^{2x} + 2(A+Bx)e^{2x} + xe^{2x} + x^2e^{2x}$. (ii)

When $x = 0$, $y = 0$.

$\therefore$ from Eqn. (i), $A = 0$.

Again, when $x = 0$, $\dfrac{dy}{dx} = 1$.

$\therefore$ from Eqn. (ii), $B + 2A = 1$, $\therefore$ $B = 1$.

So, the particular solution of the given differential equation is

$$y = xe^{2x} + \frac{1}{2}x^2e^{2x} \quad \text{or,} \quad \boldsymbol{y = \frac{1}{2}x(2+x)e^{2x}}$$

Example 16.10. *Solve the following differential equation :* $\dfrac{d^2y}{dx^2} + 4y = 2\sin 2t$.

Solution : The given differential equation may be written as $(D^2+4)y = 2\sin 2t$.

The A.E. is $f(D) = D^2 + 4 = 0$. Roots of $f(D) = 0$ are $\pm 2i$.

$\therefore$ C.F. $= A\cos 2t + B\sin 2t$.

$$\text{P.I.} = \frac{1}{D^2+4}2\sin 2t = 2t\frac{1}{2D}\sin 2t \quad [\because f(-a^2) = f(-4) = 0]$$

$$= 2t\frac{D(\sin 2t)}{2D^2} = \frac{-2t\cos 2t}{4} \quad [\because f(0) = 2D]$$

$$= \frac{-t\cos 2t}{2}.$$

$\therefore$ the general solution is $\boldsymbol{y = A\cos 2t + B\sin 2t - \frac{t}{2}\cos 2t}$.

Example 16.11. *Solve :* $\dfrac{d^2y}{dx^2} - 6\dfrac{dy}{dx} + 13y = 8e^{3x}\sin 4x + 2^x$

Solution : The given differential equation may be written as $(D^2-6D+13)y = 8e^{3x}\sin 4x + 2^x$.

Taking $y = e^{mx}$ as a trial solution, the A.E. becomes $m^2 - 6m + 13 = 0$. $\therefore$ $m = 3 \pm 2i$.

C.F. $= e^{3x}(A\cos 2x + B\sin 2x)$.

P.I. is given by

$$\text{P.I.} = \frac{1}{D^2-6D+13}(8e^{3x}\sin 4x + 2^x) = \frac{1}{D^2-6D+13}(8e^{3x}\sin 4x) + \frac{1}{D^2-6D+13}e^{x\log 2}$$

$$= 8e^{3x}\frac{1}{(D+3)^2 - 6(D+3) + 13}\sin 4x + \frac{e^{x\log 2}}{(\log 2)^2 - 6\log 2 + 13}$$

$$= 8e^{3x}\frac{1}{D^2+4}\sin 4x + \frac{2^x}{(\log 2)^2 - 6\log 2 + 13} = -\frac{2}{3}e^{3x}\sin 4x + \frac{2^x}{(\log 2)^2 - 6\log 2 + 13}.$$

The general solution is

$$\boldsymbol{y = \text{C.F.} + \text{P.I.} = e^{3x}(A\cos 2x + B\sin 2x) - \frac{2}{3}e^{3x}\sin 4x + \frac{2^x}{(\log 2)^2 - 6\log 2 + 13}}.$$

Example 16.12. *Solve* : $\frac{d^2y}{dx^2} - 4y = x\sinh x.$ **[Kerala U 1983S]**

Solution : The given differential equation can be written as

$$(D^2-4)y = x\sinh x, \quad \text{where } D \equiv \frac{d}{dx}.$$

A.E. is $D^2 - 4 = 0$, $\therefore$ $D = \pm 2$.

Then C.F. $= Ae^{2x} + Be^{-2x}$.

$$\text{P.I.} = \frac{1}{D^2-4}x\sinh x = \frac{1}{D^2-4}x\left(\frac{e^x - e^{-x}}{2}\right)$$

$$= \frac{1}{2}\left[\frac{1}{D^2-4}xe^x - \frac{1}{D^2-4}e^{-x}x\right] = \frac{1}{2}\left[e^x\frac{1}{(D+1)^2-4}x - e^{-x}\frac{1}{(D-1)^2-4}x\right]$$

$$= \frac{1}{2}\left[e^x\frac{1}{D^2+2D-3}x - e^{-x}\frac{1}{D^2-2D-3}x\right] = \frac{1}{2}\left[e^x\frac{1}{D^2+2D-3}x - e^{-x}\frac{1}{D^2-2D-3}x\right]$$

$$= \frac{1}{2}\left[\frac{e^x}{-3}\left\{1 - \left(\frac{2D}{3} + \frac{D^2}{3}\right)\right\}^{-1}x - \frac{e^{-x}}{-3}\left\{1 + \left(\frac{2D}{3} - \frac{D^2}{3}\right)\right\}^{-1}x\right]$$

$$= -\frac{1}{6}\left[e^x\left(1 + \frac{2D}{3} + \cdots\right)x - e^{-x}\left(1 - \frac{2D}{3} + \cdots\right)x\right]$$

$$= -\frac{1}{6}\left[e^x\left(x + \frac{2}{3}\right) - e^{-x}\left(x - \frac{2}{3}\right)\right] = -\frac{x}{3}\left(\frac{e^x - e^{-x}}{2}\right) - \frac{2}{9}\left(\frac{e^x + e^{-x}}{2}\right)$$

$$= -\frac{x}{3}\sinh x - \frac{2}{9}\cosh x.$$

$\therefore$ the complete solution is

$$\boldsymbol{y = Ae^{2x} + Be^{-2x} - \frac{x}{3}\sinh x - \frac{2}{9}\cosh x}, \quad \text{where } A \text{ and } B \text{ are arbitrary constants}$$

Example 16.13. *Solve* : $\frac{d^2y}{dx^2} - 2\frac{dy}{dx} + y = xe^x\sin x.$

[Kerala U 1984, Karnataka U 1988, Ranchi U 1990]

Solution : The given differential equation can be written as $(D^2 - 2D + 1)y = xe^x\sin x$.

A.E. is $D^2 - 2D + 1 = 0$, i.e., $(D-1)^2 = 0$.

The complementary function (C.F.) $= (A + Bx)e^x$.

The particular integral (P.I.) is

$$\text{P.I.} = \frac{1}{(D-1)^2}e^x x\sin x = e^x\frac{1}{(D+1-1)^2}x\sin x = e^x\frac{1}{D^2}x\sin x = e^x\frac{1}{D}\int x\sin x\,dx$$

$$= e^x \frac{1}{D}\left[x(-\cos x) - \int 1\cdot(-\cos x)dx\right] = e^x \int[-x\cos x + \sin x]dx$$

$$= e^x\left[-\left\{x\sin x - \int 1\cdot \sin x\,dx\right\} - \cos x\right] = e^x[-x\sin x - \cos x - \cos x]$$

$$= -e^x(x\sin x + 2\cos x).$$

Hence, the complete solution of the given differential equation is

$$\boldsymbol{y = (A + Bx)e^x - e^x(x\sin x + 2\cos x)},\quad \text{where } A \text{ and } B \text{ are arbitrary constants.}$$

Example 16.14. *Solve* : $\dfrac{d^2y}{dt^2} + 2\dfrac{dy}{dt} + 5y = \sin\omega t.$

Solution : The given equation may be written as $(D^2 + 2D + 5)y = \sin\omega t.$

A.E. is $D^2 + 2D + 5 = 0.$ $\quad\therefore\ D = 1 \pm 2i.$

$\therefore$ C.F. $= Ae^{-t}\cos(2t + \phi)$, where A and ϕ are two arbitrary constants.

The particular integral is

$$\text{P.I.} = \frac{1}{D^2 + 2D + 5}\sin\omega t = \frac{1}{-\omega^2 + 2D + 5}\sin\omega t = \frac{2D - (5 - \omega^2)}{4D^2 - (5 - \omega^2)^2}\sin\omega t$$

$$= \frac{5 - \omega^2 - 2D}{-4(-\omega^2) + (5 - \omega^2)^2}\sin\omega t = \frac{5 - \omega^2 - 2D}{-6\omega^2 + \omega^4 + 25}\sin\omega t$$

$$= \frac{(5 - \omega^2)\sin\omega t - 2\omega\cos\omega t}{\omega^4 - 6\omega^2 + 25}\sin\omega t.$$

$\therefore$ the complete solution is

$$\boldsymbol{y = Ae^{-t}\cos(2t + \phi) + \frac{(5 - \omega^2)\sin\omega t - 2\omega\cos\omega t}{\omega^4 - 6\omega^2 + 25}}.$$

Example 16.15. *Solve* : $(D^2 - 5D + 6)y = e^x\cos 2x.$ **[AMIETE(Winter) 1990, 1991]**

Solution : The given equation is $(D^2 - 5D + 6)y = e^x\cos 2x.$

A.E. is $D^2 - 5D - 6 = 0$ or, $(D - 2)(D - 3) = 0,$ $\quad\therefore\ D = 2, 3$

C.F. $= Ae^{2x} + Be^{3x}$, where A and B are arbitrary constants.

$$\text{P.I.} = \frac{1}{D^2 - 5D + 6}e^x\cos 2x = e^x\frac{1}{(D+1)^2 - 5(D+1) + 6}\cos 2x \quad [\text{replacing } D \text{ by } (D+1)]$$

$$= e^x\frac{1}{D^2 - 3D + 2}\cos 2x = e^x\frac{1}{-4 - 3D + 2}\cos 2x = -e^x\frac{1}{3D + 2}\cos 2x$$

$$= -e^x\frac{3D - 2}{9D^2 - 4}\cos 2x = -e^x\frac{3D - 2}{9(-4) - 4}\cos 2x = \frac{e^x}{40}(3D - 2)\cos 2x$$

$$= \frac{e^x}{40}(-6\sin 2x - 2\cos 2x) = -\frac{e^x}{20}(3\sin 2x + \cos 2x)$$

$\therefore$ the complete solution of the given differential equation is

$$\boldsymbol{y = Ae^{2x} + Be^{3x} - \frac{e^x}{20}(3\sin 2x + \cos 2x)},\quad \text{where } A \text{ and } B \text{ are two arbitrary constants.}$$

Example 16.16. *Solve* : $\dfrac{d^2y}{dx^2} + a^2y = \sec ax.$

Solution : The given differential equation may be written as $(D^2 + a^2)y = \sec ax.$

$\therefore$ A.E. is $D^2 + a^2 = 0$ or, $D = \pm ia.$

The C.F. $= A_1 \cos ax + A_2 \sin ax$, where A_1 and A_2 are two arbitrary constants.

$$\text{Now, P.I.} = \frac{1}{D^2 + a^2}\sec ax = \frac{1}{(D + ia)(D - ia)}\sec ax = \frac{1}{2ia}\left[\frac{1}{D - ia} - \frac{1}{D + ia}\right]\sec ax.$$

$$= \frac{1}{2ia}\left[\frac{1}{D - ia}\sec ax - \frac{1}{D + ia}\sec ax\right]$$

$$\text{Now, } \frac{1}{D - ia}\sec ax = e^{iax}\int \sec ax \cdot e^{-iax}dx \qquad \left[\because \frac{1}{D - a}X = e^{ax}\int Xe^{-ax}dx\right]$$

$$= e^{iax}\int \frac{\cos ax - i\sin ax}{\cos ax}dx = e^{iax}\int(1 - i\tan ax)dx = e^{iax}\left(x + \frac{i}{a}\log\cos ax\right).$$

Again, changing i to $-i$, we get

$$\frac{1}{D + ia}\sec ax = e^{-iax}\left\{x - \frac{i}{a}\log\cos ax\right\}.$$

$$\therefore \quad \text{P.I.} = \frac{1}{2ia}\left[e^{iax}\left\{x + \frac{i}{a}\log\cos ax\right\} - e^{-iax}\left\{x - \frac{i}{a}\log\cos ax\right\}\right]$$

$$= \frac{x}{a}\frac{e^{iax} - e^{-iax}}{2i} + \frac{1}{a^2}\log\cos ax \cdot \frac{e^{iax} + e^{-iax}}{2} = \frac{x}{a}\sin ax + \frac{1}{a^2}\log\cos ax \cdot \cos ax.$$

Hence, the complete solution of the differential equation is

$$\boldsymbol{y = A_1 \cos ax + A_2 \sin ax + \frac{x}{a}\sin ax + \frac{1}{a^2}\log\cos ax \cdot \cos ax.}$$

Example 16.17. *Solve* : $\dfrac{d^3y}{dx^3} - 2\dfrac{d^2y}{dx^2} - 5\dfrac{dy}{dx} + 6y = e^{3x}.$

Solution : The given differential equation may be written as $(D^3 - 2D^2 - 5D + 6)y = e^{3x}.$

A.E. is $D^3 - 2D^2 - 5D + 6 = 0$ or, $(D - 1)(D - 3)(D + 2) = 0$, $\therefore$ $D = 1, 3, -2.$

C.F. is $y = Ae^x + Be^{3x} + Ce^{-2x}.$

Now, the particular integral is P.I. $= \dfrac{1}{(D - 1)(D - 3)(D + 2)}e^{3x}.$

Here, $f(a) = f(3) = 0$. So, the method to find P.I. fails.

$$\therefore \quad \text{P.I.} = \frac{1}{D - 3}\left[\frac{1}{(D - 1)(D + 2)}e^{3x}\right] = \frac{1}{D - 3}\left[\frac{1}{(3 - 1)(3 + 2)}e^{3x}\right]$$

$$= \frac{1}{D - 3}\cdot\frac{1}{10}e^{3x} = \frac{1}{10}\frac{1}{D - 3}e^{3x}.$$

Now, let $\frac{1}{D-3}e^{3x} = z$, then $(D - 3)z = e^{3x}$ or, $\dfrac{dz}{dx} - 3z = e^{3x}.$

The integrating factor of this first order linear differential equation is $e^{\int -3dx} = e^{-3x}$.

Multiplying both sides by e^{-3x} and integrating, we get

$$\therefore \quad e^{-3z} \cdot z = \int dx = x \quad \text{or,} \quad z = xe^{3x}.$$

$$\text{P.I. is} = \frac{1}{10} xe^{3x}.$$

Now the complete solution of the given differential equation is

$$\boldsymbol{y = Ae^{x} + Be^{3x} + Ce^{-3x} + \frac{1}{10}xe^{3x}}, \text{ where } A \text{ and } B \text{ are arbitrary constants.}$$

Example 16.18. *Solve* : $\dfrac{d^2y}{dx^2} - y = 4xe^x$.

Solution : The given differential equation may be written as

$$(D^2-1)y = 4xe^x \quad \text{or,} \quad (D-1)(D+1)y = 4xe^x \left[D \equiv \frac{d}{dx}\right]$$

A.E. is $(D-1)(D+1) = 0$ or, $D = 1, -1$.

$\therefore$ C.F. $= Ae^{-x} + Be^x$.

P.I. is given by

$$\text{P.I.} = \frac{1}{(D-1)(D+1)}4xe^x = \frac{1}{2}\left(\frac{1}{(D-1)} - \frac{1}{(D+1)}\right)4xe^x = 2\frac{1}{D-1}xe^x - 2\frac{1}{D+1}xe^x$$

$$= 2e^x \int xe^x \cdot e^{-x} dx - 2e^{-x}\int xe^x \cdot e^x dx = 2e^x \int x\, dx - 2e^{-x}\int xe^{2x} dx$$

$$= 2e^x \cdot \frac{x^2}{2} - 2e^{-x}\left[x\int e^{2x}dx - \int\left\{\left(\frac{dx}{dx}\right)\int e^{2x}dx\right\}dx\right]$$

$$= x^2e^x - 2e^{-x}\left[\frac{xe^{2x}}{2} - \frac{e^{2x}}{4}\right] = e^x\left(x^2 - x + \frac{1}{2}\right).$$

$\therefore$ the general solution of the given differential equation is

$$\boldsymbol{y = Ae^{-x} + Be^x + e^x\left(x^2 - x + \frac{1}{2}\right)}, \text{ where } A \text{ and } B \text{ are two arbitrary constants.}$$

Example 16.19. *Solve* : $\dfrac{d^2y}{dx^2} - 3\dfrac{dy}{dx} + 2y = e^x$, if $y = 3$ and $\dfrac{dy}{dx} = 3$, when $x = 0$.

Solution : The given equation may be written as $(D^2 - 3D + 2)y = e^x$, $D \equiv \dfrac{d}{dx}$.

A.E. is $(D^2 - 3D + 2 = 0$ or, $(D-1)(D-2) = 0$ or, $D = 1,\ 2$.

C.F. is $y = Ae^x + Be^{2x}$, where A and B are two arbitrary constants.

$$\text{Now, P.I.} = \frac{1}{(D-1)(D-2)}e^x = -\frac{1}{D-1}e^x = -e^x\int e^x \cdot e^{-x}dx = -e^{-x}\int dx = -xe^x.$$

$\therefore$ general solution is $\boldsymbol{y = Ae^x + Be^{2x} - xe^x}$.

$\therefore \quad 3 = A + B$ or, $A + B = 3$. (i)

Also, $\dfrac{dy}{dx} = Ae^x + 2Be^{2x} - e^x - xe^x = (A-1-x)e^x + 2Be^{2x}$.

Putting $\dfrac{dy}{dx} = 3$ for $x = 0$,

$$3 = (A-1-0) + 2B \quad \text{or,} \quad A - 1 + 2B = 3. \tag{ii}$$

Solving eqns. (i) and (ii), we get $A = 2$, $B = 1$.

$\therefore$ the required general equation is $\boldsymbol{y = (1-x)e^x + 2e^{2x}}$.

Example 16.20. *Apply the method of variation of parameters to solve,* $\dfrac{d^2y}{dx^2} + 4y = 4\sec^2 x$.

[WBUT 2006]

Solution : The given differential equation may be written as $(D^2+4)y = 4\sec^2 2x$, $D \equiv \dfrac{d}{dx}$.

C.F. is found from $(D^2+4)y = 0$.

Let $y = e^{mx}$ be a trial solution, then the A.E. becomes $m^2 + 4 = 0$ or, $m = \pm 2i$.

$\therefore$ C.F. $= A\cos 2x + B\sin 2x$.

To find the P.I., let $y_1 = \cos 2x$, $y_2 = \sin 2x$ and $X = 4\sec^2 2x$.

$$\therefore \quad W = \begin{vmatrix} y_1 & y_2 \\ y_1' & y_2' \end{vmatrix} = \begin{vmatrix} \cos 2x & \sin 2x \\ -2\sin 2x & 2\cos 2x \end{vmatrix} = 2 \neq 0.$$

$\therefore$ y_1 and y_2 are two linearly independent solutions of the corresponding homogeneous equation of the given differential equation.

$$\begin{aligned}
\therefore \quad \text{P.I.} &= -y_1 \int \frac{y_2 X}{W}\,dx + y_2 \int \frac{y_1 X}{W}\,dx \\
&= -\cos 2x \int \frac{1}{2}(\sin 2x) 4\sec^2 2x\,dx + \sin 2x \int \frac{1}{2}(\cos 2x) 4\sec^2 2x\,dx \\
&= -\cos 2x \int 2\sec 2x \tan 2x\,dx + \sin 2x \int 2\sec^2 2x\,dx \\
&= -\cos 2x \sec 2x + \sin 2x \log(\sec 2x + \tan 2x) \\
&= -1 + \sin 2x \log(\sec 2x + \tan 2x).
\end{aligned}$$

$\therefore$ the general solution is

$$\boldsymbol{y = \text{C.F.} + \text{P.I.} = A\cos 2x + B\sin 2x - 1 + \sin 2x \log(\sec 2x + \tan 2x)},$$

where A and B are two arbitrary constants

Example 16.21. *Solve by the method of variation of parameters :* $\dfrac{d^2y}{dx^2} - y = \dfrac{2}{1+e^x}$.

Solution : We have $\dfrac{d^2y}{dx^2} - y = \dfrac{2}{1+e^x}$ or, $(D^2-1)y = \dfrac{2}{1+e^x}$, $D \equiv \dfrac{d}{dx}$.

$\therefore$ A.E. is $(D^2 - 1) = 0$ or $D^2 = 1$ or, $D = \pm 1$.

C.F. $= Ae^x + Be^{-x} = Ay_1 + By_2$; A, B are arbitrary constants,

where $y_1 = e^x$ and $y_2 = e^{-x}$ and $y_1y_2' - y_1'y_2 = -e^x e^{-x} - e^x e^{-x} = -2$.

$$u=\int\frac{-y_2X}{y_1y_2'-y_1'y_2}dx=-\int\frac{e^{-x}}{-2}\cdot\frac{2}{1+e^x}dx=\int\frac{e^{-x}}{1+e^x}dx=\int\frac{dx}{e^x(1+e^x)}$$

$$=\int\left(\frac{1}{e^x}-\frac{1}{1+e^x}\right)dx=\int e^{-x}dx-\int\frac{e^{-x}}{e^{-x}+1}dx=-e^{-x}+\log\left(e^{-x}+1\right).$$

$$v=\int\frac{y_1X}{y_1y_2'-y_1'y_2}dx=\int\frac{e^x}{-2}\cdot\frac{2}{1+e^x}dx=-\int\frac{e^x}{1+e^x}dx=-\log(1+e^x)$$

$$\text{P.I.}=uy_1+vy_2=\left[-e^{-x}+\log(e^{-x}+1)\right]e^x-e^{-x}\log(1+e^x)$$

$$=1+e^x\log(e^{-x}+1)-e^{-x}\log(e^x+1).$$

Hence, the general solution is

$$\boldsymbol{y=Ae^x+Be^{-x}-1+e^x\log(e^{-x}+1)-e^{-x}\log(e^x+1)}.$$

Example 16.22. *Solve using the method of variation of parameters :*

$$\frac{d^2y}{dx^2}-3\frac{dy}{dx}+2y=\frac{e^x}{1+e^x}.$$

Solution : The given differential equation may be written as

$$(D^2-3D+2)y=\frac{e^x}{1+e^x},\quad\text{where } D\equiv\frac{d}{dx}.$$

A.E. is $D^2-3D+2=0$ or, $D=1,2$.

$\therefore$ C.F. $=Ae^x+Be^{2x}$, A, B are arbitrary constants.

Now, let $y_1=e^x$, $y_2=e^{2x}$ and $X=\frac{e^x}{1+e^x}$.

$$\text{Wronskian, } W=\begin{vmatrix}y_1 & y_2\\ y_1' & y_2'\end{vmatrix}=\begin{vmatrix}e^x & e^{2x}\\ e^x & 2e^{2x}\end{vmatrix}=e^{3x}\neq 0.$$

$\therefore$ y_1 and y_2 are two linearly independent solutions of the corresponding homogeneous equations of the given differential equation.

$$\text{Now, P.I.}=-y_1\int\frac{y_2X}{W}dx+y_2\int\frac{y_1X}{W}dx$$

$$=-e^x\int\frac{e^{2x}}{e^{3x}}\cdot\frac{e^x}{(1+e^x)}dx+e^{2x}\int\frac{e^x}{e^{3x}}\cdot\frac{e^x}{(1+e^x)}dx=-e^x\int\frac{dx}{1+e^x}+e^{2x}\int\frac{dx}{e^x(1+e^x)}$$

$$=-e^x\int\frac{dx}{1+e^x}+e^{2x}\int\frac{(1+e^x)-e^x}{e^x(1+e^x)}dx=-(e^x+e^{2x})\int\frac{dx}{1+e^x}+e^{2x}\int e^{-x}dx$$

$$=-(e^x+e^{2x})\int\frac{e^{-x}}{e^{-x}+1}dx-e^{2x}\cdot e^{-x}=(e^x+e^{2x})\log(e^{-x}+1)-e^x.$$

$\therefore$ the general solution is

$$y=\text{C.F.}+\text{P.I.}=\boldsymbol{Ae^x+Be^{2x}+(e^x+e^{2x})\log(e^{-x}+1)-e^x}.$$

Example 16.23. *Apply the method of variation of parameters to solve the equation*

$$\frac{d^2y}{dx^2}+y=\sec^3x\tan x.$$

[WBUT 2007]

Solution : The given equation may be written as $(D^2+1)y=0$, where $D \equiv \dfrac{d}{dx}$.

The A.E. equation is $D^2+1=0$ or, $D=\pm i$.

$\therefore$ C.F. $= A\cos x + B\sin x$, where A and B are arbitrary constants.

To determine P.I., let $y_1=\cos x$, $y_2=\sin x$ and $X=\sec^3 x\tan x$.

The Wronskian, $W=\begin{vmatrix} y_1 & y_2 \\ y_1' & y_2' \end{vmatrix} = \begin{vmatrix} \cos x & \sin x \\ -\sin x & \cos x \end{vmatrix} = 1 \neq 0.$

$\therefore$ y_1 and y_2 are two linearly independent solutions of the corresponding homogeneous equation of the given differential equation.

$$\therefore \quad \text{P.I.} = -y_1\int \frac{y_2X}{W}dx + y_2\int\frac{y_1X}{W}dx$$

$$= -\cos x\int \sin x\sec^3 x\tan x\,dx + \sin x\int\cos x\sec^3 x\tan x\,dx$$

$$= -\cos x\int\sec^2 x\tan^2 x\,dx + \sin x\int \sec^2 x\tan x\,dx$$

$$= -\frac{1}{3}\cos x\tan^3 x + \frac{1}{2}\sin x\tan^2 x = -\frac{1}{3}\sin x\tan^2 x + \frac{1}{2}\sin x\tan^2 x$$

$$= \frac{1}{6}\sin x\tan^2 x.$$

$\therefore$ the complete solution is $\boldsymbol{y}=$ C.F. + P.I. $= \boldsymbol{A\cos x + B\sin x + \frac{1}{6}\sin x\tan^2 x}$.

Example 16.24. *Solve :* $\dfrac{dx}{dt}+3x+y=e^t$, $\dfrac{dy}{dt}-x+y=e^{2t}$. **[WBUT 2004]**

Solution : The given equations can be written as

$$(D+3)x+y=e^t \qquad \text{(i)}$$

$$-x+(D+1)y=e^{2t}, \qquad \text{(ii)}$$

where $D\equiv\dfrac{d}{dt}$.

Operating both sides of Eqn. (i) and then subtracting from Eqn. (ii), we get

$$-\{(D+1)(D+3)+1\}x = e^{2t}-(D+1)e^t$$

$$\text{or,} \quad (D+2)^2x = 2e^t - e^{2t}. \qquad \text{(iii)}$$

The C.F. of Eqn. (iii) is $(D+2)^2=0$, $\therefore$ $D=-2,-2$.

C.F. $=(A+Bt)e^{-2t}$, A and B are arbitrary constants.

$$\text{Now, P.I.} = \frac{1}{(D+2)^2}(2e^t-e^{2t}) = \frac{1}{(D+2)^2}(2e^t) - \frac{1}{(D+2)^2}(e^{2t})$$

$$= 2\cdot\frac{1}{(1+2)^2}e^t - \frac{1}{(2+2)^2}e^{2t} = \frac{2}{9}e^t - \frac{1}{16}e^{2t}.$$

$$\therefore \quad x=(A+Bt)e^{-2t}+\frac{2}{9}e^t-\frac{1}{16}e^{2t}. \qquad \text{(iv)}$$

Using the value of x from Eqn. (iv) in Eqn. (i), we get

$$y = e^t - (D+3)x = e^t - (D+3)\left\{(A+Bt)e^{-2t} + \frac{2}{9}e^t - \frac{1}{16}e^{2t}\right\}$$

$$= e^t - \left\{Be^{-2t} - 2(A+Bt)e^{-2t} + \frac{2}{9}e^t + \frac{1}{8}e^{2t} + 3(A+Bt)e^{-2t} + \frac{6}{9}e^t - \frac{3}{16}e^{2t}\right\}$$

$$= -Be^{-2t} - (A+Bt)e^{-2t} + \frac{1}{9}e^t + \frac{5}{16}e^{2t}.$$

Hence, the required solution is

$$\boldsymbol{x = (A+Bt)e^{-2t} + \frac{2}{9}e^t - \frac{1}{16}e^{2t}}$$

and $\boldsymbol{y = -Be^{-2t} - (A+Bt)e^{-2t} + \frac{e^t}{9} + \frac{5e^{2t}}{16}}.$

Example 16.25. *The functions $x(t)$ and $y(t)$ are governed by $(D-3)x + 2(D+2)y = e^{2t}$ and $2(D+1)x + (D-1)y = 0$, where $D \equiv \frac{d}{dt}$. Determine $x(t)$.* **[AMIE(Winter) 1991]**

Solution : We have $(D-3)x + 2(D+2)y = e^{2t}$ (i)

$$2(D+1)x + (D-1)y = 0. \quad \text{(ii)}$$

We multiply Eqn. (i) by $(D-1)$ and Eqn. (ii) by $2(D+2)$ and we get

$$(D-1)(D-3)x + 2(D-1)(D+2)y = (D-1)e^{2t} \quad \text{(iii)}$$

or, $4(D+1)(D+2)x + 2(D-1)(D+2)y = 0.$ (iv)

Now, subtracting Eqn. (iv) from (iii), we get

$$[(D-1)(D-3) - 4(D+1)(D+2)]x = (D-1)e^{2t}$$

or, $[D^2 - 4D + 3 - 4D^2 - 12D - 8]x = 2e^{2t} - e^{2t}$

or, $[-3D^2 - 16D - 5]x = e^{2t}$ or, $[3D^2 + 16D + 5]x = -e^{2t}$.

A.E. is $3D^2 + 16D + 5 = 0$.

$$\therefore \quad D = \frac{-16 \pm \sqrt{256-60}}{2\cdot 3} = \frac{-16 \pm \sqrt{196}}{6} = \frac{-16 \pm 14}{6} = -\frac{2}{6}, -\frac{30}{6} = -\frac{1}{3}, -5.$$

$\therefore$ C.F. is $Ae^{-\frac{t}{3}} + Be^{-5t}$, where A and B are arbitrary constants.

The particular integral is P.I. $= \frac{1}{3D^2 + 16D + 5}(e^{-2t}) = -\frac{1}{3(2)^2 + 16(2) + 5}(e^{2t}) = -\frac{e^{-2t}}{49}.$

Hence the general solution is $\boldsymbol{x = Ae^{-\frac{t}{3}} + Be^{-5t} - \frac{e^{-2t}}{49}}.$

16.9 Applications of Differential Equations

Example 16.26. *A particle of mass m is projected vertically upwards from a horizontal level with a velocity u. Assuming the resistance of the air to be proportional to velocity find the maximum height reached by the particle.*

Solution : Let x be the height attained by the particle from the point of projection at any instant t. The forces acting on the particle are : (i) the gravitational force mg and (ii) the air resistance λmv, where λ being a constant of proportionality per unit mass.

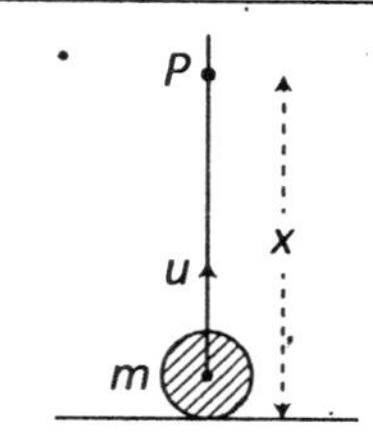

Fig. 16.1: Position of the particle at time t

$\therefore$ the differential equation of motion is

$$m\frac{d^2x}{dt^2} = -mg - \lambda m\frac{dx}{dt} \qquad \text{(i)}$$

or, $$\frac{d^2x}{dt^2} + \lambda\frac{dx}{dt} = -g. \qquad \text{(ii)}$$

$$\text{A.E. is } D^2 + \lambda D = 0,\ D \equiv \frac{d}{dx}, \qquad \therefore\ D = 0,\ -\lambda.$$

The complementary function C.F. $= A + Be^{-\lambda t}$, where A and B are particular arbitrary constants.

$$\text{P.I.} = \frac{1}{D^2+\lambda D}(-g) = (-g)\frac{1}{\lambda D}\left(1+\frac{D}{\lambda}\right)^{-1}(1) = -\frac{g}{\lambda}\frac{1}{D}(1) = -\frac{gt}{\lambda}.$$

Hence, the general solution is $\boldsymbol{x = A + Be^{-\lambda t} - \frac{gt}{\lambda}}$. (iii)

At $t = 0$, $x = 0$,

$\therefore$ $$0 = A + B. \qquad \text{(iv)}$$

Again, $\dfrac{dx}{dt} = -Be^{-\lambda t} - \dfrac{g}{\lambda}$.

At $t = 0$, $\frac{dx}{dt} = u$,

$\therefore$ $$u = -B\lambda - \frac{g}{\lambda} \quad \text{or,} \quad B = -\frac{g+u\lambda}{\lambda^2}. \qquad \text{(v)}$$

From Eqn. (iv), $A = \dfrac{g+u\lambda}{\lambda^2}$. (vi)

So, $$x = \frac{g+u\lambda}{\lambda^2}(1-e^{-\lambda t}) - \frac{gt}{\lambda}.$$

$\therefore$ $$\frac{dx}{dt} = -B\lambda e^{-\lambda t} - \frac{g}{h} = 0. \qquad \text{(vii)}$$

But for maximum height x_m, $\frac{dx}{dt} = 0$.

$\therefore$ from Eqn. (vii), $e^{-\lambda t} = -\dfrac{g}{B\lambda^2} = \dfrac{g}{A\lambda^2}$

or, $$t = -\frac{1}{\lambda}\log\left(\frac{g}{A\lambda^2}\right) = -\frac{1}{\lambda}\log\left(\frac{g}{g+u\lambda}\right). \qquad \text{(viii)}$$

Also, $$\frac{d^2x}{dt^2} = B\lambda^2 e^{-\lambda t} = -A\lambda^2 e^{-\lambda t} = -g < 0.$$

$\therefore$ $x = x_m$, the maximum value of x for time t given by Eqn. (viii).

$\therefore$ $$x_m = \left(\frac{g+u\lambda}{\lambda^2}\right)\left(1 - \frac{g}{g+u\lambda}\right) + \frac{g}{\lambda^2}\log\left(\frac{g}{g+u\lambda}\right)$$

or, $$\boldsymbol{x_m = \frac{u}{\lambda} + \frac{g}{\lambda^2}\log\left(\frac{g}{g+u\lambda}\right)}. \qquad \text{(ix)}$$

Example 16.27. *A beam of elastic material supported at the ends is subjected to a concentrated load W at the middle. If Y be the Young's modulus of the material of elastic constant E, then show that*

$$\text{(a) } M = EI\frac{d^2y}{dx^2} \quad \text{(b) } EI\frac{d^3y}{dx^3} = V(x) \quad \text{(c) } EI\frac{d^4y}{dx^4} = \omega(x),$$

where M is the bending moment and I, the geometrical moment of inertia of cross section about the neutral axis of the bent beam.

Solution : Fig. 16.2 shows the configuration of the bent beam.

(a) During bending the upper half becomes concave and the lower half convex. The filaments in the upper half are compressed while in the lower half the filaments are stretched. But in the middle region there is a surface in which the filaments are neither compressed nor stretched. This surface is called neutral surface. The intersection of the neutral surface with the plane of cross section is known as neutral axis. $M'N'$ is the neutral axis and the line RS, which was initially along the X-axis, now lies on the neutral plane, is called the elastic curve.

If R be the radius of curvature of the elastic curve at the point $P(x, y)$, then the bending moment M about $M'N'$ is given by

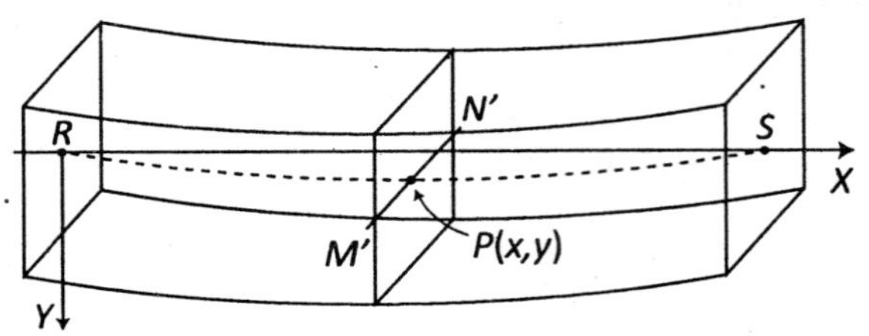

Fig. 16.2: Bending of a beam

$$M = \frac{EI}{R}.$$

$$\text{But, } R = \frac{\left[1 + \left(\frac{dy}{dx}\right)^2\right]^{\frac{3}{2}}}{\frac{d^2y}{dx^2}} = \frac{1}{\frac{d^2y}{dx^2}}$$

as for small bending, $\frac{dy}{dx} = 0$.

So, $$\boldsymbol{M = EI\frac{d^2y}{dx^2}}. \qquad \text{(i)}$$

(b) Again, shearing force $= \frac{dM}{dx}$ and intensity of loading $= \frac{d^2y}{dx^2}$, under transverse load.

$$\frac{dM}{dx} = \frac{d}{dx}\left(EI\frac{d^2y}{dx^2}\right) = V(x)$$

or, $$\boldsymbol{EI\frac{d^3y}{dx^3} = V(x)}. \qquad \text{(ii)}$$

Again, $$\frac{d^2M}{dx^2} = \frac{d^2}{dx^2}\left(EI\frac{d^2y}{dx^2}\right) = w(x)$$

or, $$\boldsymbol{EI\frac{d^4y}{dx^4} = w(x)}. \qquad \text{(iii)}$$

$w(x)$ is the load per unit length and is a constant, $V(x)$ is the shear which is positive along the negative Y-axis.

Example 16.28. *A beam of uniform stiffness EI is simply supported at its ends $x = 0$ and $x = a$. It carries a uniform load of 1200* kg *per unit length.* Find the deflection y at any point of the beam and show that its maximum deflection at $x = \frac{a}{2}$.

Solution : The equation of motion of the beam is

$$EI\frac{d^4y}{dx^4} = w(x) = 1200$$

or, $$\lambda\frac{d^4y}{dx^4} = 1200 \quad [\text{where } \lambda = EI \text{ is a constant}] \qquad \text{(i)}$$

Integrating successively w.r.t. x, we get

$$\lambda y = 5xx^4 + \frac{A}{6}x^3 + \frac{B}{2}x^2 + Cx + D,$$ where A, B, C, D are constants of integration.

Boundary conditions are :

$$y = 0 \text{ at } x = 0 \qquad \text{(ii)}$$

$$y = 0 \text{ at } x = a \qquad \text{(iii)}$$

$$\frac{d^2y}{dx^2} = 0 \text{ at } x = 0 \qquad \text{(iv)}$$

$$\frac{d^2y}{dx^2} = 0 \text{ at } x = a. \qquad \text{(v)}$$

Applying boundary conditions (ii), we have

$$D = 0. \qquad \text{(vi)}$$

By condition (iii),

$$\lambda \cdot 0 = \frac{A}{6}a^3 + \frac{B}{2}a^2 + Ca$$

or, $$\frac{A}{6}a^2 + \frac{Ba}{2} + C = 0. \qquad \text{(vii)}$$

Again, $$\lambda\frac{d^2y}{dx^2} = 600x^2 + Ax + B.$$

Using boundary condition (iv) and (v),

$$B = 0 \qquad \text{(viii)}$$

and $$600a^2 + Aa + B = 0. \qquad \text{(ix)}$$

Thus, from Eqns. (vi) to (ix), we get $B = 0$, $D = 0$, $A = -600a$, $C = 50a^3$.

$$\therefore \quad y = \frac{1}{\lambda}[50x^4 - 100x^3 + 50a^3x].$$

Again, $\frac{dy}{dx} = 0$ for $200x^3 - 300ax^2 + 50a^3 = 0$ or, $4x^3 - 6ax^2 + a^3 = 0$.

$$\therefore \quad x = \frac{a}{2}.$$

But at $\frac{a}{2}$, $\frac{d^2y}{dx^2} = -\frac{150}{\lambda}a^2 < 0.$

Hence, **the maximum deflection of the beam occurs at $x = \frac{a}{2}$.**

16.10 Free Oscillation

We consider a body of mass m free to move along y-axis. The mass is attracted towards the origin with a force proportional to displacement y at any instant of time.

The equation of motion of the body is

$$m\frac{d^2y}{dt^2} = -\lambda y, \quad \text{where } \lambda \text{ is } \textbf{stiffness constant}. \tag{i}$$

Eqn. (i) can be written as

$$\frac{d^2y}{dt^2} + \omega^2 y = 0, \quad \text{where } \omega = \sqrt{\frac{\lambda}{m}}. \tag{ii}$$

The general solution of Eqn. (ii) is

$$y = A\cos\omega t + B\sin\omega t, \quad \text{where } A \text{ and } B \text{ are arbitrary constants.} \tag{iii}$$

Now, $$v = \frac{dy}{dt} = -\omega(A\sin\omega t + B\cos\omega t). \tag{iv}$$

If the motion be started with initial displacement y_0 from rest, then $v = 0$.

$\therefore \quad A = y_0 \quad \text{and} \quad B = 0.$

So, $$y = y_0\cos\omega t. \tag{v}$$

This represents an **undamped** simple harmonic motion of constant amplitude y_0.

The time period of motions is $$\boldsymbol{T = \frac{2\pi}{\omega} = 2\pi\sqrt{\frac{m}{\lambda}}} \tag{vi}$$

and the frequency of oscillation is $$\boldsymbol{f = \frac{1}{T} = \frac{1}{2\pi}\sqrt{\frac{\lambda}{m}}} \tag{vii}$$

16.11 Damped Oscillation

If the medium in which the body moves offers a resistance to the motion, then the motion of the body is **damped**. The motion of the body is then called **damped oscillation**. The damping force is proportional to velocity $\frac{dy}{dt}$ and is equal to $-k\frac{dy}{dt}$, where k is the damping force per unit velocity and is known as **damping constant**.

The equation of motion of the damped oscillator is

$$m\frac{d^2y}{dt^2} + k\frac{dy}{dt} + \lambda y = 0 \tag{i}$$

or, $$m\frac{d^2y}{dt^2} + 2b\frac{dy}{dt} + \omega^2 y = 0, \tag{ii}$$

where $b = \frac{k}{2m}$ and $\omega = \sqrt{\frac{\lambda}{m}}$.

The auxiliary equation is

$$n^2 + 2bn + \omega^2 = 0 \quad [\text{Put } y = e^{nt} \text{ as the trial solution of Eqn. (ii)}]$$

or, $$n = \frac{-2b \pm \sqrt{4b^2 - \omega^2}}{2} = -2b \pm \sqrt{4b^2 - \omega^2}.$$

The general solution depends on the nature of the roots n_1 and n_2.

Case I : Over-damped motion

If the damping force is greater than stiffness, i.e., $b > \omega$, then the general solution is

$$y = Ae^{m_1 t} + Be^{m_2 t}. \quad \text{(iii)}$$

If the motion be started with initial displacement y_0 but no initial velocity, i.e., $v = 0$ then after evaluating the values of the constants A and B, we get

$$\boldsymbol{y = \frac{y_0}{n_1 - n_2}(n_1 e^{n_2 t} - n_2 e^{n_1 t})}. \quad \text{(iv)}$$

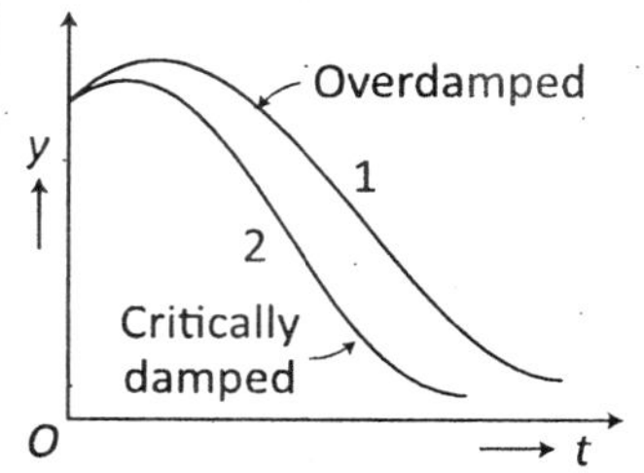

Fig. 16.3

Eqn. (iv) represents an **over-damped** motion where no oscillation occurs (Fig. 16.3).

Case II : Damped oscillatory motion

If $b < \omega$, then n_1 and n_2 form a complex-conjugate pair, $-b \pm ia$, where $a = \sqrt{\omega^2 - b^2}$. The general solution of Eqn. (ii) is

$$y = (A\cos at + B\sin at)e^{-bt} \quad \text{(vii)}$$

With the boundary conditions as stated earlier, we get

$$\boldsymbol{y = \frac{y_0}{a}(a\cos at + b\sin at)e^{-bt} = \left(\frac{y_0\sqrt{a^2+b^2}}{a}\right)\cos(at - \theta)e^{-bt}}. \quad \text{(viii)}$$

where $\theta = \tan^{-1}\left(\frac{b}{a}\right)$.

Eqn. (viii) shows that the body oscillates with an amplitude $A = \frac{y_0\sqrt{a^2+b^2}}{a}$, which falls of exponentially with time as shown in Fig. 16.4. Strictly, the motion is not periodic, but the graph crosses equilibrium position, $y = 0$ at regular interval of time. The '**period**' T is given by

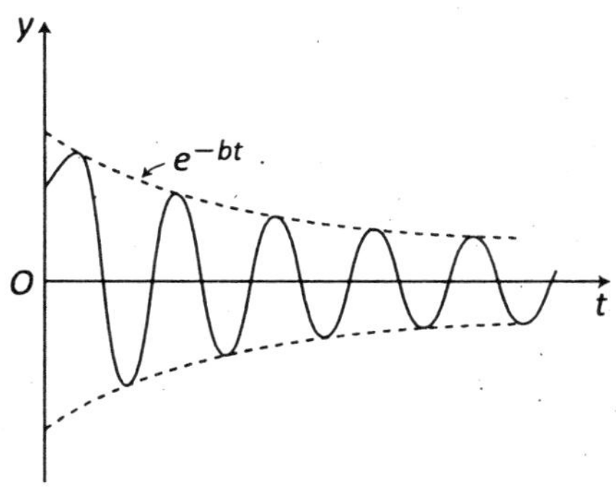

Fig. 16.4

$$\boldsymbol{T = \frac{2\pi}{a} = \frac{2\pi}{\sqrt{\omega^2 - b^2}} = \frac{2\pi}{\sqrt{\left(\frac{\lambda}{m}\right) - \left(\frac{k^2}{4m^2}\right)}}}. \quad \text{(ix)}$$

The frequency of oscillation, f, is

$$\boldsymbol{f = \frac{1}{T} = \frac{1}{2\pi}\sqrt{\omega^2 - b^2} = \frac{1}{2\pi}\sqrt{\frac{\lambda}{m} - \frac{k^2}{4m^2}}} \quad \text{(x)}$$

and is called the **natural frequency** of oscillation.

16.12 Forced Vibration with Damping

In addition to the force restitution $(-\lambda y)$ and damping force $\left(-k\frac{dy}{dt}\right)$, if a periodic force $F\cos\omega t$ be applied to the system, then the equation of motions becomes

$$m\frac{d^2y}{dt^2} + k\frac{dy}{dt} + \lambda y = F\cos\omega t. \quad \text{(i)}$$

The auxiliary equation becomes $m\alpha^2 + k\alpha + \lambda = 0$,

$$\text{i.e.,} \quad \alpha = \frac{-k \pm \sqrt{-k^2 - 4\lambda m}}{2m}. \quad \text{(ii)}$$

Taking $\frac{k}{m} = 2b$ and $\frac{\lambda}{m} = \omega_0^2$, we have

$$\alpha = -b \pm \sqrt{b^2 - \omega_0^2}.$$

$$\therefore \quad \text{C.F.} = e^{-bt}\left[Ae^{(\sqrt{b^2-\omega_0^2})t} + Be^{-(\sqrt{b^2-\omega_0^2})t}\right] \tag{iii}$$

and
$$\text{P.I.} = \left(\frac{F}{m}\right)\frac{1}{D^2 + 2bD + \omega_0^2}\cos\omega t. \tag{iv}$$

For $\omega_0 \neq \omega$,

$$\text{P.I.} = \left(\frac{F}{m}\right)\frac{1}{2bD + (\omega_0^2 - \omega^2)}\cos\omega t = \left(\frac{F}{m}\right)\frac{2bD + (\omega^2 - \omega_0^2)}{4b^2D^2 - (\omega_0^2 - \omega^2)}\cos\omega t$$

$$= \left(-\frac{F}{m}\right)\frac{[2bD + (\omega^2 - \omega_0^2)]}{(\omega_0^2 - \omega^2)^2 + 4b^2\omega^2}\cos\omega t = \left(\frac{F}{m}\right)\frac{(\omega_0^2 - \omega^2)^2\cos\omega t + 2\omega b\sin\omega t}{(\omega_0^2 - \omega^2)^2 + 4b^2\omega^2}.$$

The solution of Eqn. (i) is

$$y = e^{-bt}\left[Ae^{(\sqrt{b^2-\omega_0^2})t} + Be^{-(\sqrt{b^2-\omega_0^2})t}\right] + \left(\frac{F}{m}\right)\frac{(\omega_0^2 - \omega^2)^2\cos\omega t + 2\omega b\sin\omega t}{(\omega_0^2 - \omega^2)^2 + 4b^2\omega^2}. \tag{v}$$

For $\omega = \omega_0$,

$$\text{P.I.} = \left(\frac{F}{m}\right)\frac{1}{D^2 + 2bD + \omega^2}\cos\omega t = \left(\frac{F}{m}\right)\frac{D}{2bD^2}\cos\omega t = \frac{F}{2b\omega^2 m}\omega\sin\omega t = \frac{F\sin\omega t}{2bm\omega}. \tag{vi}$$

Hence, the solution of Eqn. (i) is

$$\boldsymbol{y = e^{-bt}\left[Ae^{(\sqrt{b^2-\omega_0^2})t} + Be^{-(\sqrt{b^2-\omega_0^2})t}\right] + \frac{F\sin\omega t}{2bm\omega}}. \tag{vii}$$

16.13 Forced Oscillation Without Damping

If damping is absent, then $k = 0$.

$\therefore$ Eqn. (i) of (16.12) becomes

$$m\frac{d^2y}{dt^2} + \lambda y = F\cos\omega t \tag{i}$$

or,
$$\frac{d^2y}{dt^2} + \omega_0^2 y = \frac{F}{m}\cos\omega t \qquad \left[\because\ \omega_0^2 = \frac{\lambda}{m}\right]$$

$$\therefore \quad \text{C.F.} = A\cos\omega t + B\sin\omega t$$

$$\text{P.I.} = \frac{F}{m}\frac{\cos\omega t}{(\omega_0^2 - \omega^2)} \qquad [\text{for } \omega_0 \neq \omega]$$

$\therefore$ general solution is

$$y = A\cos\omega_0 t + B\sin\omega_0 t + \frac{F}{m}\frac{\cos\omega t}{(\omega_0^2 - \omega^2)}, \quad (\omega_0 \neq \omega) \tag{ii}$$

and
$$\boldsymbol{y = A\cos\omega_0 t + B\sin\omega_0 t + \frac{Ft\sin\omega t}{2m\omega}}, \quad (\omega_0 \neq \omega). \tag{iii}$$

Eqn. (ii) shows the oscillation is the superposition of two oscillations of periods $\frac{2\pi}{\omega_0}$ and $\frac{2\pi}{\omega}$. When $\omega_0 = \omega$, the frequency of the periodic force is the same as that of the system, i.e., $\omega = \omega_0$ which is a case of resonance.

16.14 LCR Circuit with an Alternating emf

Let an alternating emf $E_0 \sin \omega t$ be applied to a series combination of an inductance L, capacitance C and a resistance R as shown in Fig. 16.5.

In the steady state if q be the change on the capacitor and I, the current at any instant t, then the ohmic equation of the circuit is

$$L\frac{dI}{dt} + RI + \frac{q}{C} = E_0 \sin \omega t. \quad \text{(i)}$$

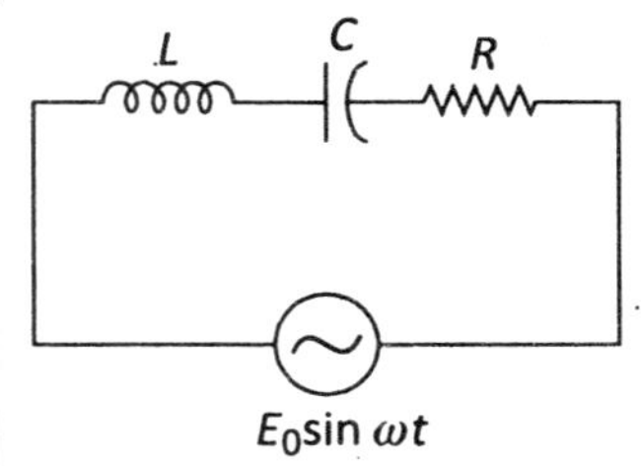

Fig. 16.5

Differentiating with respect to t,

$$L\frac{d^2I}{dt^2} + R\frac{dI}{dt} + \frac{I}{C} = E_0\omega \cos \omega t \quad \left[I = \frac{dq}{dt}\right]$$

or, $$\left(LD^2 + RD + \frac{1}{C}\right) = E_0 \cos \omega t. \quad \text{(ii)}$$

The steady state solution is the particular integral of Eqn. (i) and is given by

$$I = \frac{E_0\omega \cos \omega t}{LD^2 + RD + \frac{1}{C}} = \frac{E_0\omega \cos \omega t}{-L\omega^2 + RD + \frac{1}{C}} = \frac{E_0\omega \cos \omega t}{RD - \left(\omega L - \frac{1}{\omega C}\right)\omega}$$

$$= \frac{E_0\omega \cos \omega t}{RD - X\omega}, \quad \text{where } X = \omega L - \frac{1}{\omega C}$$

$$= E_0 \frac{\omega(RD + X\omega)}{R^2D^2 - X^2\omega^2} \cos \omega t = \frac{E_0\omega}{-R^2\omega^2 - X^2\omega^2}(-R\omega \sin \omega t + X\omega \cos \omega t)$$

$$= \frac{E_0}{R^2 + X^2}(A \cos \omega t \cos \phi - X \cos \omega t \sin \phi)$$

$$\left[\text{where } R = A\cos\phi,\ X = A\sin\phi, \text{ so that } A = \sqrt{R^2 + X^2} \text{ and } \phi = \tan^{-1}\left(\frac{X}{R}\right)\right]$$

$$= \frac{E_0 A}{R^2 + X^2} \sin(\omega t - \phi)$$

or, $$\boldsymbol{I = I_0 \sin(\omega t - \phi)}, \quad \text{(iii)}$$

where $I_0 = \dfrac{E_0}{R^2 + \left(\omega L - \frac{1}{\omega C}\right)^2}$ is the amplitude value or maximum value of current. (iv)

$$Z = \sqrt{R^2 + X^2} = \sqrt{R^2 + \left(\omega L - \frac{1}{\omega C}\right)^2} \text{ is the impedance of the circuit}$$

and $X = \omega L - \dfrac{1}{\omega C}$ is the effective resistance.

For $X = 0$, $\omega L = \dfrac{1}{\omega C}$, the value of I_0 is the maximum and the circuit is said to be in **(series) resonance**. The condition for resonance is

$$\omega L = \frac{1}{\omega C} \quad \text{or,} \quad \boldsymbol{\omega = \frac{1}{\sqrt{LC}}} = \omega_0 \text{ (say)}$$

or, $\quad f_0 = \frac{1}{2\pi}\sqrt{\frac{1}{LC}}$ is the resonant frequency

$\phi = \tan^{-1}\left(\frac{x}{R}\right)$ is the phase difference between emf and the current.

Example 16.29. *A condenser of capacity C discharged through an inductance L and resistance R in series and the charge q at any time t satisfies the equation $L\frac{d^2q}{dt^2} + R\frac{dq}{dt} + \frac{q}{C} = 0$.*

Given, $L = 0.25$ H, $R = 250\ \Omega$ and $C = 2 \times 10^{-6}$ F and that when $t = 0$, $q = 0.002$ C and $\frac{dq}{dt} = 0$, obtain the value of q in terms of t. **[BU(H),1999]**

Solution : The ohmic equation of the LCR circuit is

$$L\frac{d^2q}{dt^2} + R\frac{dq}{dt} + \frac{q}{C} = 0 \quad \text{or,} \quad \frac{d^2q}{dt^2} + \frac{R}{L}\frac{dq}{dt} + \frac{1}{LC}q = 0$$

or, $$\frac{d^2q}{dt^2} + 2\lambda\frac{dq}{dt} + \omega^2 q = 0, \quad \text{where } \lambda = \frac{R}{2L},\ \omega = \frac{1}{\sqrt{LC}}.$$

Here, $\lambda = \frac{R}{2L} = \frac{250}{2 \times 0.25} = 500$ and $\omega^2 = \frac{1}{LC} = \frac{1}{0.25 \times 2 \times 10^{-6}} = 2 \times 10^6$.

If α be the root of the auxiliary equation $D^2 + 2\lambda D + \mu^2 = 0$, then

$$\alpha = \frac{-2\lambda \pm \sqrt{4\lambda^2 - 4\mu^2}}{2} = -\lambda \pm \sqrt{\lambda^2 - \mu^2}.$$

As $\lambda < \mu$, $\alpha = -\lambda \pm \sqrt{\mu^2 - \lambda^2} = -500 \pm \sqrt{2 \times 10^6 - 25 \times 10^4} = -500 \pm 500\sqrt{7}t$.

$\therefore$ the general equation for q is $\boldsymbol{q = e^{-500t}\left(A\cos 500\sqrt{7}t + B\sin 500\sqrt{7}t\right)}$.

For $t = 0$, $q = 0.002$, $\quad \therefore\ A = 0.002$.

For $t = 0$, $\frac{dq}{dt} = 0$, $\quad \therefore\ B = \frac{A}{\sqrt{7}} = \frac{0.002}{\sqrt{7}}$.

$\therefore$ the particular solution is $\boldsymbol{q = 0.002e^{-500t}\left[\cos 500\sqrt{7}t + \frac{1}{\sqrt{7}}\sin 500\sqrt{7}t\right]}$.

16.15 Motion of a Charged Particle in Electromagnetic Field

If a charged particle of mass m and charge q moves in an electromagnetic field with velocity $\vec{v}$, then the **Lorentz force** acting on the particle is

$$\vec{F} = q(\vec{E} + \vec{v} \times \vec{B}), \tag{i}$$

where $\vec{E}$ and $\vec{B}$ are the electric and magnetic field intensity respectively.

The equation of motion of the particle is

$$m\frac{d\vec{v}}{dt} = q(\vec{E} + \vec{v} \times \vec{B}). \tag{ii}$$

Case I : When $\vec{E}$ is constant

Let the particle be initially at rest at (x_0, y_0, z_0) and the constant electric force acts along the X-axis, i.e., $E = (E, 0, 0)$, then the equations of motion of the particle are :

$$m\ddot{x} = qE$$
$$m\ddot{y} = 0$$
$$m\ddot{z} = 0.$$

The solutions of the above equations are

$$\left.\begin{aligned} x &= x_0 + \frac{1}{2}\frac{qE}{m}t^2 \\ y &= y_0 \\ z &= z_0. \end{aligned}\right\} \tag{iii}$$

So, the acceleration of the particle at any instant of times is

$$a = \frac{qE}{m}. \tag{iv}$$

If the particle be acted upon by a potential difference of V volts, then its potential energy is qV which is converted into kinetic energy $\frac{1}{2}mv^2$.

$$\therefore \quad \frac{1}{2}mv^2 = qV \quad \text{or,} \quad \boldsymbol{v} = \sqrt{\frac{\boldsymbol{2qV}}{\boldsymbol{m}}}. \tag{v}$$

Case II : When $\vec{B}$ is constant

The equation of motion of the particle in the present case is

$$m\frac{d\vec{v}}{dt} = q\vec{v} \times \vec{B}.$$

Taking dot product with $\vec{v}$ on both sides, we get

$$\left(m\frac{d\vec{v}}{dt}\right) \cdot \vec{v} = (q\vec{v} \times \vec{B}) \cdot \vec{v} \quad \text{or,} \quad \frac{d}{dt}\left(\frac{1}{2}mv^2\right) = q(\vec{v} \times \vec{v}) \cdot \vec{B} = 0,$$

i.e., $$\frac{\mathbf{1}}{\mathbf{2}}\boldsymbol{mv^2} = \textbf{constant}. \tag{vi}$$

This shows that the kinetic energy as well as the speed of the particle remain constant. Due to the action of $\vec{E}$, however, the direction of motion of the particle will be changed.

Now we consider that $\vec{B}$ is constant and acts along X-direction. If $\vec{v_1}$ and $\vec{v_2}$ be the velocities of the charged particles along perpendicular and parallel direction of $\vec{B}$, then the corresponding components of forces acting on the particle are

$$\left.\begin{aligned} \vec{F_1} &= q\vec{v_1} \times \vec{B} \\ \text{and} \quad \vec{F_2} &= q\vec{v_2} \times \vec{B} = 0. \end{aligned}\right\} \tag{vii}$$

$$\therefore \quad \vec{F_2} = m\frac{d\vec{v_2}}{dt} = 0,$$

i.e., $$\vec{v_2} = \text{constant}. \tag{viii}$$

$\because$ $v^2 = v_1^2 + v_2^2$ is constant [eqn. (vi)], then it follows from equation (viii) that v_1 is also constant.

From eqn. (ii), $m\dfrac{d\vec{v_1}}{dt} = q\vec{v_1} \times \vec{B}.$

So, the magnitude of $\vec{v_1}$ is constant and its direction changes continuously.

Trajectory of the particle : If $\overrightarrow{B} = \hat{i}B$, then

$$m\frac{d\overrightarrow{v}}{dt} = q\overrightarrow{v} \times \hat{i}B$$

or, $$\frac{d\overrightarrow{v}}{dt} = \frac{qB}{m}(\overrightarrow{v} \times \hat{i}) = \xi(\overrightarrow{v} \times \hat{i}), \quad \text{where } \xi = \frac{qB}{m}. \tag{ix}$$

$\because \overrightarrow{v} = \hat{i}v_x + \hat{j}v_y + \hat{k}v_z$, the above equation can be expressed in component form as

$$\frac{dv_x}{dt} = 0, \quad \frac{dv_y}{dt} = \xi v_x, \quad \frac{dv_z}{dt} = -\xi v_y.$$

$\therefore \quad v_x = \text{constant} = v_2.$

So, $$x = x_0 + vt. \tag{x}$$

Now eliminating v_y from the last two equations in the above, we get

$$\frac{d}{dt}\left(-\frac{1}{\xi}\frac{dv_2}{dt}\right) = \xi v_x \quad \text{or,} \quad \frac{d^2v_2}{dt^2} + \xi^2 v_x = 0. \tag{xi}$$

The solution of eqn. (xi) is

$$v_2 = A\cos\xi t + B\sin\xi t. \tag{xii}$$

$$\therefore \quad v_y = -\frac{1}{\xi}\frac{dv_z}{dt} = -A\sin\xi t + B\cos\xi t. \tag{xiii}$$

Integrating equations (xii) and (xiii), we get

$$\left.\begin{aligned} \xi(y - y_0) &= A\cos\xi t + B\sin\xi t \\ \text{and} \quad \xi(z - z_0) &= A\sin\xi t + B\cos\xi t \end{aligned}\right\} \tag{xiv}$$

Here, y_0 and z_0 are the constants of integration and do not correspond to the position at $t = 0$.

Now, $[\xi(y - y_0)]^2 + [\xi(z - z_0)]^2 = (A^2 + B^2)$

or, $$\boldsymbol{(y - y_0)^2 + (z - z_0)^2 = \frac{1}{\xi^2}(A^2 + B^2) = \text{constant}.} \tag{xv}$$

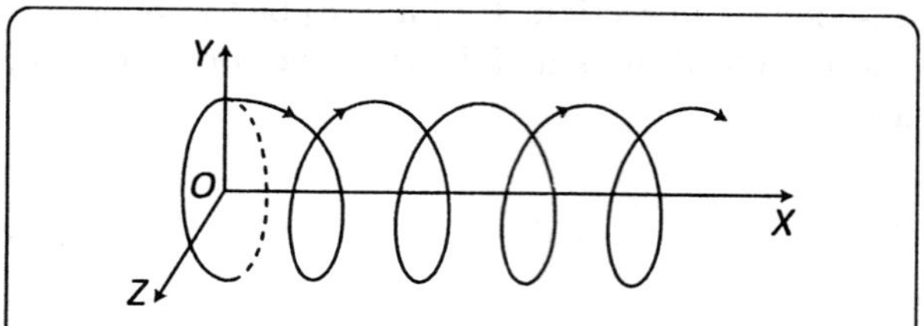

Fig. 16.6: Trajectory of a charged particle in an electromagnetic field when magnetic field is constant

This represents a circle with centre at (y_0, z_0).

Thus, the particle will move in a circular path about X-axis and at the same time it will move with a constant velocity along the X-axis. The trajectory of the charged particle under this condition is thus a **helix** as shown in Fig. 16.6.

Case III : Motion in combined field

Let the velocity components of the particle at $t = 0$ be (u_0, v_0, w_0), when the particle was at $(0, 0, 0)$ of a Cartesian coordinate system. If $\overrightarrow{B}$ be along Y-direction and $\overrightarrow{E}$ in the XY-plane inclined at an angle with the X-axis, then $B = (0, B, 0)$ and $E = E(E_x, E_y, 0)$.

Then the equations of motion are

$$\left.\begin{aligned} \frac{du}{dt} &= a_x - \xi w \\ \frac{dv}{dt} &= a_y \\ \text{and} \quad \frac{dw}{dt} &= \xi u \end{aligned}\right\} \tag{xvi}$$

$$\left.\begin{aligned} \text{where} \quad a_x &= \frac{qE_x}{m} \\ a_y &= \frac{qE_y}{m} \\ \text{and} \quad \xi &= \frac{qB}{m} \end{aligned}\right\} \tag{xvii}$$

From first and third equations of (xvi), we have

$$\frac{d^2u}{dt^2} = -\xi\frac{dw}{dt} = -\xi^2 u \quad \text{or,} \quad \frac{d^2u}{dt^2} + \xi^2 u = 0,$$

solution of the above equation being $u = C\cos\xi t + A\sin\xi t$.

Assuming at $t = 0$, $u_0 = 0$, we get $C = 0$. Then,

$$u = A\sin\xi t. \tag{xviii}$$

From the third equation of (xvi), $\dfrac{dw}{dt} = \xi A\sin\xi t$.

Integrating, $w - w_0 = \xi A\displaystyle\int_0^t \sin\xi t$

$$\text{or,} \quad w - w_0 = A(1 - \cos\xi t). \tag{xix}$$

Again integrating equations (xviii) and (xix), we get

$$x = \int_0^t A\sin\xi t\,dt = \frac{A}{\xi}(1 - \cos\xi t) \tag{xx}$$

$$\text{and} \quad z = w_0 t + \frac{A}{\xi}(\xi t - \sin\xi t). \tag{xxi}$$

Eqns. (xx) and (xxi) represent a cycloid.

Now from the equation $\frac{dv}{dt} = a_y$, we get $v = v_0 + a_y t$ so that,

$$y = v_0 t + \frac{1}{2}a_y t^2. \tag{xxii}$$

Hence, due to the electric component E_y and the initial velocity component, v_0, the particle moves in the y-direction.

If now $E_y = 0$, then $a_y = 0$, the particle moves along y-axis with constant velocity v_0 and the distance of the particle at time t is $v_0 t$.

From eqns. (xvi) and (xviii),

$$\frac{du}{dt} = a_x - \xi w \quad \text{or,} \quad A\xi\cos\xi t = a_x - \xi w.$$

At $t = 0$, $A\xi = a_x - \xi w_0$.

$$\therefore \quad A = \frac{1}{\xi}(a_x - \xi w_0) = \left(\frac{E_x}{B} - w_0\right) \quad \left[\because \xi = \frac{qB}{m}\right]$$

Then Eqn. (xviii) takes the form : $\boldsymbol{u = \left(\frac{E_x}{B} - w_0\right) \sin \xi t}$. (xxiii)

From the above equation we note that if $w_0 = \frac{E_x}{B}$, then $u = 0$, i.e., the particle does not move along the x-axis. Such particles are called to be filtered out.

Hence, by choosing $\vec{E}$ and $\vec{B}$, particles of desired velocity may be selected and those of undesired velocity may be filtered out. Such a combination of cross electric field $\vec{E}$ and magnetic field $\vec{B}$, forms the basic principles of velocity filter of charged particles.

Exercises

16(A) : Objective Type Questions

Select the correct answer

16.1 The complementary function of $\frac{d^2y}{dx^2} + 4\frac{dy}{dx} + 5y = 65 \sin 2x$ is

(a) $e^{-x}(C_1 \cos x + C_2 \sin x)$ **(b)** $C_1 \cos x + C_2 \sin x$

(c) $e^{-2x}(C_1 \cos x + C_2 \sin x)$ **(d)** $e^{-2x} C_1 \cos x$.

16.2 The general solution of the ordinary differential equation $\frac{d^2y}{dx^2} + 4y = 0$ is

(a) $Ae^{2x} + Be^{-2x}$ **(b)** $(A + Bx)e^{-2x}$ **(c)** $A \cos 2x + B \sin 2x$ **(d)** $(A + Bx) \cos 2x$.

[WBUT 2006]

16.3 The general solution of the ordinary differential equation $D^2y + 9y = 0$ is

(a) $Ae^{3x} + Be^{-3x}$ **(b)** $(A + Bx)e^{3x}$ **(c)** $A \cos 3x + B \sin 3x$ **(d)** $(A + Bx) \sin 3x$.

[WBUT 2008]

16.4 $\frac{1}{D^2 - 1} x^2$ is equal to

(a) $x^2 + 2x + 2$ **(b)** $-(x^2 + 2x + 2)$ **(c)** $2x - x^2$ **(d)** $-(2x - x^2)$. **[WBUT 2008, '10]**

16.5 The general solution of $(D - 1)^2 y = 0$, $D \equiv \frac{d}{dx}$ is

(a) $y = C_1 + C_2 x$ **(b)** $y = C_1 e^x$ **(c)** $y = (C_1 + C_2 x)e^x$ **(d)** none of these.

16.6 The solution of $\frac{d^2y}{dx^2}(e^x + 1) + y' = 0$ is

(a) $y = C_1 x + C_2 e^{-x}$ **(b)** $y = C_1(x - e^{-x}) + C_2$

(c) $y = C_1(x + e^{-x})$ **(d)** $y = C_1(x + e^{-x}) + C_2$.

16.7 $\frac{1}{D^2 + D + 1} \cos x$ is equal to

(a) $-\sin x$ **(b)** $\cos x$ **(c)** $\sin x$ **(d)** $-\cos x$.

16.8 $\dfrac{1}{(D^2-1)^2}e^x$ is equal to

(a) $\dfrac{e^x}{x}$ (b) e^x (c) x^2e^x (d) $\dfrac{e^x}{2}x^2$.

16.9 $\dfrac{1}{D-a}X$ is equal to

(a) $\int X\,dx$ (b) $\int Xe^{-ax}dx$ (c) $e^{ax}\int Xe^{-ax}dx$ (d) $e^{-ax}\int Xe^{ax}dx$.

16.10 Solution of $\dfrac{d^2y}{dx^2}\omega^2 y = 0$ is

(a) $y = A\cos\omega x$ (b) $y = B\sin\omega x$ (c) $y = A\cos\omega x + B\sin\omega x$ (d) none of these.

16.11 $\dfrac{1}{f(D^2)}\sin ax$ is equal to

(a) $\dfrac{\sin x}{f(-a^2)}$ (b) $\dfrac{\sin ax}{f(-a^2)}$ (c) $\dfrac{\sin ax}{f(a^2)}$ (d) $\dfrac{\cos ax}{f(-a^2)}$.

16.12 Solution of $(D^2+a^2)y = \sin ax$ is

(a) $C_1\cos ax + C_2\sin ax$ (b) $C_1\cos ax - C_2\sin ax$

(c) $C_1\cos ax + C_2\sin ax - \dfrac{x}{2a}\cos ax$ (d) $C_1\cos ax + C_2\sin ax + \dfrac{x}{2a}\cos ax$.

16.13 Solution of the equations $\dfrac{dx}{dt} = -\omega y$ and $\dfrac{dy}{dt} = \omega x$ is

(a) $x = -A\sin\omega t$ (b) $x = -B\cos\omega t$ (c) $x = -A\sin\omega t + B\cos\omega t$ (d) none of these.

16.14 Solution of $\dfrac{d^2y}{dx^2} - 5\dfrac{dy}{dt} + 6y = 2e^x$ when $x = 0$, $y = 1$ and $\dfrac{dy}{dx} = 1$ is

(a) $y = Ae^x + Be^{-x}$ (b) $y = e^x$ (c) $y = e^{-x}$ (d) $y = e^{a-x}$.

16.15 The general solution of the differential equation $y\dfrac{d^2y}{dx^2} = \left(\dfrac{dy}{dx}\right)$ is

(a) $y = Ae^{-Cx}$ (b) $y = Ae^{Cx}$ (c) $y = Ae^{-x}$ (d) $y = Ae^x$.

Answers

16.1 (c) **16.2** (c) **16.3** (c) **16.4** (b) **16.5** (c) **16.6** (b) **16.7** (c) **16.8** (d)
16.9 (c) **16.10** (c) **16.11** (b) **16.12** (c) **16.13** (c) **16.14** (b) **16.15** (b).

16(B) : Short Answer Type Questions

16.1 Explain what do you mean by Complementary Function (C.F.) and Particular Integral (P.I.) of a linear differential equation with constant coefficients.

16.2 What is an inverse operator?

Show that, $(D-k)^{-1} = -\dfrac{1}{k}\left(1 + \dfrac{D}{k} + \dfrac{D^2}{k^2} + \cdots\right)$, where $D \equiv \dfrac{d}{dx}$ and k is a constant.

16.3 If $y_1(x)$ and $y_2(x)$ be two linearly independent solutions of homogeneous differential equation, then show that its general solution is given by $y(x) = C_1y_1(x) + C_2y_2(x)$, where C_1 and C_2 are two arbitrary constants.

16.4 If $u(x)$ and $v(x)$ be respectively the general solution and particular solution of a homogeneous differential equation $\frac{d^2y}{dx^2} + P_1\frac{dy}{dx} + P_2y = X$, then its general solution is given by $y = u(x) + v(x)$.

16.5 Prove that, $\frac{1}{f(D^2)}\sin(ax+b) = \frac{1}{f'(-a^2)}\sin(ax+b)$ provided $f'(a^2) \neq 0$.

16.6 Prove that, $\frac{1}{f(D)}(e^{ax}V) = e^{ax}\frac{1}{f(D+a)}V$.

16.7 Discuss the method of variation of parameters as applied to find the particular solution of a linear non-homogeneous differential equation of second order with variable coefficients.

16.8 Find whether $d\phi$ is an exact differential where,

(i) $d\phi = (x^2 - y)dx + xdy$, **[CU(H) 2009]**

(ii) $d\phi = (x^3 - 4xy)dx + (y^2 - 2x)dy$. **[CU(H) 2007]**

16(C) : Broad Answer Type Questions

16.1 Solve the following differential equations :

(i) $\frac{d^2y}{dx^2} + 4y = x^2 + \cos 2x$ **[Ans:** $y = C_1\cos 2x + C_2\sin 2x + \frac{x^2}{4} - \frac{1}{8} - \frac{1}{4}x\cos 2x$**]**

(ii) $\frac{d^2y}{dx^2} - 2b\frac{dy}{dx} + b^2y = 0$ **[Ans:** $y = (A + Bx)e^{bx}$**]**

(iii) $\frac{d^2y}{dx^2} + \frac{dy}{dx} - 30y = 0$ **[Ans:** $y = C_1e^{5x} + C_2e^{-6x}$**]**

(iv) $\frac{d^2y}{dx^2} + 2\frac{dy}{dx} + y = 0,\ y(0) = 1,\ y'(0) = -1$

[AMIETE(Winter) 1997] **[Ans:** $y = e^{-x}$**]**

(v) $\frac{d^2y}{dx^2} - y = x^2\cos x$ **[Ans:** $y = C_1e^x + C_2e^{-x} + \frac{1}{2}(1 - x^2)\cos x + x\sin x$**]**

(vi) $\frac{d^2y}{dx^2} - 6\frac{dy}{dx} + 9y = 0$ **[Ans:** $y = (A + Bx)e^{3x}$**]**

(vii) $\frac{d^2y}{dx^2} + 2\frac{dy}{dx} = 1 + x^2$ **[Ans:** $y = C_1 + C_2e^{-2x} + \frac{x}{12}(2x^2 - 3x + 9)$**]**

(viii) $D^2(D^2 + 4)y = 96x^2$ **[Ans:** $y = C_1 + C_2x + C_3\cos 2x + C_4\sin 2x + 2x^2(x^2 - 3)$**]**

(ix) $\frac{d^2y}{dx^2} - 2\frac{dy}{dx} + 4y = e^x\cos x$ **[Ans:** $y = e^x(C_1\cos\sqrt{3}x + C_2\sin\sqrt{3}x + \frac{1}{2}e^x\cos x$**]**

(x) $\frac{d^2y}{dx^2} - 7\frac{dy}{dx} + 6y = 2e^{3x}$, given $y = 1,\ \frac{dy}{dx} = 0$ when $x = 0$

[Ans: $y = \frac{7}{5}e^x - \frac{1}{15}e^{6x} - \frac{1}{3}e^{3x}$**]**

(xi) $\dfrac{d^2y}{dx^2} + y = 6,\ \dfrac{dy}{dt} = 0$ when $t = \pi$; $y = 12$ when $t = 0$ **[Ans:** $y = 6(1 + \cos t)$]

(xii) $x^2\dfrac{d^2y}{dx^2} + x\dfrac{dy}{dx} + a^2y = 0$ **[CU(H) 2012]**

(xiii) $yy' + y'^2 + 4 = 0$ for $x = 1 \to 3,\ y' = 0$ **[CU(H) 2009]**

16.2 Solve by the method of variation of parameters :

(i) $\dfrac{d^2y}{dx^2} - 2\dfrac{dy}{dx} = e^x \sin x$ **[Ans:** $y = C_1 + C_2e^{2x} - \frac{1}{2}e^x \sin x$]

(ii) $\dfrac{d^2y}{dx^2} - 2\dfrac{dy}{dx} + 2y = e^x \tan x$

[Ans: $y = e^x(C_1 \cos x + C_2 \sin x) - e^x \cos x \log(\sec x + \tan x)$]

(iii) $\dfrac{d^2y}{dx^2} + 4y = 4\tan 2x$ **[WBUT 2002]**

[Ans: $y = C_1 \cos 2x + C_2 \sin 2x - \cos 2x \log(\sec 2x + \tan 2x)$]

(iv) $\dfrac{d^2y}{dx^2} + 9y = \sec 3x$ **[WBUT 2005, '08, '09]**

[Ans: $y = C_1 \cos 3x + C_2 \sin 3x + \frac{x}{3}\sin 3x + \frac{1}{9}\cos 3x \log\cos 3x$]

(v) $\dfrac{d^2y}{dx^2} + y = \sec x \tan x$ **[Ans:** $C_1 \cos x + C_2 \sin x + x\cos x + \sin x \log \sec x - \sin x$]

(vi) $\dfrac{d^2y}{dx^2} + y = \sec x$ **[AMIETE(Winter) 1990]**

[Ans: $y = C_1 \cos x + C_2 \sin x + x \sin x + \cos x \log \cos x$]

(vii) $\dfrac{d^2y}{dx^2} - 3\dfrac{dy}{dx} + 2y = \sin x$ **[AMIETE(Winter) 1992]**

[Ans: $C_1e^x + C_2e^{2x} + \frac{1}{10}(3\cos x + \sin x)$]

(viii) $y'' - 2y' + 2y = e^x \tan x$

[Ans: $y = e^x(C_1 \cos x + C_2 \sin x) - e^x \cos x \log(\sec x + \tan x)$]

(ix) $\dfrac{d^2y}{dx^2} + y = \dfrac{1}{1 + \sin x}$

[Ans: $y = C_1 \cos x + C_2 \sin x + \sin x \log(1 + \sin x) - x\cos x - 1$]

(x) $\dfrac{d^2y}{dx^2} + a^2y = \tan ax\ (a \neq 0)$.

[Ans: $y = C_1 \cos ax + C_2 \sin ax - \frac{1}{a^2}\cos ax \log|\sec ax + \tan ax|$]

16.3 Solve the following systems of simultaneous equations :

(i) $\dfrac{dy}{dx} + 4x + 2y = t,\ \dfrac{dy}{dt} + 2x + 5y = e^t$

[Ans: $x = C_1e^{-2t} + C_2e^{-7t} - \frac{31}{196} + \frac{5}{14}t - \frac{1}{8}e^t$

$y = -\frac{2}{3}C_1e^{-2t} + C_2e^{-7t} + \frac{9}{98} - \frac{1}{7}t + \frac{5}{24}e^t$]

(ii) $\dfrac{dy}{dx} + y = \sin t,\ \dfrac{dy}{dx} + x = \cos t$ given that $x = 2$ and $y = 0$ at $t = 0$.

[Ans: $x = e^t + e^{-t},\ y = e^{-t} + e^t + \sin t$]

(iii) $\dfrac{dx}{dt} + 2y + \sin t = 0,\ \dfrac{dy}{dt} - 2x - \cos t = 0$, given $x = 0,\ y = 1$ at $t = 0$.

[Ans: $x = \cos 2t - \sin 2t - \cos t,\ y = \cos 2t + \sin 2t - \sin t$]

(iv) $\dfrac{dx}{dt} + 2x - 3y = 5t,\ \dfrac{dy}{dt} - 3x + 2y = 2e^{2t}$

[**Ans:** $x = C_1e^t + C_2e^{-5t} + \frac{6}{7}e^{2t} - 2t - \frac{13}{5}$
$y = C_1e^t - C_2e^{-5t} - \frac{8}{7}e^{2t} - 3t - \frac{12}{5}$]

(v) $\dfrac{dx}{dt} - 5x - y = 0,\ \dfrac{dy}{dt} - y + 4x = 0$

[**Ans:** $x = (C_1 + C_2x)e^{3x},\ y = (1 - 2x)(C_2 - 2C_1)e^{3x}$]

(vi) $\dfrac{dx}{dt} + 5x - 2y = t,\ \dfrac{dy}{dt} + 2x + y = 0$ so that $x = y = 0$ at $t = 0$

[AMIETE(Summer) 1993]

[**Ans:** $x = -\frac{1}{27}(1 + 6t)e^{-3t} + \frac{1}{27}(1 + 3t),\ y = -\frac{2}{27}(2 + 3t)e^{-3t} + \frac{2}{27}(2 - 3t)$]

(vii) $\dfrac{dy}{dx} - y = t,\ \dfrac{dy}{dt} = t^2 - x$

[**Ans:** $x = C_1 \cos t + C_2 \sin t + t^2 - 1,\ y = -C_1 \sin t + C_2 \cos t + t$]

16.4 Solve the following differential equations and evaluate from the given data, the arbitrary constants of integration :

(i) $\dfrac{d^2y}{dx^2} + y = 6,\ \dfrac{dy}{dt} = 0$, when $t = \pi$, $y = 12$ when $t = 0$. [**Ans:** $y = 6(1 + \cos t)$]

(ii) $\dfrac{d^2y}{dx^2} = -9x - 4\cos x;\ y = 2 - \dfrac{\pi}{2}$, when $x = \frac{\pi}{2}$ and $y = 1$, when $x = 0$.

[**Ans:** $y = 9 \sin x + 4 \cos x - \left(\frac{8}{\pi} + 1\right)x - 3$]

(iii) $\dfrac{d^2y}{dx^2} + 6y + 10y = \cos 2x,\ y(0) = 0$ and $\dfrac{dy}{dx}(0) = 0$

[**Ans:** $y = \frac{e^{-3x}}{18}(\cos x - \sin x) - \frac{1}{18}\cos 2x + \frac{1}{9}\sin 2x$]

(iv) $\dfrac{d^2y}{dx^2} + 4\dfrac{dy}{dx} + 5y = x^2 + 5,\ y(0) = 0$ and $\dfrac{dy}{dx}(0) = 0$

[**Ans:** $y = Ae^{4x} + Be^{-4x} + \frac{1}{8}xe^{4x}$]

(v) $\dfrac{d^2x}{dt^2} = g$ (a constant), $x = x_0$ and $\dfrac{dx}{dt} = w_0$ at $t = 0$ [**Ans:** $x = x_0 + w_0t + \frac{1}{2}gt^2$]

(vi) $y'' - 4y' + 3y = 10e^{-2x}$ subject to the conditions $y(0) = 1,\ y'(0) = -3$

[BU(H) 1995] [**Ans:** $y = \frac{11}{6}e^x - \frac{7}{6}e^{3x} + \frac{2}{3}e^{-2x}$]

(vii) $\dfrac{d^2x}{dt^2} + \dfrac{dy}{dt} + 3x = e^{-1}$ and $\dfrac{d^2y}{dt^2} - \dfrac{dx}{dt} + 3y = \sin 2t$. **[VU(H) 2005, 2012]**

16.5 Find the general solutions of the following differential equations :

(i) $\dfrac{d^2y}{dx^2} + \dfrac{dy}{dx} - 12y = e^x$ [**Ans:** $y = Ae^{3x} + Be^{-4x} - \frac{1}{10}e^x$]

(ii) $\dfrac{d^2y}{dx^2} + 1 = \sin x$ [**Ans:** $y = -\frac{x^2}{2} - \sin x + Ax + B$]

(iii) $(x^2D^2 - xD + 1)y = \log x$ [**Ans:** $y = (A + B\log x)x + \log x + 2$]

[**Hint :** Put $x = e^z$]

(iv) $(D^2 - 3D + 2)y = 3e^{2x}$ [**Ans:** $y = e^x + Be^{2x} + xe^{2x}$]

(v) $\dfrac{d^2y}{dx^2} + 3\dfrac{dy}{dx} + 2y = 3e^{2x}$ [**Ans:** $x = Ae^{-x} + Be^{-2x} + \frac{1}{4}e^{2x}$]

(vi) $\dfrac{d^2y}{dx^2} + 2\dfrac{dy}{dx} + 2y = xe^{-x}$ [**Ans:** $y = e^{-x}(A\cos x + B\sin x + xe^{-x}$]

(vii) $(D^2 + 1)y = e^x \cos x$ [**Ans:** $y = A\cos x + B\sin x + \frac{e^x}{5}(\cos x + 2\sin x)$]

(viii) $(D^2 + kD)y = x$, where k is a constant. [**Ans:** $y = A + Be^{-kx} + \frac{x^2}{2k} - \frac{x}{k^2} + \frac{1}{k^3}$]

(ix) Solve the differential equation $y'' - 3y' + 2y = x^3$. **[VU(H) 2013]**

(x) $x^2y'' - xy' + y = \ln x$ **[VU(H) 2011]**

16.6 Solve the following equations of Clairaut and Lagrange :

(i) $y = x\left(\dfrac{dy}{dx}\right)^2 - 2\left(\dfrac{dy}{dx}\right)^3$ [**Ans:** $x = C(p-1)^{-2} + 2p + 1$]

(ii) $y + x\dfrac{dy}{dx} = 4\sqrt{\dfrac{dy}{dx}}$ [**Ans:** $x\sqrt{p} = \ln p + C$, $y = \sqrt{p}(4 - \ln p - C)$].

16.7 A solution of a differential equation $\dfrac{d^2y}{dx^2} + 2\dfrac{dy}{dx} + y = 4e^{-x}$ takes the value 1 when $x = 0$ and a value $\dfrac{1}{e}$ when $x = 1$. What is its value when $x = 2$? **[BU(H) 2013]**

16(D) : Applications of Differential Equations : Physical Problems

16.1 Motion of a particle along a straight line along X-axis is determined by the equation

$$\frac{d^2x}{dt^2} + 4\frac{dx}{dt} + 8x = 20\cos 2t.$$

The particle starts from rest at $x = 0$. Find x as a function of t. **[BU(H) 1996]**

[**Ans:** $y = \cos 2t + 2\sin 2t - e^{-2t}(\cos 2t + 3\sin 2t)$]

16.2 An inductance L, a capacitance C is discharged through R so that the instantaneous charge $q(t)$ satisfies the equation $L\dfrac{d^2q}{dt^2} + R\dfrac{dq}{dt} + \dfrac{q}{C} = 0$.

Solve the equation and find $q(t)$ for $t > 0$, subject to the condition that at $t = 0$, $q = 0.001$ C and $\dfrac{dq}{dt} = 0$. [Given : $L = 0.25$ H, $R = 250\ \Omega$ and $C = 2\ \mu$F]

[BU(H) 1999] [**Ans:** $q = 0.001e^{-500t}\left\{\cos 500\sqrt{7}t + \frac{1}{\sqrt{7}}\sin 500\sqrt{7}t\right\}$]

16.3 A force on a body of mass m moving in a straight line is directed towards a fixed point O in the line and is of magnitude km/x^2, where x is the distance from O at time t. If E be the kinetic energy at $x = a$, find the kinetic energy at any distance x.

16.4 If $a\dfrac{d\theta}{dt} + 2b\theta + u = 0$ and $\theta = \dfrac{dx}{dt}$ and $x = ku$, find the differential equation connecting x, t and the constant a, b and k. Express the relation between a, b and k so that if x denotes a displacement and t the time, the motion may just be non-oscillatory. On this supposition show that,

$$x = e^{-\frac{b}{a}t}\left\{x_0 + \frac{b}{a}x_0 t + \theta_0 t\right\},$$

where x_0 and θ_0 are the values of x and θ when $t = 0$.

16.5 If the resistance varies as the fourth power of the velocity, show that the kinetic energy of a particle of mass m falling freely under gravity at a distance x is

$$E \tanh\left(\frac{mgx}{E}\right), \text{ where } E \text{ is the terminal energy.}$$

16.6 A particle of unit mass is projected with velocity u at an angle α with the horizon in a medium whose resistance is μ times the velocity. Show that its direction will again make an angle α with the horizon after a time $\frac{1}{\mu}\log\left[1 + \frac{2\mu}{g}V\right]$, where V is the initial velocity in vertical direction.

16.7 Two particles each of mass m gm are suspended from two springs of same stiffness k as in adjoining figure. After the system comes to rest, the lower mass is filled l cm downwards and released. Discuss their motion.

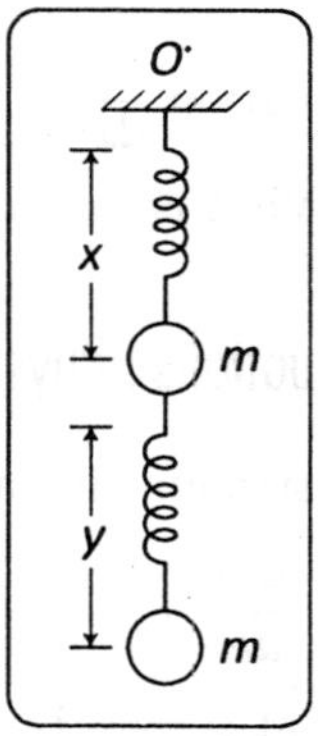

16.8 The voltage V and the current I at a distance x from the sending end of the transmission line satisfy the equation :

$$-\frac{dV}{dx} = RI, \quad -\frac{dI}{dx} = GV, \text{ where } R \text{ and } G \text{ are constants.}$$

If $V = V_0$ at the sending end ($x = 0$) and $V = 0$ at the receiving end ($x = l$), show that,

$$V = V_0\left[\frac{\sinh n(l - x)}{\sin nl}\right], \quad \text{where } n^2 = RG.$$

16.9 Two equal masses are suspended at the end of a weightless elastic spring of stiffness constant k. One mass falls off. Solve the differential equation of motion of the remaining mass.

16.10 A mass m is projected vertically upwards from a point O with initial velocity v_0. Show that the maximum height attained assuming that the resistance of air is proportional to the velocity, is given by

$$H = \frac{1}{k}\left[v_0 - \frac{g}{k}\ln\left(\frac{9 + kv_0}{9}\right)\right].$$

16.11 A mass m is projected vertically upwards from a point O with initial velocity v_0. Find the maximum height reached assuming that the resistance of air is proportional to the velocity. **[VU(H) 1990]**

16.12 The equation of motion of a small particle of mass m is given by $m\ddot{x} + y\dot{x} - mg = 0$. Assuming the initial speed of the mass to be u what is its terminal speed? **[BU(H) 2012]**

16.13 Solve the differential equation : $\dfrac{dx^2}{dt^2} + \dfrac{g}{l}x = \dfrac{g}{l}L$,

where g, l and L are constants, subject to the initial conditions $x = a$ and $\frac{dx}{dt} = 0$ at $t = 0$. **[BU(H) 2011]**

16.14 (a) State the order and degree of the following differential equation :

$$\frac{d^3y}{dx^3} + \left(\frac{dy}{dx}\right)^2 + xy = 0.$$ **[CU(H) 2014, '16]**

(b) Solve the equation, $(1 - x^2)\dfrac{d^2y}{dx^2} - 2x\dfrac{dy}{dx} + 6y = 0$ around the point $x = 0$. **[CU(H) 2008]**

CHAPTER 17

Series Solutions of Differential Equations

17.1 Introduction

In many physical problems, we need to solve linear differential equations with variable coefficients which do not permit a general solution in terms of known functions. Though numerical methods provide a way to solve such equations, series solution methods are more convenient.

17.2 Solution in the Form of a Series

We consider a differential equation of the form

$$P_0(x)\frac{d^2y}{dx^2} + P_1(x)\frac{dy}{dx} + P_2(x)y = 0, \quad \text{where } P_0(0) \neq 0. \tag{17.1}$$

Eqn. (17.1) can be solved by assuming y as an infinite series, like

$$y = a_0 + a_1x + a_2x^2 + \cdots + a_nx^n + \cdots. \tag{17.2}$$

The series solution of some differential equations gives rise to special functions, like Bessel function, Legendre's polynomials, Lagurre's polynomial, Hermite polynomial, Chebyshev polynomials, etc. which have important applications in science and engineering.

Calculating $\frac{dy}{dx}, \frac{d^2y}{dx^2}, \ldots$ from equation (17.2) and putting in (17.1), we get a series of relation among the coefficients, called **recurrence relation**, by equating to zero the coefficient of x^n.

Then using the values of $a_0, a_1, \ldots, a_n$ in eqn. (17.2), we get the series solution of eqn. (17.1).

17.3 Validity of the Series Solutions

The validity of the series solution of the differential equation (17.1),

$$P_0(x)\frac{d^2y}{dx^2} + P_1(x)\frac{dy}{dx} + P_2(x)y = 0 \quad \text{where } P_0(0) \neq 0 \tag{17.1}$$

can be determined under conditions :

1. If $P_0(a) \neq 0$, then $x = a$ is called an **ordinary point** of Eqn. (17.1).
2. If $P_0(a) = 0$, then $x = a$ is called a **singular point** of Eqn. (17.1).
3. If for a singular point $x = a$ of eqn. (17.1), it can be expressed as

$$\frac{d^2y}{dx^2} + \frac{Q_1(x)}{(x-a)}\frac{dy}{dx} + \frac{Q_2(x)}{(x-a)^2}y = 0, \tag{17.3}$$

where $Q_1(x)$ and $Q_2(x)$ possess derivatives of all orders about the point $x = a$, then the point $x = a$ is called **regular**, otherwise **irregular**.

Frobenius-Fuch's Theorem

1. If $x = a$ be an ordinary point of eqn. (17.1), then its solution can be expressed in the form,

$$y = a_0 + a_1(x-a) + a_2(x-a)^2 + \cdots + a_n(x-a)^n + \cdots. \tag{17.4}$$

2. If $x = a$ be a regular singular point of eqn. (17.1), then its solution can be expressed in the form,

$$y = (x-a)^m[a_0 + a_1(x-a) + a_2(x-a)^2 + \cdots + a_n(x-a)^n + \cdots]. \quad (17.5)$$

3. The series (17.4) and (17.5) are convergent at every point $x = a$.

A series solution is valid only if the series is convergent.

Example 17.1. *Find the solution of* $x\dfrac{dy}{dx} - y = x + 1$ *in powers of* $(x-1)$.

Solution : We have $x\dfrac{dy}{dx} - y = x + 1.$ (i)

Put $z = x - 1, \quad \therefore \quad \dfrac{dz}{dx} = 1.$

$$\frac{dy}{dx} = \frac{dy}{dz}\frac{dz}{dx} = \frac{dy}{dz} \qquad \left[\because \frac{dz}{dx} = 1\right]$$

Eqn. (i) then reduced to the form,

$$(z+1)\frac{dy}{dz} - y = z + 2. \quad \text{(ii)}$$

Let $y = a_0 + a_1 z + a_2 z^2 + \cdots + a_n z^n + \cdots$.

$$\therefore \quad \frac{dy}{dz} = a_1 + 2a_2 z + 3a_3 z^2 + \cdots + na_n z^{n-1} + \cdots.$$

Substituting this value of $\frac{dy}{dz}$ in Eqn. (ii), we get

$$(z+1)[a_1 + 2a_2 z + 3a_3 z^2 + \cdots + na_n z^{n-1} + \cdots] - [a_0 + a_1 z + a_2 z^2 + \cdots + a_n z^n + \cdots] = z + 2.$$

Equating the coefficients of z, we get

$$a_1 - a_0 = 2 \quad \text{or,} \quad a_1 = (a_0 + 2)$$

$$a_1 + 2a_2 - a_1 = 1 \quad \text{or,} \quad a_2 = \frac{1}{2}$$

$$3a_3 + 2a_2 - a_2 = 0 \quad \text{or,} \quad a_3 = -\frac{a_2}{3} = -\frac{1}{3\cdot 2}.$$

In general, $na_n + (n+1)a_{n+1} - a_n = 0$.

$$\therefore \quad a_{n+1} = -\frac{(n-1)}{(n+1)}a_n, \quad n > 1.$$

$$\therefore \quad a_n = -\frac{(n-2)}{n}a_{n-1}$$

$$= (-1)^n \frac{(n-2)(n-3)}{n(n-1)}a_{n-2}$$

$$= (-1)^3 \frac{(n-2)(n-3)(n-4)}{n(n-1)(n-2)}a_{n-3}$$

$$\cdots\cdots\cdots\cdots\cdots\cdots$$

$$= (-1)^n \frac{1}{n(n-1)}, \quad n > 1.$$

From Eqn. (i),

$$y = a_0 + (a_0 + 2)z + \sum_{n=2}^{\infty}(-1)^n \frac{z^n}{n(n-1)} = a_0 + (a_0 + 2)(x-1) + \sum_{n=2}^{\infty}(-1)^n \frac{(x-1)^n}{n(n-1)}$$

or, $$y = a_0 + 2(x-1) + \sum_{n=2}^{\infty} \frac{(-1)^n (x-1)^n}{n(n-1)}.$$

17.4 Series Solution when $x = 0$ is a Regular Singularity of the Equation

Let us consider the equation (17.1),

$$P_0(x)\frac{d^2y}{dx^2} + P_1(x)\frac{dy}{dx} + P_2(x)y = 0, \quad \text{where } P_0(0) \neq 0. \tag{17.1}$$

(i) The series solution of eqn. (17.1) is

$$y = x^m(a_0 + a_1x + a_2x^2 + a_3x^3 + \cdots) = \sum_{k=0}^{\infty} a_k x^{m+k}. \tag{17.2}$$

(ii) Substituting for y, $\frac{dy}{dx}$, $\frac{d^2y}{dx^2}$ from eqn. (17.2) in eqn. (17.1) we get the identity.

(iii) The lowest degree term in x is then equated to zero. Then a **quadratic equation**, known as the **indicial equation**, is obtained.

(iv) From the coefficients of the other powers of x, we find the values of a_1, a_2, a_3, ... in terms of a_0.

The series solution of eqn. (17.1) will depend on the nature of the indicial equation.

Now, the roots of the quadratic equation (the indicial equation), say, m_1 and m_2 may be, such that,

(i) roots m_1 are distinct and do not differ by an integer,

(ii) roots m_1 and m_2 are distinct and equal,

(iii) roots m_1 and m_2 are distinct and differ by an integer.

Specific method of solution are to be followed for each of the above three cases. These methods are known as **Frobenius method** after the name of the German mathematician **Ferdinand Georg Frobenius** (1849–1917).

Solution of the differential equation are as discussed below :

Case I : When m_1 and m_2 are distinct and not differing by an integer, the complete solution is

$$y = C_1(y)_{m_1} + C_2(y)_{m_2}.$$

Case II : When $m_1 = m_2$, then solution is

$$y = C_1(y)_{m_1} + C_2\left(\frac{\partial y}{\partial m}\right)_{m_1}$$

If some of the coefficients of y series become infinite when $m = m_1$, to overcome this difficulty, we replace a_0 by $b_0(m - m_2)$. Then the complete solution is also given by

$$y = C_1(y)_{m_1} + C_2\left(\frac{\partial y}{\partial m}\right)_{m_2}$$

Case III : When the roots are distinct and differ by an integer, the complete solution is given by

$$y = C_1(y)_{m_1} + C_2(y)_{m_2},$$

if the coefficients are not infinite when $m = m_2$.

Example 17.2. **(Roots are not differing by an integer)** *Solve in series the equation,*

$$9x(1-x)\frac{d^2y}{dx^2} - 12\frac{dy}{dx} + 4y = 0. \qquad \text{(i)}$$

[Madras U 2006, Roorkee U 2000]

Solution : The coefficient of $\frac{d^2y}{dx^2}$, i.e., $9x(1-x) = 0$ at $x = 0$. So $x = 0$ is a singular point.

Let $y = a_0x^m + a_1x^{m+1} + a_2x^{m+2} + \cdots$

$$\therefore \quad \frac{dy}{dx} = ma_0x^{m-1} + (m+1)a_1x^m + (m+2)a_2x^{m+1} + \cdots.$$

$$\text{and} \quad \frac{d^2y}{dx^2} = m(m-1)a_0 + (m+1)ma_1x^{m-1} + (m+2)(m+1)a_2x^m + \cdots$$

Using y, $\frac{dy}{dx}$ and $\frac{d^2y}{dx^2}$ in eqn. (i), we get

$$9x(x-1)[m(m-1)a_0x^{m-2} + (m+1)ma_1x^{m-1} + (m+2)(m+1)a_2x^m + \cdots]$$
$$- 12[max^{m-1} + (m+1)a_1x^m + (m+2)a_2x^{m+1} + \cdots]$$
$$+ 4[a_0x^m + a_1x^{m+1} + a_2x^{m+2} + \cdots] = 0. \qquad \text{(ii)}$$

The lowest power of x in eqn. (ii) is x^{m-1}. Equating the coefficients of x^{m+1} to zero, we get

$$a_0[9m(m-1) - 12m] = 0 \quad \text{or,} \quad m(3m-7) = 0 \qquad [\because a_0 \neq 0]$$

The roots of the individual equation are thus $m = 0$ and $\frac{7}{3}$. So the roots are distinct and do not differ by an integer.

Now, equating the coefficient of x^m to zero, we get

$$a_1[9(m+1)m - 12(m+1) + a_0(4 - 9m(m-1))] = 0$$

$$\text{or,} \quad 3a_1(3m-4)(m+1) - a_0(3m-4)(3m+1) = 0$$

$$\text{or,} \quad 3a_1(m+1) = a_0(3m+1).$$

Similarly, $3a_2(m+2) = a_1(3m+4)$, $3a_2(m+3) = a_2(5m+7)$ and so on.

$$\therefore \quad a_1 = \frac{3m+1}{3(m+1)}a_0, \qquad a_2 = \frac{(3m+4)a_1}{3(m+2)} = \frac{(3m+4)(3m+1)}{3^2(m+2)(m+1)}a_0,$$

$$a_3 = \frac{(3m+7)(3m+4)(3m+1)}{3^2(m+3)(m+2)(m+1)}a_0, \text{ etc.}$$

When $m = 0$, $a_1 = \frac{1}{3}a_0$, $\quad a_2 = \frac{1}{3}\cdot\frac{4}{6}a_0$, $\quad a_3 = \frac{1}{3}\cdot\frac{4}{6}\cdot\frac{7}{9}a_0$, etc.

$\therefore$ the particular solution for $m = 0$ is

$$y_1 = a_0\left[1 + \frac{1}{3}x + \frac{1}{3}\cdot\frac{4}{6}x^2 + \frac{1}{3}\cdot\frac{4}{6}\cdot\frac{7}{9}x^3 + \cdots\right].$$

When $m = \frac{7}{3}$, the particular solution is

$$y_2 = a_0x^{\frac{7}{3}}\left[1 + \frac{8}{10}x + \frac{8}{10}\cdot\frac{11}{13}x^2 + \frac{8}{10}\cdot\frac{11}{13}\cdot\frac{14}{16}x^3 + \cdots\right].$$

Hence, the complete solution of eqn. (i) is

$$\boldsymbol{y = c_1y_1 + c_2y_2}$$

$$= C_1\left[1+\frac{1}{3}x+\frac{1}{3}\cdot\frac{4}{6}x^2+\frac{1}{3}\cdot\frac{4}{6}\cdot\frac{7}{9}x^3+\cdots\right]$$

$$+ C_2x^{\frac{7}{3}}\left[1+\frac{8}{10}x+\frac{8}{10}\cdot\frac{11}{13}x^2+\frac{8}{10}\cdot\frac{11}{13}\cdot\frac{14}{16}x^3+\cdots\right]$$

[where $C_1 = c_1a_0$, $C_2 = c_2a_0$]

Example 17.3. **(Roots are equal)**

Find the solution in generalised series about $x = 0$, of the differential equation,

$$x\frac{d^2y}{dx^2}+\frac{dy}{dx}+xy=0.$$ **[AMIETE(Summer) 1995, SVTU 2007]**

Solution : The given equation is $x\dfrac{d^2y}{dx^2}+\dfrac{dy}{dx}+xy=0.$ (i)

As $x = 0$ is a regular singular point, the solution may be assumed in the form

$$y=\sum_{k=0}^{\infty}a_kx^{m+k}.$$

$$\therefore\quad \frac{dy}{dx}=\sum_{k=0}^{\infty}a_k(m+k)x^{m+k-1} \quad\text{and}\quad \frac{d^2y}{dx^2}=\sum_{k=0}^{\infty}a_k(m+k)(m+k-1)x^{m+k-2}.$$

Substituting for y, $\frac{dy}{dx}$, $\frac{d^2y}{dx^2}$ in Eqn. (i), we get

$$\sum_k a_k(m+k)(m+k-1)x^{m+k-1}+\sum_k a_k(m+k)x^{m+k-1}+\sum_k a_kx^{m+k-1}=0$$

$$\text{or,}\quad \sum_k a_k[(m+k)(m+k-1)+(m+k)x^{m+k-1}]+\sum_k a_kx^{m+k+1}=0$$

$$\text{or,}\quad \sum_k a_k(m+k)(m+k)x^{m+k-1}+\sum_k a_kx^{m+k+1}=0.$$

The lowest degree term is x^{m-1}. Its coefficient is obtained by putting $k = 0$ in the first term and equating it to zero.

$$a_0m^2=0 \quad [\because\ a\neq 0] \quad \therefore\ m=0,0$$

The coefficient of x^m next lowest degree term is obtained by putting $k = 1$ in the first term only and then equating to zero,

$$a_1(m+1)(m+1)=0, \quad \therefore\ \mathbf{a_1=0.}$$

Equating the coefficient of x^{m+k+1}, the recurrence relation is given by

$$a_{k+2}(m+k+2)^2+a_k=0 \quad\text{or,}\quad a_{k+2}=\frac{-1}{(m+k+2)^2}a^k.$$

$$\therefore\quad a_2=\frac{-1}{(m+2)^2}a_0,\ a_3=\frac{-1}{(m+3)^2}a_1=0,\ a_4=\frac{-1}{(m+4)^2}a_2=\frac{1}{(m+2)^2(m+4)^2}a_0,$$

$$a_5=\frac{-1}{(m+5)}a_3=\frac{1}{(m+3)^2(m+5)^2}a_1=0,$$

$$a_6 = \frac{-a_4}{(m+6)^2} = \frac{a_0}{(m+6)^2(m+4)^2(m+2)^2}.$$

$$\therefore \quad y = a_0 x^m \left[1 - \frac{x^2}{(m+2)^2} + \frac{x^4}{(m+2)^2(m+4)^2} - \frac{x^6}{(m+2)^2(m+4)^2(m+6)^2} + \cdots\right].$$

For $m = 0, 0$ we get only one solution instead of two. The second solution is given by

$$y_2 = \left(\frac{dy}{dm}\right)_{m=0} = a_0 x^m \log x \left[1 - \frac{x^2}{(m+2)^2} + \frac{x^4}{(m+2)^2(m+4)^2} + \cdots\right]$$
$$+ a_0 x^m \left[\frac{2x^2}{(m+2)^3} + \frac{-2x^4}{(m+2)^2(m+4)^2} + \frac{-2x^4}{(m+2)^2(m+4)^3} + \cdots\right]$$

$$\text{or,} \quad y_2 = y_1 \log x + a_0 x^m \left[\frac{x^2}{(m+2)^2}\frac{2}{m+2} - \frac{x^4}{(m+2)^2(m+4)^2}\left\{\frac{2}{(m+2)} + \frac{2}{m+4}\right\} + \cdots\right].$$

For $m = 0$,

$$y_2 = y_1 \log x + a_0 \left[\frac{x^2}{2^2} - \frac{1}{2^2}\frac{1}{4^2}\left(1 + \frac{1}{2}\right)x^4 + \frac{1}{2^2 4^2 6^2}\left\{1 + \frac{1}{2} + \frac{1}{3}\right\}x^6 \cdots\right].$$

$\therefore$ the complete solution is

$$y = C_1 y_1 + C_2 y_2$$

$$\text{or,} \quad y = C_1\left[1 + \frac{1}{2^2}x^2 + \frac{1}{2^2 4^2}x^4 - \frac{1}{2^2 4^2 6^2}x^6 + \cdots\right]$$
$$+ C_2\left[y_1 \log x + a_0\left\{\frac{1}{2^2}x^2 - \frac{1}{2^2 4^2}\left(1 + \frac{1}{2}\right)x^4 + \frac{1}{2^2 4^2 6^2}\left(1 + \frac{1}{2} + \frac{1}{3}\right)x^6 + \cdots\right\}\right].$$

Example 17.4. **(Roots differ by an integer)**

Solve : $x^2\dfrac{d^2y}{dx^2} + x\dfrac{dy}{dx} + (x^2 - 4)y = 0$ **[Rajasthan U 2003, AMIETE(Summer) 1997]**

Solution : The coefficient of $\frac{d^2y}{dx^2} = 0$ for $x = 0$. So $x = 0$ is a singularity of the given equation.

Let $y = \sum a_k x^{m+k}$,

$$\therefore \quad \frac{dy}{dx} = \sum a_k (m+k) x^{m+k-1} \quad \text{and} \quad \frac{d^2y}{dx^2} = \sum a_k (m+k)(m+k-1) x^{m+k-2}.$$

Substituting the values of y, $\frac{dy}{dx}$ and $\frac{d^2y}{dx^2}$ in the given differential equation, we get

$$x^2 \sum a_k (m+k)(m+k-1)x^{m+k-2} + x\sum a_k (m+k)x^{m+k-1} + (x^2 - 4)\sum a_k x^{m+k} = 0$$

$$\text{or,} \quad \sum a_k[(m+k)(m+k-1) + (m+k) - 4]x^{m+k} + \sum a_k x^{m+k+2} = 0$$

$$\text{or,} \quad \sum a_k (m+k+2)(m+k-2)x^{m+k} + \sum a_k x^{m+k+2} = 0.$$

The coefficient of lowest degree term x^m is obtained by putting $k = 0$ in the first term only and then equating it to zero.

$\therefore \quad a_0(m+2)(m-2)=0$ or, $m=2,-2$ $[\because a_0 \neq 0]$.

Equating the coefficient of x^{m+k+2} to zero,

$$a_{k+2}(m+k+4)(m+k)+a_k=0 \quad \text{or,} \quad a_{k+2}=-\frac{a_k}{(m+k+4)(m+k)}.$$

$$a_1=a_3=a_5=\cdots=0$$

$$a_2=-\frac{a_0}{m(m+4)}$$

$$a_4=-\frac{a_2}{(m+2)(m+6)}=-\frac{a_0}{m(m+2)(m+4)(m+6)}$$

$$a_6=-\frac{a_4}{(m+4)(m+8)}=-\frac{a_0}{m(m+2)(m+4)^2(m+6)(m+8)}.$$

$$\therefore \quad y=a_0x^m\left[1-\frac{x^2}{m(m+4)}+\frac{x^4}{m(m+2)(m+4)(m+6)}-\frac{x^6}{m(m+2)(m+4)^2(m+6)(m+8)}+\cdots\right] \qquad \text{(i)}$$

Putting $m=2$ in eqn. (i),

$$y=a_0x^2\left[1-\frac{x^2}{2\times 6}+\frac{x^4}{2\times 4\times 6\times 8}-\frac{x^6}{2\times 4\times 6^2\times 8\times 10}+\cdots\right]. \qquad \text{(ii)}$$

For $m=-2$, the coefficients of x^4, x^6, ...in Eqn. (i) becomes infinite. To overcome this difficulty we put $a_0=b_0(m+2)$ in Eqn. (i) and get

$$y=b_0x^m\left[(m+2)-\frac{(m+2)x^2}{m(m+4)}+\frac{x^4}{m(m+4)(m+6)}-\frac{x^6}{m(m+4)^2(m+6)(m+8)}+\cdots\right]. \qquad \text{(iii)}$$

Differing Eqn. (iii) w.r.t. m, we get

$$\frac{\partial y}{\partial m}=b_0(x^m\cdot\log x)\left[(m+2)-\frac{(m+2)x^2}{m(m+4)}+\frac{x^4}{m(m+4)(m+6)}-\frac{x^6}{m(m+4)^2(m+6)(m+8)}+\cdots\right]$$
$$+b_0x^m\left[1-\frac{(m+2)x^2}{m(m+4)}\left(\frac{1}{m+2}-\frac{1}{m}-\frac{1}{m+4}\right)+\frac{x^4}{m(m+4)(m+6)}\left(-\frac{1}{m}-\frac{1}{m+4}-\frac{1}{m+6}\right)+\cdots\right].$$

On replacing m by -2, we get

$$\left(\frac{\partial y}{\partial m}\right)_{m=-2}=(b_0x^{-2}\log x)\left[0-0+\frac{x^4}{(-2)(2)(4)}-\frac{x^6}{(-2)(2)^2(4)(6)}+\cdots\right]$$
$$+b_0x^{-2}\left[1+\frac{x^2}{2^2}+\frac{x^4}{2^2\times 4}\left(\frac{1}{4}\right)+\cdots\right]$$

$$y_2 = b_0 x^2 \log x \left[-\frac{1}{2^2 \times 4} + \frac{x^2}{2^3 \times 4 \times 6} - \frac{x^4}{2^3 \times 4^2 \times 6 \times 8} + \cdots \right] + b_0 x^{-2} \left[1 + \frac{x^2}{2^2} + \frac{x^4}{2^2 \times 4^2} + \cdots \right].$$

Hence, the required general solution is

$$y = C_1 x^2 \left[1 - \frac{x^2}{2 \times 6} + \frac{x^4}{2 \times 4 \times 6 \times 8} - \frac{x^6}{2 \times 4 \times 6^2 \times 8 \times 10} + \cdots \right] + C_2 \left[x^2 \log x \left(1 - \frac{1}{2^2 \times 4} + \frac{x^4}{2^3 \times 4 \times 6} - \frac{x^4}{2^3 \times 4^2 \times 6 \times 8} + \cdots \right) + x^{-2} \left(1 + \frac{x^2}{2^2} + \frac{x^4}{2^2 \times 4} + \cdots \right) \right].$$

17.5 Solved Problems

Example 17.1. *Find power series solution of* $\dfrac{d^2y}{dx^2} + x\dfrac{dy}{dx} + y = 0.$

Solution : We assume the solution to be of the form

$$y = \sum_{k=0}^{\infty} a_k x^k. \tag{i}$$

Then $\dfrac{dy}{dx} = \displaystyle\sum_{k=1}^{\infty} k a_k x^{k-1}$ and $\dfrac{d^2y}{dx^2} = \displaystyle\sum_{k=2}^{\infty} k(k-1) a_k x^{k-2}.$

Substituting the values of y, $\frac{dy}{dx}$ and $\frac{d^2y}{dx^2}$ in the given equation, we get

$$\sum_{k=2}^{\infty} k(k-1) a_k x^{k-2} + x \sum_{k=1}^{\infty} k a_k x^{k-1} + \sum_{k=0}^{\infty} a_k x^k = 0.$$

We now collect all the coefficients of x^k. Since the first series starts from $k = 2$, the coefficient of x^k will be given for $k \to k+2$. Similarly the second series starts from $k = 1$, but the series is multiplied by x, the coefficient of x^k will be given for $k \to k$. Same argument holds for the third series.

$\therefore$ the coefficient of x^k is

$$(k+2)(k+1)a_{k+2} + k a_k + a_k = 0 \quad \text{or,} \quad a_{k+2} = -\frac{a_k}{k+2}.$$

Replacing $(k+2)$ by k, we get

$$a_k = -\frac{1}{k} a_{k-2}. \tag{ii}$$

Eqn (ii) is a recursion formula. Starting from $k = 2$, we find

$$a_2 = -\frac{1}{2} a_0, \qquad a_3 = -\frac{1}{3} a_1, \qquad a_4 = -\frac{1}{4} a_2 = \frac{1}{2 \times 4} a_0,$$

$$a_5 = -\frac{1}{5} a_3 = \frac{1}{3 \times 5} a_1, \quad a_6 = -\frac{1}{6} a_4 = -\frac{1}{2 \times 4 \times 6} a_0, \quad a_7 = -\frac{1}{7} a_5 = \frac{1}{3 \times 5 \times 7} a_1.$$

$\therefore$ the solution of the given differential equation is

$$y = a_0\left(1 - \frac{x^2}{2} + \frac{x^4}{2\times 4} - \frac{x^6}{2\times 4\times 6}\right) + a_1\left(x - \frac{x^3}{3} + \frac{x^5}{3\times 5} - \frac{x^7}{3\times 5\times 7} + \cdots\right)$$

or, $y = a_0 y_1 + a_1 y_2$.

The Wronskian of the solutions at $x = 0$ is

$$W = \begin{vmatrix} y_1(0) & y_2(0) \\ y_1'(0) & y_2'(0) \end{vmatrix} = \begin{vmatrix} 1 & 0 \\ 0 & 1 \end{vmatrix} = 1.$$

Since $W \neq 0$, the solutions y_1 and y_2 are the pair of basic solutions of the given equation.

Example 17.2. *Solve in series the equation,* $\dfrac{d^2y}{dx^2} + x^2 y = 0.$ **[AMIETE(Winter) 1995]**

Solution : Let $y = a_0 + a_1x + a_2x^2 + a_3x^3 + a_4x^4 + a_5x^5 + a_6x^6$

$$+ a_7x^7 + a_8x^8 + \cdots + a_kx^k + \cdots \qquad \text{(i)}$$

$$= \sum_{k=0}^{\infty} a_k x^k \quad (\because\ x = 0 \text{ is an ordinary point of the given equation}).$$

$$\therefore \quad \frac{dy}{dx} = a_1 + 2a_2x + 3a_3x^2 + 4a_2x^3 + 5a_5x^4 + 6a_6x^5 + 7a_7x^6 + 8a_8x^7 + \cdots$$

and $$\frac{d^2y}{dx^2} = 2a_2 + 2\cdot 3a_3x + 3\cdot 4a_4x^2 + 4\cdot 5a_5x^3 + 5\cdot 6a_6x^4 + 6\cdot 7a_7x^5 + 7\cdot 8a_8x^6 + \cdots.$$

Substituting the values of y, $\frac{dy}{dx}$ and $\frac{d^2y}{dx^2}$ in the given equation, we get

$$2a_2 + 2\cdot 3a_3x + 3\cdot 4x^2 + 4\cdot 5a_5x^3 + 5\cdot 6a_6x^4 + 6\cdot 7a_7x^5 + 7\cdot 8a_8x^6 + \cdots$$
$$+ x^2(a_0 + a_1x + a_2x^2 + a_3x^3 + a_4x^4 + a_5x^5 + \cdots) = 0$$

or, $$2a_2 + 6a_3x + (a_0 + 12a_4)x^2 + (a_1 + 20a_5)x^3 + (a_2 + 30a_6)x^4 + \cdots$$
$$+ [a_{k-2} + (k+2)(k+1)a_{k+2}]x^k + \cdots = 0.$$

Equating the various powers of x to zero, we get $\boldsymbol{a_2 = 0,\ a_3 = 0}$.

$$a_0 + 12a_4 = 0, \quad \therefore\ a_4 = -\frac{1}{12}a_0$$

$$a_1 + 20a_5 = 0, \quad \therefore\ a_5 = -\frac{1}{20}a_1$$

$$a_3 + 30a_6 = 0, \quad \therefore\ a_6 = -\frac{1}{30}a_2 \quad (\because\ a_2 = 0)$$

and so on.

In general, $a_{k-2} + (k+2)(k+1)a_{k+2} = 0$ or, $a_{k+2} = -\dfrac{a_{k-2}}{(k+1)(k+2)}$.

For $k = 5,\quad a_7 = -\dfrac{a_3}{6\cdot 7} = 0$

$k = 6,\quad a_8 = -\dfrac{a_4}{7\cdot 8} = \dfrac{a_0}{12\cdot 7\cdot 8}$

$k = 7,\quad a_9 = -\dfrac{a_5}{8\cdot 9} = \dfrac{a_1}{20\cdot 8\cdot 9}$

$$k = 8, \quad a_{10} = -\frac{a_6}{9 \cdot 10} = 0 \quad [\because a_6 = 0]$$

$$k = 9, \quad a_{11} = \frac{a_7}{11 \cdot 10} = 0 \quad [\because a_7 = 0]$$

$$k = 10, \; a_{12} = \frac{a_8}{12 \cdot 11} = -\frac{a_0}{12 \cdot 8 \cdot 7 \cdot 11 \cdot 12}.$$

Substituting these values in equation (i), we get

$$y = a_0 + a_1 x - \frac{1}{12} a_0 x^4 - \frac{a_1}{20} x^5 + \frac{a_0}{12 \cdot 7 \cdot 8} x^8 + \frac{a_1}{20 \cdot 8 \cdot 9} x^9 - \frac{a_0}{12 \cdot 8 \cdot 7 \cdot 11 \cdot 12} x^{12} + \cdots$$

$$= a_0 \left[1 - \frac{1}{12} x^4 + \frac{x^8}{12 \cdot 7 \cdot 8} - \frac{x^{12}}{12 \cdot 8 \cdot 7 \cdot 11 \cdot 12} + \cdots \right] + a_1 \left[x - \frac{x^5}{20} + \frac{x^9}{20 \cdot 8 \cdot 9} \cdots \right].$$

Example 17.3. *Solve the equation in series,* $\frac{d^2y}{dx^2} + xy = 0.$

Solution : Since the coefficients of $\frac{d^2y}{dx^2} \neq 0$ at $x = 0$, here $x = 0$ is an ordinary point.

Let the solution of the given equation be

$$y = a_0 + a_1 x + a_2 x^2 + a_3 x^3 + \cdots + a_k x^k + \cdots. \tag{i}$$

$$\therefore \quad \frac{dy}{dx} = a_1 + 2a_2 x + 3a_3 x^2 + \cdots + k a_k x^{k-1} + \cdots$$

and
$$\frac{d^2y}{dx^2} = 2 \cdot 1 a_2 + 3 \cdot 2 a_3 x + 4 \cdot 3 a_4 x^2 + \cdots + k(k-1) a_k x^{k-2} + \cdots.$$

Substituting the values of y, $\frac{dy}{dx}$, $\frac{d^2y}{dx^2}$ in the given equation, we get

$$2 \cdot 1 a_2 + 3 \cdot 2 a_3 x + 4 \cdot 3 a_4 x^2 + \cdots + k(k-1) a_k x^{k-2} + \cdots + x(a_0 + a_1 x + a_2 x^2 + \cdots + a_k x^k + \cdots) = 0$$

$$2 \cdot 1 a_2 + (3 \cdot 2 a_3 x + a_0) x + (4 \cdot 3 a_4 + a_1) x^2 + (5 \cdot 4 a_5 + a_2) x^3 + \cdots + [(k+2)(k+1) a_{k+2} + a_{k-1}] x^k + \cdots = 0.$$

Equating the coefficients of various powers of x to zero, $a_2 = 0$.

$$3 \cdot 2 a_3 + a_0 = 0, \quad \therefore \; a_3 = -\frac{a_0}{3!}$$

$$4 \cdot 3 a_4 + a_1 = 0, \quad \therefore \; a_4 = -\frac{2a_1}{4!}$$

$$5 \cdot 3 a_5 + a_2 = 0, \quad \therefore \; a_5 = -\frac{a_2}{5 \cdot 4} = 0 \text{ and so on.}$$

In general, $(k+2)(k+1) a_{k+2} + a_{k-1} = 0$.

$$\therefore \quad a_{k+2} = -\frac{a_{k-1}}{(k+2)(k+1)}. \tag{ii}$$

Putting $k = 4, 5, 6, \ldots$ in Eqn. (ii) successively,

$$a_6 = -\frac{a_3}{6\cdot 5} = \frac{4a_0}{6!}, \quad a_7 = -\frac{a_4}{7\cdot 6} = \frac{5\cdot 2a_1}{7!}, \quad a_8 = -\frac{a_5}{8\cdot 7} = 0,$$

$$a_9 = -\frac{a_6}{9\cdot 8} = -\frac{7\cdot 4a_0}{9!} \text{ and so on.}$$

Substituting these values in Eqn. (i), we get

$$y = a_0\left(1 - \frac{x^3}{3!} + \frac{1\cdot 4x^6}{6!} - \frac{1\cdot 4\cdot 7x^9}{9!} + \cdots\right) + a_1\left(x - \frac{2x^4}{4!} + \frac{2\cdot 5x^7}{7!} - \cdots\right).$$

Example 17.4. *Solve the differential equation,* $(1+x^2)\frac{d^2y}{dx^2} + x\frac{dy}{dx} - y = 0.$

Solution : Since $x = 0$ is an ordinary point of the given equation, we assume the solution of the given equation in the form,

$$y = a_0 + a_1x + a_2x^2 + a_3x^3 + \cdots. \tag{i}$$

$$\therefore \quad \frac{dy}{dx} = a_1 + 2a_2x + 3a_3x^2 + 4a_4x^3 + \cdots$$

and $$\frac{d^2}{dx^2} = 2a_2 + 3\cdot 2a_3x + 4\cdot 3x^2 + 5\cdot 4x^3 + \cdots.$$

Substituting the values of y, $\frac{dy}{dx}$, $\frac{d^2y}{dx^2}$ in the given equation, we get

$$(1+x^2)(2a_2 + 6a_3x + 12a_4x^2 + 20a_5x^3 + \cdots) + x(a_1 + 2a_2x + 3a_3x^2 + 4a_4x^3 + \cdots) - (a_0 + a_1x + a_2x^2 + \cdots) = 0$$

or, $$(2a_2 - a_0) + (6a_3 + a_1 - a_1)x + (2a_2 + 12a_4 + a_2)x^2 + (20a_5 + 6a_3 + 3a_3 - a_3)x^3 + \cdots = 0.$$

Equating the coefficients of various powers of x to zero, we get

$$2a_2 - a_0 = 0, \quad \therefore a_2 = -\frac{a_0}{2}$$

$$6a_3 = 0, \quad \therefore a_3 = 0$$

$$12a_4 + 3a_2 = 0, \quad \therefore a_4 = -\frac{a_2}{4} = -\frac{a_0}{8}$$

$$20a_5 + 8a_3 = 0, \quad \therefore a_5 = -\frac{8a_3}{20} = 0 \quad (\because a_3 = 0).$$

Substituting the values in Eqn. (i) we get the series solution of the given differential equation,

$$y = a_0\left[1 + \frac{x^2}{2} - \frac{x^4}{8} + \cdots\right] + a_1x.$$

Example 17.5. *Solve in series,* $(1-x^2)\frac{d^2y}{dx^2} - x\frac{dy}{dx} + 4y = 0.$ **[Rajasthan U 2006]**

Solution : As the coefficient of $\frac{d^2y}{dx^2}$ is not equal to zero at $x = 0$, the point $x = 0$ is an ordinary point.

Let the solution of the given differential equation be

$$y = a_0 + a_1 x + a_2 x^2 + a_3 x^3 + \cdots + a_k x^k + \cdots.$$

$$\therefore \quad \frac{dy}{dx} = a_1 + 2a_2 x + 3a_3 x^2 + \cdots + k a_k x^{k-1} + \cdots$$

and $$\frac{d^2y}{dx^2} = 2a_2 + 3 \cdot 2a_3 x + \cdots + k(k-1)x^{k-2} + \cdots.$$

Substituting the values of y, $\frac{dy}{dx}$, $\frac{d^2y}{dx^2}$ in the given equation, we get

$$(1 - x^2)[2a_2 + 3 \cdot 2a_3 x + 4 \cdot 3a_4 x^2 + \cdots + k(k-1)a_k x^{k-2} + \cdots]$$
$$- x[a_1 + 2a_2 x + 3a_3 x^2 + \cdots + k a_k x^{k-1} + \cdots]$$
$$- 4[a_0 + a_1 x + a_2 x^2 + \cdots + a_k x^k + \cdots] = 0.$$

Equating the coefficients of various powers of x to zero, we get

$$2a_2 + 4a_0 = 0, \qquad \therefore \mathbf{a_2 = -a_0}$$

$$3 \cdot 2a_3 - a_1 + 4a_1 = 0, \qquad \therefore \mathbf{a_3 = -\frac{1}{2}a_1}$$

$$4 \cdot 3a_4 - 2a_2 - 2a_2 + 4a_2 = 0, \qquad \therefore \mathbf{a_4 = 0}$$

$$5 \cdot 4a_5 - 3 \cdot 2a_3 - 3a_3 + 4a_3 = 0, \text{ i.e., } 20a_5 - 5a_3 = 0.$$

$$\therefore \quad \mathbf{a_5 = -\frac{a_1}{8}} \textbf{ and so on.}$$

Putting $k = 4, 5, 6, 7, \ldots$ in Eqn. (ii) successively, we get

$$\mathbf{a_6 = 0, \quad a_7 = \frac{3}{6}a_5 = -\frac{3}{6} \cdot \frac{a_1}{8}, \quad a_8 = 0, \quad a_9 = -\frac{5}{8} \cdot \frac{3}{6} \cdot \frac{a_1}{8} \cdots.}$$

Substituting these values in Eqn. (i), we get

$$\mathbf{y = a_0(1 - 2x^2) + a_1 x \left(1 - \frac{x^2}{2} - \frac{x^4}{8} - \frac{3}{6} \cdot \frac{x^6}{8} - \frac{5 \cdot 3}{8 \cdot 6} \frac{x^8}{8} - \cdots\right)}.$$

Example 17.6. *Solve the following differential equations by Frobenius method : (Roots are not differing by an integer)*

(i) $2x^2 \frac{d^2y}{dx^2} - x\frac{dy}{dx} + (1 - x^2)y = 0$ (ii) $3x\frac{d^2y}{dx^2} + 2\frac{dy}{dx} + y = 0$

(iii) $(2x + x^3)\frac{d^2y}{dx^2} + 3\frac{dy}{dx} - 2xy = 0.$

Solution : (i) Here $x = 0$ is a regular singularity, the solution of the given differential equation may be written as

$$y = a_0 x^m + a_1 x^{m+1} + \cdots.$$

$$\therefore \quad \frac{dy}{dx} = m a_0 x^{m-1} + (m+1)a_1 x^m + \cdots$$

and $$\frac{d^2y}{dx^2} = m(m-1)a_0 x^{m-2} + (m+1)m a_1 x^{m-1} + \cdots.$$

Substituting the values of y, $\frac{dy}{dx}$, $\frac{d^2y}{dx^2}$ in the given equation, we get

$$2[m(m-1)a_0x^m + (m+1)ma_1x^{m+1} + \cdots] - [ma_0x^m + (m+1)a_1x^{m+1} + \cdots] + (1-x^2)[a_0x^m + a_1x^{m+1} + \cdots] = 0.$$

Equating the coefficient of lowest power of x (here x^m) to zero, we get

$$[2m(m-1) - m + 1] = 0 \quad \text{or,} \quad 2m^2 - 3m + 1 = 0 \quad (\because\ a_0 \neq 0).$$

This is indicial equation whose roots are $m = 1, \dfrac{1}{2}$.

Now equating the coefficient of x^{m+1} to zero,

$$2(m+1)ma_1 - (m+1)a_1 + a_1 = 0 \quad \text{or,} \quad m(m+2)a_1 = 0.$$

$$\therefore \quad \boldsymbol{a_1 = 0}.$$

Then equating the coefficient of x^{m+2} to zero, we get

$$2(m+2)(m+1)a_2 - (m+2)a_2 + a_2 - a_0 = 0 \quad \text{or,} \quad (2m^2 + 5m + 3)a_2 = a_0.$$

$$\therefore \quad a_2 = \frac{a_0}{(2m+3)(m+1)}.$$

Similarly, $a_3 = \dfrac{a_1}{(2m+5)(m+2)} = 0$

$$a_4 = \frac{a_2}{(2m+7)(m+3)} = \frac{a_0}{(2m+3)(2m+7)(m+1)(m+3)}$$

$$a_5 = 0$$

$$a_6 = \frac{a_4}{(2m+11)(m+5)} = \frac{a_4}{(2m+3)(2m+7)(2m+11)(m+1)(m+3)(m+5)}.$$

When $m = 1$,

$$a_2 = \frac{a_0}{2 \cdot 5} a_4 = \frac{a_4}{2 \cdot 4 \cdot 5 \cdot 9}, \quad a_6 = \frac{a_0}{2 \cdot 4 \cdot 5 \cdot 9 \cdot 13} \text{ and so on}$$

$$a_1 = a_3 = \cdots = 0.$$

$$\therefore \text{ one solution is : } y_1 = a_0\left(x + \frac{x^3}{2 \cdot 5} + \frac{x^5}{2 \cdot 4 \cdot 5 \cdot 9} + \frac{x^7}{2 \cdot 4 \cdot 6 \cdot 5 \cdot 9 \cdot 13} + \cdots\right).$$

When $m = \dfrac{1}{2}$,

$$a_1 = a_3 = a_5 = \cdots = 0$$

$$a_2 = \frac{a_0}{2 \cdot 3} a_4 = \frac{a_0}{2 \cdot 4 \cdot 3 \cdot 7}, \quad a_6 = \frac{a_0}{2 \cdot 4 \cdot 6 \cdot 3 \cdot 7 \cdot 11}.$$

$$\therefore \text{ the other solution is : } y_2 = a_0\left[x^{\frac{1}{2}} + \frac{x^{\frac{5}{2}}}{2 \cdot 5} + \frac{x^{\frac{9}{2}}}{2 \cdot 4 \cdot 3 \cdot 7} + \frac{x^{\frac{13}{2}}}{2 \cdot 4 \cdot 6 \cdot 3 \cdot 7 \cdot 11} + \cdots\right].$$

So the complete solution is

$$y = C_1y_1 + C_2y_2 = C_1(y)_{m_1} + C_2(y)_{m_2}$$

$$= A\left(x + \frac{x^3}{2 \cdot 5} + \frac{x^5}{2 \cdot 4 \cdot 5 \cdot 9} + \cdots\right) + B\sqrt{x}\left(1 + \frac{x^2}{2 \cdot 3} + \frac{x^4}{2 \cdot 4 \cdot 3 \cdot 7} + \cdots\right).$$

(ii) $3x\frac{d^2y}{dx^2}+2\frac{dy}{dx}+y=0$ (i)

Here, $x=0$ is a regular singular point, the solution may be assumed to be

$$y=\sum_{k=0}^{\infty}a_kx^{m+k}.$$

$$\therefore \quad \frac{dy}{dx}=\sum_{k=0}^{\infty}a_k(m+k)x^{m+k-1} \quad \text{and} \quad \frac{d^2y}{dx^2}=\sum_{k=0}^{\infty}a_k(m+k)(m+k-1)x^{m+k-2}.$$

Substituting values in Eqn. (i),

$$3\sum a_k(m+k)(m+k-1)x^{m+k-1}+2\sum a_k(m+k)x^{m+k-1}+\sum a_kx^{m+k}=0$$

$$\text{or,} \quad a_k[3(m+k)(m+k-1)+2(m+k)]x^{m+k-1}+\sum a_kx^{m+k}=0.$$

The coefficient of the lowest degree term x^{m-1} is obtained by putting $k=0$ in the final term only and equating to zero,

$$a_0[3m(m-1)+2m]=0 \quad \text{or,} \quad a_0[3m^2-m]=0 \quad \text{or,} \quad a_0m(3m-1)=0.$$

$$\because \; a_0\neq 0, \quad m=0 \quad \text{or} \quad \frac{1}{3}.$$

The coefficient of the next lowest degree term x^m is obtained by putting $k=1$ in the first term and $k=0$ in the second term and equating it to zero.

$$a_1[3(m+1)m+2(m+1)]+a_0=0 \quad \text{or,} \quad a_1[3m^2+5m+2]+a_0=0$$

$$\text{or,} \quad a_1(3m+2)(m+1)+a_0=0 \quad \text{or,} \quad a_1=-\frac{a_0}{(3m+2)(m+1)}.$$

Equating to zero, the coefficient of x^{m+k}, the recurrence relation is

$$a_{k+1}[3(m+k+1)(m+k)+2(m+k+1)]+a_k=0$$

$$\text{or,} \quad a_{k+1}(m+k+1)(3m+3k+2)a_k=0 \quad \text{or,} \quad a_{k+1}=\frac{-1}{(m+k+1)(3m+3k+2)}a_k.$$

For $k=0$,

$$\boldsymbol{a_1=\frac{-1}{(m+1)(3m+2)}a_0}$$

$$\boldsymbol{a_2=\frac{-1}{(m+2)(3m+5)}a_1=\frac{1}{(m+1)(m+2)(3m+2)(3m+5)}a_0.}$$

For $k=2$,

$$\boldsymbol{a_3=\frac{-1}{(m+3)(3m+8)}a_2=\frac{-1}{(m+1)(m+2)(m+3)(3m+2)(3m+5)(3m+8)}a_0.}$$

For $m=0$, $\boldsymbol{a_1=-\frac{a_0}{2}}$, $\boldsymbol{a_2=-\frac{a_0}{20}}$, $\boldsymbol{a_3=-\frac{1}{480}a_0}$.

Hence, for $m=0$, $\boldsymbol{y_1=a_0\left(1-\frac{x}{2}+\frac{x^2}{20}-\frac{x^3}{480}+\cdots\right)}$.

For $m = \frac{1}{3}$, $a_1 = -\frac{a_0}{4}$, $a_2 = -\frac{a_0}{56}$, $a_3 = -\frac{1}{1680}a_0$.

Hence, for $m = \frac{1}{3}$, $y_2 = a_0\left(x^{\frac{1}{3}} - \frac{x^{\frac{4}{3}}}{4} + \frac{x^{\frac{7}{3}}}{56} - \frac{1}{1680}x^{\frac{10}{3}} + \cdots\right)$.

$\therefore$ the complete solution is

$$y = Ay_1 + By_2 = a_0\left(1 + \frac{x}{2} + \frac{x^2}{20} - \frac{x^3}{480} + \cdots\right) + b_0 x^{\frac{1}{3}}\left(1 - \frac{x}{4} + \frac{x^2}{56} - \frac{x^3}{1680} + \cdots\right).$$

(iii) $(2x + x^3)\frac{d^2y}{dx^2} + 3\frac{dy}{dx} - 2xy = 0$ (i)

The given equation has regular point at $x = 0$. Let the solution of the Eqn. (i) be

$$y = \sum_{k=0}^{\infty} a_k x^{m+k}.$$

$$\therefore \quad \frac{dy}{dx} = \sum_{k=0}^{\infty} a_k(m+k)x^{m+k-1} \quad \text{and} \quad \frac{d^2y}{dx^2} = \sum_{k=0}^{\infty} a_k(m+k)(m+k-1)x^{m+k-2}.$$

Substituting these values in Eqn. (i), we get

$$\sum_{k=0}^{\infty} a_k[(m+k)(m+k-1) - 2]x^{m+k+1} + \sum_{k=0}^{\infty} a_k[2(m+k)(m+k-1) + 3(m+k)]x^{m+k+1} = 0.$$

Equating the coefficient of lowest power of x (i.e., x^{k-1}), we get

$$[2k(k-1) + k]a_0 = 0 \quad \text{or,} \quad 2k(k-1) + k = 0 \quad [\because\ a_0 \neq 0],$$

$$\therefore \quad k = 0 \quad \text{or} \quad \frac{1}{2}.$$

Now equating to zero, the coefficients of x^k, x^{k+1}, ..., etc. :

x^k: $[2k(k+1) + k + 1]a_1 + (1-k)a_0 = 0$

x^{k+1}: $[2(k+1)(k+2) + k + 2]a_2 + [1 - (k+1)]a_1 = 0$

x^{k+2}: $[2(k+2)(k+3) + k + 3]a_3 + [1 - (k+2)]a_2 = 0$

....

x^{m+k}: $[2(k+m)(k+m+1) + (k+m+1)]a_{m+1} + [1 - (k+m)]a_m = 0.$

So, the recurrence relation is

$$a_{m+1} = -\frac{a_m(1-k-m)}{f(m,k)} = \frac{k+m-1}{f(m,k)}a_m,$$

where $f(m,k) = 2(k+m)(k+m+1) + (k+m+1)$.

For $k = 0$, $a_{m+1} = -\frac{(1-m)a_m}{f(m,0)} = \frac{m-1}{f(m,0)}a_m$, (ii)

where $f(m,0) = 2m(m+1) + (m+1) = (m+1)(2m+1)$.

For $k=\frac{1}{2}$, $a_{m+1}=\frac{-\left(\frac{1}{2}-m\right)a_m}{f\left(m,\frac{1}{2}\right)}=\frac{m-\frac{1}{2}}{f\left(m,\frac{1}{2}\right)}a_m,$ (iii)

where $f\left(m,\frac{1}{2}\right)=\left(m+\frac{3}{2}\right)(2m+2)=(2m+3)(m+1).$

From Eqn. (ii), for $k=0$,

$$a_1=-a_0, \quad a_2=a_3=\cdots=a_m=0.$$

$\therefore$ the solution for $k=0$ is $y_1=a_0(1-x)$.

From Eqn. (iii), the coefficient for $k=\frac{1}{2}$,

$$a_1=-\frac{1}{2}\frac{a_0}{3\cdot 1}=-\frac{a_0}{3!}, \quad a_2=\frac{a_1}{2\cdot 5\cdot 2}=-\frac{1}{2^2}\frac{a_0}{2\cdot 3\cdot 5}=-\frac{a_0}{5!}$$

$$a_3=\frac{3}{2}\cdot\frac{a_2}{7\cdot 3}=-\frac{3}{2^2}\frac{a_0}{2\cdot 3\cdot 5\cdot 7\cdot 3}=-\frac{3a_0}{7!}.$$

So, the solution for $k=\frac{1}{2}$ is,

$$y_2=a_0x^{\frac{1}{2}}\left[1-\frac{x}{3!}-\frac{x^2}{5!}-\frac{3x^3}{7!}-\cdots\right].$$

Hence, the general solution is,

$$y=y_1+y_2=a_0(1-x)+b_0x^{\frac{1}{2}}\left(1-\frac{x}{3!}-\frac{x^2}{5!}-\frac{3x^3}{7!}-\cdots\right).$$

Example 17.7. **Roots are equal.** *Solve*: (i) $x\frac{d^2y}{dx^2}+\frac{dy}{dx}+2y=0$ (ii) $x\frac{d^2y}{dx^2}+\frac{dy}{dx}+x^2y=0.$

Solution : (i) The given equation is $x\frac{d^2y}{dx^2}+\frac{dy}{dx}+2y=0.$ (i)

Let the solution of Eqn. (i) be $y=\sum_{n=0}^{\infty}a_nx^{n+r}$.

$$\therefore \quad \frac{dy}{dx}=\sum_{n=0}^{\infty}a_n(n+r)x^{n+r-1} \quad \text{and} \quad \frac{d^2y}{dx^2}=\sum_{n=0}^{\infty}a_n(n+r)(n+r-1)x^{n+r-2}.$$

Substituting the values of y, $\frac{dy}{dx}$, $\frac{d^2y}{dx^2}$ in Eqn. (i), we get

$$\sum_{n=0}^{\infty}a_n(n+r)(n+r-1)x^{n+r-1}+\sum_{n=0}^{\infty}a_n(n+r)x^{n+r-1}+\sum_{n=0}^{\infty}2a_nx^{n+r}=0. \quad \text{(ii)}$$

Equating to zero the coefficient of lowest degree term (i.e., x^{r-1}), we get

$$r(r-1)+r=0 \quad \text{or,} \quad r^2=0, \quad \therefore r=0,0.$$

Thus, the roots of the indicial equation are equal. The recursion formula for finding a_1, a_2, ... is obtained from Eqn. (ii).

$$[(n+r)(n+r-1)+(n+r)]a_n=-2a_{n-1} \quad \text{or,} \quad (n+r)^2a_n=-2a_{n-1}.$$

Putting, $n = 1, 2, \ldots$, we get

$$(r+1)^2 a_1 = -2a_0, \quad \therefore\ a_1 = -\frac{2a_0}{(r+1)^2}$$

$$(r+2)^2 a_2 = -2a_1, \quad \therefore\ a_2 = -\frac{2a_1}{r+2} = \frac{2^2 a_0}{(r+2)^2(r+1)^2}$$

$$(r+3)^2 a_3 = -2a_2, \quad \therefore\ a_3 = -\frac{2^3 a_0}{(r+3)^2(r+2)^2(r+1)^2} \text{ and so on.}$$

Taking $a_0 = 1$, we get one solution of the equation as

$$y_1(x) = x^r \left[1 - \frac{2x}{(r+1)^2} + \frac{2^2x^2}{(r+2)^2(r+1)^2} - \frac{2^3 2x^3}{(r+3)^2(r+2)^2(r+1)^2} + \cdots\right] \quad \text{(iii)}$$

For $r = 0$,

$$y_1(x) = \left[1 - \frac{2x}{1^2} + \frac{2^2x^2}{2^2 \cdot 1^2} - \frac{2^3x^3}{3^2 \cdot 2^2 \cdot 1^2} + \cdots\right]. \quad \text{(iv)}$$

The second independent solution is $y_2(x) = \dfrac{\partial}{\partial r}[y_1(x)_r]$ for $r = 0$.

But, $y_1(x) = x^r F(r)$, where $F(r) = 1 - \dfrac{2x}{(r+1)^2} + \dfrac{2^2x^2}{(r+2)^2(r+1)^2} - \dfrac{2^3x^3}{(r+3)^2(r+2)^2(r+1)^2} - \cdots$

$\therefore \quad \log y_1 = r \log x + \log F.$

Differentiating partially w.r.t. r,

$$\frac{1}{y_1}\frac{\partial y_1}{\partial r} = r \log x + \frac{1}{F}\frac{\partial F}{\partial r}.$$

$$\therefore \quad \frac{\partial y_1}{\partial r} = y_1 \log x + \frac{y_1}{F}\frac{\partial F}{\partial r} = y_1 \log x + x^r \frac{\partial F}{\partial r} \quad [\because\ y_1(x)_r = x^r F(r)] \quad \text{(v)}$$

$$\text{Now, } \frac{\partial F}{\partial r} = \left[\frac{2x}{(r+1)^2} \cdot \frac{2}{(r+1)} - \frac{2^2x^2}{(r+2)^2(r+1)^2}\left(\frac{2}{(r+1)} + \frac{2}{(r+2)}\right)\right.$$
$$\left. + \frac{2^3x^3}{(r+1)^2(r+2)^2(r+3)^2}\left(\frac{2}{(r+1)} + \frac{2}{(r+2)} + \frac{2}{(r+3)}\right) + \cdots\right]$$

$$\therefore \quad \left(\frac{\partial F}{\partial r}\right)_{r=0} = 2\left[\frac{2x}{1^2} - \frac{2^2x^2}{1^2 \cdot 2^2}\left(1 + \frac{1}{2}\right) + \frac{2^3x^3}{1^2 \cdot 2^2 \cdot 3^2}\left(1 + \frac{1}{2} + \frac{1}{3}\right) - \cdots\right] = p(x), \text{ say.} \quad \text{(vi)}$$

From Eqn. (v), $\left(\dfrac{\partial y_1}{\partial r}\right)_{r=0} = y_1 \log x + p(x).$

Hence, the complete solution of the given differential equation is

$$y(x) = Ay_1(x) + B[y_1(x) \log x + p(x)]$$
$$= (A + B \log x)\left[1 - \frac{2x}{1^2} + \frac{2^2x^2}{1^2 \cdot 2^2} - \frac{2^3x^3}{1^2 \cdot 2^2 \cdot 3^2} + \cdots\right]$$
$$+ B\left[\frac{2x}{1^2} - \frac{2^2x^2}{1^2 \cdot 2^2}\left(1 + \frac{1}{2}\right) + \frac{2^3x^3}{1^2 \cdot 2^2 \cdot 3^2}\left(1 + \frac{1}{2} + \frac{1}{3}\right) \cdots\right].$$

(ii) The given differential equation is $x\dfrac{d^2y}{dx^2}+\dfrac{dy}{dx}+x^2y=0.$ (i)

Here, $x=0$ is a singularity. So, the solution my be taken as

$$y=\sum_{k=0}^{\infty}a_kx^{m+k}=a_0x^m+a_1x^{m+1}+a_2x^{m+2}+\cdots. \quad \text{(ii)}$$

$$\therefore \quad \frac{dy}{dx}=a_0mx^{m-1}+a_1(m+1)x^m+a_2(m+2)x^{m+1}+\cdots$$

and $$\frac{d^2y}{dx^2}=a_0m(m-1)x^{m-2}+a_1(m+1)mx^{m-1}+a_2(m+2)(m+1)x^m+\cdots.$$

Substituting the values of y, $\frac{dy}{dx}$, $\frac{d^2y}{dx^2}$ in Eqn. (ii),

$$x[m(m-1)a_0x^{m-2}+(m+1)ma_1x^{m-1}+(m+2)(m+1)a_2x^m+\cdots]$$
$$+[ma_0x^{m-1}+(m+1)a_1x^m+(m+2)a_2x^{m+1}+\cdots]$$
$$+x^2[a_0x^m+a_1x^{m+1}+a_2x^{m+2}+\cdots]=0.$$

Equating the coefficient of the lowest power of x, i.e., x^{m-1} to zero, we get

$$[m(m-1)+m]a_0=0.$$

$$\therefore \quad m^2=0 \quad \text{or,} \quad m=0,0 \quad [\because\ a_0\neq 0].$$

Now taking the coefficients of x^m, x^{m+1}, ..., we get

$(m+1)ma_1+(m+1)a_1=0,$ $\quad \therefore\ \boldsymbol{a_1=0}$

$(m+2)(m+1)a_2+(m+2)a_2=0,$ $\quad \therefore\ \boldsymbol{a_2=0}$

$(m+3)(m+2)a_3+(m+3)a_3+a_0=0,$ $\quad \therefore\ \boldsymbol{a_3=-\dfrac{a_0}{(m+3)^2}}$

$(m+4)(m+3)a_4+(m+4)a_4+a_1=0,$ $\quad \therefore\ \boldsymbol{a_4=0}$

$(m+5)(m+4)a_5+(m+5)a_5+a_2=0,$ $\quad \therefore\ \boldsymbol{a_5=0}$

$(m+6)(m+5)a_6+(m+6)a_6+a_3=0,$ $\quad \therefore\ \boldsymbol{a_6=(-1)^3\dfrac{a_0}{(m+3)^2(m+6)^2}}$ **and so on.**

Now, for $m=0$, $a_0\neq 0$, $a_1=0$, $a_2=0$, $a_3=-\dfrac{a_0}{3^2}$, $a_4=0$, $a_5=0$, $a_6=(-1)^3\dfrac{a_0}{3^2\cdot 6^2}$.

$\therefore$ one solution of the given differential equation is $\boldsymbol{y_1=a_0\left[1-\dfrac{x^3}{3^2}+\dfrac{x^6}{3^2\cdot 6^2}-\cdots\right]}$.

The second solution is $y_2=\left(\dfrac{\partial y_1}{\partial m}\right)_{m=0}$.

$$\text{Now,}\ \frac{\partial y_1}{\partial m}=\frac{\partial}{\partial m}\left[a_0x^m-a_0\frac{x^{m+3}}{(m+3)^2}+a_0\frac{x^{m+6}}{(m+3)^2(m+6)^2}-\cdots\right]$$

$$=\frac{\partial}{\partial m}\left[a_0x^m\left\{1-\frac{x^3}{(m+3)^2}+\frac{x^6}{(m+3)^2(m+6)^2}-\cdots\right\}\right]$$

$$=\frac{\partial}{\partial m}[a_0x^mA],\quad \text{where } A=\left\{1-\frac{x^3}{(m+3)^2}+\frac{x^6}{(m+3)^2(m+6)^2}-\cdots\right\}.$$

Let $y_1 = a_0 x^m A$.

$$\therefore \quad \log y_1 = \log a_0 + m \log x + \log A.$$

Differentially partially w.r.t. m, we get

$$\frac{1}{y_1}\frac{\partial y_1}{\partial m} = \log x + \frac{1}{A}\left(\frac{\partial A}{\partial m}\right).$$

$$\therefore \quad \frac{\partial y_1}{\partial m} = y_1 \log x + \frac{y_1}{A}\left(\frac{\partial A}{\partial m}\right)$$

$$= y_1 \log x + a_0 x^m \frac{\partial}{\partial m}\left[1 - \frac{x^2}{(m+3)^2} + \frac{x^6}{(m+3)^2(m+6)^2} - \cdots\right]$$

$$= y_1 \log x + a_0 x^m \left[\frac{x^3}{(m+3)^2}\left(\frac{2}{m+3}\right) - \frac{x^6}{(m+3)^2(m+6)^2}\left(\frac{2}{m+3} + \frac{2}{m+6}\right) + \cdots\right]$$

$$\therefore \quad \left(\frac{\partial y_1}{\partial m}\right)_{m=0} = y_1 \log x + 2a_0\left[\frac{x^3}{3^3} - \frac{x^6}{3^3 \cdot 6^3}\left(1 + \frac{1}{2}\right) + \frac{x^9}{3^3 \cdot 6^3 \cdot 9^3}\left(1 + \frac{1}{2} + \frac{1}{3}\right) - \cdots\right]$$

$$= y_2 \text{ (say)}.$$

Hence, the complete solution is

$$\boldsymbol{y = Ay_1 + By_2 = Ay_1 + B\left(\frac{\partial y_1}{\partial m}\right)_{m=0}}$$

$$\boldsymbol{= A\left[1 - \frac{x^3}{3^3} + \frac{x^6}{3^2 \cdot 6^2} - \cdots\right] + B \log x \left[1 - \frac{x^3}{3^3} + \frac{x^6}{3^2 \cdot 6^2} - \cdots\right]}$$

$$\boldsymbol{+ 2B\left[\frac{x^3}{3^3} - \frac{x^6}{3^3 \cdot 6^3}\left(1 + \frac{1}{2}\right) + \frac{x^9}{3^3 \cdot 6^3 \cdot 9^3}\left(1 + \frac{1}{2} + \frac{1}{3}\right) - \cdots\right]}$$

$$\boldsymbol{= (A + B \log x)\left[1 - \frac{x^3}{3^2} + \frac{x^6}{3^4 \cdot (2!)^2} - \cdots\right]}$$

$$\boldsymbol{+ 2B\left[\frac{x^3}{3^3} - \frac{x^6}{3^6 \cdot (2!)^3}\left(1 + \frac{1}{2}\right) + \frac{x^9}{3^9 \cdot (3!)^3}\left(1 + \frac{1}{2} + \frac{1}{3}\right) - \cdots\right].}$$

Example 17.8. *Roots differing by an integer.*

(i) *Solve in series* $x\frac{d^2y}{dx^2} + 2\frac{dy}{dx} + xy = 0$ **[UPTU 2003]**

(ii) $x^2\frac{d^2y}{dx^2} + 4x\frac{dy}{dx} + (x^2 + 2)y = 0.$ **[AMIETE(Summer) 1998]**

Solution : (i) The given differential equation is $x\frac{d^2y}{dx^2} + 2\frac{dy}{dx} + xy = 0.$ (i)

Since the coefficient of $\frac{d^2y}{dx^2}$ is zero at $x = 0$, $x = 0$ is a singular point.

The solution of Eqn. (i) is of the form,

$$y = a_0 x^m + a_1 x^{m+1} + a_2 x^{m+2} + a_3 x^{m+3} + \cdots = \sum_{k=0}^{\infty} a_k x^{m+k}. \quad \text{(ii)}$$

$$\therefore \quad \frac{dy}{dx} = ma_0 x^{m-1} + (m+1)a_4 x^m + (m+2)a_2 x^{m+1} + (m+3)a_3 x^{m+2} + \cdots$$

and $\dfrac{d^2y}{dx^2} = m(m-1)a_0x^{m-2} + (m+1)ma_1x^{m-1} + (m+2)(m+1)a_2x^m + \cdots.$

Substituting y, $\frac{dy}{dx}$ and $\frac{d^2y}{dx^2}$ in Eqn. (i), we get

$$x[m(m-1)a_0x^{m-2} + (m+1)ma_1x^{m-1} + (m+2)(m+1)a_2x^m + \cdots]$$
$$+ 2[ma_0x^{m-1} + (m+1)a_1x^m + (m+2)a_2x^{m+1} + (m+3)a_3x^{m+2} + \cdots]$$
$$+ x[a_0x^m + a_1x^{m+1} + a_2x^{m+2} + a_3x^{m+3} + \cdots] = 0.$$

Equating to zero, the coefficient of lowest power of x, we get

$$m(m-1)a_0 + 2ma_0 = 0 \quad \text{or,} \quad m(m-1) + 2m = 0 \quad [\because a_0 \neq 0]$$

$\therefore \quad m = 0, \ -1.$

Thus, the roots are distinct and differ by an integer.

Equating the coefficient of x^m to zero, we get

$$(m+1)ma_1 + 2(m+1)a_1 = 0 \quad \text{or,} \quad (m+1)(m+2)a_1 = 0$$

or, $\quad (m+1)a_1 = 0 \quad [\because m+1 \neq 0].$

When $m = -1$, $a_1 = \dfrac{0}{0}$, i.e., indeterminate.

Hence, the complete solution will be obtained by putting $m = -1$ in y itself (which contains arbitrary constants a_0 and a_1).

Equating to zero the coefficient of x^{m+1}, x^{m+2}, x^{m+3}, x^{m+4}, etc. successively, we get

$$(m+2)(m+3)a_2 + a_0 = 0, \quad (m+3)(m+4)a_3 + a_1 = 0, \quad (m+4)(m+5)a_4 + a_2 = 0$$
$$(m+5)(m+6)a_5 + a_3 = 0 \text{ and so on.}$$

$$\therefore \quad a_2 = -\frac{a_0}{(m+2)(m+3)}, \quad a_3 = -\frac{a_1}{(m+3)(m+4)}, \quad a_4 = \frac{a_0}{(m+2)(m+3)(m+4)(m+5)},$$

$$a_5 = \frac{a_1}{(m+3)(m+4)(m+5)(m+6)} \text{ and so on.}$$

Substituting these values in Eqn. (ii), we get

$$y = x^m \left[a_0 + a_1x - \frac{a_0}{(m+2)(m+3)}x^2 - \frac{a_1}{(m+3)(m+4)}x^3 + \frac{a_0}{(m+2)(m+3)(m+4)(m+5)}x^4 + \frac{a_1}{(m+3)(m+4)(m+5)(m+6)}x^5 - \cdots\right].$$

Putting $m = -1$, the complete solution is

$$y = x^{-1}\left[a_0\left(1 - \frac{x^2}{1\cdot 2} + \frac{x^4}{1\cdot 2\cdot 3\cdot 4} - \cdots\right) + a_1\left(x - \frac{x^3}{1\cdot 2\cdot 3} + \frac{x^5}{1\cdot 2\cdot 3\cdot 4\cdot 5} - \cdots\right)\right]$$
$$= x^{-1}(a_0 \cos x + a_1 \sin x).$$

(ii) The given differential equation is, $x^2\dfrac{d^2y}{dx^2} + 4x\dfrac{dy}{dx} + (x^2+2)y = 0.$ (i)

The solution of Eqn. (i) may be written in the form, $y = \displaystyle\sum_{k=0}^{\infty} a_kx^{m+k}.$

Then $\frac{dy}{dx} = \sum_k a_k(m+k)x^{m+k-1}$ and $\frac{d^2y}{dx^2} = \sum_k a_k(m+k)(m+k-1)x^{m+k-2}$.

Substituting the values of y, $\frac{dy}{dx}$, $\frac{d^2y}{dx^2}$ in Eqn. (i), we get

$$x^2\sum_k a_k(m+k)(m+k-1)x^{m+k-2} + 4x\sum_k a_k(m+k)x^{m+k-1} + (x^2+2)\sum_k a_k x^{m+k} = 0$$

or, $$\sum_k a_k(m+k)(m+k-1)x^{m+k} + 4\sum_k a_k(m+k)x^{m+k} + \sum_k a_k x^{m+k+2} + \sum_k 2a_k x^{m+k} = 0$$

or, $$\sum_k a_k[(m+k)(m+k-1) + 4(m+k) + 2]x^{m+k} + \sum_k a_k x^{m+k+2} = 0$$

or, $$\sum_k a_k[(m+k)^2 + 3(m+k) + 2]x^{m+k} + \sum_k a_k x^{m+k+2} = 0.$$

The coefficient of lowest degree term x^m is obtained by putting $k = 0$ in the first term only and then equating to zero, we get,

$$a_0(m^2+3m+2) = 0$$

or, $m^2 + 3m + 2 = 0 \quad \because \quad a_0 \neq 0$

or, $(m+1)(m+2) = 0$ i.e., $m = -1, -2$.

Putting $k = 1$ in the first term only and then equating the coefficient of x^{m+1} to zero,

$$a_1[m^2+5m+6] = 0 \text{ or, } a_1(m+2)(m+3) = 0, \quad \because m = -2,\ m \neq -3,\ a_1 \neq 0.$$

Equating the coefficient of x^{m+k+2},

$$a_{k+2}[m^2 + (2k+4+3)m + (k+2)^2 + 3(k+2) + 2] + a_k = 0$$

or, $$a_{k+2}[m^2 + (2k+7)m + k^2 + 7k + 12] + a_k = 0.$$

$$\therefore \quad a_{k+2} = -\frac{1}{m^2 + (2k+7)m + k^2 + 7k + 12}a_k.$$

For, $k = 0$, $a_2 = -\frac{a_0}{m^2+7m+12} = -\frac{a_0}{(m+3)(m+4)}$

$k = 1$, $a_3 = -\frac{a_1}{m^2+9m+20} = -\frac{a_1}{(m+4)(m+5)}$

$k = 2$, $a_4 = -\frac{a_2}{m^2+11m+30} = \frac{a_0}{(m+3)(m+4)(m+5)(m+6)}$

$k = 3$, $a_5 = -\frac{a_3}{m^2+13m+42} = \frac{a_0}{(m+4)(m+5)(m+6)(m+7)}$.

For $m = -1$, $a_2 = -\frac{a_0}{6}$, $a_3 = -\frac{a_1}{12}$, $a_4 = \frac{a_0}{120}$, $a_5 = \frac{a_5}{360}$.

Hence, for $m = -1$,

$$y_1 = x^{-1}\left[1 - \frac{x^2}{6} + \frac{x^4}{120} + \cdots\right]a_0 + \left[1 - \frac{x^2}{12} + \frac{x^4}{360} + \cdots\right]a_1.$$

For $m = -2$,

$$y_2 = x^{-2}\left[1 - \frac{x^2}{2} + \frac{x^4}{24} + \cdots\right]a_0 + \left[\frac{1}{x} - \frac{x}{6} + \frac{x^3}{120} + \cdots\right]a_1$$

$$= x^{-2}\left[\left\{1 - \frac{x^2}{2!} + \frac{x^4}{4!} + \cdots\right\}a_0 + \left\{x - \frac{x^3}{3!} + \frac{x^5}{5!} + \cdots\right\}a_1\right]$$

$$= x^{-2}[a_0 \cos x + a_1 \sin x].$$

Exercises

17(A) : Objective Type Questions

Select the correct answer

17.1 Power series solution is applicable to differential equations which are

(a) second order of degree n (b) partial differential equations
(c) linear homogeneous (d) none of these.

17.2 For the differential equations $x^2\dfrac{d^2y}{dx^2} + (x^2 - x)\dfrac{dy}{dx} + 2y = 0$, power series method

(a) may be applied (b) may not be applied (c) none of these.

17.3 Consider the differential equation, $P_0(x)\dfrac{d^2y}{dx^2} + P_1(x)\dfrac{dy}{dx} + P_2y = 0$. If $P_0(a) \neq 0$, then $x = a$ is called

(a) an ordinary point (b) a singular point (c) a regular singularity (d) none of these.

17.4 If $P_0(a) = 0$, in equation of 17.3, then $x = a$ is called

(a) an ordinary point (b) a singular point (c) a regular singularity (d) none of these.

17.5 If the differential equation in 17.3 has singularity at $x = a$, be expressed as

$$\frac{d^2y}{dx^2} + \frac{Q_1(x)}{(x-a)}\frac{dy}{dx} + \frac{Q_2(x)}{(x-a)^2}y = 0,$$

where $Q_1(x)$ and $Q_2(x)$ have derivatives of all orders about the point $x = a$, then the point $x = a$ is called

(a) an irregular singular (b) a regular singular (c) ordinary (d) none of these.

17.6 A function $f(x)$ is said to be analytic or regular, if

$$\sum_n \frac{f^n(a)}{n!}(x-a)^n$$

(a) exists and converges to $f(x)$ in some open interval $b < a < c$

(b) does not exist but converges to $f(x)$ in $b < a < c$

(c) exists but does not converge to $f(x)$ in $b < a < c$

(d) none of these.

17.7 For the differential equation $(1-x^2)\frac{d^2y}{dx^2} - 2x\frac{dy}{dx} + 2y = 0$, $x = 0$ is

(a) a singular point (b) an ordinary point
(c) a regular singular point (d) an irregular singular point.

17.8 The number of singularity of equation, $\frac{d^2y}{dx^2} + \frac{10}{3x(2-x)}\frac{dy}{dx} + \frac{11}{4x}y = 0$ is equal to

(a) two (at $x = 0, x = 2$) (b) one at $x = 0$ (b) one at $x = 2$ (d) none of these.

17.9 The series solution of a differential equation having no singularity can be written as

(a) $y = x^m \sum_{k=0}^{\infty} a_k x^k$ (b) $y = \sum_{k=0}^{\infty} a_k x^k$ (c) $y = a^m \sum_{k=0}^{\infty} x^{k+m}$ (d) none of these.

17.10 The series solution of a differential equation having point of singularity can be written as

(a) $y = x^m \sum_{k=0}^{\infty} a_k x^k$ (b) $y = \sum_{k=0}^{\infty} a_k x^k$ (c) $y = \sum_{k=0}^{\infty} x^{k+m}$ (d) none of these.

Answers

17.1 (c) **17.2** (b) **17.3** (a) **17.4** (b) **17.5** (b) **17.6** (a) **17.7** (b) **17.8** (a) **17.9** (a) **17.10** (a)

17(B) : Broad Answer Type Questions and Problems

Solve the following differential equations series :

17.1 $(x+1)\frac{dy}{dx} - (x+2)y = 0$ **[Ans:** $y = a_0\left(1 + 2x + \frac{3}{2}x^2 + \frac{2}{3}x^3 + \frac{5}{24}x^4 + \cdots\right)$**]**

17.2 $(1-x)^2\frac{d^2y}{dx^2} - x\frac{dy}{dx} + 2y = 0$

[UPTU 2004] **[Ans:** $y = a_0\left(1 - x^2 + \frac{1}{3}x^4 - \frac{1}{5}x^6 + \cdots\right) + a_1 x$**]**

17.3 $(1+x)^2\frac{d^2y}{dx^2} + x\frac{dy}{dx} - y = 0$

[DU 2002] **[Ans:** $y = a_0\left(1 + \frac{x^2}{2} - \frac{x^4}{8} + \frac{x^6}{16} - \frac{5x^8}{128} + \cdots\right) + a_1 x$**]**

17.4 (a) Write down the general form of a linear homogeneous, second order ordinary differential equation. When do you expect to get the solution of such an equation by the method of series substitution?

(b) Solve the equation $(1-x)^2\frac{d^2y}{dx^2} - 2x\frac{dy}{dx} + 2y = 0$ by the method of series substitution. **[BU(H) 2003]**

17.5 Solve the equation $x(1-x)\frac{d^2y}{dx^2} - 3\frac{dy}{dx} + 2y = 0$ in power series. **[BU(H) 2005]**

[Ans: $y = a_0\left(1 - \frac{2}{3}x + \frac{1}{3}x^2 + \cdots\right) + a_1 x^4(1 - 2x + 3x^2 - 4x^3 + \cdots)$**]**

17.6 Solve : $(1-x^2)y'' + 2y = 0$ in series, given $y(0) = 4$, $y'(0) = 5$. **[PTU 2006]**

17.7 Solve : $(x+1)y' - (2x+3)y = 0$.

[PTU 2006] **[Ans:** $y = a_0\left[1 + 3x + 4x^2 + \frac{10}{3}x^3 + 2x^4 \cdots\right]$**]**

17.8 $(x+1)\dfrac{dy}{dx}-(x+2)y=0$ **[Ans:** $y=a_0\left(1+2x+\frac{3}{2}x^2+\frac{2}{3}x^3+\frac{5}{24}x^4+\cdots\right)$**]**

17.9 Solve : $\dfrac{d^2y}{dx^2}+x\dfrac{dy}{dx}+y=0.$

[Ans: $y=a_0\left(1-\frac{x^2}{2}+\frac{x^2}{2\cdot4}-\frac{x^6}{2\cdot4\cdot6}+\cdots\right)+a_1(x-\frac{x^3}{3}+\frac{x^5}{3\cdot5}-\frac{x^7}{3\cdot5\cdot7}+\cdots)$**]**

17.10 Solve the following equation in power series : $x\dfrac{d^2y}{dx^2}+\dfrac{dy}{dx}-y=0.$

[Ans: $y=(C_1+C_2\log x)\left[1+x+\frac{x^2}{(2!)^2}+\frac{x^3}{(3!)^3}+\cdots\right]$

$-2C_2\left[x+\frac{1}{(2!)^2}\left(1+\frac{1}{2}\right)x^2+\frac{1}{(3!)^2}\left(1+\frac{1}{2}+\frac{1}{3}\right)x^3+\cdots\right]$**]**

17.11 Solve in power series method : $x\dfrac{d^2y}{dx^2}+(1+x)\dfrac{dy}{dx}+2y=0.$ **[JNTU 2006]**

[Ans: $y=a_0\left(1-2x+\frac{3}{2!}x^2-\frac{4}{3!}x^3+\cdots\right)+a_1\left[y_1\log x+a_0\left(3x-\frac{13}{4}x^2+\cdots\right)\right]$**]**

17.12 Solve : $x\dfrac{d^2y}{dx^2}+\dfrac{dy}{dx}+xy=0.$

[Ans: $y=(A+B\log x)\left[1-\frac{x^2}{2^2}+\frac{x^2}{2^2.4^2}-\frac{x^6}{2^2.4^2.6^2}+\cdots\right]$

$+B\left[\frac{x^2}{2^2}-\frac{x^4}{2^2.4^2}\left(1+\frac{1}{2}\right)+\frac{x^6}{2^2.4^2.6^2}\left(1+\frac{1}{2}+\frac{1}{3}\right)-\cdots\right]$**]**

17.13 Solve in power series method : $x(1-2x)\dfrac{d^2y}{dx^2}-(6x+1)\dfrac{dy}{dx}-2y=0.$

17.14 By Frobenius method, one solution of $xy''-(1+2x)y'+(1+x)y=0$ is obtained as $y=Ae^x$. Find the second solution. **[Ans:** $y_2=xe^x$**]**

17.15 **(a)** What do you mean by (i) ordinary point (i) regular and irregular singular points of a homogeneous linear second order differential equation?

(b) Obtain the complete series solution of the following differential equation :

$$4xy''+2(1-x)y'-y=0.$$ **[BU(H) 2001]**

17.16 Solve by the Frobenius method of the equation :

$$x(x-1)y''+(3x-1)y'+y=0.$$ **[BU(H) 1994]**

17.17 Obtain the series solution of the differential equation

$$x(1-x)\frac{d^2y}{dx^2}-(1+3x)\frac{dy}{dx}+y=0.$$ **[BU(H) 2000]**

17.18 Obtain the series solution of the following equation by Frobenius method :

$$2x(1-x)\frac{d^2y}{dx^2}+(1-x)\frac{dy}{dx}+3y=0.$$ **[BU(H) 1993]**

17.19 **(a)** Apply Frobenius method to the equation $\dfrac{d^2y}{dx^2}+\omega^2y=0$ setting $\sum_{\lambda=0}^{\infty}a_\lambda x^{k+\lambda}$ with $a_0\neq0$.

(b) Verify that the indicial equation is $k(k-1)=0$.

(c) Find the recurrence relation.

(d) Using the recurrence relation, show that $k = 1$ leads to the solution

$$y(x) = \frac{a_0}{\omega} \sin \omega x. \qquad \textbf{[CU(H) 2014]}$$

17.20 Solve the equation $x^2 \frac{d^2y}{dx^2} + x\frac{dy}{dx} - a^2y = 0$ by the Frobenius method. **[CU(H) 2014]**

17.21 (a) Consider the equation $x^2 \frac{d^2y}{dx^2} + \omega^2 y = 0$. Since zero is a ordinary point dx of this differential equation, obtain a power series solution around zero.

(b) Consider the equation $x^2 \frac{d^2y}{dx^2} + \omega^2 y = 0$. Since zero is a regular singular point, assume a solution $y = \sum a_i \cdot x^{i+\lambda}$. Show that the possible solutions of λ are -2 and 3. Show that if we choose $\lambda = 3$ only $a_0 = 0$ and if we choose $\lambda = -2$ then only a_0 and a_5 are non zero. Hence, show that $y = C_1x^3 + C_2x^{-2}$ is a solution. **[CU(H) 2011]**

17.22 (a) Use Frobenius method to solve the differential equation for Hermite polynomial around $x = 0$.

$$\frac{d^2y}{dx^2} - 2x\frac{dy}{dx} + 2xy = 0, \; n \text{ is a non-negative integer.} \qquad \textbf{[CU(H) 2010]}$$

(b) What is the nature of singularity of the following differential equation $y'' - \frac{6}{x^2}y = 0$.

Solve by Frobenius method : $\ddot{x} + \omega^2 x = 0$ **[VU(H) 2014]**

17.23 (a) Solve the differential equation $y'' - 3y' + 2y + x^3 = 0$

(b) A differential equation is given by $(1 - x^2)\frac{d^2y}{dx^2} - 2x\frac{dy}{dx} + \lambda y = 0$

(i) Find the singular points.

(ii) Check whether the singular points are essential or non-essential.

(iii) Comment whether series solution of this equation is possible or not. **[VU(H) 2011]**

17.24 (a) Use Frobenius method to solve $2x^2\frac{d^2y}{dx^2} + 3x\frac{dy}{dx} + (x - 1)y = 0$ at $x = 0$. **[VU(H) 2007]**

(b) Solve the differential equation $(1 + x^2)y^n + xy' - y = 0$.

(c) Define ordinary and singular points of a differential equation. Locate and classify the singular points of the following equation

$$(1 - x^2)\frac{d^2y}{dx^2} - 2x\frac{dy}{dx} + 2y = 0 \qquad \textbf{[NBU(H) 2008]}$$

17.25 (a) What is meant by the words "homogeneous" and linear in connection with second order differential equations?

(b) Find two independent power series solutions of the following equation around the origin : $y'' - 3y' + 2y = 0$. **[BU(H) 2013]**

17.26 (a) Find whether $x = 0$ is an ordinary point, a regular singular point or an irregular singular point of the differential equation $x^2\frac{d^2y}{dx^2} + x\frac{dy}{dx} + y = 0$.

(b) Let $y_1(x)$ and $y_2(x)$ be two linearly independent solutions of the differential equation $\frac{d^2y}{dx^2} + 3\frac{dy}{dx} + f(x)y = 0$ and let $W(x) = y_1\frac{dy_2(x)}{dx} - y_2(x)\frac{dy_1(x)}{dx}$.

If $W(0) = 1$, find the value of $W(1)$.

(c) Solve the following differential equation by the method of power series method :

$$\frac{d^2y}{dx^2} + \omega^2 y = 0.$$ **[BU(H) 2012]**

17.27 (a) Solve the differential equation $4x\frac{d^2y}{dx^2} + 2\frac{dy}{dx} + y = 0$ in power series. **[BU(H) 2010]**

(b) The functions $y_1 = \sin x$ and $y_2 = \cos x$ are both solutions of the equation $y^2 + y = 0$. Evaluate the Wronskian of these two solutions and hence show that they are linearly independent. **[BU(H) 2009]**

17.28 (a) What is meant by singular points of a second order differential equation?

(b) Solve the following differential equation by Frobenius power series method :

$$y'' + \omega^2 y = 0.$$

17.29 (a) Obtain the general solution of $(1 - x^2)y'' - 2xy' + n(n+1)y = 0$ about an ordinary point using power series method. **[Gujarat U 2017]**

(b) Solve Hermite differential equation by Frobenius method $y'' - 2xy' + 2xy = 0$: given n is an integer.

(c) Discuss whether each of the points $x = +1, 0, -1$ is ordinary regular singular point or regular singular point of the differential equation

$$(1 - x^2)y'' + x(1 - x)y' + (1 + x)y = 0.$$

17.30 (a) Apply Frobenius method to solve the equation :

$$\frac{d^2y}{dx^2} + y = 0, \text{ setting } y(x) = \sum_{\lambda=0}^{\infty} a_\lambda x^{k+\lambda} \text{ with } a_0 = 0.$$

(i) Verify that the indicial equation is $k(k - 1) = 0$.

(ii) For $k = 1$, show that a_1 is necessarily zero.

(iii) Find the recurrence relation.

(iv) Hence, show that $k = 1$, the solution is $y = a_0 \sin x$. **[CU(H) 2017]**

(b) Solve the equation $(1 - x^2)\frac{d^2y}{dx^2} - 2x\frac{dy}{dx} + 6y = 0$ around the point $x = 0$. **[CU(H) 2008]**

17.31 (a) Consider the equation $\frac{d^2y}{dx^2} + \omega^2 y = 0$. Since zero is an ordinary point dx of this differential equation, obtain power series solution.

(b) Consider the equation $x^2\frac{d^2y}{dx^2} - 6y = 0$. Since zero is a regular singular point, assume a solution $y = \sum a_i x^{i+\lambda}$. Show that the possible solutions of λ are -2 and 3. Show that if we choose $\lambda = 3$ only, $a_0 = 0$ and if we choose $\lambda = -2$, then only a_0 and a_5 are non-zero. Show that $y = C_1x^3 + C_2x^{-2}$ is a solution.

CHAPTER 18

Special Functions

18.1 Introduction

Special functions like Bessel function, Legendre function are the solutions of Bessel equation and Legendre equation respectively which have important applications in many physical problems. For example, Bessel function has wide applications in physical problems involving vibration of circular membrane, heat conduction in cylindrical regions, diffraction at a circular aperture, stability of struts, etc. Legendre function finds important application in the solution of Laplace's equation in spherical polar coordinates, etc.

18.2 Legendre's Differential Equation

The differential equation is of the form

$$(1-x^2)\frac{d^2y}{dx^2} - 2x\frac{dy}{dx} + n(n+1)y = 0, \tag{18.1}$$

where n is a real constant, is known as **Legendre's differential equation** named after the French mathematician **Adrien-Marie Legendre** (1752–1833).

In most physical problems only the integral values of n are required.

Eqn. (18.1) can also be expressed in the form,

$$\frac{d}{dx}\left[(1-x^2)\frac{dy}{dx}\right] + n(n+1)y = 0. \tag{18.2}$$

The solution of Eqn. (18.1) is known as **Legendre function**.

18.3 General Solution

Eqn. (18.1) can be solved (i.e., integrated) in series of ascending or descending powers of x. Eqn. (18.1) has no singularity at $x = 0$ and hence a power series solution about $x = 0$ can be developed.

Let the series in descending power of x be $y = \sum_{r=0}^{\infty} a_r x^{m-r}$.

$$\therefore \quad \frac{dy}{dx} = \sum_{r=0}^{\infty} a_r(m-r)x^{m-r-1} \quad \text{and} \quad \frac{d^2y}{dx^2} = \sum_{r=0}^{\infty} a_r(m-r-1)x^{m-r-2}.$$

Substituting these values in eqn. (18.1), we get

$$(1-x^2)\sum_{r=0}^{\infty} a_r(m-r)(m-r-1)x^{m-r-2} - 2x\sum_{r=0}^{\infty} a_r(m-r)x^{m-r-1} + n(n+1)\sum_{r=0}^{\infty} x^{m-r}a_r = 0$$

$$\text{or,} \quad \sum_{r=0}^{\infty} a_r(m-r)(m-r-1)x^{m-r-2} + [n(n+1) - 2(m-r) - (m-r)(m-r-1)]x^{m-r}a_r = 0$$

or, $$\sum_{r=0}^{\infty}[(m-r)(m-r-1)x^{m-r-2}+\{n(n+1)-(m-r)(m-r+1)\}x^{m-r}]a_r=0. \quad (18.3)$$

Since eqn. (18.3) is an identity, the coefficients of various powers of x must be zero.

Equating the coefficient of the highest power of x, i.e., x^m $(r=0)$ to zero, we get from Eqn. (18.3),

$$a_0[n(n+1)-m(m+1)]=0 \quad (18.4)$$

or, $n(n+1)-m(m+1)=0 \quad [\because\ a_0 \neq 0]$

or, $(n^2-m^2)+(n-m)=0$ or, $(n-m)(n+m+1)=0.$

$\therefore \quad m=n$ or, $m=-n-1.$

Eqn. (18.4) determines the index and hence called **indicial equation.**

Next, equating to zero, the coefficient of x^{m-1} by putting $r=1$,

$a_1[n(n+1)-(m-1)m]=0$ or, $a_1[(m+n)(m-n-1)]=0.$

$$\therefore \quad a_1=0 \quad [\because\ (m+n)(m-n-1)\neq 0] \quad (18.5)$$

Now, equating the coefficients of x^{m-r-2} to zero, we get

$$(m-r)(m-r-1)a_r+[n(n+1)-(m-r-2)(m-r-1)]a_{r+2}=0.$$

But, $$\begin{aligned} n(n+1)-(m-r-2)(m-r-1) &= n^2+n-(m-r-1-1)(m-r-1) \\ &= -[(m-r-1)^2-(m-r-1)-n^2-n] \\ &= -[(m-r-1+n)(m-r-1-n)-(m-r-1+n)] \\ &= -[(m-r-1+n)(m-r-1-n-1)] \\ &= (m-r+n-1)(m-r+n-2) \end{aligned}$$

or, $(m-r)(m-r-1)a_r-(m-r)-(m-r+n-1)(m-r-n-2)a_{r+2}=0$

$$\therefore \quad a_{r+2}=-\frac{(m-r)(m-r-1)}{(m-r+n-1)(m-r+n-2)}a_r. \quad (18.6)$$

But $a_1=0$,

$$\therefore \quad \boldsymbol{a_3=a_5=a_7=\cdots=0.} \quad (18.7)$$

For two values of m viz., $m=n$ and $m=-n-1$, the following two cases arise :

Case I : When $m=n$

$$a_{r+2}=-\frac{(n-r)(n-r-1)}{(2n-r-1)(r+2)}$$

so that, $$a_2=-\frac{n(n-1)}{(2n-1)\cdot 2}a_0,$$

$$a_4=-\frac{(n-2)(n-3)}{(2n-3)}a_2=\frac{n(n-1)(n-2)(n-3)}{(2n-1)(2n-3)\cdot 2\cdot 4}a_0 \text{ and so on.}$$

Hence, the series in Eqn. (18.2) becomes

$$\boldsymbol{y=a_0\left[x^n-\frac{n(n-1)}{(2n-1)\cdot 2}x^{n-2}+\frac{n(n-1)(n-2)(n-3)}{(2n-1)(2n-3)\cdot 2\cdot 4}x^{n-4}-\cdots\right].} \quad (18.8)$$

Case II : When $m = -(n+1)$

$$a_{r+2} = \frac{(n+r+1)(n+r+2)}{(r+2)(2n+r+3)} a_r.$$

$$\therefore \quad a_2 = \frac{(n+1)(n+2)}{2(2n+3)} a_0,$$

$$a_4 = \frac{(n+3)(n+4)}{4(2n+5)} a_2 = \frac{(n+1)(n+2)(n+3)(n+4)}{2 \cdot 4 \cdot (2n+3)(2n+5)} a_0 \text{ and so on.}$$

$\therefore$ the series of Eqn. (18.2) becomes

$$y = a_0 \left[x^{-n-1} + \frac{(n+1)(n+2)}{2(2n+3)} x^{-n-3} + \frac{(n+1)(n+2)(n+3)(n+4)}{2 \cdot 4 \cdot (2n+3)(2n+5)} x^{-n-5} + \cdots \right]. \quad (18.9)$$

This is another solution of Eqn. (18.1).

18.4 Legendre Polynomial $P_n(x)$

The solution of the Legendre's equations $(1-x^2)\frac{d^2y}{dx^2} - 2x\frac{dy}{dx} + n(n+1)y = 0$ in series of descending power of x is

$$y = a_0 \left[x^n - \frac{n(n-1)}{(2n-1)2} x^{n-2} + \frac{n(n-1)(n-2)(n-3)}{(2n-1)(2n-3) \cdot 2 \cdot 4} x^{n-4} - \cdots \right],$$

where a_0 is an arbitrary constant.

If now n is a +ve integer and $a_0 = \frac{1 \cdot 3 \cdot 5 \cdots (2n-1)}{n!}$, then the above solution is $P_n(x)$, where

$$P_n(x) = \frac{1 \cdot 3 \cdot 5 \cdots (2n-1)}{n!} \left[x^n - \frac{n(n-1)}{(2n-1)2} x^{n-2} + \cdots \right]. \quad (18.10)$$

$P_n(x)$ is a **terminating series** containing $\left(\frac{n}{2}+1\right)$ terms when n is even and $\frac{1}{2}(n+1)$ terms when n is odd. $P_n(x)$ is called the **Legendre's function of the first kind**.

If, however, a_0 of the other solution (18.9) be, $a_0 = \frac{n!}{1 \cdot 3 \cdot 5 \cdots (2n+1)}$, then the solution (18.9) of the Legendre's differential equation is $y = Q_n(x)$, where

$$Q_n(x) = \frac{n!}{1 \cdot 3 \cdot 5 \cdots (2n+1)} \left[x^{-n-1} + \frac{(n+1)(n+2)}{2(2n+3)} x^{-n-3} + \cdots \right]. \quad (18.11)$$

$Q_n(x)$ is **non-terminating** series and is called the **Legendre function of the second kind.**

18.5 General Solution of Legendre's Equation

The general solution of Eqn. (18.1) in terms of $P_n(x)$ and $Q_n(x)$, the two independent solutions of the same equation is

$$y = AP_n(x) + BQ_n(x),$$

where A and B are two arbitrary constants.

Note:

(i) If n is a positive integer, the general solution consists of a **polynomial solution** when n is even and an **infinite series solution** when n is an odd integer.

(ii) If a_0 or a_1 or the polynomial solution be so chosen that the value of the polynomial is 1 for $x = 1$, then these are called the **Legendre's polynomials of order n.**

18.6 Rodrigue's Formula for Legendre Polynomials

Rodrigue's formula for Legendre polynomial is given by

$$P_n(x) = \frac{1}{2^n n!}\frac{d^n}{dx^n}(x^2 - 1)^n. \tag{18.12}$$

Proof : Let $v = (x^2 - 1)^n$. (18.13)

Then $\dfrac{dv}{dx} = n(x^2 - 1)^{n-1}(2x)$.

Multiplying both sides by $(x^2 - 1)$, we get

$$(x^2 - 1)\frac{dv}{dx} = 2n(x^2 - 1)^n x$$

or, $$(x^2 - 1)\frac{dv}{dx} = 2nvx. \tag{18.14}$$

Differentiating Eqn. (18.14) $(n + 1)$ times by **Leibnitz theorem**, we get

$$(x^2 - 1)\frac{d^{n+2}v}{dx^{n+2}} + {}^{n+1}C_1(2x)\frac{d^{n+1}v}{dx^{n+1}} + {}^{n+1}C_2(2)\frac{d^n v}{dx^n} = 2n\left[x\frac{d^{n+1}v}{dx^{n+1}} + {}^{n+1}C_1(1)\frac{d^n v}{dx^n}\right]$$

or, $$(x^2 - 1)\frac{d^{n+2}v}{dx^{n+2}} + 2x\left[{}^{n+1}C_1 - n\right]\frac{d^{n+1}v}{dx^{n+1}} + 2\left[{}^{n+1}C_2 - n^{n+1}C_1\right]\frac{d^n v}{dx^n} = 0$$

or, $$(x^2 - 1)\frac{d^{n+2}v}{dx^{n+2}} + 2x\frac{d^{n+1}v}{dx^{n+1}} - n(n + 1)\frac{d^n v}{dx^n} = 0. \tag{18.15}$$

Putting $\frac{d^n v}{dx^n} = y$, Eqn. (18.5) becomes

$$(x^2 - 1)\frac{d^2y}{dx^2} + 2x\frac{dy}{dx} - n(n + 1)y = 0$$

or, $$(x^2 - 1)\frac{d^2y}{dx^2} - 2x\frac{dy}{dx} + n(n + 1)y = 0, \quad \text{[Legendre's equation]}.$$

This shows that $y = \dfrac{d^n v}{dx^n}$ is a solution of **Legendre's equation**.

$$\therefore \quad C\frac{d^n v}{dx^n} = P_n(x), \quad \text{where } C \text{ is a constant.} \tag{18.16}$$

But, $v = (x^2 - 1)^n = (x + 1)^n(x - 1)^n$ shows that,

$$\frac{d^n v}{dx^n} = (x + 1)^n\frac{d^n}{dx^n}(x - 1)^n + {}^nC_1 \cdot n(x + 1)^{n-1}\frac{d^{n-1}}{dx^{n-1}}(x - 1)^n + \cdots + (x - 1)\frac{d^n}{dx^n}(x + 1)^n = 0.$$

When $x = 1$, $\dfrac{d^n v}{dx^n} = 2^n \cdot n!$.

Other terms, except the first, disappear as $(x - 1)$ is a factor in every term except first.

$\therefore$ when $x = 1$ Eqn. (18.16) gives

$$C \cdot 2^n \cdot n! = P_n(1) = 1 \quad [\because \ P_n(1) = 1]$$

$$\therefore \qquad C = \frac{1}{2^n \cdot n!}.$$

Now, from Eqn. (18.16),

$$P_n(x) = \frac{1}{2^n \cdot n!} \frac{d^n v}{dx^n}$$

or, $$\boldsymbol{P_n(x) = \frac{1}{2^n \cdot n!} \frac{d^n v}{dx^n}(x^n - 1)^n} \quad [\because \ v = (x^2 - 1)^n].$$

18.7 Legendre's Polynomials

Rodrigue's formula is $\boldsymbol{P_n(x) = \frac{1}{2^n \cdot n!} \cdot \frac{d^n}{dx^n}(x^n - 1)^n}$.

Putting $n = 0$, $P_0(x) = \frac{1}{2^0 0!} = 1$

$$n = 1, \quad P_1(x) = \frac{1}{2^1 1!} \frac{d}{dx}(x^2 - 1) = \frac{1}{2} 2x = x$$

$$n = 2, \quad P_2(x) = \frac{1}{2^2 2!} \frac{d}{dx^2}(x^2 - 1)^2 = \frac{1}{8} \frac{d}{dx}[2(x^2 - 1)(2x)] = \frac{1}{2}[(x^2 - 1) \cdot 1 + 2x \cdot x]$$

$$= \frac{1}{2}(3x^2 - 1).$$

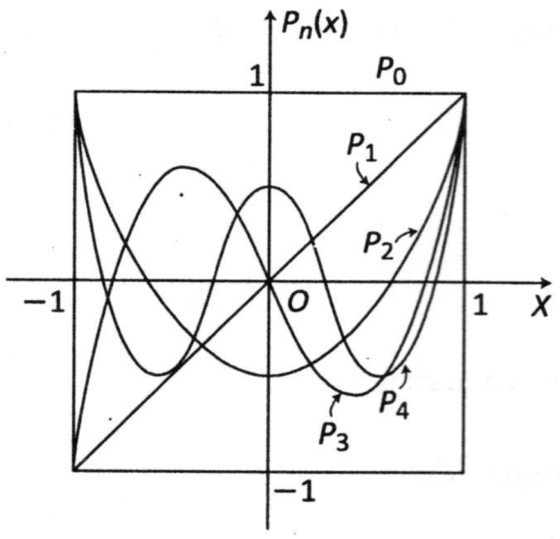

Fig. 18.1: Plot of $P_n(x)$ against x

Similarly, $P_3(x) = \frac{1}{2}(5x^3 - 3x)$

$$P_4(x) = \frac{1}{8}(35x^4 - 30x^2 + 3)$$

$$P_5(x) = \frac{1}{8}(63x^5 - 70x^3 + 15x)$$

$$P_6(x) = \frac{1}{16}(231x^6 - 315x^4 + 105x^2 - 5)$$

....

$$P_n(x) = \sum_{r=0}^{N} \frac{(-1)^r (2n - 2r)!}{2^n \cdot r!(n - r)!(n - 2r)!} x^{n-2r}$$

where $N = \frac{n}{2}$, if n is even

$N = \frac{1}{2}(n - 1)$, if n is odd.

The plot of $P_n(x)$ against x for different n-values is shown in Fig. 18.1.

Example 18.1. *Express $f(x) = x^4 - 2x^3 + 3x^2 + 5x - 9$ in terms of Legendre polynomials.*

Solution : We know that, $P_4(x) = \frac{1}{8}(35x^4 - 30x^2 + 3)$.

$$\therefore \qquad x^4 = \frac{8}{35} P_4(x) + \frac{6}{7} x^2 - \frac{3}{35} \quad \text{and} \quad P_3(x) = \frac{1}{2}(5x^3 - 3x).$$

$$\therefore \quad x^3 = \frac{2}{5}P_3(x) + \frac{3}{5}x \quad \text{and} \quad P_2(x) = \frac{1}{2}(3x^2 - 1).$$

$$\therefore \quad x^2 = \frac{2}{3}P_2(x) + \frac{1}{3} \quad \text{and} \quad x = P_1(x).$$

Hence, $f(x) = x^4 - 2x^3 + 3x^2 + 5x - 9 = \frac{8}{35}P_4(x) + \frac{6}{7}x^2 - \frac{3}{35} - \frac{4}{3}P_3(x) - \frac{6}{5}x + 3x^2 + 5x - 9$

$$= \frac{8}{35}P_4(x) + \frac{27}{7}\left[\frac{2}{3}P_2(x) + \frac{1}{3}\right] + \frac{19}{5}P_1(x) - \frac{4}{3}P_3(x) - \left(9 + \frac{3}{35}\right)$$

$$= \mathbf{\frac{8}{35}P_4(x) - \frac{4}{5}P_3(x) + \frac{18}{7}P_2(x) + \frac{19}{5}P_1(x) - \frac{35}{5}}.$$

18.8 Generating Function for $P_n(x)$

Legendre polynomial $P_n(x)$ is the coefficient of t^n in the expansion of $(1 - 2xt + t^2)^{-\frac{1}{2}}$, i.e.,

$$\mathbf{(1 - 2xt + t^2)^{-\frac{1}{2}} = \sum_{n=0}^{\infty} t^n P_n(x).} \tag{18.17}$$

Proof : We have $[1 - (2xt - t^2)]^{-\frac{1}{2}} = [1 - t(2x - t)]^{-\frac{1}{2}}$

$$= 1 + \frac{t}{2}(2x - t) + \frac{1 \cdot 3}{2 \cdot 4}t^2(2x - t)^2 + \frac{1 \cdot 3 \cdot 5}{2 \cdot 4 \cdot 6}(2x - t)^3 + \cdots + \frac{1 \cdot 3 \cdot 5 \cdots (2n-1)}{2 \cdot 4 \cdot 6 \cdots 2n}t^n(2x - t)^n.$$

Now, collecting the coefficient of t^n, we get

$$\frac{1 \cdot 3 \cdot 5 \cdots (2n-1)}{2 \cdot 4 \cdot 6 \cdots 2n}(2x)^n - \frac{1 \cdot 3 \cdot 5 \cdots (2n-3)}{2 \cdot 4 \cdot 6 \cdots (2n-2)}(2x)^{n-2}\,{}^{n-1}C_1 + \frac{1 \cdot 3 \cdot 5 \cdots (2n-5)}{2 \cdot 4 \cdot 6 \cdots (2n-4)}(2x)^{n-4}\,{}^{n-2}C_2 - \cdots$$

$$= \frac{1 \cdot 3 \cdot 5 \cdots (2n-1)}{n!}\left[x^n - \frac{n(n-1)}{2(2n-1)}x^{n-2} + \frac{n(n-1)(n-2)(n-3)}{2 \cdot 4 \cdot (2n-1)(2n-3)}x^{n-4} - \cdots\right] = P_n(x).$$

$$\therefore \quad \mathbf{(1 - 2xt + t^2)^{-\frac{1}{2}} = \sum_{n=0}^{\infty} t^n P_n(x).}$$

Thus, $(1 - 2xt + t^2)^{-\frac{1}{2}}$ is the generating function of Legendre polynomial $P_n(x)$.

18.8.1 Some Useful Relations

(1) $\mathbf{P_n(1) = 1}$ **[VH(H) '99, 2006]**

Putting $x = 1$ in (18.17), we get

$$(1 - 2t + t^2)^{-\frac{1}{2}} = \sum_{n=0}^{\infty} t^n P_n(1)$$

or, $$\sum_{n=0}^{\infty} t^n P_n(1) = (1 - t)^{-1} = 1 + t + t^2 + \cdots = \sum_{n=0}^{\infty} t^n.$$

$$\therefore \quad \mathbf{P_n(1) = 1.} \tag{18.18}$$

(2) $P_n(-1)=(-1)^n$

Putting $x=-1$ in Eqn. (18.7),

$$\sum_{n=0}^{\infty} t^n P_n(-1)=(1+2t+t^2)^{-\frac{1}{2}}=(1+t)^{-1}=\sum_{n=0}^{\infty}(-1)^n t^n.$$

$$\therefore \quad P_n(-1)=(-1)^n. \tag{18.19}$$

(3) $P_n(-x)=(-1)^n P_n(x)$ **[VH(H) 2006, CU(H) 2007, BU(H) 2003]**

In eqn. (18.17) replacing t by $-t$ and x by $-x$, we get

$$[1-2(-t)(-x)+(-t)^2]^{-\frac{1}{2}}=\sum_{n=0}^{\infty} P_n(-x)(-t)^n$$

or, $$(1-2tx+t^2)^{-\frac{1}{2}}=\sum_{n=0}^{\infty}(-1)^n P_n(-x)t^n$$

or, $$\sum_{n=0}^{\infty} t^n P_n(x)=\sum_{n=0}^{\infty} t^n(-1)^n P_n(-x) \quad \text{or,} \quad P_n(x)=(-1)^n P_n(-x)$$

or, $$P_n(-x)=(-1)^n P_n(x). \tag{18.20}$$

(4) $P_{2n}(0)=(-1)^n\dfrac{1\cdot 3\cdot 5\cdots(2n-1)}{2\cdot 4\cdot 6\cdots(2n)}$ and $P_{2n+1}(0)=0$.

We have $\sum_{n=0}^{\infty} t^n P_n(x)=(1-2xt+t^2)^{-\frac{1}{2}}$.

For $x=0$,

$$\sum_{n=0}^{\infty} t^n P_n(0)=(1+t^2)^{-\frac{1}{2}}=1-\frac{1}{2}t^2+\frac{1}{2}\cdot\frac{3}{4}t^4-\cdots+(-1)^r\frac{1\cdot 3\cdots(2r-1)x^{2r}}{2\cdot 4\cdots(2r)}x^{2r}-\cdots$$

Equating the coefficient of x^{2n} from both sides,

$$P_{2n}(0)=(-1)^n\frac{1\cdot 3\cdot 5\cdots(2n-1)}{2\cdot 4\cdot 6\cdots(2n)}. \tag{18.21}$$

Coefficient of $t^{2n+1}=P_{2n+1}(0)=0.$ (18.22)

(5) If (r_1,θ_1) and (r_2,θ_2) be the polar coordinates of any two points and $\theta=\theta_1-\theta_2$, then the reciprocal of the distance between the points can be expressed in terms of Legendre polynomial.

Now, the reciprocal of the distance between two points (r_1,θ_1) and (r_2,θ_2) is

$$\frac{1}{(r_1^2-2r_1r_2\cos\theta+r_2^2)}=(r_1^2-2r_1r_2\cos\theta+r_2^2)^{-\frac{1}{2}}=\frac{1}{r_1}\left[1-2\frac{r_2}{r_1}\cos\theta+\left(\frac{r_2}{r_1}\right)^2\right]^{-\frac{1}{2}}$$

$$=\frac{1}{r_1}\sum_{n=0}^{\infty}P_n(\cos\theta)\left(\frac{r_2}{r_1}\right)^n \quad \left[\text{putting } t=\frac{r_2}{r_1}\right]$$

$$=\sum_{n=0}^{\infty}\frac{r_2^n}{r_1^{n+1}}P_n(\cos\theta). \tag{18.23}$$

18.9 Recurrence Formula for $P_n(x)$

The following recurrence formulae can be derived from the general function for $P_n(x)$:

(1) $(n+1)P_{n+1}(x) = (2n+1)xP_n(x) - nP_{n-1}(x)$

[CU(H) 2007, BU(H) 1997, '94, NBU(H) 2006, SVTU 2007, VTU 2003, VU(H) 2002]

(2) $nP_n(x) = xP'_n(x) - P'_{n-1}(x)$ [VTU 2006, JNTU 2006, BU(H) 2001]

(3) $(2n+1)P_n(x) = P'_{n+1}(x) - P'_{n-1}(x)$ [Madras U 2006]

(4) $P'_n(x) - xP'_{n-1}(x) = nP_{n-1}(x)$

(5) $(1-x^2)P'_n(x) = n[P_{n-1}(x) - xP_n(x)$

(6) $nP_n(x) = (2n-1)xP_{n-1}(x) - (n-1)P_{n-2}(x)$

(7) $(x^2-1)P'_n(x) = (n+1)[P_{n+1}(x) - xP_n(x)]$

(8) **Christoffel's expansion :**

$$P'_n(x) = (2n-1)P_n(x) + (2n-5)P_{n-3}(x) + (2n-9)P_{n-5}(x) + \cdots$$

Proof of (1) : We have, $(1-2xt+t^2)^{-\frac{1}{2}} = \sum_{n=0}^{\infty} P_n(x)t^n$. (i)

Differentiating partially with respect to t, we get

$$-\frac{1}{2}(1-2xt+t^2)^{-\frac{3}{2}}(-2x+2t) = \sum nP_n(x)t^{n-1}$$

or, $(x-t)(1-2xt+t^2)^{-\frac{1}{2}} = (1-2xt+t^2)\sum nP_n(x)t^{n-1}$

or, $(x-t)\sum P_n(x)t^n = (1-2xt+t^2)\sum nP_n(x)t^{n-1}$.

Equating the coefficient of t^n from both sides, we get

$$xP_n(x) - P_{n-1}(x) = (n+1)P_{n+1}(x) - 2nxP_n(x) + (n-1)P_{n-1}(x)$$

or, $(n+1)P_{n+1}(x) = (2n+1)xP_n(x) - nP_{n-1}(x)$ **(Proved).**

Proof of (2) : Differentiating partially Eqn. (i) with respect to x, we get

$$-\frac{1}{2}(1-2xt+t^2)^{-\frac{3}{2}}(-2t) = \sum P'_n(x)t^n$$

or, $t(1-2xt+t^2)^{-\frac{3}{2}} = \sum P'_n(x)t^n$. (ii)

Again differentiating partially Eqn. (i) w.r.t. t, we get

$$(x-t)(1-2xt+t^2)^{-\frac{3}{2}} = \sum nP_n(x)t^{n-1}. \quad \text{(iii)}$$

Dividing Eqn. (iii) by (ii), we get

$$\frac{x-t}{t} = \frac{\sum nP_n(x)t^{n-1}}{\sum P'_n(x)t^n} \quad \text{or,} \quad \sum nP_n(x)t^n = (x-t)\sum P'_n(x)t^n.$$

Now, equating the coefficients of t^n from both sides, we get

$$nP_n(x) = xP'_n(x) - P'_{n-1}(x) \quad \textbf{(Proved).}$$

Proof of (3) : We know that, $(1-2xt+t^2)^{-\frac{1}{2}} = \sum t^n P_n(x)$.

Differentiating w.r.t. t, we get

$$-\frac{1}{2}(1-2xt+t^2)^{-\frac{3}{2}}(-2x+2t) = \sum nt^{n-1}P_n(x).$$

Multiplying both sides by $(1-2xt+t^2)$, we get

$$(1-2xt+t^2)^{-\frac{1}{2}}(x-t) = (1-2xt+t^2)\sum nt^{n-1}P_n(x)$$

or, $$(x-t)\sum t^n P_n(x) = (1-2xt+t^2)\sum nt^{n-1}P_n(x) \quad \text{(i)}$$

$$\left[\because (1-2xt+t^2)^{-\frac{1}{2}} = \sum t^n P_n(x)\right]$$

Equating the coefficients of t^{n-1} from both sides, we get

$$xP_{n-1}(x) - P_{n-2}(x) = nP_n(x) - 2x(n-1)P_{n-1} + (n-2)P_{n-2}(x)$$

or, $$\boldsymbol{nP_n(x) = (2n-1)xP_{n-1}(x) - (n-1)P_{n-2}(x).} \quad \text{(ii)}$$

Now replacing n by $(n+1)$,

$$(n+1)P_{n+1}(x) = (2n+2-1)xP_n(x) - nP_{n-1}$$

or, $$(n+1)P_{n+1}(x) = (2n+1)xP_n(x) - nP_{n-1}. \quad \text{(iii)}$$

Differentiating eqn. (iii) w.r.t., x we get

$$(n+1)P'_{n+1}(x) = (2n+1)P_n(x) + (2n+1)xP'_n(x) - nP'_{n-1}(x) \quad \text{(iv)}$$

Again $xP'_n(x) - P'_{n-1}(x) = nP_n$ [by recurrence formula (2)] (v)

Substituting the value of $xP'_n(x)$ from Eqn. (v) into Eqn. (iv), we get

$$(n+1)P'_{n+1}(x) = (2n+1)P_n(x) + (2n+1)[nP_n(x) + P'_{n+1}(x)] - nP'_{n-1}(x)$$

or, $$(n+1)P'_{n+1}(x) - (n+1)P'_{n-1}(x) = (2n+1)(1+n)P_n(x)$$

or, $$\boldsymbol{P'_{n+1}(x) - P'_{n-1}(x) = (2n+1)P_n(x)}$$

or, $$\boldsymbol{(2n+1)P_n(x) = P'_{n+1}(x) - P'_{n-1}(x)} \quad \textbf{(Proved).}$$

Proof of (4) : We have $nP_n(x) = (2n-1)xP_{n-1}(x) - (n-1)P_{n-2}(x)$ [recurrence formula (6)]

Differentiating the above formula w.r.t. x, we get

$$nP'_n(x) = (2n-1)P_{n-1}(x) + (2n-1)xP'_{n-1}(x) - (n-1)P'_{n-2}(x)$$

or, $n[P'_n(x) - xP'_{n-1}(x)] - (n-1)[xP'_{n-2}(x)] = (2n-1)P_{n-1}(x)$

or, $n[P'_n(x) - xP'_{n-1}(x)] - (n-1)[(n-1)P_{n-1}(x)] = (2n-1)P_{n-1}(x)$ [by formula (2)]

or, $n[P'_n(x) - xP'_{n-1}(x)] = [(n-1)^2 + (2n-1)]P_{n-1}(x) = n^2 P_{n-1}(x)$

or, $\boldsymbol{P'_n(x) - xP'_{n-1}(x) = nP_{n-1}(x).}$

Proof of (5) : From (2) and (4) we can write,

$$xP'_n(x) - P'_{n-1}(x) = nP_n(x) \quad \text{(i)}$$

$$P'_n(x) - xP'_{n-1}(x) = nP_{n-1}(x). \quad \text{(ii)}$$

Multiplying Eqn. (i) by x and then subtracting from (ii), we get

$$(1-x^2)P_n'(x) = n[P_{n-1}(x) - xP_n(x)] \quad \textbf{(Proved)}.$$

Proof of (6) : We know that, $(1-2xt+t^2)^{-\frac{1}{2}} = \sum t^n P_n(x)$.

Differentiating w.r.t. t, we get

$$-\frac{1}{2}(1-2xt+t^2)^{-\frac{3}{2}}(-2x+2t) = \sum nt^{n-1}P_n(x).$$

Multiplying both sides by $(1-2xt+t^2)$, we get

$$(1-2xt+t^2)^{-\frac{1}{2}}(x-t) = (1-2xt+t^2)\sum nt^{n-1}P_n(x)$$

or, $(x-t)\sum t^n P_n(x) = (1-2xt+t^2)\sum nt^{n-1}P_n(x)$.

Equating the coefficients of t^{n-1} from both sides,

$$xP_{n-1}(x) - P_{n-2}(x) = nP_n(x) - 2x(n-1)P_{n-1}(x) + (n-2)P_{n-2}$$

or, $\boldsymbol{nP_n(x) = (2n-1)xP_{n-1}(x) - (n-1)P_{n-2}(x)}$ **(Proved)**.

Proof of (7) : We have $nP_n(x) = (2n-1)xP_{n-1}(x) - (n-1)P_{n-2}(x)$, [by formula (6)]

Replacing n by $(n+1)$ we get,

$$(n+1)P_{n+1}(x) = (2n+2-1)xP_n(x) - nP_{n-1}(x)$$

or, $(n+1)P_{n+1}(x) = (2n+1)xP_n(x) - nP_{n-1}(x)$ (i)

which can be written as,

$$(n+1)[P_{n+1}(x) - xP_n(x)] = n[xP_n(x) - P_{n-1}(x)] \quad \text{(i)}$$

But, $(x^2-1)P_n'(x) = n[xP_{n-1}(x) - P_{n-1}(x)]$ (ii)

From (i) and (ii) we get,

$$(x^2-1)P_n'(x) = (n+1)[P_{n+1}(x) - xP_n(x)] \quad \textbf{(Proved)}.$$

Proof of (8) : From relation (3), $P_{n+1}'(x) - P_{n-1}'(x) = (2n+1)P_n(x)$.

Replacing n by $(n-1)$, we get

$$P_n'(x) = (2n-1)P_{n-1}(x) + P_{n-2}'(x) = (2n-1)P_n(x) + (2n-5)P_{n-3}(x) + P_{n-4}'(x) + \cdots$$

Proceeding successively, we get

$$P_n'(x) = (2n-1)P_n(x) + (2n-5)P_{n-3}(x) + (2n-9)P_{n-5}(x) + \cdots \quad \textbf{(Proved)}.$$

The last term may be either $3P_1(x)$ or $P_0(x)$ according as n is even or odd.

18.10 Orthogonality of Legendre Polynomials

Due to the following property, the set of Legendre polynomials is said to be orthogonal in the interval $(-1, 1)$.

$$\int_{-1}^{1} P_m(x)P_n(x)dx = 0, \quad [m \neq n]$$

[BU(H) 1993, Madras U 2006, VTU 2006, AMIETE(Winter) 1997, (Summer) 1996]

$$\int_{-1}^{1} P_m(x)P_n(x)dx = \frac{2}{2m+1}, \quad [m=n].$$ **[BU(H) 1993, Madras U 2006, VTU 2006]**

Proof : $P_n(x)$ is a solution of $(1-x^2)\frac{d^2y}{dx^2} - 2x\frac{dy}{dx} + n(n+1)y = 0.$ (18.24)

$P_m(x)$ is a solution of $(1-x^2)\frac{d^2z}{dx^2} - 2x\frac{dz}{dx} + m(m+1)z = 0.$ (18.25)

Multiplying (18.24) by z and (18.25) by y and then subtracting, we get

$$(1-x^2)\left[z\frac{d^2y}{dx^2} - y\frac{d^2z}{dx^2}\right] - 2x\left[z\frac{dy}{dx} - y\frac{dz}{dx}\right] + [n(n+1) - m(m+1)]yz = 0$$

or, $$\frac{d}{dx}\left[(1-x^2)\left(z\frac{dy}{dz} - y\frac{dz}{dz}\right)\right] + (n-m)(n+m+1)yz = 0.$$

Now, integrating from -1 to 1, we get

$$\left[(1-x^2)\left(z\frac{dy}{dx} - y\frac{dz}{dx}\right)\right]^{-1} + (n-m)(n+m-1)\int_{-1}^{+1} yz\,dz = 0$$

or, $$0 + (n-m)(n+m+1)\int_{-1}^{+1} yz\,dx = 0$$

or, $$\int_{-1}^{+1} P_m(x)P_n(x)dx = 0 \quad [\text{if } n \neq m]$$

$\therefore$ $$\int_{-1}^{+1} P_m(x)P_n(x)dx = 0 \quad \textbf{(Proved)}.$$ (18.26)

Again, to prove the second relation, we know that

$$(1-2xt+t^2)^{-1} = \sum t^n P_n(x).$$

Integrating both sides from -1 to 1, we get

$$\int_{-1}^{1}\sum t^{2n}P_n^2(x)dx + \int_{-1}^{1} 2\sum t^{m+n}P_m(x)P_n(x)dx = \int_{-1}^{1}(1-2xt+t^2)dx$$

or, $$\int_{-1}^{1}\sum t^{2n}P_n^2(x)dx + 0 = \int_{-1}^{1}\frac{1}{(1-2xt+t^2)}dx$$

or, $$\sum t^{2n}\int_{-1}^{1}P_n^2(x)dx = -\frac{1}{2t}\left[\log(1-2xt+t^2)\right]_{-1}^{1} = -\frac{1}{2t}\log\frac{1-2t+t^2}{1+2t+t^2}$$

$$= -\frac{1}{2t}\log\left(\frac{1-z}{1+z}\right)^2 = \frac{1}{t}\log\frac{1+t}{1-t} = \frac{1}{t}[\log(1+t) - \log(1-t)]$$

$$= \frac{1}{t}\left[\left(t - \frac{t^2}{2} + \frac{t^3}{3} - \frac{t^4}{4} + \cdots + \frac{t^{2n+1}}{2n+1} + \cdots\right)\right.$$

$$\left. - \left(-t - \frac{t^2}{2} - \frac{t^3}{3} - \frac{t^4}{4} - \cdots - \frac{t^{2n+1}}{2n+1}\right)\right]$$

$$= \frac{2}{t}\left[t + \frac{t^3}{3} + \frac{t^5}{5} + \cdots + \frac{t^{2n+1}}{2n+1} + \cdots\right]$$

$$= 2\left[1 + \frac{t^2}{3} + \frac{t^4}{5} + \cdots + \frac{t^{2n}}{2n+1} + \cdots\right].$$

Equating the coefficients of t^{2n} from both sides of the above equation, we get

$$\int_{-1}^{1} [P_n(x)]^2 dx = \frac{2}{2n+1}, \quad n = 0, 1, 2, \ldots$$

i.e., $$\int_{-1}^{1} P_m(x)P_n(x)\, dx = \frac{2}{2m+1} \quad [\text{for } m = n]. \tag{18.27}$$

Note : Equations (18.26) and (18.27) can be combined into one as below :

$$\int_{-1}^{1} P_m(x)P_n(x) = \frac{2}{2m+1}\delta_{m,n} \tag{18.28}$$

where $\delta = \begin{cases} 0, & \text{for } m \neq n \\ 1, & \text{for } m = n. \end{cases}$

which is the required condition of orthogonality.

18.10.1 Fourier-Legendre Expansion of $f(x)$

If $f(x)$ is a function defined from $x = -1$ to $x = 1$, we can write,

$$f(x) = C_0P_0(x) + C_1P_1(x) + \cdots + C_nP_n(x) = \sum_{n=0}^{\infty} C_nP_n(x), \tag{18.29}$$

where $C_0, C_1, \ldots, C_n$ are all constants, then we can show that,

$$C_m = \frac{2m+1}{2}\int_{-1}^{+1} f(x)P_m(x)dx.$$ **[AMIETE(Summer) 1997]**

Solution : Multiplying both sides of Eqn. (18.29) by $P_m(x)$, we get

$$P_m(x)f(x) = C_0P_0(x)P_m(x) + C_1P_1(x)P_m(x) + C_2P_2(x)P_m(x) + \cdots + C_mP_m^2(x) + \cdots.$$

$$\therefore \int_{-1}^{1} f(x)P_m(x)dx = \int_{-1}^{1}[C_0P_0(x)P_m(x) + C_1P_1(x)P_m(x) + C_2P_2(x)P_m(x) + \cdots + C_mP_m^2(x) + \cdots]dx$$

$$= \left[0 + 0 + \cdots + C_m\frac{2}{2m+1} + \cdots\right] = \frac{2C_m}{2m+1}.$$

$$\therefore C_m = \frac{2m+1}{2}\int_{-1}^{+1} f(x)P_m(x)dx \quad \textbf{(Proved)}.$$

Example 18.2. *Show that,* $\displaystyle\int_{-1}^{+1} x^2P_{n-1}(x)P_{n+1}(x)dx = \frac{2n(n+1)}{(2n-1)(2n+1)(2n+3)}.$

Solution : We have the recurrence relation,

$$(2n+1)xP_n(x) = (n+1)P_{n+1}(x) + nP_{n-1}(x).$$

$$\therefore \quad xP_{n-1}(x) = \frac{1}{2n-1}[nP_n(x) + (n-1)P_{n-1}(x)]$$

and $$xP_{n+1}(x) = \frac{1}{2n+3}[(n+2)P_{n+2}(x) + (n+1)P_n(x)].$$

$$\therefore \quad x^2 P_{n-1}(x)P_{n+1}(x) = \frac{1}{(2n-1)(2n+3)}[n(n+2)P_n(x)P_{n+2}(x) + n(n+1)P_n^2(x)$$
$$+ (n-1)(n+2)P_{n-2}(x)P_{n+2}(x) + (n^2-1)P_n(x)P_{n-2}(x)].$$

Now integrating both sides from -1 to 1 and using orthogonality of Legendre polynomials, we get

$$\int_{-1}^{1} x^2 P_{n-1}(x)P_{n+1}(x)dx = \frac{n(n+1)}{(2n-1)(2n+3)}\int_{-1}^{1} P_n^2(x)dx$$

$$= \frac{2n(n+1)}{(2n-1)(2n+1)(2n+3)} \quad \textbf{(Proved)}.$$

18.11 Associated Legendre's Equation

The associated Legendre equation is given by

$$(1-x^2)\frac{d^2v}{dx^2} - 2x\frac{dv}{dx} + \left[n(n+1) - \frac{m^2}{1-x^2}\right]y = 0. \tag{18.30}$$

The above equation can be solved by the power series method. The particular solution of (18.30) having physical significance is given by,

$$v = (1-x^2)^{\frac{m}{2}} \frac{d^m}{dx^m}[P_n(x)] \tag{18.31}$$

Proof : The Legendre differential equation is $(1-x^2)\frac{d^2y}{dx^2} - 2x\frac{dy}{dx} + n(n+1)y = 0.$

Differentiating successively m times with the help of Leibnitz theorem, we get

$$(1-x^2)y_{m+2} - 2(m+1)xy_{m+1} + [n(n+1) - m(m+1)]y_m = 0. \tag{18.32}$$

Using transformation,

$$v = (1-x^2)^{\frac{m}{2}} y_m, \tag{18.33}$$

where y_m is the mth differentiation w.r.t. x,

$$\frac{dv}{dx} = (1-x^2)^{\frac{m}{2}} y_{m+1} - mx(1-x^2)^{\frac{m}{2}-1} y_m \tag{18.34}$$

$$\frac{d^2v}{dx^2} = (1-x^2)^{\frac{m}{2}} y_{m+2} - 2mx(1-x^2)^{\frac{m}{2}-1} y_{m+1} - m(1-x^2)^{\frac{m}{2}-1} y_m$$
$$+ 2m\left(\frac{m}{2}-1\right)x^2(1-x^2)^{\frac{m}{2}-1} y_m. \tag{18.35}$$

Using Eqn. (18.33), (18.34) and (18.35) in Eqn. (18.32), we get

$$(1-x^2)\frac{d^2v}{dx^2} - 2x\frac{dv}{dx} + \left[n(n+1) - \frac{m}{1-n^2}\right]v = 0 \tag{18.36}$$

Eqn. (18.36) is the **associated Legendre's equation.**

If $y = P_n(x)$ and $y = Q_n(x)$ be the solutions of the Legendre's equation, then from Eqn. (18.33) and (18.36), we can write,

$$t_n^m = (1-x^2)^{\frac{m}{2}} \frac{d^m P_n(x)}{dx^m} \tag{18.37}$$

and $$Q_n^m = (1-x^2)^{\frac{m}{2}} \frac{d^m Q_n(x)}{dx^m}. \tag{18.38}$$

which are the two solutions of (18.36). $P_n^m(x)$ and $Q_n^m(x)$ are the **associated Legendre functions or Legendre polynomials.**

From Eqn. (18.37),

$$\boldsymbol{P_1^1(x) = (1-x^2)^{\frac{1}{2}}, \quad P_2^1(x) = 3x(1-x^2)^{\frac{1}{2}}, \quad P_3^2(x) = 15(1-x^2),}$$

$$\boldsymbol{P_3^3(x) = 15(1-x^2)^{\frac{3}{2}}, \ldots} \textbf{ and so on.}$$

Example 18.3. *Prove that* $P_n^m(-x) = (-1)^{n+m} P_n^m(x)$.

Solution : We have $P_n^m(x) = \left(1-x^2\right)^{\frac{m}{2}} \dfrac{d^m}{dx^m} P_n(x)$.

$$\therefore \quad P_n^m(-x) = \left(1-x^2\right)^{\frac{m}{2}} \frac{d^m}{dx^m} P_n(-x). \tag{i}$$

Now, $\dfrac{d^m}{dx^m}[P_n(-x)] = (-1)^m \dfrac{d^m}{dx^m}[P_n(-x)]$.

Since, in each differentiation (-1) is coming as a factor,

$$\frac{d^m}{dx^m}[P_n(-x)] = (-1)^m(-1)^n \frac{d^m}{dx^m}[P_n(x)].$$

From Eqn. (i),

$$\boldsymbol{P_n^m(-x) = (-1)^{m+n} \frac{d^m}{dx^m} P_n(x) = (-1)^{m+n} P_n^m(x)} \quad \textbf{(Proved).}$$

18.12 Bessel's Equation

The differential equation

$$\boldsymbol{x^2 \frac{d^2y}{dx^2} + x\frac{dy}{dx} + (x^2 - n^2)y = 0} \tag{18.39}$$

is called the **Bessel's differential equation** after the name of German mathematician and astronomer **Friedrich Wilhelm Bessel** (1784–1846). The particular solutions of the equation are called the **Bessel's function of order *n*.**

18.13 General Solution of Bessel Equation

Equation (18.39) has a regular singularity at $x = 0$. So, power series solution can be developed about the point $x = 0$.

Let the solution be $y = \sum_r a_r x^{m+r}$

or, $$y = a_0 x^m + a_1 x^{m+1} + a_2 x^{m+2} + \cdots. \tag{18.40}$$

$$\therefore \quad \frac{dy}{dx} = \sum_r a_r(m+r)x^{m+r-1} \quad \text{and} \quad \frac{d^2y}{dx^2} = \sum_r a_r(m+r)(m+r-1)x^{m+r-2}.$$

Substituting the values of y, $\frac{dy}{dx}$ and $\frac{d^2y}{dx^2}$ in Eqn. (18.40), we get

$$x^2\sum a_r[(m+r)(m+r-1)+(m+r)-n^2]x^{m+r}+\sum a_r x^{m+r+2}=0$$

or, $$\sum a_r[(m+r)^2-n^2]x^{m+r}+\sum a_r x^{m+r+2}=0.$$

Equating the coefficient of x^m to zero we get the **indicial equation,**

$$a_0[(m+0)^2-n^2]=0 \quad [\because r=0]$$

or, $$m^2=n^2,\ \text{i.e.,}\ m=\pm n \quad [\because a_0\neq 0]$$

Equating the coefficient of x^{m+1} to zero, we get

$$a_1[(m+1)^2-n^2]=0$$

or, $$\boldsymbol{a_1=0} \quad [\because (m+1)^2-n^2]\neq 0$$

Equating the coefficient of x^{m+2} to zero, we get

$$a_{m+2}[(m+r+2)^2-n^2]+a_r=0$$

or, $$\boldsymbol{a_{r+2}=\frac{1}{(m+r+2)^2-n^2}}. \tag{18.41}$$

This is the **recurrence relation.**

Since $a_1=0$, we have from Eqn. (18.41), $\boldsymbol{a_1=a_3=a_5=\cdots=0}$.

If $r=0$, $$a_2=\frac{1}{[(m+2)^2-n^2]}a_0$$

$r=2$, $$a_4=\frac{1}{[(m+4)^2-n^2]}a_2=\frac{1}{[(m+2)^2-n^2][(m+4)^2-n^2]}a_0.$$

Substituting these values in Eqn. (18.40), we get

$$\boldsymbol{y_1=a_0x^n\left[1-\frac{x^2}{2(2n+2)}+\frac{x^4}{2\cdot 4(2n+2)(2n+4)}-\cdots\right]}$$

for $m=n$, if n is not a negative integer.

and $$\boldsymbol{y_2=a_0x^{-n}\left[1-\frac{x^2}{2(-2n+2)}+\frac{x^4}{2\cdot 4(-2n+2)(-2n+4)}-\cdots\right]} \quad \text{for } m=-n.$$

18.14 Bessel Functions, $J_n(x)$

Case I: When n is not an integral or zero

We have the solution of the Bessel's equation, $\boldsymbol{x^2\frac{d^2y}{dx^2}+x\frac{dy}{dx}+(x^2-n^2)y=0}$ as

$$y=a_0x^n\left[1-\frac{x^2}{2\cdot 2(n+1)}+\frac{x^4}{2\cdot 4\cdot 2^2(n+1)(n+2)}-\cdots\right.$$
$$\left.+(-1)^r\frac{x^{2r}}{(2^r\cdot r!)2^r(n+1)(n+2)\cdots(n+r)}+\cdots\right]$$

$$=a_0x^n\sum_{r=0}^{\infty}(-1)^r\frac{x^{2r}}{(2^r\cdot r!)(n+1)(n+2)\cdots(n+r)},$$ a_0 being an arbitrary constant.

Putting $a_0 = \frac{1}{2^n\Gamma(n+1)}$ in the above equation, we get

$$y = J_n(x) = \frac{1}{2^n\Gamma(n+1)}\sum_r (-1)^r \frac{x^{n+2r}}{2^r \cdot r!(n+1)(n+2)\cdots(n+r)}.$$

or, $$J_n(x) = \sum_{r=0}^{\infty} \frac{(-1)^r}{r!(n+r)!}\left(\frac{x}{2}\right)^{n+2r} \qquad (18.42)$$

$J_n(x)$ is called the **Bessel function of the first kind of order n.**

For $m = -n$, we replace n by $-n$ in Eqn. (18.42) and get

$$J_{-n}(x) = \sum_{r=0}^{\infty} \frac{(-1)^r}{r!\Gamma(-n+r+1)}\left(\frac{x}{2}\right)^{-n+2r}. \qquad (18.43)$$

$J_{-n}(x)$ is called the **Bessel function of the first kind of order $-n$.**

$\therefore$ the complete solution of the Bessel's equation is

$$y = AJ_n(x) + BJ_{-n}(x), \quad \text{where } A \text{ and } B \text{ are arbitrary constants.} \qquad (18.44)$$

For $n = 0$, we get from Eqn. (18.44)

$$J_0(x) = \sum \frac{(-1)^r}{(r!)^2}\left(\frac{x}{2}\right)^{2r} = 1 - \frac{x^2}{2^2} + \frac{x^4}{2^2 \cdot 4^2} - \frac{x^6}{2^2 \cdot 4^2 \cdot 6^2} + \cdots. \qquad (18.45)$$

For $n = 1$,

$$J_1(x) = \frac{x}{2} - \frac{x^3}{2^2 \cdot 4} + \frac{x^5}{2^2 \cdot 4^2 \cdot 6} - \cdots.$$

The plot of $J_0(x)$ and $J_1(x)$ against x is shown in Fig. (18.2). It is clear that both the functions are oscillatory with a varying period and decreasing amplitude.

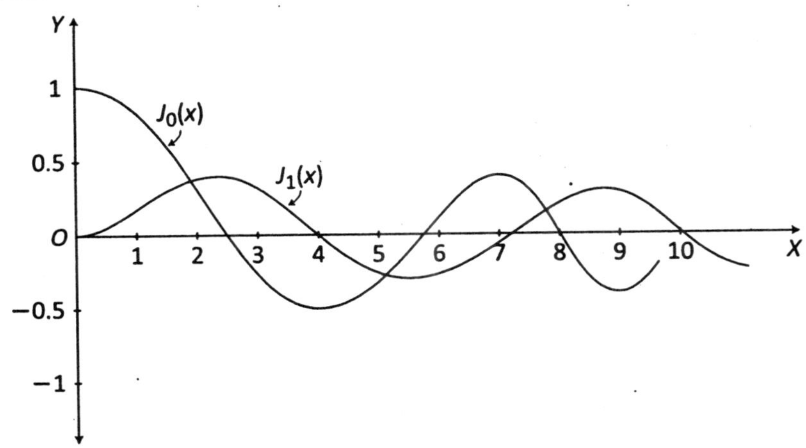

Fig. 18.2: Plot of $J_0(x)$ and $J_1(x)$ against x

Case II : When n is integral

When n is integral, y_2 cannot give a solution for the positive values of n. The first n terms of the series would then vanish as the Γ-function of negative integers is infinite.

Let $k = r - n$, where k is a positive integer, then

$$J_{-n}(x) = \sum_{k=0}^{\infty}(-1)^{k+n}\frac{\left(\frac{x}{2}\right)^{n+2k}}{(k+n)!\Gamma(k+1)} = (-1)^{k+n}\sum(-1)^k\frac{\left(\frac{x}{2}\right)^{n+2k}}{k!\Gamma(k+n+1)}$$

$$= (-1)^n J_n(x). \tag{18.46}$$

$\therefore$ **for $n = 0, 1, 2, 3, \ldots, J_{-n}(x)$ does not represent the second independent solution Eqn. (18.39).**

Another independent integral of Bessel equation (18.39) is required to form its general solution when n is an integer.

Let $y = u(x)J_n(x)$ be a solution of (18.29).

$$\therefore \quad \frac{dy}{dx} = \frac{du}{dx}J_n + u\frac{dJ_n}{dx}$$

$$\frac{d^2y}{dx^2} = \frac{d^2u}{dx^2}J_n + 2\frac{du}{dx}\frac{dJ_n}{dx} + u\frac{d^2J_n}{dx^2}.$$

Substituting these values in Bessel's equation, we get

$$x^2\left(\frac{d^2u}{dx^2}J_n + 2\frac{du}{dx}\frac{dJ_n}{dx} + u\frac{d^2J_n}{dx^2}\right) + x\left(\frac{du}{dx}J_n + u\frac{dJ_n}{dx}\right) + (x^2 - n^2)uJ_n = 0$$

or, $$u\left[x^2\frac{d^2J_n}{dx^2} + x\frac{dJ_n}{dx} + (x^2 - n^2)J_n\right] + x^2\frac{d^2u}{dx^2}J_n + 2x^2\frac{du}{dx}\frac{dJ_n}{dx} + x\frac{du}{dx}J_n = 0.$$

$\because$ J_n is a solution of eqn. (18.39), we have

$$x^2\frac{d^2u}{dx^2}J_n + 2x^2\frac{du}{dx}\frac{dJ_n}{dx} + x\frac{du}{dx}J_n = 0.$$

Dividing by $x^2\frac{du}{dx}J_n$, we get

$$\frac{\frac{d^2u}{dx^2}}{\frac{du}{dx}} + 2\frac{\frac{dJ_n}{dx}}{J_n} + \frac{1}{x} = 0 \quad \text{or,} \quad \frac{d}{dx}\left(\ln\frac{du}{dx}\right) + 2\frac{d}{dx}(\ln J_n) + \frac{d}{dx}(\ln x) = 0$$

or, $$\frac{d}{dx}\left[\ln\left(\frac{du}{dx}J_n^2 x\right)\right] = 0.$$

Integrating, $x\frac{du}{dx}J_n^2 = B$ (B is a constant)

or, $$\frac{du}{dx} = \frac{B}{xJ_n^2} \quad \text{or,} \quad u = B\int\frac{dx}{xJ_n^2} + A.$$

$$\therefore \quad y = AJ_n(x) + BJ_n(x)\int\frac{dx}{x[J_n(x)]^2}.$$

Hence, the complete solution of Bessel's equation is

$$\boldsymbol{y = AJ_n(x) + BY_n(x)}, \tag{18.47}$$

where $$\boldsymbol{Y_n(x) = J_n(x)\int\frac{dx}{x[J_n(x)]^2}}. \tag{18.48}$$

$Y_n(x)$ is called the **Bessel function of the second kind of order n.**

18.15 Values of $J_{\frac{1}{2}}(x)$

We have $J_n(x) = \sum_{r=0}^{\infty} \frac{(-1)^r}{r!\Gamma(n+r+1)} \left(\frac{x}{2}\right)^{n+2r} = \left(\frac{x}{2}\right)^n \sum_{r=0}^{\infty} \frac{(-1)^r}{r!\Gamma(n+r+1)} \left(\frac{x}{2}\right)^{n+2r}$.

Substituting $n = \frac{1}{2}$ in the above relation, we get

$$J_{\frac{1}{2}}(x) = \left(\frac{x}{2}\right)^{\frac{1}{2}} \left[\frac{1}{\Gamma\left(\frac{3}{2}\right)} - \frac{1}{1!\Gamma\left(\frac{5}{2}\right)}\left(\frac{x}{2}\right)^2 + \frac{1}{2!\Gamma\left(\frac{7}{2}\right)}\left(\frac{x}{2}\right)^4 - \cdots\right]$$

$$= \left(\frac{x}{2}\right)^{\frac{1}{2}} \left[\frac{1}{\frac{1}{2}\Gamma\left(\frac{1}{2}\right)} - \frac{1}{\frac{3}{2}\cdot\frac{1}{2}\Gamma\left(\frac{1}{2}\right)}\left(\frac{x}{2}\right)^2 + \frac{1}{2\cdot\frac{5}{2}\cdot\frac{3}{2}\cdot\frac{1}{2}\Gamma\left(\frac{1}{2}\right)}\left(\frac{x}{2}\right)^4 - \cdots\right]$$

$$= \frac{\sqrt{x}}{\sqrt{2}\sqrt{\pi}}\left[\frac{2}{1!} - \frac{2x^3}{3!} + \frac{2x^4}{5!} - \cdots\right] \quad \left[\because \Gamma\left(\frac{1}{2}\right) = \sqrt{\pi}\right] \tag{18.49}$$

Multiplying the expression in the bracket by $\frac{x}{2}$ and outside by $\frac{2}{x}$, we get

$$J_{\frac{1}{2}}(x) = \frac{\sqrt{2}}{\sqrt{x}\sqrt{\pi}}\left[\frac{x}{1!} - \frac{x^3}{3!} + \frac{x^5}{5!} - \cdots\right]$$

or, $$\boldsymbol{J_{\frac{1}{2}}(x) = \sqrt{\frac{2}{\pi x}}\sin x.}$$

Again putting $n = -\frac{1}{2}$ in the expression on for $J_n(x)$, we get

$$J_{-\frac{1}{2}}(x) = \frac{x^{-\frac{1}{2}}}{2^{-\frac{1}{2}}\Gamma\left(-\frac{1}{2}+1\right)}\left[1 - \frac{x^2}{2\cdot2\cdot\left(-\frac{1}{2}+1\right)} + \frac{x^4}{2\cdot4\cdot2^3\left(-\frac{1}{2}+1\right)\left(-\frac{1}{2}+2\right)} - \cdots\right].$$

$$= \frac{\sqrt{2}}{\sqrt{x}\Gamma\left(\frac{1}{2}\right)}\left[1 - \frac{x^2}{2!} + \frac{x^4}{4!} - \cdots\right]$$

$$\therefore \quad \boldsymbol{J_{-\frac{1}{2}}(x) = \sqrt{\frac{2}{\pi x}}\cos x} \quad \left[\because \Gamma\left(\frac{1}{2}\right) = \sqrt{\pi}\right] \tag{18.50}$$

18.16 Recurrence Formulae for Bessel Functions

1. $\boldsymbol{xJ_n' = nJ_n - xJ_{n+1}}$
2. $\boldsymbol{xJ_n' = -nJ_n + xJ_{n-1}}$
3. $\boldsymbol{2J_n' = J_{n-1} - J_{n+1}}$
4. $\boldsymbol{2nJ_n = x(J_{n-1} + J_{n+1})}$

[AMIETE(Summer) 1997, Anna U 2005S, VTU 2003S, BU(H) 1996]

5. $\boldsymbol{\frac{d}{dx}(x^{-n}J_n) = -x^{-n}J_{n+1}}$
6. $\boldsymbol{\frac{d}{dx}(x^nJ_n) = x^nJ_{n-1}}$. **[PTU 2006, BPTU 2005]**

Proof of (1) : We know that $J_n = \sum_{r=0}^{\infty} \frac{(-1)^r}{r!\Gamma(n+r+1)} \left(\frac{x}{2}\right)^{n+2r}$.

Differentiating w.r.t. x, we get

$$J_n' = \sum \frac{(-1)^r(n+2r)}{r!\Gamma(n+r+1)} \left(\frac{x}{2}\right)^{n+2r-1}$$

$$\text{or,}\quad xJ_n' = n\sum \frac{(-1)^r}{r!\Gamma(n+r+1)} \left(\frac{x}{2}\right)^{n+2r} + x\sum \frac{(-1)^r \cdot 2r}{2 \cdot r!\Gamma(n+r+1)} \left(\frac{x}{2}\right)^{n+2r-1}$$

$$= nJ_n + x\sum_{r=1}^{\infty} \frac{(-1)^n}{(r-1)!\Gamma(n+r+1)} \left(\frac{x}{2}\right)^{n+2r-1}$$

$$= nJ_n + x\sum_{s=0}^{\infty} \frac{(-1)^{s+1}}{s!\Gamma(n+s+2)} \left(\frac{x}{2}\right)^{n+2r+1} \qquad [\text{where } r-1=s]$$

$$= nJ_n + \sum_{s=0}^{\infty} \frac{(-1)^s}{s!\Gamma(n+1+s+1)} \left(\frac{x}{2}\right)^{(n+1)+2s}$$

$$\text{or,}\quad \boldsymbol{xJ_n' = nJ_n - xJ_{n+1}} \quad \textbf{(Proved)}. \tag{18.51}$$

Proof of (2) : We know that $J_n = \sum_{r=0}^{\infty} \frac{(-1)^r}{r!\Gamma(n+r+1)} \left(\frac{x}{2}\right)^{n+2r}$

Differentiating w.r.t. x, we get

$$J_n' = \sum_{r=0}^{\infty} \frac{(-1)^r(n+2r)}{r!\Gamma(n+r+1)} \left(\frac{x}{2}\right)^{n+2r-1} \cdot \frac{1}{2}$$

$$\text{or,}\quad xJ_n' = \sum_{r=0}^{\infty} \frac{(-1)^r[(2n+2r)-n]}{r!\Gamma(n+r+1)} \left(\frac{x}{2}\right)^{n+2r}$$

$$= \sum \frac{(-1)^r[(2n+2r-n)-n]}{r!\Gamma(n+r+1)} \left(\frac{x}{2}\right)^{n+2r} - n\sum \frac{(-1)^r}{r!\Gamma(n+r+1)} \left(\frac{x}{2}\right)^{n+2r}$$

$$= \sum \frac{(-1)^r 2}{r!\Gamma(n+1)} \left(\frac{x}{2}\right)^{n+2r} - nJ_n = x\sum \frac{(-1)^r}{r!\Gamma(n-1+r+1)} \left(\frac{x}{2}\right)^{n-1+2r} - nJ_n$$

$$\text{or,}\quad \boldsymbol{xJ_n' = -nJ_n + xJ_{n-1}} \quad \textbf{(Proved)} \tag{18.52}$$

Proof of (3) : We have from (18.51) and (18.52),

$$xJ_n' = nJ_n - xJ_{n+1}$$

$$xJ_n' = -nJ_n + xJ_{n-1}.$$

Adding, we get

$$2xJ_n' = -xJ_{n+1} + xJ_{n-1}$$

$$\text{or,}\quad \boldsymbol{2J_n' = J_{n-1} - J_{n+1}} \quad \textbf{(Proved)}. \tag{18.53}$$

Proof of (4) : We have from (18.51) and (18.52),

$$xJ_n' = nJ_n - xJ_{n+1}$$

$$xJ_n' = -nJ_n + xJ_{n-1}$$

Subtracting,

$$0 = 2nJ_n - xJ_{n+1} - xJ_{n-1}$$

or, $\boldsymbol{2nJ_n = x(J_{n-1} + J_{n+1})}$ **(Proved).** (18.54)

Proof of (5) : We have from (18.51), $xJ_n' = nJ_n - xJ_{n+1}$.

Multiplying by x^{-n-1}, we get

$$x^{-n}J_n' = nx^{-n-1}J_n - x^{-n}J_{n+1} \quad \text{or,} \quad x^{-n}J_n' - nx^{-n-1}J_n = -x^{-n}J_{n+1}$$

or, $\boldsymbol{\frac{d}{dx}(x^{-n}J_n) = -x^{-n}J_{n+1}}$ **(Proved).** (18.55)

Proof of (6) : From (18.52), $xJ_n' = -nJ_n + xJ_{n-1}$.

Multiplying by x^{n-1}, we get

$$x^nJ_n' = -nx^{n-1}J_n + x^nJ_{n-1} \quad \text{or,} \quad x^nJ_n' + nx^{n-1}J_n = x^nJ_{n-1}$$

or, $\boldsymbol{\frac{d}{dx}(x^nJ_n) = x^nJ_{n-1}}$ **(Proved).** (18.56)

Example 18.4. *Find the values of* $J_{\frac{3}{2}}(x)$ *and* $J_{-\frac{3}{2}}(x)$.

Solution : We have $2nJ_n(x) = x[J_{n-1}(x) + J_{n+1}(x)]$. (i)

Putting $n = \frac{1}{2}$,

$$J_{\frac{1}{2}}(x) = x[J_{-\frac{1}{2}}(x) + J_{\frac{3}{2}}(x)]$$

$$\therefore \quad J_{\frac{3}{2}}(x) = \frac{1}{x}J_{\frac{1}{2}}(x) - J_{-\frac{1}{2}}(x)$$

$$= \frac{1}{x}\left[\sqrt{\frac{2}{\pi x}}\sin x\right] - \left[\sqrt{\frac{2}{\pi x}}\cos x\right], \quad \left[\because J_{\frac{1}{2}}(x) = \sqrt{\frac{2}{\pi x}}\sin x \text{ and } J_{-\frac{1}{2}}(x) = \sqrt{\frac{2}{\pi x}}\cos x\right]$$

$$= \sqrt{\frac{2}{\pi x}}\left[\frac{\sin x}{x} - \cos x\right] = \sqrt{\frac{2}{\pi x}}\left[\frac{1}{x}\left(x - \frac{x^3}{3!} + \frac{x^5}{5!} - \cdots\right) - \left(1 - \frac{x^2}{2!} + \frac{x^4}{4!} - \cdots\right)\right]$$

$$= \sqrt{\frac{2}{\pi x}}\left[\frac{2x^2}{3!} - \frac{4x^4}{5!} + \frac{6x^6}{7!} - \cdots\right] = \sqrt{\frac{2}{\pi x}}\sum_{n=1}^{\infty}(-1)^{n-1}\frac{2nx^{2n}}{(2n+1)!}.$$

Putting $n = -\frac{1}{2}$ in (i), we get

$$-J_{-\frac{1}{2}}(x) = x[J_{-\frac{3}{2}}(x) + J_{\frac{1}{2}}(x)]$$

or, $\boldsymbol{J_{-\frac{3}{2}}(x) = -\frac{1}{x}[J_{-\frac{1}{2}}(x) - J_{\frac{1}{2}}(x)]}$.

Using the values of $J_{\frac{1}{2}}(x)$ and $J_{-\frac{1}{2}}(x)$, we get

$$\boldsymbol{J_{-\frac{3}{2}}(x)} = -\sqrt{\frac{2}{\pi x}}\left[\frac{\cos x}{x} + \sin x\right] = -\sqrt{\frac{2}{\pi x}}\left[\frac{1}{x} + \left(\frac{1}{2!}x - \frac{3}{4!}x^3 + \frac{5}{6!}x^5 - \cdots\right)\right]$$

$$= \boldsymbol{-\sqrt{\frac{2}{\pi x}}\left[\frac{1}{x} + \sum_{n=0}^{\infty}\frac{(-1)^n(2n+1)x^{2n+1}}{(2n+2)!}\right]}.$$

Example 18.5. *Prove that,* $\frac{d}{dx}[x^n J_n(x)] = x^n J_{n-1}(x)$.

Solution : We have $x^n J_n(x) = x^n \sum_{r=0}^{\infty} \frac{(-1)^r}{r!\Gamma(n+r+1)} \left(\frac{x}{2}\right)^{n+2r} = \sum \frac{(-1)^r x^{2n+2r}}{r!\Gamma(n+r+1)\cdot 2^{n+2r}}$.

$$\therefore \quad \frac{d}{dx}[x^n J_n(x)] = \sum \frac{(-1)^r (2n+2r) x^{2n+2r-1}}{r!\Gamma(n+r+1)\cdot 2^{n+2r}} = x^n \sum \frac{(-1)^r (n+r)}{r!\Gamma(n+r+1)} \left(\frac{x}{2}\right)^{n+2r-1}$$

$$= x^n \sum \frac{(-1)^r}{r!\Gamma(n+r)} \left(\frac{x}{2}\right)^{n+2r-1} = x^n \sum \frac{(-1)^r}{r!\Gamma(n-1+r+1)} \left(\frac{x}{2}\right)^{(n-r)+2r}$$

or, $\boldsymbol{\frac{d}{dx}[x^n J_n(x)] = x^n J_{n-1}(x)}$ **(Proved).**

18.17 Equations Reducible to Bessel's Equations

In many areas of science and technology we come across differential equations which can be reduced to Bessel's equations by proper substitution which can be solved by Bessel functions.

Example 18.6. *Reduce the following differential equation to Bessel form :*

$$\boldsymbol{x^2 \frac{d^2y}{dx^2} + x\frac{dy}{dx} + (k^2x^2 - n^2)y = 0.} \qquad \text{(i)}$$

Solution : Put $t = kx$.

$$\therefore \quad \frac{dt}{dx} = k, \quad \frac{dy}{dx} = \frac{dy}{dt}\frac{dt}{dx} = k\frac{dy}{dt} \quad \text{and} \quad \frac{d^2y}{dx^2} = \frac{d}{dx}\left(k\frac{dy}{dt}\right) = \frac{d}{dt}\left(k\frac{dy}{dt}\right)\frac{dt}{dx} = k^2\frac{d^2y}{dx^2}.$$

Now, using the values of y, $\frac{dy}{dx}$ and $\frac{d^2y}{dx^2}$ in Eqn. (i),

$$\frac{t^2}{k^2}\left(k^2\frac{d^2y}{dt^2}\right) + \frac{t}{k}\left(k\frac{dy}{dt}\right) + (t^2 - n^2)y = 0$$

or, $\boldsymbol{t^2\frac{d^2y}{dx^2} + t\frac{dy}{dx} + (t^2 - n^2)y = 0.}$

Solution of the above equation is $y = C_1 J_1(t) + C_2 J_{-n}(t)$.

$\therefore$ solution of Eqn. (i) is $\boldsymbol{y = C_1 J_n(kx) + C_2 J_{-n}(kx)}$.

18.18 Spherical Bessel's Functions

The spherical Bessel function is defined as

$$\boldsymbol{J_n(x) = \sqrt{\frac{\pi}{2x}} J_{n+\frac{1}{2}}(x),}$$

which is the solution of the differential equation

$$\boldsymbol{x^2\frac{d^2R}{dx^2} + 2x\frac{dR}{dx} + [x^2 - n(n+1)]R = 0.}$$

This equation can be transformed to Bessel form. **[BU(H) 1996]**

We consider the differential equation,

$$x^2\frac{d^2R}{dx^2} + 2x\frac{dR}{dx} + [x^2 - n(n+1)]R = 0. \qquad \text{(i)}$$

Put $R(x) = \frac{z(x)}{\sqrt{x}}$, then Eqn. (i) reduces to

$$x^2\frac{d^2z}{dx^2} + x\frac{dz}{dx} + \left[x^2 - \left(n + \frac{1}{2}\right)^2\right] z = 0. \tag{ii}$$

Eqn. (ii) is a Bessel's equation of order $\left(n + \frac{1}{2}\right)$.

Solution of Eqn. (ii) is $J_{n+\frac{1}{2}}(x)$. (iii)

The normalisation factor $\sqrt{\pi/2}$ is so chosen that, $J_0(x) = \dfrac{\sin x}{x}$.

$$\therefore \qquad J_n(x) = R(x) = \sqrt{\frac{\pi}{2x}} J_{n+\frac{1}{2}}(x). \tag{iv}$$

Again, $\because$ $J_{n+\frac{1}{2}}(x) = \displaystyle\sum_{r=0}^{\infty} \frac{(-1)^r}{\Gamma(r+1)\Gamma\left(n+r+\frac{3}{2}\right)} \left(\frac{x}{2}\right)^{n+2r+\frac{1}{2}}$

$$= \sum_{r=0}^{\infty} \frac{(-1)^r 2^{2n+2r+1}(n+r)!}{\sqrt{\pi}(2n+2r+1)!k!} \left(\frac{x}{2}\right)^{n+2r+\frac{1}{2}}.$$

$$\therefore \qquad J_n(x) = 2^n x^n \sum_{r=0}^{\infty} \frac{(-1)^r (n+r)!}{r!(2n+2r+1)!} x^{2r}. \tag{v}$$

18.19 Orthogonality of Bessel Functions

Let α and β be different roots of $J_n(x) = 0$, then

$$\int_0^1 xJ_n(\alpha x)J_n(\beta x)dx = 0, \qquad \text{if } \alpha \neq \beta$$
$$= \frac{1}{2}[J_{n+1}(\alpha)^2]^2, \qquad \text{if } \alpha = \beta.$$

Proof : $J_n(x)$ is the solution of $x^2\dfrac{d^2y}{dx^2} + x\dfrac{dy}{dx} + (x^2 - n^2)y = 0.$ (i)

Put $x = \alpha r$, $y = u = J_n(\alpha r)$, then

$$\frac{dy}{dx} = \frac{du}{dx} = \frac{du}{dr}\frac{dr}{dx} = \frac{1}{a}\frac{du}{dr}$$

and $$\frac{d^2y}{dx^2} = \frac{d}{dx}\left(\frac{dy}{dx}\right) = \frac{d}{dr}\left(\frac{dy}{dx}\right)\frac{dr}{dx} = \frac{1}{\alpha}\frac{d}{dr}\left(\frac{dy}{dx}\right) = \frac{1}{\alpha}\frac{d}{dr}\left(\frac{1}{\alpha}\frac{du}{dr}\right) = \frac{1}{\alpha^2}\frac{d^2u}{dx^2}.$$

Using these values in Eqn. (i), we get

$$r^2\frac{d^2u}{dr^2} + r\frac{du}{dr} + (\alpha^2r^2 - n^2)u = 0. \tag{ii}$$

Similarly, putting $x = \beta r$, $y = v = J_n(\beta r)$, Eqn. (i) reduces to

$$r^2\frac{d^2v}{dr^2} + r\frac{dv}{dr} + (\beta^2r^2 - n^2)v = 0. \tag{iii}$$

Multiplying Eqn. (ii) by $\frac{v}{r}$ and (iii) by $\frac{u}{r}$ and subtracting,

$$r[vu'' - uv''] + [vu' - uv'] + (\alpha^2 - \beta^2)ruv = 0 \quad \text{or,} \quad \frac{d}{dr}[r(vu' - uv')] = (\beta^2 - \alpha^2)ruv.$$

Integrating, $r(vu' - uv') = (\beta^2 - \alpha^2)\int ruv\,dr$.

$$\therefore \quad \int_0^1 ruv\,dr = \left[\frac{r(vu' - uv')}{\beta^2 - \alpha^2}\right]_0^1$$

$$\text{or,} \quad \int_0^1 rJ_n(\alpha r)J_n(\beta r)dr = \left[\frac{r\{\alpha J_n'(\alpha r)J_n(\beta r) - \beta J_n(\alpha r)J_r'(\beta r)\}}{\beta^2 - \alpha^2}\right]$$

$$= \frac{\alpha J_n'(\alpha)J_n(\beta) - \beta J_n(\alpha)J_n'(\beta)}{\beta^2 - \alpha^2}. \tag{iv}$$

$\because$ α and β are the roots of $J_n(x) = 0$, $\therefore$ $J_n(\alpha) = J_n(\beta) = 0$, [if $\alpha \neq \beta$]

$$\therefore \quad \int_0^1 rJ_n(\alpha r)J_n(\beta r)dr = 0$$

$$\text{or,} \quad \int_0^1 \boldsymbol{xJ_n(\alpha x)J_n(\beta x)dx = 0} \quad [\text{if } \alpha \neq \beta]. \tag{v}$$

This is the condition of orthogonality of Bessel function.

Again, if $\beta = \alpha$, the RHS of Eqn. (v) is indeterminant.

In this case, let α be a root of $J_n(x) = 0$, i.e., $J_n(a) = 0$ and $\beta \to \alpha$.

Then Eqn. (v) becomes

$$\underset{\beta\to\alpha}{\text{Lt}} \int_0^1 rJ_n(\alpha r)J_n(\beta r)dr = \underset{\beta\to\alpha}{\text{Lt}} \frac{\alpha J_n'(\alpha)J_n(\beta)}{\beta^2 - \alpha^2} \left[\text{which is of the form } \left(\frac{0}{0}\right) \text{ as } J_n(\alpha) = 0\right].$$

Applying **L Hospital's** rule on the RHS of the above equation, we get

$$\underset{\beta\to\alpha}{\textbf{Lt}} \int_0^1 \boldsymbol{rJ_n(\alpha r)J_n(\beta r)dr} = \underset{\beta\to\alpha}{\text{Lt}} \frac{\alpha J_n'(\alpha)J_n'(\beta)}{2\beta} \tag{vi}$$

$$= \frac{1}{2}[J_n'(a)]^2 \quad [\text{if } \alpha = \beta].$$

This is **another condition of orthogonality of $J_n(x)$ when two roots of $J_n(x) = 0$ are equal.**

18.20 Generating Function for $J_n(x)$

We can show that $J_n(x)$ is the coefficient of t^n in the expansion of $e^{\frac{x}{2}}\left(t - \frac{1}{t}\right)$.

Proof : We have $e^{\frac{x}{2}\left(t-\frac{1}{t}\right)} = e^{\frac{xt}{2}} \times e^{-\frac{x}{2t}}$

$$= \left[1 + \frac{xt}{2} + \frac{1}{2!}\left(\frac{xt}{2}\right)^2 + \frac{1}{3!}\left(\frac{xt}{2}\right)^3 + \cdots\right]$$

$$\times \left[1 - \left(\frac{x}{2t}\right) + \frac{1}{2!}\left(\frac{x}{2t}\right)^2 - \frac{1}{3!}\left(\frac{x}{2t}\right)^3 + \cdots\right]. \tag{i}$$

The coefficient of t^n in the product of Eqn. (i) is

$$\frac{1}{n!}\left(\frac{x}{2}\right)^n - \frac{1}{(n+1)!}\left(\frac{x}{2}\right)^{n+2} + \frac{1}{2!(n+2)!}\left(\frac{x}{2}\right)^{n+4} - \cdots = J_n(x).$$

Similarly, the coefficient of t^{-n} in the product of Eqn. (i) is $J_{-n}(x)$.

$$\therefore \quad e^{\frac{x}{2}\left(t-\frac{1}{t}\right)} = J_0 + tJ_1 + t^2J_2 + t^3J_3 + \cdots + t^{-1}J_{-1} + t^{-2}J_{-2} + t^{-3}J_{-3} + \cdots = \sum_{n=-\infty}^{\infty} t^n J_n(x).$$

For this reason $e^{\frac{x}{2}\left(t-\frac{1}{t}\right)}$ is called the generating function of Bessel function.

Note : The coefficient of t^0 from Eqn. (i) is

$$1 - \frac{x^2}{2^2} + \frac{x^4}{2^2 \cdot 4^2} - \cdots$$

or, $$J_0(x) = 1 - \frac{x^2}{2^2} + \frac{x^4}{2^2 \cdot 4^2} - \cdots.$$

18.21 BER and BEI Functions

Let us consider the differential equation, $x\dfrac{d^2y}{dx^2} + \dfrac{dy}{dx} - ixy = 0.$ (18.57)

This equation is obtained by putting $n = 0$, $k^2 = -i$ in

$$x^2\frac{d^2y}{dx^2} + x\frac{dy}{dx} + (k^2x^2 - n^2)y = 0.$$

$\therefore$ The particular solution of Eqn. (18.57) is

$$y = J_0(kx) = J_0[(-i)^{\frac{1}{2}}x] = J_0(i^{\frac{3}{2}}x).$$

Replacing $(i^{\frac{3}{2}}x)$ in the series for $J_0(x)$, we get

$$y = 1 - \frac{i^3x^3}{2^2} + \frac{i^6x^4}{(2!)^2 2^4} - \frac{i^9x^6}{(3!)^2 2^6} + \frac{i^{12}x^8}{(4!)^2 2^8} - \cdots$$

$$= \left[1 - \frac{x^4}{2^2 \cdot 4^2} + \frac{x^8}{2^2 \cdot 4^2 \cdot 6^2 \cdot 8^2} - \cdots\right]$$

$$+ i\left[\frac{x^2}{2^2} - \frac{x^6}{2^2 \cdot 4^2 \cdot 6^2} + \frac{x^{10}}{2^2 \cdot 4^2 \cdot 6^2 \cdot 8^2 \cdot 10^2} - \cdots\right] \quad (18.58)$$

Thus, y is complex if x is real.

or, $$y = \left[1 + \sum_{m=1}^{\infty}(-1)^m \frac{x^{4m}}{2^2 \cdot 4^2 \cdot 6^2 \cdots (4m)^2}\right] + i\left[-\sum_{m=1}^{\infty}(-1)^m \frac{x^{4m}}{2^2 \cdot 4^2 \cdot 6^2 \cdots (4m-2)^2}\right]$$

$$= \textbf{\textit{Ber}}\ x + i\textbf{\textit{Bei}}\ x, \quad (18.59)$$

where $$Ber\ x = 1 + \sum_{m=1}^{\infty}(-1)^m \frac{x^{4m}}{2^2 \cdot 4^2 \cdot 6^2 \cdots (4m)^2} \quad (18.60)$$

$$Bei\ x = -\sum_{m=1}^{\infty}(-1)^m \frac{x^{4m-2}}{2^2 \cdot 4^2 \cdot 6^2 \cdots (4m-2)^2}. \quad (18.61)$$

Here, *Ber* x and *Bei* x are respectively **Bessel real and Bessel imaginary functions.**

18.22 Bessel Function $J_n(x)$ can be Represented in the form of Integral

To prove that (a) $J_0(x) = \dfrac{1}{\pi}\displaystyle\int_0^{\pi} \cos(x\sin\theta)d\theta$ (b) $J_n(x) = \dfrac{1}{\pi}\displaystyle\int_0^{\pi} \cos(n\theta - \pi\sin\theta)d\theta$.

Proof : We have $\cos(x\sin\theta) = J_0 + 2J_2\cos 2\theta + 2J_4\cos 4\theta + \cdots$ (i)

$$\sin(x\sin\theta) = 2J_1\sin\theta + 2J_3\sin 3\theta + 2J_5\sin 5\theta + \cdots \quad \text{(ii)}$$

(a) Integrating (i) between the limits 0 and π, we get

$$\int_0^\pi \cos(x\sin\theta)d\theta = \int_0^\pi (J_0 + 2J_2\cos 2\theta + 2J_4\cos 4\theta + \cdots)d\theta$$

$$= J_0\int_0^\pi d\theta + 2J_2\int_0^\pi \cos 2\theta d\theta + 2J_4\int_0^\pi \cos 4\theta\, d\theta + \cdots$$

$$= J_0\pi + 0 + 0.$$

$$\therefore \quad J_0 = \frac{1}{\pi}\int_0^\pi \cos(x\sin\theta)d\theta \quad \textbf{(Proved).} \tag{18.53}$$

(b) Multiplying eqn. (i) by $\cos n\theta$ and integrating between the limits 0 and π, we have

$$\int_0^\pi \cos(x\sin\theta)\cos n\theta\, d\theta = \int_0^\pi [J_0\cos n\theta + 2J_0\cos 2\theta\cos n\theta + 2J_4\cos 4\theta\cos n\theta + \cdots]d\theta$$

$$= 2J_0\int_0^\pi \cos n\theta\, d\theta + 2J_2\int_0^\pi \cos 2\theta\cos n\theta\, d\theta + \cdots$$

$$= 0, \quad \text{if } n \text{ is odd} \quad \text{(iii)}$$

$$= \pi J_n, \quad \text{if } n \text{ is even.} \quad \text{(iv)}$$

Again multiplying Eqn. (ii) by $\sin n\theta$ and integrating between the limits 0 and π, we get

$$\int_0^\pi \sin(x\sin\theta)\sin n\theta\, d\theta = \int_0^\pi 2J_1\sin\theta\sin n\theta + 2J_3\sin 3\theta\sin n\theta + \cdots]d\theta$$

$$= 2J_1\int_0^\pi \sin\theta\sin n\theta\, d\theta + 2J_3\int_0^\pi \sin 3\theta\sin n\theta\, d\theta + \cdots$$

$$= 0, \quad \text{if } n \text{ is even} \quad \text{(v)}$$

$$= \pi J_n, \quad \text{if } n \text{ is odd.} \quad \text{(vi)}$$

Adding (iv) and (v), we get

$$\int_0^\pi [\cos(x\sin\theta)\cos n\theta + \sin(x\sin\theta)\sin n\theta]d\theta = \pi J_n \quad \text{or,} \quad \int_0^\pi \cos(n\theta - x\sin\theta)d\theta = \pi J_n$$

$$\text{or,} \quad J_n = \frac{1}{\pi}\int_0^\pi \cos(n\theta - x\sin\theta)d\theta \quad \textbf{(Proved).} \tag{18.63}$$

18.23 Hermite's Equation

Hermite differential equation is

$$\frac{d^2y}{dx^2} - 2x\frac{dy}{dx} + 2xy = 0, \tag{18.64}$$

where n is a **non-negative integer**. Since there is no singularity in the finite plane Eqn. (18.64) can be solved by power series method.

The power series solution is $y = \sum_{r=0}^{\infty} a_r x^{m+r}$. (18.65)

$\therefore \quad \dfrac{dy}{dx} = \sum_{r=0}^{\infty} a_r(m+r)x^{m+r-1}$ and $\dfrac{d^2y}{dx^2} = \sum_{r=0}^{\infty} a_r(m+r)(m+r-1)x^{m+r-2}.$

Substituting in Eqn. (18.64), we get

$$\sum_{r=0}^{\infty}[(m+r)(m+r-1)x^{m+r-2} - 2(m+r)x^{m+r} + 2nx^{m+r}]a_r = 0$$

or, $$\sum_{r=0}^{\infty}[(m+r)(m+r-1)x^{m+r-2} - 2(m+r-n)x^{m+r}]a_r = 0. \tag{18.66}$$

Equating the coefficient of the lowest power of x, i.e., x^{m-2} to zero, we get

$$a_0[m(m-1)] = 0, \tag{18.67}$$

$\because \quad a_0 \neq 0 \quad \therefore \quad m = 0 \quad$ or, $\quad m = 1.$

Again, equating the coefficient of x^{m-1} to zero, we get

$$a_1[(m+1)m] = 0. \tag{18.68}$$

Now, equating the coefficient of the general term x^{m+r} in Eqn. (18.66) to zero,

$$a_{r+2}(m+r+2)(m+r+1) - 2a_r(m+r-n) = 0$$

or, $$\boldsymbol{a_{r+2} = \frac{2(m+r-n)}{(m+r+2)(m+r+1)}a_r = \frac{2(m+r)-2n}{(m+r+2)(m+r+1)}a_r.} \tag{18.69}$$

This is the **recursion relation** between the coefficients.

Case I : When $m = 0$

From relation (19.69) for $m = 0$,

$$a_{r+2} = \frac{2r-2n}{(r+2)(r+1)}a_r. \tag{18.70}$$

$$\therefore \quad a_2 = -\frac{2n}{2!}a_0, \quad a_4 = \frac{4-2n}{4\cdot 3}a_2 = -\frac{(4-2n)\cdot 2n}{4\cdot 3\cdot 2!}a_0 = \frac{2^2(-2+n)n}{4!}a_0,$$

$$a_{2m} = \frac{(-2)^k n(n-2)\cdots(n-2k+2)}{(2k)!}a_0.$$

Similarly, $a_3 = \dfrac{2-2n}{3\cdot 2}a_1 = -\dfrac{2(n-1)}{3!}a_1$

$$a_5 = \frac{6-2n}{5\cdot 4}a_3 = -\frac{2(n-1)(6-2n)}{5\cdot 4\cdot 3!} = (-2)^2\frac{(n-1)(n-3)}{5!}a_1, \text{ etc.}$$

The general term is,

$$a_{2k+1} = \frac{(-2)^k(n-1)(n-3)\cdots(n-2k+1)}{(2k+1)!}.$$

$$\therefore \quad y = \sum_{r=0}^{\infty} a_r x^{m+r} = \sum_{r=0}^{\infty} a_r x^r \quad (\because\ m = 0)$$

$$= a_0 + a_1x + a_2x^2 + a_3x^3 + \cdots$$

$$= a_0\left[1 - \frac{2n}{2!}x^2 + \frac{2^2 n(n-2)}{4!}x^4 + \cdots + \frac{(-2)^k n(n-2)\cdots(n-2k+2)}{2k!}x^{2k}\right]$$

$$+ a_1\left[x - \frac{2(n-1)}{3!}x^3 + \frac{2^2(n-1)(n-3)}{5!}x^5 + \cdots\right] \quad (\because\ a_1 \neq 0) \qquad (18.71)$$

and $$y = a_0\left[1 - \frac{2n}{2!}x^2 + \frac{2^2 n(n-2)}{4!}x^4 + \cdots\right] \qquad [\text{for } a_1 = 0]$$

$$= u \text{ (say)}. \qquad (18.72)$$

Case II : When $m = 1$

In this case, $a_{r+2} = \dfrac{2(r+1) - 2n}{(r+3)(r+2)} a_r$.

Putting $r = 1, 3, 5, \ldots$, we get from the above relation

$$a_3 = a_5 = \cdots = 0 \quad (\because\ a_1 = 0)$$

Also putting $r = 0, 2, 4, \ldots$, we get

$$a_2 = \frac{2-2n}{3\cdot 2}a_0 = -\frac{2(n-1)}{3!}a_0, \quad a_4 = \frac{6-2n}{5\cdot 4}a_2 = \frac{2^2(n-1)(n-3)}{5!}a_0 \text{ and so on.}$$

The general term is given by

$$a_{2k} = \frac{(-1)^k(n-1)(n-3)\cdots(n-2k+1)}{(2k+1)!}.$$

$$\therefore \quad \sum_{r=0}^{\infty} a_r x^{r+1} = a_0 x + a_2 x^3 + a_4 x^5 + \cdots + a_{2k}x^{2k+1}$$

or, $$y = a_0\left[x - \frac{2(n-1)}{3!}x^3 + \frac{2^2(n-1)(n-3)}{5!}x^5 + \cdots\right]$$

$$= v \text{ (say)}. \qquad (18.73)$$

Solution (18.72) is a part of the solution (18.71) but both being the solutions of the same differential equation, (18.72) cannot be a part of the solution (18.71).

So, $a_4 = 0$ and the solution for $m = 0$ must be given by (18.72).

$\therefore$ the general solution is $\boldsymbol{y = Au + Bv}$.

18.24 Hermite Polynomial $H_n(x)$

Taking $a_0 = (-1)^{\frac{n}{2}}.\frac{n!}{(\frac{n}{2})!}$ and n even the coefficient of x^n in Eqn. (18.71) becomes

$$(-1)^{\frac{n}{2}}\frac{n!}{(\frac{n}{2})!}\frac{(-2)^{\frac{n}{2}}n(n-2)\cdots(n-n+2)}{n!} = 2^n\frac{\frac{n}{2}\left(\frac{n}{2}-1\right)\cdots 1}{(\frac{n}{2})!} = 2^n.$$

Similarly, the coefficient of x^{n-2} is given by

$$(-1)^{\frac{n}{2}}\frac{n!}{(\frac{n}{2})!} - \frac{n(n-1)}{1!}(2x)^{n-2} + \frac{n(n-1)(n-2)(n-3)}{2!}(2x)^{n-4} + \cdots + (-1)^{\frac{n}{2}}\frac{n!}{(\frac{n}{2})!}$$

$$= \boldsymbol{H_n(x)} \textbf{ (say)}.$$

$H_n(x)$ is the polynomial y_n and is called **Hermite polynomial of degree n.**

$$\therefore \quad H_n(x) = \begin{cases} \sum_{r=0}^{\infty}(-1)^r \frac{n!}{r!(n-2r)!}(2x)^{n-2r}, & (n \text{ even}) \\ \sum_{r=0}^{\frac{n}{2}-1}(-1)^r \frac{n!}{r!(n-2r)!}(2x)^{n-2r}, & (n \text{ odd}) \end{cases} \tag{18.73}$$

18.25 Generating Function for $H_n(x)$

The generating function for $H_n(x)$ for x real and n integral is given by

$$e^{2tx-t^2} = \sum_{n=0}^{\infty} \frac{t^n}{n!} H_n(x). \tag{18.74}$$

Proof : Here, $e^{2tx-t^2} = e^{2tx} \cdot e^{-t^2} = \sum_{r=0}^{\infty} \frac{(2tx)^r}{r!} \sum_{s=0}^{\infty} \frac{(-t^2)^s}{s!} = \sum_{r,s=0}^{\infty} (-1)^s \frac{(2x)^r}{r!s!} t^{r+2s}$.

Coefficient of t^n for fixed s value is

$(-1)^s \frac{(2x)^{n-2s}}{(n-2s)!s!}$ [putting, $r + 2s = n$].

The total coefficient of t^n can be obtained by summing over all the allowed values of s, since

$r = n - 2s \geq 0$ or, $s \leq \frac{n}{2}$.

So, if n is even, $s \to 0$ to $\frac{\pi}{2}$;

if n is odd, $s \to 0$ to $\frac{1}{2}(n-1)$.

So, the required coefficient of t^n is

$$\sum_{s=0}^{\frac{n}{2}} (-1)^s \frac{(2x)^{n-2s}}{(n-2s)!s!} = \frac{H_n(x)}{n!} \quad \text{(by Eqn. 18.73)}.$$

$$\therefore \quad e^{2tx-t^2} = \sum_{n=0}^{\infty} \frac{t^n}{n!} H_n(x) \tag{18.75}$$

$$\text{or,} \quad e^{x^2-(t-x)^2} = \sum_{n=0}^{\infty} \frac{t^n}{n!} H_n(x). \tag{18.76}$$

18.26 Rodrigue's Formula for $H_n(x)$

Differentiating both sides of Eqn. (18.76) n times w.r.t. t and putting $t = 0$, we get

$$\frac{H_n(x)}{n!} n! = \left[\frac{\partial^n}{\partial t^n} e^{-(t-x)^2}\right]_{t=0} e^{x^2}.$$

Putting $u = t - x$ when $x = -u$ as $t = 0$, so that $\frac{\partial}{\partial t} \equiv \frac{\partial}{\partial u}$, we get

$$\left[\frac{\partial^n}{\partial t^n} e^{-(t-x)^2}\right]_{t=0} = \frac{\partial^n}{\partial u^n}(e^{-u^2}) = (-1)^n \frac{\partial^n}{\partial x^n}(e^{-x^2}) = (-1)^n \frac{d^n}{dx^n}(e^{-x^2}).$$

$$\therefore \quad H_n(x) = (-1)^n e^{x^2} \frac{d^n}{dx^n}(e^{-x^2}) \tag{18.77}$$

Equating (18.77) is the differential form of Hermite polynomial and is called **Rodrigue's formula.**

Putting $n = 0, 1, 2, \ldots$ in eqn. (18.77), we get

$$H_0(x) = e^{x^2} \cdot e^{-x^2} = 1,$$

$$H_1(x) = (-1)e^{x^2} \frac{d}{dx}\left(e^{-x^2}\right) = 2x,$$

$$H_2(x) = (-1)e^{x^2} \frac{d^2}{dx^2}\left(e^{-x^2}\right) = e^{x^2} \frac{d}{dx}\left(-2xe^{-x^2}\right) = 4x^2 - 2,$$

$$H_3(x) = 8x^3 - 12x,$$

$H_4(x) = 16x^4 - 48x^2 + 12$ and so on.

18.27 Recurrence Formulae for $H_n(x)$

We have from Eqn. (18.75), $e^{2tx-t^2} = \sum_{n=0}^{\infty} \frac{t^n}{n!} H_n(x)$.

Differentiating w.r.t. x,

$$2te^{2tx-t^2} = \sum_{n=0}^{\infty} H_n'(x)\frac{t^n}{n!} \quad \text{or,} \quad 2t\sum_{n=0}^{\infty} H_n(x)\frac{t^n}{n!} = \sum_{n=0}^{\infty} H_n'(x)\frac{t^n}{n!}.$$

Equating the coefficient of $\frac{t^n}{n!}$ from both sides,

$$2\frac{n!}{(n-1)!}H_{n-1}(x) = H_n'(x)$$

or, $$2nH_{n-1}(x) = H_n'(x) \tag{18.78}$$

18.28 Orthogonality of Hermite Polynomials

As $H_n(x)$ is the solution of Hermite equation, we write

$$\frac{d^2H_n}{dx^2} - 2x\frac{dH_n}{dx} + 2nH_n(x) = 0. \tag{18.79}$$

Multiplying both sides by e^{-x^2},

$$\frac{d}{dx}\left[e^{-x^2}\frac{dH_n}{dx}\right] = -2n\left[e^{-x^2}H_n(x)\right]. \tag{18.80}$$

Also, $$\frac{d}{dx}\left[e^{-x^2}\frac{dH_n}{dx}\right] = -2m\left[e^{-x^2}H_m(x)\right] \quad [\text{for } m \neq n] \tag{18.81}$$

Now, multiplying Eqn. (18.80) by H_m and (18.78) by H_n and then subtracted. The resulting equation is then integrated w.r.t. x from $-\infty$ and ∞ to get

$$\int_{-\infty}^{+\infty}\left\{H_m(x)\frac{d}{dx}\left(e^{-x^2}\frac{dH_n}{dx}\right) - H_n(x)\frac{d}{dx}\left(e^{-x^2}\frac{dH_m}{dx}\right)\right\}dx \cdot$$

$$= 2(m-n)\int_{-\infty}^{+\infty} e^{-x^2}H_m(x)H_n(x)dx. \tag{18.82}$$

The LHS of the above equation is

$$\int_{-\infty}^{+\infty} \frac{d}{dx}\left\{H_m(x)\left(e^{-x^2}\frac{dH_n}{dx}\right)\right\}dx - \int_{-\infty}^{+\infty} \frac{d}{dx}\left\{H_n(x)\left(e^{-x^2}\frac{dH_m}{dx}\right)\right\}dx = 0.$$

$$\therefore \quad \int_{-\infty}^{+\infty} e^{-x^2} H_m(x)H_n(x)dx = 0, \quad m \neq n. \tag{18.83}$$

So, Hermite polynomials are orthogonal to e^{-x^2}.

18.29 Prove that, $\int_{-\infty}^{+\infty} e^{-x^2} H_n^2(x)dx = 2^n\sqrt{\pi}\cdot n!$.

Proof : We have the generating function, $e^{2tx-t^2} = \sum_{n=0}^{\infty} H_n(x)\frac{t^n}{n!}$.

Differentiating partially w.r.t. x,

$$2(x-t)e^{2tx-t^2} = \sum_{n=1}^{\infty} H_n(x)\frac{t^{n-1}}{(n-1)!}$$

or, $$2x\sum_{n=0}^{\infty} H_n(x)\frac{t^n}{n!} - 2\sum_{n=0}^{\infty} H_n(x)\frac{t^{n+1}}{n!} = \sum_{n=1}^{\infty} H_n(x)\frac{t^{n+1}}{(n-1)!}.$$

Equating the coefficient of t^n from both sides, we get

$$2x\frac{H_n(x)}{n!} - 2\frac{H_{n-1}(x)}{(n-1)!} = \frac{H_{n+1}(x)}{n!}$$

or, $$H_{n+1} - 2xH_n + 2nH_{n-1} = 0. \tag{18.84}$$

Replacing n by $n-1$,

$$H_n - 2xH_{n-1} + 2(n-1)H_{n-2} = 0$$

or, $$H_n^2 - 2xH_nH_{n-1} + 2(n-1)H_nH_{n-2} = 0. \tag{18.85}$$

Multiplying Eqn. (18.84) by H_{n-1},

$$H_{n-1}H_{n+1} - 2xH_nH_{n-1} + 2nH_{n-1}^2 = 0. \tag{18.86}$$

Subtracting Eqn. (18.86) from (18.85), we get

$$H_n^2 - H_{n-1}H_{n+1} + 2(n-1)H_nH_{n-2} - 2nH_{n-1}^2 = 0. \tag{18.87}$$

Multiplying Eqn. (18.87) by e^{-x^2} and then integrating with respect to x from $-\infty$ to ∞, we get

$$\int_{-\infty}^{+\infty} e^{-x^2}H_n^2\,dx - \int_{-\infty}^{+\infty} e^{-x^2}H_{n+1}H_{n-1}\,dx + 2(n-1)\int_{-\infty}^{+\infty} e^{-x^2}H_nH_{n-2}\,dx - 2n\int_{-\infty}^{+\infty} e^{-x^2}H_{n-1}^2\,dx = 0. \tag{18.88}$$

Using the orthogonal property (Eqn. 18.83), we get

$$\int_{-\infty}^{+\infty} e^{-x^2}H_n^2(x)dx = 2n\int_{-\infty}^{+\infty} e^{-x^2}H_{n-1}^2(x)dx = 2^2n(n-1)\int_{-\infty}^{+\infty} e^{-x^2}H_{n-2}^2(x)dx$$

$$\cdots\cdots\cdots\cdots\cdots\cdots\cdots\cdots\cdots\cdots\cdots\cdots\cdots$$

$$= 2^n \cdot n! \int_{-\infty}^{+\infty} e^{-x^2} H_0^2(x) dx = 2^n \cdot n! \int_{-\infty}^{+\infty} e^{-x^2} dx = 2^n \cdot n! 2 \frac{\sqrt{\pi}}{2}$$

$$= 2^n \cdot n! \sqrt{\pi}. \tag{18.89}$$

Eqn. (18.83) and (18.89) may be combined to the following form :

$$\int_{-\infty}^{+\infty} e^{-x^2} H_n(x) H_m(x) dx = 2^n n! \sqrt{\pi} \delta_{m,n}, \tag{18.90}$$

where $\delta_{m,n} = \begin{cases} 0, & \text{if } n \neq m \\ 1, & \text{if } n = m. \end{cases}$

18.30 Laguerre Equation

Laguerre differential equation is given by

$$x \frac{d^2y}{dx^2} + (1 - x) \frac{dy}{dx} + ny = 0, \quad \text{where } n \text{ is an integer.} \tag{18.91}$$

Since Eqn. (18.91) has non-essential singularity at $x = 0$, method of series can be applied to find its solution.

Let us assume, $x = \frac{1}{z}$. $\therefore$ $z = \frac{1}{x}$, so that

$$\frac{dy}{dx} = -z^2 \frac{dy}{dz}; \qquad \frac{d^2y}{dx^2} = z^4 \frac{d^2y}{dz^2} + 2x^3 \frac{dy}{dz}.$$

Substituting these values in Eqn. (18.91), we get

$$z^3 \frac{d^2y}{dz^2} + z^2 \frac{dy}{dz} + z \frac{dy}{dz} + ny = 0. \tag{18.92}$$

Let the solution of Eqn. (18.92) be $y = \sum_{r=0}^{\infty} a_r x^{m+r}$.

$$\therefore \quad \frac{dy}{dz} = \sum_{r=0}^{\infty} a_r (m + r) z^{m+r-1}; \qquad \frac{d^2y}{dz^2} = \sum_{r=0}^{\infty} a_r (m + r)(m + r - 1) z^{m+r-2}.$$

Substituting these values in Eqn. (18.92), we get

$$\sum_{r=0}^{\infty} a_r (m + r)(m + r - 1) z^{m+r-2} + \sum_{r=0}^{\infty} a_r (m + r) z^{m+r-1} + \sum_{r=0}^{\infty} a_r (m + r) z^{m+r} + n \sum_{r=0}^{\infty} a_r z^{m+r} = 0$$

$$\sum_{r=0}^{\infty} a_r (m + r)^2 z^{m+r+1} + \sum_{r=0}^{\infty} a_r (m + r + n) z^{m+r} = 0. \tag{18.93}$$

Equating the coefficient of z^{m+r} to zero,

$$a_{r-1}(m + r - 1)^2 + a_r(m + r + n) = 0 \quad \text{or,} \quad a_r = \frac{(m + r - 1)^2}{m + n + r} a_r - 1.$$

$$\therefore \quad a_1 = -\frac{m^2}{m + n + 1} a_0$$

$$a_2 = -\frac{(m + n)^2}{m + n + 2} a_1 = \frac{m^2 (m + 1)^2}{(m + n + 1)(m + n + 2)} a_0 \cdots .$$

$$\therefore \quad y = a_0 z^n \left[1 + \frac{a_1}{a_0} z + \frac{a_2}{a_0} z^2 + \cdots \right]$$

or, $$y = a_0 z^m \left[1 - \frac{m^2}{m+n+1} z + \frac{m^2(m+1)^2}{(m+n+1)(m+n+2)} z^2 + \cdots \right]. \quad (18.94)$$

Using the above equation in (18.92), we get

$$z^3 \frac{d^2y}{dz^2} + z^2 \frac{dy}{dz} + z\frac{dy}{dz} + ny = a_0(m+n)z^m.$$

For $m = -n$, the RHS of the above equation becomes zero.

$\therefore$ the solution of Laguerre equation is

$$y = a_0 x^{-n} \left[1 - \frac{n^2}{1!}\frac{1}{x} + \frac{n^2(n-1)^2}{2!}\frac{1}{x^2} - \cdots \right]. \quad (18.95)$$

18.31 Laguerre Polynomial

For $n = +$ve integer, the solution of Laguerre differential equation is

$$y = a_0 \left[x^n - \frac{n^2}{1!}x^{n-1} + \frac{n^2(n-1)^2}{2!}x^{n-2} + \cdots + (-1)^n n!\right].$$

Putting $a_0 = (-1)^n$, we get the Laguerre polynomial as

$$L_n(x) = (-1)^n \left[x^n - \frac{n^2}{1!}x^{n-1} + \frac{n^2(n-1)^2}{2!}x^{n-2} + \cdots + (-1)^n n!\right]. \quad (18.96)$$

Putting $n = 0, 1, 2, \ldots$, we get

$$L_0(x) = 1,\; L_1(x) = 1 - x,\; L_2(x) = 1 - 2x + \frac{x^2}{2},\; L_3(x) = 1 - 3x + \frac{3}{2}x^2 - \frac{x^3}{6},$$

$$L_4(x) = 1 - 4x + 3x^2 + \frac{2x^3}{3} + \frac{x^4}{24} \text{ and so on.}$$

18.32 Generating Function of $L_x(x)$

We have $$\frac{1}{1-t}e^{-\frac{xt}{1-t}} = \frac{1}{1-t}\left[1 - \frac{xt}{1-t} + \frac{x^2t^2}{2!(1-t)^2} + \cdots + (-1)^n \frac{x^n t^n}{n!(1-t)^n} + \cdots \right]$$

$$= (1-t)^{-1} - xt(1-t^2) + \frac{1}{2!}x^2t^2(1-t)^{-3} + \cdots$$

$$+ (-1)^n \frac{x^n t^n}{n!}(1-t)^{-n+1} + \cdots$$

$$\therefore \text{ coefficient of } t^n = \frac{(-1)^n x^n}{n!} + \frac{(-1)^{n-1}x^{n-1}}{(n-1)!}n + \frac{(-1)^{n-2}x^{n-2}}{(n-2)!}\frac{n(n-1)}{2} + \cdots$$

$$= \frac{(-1)^n}{n!}\left[x^n - n^2x^{n-1} + \frac{n^2(n-1)^2}{2!}x^{n-2} + \cdots \right] = \frac{L_n(x)}{n!}.$$

$$\therefore \quad \frac{1}{1-t}e^{-\frac{xt}{1-t}} = \sum_{n=0}^{\infty} \frac{L_n(x)}{n!}t^n. \quad (18.97)$$

which is the generating function of $L_n(x)$.

18.33 Rodrigue's Formula for $L_n(x)$

$$\frac{d^n}{dx^n}(x^n e^{-x}) = (-1)^n e^{-x} x^n + (-1)^{n-1} {}^nC_1 e^{-x} n x^{n-1} + (-1)^{n-2} {}^nC_2 e^{-x} n(n-1) x^{n-2} + \cdots$$

$$= (-1)^n e^{-x}\left[x^n - n^2 x^{n-1} + \frac{n^2(n-1)^2}{2!} x^{n-2} - \cdots\right] = e^{-x} L_n(x).$$

$$\therefore \quad L_n(x) = e^x \frac{d^n}{dx^n}(x^n e^{-x}). \tag{18.98}$$

This is the **Rodrigue's formula for $L_n(x)$**.

From Eqn. (18.98), putting $n = 0, 1, 2, \ldots$, we get

$$L_0(x) = e^x(x^0 e^{-x}) = 1, \quad L_1(x) = e^x \frac{d}{dx}(xe^{-x}) = e^x[x(-e^{-x}) + e^{-x}(1)] = 1 - x.$$

$$L_2(x) = e^x \frac{d^2}{dx^2}(x^2 e^{-x}) = e^x[x^2 e^{-x} - 4xe^{-x} + 2e^{-x}] = x^2 - 4x + 2 \text{ and so on.}$$

18.34 Recurrence Relation for $L_n(x)$

We have $(1-t)\sum \frac{L_n t^n}{n!} = e^{-\frac{xt}{1-t}} = \phi(x, t)$, say. $\quad$ (18.99)

$$\therefore \quad \ln \phi = -\frac{xt}{1-t} = \frac{x(1-t) - x}{1-t} = x - \frac{x}{1-t}.$$

Differentiating w.r.t. x and t successively

$$\frac{1}{\phi}\frac{\partial \phi}{\partial x} = 1 - \frac{1}{1-t} = -\frac{t}{1-t} \tag{18.100}$$

$$\frac{1}{\phi}\frac{\partial \phi}{\partial t} = -\frac{x}{(1-t)^2}. \tag{18.101}$$

Now, from (18.99), $\frac{\partial \phi}{\partial x} = (1-t)\sum_{n=0}^{\infty} \frac{L_n' t^n}{n!}$

and from (18.100), $\frac{\partial \phi}{\partial x} = -\frac{\phi t}{1-t} = -t\sum_{n=0}^{\infty} \frac{L_n t^n}{n!}.$

$$\therefore \quad -t\sum_{n=0}^{\infty} \frac{L_n t^n}{n!} = (1-t)\sum_{n=0}^{\infty} \frac{L_n' t^n}{n!}. \tag{18.102}$$

Equating the coefficient of t^n from both sides of (18.102),

$$-\frac{L_{n-1}}{(n-1)!} = \frac{L_n'}{n!} - \frac{L_{n-1}'}{(n-1)!}$$

or, $\quad \boldsymbol{L_n' - nL_{n-1}' + nL_{n-1} = 0}.$ $\quad$ (18.103)

Again, from (18.99), $\frac{\partial \phi}{\partial t} = \sum_{n=0}^{\infty} \frac{nL_n t^{n-1}}{n!} = -\sum_{n=0}^{\infty} \frac{(n+1)L_n t^n}{n!}$

and from (18.101), $\frac{\partial \phi}{\partial t} = -\frac{x}{(1-t)^2}\omega = -\frac{x}{1-t}\sum_{n=0}^{\infty} \frac{L_n t^n}{n!}.$

$$\therefore \quad -x\sum_{n=0}^{\infty}\frac{L_n}{n!}t^n = (1-t)\sum_{n=0}^{\infty}\frac{L_n t^{n-1}}{(n-1)!} - (1-t)\sum_{n=0}^{\infty}\frac{L_n t^n(n+1)}{n!}. \tag{18.104}$$

Equating the coefficients of t^n,

$$-x\frac{L_n}{n!} = \frac{L_{n+1}}{n!} - \frac{L_n}{(n-1)!} - \frac{(n+1)}{n!}L_n + \frac{L_{n-1}}{(n-1)!}n = \frac{1}{n!}[L_{n+1} - (2n+1)L_n + n^2 L_{n-1}]$$

or, $$\boldsymbol{L_{n+1} + (x - 2n - 1)L_n + n^2 L_{n-1} = 0.} \tag{18.105}$$

18.35 Orthogonal Properties of $L_n(x)$

The orthogonal properties of $L_n(x)$ are expressed as

1. $\displaystyle\int_0^\infty e^{-x}L_m(x)L_n(x)dx = 0$, **when** $m \neq n$

2. $\displaystyle\int_0^\infty e^{-x}L_n^2(x)dx = [n!]^2$, **when** $m = n$.

Proof of (1) : If $y = L_n(x)$ be the solution of Laguerre equation, then

$$x\frac{d^2L_n}{dx^2} + (1-x)\frac{dL_n}{dx} + nL_n = 0 \quad \text{or,} \quad xe^{-x}\frac{d^2L_n}{dx^2} + (1-x)e^{-x}\frac{dL_n}{dx} + ne^{-x}L_n = 0$$

or, $$\frac{d}{dx}\left[e^{-x}x\frac{dL_n}{dx}\right] + ne^{-x}L_n = 0. \tag{18.106}$$

Similarly, $$\frac{d}{dx}\left[e^{-x}x\frac{dL_m}{dx}\right] + me^{-x}L_m = 0. \tag{18.107}$$

Multiplying (18.106) by L_m and (18.107) by L_n and subtracting and then integrating the result w.r.t. x from $-\infty$ to $+\infty$, we get

$$\int_0^\infty L_m\frac{d}{dx}\left(xe^{-x}\frac{dL_n}{dx}\right)dx - \int_0^\infty L_n\frac{d}{dx}\left(xe^{-x}\frac{dL_m}{dx}\right)dx + (n-m)\int_0^\infty e^{-x}L_mL_n\,dx = 0 \tag{18.108}$$

or, $$(n-m)\int_0^\infty e^{-x}L_mL_n\,dx = 0.$$

$$\left[\because \int_0^\infty L_m\frac{d}{dx}\left(xe^{-x}\frac{dL_n}{dx}\right)dx = L_mxe^{-x}\frac{dL_n}{dx} - \int_0^\infty xe^{-x}\frac{dL_m}{dx}\frac{dL_n}{dx} = 0 - \int_0^\infty xe^{-x}\frac{dL_m}{dx}\frac{dL_n}{dx}\right.$$

and $$\left.\int_0^\infty L_n\frac{d}{dx}\left(xe^{-x}\frac{dL_m}{dx}\right)dx = -\int_0^\infty xe^{-x}\frac{dL_m}{dx}\frac{dL_n}{dx}\right]$$

$$\therefore \quad \int_0^\infty \boldsymbol{e^{-x}L_mL_n\,dx = 0}, \quad \text{if } m \neq n. \tag{18.109}$$

Proof of (2) : We have $L_{n+1} + (x-2n-1)L_n + n^2L_{n-1} = 0$ [18.105]

Replacing n by $(n-1)$, we get

$$L_n + (x-2n-1)L_{n-1} + (n-1)^2L_{n-2} = 0. \tag{18.110}$$

Multiplying Eqn. (18.110) by L_n and (18.105) by L_{n-1}, we get

$$L_n^2 + (x - 2n + 1)L_n L_{n-1} + (n-1)^2 L_n L_{n-2} = 0 \tag{18.111}$$

$$L_{n+1}L_{n-1} + (x - 2n - 1)L_n L_{n-1} + n^2 L_{n-1}^2 = 0. \tag{18.112}$$

Subtracting, Eqn. (18.112) from (18.111), we get

$$L_n^2 - L_{n+1}L_n + 2L_n L_{n-1} + (n-1)^2 L_n L_{n-2} - n^2 L_{n-1}^2 = 0.$$

Multiplying by e^{-x} and integrating w.r.t. x from 0 to ∞ and using the orthogonal property (1), we get

$$\int_0^\infty e^{-x} L_n^2 \, dx = n^2 \int_0^\infty e^{-x} L_{n-1}^2 \, dx = n^2 (n-1)^2 \int_0^\infty e^{-x} L_{n-2}^2 \, dx = \cdots$$

$$= n^2(n-1)^2 \cdots 1^2 \int_0^\infty e^{-x} L_0^2 \, dx = (n!)^2 \int_0^\infty e^{-x} dx \qquad (\because \; L_0 = 1)$$

$$= (n!)^2.$$

$$\therefore \quad \int_0^\infty \boldsymbol{e^{-x} L_n^2 dx = (n!)^2}. \tag{18.113}$$

18.36 Associated Laguerre's Equation

We consider the equation, $x\dfrac{d^2 z}{dx^2} + (k + 1 - x)\dfrac{dz}{dx} + (n - k)z = 0,$ (18.114)

where $z = L_n^k(x) = \dfrac{d^k}{dx^k} L_n(x)$ is known as the **associated Laguerre equation**. The function $L_n^k(x)$ is called the **associated Laguerre's** polynomials of degree $(n - k)$.

The **generating function** for associated Laguerre's polynomial is

$$\frac{(-t)^k}{(1-t)^{k+1}} e^{-\frac{tx}{1-t}} = \sum_{n=0}^{\infty} L_n^k(x) t^n. \tag{18.115}$$

The **Rodrigue's formula** for associated Laguerre's polynomial is

$$L_n^k(x) = \frac{1}{n!} e^x x^{-k} \frac{d^n}{dx^n}(e^{-x} x^{n+k}). \tag{18.116}$$

The function, $\boldsymbol{Y_{n,k}(x) = e^{-\frac{x}{2}} x^{\frac{k}{2}} L_n^k(x)}$ (18.117)

is called **associated Laguerre's function.**

The orthogonality conditions are

$$\int_0^\infty \boldsymbol{Y_{n,k}(x) Y_{m,k}(x) dx} = \begin{cases} \boldsymbol{0} & \textbf{, for } \boldsymbol{m \neq n} \\ \boldsymbol{\dfrac{(k+n)!}{n!}} & \textbf{, for } \boldsymbol{m = n} \end{cases} \tag{18.118}$$

A piece-wise continuous function $f(x)$ in $(0, \infty)$ can be expanded in terms Laguerre's function as

$$f(x) = \sum_{n=0}^{\infty} a_n L_n(x), \tag{18.119}$$

$$\text{where } a_n = \int_0^\infty L_n(x) f(x) dx. \tag{18.120}$$

18.37 Chebyshev Polynomials

We consider the differential equation,

$$(1-x^2)\frac{d^2y}{dx^2} - x\frac{dy}{dx} + n^2y = 0, \tag{18.121}$$

where n is **a positive integer**. The solution of the above equation can be found out by putting $x = \cos\phi$. Then the above equation reduces to

$$\frac{d^2y}{d\phi^2} + n^2y = 0, \tag{18.122}$$

whose solutions are :

$$\begin{aligned} y_1 &= \cos n\phi = \cos(n\cos^{-1}x) \\ &= x^n - {}^2C_n x^{n-2}(1-x^2) + {}^4C_n x^{n-4}(1-x^2)^2 - \cdots \end{aligned} \tag{18.123}$$

$$\begin{aligned} y_2 &= \sin n\phi = \sin(n\cos^{-1}x) \\ &= {}^1C_n x^{n-1}(1-x^2)^{\frac{1}{2}} - {}^3C_n x^{n-3}(1-x^2)^{\frac{3}{2}} + \cdots \end{aligned} \tag{18.124}$$

These are polynomials of degree n and called **Chebyshev's polynomials of first kind**. Thus,

$$T_n(x) = \cos(n\cos^{-1}x) \tag{18.125}$$

and $$T_n(\cos\phi) = \cos n\phi. \tag{18.126}$$

Rodrigue's formula for Chebyshev formula of first kind is

$$T_n(x) = \frac{2^n n!}{(2n)!}\sqrt{n^2-1}\frac{d^n}{dx^n}[(x^2-1)^{n-\frac{1}{2}} \tag{18.127}$$

$$\therefore \quad \left.\begin{aligned} T_n(-x) &= (-1)^n T_n(x) \\ T_n(1) &= 1 \\ T_n(-1) &= (-1)^n \end{aligned}\right\} \tag{18.128}$$

So, $T_0(x) = 1,\ T_1(x) = x,\ T_2(x) = 2x^2 - 1,\ T_3(x) = 4x^3 - 3x,$
$T_4(x) = 8x^4 - 8x^2 + 1,\ T_5(x) = 16x^5 - 20x^3 + 5x.$

The orthogonal property is $\displaystyle\int_{-1}^{+1}(1-x^2)T_m(x)T_n(x)dx = \begin{cases} 0, & m \neq n \\ \frac{\pi}{2}, & m = n \\ \pi, & m = n = 0\,. \end{cases}$

18.38 Solved Problems

Example 18.1. (a) *Using the generating function for the Legendre polynomials*

$$G(x,t) = (1-2xt+t^2)^{-\frac{1}{2}} = \sum_0^\infty P_n(x)t^n, \quad -1 \le x \le 1,\ t < 1.$$

Show that, $P_n(-x) = (-1)^n P_n(x)$ *and* $P_n(1) = 1$. **[CU(H) 2007, BU(H) 2005]**

(b) *Prove that,* $\displaystyle\int_{-1}^{1} P_m(x)P_n(x)dx = 0$, *for* $m \neq n$.

[VU(H) 2009, '06; BU(H) 1993; CU(H) 2006]

Solution : (a) **See Art. 18.8.1 (p. 674 and p. 675)**
(b) **See Art. 18.10 (p. 678)**

Example 18.2. (a) *Show that for the Legendre polynomial* $P_n(x)$

$$\int_{-1}^{+1} [P_n(x)]^2 dx = \frac{2}{2n+1} \text{ (when } n \text{ is zero or a positive integer).}$$ **[VU(H) 2004]**

(b) *Consider the following generating relation for Legendre polynomials,*

$$(1 - 2xt + t^2)^{-\frac{1}{2}} = \sum_0^\infty P_n(x)t^n.$$

Using the above derive the following recursion relation,

$$(n+1)P_{n+1}(x) = (2n+1)xP_n(x) - nP_{n-1}(x).$$

[cf. VU(H) 2010, VU(H) 2002, '05; CU(H) 2007]

Solution : (a) **See Art. 18.10 (p. 678)**
(b) **See Art. 18.9 (p. 676)**

Example 18.3. *Prove that,* $P_l(x) = \dfrac{1}{2^l l!}\dfrac{d^l}{dx^l}(x^2-1)^l$ *satisfies the Legendre differential equation.* **[BU(H) 2000, VU(H) 2001]**

Solution : Let $u = (x^2-1)^l$.

Then, $\dfrac{du}{dx} = 2lx(x^2-1)^{l-1}$ or, $(1-x^2)\dfrac{du}{dx} + 2lxu = 0.$

Differentiating w.r.t. x, we get

$$(1-x^2)\frac{d^2u}{dx^2} - 2x\frac{du}{dx} + 2lx\frac{du}{dx} + 2lu = 0$$

or, $$(1-x^2)\frac{d^2u}{dx^2} + 2(l-1)x\frac{du}{dx} + 2lu = 0. \quad \text{(i)}$$

Differentiating successively n times with the help of Leibnitz's theorem,

$$(1-x^2)\frac{d^{n+2}}{dx^{n+2}} + 2(l-n-1)x\frac{d^{n+1}}{dx^{n+1}} + (n+1)(2l-n)\frac{d^n u}{dx^n} = 0.$$

Putting $\dfrac{du}{dx} = u_n$, we get $(1-x^2)\dfrac{d^2}{dx^2}(u_n) + 2(l-n-1)x\dfrac{d}{dx}(u_n) + (n+1)(2l-n)u_n = 0.$

Now, if $n = l$, then

$$(1-x^2)\frac{d^2}{dx^2}(u_l) - 2x\frac{d}{dx}(u_l) + l(l+1) = 0. \quad \text{(ii)}$$

This is the Legendre's differential equation in u_l, where $u_l = \dfrac{d^l}{dx^l}(x^2-1)^l$.

$\therefore$ $\quad \boldsymbol{P_l(x) = Cu_l}$, **where** $\boldsymbol{C = \dfrac{1}{2^l l!}}$

i.e. $\quad P_l(x) = \dfrac{1}{2^l l!}\dfrac{d^l}{dx^l}(x^2-1)^l.$

Hence, u_l and $P_l(x)$ satisfies Legendre differential equation.

Example 18.4. *Prove that $P_l(x)$ satisfies the differential equation*

$$(1-x^2)y''-2xy'+l(l+1)y=0. \quad \text{[BU(H) 1992]}$$

Solution : We have the relation, $(x^2-1)P_l'(x)=l[xP_l(x)-P_{l-1}(x)]$. (i)

Differentiating w.r.t. x, we get

$$(x^2-1)P_l''(x)+2xP_l'(x)=l[P_l(x)+xP_l'(x)-P_{l-1}'(x)]. \quad \text{(ii)}$$

Again, we have the relation,

$$P_{l-1}'(x)+lP_l(x)=xP_l'(x) \quad \text{or,} \quad lP_{l-1}'(x)+l^2P_l(x)=xlP_l'(x). \quad \text{(iii)}$$

Adding (ii) and (iii), we get

$$(1-x^2)P_l''(x)-2xP_l'(x)+l^2P_l(x)=lxP_l'(x)-lP_l(x)-lxP_l'(x)$$

or, $$\boldsymbol{(1-x^2)P_l''(x)-2xP_l'(x)+l(l+1)P_l(x)=0.} \quad \text{(iv)}$$

Hence, $P_l(x)$ satisfies Legendre equation.

Example 18.5. *Using Rodrigue's formula evaluate $P_1(x)$, $P_2(x)$, $P_3(x)$.* [BU(H) 1994]

Solution : We know the Rodrigue's formula, $P_n(x)=\dfrac{1}{2^n\cdot n!}\dfrac{d^n}{dx^n}(x^2-1)^n$.

For $n=1$, $P_1(x)=\dfrac{1}{2\cdot 1}\dfrac{d}{dx}(x^2-1)=\dfrac{2x}{2}=x$.

$\therefore$ $\boldsymbol{P_1(x)=x}$

For $n=2$, $P_2(x)=\dfrac{1}{2^2\cdot 2!}\dfrac{d^2}{dx^2}(x^2-1)^2=\dfrac{1}{8}\dfrac{d^2}{dx^2}(x^4-2x^2+1)=\dfrac{1}{8}\dfrac{d}{dx}(4x^3-4x)=\dfrac{1}{2}\dfrac{d}{dx}(x^3-x)$.

$\therefore$ $\boldsymbol{P_2(x)=\frac{1}{2}(3x^2-1)}$.

For $n=3$, $P_3(x)=\dfrac{1}{2^3\cdot 3!}\dfrac{d^3}{dx^3}(x^2-1)^3=\dfrac{1}{48}\dfrac{d^3}{dx^3}[3(x^2-1)2x]=\dfrac{1}{8}\dfrac{d^2}{dx^2}(x^5-2x^3+x)$.

$\therefore$ $\boldsymbol{P_3(x)=\frac{1}{2}(5x^3-3x)}$.

Example 18.6. *Prove that,* $\displaystyle\int_{-1}^{+1}xP_n(x)P_{n-1}(x)dx=\frac{2n}{4n^2-1}$. [VU(H) 2012, NBU(H) 2007]

Solution : From the recurrence relation,

$$(2n+1)xP_n(x)=(n+1)P_{n+1}(x)+nP_{n-1}(x)$$

or, $(2n-1)xP_{n-1}(x)=nP_n(x)+(n-1)P_{n-2}(x)$ [replacing n by $n-1$]

Multiplying both sides by $P_n(x)$,

$$xP_n(x)P_{n-1}(x)=\frac{1}{2n-1}\int_{-1}^{+1}P_n^2dx+\frac{n-1}{2n-1}\int_{-1}^{+1}P_n(x)dxP_{n-2}(x)dx$$

$$=\frac{n}{2n-1}\left(\frac{2}{2n+1}\right)+\frac{n}{2n-1} \quad \text{[by orthogonality of property of } P_n(x)]$$

$$\therefore \quad \boldsymbol{\int_{-1}^{+1}xP_n(x)P_{n-1}(x)dx=\frac{2n}{4n^2-1}} \quad \textbf{(Proved).}$$

Example 18.7. *Using the Rodrigue's formula for Legendre functions, prove that*

$$\int_{-1}^{+1} x^m P_n(x)dx = 0,$$

where m, n are positive integers and $m < n$. **[AMIETE(Summer) 1997]**

Solution : $\displaystyle\int_{-1}^{+1} x^m P_n(x)dx = \int_{-1}^{+1} x^m \frac{1}{2^n \cdot n!}\frac{d^n}{dx^n}(x^2-1)^n dx = \frac{1}{2^n \cdot n!}\int_{-1}^{+1} x^m \frac{d^n}{dx^n}(x^2-1)^n dx.$

Integrating by parts,

$$= \frac{1}{2^n \cdot n!}\left[\left\{x^m \frac{d^{n-1}}{dx^{n-1}}(x^2-1)^n\right\}_{-1}^{+1} - \int_{-1}^{+1} mx^{m-1}\frac{d^{n-1}}{dx^{n-1}}(x^2-1)^n dx\right]$$

$$= 0 - \frac{m}{2^n \cdot n!}\int_{-1}^{+1} x^{m-1}\frac{d^{n-1}}{dx^{n-1}}(x^2-1)^n dx$$

$$\therefore \int_{-1}^{+1} x^m P_n(x)dx = -\frac{m}{2^n \cdot n!}\int_{-1}^{+1} x^{m-1}\frac{d^{n-1}}{dx^{n-1}}(x^2-1)^n dx$$

$$= -(-1)^n \frac{m(m-1)}{2^n \cdot n!}\int_{-1}^{+1} x^{m-2}\frac{d^{n-2}}{dx^{n-2}}(x^2-1)^n dx.$$

Integrating $(m-2)$ times, we get

$$= -(-1)^m \frac{m(m-1)\cdots 1}{2^n \cdot n!}\int_{-1}^{+1} x^{n-m}\frac{d^{n-m}}{dx^{n-1}}(x^2-1)^n dx$$

$$= \frac{(-1)^m m!}{2^n \cdot n!}\int_{-1}^{+1} x^{n-1}\frac{d^{n-1}}{dx^{n-m}}(x^2-1)^n dx$$

$$\therefore \int_{-1}^{+1} \boldsymbol{x^m P_n(x)dx} = \frac{(-1)^m m!}{2^n \cdot n!}\left[\frac{d^{n-m-1}}{dx^{n-m-1}}(x^2-1)^m\right]_{-1}^{+1} = 0 \textbf{(Proved)}.$$

Example 18.8. *Starting from the generating function for Legendre's polynomials prove that,*

$$P_{2n}(0) = (-1)^n \frac{1\cdot 3\cdot 5\cdots(2n-1)}{2\cdot 4\cdot 6\cdots 2n}.$$ **[BU(H) 2004, AMIE(Winter) 1988]**

Solution : We have the generating function for Legendre's polynomials

$$\sum t^{2n} P_{2n}(x) = (1-2xt+t^2)^{-\frac{1}{2}}$$

or, $\displaystyle\sum t^{2n} P_{2n}(0) = (1+z^2)^{-\frac{1}{2}}$ [for $x = 0$]

$$= 1 + \left(-\frac{1}{2}\right)t^2 + \frac{\left(-\frac{1}{2}\right)\left(-\frac{3}{2}\right)}{2!}(t^2)^2 + \frac{\left(-\frac{1}{2}\right)\left(-\frac{3}{4}\right)\left(-\frac{5}{2}\right)}{3!}(t^2)^3$$

$$+ \frac{\left(-\frac{1}{2}\right)\left(-\frac{3}{4}\right)\left(-\frac{5}{2}\right)\cdots\left(-\frac{1}{2}-n+1\right)}{n!}(t^2)^n + \cdots.$$

Equating the coefficients of t^{2n} from both sides, we get

$$\boldsymbol{P_{2n}} = \frac{\left(-\frac{1}{2}\right)\left(-\frac{3}{4}\right)\left(-\frac{5}{2}\right)\cdots\left(-\frac{1}{2}-n+1\right)}{n!} = (-1)^n\frac{1\cdot 3\cdot 5\cdots(2n-1)}{2\cdot 4\cdot 6\cdot 8\cdots 2n} \textbf{ (Proved)}.$$

Example 18.9. *Starting from the generating function of Legendre polynomials show that,*

$$P_l(-1) = (-1)^l.$$

Solution : **See Art. 18.8.1 (p. 674)**

Example 18.10. *Show that,* $\displaystyle\int_{-1}^{+1} x^2 P_{n-1}P_{n+1}dx = \frac{2n(n+1)}{(2n-1)(2n+1)(2n+3)}.$

[Kerala U M.Tech 2005, JNTU 2005]

Solution : We know the recurrence relation, $(2n+1)xP_n = (n+1)P_{n+1} + nP_{n-1}$.

$$xP_{n-1} = \frac{1}{2n-1}\{nP_n + (n-1)P_{n-2}\}.$$

$$\therefore \quad xP_{n+1} = \frac{1}{2n+3}\{(n+2)P_{n+2} + (n+1)P_n\}.$$

$$\therefore \quad x^2 P_{n-1}P_{n+1} = \frac{1}{(2n+3)(2n-1)}\{n(n+2)P_nP_{n+2} + n(n+1)P_n^2 + (n-1)(n+2)P_{n-2}P_{n+2} + (n^2-1)P_nP_{n-2}\}.$$

Integrating both sides from -1 to $+1$ and using orthogonality of Legendre polynomials, we get

$$\int_{-1}^{+1} x^2 P_{n-1}P_{n+1}\,dx = \frac{n(n+1)}{(2n-1)(2n+3)}\int_{-1}^{+1} P^2 n\,dx = \frac{2n(n+1)}{(2n-1)(2n+1)(2n+3)}.$$

Example 18.11. *If $f(x) = |x|$ defined for $-1 \le x \le 1$, expand $f(x)$ in terms of $P_n(x)$.*

Solution : Let $f(x) = \displaystyle\sum_{x=0}^{\infty} C_n P_n(x)$, then $C_n = \left(n + \frac{1}{2}\right)\displaystyle\int_{-1}^{+1} f(x)P_n(x)dx.$

Here, $f(x) = -x, \quad -1 \le x < 0$

$\qquad\qquad = 0, \quad 0 \le x \le 1$

$$\therefore \quad C_0 = \frac{1}{2}\int_{-1}^{1} f(x)P_0(x) = \frac{1}{2}\left[\int_{-1}^{1} x\,dx + \int_{0}^{1} x\,dx\right] = \frac{1}{2}$$

$$C_1 = \frac{3}{2}\int_{-1}^{1} f(x)P_1(x) = \frac{3}{2}\left[\int_{-1}^{0} -x^2dx + \int_{0}^{1} x^2dx\right] = 0$$

$$C_2 = \frac{5}{2}\int_{-1}^{1} f(x)P_2(x) = \frac{5}{3}\left[\int_{-1}^{0} -x\frac{1}{2}(3x^2-1)dx + \int_{0}^{1} x\frac{1}{2}(3x^2-1)dx\right]$$

$$= \frac{5}{4}\left[\int_{-1}^{0} (x-3x^3)dx + \int_{0}^{1} (3x^2-x)dx\right] = \frac{5}{8}$$

$$C_3 = \frac{7}{2}\int_{-1}^{1} f(x)P_3(x) = \frac{7}{2}\left[\int_{-1}^{0} -x\frac{1}{2}(5x^3-3x)dx + \int_{0}^{1} x(5x^3-3x)dx\right]$$

$$= \frac{7}{4}\left[\int_{-1}^{0} (3x^2-5x^4)dx + \int_{0}^{1} (5x^4-3x^2)dx\right] = 0$$

$$C_4 = \frac{9}{2}\int_{-1}^{1} f(x)P_4(x) = \frac{9}{2}\int_{-1}^{0} -\frac{x}{8}(35x^4 - 30x + 3)dx + \frac{9}{2}\int_{0}^{1}\frac{x}{8}(35x^4 - 30x^2 + 3)dx$$

$$= -\frac{3}{16}.$$

$$\therefore \quad f(x) = \frac{1}{2}P_0(x) + \frac{5}{8}P_2(x) - \frac{3}{16}P_4(x) + \cdots.$$

Example 18.12. *Bessel differential equation is given by $x^2y'' + xy' + (x^2 - n^2)y = 0$.*

Find by the method of series substitution, two independent solutions of this equation when n is not an integer.

Is your second solution independent of the first when n is an integer? **[BU(H) 1999]**

Solution : See Art 18.14 (p. 683)

For $n =$ an integer, i.e., $n = 0, 1, 2, 3, \ldots J_{-n}(x)$, the second solution is independent of the first.

Example 18.13. *Using the following recurrence relations satisfied by Bessel's functions of first kind* (i) $J_{n-1}(x) + J_{n+1}(x) = \frac{2n}{x}J_n(x)$ (ii) $J_{n-1}(x) - J_{n+1}(x) = 2\frac{d}{dx}J_n(x)$.

Show that $\left(\frac{1}{x}\frac{d}{dx}\right)^m [x^n J_n(x)] = x^{n-m}J_{n-m}(x)$

and $\left(\frac{1}{x}\frac{d}{dx}\right)^m [x^{-n} J_n(x)] = (-1)^m x^{-(n+m)}J_{n+m}(x)$. **[BU(H) 1996]**

Solution : We have $\frac{d}{dx}[x^n J_n(x)] = x^n J_n'(x) + nx^{n-1}J_n(x) = x^n\left[J_n'(x) + \frac{n}{x}J_n(x)\right]$

$$= x^n\left[\frac{1}{2}J_{n-1}(x) - \frac{1}{2}J_{n+1}(x) + \frac{1}{2}J_{n-1}(x) + \frac{1}{2}J_{n+1}(x)\right]$$

[from (i)]

$$= x^n J_{n-1}(x).$$

$$\therefore \quad \left(\frac{1}{x}\frac{d}{dx}\right)^1 [x^n J_n(x)] = x^{n-1}J_{n-1}(x).$$

If it holds good for the mth differentiation,

$$\left(\frac{1}{x}\frac{d}{dx}\right)^m [x^m J_n(x)] = x^{n-m}J_{n-m}(x)$$

$$\left(\frac{1}{x}\frac{d}{dx}\right)^{m+1} [x^n J_n(x)] = \left(\frac{1}{x}\frac{d}{dx}\right)\left(\frac{1}{x}\frac{d}{dx}\right)^m [x^n J_n(x)] = \frac{1}{x}\frac{d}{dx}[x^{n-m}J_{n-m}(x)]$$

$$= \frac{1}{x}[x^{n-m}J_{n-m-1}(x)] = x^{n-(m+1)}J_{n-(m+1)}(x).$$

So, if it is true for m, then it is also true for $(m+1)$th differentiation. But it is true for the first differentiation. So, it is true for 2, 3, 4, ... differentiation. Hence **proved**.

$$\therefore \quad \left(\frac{1}{x}\frac{d}{dx}\right)^m [x^n J_n(x)] = x^{n-m}J_{n-m}(x) \quad \textbf{(Proved)}.$$

Again, $\frac{d}{dx}[x^{-n}J_n(x)] = x^{-n}J_n'(x) - nx^{-n-1}J_n(x) = x^{-n}\left[J_n'(x) - \frac{n}{x}J_n(x)\right]$

$$= x^{-n}\left[\frac{1}{2}J_{n-1}(x) - \frac{1}{2}J_{n+1}(x) - \frac{1}{2}J_{n-1}(x) - \frac{1}{2}J_{n+1}(x)\right]$$

$$= -x^{-n}J_{n+1}(x).$$

$$\therefore \quad \left(\frac{1}{x}\frac{d}{dx}\right)[x^{-n}J_n(x)] = -x^{-n-1}J_{n+1}(x)$$

As before we can prove that,

$$\left(\frac{1}{x}\frac{d}{dx}\right)^m [x^{-n}J_n(x)] = (-1)^m x^{-(n+m)}J_{n+m}(x) \quad \textbf{(Proved).}$$

Example 18.14. *Prove that,* (i) $J_{n-1}(x) - J_{n+1}(x) = 2J_n'(x)$

(ii) $J_{n-1}(x) + J_{n-1}(x) = \frac{2n}{x}J_n(x)$, where the symbols have their usual meanings.

[BU(H) 2003]

Solution : (i) We have the recurrence relation

$$xJ_n'(x) = nJ_n(x) - xJ_{n+1}(x) \qquad \text{(i)}$$

$$xJ_n'(x) = -nJ_n(x) + xJ_{n-1}(x). \qquad \text{(ii)}$$

Adding Eqns. (i) and (ii), we get

$$2xJ_n'(x) = -xJ_{n+1}(x) + xJ_{n-1}(x)$$

or, $\boldsymbol{2J_n'(x) = J_{n-1}(x) - J_{n+1}(x)}$ **(Proved).**

(ii) We have, $xJ_n'(x) = -nJ_n(x) + xJ_{n-1}(x)$

$$xJ_n'(x) = -nJ_n(x) + xJ_{n+1}(x).$$

Subtracting, $0 = 2nJ_n(x) - xJ_{n+1}(x) - xJ_{n-1}(x)$

or, $\boldsymbol{J_{n-1}(x) + J_{n+1}(x) = \frac{2n}{x}J_n(x)}$ **(Proved).**

Example 18.15. *Starting from the generating function for Bessel's function prove the following recurrence relation :*

$$J_n(x) = \frac{x}{2n}[J_{n-1}(x) + J_{n+1}(x)] \quad \text{and} \quad J_n'(x) = \frac{1}{2}[J_{n-1}(x) - J_{n+1}(x)].$$ [BU(H) 2000]

Solution : See Art. 18.16 (p. 686)

Example 18.16. *Obtain the complete series solution of the Bessel's differential equation :* $x^2y'' + xy' + (x^2 - n^2)y = 0$, where n is an integer.

Prove that, $J_{-n}(x) = (-1)^n J_n(x)$. [BU(H) 2001]

Solution : See Art. 18.13 (p. 682); Case II : (p. 684)

Example 18.17. *Given the recurrence relations for Bessel functions :*

$$J_{n-1}(x) + J_{n+1}(x) = \frac{2n}{x} J_n(x) \quad \text{and} \quad J_{n-1}(x) - J_{n+1}(x) = 2J_n'(x),$$

where the prime denotes differentiation with respect to x, prove the following relations :

(i) $\frac{d}{dx}[x^n J_n(x)] = x^n J_{n-1}(x)$ (ii) $\frac{d}{dx}[x^{-n} J_n(x)] = -x^{-n} J_{n+1}(x)$. **[BU(H) 1998]**

Solution : See Art. 18.16 (p. 686).

Example 18.18. (a) *If y_1 and y_2 are two linearly independent solutions of the Bessel differential equation $x^2\frac{d^2y}{dx^2}+x\frac{dy}{dx}+(x^2-n^2)y = 0$, show that, the Wronskian of y_1 and y_2 is proportional to $\frac{1}{x}$.*

(b) *Define Bessel function of the first kind of order n denoted by $J_n(x)$.*

Prove the following recurrence relation : $J_{n-1}(x) + J_{n+2}(x) = \frac{2n}{x} J_n(x)$.

Show that, $J_{\frac{1}{2}}(x) = \sqrt{\frac{\pi}{2x}} \sin x$. **[Anna U 2005S; VTU 2003S]**

Solution : (a) If y_1 and y_2 be two solutions of Bessel's equation,

$$x^2\frac{d^2y}{dx^2} + x\frac{dy}{dx} + (x^2 - n^2)y = 0, \quad \text{i.e.} \quad \frac{d}{dx}\left(x\frac{dy}{dx}\right) + \left(x - \frac{n^2}{x}\right) y = 0,$$

then $$\frac{d}{dx}(xy_1') + \left(x - \frac{n^2}{x}\right) y_1 = 0 \tag{i}$$

and $$\frac{d}{dx}(xy_2') + \left(x - \frac{n^2}{x}\right) y_2 = 0. \tag{ii}$$

Multiplying Eqn. (i) by $-y_2$ and (ii) by y_1 and then adding,

$$\frac{d}{dx}[x(y_1y_2' - y_2y_1')] = 0.$$

Integrating w.r.t. x,

$$x(y_1y_2' - y_2y_1') = C \text{ (constant)} \quad \text{or,} \quad \begin{vmatrix} y_1 & y_2 \\ y_1' & y_2' \end{vmatrix} \propto \frac{1}{x} \quad \textbf{(Proved)}.$$

(b) See Art. 18.14 (p. 683); Art. 18.16 (p. 686); Art. 18.15 (p. 686).

Example 18.19. *Show that the equation, $x^2\frac{d^2y}{dx^2} + 2x\frac{dy}{dx} + [x^2 - n(n+1)]y = 0$, where n is a positive integer, can be transferred to Bessel's equation of order $\left(n + \frac{1}{2}\right)$ by substitution $y(x) = \frac{z(x)}{x^{\frac{1}{2}}}$.* **[BU(H) 1996]**

Solution : Here $y = \frac{z}{\sqrt{x}} = zx^{-\frac{1}{2}}$.

$$\frac{dy}{dz} = x^{-\frac{1}{2}}\frac{dz}{dx} - \frac{1}{2}x^{-\frac{3}{2}}z \quad \text{and} \quad \frac{d^2z}{dx^2} = x^{-\frac{1}{2}}\frac{d^2z}{dx^2} - x^{-\frac{3}{2}}\frac{dz}{dx} + \frac{3}{4}zx^{-\frac{5}{2}}.$$

Substituting the values of y, $\frac{dy}{dx}$, $\frac{d^2z}{dx^2}$ in the given equation, we get

$$x^{\frac{3}{2}}\frac{d^2z}{dx^2} + x^{\frac{1}{2}}\frac{dz}{dx} - \frac{1}{2}x^{-\frac{1}{2}}z + [x^2 - (n^2+n)]zx^{-\frac{1}{2}} = 0$$

or, $$\boldsymbol{x^2\frac{d^2z}{dx^2} + x\frac{dz}{dz} + \left[x^2 - \left(n + \frac{1}{2}\right)\right]z = 0.}$$

This is Bessel's equation of order $(n + \frac{1}{2})$.

Example 18.20. *Solve the differential equation* $x\frac{d^2y}{dx^2} + C\frac{dy}{dx} + m^2xy = 0$, *substituting* $y = x^nz$.

Solution : Since $y = x^nz$,

$$\therefore \quad \frac{dy}{dx} = nx^{n-1}z + x^n\frac{dz}{dx} \quad \text{and} \quad \frac{d^2y}{dx^2} = n(n-1)x^{n-2}z + 2nx^{n-1}\frac{dz}{dx} + x^n\frac{d^2z}{dx^2}.$$

Substituting the values of y, $\frac{dy}{dx}$, $\frac{d^2y}{dx^2}$ in the given equation, we get

$$x^{n+1}\frac{d^2z}{dx^2} + 2nx^n\frac{dz}{dx} + n(n-1)x^{n-1}z + nx^{n-1}Cz + Cx^n\frac{dz}{dx} + m^2x^{n+1}z = 0$$

or, $$x^{n+1}\frac{d^2z}{dx^2} + (2n+C)x^n\frac{dz}{dx} + x^{n-1}(m^2x^2 + n^2 + nC - n)z = 0.$$

Dividing by x^{n-1} and putting $2n + C = 1$, we get

$$x^2\frac{d^2z}{dx^2} + x\frac{dz}{dx} + [m^2x^2 + n^2 + n(C-1)]z = 0$$

or, $$\boldsymbol{x^2\frac{d^2z}{dx^2} + x\frac{dz}{dx} + [m^2x^2 - n^2]z = 0.}$$

This is a Bessel's equation. The solution of the equation is

$\boldsymbol{z = AJ_n(mx) + BJ_{-n}(mx)}$, **where $\boldsymbol{n}$ is not integer**

and $\boldsymbol{z = AJ_n(mx) + BY_n(mx)}$ **for integral $\boldsymbol{n}$.**

Example 18.21. *Reduce the given differential equation to Bessel's equation and then solve it.*

$$x\frac{d^2y}{dx^2} + \alpha\frac{dy}{dx} + k^2xy = 0.$$ **[Madras U 2006]**

Solution : Let $y = x^nz$, then

$$\frac{dy}{dx} = x^n\frac{dz}{dx} + nx^{n-1}z \quad \text{and} \quad \frac{d^2y}{dx^2} = x^n\frac{d^2z}{dx^2} + 2nx^{n-1}\frac{dz}{dx} + n(n-1)x^{n-2}z.$$

Substituting the values of y, $\frac{dy}{dx}$, $\frac{d^2y}{dx^2}$ in the given equation, we get

$$x^{n+1}\frac{d^2z}{dx^2} + (2n+\alpha)x^n\frac{dz}{dx} + [k^2x^2 + n^2 + (\alpha-1)n]x^{n-1}z = 0.$$

Dividing throughout by x^{n-1} and putting $2n+\alpha=1$, we get

$$x^2\frac{d^2z}{dx^2}+x\frac{dz}{dx}+(k^2x^2-n^2)z=0.$$

It is a Bessel's equation whose solution is

$$\boldsymbol{z=C_1J_n(kx)+C_2J_{-n}(kx),\ n \text{ is not an integer}}$$

and $\boldsymbol{z=C_1J_n(kx)+C_2Y_n(kx),\ n}$ **is an integer**

Example 18.22. *Prove the following recurrence relations :*

(i) $2xH_n(x)=2nH_{n-1}(x)-H_{n+1}(x)$

(ii) $H_n'(x)=2xH_n(x)-H_{n+1}(x)$

(iii) $H_n''(x)-2xH_n'(x)+2nH_n(x)=0.$

Solution : (i) We have $e^{2tx-t^2}=\sum_{n=0}^{\infty}H_n(x)\frac{t^n}{n!}$.

Differentiating partially with respect to t,

$$2(x-t)e^{2tx-t^2}=\sum_{n=0}^{\infty}H_n(x)\frac{t^{n-1}}{(n-1)!}=2x\sum_{n=0}^{\infty}H_n(x)\frac{t^n}{n!}-2\sum_{n=0}^{\infty}H_n(x)\frac{t^{n+1}}{n!}$$

$$=\sum_{n=0}^{\infty}H_n(x)\frac{t^{n+1}}{(n-1)!}.$$

Equating the coefficient of t^n from both sides, we get

$$2x\frac{H_n(x)}{n!}-2\frac{H_{n-1}(x)}{(n-1)!}=\frac{H_{n+1}(x)}{n!}.$$

or, $\boldsymbol{2xH_n(x)=2nH_{n-1}(x)+H_{n+1}(x)}$ **(Proved).** (i)

(ii) Using the recurrence relation

$$H_n'(x)=2nH_{n-1}(x) \tag{ii}$$

in Eqn. (i) above, we get

$$2xH_n(x)=H_n'(x)+H_{n+1}(x)$$

or, $\boldsymbol{H_n'(x)=2xH_n(x)-H_{n+1}(x)}$ **(Proved).** (iii)

(iii) Differentiating Eqn. (ii) w.r.t. x, we get

$$H_n''(x)=2xH_n'(x)+2H_n(x)-H_{n+1}(x).$$

Replacing n by $(n+1)$ in Eqn. (ii), we get

$$H_{n+1}'(x)=2(n+1)H_n(x).$$

$\therefore\quad H_n''(x)=2xH_n'(x)+2H_n(x)-2(n+1)H_n(x)$

or, $\boldsymbol{H_n''(x)-2xH_n'(x)+2nH_n(x)=0}$ **(Proved).** (iv)

Example 18.23. *Prove that, $H_n(-x) = (-1)^n H_n(x)$.*

Solution : We have the Hermite polynomial of degree n denoted by $H_n(x)$,

$$H_n(x) = \sum_{n=0}^{N}(-1)^m \frac{n!}{m!(n-2m)!}(2x)^{n-2m}.$$

Then, $H_n(-x) = \sum_{n=0}^{N}(-1)^m \frac{n!}{m!(n-2m)!}(-2x)^{n-2m} = \sum_{n=0}^{N}(-1)^m(-1)^{n-2m}\frac{n!}{m!(n-2m)!}(2x)^{n-2m}$

$$= (-1)^n \sum_{n=0}^{N}(-1)^m \frac{n!}{m!(n-2m)!}(2x)^{n-2m}.$$

$\therefore \quad \boldsymbol{H_n(-x) = (-1)^n H_n(x)}$ **(Proved).**

Example 18.24. *Expand the function* $f(x) = \begin{cases} -1, & \text{for } x < 0 \\ 0, & \text{for } x = 0 \\ 1, & \text{for } x > 1 \end{cases}$

in terms of Hermite polynomials.

Solution : Let $f(x) = \sum_{n=0}^{\infty} A_n H_n(x)$.

Then, $A_n = \dfrac{1}{2^n(n!)\sqrt{\pi}}\displaystyle\int_{-\infty}^{+\infty} e^{-x^2} H_n(x) f(x) dx.$

Since, $f(x)$ is odd, $A_{2n} = 0$.

$$\therefore \quad A_{2n+1} = \frac{1}{2^{2n}(2n+1)!\sqrt{\pi}}\int_0^{\infty} e^{-x^2} H_{2n+1}(x)dx. \tag{i}$$

Using the relation,

$$H_{n+1}(x) + H_n'(x) - 2xH_n(x) = 0, \tag{ii}$$

we get $A_{2n+1} = \dfrac{1}{2^{2n}(2n+1)!\sqrt{\pi}}\left[2\displaystyle\int_0^{\infty} xe^{-x^2} H_{2n}(x)dx - \int_0^{\infty} e^{-x^2} H_{2n}'(x)dx\right].$ (iii)

Now, $\displaystyle\int_0^{\infty} xe^{-x^2} H_{2n}(x)dx = H_{2n}(x)\int_0^{\infty} 2xe^{-x^2}dx + \int_0^{\infty} H_{2n}'(x)e^{-x^2}dx$

$$= \left[-e^{-x^2} H_{2n}(x)\right]_0^{\infty} + \int_0^{\infty} H_{2n}'(x)e^{-x^2}dx. \tag{iv}$$

From Eqns. (ii) and (iv), we get

$$A_{2n+1.} = \frac{H_{2n}(0)}{2^{2n}(2n+1)!\sqrt{\pi}}. \tag{v}$$

Again, $e^{-t^2+2tx} = \sum_{n=0}^{\infty} \frac{1}{n!} H_n(x)t^n.$

$$\text{Putting } x = 0, \quad e^{-t^2} = \sum_{n=0}^{\infty} \frac{1}{n!} H_n(0)t^n. \tag{vi}$$

Comparing the coefficients of t^{2n} on both sides, we get $H_{2n}(0) = (-1)^n \dfrac{2n!}{n!}$.

$$\therefore \quad A_{2n+1} = (-1)^n \frac{1}{2^{2n}(2n+1)n!\sqrt{\pi}}. \tag{vi}$$

$$\therefore \quad f(x) = \frac{1}{\sqrt{\pi}} \sum_{n=0}^{\infty} \frac{(-1)^n}{2^{2n}(2n+1)n!} H_{2n+1}(x)$$

$$= \frac{1}{\sqrt{\pi}} \left[H_1(x) - \frac{1}{12} H_3(x) + \frac{1}{160} H_5(x) - \cdots \right].$$

Example 18.25. *Expand the function $f(x) = e^{-\alpha x}$ in terms of Laguerre's polynomial.*

Solution : Let $e^{-\alpha x} = \sum A_n L_n(x)$, then

$$A_n = \int_{-\infty}^{+\infty} e^{-x} L_n(x) e^{-\alpha x} dx$$

$$= \frac{1}{n!} \cdot \int_{-\infty}^{+\infty} e^{-\alpha x} \frac{d^n}{dx^n}(x^n e^{-x}) dx, \quad \left[\because L_n(x) = \frac{e^x}{n!} \frac{d^n}{dx^n}(x^n e^{-x}) \right]$$

$$= \frac{1}{n!} \int_0^{\infty} x^n e^{-(\alpha+1)x} dx, \quad \text{(integrating by parts)}.$$

Putting $(\alpha + 1)x = y$, we get

$$a_n = \frac{\alpha^n}{n!(\alpha+1)^{n+1}} \cdot \int_0^{\infty} y^n e^{-y} dy = \frac{\Gamma(n+1)\alpha^n}{n!(\alpha+1)^{n+1}}.$$

Thus, $e^{-\alpha x} = \dfrac{1}{\alpha+1} \displaystyle\sum_{n=0}^{\infty} 1 \left(\frac{\alpha}{\alpha+1} \right)^n L_n(x)$.

For $\alpha = 1$,

$$e^{-x} = \frac{1}{2} \sum_{n=0}^{\infty} \frac{1}{2^n} L_n(x) = \frac{1}{2} \left[L_0(x) + \frac{1}{2} L_1(x) + \frac{1}{4} L_2(x) + \frac{1}{8} L_3(x) + \cdots \right].$$

Exercises

18(A) : Objective Type Questions

Select the correct answer

18.1 The Legendre polynomial $P_n(x)$ has

(a) n real zeros between 0 and 1

(b) n zeros of which only one is between -1 and 1

(c) $(2n - 1)$ real zeros between -1 and 1

(d) none of these.

18.2 The value of $\int_{-1}^{1}(2x+1)P_3(x)dx$, where $P_3(x)$ is the third degree Legendre polynomial is

(a) 1 (b) -1 (c) 2 (d) 0.

18.3 If $P_n(x)$ is the Legendre polynomial, then $P_n'(-x)$ is

(a) $(-1)^{n+1}P_n'(x)$ (b) $(-1)^n P_n'(x)$ (c) $(-1)^n P_n(x)$ (d) $P_n''(x)$.

18.4 If $P_m(x)$ and $P_n(x)$ are the Legendre polynomials, then for $m \neq n$, $\int_{-1}^{1} P_m(x)P_n(x)$ is equal to

(a) $\frac{2}{2n+1}$ (b) 0 (c) -1 (d) ∞.

18.5 If $P_n(x)$ be the Legendre polynomial, then $P_n'(1)$ is equal to

(a) 0 (b) 1 (c) $\frac{n(n+1)}{2}$ (d) $\frac{2n}{(n+1)}$.

18.6 If Rodrigue's formula for $P_n(x)$, the Legendre polynomial of degree n is

$$P_n(x) = k\frac{d^n|(x^2-1)^n|}{dx^n},$$

where

(a) $k = \frac{n!}{2^n}$ (b) $k = \frac{2^n}{n!}$ (c) $k = \frac{1}{2^n n!}$ (d) $\frac{1}{2^n (n!)}$.

18.7 If $P_3^2(x)$ be the associated Legendre polynomial of rank 2, then $P_3^2(x)$ is equal to

(a) $15(1-x^2)^{\frac{3}{2}}$ (b) $15x(1-x^2)$ (c) $15(1-x^2)$ (d) $15x(1-x^2)^{\frac{1}{2}}$.

18.8 If $P_n(x)$ is the Legendre polynomial of order n, then $3x^2+3x+1$ is equal to

(a) $2P_2+3P_1$ (b) $4P_2+2P_1+P_0$ (c) $3P_2+3P_1+P_0$ (d) $2P_2+3P_1+2P_0$.

18.9 If $P_{2n+1}(x)$ be the Legendre polynomial of order n, then $P_{2n+1}(0)$ is equal to

(a) -1 (b) 1 (c) 0 (d) ∞.

18.10 If $P_n(x)$ be a Legendre polynomial of order n, then $\left[\frac{d}{dx}P_n(x)\right]_{x=1}$ is

(a) $n(n+1)$ (b) $\frac{n(n+1)}{2}$ (c) $\frac{1}{2}(n+1)$ (d) $\frac{(n+1)}{2n}$.

18.11 $J_{\frac{1}{2}}(x)$ is given by

(a) $\sqrt{\frac{2\pi}{x}}\sin x$ (b) $\sqrt{\frac{2\pi}{x}}\cos x$ (c) $\sqrt{\frac{\pi}{2x}}\cos x$ (d) $\sqrt{\frac{\pi}{2x}}\sin x$.

18.12 The value of the integral $\int_0^{\pi}[J_{-2}(x)-J_2(x)]dx$, where $J_n(x)$ is the Bessel's function of first kind of order n, is equal to

(a) 0 (b) -2 (c) 2 (d) 1.

18.13 If J_0 and J_1 are Bessel's function, then $J_1'(x)$ is given by

(a) $J_0(x) - \frac{1}{x}J_1(x)$ (b) $-J_0(x)$ (c) $J_0(x)+\frac{1}{x}J_1(x)$ (d) $J_0(x)-\frac{1}{x^2}J_1(x)$.

18.14 $\frac{1}{\pi}\int_0^\pi \sin(x\sin\theta)\sin m\theta$ for m odd is equal to

(a) 0 (b) 1 (c) -1 (d) J_m.

18.15 If $J_n(x)$ be the Bessel's function, then $\left[J_{n-1}(x) - \frac{n}{x}J_n(x)\right]$ is equal to

(a) $J_n(x)$ (b) $J_n'(x)$ (c) $J_n''(x)$ (d) $J_n(x)J_n'(x)$.

18.16 If $H_n(x)$ be the Hermite polynomial, then $[2nH_{n-1}(x) + H_{n+1}(x)]$ is equal to

(a) $2H_n(x)$ (b) $2xH_n(x)$ (c) $xH_n(x)$ (d) $\frac{xH_n(x)}{2}$.

18.17 The value of the integral $\int_{-\infty}^{\infty} e^{-x^2}H_m(x)H_n(x)dx$ for $m = n$ is

(a) 0 (b) 1 (c) $2^n n!\pi$ (d) $2^n n!\sqrt{\pi}$.

18.18 If $H_n(x) = (-1)^n e^{x^2}\frac{d^n}{dx^n}(e^{-x^2})$, then $H_2(x)$ is equal to

(a) $2x$ (b) $4x^2 - 2$ (c) $8x^3 - 12x$ (d) $4x^2 - 8$.

18.19 The value of the integral $\int_0^\infty e^{-x}L_m(x)L_n(x)dx$ for $m \neq n$ is equal to

(a) 0 (b) $(n!)^2$ (c) $\frac{n}{2n+1}$ (d) $\frac{2n+1}{n}$.

18.20 If $L_n(x) = e^x\frac{d^n}{dx^n}(x^n e^{-x})$, then $L_1(x)$ is equal to

(a) 1 (b) $(1-x)$ (c) $(2 - 4x + x^2)$ (d) x.

18.21 The value of the integral $\int_0^\infty e^{-x}L_mL_n\,dx$ for $m \neq n$ is equal to

(a) 0 (b) 1 (c) $\frac{(m+n)}{2x}$ (d) $\frac{mn}{2x}$.

Answers

18.1 (a) **18.2** (d) **18.3** (a) **18.4** (b) **18.5** (c) **18.6** (c) **18.7** (b) **18.8** (d)
18.9 (c) **18.10** (b) **18.11** (d) **18.12** (a) **18.13** (a) **18.14** (d) **18.15** (b) **18.16** (b)
18.17 (d) **18.18** (b) **18.19** (b) **18.20** (b) **18.21** (a).

18(B) : Short Answer Type Questions

18.1 Write down Legendre differential equation. State its solution for $m = n$ and $m = -(n+1)$, where m is the root of the indicial equation and n is a real constant.

18.2 Write down Legendre's function of the first kind $P_n(x)$ and Legendre's function of the second kind $Q_n(x)$. What is the general solution of Legendre's differential equation in terms of $P_n(x)$ and $Q_n(x)$?

18.3 If $P_n(x) = \frac{1}{2^n n!}\frac{d^n}{dx^n}(x^2-1)^n$, then find the values of $P_2(x)$ and $P_3(x)$.

18.4 Starting from the generating function $(1-2xt+t^2)^{-\frac{1}{2}} = \sum_{n=0}^{\infty} t^n P_n(x)$, prove that

$$P_n(-x) = (-1)^n P_n(x).$$

18.5 Prove that, $P_{2n}(0) = (-1)^n \dfrac{1\cdot 3\cdot 5\cdots(2n-1)}{2\cdot 4\cdot 6\cdots(2n)}$ and hence $P_{2n+1}(0) = 0$.

18.6 Write the condition of orthogonality of Legendre polynomials. State Fourier-Legendre expansion of $f(x)$.

18.7 Find the values of the associated Legendre polynomials $P_1'(x)$ and $P_2'(x)$.

18.8 Write the Bessel's differential equation and find the values of $J_0(x)$.

18.9 Prove the recurrence relation, $2nJ_n(x) = x[J_{n-1}(x) + J_{n+1}(x)]$.

18.10 Show that under the transformation $y = \dfrac{u}{\sqrt{x}}$, the Bessel's equation $x^2\dfrac{d^2y}{dx^2} + x\dfrac{dy}{dx} + (x^2-n^2)y = 0$ reduces to $\dfrac{d^2u}{dx^2} + \left(1 + \dfrac{1-4x^2}{4x^2}\right)u = 0$.

18.11 Write the condition of normality of Bessel's function. What is its generating function?

18.12 Write the expressions of BER and BEI functions.

18.13 Prove that $J_0(x) = \dfrac{1}{\pi}\displaystyle\int_0^{\pi} \cos(x\sin\theta)d\theta$.

18.14 Write down Hermite differential equation and Hermite polynomial.

18.15 Starting from Rodrigue's formula find the values of $H_0(x)$ and $H_1(x)$.

18.16 Show that, $\displaystyle\int_{-\infty}^{\infty} e^{-x} H_m(x) H_n(x) dx = 0$ for $m \neq n$.

18.17 What is Laguerre's differential equation? Write down its solution.

18.18 Starting from the generating function of Laguerre's function $L_n(x)$ show that,

$$L_{n+1}(x) + (x-2n-1)L_n(x) + n^2 L_{n-1}(x) = 0.$$

18.19 Write down associated Laguerre's equation. State its orthogonality condition.

18.20 What is Chebyshev differential equation. State Chebyshev polynomial of first kind.

18.21 Starting from the Rodrigue's formula for Chebyshev polynomial, prove that (i) $T_1(x) = 1$ and (ii) $T_3(x) = 4x^3 - 3x$.

18.22 The Legendre polynomials are defined in terms of its generating function $F(x,t) = (1-2xt+t^2)^{-\frac{1}{2}}$ as follows : $F(x,t) = \sum_{n=0}^{\infty} P_n(x)t^n$.

Find the value of $P_3(-1)$. **[BU(H) 2012]**

18(C) : Broad Answer Type Questions

18.1 Identify the singularity of the Legendre differential equation,

$$(1-x^2)\frac{d^2y}{dx^2} - 2x\frac{dy}{dx} + n(n+1) = 0, \text{ where } n \text{ is a constant}$$

Obtain the general series solution near $x = 0$. Discuss the convergence of the series solution.

When do you have a polynomial solution? **[BU(H) 1996, CU(H) 2003]**

18.2 (a) Prove that, $\int_0^{\pi} P_l(\cos\theta)P_m(\cos\theta)\sin\theta\, d\theta = \frac{2}{2l+1}\delta_{lm}$

where $P_l(\cos\theta)$ is a Legendre polynomial and δ_{lm} is a Kronecker delta. **[BU(H) 1995]**

(b) Show that, $\int_{-1}^{+1} \frac{P_n(x)dx}{\sqrt{1-2xh+h^2}} = \frac{2h^n}{2n+1}$.

18.3 (a) Obtain by Frobenins' method the series solution of Legendre equation :

$$(1-x^2)\frac{d^2y}{dx^2} - 2x\frac{dy}{dx} + l(l+1)y = 0.$$

Under what condition, you get polynomial solution?

(b) Prove that, $P_l(x) = \frac{1}{2^l l!}\frac{d^l}{dx^l}(x^2-1)^l$ satisfies Legendre equation. **[BU(H) 2000]**

18.4 (a) Starting from the generating function for Legendre polynomials, prove the following recurrence relations :

(i) $(l+1)P_{l+1}(x) - (2l+1)xP_l(x) + lP_{l-1}(x) = 0$ **[BU(H) 1997]**

(ii) $lP_l(x) = xP'_l(x) - P'_{l-1}(x)$.

(b) Prove that, $\int_0^{\pi} P_l^2(\cos\theta)\sin\theta\, d\theta = \frac{2}{2l+1}$. **[BU(H) 2001]**

18.5 (a) For the Legendre polynomials, prove the following integrals :

(i) $\int_{-1}^{+1} P_l(x)P_m(x)dx = 0 \quad (l \neq m)$ (ii) $\int_{-1}^{+1} [P_l(x)]^2 dx = \frac{2}{2l+1} \quad (l = m)$

(b) Show that under the transformation $y = \frac{u}{\sqrt{x}}$, the Bessel's equation

$$x^2\frac{d^2y}{dx^2} + x\frac{dy}{dx} + (x^2 - n^2)y = 0 \text{ becomes } \frac{d^2u}{dx^2} + \left(1 + \frac{1-4n^2}{4x^2}\right)u = 0.$$

Hence, deduce that for $x \to \infty$, the solution of Bessel's equation has the form :

$$\frac{C_1 \sin x}{\sqrt{x}} + \frac{C_2 \cos x}{\sqrt{x}}.$$

[BU(H) 1993]

18.6 Legendre polynomials $P_n(x)$ are defined by the generating function

$$g(x,t) = (1 - 2xt + t^2)^{-\frac{1}{2}} = \sum_{n=0}^{\infty} P_n(x)t^n, \quad |t| < 1.$$

Hence, find the recurrence relations :

(i) $(2n+1)xP_n(x) = (n+1)P_{n+1}(x) + nP_{n-1}(x)$ **[NBU(H) 2000]**

(ii) $P'_{n+1}(x) + P'_{n-1}(x) = 2xP'_n(x) + P_n(x); \quad n = 1, 2, 3, \ldots$

Prime denotes differentiation w.r.t. x.

Also prove that $P_n(1) = 1$ for all n, $P_n(-1) = (-1)^n$ and if $x = \cos\theta$, $|P_n(x)| \le P_n(1)$.
[CU(H) 2014, BU(H) 1999]

18.7 **(a)** If $I = \int_{-1}^{+1} x^3 P_2(x)$, evaluate I, without actually performing integration and justify your answer. **[NBU(H) 2008]**

(b) Prove the orthogonality condition of Legendre's polynomials.

(c) Find the value of $\int_{-1}^{+1} P_3(x)dx$. **[NBU(H) 2009]**

18.8 **(a)** Using the generating function for the Legendre polynomials,

$$G(x,t) = (1 - 2xt + t^2)^{-\frac{1}{2}} = \sum_{n=0}^{\infty} P_n(x)t^n, \quad -1 \le x \le 1,\ t < 1,$$

show that, $P_n(-x) = (-1)^n P_n(x)$ and $P_n(1) = 1$. **[CU(H) 2014, VU(H) 2009, '06]**

(b) Show that, $\int_{-1}^{+1} P_l(x)P_m(x)dx = 0$ for $l \ne m$; $P_l(x)$ is the Legendre polynomial of order l.

18.9 Prove that :

(i) $\int_{-1}^{+1} [P_2(x)]^2 dx$ **[PTU 2002]**

(ii) $\int_{-1}^{+1} P_n(x)]dx = 0 \quad (n \ne 0)$

(iii) $\int_{-1}^{+1} (1 - x^2)P'_m(x)P'_n(x)dx = \begin{cases} 0 & \text{, for } m \ne n \\ \dfrac{2n(n+1)}{2n+1}, & \text{for } m = n \end{cases}$

where $P_m(x)$ is Legendre polynomial of order m.

18.10 Express in terms of Legendre polynomials

(i) $5x^3 + x$ **[Ans: $2P_3 + 4P_1$]**

(ii) $1 + x - x^2$ **[Ans: $-\frac{2}{3}P_2(x) + P_1(x) + \frac{2}{3}P_0(x)$]**

(iii) $x^4 - 2x^3 + 3x^2 + 5x - 9$. **[Ans: $\frac{8}{35}P_4(x) - \frac{4}{3}P_3(x) + \frac{18}{7}P_2(x) + \frac{19}{5}P_1(x) - \frac{39}{5}$]**

18.11 Prove that : **(a)** $P'_{2n+1}(0) = \dfrac{(-1)^n(2n+1)!}{2^{2n}(n!)^2}$ **(b)** $P'_{2n}(0) = 0$.

18.12 Expand the following functions in terms of Legendre polynomials in the interval, $-1 < x < 1$.

(i) $f(x) = x^4 + x^3 + 2x^2 - x - 3$

(ii) If $f(x) = 0, \quad -1 < x < 0$

$\qquad = 1, \quad 0 < x < 1$.

Show that $f(x) = \frac{1}{2}P_0(x) + \frac{3}{4}P_1(x) - \frac{7}{16}P_3(x) + \cdots$.

18.13 Show that under the transformation $y = \dfrac{u}{\sqrt{x}}$, the Bessel's equation

$$x^2\frac{d^2y}{dx^2} + x\frac{dy}{dx} + (x^2 - n^2)y = 0 \text{ becomes } \frac{d^2u}{dx^2} + \left(1 + \frac{1-4x^2}{4x^2}\right)u = 0.$$

Hence, deduce that for $x \to \infty$, the solution of Bessel's equation has the form :

$$\frac{1}{\sqrt{x}}[C_1 \sin x + C_2 \cos x].$$

[BU(H) 1993]

18.14 Obtain the complete series solution of the Bessel's differential equation :

$$x^2y'' + xy' + (x^2 - n^2)y = 0, \text{ where } n \text{ is an integer.}$$

If n is an integer, prove that, $J_{-n}(x) = (-1)^n J_n(x)$. **[BU(H) 2001]**

18.15 Bessel's differential equation is given by $x^2y'' + xy' + (x^2 - n^2)y = 0$.

Find by the method of series substitution two independent solutions of this equation when n is not an integer. Is your second solution is independent of the first when n is an integer? **[BU(H) 1999]**

18.16 Show that under the equation, $x^2\dfrac{d^2y}{dx^2} + 2x\dfrac{dy}{dx} + [x^2 - n(n+1)]y = 0$, where n is a positive integer can be transformed to Bessel's equation of the order $\left(n + \dfrac{1}{2}\right)$ by the substitution $y(x) = \dfrac{z(x)}{x^{\frac{1}{2}}}$. **[BU(H) 1996]**

18.17 Find the general solution of the Bessel's equation $x^2\dfrac{d^2y}{dx^2} + x\dfrac{dy}{dx} + (x^2 - n^2)y = 0$, when n is not an integer. **[BU(H) 2002]**

If n is an integer, show that, $J_{-n}(x) = (-1)^n J_n(x)$.

How would you find the second solution when n is an integer. **[BU(H) 2004]**

18.18 Prove the following recurrence relations satisfied by Bessel's functions :

(i) $J_n(x) = \dfrac{x}{2n}[J_{n-1}(x) + J_{n+1}(x)]$ **[BU(H) 2000, '95, '03]**

(ii) $J_n(x) = \dfrac{1}{2}[J_{n-1}(x) - J_{n+1}(x)]$ **[BU(H) 1995, 2000]**

(iii) $J_{n-1}(x) + J_{n+1}(x) = \dfrac{2n}{x}J_n(x)$

(iv) $J_{n-1}(x) - J_{n+1}(x) = 2J_n'(x)$ **[BU(H) 1998, 2000]**

(v) $\dfrac{d}{dx}[x^n J_n(x)] = x^n J_{n-1}(x)$

(vi) $\dfrac{d}{dx}[x^{-n} J_n(x)] = -x^{-n} J_{n+1}(x)$. **[BU(H) 1998]**

18.19 The generating function of the Hermite polynomial is $g(x,t) = e^{-t+2tx} = \sum_{n=0}^{\infty} H_n(x)\dfrac{t^n}{n!}$

Use above generating function to prove that

(i) $H_n'(x) = 2nH_{n-1}(x)$ (ii) $H_{2n}(0) = (-1)^n\dfrac{(2n)!}{n!}$. **[VU(H) 2006]**

18.20 Given, $e^{-t^2+2tx} = \sum_{n=0}^{\infty} H_n(x)\frac{t^n}{n!}$; $H_n(x)$ being Hermite polynomials. Prove that

(i) $H_{n+1}(x) = 2xH_n(x) - 2nH_{n-1}(x)$ (ii) $H_n(x) = (-1)^n H_n(-x)$. **[VU(H) 2009]**

18.21 (a) Prove the following recurrence formulae :

(i) $H_n'(x) = 2xH_n(x) - H_{n+1}(x)$ **[NBU(H) 2006]**

(ii) $H_n''(x) - 2xH_n'(x) + 2nH_n(x) = 0$

(iii) $2nH_{n-1}(x) = H_n'(x)$.

(b) Prove that :

$$\int_{-1}^{+1} e^{-x^2} H_n(x)H_m(x)dx = 2^n \cdot n!\sqrt{\pi} \cdot \delta_{mn}, \quad \text{where } \delta_{mn} = \begin{cases} 0, \text{ for } n \neq m \\ -1, \text{ for } n = m. \end{cases}$$

18.22 (a) Solve the Laguarre's differential equation $x\frac{d^2y}{dx^2} + (1-x)\frac{dy}{dx} + ny = 0$ by the method of power series.

(b) Use Rodrigue's formula for $L_n(x)$, i.e., $L_n(x) = e^x \frac{d^m}{dx^n}(x^n e^{-x})$ to find out the expression for $L_0(x)$, $L_1(x)$, $L_2(x)$.

18.23 (a) Given the generating function $(1-t)\sum L_n(x)\frac{t^n}{n!} = e^{-\frac{xt}{1-t}}$, where $L_n(x)$ is the Laguerre's polynomial of order n. Prove the following recurrence relations :

(i) $L_n'(x) - nL_{n-1}'(x) + nL_{n-1} = 0$

(ii) $L_{n+1}(x) + (x - 2n - 1)L_n(x) + n^2 L_{n-1}(x) = 0$.

(b) Prove that :

(i) $\int_0^{\infty} e^{-x} L_m(x)L_n(x)dx = 0$, when $m \neq n$

(ii) $\int_0^{\infty} e^{-x} L_n^2(x)dx = (n!)^2$, when $m = n$.

18.24 Starting from Hermite's equation $y'' - 2xy' + 2xy = 0$ (n = a constant integer) derive by Frobenius method the standard form of Hermite polynomial $H_n(x)$. Hence, prove that, $H_n(x) = 2nH_{n-1}(x)$. **[VU(H) 2005]**

18.25 (a) Solve Legendre's equation $(1 - x^2)y'' - 2xy' + n(n-1)y = 0$ by Frobenius method by taking $y(x) = \sum_{\lambda=0}^{\infty} a_\lambda x^{k+\lambda}$

(i) Verify that the indicial equation is $k(k-1) = 0$.

(ii) Obtain the recurrence relation relating a_{j+1} to a_j when $k = 0$.

(iii) Repeat (ii) when $k = 1$.

Show that by appropriate choice of n, one series at time may be converted into a polynomial.

(b) Hermite polynomial may be expressed as, $e^{-t^2+2tx} = \sum_{n=0}^{\infty} H_n(x)\frac{t^n}{n!}$.

Use this to show that, $H_n(-x) = (-1)^n H_n(x)$. **[CU(H) 2013, 2011]**

18.26 (a) Legendre polynomial may be expressed as $(1-2tx+t^2)^{-\frac{1}{2}} = \sum_{n=0}^{\infty} P_n(x)t^n$.

Use this to prove that, $P_n(-x) = (-1)^n P_n(x)$. **[CU(H) 2010]**

(b) Show that, $\int_{-1}^{+1} P_l(x)P_m(x)dx = 0$ for $l \neq m$; $P_1(x)$ is the Legendre polynomial of order 1. **[CU(H) 2006]**

18.27 Prove the recurrence relation for Legendre polynomials, $(1-x^2)P_L'(x) = (L+1)xP_L(x) - (L+1)P_{L+1}(x)$ hence evaluate the integral $\int_{-1}^{+1}(1-x^2)P_L'(x)P_m(x)dx$. **[VU(H) 2013]**

18.28 (a) Prove that $H_n(x) = (-1)^n e^{x^2}\frac{d^n}{dx^n}(e^{-x^2})$

(b) Show that, $\int_{-\infty}^{+\infty}[H_2(x)]^2 = 8\sqrt{\pi}$. **[VU(H) 2012]**

18.29 Generating function for Hermite polynomial is given by $e^{2tx-t^2} = \sum_{n=0}^{\infty}\frac{H_n(x)}{n!}t^n$

Prove : (i) $H_n'(x) = 2nH_{n-1}(x)$ (ii) $H_n(-x) = (-1)^n H_n(x)$ **[VU(H) 2011]**

18.30 Legendre polynomial may be expressed as $(1-2xt+t^2)^{-\frac{1}{2}} = \sum_{n=0}^{\infty} P_n(x)t^n$.

Use the above relation to show that

(i) $P_n(x) = (-1)^n P_n(x)t^n$

(ii) $(l+1)P_{l+1}(x) = (2l+1)xP_l(x) - lP_{l-1}(x)$. **[VU(H) 2010]**

18.31 (a) State the conditions for which $x = x_0$ is a regular singular point of the differential equation :

$$P_1(x)y'' + P_2(x)y' + P_3(x)y = 0.$$

What can be said about the solution of the differential equation when $x = x_0$ is a singular point?

(b) Let $J_n(x)$ denote the Bessel function of the first kind of order n. If $J_{\frac{1}{2}} = \sqrt{\frac{2}{\pi x}}\sin x$ and $J_{-\frac{1}{2}}(x) = \sqrt{\frac{2}{\pi x}}\cos x$ find the value of $J_{\frac{5}{2}}(x)$. **[BU(H) 2014]**

18.32 (a) Show that the Legendre polynomials can be represented by Rodrigue's formula

$$P_n(x) = \frac{1}{2^n!}\frac{d^n}{dx^n}(x^2-1)^n$$

(b) Obtain the explicit form of $P_3(x)$.

(c) Using Rodrigue's formula show that $\int_{-1}^{+1} P_n^2(x)dx = \frac{2}{2n+1}$. **[BU(H) 2014]**

18.33 (a) Write the Legendre differential equation.

What is the form of the generating function for the Legendre's polynomials?

(b) Prove the recurrence formula $(n+1)P_{n+1}(x) - (2n+1)xP_n + nP_{n-1}(x) = 0$

[BU(H) 2013]

18.34 (a) Show that under the transformation $y = \frac{u}{\sqrt{x}}$, Bessel's equation $x^2\frac{d^2y}{dx^2} + x\frac{dy}{dx} + (x^2 - n^2)y = 0$ reduces to $u'' + \left(1 + \frac{1-4n^2}{4x^2}\right)u = 0$.

Hence, show that for $x \to \infty$, the solution of Bessel's equation has the form :

$$C_1\frac{\sin x}{\sqrt{x}} + C_2\frac{\cos x}{\sqrt{x}},$$ where C_1 and C_2 are arbitrary constants.

(b) Prove the following recurrence relations involving Bessel's functions of the first kind :

(i) $J_n'(x) = \frac{n}{x}J_0(x) - J_{n+1}(x)$

(ii) $J_{n+1}(x) = \frac{2n}{x}J_n(x) - J_{n-1}(x)$. **[BU(H) 2011]**

18.35 (a) Legendre equation has the form : $(1 - x^2)y'' - 2xy' + n(n+1)y = 0$, where n is a constant. Examine and explain the nature of the points at $x = 0$ and

(b) Starting from the generating function of Legendre polynomials show that

$$P_l(-1) = (-1)^l.$$ **[BU(H) 2009]**

18.36 (a) Find the indicial equation and its roots for the given differential equation $x^2y'' + 3xy' + 4xy = 0$.

(b) Find the value of $P(-1)$.

(c) Express $f(x)$ as a linear combination of Legendre Polynomials, where

$$f(x) = 5x^3 + 2x^2 - 7x + 4.$$

18.37 (a) Solve the given differential equation $y'' - 2xy' + 2\lambda y = 0$, λ is a real constant. For $\lambda = 0, 1, 2, 3\cdots$ Show that one of the solution is a polynomial of order n. Name the polynomial.

(b) Show that, $J_{\frac{1}{2}}(x) = \sqrt{\frac{2}{\pi x}}\sin(x)$.

(c) Show that $\frac{d}{dx}\{x^n J_n(x)\} = x^n J_{n-1}(x)$.

18.38 (a) Solve the differential equation $x^2y'' + xy' + (x^2 - 4)y = 0$.

(b) For Legendre polynomial $P_n(x)$, prove that, $nP_n = (2n-1)xP_{n-1} - (n-1)P_{n-2}$

18.39 (a) What is the generating function of Hermite polynomial?

[Mahatma Gandhi (U) 2012]

(b) Evaluate $J_{\frac{3}{2}}(x)$ and $J_{(-\frac{3}{2})}(x)$.

18.40 (a) Verify orthogonality property of Legendre's polynomials.

(b) What is Bessel's differential equation? Using Frobenius method find the solution of Bessel equation. **[Mahatma Gandhi, U. 2012]**

18.41 (a) Obtain values of $P_n(1)$ and $P_n(-1)$.

(b) Write two recurrence relations for the Bessel function.

(c) Show that $\cos x = J_0(x) + 2\sum_{n=1}^{\infty}(-1)^n J_{2n}(x)$. **[Mahatma Gandhi U(MSc) 2012]**

18.42 (a) Derive the orthogonality relation for Legendre polynomial .

$$\int_{-1}^{1} P_n(x)P_m(x)dx = \frac{2}{2n+1}\delta_{mn}.$$

(b) Discuss the Rodrigue's formula orthogonality relation and recurrence relations of hermite polynomial. **[Mahatma Gandhi U.(M.Sc.) 2012]**

18.43 (a) Prove that $J_n(x) = (-1)^n J_n(x)$.

(b) Prove that $P_n(-1) = (-1)^n$.

(c) Show that Bessel's equation $x^2y'' + xy' + (x^2 - x^2)y = 0$ is transferred into $x^2y'' + xy' + (\lambda^2x^2 - n^2)y = 0$, if x is replaced by λx, where λ is a constant.

18.44 (a) Solve Hermite differential equation by Frobenius method, $y'' - 2xy' + 2xy = 0$, given n is an integer.

(b) Obtain Rodrigue's formula for Legendre's polynomials, $P_n(x) = \frac{1}{n!}\frac{d^n}{dx^n}(x^2-1)^n$. Evaluate the value of $P_3(x)$ using Rodrigue's formula.

(c) Expand $f(x) = x^2 + 5x - 2$ in a series of the form : $\sum_{n=0}^{\infty} a_n p_n(x)$.

[Given, $P_0(x) = 1,\ P_1(x) = x,\ P_2(x) = \frac{3x^2-1}{2}$]

18.45 (a) Using generating function of Bessel's function verify that,

(i) $\cos(x\sin\theta) = J_0(x) + 2J_2(x)\cos 2\theta + 2J_4(x)\cos 4\theta$.

(ii) $\sin(x\sin\theta) = 2J_1(x)\sin\theta + 2J_3(x)\sin 3\theta + 2J_5(x)\sin 5\theta + \cdots$

(b) Show that, $J_3(x) = \sqrt{\frac{2}{\pi x}}\left(\frac{\sin x - \cos x}{x}\right)$.

18.46 (a) Deduce the generating function for the Legendre polynomials. Also obtain the recursion relation $(n+1)P_{n+1}(z) - (2n+1)zP_n(z) + nP_{n-1}(2) = 0$.

(b) Discuss orthogonality and normalisation properties of the Legendre polynomials.

(c) Find $P_n(x)$ for $n = 4$ and 5 **[Gujrat U(MSc) 2017]**

18.47 (a) For Hermite polynomial $H_n(x)$, prove that, $H_n(x) = 2nH_{n-1}(x)$.

(b) Prove that, $(n+1)L_{n+1}(x) = (2n+1-x)nL_n(x) - L_{n-1}(x)$.

(c) Prove that, $L_n(0) = 1$.

18.48 (a) Prove the recurrence relation : $xJ_n(x) = nJ_n(x) - xJ_{n+1}(x)$.

(b) Express $J_2(x)$ in terms of $J_0(x)$ and $J_1(x)$.

18.49 (a) Determine Laguerre polynomials $L_0(x), L_1(x)$ and $L_2(x)$ by using Rodrigue's formula.

(b) Prove the recurrence relation for Legendre polynomial $P_{n+1}(x) + P'_{n-1}(x) = 2xP'_n(x) + P_n(x)$.

(c) For the Bessel function of first kind show that $J'_n(x) = \frac{1}{2}[J_{n-1}(x) - J_{n+1}(x)]$.

18.50 (a) Prove the recurrence relation for Hermite polynomials

$$H_{n+1}(x) = 2xH_n(x) - 2nH_{n-1}(x).$$

(b) For Bessel function of the first kind show that, $J_n(x) = (-1)^n J_n(-x)$.

[Pune U(MSc) 2013]

18.51 (a) Prove that; $H_n(x) = 2nH_n(x) - 2nH_{n-1}(x)$.
(b) State and prove orthogonality property of Hermite functions.
(c) Prove that, $L_{n+1}(x) = 2(n+1-x)L_n(x) - n^2L_{n-1}(x)$. [Pune U(MSc) 2008]

18.52 (a) Write down generating function for Hermite polynomial and derive two basic recurrence relations.
(b) Write expression for Legendre function and determine its parity.
(c) For Legendre function prove the orthogonality condition. [Pune U(MSc) 2013]

18.53 (a) Obtain the Associated Legendre function $P_2^3(x)$.
(b) Prove that, $H_n'(x) = 2nH_{n-1}(x)$.
(c) Prove that, $J_{n+1}(x) = \frac{2n}{x}J_n(x) - J_{n-1}(x)$. [Pune U(MSc) 2014]

18.54 (a) Consider the Legendre equation $(1-x^2)y'' - 2xy' + n(n+1)y = 0$. Assume a power series solution $y(x) = \sum_{\lambda=0}^{\infty} a_K x^{K+\lambda}$.
(i) Verify that indicial equation is $k(k-1) = 0$.
(ii) Obtain the recurrence relation relating a_{j+2} to a_j, when $k = 0, 2$.
(b) The generating function for the Hermite polynomial is

$$g(x,t) = e^{-t^2+2tx} = \sum_{n=0}^{\infty} \frac{H_n(x)}{n!} t^n.$$

Derive the recurrence relations :
(i) $H_n' = 2_n H_{n-1}$ (ii) $H = 2xH_n - 2nH_{n-1}$. [CU(H) 2016]

18.55 (a) Using the generating function $\phi(x,h) = (1-2xh+h^2)^{-\frac{1}{2}} = \sum_{l=0}^{\infty} h^l P_L(x)(|h| < l)$, find the Legendre polynomial $P_2(x)$.
(b) Given that, $e^{-t^2+2tx} = \sum_{n=0}^{\infty} H_n(x)\frac{t^n}{n!}$, where H is the Hermite polynomial. Show that $H_n(x) = (-1)^n H_n(-x)$. [C.U.(H), 2017]

18.56 (a) Consider Hermite equation $\frac{d^2y}{dx^2} - 2x\frac{dy}{dx} + 2ay = 0$ and assume a series solution $Y(x) = \sum_{\lambda=0}^{\infty} a_\lambda - x^{k+\lambda}$.
(i) Find the indicial equation.
(ii) Find recurrence relation.
(iii) Find the condition on 'α' so that the infinite series solution becomes a polynomial.
(b) Solve : $x^2\frac{d^2y}{dx^2} - 6y = 0$, by the method of Frobenius. [CU(H) 2015]

18.57 (a) Apply Frobenius method to equation $\frac{d^2y}{dx^2} + \omega^2 y = 0$ and setting $y(x) = \sum_{\lambda=0}^{\infty} a_\lambda x^{k+\lambda}$ with $a_0 \neq 0$.

(i) Verify that indicial equation is $k(k-1)=0$.

(ii) For $k=1$, a_λ is necessarily zero.

(iii) Find the recurrence relation.

(iv) Using the recurrence relation, show that $k=1$ leads to the solution $y(x) = \frac{a_0}{\omega}\sin\omega x$.

(b) Legendre polynomial $P_n(x)$ may be expressed as $(1-2xt+t^2)^{\frac{t}{2}} = \sum_{n=0}^{\infty} P_n(x)t^n$. Use this to show that $P_n(-x) = (-1)^n P_n(x)$.

[CU(H) 2014]

CHAPTER 19

Partial Differentiation

19.1 Introduction

In science and technology we often come across many physical quantities whose values depend on more than one variable. For example, the area of a rectangle of length x and breadth y is $A = xy$. A has a definite value of a given pair of values of x and y. So, the area A of the rectangle is expressed as $u = f(x, y)$ where u is a function of two independent variables, x and y. Here u has a unique value for every pair of values of x and y.

In fact, y may be a function of more than two independent variables expressed as

$$u = f(x, y, z, t, \ldots)$$

where x, y, z, t, $\ldots$ are a set of independent variables.

19.2 Limit and Continuity

A function $f(x, y)$ is said to tend to the limit l as $x \to a$, $y \to b$, if and only if the limit is independent of path followed by the point $P(x, y)$ as $x \to a$ and $y \to b$, i.e.,

$$\underset{\substack{x\to a\\ y\to b}}{\text{Lt}}\ f(x) = l$$

In terms of circular neighbourhood, the function $f(x, y)$ defined in a region R is said to have the limit l as $x \to a$, $y \to b$, if and only if corresponding to a positive number ϵ, there exists another positive number δ, such that

$$|f(x, y) - l| < \epsilon \text{ for } 0 < (x - a)^2 + (y - b)^2 < \delta^2$$

for every point (x, y) in region R.

Continuity : A function $f(x, y)$ is said to be continuous at the point (a, b), if

$$\underset{\substack{x\to a\\ y\to a}}{\text{Lt}}\ f(x, y) \text{ exists and } = f(a, b).$$

If $f(x, y)$ is continuous at all points in R, then it is continuous in R. If $f(x, y)$ is not continuous at any point (x, y), then it is discontinuous at that point.

19.3 Partial Derivatives

Let $f(x, y)$ be a continuous function of variables x and y. If y be treated as constant, then the ordinary differentiation of $f(x, y)$ with respect to x, is called the partial derivative of $f(x, y)$ with respect to x and is denoted by $\frac{\partial f}{\partial x}$, $\left(\frac{\partial f}{\partial x}\right)_x$ or f_x.

$$\therefore\ \frac{\partial f}{\partial x} = \underset{\Delta x\to 0}{\text{Lt}} = \frac{f(x + \Delta x) - f(x, y)}{\Delta x} \tag{19.1}$$

Similarly, treating x as constant, we get $\frac{\partial f}{\partial y}$, $\left(\frac{\partial f}{\partial y}\right)_x$ or f_y, where

$$\frac{\partial f}{\partial y} = \underset{\Delta y \to 0}{\text{Lt}} = \frac{f(x, y + \Delta y) - f(x, y)}{\Delta y}. \tag{19.2}$$

In a similar way the second order partial derivative of $f(x, y)$ may be found out as follows : Since $f_x(x, y)$ and $f_y(x, y)$ are simple functions of x and y, we have

$$f_{xx} = \frac{\partial^2 f}{\partial x^2} = \underset{\Delta x \to 0}{\text{Lt}} \frac{f_x(x + \Delta x, y) - f_x(x, y)}{\Delta x} \tag{19.3}$$

and $$f_{yy} = \frac{\partial^2 f}{\partial y^2} = \underset{\Delta y \to 0}{\text{Lt}} \frac{f_y(x, y + \Delta y) - f_y(x, y)}{\Delta y}. \tag{19.4}$$

$$f_{xy} = \frac{\partial^2 f}{\partial x \partial y} = \frac{\partial}{\partial x}\left(\frac{\partial f}{\partial y}\right) = \underset{\Delta x \to 0}{\text{Lt}} \frac{f_y(x + \Delta x, y) - f_y(x, y)}{\Delta x} \tag{19.5}$$

and $$f_{yx} = \frac{\partial^2 f}{\partial y \partial x} = \frac{\partial}{\partial y}\left(\frac{\partial f}{\partial x}\right) = \underset{\Delta y \to 0}{\text{Lt}} \frac{f_x(x, y + \Delta y) - f_x(x, y)}{\Delta y}. \tag{19.6}$$

Example 19.1. *If* $u = x^2 \tan^{-1} \frac{y}{x} - y^2 \tan^{-1} \frac{x}{y}$, *show that*

$$\frac{\partial^2 u}{\partial x \partial y} = \frac{x^2 - y^2}{x^2 + y^2} \quad \text{and} \quad \frac{\partial^2 u}{\partial x \partial y} = \frac{\partial^2 u}{\partial y \partial x}.$$ **[AMIETE(Summer) 1990]**

Solution : Since $u = x^2 \tan^{-1} \frac{y}{x} - y^2 \tan^{-1} \frac{x}{y}$,

$$\therefore \quad \frac{\partial u}{\partial y} = x^2 \cdot \frac{1}{1 + \left(\frac{y}{x}\right)^2} \cdot \frac{1}{x} - \left\{ 2y \tan^{-1} \frac{y}{x} + y^2 \cdot \frac{1}{1 + \left(\frac{x}{y}\right)^2} \cdot \left(-\frac{x}{y}\right) \right\}$$

$$= \frac{x^3}{x^2 + y^2} - 2y \tan^{-1} \frac{x}{y} + \frac{xy^2}{x^2 + y^2} = x - 2y \tan^{-1} \frac{x}{y}.$$

$$\therefore \quad \frac{\partial^2 u}{\partial x \partial y} = \frac{\partial}{\partial x}\left[x - 2y \tan^{-1} \frac{x}{y}\right] = 1 - 2y \cdot \frac{1}{1 + \left(\frac{x}{y}\right)^2} \cdot \frac{1}{y} = 1 - \frac{2y^2}{x^2 + y^2} = \frac{x^2 - y^2}{x^2 + y^2}$$

Similarly, $\frac{\partial u}{\partial x} = 2x \tan^{-1} \frac{y}{x} - y$ and $\frac{\partial^2 u}{\partial y \partial x} = \frac{\partial}{\partial y}\left\{2x \tan^{-1} \frac{y}{x} - y\right\} = \frac{x^2 - y^2}{x^2 + y^2}$

Hence, $\frac{\partial^2 u}{\partial x \partial y} = \frac{\partial^2 u}{\partial y \partial x}$.

19.4 Differentials

Let $y = f(x)$. If x changes to $x + \Delta x$, where Δx is small, then y changes to $y + \Delta y$ and $\frac{dy}{dx} = \underset{\Delta x \to 0}{\text{Lt}} \frac{\Delta y}{\Delta x}$.

The relation, $\frac{\Delta y}{\Delta x} = \frac{dy}{dx}$ is approximately true.

For smaller Δx, the relation becomes true.

$$\therefore \quad \Delta y = \frac{dy}{dx}\Delta x. \tag{19.7}$$

If $\Delta x \to 0$ and $\Delta y \to 0$, then

$$dy = \frac{dy}{dx}dx. \tag{19.8}$$

This is exactly true. Under such condition dx and dy are called the differentials of x and y respectively.

19.5 Total Differential

We consider a function $Z = f(x, y)$ of two variables x and y. If x changes to $x + \Delta x$ and y changes to $y + \Delta y$, then z changes to $z + \Delta z$, so that

$$\begin{aligned}\Delta z &= f(x + \Delta x, y + \Delta y) - f(x, y) \\ &= f(x + \Delta x, y + \Delta y) - f(x, y + \Delta y) + f(x, y + \Delta y) - f(x, y).\end{aligned} \tag{19.9}$$

By Lagrange's mean value theorem,

$$f(x + \Delta x) - f(x) = \Delta x f'(x + \theta\Delta x), \ (0 < \theta < 1). \tag{19.10}$$

Again by mean value theorem,

$$f(x + \Delta x, y + \Delta y) - f(x, y + \Delta y) = \Delta x f_x(x + \theta_1\Delta x, y + \Delta y), \ (0 < \theta_1 < 1) \tag{19.11}$$

and $f(x, y + \Delta y) - f(x, y) = \Delta y f_y(x, y + \theta_2\Delta y), \ (0 < \theta_2 < 1).$ (19.12)

Since $f_x(x, y)$ and $f_y(x, y)$ are continuous,

$$\therefore \quad f_x(x + \theta_1\Delta x, y + \Delta y) = f_x(x, y) + C_1 \tag{19.13}$$

and $f_y(x, y + \theta_2\Delta y) = f_y(x, y) + C_2,$ (19.14)

where both C_1 and C_2 tend to zero as Δx and Δy tend to zero.

Now, from Eqns. (19.11) and (19.12),

$$f(x, y + \Delta y) - f(x, y) = [f_y(x, y) - C_2]\Delta y.$$

$\therefore$ from eqn. (19.9),

$$\Delta z = f(x, y)\Delta x + f_y(x, y) + C_1\Delta x + C_2\Delta y. \tag{19.15}$$

Now, the differential of z is

$$dz = \frac{\partial f}{\partial x}\Delta x + \frac{\partial f}{\partial y}\Delta y. \tag{19.16}$$

Using Eqn. (19.16) in (19.15), we get

$$\Delta z = dz + C_1\Delta x + C_2\Delta y. \tag{19.17}$$

But for $\Delta x \to 0$, $C_1 \to 0$ and $\Delta y \to 0$, $C_2 \to 0$.

$$\therefore \quad \Delta z = dz. \tag{19.18}$$

Replacing Δx by dx and Δy by dy in Eqn. (19.16), we get

$$\boldsymbol{dz = \frac{\partial f}{\partial x}dx + \frac{\partial f}{\partial y}dy}. \tag{19.19}$$

dz is called the **total differential** of $z = f(x, y)$.

N.B. If $z = f(x_1, x_2, \dots x_i, \dots x_n)$, then

$$dz = \frac{\partial z}{\partial x_1}dx_1 + \frac{\partial z}{\partial x_2}dx_2 + \cdots + \frac{\partial z}{\partial x_i}dx_i + \cdots = \sum_{i=1}^{n} \frac{\partial z}{\partial x_i}dx_i. \tag{19.20}$$

19.5.1 Error and Approximation

We consider a continuous function $f(x, y)$. If δx and δy be the slight increments in x and y respectively, then the increments in $f(x, y)$ is

$$\delta f = f(x + \delta x, y + \delta y) - f(x, y).$$

Now, by **Taylor's theorem,**

$$f(x + \delta x, y + \delta y) = f(x, y) + \frac{\partial f}{\partial x}\delta x + \frac{\partial f}{\partial y}\delta y + \text{higher order terms.}$$

As δx and δy are considerably small the higher order terms are negligible.

So, $$\delta f = \frac{\partial f}{\partial x}\delta x + \frac{\partial f}{\partial y}\delta y. \tag{19.21}$$

Here, δx, δy are actual errors in x and y respectively and δf is the approximate error in $f(x, y)$.

Now $\frac{\delta f}{f}$ is called the **proportional error** in $f(x, y)$ and $\frac{\delta f}{f} \times 100$ is called the **percentage error**.

$\therefore$ **% error in** $\boldsymbol{f(x, y) = \frac{\delta f}{f} \times 100}$.

Example 19.2. *The period of a simple pendulum is $T = 2\pi\sqrt{\frac{l}{g}}$, find the maximum percentage error in T due to the possible error up to 1% in l and 2.5% in g.*

Solution : Since $T = 2\pi\sqrt{\frac{l}{g}}$,

$$\therefore \quad \log T = \log 2\pi + \frac{1}{2}\log l - \frac{1}{2}\log g \quad \text{or,} \quad \frac{\delta T}{T} = 0 + \frac{1}{2}\frac{\delta l}{l} - \frac{1}{2}\frac{\delta g}{g}.$$

$$\therefore \quad \frac{\delta T}{T} \times 100 = \frac{1}{2}\frac{\delta l}{l} \times 100 - \frac{1}{2}\frac{\delta g}{g} = \frac{1}{2}(1 \pm 2.5) \quad \text{or,} \quad -0.75.$$

$\therefore$ the maximum % error in $T =$ **1.75%**.

19.5.2 Total Derivative

If $u = f(x, y)$ where $x = \phi(t)$ and $y = \psi(t)$, then u may be expressed as a function of t alone by substituting the values of x and y in $f(x, y)$. Then we can find the total derivative $\frac{du}{dt}$, called the total derivative of u, to distinguish it from the partial derivatives $\frac{\partial u}{\partial x}$ and $\frac{\partial u}{\partial z}$.

We can find out $\frac{du}{dt}$ without actually substituting the values of x and y in (x, y).

Here we shall prove two important formulae as follows :

1. $\frac{du}{dt} = \frac{\partial u}{\partial t}\frac{dx}{dt} + \frac{\partial u}{\partial y}\frac{du}{dt}$ and 2. $\frac{dy}{dx} = -\frac{\frac{\partial f}{\partial x}}{\frac{\partial f}{\partial y}}$.

Proof : Since $u = f(x, y)$, the differential du is given by

$$du = \frac{\partial u}{\partial x}dx + \frac{\partial u}{\partial y}dy. \tag{19.22}$$

Again, from the definition of differentials,

$$du = \frac{du}{dt}dt, \quad dx = \frac{dx}{dt}dt, \quad dy = \frac{dy}{dt}dt.$$

Substituting these values in Eqn. (19.22), we get

$$\frac{du}{dt}dt = \frac{\partial u}{\partial x}\frac{dx}{dt}dt + \frac{\partial u}{\partial y}\frac{dy}{dt}dt$$

or, $$\boldsymbol{\frac{du}{dt} = \frac{\partial u}{\partial x}\frac{dx}{dt} + \frac{\partial u}{\partial y}\frac{dy}{dt}}. \tag{19.23}$$

N.B. If $u = f(x, y, z)$ and $x = \phi(t)$, $y = \psi(t)$, $z = \theta(t)$, then

$$\boldsymbol{\frac{du}{dt} = \frac{\partial u}{\partial x}\frac{dx}{dt} + \frac{\partial u}{\partial y}\frac{dy}{dt} + \frac{\partial u}{\partial z}\frac{dz}{dt}}. \tag{19.24}$$

Differentiation of Implicit Functions

If $u = f(x, y) = C$, a constant, then y is an implicit function of x and $\frac{du}{dx} = 0$. Putting $t = x$ in Eqn. (19.23), we get

$$\frac{du}{dx} = \frac{\partial u}{\partial x}\frac{dx}{dx} + \frac{\partial y}{\partial y}\frac{dy}{dx} \quad \text{or,} \quad \frac{\partial u}{\partial x} + \frac{\partial u}{\partial y}\frac{dy}{dx} = 0 \quad \text{or,} \quad f_x + f_y\frac{dy}{dx} = 0$$

or, $$\boldsymbol{\frac{dy}{dx} = -\frac{f_x}{f_y}}. \tag{19.25}$$

Example 19.3. *If $u = x\log(x, y)$, where $x^3 + y^3 + 3xy = 1$, find $\frac{du}{dx}$.*

Solution : From $f(x, y) = x^3 + y^3 + 3xy - 1$, we get

$$\frac{dy}{dx} = -\frac{\frac{\partial f}{\partial x}}{\frac{\partial f}{\partial y}} = -\frac{3x^2 + 3y}{3y^2 + 3x} = -\frac{x^2 + y}{y^2 + x}.$$

Again, $$\frac{du}{dx} = \frac{\partial u}{\partial x} + \frac{\partial u}{\partial y} + \frac{\partial u}{\partial z} = \left(1 \cdot \log xy + x \cdot \frac{1}{x}\right) + \left(\frac{x}{y}\right)\frac{dy}{dx}.$$

$\therefore$ $$\boldsymbol{\frac{du}{dx} = 1 + \log(xy) - \frac{x(x^2 + y)}{y(y^2 + x)}}.$$

19.5.3 Change of Variables

Let $u = f(x, y)$ be a continuous and differentiable function of x and y and $x = \phi(s, t)$ and $y = \psi(s, t)$. Then u is a function of x and y and also of s and t.

Now keeping t constant and differentiating u with respect to s we get from Eqn. (19.23),

$$\left(\frac{\partial u}{\partial s}\right)_t = \frac{\partial u}{\partial x}\left(\frac{\partial x}{\partial s}\right)_t + \frac{\partial u}{\partial y}\left(\frac{\partial y}{\partial s}\right)_t. \tag{19.26}$$

Again keeping t constant and differentiating u with respect to s we get from Eqn. (19.23),

$$\left(\frac{\partial u}{\partial t}\right)_s = \frac{\partial u}{\partial x}\left(\frac{\partial x}{\partial t}\right)_s + \frac{\partial u}{\partial y}\left(\frac{\partial y}{\partial t}\right)_s. \tag{19.27}$$

Solving Eqns. (19.26) and (19.27) we can find out the values of $\frac{\partial u}{\partial x}$ and $\frac{\partial u}{\partial y}$ in terms of $\frac{\partial u}{\partial s}$, $\frac{\partial u}{\partial t}$, u, s, t.

If $s = g(x, y)$ and $t = h(x, y)$, then

$$\frac{\partial u}{\partial x} = \frac{\partial u}{\partial x}\frac{\partial s}{\partial x} + \frac{\partial u}{\partial t}\frac{\partial t}{\partial x} \tag{19.28}$$

and $$\frac{\partial u}{\partial y} = \frac{\partial u}{\partial s}\frac{\partial s}{\partial y} + \frac{\partial u}{\partial t}\frac{\partial t}{\partial y}. \tag{19.29}$$

Example 19.4. *If $u = f(x-y, y-z, z-x)$, then show that,* $\frac{\partial u}{\partial x} + \frac{\partial u}{\partial y} + \frac{\partial u}{\partial z} = 0$.

Solution : Let $x - y = r$, $y - z = s$, $y - z = t$.

$\therefore \quad u = f(r, s, t)$.

$$\therefore \quad \frac{\partial u}{\partial x} = \frac{\partial u}{\partial r}\cdot\frac{\partial r}{\partial x} + \frac{\partial u}{\partial s}\cdot\frac{\partial s}{\partial x} + \frac{\partial u}{\partial t}\cdot\frac{\partial t}{\partial x} = \frac{\partial u}{\partial r}(1) + \frac{\partial u}{\partial s}(0) + \frac{\partial u}{\partial t}(-1) = \frac{\partial u}{\partial r} - \frac{\partial u}{\partial t}.$$

$$\frac{\partial u}{\partial y} = \frac{\partial u}{\partial r}\cdot\frac{\partial r}{\partial y} + \frac{\partial u}{\partial s}\cdot\frac{\partial s}{\partial y} + \frac{\partial u}{\partial t}\cdot\frac{\partial t}{\partial y} = \frac{\partial u}{\partial r}(-1) + \frac{\partial u}{\partial s}(1) + \frac{\partial u}{\partial t}(0) = \frac{\partial u}{\partial s} - \frac{\partial u}{\partial r}.$$

$$\frac{\partial u}{\partial z} = \frac{\partial u}{\partial r}\cdot\frac{\partial r}{\partial z} + \frac{\partial u}{\partial s}\cdot\frac{\partial s}{\partial z} + \frac{\partial u}{\partial t}\cdot\frac{\partial t}{\partial y} = \frac{\partial u}{\partial r}(0) + \frac{\partial u}{\partial s}(-1) + \frac{\partial u}{\partial t}(1) = \frac{\partial u}{\partial t} - \frac{\partial u}{\partial s}.$$

$$\therefore \quad \frac{\partial u}{\partial x} + \frac{\partial u}{\partial y} + \frac{\partial u}{\partial z} = 0.$$

19.6 Homogeneous Functions : Euler's Theorem

Definition : A function $f(x, y)$ which can be expressed as

$$x^n \phi\left(\frac{y}{x}\right)$$

is called a **homogeneous function** of degree n in x and y.

We consider the expression,

$$f(x, y) = a_0 x^n + a_1 x^{n-1} y + a_2 x^{n-2} y^2 + \cdots + a_n y^n, \tag{19.30}$$

whose every term is of degree n. Such a function is a homogeneous function of degree n.

Above expression may be put in the form,

$$f(x, y) = x^n \left[a_0 + a_1 \frac{y}{x} + a_2 \left(\frac{y}{x}\right)^2 + \cdots + a_n \left(\frac{y}{x}\right)^n\right]$$

or, $$u = x^n \phi\left(\frac{y}{x}\right). \tag{19.31}$$

N.B. If the function $f(x, y, z, t, \ldots)$ be a homogeneous function of degree n in x, y, z, t, ... then it can be expressed as

$$x^n \phi\left(\frac{y}{x}, \frac{z}{x}, \frac{t}{x}, \ldots\right). \tag{19.32}$$

Euler's Theorem : If $u = f(x, y)$ be a homogeneous function of degree n in x and y, then according to Euler's theorem,

$$x\frac{\partial u}{\partial x} + y\frac{\partial u}{\partial y} = nu. \tag{19.33}$$

Proof : Here, $u = f(x, y)$ is a homogeneous function of degree n in x and y,

$$\therefore \quad u = x^n \phi\left(\frac{y}{x}\right).$$

$$\therefore \quad \frac{\partial u}{\partial x} = nx^{n-1}\phi\left(\frac{y}{x}\right) + x^n\phi'\left(\frac{y}{x}\right)\cdot y\left(-\frac{1}{x^2}\right) = nx^{n-1}\phi\left(\frac{y}{x}\right) - yx^{n-2}\phi'\left(\frac{y}{x}\right)$$

and $\dfrac{\partial u}{\partial y} = x^n\phi'\left(\dfrac{y}{x}\right)\cdot\dfrac{1}{x} = x^{n-1}\phi'\left(\dfrac{y}{x}\right).$

$$\therefore \quad x\frac{\partial u}{\partial x} + y\frac{\partial u}{\partial y} = nx^n\phi\left(\frac{y}{x}\right) = nu \quad \textbf{(Proved)}$$

N.B. If $u = f(x, y, z, t, \ldots)$, then

$$x\frac{\partial u}{\partial x} + y\frac{\partial u}{\partial y} + z\frac{\partial u}{\partial z} + t\frac{\partial u}{\partial t} = nu. \tag{19.34}$$

19.6.1 Deductions from Euler's Theorem

Following deductions may be made from Euler's theorem :

(1) $x\dfrac{\partial u}{\partial x} + y\dfrac{\partial u}{\partial y} = n\dfrac{f(u)}{f'(u)}$, if z is a homogeneous function of degree n and $z = f(u)$.

(2) $x^2\dfrac{\partial^2 u}{\partial x^2} + 2xy\dfrac{\partial^2 u}{\partial x\partial y} + y^2\dfrac{\partial^2 u}{\partial y^2} = g(u)[g'(u) - 1]$, where $g(u) = n\dfrac{f(u)}{f'(u)}$.

Proof of (1) : As z is a homogeneous function of degree n, we have from Euler's theorem

$$x\frac{\partial u}{\partial x} + y\frac{\partial u}{\partial y} = nz,$$

$$\because \quad z = f(u) \quad \therefore \quad \frac{\partial z}{\partial x} = f'(u)\frac{\partial u}{\partial x}; \quad \frac{\partial z}{\partial y} = f'(u)\frac{\partial u}{\partial y}.$$

$$\therefore \quad xf'(u)\frac{\partial z}{\partial x} + y'f'(u)\frac{\partial u}{\partial y} = nf(u)$$

$$\text{or,} \quad x\frac{\partial u}{\partial x} + y\frac{\partial u}{\partial y} = n\frac{f(u)}{f'(u)}. \tag{19.35}$$

Proof of (2) : Using $g(u) = n\frac{f(u)}{f'(u)}$ in Eqn. (19.35), we get

$$x\frac{\partial u}{\partial x} + y\frac{\partial u}{\partial y} = n\frac{f(u)}{f'(u)} = g(u).$$

Differentiating partially with respect to x,

$$\left(x\frac{\partial^2 u}{\partial x^2} + \frac{\partial u}{\partial x}\right) + y\frac{\partial^2 u}{\partial x\partial y} = g'(u)\frac{\partial u}{\partial x}$$

$$\text{or,} \quad x^2\frac{\partial^2 u}{\partial x^2} + y\frac{\partial^2 u}{\partial x\partial y} = [g'(u) - 1]\frac{\partial u}{\partial x}. \tag{19.36}$$

Again, differentiating partially with respect to y,

$$y\frac{\partial^2 u}{\partial y^2} + x\frac{\partial^2 u}{\partial y\partial x} = [g'(u) - 1]\frac{\partial u}{\partial x}. \tag{19.37}$$

Now multiplying (19.36) by x and (19.37) by y and then adding the results,

$$x^2\frac{\partial^2 u}{\partial x^2}+2xy\frac{\partial^2 u}{\partial x\partial y}+y^2\frac{\partial^2 u}{\partial y^2}=[g'(u-1)]\left[x\frac{\partial u}{\partial x}+y\frac{\partial u}{\partial y}\right]=g(u)[g'(u)-1]$$

or, $$x^2\frac{\partial^u}{\partial x^2}+2xy\frac{\partial^2 y}{\partial x\partial y}+y^2\frac{\partial^2 u}{\partial y^2}=g(u)[g'(u)-1]. \tag{19.38}$$

Example 19.5. *If z is a homogeneous function of degree n in x and y, show that,*

$$x^2\frac{\partial^2 z}{\partial x^2}+2xy\frac{\partial^2 z}{\partial x\partial y}+y^2\frac{\partial^2 z}{\partial y^2}=n(n-1)z.$$

Solution : According to **Euler's theorem**,

$$x\frac{\partial z}{\partial x}+y\frac{\partial z}{\partial y}=nz. \tag{19.39}$$

Differentiating with respect to x,

$$x\frac{\partial^2 y}{\partial x^2}+\frac{\partial z}{\partial x}+y\frac{\partial^2 z}{\partial x\partial y}=n\frac{\partial z}{\partial x}. \tag{19.40}$$

Differentiating again with respect to y,

$$x\frac{\partial^2 y}{\partial y\partial x}+\frac{\partial z}{\partial y}+y\frac{\partial^2 z}{\partial y^2}=n\frac{\partial z}{\partial y}. \tag{19.41}$$

Multiplying Eqn. (19.40) by x and (19.41) by y and then adding,

$$x^2\frac{\partial^2 y}{\partial x^2}+2xy\frac{\partial^2 z}{\partial x\partial y}+y^2\frac{\partial^2 z}{\partial y^2}=(n-1)\left(x\frac{\partial z}{\partial x}+y\frac{\partial z}{\partial y}\right)=n(n-1)z.$$

19.7 Jacobian

If $u=u(x,y)$ and $v=v(x,y)$, then the Jacobian of u and v with respect to x and y is denoted by $J\left(\frac{u,v}{x,y}\right)$ or $\frac{\partial(u,v)}{\partial(x,y)}$ and is defined as

$$\frac{\partial(u,v)}{\partial(x,y)}=\begin{vmatrix}\frac{\partial u}{\partial x} & \frac{\partial u}{\partial y}\\ \frac{\partial v}{\partial x} & \frac{\partial v}{\partial y}\end{vmatrix}. \tag{19.42}$$

It is also known as **functional determinant**.

N.B. The name Jacobian is derived from the name of **Carl Gustav Jacob Jacobi** (1804–1851).

In a similar way, if $u=u(x,y)$, $v=v(x,y)$ and $w=w(x,y)$ then the Jacobian of u, v, w with respect to the variables x, y and z is given by

$$\frac{\partial(u,v,w)}{\partial(x,y,z)}=\begin{vmatrix}\frac{\partial u}{\partial x} & \frac{\partial u}{\partial y} & \frac{\partial u}{\partial z}\\ \frac{\partial v}{\partial x} & \frac{\partial v}{\partial y} & \frac{\partial v}{\partial z}\\ \frac{\partial w}{\partial x} & \frac{\partial w}{\partial y} & \frac{\partial w}{\partial z}\end{vmatrix}$$

19.7.1 Properties of Jacobians

(i) If $J = \dfrac{\partial(u,v)}{\partial(x,y)}$ and $J' = \dfrac{\partial(x,y)}{\partial(u,v)}$, then $JJ' = 1$.

Proof : Let $u = f(x,y)$ and $v = g(x,y)$. Solving for x and y, let $x = \phi(u,v)$ and $y = \psi(u,v)$.

Then, $\dfrac{\partial u}{\partial u} = 1 = \dfrac{\partial u}{\partial x}\dfrac{\partial x}{\partial u} + \dfrac{\partial u}{\partial y}\dfrac{\partial y}{\partial u}$, $\dfrac{\partial u}{\partial v} = 0 = \dfrac{\partial u}{\partial x}\dfrac{\partial x}{\partial v} + \dfrac{\partial u}{\partial y}\dfrac{\partial y}{\partial v}$,

$$\frac{\partial v}{\partial x} = 0 = \frac{\partial v}{\partial x}\frac{\partial x}{\partial u} + \frac{\partial v}{\partial y}\frac{\partial y}{\partial u} \text{ and } \frac{\partial v}{\partial v} = 1 = \frac{\partial v}{\partial x}\frac{\partial x}{\partial v} + \frac{\partial v}{\partial y}\frac{\partial y}{\partial v}.$$

$$\therefore \quad JJ' = \begin{vmatrix} \frac{\partial u}{\partial x} & \frac{\partial u}{\partial y} \\ \frac{\partial v}{\partial x} & \frac{\partial v}{\partial y} \end{vmatrix} \times \begin{vmatrix} \frac{\partial x}{\partial u} & \frac{\partial x}{\partial v} \\ \frac{\partial y}{\partial u} & \frac{\partial y}{\partial v} \end{vmatrix} = \begin{vmatrix} \frac{\partial u}{\partial x} & \frac{\partial u}{\partial y} \\ \frac{\partial v}{\partial x} & \frac{\partial v}{\partial y} \end{vmatrix} \times \begin{vmatrix} \frac{\partial x}{\partial u} & \frac{\partial y}{\partial u} \\ \frac{\partial x}{\partial v} & \frac{\partial y}{\partial v} \end{vmatrix}$$

[interchanging rows and columns of the second determinant].

$$= \begin{vmatrix} 1 & 0 \\ 0 & 1 \end{vmatrix} = 1 \quad \textbf{(Proved)}. \tag{19.43}$$

Similarly we can prove that, $\dfrac{\partial(u,v,w)}{\partial(x,y,z)} \times \dfrac{\partial(x,y,z)}{\partial(u,v,w)} = 1$.

(ii) **Chain Rule for Jacobians :** If u and v be functions of r and s where r and s are functions of x and y, then

$$\frac{\partial(u,v)}{\partial(x,y)} = \frac{\partial(u,v)}{\partial(r,s)} \times \frac{\partial(r,s)}{\partial(x,y)}.$$

Proof : $$\frac{\partial(u,v)}{\partial(r,s)} \times \frac{\partial(r,s)}{\partial(x,y)} = \begin{vmatrix} \frac{\partial u}{\partial r} & \frac{\partial u}{\partial s} \\ \frac{\partial v}{\partial r} & \frac{\partial v}{\partial s} \end{vmatrix} \times \begin{vmatrix} \frac{\partial r}{\partial x} & \frac{\partial r}{\partial y} \\ \frac{\partial s}{\partial x} & \frac{\partial s}{\partial y} \end{vmatrix} = \begin{vmatrix} \frac{\partial u}{\partial r} & \frac{\partial u}{\partial s} \\ \frac{\partial v}{\partial r} & \frac{\partial v}{\partial s} \end{vmatrix} \times \begin{vmatrix} \frac{\partial r}{\partial x} & \frac{\partial s}{\partial x} \\ \frac{\partial r}{\partial y} & \frac{\partial s}{\partial y} \end{vmatrix}$$

[interchanging rows and columns in the second determinant].

$$= \begin{vmatrix} \frac{\partial u}{\partial r}\frac{\partial r}{\partial x} + \frac{\partial u}{\partial s}\frac{\partial s}{\partial x} & \frac{\partial u}{\partial r}\frac{\partial r}{\partial y} + \frac{\partial u}{\partial s}\frac{\partial s}{\partial y} \\ \frac{\partial v}{\partial r}\frac{\partial r}{\partial x} + \frac{\partial v}{\partial s}\frac{\partial s}{\partial x} & \frac{\partial v}{\partial r}\frac{\partial r}{\partial y} + \frac{\partial v}{\partial s}\frac{\partial s}{\partial y} \end{vmatrix} = \begin{vmatrix} \frac{\partial u}{\partial x} & \frac{\partial u}{\partial y} \\ \frac{\partial v}{\partial x} & \frac{\partial v}{\partial y} \end{vmatrix}$$

$$= \frac{\partial(u,v)}{\partial(x,y)} \quad \textbf{(Proved)}. \tag{19.44}$$

Similarly, $\dfrac{\partial(u,v,w)}{\partial(x,y,z)} = \dfrac{\partial(u,v,w)}{\partial(r,s,t)} \times \dfrac{\partial(r,s,t)}{\partial(x,y,z)}$.

(iii) If u, v, w be the functions of three independent variables x, y, z, which are not independent, then

$$\frac{\partial(u,v,w)}{\partial(x,y,z)} = 0.$$

Proof : Since u, v, w are not independent, $f(u, v, w) = 0$.

Differentiating with respect to x, y, z, we get

$$\left.\begin{aligned}\frac{\partial f}{\partial u}\cdot\frac{\partial u}{\partial x}+\frac{\partial f}{\partial v}\cdot\frac{\partial v}{\partial x}+\frac{\partial f}{\partial w}\cdot\frac{\partial w}{\partial x}=0\\ \frac{\partial f}{\partial u}\cdot\frac{\partial u}{\partial y}+\frac{\partial f}{\partial v}\cdot\frac{\partial v}{\partial y}+\frac{\partial f}{\partial w}\cdot\frac{\partial w}{\partial y}=0\\ \frac{\partial f}{\partial u}\cdot\frac{\partial u}{\partial z}+\frac{\partial f}{\partial v}\cdot\frac{\partial v}{\partial z}+\frac{\partial f}{\partial w}\cdot\frac{\partial w}{\partial z}=0\end{aligned}\right\} \tag{19.45}$$

Now eliminating $\frac{\partial f}{du}$, $\frac{\partial f}{dv}$ and $\frac{\partial f}{dw}$ from Eqn. (19.45), we get

$$\begin{vmatrix}\frac{\partial u}{\partial x} & \frac{\partial v}{\partial x} & \frac{\partial w}{\partial x}\\ \frac{\partial u}{\partial y} & \frac{\partial v}{\partial y} & \frac{\partial w}{\partial y}\\ \frac{\partial u}{\partial z} & \frac{\partial v}{\partial z} & \frac{\partial w}{\partial z}\end{vmatrix}=0 \quad \text{or,} \quad \begin{vmatrix}\frac{\partial u}{\partial x} & \frac{\partial u}{\partial y} & \frac{\partial u}{\partial z}\\ \frac{\partial v}{\partial x} & \frac{\partial v}{\partial y} & \frac{\partial v}{\partial z}\\ \frac{\partial w}{\partial x} & \frac{\partial w}{\partial y} & \frac{\partial w}{\partial z}\end{vmatrix}=0 \quad \text{[interchanging rows and columns]}$$

$$\text{or,} \quad \frac{\partial(u, v, w)}{\partial(x, y, z)} = 0 \quad \textbf{(Proved).} \tag{19.46}$$

Example 19.6. *If $y_1 = \frac{x_2 x_3}{x_1}$, $y_2 = \frac{x_3 x_1}{x_2}$, $y_3 = \frac{x_1 x_2}{x_3}$, show that, the Jacobian of y_1, y_2, y_3 with respect to x_1, x_2, x_3 is 4.* **[UPTU 2006]**

Solution : We have $\frac{\partial y_1}{\partial x_1} = -\frac{x_3 x_2}{x_1^2}$, $\frac{\partial y_1}{\partial x_2} = \frac{x_3}{x_1}$ and $\frac{\partial y_1}{\partial x_3} = \frac{x_2}{x_1}$

$$\frac{\partial y_2}{\partial x_1} = -\frac{x_3}{x_2}, \quad \frac{\partial y_2}{\partial x_2} = -\frac{x_3 x_1}{x_2^2} \quad \text{and} \quad \frac{\partial y_2}{\partial x_3} = \frac{x_1}{x_2}$$

$$\frac{\partial y_3}{\partial x_1} = \frac{x_2}{x_3}, \quad \frac{\partial y_3}{\partial x_2} = \frac{x_1}{x_3} \quad \text{and} \quad \frac{\partial y_3}{\partial x_3} = -\frac{x_1 x_2}{x_3^2}.$$

$$\therefore \quad \frac{\partial(y_1, y_2, y_3)}{\partial(x, y, z)} = \begin{vmatrix}\frac{\partial y_1}{\partial x_1} & \frac{\partial y_1}{\partial x_2} & \frac{\partial y_1}{\partial x_3}\\ \frac{\partial y_2}{\partial x_1} & \frac{\partial y_2}{\partial x_2} & \frac{\partial y_2}{\partial x_3}\\ \frac{\partial y_3}{\partial x_1} & \frac{\partial y_3}{\partial x_2} & \frac{\partial y_3}{\partial x_3}\end{vmatrix} = \begin{vmatrix}-\frac{x_2 x_3}{x_1^2} & \frac{x_3}{x_1} & \frac{x_2}{x_1}\\ \frac{x_3}{x_2} & -\frac{x_3 x_1}{x_2^2} & \frac{x_1}{x_2}\\ \frac{x_2}{x_3} & \frac{x_1}{x_3} & -\frac{x_1 x_3}{x_3^2}\end{vmatrix}$$

$$= -\frac{1}{x_1^2 x_2^2 x_3^2}\begin{vmatrix}-x_2 x_3 & x_2 x_1 & x_1 x_2\\ x_2 x_3 & -x_3 x_1 & x_1 x_2\\ x_2 x_3 & x_3 x_1 & -x_1 x_2\end{vmatrix} = -\frac{x_1^2 x_2^2 x_3^2}{x_1^2 x_2^2 x_3^2}\begin{vmatrix}-1 & 1 & 1\\ 1 & -1 & 1\\ 1 & 1 & -1\end{vmatrix}$$

$$= -1(1-1) - 1(-1-1) + 1(1+1) = 0 + 2 + 2 = 4 \quad \textbf{(Proved).}$$

19.7.2 Jacobian of Implicit Functions

If x, y, u and v are connected by the implicit functions,

$$f_1(x, y, u, v) = 0 \tag{19.47}$$

$$f_2(x, y, u, v) = 0. \tag{19.48}$$

Here u and v are implicit functions of x and y.

Differentiating Eqns. (19.47) and (19.48) with respect to x and y in turn respectively, we get

$$\frac{\partial f_1}{\partial x} + \frac{\partial f_1}{\partial u}\cdot\frac{\partial u}{\partial x} + \frac{\partial f_1}{\partial v}\cdot\frac{\partial v}{\partial x} = 0 \tag{19.49}$$

$$\frac{\partial f_1}{\partial y} + \frac{\partial f_1}{\partial u}\cdot\frac{\partial u}{\partial y} + \frac{\partial f_1}{\partial v}\cdot\frac{\partial v}{\partial y} = 0 \tag{19.50}$$

and $$\frac{\partial f_2}{\partial x} + \frac{\partial f_2}{\partial u}\cdot\frac{\partial u}{\partial x} + \frac{\partial f_2}{\partial v}\cdot\frac{\partial v}{\partial x} = 0 \tag{19.51}$$

$$\frac{\partial f_2}{\partial y} + \frac{\partial f_2}{\partial u}\cdot\frac{\partial u}{\partial y} + \frac{\partial f_2}{\partial v}\cdot\frac{\partial v}{\partial y} = 0. \tag{19.52}$$

$$\therefore \quad \frac{\partial(f_1, f_2)}{\partial(u, v)} \times \frac{\partial(u, v)}{\partial(x, y)} = \begin{vmatrix} \frac{\partial f_1}{\partial u} & \frac{\partial f_1}{\partial v} \\ \frac{\partial f_2}{\partial u} & \frac{\partial f_2}{\partial v} \end{vmatrix} \times \begin{vmatrix} \frac{\partial u}{\partial x} & \frac{\partial u}{\partial y} \\ \frac{\partial v}{\partial x} & \frac{\partial v}{\partial y} \end{vmatrix}$$

$$= \begin{vmatrix} \frac{\partial f_1}{\partial u}\cdot\frac{\partial u}{\partial x} + \frac{\partial f_1}{\partial v}\cdot\frac{\partial v}{\partial x} & \frac{\partial f_1}{\partial u}\cdot\frac{\partial u}{\partial y} + \frac{\partial f_1}{\partial v}\cdot\frac{\partial v}{\partial y} \\ \frac{\partial f_2}{\partial u}\cdot\frac{\partial u}{\partial x} + \frac{\partial f_2}{\partial v}\cdot\frac{\partial v}{\partial x} & \frac{\partial f_2}{\partial u}\cdot\frac{\partial u}{\partial y} + \frac{\partial f_2}{\partial v}\cdot\frac{\partial v}{\partial y} \end{vmatrix} = \begin{vmatrix} -\frac{\partial f_1}{\partial x} & -\frac{\partial f_1}{\partial y} \\ -\frac{\partial f_2}{\partial x} & -\frac{\partial f_2}{\partial y} \end{vmatrix}$$

$$= (-1)^2 \frac{\partial(f_1, f_2)}{\partial(x, y)},$$

using relations (19.49) to (19.52),

$$\therefore \quad \boldsymbol{\frac{\partial(u, v)}{\partial(x, y)} = (-1)^2 \frac{\frac{\partial(f_1, f_2)}{\partial(x, y)}}{\frac{\partial(f_1, f_2)}{\partial(u, v)}}}. \tag{19.53}$$

Example 19.7. *If $u = x^2 - y^2$, $v = 2xy$ and $x = r\cos\theta$, $y = r\sin\theta$, find $\dfrac{\partial(u, v)}{\partial(r, \theta)}$.*

[VTU 2007, Madras U 2006]

Solution : We know that, $\dfrac{\partial(u, v)}{\partial(r, \theta)} = \dfrac{\partial(u, v)}{\partial(x, y)} \times \dfrac{\partial(x, y)}{\partial(r, \theta)}$.

$\because \; u = x^2 - y^2$, $v = 2xy$.

$$\therefore \quad \frac{\partial(u, v)}{\partial(x, y)} = \begin{vmatrix} \frac{\partial u}{\partial x} & \frac{\partial u}{\partial y} \\ \frac{\partial v}{\partial x} & \frac{\partial v}{\partial y} \end{vmatrix} = \begin{vmatrix} 2x & -2y \\ 2y & 2x \end{vmatrix} = 4(x^2 + y^2).$$

Again, $\because\ x = r\cos\theta,\ y = r\sin\theta$,

$$\therefore \qquad \frac{\partial(x,y)}{\partial(r,\theta)} = \begin{vmatrix} \dfrac{\partial x}{\partial r} & \dfrac{\partial x}{\partial \theta} \\ \dfrac{\partial y}{\partial r} & \dfrac{\partial y}{\partial \theta} \end{vmatrix} = \begin{vmatrix} \cos\theta & -r\sin\theta \\ \sin\theta & r\cos\theta \end{vmatrix} = r.$$

Hence, $\dfrac{\partial(u,v)}{\partial(r,\theta)} = 4(x^2+y^2)\cdot r = 4(r^2\cos^2\theta + r^2\sin^2\theta)\cdot r = 4r^3.$

19.8 Taylor's Theorem for Functions of Two Variables

If (x, y) and all its partial derivatives up to nth order are finite and continuous at all points (x, y), then

$$f(x+h, y+k) = f(x,y) + \left(h\frac{\partial}{\partial x} + k\frac{\partial}{\partial y}\right) f + \frac{1}{2!}\left(h\frac{\partial}{\partial x} + k\frac{\partial}{\partial y}\right)^2 f + \cdots.$$

This is **Taylor's theorem** for a function of two variables.

Proof : Considering $f(x+h, y+k)$ as a function of a single variable x only, we have by Taylor's theorem,

$$f(x+h, y+k) = f(x, y+k) + h\frac{\partial f(x,y+k)}{\partial x} + \frac{h^2}{2!}\frac{\partial^2 f(x,y+k)}{\partial x^2} + \cdots. \qquad (19.54)$$

Also, $f(x, y+k) = f(x,y) + k\dfrac{\partial f(x,y)}{\partial y} + \dfrac{k^2}{2!}\dfrac{\partial^2 f(x,y)}{\partial y^2} + \cdots. \qquad (19.55)$

Substituting Eqn. (19.55) in (19.54), we get

$$\begin{aligned} f(x+h, y+k) &= f(x,y) + k\frac{\partial f(x,y)}{\partial y} + \frac{k^2}{2!}\frac{\partial^2 f(x,y)}{\partial y^2} + \cdots \\ &\quad + h\frac{\partial}{\partial x}\left[f(x,y) + k\frac{\partial f(x,y)}{\partial y} + \frac{k^2}{2!}\frac{\partial^2 f(x,y)}{\partial y^2} + \cdots\right] \\ &\quad + \frac{h^2}{2!}\frac{\partial}{\partial x^2}\left[f(x,y) + k\frac{\partial f(x,y)}{\partial y} + \frac{k^2}{2!}\frac{\partial^2 f(x,y)}{\partial y^2} + \cdots\right] + \cdots \\ &= f(x,y) + \left(h\frac{\partial}{\partial x} + k\frac{\partial}{\partial y}\right) + \frac{1}{2!}\left(h^2\frac{\partial^2 f}{\partial x^2} + 2hk\frac{\partial^2 f}{\partial x\partial y} + k^2\frac{\partial^2 f}{\partial y^2}\right) + \cdots \end{aligned}$$

or,
$$\begin{aligned} f(x+h, y+k) &= f(x,y) + \left(h\frac{\partial}{\partial x} + k\frac{\partial}{\partial y}\right) f \\ &\quad + \frac{1}{2!}\left(h\frac{\partial}{\partial x} + k\frac{\partial}{\partial y}\right)^2 f + \cdots. \end{aligned} \qquad (19.56)$$

This proves Taylor's theorem for a function of two variables.

N.B. (1) Putting $x = a$ and $y = b$, we get from Eqn. (19.56),

$$\begin{aligned} f(a+h, b+k) &= f(a,b) + [hf_x(a,b) + kf_y(a,b)] \\ &\quad + \frac{1}{2!}[h^2 f_{xx}(a,b) + 2hk f_{xy}(a,b) + k^2 f_{yy}(a,b)] + \cdots. \end{aligned} \qquad (19.57)$$

Taylor's expansion of $f(x,y)$ in powers of $(x-a)$ and $(y-b)$ can be obtained by putting $a+h=x$ and $b+k=y$.

$$\therefore \quad f(x,y) = f(a,b) + [(x-a)f_x(a,b) + (y-b)f_y(a,b)]$$
$$+ \frac{1}{2!}[(x-a)^2 f_{xx}(a,b) + 2(x-a)(y-b)f_{xy}(a,b) + (y-b)^2 f_{yy}(a,b)]. \quad (19.58)$$

(2) If $a=0$, $b=0$, then Eqn. (19.58) reduces to

$$f(x,y) = f(0,0) + [xf_x(0,0) + yf_y(0,0)]$$
$$+ \frac{1}{2!}[x^2 f_{xx}(a,b) + 2xyf_{xy}(0,0) + y^2 f_{yy}(0,0)] + \cdots. \quad (19.59)$$

This is **Maclaurin's expansion** of (x,y).

Example 19.8. *Expand $e^x \log(1+y)$ in powers of x and y in Taylor's series in the neighbourhood of $(0,0)$ as far as the first ten terms.* **[JNTU 2006, PTU 2006]**

Solution : We have $f(x,y) = e^x \log(1+y)$, $\quad \therefore \; f(0,0) = 0$

$$\frac{\partial f}{\partial x} = e^x \log(1+x), \qquad \therefore \; f_x(0,0) = 0$$

$$\frac{\partial f}{\partial y} = \frac{e^x}{1+y}, \qquad \therefore \; f_y(0,0) = 1$$

$$\frac{\partial^2 f}{\partial x^2} = e^x \log(1+x), \qquad \therefore \; f_{xx}(0,0) = 0$$

$$\frac{\partial^2 f}{\partial y^2} = -\frac{e^x}{(1+y)^2}, \qquad \therefore \; f_{yy}(0,0) = -1$$

$$\frac{\partial^2 f}{\partial x \partial y} = \frac{e^x}{1+y}, \qquad \therefore \; f_{xy}(0,0) = 1$$

$$\frac{\partial^3 f}{\partial x^3} = e^x \log(1+y), \qquad \therefore \; f_{xxx}(0,0) = 0$$

$$\frac{\partial^3 f}{\partial y \partial x \partial x} = \frac{e^x}{1+y}, \qquad \therefore \; f_{xxy}(0,0) = 1$$

$$\frac{\partial^3 f}{\partial y \partial y \partial x} = -\frac{e^x}{(1+y)^2}, \qquad \therefore \; f_{xyy}(0,0) = -1$$

and
$$\frac{\partial^3 f}{\partial y^3} = 2e^x(1+y)^{-3}, \qquad \therefore \; f_{yyy}(0,0) = 2.$$

Now, according to Maclaurin's expansion of $f(x,y)$,

$$f(x,y) = f(0,0) + xf_x(0,0) + yf_y(0,0) + \frac{1}{2!}[x^2 f_{xx}(0,0) + 2xyf_{xy}(0,0) + y^2 f_{yy}(0,0)]$$
$$+ \frac{1}{3!}[x^3 f_{xxx}(0,0) + 3x^2 y f_{xxy}(0,0) + 3xy^2 f_{xyy}(0,0) + y^3 f_{yyy}(0,0)] + \cdots.$$

$$\therefore \quad e^x \log(1+x) = 0 + x(0) + y(1) + \frac{1}{2!}[x^2(0) + 2xy(1) + y^2(-1)]$$
$$+ \frac{1}{3!}[x^3(0) + 3x^2y(1) + 3xy^2(-1) + y^3(2)] + \cdots$$

or, $$e^x \log(1+x) = y + xy - \frac{1}{2}y^2 + \frac{1}{2}(x^2y - xy^2) + \frac{1}{3}y^3 + \cdots.$$

19.9 (1) Geometrical Interpretation

To understand the geometrical interpretation of partial differential coefficient, we consider Fig. 19.1. Let OX, OY, OZ be three mutually perpendicular axes with respect to which the equation $z = f(x, y)$ represents a surface and $P(x, y, z)$ represents any point on the surface.

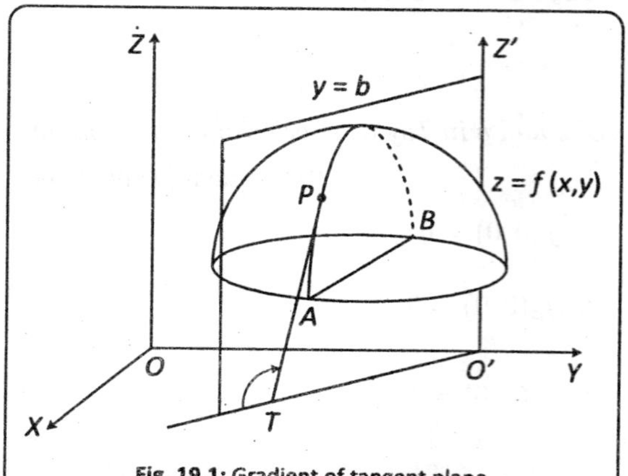

Fig. 19.1: Gradient of tangent plane

Let a plane $y = b$ parallel to xz-plane intersect the surface $z = f(x, y)$ along the curve $AP'B$ given by $z = f(x, b)$.

As y remains equal to b and x varies then P moves along the curve APB and $\frac{\partial z}{\partial x}$ is the ordinary derivative of $f(x, b)$ w.r.t. x,

so, $\frac{\partial z}{\partial x}$ **at P is the gradient to the tangent to the curve of intersection of the surface $z = f(x, y)$ and a plane parallel to xz-plane.**

Similarly, $\frac{\partial z}{\partial y}$ at P is the gradient of the tangent to the curve of intersection of the surface $z = f(x, y)$ and the plane parallel to yz-plane.

(2) Tangent Plane and Normal to the Surface

Let $P(x, y, z)$ and $Q(x + \Delta x, y + \Delta y, z + \Delta z)$ be two neighbouring points on the surface $f(x, y, z) = 0$. Let arc $PQ = \Delta s$ and chord $PQ = \Delta l$.

If $Q \to P$, $\Delta S \to \Delta P$.

then, $\text{Lt}\left(\frac{\Delta s}{\Delta l}\right) = 1$ as $Q \to P$.

The direction cosines of PQ are

$$\frac{\Delta x}{\Delta l}, \frac{\Delta y}{\Delta l}, \frac{\Delta z}{\Delta l} = \frac{\Delta x}{\Delta s}\frac{\Delta s}{\Delta l}, \frac{\Delta y}{\Delta s}\frac{\Delta s}{\Delta l}, \frac{\Delta z}{\Delta s}\frac{\Delta s}{\Delta l}.$$

Now, as $Q \to P$, $\Delta l \to 0$ and the chord PQ becomes the tangent PT, so, the direction cosines of the tangent PT are

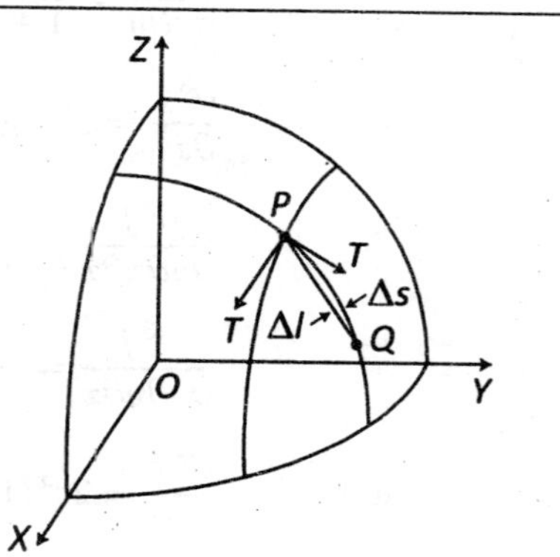

Fig. 19.2: Tangent plane and normal to the surface

$$\underset{\Delta s \to 0}{\text{Lt}} \frac{\Delta x}{\Delta s}, \quad \underset{\Delta l \to 0}{\text{Lt}} \frac{\Delta s}{\Delta l}, \quad \underset{\Delta s \to 0}{\text{Lt}} \frac{\Delta y}{\Delta s}, \quad \underset{\Delta l \to 0}{\text{Lt}} \frac{\Delta s}{\Delta l},$$
$$\underset{\Delta s \to 0}{\text{Lt}} \frac{\Delta z}{\Delta s}, \quad \underset{\Delta l \to 0}{\text{Lt}} \frac{\Delta s}{\Delta l}$$

or, $$\frac{dx}{ds}, \frac{dy}{ds}, \frac{dz}{ds}. \qquad (19.60)$$

Differentiating $f(x, y, z) = 0$ with respect to s,

$$\frac{\partial f}{\partial x} \cdot \frac{dx}{ds} + \frac{\partial f}{\partial y} \cdot \frac{dy}{ds} + \frac{\partial f}{\partial z} \cdot \frac{dz}{ds} = 0. \tag{19.61}$$

This shows that the tangent to the curve is perpendicular to the line whose direction cosines are given by

$$\frac{\partial f}{\partial x}, \frac{\partial f}{\partial y}, \frac{\partial f}{\partial z}. \tag{19.62}$$

A number of curves along this surface may join P and Q giving several tangents perpendicular to the line having direction cosines $\frac{\partial f}{\partial x}, \frac{\partial f}{\partial y}, \frac{\partial f}{\partial z}$. Since, all the tangents lie on a single plane, they form a tangent plane at P. This tangent plane possesses a normal with the direction cosines as above.

$\therefore$ the equation to the tangent plane is $\frac{\partial f}{\partial x}(X - x) + \frac{\partial f}{\partial y}(Y - y) + \frac{\partial f}{\partial z}(Z - z) = 0.$ (19.63)

For a particular point $P(x_0, x_0, z_0)$ on the surface, the equation to the tangent plane is

$$A(x - x_0) + B(y - y_0) + C(z - z_0) = 0,$$

where $A = \left(\frac{\partial f}{\partial x}\right)_{(x_0, y_0)}$; $B = \left(\frac{\partial f}{\partial y}\right)_{(x_0, y_0)}$; $C = \left(\frac{\partial f}{\partial z}\right)_{(x_0, y_0)}$. (19.64)

The equation to the normal to the surface at $P(x, y, z)$ is

$$\frac{X - x}{\left(\frac{\partial f}{\partial x}\right)} = \frac{Y - y}{\left(\frac{\partial f}{\partial y}\right)} = \frac{Z - z}{\left(\frac{\partial f}{\partial z}\right)}. \tag{19.65}$$

Hence, normal to the surface at $P(x_0, y_0, z_0)$ is

$$\frac{x - x_0}{A} = \frac{y - y_0}{B} = \frac{z - z_0}{C}, \tag{19.66}$$

where A, B and C are as given above.

19.10 Maxima and Minima of a Function of Two Variables

1. Definition : A function $f(x, y)$ of two variables x and y is said to have a maximum at (a, b), if (i) $[f(a + h, b + k) - f(a, b)] < 0$ or, a minima at (a, b), if (ii) $[f(a + h, b + k) - f(a, b)] > 0$ for all positive or negative small values of h and k.

In other words, if $\Delta = [f(a + h, b + k) - f(a, b)]$ is of the same sign for all small values of h and k and if the sign is **negative**, then $f(a, b)$ is **maximum**. If this sign is **positive** $f(a, b)$ is a **minimum**.

2. Condition for $f(x, y)$ to the maximum or minimum : By Taylor's theorem,

$$f(a+h, b+k) - f(a, b) = \left(h\frac{\partial f}{\partial x} + k\frac{\partial f}{\partial y}\right)_{a,b} + \frac{1}{2!}\left(h^2\frac{\partial^2 f}{\partial x^2} + 2hk\frac{\partial^2 f}{\partial x \partial y} + k^2\frac{\partial^2 f}{\partial y^2}\right) + \cdots, \tag{19.67}$$

neglecting the higher order terms which are still smaller for small h and k.

Thus, sign of Δ = sign of $[hf_x(a, b) + kf_y(a, b)]$. (19.68)

Putting $h = 0$, the right-hand side changes sign when k changes sign. So, $f(x, y)$ cannot be maximum or minimum at (a, b) unless $f_y(a, b) = 0$. Similarly, putting $h = 0$, we find that $f(x, y)$ cannot have a maximum or minimum at (a, b) unless $f_x(a, b) = 0$.

$\therefore$ the necessary conditions for $f(x, y)$ to have a maximum or minimum at (a, b) are :

$$\left.\begin{aligned} f_x(a, b) = 0 \\ f_y(a, b) = 0 \end{aligned}\right\}. \tag{19.69}$$

If the above conditions are satisfied, then from Eqn. (19.67) we have for small values of h and k,

$$\text{sign of } \Delta = \text{sign of } \left[\frac{1}{2!}(h^2r + 2hks + k^2t)\right],$$

where $r = f_{xx}(a, b)$, $s = f_{xy}(a, b)$, $t = f_{yy}(a, b)$.

Again, $h^2r + 2hks + k^2t = \dfrac{1}{r}[h^2r^2 + 2hkrs + k^2rt] = \dfrac{1}{r}[(hr + ks)^2 + h^2(rt - s^2)]$.

So, $$\text{sign of } \Delta = \text{sign of } \frac{1}{2r}[(hr + ks)^2 + k^2(rt - s^2). \tag{19.70}$$

Eqn. (19.70) shows that $(hr + ks)^2$ is always positive and $k^2(rt - s^2)$ will be positive if $(rt - s^2) > 0$. Then Δ will have the same sign as that of r for all values of h and k.

Hence, 1. if $(rt - s^2) > 0$, $f(x, y)$ has a maximum or minimum at (a, b) according as $r < 0$ or $r > 0$

2. if $(rt - s^2) = 0$, further investigation is needed to find whether $f(x, y)$ has a maximum or minimum at (a, b).

N.B. (i) **Stationary value :** If $f_x(a, b) = 0$ and $f_y(a, b) = 0$, then the function $f(a, b)$ is said to be a stationary value of $f(x, y)$ at (a, b).

(ii) The points where $f_x(x, y) = 0$ and $f_y(x, y)$ but $f(x, y)$ does not possess maxima or minima are known as **saddle point**.

Finally, the **conditions** for $f(x, y)$ to have a maximum or minimum at (a, b) are :

1. **$f_x(a, b) = 0$, $f_y(a, b) = 0$**
2. **If $(rt - s^2) > 0$, then $f(x, y)$ has a maximum at (a, b) provided $r < 0$.**
3. **If $(rt - s^2) > 0$, then $f(x, y)$ has a minimum at (a, b) provided $r > 0$.**
4. **If $rt - s^2 = 0$, then further investigation is necessary to find whether $f(x, y)$ has a maximum or minimum at (a, b).**

Example 19.9. *Examine $f(x, y) = x^3 + y^3 - 3xy$ for maximum and minimum values.*

Solution : Given that $f(x, y) = x^3 + y^3 - 3xy$. (i)

$$\therefore \quad \frac{\partial f}{\partial x} = 3x^2 - 3y, \ \frac{\partial f}{\partial y} = 3y^2 - 3x, \ r = \frac{\partial^2 f}{\partial x^2} = 6x, \ s = \frac{\partial^2 f}{\partial x \partial y} = -3 \text{ and } t = 6y.$$

For extremum (i.e., maximum or minimum),

$$\frac{\partial f}{\partial x} = 0 \text{ and } \frac{\partial f}{\partial y} = 0 \text{ or, } 3x^2 - 3x = 0 \text{ and } 3y^2 - 3x = 0$$

$$\text{or,} \quad x^2 = y \tag{ii}$$

$$\text{and} \quad y^2 = x. \tag{iii}$$

Putting $y = x^2$ in $y^2 = x$, we get

$$x^4 = x \quad \text{or,} \quad x(x^3 - 1) = 0 \quad \text{or,} \quad x(x-1)(x^2 + x + 1) = 0.$$

$\therefore \quad x = 0,\ 1.$

When $x = 0$, Eqn. (ii) gives $y = 0$

and when $x = 1$, Eqn. (i) gives $y = 1$.

$\therefore$ $(0, 0)$ and $(1, 1)$ are the stationary points.

Now, at $(0, 0)$, $r = 6 \times 0 = 0$, $s = -3$, $t = 6 \times 0 = 0$.

$\therefore \quad rt - s^2 = 0 - 9 = -9$, which is negative.

So at $(0, 0)$ there is no extremum value.

But at $(1, 1)$, $r = 6$, $s = \dfrac{\partial^2 f}{\partial x \partial y} = -3$, $t = 6$.

$\therefore \quad rt - s^2 = 6 \times 6 - 9 = 36 - 9 = 27$ and $r = \pm 6$.

Hence, **the point $(1, 1)$ is a minimum.**

The minimum value of $f(x, y)$ at $(1, 1)$ is

$$f(1, 1) = 1 + 1 - 3 = -1.$$

19.11 Lagrange's Method of Undetermined Multiplier

If the variables of a function are connected by a relation and not all independent, then the stationary points and stationary values of the function can be determined by **Lagrange's method of undetermined multiplier**.

Let $u = f(x, y, z)$ be a function of three variables and the variables are connected by

$$\phi(x, y, z) = 0. \tag{19.71}$$

Thus, the variables x, y, z are not independent. Since there is one equation of constraint, only two of the three variables are independent.

For the conditions for stationary points in $u = f(x, y, z)$, we have

$$\frac{\partial u}{\partial x} = 0, \quad \frac{\partial u}{\partial y} = 0, \quad \frac{\partial u}{\partial z} = 0. \tag{19.72}$$

$$\therefore \quad du = \frac{\partial u}{\partial x} dx + \frac{\partial u}{\partial y} dy + \frac{\partial u}{\partial z} dz = 0. \tag{19.73}$$

Also from Eqn. (19.71),

$$\frac{\partial \phi}{\partial x} dx + \frac{\partial \phi}{\partial y} dy + \frac{\partial \phi}{\partial z} dz = 0. \tag{19.74}$$

Now multiplying Eqn. (19.74) by a parameter λ and then adding the result with eqn. (19.73), we get

$$\left(\frac{\partial u}{\partial x} + \lambda \frac{\partial \phi}{\partial x}\right) dx + \left(\frac{\partial u}{\partial y} + \lambda \frac{\partial \phi}{\partial y}\right) dy + \left(\frac{\partial u}{\partial z} + \lambda \frac{\partial \phi}{\partial z}\right) dz = 0. \tag{19.75}$$

$\therefore$ Eqn. (19.75) is satisfied, if

$$\left.\begin{aligned} \frac{\partial u}{\partial x} + \lambda\frac{\partial \phi}{\partial x} &= 0 \\ \frac{\partial u}{\partial y} + \lambda\frac{\partial \phi}{\partial y} &= 0 \\ \frac{\partial u}{\partial z} + \lambda\frac{\partial \phi}{\partial z} &= 0 \end{aligned}\right\}. \quad (19.76)$$

The values of x, y, z and λ can be determined from Eqns. (19.71) and (19.76). Finally, the stationary values of $u = f(x, y, z)$, which may be maximum or minimum, is obtained.

Working rule :

(i) Write $F = f(x, y, z) + \lambda\phi(x, y, z)$.

(ii) Obtain the equations, $\frac{\partial F}{\partial x} = 0, \frac{\partial F}{\partial y} = 0, \frac{\partial F}{\partial z} = 0$.

(iii) Solve the above equations together with $\phi(x, y, z)$.

N.B. Lagrange's method cannot determine the nature of the stationary points. From physical consideration of the problem this can, however, be determined.

Example 19.10. *Using the Lagrange's method of undetermined multiplier, find the maximum of $F = 4xyz$ subject to the constraint $x^2 + y^2 + z^2 - a^2 = 0$.* **[BU(H) 1997]**

Solution : Here, the equation of the constraint is

$$x^2 + y^2 + z^2 - a^2 = 0. \quad \text{(i)}$$

We consider $f(x, y, z) = 4xyz + \lambda(x^2 + y^2 + z^2 - a^2)$. (ii)

$$\therefore \quad \frac{\partial f}{\partial x} = 4yz + 2\lambda x = 0 \quad \text{(iii)}$$

$$\frac{\partial f}{\partial y} = 4zx + 2\lambda y = 0 \quad \text{(iv)}$$

$$\frac{\partial f}{\partial z} = 4xy + 2\lambda z = 0. \quad \text{(v)}$$

Comparing the λ values in (iii) and (iv), we get $\lambda = -\frac{2yz}{x} = -\frac{2zx}{y}$.

$\therefore \quad y^2 = x^2$, i.e., $y = x$.

Similarly, from (iv) and (v), $x = z$.

So, $x = y = z$. (vi)

From Eqn. (i),

$$x = y = z = \pm\frac{9}{\sqrt{3}}. \quad \text{(vii)}$$

Again, the value of $F = 4xyz$ which is between $-\frac{4a}{3\sqrt{3}}$ to $\frac{4a}{3\sqrt{3}}$.

$\therefore$ the maximum value of F is $\frac{4a}{3\sqrt{3}}$.

19.12 Differentiation under the Integral Sign : Leibnitz Rule

Leibnitz rule is applied to evaluate definite integrals which are not integrable otherwise.

If $f(x,\alpha)$ and $\frac{\partial f(x,\alpha)}{\partial x}$ be continuous functions of x and α, then according to Leibnitz rule,

$$\frac{d}{dx}\left[\int_a^b f(x,\alpha)d\alpha\right] = \int_a^b \frac{\partial f(x,\alpha)}{\partial x}dx,$$

where a and b are independent of α. **[BU(H) 1996]**

Proof : Let $F(\alpha) = \int_a^b f(x,\alpha)dx$.

$$\therefore \quad F(\alpha+\delta\alpha) = \int_a^b f(x,\alpha+\delta\alpha)dx.$$

$$\therefore \quad \frac{F(\alpha+\delta\alpha)-F(\alpha)}{\delta\alpha} = \int_a^b \frac{f(x,\alpha+\delta)-f(x,\alpha)}{\delta\alpha}dx$$

$$= \int_a^b \frac{f(x,\alpha+\theta\delta\alpha)}{\delta\alpha}dx \quad (\text{where } 0<\theta<1) \quad [\text{by Mean Value Theorem}]$$

$$\therefore \quad \operatorname*{Lt}_{\delta\alpha\to 0}\frac{F(\alpha+\delta\alpha)-F(\alpha)}{\delta\alpha} = \int_a^b \frac{f(x,\alpha)}{\delta\alpha}dx \quad \text{or,} \quad \frac{d}{d\alpha}[F(\alpha)] = \int_a^b \frac{f(x,\alpha)}{\delta\alpha}dx$$

$$\text{or,} \quad \frac{d}{d\alpha}\left[\int_a^b f(x,\alpha)dx\right] = \int_a^b \frac{f(x,\alpha)}{\delta\alpha}dx. \qquad (19.77)$$

Example 19.11. *Evaluate the integral* $\int_0^\infty e^{-x}\frac{\sin\alpha x}{x}dx$. **[BU(H)(Old) 2000]**

Solution : Let $F(\alpha) = \int_0^\infty e^{-x}\frac{\sin\alpha x}{x}dx$.

$$\therefore \quad \frac{d}{d\alpha}F(\alpha) = \int_0^\infty \frac{\partial}{\partial\alpha}\left[e^{-x}\frac{\sin\alpha x}{x}\right]dx = \int_0^\infty \frac{xe^{-x}\cos\alpha x}{x}dx = \int_0^\infty e^{-x}\cos\alpha x\,dx.$$

Now, let $I = \int_0^\infty e^{-x}\cos\alpha x\,dx = \frac{1}{\alpha}e^{-x}\sin\alpha x + \frac{1}{\alpha}e^{-x}\sin\alpha x\,dx$

$$= \frac{1}{\alpha}e^{-x}\sin\alpha x - \frac{1}{\alpha^2}\left[e^{-x}\cos\alpha x + \int e^{-x}\cos\alpha x\,dx\right].$$

$$\therefore \quad \left(1+\frac{1}{\alpha^2}\right)I = \frac{1}{\alpha}e^{-x}\sin\alpha x - \frac{1}{\alpha^2}e^{-x}\cos\alpha x \quad \text{or,} \quad I = \frac{1}{(1+\alpha^2)}[\alpha e^{-x}\sin\alpha x - e^{-x}\cos\alpha x].$$

$$\therefore \quad \frac{d}{d\alpha}[F(\alpha)] = \frac{1}{1+\alpha^2}\left[ae^{-x}\sin\alpha x - e^{-x}\cos\alpha x\right]_0^\infty = \frac{1}{1+\alpha^2}.$$

$$\therefore \quad F(\alpha) = \tan^{-1}(\alpha) + C.$$

But $F(0) = 0$, $\therefore C = 0$.

Hence, $\boldsymbol{\int_0^\infty e^{-x}\frac{\sin\alpha x}{x}dx = \tan^{-1}x}$.

19.13 Important Relations

If the variables x, y, z of a function $f(x,y,z)=0$ are related by functional equations, so that each variable can be expressed as a function of the other two, then

$$1.\ \left(\frac{\partial z}{\partial x}\right)_y=\frac{1}{\left(\frac{\partial x}{\partial z}\right)_y},\quad 2.\ \left(\frac{\partial x}{\partial y}\right)_z\left(\frac{\partial y}{\partial z}\right)_x\left(\frac{\partial z}{\partial y}\right)_y=-1.$$

Proof : We have $f(x,y,z)=0$, $x=x(y,z)$, $z=z(x,y)$.

$$\therefore\quad dx=\left(\frac{\partial x}{\partial y}\right)_z dy+\left(\frac{\partial x}{\partial z}\right)_y dz \tag{19.78}$$

and
$$dz=\left(\frac{\partial z}{\partial x}\right)_y dx+\left(\frac{\partial z}{\partial y}\right)_x dy. \tag{19.79}$$

Eliminating dz from Eqn. (19.78) by Eqn. (19.79), we get

$$dx=\left(\frac{\partial x}{\partial y}\right)_z dy+\left(\frac{\partial x}{\partial z}\right)_y\left[\left(\frac{\partial z}{\partial x}\right)_y dx+\left(\frac{\partial z}{\partial y}\right)_z dy\right]. \tag{19.80}$$

As dx and dy are independent of each other, equating the coefficients of dx from both sides of Eqn. (19.80), we get

$$\left(\frac{\partial x}{\partial z}\right)_y\left(\frac{\partial z}{\partial x}\right)_y=1\quad\text{or,}\quad\left(\frac{\partial z}{\partial x}\right)_y=\frac{1}{\left(\frac{\partial x}{\partial z}\right)_y}.$$

Again, equating the coefficients of dy,

$$\left(\frac{\partial x}{\partial y}\right)_z+\left(\frac{\partial x}{\partial z}\right)_y\left(\frac{\partial z}{\partial y}\right)_x=0\quad\text{or,}\quad\left(\frac{\partial y}{\partial z}\right)_x\left(\frac{\partial z}{\partial x}\right)_y\left(\frac{\partial x}{\partial y}\right)_x+1=0$$

or,
$$\left(\frac{\partial x}{\partial y}\right)_z\left(\frac{\partial y}{\partial z}\right)_x\left(\frac{\partial z}{\partial x}\right)_x=-1.$$

19.14 Solved Problems

Example 19.1. *If* $u=\log(x^3+y^3+z^3-3xyz)$, *show that*

$$\left(\frac{\partial}{\partial x}+\frac{\partial}{\partial y}+\frac{\partial}{\partial z}\right)^2 u=\frac{-9}{(x^2+y^2+z^2)^2}.$$ **[KU 2006, UPTU 2006]**

Solution : Since, $u=\log(x^3+y^3+z^3-3xyz)$,

$$\therefore\quad \frac{\partial u}{\partial x}=\frac{3x^2-3yz}{x^3+y^3+z^3-3xyz},\quad \frac{\partial u}{\partial y}=\frac{3y^2-3zx}{x^3+y^3+z^3-3xyz},\quad \frac{\partial u}{\partial z}=\frac{3z^2-3xy}{x^3+y^3+z^3-3xyz}.$$

$$\therefore\quad \frac{\partial u}{\partial x}+\frac{\partial u}{\partial y}+\frac{\partial u}{\partial z}=\frac{3(x^2+y^2+z^2-xy-yz-zx)}{(x+y+z)(x^2+y^2+z^2-xy-yz-zx)}=\frac{3}{x+y+z}.$$

$$\therefore\quad \left(\frac{\partial}{\partial x}+\frac{\partial}{\partial y}+\frac{\partial}{\partial z}\right)^2 u=\left(\frac{\partial}{\partial x}+\frac{\partial}{\partial y}+\frac{\partial}{\partial z}\right)\left(\frac{\partial u}{\partial x}+\frac{\partial u}{\partial y}+\frac{\partial u}{\partial z}\right)$$

$$=\left(\frac{\partial}{\partial x}+\frac{\partial}{\partial y}+\frac{\partial}{\partial z}\right)\left(\frac{3}{x+y+z}\right)$$

$$= -\frac{3}{(x+y+z)^2} - \frac{3}{(x+y+z)^2} - \frac{3}{(x+y+z)^2}$$

$$= -\frac{9}{(x+y+z)^2} \quad \textbf{(Proved)}.$$

Example 19.2. *If $z = e^{ax+by} f(ax - by)$, prove that, $b\frac{\partial z}{\partial x} + a\frac{\partial z}{\partial y} = 2abz$.*

Solution : We have $z = e^{ax+by} f(ax - by)$.

$$\therefore \quad \frac{\partial z}{\partial x} = ae^{ax+by} \cdot f(ax - by) + e^{ax+by} af'(ax - by).$$

$$\therefore \quad b\frac{\partial z}{\partial x} = abe^{ax+by} \cdot f(ax - by) + abe^{ax+by} \cdot f'(ax - by). \quad \text{(i)}$$

$$\therefore \quad \frac{\partial z}{\partial y} = be^{ax+by} \cdot f(ax - by) + e^{ax+by}(-b)f'(ax - by).$$

$$\therefore \quad a\frac{\partial z}{\partial y} = abe^{ax+by} \cdot f(ax - by) - abe^{ax+by} \cdot f'(ax - by). \quad \text{(ii)}$$

Adding equations (i) and (ii), we get

$$b\frac{\partial z}{\partial x} + a\frac{\partial z}{\partial y} = 2abe^{ax+by} f(ax - by) \quad \text{or,} \quad \boldsymbol{b\frac{\partial z}{\partial x} + a\frac{\partial z}{\partial y} = 2abz} \quad \textbf{(Proved)}.$$

Example 19.3. *If $z = f(x + ct) + \phi(x - ct)$, prove that $\frac{\partial^2 z}{\partial t^2} = c^2\frac{\partial^2 z}{\partial x^2}$, where f and ϕ are assumed to be twice differentiable and c is any constant.*

[VU(H) 2001, Pune U May 1980, AMIETE(Winter) 1994, KU 2006, UPTU 2006]

Solution : We have $z = f(x + ct) + \phi(x - ct)$.

$$\therefore \quad \frac{\partial z}{\partial x} = f'(x + ct) \cdot \frac{\partial}{\partial x}(x + ct) + \phi(x - ct)\frac{\partial}{\partial x}(x - ct) = f'(x + ct) + \phi'(x - ct)$$

$$\text{and} \quad \frac{\partial^2 z}{\partial x^2} = f''(x + ct) + \phi''(x - ct). \quad \text{(i)}$$

$$\text{Again,} \quad \frac{\partial z}{\partial t} = f'(x + ct) \cdot \frac{\partial}{\partial t}(x + ct) + \phi'(x - ct)\frac{\partial}{\partial t}(x - ct) = cf'(x + ct) - c\phi'(x - ct)$$

$$\text{and} \quad \frac{\partial^2 z}{\partial t^2} = c^2[f''(x + ct) + \phi''(x - ct)]. \quad \text{(ii)}$$

From Eqns. (i) and (ii),

$$\boldsymbol{\frac{\partial^2 z}{\partial t^2} = c^2\frac{\partial^2 z}{\partial x^2}} \quad \textbf{(Proved)}.$$

Example 19.4. *If $u = x^y$, show that $\frac{\partial^3 u}{\partial x^2 \partial y} = \frac{\partial^3 u}{\partial x \partial y \partial z}$.* **[AMIETE(Summer) 1995]**

Solution : Since $u = x^y$,

$$\therefore \quad \frac{du}{dy} = x^y \log x \quad \text{and} \quad \frac{d^2 u}{dx dy} = yx^{y-1}\log x + x^y \cdot \frac{1}{x} = x^{y-1}(y \log x + 1).$$

$$\therefore \quad \frac{\partial^3 u}{\partial x^2 \partial y} = \frac{\partial}{\partial x}[x^{y-1}(y\log x + 1)]. \qquad \text{(i)}$$

Again, $\dfrac{du}{dx} = yx^{y-1}$ and $\dfrac{\partial^2 u}{\partial x^2} = 1\cdot x^{y-1} + y(\frac{1}{x}x^y \log x) = x^{y-1}(1 + y\log x)$.

$$\text{and} \quad \frac{\partial^3 u}{\partial x \partial y \partial z} = \frac{\partial}{\partial x}[x^{y-1}(y\log x + 1)] \qquad \text{(ii)}$$

From Eqns. (i) and (ii),

$$\boldsymbol{\frac{\partial^3 u}{\partial x^2 \partial y} = \frac{\partial^3 u}{\partial x \partial y \partial z}} \quad \textbf{(Proved).}$$

Example 19.5. *If $\theta = t^n e^{-\frac{r^2}{4t}}$, find out what value of n will make $\dfrac{1}{r^2}\dfrac{\partial}{\partial r}\left(r^2\dfrac{\partial\theta}{\partial r}\right) = \dfrac{\partial\theta}{\partial r}$.*

[KU 2006, UPTU 2006]

Solution : Since $\theta = t^n e^{-\frac{r^2}{4t}}$,

$$\therefore \quad \frac{\partial\theta}{\partial r} = t^n e^{-\frac{r^2}{4t}}\left(-\frac{2t}{4t}\right) = -\frac{1}{2}rt^{n-1}e^{-\frac{r^2}{4t}} \quad \text{and} \quad r^2\frac{\partial\theta}{\partial r} = -\frac{r^3}{2}t^{n-1}e^{-\frac{r^2}{4t}}.$$

$$\therefore \quad \frac{\partial}{\partial r}\left(r^2\frac{\partial\theta}{\partial r}\right) = -\frac{3r^2}{2}t^{n-1}\cdot e^{-\frac{r^2}{4t}} - \frac{r^3}{2}t^{n-1}\cdot e^{-\frac{r^2}{4t}} \times \left(-\frac{2r}{4t}\right)$$

$$\text{and} \quad \frac{1}{r^2}\frac{\partial}{\partial r}\left(r^2\frac{\partial\theta}{\partial r}\right) = \left(-\frac{3}{2}t^{n-1} + \frac{r^2}{4}t^{n-2}\right)e^{-\frac{r^2}{4t}} = e^{-\frac{r^2}{4t}}\left[nt^{n-1} + \frac{1}{4}r^2t^{n-2}\right]. \qquad \text{(i)}$$

$$\text{Again,} \quad \frac{\partial\theta}{\partial t} = nt^{n-1}e^{-\frac{t^2}{4t}} + t^n e^{-\frac{t^2}{4t}}\left(\frac{r^2}{4t^2}\right) = e^{-\frac{t^2}{4t}}\left[nt^{n-1} + \frac{1}{4}r^2t^{n-2}\right]. \qquad \text{(ii)}$$

From Eqns. (i) and (ii),

$$e^{-\frac{t^2}{4t}}\left(-\frac{3}{2}t^{n-1} + \frac{1}{4}r^2t^{n-2}\right) = e^{-\frac{t^2}{4t}}\left(nt^{n-1} + \frac{1}{4}r^2t^{n-2}\right) \quad \text{or,} \quad \left(n + \frac{3}{2}\right)t^{n-1}\cdot e^{-\frac{r^2}{2t}} = 0.$$

$$\therefore \quad \boldsymbol{n = -\frac{3}{2}}.$$

Example 19.6. *If $u = (1 - 2xy + y^2)^{-\frac{1}{2}}$, prove that, $\dfrac{\partial}{\partial x}\left[(1 - x^2)\dfrac{\partial u}{\partial x}\right] + \dfrac{\partial}{\partial y}\left(r^2\dfrac{\partial u}{\partial y}\right) = 0$.*

Solution : Since $u = (1 - 2xy + y^2)^{-\frac{1}{2}}$, (i)

$$\therefore \quad \frac{\partial u}{\partial x} = -\frac{1}{2}(1 - 2xy + y^2)^{-\frac{3}{2}}(-2y) = \frac{y}{(1 - 2xy + y^2)^{\frac{3}{2}}}.$$

$$\therefore \quad (1 - x^2)\frac{\partial u}{\partial x} = \frac{(1 - x^2)y}{(1 - 2xy + y^2)^{\frac{3}{2}}}.$$

$$\text{So,} \quad \frac{\partial}{\partial x}\left[(1 - x^2)\frac{\partial u}{\partial x}\right] = \frac{(1 - 2xy + y^2)^{\frac{3}{2}}(-2xy) - (1 - x^2)y^{\frac{3}{2}}(1 - 2xy + y^2)^{\frac{1}{2}}(-2y)}{(1 - 2xy + y^2)^3}$$

$$= \frac{(1 - 2xy + y^2)(-2xy) + 3(1 - x^2)y^2}{(1 - 2xy + y^2)^{\frac{5}{2}}}$$

$$= \frac{-2xy + 4x^2y^2 - 2xy^3 + 3y^2 - 3x^2y^2}{(1-2xy+y^2)^{\frac{5}{2}}}$$

$$= \frac{x^2y^2 - 2xy^3 - 2xy + 3y^2}{(1-2xy+y^2)^{\frac{5}{2}}}. \tag{ii}$$

Now, from Eqn. (i),

$$\frac{\partial u}{\partial y} = -\frac{1}{2}(1-2xy+y^2)^{-\frac{3}{2}}(-2x+2y) = \frac{x-y}{(1-2xy+y^2)^{\frac{3}{2}}}.$$

$$y^2\frac{\partial u}{\partial y} = \frac{xy^2 - y^3}{(1-2xy+y^2)^{\frac{3}{2}}}$$

$$\frac{\partial}{\partial y}\left(y^2\frac{\partial u}{\partial x}\right) = \frac{(1-2xy+y^2)^{\frac{3}{2}}(2xy-3y^2) - (xy^2-y^3)\frac{3}{2}(1-2xy+y^2)^{\frac{1}{2}}(-2x+2y)}{(1-2xy+y^2)^3}$$

$$= \frac{(1-2xy+y^2)(2xy-3y^2) + (xy^2-y^3)3(x-y)}{(1-2xy+y^2)^{\frac{5}{2}}}$$

$$= \frac{2xy - 4x^2y^2 + 2xy^3 - 3y^2 + 6xy^3 - 3y^4 + 3x^2y^2 - 3xy^3 - 3xy^3 + 3y^4}{(1-2xy+y^2)^{\frac{5}{2}}}$$

$$= \frac{-x^2y^2 + 2xy^3 + 2xy - 3y^2}{(1-2xy+y^2)^{\frac{5}{2}}}. \tag{iii}$$

Adding Eqns. (ii) and (iii), we get

$$\frac{\partial}{\partial x}\left[(1-x^2)\frac{\partial u}{\partial x}\right] + \frac{\partial}{\partial x}\left(y^2\frac{\partial u}{\partial y}\right) = \frac{x^2y^2 - 2xy^3 - 2xy + 3y^2}{(1-2xy+y^2)^{\frac{5}{2}}} + \frac{-x^2y^2 + 2xy^3 + 2xy - 3y^2}{(1-2xy+y^2)^{\frac{5}{2}}}$$

$$= 0 \quad \textbf{(Proved)}.$$

Example 19.7. *If* $u = f\left(\frac{x}{y}, \frac{y}{z}, \frac{z}{x}\right)$, *prove that,* $x\frac{\partial u}{\partial x} + y\frac{\partial u}{\partial y} + z\frac{\partial u}{\partial z} = 0.$

Solution : Let $\frac{x}{y} = p$, $\frac{y}{z} = q$, $\frac{z}{x} = r$; then $u = f\left(\frac{x}{y}, \frac{y}{z}, \frac{z}{x}\right) = f(p,q,r)$.

$$\therefore \quad \frac{\partial u}{\partial x} = \frac{\partial u}{\partial p}\frac{\partial p}{\partial x} + \frac{\partial u}{\partial q}\frac{\partial q}{\partial x} + \frac{\partial u}{\partial r}\frac{\partial r}{\partial x} = \frac{\partial u}{\partial p}\left(\frac{1}{y}\right) + \frac{\partial u}{\partial q}\cdot 0 + \frac{\partial u}{\partial r}\left(-\frac{z}{x^2}\right)$$

or, $$x\frac{\partial u}{\partial x} = \frac{\partial u}{\partial p}\left(\frac{x}{y}\right) - \frac{\partial u}{\partial r}\left(\frac{z}{x}\right) = p\frac{\partial u}{\partial p} - r\frac{\partial u}{\partial r}. \tag{i}$$

Again, $$\frac{\partial u}{\partial y} = \frac{\partial u}{\partial p}\frac{\partial p}{\partial y} + \frac{\partial u}{\partial q}\frac{\partial q}{\partial y} + \frac{\partial u}{\partial r}\frac{\partial r}{\partial y} = \frac{\partial u}{\partial p}\left(-\frac{x}{y^2}\right) + \frac{\partial u}{\partial q}\left(\frac{1}{z}\right) + \frac{\partial u}{\partial r}\cdot 0$$

or, $$y\frac{\partial u}{\partial y} = \frac{\partial u}{\partial p}\left(-\frac{x}{y}\right) + \frac{\partial u}{\partial q}\left(\frac{y}{z}\right) = q\frac{\partial u}{\partial q} - p\frac{\partial u}{\partial p}. \tag{ii}$$

Similarly, $$z\frac{\partial u}{\partial z} = r\frac{\partial u}{\partial r} - q\frac{\partial u}{\partial q}. \tag{iii}$$

Adding Eqns. (i), (ii) and (iii), we get

$$\boldsymbol{x\frac{\partial u}{\partial x}+y\frac{\partial u}{\partial y}+z\frac{\partial u}{\partial z}=0} \quad \textbf{(Proved).}$$

Example 19.8. *Express the two-dimensional Laplace's equation* $\frac{\partial^2 u}{\partial x^2}+\frac{\partial^2 u}{\partial y^2}=0$ *in polar coordinates.* **[BU(H) 1999, 2003]**

Solution : Let the Cartesian coordinates of a point (x, y) be transferred to polar form (r, θ).

Then $x = r\cos\theta$, $y = r\sin\theta$.

$$\therefore \quad \left.\begin{aligned} r &= \sqrt{x^2+y^2} \\ \theta &= \tan^{-1}\left(\frac{y}{x}\right) \end{aligned}\right\} \tag{i}$$

$$\left.\begin{aligned} \text{Now,}\quad \frac{\partial r}{\partial x} &= \frac{2x}{2\sqrt{x^2+y^2}} = \frac{x}{\sqrt{x^2+y^2}} = \cos\theta \\ \text{and}\quad \frac{\partial r}{\partial y} &= \frac{y}{\sqrt{x^2+y^2}} = \sin\theta \end{aligned}\right\} \tag{ii}$$

$$\left.\begin{aligned} \frac{\partial\theta}{\partial x} &= \frac{1}{1+\frac{y^2}{x^2}}\cdot\left(-\frac{y}{x^2}\right) = -\frac{y}{x^2+y^2} = -\frac{\sin\theta}{r} \\ \frac{\partial\theta}{\partial y} &= \frac{1}{1+\frac{y^2}{x^2}}\cdot\left(\frac{1}{x}\right) = \frac{x}{x^2+y^2} = \frac{\cos\theta}{r} \end{aligned}\right\} \tag{iii}$$

By chain rule,

$$\left.\begin{aligned} \frac{\partial}{\partial x} &= \frac{\partial}{\partial r}\cdot\frac{\partial r}{\partial x}+\frac{\partial}{\partial\theta}\cdot\frac{\partial\theta}{\partial x} = \cos\theta\frac{\partial}{\partial r}-\frac{\sin\theta}{r}\frac{\partial}{\partial\theta} \\ \text{and}\quad \frac{\partial}{\partial y} &= \frac{\partial}{\partial r}\cdot\frac{\partial r}{\partial y}+\frac{\partial}{\partial\theta}\cdot\frac{\partial\theta}{\partial y} = \sin\theta\frac{\partial}{\partial\theta}+\frac{\cos\theta}{r}\frac{\partial}{\partial\theta} \end{aligned}\right\} \tag{iv}$$

$$\therefore \quad \frac{\partial^2}{\partial x^2} = \frac{\partial}{\partial x}\left(\frac{\partial}{\partial x}\right) = \left(\cos\theta\frac{\partial}{\partial r}-\frac{\sin\theta}{r}\frac{\partial}{\partial\theta}\right)\left(\cos\theta\frac{\partial}{\partial r}-\frac{\sin\theta}{r}\frac{\partial}{\partial\theta}\right)$$

$$= \cos^2\theta\frac{\partial^2}{\partial r^2}-\frac{2\sin\theta\cos\theta}{r}\frac{\partial^2}{\partial r\partial\theta}+\frac{\sin^2\theta}{r^2}\frac{\partial^2}{\partial\theta^2}$$

$$+\frac{\sin^2\theta}{r}\frac{\partial}{\partial\theta}+\frac{2\sin\theta\cos\theta}{r^2}\frac{\partial}{\partial\theta}\cdots \tag{v}$$

$$\text{and}\quad \frac{\partial^2}{\partial y^2} = \frac{\partial}{\partial y}\left(\frac{\partial}{\partial y}\right) = \left(\sin\theta\frac{\partial}{\partial r}+\frac{\cos\theta}{r}\frac{\partial}{\partial\theta}\right)\left(\sin\theta\frac{\partial}{\partial r}+\frac{\cos\theta}{r}\frac{\partial}{\partial\theta}\right)$$

$$= \sin^2\theta\frac{\partial^2}{\partial r^2}+\frac{2\sin\theta\cos\theta}{r}\frac{\partial^2}{\partial r\partial\theta}+\frac{\cos^2\theta}{r^2}\frac{\partial^2}{\partial\theta^2}$$

$$+\frac{\cos^2\theta}{r}\frac{\partial}{\partial r}-\frac{2\sin\theta\cos\theta}{r^2}\frac{\partial}{\partial\theta}\cdots. \tag{vi}$$

Adding Eqns. (v) and (vi), we get

$$\Delta^2 = \frac{\partial^2}{\partial x^2}+\frac{\partial^2}{\partial y^2} \quad \text{or,} \quad \boldsymbol{\Delta^2 = \frac{\partial^2}{\partial r^2}+\frac{1}{r}\frac{\partial^2}{\partial\theta^2}+\frac{1}{r}\frac{\partial}{\partial r}}. \tag{vii}$$

Example 19.9. *If $u = f(r)$ and $x = r\cos\theta$, $y = r\sin\theta$, prove that,*

$$\frac{\partial^2 u}{\partial x^2} + \frac{\partial^2 u}{\partial y^2} = f''(r) + \frac{1}{r}f'(r).$$ **[BU(H) 1995, AMIETE 1994]**

Solution : We have $u = f(r)$, $x = r\cos\theta$, $y = r\sin\theta$.

$\therefore \quad r^2 = x^2 + y^2$ so that $\dfrac{\partial r}{\partial x} = \dfrac{x}{r}$.

$$\frac{\partial u}{\partial x} = \frac{df}{dr}\cdot\frac{\partial f}{\partial r} = \frac{df}{dx}\cdot\frac{x}{r}.$$

Differentiating again w.r.t. x,

$$\frac{\partial^2 u}{\partial x^2} = \left(\frac{d^2 f}{dr^2}\frac{\partial r}{\partial x}\right)\cdot\frac{x}{r} + \frac{df}{dr}\left[\frac{r\cdot 1 - x\frac{\partial r}{\partial x}}{r^2}\right] = \left(\frac{d^2 f}{dr^2}\frac{x}{r}\right)\cdot\frac{x}{r} + \frac{df}{dr}\left[\frac{r\cdot 1 - x\cdot\frac{x}{r}}{r^2}\right]$$

$$= \frac{d^2 f}{dr^2}\frac{x^2}{r^2} + \frac{df}{dr}\frac{r^2 - x^2}{r^3} = \frac{d^2 f}{dr^2}\frac{x^2}{r} + \frac{df}{dr}\frac{y^2}{r^3}. \qquad \text{(i)}$$

Similarly, $\dfrac{\partial^2 u}{\partial y^2} = \dfrac{d^2 f}{dr^2}\dfrac{y^2}{r^2} + \dfrac{df}{dr}\dfrac{x^2}{r^3}$. (ii)

Adding Eqns. (i) and (ii), we get

$$\frac{\partial^2 u}{\partial x^2} + \frac{\partial^2 u}{\partial y^2} = \frac{d^2 f}{dr^2}\frac{x^2 + y^2}{r^2} + \frac{df}{dr}\frac{x^2 + y^2}{r^3} = \frac{d^2 f}{dr^2} + \frac{df}{dr}\frac{1}{r}$$

or, $\boldsymbol{\dfrac{\partial^2 u}{\partial x^2} + \dfrac{\partial^2 u}{\partial y^2} = f''(r) + \dfrac{1}{r}f'(r)}$ **(Proved)**.

Example 19.10. *If $u = \dfrac{1}{r}$, where $r = \sqrt{x^2 + y^2 + z^2}$, show that, $\dfrac{\partial^2 u}{\partial x^2} + \dfrac{\partial^2 u}{\partial y^2} + \dfrac{\partial^2 u}{\partial z^2} = 0$.*

[BU(H) 1993, VTU 2006]

Solution : Since $u = \dfrac{1}{r} = (x^2 + y^2 + z^2)^{-\frac{1}{2}}$,

$$\therefore \quad \frac{\partial u}{\partial x} = -\frac{1}{2}(x^2 + y^2 + z^2)^{-\frac{3}{2}}\cdot 2x = -x(x^2 + y^2 + z^2)^{-\frac{3}{2}}$$

and $$\frac{\partial^2 u}{\partial x^2} = -\left[1\cdot(x^2 + y^2 + z^2)^{-\frac{3}{2}} + x\left(-\frac{3}{2}\right)(x^2 + y^2 + z^2)^{-\frac{5}{2}}\cdot 2x\right]$$

$$= -(x^2 + y^2 + z^2)^{-\frac{5}{2}}[x^2 + y^2 + z^2 - 3x^2] = (x^2 + y^2 + z^2)^{-\frac{5}{2}}(2x^2 - y^2 - z^2).$$

Similarly, $\dfrac{\partial^2 u}{\partial y^2} = (x^2 + y^2 + z^2)^{-\frac{5}{2}}(-x^2 + 2y^2 - z^2)$

and $\dfrac{\partial^2 u}{\partial z^2} = (x^2 + y^2 + z^2)^{-\frac{5}{2}}(-x^2 - y^2 + 2z^2)$.

$$\therefore \quad \boldsymbol{\frac{\partial^2 u}{\partial x^2} + \frac{\partial^2 u}{\partial y^2} + \frac{\partial^2 u}{\partial z^2} = (x^2 + y^2 + z^2)^{-\frac{5}{2}}\cdot 0 = 0} \quad \textbf{(Proved)}.$$

Example 19.11. *If* $u = \sin^{-1}\dfrac{x+y}{\sqrt{x}+\sqrt{y}}$, *prove that,*

$$x\frac{\partial u}{\partial x} + y\frac{\partial u}{\partial y} = \frac{1}{2}\tan u$$ **[RU 2006, Calicut 2005]**

and $$x^2\frac{\partial^2 u}{\partial x^2} + 2xy\frac{\partial^2 u}{\partial x \partial y} + y^2\frac{\partial^2 u}{\partial y^2} = -\frac{\sin u \cos 2u}{4\cos^3 u}.$$ **[PTU 2006]**

Solution : We note that, u is not a homogeneous function but $z = \sin u = \frac{x+y}{\sqrt{x}+\sqrt{y}}$ is a homogeneous function of degree $\frac{1}{2}$ in x and y.

By Euler's theorem,

$$x\frac{\partial z}{\partial x} + y\frac{\partial z}{\partial y} = \frac{1}{2}z \quad \text{or,} \quad x\cos u\frac{\partial u}{\partial x} + y\cos u\frac{\partial u}{\partial y} = \frac{1}{2}\sin u.$$

Thus, $$\boldsymbol{x\frac{\partial u}{\partial x} + y\frac{\partial u}{\partial y} = \frac{1}{2}\tan u.} \qquad \text{(i)}$$

Differentiating Eqn. (i) partially w.r.t. x,

$$x\frac{\partial^2 u}{\partial x^2} + \frac{\partial u}{\partial x} + y\frac{\partial^2 u}{\partial x \partial y} = \frac{1}{2}\sec^2 u\frac{\partial u}{\partial x} \qquad \text{(ii)}$$

or, $$x\frac{\partial^2 u}{\partial x^2} + y\frac{\partial^2 u}{\partial x \partial y} = \left(\frac{1}{2}\sec^2 u - 1\right)\frac{\partial u}{\partial x}.$$

Differentiating again partially w.r.t. y,

$$x\frac{\partial^2 u}{\partial y \partial x} + y\frac{\partial^2 u}{\partial y^2} + \frac{\partial u}{\partial y} = \frac{1}{2}\sec^2 u\frac{\partial u}{\partial y}$$

or, $$x\frac{\partial^2 u}{\partial x \partial y} + y\frac{\partial^2 u}{\partial y^2} = \left(\frac{1}{2}\sec^2 u - 1\right)\frac{\partial u}{\partial y}. \qquad \text{(iii)}$$

Now, multiplying Eqn. (ii) by x and (iii) by y and then adding, we get

$$x^2\frac{\partial^2 u}{\partial x^2} + 2xy\frac{\partial^2 u}{\partial x \partial y} + y^2\frac{\partial^2 u}{\partial y^2} = \left(\frac{1}{2}\sec^2 u - 1\right)\left(x\frac{\partial u}{\partial x} + y\frac{\partial u}{\partial y}\right)$$

or, $$x^2\frac{\partial^2 u}{\partial x^2} + 2xy\frac{\partial^2 u}{\partial x \partial y} + y^2\frac{\partial^2 u}{\partial y^2} = \left(\frac{1}{2}\sec^2 u - 1\right)\left(\frac{1}{2}\tan u\right) = \frac{1}{4}\frac{\sin u}{\cos^3 u} - \frac{1}{2}\frac{\sin u}{\cos u}$$

$$= -\frac{\sin u(2\cos^2 u - 1)}{4\cos^3 u}.$$

Hence, $$\boldsymbol{x^2\frac{\partial^2 u}{\partial x^2} + 2xy\frac{\partial^2 u}{\partial x \partial y} + y^2\frac{\partial^2 u}{\partial y^2} = -\frac{\sin u \cos 2u}{4\cos^3 u}} \quad \textbf{(Proved)}.$$

Example 19.12. *If* $u = x\log xy$, *where* $x^3 + y^3 + 3xy = 1$, *find* $\dfrac{du}{dx}$.

Solution : We have $f(x, y) = x^3 + y^3 + 3xy - 1$.

$$\therefore \quad \frac{dy}{dx} = -\frac{\frac{\partial f}{\partial x}}{\frac{\partial f}{\partial y}} = -\frac{3x^2 + 3y}{3y^2 + 3x} = -\frac{x^2 + y}{y^2 + x}$$

and $\dfrac{du}{dx} = \dfrac{\partial u}{\partial x} + \dfrac{\partial u}{\partial y}\dfrac{dy}{dx} = \left(1 \cdot \log xy + x \cdot \dfrac{1}{x}\right) + \left(\dfrac{x}{y}\right) \cdot \dfrac{dy}{dx}.$

$$\therefore \quad \boldsymbol{\frac{du}{dx} = 1 + \log xy - \frac{x(x^2+y)}{y(y^2+x)}}.$$

Example 19.13. *If $x = r\cos\theta$, $y = r\sin\theta$, evaluate $\dfrac{\partial(x,y)}{\partial(r,\theta)}$ and $\dfrac{\partial(r,\theta)}{\partial(x,y)}$.*

[VTU 2004, AU 2000, AMIETE(Summer) 1995]

Solution : We have $x = r\cos\theta$ and $y = r\sin\theta$.

$$\therefore \quad \frac{\partial x}{\partial r} = \cos\theta, \quad \frac{\partial y}{\partial r} = 1 \cdot \sin\theta, \quad \frac{\partial x}{\partial \theta} = -r\sin\theta, \quad \frac{\partial y}{\partial \theta} = r\cos\theta.$$

$$\therefore \quad \frac{\partial(x,y)}{\partial(r,\theta)} = \begin{vmatrix} \dfrac{\partial x}{\partial r} & \dfrac{\partial x}{\partial \theta} \\ \dfrac{\partial y}{\partial r} & \dfrac{\partial y}{\partial \theta} \end{vmatrix} = \begin{vmatrix} \cos\theta & -r\sin\theta \\ \sin\theta & r\cos\theta \end{vmatrix} = r\cos^2\theta + r\sin^2\theta = r(\cos^2\theta + \sin^2\theta) = r.$$

$$\therefore \quad \boldsymbol{\frac{\partial(x,y)}{\partial(r,\theta)} = r}.$$

Again, $r^2 = x^2 + y^2, \quad \theta = \tan^{-1}\dfrac{y}{x}.$

$$\therefore \quad \frac{\partial r}{\partial x} = \frac{x}{r}, \quad \frac{\partial \theta}{\partial x} = \frac{-y}{x^2+y^2} = -\frac{y}{r^2},$$

$$\frac{\partial r}{\partial y} = \frac{y}{r}, \quad \frac{\partial \theta}{\partial y} = \frac{x}{x^2+y^2} = \frac{x}{r^2}.$$

$$\therefore \quad \boldsymbol{\frac{\partial(r,\theta)}{\partial(x,y)}} = \begin{vmatrix} \dfrac{\partial r}{\partial x} & \dfrac{\partial r}{\partial y} \\ \dfrac{\partial \theta}{\partial x} & \dfrac{\partial \theta}{\partial y} \end{vmatrix} = \begin{vmatrix} \dfrac{x}{r} & \dfrac{y}{r} \\ -\dfrac{y}{r^2} & \dfrac{x}{r^2} \end{vmatrix} = \frac{x^2}{r^3} + \frac{y^2}{r^3} = \frac{x^2+y^2}{r^3} = \frac{r^2}{r^3} = \boldsymbol{\frac{1}{r}}.$$

$$\therefore \quad \boldsymbol{\frac{\partial(x,y)}{\partial(r,\theta)} \times \frac{\partial(r,\theta)}{\partial(x,y)} = r \cdot \frac{1}{r} = 1}.$$

Example 19.14. *Find the Jacobian of the transformation from rectangular Cartesian coordinates (x, y, z) to spherical polar coordinates (r, θ, ϕ).* **[Rohtak U 2003, BU(H) 1998, 2004]**

Also evaluate the integral, $\iiint_V (x^2 + y^2 + z^2)\,dxdydz$ where V is the volume of a sphere with centre at the origin and radius R. **[BU(H) 1992]**

Solution : The transformation from Cartesian coordinates (x, y, z) to spherical polar coordinates (r, θ, ϕ) is

$$x = r\sin\theta\cos\phi, \qquad y = r\sin\theta\cos\phi, \qquad z = r\cos\theta.$$

$$\therefore \quad \frac{\partial x}{\partial r} = \sin\theta\cos\phi, \qquad \frac{\partial y}{\partial r} = \sin\theta\cos\phi, \qquad \frac{\partial z}{\partial r} = \cos\theta,$$

$$\frac{\partial x}{\partial \theta} = r\cos\theta\cos\phi, \qquad \frac{\partial y}{\partial \theta} = r\cos\theta\sin\phi, \qquad \frac{\partial z}{\partial \theta} = -r\sin\theta,$$

$$\frac{\partial x}{\partial \phi} = -r\sin\theta\sin\phi, \qquad \frac{\partial y}{\partial \phi} = r\sin\theta\cos\phi, \qquad \frac{\partial z}{\partial \phi} = 0.$$

$$\therefore \quad J = \frac{\partial(x, y, z)}{\partial(r, \theta, \phi)} = \begin{vmatrix} \sin\theta\cos\phi & r\cos\theta\cos\phi & -r\sin\theta\sin\phi \\ \sin\theta\sin\phi & r\cos\theta\sin\phi & r\sin\theta\cos\phi \\ \cos\theta & -r\sin\theta & 0 \end{vmatrix}$$

$$= \cos\theta \begin{vmatrix} r\cos\theta\cos\phi & -r\sin\theta\sin\phi \\ r\cos\theta\sin\phi & r\sin\theta\cos\phi \end{vmatrix} + r\sin\theta \begin{vmatrix} \sin\theta\cos\phi & -r\sin\theta\sin\phi \\ \sin\theta\sin\phi & r\sin\theta\cos\phi \end{vmatrix}$$

$$= r^2\sin\theta\cos^2\theta \begin{vmatrix} \cos\phi & -\sin\phi \\ \sin\phi & \cos\phi \end{vmatrix} + r^2\sin^3\theta \begin{vmatrix} \cos\phi & -\sin\phi \\ \sin\phi & \cos\phi \end{vmatrix}$$

$$= r^2\sin\theta\cos^2\theta + r^2\sin^3\theta = r^2\sin\theta(\cos^2\theta + \sin^2\theta) = r^2\sin\theta.$$

Now, $$\iiint\limits_V (x^2 + y^2 + z^2)dxdydz = \int_{r=0}^{R}\int_{\theta=0}^{\pi}\int_{\phi=0}^{2\pi} r^2 \cdot r^2\sin\theta \, dr\, d\theta\, d\phi$$

$$= \int_{\phi=0}^{2\pi}\left[\int_{\theta=0}^{\pi}\left\{\int_{r=0}^{R} r^4 dr\right\} r^2\sin\theta\, d\theta\right] d\phi$$

$$= \frac{R^5}{5}\int_{\phi=0}^{2\pi}\int_{\theta=0}^{\pi}\sin\theta\, d\theta d\phi = -\frac{R^5}{5}\int_{\phi=0}^{2\pi}[\cos\theta]_0^{\pi}\, d\theta$$

$$= \frac{2R^5}{5}\int_{\phi=0}^{2\pi} d\phi = \frac{\mathbf{4\pi R^5}}{\mathbf{5}}.$$

Example 19.15. *In cylindrical coordinate, $x = \rho\cos\phi$, $y = \rho\sin\phi$ and $z = z$, show that,*

$$\frac{\partial(x, y, z)}{\partial(\rho, \phi, z)} = \rho.$$

Solution : Given, $x = \rho\cos\phi$, $y = \rho\sin\phi$ and $z = z$.

$$\therefore \quad \frac{\partial x}{\partial \rho} = \cos\phi, \qquad \frac{\partial x}{\partial \phi} = -\rho\sin\phi, \qquad \frac{\partial x}{\partial z} = 0$$

$$\frac{\partial y}{\partial \rho} = \sin\phi, \qquad \frac{\partial y}{\partial \phi} = \rho\cos\phi, \qquad \frac{\partial y}{\partial z} = 0$$

and $$\frac{\partial z}{\partial \rho} = 0, \qquad \frac{\partial z}{\partial \phi} = 0, \qquad \frac{\partial z}{\partial z} = 1.$$

$$\therefore \quad \frac{\boldsymbol{\partial(x, y, z)}}{\boldsymbol{\partial(\rho, \phi, z)}} = \begin{vmatrix} \cos\phi & -\rho\sin\phi & 0 \\ \sin\phi & \rho\cos\phi & 0 \\ 0 & 0 & 1 \end{vmatrix} = \boldsymbol{\rho} \quad \textbf{(Proved)}.$$

Example 19.16. *Verify if the following functions are functionally dependent and if so, find the relation between them : $u = \dfrac{x+y}{1-xy}$ and $v = \tan^{-1}x + \tan^{-1}y$.*

Solution : We have $\dfrac{\partial(u,v)}{\partial(x,y)} = \begin{vmatrix} \dfrac{\partial u}{\partial x} & \dfrac{\partial u}{\partial y} \\ \dfrac{\partial v}{\partial x} & \dfrac{\partial v}{\partial y} \end{vmatrix}$.

But $\dfrac{\partial u}{\partial x} = \dfrac{1+y^2}{(1-xy)^2}$, $\dfrac{\partial u}{\partial y} = \dfrac{1+x^2}{(1-xy)^2}$, $\dfrac{\partial v}{\partial x} = \dfrac{1}{1+x^2}$ and $\dfrac{\partial v}{\partial y} = \dfrac{1}{1+y^2}$.

$$\therefore \quad \frac{\partial(u,v)}{\partial(x,y)} = \begin{vmatrix} \dfrac{1+y^2}{(1-xy)^2} & \dfrac{1+x^2}{(1-xy)^2} \\ \dfrac{1}{1+x^2} & \dfrac{1}{1+y^2} \end{vmatrix} = \frac{1}{(1-xy)^2} - \frac{1}{(1-xy)^2} = 0.$$

So, the functions u and v are functionally dependent.

$$v = \tan^{-1} x + \tan^{-1} y = \tan^{-1} \frac{x+y}{1-xy} = \tan^{-1} u \quad \text{or,} \quad \boldsymbol{u = \tan v}.$$

Example 19.17. *If $x+y+z=u$, $y+z=uv$, $z=uvw$, show that,* $\dfrac{\partial(x,y,z)}{\partial(u,v,w)} = u^2v$.

Solution : Let $f_1 = x+y+z-u$, $f_2 = y+z-uv$, $f_3 = z-uvw$.

$$\therefore \quad \frac{\partial(f_1,f_2,f_3)}{\partial(x,y,z)} = \begin{vmatrix} \dfrac{\partial f_1}{\partial x} & \dfrac{\partial f_1}{\partial y} & \dfrac{\partial f_1}{\partial z} \\ \dfrac{\partial f_2}{\partial x} & \dfrac{\partial f_2}{\partial y} & \dfrac{\partial f_2}{\partial z} \\ \dfrac{\partial f_3}{\partial x} & \dfrac{\partial f_3}{\partial y} & \dfrac{\partial f_3}{\partial z} \end{vmatrix} = \begin{vmatrix} 1 & 1 & 1 \\ 0 & 1 & 1 \\ 0 & 0 & 1 \end{vmatrix} = 1.$$

$$\text{Again,} \quad \frac{\partial(f_1,f_2,f_3)}{\partial(u,v,w)} = \begin{vmatrix} \dfrac{\partial f_1}{\partial u} & \dfrac{\partial f_1}{\partial v} & \dfrac{\partial f_1}{\partial w} \\ \dfrac{\partial f_2}{\partial u} & \dfrac{\partial f_2}{\partial v} & \dfrac{\partial f_2}{\partial w} \\ \dfrac{\partial f_3}{\partial u} & \dfrac{\partial f_3}{\partial v} & \dfrac{\partial f_3}{\partial w} \end{vmatrix} = \begin{vmatrix} -1 & 0 & 0 \\ -v & -u & 0 \\ -vw & -uv & -uv \end{vmatrix} = -u^2v.$$

$$\text{But,} \quad \frac{\partial(x,y,z)}{\partial(u,v,w)} = (-1)^3 \frac{\frac{\partial(f_1,f_2,f_3)}{\partial(u,v,w)}}{\frac{\partial(f_1,f_2,f_3)}{\partial(x,y,z)}} = -\frac{-u^2v}{1} = \boldsymbol{u^2v}.$$

Example 19.18. *If $u = x+3y^2-z^3$, $v = 4x^2yz$, $w = 2z^2-xy$, evaluate* $\dfrac{\partial(u,v,w)}{\partial(x,y,z)}$ *at* $(1,-1,0)$. **[VTU 2006]**

$$\textbf{Solution :} \quad \frac{\partial(u,v,w)}{\partial(x,y,z)} = \begin{vmatrix} \dfrac{\partial u}{\partial x} & \dfrac{\partial u}{\partial y} & \dfrac{\partial u}{\partial z} \\ \dfrac{\partial v}{\partial x} & \dfrac{\partial v}{\partial y} & \dfrac{\partial v}{\partial z} \\ \dfrac{\partial w}{\partial x} & \dfrac{\partial w}{\partial y} & \dfrac{\partial w}{\partial z} \end{vmatrix} = \begin{vmatrix} 1 & 6y & -3z^2 \\ 8xyz & 4x^2z & 4x^2y \\ -y & -x & 4z \end{vmatrix}.$$

$$\text{At } (1,-1,0),\ \frac{\partial(u,v,w)}{\partial(x,y,z)} = \begin{vmatrix} 1 & -6 & 0 \\ 0 & 0 & -4 \\ 1 & -1 & 0 \end{vmatrix} = 4(-1+6) = \mathbf{20}$$

Example 19.19. *Expand $e^x \cos y$ near the point $\left(1, \frac{\pi}{4}\right)$ by Taylor's theorem.*

Solution : We have by Taylor's theorem,

$$f(x+h, y+k) = f(x,y) + \left(h\frac{\partial}{\partial x} + k\frac{\partial}{\partial y}\right) f + \frac{1}{2!}\left(h\frac{\partial}{\partial x} + k\frac{\partial}{\partial y}\right)^2 f + \frac{1}{3!}\left(h\frac{\partial}{\partial x} + k\frac{\partial}{\partial y}\right)^3 f + \cdots.$$

Let $f(x,y) = e^x \cos y$.

$$\therefore \quad e^x \cos y = f\left[1 + (x-1), \frac{\pi}{4} + \left(y - \frac{\pi}{4}\right)\right]$$

$$= f\left(1+h, \frac{\pi}{4} + k\right) \quad \left[\text{where } a = 1,\ b = \frac{\pi}{4},\ h = x-1,\ k = y - \frac{\pi}{4}\right]$$

$$= f\left(1, \frac{\pi}{4}\right) + \left[(x-1)\frac{\partial f}{\partial x} + \left(y - \frac{\pi}{4}\right)\frac{\partial f}{\partial y}\right] + \frac{1}{2!}\left[(x-1)^2\frac{\partial^2 f}{\partial x^2} + 2(x-1)\left(y-\frac{\pi}{4}\right)\frac{\partial^2 f}{\partial x \partial y} + \left(y - \frac{\pi}{4}\right)\frac{\partial^2 f}{\partial y^2}\right] + \cdots \quad \text{(i)}$$

$$\because \quad f(x,y) = e^x \cos y \quad \text{and} \quad f\left(1, \frac{\pi}{4}\right) = \frac{e}{\sqrt{2}},$$

$$\frac{\partial f}{\partial x} = e^x \cos y, \qquad \therefore \quad \frac{\partial f}{\partial x}\left(1, \frac{\pi}{4}\right) = \frac{e}{\sqrt{2}}$$

$$\frac{\partial f}{\partial y} = -e^x \sin y, \qquad \therefore \quad \frac{\partial f}{\partial y}\left(1, \frac{\pi}{4}\right) = \frac{-e}{\sqrt{2}}$$

$$\frac{\partial^2 f}{\partial x^2} = e^x \cos y, \qquad \therefore \quad \frac{\partial^2 f}{\partial x^2}\left(1, \frac{\pi}{4}\right) = \frac{e}{\sqrt{2}}$$

$$\frac{\partial^2 f}{\partial y^2} = -e^x \cos y, \qquad \therefore \quad \frac{\partial^2 f}{\partial x^2}\left(1, \frac{\pi}{4}\right) = \frac{-e}{\sqrt{2}}$$

$$\frac{\partial^2 f}{\partial x \partial y} = -e^x \sin y, \qquad \therefore \quad \frac{\partial^2 f}{\partial x \partial y}\left(1, \frac{\pi}{4}\right) = \frac{-e}{\sqrt{2}}.$$

Using these values in Eqn. (i), we get

$$e^x \cos y = \frac{e}{\sqrt{2}} + \left[(x-1)\frac{e}{\sqrt{2}} + \left(y - \frac{\pi}{4}\right)\left(\frac{-e}{\sqrt{2}}\right)\right] + \frac{1}{2!}\left[(x-1)^2\frac{e}{\sqrt{2}} + 2(x-1)\left(y - \frac{\pi}{4}\right)\left(\frac{-e}{\sqrt{2}}\right) + \left(y - \frac{\pi}{4}\right)^2\left(\frac{-e}{\sqrt{2}}\right)\right] + \cdots$$

$$\therefore \quad \mathbf{e^x \cos y = \frac{e}{\sqrt{2}} + \left[1 + (x-1) - \left(y - \frac{\pi}{4}\right) + \frac{(x-1)^2}{2} - (x-1)\left(y - \frac{\pi}{4}\right) - \frac{1}{2}\left(y - \frac{\pi}{4}\right)^2 + \cdots\right]}.$$

Example 19.20. *If $f(x, y) = \tan^{-1}(xy)$, compute an approximate value of* $(0.9, -1.2)$.

Solution : $f(x, y) = \tan^{-1}(xy)$.

We expand $f(x, y)$ near the point $(1, -1)$.

$$f(0.9, -1.2) = f(1 - 0.1, -1 - 0.2)$$

$$= f(1, 1) + \left[(-0.1)\frac{\partial f}{\partial x} + (-0.2)\frac{\partial f}{\partial y}\right]$$

$$+ \frac{1}{2!}\left[(-0.1)^2\frac{\partial^2 f}{\partial x^2} + 2(-0.1)(-0.2)\frac{\partial^2 f}{\partial x \partial y} + (-0.2)^2\frac{\partial^2 f}{\partial y^2}\right] + \cdots. \quad \text{(i)}$$

$\because\ f(x, y) = \tan^{-1}(x, y), \quad f(1, -1) = \tan^{-1}(-1) = -\frac{\pi}{4},$

$$\frac{\partial f}{\partial x} = \frac{y}{1 + x^2y^2}, \qquad \therefore\ \frac{\partial f}{\partial x}(1, -1) = \frac{-1}{1+1} = -\frac{1}{2}$$

$$\frac{\partial f}{\partial y} = \frac{x}{1 + x^2y^2}, \qquad \therefore\ \frac{\partial f}{\partial y}(1, -1) = \frac{1}{1+1} = \frac{1}{2}$$

$$\frac{\partial^2 f}{\partial x^2} = -\frac{(2x)y}{(1 + x^2y^2)^2}, \qquad \therefore\ \frac{\partial^2 f}{\partial x^2}(1, -1) = \frac{(-2)(-1)}{(1+1)^2} = \frac{1}{2}$$

$$\frac{\partial^2 f}{\partial x \partial y} = \frac{1 + x^2y^2 - x(2xy^2)}{1 + x^2y^2} = \frac{1 - x^2y^2}{(1 + x^2y^2)^2}, \qquad \therefore\ \frac{\partial^2 f}{\partial x \partial y}(1, -1) = \frac{1 - 1}{(1+1)^2} = 0$$

$$\frac{\partial^2 f}{\partial y^2} = \frac{-x(2x^2y)}{(1 + x^2y^2)}, \qquad \therefore\ \frac{\partial^2 f}{\partial y^2}(1, -1) = \frac{2}{(1+1)^2} = \frac{1}{2}.$$

Substituting these values in Eqn. (i), we get

$$f(0.9, -1.2) = -\frac{\pi}{4} + (-0.1)\left(-\frac{1}{2}\right) + (-0.2)\left(\frac{1}{2}\right)$$

$$+ \frac{1}{2}\left[(-0.1)^2\left(\frac{1}{2}\right) + 2(-0.1)(-0.2)0 + (-0.2)^2\left(\frac{1}{2}\right)\right] + \cdots$$

$$= -\frac{\pi}{4} + 0.05 - 0.1 + \frac{1}{2}(0.005 + 0.04) = -\frac{\pi}{4} + 0.05 - 0.1 + 0.0225$$

$\therefore$ $\boldsymbol{f(0.9, -1.2) = -0.8229}$.

Example 19.21. *Expand $f(x) = \sin xy$ in powers of $(x - 1)$ and $\left(y - \frac{\pi}{2}\right)$ up to second order terms* **[BU(H) 2002]**

Solution : We have Taylor's theorem of $f(x, y)$ in powers of $(x - a)$ and $(y - b)$ as

$$f(x, y) = f(a, b) + [(x - a)f_x(a, b) + (y - b)f_y(a, b)]$$

$$+ \frac{1}{2!}[(x - a)^2 f_{xx}(a, b) + 2(x - a)(y - b)f_{xy}(a, b) + (y - b)^2 f_{yy}(a, b)] + \cdots. \quad \text{(i)}$$

Now since, $f(x, y) = \sin xy, \qquad \therefore\ f\left(1, \frac{\pi}{2}\right) = 1$

$$f_x(x, y) = y \cos xy, \qquad \therefore\ f_x\left(1, \frac{\pi}{2}\right) = 0$$

$$f_y(x, y) = x\cos xy, \qquad \therefore\ f_y\left(1, \frac{\pi}{2}\right) = 0$$

$$f_{xx}(x, y) = -y^2 \sin xy, \qquad \therefore\ f_{xx}\left(1, \frac{\pi}{2}\right) = -\frac{\pi^2}{4}$$

$$f_{xy}(x, y) = \cos xy - xy\sin xy, \qquad \therefore\ f_{xy}\left(1, \frac{\pi}{2}\right) = -\frac{\pi}{2}$$

$$f_{yy}(x, y) = -x^2 \sin xy, \qquad \therefore\ f_{yy}\left(1, \frac{\pi}{2}\right) = -1.$$

Substituting these values in Eqn. (i), we get

$$\boldsymbol{f(x, y) = 1 - \frac{\pi^2}{8}(x-1)^2 - \frac{\pi}{2}(x-1)^2\left(y - \frac{\pi}{2}\right) - \frac{1}{2}\left(y - \frac{\pi}{2}\right)^2 + \cdots.}$$

Example 19.22. *Expand by Taylor's theorem* $\tan^{-1}\left(\frac{y}{x}\right)$ *about* $(1, 1)$ *up to and including second degree terms.* **[VU(H) 2008]**

Solution : Taylor's theorem is

$$f(x, y) = f(a, b) + \left[(x-a)\frac{\partial f}{\partial x} + (y-b)\frac{\partial f}{\partial y}\right] + \frac{1}{2!}\left[(x-a)^2\frac{\partial^2 f}{\partial x^2} + 2(x-a)(y-b)\frac{\partial^2 f}{\partial x \partial y} + (y-b)\frac{\partial^2 f}{\partial y^2}\right] + \cdots. \quad \text{(i)}$$

Now $f(x, y) = \tan^{-1}\left(\frac{y}{x}\right)$, $\qquad \therefore\ f(1, 1) = \frac{\pi}{4}$

$$\frac{\partial f}{\partial x} = -\frac{y}{x^2 + y^2}, \qquad \therefore\ \frac{\partial f}{\partial x}(1, 1) = -\frac{1}{2}$$

$$\frac{\partial f}{\partial y} = \frac{x}{x^2 + y^2}, \qquad \therefore\ \frac{\partial f}{\partial y}(1, 1) = \frac{1}{2}$$

$$\frac{\partial^2 f}{\partial x^2} = \frac{y(2x)}{(x^2+y^2)^2} = \frac{2xy}{(x^2+y^2)^2}, \qquad \therefore\ \frac{\partial^2 f}{\partial x^2}(1, 1) = \frac{1}{2}$$

$$\frac{\partial^2 f}{\partial x \partial y} = \frac{(x^2+y^2) - x(2x)}{(x^2+y^2)^2} = \frac{y^2 - x^2}{(x^2+y^2)^{\frac{1}{2}}}, \qquad \therefore\ \frac{\partial^2 f}{\partial x \partial y}(1, 1) = 0$$

$$\frac{\partial^2 f}{\partial y^2} = \frac{-2xy}{(x^2+y^2)^2}, \qquad \therefore\ \frac{\partial^2 f}{\partial y^2}(1, 1) = -\frac{1}{2}$$

Here, $f(x, y) = \tan^{-1}\left(\frac{y}{x}\right)$, $a = 1$, $b = 1$.

$$\therefore \quad \boldsymbol{\tan^{-1}\left(\frac{y}{x}\right)} = \frac{\pi}{4} + (x-1)\left(-\frac{1}{2}\right) + (y-1)\left(\frac{1}{2}\right) + \frac{1}{2!}\left[(x-1)^2\left(\frac{1}{2}\right) + 2(x-1)(y-1)(0) + (y-1)^2\left(-\frac{1}{2}\right)\right] + \cdots$$

$$\boldsymbol{= \frac{\pi}{4} - \frac{1}{2}(x-1) + \frac{1}{2}(y-1) + \frac{1}{4}(x-1)^2 - \frac{1}{4}(y-1)^2 + \cdots.}$$

Example 19.23. *The time period of oscillation of a simple pendulum is given by the expression $T = 2\pi\sqrt{\frac{l}{g}}$, where the symbols have their usual meaning. In an experiment carried out to find the value of g, error of 1% and 0.5% are possible in the values of l and T. Find the maximum error in the calculation of T.* **[BU(H) 1993]**

Solution : Given, $T = 2\pi\sqrt{\frac{l}{g}}$.

$\therefore \quad \log T = \log 2\pi + \frac{1}{2}\log l - \frac{1}{2}\log g \quad$ or, $\quad \frac{\delta T}{T} = 0 + \frac{1}{2}\frac{\delta l}{l} - \frac{1}{2}\frac{\delta g}{g}$

or, $\quad \frac{\delta T}{T} \times 100 = \frac{1}{2}\left(\frac{\delta l}{l} \times 100 - \frac{\delta g}{g} \times 100\right) = \frac{1}{2}(1 \pm 0.5) = \frac{1.5}{2} = 0.75.$

$\therefore$ maximum error in $\boldsymbol{T} = \mathbf{0.75\%}$.

Example 19.24. *The length of a simple pendulum is increased by 2%. By what percentage will the frequency increase or decrease?* **[BU(H) 2003]**

Solution : The frequency of oscillation of a simple pendulum is given by

$$f = \frac{1}{2\pi}\sqrt{\frac{l}{g}},$$

where g and l are the acceleration due to gravity and effective length respectively.

$\therefore \quad \log f = \log\left(\frac{1}{2\pi}\right) + \frac{1}{2}\log g - \frac{1}{2}\log l \quad$ or, $\quad \frac{df}{f} = 0 + 0 - \frac{1}{2}\frac{\delta l}{l} = -\frac{1}{2}\frac{\delta l}{l}$

or, $\quad \frac{df}{f} \times 100 = -\frac{1}{2}\frac{\delta l}{l} \times 100 = -\frac{1}{2} \times 2 = -1.$

$\therefore$ the frequency of the simple pendulum will decrease by **1%**.

Example 19.25. *The rate of flow (V) of a liquid through a capillary tube of radius r and length l at a pressure difference P between its ends is given by $V = \frac{\pi P r^4}{8\eta l}$, where η is the coefficient of viscosity of the liquid. In an experiment the errors in the measurement of P, r, l and V are 1%, 5%, 0.5% and 2% respectively. Evaluate the error in the measurement of η.* **[BU(H) 1995]**

Solution : Given, $V = \frac{\pi P r^4}{8\eta l}$, $\quad \therefore \ \eta = \frac{\pi P r^4}{8Vl}$.

So, $\log \eta = \log \pi + \log P + \log r^4 - \log 8 - \log V - \log r$

or, $\quad \frac{\delta\eta}{\eta} = 0 + \frac{\delta P}{P} + 4\frac{\delta r}{r} - 0 - \frac{\delta V}{V} - \frac{\delta r}{r}.$

For maximum proportional error,

$$\frac{\delta\eta}{\eta} = \frac{\delta P}{P} + 4\frac{\delta r}{r} + \frac{\delta V}{V} + \frac{\delta r}{r}$$

$$\text{or,}\quad \frac{\delta\eta}{\eta}\times 100 = \frac{\delta P}{P}\times 100 + 4\frac{\delta r}{r}\times 100 + \frac{\delta V}{V}\times 100 + \frac{\delta r}{r}\times 100$$

$$= 1 + 4\times 1.5 + 0.5 + 2 = 1 + 6 + 0.5 + 2 = 9.5.$$

$\therefore$ maximum percentage error in η is **9.5%**.

Example 19.26. *Show that g is reduced by 1%, corresponding to an increase of 0.5% in altitude, given that at an external point g is given by $k\left(\frac{r}{x}\right)^2$, where r is the radius of the earth and x is the distance from the centre of the earth.* **[AMIETE(Summer) 1990]**

Solution : Given, $k\left(\frac{r}{x}\right)^2$.

$$\therefore \quad \log g = \log k + 2\log r - 2\log x.$$

$$\text{Differentiating,}\quad \frac{\delta g}{g} = \frac{\delta k}{k} + 2\frac{\delta r}{r} - 2\frac{\delta x}{x} = 0 + 0 - 2\frac{\delta x}{x}$$

$$\text{or,}\quad \frac{\delta g}{g}\times 100 = -2\frac{\delta x}{x}\times 100 = -2(0.5) = -1.$$

$\therefore$ **g is reduced by 1%.**

Example 19.27. *What is the percentage error in the density of the material of a cube if the percentage error in the measurement of mass is 0.5 and that of length is 1.* **[BU(H) 2002]**

Solution : If a be the length of a side and μ the mass of the cube, then $M = \rho a^3$, where ρ is the density of its material.

$$\therefore \quad \log M = \log\rho + 3\log a \quad \text{or,}\quad \log\rho = \log M - 3\log a \quad \text{or,}\quad \frac{\delta\rho}{\rho} = \frac{\delta M}{M} - 3\times\frac{\delta a}{a}.$$

$$\text{For maximum error,}\quad \frac{\delta\rho}{\rho} = \frac{\delta M}{M} + 3\frac{\delta a}{a}.$$

$$\therefore \quad \frac{\delta\rho}{\rho}\times 100 = \frac{\delta M}{M}\times 100 + 3\frac{\delta a}{a}\times 100 = 0.5 + 3\times 1 = 3.5.$$

$\therefore$ **% error in the density of the material is 3.5%.**

Example 19.28. *Find the maxima and minima of the function*

$$f(x,y) = x^2y^2 - 5x^2 - 8xy - 5y^2.$$

Solution : Since $(x,y) = x^2y^2 - 5x^2 - 8xy - 5y^2$.

$$\therefore \quad f_x = 2xy^2 - 10x - 8y$$

$$f_y = 2x^2y - 8x - 10y.$$

For $f_x = 0$ and $f_y = 0$,

$$xy^2 - 5x - 4y = 0 \qquad \text{(i)}$$

$$x^2y - 4x - 5y = 0. \qquad \text{(ii)}$$

Multiplying Eqn. (i) by x and (ii) by y and then subtracting, we get

$$x^2 = y^2, \quad \text{i.e.,}\quad x = \pm y.$$

Using in Eqns. (i) and (ii) we get the stationary points $(0,0)$, $(1,-1)$, $(-1,1)$, $(3,3)$, $(-3,-3)$.

Again, $r = f_{xx} = 2y^2 - 10, \quad s = f_{xy} = 4xy - 8, \quad t = f_{yy} = 2x^2 - 10.$

At $(0,0)$, $r = -10$, $s = -8$, $t = -10$.

$\therefore \quad rt - s^2 = 100 - 64 = 36 > 0, \; r < 0.$

So, $f(x,y)$ is maximum at $(0,0)$.

At $(1,-1)$, $r = -8$, $s = -12$, $t = -8$.

$\therefore \quad rt - s^2 = 64 - 144 = -8 < 0.$

$\therefore$ $(1,-1)$ is a saddle point.

Similarly, $(-1,1)$, $(3,3)$ and $(-3,-3)$ are also saddle points.

Example 19.29. *Find the absolute maximum and minimum values of $f(x,y) = 2 + 2x + 2y - x^2 - y^2$ on triangular plate in the first quadrant bounded by lines $x = 0$, $y = 0$ and $y = 9 - x$.*

Solution : Given $f(x,y) = 2 + 2x + 2y - x^2 - y^2$.

$\therefore \quad f_x = 2 - 2x, \; f_y = 2 - 2y, \; f_{xx} = -2, \; f_{yy} = -2, \; f_{xy} = 0.$

For maxima and minima, $f_x = 0$, $f_y = 0$.

$\therefore \quad 2 - 2x = 0, \; x = 1 \quad$ and $\quad 2 - 2y = 0, \; y = 1.$

At $(1,1)$, $rt - s^2 = (-2)(-2) = 4$.

$\therefore$ $f(x,y)$ is maximum at $(1,1)$.

$\therefore$ maximum value of $f(x,y) = 2 + 2 + 2 - 1 - 1 = 4$.

Example 19.30. *In a plane triangle ABC, find the maximum value of $\cos A \cos B \cos C$.*

Solution : Let
$$\begin{aligned} f(A,B) &= \cos A \cos B \cos C \\ &= \cos A \cos B \cos[\pi - (A+B)] \qquad [\because A + B + C = \pi] \\ &= -\cos A \cos B \cos(A+B). \end{aligned}$$

$$\therefore \quad \frac{\partial f}{\partial A} = \sin A \cos B \cos(A+B) + \cos A \cos B \cos(A+B) = \cos B \sin(A+B+C)$$
$$= \cos B \sin(2A+B).$$
$$\frac{\partial f}{\partial B} = -\cos A[-\sin B \cos(A+B) - \cos B \sin(A+B)] = \cos A \sin(A+B+B)$$
$$= \cos A \sin(A+2B).$$

For maxima and minima,
$$\frac{\partial f}{\partial A} = 0 \quad \text{or,} \quad \cos B \sin(2A+B) = 0 \quad \text{or,} \quad \sin(2A+B) = 0$$
or, $$2A + B = \pi; \tag{i}$$
and $$\frac{\partial f}{\partial B} = 0 \quad \text{or,} \quad \cos A \sin(A+2B) = 0 \quad \text{or,} \quad \sin(A+2B) = 0$$
or, $$A + 2B = \pi. \tag{ii}$$

Solving Eqns. (i) and (ii), we get $A = B = \dfrac{\pi}{3}$.

Again, $\dfrac{\partial^2 f}{\partial A^2} = 2B\cos(2A+B)$

$$\frac{\partial^2 f}{\partial A \partial B} = \cos B\cos(2A+B) - \sin B\sin(2A+B) = \cos(2A+2B)$$

$$\frac{\partial^2 f}{\partial B^2} = 2\cos A\cos(A+2B).$$

At $A = B = \dfrac{\pi}{3}$,

$$\frac{\partial^2 f}{\partial A^2} = 2\cos\frac{\pi}{3}\cos\pi = -1$$

$$\frac{\partial^2 f}{\partial A \partial B} = \cos\left(\frac{4\pi}{3}\right) = -\frac{1}{2}$$

$$\frac{\partial^2 f}{\partial B^2} = 2\cos\frac{\pi}{3}\cos\pi = -1.$$

Now, (1) $\dfrac{\partial^2 f}{\partial A^2} \times \dfrac{\partial^2 f}{\partial B^2} > \dfrac{\partial^2 f}{\partial A \partial B}$ (2) $\dfrac{\partial^2 f}{\partial A^2}$ is $-$ve.

$\therefore$ $f(A, B)$ is maximum for $A = B = C = \dfrac{\pi}{3}$.

The maximum value of

$\therefore$ $$\mathbf{\cos A\cos B\cos C = \cos\frac{\pi}{3}\cos\frac{\pi}{3}\cos\frac{\pi}{3} = \frac{1}{2}\times\frac{1}{2}\times\frac{1}{2} = \frac{1}{8}}.$$

Example 19.31. *The temperature T at any point (x, y, z) in space is $T = 400xyz^2$. Find the highest temperature at the surface of a unit sphere $x^2 + y^2 + z^2 = 1$.*

Solution : Given, $T = 400xyz^2$, $\phi(x, y, z) = x^2 + y^2 + z^2 - 1$.

$\therefore$ by Lagrange's method of undetermined multiplier,

$$\frac{\partial T}{\partial x} + \lambda\frac{\partial \phi}{\partial x} \quad \text{or,} \quad 400yz^2 + \lambda(2x) = 0 \tag{i}$$

$$\frac{\partial T}{\partial y} + \lambda\frac{\partial \phi}{\partial y} \quad \text{or,} \quad 400xz^2 + \lambda(2y) = 0 \tag{ii}$$

$$\frac{\partial T}{\partial z} + \lambda\frac{\partial \phi}{\partial z} \quad \text{or,} \quad 800xyz + \lambda(2z) = 0. \tag{iii}$$

Now, multiplying Eqn. (i) by x, (ii) by y and (iii) by z and then adding,

$$1600xyz^2 + 2\lambda(x^2+y^2+z^2) = 0 \quad \text{or,} \quad 1600xyz^2 + 2\lambda(1) = 0 \quad \text{or,} \quad \lambda = -800xyz^2.$$

Putting the value of λ in Eqn. (i), we get

$$400yz^2 + 2x(-800xyz^2) = 0 \quad \text{or,} \quad 1 - 4x^2 = 0 \quad \text{or,} \quad x = \pm\frac{1}{2}.$$

Similarly, $y = \pm\dfrac{1}{2}$.

Putting the value of λ in Eqn. (iii), we get

$$800xyz - 1600xyz^2 \quad \text{or,} \quad 1 - 2z^2 = 0.$$

$\therefore \quad Z = \pm\frac{1}{\sqrt{2}}.$

Now, using the values of x, y, z in T, we get

$$\boldsymbol{T} = 400 \times \frac{1}{2} \times \frac{1}{2} \times \frac{1}{2} = \mathbf{50}.$$

Example 19.32. *Using the method of Lagrange multiplier find the maximum of $F = 4xyz$ subject to the constraint $x^2 + y^2 + z^2 - a^2 = 0$.* **[BU(H) 1997]**

Solution : Let $\phi(x, y, z) = x^2 + y^2 + z^2 - a^2 = 0.$ (i)

We consider, $f(x, y, z) = F(x, y, z) + \lambda\phi(x) = 4xyz - \lambda(x^2 + y^2 + z^2 - a^2).$ (ii)

$$\therefore \quad \frac{\partial f}{\partial x} = 4yz + 2\lambda x = 0 \qquad \text{(iii)}$$

$$\frac{\partial f}{\partial y} = 4zx + 2\lambda y = 0 \qquad \text{(iv)}$$

$$\frac{\partial f}{\partial z} = 4xy + 2\lambda z = 0. \qquad \text{(v)}$$

From Eqns. (iii) and (iv), we get

$$\lambda = -\frac{2yz}{x} = -\frac{2zx}{y}, \quad \text{i.e.,} \quad y^2 = x^2 \quad \text{or,} \quad y = x.$$

Similarly, from Eqns. (iv) and (v), we get $x = z$.

$\therefore \quad x = y = z.$ (vi)

From Eqn. (i), $x = y = z = \pm\frac{a}{\sqrt{3}}.$ (vii)

The value of $F = 4xyz$ will lie within $-\frac{4a}{3\sqrt{3}}$ to $\frac{4a}{3\sqrt{3}}$.

So, the maximum value of F will be $\boldsymbol{\frac{4a}{3\sqrt{3}}}$.

Example 19.33. *Find the height of a right cylinder with greatest volume that can be inscribed in a sphere of radius R.*

Solution : In Fig. 19.3, $ABCD$ is the cylinder of greatest volume that can be inscribed in the sphere of radius $OP = R$.

If $OQ = r$ be the radius of the cylinder, then $r^2 = x^2 + y^2$.

Volume of the cylinder $V = \pi r^2 2z$

or, $V(x, y, z) = 2\pi(x^2 + y^2)z$ (i)

Equation of the sphere is $x^2 + y^2 + z^2 - R^2 = 0$

or, $\phi(x, y, z) = x^2 + y^2 + z^2 - R^2 = 0.$ (ii)

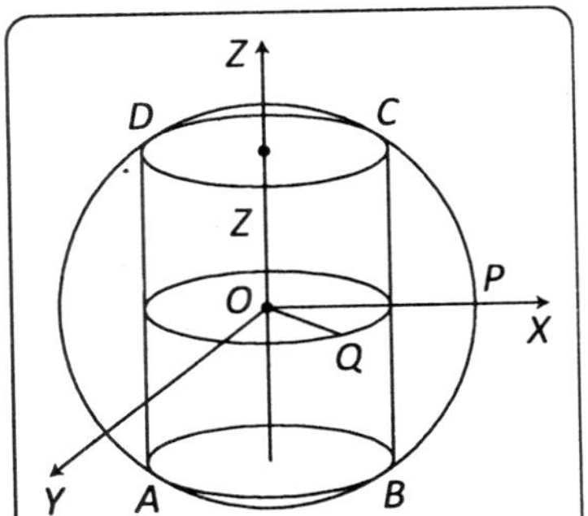

Fig. 19.3

By Lagrange's method,

$$f(x, y, z) = V + \lambda\phi = 2\pi(x^2 + y^2)z + \lambda(x^2 + y^2 + z^2 - R^2)$$

$$\frac{\partial f}{\partial x} = 4\pi xz + 2\lambda x = 0, \quad \therefore \quad \lambda = -2\pi z$$

$$\frac{\partial f}{\partial y} = 4\pi yz + 2\lambda y = 0, \quad \therefore \quad \lambda = -2\pi z$$

$$\frac{\partial f}{\partial z} = 4\pi(x^2 + y^2) + 2\lambda z = 0, \quad \therefore \quad \lambda = -\frac{2\pi(x^2+y^2)}{z}.$$

$$\therefore \quad \frac{-2\pi(x^2+y^2)}{z} = -2\pi z \quad \text{or,} \quad x^2 + y^2 = z^2. \tag{iii}$$

From Eqn. (ii),

$$x^2 + y^2 + z^2 - R^2 = 0 \quad \text{or,} \quad 2z^2 - R^2 = 0 \quad \text{or,} \quad z = \frac{R}{\sqrt{2}}.$$

$\therefore$ the height of the cylinder with maximum volume is $2z = 2\frac{R}{\sqrt{2}} = \sqrt{2}R$.

$$V_{\text{max}} = 2\pi(x^2+y^2)z = 2\pi z^2 \times z = 2\pi\left(\frac{R}{\sqrt{2}}\right)^2 \times \frac{R}{\sqrt{2}} = 2\pi\frac{R^2}{2} \times \frac{R}{\sqrt{2}} = \frac{\pi R^3}{\sqrt{2}}.$$

Example 19.34. *Show that the rectangular solid of maximum volume that can be inscribed in a sphere is a cube.* **[AMIETE(Summer) 1997, '94, '92]**

Solution : Let R be the radius of the sphere and $2x$, $2y$, $2z$, the length, breadth and height of the rectangular solid.

Then volume of the solid is

$$V(x, y, z) = 8xyz \tag{i}$$

and the equation of the sphere is

$$x^2 + y^2 + z^2 - R^2 = 0 \quad \text{or,} \quad \phi(x, y, z) = x^2 + y^2 + z^2 - R^2. \tag{ii}$$

$$\therefore \quad f(x, y, z) = V(x, y, z) + \lambda\phi(x, y, z) = 8xyz + \lambda(x^2 + y^2 + z^2 - R^2).$$

$$\therefore \quad \frac{\partial f}{\partial x} = 8yz + \lambda(2x) = 0. \tag{iii}$$

$$\frac{\partial f}{\partial y} = 8xz + \lambda(2y) = 0 \tag{iv}$$

$$\text{and} \quad \frac{\partial f}{\partial z} = 8xy + \lambda(2z) = 0. \tag{v}$$

From Eqn. (iii), $2\lambda x = -8yz$ or, $2\lambda x^2 = -8xyz$.

From Eqn. (iv), $2\lambda y^2 = -8xyz$.

From Eqn. (v), $2\lambda z^2 = -8xyz$.

$$\therefore \quad 2\lambda x^2 = 2\lambda y^2 = 2\lambda z^2 \quad \text{or,} \quad x^2 = y^2 = z^2. \quad \therefore \quad x = y = z.$$

Hence, the rectangular solid is a **cube**.

Example 19.35. *Find the volume of the greatest rectangular parallelopiped that can be inscribed in the ellipsoid* $\frac{x^2}{a^2} + \frac{y^2}{b^2} + \frac{z^2}{c^2} = 1$. **[Madras U, 2006]**

Solution : Let $2x$, $2y$, $2z$ be the edges of the parallelopiped whose edges are parallel to the X-, Y- and Z-axes respectively.

Then volume of the parallelopiped is $V = 8xyz$.

Let $\phi(x, y, z) = \dfrac{x^2}{a^2} + \dfrac{y^2}{b^2} + \dfrac{z^2}{c^2} = 1$.

By Lagrange's method of undetermined multiplier,

$$F(x, y, z) = V + \lambda\phi(x, y, z) \tag{i}$$

or, $$F(x, y, z) = 8xyz + \lambda\left(\frac{x^2}{a^2} + \frac{y^2}{b^2} + \frac{z^2}{c^2} - 1\right).$$

$\therefore$ $$\frac{\partial F}{\partial x} = 8xyz + \lambda\left(\frac{2x}{a^2}\right) = 0 \tag{ii}$$

$$\frac{\partial F}{\partial y} = 8xz + \lambda\left(\frac{2y}{b^2}\right) = 0 \tag{iii}$$

and $$\frac{\partial F}{\partial z} = 8xy + \lambda\left(\frac{2z}{c^2}\right) = 0. \tag{iv}$$

From Eqn. (ii), $\lambda = -\dfrac{8yz}{\left(\frac{2x}{a^2}\right)}$ and from Eqn. (iii), $\lambda = -\dfrac{8xz}{\left(\frac{2z}{c^2}\right)}$.

$\therefore$ $$\frac{8yza^2}{2x} = \frac{8xzc^2}{2z} \quad \text{or,} \quad \frac{x^2}{a^2} = \frac{y^2}{b^2}.$$

Similarly, from (iii) and (iv), we get $\dfrac{y^2}{b^2} = \dfrac{z^2}{c^2}$.

$\therefore$ $$\frac{x^2}{a^2} = \frac{y^2}{b^2} = \frac{z^2}{c^2}.$$

Substituting in Eqn. (i), we get $\dfrac{x^2}{a^2} = \dfrac{x^2}{a^2} = \dfrac{x^2}{a^2} = 1$ or, $\dfrac{x^2}{a^2} = \dfrac{1}{3}$.

$\therefore$ $$\frac{x^2}{a^2} = \frac{y^2}{b^2} = \frac{z^2}{c^2} = \frac{1}{3}.$$

Hence, $x = \dfrac{a}{\sqrt{3}}$, $y = \dfrac{b}{\sqrt{3}}$, $z = \dfrac{c}{\sqrt{3}}$.

Thus, the volume of the largest rectangular parallelopiped is

$$\boldsymbol{V} = 8xyz = 8\frac{a}{\sqrt{3}}\frac{b}{\sqrt{3}}\frac{c}{\sqrt{3}} = \frac{\mathbf{8abc}}{\mathbf{3\sqrt{3}}}.$$

Example 19.36. *A tent of a given volume has a square base of side $2a$ having its four sides vertical of length b and is surrounded by a regular pyramid of height h. Find the values of a and b in terms of h such that canvas required for its construction is minimum.*

Solution : The volume of the tent, $V = 4a^2b + \dfrac{1}{3}(4a^2)h$.

The surface area of the tent, $S = 8ab + 4a\sqrt{a^2 + h^2}$.

According to Lagrange's method of undetermined multiplier,

$$F(a,b,h) = S + \lambda V \quad \text{or,} \quad F(a,b,h) = 8ab + 4a\sqrt{a^2+h^2} + 4a^2b + \frac{1}{3}(4a^2)h.$$

$$\therefore \quad \frac{\partial F}{\partial a} = 8b + 4a\sqrt{a^2+h^2} + \frac{4a^2}{\sqrt{a^2+h^2}} = \lambda\left(8ab + \frac{8ah}{3}\right) = 0 \qquad \text{(i)}$$

$$\frac{\partial F}{\partial b} = 8b + 4\lambda a^2 \qquad \text{(ii)}$$

$$\frac{\partial F}{\partial h} = \frac{4ah}{\sqrt{a^2+h^2}} + \frac{4}{3}\lambda a^2 = 0. \qquad \text{(iii)}$$

From Eqn. (ii), $\lambda a + 2 = 0.$ (iv)

From Eqn. (iii), $12ah + 4\lambda a^2\sqrt{a^2+h^2} = 0$

or, $3h + \lambda a\sqrt{a^2+h^2} = 0$ (v)

or, $3h + (-2)\sqrt{a^2+h^2} = 0$ $[\because \lambda a = -2$ from Eqn. (iv)$]$

or, $9h^2 = 4a^2 + 4h^2$ or, $a = \dfrac{\sqrt{5}h}{2}.$

Now putting $\lambda a = -2$ and $a = \dfrac{\sqrt{5}}{2}h$ in Eqn. (i), we get after simplifying, $b = \dfrac{h}{2}.$

Thus, for stationary value of S, $\boldsymbol{a = \dfrac{\sqrt{5}}{2}h}$ **and** $\boldsymbol{b = \dfrac{h}{2}}.$

Example 19.37. *Using methods of Lagrange's multiplier, find the area of the largest rectangle which can be inscribed in a semi-circle of radius R with one of the largest sides of the rectangle coinciding with the diameter.* **[BU(H) 1996]**

Solution : The rectangle inscribed within the circle must be symmetric about Y-axis.

If $2x$ and y be the length and breadth of the rectangle respectively, then its area is

$$S = 2xy. \qquad \text{(i)}$$

Also, $x^2 + y^2 = R^2.$ (ii)

Let $F(x,y) = 2xy + \lambda(x^2+y^2-R^2).$

$$\therefore \quad \frac{\partial F}{\partial x} = 2y + 2\lambda x \quad \text{and} \quad \frac{\partial F}{\partial y} = 2x + 2\lambda y.$$

For $\dfrac{\partial F}{\partial x} = 0,\ 2y + 2\lambda x = 0$ or, $y + \lambda x = 0.$ (iii)

For $\dfrac{\partial F}{\partial y} = 0,\ 2x + 2\lambda y = 0$ or, $x + \lambda y = 0.$ (iv)

Equating λ from (iii) and (iv), we get $-\dfrac{y}{x} = -\dfrac{x}{y}$ or, $x^2 = y^2$ i.e., $x = y.$

$\therefore$ from Eqn. (ii), $x = \dfrac{R}{\sqrt{2}} = y.$

From Eqn. (i), $S = 2\left(\dfrac{R}{\sqrt{2}}\right)\left(\dfrac{R}{\sqrt{2}}\right) = R^2.$

So, when $y = 0$, $S = 0$. As y increases, S also increases. Again, when y increases sufficiently, x becomes so small that the area becomes zero. In these two stages at a certain level given by eqn. (iv) S is maximum. So maximum area $\boldsymbol{S_{\max} = R^2}$.

Example 19.38. *Find the point upon the plane $ax + by + cz = p$ at which the function $f(x, y, z) = x^2 + y^2 + z^2$ has a minimum value and find this minimum $f(x, y, z)$.*

Solution : Given, $f(x, y, z) = x^2 + y^2 + z^2$ (i)

and $ax + by + cz = p$ or, $\phi(x, y, z) = ax + by + cz - p$ (ii)

$\therefore \quad F(x, y, z) = f(x, y, z) + \lambda\phi(x, y, z)$.

Differentiating partially,

$$\frac{\partial F}{\partial x} = \frac{\partial f}{\partial x} + \lambda\frac{\partial \phi}{\partial x} = 2x + \lambda a = 0 \quad \text{or,} \quad x = \frac{-\lambda a}{2}$$

$$\frac{\partial F}{\partial y} = \frac{\partial f}{\partial y} + \lambda\frac{\partial \phi}{\partial y} = 2y + \lambda b = 0 \quad \text{or,} \quad y = \frac{-\lambda b}{2}$$

$$\frac{\partial F}{\partial z} = \frac{\partial f}{\partial z} + \lambda\frac{\partial \phi}{\partial z} = 2z + \lambda c = 0 \quad \text{or,} \quad y = \frac{-\lambda c}{2}.$$

Substituting the values of x, y, z in Eqn. (ii), we get

$$a\left(-\frac{\lambda a}{2}\right) + b\left(-\frac{\lambda b}{2}\right) + c\left(-\frac{\lambda c}{2}\right) = p \quad \text{or,} \quad \lambda(a^2 + b^2 + c^2) = -2p \quad \text{or,} \quad \lambda = \frac{-2p}{a^2 + b^2 + c^2}.$$

$$\therefore \quad \boldsymbol{x = \frac{ap}{x^2 + y^2 + z^2}, \quad y = \frac{bp}{a^2 + b^2 + c^2}, \quad z = \frac{cp}{a^2 + b^2 + c^2}}.$$

So the function (x, y, z) will have maximum value at (x, y, z).

$$\textbf{The minimum value of } \boldsymbol{f(x, y, z)} = \frac{a^2p^2}{(a^2 + b^2 + c^2)^2} + \frac{b^2p^2}{(a^2 + b^2 + c^2)^2} + \frac{c^2p^2}{(a^2 + b^2 + c^2)^2}$$

$$= \frac{p^2(a^2 + b^2 + c^2)}{(a^2 + b^2 + c^2)^2} = \boldsymbol{\frac{p^2}{a^2 + b^2 + c^2}}.$$

Example 19.39. *Evaluate the integral* $\int_0^\infty e^{-x}\frac{\sin \alpha x}{x}dx$. **[BU(H)(Old) 2000]**

Solution : Let $F(\alpha) = \int_0^\infty e^{-x}\frac{\sin \alpha x}{x}dx$.

$$\therefore \quad \frac{dF(\alpha)}{d\alpha} = F'(\alpha) = \int_0^\infty \frac{\partial}{\partial x}\left[e^{-x}\frac{\sin \alpha x}{x}\right]dx = \int_0^\infty \frac{xe^{-x}\cos \alpha x}{x}dx = \int_0^\infty e^{-x}\cos \alpha x\, dx$$

$$= I \text{ (say)}.$$

Then $I = \frac{1}{\alpha}e^{-x}\sin \alpha x + \frac{1}{\alpha}\int e^{-x}\sin \alpha x\, dx = \frac{1}{\alpha}e^{-x}\sin \alpha x - \frac{1}{\alpha^2}\left[e^{-x}\cos \alpha x + \int e^{-x}\cos \alpha x\, dx\right]$.

$$\therefore \quad \left(1 + \frac{1}{\alpha^2}\right)I = \frac{1}{\alpha}e^{-x}\sin \alpha x - \frac{1}{\alpha^2}e^{-x}\cos \alpha x$$

$$\therefore \quad I = \frac{1}{\alpha^2+1}\left[\alpha e^{-x}\sin\alpha x - e^{-x}\cos\alpha x\right].$$

So, $\quad F'(\alpha) = \dfrac{1}{\alpha^2+1}\left[\alpha e^{-x}\sin\alpha x - e^{-x}\cos\alpha x\right]_0^\infty \quad$ or, $\quad \dfrac{dF(\alpha)}{d\alpha} = \dfrac{1}{\alpha^2+1}$

or, $\quad dF(\alpha) = \dfrac{1}{\alpha^2+1}\,d\alpha.$

Integrating, $F(\alpha) = \tan^{-1}\alpha + C$.

But, $F(0) = 0, \quad \therefore \; C = 0$.

Hence, $\displaystyle\int_0^\infty \boldsymbol{e^{-x}\frac{\sin\alpha x}{x}\,dx = \tan^{-1}(\alpha)}.$

Example 19.40. Evaluate $\displaystyle\int_0^1 \frac{x^\alpha - 1}{\log x}dx,\ \alpha > 0.$ **[BU(H) 1998, 2002]**

Solution : Let $F(\alpha) = \displaystyle\int_0^1 \frac{x^\alpha-1}{\log x}\,dx.$ (i)

$$\therefore \quad \frac{dF(\alpha)}{d\alpha} = \int_0^1 \frac{\partial}{\partial\alpha}\left(\frac{x^\alpha-1}{\log x}\right)dx = \int_0^1 \frac{x^\alpha \log x}{\log x}dx$$

$$= \int_0^1 x^\alpha\,dx \qquad \left[\because\ \frac{d}{d\alpha}(x^\alpha) = x^\alpha \log x\right]$$

$$= \left[\frac{x^{\alpha+1}}{\alpha+1}\right] = \frac{1}{\alpha+1}.$$

Integrating both sides w.r.t. α, we get

$F(\alpha) = \log(1+\alpha) + C.$ (ii)

But for $\alpha = 0$, $F(\alpha) = 0$.

From Eqn. (ii), $F(0) = \log(1+0) + C$, i.e., $C = 0$.

Hence, $\boldsymbol{F(\alpha) = \log(1+\log\alpha)}$. (iii)

Example 19.41. Evaluate $\displaystyle\int_0^\pi \log(1+\alpha\cos x)dx$ *using the method of differentiation under sign of integration.* **[BUM(H) 1981]**

Solution : Let $F(\alpha) = \displaystyle\int_0^\pi \log(1+\alpha\cos x)dx$, then

$$\frac{dF(\alpha)}{d\alpha} = \int_0^\pi \frac{\partial}{\partial x}[\log(1+\alpha\cos x)]dx = \int_0^\pi \frac{\cos x}{1+\alpha\cos x}dx = \frac{1}{\alpha}\int_0^\pi \frac{1+\alpha\cos x - 1}{1+\cos x}dx$$

$$= \frac{1}{\alpha}\int_0^\pi dx - \frac{1}{\alpha}\int_0^\pi \frac{dx}{1+\alpha\cos x}dx = \frac{\pi}{\alpha} - \frac{1}{\alpha}\frac{2}{\sqrt{1-\alpha^2}}\left[\tan^{-1}\left(\sqrt{\frac{1-\alpha}{1+\alpha}}\tan\frac{x}{\alpha}\right)\right]_0^\pi$$

$$= \frac{\pi}{\alpha} - \frac{\pi}{\alpha\sqrt{1-\alpha^2}} = \frac{\pi(\sqrt{1-\alpha^2}-1)}{\alpha\sqrt{1-\alpha^2}}.$$

$$\therefore \quad F(\alpha) = \pi \int_0^{\pi} \frac{\sqrt{1-\alpha^2}-1}{\alpha\sqrt{1-\alpha^2}}\, d\alpha + C.$$

Putting, $\alpha = \sin\theta$, $d\alpha = \cos\theta\, d\theta$,

$$\therefore \quad F(\alpha) = \pi \log(1+\sqrt{1-\alpha^2}) + C \quad \text{or,} \quad F(0) = \pi \log 2 + C.$$

$$\therefore \quad \boldsymbol{F(\alpha) = \pi \log \frac{1}{2}(1+\sqrt{1-\alpha^2})}.$$

Example 19.42. *Using Leibnitz's rule for differentiation under the integral sign, find* $I(\alpha) = \int_0^{\infty} e^{-\alpha x} \frac{\sin x}{x}\, dx$ $(a \geq 0)$ *and hence evaluate the integral* $\int_0^{\infty} \frac{\sin x}{x}\, dx$. **[BU(H) 1992]**

Solution : We have $I(\alpha) = \int_0^{\infty} \frac{e^{-\alpha x}\sin x}{x}\, dx.$ (i)

$$\therefore \quad \frac{dI(\alpha)}{d\alpha} = -\int_0^{\infty} e^{-\alpha x} \sin x\, dx = \frac{1}{1+\alpha^2}.$$

Integrating w.r.t. α, we get

$$I(\alpha) = -\tan^{-1}\alpha + C.$$

But, as $\alpha \to \infty$, $I(\alpha) \to 0$, $\quad \therefore \; C = \frac{\pi}{2}$.

$$\therefore \quad I(\alpha) = \int_0^{\infty} \frac{e^{-\alpha x}\sin x}{x}\, dx = -\tan^{-1}\alpha + \frac{\pi}{2}. \qquad \text{(ii)}$$

$$\boldsymbol{I(\alpha) = \cot^{-1}\alpha}.$$

Putting $\alpha = 0$, we get from Eqn. (ii),

$$\boldsymbol{\int_0^{\infty} \frac{\sin x}{x}\, dx = \frac{\pi}{2}}.$$

Example 19.43. *Given* $\int_0^{\pi} \frac{dx}{a+b\cos x} = \frac{\pi}{\sqrt{a^2-b^2}}$ $(a > b)$

Show that, $\int_0^{\pi} \frac{dx}{(a+b\cos x)^2} = \frac{\pi a}{(a^2-b^2)^{\frac{3}{2}}}$.

Solution : Given, $\int_0^{\pi} \frac{dx}{a+b\cos x} = \frac{\pi}{\sqrt{a^2-b^2}}$.

Differentiating w.r.t. a,

$$\int_0^{\pi} \frac{\partial}{\partial a}\left(\frac{1}{a+b\cos x}\right) dx = \frac{\partial}{\partial a}\left[\frac{\pi}{\sqrt{a^2-b^2}}\right]$$

$$\text{or,} \quad \int_0^{\pi} \frac{dx}{(a+b\cos x)^2} = \pi\left(-\frac{1}{2}\right)(a^2-b^2)^{-\frac{3}{2}} 2a$$

$$\text{or,} \quad \boldsymbol{\int_0^{\pi} \frac{dx}{(a+b\cos x)^2} = \frac{\pi a}{(a^2-b^2)^{\frac{3}{2}}}} \quad \textbf{(Proved)}.$$

Exercises

19(A) : Objective Type Questions

Select the correct answer

19.1 If $u = x^y$, then $\dfrac{\partial u}{\partial x}$ is

(a) 0 (b) yx^{y-1} (c) $x^y \log x$ (d) none of these.

19.2 If $u = x^2 + y^2$, then $\dfrac{\partial^2 u}{\partial x \partial y}$ is

(a) 2 (b) 0 (c) $2x + 2y$ (d) $x + y$.

19.3 If $u = f\left(\dfrac{y}{x}\right)$, then

(a) $x\dfrac{\partial u}{\partial x} - y\dfrac{\partial u}{\partial y} = 0$ (b) $x\dfrac{\partial u}{\partial x} + y\dfrac{\partial u}{\partial y} = 0$ (c) $x\dfrac{\partial u}{\partial x} + y\dfrac{\partial u}{\partial y} = 2u$ (d) $x\dfrac{\partial u}{\partial x} + y\dfrac{\partial u}{\partial y} = 1$.

19.4 If $u = \log \dfrac{x^2}{y}$, then $x\dfrac{\partial u}{\partial x} + y\dfrac{\partial u}{\partial y}$ is equal to

(a) $2u$ (b) u (c) 0 (d) 1.

19.5 If $x = r\cos\theta$, $y = r\sin\theta$, then $\dfrac{\partial(x, y)}{\partial(r, \theta)}$ is equal to

(a) 1 (b) r (c) $\dfrac{1}{r}$ (d) 0.

19.6 If $u = f(x, y)$ be a homogeneous function of degree n in x, y, then $x\dfrac{\partial u}{\partial x} + y\dfrac{\partial u}{\partial y}$ is equal to

(a) u (b) 0 (c) nu (d) $\dfrac{u}{n}$.

19.7 If $J = \dfrac{\partial(u, v)}{\partial(x, y)}$ and $J' = \dfrac{\partial(x, y)}{\partial(u, v)}$, then JJ' is equal to

(a) 0 (b) -1 (c) ∞ (d) 1.

19.8 If u and v be functions of three independent variables x, y, z where (u, v, w) are not independent, then $\dfrac{\partial(u, v, w)}{\partial(x, y, z)}$ is equal to

(a) 1 (b) 0 (c) -1 (d) 2.

19.9 Fill in the blanks :

If $x = r\cos\theta$, $y = r\sin\theta$, then

(i) $\dfrac{\partial(x, y)}{\partial(r, \theta)} =$ ______.

(ii) $\dfrac{\partial(x, y)}{\partial(r, \theta)} \times \dfrac{\partial(r, \theta)}{\partial(x, y)} =$ ______.

(iii) If $u = x^2$, $v = y^2$, then the value of the Jacobian $\dfrac{\partial(u, v)}{\partial(x, y)} =$ ______.

19.10 If $r = f_{xx}(a,b)$, $s = f_{xy}(a,b)$, $t = f_{yy}(a,b)$, then $f(x,y)$ will have a maximum, if

(a) $f_x = 0$, $f_y = 0$, $(rt - s^2) > 0$ and $r < 0$
(b) $f_x = 0$, $f_y = 0$, $(rt - s^2) = 0$ and $r > 0$
(c) $f_x = 0$, $f_y = 0$, $(rt - s^2) > 0$ and $r > 0$
(d) $f_x = 0$, $f_y = 0$, $(rt - s^2) = 0$ and $r < 0$.

19.11 If an error of 1% is made in measuring its length and breadth, the percentage error in the area of a rectangle is

(a) 0.2% (b) 0.02% (c) 2% (d) 1%.

19.12 $\dfrac{\sqrt{x}-\sqrt{y}}{\sqrt{x}+\sqrt{y}}$ is a homogeneous function of degree

(a) $\dfrac{1}{2}$ (b) 1 (c) 0 (d) 2.

19.13 If the variables x, y, z of a function $f(x,y,z)$ are related by functional equations so that each variable can be expressed as a function of the other two, then

$\left(\dfrac{\partial x}{\partial y}\right)_z \left(\dfrac{\partial y}{\partial z}\right)_x \left(\dfrac{\partial z}{\partial x}\right)_y$ is equal to

(a) 0 (b) 1 (c) ∞ (d) -1.

19.14 If $f(x,y) = e^x \log(1+y)$, then f_{xy} is equal to

(a) 0 (b) -1 (c) 0 (d) none of these.

19.15 The point of minimum of the function $f(x,y) = x^3y^2(1-x-y)$ is

(a) $(0,0)$ (b) $(2,-1)$ (c) $\left(-\dfrac{1}{2},\dfrac{1}{3}\right)$ (d) $\left(\dfrac{1}{2},\dfrac{1}{3}\right)$. **[AMIETE(Winter) 1995]**

19.16 The rectangular solid that can be inscribed in a sphere is a

(a) parallelopiped (b) cube (c) a rectangular plate (d) a singular plate.

19.17 If $\displaystyle\int_0^\pi \frac{dx}{a+b\cos x} = \frac{x}{\sqrt{a^2-b^2}}$ $(a > b)$, then $\displaystyle\int_0^\pi \frac{\cos x}{(a+b\cos x)}dx$ is equal to

(a) $\dfrac{ab}{(a^2-b^2)^{\frac{3}{2}}}$ (b) $\dfrac{\pi a}{(a^2-b^2)^{\frac{3}{2}}}$ (c) $\dfrac{\pi b}{(a^2-b^2)^{\frac{3}{2}}}$ (d) $\dfrac{\pi ab}{\sqrt{a^2+b^2}}$.

Answers

19.1 (b) **19.2** (b) **19.3** (b) **19.4** (d) **19.5** (b) **19.6** (c) **19.7** (d) **19.8** (b) **19.9** **(i)** r **(ii)** 1 **(iii)** $4xy$ **19.10** (a) **19.11** (c) **19.12** (c) **19.13** (d) **19.14** (c) **19.15** (d) **19.16** (b) **19.17** (c).

19(B) : Short Answer Type Questions

19.1 If $u = e^{r\cos\theta} \cdot \cos(r\sin\theta)$, find $\dfrac{\partial u}{\partial r}$ and $\dfrac{\partial u}{\partial \theta}$.

[Ans: $e^{r\cos\theta}\cos(r\sin\theta+\theta)$, $-re^{r\cos\theta}\sin(r\sin\theta+\theta)$**]**

19.2 If $z = x^3 + y^3 - 3axy$, then find $\dfrac{\partial^2 z}{\partial x^2}$. **[Ans: $6x$]**

19.3 If $v = \dfrac{1}{t^{\frac{1}{2}}} e^{-\frac{x^2}{4a^2 t}}$, then prove that $\dfrac{\partial v}{\partial t} = a^2 \dfrac{\partial^2 v}{\partial x^2}$.

19.4 If $u = e^{xyz}$, find the value of $\dfrac{\partial^3 u}{\partial x \partial y \partial z}$. **[RU 2005, Gauhati U 1999, AMIE(Summer) 1995]**

[Ans: $e^{xyz}(x^2 + y^2 + z^2 + 3xyz + 1)$]

19.5 If $u = \sin^{-1} \dfrac{x^2 + y^2}{x + y}$, show that $x\dfrac{\partial u}{\partial x} + y\dfrac{\partial u}{\partial y} = \tan u$. **[AMIETE(Winter) 1997]**

19.6 If $u = \tan^{-1}\left(\dfrac{y^2}{x}\right)$, find $x\dfrac{\partial u}{\partial x} + y\dfrac{\partial u}{\partial y}$. **[Ans: $\frac{xy^2}{x^2+y^4}$]**

19.7 Verify Euler's theorem on homogeneous function for $u = \dfrac{x + y + z}{\sqrt{x} + \sqrt{y} + \sqrt{z}}$.

19.8 If $x = r\cos\theta,\ y = r\sin\theta$ show that,

(i) $\dfrac{\partial r}{\partial x} = \dfrac{\partial x}{\partial r}$ (ii) $\dfrac{1}{r}\dfrac{\partial x}{\partial \theta} = r\dfrac{\partial \theta}{\partial x}$ (iii) $\left(\dfrac{\partial r}{\partial x}\right)^2 + \left(\dfrac{\partial r}{\partial y}\right)^2 = 1$. **[BU(H) 2003]**

19.9 Transform the equation, $\dfrac{\partial^2 u}{\partial x^2} + \dfrac{\partial^2 u}{\partial y^2} = 0$ into polar coordinates.

[Ans: $\frac{\partial^2 u}{\partial r^2} + \frac{1}{r}\frac{\partial u}{\partial r} + \frac{1}{r^2}\frac{\partial^2 u}{\partial \theta^2} = 0$]

19.10 If the sides and angles of a triangle ABC vary in such a manner that its circumradius remains constant, prove that, $\dfrac{\partial a}{\cos A} + \dfrac{\partial b}{\cos B} + \dfrac{\partial c}{\cos C} = 0$. **[AMIE(Summer) 1990]**

[**Hint :** Let O be the centre and R the radius of the circumcircle of $\triangle ABC$. OD is the right bisector of BC.

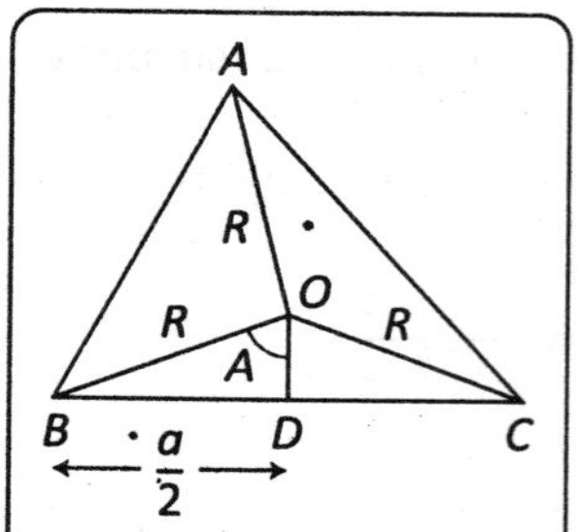

Fig. 19.4

$$\angle BOD = \frac{1}{2}\angle BOC = \frac{1}{2}(2\angle A) = \angle A.$$

Again, $BD = BO\sin A$ or, $\dfrac{a}{2} = 2\sin A$ or, $a = 2R\sin A$.

$$\therefore \quad \partial a = 2R\cos A\,\partial A \quad \text{or,} \quad \partial A = \frac{\partial a}{2R\cos A}.$$

Similarly, $\partial B = \dfrac{\partial b}{2R\cos B}$ and $\partial C = \dfrac{\partial c}{2R\cos C}$.

But, $A + B + C = \pi$,

$$\therefore \quad \partial A + \partial B + \partial C = 0 \quad \text{or,} \quad \frac{\partial a}{2R\cos A} + \frac{\partial b}{2R\cos B} + \frac{\partial c}{2R\cos C} = 0$$

$$\text{or,} \quad \frac{\partial a}{\cos A} + \frac{\partial b}{\cos B} + \frac{\partial c}{\cos C} = 0$$]

19.11 If $v = \dfrac{1}{\sqrt{x^2 + y^2 + z^2}}$, then prove that $\dfrac{\partial^2 v}{\partial x^2} + \dfrac{\partial^2 v}{\partial y^2} + \dfrac{\partial^2 v}{\partial z^2} = 0$. **[VTU 2006]**

19.12 If the three thermodynamic variables P, V, T are connected by a relation $f(P, V, T) = 0$, show that $\left(\frac{\partial P}{\partial T}\right)_V \left(\frac{\partial T}{\partial V}\right)_P \left(\frac{\partial V}{\partial P}\right)_T = -1$.

Solution : We have $f(P, V, T) = 0$, i.e., $P = P(V, T)$ and $T = T(P, V)$.

$$\therefore \quad dP = \left(\frac{\partial P}{\partial V}\right)_T dV + \left(\frac{\partial P}{\partial T}\right)_V dT \quad \text{and} \quad dT = \left(\frac{\partial T}{\partial P}\right)_V dP + \left(\frac{\partial T}{\partial V}\right)_P dV.$$

$$\text{So, } dP = \left(\frac{\partial P}{\partial V}\right)_T dV + \left(\frac{\partial P}{\partial T}\right)_V \left[\left(\frac{\partial T}{\partial P}\right)_V dP + \left(\frac{\partial T}{\partial V}\right)_P dV\right].$$

$\because$ dP and dV are independent of each other; equating the coefficients of dP from both sides of the above equation, we get

$$\left(\frac{\partial P}{\partial T}\right)_V \left(\frac{\partial T}{\partial P}\right)_V = 1 \quad \text{or,} \quad \left(\frac{\partial T}{\partial P}\right)_V = \frac{1}{\left(\frac{\partial P}{\partial T}\right)_V} \qquad \text{(i)}$$

Similarly, equating the coefficients of dV from both sides, we get

$$\left(\frac{\partial P}{\partial V}\right)_T + \left(\frac{\partial P}{\partial T}\right)_V \left(\frac{\partial T}{\partial V}\right)_P = 0$$

$$\text{or, } \left(\frac{\partial V}{\partial T}\right)_P \left(\frac{\partial T}{\partial P}\right)_V \left(\frac{\partial P}{\partial V}\right)_T + 1 = 0 \qquad \text{(ii)}$$

From Eqns. (i) and (ii), we get

$$\left(\frac{\partial P}{\partial T}\right)_V \left(\frac{\partial T}{\partial V}\right)_P \left(\frac{\partial V}{\partial P}\right)_T = -1 \quad \textbf{(Proved)}.$$

19.13 If $u = \frac{y - x}{1 + xy}$ and $v = \tan^{-1} y - \tan^{-1} x$, find $\frac{\partial(u, v)}{\partial(x, y)}$. **[Ans:** 0**]**

19.14 If $u = x^2 + y^2 + z^2$, $v = xyz$, find $\frac{\partial x}{\partial u}$. **[Ans:** $\frac{x}{2(2x^2 - y^2)}$**]**

19.15 Expand $f(x, y)$, compute $f(0.9, -1.2)$ approximately. **[Ans:** -0.8232**]**

19.16 If $PV^2 = k$ and the relative errors in P and V are respectively 0.05 and 0.025, show that the error in k is 10%. **[M.U. 1999]**

19.17 Find the stationary points of the function $f(x, y, z) = x^3 y^2 (1 - x - y)$.
[AMIE(Winter) 1995] **[Ans:** $\left(\frac{1}{2}, \frac{1}{3}\right)$, Maxima**]**

19.18 Using Lagrange's method (of multiplier), find critical (stationary values) of the function $f(x, y, z) = x^2 + y^2 + z^2$, given that, $z^2 = xy + 1$.
[AMIE(Winter) 1999] **[Ans:** $(0, 0, -1)$, $(0, 0, 1)$**]**

19(C) : Broad Answer Type Questions

19.1 (i) What is 'partial derivatives'. Give its geometrical significance.

(ii) If $u(x, t) = ae^{-gx} \sin(nt - gx)$, satisfy the equation $\frac{\partial u}{\partial x} = a^2 \frac{\partial^2 u}{\partial x^2}$, show that $ag = \sqrt{\frac{n}{2}}$.

(iii) If $z = yf(x^2 - y^2)$, show that, $y\frac{\partial z}{\partial x} + x\frac{\partial z}{\partial y} + \frac{xz}{y}$.

(iv) If $x^x y^y z^z = c$, show that at $x = y = z$, $\frac{\partial^2 z}{\partial x \partial y} = -(x \log_e x)^{-1}$.

[Pune U (May) 1986]

(v) If $z(x+y) = x^2 + y^2$, show that $\left(\frac{\partial z}{\partial x} - \frac{\partial z}{\partial y}\right)^2 = 4\left(1 - \frac{\partial z}{\partial x} - \frac{\partial z}{\partial y}\right)$. **[VTU 2003]**

(vi) If $r^2 = x^2 + y^2 + z^2$ and $V = r^m$, prove that, $V_{xx} + V_{yy} + V_{zz} = m(m+1)r^{m-2}$.

[Raipur U 2005]

(vii) If $v = x^y \cdot y^x$, prove that $x\frac{\partial v}{\partial x} + y\frac{\partial v}{\partial y} = v(x + y + \log v)$. **[Anna U 2005]**

19.2 (i) Define homogeneous function of two variables of order n. Establish Euler's theorem in this regard.

(ii) If $u = \log(\tan x + \tan y)$, prove that $\sin 2x\frac{\partial u}{\partial x} + \sin 2y\frac{\partial u}{\partial y} = 2$. **[Gujarat U 1991]**

(iii) If $u = \sin^{-1}\left[\frac{x + 2y + 3z}{x^8 + y^8 + z^8}\right]$, find the value of $x\frac{\partial u}{\partial x} + y\frac{\partial u}{\partial y} + z\frac{\partial u}{\partial z}$.

[UPTU 2004] **[Ans: $-7 \tan u$]**

(iv) Verify Euler's theorem for the function $u = \sin^{-1}\frac{x}{y} + \tan^{-1}\frac{y}{x}$.

[AMIE(Winter) 1993]

(v) If $f(x,y) = \frac{1}{x^2} + \frac{1}{xy} + \frac{\log x - \log y}{x^2 + y^2}$, prove that $x\frac{\partial f}{\partial x} + y\frac{\partial f}{\partial y} + 2f = 0$.

(vi) If z be a homogeneous function of degree n, show that

(a) $x\frac{\partial^2 z}{\partial x^2} + y\frac{\partial^2 z}{\partial x \partial y} = (n-1)\frac{\partial z}{\partial x}$

(b) $x\frac{\partial^2 z}{\partial x \partial y} + y\frac{\partial^2 z}{\partial y^2} = (n-1)\frac{\partial z}{\partial y}$

(c) $x^2\frac{\partial^2 z}{\partial x^2} + 2xy\frac{\partial^2 z}{\partial x \partial y} + y^2\frac{\partial^2 z}{\partial y^2} = n(n-1)z$. **[VTU 2007, UPTU 2006]**

(vii) If $g(u) = n\frac{f(u)}{f'(u)}$, then show that, $x^2\frac{\partial^2 u}{\partial x^2} + 2xy\frac{\partial^2 u}{\partial x \partial y} + y^2\frac{\partial^2 u}{\partial y^2} = g(u)[g'(u) - 1]$.

(viii) (a) If $z = f(x,y)$ show that the total differential of z is given by the following relation

$$dz = \frac{\partial f}{\partial x}dx + \frac{\partial f}{\partial y}dy.$$

(b) If $f(x,y) = 0$, show that $\frac{dy}{dx} = -\frac{\frac{\partial f}{\partial x}}{\frac{\partial f}{\partial y}}$.

(ix) (a) If u is a homogeneous function of nth degree in x, y, z, prove that

$$x\frac{\partial u}{\partial x} + y\frac{\partial u}{\partial y} + z\frac{\partial u}{\partial z} = nu.$$

(b) If $u = \frac{x}{y+z} + \frac{y}{z+x} + \frac{z}{x+y}$, show that $x\frac{\partial u}{\partial x} + y\frac{\partial u}{\partial y} + z\frac{\partial u}{\partial z} = 0$.

[VTU 2008S]

(x) Verify Euler's theorem, when

(a) $f(x, y) = ax^2 + 2hxy + by^2$

(b) $f(x, y) = \frac{x^2(x^2 - y^2)^3}{(x^2 + y^2)^3}$ [JNTU 1999]

(c) $f(x, y) = 3x^2yz + 5xy^2z + 4z^4$.

(xi) If $u = \tan^{-1}\frac{x^3 + y^3}{x + y}$, then find the value of $x^2\frac{\partial^2 u}{\partial x^2} + 2xy\frac{\partial^2 u}{\partial x \partial y} + y^2\frac{\partial^2 u}{\partial y^2}$.

[Ans: $\sin 2u(2\cos 2u - 1)$]

19.3 (i) If $x = e^u \cos v$, $y = e^u \sin v$, find $J = \frac{\partial(u, v)}{\partial(x, y)}$, $J = \frac{\partial(x, y)}{\partial(u, v)}$. Hence, show that $JJ' = 1$. [VTU 2005]

(ii) If $F = xu + v - y$, $G = u^2 + uv + w$, $H = zu - v + uw$. Compute $\frac{\partial(F, G, H)}{\partial(u, v, w)}$.

[Ans: $x(yv + 1 - w) + z - 2uv$]

(iii) If $u = xyz$, $v = x^2 + y^2 + z^2$, $w = x + y + z$, find $J = \frac{\partial(x, y, z)}{\partial(u, v, w)}$.

[Pune U 1990] [Ans: $-2(x - y)(y - z)(z - x)$]

(iv) Verify that $u = \frac{x - y}{x + y}$, $v = \frac{x + y}{x}$ are functionally dependent and if so, find the relation between them. [Ans: $u = \frac{2-v}{v}$]

(v) Are $x + y - z$, $x - y + z$, $x^2 + y^2 + z^2 - 2yz$ functionally dependent? If so, find a relation between them. [Ans: $2w = u(3v - u^2)$]

(vi) Use Jacobian to find $\left(\frac{\partial u}{\partial x}\right)$, if $u^2 + xv^2 - xy = 0$ and $u^2 + xyv + v^2 = 0$.

[Ans: $\frac{xyv^2 - 2v^3 + xy^2 + 2yv}{2uxy + 4uv - 4xuv}$]

(vii) If $u^2 + xv^2 = x + y$, $v^2 + yu^2 = x - y$, find $\frac{\partial u}{\partial x}, \frac{\partial v}{\partial y}$. [Ans: $\frac{1-x-v^2}{2u(1-xy)}, \frac{1+y+u^2}{-2u(1-xy)}$]

19.4 (i) State and prove Taylor's theorem for a function of two variables.

(ii) Expand e^{xy} at $(1, 1)$ up to three terms.

[Ans: $e\left[1 + (x - 1) + (y - 1) + \frac{1}{2!}\left\{(x - 1)^2 + 4(x - 1)(y - 1) + (y - 1)^2\right\}\right]$]

(iii) Expand the following functions in powers of x and y as far as terms of third degree : (i) $\sin x \sin y$ (ii) $e^x \sin y$ [BU(H) 2003, Kottayam U 2005]

19.5 (i) Establish the condition for a function $f(x, y)$ to be extremum. [Use Taylor's theorem]

(ii) Find the extremum value of the following functions :

(a) $f(x, y) = x^2 + y^2 + xy + x - 4xy + 5$ [AMIETE(Winter) 1996]

[Ans: Minimum value of $f(x, y)$ at $(-2, 3)$ is -2]

(b) $f(x,y) = \sin x \sin y \sin(x+y)$ **[Ans:** Max at $(\frac{\pi}{3}, \frac{\pi}{3})$, Min at $(\frac{2\pi}{3}, \frac{2\pi}{3})$]

(c) $f(x,y) = 2(x^2 - y^2) - x^4 + y^4$

[Osmania U 2003] **[Ans:** Max at $(\pm 1, 0)$, Min at $(0, \pm 1)$]

(d) $f(x,y) = x^2y^2 - 5x^2 - 8xy - 5y^2$

[Ans: Max at $(0,0)$; $(-1,1)$, $(3,3)$, $(-3,-3)$ are saddle points]

(e) Find the critical points of the following functions and test for minima and maxima : **(i)** $z = 1 + x^2 + y^2$ **(ii)** $z = 2x^2 - xy - 3y^2 - 3x + 7y$

[BU(H) 1994] **[Ans:** (i) Minimum at $(0,0)$, (ii) Saddle point $(1,1)$]

(f) Test the following function for maximum and minimum :

$$z = x^2 - xy + y^2 + 3x - 2y + 1.$$

Find also the maximum or minimum values.

[BU(H) 1992] **[Ans:** Max at $(-\frac{4}{3}, -\frac{2}{3})$; $\frac{13}{9}$]

(g) Without using any table or calculator find an approximate value of

$$\frac{\tan^{-1}(1.02)}{3.05}.$$ **[BU(H)(Old) 1995]** **[Ans:** $(45.06°)$]

19.6 **(a)** Discuss Leibnitz rule for differentiation under integral sign.

(b) **(i)** Show that $\displaystyle\int_0^\infty e^{-ax} \cos bx \, dx = \frac{a}{a^2 + b^2}$.

Using this result, evaluate the integral $\displaystyle\int_0^\infty e^{-ax} \sin bx \; x \, dx$.

(ii) Find the value of the integral, $\displaystyle\int_0^\infty \frac{e^{-x} - e^{-\lambda x}}{x} dx$, $\lambda > 0$.

[BU(H)(Old) 1995] **[Ans:** $\frac{1}{\lambda}$]

19.7 **(a)** Establish Lagrange's method of undetermined multiplier. Indicate its field of application.

(b) **(i)** A rectangular box open at the top is to be constructed having a volume 108 cubic feet. Find the dimensions of the box requiring least material for its construction. **[BU(H) 1995]** **[Ans:** 6 ft, 6 ft, 12 ft]

(ii) Divide 24 into three parts such that the continued of the first, the square of the second and the cube of the third may be a maximum. **[Ans:** 4, 8, 12]

(iii) A tapering log has a square cross section whose side varies uniformly and is equal to a at the top and b $\left(b > \dfrac{3a}{2}\right)$ at the bottom. Show that the volume of the greatest circular section that can be obtained from the log is $\dfrac{\pi b^3 l}{27(b-a)}$, where l is the length of the rod.

(iv) The beds of two rivers approximately represent a parabola $y = x^2$ and a straight line $y = x - 2$. It is required to connect these rivers by a straight canal of least length. Use the method of Lagrange multiplier, to find the points through which it passes. What is the least length?

[AMIETE(Winter) 1990] **[Ans:** $\frac{7\sqrt{2}}{8}$]

(v) A pyramid with vertex at $(0,0,0)$ is inscribed inside an ellipsoid $\frac{x^2}{a^2}+\frac{y^2}{b^2}+\frac{z^2}{c^2}=1$. Its base is parallel to XY-plane and the sides of the base are parallel to the principal axes of the elliptic section. Find the dimensions and the volume of the pyramid, when the volume is maximum.

[AMIETE(Winter) 1988] [**Ans:** $x=\frac{2a}{3}$, $y=\frac{2b}{3}$, $z=\frac{c}{3}$, volume $\frac{64}{81}abc$]

(vi) Find the maximum and minimum distances of the point $(3,4,12)$ from the sphere $x^2+y^2+z^2=1$.

[PTU 2005] [**Ans:** Max. dist. $=14$ and Min. dist. $=12$]

(vii) Find the dimensions of the rectangular box, open at the top of maximum capacity whose surface is 432 sq. cm.

[Madras U 2000S] [**Ans:** $12\times 12\times 6$ cm]

(viii) An open tank with a square base and vertical sides to be constructed from a metal sheet to hold a given quantity of water. Show that the cost of the material will be the least when the depth is half the width. **[BU(H) 2002]**

CHAPTER 20

Partial Differential Equations

20.1 Introduction

Partial differential equations find very important applications in physics and engineering in the domain of solid mechanics, fluid dynamics, electricity, quantum mechanics, heat flow, etc. In fact, most of the phenomena in mathematical physics and engineering science can be described in terms of partial differential equations. Though simplification may reduce the partial differential equation to ordinary differential equation, still for the complete description of the system partial differential equation is essential.

20.2 Definitions

A **partial differential equation (PDE)** is an equation **which contains partial derivatives** or differentials and **as such involves two or more independent variables.** This means, if z is a function of two independent variables x and y, then $z = f(x, y)$ and the partial derivatives are

$$p = \frac{\partial z}{\partial x}, \quad q = \frac{\partial z}{\partial y}, \quad r = \frac{\partial^2 z}{\partial x^2}, \quad s = \frac{\partial^2 z}{\partial x \partial y} \quad \text{and} \quad t = \frac{\partial^2 z}{\partial y^2}.$$

Order of an equation : The order of the highest derivative of the dependent variable in a partial differential equation is called the **order** of the equation.

Linear equation : A PDE is said to be **linear** if all derivatives appearing in it are of first degree.

A linear PDE is said to be **homogeneous** if each term of it contains either the dependent variable or one of its derivatives, otherwise it is **inhomogeneous**.

Example *If $z = f(x, y)$ where x and y are independent variables, then of the PDEs,*

(i) $\dfrac{\partial z}{\partial x} + y\dfrac{\partial z}{\partial y} = z,$

(ii) $\dfrac{\partial^2 z}{\partial x^2} + z\dfrac{\partial^2 z}{\partial y^2} = 0,$

(iii) $\dfrac{\partial^2 z}{\partial x^2} + 2\dfrac{\partial^2 z}{\partial x \partial y} + \dfrac{\partial^2 z}{\partial y^2} = e^z.$

(i) is of first, (ii) and (iii) are 2nd order partial differential equations.

PDEs (i) and (ii) are linear and homogeneous but (iii) is linear but inhomogeneous because it contains z and its partial derivatives in first degree and the product of z and its derivatives are absent.

20.3 Method of Forming Partial Differential Equations

Partial differential equations can be formed in any of the following methods :

(i) By the elimination of arbitrary constants

Let $f(x, y, z, a, b) = 0$ (20.1)

represent a family of surfaces where a and b are arbitrary constants. Here, f is a differentiable function of x and y, and z is dependent on x, y.

To eliminate a, b from Eqn. (20.1) we require two more equations containing them. These are obtained as follows :

Differentiating eqn. (20.1) w.r.t. x and y, we get

$$\frac{\partial f}{\partial x} + \frac{\partial f}{\partial y} \cdot p = 0 \tag{20.2}$$

$$\frac{\partial f}{\partial y} + \frac{\partial f}{\partial x} \cdot q = 0. \tag{20.3}$$

Eliminating a, b from eqns. (20.1), (20.2) and (20.3), we get the general relation of the type

$$F(x, y, z, p, q) = 0. \tag{20.4}$$

Eqn. (20.4) is a first order PDE. The equation formed will be of higher order if the number of constants be greater than the number of variables.

Example 20.1. *Form a partial differential equation from*

$$x^2 + y^2 + (z-c)^2 = a^2. \tag{i}$$

Solution : Given $x^2 + y^2 + (z-c)^2 = a^2$.

Eqn. (i) contains two arbitrary constants a and c.

Differentiating (i) partially w.r.t. x,

$$2x + 2(z-c)\frac{\partial z}{\partial x} = 0 \quad \text{or,} \quad x + (z-c)p = 0. \tag{ii}$$

Differentiating Eqn. (i) partially w.r.t. y,

$$2y + 2(z-c)\frac{\partial z}{\partial y} = 0 \quad \text{or,} \quad y + (z-c)\frac{\partial z}{\partial y} = 0. \tag{iii}$$

Now from Eqn. (ii), $z - c = -\dfrac{x}{p}$.

$\therefore$ from Eqn. (iii), we get

$$y - \frac{x}{p}q = 0 \quad \text{or,} \quad yp - xq = 0.$$

(ii) By the method of elimination of arbitrary functions

Example 20.2. *Form the partial differential equation from*

$$z = \phi(x+cy) + \psi(x-cy). \tag{i}$$

Solution : Differentiating twice the given relation partially w.r.t. x and y, we get

$$\frac{\partial^2 z}{\partial x^2} = \phi''(x+cy) + \psi''(x-cy) \tag{ii}$$

and $$\frac{\partial^2 z}{\partial y^2} = c^2\{\phi''(x+cy) + \psi''(x-cy)\}. \tag{iii}$$

Eliminating ϕ and ψ from Eqns. (ii) and (iii), we get

$$\frac{\partial^2 z}{\partial y^2} = c^2 \frac{\partial^2 z}{\partial x^2}.$$

This is a linear PDE of second order.

20.4 Solution of PDEs by Direct Integration

Example 20.3. *Solve* : $\dfrac{\partial^2 z}{\partial x \partial y} = x^2 y$ subject to the condition $z(x, 0) = x^2$ and $z(1, y) = \cos y$.

Solution : Integrating the given PDE w.r.t. x, we get

$$\frac{\partial z}{\partial y} = \frac{x^3}{3} y + f(y).$$

Integrating w.r.t. y, we get

$$z = \frac{x^3}{3} \frac{y^2}{2} + \int f(y) dy + g(x)$$

$$\text{or,} \quad z = \frac{x^3 y^2}{6} + F(x, y) + g(x), \quad \left[\text{where } F(x, y) = \int f(y) dy\right] \tag{i}$$

Putting $z = x^2$ and $y = 0$ in Eqn. (i), we obtain

$$x^2 = 0 + F(0) + g(x).$$

Putting the value of $g(x)$ in Eqn. (i)

$$z = \frac{x^3 y^2}{6} + F(y) + x^2 - F(0). \tag{ii}$$

Again, since $z(1, y) = \cos y$, putting $x = 1$ and $z = \cos y$ in Eqn. (ii), we get

$$\cos y = \frac{y^3}{6} + F(y) + 1 - F(0).$$

Putting the value of $F(y)$ in Eqn. (ii), we get

$$z = \frac{1}{6} x^3 y^2 + \cos y - \frac{1}{6} y^2 - 1 + F(0) + x^2 - F(0)$$

$$\text{or,} \quad \mathbf{z = \frac{1}{6} x^3 y^2 + \cos y - \frac{1}{6} y^2 - 1 + x^2.}$$

20.5 First Order Linear PDE : Lagrange's Equation

An equation of the type

$$Pp + Qq = R, \tag{20.4}$$

where P, Q and R are the functions of x, y, z and $p = \frac{\partial z}{\partial x}$, $q = \frac{\partial z}{\partial y}$ are known as the **first order linear PDE** or **Lagrange's equation**.

Solution : $Pp + Qq = R$ (1)

Such equations are obtained by eliminating an arbitrary function f from

$$f(u, v) = 0. \tag{20.5}$$

Differentiating equation (2) partially w.r.t. x, y,

$$\frac{\partial f}{\partial u}\left(\frac{\partial u}{\partial x}+\frac{\partial u}{\partial z}\frac{\partial z}{\partial x}\right)+\frac{\partial f}{\partial v}\left(\frac{\partial v}{\partial x}+\frac{\partial v}{\partial z}\frac{\partial z}{\partial x}\right)=0 \tag{20.6}$$

and $$\frac{\partial f}{\partial u}\left(\frac{\partial u}{\partial y}+\frac{\partial u}{\partial z}\frac{\partial z}{\partial y}\right)+\frac{\partial f}{\partial v}\left(\frac{\partial v}{\partial y}+\frac{\partial v}{\partial z}\frac{\partial z}{\partial y}\right)=0. \tag{20.7}$$

We now eliminate $\frac{\partial f}{\partial u}$ and $\frac{\partial f}{\partial v}$ from Eqns. (20.6) and (20.7).

From (20.6), $$\frac{\partial f}{\partial u}\left[\frac{\partial u}{\partial x}+\frac{\partial u}{\partial z}p\right]=-\frac{\partial f}{\partial v}\left[\frac{\partial v}{\partial x}+\frac{\partial v}{\partial z}p\right] \tag{20.8}$$

From (20.7), $$\frac{\partial f}{\partial u}\left[\frac{\partial u}{\partial y}+\frac{\partial u}{\partial z}q\right]=-\frac{\partial f}{\partial v}\left[\frac{\partial v}{\partial y}+\frac{\partial v}{\partial z}q\right]. \tag{20.9}$$

Dividing Eqn. (20.9) by (20.8), we get

$$\frac{\left[\frac{\partial u}{\partial x}+\frac{\partial u}{\partial z}p\right]}{\left[\frac{\partial u}{\partial y}+\frac{\partial u}{\partial z}q\right]}=\frac{\left[\frac{\partial v}{\partial x}+\frac{\partial v}{\partial z}p\right]}{\left[\frac{\partial v}{\partial y}+\frac{\partial v}{\partial z}q\right]} \quad \text{or,} \quad \left[\frac{\partial u}{\partial x}+\frac{\partial u}{\partial z}p\right]\left[\frac{\partial v}{\partial y}+\frac{\partial v}{\partial z}q\right]=\left[\frac{\partial u}{\partial y}+\frac{\partial u}{\partial z}q\right]\left[\frac{\partial v}{\partial x}+\frac{\partial v}{\partial z}p\right]$$

or, $$\frac{\partial u}{\partial x}\times\frac{\partial v}{\partial y}+\frac{\partial u}{\partial x}\times\frac{\partial v}{\partial z}q+\frac{\partial u}{\partial z}\times p\times\frac{\partial v}{\partial y}+\frac{\partial u}{\partial z}\times\frac{\partial v}{\partial z}pq$$
$$=\frac{\partial u}{\partial y}\times\frac{\partial v}{\partial x}+\frac{\partial u}{\partial y}\times\frac{\partial v}{\partial z}p+\frac{\partial u}{\partial z}\cdot q+\frac{\partial v}{\partial x}+\frac{\partial u}{\partial z}\times\frac{\partial v}{\partial z}\cdot pq$$

or, $$\left[\frac{\partial u}{\partial y}\times\frac{\partial v}{\partial z}-\frac{\partial u}{\partial z}\times\frac{\partial v}{\partial y}\right]p+\left[\frac{\partial u}{\partial z}\times\frac{\partial v}{\partial x}-\frac{\partial u}{\partial x}\times\frac{\partial v}{\partial z}\right]q=\frac{\partial u}{\partial x}\times\frac{\partial v}{\partial y}-\frac{\partial u}{\partial y}\times\frac{\partial v}{\partial x}. \tag{20.10}$$

Equating the coefficients of p and q in Eqn. (20.4) and (20.10), we get

$$\left.\begin{aligned}P&=\frac{\partial u}{\partial y}\times\frac{\partial v}{\partial z}-\frac{\partial u}{\partial z}\times\frac{\partial v}{\partial y}\\Q&=\frac{\partial u}{\partial z}\times\frac{\partial v}{\partial x}-\frac{\partial u}{\partial x}\times\frac{\partial v}{\partial z}\\R&=\frac{\partial u}{\partial x}\times\frac{\partial v}{\partial y}-\frac{\partial u}{\partial y}\times\frac{\partial v}{\partial x}\end{aligned}\right\} \tag{20.11}$$

Let $u=c_1$ and $v=c_2$ be two solutions and a, b are constants.

Differentiating $u=c_1$ and $v=c_2$,

$$\frac{\partial u}{\partial x}dx+\frac{\partial u}{\partial y}dy+\frac{\partial u}{\partial z}dz=0 \tag{20.12}$$

and $$\frac{\partial v}{\partial x}dx+\frac{\partial v}{\partial y}dy+\frac{\partial v}{\partial z}dz=0. \tag{20.13}$$

Solving Eqns. (20.12) and (20.13), we get

$$\frac{dx}{\frac{\partial u}{\partial y}\times\frac{\partial v}{\partial z}-\frac{\partial u}{\partial z}\times\frac{\partial v}{\partial y}}=\frac{dy}{\frac{\partial u}{\partial z}\times\frac{\partial v}{\partial x}-\frac{\partial u}{\partial x}\times\frac{\partial v}{\partial z}}=\frac{dz}{\frac{\partial u}{\partial x}\times\frac{\partial v}{\partial y}-\frac{\partial u}{\partial y}\times\frac{\partial v}{\partial x}}. \tag{20.14}$$

From Eqns. (20.11), $\dfrac{dx}{P}=\dfrac{dy}{Q}=\dfrac{dz}{R}$.

Solutions of these equations are : $u = c_1$ and $v = c_2$.

$\therefore$ **$f(u, v)$ is the desired solution of (20.4).**

Working rule : To solve the equation $Pp + Qq = R$,

(i) form the subsidiary equations $\dfrac{dx}{P} = \dfrac{dy}{Q} = \dfrac{dz}{R}$

(ii) solve the above simultaneous equations. Let two solutions be $u = c_1$ and $v = c_2$

(iii) write the complete solution as $f(u, v)$ or, $u = \phi(v)$.

Example 20.4. *Solve :* $(x^2 - yz)p + (y^2 - zx)q = z^2 - xy$. **[AMIE(Summer) 2001]**

Solution : $(x^2 - yz)p + (y^2 - zx)q = z^2 - xy$ (i)

The auxiliary equations are

$$\frac{dx}{x^2 - yz} = \frac{dy}{y^2 - zx} = \frac{dy}{z^2 - xy}$$

or, $$\frac{dx - dy}{x^2 - yz - y^2 + zx} = \frac{dy - dz}{y^2 - zx - z^2 + xy} = \frac{dz - dx}{z^2 - xy - x^2 + yz}$$

or, $$\frac{dx - dy}{(x - y)(x + y + z)} = \frac{dy - dz}{(x + y + z)(y - z)} = \frac{dz - dx}{(x + y + z)(z - x)}$$

or, $$\frac{dx - dy}{x - y} = \frac{dy - dz}{y - z} = \frac{dz - dx}{z - x}. \qquad \text{(ii)}$$

Integrating first two equations of (ii), we get

$$\log(x - y) = \log(y - z) + \log c_1 \quad \text{or,} \quad \log\frac{x - y}{y - z} = \log c_1 \quad \text{or,} \quad \frac{x - y}{y - z} = c_1.$$

Similarly, from the last two equations of (ii), $\dfrac{y - z}{z - x} = c_2$.

$\therefore$ the required solution is $\boldsymbol{f\left[\dfrac{x - y}{y - z}, \dfrac{y - z}{z - x}\right] = 0}$.

20.6 Method of Multipliers

Consider the auxiliary equation $\dfrac{dx}{P} = \dfrac{dy}{Q} = \dfrac{dz}{R}$.

If l, m and n be constants and functions of x, y, z, then

$$\frac{dx}{P} = \frac{dy}{Q} = \frac{dz}{R} = \frac{l\,dx + m\,dy + n\,dz}{lP + mQ + nR}. \qquad (20.15)$$

l, m and n are so chosen that,

$$lP + mQ + nR = 0. \qquad (20.16)$$

Thus, $l\,dx + m\,dy + n\,dz = 0$.

Solve this differential equation if the solution is $u = c_1$.

Similarly, choosing another set of multipliers (l_1, m_1, n_1) and if the second solution is $v = c_2$, then the required solution is $\boldsymbol{f(u, v) = 0}$.

Example 20.5. *Solve* : $x^2\dfrac{\partial z}{\partial x}+y^2\dfrac{\partial z}{\partial y}=(x+y)z.$

Solution : $x^2\dfrac{\partial z}{\partial x}+y^2\dfrac{\partial z}{\partial y}=(x+y).$ (i)

The auxiliary equation of (i) is $\dfrac{dx}{x^2}=\dfrac{dy}{y^2}=\dfrac{dz}{(x+y)z}.$ (ii)

Integrating the first two members of (ii),

$$-\frac{1}{x}=-\frac{1}{y}+c \quad \text{or,} \quad \frac{1}{x}-\frac{1}{y}=c_1. \tag{iii}$$

Eqn. (ii) can be written as

$$\frac{\frac{dx}{x}}{x}=\frac{\frac{dy}{y}}{y}-\frac{\frac{dz}{z}}{z}=\frac{\frac{dx}{x}+\frac{dy}{y}+\frac{dz}{z}}{(x+y)-(x-y)} \quad \text{or,} \quad \frac{dx}{x}+\frac{dy}{y}-\frac{dz}{z}=0.$$

On integrating,

$$\log x+\log y-\log z=\log c_2 \quad \text{or,} \quad \log\frac{xy}{z}=\log c_2 \quad \text{or,} \quad \frac{xy}{z}=c_2. \tag{iv}$$

From (iii) and (iv), $\boldsymbol{f\left[\frac{1}{x}-\frac{1}{y},\frac{xy}{z}\right]=0}.$

20.7 Non-linear Partial Differential Equations of the First Order

The partial differential equations in which p and q occur other than in the first degree are called **non-linear partial differential equations of the first order**. The methods of solving such equations in certain standard form will be discussed.

Form I : Equations of the type $f(p,q)=0$, i.e., equations containing p and q only.

Method : Let the required solution of the equation be $z=ax+by+c.$ (20.17)

$$\therefore \quad \frac{dz}{dx}=a, \quad \frac{dz}{dy}=b.$$

Putting these values in $f(p,q)=0$, we get $f(a,b)=0$. Finding the value of b in terms of a and substituting the value of b in equation (20.17), we get the required solution.

Example 20.6. *Solve* : $\sqrt{p}+\sqrt{q}=1.$ (i)

Solution : We have $\sqrt{p}+\sqrt{q}=1.$

The complete solution is $z=ax+by+c,$ (ii)

where $\sqrt{a}+\sqrt{b}=1$ or, $b=(1-\sqrt{a})^2.$

$\therefore$ Eqn. (ii) becomes, $\boldsymbol{z=ax+(1-\sqrt{a})^2y+c}.$

This is the desired solution of Eqn. (i).

Example 20.7. *Solve* : $x^2p^2+y^2q^2=z^2.$ **[Kerala U 2005, Madras U 2003]**

Solution : The given equation can be reduced to

$$\frac{x^2}{z^2}p^2+\frac{y^2}{z^2}q^2=1 \quad \text{or,} \quad \left(\frac{x}{z}\frac{\partial z}{\partial x}\right)^2+\left(\frac{y}{z}\frac{\partial z}{\partial y}\right)^2=1$$

or, $$\left[\frac{\left(\frac{\partial z}{z}\right)}{\left(\frac{\partial x}{x}\right)}\right]^2 + \left[\frac{\left(\frac{\partial z}{z}\right)}{\left(\frac{\partial y}{y}\right)}\right]^2 = 1. \quad \text{(i)}$$

Now let $\frac{\partial z}{z} = \partial Z$, $\frac{\partial x}{x} = \partial X$, $\frac{\partial y}{y} = \partial Y$, then

$$\log z = Z, \quad \log x = X, \quad \log y = Y.$$

$\therefore$ Eqn. (i) can be written as $\left(\frac{\partial Z}{\partial X}\right)^2 + \left(\frac{\partial Z}{\partial Y}\right)^2 = 1$

or, $$P^2 + Q^2 = 1. \quad \text{(ii)}$$

Let the required solution of Eqn. (ii) be

$$Z = aX + bY + c, \quad P = \frac{\partial Z}{\partial X} = a, \quad Q = \frac{\partial Z}{\partial Y} = b.$$

From Eqn. (ii), $a^2 + b^2 = 1$ or, $b = \sqrt{1 - a^2}$.

$\therefore$ $Z = aX + \sqrt{1 - a^2}Y + c$ or, $\boldsymbol{\log z = a \log x + \sqrt{1 - a^2} \log y + c}$.

Form II : Equation of the type $f(z, p, q) = 0$, i.e., equations not containing x and y.

Example 20.8. *Solve* : $q^2 = z^2p^2(1 - p^2)$. **[JNTU 2005]**

Solution : Let $u = y + ax$ and $z = f(u)$, then

$$p = \frac{\partial z}{\partial x} = \frac{dz}{du} \cdot \frac{\partial u}{\partial x} = a\frac{dz}{du} \quad \text{and} \quad q = \frac{dz}{du} \cdot \frac{\partial u}{\partial y} = \frac{dz}{du}.$$

The given equation becomes

$$\left(\frac{\partial u}{\partial z}\right)^2 = a^2z^2\left(\frac{dz}{du}\right)^2\left\{1 - a^2\left(\frac{dz}{du}\right)^2\right\} \quad \text{(i)}$$

or, $$a^4z^2\left(\frac{dz}{du}\right)^2 = a^2z^2 - 1 \quad \text{or,} \quad \frac{dz}{du} = \frac{\sqrt{a^2z^2 - 1}}{a^2z}.$$

Integrating, we get

$$\int \frac{a^2z}{\sqrt{a^2z^2 - 1}}dz = \int du + c \quad \text{or,} \quad \sqrt{(a^2z^2 - 1)} = u + c$$

or, $\boldsymbol{a^2z^2 = (y + ax + c)^2 + 1}$ $[\because u = z + ax]$.

Another form of Eqn. (i) is $\frac{dz}{du} = 0$ whose solution is $z = c$.

Form III : Equation of the type $f(x, p) = F(y, q)$, i.e., equation in which z is absent and the terms containing x and p can be separated from those containing y and q.

As a trial solution we assume that $f(x, p) = F(y, q) = a$ (say)

Then solving for p we get, $p = \phi(x)$ and solving for q we get, $q = \psi(y)$

$$\because dz = \frac{\partial z}{\partial x}dx + \frac{\partial z}{\partial y}dz = pdx + qdy \quad \therefore dz = \phi(x)dx + \psi(y)dy$$

Integrating $z = \int \phi(x)dx + \int \psi(y)dy + b$ is the complete solution containing two constants a and b.

Example 20.9. *Solve* : $p^2 + q^2 = z^2(x+y)$.

Solution : The given equation can be written as, $\left(\frac{p}{z}\right)^2 + \left(\frac{q}{z}\right)^2 = (x+y)$

$$\text{or,}\quad \left(\frac{1}{z}\frac{\partial z}{\partial x}\right)^2 + \left(\frac{1}{z}\frac{\partial z}{\partial y}\right)^2 = x+y \quad \text{or,}\quad \left[\frac{\left(\frac{\partial z}{z}\right)}{\partial x}\right]^2 + \left[\frac{\left(\frac{\partial z}{z}\right)}{\partial y}\right]^2 = x+y$$

$$\text{or,}\quad \left(\frac{\partial Z}{\partial x}\right)^2 + \left(\frac{\partial Z}{\partial y}\right)^2 = (x+y), \quad \text{where } \frac{\partial z}{z} = \partial Z \text{ or, } \log z = Z$$

or, $P^2 + Q^2 = x + y$ or, $p^2 - x = y - Q^2 = a$ (say).

$\therefore$ $p^2 - x = a$ or, $P = \sqrt{a+x}$

$y - Q^2 = a$ or, $Q = \sqrt{y-a}$.

$\therefore$ from the equation $dZ = P\,dx + Q\,dy$, we get

$$dZ = \sqrt{a+x}dx + \sqrt{y-a}dy \quad \text{or,}\quad Z = \int \sqrt{(a+x)}dx + \int \sqrt{(y-a)}dy + c$$

$$\text{or,}\quad \log z = \frac{2}{3}(a+x)^{\frac{3}{2}} + \frac{2}{3}(y-a)^{\frac{3}{2}} + c.$$

Form IV : $z = px + qy + f(p,q)$: **This is analogous of Clairaut's equation.**

The complete solution of the above equation is

$$Z = px + qy + f(p,q), \quad \text{where } f(p,q) = \sqrt{1+p^2+q^2}.$$

$\therefore$ the complete solution is $Z = ax + by + \sqrt{1+p^2+q^2}$.

20.8 Charpit's Method

Charpit's method provides in general way for solving non-linear PDEs of order one.

Let us consider the equation,

$$f(x,y,z,p,q) = 0. \tag{20.18}$$

Since z depends on x and y, we have

$$dz = \frac{\partial z}{\partial x}dx + \frac{\partial z}{\partial y}dy = p\,dx + q\,dy. \tag{20.19}$$

If $\phi(x,y,z,p,q) = 0$ (20.20)

be another relation connecting x, y, z, p, q, then we can find out p and q by solving Eqns. (20.19) and (20.20) and substitute in Eqn. (20.19). This will give the solution of Eqn. (20.19) provided (20.19) is integrable.

To determine ϕ, we differentiate Eqns. (20.18) and (20.20) w.r.t. x and y as below :

$$\frac{\partial f}{\partial x} + \frac{\partial f}{\partial z}p + \frac{\partial f}{\partial p}\frac{\partial p}{\partial x} + \frac{\partial f}{\partial q}\frac{\partial q}{\partial x} = 0 \tag{20.21}$$

$$\frac{\partial \phi}{\partial x}+\frac{\partial \phi}{\partial x}p+\frac{\partial \phi}{\partial p}\frac{\partial p}{\partial x}+\frac{\partial \phi}{\partial q}\frac{\partial q}{\partial x}=0 \tag{20.22}$$

$$\frac{\partial f}{\partial y}+\frac{\partial f}{\partial z}q+\frac{\partial f}{\partial p}\frac{\partial p}{\partial y}+\frac{\partial f}{\partial q}\frac{\partial q}{\partial y}=0 \tag{20.23}$$

$$\frac{\partial \phi}{\partial y}+\frac{\partial \phi}{\partial z}q+\frac{\partial \phi}{\partial p}\frac{\partial p}{\partial y}+\frac{\partial \phi}{\partial q}\frac{\partial q}{\partial y}=0. \tag{20.24}$$

Eliminating $\frac{\partial p}{\partial x}$ between equations (20.21) and (20.22), we get

$$\left(\frac{\partial f}{\partial x}\frac{\partial \phi}{\partial p}-\frac{\partial \phi}{\partial x}\frac{\partial f}{\partial p}\right)+\left(\frac{\partial f}{\partial z}\frac{\partial \phi}{\partial p}-\frac{\partial \phi}{\partial z}\frac{\partial f}{\partial p}\right)p+\left(\frac{\partial f}{\partial q}\frac{\partial \phi}{\partial p}-\frac{\partial \phi}{\partial q}\frac{\partial f}{\partial p}\right)\times\frac{\partial q}{\partial x}=0. \tag{20.25}$$

Eliminating $\frac{\partial q}{\partial y}$ between equations (20.23) and (20.24), we get

$$\left(\frac{\partial f}{\partial y}\frac{\partial \phi}{\partial q}-\frac{\partial \phi}{\partial y}\frac{\partial f}{\partial q}\right)+\left(\frac{\partial f}{\partial z}\frac{\partial \phi}{\partial q}-\frac{\partial \phi}{\partial z}\frac{\partial f}{\partial q}\right)q+\left(\frac{\partial f}{\partial p}\frac{\partial \phi}{\partial q}-\frac{\partial \phi}{\partial p}\frac{\partial f}{\partial q}\right)\frac{\partial p}{\partial y}=0. \tag{20.26}$$

Adding Eqns. (20.25) and (20.26) and using $\frac{\partial q}{\partial x}=\frac{\partial^2 z}{\partial x\partial y}=\frac{\partial p}{\partial y}$, the last terms in both cancel and the other terms on rearrangement give

$$\left(\frac{\partial f}{\partial x}+p\frac{\partial f}{\partial z}\right)\frac{\partial \phi}{\partial p}+\left(\frac{\partial f}{\partial y}+q\frac{\partial f}{\partial z}\right)\frac{\partial \phi}{\partial q}+\left(-p\frac{\partial f}{\partial p}-q\frac{\partial f}{\partial q}\right)\frac{\partial \phi}{\partial z}$$
$$+\left(-\frac{\partial f}{\partial p}\right)\frac{\partial \phi}{\partial x}+\left(-\frac{\partial f}{\partial q}\right)\frac{\partial \phi}{\partial y}=0 \tag{20.27}$$

or, $$\left(-\frac{\partial f}{\partial p}\right)\frac{\partial \phi}{\partial x}+\left(\frac{-\partial f}{\partial q}\right)\frac{\partial \phi}{\partial y}+\left(-p\frac{\partial f}{\partial p}-q\frac{\partial f}{\partial q}\right)\frac{\partial \phi}{\partial z}$$
$$+\left(\frac{\partial f}{\partial x}+p\frac{\partial f}{\partial z}\right)\frac{\partial \phi}{\partial p}+\left(\frac{\partial f}{\partial y}+q\frac{\partial f}{\partial z}\right)\frac{\partial \phi}{\partial q}=0. \tag{20.28}$$

This is Lagrange's linear equation with x, y, z, p, q as independent variables and ϕ as dependent variable. Its solution will depend on the solution of the auxiliary equations.

$$\frac{dx}{-\frac{\partial f}{\partial p}}=\frac{dy}{-\frac{\partial f}{\partial q}}=\frac{dz}{-p\frac{\partial f}{\partial p}-p\frac{\partial f}{\partial q}}=\frac{dp}{\frac{\partial f}{\partial x}+p\frac{\partial f}{\partial z}}=\frac{dq}{\frac{\partial f}{\partial y}+q\frac{\partial f}{\partial z}}=\frac{d\phi}{0}. \tag{20.29}$$

Any integral of these equations involving p or q or both can be taken for the required second relation (20.19) which along with (20.18) will give the values of p and q to make (20.19) integrable. Of course we should take the simplest of the integrals so that it may be easier to solve for p and q.

Example 20.10. *Solve* : $2xz-px^2-2qxy+pq=0.$ **[Rajasthan U 2006]**

Solution : Let $f(x,y,z,p,q)=2xz-px^2-2qxy+pq=0.$ (i)

The Charpit's auxiliary equations are

$$\frac{dx}{x^2-q^2}=\frac{dy}{2xy-p}=\frac{dx}{px^2-2pq+2qxy}=\frac{dp}{2x-2qy}=\frac{dq}{0}.$$

$\therefore \quad dq=0 \quad$ or, $\quad q=a.$

Putting $q = a$ in Eqn. (i), we get $p = \dfrac{2x(x-ay)}{x^2-a}$.

$$\therefore \quad dz = p\,dx + q\,dy = \frac{2x(x-ay)}{x^2-a}dx + dy \quad \text{or,} \quad \frac{dz - a\,dy}{x-ay} = \frac{2x}{x^2-a}dx.$$

Integrating the above equation, we get

$$\log(z-ay) = \log(x^2-a) + \log b = \log b(x^2-a) \quad \text{or,} \quad z - ay = b(x^2-a)$$

or, $\boldsymbol{z = ay + b(x^2 - a)}$ is the required complete solution.

20.9 Homogeneous Linear PDEs with Constant Coefficient

An equation of the form,

$$\frac{\partial^n z}{\partial x^n} + k_1\frac{\partial^n z}{\partial x^{n-1}\partial y} + \cdots + k_n\frac{\partial^n z}{\partial y^n} = F(x,y) \tag{20.30}$$

is a homogeneous linear PDE of nth order with constant coefficients. Since, all the terms of equation (20.30) contain derivatives of the same order, it is homogeneous.

Eqn. (20.30) can be written as

$$(k_0D^n + k_1D^{n-1}D' + \cdots + k_nD'^n)z = F(x,y) \quad \text{or,} \quad f(D,D')z = F(x,y),$$

where $\frac{\partial}{\partial x} = D$ and $\frac{\partial}{\partial y} = D'$.

20.10 Rules for Finding the Complementary Function

Let $\boldsymbol{k_0\dfrac{\partial^2 z}{\partial x^2} + k_1\dfrac{\partial^2 z}{\partial x\partial y} + k_2\dfrac{\partial^2 z}{\partial y^2} = 0}$ or, $\boldsymbol{(k_0D^2 + k_1DD' + k_2D'^2)z = 0}$.

Now putting $D = m$ and $D' = 1$,

$$\boldsymbol{k_0m^2 + k_1m + k_2 = 0.}$$

This is the **auxiliary** equation.

Case I : If the roots of the auxiliary equation are real and different, say m_1 and m_2, then the complementary function **C.F.** = $\boldsymbol{f_1(y+m_1x) + f_2(y+m_2x)}$.

Case II : If the roots are equal (i.e., $m_1 = m_2 = m$), then **C.F.** = $\boldsymbol{f_1(y+mx) + xf_2(y+mx)}$.

Example 20.11. *Solve* : $\dfrac{\partial^2 z}{\partial x^2} + 4\dfrac{\partial^2 z}{\partial x\partial y} - 5\dfrac{\partial^2 z}{\partial y^2} = 0.$

Solution : $(D^2 + 4DD' - 5D'^2)z = 0$

$$D^2 + 4D\cdot 1 - 5\cdot 1 = 0, \quad \left[\text{putting } \frac{\partial^2}{\partial x^2} = D = m \text{ and } \frac{\partial^2 z}{\partial x\partial y^2} = D' = 1\right]$$

or, the auxiliary equation is $\boldsymbol{m^2 + 4m - 5 = 0}$ or, $(m-1)(m+5) = 0$.

$\therefore$ $\boldsymbol{m = 1, -5}$.

The required solution is $\boldsymbol{Z = f_1(y+x) + f_2(y-5x)}$.

Example 20.12. *Solve* : $2\dfrac{\partial^2 z}{\partial x^2} + 5\dfrac{\partial^2 z}{\partial x\partial y} + 2\dfrac{\partial^2 z}{\partial y^2} = 0.$

Solution : The given equation is $(2D^2 + 5DD' + 2D^2)z = 0$.

Its auxiliary equation is $2m^2 + 5m + 2 = 0$, where $m = \frac{D}{D'}$.

$\therefore \quad m = -2, -\frac{1}{2}.$

Hence, the complete solution is $Z = f_1(y - 2x) + f_2\left(y - \frac{x}{2}\right)$.

20.11 Rules for Finding the Particular Integral

Let us consider the PDE,

$$(k_0D^2 + k_1DD' + k_2D^2)z = F(x, y) \quad \text{or,} \quad f(D, D')z = F(x, y).$$

$$\therefore \quad \text{P.I.} = \frac{1}{f(D, D')}F(x, y).$$

Case I: When $F(x, y) = e^{ax+by}$

$$\text{P.I.} = \frac{1}{f(D, D')}e^{ax+by} = \frac{e^{ax+by}}{f(a, b)} \quad [\text{Put } D = a,\ D' = b]$$

Example 20.13. *Solve* : $\frac{\partial^2 z}{\partial x^2} - \frac{\partial^2 z}{\partial y^2} = e^{x+2y}$.

Solution : The given equation can be written as $(D^2 - D'^2)z = e^{x+2y}$.

The auxiliary equation is $(m^2 - 1) = 0$ [Put $D = m$, $D' = 1$]

$\therefore \quad m = 1, -1.$

$$\text{C.F.} = f_1(y + x) + f_2(y - x)$$

$$\text{P.I.} = \frac{1}{D^2 - D'^2}e^{x+2y}$$

$$= \frac{1}{1^2 - 2^2}e^{x+2y} \quad [\text{Putting } D = 1,\ D' = 2]$$

$$= -\frac{1}{3}e^{x+2y}.$$

$$\therefore \quad \text{C.S. is } Z = f_1(y + x) + f_2(y - x) - \frac{1}{3}e^{x+2y}.$$

Case II: When $F(x, y) = \sin(ax + by)$ or $\cos(ax + by)$

$$\text{P.I.} = \frac{1}{f(D^2, DD', D'^2)}\sin(ax + by) \quad \text{or,} \quad \cos(ax + by).$$

Put $D^2 = -a^2$ or, $DD' = -ab$ or, $D^2 = -b^2$.

$$\therefore \quad \text{P.I.} = \frac{\sin(ax + by) \text{ or, } \cos(ax + by)}{f(-a^2, -ab, -b^2)}$$

Example 20.14. *Solve* : $\frac{\partial^3 z}{\partial x^3} - 4\frac{\partial^3 z}{\partial x^2 \partial y} + 4\frac{\partial^3 z}{\partial x \partial y^2} = 2\sin(3x + 2y)$

[AMIE Winter 1993, Summer 1991]

Solution : $\frac{\partial^3 z}{\partial x^3} - 4\frac{\partial^3 z}{\partial x^2 \partial y} + 4\frac{\partial^3 z}{\partial x \partial y^2} = 2\sin(3x + 2y).$

Putting, $\frac{\partial}{\partial x} = D$, $\frac{\partial}{\partial y} = D'$

$$D^3z - 4D^2D'z + 4DD'^2z = 2\sin(3x+2y)$$

or, $(D^3 - 4D^2D' + 4DD'^2)z = 2\sin(3x+2y)$

A.E. is $D^3 - 4D^2D' + 4DD'^2 = 0$ or, $D(D^2 - 4DD' + 4D'^2) = 0$

Put, $D = m$, $D' = 1$ then,

$$m(m^2 - 4m + 4) = 0 \quad \text{or,} \quad m(m-2)^2 = 0 \quad \text{or,} \quad m = 0, 2, 2$$

$\therefore$ **C.F. is $f_1(y) + f_2(y+2x) + xf_3(y+2x)$**

and **P.I.** $= \dfrac{1}{D^3 - 4D^2D' + 4DD'^2} 2\sin(3x+2y) = 2\dfrac{1}{D(D^2 - 4DD' + 4D'^2)}\sin(3x+2y)$

$$= 2\frac{1}{[-9-4(-6)+4(-4)]}\sin(3x+2y) = -\frac{2}{D}\sin(3x+2y)$$

$$= -\frac{2}{3}[-\cos(3x+2y)] = \frac{2}{3}\cos(3x+2y)$$

$\therefore$ **General solution is $z = f_1(y) + f_2(y+2x) + xf_3(y+2x) + \frac{2}{3}\cos(3x+2y)$.**

Case III : When $F(x,y) = x^m y^n$

Example 20.15. *Solve :* $\dfrac{\partial^3 z}{\partial x^3} - 2\dfrac{\partial^2 z}{\partial x^2 \partial y} = 2e^{2x} + 3x^2y$.

[AMIE(Winter) 1993, 1990, SVTU 2007]

Solution : The given equation can be written as $(D^3 - 2D^2D')Z = 2e^x + 3x^2y$.

The auxiliary equation is $m^3 - 2m^2 = 0$ or, $m^2(m-2) = 0$, $\therefore$ $m = 0, 0, 2$.

$\therefore$ **C.F.** $= f_1(y+0) + f_2(y+0) + f_3(y+2x) = \mathbf{f_1(y) + f_2(y) + f_3(y+2x)}$

P.I. $= \dfrac{1}{D^3 - 2D^2D'}(2e^{2x} + 3x^2y) = \dfrac{1}{D^3 - 2D^2D'}2e^{2x} + \dfrac{1}{D^3 - 2D^2D'}3x^2y$

$$= 2\cdot\frac{e^{2x}}{(2)^3 - 2(2)^3(0)} + 3\cdot\frac{x^2y}{D^3\left(1 - \frac{2D'}{D}\right)} = \frac{2e^{2x}}{8} + \frac{3}{D^3}\left(1 - \frac{2D'}{D}\right)^{-1}x^2y$$

$$= \frac{e^{2x}}{4} + \frac{3}{D^3}\left(1 + \frac{2D'}{D}\cdots\right)^{-1}x^2y = \frac{e^{2x}}{4} + \frac{3}{D^3}\left(x^2y + \frac{2}{D}x^2\right) = \frac{e^{2x}}{4} + \frac{3}{D^3}\left(x^2y + \frac{2x^3}{3}\right)$$

$$= \frac{e^{2x}}{4} + 3y\frac{1}{D^3}x^2 + \frac{2}{D^3}x^3 = \frac{e^{2x}}{4} + 3y\frac{x^5}{3\cdot4\cdot5} + 2\frac{x^6}{4\cdot5\cdot6} = \frac{e^{2x}}{4} + \frac{x^5y}{20} + \frac{x^6}{60}$$

$$= \frac{1}{60}(15e^{2x} + 3x^5y + x^6).$$

$\therefore$ **C.S. is $Z = f_1(y) + f_2(y) + f_3(y+2x) + \frac{1}{60}(15e^{2x} + 3x^5y + x^6)$.**

Case IV : When $F(x,y)$ is any function of x and y.

Example 20.16. *Solve :* $\dfrac{\partial^2 z}{\partial x^2} + \dfrac{\partial^2 z}{\partial x \partial y} - 6\dfrac{\partial^2 z}{\partial y^2} = \cos(2x+y)$ **[AMIE(Winter) 1993]**

Solution : In symbolic form, the given equation can be written as

$$(D^2 + DD' - 6D'^2)Z = \cos(2x + y)$$

Its auxiliary equation is

$$(m^2 + m - 6) = 0 \quad \text{or,} \quad (m+3)(m-2) = 0, \quad \text{i.e,} \quad m = 2, -3.$$

$$\therefore \quad \textbf{C.F.} = \boldsymbol{f_1(y + 2x) + f_2(y - 3x)}$$

$$\textbf{P.I.} = \frac{1}{D^2 + DD' - 6D'^2}\cos(2x + y) = \frac{1}{(D - 2D')(D + 3D')}\cos(2x + y)$$

$$= \frac{1}{D + 3D'}\left[\int \cos\{2x + (c - 2x)\}dx\right]_{c \to y+2x}$$

$$= \frac{1}{D + 3D'}\left[\int \cos c\, dx\right]_{c \to y+2x} \qquad [\because\ y = c - mx = c - 2x]$$

$$= \frac{1}{D + 3D'} x\cos(y + 2x) = \left[\int x\cos\{(c + 3x) + 2x\}dx\right]_{c \to y-3x}$$

$$= \left[\int x\cos(5x + c)dx\right]_{c \to y-3x}$$

$$= \left[\frac{x\sin(5x + c)}{5} + \frac{\cos(5x + c)}{25}\right]_{c \to y-3x} \qquad \text{[integrating by parts]}$$

$$= \frac{x}{5}\sin[5x + (y - 3x)] + \frac{1}{25}\cos[5x + (y - 3x)] = \boldsymbol{\frac{x}{5}\sin(2x + y) + \frac{1}{25}\cos(2x + y)}.$$

Hence, the complete solution (C.S.) is

$$\boldsymbol{Z = f_1(y - 3x) + f_2(y + 2x) + \frac{x}{5}\sin(2x + y) + \frac{1}{25}\cos(2x + y)}.$$

20.12 Solved Problems

Example 20.1. *Form a PDE by eliminating a, b from $(x - a)^2 + (y - b)^2 + z^2 = d^2$.*

Solution : Differentiating partially the given equation w.r.t. x and y, we get

$$2(x - a) + 2x\frac{\partial z}{\partial x} = 0 \quad \text{or,} \quad (x - a) = -pz \tag{i}$$

and $$2(y - b) + 2z\frac{\partial z}{\partial y} = 0 \quad \text{or,} \quad (y - b) = -qz. \tag{ii}$$

Eliminating a and b from the given equation and equations (i) and (ii), we get

$$(p^2 + q^2)z^2 + z^2 = d^2 \quad \text{or,} \quad \boldsymbol{(p^2 + q^2 + 1)z^2 = d^2}.$$

Example 20.2. *Solve the partial differential equation, $yq - xp = z$, where $p = \dfrac{\partial z}{\partial x}$, $q = \dfrac{\partial z}{\partial y}$.*

Solution : Given $yq - xp = z$.

Auxiliary equations are, $\dfrac{dx}{-x} = \dfrac{dy}{y} = \dfrac{dz}{z}$.

$\therefore \quad -\log x = \log y - \log a$ [from the first two equations]

or, $\quad xy = a.$ (i)

$\log y = \log z + \log b$ [from the last two equations]

or, $\quad \frac{y}{z} = b.$ (ii)

From Eqns. (i) and (ii), the solution is $\boldsymbol{f\left(xy, \frac{y}{z}\right) = 0}$.

Example 20.3. *Solve :* $\frac{\partial^2 z}{\partial x \partial y} = \sin x \sin y$, for which $\frac{\partial z}{\partial y} = -2\sin y$, when $x = 0$, $z = 0$ and y is an odd multiple of $\frac{\pi}{2}$. **[VJU 2004]**

Solution : $\frac{\partial^2 z}{\partial x \partial y} = \sin x \sin y$

Integrating w.r.t. x keeping y constant,

$$\frac{\partial z}{\partial y} = -\cos x \sin y + f(y). \quad \text{(i)}$$

When $x = 0$, $\frac{\partial z}{\partial y} = -2\sin y$,

$\therefore \quad -2\sin y = -\sin y + f(y)$ or, $f(y) = -\sin y$.

$\therefore$ from Eqn. (i), $\frac{\partial z}{\partial y} = -\cos x \sin y - \sin y.$

Now, integrating the above equation w.r.t. y, keeping x constant,

$$z = \cos x \cos y + \cos y + g(x).$$

When y is an odd multiple of $\frac{\pi}{2}$, $z = 0$.

$\therefore \quad 0 = 0 + 0 + g(x)$ or, $g(x) = 0 \quad \left[\because \cos(2n+1)\frac{\pi}{2} = 0\right]$

Hence, the complete solution is $\boldsymbol{Z = (1 + \cos x)\cos y}$.

Example 20.4. *Solve :* $x^2(y-z)p + y^2(z-x)q = z^2(x-y)$. **[SVTU 2007]**

Solution : The auxiliary equations are $\frac{dx}{x^2(y-z)} = \frac{dy}{y^2(z-x)} = \frac{dz}{z^2(x-y)}$.

Using the multipliers $\frac{1}{x}, \frac{1}{y}, \frac{1}{z}$, we get each fraction $= \frac{\frac{dx}{x^2} + \frac{dy}{y^2} + \frac{dz}{z^2}}{0}$

$$\therefore \quad \frac{dx}{x^2} + \frac{dy}{y^2} + \frac{dz}{z^2} = 0.$$

Integrating, we get

$$\log x + \log y + \log z = \log a \quad \text{or,} \quad xyz = a \quad \text{(i)}$$

Using the multipliers $\frac{1}{x^2}, \frac{1}{y^2}, \frac{1}{z^2}$ each fraction $= \frac{\frac{dx}{x^2} + \frac{dy}{y^2} + \frac{dz}{z^2}}{0}$.

$\therefore \quad \dfrac{dx}{x^2} + \dfrac{dy}{y^2} + \dfrac{dz}{z^2} = 0.$

Integrating, $\dfrac{1}{x} + \dfrac{1}{y} + \dfrac{1}{z} = 0.$ (ii)

Hence, from Eqns. (i) and (ii), we get $\boldsymbol{xyz = f\left(\dfrac{1}{x} + \dfrac{1}{y} + \dfrac{1}{z}\right)}$.

Example 20.5. *Solve* : $(mz - ny)\dfrac{\partial z}{\partial x} + (nx - lz)\dfrac{\partial z}{\partial y} = ly - mx.$

[AMIETE, 1986; AMIE, Summer, 1995; Winter 1993; VTU 2009]

Solution : The auxiliary equations in this case is $\dfrac{dx}{mz - ny} = \dfrac{dy}{nx - lz} = \dfrac{dz}{ly - mx}$.

Using the multipliers x, y, z each fraction is $\dfrac{x\,dx + y\,dy + z\,dz}{0}$.

$\therefore \quad x\,dx + y\,dy + z\,dz = 0.$

Integrating, $x^2 + y^2 + z^2 = a.$ (i)

Using the multipliers l, m, n, each fraction is $\dfrac{ldx + mdy + ndz}{0}$ or, $l\,dx + m\,dy + n\,dz = 0.$

Integrating, $lx + my + nz = b.$ (ii)

Hence from eqns. (i) and (ii) we get the solution as

$$\boldsymbol{x^2 + y^2 + z^2 = f(lx + my + nz)}.$$

Example 20.6. *Solve* : $\dfrac{y - z}{yz}p + \dfrac{z - x}{zx}q = \dfrac{x - y}{xy}$ **[AMIE(Winter) 1991]**

Solution : $\dfrac{y - z}{yz}p + \dfrac{z - x}{zx}q = \dfrac{x - y}{xy}$ or, $x(y - z)p + y(z - x)q = z(x - y).$

The auxiliary equations are $\dfrac{dx}{x(y - z)} = \dfrac{dy}{y(z - x)} = \dfrac{dz}{z(x - y)}$. (i)

Each fraction $= \dfrac{dx + dy + dz}{0}$ or, $dx + dy + dz = 0.$

Integrating, $x + y + z = a.$ (ii)

Again, Eqn. (i) can be written as

$$\frac{\frac{dx}{x}}{y - z} = \frac{\frac{dy}{y}}{z - x} = \frac{\frac{dz}{z}}{x - y} = \frac{\frac{dx}{x} + \frac{dy}{y} + \frac{dz}{z}}{0} \quad \text{or,} \quad \frac{dx}{x} + \frac{dy}{y} + \frac{dz}{z} = 0.$$

Integrating, $\log x + \log y + \log z = \log b$ or, $\log xyz = \log b$

or, $\quad xyz = b.$ (iii)

From Eqns. (ii) and (iii), we get $\boldsymbol{xyz = f(x + y + z)}$, which is the required solution.

Example 20.7. *Find the general solution of* $(x^2 - y^2 - z^2)p + 2xyq = 2xz.$

[BU(H) 2003, Rajasthan U 2006]

Solution : The auxiliary equations is $\dfrac{dx}{x^2 + y^2 + z^2} = \dfrac{dy}{2xy} = \dfrac{dz}{2xz}$.

From the last two fractions, $\frac{dy}{y} = \frac{dz}{z}$.

Integrating, $\log y = \log z + \log a$ or, $\frac{y}{z} = a$. (i)

Using multipliers x, y, z, each fraction $= \frac{x\,dx + y\,dy + z\,dz}{x(x^2 + y^2 + z^2)}$.

$$\therefore \quad \frac{2x\,dx + 2y\,dy + 2z\,dz}{x^2 + y^2 + z^2} = \frac{dz}{z}.$$

Integrating, $\log(x^2 + y^2 + z^2) = \log x + \log b$ or, $\frac{x^2 + y^2 + z^2}{x} = b$. (ii)

Hence, from Eqns. (i) and (ii), the solution is $\boldsymbol{x^2 + y^2 + z^2 = zf\left(\frac{y}{z}\right)}$.

Example 20.8. *Solve :* $x^2p^2 + y^2q^2 = z^2$. **[Kerala U 2005]**

Solution : The given equation may be written as $\left(\frac{x}{z}\frac{\partial z}{\partial x}\right)^2 + \left(\frac{y}{z}\frac{\partial z}{\partial y}\right)^2 = 1$. (i)

Putting $\frac{dx}{x} = du$, $\frac{dy}{y} = dv$, $\frac{dz}{z} = dw$ so that,

$u = \log x, \quad v = \log y, \quad w = \log z$.

Then Eqn. (i) becomes $\left(\frac{\partial w}{\partial u}\right)^2 + \left(\frac{\partial w}{\partial v}\right)^2 = 1$

or, $\quad P^2 + Q^2 = 1$, where $P = \frac{\partial w}{\partial u}$ and $Q = \frac{\partial w}{\partial v}$.

Its complete solution is $\boldsymbol{w = au + bv + c}$, (ii)

where $a^2 + b^2 = 1$ or, $b = \sqrt{1 - a^2}$.

Then, from Eqn. (ii), $w = au + (\sqrt{1 - a^2})v + c$

or, $\quad \boldsymbol{\log z = a\log x + (\sqrt{1 - a^2})\log y + c}$ is the required solution.

Example 20.9. *Solve by Jacobi's method,* $2p_1x_1x_3 + 3p_2x_3^2 + p_2^2p_3 = 0$.

Solution : The given equation is $F_1 = 2p_1x_1x_3 + 3p_2x_3^2 + p_2^2p_3 = 0$. (i)

$$\frac{dx_1}{-2x_1x_3} = \frac{dx_2}{-(3x_3^2 + 2p_2p_3)} = \frac{dx_3}{-p_2^2} = \frac{dp_1}{2p_1x_3} = \frac{dp_2}{0} = \frac{dp_3}{2p_1x_1 + 6p_2x_3}$$

or, $\quad \frac{dx_1}{x_1} + \frac{dp_1}{p_1} = 0$ and $dp_2 = 0$.

Integrating, we get

$p_1x_1 = a_1$ (say) (ii)

and $\quad p_2x_2 = a_2$ (say). (iii)

Substituting the values of p_1 and p_2 in Eqn. (i), we get

$$2a_1x_3 + 3a_2x_3^2 + a_2^2p_3 = 0 \quad \text{or,} \quad p_3 = -\frac{(2a_1x_3 + 3a_2x_3^2)}{a_2^2}.$$

Now, $dz = p_1\,dx_1 + p_2\,dx_2 + p_3\,dx_3$ gives

$$dz = \frac{a_1}{x_1}dx_1 + a_2\,dx_2 - \frac{2a_1x_3 + 3a_2x_3^2}{a_2^2}dx_3.$$

Integrating, we get

$$\boldsymbol{z = a_1 \log x_1 + a_2x_2 - \frac{a_1x_3^2 + a_2x_3^2}{a_2^3} + c.}$$

This gives the required complete integral.

Example 20.10. *Solve* : $p^2 + q^2 = x + y$. **[Madras U 2003]**

Solution : Let $p^2 - x = y - q^2 = a$, then

$$p^2 - x = a, \qquad \therefore\ p = \sqrt{a + x}$$

and $\quad y - q^2 = a, \qquad \therefore\ q = \sqrt{y - a}.$

Substituting the values of p and q in $dz = p\,dx + q\,dy$, we get

$$dz = (\sqrt{a + x})dx + (\sqrt{y - a})dy.$$

Integrating, $\boldsymbol{z = \frac{2}{3}(a + x)^{\frac{3}{2}} + \frac{2}{3}(y - a)^{\frac{3}{2}} + b}$ is the required complete solution.

Example 20.11. *Solve* : $p(1 + q) = qz$.

Solution : The given PDE is $p(1 + q) = qz$. (i)

Put $u = x + ay$.

$$\therefore \quad p = \frac{dz}{du} \quad \text{and} \quad q = \frac{dz}{du}.$$

From Eqn. (i),

$$\frac{dz}{du}\left(1 + a\frac{dz}{du}\right) = a\frac{dz}{du}z \quad \text{or,} \quad 1 + a\frac{dz}{du} = az \quad \text{or,} \quad a\frac{dz}{du} = az - 1 \quad \text{or,} \quad \frac{dz}{du} = \frac{az - 1}{a}$$

$$\text{or,} \quad \frac{du}{dz} = \frac{a}{az - 1} \quad \text{or,} \quad du = \frac{a\,dz}{az - 1}.$$

Integrating, $u = \log(az - 1) + \log c$

or, $\boldsymbol{x + ay = \log c(az - 1)}$ is the required solution.

Example 20.12. *Solve* : $\dfrac{\partial^2 z}{\partial x^2} + 4\dfrac{\partial^2 z}{\partial x \partial y} + 4\dfrac{\partial^2 z}{\partial y^2} = 0.$

Solution : In symbolic form, the given equation is $(D^2 - 4DD' + 4D'^2)z = 0$.

Its auxiliary equation is

$$m^2 - 4m + 4 = 0 \quad [D = m,\ D' = 1]$$

or, $(m - 2)^2 = 0$ i.e., $m = 2, 2$.

Now, the required solution of the given equation is $\boldsymbol{z = f_1(y + 2x) + xf_2(y + 2x)}$.

Example 21.13. *Solve* : $r - t + p - q = 0$.

Solution : In symbolic form, $(D^2 - D'^2 + D - D')z = 0$

or, $[(D - D')(D + D') + 1(D - D')z = 0$ or, $(D - D')(D + D' + 1)z = 0$.

$\therefore$ the solution is $\boldsymbol{z = f_1(y + x) + e^{-x} f_2(y - x)}$.

Example 20.14. *Solve* : $\dfrac{\partial^2 z}{\partial x^2} + 4\dfrac{\partial^2 z}{\partial x \partial y} + 3\dfrac{\partial^2 z}{\partial y^2} = 0$.

Solution : In symbolic form, the given equation is $(D^2 + 4DD' + 3D'^2)z = 0$.

The auxiliary equation is

$m^2 + 4m + 3 = 0$ $[D = m,\ D' = 1]$

or, $m^2 + 3m + m + 3 = 0$ or, $m(m + 3) + 1(m + 3) = 0$

or, $(m + 1)(m + 3) = 0$ i.e., $m = -1, -3$.

$\therefore$ the required solution is $\boldsymbol{z = f_1(y - x) + f_2(m - 3x)}$.

Example 20.15. *Solve* : $(D^2 - 3DD' + 2D'^2)z = 0$.

Solution : The auxiliary solution is

$D^2 - 3DD' + 2D'^2 = 0$

or, $m^2 - 3m + 2 = 0$ [putting $D = m,\ D' = 1$]

or, $(m - 1)(m - 2) = 0$ i.e., $m = 1, 2$.

$\therefore$ the required solution is $\boldsymbol{z = f_1(y + x) + f_2(y + 2x)}$.

Example 20.16. *Solve* : $x\dfrac{\partial z}{\partial x} + \dfrac{\partial z}{\partial y} = -zy$.

Solution : The auxiliary equation is $\dfrac{dx}{x} = \dfrac{dy}{1} = \dfrac{dy}{-zy}$. (i)

Integrating the relation $dy = \frac{dx}{x}$, we get

$$y = \log\frac{x}{a} \quad \text{or,} \quad xe^{-y} = a = \text{constant}. \tag{ii}$$

Again integrating the relation $\frac{dy}{1} = \frac{dz}{-zy}$, we get

$$-\frac{z^2}{1} = \log^1\left(\frac{z}{b}\right) \quad \text{or,} \quad z = be^{-\frac{x^2}{y}}. \tag{iii}$$

From Eqns. (ii) and (iii), the required solution is $\boldsymbol{z(x, y) = f(xe^{-y})e^{-\frac{x^2}{y}}}$.

Example 20.17. *Solve* : $(D^2 - 2DD' + D'^2)z = e^{x+2y}$.

Solution : The auxiliary equation is

$D^2 - 2DD' + D'^2 = 0$

or, $m^2 - 2m + 1 = 0$ [putting $D = m,\ D' = 1$]

or, $(m - 1)^2 = 0$, i.e., $m = 1, 1$.

$\therefore$ C.F. $= f_1(y+x) + xf_2(y+x)$

$$\text{P.I.} = \frac{e^{x+2y}}{D^2 - 2DD' + D'^2} = \frac{e^{x+2y}}{1^2 - 2\cdot 1\cdot 2 + 2^2} = e^{x+2y}.$$

Hence, the complete solution is : $\boldsymbol{z = f_1(y+x) + xf_2(y+x) + e^{x+2y}}$.

Example 20.18. *Solve* : $\dfrac{\partial^3 z}{\partial x^3} - 3\dfrac{\partial^3 z}{\partial x^2 \partial y} + 4\dfrac{\partial^3 z}{\partial y^3} = e^{x+2y}$.

[BU(H) 2003, AMIE(Winter) 1990]

Solution : In symbolic form, $(D^3 - 3D^2D' + 4D'^3)z = e^{x+2y}$

Auxiliary equation is

$m^3 - 3m^2 + 4 = 0$ [putting $D = m$, $D' = 1$]

$(m+1)(m-2)(m-2) = 0$, i.e., $m = -1, 2, 2$.

$\therefore$ **C.F.** $= \boldsymbol{f_1(y-x) + f_2(y+2x) + xf_3(y+2x)}$

$$\textbf{P.I.} = \frac{1}{D^3 - 3D^2D' + 4D'^3}e^{x+2y} = \frac{1}{1-6+32}e^{x+2y} \quad [\text{putting } D = 1,\ D' = 2]$$

$$= \frac{e^{x+2y}}{27}.$$

Hence, the complete solution is, $\boldsymbol{z = f_1(y-x) + f_2(y+2x) + xf_3(y+2x) + \dfrac{e^{x+2y}}{27}}$.

Example 20.19. *Solve* : $\dfrac{\partial^2 z}{\partial x^2} + \dfrac{\partial^2 z}{\partial x \partial y} - 6\dfrac{\partial^2 z}{\partial y^2} = y\cos x$.

[Anna U 2005S, UPTU 2003, AMIE(Winter) 1992]

Solution : The given equation in symbolic form can be written as

$(D^2 + DD' - 6D'^2)z = y\cos x$, where $\dfrac{\partial}{\partial x} = D$, $\dfrac{\partial}{\partial x^2} = D$, $\dfrac{\partial}{\partial y} = D'$ and $\dfrac{\partial^2}{\partial y^2} = D'^2$.

The auxiliary equation is

$m^2 + m - 6 = 0$ or, $(m+3)(m-2) = 0$ or, $m = 2, -3$.

$\therefore$ **C.F.** $= \boldsymbol{f_1(y+2x) + f_2(y-3x)}$

$$\textbf{P.I.} = \frac{1}{D^2 + DD' - 6D'^2}y\cos x = \frac{1}{(D-2D')(D+3D')}y\cos x$$

$$= \frac{1}{(D-2D')}\int (c+3x)\cos x\,dx \quad [\text{putting } y = (c+3x)]$$

$$= \frac{1}{(D-2D')}[(c+3x)\sin x + 3\cos x]$$

$$= \frac{1}{(D-2D')}[y\sin x + 3\cos x] \quad [\text{putting } c - 2x = y]$$

$$= \int [(c-2x)(-\cos x) - 2\sin x + 3\sin x$$

$$= \boldsymbol{-y\cos x + \sin x} \quad [\text{putting } c + 3x = y]$$

$\therefore$ the complete solution is $\boldsymbol{z = f_1(y+2x) + f_2(y-3x) + \sin x - y\cos x}$.

Example 20.20. *Solve* : $(D^2 - D'^2 - 3D + 3D')z = xy + e^{x+2y}$.

Solution : The auxiliary equation is

$$(D^2 - D'^2 - 3D + 3D') = 0$$

or, $D^2 - 1 - 3D + 3 = 0$ [putting $D = m,\ D' = 1$]

or, $m^2 - 3m + 2 = 0$ or, $m^2 - m - 2m + 2 = 0$ or, $m(m-1) - 2(m-1) = 0$

or, $(m-1)(m-2) = 0$, i.e., $m = 1, 2$.

$\therefore$ **C.F.** $= \boldsymbol{f_1(y - x) + f_2(y - 2x)}$

$$\text{P.I.} = \frac{1}{(D^2 - D'^2 - 3D + 3D')}(xy + e^{x+y}) = \frac{1}{(D - D')(D + D' - 3)}(xy + e^{x+y})$$

Now, $$\frac{1}{(D - D')(D + D' - 3)}(xy) = -\frac{1}{3D}\left(1 - \frac{D'}{D}\right)^{-1}\left(1 - \frac{D + D'}{3}\right)(xy)$$

$$= -\frac{1}{3D}\left(1 + \frac{D'}{D} + \cdots\right)\left[1 + \frac{D + D'}{3} + \frac{(D + D')}{9} + \cdots\right](xy)$$

$$= -\frac{1}{3D}\left[1 + \frac{D + D'}{3} + \frac{2DD'}{9} + \frac{D'}{D} + \frac{D'}{3} + \frac{DD'}{9} + \cdots\right](xy)$$

$$= -\frac{1}{3D}\left[1 + \frac{D}{3} + \frac{2D'}{3} + \frac{D'}{D} + \frac{DD'}{3} + \cdots\right](xy)$$

$$= -\frac{1}{3D}\left[xy + \frac{y}{3} + \frac{2x}{3} + \frac{x^2}{2} + \frac{1}{3}\right]$$

$$= -\frac{1}{3}\left[\frac{x^2y}{2} + \frac{xy}{3} + \frac{x^2}{3} + \frac{x^3}{6} + \frac{x}{3}\right]$$

and $$\frac{1}{(D - D')(D + D' - 3)}e^{x+2y} = \frac{1}{(D + D' - 3)(1 - 2)}e^{x+y} = \frac{1}{(D + D' - 3)}e^{x+2y}$$

$$= -e^{x+2y}\frac{1}{(D + 1 + D' + 2 - 3)} \cdot 1 = -e^{x+2y}\frac{1}{D'}\left(1 + \frac{D}{D'}\right)^{-1} \cdot 1$$

$$= e^{-x+2y} \cdot \frac{1}{D'} \cdot 1 = -\boldsymbol{ye^{x+2y}}.$$

Hence, the general solution is : $\boldsymbol{z = f_1(y - x) + f_2(y - 2x) - ye^{x+2y}}$.

Example 20.21. *Solve* : $(D^2 - DD' + D' - 1)z = \cos(x + 2y) + e^x$

Solution : The given equation is $(D^2 - DD' + D' - 1) = \cos(x + 2y) + e^x$.

Its auxiliary equation is $(D^2 - DD' + D' - 1) = 0$

Putting $D = m,\ D' = 1$,

$$m^2 - m + 1 - 1 = 0 \quad \text{or,} \quad m(m-1) = 0, \quad \therefore\ m = 0, 1.$$

$\therefore$ **C.F.** $= f_1(y + 0x) + f_2(y - x) = \boldsymbol{f_1(y) + f_2(y - x)}$

$$\text{P.I.} = \frac{1}{(D^2 - DD' + D' - 1)}[\cos(x+2y) + e^y].$$

Now, $\dfrac{1}{(D - DD' + D' - 1)}\cos(x+2y) = \dfrac{1}{-1^2 - (-2) + D' - 1}\cos(x+2y) = \dfrac{1}{D'}\cos(x+2y)$

$$= \frac{1}{2}\sin(x+2y)$$

and $\dfrac{1}{D^2 - DD' + D' - 1}e^y = e^y \dfrac{1}{D^2 - D(D'+1) + (D'+1) - 1}\cdot 1$

$$= e^y \frac{1}{D^2 - DD' - D + D'}\cdot 1 = -\frac{1}{D}e^y\left[1 + \left(\frac{D'}{D} + D + D'\right)\right]^{-1}\cdot 1$$

$$= -\frac{1}{D}e^y \cdot 1 = -xe^y.$$

Hence, the general solution is $z = f_1(y) + f_2(y - z) - xe^y$.

Exercises

20(A) : Objective Type Questions

Select the correct answers

20.1 A PDE involves

(a) partial derivatives

(b) partial differentials

(c) partial derivatives and differentials

(d) partial derivative with respect to one variable.

20.2 A partial differential equation is linear, if

(a) all derivatives appearing in it are of first degree

(b) only one derivative is of first degree

(c) some of the derivatives are of first degree only

(d) none of these.

20.3 A linear PDE is said to be homogeneous, if

(a) each term of it contains dependent variable

(b) each term contains its derivatives

(c) each term of it contains either the dependent variable or one of its derivatives

(d) none of these.

20.4 The PDE, $\dfrac{\partial^2 z}{\partial x^2} + z\dfrac{\partial^2 z}{\partial y^2} = 0$ is

(a) linear (b) homogeneous (c) both (a) and (b) (d) linear but inhomogeneous.

20.5 PDE can be formed by eliminating

(a) arbitrary constants (b) by eliminating arbitrary functions

(c) either by (a) or (b) (d) none of these.

20.6 The PDE obtained from $x^2 + y^2 + (z-c)^2 = a^2$ is

(a) $\frac{x}{p} - \frac{y}{q} = 0$ (b) $yp - xq = 0$ (c) $xq - \frac{y}{p} = 0$ (d) $xy - pq = 0$,

where $p = \frac{\partial z}{\partial x}$ and $q = \frac{\partial z}{\partial y}$.

20.7 In non-linear PDEs of first degree, p and q occur in

(a) first degree (b) second degree (c) other than first degree (d) any degree.

20.8 The solution of a non-linear PDE of first degree of the form $f(p,q) = 0$ can be written as

(a) $z = ax^2 + by$ (b) $z = ax + by + c$ (c) $z = ax + by^2$ (d) $z = ax + by$,

where $\frac{\partial z}{\partial x} = a$ and $\frac{\partial z}{\partial y} = b$.

20.9 The solution of a non-linear PDE of first degree of the form $f(z,p,q) = 0$ can be written as $F(z,a)$ equal to

(a) $x + ay$ (b) $ay + b$ (c) $x + ay + b$ (d) $x + b$.

20.10 The solution of a non-linear PDE of first degree of the form $f(x,p) = F(y,q)$ can be written as z equal to

(a) $\int \psi(y)dy + b$ (b) $\int \phi(x)dx + a$ (c) $\int [\psi(y)dy + \phi(x)dx]$ (d) $\int \psi(y)dy$,

where $p = \phi(x)$ and $q = \psi(y)$.

20.11 The solution of a non-linear PDE of first degree of the form $z = px + qy + f(p,q)$ can be written as z equal to

(a) $ax + by + \sqrt{a^2 + b^2}$ (b) $ax + by + \sqrt{1 + a^2}$

(c) $ax + by + \sqrt{1 + b^2}$ (d) $ax + by + \sqrt{1 + a^2 + b^2}$.

20.12 If the roots (m_1, m_2) of a linear homogeneous PDE with constant coefficients $k_0\frac{\partial^2 z}{\partial x^2} + k_1\frac{\partial^2 z}{\partial x \partial y} + k_2\frac{\partial^2 z}{\partial y^2} = 0$ be real and distinct, then its C.F. is written as

(a) $z = f_1(y + m_1x) + f_2(y + m_2x)$ (b) $z = f_1(y - m_1x) + f_2(y)$

(c) $z = f_1(y) + f_2(y - m_2x)$ (d) $z = f_1(m_1y) + f_2(m_2y)$.

20.13 If the roots be equal, then the solution of the PDE in (20.12) can be written as

(a) $z = f_1(y + m_1x)$ (b) $z = f_1(y + m_1x) + f_2(y + m_2x)$

(c) $z = f_1(y + m_1x) + xf_2(y + m_2x)$ (d) $z = xf_1(y + m_1x) + yf_2(y + m_2x)$.

20.14 The solution of a non-homogeneous linear PDE of the form $(D - mD' - a)z$ is written as

(a) $z = e^{ax}f(y+mx)$ (b) $z = e^{ax}f(my)$ (c) $z = e^{ax}f(y)$ (d) $z = e^{-ax}f(y+mx)$

20.15 The particular integral (P.I.) of a linear homogeneous PDE of the form $f(D, D')z = F(x, y)$, when $F(x, y) = e^{ax+by}$ is

(a) $\frac{1}{f(a,b)}e^{ax}$ (b) $\frac{1}{f(a,b)}e^{ax+b}$ (c) $\frac{1}{f(a,b)}e^{ax+by}$ (d) $\frac{1}{f(a,b)}e^{by}$.

20.16 The P.I. of the PDE of the form as in (20.15), when $F(x, y) = \sin(mx + ny)$ is

(a) $\frac{1}{f(-m^2, -mn, -n^2)}\sin(mx + ny)$ (b) $\frac{1}{f(-m^2, mn, -n^2)}\sin(mx + ny)$

(c) $\frac{1}{f(m^2, mn, n^2)}\sin(mx + ny)$ (d) $\frac{1}{f(-m^2, -mn, +n^2)}$.

20.17 The P.I. of the PDE of the form as in (20.15), when $F(x, y) = x^m y^n$ is

(a) $[f(D, D')]^{-1}x^m y^n$ (b) $[f(D, D')]^{-1} = \frac{x^m}{y^n}$ (c) $[f(D, D')]^{-1}yx^m$ (d) $[f(D, D')]^{-1}xy^n$

20.18 The solution of PDE of the form as in (20.15), when $F(x, y)$ is any function of x and y is

(a) $\int F(x, c - mx)dx$ (b) $\int F(x, mx)dx$ (c) $\int F(x, c)dx$ (d) $\int F(x, -mx)dx$.

20.19 The solution of a non-homogeneous linear PDE of the form $f(D, D') = F(x, y)$ can be written as z equals to

(a) $e^{ax}f(y + mx)$ (b) $e^{ax}f(mx)$ (c) $f(y + mx)$ (d) none of these,

where $b = f(a)$.

20.20 The solution of $xp + yq = z$ is

(a) $f(x^2, y^2) = 0$ (b) $f(xy, yz) = 0$ (c) $f(x, y) = 0$ (d) $f\left(\frac{x}{y}, \frac{y}{z}\right) = 0$.

20.21 The solution of $\frac{\partial^2 z}{\partial x^2} = \frac{\partial^2 z}{\partial y^2}$ is

(a) $z = f_1(y + x) + f_2(y - x)$ (b) $f_1(y + x) + f_1(y - x)$

(c) $z = f(x^2 + y^2)$ (d) $f(x^2 + y^2)$.

Answers

20.1 (c) **20.2** (a) **20.3** (c) **20.4** (c) **20.5** (c) **20.6** (b) **20.7** (c) **20.8** (b)
20.9 (c) **20.10** (a) **20.11** (d) **20.12** (a) **20.13** (c) **20.14** (a) **20.15** (c) **20.16** (a)
20.17 (a) **20.18** (a) **20.19** (a) **20.20** (d) **20.21** (a).

20(B) : Short Answer Type Questions

20.1 Show that the PDE $yp + xq = 0$ can be formed from $z = f(x^2 - y^2)$.

20.2 Derive a partial differential equation by eliminating the constants from equation

$$z = \frac{1}{2}\left(\frac{x^2}{a^2} + \frac{y^2}{b^2}\right).$$

[Ans: $z = \frac{1}{2}(xp + yq)$]

20.3 Form the partial differential equation from $f(x+y+z, x^2+y^2+z^2)=0$.

[AMIE(Winter) 1995] [**Ans:** $(y-z)p+(z-x)q=x-y$]

20.4 Solve : $yq-xp=z$, where $p=\dfrac{\partial z}{\partial x}$, $q=\dfrac{\partial z}{\partial y}$. [**Ans:** $f\left(xy, \frac{y}{z}\right)=0$]

20.5 Use the method of multiplier to solve the following equations :

(a) $p+q+2xz=0$ [**Ans:** $f(x-y)=x^2+\log z$]

(b) $xp-yq+x^2-y^2=0$ **[AMIE(summer) 1990]** [**Ans:** $f(x,y)=\frac{x^2}{2}+\frac{y^2}{2}+z$]

(c) $(y-z)p+(x-y)q=z-x$. [**Ans:** $f(x+y+z), x^2+2yz$]

20.6 Obtain the complete solutions of the following PDEs :

(a) $p^2+q^2=1$ **[Osmania U 2000]** [**Ans:** $z=ax+(\sqrt{1-a^2})y+c$]

(b) $\sqrt{p}+\sqrt{q}=2x$ **[JNTU 2006]** [**Ans:** $z=\frac{1}{6}(zx+a)^3+a^2y+b$]

(c) $pq+p+q=0$ [**Ans:** $z=ax-\frac{a}{1+a}y+c$]

(d) $pq=1$. [**Ans:** $z=ax+\frac{1}{a}y+c$]

20.7 Solve the following PDEs :

(a) $p(1+q^2)=q(z-a)$ [**Ans:** $(4bz-ab-1)=(x+by+c)^2$]

(b) $x^2p^2+y^2q^2=z$ [**Ans:** $2\sqrt{z}=\frac{\log x+a\log y}{\sqrt{1+a^2}}+c$]

(c) $4r+12s+qt=e^{3x-2y}$. [**Ans:** $z=f_1(2y-3x)+xf_2(2y-3x)+\frac{1}{2}x^2e^{3x-2y}$]

20(C) : Broad Answer Type Questions

20.1 Form the PDE :

(a) $(x-h)^2+(y-k)^2+z^2=a^2$ [**Ans:** $z^2(p^2+q^2+1)=a^2$]

(b) $ax^2+by^2+z^2=1$ [**Ans:** $z(px+qy)=z^2-1$]

(c) $z=yf(x)+xg(y)$ **[VTU 2004]** [**Ans:** $\frac{\partial^2 z}{\partial y^2}=\frac{\partial z}{\partial y}$]

(d) $z=f_1(x)f_2(y)$ [**Ans:** $z\frac{\partial^2 z}{\partial x\partial y}=\frac{\partial z}{\partial x}\cdot\frac{\partial z}{\partial y}$]

(e) $xyz=f(x+y+z)$ [**Ans:** $x(y-z)p+y(z-x)q=z(x-y)$]

(f) $z=e^{my}f(x-y)$ **[PTU 2002]** [**Ans:** $p+q=mz$]

(g) $x^2+y^2=(z-c)^2\tan^2\theta$. [**Ans:** $yp-xq=0$]

20.2 Solve the following PDEs :

(a) $\dfrac{\partial^2 z}{\partial x^2}=xy$ [**Ans:** $z=\frac{1}{6}x^3y+xf_1(y)+f_2(y)$]

(b) $\dfrac{\partial z}{\partial x}-2\dfrac{\partial z}{\partial y}=2xe^y+1$ [**Ans:** $f(2x+y)=z-\frac{(2x+1)^2}{4}\cdot\frac{y}{2}$]

(c) $\dfrac{\partial^2 z}{\partial x\partial t}=e^{-t}\cos x$ [**Ans:** $z=e^{-t}\sin x+f_1(x)+f_2(t)$]

(d) $(x+y)\left(\dfrac{\partial z}{\partial x}+\dfrac{\partial z}{\partial y}\right)=z-1$ **[AMIE(Winter) 1992]** [**Ans:** $f(x-y)=\frac{x+y}{(z-1)^2}$]

(e) $\dfrac{\partial^2 z}{\partial y^2}=z$, given that when $y=0$, $z=e^x$ and $\dfrac{\partial z}{\partial y}=e^{-x}$

[**Ans:** $z=e^x\cosh y+e^{-x}\sinh x$]

(f) $\dfrac{\partial^2 z}{\partial x^2}=a^2 z$, given that when $x=0$, $\dfrac{\partial z}{\partial x}=a\sin y$ and $\dfrac{\partial z}{\partial y}=0$

[**Ans:** $z=\sin x+e^y\cos x$]

(g) $\dfrac{\partial^2 z}{\partial x^2}+z=0$, given that when $x=0$, $z=e^y$ and $\dfrac{\partial z}{\partial x}=1$

[**Ans:** $z=\sin x+e^y\cos x$].

20.3 Solve the following PDEs :

(a) $pyz+qzx=xy$ [**Ans:** $x^2-y^2=f(y^2-z^2)$]

(b) $y^2p+xyq=x(z-2y)$ [**Ans:** $x^2+y^2=f(y^2-yz)$]

(c) $p+3q=5z+\tan(y-3x)$ [**Ans:** $f(y-3x)=\frac{e^{5x}}{5z+\tan(y-x)}$]

(d) $px(z-2y^2)=(z-qy)(z-y^2-2x^3)$ [**Ans:** $f\left(\frac{y}{z},\frac{z}{x}-\frac{y^2}{x}+x^2\right)=0$]

(e) $p\sqrt{x}+q\sqrt{y}=\sqrt{z}$ [**Ans:** $\sqrt{x}-\sqrt{y}=f(\sqrt{x}-\sqrt{z})$]

(f) $(y+2x)p-(x+yz)q=x^2-y^2$. [**Ans:** $f(x^2+y^2-z^2)=(x-y)^2-(z+1)^2$]

20.4 Solve the following PDEs :

(a) $\dfrac{\partial^2 z}{\partial x^2}-2\dfrac{\partial^2 z}{\partial x\partial y}+\dfrac{\partial^2 z}{\partial y^2}=0$ [**Ans:** $z=f_1(y+x)+xf_2(y+x)$]

(b) $\dfrac{\partial^3 z}{\partial x^3}-4\dfrac{\partial^3 z}{\partial x^2\partial y}+4\dfrac{\partial^3 z}{\partial x\partial y^2}=0$ [**Ans:** $z=f_1(y)+f_2(y+2x)+xf_3(y+2x)$]

(c) $4\dfrac{\partial^2 z}{\partial x^2}-4\dfrac{\partial^2 z}{\partial x\partial y}+\dfrac{\partial^2 z}{\partial y^2}=0$ [**Ans:** $z=f_1\left(y+\frac{x}{2}\right)+xf_2\left(y+\frac{x}{2}\right)$]

(d) $\dfrac{\partial^2 y}{\partial t^2}-a^2\dfrac{\partial^2 y}{8x^2}=0$ [**Ans:** $y=f_1(x-at)+f_2(x+at)$]

(e) $(2DD'+D'^2-3D')z=0$. [**Ans:** $z=f_1(x)+e^{3y}f_2(2y-x)$]

20.5 Solve the following PDEs :

(a) $\dfrac{\partial^2 z}{\partial x^2}-7\dfrac{\partial^2 z}{\partial x\partial y}+12\dfrac{\partial^2 z}{\partial y^2}=e^{x-y}$ [**Ans:** $z=f_1(y+3x)+f_2(y+4x)+\frac{1}{20}e^{x-y}$]

(b) $\dfrac{\partial^2 z}{\partial x^2}-\dfrac{\partial^2 z}{\partial x\partial y}=\sin x\cos 2y$ **[AMIE(Summer) 1995, '92]**

[**Ans:** $z=f_1(y)+f_2(y+x)+\frac{1}{2}\sin(x+2y)-\frac{1}{6}\sin(x-2y)$]

(c) $\dfrac{\partial^2 z}{\partial x^2}-2\dfrac{\partial^2 z}{\partial x\partial y}+\dfrac{\partial^2 z}{\partial y^2}=\sin x$

[UPTU 2004] [**Ans:** $z=f_1(y+x)+xf_2(y+x)-\sin x$]

(d) $(D^2+3DD'+2D'^2)z=24xy$ [**Ans:** $z=f_1(y-x)+f_2(y-2x)+4x^3y-3x^4$]

(e) $\dfrac{\partial^2 z}{\partial x^2} + 3\dfrac{\partial^2 z}{\partial x \partial y} + 2\dfrac{\partial^2 z}{\partial y^2} = 12xy$ [**Ans:** $z = f_1(y-x) + f_2(y-2x) + 2x^3y - \frac{3x^4}{2}$]

(f) $\dfrac{\partial^2 z}{\partial x^2} - a^2\dfrac{\partial^2 z}{\partial y^2} = x^2$ [**Ans:** $z = f_1(y+ax) + f_2(y-ax) + \frac{x^4}{12}$]

(g) $(D^2 - DD' - 2D'^2)z = (2x^2 + xy - y^2)\sin xy - \cos xy$
[**Ans:** $z = f_1(y+2x) + f_2(y-x) + \sin xy$]

(h) $\dfrac{\partial^2 z}{\partial x^2} + \dfrac{\partial^2 z}{\partial x \partial y} + \dfrac{\partial z}{\partial y} - z = e^{-x}$ [**Ans:** $z = e^{-x}f_1(y) + e^{x}f_2(y-x) - \frac{xe^{-x}}{2}$]

(i) $p^2 + 2xp + q = e^{2y}$ [**Ans:** $z = -x^2 + \frac{e^{2y}}{2} + c$]

(j) $pxy + pq + qy = zy$ [**Ans:** $z = ax + be^{y}(y+a)^{-a}$]

(k) $r - t\cos^2 x + p\tan x = 0$ [**Ans:** $z = f_1(y + \sin x) + f_2(y - \sin x)$]

(l) $y^2 r - 2ys + t = p + 6y$
[Osmania U 2002] [**Ans:** $z = y^3 - yf_1(y^2+2x) + f_2(y^2+2x)$]

(m) $\log s = x + y$ [**Ans:** $z = e^{x+y} + f_1(y) + f_2(x)$]

(n) $xr + p = 9x^2y^2$ [**Ans:** $z = x^3y^3 + f_1(y)\log x + f_2(y)$]

(o) $s - t = \dfrac{x}{y^2}$ [**Ans:** $z = f_1(x) + (x+y)\log y + F(x+y)$]

(p) $r = a^2 t$ [**Ans:** $z = f_1(y+ax) + f_2(y-ax)$]

(q) $(x^2D^2 - 4xyDD' + 4y^2D'^2 + 4yD' + xD)z = x^2y$
[**Ans:** $z = f_1(\log x^2y) + \log x f_2(\log x^2y) + \frac{1}{2}(\log x)^2x^2y^2$]

(r) $(D^2 - D')z = xe^{ax+a^2y}$. [**Ans:** $z = \sum_r A_r e^{(a_r x + a_r^2 y)} + \left(\dfrac{x^2}{4a} - \dfrac{x}{4a^2}\right)e^{(ax+a^2y)}$]

CHAPTER 21

Applications of Partial Differential Equations

21.1 Types of Partial Differential Equations (i.e., Elliptic, Parabolic and Hyperbolic Equations)

A linear second order PDE in two variables x and y, with constant coefficients is given by

$$A\frac{\partial^2 u}{\partial x^2} + 2H\frac{\partial^2 u}{\partial x \partial y} + B\frac{\partial^2 u}{\partial y^2} + 2F\frac{\partial u}{\partial x} + 2G\frac{\partial u}{\partial y} + Cu = 0, \qquad (21.1)$$

where A, H, B, F, G and C are constant coefficients.

Again, the general equation of a conic is given by

$$Ax^2 + 2Hxy + By^2 + 2Fx + 2Gy + C = 0. \qquad (21.2)$$

Now comparing Eqns. (21.1) and (21.2) we can say that the Eqns. (21.1) is of

(i) **elliptic type**, if $AB - H^2 > 0$

(ii) **parabolic type**, if $AB - H^2 = 0$, and

(iii) **hyperbolic type**, if $AB - H^2 < 0$.

Example 21.1. *State the type of the following PDEs :*

(a) $\frac{\partial^2 u}{\partial x^2} + \frac{\partial^2 u}{\partial y^2} = 0$ (b) $\frac{\partial^2 u}{\partial x^2} - k^2\frac{\partial^2 u}{\partial y^2} = 0$ (c) $\frac{\partial^2 u}{\partial x^2} - k\frac{\partial u}{\partial x} = 0$. **[BU(H) 2001]**

Solution : In Eqn. (a), $A = 1$, $H = 0$, $B = 1$, $F = G = C = 0$.

$\therefore \quad AB - H^2 = 1 - 0 = 1 > 0$

So, the Eqn. (a) is of **elliptic type**.

In Eqn. (b), $A = 1$, $H = 0$, $B = -k$, $F = G = C = 0$.

$\therefore \quad AB - H^2 = -k^2 < 0.$

So, the Eqn. (b) is of **hyperbolic type**.

In Eqn. (c), $A = 1$, $H = 0$, $B = 0$, $G = -\frac{k}{2}$ and $F = C = 0$.

$\therefore \quad AB - H^2 = 0 - 0 = 0.$

So, the Eqn. (c) is of **parabolic type**.

Example 21.2. *State the type (parabolic, hyperbolic or elliptic) of the following partial differential equations :*

(i) $\frac{\partial^2 u}{\partial t^2} = c^2\frac{\partial^2 u}{\partial x^2}$ (c-velocity) (ii) $9\frac{\partial^2 u}{\partial x^2} + 6\frac{\partial u}{\partial x}\frac{\partial u}{\partial y} + \frac{\partial^2 u}{\partial y^2} = 3x + 4y + 1$ **[BU(H) 1992]**

Solution : In Eqn. (i), $A = 1,\ H = 0,\ B = -c^2,\ F = G = c = 0$

$\therefore \quad AB - H^2 = -c^2 - 0 = -c^2 < 1.$

So Eqn. (i) is of **hyperbolic type**.

In Eqn. (ii), $A = 9, H = 3,\ B = 1,\ F = G = c = 0$

$\therefore\ \ AB - H^2 = 9 - 9 = 0.$

So, Eqn. (ii) is of **parabolic type.**

21.2 The Partial Differential Equations Used in Practical Problems

The following types of partial differential equations are generally used in practical problems :

(i) **One-dimensional wave equation :** $\dfrac{\partial^2 u}{\partial x^2} = c^2\dfrac{\partial^2 u}{\partial x^2}$

(ii) **One-dimensional heat flow equation :** $\dfrac{\partial u}{\partial t} = c^2\dfrac{\partial^2 u}{\partial x^2}$

(iii) **Two-dimensional heat flow equation :** $\dfrac{\partial^2 u}{\partial x^2} + \dfrac{\partial^2 u}{\partial y^2} = 0$

(iv) **Radio equations :** $-\dfrac{\partial V}{\partial x} = L\dfrac{\partial i}{\partial t},\ -\dfrac{\partial i}{\partial x} = C\dfrac{\partial V}{\partial t}$

(v) **Three-dimensional Poisson's equation :** $\dfrac{\partial^2 u}{\partial x^2} + \dfrac{\partial^2 u}{\partial y^2} + \dfrac{\partial^2 u}{\partial z^2} = 4\pi\rho$

and Laplace's equation : $\dfrac{\partial^2 u}{\partial x^2} + \dfrac{\partial^2 u}{\partial y^2} + \dfrac{\partial^2 u}{\partial z^2} = 0.$

21.3 Method of Solving Partial Differential Equations

Different methods to solve PDEs with given boundary conditions are as follows :

1. By inspection and trial 2. By method of Green function 3. By separation of variables.

21.4 Solved Problems

Example 21.1. *Solve by the method of separation of variable,* $\dfrac{\partial u}{\partial x} = 2\dfrac{\partial u}{\partial t} + u$ *under the condition* $u(x, 0) = 6e^{-3x}$. **[BU(H) 1995, AMIE(Winter) 1997, Kurukshetra U 2006]**

Solution : We assume the solution, $u = X(x)T(t)$, (i)

where X is a function of x and T is s function of t only.

$\therefore\ \dfrac{\partial u}{\partial x} = 2\dfrac{\partial u}{\partial t} + u$ becomes

$$\frac{\partial(X,T)}{\partial x} = 2\frac{\partial(X,T)}{\partial t} + X\cdot T \quad \text{or,}\quad T\frac{dX}{dx} = 2X\frac{dT}{dt} + X\cdot T \quad \text{or,}\quad TX' = 2XT' + XT$$

$$\text{or,}\quad \frac{X'}{X} = 2\frac{T'}{T} + 1 = C \text{ (say)}.$$

(a) $\dfrac{X'}{X} = C \quad \text{or,}\quad \dfrac{1}{X}\dfrac{dX}{dx} = C \quad \text{or,}\quad \dfrac{dX}{X} = C\,dx.$

Integrating,

$$\log x = Cx + \log a \quad \text{or,}\quad \log\frac{X}{a} = Cx \quad \text{or,}\quad \frac{X}{a} = e^{Cx} \quad \text{or,}\quad X = ae^{Cx}.$$

(b) $\dfrac{2T'}{T} + 1 = C \quad \text{or,}\quad \dfrac{T'}{T} = \dfrac{1}{2} = (C-1) \quad \text{or,}\quad \dfrac{1}{T}\dfrac{dT}{dt} = \dfrac{1}{2}(C-1) \quad \text{or,}\quad \dfrac{dT}{T} = \dfrac{1}{2}(C-1)dt.$

Integrating,

$$\log T = \frac{1}{2}(C-1)t + \log b \text{ or, } \log\frac{T}{b} = \frac{1}{2}(C-1)t \text{ or, } \frac{T}{b} = e^{\frac{1}{2}(C-1)t} \text{ or, } T = be^{\frac{1}{2}(C-1)t}.$$

Now putting the values of X and T in Eqn. (i), we get

$$u = ae^{Cx}\cdot be^{\frac{1}{2}(C-1)t} \quad \text{or,} \quad u = abe^{Cx+\frac{1}{2}(C-1)t} \tag{ii}$$

or, $u(x,0) = abe^{Cx}$.

But $u(x,0) = 6e^{-3x}$, $\therefore$ $abe^{Cx} = 6e^{-3x}$ or, $ab = 6$ and $C = -3$.

Using these values of ab and C in Eqn. (ii), we get the desired solution as

$$\boldsymbol{u = 6e^{-3x+\frac{1}{2}(-3-1)t} = 6e^{-3x-2t}}.$$

Example 21.2. *Solve the following boundary value problem* $\frac{\partial u}{\partial x} = 4\frac{\partial u}{\partial y}$, *given* $u(0,y) = 8e^{-3y}$, *by the method of separation of variables.* **[NBU(H) 2009, VU(H) 2008]**

Solution : Let the solution of the given equation be

$u = X(x)Y(y)$, where X is a function of x and Y is a function of y only.

$$\therefore \quad \frac{\partial u}{\partial x} = Y\frac{dX}{dx};\quad \frac{\partial u}{\partial y} = X\frac{dY}{dy}.$$

Substituting these values in the given equation, we get

$$Y\frac{dX}{dx} = 4X\frac{dY}{dy} \quad \text{or,} \quad \frac{1}{X}\frac{dX}{dx} = \frac{4}{Y}\frac{dY}{dy}. \tag{i}$$

Since, the LHS of Eqn. (i) is a function of x only and the RHS is a function of y only, each must be equal to a given constant C (say).

$$\therefore \quad \frac{1}{X}\frac{dX}{dx} = C \quad \text{or,} \quad \frac{dX}{X} = C\,dx.$$

Integrating, $\log X = Cx + \log a$ or, $X = ae^{Cx}$.

Again, $\frac{4}{Y}\frac{dY}{dy} = C$ or, $\frac{dY}{y} = \frac{C}{4}dy$.

Integrating, $\log Y = \frac{C}{4}y + \log b$

$$\therefore \quad Y = be^{\frac{Cy}{4}} = be^{ny} \quad \left[\text{where } n = \frac{C}{4}\right]$$

$$\therefore \quad u = XY = ae^{Cx}\cdot be^{ny} = abe^{Cx}e^{ny}.$$

But by the condition of the problem

$$8e^{-3y} = u = ke^{ny} \quad [\text{where } k = ab]$$

or, $\quad k = 8e^{-3y-ny} = 8e^{-y(3+n)}$.

$$\therefore \quad u = ke^{mx}e^{ny} = 8e^{-y(3+n)}e^{mx}e^{ny} \quad \text{or,} \quad \boldsymbol{u = 8e^{mx-3y}}.$$

Example 21.3. *Use the method of separation of variables to find the solution of the equation* $\frac{\partial^2 u}{\partial x^2} + \frac{\partial^2 u}{\partial y^2} = 0$ *in the region* $0 \le x \le a$, $0 \le y < \infty$, *satisfying the conditions* $u(0, y) = 0$, $u(a, y) = 0$, $u(x, 0) = A\left(1 - \frac{x}{a}\right)$ *and* $u(x, \infty) = 0$. **[BU(H) 1993, VU(H) 2007, 2003]**

Solution : Let $u = X(x)Y(y)$. (i)

Putting the values of $\frac{\partial^2 u}{\partial x^2}$ and $\frac{\partial^2 u}{\partial y^2}$ in the given equation, we get

$$Y\frac{d^2X}{dx^2} + X\frac{d^2Y}{dy^2} = 0 \quad \text{or,} \quad \frac{1}{Y}\frac{d^2X}{dx^2} = -\frac{1}{Y}\frac{d^2Y}{dy^2} = -p^2 \text{ (say)}$$

$$\therefore \quad \frac{d^2X}{dx^2} = -p^2X \quad \text{or,} \quad \frac{d^2X}{dx^2} + p^2X = 0 \qquad \text{(ii)}$$

$$\text{and} \quad \frac{d^2Y}{dy^2} - p^2y = 0. \qquad \text{(iii)}$$

A.E. of (ii) is $m^2 + p^2 = 0$ or, $m = \pm ip$.

$\therefore \quad X = C_1 \cos px + C_2 \sin py.$

A.E. of (iii) is $m^2 - p^2 = 0$, $\therefore$ $m = \pm p$.

$\therefore \quad Y = C_3e^{py} + C_4e^{-py}.$

$$\therefore \quad u = XY = (C_1 \cos px + C_2 \sin py)(C_3e^{px} + C_4e^{-py}). \qquad \text{(iv)}$$

By the condition $u(0, y) = 0$,

$$C_1(C_3e^{px} + C_4e^{-py}) = 0, \quad \therefore C_1 = 0.$$

$\therefore$ Eqn. (iv) is reduced to

$$u = C_2 \sin px(C_3e^{px} + C_4e^{-px}). \qquad \text{(v)}$$

Again, by the condition $u(a, y) = 0$,

$$B \sin pa(C_3e^{px} + C_4e^{-py}) = 0 \quad \text{or,} \quad \sin pa = 0 \quad [\because C_2 \neq 0]$$

$$\therefore \quad pa = n\pi, \quad n = 0, \pm 1, \pm 2, \ldots$$

$$\text{or,} \quad p = \frac{n\pi}{a}.$$

Then Eqn. (i) reduces to

$$u(x, y) = C_2 \sin \frac{n\pi x}{a}\left(C_3e^{\frac{n\pi}{a}y} + C_4e^{-\frac{n\pi}{a}y}\right).$$

Further by the condition, $u(x, \infty) = 0$,

$$u(x, \infty) = C_2C_3 \sin \frac{n\pi x}{a}, \quad \text{i.e.,} \quad C_3 = 0.$$

$\therefore$ Eqn. (i) becomes $u(x, y) = C_2C_4 \sin \frac{n\pi x}{a} e^{-\frac{n\pi}{a}y}$.

Again, $\because$ $u(x, 0) = A\left(1 - \frac{x}{a}\right)$,

$$\therefore \quad u(x, 0) = C_2C_4 \sin \frac{n\pi x}{a} = b_n \sin \frac{n\pi x}{a} \quad [\text{where } C_2C_4 = b_n]$$

Adding all solutions,

$$u(x,0)=\sum_{n=1}^{\infty} b_n \sin\frac{n\pi x}{a}=A\left(1-\frac{x}{a}\right).$$

Expanding $A\left(1-\frac{x}{a}\right)$ in Fourier half-sine series in $(0,a)$ we get,

$$b_n=\frac{2A}{a}\int_0^a\left(1-\frac{x}{a}\right)\sin\frac{n\pi x}{a}dx=-\frac{2A}{a}\left[\left(1-\frac{x}{a}\right)\frac{a}{n\pi}\cos\frac{n\pi x}{a}+\frac{a}{n^2\pi^2}\sin\left(\frac{n\pi x}{a}\right)\right]_0^a$$

$$=\frac{2A}{n\pi}.$$

Hence, the required solution is $\boldsymbol{u(x,y)=\frac{2A}{\pi}\sum_{n=1}^{\infty}\frac{1}{n}\sin\left(\frac{n\pi x}{a}\right)e^{-\frac{n\pi y}{a}}}$.

Example 21.4. *Solve Laplace's equation in*

(i) Two-dimensional Cartesian form, (ii) Two-dimensional polar form.

Solution : (i) Laplace's equation in the two-dimensional Cartesian form

Laplace's equation in two dimensional Cartesian form is

$$\frac{\partial^2 u}{\partial x^2}+\frac{\partial^2 u}{\partial y^2}=0. \tag{i}$$

Assuming the variables to be separable,

let $\quad u(x,y)=X(x)Y(y), \tag{ii}$

where X is a function of x only and Y is a function of y only.

Substituting equation (ii) in equation (i), we get

$$Y\frac{d^2X}{dx^2}+X\frac{d^2Y}{dy^2}=0. \tag{iii}$$

Dividing Eqn. (iii) by XY, we get

$$\frac{1}{X}\frac{d^2X}{dx^2}+\frac{1}{Y}\frac{d^2Y}{dy^2}=0.$$

$$\because\quad \frac{d}{dx}\left(\frac{1}{X}\frac{d^2X}{dx^2}\right)=\frac{d}{dx}\left(-\frac{1}{Y}\frac{d^2Y}{dy^2}\right)=0 \quad\text{and}\quad \frac{d}{dy}\left(\frac{1}{X}\frac{d^2X}{dx^2}\right)=\frac{d}{dy}\left(-\frac{1}{Y}\frac{d^2Y}{dy^2}\right)=0,$$

$$\therefore\quad \frac{1}{X}\frac{d^2X}{dx^2}=-\frac{1}{Y}\frac{d^2Y}{dy^2}=\lambda^2 \text{ (say)}, \tag{iv}$$

where λ^2 is a constant independent of x and y and is called the **separation constant.**

$$\therefore\quad \frac{d^2X}{dx^2}-\lambda^2X=0 \tag{v}$$

and $$\frac{d^2Y}{dy^2}+\lambda^2Y=0. \tag{vi}$$

The solutions of Eqns. (v) and (vi) are

$$\boldsymbol{X=C_1e^{\lambda x}+C_2e^{-\lambda x}} \quad\text{and}\quad \boldsymbol{Y=C_3\cos\lambda y+C_4\sin\lambda y}.$$

$\therefore$ solution of Eqn. (i) is

$$\boldsymbol{u = XY = (C_1e^{\lambda x} + C_2e^{-\lambda x})(C_3 \cos \lambda y + C_4 \sin \lambda y)}, \quad \text{(vii)}$$

where C_1, C_2, C_3, C_4 and λ are constants.

If the separation constant be taken as $-\lambda^2$, then

$$\boldsymbol{u = (C_1 \cos \lambda x + C_2 \sin \lambda x)(C_3e^{\lambda y} + C_4e^{-\lambda y})}. \quad \text{(viii)}$$

(ii) Solution of Laplace's equation in the two-dimensional Polar form

Laplace's equation in two dimensional polar form is

$$\frac{\partial^2 u}{\partial r^2} + \frac{1}{r}\frac{\partial u}{\partial r} + \frac{1}{r^2}\frac{\partial^2 u}{\partial \theta^2} = 0. \quad \text{(i)}$$

Let the solution of Eqn. (i) be

$$u(r, \theta) = R(r)\Theta(\theta). \quad \text{(ii)}$$

Substituting Eqn. (ii) in Eqn. (i), we get

$$\Theta\left(\frac{d^2R}{dr^2} + \frac{1}{r}\frac{dR}{dr}\right) + \frac{R}{r^2}\frac{d^2\Theta}{d\theta^2} = 0.$$

$$\therefore \quad \frac{r^2}{R}\left(\frac{d^2R}{dr^2} + \frac{1}{r}\frac{dR}{dr}\right) = -\frac{1}{\Theta}\frac{d^2\Theta}{d\theta^2} = \lambda^2 \text{ (say)} \quad \text{(iii)}$$

$$\text{or,} \quad \frac{d^2\Theta}{d\theta^2} + \lambda^2\Theta = 0 \quad \text{(iv)}$$

$$\text{and} \quad r^2\frac{d^2R}{dr^2} + r\frac{dR}{dr} - \lambda^2 R = 0. \quad \text{(v)}$$

Now, the solution of Eqn. (iv) is

$$\Theta = C_1 \cos \lambda\theta + C_2 \sin \lambda\theta.$$

To find the solution of Eqn. (iv), let $R = r^m$ $(m \neq 0)$.

$$\therefore \quad \frac{dR}{dr} = mr^{m-1} \quad \text{and} \quad \frac{d^2R}{dr^2} = m(m-1)r^{m-2}.$$

Substituting these values in Eqn. (iv), we get

$$m(m-1)r^m + mr^m - \lambda^2 r^m = 0 \quad \text{or,} \quad m^2 - \lambda^2 = 0. \quad \therefore \quad m = \pm\lambda.$$

$$\therefore \quad \boldsymbol{u = R\Theta = (C_3r^{\lambda} + C_4r^{-\lambda})(C_1 \cos \lambda\theta + C_2 \sin \lambda\theta)}. \quad \text{(vi)}$$

Sometimes solution extending up to the origin is required. Since u must be finite at the origin, we must have $C_4 = 0$. So, the solution becomes

$$\boldsymbol{u = (C_1' \cos \lambda\theta + C_2' \sin \lambda\theta)r^{\lambda}}, \quad \text{(vii)}$$

where $C_1' = C_3C_1$ and $C_2' = C_3C_2$.

Example 21.5. *Find the solution of Laplace's equation in*

(i) *Three-dimensional Cartesian form*

(ii) *Three-dimensional Cylindrical form*

(iii) *Spherical form.*

Solution : (i) Solution of Laplace's equation in the three dimensional Cartesian form

Laplace's equation in two-dimensional Cartesian form is

$$\frac{\partial^2 u}{\partial x^2} + \frac{\partial^2 u}{\partial y^2} + \frac{\partial^2 u}{\partial z^2} = 0. \tag{i}$$

Let $u(x, y, z) = X(x)Y(y)Z(z)$. (ii)

Substituting Eqn. (ii) in Eqn. (i), we get

$$\frac{1}{X}\frac{d^2X}{dx^2} + \frac{1}{Y}\frac{d^2Y}{dy^2} + \frac{1}{Z}\frac{d^2Z}{dz^2} = 0 \quad \text{or,} \quad f_1(x) + f_2(y) + f_3(z) = 0.$$

This relation is true if f_1, f_2 and f_3 are constant functions. If the constants be k^2, l^2 and $-(k^2 + l^2)$ respectively, then

$$\frac{d^2X}{dx^2} - k^2x = 0, \quad \frac{d^2Y}{dy^2} - l^2y = 0 \quad \text{and} \quad \frac{d^2Z}{dz^2} + (k^2 + l^2)Z = 0,$$

where k^2 and l^2 may be real or complex.

The solutions of the above set of equations are :

$$X = C_1e^{kx} + C_2e^{-kx}, \quad Y = C_3e^{ly} + C_4e^{-ly}$$

$$Z = C_5\cos(\sqrt{k^2 + l^2})Z + C_6\sin(\sqrt{k^2 + l^2})Z.$$

$\therefore$ the solution of Eqn. (i) is

$$u = XYZ$$

$$= (C_1e^{kx} + C_2e^{-kx})(C_3e^{ly} + C_4e^{-ly})[C_5\cos(\sqrt{k^2 + l^2})Z + C_6\sin(\sqrt{k^2 + l^2})Z]. \tag{iii}$$

(ii) Solution of Laplace's equation in three dimensional cylindrical form

[VU(H) 2001]

Laplace's equation in cylindrical form is

$$\frac{\partial^2 u}{\partial r^2} + \frac{1}{r}\frac{\partial u}{\partial r} + \frac{1}{r^2}\frac{\partial^2 u}{\partial r^2} + \frac{\partial^2 u}{\partial z^2} = 0. \tag{i}$$

Let the solution be $u = (r, \theta, z) = R(r)\Theta(\theta)Z(z)$. (ii)

Substituting Eqn. (ii) in Eqn. (i), we get

$$\frac{1}{R}\left(\frac{d^2R}{dr^2} + \frac{1}{r}\frac{dR}{dr}\right) + \frac{1}{r^2\Theta}\frac{d^2\Theta}{d\theta^2} + \frac{1}{Z}\frac{d^2Z}{dz^2} = 0. \tag{iii}$$

The last term must be a constant (say k^2), then

$$\frac{d^2Z}{dz^2} = k^2Z. \tag{iv}$$

$$\therefore \quad Z = C_5e^{kz} + C_6e^{-kz}.$$

Now, the Eqn. (ii) reduces to

$$\frac{1}{R}\left(\frac{d^2R}{dr^2} + \frac{1}{r}\frac{dR}{dr}\right) + \frac{1}{r^2\Theta}\frac{d^2\Theta}{d\theta^2} + k^2 = 0.$$

Multiplying both sides by r^2R, we get

$$r^2\frac{d^2R}{dr^2}+r\frac{dR}{dr}+\left(\frac{1}{\Theta}\frac{d^2\Theta}{d\theta^2}+k^2r^2\right)R=0.$$

The above equation is separated as

$$\frac{1}{\Theta}\frac{d^2\Theta}{d\theta^2}=-n^2\text{ (say)} \tag{v}$$

and $$r^2\frac{d^2R}{dr^2}+r\frac{dR}{dr}+(k^2r^2-n^2)R=0. \tag{vi}$$

From Eqn. (v), we write,

$$\frac{d^2\Theta}{d\theta^2}+n^2\Theta=0.$$

$$\therefore \quad \boldsymbol{\Theta=C_3\cos n\theta+C_4\sin n\theta.} \tag{vii}$$

Eqn. (vi) is the **Bessel's equation** and its solutions are called Bessel's functions. The solution of Eqn. (vi) is

$$\boldsymbol{R=C_1J_n(kr)+C_2Y_n(kr).} \tag{viii}$$

$\therefore$ the solution of Eqn. (i) is

$$\begin{aligned}\boldsymbol{u} &= \boldsymbol{R\Theta Z}\\ &= \boldsymbol{[C_1J_n(kr)+C_2Y_n(kr)](C_3\cos n\theta+C_4\sin n\theta)(C_5e^{kz}+C_6e^{-kz}).}\end{aligned} \tag{ix}$$

This solution of Bessel's equation is called **cylindrical harmonics**.

(iii) Solution of Laplace's equation in spherical form **[VU(H) 1997]**

Laplace's equation in spherical form is

$$\boldsymbol{\frac{\partial^2u}{\partial r^2}+\frac{2}{r}\frac{\partial u}{\partial r}+\frac{1}{r^2}\frac{\partial^2u}{\partial\theta^2}+\frac{\cot\theta}{r^2}\frac{\partial u}{\partial\theta}+\frac{1}{r^2\sin^2\theta}\frac{\partial^2u}{\partial\phi^2}=0.} \tag{i}$$

Let the solution be $u(r,\theta,\phi)=R(r)\Theta(\theta)\Phi(\phi)$. (ii)

Substituting Eqn. (ii) in Eqn. (i) and then dividing throughout by $R\Theta\Phi(\phi)$, we get

$$\frac{1}{R}\left(r^2\frac{d^2R}{dr^2}+2r\frac{dR}{dr}\right)+\frac{1}{\Theta}\left(\frac{d^2\Theta}{d\theta^2}+\cot\theta\frac{d\Theta}{d\theta}\right)+\frac{1}{\Phi\sin^2\theta}\frac{d^2\Phi}{d\phi^2}=0.$$

$$\therefore \quad \frac{1}{R}\left(r^2\frac{d^2R}{dr^2}+2r\frac{dR}{dr}\right)=n(n+1) \tag{iii}$$

and $$\frac{1}{\Phi}\frac{d^2\Phi}{d\phi^2}=-m^2, \tag{iv}$$

where $-m^2$ and $n(n+1)$ are two separation constants.

$$\therefore \quad \frac{d^2\Theta}{d\theta^2}+\cot\theta\frac{d\Theta}{d\theta}+\left[n(n+1)-\frac{m^2}{\sin^2\theta}\right]\Theta=0. \tag{v}$$

Since ϕ is an angular coordinate,

$$\Phi(\phi)=\Phi(2\pi+\theta),\quad \text{if } m \text{ is an integer.}$$

Now, the solution of Eqn. (v) will remain finite for all values of θ if n is a positive integer and $n \geq m$.

Eqn. (v) is the well-known **associated Legendre's equation** whose solution is

$$\Theta = C_1 P_n^m(\cos\theta) + C_2 Q_n^m(\cos\theta). \quad \text{(vi)}$$

Solution of Eqn. (iv) is

$$\Phi(\phi) = C_3 \cos m\phi + C_4 \sin m\phi.$$

To solve Eqn. (iii) we assume, $R = r^k$.

$$\therefore \quad k(k-1) + 2k = n(n-1).$$

This gives $k = n$ or, $k = -(n+1)$.

$$\therefore \quad R = C_5 r^n + C_6 r^{-(n+1)}. \quad \text{(vii)}$$

Hence, the solution of Eqn. (i) is

$$u = R\Theta\Phi = \sum_{n=0}^{\infty}\sum_{m=0}^{\infty}\left(C_5 r^n + \frac{C_6}{r^{n+1}}\right)(C_3 \cos m\phi + C_4 \sin m\phi)$$
$$[C_1 P_n^m(\cos\theta) + C_2 Q_n^m(\cos\theta)]. \quad \text{(viii)}$$

N.B.

(i) Solution of Laplace's equation of the form $u = R(r)\Theta(\theta)\Phi(\phi)$ are called **spherical harmonics**.

(ii) Solution of the form $\Theta(\theta)\Phi(\phi)$ is called **surface harmonic**.

Example 21.6. (a) *Assume separation of* $u = R(r)\Theta(\theta)\Phi(\phi)$.

Hence, show that the Laplace's equation can be decomposed into three total differential equations.

(b) *Find the solution of* Φ*-equation subject to the condition* $\Phi(\phi + 2\pi) = \Phi(\phi)$.

[CU(H) 2002]

Solution : (a) Laplace's equation in spherical form is

$$\frac{\partial^2 u}{\partial r^2} + \frac{2}{r}\frac{\partial u}{\partial r} + \frac{1}{r^2}\frac{\partial^2 u}{\partial\theta^2} + \frac{\cot\theta}{r^2}\frac{\partial u}{\partial\theta} + \frac{1}{r^2\sin^2\theta}\frac{\partial^2 u}{\partial\theta^2} = 0. \quad \text{(i)}$$

Putting $u = R(r)\Theta(\theta)\Phi(\phi)$ in Eqn. (i), we get

$$\Theta\Phi\frac{d^2R}{dr^2} + \frac{2}{r}\Theta\Phi\frac{dR}{dr} + \frac{1}{r^2}R\Phi\frac{d^2\Theta}{d\theta^2} + \frac{\cot\theta}{r^2}R\Phi\frac{d\Theta}{d\theta} + \frac{1}{r^2\sin^2\theta}R\Theta\frac{d^2\Phi}{d\phi^2} = 0.$$

Dividing throughout by $R\Theta\Phi$, we get

$$\frac{1}{R}\frac{d^2R}{dr^2} + \frac{2}{r}\frac{1}{R}\frac{dR}{dr} + \frac{1}{r^2}\frac{1}{\Theta}\frac{d^2\Theta}{d\theta^2} + \frac{\cot\theta}{r^2}\frac{1}{\Theta}\frac{d\Theta}{d\theta} + \frac{1}{r^2\sin^2\theta}\frac{1}{\Phi}\frac{d^2\Phi}{d\phi^2} = 0$$

or, $$\frac{1}{R}\left(r^2\frac{d^2R}{dr^2} + 2r\frac{dR}{dr}\right) + \frac{1}{\Theta}\left(\frac{d^2\Theta}{d\theta^2} + \cot\theta\frac{d\Theta}{d\theta}\right) + \frac{1}{\Phi\sin^2\theta}\frac{d^2\Phi}{d\phi^2} = 0.$$

The above equation is possible if each term is independently equal to a constant.

$$\therefore \quad \frac{1}{R}\left(r^2\frac{d^2R}{dr^2} + 2r\frac{dR}{dr}\right) = n(n+1) \quad \text{(ii)}$$

and $\dfrac{1}{\Phi}\dfrac{d^2\Phi}{d\phi^2} = -m^2,$ (iii)

where $-m^2$ and $n(n+1)$ are two separation constants.

$$\therefore \quad \frac{d^2\Theta}{d\theta^2} + \cot\theta\frac{d\Theta}{d\theta} + \left[n(n+1) - \frac{m^2}{\sin^2\theta}\right]\Theta = 0. \qquad \text{(iv)}$$

Hence, **Laplace's equation can be decomposed into three total differential equation viz., Eqns. (ii), (iii) and (iv).**

(b) To find the solution of Θ-eqn., we have

$$\frac{1}{\Phi}\frac{d^2\Phi}{d\phi^2} = -m^2 \quad \text{or,} \quad \frac{d^2\Phi}{d\phi^2} + m^2\Phi = 0. \qquad \text{(v)}$$

Now, putting D^2 for $\frac{d^2}{d\phi^2}$, we get

$$(D^2 + m^2)\Phi = 0.$$

$\therefore$ A.E. is $D^2 = -m^2$, $\therefore$ $D = \pm im$.

Hence, the solution of Eqn. (v) is $\boldsymbol{\Phi = C_1\cos m\phi + c_2\sin m\phi}$.

Example 21.7. *Find the form of two-dimensional Laplace's equation $\nabla^2 u = 0$ in plane polar coordinates and solve the resulting equation, subject to the boundary conditions :*

$$u(r,0) = 0, \quad 0 \le r \le a$$
$$u(r,\pi) = 0 \quad 0 \le r \le a$$
$$u(a,\theta) = c.$$

[BU(H)(Old) 1995]

Solution : Laplace's equation in polar coordinates is given by

$$r^2\frac{\partial^2 u}{\partial r^2} + r\frac{\partial u}{\partial r} + \frac{\partial^2 u}{\partial\theta^2} = 0. \qquad \text{(i)}$$

For solution by separation of variables,

let $\quad u(r,\theta) = R(r)\Theta(\theta).$ (ii)

Substituting Eqn. (ii), we get

$$r^2\frac{d^2R}{dr^2}\Theta + r\frac{dR}{dr}\Theta + R\frac{d^2\Theta}{d\theta^2} = 0 \quad \text{or,} \quad \Theta\left(r^2\frac{d^2R}{dr^2} + r\frac{dR}{dr}\right) = -R\frac{d^2\Theta}{d\theta^2}$$

$$\text{or,} \quad \frac{\left(r^2\frac{d^2R}{dr^2} + r\frac{dR}{dr}\right)}{R} = -\frac{\left(\frac{d^2\Theta}{d\theta^2}\right)}{\Theta} = k \text{ (say).} \qquad \text{(iii)}$$

$$\therefore \quad r^2\frac{d^2R}{dr^2} + r\frac{dR}{dr} - kR = 0 \qquad \text{(iv)}$$

$$\text{and} \quad \frac{d^2\Theta}{d\theta^2} + k\Theta = 0. \qquad \text{(v)}$$

Putting $r = e^z$, Eqn. (iv) reduces to

$$[\delta(\delta-1) + \delta - k]R = 0, \quad \text{where } \delta = \frac{d}{dz}$$

or, $(\delta^2 - k)R = 0$ or, $\dfrac{d^2R}{dz^2} - kR = 0.$

For $k > 0$, let $k = p^2$, then

$$R = C_1 e^{pz} + C_2 e^{-pz} \quad \text{or,} \quad R = C_1 r^p + C_2 e^{-p}$$

and $\Theta = C_3 \cos p\theta + C_4 \sin p\theta.$

For $k < 0$, let $k = -p^2$, then

$$\boldsymbol{R = C_5 \cos pz + C_6 \sin pz = C_5 \cos(p \log r) + C_6 \sin(p \log r)}$$

and $\boldsymbol{\Theta = C_7 e^{p\theta} + C_8 e^{-p\theta}}.$

For $k = 0$, $R = lz + m = l(\log r) + m$ and $\Theta = n\theta + q.$

Thus, there are three sets of solution of which the solution consistent with the boundary conditions $(r = 0, u = 0)$ will be

$$\boldsymbol{u(r,\theta) = (\alpha r^p + \beta r^{-p})(\gamma \cos p\theta + \delta \sin p\theta)}. \tag{vi}$$

where α, β, γ, δ are arbitrary constants to be determined from the conditions of the problem.

Using condition $u(r,\theta) = 0$, $\gamma(\alpha r^p + \beta r^{-p}) = 0$ i.e., $\gamma = 0$.

$$\therefore \quad u(r,\theta) = \delta \sin p\theta (ar^p + \beta r^{-p}). \tag{vii}$$

Using condition $u(r,\pi) = 0$, $\delta \sin p\pi(\alpha r^p + \beta r^{-p}) = 0$.

$\because$ $\delta \neq 0$, $\sin p\pi = 0 = \sin n\pi$ $[n = \pm 1, \pm 2, \ldots]$,

$\therefore$ $p = n$.

$$\text{So,} \quad u(r,\theta) = \delta \sin n\theta(\alpha r^n + \beta r^{-n}). \tag{viii}$$

But $u(r,\theta) = 0$, when $r = 0$.

$\therefore$ from (viii), $\beta = 0$.

So, $u(r,\theta) = \alpha\delta r^n \sin n\theta = b_n r^n \sin n\theta.$

$$\therefore \text{ the general solution of Eqn. (i) is } u(r,\theta) = \sum_{n=1}^{\infty} b_n r^n \sin n\theta. \tag{ix}$$

By the condition, $u(r,\theta) = \displaystyle\sum_{n=1}^{\infty} b_n a^n \sin n\theta = C.$

The constant b_n may be found out by Fourier equation

$$b_n = \frac{2}{\pi r^n} \int_0^{\pi} C \sin n\theta \, d\theta = \frac{2C}{n\pi a^n}(1 - \cos n\pi) = \frac{2C}{n\pi a^n}[1 - (-1)^n].$$

$\therefore$ if n is even, $b_n = 0$ and if n is odd, $b_n = \dfrac{4C}{n\pi a^n}.$

Finally the required solution of Eqn. (i) is

$$u(r,\theta) = \frac{4C}{\pi}\left[\left(\frac{r}{a}\right)\sin\theta + \frac{1}{3}\left(\frac{r}{a}\right)^3 \sin 3\theta + \frac{1}{5}\left(\frac{r}{a}\right)^5 \sin 5\theta + \cdots\right]$$

$$\text{or,} \quad \boldsymbol{u(r,\theta) = \frac{4C}{\pi}\sum_{n=1}^{\infty} \frac{1}{2n-1} \cdot \left(\frac{r}{a}\right)^{2n-1} \sin(2n-1)\theta}. \tag{x}$$

Example 21.8. *Solve,* $\dfrac{\partial^2 u}{\partial x^2} + \dfrac{\partial^2 u}{\partial y^2} = 0$, *which satisfies the conditions :*

$$u(0, y) = u(l, y) = u(x, 0) = 0 \quad \text{and} \quad u(x, a) = \sin\frac{n\pi x}{l}.$$

[AMIETE(Winter) 1995, 1992; JNTU 2006; Kerala M.Tech 2005]

Solution : The given equation is $\dfrac{\partial^2 u}{\partial x^2} + \dfrac{\partial^2 u}{\partial y^2} = 0.$ (i)

Let $u(x, y) = X(x)Y(y)$ (ii)

where X and Y are functions of the variables x and y respectively.

$$\therefore \quad \frac{\partial^2 u}{\partial x^2} = \frac{\partial^2 X}{\partial x^2}Y \quad \text{and} \quad \frac{\partial^2 u}{\partial y^2} = X\frac{\partial^2 Y}{\partial y^2}$$

Using these values in Eqn. (i), we get

$$\frac{d^2X}{dx^2}Y + X\frac{d^2Y}{dy^2} = 0 \quad \text{or,} \quad \frac{1}{X}\frac{d^2X}{dx^2} = -\frac{1}{Y}\frac{d^2Y}{dy^2} = -\lambda^2 \text{ (say)}.$$

$$\therefore \quad \frac{d^2X}{dx^2} = -\lambda^2 X \quad \text{or,} \quad \frac{d^2X}{dx^2} + \lambda^2 X = 0 \qquad \text{(iii)}$$

$$\text{and} \quad \frac{d^2Y}{dy^2} = \lambda^2 Y \quad \text{or,} \quad \frac{d^2Y}{dy^2} - \lambda^2 Y = 0. \qquad \text{(iv)}$$

Now, A.E. of (iii) is $m^2 + \lambda^2 = 0$ or, $m = \pm i\lambda$.

$$\therefore \quad \boldsymbol{X = C_1 \cos \lambda x + C_2 \sin \lambda x}.$$

Again, A.E. of (iv) is $m^2 - \lambda^2 = 0$ or, $m = \pm\lambda$.

$$\therefore \quad \boldsymbol{Y = C_3 e^{\lambda y} + C_4 e^{-\lambda y}}.$$

Putting the values of X and Y in Eqn. (ii), we get

$$\boldsymbol{u(x, y) = (C_1 \cos \lambda x + C_2 \sin \lambda x)(C_3 e^{\lambda y} + C_4 e^{-\lambda y})}. \qquad \text{(v)}$$

For $x = 0$, $u = 0$ in Eqn. (v),

$$0 = C_1(C_3 e^{\lambda y} + C_4 e^{-\lambda y}), \quad \therefore \; C_1 = 0.$$

Then Eqn. (v) becomes

$$u = C_2 \sin \lambda x (C_3 e^{\lambda y} + C_4 e^{-\lambda y}). \qquad \text{(vi)}$$

Again, for $x = l$, $u = 0$, we get

$$0 = C_2 \sin \lambda l (C_3 e^{\lambda y} + C_4 e^{-\lambda y}).$$

Since, $C_2 \neq 0$, $\quad \therefore \; \sin \lambda l = 0 = \sin n\pi \quad$ or, $\; nl = n\pi$, i.e., $\lambda = \dfrac{n\pi}{l}$.

Then Eqn. (vi) becomes

$$u = C_3 \sin\frac{n\pi x}{l}\left(C_3 e^{\frac{n\pi y}{l}} + C_4 e^{-\frac{n\pi y}{l}}\right). \qquad \text{(vii)}$$

Now, putting $u = 0$ and $y = 0$ in Eqn. (vii),

$$0 = C_2 \sin\frac{n\pi x}{l}(C_3 + C_4).$$

$\therefore \quad C_3 + C_4 = 0 \quad \text{or,} \quad C_4 = -C_3.$

Then Eqn. (vii) becomes

$$u = C_2C_3 \sin\frac{n\pi x}{l}\left(e^{\frac{n\pi y}{l}} - e^{-\frac{n\pi y}{l}}\right). \tag{viii}$$

Putting $y = a$ and $u = \sin\frac{n\pi x}{l}$ in Eqn. (viii),

$$\sin\frac{n\pi x}{l} = C_2C_3 \cdot \sin\frac{n\pi x}{l}\left(e^{\frac{n\pi a}{l}} - e^{-\frac{n\pi a}{l}}\right) \quad \text{or,} \quad C_2C_3 = \frac{1}{e^{\frac{n\pi a}{l}} - e^{-\frac{n\pi a}{l}}}.$$

Putting this value of C_2C_3 in Eqn. (viii), we get

$$u = \sin\frac{n\pi x}{l}\frac{e^{\frac{n\pi y}{l}} - e^{-\frac{n\pi y}{l}}}{e^{\frac{n\pi a}{l}} - e^{-\frac{n\pi a}{l}}} \quad \text{or,} \quad \boldsymbol{u = \sin\left(\frac{n\pi x}{l}\right)\cdot\frac{\sinh\left(\frac{n\pi y}{l}\right)}{\sinh\left(\frac{n\pi a}{l}\right)}}.$$

Example 21.9. *The function $v(x, y)$ satisfies the Laplace's equation in rectangular coordinates (x, y), and for points within the rectangle $x = 0$, $x = a$, $y = 0$, $y = b$ it satisfies the condition $v(0, y) = v(a, y) = v(x, b)$ and $v(x, 0) = x(a - x)$, $0 < x < a$.*

Show that $v(x, y)$ is given by

$$v(x,y) = \frac{8a^2}{\pi^3}\sum_{n=0}^{\infty}\frac{\sin\frac{(2n+1)\pi x}{a}}{(2n+1)^3}\cdot\frac{\sinh\frac{(2n+1)\pi(b-y)}{a}}{\sinh\frac{(2n+1)\pi b}{a}}.$$

[Madras U 2003]

Solution : Laplace's equation in rectangular coordinates (x, y) is

$$\frac{\partial^2 v}{\partial x^2} + \frac{\partial^2 v}{\partial y^2} = 0. \tag{i}$$

The only possible solution of Eqn. (i) is of the form,

$$v(x,y) = (C_1\cos\lambda x + C_2\sin\lambda x)(C_3e^{\lambda y} + C_4e^{-\lambda y}) \tag{ii}$$ [See problem (21.4), page 800]

The boundary conditions are :

$$\left.\begin{aligned} v(0,y) &= 0 \\ v(a,y) &= 0 \end{aligned}\right\} \tag{iii}$$

$$v(x,b) = 0 \tag{iv}$$

$$v(x,0) = x(a-x), \quad 0 < x < a. \tag{v}$$

Using Eqn. (iii) in (ii), we get

$$v(0,y) = C_1(C_3e^{\lambda y} + C_4e^{-\lambda y}) = 0, \quad \text{i.e.,} \quad C_1 = 0.$$

$\therefore$ Eqn. (ii) becomes

$$v(x,y) = C_3\sin\lambda x(C_3e^{\lambda y} + C_4e^{-\lambda y}). \tag{vi}$$

Using Eqn. (iii) in (ii), we get

$$v(a,y) = C_2\sin\lambda a(C_3e^{\lambda y} + C_4e^{-\lambda y}) = 0.$$

$$\therefore \quad \sin\lambda a = 0, \quad \text{i.e.,} \quad \lambda a = n\pi \quad \text{or,} \quad \lambda = \frac{n\pi}{a}.$$

Now, Eqn. (vi) becomes

$$v(x,y) = C_2\sin\frac{n\pi x}{a}\left(C_3e^{\frac{n\pi y}{a}} + C_4e^{-\frac{n\pi y}{a}}\right)$$

or, $$v(x,y)=\sin\frac{n\pi x}{a}\left(\alpha e^{\frac{n\pi y}{a}}+\beta e^{-\frac{n\pi y}{a}}\right),\quad \text{where } \alpha=C_2C_3,\ \beta=C_2C_4. \tag{iv}$$

Using (iv) in (ii),

$$v(x,b)=\sin\frac{n\pi x}{a}\left(\alpha e^{\frac{n\pi b}{a}}+\beta e^{-\frac{n\pi b}{a}}\right)=0 \quad\text{or,}\quad \alpha e^{\frac{n\pi b}{a}}+\beta e^{-\frac{n\pi b}{a}}=0$$

or, $$\alpha e^{\frac{n\pi b}{a}}-\beta e^{-\frac{n\pi b}{a}}=-\frac{1}{2}b_n \quad \text{(say)}.$$

Now, Eqn. (vii) becomes

$$v(x,y)=\sin\frac{n\pi x}{a}\cdot\frac{1}{2}b_n\left[e^{\frac{n\pi(b-y)}{a}}-e^{-\frac{n\pi(b-y)}{a}}\right]=b_n\sin\frac{n\pi x}{a}\sinh\frac{n\pi(b-y)}{a}.$$

Hence, the most general solution of Eqn. (i) is

$$v(x,y)=\sum_{n=1}^{\infty}b_n\sin\frac{n\pi x}{a}\sinh\frac{n\pi(b-y)}{a}. \tag{viii}$$

Using Eqn. (v) in Eqn. (ii), we get

$$x(a-x)=v(x,0)=\sum_{n=1}^{\infty}b_n\sinh\frac{n\pi b}{a}\sin\frac{n\pi x}{a},$$

where $$b_n\sin h\left(\frac{n\pi b}{a}\right)=\frac{2}{a}\int_0^a x(a-x)\sin\left(\frac{n\pi x}{a}\right)dx$$

$$=\frac{2}{a}\left[(ax-x^2)\left(\frac{-\cos\frac{n\pi x}{a}}{\frac{n\pi}{a}}\right)-(a-2x)\left(-\frac{\sin\frac{2\pi x}{a}}{\left(\frac{n\pi}{a}\right)}^2\right)+(-2)\left\{\frac{\cos\frac{n\pi x}{a}}{\left(\frac{n\pi}{a}\right)^3}\right\}\right]_0^a$$

$$=0-0+\frac{4a^2}{n^3\pi^3}(1-\cos n\pi)$$

$$=\frac{8a^2}{n^3\pi^3},\quad \text{when } n \text{ is odd; or zero when } n \text{ is even.}$$

Hence, from Eqn. (viii), the required solution of Eqn. (i) is

$$v(x,y)=\frac{8a^2}{\pi^3}\sum_{n=1,3,5}^{\infty}\frac{\sinh\frac{n\pi(b-y)}{a}}{n^3\sinh\left(\frac{n\pi b}{a}\right)}\sin\left(\frac{n\pi x}{a}\right)$$

or, $$\boldsymbol{v(x,y)=\frac{8a^2}{\pi^3}\sum_{n=0}^{\infty}\frac{\sin\left\{\frac{(2n+1)\pi(b-y)}{a}\right\}}{(2n+1)^3\sinh\left\{\frac{(2n+1)\pi b}{a}\right\}}\sin\frac{(2n+1)\pi x}{a}}.$$

21.5 Vibration of a Stretched String : Wave Equation [BU(H) 2000]

Let a flexible elastic string of length l be fixed at A and B under a constant tension T as shown in Fig. (21.1). We assume that,

(i) T is so large that weight of the string due to gravity is negligible,

(ii) Each point on the string vibrates at right angles to the equilibrium position AB of the string entirely in one plane.

Let the end A of the string be at the origin, AB along the X-axis and AY perpendicular to it as the Y-axis. Then the motion is confined entirely to the XY-plane. We consider an element PQ of the vibrating string at time t. The tangents at P and Q make angles θ_1 and θ_2 with the X-axis (Fig. 21.1).

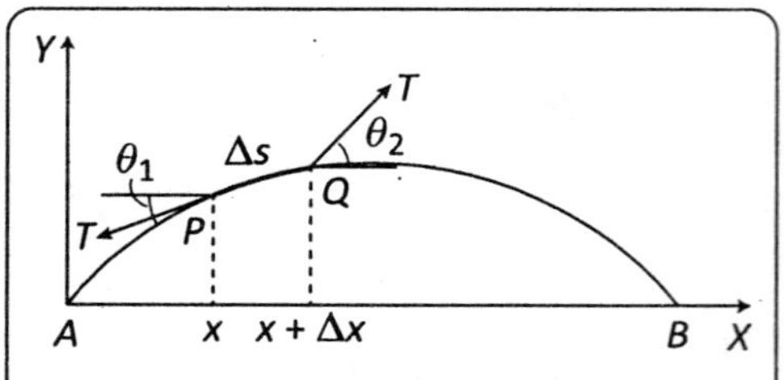

Fig. 21.1

If at any time t the vertical displacement at x be $u(x,t)$, then the acceleration is $\frac{\partial^2 u(x,t)}{\partial t^2}$.

The equation of motion of the element $PQ = \Delta x$ is then

$$T_2 \sin\theta_2 - T_1 \sin\theta_1 = m\Delta s \frac{\partial^2 u}{\partial t^2} \quad \text{(i)}$$

and $T_2 \cos\theta_2 - T_1 \cos\theta_1 = 0,$ (ii)

where m is the mass per unit length of the string.

As there is no horizontal acceleration of any point of the string,

$$T_2 \cos\theta_2 - T_1 \cos\theta_1 = T \text{ (say)}. \quad \text{(iii)}$$

Now Eqn. (i) becomes

$$T\tan\theta_2 - T\tan\theta_1 = m\Delta s \frac{\partial^2 u}{\partial t^2} \quad \text{[by using the values of } T_1 \text{ and } T_2 \text{ from Eqn. (iii)]}$$

or, $$\frac{\partial^2 u}{\partial t^2} = \frac{T}{m} \cdot \frac{1}{\Delta s}(\tan\theta_2 - \tan\theta_1). \quad \text{(iv)}$$

But $\tan\theta_2 = \left(\frac{\partial u}{\partial x}\right)_{x+\Delta x}$, $\tan\theta_1 = \left(\frac{\partial u}{\partial x}\right)_x$.

Since, the displacement $u(x,t)$ is small we can assume, $\Delta s = \Delta x$.

Eqn. (iv) now becomes

$$\frac{\partial^2 u}{\partial t^2} = \frac{T}{m} \cdot \frac{1}{\Delta x}\left[\left(\frac{\partial u}{\partial x}\right)_{x+\Delta x} - \left(\frac{\partial u}{\partial x}\right)_x\right].$$

Since $\Delta x \to 0$, $\therefore$ $$\frac{\partial^2 u}{\partial t^2} = \frac{T}{m}\frac{\partial^2 u}{\partial x^2}. \quad \text{(v)}$$

Putting $c^2 = \frac{T}{m}$, in the above Eqn. (v), we get

$$\boldsymbol{\frac{\partial^2 u}{\partial t^2} = c^2 \frac{\partial^2 u}{\partial x^2}}. \quad \text{(vi)}$$

Eqn. (vi) is the **wave equation** (one-dimensional) of transversely vibrating string under tension T.

21.6 Solution of the Wave Equation [AMIETE(Winter) 1991]

To solve Eqn. (vi) by the method of separation of variables, let

$$u(x,t) = X(x)T(t). \quad \text{(vii)}$$

Using Eqn. (vii) in Eqn. (vi), we get

$$\frac{d^2T}{dx^2}X = c^2 \frac{d^2X}{dt^2}T.$$

or, $$\frac{\left(\frac{d^2T}{dx^2}\right)}{c^2T} = \frac{\left(\frac{d^2X}{dt^2}\right)}{X} = \lambda \text{ (say)} \quad [\because x \text{ and } t \text{ are independent variables}]$$

(i) For λ positive, let $\lambda = p^2$. Then the solutions are

$$X = C_1 e^{px} + C_2 e^{-px}$$
$$T = C_3 e^{cpt} + C_4 e^{-cpt}.$$

(ii) For λ negative, let $\lambda = -p^2$, then

$$X = C_5 \cos px + C_6 \sin px$$
$$T = C_7 \cos cpt + C_8 \sin cpt.$$

(iii) For $\lambda = 0$,

$$X = m_1 x + n_1$$
$$T = p_1 t + q_1.$$

So the possible solutions of Eqn. (vi) are

$$u(x,t) = (C_1 e^{pt} + C_2 e^{-pt})(C_3 e^{cpt} + C_4 e^{-cpt}) \qquad \text{(viii)}$$
$$= (C_5 \cos px + C_6 \sin px)(C_7 \cos cpt + C_8 \sin cpt) \qquad \text{(ix)}$$
$$= (m_1 x + n_1)(p_1 t + q_1). \qquad \text{(x)}$$

Since, the displacement is periodic in this case the accepted solution of Eqn. (vi) is

$$\boldsymbol{u(x,t) = (A \cos px + B \sin px)(C \cos cpt + D \sin cpt)} \qquad \text{(xi)}$$

Example 21.10. *Solve the wave equation* $\frac{\partial^2 u}{\partial t^2} = c^2 \frac{\partial^2 u}{\partial x^2}$ *under the following conditions :*

$$u(0,t) = 0; \quad u(l,t) = 0; \quad u(x,0) = a \sin \frac{\pi x}{l}; \quad \frac{\delta u(x,0)}{\partial t} = 0.$$ **[BU(H) 2001]**

Solution : The given equation is $\frac{\partial^2 u}{\partial t^2} = c^2 \frac{\partial^2 u}{\partial x^2}$. (i)

The solution of Eqn. (i) is given by

$$u(x,t) = (C_1 \cos px + C_2 \sin px)(C_3 \cos cpt + C_4 \sin cpt). \qquad \text{(ii)}$$

Using condition $u(0,t) = 0$ in Eqn. (ii),

$$C_1 (C_3 \cos cpt + C_4 \sin cpt) = 0, \quad \therefore \; C_1 = 0.$$

$\therefore$ Eqn. (ii) becomes

$$u(x,t) = C_2 \sin px (C_3 \cos cpt + C_4 \sin cpt). \qquad \text{(iii)}$$

Using condition $u(l,t) = 0$,

$$C_2 \sin pt (C_3 \cos cpt + C_4 \sin cpt) = 0.$$

$\because \; C_2 \neq 0, \;\; \sin pt = 0 = \sin n\pi \; [n = 1, 2, \ldots].$

$$\therefore \quad p = \frac{n\pi}{l}.$$

So, Eqn. (iii) becomes

$$u(x,t) = C_2 \sin \frac{n\pi x}{l} \left(C_3 \cos \frac{n\pi c}{l} t + C_4 \cos \frac{n\pi c}{l} t \right) \qquad \text{(iv)}$$

$$\therefore \quad \frac{\partial u}{\partial t} = C_2 \sin \left(\frac{n\pi x}{l} \right) \left[\frac{n\pi c}{l} \left\{ -C_3 \sin \left(\frac{n\pi c}{l} t \right) + C_4 \cos \left(\frac{n\pi c}{l} t \right) \right\} \right].$$

Using the condition, $\frac{\partial u(x,0)}{\partial t} = 0$,

$$C_2C_4\frac{n\pi c}{l}\sin\frac{n\pi x}{l} = 0, \text{ i.e., } C_4 = 0 \quad [\because C_2 \neq 0]$$

Hence, Eqn. (iv) becomes

$$u(x,t) = C_2C_3\sin\left(\frac{n\pi x}{l}\right)\cos\left(\frac{n\pi c}{l}t\right).$$

Now, $u(x,0) = C_2C_3\sin\left(\frac{n\pi x}{l}\right) = a\sin\frac{\pi x}{l}$.

This is possible only if $C_2C_3 = a$ and $n = 1$.

$\therefore$ Eqn. (v) becomes

$$\boldsymbol{u(x,t) = a\sin\left(\frac{\pi x}{l}\right)\cos\left(\frac{\pi c}{l}t\right)} \tag{v}$$

is the required solution.

Example 21.11. *A string is stretched and fastened to two points a distance l apart. Motion is started by giving it a displacement y in the form $y = a\sin\left(\frac{\pi x}{l}\right)$ from which it is released at $t = 0$. Show that the displacement of any point at a distance x from one end at time t is given by $y(x,t) = a\sin\left(\frac{\pi x}{l}\right)\cos\left(\frac{\pi ct}{l}t\right)$.*

[BU(H) 2000; SVTU 2007; Kerala U 2005; UPTA 2004]

Solution : The transverse vibration of a stretched string is given by

$$\frac{\partial^2 y}{\partial t^2} = c^2\frac{\partial^2 y}{\partial x^2}. \tag{i}$$

Since, the end points of the string are fixed, for all value of t,

$$y(0,t) = 0 \tag{ii}$$

$$y(l,t) = 0. \tag{iii}$$

As the initial transverse vibration of the string is zero,

$$\left(\frac{\partial y}{\partial t}\right)_{t=0} = 0. \tag{iv}$$

Also, $y(x,0) = a\sin\left(\frac{\pi x}{l}\right)$. (v)

Since, the vibration of the string is periodic, the solution of Eqn. (i) is of the form

$$y(x,t) = (C_1\cos px + C_2\sin px)(C_3\cos cpt + C_4\sin cpt). \tag{vi}$$

By the condition (ii), $y(0,t) = C_1(C_3\cos cpt + C_4\sin cpt) = 0$.

This is to be true for all t, $C_1 = 0$.

Hence, $y(x,t) = C_2\sin px(C_3\cos cpt + C_4\sin cpt)$ (vii)

and $\frac{\partial y}{\partial t} = C_2\sin px[C_3(-cp\sin cpt) + C_4(cp\cdot\cos cpt)]$.

By condition, $\left(\frac{\partial y}{\partial t}\right)_{t=0} = C_2\sin px(C_4cp) = 0$, where $C_2C_4cp = 0$.

If $C_2 = 0$, then the solution of Eqn. (vii) will be $y(x,t) = 0$.

$\therefore$ the only possibility is that $C_4 = 0$.

Thus, Eqn. (vii) becomes

$$y(x,t) = C_2C_3 \sin px = 0, \text{ for all } t.$$

Since, $C_2 \neq 0$, $C_3 \neq 0$, we have $\sin pl = 0$.

$$\therefore \quad pl = n\pi \quad \text{or,} \quad p = \frac{n\pi}{l} \quad [\text{where } n = 1, 2, \ldots].$$

Hence, the solution of Eqn. (i) becomes

$$y(x,t) = C_2C_3 \sin\left(\frac{n\pi x}{l}\right) \cos\left(\frac{n\pi ct}{l}\right).$$

Now by condition (v), we get

$$y(x,0) = C_2C_3 \sin\left(\frac{n\pi x}{l}\right) = a \sin\frac{\pi x}{l}, \text{ where } C_2C_3 = a \text{ and } n = 1.$$

Hence, the required solution is $\boldsymbol{y(x,t) = a \sin\left(\frac{\pi x}{l}\right) \cos\frac{\pi ct}{l}}$. (viii)

Example 21.12. *The transverse vibrations of an elastic string are governed by the equation,*

$$\frac{\partial^2 u}{\partial t^2} = c^2 \frac{\partial^2 u}{\partial x^2},$$

where $u(x,t)$ is the deflection of the string. The string is fixed at both ends and the initial conditions are $u(x,0) = f(x)$ and $\frac{\partial u}{\partial t} = g(x)$ at $t = 0$.

Find the solution of the equation satisfying the above conditions. **[BU(H) 1994]**

Solution : The differential equation governing the transverse vibration of the string is

$$\frac{\partial^2 u}{\partial t^2} = c^2 \frac{\partial^2 u}{\partial x^2}. \tag{i}$$

The general solution of this equation is

$$u(x,t) = (C_1 \cos px + C_2 \sin px)(C_3 \cos cpt + C_4 \sin cpt). \tag{ii}$$

The string of length l is fixed at both ends,

$$\therefore \quad u(0,t) = u(l,t) = 0. \tag{iii}$$

Using Eqn. (iii) in Eqn. (ii), we get

$$u(x,t) = \sin\frac{n\pi x}{l}\left[C_2C_3 \cos\left(\frac{n\pi ct}{l}\right) + C_2C_4 \sin\left(\frac{n\pi ct}{l}\right)\right]. \tag{iv}$$

Now, $$\frac{\partial u(x,t)}{\partial t} = \frac{n\pi c}{l} \sin\left(\frac{n\pi x}{l}\right)\left[-C_2C_3 \sin\frac{n\pi c}{l}t + C_2C_4 \cos\frac{n\pi c}{l}t\right].$$

By the condition, $\frac{\partial u(x,0)}{\partial t} = g(x)$, we get

$$C_2C_4 \frac{n\pi c}{l} \sin\left(\frac{n\pi x}{l}\right) = g(x) \quad \text{or,} \quad a_n \frac{n\pi c}{l} \sin\left(\frac{n\pi x}{l}\right) = g(x).$$

b_n can be found out by Fourier expansion

$$\frac{n\pi c}{l}a_n = \frac{2}{l}\int_0^l g(x)\sin\left(\frac{n\pi x}{l}\right)dx \quad \text{or,} \quad a_n = \frac{2}{n\pi c}\int_0^l g(x)\sin\left(\frac{n\pi x}{l}\right)dx. \tag{v}$$

Using the condition $u(x,0) = f(x)$ in Eqn. (iv),

$$C_2C_3\sin\frac{n\pi x}{l} = f(x) \quad \text{or,} \quad b_n\sin\frac{n\pi x}{l} = f(x),$$

$$\text{where} \quad b_n = \frac{2}{l}\int_0^l f(x)\sin\left(\frac{n\pi x}{l}\right)dx. \tag{vi}$$

Using the values of a_n and b_n and then adding all solutions, we get

$$\boldsymbol{u(x,t) = \sum_{n=1}^{\infty}\left[a_n\sin\frac{n\pi c}{l}t + b_n\cos\frac{n\pi c}{l}t\right]\sin\frac{n\pi x}{l}}. \tag{vii}$$

Example 21.13. *A tightly stretched string with fixed end points $x = 0$ and $x = l$ is initially in position given by $y = y_0\sin^3\left(\frac{\pi x}{l}\right)$. If it is released from rest from this position find the displacement $y(x,t)$.* **[AMIE, Summer, 1994; Rajasthan U 2006; JNTU 2002]**

Solution : The equation of motion of the transversely vibrating string is

$$\frac{\partial^2 y}{\partial t^2} = c^2\frac{\partial^2 y}{\partial x^2}. \tag{i}$$

Here, the boundary conditions are

$$\left.\begin{array}{l} y(0,t) = 0 \\ y(l,t) = 0 \end{array}\right\} \tag{ii}$$

and the initial condition is

$$y(x,0) = y_0\sin^3\left(\frac{\pi x}{l}\right). \tag{iii}$$

The vibration of the string being periodic, the solution of Eqn. (i) can be written as

$$y(x,t) = (C_1\cos px + C_2\sin px)(C_3\cos cpt + C_4\sin cpt) = 0.$$

By condition (ii), $y(0,t) = C_1(C_3\cos cpt + C_4\sin cpt) = 0.$

This is true for all values of t, $\therefore\ C_1 = 0$.

$$\therefore \quad y(x,t) = C_2\sin px(C_3\cos cpt + C_4\sin cpt).$$

Also by condition (ii), $y(l,t) = C_2\sin pl(C_3\cos cpt + C_4\sin cpt) = 0$ for all values of t.

This gives $pl = n\pi$ or, $p = \frac{n\pi}{l}$, n being an integer.

$$\text{Thus,} \quad y(x,t) = C_2\sin\frac{n\pi x}{l}\left[C_3\cos\left(\frac{n\pi ct}{l}\right) + C_4\sin\left(\frac{n\pi ct}{l}\right)\right]. \tag{v}$$

$$\therefore \quad \frac{\partial y}{\partial t} = \left(C_2\sin\frac{n\pi x}{l}\right)\frac{n\pi c}{l}\left[-C_3\sin\left(\frac{n\pi ct}{l}\right) + C_4\cos\left(\frac{n\pi ct}{l}\right)\right].$$

Using condition (iv) in the above equation,

$$\left(\frac{\partial y}{\partial t}\right)_{t=0} = C_3 \sin\left(\frac{n\pi x}{l}\right)\frac{n\pi c}{l}C_4 = 0, \quad \therefore \quad C_4 = 0.$$

Now, Eqn. (v) becomes

$$y(x,t) = C_2C_3 \sin\left(\frac{n\pi x}{l}\right)\cos\left(\frac{n\pi ct}{l}\right), \quad y = b_n \sin\left(\frac{n\pi x}{l}\right)\cos\left(\frac{n\pi ct}{l}\right) \quad [b_n = C_2C_3]$$

By adding all such solutions we get the general solution of Eqn. (i) in the form

$$y(x,t) = \sum_{n=1}^{\infty} b_n \sin\left(\frac{n\pi x}{l}\right)\cos\left(\frac{n\pi ct}{l}\right). \tag{vi}$$

Again, from Eqn. (iii), we get

$$y_0 \sin^3\frac{\pi x}{l} = \sum_{n=1}^{\infty} b_n \sin\frac{n\pi x}{l}$$

or, $$y_0\left[\frac{3\sin\frac{\pi x}{l} - \sin\frac{3\pi x}{l}}{4}\right] = b_1 \sin\frac{\pi x}{l} + b_2\sin\frac{2\pi x}{l} + b_3\sin\frac{3\pi x}{l}.$$

Comparing similar terms from both sides, we get

$$b_1 = \frac{3}{4}y_0, \quad b_2 = 0, \quad b_3 = -\frac{1}{4}y_0, \quad b_4 = b_5 = \cdots = 0.$$

Hence, from Eqn. (vi) the solution of Eqn. (i) is

$$\boldsymbol{y(x,t) = \frac{3y_0}{4}\sin\frac{\pi x}{l}\cos\frac{\pi ct}{l} - \frac{y_0}{4}\sin\frac{3\pi x}{l}\cos\frac{3\pi ct}{l}}. \tag{vii}$$

Example 21.14. *Find the deflection $u(x,t)$ of a vibrating string of length l and the ends fixed corresponding to zero initial velocity and initial deflection,*

$$f(x) = \begin{cases} \frac{2k}{l}x & \text{for } 0 < x < \frac{l}{2} \\ \frac{2k}{l}(l-x) & \text{for } \frac{l}{2} < x < l. \end{cases}$$

Solution : The differential equation governing the vibration of a stretched string is

$$\frac{\partial^2 u}{\partial t^2} = c^2\frac{\partial^2 u}{\partial x^2}. \tag{i}$$

The solution of the above equation is

$$u = (C_3\cos kx + C_4\sin kx)(C_1\cos kct + C_2\sin kct). \tag{ii}$$

Here the boundary conditions are : (i) $u = 0$, $x = 0$ and (ii) $u = 0$, $x = l$.

Using the above boundary conditions in equation (ii), we get

$$0 = C_3(C_1\cos ckt + C_2\sin ckt) \quad \text{and} \quad 0 = (C_3\cos kl + C_4\sin kl)(C_1\cos ckt + C_2\sin ckt).$$

$\therefore$ $\quad C_3 = 0$ and $\sin kl = 0 = \sin n\pi$ or, $kl = n\pi$ or, $k = \frac{n\pi}{l}$, $n =$ an integer.

Substituting the values of C_3 and k in Eqn. (ii), we get after adding the solutions for different n values

$$u = \sum_{n=1}^{\infty} \sin \frac{n\pi x}{l} \left(A_n \cos \frac{n\pi ct}{l} + B_n \sin \frac{n\pi ct}{l} \right). \tag{iii}$$

Applying the boundary conditions, $u = f(x)$ and $\frac{\partial u}{\partial t} = 0$ at $t = 0$, we get

$$f(x) = \sum_{n=1}^{\infty} A_n \sin \frac{n\pi x}{l} \tag{iv}$$

and $$0 = \sum_{n=1}^{\infty} \frac{n\pi c}{l} B_n \sin \frac{n\pi x}{l}, \quad \text{i.e., } B_n = 0 \text{ for all } n. \tag{v}$$

The RHS of Eqn. (iii) is a Fourier expansion of $f(x)$.

$$\therefore \quad A_n = \frac{2}{l} \int_0^l f(x) \sin \frac{n\pi x}{l} dx = \frac{2}{l} \left[\int_0^{\frac{l}{2}} \frac{2k}{l} x \sin \frac{n\pi x}{l} dx + \int_{\frac{l}{2}}^{0} \frac{2k}{l} (l-x) \sin \frac{n\pi x}{l} dx \right]$$

$$= \frac{4k}{l^2} \left[x \left(-\cos \frac{n\pi x}{l} \right) \frac{l}{n\pi} - \left(-\sin \frac{n\pi x}{l} \right) \frac{l^2}{n^2\pi^2} \right]_0^{\frac{l}{2}}$$

$$+ \frac{4k}{l^2} \left[(l-x) \left(-\cos \frac{n\pi x}{l} \right) \frac{l}{n\pi} - (-1) \left(-\sin \frac{n\pi x}{l} \right) \frac{l^2}{n^2\pi^2} \right]_{\frac{l}{2}}^{l}$$

$$= \frac{4k}{l^2} \frac{l}{n\pi} \left[-\frac{1}{2} l \cos \frac{n\pi}{2} + \frac{l}{n\pi} \sin \frac{n\pi}{2} \right] + \frac{4k}{l^2} \frac{l}{n\pi} \left[\frac{1}{2} l \cos \frac{n\pi}{2} + \frac{l}{n\pi} \sin \frac{n\pi}{2} \right]$$

$$= \frac{8k}{n^2\pi^2} \sin \frac{n\pi}{2} = \frac{8k}{n^2\pi^2} \quad \text{or,} \quad -\frac{8k}{n^2\pi^2} \quad \text{or,} \quad 0, \tag{vi}$$

according as n is of the form $(4m+1)$, $(4m+3)$ or $2m$.

Using the values of A_n and B_n in Eqn. (iii), we get the desired solution in the form,

$$\boldsymbol{u = \frac{8k}{\pi^2} \left(\sin \frac{\pi x}{l} \cos \frac{\pi ct}{l} - \frac{1}{3^2} \sin \frac{3\pi x}{l} \cos \frac{3\pi ct}{l} + \cdots \right).}$$

Example 21.15. *A uniform flexible string of length L and linear density m is stretched between its ends with a tension T. The string is plucked at $x = \frac{l}{3}$ with an amplitude h and then released. Solve the equation for the subsequent motion of the string.* **[BU(H) 1994]**

Solution : The differential equation of transverse vibration of the string is given by

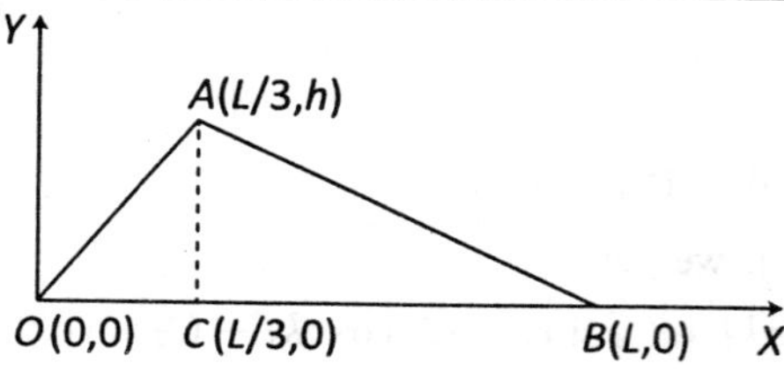

Fig. 21.2

$$\frac{\partial^2 y}{\partial t^2} = c^2 \frac{\partial^2 y}{\partial x^2}. \tag{i}$$

The boundary conditions are :

$$y(0,t) = 0 \tag{ii}$$

$$y(L,t) = 0 \tag{iii}$$

$$\frac{\partial y}{\partial t} = 0. \tag{iv}$$

The equation to the line OA is $y = \frac{3h}{L}x$ and that of AB is $y = \frac{3h}{2L}(L - x)$.
So, the boundary condition is

$$y(x, 0) = \frac{3h}{L}x, \qquad 0 \le x \le \frac{L}{3} \tag{v}$$

$$= \frac{3h}{2L}(L - x), \quad \frac{L}{3} \le x \le L. \tag{vi}$$

It can be easily shown that,

$$y(x, t) = b_n \sin\frac{n\pi x}{l}\cos\frac{n\pi ct}{L}. \tag{vii}$$

Adding all solutions,

$$y(x, t) = \sum b_n \sin\frac{n\pi x}{l}\cos\frac{n\pi ct}{l}.$$

For $t = 0$, $y(x, 0) = \sum b_n \sin\frac{n\pi x}{l}$.

The coefficient b_n can be found out with the help of Fourier half-range sine series.

$$\therefore \qquad b_n = \frac{2}{L}\left[\int_0^{\frac{L}{3}} \frac{3h}{L}x \sin\frac{n\pi x}{L}dx + \int_{\frac{L}{3}}^{L} \frac{3h}{2L}(L - x)\sin\frac{n\pi x}{L}dx\right]$$

$$= \frac{6h}{L^2}\left[\int_0^{\frac{L}{3}} x\sin\frac{n\pi x}{l}dx + \frac{1}{2}\int_{\frac{L}{3}}^{L}(L - x)\sin\frac{n\pi x}{l}dx\right]$$

$$= \frac{6h}{L^2}\left[\left(-\frac{Lx}{n\pi}\cos\frac{n\pi x}{L} + \frac{1}{2} + \frac{L^2}{n^2\pi^2}\sin\frac{n\pi x}{L}\right)_0^{\frac{L}{3}} - \frac{L^2}{2n\pi}\left(\cos\frac{n\pi x}{L}\right)_{\frac{L}{3}}^{L} - \frac{1}{2}\left(-\frac{Lx}{n\pi}\cos\frac{n\pi x}{L} + \frac{L^2}{n^2\pi^2}\sin\frac{n\pi x}{L}\right)_{\frac{L}{3}}^{L}\right]$$

$$= gh\sin\frac{n\pi}{3}.$$

Hence, the desired solution is $\boldsymbol{y(x, t) = gh\sum_{n=1}^{\infty} \sin\frac{n\pi}{3}\sin\frac{n\pi x}{L}\cos\frac{n\pi x}{l}}$.

Example 21.16. *Obtain the harmonic wave solution of the equation* $\nabla^2\psi = \frac{1}{c^2}\frac{\partial^2\psi}{\partial t^2}$.

Show that the solution represents a travelling wave whose amplitude falls off inversely as distance. **[CU(H) 2007, 2002]**

Solution : The given equation is a three-dimensional wave equation and can be written as

$$\frac{\partial^2\psi}{\partial t^2} = c^2\left(\frac{d^2\psi}{\partial x^2} + \frac{d^2\psi}{\partial y^2} + \frac{d^2\psi}{\partial z^2}\right). \tag{i}$$

The solution of Eqn. (i) may be written as

$$\psi(x, y, z, t) = X(x)Y(y)Z(z)T(t). \tag{ii}$$

Substituting Eqn. (ii) in Eqn. (i) and then dividing throughout by $XYZT$, we get

$$\frac{1}{c^2T}\frac{d^2T}{dt^2} = \frac{1}{X}\frac{d^2X}{dx^2} + \frac{1}{Y}\frac{d^2Y}{dy^2} + \frac{1}{Z}\frac{d^2Z}{dz^2}. \tag{iii}$$

The LHS of Eqn. (iii) is a function of t only and the RHS is of the form

$$f_1(x) + f_2(y) + f_3(z).$$

x, y, z, t being independent, Eqn. (iii) will be true only if each term of it is a constant.

$$\therefore \quad \frac{1}{X}\frac{d^2X}{dx^2} = -k_1^2 \tag{iv}$$

$$\frac{1}{Y}\frac{d^2Y}{dy^2} = -k_2^2 \tag{v}$$

$$\frac{1}{Z}\frac{d^2Z}{dz^2} = -k_3^2 \tag{vi}$$

and
$$\frac{1}{c^2T}\frac{d^2T}{dt^2} = -k^2. \tag{vii}$$

Eqn. (iv) can be written as $\frac{d^2X}{dx^2} + k_1^2X = 0$.

The solution of the above equation is $X = A_1 \cos k_1x + B_1 \sin k_1x$. (viii)

Similarly, the Eqns. (v) and (vi) give $\frac{d^2Y}{dy^2} + k_2^2Y = 0$ and $\frac{d^2Z}{dz^2} + k_3^2Z = 0$.

The solutions of the above equations are

$$Y = A_2 \cos k_2x + B_2 \sin k_2x \tag{ix}$$

$$Z = A_3 \cos k_3x + B_3 \sin k_3x \tag{x}$$

The Eqn. (vii) gives

$$\frac{d^2T}{dt^2} + c^2k^2T = 0 \quad \text{or,} \quad \frac{d^2T}{dt^2} + \omega^2T = 0, \quad \text{where, } \omega = ck.$$

So, the solution of Eqn. (i) is $T = A_4 \cos \omega t + B_4 \sin \omega t$. (xi)

So the solution of Eqn. (i) is

$$\boldsymbol{\psi = XYZT = (A_1 \cos k_1x + B_1 \sin k_1x)(A_2 \cos k_2x + B_2 \sin k_2x)}$$
$$\boldsymbol{\times (A_3 \cos k_3x + B_3 \sin k_3x)(A_4 \cos \omega t + B_4 \sin \omega t)}, \tag{xii}$$

where A's and B's are arbitrary constants

and
$$k_1^2 + k_2^2 + k_3^2 = k^2 = \frac{\omega^2}{c^2}. \tag{xiii}$$

The solution (xii) may also be written as

$$\boldsymbol{\psi = Ae^{\pm i(\omega t \pm k_1x \pm k_2x \pm k_3x)}}. \tag{xiv}$$

The positive sign in the exponential of Eqn. (xiv) leads to absurdity as $\psi \to \infty$ as x, y, z, $t \to 0$. **Hence, the solution (xiv) represents a travelling wave whose amplitude falls of inversely as distance.**

Example 21.17. *Consider a stretched uniform flexible string fixed at both ends.*

(i) *Obtain the differential equation for transverse vibrations of the string.*

(ii) *Solve the equation to find displacement for the nth mode.*

(iii) *Obtain an expression for the frequency in the nth mode.*

(iv) *The energy in the nth mode is given by* $E_n = \frac{1}{4} m A_n^2 \omega_n^2$, *where m is the mass of the string,* A_n *is the amplitude and* ω_n *the circular frequency of the nth mode.*

Express the above result in terms of the maximum velocity at an antinode.

[BU(H) 1993]

Solution : (i) See Art. 21.5, Eqn. (vi)

(ii) See Art. 21.6, Eqn. (ix)

Applying the boundary conditions

(i) $u = 0$ at $x = 0$ and (ii) $u = 0$ at $x = l$ to the equation

$$u = (A \cos px + B \sin px)(C \cos cpt + D \sin cpt), \tag{i}$$

we get $0 = A(C \cos pt + D \sin pt)$ and $0 = (A \cos pl + B \sin pl)(C \cos cpt + D \sin cpt)$.

$$\therefore \quad A = 0 \text{ and } \sin pl = 0 = \sin n\pi \text{ or, } p = \frac{n\pi}{l}, \ n \text{ is an integer.}$$

Using $A = 0$ and $p = \frac{n\pi}{l}$ in Eqn. (i) above we get the displacement for the nth mode of vibration as

$$u = \sum_{n=1}^{\infty} \left(C_n \cos \frac{n\pi ct}{l} + D_n \sin \frac{n\pi ct}{l} \right) B_n \sin \frac{n\pi x}{l} \tag{ii}$$

$$\text{or,} \quad u = B_n (C_n \cos \omega_n t + D_n \sin \omega_n t) \sin \frac{n\pi x}{l}, \text{ where } \left(\omega_n = \frac{n\pi c}{l} \right). \tag{iii}$$

Each $u_n(x, t)$ harmonic motion with period $\frac{2\pi}{\omega_n}$, i.e., each vibration has frequency $f_n = \frac{\omega_n}{2\pi} = \frac{nc}{2l}$ cycles/unit time in the nth mode.

(ii) Eqn. (iii) can be expressed in the form $u_n = A_n \sin p_n x \cos(\omega_n t + \phi_n)$.

At antinodes, $\sin p_n x = 1$.

$$\therefore \quad u_n = A_n \cos(\omega_n t + \phi_n).$$

The particle velocity $V_n = -\frac{\partial u_n}{\partial t}$ or, $V_n = -A_n \omega_n \sin(\omega_n t + \phi_n) = -V_{\max} \sin(\omega_n t + \phi_n)$.

Putting $A_n \omega_n = V_{\max}$, the velocity at the antinode, the energy of the vibrating string,

$$\boldsymbol{E_n = \frac{1}{4} m A_n^2 \omega_n^2 = \frac{1}{4} m V_{\max}^2}.$$

Example 21.18. *Solve the one-dimensional wave equation using D'Alembert's method.*

Solution : The one-dimensional wave equation is given by $\frac{\partial^2 y}{\partial t^2} = c^2 \frac{\partial^2 y}{\partial x^2}$. (i)

According to D'Alembert, here we introduce two variables u and v independent of each other as

$$\left.\begin{aligned} u &= x + ct \\ \text{and} \quad v &= x - ct \end{aligned}\right\} \tag{ii}$$

$$\therefore \quad \frac{\partial y}{\partial x} = \frac{\partial y}{\partial u}\frac{\partial u}{\partial x} + \frac{\partial y}{\partial v}\frac{\partial v}{\partial x} = \frac{\partial y}{\partial u}\cdot 1 + \frac{\partial y}{\partial v}\cdot 1 = \frac{\partial y}{\partial u} + \frac{\partial y}{\partial v}$$

and
$$\frac{\partial^2 y}{\partial x^2} = \frac{\partial}{\partial u}\left(\frac{\partial y}{\partial x}\right)\frac{\partial u}{\partial x} + \frac{\partial}{\partial v}\left(\frac{\partial y}{\partial x}\right)\frac{\partial v}{\partial x} = \frac{\partial}{\partial u}\left(\frac{\partial y}{\partial u} + \frac{\partial y}{\partial v}\right) + \frac{\partial}{\partial v}\left(\frac{\partial y}{\partial u} + \frac{\partial y}{\partial v}\right)$$
$$= \frac{\partial^2 y}{\partial u^2} + 2\frac{\partial^2 y}{\partial u \partial v} + \frac{\partial^2 y}{\partial v^2}. \qquad \text{(iii)}$$

Again,
$$\frac{\partial y}{\partial t} = \frac{\partial y}{\partial u}\frac{\partial u}{\partial t} + \frac{\partial y}{\partial v}\frac{\partial v}{\partial t} = \frac{\partial y}{\partial u}(c) + \frac{\partial y}{\partial v}(-c) = c\left(\frac{\partial y}{\partial u} - \frac{\partial y}{\partial v}\right)$$

and
$$\frac{\partial^2 u}{\partial t^2} = \frac{\partial}{\partial u}\left(\frac{\partial y}{\partial t}\right)\frac{\partial u}{\partial t} + \frac{\partial}{\partial v}\left(\frac{\partial v}{\partial t}\right)\frac{\partial v}{\partial t} = c^2\left[\frac{\partial}{\partial u}\left(\frac{\partial y}{\partial u} - \frac{\partial y}{\partial v}\right) - \frac{\partial}{\partial v}\left(\frac{\partial y}{\partial u} - \frac{\partial y}{\partial v}\right)\right]$$
$$= c^2\left[\frac{\partial^2 y}{\partial u^2} - 2\frac{\partial^2 y}{\partial u \partial v} + \frac{\partial^2 y}{\partial v^2}\right]. \qquad \text{(iv)}$$

Substituting the values of $\frac{\partial^2 u}{\partial x^2}$ and $\frac{\partial^2 u}{\partial t^2}$ in Eqn. (i), we get

$$\frac{\partial^2 y}{\partial u \partial v} = 0. \qquad \text{(v)}$$

Integrating Eqn. (v) w.r.t. v, we get

$$\frac{\partial y}{\partial u} = f(u), \quad \text{where } f(u) \text{ is an arbitrary function of } u.$$

Integrating Eqn. (vi) w.r.t. u, we get

$$y = \int f(u)du + \psi(v), \qquad \text{(vii)}$$

where $\psi(v)$ is an arbitrary function of v. Denoting $\int f(u)du$ by $\phi(u)$,

$$\boldsymbol{y = \phi(u) + \psi(v)} \quad \text{or,} \quad \boldsymbol{y = \phi(x+ct) + \psi(x-ct)}. \qquad \text{(viii)}$$

This is the general solution of Eqn. (i)

To determine ϕ and ψ, suppose the conditions are $u(x,0) = f(x)$ and $\dfrac{\partial u(x,0)}{\partial t} = 0.$

Now differentiating Eqn. (viii) w.r.t. t, we get

$$\frac{\partial y}{\partial t} = c\phi'(x+ct) - cv(x-ct).$$

At $t = 0$, $\phi'(x) = \psi'(x)$ (ix)

and $y(x,0) = \phi(x) + \psi(x) = f(x).$ (x)

From Eqn. (ix), we get $\phi(x) = \psi(x) + k$.

$\therefore$ Eqn. (x) becomes

$$2\psi(x) + k = f(x) \quad \text{or,} \quad \psi(x) = \frac{1}{2}[f(x) - k] \quad \text{and} \quad \phi(x) = \frac{1}{2}[f(x) + k].$$

Then Eqn. (viii) becomes

$$y(x,t) = \frac{1}{2}[f(x+ct) + k] + \frac{1}{2}[f(x-ct) - k]$$

and $\quad \boldsymbol{y(x,t) = f(x+ct) + f(x-ct)}.$ (xi)

This is the D'Alembert's solution of Eqn. (i).

Example 21.19. *Show that one-dimensional wave equation* $\frac{\partial^2 y}{\partial x^2} = \frac{1}{c^2}\frac{\partial^2 y}{\partial t^2}$ *has a solution of the form* $y = f_1(x-ct) + f_2(x+ct)$ *where* f_1 *and* f_2 *are twice differentiable functions.*

[Calicut U 1991; Kerala U 1990S; Mysore U 1987(S); VU(H) 2001]

Solution : Let us introduce variables u and v which are independent of each other such that, $u = (x-ct)$ and $v = (x+ct)$.

$$\therefore \quad \frac{\partial y}{\partial x} = \frac{\partial y}{\partial u}\frac{\partial u}{\partial x} + \frac{\partial y}{\partial v}\frac{\partial v}{\partial x} = \frac{\partial y}{\partial u}\cdot 1 + \frac{\partial y}{\partial v}\cdot 1 = \frac{\partial y}{\partial u} + \frac{\partial y}{\partial v} \tag{i}$$

and
$$\frac{\partial^2 y}{\partial x^2} = \frac{\partial}{\partial u}\left(\frac{\partial y}{\partial x}\right)\frac{\partial u}{\partial x} + \frac{\partial}{\partial v}\left(\frac{\partial y}{\partial x}\right)\frac{\partial v}{\partial x} = \frac{\partial}{\partial u}\left(\frac{\partial y}{\partial u} + \frac{\partial y}{\partial v}\right) + \frac{\partial}{\partial v}\left(\frac{\partial y}{\partial u} + \frac{\partial y}{\partial v}\right)$$

$$= \frac{\partial^2 y}{\partial u^2} + 2\frac{\partial^2 y}{\partial u \partial v} + \frac{\partial^2 y}{\partial v^2}. \tag{ii}$$

Also,
$$\frac{\partial y}{\partial t} = \frac{\partial y}{\partial u}\frac{\partial u}{\partial t} + \frac{\partial y}{\partial v}\frac{\partial v}{\partial t} = \frac{\partial y}{\partial u}(-c) + \frac{\partial y}{\partial v}(c) = -c\left[\frac{\partial y}{\partial u} - \frac{\partial y}{\partial v}\right]$$

and
$$\frac{\partial^2 y}{\partial t^2} = \frac{\partial}{\partial u}\left(\frac{\partial y}{\partial t}\right)\frac{\partial u}{\partial t} + \frac{\partial}{\partial v}\left(\frac{\partial y}{\partial t}\right)\frac{\partial v}{\partial t} = -c^2\left[\frac{\partial}{\partial u}\left(\frac{\partial y}{\partial u} - \frac{\partial y}{\partial v}\right) - \frac{\partial y}{\partial v}\left(\frac{\partial y}{\partial u} - \frac{\partial y}{\partial v}\right)\right]$$

$$= c^2\left[\frac{\partial^2 y}{\partial u^2} - 2\frac{\partial^2 y}{\partial u \partial v} + \frac{\partial^2 y}{\partial v^2}\right]. \tag{iii}$$

Substituting the values of $\frac{\partial^2 y}{\partial x^2}$ and $\frac{\partial^2 y}{\partial t^2}$ in Eqn. (iv) $\frac{\partial^2 y}{\partial x^2} = c^2\frac{\partial^2 y}{\partial t^2}$, we get

$$\frac{\partial^2 y}{\partial u \partial v} = 0 \tag{iv}$$

Integrating Eqn. (iv) w.r.t. v, we get

$$\frac{\partial y}{\partial u} = f(u), \quad \text{where } f(u) \text{ is an arbitrary function of } u. \tag{v}$$

Now integrating Eqn. (v) w.r.t. u, we get

$$y = \int f(u)du + \psi(v)$$

where $\psi(v)$ is an arbitrary function of v. Since the integral is a function of u alone we may denote it by $\phi(u)$. Thus

$$y = \phi(u) + \psi(v)$$

i.e. $\quad \boldsymbol{y(x,t) = \phi(x+ct) + \psi(x-ct)}$ (vi)

This is d'Alembert's solution of the given wave equation.

Now, to determine ϕ and ψ suppose initially $u(x,0) = f(x)$ and $\frac{\partial(x,0)}{\partial t} = 0$

Differentiating (vi) w.r.t. t, we get $\frac{\partial y}{\partial t} = c\phi'(x+ct) - c\psi'(x-ct)$

At $t = 0$, $\phi'(x) = \psi(x)$ (vii)

and $\quad y(x,0) = \phi(x) + \psi(x) = f(x)$ (viii)

Eqn. (vii) gives, $\phi(x) = \psi(x) + k$

$\therefore$ Eqn. (viii) becomes

$$2\psi(x) + k = f(x) \quad \text{or,} \quad \psi(x) = \frac{1}{2}[f(x) - k] \quad \text{and} \quad \phi(x) = \frac{1}{2}[f(x) + k]$$

Hence Eqn. (vi) takes the form,

$$\boldsymbol{y(x,t) = \frac{1}{2}[f(x+ct) + k] + \frac{1}{2}[f(x-ct) - k] = f_1(x-ct) + f_2(x+ct)} \quad \text{(ix)}$$

where f_1 and f_2 are twice differentiable functions.

Hence, **Eqn. (ix) is a solution of the given wave equation.**

Example 21.20. *A tightly stretched string with fixed end points $x = 0$ and $x = l$ is initially at rest. Find the displacement of the transversely vibrating string at any distance x from one end at any time t.* **[UPTU 2002]**

Solution : The equation of motion of a transversely vibrating string at any distance x and time t is given by

$$\frac{\partial^2 y}{\partial t^2} = c^2 \frac{\partial^2 y}{\partial x^2}. \quad \text{(i)}$$

The boundary conditions are :

$$\text{(i) } y = (0,t) = 0,\ y(l,t) = 0,\ y(x,0) = 0 \ \text{ and } \ \text{(ii) } \left(\frac{dy}{dt}\right)_{t=0} = \lambda x(l-x).$$

The vibration of the string being periodic, the solution of Eqn. (i) is

$$y(x,t) = (C_1 \cos px + C_2 \sin px)(C_3 \cos cpt + C_4 \sin cpt). \quad \text{(ii)}$$

By the condition, $y(0,t) = 0$, we have

$$y(0,t) = C_1(C_3 \cos cpt + C_4 \sin cpt) = 0.$$

Since, this is true for all values of t, $\therefore C_1 = 0$.

$$\therefore \quad y(x,t) = C_2 \sin px (C_3 \cos cpt + C_4 \sin cpt).$$

Also $\quad y(l,t) = C_2 \sin pl (C_3 \cos cpt + C_4 \sin cpt) = 0$ for all values of t.

This gives $pl = n\pi$ or, $p = \dfrac{n\pi}{l}$ [n is an integer]

$$\therefore \quad y(x,t) = C_2 \sin \frac{n\pi x}{l}\left(C_3 \cos \frac{cn\pi}{l}t + C_4 \sin \frac{cn\pi}{l}t\right).$$

By the condition $y(x,0) = 0$,

$$0 = C_2 C_3 \sin \frac{n\pi x}{l} \text{ for all } x, \quad \therefore C_2 C_3 = 0.$$

$$\therefore \quad y(x,t) = b_n \sin \frac{n\pi x}{l} \sin \frac{cn\pi t}{l}, \quad \text{where } b_n = C_2 C_4.$$

So, the general solution of Eqn. (i) for all values of n is

$$y(x,t) = \sum_{n=1}^{\infty} b_n \sin \frac{n\pi x}{l} \sin \frac{n\pi ct}{l}. \quad \text{(iii)}$$

$$\therefore \quad \frac{\partial y}{\partial t} = \sum_{n=1}^{\infty} b_n \sin\frac{n\pi x}{l}\cos\frac{n\pi ct}{l}\cdot\left(\frac{n\pi c}{l}\right).$$

By the condition, $\left(\dfrac{\partial y}{\partial t}\right)_{t=0} = \lambda x(l-x)$.

$$\therefore \quad \lambda x(l-x) = \frac{\pi c}{l}\sum_{n=1}^{\infty} nb_n \sin\frac{n\pi x}{l}$$

$$\text{or,} \quad \frac{\pi c}{l}nb_n = \frac{2}{l}\int_0^l \lambda x(l-x)\sin\frac{n\pi x}{l}dx$$

$$= \frac{2\lambda}{l}\left[(lx-x^2)\left(-\frac{l}{n\pi}\cos\frac{n\pi x}{l}\right) - (l-2x)\left(-\frac{l^2}{n^2\pi^2}\sin\frac{n\pi x}{l}\right) + (-2)\left(\frac{l^3}{n^3\pi^3}\cos\frac{n\pi x}{l}\right)\right]_0^l.$$

$$= \frac{4\lambda l^3}{n^3\pi^3}(1-\cos nx) = \frac{4\lambda l^2}{n^3\pi^3}[1-(-1)^n].$$

$$\therefore \quad b_n = \frac{4\lambda l^3}{c\pi^4 n^4}[1-(-1)^n] = \frac{8\lambda l^3}{c\pi^4(2n-1)^4}, \quad \text{where } n = 2m-1.$$

Hence, the solution is $\boldsymbol{y = \dfrac{8\lambda l^3}{c\pi^4}\displaystyle\sum_{n=1}^{\infty}\frac{1}{(2m-1)^4}\sin\frac{(2m-1)\pi x}{l}\sin\frac{(2m-1)\pi ct}{l}}$.

Example 21.21. (a) *Obtain one-dimensional heat flow equation and* (b) *solve it.*

Solution : (a) We consider homogeneous bar of uniform cross-section α and the bar is insulated so the radiation of heat through its sides is negligibly small.

Let the left end of the bar be at the origin and it lies along the X-axis so that heat flow takes place along the X-axis (Fig. 21.3). We consider a slice of width Δx and cross-sectional area α. The temperature at any point of the bar is a function of x and t.

The amount of heat flowing through the section A in time Δt is

$$Q_1 = -k\alpha\left(\frac{\partial u}{\partial x}\right)_x \Delta t,$$

where $\left(\frac{\partial u}{\partial x}\right)_x$ is the temperature gradient at A and k is the thermal conductivity of the material of the rod. The negative sign on the right side of the above equation represents that $u(x,t)$ decreases with x.

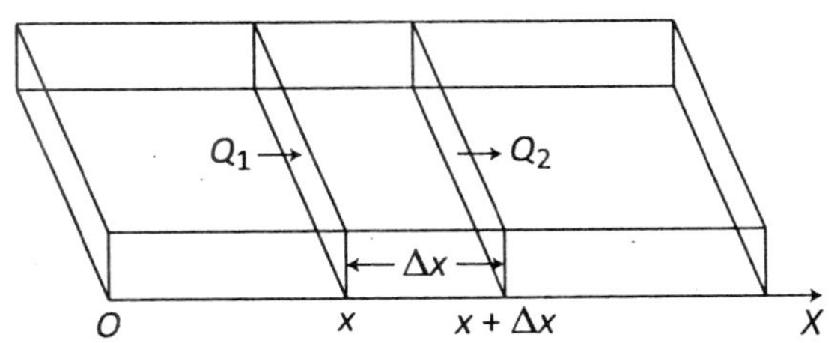

Fig. 21.3

The amount of heat flowing out through the section B is

$$Q_2 = -k\alpha\left(\frac{\partial u}{\partial x}\right)_{x+\Delta x} \Delta x.$$

$\therefore$ heat absorbed by the element is

$$Q = Q_1 - Q_2 = -k\alpha\left[\left(\frac{\partial u}{\partial x}\right)_x - \left(\frac{\partial u}{\partial x}\right)_{x+\Delta x}\right]\Delta x. \qquad \text{(i)}$$

Let Δu be the rise of temperature of the slice due to the absorption of heat Q.

If ρ be the density and s the specific heat of the material, then

$$Q = \rho\alpha\Delta x s\Delta u. \tag{ii}$$

From Eqns. (i) and (ii), we get

$$-k\alpha\left[\left(\frac{\partial u}{\partial x}\right)_x - \left(\frac{\partial u}{\partial x}\right)_{x+\Delta x}\right]\Delta t = \rho\alpha\Delta x s\Delta u.$$

$$\therefore \quad k \underset{\Delta x\to 0}{\text{Lt}} \frac{1}{\Delta x}\left[\left(\frac{\partial u}{\partial x}\right)_{x+\Delta x} - \left(\frac{\partial u}{\partial x}\right)_x\right] = \rho s \underset{\Delta t\to 0}{\text{Lt}} \frac{\Delta u}{\Delta t}$$

$$\text{or,} \quad k\frac{\partial^2 u}{\partial x^2} = \rho s\frac{\partial u}{\partial t} \quad \text{or,} \quad \frac{\partial u}{\partial t} = \left(\frac{k}{\rho s}\right)\frac{\partial^2 u}{\partial x^2}$$

or, $\boldsymbol{\dfrac{\partial u}{\partial t} = c^2\dfrac{\partial^2 u}{\partial x^2}}$, where $c^2 = \dfrac{k}{\rho s}$ is known as the **diffusivity**. (iii)

Eqn. (iii) is the equation governing one-dimensional heat flow of a longitudinal bar.

(b) Solution of the equation $\dfrac{\partial u}{\partial t} = c^2\dfrac{\partial^2 u}{\partial x^2}$. **[VU(H) 2001]**

We shall solve the above equation by using the method of separation of variables.

Let $u(x,t) = X(x)T(t)$, where $X(x)$ and $T(t)$ are the functions of only x and t respectively.

$$\therefore \quad \frac{\partial^2 u}{\partial x^2} = \frac{d^2X}{dx^2}T, \quad \frac{\partial u}{\partial t} = X\frac{dT}{dt}.$$

Using these values in $\frac{\partial u}{\partial t} = c^2\frac{\partial^2 u}{\partial x^2}$, we get

$$\frac{d^2X}{dx^2}T = \frac{1}{c^2}\frac{dT}{dt}x.$$

$\because$ x and t are independent, each side of the above equation must be a constant.

$$\therefore \quad \frac{1}{X}\frac{d^2X}{dx^2} = \frac{1}{c^2T}\frac{\partial T}{\partial t} = \lambda \text{ (say)}. \tag{i}$$

Now when $\lambda = p^2$, we have

$$\frac{d^2X}{dx^2} - p^2X = 0 \quad \text{and} \quad \frac{dT}{dt} - c^2T\lambda = 0.$$

$$\therefore \quad X = C_1e^{px} + C_1e^{-px} \quad \text{and} \quad T = C_3e^{c^2p^2t}.$$

For $\lambda = -p^2$, we get

$$\frac{d^2X}{dx^2} + p^2X = 0 \quad \text{and} \quad \frac{dT}{dt} + c^2p^2T = 0.$$

$$\therefore \quad X = C_4\cos px + C_5\sin px \quad \text{and} \quad T = a_6e^{-c^2p^2t}.$$

For $\lambda = 0$, $\dfrac{d^2X}{dt^2} = 0$ and $\dfrac{dT}{dt} = 0$.

$$\therefore \quad X = C_7x + C_8 \quad \text{and} \quad T = C_8.$$

So the possible solutions of the equation are :

$$u(x,t) = (C_1 e^{px} + C_2 e^{-px}) a_3 e^{c^2 p^2 t} \quad \text{(ii)}$$

$$= (C_4 \cos px + C_3 \sin px) e^{-c^2 p^2 t} \quad \text{(iii)}$$

$$= (C_7 x + C_8) C_9. \quad \text{(iv)}$$

Practically heat flow falls of with time and hence equation (iii) is consistent with physical situation.

Putting $C_4 C_6 = A$ and $C_5 C_6 = B$ we have the solution,

$$\boldsymbol{u(x,t) = (A \cos pt + B \sin pt) e^{-c^2 p^2 t}}.$$

Example 21.22. *One-dimensional differential equation of heat flow in an insulated metal bar is* $\frac{\partial \theta}{\partial t} = h \frac{\partial^2 \theta}{\partial x^2}$, *where h is a constant of the material of the rod.*

(i) *Find a general solution of the above equation by the method of separation of variables.*

(ii) *In steady state the temperature of the two ends of the bar are known. Find the nature of variation of temperature along the length of the rod. Plot its variation.* **[VU(H) 2006]**

Solution : (i) See answer of problem 22.21(b).

(ii) In the steady state $\frac{\partial \theta}{\partial t} = 0$, then the given equation reduces to

$$h \frac{\partial^2 \theta}{\partial x^2} = 0$$

or, $$\frac{\partial^2 \theta}{\partial x^2} = 0 \quad (\because h \neq 0) \quad \text{(i)}$$

Integrating eqn. (i) twice w.r.t. x to get

$$\theta = C_1 x + C_2. \quad \text{(ii)}$$

Let (i) $\theta = \theta_0$ at $x = 0$ and (ii) $\theta = \theta_1$ at $x = l$.

Using condition (i) in Eqn. (ii), $\theta_0 = C_2$

$$\therefore \quad \theta = C_1 x + \theta_0.$$

Using condition (ii), we get $\theta_1 = C_1 l + \theta_0$.

$$\therefore \quad C_1 = \frac{(\theta_1 - \theta_0)}{l} = -\frac{(\theta_0 - \theta_1)}{l}.$$

Using the values of C_1 and C_2 in eqn. (ii),

$$\boldsymbol{\theta = \theta_0 - \frac{(\theta_0 - \theta_1)}{l} x.} \quad \text{(iii)}$$

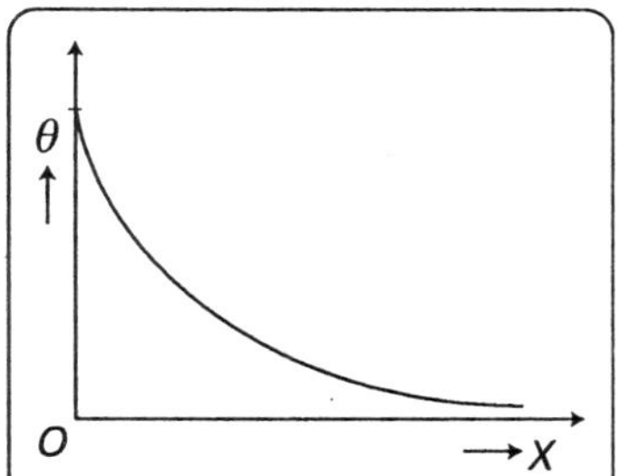

Fig. 21.4

Eqn. (iii) is the desired equation representing the variation of temperature θ along the length of the bar.

Fig. 21.4 shows the variation of temperature of the bar along the length.

Example 21.23. *Apply the method of separation of variables to solve the one-dimensional heat flow equation* $\frac{\partial \theta(x,t)}{\partial t} = h^2 \frac{\partial^2 \theta(x,t)}{\partial x^2}$, $0 < x < t$ *subject to the following conditions :*

(i) θ is finite when $t \to \infty$ (ii) $\left.\dfrac{\partial\theta(x,t)}{\partial t}\right|_{x=0} = \left.\dfrac{\partial\theta(x,t)}{\partial t}\right|_{x=l} = 0$ (iii) $\theta(x,\theta) = lx - x^2$.

[BU(H) 1996; Assam 1999]

Solution : The given equation is $\dfrac{\partial\theta(x,t)}{\partial t} = h^2\dfrac{\partial^2\theta(x,t)}{\partial x^2}, 0 < x < l.$ (i)

According to the solution of problem 21.21(b) the general solution of Eqn. (i) is

$$\theta(x,t) = (A\cos px + B\sin px)e^{-h^2p^2t}. \quad \text{(ii)}$$

$$\therefore \quad \frac{\partial\theta}{\partial t} = (Bp\cos px - Ap\sin px)e^{-h^2p^2t}. \quad \text{(iii)}$$

Using condition $\left(\frac{\partial\theta}{\partial x}\right)_{x=0} = 0$ in Eqn. (iii), we get

$$Bpe^{-h^2p^2t} = 0, \quad \therefore \; B = 0.$$

Eqn. (ii) then becomes

$$\theta(x,t) = A\cos px e^{-h^2p^2t}. \quad \text{(iv)}$$

Again, $\left(\frac{\partial\theta}{\partial x}\right)_{x=l} = 0$ gives

$$-Ap\sin pl e^{-h^2p^2t} = 0 \quad \text{or,} \quad \sin pl = 0 = \sin n\pi, \quad \text{i.e.,} \quad p = \frac{n\pi}{l} \quad (n = 1, 2, 3, \ldots).$$

Eqn. (iv) reduces to

$$\theta(x,t) = A\cos\frac{n\pi x}{l}e^{\left[\frac{-h^2n^2\pi^2t}{l}\right]} = A_n\cos\frac{n\pi x}{l}e^{\left[\frac{-h^2n^2\pi^2t}{l}\right]}.$$

Adding all solutions for different n values, we get

$$\theta(x,t) = \sum_{n=1}^{\infty} A_n\cos\left(\frac{n\pi x}{l}\right)e^{\left[\frac{-h^2n^2\pi^2t}{l}\right]}. \quad \text{(v)}$$

$$\therefore \quad \theta(x,\theta) = \sum_{n=1}^{\infty} A_n\cos\left(\frac{n\pi x}{l}\right) = lx - x^2 \text{ by the condition (iii).}$$

Now by Fourier theorem,

$$A_0 = \frac{2}{l}\int_0^l (lx - x^2)dx = \frac{l^2}{3}$$

and
$$A_n = \frac{2}{l}\int_0^l (lx - x^2)\cos\frac{n\pi x}{l}dx = -\frac{4l^2}{\pi^2n^2}, \quad \text{if } n \text{ is even}$$

$$= 0, \quad \text{if } n \text{ is odd.}$$

$$\text{Hence, } \theta(x,t) = \frac{A_0}{2} + \sum_{n=1}^{\infty} A_n\cos\frac{n\pi x}{l}e^{-\frac{h^2n^2\pi^2t}{l^2}}$$

$$= \frac{l^2}{6} - \frac{4l^2}{\pi^2}\sum_{n=2}^{\infty}\frac{1}{n^2}\cos\frac{n\pi x}{l}e^{-\frac{h^2n^2\pi^2t}{l^2}} \quad \text{[if } n \text{ is even]}$$

$$\text{or,} \quad \boldsymbol{\theta(x,t) = \frac{l^2}{6} - \frac{l^2}{\pi^2}\sum_{n=1}^{\infty}\cos\frac{2n\pi x}{l}e^{-\frac{h^2n^2\pi^2t}{l^2}}}.$$

Example 21.24. *Using the method of separation of variables, solve the one-dimensional heat flow equation :* $\frac{\partial^2 u}{\partial x^2} = \frac{1}{k}\frac{\partial u}{\partial t}$ under the following boundary conditions :

(i) $u(0,t) = u(l,t) = 0,\ t > 0$ (ii) $u(x,0) = \frac{x(l-x)}{l^2},\ 0 < x < l$ **[BU(H) 1992]**

Solution : Given, $\frac{\partial^2 u}{\partial x^2} = \frac{1}{k}\frac{\partial u}{\partial t}$. (i)

The general solution of the equation is given in problem 21.21(b) in the form

$$u(x,t) = (A\cos px + B\sin px)e^{-kp^2t}. \tag{ii}$$

By condition (i), $Ae^{-kp^2t} = 0, \quad \therefore\ A = 0.$

Then Eqn. (ii) reduces to

$$u(x,t) = B\sin px e^{-kp^2t}. \tag{iii}$$

Again, $\because\ u(l,t) = 0, \quad \therefore\ B\sin pl e^{-kp^2t} = 0.$

But $B \neq 0$,

$$\therefore \quad \sin pl = 0 = \sin n\pi \ [n = 1, 2, 3, \ldots]$$

or, $\quad pl = n\pi \quad \text{and} \quad p = \frac{n\pi}{l}.$

Hence, Eqn. (iii) reduces to

$$u(x,t) = B\sin\frac{n\pi x}{l}e^{\frac{-kn^2\pi^2t}{l^2}} = b_n\sin\frac{n\pi x}{l}e^{\frac{-kn^2\pi^2t}{l^2}}. \tag{iv}$$

Summing up all such solutions, we get

$$\boldsymbol{u(x,t) = \sum_{n=1}^{\infty} b_n \sin\frac{n\pi x}{l}e^{\frac{-kn^2\pi^2t}{l^2}}}.$$

Substituting these values in Eqn. (i) and then dividing throughout by $X(x)Y(y)T(t)$, we get

$$h^2\left(\frac{1}{X}\frac{d^2X}{dx^2} + \frac{1}{Y}\frac{d^2Y}{dy^2}\right) = \frac{1}{T}\frac{dT}{dt}. \tag{ii}$$

$$\therefore \quad \frac{1}{h^2}\frac{1}{T}\frac{dT}{dt} = \frac{1}{X}\frac{d^2X}{dx^2} + \frac{1}{Y}\frac{d^2Y}{dy^2} = -k^2 \text{ (say), where } -k^2 \text{ is a separation constant.}$$

$$\therefore \quad \frac{dT}{dt} = -k^2hT \tag{iii}$$

and $\quad \frac{1}{X}\frac{d^2X}{dx^2} + \frac{1}{Y}\frac{d^2Y}{dy^2} = -k^2 \quad \text{or,} \quad \frac{1}{X}\frac{d^2X}{dx^2} + k^2 = -\frac{1}{Y}\frac{d^2Y}{dy^2} = m^2 \text{ (say)}$

$$\text{or,} \quad \frac{d^2X}{dx^2} + (k^2 - m^2)X = 0 \tag{iv}$$

$$\text{or,} \quad \frac{d^2X}{dx^2} + q^2X = 0 \text{ (where } q^2 = k^2 - m^2) \tag{v}$$

$$\text{and} \quad \frac{d^2Y}{dy^2} + m^2Y = 0. \tag{vi}$$

Now solutions of (v) and (vi) are

$$X = C_1 \sin qx + C_2 \cos qx \tag{vii}$$

and $$Y = C_3 \sin my + C_4 \cos my. \tag{viii}$$

Solution of Eqn. (iii) is $T = Ce^{-k^2h^2t}$. (ix)

Hence, the complete solution of the two-dimensional heat conduction equation is

$$u(x, y, t) = XYT = (C_1 \sin qx + C_2 \cos qx)(C_3 \sin my + C_4 \cos my)Ce^{-k^2h^2t}$$

or, $$\boldsymbol{u(x, y, t) = [C_1 \sin(\sqrt{k^2 - m^2})x + C_2 \cos(\sqrt{k^2 - m^2})x]}$$
$$\boldsymbol{\times (C_3 \sin my + C_4 \cos my)Ce^{-k^2h^2t}}.$$

(b) The given equation, $\frac{\partial^2 u}{\partial x^2} + \frac{\partial u}{\partial y^2} = 0.$ (i)

$$\therefore \quad u(x, 0) = \sum_{n=1}^{\infty} b_n \sin \frac{n\pi x}{l}.$$

But by the condition (ii) $u(x, 0) = \frac{x(l-x)}{l^2}$, $0 < x < l$, we get

$$\frac{x(l-x)}{l^2} = \sum_{n=1}^{\infty} b_n \sin \frac{n\pi x}{l}.$$

Applying the condition of Fourier constants,

$$b_n = \frac{2}{l} \int_0^l \frac{x(l-x)}{l^2} \sin \frac{n\pi x}{l} dx = \frac{2}{l^3} \int_0^l \sin \frac{n\pi x}{l} dx.$$

Hence, the desired solution is

$$u(x, t) = \frac{4l}{\pi^3} \sum_{n=1}^{\infty} \frac{1}{n^3} [1 - (-1)^n] \sin \frac{n\pi x}{l} e^{-\frac{kn^2\pi^2 t}{l^2}}$$

or, $$\boldsymbol{u(x, t) = \frac{4l}{\pi^3} \sum_{n=1}^{\infty} \frac{(-1)^{n+1}}{n^3} \sin \frac{n\pi x}{l} e^{-\frac{kn^2\pi^2 t}{l^2}}}.$$

Example 21.25. (a) *Suppose the following differential equation refers to a problem of two dimensional conduction of heat in Cartesian coordinates* $\frac{\partial u}{\partial t} = h^2 \left(\frac{\partial^2 u}{\partial x^2} + \frac{\partial^2 u}{\partial y^2}\right)$. *Find a general solution of the equation by separation of variables.*

(b) If (i) $u(0, y) = 0$; $T(x, \infty) = 0$ and (ii) $u(0, y) = 0$; $u(x, 0) = \sin \frac{\pi x}{a}$, then find the solution of $\frac{\partial^2 u}{\partial x^2} + \frac{\partial^2 u}{\partial y^2} = 0.$

Solution : (a) Let $u = X(x)Y(y)T(t)$. (i)

where $X(x)$ and $T(t)$ are functions of only x and t respectively.

$$\therefore \quad \frac{\partial u}{\partial t} = \frac{dT}{dt} X(x)Y(y), \quad \frac{\partial^2 u}{\partial x^2} = \frac{d^2 X}{dx^2} Y(x)T(t) \quad \text{and} \quad \frac{\partial^2 u}{\partial y^2} = X(x)\frac{d^2 Y}{dx^2} T(t)$$

refer to two-dimensional heat conduction at steady state, i.e., $\frac{\partial u}{\partial t} = 0$.

By the method of separation of variables, $u = f_1(x)f_2(y)$.

$\therefore \quad \dfrac{\partial u}{\partial x} = \dfrac{df_1}{dx} f_2(y)$

and $\quad \dfrac{\partial^2 u}{\partial x^2} = \dfrac{d^2 f_1}{dx^2} f_2(y)$. (ii)

Similarly, $\dfrac{\partial^2 u}{\partial y^2} = f_1(x)\dfrac{d^2 f_2}{dy^2}$. (iii)

$\because \quad \dfrac{\partial^2 u}{\partial x^2} + \dfrac{\partial^2 u}{\partial y^2} = 0,$

$\therefore \quad \dfrac{d^2 f_1(x)}{dx^2} f_2(y) = -f_1(x)\dfrac{d^2 f_2(y)}{dy^2}$ (iv)

or, $\quad \dfrac{1}{f_1(x)}\dfrac{d^2 f_1(x)}{dx^2} = -\dfrac{1}{f_2(x)}\dfrac{d^2 f_2(y)}{dy^2} = -k^2$ (say), where $-k^2$ is the separation constant.

$\therefore \quad \dfrac{d^2 f_1(x)}{dx^2} = -k^2 f_1(x)$ (v)

and $\quad \dfrac{d^2 f_2(y)}{dy^2} = k^2 f_2(y)$. (vi)

Solutions of Eqns. (v) and (vi) are respectively,

$f_1(x) = C_1 \sin kx + C_2 \cos kx$ (vii)

and $\quad f_2(x) = C_3 e^{ky} + C_4 e^{-ky}$. (viii)

$\therefore$ the solution of Eqn. (i) is

$$u = f_1(x)f_2(y) = (C_1 \sin kx + C_2 \cos kx)(C_3 e^{ky} + C_4 e^{-ky})$$
$$= e^{-ky}(A \sin kx + B \cos kx) + e^{ky}(C \sin kx + D \cos kx), \quad \text{(ix)}$$

where $A_1B_1 = A$, $B_1B_2 = B$, $A_1A_2 = C$ and $B_1A_2 = D$.

Now from the boundary condition, $u = 0$ at $y = \infty$.

$\therefore \quad C = D = 0$ [from Eqn. (ix)]

and $\quad u = 0$ at $x = 0$.

$\therefore \quad u = e^{-ky} A \sin ky$ [from Eqn. (ix)].

Again, $u = 0$ at $x = a$,

$\therefore \quad A \sin ka = 0 \quad$ or, $\quad ka = r\pi \quad (r = 0, 1, 2, \ldots)$

$\therefore \quad u_r = A_r e^{\frac{-r\pi y}{a}} \sin \dfrac{n\pi x}{a}$. (x)

$\because \; u_r = \sin \dfrac{\pi x}{a}$ at $y = 0$, $\quad \therefore \; \sin \dfrac{\pi x}{a} = A_r \sin \dfrac{n\pi x}{a} \quad$ or, $\quad A_r = 1,\ r = 1$.

From eqn. (x), $u_r = \sin\left(\dfrac{\pi x}{a}\right) e^{\frac{-\pi y}{a}}$.

Hence, the required solution is $\boldsymbol{u = \sin\left(\dfrac{\pi x}{a}\right) e^{\frac{-\pi y}{a}}}$. (xi)

Example 21.26. *Solve the equation* $\dfrac{\partial u}{\partial t} = \dfrac{\partial^2 u}{\partial x^2}$ *with boundary conditions*

$$u(x,0) = 3\sin n\pi x,\ u(0,t) = 0,\ u(1,t) = 0, \text{ where } 0 < x < 1,\ t > 0.$$

Solution : We have the solution of the equation $\dfrac{\partial u}{\partial t} = \dfrac{\partial^2 u}{\partial x^2}$ (i)

can be found out by the method of separation of variables in the form

$$u(x,t) = (C_1 \cos px + C_2 \sin px)e^{-p^2 t}. \tag{ii}$$

When $x = 0$, $u(0,t) = C_1 e^{-p^2 t} = 0$, i.e., $C_1 = 0$.

$\therefore$ Equation (ii) becomes $u(x,t) = C_2 \sin pxe^{-k^2 t}$. (iii)

When $x = 1$, $u(1,t) = C_2 \sin pe^{-p^2 t} = 0$ or, $p = 0$, i.e., $p = n\pi$.

Equation (iii) becomes $u(x,t) = b_n e^{-n^2\pi^2 t}$, where $b_n = C_2$.

Hence, the general solution of Eqn. (i) is $\boldsymbol{u(x,t) = \sum_{n=1}^{\infty} b_n e^{-n^2\pi^2 t} \sin n\pi x}$. (iv)

By the condition, for $t = 0$, $u(x,0) = 3\sin n\pi x$. (v)

Also from (iv), $u(x,0) = \sum_{n=1}^{\infty} b_n \sin n\pi x$. (vi)

Comparing Eqns. (v) and (vi), we get $b_n = 3$.

Hence, $\boldsymbol{u(x,t) = 3\sum_{n=1}^{\infty} b_n e^{-n^2\pi^2 t} \sin n\pi x}$. (vii)

Example 21.27. *The differential equation of motion of a rod fixed at the end* $x = 0$ *and free at the end* $x = l$ *is given by* $\dfrac{\partial^2 \xi}{\partial t^2} = v^2 \dfrac{\partial^2 \xi}{\partial x^2}$, *where* $v^2 = \dfrac{Y}{\rho}$, Y *being the Young's modulus and* ρ, *the density of the material of the rod. If the free end be subjected to a sinusoidal force* $F = F_0 \sin \omega t$, *then show that the steady state displacement for forced vibration of the rod is* $\xi(x,t) = \dfrac{F_0 v}{\alpha Y \omega} \sec\left(\dfrac{\omega l}{v}\right) \sin\left(\dfrac{\omega x}{v}\right) \sin \omega t$, α *being the area of cross section of the rod.*

Solution : The steady state displacement of the forced vibration is given by

$$\xi(x,t) = (C_1 \cos \lambda x + C_2 \sin \lambda x) \sin \omega x, \tag{i}$$

where $\lambda = \dfrac{\omega}{v} = \omega\sqrt{\dfrac{\rho}{Y}}$.

By the given condition, $\xi(0,t) = 0$.

$\therefore$ from Eqn. (i), $A \sin \omega t = 0$, $\therefore$ $A = 0$.

So, $\xi(x,t) = B \sin \lambda x \sin \omega t$. (ii)

Also $x = l$, the strain due to longitudinal vibration $= \dfrac{\partial \xi}{\partial x}$.

Then the initial force at $x = l$, is $\alpha Y \left(\dfrac{\partial \xi}{\partial x}\right)_{x=l}$.

Again, since the initial force is $F = F_0 \sin \omega t$,

$$\therefore \quad \alpha Y \left(\frac{\partial \xi}{\partial x}\right)_{x=l} = F_0 \sin \omega t. \qquad \text{(iii)}$$

From Eqn. (iii),

$$\left(\frac{\partial \xi}{\partial x}\right)_{x=l} = \alpha Y C_2 \lambda (\cos \lambda x \sin \omega t)_{x=l} = \alpha Y C_2 \lambda \cos \lambda l \sin \omega t.$$

$$\therefore \quad F_0 \sin \omega t = \alpha Y C_2 \lambda \cos \lambda l \sin \omega t.$$

$$\therefore \quad C_2 = \frac{F_0}{\alpha Y \lambda \cos \lambda l} = \frac{F_0 v}{\alpha Y \omega} = \sec\left(\frac{\omega l}{v}\right).$$

Hence, from Eqn. (ii), $\boldsymbol{\xi(x,t) = \dfrac{F_0}{\alpha Y \omega} \sec\left(\dfrac{\omega l}{v}\right) \sin\left(\dfrac{\omega x}{v}\right) \sin \omega t}$ **(Proved).** (iv)

Example 21.28. *Neglecting the effect of resistance and leakage find the emf $v(x,t)$ in a line of length l, t seconds after the ends were suddenly grounded.*

Given that $i(x,0) = i_0$ and $v(x,0) = e_1 \sin \dfrac{\pi x}{l} + e_5 \sin \dfrac{5\pi x}{l}$. **[SVTU 2007]**

Solution : Neglecting resistance and leakage, the radio equation is $\dfrac{\partial^2 v}{\partial x^2} = LC \dfrac{\partial^2 v}{\partial t^2}$. (i)

As the ends are suddenly grounded, we have

$$\left.\begin{aligned} v(0,t) &= 0 \\ v(l,t) &= 0 \end{aligned}\right\} \qquad \text{(ii)}$$

Also the initial conditions are

$$i(x,0) = i_0 \qquad \text{(iii)}$$

$$\text{and} \quad v(x,0) = e_1 \sin \frac{\pi x}{l} + e_5 \sin \frac{5\pi x}{l}. \qquad \text{(iv)}$$

$$\therefore \quad \frac{\partial i}{\partial x} = -c \frac{\partial v}{\partial t}.$$

$$\text{So,} \quad \frac{\partial v(x,0)}{\partial t} = 0. \qquad \text{(v)}$$

Now let the solution of Eqn. (i) be $v = X(x)T(t)$, then from Eqn. (i),

$$\frac{d^2 X}{dx^2} T = LCX \frac{d^2 T}{dt^2} \quad \text{or,} \quad \frac{1}{X}\frac{d^2 X}{dx^2} = LC \frac{1}{T}\frac{d^2 T}{dt^2} = -k^2 \text{ (say)}.$$

$$\therefore \quad \frac{d^2 X}{dx^2} + k^2 X = 0 \quad \text{and} \quad \frac{d^2 T}{dt^2} + \frac{k^2}{LC} T = 0.$$

The solutions of the above equations are respectively,

$$X(x) = C_1 \cos kx + C_2 \sin kx \quad \text{and} \quad T(t) = C_3 \cos \frac{k}{\sqrt{LC}} t + C_4 \sin \frac{k}{\sqrt{LC}} t.$$

$\therefore$ the solution of Eqn. (i) is

$$v = XT = (C_1 \cos kx + C_2 \sin kx)\left(C_3 \cos \frac{k}{\sqrt{LC}} t + C_4 \sin \frac{k}{\sqrt{LC}} t\right). \qquad \text{(vi)}$$

By boundary condition (ii), we get $C_1 = 0$ and $k = \frac{n\pi}{l}$.

$$\therefore \quad v = \sin\frac{n\pi x}{l}\left[a_n \cos\left(\frac{n\pi}{l\sqrt{LC}}t\right) + b_n \sin\left(\frac{n\pi}{l\sqrt{LC}}t\right)\right].$$

By the initial condition (iv), we get $b_n = 0$.

$$\therefore \quad v = a_n \sin\frac{n\pi x}{l}\cos\left(\frac{n\pi}{l\sqrt{LC}}t\right).$$

Thus, the general solution of (i) is $v = \sum_{n=1}^{\infty} a_n \sin\frac{n\pi x}{l}\cos\left(\frac{n\pi}{l\sqrt{LC}}t\right)$.

By the initial condition (iv), we get

$$e_1 \sin\frac{\pi x}{l} + e_5 \sin\frac{5\pi x}{l} = \sum_{n=1}^{\infty} a_n \sin\frac{n\pi x}{l}.$$

$$\therefore \quad a_1 = e_1 \quad \text{and} \quad a_5 = e_5 \quad \text{and} \quad a_2 = a_3 = 0.$$

Hence, the final solution of the Eqn. (i) is

$$\boldsymbol{v = e_1 \sin\frac{\pi x}{l}\cos\frac{\pi t}{l\sqrt{LC}} + e_5 \sin\frac{5\pi x}{l}\cos\frac{5\pi ct}{l\sqrt{LC}}}. \tag{vii}$$

Exercises

21(A) : Objective Type Questions

Select the correct answer

21.1 The linear second order PDE in two variables x and y is

$$A\frac{\partial^2 u}{\partial x^2} + 2H\frac{\partial^2 u}{\partial x \partial y} + B\frac{\partial^2 u}{\partial y^2} + 2F\frac{\partial u}{\partial x} + 2G\frac{\partial u}{\partial y} + cu = 0.$$

The equation is of parabolic type, if

(a) $AB - H^2 > 0$ **(b)** $AB - H^2 = 0$ **(c)** $AB - H^2 < 0$ **(d)** $AB - A^2 = 0$.

21.2 The type of the equation $\frac{\partial^2 u}{\partial x^2} - k\frac{\partial u}{\partial x} = 0$ is

(a) parabolic **(b)** hyperbolic **(c)** elliptic **(d)** none of these.

21.3 The partial differential equation $\frac{\partial^2 u}{\partial t^2} - c^2\frac{\partial^2 u}{\partial x^2} = 0$ is

(a) wave equation **(b)** heat equation **(c)** Laplace's equation **(d)** none of these.

[AMIE(Winter) 1991]

21.4 Laplace's equation in Cartesian coordinates x and y is

(a) $\frac{\partial^2 u}{\partial x^2} + \frac{\partial u}{\partial y} = 0$ **(b)** $\frac{\partial^2 u}{\partial x^2} - \frac{\partial^2 u}{\partial y^2} = 0$ **(c)** $\frac{\partial^2 u}{\partial x^2} + \frac{\partial^2 u}{\partial y^2} = 0$ **(d)** $\frac{\partial u}{\partial x} + \frac{\partial^2 u}{\partial y^2} = 0$.

21.5 Laplace's equation in two-dimensional polar coordinates is

(a) $\frac{\partial^2 u}{\partial r^2}+\frac{\partial u}{\partial r}+\frac{1}{r^2}\frac{\partial^2 u}{\partial \theta^2}=0$ (b) $\frac{\partial^2 u}{\partial r^2}+\frac{1}{r}\frac{\partial u}{\partial r}+\frac{1}{r^2}\frac{\partial^2 u}{\partial \theta^2}=0$

(c) $\frac{\partial^2 u}{\partial r^2}+\frac{1}{r^2}\frac{\partial^2 u}{\partial \theta^2}=0$ (d) $\frac{\partial u}{\partial r}+\frac{1}{r^2}\frac{\partial^2 u}{\partial \theta^2}=0.$

21.6 The solution of the equation $\frac{\partial^2 u}{\partial x^2}+\frac{\partial^2 u}{\partial y^2}=0$ is

(a) $u=(C_1\cos\lambda x+C_2\sin\lambda x)(C_3e^{\lambda y}+C_4e^{-\lambda y})$ (b) $u=(C_3e^{\lambda y}+C_4e^{-\lambda y})$

(c) $u=(C_1\cos\lambda x+C_2\sin\lambda x)$ (d) $u=C_1\cos\lambda x(C_3e^{\lambda y}+C_4e^{-\lambda y})$.

21.7 The solution of $\frac{\partial u}{\partial t}=2\frac{\partial^2 u}{\partial x^2}$ if $u(0,t)=u(3,t)=0$ and $u(x,0)=5\sin 4\pi x-3\sin 8\pi x$ is

(a) $C_1\sin pt(C_3\cos pt+C_4\sin pt)$ (b) $-\frac{1}{y^2}\sin(xy)+xf_1(y)+f_2(y)$

(c) $(5\sin 4\pi x-3\sin 8\pi x)e^{-\frac{2\pi^2x^2}{9}t}$ (d) $(\sin 4\pi x+3\sin 8\pi x)e^{-\frac{2\pi^2x^2}{9}t}$.

[AMIE(Summer) 1993]

21.8 The solution of the equation $\frac{\partial^2 z}{\partial x\partial y}=0$ is

(a) $f_1(x)+f_2(y)$ (b) $xf_1(x)+yf_2(y)$ (c) $f_1(x)-yf_2(y)$ (d) $xf_1(x)-yf_2(y)$.

[AMIE(Summer) 1993]

21.9 The solution of $\frac{\partial^2 z}{\partial x^2}=\sin(xy)$ is

(a) $\sin xy+xf_1(y)+f_2(y)$ (b) $-\frac{1}{y^2}\sin xy+xf_1(y)+f_2(y)$

(c) $-\frac{1}{y^2}\sin xy+f_2(y)$ (d) $\sin xy+f_1(y)+y^2f_2(y)$.

21.10 The D'Alembert's solution of the wave equation $\frac{\partial^2 y}{\partial t^2}=c^2\frac{\partial^2 y}{\partial x^2}$ is $y(x,t)=$

(a) $f_1(x-ct)+f_2(x+ct)$ (b) $f_1(x)+f_2(x+ct)$

(c) $f_1(ct)+f_2(ct+x)$ (d) $f_1(x-ct)+f_2(ct)$.

21.11 The one-dimensional heat flow equations is

(a) $\frac{\partial u}{\partial t}=c^2\frac{\partial^2 u}{\partial x^2}$ (b) $\frac{\partial^2 u}{\partial t^2}=c^2\frac{\partial u}{\partial x}$ (c) $\frac{\partial u}{\partial t}=c^2\frac{\partial u}{\partial x}$ (d) $\frac{\partial u}{\partial t}=c^2\frac{\partial^2 u}{\partial x^2}$.

21.12 The solution of the equation $\frac{\partial u}{\partial t}=c^2\frac{\partial^2 u}{\partial x^2}$, $u(x,t)$ is

(a) $A\cos pte^{-c^2p^2t}$ (b) $(A\cos pt+B\sin pt)e^{-c^2p^2t}$

(c) $(A\cos pt+B\sin pt)$ (d) $(A+B)e^{c^2p^2t}$.

21.13 The solution of the equation $\frac{\partial^2 u}{\partial x^2} = k^2 \frac{\partial^2 u}{\partial t^2}$ for which $u(0,t) = u(l,t)$, $u(x,0) = \sin \frac{\pi x}{l} u(x,t)$ is,

(a) $\sin \frac{\pi x}{l} e^{\frac{\pi^2 t}{h^2 l^2}}$ (b) $\sin \frac{\pi x}{l} e^{-\frac{\pi^2 t}{h^2 l^2}}$ (c) $\sin \frac{\pi x}{l} e^{\frac{\pi^2 h^2}{l^2}}$ (d) $\sin \frac{\pi x}{l} e^{-\frac{\pi^2 l^2}{h}}$.

21.14 The energy of a vibrating string in terms of maximum velocity at the antinode is

(a) $\frac{1}{2} m V_m^2$ (b) $m V_m^2$ (c) $\frac{1}{4} m V_m^2$ (d) $\frac{1}{4} m^2 V_m^2$,

where m is the mass/length of the wine and V_m is the maximum velocity at an antinode.

21.15 The solution of the equation $\frac{\partial u}{\partial x} = 2\frac{\partial u}{\partial t} + u$, where $u(x,0) = 6e^{-3x}$, is $u(x,t) =$

(a) $6e^{-3x+2t}$ (b) $6e^{-3x-2t}$ (c) $12e^{-3x-2t}$ (d) $3e^{-3x-2t}$.

Answers

21.1 (b) **21.2** (b) **21.3** (a) **21.4** (c) **21.5** (b) **21.6** (a) **21.7** (c) **21.8** (a) **21.9** (b) **21.10** (a) **21.11** (a) **21.12** (b) **21.13** (b) **21.14** (c) **21.15** (b).

21(B) : Short Answer Type Questions

21.1 For what values of x and y the following equations are hyperbolic, parabolic or elliptic?

(i) $\frac{\partial^2 u}{\partial x^2} + 2x\frac{\partial^2 u}{\partial x \partial y} + \frac{\partial^2 u}{\partial y^2} = x + y$ (ii) $(x+1)\frac{\partial^2 u}{\partial x^2} + 2x\frac{\partial^2 u}{\partial x \partial y} + y\frac{\partial^2 u}{\partial y^2} = u$. **[BU(H) 1999]**

[**Ans:** (i) $|x| < 1$ (elliptic), $x = 1$ (parabolic), $|x| > 1$ (hyperbolic); (ii) parabolic on the hyperbola $x^2 - xy - y = 0$, elliptic in the interior of the hyperbola, parabolic outside the hyperbola]

21.2 Find the solution of the equation $\frac{\partial u}{\partial x} = 4\frac{\partial u}{\partial y}$.

[AMIE(Summer) 1997] [**Ans:** $u = 8e^{mx-3y}$]

21.3 Find the solution of the following differential equation $\frac{\partial^2 \psi}{\partial x^2} + \frac{\partial^2 \psi}{\partial y^2} = 0$ inside an annulus bounded by circles $x^2 + y^2 = r_1^2$ and $x^2 + y^2 = r_2^2$, which satisfies the conditions :

$\psi = \psi_2$ at $r = r_2$; $\left(\frac{\partial \psi}{\partial r}\right) = \frac{k}{r_1}$ at $r = r_1$. [**Ans:** $\psi = k \ln\left(\frac{r}{r_2}\right) + \psi_2$]

21.4 Solve : $\frac{\partial^2 u}{\partial x^2} = 2\frac{\partial u}{\partial t}$, if $u(x,0) = x(4-x)$. [**Ans:** $u = k(4-x)e^{-\frac{p^2 t}{2}}$]

21.5 Solve the one-dimensional wave equation $\frac{\partial^2 u}{\partial t^2} = c^2\frac{\partial^2 u}{\partial x^2}$ under the conditions :

$$u = 0, \quad \text{when } x = 0 \text{ and } x = \pi$$

$$\frac{\partial u}{\partial t} = 0, \quad \text{when } t = 0 \text{ and } u(x,0) = x,\ 0 < x < \pi.$$

[**Ans:** $u = 2\sum \frac{(-1)^{n+1}}{n} \sin nx \cos nct$]

21.6 What is one-dimensional heat equation? Give main steps in the solution of such a partial differential equation. **[AMIE(Summer) 1991]** [**Ans:** $\frac{\partial u}{\partial t} = c^2 \frac{\partial^2 u}{\partial x^2}$]

21.7 The equation $\dfrac{\partial u}{\partial t} = \mu \dfrac{\partial^2 u}{\partial x^2}$ refers to the condition of heat along a bar without radiation. Show that if $u = Ae^{-gx} \sin(nt - gx)$, where A, g and n are positive constant, then

$$g = \sqrt{\frac{n}{2\mu}}.$$

21.8 Find the solution of $\dfrac{\partial V}{\partial t} = k \dfrac{\partial^2 V}{\partial x^2}$, having given that, $V = V_0 \sin nt$, when $x = 0$, for all values of t, and $V = 0$ when $x \to \infty$. [**Ans:** $V = V_0 e^{-\sqrt{\frac{n}{2k}}x} \sin\left[nt - \sqrt{\frac{n}{2k}}x\right]$]

21.9 Determine the distribution of temperature in the semi-infinite medium $x \geq 0$ when the end $x = 0$ is maintained at zero temperature and the initial distribution of temperature is $f(x)$. **[Osmania U 2003]** [**Ans:** $u(x,t) = \frac{2}{\pi} \int_0^\infty f_s(s) e^{-c^2 s^2 t} \sin xs \, ds$]

21.10 A tightly stretched string of 10 cm attached at both ends is set in motion with the following conditions :

$$u(x,0) = 0, \quad \frac{\partial u(x,0)}{\partial t} = 6 \sin^3\left(\frac{\pi x}{10}\right).$$

Determine the motion. [**Ans:** $u(x,t) = \frac{45}{\pi} \sin \frac{\pi x}{10} \sin \frac{\pi ct}{10} - \frac{5}{\pi c} \sin \frac{3\pi x}{10} \sin \frac{3\pi ct}{10}$]

21(C) : Broad Answer Type Questions

21.1 Using the method of separation of variables solve the following partial differential equations :

(a) $\dfrac{\partial u}{\partial t} = 4 \dfrac{\partial^2 u}{\partial x^2}$, when $u(x,0) = 4x - \dfrac{x^2}{2}$. [**Ans:** $u = \left(4x - \frac{x^2}{2}\right) e^{-p^2 t}$]

(b) Solve by the method of separation of variables $3\dfrac{\partial u}{\partial x} + 2\dfrac{\partial u}{\partial y}$, $u(x,0) = 4e^{-x}$.

[CU(H) 2006] [**Ans:** $u(x,y) = 3e^{-5x-3x} + 2e^{-3x-2y}$]

(c) Find a solution of the equation $\dfrac{\partial^2 u}{\partial x^2} = \dfrac{\partial u}{\partial y} + 2y$ in the form $u = f(x)g(y)$. Solve the equation subject to the conditions $u = 0$ and $\dfrac{\partial u}{\partial x} = 1 + e^{-3y}$, when $x = 0$, for all values of y. **[Andhra U 2000]** [**Ans:** $u = \frac{1}{\sqrt{2}} \sinh \sqrt{2}x + e^{-3y} \sin x$]

(d) Obtain the general solution of the following partial differential equation by the method of separation of variables,

$$\frac{\partial^2 T}{\partial x^2} = \alpha^2 \frac{\partial T}{\partial t}.$$

[CU(H) 2005] [**Ans:** $u = (C_2 \cos px + C_3 \sin px) C_1 e^{-\frac{p^2 t}{\alpha^2}}$]

(e) $x^2 \dfrac{\partial u}{\partial x} + y^2 \dfrac{\partial u}{\partial y} = 0.$ **[VTU 2006]** [**Ans:** $u = Ce^{k\left(\frac{1}{y} - \frac{1}{x}\right)}$]

21.2 Solve the following Laplace's equations with the given conditions :

(a) Solve the Laplace's equation $\frac{\partial^2 u}{\partial x^2} + \frac{\partial^2 u}{\partial y^2} = 0$ subject to the conditions :

$$u(0, y) = u(l, y) = u(x, 0) = 0 \quad \text{and} \quad u(x, a) = \sin\left(\frac{n\pi x}{l}\right).$$

[JNTU 2006] [**Ans:** $u(x, y) = \frac{\sinh\left(\frac{n\pi y}{l}\right)}{\sinh\left(\frac{n\pi a}{l}\right)} \sin \frac{n\pi x}{l}$]

(b) Find the solution of the Laplace's equation $\frac{\partial^2 u}{\partial x^2} + \frac{\partial^2 u}{\partial y^2} = 0$ under the following boundary conditions :

$$u(0, y) = 0, \text{ for all values of } y$$
$$u(x, y) = 0, \text{ for all values of } y$$
$$u(x, \infty) = 0, \text{ in } 0 < x < \pi$$
$$u(x, 0) = u_0, \text{ in } 0 < x < \pi$$

[AMIETE 2005S]

[**Ans:** $u(x, y) = \frac{4u_0}{\pi}[e^{-y} \sin x + \frac{1}{3}e^{-3y} \sin 3x + \frac{1}{5}e^{-5y} \sin 5x + \cdots]$]

(c) Laplace's equation in two dimensions $\frac{\partial^2 \phi}{\partial x^2} + \frac{\partial^2 \phi}{\partial y^2} = 0$ has a solution

$$\phi(x, y) = (A_1 e^{kx} + A_2 e^{-kx})(B_1 \cos ky + B_2 \sin ky), \tag{1}$$

where A_1, A_2, B_1, B_2 and k are constants.

Now consider the following boundary conditions :

(i) $\phi(x, 0) = 0$ **(ii)** $\phi(x, a) = 0$ **(iii)** $\phi(x \to \infty, y) = 0$ **(iv)** $\phi(x, y) = F(x, y)$.

Show that the above equation (1) gives general solution of the form :

$$\phi(x, y) = \sum_{n=1}^{\infty} B_n \sin \frac{n\pi y}{a} e^{-\frac{n\pi x}{a}} = \sum_{n=1}^{\infty} \left[\frac{2}{a} \int_0^a F(x, y) \sin \frac{n\pi y}{a} dy\right] \times \sin \frac{n\pi y}{a} e^{-\frac{n\pi x}{a}}.$$

[CU(H) 2004, 1997]

(d) Solve the Laplace's equation in polar coordinates by the method of separation of variables.

$$\frac{\partial^2 u}{\partial r^2} + \frac{1}{r}\frac{\partial u}{\partial r} + \frac{1}{r^2}\frac{\partial^2 u}{\partial \theta^2} = 0.$$

[AMIE(Summer) 1992]

[**Ans:** The three possible solutions are :

$$u = (C_1 r^k + C_2 r^{-k})[C_3 \cos(k\theta) + C_4 \sin(k\theta)]$$
$$u = (C_5 + C_6 \log r)(C_7 + C_8 \theta)$$
$$u = [C_9 \cos(p \log r) + C_{10} \sin(p \log r)] \times [C_{11} e^{p\theta} + C_{12} e^{-p\theta}]]$$

(e) Solve the three-dimensional Laplace's equation

$$\nabla^2 u = 0, \text{ i.e, } \frac{\partial^2 u}{\partial x^2} + \frac{\partial^2 u}{\partial y^2} + \frac{\partial^2 u}{\partial z^2} = 0$$

in Cartesian coordinates.

[**Ans:** $u = (Ae^{kx} + Be^{-kx})(Ce^{ly} + De^{-ly})[E \cos(\sqrt{k^2 + l^2})z + F \sin(\sqrt{k^2 + l^2})z]$]

21.3 Solve the following wave equations using the given boundary conditions :

(a) Find the solution of the equation $\frac{\partial^2 y}{\partial t^2} = a^2 \frac{\partial^2 y}{\partial x^2}$ which satisfies the boundary value conditions :

$$y(0,t) = 0, \quad y(l,t) = 0, \quad y(x,0) = f(x) \text{ and } \frac{\partial y}{\partial x} \text{ at } t = 0 \text{ is } \phi(x).$$

[AMIE(Winter) 1992; BU(H) 1994] [**Ans:** $y = f(x)\cos\left(\frac{n\pi at}{l}\right) + \frac{l\phi(x)}{n\pi a}\frac{\sin\left(\frac{n\pi at}{l}\right)}{\sin\left(\frac{n\pi x}{l}\right)}$]

(b) Solve the one-dimensional wave equation $\frac{\partial^2 y}{\partial t^2} = a^2 \frac{\partial^2 y}{\partial x^2}$ (a being a constant) which satisfies the conditions :

(i) $y(0,t) = 0$, **(ii)** $y(l,t) = 0$, **(iii)** $y(x,0) = \phi(x)$ and **(iv)** $\frac{\partial y(x,0)}{\partial t} = 0$.

$$\text{where } \phi(x) = \frac{2hx}{l} \quad \text{for } 0 \le x \le \frac{l}{2}$$
$$= \frac{2h}{l}(l - x) \quad \text{for } \frac{l}{2} \le x \le l.$$

[BU(H) 2004; Kerala U ME 2005]

[**Ans:** $y = \frac{8h}{\pi^2}\left[\sin\frac{\pi x}{l}\cos\frac{\pi ct}{l} - \frac{1}{3^2}\sin\frac{3\pi x}{l}\cos\frac{3\pi ct}{l}\right]$]

(c) A tightly stretched string of length l with fixed ends is initially in equilibrium position. It is set vibrating by giving each point a velocity $v_0 \sin^3\left(\frac{\pi x}{l}\right)$. Find the displacement $y(x,t)$.

[UPTU 2003] [**Ans:** $y = \frac{lv_0}{12c\pi}\left[9\sin\frac{\pi x}{l}\sin\frac{c\pi t}{l} - \sin\frac{3\pi x}{l}\sin\frac{3\pi ct}{l}\right]$]

(d) The vibrations of an elastic string is governed by the partial differential equation

$$\frac{\partial^2 u}{\partial t^2} = \frac{\partial^2 u}{\partial x^2}.$$

The length of the string is π and the ends are fixed. The initial velocity is zero and the initial deflection is $u(x,0) = 2(\sin x + \sin 3x)$. Find the deflection $u(x,t)$ of the vibrating string for $t > 0$.

[AMIETE(Winter) 1987] [**Ans:** $u(x,t) = 4\cos x \sin 2x \cos 2t$]

(e) Using the transformations $v = x + ct$ and $z = x - ct$, solve the following equation

$$\frac{\partial^2 u}{\partial t^2} = c^2\frac{\partial^2 u}{\partial x^2}; \quad \frac{\partial u}{\partial t}(x,0) = 0; \quad u(x,0) = f(x).$$

[**Ans:** $u(x,0) = \sum_{n=1}^{\infty} b_n \cos\frac{n\pi ct}{l}\sin\frac{n\pi x}{l}$]

(f) Using D'Alembert's method find the deflection of a vibrating string of unit length having fixed ends, with initial velocity zero and initial deflection

(i) $f(x) = a(x - x^2)$ **[Kerala, M.Tech 2006]** **(ii)** $f(x) = a\sin^2 \pi x$.

[**Ans:** (i) $y(x,t) = a(x - x^2 - c^2t^2)$ (ii) $y(x,t) = \frac{a}{2}(1 - \cos 2\pi x \cos 2\pi ct)$]

(g) A uniform flexible string of length l and linear density is stretched with a tension T between the fixed end points $x = 0$ and $x = l$. Obtain an expression for the transverse displacement $y(x,t)$ of the string under the following boundary conditions :

(i) initial velocity, $v(x,0) = 0$

(ii) initial displacement, $y(x,0) = \dfrac{3bx}{l}, \quad \text{for } 0 \le x \le \dfrac{l}{3}$

$$= \frac{3b(l-x)}{2l}, \quad \text{for } \frac{l}{3} \le x \le l.$$ [BU(H) 1992]

[**Ans:** $y(x,t) = \dfrac{9b}{\pi^2}\displaystyle\sum_{n=1}^{\infty} \dfrac{1}{n^2} \sin\dfrac{n\pi}{3} \sin\dfrac{n\pi x}{l} \sin\dfrac{n\pi ct}{l}$]

(h) A uniform string of length l is stretched between the fixed ends $x = 0$, $x = l$. The string is then plucked at $x = 0$, $0 < c < l$ through a transverse height b. Show that the initial displacement is given by

$$y(x,0) = \frac{bx}{c}, \qquad 0 \le x \le c$$

$$= \frac{b(l-x)}{(l-c)}, \quad 0 \le x \le l.$$

(i) Find the displacement $y(x,t)$ of the string if it is set into vibration by striking at the centre so as to give it a velocity which varies linearly from zero at $x = 0$ to v_0 at $x = \dfrac{L}{2}$, keeping the remaining portion of the string initially at rest.

[BU(H) 1995] [**Ans:** $y(x,t) = \dfrac{4V_0 l}{\pi^3 c}\displaystyle\sum_{n=1}^{\infty} \dfrac{1}{n^3} \sin\left(\dfrac{n\pi}{2}\right) \sin\dfrac{n\pi x}{L} \sin\dfrac{n\pi ct}{L}$]

21.4 Solve the following heat flow equations with the given boundary conditions :

(a) Solve : $\dfrac{\partial u}{\partial t} = a^2 \dfrac{\partial^2 u}{\partial x^2}$ under the conditions :

(i) $u(0,t) = 0$, $t > 0$ (ii) $u(\pi,t) = 0$ (iii) $u(x,0) = x^2$, $0 < x < \pi$.

[**Ans:** $u(x,t) = \dfrac{\pi^2}{3} + 4\displaystyle\sum_{n=1}^{\infty} \dfrac{(-1)^n}{n^2} \cos nx \times e^{-a^2 n^2 t}$]

(b) Show that the three-dimensional equation of heat flow in polar coordinates has the solution

$$u(r) = \frac{Q}{4\pi kr} + B,$$

where Q is the rate of isothermal heat flow from a point source at $r = 0$ and B is a constant.

If the temperature of two concentric sphere of radii r_1 and r_2 with the point source at the centre be u_1 and u_2 respectively, then show that,

$$Q = \frac{4\pi k(u_1 - u_2)}{(r_1 - r_2)} r_1 r_2.$$

(c) Find the solution of the equation $\dfrac{\partial^2 u}{\partial x^2} = \dfrac{1}{k}\dfrac{\partial u}{\partial t}$ that remains finite, when $t \to \infty$.

[VU(H) 2001]

(d) Solve the partial differential equation $\frac{\partial u}{\partial t} = \frac{\partial^2 u}{\partial x^2}$ by the method of separation of variables in the region $0 < x < 4$ for $t > 0$. Given $u(0,t) = 0$, $u(4,t) = 0$, $u(x,0) = 2x$. **[CU(H) 2008]**

(e) Solve the following boundary value problem which arises in the heat conduction in a rod :

$$\frac{\partial u}{\partial t} = c^2 \frac{\partial^2 u}{\partial x^2}.$$

$u(0,t) = u(l,t) = 0, \quad u(x,0) = 100\frac{x}{l}.$

[**Ans:** $u(x,t) = \frac{200}{\pi} \sum_{n=1}^{\infty} \frac{(-1)^{n+1}}{n} \sin \frac{n\pi x}{l} e^{-\frac{c^2 n^2 \pi^2}{l^2} t}$]

21.5 (a) Write down the two-dimensional Laplace's equation in plane polar coordinates.

(b) Find the general solution of the above equation by the method of separation of variables.

(c) The circumference of a semicircular metal plate of radius R is maintained at a temperature $T = T_0$, while the base of the same is kept at $T = 0$. Determine the steady state temperature T at any point $P(r,\theta)$ within the area enclosed by the plate, using the solution of two-dimensional Laplace's equation with boundary conditions $T(r,0) = 0$ for $0 < r < R$, $T(r,\theta) =$ finite, when $r \to 0$ and $T(R,\theta) = T_0$ for $0 < \theta < \pi$. **[BU(H) 2014]**

21.6 The wave equation describing the transverse vibrations of a stretched membrane under tension T and having a uniform surface density ρ is

$$T\left(\frac{\partial^2 u}{\partial x^2} + \frac{\partial^2 u}{\partial y^2}\right) = \rho \frac{\partial^2 u}{\partial t^2}.$$

Find a separable solution appropriate to a membrane stretched on a frame of length a and width b. Show that the natural angular frequencies of such a membrane are given by,

$\omega^2 = \frac{\pi^2 T}{\rho}\left(\frac{n^2}{a^2} + \frac{m^2}{b^2}\right)$ where n and m are positive integers. **[BU(H) 2013]**

21.7 Show that Laplace's equation in two-dimensions can be written in the polar form as

$$\frac{\partial^2 u}{\partial r^2} + \frac{1}{r}\frac{\partial u}{\partial r} + \frac{1}{r^2}\frac{\partial^2 u}{\partial \theta^2} = 0.$$

Solve the above equation by the method of separation of variables with the following boundary conditions.

(i) $u(r,0) = 0;\ 0 \le r \le a$ **(ii)** $u(r,\pi) = 0;\ 0 \le r \le a$ and **(iii)** $u(a,\theta) = T$

21.8 (a) Use the method of separation of variables for the equation $(\nabla^2 + k)\psi(r,\theta,\phi) = 0$ to solve the radial part of the equation only. **[NBU(H) 2008]**

(b) Solve the equation $\frac{\partial u}{\partial t} = \frac{\partial^2 u}{\partial x^2}$, $t > 0$, $0 < x < L$,

using the method of separation of variables, subject to the boundary condition $u(0,t) = 0 = u(L,t)$ for all t. **[CU(H) 2014]**

21.9 **(a)** Use the method of separation of variables to solve $\frac{\partial u}{\partial x} + 4\frac{\partial u}{\partial y}$.

Given, $u(0, y) = 8e^{-3y}$. **[CU(H) 2011; VU(H) 2008]**

(b) Laplace's equation in spherical polar coordinates with azimuthal symmetry is given by

$$\frac{1}{r^2}\frac{\partial}{\partial r}\left(r^2\frac{\partial v}{\partial r}\right) + \frac{1}{r^2\sin^2\theta}\left(\sin\theta\frac{\partial v}{\partial \theta}\right) = 0.$$

Assume $v(r, \theta) = R(r)\Theta(\theta)$. Show that $\frac{1}{R}\frac{d}{dr}\left(r^2\frac{dR}{dr}\right) = \frac{1}{\Theta\sin\theta}\frac{d}{d\theta}\left(\sin\theta\frac{d\Theta}{d\theta}\right)$

for every r and θ. Why each side of the last equation is a constant? Assume the constant to be of the form $l(l+1)$, we get the radial part of the equation.

$$r^2\frac{d^2R}{dr^2} + 2r\frac{dR}{dr} - l(l+1)R = 0$$

to solve the equation substitute $r = e^x$ and show that $\frac{d^2R}{dx^2} + \frac{dR}{dx} - l(l+1)R = 0.$

Show that the solution of R in the variable x is given by $R = Ae^{lx} + Be^{-(l+1)x}$. Hence, get the solution of R in terms of r.

(c) Laplace's equation in cylindrical coordinate (with z-symmetry) is given by

$$\frac{1}{\rho}\frac{\partial}{\partial \rho}\left(\rho\frac{\partial v}{\partial \rho}\right) + \frac{1}{\rho^2}\frac{\partial^2 v}{\partial \phi^2} = 0.$$

Assume $v(\rho, \phi) = R(\rho)\Phi(\phi)$. Show that, $\frac{\rho}{R}\frac{d}{d\rho}\left(\rho\frac{dR}{d\phi}\right) = -\frac{1}{\phi}\frac{d^2\Phi}{d\phi^2}$.

Show that we have to assume $\frac{1}{\phi}\frac{d^2\Phi}{d\phi} = -m^2$ (m is an integer) to solve Φ.

[CU(H) 2011]

21.10 Solve the following equation using the method of separation of variables,

$$\frac{\partial^2 V}{\partial x^2} + \frac{\partial^2 V}{\partial y^2} = 0$$

with $V = 0$ when $y = 0$, $V = 0$ when $y = \pi$, $V = V_0$ when $x = 1$, $V(-x, y) = V(x, y)$.

[CU(H) 2007]

21.11 **(a)** A uniform flexible string of length l and linear density ρ is stretched between its ends with a tension T. The string is plucked at $x = \frac{l}{3}$ with an amplitude h and released. Solve the equation for subsequent motion of the string. (Solution of differential wave equation for a stretched string can be assumed. **[VU(H) 2010]**

(b) Solve : $\frac{\partial^2 u}{\partial x^2} + \frac{\partial^2 u}{\partial y^2} = 0$, for $\begin{cases} 0 < x < \pi \\ 0 < y < \pi \end{cases}$

with the conditions : $u(0, y) = u(\pi, y) = u(x, 0) = \sin^2 x$.

21.12 **(a)** What is an exact differential equation? Write down necessary and sufficient condition for an exact differential equation.

(b) Check $(x + y)dx + xdy = 0$ is exact or not. **[Gujrat U(MSc) 2017]**

21.13 **(a)** Find the amplitude $u(\rho, \phi, t)$ of a vibrating circular membrane of radius 'a' satisfying the wave equation :

$$\nabla^2 u - \frac{1}{v^2}\frac{\delta^2 u}{\delta t^2} = 0,$$

where v is the phase velocity of the wave fixed by elastic constants.

(b) Separate the partial differential equation $\nabla^2\psi(\rho,\phi,z) + k^2\psi(\rho,\phi,z) = 0$ into three ordinary differential equations.

(c) Obtain general solution of Laplace's equation in cylindrical co-ordinates.

21.14 **(a)** Find whether $d\phi$ is an exact differential, where $d\phi = (x^3 - 4xy)dx + (v^2 - 2x)dy$.

(b) Solve $\frac{\partial U}{\partial x} + 4\frac{\partial U}{\partial y}$ by the method of separation of variables, given, $U(0,y) = 8e^{-3v}$. **[CU(H) 2017]**

21.15 **(a)** Given the equation $\frac{\partial^2 f}{\partial x^2} = \frac{1}{k^2}\frac{\partial f}{\partial t}$, where k is a constant and the boundary conditions are $f(0,t) = f(L,t) = 0$. Show that there exists a particular solution of the form

$$f_n(x,t) = B_n \sin\frac{n\pi x}{t} e^{-\frac{t}{T_n}},$$

where B_n and T_n are constants related to the integer n. Find the expression for T_n.

(b) Laplace's equation in spherical polar co-ordinates for problem with azimuthal symmetry is given by

$$\frac{1}{r^2}\frac{\partial}{\partial r}\left(r^2\frac{\partial V}{\partial r}\right) + \frac{1}{r^2\sin\theta}\frac{\partial}{\partial\theta}\left(\sin\theta\frac{\partial V}{\partial\theta}\right) = 0.$$

Let $V(r,\theta) = R(r),\ T(\theta)$. Taking the separation constant to be $l(l+1)$, solve for $R(r)$. Also show that substitution of $W = \cos\theta$ in the angular part leads to Legendre's equation for $T(\theta) = \rho(W)$. **[CU(H) 2011]**

21.16 **(a)** Solve the following equation using the method of separation of variables

$$\frac{\partial^2 V}{\partial x^2} + \frac{\partial^2 V}{\partial y^2} = 0,$$

where $V = 0,$ for $y = 0$ and $y = \pi$

$V = V_0,$ for $x = 1$

$V(-x,y) = V(x,y).$ **[CU(H) 2015]**

(b) Solve the equation $\frac{\partial U}{\partial t} = \frac{\partial^2 U}{\partial x^2},\ t > 0,\ 0 < x < L.$ Using the method of separation of variables, subject to the boundary condition $U(0,t) = 0 = U(L,t)$ for all t.

21.17 **(a)** Solve by the method of separation of variables $3\frac{\partial u}{\partial x} + 2\frac{\partial u}{\partial y} = 0,\ u(x,0) = 4e^{-x}.$

(b) Consider the differential equation $\frac{1}{\sin\theta}\frac{d}{d\theta}\left(\sin\theta\frac{dT}{d\theta}\right) + l(l+1)T = 0$ for $T(\theta) = 0.$ Show that the substitution of $W = \cos\theta$ converts it into Legendre equation for $P(W) = T(\theta)$. **[CU(H) 2013]**

CHAPTER 22

Multiple Integral

22.1 Double Integrals

The symbol $\int_{y_1}^{y_2}\int_{x_1}^{x_2} f(x,y)dx\,dy$ is called a double integral. This means that the function $f(x,y)$ is to be integrated first with respect to x, treating y as constant, between the limits x_1 and x_2. The resulting expression is then to be integrated with respect to y between the limits y_1 and y_2.

The order of integration is denoted by

$$\boxed{I = \int_{y_1}^{y_2}\boxed{\int_{x_1}^{x_2} f(x,y)dx}\,dy} \qquad (22.1)$$

where the integration is to be performed from the inner rectangular to outer one.

The integration may also be carried out in altered order as indicated below. Here, first the integration is to be performed with rest to y, treating x as constant between y_1 and y_2. Then the resulting expression is to be integrated with respect to x between the limits x_1 and x_2.

$$\boxed{I = \int_{x_1}^{x_2}\boxed{\int_{y_1}^{y_2} f(x,y)dy}\,dx} \qquad (22.2)$$

22.2 Evaluation of Double Integral

(i) When y_1, y_2 are functions of x and x_1, x_2 are constants

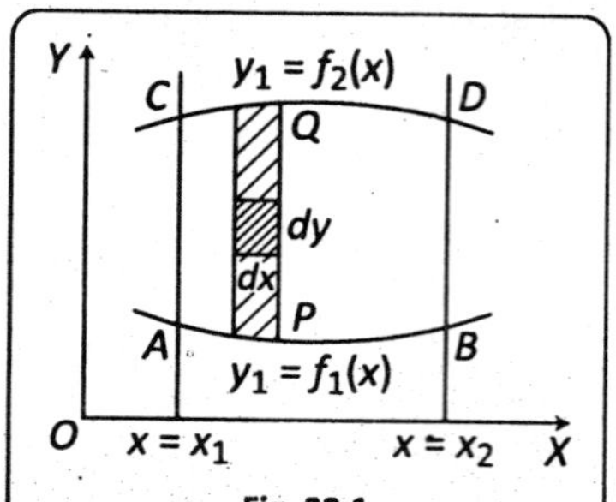

Fig. 22.1

$f(x,y)$ is first integrated with respect to y treating x as constant between the limits y_1 and y_2. Then the result is integrated with respect to x between the limits x_1 and x_2 [Eqn. (22.2)].

Fig. 22.1 shows two curves AB having equation $y_1 = f_1(x)$ and CD having equation $y_2 = f_2(x)$. The inner integration is to be carried out along one edge of the strip PQ (of width dx) from P to Q and the outer integration indicates sliding PQ from AC to BD covering the area $ABCD$.

(ii) When x_1, x_2 are functions of y and y_1, y_2 are constants

In this case $f(x,y)$ is to be first integrated with respect to x, treating y as constant between the limits x_1 and x_2 and then the resulting expression is integrated with respect to y between the given limit.

Fig. 22.2 shows the procedure corresponding to eqn. (22.1). Here the curve AB corresponds to $x_1 = f_1(y)$ and CD to $x_2 = f_2(y)$. The horizontal strip PQ of width dy and

inner integration is to be carried along one edge of the strip from P to Q and the outer one corresponds to sliding the edge from AC to BD covering the area $ABCD$.

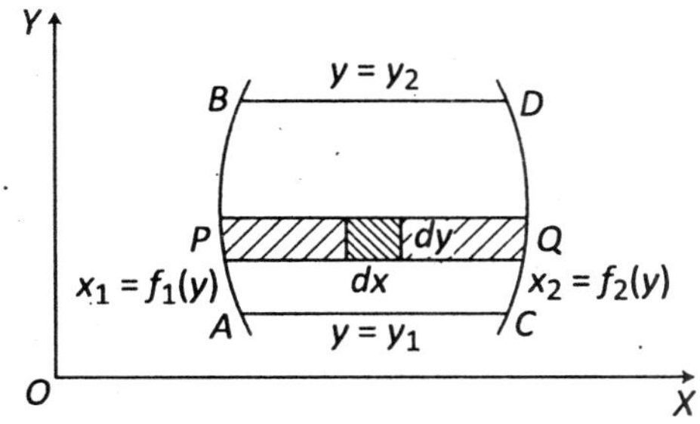

Fig. 22.2

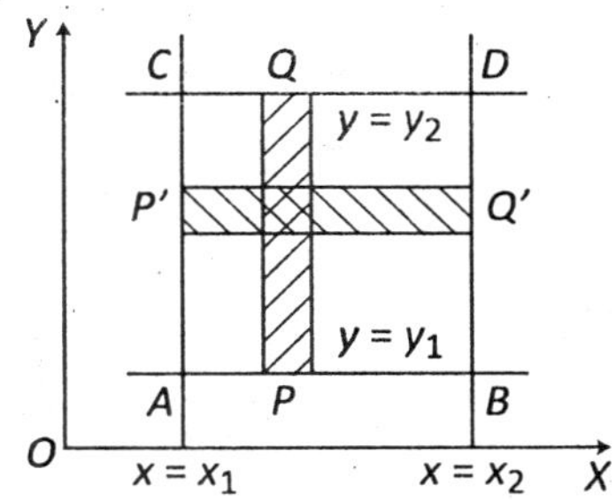

Fig. 22.3

(iii) When both x_1, x_2 and y_1, y_2 are constants

Here, we may integrate along PQ and then slide it from AC to BD (Fig. 22.3).

Or, we may integrate along $P'Q'$ and then slide it from AB to CD.

In both cases the covered area is $ABCD$.

N.B. If the limits are constant, then it is immaterial, whether the integration be carried out first with respect to x and then with respect to y or, conversely.

Example 22.1. *Evaluate $\iint_A xy\,dx\,dy$, where A is the domain bounded by X-axis, ordinate $x = 2a$ and the curve $x^2 = 4ay$.*

Solution : First method : The line $x = 2a$ and the parabola $x^2 = 4ay$ intersect at $B(2a, a)$. The domain of integration is OAB (Fig. 22.4). We first integrate w.r.t. x. The limits of integration are $x = 0$ to $2a$ and $y = 0$ to $\frac{x^2}{4a}$.

$$\therefore \iint_A \boldsymbol{xy\,dx\,dy} = \int_0^{2a}\int_0^{\frac{x^2}{4a}} xy\,dx\,dy = \int_0^{2a} x\,dx \int_0^{\frac{x^2}{4a}} y\,dy = \int_0^{2a} x\,dx\left[\frac{y^2}{2}\right]_0^{\frac{x^2}{4a}} = \int_0^{2a} x\frac{x^4}{32a^2}dx$$

$$= \frac{1}{32a^2}\int_0^{2a} x^5\,dx = \frac{1}{32a^2}\left[\frac{x^6}{6}\right]_0^{2a} = \frac{1}{192a^2}[(2a)^6 - 0] = \frac{64a^6}{192a^2} = \boldsymbol{\frac{a^4}{3}}.$$

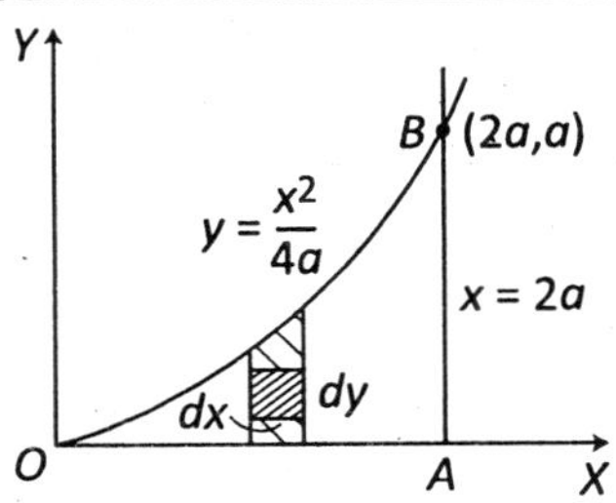

Fig. 22.4

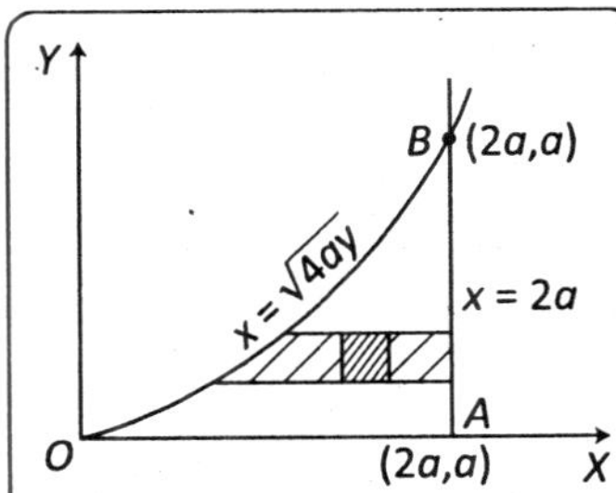

Fig. 22.5

Second method : Here we first integrate w.r.t. x and then w.r.t. y. The limits of integrations are $x = \sqrt{4ay}$ to $2a$ and $y = 0$ to a.

$$\therefore \iint_A xy\,dx\,dy = \int_0^a y\,dy \int_{\sqrt{4ay}}^{2a} x\,dx = \int_0^a y\,dy \left[\frac{x^2}{2}\right]_{\sqrt{4ay}}^{a} = \int_0^a (2a^2 - 2ay)dy$$

$$= 2a\int_0^a (ay - y^2)dy = 2a\left[\frac{ay^2}{2} - \frac{y^3}{3}\right] = 2a\left[\frac{a^2}{2} - \frac{a^3}{3}\right] = \frac{a^4}{3}.$$

22.3 Change of Order of Integration

In case of double integral with variable limit, the change of order of integration leads to a change of limits of integration. The double integrals which seems to be complicated can be solved easily by splitting the region of integration and the given integral expressed as the sum of a number of double integrals.

To find the new limits of integration, we draw a rough sketch of the region of integration.

Example 22.2. *Change the order of integration in the integral,* $I = \int_0^1 \int_{x^2}^{2-x} xy\,dy\,dx$ *and hence, evaluate the integral.* **[BU(H) 2001; SVTU 2007; UPTU 2005; Cochin U 2005]**

Solution : Here the region of integration is the area AOB which is bounded by the parabola $y = x^2$ and the line $y = 2 - x$. The point of intersection of the parabola $y = x^2$ and the line $y = 2 - x$ is $(1, 1)$.

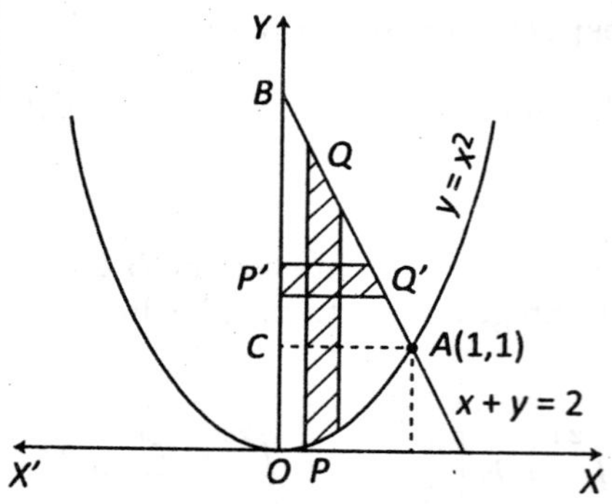

Fig. 22.6

In Fig. 22.6, the strip PQ is taken parallel to y-axis and the integration is first w.r.t. y along the strip. The order of integration is

$$\int_0^1 x\,dx \int_{x^2}^{2-x} y\,dy.$$

Changing the order of integration, first we integrate w.r.t. x along a horizontal strip $P'Q'$ parallel to X-axis. This is done by splitting the region OAB into (i) curvilinear ΔOAC and (ii) ΔABC. For ΔOAC, the limits are $x = 0$ to $x = \sqrt{y}$ and for y the limits are $y = 0$ to $y = 1$. Again for ΔABC, the limits for x are $x = 0$ to $x = 2 - y$ and those for y are $y = 1$ to $y = 2$.

$$\text{Thus, } I = \int_0^1 y\,dy \int_0^{\sqrt{y}} x\,dx + \int_1^2 y\,dy \int_0^{2-y} x\,dx = \int_0^1 y\,dy \left[\frac{x^2}{2}\right]_0^{\sqrt{y}} + \int_1^2 y\,dy \left[\frac{x^2}{2}\right]_0^{2-y}$$

$$= \frac{1}{2}\int_0^1 y^2 dy + \frac{1}{2}\int_1^2 y(2-y)^2 dy = \frac{1}{2}\left[\frac{y^3}{3}\right]_0^1 + \frac{1}{2}\int_1^2 (4y - 4y^2 + y^3)dy$$

$$= \frac{1}{6} + \frac{1}{2}\left[2y^2 - \frac{4}{3}y^3 + \frac{y^4}{4}\right]_1^2 = \frac{1}{6} + \frac{1}{2}\left[8 - \frac{32}{3} + 4 - 2 + \frac{4}{3} - \frac{1}{4}\right] = \frac{1}{6} + \frac{5}{24} = \frac{9}{24} = \frac{3}{8}.$$

22.4 Application of the Double Integrals

(a) Double integrals in Cartesian Coordinates

Area in Cartesian coordinates is given by

$$\text{Area } A = \int_{x_1}^{x_2} \int_{f_1(x)}^{f_2(x)} dx\,dy \quad \text{or,} \quad A = \int_{y_1}^{y_2} \int_{f_1(y)}^{f_2(y)} dx\,dy.$$

In Fig. 22.7, let the curves AB and CD be $y_1 = f_1(x)$ and $y_2 = f_2(x)$ and the ordinates AD and BC be $x = x_1$ and $x = x_2$. The area enclosed by the curves $y_1 = f_1(x)$ and $y_2 = f_2(x)$ and $x = x_1$ and $x = x_2$ is $ABCD$.

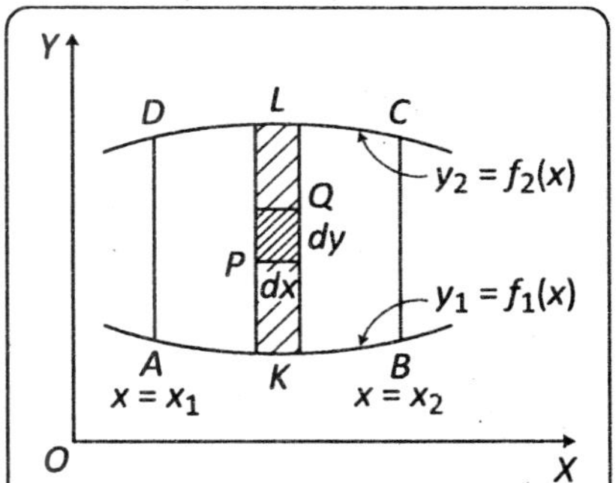

Fig. 22.7

The area $ABCD$ is divided into a number of vertical strips of width dx. Let $P(x, y)$ and $Q(x + \delta x, y + \delta y)$ be two neighbouring points; then the area of the small strip $PQ = \delta x \, \delta y$.

$$\therefore \text{ area of the strip } KL = \operatorname*{Lt}_{\delta y \to 0} \sum \delta x \delta y = \delta x \sum_{y_1}^{y_2} \delta y$$

$$= \delta x \int_{y_1}^{y_2} \delta y.$$

Adding the area of all the strips from $x = x_1$ to $x = x_2$, we get

$$\textbf{area } ABCD = \operatorname*{Lt}_{\delta x \to 0} \sum_{x_1}^{x_2} \int_{y_1}^{y_2} dy = \int_{x_1}^{x_2} dx \int_{y_1}^{y_2} dy.$$

$$\therefore \quad \textbf{area } A = \sum_{x_1}^{x_2} \sum_{y_1}^{y_2} dx\, dy.$$

(b) Double integral in polar coordinates

We consider the area enclosed by the curve $r = f(\theta)$. Let $P(r, \theta)$ and $Q(r + \delta r, \theta + \delta\theta)$ be two neighbouring points. We now draw circular areas of radii r and $r + dr$ which cut OQ at R and OP (produced) at S (Fig. 22.8).

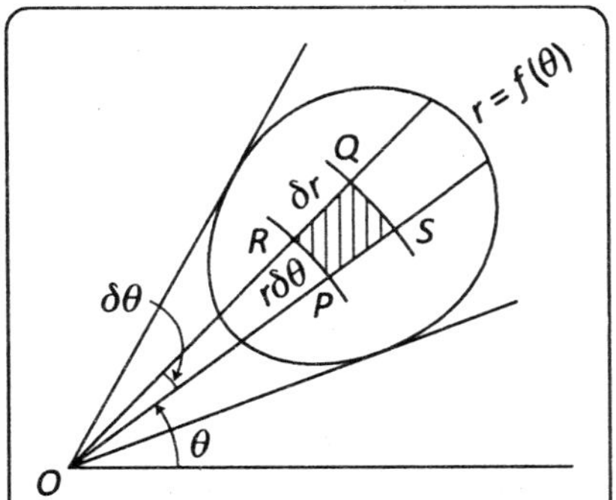

Fig. 22.8

Then, $PQ = r\delta\theta$, $PS = \delta r$.

The area of the rectangle $PQRS = r\delta\theta \cdot \delta r$.

The whole area is divided into a large number of such small rectangles.

Hence, the area $A = \operatorname*{Lt}_{\substack{\delta r \to 0 \\ \delta\theta \to 0}} = \sum \sum r\delta\theta \cdot \delta r$

$$= \iint r\, d\theta\, dr$$

Example 22.3. *Show that the area enclosed between the parabolas $y^2 = 4ax$ and $x^2 = 4ay$ is $\frac{16a^2}{3}$.*

Solution : The given parabolas are represented by the equations,

$$y^2 = 4ax \qquad \text{(i)}$$

and $\quad x^2 = 4ay. \qquad$ (ii)

Solving eqns. (i) and (ii) we get the point of intersection $(4a, 4a)$.

The area enclosed by the parabolas is divided into horizontal strips of width δy, x varies from P, $\frac{y^2}{4a}$ to Q, $\sqrt{4ay}$ and then y varies from $O(y = 0)$ to $y(x = 4a)$.

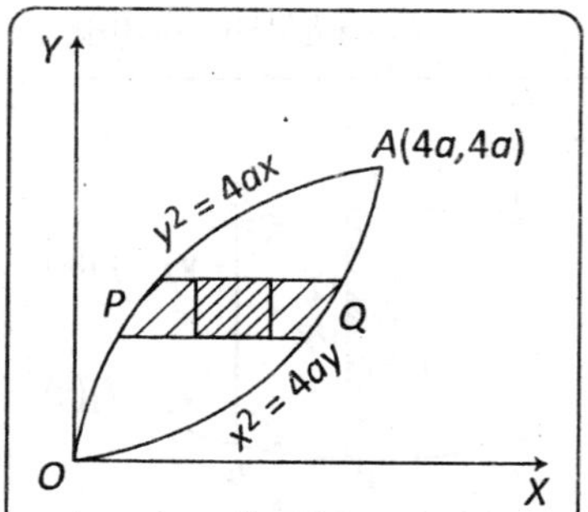

Fig. 22.9

$$\therefore \text{ the required area} = \int_0^{4a} dy \int_{\frac{y^2}{4a}}^{\sqrt{4ay}} dx = \int_0^{4a} dy [x]_{\frac{y^2}{4a}}^{\sqrt{4ay}}$$

$$= \int_0^{4a} \left[\sqrt{4ay} - \frac{y^2}{4a}\right] = \left[\sqrt{4}\frac{y^{\frac{3}{2}}}{\frac{3}{2}} - \frac{y^3}{12a}\right]_0^{4a}$$

$$= \left[\frac{4\sqrt{a}}{3}(4a)^{\frac{3}{2}} - \frac{(4a)^3}{12a}\right] = \left[\frac{32}{3}a^2 - \frac{16}{3}a^2\right]$$

$$= \frac{\mathbf{16}}{\mathbf{3}}\boldsymbol{a}^{\mathbf{2}}.$$

Example 22.4. *Calculate by double integration the volume generated by the revolution of the cardioid $r = a(1 - \cos\theta)$ about its axis.*

Solution : The volume V generated by the revolution of the cardioid about its axis (Fig. 22.10) is

$$V = 2\pi \iint y\,dx\,dy = 2\pi \iint (r\,d\theta\,dr)y = 2\pi \int d\theta \int r\,dr(r\sin\theta)$$

$$= 2\pi \int_0^{\pi} \sin\theta\,d\theta \int_0^{a(1-\cos\theta)} r^2 dr = 2\pi \int_0^{\pi} \sin\theta\,d\theta \left[\frac{r^3}{3}\right]_0^{a(1-\cos\theta)}$$

$$= \frac{2\pi}{3}\int_0^{\pi} a^3(1-\cos\theta)^3 \sin\theta\,d\theta = \frac{2\pi a^3}{3}\left[\frac{(1-\cos\theta)^4}{4}\right]_0^{\pi}$$

$$= \frac{2\pi a^3}{12}(16) = \frac{\mathbf{8}}{\mathbf{3}}\boldsymbol{\pi a^3}.$$

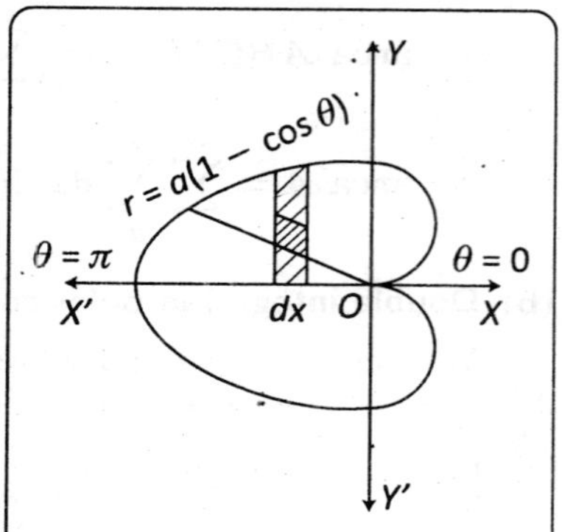

Fig. 22.10

22.5 Triple Integrals

We consider a finite region of volume V where a function $f(x, y, z)$ is defined at every point of it. If V be divided into elementary volumes δV_1, δV_2, ..., δV_n and (x_i, y_i, z_i) be any point in the ith element of volume δV_i, then the limit of the sum

$$\sum_{i=1}^{\infty} f(x_i, y_i, z_i)\delta V_i$$

if it exists, as $n \to \infty$ and $\delta V_i \to 0$, then

$$\iiint_V f(x, y, z) dV = \int_{x_1}^{x_2}\int_{y_1}^{y_2}\int_{z_1}^{z_2} f(x, y, z)\,dx\,dy\,dz \tag{22.3}$$

is known as the **triple integral** of $f(x, y, z)$ over the volume V.

Evaluation : If x_1, x_2 are constants, y_1, y_2 are either constants or functions of x and z_1, z_2 are also either constants or functions of x and y, then the above triple integral may be evaluated as described below. .

First, the integral $\int_{z_1}^{z_2} f(x, y)dz$ is integrated with respect to z between the limits z_1 and z_2 treating x and y constant. The resulting expression is then integrated with respect to y between the limits y_1 and y_2 treating x constant. Finally, the result obtained is integrated with respect to x between the limits x_1 and x_2. The procedure of evaluation is represented as follows :

$$I = \boxed{\int_{x_1}^{x_2} \boxed{\int_{y_1}^{y_2} \boxed{\int_{z_1}^{z_2} f(x, y, z) dz\, dy\, dx}}} \qquad (22.4)$$

The integration is to be carried out from the innermost rectangle to the outermost rectangle.

Example 22.5. *Evaluate the integral* $\iiint_R (x - 2y + z) dx\, dy\, dz$, *where* R *is defined as* $0 \le x \le 1$, $o \le y \le y^2$ *and* $0 \le z \le x + y$. **[AMIE(Winter) 1991]**

Solution : The integral is

$$I = \iiint_R (x - 2y + z) dx\, dy\, dz = \int_0^1 dx \int_0^{x^2} dy \int_0^{x+y} (x - 2y + z) dz$$

$$= \int_0^1 dx \int_0^{x^2} dy \left[\frac{(x - 2y + z)^2}{2}\right]_0^{x+y} = \int_0^1 dx \int_0^{x^2} dy \left[\frac{(x - 2y + x + y)^2}{2} - \frac{(x - 2y)^2}{2}\right]$$

$$= \frac{1}{2}\int_0^1 dx \int_0^{x^2} [(2x - y)^2 - (x - 2y)^2] dy = \frac{1}{2}\int_0^1 dx \int_0^{x^2} (3x^2 - 3y^2) dy$$

$$= \frac{3}{2}\int_0^1 dx \left[x^2 y - \frac{y^3}{3}\right]_0^{x^2} = \frac{3}{2}\int_0^1 dx \left(x^4 - \frac{x^6}{3}\right) = \frac{3}{2}\left[\frac{x^5}{5} - \frac{x^7}{21}\right]_0^1 = \mathbf{\frac{8}{35}}.$$

Example 22.6. *Evaluate :* $\int_0^{2\pi}\int_0^{\frac{\pi}{4}}\int_0^a r^2 \sin\theta\, dr\, d\theta\, d\phi$. **[AMIE(Summer) 1991]**

Solution : The given integral in spherical polar coordinates is

$$I = \int_0^{2\pi}\int_0^{\frac{\pi}{4}}\int_0^a r^2 \sin\theta\, dr\, d\theta\, d\phi = \int_0^{2\pi} d\phi \int_0^{\frac{\pi}{4}} \sin\theta\, d\theta \int_0^a r^2\, dr$$

$$= \int_0^{2\pi} d\phi \int_0^{\frac{\pi}{4}} \sin\theta\, d\theta \left[\frac{r^3}{3}\right]_0^a = \frac{a^3}{3}\int_0^{2\pi} d\phi \int_0^{\frac{\pi}{4}} \sin\theta\, d\theta = \frac{a^3}{3}\int_0^{2\pi} d\phi\, [-\cos\theta]_0^{\frac{\pi}{4}}$$

$$= \frac{a^3}{3}\int_0^{2\pi} d\phi \left(-\cos\frac{\pi}{4} + \cos 0\right) = \frac{a^3}{3}\left(1 - \frac{1}{\sqrt{2}}\right)\int_0^{2\pi} d\phi = \frac{a^3}{3}\left(1 - \frac{1}{\sqrt{2}}\right)[\phi]_0^{2\pi}$$

$$= \frac{a^3}{3}\left(1 - \frac{1}{\sqrt{2}}\right) 2\pi = \mathbf{\frac{2\pi a^3}{3\sqrt{2}}(\sqrt{2} - 1)}.$$

22.6 Volume of Solids

1. Volume as double integral

Let $z = f(x, y)$ represent a surface and S be the orthogonal projection of a part S' of it as shown in Fig. 22.11. Then the volume of the solid cylinder with S as its base and bounded by the given surface, with generators parallel to Z-axis, is

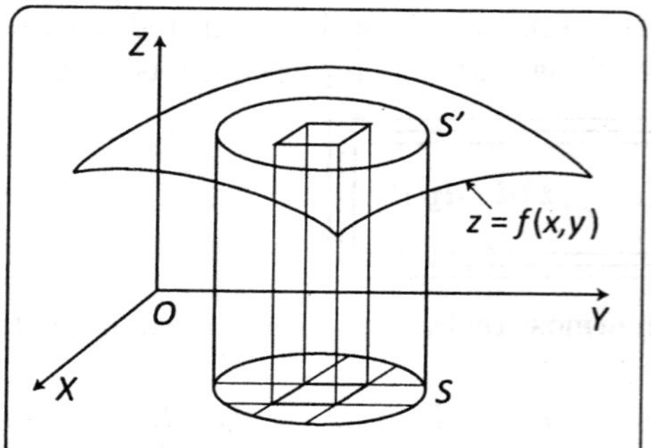

Fig. 22.11

$$V = \underset{\substack{\delta x \to 0 \\ \delta y \to 0}}{\text{Lt}} = \sum z\,\delta x\,\delta y = \iint z\,dx\,dy$$

or, $$V = \iint f(x, y)dx\,dy.$$

The integration is over the surface area.

In polar coordinates $dx\,dy$ is given by

$r\,d\theta\,dr$.

$$\therefore \quad \boldsymbol{V = \iint zr\,d\theta\,dr.} \tag{22.5}$$

2. Volume as triple integral

If an element of volume $\delta V = \delta x\,\delta y\,\delta z$ (Fig. 22.12), then the whole volume of the entire solid is obtained by evaluating the triple integral

$$\boldsymbol{V = \iiint \delta V = \iiint dx\,dy\,dz.} \tag{22.6}$$

If ρ be the density of the solid, then its mass

$$\boldsymbol{M = V \times \rho = \iiint \rho dx\,dy\,dz.} \tag{22.7}$$

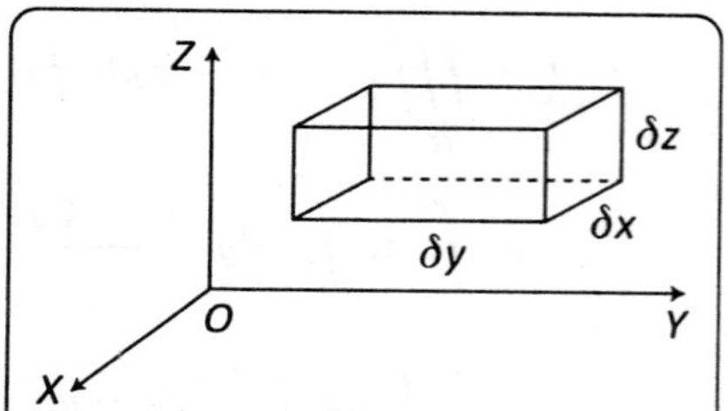

Fig. 22.12

3. Volume of solids of revolution

Let us consider an element of area $\delta x\,\delta y$ at $P(x, y)$ on a plane area A as shown in Fig. 22.13. The revolution of $\delta x\,\delta y$ about the X-axis will generate a ring of volume

$$\delta V = \pi[(y + \delta y)^2 - y^2]\delta x \simeq 2\pi y\,\delta x\,\delta y.$$

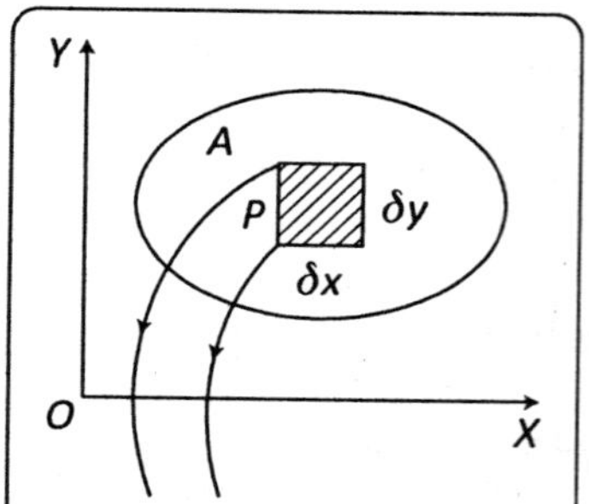

Fig. 22.13

$(\delta y)^2$ has been neglected due to smallness.

Then the whole volume generated by the entire area A about the X-axis is

$$\boldsymbol{V_1 = \iint_A 2\pi y\,dx\,dy.}$$

The volume generated by the revolution of A about the Y-axis is

$$\boldsymbol{V_2 = \iint_A 2\pi x\,dx\,dy.}$$

In polar coordinates the volume is

$$\boldsymbol{V_3 = \iint_A 2\pi r \sin\theta \cdot r\,d\theta\,dr = \iint_A 2\pi r^2 \sin\theta\,d\theta\,dr.} \tag{22.8}$$

Example 22.7. *Calculate the volume of the solid bounded by the surface $x = 0$, $y = 0$, $x + y + z = 1$ and $z = 0$.* **[UPTU 2005; AMIETE(Summer) 1993]**

Solution : Given $x = 0$, $y = 0$, $x + y + z = 1$ and $z = 0$.

$\therefore$ z varies from 0 to $1-x-y$, y varies from 0 to $1-x$ and x varies from 0 to 1.
$\therefore$ the required volume of the solid is

$$V = \iiint dx\,dy\,dz = \int_0^1 dx \int_0^{1-y} dy \int_{z=0}^{1-x-y} dz = \int_0^1 dx \int_0^{1-y} dy[z]_0^{1-x-y}$$
$$= \int_0^1 dx \int_0^{1-y} dy(1-x-y) = \int_0^1 dx \left[y - xy - \frac{y^2}{2}\right]^{1-y}$$
$$= \int_0^1 dx \left[1-x-x(1-x) - \frac{1}{2}(1-x)^2\right] = \int_0^1 \left[1-x-x-x^2-\frac{1}{2}+x-\frac{x^2}{2}\right] dx$$
$$= \int_0^1 \left[\frac{1}{2} - x + \frac{x^2}{2}\right] dx = \left[\frac{x}{2} - \frac{x^2}{2} + \frac{x^3}{6}\right]_0^1 = \frac{1}{2} - \frac{1}{2} + \frac{1}{6} = \mathbf{\frac{1}{6}}.$$

Example 22.8. *Express the volume between the surfaces whose equations are $x^2+y^2+z^2 = a^2$, $x^2+y^2 = a^2$, $z = a$ and the coordinate planes in the form : $V = \iint z\,dx\,dy$* and determine the value of V.

Solution : The required volume lies in the positive octant bounded by the sphere $x^2+y^2+z^2 = a^2$, the cylinder $x^2+y^2 = a^2$ and $z = a$ (Fig. 22.14).

We consider a strip raised on the element $\delta x\,\delta y$ parallel to Z-axis. If z be the length of the strip (z varies from the surface of the sphere to $z = a$, then

$$V = \iint z\,dx\,dy.$$

Here x varies from $\sqrt{a^2-x^2-y^2}$ to a,

y varies from 0 to $\sqrt{a^2-x^2}$

and z varies from 0 to a.

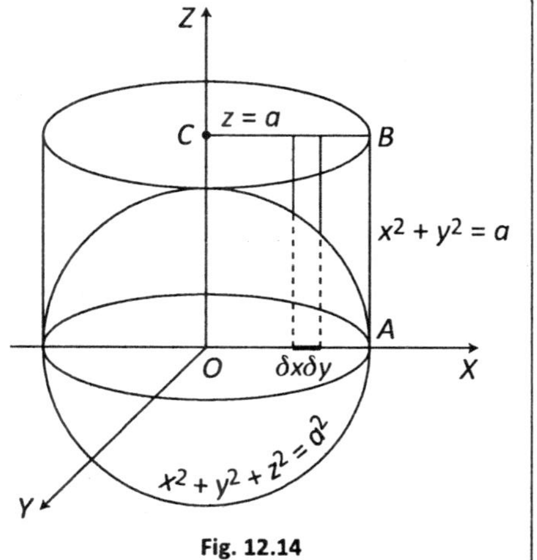

Fig. 12.14

$$\therefore \quad V = \iint [a - \sqrt{a^2-x^2-y^2}]dx\,dy$$
$$= \int \left[ay - \left\{\frac{1}{2}y\sqrt{a^2-a^2-y^2} + \frac{1}{2}(a^2-x^2)\sin^{-1}\frac{y}{\sqrt{a^2-x^2}}\right\}\right]_{y=0}^{y-\sqrt{a^2-x^2}} dx$$
$$= \int_0^a \left[a\sqrt{a^2-x^2} - \frac{1}{4}\pi(a^2-x^2)\right] dx$$
$$= \left[a\left\{\frac{1}{2}x\sqrt{a^2-x^2} + \frac{1}{2}a^2\sin^{-1}\frac{x}{a}\right\} - \frac{1}{2}\pi\left(a^2x - \frac{x^3}{3}\right)\right]_0^a$$
$$= \left[\frac{a^3}{2}\cdot\frac{\pi}{2} - \frac{1}{4}\pi\left(a^3 - \frac{a^3}{3}\right)\right] = \mathbf{\frac{\pi a^3}{12}}.$$

22.7 Change of Variables

Evaluation of double or triple integral becomes often easier by a change to appropriate choice of variables.

1. Double Integral : Suppose the variables x, y be changed to the new variables u and v by the transformation equations $x = x(u, v)$ and $y = y(u, v)$ are continuous differential functions in region R_{uv} in the uv-plane corresponding to the region R_{xy} in the xy-plane. Then,

$$\iint\limits_{R_{xy}} f(x, y)dx\,dy = \iint\limits_{R_{xy}} f[x(u, v), y(u, v)]|J|du\,dv, \tag{22.9}$$

where $|J| = \dfrac{\partial(x, y)}{\partial(u, v)} \neq 0$ is the Jacobian transformation from (x, y) to (u, v) coordinates.

2. Triple Integrals : For triple integral the corresponding formula is

$$\iiint\limits_{R_{xyz}} f(x, y, z)dx\,dy\,dz = \iiint\limits_{R_{uvw}} f[x(u, v, w), y(u, v, w)] \times |J|du\,dv\,dw, \tag{22.10}$$

where $|J| = \dfrac{\partial(x, y, z)}{\partial(u, v, w)} \neq 0$ is the Jacobian transformation from (x, y, z) to (u, v, w).

Special Cases :

(i) **Cartesian to polar coordinates :** To change from Cartesian coordinates (x, y) to polar coordinates (r, θ), we have $x = r\cos\theta$, $y = r\sin\theta$ and $J = \frac{\partial(x,y)}{\partial(r,\theta)} = r$.

$$\therefore \quad \iint\limits_{R_{xy}} \boldsymbol{f(x, y)dx\,dy = \iint\limits_{R_{r\theta}} f(r\cos\theta, r\sin\theta)r\,dr\,d\theta}. \tag{22.11}$$

(ii) **Rectangular to cylindrical coordinates :** Here, $x = \rho\cos\phi$, $y = \rho\sin\phi$, $z = z$.

$$\therefore \quad J = \frac{\partial(x, y, z)}{\partial(\rho, \phi, z)} = \rho.$$

$$\therefore \quad \iiint\limits_{R_{xyz}} \boldsymbol{f(x, y, z)dx\,dy\,dz = \iiint\limits_{R_{\rho\phi z}} f[(\rho\cos\phi, \rho\sin\phi, z)\rho\,d\rho\,d\phi\,dz}. \tag{22.12}$$

(iii) **Rectangular to spherical polar coordinates :** Here, $x = r\sin\theta\cos\phi$, $y = r\sin\theta\sin\phi$, $z = r\cos\theta$.

$$\therefore \quad J = \frac{\partial(x, y, z)}{\partial(\rho, \theta, \phi)} = r^2\sin\theta.$$

$$\therefore \quad \iiint\limits_{R_{xyz}} \boldsymbol{f(x, y, z)dx\,dy\,dz}$$

$$= \iiint\limits_{R_{r\theta\phi}} \boldsymbol{f[(r\sin\theta\cos\phi), (r\sin\theta\sin\phi), (r\cos\theta)] \times r^2\sin\theta\,dr\,d\theta\,d\phi}. \tag{22.13}$$

Example 22.9. *Rewrite the following integral in polar coordinates and then integrate :*

$$\int_0^\infty \int_0^\infty e^{-(x^2+y^2)}dx\,dy.$$ **[BU(H) 2000, Anna U 2003]**

Solution : The region of integration is the first quadrant of the xy-plane and as such r varies from 0 to ∞ and θ varies from 0 to $\frac{\pi}{2}$.

$$\therefore \quad I = \int_0^\infty \int_0^\infty e^{-(x^2+y^2)} dx\, dy = \int_{\theta=0}^{\frac{\pi}{2}} \int_{r=0}^{\infty} e^{-r^2} r\, dr\, d\theta$$

$$\left[\because |J| = \frac{\partial(x,y)}{\partial(r,\theta)} = r \text{ and } x^2+y^2 = r^2\cos^2\theta + r^2\sin^2\theta, \ \therefore \ x^2+y^2 = r^2\right]$$

$$= -\frac{1}{2}\int_0^{\frac{\pi}{2}} \left\{\int_0^\infty e^{-r^2}(-2r)dr\right\} d\theta = -\frac{1}{2}\int_0^{\frac{\pi}{2}} \left|e^{-r^2}\right|_0^\infty d\theta = \frac{1}{2}\int_0^{\frac{\pi}{2}} d\theta = \frac{1}{2}\cdot\frac{\pi}{2} = \frac{\boldsymbol{\pi}}{\mathbf{4}}.$$

Example 22.10. *Evaluate the integral,* $\iiint (x^2+y^2+z^2)dx\, dy\, dz$ *taken over the volume enclosed by the sphere* $x^2+y^2+z^2=1$.

Solution : The integral can be evaluated by changing the variables from Cartesian (x,y,z) to polar (r,θ,ϕ) as

$$x = r\sin\theta\cos\phi, \quad y = r\sin\theta\sin\phi, \quad z = r\cos\theta.$$

Here, limits of r varies from 0 to 1, limits of θ varies from 0 to π and limits of ϕ varies from 0 to 2π.

Now, $|J| = \dfrac{\partial(x,y,z)}{\partial(\rho,\theta,\phi)} = r^2\sin\theta.$

$$\therefore \quad I = \iiint (x^2+y^2+z^2)dx\, dy\, dz = \int_{\theta=0}^{\pi}\int_{\phi=0}^{2\pi}\int_{r=0}^{1} r^2(r^2\sin\theta)d\theta\, d\phi\, dr$$

$$= \int_0^{2\pi} d\phi \int_0^{\pi} \sin\theta\, d\theta \int_0^1 r^4 dr = \int_0^{2\pi} d\phi \int_0^{\pi}\sin\theta\, d\theta \times \frac{1}{5} = \frac{1}{5}\int_0^{2\pi} d\phi\,[-\cos\theta]_0^\pi$$

$$= \frac{2}{5}\int_0^{2\pi} d\phi = \frac{2}{5}\times 2\pi = \frac{\mathbf{4\pi}}{\mathbf{5}}.$$

22.8 Area of Curved Surface

We consider a curved surface area S represented by $z = f(x,y)$, whose projection on the XY-plane is A. Then we erect a strip on $\delta x\delta y$ in region A whose generator is parallel to Z-axis and meeting the surface S in an element of area δS.

$\therefore \quad \delta x\delta y = dS\cos\gamma$ (Fig. 22.15),

where γ is the angle between the XY-plane and the tangent plane to S at P, i.e., the angle between the Z-axis and the normal to S at P.

The directions cosines of the normal to the surface $F(x,y,z) = 0$ are proportional to $\frac{\partial F}{\partial x}, \frac{\partial F}{\partial y}, \frac{\partial F}{\partial z}$.

$\therefore$ the direction of the normal to $S[F = f(x,y) - 1]$ are proportional to $-\frac{\partial z}{\partial x}, -\frac{\partial z}{\partial y}$, 1 and those of Z-axis are 0, 0, 1.

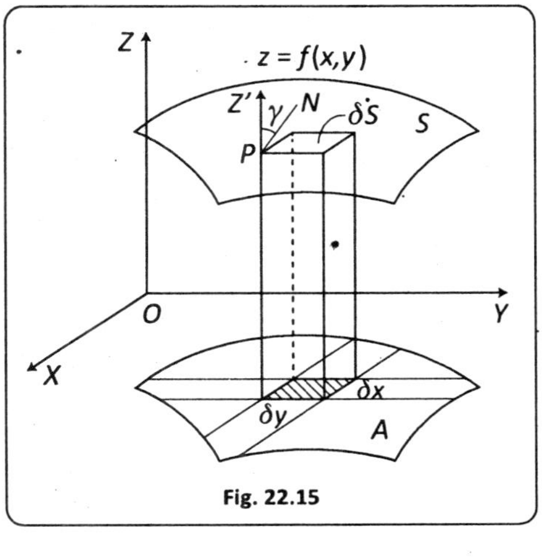

Fig. 22.15

$$\text{The direction cosines} = \frac{-\frac{\partial z}{\partial x}}{\sqrt{\left(\frac{\partial z}{\partial x}\right)^2 + \left(\frac{\partial z}{\partial y}\right)^2 + 1}}, \frac{-\frac{\partial z}{\partial y}}{\sqrt{\left(\frac{\partial z}{\partial x}\right)^2 + \left(\frac{\partial z}{\partial y}\right)^2 + 1}}, \frac{1}{\sqrt{\left(\frac{\partial z}{\partial x}\right)^2 + \left(\frac{\partial z}{\partial y}\right)^2 + 1}}.$$

$$\therefore \quad \cos\gamma = \frac{1}{\sqrt{\left(\frac{\partial z}{\partial x}\right)^2 + \left(\frac{\partial z}{\partial y}\right)^2 + 1}}.$$

$$\therefore \quad \delta S = \frac{\delta x \delta y}{\cos\gamma} = \sqrt{\left(\frac{\partial z}{\partial x}\right)^2 + \left(\frac{\partial z}{\partial y}\right)^2 + 1}\, \delta x \delta y.$$

$$\therefore \quad \boldsymbol{S} = \underset{\delta S \to 0}{\mathbf{Lt}} \sum \delta S = \iint_A \left[\left(\frac{\partial z}{\partial x}\right)^2 + \left(\frac{\partial z}{\partial y}\right)^2 + 1\right] dx\, dy. \tag{22.14}$$

Example 22.11. *Find the area of the portion of the cylinder $x^2 + z^2 = 4$ lying inside the cylinder $x^2 + y^2 = 4$.*

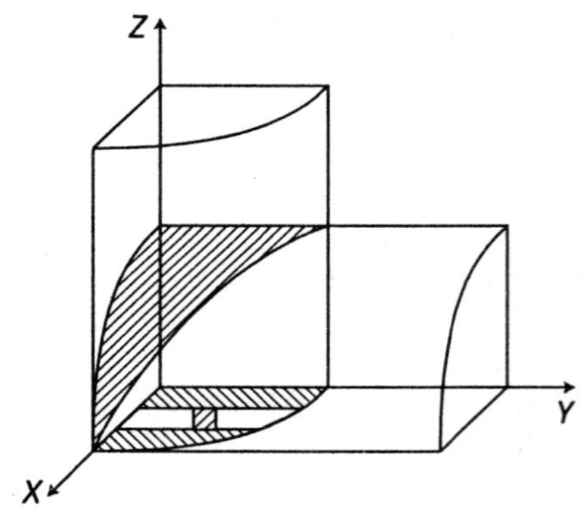

Fig. 22.16

Solution : Fig. 22.16 shows the $\frac{1}{8}$th of the required area. The projection of the said area on the XY-plane is a quadrant circle $x^2 + y^2 = 4$.

For $x^2 + z^2 = 4$, we have $\frac{\partial z}{\partial x} = -\frac{x}{z}$, $\frac{\partial z}{\partial y} = 0$.

$$\therefore \quad \left(\frac{\partial z}{\partial x}\right)^2 + \left(\frac{\partial z}{\partial y}\right)^2 + 1 = \frac{x^2 + z^2}{z^2} = \frac{4}{4 - x^2}.$$

$\therefore$ the required surface area $= 8 \times$ surface area of the upper portion of the cylinder $x^2 + z^2 = 4$ lying within the cylinder $x^2 + z^2 = 4$ in the first octant.

$$= 8 \times \int_0^2 \int_0^{\sqrt{4-x^2}} \frac{2}{\sqrt{4 - x^2}}\, dx\, dy$$

$$= 16 \int_0^2 \frac{1}{\sqrt{4 - x^2}} [y]_0^{\sqrt{4-x^2}} dx = 16 \int_0^2 \frac{1}{\sqrt{4 - x^2}} \times \sqrt{4 - x^2}\, dx = 16 \int_0^2 dx = 16[x]_0^2 = \mathbf{32}.$$

22.9 Calculation of Mass

1. For a plane lamina : Let at any point $P(x, y)$ of a plane lamina the surface density be $\rho(x, y)$ then elemental mass δm at P is

$$\delta m = \rho \delta x\, \delta y.$$

$$\text{Total mass of the lamina is } M = \iint \rho \delta x\, \delta y, \tag{22.15}$$

where the integration covers the whole area of the lamina.

$$\text{In polar coordinates the mass of the lamina is } \boldsymbol{M} = \iint \boldsymbol{\rho r\, \delta\theta\, dr}, \tag{22.16}$$

where $\rho(r, \theta)$ is the surface density at a point $P(r, \theta)$ of the lamina.

2. For solid : If $\rho(x, y, z)$ be the volume density at a point $P(x, y, z)$ of a solid, then its total mass

$$M = \iiint \rho \, dx \, dy \, dz, \tag{22.17}$$

the integration covers the whole volume of the solid.

22.10 Centre of Mass

1. Plane lamina : If $\rho(x, y)$ be the surface density at a point $P(x, y)$ of a plane lamina, then its elemental mass $\delta m = \rho \delta x \, \delta y$. Then its C.M. is given by

$$\left.\begin{aligned} \overline{x} &= \frac{\iint x\rho \, dx \, dy}{\iint \rho \, dx \, dy} \\ \text{and} \quad \overline{y} &= \frac{\iint y\rho \, dx \, dy}{\iint \rho \, dx \, dy} \end{aligned}\right\} \tag{22.18}$$

$\therefore$ the centre of mass is $(\overline{x}, \overline{y})$. The integration covers the whole lamina.

In polar coordinates $x = r\cos\theta$, $y = r\sin\theta$.

The elemental mass of the lamina is $\delta m = \rho r \, \delta\theta \, \delta r$ at $P(r, \theta)$ and the C.M. $(\overline{x}, \overline{y})$ is given by

$$\left.\begin{aligned} \overline{x} &= \frac{\iint r\cos\theta \rho r \, d\theta \, dr}{\iint \rho r \, d\theta \, dr} \\ \overline{y} &= \frac{\iiint r\sin\theta \rho r \, d\theta \, dr}{\iiint \rho r \, d\theta \, dr} \end{aligned}\right\} \tag{22.19}$$

2. Solid body : If $F(x, y, z)$ be the volume density of a solid at $P(x, y, z)$ then its C.M. $(\overline{x}, \overline{y}, \overline{z})$ is given by

$$\left.\begin{aligned} \overline{x} &= \frac{\iiint x\rho \, dx \, dy \, dz}{\iiint \rho \, dx \, dy \, dz} \\ \overline{y} &= \frac{\iiint y\rho \, dx \, dy \, dz}{\iiint \rho \, dx \, dy \, dz} \\ \overline{z} &= \frac{\iiint z\rho \, dx \, dy \, dz}{\iiint \rho \, dx \, dy \, dz} \end{aligned}\right\} \tag{22.20}$$

The range of integration covers the whole volume of the solid.

Example 22.12. *Determine the position of the centroid of the volume contained in the positive octant by* $\left(\frac{x}{a}\right)^{\frac{2}{3}} + \left(\frac{y}{b}\right)^{\frac{2}{3}} + \left(\frac{z}{c}\right)^{\frac{2}{3}} = 1.$

Solution : Let $\left(\frac{x}{a}\right)^{\frac{2}{3}} = u, \quad \therefore\ x = au^{\frac{3}{2}}, \quad dx = \frac{3}{2}au^{\frac{1}{2}}du$

$$\left(\frac{y}{b}\right)^{\frac{2}{3}} = v, \qquad y = bv^{\frac{3}{2}}, \quad dy = \frac{3}{2}bv^{\frac{1}{2}}dv$$

$$\left(\frac{z}{c}\right)^{\frac{2}{3}} = w, \qquad y = cw^{\frac{3}{2}}, \quad dz = \frac{3}{2}cw^{\frac{1}{2}}dw.$$

$$\therefore\ \overline{x} = \frac{\iiint x\,dx\,dy\,dz}{\iiint dx\,dy\,dz} = \frac{\iiint au^{\frac{3}{2}}u^{\frac{1}{2}}v^{\frac{1}{2}}w^{\frac{1}{2}}\,du\,dv\,dw}{\iiint u^{\frac{1}{2}}v^{\frac{1}{2}}w^{\frac{1}{2}}\,du\,dv\,dw},\ u+v+w=1$$

$$= a\frac{\iiint au^{3-1}v^{\frac{3}{2}-1}w^{\frac{3}{2}-1}\,du\,dv\,dw}{\iiint u^{\frac{3}{2}-1}v^{\frac{3}{2}-1}w^{\frac{3}{2}-1}\,du\,dv\,dw} = a\frac{\Gamma(3)\Gamma\left(\frac{3}{2}\right)\Gamma\left(\frac{3}{2}\right)}{\Gamma\left(3+\frac{3}{2}+\frac{3}{2}\right)} \times \frac{\Gamma\left(\frac{3}{2}+\frac{3}{2}+\frac{3}{2}+1\right)}{\Gamma\left(\frac{3}{2}\right)\Gamma\left(\frac{3}{2}\right)\Gamma\left(\frac{3}{2}\right)}$$

$$= a\frac{\Gamma(3)\Gamma\left(\frac{11}{2}\right)}{\Gamma(7)\Gamma\left(\frac{3}{2}\right)} = \frac{21}{128}a.$$

Similarly, $\overline{y} = \frac{21}{128}b$ and $\overline{z} = \frac{21}{128}c.$

Hence, the required position of the centroid is $\left(\frac{\mathbf{21}}{\mathbf{128}}\boldsymbol{a}, \frac{\mathbf{21}}{\mathbf{128}}\boldsymbol{b}, \frac{\mathbf{21}}{\mathbf{128}}\boldsymbol{c}\right)$.

Example 22.13. *Find the mass of the tetrahedron bounded by the coordinate planes and the plane* $\frac{x}{a}+\frac{y}{b}+\frac{z}{c}=1$, *the variable density* $\rho = kxyz$. **[Rohtak U 2003, UPTU 2003]**

Solution : In Fig. 22.17, the elemental mass of the elementary volume $\delta x\delta y\delta z$ at $P(x,y,z)$ is

$$\delta M = \rho\,dx\,dy\,dz = kxyz\,dx\,dy\,dz.$$

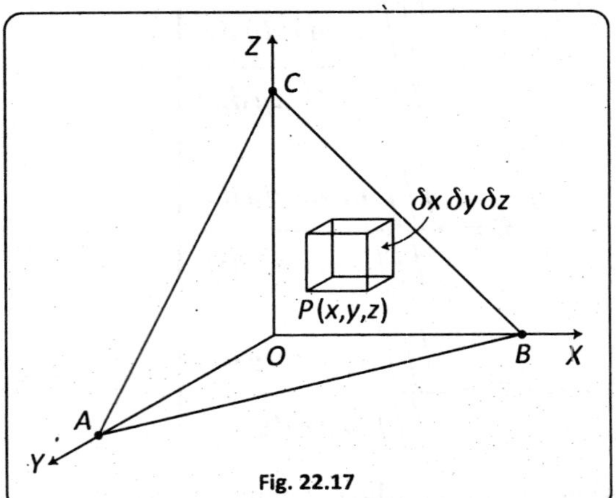

Fig. 22.17

$\therefore$ the whole mass of the tetrahedron is $M = \iiint kxyz\, dx\, dy\, dz.$

The integral covers the whole volume $OABC$ as shown in Fig. 22.17.

Here, the limits for z are : $z = 0$ to $z = c\left(1 - \frac{x}{a} - \frac{y}{b}\right)$

for y are : $y = 0$ to $y = b\left(1 - \frac{x}{a}\right)$

for x are : $x = 0$ to $x = a$.

$$\therefore \quad M = \int_0^a \int_0^{b\left(1-\frac{x}{a}\right)} \int_0^{c\left(1-\frac{x}{a}-\frac{y}{b}\right)} kxyz\, dx\, dy\, dz$$

$$= k\int_0^a \int_0^{b\left(1-\frac{x}{a}\right)} xy\left[\frac{z^2}{2}\right]_0^{c\left(1-\frac{x}{a}-\frac{y}{b}\right)} dy\, dz = k\int_0^a \int_0^{b\left(1-\frac{x}{a}\right)} xy\frac{c^2}{2}\left(1 - \frac{x}{a} - \frac{y}{b}\right)^2 dy\, dx$$

$$= \frac{kc^2}{2}\int_0^a \int_0^{b\left(1-\frac{x}{a}\right)} x\left[\left(1-\frac{x}{a}\right)^2 y - 2\left(1-\frac{x}{a}\right)^2\frac{y^2}{b} + \frac{y^3}{b^2}\right] dy\, dx$$

$$= \frac{kc^2}{2}\int_0^a x\left[\left(1-\frac{x}{a}\right)^2\frac{y^2}{2} - 2\left(1-\frac{x}{a}\right)\frac{y^3}{3b} + \frac{y^4}{4b^2}\right]_0^{b\left(1-\frac{x}{a}\right)} dx$$

$$= \frac{kc^2}{2}\int_0^a x\left[\frac{b^2}{2}\left(1-\frac{x}{a}\right)^4 - \frac{2b^2}{3}\left(1-\frac{x}{a}\right)^4 + \frac{b^2}{4}\left(1-\frac{x}{a}\right)^4\right] dx$$

$$= \frac{kb^2c^2}{24}\int_0^a x\left(1-\frac{x}{a}\right)^4 dx = \mathbf{\frac{ka^2b^2c^2}{720}}.$$

22.11 Solved Problems

Example 22.1. *Change the order of integration in* $\int_0^\infty \int_x^\infty \frac{e^{-y}}{y} dx\, dy$ *and hence find its value.* **[AMIE, Winter 1994; Summer 1992]**

Solution : Here, the region of integration is bounded by lines $y = x$, $x = 0$ and and infinite boundary.

We take a strip parallel to X-axis to change the order of integration. The extremities of the strip lies on $x = 0$ and $y = x$.
$\therefore$ limits of x are from 0 to y and limits of y are from 0 to ∞.

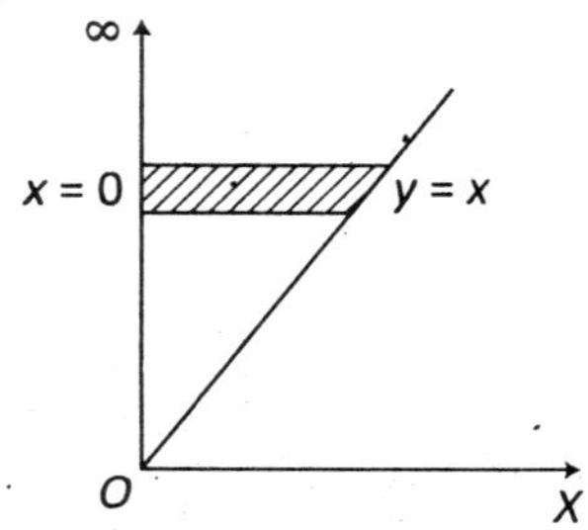

Fig. 22.18

$$\therefore \quad \int_0^\infty \int_x^\infty \frac{e^{-y}}{y} dx\, dy = \int_0^\infty \int_0^y dy\, dx = \int_0^\infty \frac{e^{-y}}{y}[x]_0^y dy$$

$$= \int_0^\infty e^{-y} dy = \left[-e^{-y}\right]_0^\infty = \mathbf{1}.$$

Example 22.2. *Change the order of integration in* $\int_0^a \int_{\frac{1}{2}\sqrt{a^2-x^2}}^{\sqrt{a^2-x^2}} V\, dx\, dy$, *where* V *is a function of* x *and* y.

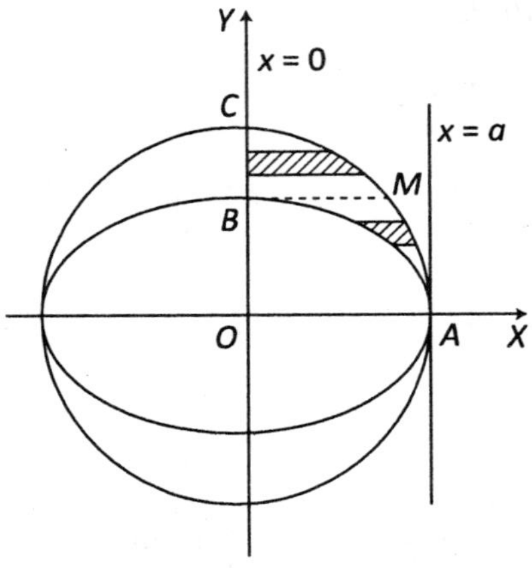

Fig. 22.19

Solution : Here, the region of integration is bounded by $y = \frac{1}{2}\sqrt{a^2 - x^2}$, i.e., $x^2 + 4y^2 = a^2$ (an ellipse) and the circle $x^2 + y^2 = a^2$, $x = 0$ (y-axis) and the line $x = a$ in the positive quadrant. Fig. 22.19 shows the region ABC.

The strip drawn parallel to X-axis changes its character at B, where $y = \frac{a}{2}$.

The strip in region ABM has its extremities on the ellipse $x^2 + 4y^2 = a^2$. So, the limits of x are from $\sqrt{a^2 - 4y^2}$ to $\sqrt{a^2 - y^2}$. Limits of y are from 0 to $\frac{a}{2}$ as B is the point $(0, \frac{a}{2})$.

In the region BCM, a strip parallel to X-axis has its extremities on $x = 0$ and the circle $x^2 + y^2 = a^2$. Hence, the limits of x are from 0 to $\sqrt{a^2 - y^2}$. Also, limits of y are from $\frac{1}{2}a$ to a.

$$\therefore \quad \int_0^a \int_{\frac{1}{2}\sqrt{a^2-x^2}}^{\sqrt{a^2-x^2}} V\, dx\, dy = \int_0^{\frac{1}{2}a} \int_{\sqrt{a^2-4x^2}}^{\sqrt{a^2-x^2}} V\, dx\, dy + \int_{\frac{1}{2}a}^{a} \int_0^{\sqrt{a^2-y^2}} V\, dx\, dy.$$

Example 22.3. *Change the order of integration in* $\int_0^1 \int_x^{x(2-x)} f(x, y)\, dx\, dy$.

Solution : Here the region of integration is bounded by $y = x$ and parabola $y = x(2 - x)$, i.e., $x^2 - 2x + y = 0$ or, $(x - 1)^2 = -(y - 1)$ and lines $x = 0$ to $x = 1$.

The parabola has its vertex at $(1, 1)$ and concavity downwards.

Now, we consider a strip parallel to axis of X. The strip has its extremities on parabola (Fig. 22.20), $(x-1)^2 = (1-y)$ and line $x = y$.

$\therefore$ limits of x are from $1 - \sqrt{1-y}$ to y (value $1 + \sqrt{1-y}$) corresponding to portion on the other side of $x = 1$.

So, y varies from 0 to A, i.e., 0 to 1.

Hence, the given integral is $\displaystyle\int_0^1 \int_{1-\sqrt{1-y}}^{y} f(x, y)\, dx\, dy.$

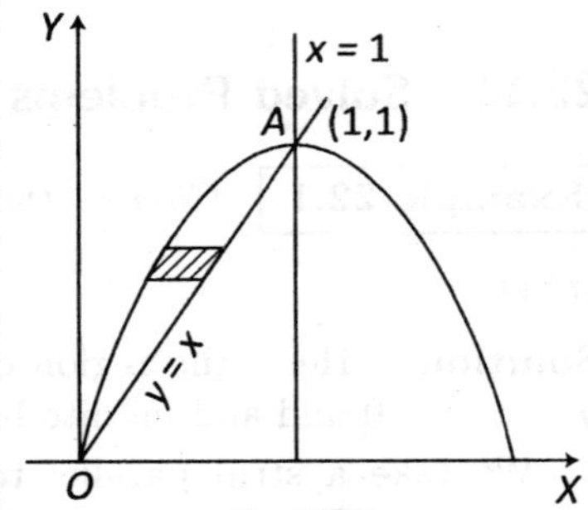

Fig. 22.20

Example 22.4. *Evaluate the integral* $\iiint_V (x^2 + y^2 + z^2)\, dx\, dy\, dz$, *where* V *is the volume of the sphere with centre at the origin and radius* R. **[BU(H) 1993]**

Solution : We convert the given integral in spherical polar coordinates

$$x = r \sin\theta \cos\phi, \quad y = r \sin\theta \sin\phi, \quad z = r\cos\theta.$$

$$\therefore \quad \iiint_V (x^2 + y^2 + z^2)\, dx\, dy\, dz = \int_{\theta=0}^{\pi} \int_{\phi=0}^{2\pi} \int_{r=0}^{R} r^2 (r^2 \sin\theta\, d\theta\, d\phi\, dr)$$

$$= \int_0^{2\pi} d\phi \int_0^{\pi} \sin\theta\, d\theta \int_0^R r^4 dr = \int_0^{2\pi} d\phi \int_0^{\pi} \sin\theta\, d\theta \left[\frac{r^5}{5}\right]_0^R$$

$$= \frac{R^5}{5}\int_0^{2\pi} d\phi [1-\cos\theta]_0^{\pi} = \frac{2R^5}{5}\int_0^{2\pi} d\phi = \frac{4\pi R^5}{5}.$$

Example 22.5. *Show that* $\int_0^{\infty}\int_0^{\infty} e^{-(x^2+y^2)} dx\, dy = \int_0^{\infty}\int_0^{\infty} e^{-r^2} r\, dr \frac{dt}{1+t^2}$, *where* $y = tx$, $x^2+y^2=r^2$. *Hence, evaluate the integral.*

Solution : Here we have $x^2+y^2=r^2$, $y=tx$.

Let $x = r\cos\theta$, $y = r\sin\theta$, then $t = \dfrac{x}{y} = \tan\theta$, $r^2 = x^2+y^2$.

$$\therefore \quad dt = \sec^2\theta\, d\theta, \quad \text{i.e.,} \quad d\theta = \frac{dt}{\sec^2\theta} = \frac{dt}{1+t^2}$$

$$\therefore \quad dx\, dy = r\, d\theta\, dr = r\frac{dt}{1+t^2}dt.$$

The limits of x and y are from 0 to ∞. The region of integration covers the positive quadrant in which r varies from 0 to ∞ and t varies from 0 to ∞ ($\because$ θ varies from 0 to $\frac{\pi}{2}$, $\tan\theta$ varies from 0 to ∞).

Thus the integral after transformation becomes

$$\int_0^{\infty}\int_0^{\infty} e^{-r^2} dr \frac{dt}{1+t^2} = \frac{1}{2}\int_0^{\infty}\left[e^{-r^2}\right]_0^{\infty}\frac{dt}{1+t^2} = \frac{1}{2}\left[\tan^{-1}\right]_0^{\infty} = \frac{1}{2}\left(\frac{\pi}{2}-0\right) = \frac{\pi}{4}.$$

Example 22.6. *Evaluate :* $\iiint\limits_S \sqrt{x^2+y^2}\, dx\, dy\, dz$, *where S is the solid bounded by the surfaces* $x^2+y^2=z^2$, $z=0$, $z=1$.

Solution : The given integral is $I = \iiint \sqrt{x^2+y^2}\, dx\, dy\, dz$

$$= \int_{-1}^{+1} dx \int_{-\sqrt{1-x^2}}^{\sqrt{1-x^2}} \sqrt{x^2+y^2} dy \int_{\sqrt{x^2+y^2}}^{1} dz$$

$$= \int_{-1}^{1} dx \int_{-\sqrt{1-x^2}}^{\sqrt{1-x^2}} \sqrt{x^2+y^2}\, dy\, [z]_{\sqrt{x^2+y^2}}^{1}$$

$$= \int_{-1}^{1} dx \int_{-\sqrt{1-x^2}}^{\sqrt{1-x^2}} dy \sqrt{x^2+y^2}\,[1-\sqrt{x^2+y^2}]$$

$$= \int_{-1}^{1} dx \int_{-\sqrt{1-x^2}}^{\sqrt{1-x^2}} [\sqrt{x^2+y^2} - x^2 - y^2] dy.$$

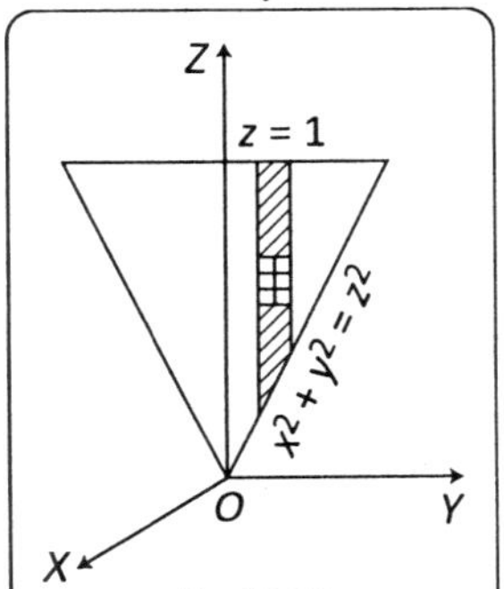

Fig. 22.21

We now change the integral into polar coordinates keeping the region of integration in the first quadrant of the circle $x^2+y^2=1$, $z=1$.

$$\therefore\ I = 4\int_0^{\frac{\pi}{2}}\int_0^1 (r - r^2)(r\,d\theta\,dr)\quad [\because\ x^2 + y^2 = r^2 \text{ and } dx\,dy = r\,d\theta\,dr]$$

$$= 4\int_0^{\frac{\pi}{2}}\left[\frac{r^3}{3} - \frac{r^4}{4}\right]_0^1 d\theta = 4\int_0^{\frac{\pi}{2}} d\theta\left(\frac{1}{3} - \frac{1}{4}\right) = \frac{4}{12}[\theta]_0^{\frac{\pi}{2}} = \frac{1}{3}\left(\frac{\pi}{2}\right) = \frac{\boldsymbol{\pi}}{\mathbf{6}}.$$

Example 22.7. *Evaluate :* $\int_0^2\int_0^{\sqrt{2x-x^2}} (x^2 + y^2)\,dy\,dx$ **[AMIE, Summer 1993; Winter 1994]**

Solution : The given integral is $I = \int_0^2\int_0^{\sqrt{2x-x^2}} (x^2 + y^2)\,dy\,dx$

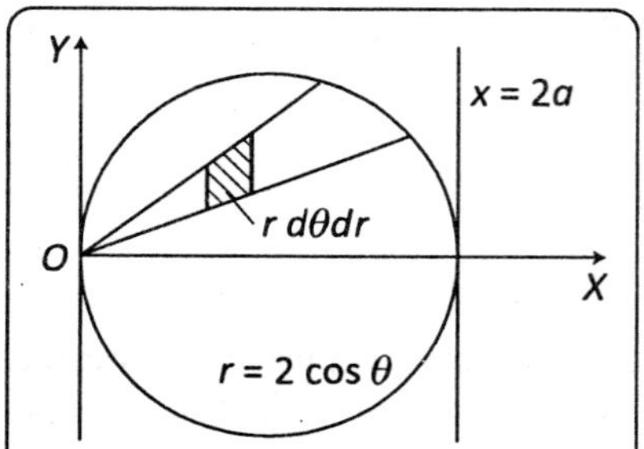

Fig. 22.22

or, $y = \sqrt{2x - x^2}$ or, $y^2 = 2x - x^2$

or, $x^2 + y^2 = 2x$ (i)

represents a circle with centre at $(1, 0)$ and radius $= 1$ (Fig. 22.22). Lower limit of $y = 0$, i.e., X-axis, and region of integration is the upper half-circle.

To convert Eqn. (i) into polar coordinates we have $x = r\cos\theta$, $y = r\sin\theta$.

$$r^2 - 2r\cos\theta = 0 \quad \text{or,} \quad r = 2\cos\theta.$$

Limits of r are 0 to $2\cos\theta$ and limits of θ are 0 to $\frac{\pi}{2}$.

$$\therefore\quad I = \int_0^{\frac{\pi}{2}}\int_0^{\sqrt{2x-x^2}} (x^2 + y^2)\,dy\,dx = \int_{\theta=0}^{\frac{\pi}{2}}\int_{r=0}^{2\cos\theta} r^2(r\,d\theta\,dr) = \int_0^{\frac{\pi}{2}} d\theta\int_0^{2\cos\theta} r^3\,dr$$

$$= \int_0^{\frac{\pi}{2}} d\theta\left[\frac{r^4}{4}\right]_0^{2\cos\theta} = 4\int_0^{\frac{\pi}{2}}\cos^4\theta\,d\theta = 4\cdot\frac{3\times 1\times\pi}{4\times 2\times 2} = \frac{\mathbf{3}\boldsymbol{\pi}}{\mathbf{4}}.$$

Example 22.8. *Evaluate :* $\iint_R (x+y)^2\,dx\,dy$, *where R is the parallelogram in the XY-plane with vertices $(1,0)$, $(3,1)$, $(2,2)$ and $(0,1)$ using the transformation, $u = x + y$, $v = x - 2y$.*

[BU(H) 1997; UPTA 2004]

Solution : Here, the region of integration R, the parallelogram $ABCD$ in XY-plane becomes the region R', i.e., the rectangle $A'B'C'D'$ in the uv-plane (Fig. 22.23).

$\because\ u = x + y,\ v = u - 2y.$

$$\therefore\quad x = \frac{1}{3}(2u + v),\quad y = \frac{1}{3}(u - v).$$

$$\text{Now, } \frac{\partial(x,y)}{\partial(u,v)} = \begin{vmatrix} \frac{\partial x}{\partial u} & \frac{\partial y}{\partial u} \\ \frac{\partial x}{\partial v} & \frac{\partial y}{\partial v} \end{vmatrix} = -\frac{1}{3}.$$

$$\therefore \text{ the given integral is } \iint_{R'} u^2|J|\,du\,dv = \int_{u=1}^{4}\int_{v=-2}^{1} u^2\frac{1}{3}du\,dv = \frac{1}{3}\left[\frac{u^3}{3}\right]_1^4 [v]_{-2}^1 = \mathbf{21}.$$

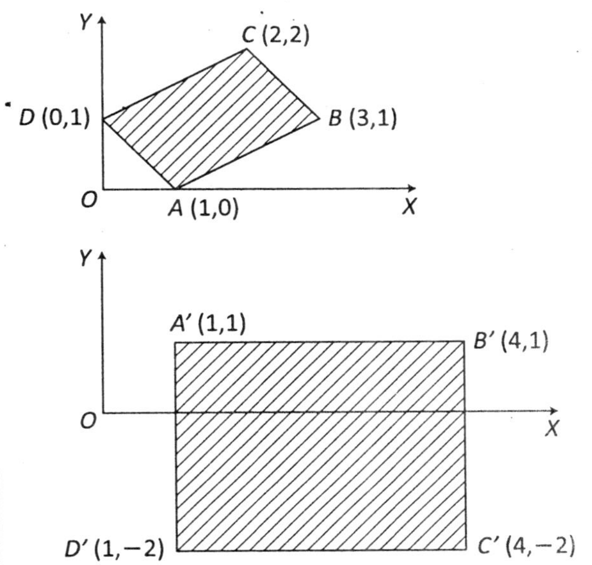

Fig. 22.23

Example 22.9. *Using the transformation $x + y = u$, $y = uv$, show that,*

$$\iint [xy(1-x-y)]^{\frac{1}{2}}\,dx\,dy = \frac{2\pi}{150},$$

integration being taken over the area of this triangle bounded by the lines, $x = 0$, $y = 0$, $x + y = 1$. [AMIE(Winter) 1996]

Solution : Given, $x + y = u$, $y = uv$.

$\therefore \quad x = u(1-v), \quad y = uv.$

$$\therefore \quad dx\,dy = \begin{vmatrix} 1-v & -u \\ v & u \end{vmatrix} du\,dv = u\,du\,dv\,.$$

and $[xy(1-x-y)]^{\frac{1}{2}} = [u^2v(1-v)(1-u)]^{\frac{1}{2}} = uv^{\frac{1}{2}}(1-v)^{\frac{1}{2}}(1-u)^{\frac{1}{2}}$.

$\therefore$ the limits of u are 0 to 1 and the limits of $v = 0$, $v = 1$ $\left[\because\ u = x + y \text{ and } v = \dfrac{v}{x+y}\right]$

$$\text{Hence, the transformed integral} = \int_0^1\int_0^2 uv^{\frac{1}{2}}(1-v)^{\frac{1}{2}}(1-u)^{\frac{1}{2}}\,u\,du\,dv$$

$$= \int_0^1\int_0^1 u^2v^{\frac{1}{2}}(1-v)^{\frac{1}{2}}(1-u)^{\frac{1}{2}}\,du\,dv.$$

Now, we put $u = \sin^2\theta$ and $v = \sin^2\phi$, then

$$\iint [xy(1-x-y)^{\frac{1}{2}}]du\,dv = 4\int_0^{\frac{\pi}{2}}\int_0^{\frac{\pi}{2}} \sin^5\theta\cos^2\theta\sin^2\phi\cos^2\phi\,d\theta\,d\phi$$

$$= \frac{4\Gamma(3)\Gamma\left(\frac{3}{2}\right)}{2\Gamma\left(\frac{9}{2}\right)}\cdot\frac{4\Gamma\left(\frac{3}{2}\right)\Gamma\left(\frac{3}{2}\right)}{2\Gamma(3)} = \frac{1}{\frac{7}{2}\cdot\frac{5}{2}\cdot\frac{3}{2}}\times\frac{\frac{\sqrt{\pi}}{2}\cdot\frac{\sqrt{\pi}}{2}}{1} = \mathbf{\frac{2\pi}{205}}.$$

Example 22.10. *Calculate* $\iint r^3\, dr\, d\theta$ *over the area bounded by the circle* $r = 2\sin\theta$ *and* $r = 4\sin\theta$.

Solution : The circles are : $r = 2\sin\theta$ and $r = 4\sin\theta$ as shown in Fig. 22.24.

The area enclosed by the two circles is the required area of integration.

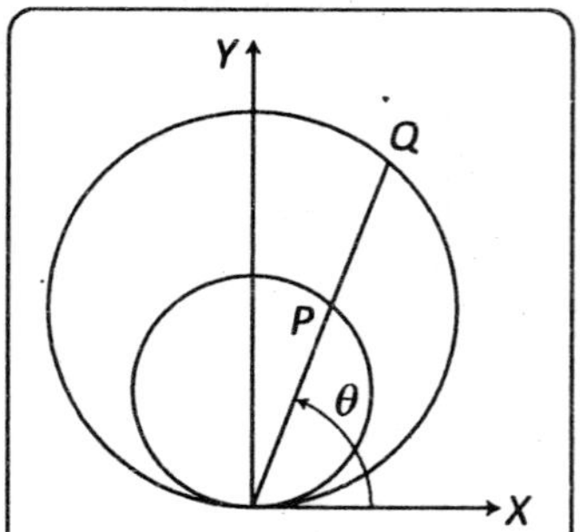

Fig. 22.24

The limits of integration of r are from $P(r = 2\sin\theta)$ to $Q(r = 4\sin\theta)$. The limits of integration of θ are from $\theta = 0$ to 2π.

$\therefore$ the required area is

$$\int_{\phi=0}^{\pi}\int_{r=2\sin\theta}^{4\sin\theta} r^3\, dr = \int_0^{\pi} d\theta\left[\frac{r^4}{4}\right]_{2\sin\theta}^{4\sin\theta} = 60\int_0^{\pi}\sin^4\theta\, d\theta$$

$$= 60\times 2\int_0^{\frac{\pi}{2}}\sin^4\theta\, d\theta = 120\times\frac{3}{4}\,\frac{1}{2}\,\frac{\pi}{2}$$

$$= \mathbf{22.5\pi}.$$

Example 22.11. *Transform the integral* $\iiint V\, dx\, dy\, dz$ *into new variables* u, v, w, *where* $x + y + z = u$, $y + z = uv$, $z = uvw$.

Solution : $\because\quad x + y + z = u \quad \therefore\quad x = u - y - z$

$$y + z = uv \quad \therefore\quad y = uv - z$$

$$z = uvw$$

$$\therefore\quad \frac{\partial x}{\partial u} = 1,\quad \frac{\partial y}{\partial v} = u,\quad \frac{\partial z}{\partial w} = uv$$

and $\quad dx\, dy\, dz = \dfrac{dx}{du}\cdot\dfrac{dy}{dv}\cdot\dfrac{dz}{dw}\, du\, dv\, dw = 1\cdot u\cdot uv\, du\, dv\, dw = u^2 v\, du\, dv\, dw.$

If the transformed volume be V', then the transformed integral is $\iiint \mathbf{V'u^2v\, du\, dv\, dw}$.

Example 22.12. *Find the volume of the torus generated by revolving the circle* $x^2 + y^2 = 4$ *about the line* $x = 3$.

Solution : Fig. 22.25 shows the required volume V of the torus.

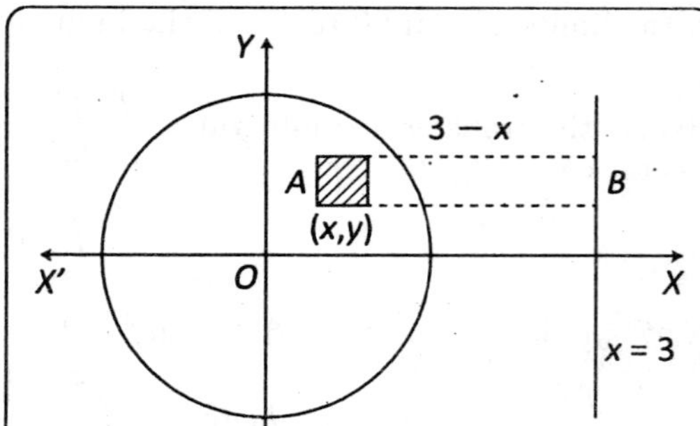

Fig. 22.25

Here, $V = \iint 2\pi(AB)dx\, dy$

$$= 2\pi\iint(3 - x)\, dx\, dy \quad [\because\ PQ = 3 - x]$$

or, $\quad V = 2\pi\int_{-2}^{2} dx\int_{-\sqrt{4-x^2}}^{\sqrt{4-x^2}}(3 - x)\, dy$

$$= 2\int_{-2}^{2} dx[3y - xy]_{-\sqrt{4-x^2}}^{\sqrt{4-x^2}}$$

$$= 2\pi \int_{-2}^{2} dx[3\sqrt{4-x^2} - x\sqrt{4-x^2} + 3\sqrt{4-x^2} - x\sqrt{4-x^2}]$$

$$= 4\pi \int_{-2}^{2} dx[3\sqrt{4-x^2} - x\sqrt{4-x^2}]dx$$

$$= 4\pi \left[3 \times \frac{x}{2}\sqrt{4-x^2} + 3 \times \frac{4}{2}\sin^{-1}\frac{\pi}{2} + \frac{1}{3}(4-x^2)^{\frac{3}{2}}\right]_{-2}^{2} = 4\pi\left[6 \times \frac{\pi}{2} + 6 \times \frac{\pi}{2}\right]$$

$$= \mathbf{24\pi^2}.$$

Example 22.13. *Find the volume bounded by the cylinder $x^2 + y^2 = 4$ and the plane $y + z = 3$ and $z = 0$.* **[SVTU 2007; Cochin U 2005]**

Solution : We have $x^2 + y^2 = 4, \quad \therefore \quad y = \pm\sqrt{4-x^2}$

$$y + z = 4, \quad \therefore \quad z = 4 - y$$

$$z = 0.$$

Thus, x varies from -2 to $+2$, y varies from $-\sqrt{4-x^2}$ to $\sqrt{4-x^2}$ and z varies from 0 to $4 - y$ (Fig. 22.26).

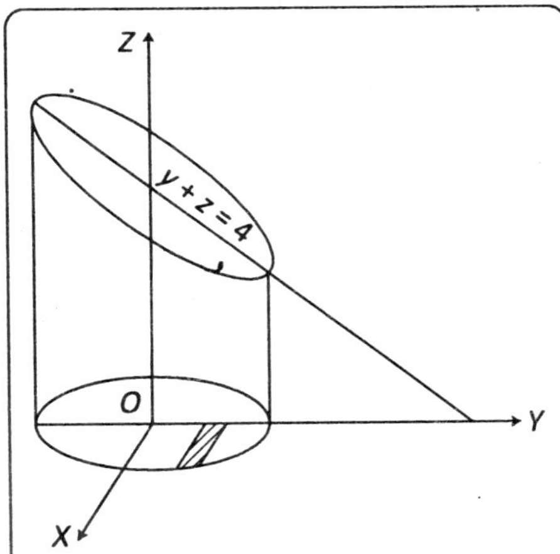

Fig. 22.26

Now, the required volume is given by

$$V = \int_{-2}^{+2} dx \int_{-\sqrt{4-x^2}}^{\sqrt{4-x^2}} dy \int_{0}^{4-y} dz$$

$$= \int_{-2}^{+2} dx \int_{-\sqrt{4-x^2}}^{\sqrt{4-x^2}} dy\, [z]_0^{4-y}$$

$$= \int_{-2}^{+2} dx \int_{-\sqrt{4-x^2}}^{\sqrt{4-x^2}} (4-y)\, dy$$

$$= \int_{-2}^{+2} dx \left[4y - \frac{y^2}{2}\right]_{-\sqrt{4-x^2}}^{\sqrt{4-x^2}}$$

$$= \int_{-2}^{+2} \left[4\sqrt{4-x^2} - \frac{1}{2}(4-x^2) + 4\sqrt{4-x^2} + \frac{1}{2}(4-x^2)\right] dx = 8\int_{-2}^{+2} \sqrt{4-x^2}$$

$$= 8\left[\frac{x}{2}\sqrt{4-x^2} + \frac{4}{2}\sin^{-1}\frac{x}{2}\right]_{-2}^{2} = \mathbf{16\pi}.$$

Example 22.14. *Evaluate the integral, $\iiint_S xyz\, dx\, dy\, dz$, where $S = f(x, y, z); x^2+y^2+z^2 \le 1$, $x \ge a$, $y \ge 0$, $z \ge 0$.*

Solution : Here S represents a sphere $x^2 + y^2 + z^2 = 1$. The positive octant of the sphere is to be considered since x, y, z are all positive.

$\because \; x^2 + y^2 + z^2 = 1, \quad \therefore \; z = \sqrt{1 - x^2 - y^2}.$

Now, the projection of the sphere on the circle is a circle $x^2 + y^2 = 1$.

Here y varies from 0 to $\sqrt{1-x^2}$ and x varies from 0 to 1.

$\therefore$ the given integral is

$$\iint xyz\,dx\,dy\,dz = \int_0^1 x\,dx \int_0^{\sqrt{1-x^2}} y\,dy \int_0^{\sqrt{1-x^2-y^2}} z\,dz$$

$$= \int_0^1 x\,dx \int_0^{\sqrt{1-x^2}} y\,dy \left[\frac{z^2}{2}\right]_0^{\sqrt{1-x^2-y^2}} = \int_0^1 x\,dx \int_0^{\sqrt{1-x^2}} dy \left[\frac{1-x^2-y^2}{2}\right]$$

$$= \frac{1}{2}\int_0^1 x\,dx \int_0^{\sqrt{1-x^2}} (y - x^2 y - y^3)dy = \frac{1}{2}\int_0^1 x\,dx \left[\frac{y^2}{2} - \frac{x^2y^2}{2} - \frac{y^4}{4}\right]_0^{\sqrt{1-x^2}}$$

$$= \frac{1}{4}\int_0^1 x\,dx \left[1 - x^2 - x^2(1-x^2) - \frac{(1-x^2)^2}{2}\right]$$

$$= \frac{1}{8}\int_0^1 x\,dx[2 - 2x^2 - 2x^2 + 2x^4 - 1 - x^4 + 2x^2] = \frac{1}{8}\int_0^1 (x - 2x^3 + x^5)dx$$

$$= \frac{1}{8}\left[\frac{x^2}{2} - \frac{x^4}{4} + \frac{x^6}{6}\right]_0^1 = \frac{1}{8}\left(\frac{1}{2} - \frac{1}{2} + \frac{1}{6}\right) = \mathbf{\frac{1}{48}}.$$

Example 22.15. *Transform to polar coordinates and integrate* $\iint \sqrt{\frac{1-x^2+y^2}{1+x^2+y^2}}\,dx\,dy$.

The integral being extended over all positive values of x and y subject to $x^2+y^2 \le 1$.

Solution : To transform to polar coordinates we have $x = r\cos\theta$, $y = r\sin\theta$.

$$\therefore \quad dx\,dy = \frac{\partial(x,y)}{\partial(r,\theta)}\,dr\,d\theta = \begin{vmatrix} \cos\theta & -r\sin\theta \\ \sin\theta & r\cos\theta \end{vmatrix} dr\,d\theta = r\,d\theta\,dr.$$

In the positive octant of $x^2 + y^2 \le 1$, r varies from 0 to 1, θ varies from 0 to $\frac{\pi}{2}$.

$$\therefore \quad \iint \sqrt{\frac{1-x^2+y^2}{1+x^2+y^2}}\,dx\,dy, \quad (x^2+y^2 \le 1)$$

$$= \int_{r=0}^{1}\int_{\theta=0}^{\frac{\pi}{2}} \sqrt{\frac{1-r^2}{1+r^2}}\,r\,dr\,d\theta = \frac{\pi}{2}\int_0^1 \sqrt{\frac{1-r^2}{1+r^2}}\,r\,dr\,d\theta$$

$$= \frac{\pi}{2}\int_{\frac{\pi}{2}}^{0} \sqrt{\frac{1-\cos t}{1+\cos t}}\cdot\frac{1}{2}(-\sin t)\,dt \quad [\text{putting } r^2 = \cos t]$$

$$= \frac{\pi}{4}\int_0^{\frac{\pi}{2}} \frac{\frac{1}{2}\sin\frac{t}{2}\cdot 2\sin\frac{t}{2}\cos\frac{t}{2}}{\cos\frac{t}{2}}\,dt = \frac{\pi}{4}\left[\frac{\pi}{2} - 1\right] = \mathbf{\frac{1}{8}\pi^2 - \frac{1}{4}\pi}.$$

Example 22.16. *Transform the integral* $\int_{-1}^{1}\int_{x}^{\frac{1}{x}} V\,dx\,dy\,dz$ *by the substitution* $u = 1 + x$ *and* $v = xy$.

Solution : Given, $u = 1 + x$ and $v = xy$.

$$\therefore \quad x = u - 1, \quad y = \frac{v}{u-1}.$$

$$\therefore \quad \frac{\partial(x,y)}{\partial(u,v)} = \begin{vmatrix} 1 & 0 \\ -\dfrac{u}{(u-1)^2} & \dfrac{1}{u-1} \end{vmatrix} = \frac{1}{u-1}.$$

So, $dx\,dy = |J|\,du\,dv = \dfrac{1}{u-1}\,du\,dv.$

Now, the limits of y are x to $\dfrac{1}{x}$.

Also $v = xy$, so v varies from $x \cdot x$ to $x \cdot \dfrac{1}{x}$, i.e., x^2 to 1.

$\therefore$ Limits of x vary from $(u-1)^2$ to 1

Limits of x are from -1 to 1

Limits of u are from $1-1$ to $1+1$, i.e., 0 to 2.

Hence, $\displaystyle\int_{-1}^{1}\int_{x}^{\frac{1}{x^2}} \boldsymbol{V\,dx\,dy} = \int_{0}^{2}\int_{(u-1)^2}^{1} \boldsymbol{V'}\frac{\mathbf{1}}{\boldsymbol{u-1}}\,\boldsymbol{du\,dv},$

where V' is the function of V changed in u and v.

Example 22.17. *If r, r' be the distances of a point in the plane of reference from two fixed points at a distance $2c$ apart on the axis of x, then show that, between corresponding limits of integration $\displaystyle\iint 2cy\,dx\,dy = \iint rr'\,dr\,dr'$.*

Solution : Fig. 22.27 shows that the point $P(x,y)$ is at a distance r and r' from O and A respectively, where $OA = 2c$.

$$r^2 = x^2 + y^2 \quad \text{and} \quad r'^2 = (2c-x)^2 + y^2$$

Now, $dx\,dy = \dfrac{\partial(x,y)}{\partial(r,r')}\,dr\,dr'.$

But, $\dfrac{\partial(x,y)}{\partial(r,r')} = \begin{vmatrix} \dfrac{x}{r} & \dfrac{y}{r} \\ -\dfrac{(2c-x)}{r'} & \dfrac{y}{r'} \end{vmatrix} = \dfrac{2cy}{rr'}.$

$\therefore \quad dr\,dr' = \dfrac{2cy}{rr'}\,dx\,dy$, i.e., $2cy\,dx\,dy = rr'\,dr\,dr'$.

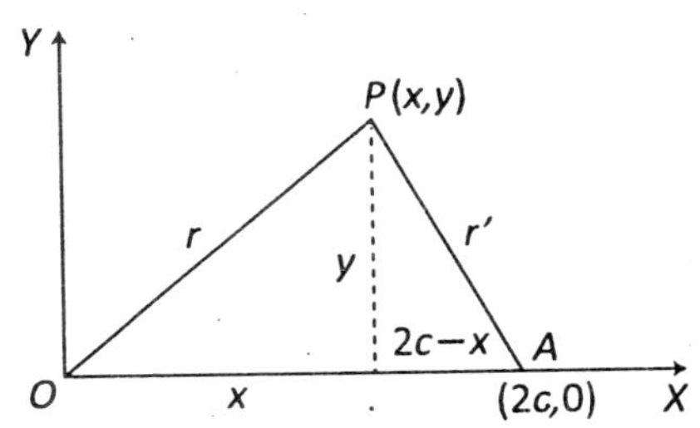

Fig. 22.27

Hence, $\displaystyle\iint \boldsymbol{2cy\,dx\,dy} = \iint \boldsymbol{rr'\,dr\,dr'}$ **(Proved).**

Example 22.18. *Show that,* $\displaystyle\int_{-\infty}^{\infty}\int_{-\infty}^{\infty} \frac{a\,dx\,dy}{(x^2+y^2+a^2)^{\frac{3}{2}}(x^2+y^2+b^2)^{\frac{1}{2}}} = \frac{2\pi}{a+b}$ *after transforming the integral in to polar coordinates.*

Solution : The region of integration in the whole xy plane as both x and y varies from $-\infty$ to ∞.

$\therefore$ the limits of r are from $-\infty$ to ∞ and limits of θ are from 0 to π, for the whole r-θ plane.

$\therefore \quad dx\,dy = r\,dr\,d\theta \quad [\text{as } x = r\cos\theta \text{ and } y = r\sin\theta]$

$\therefore$ the given integral is

$$a\int_{-\infty}^{\infty}\int_0^{\pi}\frac{r\,dr\,d\theta}{(r^2+a^2)^{\frac{3}{2}}(r^2+b^2)^{\frac{1}{2}}} = 4a\int_{-\infty}^{\infty}\int_0^{\frac{\pi}{2}}\frac{r\,dr\,d\theta}{(r^2+a^2)^{\frac{3}{2}}(r^2+b^2)^{\frac{1}{2}}}$$

$$= 2a\pi\int_0^{\infty}\frac{r\,dr}{(r^2+a^2)^{\frac{3}{2}}(r^2+b^2)^{\frac{1}{2}}}. \quad \text{(i)}$$

Now, let $r^2+a^2=\frac{1}{u}$. $\therefore$ $\log(r^2+a^2)=\log\left(\frac{1}{u}\right)$.

Differentiating, $\frac{2r}{(r^2+a^2)}dr = -\frac{du}{u}$.

From Eqn. (i), $\pi a\int_0^{\frac{1}{a^2}}\frac{du}{u\sqrt{\left[\frac{1}{u}\left(\frac{1}{u}-a^2+b^2\right)\right]}} = \pi a\int_0^{\frac{1}{a^2}}\frac{du}{\sqrt{[1+(b^2-a^2)u]}}$

$$= \pi\left[\left\{1+(b^2-a^2)u\right\}^{\frac{1}{2}}\right]_0^{\frac{1}{a^2}}\frac{1}{b^2-a^2}\cdot\frac{2}{1}$$

$$= \frac{2\pi a}{b^2-a^2}\left[\frac{b}{a}-1\right] = \frac{2\pi}{a+b} \quad \textbf{(Proved)}$$

Example 22.19. *Find the area of the surface of the sphere $x^2+y^2+z^2=a^2$ included within the cylinder $x^2+y^2=ax$.*

Solution : We have $x^2+y^2+z^2=a^2$.

$$\frac{\partial z}{\partial x}=-\frac{x}{z} \text{ and } \frac{\partial z}{\partial y}=-\frac{y}{z}.$$

$$\therefore \quad \sec\gamma=\sqrt{1+\left(\frac{\partial z}{\partial x}\right)^2+\left(\frac{\partial z}{\partial y}\right)^2}=\left(1+\frac{x^2}{z^2}+\frac{y^2}{z^2}\right)=\frac{a}{z}=\frac{a}{\sqrt{a^2-x^2-y^2}}$$

$$=\frac{a}{\sqrt{a^2-r^2}} \quad [\text{where } x=r\cos\theta,\ y=r\sin\theta]$$

Again, $x^2+y^2=ax$ becomes $r^2=ar\cos\theta$ or, $r=a\cos\theta$.

$\therefore$ required surface area is $= 4\int_0^{\frac{\pi}{2}}\int_0^{a\cos\theta}\frac{ar\,d\theta\,dr}{\sqrt{a^2-r^2}} = 4a\int_0^{\frac{\pi}{2}}\left[-\sqrt{a^2-r^2}\right]_0^{a\cos\theta}d\theta$

$$=4a^2\int_0^{\frac{\pi}{2}}(1-\sin\theta)\,d\theta = 4a^2[\theta+\cos\theta]_0^{\frac{\pi}{2}} = 4a^2\left(\frac{\pi}{2}-1\right)$$

$$= \mathbf{2a^2(\pi-2)}.$$

Example 22.20. *Prove that the area of the surface $z^2=2xy$ included between the planes $x=0$, $x=a$, $y=0$, $y=b$ is $\frac{4\sqrt{ab}}{3\sqrt{2}}(a+b)$.*

Solution : $\because$ $z^2=2xy$, $\therefore$ $\frac{\partial z}{\partial x}=\frac{y}{z}$ and $\frac{\partial z}{\partial y}=\frac{x}{z}$.

$$\therefore \quad \sec\gamma = \sqrt{1+\left(\frac{\partial z}{\partial x}\right)^2+\left(\frac{\partial z}{\partial y}\right)^2} = \sqrt{\left(1+\frac{y^2}{z^2}+\frac{x^2}{z^2}\right)} = \sqrt{\left(\frac{x^2+y^2+z^2}{z^2}\right)}$$

$$= \sqrt{\left(\frac{x^2+y^2+2xy}{z^2}\right)} = \frac{x+y}{2} = \frac{x+y}{\sqrt{2xy}} = \frac{1}{\sqrt{2}}\left[\sqrt{\frac{x}{y}}+\sqrt{\frac{y}{x}}\right].$$

Hence, the required surface area $= \dfrac{1}{\sqrt{2}}\displaystyle\int_0^a\int_0^b\left[\sqrt{\frac{x}{y}}+\sqrt{\frac{y}{x}}\right]dx\,dy$

$$= \frac{1}{\sqrt{2}}\left[\left\{\frac{2}{3}x^{\frac{3}{2}}\right\}_0^a \times \left\{2y^{\frac{1}{2}}\right\}_0^b + \left(\frac{2}{3}y^{\frac{3}{2}}\right)_0^b\left(2x^{\frac{1}{2}}\right)_0^a\right]$$

$$= \frac{\mathbf{4\sqrt{ab}(a+b)}}{\mathbf{3\sqrt{2}}} \quad \textbf{(Proved)}.$$

Example 22.21. *Evaluate :* $\displaystyle\int_0^1\int_0^{\sqrt{1-x^2}}\int_{\sqrt{x^2+y^2}}^1 \frac{dx\,dy\,dz}{\sqrt{x^2+y^2+z^2}}.$

Solution : To change the integral into spherical polar coordinates we have

$x = r\sin\theta\cos\phi, \quad y = r\sin\theta\sin\phi, \quad z = r\cos\theta.$

$\therefore \quad J = r^2\sin\theta, \quad x^2+y^2+z^2 = r.$

It is evident that the region of integration is common to the cone $z^2 = x^2+y^2$ and the cylinder $x^2+y^2 = 1$, bounded by the plane $z = 1$ in the positive quadrant as shown in Fig. 22.28.

Hence, θ varies from 0 to $\frac{\pi}{4}$, r varies from 0 to $\sec\theta$ and ϕ varies from 0 to $\frac{\pi}{2}$.

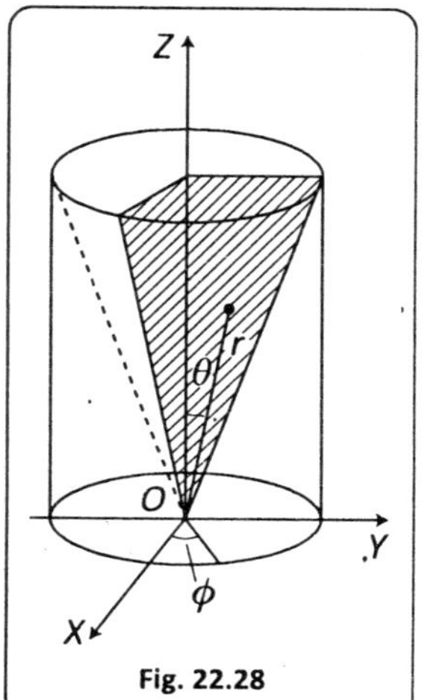

Fig. 22.28

$$\therefore \int_0^1\int_0^{\sqrt{1-x^2}}\int_{\sqrt{x^2+y^2}}^1 \frac{dz\,dy\,dx}{\sqrt{x^2+y^2+z^2}} = \int_0^{\frac{\pi}{2}}\int_0^{\frac{\pi}{4}}\int_0^{\sec\theta}\frac{1}{r}\cdot r^2\sin\theta\,dr\,d\theta\,d\phi$$

$$= \int_0^{\frac{\pi}{2}} d\phi\int_0^{\frac{\pi}{4}}\left[\frac{r^2}{2}\right]_0^{\sec\theta}\sin\theta\,d\theta$$

$$= \frac{\pi}{2}\int_0^{\frac{\pi}{4}}\frac{\sec^2\theta}{2}\sin\theta\,d\theta$$

$$= \frac{\pi}{4}\int_0^{\frac{\pi}{4}}\sec\theta\tan\theta\,d\theta = \frac{\pi}{4}[\sec\theta]_0^{\frac{\pi}{4}}$$

$$= \frac{\mathbf{(\sqrt{2}-1)\pi}}{\mathbf{4}}.$$

Example 22.22. *Find the volume of the ellipsoid* $\dfrac{x^2}{a^2}+\dfrac{y^2}{b^2}+\dfrac{z^2}{c^2} = 1.$

[PTU 2006; Kottayam U 2005]

Solution : Fig. 22.29 shows the positive octant of the given ellipsoid bounded by the planes $OAB(z=a)$, $OBC(x=0)$, $OCA(y=1)$ and the surface ABC, i.e., $\dfrac{x^2}{a^2}+\dfrac{y^2}{b^2}+\dfrac{z^2}{c^2} = 1.$

An element of volume $\delta x\,\delta y\,\delta z$ at point $P(x, y, z)$ is shown in Fig. 22.29.

The volume of the ellipsoid $= 8\iiint_R xy\,dx\,dy$ in the region R.

Here, (i) z varies from 0 to MN, $MN = c\left(\sqrt{1-\frac{x^2}{a^2}-\frac{y^2}{b^2}}\right)$

(ii) y varies from 0 to EF, $EF = b\left(\sqrt{1-\frac{x^2}{a^2}}\right)$

(iii) x varies from 0 to $OA = a$.

Now, the volume of the ellipsoid is

$$8\int_0^a\int_0^{b\sqrt{1-\frac{x^2}{a^2}}}\int_0^{c\sqrt{1-\frac{x^2}{a^2}-\frac{y^2}{b^2}}} dx\,dy\,dz = 8\int_0^a dx\int_0^{b\sqrt{1-\frac{x^2}{a^2}}} dy\,[z]_0^{c\sqrt{1-\frac{x^2}{a^2}-\frac{y^2}{b^2}}} dz$$

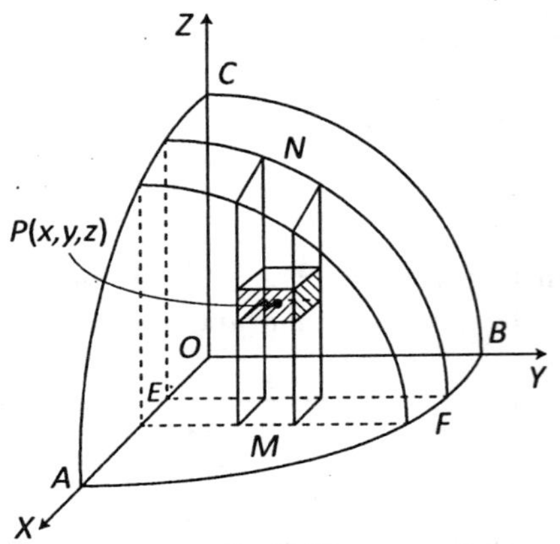

Fig. 22.29

$$= 8c\int_0^a dx\int_0^{b\sqrt{1-\frac{x^2}{a^2}}}\sqrt{1-\frac{x^2}{a^2}-\frac{y^2}{b^2}}\,dy$$

$$= \frac{8c}{b}\int_0^a dx\int_0^{\rho}\sqrt{\rho^2-y^2}\,dy, \quad \left[\rho = b\sqrt{1-\frac{x^2}{a^2}}\right]$$

$$= \frac{8c}{b}\int_0^a dx\left[\frac{y\sqrt{\rho^2-y^2}}{2}+\frac{\rho^2}{2}\sin^{-1}\frac{y}{\rho}\right]_0^{\rho}$$

$$= \frac{8c}{b}\int_0^a \frac{b^2}{2}\left(1-\frac{x^2}{a^2}\right)\frac{\pi}{2}\,dx$$

$$= 2\pi bc\int_0^a\left(1-\frac{x^2}{a^2}\right)dx = 2\pi bc\left[x-\frac{x^3}{3a^2}\right]_0^a$$

$$= \frac{\mathbf{4\pi abc}}{\mathbf{3}}.$$

Exercises

22(A) : Objective Type Questions

Select the correct answer

22.1 $\iint_R (x^2+y^2)dx\,dy$, where R is the region in the XY-plane bounded by the ellipse $\frac{x^2}{a^2}+\frac{y^2}{b^2} = 1$ is

(a) $\pi ab(a^2+b^2)+a^2b^2$ **(b)** $\frac{1}{4}\pi ab(a^2+b^2)$ **(c)** $\frac{1}{4}\pi ab(a^2+b^2)+a^2b^2$ **(d)** none of these.

22.2 $\int_0^{2\pi}\int_0^{\sqrt{2ax-x^2}} x^2\,dx$ is equal to

(a) $\frac{3}{5}\pi a^4$ **(b)** $\frac{5\pi a^4}{8}$ **(c)** $\frac{5\pi a^3}{8}$ **(d)** $\frac{8}{5}\pi a^3$.

22.3 $\iint_R x^2\,dx\,dy$, where R is the two-dimensional region bounded by the curves $y = x$ and $y = x^2$ is

(a) $\frac{1}{20}$ (b) $\frac{1}{18}$ (c) 18 (d) none of these.

22.4 $\int_0^l \int_{x^2}^{2-x} xy\,dy\,dy$ is equal to

(a) $\frac{3}{4}$ (b) $\frac{3}{8}$ (c) $\frac{5}{8}$ (d) $\frac{7}{8}$.

22.5 The area between the parabolas $y^2 = 4ax$ and $x^2 = 4ay$ is

(a) $16a^3$ (b) $\frac{16a^2}{3}$ (c) $\frac{16a^3}{3}$ (d) $\frac{3}{16}a^2$.

22.6 The area enclosed by the leminiscate $r^2 = a^2 \cos 2\theta$ is equal to

(a) a^2 (b) $\frac{a^3}{3}$ (c) $\frac{a^2}{2}$ (d) $\frac{a^4}{16}$.

22.7 $\int_0^1 dx \int_0^2 dy \int_1^2 x^2yz\,dz$ is equal to

(a) 5 (b) 2 (c) 1 (d) 0.

22.8 The volume bounded by the coordinate planes and the plane $\frac{x}{a} + \frac{y}{b} + \frac{z}{c} = 1$ is

(a) $\frac{1}{3}abc$ (b) $\frac{1}{6}abc$ (c) $\frac{1}{4}abc$ (d) $4abc$.

22.9 The volume of a cone of height h and base radius r is

(a) $\pi r^2 h$ (b) $\frac{1}{3}\pi r^2 h$ (c) $\frac{1}{4}\pi r^2 h$ (d) $\frac{1}{3}\pi r h$.

22.10 On changing the order of integration $\int_0^{2a} dx \int_0^{x} e^x\,dx\,dy$ becomes

(a) $\int_0^a dy \int_{\sqrt{4ay}}^{2a} (x+y)^3 dx$

(b) $\int_{-a}^a dy \int_{4ay}^{2a} (x+y)^3 dx$

(c) $\int_0^a dy \int_{\sqrt{4ay}}^{2a} (x+y)\,dx$

(d) $\int_0^a \int_{-4ax}^{2a} (x+y)^2 dx$.

22.11 On converting into polar coordinates $\int_0^{2a} \int_0^{\sqrt{2ax-y}} dx\,dy$ becomes

(a) $\int_0^{\frac{\pi}{2}} \int_0^{2a\cos\theta} d\theta\,dr$ (b) $\int_0^{\frac{\pi}{2}} \int_0^{2a\cos\theta} r\,d\theta\,dr$ (c) $\int_0^{\frac{\pi}{2}} \int_0^{2a} r\,d\theta\,dr$ (d) $\int_0^{\pi} \int_0^{2a\cos\theta} r\,d\theta\,dr$.

22.12 The centroid of the region in the first octant bounded by $\frac{x}{a} + \frac{y}{b} + \frac{z}{c} = 1$ is

(a) $\left(\frac{a}{2}, \frac{b}{2}, \frac{0}{2}\right)$ (b) $\left(\frac{a}{3}, \frac{b}{3}, \frac{c}{3}\right)$ (c) $\left(\frac{a}{4}, \frac{b}{4}, \frac{c}{4}\right)$ (d) none of these.

22.13 The centroid of the region bounded by $z = 1 - x^2 - y^2$ and xy-plane is

(a) $\left(0, 0, \frac{4}{3}\right)$ (b) $\left(0, \frac{4}{3}, 0\right)$ (c) $\left(\frac{4}{3}, 0, 0\right)$ (d) $\left(0, 0, \frac{2}{3}\right)$.

22.14 The volume cut by the paraboloid $4z = x^2 + y^2$ by the plane $x = 4$ is

(a) $\frac{\pi^2}{4}$ (b) $\frac{\pi^2}{8}$ (c) $\frac{\pi^2}{2}$ (d) $\frac{\pi^2}{6}$

22.15 The triple integral gives the volume of a sphere of radius R equal to

(a) $\frac{4}{3}\pi R^2$ (b) $\frac{4}{3}\pi R^3$ (c) $\frac{4\pi}{3R^2}$ (d) $\frac{4\pi R^5}{3}$.

Answers

22.1 (c) **22.2** (b) **22.3** (a) **22.4** (b) **22.5** (b) **22.6** (a) **22.7** (c) **22.8** (b)
22.9 (b) **22.10** (a) **22.11** (b) **22.12** (c) **22.13** (a) **22.14** (b) **22.15** (b).

22(B) : Short Answer Type Questions

22.1 Evaluate $\iint (x^2 + y^2) x \, dx \, dy$ over the positive quadrant of the circle $x^2 + y^2 = a$.

[Ans: $\frac{a^5}{5}$]

22.2 Evaluate $\int_0^1 \int_x^{\sqrt{x}} (x^2 + y^2) \, dx \, dy$. **[VTU 2000] [Ans: $\frac{3}{25}$]**

22.3 Evaluate $\int_0^\infty \int_0^x x e^{-\frac{x^2}{y}} \, dy \, dx$. **[SVTU 2006; VTU 2006]**

22.4 Evaluate $\iint \frac{r \, dr \, d\theta}{\sqrt{a^2 + r^2}}$ over the loop of the leminiscate $r^2 = a^2 \cos 2\theta$.

[Rohtak U 2006S] [Ans: $\frac{4}{3}a^2$]

22.5 Evaluate $\int_0^a \int_0^x \int_0^{x+y} e^{x+y+z} dz \, dy \, dx$. **[JNTU 2006] [Ans: $\frac{1}{8}e^{4a} - \frac{3}{4}e^{2a} + e^a - \frac{3}{8}$]**

22.6 $\int_0^{\frac{\pi}{2}} \int_0^{a \sin\theta} \int_0^{\frac{a^2 - r^2}{a}} r \, dz \, dr \, d\theta$. **[VTU 2007] [Ans: $\frac{5}{64}\pi a^3$]**

22.7 By using the transformation $x + y = u$, $y = uv$, show that

$$\int_0^1 \int_0^{1-x} e^{\frac{y}{x+y}} \, dy \, dx = \frac{1}{2}(e - 1).$$ **[PTU 2003]**

22.8 Change the order of integration in $\int_0^a \int_0^x \frac{f'(y) dx \, dy}{\sqrt{(a - x)(x - y)}}$.

[Ans: $\int_0^a \int_0^x \frac{f'(y) dy \, dx}{\sqrt{(a - x)(x - y)}}$]

22.9 Change the order of integration and evaluate $\int_0^b \int_0^{\frac{a}{b}\sqrt{b^2-y^2}} xy\,dx\,dy.$

[**Ans:** $\int_0^a x\,dx \int_0^{\frac{b}{a}\sqrt{a^2-x^2}} y\,dy;\ \frac{a^2b^2}{8}$]

22.10 Find the volume bounded by the cylinders $x^2+y^2=2ax$ and $x^2=2ax$. [**Ans:** $\frac{128}{15}a^2$]

22.11 Compute the area of a region bounded by the ellipse $x=a\cos\theta,\ y=b\sin\theta$. [**Ans:** πab]

22(C) : Broad Answer Type Questions

22.1 Evaluate the following integrals :

(i) $\int_0^a \int_0^{\sqrt{a^2-x^2}} dx\,dy.$ [**Ans:** $\frac{\pi a^2}{4}$]

(ii) $\iint_A (5-2x-y)\,dx\,dy$, where A is given by $y=0,\ x+2y=3,\ x=y^2$. [**Ans:** $\frac{217}{60}$]

(iii) $\int_0^a \int_{\frac{x}{a}}^{\sqrt{\frac{x}{a}}} (x^2+y^2)dx\,dy.$ [**Ans:** $\left(\frac{a^2}{28}+\frac{a}{20}\right)$]

(iv) $\int_0^\infty \int_0^x xe^{-\frac{x^2}{y}}\,dx\,dy.$ **[SVTU 2006; UTPA 2005]** [**Ans:** 1]

(v) $\iint_R x^2\,dx\,dy$, R is the two-dimensional region bounded by the curves $y=x$ and $y=x^2$. **[AMIETE(Winter) 1996]** [**Ans:** $\frac{1}{20}$]

(vi) $\int_0^2 \int_0^{x^2} e^{\frac{y}{x}}\,dy\,dx.$ [**Ans:** (e^2-1)]

(vii) $\iint xy(x+y)dx\,dy$ over the area between $y=x^2$ and $y=x$. [**Ans:** $\frac{3}{58}$]

22.2 Evaluate the integrals by changing the order of integration :

(i) Change the order of integration of the following integral :

$$I=\int_{-a}^a \int_0^{\sqrt{a^2-y^2}} f(x,y)dx\,dy.$$

[**Ans:** $\int_0^a dx \int_{-\sqrt{a^2-x^2}}^{\sqrt{a^2-x^2}} f(x,y)dy$]

(ii) Change the order of integration of the following integral :

$$\int_{-1}^{+1}\left[\int_0^{\sqrt{1-x^2}} f(x,y)dy\right]dx.$$

[**Ans:** $\int_0^1 dy \int_{-\sqrt{1-y^2}}^{\sqrt{1-y^2}} f(x,y)dx$]

(iii) Change the order of integration in $\int_0^{\frac{\pi}{3}} \int_0^{2a\cos\theta} f(r,\theta)r\,d\theta\,dr.$

[**Ans:** $\int_0^{2a} \int_0^{\cos^{-1}\left(\frac{r}{2a}\right)} f(r,\theta)r\,dr\,d\theta$]

(iv) Change the order of integration in $\int_0^{\frac{a}{2}}\int_{\frac{x^2}{a}}^{x-\frac{x^2}{a}} V\,dx\,dy$.

[Ans: $\int_0^{\frac{a}{4}}\int_{x-\sqrt{a^2-4ay}}^{\sqrt{ay}} V\,dy\,dx$]

(v) Change the order of integration in $\int_0^{2a}\int_{\sqrt{2ax-x^2}}^{\sqrt{2ax}} V\,dx\,dy$.

[Ans: $\int_0^{a}\int_{\frac{y^2}{2a}}^{a-\sqrt{a^2-y^2}} V\,dy\,dx + \int_0^{a}\int_{a+\sqrt{a^2-y^2}}^{2a} V\,dy + \int_0^{2a}\int_{\frac{y^2}{2a}}^{2a} V\,dy\,dx$]

(vi) Change the order of integration $\int_0^1\int_{x^2}^{x} (x^2+y^2)^{-\frac{1}{2}}\,dy\,dx$.

[Ans: $\int_0^1 dy\int_y^{\sqrt{y}} (x^2+y^2)^{-\frac{1}{2}}\,dx$]

(vii) Change the order of integration and then evaluate $\int_0^1\int_0^{1-y^2} x^3y\,dx\,dy$.

[VTU 2006] [Ans: $\frac{1}{24}$]

22.3 Evaluate the following integrals by changing to polar coordinates :

(i) $\int_0^{4a}\int_{\frac{y^2}{4a}}^{y} \dfrac{x^2-y^2}{x^2+y^2}\,dx\,dy$. [Ans: $8\left(\frac{\pi}{2}-\frac{5}{3}\right)a^2$]

(ii) $\iint xy(x^2+y^2)^{\frac{n}{2}}\,dx\,dy$ over the positive quadrant of $x^2+y^2=4$, supposing $n+3>0$. [SVTU 2007] [Ans: $\frac{2^{n+3}}{n+4}$]

(iii) $\iiint \left[\dfrac{1-x^2-y^2}{1+x^2+y^2}\right]^{\frac{1}{2}} dx\,dy$, the integral being extended over all positive values of x and y subject to $x^2+y^2=1$. [PTU 2005] [Ans: 0]

(iv) $\int_0^a dy\int_y^a \dfrac{dx}{\sqrt{a^2x^2+b^2y^2}}$. [Ans: $\frac{a}{b}\sin^{-1}\frac{b}{a}$]

22.4 Transform to Cartesian form and then integrate $\int_0^{\pi}\int_0^{a} r^2\sin\theta\cos\theta\,dr\,d\theta$

[PTU 2005] [Ans: 0]

22.5 Change the following integrals into spherical coordinates and then evaluate

(i) $\int_0^1\int_0^{\sqrt{1-x^2}}\int_0^{\sqrt{1-x^2-y^2}} \dfrac{dx\,dy\,dz}{\sqrt{1-x^2-y^2-z^2}}$. [VTU 2006] [Ans: $\frac{\pi^2}{8}$]

(ii) $\iiint (x^2+y^2+z^2)dx\,dy\,dz$ taken over the volume enclosed by the sphere $x^2+y^2+z^2=1$. [Ans: $\frac{4\pi}{5}$]

(iii) Find the volume bounded by xy-plane, the paraboloid $zx=x^2+y^2$ and the cylinder $x^2+y^2=4$. [Raipur U 2006] [Ans: 4π]

(iv) Find the volume bounded by the xy-plane, the cylinder $x^2+y^2=1$ and the plane $x+y+z=3$. [ISM 2001] [Ans: 3π]

22.6 Evaluate the integral $I = \int_{x=0}^{\frac{1}{2}} \int_{y=x}^{1-x} \frac{(x-y)^2}{(x+y)^2} dy\, dx$ changing to variables r and s such that, $x = \frac{1}{2}(r-s)$ and $y = \frac{1}{2}(r+s)$. **[BU(H) 1994]**

22.7 Compute the double integral $\iint_D (y-x)\, dx\, dy$ over, the region D in the xy-plane bounded by the straight lines, $y = x+1$, $y = x-3$, $y = -\frac{x}{3} + \frac{7}{3}$ and $y = -\frac{x}{3} + 5$, by changing to new variables $u = y - x$ and $v = y + \frac{\pi}{3}$.

22.8 If $x + y + z = u$, $x + y = uv$, $y = uvw$, prove that

$$\int_0^\infty \int_0^\infty \int_0^\infty V\, dx\, dy\, dz = \int_0^\infty \int_0^1 \int_0^1 V' u^2 v\, du\, dv\, dw.$$

22.9 By means of the substitution $x + y + z = u$, $v + z = uv$, $z = vn$, prove that the value of $\iiint x^{-\frac{1}{2}} + y^{-\frac{1}{2}} + z^{-\frac{1}{2}}(1 - x - y - z) dx\, dy\, dz$ taken over the volume bounded by $x = 0$, $y = 0$, $z = 0$, $x + y + z = 1$ is $\frac{\pi^2}{4}$.

22.10 Transform $\iiint (x-y)(y-z)(z-x) dx\, dy\, dz$ into one in which u, v, w are independent variables, where $u^3 = xyz$, $\frac{1}{v} = \frac{1}{x} + \frac{1}{y} + \frac{1}{z}$ and $w^2 = x^2 + y^2 + z^2$.

[Ans: $3 \iiint \left[\frac{v^2}{wu^2} \sqrt{\left(w^2 + \frac{2u^2}{v} \right)} \right]^{-1} du\, dv\, dw$**]**

22.11 Four parabolas $y^2 = 4ax$, $y^2 = 4bx$, $x^2 = cy$, $x^2 = 4by$ intersect and form a quadrilateral space. Find the area of the sphere thus enclosed. **[Ans:** $\frac{16}{3}(b-a)(d-c)$**]**

22.12 Evaluate the following triple integrals :

(i) $\int_0^a \int_0^x \int_0^{x+y} e^{x+y+z}\, dx\, dy\, dz$ **[Ans:** $\frac{1}{8}e^{4a} - \frac{3}{4}e^{2a} + e^a - \frac{3}{8}$**]**

(ii) $\iiint \frac{dx\, dy\, dz}{\sqrt{1 - x^2 - y^2 - z^2}}$, taken throughout the volume of the sphere $x^2 + y^2 + z^2 = 1$ of radius 1 lying in the first octant. **[AMIE(Summer) 1994]** **[Ans:** $\frac{\pi^2}{8}$**]**

(iii) $\iiint_R (x - y + z) dx\, dy\, dz$, where R is bounded by $x + y + z = a$, $x = 0$, $y = 0$, $z = 0$. **[AMIE(Summer) 1994]** **[Ans:** $\frac{a^4}{8}$**]**

(iv) $\int_{-c}^{c} \int_{-b}^{b} \int_{-a}^{a} (x^2 + y^2 + z^2)\, dx\, dy\, dz$. **[VTU 2000]** **[Ans:** $\frac{8}{3}abc(a^2 + b^2 + c^2)$**]**

(v) $\int_0^{\frac{\pi}{2}} \int_0^{a \sin\theta} \int_0^{\frac{a^2 - r^2}{a}} r\, dz\, dr\, d\theta$. **[VTU 2007]** **[Ans:** $\frac{5\pi a^3}{64}$**]**

22.13 (i) Find the portion of the sphere $x^2 + y^2 + z^2 = a^2$, lying inside the cylinder $x^2 + y^2 = ax$. **[Ans:** $2(\pi - 2a^2)$**]**

(ii) Find the surface area of the sphere, $x^2 + y^2 + z^2 = 16$. **[Ans:** 64π**]**

(iii) Prove that the area of the surface $az = xy$ that lies inside the cylinder $(x^2+y^2)^2 = 2a^2xy$ is $\frac{1}{9}(20 - 3\pi)a^2$.

(iv) Prove that the area of the surface $z^2 = 2xy$ included between the planes $x = 0$, $x = a$, $y = 0$, $y = b$ is $4\sqrt{ab}\frac{(a+b)}{3\sqrt{2}}$.

(v) Show that the area of the surface of the sphere $x^2+y^2+z^2 = 1$, that lies inside the cylinder $2x^2(x^2+y^2) = 3(x^2 - y^2)$ is $2\pi - 4\sqrt{2}[3\log(\sqrt{3}+\sqrt{2}) = 2\log(1+\sqrt{2})]$.

22.14 (i) Find the volume by the xy-plane, the cylinder $x^2 + y^2 = 1$ and the plane $x + y + z = 3$. **[Ans:** 3π**]**

(ii) Prove that the volume V enclosed by the sphere $x^2+y^2+z^2 = a^2$ and the cylinders $x^2 + y^2 = ay$ is given by $V = 2\iint(\sqrt{a^2 - x^2 - y^2})dx\,dy$, the integral being taken over the circle $x^2 + y^2 = ay$. Transform the integral into polar coordinates and hence show that, $V = \frac{4}{3}a^3\left(\frac{\pi}{2} - \frac{l}{3}\right)$.

(iii) Find the volume cut off from the paraboloid $x^2 + \frac{y^2}{4} + z^2 = 1$ by the plane $z = 0$. **[Ans:** π**]**

22.15 (i) Compute the mass of a hemisphere of radius a with centre at the origin, if the density ρ of its substance at any point of the hemisphere is proportional to the height of the point from the base (i.e., $\rho = kz$). **[BU(H) 1994]**

(ii) Determine the mass of a sphere of radius R with centre at the origin, if the density of its material at each point is proportional to the distance of the point from the surface. **[BU(H) 2001]**

22.16 (a) Evaluate double integral $\iint e^{-(x^2+y^2)}dxdy$ in the region bounded by $x^2 + y^2 = 1$. **[CU(H) 2014]**

(b) Evaluate $\int_S U(x, y, z)dS$, where $U(x, y, z) = x^2+y^2$ and S is the surface of the cone $z^2 - x^2 - y^2 = 0$ of both height and base radius a. Give physical interpretation of the above integral. **[VU(H) 2014]**

22.17 Prove that the volume enclosed by the sphere $x^2 + y^2 + z^2 = a^2$ and the cylinder $x^2 + y^2 = ay$ is given by

$$V = 2\iint(a^2 - x^2 - y^2)dx\,dy$$

the integration being taken over the circle $x^2 + y^2 = ay$. Transform the integral into polar coordinates and hence show that

$$V = \frac{4}{3}a^3\left[\frac{\pi}{2} - \frac{2}{3}\right].$$

22.18 (a) Evaluate $\iiint_V (x^2 + y^2)dx\,dy\,dz$, where V is the region bounded by $z = x^2 + y^2$ and $z = 8 - x^2 - y^2$.

(b) Evaluate $\iint_R \sqrt{x^2 + y^2}dydx$, where R is the region in xy-plane bounded by $x^2+y^2 = 4$ and $x^2 + y^2 = 9$. **[CU(H) 2016]**

22.19 (a) Evaluate the double integral $\iint_A xy\,dxdy$, where A is the rectangle bounded by $x = 0, x = 2, y = 0$ and $y = 1$. **[CU(H) 2015]**

(b) Evaluate the double integral $\iint_R (1+\sqrt{x^2 + y^2})dxdy$, where R is the region bounded by the circle $x^2 + y^2 = 1$. **[CU(H) 2009]**

CHAPTER 23

Probability

23.1 Some Important Concepts

To understand the principles of statistics it is essential to realise some of the basic concepts as discussed below :

1. **Data :** Data are the raw material of statistics which are numbers arising out of **measurement** or counting. For example, measure the length of 10 rods supplied to you in metre. You get five numbers, say, 5, 10, 7, 15 and 11. Each of these numbers is a data and the collection of all the five constitutes **data**.

2. **Statistics :** It is that branch of science which deals with the collection of data, organising, summarising, presenting and analysing data and then drawing valid conclusion on the basis of such analysis.

3. **Variable :** If we take weight of the students in a particular class of a school, we find that it takes different values and we refer to the **characteristic** as variable.

   Some **characteristics** are, however, not measurable, e.g., people belonging to different caste. Variables of this type are called **qualitative**.

4. **Random variables :** If the values x_i (i.e., x_1, x_2, x_3, ..., x_n) of a characteristic or variable X are obtained as a chance factor which could not be predicted before, then the variable is called a **random variable** or **stochastic variable** or **variate**.

   For example, if in tossing a pair of dice the sum of two numbers X has values $x_i = 0, 1, 2, 3, \ldots, 12$ depending purely on chance, then X is a random variable.

   Random variables may be either **discrete** or **continuous**. Discrete variables are those in which there are **gaps** or **interruptions** in the set of values. In the dice experiment above X cannot have values 2.3, 3.5, etc. which is a set of discrete random variable.

   If there is no gaps or interruptions in the values of X and X can have infinite number of values, the variables are called **continuous random variables**. For example, the values obtained in height, weight, etc. on individuals.

5. **Population :** It is defined in statistics as the largest collection of values of random variable for which we have an interest at a particular time.

   If a population has a fixed number of values of variable, then it is called **finite**. On the other hand if it consists of an endless succession of values, then the population is **infinite**.

6. **Success :** If is a **part of a population**.

   If the population consists of the heights of all the students in a class of a school, then the height of only a fraction of the students is a **sample**.

7. **Simple random sample :** If a sample of size n be drawn from a population of size N ($N \geq n$) in such a way that every possible sample of size n has the same chance of being selected, then the sample is called a **simple random sample**. Drawing a sample satisfying the above condition is called **simple random sampling**.

23.2 Measurement of Central Tendency

The central value or the average value is of importance as it gives a most representative value of a set of data. Average value is also called the **central tendency**. There are five types of averages viz., **Arithmetic mean**, **Median**, **Mode**, **Geometric mean** and **Harmonic mean**.

(1) **Arithmetic mean :** (a) If $x_1, x_2, \ldots, x_n$ are numbers, then their arithmetic mean (A.M.) is defined by

$$\textbf{A.M. } (\overline{x}) = \frac{x_1 + x_2 + \cdots + x_n}{n} = \frac{\sum_{i=1}^{n} x_i}{n}. \tag{23.1}$$

If the numbers $x_1, x_2, \ldots, x_n$ occur $f_1, f_2, \ldots, f_n$ times respectively, then

$$\textbf{A.M. } (\overline{x}) = \frac{f_1x_1 + f_2x_2 + \cdots + f_nx_n}{f_1 + f_2 + \cdots + f_n} = \frac{\sum_{i=1}^{n} f_ix_i}{\sum_{i=1}^{n} f_i}, \tag{23.2}$$

where $\sum_{i=1}^{n} f_i = f_1 + f_2 + \cdots + f_n =$ total frequency.

Example 23.1. *Find the mean of the following :*

Numbers :	7	10	15	25
Frequency :	3	5	5	2

Solution : Here, $\sum_i f_ix_i = 7 \times 3 + 10 \times 5 + 15 \times 5 + 25 \times 2$

$= 21 + 50 + 75 + 50 = 196$

$\sum_i f_i = 3 + 5 + 5 + 2 = 15.$

$$\therefore \quad \overline{x} = \frac{\sum_i f_ix_i}{\sum_i f_i} = \frac{196}{15} = \mathbf{13.07}.$$

(b) **Short cut method :** If A be the assumed mean and d, the deviation of the variate x from A, then

$$\frac{\sum_{i=1}^{n} f_id_i}{\sum_{i=1}^{n} f_i} = \frac{\sum_{i=1}^{n} f_i(x_i - A)}{\sum_{i=1}^{n} f_i} = \frac{\sum_{i=1}^{n} f_ix_i}{\sum_{i=1}^{n} f_i} - \frac{\sum_{i=1}^{n} f_id_i}{\sum_{i=1}^{n} f_i} = \overline{x} - \frac{d\sum_{i=1}^{n} f_i}{\sum_{i=1}^{n} f_i} = \overline{x} - A.$$

$$\therefore \quad \overline{x} = A + \frac{\sum_{i=1}^{n} f_id_i}{\sum_{i=1}^{n} f_i}. \tag{23.3}$$

Example 23.2. *Find the arithmetic mean for the following distributions :*

Class :	0–10	10–20	20–30	30–40	40–50
Frequency :	7	8	20	10	5

Solution : Suppose the assumed mean $A = 25$

Class	Mid-Value (x)	Frequency (y)	$x - 25 = d$	$f_i d_i$
0–10	5	7	−20	−140
10–20	15	8	−10	−80
20–35	25	20	0	0
30–40	35	10	+10	+100
40–45	45	5	−20	+100
Total		50		−20

$$\therefore \quad \overline{x} = A + \frac{\sum_i f_i d_i}{\sum_i f_i} = 25 + \frac{-20}{50} = \mathbf{24.6.}$$

(c) **Step deviation method :** If A be the assumed mean, i the class interval and $D = \frac{x-A}{i}$, then

$$\overline{\boldsymbol{x}} = \boldsymbol{A} + \frac{\sum_i \boldsymbol{f_i D}}{\sum_i \boldsymbol{f_i}} \times \boldsymbol{i}. \tag{23.4}$$

Example 23.3. *Use step deviation method to find the arithmetic mean of the data given in example* 23.2 :

$$D = \frac{x - A}{i} = -2, -1, 0, +1, +2 \quad \text{and} \quad f_i D_i = -14, -8, 0, +10, +10$$

$$\therefore \quad \text{total } f = 7 + 8 + 20 + 10 + 5 = 50 \quad \text{and} \quad f_i D_i = (-14) + (-8) + 0 + 10 + 10 = 2.$$

$$\therefore \quad \overline{x} = A + \frac{\sum_i f_i D_i}{\sum_i f_i} \times i = 25 + \frac{-2}{50} \times 10 = \mathbf{24.6.}$$

(2) **Median :** If the values of a variable are arranged in the ascending order of magnitude, the median is the middle item if the number is odd and is the mean of the two middle items if the number is even. Thus, the median is equal to the mid-value, i.e., the value which divided the total frequency into two equal parts.

In other words, if the total number of items is odd and equal to n (say), then the value of $\left(\frac{n+1}{2}\right)$th item is the median.

If the total number of the frequencies is even, say n, then there are two middle items and so the mean of the values of $\left(\frac{n}{2}\right)$th and $\left(\frac{n}{2} + 1\right)$th item is the median.

Example 23.4. (a) *Find the median of* 18, 20, 3, 5, 25, 8, 16.

(b) *Find the median of* 20, 24, 6, 9, 15, 8, 13, 16.

Solution : (a) By arranging in ascending order, we get 3, 5, 8, 16, 18, 20, 25.

So, $n = 7$ and $\left(\frac{n+1}{2}\right)$th term, i.e., the 4th term is the median = **16**.

(b) By arranging in ascending order, we get 6, 8, 9, 13, 15, 16, 20, 24.

Here, $n = 0$ and $\left(\frac{n}{2}\right)$th and $\left[\left(\frac{n}{2}\right) + 1\right]$th i.e., 4th and 5th terms are the middle terms corresponding to 13 and 15.

Hence, **the median** $= \frac{13+15}{2} = \mathbf{14}$.

For group data, **Median** $= \boldsymbol{l + \frac{\frac{1}{2}N - F}{f} \times i}$, (23.5)

where l = lower limit of the median class, f = frequency of the median class, i = width of the class interval, F = total of all the frequencies before the median class and N = total frequency of data.

(3) **Mode : Mode is defined as that value of the variable which occurs most frequently, i.e., the value of the maximum frequency.**

Example 23.5. *Find the mode of the following items :*

(i) 3, 5, 14, 7, 9, 11, 13, 11, 18, 11, 22, 11, 19, 28, 11

(ii) 3, 5, 14, 7, 9, 11, 13, 18, 19, 28

(iii) 3, 3, 5, 7, 7, 11, 13, 7, 11, 19, 28, 11.

Solution : (i) Here, 11 occurs 5 times or more. So, the mode of the distribution is 11.

(ii) As the values of the given set are all different, the distribution has mode.

(iii) Here 7 and 11 occur thrice. So, the two modes viz., 7 and 11 in the given distribution.

For grouped data, **Mode** $= \boldsymbol{l + \frac{f - f_{-1}}{2f - f_{-1} - f_1} \times i}$, (23.6)

where l = lower limit of the median class, f = frequency of the class, i = the width of the class, f_1 & f_2 = frequency before and after the nodal class respectively.

Empirical formula : Mean − Mode = 3(Mean − Median)

(4) **Geometric Mean :** If $x_1, x_2, x_3, \ldots, x_n$ be n values of variates, then the geometric mean

$$\boldsymbol{G = (x_1 \times x_2 \times x_3 \times \cdots \times x_n)^{\frac{1}{n}}}. \quad (23.7)$$

Example 23.6. *Find the geometric mean of* 2, 4, 8.

Solution : $G = (2 \times 4 \times 8)^{\frac{1}{3}} = \mathbf{4}$.

(5) **Harmonic mean :** If $x_1, x_2, x_3, \ldots, x_n$ be a set of n observations, then the harmonic mean is the reciprocal of the arithmetic mean of the reciprocals of the quantities.

$\therefore$ Harmonic mean $\boldsymbol{\frac{1}{H} = \frac{1}{n}\left[\frac{1}{x_1} + \frac{1}{x_2} + \frac{1}{x_3} + \cdots + \frac{1}{x_n}\right]}$ (23.8)

Example 23.7. *Calculate the harmonic mean of 2, 4, 8.*

Solution : $H = \frac{1}{3}\left[\frac{1}{2}+\frac{1}{4}+\frac{1}{8}\right] = \frac{1}{3}\times\frac{7}{8} = \frac{7}{24}$. $\therefore$ $\boldsymbol{H = \frac{24}{7} = 3.434}$.

23.3 Dispersion

Though the measures of central tendency gives an idea about the central value of the data-set of a variable, they fail to give any idea as to how the individual values differ from the central value, i.e., whether they are closely packed around the central value or widely scattered from it.

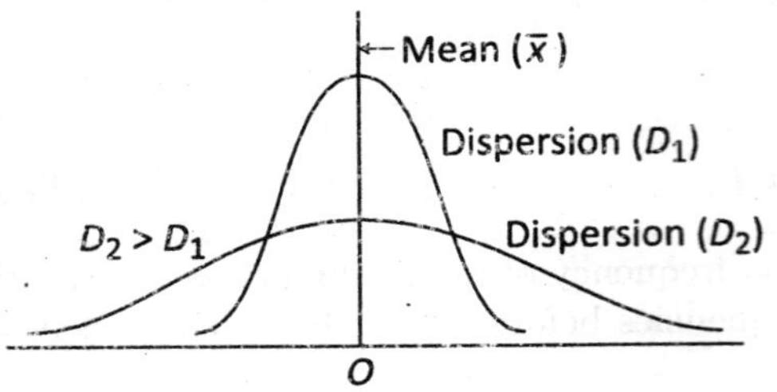

Fig. 23.1

Dispersion of a set of observation gives the variety that data-set exhibits.

Dispersion is actually the deviation of the values of a variate from its central value.

Fig. 23.1 shows that two populations may have the same mean but different dispersion where one's data may spread out from the mean more than the other. The magnitude of such variation is called **dispersion.**

(1) **Range :** It is the simplest measure of dispersion and is given by the difference between the largest value (x_l) to smallest value (x_s) of the set.

So, **Range** $\boldsymbol{R = x_l - x_s}$. (23.9)

Example 23.8. *The weekly wages of a group of labourers are :*

Rs. 85, 83, 86, 80, 89, 86, 87

What is the range?

Solution : The range R = Maximum value − Minimum value = 89 − 80 = **Rs. 9**.

(2) **Quartile deviation or semi-interquartile range :** Quartile deviation is the one-half of the interquartile range. If Q_1 and Q_3 are the first and third quartiles, the semi-interquartile range is

$$\boldsymbol{Q = \frac{1}{2}(Q_3 - Q_1)}. \tag{23.10}$$

(3) **Mean deviation : It is defined as the mean of the absolute values of the deviations of a given set of numbers from their arithmetic mean.**

If $x_1, x_2, \ldots, x_n$ be a set of n numbers, then for simple distribution, the mean deviation (M.D.) about the mean value is

$$\textbf{M.D.} = \frac{1}{n}\sum_{i=1}^{n}|x_i - \bar{x}|. \tag{23.11}$$

For frequency distribution, $\textbf{M.D.} = \frac{1}{n}\sum_{i=1}^{n} f_i|x_i - \bar{x}|$, (23.12)

where $\bar{x} = \frac{1}{n}\sum_{i=1}^{n} f_i x_i$ and $n = \sum_{i=1}^{n} f_i$.

(4) **Standard deviation :** Standard deviation is defined as the square root of the mean of the square of the deviation from the arithmetic mean.

$$\therefore \textbf{standard deviation (S.D.)} = \sigma = \sqrt{\frac{\sum_i f_i(x_i - \overline{x})^2}{\sum_i f_i}}. \tag{23.13}$$

The square of the standard deviation σ^2 is called variance, i.e.,

$$\textbf{Variance } \sigma^2 = \frac{\sum_i f_i(x_i - \overline{x})^2}{\sum_i f_i}. \tag{23.14}$$

σ^2 is called the second moment about the mean and is denoted by μ_2.

Shortest method of calculating standard deviation :

We already have

$$\sigma^2 = \frac{1}{N}\sum_i f_i(x_i - \overline{x})^2 = \frac{1}{N}\sum_i f_i(x_i - A - \overline{x - A})^2$$

$$= \frac{1}{N}\sum_i f_i(d_i - \overline{x - A})^2 \quad [\because\ x_i - A = d_i]$$

$$= \frac{1}{N}\sum_i f_i d_i^2 - 2(\overline{x} - A)\frac{1}{N}\sum_i f_i d_i + (\overline{x} - A)^2\frac{1}{N}\sum_i f_i \quad \left[\sum_i f_i = N\right]$$

$$= \frac{1}{N}\sum_i f_i d_i^2 - 2(\overline{x} - A)\frac{1}{N}\sum_i f_i d_i + (\overline{x} - A)^2.$$

Now, $\overline{x} = A + \frac{\sum_i f_i d_i}{N}$ or, $\overline{x} - a = \frac{\sum_i f_i d_i}{N}$.

$$\therefore \quad \sigma^2 = \frac{1}{N}\sum_i f_i d_i^2 - 2\left(\frac{\sum_i f_i d_i}{N}\right) + \left(\frac{1}{N}\sum_i f_i d_i\right) + \left(\frac{\sum_i f_i d_i}{N}\right)^2$$

$$= \frac{1}{N}\sum_i f_i d_i^2 - \left(\frac{\sum_i f_i d_i}{N}\right)^2.$$

$$\therefore \quad \textbf{S.D.} = \sigma = \sqrt{\frac{\sum_i f_i d_i^2}{N} - \left(\frac{\sum_i f_i d_i}{N}\right)^2}. \tag{23.15}$$

Coefficient of variation : The ratio of standard deviation to the mean is known as the coefficient of variation and is often expressed as

$$\textbf{coefficient of variation} = \frac{\sigma}{\overline{x}} \times 100. \tag{23.16}$$

Example 23.9. *Calculate the mean and standard deviation for the following :*

Size of item :	6	7	8	9	10	11	12
Frequency :	3	6	9	13	8	5	4

[VTU 2001]

Solution :

Size of item x	Frequency f	Deviation $d = x - 9$	$f \times d$	$f \times d^2$
6	3	−3	−9	27
7	6	−2	−12	24
8	9	−1	−9	9
9	13	0	0	0
10	8	1	8	8
11	5	2	10	20
12	4	3	12	36
	$\sum f = 48$		$\sum fd = 0$	$\sum fd^2 = 124$

$$\therefore \quad \textbf{Mean} = 9 + \frac{\sum fd}{\sum f} = 9$$

$$\sigma = \sqrt{\left[\frac{\sum f(x-\overline{x})^2}{\sum f}\right]} = \sqrt{\frac{\sum fd^2}{\sum f}} = \sqrt{\frac{124}{48}} = \mathbf{1.607}.$$

23.4 Introduction : Probability and Distribution

The study of probability provides a mathematical tool for some assertions (like, there is a fair chance that 'X' will pass the examination) and is essential in every decision making process.

23.5 Basic Terminology

1. **Exhaustive events :** If a set of events includes all possible events, then it is called an exhaustive events. In case of tossing a coin the number of exhaustive event is only two-head or tail.

2. **Mutually exclusive events :** In case of two events, if the occurrence of one excludes the occurrence of the other, then these two events are said to be mutually exclusive. For example, in tossing a coin, either head or tail occurs but not both at the same time.

3. **Equally likely events :** If one event cannot be expected to happen in preference to another, then the events are said to be '**equally likely**'. For instance, the appearance of **head** or **tail** is equally likely events.

4. **Compound event :** If two or more events occur simultaneously in a single trial, the event is called a **compound event**. When a die is thrown, getting 5 or 6 is a compound event.

5. **Odds in favour of an event :** If m be the number of ways favourable to an event A, and n, the number of ways not favourable to A, then odds in favour of A is $\frac{m}{n}$ and odds against A is $\frac{n}{m}$.

6. **Random experiment :** If the result of the experiments be altogether different, even though they are performed under identical conditions, then they are termed as **random experiments**. For example, tossing a coin is a random experiment.

23.6 Probability

The concept of probability may be of two types as follows :

(i) Classical probability or a priori probability

(ii) Relative frequency probability or a posteriori probability.

(i) Classical probability

Definition : If an event can occur in n mutually exclusive and equally likely ways of which m are favourable to an event A, then the probability of p of the happening of A is given by

$$p = P(A) = \frac{m}{n}. \tag{23.17}$$

Obviously, the probability of non-occurrence of A is $\left(\frac{n-m}{n}\right)$ and is written as

$$q = P(A') = \frac{n-m}{n} = 1 - p, \quad \text{i.e.,} \quad P(A') = 1 - P(A).$$

$$\therefore \quad P(A) + P(A') = 1.$$

This means if an event is certain to happen, then its probability is unity, but if it is certain not to happen, then its probability is zero.

N.B. The above definition of probability fails if (a) the number of outcomes is infinite (not exhaustive) and (b) outcomes are not equally likely.

(ii) Relative frequency probability

Definition : If some process (trial) is repeated a large number of times (say, n) under the same conditions and if out of these some event with trait A occurs m times, then the relative frequency of occurrence of A is $\frac{m}{n}$ which will approximate to the probability of A, i.e.,

$$P(A) = \underset{n\to\infty}{\text{Lt}} \frac{m}{n}. \tag{23.18}$$

Example 23.10. *What is the probability of throwing 3 with an ordinary 6-faced die?*

Solution : There are six possible ways in which the die can fall and there is only one way of throwing 3.

$$\therefore \text{ the required probability} = \frac{\text{No. of favourable ways}}{\text{Total no. of equally likely ways}} = \frac{1}{6}.$$

Example 23.11. *What is the chance that a leap year selected at random will contain 53 Sundays?* **[Madras U 2003]**

Solution : There are $365 + 1 = 366$ days in a leap year and 52 full weeks, i.e., 52 Sundays and 2 extra days. These two extra days may be

(i) Monday, Tuesday (ii) Tuesday, Wednesday (iii) Wednesday, Thursday

(iv) Thursday, Friday (v) Friday, Saturday (vi) Saturday, Sunday

(vii) Sunday, Monday

The case (vii) is favourable in the present case.

$\therefore$ the required probability is $\frac{2}{7}$.

23.7 Probability and Set Notations

(i) Probability of occurrence of events A or B is written as $P(A+B)$ or $P(A \cup B)$.

(ii) Probability of occurrence of **both the events A and B** is written as $P(AB)$ or $P(A \cap B)$.

(iii) Event A implies ($\Rightarrow$) event B is written as $A \subset B$.

(iv) 'Event A and B are mutually exclusive' is written as $A \cap B = \phi$.

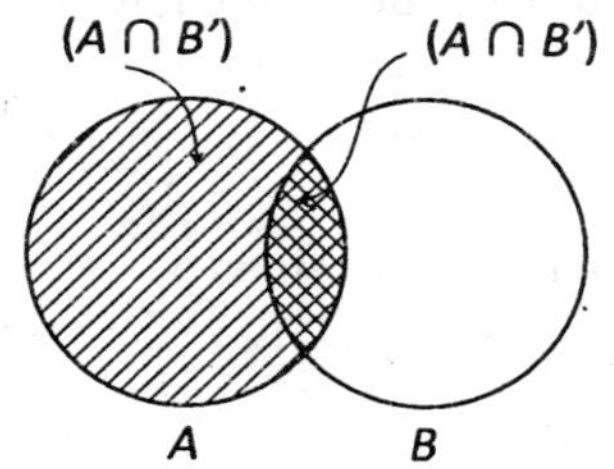

Fig. 23.2

Prove that for any two events A and B,

$$P(A \cap B') = P(A) - P(A \cap B).$$

Proof : $(A \cap B') \cup (A \cap B) = A$ [from Fig. (23.2)]

$\therefore$ $P[(A \cap B') \cup (A \cap B)] = P(A)$

or, $[P(A \cap B') + P(A \cap B)] = P(A)$

or, $P(A \cap B') = P(A) - P(A \cap B)$.

Similarly, we can prove that,

$$P(A' \cap B) = P(B) - P(A \cap B) \tag{23.19}$$

23.8 Important Properties of Probability

(1) **If some process processes n mutually exclusive outcomes (events) A_1, A_2, ..., A_n, then the probability of occurrence of any event A_i, i.e., $P(A_i)$ is assigned a non-negative number,**

$$P(A_i) \geq 0.$$

Actually, $0 \leq P(A_i) \leq 1$.

(2) **The sum of the probabilities of mutually exclusive events is equal to 1, i.e.,**

$$P(A_1) + P(A_2) + \cdots + P(A_n) = 1 \Rightarrow \sum_{i=1}^{n} P(A_n) = 1.$$

(3) **If A_i and A_j be two mutually exclusive events, then the probability of occurrence of either A_i or A_j is equal to the sum of their individual probabilities.**

$$\therefore\ P(A_i \cup A_j) = P(A_i) + P(A_j) \quad [U \equiv \text{'Unior' or 'or'}]. \tag{23.20}$$

23.8.1 Conditional Probability

The probability of happening an event A, such that event B has happened, is called the conditional probability of happening A on the condition that B has already happened.

Conditional probability is denoted by $P\left(\frac{A}{B}\right)$.

Example 23.12. *If one card is drawn at random from a pack of 52, what is the probability of getting a king?*

Solution : A king can be chosen in 4 ways.

But a card can be drawn in 52 ways.

$\therefore$ the probability of getting a king is $\frac{4}{52} = \mathbf{\frac{1}{13}}$.

23.8.2 Joint Probability

Sometimes we require to find the probability that a subject picked up at random from a group have simultaneously two characteristics. The probability in question is called **joint probability**.

Example 23.13. *What is the probability that a person picked up at random among* 100 *persons will be at the same time a male* (M) *and a drug addict* (D)?

Solution : The required probability is $P(M \cap D)$, where $\cap \equiv$ 'and' or 'intersection' so that $M \cap D$ implies the joint occurrence of M and D.

$\therefore \quad P(M \cap D) = \frac{20}{100} = \mathbf{0.2}$, where 20 is the number of drug addicted males.

23.9 Addition Law of Probability (Theorem of Total Probability)

If $P_1, P_2, \ldots, P_n$ be the separate probabilities of mutually exclusive events, then the probability P, that any one of these events will occur is given by

$$\boldsymbol{P = P_1 + P_2 + \cdots + P_n = \sum_{i=1}^{n} P_i.}$$

Proof : Let $P(A)$, $P(B)$, $P(C)$, $\ldots$, etc. be the probabilities of happening events $A, B, C, \ldots$, etc. respectively.

Let n be the total number of favourable cases to either A or B or C, $\ldots$, etc. and $m_1, m_2, m_3, \ldots$, etc. be the number of favourable cases.

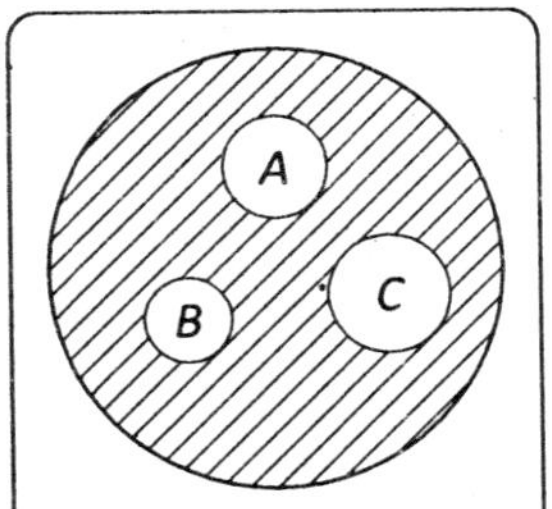

Fig. 23.3

$$\text{Hence, } P(A + B + C + \cdots) = \frac{m_1 + m_2 + \cdots + m_n}{n}$$
$$= \frac{m_1}{n} + \frac{m_2}{n} + \cdots + \frac{m_n}{n}$$
$$= P(A) + P(B) + P(C) + \cdots.$$

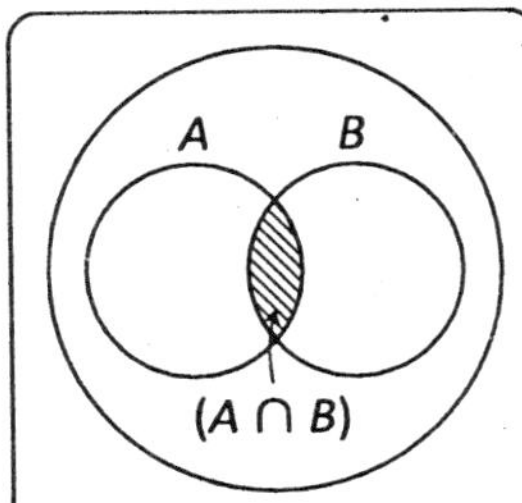

Fig. 23.4

$$\boldsymbol{\therefore \quad P = P(A) + P(B) + P(C) + \cdots = P_1 + P_2 + \cdots + P_n}$$
$$\boldsymbol{= \sum_{i=1}^{n} P_i.} \tag{23.21}$$

(i) If A and B be two mutually exclusive events, then

$$\boldsymbol{P = P(A + B) = P(A) + P(B)} \tag{23.22}$$

(ii) If two events A and B are not mutually exclusive, then the probability that event A or B or both occur is given by

$$\boldsymbol{P(A \cup B) = P(A) + P(B) - P(A \cap B).} \tag{23.23}$$

Example 23.14. *An urn contains 10 black balls and 10 red balls. What is the probability of drawing two balls of the same colour?*

Solution : There are 10 black and 10 red balls in the urn.

$\therefore$ total number of balls $= 10 + 10 = 20$.

So, the probability of drawing 2 black balls $= \dfrac{{}^{10}C_2}{{}^{20}C_2}$.

Similarly, the probability of drawing 2 red balls $= \dfrac{{}^{10}C_2}{{}^{20}C_2}$.

$\therefore$ the probability of drawing two balls of the same colour is

$$2 \times \frac{{}^{10}C_2}{{}^{20}C_2} = 2 \times \frac{\frac{10!}{(10-2)!}}{\frac{20!}{(20-2)!}} = 2 \times \frac{\frac{10\times9\times8!}{8!}}{\frac{20\times19\times18!}{18!}} = 2 \times \frac{10 \times 9}{20 \times 19} = \frac{9}{20}.$$

23.10 Multiplication Law of Probability

If there are two independent events, the respective probabilities of which are known, then the probability that both will occur is the product of the probabilities of their occurrence respectively.

$\therefore$ $P(AB) = P(A) \times P(B)$.

Proof : Let A and B be two independent events where A occurs in m_1 ways and fails to occur in n_1 ways, then

$$P(A) = \frac{m_1}{m_1 + n_1}.$$

Also let B occur in m_2 ways and fail to occur in n_2 ways, then

$$P(B) = \frac{m_2}{m_2 + n_2}.$$

Now the following four possibilities arise :

(i) A and B both occur, the number of ways $= m_1 \cdot m_2$

(ii) A may occur and B may occur, the number of ways $= m_1 \cdot n_2$

(iii) A may fail and B may occur, the number of ways $= n_1 \cdot m_2$

(iv) A and B both may fail, the number of ways $= n_1 \cdot n_2$

Now, the total number of ways $= m_1m_2 + m_1n_2 + n_1m_2 + n_1n_2 = (m_1 + n_1)(m_2 + n_2)$.

Hence, probability that both the events A and B will occur is given by

$$P(A \cap B) = \frac{m_1m_2}{(m_1+n_1)(m_2+n_2)} = \frac{m_1}{(m_1+n_1)} \times \frac{m_2}{(m_2+n_2)}.$$

$$\therefore \quad P(A \cap B) = P(A) \times P(B) \quad \text{(Proved)}. \tag{23.24}$$

Independent events : Two events A and B are said to be independent, if the probability of occurrence of A is the same regardless of whether or not B occurs.

N.B. It is to be kept in mind that 'mutually exclusive' and 'independent' do not have same meaning.

Complementary events : Events A and A' are said to be **complementary** if the probability of occurrence of event A is equal to 1 minus the probability of occurrence of event $\overline{A}$, called the **complement of A**, i.e., if

$$P(A) = 1 - P(\overline{A}) \Rightarrow P(\overline{A}) = 1 - P(A).$$

If two events are dependent, then

$$P(A_1 \cup A_2) = P(A_1) \times P\left(\frac{A_2}{A_1}\right), \tag{23.25}$$

where $\frac{A_2}{A_1}$ represents conditional probability of A_2 on the supposition that, A_1 has already occurred.

Example 23.15. *A problem in a machine is given to three students A, B and C whose chances of solving it are $\frac{1}{2}$, $\frac{1}{3}$ and $\frac{1}{4}$ respectively. What is the probability that the problem will be solved?* **[VTU 2004]**

Solution : Given that, the probability that A can solve the problem $= \frac{1}{2}$,

the probability that A cannot solve the problem $= 1 - \frac{1}{2}$.

Similarly, the probabilities that B and C cannot solve the problem are $\left(1 - \frac{1}{3}\right)$ and $\left(1 - \frac{1}{4}\right)$.

$\therefore$ the probability that A, B and C cannot solve the problem $= \left(1 - \frac{1}{2}\right)\left(1 - \frac{1}{3}\right)\left(1 - \frac{1}{4}\right)$.

Hence, the probability that, the problem will be solved i.e., at least one student will solve it is

$$= 1 - \left[\left(1 - \frac{1}{2}\right)\left(1 - \frac{1}{3}\right)\left(1 - \frac{1}{4}\right)\right] = 1 - \left[\frac{1}{2} \times \frac{2}{3} \times \frac{3}{4}\right] = 1 - \frac{1}{4} = \frac{3}{4}.$$

23.11 Probability Distribution

Probability distribution can be classified in two categories, viz., (1) Discrete probability distribution and (2) Continuous probability distribution.

(1) **Discrete probability distribution :** If a random variable takes a finite set of values, then it is a **discrete variate**. But if it assumes an infinite number of uncountable values, then it is a **continuous variate**.

Let a discrete variable X be the outcome of some experiment. If the probability that X takes the values x_i, is p_i, then

$$P(x = x_i) = p_i \quad \text{or} \quad p(x_i), \quad i = 1, 2, 3, \ldots,$$

where (i) $p(x_i) \geq 0$, for all values of i, (ii) $\sum p(x_i) = 1$.

The set of values of x_i with their probabilities p_i, constitute a **discrete probability distribution**.

Distribution function : For all the discrete variate X, the distribution function $F(x)$ is defined as

$$F(x) = P(X \leq x) = \sum_{i=1}^{x} p(x_i), \tag{23.26}$$

where x is any integer.

$F(x)$ is sometimes called **cumulative distribution** function.

Example 23.16. *The probability density function of a variable X is*

X :	0	1	2	3	4	5	6
$p(X)$:	k	$3k$	$5k$	$7k$	$9k$	$11k$	$13k$

(i) *Find $P(X < 4)$, $P(x \geq 5)$, $P(3 < X \leq 6)$.*

(ii) *What will be the maximum value of k, so that $P(X \leq 2) > 3$?* **[VTU 2004]**

Solution : (i) Let X be a random variable, then

$$\sum_{i=0}^{6} p(x_i) = 1 \quad \text{or,} \quad k + 3k + 5k + 7k + 9k + 11k + 13k = 1 \quad \text{or,} \quad 49k = 1 \quad \text{i.e.,} \quad k = \frac{1}{49}.$$

$$P(X < 4) = k + 3k + 5k + 7k = 16k \quad \text{or,} \quad \boldsymbol{P(X < 4)} = 16 \times \frac{1}{49} = \frac{\mathbf{16}}{\mathbf{49}}.$$

$$\boldsymbol{P(X \geq 5)} = 11k + 13k = 24k = \frac{\mathbf{24}}{\mathbf{49}}$$

and $\boldsymbol{P(3 < X \leq 6)} = 9k + 11k + 13k = 33k = \dfrac{\mathbf{33}}{\mathbf{49}}.$

(ii) $\boldsymbol{P(X \leq 2)} = k + 3k + 5k = \boldsymbol{9k > 0.3}.$

23.12 Theoretical Probability Distribution

The probability distribution of many variables of interest can be developed from purely theoretical considerations. Some of such distributions having wide range of applications are as below :

1. Binomial distribution, 2. Poisson distribution and 3. Normal distribution.

23.12.1 Binomial Distribution

Binomial distribution concerned with trial of a repetitive nature in which only the occurrence or non-occurrence, success or failure, acceptance or rejection, yes or no of a particular event is of interest. It has wide applications in statistics. This is also known as **Bernoulli's distribution.**

Expression for Binomial distribution $P(r)$: This means to find the probability of happening of an event once, twice, thrice, ..., r times exactly in $n(n > r)$ number of trials.

Suppose, the probability of happening an event A in one trial be p and the probability of not happening be $1 - p = q$. As there are n trials and the probability of not happening is $n - r$ times, we may represent the situation as :

$$\underset{r\text{-times}}{AAA \cdots\cdots\cdots A}\overline{A}\,\underset{(n-r)\text{-times}}{\overline{A}\,\overline{A} \cdots\cdots\cdots\cdots \overline{A}\,\overline{A}}, \tag{23.27}$$

where A and $\overline{A}$ represent respectively happening and not happening respectively.

Now, the order of arrangement in (23.27) has the probability

$$\underset{r\text{-times}}{ppp \cdots\cdots\cdots p}\,\underset{(n-r)\text{-times}}{qq \cdots\cdots\cdots\cdots q}p = p^r q^{n-r}. \tag{23.28}$$

Now, (23.27) is only one order of arrangement rA's.

The probability of (23.27) $= p^r q^{n-r} \times$ number of different arrangements of rA's.

The number of different arrangements of rA's and $(n - r)\overline{A}$'s $= {}^nC_r$.

$\therefore$ probability of happening of an event r-times $= {}^nC_r . p^r q^{n-r}$

or, $P(r) = {}^nC_r p^r q^{n-r} \quad [r = 0, 1, 2, \ldots, n]$

i.e., $\boldsymbol{P(r) = (r + 1)}$**th term of** $\boldsymbol{(q + p)^n}$. (23.29)

Eqn. (23.29) gives the **binomial distribution.**

Special Cases :

(i) If $r = 0$, then $P(0) = {}^nC_0 p^0 q^n = q^n$.

Thus the probability of happening an event A 0-times in n trial is q^n, the outcome being a failure.

(ii) If $r = 1$, $P(1) = {}^nC_1 pq^{n-1}$ = the probability of happening A in 1 time.

(iii) If $r = 2$, $P(2) = {}^nC_2 p^2 q^{n-2}$ = the probability of happening A 2 times.

(iv) If $r = 3$, $P(3) = {}^nC_3 p^3 q^{n-3}$ = the probability of happening A 3 times.

The above terms are the successive terms in the expansion $(p+q)^n$ and hence the name **binomial distribution**.

Example 23.17. *If 10% of bolts produced by a machine are defective, determine the probability that out of* 10 *bolts chosen at random* (i) 1 (ii) none (iii) at most two bolts will be defective.

[AMIETE(Winter) 1989, 1988]

Solution : By the problem, probability of defective bolts $= p = 10\% = 0.1$.

Probability of undefective bolts $= q = 1 - p = 1 - 0.1 = 0.9$.

Total number of bolts, $n = 10$.

(i) Probability of one defective bolt $= {}^{10}C_1 (0.1)^1 (0.1)^9 = \mathbf{0.3874}$.

(ii) Probability that none is defective = probability of 0 defective bolts $= P(0)$

$$= {}^{10}C_0 (0.1)^0 (0.9)^{10} = \mathbf{3487}.$$

(iii) Probability of 2 defective bolts $= {}^{10}C_2 (0.1)^2 (0.9)^8 = \mathbf{0.1937}$.

Now, probability of at most 2 defective bolts $= P(0 \text{ or } 1 \text{ or } 2) = P(0) + P(1) + P(2)$

$$= 0.3487 + 0.3874 + 0.1937$$

$$= \mathbf{0.9298}.$$

23.12.2 Mean Binomial Distribution

For the binomial distribution

$$P(r) = {}^nC_r p^r q^{n-r},$$

where r = the number of times in which the success occurs, f = the corresponding frequency.

Successes (r)	Frequency (f)	Product (rf)
0	q^n	0
1	npq^{n-1}	npq^{n-1}
2	$\frac{n(n-1)}{2} p^2 q^{n-2}$	$n(n-1)p^2q^{n-2}$
$\cdots$	$\cdots$	$\cdots$
n	p^n	np^n

$$\therefore \quad \sum fr = npq^{n-1} + n(n-1)p^2q^{n-2} + \frac{n(n-1)(n-2)}{2!}p^3q^{n-3} + \cdots + np^n$$

$$= np\left[q^{n-1} + \frac{(n-1)}{1!}pq^{n-2} + \frac{(n-1)(n-2)}{2!}p^2q^{n-3} + \cdots + p^{n-1}\right] = np(q+p)^{n-1}$$

$$= np, \quad [\because\ q + p = 1]$$

and $\sum f = q^n + npq^{n-1} + \frac{n(n-1)}{2!}p^2q^{n-2} + \cdots + p^n \doteq (q+p)^n = 1.$

$$\therefore \quad \textbf{Mean binomial distribution, } \mu = \frac{\sum fr}{\sum f} = \boldsymbol{np}. \tag{23.30}$$

23.12.3 Standard Deviation of Binomial Distribution

To derive the standard deviation (σ) of binomial distribution we refer to the following table :

Successes (r)	Frequency (f)	($r^2 f$) value
0	q^n	0
1	npq^{n-1}	npq^{n-1}
2	$\frac{n(n-1)}{2!}p^2q^{n-2}$	$2n(n-1)p^2q^{n-2}$
...	...	...
n	p^n	n^2p^n

$\therefore$ total frequency $= \sum f = 1.$

$$\sum r^2 f = npq^{n-1} + 2n(n-1)p^2q^{n-2} + \frac{3n(n-1)(n-2)}{2!}p^3q^{n-3} + \cdots + n^2p^n$$

$$= np\left[q^{n-1} + \frac{2n(n-1)}{1!}pq^{n-2} + \frac{3n(n-1)(n-2)}{2!}p^2q^{n-3} + \cdots + np^{n-1}\right]$$

$$= np\left[q^{n-1} + \frac{(n-1)pq^{n-2}}{1!} + \frac{(n-1)(n-2)}{2!}p^2q^{n-3} + \cdots + p^{n-1}\right.$$
$$\left. + \frac{(n-1)pq^{n-2}}{1!} + \frac{2(n-1)(n-2)}{2!}p^2q^{n-3} + \cdots + (n-1)p^{n-1}\right]$$

$$= np\left[q^{n-1} + (n-1)pq^{n-2} + \frac{(n-1)(n-2)}{2!}p^2q^{n-3} + \cdots + p^{n-1}\right.$$
$$\left. + (n-1)p\left\{q^{n-2} + (n-2)pq^{n-3} + \frac{(n-2)(n-3)}{2!}p^2q^{n-4} + \cdots + p^{n-2}\right\}\right]$$

$$= np[(p+q)^{n-1} + (n-1)p(p+q)^{n-2}] = np[1 + (n-1)p] = np[np + (1-p)]$$

$$= np[np + q] \quad [\because\ 1 - p = q]$$

$$= \boldsymbol{n^2p^2 + npq}$$

$$\therefore \quad \textbf{Variance, } \sigma^2 = \frac{\sum r^2 f}{\sum f} - \left(\frac{\sum rf}{\sum f}\right)^2 = n^2p^2 + npq - (np)^2 = \boldsymbol{npq}. \tag{23.31}$$

$$\therefore \quad \textbf{Standard deviation, } \boldsymbol{\sigma = \sqrt{npq}}. \tag{23.32}$$

Thus, for the binomial distribution, Mean $= np$

$\mu_2 = \sigma^2 = npq.$

Likewise, the following moment generating functions can be obtained :

$$\mu_3 = npq(q-p)$$

$$\mu_4 = npq[1+3pq(n-2)]$$

Here, the moment coefficient of skewness is

$$\beta_1 = \frac{3^2}{\mu_2^3} = \frac{[npq(q-p)]^2}{(npq)^3} = \frac{(q-p)^2}{npq}.$$

$$r_1 = \sqrt{\beta_1} = \frac{q-p}{\sqrt{npq}}.$$

Coefficient of Kurtosis is given by

$$\beta_2 = \frac{\mu_4}{\mu_2^2} = \frac{npq[1+3pq(n-2)]}{(npq)^2} = 3 + \frac{1-6pq}{npq}.$$

$$r_2 = \beta_2 - 3 = \frac{1-6pq}{npq}.$$

Example 23.18. *A coin is tossed thrice. A success is getting* 1 *or* 6 *on a toss. Find the mean and variance of the number of success.* **[AMIE(Summer) 1991]**

Solution : Here, $n = 3$, $p = \frac{1}{3}$, $q = \frac{2}{3}$.

Mean $= np = 3 \times \frac{1}{3} = \mathbf{1}$, Variance $\sigma^2 = npq = 3 \times \frac{1}{3} \times \frac{2}{3} = \mathbf{\frac{2}{3}}$.

Standard deviation, $\boldsymbol{\sigma} = \sqrt{npq} = \sqrt{\frac{2}{3}}$.

23.13 Poisson Distribution

Poisson's distribution is related to the probability of events which are extremely rare, but which have a large number of independent opportunities for occurrence.

In physical problem like **radioactive disintegration**, this distribution finds important application.

Thus, in Poisson's distribution the probability of success $p \to 0$ and the number of trials $n \to \infty$ and np remains finite. With the above consideration, Poisson's distribution can be derived as a limiting case of binomial distribution.

The probability of r successes in binomial distribution is given by

$$P(r) = {}^nC_r p^r q^{n-r} = \frac{n!}{r!(n-r)!} p^r (1-p)^{n-r} \qquad [\because\ p+q=1]$$

$$= \frac{1}{r!}[(n-1)(n-2)\cdots(n-r+1)]\left(\frac{m}{n}\right)^r\left(1-\frac{m}{n}\right)^{n-r}$$

[where np has been replaced by m, and p by $\left(\frac{m}{n}\right)$]

$$= \frac{n(n-1)(n-2)\cdots(n-r+1)m^r}{n\cdot n\cdot n\cdot n\cdots nr!}\left(1-\frac{m}{n}\right)^n\left(1-\frac{m}{n}\right)^{-r}$$

$$= \left(1-\frac{1}{n}\right)\left(1-\frac{2}{n}\right)\cdots\left(1-\frac{r-1}{n}\right)\left(1-\frac{m}{n}\right)^n\left(1-\frac{m}{n}\right)^{-r}\frac{m^r}{r!}.$$

If now $n \to \infty$, then

$$P(r) = (1-0)(1-0)\cdots(1-0)e^{-m} \times (1-0)^{-r}\frac{m^r}{r!}$$

or, $$\boldsymbol{P(r) = \frac{m^r e^{-m}}{r!}} \quad \left[\because \underset{n\to\infty}{\text{Lt}} \left(1 - \frac{m}{n}\right)^n = e^{-m}\right] \tag{23.32}$$

Eqn. (23.32) gives the **Poisson's distribution.**

Poisson's distribution may be defined as :

If x be the number of occurrences of some of events in an interval of time and space, probability that x will occur is

$$\boldsymbol{f(x) = \frac{e^{-\lambda}\lambda^x}{x!}, \quad x = 0, 1, 2, \ldots,}$$

where λ is a parameter being the number of occurrences of the random events in the interval.

N.B. From eqn. (23.32), the probabilities of 0, 1, 2, ..., r, ... successes in a Poisson's distribution are :

$$\boldsymbol{P(0) = e^{-m}} \qquad [\text{for } r = 0]$$

$$\boldsymbol{P(1) = me^{-m}} \qquad [\text{for } r = 1]$$

$$\boldsymbol{P(2) = \frac{m^2 e^{-m}}{2!}} \qquad [\text{for } r = 2]$$

$$\boldsymbol{P(3) = \frac{m^3 e^{-m}}{3!}} \qquad [\text{for } r = 3]$$

$$\cdots\cdots\cdots\cdots \qquad \cdots\cdots\cdots$$

$$\boldsymbol{P(r) = \frac{m^r e^{-m}}{r!}} \qquad [\text{for } r = n]$$

23.13.1 Mean and Standard Deviation of Poisson's Distribution

We have $P(r) = \dfrac{e^{-m}m^r}{r!}$.

Successes (r)	Frequency (f)	Product ($f \cdot r$)
0	$\frac{e^{-m}m^0}{0!}$	0
1	$\frac{e^{-m}m^1}{1!}$	$e^{-m} \cdot m$
2	$\frac{e^{-m}m^2}{2!}$	$e^{-m} \cdot m^2$
3	$\frac{e^{-m}m^3}{3!}$	$\frac{e^{-m} \cdot m^3}{2!}$
...	...	...
r	$\frac{e^{-m}m^r}{r!}$	$\frac{e^{-m} \cdot m^r}{(r-1)!}$

$\therefore$ mean of Poisson's distribution is given by

$$\mu = \sum pr = p_1r_1 + p_2r_2 + p_3r_3 + \cdots$$
$$= e^{-m}\cdot 0 + me^{-m}\cdot 1 + \frac{m^2}{2!}e^{-m}\cdot 2 + \frac{m^3}{3!}e^{-m}\cdot 3 + \cdots + e^{-m}\frac{m^r}{(r-1)!} + \cdots$$
$$= e^{-m}\cdot m\left[1 + \frac{m}{1!} + \frac{m^2}{2!} + \cdots + \frac{m^{r-1}}{(r-1)!} + \cdots\right] = me^{-m}[e^m] = m.$$

$\therefore$ **Mean $\mu = m$.** (23.34)

The variance of Poisson's distribution is $\sigma^2 = \sum pr^2 - \mu^2$.

Here, $\sum pr^2 = p_1r_1^2 + p_2r_2^2 + p_3r_3^2 + \cdots = e^{-m}\cdot 0 + me^{-m}\cdot 1^2 + \frac{m^2}{2!}e^{-m}\cdot 2^2 + \frac{m^3}{3!}e^{-m}\cdot 3^2 + \cdots$

$$= me^{-m}\left[1 + 2m + \frac{3m^2}{2!} + \frac{4m^3}{3^2} + \cdots\right]$$
$$= me^{-m}\left[\left(1 + m + \frac{m^2}{2!} + \cdots\right) + \left(m + \frac{2m^2}{2!} + \frac{3m^3}{3!} + \cdots\right)\right]$$
$$= me^{-m}\left[e^m + m\left(1 + m + \frac{2m^2}{2!} + \cdots\right)\right] = me^{-m}[e^m + me^m] = m(1+m).$$

$\therefore$ $\boldsymbol{\sigma^2} = m(1+m) - m^2 = \sum pr^2 - \sum pr = \boldsymbol{m}$ (23.35)

The standard deviation of Poisson's distribution is **S.D.** $= \boldsymbol{\sigma} = \boldsymbol{\sqrt{m}}$ (23.36)

23.13.2 In Poisson's Distribution with Unit Mean and Mean Deviation about the Mean is $\left(\frac{2}{e}\right)$ × standard deviation (S.D.)

Proof : We have $P(r) = \frac{m^r}{r!}e^{-m}$.

But, mean = 1, i.e., $m = 1$ and S.D. $= \sqrt{m} = 1$.

$\therefore$ $P(r) = \frac{e^{-1}}{r!} = \frac{1}{e}\cdot\frac{1}{r!}$.

(r)	$P(r)$	$\|r-1\|$	$P(r)\|r-1\|$	(r)	$P(r)$	$\|r-1\|$	$P(r)\|r-1\|$
0	$\frac{1}{r}$	1	$\frac{1}{e}$	4	$\frac{1}{e}\frac{1}{4!}$	3	$\frac{1}{e}\frac{3}{4!}$
1	$\frac{1}{e}$	0	0	$\cdots$	$\cdots$	$\cdots$	$\cdots$
2	$\frac{1}{e}\frac{1}{2!}$	1	$\frac{1}{e}\frac{1}{2!}$	r	$\frac{1}{e}\frac{1}{r!}$	$r-1$	$\frac{1}{e}\frac{r-1}{r!}$
3	$\frac{1}{e}\frac{1}{3!}$	2	$\frac{1}{e}\frac{2}{3!}$	$\cdots$	$\cdots$	$\cdots$	$\cdots$

$$\therefore \sum P(r)|r-1| = \frac{1}{e} + 0 + \frac{1}{e}\frac{1}{2!} + \frac{1}{e}\frac{2}{3!} + \frac{1}{e}\frac{3}{4!} + \cdots + \frac{1}{e}\frac{r-1}{r!} + \cdots$$
$$= \frac{1}{e}\left[1 + 0 + \frac{1}{2!} + \frac{2}{3!} + \frac{3}{4!} + \cdots + \frac{r-1}{r!} + \cdots\right]$$

$$= \frac{1}{e}\left[1 + \left(\frac{1}{1!} - \frac{1}{1!}\right) + \left(\frac{2}{2!} - \frac{1}{2!}\right) + \left(\frac{3}{3!} - \frac{1}{3!}\right) + \left(\frac{4}{4!} - \frac{1}{4!}\right) + \cdots + \left(\frac{r}{r!} - \frac{1}{r!}\right) + \cdots\right]$$

$$= \frac{1}{e}\left[1 + \frac{1}{1!} + \frac{2}{2!} + \frac{3}{3!} + \frac{4}{4!} + \cdots + \frac{r}{r!} + \cdots + \frac{1}{1!} - \frac{1}{2!} - \frac{1}{3!} - \frac{1}{4!} \cdots - \frac{1}{r!} \cdots\right]$$

$$= \frac{1}{e}\left[1 + \left\{1 + \frac{1}{1!} + \frac{2}{2!} + \frac{3}{3!} + \frac{4}{4!} + \cdots + \frac{1}{(r-1)!} + \cdots\right\} - \left\{1 + \frac{1}{1!} + \frac{1}{2!} + \frac{1}{3!} + \frac{1}{4!} + \cdots + \frac{1}{r!} + \cdots\right\} + 1\right]$$

$$= \frac{1}{e}[1 + e - e + 1] = \frac{2}{e} = \frac{2}{e}(1) = \frac{2}{e} \times \text{(S.D.)}. \tag{23.37}$$

23.13.3 Recurrence Formula for Poisson's Distribution

We know that in Poisson's distribution,

$$P(r) = \frac{e^{-m}m^r}{r!}. \tag{i}$$

$$\therefore \quad P(r+1) = \frac{e^{-m}m^{r+1}}{(r+1)!}. \tag{ii}$$

Dividing Eqn. (ii) by Eqn. (i), we get

$$\frac{P(r+1)}{P(r)} = \frac{e^{-m}m^{r+1}}{(r+1)!} \cdot \frac{r!}{e^{-m}m^r} = \frac{m}{r+1}.$$

$$\therefore \quad \boldsymbol{P(r+1) = \frac{m}{r+1}P(r)}. \tag{23.38}$$

Example 23.19. *In a certain factory producing cycle tyres there is a small chance of* 1 *in* 500 *tyres to be defective. The tyres are supplied in lots of* 10. *Using Poisson's distribution, calculate the approximate number of lots containing* (i) no defective; (ii) one defective; (iii) two defective tyres respectively, in consignment of 10,000 lots.

[AMIETE(Summer) 1994; AMIE(Summer) 1994; WBUT 2004]

Solution : Here, $p = \frac{1}{500}$, $n = 10$, $m = np = 10 \times \frac{1}{500}$ or, $m = \frac{1}{50} = 0.02$.

$$\therefore \quad P(r) = \frac{e^{-m}m^r}{r!}.$$

(i) Probability of no defective tyres, i.e., $P(0) = \frac{e^{-0.02}(0.02)^0}{0!} = e^{-0.02} = \mathbf{0.9802}$.

Number of lots containing no defective tyres $= 10,000 \times 0.9802 =$ **9802** lots.

(ii) Probability of one defective tyre $P(1) = \frac{e^{-0.02}(0.02)}{1!} = 0.9802 \times 0.02 = \mathbf{0.019604}$.

Number of lots containing 1 defective tyres $= 10,000 \times 0.019604 =$ **196** lots.

(iii) Probability of two defective tyres $P(2) = \dfrac{e^{-0.02}(0.02)^2}{2!} = 0.9802 \times 0.0002$

$= \mathbf{0.00019604}.$

Number of lots containing two defective tyres $= 10,000 \times 0.00019604 = \mathbf{2\ lots}.$

23.14 Normal Distribution

Normal distribution is a continuous probability distribution. Any quantity whose variation depends on random causes will be distributed according to normal distribution.

Normal distribution can be derived from binomial distribution in the limiting case when the number of trial n is very large and the probability of success is close to $\frac{1}{2}$. The normal (or, Gaussian) distribution is defined as

$$f(x) = \frac{1}{\sqrt{2\pi\sigma}} e^{-\frac{(x-\mu)}{2a^2}}, \quad -\infty < x < \infty, \tag{23.40}$$

where the mean μ and the standard deviation $\sigma > 0$ are arbitrary constants. x is the normal variate and $f(x)$ is the probability density function of the normal distribution.

The graph of normal distribution is called the normal curve which is bell-shaped and symmetrical about the mean μ as shown in Fig. 23.5. The area under the normal curve between any two coordinates $x = x_1$ and $x = x_2$ is the probability of values falling into the given interval. The total area under the curve above the X-axis is

$$\int_{-\infty}^{\infty} f(x)dx = 1. \tag{23.41}$$

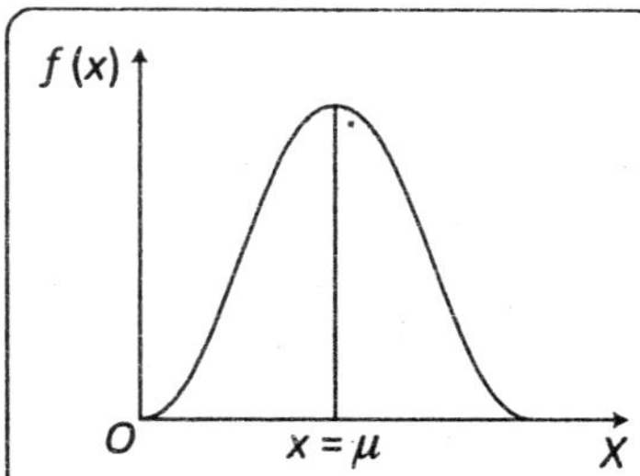

Fig. 23.5

N.B. $f(x)$ is called probability function, if

(i) $f(x) = 0$ **for every value of** x, (ii) $\int_{x_1}^{x_2} f(x)dx = P\ (x_1 < x < x_2)$

and (iii) $\int_{-\infty}^{\infty} f(x)dx = 1.$

Example 23.20. *A function $f(x)$ is defined as* $\begin{cases} 0, & x < 2 \\ \frac{1}{18}(2x+3), & 2 \le x \le 4 \\ 0, & x > 0. \end{cases}$

Show that it is a probability density function. **[AMIETE(Winter) 1994]**

Solution : Here $f(x) = 0, \quad x < 2$

$= \frac{1}{18}(2x+3), \quad 2 \le x \le 4$

$= 0, \quad x > 0.$

If $f(x)$ be a probability density function, then $\int_{-\infty}^{a} f(x)dx = 1.$

$$\therefore \int_2^4 \frac{1}{18}(2x+3)dx = \frac{1}{18}[x^2+3x]_2^4 = \frac{1}{18}[16+12-4-6] = 1.$$

$\therefore$ the given function is a probability density function.

23.14.1 Standard Form of Normal Distribution

The standard normal (or unit normal) distribution is characterised by a mean $\mu = 0$ and standard deviation $\sigma = 1$. It can be obtained by substituting the random variable $z = \frac{(x-\mu)}{\sigma}$ in equation

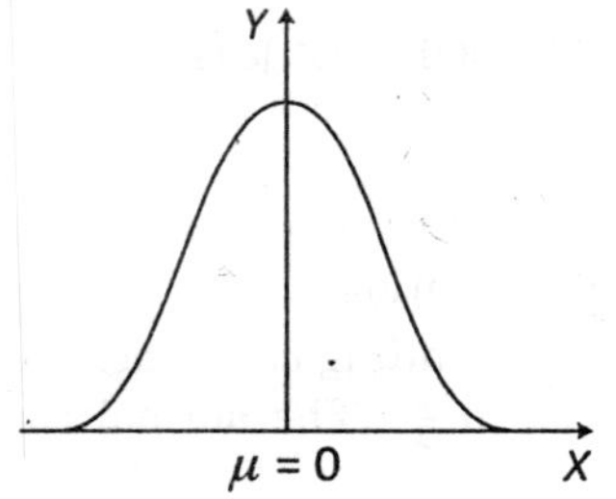

Fig. 23.6

$$f(x) = \frac{1}{\sqrt{2\pi\sigma}} e^{-\frac{(x-\mu)^2}{2\sigma^2}}, \quad -\infty < x < \infty.$$

Then the equation of normal distribution can be written as

$$f(z) = \frac{1}{\sqrt{2\pi}} e^{-\frac{z^2}{2}}, \quad -\infty < x < \infty. \tag{23.42}$$

The graph of the standard normal distribution is shown in Fig. 23.6.

(i) The probability of x lying between x_1 and x_2 is given by the area under the normal curve from x_1 to x_2.

$$\therefore \; P(x_1 \le x \le x_2) = \frac{1}{\sqrt{2\pi x}} \int_{x_1}^{x_2} e^{\frac{(x-\mu)^2}{2\sigma^2}} dx$$

$$= \frac{1}{\sqrt{2\pi}} \int_{z_1}^{z_2} e^{-\frac{z^2}{2}} dz \quad \left[\because \; z = \frac{(x-\mu)}{\sigma} \quad \text{or,} \; dz = \frac{dx}{\sigma}\right]$$

$$= \frac{1}{\sqrt{2\pi}} \left[\int_0^{z_2} e^{-\frac{z^2}{2}} dz - \int_0^{z_1} e^{-\frac{z^2}{2}} dz\right]$$

$$= P_2(z) - P_1(z). \tag{23.42}$$

The values of the above integrals are available in the table to the statistician.

(ii) The integral $\frac{1}{\sqrt{2\pi}} \int_0^z e^{-\frac{z^2}{2}} dz$ is called the **probability integral** or **the error function represented by $f(z)$.**

23.14.2 Mean for Normal Distribution

$$\text{Mean} = \int_{-\infty}^{\infty} \frac{1}{\sigma\sqrt{2\pi}} e^{-\frac{x^2}{2\sigma^2}} x \, dx$$

$$= \frac{1}{\sigma\sqrt{2\pi}} \int_{-\infty}^{\infty} e^{-\frac{t^2}{2}} (t\sigma)(\sigma \, dt) \quad \left[\text{putting } t = \frac{x}{\sigma}\right]$$

$$= \frac{\sigma}{\sqrt{2\pi}} \int_{-\infty}^{\infty} te^{-\frac{t^2}{2}} dt = \frac{\sigma}{\sqrt{2\pi}} \left[e^{-\frac{t^2}{2}}\right]_{-\infty}^{\infty} = \frac{\sigma}{\sqrt{2\pi}}(0) = 0. \tag{23.43}$$

23.14.3 Standard Deviation for Normal Distribution

$$\mu_2' = \int x^2 f(x) dx = \int_{-\infty}^{\infty} x^2 \frac{1}{\sigma\sqrt{2\pi}} e^{-\frac{x^2}{2\sigma^2}} dx.$$

Put $\frac{x^2}{2a^2} = 1$ or, $x = \sqrt{2}\sigma t^{\frac{1}{2}}$, $dx = \frac{\sqrt{2}\sigma \, dx}{2t^{\frac{1}{2}}}$.

$$\therefore \quad \mu_2' = \int_{-\infty}^{\infty} (2\sigma^2 t) \frac{1}{\sqrt{2\pi}\sigma} e^{-t\left(\frac{\sqrt{2}\sigma\, dt}{2t^{\frac{1}{2}}}\right)} = \frac{2\sigma^2}{\sigma\sqrt{2\pi}} \frac{\sqrt{2}\sigma}{2} \int_{-\infty}^{\infty} t^{\frac{3}{2}-1} e^{-t} dt$$

$$= \frac{\sigma^2}{\sqrt{\pi}} \cdot 2 \cdot \Gamma\left(\frac{3}{2}\right), \quad \left[\because \int_0^{\infty} x^{n-1} e^{-x} dx = \Gamma(n)\right]$$

$$= 2\frac{\sigma^2}{\sqrt{\pi}} \frac{1}{2} \Gamma\left(\frac{1}{2}\right) = \frac{\sigma^2}{\sqrt{\pi}} \sqrt{\pi} = \sigma^2.$$

$$\therefore \quad \mu_2 = \mu_2' - (\mu_1)^2 = \sigma^2 - (0)^2 = \sigma^2.$$

$$\textbf{S.D.} = \boldsymbol{\sigma}. \tag{23.44}$$

23.14.4 Variance and Standard Deviation for Normal Distribution

By definition of variance,

$$\text{Var } (x) = \int_{-\infty}^{\infty} (x-\mu)^2 f(x) dx = \frac{1}{\sqrt{2\pi}\sigma} \int_{-\infty}^{\infty} (x-\mu)^2 e^{-\frac{1}{2}\left(\frac{x-\mu}{\sigma}\right)^2} dx$$

$$= \frac{1}{\sqrt{2\pi}\sigma} \int_{-\infty}^{\infty} \sigma^2 z^2 e^{-\frac{z^2}{2}} \sigma\, dz, \quad \left[\text{putting } \frac{x-\mu}{\sigma} = z, \quad dx = \sigma\, dz\right]$$

$$= \frac{\sigma^2}{\sqrt{2\pi}} \cdot 2 \int_0^{\infty} z^2 e^{-\frac{z^2}{2}} dx = \frac{2\sigma^2}{\sqrt{2}} \frac{2(2)}{\sqrt{2}} \int_0^{\infty} t e^{-t} t^{-\frac{1}{2}} dt$$

$$= \frac{2\sigma^2}{\sqrt{\pi}} \int_0^{\infty} e^{-t} t^{\frac{3}{2}-1} dt, \quad \left[z = \sqrt{2} t^{\frac{1}{2}}, \ \therefore\ dz = \frac{1}{\sqrt{2}} t^{-\frac{1}{2}} dt\right]$$

$$= \frac{2\sigma^2}{\sqrt{\pi}} \Gamma\left(\frac{3}{2}\right), \quad \left[\because \Gamma(n) = \int_0^{\infty} e^{-x} x^{n-1} dx\right]$$

$$= \frac{2\sigma^2}{\sqrt{\pi}} \frac{1}{2} \Gamma\left(\frac{1}{2}\right) = \frac{2\sigma^2}{\sqrt{\pi}} \frac{1}{2} \sqrt{\pi} = \sigma^2. \tag{23.45}$$

$\therefore$ **Var** $(x) = \sigma^2$ and σ **is the standard deviation.**

Example 23.21. *In a normal distribution 31% of items are under 45 and 8% are over 64. Find the mean standard deviation of the distribution.*

[AMIE(Summer) 1991; WBUT 2003; JNTU 2003; PTU 2003; UPTU 2003]

Solution : Let the mean be m and σ, the standard deviation.

If 31% of the item are under 45, then the area to the left of the ordinate $x = 45$ is 0.31 (Fig. 23.7).

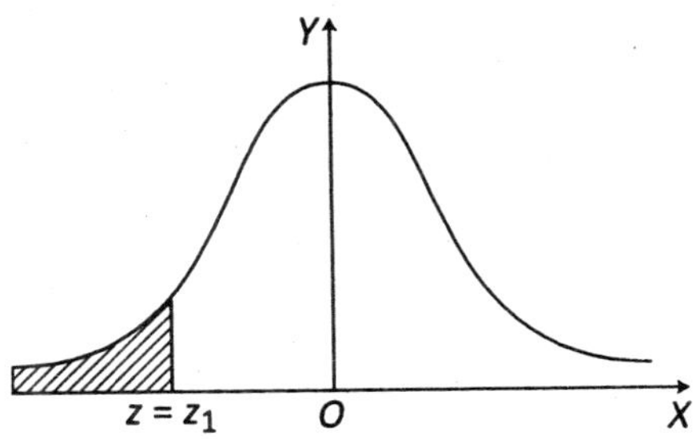

Fig. 23.7

Assuming $z = z_1$ for $x = 45$, then

$$P(z_1 < z < 0) = 0.5 - 0.31 = 1.9.$$

From the tables, corresponding $P = 0.19$ is 0.5.

$$\therefore \quad z = -0.5.$$

When $x = 64$, $z = z_2$,

$$P(0 < z < z_2) = 0.5 - 0.08 = 0.42.$$

Since 8% of the items are over 64, i.e., area to the right of the ordinate $x = 64$ is 0.08. From the tables, the value of z corresponding to this area is 1.4.

$\therefore \quad z_2 = 1.4.$

Since, $z = \dfrac{x-m}{\sigma}$, $\quad \therefore \quad -0.5 = \dfrac{45-m}{\sigma}$ and $1.45 = \dfrac{64-m}{\sigma}$.

Solving, we get $\boldsymbol{m} = \mathbf{50}$ and $\boldsymbol{\sigma} = \mathbf{10}$.

23.14.5 Table of Area $z = \frac{x}{\sigma}$

$z = \frac{x}{\sigma}$	0.00	0.01	0.02	0.03	0.04	0.05	0.06	0.07	0.08	0.09
0.0	0.5000	0.5040	0.5080	0.5120	0.5160	0.5199	0.5239	0.5279	0.5319	0.5359
0.1	0.5398	0.5438	0.5478	0.5517	0.5570	0.5596	0.5636	0.5675	0.5714	0.5753
0.2	0.5793	0.5832	0.5871	0.5910	0.5948	0.5987	0.6026	0.6064	0.6103	0.6141
0.3	0.6179	0.6217	0.6255	0.6293	0.6331	0.6368	0.6406	0.6443	0.6480	0.6517
0.4	0.6554	0.6591	0.6628	0.6664	0.6700	0.6736	0.6772	0.6808	0.6844	0.6879
0.5	0.6915	0.6950	0.6985	0.7019	0.7054	0.7088	0.7123	0.7157	0.7190	0.7224
0.6	0.7257	0.7291	0.7324	0.7357	0.7389	0.7422	0.7454	0.7486	0.7517	0.7549
0.7	0.7580	0.7611	0.7642	0.7673	0.7703	0.7734	0.7764	0.7793	0.7823	0.7852
0.8	0.7881	0.7910	0.7939	0.7697	0.7995	0.8023	0.8051	0.8078	0.8106	0.8133
0.9	0.8159	0.8186	0.8212	0.8238	0.8264	0.8289	0.8315	0.8340	0.8365	0.8389
1.0	0.8413	0.8438	0.8461	0.8485	0.8508	0.8531	0.8554	0.8577	0.8599	0.8621
1.1	0.8643	0.8665	0.8686	0.8708	0.8729	0.8749	0.8770	0.8790	0.8810	0.8830
1.2	0.8849	0.8869	0.8888	0.8906	0.8925	0.8943	0.8962	0.8980	0.8907	0.9015
1.3	0.9032	0.9049	0.9066	0.9082	0.9099	0.9115	0.9131	0.9147	0.9102	0.9177
1.4	0.9192	0.9207	0.9222	0.9236	0.9251	0.2650	0.9279	0.9292	0.9316	0.9319
1.5	0.9332	0.9345	0.9357	0.9370	0.9382	0.9394	0.9406	0.9418	0.9409	0.9441
1.6	0.9452	0.9463	0.9474	0.9494	0.9495	0.9505	0.9515	0.9525	0.9505	0.9545

23.15 Applications of Normal Distributions

The normal distribution finds important applications in the following :

(i) Calculation of errors made by chance in experimental measurement

(ii) Computation of heat probability of a shot

(iii) Statistical interference in almost every branch of science.

23.16 Solved Problems

Example 23.1. *Three coins are tossed simultaneously. What is the probability that at least two tails occur.*

Solution : Total number of outcome (sample space) is elementary events.

$\therefore$ sample space S is

$$S = (HHH, HHT, HTH, THH, HTT, THT, TTH, TTT)$$

where H = head and T = tail.

We define the event as A. A : at least two tails occur.

Here, the favourable cases are HTT, THT, TTH and TTT.

$\therefore$ the probability that at least two tails occur, $\boldsymbol{P(A)} = \frac{4}{8} = \frac{\mathbf{1}}{\mathbf{2}}$.

Example 23.2. *Calculate the probability of obtaining 4 heads in six tosses, using an unbiased coil.*

Solution : In this problem, $p = \frac{1}{2}$, $q = \frac{1}{2}$, $n = 6$ and $r = 4$.

$$\boldsymbol{P(r)} = {}^nC_r p^r q^{n-r} = \frac{6 \times 5}{1 \times 2}\left(\frac{1}{2}\right)^4\left(\frac{1}{2}\right)^2 = 15 \times \left(\frac{1}{2}\right)^6 = \frac{\mathbf{15}}{\mathbf{64}}.$$

Example 23.3. *The overall percentage of failures in an examination is 20%. If 6 candidates appear in the examination, what is the probability that at least 5 pass?*

Solution : By the problem, the probability of failures $q = \frac{20}{100} = \frac{1}{5}$.

$\therefore$ probability of pass $p = 1 - q = 1 - \frac{1}{5} = \frac{4}{5}$.

$\therefore$ probability of at least 5 pass, i.e., 5 or 6 $= {}^6C_5 p^5 q + {}^6C_6 p^6 q^0 = 6\left(\frac{4}{5}\right)^5\left(\frac{1}{5}\right) + \left(\frac{4}{5}\right)^6$

$$= \left(\frac{4}{5}\right)^6\left(\frac{6}{5} + \frac{4}{5}\right) = 2\left(\frac{4}{5}\right)^6 = \mathbf{0.6553}.$$

Example 23.4. *From a pack of well-shuffled cards, one card is drawn. Find the probability that this card is either a king or and ace.*

Solution : We define the events as A : card drawn is a king, B : card drawn is an ace.

The events A and B are mutually exclusive.

In a pack there are 52 cards, out of which 4 kings and 4 aces.

$\therefore$ $P(A) = \frac{4}{52} = \frac{1}{13}$, $\quad P(B) = \frac{4}{52} = \frac{1}{13}$.

As A and B are mutually exclusive, the probability,

P (either a king or an ace) $= P(A \cup B) = P(A) + P(B) = \frac{1}{13} + \frac{1}{13} = \frac{\mathbf{2}}{\mathbf{13}}$.

Example 23.5. *If A, B, C are mutually exclusive and exhaustive events associated with a random experiment and $P(B) = 0.6$, $P(C) = 0.2P(A)$, then find $P(A)$.*

Solution : By the condition of the problem, A, B, C are mutually exclusive and exhaustive events. We have

$$P(A) + P(B) + P(C) = 1 \quad \text{or,} \quad P(A) + 0.6P(A) + 0.2P(A) = 1 \quad \text{or,} \quad 1.8P(A) = 1.$$

$\therefore$ $\quad \boldsymbol{P(A)} = \frac{1}{1.8} = \frac{\mathbf{5}}{\mathbf{9}}$.

Example 23.6. *A bag contains 8 white and 6 red balls. Find the probability of drawing two balls of the same colour.*

Solution : 2 balls out of $8+6 = 14$ balls can be drawn in ${}^{14}C_2$ ways. This is the total outcome.

2 white balls out of 8 can be drawn in $^8C_2$ ways which is the total number of outcomes.

$\therefore$ probability of drawing 2 white balls out of 8 $= \dfrac{^8C_2}{^{14}C_2} = \dfrac{28}{91}$.

Similarly, 2 red balls out of 6 can be drawn in $^6C_2$ ways.

$\therefore$ probability of drawing 2 red balls out of 6 $= \dfrac{^6C_2}{^{14}C_2} = \dfrac{15}{91}$.

Hence, the probability of drawing 2 balls of the same colour (either both white or both red)

$$= \frac{28}{91} + \frac{15}{91} = \mathbf{\frac{43}{91}}.$$

Example 23.7. *Two dice are tossed once. Find the probability of getting an even numbers on the first die or a total* 8.

Solution : Let us define the events as :

A : getting an even number on the first die, B : getting a total of 8

$\therefore$ the set of elementary events are as follows :

$A : [(2,1),\ldots,(2,6),(4,1)\cdots(4,6),(6.1),\ldots(6,6)]$, $\quad B : [(2,6),(3,5),(4,4),(5,3),(6,2)]$.

$A \cap B : [(2,6),(4,4),(6,2)]$

$\therefore\ P(A) = \dfrac{10}{20} = \dfrac{1}{2}, \quad P(B) = \dfrac{5}{36} \quad$ and $\quad P(A\cap B) = \dfrac{3}{36} = \dfrac{1}{12}$.

$\therefore$ **P (even number on the first die or a total of 8)** $= P\cup B = P(A) + P(B) - P(A\cap B)$

$$= \frac{1}{2} + \frac{5}{36} - \frac{1}{12} = \mathbf{\frac{5}{9}}.$$

Example 23.8. *A box contains 2 white and 4 black balls. Another box B contains 5 white and 7 black balls. A ball is transferred from the box A to the box B. When a ball is drawn from the box B, find the probability that it is white.* **[VTU 2004; Cochin 1999]**

Solution : Here the required probability will depend on whether the transferred ball from box A to B is black or white.

If a black ball is transferred, then the probability $= \frac{4}{6}$.

After transference, box B will contain 5 white and 8 black balls. So, the probability of drawing a white ball from box B is $\frac{5}{13}$.

$\therefore$ the probability of drawing a white ball from B, if the transferred ball is black

$$= \frac{4}{6} \times \frac{5}{13} = \frac{10}{39}.$$

Similarly, the probability of drawing a white ball from B, if the transferred ball is white

$$= \frac{2}{6} \times \frac{6}{13} = \frac{2}{13}.$$

$\therefore$ **the required probability** $= \dfrac{10}{39} + \dfrac{2}{13} = \mathbf{\dfrac{16}{39}}$.

Example 23.9. *In shuffling a pack of cards 4 are accidentally dropped. Find the chance that the missing cards should be one from each suit.*

Solution : Here the probability of 4 missing cards from different suits

$$= \frac{{}^{13}C_1 \times {}^{13}C_1 \times {}^{13}C_1 \times {}^{13}C_1}{{}^{52}C_4} = \frac{13 \times 13 \times 13 \times 13 \times 4 \times 3 \times 2 \times 1}{52 \times 51 \times 50 \times 49}$$

$$= \frac{1}{4} \times \frac{13}{51} \times \frac{13}{50} \times \frac{13}{49} \times 24 = \mathbf{\frac{2197}{20825}}.$$

Example 23.10. *The probability that a pen manufactured by a company will be defective is $\frac{1}{10}$. If 12 such pens are manufactured, find the probability that,*

(a) *exactly two will be defective.* (b) *at least two will be defective.* (c) *none will be defective.*

[BU(H) 2003; VTU 2004]

Solution : Given that the probability of defective pen $= \frac{1}{10} = 0.1$.

$\therefore$ the probability of non-defective pen $= 1 - 0.1 = \mathbf{0.9}$.

(a) The probability that exactly two pens will be defective $= {}^{12}C_2(0.1)^2(0.9)^{10} = \mathbf{0.2301}$.

(b) The probability that at least two pens will be defective

$$= 1 - 1(\text{probability that none or one is non-defective})$$

$$= 1 - [{}^{12}C_0(0.9)^{12} + {}^{12}C_1(0.1)(0.9)^{11}] = \mathbf{0.3412}.$$

(c) The probability that none will be defective $= {}^{12}C_{12}(0.9)^{12} = \mathbf{0.2833}$.

Example 23.11. *Assume that on the average one telephone number out of fifteen called between 2 PM and 3 PM on weekdays is busy. What is the probability that if randomly selected telephone numbers are called (i) not more than three (ii) at least three of them will be busy.*

Solution : The probability of a telephone number being busy between 2 pm and 3 pm on weekdays is $p = \frac{1}{15}$.

$$\therefore\ q = 1 - p = 1 - \frac{1}{15} = \frac{14}{15},\ n = 6.$$

$$\therefore\ \text{Binomial distribution} = \frac{14}{15} + \frac{1}{15}.$$

(i) The probability that not more than three telephone numbers will be busy

$$= P(0) + P(1) + P(2) + P(3)$$

$$= {}^6C_0\left(\frac{14}{15}\right)^6\left(\frac{1}{15}\right)^0 + {}^6C_1\left(\frac{14}{15}\right)^5\left(\frac{1}{15}\right)^1 + {}^6C_2\left(\frac{14}{15}\right)^4\left(\frac{1}{15}\right)^2 + {}^6C_3\left(\frac{14}{15}\right)^3\left(\frac{1}{15}\right)^3$$

$$= \frac{14^3}{15^6}[2744 + 1176 + 210 + 20] = \frac{2744 \times 4150}{(15)^6} = \mathbf{0.9997}.$$

(ii) The probability that at least three of them will be busy

$$= P(3) + P(4) + P(5) + P(6)$$

$$= {}^6C_3\left(\frac{14}{15}\right)^3\left(\frac{1}{15}\right)^3 + {}^6C_4\left(\frac{14}{15}\right)^2\left(\frac{1}{15}\right)^4 + {}^6C_5\left(\frac{14}{15}\right)\left(\frac{1}{15}\right)^5 + {}^6C_6\left(\frac{1}{15}\right)^6$$

$$= \mathbf{0.005}.$$

Example 23.12. *A problem of physics is given to three students X, Y and Z having chances of solving it are $\frac{1}{3}$, $\frac{1}{2}$ and $\frac{2}{3}$ respectively. What is the probability that the problem will be solved.*

Solution : The probability that X can solve the problem $= \dfrac{1}{3}$.

$\therefore$ the probability that X cannot solve the problem $= \left(1 - \dfrac{1}{3}\right)$.

Similarly, the probabilities that Y and Z cannot solve the problem are $(1 - \frac{1}{2})$ and $(1 - \frac{2}{3})$ respectively.

$\therefore$ the probability that X, Y, Z cannot solve the problem

$$= \left(1 - \frac{1}{3}\right) \times \left(1 - \frac{1}{2}\right) \times \left(1 - \frac{2}{3}\right) = \frac{2}{3} \times \frac{1}{2} \times \frac{1}{3} = \frac{1}{9}.$$

Hence, the probability that the problem can be solved $= 1 - \dfrac{1}{9} = \mathbf{\dfrac{8}{9}}$.

Example 23.13. *A box containing 10 articles is likely to have 2 defectives. In a consignment of 100 boxes, how many of them are expected to have three or less defectives?*

Solution : The probability that a box will contain defective articles is $p = \dfrac{2}{10} = \dfrac{1}{5}$.

The probability of non-defective articles is $q = (1 - p) = 1 - \dfrac{1}{5} = \dfrac{4}{5}$.

$\therefore$ the probability that box will contain three or less defective articles is

$$= P(0) + P(1) + P(2) + P(3)$$

$$= {}^{10}C_0 \left(\frac{1}{5}\right)^0 \left(\frac{4}{5}\right)^{10} + {}^{10}C_1 \left(\frac{1}{5}\right)^2 \left(\frac{4}{5}\right)^9 + {}^{10}C_2 \left(\frac{1}{5}\right)^2 \left(\frac{4}{5}\right)^8 + {}^{10}C_3 \left(\frac{1}{5}\right)^3 \left(\frac{4}{5}\right)^7$$

$$= 0.10738 + 0.2684 + 0.3020 + 0.2013 = 0.87908$$

$\therefore$ the number of boxes expected to contain three or less defective articles

$$= 0.87908 \times 100 \simeq \mathbf{88}.$$

Example 23.14. *In 256 sets of 12 tosses of a coin, in how many cases one can expect 8 heads and 4 tails?* **[JNTU 2003]**

Solution : Here probability of head $P(H) = \frac{1}{2}$ and probability of tail $P(T) = \frac{1}{2}$.

According to binomial distribution, probability of 8 heads and 4 tails in 12 trials is

$$P(X = 8) = {}^{12}C_8 \left(\frac{1}{2}\right)^8 \left(\frac{1}{2}\right)^4 = \frac{12!}{8!4!} \cdot \frac{1}{2^{12}} = \frac{495}{4096}.$$

Hence, expected number of such cases in 256 sets is

$$\mathbf{P(X = 8) \times 256} = \frac{495}{4096} \times 256 = \mathbf{31}.$$

Example 23.15. *If the probability that an individual suffers a bad reaction from a certain injection is 0.001, determine the probability that out of 2000 individuals (i) exactly 3 (ii) more than 2 individuals (iii) none (iv) more than one individual will suffer a bad reaction.* **[cf. VTU 2003]**

Solution : Here the probability of occurrence is 0.001 which is very small. So Poisson's distribution is to be applied.

Here, $p = 0.001$, $n = 2000$.

$\therefore \quad m = np = (2000) \times 0.001 = 2.$

$$P(r) = \frac{e^{-m} \cdot m_r}{r!} = \frac{e^{-2}(2)^r}{r!} = \frac{2^r}{r!e^2}.$$

(i) $P(3) = \dfrac{1}{3!}\dfrac{(2)^3}{e^2} = \dfrac{4}{3} \times (0.135) = \mathbf{0.18}.$

(ii) $P(>2) = P(3) + P(4) + \cdots + P(2000) = 1 - [P(0) + P(1) + P(2)]$

$$= 1 - \left[\frac{(2)^0}{0!e^2} + \frac{2^1 1!}{e^2} + \frac{2^2}{2!e^2}\right] = 1 - \frac{1}{e^2}[1+2+2] = 1 - \frac{5}{e^2}$$

$$= 1 - 5(0.135) = 1 - 0.675 = \mathbf{0.325}.$$

(iii) $P(\text{none}) = P(0) = \dfrac{e^{-2}2^0}{0!} = \mathbf{0.325}.$

(iv) $P(>1) = P(2) + P(3) + \cdots + P(2000) = 1 - [P(0) + P(1)]$

$$= 1 - \left[\frac{e^{-2}2^0}{0!} + \frac{e^{-2}2^1}{1!}\right] = 1 - \frac{3}{e^2} = 1 - 3(0.135)$$

$$= 1 - 0.405 = \mathbf{0.595}.$$

Example 23.16. *Using Poisson's distribution, find the probability that ace of spades will be drawn from a pack of well-shuffled cards at least once in 104 consecutive trials.*

Solution : For ace of spades, probability $P = \dfrac{1}{52}$, $n = 104$.

$$m = np = 104 \times \frac{1}{52} = 2$$

$$P(r) = \frac{e^{-m}m^r}{r!} = e^{-2} \cdot \frac{2^r}{r!} = \frac{1}{e^2} \cdot \frac{2^r}{r!}.$$

P (at least once) $= P(1) + P(2) + P(3) + \cdots + P(104) = 1 - P(0)$

$$= 1 - \frac{1}{e^2} \times \frac{2^0}{0!} = 1 - \frac{1}{e^2} = 1 - 0.135 = \mathbf{0.863}.$$

Example 23.17. *3% of bolts made by a machine are defective, the defect occurred at random during production. If bolts are packaged 50 per box, find (i) exact probability (ii) Poisson approximation to it, that a given box will contain 5 defectives.*

Solution : Here, $p = \dfrac{3}{100} = 0.03.$

(i) $q = 1 - p = 1 - 0.03 = 0.97.$

$\therefore$ probability of 5 defective balls in a lot of 50 $= {}^{50}C_5(0.03)^5(0.97)^{45} = \mathbf{0.013074}.$

(ii) For Poisson distribution $m = np = 50 \times 0.03 = 1.5.$

$\therefore$ required Poisson distribution $P(r) = \dfrac{e^{-m}m^r}{r!} = \dfrac{e^{-1.5}(1.5)^5}{5!} = \mathbf{0.01412}.$

Example 23.18. *An insurance company found that only 0.01% of the population is involved in a certain type of accident each year. If its 1000 policy holders were randomly selected from the population what is the probability that not more than two of its clients one involved in such an accident next year? Given* $e^{-0.1} = 0.9048$ **[AMIETE(Summer) 1993]**

Solution : Here $p = 0.01\% = \frac{1}{100} \times \frac{1}{100} = \frac{1}{10000}$, $n = 1000$.

$$\therefore \quad m = np = 1000 \times \frac{1}{10000} = \frac{1}{10} = 0.1.$$

$$P(r) = \frac{e^{-m} m^r}{r!}$$

$$P \text{ (not more than 2)} = P \text{ (0.1 and 2)} = P(0) + P(1) + P(2)$$

$$= \frac{e^{-0.1}(0.1)^0}{0!} + \frac{e^{-0.1}(0.1)^1}{1!} + \frac{e^{-0.1}(0.1)^2}{2!} = e^{-0.1}\left(1 + 0.1 + \frac{0.01}{2}\right)$$

$$= 0.9048 \times 1.05 = \mathbf{0.9998}.$$

Example 23.19. *A manufacturer knows that the razor blades he makes contain on the average 0.5% defectives. He packs them in packets of 5. What is the probability that a packet picked up at random will contain 3 or more faulty blades?*

Solution : Here $p = 0.5 = \frac{5}{100} = 0.005$, $n = 5$, $m = np = 5 \times 0.005 = 0.025$.

Then $P(r) = \frac{e^{-m} m^r}{r!}$.

$$\therefore \quad P(r \geq 3) = P(3) + P(4) + P(5) = \frac{e^{-0.025}(0.025)^3}{3!} + \frac{e^{-0.025}(0.025)^5}{4!} + \frac{e^{-0.025}(0.025)^5}{5!}$$

$$= \frac{e^{-0.025}}{5!} \frac{(0.025)^3}{5!}[20 + .5(0.025) + (0.025)^2]$$

$$= \frac{0.975 \times 0.000015625 \times 20.125625}{120} = \mathbf{0.000002555}.$$

Example 23.20. *A car-hire firm has two cars which it hires out day by day. The number of demands for a car on each day is distributed as a Poisson distribution with mean 1.5. Calculate the proportion of days on which a car is not used and the proportion of days on which some demand is refused.* **[AMIE(Summer) 1995]**

Solution : Here $m = np = 1.5$.

(i) If the car is not used, then demand $(r) = 0$.

Then $P(r) = \frac{e^{-m} m^r}{r!}$.

$$\therefore \quad P(0) = \frac{e^{-m} \cdot m^0}{0!} = e^{-1.5} = \mathbf{0.223}.$$

(ii) Some demand is refused if the number of demands exceeds 2 (i.e., $r > 2$)

$$\therefore \quad P(r > 2) = P(3) + P(4) + \cdots = 1 - [P(0) + P(1) + P(2)]$$

$$= 1 - \left[\frac{e^{-1.5}(1.5)^0}{0!} + \frac{e^{-1.5}(1.5)^1}{1!} + \frac{e^{-1.5}(1.5)^2}{2!}\right]$$

$$= 1 - e^{-1.5}(1 + 1.5 + 1.125) \doteq 1 - 0.2231 \times 3.625$$

$$= 1 - 0.8087375 = \mathbf{0.1912625}.$$

Example 23.21. *A manufacturer produces airmail envelope whose weight is normally distributed with mean $\mu = 1.95$ g and standard deviation $\sigma = 0.05$. The envelopes are solid in lots of 1000. How many envelopes in a lot will be heavier than 2 g? Use the fact $\frac{1}{\sqrt{2\pi}}\int_0^1 e^{-\frac{x^2}{2}}\,dx = 0.3413$.*

Solution : Given, $P(0 < z < 1) = \frac{1}{\sqrt{2\pi}}\int_0^1 e^{-\frac{x^2}{2}}\,dx = 0.3413.$

Also, $m = 1.95$, $\sigma = 0.05$ and $x = 2$.

$$\therefore \quad z = \frac{x-m}{\sigma} = \frac{2-1.95}{0.05} = 1.$$

$\therefore \quad P(z > 1) =$ area of the curve right to $z = 1 = 0.5 - 0.3413.$

Hence, number of envelopes heavier than 2 g in a lot of 1000 g is

$$N = 1000 \times 0.1587 = 158.7 \simeq \mathbf{159}.$$

Example 23.22. *In an examination the candidates are awarded the following grades depending on the marks scored by them :*

distinction :	$\geq 80\%$
second class :	$45\% \leq$ *marks* $\leq 80\%$
third class :	$30\% \leq$ *marks* $\leq 45\%$
fail :	$< 30\%$

It was found that 8% of the students failed and 8% have scored distinction. Find the average marks obtained by the candidates. Deduce the percentage of students placed in second class. Assume normal distribution of marks.

Solution : Suppose, the marks obtained by the candidates in the examination $= X$.

Let $X \sim N(m, \sigma^2)$.

We have, $P(X < 30) = 0.08$ and $P(X \geq 80) = 0.08$ as shown in Fig. 23.8.

The standard normal state is $Z = \frac{X-m}{\sigma}$.

Here, $X = 30$ is to the left of $X = m$ and $x = 80$ is to the right of $X = m$.

For $X = 30$, $z = \frac{30-m}{\sigma} = -z_1$.

For $X = 80$, $z = \frac{80-m}{\sigma} = z_2$.

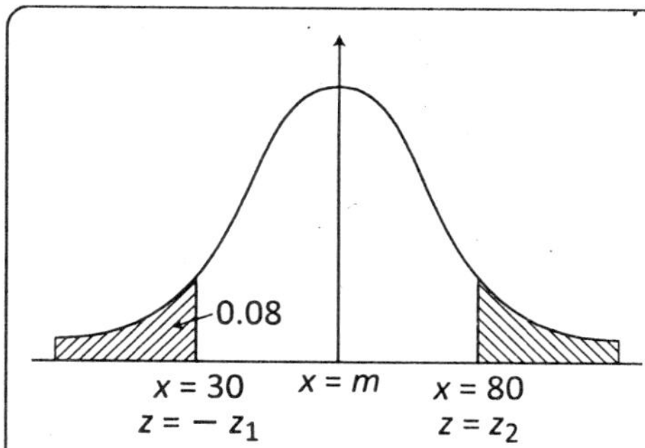

Fig. 23.8

$\because$ $P(X < 30) = 0.08$, we have

$$P(z < -z_1) \quad \text{or,} \quad P(0 < z < z_1) = 0.50 - 0.08 = 0.42.$$

From table we get $z_1 = 1.405 = z_2$.

$$\therefore \quad \frac{30 - m}{\sigma} = -1.405 \qquad \text{(i)}$$

$$\text{and} \quad \frac{80 - m}{\sigma} = 1.405. \qquad \text{(ii)}$$

Adding Eqns. (i) and (ii), we get

$$110 - 2m = 0 \quad \text{or,} \quad m = 55$$

$$\text{and} \quad \sigma = \frac{80 - m}{1.405} = \frac{25}{1.405} = \frac{25}{1.405} = \mathbf{17.79}.$$

If p be the probability that a candidate is placed in second class, then

$$P = P(45 < x < 60) = P(-0.56 \leq z < 0.28) = P(-0.56 \leq z < 0) + P(0 < z < 0.28)$$

$$= 0.2123 + 0.1103 = \mathbf{0.3226}.$$

Hence, approximately 32% candidates obtained second class.

Example 23.23. *In a normal distribution, 31% of the items are under 45 and 8% are over 64. Find the mean and the standard deviation of the distribution.* **[WBUT 2003]**

Solution : If $x = 45$, then we have, $z = \frac{(x - m)}{\sigma} = \frac{45 - m}{\sigma}$,

where $m =$ the mean and $\sigma =$ the standard deviation of the distribution.

If $x = 46$, then $z = \frac{(64 - m)}{\sigma}$.

Now, the area between $z = 0$ and $z = \frac{(45 - m)}{\sigma} = (0.50 - 0.31) = 0.19$.

From table, $z = -0.496$ for area 0.19.

Area between $z = 0$ and $z = \frac{(64 - m)}{\sigma} = 0.50 - 0.08 = 0.42$.

From table, $z = 1.405$ for area 0.42

$$\therefore \quad \frac{64 - m}{\sigma} = 1.405, \qquad \frac{45 - m}{\sigma} = -0.496.$$

Solving the above two equations we get $\boldsymbol{m = 50}$ and $\boldsymbol{\sigma = 10}$.

Example 23.24. *A manufacturer knows from experience that the resistance of resistors he produces is normal with mean $m =$ 100 ohms and standard deviation $\sigma =$ 2 ohms. What percentage of resistors will have resistance between 98 ohms and 102 ohms?*

Solution : Given $m = 100$ and $\sigma = 2$ ohms; $x_1 = 98$ and $x_2 = 102$.

$$\therefore \quad z = \frac{x - m}{\sigma} \text{ gives } z_1 = \frac{98 - 100}{2} = 1 \text{ and } z_2 = \frac{102 - 100}{2} = +1.$$

Area between $z_1 = -1$ and $z_2 = +1$ [from (Fig. 23.9)]

$= $ (area between $z = 0$ and $z = -1$)

$\qquad + $ (area between $z = 0$ and $z = +1$)

$= 2 \times$ (area between $z = 0$ and $z = +1$)

$= 2 \times 0.3413 = 0.6826.$

Hence, percentage of resistors having resistance between 98 ohms and 102 ohms = **68.26**.

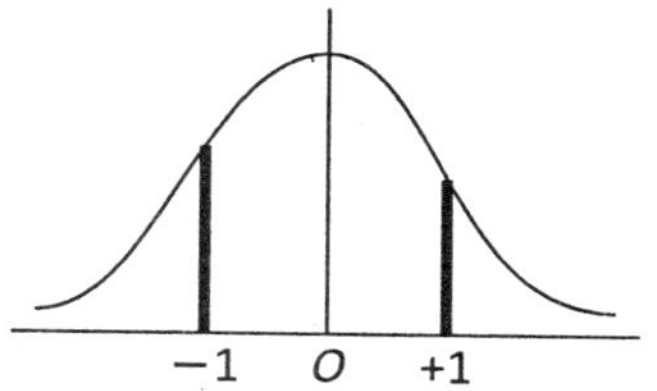

Fig. 23.9

Example 23.25. *A normal distribution has a mean* 15.73 *and a standard deviation of* 2.08. *Find the percentage of cases that fall between* 17.81 *and* 13.6. *Find also the percentage of cases lying above* 18.85.

Solution : In the problem, $m = 15.73$, $\sigma = 2.08$.

In the first case,

$$\text{when } x_1 = 13.65, \quad z_1 = \frac{x - m}{\sigma} = \frac{13.65 - 15.73}{2.08} = -\frac{2.08}{2.08} = 1$$

$$\text{and} \quad \text{when } x_2 = 17.81, \quad z_2 = \frac{17.81 - 15.73}{2.08} = -\frac{2.08}{2.08} = 1.$$

$\therefore$ the required probability, $P(13.65 < x < 17.81) = P(0 < z_1 < -1) + P(0 < z_2 < 1)$

$$= 0.34134 + 0.34134 = 0.68268.$$

Hence, the required percentage $= 100 \times 0.68268 = $ **68.268**.

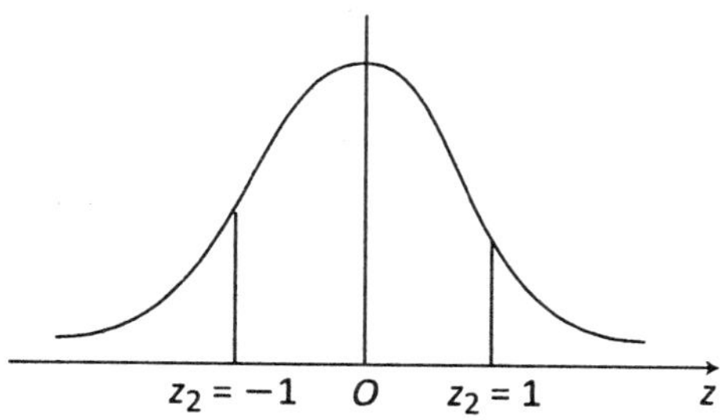

Fig. 23.10

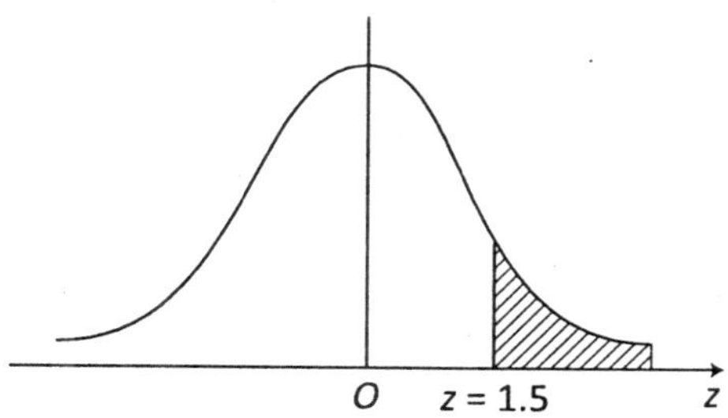

Fig. 23.11

$$\text{Again, when } x = 18.85, \quad z = \frac{18.85 - 15.73}{2.08} = \frac{3.12}{2.08} = 1.5.$$

Probability $P(z > 15) = $ area of right to $z = 1.5 (z \geq 1.5)$

$= P(0 < z < \infty) - P(0 < z < 1.5) = 0.5 - 0.43349$

$= $ **0.6651**.

[$\because$ $z = 1.5$, the probability (from table) $= 0.43349$]

and [$\because$ the total area from the mean to the right $P(0 < z < \infty)$ is 0.5]

Hence, the percentage of cases above $18.85 = 100 \times 0.0665 = $ **6.651**.

Example 23.26. *It is given that X and Y are normal variates with means 2, 5 and variances 4, 9 respectively.*

Find the value of K such that, $P(X+Y \leq K) = P(3X - Y \leq 3K)$.

Solution : We have $X \sim N(2,4)$, $Y \sim N(5,9)$ and X, Y are independent.

We define $U = X + Y$ and $V = 3X - Y$. Then U and V are also normal variates.

$$U = X + Y \sim N(2+5, 4+9) = N(7, 13)$$

$$V = 3X - Y \sim N(3(2) - 5, 9(4) + 9) = N(1, 45).$$

Let us define $z = \dfrac{U-7}{\sqrt{13}}$.

Now, $z \sim N(0,1)$ and $U = K$,

$$z = \frac{V-1}{\sqrt{45}}.$$

Again, $z \sim N(0,1)$ and $V = 3K$.

$$z = \frac{3K-1}{\sqrt{45}} = z_2.$$

We require $P(X+Y \leq K) = P(3X - Y \geq 3K)$.

$$P(z \leq z_1) = P(z \geq z_2).$$

As the normal probability curve is symmetric about $z = 0$, this implies that $z_1 = -z_2$.

$$\therefore \quad \frac{K-7}{\sqrt{13}} = -\left[\frac{3K-1}{\sqrt{45}}\right] \quad \text{or,} \quad \sqrt{45}(K-7) = \sqrt{13}(1-3K).$$

$$\therefore \quad K = \frac{7\sqrt{45} + \sqrt{13}}{\sqrt{45} + 3\sqrt{13}} = \mathbf{2.885}.$$

Example 23.27. *In a referendum 60% of voters cast vote in favour. A random sample of 200 voters was selected. Find the probability that in the sample, (i) more than 130 voted in favour (ii) between 105 and 130 inconclusive voted in favour (iii) 120 voted in favour.*

Solution : In the present problem, $n = 200$, $p = 0.6$, $q = 0.4$.

$$m = np = 200 \times 0.6 = 120, \quad \sigma = \sqrt{npq} = \sqrt{48} = 6.928.$$

(i) $P(X > 130) = P(x > 130.5) = P\left(X > \dfrac{130.5 - 120}{\sqrt{48}}\right) \doteq P(z > 1.516) = \mathbf{0.0648}$.

(ii) $P(105 < x < 130) = P(105.5 < x < 129.5) = P\left(\dfrac{105.5 - 120}{\sqrt{48}} < z < \dfrac{129.5 - 120}{\sqrt{48}}\right)$

$$= P(-2.09 < z < 1.37) = \mathbf{0.8964}.$$

(iii) $P(x = 120) = P(119.5 < x < 120.5) = P(-0.072 < z < 0.072) = \mathbf{0.0575}$.

Exercises

23(A) : Objective Type Questions

Select the correct answer

23.1 The probability that A passes a test is $\frac{2}{3}$ and the probability that B passes the same test is $\frac{3}{5}$. The probability that only one of them passes is

(a) $\frac{2}{5}$ (b) $\frac{4}{15}$ (c) $\frac{2}{15}$ (d) $\frac{7}{15}$.

23.2 On tossing a fair die, the probability of getting an odd number or a number less than 4 is

(a) 2 (b) $\frac{1}{2}$ (c) $\frac{2}{3}$ (d) $\frac{3}{4}$. **[AMIETE(Summer) 1997]**

23.3 From a pack of 52 cards one is thrown at random. Find the probability of getting a king.

(a) $\frac{3}{52}$ (b) $\frac{1}{13}$ (c) $\frac{2}{13}$ (d) $\frac{5}{52}$.

23.4 The probability that if a card is drawn at random from one ordinary pack is a diamond is

(a) $\frac{1}{3}$ (b) $\frac{2}{3}$ (c) $\frac{1}{4}$ (d) $\frac{3}{4}$.

23.5 The probability that a leap year selected at random will contain 53 Sundays is

(a) $\frac{1}{7}$ (b) $\frac{2}{7}$ (c) $\frac{5}{7}$ (d) none of these.

23.6 In a class of 12 students 5 are boys and the rest are girls; the probability that a student selected will be a girl is

(a) $\frac{5}{12}$ (b) $\frac{3}{12}$ (c) $\frac{7}{12}$ (d) $\frac{11}{12}$.

23.7 If for two events A and B, $P(A) = \frac{1}{2}$, $P(B) = \frac{1}{3}$ and $P(A \cap B) = \frac{1}{4}$, then $P\left(\frac{A}{B}\right)$ is

(a) $\frac{3}{4}$ (b) $\frac{1}{4}$ (c) $\frac{4}{3}$ (d) 1.

23.8 In a single throw of two dice the chance of throwing two aces is

(a) $\frac{1}{36}$ (b) $\frac{5}{36}$ (c) $\frac{7}{36}$ (d) $\frac{2}{9}$.

23.9 A coin is tossed four times the probability of getting one head is

(a) $\frac{3}{4}$ (b) $\frac{1}{4}$ (c) $\frac{1}{2}$ (d) $\frac{1}{3}$.

23.10 The probability that at least one of the events A and B occurs in 0.8 and the probability that both the events occur simultaneously is 0.25. The probability $P(A) + P(B)$ is

(a) 0.65 (b) 0.75 (c) 0.85 (d) 0.95. **[AMIETE(Summer) 1997]**

23.11 An unbiased coin is tossed five times. Given that heads were obtained in two of the tosses, the probability that these were obtained in the first two tosses is

(a) $\frac{1}{10}$ (b) $\frac{1}{4}$ (c) $\frac{1}{32}$ (d) none of these. **[AMIETE(Summer) 1996]**

23.12 Four cards are drawn from a pack of 52 cards. The chance that no two cards are of equal value is

(a) $\frac{2815}{3165}$ (b) $\frac{2816}{4165}$ (c) $\frac{1848}{4263}$ (d) none of these.

23.13 Five men in a company are graduates. If three men are picked up out of 20 at random, then the probability that they are graduates is

(a) $\frac{1}{114}$ (b) $\frac{3}{114}$ (c) $\frac{5}{114}$ (d) $\frac{121}{116}$.

23.14 The probability that A can solve a problem is $\frac{1}{4}$ and the probability that B can solve it is $\frac{2}{3}$. Then the probability that both of them can solve the problem is

(a) $\frac{3}{5}$ (b) $\frac{2}{7}$ (c) $\frac{1}{6}$ (d) $\frac{5}{6}$.

23.15 50 tickets are serially numbered 1 to 50. One ticket is drawn from there at random; the probability of its being multiple of 3 or 4 is

(a) $\frac{12}{25}$ (b) $\frac{14}{25}$ (c) $\frac{2}{5}$ (d) $\frac{3}{5}$.

23.16 One of the two events must occur. If the chance of one is $\frac{2}{3}$ of the other, then odds in favour of the other are

(a) 1 : 3 (b) 2 : 3 (c) 3 : 1 (d) none of these.

23.17 The random variable X has $N(1, 4)$ distribution, then

(a) $P(x > 3) > P(x > 1)$ (b) $P(x > 3) < P(x < 1)$

(c) $P(x < 3) < P(x > 1)$ (d) $P(x < 3) < P(x < 1)$. **[AMIETE(Summer) 1988]**

23.18 A bag contains 5 black and 4 white balls. Two balls are drawn at random. The probability that they match is

(a) $\frac{7}{12}$ (b) $\frac{5}{8}$ (c) $\frac{5}{9}$ (d) $\frac{4}{9}$.

23.19 The probability of having at least one tail in 4 throws with a coin is

(a) $\frac{15}{16}$ (b) $\frac{1}{16}$ (c) $\frac{1}{4}$. (d) 1.

23.20 A coin is tossed 6 times in succession, the probability of getting at least one head is

(a) $\frac{1}{64}$ (b) $\frac{3}{32}$ (c) $\frac{63}{64}$ (d) $\frac{1}{2}$.

23.21 A buys a lottery ticket in which the chance of winning is $\frac{1}{10}$; B has a ticket in which his chance of winning is $\frac{1}{20}$. The chance that at least one of them wins is

(a) $\frac{1}{200}$ (b) $\frac{29}{200}$ (c) $\frac{30}{200}$ (d) $\frac{170}{200}$.

23.22 The mean of the binomial distribution with n observations and probability of success p is

(a) pq (b) np (c) $\sqrt{np}$ (d) $\sqrt{pq}$.

23.23 The mean of Poisson's distribution is m, then the S.D. of this distribution is

(a) m^2 (b) $\sqrt{m}$ (c) m (d) none of these.

23.24 The S.D. of the binomial distribution is

(a) $\sqrt{npq}$ (b) $\sqrt{np}$ (c) npq (d) pq.

23.25 In a Poisson's distribution, $2P(x = 1) = P(x = 2)$, then the variance is

(a) 0 (b) -1 (c) 4 (d) 2.

23.26 The random variable x has a Poisson's distribution. If $P(x = 3) = \frac{1}{6}$, $P(x = 2) = \frac{1}{3}$, then $P(x = 0)$ is

(a) $e^{-\frac{3}{2}}$ (b) $e^{\frac{3}{2}}$ (c) e^{-3} (d) $e^{-\frac{1}{2}}$.

23.27 Let X be a Poisson's random variable, such that $2P(x = 0) = P(x = 2)$, then S.D. of x is

(a) 4 (b) 2 (c) $-\sqrt{2}$ (d) $\sqrt{2}$. **[AMIETE(Summer) 1992]**

23.28 In the standard normal curve, the area between $z = -1$ and $z = 1$ is nearly equal to

(a) 50% (b) 75% (c) 68% (d) 80%.

23.29 The marks obtained were found normally distributed with mean 75 and variance 100. The percentage of students who scored more than 75 marks is

(a) 50% (b) 75% (c) 40% (d) 25%.

23.30 The probability density function $f(x)$ of a continuous random variable x is defined by

$$\begin{aligned} f(x) &= \frac{A}{x^5}, \quad 5 \le x \le 10 \\ &= 0, \quad \text{otherwise.} \end{aligned}$$

The value of A is

(a) 1 (b) 50 (c) -200 (d) $\frac{200}{3}$.

23.31 If $f(x) = kxe^{-x}$, $x > 0$ is p.d.f. of x, then k is
$= 0,$ otherwise

(a) 5 (b) 1 (c) 10 (d) 150.

Answers

23.1 (d)	**23.2** (c)	**23.3** (b)	**23.4** (c)	**23.5** (b)	**23.6** (c)	**23.7** (a)	**23.8** (a)
23.9 (b)	**23.10** (b)	**23.11** (a)	**23.12** (b)	**23.13** (a)	**23.14** (c)	**23.15** (a)	**23.16** (d)
23.17 (b)	**23.18** (d)	**23.19** (a)	**23.20** (c)	**23.21** (b)	**23.22** (b)	**23.23** (b)	**23.24** (a)
23.25 (c)	**23.26** (a)	**23.27** (d)	**23.28** (c)	**23.29** (a)	**23.30** (d)	**23.31** (b).	

23(B) : Short Answer Type Questions

23.1 If one card is drawn at random from a pack of 52 cards, what is the probability of getting a queen? **[Ans: $\frac{1}{3}$]**

23.2 What is the probability of drawing a king or a queen or both from a pack of 52 cards?

[Ans: $\frac{1}{13}, \frac{1}{4}, \frac{4}{13}$]

23.3 A box A contains 2 white and 4 black balls. Another box B contains 5 white and 7 black balls. A ball is transferred from the box A to the box B. Find the probability that it is a white ball? **[Ans:** $\frac{16}{39}$**]**

23.4 An urn contains 13 balls numbering 1 to 13. Find the probability that a ball selected at random is a ball with number that is a multiple of 3 or 4.

[AMIE(Summer) 1992] **[Ans:** $\frac{6}{13}$**]**

23.5 Of the cigarette smoking population, 70% are men and 30% are women, 10% of these men and 20% of the them smoke 'WILLS'. What is the probability that a person seen smoking a 'WILLS'? What is the probability that a person seen smoking a 'WILLS' will be a man?

23.6 One bag contains four white and two black beads and another contains three of each colour. A bead is drawn from each bag. What is the probability that one is white and one is black? **[Ans:** $\frac{1}{2}$**]**

23.7 What is the probability of throwing '3' in a single die? **[Ans:** $\frac{1}{6}$**]**

23.8 Four persons are chosen at random from a group containing 3 men, 2 women and 4 children. Show that the probability that exactly two of them will be children is $\frac{10}{21}$.

23.9 What is the probability of getting a total of 5 or 8 in a single throw with two dice?

[Ans: $\frac{1}{4}$**]**

23.10 If 15 persons take seats at random at a round table, find the probability that two specified persons are seated next to each other. **[Ans:** $\frac{1}{7}$**]**

23.11 What is the probability of throwing a number greater than 3 in a single throw of one dice? **[Ans:** $\frac{1}{2}$**]**

23.12 A has one share in a lottery in which there is 1 prize and 2 blancks, B has three shares in a lottery in which there are 3 prizes and 6 blancks. Compare the probability of A's success to that of B's. **[Ans:** $7:16$**]**

23.13 Given, $P(A) = \frac{1}{2}$, $P(B) = \frac{1}{3}$ and $P(AB) = \frac{1}{4}$, find the value of $P(A+B)$.

[BU(H) 2003] **[Ans:** $\frac{7}{12}$**]**

23.14 Find the probability that a non-leap year should have 53 weeks.

[Madras U 2003] **[Ans:** $\frac{1}{7}$**]**

23.15 When a coin is tossed four times the probability of getting **(i)** exactly one head **(ii)** at most three heads and **(iii)** at least two heads. **[VTU 2000S]** **[Ans:** (i) $\frac{1}{4}$ (ii) $\frac{7}{8}$ (iii) $\frac{11}{16}$**]**

23.16 A purse contains 4 copper coins, 3 silver coins, the second purse contains 6 copper coins and 2 silver coins. If a coin is taken out of any purse, then show that the probability of it being a copper coin is $\frac{37}{56}$.

23.17 In rolling two fair dice show that, the probability of getting equal numbers or numbers with an even product is $\frac{30}{36}$.

23.18 Find the probability that in five tossings of an unbiased coin, 'head' fails at least three times in succession. **[Ans:** $\frac{1}{4}$**]**

23.19 The word ASSASIN is given. It is arranged so that S's come consecutively. Find the probability of this event. **[Ans:** $\frac{1}{7}$**]**

23.20 Of the items manufactured in a factory, 2% is defective. What is the probability that there are 3 defective items in a sample of 100?

23.21 The frequency distribution of a measurable characteristic varying between 0 and 2 is as follows :

$$f(x) = x^3, \qquad 0 \le x \le 1$$
$$= (2-x)^2, \quad 1 \le x \le 2.$$

Calculate the standard deviation and also the mean deviation about the mean.

[**Ans:** $\frac{1}{\sqrt{16}}, \frac{1}{5}$]

23(C) : Broad Answer Type Questions

23.1 **(a)** Develop the concept of probability and give its definition, both a **priori** and a **posterori**.

(b) Two balls are to be drawn simultaneously from a set of 3 red and 2 white balls. What is the number of elementary events? [**Ans:** 10 elementary events]

23.2 **(a)** State the addition rule of probability of mutually exclusive events. What happens to the rule for two events A and B if they are not mutually exclusive?

(b) A die is tossed 2 times. Find the probability of getting 4, 5 or 6 in the first toss and 3, 4 or 5 in the second toss. [**Ans:** $\frac{1}{2}, \frac{1}{2}$]

23.3 Three machines A, B and C manufacture respectively 0.4, 0.5 and 0.1 of the total production. The percentage of defective items produced by A, B and C is 2, 4 and 1 per cent respectively. For an item chosen at random, what is the probability that it is defective? [**Ans:** 0.029]

23.4 **(a)** State and prove the multiplication law of probability.

(b) A husband and a wife appear in an interview for two vacancies in the same post. The probability of husband's selection is $\frac{1}{7}$ and that of wife's selection is $\frac{1}{5}$. What is the probability that

(i) both of them will be selected?

(ii) only one of them will be selected and

(iii) none of them will be selected? [**Ans:** (i) $\frac{1}{35}$ (ii) $\frac{2}{7}$ (iii) $\frac{24}{35}$]

23.5 From 7 consonants and 5 vowels, how many words can be formed consisting of 4 different consonants and 3 different vowels? The words need not have meaning. [**Ans:** 1,764,000]

23.6 Out of 5 mathematicians and 7 physicists, a committee consisting of 2 mathematicians and 3 physicists is to be formed. In how many ways can this be done, if **(a)** any mathematician and any physicist can be included, **(b)** one particular physicist must be on the committee, **(c)** two particular mathematician cannot be on the committee?

[**Ans:** (a) 350 (b) 150 (c) 105]

23.7 A shelf has 6 mathematics books and 4 physics books. Find the probability that 3 particular mathematics books will be together. [**Ans:** $\frac{1}{15}$]

23.8 The probabilities that a husband and a wife will be alive 20 years from now are given by 0.8 and 0.9 respectively. Find the probability that in 20 years **(a)** both **(b)** neither **(c)** at least one will be alive. [**Ans:** (a) 0.7 (b) 0.02 (c) 0.98]

23.9 India plays two matches each with West Indies and Australia. In many matches the probabilities of India getting 0, 1 and 2 are 0.45, 0.05 and 0.50 respectively. Assuming that the outcomes are independent, show that the probability of India getting at least 7 points is 0.0875.

23.10 A speaks the truth in 75% cases and B in 80% of the cases. In what percentage of cases are they likely to contradict each other in stating the same fact?

[VTU 2000S] **[Ans:** $\frac{7}{20}$]

23.11 From 5 statistician and 6 economists, a committee consisting of 3 statisticians and two economists is to be formed. How many different committees can be formed, if **(a)** no restrictions are imposed **(b)** two particular statisticians can be in the committee **(c)** one particular economist cannot be in the committee. **[Ans:** (a) 150 (b) 45 (c) 100]

23.12 The overall percentage of failures in a certain examination is 20%. If six candidates appear in the exàmination, what is the probability, that at least five pass the examination? **[AMIETE(Winter) 1994]** **[Ans:** 0.65536]

23.13 The probability that a man aged 60 will live to be 70 is 0.065. What is the probability that out of 10 men, now 60, at least 7 will live to be 70?

[AMIETE(Winter) 1990] **[Ans:** 0.5137]

23.14 In a hurdle reace, a player has to cross 10 hurdles. The probability that he will clear each hurdle is $\frac{5}{6}$. What is the probability that he will knock down less than 2 hurdles?

[AMIETE(Summer) 1985] **[Ans:** $\frac{8}{3}\left(\frac{5}{6}\right)^9$]

23.15 In 256 sets of 12 tosses of a coin, in how many cases one can expect 8 heads and 4 tails? **[JNUT 2003]**

Solution : Probability of head $P(H) = \frac{1}{2}$, probability of tail $P(T) = \frac{1}{2}$.

By binomial distribution, the probability of $8H$ and $4T$ in 12 trials is

$$P(x = 8) = {}^{12}C_8 \left(\frac{1}{8}\right)^8 \left(\frac{1}{4}\right)^4 = \frac{12!}{8!4!}\frac{1}{2^{12}} = \frac{495}{4096}.$$

The expected number of such cases in 256 sets

$$= 256 \times P(x = 8) = 256 \times \frac{495}{4096} = 30.9 \simeq \mathbf{31}.$$

23.16 During war, one ship out of nine was sunk on an average in making a certain voyage. What was the probability that exactly 3 out of a convoy of 6 ships would arrive safely?

[Ans: $\frac{10240}{6}$]

23.17 A sortie of 20 aeroplanes is sent on an operational flight. The chances that an aeroplane fails to return is 5%. Find the probability that **(i)** one plane does not return **(ii)** at most 5 planes do not return. And **(iii)** what is the most probable number of return?

[Hissar U 2007] **[Ans:** (i) ${}^{20}C_1\left(\frac{1}{20}\right)\left(\frac{19}{20}\right)^9$, (ii) $\sum_{r=0}^{0} {}^{20}C_r \left(\frac{1}{20}\right)^r \left(\frac{19}{20}\right)^{20-r}$, (iii) 19]

23.18 Six dice are drawn 729 times. How many times do you expect that at least three dice to show a five or six? **[Ans:** 233]

23.19 Two persons A and B toss an unbiased coin alternately with the understanding that one who gets head first wins. If A starts the game, find the respective chance of winning?

[Ans: $\frac{2}{3}, \frac{1}{3}$]

23.20 Assume that the diameters of 1000 brass taken consecutively from a machine from a normal distribution with mean 0.7515 cm and standard deviation 0.020 cm. How many plugs are likely to be rejected if the approved diameter is 0.752 ± 0.004 cm? **[Ans:** 52]

23.21 Fit a binomial distribution following the frequency distribution :

x	0	1	2	3	4	5	6
f	13	25	52	58	32	16	4

[SVTU 2007; Madras U 2000S] [**Ans:** $200(0.554 + 0.446)^6$]

23.22 In a bombing action there is 54% chance that any bomb will strike the target. Two direct hits are needed to destroy the target completely. How many bombs are required to be dropped to give a 99% chance or better of completely destroying the target.

[VTU 2003S] [**Ans:** 11]

23.23 If on an average 1 vessel is wrecked, find the probability that out of 5 vessels expected to arrive, at least 4 will arrive safely. **[PTU 2005]** [**Ans:** $\frac{45927}{50000}$]

23.24 In a certain factory turning out razor blades, there is a small chance of 0.002 for any blade to be defective. The blades are supplied in packets of 10. Use Poisson's distribution to calculate the approximate number of packets containing **(a)** no defective blade **(b)** one defective blade and **(c)** two defective blades respectively in a consignment of 10,000 packets. [**Ans:** (a) 9802, (b) 196, (c) 2]

23.25 A certain screw making machine produces on an average two defective screws out of 100 and packs them in boxes of 500. Find the probability that a box contains 15 defective screws. **[Rohtak U 2006]** [**Ans:** 0.035]

23.26 A source of liquid is known to contain bacteria with the mean number of bacteria per cubic centimetre equal to 3. Ten 1 cc test tubes are filled with the liquid. Assuming that the Poisson's distribution is applicable, calculate the probability that all the test tubes will show growth, i.e., contain at least one bacterium each.

[VTU 2003] [**Ans:** 0.6]

23.27 Let x be the number of cars passing a certain point per minute at a particular time. Assuming that x has a Poisson's distribution with mean 0.5, find the probability of observing 3 or a few cars during any given minute.

[AMIETE(Summer) 1997] [**Ans:** 0.998]

23.28 Wireless sets are manufactured with 25 soldered joints each. On the average 1 joint in 500 is defective. How many sets can be expected to be free from defective joints in a consignment of 10000 sets? [**Ans:** 9512]

23.29 According to National Office of Vital Statistics of US, Department of Health and Human Services, the average number of accidental drowning per year in the United States is 3.0 per 100000 population. Find the probability that in a city of population 200000 there will be **(a)** 0 **(b)** 2 **(c)** 6 **(d)** 8 **(e)** between 4 and 8 **(f)** fewer than 3, accidental drowning per year. [**Ans:** (a) 0.0024, (b) 0.04462, (c) 0.1607, (d) 0.1033, (e) 0.6964, (f) 0.0620]

23.30 The length of bolts produced by a machine is normally distributed with mean 4 and S.D. of 0.5. A bolt is defective if its length does not lie in the interval $(3.8, 4.3)$. Find the percentage of defective bolts produced by the machine.

Given, $\frac{1}{\sqrt{2\pi}}\int_{-\infty}^{0.6} e^{-\frac{t^2}{2}}\,dt = 0.7257$; $\frac{1}{\sqrt{2\pi}}\int_{-\infty}^{0.4} e^{-\frac{t^2}{2}}\,dt = 0.6554$. **[WBUT 2004]**

23.31 The weight of students in a college is normally distributed with mean 40 kg, $\sigma = 5$ kg. Find the percentage of the students that have weight **(i)** greater than 40 kg **(ii)** greater than 50 kg **(iii)** between 38 and 52 kg.

Given, $\frac{1}{\sqrt{2\pi}}\int_{-\infty}^{2} e^{-\frac{t^2}{2}}dt = 0.9772;\quad \frac{1}{\sqrt{2\pi}}\int_{-\infty}^{2.4} e^{-\frac{t^2}{2}}dt = 0.9918;$

$$\frac{1}{\sqrt{2\pi}}\int_{-\infty}^{0.4} e^{-\frac{t^2}{2}}dt = 0.6554.$$

23.32 A random variable X is normally distributed with mean 68 cm and S.D. 2.5 m. What should be the size of the sample whose mean shall not differ from population mean by more than 1 cm with probability 0.95? [Given that area under the standard normal curve to the right of the ordinate at 1.6 is 0.025] **[WBUT 2003]**

23.33 In a male population of 1000, the mean height is 68.16 inches and standard deviation is 3.2 inches. How many men may be more than 6 feet? [$\phi(1.15) = 0.8749$; $\phi(1.2) = 0.8849$; $\phi(1.25) = 0.8944$; where the argument is the standard normal variable.]

[AMIETE(Winter) 1997] [Ans: 115]

23.34 The breaking strength X of a cotton fabric is normally distributed with $E(x) = 16$ and $\sigma(x) = 1$. The fabric is said to be good if $X \geq 14$. What is the probability that a fabric chosen at random is good? [Given, $\phi(2) = 0.9772$] **[Ans: 0.9772]**

23.35 The income of a group of 10,000 persons was found to be normally distributed with mean Rs. 750 pm, and standard deviation of Rs. 50. Show that of this group about 95% had income exceeding Rs. 688 and only 5% had income exceeding Rs. 832. Also find the lowest income among the richest 100. **[UPTU 2004S]**

23.36 **(a)** If $f(x) = Ae^{-\frac{(x-\mu)^2}{2\alpha^2}}$ is a probability density function, show that $A = \frac{1}{\sigma\sqrt{2\pi}}$.

[Use $\int_{-\infty}^{\infty} e^{-x^2}\,dx = \sqrt{\pi}$] **[CU(H) 2016]**

(b) Show that for Poisson's distribution $P(k) = \frac{e^{-\lambda}\lambda^k}{k!}(k = 0, 1, 2\cdots)$

variance $(x) = E(x^2) - (E(X)) = \lambda$, where $E(X) = \sum_{k=0}^{\infty} kP(x)$. **[CU(H) 2016]**

23.37 **(a)** The Gaussian distribution, i.e., normal probability distribution is given by

$$P_0(x, \mu, \sigma) = \frac{1}{\sqrt{2\pi\sigma}}\exp\left[-\frac{1}{2}\left(\frac{x-\mu}{\sigma}\right)^2\right] (-\infty < x < \propto).$$

Find out the mean and variate of this distribution. **[CU(H) 2015]**

(b) Six coins are tossed simultaneously. What is the probability of getting (i) 2 heads (ii) at least 2 heads. **[CU(H) 2013]**

(c) Sketch the two Gaussian probability distribution functions $f_1(x)$ and $f_2(x)$ with the same mean $x = 0$, but with two different standard deviations σ_1 and σ_2, where $\sigma_2 > \sigma_1$ **[CU(H) 2009, '13]**

23.38 **(a)** For free paths of length x during which a molecule does not suffer a collision with another molecule in dilute gas, one uses the exponential distribution :

$$P_E(x, \lambda) = \frac{1}{\lambda}e^{-\frac{x}{\lambda}}\ (0 \leq x \leq \alpha).$$

Show that the average value of x is λ in the above distribution. **[CU(H) 2013]**

(b) The Gaussian probability distribution is given by

$$P_G(x,\mu,\sigma) = \frac{1}{\sqrt{2\pi\sigma}} \exp\left[-\frac{1}{2}\left(\frac{x-\mu}{\sigma}\right)^2\right], -\infty < x < \infty$$

in usual notation. Show that it has two points of inflexion at $x = \mu \pm \infty$.

[CU(H) 2012]

23.39 **(a)** The weather record shows that on the average, Kolkata has 5 norwesters in 30 days of April. What is the binomial probability that a particular April has at least 4 norwesters? [C.U.(H), 2008]

(b) A random variable X being the interval from σ to β is expressed by the formula

$$f(x) = \frac{1}{\sigma\sqrt{2\pi}} e^{-\frac{(x-\mu)^2}{2\sigma^2}}, -\infty < x < \infty.$$

Show that the probability of random variable x being in the interval from α to β is expressed by the formula,

$$P(\alpha < x < \beta) = \phi\left(\frac{\beta-\mu}{\sigma}\right) - \phi\left(\frac{\alpha-\mu}{\sigma}\right)$$

where the function $\phi(x) = \frac{1}{\sqrt{2\pi}} \int_0^x e^{-\frac{t^2}{2}}$.

23.40 **(a)** Colloidal particles each of mass m are suspended in air under the force of gravitation. Show that the number density of particles decreases exponentially with the height. One can neglect the density of air with respect to the density of the particles. Suppose, these particles are positively charged and an electric field is applied along the vertical, in a direction opposite to the gravitational force. How the above distribution will be modified? [CU(H) 2009]

(b) Write down the expression for standard normal distribution function. Calculate the standard derivation for this distribution. [CU(H) 2007]

23.41 **(a)** A random variable X takes the values $0, 1, 2, 3$ with probabilities $\frac{1}{8}, \frac{3}{8}, \frac{3}{8}, \frac{1}{8}$ respectively. What will be the expectation value of x? [CU(H) 2016]

(b) A disc is thrown 8 times. What is the probability that 4 will be seen **(i)** exactly twice **(ii)** at least 6 times? [CU(H) 2017]

Appendix–A

Stirling's Approximations

Stirling's approximation has wide applications in areas of statistical physics according to which

$$\ln(n!) = n \ln - n, \tag{1}$$

where n is very large.

Proof : From gamma function,

$$n! = \Gamma(n+1) = \int_0^\infty x^n e^{-x} dx$$

$$= \int_0^\infty e^{n \ln x - x} dx. \tag{2}$$

We substitute, $x = n + \sqrt{n}y$.

$\therefore \qquad dx = \sqrt{n}dy.$

Then from eqn. (2),

$$n! = \sqrt{n} \int_{-\sqrt{n}}^\infty e^{[n \ln(n+\sqrt{n}y) - (n+\sqrt{n}y)]} dy \tag{3}$$

Now, $\qquad \ln(n\sqrt{n}y) = \ln n + \ln\left(1 + \frac{y}{\sqrt{n}}\right)$

$$= \ln n + \frac{y}{\sqrt{n}} - \frac{y^2}{2n} \quad \text{[for large } n\text{]}. \tag{4}$$

Again, for large n, we may replace the lower limit of the integral of eqn. (3) by $-\infty$. So, from eqns. (3) and (4) we may write,

$$n! = \sqrt{n}n^n e^{-n} \int_{-\infty}^\infty e^{-y^2/2} dy$$

$$= n^{n+\frac{1}{2}} e^{-n} \sqrt{2\pi} \left(\because \int_{-\infty}^\infty e^{-ax^2} dx = \sqrt{\frac{\pi}{a}}, \text{ by eqn. (2)}\right). \tag{5}$$

$$\therefore \qquad \ln(n!) = \left(n + \frac{1}{2}\right) \ln n - n + \frac{1}{2} \ln(2\pi). \tag{6}$$

We may neglect $\frac{1}{2}\ln(2\pi)$ and $\frac{1}{2}n(n+\frac{1}{2})$ for very large n.

Then, $\qquad \mathbf{\ln(n!) = n \ln n - n}.$

This is Stirling's approximation.

Appendix–B

B(1) : Dirac Delta Function :

The one-dimensional dirac delta function is defined as :

$$\delta(x) = 0, \text{ for } x \neq 0 \tag{1}$$

$$\int_{-\infty}^{\infty} f(x)\delta(x)dx = f(0) \tag{2}$$

where $f(x)$ is any well-behaved function and the integration operation includes the origin.

A special case of eqn. (2) is

$$\int_{-\infty}^{\infty} \delta(x)dx = 1. \tag{3}$$

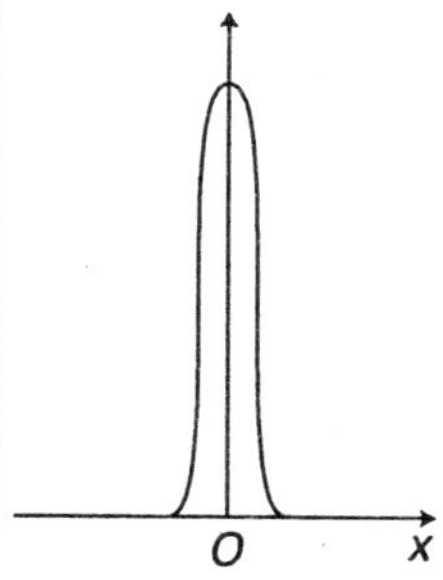

Fig. 1

Shifting the origin of co-ordinates, eqn. (2) becomes

$$\int_{-\infty}^{\infty} f(x)\delta(x-a)dx = f(a), \tag{4}$$

which is valid for any continuous function.

Eqn. (2) shows that $\delta(x)$ must be infinitely high, infinitely thin spike at $x = 0$ (Fig. 1).

B(2) : Different expressions for delta function :

Following are different forms of delta function :

1. $\delta(x) = \dfrac{1}{2\pi}\displaystyle\int_{-\infty}^{\infty} e^{i\theta x} d\theta$
2. $\sigma(x) = \underset{\sigma \to 0}{\text{Lt}} \dfrac{1}{\sigma\sqrt{2\pi}} e^{-x^2/2\pi}, \ \sigma > 0$
3. $\xi(x) = 0, \quad \text{for } x < 0$

   $ 1, \quad \text{for } x > 0$
4. $\delta(x) = \underset{\lambda \to 0}{\text{Lt}} \dfrac{\sin \lambda x}{\pi x}.$

B(3) : Equations satisfied by delta function :

Delta function satisfies the following equations.

1. $\delta^*(x) = \delta(x)$
2. $\delta(-x) = \delta(x)$
3. $d\delta(x) = 0$
4. $\delta(ax) = \dfrac{1}{a}\delta(x), \ a > 0$
5. $\delta(x^2 - a^2) = \dfrac{1}{2a}[\delta(x+a) + \delta(x-a)], \ a > 0$
6. $f(x)\delta(x-a) = f(a)\delta(x-a)$
7. $\displaystyle\int \delta(x-a)\delta(x-a)dx = \delta(a-b)$

Appendix–C

Greek Letters used

α – alpha	μ – mu
β – beta	ν – nu
γ – gamma	π – pi
δ – delta	ρ – rho
ϵ – epsilon	σ – sigma
ι – iota	τ – tau
θ – theta	χ – chi
ϕ – phi	ω – omega
ψ – psi	Γ – cap. gamma
ξ – xi	Δ – cap. delta
η – eta	Σ – cap. sigma
ζ – zeta	Θ – cap. theta
λ – lambda	Φ – cap. phi
	Λ – cap. lambda

Bibliography

1. **Mathematical Methods for Physics** by G. Arfken.
2. **Determinants and Matrices** by A. C. Aitken.
3. **Mathematics for Engineers and Scientists** by E. Suhir.
4. **Mathematics for Scientists and Engineers** by H. Cohen.
5. **Theory and applications of infinite series** by K. Knopp.
6. **Methods of Mathematical Physics** by Liu Lianshou and W. Zhengqing.
7. **Mathematical Physics** by A. K. Ghatak, I. C. Goyal and S. J. Chua.
8. **Vector Methods** by D. E. Rutherford.
9. **Mathematics in Physics and Engineering** by J. Irving and N. Mullineux.
10. **Introduction to the theory of Fourier Series and integrals** by H. S. Carslaw.
11. **Higher engineering mathematics** by B. S. Grewal.
12. **Mathematical methods for science students** by G. Stephenson.
13. **The mathematics of Physics and Chemistry** by H. Margenau and G. M. Murphy.
14. **Methods of Mathematical Physics** by R. Courant and D. Hilbert.
15. **Mathematical Physics** by E. Butkof.
16. **Introduction to applied mathematics** by F. D. Murnaghan.
17. **The use of integral transforms** by I. H. Sneddon.
18. **What is Mathematics?** by R. Courant and H. Robbins.
19. **Ordinary differential equations** by E. I. Ince.
20. **Mathematical Physics** by E. Butkov.
21. **Bessel functions for engineers** by N. W. McLachlan.
22. **Partial differential equations of Mathematical Physics** by A. G. Webster.
23. **Partial differential equations and boundary-value problems with applications** by M. A. Pinsky.
24. **Theory of Bessel functions** by G. N. Watson.
25. **Foundations of the theory of probability** by A. Kolmogorov.
26. **Applied probability for engineers and scientists** by E. Suhir.
27. **Higher Mathematics Vol 3** (Translated from Russian) by V. Shokurov.
28. **Complex analysis with applications** by R. A. Silverman and N. J. Englewood Cliffs.
29. **Fourier Series** by I. N. Sneddon.
30. **Advance calculus** by M. R. Spiegel.
31. **Laplace transform** by M. R. Spiegel.
32. **Theory and problems of vector analysis** by M. R. Spiegel.
33. **Integral equations** by F. G. Tricomi.
34. **Complex integration and Cauchy's theorem** by G. N. Watson.
35. **Special functions** by D. C. Kuznetsov.
36. **Vector calculus** by B. W. Lindgren.
37. **Theory of functions of a complex variable** by E. T. Copson.

Index